Plant Form and Function

History of Life

Animal Form and Function

Evolution

Ecology

CAMPBELL
BIOLOGY
IN FOCUS
Second Edition

Intro

Genetics

Chemistry and Cells

27 28 29 30 31 32 33 34 35 36 37 38 39 40 41 42 43 1 2 3 4 5 6 7 8 9 10 11 12 13 14 15 16 17 18 19 20 21 22 23 24 25 26

# CAMPBELL
# BIOLOGY IN FOCUS

## SECOND EDITION

**Lisa A. Urry**
Mills College, Oakland, California

**Michael L. Cain**
Bowdoin College, Brunswick, Maine

**Steven A. Wasserman**
University of California, San Diego

**Peter V. Minorsky**
Mercy College, Dobbs Ferry, New York

**Jane B. Reece**
Berkeley, California

**PEARSON**

Editor-in-Chief: *Beth Wilbur*

Executive Editor: *Josh Frost*

Executive Editorial Manager: *Ginnie Simione Jutson*

Supervising Editors: *Beth N. Winickoff and Pat Burner*

Senior Developmental Editors: *John Burner and Mary Ann Murray*

Developmental Artist: *Andrew Recher, Lachina*

Editorial Assistant: *Alexander Helmintoller*

Director, Product Management Services: *Erin Gregg*

Program Management Team Lead: *Michael Early*

Program Manager: *Leata Holloway*

Project Management Team Lead: *David Zielonka*

Project Manager: *Lori Newman*

Production Management and Composition: *S4Carlisle Publishing Services*

Illustrations: *Lachina*

Design Manager: *Marilyn Perry*

Cover and Text Design: *Hespenheide Design*

Rights & Permissions Manager: *Rachel Youdelman*

Photo Permissions Project Manager: *Donna Kalal*

Photo Researcher: *Maureen Spuhler*

Photo Permissions Specialist: *Lumina Datamatics*

Text Permissions Project Manager: *Tim Nicholls*

Text Permissions Specialist: *Lumina Datamatics*

Director of Content Development, MasteringBiology®: *Natania Mlawer*

Senior Developmental Editor, MasteringBiology®: *Sarah Jensen*

Senior Editorial Content Producer: *Lee Ann Doctor*

Executive Mastering® Media Producer: *Katie Foley*

Associate Mastering® Media Producer: *Charles Hall*

Director of Marketing: *Christy Lesko*

Executive Marketing Manager: *Lauren Harp*

Senior Marketing Manager: *Amee Mosley*

Manufacturing Buyer: *Stacey Weinberger*

Text Printer: *RR Donnelley / Willard*

Cover Printer: *Phoenix Color/Hagerstown*

Cover Photo Credit: *Klaus Honal/AGE Fotostock*

**Library of Congress Cataloging-in-Publication Data**

Campbell biology in focus. – Second edition / Lisa A. Urry, Michael L. Cain, Steve A. Wasserman, and Peter V. Minorsky.

   pages cm

   Includes index.

   ISBN 978-0-321-96275-1

   ISBN 0-321-96275-3

   1. Biology. I. Urry, Lisa A. II. Cain, Michael L. (Michael Lee), 1956- III. Wasserman, Steven Alexander IV. Minorsky, Peter V. V. Title: Biology in focus.

   QH308.2C347 2016

   570—dc23

          2015005641

ISBN 10: 0-321-96275-3; ISBN 13: 978-0-321-96275-1 (Student Edition)

ISBN 10: 0-134-20314-3; ISBN 13: 978-0-134-20314-0 (Books a la Carte Edition)

4   17

# Preface

The snow leopard (*Panthera uncia*) that peers intently from the cover of this book has a suite of evolutionary adaptations that enable it to spot, track, and ambush its prey. The snow leopard's keen eye is a metaphor for our goal in writing this text: to focus with high intensity on the core concepts that biology majors need to master in the introductory biology course.

The current explosion of biological information, while exhilarating in its scope, poses a significant challenge—how best to teach a subject that is constantly expanding its boundaries. In particular, instructors have become increasingly concerned that their students are overwhelmed by a growing volume of detail and are losing sight of the big ideas in biology. In response to this challenge, various groups of biologists have initiated efforts to refine and in some cases redesign the introductory biology course. In particular, the report *Vision and Change in Undergraduate Biology Education: A Call to Action*\* advocates focusing course material and instruction on key ideas while transforming the classroom through active learning and scientific inquiry. Many instructors have embraced such approaches and have changed how they teach. Cutting back on the amount of detail they present, they focus on core biological concepts, explore select examples, and engage in a rich variety of active learning exercises.

We were inspired by these ongoing changes in biology education to write the first edition of *Campbell BIOLOGY IN FOCUS*, a new, shorter textbook that was received with widespread excitement by instructors. Guided by their feedback, we honed the Second Edition so that it does an even better job of helping students explore the key questions, approaches, and ideas of modern biology.

## New to This Edition

Here we briefly describe the new features that we have developed for the Second Edition, but we invite you to explore pages xii–xxvi for more information and examples.

### New in the Text

- The impact of **genomics** across biology is explored throughout the Second Edition with examples that reveal how our ability to rapidly sequence DNA and proteins on a massive scale is transforming all areas of biology, from molecular and cell biology to phylogenetics, physiology, and ecology. Illustrative examples are distributed throughout the text.
- The Second Edition provides increased coverage of the urgent issue of **global climate change**. Starting with a new figure (Figure 1.11) and discussion in Chapter 1 and concluding with significantly expanded material on causes and effects of climate change in Chapter 43, including a new Make Connections Figure (Figure 43.28), the text explores the impact of climate change at all levels of the biological hierarchy.
- Ten **Make Connections Figures** pull together content from different chapters to assemble a visual representation of "big-picture" relationships. By reinforcing fundamental conceptual connections throughout biology, these figures help overcome students' tendencies to compartmentalize information.
- **Interpret the Data Questions** throughout the text engage students in scientific inquiry by asking them to analyze data presented in a graph, figure, or table. The Interpret the Data Questions can be assigned and automatically graded in MasteringBiology.®
- **Synthesize Your Knowledge Questions** at the end of each chapter ask students to synthesize the material in the chapter and demonstrate their big-picture understanding. A striking, thought-provoking photograph leads to a question that helps students realize that what they have learned in the chapter connects to their world and provides understanding and insight into natural phenomena.
- Scannable QR codes and URLs at the end of every chapter give students quick access to **Vocabulary Self-Quizzes** and **Practice Tests** that students can use on a smartphone, tablet, or computer.
- Detailed information about the organization of the text and new content in the Second Edition is provided on pages vi–ix, following this Preface.

---

\* Copyright 2011 American Association for the Advancement of Science. See also *Vision and Change in Undergraduate Biology Education: Chronicling Change, Inspiring the Future*, copyright 2015 American Association for the Advancement of Science. For more information, see www.visionandchange.org

## New in MasteringBiology®

- **Ready-to-Go Teaching Modules** in the Instructor Resources area help instructors efficiently make use of the available teaching tools for many key topics in introductory biology. Before-class assignments, in-class activities, and after-class assignments are provided for ease of use. Instructors can incorporate active learning into their course with the suggested activity ideas and clicker questions or Learning Catalytics questions.
- New **MasteringBiology tutorials** extend the power of MasteringBiology:
  - **Interpret the Data Questions** ask students to analyze a graph, figure, or table.
  - **Solve It Tutorials** engage students in a multistep investigation of a "mystery" or open question in which they must analyze real data.
  - **HHMI Short Films,** documentary-quality movies from the Howard Hughes Medical Institute, engage students in topics from the discovery of the double helix to evolution, with assignable questions.
  - **Video Field Trips** allow students to study ecology by taking virtual field trips and answering follow-up questions.

## Our Guiding Principles

Our key objective in creating CAMPBELL BIOLOGY IN FOCUS was to produce a shorter text by streamlining selected material, while emphasizing conceptual understanding and maintaining clarity, proper pacing, and rigor. Here, briefly, are the five guiding principles of our approach:

### 1. Focus on Core Concepts

We developed this text to help students master the fundamental content and scientific skills they need as college biology majors. In structuring the text, we were guided by discussions with biology professors across the country, analysis of hundreds of syllabi, study of the debates in the literature of scientific pedagogy, and our experience as instructors at a range of institutions. The result is a **briefer book for biology majors** that informs, engages, and inspires.

### 2. Establish Evolution as the Foundation of Biology

Evolution is the central theme of all biology, and it is the core theme of this text, as exemplified by the various ways that evolution is integrated into the text:

- Every chapter explicitly addresses the topic of evolution through an **Evolution section** that leads students to consider the material in the context of natural selection and adaptation.
- Each Chapter Review includes a **Focus on Evolution Question** that asks students to think critically about how an aspect of the chapter relates to evolution.

- Evolution is the unifying idea of **Chapter 1, Introduction: Evolution and the Foundations of Biology**, which devotes Concept 1.2 to the core theme of evolution, providing students with a foundation in evolution early in their study of biology.
- Following the in-depth coverage of **evolutionary mechanisms in Unit 3**, evolution also provides the storyline for the novel approach to presenting biological diversity in **Unit 4, The Evolutionary History of Life**. Focusing on landmark events in the history of life, Unit 4 highlights how key adaptations arose within groups of organisms and how evolutionary events led to the diversity of life on Earth today.

### 3. Engage Students in Scientific Thinking

Helping students learn to "think like a scientist" is a nearly universal goal of introductory biology courses. Students need to understand how to formulate and test hypotheses, design experiments, and interpret data. Scientific thinking and data interpretation skills top lists of learning outcomes and foundational skills desired for students entering higher-level courses. CAMPBELL BIOLOGY IN FOCUS, Second Edition, meets this need in several ways:

- **Scientific Skills Exercises** in every chapter use real data to build skills in graphing, interpreting data, designing experiments, and working with math—skills essential for students to succeed in biology. These exercises can also be assigned and automatically graded in MasteringBiology.
- New **Interpret the Data Questions** ask students to analyze a graph, figure, or table. These questions are also assignable in MasteringBiology.
- **Scientific Inquiry Questions** in the end-of-chapter material give students further practice in scientific thinking.
- **Inquiry Figures** and **Research Method Figures** reveal *how* we know *what* we know and model the process of scientific inquiry.

### 4. Use Outstanding Pedagogy to Help Students Learn

CAMPBELL BIOLOGY IN FOCUS, Second Edition, builds on our hallmarks of clear and engaging text and superior pedagogy to promote student learning:

- In each chapter, a framework of carefully selected **Key Concepts** helps students distinguish the "forest" from the "trees."
- Questions throughout the text catalyze learning by encouraging students to **actively engage with and synthesize key material.** Active learning questions include Concept Check Questions, Make Connections Questions, What If? Questions, Figure Legend Questions, Draw It Exercises, Summary Questions, and the new Synthesize Your Knowledge and Interpret the Data Questions.

- **Test Your Understanding Questions** at the end of each chapter are organized into three levels based on **Bloom's Taxonomy.**

## 5. Create Art and Animations That Teach

Biology is a visual science, and students learn from the art as much as the text. Therefore, we have developed our art and animations to teach with clarity and focus. Here are some of the ways our art and animations serve as superior teaching tools:

- The ten new **Make Connections Figures** help students see connections between topics across the entire introductory biology course.
- Each unit in *CAMPBELL BIOLOGY IN FOCUS*, Second Edition, opens with a **visual preview** that tells the story of the chapters' contents, showing how the material in the unit fits into a larger context.
- **BioFlix® 3-D Animations** help students visualize biology with movie-quality animations that can be shown in class and reviewed by students in the Study Area. **BioFlix Tutorials** use the animations as a jumping-off point for MasteringBiology coaching assignments with feedback.
- By integrating text, art, and photos, **Exploring Figures** help students access information efficiently.
- **Guided Tour Figures** use descriptions in blue type to walk students through complex figures as an instructor would, pointing out key structures, functions, and steps of processes.
- Because text and illustrations are equally important for learning biology, the **page layouts** are carefully designed to place figures together with their discussions in the text.
- **PowerPoint®** slides are painstakingly developed for optimum presentation in lecture halls, with enlarged editable labels, art broken into steps, and links to animations and videos.
- Many **Tutorials** and **Activities** in MasteringBiology integrate art from the text, providing a unified learning experience.

## MasteringBiology®

**MasteringBiology** is the most widely used online assessment and tutorial program for biology, providing an extensive library of homework assignments that are graded automatically. **Self-paced tutorials provide individualized coaching with specific hints and feedback** on the most difficult topics in the course. In addition to the new tutorials already mentioned, MasteringBiology includes hundreds of online exercises that can be assigned. For example:

- The **Scientific Skills Exercises** from the text can be assigned and automatically graded in MasteringBiology.
- **BioFlix® Tutorials** use 3-D animations to help students master tough topics.
- **Make Connections Tutorials** help students connect what they are learning in one chapter with material they have learned in another chapter.

- **BLAST Data Analysis Tutorials** teach students how to work with real data from the BLAST database.
- **Experimental Inquiry Tutorials** allow students to replicate a classic biology experiment and learn the conceptual aspects of experimental design.
- **Reading Quiz Questions** and approximately 3,000 **Test Bank Questions** are available for assignment.
- Optional **Adaptive Follow-up Assignments** are based on each student's performance on the original MasteringBiology assignment and provide additional coaching and practice as needed.

Every assignment is automatically graded and entered into a **gradebook**. Instructors can check the gradebook to see what topics students are struggling with and then address those topics in class.

The following resources are also available in MasteringBiology:

- The **Instructor Resources** area provides everything needed to teach the course, including the new **Ready-to-Go Teaching Modules**.
- **Learning Catalytics™** allows students to use their smartphones, tablets, or laptops to respond to questions in class.
- **Dynamic Study Modules** provide students with multiple sets of questions with extensive feedback so that they can test, learn, and retest until they achieve mastery of the textbook material. Students can use these modules on their smartphones on their own or the modules can be assigned.
- Students can read the **eText** and use the self-study resources in the **Study Area**.

MasteringBiology and the text work together to provide an unparalleled learning experience. For more information about MasteringBiology, see pages xv–xvi and xx–xxiv.

see pages xv–xvi and xx–xxiv.

\* \* \*

Our overall goal in developing and revising this text was to assist instructors and students in their exploration of biology by emphasizing essential content and skills while maintaining rigor. Although this Second Edition is now completed, we recognize that *CAMPBELL BIOLOGY IN FOCUS*, like its subject, will evolve. As its authors, we are eager to hear your thoughts, questions, comments, and suggestions for improvement. We are counting on you—our teaching colleagues and all students using this book—to provide us with this feedback, and we encourage you to contact us directly by e-mail:

Lisa Urry (Chapter 1, Units 1 and 2): lurry@mills.edu

Michael Cain (Chapter 1, Units 3, 4, and 7): mcain@bowdoin.edu

Peter Minorsky (Chapter 1, Unit 5): pminorsky@mercy.edu

Steven Wasserman (Chapter 1, Unit 6): stevenw@ucsd.edu

Jane Reece: janereece@cal.berkeley.edu

CAMPBELL *BIOLOGY IN FOCUS*, Second Edition, is organized into an introductory chapter and seven units that cover core concepts of biology at a thoughtful pace. When we adapted CAMPBELL *BIOLOGY* to write the first edition of this text, we made informed choices about how to design each chapter of CAMPBELL *BIOLOGY IN FOCUS* to meet the needs of instructors and students. In some chapters, we retained most of the material; in other chapters, we pruned material; and in still others, we completely reconfigured the material. In creating the Second Edition, we solicited feedback from reviewers and used their thoughtful critiques to further fine-tune the content and pedagogy. We have also updated the content wherever appropriate, and in a few cases reintroduced material. Here, we present synopses of the seven units and highlight the major revisions made to the Second Edition of CAMPBELL *BIOLOGY IN FOCUS*.

### CHAPTER 1  Introduction: Evolution and the Foundations of Biology

Chapter 1 introduces the **five biological themes** woven throughout the text: the core theme of **Evolution**, together with **Organization**, **Information**, **Energy and Matter**, and **Interactions**. Chapter 1 also explores the process of scientific inquiry through a case study describing experiments on the evolution of coat color in the beach mouse. The chapter concludes with a discussion of the importance of diversity within the scientific community.

In the Second Edition, a new figure (Figure 1.8) on gene expression uses lens cells in the eye as an example of DNA → RNA → protein and introduces the terms transcription and translation. This new figure and text equip students from the outset with an understanding of how gene sequences determine an organism's characteristics. New text and a new photo (Figure 1.11) inform students about the effects of climate change in general, and global warming in particular, on species survival and diversity. Concept 1.3 has been thoroughly revised to more realistically reflect the process of science. A new section has been added on the Flexibility of the Scientific Process, accompanied by a new Figure 1.19 that depicts the more realistic and complex process of science. The text now discusses searching the scientific literature, and a new question in the Chapter Review asks students to use PubMed.

## UNIT 1  Chemistry and Cells

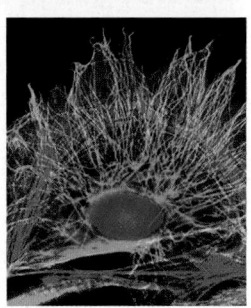

A succinct, two-chapter treatment of basic chemistry (Chapters 2 and 3) provides the foundation for this unit focused on cell structure and function. The related topics of cell membranes and cell signaling are consolidated into one chapter (Chapter 5). Due to the importance of the fundamental concepts in Units 1 and 2, much of the material in the rest of these two units has been retained from CAMPBELL *BIOLOGY*.

For the Second Edition, a new table has been added to Chapter 2 detailing the elements in the human body, with an associated Interpret the Data question. Chapter 3 includes a new section on isomers, with an accompanying figure (Figure 3.5), and ends with a new Concept 3.7 that includes cutting-edge coverage of DNA sequencing and introduces genomics and proteomics, as well as bioinformatics. A new Make Connections Figure (Figure 3.30) entitled "Contributions of Genomics and Proteomics to Biology" provides an overview of areas in which genomics and proteomics have had significant impacts—including evolution, conservation biology, paleontology, medical science, and species interactions—with the aim of inspiring and motivating students. A striking photo of thermophilic cyanobacteria has been added to Figure 6.16 on environmental factors affecting enzyme activity. In Chapter 7, a computer model of ATP synthase has been added to Figure 7.13. The icon for this enzyme in Chapters 7 and 8 has been re-drawn to more closely represent its structure. A new Make Connections Figure (Figure 8.20, "The Working Cell") integrates all the cellular activities covered in Chapters 3–8 in the context of a single working plant cell.

## UNIT 2  Genetics

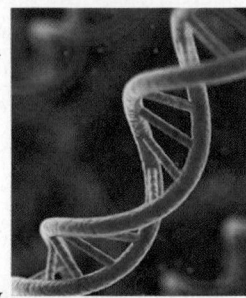

Topics in this unit include meiosis and classical genetics as well as the chromosomal and molecular basis for genetics and gene expression (Chapters 10–14). We also include a chapter on the regulation of gene expression (Chapter 15) and one on the role of gene regulation in development, stem cells, and cancer (Chapter 16). Methods in biotechnology are integrated into appropriate chapters. The stand-alone chapter on viruses (Chapter 17) can be taught at any point in the course. The final chapter in the unit, on genome evolution (Chapter 18), provides both a capstone for the study of genetics and a bridge to the evolution unit.

Chapter 10 of the Second Edition includes a new section on "Crossing Over and Synapsis During Prophase I" that explains the events of prophase I in more detail, supported by new Figure 10.9, which clearly shows and describes these events. In Chapter 11, to incorporate more molecular biology into the discussion of Mendelian genetics, Figure 11.4 on alleles has been enhanced and a new Figure 11.16 on sickle-cell disease has been added. Chapter 13 includes new text and two new figures (Figures 13.29 and 13.30) covering advances in sequencing technology. Also in this chapter, a new section, including new Figure 13.31, describes gene editing using the CRISPR-Cas9 system. In Chapter 15, the section on noncoding RNAs has been updated, and Figure 15.14 on *in situ* hybridization has been expanded and enhanced to help students understand this important technique. Chapter 16 includes a new Inquiry Figure (Figure 16.16) on induced pluripotent stem cells (iPS cells). Material on embryonic stem cells and induced pluripotent stem cells has been significantly updated. A new Make Connections Figure (Figure 16.21), "Genomics, Cell Signaling, and Cancer," illustrates recent research on subtypes of breast cancer, connecting content that students have learned in Chapters 5, 9, and 16. It also addresses treatment for one subtype of breast cancer as an example. In Chapter 17, the discussion of the importance of cell-surface proteins in determining host range has been enhanced. A new figure (Figure 17.9) presents the example of the receptor and co-receptor proteins for HIV. Coverage of the CRISPR system, as a bacterial "immune" system, has been added, supported by new Figure 17.6. Coverage of recent epidemics has been inserted (Ebola) or updated (H5N1). Chapter 18 has been significantly updated to reflect recent sequencing advances, including a discussion of the results of the ENCODE project, information on the bonobo genome, and use of high-throughput techniques to address the problem of cancer. Regarding protein structure, the discussion of BLAST searches has been enhanced, and computer models of lysozyme and α-lactalbumin have been added to support the discussion of the evolution of genes with novel functions.

## UNIT 3 Evolution

This unit provides in-depth coverage of essential evolutionary topics, such as mechanisms of natural selection, population genetics, and speciation. Early in the unit, Chapter 20 introduces "tree thinking" to support students in interpreting phylogenetic trees and thinking about the big picture of evolution. Chapter 23 focuses on mechanisms that have influenced long-term patterns of evolutionary change. Throughout the unit, new discoveries in fields ranging from paleontology to phylogenomics highlight the interdisciplinary nature of modern biology.

Revisions in the Second Edition aim to strengthen connections among fundamental evolutionary concepts. For example, Concept 20.5 includes new text on horizontal gene transfer among eukaryotes, reinforcing the overall discussion of how horizontal gene transfer has played an important role in the evolutionary history of life. Also in Concept 20.5, a new Scientific Skills Exercise walks students through the process of comparing and interpreting amino acid sequences to determine whether horizontal gene transfer may have occurred in certain organisms. Chapter 20 also includes more discussion of tree thinking, as well as a new figure (Figure 20.11) that distinguishes between paraphyletic and polyphyletic taxa. New material in Chapter 21 clarifies the interplay between mutation, genetic variation, and natural selection. A new Make Connections Figure (Figure 21.15, "The Sickle-Cell Allele") integrates material from chapters across the book in exploring the sickle-cell allele and its impact from the molecular and cellular levels to the allele's global distribution in the human population. Other changes in the unit include new examples and figures that reinforce evolutionary concepts. For example, a new introduction to Chapter 23 tells the story of the discovery of whale fossils from the Sahara Desert, striking evidence of how organisms in the past differed from organisms living today. In Chapter 22, a new figure (Figure 22.11) has been added to support the expanded text discussion of allopolyploid speciation in *Tragopogon* in the Pacific Northwest. Dates have also been revised in the text, Table 23.1 (The Geologic Record), and figures in Chapter 23 and throughout the Second Edition to reflect the International Commission on Stratigraphy 2013 revision of the Geologic Time Scale.

## UNIT 4 The Evolutionary History of Life

This unit employs a novel approach to studying the evolutionary history of biodiversity. Each chapter focuses on one or more major steps in the history of life, such as the origin of cells or the colonization of land. Likewise, the coverage of natural history and biological diversity emphasizes the evolutionary process—how factors such as the origin of key adaptations have influenced the rise and fall of different groups of organisms over time.

In the Second Edition, we have expanded our coverage of genomic and other molecular studies. Examples include a new figure (Figure 24.25) and text on the potential use and significance of CRISPR-Cas systems, a new Scientific Skills Exercise in Chapter 26 on genomic analyses of mycorrhizal and nonmycorrhizal fungi, and a new figure (Figure 27.36) and text related to evidence of gene flow between Neanderthals and modern humans. In addition, many phylogenies have been revised to reflect recent miRNA and genomic data. The unit also includes more connections to other chapters. For instance, a new Make Connections Question in Figure 24.4 asks students to apply material from Chapter 3 to explain how a membrane-like bilayer can self-assemble and form a vesicle, and a new Make Connections Figure (Figure 26.14) explores the diverse structural solutions for maximizing surface area that have evolved in cells, organ systems, and whole organisms. Other changes enhance the evolutionary storyline of the unit. For example, in Chapter 26, the chapter title, Figure 26.2, Key Concept 26.2, and text in Concepts 26.1 and 26.2 have all been revised to emphasize and explain that fungi are not closely related to plants, although they likely played a role in facilitating the colonization of land by plants, and that fungi possess their own novel adaptations for terrestrial life. Likewise, in Chapter 27, the discussion of the evolutionary impact of animals has been expanded, and new text and four new figures (Figures 27.12, 27.13, 27.30, and 27.31) on molluscs, birds, and mammals have been added. The chapter also includes expanded coverage of human evolution, including three new figures (Figures 27.34, 27.35, and 27.36). Supporting the extensive revision of Chapter 27, the number of Key Concepts in this chapter has increased from five to seven.

## UNIT 5 Plant Form and Function

The form and function of higher plants are often treated as separate topics, thereby making it difficult for students to make connections between the two. In Unit 5, plant anatomy (Chapter 28) and the acquisition and transport of resources (Chapter 29) are bridged by a discussion of how plant architecture influences resource acquisition. Chapter 30 provides an introduction to plant reproduction and examines controversies surrounding the genetic engineering of crop plants. The final chapter (Chapter 31) explores how plants respond to environmental challenges and opportunities and how the integration of this diverse information by plant hormones influences plant growth and reproduction.

In the Second Edition, a new micrograph of parenchyma cells and new information relating to root hair density, length, and function have been added to Chapter 28. In Chapter 29, a new Make Connections Figure (Figure 29.10, "Mutualism Across Kingdoms and Domains") enables students to integrate what they have learned about plant mutualisms with other examples across the natural realm. A new Inquiry Figure (Figure 29.11) examines the metagenomics of soil bacteria. A discussion on mycorrhizae and plant evolution has also been added in Chapter 29. In Chapter 30, the angiosperm life cycle figure and related text are more closely integrated, with all the numbered steps now identified in the text. Also, a discussion of coevolution of flowers and pollinators has been added. The in-depth discussion of the development from seed to flowering plant has been expanded to include the transition from vegetative growth to reproductive growth, making a connection to what students learned about development in Chapter 28. In addition, the depictions of the structure of maize root systems and raspberry fruit development have been improved. The information in Concept 31.4 concerning plant defenses against disease has been thoroughly revised and updated to reflect rapid advances in our understanding of plant immunity. Updated information relates to the two types of plant immunity: PAMP-triggered immunity and effector-triggered immunity. New Figure 31.23 highlights examples of physical, chemical, and behavioral defenses against herbivory.

## UNIT 6 Animal Form and Function

In this unit, a focused exploration of animal physiology and anatomy applies a comparative approach to a limited set of examples to bring out fundamental principles and conserved mechanisms. Students are first introduced to the closely related topics of endocrine signaling and homeostasis in an integrative introductory chapter (Chapter 32). Additional melding of interconnected material is reflected in chapters that combine treatment of circulation and gas exchange, reproduction and development, neurons and nervous systems, and motor mechanisms and behavior.

In the Second Edition, we re-envisioned the introductory chapter of this unit (Chapter 32), as conveyed by its new title, "The Internal Environment of Animals: Organization and Regulation." Endocrine signaling and the integration of nervous and endocrine system function now precede the introduction of homeostasis and the consideration of the two major examples: thermoregulation and osmoregulation. Figures on simple hormone and neurohormone pathways (Figures 32.6 and 32.7) and hormone cascades (Figure 32.8) have been substantially revised to provide clear and consistent presentation of hormone function and of the regulation of hormone secretion. The presentation of the mechanism for filtrate processing in the kidney has been substantially revised, with a single figure (Figure 32.22) in place of two and with the accompanying numbered text walking students through a carefully paced tour of the nephron. In this chapter and throughout the unit, figures illustrating homeostatic regulation have been revised to highlight the common principles and features of homeostatic mechanisms. The unit includes two new Make Connections Figures: Figure 32.3 illustrates shared and divergent solutions to fundamental challenges common to plants and animals, and Figure 37.8, on ion movements and gradients, explores the fundamental role of concentration gradients in life processes ranging from osmoregulation and gas exchange to locomotion. Also in Chapter 37, the treatments of synaptic signaling, summation, modulating signaling, and neurotransmitters have been revised to highlight key ideas, ensuring appropriate pacing and helping students focus on fundamental principles rather than memorization. Updates in Unit 6 informed by current research include new Figure 33.15 and text highlighting the explosion of interest in and understanding of the microbiome. Chapter 38 opens with a new photograph and introductory text that showcase the "brainbow" technique for labeling individual brain neurons.

## UNIT 7 Ecology

This unit applies the key themes of the text, including evolution, interactions, and energy and matter, to help students learn ecological principles. Chapter 40 integrates material on population growth and Earth's environment, highlighting the importance of both biological and physical processes in determining where species are found. Chapter 43 ends the book with a focus on global ecology and conservation biology. This chapter illustrates the threats to all species from increased human population growth and resource use. It begins with local factors that threaten individual species and ends with global factors that alter ecosystems, landscapes, and biomes.

The increased emphasis throughout the Second Edition on global climate change is capped by new discussions and figures in Unit 7. Chapter 43, for example, includes a new figure on the greenhouse effect (Figure 43.26) as well as new text examining aspects of climate change other than global warming. The chapter explores documented examples of the impacts to organisms in a new section on "Biological Effects of Climate Change" and a new Make Connections Figure (Figure 43.28, "Climate Change Has Effects at All Levels of Biological Organization"). Throughout the unit, the presentation of several other key topics has been revised. For example, in Chapter 40, the discussion of each of the following concepts or models was revised to standardize and clarify their meaning: life tables, per capita population growth, the per capita rate of increase ($r$), exponential population growth, and logistic population growth. The discussion of species interactions in Chapter 41 was modified to group species interactions according to whether they have positive (+) or negative (−) effects on survival and reproduction; as a result, there is a new section on "Exploitation" (which includes predation, herbivory, and parasitism) and another new section on "Positive Interactions" (which includes mutualism and commensalism). Material throughout Chapter 42 was revised to reinforce the fact that energy flows through ecosystems, whereas chemical elements cycle within ecosystems. New Figure Legend Questions give students practice in actively interpreting results; see, for example, the new questions with Figure 43.22 (biological magnification of PCBs) and Figure 43.31 (a new figure on per capita ecological footprints). The unit also includes a new Make Connections Figure (Figure 42.18, "The Working Ecosystem") that ties together population, community, and ecosystem processes in the arctic tundra.

# About the Authors

The author team's contributions reflect their biological expertise as researchers and their teaching sensibilities gained from years of experience as instructors at diverse institutions. They are also experienced textbook authors, having written CAMPBELL BIOLOGY in addition to CAMPBELL BIOLOGY IN FOCUS.

## Lisa A. Urry

Lisa Urry (Chapter 1 and Units 1 and 2) is Professor of Biology and Chair of the Biology Department at Mills College in Oakland, California, and a Visiting Scholar at the University of California, Berkeley. After graduating from Tufts University with a double major in biology and French, Lisa completed her Ph.D. in molecular and developmental biology at Massachusetts Institute of Technology (MIT) in the MIT/Woods Hole Oceanographic Institution Joint Program. She has published a number of research papers, most of them focused on gene expression during embryonic and larval development in sea urchins. Lisa has taught a variety of courses, from introductory biology to developmental biology and senior seminar. As a part of her mission to increase understanding of evolution, Lisa also teaches a nonmajors course called Evolution for Future Presidents and is on the Teacher Advisory Board for the Understanding Evolution website developed by the University of California Museum of Paleontology. Lisa is also deeply committed to promoting opportunities for women and underrepresented minorities in science.

## Michael L. Cain

Michael Cain (Chapter 1 and Units 3, 4, and 7) is an ecologist and evolutionary biologist who is now writing full-time. Michael earned a joint degree in biology and math at Bowdoin College, an M.Sc. from Brown University, and a Ph.D. in ecology and evolutionary biology from Cornell University. As a faculty member at New Mexico State University and Rose-Hulman Institute of Technology, he taught a wide range of courses, including introductory biology, ecology, evolution, botany, and conservation biology. Michael is the author of dozens of scientific papers on topics that include foraging behavior in insects and plants, long-distance seed dispersal, and speciation in crickets. In addition to his work on CAMPBELL BIOLOGY IN FOCUS, Michael is also the lead author of an ecology textbook.

## Steven A. Wasserman

Steve Wasserman (Chapter 1 and Unit 6) is Professor of Biology at the University of California, San Diego (UCSD). He earned his A.B. in biology from Harvard University and his Ph.D. in biological sciences from MIT. Through his research on regulatory pathway mechanisms in the fruit fly *Drosophila*, Steve has contributed to the fields of developmental biology, reproduction, and immunity. As a faculty member at the University of Texas Southwestern Medical Center and UCSD, he has taught genetics, development, and physiology to undergraduate, graduate, and medical students. He currently focuses on teaching introductory biology. He has also served as the research mentor for more than a dozen doctoral students and more than 50 aspiring scientists at the undergraduate and high school levels. Steve has been the recipient of distinguished scholar awards from both the Markey Charitable Trust and the David and Lucille Packard Foundation. In 2007, he received UCSD's Distinguished Teaching Award for undergraduate teaching.

## Peter V. Minorsky

Peter Minorsky (Chapter 1 and Unit 5) is Professor of Biology at Mercy College in New York, where he teaches introductory biology, evolution, ecology, and botany. He received his A.B. in biology from Vassar College and his Ph.D. in plant physiology from Cornell University. He is also the science writer for the journal *Plant Physiology*. After a postdoctoral fellowship at the University of Wisconsin at Madison, Peter taught at Kenyon College, Union College, Western Connecticut State University, and Vassar College. His research interests concern how plants sense environmental change. Peter received the 2008 Award for Teaching Excellence at Mercy College.

## Jane B. Reece

The head of the author team for recent editions of *CAMPBELL BIOLOGY*, Jane Reece was Neil Campbell's longtime collaborator. Earlier, Jane taught biology at Middlesex County College and Queensborough Community College. She holds an A.B. in biology from Harvard University, an M.S. in microbiology from Rutgers University, and a Ph.D. in bacteriology from the University of California, Berkeley. Jane's research as a doctoral student and postdoctoral fellow focused on genetic recombination in bacteria. Besides her work on the Campbell textbooks for biology majors, she has been an author of *Campbell Biology: Concepts & Connections*, *Campbell Essential Biology*, and *The World of the Cell*.

## Neil A. Campbell

Neil Campbell (1946–2004) combined the investigative nature of a research scientist with the soul of an experienced and caring teacher. He earned his M.A. in zoology from the University of California, Los Angeles, and his Ph.D. in plant biology from the University of California, Riverside, where he received the Distinguished Alumnus Award in 2001. Neil published numerous research articles on desert and coastal plants and how the sensitive plant (*Mimosa*) and other legumes move their leaves. His 30 years of teaching in diverse environments included introductory biology courses at Cornell University, Pomona College, and San Bernardino Valley College, where he received the college's first Outstanding Professor Award in 1986. He was a visiting scholar in the Department of Botany and Plant Sciences at the University of California, Riverside. Neil was the lead author of *Campbell Biology: Concepts & Connections*, *Campbell Essential Biology*, and *CAMPBELL BIOLOGY*, upon which this book is based.

**NEW! Ten Make Connections Figures** integrate content from different chapters and provide a visual representation of "big picture" relationships.

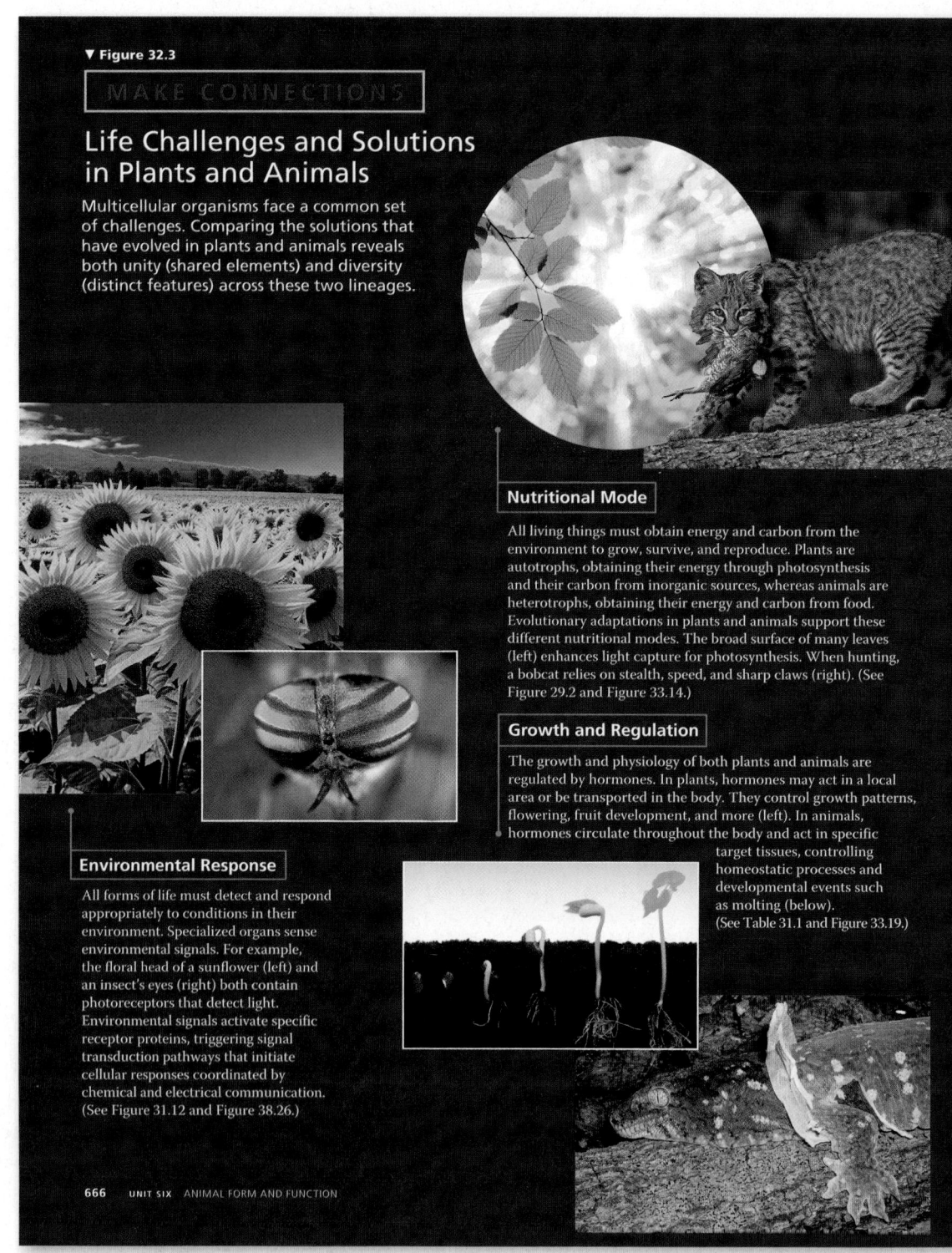

▼ Figure 32.3

MAKE CONNECTIONS

## Life Challenges and Solutions in Plants and Animals

Multicellular organisms face a common set of challenges. Comparing the solutions that have evolved in plants and animals reveals both unity (shared elements) and diversity (distinct features) across these two lineages.

### Nutritional Mode

All living things must obtain energy and carbon from the environment to grow, survive, and reproduce. Plants are autotrophs, obtaining their energy through photosynthesis and their carbon from inorganic sources, whereas animals are heterotrophs, obtaining their energy and carbon from food. Evolutionary adaptations in plants and animals support these different nutritional modes. The broad surface of many leaves (left) enhances light capture for photosynthesis. When hunting, a bobcat relies on stealth, speed, and sharp claws (right). (See Figure 29.2 and Figure 33.14.)

### Growth and Regulation

The growth and physiology of both plants and animals are regulated by hormones. In plants, hormones may act in a local area or be transported in the body. They control growth patterns, flowering, fruit development, and more (left). In animals, hormones circulate throughout the body and act in specific target tissues, controlling homeostatic processes and developmental events such as molting (below). (See Table 31.1 and Figure 33.19.)

### Environmental Response

All forms of life must detect and respond appropriately to conditions in their environment. Specialized organs sense environmental signals. For example, the floral head of a sunflower (left) and an insect's eyes (right) both contain photoreceptors that detect light. Environmental signals activate specific receptor proteins, triggering signal transduction pathways that initiate cellular responses coordinated by chemical and electrical communication. (See Figure 31.12 and Figure 38.26.)

666 UNIT SIX ANIMAL FORM AND FUNCTION

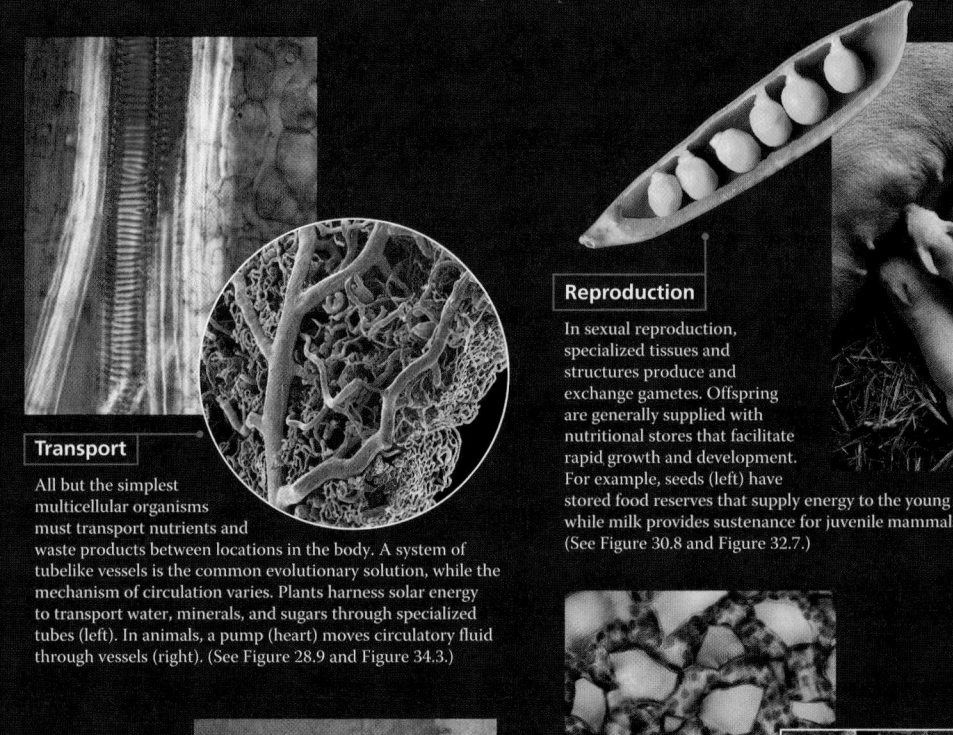

## Transport

All but the simplest multicellular organisms must transport nutrients and waste products between locations in the body. A system of tubelike vessels is the common evolutionary solution, while the mechanism of circulation varies. Plants harness solar energy to transport water, minerals, and sugars through specialized tubes (left). In animals, a pump (heart) moves circulatory fluid through vessels (right). (See Figure 28.9 and Figure 34.3.)

## Reproduction

In sexual reproduction, specialized tissues and structures produce and exchange gametes. Offspring are generally supplied with nutritional stores that facilitate rapid growth and development. For example, seeds (left) have stored food reserves that supply energy to the young seedling, while milk provides sustenance for juvenile mammals (right). (See Figure 30.8 and Figure 32.7.)

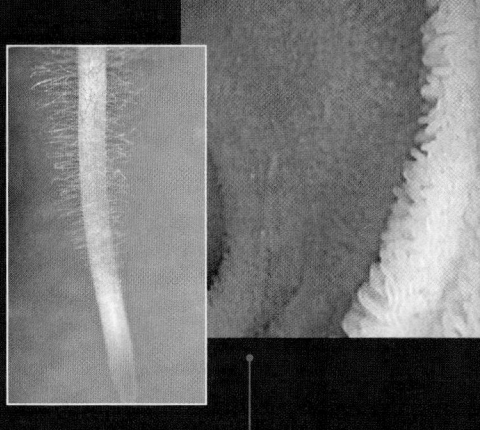

## Gas Exchange

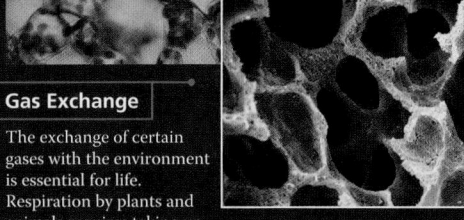

The exchange of certain gases with the environment is essential for life. Respiration by plants and animals requires taking up oxygen ($O_2$) and releasing carbon dioxide ($CO_2$). In photosynthesis, net exchange occurs in the opposite direction: $CO_2$ uptake and $O_2$ release. In both plants and animals, highly convoluted surfaces that increase the area available for gas exchange have evolved, such as the spongy mesophyll of leaves (left) and the alveoli of lungs (right). (See Figure 28.17 and Figure 34.20.)

## Absorption

Organisms need to absorb nutrients. The root hairs of plants (left) and the villi (projections) that line the intestines of vertebrates (right) increase the surface area available for absorption. (See Figure 28.4 and Figure 33.10.)

**MAKE CONNECTIONS** *Compare the adaptations that enable plants and animals to respond to the challenges of living in hot and cold environments.* See Concepts 31.3 and 32.3.

 **ANIMATION** Visit the Study Area in **MasteringBiology** for the BioFlix® 3-D Animations on Water Transport in Plants (Chapter 29), Homeostasis: Regulating Blood Sugar (Chapter 33), and Gas Exchange (Chapter 34).

◀ **Make Connections Questions** ask students to relate content in the chapter to material presented earlier in the course.

# Practice Scientific Skills

**Scientific Skills Exercises** in every chapter use real data to build key skills needed for biology, including **data analysis**, **graphing**, **experimental design**, and **math skills**.

Each Scientific Skills ▶ Exercise is based on an experiment related to the chapter content.

### Scientific Skills Exercise

## Making and Testing Predictions

**Can Predation Result in Natural Selection for Color Patterns in Guppies?** What we know about evolution changes constantly as new observations lead to new hypotheses—and hence to new ways to test our understanding of evolutionary theory. Consider the wild guppies (*Poecilia reticulata*) that live in pools connected by streams on the Caribbean island of Trinidad. Male guppies have highly varied color patterns that are controlled by genes that are only expressed in adult males. Female guppies choose males with bright color patterns as mates more often than they choose males with drab coloring. But the bright colors that attract females also can make the males more conspicuous to predators. Researchers observed that in pools with few predator species, the benefits of bright colors appear to "win out," and males are more brightly colored than in pools where predation is more intense.

One guppy predator, the killifish, preys on juvenile guppies that have not yet displayed their adult coloration. Researchers predicted that if adult guppies with drab colors were transferred to a pool with only killifish, eventually the descendants of these guppies would be more brightly colored (because of the female preference for brightly colored males).

**How the Experiment Was Done** Researchers transplanted 200 guppies from pools containing pike-cichlid fish, intense predators of adult guppies, to pools containing killifish, less active predators that prey mainly on juvenile guppies. They tracked the number of bright-colored spots and the total area of those spots on male guppies in each generation.

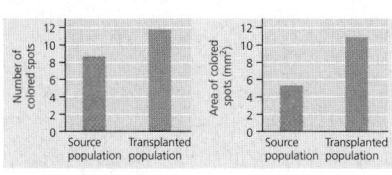

Guppies transplanted

Pools with pike-cichlids and guppies

Pools with killifish, but no guppies prior to transplant

**Data from the Experiment** After 22 months (15 generations), researchers compared the color pattern data for guppies from the source and transplanted populations.

**Data from** J. A. Endler, Natural selection on color patterns in *Poecilia reticulata*, *Evolution* 34:76–91 (1980).

#### INTERPRET THE DATA

1. Identify the following elements of hypothesis-based science in this example: (a) question, (b) hypothesis, (c) prediction, (d) control group, and (e) experimental group. (For additional information about hypothesis-based science, see Chapter 1 and the Scientific Skills Review in Appendix F and the Study Area of MasteringBiology.)
2. Explain how the types of data the researchers chose to collect enabled them to test their prediction.
3. What conclusion do you draw from the data presented above?
4. Predict what would happen if, after 22 months, guppies from the transplanted population were returned to the source pool. Describe an experiment to test your prediction.

(MB) A related version of this Scientific Skills Exercise can be assigned in MasteringBiology.

◀ Most Scientific Skills Exercises use **data from published research**, cited in the exercise.

◀ **Questions build in difficulty**, walking students through new skills step by step and providing opportunities for higher-level critical thinking.

## Every chapter has a Scientific Skills Exercise:

Each Scientific Skills Exercise from the text also has an **assignable, interactive tutorial version in MasteringBiology** that is automatically graded and includes coaching feedback.

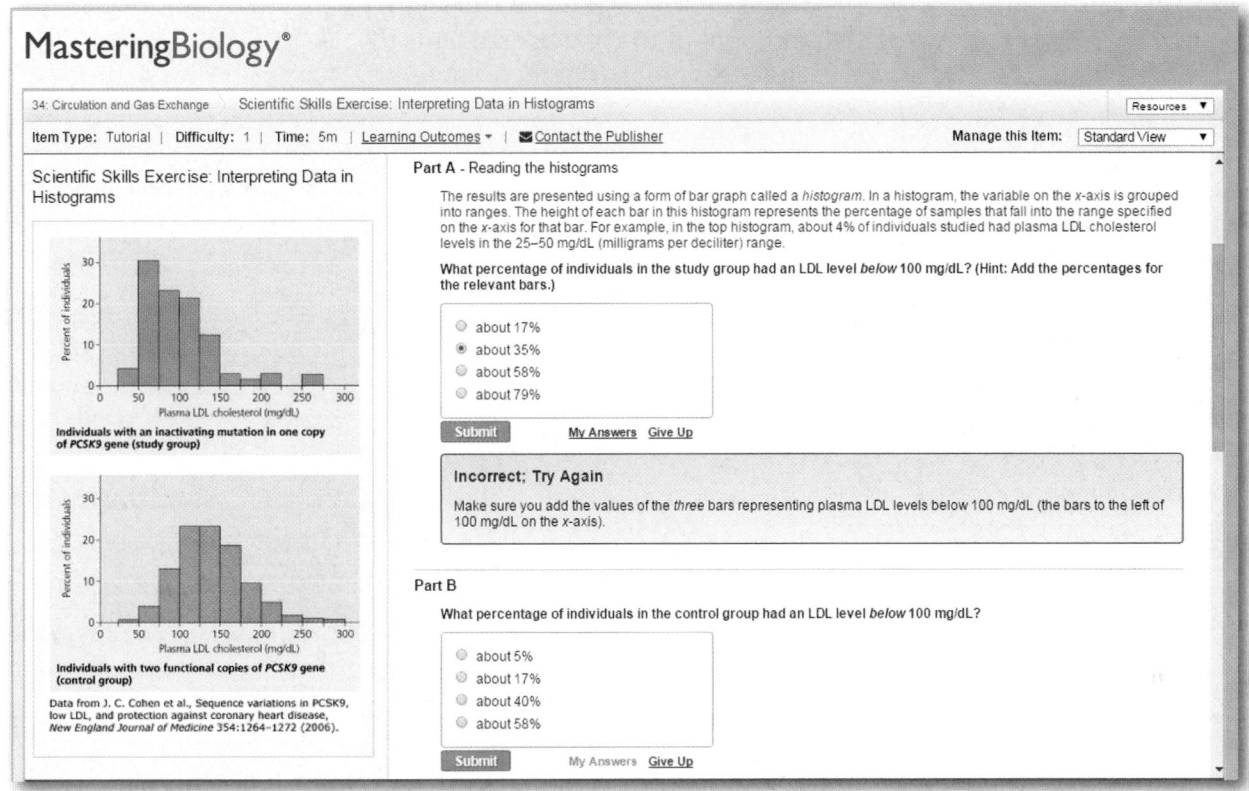

# MasteringBiology®

To learn more, visit www.masteringbiology.com

# Interpret Data

CAMPBELL *BIOLOGY IN FOCUS*, Second Edition, and MasteringBiology offer a wide variety of ways for students to move beyond memorization and to **think like a scientist.**

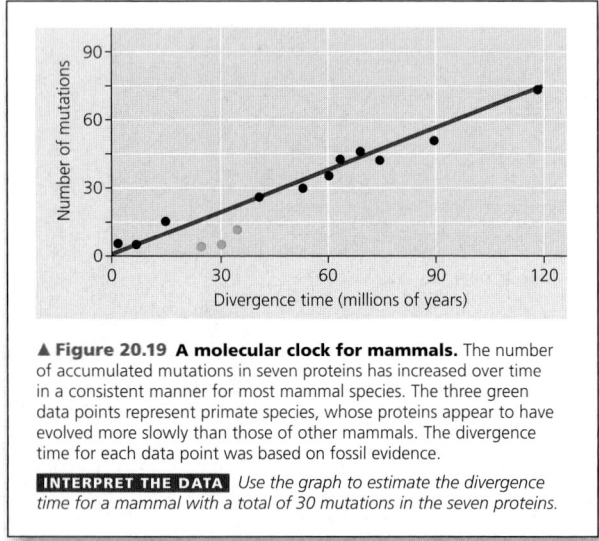

▲ **Figure 20.19 A molecular clock for mammals.** The number of accumulated mutations in seven proteins has increased over time in a consistent manner for most mammal species. The three green data points represent primate species, whose proteins appear to have evolved more slowly than those of other mammals. The divergence time for each data point was based on fossil evidence.

**INTERPRET THE DATA** *Use the graph to estimate the divergence time for a mammal with a total of 30 mutations in the seven proteins.*

◄ **NEW! Interpret the Data Questions** throughout the text ask students to analyze a graph, figure, or table.

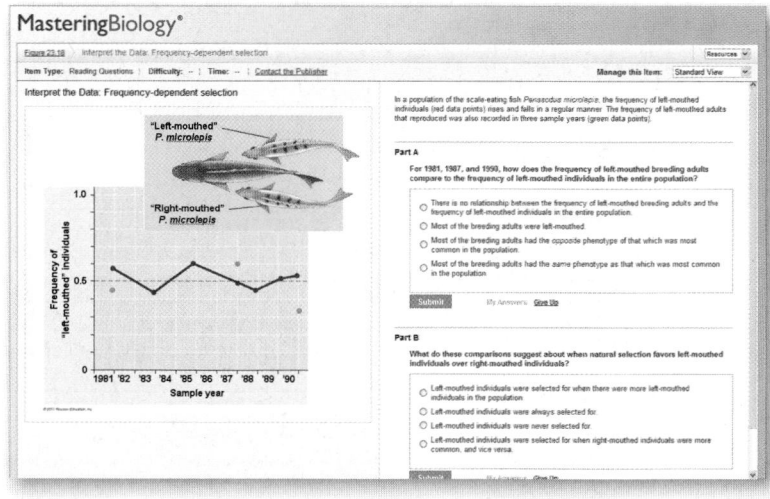

▲ **NEW! Every Interpret the Data** question from the text is **assignable in MasteringBiology.**

**MasteringBiology®**

**Learn more at**
**www.masteringbiology.com**

◄ **NEW! Solve It Tutorials** engage students in a multi-step investigation of a "mystery" or open question in which they must analyze real data. These are assignable in MasteringBiology. Topics include:

- Which Biofuel Has the Most Potential to Reduce Our Dependence on Fossil Fuels?

- Is It Possible to Treat Bacterial Infections Without Traditional Antibiotics?

- Which Insulin Mutations May Result in Disease?

- Are You Getting the Fish You Paid For?

- Why Are Honey Bees Vanishing?

- What Is Causing Episodes of Muscle Weakness in a Patient?

- How Can the Severity of Forest Fires Be Reduced?

# Keep Current with New Scientific Advances

**NEW!** The Second Edition incorporates **up-to-date content** on genomics, gene editing, human evolution, microbiomes, climate change, and more.

**NEW!** The Second Edition shows students how our ability to **sequence DNA and proteins rapidly and inexpensively** is transforming every subfield of biology, from cell biology to physiology to ecology. For instance, the examples in this figure from Chapter 3 are explored in greater depth later in the text.

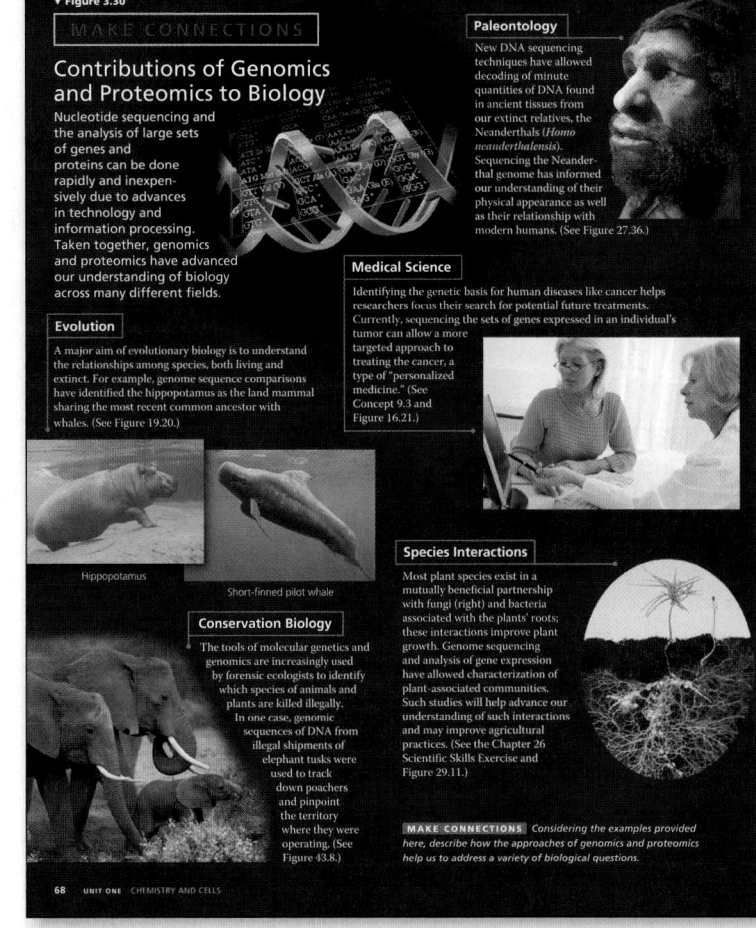

▼ Figure 3.30

MAKE CONNECTIONS

## Contributions of Genomics and Proteomics to Biology

Nucleotide sequencing and the analysis of large sets of genes and proteins can be done rapidly and inexpensively due to advances in technology and information processing. Taken together, genomics and proteomics have advanced our understanding of biology across many different fields.

**Paleontology**

New DNA sequencing techniques have allowed decoding of minute quantities of DNA found in ancient tissues from our extinct relatives, the Neanderthals (*Homo neanderthalensis*). Sequencing the Neanderthal genome has informed our understanding of their physical appearance as well as their relationship with modern humans. (See Figure 27.36.)

**Evolution**

A major aim of evolutionary biology is to understand the relationships among species, both living and extinct. For example, genome sequence comparisons have identified the hippopotamus as the land mammal sharing the most recent common ancestor with whales. (See Figure 19.20.)

**Medical Science**

Identifying the genetic basis for human diseases like cancer helps researchers focus their search for potential future treatments. Currently, sequencing the sets of genes expressed in an individual's tumor can allow a more targeted approach to treating the cancer, a type of "personalized medicine." (See Concept 9.3 and Figure 16.21.)

Hippopotamus

Short-finned pilot whale

**Conservation Biology**

The tools of molecular genetics and genomics are increasingly used by forensic ecologists to identify which species of animals and plants are killed illegally. In one case, genomic sequences of DNA from illegal shipments of elephant tusks were used to track down poachers and pinpoint the territory where they were operating. (See Figure 43.8.)

**Species Interactions**

Most plant species exist in a mutually beneficial partnership with fungi (right) and bacteria associated with the plants' roots; these interactions improve plant growth. Genome sequencing and analysis of gene expression have allowed characterization of plant-associated communities. Such studies will help advance our understanding of such interactions and may improve agricultural practices. (See the Chapter 26 Scientific Skills Exercise and Figure 29.11.)

MAKE CONNECTIONS *Considering the examples provided here, describe how the approaches of genomics and proteomics help us to address a variety of biological questions.*

68    UNIT ONE   CHEMISTRY AND CELLS

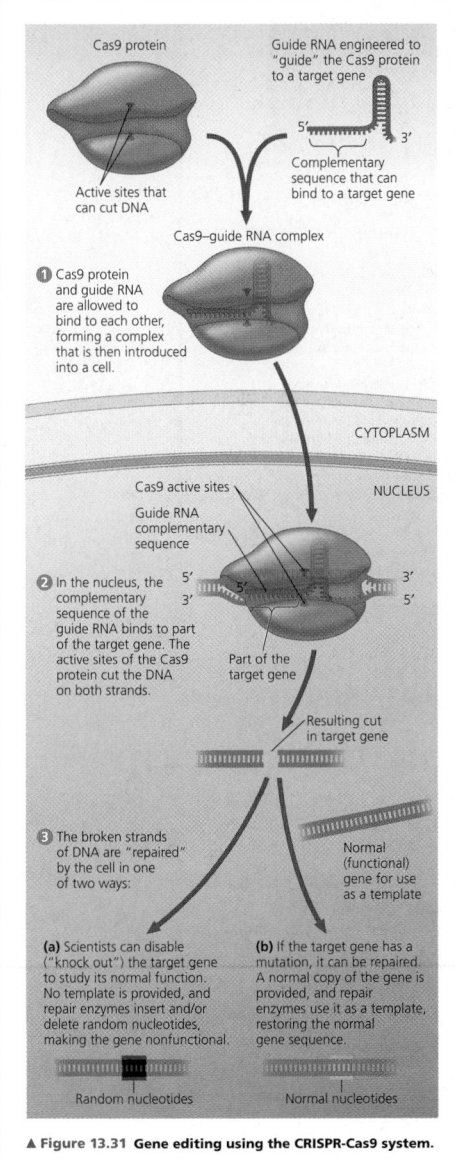

Cas9 protein

Guide RNA engineered to "guide" the Cas9 protein to a target gene

Active sites that can cut DNA

5′   3′

Complementary sequence that can bind to a target gene

Cas9–guide RNA complex

❶ Cas9 protein and guide RNA are allowed to bind to each other, forming a complex that is then introduced into a cell.

CYTOPLASM

Cas9 active sites

NUCLEUS

Guide RNA complementary sequence

❷ In the nucleus, the complementary sequence of the guide RNA binds to part of the target gene. The active sites of the Cas9 protein cut the DNA on both strands.

5′   3′
3′   5′

Part of the target gene

Resulting cut in target gene

❸ The broken strands of DNA are "repaired" by the cell in one of two ways:

Normal (functional) gene for use as a template

**(a)** Scientists can disable ("knock out") the target gene to study its normal function. No template is provided, and repair enzymes insert and/or delete random nucleotides, making the gene nonfunctional.

Random nucleotides

**(b)** If the target gene has a mutation, it can be repaired. A normal copy of the gene is provided, and repair enzymes use it as a template, restoring the normal gene sequence.

Normal nucleotides

▲ Figure 13.31 **Gene editing using the CRISPR-Cas9 system.**

◀ **NEW!** Chapter 13 describes **gene editing using the CRISPR-Cas9 system**, and Chapter 17 describes the basic biology of this system in bacteria.

**NEW!** Chapter 27 includes new material on human origins, including how sequencing DNA extracted from this fossil jawbone recently revealed **evidence of human-Neanderthal interbreeding.** ▶

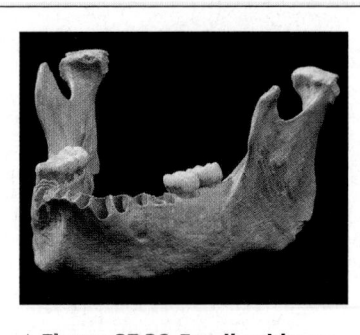

▲ **Figure 27.36 Fossil evidence of human-Neanderthal interbreeding.**

# Focus on the Key Concepts

Each chapter is organized around a framework of **3 to 6 Key Concepts** that focus on the big picture and provide a context for the supporting details.

**The list of Key Concepts ▶ introduces the big ideas covered in the chapter.**

CHAPTER

## 14 Gene Expression: From Gene to Protein

**KEY CONCEPTS**

14.1 Genes specify proteins via transcription and translation

14.2 Transcription is the DNA-directed synthesis of RNA: *a closer look*

14.3 Eukaryotic cells modify RNA after transcription

14.4 Translation is the RNA-directed synthesis of a polypeptide: *a closer look*

14.5 Mutations of one or a few nucleotides can affect protein structure and function

▲ **Figure 14.1** How does a single faulty gene result in the dramatic appearance of an albino donkey?

**Every chapter opens with a visually dynamic photo accompanied by an intriguing question that invites students into the chapter.**

### The Flow of Genetic Information

The island of Asinara lies off the coast of Sardinia, an Italian island. The name Asinara probably originated from the Latin word *sinuaria*, which means "sinus-shaped." A second meaning of Asinara is "donkey-inhabited," which is particularly appropriate because Asinara is home to a wild population of albino donkeys **(Figure 14.1)**. The donkeys were brought to Asinara in the early 1800s and abandoned there in 1885 when the 500 residents were forced to leave the island so it could be used as a penal colony. What is responsible for the phenotype of the albino donkey, strikingly different from its pigmented relative?

Inherited traits are determined by genes, and the trait of albinism is caused by a recessive allele of a pigmentation gene (see Concept 11.4). The information content of genes is in the form of specific sequences of nucleotides along strands of DNA, the genetic material. But how does this information determine an organism's traits? Put another way, what does a gene actually say? And how is its message

translated by cells into a specific trait, such as brown hair, type A blood, or, in the case of an albino donkey, a total lack of pigment? The albino donkey has a faulty version of a key protein, an enzyme required for pigment synthesis, and this protein is faulty because the gene that codes for it contains incorrect information.

This example illustrates the main point of this chapter: The DNA inherited by an organism leads to specific traits by dictating the synthesis of proteins and of RNA molecules involved in protein synthesis. In other words, proteins are the link between genotype and phenotype. **Gene expression** is the process by which DNA directs the synthesis of proteins (or, in some cases, just RNAs). The expression of genes that code for proteins includes two stages: transcription and translation. This chapter describes the flow of information from gene to protein and explains how genetic mutations affect organisms through their proteins. Understanding the processes of gene expression, which are similar in all three domains of life, will allow us to revisit the concept of the gene in more detail at the end of the chapter.

278

**After reading a Key Concept section, ▶ students can check their understanding using the Concept Check questions:**

**Make Connections questions** ask students to relate content in the chapter to material presented earlier in the course.

**What if? questions** ask students to apply what they've learned.

**Draw It Exercises** ask students to put pencil to paper and draw a structure, annotate a figure, or graph experimental data.

---

**CONCEPT CHECK 14.5**

1. What happens when one nucleotide pair is lost from the middle of the coding sequence of a gene?

2. **MAKE CONNECTIONS** Individuals heterozygous for the sickle-cell allele show effects of the allele under some circumstances (see Concept 11.4). Explain in terms of gene expression.

3. **WHAT IF?** **DRAW IT** The template strand of a gene includes this sequence: 3'-TACTTGTCCGATATC-5'. It is mutated to 3'-TACTTGTCCAATATC-5'. For both versions, draw the DNA, the mRNA, and the encoded amino acid sequence. What is the effect on the amino acid sequence?

For suggested answers, see Appendix A.

The **Summary of Key Concepts** refocuses students on the main points of the chapter.

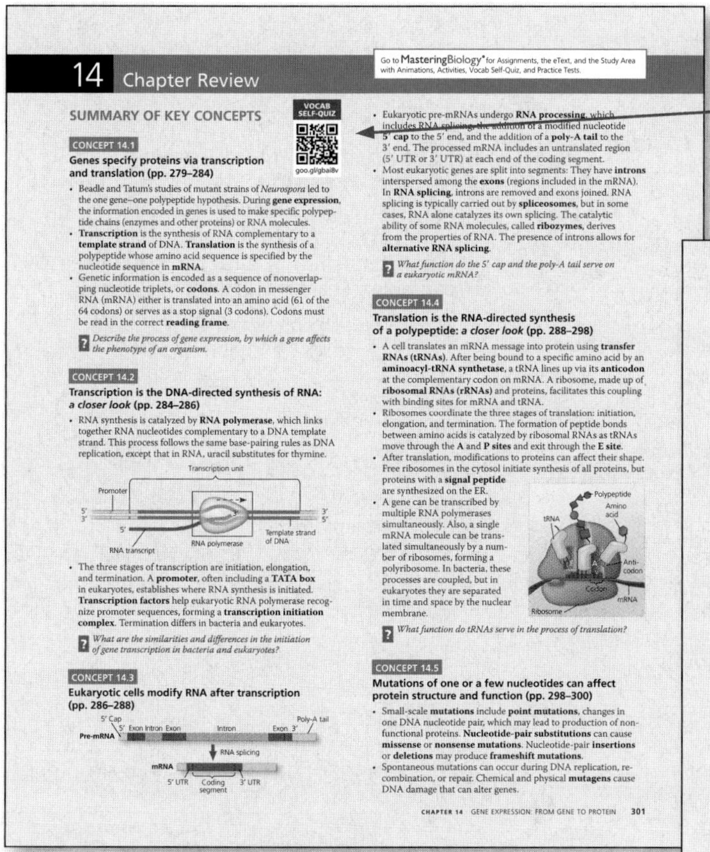

14 Chapter Review

Go to MasteringBiology® for Assignments, the eText, and the Study Area with Animations, Activities, Vocab Self-Quiz, and Practice Tests.

SUMMARY OF KEY CONCEPTS

VOCAB SELF-QUIZ goo.gl/gbai8v

CONCEPT 14.1

**Genes specify proteins via transcription and translation (pp. 279–284)**

- Beadle and Tatum's studies of mutant strains of *Neurospora* led to the one gene–one polypeptide hypothesis. During **gene expression**, the information encoded in genes is used to make specific polypeptide chains (enzymes and other proteins) or RNA molecules.
- **Transcription** is the synthesis of RNA complementary to a **template strand** of DNA. **Translation** is the synthesis of a polypeptide whose amino acid sequence is specified by the nucleotide sequence in **mRNA**.
- Genetic information is encoded as a sequence of nonoverlapping nucleotide triplets, or **codons**. A codon in messenger RNA (mRNA) either is translated into an amino acid (61 of the 64 codons) or serves as a stop signal (3 codons). Codons must be read in the correct **reading frame**.

? *Describe the process of gene expression, by which a gene affects the phenotype of an organism.*

CONCEPT 14.2

**Transcription is the DNA-directed synthesis of RNA: *a closer look* (pp. 284–286)**

- RNA synthesis is catalyzed by **RNA polymerase**, which links together RNA nucleotides complementary to a DNA template strand. This process follows the same base-pairing rules as DNA replication, except that in RNA, uracil substitutes for thymine.


Transcription unit
Promoter
5′
3′
5′
3′
5′
3′
Template strand of DNA
RNA polymerase
RNA transcript


- The three stages of transcription are initiation, elongation, and termination. A **promoter**, often including a **TATA box** in eukaryotes, establishes where RNA synthesis is initiated. **Transcription factors** help eukaryotic RNA polymerase recognize promoter sequences, forming a **transcription initiation complex**. Termination differs in bacteria and eukaryotes.

? *What are the similarities and differences in the initiation of gene transcription in bacteria and eukaryotes?*

CONCEPT 14.3

**Eukaryotic cells modify RNA after transcription (pp. 286–288)**


5′ Cap
Poly-A tail
5′ Exon Intron Exon
Intron
Exon 3′
Pre-mRNA
RNA splicing
mRNA
5′ UTR
Coding segment
3′ UTR


- Eukaryotic pre-mRNAs undergo **RNA processing**, which includes RNA splicing, the addition of a modified nucleotide **5′ cap** to the 5′ end, and the addition of a **poly-A tail** to the 3′ end. The processed mRNA includes an untranslated region (5′ UTR or 3′ UTR) at each end of the coding segment.
- Most eukaryotic genes are split into segments: They have **introns** interspersed among the **exons** (regions included in the mRNA). In **RNA splicing**, introns are removed and exons joined. RNA splicing is typically carried out by **spliceosomes**, but in some cases, RNA alone catalyzes its own splicing. The catalytic ability of some RNA molecules, called **ribozymes**, derives from the properties of RNA. The presence of introns allows for **alternative RNA splicing**.

? *What function do the 5′ cap and the poly-A tail serve on a eukaryotic mRNA?*

CONCEPT 14.4

**Translation is the RNA-directed synthesis of a polypeptide: *a closer look* (pp. 288–298)**

- A cell translates an mRNA message into protein using **transfer RNAs (tRNAs)**. After being bound to a specific amino acid by an **aminoacyl-tRNA synthetase**, a tRNA lines up via its **anticodon** at the complementary codon on mRNA. A ribosome, made up of **ribosomal RNAs (rRNAs)** and proteins, facilitates this coupling with binding sites for mRNA and tRNA.
- Ribosomes coordinate the three stages of translation: initiation, elongation, and termination. The formation of peptide bonds between amino acids is catalyzed by ribosomal RNAs as tRNAs move through the **A** and **P sites** and exit through the **E site**.
- After translation, modifications to proteins can affect their shape. Free ribosomes in the cytosol initiate synthesis of all proteins, but proteins with a **signal peptide** are synthesized on the ER.
- A gene can be transcribed by multiple RNA polymerases simultaneously. Also, a single mRNA molecule can be translated simultaneously by a number of ribosomes, forming a polyribosome. In bacteria, these processes are coupled, but in eukaryotes they are separated in time and space by the nuclear membrane.


Polypeptide
Amino acid
tRNA
Anticodon
Codon
mRNA
Ribosome


? *What function do tRNAs serve in the process of translation?*

CONCEPT 14.5

**Mutations of one or a few nucleotides can affect protein structure and function (pp. 298–300)**

- Small-scale **mutations** include **point mutations**, changes in one DNA nucleotide pair, which may lead to production of nonfunctional proteins. **Nucleotide-pair substitutions** can cause **missense** or **nonsense mutations**. Nucleotide-pair **insertions** or **deletions** may produce **frameshift mutations**.
- Spontaneous mutations can occur during DNA replication, recombination, or repair. Chemical and physical **mutagens** cause DNA damage that can alter genes.

CHAPTER 14 GENE EXPRESSION: FROM GENE TO PROTEIN 301

VOCAB SELF-QUIZ

goo.gl/gbai8v

**NEW!** QR Codes and URLs at the end of every chapter give students quick access to **Vocabulary Self-Quizzes** and **Practice Tests** on their smartphones, tablets, and computers.

PRACTICE TEST

goo.gl/CRZjvS

- **Summary of Key Concepts questions** check students' understanding of a key idea from each concept.
- **Summary figures** recap key information visually.

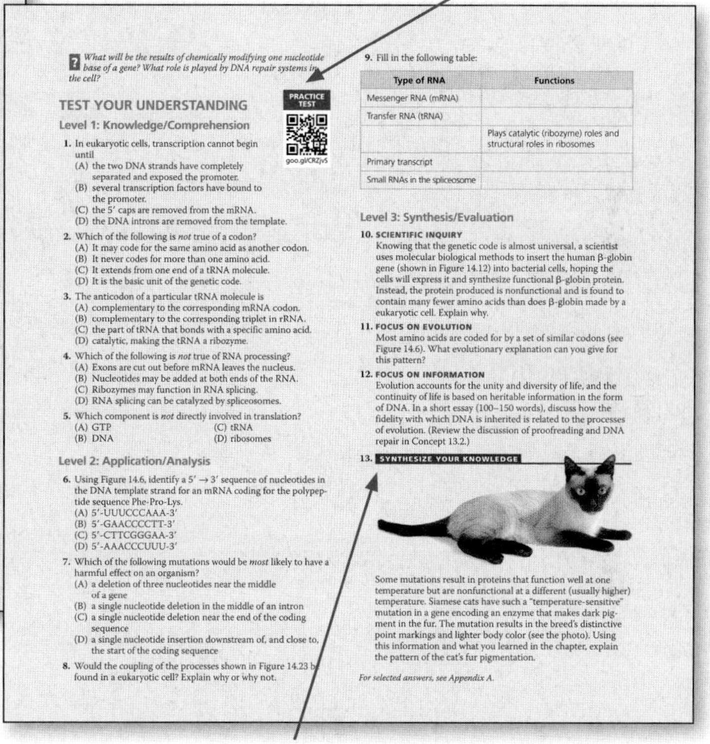

? *What will be the results of chemically modifying one nucleotide base of a gene? What role is played by DNA repair systems in the cell?*

TEST YOUR UNDERSTANDING

PRACTICE TEST goo.gl/CRZjvS

Level 1: Knowledge/Comprehension

1. In eukaryotic cells, transcription cannot begin until
   (A) the two DNA strands have completely separated and exposed the promoter.
   (B) several transcription factors have bound to the promoter.
   (C) the 5′ caps are removed from the mRNA.
   (D) the DNA introns are removed from the template.
2. Which of the following is *not* true of a codon?
   (A) It may code for the same amino acid as another codon.
   (B) It never codes for more than one amino acid.
   (C) It extends from one end of a tRNA molecule.
   (D) It is the basic unit of the genetic code.
3. The anticodon of a particular tRNA molecule is
   (A) complementary to the corresponding mRNA codon.
   (B) complementary to the corresponding triplet in rRNA.
   (C) the part of tRNA that bonds with a specific amino acid.
   (D) catalytic, making the tRNA a ribozyme.
4. Which of the following is *not* true of RNA processing?
   (A) Exons are cut out before mRNA leaves the nucleus.
   (B) Nucleotides may be added at both ends of the RNA.
   (C) Ribozymes may function in RNA splicing.
   (D) RNA splicing can be catalyzed by spliceosomes.
5. Which component is *not* directly involved in translation?
   (A) GTP (C) tRNA
   (B) DNA (D) ribosomes

Level 2: Application/Analysis

6. Using Figure 14.6, identify a 5′ → 3′ sequence of nucleotides in the DNA template strand for an mRNA coding for the polypeptide sequence Phe-Pro-Lys.
   (A) 5′-UUUCCCAAA-3′
   (B) 5′-GAACCCCTT-3′
   (C) 5′-CTTCGGGAA-3′
   (D) 5′-AAACCCUUU-3′
7. Which of the following mutations would be *most* likely to have a harmful effect on an organism?
   (A) a deletion of three nucleotides near the middle of a gene
   (B) a single nucleotide deletion in the middle of an intron
   (C) a single nucleotide deletion near the end of the coding sequence
   (D) a single nucleotide insertion downstream of, and close to, the start of the coding sequence
8. Would the coupling of the processes shown in Figure 14.23 be found in a eukaryotic cell? Explain why or why not.
9. Fill in the following table:

| Type of RNA | Functions |
|---|---|
| Messenger RNA (mRNA) | |
| Transfer RNA (tRNA) | |
| | Plays catalytic (ribozyme) roles and structural roles in ribosomes |
| Primary transcript | |
| Small RNAs in the spliceosome | |

Level 3: Synthesis/Evaluation

10. SCIENTIFIC INQUIRY
    Knowing that the genetic code is almost universal, a scientist uses molecular biological methods to insert the human β-globin gene (shown in Figure 14.12) into bacterial cells, hoping the cells will express it and synthesize functional β-globin protein. Instead, the protein produced is nonfunctional and is found to contain many fewer amino acids than does β-globin made by a eukaryotic cell. Explain why.
11. FOCUS ON EVOLUTION
    Most amino acids are coded for by a set of similar codons (see Figure 14.6). What evolutionary explanation can you give for this pattern?
12. FOCUS ON INFORMATION
    Evolution accounts for the unity and diversity of life, and the continuity of life is based on heritable information in the form of DNA. In a short essay (100–150 words), discuss how the fidelity with which DNA is inherited is related to the processes of evolution. (Review the discussion of proofreading and DNA repair in Concept 13.2.)
13. SYNTHESIZE YOUR KNOWLEDGE

    Some mutations result in proteins that function well at one temperature but are nonfunctional at a different (usually higher) temperature. Siamese cats have such a "temperature-sensitive" mutation in a gene encoding an enzyme that makes dark pigment in the fur. The mutation results in the breed's distinctive point markings and lighter body color (see the photo). Using this information and what you learned in the chapter, explain the pattern of the cat's fur pigmentation.

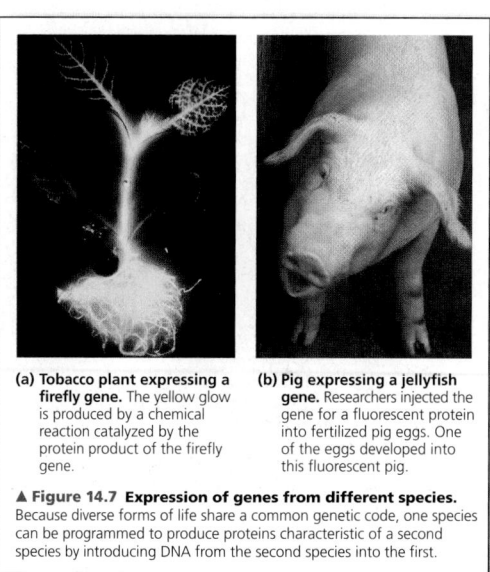

*For selected answers, see Appendix A.*

**NEW! Synthesize Your Knowledge questions** ask students to apply their understanding of the chapter content to explain an intriguing photo.

## **Evolution**, the fundamental theme of biology, is emphasized throughout. For example:

- Every Chapter Review includes a "Focus on Evolution" question (shown above right).
- Every chapter has a section explicitly relating the chapter content to evolution (shown at right).

### *Evolution of the Genetic Code*

EVOLUTION The genetic code is nearly universal, shared by organisms from the simplest bacteria to the most complex plants and animals. The mRNA codon CCG, for instance, is translated as the amino acid proline in all organisms whose genetic code has been examined. In laboratory experiments, genes can be transcribed and translated after being transplanted from one species to another, sometimes with quite striking results, as shown in **Figure 14.7**. Bacteria can be programmed by the insertion of human genes to synthesize certain human proteins for medical use, such as insulin. Such applications have produced many exciting developments in the area of genetic engineering (see Concept 13.4).

Despite a small number of exceptions in which a few codons differ from the standard ones, the evolutionary significance of the code's *near* universality is clear. A language shared by all living things must have been operating very early in the

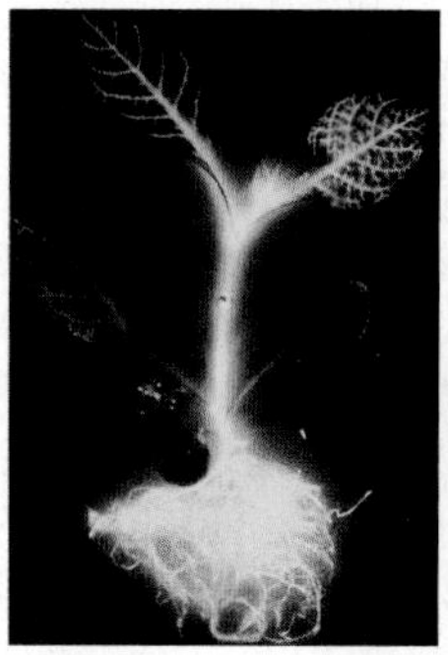

**(a) Tobacco plant expressing a firefly gene.** The yellow glow is produced by a chemical reaction catalyzed by the protein product of the firefly gene.

**(b) Pig expressing a jellyfish gene.** Researchers injected the gene for a fluorescent protein into fertilized pig eggs. One of the eggs developed into this fluorescent pig.

▲ **Figure 14.7 Expression of genes from different species.** Because diverse forms of life share a common genetic code, one species can be programmed to produce proteins characteristic of a second species by introducing DNA from the second species into the first.

# NEW! Ready-to-Go Teaching Modules

**Ready-to-Go Teaching Modules** help instructors efficiently make use of the best teaching tools before, during, and after class.

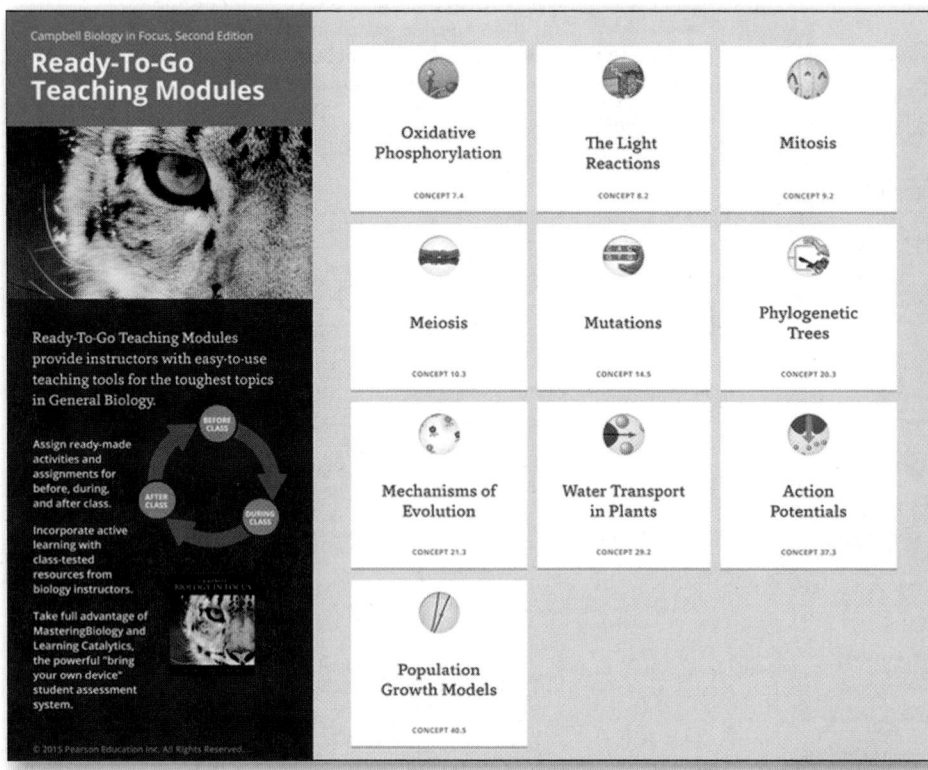

Campbell Biology in Focus, Second Edition

**Ready-To-Go Teaching Modules**

Ready-To-Go Teaching Modules provide instructors with easy-to-use teaching tools for the toughest topics in General Biology.

Assign ready-made activities and assignments for before, during, and after class.

Incorporate active learning with class-tested resources from biology instructors.

Take full advantage of MasteringBiology and Learning Catalytics, the powerful "bring your own device" student assessment system.

© 2015 Pearson Education Inc. All Rights Reserved.

| | | |
|---|---|---|
| Oxidative Phosphorylation — CONCEPT 7.4 | The Light Reactions — CONCEPT 8.2 | Mitosis — CONCEPT 9.2 |
| Meiosis — CONCEPT 10.3 | Mutations — CONCEPT 14.5 | Phylogenetic Trees — CONCEPT 20.3 |
| Mechanisms of Evolution — CONCEPT 21.3 | Water Transport in Plants — CONCEPT 29.2 | Action Potentials — CONCEPT 37.3 |
| Population Growth Models — CONCEPT 40.5 | | |

◀ **NEW!** The **Ready-to-Go Teaching Modules** incorporate the best that the text, MasteringBiology, and Learning Catalytics have to offer, along with new ideas for in-class activities. The modules can be accessed through the Instructor Resources area of MasteringBiology.

Instructors can easily incorporate **active learning** into their courses using suggested activity ideas and questions for use with classroom response systems, including Learning Catalytics. ▶

▼ **Learning Catalytics™** allows students to use their smartphone, tablet, or laptop to respond to questions in class. Visit www.learningcatalytics.com.

The following conditions were detected in a mutant cell:

The cell is running out of ATP, while ADP is building up to very high levels. NADH is building up to very high levels, while the level of NAD⁺ is becoming very low.

The amount of protons in the intermembrane space and in the matrix is becoming more equal (the strength of the proton gradient is decreasing/wea

Use this information to predict which stage of cellular respiration is not functioning normally in this mutant cell.

# More Instructor Resources

**Instructor's Resource DVD (IRDVD) Package**
0134100085 / 9780134100081
The instructor resources for *Campbell BIOLOGY IN FOCUS*, **Second Edition,** are combined into one chapter-by-chapter resource that includes DVDs of all chapter visual resources. Assets include:

- Editable figures (art and photos) and tables from the text in PowerPoint®
- Prepared PowerPoint® Lecture Presentations for each chapter, with lecture notes, editable figures, tables, and links to animations and videos
- 250+ Instructor Animations and Videos, including BioFlix® 3-D Animations and *ABC News* Videos
- JPEG Images, including labeled and unlabeled art, photos from the text, and extra photos
- Digital Transparencies
- Clicker Questions in PowerPoint®
- Quick Reference Guide
- Test Bank questions in TestGen® software and Microsoft® Word

**Instructor Resources Area in MasteringBiology**
This area includes:

- Ready-to-Go Teaching Modules
- Art and Photos in PowerPoint®
- Lecture Presentations in PowerPoint®
- Animations and Videos, including BioFlix®
- JPEG Images
- Digital Transparencies
- Clicker Questions
- IRDVD Quick Reference Guide
- Test Bank Files
- Learning Objectives
- Instructor Answers for Scientific Skills Exercises, Interpret the Data Questions, and Short-Answer Essay Questions; also includes a rubric and tips for grading short-answer essays
- Instructor Guides for Supplements
- Video Field Trips

▲ **Customizable PowerPoints®** provide a jumpstart for each lecture.

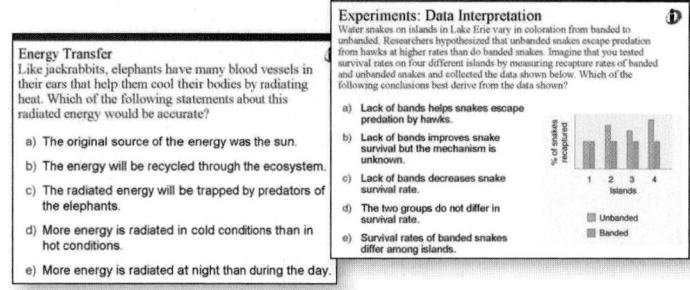

▲ **All of the art, graphs, and photos from the book are provided with customizable labels.** More than 1,600 photos from the text and other sources are included.

▲ **Clicker Questions** can be used to stimulate effective classroom discussions (for use with or without clickers).

---

**Test Bank**
0134100468 / 9780134100463
This invaluable resource contains more than 3,000 questions, including scenario-based questions and art, graph, and data interpretation questions. The Test Bank is available electronically in MasteringBiology and on the Instructor's Resource DVD Package.

**Course Management Systems**
Content is available in **Modified MasteringBiology,** which offers the usual Mastering features plus:

- Integration with other learning management systems using single sign-on
- Temporary access (grace period)
- Discussion boards
- Email
- Chat and class live (synchronous whiteboard presentation)
- Submissions (Dropbox)

# Personalized Coaching in MasteringBiology®

Instructors can assign **self-paced MasteringBiology® tutorials** that provide students with individualized coaching with specific **hints and feedback** on the toughest topics in the course.

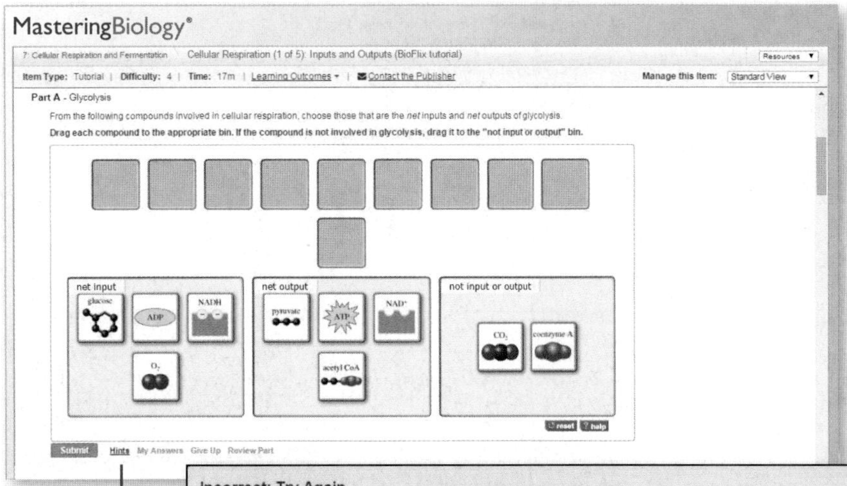

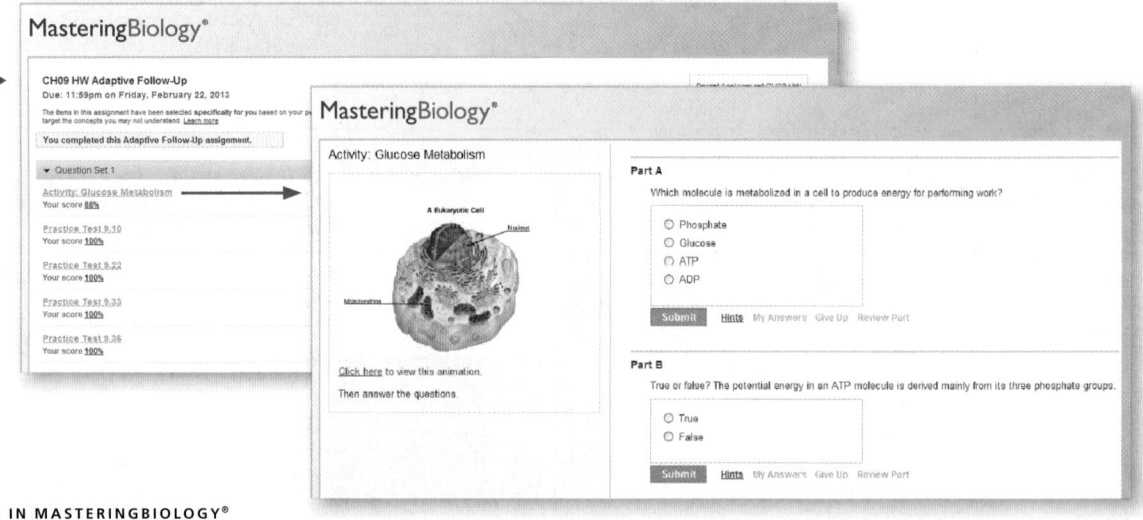

## MasteringBiology®

**Learn more at**
**www.masteringbiology.com**

1. If a student gets stuck...

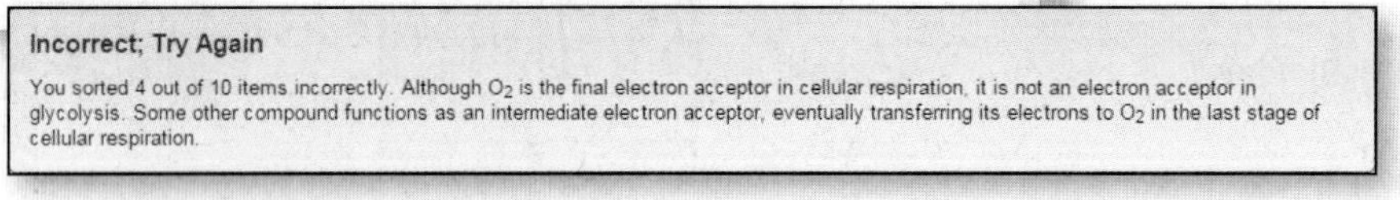

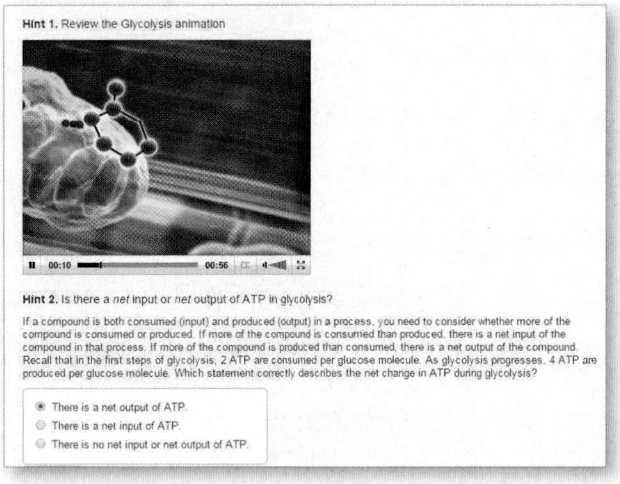

2. specific wrong-answer **feedback** appears in the purple feedback box.

3. **Hints** coach the student to the correct response.

4. Optional **Adaptive Follow-Up Assignments** are based on each student's performance on the original homework assignment and provide additional coaching and practice as needed.

Question sets in the Adaptive Follow-Up Assignments **continuously adapt** to each student's needs, making efficient use of study time.

▼ MasteringBiology® offers a wide variety of **tutorials** that can be assigned as homework. Some examples are shown below.

**BioFlix® Tutorials** use 3-D, movie-quality animations and **coaching exercises** to help students master tough topics outside of class. Animations can also be shown in class.

**NEW! HHMI Short Films**, documentary-quality movies from the Howard Hughes Medical Institute, engage students in topics from the discovery of the double helix to evolution, with assignable questions.

**Video Field Trips** allow students to study ecology by taking virtual field trips and answering follow-up questions.

▼ The **MasteringBiology® Gradebook** provides instructors with quick results and easy-to-interpret insights into student performance. Every assignment is automatically graded. Shades of red highlight vulnerable students and challenging assignments.

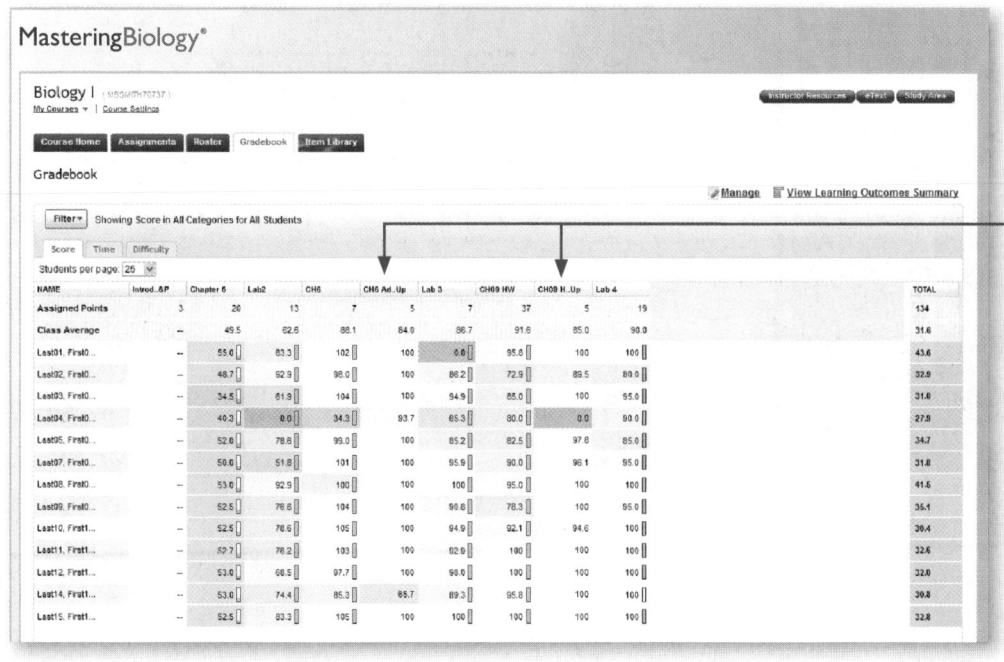

**Student scores** on the optional **Adaptive Follow-Up Assignments** are recorded in the gradebook and offer additional diagnostic information for instructors to monitor learning outcomes and more.

# Self-Study Tools in MasteringBiology®

## eTEXT

**Access the complete textbook online!**

▲ The Pearson eText gives students access to an electronic version of the text. The eText includes powerful interactive and customization functions, such as instructor and student note-taking, highlighting, bookmarking, search, and links to glossary terms, media, and quizzes.

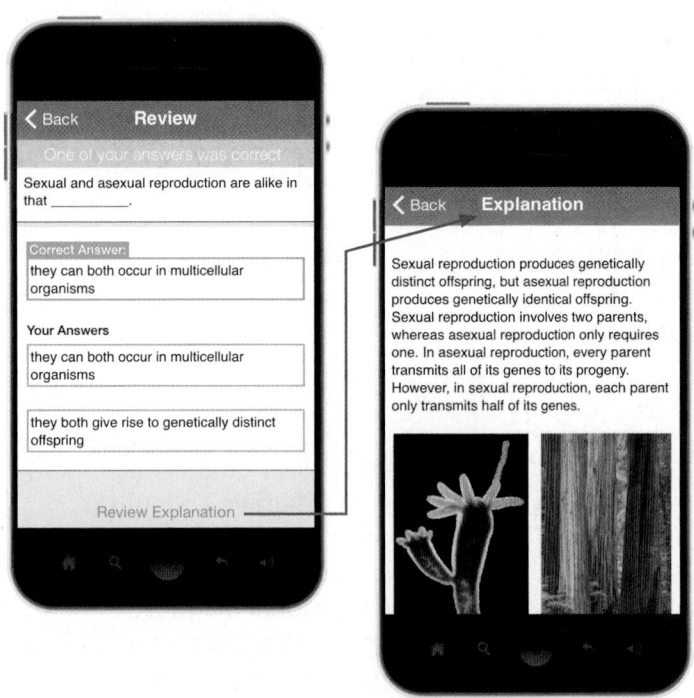

▲ **NEW! Dynamic Study Modules** help students acquire, retain, and recall information faster and more efficiently than ever before. The modules are available as a self-study tool or can be assigned by the instructor.

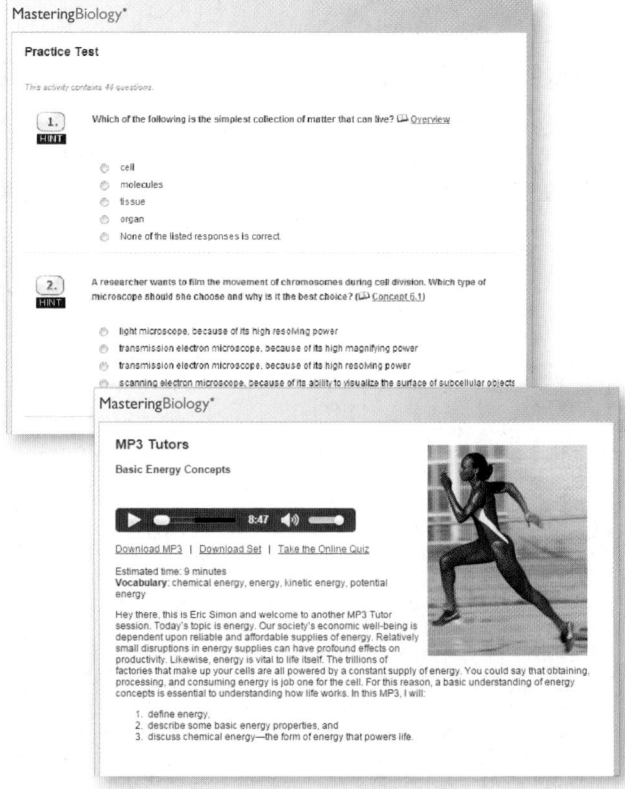

▲ **Study Area**
The Study Area includes:
- Practice Tests
- MP3 Tutors
- Animations
- Videos
- Activities
  and much more!

# Student and Lab Supplements

## STUDENT SUPPLEMENTS

*Inquiry in Action: Interpreting Scientific Papers*, Third Edition*
by Ruth Buskirk, *University of Texas at Austin*,
and Christopher M. Gillen, *Kenyon College*
9780321834171 / 0321834178
This guide helps students learn how to read and understand primary research articles. Part A presents complete articles accompanied by questions that help students analyze the article. Related Inquiry Figures are included in the supplement. Part B covers every part of a research paper, explaining the aim of the sections and how the paper works as a whole.

*Practicing Biology: A Student Workbook*, Fifth Edition*
by Jean Heitz and Cynthia Giffen, *University of Wisconsin, Madison*
9780321877055 / 0321877055
This workbook offers a variety of activities to suit different learning styles. Activities such as modeling and concept mapping allow students to visualize and understand biological processes. Other activities focus on basic skills, such as reading and drawing graphs.

*Biological Inquiry: A Workbook of Investigative Cases*,
Fourth Edition*
by Margaret Waterman, *Southeast Missouri State University*, and Ethel Stanley, *BioQUEST Curriculum Consortium and Beloit College*
9780321833914 / 0321833910
This workbook offers ten investigative cases. Each case study requires students to synthesize information from multiple chapters of the text and apply that knowledge to a real-world scenario as they pose hypotheses, gather new information, analyze evidence, graph data, and draw conclusions. A link to a student website is in the Study Area in MasteringBiology.

*Study Card*
9780321834157 / 0321834151
This quick-reference card provides students with an overview of the entire field of biology, helping them see the connections among topics.

*Spanish Glossary*
by Laura P. Zanello, *University of California, Riverside*
9780321834980 / 0321834984
This resource provides definitions in Spanish for glossary terms.

*Into the Jungle: Great Adventures in the Search for Evolution*
by Sean B. Carroll, *University of Wisconsin, Madison*
9780321556714 / 0321556712
These nine short tales vividly depict key discoveries in evolutionary biology and the excitement of the scientific process. Online resources are available at www.aw-bc.com/carroll.

*Get Ready for Biology*
9780321500571 / 0321500571
This engaging workbook helps students brush up on important math and study skills and get up to speed on biological terminology and the basics of chemistry and cell biology. *Get Ready for Biology* is also available in the Study Area of MasteringBiology.

*A Short Guide to Writing About Biology*, Ninth Edition
by Jan A. Pechenik, *Tufts University*
9780134143736 / 0134143736
This best-selling writing guide teaches students to think as biologists and to express ideas clearly and concisely through their writing.

*An Introduction to Chemistry for Biology Students*, Ninth Edition
by George I. Sackheim, *University of Illinois, Chicago*
9780805395716 / 0805395717
This text/workbook helps students review and master all the basic facts, concepts, and terminology of chemistry that they need for their life science course.

*An Instructor Guide is available for download in the Instructor Resources Area in MasteringBiology.

---

## LAB SUPPLEMENTS

*Investigating Biology Laboratory Manual*, Eighth Edition
by Judith Giles Morgan, *Emory University*, and M. Eloise Brown Carter, *Oxford College of Emory University*
9780321838995 / 0321838998

Now in full color! With its distinctive investigative approach to learning, this best-selling laboratory manual is now more engaging than ever, with full-color art and photos throughout. As always, the lab manual encourages students to participate in the process of science and develop creative and critical-reasoning skills.

The Eighth Edition includes major revisions that reflect new molecular evidence and the current understanding of phylogenetic relationships for plants, invertebrates, protists, and fungi. A new lab topic, "Fungi," has been added, providing expanded coverage of the major fungi groups. The "Protists" lab topic has been revised and expanded with additional examples of all the major clades. In the new edition, population genetics is covered in one lab topic with new problems and examples that connect ecology, evolution, and genetics.

*Annotated Instructor Edition for Investigating Biology Laboratory Manual*, Eighth Edition
by Judith Giles Morgan, *Emory University*, and M. Eloise Brown Carter, *Oxford College of Emory University*
9780321834973 / 0321834976

*Preparation Guide for Investigating Biology Laboratory Manual*,
Eighth Edition
by Judith Giles Morgan, *Emory University*, and M. Eloise Brown Carter, *Oxford College of Emory University*
9780321834454 / 0321834453
Available for download in the Instructor Resources Area in MasteringBiology.

**The Pearson Custom Laboratory Program for the Biological Sciences**
www.pearsonlearningsolutions.com/custom-library/

**MasteringBiology® Virtual Labs**
www.masteringbiology.com
This online environment promotes critical thinking skills using virtual experiments and explorations that may be difficult to perform in a wet lab environment due to time, cost, or safety concerns. Designed to supplement or substitute for existing wet labs, this product offers students unique learning experiences and critical thinking exercises in the areas of microscopy, molecular biology, genetics, ecology, and systematics.

# Featured Figures

*The Inquiry Figure, original research paper, and a worksheet to guide you through the paper are provided in *Inquiry in Action: Interpreting Scientific Papers*, Second Edition.

†A related Experimental Inquiry Tutorial can be assigned in MasteringBiology.

# Acknowledgments

The authors wish to express their gratitude to the global community of instructors, researchers, students, and publishing professionals who have contributed to the Second Edition of **CAMPBELL BIOLOGY IN FOCUS**.

As authors of this text, we are mindful of the daunting challenge of keeping up to date in all areas of our rapidly expanding subject. We are grateful to the many scientists who helped shape this text by discussing their research fields with us, answering specific questions in their areas of expertise, and sharing their ideas about biology education. We are especially grateful to the following, listed alphabetically: Monika Abedin, John Archibald, Kristian Axelsen, Daniel Boyce, Nick Butterfield, Jean DeSaix, Rachel Kramer Green, Eileen Gregory, Hopi Hoekstra, Fred Holtzclaw, Theresa Holtzclaw, Azarias Karamanlidis, Patrick Keeling, David Lamb, Brian Langerhans, Joe Montoya, Kevin Peterson, Michael Pollock, Susannah Porter, T. K. Reddy, Andrew Roger, Andrew Schaffner, Tom Silhavy, Alastair Simpson, Doug Soltis, Pamela Soltis, and George Watts. In addition, the biologists listed on pages xxviii–xxx provided detailed reviews, helping us ensure the text's scientific accuracy and improve its pedagogical effectiveness.

Thanks also to the other professors and students, from all over the world, who contacted the authors directly with useful suggestions. We alone bear the responsibility for any errors that remain, but the dedication of our consultants, reviewers, and other correspondents makes us confident in the accuracy and effectiveness of this text.

The value of **CAMPBELL BIOLOGY IN FOCUS** as a learning tool is greatly enhanced by the supplementary materials that have been created for instructors and students. We recognize that the dedicated authors of these materials are essentially writing mini (and not so mini) books. We appreciate the hard work and creativity of all the authors listed, with their creations, on page xxv. We are also grateful to Kathleen Fitzpatrick and Nicole Tunbridge (PowerPoint® Lecture Presentations); Douglas Darnowski, James Langeland, Murty Kambhampati, and Roberta Batorsky (Clicker Questions); Ford Lux, Chris Romero, Ruth Sporer, Kristen Miller, Justin Shaffer, Murty Kambhampati, and Suann Yang (Test Bank).

MasteringBiology® and the electronic media for this text are invaluable teaching and learning aids. We thank the hardworking, industrious instructors who worked on the revised and new media: Rebecca Orr (team lead), Eileen Gregory, Angela Hodgson, Karen Resendes, Allison Silveus, Maureen Leupold, Jenny Gernhart, Sara Tallarovic, Molly Jacobs, Jonathan Fitz Gerald, and Martin Kelly (Ready-to-Go Teaching Modules); Emily Carlisle (Practice Tests); Roberta Batorsky (Reading Quizzes); Jennifer Yeh (Interpret the Data Questions in Mastering); Natalie Bronstein, Linda Logdberg, Matt McArdle, Ria Murphy, Chris Romero, Justin Walguarnery, and Andy Stull (Dynamic Study Modules); and Eileen Gregory, Rebecca Orr, and Elena Pravosudova (Adaptive Follow-up Assignments). We are also grateful to the many other people—biology instructors, editors, and production experts—who are listed in the credits for these and other elements of the electronic media that accompany the text.

**CAMPBELL BIOLOGY IN FOCUS** results from an unusually strong synergy between a team of scientists and a team of publishing professionals. Our editorial team at Pearson Science again demonstrated unmatched talents, commitment, and pedagogical insights. Our Executive Editor, Josh Frost, brought publishing savvy, intelligence, and a much-appreciated level head to leading the whole team. The clarity and effectiveness of every page owes much to our extraordinary Supervising Editors Pat Burner and Beth Winickoff, who worked with top-notch Senior Developmental Editors John Burner and Mary Ann Murray. Our unsurpassed Editor-in-Chief Beth Wilbur, Executive Editorial Manager Ginnie Simione Jutson, Senior Administrative Coordinator Nick Nickel, and Editorial Assistant Alexander Helmintoller were indispensable in moving the project in the right direction. We also want to thank Robin Heyden for organizing the annual Biology Leadership Community meetings and keeping us in touch with the world of AP Biology.

You would not have this beautiful text if not for the work of the Product Management Services team: Program Management Team Lead Mike Early, Project Management Team Lead David Zielonka, Project Manager Lori Newman, and Program Manager Leata Holloway; the Rights Management team: Manager Rachel Youdelman, Project Manager Donna Kalal, Senior Project Manager Tim Nicholls, Photo Researcher Maureen Spuhler, and Project Managers Elaine Kosta and Jen Simmons at Lumina Datamatics; the Production Team: Copy Editor Joanna Dinsmore, Proofreader Pete Shanks, Project Managers Norine Strang and Roxanne Klaas and the rest of the staff at S4Carlisle Publishing Services; Art Project Managers Mimi Polk, Kristina Seymour, and Rebecca Marshall, Developmental Artist Andrew Recher, and the rest of the staff at Precision Graphics (a Lachina Company); Design Manager Marilyn Perry; Text and Cover Designer Gary Hespenheide; and Senior Procurement Specialist Stacey Weinberger. We also thank those who worked on the text's supplements: Eddie Lee, Supervising Project Manager, Instructor Media, and Sylvia Rebert and Marsha Hall at Progressive Publishing Services. And for creating the wonderful package of electronic media that accompanies the text, we are grateful to Tania Mlawer (Director of Content Development for MasteringBiology), Sarah Jensen, Lee Ann Doctor, Katie Foley, Charles Hall, Sarah Young-Dualan, Jackie Jakob, Katie Cook, Jim Hufford, Dario Wong, as well as Director of Media Development Laura Tommasi and Director of Digital Strategy Stacy Treco.

For their important roles in marketing the text and media, we thank Christy Lesko, Lauren Harp, Amee Mosley, Scott Dustan, and Jane Campbell. For her market development support, we thank Leslie Allen. We are grateful to Adam Jaworksi, Vice President of Pearson Science Editorial, for his enthusiasm, encouragement, and support.

The Pearson sales team, which represents **CAMPBELL BIOLOGY IN FOCUS** on campus, is an essential link to the users of the text. They tell us what you like and don't like about the text, communicate the features of the text, and provide prompt service. We thank them for their hard work and professionalism. For representing our text to our international audience, we thank our sales and marketing partners throughout the world. They are all strong allies in biology education.

Finally, we wish to thank our families and friends for their encouragement and patience throughout this long project. Our special thanks to Ross, Lily, and Alex (L.A.U.); Debra and Hannah (M.L.C.); Harry, Elga, Aaron, Sophie, Noah, and Gabriele (S.A.W.); Natalie (P.V.M.); and Paul, Dan, Maria, Armelle, Sean, and Jeff (J.B.R.). And, as always, thanks to Rochelle, Allison, Jason, McKay, and Gus.

Lisa A. Urry, Michael L. Cain, Steve A. Wasserman, Peter V. Minorsky, and Jane B. Reece

# Reviewers

## Reviewers of the Second Edition

Steve Abedon, *Ohio State University*
John Alcock, *Arizona State University*
Mary Allen, *Hartwick College*
Phillip Allman, *Florida Gulf Coast College*
Rodney Allrich, *Purdue University*
John Archibald, *Dalhousie University*
Mary Ashley, *University of Illinois at Chicago*
Linda Barnes, *Marshalltown Community College*
Jim Barron, *Montana State University Billings*
Aimee Bernard, *University of Colorado Denver*
James Blevins, *Salt Lake Community College*
Christopher Bloch, *Bridgewater State University*
Chris Botanga, *Chicago State University*
Jeffrey Bowen, *Bridgewater State University*
Scott Bowling, *Auburn University*
Chad Brassil, *University of Nebraska*
Paul Broady, *University of Canterbury*
Judith Bronstein, *University of Arizona*
Beverly Brown, *Nazareth College*
Lesley Bulluck, *Virginia Commonwealth University*
Tessa Burch, *University of Tennessee*
Warren Burggren, *University of North Texas*
Patrick Cafferty, *Emory University*
Michael Campbell, *Portland State University*
Mickael Cariveau, *Mount Olive College*
Jeffrey Carmichael, *University of North Dakota*
Margaret Carroll, *Framingham State University*
Sam Chapman, *Villanova University*
Bryant Chase, *Florida State University*
Mark Chiappone, *Miami Dade Homestead Campus*
Steve Christenson, *Brigham Young University Idaho*
Amy Clark, *Cape Cod Community College*
Curt Coffman, *Vincennes University*
Bill Cohen, *University of Kentucky*
Jim Colbert, *Iowa State University*
Kathy Cole, *University of Hawaii at Manoa*
Sean Coleman, *University of the Ozarks*
William Cook, *Midwestern State University*
Lewis Coons, *University of Memphis*
Ron Cooper, *University of California, Los Angeles*
Samantha Croft, *Austin Community College*
Curt Daeler, *University of Hawaii at Manoa*
Deborah Dardis, *Southeastern Louisiana University*
Douglas Darnowski, *Indiana University Southeast*
Melissa Deadmond, *Truckee Meadows Community College*
Charles Delwiche, *University of Maryland*
Jean DeSaix, *University of North Carolina*
Bill Detrich, *Northeastern University*
Kevin Dixon, *Florida State University*
Tim Dolan, *Butler University*
Uvetta Dozier, *Bowie State University*
Anna Edlund, *Lafayette College*
Rob Erdman, *Florida Gulf Coast College*
Dale Erskine, *Lebanon Valley College*

Rebecca Escamilla, *El Paso Community College*
Danilo Fernando, *SUNY College of Environmental Science and Forestry (Syracuse)*
Miriam Ferzli, *North Carolina State University*
Christina Fieber, *Horry Georgetown Technical College*
Melissa Fierke, *SUNY College of Environmental Science and Forestry (Syracuse)*
Shannon Finerty, *Bowling Green State University*
Mark Flood, *Fairmont State University*
Robert Fowler, *San Jose State University*
Valerie Franck, *Hawaii Pacific University Hawaii-Loa Windward Campus*
Jason Fuller, *Belleview College*
Rebecca Fuller, *University of Illinois at Urbana–Champaign*
Kristen Genet, *Anoka Ramsey Community College*
Michael Ghedotti, *Regis University*
James Gould, *Princeton University*
Eileen Gregory, *Rollins College*
Mark Grobner, *California State University, Stanislaus*
Becky Gullette, *Panola College*
Carla Haas, *Pennsylvania State University*
Gokhan Hacisalihoglu, *Florida A&M University*
Ken Halanych, *Auburn University*
Matt Halfhill, *St. Ambrose University*
Monica Hall-Woods, *St. Charles Community College*
Dennis Haney, *University of Florida*
Jean Hardwick, *Ithaca College*
Luke Harmon, *University of Idaho*
Chris Haynes, *Shelton State Community College*
Triscia Hendrickson, *Morehouse College*
Albert Herrera, *University of Southern California*
Bruce Heyer, *De Anza College*
Karen Hicks, *Kenyon College*
Kendra Hill, *San Diego State University*
Liz Hobson, *New Mexico State University*
Angela Hodgson, *North Dakota State University*
Rick Holloway, *Northern Arizona University*
Ken Holscher, *Iowa State University*
David Hooper, *Western Washington University*
Jodee Hunt, *Grand Valley State University*
Erin Irish, *University of Iowa*
Sally Irwin, *University of Hawaii, Maui*
Emmanuelle Javaux, *University of Liège*
Dianne Jennings, *Virginia Commonwealth University*
Jaime Jensen, *Brigham Young University*
Liesl Jones, *Lehman College*
Mark Jordan, *Indiana University-Purdue University*
Ari Jumpponen, *Kansas State University*
Leann Kanda, *Ithaca College*
Doug Kane, *Defiance College*
Jessica Kaufman, *Endicott College*
Mary Jane Keleher, *Salt Lake Community College*
Paul Kenrick, *Natural History Museum, London*
Shannon King, *North Dakota State University*

Daniel Kjar, *Elmira College*
Jennifer Kneafsey, *School Tulsa Community College*
Jacob Krans, *Western New England University*
Patrick Krug, *California State University, Los Angeles*
Barb Kuemerle, *Case Western Reserve University*
Stephen W. L'Hernault, *Emory University*
Jim Langeland, *Kalamazoo College*
Grace Lasker, *Lake Washington Institute of Technology*
Kadee Lawrence, *Highline Community College*
Terolyn Lay, *Chipola College*
Lisa Leege, *Georgia Southern University*
Jani Lewis, *State University of New York*
Tatyana Lobova, *Old Dominion University*
David Longstreth, *Louisiana State University*
Donald Lovett, *College of New Jersey*
Lisa Lyons, *Florida State University*
Dale Mabry, *Hillsborough Community College*
Nancy Magill, *Indiana University*
Mark Maloney, *Spelman College*
Michael Manson, *Texas A&M University*
Mary Martin, *Northern Michigan University*
Bryant McAllister, *University of Iowa*
Tanya McGhee, *Craven Community College*
Paul McMillan, *Capilano University*
Ana Medrano, *University of Houston*
Chris Meloce, *Metro State University*
Jenny Metzler, *Ball State University*
Marcella Meyers, *St. Catherine University*
James Mickle, *North Carolina State University*
Grace Miller, *Indiana Wesleyan University*
Iain Miller, *Ohio University*
Jonathan Miller, *Edmonds Community College*
Mill Miller, *Wright State University*
Sarah Milton, *Florida Atlantic University*
Elizabeth Morgan, *Lone Star College Kingwood*
Mark Mort, *University of Kansas*
Jeff Murray, *University of Iowa*
Barbara Musolf, *Clayton State University*
Barbara Nash, *Mercy College*
Karen Neal, *J. Sargeant Reynolds Community College*
Judy Nesmith, *University of Michigan Dearborn*
Zia Nisani, *Antelope Valley College*
Shawn Nordell, *Saint Louis University*
Kavita S. Oommen, *Georgia State University*
Rebecca Orr, *Spring Creek College*
Patrick Owen, *University of Cincinnati, Blue Ash College*
Matt Palmtag, *Florida Gulf Coast University*
Robert Patterson, *San Francisco State University*
Eric Peters, *Chicago State University*
Kevin Peterson, *Dartmouth College*
John Pleasants, *Iowa State University*
Jason Porter, *University of the Sciences, Philadelphia*
Elena Pravosudova, *University of Nevada, Reno*
Shira Rabin, *University of Louisville*

Robert Reavis, *Glendale Community College*
Erin Rehrig, *Fitchburg State University*
Wayne Rickroll, *Western Oregon University*
Andrew Rogers, *Dalhousie University*
Scott Russell, *University of Oklahoma*
Jenny Rygh, *Hawaii Community College*
Karin Scarpinato, *Georgia Southern University*
Cara Schillington, *Eastern Michigan University*
Carrie Schwartz, *Western Washington University*
David Schwartz, *Houston Community College*
Joan Sharp, *Simon Fraser University*
Alison Sherwood, *University of Hawaii at Manoa*
Brian Shmaefsky, *Lonestar University*
Eric Shows, *Jones County Junior College*
Charles Shuster, *New Mexico State University*
Michele Shuster, *New Mexico State University*
Mitchell Singer, *University of California, Davis*
Ramona Smith, *Brevard College*
Douglas Soltis, *University of Florida*
Rebecca Sperry, *Salt Lake Community College*
Clint Springer, *St. Joseph's College*
Mark Sturtevant, *Oakland University*
Diane Sweeney, *Punahou School*
Kevin Swier, *Chicago State University*
William Tanner, *Salt Lake Community College*
Kristen Taylor, *Salt Lake Community College*
Max Telford, *University College London*
Rebekah Thomas, *College of St. Joseph*
Melissa Tillack, *Salt Lake Community College*
Mike Toliver, *Eureka College*
Clint Tuberville, *Virginia Commonwealth University*
Marty Vaughan, *Indiana University – Purdue University Indianapolis*
James Wandersee, *Louisiana State University*
Rebekah Ward, *Georgia Gwinnett College*
Jim Wee, *Loyola University New Orleans*
Jennifer R. Welch, *Madisonville Community College*
Charles Wellman, *University of Sheffield*
Christopher Whipps, *State University of New York College of Environmental Science and Forestry*
Philip White, *James Hutton Institute*
Jessica White-Phillip, *Our Lady of the Lake University*
Kathy Williams, *San Diego State University*
Ruth Wrightsman, *Flathead Valley Community College*
Robert Yost, *Indiana University – Purdue University Indianapolis*
Stephanie Zamule, *Nazareth College*

## First Edition Reviewers

Ann Aguanno, *Marymount Manhattan College*
Marc Albrecht, *University of Nebraska*
John Alcock, *Arizona State University*
Eric Alcorn, *Acadia University*

Rodney Allrich, *Purdue University*
John Archibald, *Dalhousie University*
Terry Austin, *Temple College*
Brian Bagatto, *University of Akron*
Virginia Baker, *Chipola College*
Teri Balser, *University of Wisconsin, Madison*
Bonnie Baxter, *Westminster College*
Marilee Benore, *University of Michigan, Dearborn*
Catherine Black, *Idaho State University*
William Blaker, *Furman University*
Edward Blumenthal, *Marquette University*
David Bos, *Purdue University*
Scott Bowling, *Auburn University*
Beverly Brown, *Nazareth College*
Beth Burch, *Huntington University*
Warren Burggren, *University of North Texas*
Dale Burnside, *Lenoir-Rhyne University*
Ragan Callaway, *The University of Montana*
Kenneth M. Cameron, *University of Wisconsin, Madison*
Patrick Canary, *Northland Pioneer College*
Cheryl Keller Capone, *Pennsylvania State University*
Mickael Cariveau, *Mount Olive College*
Karen I. Champ, *Central Florida Community College*
David Champlin, *University of Southern Maine*
Brad Chandler, *Palo Alto College*
Wei-Jen Chang, *Hamilton College*
Jung Choi, *Georgia Institute of Technology*
Steve Christenson, *Brigham Young University, Idaho*
Reggie Cobb, *Nashville Community College*
James T. Colbert, *Iowa State University*
Sean Coleman, *University of the Ozarks*
William Cushwa, *Clark College*
Deborah Dardis, *Southeastern Louisiana University*
Shannon Datwyler, *California State University, Sacramento*
Melissa Deadmond, *Truckee Meadows Community College*
Eugene Delay, *University of Vermont*
Daniel DerVartanian, *University of Georgia*
Jean DeSaix, *University of North Carolina, Chapel Hill*
Janet De Souza-Hart, *Massachusetts College of Pharmacy & Health Sciences*
Jason Douglas, *Angelina College*
Kathryn A. Durham, *Lorain Community College*
Anna Edlund, *Lafayette College*
Curt Elderkin, *College of New Jersey*
Mary Ellard-Ivey, *Pacific Lutheran University*
Kurt Elliott, *Northwest Vista College*
George Ellmore, *Tufts University*
Rob Erdman, *Florida Gulf Coast College*
Dale Erskine, *Lebanon Valley College*
Robert C. Evans, *Rutgers University, Camden*
Sam Fan, *Bradley University*
Paul Farnsworth, *University of New Mexico*
Myriam Alhadeff Feldman, *Cascadia Community College*

Teresa Fischer, *Indian River Community College*
David Fitch, *New York University*
T. Fleming, *Bradley University*
Robert Fowler, *San Jose State University*
Robert Franklin, *College of Charleston*
Art Fredeen, *University of Northern British Columbia*
Kim Fredericks, *Viterbo University*
Matt Friedman, *University of Chicago*
Cynthia M. Galloway, *Texas A&M University, Kingsville*
Kristen Genet, *Anoka Ramsey Community College*
Phil Gibson, *University of Oklahoma*
Eric Gillock, *Fort Hayes State University*
Simon Gilroy, *University of Wisconsin, Madison*
Edwin Ginés-Candelaria, *Miami Dade College*
Jim Goetze, *Laredo Community College*
Lynda Goff, *University of California, Santa Cruz*
Roy Golsteyn, *University of Lethbridge*
Barbara E. Goodman, *University of South Dakota*
Eileen Gregory, *Rollins College*
Bradley Griggs, *Piedmont Technical College*
David Grise, *Texas A&M University, Corpus Christi*
Edward Gruberg, *Temple University*
Karen Guzman, *Campbell University*
Carla Haas, *Pennsylvania State University*
Pryce "Pete" Haddix, *Auburn University*
Heather Hallen-Adams, *University of Nebraska, Lincoln*
Monica Hall-Woods, *St. Charles Community College*
Bill Hamilton, *Washington & Lee University*
Devney Hamilton, *Stanford University (student)*
Matthew B. Hamilton, *Georgetown University*
Dennis Haney, *Furman University*
Jean Hardwick, *Ithaca College*
Luke Harmon, *University of Idaho*
Jeanne M. Harris, *University of Vermont*
Stephanie Harvey, *Georgia Southwestern State University*
Bernard Hauser, *University of Florida*
Chris Haynes, *Shelton State Community College*
Andreas Hejnol, *Sars International Centre for Marine Molecular Biology*
Albert Herrera, *University of Southern California*
Chris Hess, *Butler University*
Kendra Hill, *San Diego State University*
Jason Hodin, *Stanford University*
Laura Houston, *Northeast Lakeview College*
Sara Huang, *Los Angeles Valley College*
Catherine Hurlbut, *Florida State College, Jacksonville*
Diane Husic, *Moravian College*
Thomas Jacobs, *University of Illinois*

Kathy Jacobson, *Grinnell College*
Mark Jaffe, *Nova Southeastern University*
Emmanuelle Javaux, *University of Liege, Belgium*
Douglas Jensen, *Converse College*
Lance Johnson, *Midland Lutheran College*
Roishene Johnson, *Bossier Parish Community College*
Cheryl Jorcyk, *Boise State University*
Caroline Kane, *University of California, Berkeley*
The-Hui Kao, *Pennsylvania State University*
Nicholas Kapp, *Skyline College*
Jennifer Katcher, *Pima Community College*
Judy Kaufman, *Monroe Community College*
Eric G. Keeling, *Cary Institute of Ecosystem Studies*
Chris Kennedy, *Simon Fraser University*
Hillar Klandorf, *West Virginia University*
Mark Knauss, *Georgia Highlands College*
Charles Knight, *California Polytechnic State University*
Roger Koeppe, *University of Arkansas*
Peter Kourtev, *Central Michigan University*
Jacob Krans, *Western New England University*
Eliot Krause, *Seton Hall University*
Steven Kristoff, *Ivy Tech Community College*
William Kroll, *Loyola University*
Barb Kuemerle, *Case Western Reserve University*
Rukmani Kuppuswami, *Laredo Community College*
Lee Kurtz, *Georgia Gwinnett College*
Michael P. Labare, *United States Military Academy, West Point*
Ellen Lamb, *University of North Carolina, Greensboro*
William Lamberts, *College of St. Benedict and St. John's University*
Tali D. Lee, *University of Wisconsin, Eau Claire*
Hugh Lefcort, *Gonzaga University*
Alcinda Lewis, *University of Colorado, Boulder*
Jani Lewis, *State University of New York*
Graeme Lindbeck, *Valencia Community College*
Hannah Lui, *University of California, Irvine*
Nancy Magill, *Indiana University*
Cindy Malone, *California State University, Northridge*
Mark Maloney, *University of South Mississippi*
Julia Marrs, *Barnard College (student)*
Kathleen Marrs, *Indiana University – Purdue University Indianapolis*
Mike Mayfield, *Ball State University*
Kamau Mbuthia, *Bowling Green State University*
Tanya McGhee, *Craven Community College*
Darcy Medica, *Pennsylvania State University*
Susan Meiers, *Western Illinois University*
Mike Meighan, *University of California, Berkeley*
Jan Mikesell, *Gettysburg College*
Alex Mills, *University of Windsor*
Sarah Milton, *Florida Atlantic University*
Eli Minkoff, *Bates College*
Subhash Minocha, *University of New Hampshire*
Ivona Mladenovic, *Simon Fraser University*

Barbara Modney, *Cleveland State University*
Linda Moore, *Georgia Military College*
Courtney Murren, *College of Charleston*
Karen Neal, *Reynolds University*
Ross Nehm, *Ohio State University*
Kimberlyn Nelson, *Pennsylvania State University*
Jacalyn Newman, *University of Pittsburgh*
Kathleen Nolta, *University of Michigan*
Gretchen North, *Occidental College*
Margaret Olney, *St. Martin's University*
Aharon Oren, *The Hebrew University*
Rebecca Orr, *Spring Creek College*
Henry R. Owen, *Eastern Illinois University*
Matt Palmtag, *Florida Gulf Coast University*
Stephanie Pandolfi, *Michigan State University*
Nathalie Pardigon, *Institut Pasteur*
Cindy Paszkowski, *University of Alberta*
Andrew Pease, *Stevenson University*
Nancy Pelaez, *Purdue University*
Irene Perry, *University of Texas of the Permian Basin*
Roger Persell, *Hunter College*
Eric Peters, *Chicago State University*
Larry Peterson, *University of Guelph*
Mark Pilgrim, *College of Coastal Georgia*
Vera M. Piper, *Shenandoah University*
Deb Pires, *University of California, Los Angeles*
Crima Pogge, *City College of San Francisco*
Michael Pollock, *Mount Royal University*
Roberta Pollock, *Occidental College*
Therese M. Poole, *Georgia State University*
Angela R. Porta, *Kean University*
Jason Porter, *University of the Sciences, Philadelphia*
Robert Powell, *Avila University*
Elena Pravosudova, *University of Nevada, Reno*
Eileen Preston, *Tarrant Community College Northwest*
Terrell Pritts, *University of Arkansas, Little Rock*
Pushpa Ramakrishna, *Chandler-Gilbert Community College*
David Randall, *City University Hong Kong*
Monica Ranes-Goldberg, *University of California, Berkeley*
Robert S. Rawding, *Gannon University*
Robert Reavis, *Glendale Community College*
Sarah Richart, *Azusa Pacific University*
Todd Rimkus, *Marymount University*
John Rinehart, *Eastern Oregon University*
Kenneth Robinson, *Purdue University*
Deb Roess, *Colorado State University*
Heather Roffey, *Marianopolis College*
Suzanne Rogers, *Seton Hill University*
Patricia Rugaber, *College of Coastal Georgia*
Scott Russell, *University of Oklahoma*
Glenn-Peter Saetre, *University of Oslo*
Sanga Saha, *Harold Washington College*
Kathleen Sandman, *Ohio State University*
Louis Santiago, *University of California, Riverside*
Tom Sawicki, *Spartanburg Community College*

Andrew Schaffner, *California Polytechnic State University, San Luis Obispo*
Thomas W. Schoener, *University of California, Davis*
Patricia Schulte, *University of British Columbia*
Brenda Schumpert, *Valencia Community College*
David Schwartz, *Houston Community College*
Duane Sears, *University of California, Santa Barbara*
Brent Selinger, *University of Lethbridge*
Alison M. Shakarian, *Salve Regina University*
Joan Sharp, *Simon Fraser University*
Robin L. Sherman, *Nova Southeastern University*
Eric Shows, *Jones County Junior College*
Sedonia Sipes, *Southern Illinois University, Carbondale*
John Skillman, *California State University, San Bernardino*
Doug Soltis, *University of Florida, Gainesville*
Joel Stafstrom, *Northern Illinois University*
Alam Stam, *Capital University*
Judy Stone, *Colby College*
Cynthia Surmacz, *Bloomsburg University*
David Tam, *University of North Texas*
Yves Tan, *Cabrillo College*
Emily Taylor, *California Polytechnic State University*
Marty Taylor, *Cornell University*
Franklyn Tan Te, *Miami Dade College*
Kent Thomas, *Wichita State University*
Mike Toliver, *Eureka College*
Saba Valadkhan, *Center for RNA Molecular Biology*
Sarah Van Vickle-Chavez, *Washington University, St. Louis*
William Velhagen, *New York University*
Amy Volmer, *Swarthmore College*
Janice Voltzow, *University of Scranton*
Margaret Voss, *Penn State Erie*
Charles Wade, *C.S. Mott Community College*
Claire Walczak, *Indiana University*
Jerry Waldvogel, *Clemson University*
Robert Lee Wallace, *Ripon College*
James Wandersee, *Louisiana State University*
Fred Wasserman, *Boston University*
James Wee, *Loyola University*
John Weishampel, *University of Central Florida*
Susan Whittemore, *Keene State College*
Murray Wiegand, *University of Winnipeg*
Kimberly Williams, *Kansas State University*
Janet Wolkenstein, *Hudson Valley Community College*
Grace Wyngaard, *James Madison University*
Shuhai Xiao, *Virginia Polytechnic Institute*
Paul Yancey, *Whitman College*
Anne D. Yoder, *Duke University*
Ed Zalisko, *Blackburn College*
Nina Zanetti, *Siena College*
Sam Zeveloff, *Weber State University*
Theresa Zucchero, *Methodist University*

# Detailed Contents

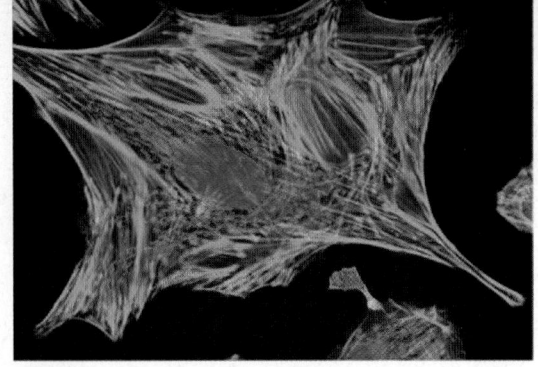

# 7 Cellular Respiration and Fermentation 141

# 8 Photosynthesis 161

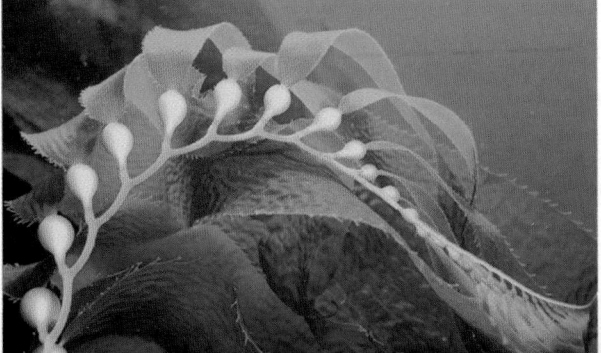

# 9 The Cell Cycle 182

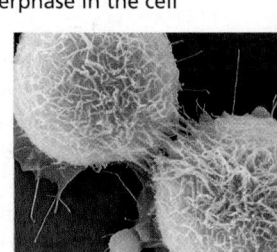

# UNIT 2 Genetics 199

# 10 Meiosis and Sexual Life Cycles 200

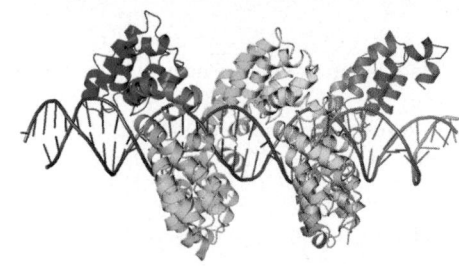

## UNIT 4 The Evolutionary History of Life 473

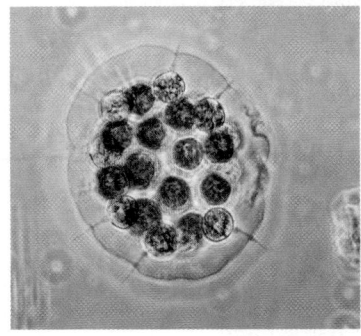

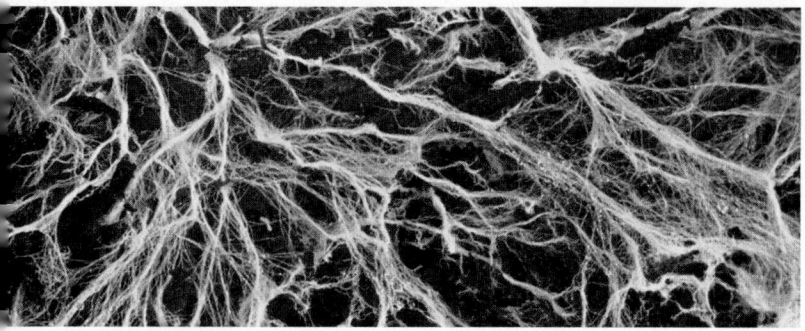

## UNIT 5 Plant Form and Function 574

## 29 Resource Acquisition, Nutrition, and Transport in Vascular Plants 593

## 30 Reproduction and Domestication of Flowering Plants 619

## 31 Plant Responses to Internal and External Signals 639

## UNIT 6 Animal Form and Function 662

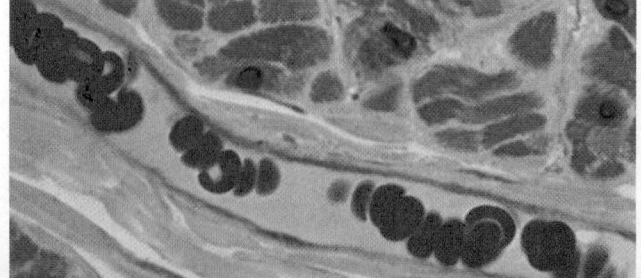

## 35 The Immune System 733

## 36 Reproduction and Development 751

## 37 Neurons, Synapses, and Signaling 772

## UNIT 7 Ecology 839

## 40 Population Ecology and the Distribution of Organisms 840

# 41 Species Interactions 867

# 42 Ecosystems and Energy 886

# 43 Global Ecology and Conservation Biology 906

# 1 Introduction: Evolution and the Foundations of Biology

▲ **Figure 1.1 What can this beach mouse (*Peromyscus polionotus*) teach us about biology?**

## KEY CONCEPTS

**1.1** The study of life reveals common themes

**1.2** The Core Theme: Evolution accounts for the unity and diversity of life

**1.3** In studying nature, scientists form and test hypotheses

## Inquiring About Life

There are few hiding places for a mouse among the sparse clumps of beach grass that dot the brilliant white sand dunes along the Florida seashore. However, the beach mice that live there have light, dappled fur, allowing them to blend into their surroundings **(Figure 1.1)**. Mice of the same species (*Peromyscus polionotus*) also inhabit nearby inland areas. These mice are much darker in color, as are the soil and vegetation where they live **(Figure 1.2)**. For both beach mice and inland mice, the close color match of coat (fur) and environment is vital for survival, since hawks, herons, and other sharp-eyed predators periodically scan the landscape for prey. How has the color of each group of mice come to be so well matched, or *adapted*, to the local background?

An organism's adaptations to its environment, such as the mouse's protective camouflage, are the result of **evolution**, the process of change over time that has

▲ **Figure 1.2 An inland mouse of the species *Peromyscus polionotus*.** This mouse has a much darker back, side, and face than mice of the same species that inhabit sand dunes.

resulted in the astounding array of organisms found on Earth. Evolution is the fundamental principle of biology and the core theme of this book.

Posing questions about the living world and seeking answers through scientific inquiry are the central activities of **biology**, the scientific study of life. Biologists' questions can be ambitious. They may ask how a single tiny cell becomes a tree or a dog, how the human mind works, or how the different forms of life in a forest interact. When questions occur to you as you observe the living world, you are thinking like a biologist.

How do biologists make sense of life's diversity and complexity? This opening chapter sets up a framework for answering this question. We begin with a panoramic view of the biological "landscape," organized around a set of unifying themes. We'll then focus on biology's core theme, evolution. Finally, we'll examine the process of scientific inquiry—how scientists ask and attempt to answer questions about the natural world.

# The study of life reveals common themes

Biology is a subject of enormous scope, and exciting new biological discoveries are being made every day. How can you organize and make sense of all the information you'll encounter as you study biology? Focusing on a few big ideas will help. Here are five unifying themes—ways of thinking about life that will still hold true decades from now:

- Organization
- Information
- Energy and Matter
- Interactions
- Evolution

In this chapter, we'll briefly define and explore each theme.

## Theme: New Properties Emerge at Successive Levels of Biological Organization

**ORGANIZATION** The study of life on Earth extends from the microscopic scale of the molecules and cells that make up organisms to the global scale of the entire living planet. As biologists, we can divide this enormous range into different levels of biological organization.

In **Figure 1.3**, we zoom in from space to take a closer and closer look at life in a mountain meadow. This journey, depicted in the figure as a series of numbered steps, highlights the hierarchy of biological organization.

Zooming in at ever-finer resolution illustrates the principle that underlies *reductionism*, an approach that reduces complex systems to simpler components that are more manageable to study. Reductionism is a powerful strategy in biology. For example, by studying the molecular structure of DNA that had been extracted from cells, James Watson and Francis Crick inferred the chemical basis of biological inheritance. Despite its importance, reductionism provides an incomplete view of life, as we'll discuss next.

### Emergent Properties

Let's reexamine Figure 1.3, beginning this time at the molecular level and then zooming out. Viewed this way, we see that novel properties emerge at each level that are absent from the preceding one. These **emergent properties** are due to the arrangement and interactions of parts as complexity increases. For example, although photosynthesis occurs in an intact chloroplast, it will not take place if chlorophyll and other chloroplast molecules are simply mixed in a test tube. The coordinated processes of photosynthesis require a specific organization of these molecules in the chloroplast. In general, isolated components of living systems—the objects of study in a reductionist approach—lack a number of significant properties that emerge at higher levels of organization.

Emergent properties are not unique to life. A box of bicycle parts won't transport you anywhere, but if they are arranged in a certain way, you can pedal to your chosen destination. Compared with such nonliving examples, however, biological systems are far more complex, making the emergent properties of life especially challenging to study.

To fully explore emergent properties, biologists complement reductionism with **systems biology**, the exploration of the network of interactions that underlie the emergent properties of a system. A single leaf cell can be considered a system, as can a frog, an ant colony, or a desert ecosystem. By examining and modeling the dynamic behavior of an integrated network of components, systems biology enables us to pose new kinds of questions. For example, how do networks of genes in our cells produce oscillations in the activity of the molecules that generate our 24-hour cycle of wakefulness and sleep? At a larger scale, how does a gradual increase in atmospheric carbon dioxide alter ecosystems and the entire biosphere? Systems biology can be used to study life at all levels.

### Structure and Function

At each level of biological organization, we find a correlation between structure and function. Consider a leaf in Figure 1.3: Its thin, flat shape maximizes the capture of sunlight by chloroplasts. Because such correlations of structure and function are common in all forms of life, analyzing a biological structure gives us clues about what it does and how it works. A good example from the animal kingdom is the hummingbird. The hummingbird's anatomy allows its wings to rotate at the shoulder, so hummingbirds have the ability, unique among birds, to fly backward or hover in place. While hovering, the birds can extend their long slender beaks into flowers and feed on nectar. Such an elegant match of form and function in the structures of life is explained by natural selection, as we'll explore shortly.

### The Cell: An Organism's Basic Unit of Structure and Function

The cell is the smallest unit of organization that can perform all activities required for life. In fact, the actions of an organism are all based on the activities of its cells. For instance, the movement of your eyes as you read this sentence results from the activities of muscle and nerve cells. Even a process that occurs on a global scale, such as the recycling of carbon atoms, is the cumulative product of cellular functions,

### ◀ 1 The Biosphere

Even from space, we can see signs of Earth's life—in the green mosaic of the forests, for example. We can also see the entire **biosphere**, which consists of all life on Earth and all the places where life exists: most regions of land, most bodies of water, the atmosphere to an altitude of several kilometers, and even sediments far below the ocean floor.

### ◀ 2 Ecosystems

Our first scale change brings us to a North American mountain meadow, which is an example of an ecosystem, as are tropical forests, grasslands, deserts, and coral reefs. An **ecosystem** consists of all the living things in a particular area, along with all the nonliving components of the environment with which life interacts, such as soil, water, atmospheric gases, and light.

### ▶ 3 Communities

The array of organisms inhabiting a particular ecosystem is called a biological **community**. The community in our meadow ecosystem includes many kinds of plants, various animals, mushrooms and other fungi, and enormous numbers of diverse microorganisms, such as bacteria, that are too small to see without a microscope. Each of these forms of life belongs to a *species*—a group whose members can only reproduce with other members of the group.

### ▶ 4 Populations

A **population** consists of all the individuals of a species living within the bounds of a specified area. For example, our meadow includes a population of lupine (some of which are shown here) and a population of mule deer. A community is therefore the set of populations that inhabit a particular area.

### ▲ 5 Organisms

Individual living things are called **organisms**. Each plant in the meadow is an organism, and so is each animal, fungus, and bacterium.

including the photosynthetic activity of chloroplasts in leaf cells.

All cells share certain characteristics, such as being enclosed by a membrane that regulates the passage of materials between the cell and its surroundings. Nevertheless, we distinguish two main forms of cells: prokaryotic and eukaryotic. The cells of two groups of single-celled microorganisms—bacteria and archaea—are prokaryotic. All other forms of life, including plants and animals, are composed of eukaryotic cells.

A **eukaryotic cell** contains membrane-enclosed organelles **(Figure 1.4)**. Some organelles, such as the DNA-containing nucleus, are found in the cells of all eukaryotes; other organelles are specific to particular cell types. For example, the chloroplast in Figure 1.3 is an organelle found only in eukaryotic cells that carry out photosynthesis. In contrast to eukaryotic cells, a **prokaryotic cell** lacks a nucleus or other membrane-enclosed organelles. Furthermore, prokaryotic cells are generally smaller than eukaryotic cells, as shown in Figure 1.4.

## ▼ 6 Organs

The structural hierarchy of life continues to unfold as we explore the architecture of a complex organism. A leaf is an example of an **organ**, a body part that is made up of multiple tissues and has specific functions in the body. Leaves, stems, and roots are the major organs of plants. Within an organ, each tissue has a distinct arrangement and contributes particular properties to organ function.

## ▼ 7 Tissues

Viewing the tissues of a leaf requires a microscope. Each **tissue** is a group of cells that work together, performing a specialized function. The leaf shown here has been cut on an angle. The honeycombed tissue in the interior of

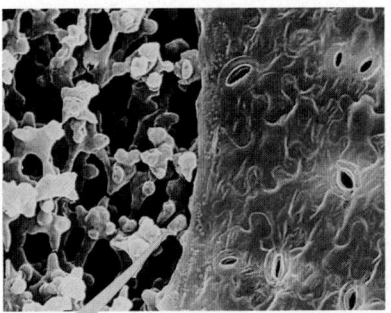

the leaf (left side of photo) is the main location of photosynthesis, the process that converts light energy to the chemical energy of sugar. The jigsaw puzzle–like "skin" on the surface of the leaf is a tissue called epidermis (right side of photo). The pores through the epidermis allow entry of the gas $CO_2$, a raw material for sugar production.

50 μm

Cell    10 μm

## ▶ 8 Cells

The **cell** is life's fundamental unit of structure and function. Some organisms consist of a single cell, which performs all the functions of life. Other organisms are multicellular and feature a division of labor among specialized cells. Here we see a magnified view of a cell in a leaf tissue. This cell is about 40 micrometers (μm) across—about 500 of them would reach across a small coin. Within these tiny cells are even smaller green structures called chloroplasts, which are responsible for photosynthesis.

## ▼ 9 Organelles

Chloroplasts are examples of **organelles**, the various functional components present in cells. The image below, taken by a powerful microscope, shows a single chloroplast.

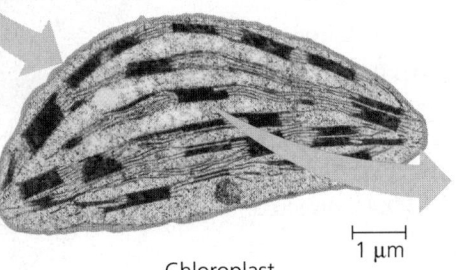

Chloroplast    1 μm

## ▼ 10 Molecules

Our last scale change drops us into a chloroplast for a view of life at the molecular level. A **molecule** is a chemical structure consisting of two or more units called atoms, represented as balls in this computer graphic of a chlorophyll molecule.

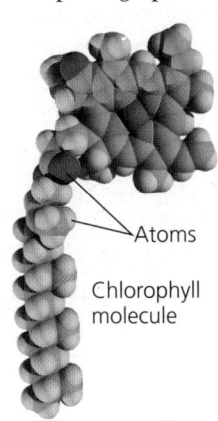

Chlorophyll is the pigment that makes a leaf green, and it absorbs sunlight during photosynthesis. Within each chloroplast, millions of chlorophyll molecules are organized into systems that convert light energy to the chemical energy of food.

Atoms

Chlorophyll molecule

---

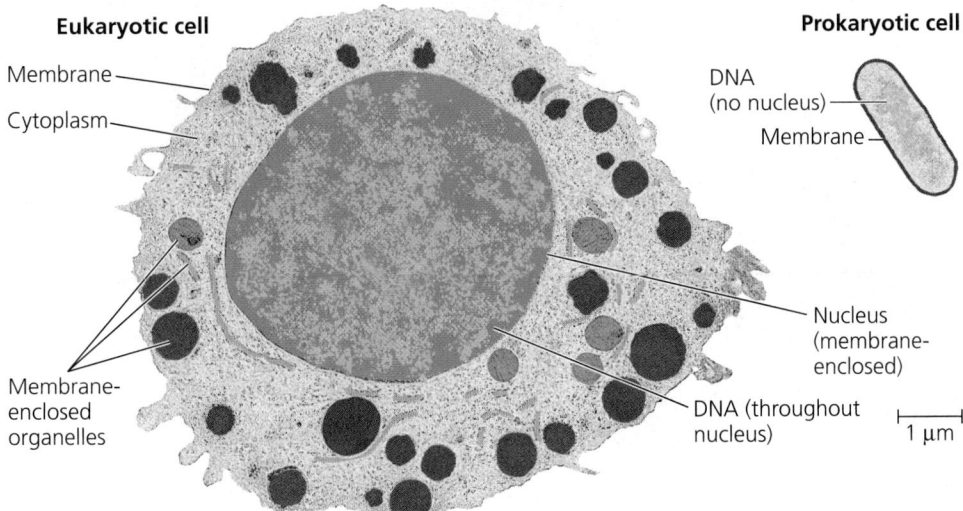

Eukaryotic cell

Membrane

Cytoplasm

Membrane-enclosed organelles

Prokaryotic cell

DNA (no nucleus)

Membrane

Nucleus (membrane-enclosed)

DNA (throughout nucleus)

1 μm

◀ **Figure 1.4 Contrasting eukaryotic and prokaryotic cells in size and complexity.** Cells vary in size, but eukaryotic cells are generally much larger than prokaryotic cells.

## Theme: Life's Processes Involve the Expression and Transmission of Genetic Information

**INFORMATION** Within cells, structures called chromosomes contain genetic material in the form of **DNA (deoxyribonucleic acid)**. In cells that are preparing to divide, the chromosomes may be made visible using a dye that appears blue when bound to the DNA **(Figure 1.5)**.

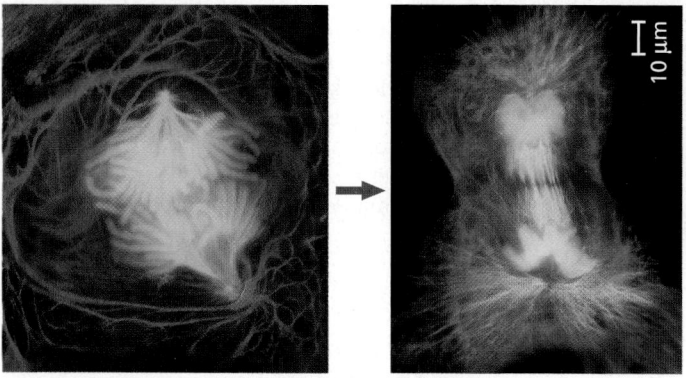

▲ **Figure 1.5 A lung cell from a newt divides into two smaller cells that will grow and divide again.**

### DNA, the Genetic Material

Each chromosome contains one very long DNA molecule with hundreds or thousands of **genes**, each a section of the DNA of the chromosome. Transmitted from parents to offspring, genes are the units of inheritance. They encode the information necessary to build all of the molecules synthesized within a cell, which in turn establish that cell's identity and function. You began as a single cell stocked with DNA inherited from your parents. The replication of that DNA during each round of cell division transmitted copies of the DNA to what eventually became the trillions of cells of your body. As the cells grew and divided, the genetic information encoded by the DNA directed your development **(Figure 1.6)**.

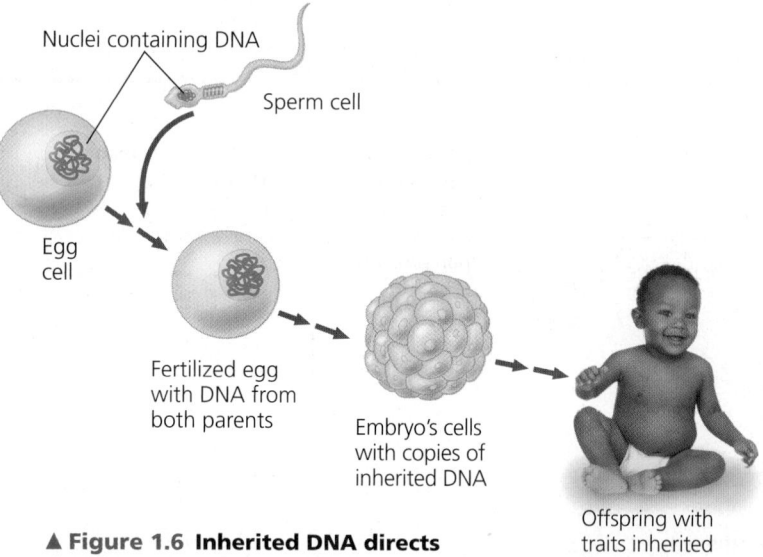

Nuclei containing DNA

Sperm cell

Egg cell

Fertilized egg with DNA from both parents

Embryo's cells with copies of inherited DNA

Offspring with traits inherited from both parents

▲ **Figure 1.6 Inherited DNA directs development of an organism.**

The molecular structure of DNA accounts for its ability to store information. A DNA molecule is made up of two long chains, called strands, arranged in a double helix. Each chain is made up of four kinds of chemical building blocks called nucleotides, abbreviated A, T, C, and G **(Figure 1.7)**. Specific sequences of these four nucleotides encode the information in genes. The way DNA encodes information is analogous to how we arrange the letters of the alphabet into words and phrases with specific meanings. The word *rat*, for example, evokes a rodent; the words *tar* and *art*, which contain the same letters, mean very different things. We can think of the set of nucleotides as a four-letter alphabet.

For many genes, the sequence provides the blueprint for making a protein. For instance, a given bacterial gene may specify a particular protein (an enzyme) required to assemble the cell membrane, while a certain human gene may denote a different protein (an antibody) that helps fight off infection. Overall, proteins are major players in building and maintaining the cell and in carrying out its activities.

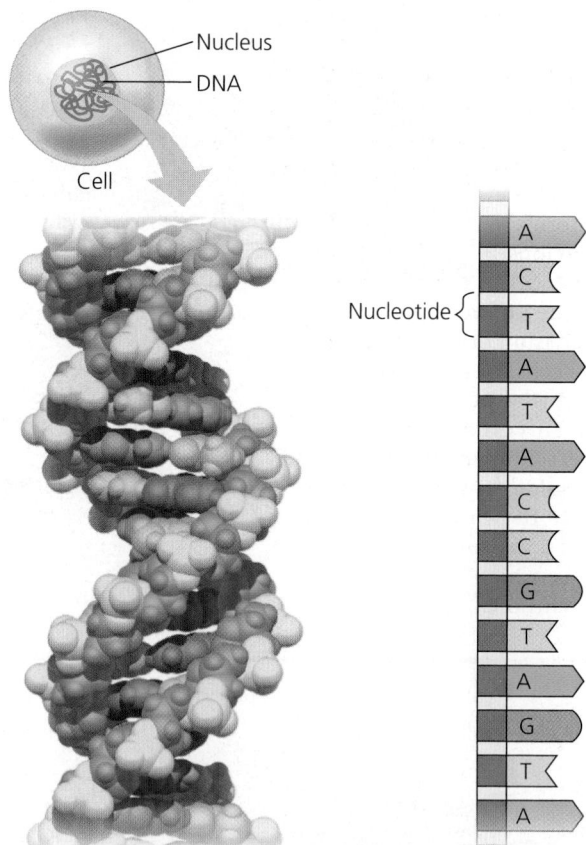

Nucleus

DNA

Cell

Nucleotide

A
C
T
A
T
A
C
C
G
T
A
G
T
A

**(a) DNA double helix.** This model shows the atoms in a segment of DNA. Made up of two long chains (strands) of building blocks called nucleotides, a DNA molecule takes the three-dimensional form of a double helix.

**(b) Single strand of DNA.** These geometric shapes and letters are simple symbols for the nucleotides in a small section of one strand of a DNA molecule. Genetic information is encoded in specific sequences of the four types of nucleotides. Their names are abbreviated A, T, C, and G.

▲ **Figure 1.7 DNA: The genetic material.**

Genes control protein production indirectly, using a related molecule called mRNA as an intermediary (**Figure 1.8**). The sequence of nucleotides along a gene is transcribed into mRNA, which is then translated into a chain of protein building blocks called amino acids. Once completed, this chain forms a specific protein with a unique shape and function. The entire process by which the information in a gene directs the production of a cellular product is called **gene expression**.

In carrying out gene expression, all forms of life employ essentially the same genetic code: A particular sequence of nucleotides says the same thing in one organism as it does in another. Differences between organisms reflect differences between their nucleotide sequences rather than between their genetic codes. This universality of the genetic code is a strong piece of evidence that all life is related. Comparing the sequences in several species for a gene that codes for a particular protein can provide valuable information both about the protein and about the evolutionary relationship of the species to each other.

The mRNA molecule in Figure 1.8 is translated into a protein, but other cellular RNAs function differently. For example, we have known for decades that some types of RNA are actually components of the cellular machinery that manufactures proteins. Recently, scientists have discovered whole new classes of RNA that play other roles in the cell, such as regulating the function of protein-coding genes. Genes also specify all of these RNAs, and their production is also referred to as gene expression. By carrying the instructions for making proteins and RNAs and by replicating with each cell division, DNA ensures faithful inheritance of genetic information from generation to generation.

### Genomics: Large-Scale Analysis of DNA Sequences

The entire "library" of genetic instructions that an organism inherits is called its **genome**. A typical human cell has two similar sets of chromosomes, and each set has approximately 3 billion nucleotide pairs of DNA. If the one-letter abbreviations for the nucleotides of one strand in a set were written in letters the size of those you are now reading, the genomic text would fill about 700 biology textbooks.

Since the early 1990s, the pace at which researchers can determine the sequence of a genome has accelerated at an astounding rate, enabled by a revolution in technology. The genome sequence—the entire sequence of nucleotides for a representative member of a species—is now known for humans and many other animals, as well as numerous plants, fungi, bacteria, and archaea. To make sense of the deluge of data from genome-sequencing projects and the growing catalog of known gene functions, scientists are applying a systems biology approach at the cellular and molecular levels. Rather than investigating a single gene at a time, researchers study whole sets of genes in one or more species—an approach called **genomics**. Likewise, the term **proteomics** refers to the study of sets of proteins and their properties. (The entire set of proteins expressed by a given cell or group of cells is called a **proteome**.)

▼ **Figure 1.8 Gene expression: Cells use information encoded in a gene to synthesize a functional protein.**

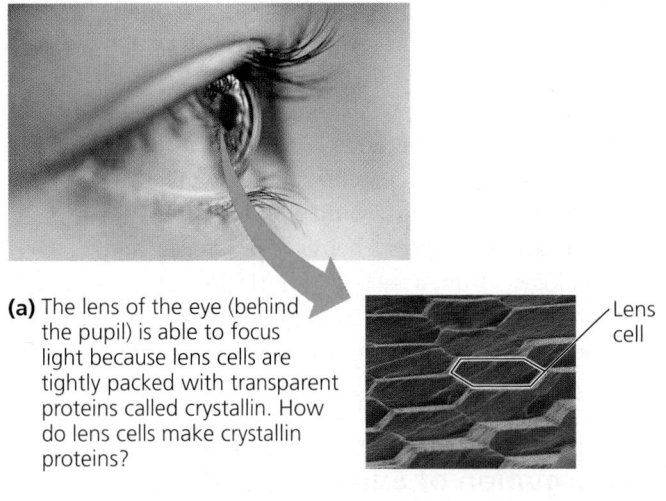

**(a)** The lens of the eye (behind the pupil) is able to focus light because lens cells are tightly packed with transparent proteins called crystallin. How do lens cells make crystallin proteins?

Lens cell

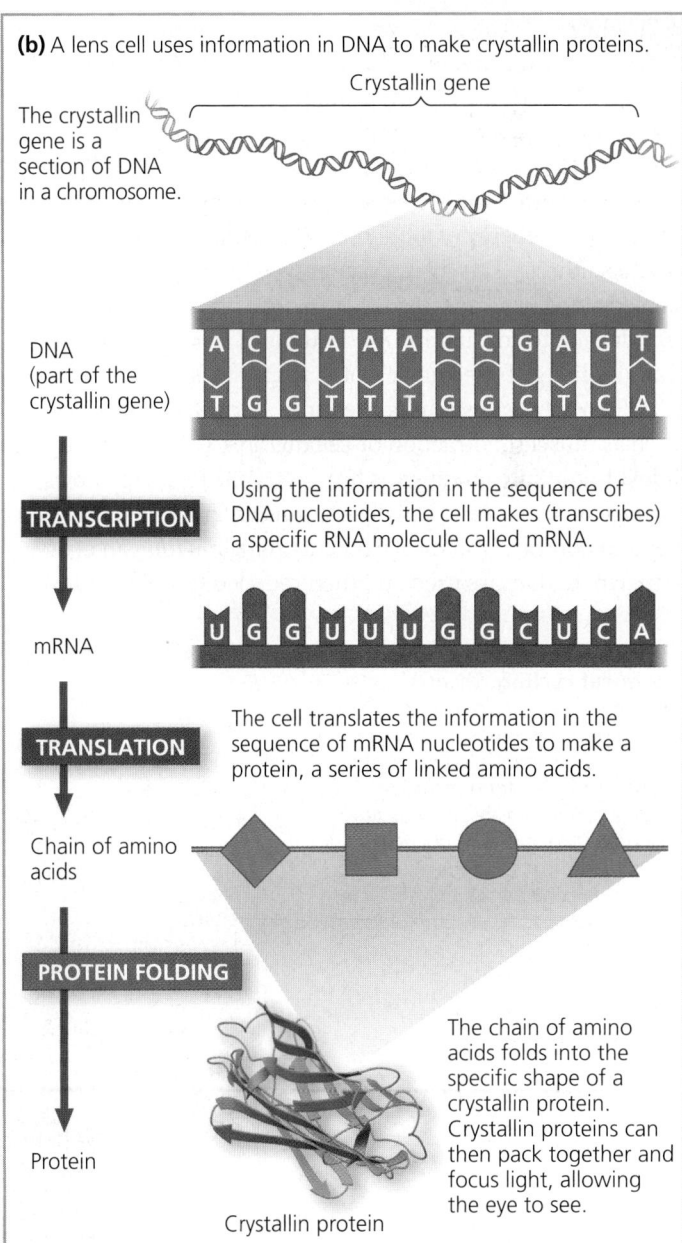

**(b)** A lens cell uses information in DNA to make crystallin proteins.

Crystallin gene

The crystallin gene is a section of DNA in a chromosome.

DNA (part of the crystallin gene)

A C C A A A C C G A G T
T G G T T T G G C T C A

**TRANSCRIPTION**

Using the information in the sequence of DNA nucleotides, the cell makes (transcribes) a specific RNA molecule called mRNA.

mRNA

U G G U U U G G C U C A

**TRANSLATION**

The cell translates the information in the sequence of mRNA nucleotides to make a protein, a series of linked amino acids.

Chain of amino acids

**PROTEIN FOLDING**

Protein

The chain of amino acids folds into the specific shape of a crystallin protein. Crystallin proteins can then pack together and focus light, allowing the eye to see.

Crystallin protein

Three important research developments have made the genomic and proteomic approaches possible. One is "high-throughput" technology, tools that can analyze many biological samples very rapidly. The second major development is **bioinformatics**, the use of computational tools to store, organize, and analyze the huge volume of data that results from high-throughput methods. The third key development is the formation of interdisciplinary research teams—groups of diverse specialists that may include computer scientists, mathematicians, engineers, chemists, physicists, and, of course, biologists from a variety of fields. Researchers in such teams aim to learn how the activities of all the proteins and RNAs encoded by the DNA are coordinated in cells and in whole organisms.

## Theme: Life Requires the Transfer and Transformation of Energy and Matter

**ENERGY AND MATTER** Moving, growing, reproducing, and the various cellular activities of life are work, and work requires energy. The input of energy, primarily from the sun, and the transformation of energy from one form to another make life possible **(Figure 1.9)**. When a plant's leaves absorb sunlight, molecules within the leaves convert the energy of sunlight to the chemical energy of food, such as sugars, in the process of photosynthesis. The chemical energy in food molecules is then passed along by plants and other photosynthetic organisms (producers) to consumers. A consumer is an organism that obtains its energy by feeding on other organisms or their remains.

When an organism uses chemical energy to perform work, such as muscle contraction or cell division, some of that energy is lost to the surroundings as heat. As a result, energy flows *through* an ecosystem, usually entering as light and exiting as heat. In contrast, chemical elements remain *within* an ecosystem, where they are used and then recycled (see Figure 1.9).

Chemicals that a plant absorbs from the air or soil may be incorporated into the plant's body and then passed to an animal that eats the plant. Eventually, these chemicals will be returned to the environment by decomposers, such as bacteria and fungi, that break down waste products, organic debris, and the bodies of dead organisms. The chemicals are then available to be taken up by plants again, thereby completing the cycle.

## Theme: Organisms Interact with Other Organisms and the Physical Environment

**INTERACTIONS** Every organism in an ecosystem interacts with other organisms. A flowering plant, for example, interacts with soil microorganisms associated with its roots, insects that pollinate its flowers, and animals that eat its leaves and petals. Interactions between organisms include those that are mutually beneficial (as when fish eat small parasites on a turtle, shown in **Figure 1.10**), and those in which one species benefits and the other is harmed (as when a lion kills and eats a zebra). In some interactions between species both are harmed (as when two plants compete for a soil resource that is in short supply).

Each organism in an ecosystem also interacts continuously with physical factors in its environment. The leaves of a flowering plant, for example, absorb light from the sun, take in carbon dioxide from the air, and release oxygen to the air. The environment is also affected by the organisms living there. For example, a plant takes up water and minerals from the soil through its roots, and its roots break up rocks, thereby contributing to the formation of soil. On a global scale, plants and other photosynthetic organisms have generated all the oxygen in the atmosphere.

Like other organisms, we humans interact with our environment. Unfortunately, our interactions sometimes have dire consequences. For example, over the past 150 years, humans have greatly increased the burning of fossil fuels (coal, oil, and gas). This practice releases large amounts of carbon dioxide

▶ **Figure 1.9 Energy flow and chemical cycling.** There is a one-way flow of energy in an ecosystem: During photosynthesis, plants convert energy from sunlight to chemical energy (stored in food molecules such as sugars), which is used by plants and other organisms to do work and is eventually lost from the ecosystem as heat. In contrast, chemicals cycle between organisms and the physical environment.

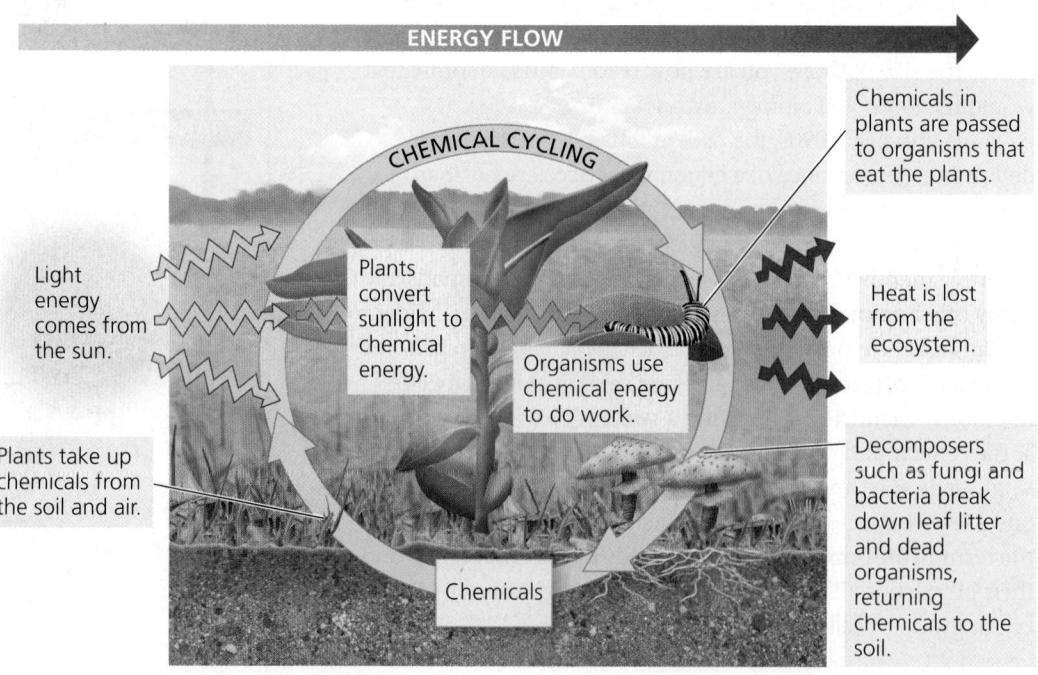

ENERGY FLOW

CHEMICAL CYCLING

Chemicals in plants are passed to organisms that eat the plants.

Light energy comes from the sun.

Plants convert sunlight to chemical energy.

Organisms use chemical energy to do work.

Heat is lost from the ecosystem.

Plants take up chemicals from the soil and air.

Chemicals

Decomposers such as fungi and bacteria break down leaf litter and dead organisms, returning chemicals to the soil.

▶ **Figure 1.10 A mutually beneficial interaction between species.** These fish feed on small organisms living on the sea turtle's skin and shell. The sea turtle benefits from the removal of parasites, and the fish gain a meal and protection from enemies. For more examples of mutually beneficial relationships (mutualisms), see Make Connections Figure 29.10.

($CO_2$) and other gases into the atmosphere. About half of this $CO_2$ stays in the atmosphere, causing heat to be trapped close to Earth's surface (see Figure 43.26). Scientists calculate that the $CO_2$ that human activities have added to the atmosphere has increased the average temperature of the planet by about 1°C since 1900. At the current rates that $CO_2$ and other gases are being added to the atmosphere, global models predict an additional rise of at least 3°C before the end of this century.

This ongoing global warming is a major aspect of **climate change**, a directional change to the global climate that lasts for three decades or more (as opposed to short-term changes in the weather). But global warming is not the only way the climate is changing: wind and precipitation patterns are also shifting, and extreme weather events such as storms and droughts are occurring more often. Climate change has already affected organisms and their habitats all over the planet. For example, polar bears have lost much of the ice platform from which they hunt, leading to food shortages and increased mortality rates. As habitats deteriorate, hundreds of plant and animal species are shifting their ranges to more suitable locations—but for some, there is insufficient suitable habitat, or they may not be able to migrate quickly enough. As a result, the populations of many species are shrinking in size or even disappearing **(Figure 1.11)**.

This trend can ultimately result in extinction, the permanent loss of a species. As we'll discuss in greater detail in Concept 43.4, the consequences of these changes for humans and other organisms may be profound.

## Evolution, the Core Theme of Biology

Having considered four of the unifying themes that run through this text (organization, information, energy and matter, and interactions), let's now turn to biology's core theme—evolution. Evolution makes sense of everything we know about living organisms. As the fossil record clearly shows, life has been evolving on Earth for billions of years, resulting in a vast diversity of past and present organisms. But along with the diversity there is also unity. For example, while sea horses, jackrabbits, hummingbirds, crocodiles, and giraffes all look very different, their skeletons are organized in the same basic way.

The scientific explanation for the unity and diversity of organisms—as well as for the adaptation of organisms to their particular environments—is **evolution**: the concept that the organisms living on Earth today are the modified descendants of common ancestors. As a result of descent with modification, two species share certain traits (unity) simply because they have descended from a common ancestor. Furthermore, we can account for differences between two species (diversity) with the idea that certain heritable changes occurred after the two species diverged from their common ancestor. An abundance of evidence of different types supports the occurrence of evolution and the theory that describes how it takes place, which we'll discuss in detail in Chapters 19–23. Meanwhile, in the next section, we'll continue our introduction to the fundamental concept of evolution.

▶ **Figure 1.11 Threatened by global warming.** A warmer environment causes lizards in the genus *Sceloporus* to spend more time in refuges from the heat, reducing the time available for foraging. The lizards' food intake drops, decreasing their reproductive success. Indeed, surveys of 200 populations of *Sceloporus* species in Mexico show that 12% of these populations have disappeared since 1975. For more examples of how climate change is affecting life on Earth, see Make Connections Figure 43.28.

**CONCEPT CHECK 1.1**

1. Starting with the molecular level in Figure 1.3, write a sentence that includes components from the previous (lower) level of biological organization, for example, "A molecule consists of *atoms* bonded together." Continue with organelles, moving up the biological hierarchy.
2. Identify the theme or themes exemplified by (a) the sharp quills of a porcupine, (b) the development of a multicellular organism from a single fertilized egg, and (c) a hummingbird using sugar to power its flight.
3. **WHAT IF?** For each theme discussed in this section, give an example not mentioned in the text.

For suggested answers, see Appendix A.

# The Core Theme: Evolution accounts for the unity and diversity of life

**EVOLUTION** Diversity is a hallmark of life. Biologists have identified and named about 1.8 million species of organisms, and estimates of the number of living species range from about 10 million to over 100 million. These remarkably diverse forms of life arose by evolutionary processes. Before exploring evolution further, however, let's first consider how biologists organize the enormous variety of life forms on this planet into manageable and informative groupings.

## Classifying the Diversity of Life

Humans have a tendency to group diverse items according to their similarities and relationships to each other. Following this inclination, biologists have long used careful comparisons of form and function to classify life-forms into a hierarchy of increasingly inclusive groups. Consider, for example, the species known as the leopard (*Panthera pardus*). Leopards belong to the same genus (*Panthera*) as tigers and lions. Bringing together several similar genera forms a family, which in turn is a component of an order and then a class. For the leopard, this means being grouped with cougars, cheetahs, and others in the family Felidae, with wolves in the order Carnivora, and with dolphins (and us) in the class Mammalia (see Figure 20.3). These animals can be classified into still broader groupings: the phylum Chordata and the kingdom Animalia.

In the last few decades, new methods of assessing species relationships, especially comparisons of DNA sequences, have led to a reevaluation of the larger groupings. Although this reevaluation is ongoing, there is consensus among biologists that the kingdoms of life, whatever their number, can be further grouped into three higher levels of classification called domains: Bacteria, Archaea, and Eukarya **(Figure 1.12)**.

▼ **Figure 1.12 The three domains of life.**

**(a) Domain Bacteria**

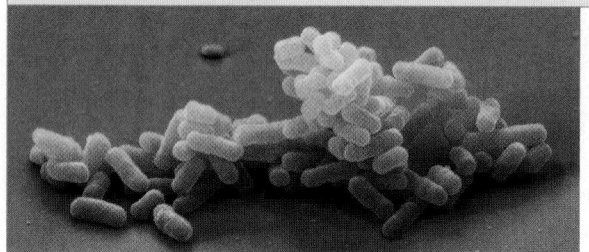

2 μm

**Bacteria** are the most diverse and widespread prokaryotes and are now classified into multiple kingdoms. Each rod-shaped structure in this photo is a bacterial cell.

**(b) Domain Archaea**

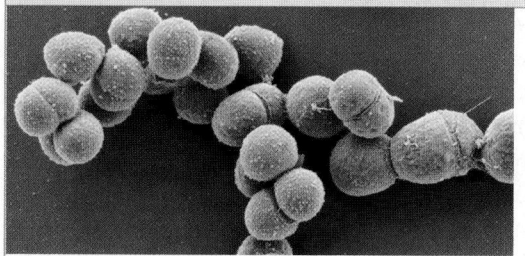

2 μm

Domain Archaea includes multiple kingdoms. Some of the prokaryotes known as **archaea** live in Earth's extreme environments, such as salty lakes and boiling hot springs. Each round structure in this photo is an archaeal cell.

**(c) Domain Eukarya**

◀ **Kingdom Animalia** consists of multicellular eukaryotes that ingest other organisms.

100 μm

▲ **Kingdom Plantae** (land plants) consists of terrestrial multicellular eukaryotes that carry out photosynthesis, the conversion of light energy to the chemical energy in food.

▶ **Kingdom Fungi** is defined in part by the nutritional mode of its members (such as this mushroom), which absorb nutrients from outside their bodies.

▶ **Protists** are mostly unicellular eukaryotes and some relatively simple multicellular relatives. Pictured here is an assortment of protists inhabiting pond water. Scientists are currently debating how to classify protists in a way that accurately reflects their evolutionary relationships.

The organisms making up two of the three domains—**Bacteria** and **Archaea**—are prokaryotic. All the eukaryotes (organisms with eukaryotic cells) are grouped in domain **Eukarya**. This domain includes three kingdoms of multicellular eukaryotes: Plantae, Fungi, and Animalia. These three kingdoms are distinguished partly by their modes of nutrition. Plants produce their own sugars and other food molecules by photosynthesis; fungi absorb dissolved nutrients from their surroundings; and animals obtain food by eating and digesting other organisms. Animalia is, of course, our own kingdom.

The most numerous and diverse eukaryotes are the mostly single-celled protists. Although protists once were placed in a single kingdom, they are now classified into several groups. One major reason for this change is the recent DNA evidence showing that some protists are less closely related to other protists than they are to plants, animals, or fungi.

## Unity in the Diversity of Life

Although diversity is apparent in the many forms of life, there is also remarkable unity. Consider, for example, the similar skeletons of different animals and the universal genetic language of DNA (the genetic code), both mentioned earlier. In fact, similarities between organisms are evident at all levels of the biological hierarchy.

How can we account for life's dual nature of unity and diversity? The process of evolution, explained next, illuminates both the similarities and differences in the world of life. It also introduces another dimension of biology: the passage of time. The history of life, as documented by fossils and other evidence, is the saga of an ever-changing Earth billions of years old, inhabited by an evolving cast of living forms (**Figure 1.13**).

▲ **Figure 1.13 Digging into the past.** Paleontologists carefully excavate the hind leg of a long-necked dinosaur (*Rapetosaurus krausei*) from rocks in Madagascar.

▶ **Figure 1.14 Charles Darwin as a young man.** His revolutionary book, which is commonly referred to as *The Origin of Species*, was first published in 1859.

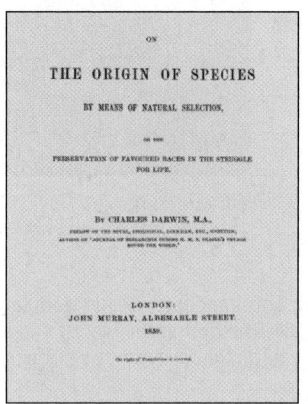

## Charles Darwin and the Theory of Natural Selection

An evolutionary view of life came into sharp focus in 1859, when Charles Darwin published one of the most important and influential books ever written, *On the Origin of Species by Means of Natural Selection* (**Figure 1.14**). *The Origin of Species* articulated two main points. The first was that species arise from a succession of ancestors that were different from them. Darwin called this process "descent with modification." This insightful phrase captured the duality of life's unity and diversity—unity in the kinship among species that descended from common ancestors and diversity in the modifications that evolved as species branched from their common ancestors (**Figure 1.15**). Darwin's second main point was his proposal that "natural selection" is a primary cause of descent with modification.

▼ **American flamingo**   ▼ **European robin**   ▼ **Gentoo penguin**

▲ **Figure 1.15 Unity and diversity among birds.** These three birds are variations on a common body plan. For example, each has feathers, a beak, and wings, although these features are highly specialized for the birds' diverse lifestyles.

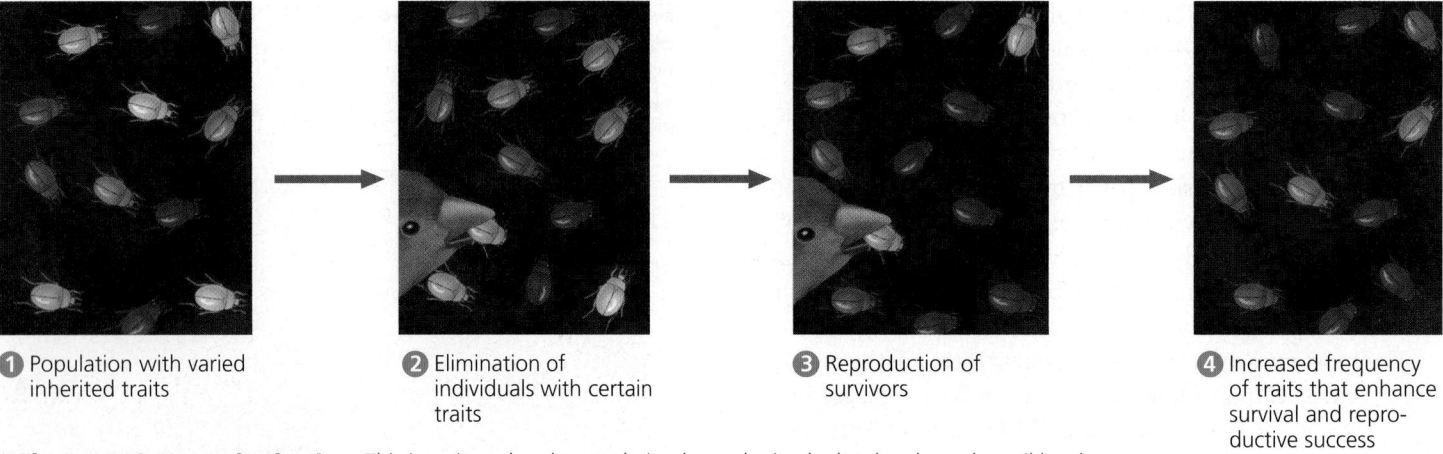

① Population with varied inherited traits

② Elimination of individuals with certain traits

③ Reproduction of survivors

④ Increased frequency of traits that enhance survival and reproductive success

▲ **Figure 1.16 Natural selection.** This imaginary beetle population has colonized a locale where the soil has been blackened by a recent brush fire. Initially, the population varies extensively in the inherited coloration of the individuals, from very light gray to charcoal. For hungry birds that prey on the beetles, it is easiest to spot the beetles that are lightest in color.

Darwin developed his theory of natural selection from observations that by themselves were neither new nor profound. However, although others had described the pieces of the puzzle, it was Darwin who saw how they fit together. His three essential observations were the following: First, individuals in a population vary in their traits, many of which seem to be heritable (passed on from parents to offspring). Second, a population can produce far more offspring than can survive to produce offspring of their own. Competition is thus inevitable. Third, species generally are suited to their environments—in other words, they are adapted to their circumstances. For instance, a common adaptation among birds that eat mostly hard seeds is an especially strong beak.

By making inferences from these three observations, Darwin arrived at his theory of evolution. He reasoned that individuals with inherited traits that are better suited to the local environment are more likely to survive and reproduce than are less well-suited individuals. Over many generations, a higher and higher proportion of individuals in a population will have the advantageous traits. Darwin called this mechanism of evolutionary adaptation **natural selection** because the natural environment consistently "selects" for the propagation of certain traits among naturally occurring variant traits in the population **(Figure 1.16)**.

## The Tree of Life

For another example of unity and diversity, consider the human arm. The bones, joints, nerves, and blood vessels in your forelimb are very similar to those in the foreleg of a horse, the flipper of a whale, and the wing of a bat. Indeed, all mammalian forelimbs are anatomical variations of a common architecture. According to the Darwinian concept of descent with modification, the shared anatomy of mammalian limbs reflects inheritance of the limb structure from a common ancestor—the "prototype" mammal from which all other mammals descended. The diversity of mammalian forelimbs results from modification by natural selection operating over millions of years in different environmental contexts.

Darwin proposed that natural selection, by its cumulative effects over long periods of time, could cause an ancestral species to give rise to two or more descendant species. This could occur, for example, if one population of organisms fragmented into several subpopulations isolated in different environments. In these separate arenas of natural selection, one species could gradually radiate into multiple species as the geographically isolated populations adapted over many generations to different environmental conditions.

The "family tree" of six finch species shown in **Figure 1.17** illustrates a famous example of the process of radiation. Darwin collected specimens of finches during his 1835 visit to the remote Galápagos Islands, 900 kilometers (km) west of South America. The Galápagos finches are believed to have descended from an ancestral finch species that reached the archipelago from South America or the Caribbean. Over time, the Galápagos finches diversified from their ancestor as populations became adapted to different food sources on their particular islands. Years after Darwin collected the finches, researchers began to sort out their evolutionary relationships, first from anatomical and geographic data and more recently using DNA sequence comparisons.

Biologists' diagrams of such evolutionary relationships generally take treelike forms, though the trees are often turned sideways, as in Figure 1.17. Tree diagrams make sense: Just as an individual has a genealogy that can be diagrammed as a family tree, each species is one twig of a branching tree of life extending back in time through ancestral species more and more remote. Species that are very similar, such as the Galápagos finches, share a relatively recent common ancestor. Through an ancestor that lived much farther back in time, finches are related to sparrows, hawks, penguins, and all other birds. Furthermore, finches and other birds are related to us through a common ancestor even more ancient. Trace life back far enough, and we reach the early prokaryotes that inhabited Earth 3.5 billion years ago. We can recognize their vestiges in our own cells—in the universal genetic code, for example. Indeed, all of life is connected through its long evolutionary history.

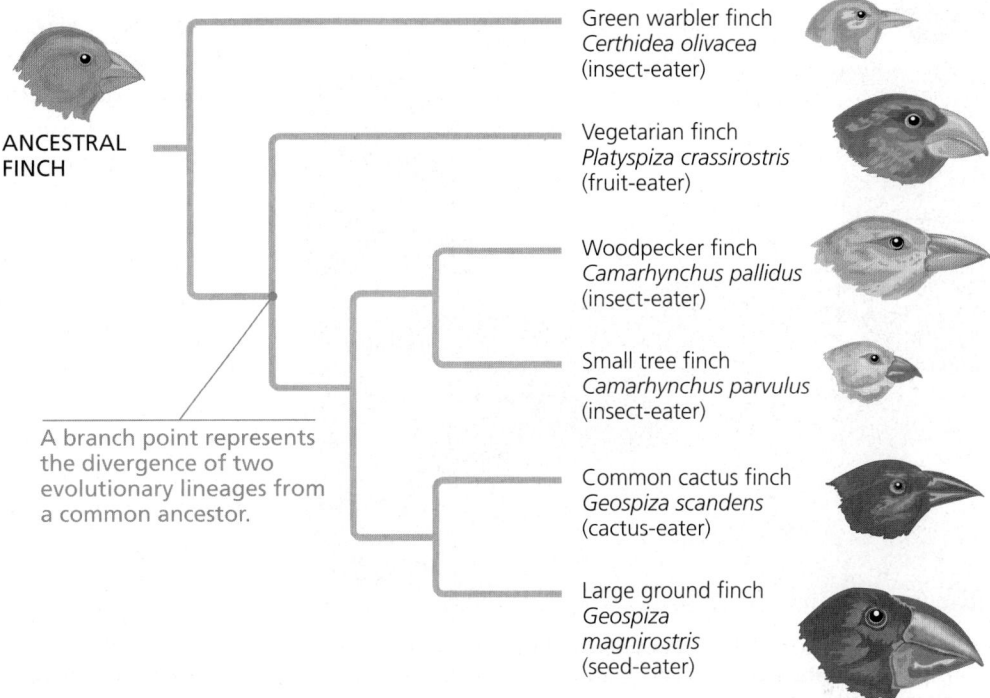

► **Figure 1.17 Descent with modification: finches on the Galápagos Islands.** This "tree" diagram illustrates a current model for the evolutionary relationships of finches on the Galápagos. Note the various beaks, which are adapted to particular food sources. For example, heavier, thicker beaks are better at cracking seeds, while more slender beaks are better at grasping insects.

ANCESTRAL FINCH

A branch point represents the divergence of two evolutionary lineages from a common ancestor.

Green warbler finch
*Certhidea olivacea*
(insect-eater)

Vegetarian finch
*Platyspiza crassirostris*
(fruit-eater)

Woodpecker finch
*Camarhynchus pallidus*
(insect-eater)

Small tree finch
*Camarhynchus parvulus*
(insect-eater)

Common cactus finch
*Geospiza scandens*
(cactus-eater)

Large ground finch
*Geospiza magnirostris*
(seed-eater)

---

**CONCEPT CHECK 1.2**

1. How is a mailing address analogous to biology's hierarchical classification system?
2. Explain why "editing" is an appropriate metaphor for how natural selection acts on a population's heritable variation.
3. **DRAW IT** Recent evidence indicates that fungi and animals are more closely related to each other than either of these kingdoms is to plants. Draw a simple branching pattern that symbolizes the proposed relationship between these three kingdoms of multicellular eukaryotes.

   For suggested answers, see Appendix A.

**CONCEPT 1.3**

# In studying nature, scientists form and test hypotheses

**Science** is a way of knowing—an approach to understanding the natural world. It developed out of our curiosity about ourselves, other life forms, our planet, and the universe.

At the heart of science is **inquiry**, a search for information and explanations of natural phenomena. There is no formula for successful scientific inquiry, no single scientific method that researchers must rigidly follow. As in all quests, science includes elements of challenge, adventure, and luck, along with careful planning, reasoning, creativity, patience, and the persistence to overcome setbacks. Such diverse elements of inquiry make science far less structured than most people realize. That said, it is possible to highlight certain characteristics that help to distinguish science from other ways of describing and explaining nature.

Scientists use a process of inquiry that includes making observations, forming logical explanations (*hypotheses*), and testing them. The process is necessarily repetitive: In testing a hypothesis, our observations may inspire revision of the original hypothesis or formation of a new one, thus leading to further testing. In this way, scientists circle closer and closer to their best estimation of the laws governing nature.

## Exploration and Discovery

Biology, like other sciences, begins with careful observation. In gathering information, biologists often use tools, such as microscopes, precision thermometers, or high-speed cameras, that extend their senses or facilitate careful measurement. Observations can reveal valuable information about the natural world. For example, a series of detailed observations have shaped our understanding of cell structure. Another set of observations is currently expanding our databases of genome sequences from diverse species and of genes whose expression is altered in diseases.

In exploring nature, biologists also rely heavily on the scientific literature, the published contributions of fellow scientists. By reading about and understanding past studies, scientists can build on the foundation of existing knowledge, focusing their investigations on observations that are original and on hypotheses that are consistent with previous findings. Identifying publications relevant to a new line of research is now easier than at any point in the past, thanks to indexed and searchable electronic databases.

## Gathering and Analyzing Data

Recorded observations are called **data**. Put another way, data are items of information on which scientific inquiry is based. Some data are *qualitative*, such as descriptions of what is observed. For example, British primate researcher Jane Goodall spent decades recording her observations of chimpanzee behavior during field research in a Tanzanian jungle

▲ **Figure 1.18 Jane Goodall collecting qualitative data on chimpanzee behavior.** Goodall recorded her observations in field notebooks, often with sketches of the animals' behavior.

**(Figure 1.18)**. She also documented her observations with photographs and movies.

In her studies, Goodall also gathered and recorded volumes of *quantitative* data, such as the frequency and duration of specific behaviors for different members of a group of chimpanzees in a variety of situations. Quantitative data are generally expressed as numerical measurements and often organized into tables or graphs. Scientists analyze their data using a type of mathematics called *statistics* to test whether their results are significant or merely due to random fluctuations. All results presented in this text have been shown to be statistically significant.

Collecting and analyzing observations can lead to important conclusions based on a type of logic called **inductive reasoning**. Through induction, we derive generalizations from a large number of specific observations. The generalization "All organisms are made of cells" was based on two centuries of microscopic observations made by biologists examining cells in diverse biological specimens. Careful observations and data analyses, along with the generalizations reached by induction, are fundamental to our understanding of nature.

## Forming and Testing Hypotheses

Our innate curiosity often stimulates us to pose questions about the natural basis for the phenomena we observe in the world. What *caused* the different chimpanzee behaviors observed in the wild? What *explains* the variation in coat color among the mice of a single species, shown in Figures 1.1 and 1.2? In science, answering such questions usually involves forming and testing logical explanations—that is, hypotheses.

In science, a **hypothesis** is an explanation, based on observations and assumptions, that leads to a testable prediction.

Said another way, a hypothesis is an explanation on trial. The hypothesis is usually a rational accounting for a set of observations, based on the available data and guided by inductive reasoning. A scientific hypothesis must lead to predictions that can be tested by making additional observations or by performing experiments. An **experiment** is a scientific test, often carried out under controlled conditions.

We all make observations and develop questions and hypotheses in solving everyday problems. Let's say, for example, that your desk lamp is plugged in and turned on but the bulb isn't lit. That's an observation. The question is obvious: Why doesn't the lamp work? Two reasonable hypotheses based on your experience are that (1) the bulb is burnt out or (2) the lamp is broken. Each of these hypotheses leads to predictions you can test with experiments. For example, the burnt-out bulb hypothesis predicts that replacing the bulb will fix the problem. Figuring things out in this way by trial and error is a hypothesis-based approach.

### Deductive Reasoning

A type of logic called deduction is also built into the use of hypotheses in science. While induction entails reasoning from a set of specific observations to reach a general conclusion, **deductive reasoning** involves logic that flows in the opposite direction, from the general to the specific. From general premises, we extrapolate to the specific results we should expect if the premises are true. In the scientific process, deductions usually take the form of predictions of results that will be found if a particular hypothesis (premise) is correct. We then test the hypothesis by carrying out experiments or observations to see whether or not the results are as predicted. This deductive testing takes the form of "*If . . . then*" logic. In the case of the desk lamp example: *If* the burnt-out bulb hypothesis is correct, *then* the lamp should work when you replace the bulb with a new one.

We can use the desk lamp example to illustrate two other key points about the use of hypotheses in science. First, one can always devise additional hypotheses to explain a set of observations. For instance, another of the many possible alternative hypotheses to explain our dead desk lamp is that the wall outlet is faulty. Although you could design an experiment to test this hypothesis, we can never test all possible explanations. Second, we can never *prove* that a hypothesis is true. The burnt-out bulb hypothesis is the most likely explanation, but testing supports that hypothesis *not* by proving that it is correct, but rather by not finding that it is false. For example, if replacing the bulb fixed the desk lamp, it might have been because the bulb we replaced was good but not screwed in properly.

Although a hypothesis can never be proved beyond the shadow of a doubt, testing it in various ways can significantly increase our confidence in its validity. Often, rounds of hypothesis formulation and testing lead to a scientific consensus—the shared conclusion of many scientists that a particular hypothesis explains the known data well and stands up to experimental testing.

### Questions That Can and Cannot Be Addressed by Science

Scientific inquiry is a powerful way to learn about nature, but there are limitations to the kinds of questions it can answer. A scientific hypothesis must be *testable*; there must be some observation or experiment that could reveal if such an idea is more likely to be true or false. For example, the hypothesis that a burnt-out bulb is the sole reason the lamp doesn't work would not be supported if replacing the bulb with a new one didn't fix the lamp.

Not all hypotheses meet the criteria of science: You wouldn't be able to test the hypothesis that invisible ghosts are fooling with your desk lamp! Because science only deals with natural, testable explanations for natural phenomena, it can neither support nor contradict the invisible ghost hypothesis, nor whether spirits or elves cause storms, rainbows, or illnesses. Such supernatural explanations are simply outside the bounds of science, as are religious matters, which are issues of personal faith.

▶ **Figure 1.19 The process of science: A realistic model.** The process of science is not linear, but instead involves backtracking, repetition, and feedback between different steps of the process. This illustration is based on a model (How Science Works) from the website Understanding Science.

### The Flexibility of the Scientific Process

The way that researchers answer questions about the natural and physical world is often idealized as the scientific method. However, very few scientific studies adhere rigidly to the sequence of steps that are typically used to describe this approach. For example, a scientist may start to design an experiment, but then backtrack after realizing that more preliminary observations are necessary. In other cases, observations remain too puzzling to prompt well-defined questions until further study provides a new context in which to view those observations. For example, scientists could not unravel the details of how genes encode proteins until *after* the discovery of the structure of DNA (an event that took place in 1953).

A more realistic model of the scientific process is shown in **Figure 1.19**. The focus of this model, shown in the central circle in the figure, is the forming and testing of hypotheses. This core set of activities is the reason that science does so well in explaining phenomena in the natural world. These activities, however, are shaped by exploration and discovery (upper circle)

**EXPLORATION AND DISCOVERY**
- Observing nature
- Asking questions
- Reading the scientific literature

**FORMING AND TESTING HYPOTHESES**

**Testing Ideas**
- Forming hypotheses
- Predicting results
- Doing experiments and/or making observations
- Gathering data
- Analyzing results

**Interpreting Results**
Data may...
- Support a hypothesis
- Contradict a hypothesis
- Inspire a revised or new hypothesis

**SOCIETAL BENEFITS AND OUTCOMES**
- Developing technology
- Informing policy
- Solving problems
- Building knowledge

**COMMUNITY ANALYSIS AND FEEDBACK**
- Feedback and peer review
- Replication of findings
- Publication
- Consensus building

and influenced by interactions with other scientists and with society more generally (lower circles). For example, the community of scientists influences which hypotheses are tested, how test results are interpreted, and what value is placed on the findings. Similarly, societal needs—such as the push to cure cancer or understand the process of climate change—may help shape what research projects are funded and how extensively the results are discussed.

Now that we have highlighted the key features of scientific inquiry—making observations and forming and testing hypotheses—you should be able to recognize these features in a case study of actual scientific research.

## A Case Study in Scientific Inquiry: Investigating Coat Coloration in Mouse Populations

Our case study begins with a set of observations and inductive generalizations. Color patterns of animals vary widely in nature, sometimes even among members of the same species. What accounts for such variation? As you may recall, the two mice depicted at the beginning of this chapter are members of the same species (*Peromyscus polionotus*), but they have different coat (fur) color patterns and reside in different environments. The beach mouse lives along the Florida seashore, a habitat of brilliant white sand dunes with sparse clumps of beach grass. The inland mouse lives on darker, more fertile soil farther inland **(Figure 1.20)**. Even a brief glance at the photographs in Figure 1.20 reveals a striking match of mouse coloration to its habitat. The natural predators of these mice, including hawks, owls, foxes, and coyotes, all use their sense of sight to hunt for prey. It was logical, therefore, for Francis B. Sumner, a naturalist studying populations of these mice in the 1920s, to hypothesize that their coloration patterns had evolved as adaptations that camouflage the mice in their native environments, protecting them from predation.

As obvious as the camouflage hypothesis may seem, it still required testing. In 2010, biologist Hopi Hoekstra of Harvard

University and a group of her students headed to Florida to test the prediction that mice with coloration that did not match their habitat would be preyed on more heavily than the native, well-matched mice. **Figure 1.21** summarizes this field experiment, introducing a format we will use throughout the book to walk through other examples of biological inquiry.

The researchers built hundreds of models of mice and spray-painted them to resemble either beach or inland mice, so that the models differed only in their color patterns. The researchers placed equal numbers of these model mice randomly in both habitats and left them overnight. The mouse models resembling the native mice in the habitat were the *control* group (for instance, light-colored mouse models in the beach habitat), while the models with the non-native coloration were the *experimental* group (for example, darker models in the beach habitat). The following morning, the team counted and recorded signs of predation events, which ranged from bites and gouge marks on some models to the outright disappearance of other models. Judging by the shape of the predators' bites and the tracks surrounding the experimental sites, the predators appeared to be split fairly evenly between mammals (such as foxes and coyotes) and birds (such as owls, herons, and hawks).

For each environment, the researchers then calculated the percentage of predator attacks that targeted camouflaged models. The results were clear-cut: Camouflaged models showed much lower predation rates than those lacking camouflage in both the dune habitat (where light mice were less vulnerable) and the inland habitat (where dark mice were less vulnerable). The data thus fit the key prediction of the camouflage hypothesis.

## Experimental Variables and Controls

In carrying out an experiment, a researcher often manipulates a factor in a system and observes the effects of this change. The mouse camouflage experiment described in Figure 1.21 is an example of a **controlled experiment**, one that is designed to compare an experimental group (the non-camouflaged models, in this case) with a control group (the camouflaged

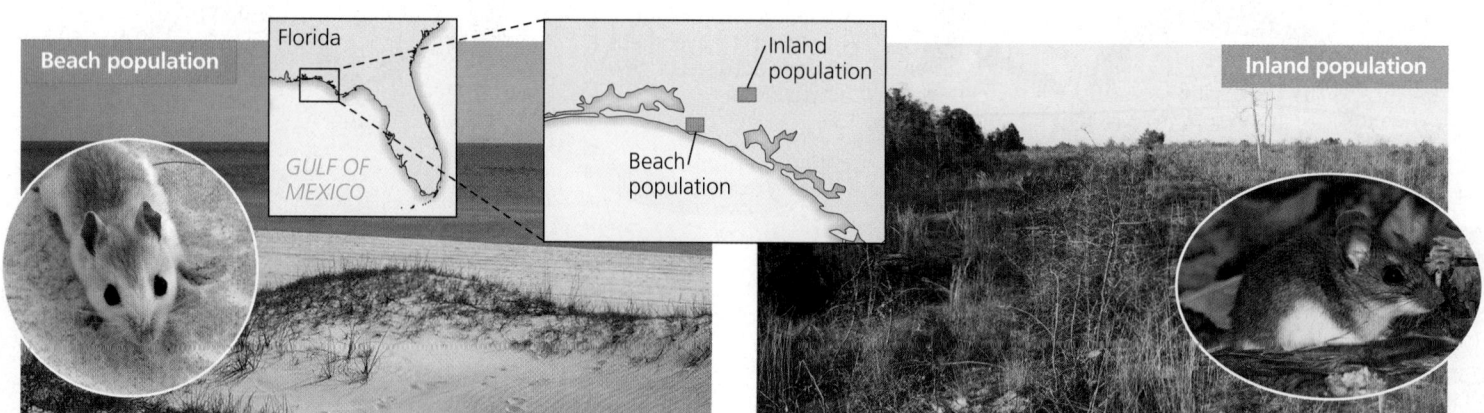

Beach mice living on sparsely vegetated sand dunes along the coast have light tan, dappled fur on their backs that allows them to blend into their surroundings, providing camouflage.

Members of the same species living about 30 km inland have dark fur on their backs, camouflaging them against the dark ground of their habitat.

▲ **Figure 1.20 Different coloration in beach and inland populations of *Peromyscus polionotus*.**

## Does camouflage affect predation rates on two populations of mice?

**Experiment** Hopi Hoekstra and colleagues tested the hypothesis that coat coloration provides camouflage that protects beach and inland populations of *Peromyscus polionotus* mice from predation in their habitats. The researchers spray-painted mouse models with light or dark color patterns that matched those of the beach and inland mice and placed models with each of the patterns in both habitats. The next morning, they counted damaged or missing models.

**Results** For each habitat, the researchers calculated the percentage of attacked models that were camouflaged or non-camouflaged. In both cases, the models whose pattern did not match their surroundings suffered much higher "predation" than did the camouflaged models.

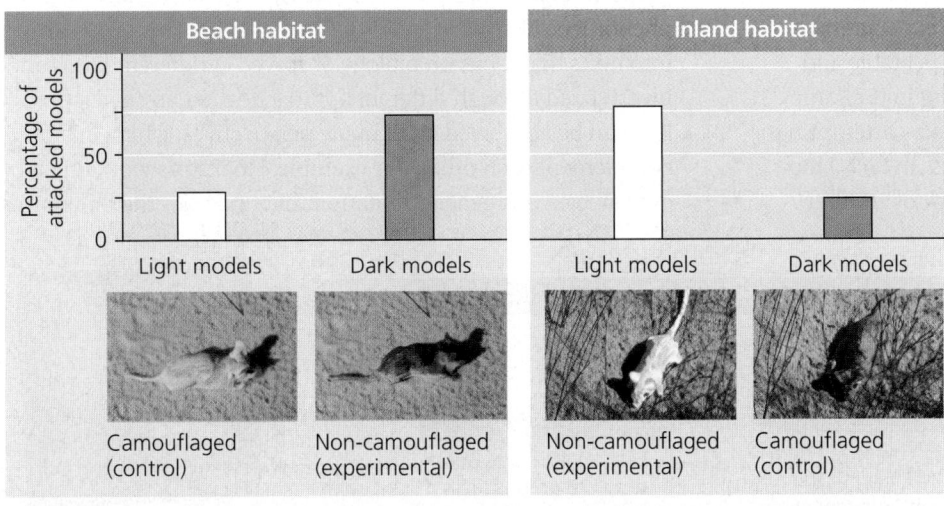

**Conclusion** The results are consistent with the researchers' prediction that mouse models with camouflage coloration would be preyed on less often than non-camouflaged mouse models. Thus, the experiment supports the camouflage hypothesis.

**Data from** S. N. Vignieri, J. G. Larson, and H. E. Hoekstra, The selective advantage of crypsis in mice, *Evolution* 64:2153–2158 (2010).

**INTERPRET THE DATA** *The bars indicate the percentage of the attacked models that were either light or dark. Assume 100 mouse models were attacked in each habitat. For the beach habitat, how many were light models? Dark models? Answer the same questions for the inland habitat.*

models). Both the factor that is manipulated and the factor that is subsequently measured are experimental **variables**—a feature or quantity that varies in an experiment. In our example, the color of the mouse model was the **independent variable**—the factor manipulated by the researchers. The **dependent variable** is the factor being measured that is predicted to be affected by the independent variable; in this case, the researchers measured the amount of predation in response to variation in color of the mouse model. Ideally, the experimental and control groups differ in only one independent variable—in the mouse experiment, color.

Without the control group of camouflaged models, the researchers would not have been able to rule out other factors as causes of the more frequent attacks on the non-camouflaged models—such as different numbers of predators or different temperatures in the various test areas. The clever experimental design left coloration as the only factor that could account for the low predation rate on models camouflaged with respect to the surrounding environment.

A common misconception is that the term *controlled experiment* means that scientists control all features of the experimental environment. But that's impossible in field research and can be very difficult even in a highly regulated laboratory setting. Researchers usually "control" unwanted variables not by *eliminating* them but by *canceling out* their effects using control groups.

### Theories in Science

"It's just a theory!" Our everyday use of the term *theory* often implies an untested speculation. But the term *theory* has a different meaning in science. What is a scientific theory, and how is it different from a hypothesis or from mere speculation?

First, a scientific **theory** is much broader in scope than a hypothesis. *This* is a hypothesis: "Coat coloration well-matched to their habitat is an adaptation that protects mice from predators." But *this* is a theory: "Evolutionary adaptations arise by natural selection." This theory proposes that natural selection accounts for an enormous variety of adaptations, of which coat color in mice is but one example.

Second, a theory is general enough to spin off many new, testable hypotheses. For example, the theory of natural selection motivated two researchers at Princeton University, Peter and Rosemary Grant, to test the specific hypothesis that the beaks of Galápagos finches evolve in response to changes in the types of available food. (For their results, see the introduction to Chapter 21.)

And third, compared to any one hypothesis, a theory is generally supported by a much greater body of evidence. Those theories that become widely adopted in science (such as the theory of natural selection or the theory of gravity) explain a great diversity of observations and are supported by a vast accumulation of evidence.

In spite of the body of evidence supporting a widely accepted theory, scientists will sometimes modify or even reject theories when new research produces results that don't fit. For example, biologists once lumped bacteria and archaea together as a kingdom of prokaryotes. When new methods for comparing cells and molecules could be used to test such relationships, the evidence led scientists to reject the theory that bacteria and archaea are members of the same kingdom. If there is "truth" in science, it is conditional, based on the weight of available evidence.

## Science as a Social Process

The great scientist Sir Isaac Newton once said: "To explain all nature is too difficult a task for any one man or even for any one age. 'Tis much better to do a little with certainty, and leave the rest for others that come after you." Anyone who becomes a scientist, driven by curiosity about nature, is sure to benefit from the rich storehouse of discoveries by others who have come before. In fact, while movies and cartoons sometimes portray scientists as loners working in isolated labs, science is an intensely social activity. Most scientists work in teams, which often include graduate and undergraduate students.

Science is continuously vetted through the expectation that observations and experiments must be repeatable and hypotheses must be testable. Scientists working in the same research field often check one another's claims by attempting to confirm observations or repeat experiments. In fact, Hopi Hoekstra's experiment benefited from the work of another researcher, D. W. Kaufman, four decades earlier. You can study the design of Kaufman's experiment and interpret the results in the **Scientific Skills Exercise**.

If scientific colleagues cannot repeat experimental findings, this failure may reflect an underlying weakness in the original claim, which will then have to be revised. In this sense, science polices itself. Adherence to high professional standards in reporting results is central to the scientific endeavor, since the validity of experimental data is key to designing further inquiry.

Biologists may approach questions from different angles. Some biologists focus on ecosystems, while others study natural phenomena at the level of organisms or cells. This text is divided into units that focus on biology observed at different levels and investigated through different approaches. Yet any given problem can be addressed from many perspectives, which in fact complement each other. For example, Hoekstra's work uncovered at least one genetic mutation that underlies the differences

---

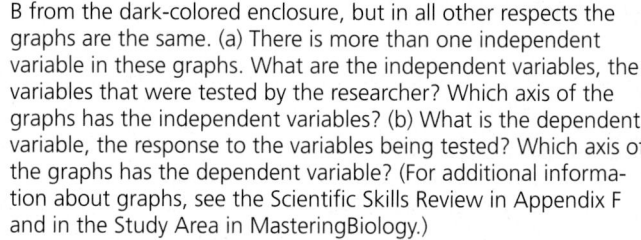

## Scientific Skills Exercise

### Interpreting a Pair of Bar Graphs

**How Much Does Camouflage Affect Predation on Mice by Owls with and without Moonlight?** D. W. Kaufman hypothesized that the extent to which the coat color of a mouse contrasted with the color of its surroundings would affect the rate of nighttime predation by owls. He also hypothesized that contrast would be affected by the amount of moonlight. In this exercise, you will analyze data from his studies of owl-mouse predation that tested these hypotheses.

**How the Experiment Was Done** Pairs of mice (*Peromyscus polionotus*) with different coat colors, one light brown and one dark brown, were released simultaneously into an enclosure that contained a hungry owl. The researcher recorded the color of the mouse that was first caught by the owl. If the owl did not catch either mouse within 15 minutes, the test was recorded as a zero. The release trials were repeated multiple times in enclosures with either a dark-colored soil surface or a light-colored soil surface. The presence or absence of moonlight during each assay was recorded.

**Data from the Experiment**

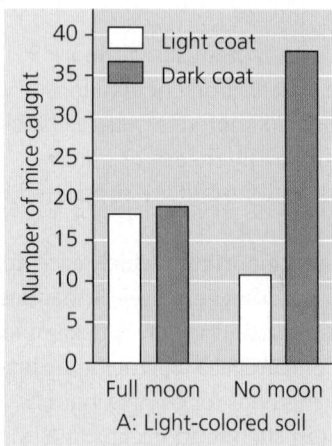 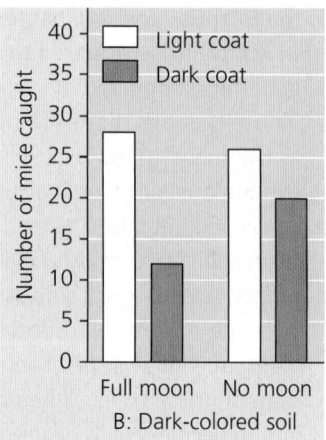

**Data from** D. W. Kaufman, Adaptive coloration in *Peromyscus polionotus*: Experimental selection by owls, *Journal of Mammalogy* 55:271–283 (1974).

**INTERPRET THE DATA**

1. First, make sure you understand how the graphs are set up. Graph A shows data from the light-colored soil enclosure and graph B from the dark-colored enclosure, but in all other respects the graphs are the same. (a) There is more than one independent variable in these graphs. What are the independent variables, the variables that were tested by the researcher? Which axis of the graphs has the independent variables? (b) What is the dependent variable, the response to the variables being tested? Which axis of the graphs has the dependent variable? (For additional information about graphs, see the Scientific Skills Review in Appendix F and in the Study Area in MasteringBiology.)

2. (a) How many dark brown mice were caught in the light-colored soil enclosure on a moonlit night? (b) How many dark brown mice were caught in the dark-colored soil enclosure on a moonlit night? (c) On a moonlit night, would a dark brown mouse be more likely to escape predation by owls on dark- or light-colored soil? Explain your answer.

3. (a) Is a dark brown mouse on dark-colored soil more likely to escape predation under a full moon or with no moon? (b) What about a light brown mouse on light-colored soil? Explain.

4. (a) Under which conditions would a dark brown mouse be most likely to escape predation at night? (b) A light brown mouse?

5. (a) What combination of independent variables led to the highest predation level in enclosures with light-colored soil? (b) What combination of independent variables led to the highest predation level in enclosures with dark-colored soil?

6. Thinking about your answers to question 5, provide a simple statement describing conditions that are especially deadly for either color of mouse.

7. Combining the data from both graphs, estimate the number of mice caught in moonlight versus no-moonlight conditions. Which condition is optimal for predation by the owl? Explain.

(MB) A version of this Scientific Skills Exercise can be assigned in MasteringBiology.

between beach and inland mouse coloration. Because the biologists in her lab have different specialties, her research group has been able not only to characterize evolutionary adaptations, but also to define the molecular basis for particular adaptations in the DNA sequence of the mouse genome.

The research community is part of society at large. The relationship of science to society becomes clearer when we add technology to the picture. The goal of **technology** is to *apply* scientific knowledge for some specific purpose. Because scientists put new technology to work in their research, science and technology are interdependent.

In centuries past, many major technological innovations originated along trade routes, where a rich mix of different cultures ignited new ideas. For example, the printing press was invented around 1440 by Johannes Gutenberg, living in what is now Germany. This invention relied on several innovations from China, including paper and ink, and from Iraq, where technology was developed for the mass production of paper. Like technology, science stands to gain much from embracing a diversity of backgrounds and viewpoints among its practitioners.

The scientific community reflects the customs and behaviors of society at large. It is therefore not surprising that until recently, women and certain racial and ethnic groups have faced huge obstacles in their pursuit to become professional scientists. Over the past 50 years, changing attitudes about career choices have increased the proportion of women in biology, and women now constitute roughly half of undergraduate majors and Ph.D. students in the field. The pace of change has been slow at higher levels in the profession, however, and women and many racial and ethnic groups are still significantly underrepresented in many branches of science. This lack of diversity hampers the progress of science. The more voices that are heard at the table, the more robust and productive the scientific conversation will be. The authors of this textbook welcome all students to the community of biologists, wishing you the joys and satisfactions of this exciting field of science.

## CONCEPT CHECK 1.3

1. Contrast inductive reasoning with deductive reasoning.
2. What qualitative observation led to the quantitative study outlined in Figure 1.21?
3. Why is natural selection called a theory?
4. How does science differ from technology?

**For suggested answers, see Appendix A.**

---

Go to **MasteringBiology®** for Assignments, the eText, and the Study Area with Animations, Activities, Vocab Self-Quiz, and Practice Tests.

# 1  Chapter Review

## SUMMARY OF KEY CONCEPTS

**VOCAB SELF-QUIZ**

goo.gl/gbai8v

### CONCEPT 1.1

**The study of life reveals common themes (pp. 3–9)**

**Theme: Organization**
- The hierarchy of life unfolds as follows: biosphere > ecosystem > community > population > organism > organ system > organ > tissue > cell > organelle > molecule > atom. With each step up, new properties emerge (**emergent properties**) as a result of interactions among components at the lower levels.
- Structure and function are correlated at all levels of biological organization. The cell is the lowest level of organization that can perform all activities required for life. Cells are either prokaryotic or eukaryotic. **Eukaryotic cells** have a DNA-containing nucleus and other membrane-enclosed organelles. **Prokaryotic cells** lack such organelles.

**Theme: Information**
- Genetic information is encoded in the nucleotide sequences of **DNA**. It is DNA that transmits heritable information from parents to offspring. DNA sequences (called **genes**) program a cell's protein production by being transcribed into mRNA and then translated into specific proteins, a process called **gene expression**. Gene expression also produces RNAs that are not translated into proteins but serve other important functions.

**Theme: Energy and Matter**
- Energy flows through an ecosystem. All organisms must perform work, which requires energy. Producers convert energy from sunlight to chemical energy, some of which is then passed on to consumers (the rest is lost from the ecosystem as heat). Chemicals cycle between organisms and the environment.

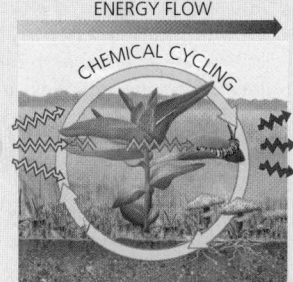

ENERGY FLOW

CHEMICAL CYCLING

**Theme: Interactions**
- Organisms interact continuously with physical factors. Plants take up nutrients from the soil and chemicals from the air and use energy from the sun. Interactions among plants, animals, and other organisms affect the participants in varying ways.

**Core Theme: Evolution**
- Evolution accounts for the unity and diversity of life and also for the match of organisms to their environments.

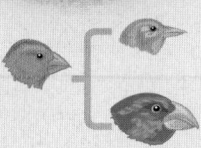

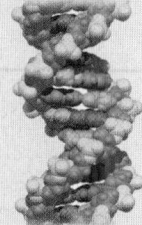

*Thinking about the muscles and nerves in your hand, how does the activity of text messaging reflect the five unifying themes of biology described in this chapter?*

## The Core Theme: Evolution accounts for the unity and diversity of life (pp. 10–13)

- Biologists classify species according to a system of broader and broader groups. Domain **Bacteria** and domain **Archaea** consist of prokaryotes. Domain **Eukarya**, the eukaryotes, includes various groups of protists as well as plants, fungi, and animals. As diverse as life is, there is also evidence of remarkable unity, which is revealed in the similarities between different kinds of organisms.
- Darwin proposed **natural selection** as the mechanism for evolutionary adaptation of populations to their environments. Natural selection is the evolutionary process that occurs when a population is exposed to environmental factors that consistently cause individuals with certain heritable traits to have greater reproductive success than do individuals with other heritable traits.
- Each species is one twig of a branching tree of life extending back in time through more and more remote ancestral species. All of life is connected through its long evolutionary history.

**?** *How could natural selection have led to the evolution of adaptations such as camouflaging coat color in beach mice?*

## In studying nature, scientists form and test hypotheses (pp. 13–19)

- In scientific **inquiry**, scientists make and record observations (collect **data**) and use **inductive reasoning** to draw a general conclusion, which can be developed into a testable **hypothesis**. **Deductive reasoning** makes predictions that can be used to test hypotheses. Scientific hypotheses must be testable.
- **Controlled experiments**, such as the investigation of coat color in mouse populations, are designed to demonstrate the effect of one **variable** by testing control groups and experimental groups differing in only that one variable.
- A scientific **theory** is broad in scope, generates new hypotheses, and is supported by a large body of evidence.
- Observations and experiments must be repeatable, and hypotheses must be testable. Biologists approach questions at different levels; their approaches complement each other. **Technology** is a method or device that applies scientific knowledge for some specific purpose that affects society as well as for scientific research. Diversity among scientists promotes progress in science.

**?** *What are the roles of gathering and interpreting data in scientific inquiry?*

## TEST YOUR UNDERSTANDING

**PRACTICE TEST**

goo.gl/CRZjvS

### Level 1: Knowledge/Comprehension

1. All the organisms on your campus make up
   - (A) an ecosystem.
   - (B) a community.
   - (C) a population.
   - (D) a taxonomic domain.

2. Which of the following best demonstrates the unity among all organisms?
   - (A) emergent properties
   - (B) descent with modification
   - (C) DNA structure and function
   - (D) natural selection

3. A controlled experiment is one that
   - (A) proceeds slowly enough that a scientist can make careful records of the results.
   - (B) tests experimental and control groups in parallel.
   - (C) is repeated many times to make sure the results are accurate.
   - (D) keeps all variables constant.

4. Which of the following statements best distinguishes hypotheses from theories in science?
   - (A) Theories are hypotheses that have been proved.
   - (B) Hypotheses are guesses; theories are correct answers.
   - (C) Hypotheses usually are relatively narrow in scope; theories have broad explanatory power.
   - (D) Theories are proved true; hypotheses are often contradicted by experimental results.

### Level 2: Application/Analysis

5. Which of the following best describes the logic of scientific inquiry?
   - (A) If I generate a testable hypothesis, tests and observations will support it.
   - (B) If my prediction is correct, it will lead to a testable hypothesis.
   - (C) If my observations are accurate, they will support my hypothesis.
   - (D) If my hypothesis is correct, I can expect certain test results.

6. **DRAW IT** With rough sketches, draw a biological hierarchy similar to the one in Figure 1.3 but using a coral reef as the ecosystem, a fish as the organism, its stomach as the organ, and DNA as the molecule. Include all levels in the hierarchy.

### Level 3: Synthesis/Evaluation

7. **SCIENTIFIC INQUIRY**
   Based on the results of the mouse coloration case study, propose a hypothesis researchers might use to further study the role of predators in the natural selection process.

8. **SCIENTIFIC INQUIRY**
   Scientists search the scientific literature by means of electronic databases such as PubMed, a free online database maintained by the National Center for Biotechnology Information. Use PubMed to find the abstract of a scientific article that Hopi Hoekstra published in 2014 or later.

9. **FOCUS ON EVOLUTION**
   In a short essay (100–150 words), describe Darwin's view of how natural selection resulted in both unity and diversity of life on Earth. Include in your discussion some of his evidence. (See tips for writing good essays and a suggested grading rubric in the Study Area of MasteringBiology under "Writing Tips and Rubric.")

10. **FOCUS ON INFORMATION**
    A typical prokaryotic cell has about 3,000 genes in its DNA, while a human cell has almost 21,000 genes. About 1,000 of these genes are present in both types of cells. (a) Based on your understanding of evolution, explain how such different organisms could have this same subset of 1,000 genes. (b) What sorts of functions might these shared genes have? Justify your choices.

11. **SYNTHESIZE YOUR KNOWLEDGE**

Can you pick out the mossy leaf-tailed gecko lying against the tree trunk in this photo? How is the appearance of the gecko a benefit in terms of survival? Given what you learned about evolution, natural selection, and genetic information in this chapter, describe how the gecko's coloration might have evolved.

*For selected answers, see Appendix A.*

# Unit 1 Chemistry and Cells

## 2 The Chemical Context of Life

The structures and functions of living organisms are based on the **chemistry** of atoms and molecules.

## 3 Carbon and the Molecular Diversity of Life

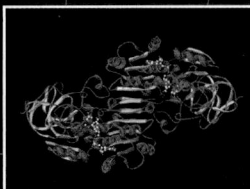

The carbon atom is the foundation of all organic molecules, and its versatility gives rise to the **molecular diversity of life.**

## 4 A Tour of the Cell

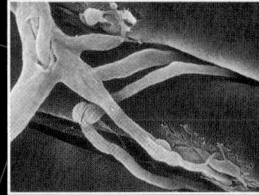

The basic structural and functional unit of life is the **cell.**

## 5 Membrane Transport and Cell Signaling

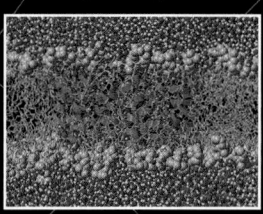

The **plasma membrane** regulates the passage of substances into and out of the cell and enables **signaling** between cells.

## 6 An Introduction to Metabolism

The cellular processes that transform matter and energy make up **cell metabolism.**

## 7 Cellular Respiration and Fermentation

Organisms obtain energy from food by breaking it down by means of **cellular respiration** or **fermentation.**

## 8 Photosynthesis

**Photosynthesis** is the basis of life on planet Earth: Photosynthetic organisms capture light energy and use it to make the food that all organisms depend on.

## 9 The Cell Cycle

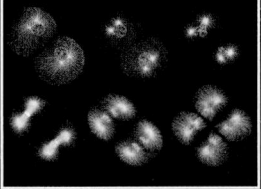

A eukaryotic cell grows and then divides in two, passing along identical genetic information to its daughter cells via **mitosis.** The **cell cycle** describes this progression.

Chemistry and Cells

2 The Chemical Context of Life
3 Carbon and the Molecular Diversity of Life
4 A Tour of the Cell
5 Membrane Transport and Cell Signaling
6 An Introduction to Metabolism
7 Cellular Respiration and Fermentation
8 Photosynthesis
9 The Cell Cycle

Ecology
Animals
Plants
History of Life
Evolution
Genetics

▲ **Figure 2.1 What weapon are these wood ants shooting into the air?**

# A Chemical Connection to Biology

Like other animals, ants have mechanisms that defend them from attack. Wood ants live in colonies of hundreds or thousands, and the colony as a whole has a particularly effective way of dealing with enemies. When threatened from above, the ants shoot volleys of formic acid into the air from their abdomens, and the acid bombards the potential predator, such as a hungry bird **(Figure 2.1)**. Formic acid is produced by many species of ants and got its name from the Latin word for ant, *formica*. In quite a few ant species, the formic acid isn't shot out, but probably serves as a disinfectant that protects the ants against microbial parasites. Scientists have long known that chemicals play a major role in insect communication, the attraction of mates, and defense against predators.

Research on ants and other insects is a good example of how relevant chemistry is to the study of life. Unlike college courses, nature is not neatly packaged into individual sciences—biology, chemistry, physics, and so forth. Biologists specialize in the study of life, but organisms and their environments are natural systems to which the concepts of chemistry and physics apply. Biology is multidisciplinary.

This unit of chapters introduces some basic concepts of chemistry that apply to the study of life. Somewhere in the transition from molecules to cells, we will cross the blurry boundary between nonlife and life. This chapter focuses on the chemical components that make up all matter, with a final section on the substance that supports all of life—water.

## CONCEPT 2.1

# Matter consists of chemical elements in pure form and in combinations called compounds

Organisms are composed of **matter**, which is anything that takes up space and has mass. Matter exists in many forms. Rocks, metals, oils, gases, and living organisms are a few examples of what seems to be an endless assortment of matter.

## Elements and Compounds

Matter is made up of elements. An **element** is a substance that cannot be broken down to other substances by chemical reactions. Today, chemists recognize 92 elements occurring in nature; gold, copper, carbon, and oxygen are examples. Each element has a symbol, usually the first letter or two of its name. Some symbols are derived from Latin or German; for instance, the symbol for sodium is Na, from the Latin word *natrium*.

A **compound** is a substance consisting of two or more different elements combined in a fixed ratio. Table salt, for example, is sodium chloride (NaCl), a compound composed of the elements sodium (Na) and chlorine (Cl) in a 1:1 ratio. Pure sodium is a metal, and pure chlorine is a poisonous gas. When chemically combined, however, sodium and chlorine form an edible compound. Water ($H_2O$), another compound, consists of the elements hydrogen (H) and oxygen (O) in a 2:1 ratio. These are simple examples of organized matter having *emergent properties*: A compound has chemical and physical characteristics different from those of its constituent elements **(Figure 2.2)**.

## The Elements of Life

Of the 92 natural elements, about 20–25% are **essential elements** that an organism needs to live a healthy life and reproduce. The essential elements are similar among organisms, but there is some variation—for example, humans need 25 elements, but plants need only 17.

Relative amounts of all the elements in the human body are listed in **Table 2.1**. Just four elements—oxygen (O), carbon (C), hydrogen (H), and nitrogen (N)—make up approximately 96% of living matter. Calcium (Ca), phosphorus (P), potassium (K), sulfur (S), and a few other elements account for most of the remaining 4% or so of an organism's mass. **Trace elements** are required by an organism in only minute quantities. Some trace elements, such as iron (Fe), are needed by all forms of life; others are required only by certain species. For example, in vertebrates (animals with backbones), the element iodine (I) is an essential ingredient of a hormone produced by the thyroid gland. A daily intake of only 0.15 milligram (mg) of iodine is adequate for normal activity of the human thyroid. An iodine deficiency in the diet causes the thyroid gland to grow to abnormal size, a condition called goiter. Consuming seafood or iodized salt reduces the incidence of goiter.

**Sodium**          **Chlorine**          **Sodium chloride**

▲ **Figure 2.2 The emergent properties of a compound.** The metal sodium combines with the poisonous gas chlorine, forming the edible compound sodium chloride, or table salt.

**Table 2.1 Elements in the Human Body**

| Element | Symbol | Percentage of Body Mass (including water) | |
|---|---|---|---|
| Oxygen | O | 65.0% | |
| Carbon | C | 18.5% | 96.3% |
| Hydrogen | H | 9.5% | |
| Nitrogen | N | 3.3% | |
| Calcium | Ca | 1.5% | |
| Phosphorus | P | 1.0% | |
| Potassium | K | 0.4% | |
| Sulfur | S | 0.3% | 3.7% |
| Sodium | Na | 0.2% | |
| Chlorine | Cl | 0.2% | |
| Magnesium | Mg | 0.1% | |

Trace elements (less than 0.01% of mass): Boron (B), chromium (Cr), cobalt (Co), copper (Cu), fluorine (F), iodine (I), iron (Fe), manganese (Mn), molybdenum (Mo), selenium (Se), silicon (Si), tin (Sn), vanadium (V), zinc (Zn)

**INTERPRET THE DATA** *Given the makeup of the human body, what compound do you think accounts for the high percentage of oxygen?*

## Evolution of Tolerance to Toxic Elements

**EVOLUTION** Some naturally occurring elements are toxic to organisms. In humans, for instance, the element arsenic has been linked to numerous diseases and can be lethal. Some species, however, have become adapted to environments containing elements that are usually toxic. For example, sunflower plants can take up lead, zinc, and other heavy metals in concentrations that would kill most organisms. This capability enabled sunflowers to be used to detoxify contaminated soils after Hurricane Katrina. Presumably, variants of ancestral sunflower species were able to grow in soils with heavy metals, and subsequent natural selection resulted in their survival and reproduction.

### CONCEPT CHECK 2.1

1. Is a trace element an essential element? Explain.
2. **WHAT IF?** In humans, iron is a trace element required for the proper functioning of hemoglobin, the molecule that carries oxygen in red blood cells. What might be the effects of an iron deficiency?

For suggested answers, see Appendix A.

**CONCEPT 2.2**

# An element's properties depend on the structure of its atoms

Each element consists of a certain type of atom that is different from the atoms of any other element. An **atom** is the smallest unit of matter that still retains the properties of an element. Atoms are so small that it would take about a million of them to stretch across the period at the end of this sentence. We symbolize atoms with the same abbreviation used for the element that is made up of those atoms. For example, the symbol C stands for both the element carbon and a single carbon atom.

## Subatomic Particles

Although the atom is the smallest unit having the properties of an element, these tiny bits of matter are composed of even smaller parts, called *subatomic particles*. Using high-energy collisions, physicists have produced more than a hundred types of particles from the atom, but only three kinds of particles are relevant here: **neutrons**, **protons**, and **electrons**. Protons and electrons are electrically charged. Each proton has one unit of positive charge, and each electron has one unit of negative charge. A neutron, as its name implies, is electrically neutral.

Protons and neutrons are packed together tightly in a dense core, or **atomic nucleus**, at the center of an atom; protons give the nucleus a positive charge. The rapidly moving electrons form a "cloud" of negative charge around the nucleus, and it is the attraction between opposite charges that keeps the electrons in the vicinity of the nucleus. **Figure 2.3** shows two commonly used models of the structure of the helium atom as an example.

The neutron and proton are almost identical in mass, each about $1.7 \times 10^{-24}$ gram (g). Grams and other conventional units are not very useful for describing the mass of objects that are so minuscule. Thus, for atoms and subatomic particles (and for molecules, too), we use a unit of measurement called the **dalton** (the same as the *atomic mass unit*, or *amu*). Neutrons and protons have masses close to 1 dalton. Because the mass of an electron is only about 1/2,000 that of a neutron or proton, we can ignore electrons when computing the total mass of an atom.

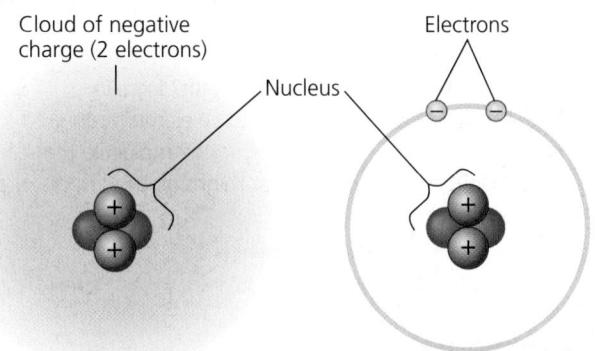

Cloud of negative charge (2 electrons)

Electrons

Nucleus

**(a)** This model shows the two electrons as a cloud of negative charge, a result of their motion around the nucleus.

**(b)** In this more simplified model, the electrons are shown as two small yellow spheres on a circle around the nucleus.

▲ **Figure 2.3 Simplified models of a helium (He) atom.** The helium nucleus consists of 2 neutrons (brown) and 2 protons (pink). Two electrons (yellow) exist outside the nucleus. These models are not to scale; they greatly overestimate the size of the nucleus in relation to the electron cloud.

## Atomic Number and Atomic Mass

Atoms of the various elements differ in their number of subatomic particles. All atoms of a particular element have the same number of protons in their nuclei. This number of protons, which is unique to that element, is called the **atomic number** and is written as a subscript to the left of the symbol for the element. The abbreviation $_2$He, for example, tells us that an atom of the element helium has 2 protons in its nucleus. Unless otherwise indicated, an atom is neutral in electrical charge, which means that its protons must be balanced by an equal number of electrons. Therefore, the atomic number tells us the number of protons and also the number of electrons in an electrically neutral atom.

We can deduce the number of neutrons from a second quantity, the **mass number**, which is the total number of protons and neutrons in the nucleus of an atom. The mass number is written as a superscript to the left of an element's symbol. For example, we can use this shorthand to write an atom of helium as $_2^4$He. Because the atomic number indicates how many protons there are, we can determine the number of neutrons by subtracting the atomic number from the mass number: The helium atom $_2^4$He has 2 neutrons. For sodium (Na):

**Mass number** = number of protons + neutrons
= 23 for sodium

$_{11}^{23}$Na

**Atomic number** = number of protons
= number of electrons in a neutral atom
= 11 for sodium

Number of neutrons = mass number − atomic number
= 23 − 11 = 12 for sodium

The simplest atom is hydrogen $_1^1$H, which has no neutrons; it consists of a single proton with a single electron.

Because the contribution of electrons to mass is negligible, almost all of an atom's mass is concentrated in its nucleus. Neutrons and protons each have a mass very close to 1 dalton, so the mass number is close to, but slightly different from, the total mass of an atom, called its **atomic mass**. For example, the mass number of sodium ($_{11}^{23}$Na) is 23, but its atomic mass is 22.9898 daltons.

## Isotopes

All atoms of a given element have the same number of protons, but some atoms have more neutrons than other atoms of the same element and thus have greater mass. These different atomic forms of the same element are called **isotopes** of the element. In nature, an element may occur as a mixture of its isotopes. As an example, the element carbon, which has the atomic number 6, has three naturally occurring isotopes. The most common isotope is carbon-12, $_6^{12}$C, which accounts for about 99% of the carbon in nature. The isotope $_6^{12}$C has 6 neutrons. Most of the remaining 1% of carbon consists of atoms of the isotope $_6^{13}$C, with 7 neutrons. A third, even rarer isotope, $^{14}_{6}C$, has 8 neutrons. Notice that all three isotopes of carbon have 6 protons; otherwise, they would not be carbon. Although the isotopes of an element have slightly different masses, they behave identically in chemical reactions. (For an element with more than one naturally occurring isotope, the atomic mass is an average of those isotopes, weighted by their abundance. Thus carbon has an atomic mass of 12.01 daltons.)

Both $^{12}C$ and $^{13}C$ are stable isotopes, meaning that their nuclei do not have a tendency to lose subatomic particles, a process called decay. The isotope $^{14}C$, however, is unstable, or radioactive. A **radioactive isotope** is one in which the nucleus decays spontaneously, giving off particles and energy. When the radioactive decay leads to a change in the number of protons, it transforms the atom to an atom of a different element. For example, when an atom of $^{14}C$ decays, it becomes an atom of nitrogen.

Radioactive isotopes have many useful applications in biology. For example, researchers use measurements of radioactivity in fossils to date these relics of past life (see Concept 23.1). Radioactive isotopes are also useful as tracers to follow atoms through metabolism, the chemical processes of an organism. Cells can use radioactive atoms just as they would use nonradioactive isotopes of the same element. The radioactive isotopes are incorporated into biologically active molecules, which can then be tracked by monitoring the radioactivity.

Radioactive tracers are important diagnostic tools in medicine. For example, certain kidney disorders can be diagnosed by injecting small doses of substances containing radioactive isotopes into the blood and then measuring the amount of tracer excreted in the urine. Radioactive tracers are also used in combination with sophisticated imaging instruments, such as PET scanners, that can monitor the growth and metabolism of cancers in the body (**Figure 2.4**).

Although radioactive isotopes are very useful in biological research and medicine, radiation from decaying isotopes also poses a hazard to life by damaging cellular molecules. The severity of this damage depends on the type and amount of radiation an organism absorbs. One of the most serious environmental threats is radioactive fallout from nuclear accidents. The doses of most isotopes used in medical diagnosis, however, are relatively safe.

▶ **Figure 2.4 A PET scan, a medical use for radioactive isotopes.** PET (positron-emission tomography) detects locations of intense chemical activity in the body. The bright yellow spot marks an area with an elevated level of radioactively labeled glucose, which in turn indicates high metabolic activity, a hallmark of cancerous tissue.

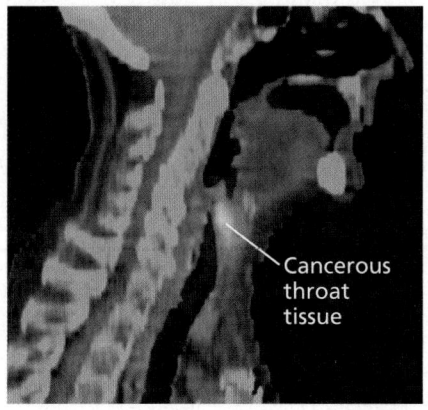
Cancerous throat tissue

## The Energy Levels of Electrons

The simplified models of the atom in Figure 2.3 greatly exaggerate the size of the nucleus relative to that of the whole atom. If an atom of helium were the size of a typical football stadium, the nucleus would be the size of a pencil eraser in the center of the field. Moreover, the electrons would be like two tiny gnats buzzing around the stadium. Atoms are mostly empty space.

When two atoms approach each other during a chemical reaction, their nuclei do not come close enough to interact. Of the three kinds of subatomic particles we have discussed, only electrons are directly involved in the chemical reactions between atoms.

An atom's electrons vary in the amount of energy they possess. **Energy** is defined as the capacity to cause change—for instance, by doing work. **Potential energy** is the energy that matter possesses because of its location or structure. For example, water in a reservoir on a hill has potential energy because of its altitude. When the gates of the reservoir's dam are opened and the water runs downhill, the energy can be used to do work, such as moving the blades of turbines to generate electricity. Because energy has been expended, the water has less energy at the bottom of the hill than it did in the reservoir. Matter has a natural tendency to move toward the lowest possible state of potential energy; in our example, the water runs downhill. To restore the potential energy of a reservoir, work must be done to elevate the water against gravity.

The electrons of an atom have potential energy due to their distance from the nucleus (**Figure 2.5**). The negatively charged

**(a)** A ball bouncing down a flight of stairs can come to rest only on each step, not between steps. Similarly, an electron can exist only at certain energy levels, not between levels.

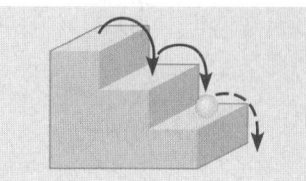

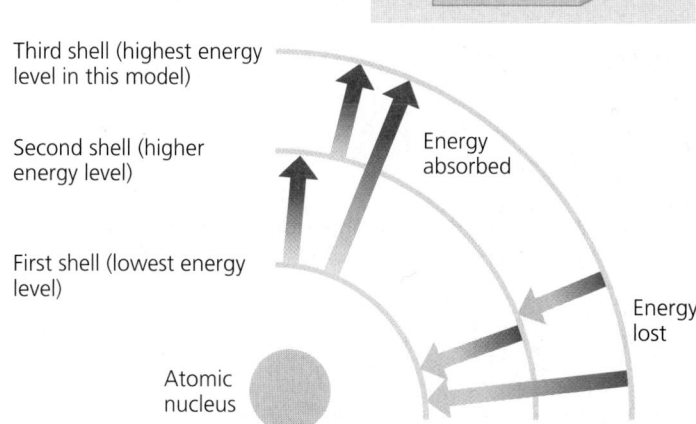

Third shell (highest energy level in this model)

Second shell (higher energy level)

First shell (lowest energy level)

Energy absorbed

Energy lost

Atomic nucleus

**(b)** An electron can move from one shell to another only if the energy it gains or loses is exactly equal to the difference in energy between the energy levels of the two shells. Arrows in this model indicate some of the stepwise changes in potential energy that are possible.

▲ **Figure 2.5 Energy levels of an atom's electrons.** Electrons exist only at fixed levels of potential energy called electron shells.

electrons are attracted to the positively charged nucleus. It takes work to move a given electron farther away from the nucleus, so the more distant an electron is from the nucleus, the greater its potential energy. Unlike the continuous flow of water downhill, changes in the potential energy of electrons can occur only in steps of fixed amounts. An electron having a certain amount of energy is something like a ball on a staircase (Figure 2.5a). The ball can have different amounts of potential energy, depending on which step it is on, but it cannot spend much time between the steps. Similarly, an electron's potential energy is determined by its energy level. An electron can exist only at certain energy levels, not between them.

An electron's energy level is correlated with its average distance from the nucleus. Electrons are found in different **electron shells**, each with a characteristic average distance and energy level. In diagrams, shells can be represented by concentric circles (Figure 2.5b). The first shell is closest to the nucleus, and electrons in this shell have the lowest potential energy. Electrons in the second shell have more energy, and electrons in the third shell even more energy. An electron can move from one shell to another, but only by absorbing or losing an amount of energy equal to the difference in potential energy between its position in the old shell and that in the new shell. When an electron absorbs energy, it moves to a shell farther out from the nucleus. For example, light energy can excite an electron to a higher energy level. (Indeed, this is the first step taken when plants harness the energy of sunlight for photosynthesis, the process that produces food from carbon dioxide and water. You'll learn more about photosynthesis in Chapter 8.) When an electron loses energy, it "falls back" to a shell closer to the nucleus, and the lost energy is usually released to the environment as heat. For example, sunlight excites electrons in the surface of a car to higher energy levels. When the electrons fall back to their original levels, the car's surface heats up. This thermal energy can be transferred to the air or to your hand if you touch the car.

## Electron Distribution and Chemical Properties

The chemical behavior of an atom is determined by the distribution of electrons in the atom's electron shells. Beginning with hydrogen, the simplest atom, we can imagine building the atoms of the other elements by adding 1 proton and 1 electron at a time (along with an appropriate number of neutrons). **Figure 2.6**, a modified version of what is called the *periodic table of the elements*, shows this distribution of electrons for the first 18 elements, from hydrogen ($_1$H) to argon ($_{18}$Ar). The elements are arranged in three rows, or *periods*, corresponding to the number of electron shells in their atoms. The

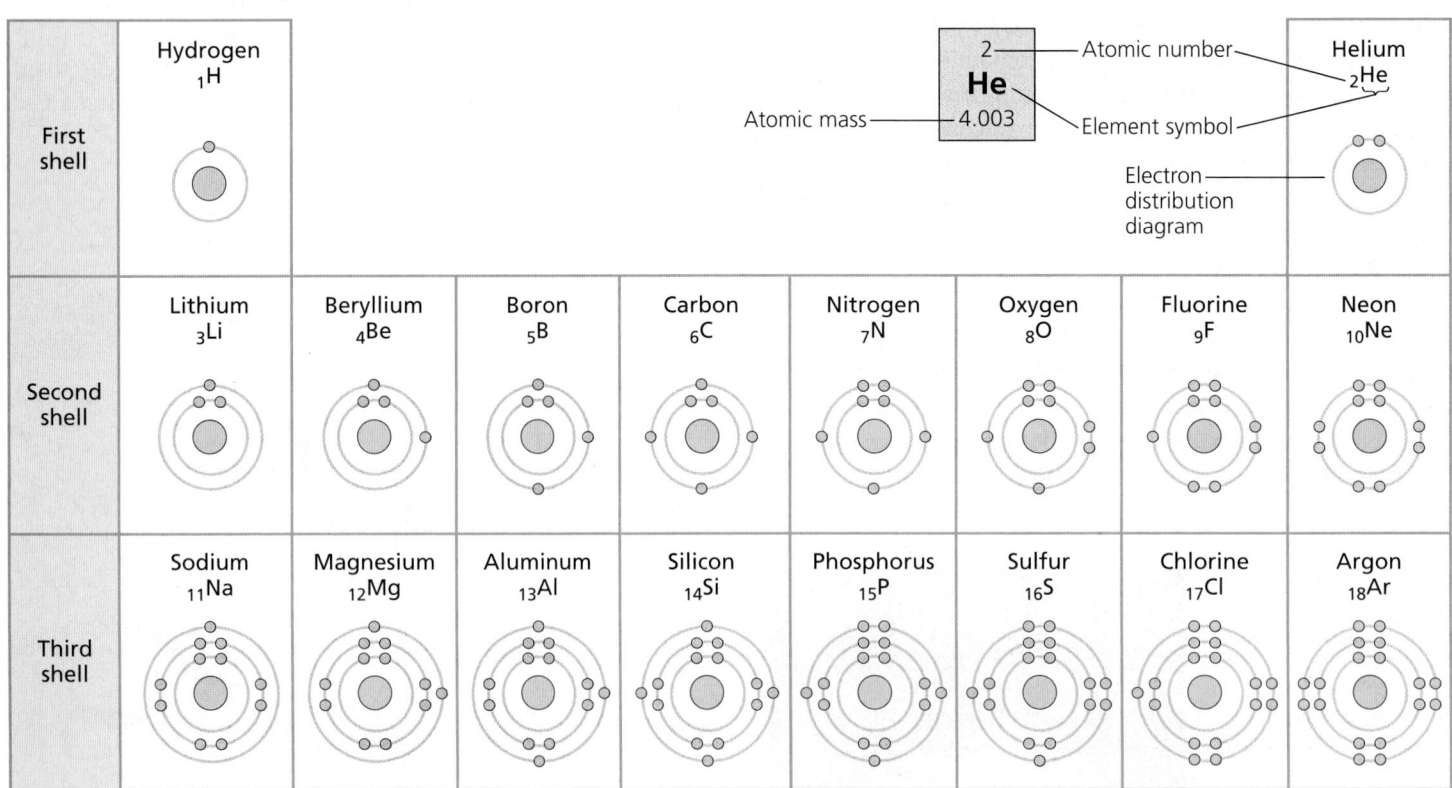

▲ **Figure 2.6 Electron distribution diagrams for the first 18 elements in the periodic table.** In a standard periodic table (see Appendix B), information for each element is presented as shown for helium in the inset. In the diagrams in this table, electrons are represented as yellow dots and electron shells as concentric circles. These diagrams are a convenient way to picture the distribution of an atom's electrons among its electron shells, but these simplified models do not accurately represent the shape of the atom or the location of its electrons. The elements are arranged in rows, each representing the filling of an electron shell. As electrons are added, they occupy the lowest available shell.

**?** *What is the atomic number of magnesium? How many protons and electrons does it have? How many electron shells? How many valence electrons?*

left-to-right sequence of elements in each row corresponds to the sequential addition of electrons and protons. (See Appendix B for the complete periodic table.)

Hydrogen's 1 electron and helium's 2 electrons are located in the first shell. Electrons, like all matter, tend to exist in the lowest available state of potential energy. In an atom, this state is in the first shell. However, the first shell can hold no more than 2 electrons; thus, hydrogen and helium are the only elements in the first row of the table. In an atom with more than 2 electrons, the additional electrons must occupy higher shells because the first shell is full. The next element, lithium, has 3 electrons. Two of these electrons fill the first shell, while the third electron occupies the second shell. The second shell holds a maximum of 8 electrons. Neon, at the end of the second row, has 8 electrons in the second shell, giving it a total of 10 electrons.

The chemical behavior of an atom depends mostly on the number of electrons in its *outermost* shell. We call those outer electrons **valence electrons** and the outermost electron shell the **valence shell**. In the case of lithium, there is only 1 valence electron, and the second shell is the valence shell. Atoms with the same number of electrons in their valence shells exhibit similar chemical behavior. For example, fluorine (F) and chlorine (Cl) both have 7 valence electrons, and both form compounds when combined with the element sodium (Na): Sodium fluoride (NaF) is commonly added to toothpaste to prevent tooth decay, and, as described earlier, NaCl is table salt (see Figure 2.2). An atom with a completed valence shell is unreactive; that is, it will not interact readily with other atoms. At the far right of the periodic table are helium, neon, and argon, the only three elements shown in Figure 2.6 that have full valence shells. These elements are said to be *inert*, meaning chemically unreactive. All the other atoms in Figure 2.6 are chemically reactive because they have incomplete valence shells.

Notice that as we "build" the atoms in Figure 2.6, the first 4 electrons added to the second and third shells are not shown in pairs; only after 4 electrons are present do the next electrons complete pairs. The reactivity of an atom arises from the presence of one or more unpaired electrons in its valence shell. As you will see in the next section, atoms interact in a way that completes their valence shells. When they do so, it is the *unpaired* electrons that are involved.

### CONCEPT CHECK 2.2

1. A nitrogen atom has 7 protons, and the most common isotope of nitrogen has 7 neutrons. A radioactive isotope of nitrogen has 8 neutrons. Write the atomic number and mass number of this radioactive nitrogen as a chemical symbol with a subscript and superscript.

2. How many electrons does fluorine have? How many electron shells? How many electrons are needed to fill the valence shell?

3. **WHAT IF?** In Figure 2.6, if two or more elements are in the same row, what do they have in common? If two or more elements are in the same column, what do they have in common?

For suggested answers, see Appendix A.

---

# The formation and function of molecules depend on chemical bonding between atoms

Now that we have looked at the structure of atoms, we can move up the hierarchy of organization and see how atoms combine to form molecules and ionic compounds. Atoms with incomplete valence shells can interact with certain other atoms in such a way that each partner completes its valence shell: The atoms either share or transfer valence electrons. These interactions usually result in atoms staying close together, held by attractions called **chemical bonds**. The strongest kinds of chemical bonds are covalent bonds and ionic bonds (when in dry ionic compounds; ionic bonds are weak when in aqueous solutions).

## Covalent Bonds

A **covalent bond** is the sharing of a pair of valence electrons by two atoms. For example, let's consider what happens when two hydrogen atoms approach each other. Recall that hydrogen has 1 valence electron in the first shell, but the shell's capacity is 2 electrons. When the two hydrogen atoms come close enough for their electron shells to overlap, they can share their electrons **(Figure 2.7)**. Each hydrogen atom is now associated with 2 electrons in what amounts to a completed valence shell. Two or more atoms held together by covalent bonds constitute a **molecule**, in this case a hydrogen molecule.

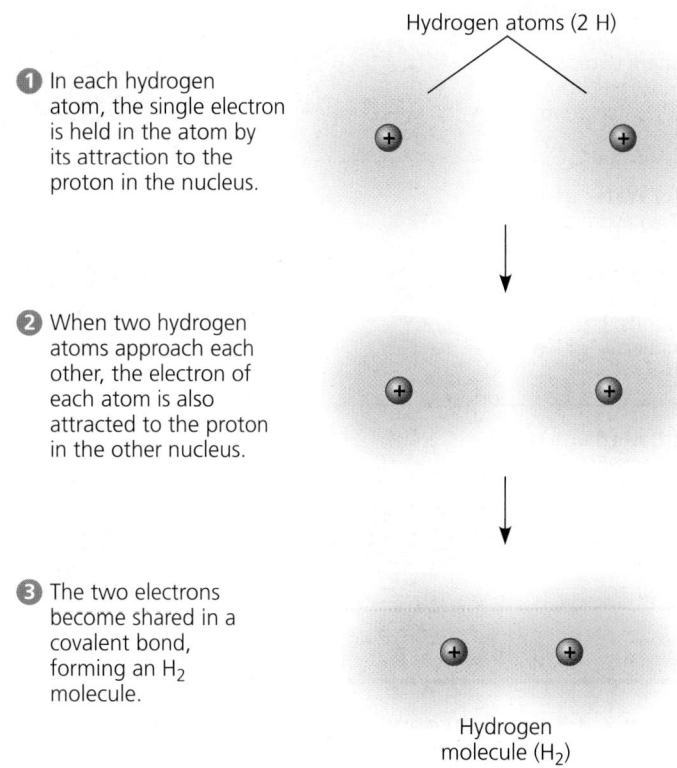

Hydrogen atoms (2 H)

❶ In each hydrogen atom, the single electron is held in the atom by its attraction to the proton in the nucleus.

❷ When two hydrogen atoms approach each other, the electron of each atom is also attracted to the proton in the other nucleus.

❸ The two electrons become shared in a covalent bond, forming an $H_2$ molecule.

Hydrogen molecule ($H_2$)

▲ **Figure 2.7 Formation of a covalent bond.**

Figure 2.8a shows several ways of representing a hydrogen molecule. Its *molecular formula*, $H_2$, simply indicates that the molecule consists of two atoms of hydrogen. Electron sharing can be depicted by an electron distribution diagram or by a *structural formula*, H—H, where the line represents a **single bond**, a pair of shared electrons. A space-filling model comes closest to representing the actual shape of the molecule.

Oxygen has 6 electrons in its second electron shell and therefore needs 2 more electrons to complete its valence shell. Two oxygen atoms form a molecule by sharing *two* pairs of valence electrons **(Figure 2.8b)**. The atoms are thus joined by a **double bond** (O=O).

Each atom that can share valence electrons has a bonding capacity corresponding to the number of covalent bonds the atom can form. When the bonds form, they give the atom a full complement of electrons in the valence shell. The bonding capacity of oxygen, for example, is 2. This bonding capacity is called the atom's **valence** and usually equals the number of electrons required to complete the atom's outermost (valence) shell. See if you can determine the valences of hydrogen,

oxygen, nitrogen, and carbon by studying the electron distribution diagrams in Figure 2.6. You can see that the valence of hydrogen is 1; oxygen, 2; nitrogen, 3; and carbon, 4. (The situation is more complicated for phosphorus, in the third row of the periodic table, which can have a valence of 3 or 5 depending on the combination of single and double bonds it makes.)

The molecules $H_2$ and $O_2$ are pure elements rather than compounds because a compound is a combination of two or more *different* elements. Water, with the molecular formula $H_2O$, is a compound. Two atoms of hydrogen are needed to satisfy the valence of one oxygen atom. **Figure 2.8c** shows the structure of a water molecule. Water is so important to life that the last section of this chapter, Concept 2.5, is devoted to its structure and behavior.

Methane, the main component of natural gas, is a compound with the molecular formula $CH_4$. It takes four hydrogen atoms, each with a valence of 1, to complement one atom of carbon, with its valence of 4 **(Figure 2.8d)**. (We will look at many other compounds of carbon in Chapter 3.)

Atoms in a molecule attract shared bonding electrons to varying degrees, depending on the element. The attraction of a particular atom for the electrons of a covalent bond is called its **electronegativity**. The more electronegative an atom is, the more strongly it pulls shared electrons toward itself. In a covalent bond between two atoms of the same element, the electrons are shared equally because the two atoms have the same electronegativity—the tug-of-war is at a standoff. Such a bond is called a **nonpolar covalent bond**. For example, the single bond of $H_2$ is nonpolar, as is the double bond of $O_2$. However, when an atom is bonded to a more electronegative atom, the electrons of the bond are not shared equally. This type of bond is called a **polar covalent bond**. Such bonds vary in their polarity, depending on the relative electronegativity of the two atoms. For example, the bonds between the oxygen and hydrogen atoms of a water molecule are quite polar **(Figure 2.9)**. Oxygen is one of the most electronegative elements, attracting shared electrons much more strongly than hydrogen does. In a covalent bond between oxygen and hydrogen, the electrons spend more time near the oxygen nucleus than they do near the hydrogen nucleus. Because electrons have a negative charge and are pulled toward oxygen in a water molecule, the oxygen atom has a partial negative charge (indicated by the Greek letter δ with a

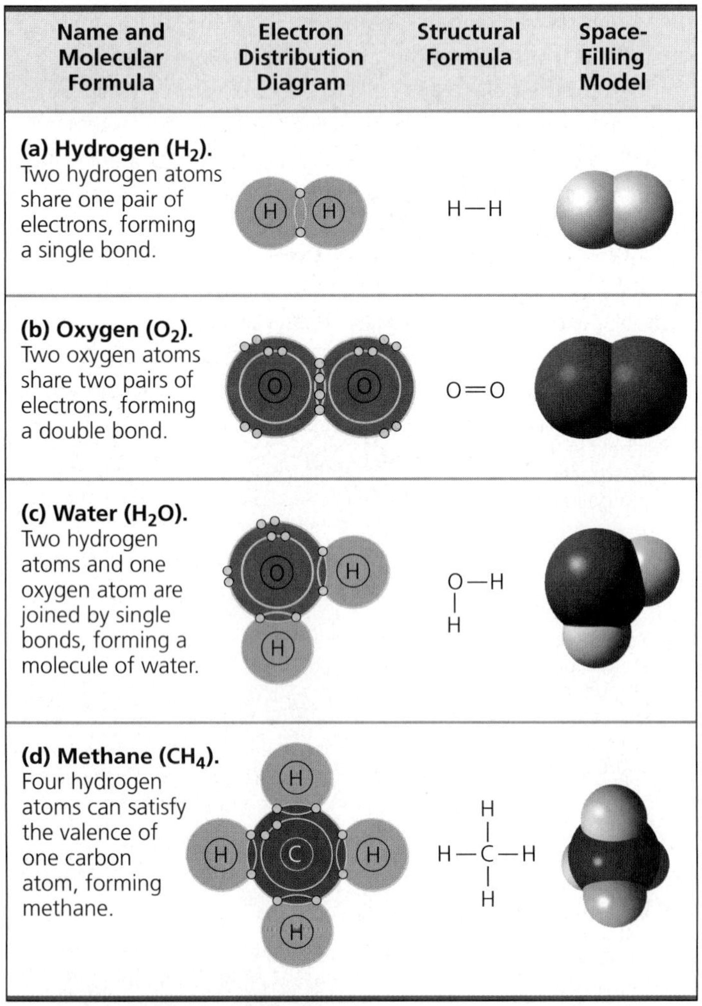

| Name and Molecular Formula | Electron Distribution Diagram | Structural Formula | Space-Filling Model |
|---|---|---|---|
| **(a) Hydrogen ($H_2$).** Two hydrogen atoms share one pair of electrons, forming a single bond. | H  H | H—H | |
| **(b) Oxygen ($O_2$).** Two oxygen atoms share two pairs of electrons, forming a double bond. | O  O | O=O | |
| **(c) Water ($H_2O$).** Two hydrogen atoms and one oxygen atom are joined by single bonds, forming a molecule of water. | O  H  H | O—H  \|  H | |
| **(d) Methane ($CH_4$).** Four hydrogen atoms can satisfy the valence of one carbon atom, forming methane. | H  H  C  H  H | H  \|  H—C—H  \|  H | |

▲ **Figure 2.8 Covalent bonding in four molecules.** The number of electrons required to complete an atom's valence shell generally determines how many covalent bonds that atom will form. This figure shows several ways of indicating covalent bonds.

Because oxygen (O) is more electronegative than hydrogen (H), shared electrons are pulled more toward oxygen.

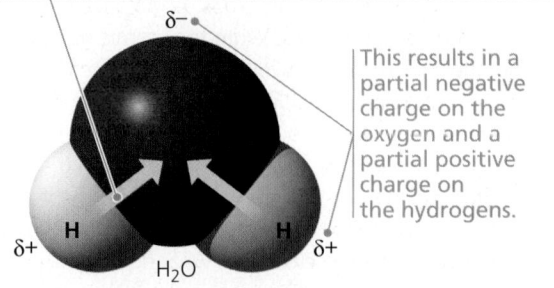

This results in a partial negative charge on the oxygen and a partial positive charge on the hydrogens.

$δ-$

$δ+$  H     H  $δ+$

$H_2O$

▲ **Figure 2.9 Polar covalent bonds in a water molecule.**

minus sign, δ−, or "delta minus"), and each hydrogen atom has a partial positive charge (δ+, or "delta plus"). In contrast, the individual bonds of methane ($CH_4$) are much less polar because the electronegativities of carbon and hydrogen are similar.

## Ionic Bonds

In some cases, two atoms are so unequal in their attraction for valence electrons that the more electronegative atom strips an electron completely away from its partner. The two resulting oppositely charged atoms (or molecules) are called **ions**. A positively charged ion is called a **cation**, while a negatively charged ion is called an **anion**. Because of their opposite charges, cations and anions attract each other; this attraction is called an **ionic bond**. Note that the transfer of an electron is not, by itself, the formation of a bond; rather, it allows a bond to form because it results in two ions of opposite charge. Any two such ions can form an ionic bond—the ions do not need to have acquired their charge by an electron transfer with each other.

This is what happens when an atom of sodium ($_{11}Na$) encounters an atom of chlorine ($_{17}Cl$) **(Figure 2.10)**. A sodium atom has a total of 11 electrons, with its single valence electron in the third electron shell. A chlorine atom has a total of 17 electrons, with 7 electrons in its valence shell. When these two atoms meet, the lone valence electron of sodium is transferred to the chlorine atom, and both atoms end up with their valence shells complete. (Because sodium no longer has an electron in the third shell, the second shell is now the valence shell.)

The electron transfer between the two atoms moves one unit of negative charge from sodium to chlorine. Sodium, now with 11 protons but only 10 electrons, has a net electrical charge of 1+; the sodium atom has become a cation. Conversely, the chlorine atom, having gained an extra electron, now has 17 protons and 18 electrons, giving it a net electrical charge of 1−; it has become a chloride ion—an anion.

▲ **Figure 2.11 A sodium chloride (NaCl) crystal.** The sodium ions ($Na^+$) and chloride ions ($Cl^-$) are held together by ionic bonds. The formula NaCl tells us that the ratio of $Na^+$ to $Cl^-$ is 1:1.

Compounds formed by ionic bonds are called **ionic compounds**, or **salts**. We know the ionic compound sodium chloride (NaCl) as table salt **(Figure 2.11)**. Salts are often found in nature as crystals of various sizes and shapes. Each salt crystal is an aggregate of vast numbers of cations and anions bonded by their electrical attraction and arranged in a three-dimensional lattice. Unlike a covalent compound, which consists of molecules having a definite size and number of atoms, an ionic compound does not consist of molecules. The formula for an ionic compound, such as NaCl, indicates only the ratio of elements in a crystal of the salt. "NaCl" by itself is not a molecule.

Not all salts have equal numbers of cations and anions. For example, the ionic compound magnesium chloride ($MgCl_2$) has two chloride ions for each magnesium ion. Magnesium ($_{12}Mg$) must lose 2 outer electrons if the atom is to have a complete valence shell, so it has a tendency to become a cation with a net charge of 2+ ($Mg^{2+}$). One magnesium cation can therefore form ionic bonds with two chloride anions ($Cl^-$).

The term *ion* also applies to entire molecules that are electrically charged. In the salt ammonium chloride ($NH_4Cl$), for instance, the anion is a single chloride ion ($Cl^-$), but the cation is ammonium ($NH_4^+$), a nitrogen atom covalently bonded to four hydrogen atoms. The whole ammonium ion has an electrical charge of 1+ because it has given up 1 electron and thus is 1 electron short.

Environment affects the strength of ionic bonds. In a dry salt crystal, the bonds are so strong that it takes a hammer and chisel to break enough of them to crack the crystal in two. If the same salt crystal is dissolved in water, however, the ionic bonds are much weaker because each ion is partially shielded by its interactions with water molecules. Most drugs are manufactured as salts because they are quite stable when dry but can dissociate (come apart) easily in water.

❶ The lone valence electron of a sodium atom is transferred to join the 7 valence electrons of a chlorine atom.

❷ Each resulting ion has a completed valence shell. An ionic bond can form between the oppositely charged ions.

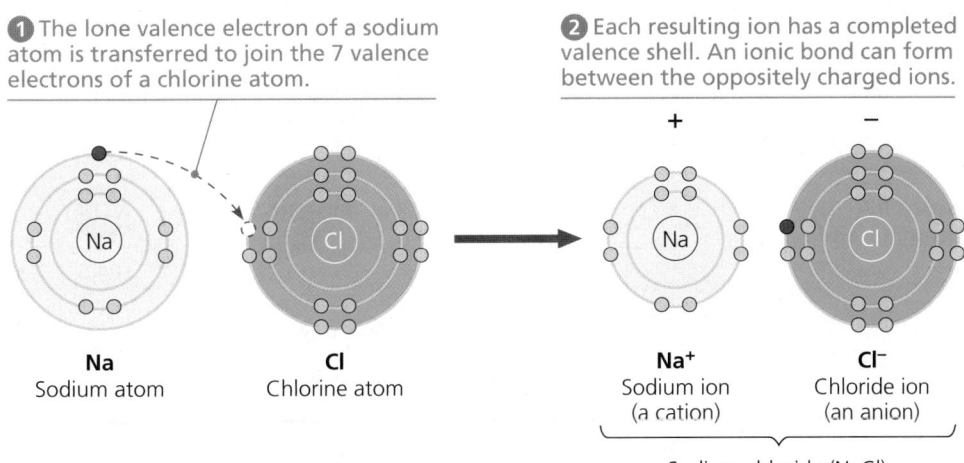

**Na**
Sodium atom

**Cl**
Chlorine atom

**Na$^+$**
Sodium ion
(a cation)

**Cl$^-$**
Chloride ion
(an anion)

Sodium chloride (NaCl)

▲ **Figure 2.10 Electron transfer and ionic bonding.** The attraction between oppositely charged atoms, or ions, is an ionic bond. An ionic bond can form between any two oppositely charged ions, even if they have not been formed by transfer of an electron from one to the other.

## Weak Chemical Bonds

In organisms, most of the strongest chemical bonds are covalent bonds, which link atoms to form a cell's molecules. But weaker bonding within and between molecules is also indispensable in the cell, contributing greatly to the emergent properties of life. Many large biological molecules are held in their functional form by weak bonds. In addition, when two molecules in the cell make contact, they may adhere temporarily by weak bonds. The reversibility of weak bonding can be an advantage: Two molecules can come together, respond to one another in some way, and then separate.

Several types of weak chemical bonds are important in organisms. One is the ionic bond as it exists between ions dissociated in water, which we just discussed. Hydrogen bonds and van der Waals interactions are also crucial to life.

### Hydrogen Bonds

Among the various kinds of weak chemical bonds, hydrogen bonds are so central to the chemistry of life that they deserve special attention. When a hydrogen atom is covalently bonded to an electronegative atom, the hydrogen atom has a partial positive charge that allows it to be attracted to a different electronegative atom nearby. This noncovalent attraction between a hydrogen and an electronegative atom is called a **hydrogen bond**. In living cells, the electronegative partners are usually oxygen or nitrogen atoms. Refer to **Figure 2.12** to examine the simple case of hydrogen bonding between water ($H_2O$) and ammonia ($NH_3$).

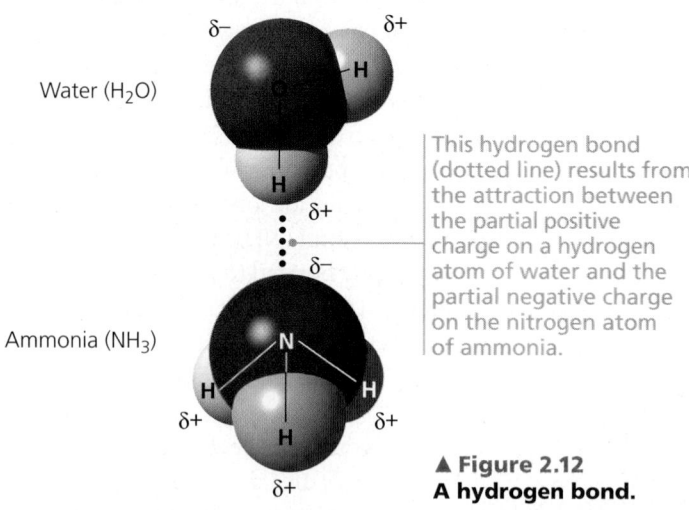

This hydrogen bond (dotted line) results from the attraction between the partial positive charge on a hydrogen atom of water and the partial negative charge on the nitrogen atom of ammonia.

▲ Figure 2.12
**A hydrogen bond.**

### Van der Waals Interactions

Even a molecule with nonpolar covalent bonds may have positively and negatively charged regions. Electrons are not always symmetrically distributed in such a molecule; at any instant, they may accumulate by chance in one part of the molecule or another. The results are ever-changing regions of positive and negative charge that enable all atoms and molecules to stick to one another. These **van der Waals interactions** are individually weak and occur only when atoms and molecules are

very close together. When many such interactions occur simultaneously, however, they can be powerful: Van der Waals interactions allow the gecko lizard, shown here, to walk straight up a wall! A gecko toe has hundreds of thousands of tiny hairs with multiple projections on each, which help to maximize surface contact with the wall. The van der Waals interactions between the molecules of the foot and those of the wall's surface are so numerous that despite their individual weakness, together they can support the gecko's body weight. This discovery has inspired development of an artificial adhesive called Geckskin: A patch the size of an index card can hold a 700-pound weight to a wall!

Van der Waals interactions, hydrogen bonds, ionic bonds in water, and other weak bonds may form not only between molecules but also between parts of a large molecule, such as a protein. The cumulative effect of weak bonds is to reinforce the three-dimensional shape of the molecule. (You will learn more about the very important biological roles of weak bonds in Concept 3.5.)

## Molecular Shape and Function

A molecule has a characteristic size and shape, which are key to its function in the living cell. A molecule consisting of two atoms, such as $H_2$ or $O_2$, is always linear, but most molecules with more than two atoms have more complicated shapes. To take a very simple example, a water molecule ($H_2O$) is shaped roughly like a V, with its two covalent bonds spread apart at an angle of 104.5° **(Figure 2.13)**. A methane molecule ($CH_4$) has a geometric shape called a tetrahedron, a pyramid with a triangular base. The carbon nucleus is inside, at the center, with its four covalent bonds radiating to hydrogen nuclei at the

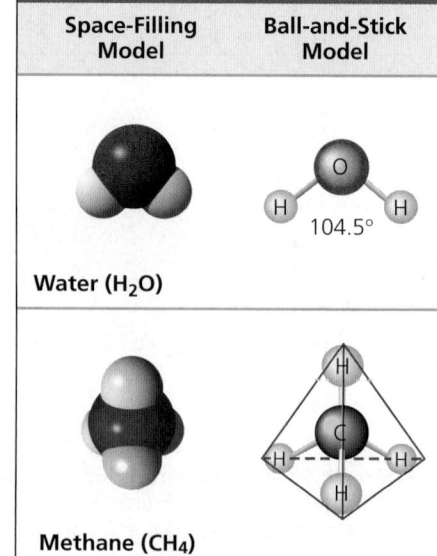

▶ Figure 2.13
**Models showing the shapes of two small molecules.**
Each of the molecules, water and methane, is represented in two different ways.

| Space-Filling Model | Ball-and-Stick Model |
| --- | --- |
| Water ($H_2O$) | |
| Methane ($CH_4$) | |

corners of the tetrahedron. Larger molecules containing multiple carbon atoms, including many of the molecules that make up living matter, have more complex overall shapes. However, the tetrahedral shape of a carbon atom bonded to four other atoms is often a repeating motif within such molecules.

Molecular shape is crucial in biology: It determines how biological molecules recognize and respond to one another with specificity. Biological molecules often bind temporarily to each other by forming weak bonds, but only if their shapes are complementary. Consider the effects of opiates (drugs derived from opium); morphine and heroin are two examples. Opiates relieve pain and alter mood by weakly binding to specific receptor molecules on the surfaces of brain cells. Why would brain cells carry receptors for opiates, compounds that are not made by the body? In 1975, the discovery of endorphins answered this question. Endorphins are signaling molecules made by the pituitary gland that bind to the receptors, relieving pain and producing euphoria during times of stress, such as intense exercise. Opiates have shapes similar to endorphins and mimic them by binding to endorphin receptors in the brain. That is why opiates and endorphins have similar effects (**Figure 2.14**).

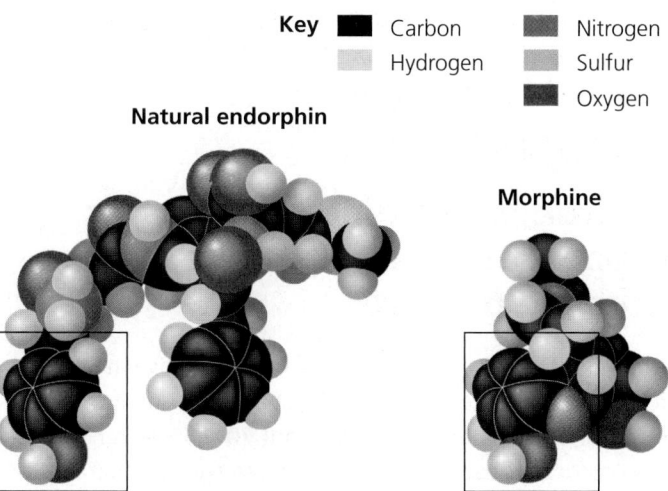

**Key** ■ Carbon ■ Nitrogen
■ Hydrogen ■ Sulfur
■ Oxygen

**Natural endorphin**

**Morphine**

**(a) Structures of endorphin and morphine.** The boxed portion of the endorphin molecule (left) binds to receptor molecules on target cells in the brain. The boxed portion of the morphine molecule (right) is a close match.

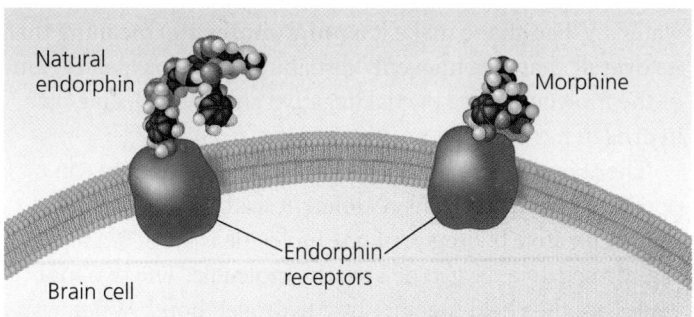

Natural endorphin

Morphine

Endorphin receptors

Brain cell

**(b) Binding to endorphin receptors.** Both endorphin and morphine can bind to endorphin receptors on the surface of a brain cell.

▲ **Figure 2.14 A molecular mimic.** Morphine affects pain perception and emotional state by mimicking the brain's natural endorphins.

The role of molecular shape in brain chemistry illustrates the match between structure and function in biological organization, one of biology's unifying themes.

**CONCEPT 2.4**

# Chemical reactions make and break chemical bonds

The making and breaking of chemical bonds, leading to changes in the composition of matter, are called **chemical reactions**. An example is the reaction between hydrogen and oxygen molecules that forms water:

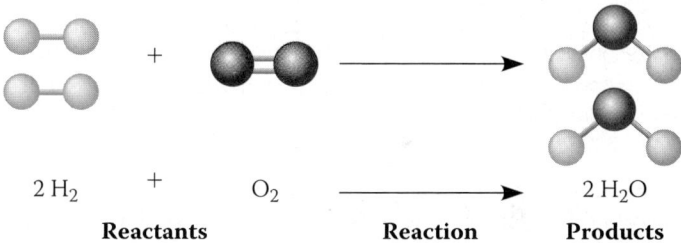

$2 H_2$ + $O_2$ ⟶ $2 H_2O$

**Reactants** **Reaction** **Products**

This reaction breaks the covalent bonds of $H_2$ and $O_2$ and forms the new bonds of $H_2O$. When we write the equation for a chemical reaction, we use an arrow to indicate the conversion of the starting materials, called the **reactants**, to the resulting materials, or **products**. The coefficients indicate the number of molecules involved; for example, the coefficient 2 in front of $H_2$ means that the reaction starts with two molecules of hydrogen. Notice that all atoms of the reactants must be accounted for in the products. Matter is conserved in a chemical reaction: Reactions cannot create or destroy atoms but can only rearrange (redistribute) the electrons among them.

Photosynthesis, which takes place within the cells of green plant tissues, is an important biological example of how chemical reactions rearrange matter. Humans and other animals ultimately depend on photosynthesis for food and oxygen, and this process is at the foundation of almost all ecosystems. The following chemical shorthand summarizes the process of photosynthesis:

$$6 CO_2 + 6 H_2O \rightarrow C_6H_{12}O_6 + 6 O_2$$

The raw materials of photosynthesis are carbon dioxide ($CO_2$), which is taken from the air, and water ($H_2O$), which is

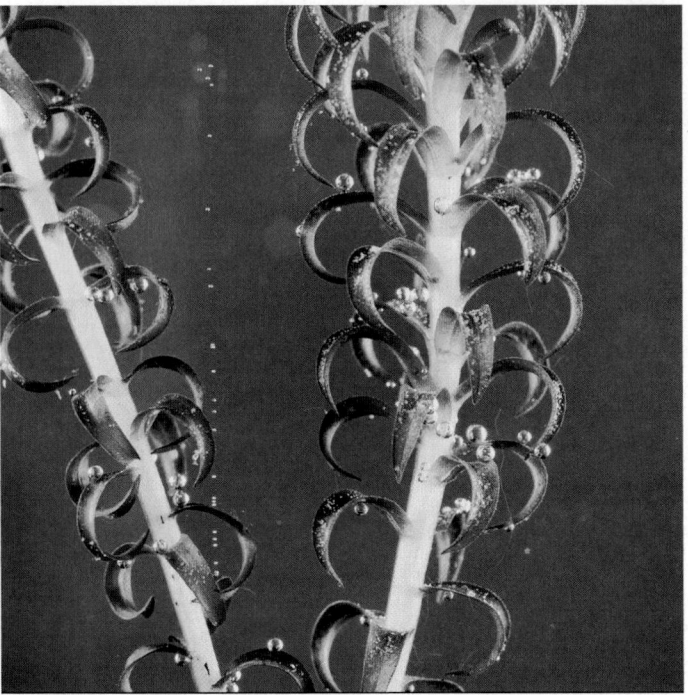

▲ **Figure 2.15 Photosynthesis: a solar-powered rearrangement of matter.** *Elodea*, a freshwater plant, produces sugar by rearranging the atoms of carbon dioxide and water in the chemical process known as photosynthesis, which is powered by sunlight. Much of the sugar is then converted to other food molecules. Oxygen gas ($O_2$) is a by-product of photosynthesis; notice the bubbles of $O_2$-containing gas escaping from the leaves submerged in water.

**?** *Explain how this photo relates to the reactants and products in the equation for photosynthesis given in the text. (You will learn more about photosynthesis in Chapter 8.)*

absorbed from the soil. Within the plant cells, sunlight powers the conversion of these ingredients to a sugar called glucose ($C_6H_{12}O_6$) and oxygen molecules ($O_2$), a by-product that the plant releases into the surroundings **(Figure 2.15)**. Although photosynthesis is actually a sequence of many chemical reactions, we still end up with the same number and types of atoms that we had when we started. Matter has simply been rearranged, with an input of energy provided by sunlight.

All chemical reactions are reversible, with the products of the forward reaction becoming the reactants for the reverse reaction. For example, hydrogen and nitrogen molecules can combine to form ammonia, but ammonia can also decompose to regenerate hydrogen and nitrogen:

$$3 H_2 + N_2 \rightleftharpoons 2 NH_3$$

The two opposite-headed arrows indicate that the reaction is reversible.

One of the factors affecting the rate of a reaction is the concentration of reactants. The greater the concentration of reactant molecules, the more frequently they collide with one another and have an opportunity to react and form products. The same holds true for products. As products accumulate, collisions resulting in the reverse reaction become more frequent. Eventually, the forward and reverse reactions occur at the same

rate, and the relative concentrations of products and reactants stop changing. The point at which the reactions offset one another exactly is called **chemical equilibrium**. This is a dynamic equilibrium; reactions are still going on, but with no net effect on the concentrations of reactants and products. Equilibrium does *not* mean that the reactants and products are equal in concentration, but only that their concentrations have stabilized at a particular ratio. The reaction involving ammonia reaches equilibrium when ammonia decomposes as rapidly as it forms. In some chemical reactions, the equilibrium point may lie so far to the right that these reactions go essentially to completion; that is, virtually all the reactants are converted to products.

To conclude this chapter, we focus on water, the substance in which all the chemical processes of organisms occur.

**CONCEPT CHECK 2.4**

1. Which type of chemical reaction occurs faster at equilibrium, the formation of products from reactants or that of reactants from products?
2. **WHAT IF?** Write an equation that uses the products of photosynthesis as reactants and the reactants of photosynthesis as products. Add energy as another product. This new equation describes a process that occurs in your cells. Describe this equation in words. How does this equation relate to breathing?

For suggested answers, see Appendix A.

**CONCEPT 2.5**

# Hydrogen bonding gives water properties that help make life possible on Earth

All organisms are made mostly of water and live in an environment dominated by water. Most cells are surrounded by water, and cells themselves are about 70–95% water. Water is so common that it is easy to overlook the fact that it is an exceptional substance with many extraordinary qualities. We can trace water's unique behavior to the structure and interactions of its molecules. As you saw in Figure 2.9, the connections between the atoms of a water molecule are polar covalent bonds. The unequal sharing of electrons and water's V-like shape make it a **polar molecule**, meaning that its overall charge is unevenly distributed: The oxygen region of the molecule has a partial negative charge ($\delta-$), and each hydrogen has a partial positive charge ($\delta+$).

The properties of water arise from attractions between oppositely charged atoms of different water molecules: The slightly positive hydrogen of one molecule is attracted to the slightly negative oxygen of a nearby molecule. The two molecules are thus held together by a hydrogen bond. When water is in its liquid form, its hydrogen bonds are very fragile, each only about 1/20 as strong as a covalent bond. The hydrogen bonds form, break, and re-form with great frequency. Each lasts only a few trillionths of a second, but the molecules are

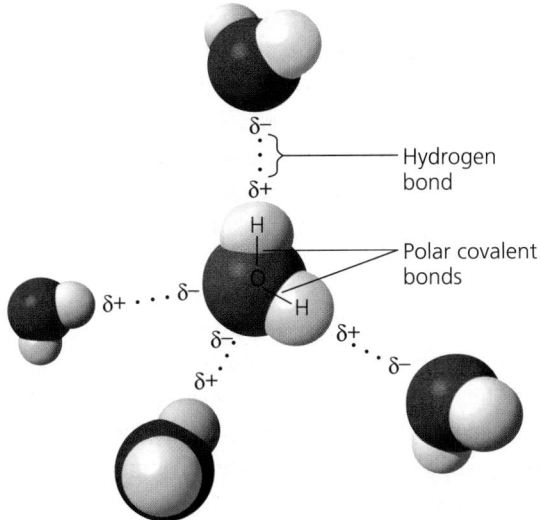

**▲ Figure 2.16 Hydrogen bonds between water molecules.** The charged regions in a water molecule are due to its polar covalent bonds. Oppositely charged regions of neighboring water molecules are attracted to each other, forming hydrogen bonds. Each molecule can hydrogen-bond to multiple partners, and these associations are constantly changing.

**DRAW IT** *Draw partial charges on all the atoms of the water molecule on the far left, and draw two more water molecules hydrogen-bonded to it.*

constantly forming new hydrogen bonds with a succession of partners. Therefore, at any instant, most of the water molecules are hydrogen-bonded to their neighbors **(Figure 2.16).** The extraordinary properties of water emerge from this hydrogen bonding, which organizes water molecules into a higher level of structural order. We will examine four emergent properties of water that contribute to Earth's suitability as an environment for life: cohesive behavior, ability to moderate temperature, expansion upon freezing, and versatility as a solvent. After that, we'll discuss a critical aspect of water chemistry—acids and bases.

## Cohesion of Water Molecules

Water molecules stay close to each other as a result of hydrogen bonding. At any given moment, many of the molecules in liquid water are linked by multiple hydrogen bonds. These linkages make water more structured than most other liquids. Collectively, the hydrogen bonds hold the substance together, a phenomenon called **cohesion**.

Cohesion due to hydrogen bonding contributes to the transport of water and dissolved nutrients against gravity in plants **(Figure 2.17).** Water from the roots reaches the leaves through a network of water-conducting cells. As water evaporates from a leaf, hydrogen bonds cause water molecules leaving the veins to tug on molecules farther down, and the upward pull is transmitted through the water-conducting cells all the way to the roots. **Adhesion**, the clinging of one substance to another, also plays a role. Adhesion of water to cell walls by hydrogen bonds helps counter the downward pull of gravity.

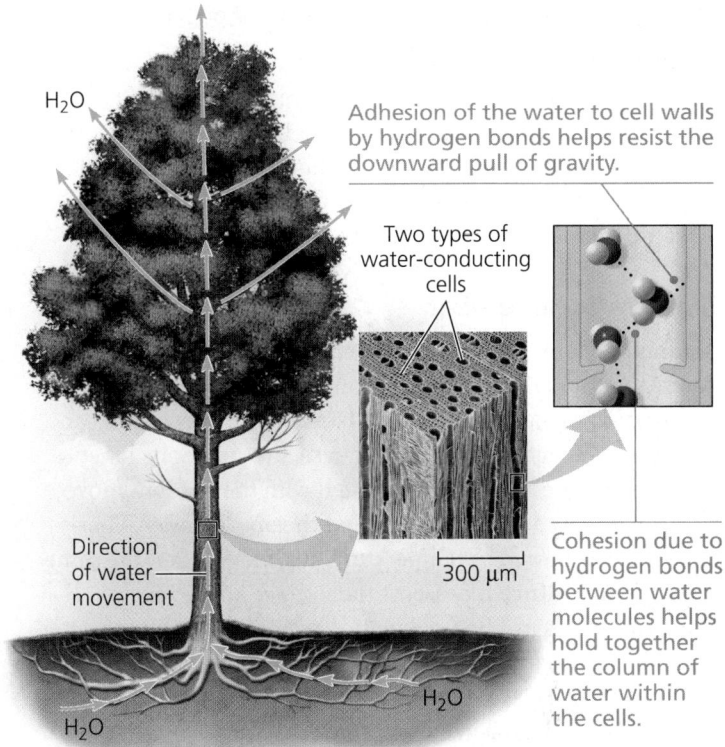

**▲ Figure 2.17 Water transport in plants.** Evaporation from leaves pulls water upward from the roots through water-conducting cells. Because of the properties of cohesion and adhesion, the tallest trees can transport water more than 100 m upward—approximately one-quarter the height of the Empire State Building in New York City.

**ANIMATION** Visit the Study Area in **MasteringBiology** for the BioFlix® 3-D Animation on Water Transport in Plants.

Related to cohesion is **surface tension**, a measure of how difficult it is to stretch or break the surface of a liquid. The hydrogen bonds in water give it an unusually high surface tension, making it behave as though it were coated with an invisible film. You can observe the surface tension of water by slightly overfilling a drinking glass; the water will stand above the rim. The spider in **Figure 2.18** takes advantage of the surface tension of water to walk across a pond without breaking the surface.

**▶ Figure 2.18 Walking on water.** The high surface tension of water, resulting from the collective strength of its hydrogen bonds, allows this raft spider to walk on the surface of a pond.

## Moderation of Temperature by Water

Water moderates air temperature by absorbing heat from air that is warmer and releasing the stored heat to air that is cooler. Water is effective as a heat bank because it can absorb or release a relatively large amount of heat with only a slight change in its own temperature. To understand this capability of water, we must first look briefly at temperature and heat.

### Temperature and Heat

Anything that moves has **kinetic energy**, the energy of motion. Atoms and molecules have kinetic energy because they are always moving, although not necessarily in any particular direction. The faster a molecule moves, the greater its kinetic energy. The kinetic energy associated with the random movement of atoms or molecules is called **thermal energy**. Thermal energy is related to temperature, but they are not the same thing. **Temperature** represents the *average* kinetic energy of the molecules in a body of matter, regardless of volume, whereas the thermal energy of a body of matter reflects the *total* kinetic energy and thus depends on the matter's volume. When water is heated in a coffeemaker, the average speed of the molecules increases, and the thermometer records this as a rise in temperature of the liquid. The total amount of thermal energy also increases in this case. Note, however, that although the pot of coffee has a much higher temperature than, say, the water in a swimming pool, the swimming pool contains more thermal energy because of its much greater volume.

Whenever two objects of different temperature are brought together, thermal energy passes from the warmer to the cooler object until the two are the same temperature. Molecules in the cooler object speed up at the expense of the thermal energy of the warmer object. An ice cube cools a drink not by adding coldness to the liquid, but by absorbing thermal energy from the liquid as the ice itself melts. Thermal energy in transfer from one body of matter to another is defined as **heat**.

One convenient unit of heat used in this book is the **calorie (cal)**. A calorie is the amount of heat it takes to raise the temperature of 1 g of water by 1°C. Conversely, a calorie is also the amount of heat that 1 g of water releases when it cools by 1°C. A **kilocalorie (kcal)**, 1,000 cal, is the quantity of heat required to raise the temperature of 1 kilogram (kg) of water by 1°C. (The "calories" on food packages are actually kilocalories.) Another energy unit used in this book is the **joule (J)**. One joule equals 0.239 cal; one calorie equals 4.184 J.

### Water's High Specific Heat

The ability of water to stabilize temperature stems from its relatively high specific heat. The **specific heat** of a substance is defined as the amount of heat that must be absorbed or lost for 1 g of that substance to change its temperature by 1°C. We already know water's specific heat because we have defined a calorie as the amount of heat that causes 1 g of water to change its temperature by 1°C. Therefore, the specific heat of water is 1 calorie per gram per degree Celsius, abbreviated as 1 cal/(g • °C). Compared with most other substances, water has an unusually high specific heat. As a result, water will change its temperature less than other liquids when it absorbs or loses a given amount of heat. The reason you can burn your fingers by touching the side of an iron pot on the stove when the water in the pot is still lukewarm is that the specific heat of water is ten times greater than that of iron. In other words, the same amount of heat will raise the temperature of 1 g of the iron much faster than it will raise the temperature of 1 g of the water. Specific heat can be thought of as a measure of how well a substance resists changing its temperature when it absorbs or releases heat. Water resists changing its temperature; when it does change its temperature, it absorbs or loses a relatively large quantity of heat for each degree of change.

We can trace water's high specific heat, like many of its other properties, to hydrogen bonding. Heat must be absorbed in order to break hydrogen bonds; by the same token, heat is released when hydrogen bonds form. A calorie of heat causes a relatively small change in the temperature of water because much of the heat is used to disrupt hydrogen bonds before the water molecules can begin moving faster. And when the temperature of water drops slightly, many additional hydrogen bonds form, releasing a considerable amount of energy in the form of heat.

What is the relevance of water's high specific heat to life on Earth? A large body of water can absorb and store a huge amount of heat from the sun in the daytime and during summer while warming up only a few degrees. At night and during winter, the gradually cooling water can warm the air. This capability of water serves to moderate air temperatures in coastal areas **(Figure 2.19)**. The high specific heat of water also tends to stabilize ocean temperatures, creating a favorable environment for marine life. Thus, because of its high specific heat, the water that covers most of Earth keeps temperature fluctuations on land and in water within limits that permit life. Also, because organisms are made primarily of water, they are better able to resist changes in their own temperature than if they were made of a liquid with a lower specific heat.

▲ **Figure 2.19 Temperatures for the Pacific Ocean and Southern California on an August day.**

INTERPRET THE DATA *Explain the pattern of temperatures shown in this diagram.*

### Evaporative Cooling

Molecules of any liquid stay close together because they are attracted to one another. Molecules moving fast enough to overcome these attractions can depart the liquid and enter the air as a gas (vapor). This transformation from a liquid to a gas is called vaporization, or *evaporation*. Recall that the speed of molecular movement varies and that temperature is the *average* kinetic energy of molecules. Even at low temperatures, the speediest molecules can escape into the air. Some evaporation occurs at any temperature; a glass of water at room temperature, for example, will eventually evaporate completely. If a liquid is heated, the average kinetic energy of molecules increases and the liquid evaporates more rapidly.

**Heat of vaporization** is the quantity of heat a liquid must absorb for 1 g of it to be converted from the liquid to the gaseous state. For the same reason that water has a high specific heat, it also has a high heat of vaporization relative to most other liquids. To evaporate 1 g of water at 25°C, about 580 cal of heat is needed—nearly double the amount needed to vaporize a gram of alcohol, for example. Water's high heat of vaporization is another emergent property resulting from the strength of its hydrogen bonds, which must be broken before the molecules can exit from the liquid in the form of water vapor.

The high amount of energy required to vaporize water has a wide range of effects. On a global scale, for example, it helps moderate Earth's climate. A considerable amount of solar heat absorbed by tropical seas is consumed during the evaporation of surface water. Then, as moist tropical air circulates poleward, it releases heat as it condenses and forms rain. On an organismal level, water's high heat of vaporization accounts for the severity of steam burns. These burns are caused by the heat energy released when steam condenses into liquid on the skin.

As a liquid evaporates, the surface of the liquid that remains behind cools down (its temperature decreases). This **evaporative cooling** occurs because the "hottest" molecules, those with the greatest kinetic energy, are the ones most likely to leave as gas. It is as if the hundred fastest runners at a college transferred to another school; the average speed of the remaining students would decline.

Evaporative cooling of water contributes to the stability of temperature in lakes and ponds and also provides a mechanism that prevents terrestrial organisms from overheating. For example, evaporation of water from the leaves of a plant helps keep the tissues in the leaves from becoming too warm in the sunlight. Evaporation of sweat from human skin dissipates body heat and helps prevent overheating on a hot day or when excess heat is generated by strenuous activity. High humidity on a hot day increases discomfort because the high concentration of water vapor in the air inhibits the evaporation of sweat from the body.

## Floating of Ice on Liquid Water

Water is one of the few substances that are less dense as a solid than as a liquid. In other words, ice floats on liquid water. While other materials contract and become denser when they solidify, water expands. The cause of this exotic behavior is, once again, hydrogen bonding. At temperatures above 4°C, water behaves like other liquids, expanding as it warms and contracting as it cools. As the temperature falls from 4°C to 0°C, water begins to freeze because more and more of its molecules are moving too slowly to break hydrogen bonds. At 0°C, the molecules become locked into a crystalline lattice, each water molecule hydrogen-bonded to four partners **(Figure 2.20)**. The hydrogen bonds keep the molecules at "arm's length," far enough apart to make ice about 10% less dense than liquid water at 4°C. When ice absorbs enough heat for its temperature to rise above 0°C, hydrogen bonds between molecules are disrupted. As the crystal collapses, the ice melts, and molecules are free to slip closer together. Water reaches its greatest density at 4°C and then begins to expand as the molecules move faster.

The ability of ice to float due to its lower density is an important factor in the suitability of the environment for life. If ice sank, then eventually all ponds, lakes, and even oceans

Hydrogen bond

**Ice:**
Hydrogen bonds are stable

**Liquid water:**
Hydrogen bonds break and re-form

◄ **Figure 2.20 Ice: crystalline structure and floating barrier.** In ice, each molecule is hydrogen-bonded to four neighbors in a three-dimensional crystal. Because the crystal is spacious, ice has fewer molecules than an equal volume of liquid water. In other words, ice is less dense than liquid water. Floating ice becomes a barrier that insulates the liquid water below from the colder air. The marine organism shown here is a type of shrimp called krill; it was photographed beneath floating ice in the Southern Ocean near Antarctica.

**WHAT IF?** *If water did not form hydrogen bonds, what would happen to the shrimp's habitat, shown here?*

would freeze solid, making life as we know it impossible on Earth. During summer, only the upper few inches of the ocean would thaw. Instead, when a deep body of water cools, the floating ice insulates the liquid water below, preventing it from freezing and allowing life to exist under the frozen surface, as shown in the photo in Figure 2.20.

## Water: The Solvent of Life

A sugar cube placed in a glass of water will dissolve. Eventually, the glass will contain a uniform mixture of sugar and water; the concentration of dissolved sugar will be the same everywhere in the mixture. A liquid that is a completely homogeneous mixture of two or more substances is called a **solution**. The dissolving agent of a solution is the **solvent**, and the substance that is dissolved is the **solute**. In this case, water is the solvent and sugar is the solute. An **aqueous solution** is one in which the solute is dissolved in water; water is the solvent.

Water is a very versatile solvent, a quality we can trace to the polarity of the water molecule. Suppose, for example, that a spoonful of table salt, the ionic compound sodium chloride (NaCl), is placed in water **(Figure 2.21)**. At the surface of each grain, or crystal, of salt, the sodium and chloride ions are exposed to the solvent. These ions and regions of the water molecules are attracted to each other due to their opposite charges. The oxygen regions of the water molecules are negatively

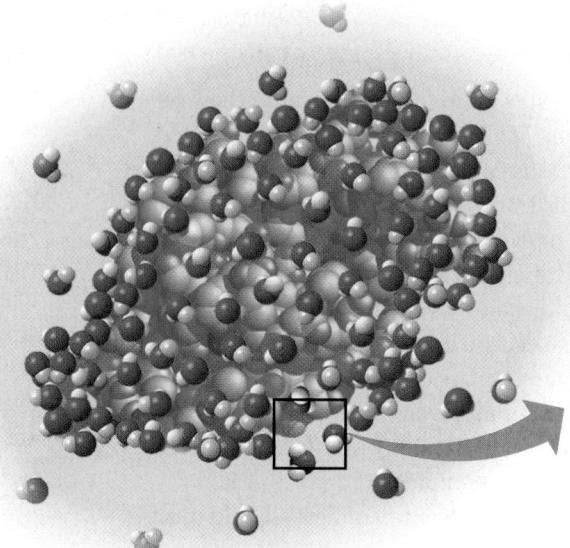
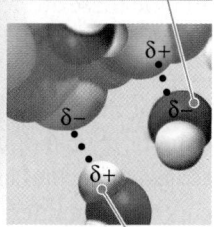

This oxygen is attracted to a slight positive charge on the lysozyme molecule.

This hydrogen is attracted to a slight negative charge on the lysozyme molecule.

▲ **Figure 2.22 A water-soluble protein.** Human lysozyme is a protein found in tears and saliva that has antibacterial action. This model shows the lysozyme molecule (purple) in an aqueous environment. Ionic and polar regions on the protein's surface attract the slightly charged regions of water molecules.

charged and are attracted to sodium cations. The hydrogen regions are positively charged and are attracted to chloride anions. As a result, water molecules surround the individual sodium and chloride ions, separating and shielding them from one another. The sphere of water molecules around each dissolved ion is called a **hydration shell**. Working inward from the surface of each salt crystal, water eventually dissolves all the ions. The result is a solution of two solutes, sodium cations and chloride anions, homogeneously mixed with water, the solvent. Other ionic compounds also dissolve in water. Seawater, for instance, contains a great variety of dissolved ions, as do living cells.

A compound does not need to be ionic to dissolve in water; many compounds made up of nonionic polar molecules, such as sugars, are also water-soluble. Such compounds dissolve when water molecules surround each of the solute molecules, forming hydrogen bonds with them. Even molecules as large as proteins can dissolve in water if they have ionic and polar regions on their surface **(Figure 2.22)**. Many different kinds of polar compounds are dissolved (along with ions) in the water of such biological fluids as blood, the sap of plants, and the liquid within all cells. Water is the solvent of life.

### Hydrophilic and Hydrophobic Substances

Any substance that has an affinity for water is said to be **hydrophilic** (from the Greek *hydro*, water, and *philos*, loving). In some cases, substances can be hydrophilic without actually dissolving. For example, some molecules in cells are so large that they do not dissolve. Another example of a hydrophilic substance that does not dissolve is cotton, a plant product. Cotton consists of giant molecules of cellulose, a compound with numerous regions of partial positive and partial negative charges that can form hydrogen bonds with water. Water adheres to the

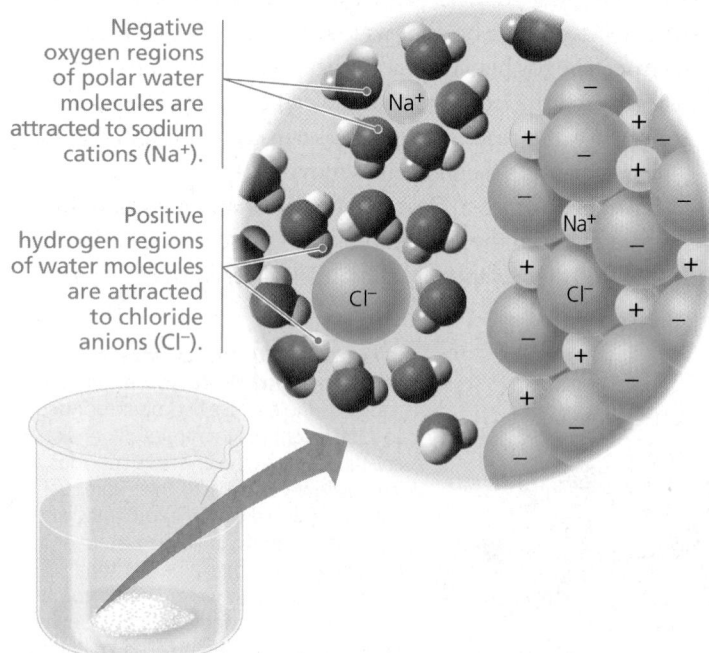

Negative oxygen regions of polar water molecules are attracted to sodium cations (Na⁺).

Positive hydrogen regions of water molecules are attracted to chloride anions (Cl⁻).

▲ **Figure 2.21 Table salt dissolving in water.** A sphere of water molecules, called a hydration shell, surrounds each solute ion.

**WHAT IF?** *What would happen if you heated this solution for a long time?*

cellulose fibers. Thus, a cotton towel does a great job of drying the body, yet it does not dissolve in the washing machine. Cellulose is also present in the walls of plant cells that conduct water; you read earlier how the adhesion of water to these hydrophilic walls helps water move up the plant against gravity.

There are, of course, substances that do not have an affinity for water. Substances that are nonionic and nonpolar (or otherwise cannot form hydrogen bonds) actually seem to repel water; these substances are said to be **hydrophobic** (from the Greek *phobos*, fearing). An example from the kitchen is vegetable oil, which, as you know, does not mix stably with water-based substances such as vinegar. The hydrophobic behavior of the oil molecules results from a prevalence of relatively nonpolar covalent bonds, in this case bonds between carbon and hydrogen, which share electrons almost equally. Hydrophobic molecules related to oils are major ingredients of cell membranes. (Imagine what would happen to a cell if its membrane dissolved!)

### Solute Concentration in Aqueous Solutions

Most of the chemical reactions in organisms involve solutes dissolved in water. To understand such reactions, we must know how many atoms and molecules are involved and be able to calculate the concentration of solutes in an aqueous solution (the number of solute molecules in a volume of solution).

When carrying out experiments, we use mass to calculate the number of molecules. We first calculate the **molecular mass**, which is simply the sum of the masses of all the atoms in a molecule. As an example, let's calculate the molecular mass of table sugar (sucrose), $C_{12}H_{22}O_{11}$, by multiplying the number of atoms by the atomic mass of each element (see Appendix B). In round numbers, sucrose has a molecular mass of $(12 \times 12) + (22 \times 1) + (11 \times 16) = 342$ daltons. Because we can't measure out small numbers of molecules, we usually measure substances in units called moles. Just as a dozen always means 12 objects, a **mole (mol)** represents an exact number of objects: $6.02 \times 10^{23}$, which is called Avogadro's number. There are $6.02 \times 10^{23}$ daltons in 1 g. Once we determine the molecular mass of a molecule such as sucrose, we can use the same number (342), but with the unit *gram*, to represent the mass of $6.02 \times 10^{23}$ molecules of sucrose, or 1 mol of sucrose. To obtain 1 mol of sucrose in the lab, therefore, we weigh out 342 g.

The practical advantage of measuring a quantity of chemicals in moles is that a mole of one substance has exactly the same number of molecules as a mole of any other substance. Measuring in moles makes it convenient for scientists working in the laboratory to combine substances in fixed ratios of molecules.

How would we make a liter (L) of solution consisting of 1 mol of sucrose dissolved in water? We would measure out 342 g of sucrose and then add enough water to bring the total volume of the solution up to 1 L. At that point, we would have a 1-molar (1 $M$) solution of sucrose. **Molarity**—the number of moles of solute per liter of solution—is the unit of concentration most often used by biologists for aqueous solutions.

## Acids and Bases

Occasionally, a hydrogen atom participating in a hydrogen bond between two water molecules shifts from one molecule to the other. When this happens, the hydrogen atom leaves its electron behind, and what is actually transferred is a **hydrogen ion** ($H^+$), a single proton with a charge of $1+$. The water molecule that lost a proton is now a **hydroxide ion** ($OH^-$), which has a charge of $1-$. The proton binds to the other water molecule, making that molecule a **hydronium ion** ($H_3O^+$):

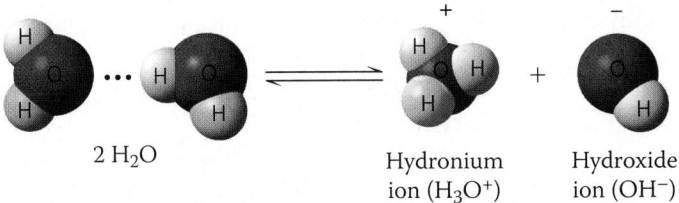

2 $H_2O$ ⇌ Hydronium ion ($H_3O^+$) + Hydroxide ion ($OH^-$)

By convention, $H^+$ (the hydrogen ion) is used to represent $H_3O^+$ (the hydronium ion), and we follow that practice here. Keep in mind, though, that $H^+$ does not exist on its own in an aqueous solution. It is always associated with a water molecule in the form of $H_3O^+$.

As indicated by the double arrows, this is a reversible reaction that reaches a state of dynamic equilibrium when water molecules dissociate at the same rate that they are being reformed from $H^+$ and $OH^-$. At this equilibrium point, the concentration of water molecules greatly exceeds the concentrations of $H^+$ and $OH^-$. In pure water, only one water molecule in every 554 million is dissociated; the concentration of each ion in pure water is $10^{-7}$ $M$ (at 25°C). This means there is only one ten-millionth of a mole of hydrogen ions per liter of pure water and an equal number of hydroxide ions.

Although the dissociation of water is reversible and statistically rare, it is exceedingly important in the chemistry of life. $H^+$ and $OH^-$ are very reactive. Changes in their concentrations can drastically affect a cell's proteins and other complex molecules. As we have seen, the concentrations of $H^+$ and $OH^-$ are equal in pure water, but adding certain kinds of solutes, called acids and bases, disrupts this balance.

What would cause an aqueous solution to have an imbalance in $H^+$ and $OH^-$ concentrations? When acids dissolve in water, they donate additional $H^+$ to the solution. An **acid** is a substance that increases the hydrogen ion concentration of a solution. For example, when hydrochloric acid (HCl) is added to water, hydrogen ions dissociate from chloride ions:

$$HCl \rightarrow H^+ + Cl^-$$

This source of $H^+$ (dissociation of water is the other source) results in an acidic solution—one having more $H^+$ than $OH^-$.

A substance that *reduces* the hydrogen ion concentration of a solution is called a **base**. Some bases reduce the $H^+$ concentration directly by accepting hydrogen ions. Ammonia ($NH_3$), for instance, acts as a base when the unshared electron pair in nitrogen's valence shell attracts a hydrogen ion from the solution, resulting in an ammonium ion ($NH_4^+$):

$$NH_3 + H^+ \rightleftharpoons NH_4^+$$

Other bases reduce the $H^+$ concentration indirectly by dissociating to form hydroxide ions, which combine with hydrogen ions and form water. One such base is sodium hydroxide (NaOH), which in water dissociates into its ions:

$$NaOH \rightarrow Na^+ + OH^-$$

In either case, the base reduces the $H^+$ concentration. Solutions with a higher concentration of $OH^-$ than $H^+$ are known as basic solutions. A solution in which the $H^+$ and $OH^-$ concentrations are equal is said to be neutral.

Notice that single arrows were used in the reactions for HCl and NaOH. These compounds dissociate completely when mixed with water, so hydrochloric acid is called a strong acid and sodium hydroxide a strong base. In contrast, ammonia is a relatively weak base. The double arrows in the reaction for ammonia indicate that the binding and release of hydrogen ions are reversible reactions, although at equilibrium there will be a fixed ratio of $NH_4^+$ to $NH_3$.

Weak acids are acids that reversibly release and accept back hydrogen ions. An example is carbonic acid:

$$\underset{\substack{\text{Carbonic} \\ \text{acid}}}{H_2CO_3} \rightleftharpoons \underset{\substack{\text{Bicarbonate} \\ \text{ion}}}{HCO_3^-} + \underset{\substack{\text{Hydrogen} \\ \text{ion}}}{H^+}$$

Here the equilibrium so favors the reaction in the left direction that when carbonic acid is added to pure water, only 1% of the molecules are dissociated at any particular time. Still, that is enough to shift the balance of $H^+$ and $OH^-$ from neutrality.

### The pH Scale

In any aqueous solution at 25°C, the *product* of the $H^+$ and $OH^-$ concentrations is constant at $10^{-14}$. This can be written

$$[H^+][OH^-] = 10^{-14}$$

In such an equation, brackets indicate molar concentration. In a neutral solution at room temperature (25°C), $[H^+] = 10^{-7}$ and $[OH^-] = 10^{-7}$. In this case, $10^{-7} \times 10^{-7} = 10^{-14}$. If enough acid is added to a solution to increase $[H^+]$ to $10^{-5}\,M$, then $[OH^-]$ will decline by an equivalent factor to $10^{-9}\,M$ (note that $10^{-5} \times 10^{-9} = 10^{-14}$). This constant relationship expresses the behavior of acids and bases in an aqueous solution. An acid not only adds hydrogen ions to a solution, but also removes hydroxide ions because of the tendency for $H^+$ to combine with $OH^-$, forming water. A base has the opposite effect, increasing $OH^-$ concentration but also reducing $H^+$ concentration by the formation of water. If enough of a base is added to raise the $OH^-$ concentration to $10^{-4}\,M$, it will cause the $H^+$ concentration to drop to $10^{-10}\,M$. Whenever we know the concentration of either $H^+$ or $OH^-$ in an aqueous solution, we can deduce the concentration of the other ion.

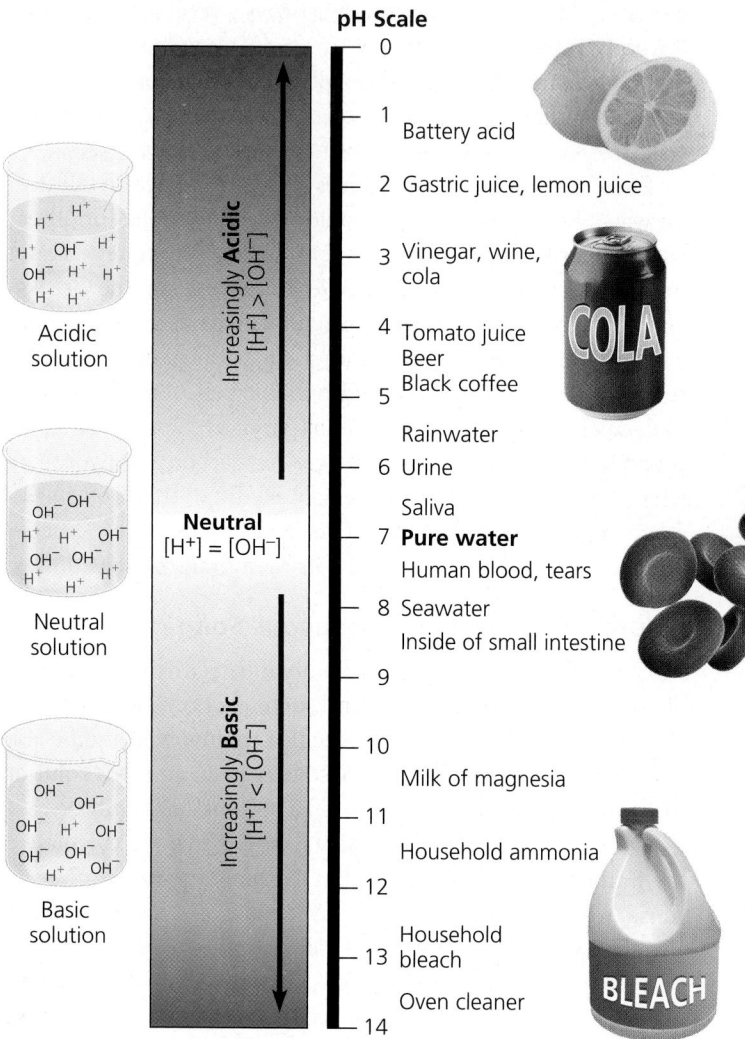

▲ **Figure 2.23 The pH scale and pH values of some aqueous solutions.**

Because the $H^+$ and $OH^-$ concentrations of solutions can vary by a factor of 100 trillion or more, scientists have developed a way to express this variation more conveniently than in moles per liter. The pH scale **(Figure 2.23)** compresses the range of $H^+$ and $OH^-$ concentrations by employing logarithms. The **pH** of a solution is defined as the negative logarithm (base 10) of the hydrogen ion concentration:

$$pH = -\log[H^+]$$

For a neutral aqueous solution, $[H^+]$ is $10^{-7}\,M$, giving us

$$-\log 10^{-7} = -(-7) = 7$$

Notice that pH *declines* as $H^+$ concentration *increases*. Notice, too, that although the pH scale is based on $H^+$ concentration, it also implies $OH^-$ concentration. A solution of pH 10 has a hydrogen ion concentration of $10^{-10}\,M$ and a hydroxide ion concentration of $10^{-4}\,M$.

The pH of a neutral aqueous solution at 25°C is 7, the midpoint of the pH scale. A pH value less than 7 denotes an acidic solution; the lower the number, the more acidic the solution. The pH for basic solutions is above 7. Most biological fluids, such as blood and saliva, are within the range of pH 6–8. There are a few exceptions, however, including the strongly acidic digestive juice of the human stomach, which has a pH of about 2.

Remember that each pH unit represents a tenfold difference in $H^+$ and $OH^-$ concentrations. It is this mathematical feature that makes the pH scale so compact. A solution of pH 3 is not twice as acidic as a solution of pH 6, but a thousand times ($10 \times 10 \times 10$) more acidic. When the pH of a solution changes slightly, the actual concentrations of $H^+$ and $OH^-$ in the solution change substantially.

## Buffers

The internal pH of most living cells is close to 7. Even a slight change in pH can be harmful because the chemical processes of the cell are very sensitive to the concentrations of hydrogen and hydroxide ions. The pH of human blood is very close to 7.4, which is slightly basic. A person cannot survive for more than a few minutes if the blood pH drops to 7 or rises to 7.8, and a chemical system exists in the blood that maintains a stable pH. If 0.01 mol of a strong acid is added to a liter of pure water, the pH drops from 7.0 to 2.0. If the same amount of acid is added to a liter of blood, however, the pH decrease is only from 7.4 to 7.3. Why does the addition of acid have so much less of an effect on the pH of blood than it does on the pH of water?

The presence of substances called buffers allows biological fluids to maintain a relatively constant pH despite the addition of acids or bases. A **buffer** is a substance that minimizes changes in the concentrations of $H^+$ and $OH^-$ in a solution. It does so by accepting hydrogen ions from the solution when they are in excess and donating hydrogen ions to the solution when they have been depleted. Most buffer solutions contain a weak acid and its corresponding base, which combine reversibly with hydrogen ions.

Several buffers contribute to pH stability in human blood and many other biological solutions. One of these is carbonic acid ($H_2CO_3$), which is formed when $CO_2$ reacts with water in blood plasma. As mentioned earlier, carbonic acid dissociates to yield a bicarbonate ion ($HCO_3^-$) and a hydrogen ion ($H^+$):

$$\underset{\substack{H^+\ donor \\ (acid)}}{H_2CO_3} \underset{\substack{\text{Response to} \\ \text{a drop in pH}}}{\overset{\substack{\text{Response to} \\ \text{a rise in pH}}}{\rightleftharpoons}} \underset{\substack{H^+\ acceptor \\ (base)}}{HCO_3^-} + \underset{\substack{\text{Hydrogen} \\ \text{ion}}}{H^+}$$

The chemical equilibrium between carbonic acid and bicarbonate acts as a pH regulator, the reaction shifting left or right as other processses in the solution add or remove hydrogen ions. If the $H^+$ concentration in blood begins to fall (that is, if pH rises), the reaction proceeds to the right and more carbonic acid dissociates, replenishing hydrogen ions. But when the $H^+$ concentration in blood begins to rise (when pH drops), the

reaction proceeds to the left, with $HCO_3^-$ (the base) removing the hydrogen ions from the solution and forming $H_2CO_3$. Thus, the carbonic acid–bicarbonate buffering system consists of an acid and a base in equilibrium with each other. Most other buffers are also acid-base pairs.

### Acidification: A Threat to Our Oceans

Among the many threats to water quality posed by human activities is the burning of fossil fuels, which releases $CO_2$ into the atmosphere. The resulting increase in atmospheric $CO_2$ levels has caused global warming (see Concept 43.4). In addition, about 25% of human-generated $CO_2$ is absorbed by the oceans. In spite of the huge volume of water in the oceans, scientists worry that the absorption of so much $CO_2$ will harm marine ecosystems.

Recent data have shown that such fears are well founded. When $CO_2$ dissolves in seawater, it reacts with water to form carbonic acid, which lowers ocean pH. This process, known as **ocean acidification**, alters the delicate balance of conditions for life in the oceans **(Figure 2.24)**. Based on measurements of $CO_2$ levels in air bubbles trapped in ice over thousands of years, scientists calculate that the pH of the oceans is 0.1 pH unit lower now than at any time in the past 420,000 years. Recent studies predict that it will drop another 0.3–0.5 pH unit by the end of this century.

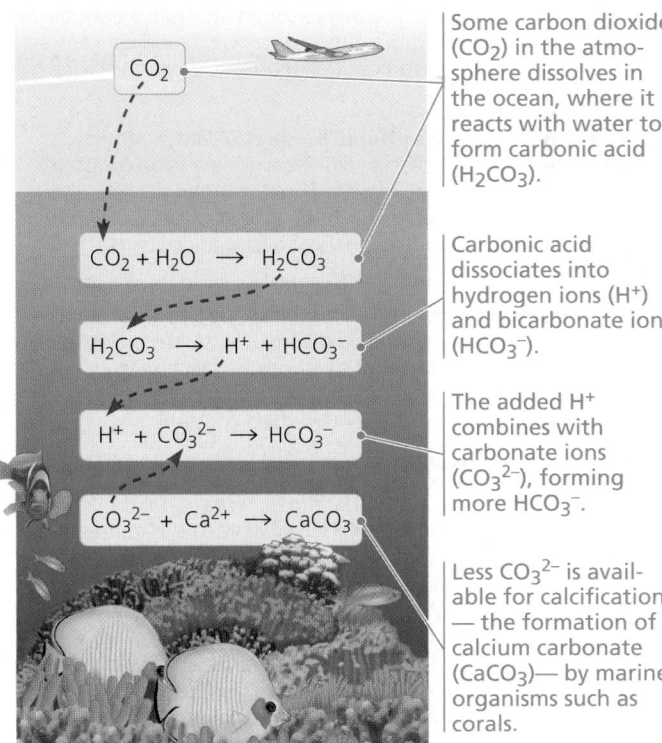

Some carbon dioxide ($CO_2$) in the atmosphere dissolves in the ocean, where it reacts with water to form carbonic acid ($H_2CO_3$).

$$CO_2 + H_2O \rightarrow H_2CO_3$$

Carbonic acid dissociates into hydrogen ions ($H^+$) and bicarbonate ions ($HCO_3^-$).

$$H_2CO_3 \rightarrow H^+ + HCO_3^-$$

The added $H^+$ combines with carbonate ions ($CO_3^{2-}$), forming more $HCO_3^-$.

$$H^+ + CO_3^{2-} \rightarrow HCO_3^-$$

$$CO_3^{2-} + Ca^{2+} \rightarrow CaCO_3$$

Less $CO_3^{2-}$ is available for calcification — the formation of calcium carbonate ($CaCO_3$)— by marine organisms such as corals.

▲ **Figure 2.24 Atmospheric $CO_2$ from human activities and its fate in the ocean.**

**WHAT IF?** *Would lowering the ocean's carbonate concentration have any effect, even indirectly, on organisms that don't form $CaCO_3$? Explain.*

As seawater acidifies, the extra hydrogen ions combine with carbonate ions ($CO_3^{2-}$) to form bicarbonate ions ($HCO_3^-$), thereby reducing the carbonate ion concentration (see Figure 2.24). Scientists predict that ocean acidification will cause the carbonate ion concentration to decrease by 40% by the year 2100. This is of great concern because carbonate ions are required for calcification, the production of calcium carbonate ($CaCO_3$), by many marine organisms, including reef-building corals and animals that build shells. The **Scientific Skills Exercise** allows you to work with data from an experiment examining the effect of carbonate ion concentration on coral reefs. Coral reefs are sensitive ecosystems that act as havens for a great diversity of marine life. The disappearance of coral reef ecosystems would be a tragic loss of biological diversity.

## Scientific Skills Exercise

# Interpreting a Scatter Plot with a Regression Line

**How Does the Carbonate Ion Concentration of Seawater Affect the Calcification Rate of a Coral Reef?** Scientists predict that acidification of the ocean due to higher levels of atmospheric $CO_2$ will lower the concentration of dissolved carbonate ions, which living corals use to build calcium carbonate reef structures. In this exercise, you will analyze data from a controlled experiment that examined the effect of carbonate ion concentration ($[CO_3^{2-}]$) on calcium carbonate deposition, a process called calcification.

**How the Experiment Was Done** For several years, scientists conducted research on ocean acidification using a large coral reef aquarium at Biosphere 2 in Arizona. They measured the rate of calcification by the reef organisms and examined how the calcification rate changed with differing amounts of dissolved carbonate ions in the seawater.

**Data from the Experiment** The black data points in the graph form a scatter plot. The red line, known as a linear regression line, is the best-fitting straight line for these points. These data are from one set of experiments, in which the pH, temperature, and calcium ion concentration of the seawater were held constant.

**Data from** C. Langdon et al., Effect of calcium carbonate saturation state on the calcification rate of an experimental coral reef, *Global Biogeochemical Cycles* 14:639–654 (2000).

**INTERPRET THE DATA**

1. When presented with a graph of experimental data, the first step in your analysis is to determine what each axis represents. (a) In words, explain what is being shown on the x-axis. Be sure to include the units. (b) What is being shown on the y-axis (including units)? (c) Which variable is the independent variable—the variable that was *manipulated* by the researchers? (d) Which variable is the dependent variable—the variable that responded to or depended on the treatment, which was *measured* by the researchers? (For additional information about graphs, see the Scientific Skills Review in Appendix F and in the Study Area in MasteringBiology.)

2. Based on the data shown in the graph, describe in words the relationship between carbonate ion concentration and calcification rate.

3. (a) If the seawater carbonate ion concentration is 270 µmol/kg, what is the approximate rate of calcification, and approximately

how many days would it take 1 square meter of reef to accumulate 30 mmol of calcium carbonate ($CaCO_3$)? (b) If the seawater carbonate ion concentration is 250 µmol/kg, what is the approximate rate of calcification, and approximately how many days would it take 1 square meter of reef to accumulate 30 mmol of calcium carbonate? (c) If carbonate ion concentration decreases, how does the calcification rate change, and how does that affect the time it takes coral to grow?

4. (a) Referring to the reactions in Figure 2.24, determine which step of the process is measured in this experiment. (b) Are the results of this experiment consistent with the hypothesis that increased atmospheric [$CO_2$] will slow the growth of coral reefs? Why or why not?

MB A version of this Scientific Skills Exercise can be assigned in MasteringBiology.

Go to **MasteringBiology®** for Assignments, the eText, and the Study Area with Animations, Activities, Vocab Self-Quiz, and Practice Tests.

## SUMMARY OF KEY CONCEPTS

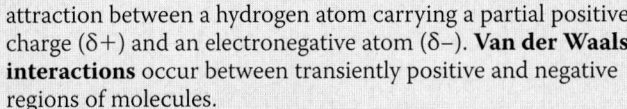
**VOCAB SELF-QUIZ**

goo.gl/gbai8v

### CONCEPT 2.1

**Matter consists of chemical elements in pure form and in combinations called compounds (pp. 22–23)**

- **Elements** cannot be broken down chemically to other substances. A **compound** contains two or more different elements in a fixed ratio. Oxygen, carbon, hydrogen, and nitrogen make up approximately 96% of living matter.

  **?** *In what way does the need for iodine or iron in your diet differ from your need for calcium or phosphorus?*

### CONCEPT 2.2

**An element's properties depend on the structure of its atoms (pp. 23–27)**

- An **atom**, the smallest unit of an element, has the following components:

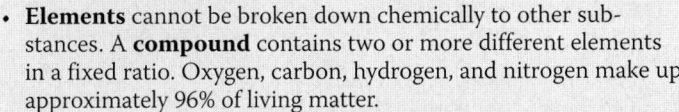
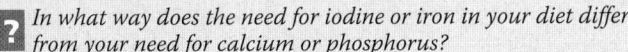

**Nucleus**

**Protons** (+ charge) determine element

**Neutrons** (no charge) determine isotope

**Electrons** (– charge) form negative cloud and determine chemical behavior

**Atom**

- An electrically neutral atom has equal numbers of electrons and protons; the number of protons determines the **atomic number**. **Isotopes** of an element differ from each other in neutron number and therefore mass. Unstable isotopes give off particles and energy as radioactivity.
- In an atom, electrons occupy specific **electron shells**; the electrons in a shell have a characteristic energy level. Electron distribution in shells determines the chemical behavior of an atom. An atom that has an incomplete outer shell, the **valence shell**, is reactive.

  **DRAW IT** *Draw the electron distribution diagrams for neon ($_{10}$Ne) and argon ($_{18}$Ar). Why are they chemically unreactive?*

### CONCEPT 2.3

**The formation and function of molecules depend on chemical bonding between atoms (pp. 27–31)**

- **Chemical bonds** form when atoms interact and complete their valence shells. **Covalent bonds** form when pairs of electrons are shared. $H_2$ has a **single bond**: H — H. A **double bond** is the sharing of two pairs of electrons, as in O═O.
- **Molecules** consist of two or more covalently bonded atoms. The attraction of an atom for the electrons of a covalent bond is its **electronegativity**. Electrons of a **polar covalent bond** are pulled closer to the more electronegative atom, such as the oxygen in $H_2O$.
- An **ion** forms when an atom or molecule gains or loses an electron and becomes charged. An **ionic bond** is the attraction between two oppositely charged ions, such as $Na^+$ and $Cl^-$.
- Weak bonds reinforce the shapes of large molecules and help molecules adhere to each other. A **hydrogen bond** is an attraction between a hydrogen atom carrying a partial positive charge ($\delta+$) and an electronegative atom ($\delta-$). **Van der Waals interactions** occur between transiently positive and negative regions of molecules.
- Molecular shape is usually the basis for the recognition of one biological molecule by another.

  **?** *In terms of electron sharing between atoms, compare nonpolar covalent bonds, polar covalent bonds, and the formation of ions.*

### CONCEPT 2.4

**Chemical reactions make and break chemical bonds (pp. 31–32)**

- **Chemical reactions** change **reactants** into **products** while conserving matter. All chemical reactions are theoretically reversible. **Chemical equilibrium** is reached when the forward and reverse reaction rates are equal.

  **?** *What would happen to the concentration of products if more reactants were added to a reaction that was in chemical equilibrium? How would this addition affect the equilibrium?*

### CONCEPT 2.5

**Hydrogen bonding gives water properties that help make life possible on Earth (pp. 32–40)**

- A **hydrogen bond** forms when the slightly negatively charged oxygen of one water molecule is attracted to the slightly positively charged hydrogen of a nearby water molecule. Hydrogen bonding between water molecules is the basis for water's properties.

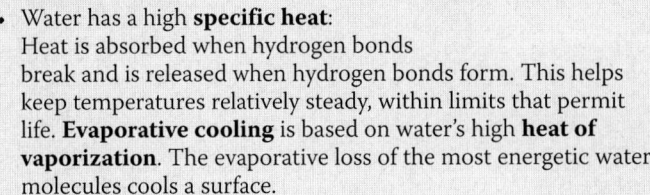

- Hydrogen bonding keeps water molecules close to each other, giving water **cohesion**. Hydrogen bonding is also responsible for water's **surface tension**.
- Water has a high **specific heat**: Heat is absorbed when hydrogen bonds break and is released when hydrogen bonds form. This helps keep temperatures relatively steady, within limits that permit life. **Evaporative cooling** is based on water's high **heat of vaporization**. The evaporative loss of the most energetic water molecules cools a surface.
- Ice floats because it is less dense than liquid water. This property allows life to exist under the frozen surfaces of lakes and seas.

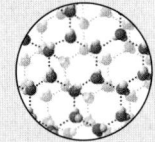
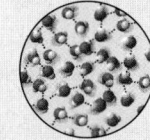

**Ice:** stable hydrogen bonds

**Liquid water:** transient hydrogen bonds

- Water is an unusually versatile **solvent** because its polar molecules are attracted to ions and polar substances that can form hydrogen bonds. **Hydrophilic** substances have an affinity for water; **hydrophobic** substances do not. **Molarity**, the number of moles of **solute** per liter of **solution**, is used as a measure of solute concentration in solutions. A **mole** is a certain number of molecules of a substance. The mass of a mole of a substance in grams is the same as the **molecular mass** in daltons.

- A water molecule can transfer an $H^+$ to another water molecule to form $H_3O^+$ (represented simply by $H^+$) and $OH^-$.
- The concentration of $H^+$ is expressed as **pH**; $pH = -\log[H^+]$. A **buffer** consists of an acid-base pair that combines reversibly with hydrogen ions, allowing it to resist pH changes.
- The burning of fossil fuels increases the amount of $CO_2$ in the atmosphere. Some $CO_2$ dissolves in the oceans, causing **ocean acidification**, which has potentially grave consequences for coral reefs.

| | |
|---|---|
| **Acidic** $[H^+] > [OH^-]$ | 0 — **Acids** donate $H^+$ in aqueous solutions. |
| **Neutral** $[H^+] = [OH^-]$ | 7 |
| **Basic** $[H^+] < [OH^-]$ | **Bases** donate $OH^-$ or accept $H^+$ in aqueous solutions. 14 |

**?** *Describe how the properties of water result from the molecule's polar covalent bonds and how these properties contribute to Earth's suitability for life.*

## TEST YOUR UNDERSTANDING

**PRACTICE TEST**

goo.gl/CRZjvS

### Level 1: Knowledge/Comprehension

1. The reactivity of an atom arises from
   (A) the average distance of the outermost electron shell from the nucleus.
   (B) the existence of unpaired electrons in the valence shell.
   (C) the sum of the potential energies of all the electron shells.
   (D) the potential energy of the valence shell.

2. Which of the following statements correctly describes any chemical reaction that has reached equilibrium?
   (A) The concentrations of products and reactants are equal.
   (B) The reaction is now irreversible.
   (C) Both forward and reverse reactions have halted.
   (D) The rates of the forward and reverse reactions are equal.

3. Many mammals control their body temperature by sweating. Which property of water is most directly responsible for the ability of sweat to lower body temperature?
   (A) water's change in density when it condenses
   (B) water's ability to dissolve molecules in the air
   (C) the release of heat by the formation of hydrogen bonds
   (D) the absorption of heat by the breaking of hydrogen bonds

4. We can be sure that a mole of table sugar and a mole of vitamin C are equal in their
   (A) mass.          (C) number of atoms.
   (B) volume.        (D) number of molecules.

5. Measurements show that the pH of a particular lake is 4.0. What is the hydrogen ion concentration of the lake?
   (A) $4.0\,M$   (B) $10^{-4}\,M$   (C) $10^4\,M$   (D) $10^{-10}\,M$

### Level 2: Application/Analysis

6. The atomic number of sulfur is 16. Sulfur combines with hydrogen by covalent bonding to form a compound, hydrogen sulfide. Based on the number of valence electrons in a sulfur atom, predict the molecular formula of the compound.
   (A) HS   (B) $HS_2$   (C) $H_2S$   (D) $H_3S_2$

7. What coefficients must be placed in the following blanks so that all atoms are accounted for in the products?
$$C_6H_{12}O_6 \rightarrow \underline{\quad} C_2H_6O + \underline{\quad} CO_2$$
   (A) 1; 2   (B) 3; 1   (C) 1; 3   (D) 2; 2

8. A slice of pizza has 500 kcal. If we could burn the pizza and use all the heat to warm a 50-L container of cold water, what would be the approximate increase in the temperature of the water? (*Note:* A liter of cold water weighs about 1 kg.)
   (A) 50°C   (B) 5°C   (C) 1°C   (D) 10°C

9. **DRAW IT** Draw the hydration shells that form around a potassium ion and a chloride ion when potassium chloride (KCl) dissolves in water. Label the positive, negative, and partial charges on the atoms.

10. **MAKE CONNECTIONS** What do climate change (see Concept 1.1) and ocean acidification have in common?

### Level 3: Synthesis/Evaluation

11. **SCIENTIFIC INQUIRY**
Female luna moths (*Actias luna*) attract males by emitting chemical signals that spread through the air. A male hundreds of meters away can detect these molecules and fly toward their source. The sensory organs responsible for this behavior are the comblike antennae visible in the photograph shown here. Each filament of an antenna is equipped with thousands of receptor cells that detect the sex attractant. (a) Based on what you learned in this chapter, propose a hypothesis to account for the ability of the male moth to detect a specific molecule in the presence of many other molecules in the air. (b) Describe predictions your hypothesis enables you to make. (c) Design an experiment to test one of these predictions.

12. **FOCUS ON EVOLUTION**
The percentages of naturally occurring elements making up the human body are similar to the percentages of these elements found in other organisms. How could you account for this similarity among organisms? Explain your thinking.

13. **FOCUS ON ORGANIZATION**
Several emergent properties of water contribute to the suitability of the environment for life. In a short essay (100–150 words), describe how the ability of water to function as a versatile solvent arises from the structure of water molecules.

14. **SYNTHESIZE YOUR KNOWLEDGE**

How do cats drink? Scientists using high-speed video have shown that cats use an interesting technique to drink aqueous substances like water and milk. Four times a second, the cat touches the tip of its tongue to the water and draws a column of water up into its mouth (as you can see in the photo), which then shuts before gravity can pull the water back down. Describe how the properties of water allow cats to drink in this fashion, including how water's molecular structure contributes to the process.

*For selected answers, see Appendix A.*

# 3 Carbon and the Molecular Diversity of Life

## KEY CONCEPTS

**3.1** Carbon atoms can form diverse molecules by bonding to four other atoms

**3.2** Macromolecules are polymers, built from monomers

**3.3** Carbohydrates serve as fuel and building material

**3.4** Lipids are a diverse group of hydrophobic molecules

**3.5** Proteins include a diversity of structures, resulting in a wide range of functions

**3.6** Nucleic acids store, transmit, and help express hereditary information

**3.7** Genomics and proteomics have transformed biological inquiry and applications

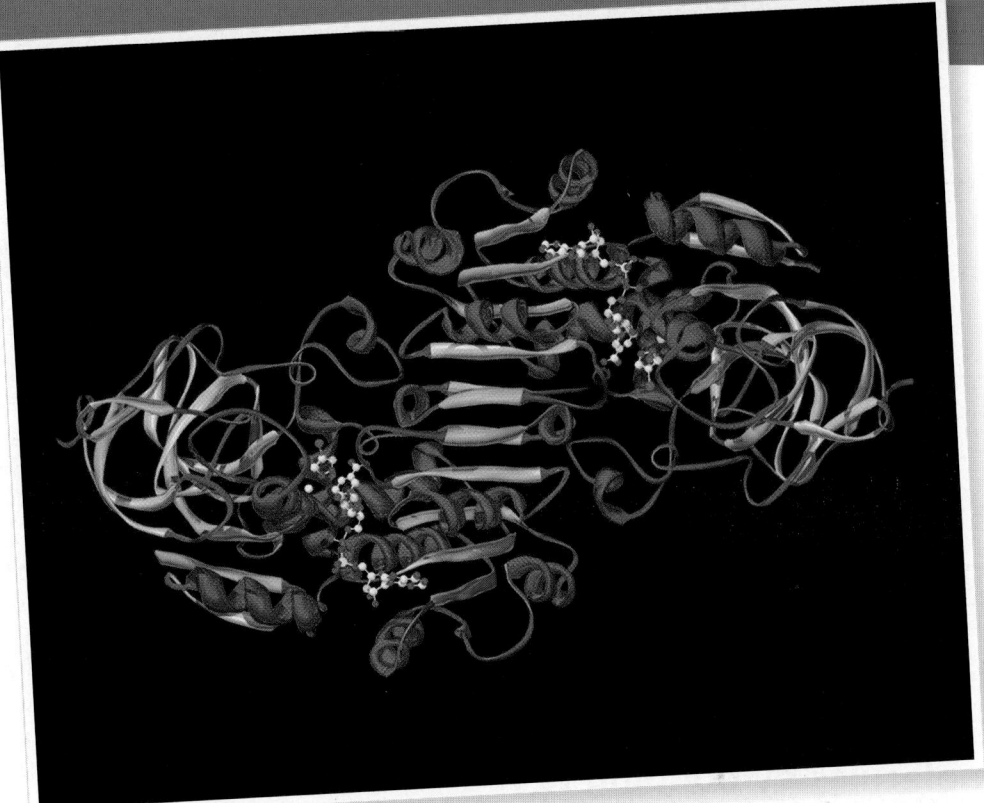

▲ **Figure 3.1 Why is the structure of a protein important for its function?**

## Carbon Compounds and Life

Water is the universal medium for life on Earth, but water aside, living organisms are made up of chemicals based mostly on the element carbon. Of all chemical elements, carbon is unparalleled in its ability to form molecules that are large, complex, and varied. Hydrogen (H), oxygen (O), nitrogen (N), sulfur (S), and phosphorus (P) are other common ingredients of these compounds, but it is the element carbon (C) that accounts for the enormous variety of biological molecules. For historical reasons, a compound containing carbon is said to be an **organic compound**; furthermore, almost all organic compounds associated with life contain hydrogen atoms in addition to carbon atoms. Different species of organisms and even different individuals within a species are distinguished by variations in their large organic compounds.

Given the rich complexity of life on Earth, it may surprise you to learn that the critically important large molecules of all living things—from bacteria to elephants—fall into just four

main classes: carbohydrates, lipids, proteins, and nucleic acids. On the molecular scale, members of three of these classes—carbohydrates, proteins, and nucleic acids—are huge and are therefore called **macromolecules**. For example, a protein may consist of thousands of atoms that form a molecular colossus with a mass well over 100,000 daltons. Considering the size and complexity of macromolecules, it is noteworthy that biochemists have determined the detailed structure of so many of them. The image in **Figure 3.1** is a molecular model of a protein called alcohol dehydrogenase, which breaks down alcohol in the body. The structures of macromolecules can provide important information about their functions.

In this chapter, we'll first investigate the properties of small organic molecules and then go on to discuss the larger biological molecules. After considering how macromolecules are built, we'll examine the structure and function of all four classes of large biological molecules. Like small molecules, large biological molecules exhibit unique emergent properties arising from the orderly arrangement of their atoms.

| Molecule and Molecular Shape | Molecular Formula | Structural Formula | Ball-and-Stick Model (molecular shape in pink) | Space-Filling Model |
|---|---|---|---|---|
| **(a) Methane.** When a carbon atom has four single bonds to other atoms, the molecule is tetrahedral. | $CH_4$ | H—C—H with H above and H below | | |
| **(b) Ethane.** A molecule may have more than one tetrahedral group of single-bonded atoms. (Ethane consists of two such groups.) | $C_2H_6$ | H—C—C—H with H above and below each C | | |
| **(c) Ethene (ethylene).** When two carbon atoms are joined by a double bond, all atoms attached to those carbons are in the same plane, and the molecule is flat. | $C_2H_4$ | H₂C=CH₂ | | |

## CONCEPT 3.1

# Carbon atoms can form diverse molecules by bonding to four other atoms

The key to an atom's chemical characteristics is its electron configuration. This configuration determines the kinds and number of bonds an atom will form with other atoms, and it is the source of carbon's versatility.

## The Formation of Bonds with Carbon

Carbon has 6 electrons, with 2 in the first electron shell and 4 in the second shell; thus, it has 4 valence electrons in a shell that can hold up to 8 electrons. A carbon atom usually completes its valence shell by sharing its 4 electrons with other atoms so that 8 electrons are present. Each pair of shared electrons constitutes a covalent bond (see Figure 2.8d). In organic molecules, carbon usually forms single or double covalent bonds. Each carbon atom acts as an intersection point from which a molecule can branch off in as many as four directions. This enables carbon to form large, complex molecules.

When a carbon atom forms four single covalent bonds, the bonds angle toward the corners of an imaginary tetrahedron. The bond angles in methane ($CH_4$) are 109.5° **(Figure 3.2a)**, and they are roughly the same in any group of atoms where carbon has four single bonds. For example, ethane ($C_2H_6$) is shaped like two overlapping tetrahedrons **(Figure 3.2b)**.

In molecules with more carbons, every grouping of a carbon bonded to four other atoms has a tetrahedral shape. But when two carbon atoms are joined by a double bond, as in ethene ($C_2H_4$), the atoms joined to those carbons are in the same plane as the carbons **(Figure 3.2c)**. We find it convenient to write molecules as structural formulas, as if the molecules being represented are two-dimensional, but keep in mind that molecules are three-dimensional and that the shape of a molecule is central to its function.

The electron configuration of carbon gives it covalent compatibility with many different elements. **Figure 3.3** shows electron distribution diagrams for carbon and its most frequent partners—hydrogen, oxygen, and nitrogen. These are the four main atoms in organic molecules. The number of unpaired electrons in the valence shell of an atom is generally equal to the atom's **valence**, the number of covalent bonds it can form. Let's consider how valence and the rules of covalent bonding

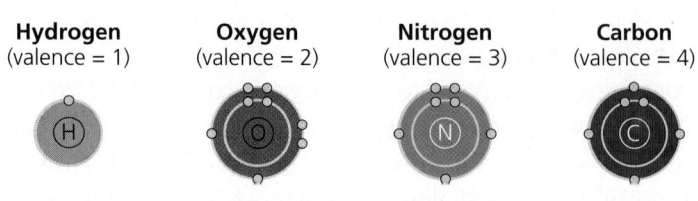

| Hydrogen (valence = 1) | Oxygen (valence = 2) | Nitrogen (valence = 3) | Carbon (valence = 4) |
|---|---|---|---|

▲ **Figure 3.3 Valences of the major elements of organic molecules.** Valence is the number of covalent bonds an atom can form. It is generally equal to the number of electrons required to complete the valence (outermost) shell (see Figure 2.6). Note that carbon can form four bonds.

apply to carbon atoms with partners other than hydrogen. We'll first look at the simple example of carbon dioxide.

In the carbon dioxide molecule ($CO_2$), a single carbon atom is joined to two atoms of oxygen by double covalent bonds. The structural formula for $CO_2$ is shown here:

$$O=C=O$$

Each line in a structural formula represents a pair of shared electrons. Thus, the two double bonds in $CO_2$ have the same number of shared electrons as four single bonds. The arrangement completes the valence shells of all atoms in the molecule:

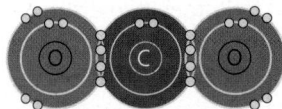

Because $CO_2$ is a very simple molecule and lacks hydrogen, it is often considered inorganic, even though it contains carbon. Whether we call $CO_2$ organic or inorganic, however, it is clearly important to the living world as the source of carbon for all organic molecules in organisms.

Carbon dioxide is a molecule with only one carbon atom. But a carbon atom can also use one or more valence electrons to form covalent bonds to other carbon atoms, linking the atoms into chains, as shown here for $C_3H_8$:

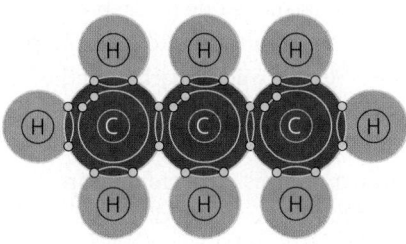

## Molecular Diversity Arising from Variation in Carbon Skeletons

Carbon chains form the skeletons of most organic molecules. The skeletons vary in length and may be straight, branched, or arranged in closed rings **(Figure 3.4)**. Some carbon skeletons have double bonds, which vary in number and location. Such variation in carbon skeletons is one important source of the molecular complexity and diversity that characterize living matter. In addition, atoms of other elements can be bonded to the skeletons at available sites.

### *Hydrocarbons*

All of the molecules shown in Figures 3.2 and 3.4 are **hydrocarbons**, organic molecules consisting of only carbon and hydrogen. Atoms of hydrogen are attached to the carbon skeleton wherever electrons are available for covalent bonding. Hydrocarbons are the major components of petroleum, which is called a fossil fuel because it consists of the partially decomposed remains of organisms that lived millions of years ago. Although hydrocarbons are not prevalent in most living

▼ **Figure 3.4 Four ways that carbon skeletons can vary.**

**(a) Length**

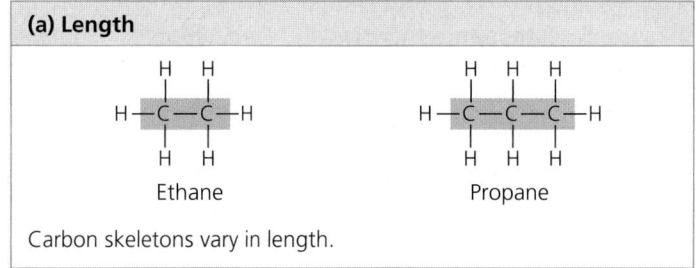

Ethane      Propane

Carbon skeletons vary in length.

**(b) Branching**

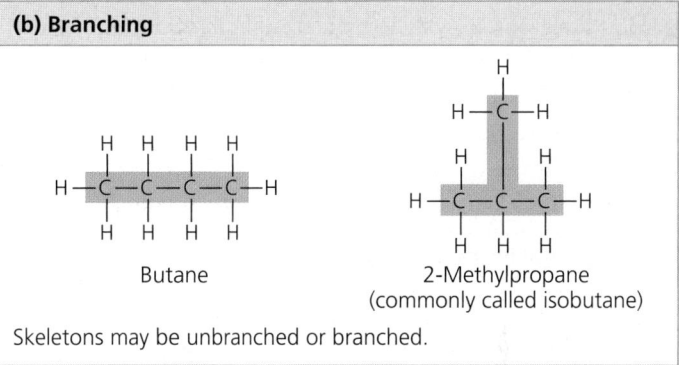

Butane     2-Methylpropane (commonly called isobutane)

Skeletons may be unbranched or branched.

**(c) Double bond position**

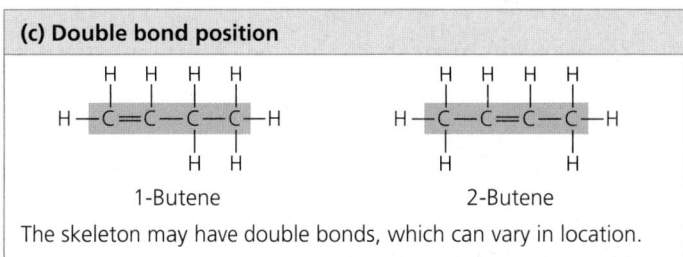

1-Butene     2-Butene

The skeleton may have double bonds, which can vary in location.

**(d) Presence of rings**

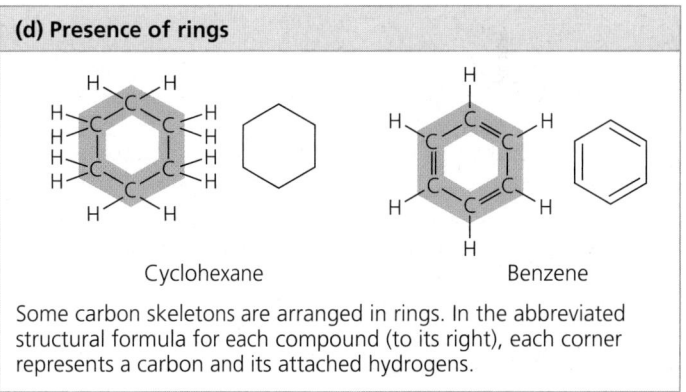

Cyclohexane     Benzene

Some carbon skeletons are arranged in rings. In the abbreviated structural formula for each compound (to its right), each corner represents a carbon and its attached hydrogens.

organisms, many of a cell's organic molecules have regions consisting of only carbon and hydrogen. For example, the molecules known as fats have long hydrocarbon tails attached to a nonhydrocarbon component (as you will see in Figure 3.13). Neither petroleum nor fat dissolves in water; both are hydrophobic compounds because the great majority of their bonds are relatively nonpolar carbon-to-hydrogen linkages. Another characteristic of hydrocarbons is that they can undergo reactions that release a relatively large amount of energy. The gasoline that fuels a car consists of hydrocarbons, and the hydrocarbon tails of fats serve as stored fuel for plant embryos (seeds) and animals.

## Isomers

Variation in the architecture of organic molecules can be seen in **isomers**, compounds that have the same numbers of atoms of the same elements but different structures and hence different properties. We will examine three types of isomers: structural isomers, *cis-trans* isomers, and enantiomers.

**Structural isomers** differ in the covalent arrangements of their atoms. Compare, for example, the two five-carbon compounds in **Figure 3.5a.** Both have the molecular formula $C_5H_{12}$, but they differ in the covalent arrangement of their carbon skeletons. The skeleton is straight in one compound but branched in the other. The number of possible isomers increases tremendously as carbon skeletons increase in size.

---

▼ **Figure 3.5 Three types of isomers, compounds with the same molecular formula but different structures.**

**(a) Structural isomers**

Pentane        2-Methylbutane

Structural isomers differ in covalent partners, as shown in this example of two isomers of $C_5H_{12}$.

**(b) *Cis-trans* isomers**

*cis* isomer: The two Xs are on the same side.     *trans* isomer: The two Xs are on opposite sides.

*Cis-trans* isomers differ in arrangement about a double bond. In these diagrams, X represents an atom or group of atoms attached to a double-bonded carbon.

**(c) Enantiomers**

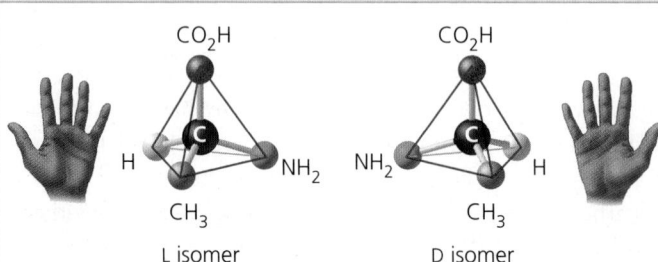

L isomer        D isomer

Enantiomers differ in spatial arrangement around an asymmetric carbon, resulting in molecules that are mirror images, like left and right hands. The two isomers here are designated the L and D isomers from the Latin for "left" and "right" (*levo* and *dextro*). Enantiomers cannot be superimposed on each other.

---

**DRAW IT** *There are three structural isomers of $C_5H_{12}$ (two of which are shown in Figure 3.5a); draw the one not shown in (a).*

---

There are only three forms of $C_5H_{12}$ (two of which are shown in Figure 3.5a), but there are 18 variants of $C_8H_{18}$ and 366,319 possible structural isomers of $C_{20}H_{42}$. Structural isomers may also differ in the location of double bonds.

In ***cis-trans*** isomers, carbons have covalent bonds to the same atoms, but these atoms differ in their spatial arrangements due to the inflexibility of double bonds. Single bonds allow the atoms they join to rotate freely about the bond axis without changing the compound. In contrast, double bonds do not permit such rotation. If a double bond joins two carbon atoms, and each C also has two different atoms (or groups of atoms) attached to it, then two distinct *cis-trans* isomers are possible. Consider a simple molecule with two double-bonded carbons, each of which has an H and an X attached to it **(Figure 3.5b)**. The arrangement with both Xs on the same side of the double bond is called a *cis isomer*, and that with the Xs on opposite sides is called a *trans isomer*. The subtle difference in shape between such isomers can dramatically affect the biological activities of organic molecules. For example, the biochemistry of vision involves a light-induced change of retinal, a chemical compound in the eye, from the *cis* isomer to the *trans* isomer (see Figure 38.26). Another example involves *trans* fats, which are discussed in later in this chapter.

**Enantiomers** are isomers that are mirror images of each other and that differ in shape due to the presence of an *asymmetric carbon*, one that is attached to four different atoms or groups of atoms. (See the middle carbon in the ball-and-stick models shown in **Figure 3.5c**.) The four groups can be arranged in space around the asymmetric carbon in two different ways that are mirror images. Enantiomers are, in a way, left-handed and right-handed versions of the molecule. Just as your right hand won't fit into a left-handed glove, a "right-handed" molecule won't fit into the same space as the "left-handed" version. Usually, only one isomer is biologically active because only that form can bind to specific molecules in an organism.

The concept of enantiomers is important in the pharmaceutical industry because the two enantiomers of a drug may not be equally effective, as is the case for both ibuprofen and the asthma medication albuterol. Methamphetamine also occurs in two enantiomers that have very different effects. One enantiomer is the highly addictive stimulant drug known as "crank," sold illegally in the street drug trade. The other has a much weaker effect and is the active ingredient in an over-the-counter vapor inhaler for treatment of nasal congestion. The differing effects of enantiomers in the body demonstrate that organisms are sensitive to even the most subtle variations in molecular architecture. Once again, we see that molecules have emergent properties that depend on the specific arrangement of their atoms.

## The Chemical Groups Most Important to Life

The distinctive properties of an organic molecule depend not only on the arrangement of its carbon skeleton but also on the various chemical groups attached to that skeleton **(Figure 3.6)**.

▼ Figure 3.6 **Some biologically important chemical groups.**

| Chemical Group | Compound Name | Examples |
|---|---|---|
| **Hydroxyl group** (—OH) 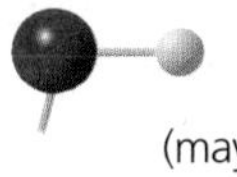<br><br>—OH<br>(may be written HO—) | **Alcohol**<br>(The specific name usually ends in -ol.) | 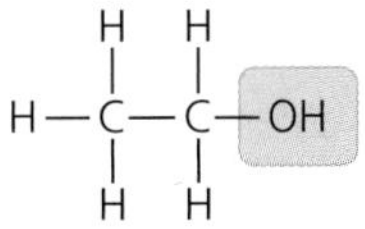 <br>**Ethanol**, the alcohol present in alcoholic beverages |
| **Carbonyl group** (⟩C═O)  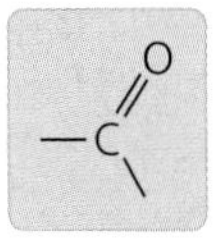<br> | **Ketone** if the carbonyl group is within a carbon skeleton<br>**Aldehyde** if the carbonyl group is at the end of a carbon skeleton | 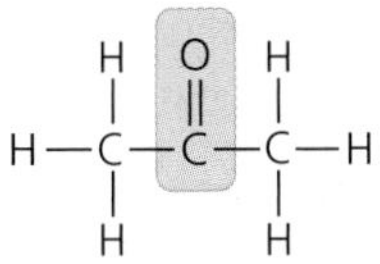 <br>**Acetone**, the simplest ketone 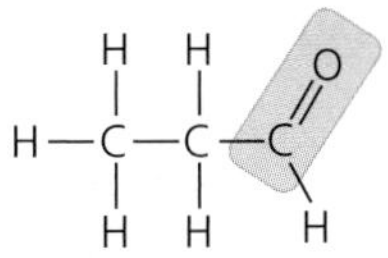     **Propanal**, an aldehyde |
| **Carboxyl group** (—COOH) 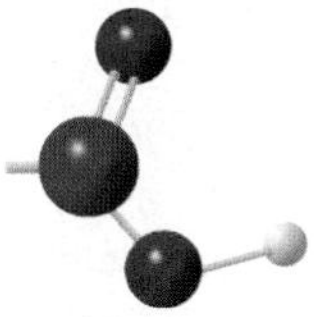 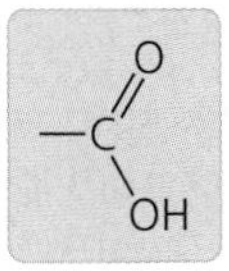<br> | **Carboxylic acid**, or **organic acid** | 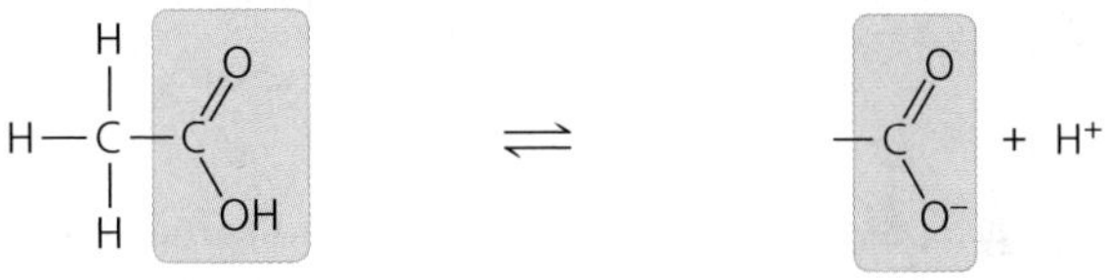 <br>**Acetic acid**, which gives vinegar its sour taste    Ionized form of —COOH (carboxylate ion), found in cells |
| **Amino group** (—NH₂)<br> 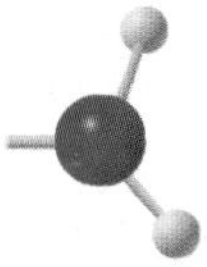 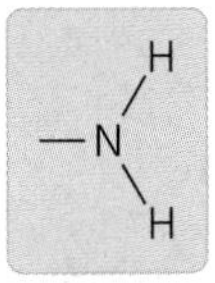 | **Amine** | 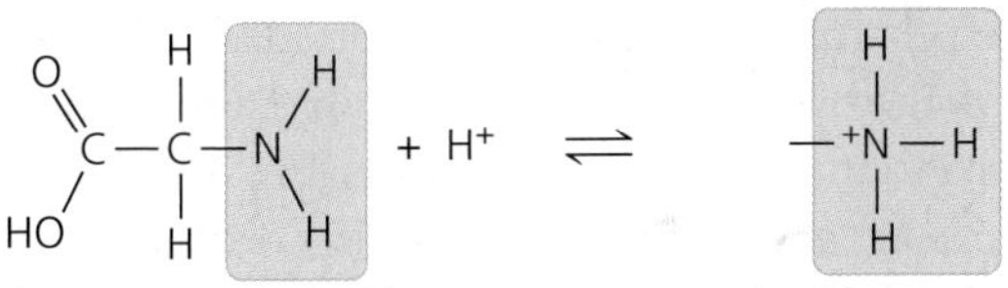 <br>**Glycine**, an amino acid (note its carboxyl group)    Ionized form of —NH₂, found in cells |
| **Sulfhydryl group** (—SH)<br> 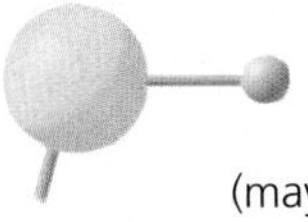<br>—SH<br>(may be written HS —) | **Thiol** | 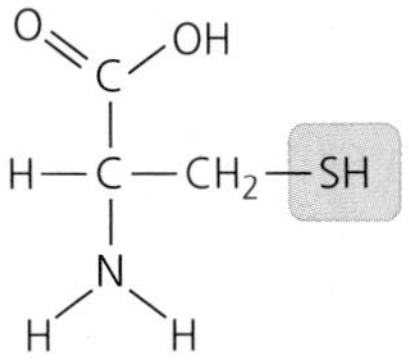 <br>**Cysteine**, a sulfur-containing amino acid |
| **Phosphate group** (—OPO₃²⁻)<br> 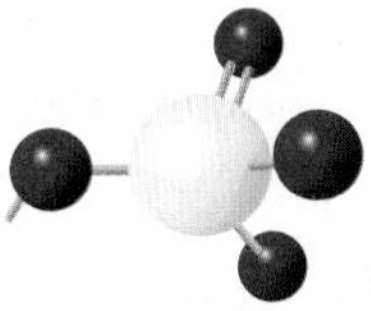 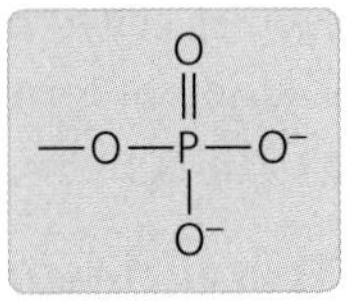 | **Organic phosphate** | 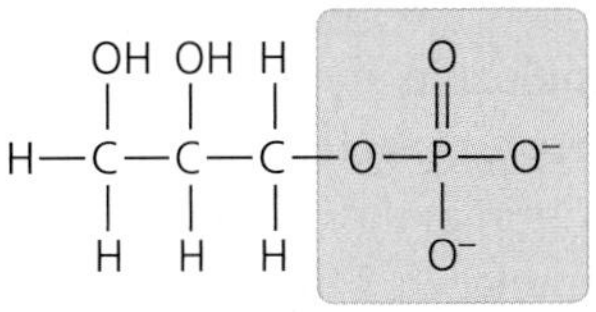 <br>**Glycerol phosphate**, which takes part in many important chemical reactions in cells |
| **Methyl group** (—CH₃)<br> 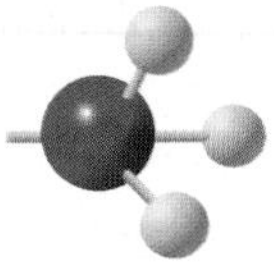 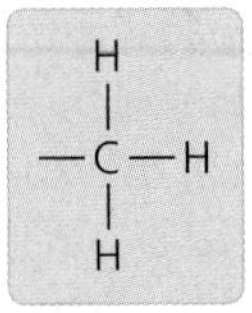 | **Methylated compound** | 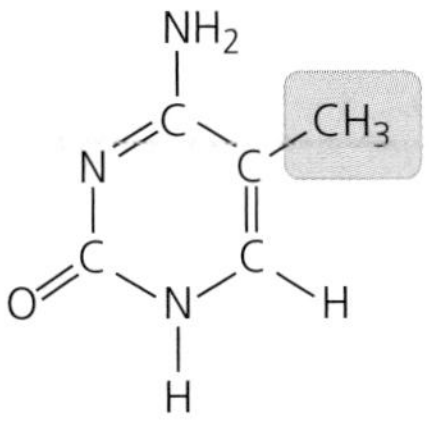 <br>**5-Methylcytosine**, a component of DNA that has been modified by addition of a methyl group |

We can think of hydrocarbons, the simplest organic molecules, as the underlying framework for more complex organic molecules. A number of chemical groups can replace one or more of the hydrogens bonded to the carbon skeleton of the hydrocarbon. The number and arrangement of chemical groups help give each organic molecule its unique properties.

In some cases, chemical groups contribute to function primarily by affecting the molecule's shape. This is true for the steroid sex hormones estradiol (a type of estrogen) and testosterone, which differ in attached chemical groups and act to produce the contrasting features of male and female vertebrates.

**Estradiol**

**Testosterone**

In other cases, the chemical groups affect molecular function by being directly involved in chemical reactions; these important chemical groups are known as **functional groups**. Each functional group participates in chemical reactions in a characteristic way.

The seven chemical groups most important in biological processes are the hydroxyl, carbonyl, carboxyl, amino, sulfhydryl, phosphate, and methyl groups (see Figure 3.6). The first six groups can act as functional groups; also, except for the sulfhydryl group, they are hydrophilic and thus increase the solubility of organic compounds in water. The last group, the methyl group, is not reactive, but instead often serves as a recognizable tag on biological molecules. Study Figure 3.6 to become familiar with these biologically important chemical groups. At normal cellular pH, the carboxyl group and amino group are ionized, as shown at the right.

### ATP: An Important Source of Energy for Cellular Processes

The "Phosphate group" row in Figure 3.6 shows a simple example of an organic phosphate molecule. A more complicated organic phosphate, **adenosine triphosphate**, or **ATP**, is worth mentioning here because its function in the cell is so important. ATP consists of an organic molecule called adenosine attached to a string of three phosphate groups:

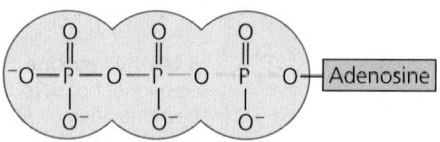

Where three phosphates are present in series, as in ATP, one phosphate may be split off as a result of a reaction with water.

This inorganic phosphate ion, $HOPO_3^{2-}$, is often abbreviated ℗$_i$ in this book, and a phosphate group in an organic molecule is often written as ℗. Having lost one phosphate, ATP becomes adenosine *di*phosphate, or ADP. Although ATP is sometimes said to store energy, it is more accurate to think of it as storing the potential to react with water. This reaction releases energy that can be used by the cell. (You will learn about this in more detail in Chapter 6.)

Reacts with $H_2O$

℗–℗–℗–Adenosine ⟶ ℗$_i$ + ℗–℗–Adenosine + Energy

ATP        Inorganic        ADP
          phosphate

### CONCEPT CHECK 3.1

1. How are gasoline and fat chemically similar?
2. Which molecules in Figure 3.4 are isomers? For each pair, identify the type of isomer.
3. What does the term *amino acid* signify about the structure of such a molecule?
4. **DRAW IT** Suppose you had an organic molecule such as cysteine (see Figure 3.6, sulfhydryl group example), and you chemically removed the —$NH_2$ group and replaced it with —COOH. Draw this structure. How would this change the chemical properties of the molecule? Is the central carbon asymmetric before the change? After?

For suggested answers, see Appendix A.

## CONCEPT 3.2

# Macromolecules are polymers, built from monomers

The macromolecules in three of the four classes of life's organic compounds—carbohydrates, proteins, and nucleic acids—are chain-like molecules called polymers (from the Greek *polys*, many, and *meros*, part). A **polymer** is a long molecule consisting of many similar or identical building blocks linked by covalent bonds, much as a train consists of a chain of cars. The repeating units that serve as the building blocks of a polymer are smaller molecules called **monomers** (from the Greek *monos*, single). In addition to forming polymers, some monomers have functions of their own.

### The Synthesis and Breakdown of Polymers

Although each class of polymer is made up of a different type of monomer, the chemical mechanisms by which cells make and break down polymers are basically the same in all cases. In cells, these processes are facilitated by **enzymes**, specialized macromolecules (usually proteins) that speed up chemical reactions. Monomers are connected by a reaction in which two molecules are covalently bonded to each other, with the loss of a water molecule; this is known as a **dehydration reaction**

**(Figure 3.7a)**. When a bond forms between two monomers, each monomer contributes part of the water molecule that is released during the reaction: One monomer provides a hydroxyl group (—OH), while the other provides a hydrogen (—H). This reaction is repeated as monomers are added to the chain one by one, making a polymer (also called polymerization).

Polymers are disassembled to monomers by **hydrolysis**, a process that is essentially the reverse of the dehydration reaction **(Figure 3.7b)**. Hydrolysis means water breakage (from the Greek *hydro*, water, and *lysis*, break). The bond between the monomers is broken by the addition of a water molecule, with a hydrogen from water attaching to one monomer and the hydroxyl group attaching to the other. An example of hydrolysis working within our bodies is the process of digestion. The bulk of the organic material in our food is in the form of polymers that are much too large to enter our cells. Within the digestive tract, various enzymes attack the polymers, speeding up hydrolysis. Released monomers are then absorbed into the bloodstream for distribution to all body cells. Those cells can then use dehydration reactions to assemble the monomers into new, different polymers that can perform specific functions required by the cell.

## The Diversity of Polymers

A cell has thousands of different macromolecules; the collection varies from one type of cell to another even in the same organism. The inherited differences between close relatives, such as human siblings, reflect small variations in polymers, particularly DNA and proteins. Molecular differences between unrelated individuals are more extensive, and those between species greater still. The diversity of macromolecules in the living world is vast, and the possible variety is effectively limitless.

What is the basis for such diversity in life's polymers? These molecules are constructed from only 40 to 50 common monomers and some others that occur rarely. Building a huge variety of polymers from such a limited number of monomers is analogous to constructing hundreds of thousands of words from only 26 letters of the alphabet. The key is arrangement—the particular linear sequence that the units follow. However, this analogy falls far short of describing the great diversity of macromolecules because most biological polymers have many more monomers than the number of letters in even the longest word. Proteins, for example, are built from 20 kinds of amino acids arranged in chains that are typically hundreds of amino acids long. The molecular logic of life is simple but elegant: Small molecules common to all organisms are ordered into unique macromolecules.

Despite this immense diversity, molecular structure and function can still be grouped roughly by class. Let's examine each of the four major classes of large biological molecules. For each class, the large molecules have emergent properties not found in their individual building blocks.

**CONCEPT CHECK 3.2**

1. How many molecules of water are needed to completely hydrolyze a polymer that is ten monomers long?
2. **WHAT IF?** Suppose you eat a serving of fish. What reactions must occur for the amino acid monomers in the protein of the fish to be converted to new proteins in your body?

For suggested answers, see Appendix A.

**CONCEPT 3.3**

# Carbohydrates serve as fuel and building material

**Carbohydrates** include both sugars and polymers of sugars. The simplest carbohydrates are the monosaccharides, or simple sugars; these are the monomers from which more complex carbohydrates are built. Disaccharides are double sugars, consisting of two monosaccharides joined by a covalent bond. Carbohydrates also include macromolecules called polysaccharides, polymers composed of many sugar building blocks joined together by dehydration reactions.

## Sugars

**Monosaccharides** (from the Greek *monos*, single, and *sacchar*, sugar) generally have molecular formulas that are some multiple of the unit $CH_2O$. Glucose ($C_6H_{12}O_6$), the most common

---

▼ **Figure 3.7** The synthesis and breakdown of polymers.

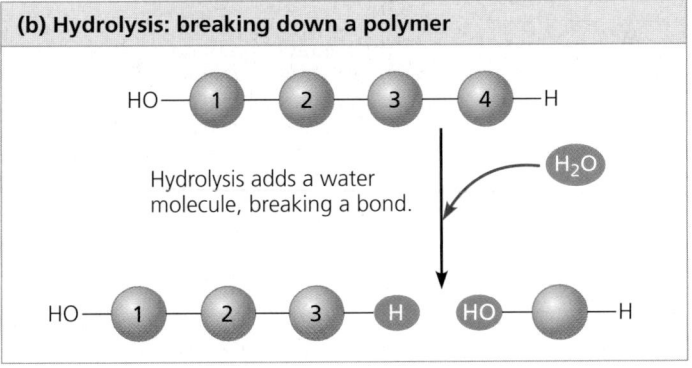

**(a) Dehydration reaction: synthesizing a polymer**

Short polymer — Unlinked monomer

Dehydration removes a water molecule, forming a new bond. → $H_2O$

Longer polymer

**(b) Hydrolysis: breaking down a polymer**

Hydrolysis adds a water molecule, breaking a bond. $H_2O$

▼ **Figure 3.8 Examples of monosaccharides.** Sugars vary in the location of their carbonyl groups (orange) and the length of their carbon skeletons.

| Triose: three-carbon sugar ($C_3H_6O_3$) | Pentose: five-carbon sugar ($C_5H_{10}O_5$) | Hexoses: six-carbon sugars ($C_6H_{12}O_6$) |
|---|---|---|
| 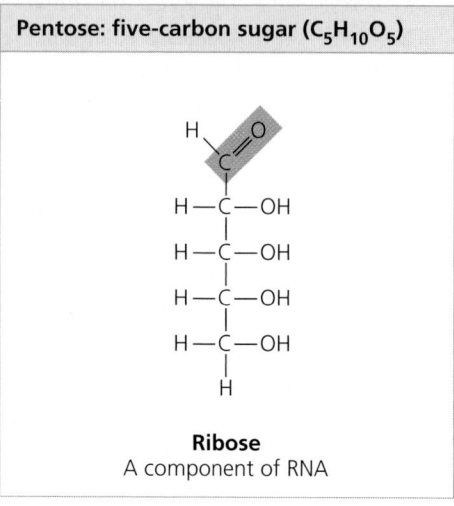 | | 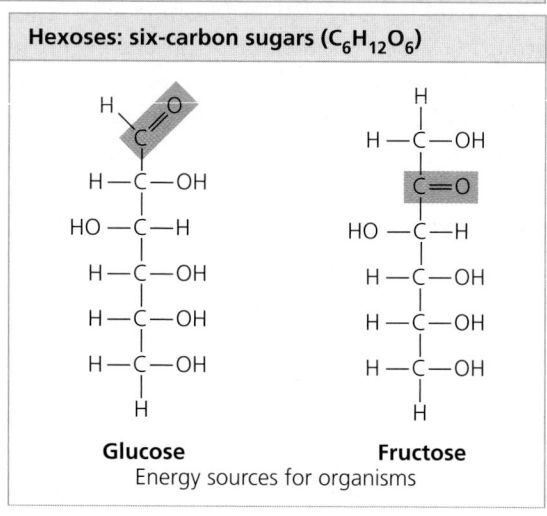 |
| **Glyceraldehyde** An initial breakdown product of glucose in cells | **Ribose** A component of RNA | **Glucose** **Fructose** Energy sources for organisms |

monosaccharide, is of central importance in the chemistry of life. In the structure of glucose, we can see the trademarks of a sugar: The molecule has a carbonyl group ($>C=O$) and multiple hydroxyl groups (—OH) **(Figure 3.8)**. The carbonyl group can be on the end of the linear sugar molecule, as in glucose, or attached to an interior carbon, as in fructose. (Thus, sugars are either aldehydes or ketones; see Figure 3.6.) The carbon skeleton of a sugar molecule ranges from three to seven carbons long. Glucose, fructose, and other sugars that have six carbons are called hexoses. Trioses (three-carbon sugars) and pentoses (five-carbon sugars) are also common. Note that most names for sugars end in -ose.

Although it is convenient to draw glucose with a linear carbon skeleton, this representation is not completely accurate. In aqueous solutions, glucose molecules (as well as most other five- and six-carbon sugars) form rings, because they are the most stable form of these sugars under physiological conditions **(Figure 3.9)**.

Monosaccharides, particularly glucose, are major nutrients for cells. In the process known as cellular respiration, cells extract energy from glucose molecules by breaking them down in a series of reactions. Also, the carbon skeletons of sugars serve as raw material for the synthesis of other types of small organic molecules, such as amino acids. Sugar molecules that are not immediately used in these ways are generally incorporated as monomers into disaccharides or polysaccharides.

A **disaccharide** consists of two monosaccharides joined by a **glycosidic linkage**, a covalent bond formed between two monosaccharides by a dehydration reaction (*glyco* refers to carbohydrate). The most prevalent disaccharide is sucrose, which is table sugar. Its two monomers are glucose and fructose **(Figure 3.10)**. Plants generally transport carbohydrates from leaves to roots and other nonphotosynthetic organs in the form of sucrose. Some other disaccharides are lactose, the sugar present in milk, and maltose, used in making beer.

**(a) Linear and ring forms.** Chemical equilibrium between the linear and ring structures greatly favors the formation of rings. The carbons of the sugar are numbered 1 to 6, as shown. To form the glucose ring, carbon 1 (magenta) bonds to the oxygen (blue) attached to carbon 5.

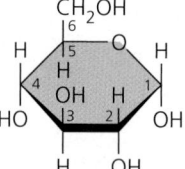

**(b) Abbreviated ring structure.** Each unlabeled corner represents a carbon. The ring's thicker edge indicates that you are looking at the ring edge-on; the components attached to the ring lie above or below the plane of the ring.

▲ **Figure 3.9 Linear and ring forms of glucose.**

**DRAW IT** *Start with the linear form of fructose (see Figure 3.8) and draw the formation of the fructose ring in two steps, as shown in (a). First, number the carbons starting at the top of the linear structure. Then draw the molecule in a ringlike orientation, attaching carbon 5 via its oxygen to carbon 2. Compare the number of carbons in the fructose and glucose rings.*

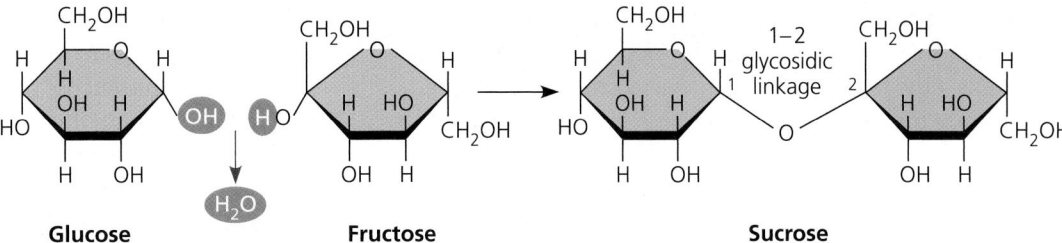

► **Figure 3.10 Disaccharide synthesis.** Sucrose is a disaccharide formed from glucose and fructose by a dehydration reaction. Notice that fructose, though a hexose like glucose, forms a five-sided ring.

**DRAW IT** *Number the carbons in each sugar (see Figure 3.9). How does the name of the linkage relate to the numbers?*

Glucose          Fructose          Sucrose

## Polysaccharides

**Polysaccharides** are macromolecules, polymers with a few hundred to a few thousand monosaccharides joined by glycosidic linkages. Some polysaccharides serve as storage material, hydrolyzed as needed to provide sugar for cells. Other polysaccharides serve as building material for structures that protect the cell or the whole organism. The structure and function of a polysaccharide are determined by its sugar monomers and by the positions of its glycosidic linkages.

### Storage Polysaccharides

Both plants and animals store sugars for later use in the form of storage polysaccharides **(Figure 3.11)**. Plants store **starch**, a polymer of glucose monomers, as granules within cells. Synthesizing starch enables the plant to stockpile surplus glucose. Because glucose is a major cellular fuel, starch represents stored energy. The sugar can later be withdrawn from this carbohydrate "bank" by hydrolysis, which breaks the bonds between the glucose monomers. Most animals, including

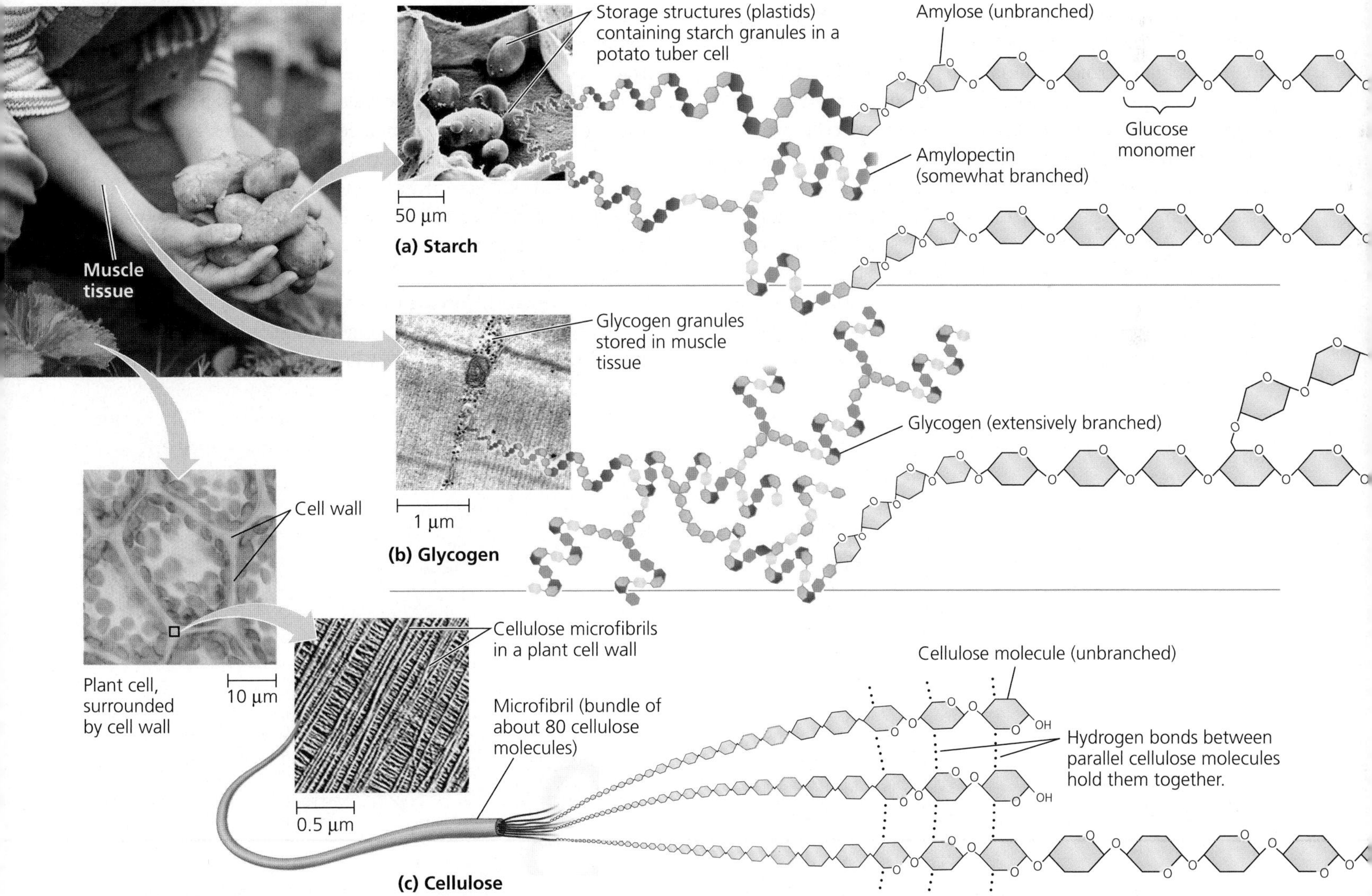

Storage structures (plastids) containing starch granules in a potato tuber cell

Amylose (unbranched)

Glucose monomer

Amylopectin (somewhat branched)

50 μm

**(a) Starch**

Muscle tissue

Glycogen granules stored in muscle tissue

Glycogen (extensively branched)

1 μm

**(b) Glycogen**

Cell wall

Plant cell, surrounded by cell wall

10 μm

Cellulose microfibrils in a plant cell wall

Microfibril (bundle of about 80 cellulose molecules)

Cellulose molecule (unbranched)

Hydrogen bonds between parallel cellulose molecules hold them together.

0.5 μm

**(c) Cellulose**

▲ **Figure 3.11 Polysaccharides of plants and animals. (a)** Starch stored in plant cells, **(b)** glycogen stored in muscle cells, and **(c)** structural cellulose fibers in plant cell walls are all polysaccharides composed entirely of glucose monomers (green hexagons). In starch and glycogen, the polymer chains tend to form helices in unbranched regions because of the angle of the 1–4 linkage between the glucose monomers. Cellulose, with a different kind of 1–4 linkage, is always unbranched.

humans, also have enzymes that can hydrolyze plant starch, making glucose available as a nutrient for cells. Potato tubers and grains—the fruits of wheat, maize (corn), rice, and other grasses—are the major sources of starch in the human diet.

Most of the glucose monomers in starch are joined by 1–4 linkages (number 1 carbon to number 4 carbon). The simplest form of starch, amylose, is unbranched, as shown in **Figure 3.11a**. Amylopectin, a more complex starch, is a branched polymer with 1–6 linkages at the branch points.

Animals store a polysaccharide called **glycogen**, a polymer of glucose that is like amylopectin but more extensively branched (see **Figure 3.11b**). Humans and other vertebrates store glycogen mainly in liver and muscle cells. Hydrolysis of glycogen in these cells releases glucose when the demand for sugar increases. This stored fuel cannot sustain an animal for long, however. In humans, for example, glycogen stores are depleted in about a day unless they are replenished by eating. This is an issue of concern in ultra-low-carbohydrate diets, which can result in weakness and fatigue.

### Structural Polysaccharides

Organisms build strong materials from structural polysaccharides. The polysaccharide called **cellulose** is a major component of the tough walls that enclose plant cells (see **Figure 3.11c**). On a global scale, plants produce almost $10^{14}$ kg (100 billion tons) of cellulose per year; it is the most abundant organic compound on Earth. Like starch, cellulose is a polymer of glucose with 1–4 glycosidic linkages, but the linkages in these two polymers differ. The difference is based on the fact that there are actually two slightly different ring structures for glucose **(Figure 3.12a)**. When glucose forms a ring, the hydroxyl group attached to the number 1 carbon is positioned either below or above the plane of the ring. These two ring forms for glucose are called alpha (α) and beta (β), respectively. In starch, all the glucose monomers are in the α configuration **(Figure 3.12b)**, the arrangement we saw in Figure 3.9. In contrast, the glucose monomers of cellulose are all in the β configuration, making every glucose monomer "upside down" with respect to its neighbors **(Figure 3.12c)**.

The differing glycosidic linkages in starch and cellulose give the two molecules distinct three-dimensional shapes. Whereas certain starch molecules are largely helical, a cellulose molecule is straight. Cellulose is never branched, and some hydroxyl groups on its glucose monomers are free to hydrogen-bond with the hydroxyls of other cellulose molecules lying parallel to it. In plant cell walls, parallel cellulose molecules held together in this way are grouped into units called microfibrils (see Figure 3.11c). These cable-like microfibrils are a strong building material for plants and an important substance for humans because cellulose is the major constituent of paper and the only component of cotton.

Enzymes that digest starch by hydrolyzing its α linkages are unable to hydrolyze the β linkages of cellulose due to the different shapes of these two molecules. In fact, few organisms possess enzymes that can digest cellulose. Almost all animals, including humans, do not; the cellulose in our food passes through the digestive tract and is eliminated with the feces. Along the way, the cellulose abrades the wall of the digestive tract and stimulates the lining to secrete mucus, which aids in the smooth passage of food through the tract. Thus, although cellulose is not a nutrient for humans, it is an important part of a healthful diet. Most fruits, vegetables, and whole grains are rich in cellulose. On food packages, "insoluble fiber" refers mainly to cellulose.

Some microorganisms can digest cellulose, breaking it down into glucose monomers. A cow harbors cellulose-digesting

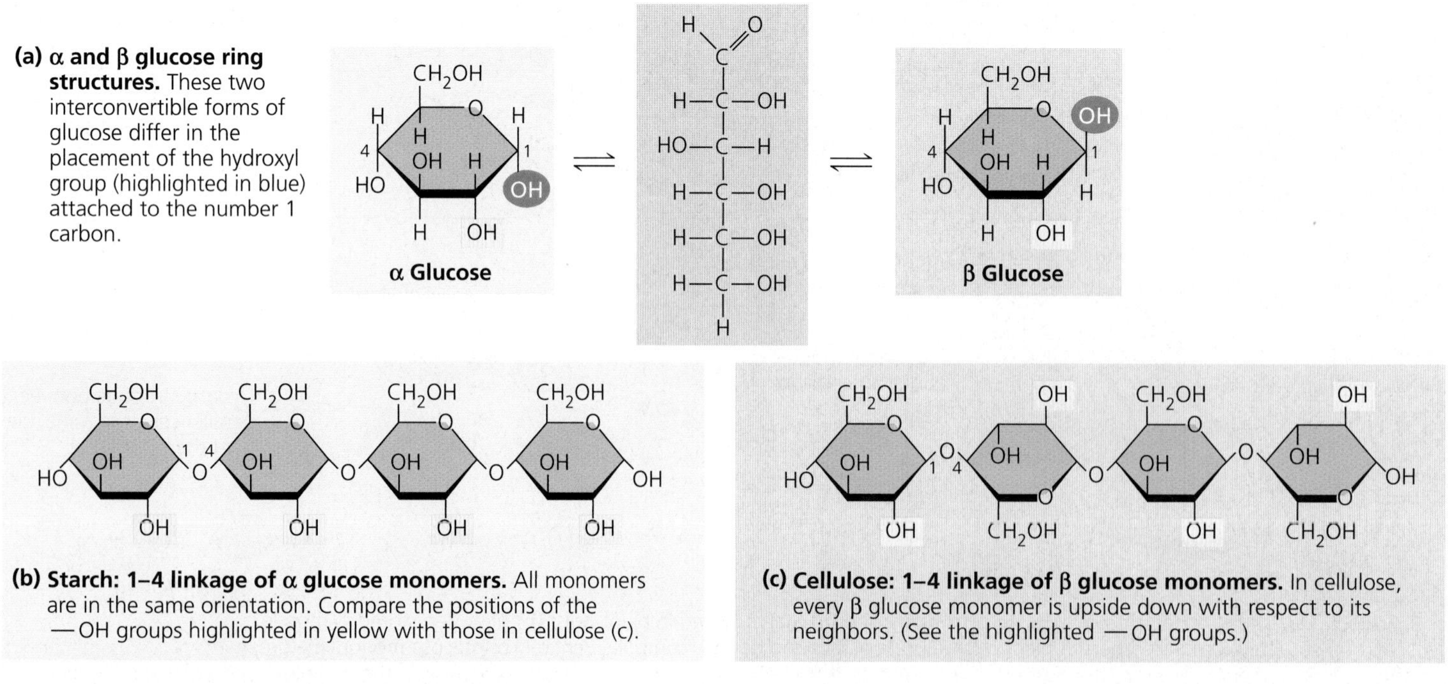

**(a) α and β glucose ring structures.** These two interconvertible forms of glucose differ in the placement of the hydroxyl group (highlighted in blue) attached to the number 1 carbon.

α Glucose

β Glucose

**(b) Starch: 1–4 linkage of α glucose monomers.** All monomers are in the same orientation. Compare the positions of the —OH groups highlighted in yellow with those in cellulose (c).

**(c) Cellulose: 1–4 linkage of β glucose monomers.** In cellulose, every β glucose monomer is upside down with respect to its neighbors. (See the highlighted —OH groups.)

▲ **Figure 3.12 Starch and cellulose structures.**

prokaryotes and protists in its gut. These microbes hydrolyze the cellulose of hay and grass and convert the glucose to other compounds that nourish the cow. Similarly, a termite, which is unable to digest cellulose by itself, has prokaryotes or protists living in its gut that can make a meal of wood. Some fungi can also digest cellulose in soil and elsewhere, thereby helping recycle chemical elements within Earth's ecosystems.

Another important structural polysaccharide is **chitin**, the carbohydrate used by arthropods (insects, spiders, crustaceans, and related animals) to build their exoskeletons—hard cases that surround the soft parts of these animals. Chitin is also found in many fungi, which use this polysaccharide as the building material for their cell walls. Chitin is similar to cellulose except that the glucose monomer of chitin has a nitrogen-containing attachment.

### CONCEPT CHECK 3.3

1. Write the formula for a monosaccharide that has three carbons.
2. A dehydration reaction joins two glucose molecules to form maltose. The formula for glucose is $C_6H_{12}O_6$. What is the formula for maltose?
3. **WHAT IF?** After a cow is given antibiotics to treat an infection, a vet gives the animal a drink of "gut culture" containing various prokaryotes. Why is this necessary?

For suggested answers, see Appendix A.

---

### CONCEPT 3.4

# Lipids are a diverse group of hydrophobic molecules

Lipids are the one class of large biological molecules that does not include true polymers, and they are generally not big enough to be considered macromolecules. The compounds called **lipids** are grouped together because they share one important trait: They mix poorly, if at all, with water. The hydrophobic behavior of lipids is based on their molecular structure. Although they may have some polar bonds associated with oxygen, lipids consist mostly of hydrocarbon regions. Lipids are varied in form and function. They include waxes and certain pigments, but we will focus on the types of lipids that are most biologically important: fats, phospholipids, and steroids.

## Fats

Although fats are not polymers, they are large molecules assembled from smaller molecules by dehydration reactions. A **fat** is constructed from two kinds of smaller molecules: glycerol and fatty acids (**Figure 3.13a**). Glycerol is an alcohol; each of its three carbons bears a hydroxyl group. A **fatty acid** has a long carbon skeleton, usually 16 or 18 carbon atoms in length. The carbon at one end of the skeleton is part of a carboxyl group, the functional group that gives these molecules the name fatty *acid*. The rest of the skeleton consists of a hydrocarbon chain. The relatively nonpolar C—H bonds in the

hydrocarbon chains of fatty acids are the reason fats are hydrophobic. Fats separate from water because the water molecules hydrogen-bond to one another and exclude the fats. This is the reason that vegetable oil (a liquid fat) separates from the aqueous vinegar solution in a bottle of salad dressing.

In making a fat, three fatty acid molecules are each joined to glycerol by an ester linkage, a bond between a hydroxyl group and a carboxyl group. The resulting fat, also called a **triacylglycerol**, thus consists of three fatty acids linked to one glycerol molecule. (Still another name for a fat is *triglyceride*, a word often found in the list of ingredients on packaged foods.) The fatty acids in a fat can be the same, or they can be of two or three different kinds, as in **Figure 3.13b**.

The terms *saturated fats* and *unsaturated fats* are commonly used in the context of nutrition. These terms refer to the structure of the hydrocarbon chains of the fatty acids. If there are no double bonds between carbon atoms composing a chain, then as many hydrogen atoms as possible are bonded to the carbon skeleton. Such a structure is said to be *saturated* with hydrogen, and the resulting fatty acid is called a **saturated fatty acid**. An **unsaturated fatty acid** has one or more double bonds, with one fewer hydrogen atom on each double-bonded carbon. Nearly every double bond in naturally occurring fatty acids is a *cis* double bond, which creates a kink

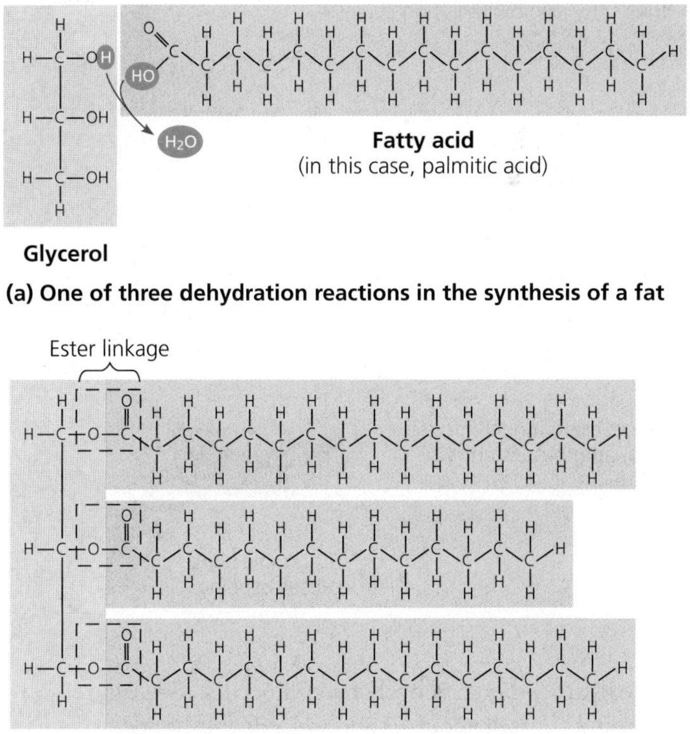

Glycerol

**Fatty acid**
(in this case, palmitic acid)

**(a) One of three dehydration reactions in the synthesis of a fat**

Ester linkage

**(b) Fat molecule (triacylglycerol)**

▲ **Figure 3.13 The synthesis and structure of a fat, or triacylglycerol.** The molecular building blocks of a fat are one molecule of glycerol and three molecules of fatty acids. **(a)** One water molecule is removed for each fatty acid joined to the glycerol. **(b)** A fat molecule with three fatty acid units, two of them identical. The carbons of the fatty acids are arranged zigzag to suggest the actual orientations of the four single bonds extending from each carbon (see Figure 3.2a).

## ▼ Figure 3.14 Saturated and unsaturated fats and fatty acids.

### (a) Saturated fat

At room temperature, the molecules of a saturated fat, such as the fat in butter, are packed closely together, forming a solid.

Structural formula of a saturated fat molecule (Each hydrocarbon chain is represented as a zigzag line, where each bend represents a carbon atom and hydrogens are not shown.)

Space-filling model of stearic acid, a saturated fatty acid (red = oxygen, black = carbon, gray = hydrogen)

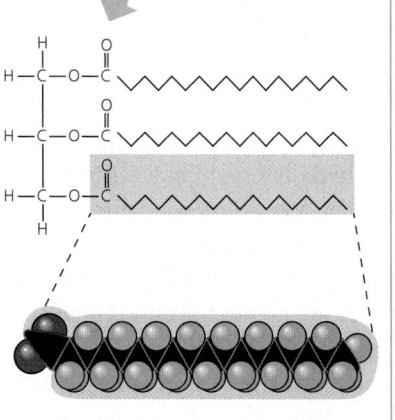

### (b) Unsaturated fat

At room temperature, the molecules of an unsaturated fat such as olive oil cannot pack together closely enough to solidify because of the kinks in some of their fatty acid hydrocarbon chains.

Structural formula of an unsaturated fat molecule

Space-filling model of oleic acid, an unsaturated fatty acid

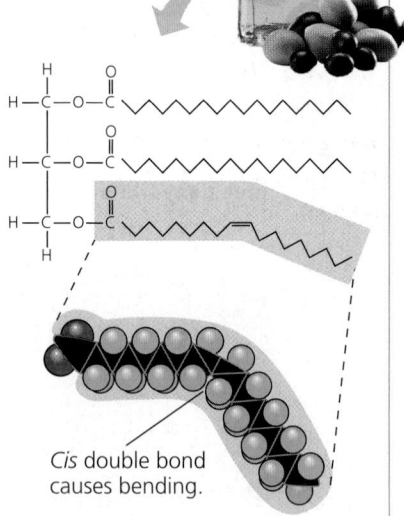

*Cis* double bond causes bending.

in the hydrocarbon chain wherever it occurs. (See Figure 3.5b to remind yourself about *cis* and *trans* double bonds.)

A fat made from saturated fatty acids is called a saturated fat. Most animal fats are saturated: The hydrocarbon chains of their fatty acids—the "tails" of the fat molecules—lack double bonds, and their flexibility allows the fat molecules to pack together tightly. Saturated animal fats—such as lard and butter—are solid at room temperature **(Figure 3.14a)**. In contrast, the fats of plants and fishes are generally unsaturated, meaning that they

are built of one or more types of unsaturated fatty acids. Usually liquid at room temperature, plant and fish fats are referred to as oils—olive oil and cod liver oil are examples **(Figure 3.14b)**. The kinks where the *cis* double bonds are located prevent the molecules from packing together closely enough to solidify at room temperature. The phrase "hydrogenated vegetable oils" on food labels means that unsaturated fats have been synthetically converted to saturated fats by adding hydrogen, allowing them to solidify. This process also produces unsaturated fats with *trans* double bonds, known as ***trans* fats**. It appears that *trans* fats can contribute to coronary heart disease (see Concept 34.4). Because *trans* fats are especially common in baked goods and processed foods, the U.S. Food and Drug Administration (FDA) requires nutritional labels to include information on *trans* fat content. The FDA has proposed a ban on *trans* fats in the U.S. food supply; some countries, such as Denmark and Switzerland, have already implemented restrictions on the level of *trans* fats in foods.

The major function of fats is energy storage. The hydrocarbon chains of fats are similar to gasoline molecules and just as rich in energy. A gram of fat stores more than twice as much energy as a gram of a polysaccharide, such as starch. Because plants are relatively immobile, they can function with bulky energy storage in the form of starch. (Vegetable oils are generally obtained from seeds, where more compact storage is an asset to the plant.) Animals, however, must carry their energy stores with them, so there is an advantage to having a more compact reservoir of fuel—fat.

## Phospholipids

Cells could not exist without another type of lipid, called phospholipids. Phospholipids are essential for cells because they are major constituents of cell membranes. Their structure provides a classic example of how form fits function at the molecular level. As shown in **Figure 3.15**, a **phospholipid** is similar to a fat molecule but has only two fatty acids attached to glycerol rather than three. The third hydroxyl group of glycerol is joined to a phosphate group, which has a negative electrical charge in the cell. Additional small molecules, which are usually charged or polar, can be linked to the phosphate group to form a variety of phospholipids.

The two ends of a phospholipid exhibit different behavior toward water. The hydrocarbon tails are hydrophobic and are excluded from water. However, the phosphate group and its attachments form a hydrophilic head that has an affinity for water. When phospholipids are added to water, they self-assemble into double-layered structures called "bilayers," shielding their hydrophobic portions from water (see Figure 3.15d).

At the surface of a cell, phospholipids are arranged in a similar bilayer. The hydrophilic heads of the molecules are on the outside of the bilayer, in contact with the aqueous solutions inside and outside of the cell. The hydrophobic tails point toward the interior of the bilayer, away from the water. The phospholipid bilayer forms a boundary between the cell and its external environment; in fact, the existence of cells depends on the properties of phospholipids.

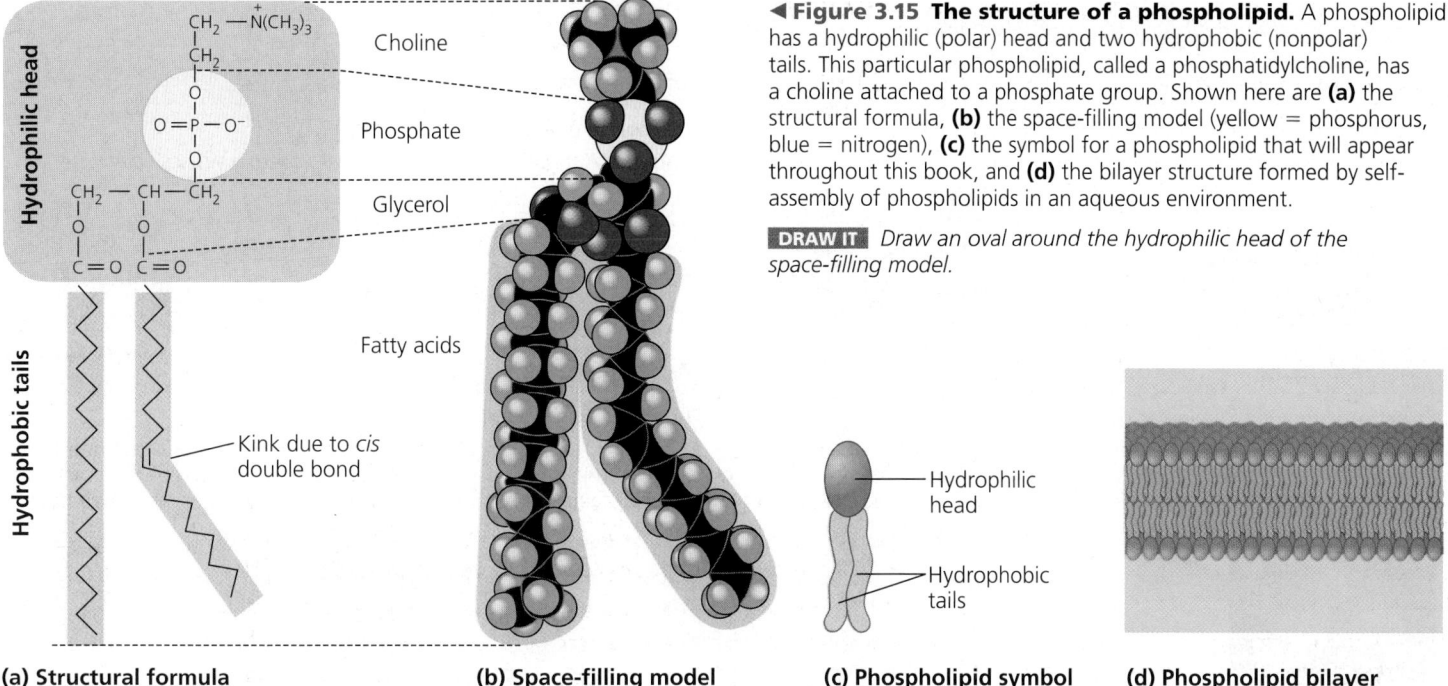

has a hydrophilic (polar) head and two hydrophobic (nonpolar) tails. This particular phospholipid, called a phosphatidylcholine, has a choline attached to a phosphate group. Shown here are **(a)** the structural formula, **(b)** the space-filling model (yellow = phosphorus, blue = nitrogen), **(c)** the symbol for a phospholipid that will appear throughout this book, and **(d)** the bilayer structure formed by self-assembly of phospholipids in an aqueous environment.

**DRAW IT** *Draw an oval around the hydrophilic head of the space-filling model.*

(a) Structural formula    (b) Space-filling model    (c) Phospholipid symbol    (d) Phospholipid bilayer

## Steroids

**Steroids** are lipids characterized by a carbon skeleton consisting of four fused rings. Different steroids are distinguished by the particular chemical groups attached to this ensemble of rings. Shown in **Figure 3.16, cholesterol** is a crucial steroid in animals. It is a common component of animal cell membranes and is also the precursor from which other steroids are synthesized, such as the vertebrate sex hormones estrogen and testosterone (see Concept 3.1). In vertebrates, cholesterol is synthesized in the liver and is also obtained from the diet. A high level of cholesterol in the blood may contribute to atherosclerosis, although some researchers are questioning the roles of cholesterol and saturated fats in development of this condition.

▶ **Figure 3.16 Cholesterol, a steroid.**

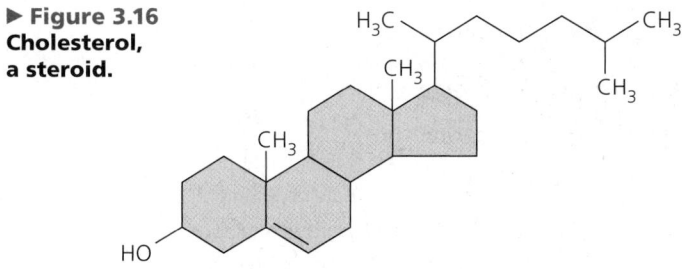

### CONCEPT CHECK 3.4

1. Compare the structure of a fat (triglyceride) with that of a phospholipid.
2. Why are human sex hormones considered lipids?
3. **WHAT IF?** Suppose a membrane surrounded an oil droplet, as it does in the cells of plant seeds. Describe and explain the form it might take.

For suggested answers, see Appendix A.

### CONCEPT 3.5

# Proteins include a diversity of structures, resulting in a wide range of functions

Nearly every dynamic function of a living being depends on proteins. In fact, the importance of proteins is underscored by their name, which comes from the Greek word *proteios*, meaning "first," or "primary." Proteins account for more than 50% of the dry mass of most cells, and they are instrumental in almost everything organisms do. Some proteins speed up chemical reactions, while others play a role in defense, storage, transport, cellular communication, movement, or structural support. **Figure 3.17** shows examples of proteins with these functions (which you'll learn more about in later chapters).

Life would not be possible without enzymes, most of which are proteins. Enzymatic proteins regulate metabolism by acting as **catalysts**, chemical agents that selectively speed up chemical reactions without being consumed in the reaction. Because an enzyme can perform its function over and over again, these molecules can be thought of as workhorses that keep cells running by carrying out the processes of life.

A human has tens of thousands of different proteins, each with a specific structure and function; proteins, in fact, are the most structurally sophisticated molecules known. Consistent with their diverse functions, they vary extensively in structure, each type of protein having a unique three-dimensional shape.

Proteins are all constructed from the same set of 20 amino acids, linked in unbranched polymers. The bond between amino acids is called a peptide bond, so a polymer of amino

**▼ Figure 3.17 An overview of protein functions.**

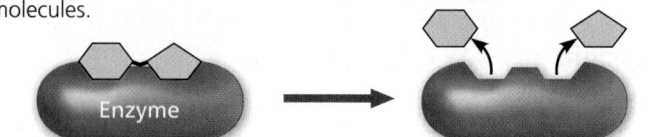

**Enzymatic proteins**

**Function:** Selective acceleration of chemical reactions

**Example:** Digestive enzymes catalyze the hydrolysis of bonds in food molecules.

Enzyme

**Defensive proteins**

**Function:** Protection against disease

**Example:** Antibodies inactivate and help destroy viruses and bacteria.

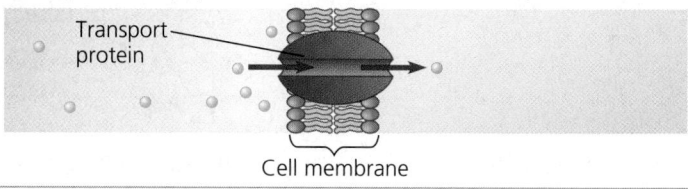

Antibodies

Virus

Bacterium

**Storage proteins**

**Function:** Storage of amino acids

**Examples:** Casein, the protein of milk, is the major source of amino acids for baby mammals. Plants have storage proteins in their seeds. Ovalbumin is the protein of egg white, used as an amino acid source for the developing embryo.

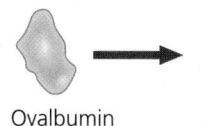

Ovalbumin — Amino acids for embryo

**Transport proteins**

**Function:** Transport of substances

**Examples:** Hemoglobin, the iron-containing protein of vertebrate blood, transports oxygen from the lungs to other parts of the body. Other proteins transport molecules across membranes, as shown here.

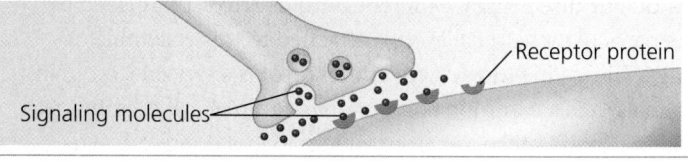

Transport protein

Cell membrane

**Hormonal proteins**

**Function:** Coordination of an organism's activities

**Example:** Insulin, a hormone secreted by the pancreas, causes other tissues to take up glucose, thus regulating blood sugar concentration.

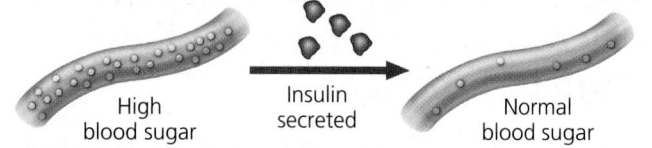

High blood sugar — Insulin secreted — Normal blood sugar

**Receptor proteins**

**Function:** Response of cell to chemical stimuli

**Example:** Receptors built into the membrane of a nerve cell detect signaling molecules released by other nerve cells.

Receptor protein

Signaling molecules

**Contractile and motor proteins**

**Function:** Movement

**Examples:** Motor proteins are responsible for the undulations of cilia and flagella. Actin and myosin proteins are responsible for the contraction of muscles.

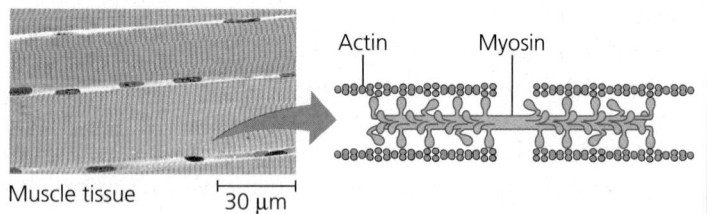

Actin    Myosin

Muscle tissue    30 μm

**Structural proteins**

**Function:** Support

**Examples:** Keratin is the protein of hair, horns, feathers, and other skin appendages. Insects and spiders use silk fibers to make their cocoons and webs, respectively. Collagen and elastin proteins provide a fibrous framework in animal connective tissues.

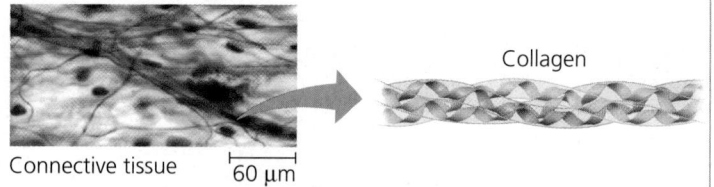

Collagen

Connective tissue    60 μm

---

acids is called a **polypeptide**. A **protein** is a biologically functional molecule made up of one or more polypeptides folded and coiled into a specific three-dimensional structure.

## Amino Acid Monomers

All amino acids share a common structure. An **amino acid** is an organic molecule with both an amino group and a carboxyl group; the small figure shows the general formula for an amino acid. At the

Side chain (R group)

R

α carbon

H   N   C   O   OH
H   H

Amino group    Carboxyl group

center of the amino acid is a carbon atom called the *alpha* (α) *carbon*. Its four different partners are an amino group, a carboxyl group, a hydrogen atom, and a variable group symbolized by R. The R group, also called the side chain, differs with each amino acid **(Figure 3.18)**.

The 20 amino acids in Figure 3.18 are the ones cells use to build their proteins. Here the amino groups and carboxyl groups are all depicted in ionized form, the way they usually exist at the pH found in a cell. The

▼ **Figure 3.18 The 20 amino acids of proteins.** The amino acids are grouped here according to the properties of their side chains (R groups) and shown in their prevailing ionic forms at pH 7.2, the pH within a cell. The three-letter and one-letter abbreviations for the amino acids are in parentheses.

## Nonpolar side chains; hydrophobic

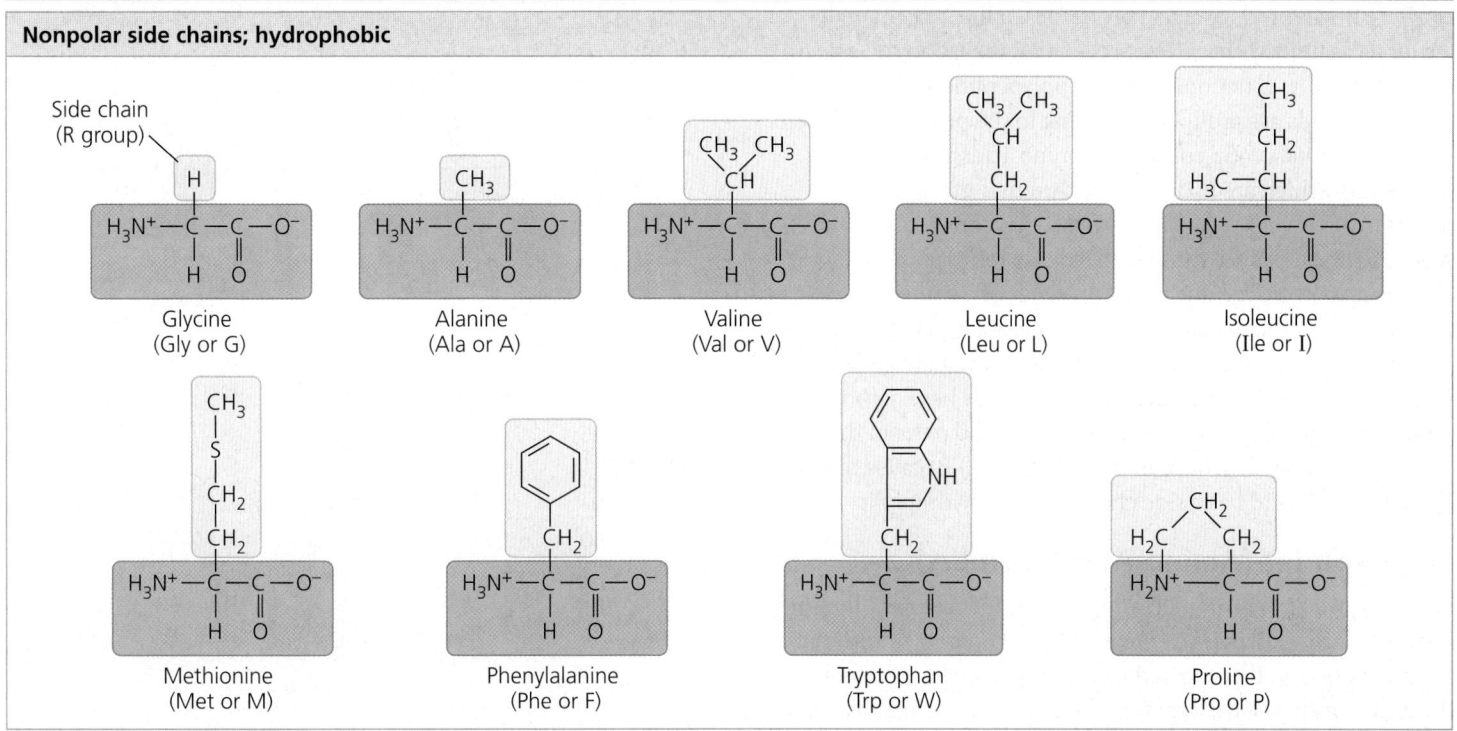

## Polar side chains; hydrophilic

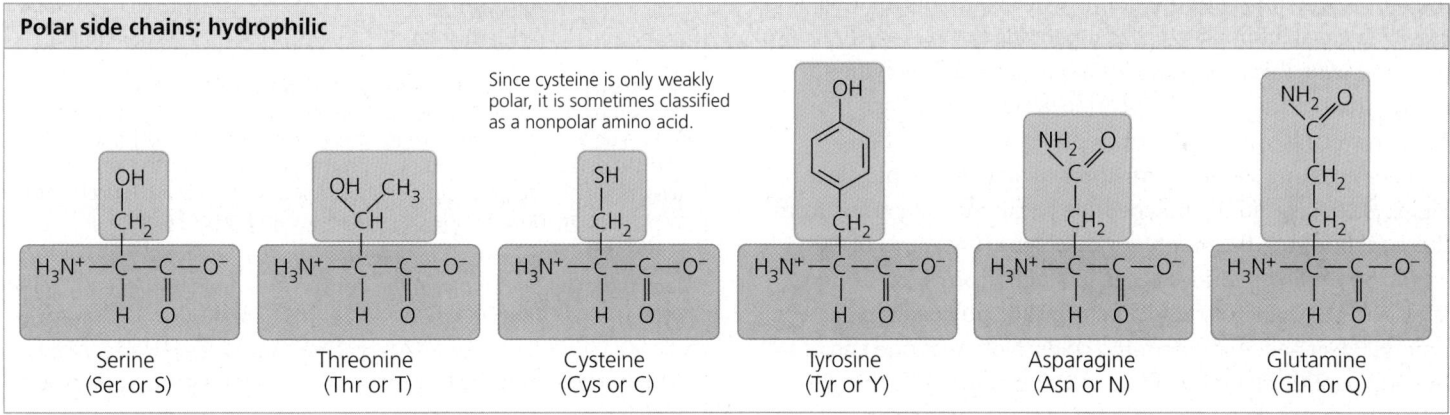

## Electrically charged side chains; hydrophilic

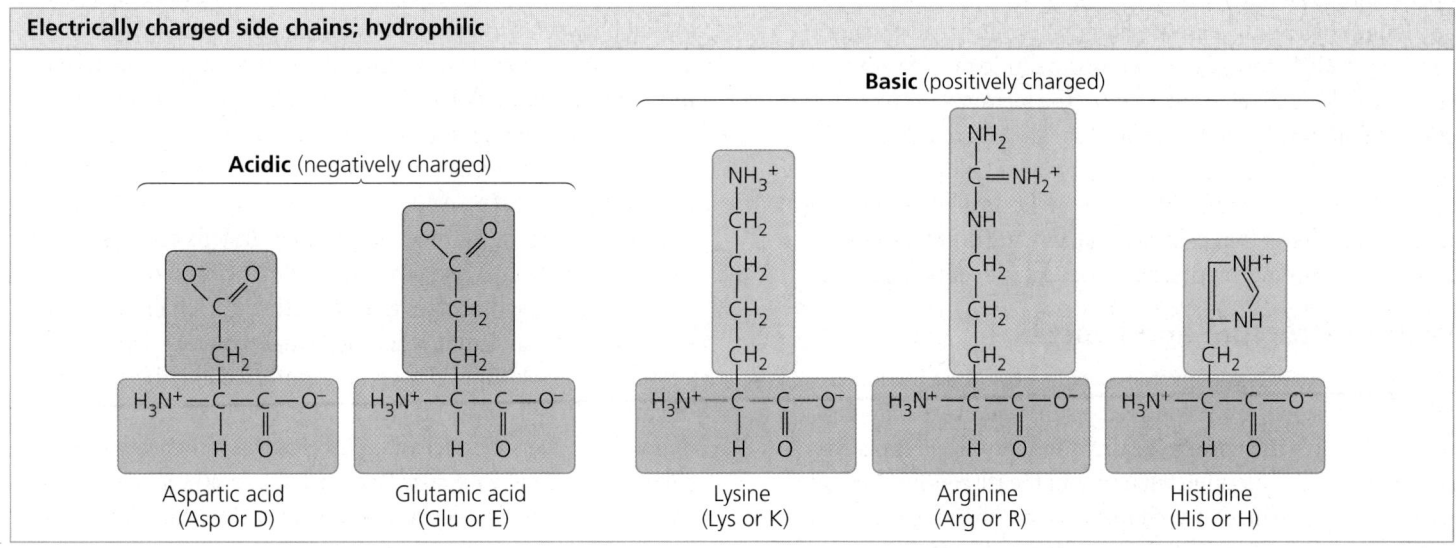

side chain (R group) may be as simple as a hydrogen atom, as in the amino acid glycine, or it may be a carbon skeleton with various functional groups attached, as in glutamine.

The physical and chemical properties of the side chain determine the unique characteristics of a particular amino acid, thus affecting its functional role in a polypeptide. In Figure 3.18, the amino acids are grouped according to the properties of their side chains. One group consists of amino acids with nonpolar side chains, which are hydrophobic. Another group consists of amino acids with polar side chains, which are hydrophilic. Acidic amino acids are those with side chains that are generally negative in charge due to the presence of a carboxyl group, which is usually dissociated (ionized) at cellular pH. Basic amino acids have amino groups in their side chains that are generally positive in charge. (Notice that *all* amino acids have carboxyl groups and amino groups; the terms *acidic* and *basic* in this context refer only to groups in the side chains.) Because they are charged, acidic and basic side chains are also hydrophilic.

## Polypeptides (Amino Acid Polymers)

Now that we have examined amino acids, let's see how they are linked to form polymers **(Figure 3.19)**. When two amino acids are positioned so that the carboxyl group of one is adjacent to the amino group of the other, they can become joined by a dehydration reaction, with the removal of a water molecule. The resulting covalent bond is called a **peptide bond**. Repeated over and over, this process yields a polypeptide, a polymer of many amino acids linked by peptide bonds. You'll learn more about how cells synthesize polypeptides in Chapter 14.

The repeating sequence of atoms highlighted in purple in Figure 3.19 is called the polypeptide backbone. Extending from this backbone are the different side chains (R groups) of the amino acids. Polypeptides range in length from a few amino acids to a thousand or more. Each specific polypeptide has a unique linear sequence of amino acids. Note that one end of the polypeptide chain has a free amino group, while the opposite end has a free carboxyl group. Thus, a polypeptide of any length has a single amino end (N-terminus) and a single carboxyl end (C-terminus). In a polypeptide of any significant size, the side chains far outnumber the terminal groups, so the chemical nature of the molecule as a whole is determined by the kind and sequence of the side chains. The immense variety of polypeptides in nature illustrates an important concept introduced earlier—that cells can make many different polymers by linking a limited set of monomers into diverse sequences.

## Protein Structure and Function

The specific activities of proteins result from their intricate three-dimensional architecture, the simplest level of which is the sequence of their amino acids. What can the amino acid sequence of a polypeptide tell us about the three-dimensional structure (commonly referred to simply as "the structure") of the protein and its function? The term *polypeptide* is not

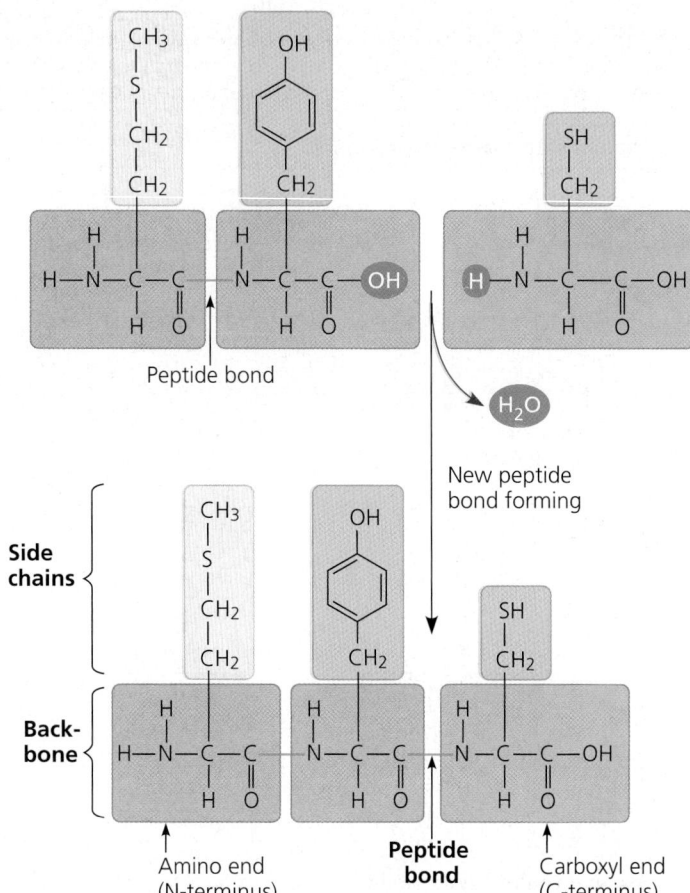

▲ **Figure 3.19 Making a polypeptide chain.** Peptide bonds are formed by dehydration reactions, which link the carboxyl group of one amino acid to the amino group of the next. The peptide bonds are formed one at a time, starting with the amino acid at the amino end (N-terminus). The polypeptide has a repetitive backbone (purple) from which the amino acid side chains (yellow and green) extend.

**DRAW IT** *At the top of the figure, circle and label the carboxyl and amino groups that will form the new peptide bond. Under each amino acid, write its three- and one-letter abbreviations (see Figure 3.18).*

synonymous with the term *protein*. Even for a protein consisting of a single polypeptide, the relationship is somewhat analogous to that between a long strand of yarn and a sweater of particular size and shape that can be knitted from the yarn. A functional protein is not *just* a polypeptide chain, but one or more polypeptides precisely twisted, folded, and coiled into a molecule of unique shape, which can be shown in several different types of models **(Figure 3.20)**. And it is the amino acid sequence of each polypeptide that determines what three-dimensional structure the protein will have under normal cellular conditions.

When a cell synthesizes a polypeptide, the chain may fold spontaneously, assuming the functional structure for that protein. This folding is driven and reinforced by the formation of various bonds between parts of the chain, which in turn depends on the sequence of amino acids. Many proteins are roughly spherical (*globular proteins*), while others are shaped like long fibers (*fibrous proteins*). Even within these broad categories, countless variations exist.

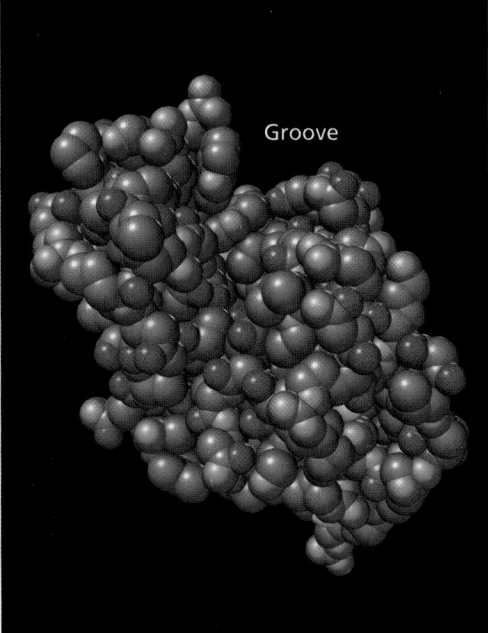

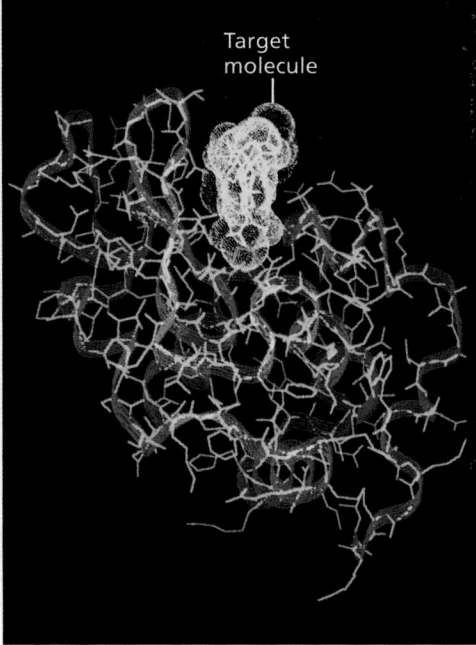

Groove

Groove

Target molecule

**(a)** A **ribbon model** shows how the single polypeptide chain folds and coils to form the functional protein. (The yellow lines represent disulfide bridges that stabilize the protein's shape.)

**(b)** A **space-filling model** shows more clearly the globular shape seen in many proteins, as well as the specific three-dimensional structure unique to lysozyme. The groove is the site that will bind to the target bacterial molecule.

**(c)** In this view, a ribbon model is superimposed on a **wireframe model**, which shows the backbone with the side chains extending from it. The yellow structure is the target molecule on the bacterial cell surface.

▲ **Figure 3.20  Structure of a protein, the enzyme lysozyme.** Present in our sweat, tears, and saliva, lysozyme is an enzyme that helps prevent infection by binding to and catalyzing the destruction of specific (target) molecules on the surface of many kinds of bacteria. The groove is the part of the protein that recognizes and binds to the target molecules on the surface of bacterial cell walls.

A protein's specific structure determines how it works. In almost every case, the function of a protein depends on its ability to recognize and bind to some other molecule. In an especially striking example of the marriage of form and function, **Figure 3.21** shows the exact match of shape between an

Antibody protein          Protein from flu virus

▲ **Figure 3.21  An antibody binding to a protein from a flu virus.** A technique called X-ray crystallography was used to generate a computer model of an antibody protein (blue and orange, left) bound to a flu virus protein (green and yellow, right). Computer software was then used to back the images away from each other, revealing the exact complementarity of shape between the two protein surfaces.

antibody (a protein in the body) and the particular foreign substance on a flu virus that the antibody binds to and marks for destruction. (In Chapter 35, you'll learn more about how the immune system generates antibodies that match the shapes of specific foreign molecules so well.)

Another example of molecules with matching shapes is endorphin molecules (produced by the body) and morphine molecules (a manufactured drug), both of which fit into receptor molecules on the surface of brain cells in humans, producing euphoria and relieving pain. Morphine, heroin, and other opiate drugs are able to mimic endorphins because they all share a similar shape with endorphins and can thus fit into and bind to endorphin receptors. This fit is very specific, something like a lock and key (see Figure 2.14). The endorphin receptor, like other receptor molecules, is a protein. The function of a protein—for instance, the ability of a receptor protein to bind to a particular pain-relieving signaling molecule—is an emergent property resulting from exquisite molecular order.

### Four Levels of Protein Structure

In spite of their great diversity, all proteins share three superimposed levels of structure, known as primary, secondary, and tertiary structure. A fourth level, quaternary structure, arises when a protein consists of two or more polypeptide chains. **Figure 3.22** describes these four levels of protein structure. Be sure to study this figure thoroughly before going on to the next section.

## Primary Structure
### Linear chain of amino acids

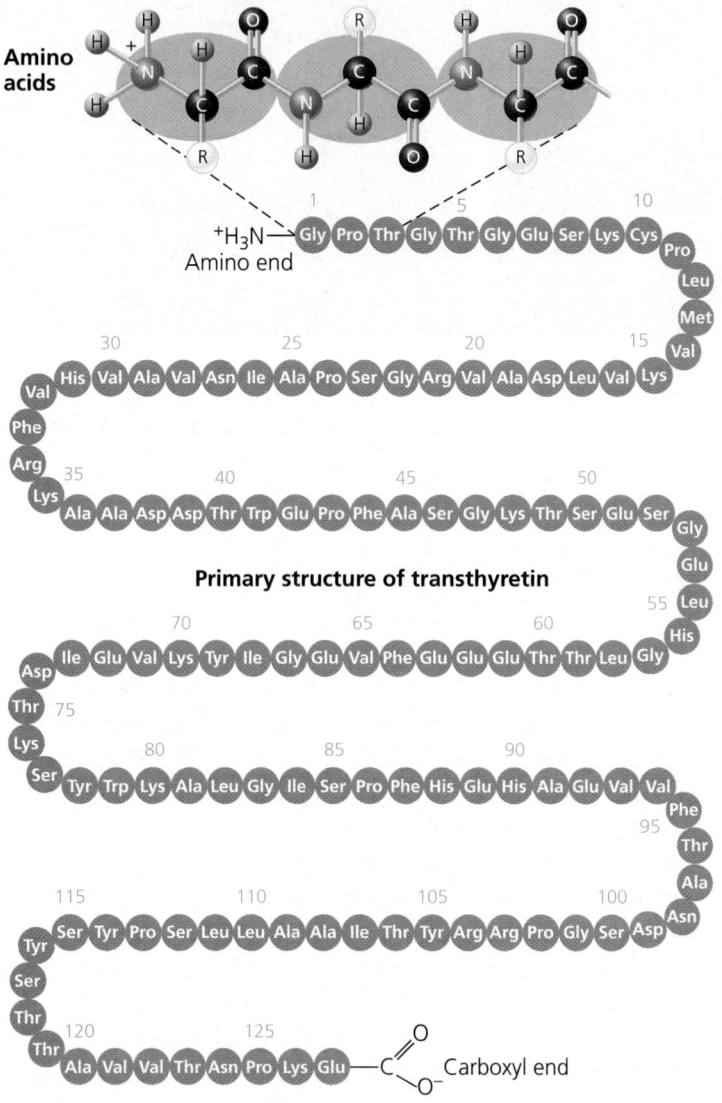

Primary structure of transthyretin

The **primary structure** of a protein is its sequence of amino acids. As an example, let's consider transthyretin, a globular blood protein that transports vitamin A and one of the thyroid hormones throughout the body. Transthyretin is made up of four identical polypeptide chains, each composed of 127 amino acids. Shown here is one of these chains unraveled for a closer look at its primary structure. Each of the 127 positions along the chain is occupied by one of the 20 amino acids, indicated here by its three-letter abbreviation.

The primary structure is like the order of letters in a very long word. If left to chance, there would be $20^{127}$ different ways of making a polypeptide chain 127 amino acids long. However, the precise primary structure of a protein is determined not by the random linking of amino acids, but by inherited genetic information. The primary structure in turn dictates secondary and tertiary structure, due to the chemical nature of the backbone and the side chains (R groups) of the amino acids along the polypeptide.

## Secondary Structure
### Regions stabilized by hydrogen bonds between atoms of the polypeptide backbone

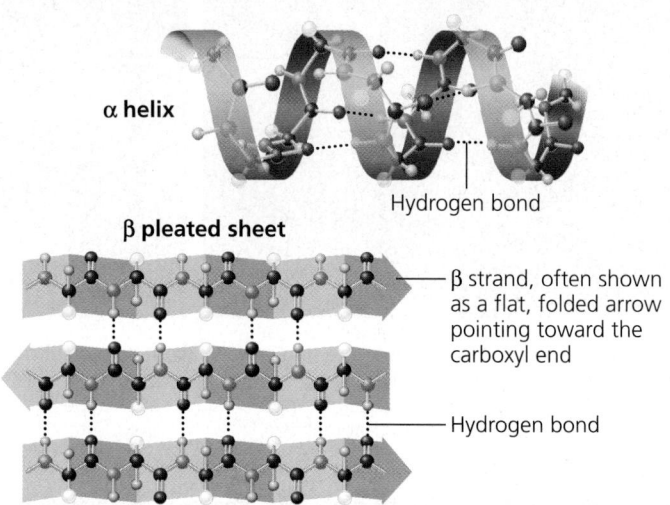

α helix

Hydrogen bond

β pleated sheet

β strand, often shown as a flat, folded arrow pointing toward the carboxyl end

Hydrogen bond

Most proteins have segments of their polypeptide chains repeatedly coiled or folded in patterns that contribute to the protein's overall shape. These coils and folds, collectively referred to as **secondary structure**, are the result of hydrogen bonds between the repeating constituents of the polypeptide backbone (not the amino acid side chains). Within the backbone, the oxygen atoms have a partial negative charge, and the hydrogen atoms attached to the nitrogens have a partial positive charge (see Figure 2.12); therefore, hydrogen bonds can form between these atoms. Individually, these hydrogen bonds are weak, but because they are repeated many times over a relatively long region of the polypeptide chain, they can support a particular shape for that part of the protein.

One such secondary structure is the **α helix**, a delicate coil held together by hydrogen bonding between every fourth amino acid, as shown here. Although each transthyretin polypeptide has only one α helix region (see the Tertiary Structure section), other globular proteins have multiple stretches of α helix separated by nonhelical regions (see hemoglobin in the Quaternary Structure section). Some fibrous proteins, such as α-keratin, the structural protein of hair, have the α helix formation over most of their length.

The other main type of secondary structure is the **β pleated sheet**. As shown here, in this structure two or more segments of the polypeptide chain lying side by side (called β strands) are connected by hydrogen bonds between parts of the two parallel segments of polypeptide backbone. β pleated sheets make up the core of many globular proteins, as is the case for transthyretin (see Tertiary Structure), and dominate some fibrous proteins, including the silk protein of a spider's web. The teamwork of so many hydrogen bonds makes each spider silk fiber stronger than a steel strand of the same weight.

▶ Spiders secrete silk fibers made of a structural protein containing β pleated sheets, which allow the spider web to stretch and recoil.

## Tertiary Structure

### Three-dimensional shape stabilized by interactions between side chains

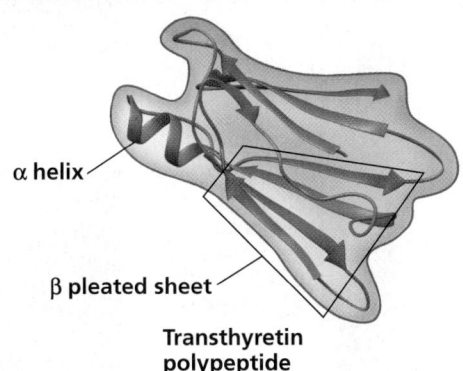

α helix

β pleated sheet

**Transthyretin polypeptide**

Superimposed on the patterns of secondary structure is a protein's tertiary structure, shown here in a ribbon model of the transthyretin polypeptide. While secondary structure involves interactions between backbone constituents, **tertiary structure** is the overall shape of a polypeptide resulting from interactions between the side chains (R groups) of the various amino acids. One type of interaction that contributes to tertiary structure is called—somewhat misleadingly—a **hydrophobic interaction**. As a polypeptide folds into its functional shape, amino acids with hydrophobic (nonpolar) side chains usually end up in clusters at the core of the protein, out of contact with water. Thus, a "hydrophobic interaction" is actually caused by the exclusion of nonpolar substances by water molecules. Once nonpolar amino acid side chains are close together, van der Waals interactions help hold them together. Meanwhile, hydrogen bonds between polar side chains and ionic bonds between positively and negatively charged side chains also help stabilize tertiary structure. These are all weak interactions in the aqueous cellular environment, but their cumulative effect helps give the protein a unique shape.

Covalent bonds called **disulfide bridges** may further reinforce the shape of a protein. Disulfide bridges form where two cysteine monomers, which have sulfhydryl groups (—SH) on their side chains (see Figure 3.6), are brought close together by the folding of the protein. The sulfur of one cysteine bonds to the sulfur of the second, and the disulfide bridge (—S—S—) rivets parts of the protein together (see yellow lines in Figure 3.20a). All of these different kinds of interactions can contribute to the tertiary structure of a protein, as shown here in a small part of a hypothetical protein:

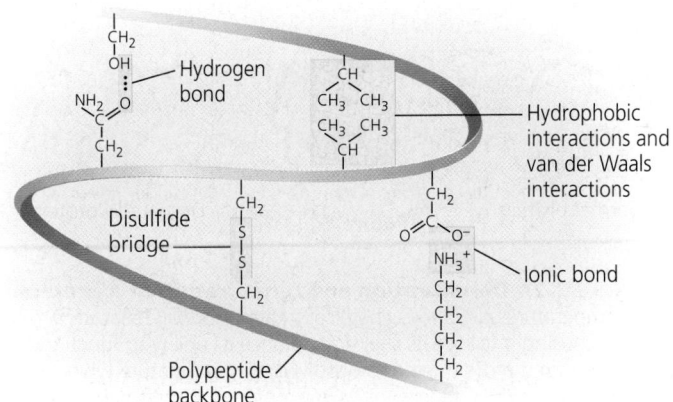

Hydrogen bond

Hydrophobic interactions and van der Waals interactions

Disulfide bridge

Ionic bond

Polypeptide backbone

## Quaternary Structure

### Association of two or more polypeptides (some proteins only)

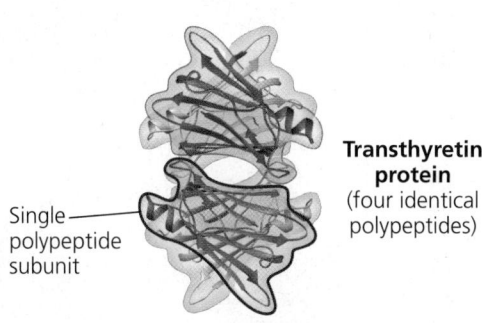

Single polypeptide subunit

**Transthyretin protein** (four identical polypeptides)

Some proteins consist of two or more polypeptide chains aggregated into one functional macromolecule. **Quaternary structure** is the overall protein structure that results from the aggregation of these polypeptide subunits. For example, shown here is the complete globular transthyretin protein, made up of its four polypeptides.

Another example is collagen, which is a fibrous protein that has three identical helical polypeptides intertwined into a larger triple helix, giving the long fibers great strength. This suits collagen fibers to their function as the girders of connective tissue in skin, bone, tendons, ligaments, and other body parts. (Collagen accounts for 40% of the protein in a human body.)

**Collagen**

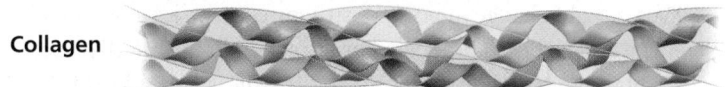

Hemoglobin, the oxygen-binding protein of red blood cells, is another example of a globular protein with quaternary structure. It consists of four polypeptide subunits, two of one kind (α) and two of another kind (β). Both α and β subunits consist primarily of α-helical secondary structure. Each subunit has a nonpolypeptide component, called heme, with an iron atom that binds oxygen.

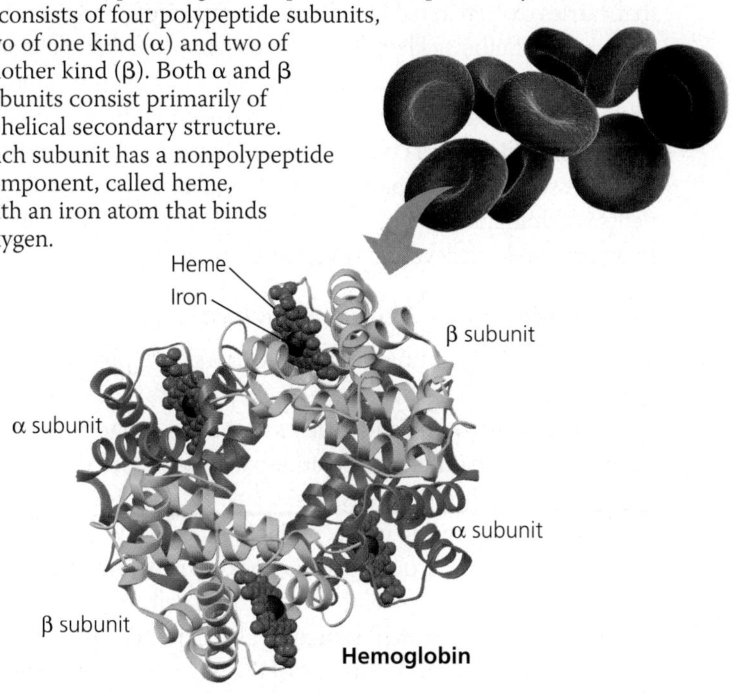

Heme

Iron

β subunit

α subunit

α subunit

β subunit

**Hemoglobin**

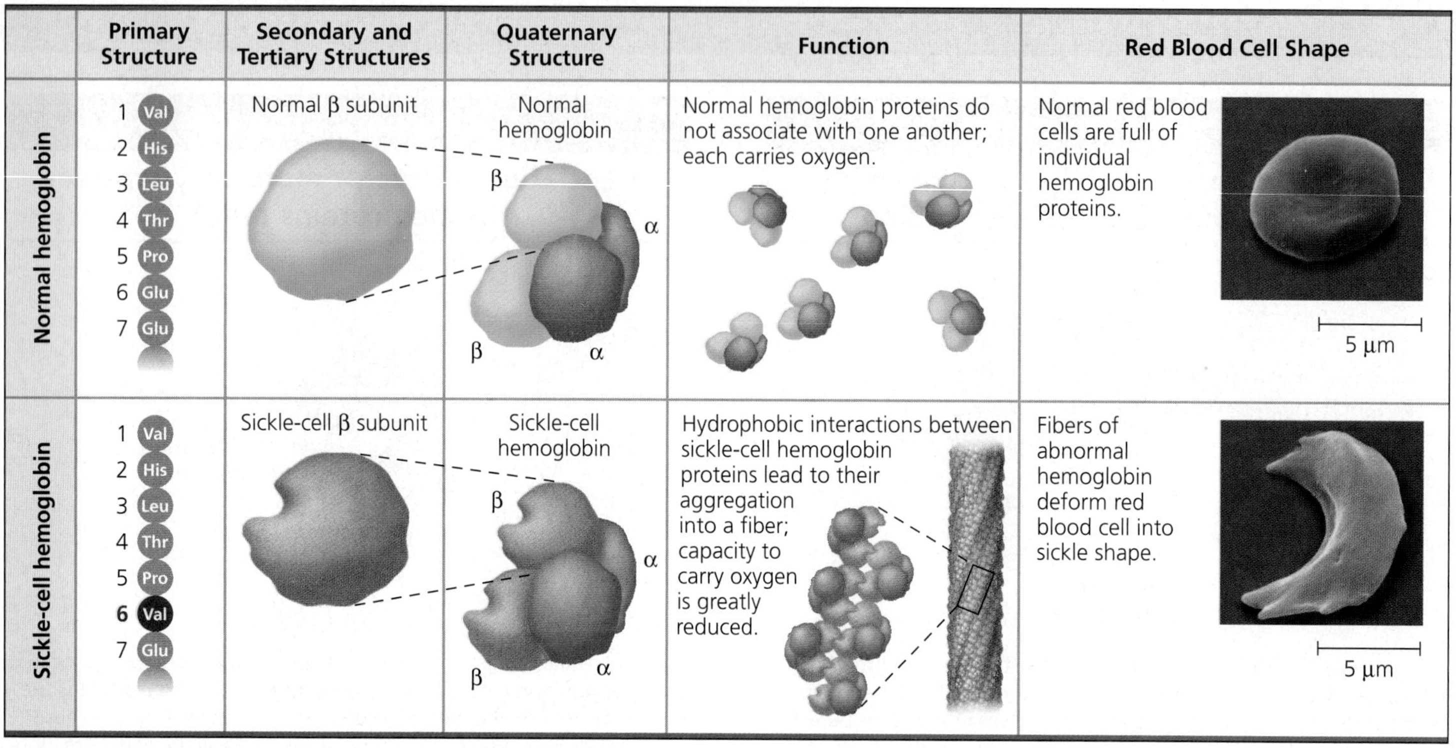

| | Primary Structure | Secondary and Tertiary Structures | Quaternary Structure | Function | Red Blood Cell Shape |
|---|---|---|---|---|---|
| **Normal hemoglobin** | 1 Val<br>2 His<br>3 Leu<br>4 Thr<br>5 Pro<br>6 Glu<br>7 Glu | Normal β subunit | Normal hemoglobin<br>β<br>α<br>β α | Normal hemoglobin proteins do not associate with one another; each carries oxygen. | Normal red blood cells are full of individual hemoglobin proteins.<br>5 μm |
| **Sickle-cell hemoglobin** | 1 Val<br>2 His<br>3 Leu<br>4 Thr<br>5 Pro<br>**6 Val**<br>7 Glu | Sickle-cell β subunit | Sickle-cell hemoglobin<br>β<br>α<br>β α | Hydrophobic interactions between sickle-cell hemoglobin proteins lead to their aggregation into a fiber; capacity to carry oxygen is greatly reduced. | Fibers of abnormal hemoglobin deform red blood cell into sickle shape.<br>5 μm |

▲ **Figure 3.23** A single amino acid substitution in a protein causes sickle-cell disease.

**MAKE CONNECTIONS** *Considering the chemical characteristics of the amino acids valine and glutamic acid (see Figure 3.18), propose a possible explanation for the dramatic effect on protein function that occurs when valine is substituted for glutamic acid.*

### Sickle-Cell Disease: A Change in Primary Structure

Even a slight change in primary structure can affect a protein's shape and ability to function. For instance, **sickle-cell disease**, an inherited blood disorder, is caused by the substitution of one amino acid (valine) for the normal one (glutamic acid) at a particular position in the primary structure of hemoglobin, the protein that carries oxygen in red blood cells. Normal red blood cells are disk-shaped, but in sickle-cell disease, the abnormal hemoglobin molecules tend to aggregate into fibers, deforming some of the cells into a sickle shape **(Figure 3.23)**. A person with the disease has periodic "sickle-cell crises" when the angular cells clog tiny blood vessels, impeding blood flow. The toll taken on such patients is a dramatic example of how a simple change in protein structure can have devastating effects on protein function.

### What Determines Protein Structure?

You've learned that a unique shape endows each protein with a specific function. But what are the key factors determining protein structure? You already know most of the answer: A polypeptide chain of a given amino acid sequence can be arranged into a three-dimensional shape determined by the interactions responsible for secondary and tertiary structure. This folding normally occurs as the protein is being synthesized in the crowded environment within a cell, aided by other proteins. However, protein structure also depends on the physical and chemical conditions of the protein's environment.

If the pH, salt concentration, temperature, or other aspects of its environment are altered, the weak chemical bonds and interactions within a protein may be destroyed, causing the protein to unravel and lose its native shape, a change called **denaturation (Figure 3.24)**. Because it is misshapen, the denatured protein is biologically inactive.

Most proteins become denatured if they are transferred from an aqueous environment to a nonpolar solvent, such as ether or chloroform; the polypeptide chain refolds so that its hydrophobic regions face outward toward the solvent.

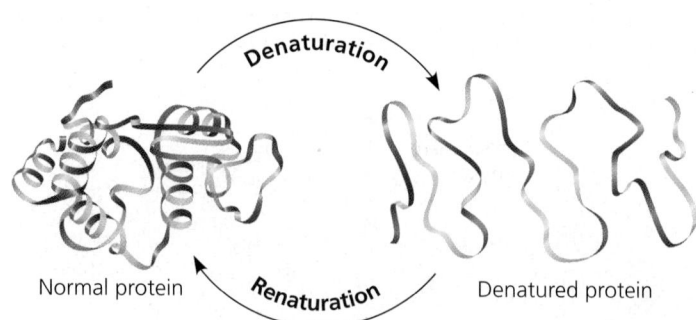

Normal protein — Denaturation → Denatured protein — Renaturation

▲ **Figure 3.24 Denaturation and renaturation of a protein.** High temperatures or various chemical treatments will denature a protein, causing it to lose its shape and hence its ability to function. If the denatured protein remains dissolved, it may renature when the chemical and physical aspects of its environment are restored to normal.

Other denaturation agents include chemicals that disrupt the hydrogen bonds, ionic bonds, and disulfide bridges that maintain a protein's shape. Denaturation can also result from excessive heat, which agitates the polypeptide chain enough to overpower the weak interactions that stabilize the structure. The white of an egg becomes opaque during cooking because the denatured proteins are insoluble and solidify. This also explains why excessively high fevers can be fatal: Proteins in the blood tend to denature at very high body temperatures.

When a protein in a test-tube solution has been denatured by heat or chemicals, it can sometimes return to its functional shape when the denaturing agent is removed. (Sometimes this is not possible: For example, a fried egg will not become lique-fied when placed back into the refrigerator!) We can conclude that the information for building a specific shape is intrinsic to the protein's primary structure. The sequence of amino acids determines the protein's shape—where an α helix can form, where β pleated sheets can exist, where disulfide bridges are located, where ionic bonds can form, and so on. But how does protein folding occur in the cell?

### Protein Folding in the Cell

Biochemists now know the amino acid sequence for nearly 50 million proteins, with about 1.7 million added each month, and the three-dimensional shape for more than 31,000. Researchers have tried to correlate the primary structure of many proteins with their three-dimensional structure to dis-cover the rules of protein folding. Unfortunately, however, the protein-folding process is not that simple. Most proteins probably go through several intermediate structures on their way to a stable shape, and looking at the mature structure does not reveal the stages of folding required to achieve that form. However, biochemists have developed methods for tracking a protein through such stages. They are still working to develop computer programs that can predict the 3-D structure of a polypeptide from its primary structure alone.

Misfolding of polypeptides is a serious problem in cells. Many diseases, such as Alzheimer's, Parkinson's, and mad cow disease, are associated with an accumulation of misfolded pro-teins. In fact, misfolded versions of the transthyretin protein featured in Figure 3.22 have been implicated in several dis-eases, including one form of senile dementia.

Even when scientists have a correctly folded protein in hand, determining its exact three-dimensional structure is not simple, for a single protein molecule has thousands of atoms. The method most commonly used to determine the 3-D shape of a protein is **X-ray crystallography**, which depends on the dif-fraction of an X-ray beam by the atoms of a crystallized mole-cule. Using this technique, scientists can build a 3-D model that shows the exact position of every atom in a protein molecule **(Figure 3.25)**. Nuclear magnetic resonance (NMR) spectros-copy and bioinformatics (see Concept 1.1) are complementary approaches to understanding protein structure and function.

▼ **Figure 3.25** **Research Method**

### X-Ray Crystallography

**Application** Scientists use X-ray crystallography to determine the three-dimensional (3-D) structure of macromolecules such as nucleic acids and proteins. As an example, we show how this method was used to determine the 3-D shape of transthyretin, the blood transport protein whose levels of structure are described in Figure 3.22.

**Technique** Researchers aimed an X-ray beam through the crystal-lized protein. The atoms of the crystal diffracted (bent) the X-rays into an orderly array that a digital detector recorded as a pattern of spots called an X-ray diffraction pattern, an example of which is shown below.

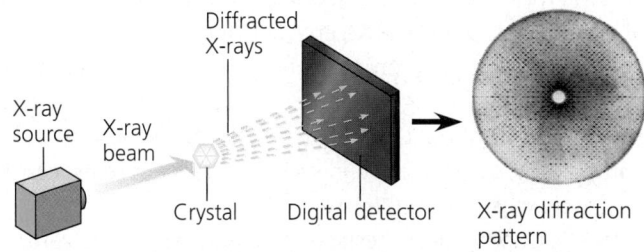

**Results** Using data from X-ray diffraction patterns, as well as the amino acid sequence determined by chemical methods, researchers built a 3-D model of the four-subunit transthyretin protein with the help of computer software.

### CONCEPT CHECK 3.5

1. Why does a denatured protein no longer function normally?
2. What parts of a polypeptide participate in the bonds that hold together secondary structure? Tertiary structure?
3. **WHAT IF?** Where would you expect a polypeptide region rich in the amino acids valine, leucine, and isoleucine to be located in a folded polypeptide? Explain.

For suggested answers, see Appendix A.

# Nucleic acids store, transmit, and help express hereditary information

If the primary structure of polypeptides determines a protein's shape, what determines primary structure? The amino acid sequence of a polypeptide is programmed by a discrete unit of inheritance known as a **gene**. Genes consist of DNA, which belongs to the class of compounds called nucleic acids. **Nucleic acids** are polymers made of monomers called nucleotides.

## The Roles of Nucleic Acids

The two types of nucleic acids, **deoxyribonucleic acid (DNA)** and **ribonucleic acid (RNA)**, enable living organisms to reproduce their complex components from one generation to the next. Unique among molecules, DNA provides directions for its own replication. DNA also directs RNA synthesis and, through RNA, controls protein synthesis; this entire process is called **gene expression (Figure 3.26)**.

DNA is the genetic material that organisms inherit from their parents. Each chromosome contains one long DNA molecule, usually carrying several hundred or more genes. When a cell reproduces itself by dividing, its DNA molecules are copied and passed along from one generation of cells to the

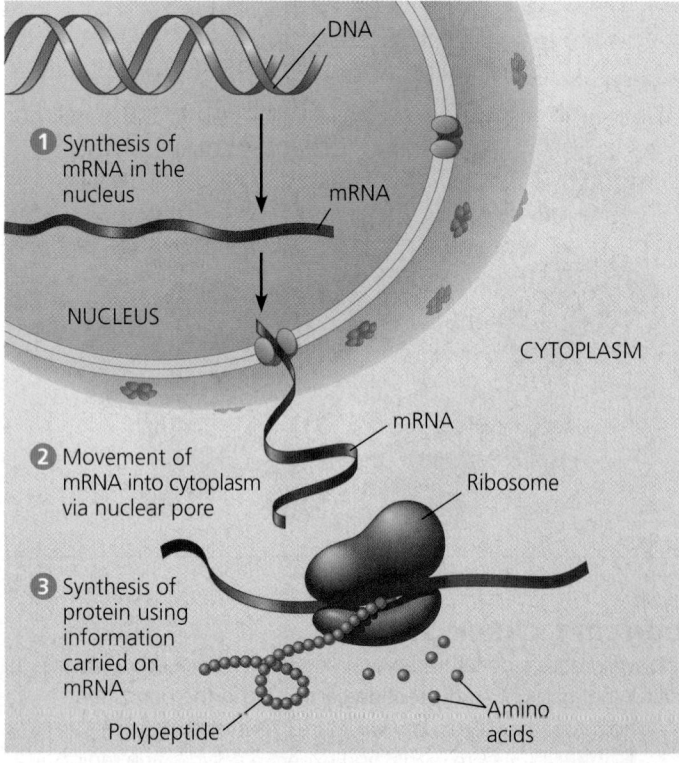

▲ **Figure 3.26 Gene expression: DNA → RNA → protein.** In a eukaryotic cell, DNA in the nucleus programs protein production in the cytoplasm by dictating synthesis of messenger RNA (mRNA).

next. Encoded in the structure of DNA is the information that programs all the cell's activities. The DNA, however, is not directly involved in running the operations of the cell, any more than computer software by itself can print a bank statement or read the bar code on a box of cereal. Just as a printer is needed to print out a statement and a scanner is needed to read a bar code, proteins are required to implement genetic programs. The molecular hardware of the cell—the tools for biological functions—consists mostly of proteins. For example, the oxygen carrier in red blood cells is the protein hemoglobin (see Figure 3.22), not the DNA that specifies its structure.

How does RNA, the other type of nucleic acid, fit into gene expression, the flow of genetic information from DNA to proteins? A given gene along a DNA molecule can direct synthesis of a type of RNA called *messenger RNA* (*mRNA*). The mRNA molecule interacts with the cell's protein-synthesizing machinery to direct production of a polypeptide, which folds into all or part of a protein. We can summarize the flow of genetic information as DNA → RNA → protein (see Figure 3.26). The sites of protein synthesis are cellular structures called ribosomes. In a eukaryotic cell, ribosomes are in the cytoplasm—the region between the nucleus and the cell's outer boundary, the plasma membrane—but DNA resides in the nucleus. Messenger RNA conveys genetic instructions for building proteins from the nucleus to the cytoplasm. Prokaryotic cells lack nuclei but still use mRNA to convey a message from the DNA to ribosomes and other cellular equipment that translate the coded information into amino acid sequences. Later in the book, you'll read about other functions of some recently discovered RNA molecules; the stretches of DNA that direct synthesis of these RNAs are also considered genes (see Concept 15.3).

## The Components of Nucleic Acids

Nucleic acids are macromolecules that exist as polymers called **polynucleotides (Figure 3.27a)**. As indicated by the name, each polynucleotide consists of monomers called **nucleotides**. A nucleotide, in general, is composed of three parts: a nitrogen-containing (nitrogenous) base, a five-carbon sugar (a pentose), and one to three phosphate groups **(Figure 3.27b)**. The beginning monomer used to build a polynucleotide has three phosphate groups, but two are lost during the polymerization process. The portion of a nucleotide without any phosphate groups is called a *nucleoside*.

To understand the structure of a single nucleotide, let's first consider the nitrogenous bases **(Figure 3.27c)**. Each nitrogenous base has one or two rings that include nitrogen atoms. (They are called nitrogenous *bases* because the nitrogen atoms tend to take up $H^+$ from solution, thus acting as bases.) There are two families of nitrogenous bases: pyrimidines and purines. A **pyrimidine** has one six-membered ring of carbon and nitrogen atoms. The members of the pyrimidine family are cytosine (C), thymine (T), and uracil (U). **Purines** are larger, with a

▼ **Figure 3.27 Components of nucleic acids. (a)** A polynucleotide has a sugar-phosphate backbone with variable appendages, the nitrogenous bases. **(b)** A nucleotide monomer includes a nitrogenous base, a sugar, and a phosphate group. Note that carbon numbers in the sugar include primes ('). **(c)** A nucleoside includes a nitrogenous base (purine or pyrimidine) and a five-carbon sugar (deoxyribose or ribose).

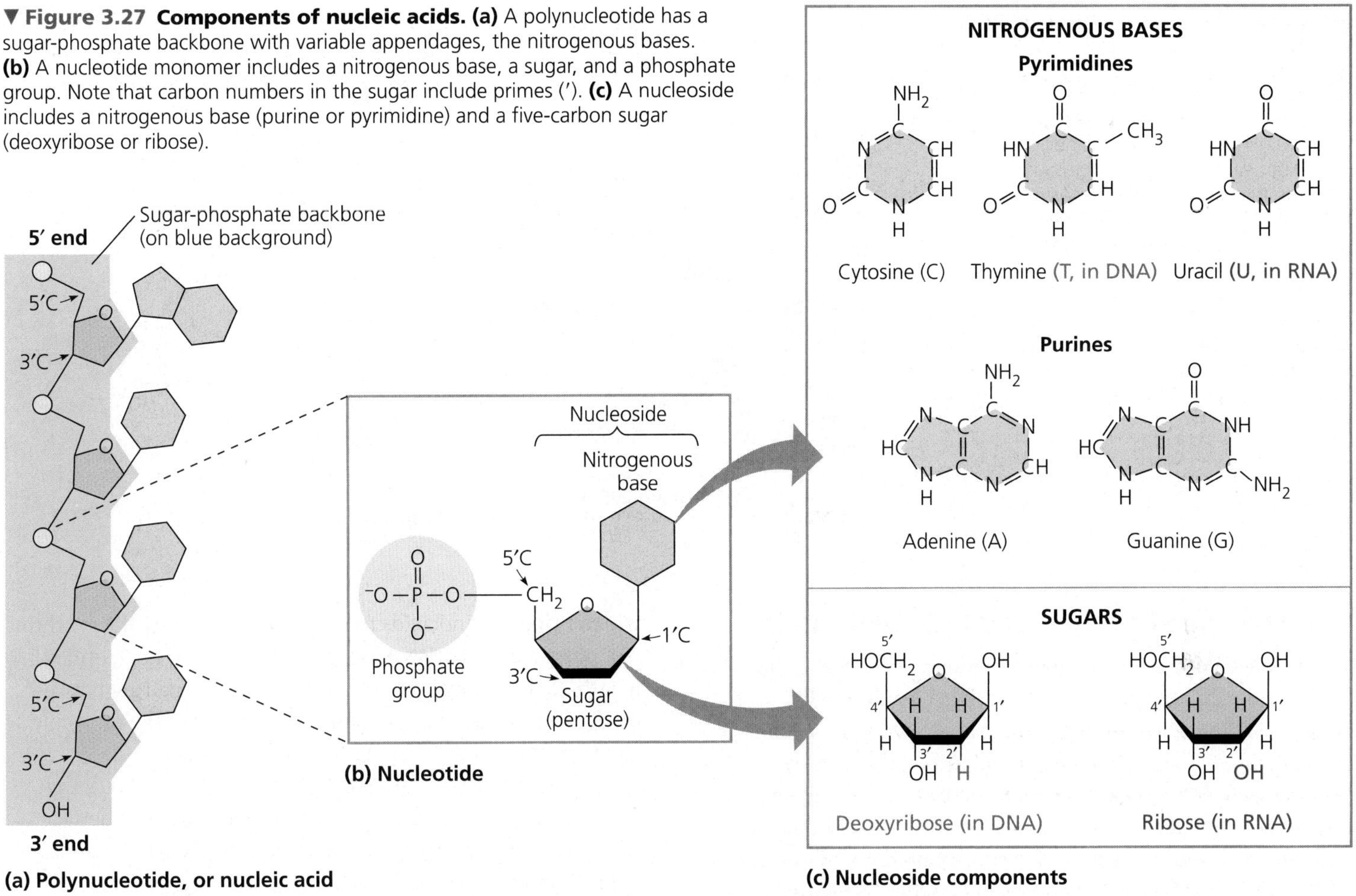

**(a) Polynucleotide, or nucleic acid**

**(b) Nucleotide**

**(c) Nucleoside components**

six-membered ring fused to a five-membered ring. The purines are adenine (A) and guanine (G). The specific pyrimidines and purines differ in the chemical groups attached to the rings. Adenine, guanine, and cytosine are found in both DNA and RNA; thymine is found only in DNA and uracil only in RNA.

Now let's add the sugar to which the nitrogenous base is attached. In DNA the sugar is **deoxyribose**; in RNA it is **ribose** (see Figure 3.27c). The only difference between these two sugars is that deoxyribose lacks an oxygen atom on the second carbon in the ring, hence the name *deoxy*ribose.

So far, we have built a nucleoside (nitrogenous base plus sugar). To complete the construction of a nucleotide, we attach a phosphate group to the 5′ carbon of the sugar (the carbon numbers in the sugar include ′, the prime symbol; see Figure 3.27b). The molecule is now a nucleoside monophosphate, more often called a nucleotide.

## Nucleotide Polymers

The linkage of nucleotides into a polynucleotide involves a dehydration reaction. (You will learn the details in Concept 13.2.) In the polynucleotide, adjacent nucleotides are joined by a phosphodiester linkage, which consists of a phosphate group that covalently links the sugars of two nucleotides. This

bonding results in a backbone with a repeating pattern of sugar-phosphate units called the *sugar-phosphate backbone* (see Figure 3.27a). (Note that the nitrogenous bases are not part of the backbone.) The two free ends of the polymer are distinctly different from each other. One end has a phosphate attached to a 5′ carbon, and the other end has a hydroxyl group on a 3′ carbon; we refer to these as the *5′ end* and the *3′ end*, respectively. We can say that a polynucleotide has a built-in directionality along its sugar-phosphate backbone, from 5′ to 3′, somewhat like a one-way street. All along this sugar-phosphate backbone are appendages consisting of the nitrogenous bases.

The sequence of bases along a DNA (or mRNA) polymer is unique for each gene and provides very specific information to the cell. Because genes are hundreds to thousands of nucleotides long, the number of possible base sequences is effectively limitless. A gene's meaning to the cell is encoded in its specific sequence of the four DNA bases. For example, the sequence 5′-AGGTAACTT-3′ means one thing, whereas the sequence 5′-CGCTTTAAC-3′ has a different meaning. (Entire genes, of course, are much longer.) The linear order of bases in a gene specifies the amino acid sequence—the primary structure—of a protein, which in turn specifies that protein's three-dimensional structure, thus enabling its function in the cell.

▶ **Figure 3.28 The structures of DNA and tRNA molecules. (a)** The DNA molecule is usually a double helix, with the sugar-phosphate backbones of the antiparallel polynucleotide strands (symbolized here by blue ribbons) on the outside of the helix. Hydrogen bonds between pairs of nitrogenous bases hold the two strands together. As illustrated here with symbolic shapes for the bases, adenine (A) can pair only with thymine (T), and guanine (G) can pair only with cytosine (C). Each DNA strand in this figure is the structural equivalent of the polynucleotide diagrammed in Figure 3.27a. **(b)** A tRNA molecule has a roughly L-shaped structure, due to complementary base pairing of antiparallel stretches of RNA. In RNA, A pairs with U.

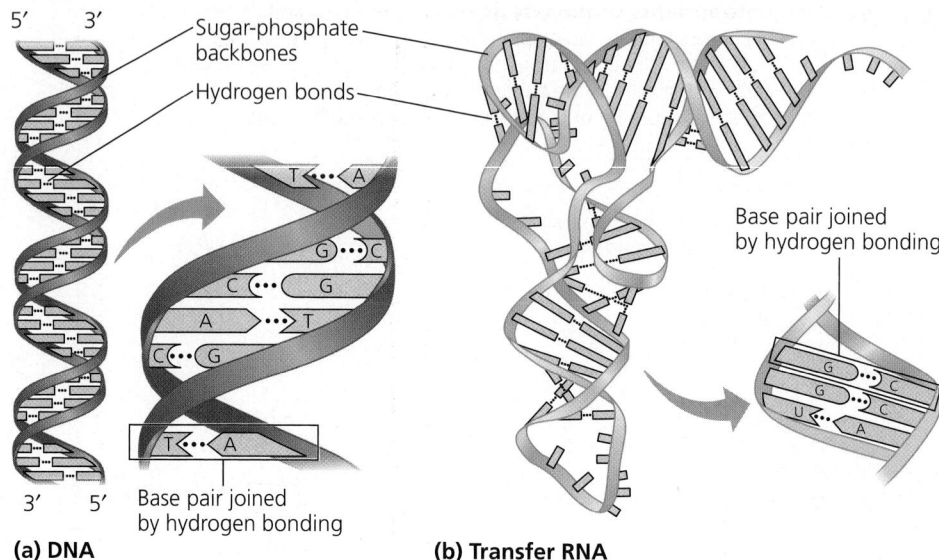

**(a) DNA**

**(b) Transfer RNA**

## The Structures of DNA and RNA Molecules

DNA molecules have two polynucleotides, or "strands," that wind around an imaginary axis, forming a **double helix (Figure 3.28a)**. The two sugar-phosphate backbones run in opposite 5′ → 3′ directions from each other; this arrangement is referred to as **antiparallel**, somewhat like a divided highway. The sugar-phosphate backbones are on the outside of the helix, and the nitrogenous bases are paired in the interior of the helix. The two strands are held together by hydrogen bonds between the paired bases (see Figure 3.28a). Most DNA molecules are very long, with thousands or even millions of base pairs. The one long DNA double helix in a eukaryotic chromosome includes many genes, each one a particular segment of the molecule.

In base pairing, only certain bases in the double helix are compatible with each other. Adenine (A) in one strand always pairs with thymine (T) in the other, and guanine (G) always pairs with cytosine (C). The two strands of the double helix are said to be *complementary*, each the predictable counterpart of the other. It is this feature of DNA that makes it possible to generate two identical copies of each DNA molecule in a cell that is preparing to divide. When the cell divides, the copies are distributed to the daughter cells, making them genetically identical to the parent cell. Thus, the structure of DNA accounts for its function of transmitting genetic information whenever a cell reproduces.

RNA molecules, by contrast, exist as single strands. Complementary base pairing can occur, however, between regions of two RNA molecules or even between two stretches of nucleotides in the *same* RNA molecule. In fact, base pairing within an RNA molecule allows it to take on the particular three-dimensional shape necessary for its function. Consider, for example, the type of RNA called *transfer RNA* (*tRNA*),

which brings amino acids to the ribosome during the synthesis of a polypeptide. A tRNA molecule is about 80 nucleotides in length. Its functional shape results from base pairing between nucleotides where complementary stretches of the molecule can run antiparallel to each other **(Figure 3.28b)**.

Note that in RNA, adenine (A) pairs with uracil (U); thymine (T) is not present in RNA. Another difference between RNA and DNA is that DNA almost always exists as a double helix, whereas RNA molecules are more variable in shape. RNAs are versatile molecules, and many biologists believe RNA may have preceded DNA as the carrier of genetic information in early forms of life (see Concept 24.1).

**CONCEPT CHECK 3.6**

1. **DRAW IT** Go to Figure 3.27a and, for the top three nucleotides, number all the carbons in the sugars (don't forget the primes), circle the nitrogenous bases, and star the phosphates.
2. **DRAW IT** In a DNA double helix, a region along one DNA strand has the following sequence of nitrogenous bases: 5′-TAGGCCT-3′. Copy this sequence, and write down its complementary strand, clearly indicating the 5′ and 3′ ends of the complementary strand.

For suggested answers, see Appendix A.

## CONCEPT 3.7

# Genomics and proteomics have transformed biological inquiry and applications

Experimental work in the first half of the 20th century established the role of DNA as the bearer of genetic information, passed from generation to generation, that specified the functioning of living cells and organisms. Once the structure of the DNA molecule was described in 1953, and the linear sequence of nucleotide bases was understood to specify the amino acid sequence of proteins, biologists sought to "decode" genes by learning their base sequences.

The first chemical techniques for *DNA sequencing*, or determining the sequence of nucleotides along a DNA strand, one by one, were developed in the 1970s. Researchers began to study gene sequences, gene by gene, and the more they learned, the more questions they had: How was expression of genes regulated? Genes and their protein products clearly interacted with each other, but how? What was the function, if any, of the DNA that is not part of genes? To fully understand the genetic functioning of a living organism, the entire sequence of the full complement of DNA, the organism's *genome*, would be most enlightening. In spite of the apparent impracticality of this idea, in the late 1980s several prominent biologists put forth an audacious proposal to launch a project that would sequence the entire human genome—all 3 billion bases of it! This endeavor began in 1990 and was effectively completed in the early 2000s.

An unplanned but profound side benefit of this project— the Human Genome Project—was the rapid development of faster and less expensive methods of sequencing. This trend has continued apace: The cost for sequencing 1 million bases in 2001, well over $5,000, has decreased to about $0.08 in 2014. And a human genome, the first of which took over 10 years to sequence, could be completed at today's pace in just a few days **(Figure 3.29)**. The number of genomes that have been fully sequenced has burgeoned, generating reams of data and prompting development of *bioinformatics*, the use of computer software and other computational tools that can handle and analyze these large data sets.

The reverberations of these developments have transformed the study of biology and related fields. Biologists often look at problems by analyzing large sets of genes or even comparing whole genomes of different species, an approach called **genomics**. A similar analysis of large sets of proteins, including their sequences, is called **proteomics**. (Protein sequences

▲ **Figure 3.29 Automatic DNA sequencing machines and abundant computing power enable rapid sequencing of genes and genomes.**

can be determined either by using biochemical techniques or by translating the DNA sequences that code for them.) These approaches permeate all fields of biology, some examples of which are shown in **Figure 3.30**.

## DNA and Proteins as Tape Measures of Evolution

**EVOLUTION** We are accustomed to thinking of shared traits, such as hair and milk production in mammals, as evidence of shared ancestry. Because DNA carries heritable information in the form of genes, sequences of genes and their protein products document the hereditary background of an organism. The linear sequences of nucleotides in DNA molecules are passed from parents to offspring; these sequences determine the amino acid sequences of proteins. As a result, siblings have greater similarity in their DNA and proteins than do unrelated individuals of the same species.

Given our evolutionary view of life, we can extend this concept of "molecular genealogy" to relationships between species: We would expect two species that appear to be closely related based on anatomical evidence (and possibly fossil evidence) to also share a greater proportion of their DNA and protein sequences than do less closely related species. In fact, that is the case. An example is the comparison of the β polypeptide chain of human hemoglobin with the corresponding hemoglobin polypeptide in other vertebrates. In this chain of 146 amino acids, humans and gorillas differ in just 1 amino acid, while humans and frogs, more distantly related, differ in 67 amino acids. In the **Scientific Skills Exercise**, you can apply this sort of reasoning to additional species. The relative sequence similarity also holds true when comparing whole genomes: The human genome is 95–98% identical to that of the chimpanzee, but only roughly 85% identical to that of the mouse, a more distant evolutionary relative. Molecular biology has added a new tape measure to the toolkit biologists use to assess evolutionary kinship.

Perhaps the most significant impact of genomics and proteomics on the field of biology has been their contributions to our understanding of evolution. To quote one of the founders of modern evolutionary theory, Theodosius Dobzhansky, "Nothing in biology makes sense except in the light of evolution." In addition to confirming evidence for evolution from the study of fossils and characteristics of currently existing species, genomics has helped us tease out relationships among different groups of organisms that had not been resolved by previous types of evidence, and thus infer their evolutionary history.

**CONCEPT CHECK 3.7**

1. How would sequencing the entire genome of an organism help scientists to understand how that organism functioned?
2. Given the function of DNA, why would you expect two species with very similar traits to also have very similar genomes?

For suggested answers, see Appendix A.

## MAKE CONNECTIONS

# Contributions of Genomics and Proteomics to Biology

Nucleotide sequencing and the analysis of large sets of genes and proteins can be done rapidly and inexpensively due to advances in technology and information processing. Taken together, genomics and proteomics have advanced our understanding of biology across many different fields.

### Paleontology

New DNA sequencing techniques have allowed decoding of minute quantities of DNA found in ancient tissues from our extinct relatives, the Neanderthals (*Homo neanderthalensis*). Sequencing the Neanderthal genome has informed our understanding of their physical appearance as well as their relationship with modern humans. (See Figure 27.36.)

### Evolution

A major aim of evolutionary biology is to understand the relationships among species, both living and extinct. For example, genome sequence comparisons have identified the hippopotamus as the land mammal sharing the most recent common ancestor with whales. (See Figure 19.20.)

### Medical Science

Identifying the genetic basis for human diseases like cancer helps researchers focus their search for potential future treatments. Currently, sequencing the sets of genes expressed in an individual's tumor can allow a more targeted approach to treating the cancer, a type of "personalized medicine." (See Concept 9.3 and Figure 16.21.)

Hippopotamus

Short-finned pilot whale

### Conservation Biology

The tools of molecular genetics and genomics are increasingly used by forensic ecologists to identify which species of animals and plants are killed illegally. In one case, genomic sequences of DNA from illegal shipments of elephant tusks were used to track down poachers and pinpoint the territory where they were operating. (See Figure 43.8.)

### Species Interactions

Most plant species exist in a mutually beneficial partnership with fungi (right) and bacteria associated with the plants' roots; these interactions improve plant growth. Genome sequencing and analysis of gene expression have allowed characterization of plant-associated communities. Such studies will help advance our understanding of such interactions and may improve agricultural practices. (See the Chapter 26 Scientific Skills Exercise and Figure 29.11.)

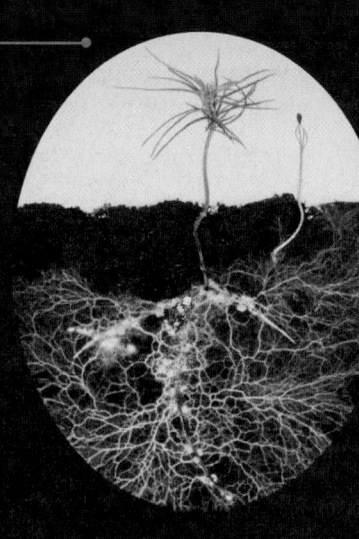

**MAKE CONNECTIONS** *Considering the examples provided here, describe how the approaches of genomics and proteomics help us to address a variety of biological questions.*

 ► Human   ► Rhesus monkey   ► Gibbon

## Analyzing Polypeptide Sequence Data

**Are Rhesus Monkeys or Gibbons More Closely Related to Humans?** DNA and polypeptide sequences from closely related species are more similar to each other than are sequences from more distantly related species. In this exercise, you will look at amino acid sequence data for the β polypeptide chain of hemoglobin, often called β-globin. You will then interpret the data to hypothesize whether the monkey or the gibbon is more closely related to humans.

**How Such Experiments Are Done** Researchers can isolate the polypeptide of interest from an organism and then determine the amino acid sequence. More frequently, the DNA of the relevant gene is sequenced, and the amino acid sequence of the polypeptide is deduced from the DNA sequence of its gene.

**Data from the Experiments** In the data below, the letters give the sequence of the 146 amino acids in β-globin from humans, rhesus monkeys, and gibbons. Because a complete sequence would not fit on one line here, the sequences are broken into three segments.

The sequences for the three different species are aligned so that you can compare them easily. For example, you can see that for all three species, the first amino acid is V (valine; see Figure 3.18) and the 146th amino acid is H (histidine).

### INTERPRET THE DATA

1. Scan along the monkey and gibbon sequences, letter by letter, circling any amino acids that do not match the human sequence. (a) How many amino acids differ between the monkey and the human sequences? (b) Between the gibbon and human?
2. For each nonhuman species, what percent of its amino acids is identical to the human sequence of β-globin?
3. Based on these data alone, state a hypothesis for which of these two species is more closely related to humans. What is your reasoning?
4. What other evidence could you use to support your hypothesis?

(MB) A version of this Scientific Skills Exercise can be assigned in MasteringBiology.

| Species | Alignment of Amino Acid Sequences of β-globin |
|---|---|
| Human | 1 VHLTPEEKSA VTALWGKVNV DEVGGEALGR LLVVYPWTQR FFESFGDLST |
| Monkey | 1 VHLTPEEKNA VTTLWGKVNV DEVGGEALGR LLLVYPWTQR FFESFGDLSS |
| Gibbon | 1 VHLTPEEKSA VTALWGKVNV DEVGGEALGR LLVVYPWTQR FFESFGDLST |
| | |
| Human | 51 PDAVMGNPKV KAHGKKVLGA FSDGLAHLDN LKGTFATLSE LHCDKLHVDP |
| Monkey | 51 PDAVMGNPKV KAHGKKVLGA FSDGLNHLDN LKGTFAQLSE LHCDKLHVDP |
| Gibbon | 51 PDAVMGNPKV KAHGKKVLGA FSDGLAHLDN LKGTFAQLSE LHCDKLHVDP |
| | |
| Human | 101 ENFRLLGNVL VCVLAHHFGK EFTPPVQAAY QKVVAGVANA LAHKYH |
| Monkey | 101 ENFKLLGNVL VCVLAHHFGK EFTPQVQAAY QKVVAGVANA LAHKYH |
| Gibbon | 101 ENFRLLGNVL VCVLAHHFGK EFTPQVQAAY QKVVAGVANA LAHKYH |

**Data from** Human: http://www.ncbi.nlm.nih.gov/protein/AAA21113.1; rhesus monkey: http://www.ncbi.nlm.nih.gov/protein/122634; gibbon: http://www.ncbi.nlm.nih.gov/protein/122616

---

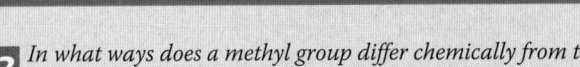

Go to **MasteringBiology**® for Assignments, the eText, and the Study Area with Animations, Activities, Vocab Self-Quiz, and Practice Tests.

# 3 Chapter Review

## SUMMARY OF KEY CONCEPTS

**VOCAB SELF-QUIZ**

goo.gl/gbai8v

### CONCEPT 3.1

**Carbon atoms can form diverse molecules by bonding to four other atoms (pp. 44–48)**

- Carbon, with a **valence** of 4, can bond to various other atoms, including O, H, and N. Carbon can also bond to other carbon atoms, forming the carbon skeletons of **organic compounds**. These skeletons vary in length and shape. **Hydrocarbons** consist of carbon and hydrogen. **Isomers** have the same molecular formula but different structures and properties.
- Chemical groups attached to the carbon skeletons of organic molecules participate in chemical reactions (as **functional groups**) or contribute to function by affecting molecular shape.
- **ATP (adenosine triphosphate)** can react with water, releasing energy that can be used by the cell.

? *In what ways does a methyl group differ chemically from the other six important chemical groups shown in Figure 3.6?*

### CONCEPT 3.2

**Macromolecules are polymers, built from monomers (pp. 48–49)**

- Large carbohydrates (polysaccharides), proteins, and nucleic acids are **polymers**, which are chains of **monomers**. The components of lipids vary. Monomers form larger molecules by **dehydration reactions**, in which water molecules are released. Polymers can disassemble by the reverse process, **hydrolysis**. An immense variety of polymers can be built from a small set of monomers.

? *What is the fundamental basis for the differences between large carbohydrates, proteins, and nucleic acids?*

| Large Biological Molecules | Components | Examples | Functions |
|---|---|---|---|
| **CONCEPT 3.3**<br><br>**Carbohydrates serve as fuel and building material (pp. 49–53)**<br><br>**?** *Compare the composition, structure, and function of starch and cellulose. What role do starch and cellulose play in the human body?* | <br>Monosaccharide monomer | **Monosaccharides:** glucose, fructose<br><br>**Disaccharides:** lactose, sucrose<br><br>**Polysaccharides:**<br>• Cellulose (plants)<br>• Starch (plants)<br>• Glycogen (animals)<br>• Chitin (animals and fungi) | Fuel; carbon sources that can be converted to other molecules or combined into polymers<br><br>• Strengthens plant cell walls<br>• Stores glucose for energy<br>• Stores glucose for energy<br>• Strengthens exoskeletons and fungal cell walls |
| **CONCEPT 3.4**<br><br>**Lipids are a diverse group of hydrophobic molecules (pp. 53–55)**<br><br>**?** *Why are lipids not considered to be polymers or macromolecules?* | 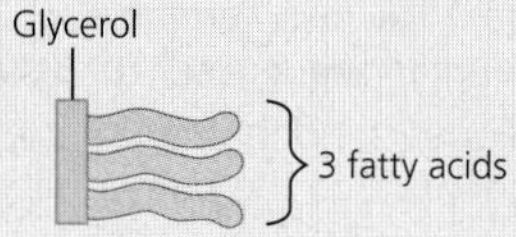 Glycerol<br>3 fatty acids | **Triacylglycerols** (fats or oils): glycerol + three fatty acids | Important energy source 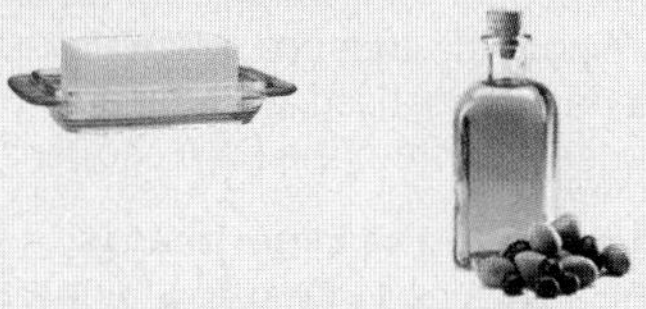 |
| | 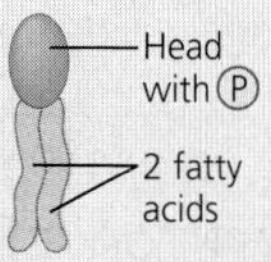 Head with (P)<br>2 fatty acids | **Phospholipids:** glycerol + phosphate group + two fatty acids | Lipid bilayers of membranes 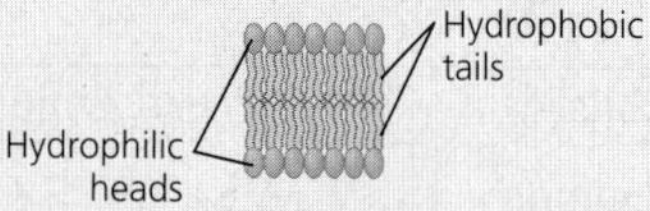 |
| | 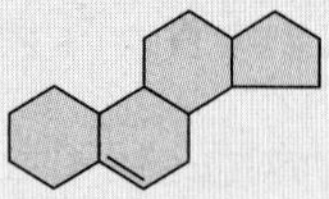 <br>Steroid backbone | **Steroids:** four fused rings with attached chemical groups | • Component of cell membranes (cholesterol)<br>• Signaling molecules that travel through the body (hormones) |
| **CONCEPT 3.5**<br><br>**Proteins include a diversity of structures, resulting in a wide range of functions (pp. 55–63)**<br><br>**?** *Explain the basis for the great diversity of proteins.* | 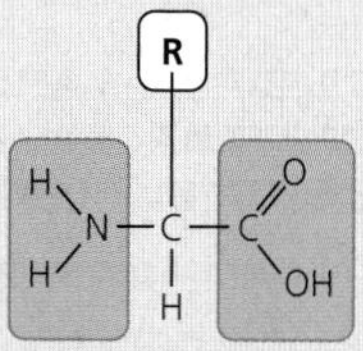 <br>Amino acid monomer<br>(20 types) | • Enzymes<br>• Structural proteins<br>• Storage proteins<br>• Transport proteins<br>• Hormones<br>• Receptor proteins<br>• Motor proteins<br>• Defensive proteins | • Catalyze chemical reactions<br>• Provide structural support<br>• Store amino acids<br>• Transport substances<br>• Coordinate organismal responses<br>• Receive signals from outside cell<br>• Function in cell movement<br>• Protect against disease |
| **CONCEPT 3.6**<br><br>**Nucleic acids store, transmit, and help express hereditary information (pp. 64–66)**<br><br>**?** *What role does complementary base pairing play in the functions of nucleic acids?* | 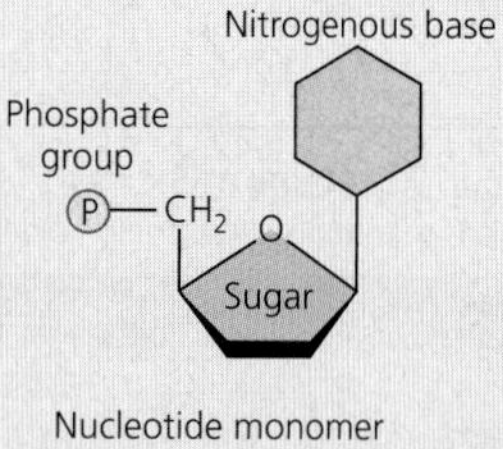 Nitrogenous base<br>Phosphate group<br><br>Nucleotide monomer | **DNA:** <br>• Sugar = deoxyribose<br>• Nitrogenous bases = C, G, A, T<br>• Usually double-stranded | Stores hereditary information |
| | | **RNA:** <br>• Sugar = ribose<br>• Nitrogenous bases = C, G, A, U<br>• Usually single-stranded | Various functions in gene expression, including carrying instructions from DNA to ribosomes |

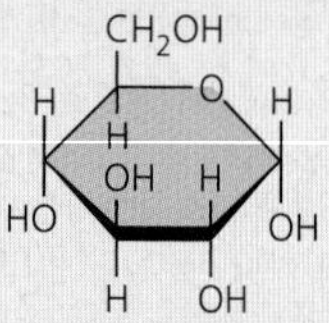

**CONCEPT 3.7**

## Genomics and proteomics have transformed biological inquiry and applications (pp. 66–69)

- Recent technological advances in DNA sequencing have given rise to **genomics**, an approach that analyzes large sets of genes or whole genomes, and **proteomics**, a similar approach for large sets of proteins. Bioinformatics is the use of computational tools and computer software to analyze these large data sets.

- The more closely two species are related evolutionarily, the more similar their DNA sequences are. DNA sequence data confirm models of evolution based on fossils and anatomical evidence.

**?** *Given the sequences of a particular gene in fruit flies, fish, mice, and humans, predict the relative similarity of the human sequence to that of each of the other species.*

# TEST YOUR UNDERSTANDING

PRACTICE TEST

goo.gl/CRZjvS

## Level 1: Knowledge/Comprehension

**1.** Choose the term that correctly describes the relationship between these two sugar molecules:

(A) structural isomers     (C) enantiomers
(B) *cis-trans* isomers     (D) isotopes

**2.** Which functional group is *not* present in this molecule?
(A) carboxyl
(B) sulfhydryl
(C) hydroxyl
(D) amino

**3.** MAKE CONNECTIONS Which chemical group is most likely to be responsible for an organic molecule behaving as a base (see Concept 2.5)?
(A) hydroxyl     (C) amino
(B) carbonyl     (D) phosphate

**4.** Which of the following categories includes all others in the list?
(A) disaccharide     (C) carbohydrate
(B) starch     (D) polysaccharide

**5.** Which of the following statements concerning *unsaturated* fats is true?
(A) They are more common in animals than in plants.
(B) They have double bonds in their fatty acid chains.
(C) They generally solidify at room temperature.
(D) They contain more hydrogen than do saturated fats having the same number of carbon atoms.

**6.** The structural level of a protein *least* affected by a disruption in hydrogen bonding is the
(A) primary level.     (C) tertiary level.
(B) secondary level.     (D) quaternary level.

**7.** Enzymes that break down DNA catalyze the hydrolysis of the covalent bonds that join nucleotides together. What would happen to DNA molecules treated with these enzymes?
(A) The two strands of the double helix would separate.
(B) The phosphodiester linkages of the polynucleotide backbone would be broken.
(C) The pyrimidines would be separated from the deoxyribose sugars.
(D) All bases would be separated from the deoxyribose sugars.

## Level 2: Application/Analysis

**8.** Which of the following hydrocarbons has a double bond in its carbon skeleton?
(A) $C_3H_8$     (C) $C_2H_4$
(B) $C_2H_6$     (D) $C_2H_2$

**9.** The molecular formula for glucose is $C_6H_{12}O_6$. What would be the molecular formula for a polymer made by linking ten glucose molecules together by dehydration reactions?
(A) $C_{60}H_{120}O_{60}$     (C) $C_{60}H_{100}O_{50}$
(B) $C_{60}H_{102}O_{51}$     (D) $C_{60}H_{111}O_{51}$

**10.** Construct a table that organizes the following terms, and label the columns and rows.

| Monosaccharides | Polypeptides | Phosphodiester linkages |
| Fatty acids | Triacylglycerols | Peptide bonds |
| Amino acids | Polynucleotides | Glycosidic linkages |
| Nucleotides | Polysaccharides | Ester linkages |

**11.** DRAW IT Copy the polynucleotide strand in Figure 3.27a and label the bases G, T, C, and T, starting from the 5′ end. Assuming this is a DNA polynucleotide, now draw the complementary strand, using the same symbols for phosphates (circles), sugars (pentagons), and bases. Label the bases. Draw arrows showing the 5′ → 3′ direction of each strand. Use the arrows to make sure the second strand is antiparallel to the first. *Hint*: After you draw the first strand vertically, turn the paper upside down; it is easier to draw the second strand from the 5′ toward the 3′ direction as you go from top to bottom.

## Level 3: Synthesis/Evaluation

**12. SCIENTIFIC INQUIRY**
Suppose you are a research assistant in a lab studying DNA-binding proteins. You have been given the amino acid sequences of all the proteins encoded by the genome of a certain species and have been asked to find candidate proteins that could bind DNA. What type of amino acids would you expect to see in the DNA-binding regions of such proteins? Explain your thinking.

**13. FOCUS ON EVOLUTION**
Comparisons of amino acid sequences can shed light on the evolutionary divergence of related species. If you were comparing two living species, would you expect all proteins to show the same degree of divergence? Why or why not? Justify your answer.

**14. FOCUS ON ORGANIZATION**
Proteins, which have diverse functions in a cell, are all polymers of the same kinds of monomers—amino acids. Write a short essay (100–150 words) that discusses how the structure of amino acids allows this one type of polymer to perform so many functions.

**15. SYNTHESIZE YOUR KNOWLEDGE**

Given that the function of egg yolk is to nourish and support the developing chick, explain why egg yolks are so high in fat, protein, and cholesterol.

*For selected answers, see Appendix A.*

# A Tour of the Cell

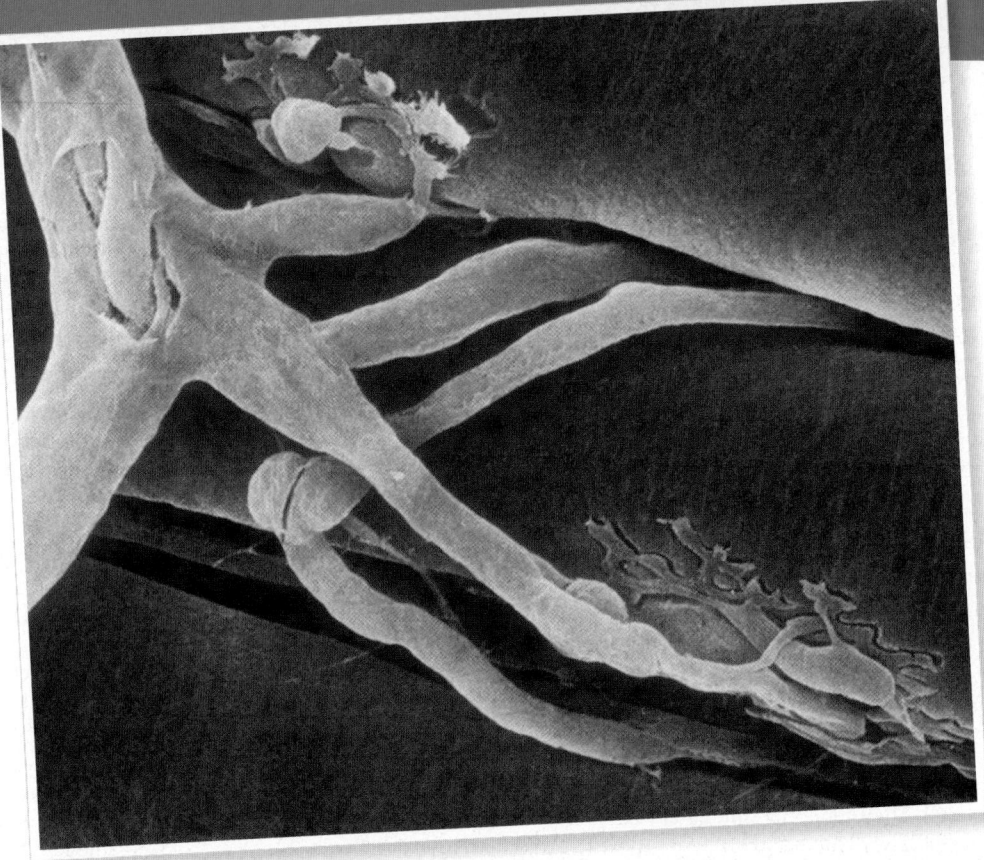

▲ **Figure 4.1** How do your cells help you learn about biology?

## The Fundamental Units of Life

Given the scope of biology, you may wonder sometimes how you will ever learn all the material in this course! The answer involves cells, which are as fundamental to the living systems of biology as the atom is to chemistry. The contraction of muscle cells moves your eyes as you read this sentence. **Figure 4.1** shows extensions from a nerve cell (orange) making contact with muscle cells (red). The words on the page are translated into signals that nerve cells carry to your brain, where they are passed on to other nerve cells. As you study, you are making cell connections like these that solidify memories and permit learning to occur.

All organisms are made of cells. In the hierarchy of biological organization, the cell is the simplest collection of matter that can be alive. Indeed,

40 μm

many forms of life exist as single-celled organisms. (You may be familiar with single-celled eukaryotic organisms that live in pond water, such as paramecia.) Larger, more complex organisms, including plants and animals, are multicellular; their bodies are cooperatives of many kinds of specialized cells that could not survive for long on their own. Even when cells are arranged into higher levels of organization, such as tissues and organs, the cell remains the organism's basic unit of structure and function.

All cells are related by their descent from earlier cells. During the long evolutionary history of life on Earth, cells have been modified in many different ways. But although cells can differ substantially from one another, they share common features. In this chapter, we'll first examine the tools and techniques that allow us to understand cells, and then tour the cell and become acquainted with its components.

◀ *Paramecium caudatum*

# CONCEPT 4.1

## Biologists use microscopes and the tools of biochemistry to study cells

How can cell biologists investigate the inner workings of a cell, usually too small to be seen by the unaided eye? Before we tour the cell, it will be helpful to learn how cells are studied.

### Microscopy

The development of instruments that extend the human senses has gone hand in hand with the advance of science. Microscopes were invented in 1590 and further refined during the 1600s. Cell walls were first seen by Robert Hooke in 1665 as he looked through a microscope at dead cells from the bark of an oak tree. But it took the wonderfully crafted lenses of Antoni van Leeuwenhoek to visualize living cells. Imagine Hooke's awe when he visited van Leeuwenhoek in 1674 and the world of microorganisms—what his host called "very little animalcules"—was revealed to him.

The microscopes first used by Renaissance scientists, as well as the microscopes you are likely to use in the laboratory, are all light microscopes. In a **light microscope (LM)**, visible light is passed through the specimen and then through glass lenses. The lenses refract (bend) the light in such a way that the image of the specimen is magnified as it is projected into the eye or into a camera (see Appendix D).

Three important parameters in microscopy are magnification, resolution, and contrast. *Magnification* is the ratio of an object's image size to its real size. Light microscopes can magnify effectively to about 1,000 times the actual size of the specimen; at greater magnifications, additional details cannot be seen clearly. *Resolution* is a measure of the clarity of the image; it is the minimum distance two points can be separated and still be distinguished as separate points. For example, what appears to the unaided eye as one star in the sky may be resolved as twin stars with a telescope, which has a higher resolving ability than the eye. Similarly, using standard techniques, the light microscope cannot resolve detail finer than about 0.2 micrometer (μm), or 200 nanometers (nm), regardless of the magnification **(Figure 4.2)**. The third parameter, *contrast*, is the difference in brightness between the light and dark areas of an image. Methods for enhancing contrast in light microscopy include staining or labeling cell components to stand out visually. **Figure 4.3** shows some different types of microscopy; study this figure as you read the rest of this section.

Until recently, the resolution barrier prevented cell biologists from using standard light microscopy when studying **organelles**, the membrane-enclosed structures within eukaryotic cells. To see these structures in any detail required the development of a new instrument. In the 1950s, the electron microscope was introduced to biology. Rather than focusing light, the **electron microscope (EM)** focuses a beam

of electrons through the specimen or onto its surface (see Appendix D). Resolution is inversely related to the wavelength of the light (or electrons) a microscope uses for imaging, and electron beams have much shorter wavelengths than visible light. Modern electron microscopes can theoretically achieve a resolution of about 0.002 nm, though in practice they usually cannot resolve structures smaller than about 2 nm across. Still, this is a 100-fold improvement over the standard light microscope.

The **scanning electron microscope (SEM)** is especially useful for detailed study of the topography of a specimen

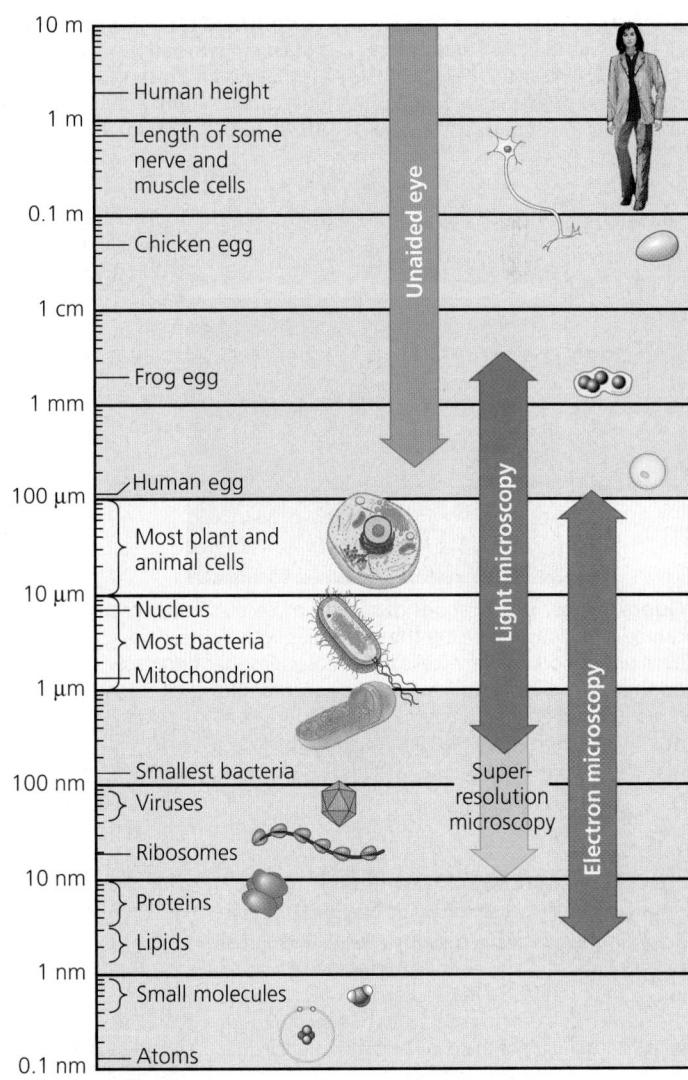

1 centimeter (cm) = $10^{-2}$ meter (m) = 0.4 inch
1 millimeter (mm) = $10^{-3}$ m
1 micrometer (μm) = $10^{-3}$ mm = $10^{-6}$ m
1 nanometer (nm) = $10^{-3}$ μm = $10^{-9}$ m

▲ **Figure 4.2 The size range of cells.** Most cells are between 1 and 100 μm in diameter (yellow region of chart), and their components are even smaller, as are viruses. Notice that the scale along the left side is logarithmic to accommodate the range of sizes shown. Starting at the top of the scale with 10 m and going down, each reference measurement marks a tenfold decrease in diameter or length. For a complete table of the metric system, see Appendix C.

## Light Microscopy (LM)

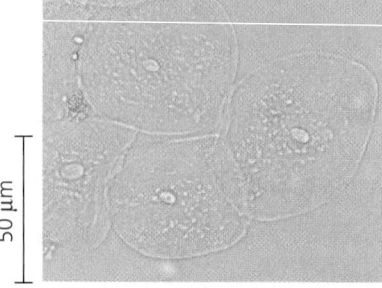  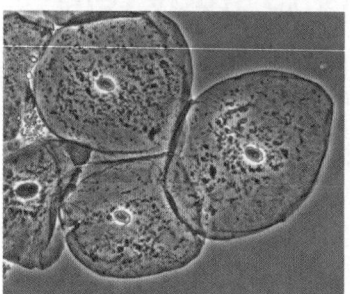 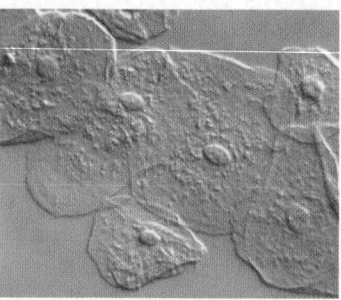

**Brightfield (unstained specimen).** Light passes directly through the specimen. Unless the cell is naturally pigmented or artificially stained, the image has little contrast.

**Brightfield (stained specimen).** Staining with various dyes enhances contrast. Most staining procedures require that cells be fixed (preserved), thereby killing them.

**Phase-contrast.** Variations in density within the specimen are amplified to enhance contrast in unstained cells; this is especially useful for examining living, unpigmented cells.

**Differential-interference contrast (Nomarski).** As in phase-contrast microscopy, optical modifications are used to exaggerate differences in density; the image appears almost 3-D.

The light micrographs above show human cheek epithelial cells; the scale bar pertains to all four micrographs.

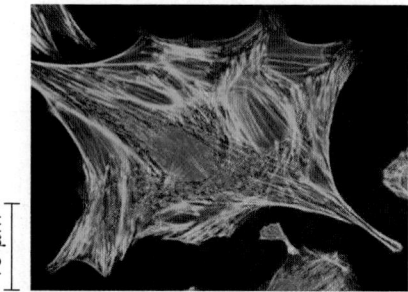 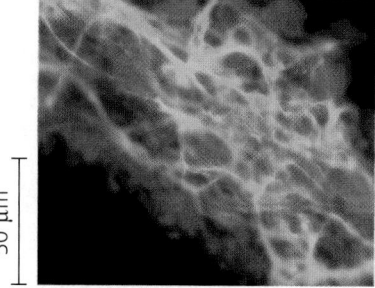 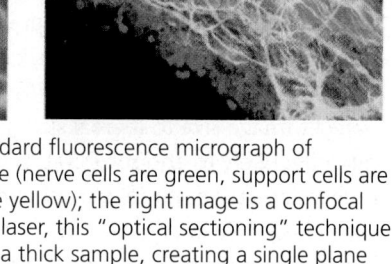

**Fluorescence.** The locations of specific molecules in the cell can be revealed by labeling the molecules with fluorescent dyes or antibodies; some cells have molecules that fluoresce on their own. Fluorescent substances absorb ultraviolet radiation and emit visible light. In this fluorescently labeled uterine cell, nuclear material is blue, organelles called mitochondria are orange, and the cell's "skeleton" is green.

**Confocal.** The left image is a standard fluorescence micrograph of fluorescently labeled nervous tissue (nerve cells are green, support cells are orange, and regions of overlap are yellow); the right image is a confocal image of the same tissue. Using a laser, this "optical sectioning" technique eliminates out-of-focus light from a thick sample, creating a single plane of fluorescence in the image. By capturing sharp images at many different planes, a 3-D reconstruction can be created. The standard image is blurry because out-of-focus light is not excluded.

## Electron Microscopy (EM)

**Scanning electron microscopy (SEM).** Micrographs taken with a scanning electron microscope show a 3-D image of the surface of a specimen. This SEM shows the surface of a cell from a trachea (windpipe) covered with cilia. (Beating of the cilia helps move inhaled debris upward toward the throat.)

Electron micrographs are black and white but are often artificially colorized to highlight particular structures, as has been done with both electron micrographs (SEM and TEM) shown here.

Abbreviations used in figure legends throughout this text:
LM = Light Micrograph
SEM = Scanning Electron Micrograph
TEM = Transmission Electron Micrograph

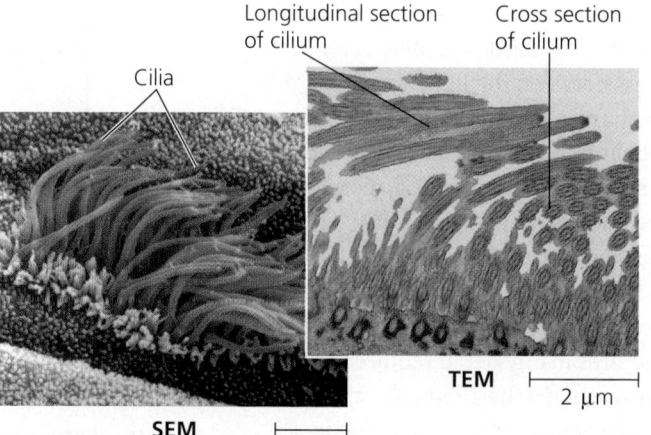

Cilia

Longitudinal section of cilium

Cross section of cilium

SEM

2 μm

TEM

2 μm

**Transmission electron microscopy (TEM).** A transmission electron microscope profiles a thin section of a specimen. This TEM shows a section through a tracheal cell, revealing its internal structure. In preparing the specimen, some cilia were cut along their lengths, creating longitudinal sections, while other cilia were cut straight across, creating cross sections.

(see Figure 4.3). The electron beam scans the surface of the sample, usually coated with a thin film of gold. The beam excites electrons on the surface, and these secondary electrons are detected by a device that translates the pattern of electrons into an electronic signal sent to a video screen. The result is an image of the specimen's surface that appears three-dimensional.

The **transmission electron microscope (TEM)** is used to study the internal structure of cells (see Figure 4.3). The TEM aims an electron beam through a very thin section of the specimen, much as a light microscope aims light through a sample on a slide. For the TEM, the specimen has been stained with atoms of heavy metals, which attach to certain cellular structures, thus enhancing the electron density of some parts of the cell more than others. The electrons passing through the specimen are scattered more in the denser regions, so fewer are transmitted. The image displays the pattern of transmitted electrons. Instead of using glass lenses, both the SEM and TEM use electromagnets as lenses to bend the paths of the electrons, ultimately focusing the image onto a monitor for viewing.

Electron microscopes have revealed many subcellular structures that were impossible to resolve with the light microscope. But the light microscope offers advantages, especially in studying living cells. A disadvantage of electron microscopy is that the methods used to prepare the specimen kill the cells. Specimen preparation for any type of microscopy can introduce artifacts, structural features seen in micrographs that do not exist in the living cell.

In the past several decades, light microscopy has been revitalized by major technical advances. Labeling individual cellular molecules or structures with fluorescent markers has made it possible to see such structures with increasing detail. In addition, confocal and other newer types of fluorescent light microscopy have produced sharper images of three-dimensional tissues and cells. Finally, a group of new techniques and labeling molecules developed in recent years have allowed researchers to "break" the resolution barrier and distinguish subcellular structures as small as 10–20 nm across. As this "super-resolution microscopy" becomes more widespread, the images we see of living cells are proving as awe-inspiring to us as van Leeuwenhoek's were to Robert Hooke 350 years ago.

Microscopes are the most important tools of *cytology*, the study of cell structure. Understanding the function of each structure, however, required the integration of cytology and *biochemistry*, the study of the chemical processes (metabolism) of cells.

## Cell Fractionation

A useful technique for studying cell structure and function is **cell fractionation**. Broken-up cells are placed in a tube that is spun in a centrifuge. The resulting force causes the largest cell components to settle to the bottom of the tube, forming a pellet. The liquid above the pellet is poured into a new tube and centrifuged at a higher speed for a longer time. This process is repeated several times, resulting in a series of pellets that consist of nuclei, mitochondria (and chloroplasts if the cells are from a photosynthetic organism), pieces of membrane, and ribosomes, the smallest components.

Cell fractionation enables researchers to prepare specific cell components in bulk and identify their functions, a task not usually possible with intact cells. For example, in one of the cell fractions resulting from centrifugation, biochemical tests showed the presence of enzymes involved in cellular respiration, while electron microscopy revealed large numbers of the organelles called mitochondria. Together, these data helped biologists determine that mitochondria are the sites of cellular respiration. Biochemistry and cytology thus complement each other in correlating cell function with structure.

### CONCEPT CHECK 4.1

1. How do stains used for light microscopy compare with those used for electron microscopy?
2. **WHAT IF?** Which type of microscope would you use to study (a) the changes in shape of a living white blood cell and (b) the details of surface texture of a hair?

For suggested answers, see Appendix A.

---

### CONCEPT 4.2

# Eukaryotic cells have internal membranes that compartmentalize their functions

Cells—the basic structural and functional units of every organism—are of two distinct types: prokaryotic and eukaryotic. Organisms of the domains Bacteria and Archaea consist of prokaryotic cells. Protists, fungi, animals, and plants all consist of eukaryotic cells. ("Protist" is an informal term referring to a group of mostly unicellular eukaryotes.)

## Comparing Prokaryotic and Eukaryotic Cells

All cells share certain basic features: They are all bounded by a selective barrier, called the *plasma membrane*. Inside all cells is a semifluid, jellylike substance called **cytosol**, in which subcellular components are suspended. All cells contain *chromosomes*, which carry genes in the form of DNA. And all cells have *ribosomes*, tiny complexes that make proteins according to instructions from the genes.

A major difference between prokaryotic and eukaryotic cells is the location of their DNA. In a **eukaryotic cell**, most of the DNA is in an organelle called the *nucleus*, which is bounded by a double membrane (see Figure 4.7). In a **prokaryotic cell**, the DNA is concentrated in a region that is not membrane-enclosed, called the **nucleoid (Figure 4.4)**. *Eukaryotic* means "true nucleus" (from the Greek *eu*, true, and *karyon*, kernel, here referring to the nucleus), and *prokaryotic* means "before nucleus" (from the Greek *pro*, before), reflecting the earlier evolution of prokaryotic cells.

The interior of either type of cell is called the **cytoplasm**; in eukaryotic cells, this term refers only to the region between the nucleus and the plasma membrane. Within the cytoplasm of a eukaryotic cell, suspended in cytosol, are a variety of organelles of specialized form and function. These membrane-bounded structures are absent in prokaryotic cells, another distinction between prokaryotic and eukaryotic cells. However, in spite of the absence of organelles, the prokaryotic cytoplasm is not a formless soup. For example, some prokaryotes contain regions surrounded by proteins (not membranes), within which specific reactions take place.

Eukaryotic cells are generally much larger than prokaryotic cells (see Figure 4.2). Size is a general feature of cell structure that relates to function. The logistics of carrying out cellular metabolism sets limits on cell size. At the lower limit, the smallest cells known are bacteria called mycoplasmas, which have diameters between 0.1 and 1.0 μm. These are perhaps the smallest packages with enough DNA to program metabolism and enough enzymes and other cellular equipment to carry out the activities necessary for a cell to sustain itself and reproduce. Typical bacteria are 1–5 μm in diameter, about ten times the size of mycoplasmas. Eukaryotic cells are typically 10–100 μm in diameter.

Metabolic requirements also impose theoretical upper limits on the size that is practical for a single cell. At the boundary of every cell, the **plasma membrane** functions as a selective barrier that allows passage of enough oxygen, nutrients, and wastes to service the entire cell **(Figure 4.5)**. For each square micrometer of membrane, only a limited amount of a particular substance can cross per second, so the ratio of surface area to volume is critical. As a cell (or any other object) increases in size, its surface area grows proportionately less than its volume. (Area is proportional to a linear dimension squared, whereas volume is proportional to the linear dimension cubed.) Thus, a smaller object has a greater ratio of surface area to volume **(Figure 4.6)**. The **Scientific Skills Exercise** for this chapter gives you a chance to calculate the volumes and surface areas of two actual cells—a mature yeast cell and a cell budding from it. To see different ways organisms maximize the surface area of cells, see Make Connections Figure 26.14.

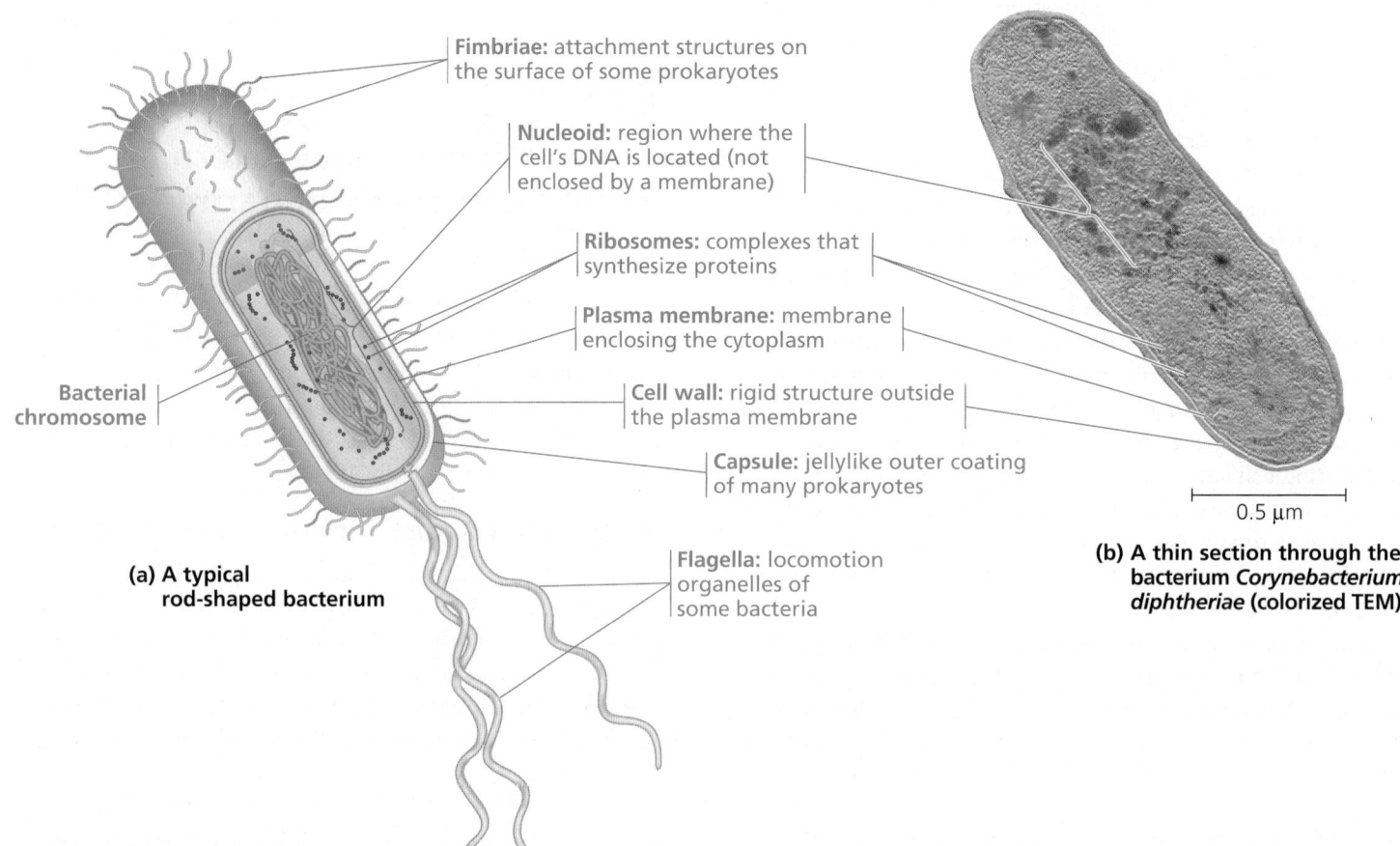

**Fimbriae:** attachment structures on the surface of some prokaryotes

**Nucleoid:** region where the cell's DNA is located (not enclosed by a membrane)

**Ribosomes:** complexes that synthesize proteins

**Plasma membrane:** membrane enclosing the cytoplasm

**Cell wall:** rigid structure outside the plasma membrane

**Capsule:** jellylike outer coating of many prokaryotes

Bacterial chromosome

**(a) A typical rod-shaped bacterium**

**Flagella:** locomotion organelles of some bacteria

0.5 μm

**(b) A thin section through the bacterium *Corynebacterium diphtheriae* (colorized TEM)**

▲ **Figure 4.4  A prokaryotic cell.** Lacking a true nucleus and the other membrane-enclosed organelles of the eukaryotic cell, the prokaryotic cell appears much simpler in internal structure. Prokaryotes include bacteria and archaea; the general cell structure of the two domains is quite similar.

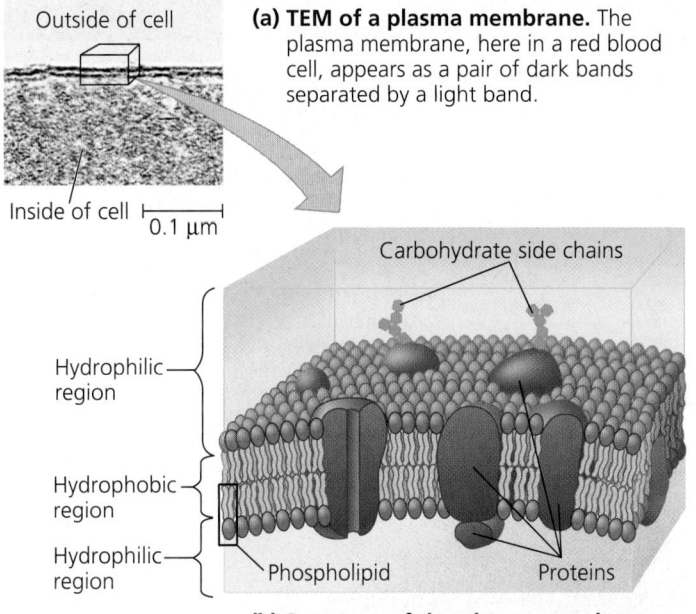

**(a) TEM of a plasma membrane.** The plasma membrane, here in a red blood cell, appears as a pair of dark bands separated by a light band.

Outside of cell

Inside of cell | 0.1 μm

Carbohydrate side chains

Hydrophilic region

Hydrophobic region

Hydrophilic region

Phospholipid

Proteins

**(b) Structure of the plasma membrane**

▲ **Figure 4.5 The plasma membrane.** The plasma membrane and the membranes of organelles consist of a double layer (bilayer) of phospholipids with various proteins attached to or embedded in it. The hydrophobic parts of phospholipids and membrane proteins are found in the interior of the membrane, while the hydrophilic parts are in contact with aqueous solutions on either side. Carbohydrate side chains may be attached to proteins or lipids on the outer surface of the plasma membrane.

**MAKE CONNECTIONS** *Review Figure 3.15 and describe the characteristics of phospholipids that allow them to function as the major components of the plasma membrane.*

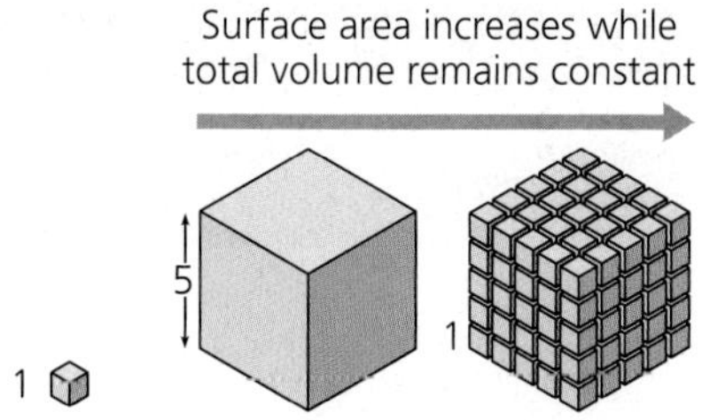

The need for a surface area sufficiently large to accommodate the volume helps explain the microscopic size of most cells and the narrow, elongated shapes of others, such as nerve cells. Larger organisms do not generally have *larger* cells than smaller organisms—they simply have *more* cells. A sufficiently high ratio of surface area to volume is especially important in cells that exchange a lot of material with their surroundings, such as intestinal cells. Such cells may have many long, thin projections from their surface called *microvilli*, which increase surface area without an appreciable increase in volume.

The evolutionary relationships between prokaryotic and eukaryotic cells will be discussed later in this chapter, and prokaryotic cells will be described in detail elsewhere (see Chapter 24). Most of the discussion of cell structure that follows in this chapter applies to eukaryotic cells.

## A Panoramic View of the Eukaryotic Cell

In addition to the plasma membrane at its outer surface, a eukaryotic cell has extensive, elaborately arranged internal membranes that divide the cell into compartments—the organelles mentioned earlier. The cell's compartments provide different local environments that support specific metabolic functions,

Surface area increases while total volume remains constant

| | 1 | 5 | 1 |
|---|---|---|---|
| **Total surface area** [sum of the surface areas (height × width) of all box sides × number of boxes] | 6 | 150 | 750 |
| **Total volume** [height × width × length × number of boxes] | 1 | 125 | 125 |
| **Surface-to-volume (S-to-V) ratio** [surface area ÷ volume] | 6 | 1.2 | 6 |

▲ **Figure 4.6 Geometric relationships between surface area and volume.** In this diagram, cells are represented as boxes. Using arbitrary units of length, we can calculate the cell's surface area (in square units, or units$^2$), volume (in cubic units, or units$^3$), and ratio of surface area to volume. A high surface-to-volume ratio facilitates the exchange of materials between a cell and its environment.

so incompatible processes can go on simultaneously inside a single cell. The plasma membrane and organelle membranes also participate directly in the cell's metabolism, because many enzymes are built right into the membranes.

The basic fabric of most biological membranes is a double layer of phospholipids and other lipids. Embedded in this lipid bilayer or attached to its surfaces are diverse proteins (see Figure 4.5). However, each type of membrane has a unique composition of lipids and proteins suited to that membrane's specific functions. For example, enzymes embedded in the membranes of the organelles called mitochondria function in cellular respiration. Because membranes are so fundamental to the organization of the cell, Chapter 5 will discuss them in more detail.

Before continuing with this chapter, examine the eukaryotic cells in **Figure 4.7**. The generalized diagrams of an animal cell and a plant cell introduce the various organelles and show the key differences between animal and plant cells. The micrographs at the bottom of the figure give you a glimpse of cells from different types of eukaryotic organisms.

**CONCEPT CHECK 4.2**

1. Briefly describe the structure and function of the nucleus, the mitochondrion, the chloroplast, and the endoplasmic reticulum.

2. **WHAT IF?** Imagine an elongated cell (such as a nerve cell) that measures 125 × 1 × 1 arbitrary units. Predict how its surface-to-volume ratio would compare with those in Figure 4.6. Then calculate the ratio and check your prediction.

**For suggested answers, see Appendix A.**

## Animal Cell (cutaway view of generalized cell)

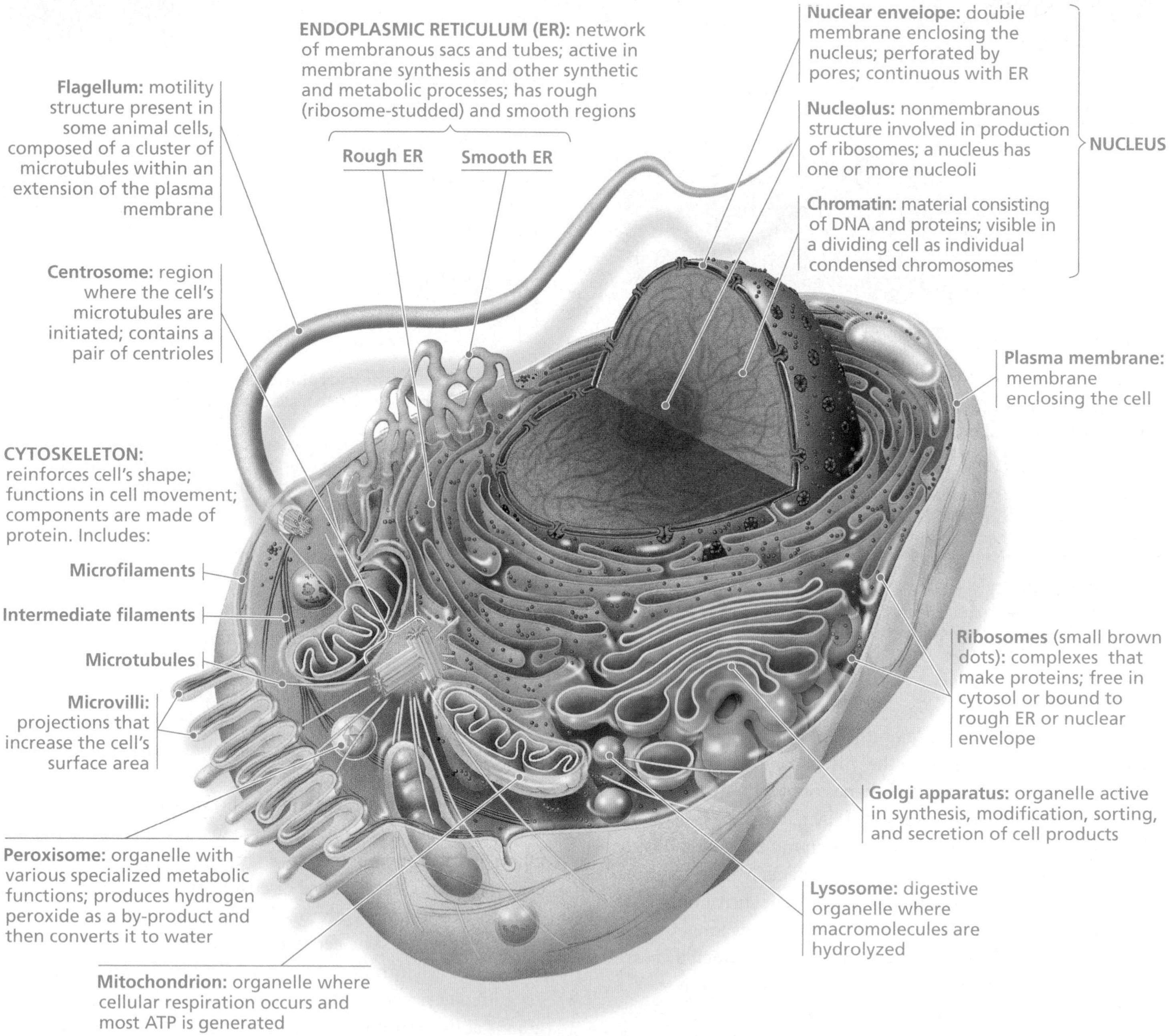

**ENDOPLASMIC RETICULUM (ER):** network of membranous sacs and tubes; active in membrane synthesis and other synthetic and metabolic processes; has rough (ribosome-studded) and smooth regions

**Rough ER**     **Smooth ER**

**Flagellum:** motility structure present in some animal cells, composed of a cluster of microtubules within an extension of the plasma membrane

**Centrosome:** region where the cell's microtubules are initiated; contains a pair of centrioles

**CYTOSKELETON:** reinforces cell's shape; functions in cell movement; components are made of protein. Includes:

**Microfilaments**

**Intermediate filaments**

**Microtubules**

**Microvilli:** projections that increase the cell's surface area

**Peroxisome:** organelle with various specialized metabolic functions; produces hydrogen peroxide as a by-product and then converts it to water

**Mitochondrion:** organelle where cellular respiration occurs and most ATP is generated

**Nuclear envelope:** double membrane enclosing the nucleus; perforated by pores; continuous with ER

**Nucleolus:** nonmembranous structure involved in production of ribosomes; a nucleus has one or more nucleoli

**NUCLEUS**

**Chromatin:** material consisting of DNA and proteins; visible in a dividing cell as individual condensed chromosomes

**Plasma membrane:** membrane enclosing the cell

**Ribosomes** (small brown dots): complexes that make proteins; free in cytosol or bound to rough ER or nuclear envelope

**Golgi apparatus:** organelle active in synthesis, modification, sorting, and secretion of cell products

**Lysosome:** digestive organelle where macromolecules are hydrolyzed

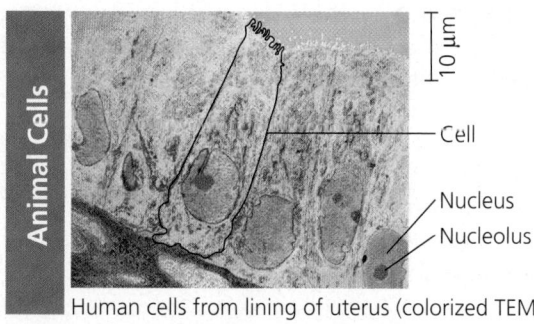

**Animal Cells**

10 μm

Cell

Nucleus

Nucleolus

Human cells from lining of uterus (colorized TEM)

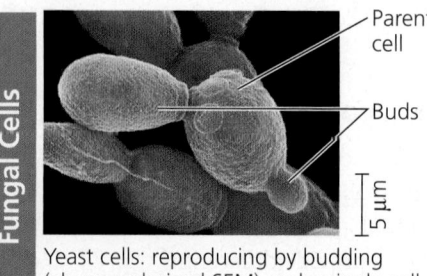

**Fungal Cells**

Parent cell

Buds

5 μm

Yeast cells: reproducing by budding (above, colorized SEM) and a single cell (right, colorized TEM)

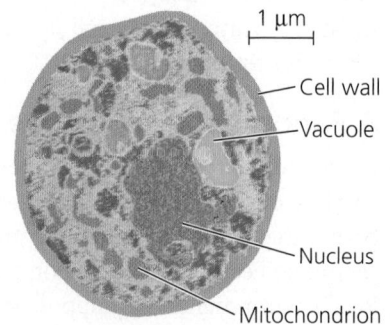

1 μm

Cell wall

Vacuole

Nucleus

Mitochondrion

## Plant Cell (cutaway view of generalized cell)

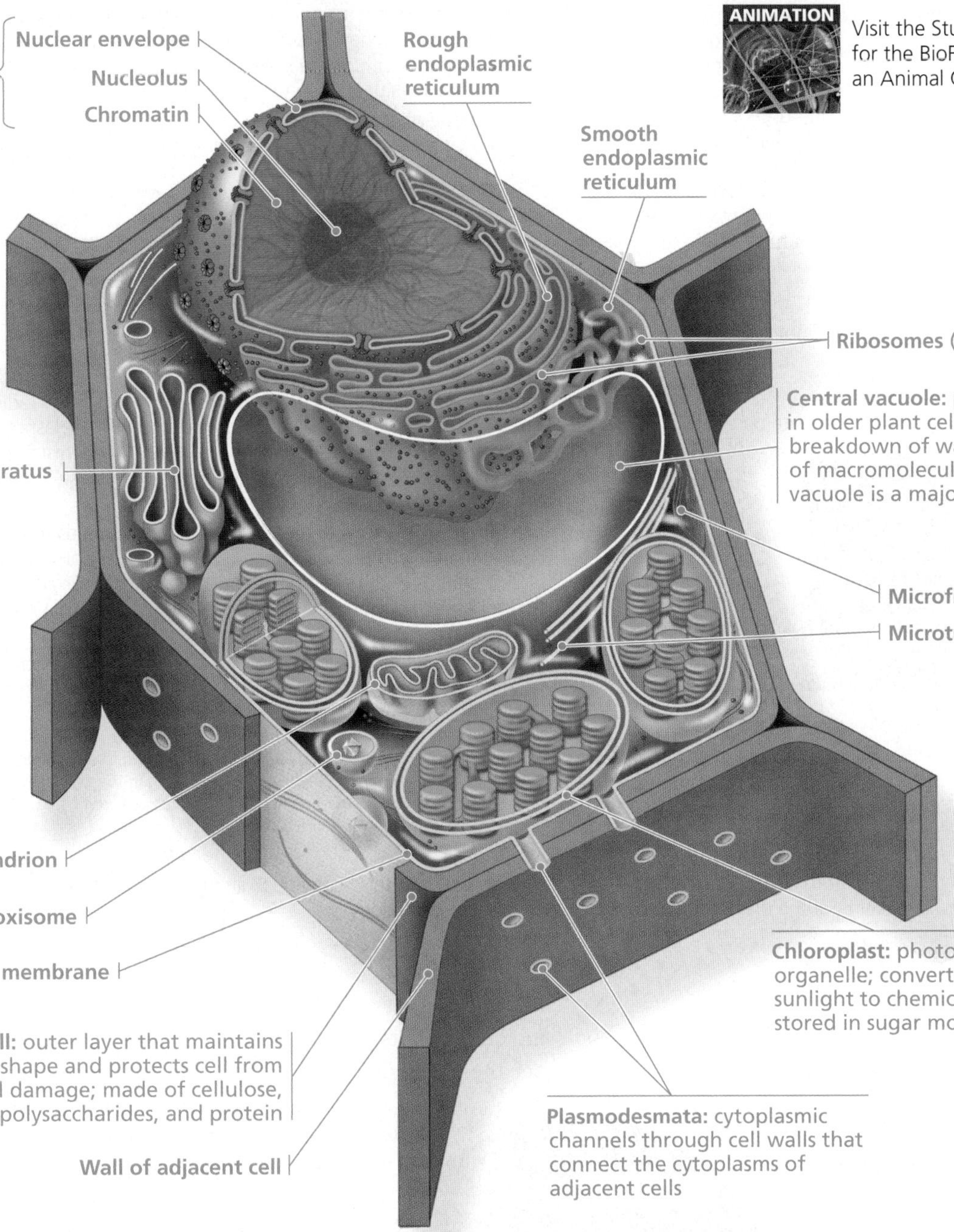

**NUCLEUS** {
- Nuclear envelope
- Nucleolus
- Chromatin

Rough endoplasmic reticulum

Smooth endoplasmic reticulum

**ANIMATION** Visit the Study Area in **MasteringBiology** for the BioFlix® 3-D Animations Tour of an Animal Cell and Tour of a Plant Cell.

**Ribosomes** (small brown dots)

**Central vacuole:** prominent organelle in older plant cells; functions include storage, breakdown of waste products, and hydrolysis of macromolecules; enlargement of the vacuole is a major mechanism of plant growth

Golgi apparatus

**Microfilaments** } **CYTOSKELETON**
**Microtubules**

Mitochondrion

Peroxisome

Plasma membrane

**Chloroplast:** photosynthetic organelle; converts energy of sunlight to chemical energy stored in sugar molecules

**Cell wall:** outer layer that maintains cell's shape and protects cell from mechanical damage; made of cellulose, other polysaccharides, and protein

Wall of adjacent cell

**Plasmodesmata:** cytoplasmic channels through cell walls that connect the cytoplasms of adjacent cells

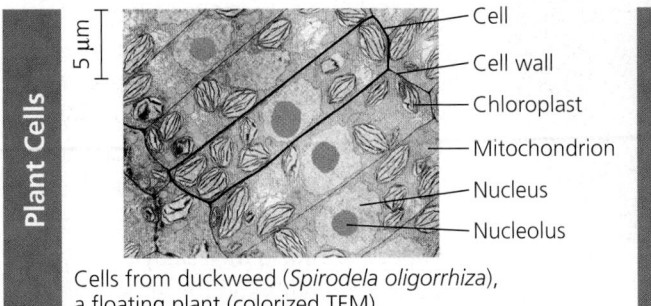

**Plant Cells**

5 μm

- Cell
- Cell wall
- Chloroplast
- Mitochondrion
- Nucleus
- Nucleolus

Cells from duckweed (*Spirodela oligorrhiza*), a floating plant (colorized TEM)

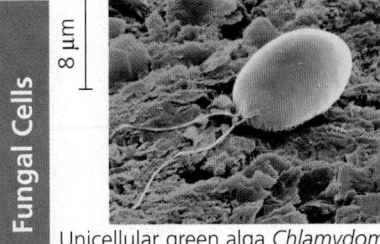

**Fungal Cells**

8 μm

Unicellular green alga *Chlamydomonas* (above, colorized SEM; right, colorized TEM)

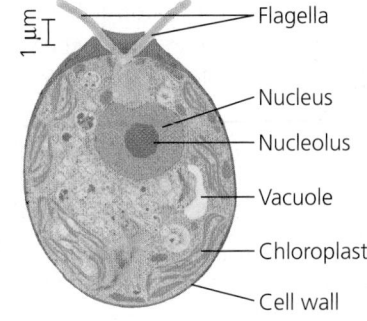

1 μm

- Flagella
- Nucleus
- Nucleolus
- Vacuole
- Chloroplast
- Cell wall

## Using a Scale Bar to Calculate Volume and Surface Area of a Cell

**How Much New Cytoplasm and Plasma Membrane Are Made by a Growing Yeast Cell?** The unicellular yeast *Saccharomyces cerevisiae* divides by budding off a small new cell that then grows to full size (see the yeast cells at the bottom of Figure 4.7). During its growth, the new cell synthesizes new cytoplasm, which increases its volume, and new plasma membrane, which increases its surface area.

In this exercise, you will use a scale bar to determine the sizes of a mature parent yeast cell and a cell budding from it. You will then calculate the volume and surface area of each cell. You will use your calculations to determine how much cytoplasm and plasma membrane the new cell needs to synthesize to grow to full size.

**How the Experiment Was Done** Yeast cells were grown under conditions that promoted division by budding. The cells were then viewed with a differential interference contrast light microscope and photographed.

**Data from the Experiment** This light micrograph shows a budding yeast cell about to be released from the mature parent cell:

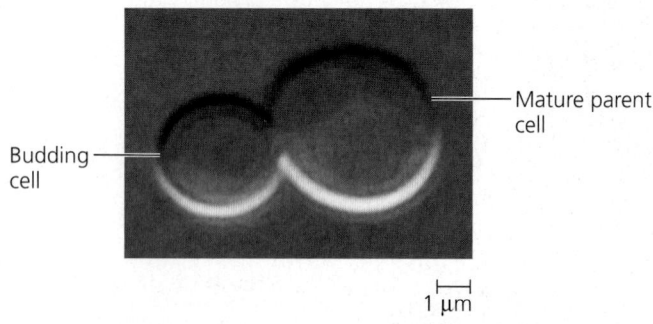

Budding cell

Mature parent cell

1 μm

INTERPRET THE DATA

1. Examine the micrograph of the yeast cells. The scale bar under the photo is labeled 1 μm. The scale bar works in the same way as a scale on a map, where, for example, 1 inch equals 1 mile. In this case the bar represents one thousandth of a millimeter. Using the scale bar as a basic unit, determine the diameter of the

mature parent cell and the new cell. Start by measuring the scale bar and the diameter of each cell. The units you use are irrelevant, but working in millimeters is convenient. Divide each diameter by the length of the scale bar and then multiply by the scale bar's length value to give you the diameter in micrometers.

2. The shape of a yeast cell can be approximated by a sphere.
   (a) Calculate the volume of each cell using the formula for the volume of a sphere:

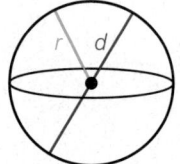

$$V = \frac{4}{3}\pi r^3$$

   Note that π (the Greek letter pi) is a constant with an approximate value of 3.14, *d* stands for diameter, and *r* stands for radius, which is half the diameter. (b) What volume of new cytoplasm will the new cell have to synthesize as it matures? To determine this, calculate the difference between the volume of the full-sized cell and the volume of the new cell.

3. As the new cell grows, its plasma membrane needs to expand to contain the increased volume of the cell. (a) Calculate the surface area of each cell using the formula for the surface area of a sphere: $A = 4\pi r^2$. (b) How much area of new plasma membrane will the new cell have to synthesize as it matures?

4. When the new cell matures, it will be approximately how many times greater in volume and how many times greater in surface area than its current size?

**Micrograph from** Kelly Tatchell, using yeast cells grown for experiments described in L. Kozubowski et al., Role of the septin ring in the asymmetric localization of proteins at the mother-bud neck in *Saccharomyces cerevisiae*, *Molecular Biology of the Cell* 16:3455–3466 (2005).

(MB) A version of this Scientific Skills Exercise can be assigned in MasteringBiology.

---

CONCEPT 4.3

# The eukaryotic cell's genetic instructions are housed in the nucleus and carried out by the ribosomes

On the first stop of our detailed tour of the eukaryotic cell, let's look at two cellular components involved in the genetic control of the cell: the nucleus, which houses most of the cell's DNA, and the ribosomes, which use information from the DNA to make proteins.

## The Nucleus: Information Central

The **nucleus** contains most of the genes in the eukaryotic cell. (Some genes are located in mitochondria

Nucleus

5 μm

and chloroplasts.) It is usually the most conspicuous organelle (see the purple structure in the cell in the fluorescence micrograph), averaging about 5 μm in diameter. The **nuclear envelope** encloses the nucleus **(Figure 4.8)**, separating its contents from the cytoplasm.

The nuclear envelope is a *double* membrane. The two membranes, each a lipid bilayer with associated proteins, are separated by a space of 20–40 nm. The envelope is perforated by pore structures that are about 100 nm in diameter. At the lip of each pore, the inner and outer membranes of the nuclear envelope are continuous. An intricate protein structure called a *pore complex* lines each pore and plays an important role in the cell by regulating the entry and exit of proteins and RNAs, as well as

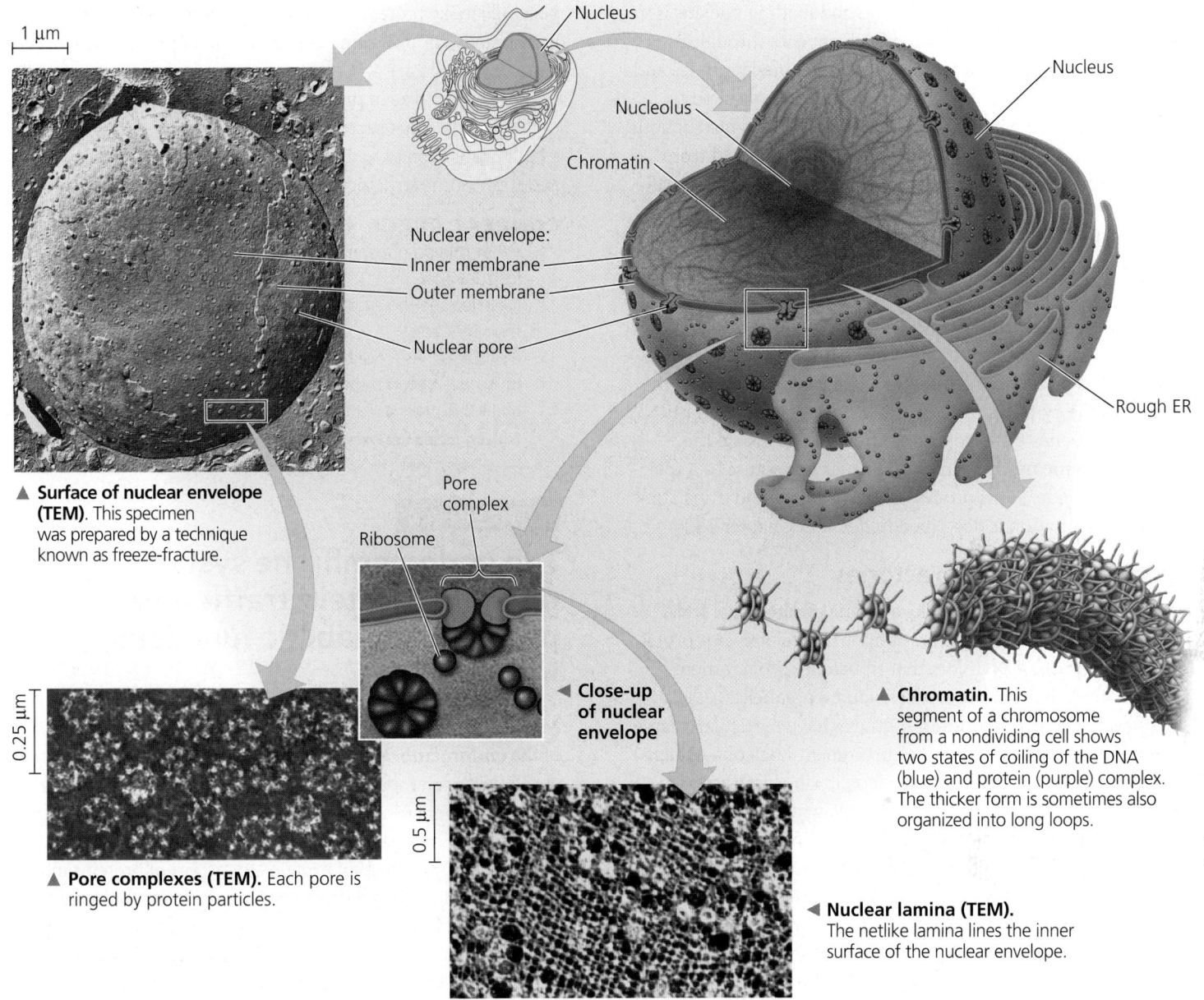

**1 μm**

Nucleus

Nucleolus

Chromatin

Nuclear envelope:
- Inner membrane
- Outer membrane

Nuclear pore

Nucleus

Rough ER

▲ **Surface of nuclear envelope (TEM).** This specimen was prepared by a technique known as freeze-fracture.

Pore complex

Ribosome

**0.25 μm**

▲ **Pore complexes (TEM).** Each pore is ringed by protein particles.

◄ **Close-up of nuclear envelope**

**0.5 μm**

◄ **Nuclear lamina (TEM).** The netlike lamina lines the inner surface of the nuclear envelope.

▲ **Chromatin.** This segment of a chromosome from a nondividing cell shows two states of coiling of the DNA (blue) and protein (purple) complex. The thicker form is sometimes also organized into long loops.

▲ **Figure 4.8 The nucleus and its envelope.** Within the nucleus are the chromosomes, which appear as a mass of chromatin (DNA and associated proteins), and one or more nucleoli (singular, *nucleolus*), which function in ribosome synthesis. The nuclear envelope, which consists of two membranes separated by a narrow space, is perforated with pores and lined by the nuclear lamina.

**MAKE CONNECTIONS** *Since the chromosomes contain the genetic material and reside in the nucleus, how does the rest of the cell get access to the information they carry? See Figure 3.26.*

large complexes of macromolecules. Except at the pores, the nuclear side of the envelope is lined by the **nuclear lamina**, a netlike array of protein filaments that maintains the shape of the nucleus by mechanically supporting the nuclear envelope.

Within the nucleus, the DNA is organized into discrete units called **chromosomes**, structures that carry the genetic information. Each chromosome contains one long DNA molecule associated with many proteins. Some of the proteins help coil the DNA molecule of each chromosome, reducing its length and allowing it to fit into the nucleus. The complex of DNA and proteins making up chromosomes is called

**chromatin**. When a cell is not dividing, stained chromatin appears as a diffuse mass in micrographs, and the chromosomes cannot be distinguished from one another, even though discrete chromosomes are present. As a cell prepares to divide, however, the chromosomes coil (condense) further, becoming thick enough to be distinguished under a microscope as separate structures. Each eukaryotic species has a characteristic number of chromosomes. For example, a typical human cell has 46 chromosomes in its nucleus; the exceptions are the sex cells (eggs and sperm), which have only 23 chromosomes in humans.

A prominent structure within the nondividing nucleus is the **nucleolus** (plural, *nucleoli*), which appears through the electron microscope as a mass of densely stained granules and fibers adjoining part of the chromatin. Here a type of RNA called *ribosomal RNA* (*rRNA*) is synthesized from instructions in the DNA. Also in the nucleolus, proteins imported from the cytoplasm are assembled with rRNA into large and small subunits of ribosomes. These subunits then exit the nucleus through the nuclear pores to the cytoplasm, where a large and a small subunit can assemble into a ribosome. Sometimes there are two or more nucleoli.

As we saw in Figure 3.26, the nucleus directs protein synthesis by synthesizing messenger RNA (mRNA) according to instructions provided by the DNA. The mRNA is then transported to the cytoplasm via the nuclear pores. Once an mRNA molecule reaches the cytoplasm, ribosomes translate the mRNA's genetic message into the primary structure of a specific polypeptide. (This process of transcribing and translating genetic information is described in detail in Chapter 14.)

## Ribosomes: Protein Factories

**Ribosomes**, which are complexes made of ribosomal RNA and protein, are the cellular components that carry out protein synthesis **(Figure 4.9)**. (Note that ribosomes are not membrane bound and thus are not considered organelles.) Cells that have high rates of protein synthesis have particularly large numbers of ribosomes as well as prominent nucleoli—which makes sense, given the role of nucleoli in ribosome assembly. For example, a human pancreas cell, which makes many digestive enzymes, has a few million ribosomes.

Ribosomes build proteins in two cytoplasmic locales. At any given time, *free ribosomes* are suspended in the cytosol, while *bound ribosomes* are attached to the outside of the endoplasmic reticulum or nuclear envelope (see Figure 4.9). Bound and free ribosomes are structurally identical, and ribosomes can alternate between the two roles. Most of the proteins made on free ribosomes function within the cytosol; examples are enzymes that catalyze the first steps of sugar breakdown. Bound ribosomes

generally make proteins that are destined for insertion into membranes, for packaging within certain organelles such as lysosomes (see Figure 4.7), or for export from the cell (secretion). Cells that specialize in protein secretion—for instance, the cells of the pancreas that secrete digestive enzymes—frequently have a high proportion of bound ribosomes. (You will learn more about ribosome structure and function in Concept 14.4.)

### CONCEPT CHECK 4.3

1. What role do ribosomes play in carrying out genetic instructions?
2. Describe the molecular composition of nucleoli, and explain their function.
3. **WHAT IF?** As a cell begins the process of dividing, its chromosomes become shorter, thicker, and individually visible in an LM. Explain what is happening at the molecular level.

For suggested answers, see Appendix A.

---

CONCEPT 4.4

# The endomembrane system regulates protein traffic and performs metabolic functions in the cell

Many of the different membranes of the eukaryotic cell are part of the **endomembrane system**, which includes the nuclear envelope, the endoplasmic reticulum, the Golgi apparatus, lysosomes, various kinds of vesicles and vacuoles, and the plasma membrane. This system carries out a variety of tasks in the cell, including synthesis of proteins, transport of proteins into membranes and organelles or out of the cell, metabolism and movement of lipids, and detoxification of poisons. The membranes of this system are related either through direct physical continuity or by the transfer of membrane segments as tiny **vesicles** (sacs made of membrane). Despite these relationships, the various membranes are not identical in structure and function. Moreover, the thickness, molecular composition, and types

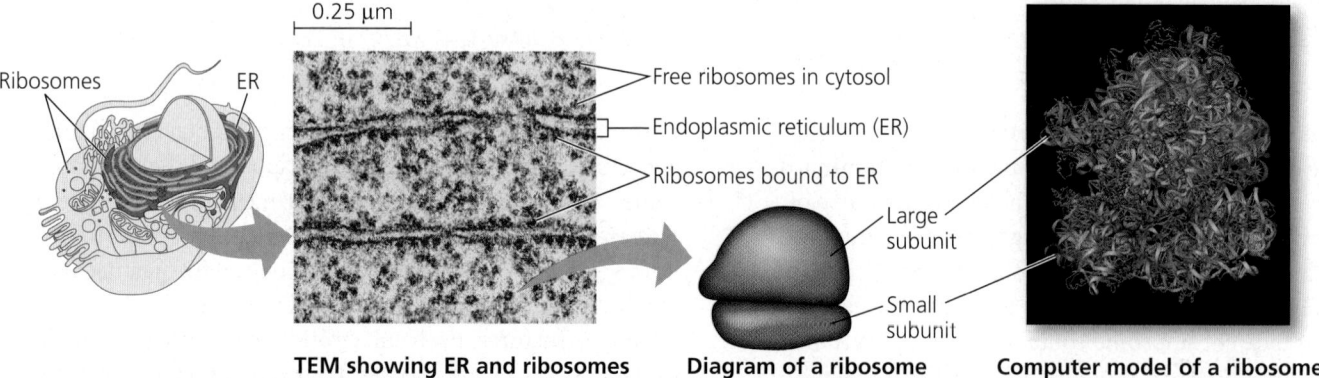

0.25 μm

Ribosomes ER

Free ribosomes in cytosol

Endoplasmic reticulum (ER)

Ribosomes bound to ER

Large subunit

Small subunit

**TEM showing ER and ribosomes**   **Diagram of a ribosome**   **Computer model of a ribosome**

▲ **Figure 4.9 Ribosomes.** This electron micrograph of part of a pancreas cell shows both free and bound ribosomes. The simplified diagram and computer model show the two subunits of a ribosome.

**DRAW IT** *After you have read the section on ribosomes, circle a ribosome in the micrograph that might be making a protein that will be secreted.*

of chemical reactions carried out in a given membrane are not fixed, but may be modified several times during the membrane's life. Having already discussed the nuclear envelope, we will now focus on the endoplasmic reticulum and the other endomembranes to which the endoplasmic reticulum gives rise.

## The Endoplasmic Reticulum: Biosynthetic Factory

The **endoplasmic reticulum (ER)** is such an extensive network of membranes that it accounts for more than half the total membrane in many eukaryotic cells. (The word *endoplasmic* means "within the cytoplasm," and *reticulum* is Latin for "little net.") The ER consists of a network of membranous tubules and sacs called cisternae (from the Latin *cisterna*, a reservoir for a liquid). The ER membrane separates the internal compartment of the ER, called the ER lumen (cavity) or cisternal space, from the cytosol. And because the ER membrane is continuous with the nuclear envelope, the space between the two membranes of the envelope is continuous with the lumen of the ER **(Figure 4.10)**.

There are two distinct, though connected, regions of the ER that differ in structure and function: smooth ER and rough ER. **Smooth ER** is so named because its outer surface lacks ribosomes. **Rough ER** is studded with ribosomes on the outer surface of the membrane and thus appears rough through the electron microscope. As already mentioned, ribosomes are also attached to the cytoplasmic side of the nuclear envelope's outer membrane, which is continuous with rough ER.

### Functions of Smooth ER

The smooth ER functions in diverse metabolic processes, which vary with cell type. These processes include synthesis of lipids, metabolism of carbohydrates, detoxification of drugs and poisons, and storage of calcium ions.

Enzymes of the smooth ER are important in the synthesis of lipids, including oils, steroids, and new membrane phospholipids. Among the steroids produced by the smooth ER in animal cells are the sex hormones of vertebrates and the various steroid hormones secreted by the adrenal glands. The cells that synthesize and secrete these hormones—in the testes and ovaries, for example—are rich in smooth ER, a structural feature that fits the function of these cells.

Other enzymes of the smooth ER help detoxify drugs and poisons, especially in liver cells. Detoxification usually involves adding hydroxyl groups to drug molecules, making them more soluble and easier to flush from the body. The sedative phenobarbital and other barbiturates are examples of drugs metabolized in this manner by smooth ER in liver cells. In fact, barbiturates, alcohol, and many other drugs induce the proliferation of smooth ER and its associated detoxification enzymes, thus increasing the rate of detoxification. This, in turn, increases tolerance to the drugs, meaning that higher doses are required to achieve a particular effect, such as sedation. Also, because some of the detoxification enzymes have relatively broad action, the proliferation of smooth ER in response to

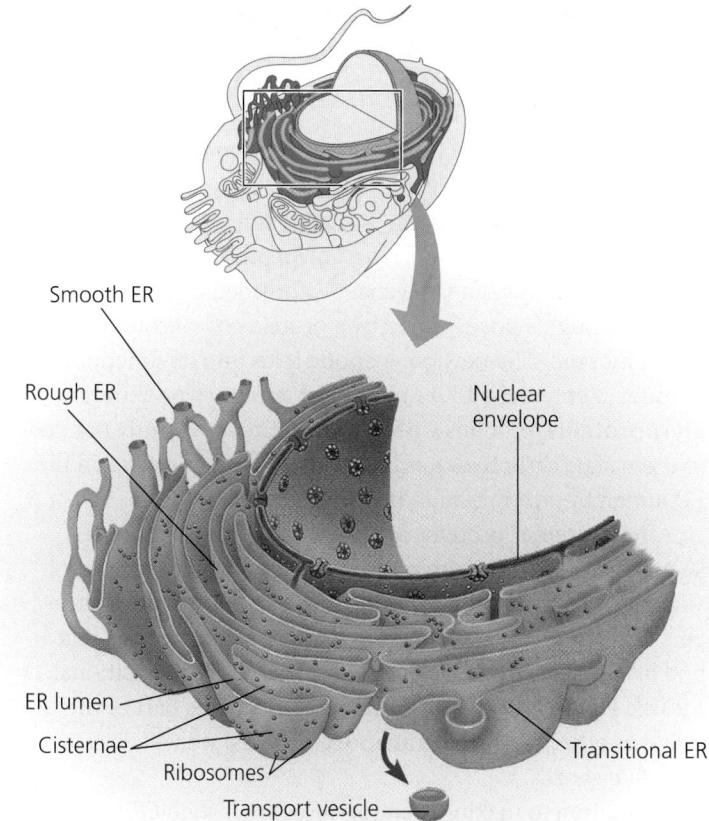

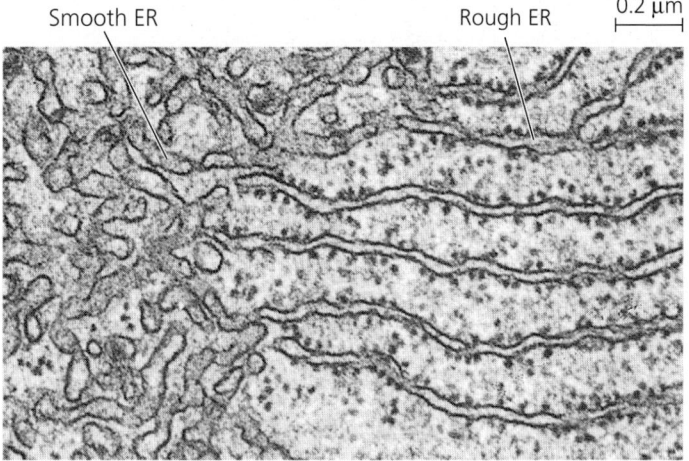

▲ **Figure 4.10 Endoplasmic reticulum (ER).** A membranous system of interconnected tubules and flattened sacs called cisternae, the ER is also continuous with the nuclear envelope, as shown in the cutaway diagram at the top. The membrane of the ER encloses a continuous compartment called the ER lumen (or cisternal space). Rough ER, which is studded on its outer surface with ribosomes, can be distinguished from smooth ER in the electron micrograph (TEM). Transport vesicles bud off from a region of the rough ER called transitional ER and travel to the Golgi apparatus and other destinations.

one drug can increase the need for higher dosages of other drugs as well. Barbiturate abuse, for example, can decrease the effectiveness of certain antibiotics and other useful drugs.

The smooth ER also stores calcium ions. In muscle cells, for example, the smooth ER membrane pumps calcium ions from the cytosol into the ER lumen. When a muscle cell is stimulated by a nerve impulse, calcium ions rush back across the ER membrane into the cytosol and trigger contraction of the muscle cell.

### Functions of Rough ER

Many cells secrete proteins that are produced by ribosomes attached to rough ER. For example, certain pancreatic cells synthesize the protein insulin in the ER and secrete this hormone into the bloodstream. As a polypeptide chain grows from a bound ribosome, the chain is threaded into the ER lumen through a pore formed by a protein complex in the ER membrane. The new polypeptide folds into its functional shape as it enters the ER lumen. Most secretory proteins are **glycoproteins**, proteins with carbohydrates covalently bonded to them. The carbohydrates are attached to the proteins in the ER lumen by enzymes built into the ER membrane.

After secretory proteins are formed, the ER membrane keeps them separate from proteins that remain in the cytosol, which are produced by free ribosomes. Secretory proteins depart from the ER wrapped in the membranes of vesicles that bud like bubbles from a specialized region called transitional ER (see Figure 4.10). Vesicles in transit from one part of the cell to another are called **transport vesicles**; we will discuss their fate shortly.

In addition to making secretory proteins, rough ER is a membrane factory for the cell; it grows in place by adding membrane proteins and phospholipids to its own membrane. As polypeptides destined to be membrane proteins grow from the ribosomes, they are inserted into the ER membrane itself and anchored there by their hydrophobic portions. Like the smooth ER, the rough ER also makes membrane phospholipids; enzymes built into the ER membrane assemble phospholipids from precursors in the cytosol. The ER membrane expands, and portions of it are transferred in the form of transport vesicles to other components of the endomembrane system.

## The Golgi Apparatus: Shipping and Receiving Center

After leaving the ER, many transport vesicles travel to the **Golgi apparatus**. We can think of the Golgi as a warehouse for receiving, sorting, shipping, and even some manufacturing. Here, products of the ER, such as proteins, are modified and stored and then sent to other destinations. Not surprisingly, the Golgi apparatus is especially extensive in cells specialized for secretion.

The Golgi apparatus consists of flattened membranous sacs—cisternae—looking like a stack of pita bread **(Figure 4.11)**. A cell may have many, even hundreds, of these stacks. The membrane of each cisterna in a stack separates its internal space from the cytosol. Vesicles concentrated in the vicinity of the Golgi apparatus are engaged in the transfer of material between parts of the Golgi and other structures.

▼ **Figure 4.11 The Golgi apparatus.** The Golgi apparatus consists of stacks of flattened sacs, or cisternae, which (unlike ER cisternae) are not physically connected, as you can see in the cutaway diagram. A Golgi stack receives and dispatches transport vesicles and the products they contain. A Golgi stack has a structural and functional directionality, with a *cis* face that receives vesicles containing ER products and a *trans* face that dispatches vesicles. The cisternal maturation model proposes that the Golgi cisternae themselves "mature," moving from the *cis* to the *trans* face while carrying some proteins along. In addition, some vesicles recycle enzymes that had been carried forward in moving cisternae, transporting them "backward" to a less mature region where their functions are needed.

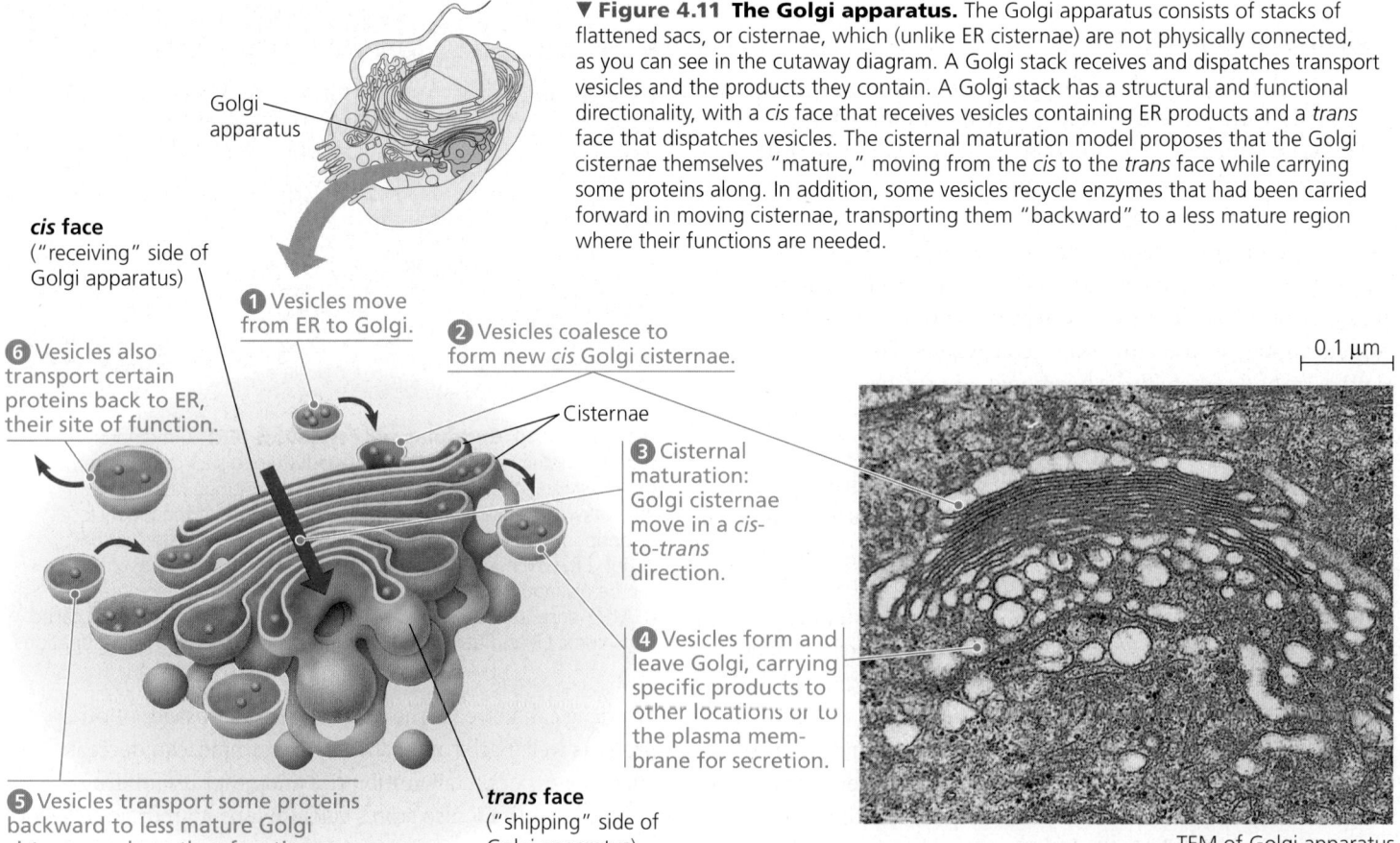

Golgi apparatus

**cis face**
("receiving" side of Golgi apparatus)

❻ Vesicles also transport certain proteins back to ER, their site of function.

❶ Vesicles move from ER to Golgi.

❷ Vesicles coalesce to form new *cis* Golgi cisternae.

Cisternae

❸ Cisternal maturation: Golgi cisternae move in a *cis*-to-*trans* direction.

❹ Vesicles form and leave Golgi, carrying specific products to other locations or to the plasma membrane for secretion.

❺ Vesicles transport some proteins backward to less mature Golgi cisternae, where they function.

**trans face**
("shipping" side of Golgi apparatus)

0.1 μm

TEM of Golgi apparatus

A Golgi stack has a distinct structural directionality, with the membranes of cisternae on opposite sides of the stack differing in thickness and molecular composition. The two sides of a Golgi stack are referred to as the *cis* face and the *trans* face; these act, respectively, as the receiving and shipping departments of the Golgi apparatus. The term *cis* means "on the same side," and the *cis* face is usually located near the ER. Transport vesicles move material from the ER to the Golgi apparatus. A vesicle that buds from the ER can add its membrane and the contents of its lumen to the *cis* face by fusing with a Golgi membrane. The *trans* face ("on the opposite side") gives rise to vesicles that pinch off and travel to other sites.

Products of the endoplasmic reticulum are usually modified during their transit from the *cis* region to the *trans* region of the Golgi apparatus. For example, glycoproteins formed in the ER have their carbohydrates modified, first in the ER itself, and then as they pass through the Golgi. The Golgi removes some sugar monomers and substitutes others, producing a large variety of carbohydrates. Membrane phospholipids may also be altered in the Golgi.

In addition to its finishing work, the Golgi apparatus also manufactures some macromolecules. Many polysaccharides secreted by cells are Golgi products. For example, pectins and certain other noncellulose polysaccharides are made in the Golgi of plant cells and then incorporated along with cellulose into their cell walls. Like secretory proteins, nonprotein Golgi products that will be secreted depart from the *trans* face of the Golgi inside transport vesicles that eventually fuse with the plasma membrane.

The Golgi manufactures and refines its products in stages, with different cisternae containing unique teams of enzymes. Until recently, biologists viewed the Golgi as a static structure, with products in various stages of processing transferred from one cisterna to the next by vesicles. While this may occur, research from several labs has given rise to a new model of the Golgi as a more dynamic structure. According to the *cisternal maturation model*, the cisternae of the Golgi actually progress forward from the *cis* to the *trans* face, carrying and modifying their cargo as they move. Figure 4.11 shows the details of this model.

Before a Golgi stack dispatches its products by budding vesicles from the *trans* face, it sorts these products and targets them for various parts of the cell. Molecular identification tags, such as phosphate groups added to the Golgi products, aid in sorting by acting like zip codes on mailing labels. Finally, transport vesicles budded from the Golgi may have external molecules on their membranes that recognize "docking sites" on the surface of specific organelles or on the plasma membrane, thus targeting the vesicles appropriately.

## Lysosomes: Digestive Compartments

A **lysosome** is a membranous sac of hydrolytic enzymes that many eukaryotic cells use to digest (hydrolyze) macromolecules

**(Figure 4.12)**. Lysosomal enzymes work best in the acidic environment found in lysosomes. If a lysosome breaks open or leaks its contents, the released enzymes are not very active because the cytosol has a near-neutral pH. However, excessive leakage from a large number of lysosomes can destroy a cell by self-digestion.

Hydrolytic enzymes and lysosomal membrane are made by rough ER and then transferred to the Golgi apparatus for further processing. At least some lysosomes probably arise by budding from the *trans* face of the Golgi apparatus (see Figure 4.11). How are the proteins of the inner surface of the lysosomal membrane and the digestive enzymes themselves spared from destruction? Apparently, the three-dimensional shapes of these lysosomal proteins protect vulnerable bonds from enzymatic attack.

Lysosomes carry out intracellular digestion in a variety of circumstances. Amoebas and many other unicellular eukaryotes eat by engulfing smaller organisms or food particles, a process called **phagocytosis** (from the Greek *phagein*, to eat, and *kytos*, vessel, referring here to the cell). The *food vacuole* formed in this way then fuses with a lysosome, whose enzymes digest the food (see Figure 4.12, bottom). Digestion products, including simple sugars, amino acids, and other monomers,

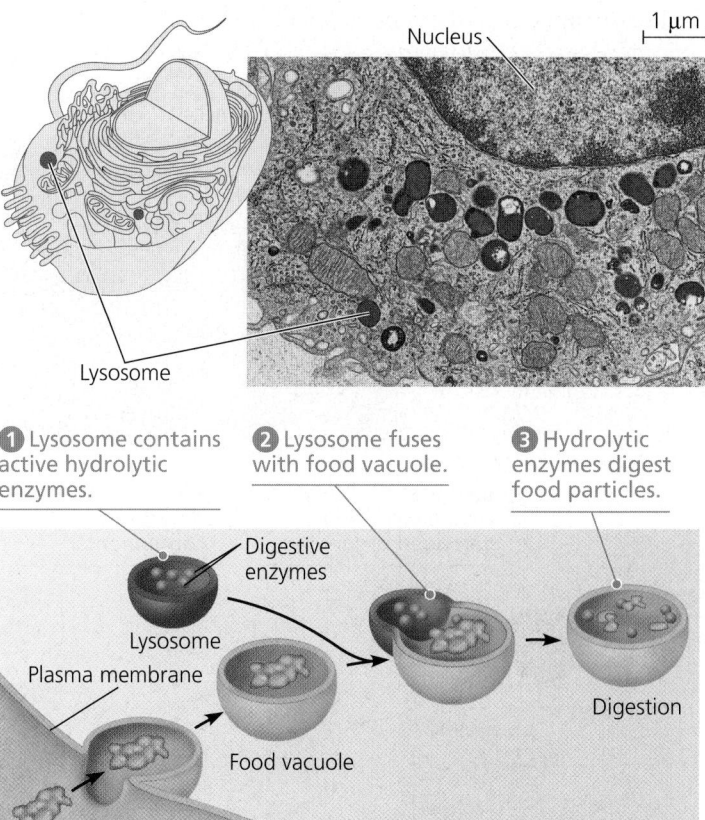

▲ **Figure 4.12 Lysosomes: Phagocytosis.** In phagocytosis, lysosomes digest (hydrolyze) materials taken into the cell and recycle intracellular materials. *Top*: In this macrophage (a type of white blood cell) from a rat, the lysosomes are very dark because of a stain that reacts with one of the products of digestion inside the lysosome (TEM). Macrophages ingest bacteria and viruses and destroy them using lysosomes. *Bottom*: This diagram shows a lysosome fusing with a food vacuole during the process of phagocytosis by a unicellular eukaryote.

pass into the cytosol and become nutrients for the cell. Some human cells also carry out phagocytosis. Among them are macrophages, a type of white blood cell that helps defend the body by engulfing and destroying bacteria and other invaders (see Figure 4.12, top, and Figure 4.28).

Lysosomes also use their hydrolytic enzymes to recycle the cell's own organic material, a process called *autophagy*. During autophagy, a damaged organelle or small amount of cytosol becomes surrounded by a double membrane, and a lysosome fuses with the outer membrane of this vesicle **(Figure 4.13)**. The lysosomal enzymes dismantle the enclosed material, and the resulting small organic compounds are released to the cytosol for reuse. With the help of lysosomes, the cell continually renews itself. A human liver cell, for example, recycles half of its macromolecules each week.

The cells of people with inherited lysosomal storage diseases lack a functioning hydrolytic enzyme normally present in lysosomes. The lysosomes become engorged with indigestible material, which begins to interfere with other cellular activities. In Tay-Sachs disease, for example, a lipid-digesting enzyme is missing or inactive, and the brain becomes impaired by an accumulation of lipids in the cells. Fortunately, lysosomal storage diseases are rare in the general population.

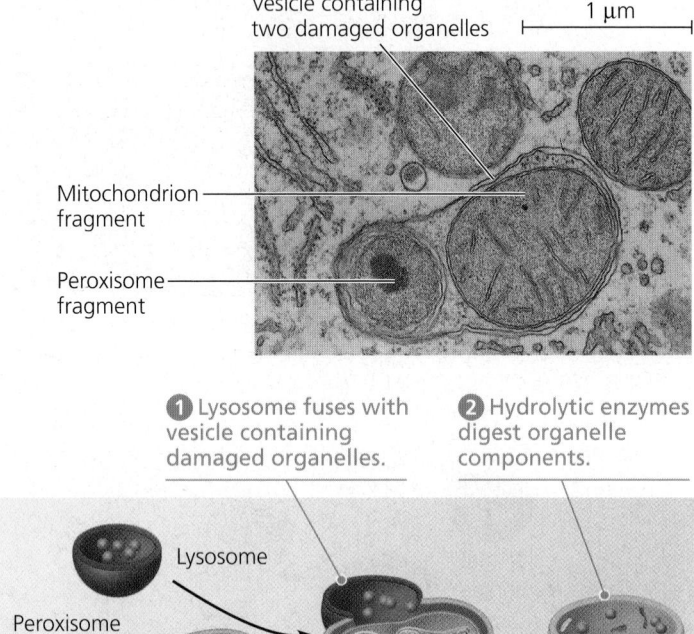

① Lysosome fuses with vesicle containing damaged organelles.

② Hydrolytic enzymes digest organelle components.

Lysosome

Peroxisome

Mitochondrion

Vesicle

Digestion

▲ **Figure 4.13 Lysosomes: Autophagy.** In autophagy, lysosomes recycle intracellular materials. *Top:* In the cytoplasm of this rat liver cell is a vesicle containing two disabled organelles; the vesicle will fuse with a lysosome in the process of autophagy (TEM). *Bottom:* This diagram shows fusion of such a vesicle with a lysosome and the subsequent digestion of the damaged organelles.

## Vacuoles: Diverse Maintenance Compartments

**Vacuoles** are large vesicles derived from the endoplasmic reticulum and Golgi apparatus. Thus, vacuoles are an integral part of a cell's endomembrane system. Like all cellular membranes, the vacuolar membrane is selective in transporting solutes; as a result, the solution inside a vacuole differs in composition from the cytosol.

Vacuoles perform a variety of functions in different kinds of cells. **Food vacuoles**, formed by phagocytosis, have already been mentioned (see Figure 4.12). Many unicellular eukaryotes living in fresh water have **contractile vacuoles** that pump excess water out of the cell, thereby maintaining a suitable concentration of ions and molecules inside the cell (see Figure 5.12). In plants and fungi, certain vacuoles carry out enzymatic hydrolysis, a function shared by lysosomes in animal cells. (In fact, some biologists consider these hydrolytic vacuoles to be a type of lysosome.) In plants, small vacuoles can hold reserves of important organic compounds, such as the proteins stockpiled in the storage cells in seeds. Vacuoles may also help protect the plant against herbivores by storing compounds that are poisonous or unpalatable to animals. Some plant vacuoles contain pigments, such as the red and blue pigments of petals that help attract pollinating insects to flowers.

Mature plant cells generally contain a large **central vacuole (Figure 4.14)**, which develops by the coalescence of smaller vacuoles. The solution inside the central vacuole, called cell sap, is the plant cell's main repository of inorganic ions, including potassium and chloride. The central vacuole plays a major role in the growth of plant cells, which enlarge as the vacuole absorbs water, enabling the cell to become larger with a minimal investment in new cytoplasm. The cytosol often occupies only a thin layer between the central vacuole and the plasma

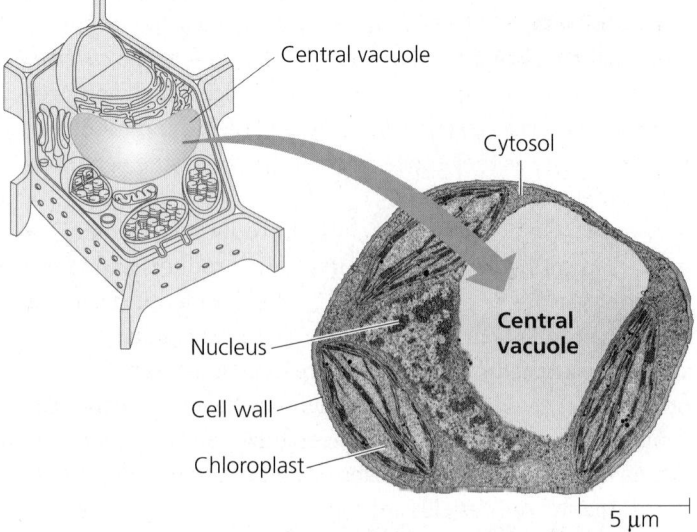

Central vacuole

Cytosol

Central vacuole

Nucleus

Cell wall

Chloroplast

▲ **Figure 4.14 The plant cell vacuole.** The central vacuole is usually the largest compartment in a plant cell; the rest of the cytoplasm is often confined to a narrow zone between the vacuolar membrane and the plasma membrane (TEM).

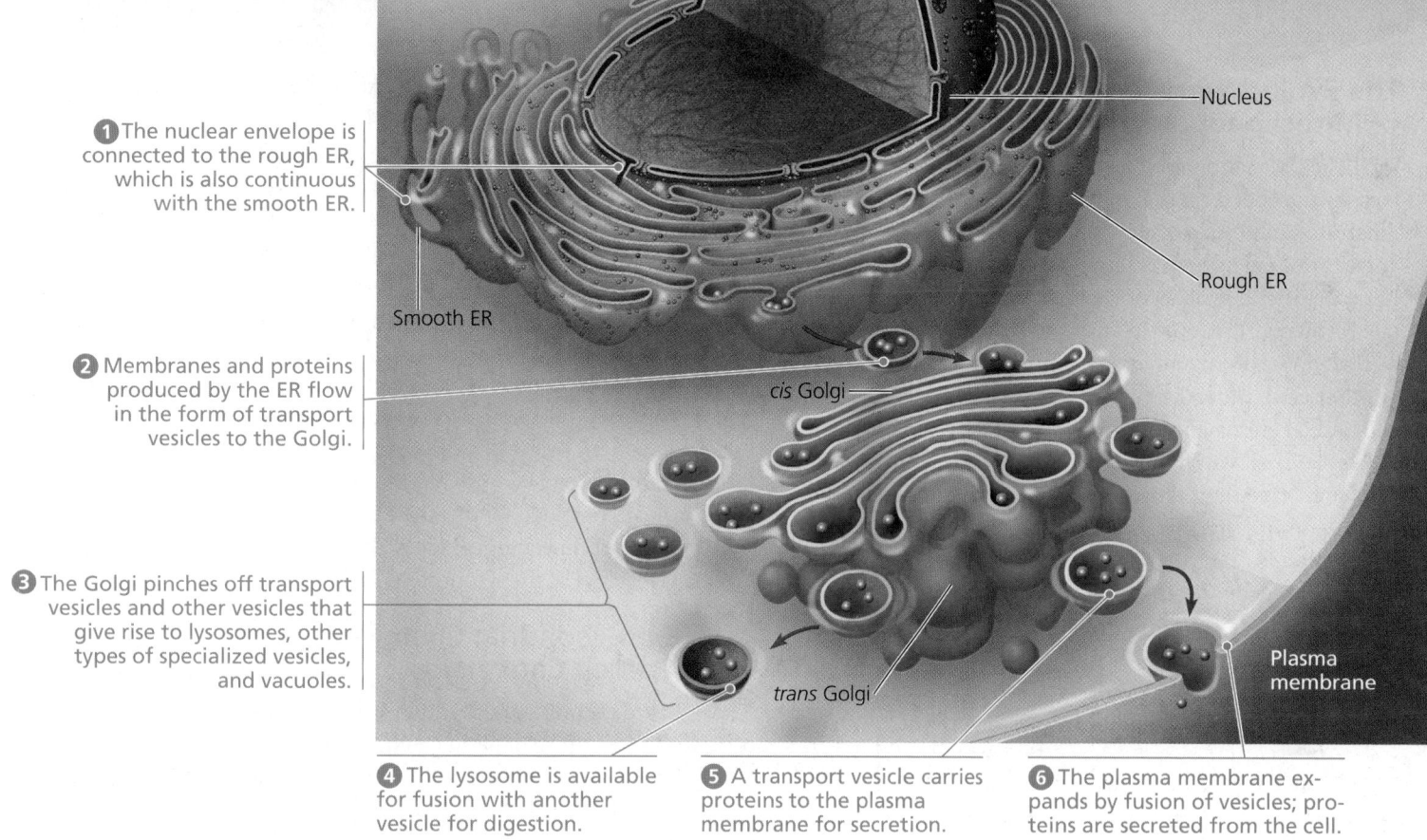

**❶** The nuclear envelope is connected to the rough ER, which is also continuous with the smooth ER.

Smooth ER

**❷** Membranes and proteins produced by the ER flow in the form of transport vesicles to the Golgi.

**❸** The Golgi pinches off transport vesicles and other vesicles that give rise to lysosomes, other types of specialized vesicles, and vacuoles.

Nucleus

Rough ER

*cis* Golgi

*trans* Golgi

Plasma membrane

**❹** The lysosome is available for fusion with another vesicle for digestion.

**❺** A transport vesicle carries proteins to the plasma membrane for secretion.

**❻** The plasma membrane expands by fusion of vesicles; proteins are secreted from the cell.

▲ **Figure 4.15 Review: relationships among organelles of the endomembrane system.** The red arrows show some of the migration pathways for membranes and the materials they enclose.

membrane, so the ratio of plasma membrane surface to cytosolic volume is sufficient, even for a large plant cell.

### The Endomembrane System: *A Review*

**Figure 4.15** reviews the endomembrane system, showing the flow of membrane lipids and proteins through the various organelles. As the membrane moves from the ER to the Golgi and then elsewhere, its molecular composition and metabolic functions are modified, along with those of its contents. The endomembrane system is a complex and dynamic player in the cell's compartmental organization.

We'll continue our tour of the cell with some organelles that are not closely related to the endomembrane system but play crucial roles in the energy transformations carried out by cells.

**CONCEPT CHECK 4.4**

1. Describe the structural and functional distinctions between rough and smooth ER.
2. Describe how transport vesicles integrate the endomembrane system.
3. **WHAT IF?** Imagine a protein that functions in the ER but requires modification in the Golgi apparatus before it can achieve that function. Describe the protein's path through the cell, starting with the mRNA molecule that specifies the protein.

For suggested answers, see Appendix A.

CONCEPT 4.5

# Mitochondria and chloroplasts change energy from one form to another

Organisms transform the energy they acquire from their surroundings. In eukaryotic cells, mitochondria and chloroplasts are the organelles that convert energy to forms that cells can use for work. **Mitochondria** (singular, *mitochondrion*) are the sites of cellular respiration, the metabolic process that uses oxygen to drive the generation of ATP by extracting energy from sugars, fats, and other fuels. **Chloroplasts**, found in plants and algae, are the sites of photosynthesis. This process in chloroplasts converts solar energy to chemical energy by absorbing sunlight and using it to drive the synthesis of organic compounds such as sugars from carbon dioxide and water.

In addition to having related functions, mitochondria and chloroplasts share similar evolutionary origins, which we'll discuss briefly before describing their structures. In this section, we will also consider the peroxisome, an oxidative organelle. The evolutionary origin of the peroxisome, as well as its relation to other organelles, is still under debate.

## The Evolutionary Origins of Mitochondria and Chloroplasts

**EVOLUTION** Mitochondria and chloroplasts display similarities with bacteria that led to the **endosymbiont theory**, illustrated in **Figure 4.16**. This theory states that an early ancestor of eukaryotic cells engulfed an oxygen-using nonphotosynthetic prokaryotic cell. Eventually, the engulfed cell formed a relationship with the host cell in which it was enclosed, becoming an *endosymbiont* (a cell living within another cell). Indeed, over the course of evolution, the host cell and its endosymbiont merged into a single organism, a eukaryotic cell with a mitochondrion. At least one of these cells may have then taken up a photosynthetic prokaryote, becoming the ancestor of eukaryotic cells that contain chloroplasts.

This is a widely accepted theory, which we will discuss in detail in Concept 25.1. This theory is consistent with many structural features of mitochondria and chloroplasts. First, rather than being bounded by a single membrane like organelles of the endomembrane system, mitochondria and typical chloroplasts have two membranes surrounding them. (Chloroplasts also have an internal system of membranous sacs.) There is evidence that the ancestral engulfed prokaryotes had two outer membranes, which became the double membranes of mitochondria and chloroplasts. Second, like prokaryotes, mitochondria and chloroplasts contain ribosomes, as well as multiple circular DNA molecules associated with their inner membranes. The DNA in these organelles programs the synthesis of some organelle proteins on ribosomes that have been synthesized and assembled there as well. Third, also consistent with their probable evolutionary origins as cells, mitochondria and chloroplasts are autonomous (somewhat independent) organelles that grow and reproduce within the cell. Next we focus on the structures of mitochondria and chloroplasts, while providing an overview of their functions.

## Mitochondria: Chemical Energy Conversion

Mitochondria are found in nearly all eukaryotic cells, including those of plants, animals, fungi, and most unicellular eukaryotes. Some cells have a single large mitochondrion, but more often a cell has hundreds or even thousands of mitochondria; the number correlates with the cell's level of metabolic activity. For example, cells that move or contract have proportionally more mitochondria per volume than less active cells.

Each of the two membranes enclosing the mitochondrion is a phospholipid bilayer with a unique collection of embedded proteins **(Figure 4.17)**. The outer membrane is smooth, but the inner membrane is convoluted, with infoldings called **cristae**. The inner membrane divides the mitochondrion into two internal compartments. The first is the intermembrane space, the narrow region between the inner and outer membranes. The second compartment, the **mitochondrial matrix**, is enclosed by the inner membrane. The matrix contains many different enzymes as well as the mitochondrial DNA and ribosomes. Enzymes in the matrix catalyze some of the steps of cellular respiration. Other proteins that function in respiration, including the enzyme that makes ATP, are built into the inner membrane. As highly folded surfaces, the cristae give the inner mitochondrial membrane a large surface area, thus enhancing the productivity of cellular respiration. This is another example of structure fitting function. (Chapter 7 discusses cellular respiration in detail.)

Mitochondria are generally in the range of 1–10 μm long. Time-lapse films of living cells reveal mitochondria moving around, changing their shapes, and fusing or dividing in two, unlike the static structures seen in electron micrographs of dead cells. These studies helped biologists understand that mitochondria form a branched tubular network (see Figure 4.17b) in a dynamic state of flux.

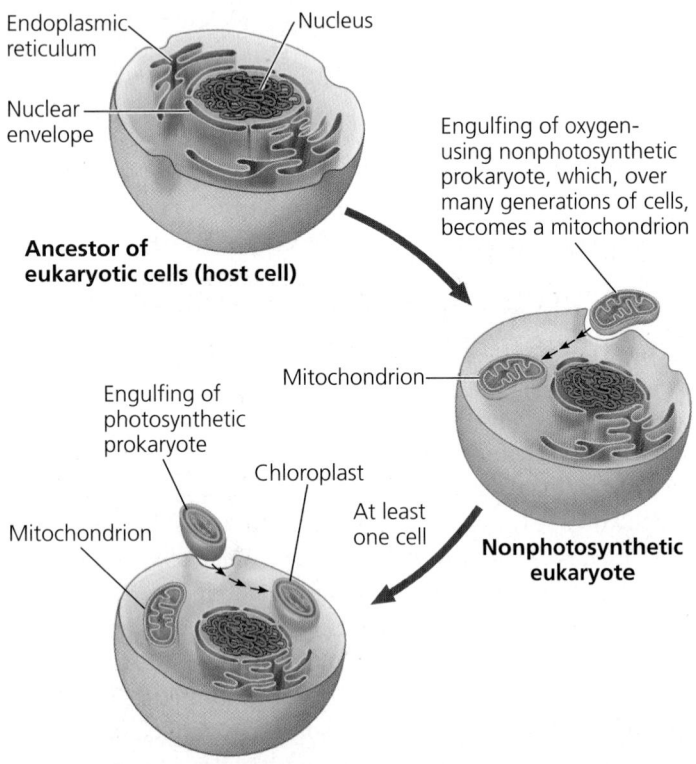

**Ancestor of eukaryotic cells (host cell)**

Endoplasmic reticulum

Nucleus

Nuclear envelope

Engulfing of oxygen-using nonphotosynthetic prokaryote, which, over many generations of cells, becomes a mitochondrion

Mitochondrion

**Nonphotosynthetic eukaryote**

Engulfing of photosynthetic prokaryote

Chloroplast

Mitochondrion

At least one cell

**Photosynthetic eukaryote**

▲ **Figure 4.16 The endosymbiont theory of the origins of mitochondria and chloroplasts in eukaryotic cells.** According to this theory, the proposed ancestors of mitochondria were oxygen-using nonphotosynthetic prokaryotes, while the proposed ancestors of chloroplasts were photosynthetic prokaryotes. The large arrows represent change over evolutionary time; the small arrows inside the cells show the process of the endosymbiont becoming an organelle, also over long periods of time.

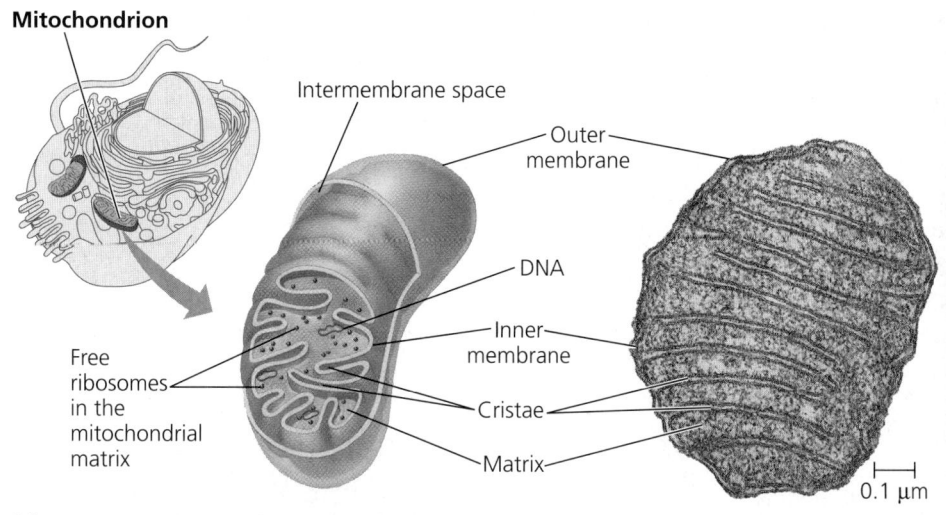

**Mitochondrion**

Intermembrane space

Outer membrane

DNA

Inner membrane

Cristae

Matrix

Free ribosomes in the mitochondrial matrix

0.1 μm

**(a) Diagram and TEM of mitochondrion**

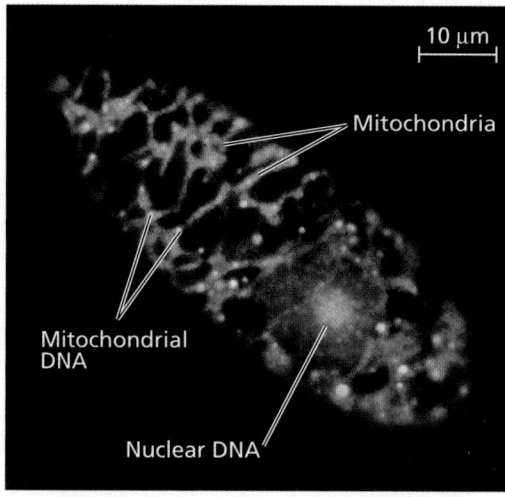

10 μm

Mitochondria

Mitochondrial DNA

Nuclear DNA

**(b) Network of mitochondria in *Euglena* (LM)**

▲ **Figure 4.17 The mitochondrion, site of cellular respiration. (a)** The inner and outer membranes of the mitochondrion are evident in the drawing and electron micrograph (TEM). The cristae are infoldings of the inner membrane, which increase its surface area. The cutaway drawing shows the two compartments bounded by the membranes: the intermembrane space and the mitochondrial matrix. Many respiratory enzymes are found in the inner membrane and the matrix. Free ribosomes are also present in the matrix. The circular DNA molecules are associated with the inner mitochondrial membrane. **(b)** The light micrograph shows an entire unicellular eukaryote (*Euglena gracilis*) at a much lower magnification than the TEM. The mitochondrial matrix has been stained green. The mitochondria form a branched tubular network. The nuclear DNA is stained red; molecules of mitochondrial DNA appear as bright yellow spots.

## Chloroplasts: Capture of Light Energy

Chloroplasts contain the green pigment chlorophyll, along with enzymes and other molecules that function in the photosynthetic production of sugar. These lens-shaped organelles, about 3–6 μm in length, are found in leaves and other green organs of plants and in algae **(Figure 4.18)**.

The contents of a chloroplast are partitioned from the cytosol by an envelope consisting of two membranes separated by a very narrow intermembrane space. Inside the chloroplast is another membranous system in the form of flattened, interconnected sacs called **thylakoids**. In some regions, thylakoids are stacked like poker chips; each stack is called a **granum**

▼ **Figure 4.18 The chloroplast, site of photosynthesis. (a)** Many plants have lens-shaped chloroplasts, as shown here in a diagram and a TEM. A chloroplast has three compartments: the intermembrane space, the stroma, and the thylakoid space. Free ribosomes are present in the stroma, as are copies of chloroplast DNA molecules. **(b)** This fluorescence micrograph, at much lower magnification than the TEM, shows a whole cell of the green alga *Spirogyra crassa*, which is named for its spiral chloroplasts. Under natural light the chloroplasts appear green, but under ultraviolet light they naturally fluoresce red, as shown here.

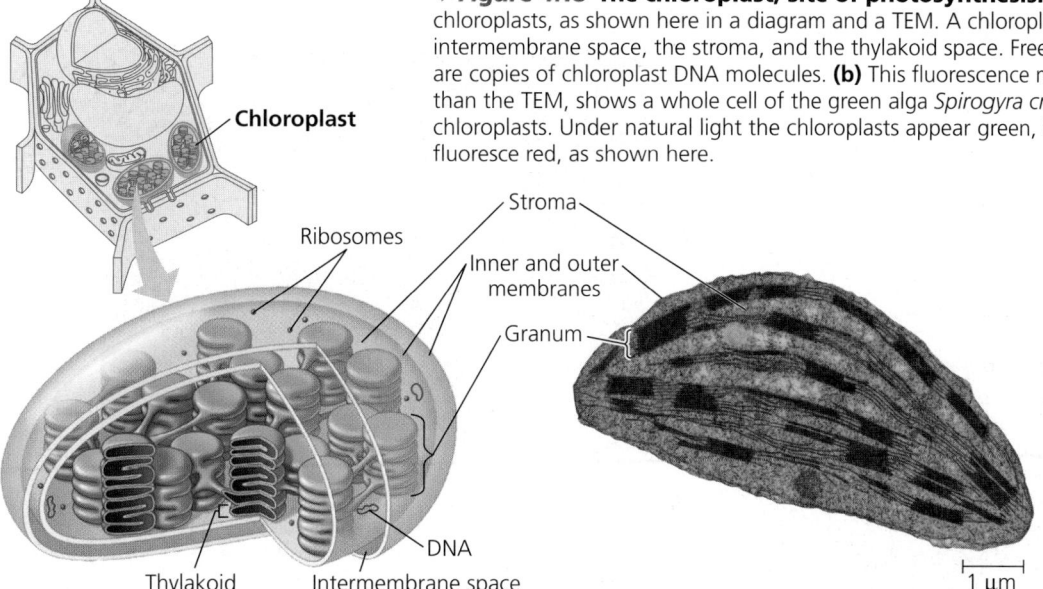

**Chloroplast**

Ribosomes

Stroma

Inner and outer membranes

Granum

DNA

Thylakoid

Intermembrane space

1 μm

**(a) Diagram and TEM of chloroplast**

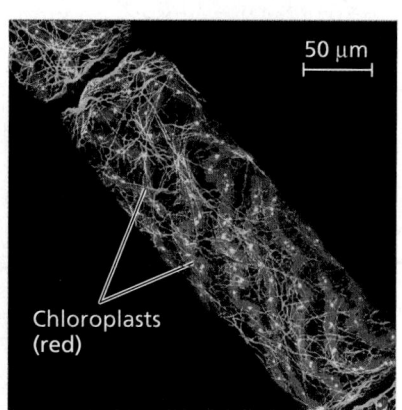

50 μm

Chloroplasts (red)

**(b) Chloroplasts in an algal cell**

(plural, *grana*). The fluid outside the thylakoids is the **stroma**, which contains the chloroplast DNA and ribosomes as well as many enzymes. The membranes of the chloroplast divide the chloroplast space into three compartments: the intermembrane space, the stroma, and the thylakoid space. This compartmental organization enables the chloroplast to convert light energy to chemical energy during photosynthesis. (You will learn more about photosynthesis in Chapter 8.)

As with mitochondria, the static and rigid appearance of chloroplasts in micrographs or schematic diagrams is not true to their dynamic behavior in the living cell. Their shape is changeable, and they grow and occasionally pinch in two, reproducing themselves. They are mobile and, as with mitochondria and other organelles, move around the cell along tracks of the cytoskeleton, a structural network we will consider in Concept 4.6.

The chloroplast is a specialized member of a family of closely related plant organelles called **plastids**. One type of plastid, the *amyloplast*, is a colorless organelle that stores starch (amylose), particularly in roots and tubers. Another is the *chromoplast*, which has pigments that give fruits and flowers their orange and yellow hues.

## Peroxisomes: Oxidation

The **peroxisome** is a specialized metabolic compartment bounded by a single membrane **(Figure 4.19)**. Peroxisomes contain enzymes that remove hydrogen atoms from certain molecules and transfer them to oxygen ($O_2$), producing hydrogen peroxide ($H_2O_2$). These reactions have many different functions. For example, peroxisomes in the liver detoxify alcohol and other harmful compounds by transferring hydrogen from the poisons to oxygen. The $H_2O_2$ formed by peroxisomes is itself toxic, but the organelle also contains an enzyme that converts $H_2O_2$ to water. This is an excellent example of how the cell's compartmental structure is crucial to its functions: The enzymes that produce $H_2O_2$ and those that dispose of this toxic compound are sequestered from other cellular components that could be damaged.

Peroxisomes grow larger by incorporating proteins made in the cytosol and ER, as well as lipids made in the ER and within the peroxisome itself. But how peroxisomes increase in number and how they arose in evolution are still open questions.

**CONCEPT CHECK 4.5**
1. Describe two characteristics shared by chloroplasts and mitochondria. Consider both function and membrane structure.
2. Do plant cells have mitochondria? Explain.
3. **WHAT IF?** A classmate proposes that mitochondria and chloroplasts should be classified in the endomembrane system. Argue against the proposal.

For suggested answers, see Appendix A.

## CONCEPT 4.6

# The cytoskeleton is a network of fibers that organizes structures and activities in the cell

In the early days of electron microscopy, biologists thought that the organelles of a eukaryotic cell floated freely in the cytosol. But improvements in both light microscopy and electron microscopy have revealed the **cytoskeleton**, a network of fibers extending throughout the cytoplasm **(Figure 4.20)**. The cytoskeleton plays a major role in organizing the structures and activities of the cell.

## Roles of the Cytoskeleton: Support and Motility

The most obvious function of the cytoskeleton is to give mechanical support to the cell and maintain its shape. This is especially important for animal cells, which lack walls. The remarkable strength and resilience of the cytoskeleton as a whole are based on its architecture. Like a dome tent, the cytoskeleton is stabilized by a balance between opposing forces exerted by its elements. And just as the skeleton of an animal helps fix the positions of other body parts, the cytoskeleton provides anchorage for many organelles and even cytosolic enzyme molecules. The cytoskeleton is more dynamic than an animal skeleton,

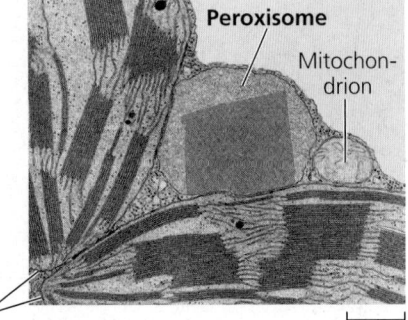

▶ **Figure 4.19 A peroxisome.** Peroxisomes are roughly spherical and often have a granular or crystalline core that is thought to be a dense collection of enzyme molecules. Chloroplasts and mitochondria cooperate with peroxisomes in certain metabolic functions (TEM).

Peroxisome

Mitochondrion

Chloroplasts

1 μm

▲ **Figure 4.20 The cytoskeleton.** As shown in this fluorescence micrograph, the cytoskeleton extends throughout the cell. The cytoskeletal elements have been tagged with different fluorescent molecules: green for microtubules and reddish-orange for microfilaments. A third component of the cytoskeleton, intermediate filaments, is not evident here. (The blue color tags the DNA in the nucleus.)

10 μm

however. It can be quickly dismantled in one part of the cell and reassembled in a new location, changing the shape of the cell.

Some types of cell motility (movement) also involve the cytoskeleton. The term *cell motility* includes both changes in cell location and more limited movements of cell parts. Cell motility generally requires the interaction of the cytoskeleton with **motor proteins**. Examples of such cell motility abound. Cytoskeletal elements and motor proteins work together with plasma membrane molecules to allow whole cells to move along fibers outside the cell. Inside the cell, vesicles and other organelles often use motor protein "feet" to "walk" to their destinations along a track provided by the cytoskeleton. For instance, this is how vesicles containing neurotransmitter molecules migrate to the tips of axons, the long extensions of nerve cells that release these molecules as chemical signals to adjacent nerve cells **(Figure 4.21)**. The cytoskeleton also manipulates the plasma membrane, bending it inward to form food vacuoles or other phagocytic vesicles.

## Components of the Cytoskeleton

Let's look more closely at the three main types of fibers that make up the cytoskeleton: *Microtubules* are the thickest, *microfilaments* (actin filaments) are the thinnest, and *intermediate filaments* are fibers with diameters in a middle range. **Table 4.1** summarizes the properties of these fibers.

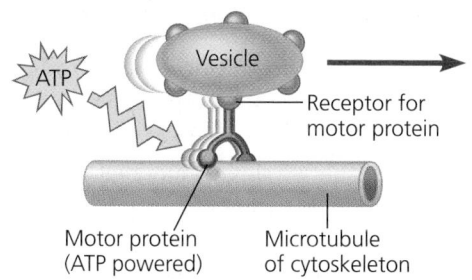

**(a)** Motor proteins that attach to receptors on vesicles can "walk" the vesicles along microtubules or, in some cases, along microfilaments.

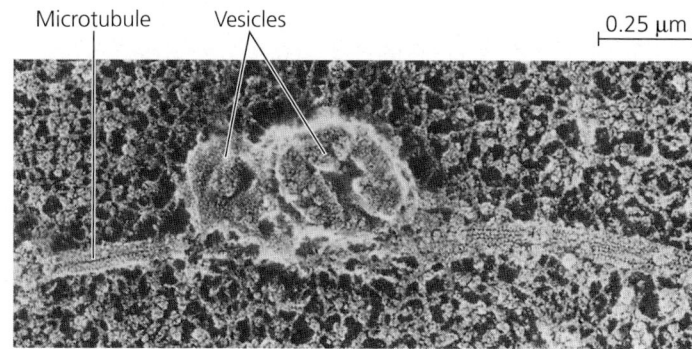

**(b)** Two vesicles containing neurotransmitters move along a microtubule toward the tip of a nerve cell extension called an axon (SEM).

▲ **Figure 4.21  Motor proteins and the cytoskeleton.**

**Table 4.1 The Structure and Function of the Cytoskeleton**

| Property | Microtubules (Tubulin Polymers) | Microfilaments (Actin Filaments) | Intermediate Filaments |
|---|---|---|---|
| **Structure** | Hollow tubes | Two intertwined strands of actin | Fibrous proteins coiled into cables |
| **Diameter** | 25 nm with 15-nm lumen | 7 nm | 8–12 nm |
| **Protein subunits** | Tubulin, a dimer consisting of α-tubulin and β-tubulin | Actin | One of several different proteins (such as keratins) |
| **Main functions** | Maintenance of cell shape; cell motility; chromosome movements in cell division; organelle movements | Maintenance of cell shape; changes in cell shape; muscle contraction; cytoplasmic streaming (plant cells); cell motility; cell division (animal cells) | Maintenance of cell shape; anchorage of nucleus and certain other organelles; formation of nuclear lamina |
| **Fluorescence micrographs of fibroblasts.** Fibroblasts are a favorite cell type for cell biology studies because they spread out flat and their internal structures are easy to see. In each, the structure of interest has been tagged with fluorescent molecules. The DNA in the nucleus has also been tagged in the first micrograph (blue) and third micrograph (orange). | 10 μm  Column of tubulin dimers  25 nm  α  β  Tubulin dimer | 10 μm  Actin subunit  7 nm | 5 μm  Keratin proteins  Fibrous subunit (keratins coiled together)  8–12 nm |

## Microtubules

All eukaryotic cells have **microtubules**, hollow rods constructed from a globular protein called tubulin. Each tubulin protein is a *dimer*, a molecule made up of two subunits. A tubulin dimer consists of two slightly different polypeptides, α-tubulin and β-tubulin. Microtubules grow in length by adding tubulin dimers; they can also be disassembled and their tubulin used to build microtubules elsewhere in the cell.

Microtubules shape and support the cell and serve as tracks along which organelles equipped with motor proteins can move (see Figure 4.21). Microtubules are also involved in the separation of chromosomes during cell division (see Figure 9.7).

**Centrosomes and Centrioles** In animal cells, microtubules grow out from a **centrosome**, a region that is often located near the nucleus and is considered a "microtubule-organizing center." These microtubules function as compression-resisting girders of the cytoskeleton. Within the centrosome is a pair of **centrioles**, each composed of nine sets of triplet microtubules arranged in a ring **(Figure 4.22)**. Although centrosomes with centrioles may help organize microtubule assembly in animal cells, many other eukaryotic cells lack centrosomes with centrioles and instead organize microtubules by other means.

**Cilia and Flagella** In eukaryotes, a specialized arrangement of microtubules is responsible for the beating of **flagella** (singular, *flagellum*) and **cilia** (singular, *cilium*), microtubule-containing extensions that project from some cells. (The bacterial flagellum,

shown in Figure 4.4, has a completely different structure.) Many unicellular eukaryotes are propelled through water by cilia or flagella that act as locomotor appendages, and the sperm of animals, algae, and some plants have flagella. When cilia or flagella extend from cells that are held in place as part of a tissue layer, they can move fluid over the surface of the tissue. For example, the ciliated lining of the trachea (windpipe) sweeps mucus containing trapped debris out of the lungs (see the EMs in Figure 4.3). In a woman's reproductive tract, the cilia lining the oviducts help move an egg toward the uterus.

Motile cilia usually occur in large numbers on the cell surface. Flagella are usually limited to just one or a few per cell, and they are longer than cilia. Flagella and cilia also differ in their beating patterns. A flagellum has an undulating motion like the tail of a fish. In contrast, cilia have alternating power and recovery strokes, like the oars of a racing crew boat.

A cilium may also act as a signal-receiving antenna for the cell. Cilia that have this function are generally nonmotile, and there is only one per cell. (In fact, in vertebrate animals, it appears that almost all cells have such a cilium, which is called a *primary cilium*.) Membrane proteins on this kind of cilium transmit molecular signals from the cell's environment to its interior, triggering signaling pathways that may lead to changes in the cell's activities. Cilium-based signaling appears to be crucial to brain function and to embryonic development.

Though different in length, number per cell, and beating pattern, motile cilia and flagella share a common structure. Each motile cilium or flagellum has a group of microtubules sheathed in an extension of the plasma membrane **(Figure 4.23a)**. Nine doublets of microtubules are arranged in a ring with two single microtubules in its center **(Figure 4.23b)**. This arrangement, referred to as the "9 + 2" pattern, is found in nearly all eukaryotic flagella and motile cilia. (Nonmotile primary cilia have a "9 + 0" pattern, lacking the central pair of microtubules.) The microtubule assembly of a cilium or flagellum is anchored in the cell by a **basal body**, which is structurally similar to a centriole, with microtubule triplets in a "9 + 0" pattern **(Figure 4.23c)**. In fact, in many animals (including humans), the basal body of the fertilizing sperm's flagellum enters the egg and becomes a centriole.

How does the microtubule assembly produce the bending movements of flagella and motile cilia? Bending involves large motor proteins called **dyneins** (red in the diagram in Figure 4.23b) that are attached along each outer microtubule doublet. A typical dynein protein has two "feet" that "walk" along the microtubule of the adjacent doublet, using ATP for energy. One foot maintains contact, while the other releases and reattaches one step farther along the microtubule (see Figure 4.21). The outer doublets and two central microtubules are held together by flexible cross-linking proteins (blue in Figure 4.23b). If the doublets were not held in place, the walking action would make them slide past each other. Instead, the movements of the dynein feet cause the microtubules—and the organelle as a whole—to bend.

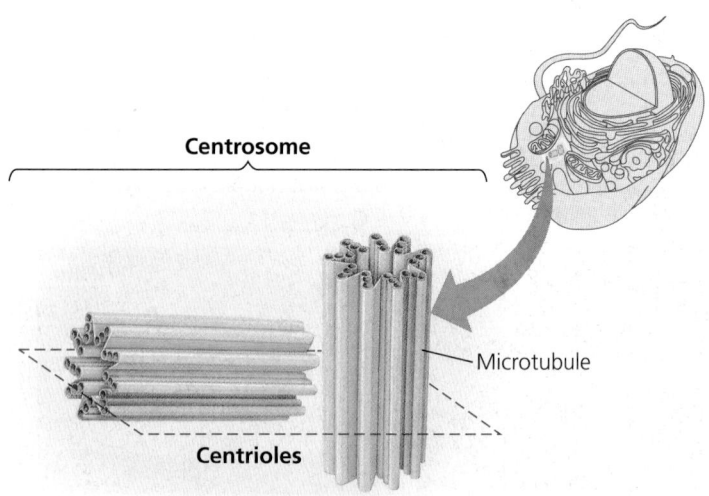

**Centrosome**

**Microtubule**

**Centrioles**

▲ **Figure 4.22 Centrosome containing a pair of centrioles.** Most animal cells have a centrosome, a region near the nucleus where the cell's microtubules are initiated. Within the centrosome is a pair of centrioles, each about 250 nm (0.25 μm) in diameter. The two centrioles are at right angles to each other, and each is made up of nine sets of three microtubules. The blue portions of the drawing represent nontubulin proteins that connect the microtubule triplets.

? *How many microtubules are in a centrosome? In the drawing, circle and label one microtubule and describe its structure. Circle and label a triplet.*

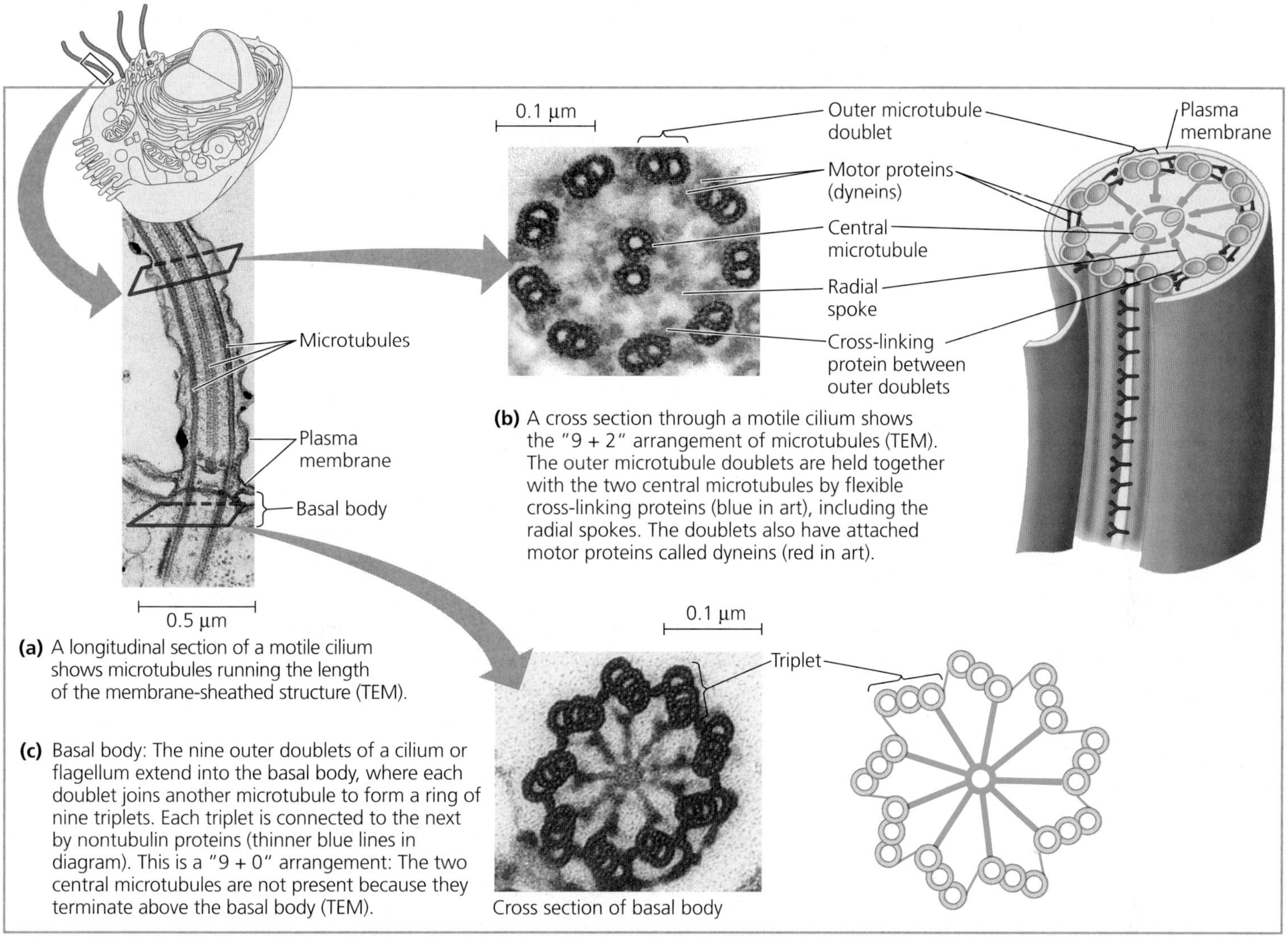

0.1 μm

Outer microtubule doublet

Motor proteins (dyneins)

Central microtubule

Radial spoke

Cross-linking protein between outer doublets

Plasma membrane

Microtubules

Plasma membrane

Basal body

0.5 μm

**(b)** A cross section through a motile cilium shows the "9 + 2" arrangement of microtubules (TEM). The outer microtubule doublets are held together with the two central microtubules by flexible cross-linking proteins (blue in art), including the radial spokes. The doublets also have attached motor proteins called dyneins (red in art).

**(a)** A longitudinal section of a motile cilium shows microtubules running the length of the membrane-sheathed structure (TEM).

0.1 μm

Triplet

**(c)** Basal body: The nine outer doublets of a cilium or flagellum extend into the basal body, where each doublet joins another microtubule to form a ring of nine triplets. Each triplet is connected to the next by nontubulin proteins (thinner blue lines in diagram). This is a "9 + 0" arrangement: The two central microtubules are not present because they terminate above the basal body (TEM).

Cross section of basal body

▲ **Figure 4.23 Structure of a flagellum or motile cilium.**

`DRAW IT` *In **(a)**, circle and label the central pair of microtubules. Show where they terminate, and explain why they aren't seen in the cross section of the basal body in **(c)**.*

## Microfilaments (Actin Filaments)

**Microfilaments** are thin solid rods. They are also called actin filaments because they are built from molecules of **actin**, a globular protein. A microfilament is a twisted double chain of actin subunits (see Table 4.1). Besides occurring as linear filaments, microfilaments can form structural networks when certain proteins bind along the side of such a filament and allow a new filament to extend as a branch.

The structural role of microfilaments in the cytoskeleton is to bear tension (pulling forces). A three-dimensional network formed by microfilaments just inside the plasma membrane helps support the cell's shape. In some kinds of animal cells, such as nutrient-absorbing intestinal cells, bundles of microfilaments make up the core of microvilli, delicate projections that increase the cell's surface area **(Figure 4.24)**.

Microfilaments are well known for their role in cell motility. Thousands of actin filaments and thicker filaments of a

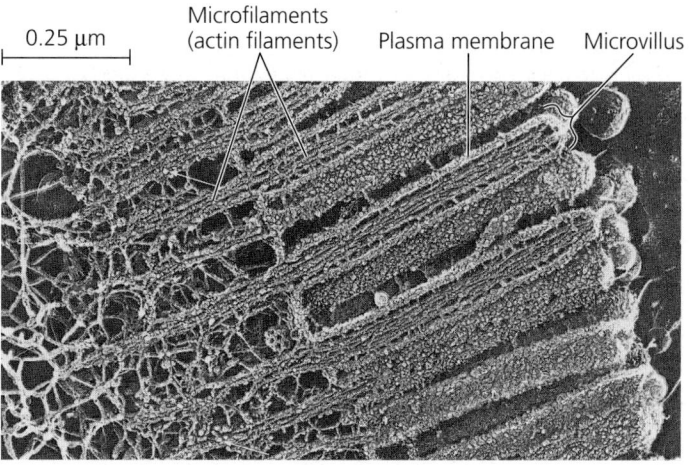

0.25 μm

Microfilaments (actin filaments)

Plasma membrane

Microvillus

▲ **Figure 4.24 A structural role of microfilaments.** The surface area of this intestinal cell is increased by its many microvilli (singular, *microvillus*), cellular extensions reinforced by bundles of microfilaments (TEM).

motor protein called **myosin** interact to cause contraction of muscle cells (described in detail in Concept 39.1). In the unicellular eukaryote *Amoeba* and some of our white blood cells, localized contractions brought about by actin and myosin are involved in the amoeboid (crawling) movement of the cells. In plant cells, actin-myosin interaction contributes to *cytoplasmic streaming,* a circular flow of cytoplasm within cells. This movement, which is especially common in large plant cells, speeds the distribution of materials within the cell.

### Intermediate Filaments

**Intermediate filaments** are named for their diameter, which is larger than the diameter of microfilaments but smaller than that of microtubules (see Table 4.1). While microtubules and microfilaments are found in all eukaryotic cells, intermediate filaments are only found in the cells of some animals, including vertebrates. Specialized for bearing tension (like microfilaments), intermediate filaments are a diverse class of cytoskeletal elements. Each type is constructed from a particular molecular subunit belonging to a family of proteins whose members include the keratins in hair and nails.

Intermediate filaments are more permanent fixtures of cells than are microfilaments and microtubules, which are often disassembled and reassembled in various parts of a cell. Even after cells die, intermediate filament networks often persist; for example, the outer layer of our skin consists of dead skin cells full of keratin filaments. Intermediate filaments are especially sturdy and play an important role in reinforcing the shape of a cell and fixing the position of certain organelles. For instance, the nucleus typically sits within a cage made of intermediate filaments. Other intermediate filaments make up the nuclear lamina, which lines the interior of the nuclear envelope (see Figure 4.8). In general, the various kinds of intermediate filaments seem to function together as the permanent framework of the entire cell.

### CONCEPT CHECK 4.6

1. How do cilia and flagella bend?
2. **WHAT IF?** Males afflicted with Kartagener's syndrome are sterile because of immotile sperm, and they tend to suffer from lung infections. This disorder has a genetic basis. Suggest what the underlying defect might be.

For suggested answers, see Appendix A.

---

### CONCEPT 4.7

# Extracellular components and connections between cells help coordinate cellular activities

Having crisscrossed the cell to explore its interior components, we complete our tour of the cell by returning to the surface of this microscopic world, where there are additional structures with important functions. The plasma membrane is usually regarded as the boundary of the living cell, but most cells synthesize and secrete materials extracellularly (to the outside of the cell). Although these materials and the structures they form are outside the cell, their study is important to cell biology because they are involved in a great many cellular functions.

## Cell Walls of Plants

The **cell wall** is an extracellular structure of plant cells **(Figure 4.25)**. This is one of the features that distinguishes plant cells from animal cells (see Figure 4.7). The wall protects the plant cell, maintains its shape, and prevents excessive uptake of water. On the level of the whole plant, the strong walls of specialized cells hold the plant up against the force of gravity. Prokaryotes, fungi, and some unicellular eukaryotes also have cell walls, as you saw in Figures 4.4 and 4.7, but we will postpone discussion of them until Chapters 24–26.

Plant cell walls are much thicker than the plasma membrane, ranging from 0.1 μm to several micrometers. The exact chemical composition of the wall varies from species to species and even from one cell type to another in the same plant, but the basic design of the wall is consistent. Microfibrils made of the polysaccharide cellulose (see Figure 3.11) are synthesized by an enzyme called cellulose synthase and secreted to the extracellular space,

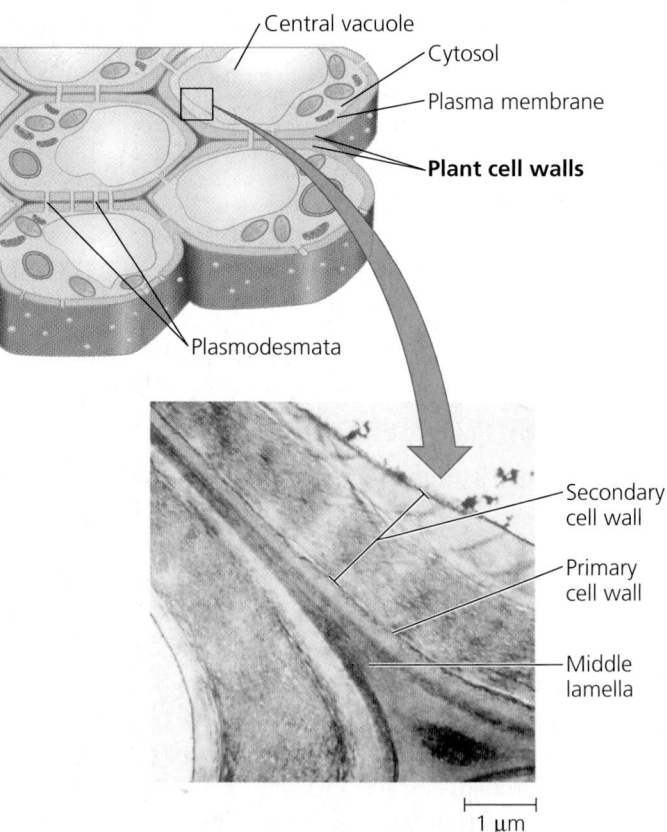

▲ **Figure 4.25 Plant cell walls.** The drawing shows several cells, each with a large vacuole, a nucleus, and several chloroplasts and mitochondria. The TEM shows the cell walls where two cells come together. The multilayered partition between plant cells consists of adjoining walls individually secreted by the cells. Plasmodesmata are channels through cell walls that connect the cytoplasm of adjacent plant cells.

where they become embedded in a matrix of other polysaccharides and proteins. This combination of materials, strong fibers in a "ground substance" (matrix), is the same basic architectural design found in steel-reinforced concrete and in fiberglass.

A young plant cell first secretes a relatively thin and flexible wall called the **primary cell wall** (see Figure 4.25). Between primary walls of adjacent cells is the **middle lamella**, a thin layer rich in sticky polysaccharides called pectins. The middle lamella glues adjacent cells together. (Pectin is used as a thickening agent in jams and jellies.) When the cell matures and stops growing, it strengthens its wall. Some plant cells do this simply by secreting hardening substances into the primary wall. Other cells add a **secondary cell wall** between the plasma membrane and the primary wall. The secondary wall, often deposited in several laminated layers, has a strong and durable matrix that affords the cell protection and support. Wood, for example, consists mainly of secondary walls. Plant cell walls are usually perforated by channels between adjacent cells called plasmodesmata, which will be discussed shortly.

## The Extracellular Matrix (ECM) of Animal Cells

Although animal cells lack walls akin to those of plant cells, they do have an elaborate **extracellular matrix (ECM)**. The main ingredients of the ECM are glycoproteins and other carbohydrate-containing molecules secreted by the cells. (Recall that glycoproteins are proteins with covalently bonded carbohydrates.) The most abundant glycoprotein in the ECM of most animal cells is **collagen**, which forms strong fibers outside the cells (see Figure 3.22, carbohydrate not shown). In fact, collagen accounts for about 40% of the total protein in the human body. The collagen fibers are embedded in a network woven of secreted **proteoglycans (Figure 4.26)**. A proteoglycan molecule consists of a small core protein with many carbohydrate chains covalently attached; it may be up to 95% carbohydrate. Large proteoglycan complexes can form when hundreds of proteoglycan molecules become noncovalently attached to a single long polysaccharide molecule, as shown in Figure 4.26. Some cells are attached to the ECM by ECM glycoproteins such as **fibronectin**. Fibronectin and other ECM proteins bind to cell-surface receptor proteins called **integrins** that are built into the plasma membrane. Integrins span the membrane and bind on their cytoplasmic side to associated proteins attached to microfilaments of the cytoskeleton. The name *integrin* is based on the word *integrate*: Integrins are in a position to transmit signals between the ECM and the cytoskeleton and thus to integrate changes occurring outside and inside the cell.

Current research is revealing the influential role of the ECM in the lives of cells. By communicating with a cell through integrins, the ECM can regulate a cell's behavior. For example, some cells in a developing embryo migrate along specific pathways by matching the orientation of their microfilaments to the "grain" of fibers in the extracellular matrix. Researchers have also learned that the extracellular matrix around a cell can influence the activity of genes in the nucleus. Information about the ECM probably reaches the nucleus by a combination of mechanical and chemical signaling pathways. Mechanical signaling involves fibronectin, integrins, and microfilaments of the cytoskeleton. Changes in the cytoskeleton may in turn trigger chemical signaling pathways inside the cell, leading to changes in the set of proteins being made by the cell and therefore changes in the cell's function. In this way, the extracellular

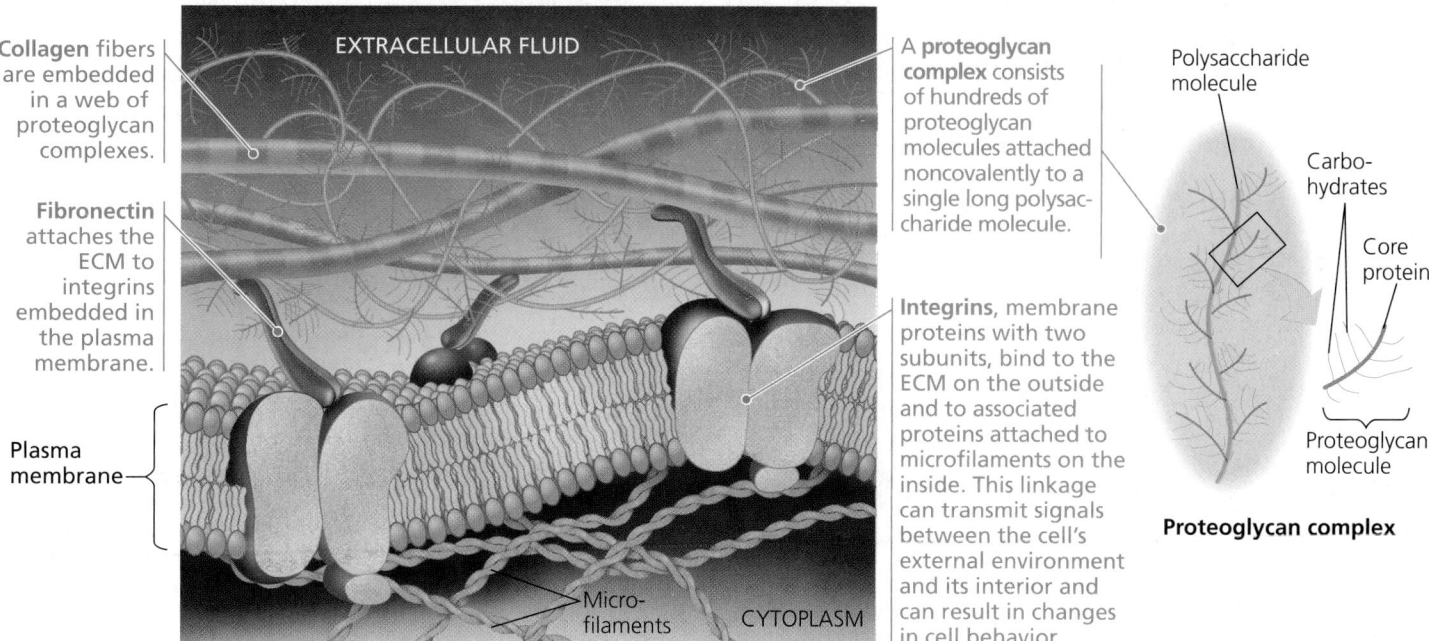

**Collagen** fibers are embedded in a web of proteoglycan complexes.

**Fibronectin** attaches the ECM to integrins embedded in the plasma membrane.

Plasma membrane

EXTRACELLULAR FLUID

A **proteoglycan complex** consists of hundreds of proteoglycan molecules attached noncovalently to a single long polysaccharide molecule.

**Integrins**, membrane proteins with two subunits, bind to the ECM on the outside and to associated proteins attached to microfilaments on the inside. This linkage can transmit signals between the cell's external environment and its interior and can result in changes in cell behavior.

Microfilaments

CYTOPLASM

Polysaccharide molecule

Carbohydrates

Core protein

Proteoglycan molecule

**Proteoglycan complex**

▲ **Figure 4.26 Extracellular matrix (ECM) of an animal cell.** The molecular composition and structure of the ECM vary from one cell type to another. In this example, three different types of ECM molecules are present: proteoglycans, collagen, and fibronectin.

matrix of a particular tissue may help coordinate the behavior of all the cells of that tissue. Direct connections between cells also function in this coordination, as we discuss next.

## Cell Junctions

Neighboring cells in an animal or plant often adhere, interact, and communicate via sites of direct physical contact.

### Plasmodesmata in Plant Cells

It might seem that the nonliving cell walls of plants would isolate plant cells from one another. But in fact, as shown in Figure 4.25, cell walls are perforated with **plasmodes-mata** (singular, *plasmodesma*; from the Greek *desma*, bond), membrane-lined channels filled with cytosol. By joining adjacent cells, plasmodesmata unify most of a plant into one living continuum. The plasma membranes of adjacent cells line the channel of each plasmodesma and thus are continuous. Water and small solutes can pass freely from cell to cell, and experiments have shown that in some circumstances, certain proteins and RNA molecules can do this as well. The macro-molecules transported to neighboring cells appear to reach the plasmodesmata by moving along fibers of the cytoskeleton.

### Tight Junctions, Desmosomes, and Gap Junctions in Animal Cells

In animals, there are three main types of cell junctions: *tight junctions*, *desmosomes*, and *gap junctions* (**Figure 4.27**). All three types are especially common in epithelial tissue, which

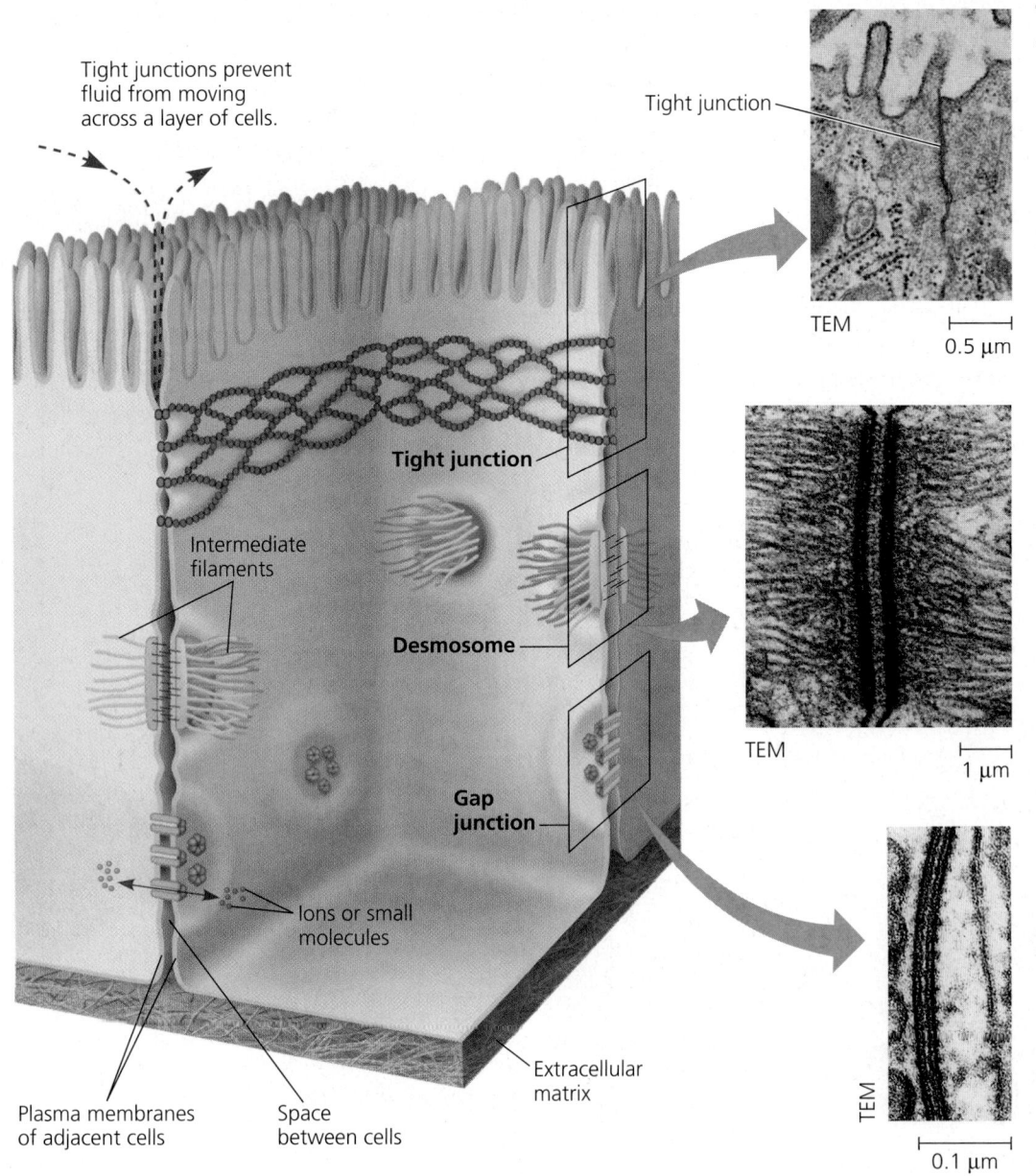

Tight junctions prevent fluid from moving across a layer of cells.

Tight junction

Intermediate filaments

Tight junction

Desmosome

Gap junction

Ions or small molecules

Plasma membranes of adjacent cells

Space between cells

Extracellular matrix

## Tight Junctions

At **tight junctions**, the plasma membranes of neighboring cells are very tightly pressed against each other, bound together by specific proteins (purple). Forming con-tinuous seals around the cells, tight junctions establish a barrier that prevents leakage of extracellular fluid across a layer of epithelial cells (see red dashed arrow). For ex-ample, tight junctions between skin cells make us watertight.

TEM    0.5 μm

## Desmosomes

**Desmosomes** (one type of *anchoring junction*) function like rivets, fastening cells together into strong sheets. Intermediate filaments made of sturdy keratin proteins an-chor desmosomes in the cytoplasm. Desmosomes attach muscle cells to each other in a muscle. Some "muscle tears" involve the rupture of desmosomes.

TEM    1 μm

## Gap Junctions

**Gap junctions** (also called *com-municating junctions*) provide cytoplasmic channels from one cell to an adjacent cell and in this way are similar in their function to the plasmodesmata in plants. Gap junctions consist of membrane pro-teins that surround a pore through which ions, sugars, amino acids, and other small molecules may pass. Gap junctions are necessary for communication between cells in many types of tissues, such as heart muscle, and in animal embryos.

TEM    0.1 μm

lines the external and internal surfaces of the body. Figure 4.27 uses epithelial cells of the intestinal lining to illustrate these junctions. (Gap junctions are most like the plasmodesmata of plants, although gap junction pores are not lined with membrane.)

### CONCEPT CHECK 4.7

1. In what way are the cells of plants and animals structurally different from single-celled eukaryotes?

2. **WHAT IF?** If the plant cell wall or the animal extracellular matrix were impermeable, what effect would this have on cell function?

3. **MAKE CONNECTIONS** The polypeptide chain that makes up a tight junction weaves back and forth through the membrane four times, with two extracellular loops, and one loop plus short C-terminal and N-terminal tails in the cytoplasm. Looking at Figure 3.18, what would you predict about the amino acids making up the tight junction protein?

For suggested answers, see Appendix A.

## The Cell: A Living Unit Greater Than the Sum of Its Parts

From our panoramic view of the cell's compartmental organization to our close-up inspection of each organelle's architecture, this tour of the cell has provided many opportunities to correlate structure with function. But even as we dissect the cell, remember that none of its components works alone. As an example of cellular integration, consider the microscopic scene in **Figure 4.28**. The large cell is a macrophage (see Figure 4.12). It helps defend the mammalian body against infections by ingesting bacteria (the smaller cells) into phagocytic vesicles. The macrophage crawls along a surface and reaches out to the bacteria with thin cell extensions called pseudopodia (specifically, filopodia). Actin filaments interact with other elements of the cytoskeleton in these movements. After the macrophage engulfs the bacteria, they are destroyed by lysosomes. The elaborate endomembrane system produces the lysosomes. The digestive enzymes of the lysosomes and the proteins of the cytoskeleton are all made on ribosomes. And the synthesis of these proteins is programmed by genetic messages dispatched from the DNA in the nucleus. All these processes require energy, which mitochondria supply in the form of ATP. To see how these processes work together in the living cell, see Make Connections Figure 8.20. Cellular functions arise from cellular order: The cell is a living unit greater than the sum of its parts.

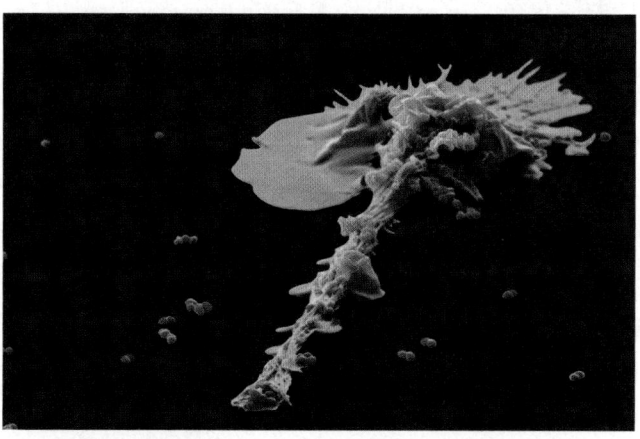

10 μm

▲ **Figure 4.28 The emergence of cellular functions.** The ability of this macrophage (brown) to recognize, apprehend, and destroy *Staphylococcus* bacteria (orange) is a coordinated activity of the whole cell. Its cytoskeleton, lysosomes, and plasma membrane are among the components that function in phagocytosis (colorized SEM).

---

# 4 Chapter Review

Go to **MasteringBiology®** for Assignments, the eText, and the Study Area with Animations, Activities, Vocab Self-Quiz, and Practice Tests.

## SUMMARY OF KEY CONCEPTS

VOCAB SELF-QUIZ

goo.gl/gbai8v

### CONCEPT 4.1

**Biologists use microscopes and the tools of biochemistry to study cells (pp. 73–75)**

- Improvements in microscopy that affect the parameters of magnification, resolution, and contrast have catalyzed progress in the study of cell structure. **Light microscopy** (LM) and **electron microscopy** (EM), as well as other types, remain important tools.
- Cell biologists can obtain pellets enriched in particular cellular components by centrifuging disrupted cells at sequential speeds, a process known as **cell fractionation**. Larger cellular components are in the pellet after lower-speed centrifugation, and smaller components are in the pellet after higher-speed centrifugation.

**?** *How do microscopy and biochemistry complement each other to reveal cell structure and function?*

### CONCEPT 4.2

**Eukaryotic cells have internal membranes that compartmentalize their functions (pp. 75–80)**

- All cells are bounded by a **plasma membrane**, a bilayer of phospholipids with their hydrophobic tails on in the interior of the membrane and their hydrophilic heads in contact with the aqueous solutions on either side.
- **Prokaryotic cells** lack nuclei and other membrane-enclosed **organelles**, while **eukaryotic cells** have internal membranes that compartmentalize cellular functions.
- The surface-to-volume ratio is an important parameter affecting cell size and shape.
- Plant and animal cells have most of the same organelles: a nucleus, endoplasmic reticulum, Golgi apparatus, and mitochondria. Some organelles are found only in plant or in animal cells. Chloroplasts are present only in cells of photosynthetic eukaryotes.

**?** *Explain how the compartmental organization of a eukaryotic cell contributes to its biochemical functioning.*

| Cell Component | Structure | Function |
|---|---|---|

**CONCEPT 4.3**

**The eukaryotic cell's genetic instructions are housed in the nucleus and carried out by the ribosomes (pp. 80–82)**

**?** *Describe the relationship between the nucleus and ribosomes.*

| Cell Component | Structure | Function |
|---|---|---|
| Nucleus (ER) | Surrounded by nuclear envelope (double membrane) perforated by nuclear pores; nuclear envelope continuous with endoplasmic reticulum (ER) | Houses chromosomes, which are made of chromatin (DNA and proteins); contains nucleoli, where ribosomal subunits are made; pores regulate entry and exit of materials |
| Ribosome  | Two subunits made of ribosomal RNA and proteins; can be free in cytosol or bound to ER | Protein synthesis |

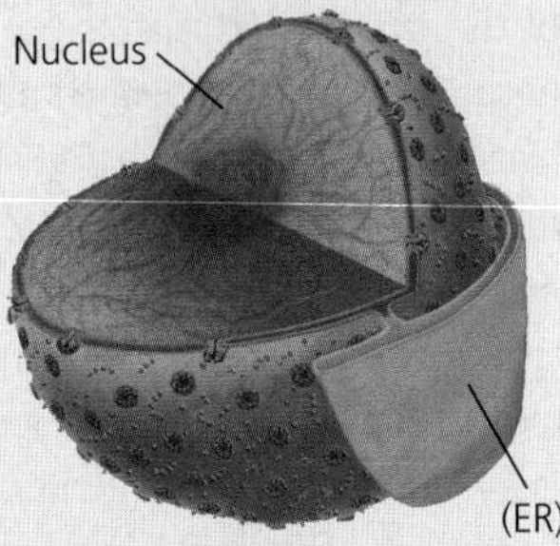

**CONCEPT 4.4**

**The endomembrane system regulates protein traffic and performs metabolic functions in the cell (pp. 82–87)**

**?** *Describe the key role played by transport vesicles in the endomembrane system.*

| Cell Component | Structure | Function |
|---|---|---|
| Endoplasmic reticulum (Nuclear envelope) | Extensive network of membrane-bounded tubules and sacs; membrane separates lumen from cytosol; continuous with nuclear envelope | Smooth ER: synthesis of lipids, metabolism of carbohydrates, $Ca^{2+}$ storage, detoxification of drugs and poisons<br>Rough ER: aids in synthesis of secretory and other proteins from bound ribosomes; adds carbohydrates to proteins to make glycoproteins; produces new membrane |
| Golgi apparatus 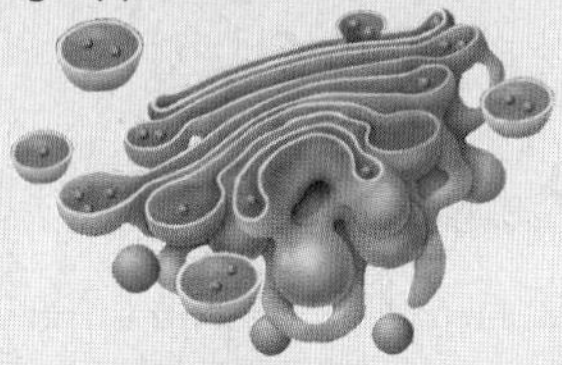 | Stacks of flattened membranous sacs; has polarity (*cis* and *trans* faces) | Modification of proteins, carbohydrates on proteins, and phospholipids; synthesis of many polysaccharides; sorting of Golgi products, which are then released in vesicles |
| Lysosome  | Membranous sac of hydrolytic enzymes (in animal cells) | Breakdown of ingested substances, cell macromolecules, and damaged organelles for recycling |
| Vacuole 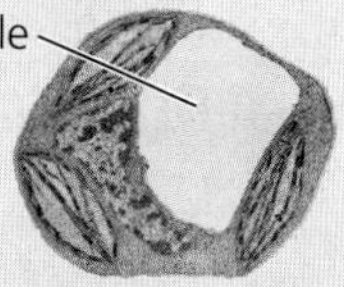 | Large membrane-bounded vesicle | Digestion, storage, waste disposal, water balance, plant cell growth and protection |

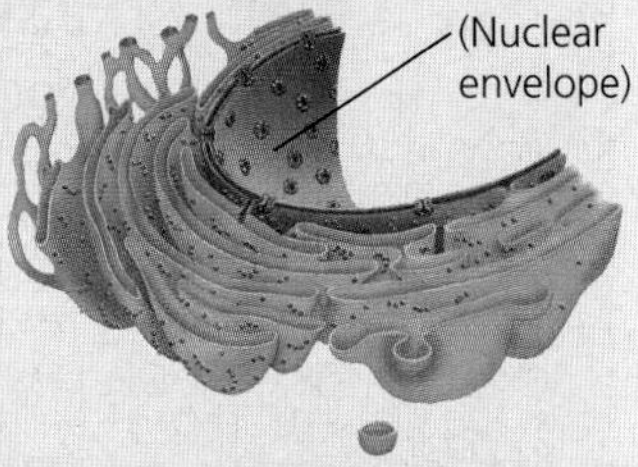

**CONCEPT 4.5**

**Mitochondria and chloroplasts change energy from one form to another (pp. 87–90)**

**?** *What is the endosymbiont theory?*

| Cell Component | Structure | Function |
|---|---|---|
| Mitochondrion | Bounded by double membrane; inner membrane has infoldings (cristae) | Cellular respiration |
| Chloroplast  | Typically two membranes around fluid stroma, which contains thylakoids stacked into grana (in cells of photosynthetic eukaryotes, including plants) | Photosynthesis |
| Peroxisome 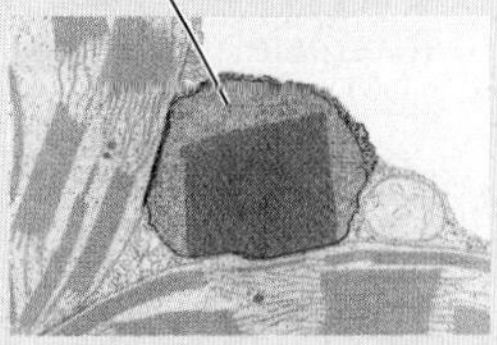 | Specialized metabolic compartment bounded by a single membrane | Contains enzymes that transfer hydrogen atoms from certain molecules to oxygen, producing hydrogen peroxide ($H_2O_2$) as a by-product; $H_2O_2$ is converted to water by another enzyme |

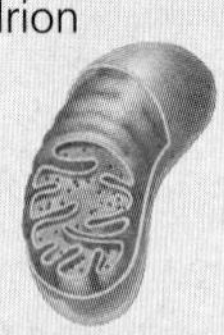

## The cytoskeleton is a network of fibers that organizes structures and activities in the cell (pp. 90–94)

- The **cytoskeleton** functions in structural support for the cell and in motility and signal transmission.
- **Microtubules** shape the cell, guide organelle movement, and separate chromosomes in dividing cells. **Cilia** and **flagella** are motile appendages containing microtubules. *Primary cilia* play sensory and signaling roles. **Microfilaments** are thin rods that function in muscle contraction, amoeboid movement, cytoplasmic streaming, and support of microvilli. **Intermediate filaments** support cell shape and fix organelles in place.

> **?** *Describe the role of motor proteins inside the eukaryotic cell and in whole-cell movement.*

## Extracellular components and connections between cells help coordinate cellular activities (pp. 94–97)

- Plant **cell walls** are made of cellulose fibers embedded in other polysaccharides and proteins.
- Animal cells secrete glycoproteins and proteoglycans that form the **extracellular matrix (ECM)**, which functions in support, adhesion, movement, and regulation.
- Cell junctions connect neighboring cells in plants and animals. Plants have **plasmodesmata** that pass through adjoining cell walls. Animal cells have **tight junctions**, **desmosomes**, and **gap junctions**.

> **?** *Compare the structure and functions of a plant cell wall and the extracellular matrix of an animal cell.*

# TEST YOUR UNDERSTANDING

## Level 1: Knowledge/Comprehension

goo.gl/CRZjvS

1. Which structure is *not* part of the endomembrane system?
   - (A) nuclear envelope
   - (B) chloroplast
   - (C) Golgi apparatus
   - (D) plasma membrane

2. Which structure is common to plant *and* animal cells?
   - (A) chloroplast
   - (B) wall made of cellulose
   - (C) mitochondrion
   - (D) centriole

3. Which of the following is present in a prokaryotic cell?
   - (A) mitochondrion
   - (B) ribosome
   - (C) nuclear envelope
   - (D) chloroplast

4. Which structure-function pair is *mismatched*?
   - (A) microtubule; muscle contraction
   - (B) ribosome; protein synthesis
   - (C) Golgi; protein trafficking
   - (D) nucleolus; production of ribosomal subunits

## Level 2: Application/Analysis

5. Cyanide binds to at least one molecule involved in producing ATP. If a cell is exposed to cyanide, most of the cyanide will be found within the
   - (A) mitochondria.
   - (B) ribosomes.
   - (C) peroxisomes.
   - (D) lysosomes.

6. What is the most likely pathway taken by a newly synthesized protein that will be secreted by a cell?
   - (A) ER → Golgi → nucleus
   - (B) nucleus → ER → Golgi
   - (C) ER → Golgi → vesicles that fuse with plasma membrane
   - (D) ER → lysosomes → vesicles that fuse with plasma membrane

7. Which cell would be best for studying lysosomes?
   - (A) muscle cell
   - (B) nerve cell
   - (C) phagocytic white blood cell
   - (D) bacterial cell

8. **DRAW IT** From memory, draw two eukaryotic cells. Label the structures listed here and show any physical connections between the internal structures of each cell: nucleus, rough ER, smooth ER, mitochondrion, centrosome, chloroplast, vacuole, lysosome, microtubule, cell wall, ECM, microfilament, Golgi apparatus, intermediate filament, plasma membrane, peroxisome, ribosome, nucleolus, nuclear pore, vesicle, flagellum, microvilli, plasmodesma.

## Level 3: Synthesis/Evaluation

9. **SCIENTIFIC INQUIRY**
   In studying micrographs of an unusual protist (single-celled eukaryote) that you found in a sample of pond water, you spot an organelle that you can't recognize. You successfully develop a method for growing this organism in liquid in the laboratory. Describe how you would go about finding out what this organelle is and what it does in the cell. Assume that you would make use of additional microscopy, cell fractionation, and biochemical tests.

10. **FOCUS ON EVOLUTION**
    Compare different aspects of cell structure. (a) What structures best reveal evolutionary unity? (b) Provide an example of diversity related to specialized modifications.

11. **FOCUS ON ORGANIZATION**
    Considering some of the characteristics that define life and drawing on your new knowledge of cellular structures and functions, write a short essay (100–150 words) that discusses this statement: Life is an emergent property that appears at the level of the cell. (Review the section on emergent properties in Concept 1.1.)

12. **SYNTHESIZE YOUR KNOWLEDGE**

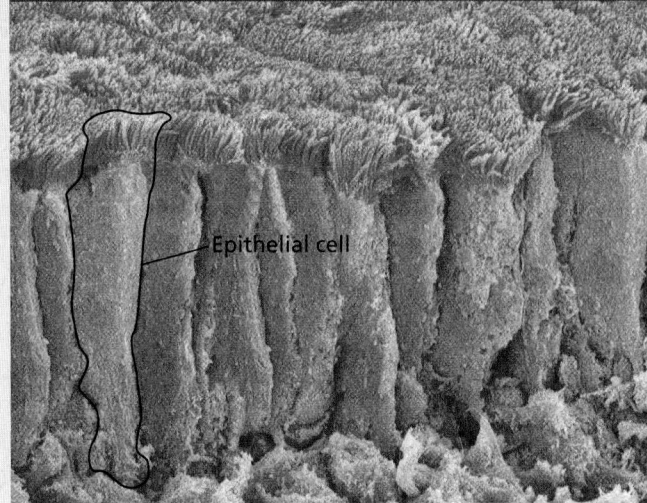

Epithelial cell

The cells in the SEM are epithelial cells from the small intestine. Discuss how their cellular structure contributes to their specialized functions of nutrient absorption and as a barrier between intestinal contents and the blood supply on the other side of the cell sheet.

*For selected answers, see Appendix A.*

# 5 Membrane Transport and Cell Signaling

## KEY CONCEPTS

**5.1** Cellular membranes are fluid mosaics of lipids and proteins

**5.2** Membrane structure results in selective permeability

**5.3** Passive transport is diffusion of a substance across a membrane with no energy investment

**5.4** Active transport uses energy to move solutes against their gradients

**5.5** Bulk transport across the plasma membrane occurs by exocytosis and endocytosis

**5.6** The plasma membrane plays a key role in most cell signaling

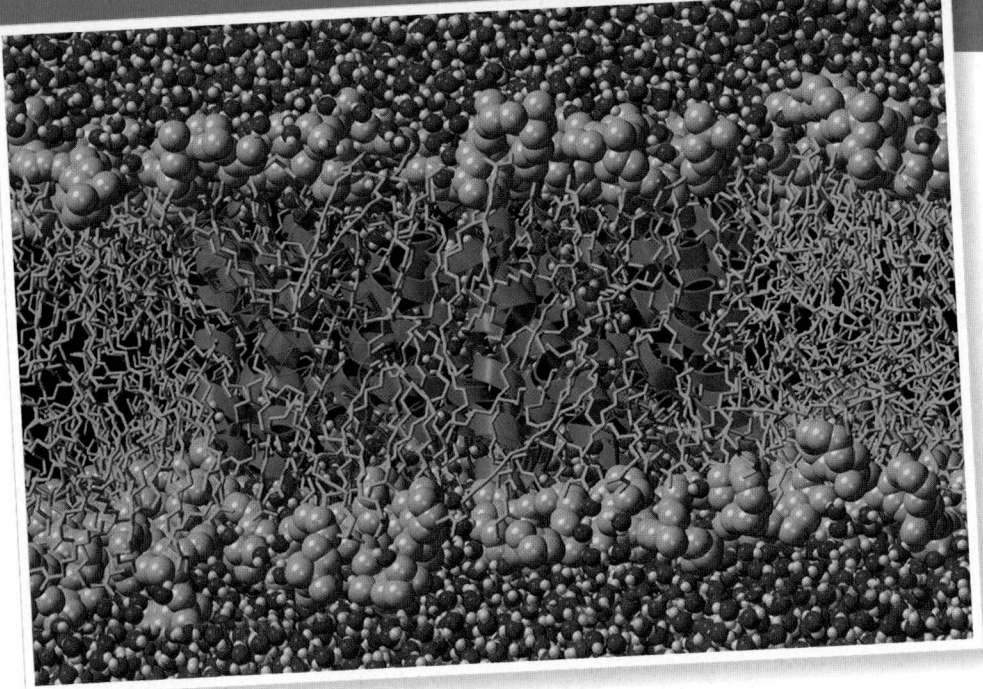

▲ **Figure 5.1** How do cell membrane proteins help regulate chemical traffic?

## Life at the Edge

The plasma membrane is the edge of life, the boundary that separates the living cell from its surroundings. A remarkable film only about 8 nm thick—it would take over 8,000 plasma membranes to equal the thickness of a piece of paper—the plasma membrane controls traffic into and out of the cell it surrounds. Like all biological membranes, the plasma membrane exhibits **selective permeability**; that is, it allows some substances to cross it more easily than others. The resulting ability of the cell to discriminate in its chemical exchanges with its environment is fundamental to life.

Most of this chapter is devoted to how cellular membranes control the passage of substances through them. **Figure 5.1** shows a computer model of water molecules (red and gray) passing through a short section of a membrane, a phospholipid bilayer (phosphates are yellow, and hydrocarbon tails are green). The blue ribbons within the lipid bilayer represent helical regions of a membrane protein called an aquaporin. One molecule of this protein enables billions of water molecules to pass through the membrane every second, many more than could cross on their own. Found in many kinds of cells, aquaporins are but one example of how the plasma membrane and its proteins enable cells to survive and function.

To understand how membranes work, we'll begin by examining their molecular structure. Then we'll describe in some detail how plasma membranes control transport into and out of cells. Finally, we'll discuss cell signaling, emphasizing the role of the plasma membrane in cell communication.

CONCEPT 5.1

## Cellular membranes are fluid mosaics of lipids and proteins

**Figure 5.2** shows the currently accepted model of the arrangement of molecules in the plasma membrane. Lipids and proteins are the staple ingredients of membranes, although carbohydrates are also important. The most abundant lipids in most membranes are phospholipids. The ability of phospholipids to form membranes is inherent in their molecular structure. A phospholipid is an **amphipathic** molecule, meaning it has both a hydrophilic region and a hydrophobic region (see Figure 3.15). A phospholipid bilayer can exist as a stable boundary between two aqueous compartments because the molecular arrangement shelters the hydrophobic tails of the phospholipids from water while exposing the hydrophilic heads to water **(Figure 5.3)**.

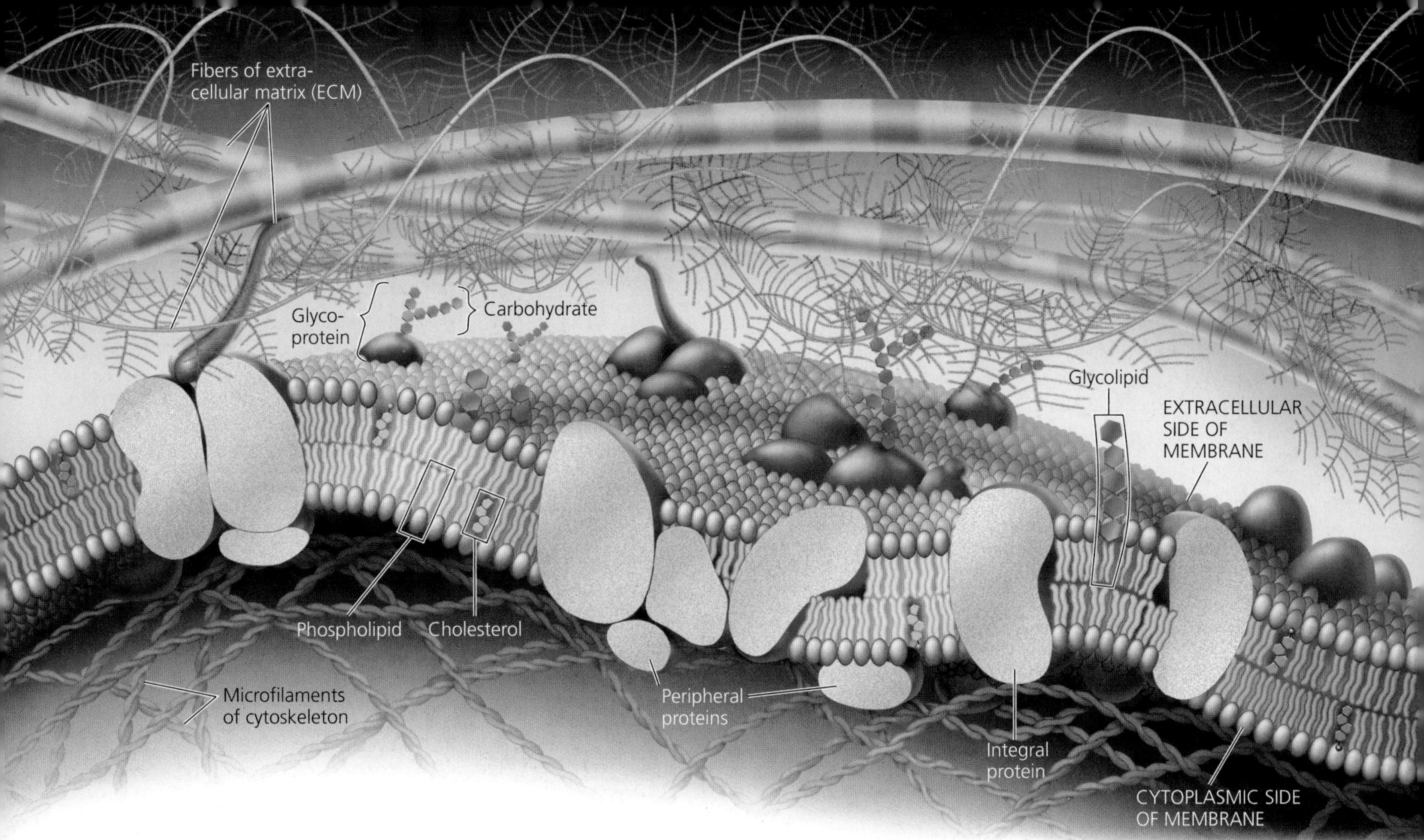

Fibers of extra-
cellular matrix (ECM)

Glyco-
protein

Carbohydrate

Glycolipid

EXTRACELLULAR
SIDE OF
MEMBRANE

Phospholipid   Cholesterol

Microfilaments
of cytoskeleton

Peripheral
proteins

Integral
protein

CYTOPLASMIC SIDE
OF MEMBRANE

▲ **Figure 5.2 Current model of an animal cell's plasma membrane (cutaway view).** Lipids are colored gray and gold, proteins purple, and carbohydrates green.

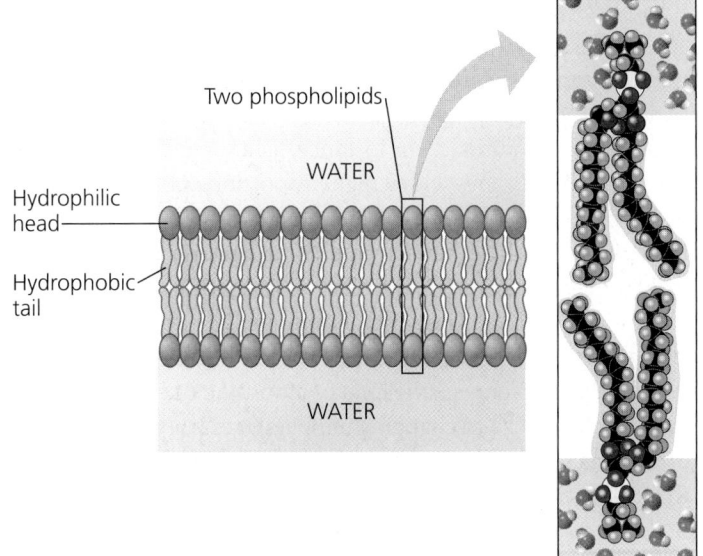

Two phospholipids

WATER

Hydrophilic
head

Hydrophobic
tail

WATER

▲ **Figure 5.3 Phospholipid bilayer (cross section).**

**MAKE CONNECTIONS** *Refer to Figure 3.15b, and then circle the hydrophilic and hydrophobic portions of the upper phospholipid on the right side of Figure 5.3. Explain what each portion contacts when the phospholipid is in the plasma membrane.*

Like membrane lipids, most membrane proteins are amphipathic. Such proteins can reside in the phospholipid bilayer with their hydrophilic regions protruding. This molecular orientation maximizes contact of the hydrophilic regions of a protein with water in the cytosol and extracellular fluid, while providing its hydrophobic parts with a nonaqueous environment.

In the **fluid mosaic model** in Figure 5.2, the membrane is a mosaic of protein molecules bobbing in a fluid bilayer of phospholipids. The proteins are not randomly distributed in the membrane, however. Groups of proteins are often associated in long-lasting, specialized patches, as are certain lipids. In some regions, the membrane may be much more packed with proteins than shown in Figure 5.2. Like all models, the fluid mosaic model is continually being refined as new research reveals more about membrane structure.

## The Fluidity of Membranes

Membranes are not static sheets of molecules locked rigidly in place. A membrane is held together primarily by hydrophobic interactions, which are much weaker than covalent bonds (see Figure 3.22). Most of the lipids and some of the proteins can shift about laterally—that is, in the plane of the membrane—like partygoers elbowing their way through a crowded room.

The lateral movement of phospholipids within the membrane is rapid. Proteins are much larger than lipids and move more slowly, but some membrane proteins do drift, as shown

## Do membrane proteins move?

**Experiment** Larry Frye and Michael Edidin, at Johns Hopkins University, labeled the plasma membrane proteins of a mouse cell and a human cell with two different markers and fused the cells. Using a microscope, they observed the markers on the hybrid cell.

**Results**

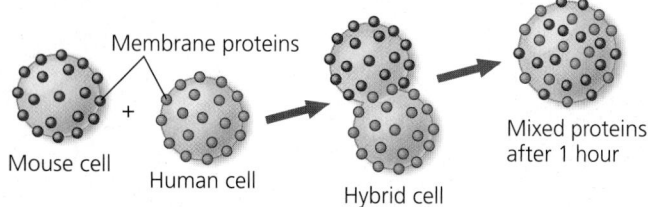

Membrane proteins

Mouse cell

Human cell

Hybrid cell

Mixed proteins after 1 hour

**Conclusion** The mixing of the mouse and human membrane proteins indicates that at least some membrane proteins move sideways within the plane of the plasma membrane.

**Data from** L. D. Frye and M. Edidin, The rapid intermixing of cell surface antigens after formation of mouse-human heterokaryons, *Journal of Cell Science* 7:319 (1970).

**WHAT IF?** Suppose the proteins did not mix in the hybrid cell, even many hours after fusion. Would you be able to conclude that proteins don't move within the membrane? What other explanation could there be?

▼ **Figure 5.5** Factors that affect membrane fluidity.

**(a) Unsaturated versus saturated hydrocarbon tails.**

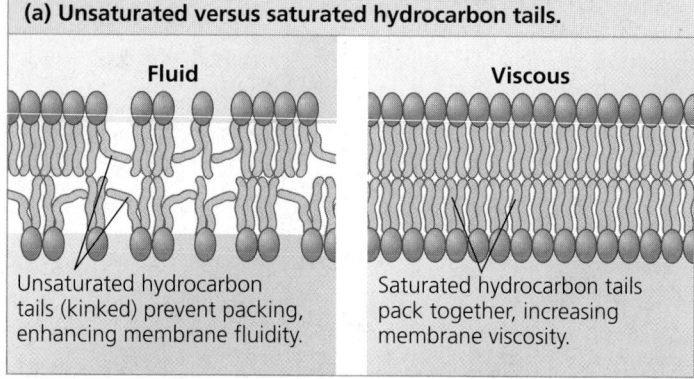

Fluid

Viscous

Unsaturated hydrocarbon tails (kinked) prevent packing, enhancing membrane fluidity.

Saturated hydrocarbon tails pack together, increasing membrane viscosity.

**(b) Cholesterol within the animal cell membrane.**

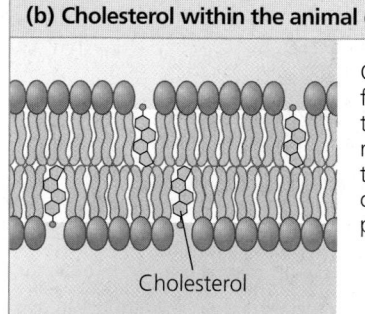

Cholesterol reduces membrane fluidity at moderate temperatures by reducing phospholipid movement, but at low temperatures it hinders solidification by disrupting the regular packing of phospholipids.

Cholesterol

---

in a classic experiment described in **Figure 5.4**. And some membrane proteins seem to move in a highly directed manner, perhaps driven along cytoskeletal fibers by motor proteins. However, many other membrane proteins seem to be held immobile by their attachment to the cytoskeleton or to the extracellular matrix (see Figure 5.2).

A membrane remains fluid as temperature decreases until the phospholipids settle into a closely packed arrangement and the membrane solidifies, much as bacon grease forms lard when it cools. The temperature at which a membrane solidifies depends on the types of lipids it is made of. The membrane remains fluid to a lower temperature if it is rich in phospholipids with unsaturated hydrocarbon tails (see Figures 3.14 and 3.15). Because of kinks in the tails where double bonds are located, unsaturated hydrocarbon tails cannot pack together as closely as saturated hydrocarbon tails, and this looseness makes the membrane more fluid **(Figure 5.5a)**.

The steroid cholesterol, which is wedged between phospholipid molecules in the plasma membranes of animal cells, has different effects on membrane fluidity at different temperatures **(Figure 5.5b)**. At relatively high temperatures—at 37°C, the body temperature of humans, for example—cholesterol makes the membrane less fluid by restraining phospholipid movement. However, because cholesterol also hinders the close packing of phospholipids, it lowers the temperature required for the membrane to solidify. Thus, cholesterol helps membranes resist changes in fluidity when the temperature changes.

Membranes must be fluid to work properly; they are usually about as fluid as salad oil. When a membrane solidifies, its

permeability changes, and enzymatic proteins in the membrane may become inactive. However, membranes that are too fluid cannot support protein function either. Therefore, extreme environments pose a challenge for life, resulting in evolutionary adaptations that include differences in membrane lipid composition.

## Evolution of Differences in Membrane Lipid Composition

**EVOLUTION** Variations in the cell membrane lipid compositions of many species appear to be evolutionary adaptations that maintain the appropriate membrane fluidity under specific environmental conditions. For instance, fishes that live in extreme cold have membranes with a high proportion of unsaturated hydrocarbon tails, enabling their membranes to remain fluid (see Figure 5.5a). At the other extreme, some bacteria and archaea thrive at temperatures greater than 90°C (194°F) in thermal hot springs and geysers. Their membranes include unusual lipids that help prevent excessive fluidity at such high temperatures.

The ability to change the lipid composition of cell membranes in response to changing temperatures has evolved in organisms that live where temperatures vary. In many plants that tolerate extreme cold, such as winter wheat, the percentage of unsaturated phospholipids increases in autumn, keeping the membranes from solidifying during winter. Some bacteria and archaea can also change the proportion of unsaturated phospholipids in their cell membranes, depending on the temperature at which they are growing. Overall, natural selection has apparently favored organisms whose mix of membrane lipids ensures an appropriate level of membrane fluidity for their environment.

## Membrane Proteins and Their Functions

Now we return to the *mosaic* aspect of the fluid mosaic model. Somewhat like a tile mosaic, a membrane is a collage of different proteins embedded in the fluid matrix of the lipid bilayer (see Figure 5.2). More than 50 kinds of proteins have been found so far in the plasma membrane of red blood cells, for example. Phospholipids form the main fabric of the membrane, but proteins determine most of the membrane's functions. Different types of cells contain different sets of membrane proteins, and the various membranes within a cell each have a unique collection of proteins.

Notice in Figure 5.2 that there are two major populations of membrane proteins: integral proteins and peripheral proteins. **Integral proteins** penetrate the hydrophobic interior of the lipid bilayer. The majority are *transmembrane proteins*, which span the membrane; other integral proteins extend only partway into the hydrophobic interior. The hydrophobic regions of an integral protein consist of one or more stretches of nonpolar amino acids (see Figure 3.18), usually coiled into α helices **(Figure 5.6)**. The hydrophilic parts of the molecule are exposed to the aqueous solutions on either side of the membrane. Some proteins also have one or more hydrophilic channels that allow passage of hydrophilic substances (even water itself; see Figure 5.1). **Peripheral proteins** are not embedded in the lipid bilayer at all; they are loosely bound to the surface of the membrane, often to exposed parts of integral proteins (see Figure 5.2).

On the cytoplasmic side of the plasma membrane, some membrane proteins are held in place by attachment to the cytoskeleton. And on the extracellular side, certain membrane proteins are attached to fibers of the extracellular matrix (see Figure 4.26). These attachments combine to give animal cells a stronger framework than the plasma membrane alone could provide.

**Figure 5.7** gives an overview of six major functions performed by proteins of the plasma membrane. A single cell may have membrane proteins carrying out several of these functions, and a single membrane protein may have multiple

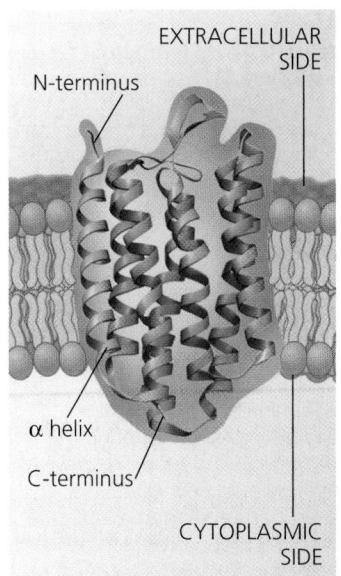

◀ **Figure 5.6 The structure of a transmembrane protein.** Bacteriorhodopsin (a bacterial transport protein) has a distinct orientation in the membrane, with its N-terminus outside the cell and its C-terminus inside. This ribbon model highlights the α-helical secondary structure of the hydrophobic parts, which lie mostly within the hydrophobic interior of the membrane. The protein includes seven transmembrane helices. The nonhelical hydrophilic segments are in contact with the aqueous solutions on the extracellular and cytoplasmic sides of the membrane.

**(a) Transport.** *Left:* A protein that spans the membrane may provide a hydrophilic channel across the membrane that is selective for a particular solute. *Right:* Other transport proteins shuttle a substance from one side to the other by changing shape. Some of these proteins hydrolyze ATP as an energy source to actively pump substances across the membrane.

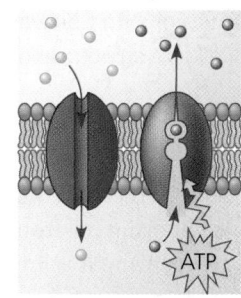

**(b) Enzymatic activity.** A protein built into the membrane may be an enzyme with its active site (where the target molecule binds) exposed to substances in the adjacent solution. In some cases, several enzymes in a membrane are organized as a team that carries out sequential steps of a metabolic pathway.

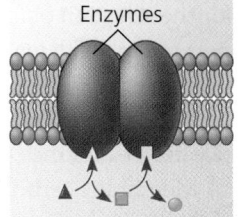

**(c) Signal transduction.** A membrane protein (receptor) may have a binding site with a specific shape that fits the shape of a chemical messenger, such as a hormone. The external messenger (signaling molecule) may cause the protein to change shape, allowing it to relay the message to the inside of the cell, usually by binding to a cytoplasmic protein.

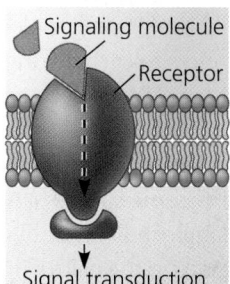

**(d) Cell-cell recognition.** Some glycoproteins serve as identification tags that are specifically recognized by membrane proteins of other cells. This type of cell-cell binding is usually short-lived compared to that shown in (e).

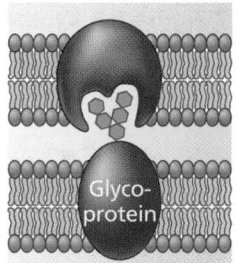

**(e) Intercellular joining.** Membrane proteins of adjacent cells may hook together in various kinds of junctions, such as gap junctions or tight junctions. This type of binding is more long-lasting than that shown in (d).

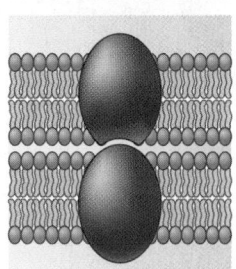

**(f) Attachment to the cytoskeleton and extracellular matrix (ECM).** Microfilaments or other elements of the cytoskeleton may be noncovalently bound to membrane proteins, a function that helps maintain cell shape and stabilizes the location of certain membrane proteins. Proteins that can bind to ECM molecules can coordinate extracellular and intracellular changes.

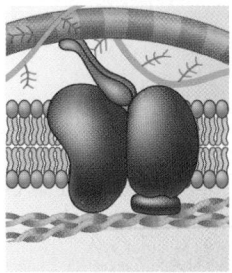

▲ **Figure 5.7 Some functions of membrane proteins.** In many cases, a single protein performs multiple tasks.

**?** *Some transmembrane proteins can bind to a particular ECM molecule and, when bound, transmit a signal into the cell. Use the proteins shown in (c) and (f) to explain how this might occur.*

functions. In this way, the membrane is a functional mosaic as well as a structural one.

## The Role of Membrane Carbohydrates in Cell-Cell Recognition

Cell-cell recognition, a cell's ability to distinguish one type of neighboring cell from another, is crucial to the functioning of an organism. It is important, for example, in the sorting of cells into tissues and organs in an animal embryo. It is also the basis for the rejection of foreign cells by the immune system, an important line of defense in vertebrate animals (see Concept 35.3). Cells recognize other cells by binding to molecules, often containing carbohydrates, on the extracellular surface of the plasma membrane (see Figure 5.7d).

Membrane carbohydrates are usually short, branched chains of fewer than 15 sugar units. Some are covalently bonded to lipids, forming molecules called **glycolipids**. (Recall that *glyco* refers to carbohydrate.) However, most are covalently bonded to proteins, which are thereby **glycoproteins**.

The carbohydrates on the extracellular side of the plasma membrane vary from species to species, among individuals of the same species, and even from one cell type to another in a single individual. The diversity of the molecules and their location on the cell's surface enable membrane carbohydrates to function as markers that distinguish one cell from another. For example, the four human blood types designated A, B, AB, and O reflect variation in the carbohydrate part of glycoproteins on the surface of red blood cells.

## Synthesis and Sidedness of Membranes

Membranes have distinct inside and outside faces. The two lipid layers may differ in lipid composition, and each protein has directional orientation in the membrane (see Figure 5.6, for example). **Figure 5.8** shows how membrane sidedness arises: The asymmetric arrangement of proteins, lipids, and their associated carbohydrates in the plasma membrane is determined as the membrane is being built by the endoplasmic reticulum (ER) and Golgi apparatus.

### CONCEPT CHECK 5.1

1. Plasma membrane proteins have carbohydrates attached to them in the ER and Golgi apparatus and then are transported in vesicles to the cell surface. On which side of the vesicle membrane are the carbohydrates?

2. **WHAT IF?** How would the membrane lipid composition of a native grass found in very warm soil around hot springs compare with that of a native grass found in cooler soil? Explain.

For suggested answers, see Appendix A.

▼ **Figure 5.8 Synthesis of membrane components and their orientation in the membrane.** The cytoplasmic (orange) face of the plasma membrane differs from the extracellular (aqua) face. The latter arises from the inside face of ER, Golgi, and vesicle membranes.

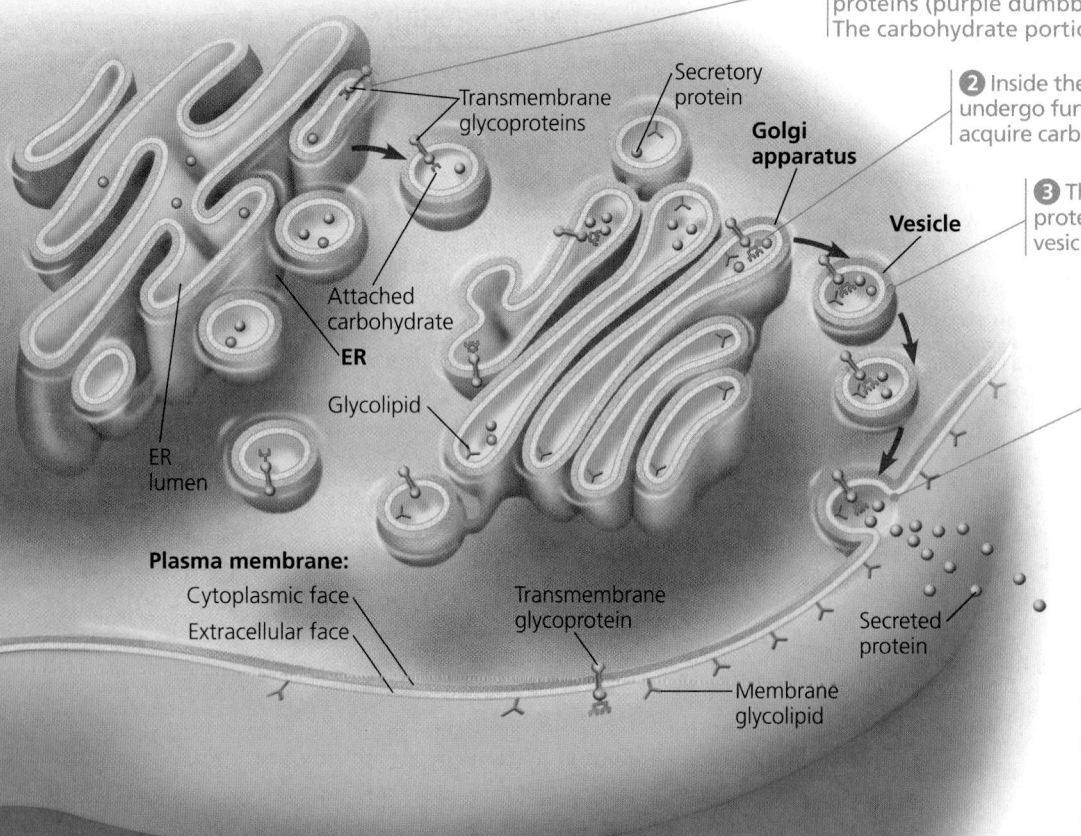

❶ Membrane proteins and lipids are synthesized in association with the endoplasmic reticulum (ER). In the ER, carbohydrates (green) are added to the transmembrane proteins (purple dumbbells), making them glycoproteins. The carbohydrate portions may then be modified.

❷ Inside the Golgi apparatus, the glycoproteins undergo further carbohydrate modification, and lipids acquire carbohydrates, becoming glycolipids.

❸ The glycoproteins, glycolipids, and secretory proteins (purple spheres) are transported in vesicles to the plasma membrane.

❹ As vesicles fuse with the plasma membrane, the outside face of the vesicle becomes continuous with the inside (cytoplasmic) face of the plasma membrane. This releases the secretory proteins from the cell, a process called *exocytosis*, and positions the carbohydrates of membrane glycoproteins and glycolipids on the outside (extracellular) face of the plasma membrane.

**DRAW IT** *Draw an integral membrane protein extending from partway through the ER membrane into the ER lumen. Next, draw the protein where it would be located in a series of numbered steps ending at the plasma membrane. Would the protein contact the cytoplasm or the extracellular fluid?*

## CONCEPT 5.2

# Membrane structure results in selective permeability

The biological membrane is an exquisite example of a supramolecular structure—many molecules ordered into a higher level of organization—with emergent properties beyond those of the individual molecules. We now focus on one of the most important of those properties: the ability to regulate transport across cellular boundaries, a function essential to the cell's existence. We will see once again that form fits function: The fluid mosaic model helps explain how membranes regulate the cell's molecular traffic.

A steady traffic of small molecules and ions moves across the plasma membrane in both directions. Consider the chemical exchanges between a muscle cell and the extracellular fluid that bathes it. Sugars, amino acids, and other nutrients enter the cell, and metabolic waste products leave it. The cell takes in $O_2$ for use in cellular respiration and expels $CO_2$. Also, the cell regulates its concentrations of inorganic ions, such as $Na^+$, $K^+$, $Ca^{2+}$, and $Cl^-$, by shuttling them one way or the other across the plasma membrane. In spite of heavy traffic through them, cell membranes are selectively permeable, and substances do not cross the barrier indiscriminately. The cell is able to take up some small molecules and ions and exclude others.

## The Permeability of the Lipid Bilayer

Nonpolar molecules, such as hydrocarbons, $CO_2$, and $O_2$, are hydrophobic. They can therefore dissolve in the lipid bilayer of the membrane and cross it easily, without the aid of membrane proteins. However, the hydrophobic interior of the membrane impedes the direct passage through the membrane of ions and polar molecules, which are hydrophilic. Polar molecules such as glucose and other sugars pass only slowly through a lipid bilayer, and even water, a very small polar molecule, does not cross rapidly. A charged atom or molecule and its surrounding shell of water (see Figure 2.21) are even less likely to penetrate the hydrophobic interior of the membrane. Furthermore, the lipid bilayer is only one aspect of the gatekeeper system responsible for a cell's selective permeability. Proteins built into the membrane play key roles in regulating transport.

## Transport Proteins

Specific ions and a variety of polar molecules can't move through cell membranes on their own. However, these hydrophilic substances can avoid contact with the lipid bilayer by passing through **transport proteins** that span the membrane.

Some transport proteins, called *channel proteins*, function by having a hydrophilic channel that certain molecules or atomic ions use as a tunnel through the membrane (see Figure 5.7a, left). For example, as you read earlier, the passage of water molecules through the plasma membrane of certain cells is greatly facilitated by channel proteins called

aquaporins (see Figure 5.1). Most aquaporin proteins consist of four identical subunits (see Figure 3.22). The polypeptide making up each subunit forms a channel that allows single-file passage of up to *3 billion* ($3 \times 10^9$) water molecules per second, many more than would cross the membrane without aquaporin. Other transport proteins, called *carrier proteins*, hold onto their passengers and change shape in a way that shuttles them across the membrane (see Figure 5.7a, right).

A transport protein is specific for the substance it translocates (moves), allowing only a certain substance (or a small group of related substances) to cross the membrane. For example, a specific carrier protein in the plasma membrane of red blood cells transports glucose across the membrane 50,000 times faster than glucose can pass through on its own. This "glucose transporter" is so selective that it even rejects fructose, a structural isomer of glucose (see Figure 3.8).

Thus, the selective permeability of a membrane depends on both the discriminating barrier of the lipid bilayer and the specific transport proteins built into the membrane. But what establishes the *direction* of traffic across a membrane? At a given time, what determines whether a particular substance will enter the cell or leave the cell? And what mechanisms actually drive molecules across membranes? We will address these questions next as we explore two modes of membrane traffic: passive transport and active transport.

### CONCEPT CHECK 5.2

1. What property allows $O_2$ and $CO_2$ to cross a lipid bilayer without the help of membrane proteins?
2. Why is a transport protein needed to move many water molecules rapidly across a membrane?
3. **MAKE CONNECTIONS** Aquaporins exclude passage of hydronium ions ($H_3O^+$), but some aquaporins allow passage of glycerol, a three-carbon alcohol (see Figure 3.13), as well as $H_2O$. Since $H_3O^+$ is closer in size to water than glycerol is, yet cannot pass through, what might be the basis of this selectivity?

For suggested answers, see Appendix A.

## CONCEPT 5.3

# Passive transport is diffusion of a substance across a membrane with no energy investment

Molecules have a type of energy called thermal energy, due to their constant motion (see Concept 2.5). One result of this motion is **diffusion**, the movement of particles of any substance so that they tend to spread out into the available space. Each molecule moves randomly, yet diffusion of a *population* of molecules may be directional. To understand this process, let's imagine a synthetic membrane separating pure water from

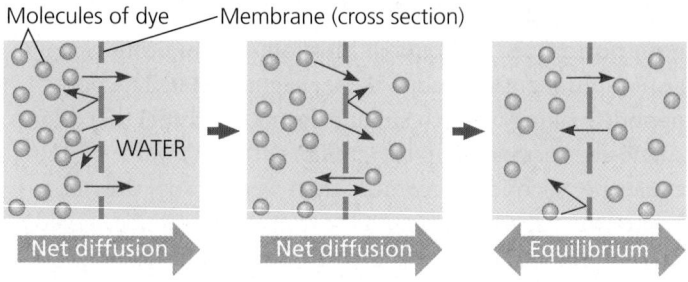

Molecules of dye — Membrane (cross section)

WATER

Net diffusion → Net diffusion → Equilibrium ↔

**(a) Diffusion of one solute.** The membrane has pores large enough for molecules of dye to pass through. Random movement of dye molecules will cause some to pass through the pores; this will happen more often on the side with more dye molecules. The dye diffuses from where it is more concentrated to where it is less concentrated (called diffusing down a concentration gradient). This leads to a dynamic equilibrium: The solute molecules continue to cross the membrane, but at roughly equal rates in both directions.

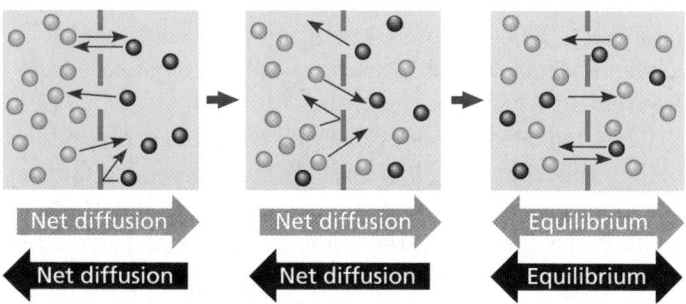

Net diffusion → Net diffusion → Equilibrium
Net diffusion ← Net diffusion ← Equilibrium

**(b) Diffusion of two solutes.** Solutions of two different dyes are separated by a membrane that is permeable to both. Each dye diffuses down its own concentration gradient. There will be a net diffusion of the purple dye toward the left, even though the *total* solute concentration was initially greater on the left side.

▲ **Figure 5.9 The diffusion of solutes across a synthetic membrane.** Each of the large arrows under the diagrams shows the net diffusion of the dye molecules of that color.

a solution of a dye in water. Study **Figure 5.9a** to appreciate how diffusion would result in both solutions having equal concentrations of the dye molecules. Once that point is reached, there will be a dynamic equilibrium, with roughly as many dye molecules crossing the membrane each second in one direction as in the other.

We can now state a simple rule of diffusion: In the absence of other forces, a substance will diffuse from where it is more concentrated to where it is less concentrated. Put another way, any substance will diffuse down its **concentration gradient**, the region along which the density of a substance increases or decreases (in this case, decreases). No work must be done to make this happen; diffusion is a spontaneous process, needing no input of energy. Note that each substance diffuses down its *own* concentration gradient, unaffected by the concentration gradients of other substances **(Figure 5.9b)**.

Much of the traffic across cell membranes occurs by diffusion. When a substance is more concentrated on one side of a membrane than on the other, there is a tendency for the substance to diffuse across the membrane down its concentration gradient (assuming that the membrane is permeable to that substance). One important example is the uptake of oxygen by a cell performing

cellular respiration. Dissolved oxygen diffuses into the cell across the plasma membrane. As long as cellular respiration consumes the $O_2$ as it enters, diffusion into the cell will continue because the concentration gradient favors movement in that direction.

The diffusion of a substance across a biological membrane is called **passive transport** because the cell does not have to expend energy to make it happen. The concentration gradient itself represents potential energy (see Concept 2.2 and Figure 6.5b) and drives diffusion. Remember, however, that membranes are selectively permeable and therefore have different effects on the rates of diffusion of various molecules. In the case of water, aquaporins allow water to diffuse very rapidly across the membranes of certain cells. As we'll see next, the movement of water across the plasma membrane has important consequences for cells.

## Effects of Osmosis on Water Balance

To see how two solutions with different solute concentrations interact, picture a U-shaped glass tube with a selectively permeable artificial membrane separating two sugar solutions **(Figure 5.10)**. Pores in this synthetic membrane are too small

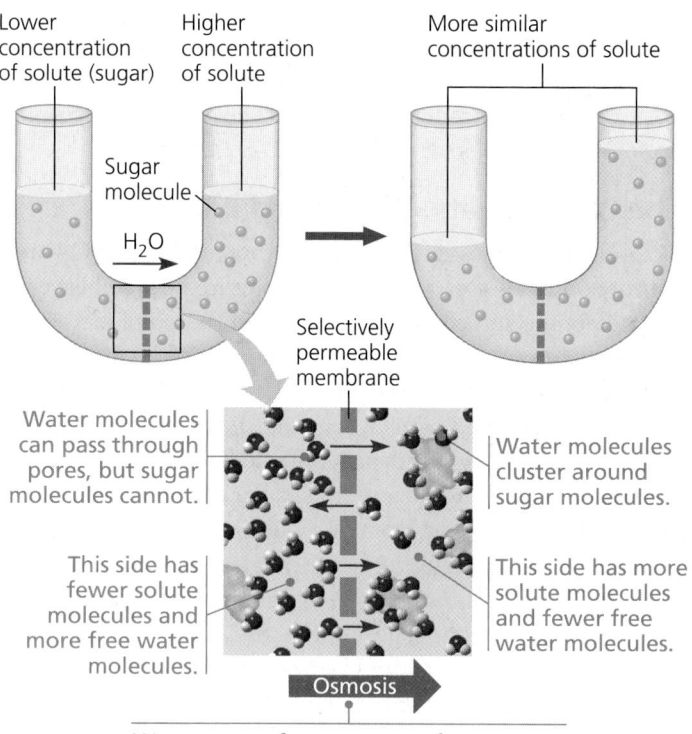

Lower concentration of solute (sugar)
Higher concentration of solute
More similar concentrations of solute

Sugar molecule

$H_2O$

Selectively permeable membrane

Water molecules can pass through pores, but sugar molecules cannot.
Water molecules cluster around sugar molecules.

This side has fewer solute molecules and more free water molecules.
This side has more solute molecules and fewer free water molecules.

Osmosis

Water moves from an area of higher to lower free water concentration (lower to higher solute concentration).

▲ **Figure 5.10 Osmosis.** Two sugar solutions of different concentrations are separated by a membrane that the solvent (water) can pass through but the solute (sugar) cannot. Water molecules move randomly and may cross in either direction, but overall, water diffuses from the solution with less concentrated solute to that with more concentrated solute. This passive transport of water, or osmosis, makes the sugar concentrations on both sides roughly equal.

**WHAT IF?** *If an orange dye capable of passing through the membrane was added to the left side of the tube above, how would it be distributed at the end of the experiment? (See Figure 5.9.) Would the final solution levels in the tube be affected?*

for sugar molecules to pass through but large enough for water molecules. However, tight clustering of water molecules around the hydrophilic solute molecules makes some of the water unavailable to cross the membrane. As a result, the solution with a higher solute concentration has a lower *free* water concentration. Water diffuses across the membrane from the region of higher free water concentration (lower solute concentration) to that of lower free water concentration (higher solute concentration) until the solute concentrations on both sides of the membrane are more nearly equal. The diffusion of free water across a selectively permeable membrane, whether artificial or cellular, is called **osmosis**. The movement of water across cell membranes and the balance of water between the cell and its environment are crucial to organisms. Let's now apply to living cells what we've learned about osmosis in this system to living cells.

**(a) Animal cell.** An animal cell fares best in an isotonic environment unless it has special adaptations that offset the osmotic uptake or loss of water.

**(b) Plant cell.** Plant cells are turgid (firm) and generally healthiest in a hypotonic environment, where the uptake of water is eventually balanced by the wall pushing back on the cell.

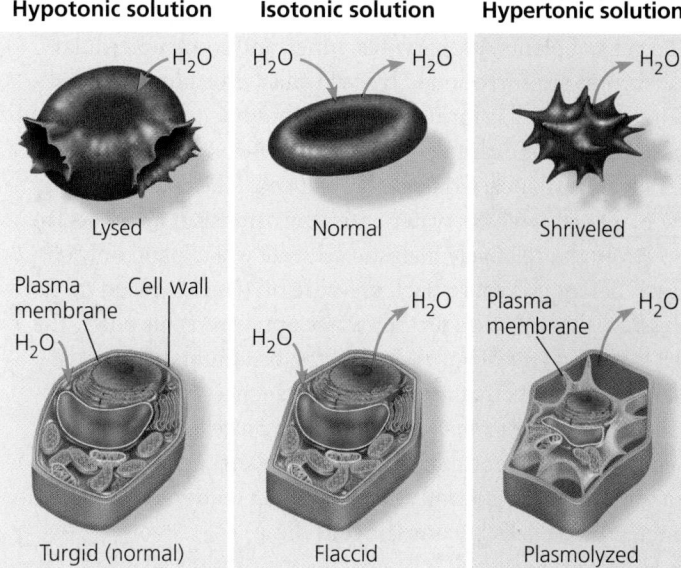

**▲ Figure 5.11 The water balance of living cells.** How living cells react to changes in the solute concentration of their environment depends on whether or not they have cell walls. **(a)** Animal cells, such as this red blood cell, do not have cell walls. **(b)** Plant cells do. (Arrows indicate net water movement after the cells were first placed in these solutions.)

### Water Balance of Cells Without Cell Walls

To explain the behavior of a cell in a solution, we must consider both solute concentration and membrane permeability. Both factors are taken into account in the concept of **tonicity**, the ability of a surrounding solution to cause a cell to gain or lose water. The tonicity of a solution depends in part on its concentration of solutes that cannot cross the membrane (nonpenetrating solutes) relative to that inside the cell. If there is a higher concentration of nonpenetrating solutes in the surrounding solution, water will tend to leave the cell, and vice versa.

If a cell without a cell wall, such as an animal cell, is immersed in an environment that is **isotonic** to the cell (*iso* means "same"), there will be no *net* movement of water across the plasma membrane. Water diffuses across the membrane, but at the same rate in both directions. In an isotonic environment, the volume of an animal cell is stable, as shown in the middle of **Figure 5.11a**.

Let's transfer the cell to a solution that is **hypertonic** to the cell (*hyper* means "more," in this case referring to nonpenetrating solutes). The cell will lose water, shrivel, and probably die (see Figure 5.11a, right). This is why an increase in the salinity (saltiness) of a lake can kill the animals there; if the lake water becomes hypertonic to the animals' cells, they might shrivel and die. However, taking up too much water can be just as hazardous as losing water. If we place the cell in a solution that is **hypotonic** to the cell (*hypo* means "less"), water will enter the cell faster than it leaves, and the cell will swell and lyse (burst) like an overfilled water balloon (see Figure 5.11a, left).

A cell without rigid cell walls can tolerate neither excessive uptake nor excessive loss of water. This problem of water balance is automatically solved if such a cell lives in isotonic surroundings. Seawater is isotonic to many marine invertebrates. The cells of most terrestrial (land-dwelling) animals are bathed in an extracellular fluid that is isotonic to the cells. In hypertonic or hypotonic environments, however, organisms that lack rigid cell walls must have other adaptations for **osmoregulation**, the control of solute concentrations and water balance. For example, the unicellular protist *Paramecium caudatum* lives in pond water, which is hypotonic to the cell. Water continually enters the cell. The *P. caudatum* cell doesn't burst because it is equipped with a contractile vacuole, an organelle that functions as a bilge pump to force water out of the cell as fast as it enters by osmosis **(Figure 5.12)**. We will examine other evolutionary adaptations for osmoregulation in Concept 32.4.

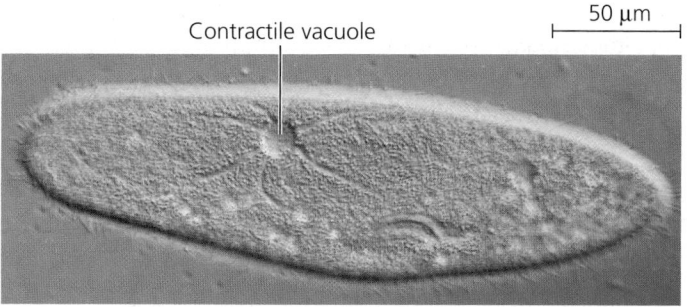

**▲ Figure 5.12 The contractile vacuole of *Paramecium caudatum.*** The vacuole collects fluid from a system of canals in the cytoplasm. When full, the vacuole and canals contract, expelling fluid from the cell (LM).

### Water Balance of Cells with Cell Walls

The cells of plants, prokaryotes, fungi, and some unicellular eukaryotes are surrounded by cell walls (see Figure 4.25). When such a cell is immersed in a hypotonic solution—bathed in rainwater, for example—the cell wall helps maintain the cell's water balance. Consider a plant cell. Like an animal cell, the plant cell swells as water enters by osmosis **(Figure 5.11b)**. However, the relatively inelastic cell wall will expand only so much before it exerts a back pressure on the cell, called *turgor pressure*, that opposes further water uptake. At this point, the cell is **turgid** (very firm), which is the healthy state for most plant cells. Plants that are not woody, such as most house-plants, depend for mechanical support on cells kept turgid by a surrounding hypotonic solution. If a plant's cells and their surroundings are isotonic, there is no net tendency for water to enter, and the cells become **flaccid** (limp).

However, a cell wall is of no advantage if the cell is immersed in a hypertonic environment. In this case, a plant cell, like an animal cell, will lose water to its surroundings and shrink. As the plant cell shrivels, its plasma membrane pulls away from the cell wall at multiple places. This phenomenon, called **plasmolysis**, causes the plant to wilt and can lead to plant death. The walled cells of bacteria and fungi also plasmolyze in hypertonic environments.

## Facilitated Diffusion: Passive Transport Aided by Proteins

Let's look more closely at how water and certain hydrophilic solutes cross a membrane. As mentioned earlier, many polar molecules and ions impeded by the lipid bilayer of the membrane diffuse passively with the help of transport proteins that span the membrane. This phenomenon is called **facilitated diffusion**. Cell biologists are still trying to learn exactly how various transport proteins facilitate diffusion. Most transport proteins are very specific: They transport some substances but not others.

As mentioned earlier, the two types of transport proteins are channel proteins and carrier proteins. Channel proteins simply provide corridors that allow specific molecules or ions to cross the membrane **(Figure 5.13a)**. The hydrophilic passageways provided by these proteins can allow water molecules or small ions to diffuse very quickly from one side of the membrane to the other. Aquaporins, the water channel proteins, facilitate the massive amounts of diffusion that occur in plant cells and in animal cells such as red blood cells. Certain kidney cells also have many aquaporin molecules, allowing them to reclaim water from urine before it is excreted. If the kidneys did not perform this function, you would excrete about 180 L of urine per day—and have to drink an equal volume of water!

Channel proteins that transport ions are called **ion channels**. Many ion channels function as **gated channels**, which open or close in response to a stimulus. For some gated channels, the stimulus is electrical. In a nerve cell, for example,

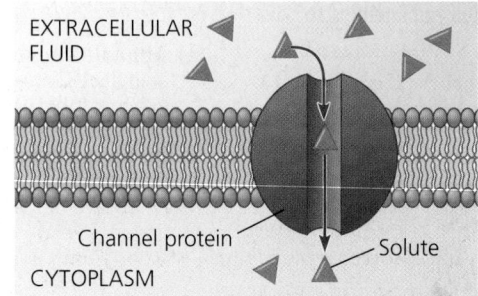

**(a)** A channel protein has a channel through which water molecules or a specific solute can pass.

EXTRACELLULAR FLUID

Channel protein    Solute

CYTOPLASM

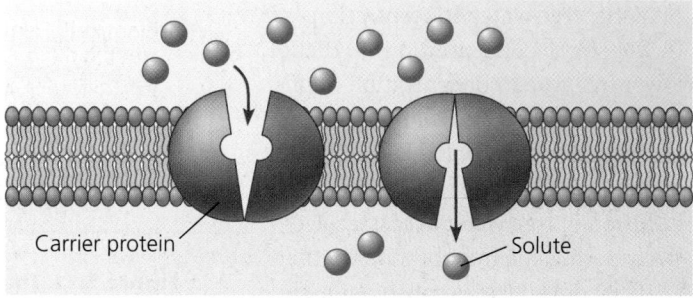

Carrier protein    Solute

**(b)** A carrier protein alternates between two shapes, moving a solute across the membrane during the shape change.

▲ **Figure 5.13 Two types of transport proteins that carry out facilitated diffusion.** In both cases, the protein can transport the solute in either direction, but the net movement is down the concentration gradient of the solute.

an ion channel opens in response to an electrical stimulus, allowing a stream of potassium ions to leave the cell. This restores the cell's ability to fire again. Other gated channels open or close when a specific substance other than the one to be transported binds to the channel. These gated channels are also important in the functioning of the nervous system (as you'll learn in Concept 37.3).

Carrier proteins, such as the glucose transporter mentioned earlier, seem to undergo a subtle change in shape that somehow translocates the solute-binding site across the membrane **(Figure 5.13b)**. Such a change in shape may be triggered by the binding and release of the transported molecule. Like ion channels, carrier proteins involved in facilitated diffusion result in the net movement of a substance down its concentration gradient. No energy input is required: This is passive transport. The **Scientific Skills Exercise** gives you an opportunity to work with data from an experiment related to glucose transport.

### CONCEPT CHECK 5.3

1. How do you think a cell performing cellular respiration rids itself of the resulting $CO_2$?
2. **WHAT IF?** If a *Paramecium caudatum* cell swims from a hypotonic to an isotonic environment, will its contractile vacuole become more active or less? Why?

For suggested answers, see Appendix A.

## Interpreting a Scatter Plot with Two Sets of Data

► 15-day-old and 1-month-old guinea pigs

**Is Glucose Uptake into Cells Affected by Age?** Glucose, an important energy source for animals, is transported into cells by facilitated diffusion using protein carriers. In this exercise, you will interpret a graph with two sets of data from an experiment that examined glucose uptake over time in red blood cells from guinea pigs of different ages. You will determine if the age of the guinea pigs affected their cells' rate of glucose uptake.

**How the Experiment Was Done** Researchers incubated guinea pig red blood cells in a 300 m*M* (millimolar) radioactive glucose solution at pH 7.4 at 25°C. Every 10 or 15 minutes, they removed a sample of cells from the solution and measured the concentration of radioactive glucose inside those cells. The cells came from either a 15-day-old or 1-month-old guinea pig.

**Data from the Experiment** When you have multiple sets of data, it can be useful to plot them on the same graph for comparison. In the graph here, each set of dots (dots of the same color) forms a *scatter plot*, in which every data point represents two numerical values, one for each variable. For each data set, a curve that best fits the points has been drawn to make it easier to see the trends. (For additional information about graphs, see the Scientific Skills Review in Appendix F and in the Study Area in MasteringBiology.)

### INTERPRET THE DATA

1. First make sure you understand the parts of the graph. (a) Which variable is the independent variable—the variable that was controlled by the researchers? (b) Which variable is the dependent variable—the variable that depended on the treatment and was measured by the researchers? (c) What do the red dots represent? (d) The blue dots?

2. From the data points on the graph, construct a table of the data. Put "Incubation Time (min)" in the left column of the table.

3. What does the graph show? Compare and contrast glucose uptake in red blood cells from a 15-day-old and a 1-month-old guinea pig.

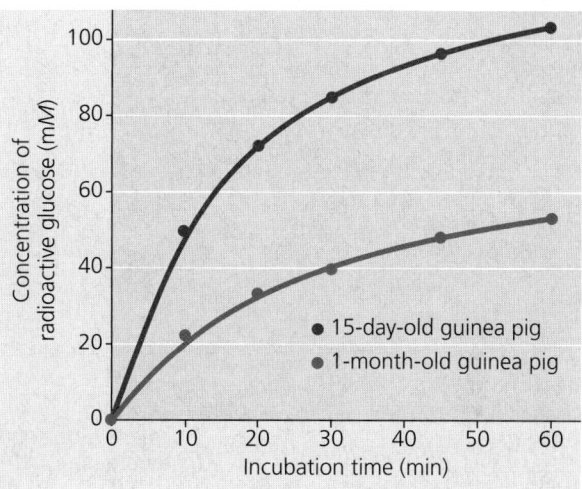

Glucose Uptake over Time in Guinea Pig Red Blood Cells

● 15-day-old guinea pig
● 1-month-old guinea pig

**Data from** T. Kondo and E. Beutler, Developmental changes in glucose transport of guinea pig erythrocytes, *Journal of Clinical Investigation* 65:1–4 (1980).

4. Develop a hypothesis to explain the difference between glucose uptake in red blood cells from a 15-day-old and a 1-month-old guinea pig. (Think about how glucose gets into cells.)

5. Design an experiment to test your hypothesis.

(MB) A version of this Scientific Skills Exercise can be assigned in MasteringBiology.

---

### CONCEPT 5.4

# Active transport uses energy to move solutes against their gradients

Despite the help of transport proteins, facilitated diffusion is considered passive transport because the solute is moving down its concentration gradient, a process that requires no energy. Facilitated diffusion speeds transport of a solute by providing efficient passage through the membrane, but it does not alter the direction of transport. Some other transport proteins, however, can move solutes against their concentration gradients, across the plasma membrane from the side where they are less concentrated (whether inside or outside) to the side where they are more concentrated.

## The Need for Energy in Active Transport

To pump a solute across a membrane against its gradient requires work; the cell must expend energy. Therefore, this type of membrane traffic is called **active transport**. The transport proteins that move solutes against their concentration gradients are all carrier proteins rather than channel proteins. This makes sense because when channel proteins are open, they merely allow solutes to diffuse down their concentration gradients rather than picking them up and transporting them against their gradients. Active transport enables a cell to maintain internal concentrations of small solutes that differ from concentrations in its environment. For example, compared with its surroundings, an animal cell has a much higher concentration of potassium ions ($K^+$) and a much lower concentration of sodium ions ($Na^+$). The plasma membrane helps maintain these steep gradients by pumping $Na^+$ out of the cell and $K^+$ into the cell.

As in other types of cellular work, ATP supplies the energy for most active transport. One way ATP can power active transport is by transferring its terminal phosphate group directly to the transport protein. This can induce the protein to change its shape in a manner that translocates a solute bound to the protein across the membrane. One transport system

► **Figure 5.14 The sodium-potassium pump: a specific case of active transport.** This transport system pumps ions against steep concentration gradients: Sodium ion concentration ($[Na^+]$) is high outside the cell and low inside, while potassium ion concentration ($[K^+]$) is low outside the cell and high inside. The pump oscillates between two shapes in a cycle that moves 3 $Na^+$ out of the cell (steps ➊ through ➌) for every 2 $K^+$ pumped into the cell (steps ➍ through ➏). The two shapes have different binding affinities for $Na^+$ and $K^+$. ATP powers the shape change by transferring a phosphate group to the transport protein (phosphorylating the protein).

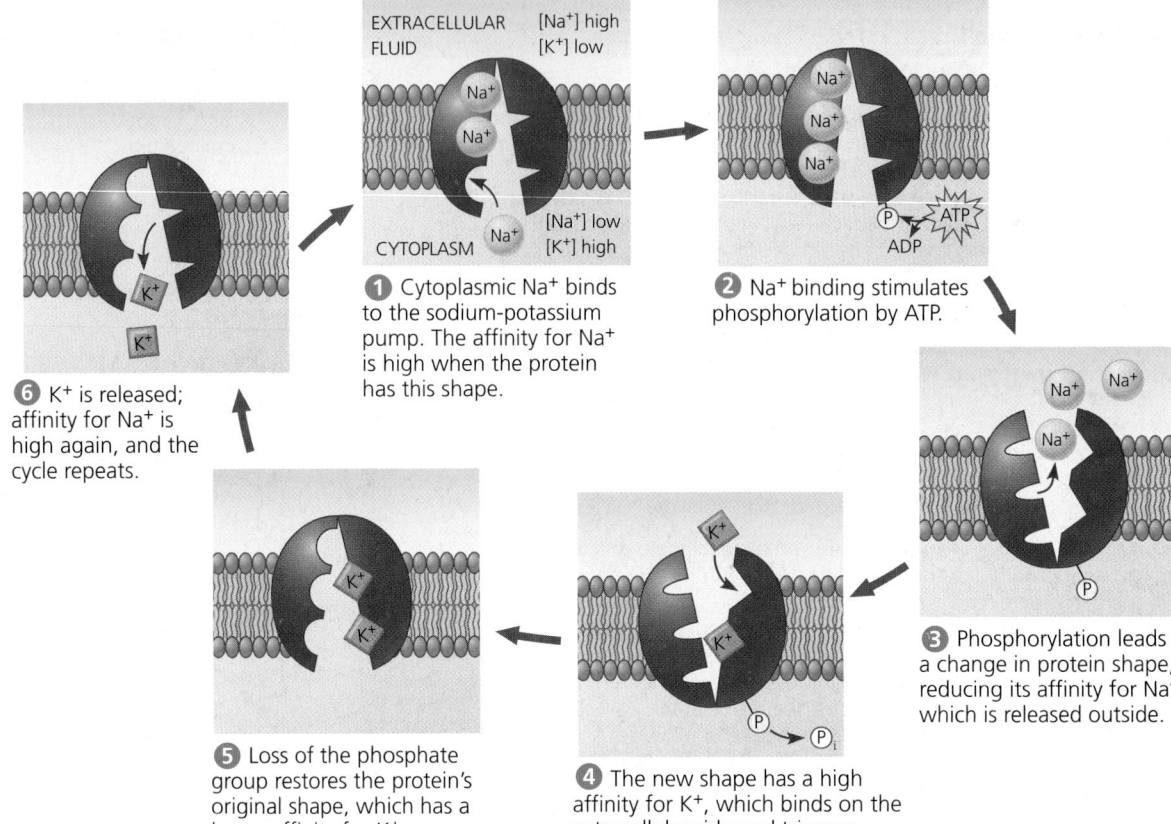

**1** Cytoplasmic $Na^+$ binds to the sodium-potassium pump. The affinity for $Na^+$ is high when the protein has this shape.

**2** $Na^+$ binding stimulates phosphorylation by ATP.

**3** Phosphorylation leads to a change in protein shape, reducing its affinity for $Na^+$, which is released outside.

**6** $K^+$ is released; affinity for $Na^+$ is high again, and the cycle repeats.

**5** Loss of the phosphate group restores the protein's original shape, which has a lower affinity for $K^+$.

**4** The new shape has a high affinity for $K^+$, which binds on the extracellular side and triggers release of the phosphate group.

that works this way is the **sodium-potassium pump**, which exchanges $Na^+$ for $K^+$ across the plasma membrane of animal cells **(Figure 5.14)**. The distinction between passive transport and active transport is reviewed in **Figure 5.15**.

## How Ion Pumps Maintain Membrane Potential

All cells have voltages across their plasma membranes. Voltage is electrical potential energy—a separation of opposite charges. The cytoplasmic side of the membrane is negative in charge relative to the extracellular side because of an unequal distribution of anions and cations on the two sides. The voltage across a membrane, called a **membrane potential**, ranges from about −50 to −200 millivolts (mV). (The minus sign indicates that the inside of the cell is negative relative to the outside.)

The membrane potential acts like a battery, an energy source that affects the traffic of all charged substances across the membrane. Because the inside of the cell is negative compared with the outside, the membrane potential favors the passive transport of cations into the cell and anions out of the cell. Thus, *two* forces drive the diffusion of ions across a membrane: a chemical force (the ion's concentration gradient) and an electrical force (the effect of the membrane potential on the ion's movement). This combination of forces acting on an ion is called the **electrochemical gradient**.

▼ **Figure 5.15 Review: passive and active transport.**

**Passive transport.** Substances diffuse spontaneously down their concentration gradients, crossing a membrane with no expenditure of energy by the cell. The rate of diffusion can be greatly increased by transport proteins in the membrane.

**Active transport.** Some transport proteins act as pumps, moving substances across a membrane against their concentration (or electrochemical) gradients. Energy for this work is usually supplied by ATP.

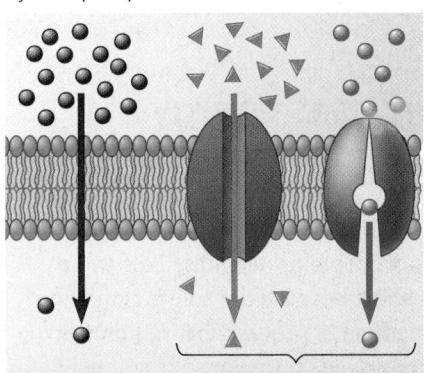

**Diffusion.** Hydrophobic molecules and (at a slow rate) very small uncharged polar molecules can diffuse through the lipid bilayer.

**Facilitated diffusion.** Many hydrophilic substances diffuse through membranes with the assistance of transport proteins, either channel proteins (left) or carrier proteins (right).

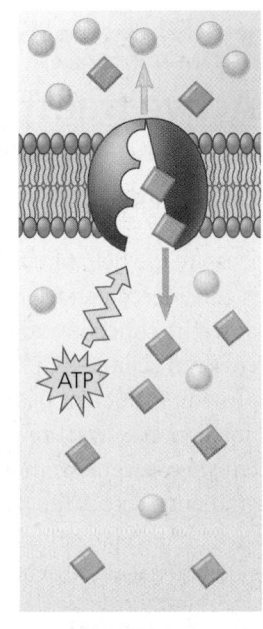

**?** *For each solute in the right panel, describe its direction of movement, and state whether it is moving with or against its concentration gradient.*

In the case of ions, then, we must refine our concept of passive transport: An ion diffuses not simply down its *concentration* gradient but, more exactly, down its *electrochemical* gradient. For example, the concentration of $Na^+$ inside a resting nerve cell is much lower than outside it. When the cell is stimulated, gated channels open that facilitate $Na^+$ diffusion. Sodium ions then "fall" down their electrochemical gradient, driven by the concentration gradient of $Na^+$ and by the attraction of these cations to the negative side (inside) of the membrane. In this example, both electrical and chemical contributions to the electrochemical gradient act in the same direction across the membrane, but this is not always so. In cases where electrical forces due to the membrane potential oppose the simple diffusion of an ion down its concentration gradient, active transport may be necessary. In Chapter 37, you'll learn about the importance of electrochemical gradients and membrane potentials in the transmission of nerve impulses.

Some membrane proteins that actively transport ions contribute to the membrane potential. An example is the sodium-potassium pump. Notice in Figure 5.14 that the pump does not translocate $Na^+$ and $K^+$ one for one, but pumps three sodium ions out of the cell for every two potassium ions it pumps into the cell. With each "crank" of the pump, there is a net transfer of one positive charge from the cytoplasm to the extracellular fluid, a process that stores energy as voltage. A transport protein that generates voltage across a membrane is called an **electrogenic pump**. The sodium-potassium pump appears to be the major electrogenic pump of animal cells. The main electrogenic pump of plants, fungi, and bacteria is a **proton pump**, which actively transports protons (hydrogen ions, $H^+$) out of the cell. The pumping of $H^+$ transfers positive charge from the cytoplasm to the extracellular solution **(Figure 5.16)**. By generating voltage across membranes, electrogenic pumps help store energy that can be tapped for cellular work. One important use of proton gradients in the cell is for ATP synthesis during cellular respiration (as you will see in Concept 7.4). Another is a type of membrane traffic called cotransport.

## Cotransport: Coupled Transport by a Membrane Protein

A solute that exists in different concentrations across a membrane can do work as it moves across that membrane by diffusion down its concentration gradient. This is analogous to water that has been pumped uphill and performs work as it flows back down. In a mechanism called **cotransport**, a transport protein (a cotransporter) can couple the "downhill" diffusion of the solute to the "uphill" transport of a second substance against its own concentration (or electrochemical) gradient. For instance, a plant cell uses the gradient of $H^+$ generated by its ATP-powered proton pumps to drive the active transport of amino acids, sugars, and several other nutrients into the cell. In the example shown in **Figure 5.17**, a cotransporter couples the return of $H^+$ to the transport of sucrose into the cell. This protein can translocate sucrose into the cell against its concentration gradient, but only if the sucrose molecule travels in the company of an $H^+$. The $H^+$ uses the transport protein as an avenue to diffuse down its own electrochemical gradient, which is maintained by the proton pump. Plants use sucrose-$H^+$ cotransport to load sucrose produced by photosynthesis into cells in the veins of leaves. The vascular tissue of the plant can then distribute the sugar to nonphotosynthetic organs, such as roots.

What we know about cotransport proteins in animal cells has helped us find more effective treatments for diarrhea, a serious problem in developing countries. Normally, sodium in waste is reabsorbed in the colon, maintaining constant levels

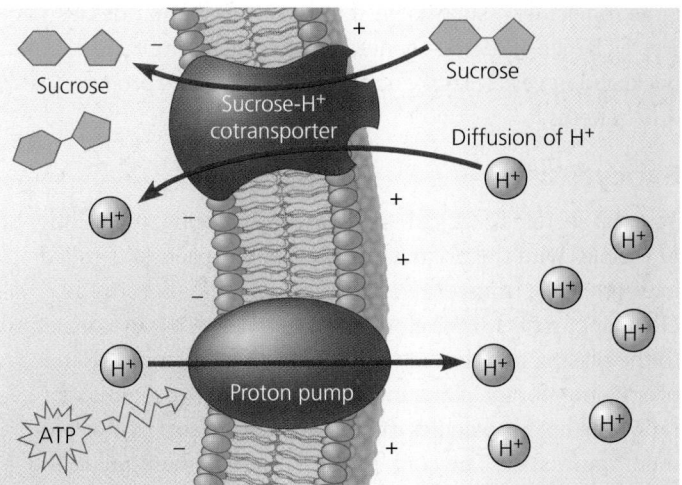

▲ **Figure 5.17 Cotransport: active transport driven by a concentration gradient.** A carrier protein, such as this sucrose-$H^+$ cotransporter in a plant cell (top), is able to use the diffusion of $H^+$ down its electrochemical gradient into the cell to drive the uptake of sucrose against its concentration gradient. (The cell wall is not shown.) Although not technically part of the cotransport process, an ATP-driven proton pump is shown here (bottom), which concentrates $H^+$ outside the cell. The resulting $H^+$ gradient represents potential energy that can be used for active transport—of sucrose, in this case. Thus, ATP indirectly provides the energy necessary for cotransport.

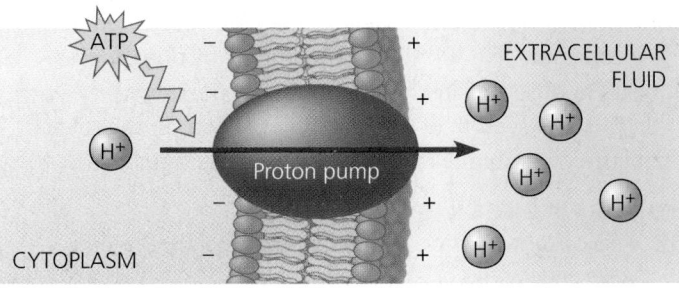

▲ **Figure 5.16 A proton pump.** Proton pumps are electrogenic pumps that store energy by generating voltage (charge separation) across membranes. A proton pump translocates positive charge in the form of hydrogen ions (that is, protons). The voltage and $H^+$ concentration gradient represent a dual energy source that can drive other processes, such as the uptake of nutrients. Most proton pumps are powered by ATP.

in the body, but diarrhea expels waste so rapidly that reabsorption is not possible, and sodium levels fall precipitously. To treat this life-threatening condition, patients are given a solution to drink containing high concentrations of salt (NaCl) and glucose. The solutes are taken up by sodium-glucose cotransporters on the surface of intestinal cells and passed through the cells into the blood. This simple treatment has lowered infant mortality worldwide.

### CONCEPT CHECK 5.4

1. Sodium-potassium pumps help nerve cells establish a voltage across their plasma membranes. Do these pumps use ATP or produce ATP? Explain.
2. Explain why the sodium-potassium pump in Figure 5.14 would not be considered a cotransporter.
3. **MAKE CONNECTIONS** Review the characteristics of the lysosome discussed in Concept 4.4. Given the internal environment of a lysosome, what transport protein might you expect to see in its membrane?

For suggested answers, see Appendix A.

---

CONCEPT 5.5

# Bulk transport across the plasma membrane occurs by exocytosis and endocytosis

Water and small solutes enter and leave the cell by diffusing through the lipid bilayer of the plasma membrane or by being moved across the membrane by transport proteins. However, large molecules—such as proteins and polysaccharides, as well as larger particles—generally cross the membrane in bulk, packaged in vesicles. Like active transport, these processes require energy.

## Exocytosis

The cell secretes certain biological molecules by the fusion of vesicles with the plasma membrane; this process is called **exocytosis**. A transport vesicle that has budded from the Golgi apparatus moves along microtubules of the cytoskeleton to the plasma membrane. When the vesicle membrane and plasma membrane come into contact, specific proteins rearrange the lipid molecules of the two bilayers so that the two membranes fuse. The contents of the vesicle then spill to the outside of the cell, and the vesicle membrane becomes part of the plasma membrane (see Figure 5.8, step 4).

Many secretory cells use exocytosis to export products. For example, the cells in the pancreas that make insulin secrete it into the extracellular fluid by exocytosis. In another example, nerve cells use exocytosis to release neurotransmitters that signal other neurons or muscle cells. When plant cells are making cell walls, exocytosis delivers proteins and carbohydrates from Golgi vesicles to the outside of the cell.

## Endocytosis

In **endocytosis**, the cell takes in molecules and particulate matter by forming new vesicles from the plasma membrane. Although the proteins involved in the two processes are different, the events of endocytosis look like the reverse of exocytosis. First, a small area of the plasma membrane sinks inward to form a pocket. Then, as the pocket deepens, it pinches in, forming a vesicle containing material that had been outside the cell. Study **Figure 5.18** carefully to understand the three types of endocytosis: phagocytosis ("cellular eating"), pinocytosis ("cellular drinking"), and receptor-mediated endocytosis.

Human cells use receptor-mediated endocytosis to take in cholesterol for membrane synthesis and the synthesis of other steroids. Cholesterol travels in the blood in particles called low-density lipoproteins (LDLs), each a complex of lipids and a protein. LDLs bind to LDL receptors on plasma membranes and then enter the cells by endocytosis. In the inherited disease familial hypercholesterolemia, characterized by a very high level of cholesterol in the blood, LDLs cannot enter cells because the LDL receptor proteins are defective or missing:

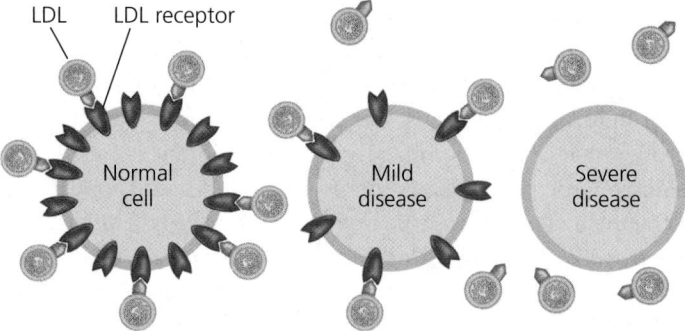

Consequently, cholesterol accumulates in the blood, where it contributes to early atherosclerosis, the buildup of lipid deposits within the walls of blood vessels. This buildup narrows the space in the vessels and impedes blood flow, potentially resulting in heart damage or stroke.

Endocytosis and exocytosis also provide mechanisms for rejuvenating or remodeling the plasma membrane. These processes occur continually in most eukaryotic cells, yet the amount of plasma membrane in a nongrowing cell remains fairly constant. The addition of membrane by one process appears to offset the loss of membrane by the other.

In the final section of this chapter, we'll look at the role of the plasma membrane and its proteins in cell signaling.

### CONCEPT CHECK 5.5

1. As a cell grows, its plasma membrane expands. Does this involve endocytosis or exocytosis? Explain.
2. **DRAW IT** Return to Figure 5.8, and circle a patch of plasma membrane that is coming from a vesicle involved in exocytosis.
3. **MAKE CONNECTIONS** In Concept 4.7, you learned that animal cells make an extracellular matrix (ECM). Describe the cellular pathway of synthesis and deposition of an ECM glycoprotein.

For suggested answers, see Appendix A.

## Phagocytosis

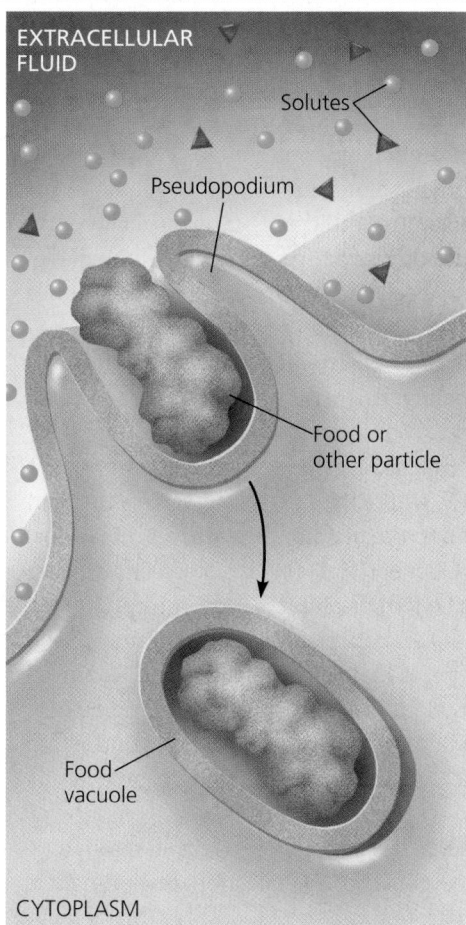

In **phagocytosis**, a cell engulfs a particle by extending pseudopodia (singular, *pseudopodium*) around it and packaging it within a membranous sac called a food vacuole. The particle will be digested after the food vacuole fuses with a lysosome containing hydrolytic enzymes (see Figure 4.12).

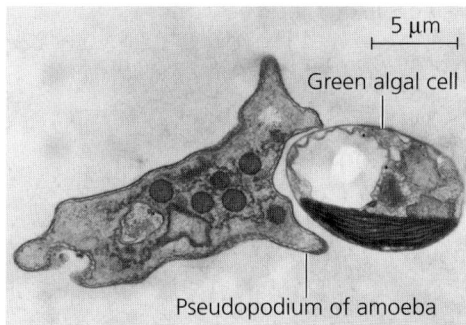

An amoeba engulfing a green algal cell via phagocytosis (TEM).

ANIMATION    Visit the Study Area in **MasteringBiology** for the BioFlix® 3-D Animation on Membrane Transport.

## Pinocytosis

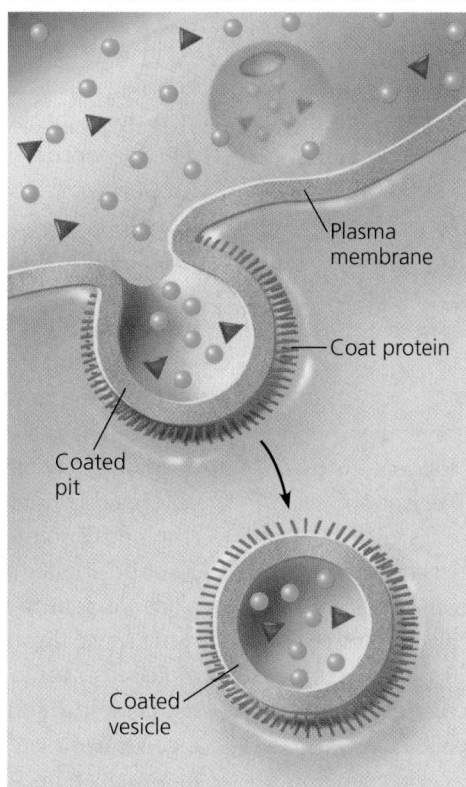

In **pinocytosis**, a cell continually "gulps" droplets of extracellular fluid into tiny vesicles, formed by infoldings of the plasma membrane. In this way, the cell obtains molecules dissolved in the droplets. Because any and all solutes are taken into the cell, pinocytosis as shown here is nonspecific for the substances it transports. In many cases, as above, the parts of the plasma membrane that form vesicles are lined on their cytoplasmic side by a fuzzy layer of coat protein; the "pits" and resulting vesicles are said to be "coated."

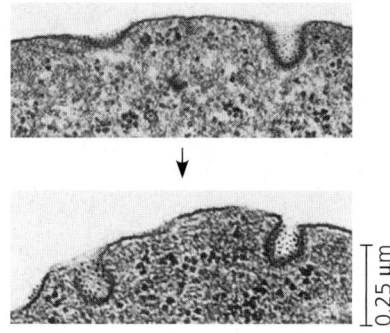

Pinocytotic vesicles forming (TEMs).

## Receptor-Mediated Endocytosis

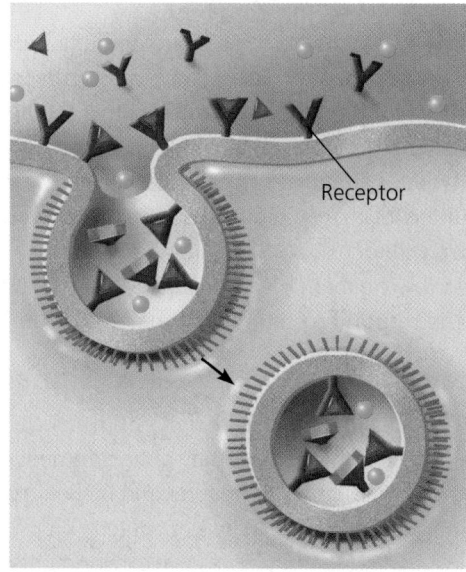

**Receptor-mediated endocytosis** is a specialized type of pinocytosis that enables the cell to acquire bulk quantities of specific substances, even though those substances may not be very concentrated in the extracellular fluid. Embedded in the plasma membrane are proteins with receptor sites exposed to the extracellular fluid. Specific solutes bind to the sites. The receptor proteins then cluster in coated pits, and each coated pit forms a vesicle containing the bound molecules. Notice that there are relatively more bound molecules (purple triangles) inside the vesicle, but other molecules (green balls) are also present. After the ingested material is liberated from the vesicle, the emptied receptors are recycled to the plasma membrane by the same vesicle (not shown).

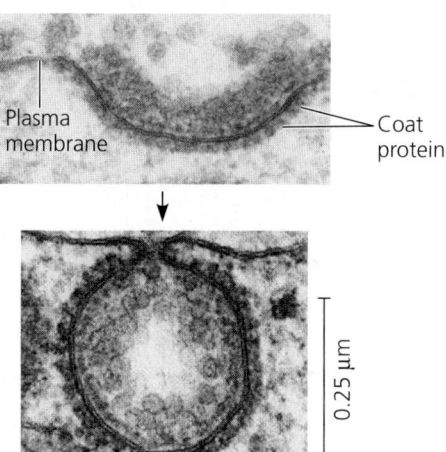

*Top*: A coated pit. *Bottom*: A coated vesicle forming during receptor-mediated endocytosis (TEMs).

# The plasma membrane plays a key role in most cell signaling

In a multicellular organism, whether a human being or an oak tree, it is cell-to-cell communication that allows the trillions of cells of the body to coordinate their activities, and the communication process usually involves the cells' plasma membranes. In fact, communication between cells is also essential for many unicellular organisms, including prokaryotes. However, here we will focus on cell signaling in animals and plants. We'll describe the main mechanisms by which cells receive, process, and respond to chemical signals sent from other cells.

## Local and Long-Distance Signaling

The signaling molecules sent out from cells are targeted for other cells that may or may not be immediately adjacent. As discussed earlier in this chapter and in Concept 4.7, eukaryotic cells may communicate by direct contact, a type of local signaling. Both animals and plants have cell junctions that, where present, directly connect the cytoplasms of adjacent cells; in animals, these are gap junctions (see Figure 4.27), and in plants, plasmodesmata (see Figure 4.25). In these cases, signaling substances dissolved in the cytosol can pass freely between adjacent cells. Also, animal cells may communicate via direct contact between membrane-bound cell-surface molecules in cell-cell recognition (see Figure 5.7d). This sort of local signaling is especially important in embryonic development and in the immune response.

In many other cases of local signaling, the signaling cell secretes messenger molecules. Some of these travel only short distances; such local regulators influence cells in the vicinity. One class of local regulators in animals, *growth factors*, are compounds that stimulate nearby target cells to grow and divide. Numerous cells can simultaneously receive and respond to the molecules of growth factor produced by a nearby cell. This type of local signaling in animals is called *paracrine signaling* **(Figure 5.19a)**. (Local signaling in plants is discussed in Concept 31.1.)

A more specialized type of local signaling called *synaptic signaling* occurs in the animal nervous system **(Figure 5.19b)**. An electrical signal moving along a nerve cell triggers the secretion of neurotransmitter molecules carrying a chemical signal. These molecules diffuse across the synapse, the narrow space between the nerve cell and its target cell (often another nerve cell), triggering a response in the target cell.

Both animals and plants use chemicals called **hormones** for long-distance signaling. In hormonal signaling in animals, also known as *endocrine signaling*, specialized cells release hormone molecules, which travel via the circulatory system to other parts of the body, where they reach target cells that can recognize and respond to the hormones **(Figure 5.19c)**. Most plant hormones (see Concept 31.1) reach distant targets via plant vascular tissues (xylem or phloem; see Concept 28.1), but some travel through the air as a gas. Hormones vary widely in molecular size and type, as do local regulators. For instance, the plant hormone ethylene, a gas that promotes fruit ripening, is a hydrocarbon of only six atoms ($C_2H_4$). In contrast, the mammalian hormone insulin, which regulates sugar levels in the blood, is a protein with thousands of atoms.

What happens when a cell encounters a secreted signaling molecule? We will now consider this question, beginning with a bit of historical background.

▼ **Figure 5.19 Local and long-distance cell signaling by secreted molecules in animals.** In both local and long-distance signaling, only specific target cells that can recognize a given signaling molecule will respond to it.

**Local signaling**

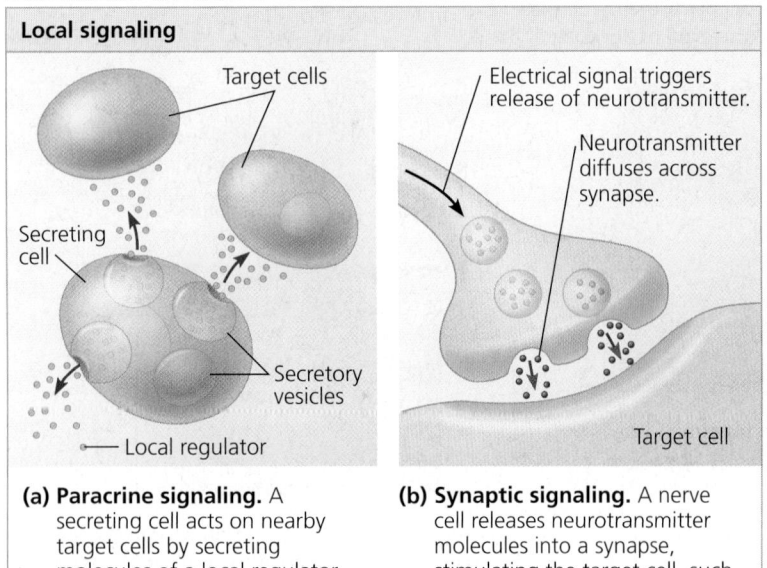

(a) **Paracrine signaling.** A secreting cell acts on nearby target cells by secreting molecules of a local regulator (a growth factor, for example).

(b) **Synaptic signaling.** A nerve cell releases neurotransmitter molecules into a synapse, stimulating the target cell, such as a muscle or nerve cell.

**Long-distance signaling**

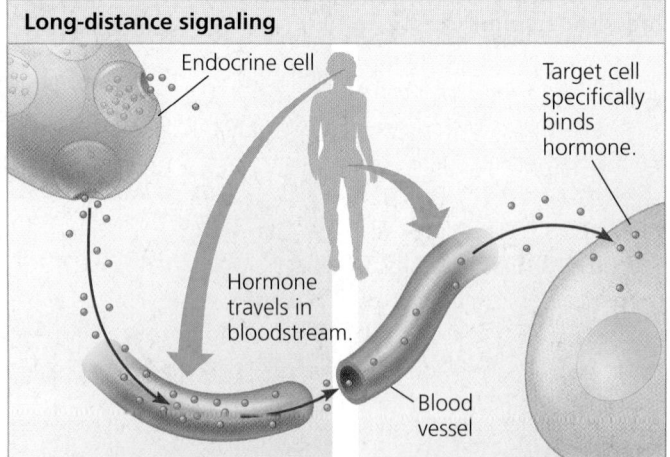

(c) **Endocrine (hormonal) signaling.** Specialized endocrine cells secrete hormones into body fluids, often blood. Hormones reach virtually all body cells, but are bound only by some cells.

## The Three Stages of Cell Signaling: *A Preview*

Our current understanding of how chemical messengers act on cells had its origins in the pioneering work of the American Earl W. Sutherland about a half-century ago. He was investigating how the animal hormone epinephrine (also called adrenaline) triggers the "fight-or-flight" response in animals by stimulating the breakdown of the storage polysaccharide glycogen within liver cells and skeletal muscle cells. Glycogen breakdown releases the sugar glucose 1-phosphate, which the cell converts to glucose 6-phosphate. The liver or muscle cell can then use this compound, an early intermediate in glycolysis, for energy production. Alternatively, the compound can be stripped of phosphate and released from the cell into the blood as glucose, which can fuel cells throughout the body. Thus, one effect of epinephrine is the mobilization of fuel reserves, which can be used by an animal to either defend itself (fight) or escape whatever elicited a scare (flight), as this impala is doing. Sutherland's research team discovered that epinephrine stimulates glycogen breakdown by activating a cytosolic enzyme (glycogen phosphorylase) while never actually entering the glycogen-containing cells. This discovery provided two insights. First, epinephrine does not interact directly with glycogen phosphorylase; an intermediate step or series of steps must be occurring in the cell. Second, the plasma membrane must somehow be involved in transmitting the signal. Sutherland's research suggested that the process going on at the receiving end of a cell-to-cell message can be divided into three stages: reception, transduction, and response **(Figure 5.20).** ❶ **Reception** is the target cell's detection of a signaling molecule coming from outside the cell. A chemical signal is "detected" when the signaling molecule binds to a receptor protein located at the cell's surface or, in some cases, inside the cell. ❷ **Transduction** is a step or series of steps that converts the signal to a form that can bring about a specific cellular response. Transduction usually requires a sequence of changes in a series of different molecules—a **signal transduction pathway**. The molecules in the pathway are often called relay molecules. ❸ In the third stage of cell signaling, the transduced signal finally triggers a specific cellular **response**. The response may be almost any imaginable cellular activity, such as catalysis by an enzyme (for example, glycogen phosphorylase),

rearrangement of the cytoskeleton, or activation of specific genes in the nucleus. The cell-signaling process helps ensure that crucial activities like these occur in the right cells, at the right time, and in proper coordination with the activities of other cells of the organism. We'll now explore the mechanisms of cell signaling in more detail.

## Reception, the Binding of a Signaling Molecule to a Receptor Protein

A radio station broadcasts its signal indiscriminately, but it can be picked up only by radios tuned to the right frequency; reception of the signal depends on the receiver. Similarly, in the case of epinephrine, the hormone encounters many types of cells as it circulates in the blood, but only certain target cells detect and react to the epinephrine molecule. A receptor protein on or in the target cell allows the cell to detect the signal and respond to it. The signaling molecule is complementary in shape to a specific site on the receptor and attaches there, like a key in a lock. The signaling molecule acts as a **ligand**, a molecule that specifically binds to another molecule, often a larger one. (LDLs, mentioned in Concept 5.5, act as ligands when they bind to their receptors, as do the molecules that bind to enzymes; see Figure 3.17.) Ligand binding generally causes a receptor protein to undergo a change in shape. For many receptors, this shape change directly activates the receptor, enabling it to interact with other cellular molecules.

Most signal receptors are plasma membrane proteins. Their ligands are water-soluble and generally too large to pass freely through the plasma membrane. Other signal receptors, however, are located inside the cell. We discuss both of these types next.

### Receptors in the Plasma Membrane

Most water-soluble signaling molecules bind to specific sites on receptor proteins that span the cell's plasma membrane. Such a transmembrane receptor transmits information from the extracellular environment to the inside of the cell by

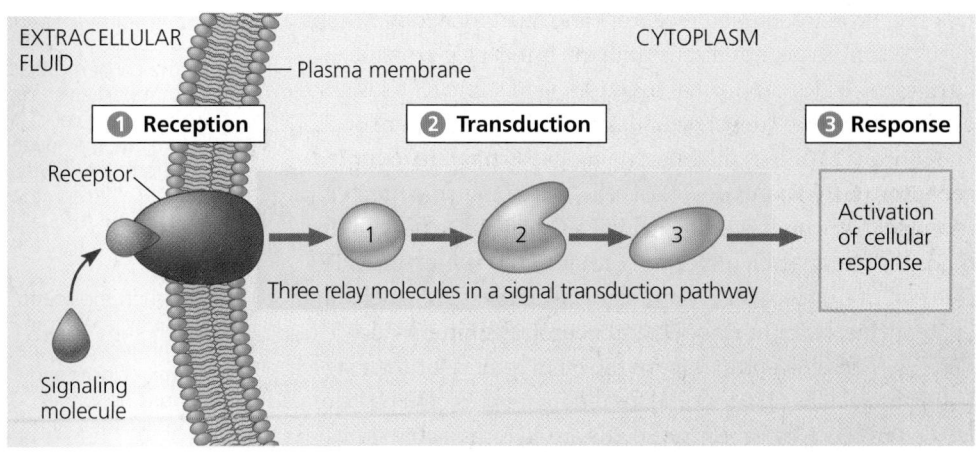

▲ **Figure 5.20 Overview of cell signaling.** From the perspective of the cell receiving the message, cell signaling can be divided into three stages: signal reception, signal transduction, and cellular response. When reception occurs at the plasma membrane, as shown here, the transduction stage is usually a pathway of several steps, with each specific relay molecule in the pathway bringing about a change in the next molecule. The final molecule in the pathway triggers the cell's response. The three stages are explained in more detail in the text.

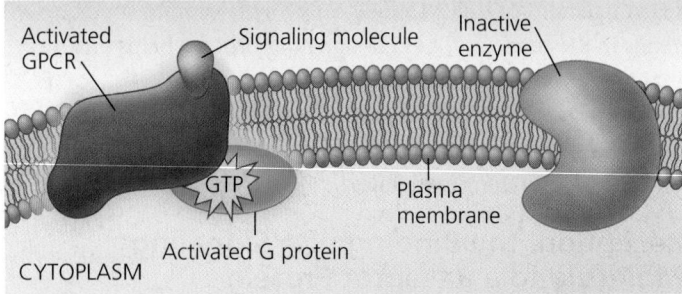

**1** When the appropriate signaling molecule binds to the extracellular side of the receptor, the receptor is activated and changes shape. Its cytoplasmic side then binds and activates a G protein. The activated G protein carries a GTP molecule.

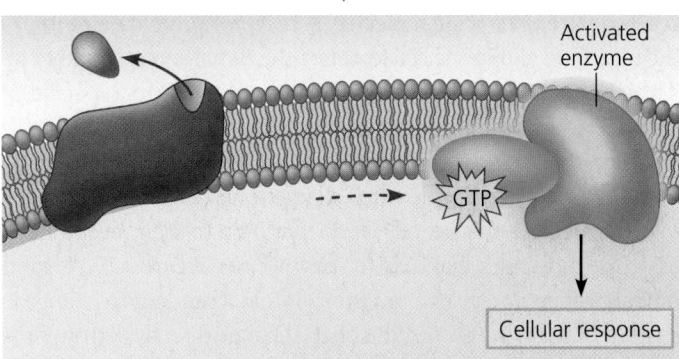

**2** The activated G protein leaves the receptor, diffuses along the membrane, and then binds to an enzyme, altering the enzyme's shape and activity. Once activated, the enzyme can trigger the next step leading to a cellular response. Binding of signaling molecules is reversible. The activating change in the GPCR, as well as the changes in the G protein and enzyme, are only temporary; these molecules soon become available for reuse.

▲ **Figure 5.21 A G protein-coupled receptor (GPCR) in action.**

changing shape when a specific ligand binds to it. We can see how transmembrane receptors work by looking at two major types: G protein-coupled receptors and ligand-gated ion channels. (A third type, not discussed here, is receptor tyrosine kinases, or RTKs. Abnormal functioning of some RTKs is associated with breast cancer; see Make Connections Figure 16.21.)

**Figure 5.21** shows the functioning of a **G protein-coupled receptor (GPCR)**. A GPCR is a cell-surface transmembrane receptor that works with the help of a **G protein**, a protein that binds the energy-rich molecule GTP, which is similar to ATP (see end of Concept 3.1). Many signaling molecules—including epinephrine, other hormones, and neurotransmitters—use GPCRs. These receptors vary in the binding sites for their signaling molecules (ligands) and for different types of G proteins inside the cell. Nevertheless, GPCRs are all remarkably similar in structure, as are many G proteins, suggesting that these signaling systems evolved very early in the history of life.

The nearly 1,000 GPCRs examined to date make up the largest family of cell-surface receptors in mammals. GPCR pathways are extremely diverse in their functions, which

include roles in embryonic development and the senses of sight, smell, and taste. They are also involved in many human diseases. For example, cholera, pertussis (whooping cough), and botulism are caused by bacterial toxins that interfere with G protein function. Up to 60% of all medicines used today exert their effects by influencing G protein pathways.

A **ligand-gated ion channel** is a membrane receptor with a region that can act as a "gate" for ions when the receptor assumes a certain shape **(Figure 5.22)**. When a signaling molecule binds as a ligand to the receptor protein, the gate opens or closes, allowing or blocking the diffusion of specific ions, such as $Na^+$ or $Ca^{2+}$, through a channel in the protein. Like other membrane receptors, these proteins bind the ligand at a specific site on their extracellular side.

Ligand-gated ion channels are very important in the nervous system. For example, the neurotransmitter molecules released at a synapse between two nerve cells (see Figure 5.19b) bind as ligands to ion channels on the receiving cell, causing

**1** Here we show a ligand-gated ion channel receptor in which the gate remains closed until a ligand binds to the receptor.

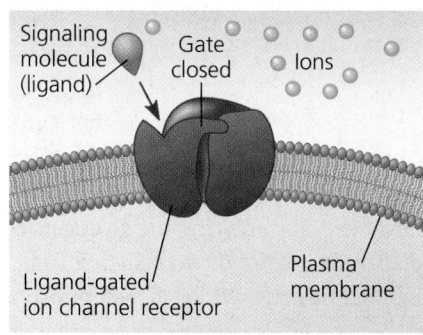

**2** When the ligand binds to the receptor and the gate opens, specific ions can flow through the channel and rapidly change the concentration of that particular ion inside the cell. This change may directly affect the activity of the cell in some way.

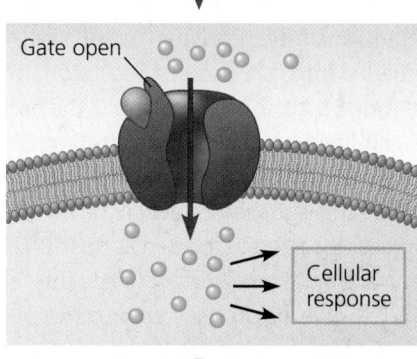

**3** When the ligand dissociates from this receptor, the gate closes and ions no longer enter the cell.

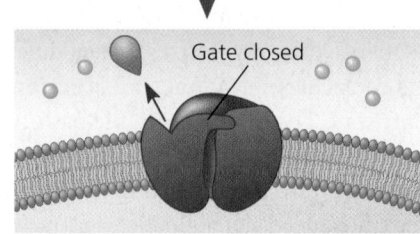

▲ **Figure 5.22 Ion channel receptor.** This is a ligand-gated ion channel, a type of receptor protein that regulates the passage of specific ions across the membrane. Whether the channel is open or closed depends on whether a specific ligand is bound to the protein.

the channels to open. The diffusion of ions through the open channels may trigger an electrical signal that propagates down the length of the receiving cell. (You'll learn more about ion channels in Chapter 37.)

### Intracellular Receptors

Intracellular receptor proteins are found in either the cytoplasm or nucleus of target cells. To reach such a receptor, a signaling molecule passes through the target cell's plasma membrane. A number of important signaling molecules can do this because they are hydrophobic enough to cross the hydrophobic interior of the membrane. These hydrophobic chemical messengers include the steroid hormones and thyroid hormones of animals. In both animals and plants, another chemical signaling molecule with an intracellular receptor is nitric oxide (NO), a gas; its very small, hydrophobic molecules can easily pass between the membrane phospholipids.

The behavior of aldosterone is representative of steroid hormones. This hormone is secreted by cells of the adrenal gland, a gland that lies over the kidney. It then travels through the blood and enters cells all over the body. However, a response occurs only in kidney cells, which contain receptor molecules for aldosterone. In these cells, the hormone binds to the receptor protein, activating it **(Figure 5.23)**. With the hormone attached, the active form of the receptor protein then enters the nucleus and turns on specific genes that control water and sodium flow in kidney cells, ultimately affecting blood volume.

How does the activated hormone-receptor complex turn on genes? Recall that the genes in a cell's DNA function by being transcribed and processed into messenger RNA (mRNA), which leaves the nucleus and is translated into a specific protein by ribosomes in the cytoplasm (see Figure 3.26). Special proteins called *transcription factors* control which genes are turned on—that is, which genes are transcribed into mRNA—in a particular cell at a particular time. When the aldosterone receptor is activated, it acts as a transcription factor that turns on specific genes.

By acting as a transcription factor, the aldosterone receptor itself carries out the transduction part of the signaling pathway. Most other intracellular receptors function in the same way, although many of them, such as the thyroid hormone receptor, are already in the nucleus before the signaling molecule reaches them. Interestingly, many of these intracellular receptor proteins are structurally similar, suggesting an evolutionary kinship.

## Transduction by Cascades of Molecular Interactions

When receptors for signaling molecules are plasma membrane proteins, like most of those we have discussed, the transduction stage of cell signaling is usually a multistep pathway involving many molecules. Steps often include activation of proteins by addition or removal of phosphate groups or release of other small molecules or ions that act as messengers. One benefit of multiple steps is the possibility of greatly amplifying a signal. If

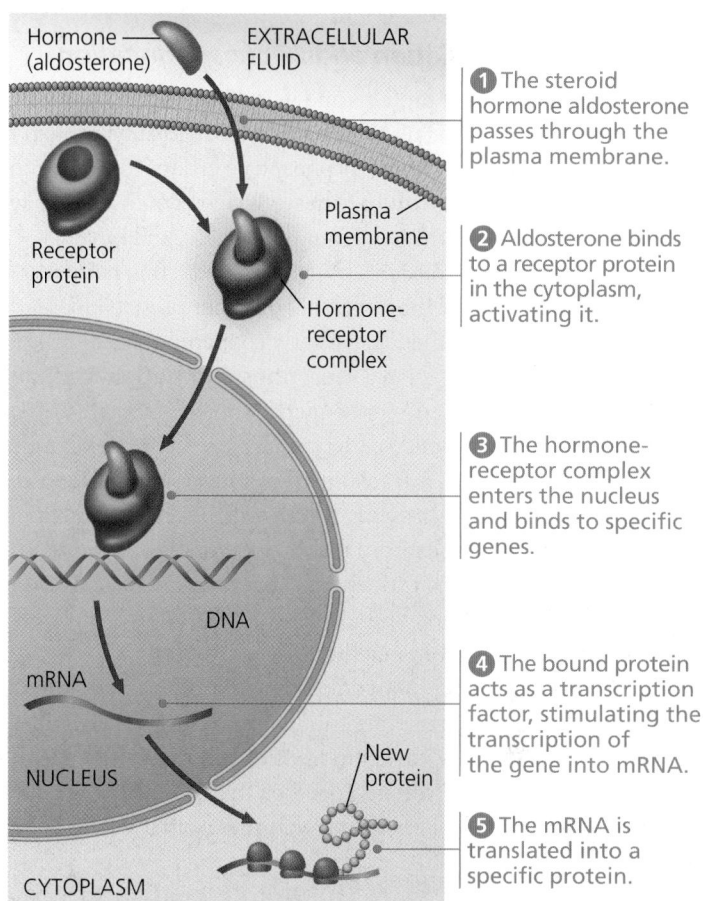

**▲ Figure 5.23 Steroid hormone interacting with an intracellular receptor.**

**?** *Why is a cell-surface receptor protein not required for this steroid hormone to enter the cell?*

each molecule in a pathway transmits the signal to numerous molecules at the next step in the series, the result is a geometric increase in the number of activated molecules by the end of the pathway. Moreover, multistep pathways provide more opportunities for coordination and control than do simpler systems.

The binding of a specific signaling molecule to a receptor in the plasma membrane triggers the first step in the chain of molecular interactions—the signal transduction pathway—that leads to a particular response within the cell. Like falling dominoes, the signal-activated receptor activates another molecule, which activates yet another molecule, and so on, until the protein that produces the final cellular response is activated. The molecules that relay a signal from receptor to response, which we call relay molecules in this book, are often proteins. The interaction of proteins is a major theme of cell signaling.

Keep in mind that the original signaling molecule is not physically passed along a signaling pathway; in most cases, it never even enters the cell. When we say that the signal is relayed along a pathway, we mean that certain information is passed on. At each step, the signal is transduced into a different form, commonly via a shape change in a protein. Very often, the shape change is brought about by phosphorylation, the addition of phosphate groups to a protein (see Figure 3.6).

## Protein Phosphorylation and Dephosphorylation

The phosphorylation of proteins and its reverse, dephosphorylation, are a widespread cellular mechanism for regulating protein activity. An enzyme that transfers phosphate groups from ATP to a protein is known as a **protein kinase**. Such enzymes are widely involved in signaling pathways in animals, plants, and fungi.

Many of the relay molecules in signal transduction pathways are protein kinases, and they often act on other protein kinases in the pathway. A hypothetical pathway containing two different protein kinases that form a short **phosphorylation cascade** is depicted in **Figure 5.24**. The sequence shown is similar to many known pathways, although typically three protein kinases are involved. The signal is transmitted by a cascade of protein phosphorylations, each bringing with it a shape change. Each such shape change results from the interaction of the newly added phosphate groups with charged or polar amino acids (see Figure 3.18). The addition of phosphate groups often changes the form of a protein from inactive to active.

The importance of protein kinases can hardly be overstated. About 2% of our own genes are thought to code for protein kinases. A single cell may have hundreds of different kinds, each specific for a different protein. Together, they probably regulate a large proportion of the thousands of proteins in a cell. Among these are most of the proteins that, in turn, regulate cell division. Abnormal activity of such a kinase can cause abnormal cell division and contribute to the development of cancer.

Equally important in the phosphorylation cascade are the **protein phosphatases** (see Figure 5.24), enzymes that can rapidly remove phosphate groups from proteins, a process called dephosphorylation. By dephosphorylating and thus inactivating protein kinases, phosphatases provide the mechanism for turning off the signal transduction pathway when the initial signal is no longer present. Phosphatases also make the protein kinases available for reuse, enabling the cell to respond again to an extracellular signal. A phosphorylation-dephosphorylation system acts as a molecular switch in the cell, turning an activity on or off, or up or down, as required. At any given moment, the activity of a protein regulated by phosphorylation depends on the balance in the cell between active kinase molecules and active phosphatase molecules.

### Small Molecules and Ions as Second Messengers

Not all components of signal transduction pathways are proteins. Many signaling pathways also involve small, nonprotein, water-soluble molecules or ions called **second messengers**. (The pathway's "first messenger" is considered to be the extracellular signaling molecule that binds to the membrane receptor.) Because they are small, second messengers can readily spread throughout the cell by diffusion. The two most common second messengers are cyclic AMP and calcium ions, $Ca^{2+}$. Here we'll limit our discussion to cyclic AMP.

In his research on epinephrine, Earl Sutherland discovered that the binding of epinephrine to the plasma membrane of a liver cell elevates the cytosolic concentration of **cyclic AMP** (**cAMP**; cyclic adenosine monophosphate). The binding of epinephrine to a G protein-coupled receptor leads, via a G protein, to activation of adenylyl cyclase, an enzyme embedded in the plasma membrane that converts ATP to cAMP **(Figure 5.25)**. Each molecule of adenylyl cyclase can catalyze the synthesis of many molecules of cAMP. In this way, the normal cellular concentration of cAMP can be boosted 20-fold in a matter of seconds. The cAMP broadcasts the signal to the cytoplasm. It does not persist for long in the absence of the hormone because a different enzyme converts cAMP to AMP. Another surge of epinephrine is needed to boost the cytosolic concentration of cAMP again.

Subsequent research has revealed that epinephrine is only one of many

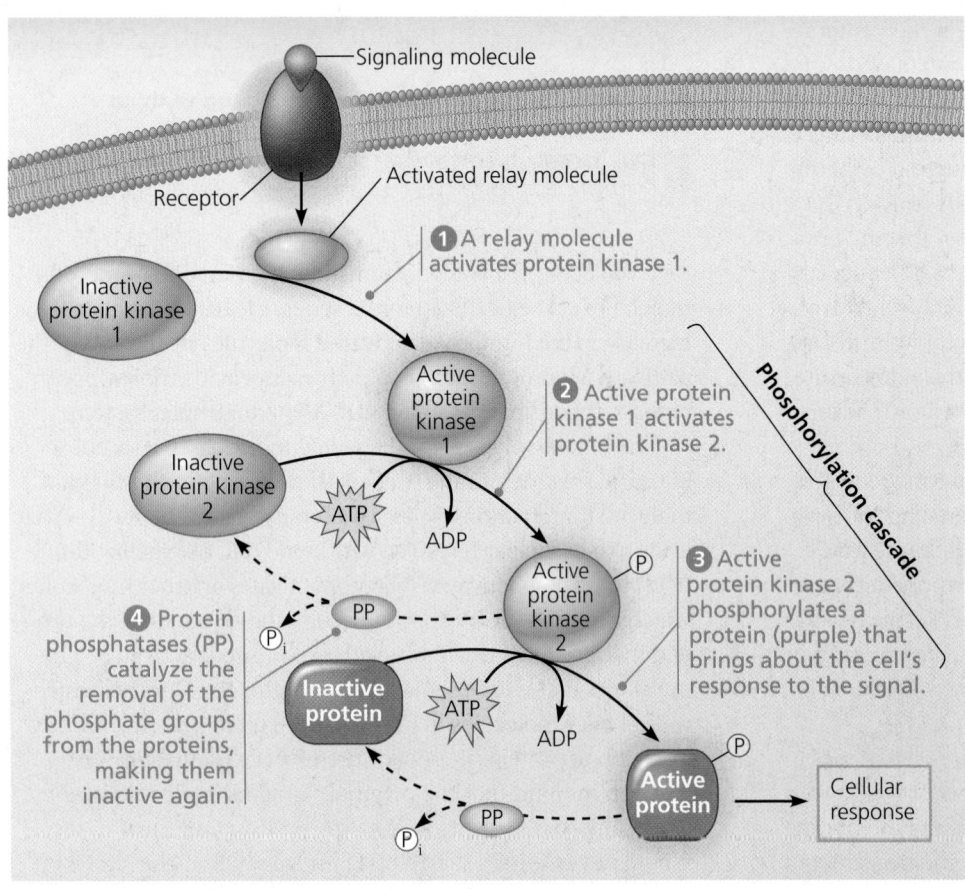

▲ **Figure 5.24 A phosphorylation cascade.** In a phosphorylation cascade, a series of different proteins in a pathway are phosphorylated in turn, each protein adding a phosphate group to the next one in line. Dephosphorylation by protein phosphatases (PP) can then return the protein to its inactive form.

**?** *Which protein is responsible for activation of protein kinase 2?*

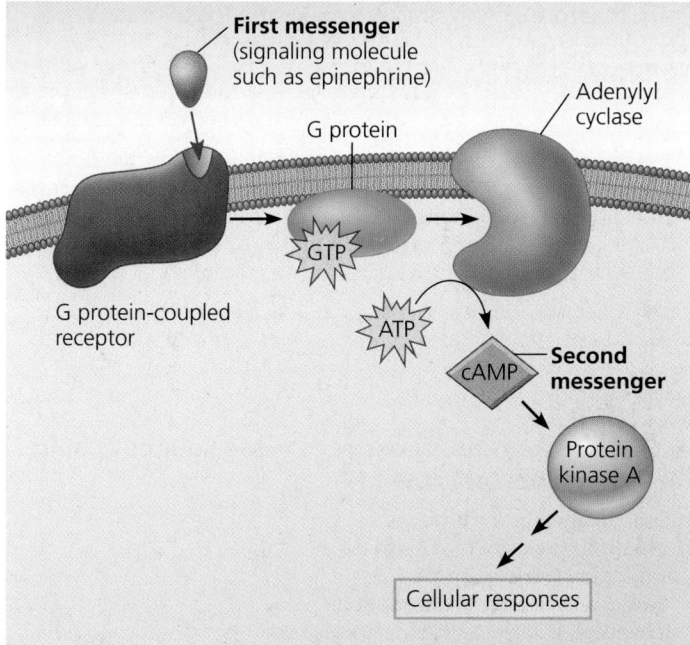

▲ **Figure 5.25 cAMP as a second messenger in a G protein signaling pathway.** The first messenger activates a G protein-coupled receptor, which activates a specific G protein. In turn, the G protein activates adenylyl cyclase, which catalyzes the conversion of ATP to cAMP. The cAMP then acts as a second messenger and activates another protein, usually protein kinase A, leading to cellular responses.

hormones and other signaling molecules that trigger the formation of cAMP. The immediate effect of cAMP is usually the activation of a protein kinase called *protein kinase A.* The activated protein kinase A then phosphorylates various other proteins.

## Response: Regulation of Transcription or Cytoplasmic Activities

What is the nature of the final step in a signaling pathway—the *response* to an external signal? Ultimately, a signal transduction pathway leads to the regulation of one or more cellular activities. The response may occur in the nucleus of the cell or in the cytoplasm.

Many signaling pathways ultimately regulate protein synthesis, usually by turning specific genes on or off in the nucleus. Like an activated steroid receptor (see Figure 5.23), the final activated molecule in a signaling pathway may function as a transcription factor. **Figure 5.26** shows an example in which a signaling pathway activates a transcription factor that turns a gene on: The response to this growth factor signal is transcription, the synthesis of one or more specific mRNAs, which will be translated in the cytoplasm into specific proteins. In other cases, the transcription factor might regulate a gene by turning it off. Often a transcription factor regulates several different genes.

Sometimes a signaling pathway may regulate the *activity* of proteins rather than causing their *synthesis* by activating gene expression. This directly affects proteins that function

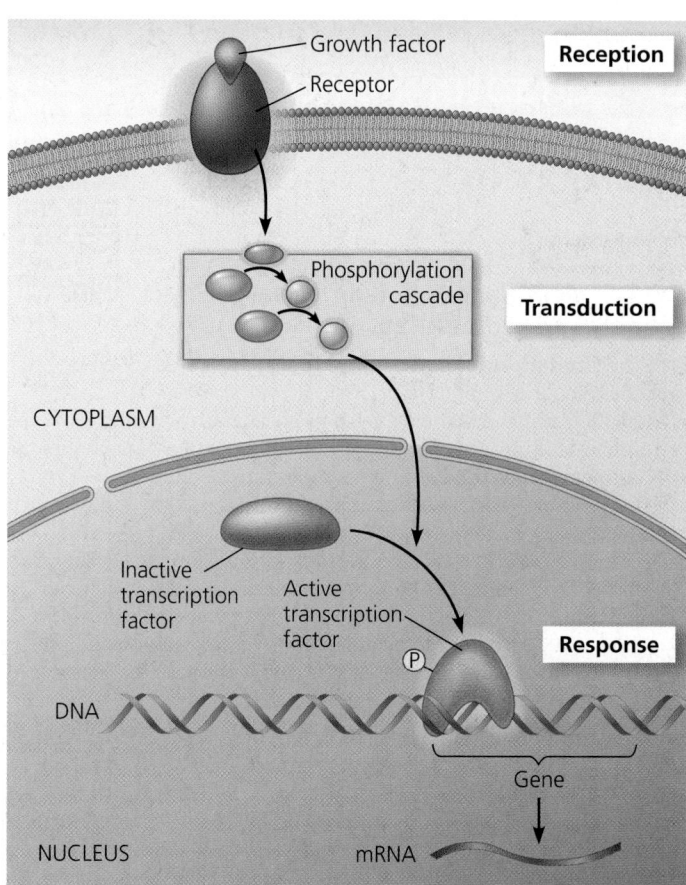

▲ **Figure 5.26 Nuclear response to a signal: the activation of a specific gene by a growth factor.** This diagram shows a typical signaling pathway that leads to the regulation of gene activity in the cell nucleus. The initial signaling molecule, a local regulator called a growth factor, triggers a phosphorylation cascade. (The ATP molecules and phosphate groups are not shown.) Once phosphorylated, the last kinase in the sequence enters the nucleus and activates a transcription factor, which stimulates transcription of a specific gene. The resulting mRNAs then direct the synthesis of a particular protein in the cytoplasm.

outside the nucleus. For example, a signal may cause the opening or closing of an ion channel in the plasma membrane or a change in cell metabolism. As we have discussed, the response of cells to the hormone epinephrine helps regulate cellular energy metabolism by affecting the activity of an enzyme: The final step in the signaling pathway that begins with epinephrine binding activates the enzyme that catalyzes the breakdown of glycogen.

### CONCEPT CHECK 5.6

1. During an epinephrine-initiated signal in liver cells, in which of the three stages of cell signaling does glycogen phosphorylase act?

2. When a signal transduction pathway involves a phosphorylation cascade, what turns off the cell's response?

3. **WHAT IF?** How can a target cell's response to a single hormone molecule result in a response that affects a million other molecules?

For suggested answers, see Appendix A.

# 5 | Chapter Review

Go to **MasteringBiology**® for Assignments, the eText, and the Study Area with Animations, Activities, Vocab Self-Quiz, and Practice Tests.

## SUMMARY OF KEY CONCEPTS

**VOCAB SELF-QUIZ**

goo.gl/gbai8v

### CONCEPT 5.1

**Cellular membranes are fluid mosaics of lipids and proteins (pp. 100–104)**

- In the **fluid mosaic model**, **amphipathic** proteins are embedded in the phospholipid bilayer.
- Phospholipids and some proteins move laterally within the membrane. The unsaturated hydrocarbon tails of some phospholipids keep membranes fluid at lower temperatures, while cholesterol helps membranes resist changes in fluidity caused by temperature changes.
- Membrane proteins function in transport, enzymatic activity, attachment to the cytoskeleton and extracellular matrix, cell-cell recognition, intercellular joining, and signal transduction. Short chains of sugars linked to proteins (in **glycoproteins**) and lipids (in **glycolipids**) on the exterior side of the plasma membrane interact with surface molecules of other cells.
- Membrane proteins and lipids are synthesized in the ER and modified in the ER and Golgi apparatus. The inside and outside faces of membranes differ in molecular composition.

**?** *In what ways are membranes crucial to life?*

### CONCEPT 5.2

**Membrane structure results in selective permeability (p. 105)**

- A cell must exchange molecules and ions with its surroundings, a process controlled by the **selective permeability** of the plasma membrane. Hydrophobic molecules are soluble in lipids and pass through membranes rapidly, whereas polar molecules and ions usually need specific **transport proteins**.

**?** *How do aquaporins affect the permeability of a membrane?*

### CONCEPT 5.3

**Passive transport is diffusion of a substance across a membrane with no energy investment (pp. 105–109)**

- **Diffusion** is the spontaneous movement of a substance down its **concentration gradient**. Water diffuses out through the permeable membrane of a cell (**osmosis**) if the solution outside has a higher solute concentration than the cytosol (is **hypertonic**); water enters the cell if the solution has a lower solute concentration (is **hypotonic**). If the concentrations are equal (**isotonic**), no net osmosis occurs. Cell survival depends on balancing water uptake and loss.

Passive transport: Facilitated diffusion

Channel protein — Carrier protein

- In **facilitated diffusion**, a transport protein speeds the movement of water or a solute across a membrane down its concentration gradient. **Ion channels** facilitate the diffusion of ions across a membrane. Carrier proteins can undergo changes in shape that translocate bound solutes across the membrane.

**?** *What happens to a cell placed in a hypertonic solution? Describe the free water concentration inside and out.*

### CONCEPT 5.4

**Active transport uses energy to move solutes against their gradients (pp. 109–112)**

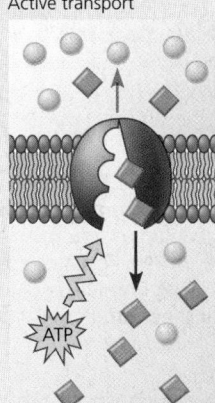

Active transport

- Specific membrane proteins use energy, usually in the form of ATP, to do the work of **active transport**.
- Ions can have both a concentration (chemical) gradient and an electrical gradient (voltage). These gradients combine in the **electrochemical gradient**, which determines the net direction of ionic diffusion.
- **Cotransport** of two solutes occurs when a membrane protein enables the "downhill" diffusion of one solute to drive the "uphill" transport of the other.

**?** *ATP is not directly involved in the functioning of a cotransporter. Why, then, is cotransport considered active transport?*

### CONCEPT 5.5

**Bulk transport across the plasma membrane occurs by exocytosis and endocytosis (pp. 112–113)**

- Three main types of **endocytosis** are **phagocytosis**, **pinocytosis**, and **receptor-mediated endocytosis**.

**?** *Which type of endocytosis involves the binding of specific substances in the extracellular fluid to membrane proteins? What does this type of transport enable a cell to do?*

### CONCEPT 5.6

**The plasma membrane plays a key role in most cell signaling (pp. 114–119)**

- Local signaling by animal cells involves direct contact or the secretion of growth factors and other signaling molecules. For long-distance signaling, animal and plant cells use **hormones**; animals also signal electrically.
- Signaling molecules that bind to membrane receptors trigger a three-stage cell-signaling pathway:

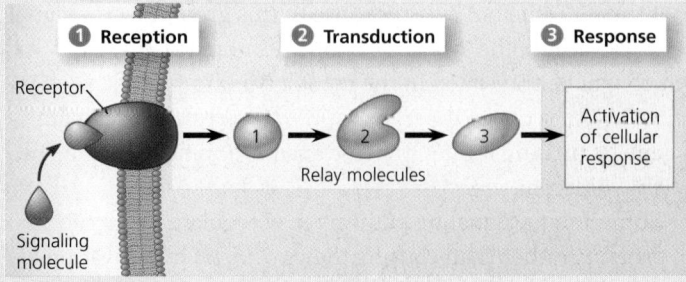

**1** Reception    **2** Transduction    **3** Response

Receptor

Relay molecules

Activation of cellular response

Signaling molecule

- In **reception**, a signaling molecule binds to a receptor protein, causing the protein to change shape. Two major types of membrane receptors are **G protein-coupled receptors (GPCRs)**, which work with the help of cytoplasmic **G proteins**, and **ligand-gated ion channels**, which open or close in response to binding by signaling molecules. Signaling molecules that are hydrophobic cross the plasma membrane and bind to receptors inside the cell.
- At each step in a **signal transduction pathway**, the signal is *transduced* into a different form, which commonly involves a change in a protein's shape. Many pathways include **phosphorylation cascades**, in which a series of **protein kinases** each add a phosphate group to the next one in line, activating it. The balance between phosphorylation and dephosphorylation, by **protein phosphatases**, regulates the activity of proteins in the pathway.
- **Second messengers**, such as the small molecule **cyclic AMP (cAMP)**, diffuse readily through the cytosol and thus help broadcast signals quickly. Many G proteins activate the enzyme that makes cAMP from ATP.
- The cell's **response** to a signal may be the regulation of transcription in the nucleus or of an activity in the cytoplasm.

**?** *What determines* whether *a cell responds to a hormone such as epinephrine? What determines* how *the cell responds?*

## TEST YOUR UNDERSTANDING

PRACTICE TEST

goo.gl/CRZjvS

### Level 1: Knowledge/Comprehension

1. In what way do the membranes of a eukaryotic cell vary?
   (A) Phospholipids are found only in certain membranes.
   (B) Certain proteins are unique to each kind of membrane.
   (C) Only certain membranes of the cell are selectively permeable.
   (D) Only certain membranes are constructed from amphipathic molecules.

2. Which of the following factors would tend to increase membrane fluidity?
   (A) a greater proportion of unsaturated phospholipids
   (B) a greater proportion of saturated phospholipids
   (C) a lower temperature
   (D) a relatively high protein content in the membrane

3. Phosphorylation cascades involving a series of protein kinases are useful for cellular signal transduction because
   (A) they are species specific.
   (B) they always lead to the same cellular response.
   (C) they amplify the original signal manyfold.
   (D) they counter the harmful effects of phosphatases.

4. Lipid-soluble signaling molecules, such as aldosterone, cross the membranes of all cells but affect only target cells because
   (A) only target cells retain the appropriate DNA segments.
   (B) intracellular receptors are present only in target cells.
   (C) only target cells have enzymes that break down aldosterone.
   (D) only in target cells is aldosterone able to initiate the phosphorylation cascade that turns genes on.

### Level 2: Application/Analysis

5. Which of the following processes includes all the others?
   (A) osmosis
   (B) diffusion of a solute across a membrane
   (C) passive transport
   (D) transport of an ion down its electrochemical gradient

6. Based on Figure 5.17, which of these experimental treatments would increase the rate of sucrose transport into a plant cell?
   (A) decreasing extracellular sucrose concentration
   (B) decreasing extracellular pH
   (C) decreasing cytoplasmic pH
   (D) adding a substance that makes the membrane more permeable to hydrogen ions

### Level 3: Synthesis/Evaluation

7. **SCIENTIFIC INQUIRY**
   An experiment is designed to study the mechanism of sucrose uptake by plant cells. Cells are immersed in a sucrose solution, and the pH of the solution is monitored. Samples of the cells are taken at intervals, and their sucrose concentration is measured. The pH is observed to decrease until it reaches a steady, slightly acidic level, and then sucrose uptake begins. (a) Evaluate these results and propose a hypothesis to explain them. (b) Predict what would happen if an inhibitor of ATP regeneration by the cell were added to the beaker once the pH was at a steady level? Explain your thinking.

8. **SCIENCE, TECHNOLOGY, AND SOCIETY**
   Extensive irrigation in arid regions causes salts to accumulate in the soil. (When water evaporates, salts that were dissolved in the water are left behind in the soil.) Based on what you have learned about water balance in plant cells, explain why increased soil salinity (saltiness) might be harmful to crops.

9. **FOCUS ON EVOLUTION**
   *Paramecium* and other unicellular eukaryotes that live in hypotonic environments have cell membranes that limit water uptake, while those living in isotonic environments have membranes that are more permeable to water. Describe what water regulation adaptations might have evolved in unicellular eukaryotes in hypertonic habitats such as the Great Salt Lake and in habitats with changing salt concentration.

10. **FOCUS ON INTERACTIONS**
    A human pancreatic cell obtains $O_2$—and necessary molecules such as glucose, amino acids, and cholesterol—from its environment, and it releases $CO_2$ as a waste product. In response to hormonal signals, the cell secretes digestive enzymes. It also regulates its ion concentrations by exchange with its environment. Based on what you have just learned about the structure and function of cellular membranes, write a short essay (100–150 words) to describe how such a cell accomplishes these interactions with its environment.

11. **SYNTHESIZE YOUR KNOWLEDGE**

In the supermarket, lettuce and other produce are often sprayed with water. Explain why this makes vegetables crisp.

*For selected answers, see Appendix A.*

# 6 An Introduction to Metabolism

## KEY CONCEPTS

**6.1** An organism's metabolism transforms matter and energy

**6.2** The free-energy change of a reaction tells us whether or not the reaction occurs spontaneously

**6.3** ATP powers cellular work by coupling exergonic reactions to endergonic reactions

**6.4** Enzymes speed up metabolic reactions by lowering energy barriers

**6.5** Regulation of enzyme activity helps control metabolism

▲ **Figure 6.1** What causes these breaking waves to glow?

## The Energy of Life

The living cell is a chemical factory in miniature, where thousands of reactions occur within a microscopic space. Sugars can be converted to amino acids that are linked together into proteins when needed, and when food is digested, proteins are dismantled into amino acids that can be converted to sugars. The process called cellular respiration drives the cellular economy by extracting the energy stored in sugars and other fuels. Cells apply this energy to perform various types of work. In an exotic example, the ocean waves shown in **Figure 6.1** are brightly illuminated from within by free-floating single-celled marine organisms called dinoflagellates. These dinoflagellates convert the energy stored in certain organic molecules to light, a process called bioluminescence. Such metabolic activities are precisely coordinated and controlled in the cell. In its complexity, its efficiency, and its responsiveness to subtle changes, the cell is peerless as a chemical factory. The concepts of metabolism that you learn in this chapter will help you understand how matter and energy flow during life's processes and how that flow is regulated.

## CONCEPT 6.1

# An organism's metabolism transforms matter and energy

The totality of an organism's chemical reactions is called **metabolism** (from the Greek *metabole*, change). Metabolism is an emergent property of life that arises from orderly interactions between molecules.

## Metabolic Pathways

We can picture a cell's metabolism as an elaborate road map of many chemical reactions, arranged as intersecting metabolic pathways. In a **metabolic pathway**, a specific molecule is altered in a series of defined steps, resulting in a product. Each step of the pathway is catalyzed by a specific enzyme:

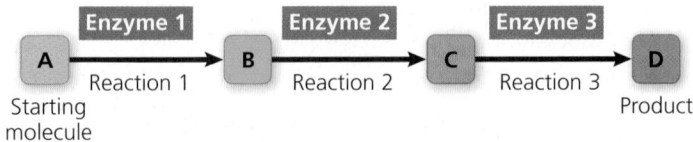

Analogous to the red, yellow, and green stoplights that control the flow of automobile traffic, mechanisms that regulate enzymes balance metabolic supply and demand.

Metabolism as a whole manages the material and energy resources of the cell. Some metabolic pathways release energy by breaking down complex molecules to simpler compounds. These degradative processes are called **catabolic pathways**, or breakdown pathways. A major pathway of catabolism is cellular respiration, in which the sugar glucose and other organic fuels are broken down in the presence of oxygen to carbon dioxide and water. Energy stored in the organic molecules becomes available to do the work of the cell, such as ciliary beating or membrane transport. **Anabolic pathways**, in contrast, consume energy to build complicated molecules from simpler ones; they are sometimes called biosynthetic pathways. Examples of anabolism are the synthesis of an amino acid from simpler molecules and the synthesis of a protein from amino acids. Catabolic and anabolic pathways are the "downhill" and "uphill" avenues of the metabolic landscape. Energy released from the downhill reactions of catabolic pathways can be stored and then used to drive the uphill reactions of anabolic pathways.

In this chapter, we will focus on mechanisms common to metabolic pathways. Because energy is fundamental to all metabolic processes, a basic knowledge of energy is necessary to understand how the living cell works. Although we will use some nonliving examples to study energy, the concepts demonstrated by these examples also apply to **bioenergetics**, the study of how energy flows through living organisms.

## Forms of Energy

**Energy** is the capacity to cause change. In everyday life, energy is important because some forms of energy can be used to do work—that is, to move matter against opposing forces, such as gravity and friction. Put another way, energy is the ability to rearrange a collection of matter. For example, you expend energy to turn the pages of this book, and your cells expend energy in transporting certain substances across membranes. Energy exists in various forms, and the work of life depends on the ability of cells to transform energy from one form to another.

Energy can be associated with the relative motion of objects; this energy is called **kinetic energy**. Moving objects can perform work by imparting motion to other matter: Water gushing through a dam turns turbines, and the contraction of leg muscles pushes bicycle pedals. **Thermal energy** is kinetic energy associated with the random movement of atoms or molecules; thermal energy in transfer from one object to another is called **heat**. Light is also a type of energy that can be harnessed to perform work, such as powering photosynthesis in green plants.

An object not presently moving may still possess energy. Energy that is not kinetic is called **potential energy**; it is energy that matter possesses because of its location or structure. Water behind a dam, for instance, possesses energy because of its altitude above sea level. Molecules possess energy because of the arrangement of electrons in the bonds between their atoms. **Chemical energy** is a term used by biologists to refer to the potential energy available for release in a chemical reaction. Recall that catabolic pathways release energy by breaking down complex molecules. Biologists say that these complex molecules, such as glucose, are high in chemical energy. During a catabolic reaction, some bonds are broken and others are formed, releasing energy and resulting in lower-energy breakdown products. This transformation also occurs in the engine of a car when the hydrocarbons of gasoline react explosively with oxygen, releasing the energy that pushes the pistons and producing exhaust. Although less explosive, a similar reaction of food molecules with oxygen provides chemical energy in biological systems, producing carbon dioxide and water as waste products. Biochemical pathways, carried out in the context of cellular structures, enable cells to release chemical energy from food molecules and use the energy to power life processes.

How is energy converted from one form to another? Consider **Figure 6.2**. The young woman climbing the ladder to the diving platform is releasing chemical energy from the food she ate for lunch and using some of that energy to perform the work of climbing. The kinetic energy of muscle movement is thus being transformed into potential energy due to her increasing height above the water. The young man diving is converting his potential energy to kinetic energy, which is then transferred to the water as he enters it. A small amount of energy is lost as heat due to friction.

Now let's consider the original source of the organic food molecules that provided the necessary chemical energy for the diver to climb the steps. This chemical energy was itself derived from light energy absorbed by plants during photosynthesis. Organisms are energy transformers.

A diver has more potential energy on the platform than in the water.

Diving converts potential energy to kinetic energy.

Climbing up converts the kinetic energy of muscle movement to potential energy.

A diver has less potential energy in the water than on the platform.

▲ **Figure 6.2 Transformations between potential and kinetic energy.**

## The Laws of Energy Transformation

The study of the energy transformations that occur in a collection of matter is called **thermodynamics**. Scientists use the word *system* to denote the matter under study; they refer to the rest of the universe—everything outside the system—as the *surroundings*. An *isolated system*, such as that approximated by liquid in a thermos bottle, is unable to exchange either energy or matter with its surroundings outside the thermos. In an *open system*, energy and matter can be transferred between the system and its surroundings. Organisms are open systems. They absorb energy—for instance, light energy or chemical energy in the form of organic molecules—and release heat and metabolic waste products, such as carbon dioxide, to the surroundings. Two laws of thermodynamics govern energy transformations in organisms and all other collections of matter.

### The First Law of Thermodynamics

According to the **first law of thermodynamics**, the energy of the universe is constant: *Energy can be transferred and or transformed, but it cannot be created or destroyed.* The first law is also known as the *principle of conservation of energy*. The electric company does not make energy, but merely converts it to a form that is convenient for us to use. By converting sunlight to chemical energy, a plant acts as an energy transformer, not an energy producer.

The brown bear in **Figure 6.3a** will convert the chemical energy of the organic molecules in its food to kinetic and other forms of energy as it carries out biological processes. What happens to this energy after it has performed work? The second law of thermodynamics helps to answer this question.

### The Second Law of Thermodynamics

If energy cannot be destroyed, why can't organisms simply recycle their energy over and over again? It turns out that during every energy transfer or transformation, some energy is converted to thermal energy and released as heat, becoming unavailable to do work. Only a small fraction of the chemical energy from the food in Figure 6.3a is transformed into the motion of the brown bear shown in **Figure 6.3b**; most is lost as heat, which dissipates rapidly through the surroundings.

A system can put thermal energy to work only when there is a temperature difference that results in the thermal energy flowing as heat from a warmer location to a cooler one. If temperature is uniform, as it is in a living cell, then the heat generated during a chemical reaction will simply warm a body of matter, such as the organism. (This can make a room crowded with people uncomfortably warm, as each person is carrying out a multitude of chemical reactions!)

A logical consequence of the loss of usable energy as heat to the surroundings is that each energy transfer or transformation makes the universe more disordered. Scientists use a quantity called **entropy** as a measure of disorder, or randomness. The more randomly arranged a collection of matter is, the greater its entropy. We can now state the **second law of thermodynamics**: *Every energy transfer or transformation increases the entropy of the universe.* Although order can increase locally, there is an unstoppable trend toward randomization of the universe as a whole.

In many cases, increased entropy is evident in the physical disintegration of a system's organized structure. For example, you can observe increasing entropy in the gradual decay of an unmaintained building. Much of the increasing entropy of the universe is less obvious, however, because it takes the form of increasing amounts of heat and less ordered forms of matter. As the bear in Figure 6.3b converts chemical energy to kinetic energy, it is also increasing the disorder of its surroundings by producing heat and small molecules, such as the $CO_2$ it exhales, that are the breakdown products of food.

The concept of entropy helps us understand why certain processes are energetically favorable and occur on their own. It turns out that if a given process, by itself, leads to an increase in entropy, that process can proceed without requiring

**(a) First law of thermodynamics:** Energy can be transferred or transformed but neither created nor destroyed. For example, chemical reactions in this brown bear will convert the chemical (potential) energy in the fish into the kinetic energy of running.

**(b) Second law of thermodynamics:** Every energy transfer or transformation increases the disorder (entropy) of the universe. For example, as the bear runs, disorder is increased around it by the release of heat and small molecules that are the by-products of metabolism. A brown bear can run at speeds up to 35 miles per hour (56 km/hr) —as fast as a racehorse.

▲ **Figure 6.3 The two laws of thermodynamics.**

an input of energy. Such a process is called a **spontaneous process**. Note that as we're using it here, the word *spontaneous* does not imply that the process would occur quickly; rather, the word signifies that it is energetically favorable. (In fact, it may be helpful for you to think of the phrase "energetically favorable" when you read the formal term "spontaneous.") Some spontaneous processes, such as an explosion, may be virtually instantaneous, while others, such as the rusting of an old car over time, are much slower.

A process that, on its own, leads to a decrease in entropy is said to be nonspontaneous: It will happen only if energy is supplied. We know from experience that certain events occur spontaneously and others do not. For instance, we know that water flows downhill spontaneously but moves uphill only with an input of energy, such as when a machine pumps the water against gravity. This understanding gives us another way to state the second law: *For a process to occur spontaneously, it must increase the entropy of the universe.*

### Biological Order and Disorder

Living systems increase the entropy of their surroundings, as predicted by thermodynamic law. It is true that cells create ordered structures from less organized starting materials. For example, simpler molecules are ordered into the more complex structure of an amino acid, and amino acids are ordered into polypeptide chains. At the organismal level as well, complex and beautifully ordered structures result from biological processes that use simpler starting materials **(Figure 6.4)**. On the other hand, an organism also takes in organized forms of matter and energy from the surroundings and replaces them with less ordered forms. For example, an animal obtains starch, proteins, and other complex molecules from the food it eats. As catabolic pathways break these molecules down, the animal releases carbon dioxide and water—small molecules that possess less chemical energy than the food did. The depletion of chemical energy is accounted for by heat

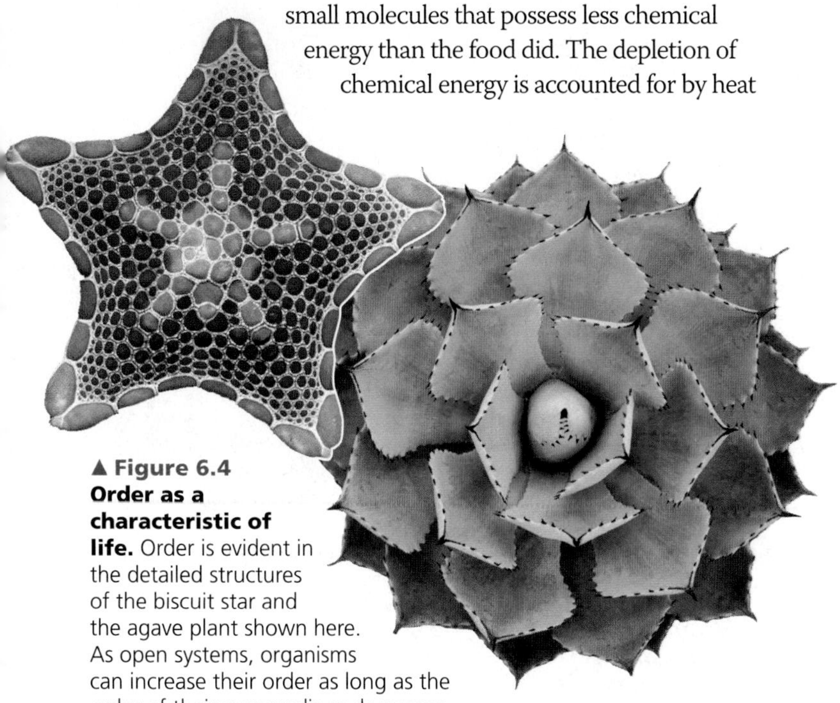

▲ Figure 6.4
**Order as a characteristic of life.** Order is evident in the detailed structures of the biscuit star and the agave plant shown here. As open systems, organisms can increase their order as long as the order of their surroundings decreases.

generated during metabolism. On a larger scale, energy flows into most ecosystems in the form of light and exits in the form of heat.

During the early history of life, complex organisms evolved from simpler ancestors. For instance, we can trace the ancestry of the plant kingdom from much simpler organisms called green algae to more complex flowering plants. However, this increase in organization over time in no way violates the second law. The entropy of a particular system, such as an organism, may actually decrease as long as the total entropy of the *universe*—the system plus its surroundings—increases. Thus, organisms are islands of low entropy in an increasingly random universe. The evolution of biological order is perfectly consistent with the laws of thermodynamics.

**CONCEPT CHECK 6.1**

1. **MAKE CONNECTIONS** How does the second law of thermodynamics help explain the diffusion of a substance across a membrane? (See Figure 5.9.)
2. Describe the forms of energy found in an apple as it grows on a tree, falls, and then is digested by someone who eats it.

   **For suggested answers, see Appendix A.**

CONCEPT 6.2

# The free-energy change of a reaction tells us whether or not the reaction occurs spontaneously

The laws of thermodynamics that we've just discussed apply to the universe as a whole. As biologists, we want to understand the chemical reactions of life—for example, which reactions occur spontaneously and which ones require some input of energy from outside. But how can we know this without assessing the energy and entropy changes in the entire universe for each separate reaction?

### Free-Energy Change (ΔG), Stability, and Equilibrium

Recall that the universe is really equivalent to "the system" plus "the surroundings." In 1878, J. Willard Gibbs, a professor at Yale, defined a very useful function called the Gibbs free energy of a system (without considering its surroundings), symbolized by the letter $G$. We'll refer to the Gibbs free energy simply as free energy. **Free energy** is the portion of a system's energy that can perform work when temperature and pressure are uniform throughout the system, as in a living cell. Biologists find it most informative to focus on the *change* in free energy ($\Delta G$) during the chemical reactions of life. $\Delta G$ represents the difference between the free energy of the final state and the free energy of the initial state:

$$\Delta G = G_{\text{final state}} - G_{\text{initial state}}$$

Using chemical methods, we can measure $\Delta G$ for any reaction. More than a century of experiments has shown that only reactions with a negative $\Delta G$ can occur with no input of energy, so the value of $\Delta G$ tells us whether a particular reaction is a spontaneous one. This principle is very important in the study of metabolism, where a major goal is to determine which reactions occur spontaneously and can be harnessed to supply energy for cellular work.

For a reaction to have a negative $\Delta G$, the system must lose free energy during the change from initial state to final state. Because it has less free energy, the system in its final state is less likely to change and is therefore more stable than it was previously. We can think of free energy as a measure of a system's instability—its tendency to change to a more stable state. Unstable systems (higher $G$) tend to change in such a way that they become more stable (lower $G$), as shown in **Figure 6.5**.

Another term that describes a state of maximum stability is chemical *equilibrium*. At equilibrium, the forward and reverse reactions occur at the same rate, and there is no further net change in the relative concentration of products and reactants. For a system at equilibrium, $G$ is at its lowest possible value in that system. We can think of the equilibrium state as a free-energy valley. Any change from the equilibrium position will have a positive $\Delta G$ and will not be spontaneous. For this reason, systems never spontaneously move away from equilibrium. Because a system at equilibrium cannot spontaneously change, it can do no work. *A process is spontaneous and can perform work only when it is moving toward equilibrium.*

## Free Energy and Metabolism

We can now apply the free-energy concept more specifically to the chemistry of life's processes.

### Exergonic and Endergonic Reactions in Metabolism

Based on their free-energy changes, chemical reactions can be classified as either exergonic ("energy outward") or endergonic ("energy inward"). An **exergonic reaction** proceeds with a net release of free energy **(Figure 6.6a)**. Because the chemical mixture loses free energy ($G$ decreases), $\Delta G$ is negative for an exergonic reaction. Using $\Delta G$ as a standard for spontaneity, exergonic reactions are those that occur spontaneously. (Remember, the word *spontaneous* implies that it is energetically favorable, not that it will occur rapidly.) The magnitude of $\Delta G$ for an exergonic reaction represents the maximum amount of work the reaction can perform (some of the free energy is released as heat and cannot do work). The greater the decrease in free energy, the greater the amount of work that can be done.

Consider the overall reaction for cellular respiration:

$$C_6H_{12}O_6 + 6\,O_2 \rightarrow 6\,CO_2 + 6\,H_2O$$
$$\Delta G = -686 \text{ kcal/mol} \ (-2{,}870 \text{ kJ/mol})$$

686 kcal (2,870 kJ) of energy are made available for work for each mole (180 g) of glucose broken down by respiration under "standard conditions" (1 $M$ of each reactant and product, 25°C, pH 7). Because energy must be conserved, the products of respiration store 686 kcal less free energy per mole than the reactants. The products are the "exhaust" of a process that tapped the free energy stored in the bonds of the sugar molecules.

It is important to realize that the breaking of bonds does not release energy; on the contrary, as you will soon see, it requires energy. The phrase "energy stored in bonds" is shorthand for the potential energy that can be released when new bonds are

- More free energy (higher $G$)
- Less stable
- Greater work capacity

In a **spontaneous change**
- The free energy of the system decreases ($\Delta G < 0$)
- The system becomes more stable
- The released free energy can be harnessed to do work

- Less free energy (lower $G$)
- More stable
- Less work capacity

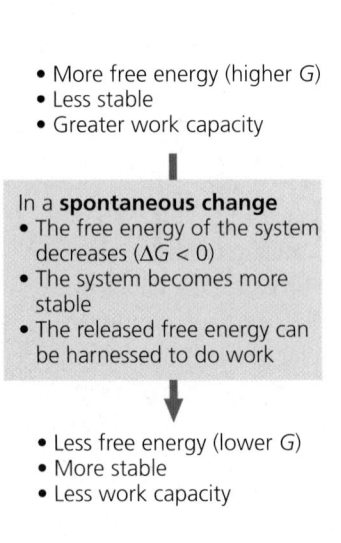

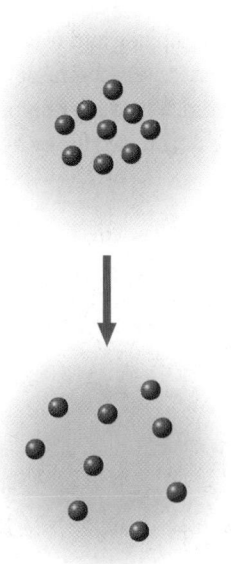

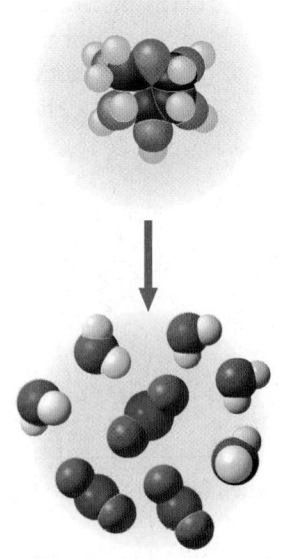

**(a) Gravitational motion.** Objects move spontaneously from a higher altitude to a lower one.

**(b) Diffusion.** Molecules in a drop of dye diffuse until they are randomly dispersed.

**(c) Chemical reaction.** In a cell, a glucose molecule is broken down into simpler molecules.

▲ **Figure 6.5 The relationship of free energy to stability, work capacity, and spontaneous change.** Unstable systems (top) are rich in free energy, $G$. They have a tendency to change spontaneously to a more stable state (bottom), and it is possible to harness this "downhill" change to perform work.

## Figure 6.6

▼ **Figure 6.6 Free energy changes ($\Delta G$) in exergonic and endergonic reactions.**

**(a) Exergonic reaction: energy released, spontaneous**

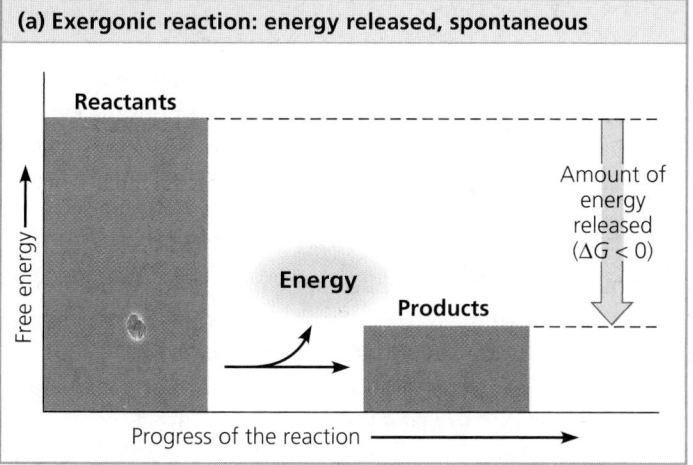

**(b) Endergonic reaction: energy required, nonspontaneous**

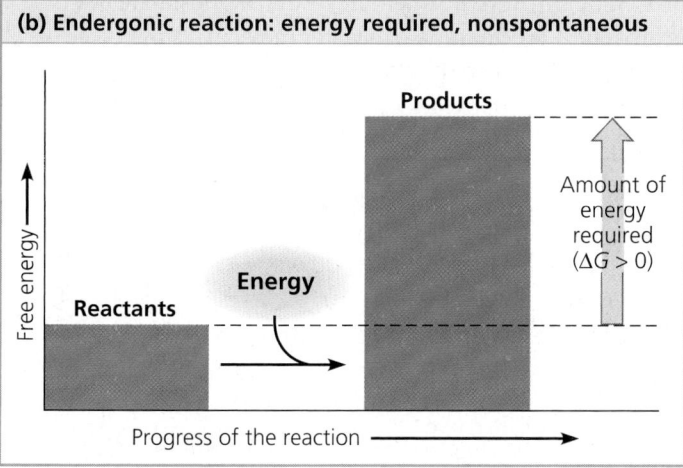

formed after the original bonds break, as long as the products are of lower free energy than the reactants.

An **endergonic reaction** is one that absorbs free energy from its surroundings **(Figure 6.6b)**. Because this kind of reaction essentially *stores* free energy in molecules ($G$ increases), $\Delta G$ is positive. Such reactions are nonspontaneous, and the magnitude of $\Delta G$ is the quantity of energy required to drive the reaction. If a chemical process is exergonic (downhill), releasing energy in one direction, then the reverse process must be endergonic (uphill), using energy. A reversible process cannot be downhill in both directions. If $\Delta G = -686$ kcal/mol for respiration, which converts glucose and oxygen to carbon dioxide and water, then the reverse process—the conversion of carbon dioxide and water to glucose and oxygen—must be strongly endergonic, with $\Delta G = +686$ kcal/mol. Such a reaction would never happen by itself.

How, then, do plants make the sugar that organisms use for energy? Plants get the required energy—686 kcal to make a mole of glucose—from the environment by capturing light and converting its energy to chemical energy. Next, in a long series of exergonic steps, they gradually spend that chemical energy to assemble glucose molecules.

## Equilibrium and Metabolism

Reactions in an isolated system eventually reach equilibrium and can then do no work, as illustrated by the isolated hydroelectric system in **Figure 6.7a**. The chemical reactions of metabolism are reversible, and they, too, would reach equilibrium if they occurred in the isolation of a test tube. Because systems at equilibrium are at a minimum of $G$ and can do no work, a cell that has reached metabolic equilibrium is dead! The fact that metabolism as a whole is never at equilibrium is one of the defining features of life.

Like most systems, a living cell is not in equilibrium. The constant flow of materials in and out of the cell keeps the metabolic pathways from ever reaching equilibrium, and the cell continues to do work throughout its life. This principle is illustrated by the open (and more realistic) hydroelectric system in **Figure 6.7b**. However, unlike this simple single-step system,

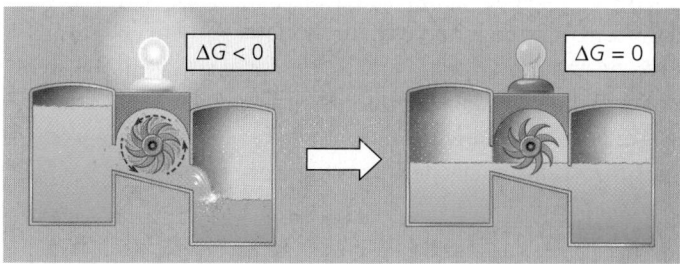

**(a) An isolated hydroelectric system.** Water flowing downhill turns a turbine that drives a generator providing electricity to a lightbulb, but only until the system reaches equilibrium.

**(b) An open hydroelectric system.** Flowing water keeps driving the generator because intake and outflow of water keep the system from reaching equilibrium.

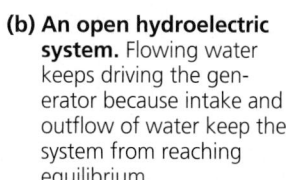

**(c) A multistep open hydroelectric system.** Cellular respiration is analogous to this system: Glucose is broken down in a series of exergonic reactions that power the work of the cell. The product of each reaction is used as the reactant for the next, so no reaction reaches equilibrium.

▲ **Figure 6.7 Equilibrium and work in isolated and open systems.**

a catabolic pathway in a cell releases free energy in a series of reactions. An example is cellular respiration, illustrated by the hydroelectric system analogy in **Figure 6.7c**. Some of the reversible reactions of respiration are constantly "pulled" in one direction—that is, they are kept out of equilibrium. The key to maintaining this lack of equilibrium is that the product of a reaction does not accumulate but instead becomes a reactant in the next step; finally, waste products are expelled from the cell. The overall sequence of reactions is kept going by the huge free-energy difference between glucose and oxygen at the top of the energy "hill" and carbon dioxide and water at the "downhill" end. As long as our cells have a steady supply of glucose or other fuels and oxygen and are able to expel waste products to the surroundings, their metabolic pathways never reach equilibrium and can continue to do the work of life.

We see once again how important it is to think of organisms as open systems. Sunlight provides a daily source of free energy for an ecosystem's plants and other photosynthetic organisms. Animals and other nonphotosynthetic organisms in an ecosystem must have a source of free energy in the form of the organic products of photosynthesis. Now that we have applied the free-energy concept to metabolism, we are ready to see how a cell actually performs the work of life.

### CONCEPT CHECK 6.2

1. Cellular respiration uses glucose and oxygen, which have high levels of free energy, and releases $CO_2$ and water, which have low levels of free energy. Is cellular respiration spontaneous or not? Is it exergonic or endergonic? What happens to the energy released from glucose?
2. How do the processes of catabolism and anabolism relate to Figure 6.5c?
3. **WHAT IF?** Some nighttime partygoers wear glow-in-the-dark necklaces. The necklaces start glowing once they are "activated" by snapping the necklace in a way that allows two chemicals to react and emit light in the form of chemiluminescence. Is this chemical reaction exergonic or endergonic? Explain your answer.

For suggested answers, see Appendix A.

---

**CONCEPT 6.3**

# ATP powers cellular work by coupling exergonic reactions to endergonic reactions

A cell does three main kinds of work:

- *Chemical work*, the pushing of endergonic reactions that would not occur spontaneously, such as the synthesis of polymers from monomers (chemical work will be discussed further in this chapter and in Chapters 7 and 8)
- *Transport work*, the pumping of substances across membranes against the direction of spontaneous movement (see Concept 5.4)

- *Mechanical work*, such as the beating of cilia (see Concept 4.6), the contraction of muscle cells, and the movement of chromosomes during cellular reproduction

A key feature in the way cells manage their energy resources to do this work is **energy coupling**, the use of an exergonic process to drive an endergonic one. ATP is responsible for mediating most energy coupling in cells, and in most cases it acts as the immediate source of energy that powers cellular work.

## The Structure and Hydrolysis of ATP

**ATP (adenosine triphosphate**; see Concept 3.1) contains the sugar ribose, with the nitrogenous base adenine and a chain of three phosphate groups bonded to it **(Figure 6.8a)**. In addition to its role in energy coupling, ATP is also one of the nucleoside triphosphates used to make RNA (see Figure 3.27).

The bonds between the phosphate groups of ATP can be broken by hydrolysis. When the terminal phosphate bond is broken by the addition of a water molecule, a molecule of inorganic phosphate ($HOPO_3^{2-}$, which is abbreviated $\textcircled{P}_i$ throughout this book) leaves the ATP. In this way, adenosine

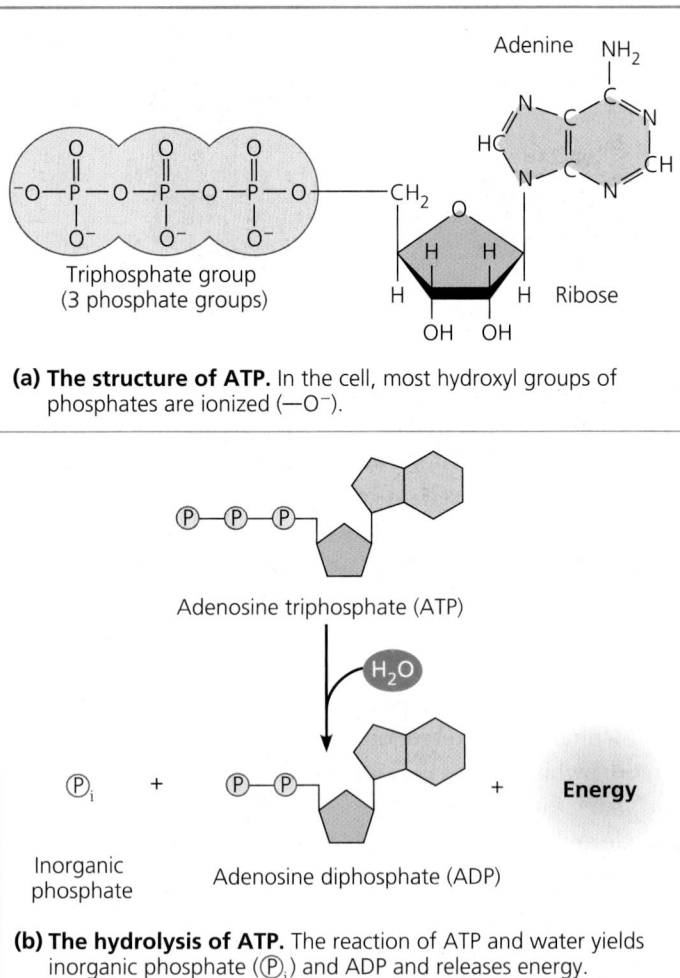

(a) **The structure of ATP.** In the cell, most hydroxyl groups of phosphates are ionized ($-O^-$).

(b) **The hydrolysis of ATP.** The reaction of ATP and water yields inorganic phosphate ($\textcircled{P}_i$) and ADP and releases energy.

▲ **Figure 6.8 The structure and hydrolysis of adenosine triphosphate (ATP).** Throughout this book, the chemical structure of the triphosphate group seen in (a) will be represented by the three joined yellow circles shown in (b).

*tri*phosphate becomes adenosine *di*phosphate, or ADP **(Figure 6.8b)**. The reaction is exergonic and releases 7.3 kcal of energy per mole of ATP hydrolyzed:

$$ATP + H_2O \rightarrow ADP + \textcircled{P}_i$$
$$\Delta G = -7.3 \text{ kcal/mol } (-30.5 \text{ kJ/mol})$$

This is the free-energy change measured under standard conditions. In the cell, conditions do not conform to standard conditions, primarily because reactant and product concentrations differ from 1 *M*. For example, when ATP hydrolysis occurs under cellular conditions, the actual $\Delta G$ is about −13 kcal/mol, 78% greater than the energy released by ATP hydrolysis under standard conditions.

Because their hydrolysis releases energy, the phosphate bonds of ATP are sometimes referred to as high-energy phosphate bonds, but the term is misleading. The phosphate bonds of ATP are not unusually strong bonds, as "high-energy" may imply; rather, the reactants (ATP and water) themselves have high energy relative to the energy of the products (ADP and $\textcircled{P}_i$). The release of energy during the hydrolysis of ATP comes from the chemical change to a state of lower free energy, not from the phosphate bonds themselves.

ATP is useful to the cell because the energy it releases on losing a phosphate group is somewhat greater than the energy most other molecules could deliver. But why does this hydrolysis release so much energy? If we reexamine the ATP molecule

in Figure 6.8a, we can see that all three phosphate groups are negatively charged. These like charges are crowded together, and their mutual repulsion contributes to the instability of this region of the ATP molecule. The triphosphate tail of ATP is the chemical equivalent of a compressed spring.

## How the Hydrolysis of ATP Performs Work

When ATP is hydrolyzed in a test tube, the release of free energy merely heats the surrounding water. In an organism, this same generation of heat can sometimes be beneficial. For instance, the process of shivering uses ATP hydrolysis during muscle contraction to warm the body. In most cases in the cell, however, the generation of heat alone would be an inefficient (and potentially dangerous) use of a valuable energy resource. Instead, the cell's proteins harness the energy released during ATP hydrolysis in several ways to perform the three types of cellular work—chemical, transport, and mechanical.

For example, with the help of specific enzymes, the cell is able to use the energy released by ATP hydrolysis directly to drive chemical reactions that, by themselves, are endergonic **(Figure 6.9)**. If the $\Delta G$ of an endergonic reaction is less than the amount of energy released by ATP hydrolysis, then the two reactions can be coupled so that, overall, the coupled reactions are exergonic. This usually involves phosphorylation, the transfer of a phosphate group from ATP to some other

**(a) Glutamic acid conversion to glutamine.** Glutamine synthesis from glutamic acid (Glu) by itself is endergonic ($\Delta G$ is positive), so it is not spontaneous.

Glutamic acid   Ammonia        Glutamine        $\Delta G_{Glu} = +3.4$ kcal/mol

**(b) Conversion reaction coupled with ATP hydrolysis.** In the cell, glutamine synthesis occurs in two steps, coupled by a phosphorylated intermediate. ❶ ATP phosphorylates glutamic acid, making it less stable. ❷ Ammonia displaces the phosphate group, forming glutamine.

Glutamic acid        Phosphorylated        Glutamine
                     intermediate

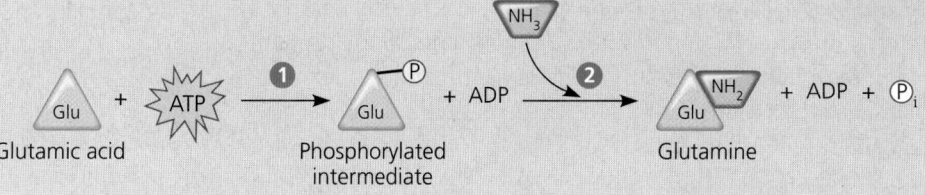

**(c) Free-energy change for coupled reaction.** $\Delta G$ for the glutamic acid conversion to glutamine (+3.4 kcal/mol) plus $\Delta G$ for ATP hydrolysis (−7.3 kcal/mol) gives the free-energy change for the overall reaction (−3.9 kcal/mol). Because the overall process is exergonic (net $\Delta G$ is negative), it occurs spontaneously.

$\Delta G_{Glu} = +3.4$ kcal/mol

$\Delta G_{ATP} = -7.3$ kcal/mol

$$\Delta G_{Glu} = +3.4 \text{ kcal/mol}$$
$$+ \Delta G_{ATP} = -7.3 \text{ kcal/mol}$$
$$\text{Net } \Delta G = -3.9 \text{ kcal/mol}$$

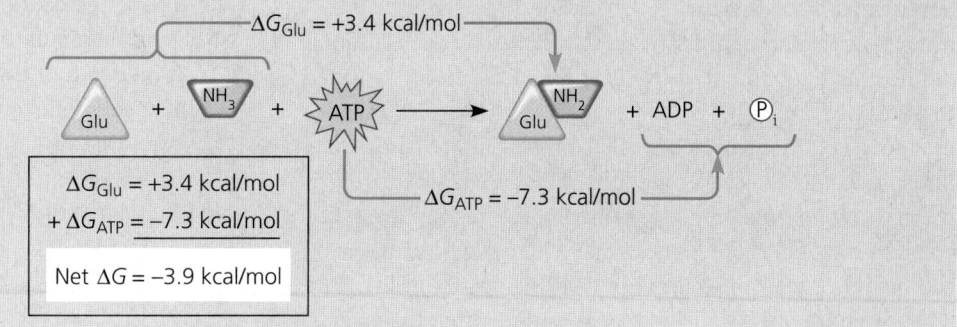

▲ **Figure 6.9 How ATP drives chemical work: energy coupling using ATP hydrolysis.** In this example, the exergonic process of ATP hydrolysis is used to drive an endergonic process—the cellular synthesis of the amino acid glutamine from glutamic acid and ammonia.

**MAKE CONNECTIONS** *Explain why glutamine is drawn as a glutamic acid (Glu) with an amino group attached. (See Figure 3.18.)*

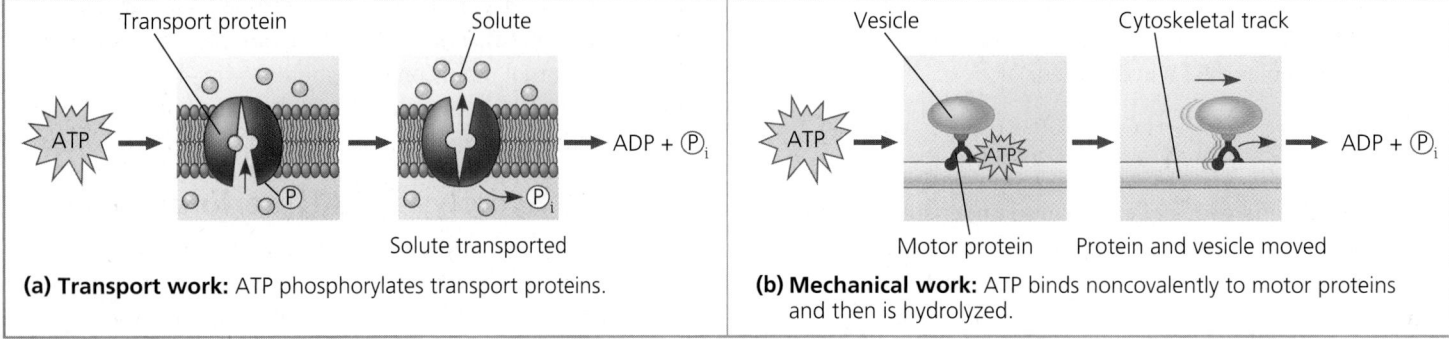

**(a) Transport work:** ATP phosphorylates transport proteins.

**(b) Mechanical work:** ATP binds noncovalently to motor proteins and then is hydrolyzed.

▲ **Figure 6.10 How ATP drives transport and mechanical work.** ATP hydrolysis causes changes in the shapes and binding affinities of proteins. This can occur either **(a)** directly, by phosphorylation, as shown for a membrane protein carrying out active transport of a solute (see also Figure 5.14), or **(b)** indirectly, via noncovalent binding of ATP and its hydrolytic products, as is the case for motor proteins that move vesicles (and other organelles) along cytoskeletal "tracks" in the cell (see also Figure 4.21).

molecule, such as the reactant (see Figure 6.9b). The recipient with the phosphate group covalently bonded to it is then called a **phosphorylated intermediate**. The key to coupling exergonic and endergonic reactions is the formation of this phosphorylated intermediate, which is more reactive (less stable) than the original unphosphorylated molecule.

Transport and mechanical work in the cell are also nearly always powered by the hydrolysis of ATP. In these cases, ATP hydrolysis leads to a change in a protein's shape and often its ability to bind another molecule. Sometimes this occurs via a phosphorylated intermediate, as seen for the transport protein in **Figure 6.10a**. In most instances of mechanical work involving motor proteins "walking" along cytoskeletal elements **(Figure 6.10b)**, a cycle occurs in which ATP is first bound noncovalently to the motor protein. Next, ATP is hydrolyzed, releasing ADP and $P_i$. Another ATP molecule can then bind. At each stage, the motor protein changes its shape and ability to bind the cytoskeleton, resulting in movement of the protein along the cytoskeletal track. Phosphorylation and dephosphorylation also promote crucial protein shape changes during cell signaling (see Figure 5.24).

## The Regeneration of ATP

An organism at work uses ATP continuously, but ATP is a renewable resource that can be regenerated by the addition of phosphate to ADP **(Figure 6.11)**. The free energy required to phosphorylate ADP comes from exergonic breakdown reactions (catabolism) in the cell. This shuttling of inorganic phosphate and energy is called the ATP cycle, and it couples the cell's energy-yielding (exergonic) processes to the energy-consuming (endergonic) ones. The ATP cycle proceeds at an astonishing pace. For example, a working muscle cell recycles its entire pool of ATP in less than a minute. That turnover represents 10 million molecules of ATP consumed and regenerated per second per cell. If ATP could not be regenerated by the phosphorylation of ADP, humans would use up nearly their body weight in ATP each day.

ATP synthesis from ADP + $P_i$ requires energy.

ATP hydrolysis to ADP + $P_i$ yields energy.

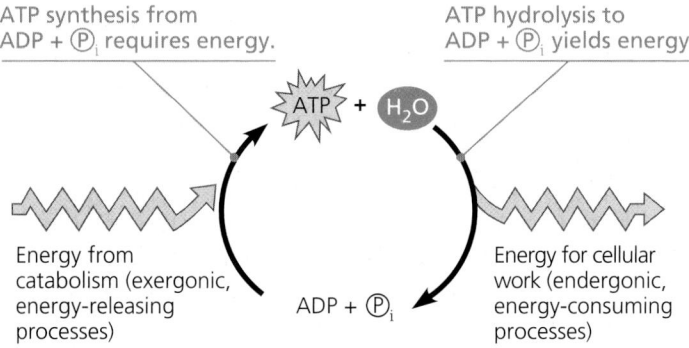

Energy from catabolism (exergonic, energy-releasing processes)

Energy for cellular work (endergonic, energy-consuming processes)

▲ **Figure 6.11 The ATP cycle.** Energy released by breakdown reactions (catabolism) in the cell is used to phosphorylate ADP, regenerating ATP. Chemical potential energy stored in ATP drives most cellular work.

Because both directions of a reversible process cannot be downhill, the regeneration of ATP from ADP and $P_i$ is necessarily endergonic:

$$ADP + P_i \rightarrow ATP + H_2O$$
$$\Delta G = +7.3 \text{ kcal/mol } (+30.5 \text{ kJ/mol}) \text{ (standard conditions)}$$

Since ATP formation from ADP and $P_i$ is not spontaneous, free energy must be spent to make it occur. Catabolic (exergonic) pathways, especially cellular respiration, provide the energy for the endergonic process of making ATP. Plants also use light energy to produce ATP. Thus, the ATP cycle is a revolving door through which energy passes during its transfer from catabolic to anabolic pathways.

**CONCEPT CHECK 6.3**

1. How does ATP typically transfer energy from exergonic to endergonic reactions in the cell?
2. Which of the following combinations has more free energy: glutamic acid + ammonia + ATP or glutamine + ADP + $P_i$? Explain your answer.
3. **MAKE CONNECTIONS** Does Figure 6.10a show passive or active transport? Explain. (See Concepts 5.3 and 5.4.)

For suggested answers, see Appendix A.

# Enzymes speed up metabolic reactions by lowering energy barriers

The laws of thermodynamics tell us what will and will not happen under given conditions but say nothing about the *rate* of these processes. A spontaneous chemical reaction occurs without any requirement for outside energy, but it may occur so slowly that it is imperceptible. For example, even though the hydrolysis of sucrose (table sugar) to glucose and fructose is exergonic, occurring spontaneously with a release of free energy ($\Delta G = -7$ kcal/mol), a solution of sucrose dissolved in sterile water will sit for years at room temperature with no appreciable hydrolysis. However, if we add a small amount of the enzyme sucrase to the solution, then all the sucrose may be hydrolyzed within seconds, as shown here:

Sucrose ($C_{12}H_{22}O_{11}$) + $H_2O$ → [Sucrase] Glucose ($C_6H_{12}O_6$) + Fructose ($C_6H_{12}O_6$)

How does the enzyme do this?

An **enzyme** is a macromolecule that acts as a **catalyst**, a chemical agent that speeds up a reaction without being consumed by the reaction. In this chapter, we are focusing on enzymes that are proteins. (Some RNA molecules, called ribozymes, can function as enzymes; these will be discussed in Concepts 14.3 and 24.1.) Without regulation by enzymes, chemical traffic through the pathways of metabolism would become terribly congested because many chemical reactions would take such a long time. In the next two sections, we will see why spontaneous reactions can be slow and how an enzyme changes the situation.

## The Activation Energy Barrier

Every chemical reaction between molecules involves both bond breaking and bond forming. For example, the hydrolysis of sucrose involves breaking the bond between glucose and fructose and one of the bonds of a water molecule and then forming two new bonds, as shown above. Changing one molecule into another generally involves contorting the starting molecule into a highly unstable state before the reaction can proceed. This contortion can be compared to the bending of a metal key ring when you pry it open to add a new key. The key ring is highly unstable in its opened form but returns to a stable state once the key is threaded all the way onto the ring. To reach the contorted state where bonds can change, reactant molecules must absorb energy from their surroundings. When the new bonds of the product molecules form, energy is released as heat, and the molecules return to stable shapes with lower energy than the contorted state.

The initial investment of energy for starting a reaction—the energy required to contort the reactant molecules so the bonds can break—is known as the *free energy of activation*, or **activation energy**, abbreviated $E_A$ in this book. We can think of activation energy as the amount of energy needed to push the reactants to the top of an energy barrier, or uphill, so that the "downhill" part of the reaction can begin. Activation energy is often supplied by heat in the form of thermal energy that the reactant molecules absorb from the surroundings. The absorption of thermal energy accelerates the reactant molecules, so they collide more often and more forcefully. It also agitates the atoms within the molecules, making the breakage of bonds more likely. When the molecules have absorbed enough energy for the bonds to break, the reactants are in an unstable condition known as the *transition state*.

**Figure 6.12** graphs the energy changes for a hypothetical exergonic reaction that swaps portions of two reactant molecules:

$$AB + CD \rightarrow AC + BD$$
Reactants     Products

The activation of the reactants is represented by the uphill portion of the graph, in which the free-energy content of the reactant molecules is increasing. At the summit, when energy

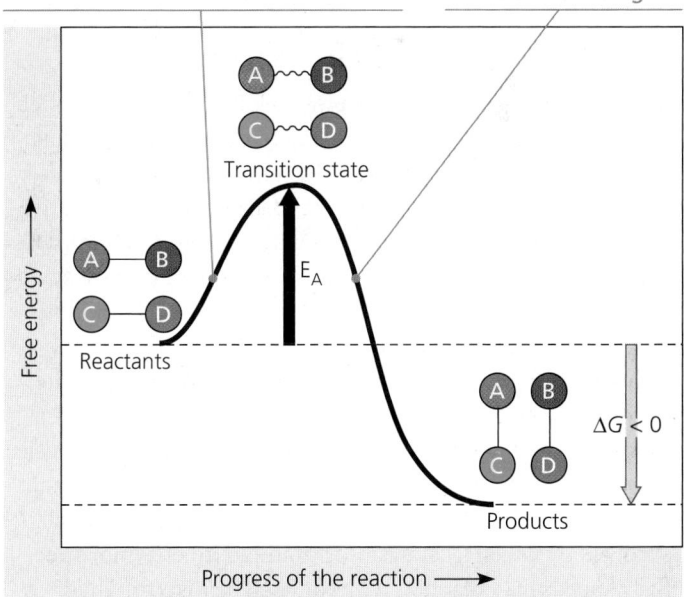

▲ **Figure 6.12 Energy profile of an exergonic reaction.** The "molecules" are hypothetical, with A, B, C, and D representing portions of the molecules. Thermodynamically, this is an exergonic reaction, with a negative $\Delta G$, and the reaction occurs spontaneously. However, the activation energy ($E_A$) provides a barrier that determines the rate of the reaction.

**DRAW IT** *Graph the progress of an endergonic reaction in which EF and GH form products EG and FH, assuming that the reactants must pass through a transition state.*

equivalent to $E_A$ has been absorbed, the reactants are in the transition state: They are activated, and their bonds can be broken. As the atoms then settle into their new, more stable bonding arrangements, energy is released to the surroundings. This corresponds to the downhill part of the curve, which shows the loss of free energy by the molecules. The overall decrease in free energy means that $E_A$ is repaid with dividends, as the formation of new bonds releases more energy than was invested in the breaking of old bonds.

The reaction shown in Figure 6.12 is exergonic and occurs spontaneously ($\Delta G < 0$). However, the activation energy provides a barrier that determines the rate of the reaction. The reactants must absorb enough energy to reach the top of the activation energy barrier before the reaction can occur. For some reactions, $E_A$ is modest enough that even at room temperature there is sufficient thermal energy for many of the reactant molecules to reach the transition state in a short time. In most cases, however, $E_A$ is so high and the transition state is reached so rarely that the reaction will hardly proceed at all. In these cases, the reaction will occur at a noticeable rate only if energy is provided, usually as heat. For example, the reaction of gasoline and oxygen is exergonic and will occur spontaneously, but energy is required for the molecules to reach the transition state and react. Only when the spark plugs fire in an automobile engine can there be the explosive release of energy that pushes the pistons. Without a spark, a mixture of gasoline hydrocarbons and oxygen will not react because the $E_A$ barrier is too high.

## How Enzymes Speed Up Reactions

Proteins, DNA, and other complex cellular molecules are rich in free energy and have the potential to decompose spontaneously; that is, the laws of thermodynamics favor their breakdown. These molecules persist only because at temperatures typical for cells, few molecules can make it over the hump of activation energy. The barriers for selected reactions must occasionally be surmounted, however, for cells to carry out the processes needed for life. Heat can increase the rate of a reaction by allowing reactants to attain the transition state more often, but this would not work well in biological systems. First, high temperature denatures proteins and kills cells. Second, heat would speed up *all* reactions, not just those that are needed. Instead of heat, organisms carry out **catalysis**, a process by which a catalyst (for example, an enzyme) selectively speeds up a reaction without itself being consumed. (You learned about catalysts in Concept 3.5.)

An enzyme catalyzes a reaction by lowering the $E_A$ barrier **(Figure 6.13)**, enabling the reactant molecules to absorb enough energy to reach the transition state even at moderate temperatures. An enzyme cannot change the $\Delta G$ for a reaction; it cannot make an endergonic reaction exergonic. Enzymes can only hasten reactions that would eventually occur anyway, but this enables the cell to have a dynamic metabolism, routing chemicals smoothly through metabolic pathways.

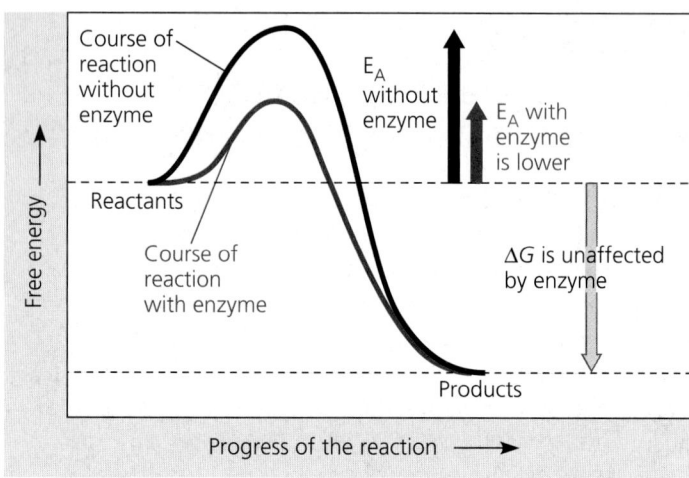

▲ **Figure 6.13 The effect of an enzyme on activation energy.** Without affecting the free-energy change ($\Delta G$) for a reaction, an enzyme speeds the reaction by reducing its activation energy ($E_A$).

Also, enzymes are very specific for the reactions they catalyze, so they determine which chemical processes will be going on in the cell at any given time.

## Substrate Specificity of Enzymes

The reactant an enzyme acts on is referred to as the enzyme's **substrate**. The enzyme binds to its substrate (or substrates, when there are two or more reactants), forming an **enzyme-substrate complex**. While enzyme and substrate are joined, the catalytic action of the enzyme converts the substrate to the product (or products) of the reaction. The overall process can be summarized as follows:

| Enzyme + | | Enzyme- | | Enzyme + |
|----------|---|---------|---|----------|
| Substrate(s) | $\rightleftharpoons$ | substrate complex | $\rightleftharpoons$ | Product(s) |

For example, the enzyme sucrase (most enzyme names end in *-ase*) catalyzes the hydrolysis of the disaccharide sucrose into its two monosaccharides, glucose and fructose (see the illustrated equation at the beginning of Concept 6.4):

| Sucrase + | | Sucrase- | | Sucrase + |
|-----------|---|----------|---|-----------|
| Sucrose + | $\rightleftharpoons$ | sucrose-$H_2O$ | $\rightleftharpoons$ | Glucose + |
| $H_2O$ | | complex | | Fructose |

The reaction catalyzed by each enzyme is very specific; an enzyme can recognize its specific substrate even among closely related compounds. For instance, sucrase will act only on sucrose and will not bind to other disaccharides, such as maltose. What accounts for this molecular recognition? Recall that most enzymes are proteins, and proteins are macromolecules with unique three-dimensional configurations. The specificity of an enzyme results from its shape, which is a consequence of its amino acid sequence.

Only a restricted region of the enzyme molecule actually binds to the substrate. This region, known as the **active site**, is typically a pocket or groove on the surface of the enzyme where catalysis occurs **(Figure 6.14a)**. Usually, the active site is formed by only a few of the enzyme's amino acids, with the rest of the protein molecule providing a framework that determines the shape of the active site. The specificity of an enzyme is attributed to a complementary fit between the shape of its active site and the shape of the substrate, like that seen in the binding of a signaling molecule to a receptor protein (see Concept 5.6).

An enzyme is not a stiff structure locked into a given shape. In fact, recent work by biochemists has shown that enzymes (and other proteins) seem to "dance" between subtly different shapes in a dynamic equilibrium, with slight differences in free energy for each "pose." The shape that best fits the substrate isn't necessarily the one with the lowest energy, but during the very short time the enzyme takes on this shape, its active site can bind to the substrate. The active site itself is also not a rigid receptacle for the substrate. As the substrate enters the active site, the enzyme changes shape slightly due to interactions between the substrate's chemical groups and chemical groups on the side chains of the amino acids that form the active site. This shape change makes the active site fit even more snugly around the substrate **(Figure 6.14b)**. This tightening of the binding after initial contact—called **induced fit**—is like a clasping handshake. Induced fit brings chemical groups of the active site into positions that enhance their ability to catalyze the chemical reaction.

## Catalysis in the Enzyme's Active Site

In most enzymatic reactions, the substrate is held in the active site by so-called weak interactions, such as hydrogen bonds and ionic bonds. The R groups of a few of the amino acids that make up the active site catalyze the conversion of substrate to product, and the product departs from the active site. The enzyme is then free to take another substrate molecule into its active site. The entire cycle happens so fast that a single enzyme molecule typically acts on about 1,000 substrate molecules per second, and some enzymes are even faster. Enzymes, like other catalysts, emerge from the reaction in their original form. Therefore, very small amounts of enzyme can have a huge metabolic impact by functioning over and over again in catalytic cycles. **Figure 6.15** shows a catalytic cycle involving two substrates and two products.

Most metabolic reactions are reversible, and an enzyme can catalyze either the forward or the reverse reaction, depending on which direction has a negative $\Delta G$. This in turn depends mainly on the relative concentrations of reactants and products. The net effect is always in the direction of equilibrium.

Enzymes use a variety of mechanisms that lower activation energy and speed up a reaction. First, in reactions involving

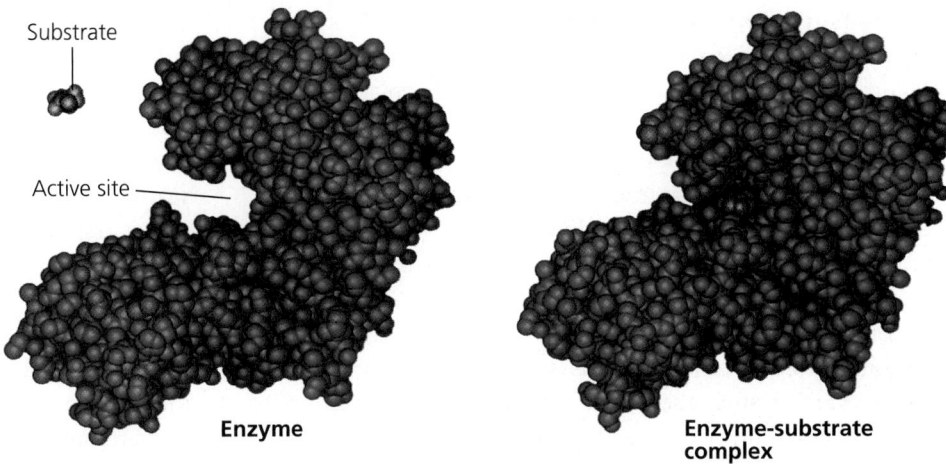

Substrate

Active site

**Enzyme**

**Enzyme-substrate complex**

**(a)** In this computer graphic model, the active site of this enzyme (hexokinase, shown in blue) forms a groove on its surface. Its substrate is glucose (red).

**(b)** When the substrate enters the active site, it forms weak bonds with the enzyme, inducing a change in the shape of the protein. This change allows additional weak bonds to form, causing the active site to enfold the substrate and hold it in place.

▲ **Figure 6.14 Induced fit between an enzyme and its substrate.**

two or more reactants, the active site provides a template on which the substrates can come together in the proper orientation for a reaction to occur between them (see Figure 6.15, step ❷). Second, as the active site of an enzyme clutches the bound substrates, the enzyme may stretch the substrate molecules toward their transition state form, stressing and bending critical chemical bonds that must be broken during the reaction. Because $E_A$ is proportional to the difficulty of breaking the bonds, distorting the substrate helps it approach

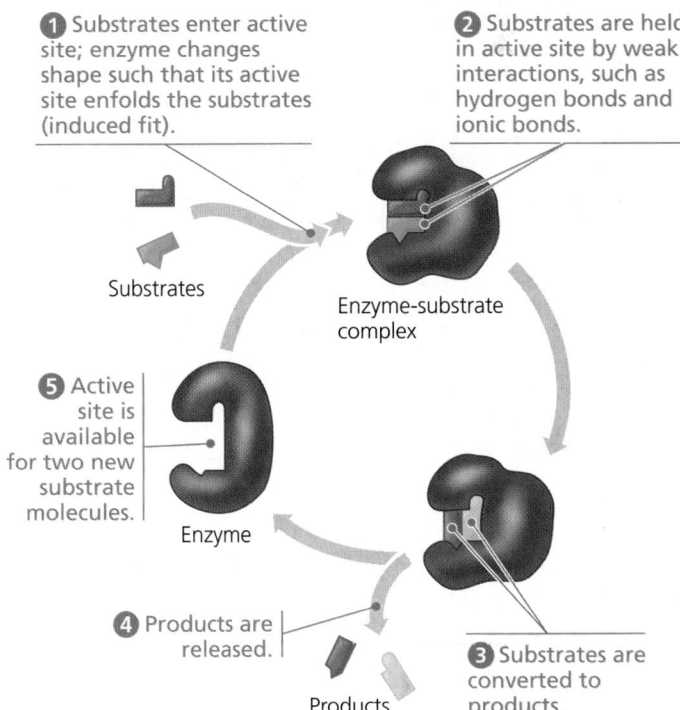

❶ Substrates enter active site; enzyme changes shape such that its active site enfolds the substrates (induced fit).

❷ Substrates are held in active site by weak interactions, such as hydrogen bonds and ionic bonds.

Substrates

Enzyme-substrate complex

❺ Active site is available for two new substrate molecules.

Enzyme

❹ Products are released.

Products

❸ Substrates are converted to products.

▲ **Figure 6.15 The active site and catalytic cycle of an enzyme.** An enzyme can convert one or more reactant molecules to one or more product molecules. The enzyme shown here converts two substrate molecules to two product molecules.

the transition state and thus reduces the amount of free energy that must be absorbed to achieve that state.

Third, the active site may also provide a microenvironment that is more conducive to a particular type of reaction than the solution itself would be without the enzyme. For example, if the active site has amino acids with acidic R groups, the active site may be a pocket of low pH in an otherwise neutral cell. In such cases, an acidic amino acid may facilitate $H^+$ transfer to the substrate as a key step in catalyzing the reaction.

A fourth mechanism of catalysis is the direct participation of the active site in the chemical reaction. Sometimes this process even involves brief covalent bonding between the substrate and the side chain of an amino acid of the enzyme. Subsequent steps of the reaction restore the side chains to their original states, so that the active site is the same after the reaction as it was before.

The rate at which a particular amount of enzyme converts substrate to product is partly a function of the initial concentration of the substrate: The more substrate molecules that are available, the more frequently they access the active sites of the enzyme molecules. However, there is a limit to how fast the reaction can be pushed by adding more substrate to a fixed concentration of enzyme. At some point, the concentration of substrate will be high enough that all enzyme molecules will have their active sites engaged. As soon as the product exits an active site, another substrate molecule enters. At this substrate concentration, the enzyme is said to be *saturated*, and the rate of the reaction is determined by the speed at which the active site converts substrate to product. When an enzyme population is saturated, the only way to increase the rate of product formation is to add more enzyme. Cells often increase the rate of a reaction by producing more enzyme molecules. You can graph the progress of an enzymatic reaction in the **Scientific Skills Exercise**.

## Scientific Skills Exercise

# *Making a Line Graph and Calculating a Slope*

**Does the Rate of Glucose 6-Phosphatase Activity Change over Time in Isolated Liver Cells?** Glucose 6-phosphatase, which is found in mammalian liver cells, is a key enzyme in control of blood glucose levels. The enzyme catalyzes the breakdown of glucose 6-phosphate into glucose and inorganic phosphate ($(P_i)$). These products are transported out of liver cells into the blood, increasing blood glucose levels. In this exercise, you will graph data from a time-course experiment that measured $(P_i)$ concentration in the buffer outside isolated liver cells, thus indirectly measuring glucose 6-phosphatase activity inside the cells.

**How the Experiment Was Done** Isolated rat liver cells were placed in a dish with buffer at physiological conditions (pH 7.4, 37°C). Glucose 6-phosphate (the substrate) was added to the dish, where it was taken up by the cells. Then a sample of buffer was removed every 5 minutes and the concentration of $(P_i)$ determined.

**Data from the Experiment**

| Time (min) | Concentration of $(P_i)$ (µmol/mL) |
|:---:|:---:|
| 0 | 0 |
| 5 | 10 |
| 10 | 90 |
| 15 | 180 |
| 20 | 270 |
| 25 | 330 |
| 30 | 355 |
| 35 | 355 |
| 40 | 355 |

**Data from** S. R. Commerford et al., Diets enriched in sucrose or fat increase gluconeogenesis and G-6-Pase but not basal glucose production in rats, *American Journal of Physiology—Endocrinology and Metabolism* 283:E545–E555 (2002).

**INTERPRET THE DATA**

1. To see patterns in the data from a time-course experiment like this, it is helpful to graph the data. First, determine which set of data goes on each axis. (a) What did the researchers intentionally vary in the experiment? This is the independent variable, which goes on the x-axis.

(b) What are the units (abbreviated) for the independent variable? Explain in words what the abbreviation stands for. (c) What was measured by the researchers? This is the dependent variable, which goes on the y-axis. (d) What does the units abbreviation stand for? Label each axis, including the units.

2. Next, you'll want to mark off the axes with just enough evenly spaced tick marks to accommodate the full set of data. Determine the range of data values for each axis. (a) What is the largest value to go on the x-axis? What is a reasonable spacing for the tick marks, and what should be the highest one? (b) What is the largest value to go on the y-axis? What is a reasonable spacing for the tick marks, and what should be the highest one?

3. Plot the data points on your graph. Match each x-value with its partner y-value and place a point on the graph at that coordinate. Draw a line that connects the points. (For additional information about graphs, see the Scientific Skills Review in Appendix F and in the Study Area in MasteringBiology.)

4. Examine your graph and look for patterns in the data. (a) Does the concentration of $(P_i)$ increase evenly through the course of the experiment? To answer this question, describe the pattern you see in the graph. (b) What part of the graph shows the highest rate of enzyme activity? Consider that the rate of enzyme activity is related to the slope of the line, $\Delta y/\Delta x$ (the "rise" over the "run"), in µmol/(mL · min), with the steepest slope indicating the highest rate of enzyme activity. Calculate the rate of enzyme activity (slope) where the graph is steepest. (c) Can you think of a biological explanation for the pattern you see?

5. If your blood sugar level is low from skipping lunch, what reaction (discussed in this exercise) will occur in your liver cells? Write out the reaction and put the name of the enzyme over the reaction arrow. How will this reaction affect your blood sugar level?

(MB) A version of this Scientific Skills Exercise can be assigned in MasteringBiology.

## Effects of Local Conditions on Enzyme Activity

The activity of an enzyme—how efficiently the enzyme functions—is affected by general environmental factors, such as temperature and pH. It can also be affected by chemicals that specifically influence that enzyme. In fact, researchers have learned much about enzyme function by employing such chemicals.

### Effects of Temperature and pH

The three-dimensional structures of proteins are sensitive to their environment (see Concept 3.5). As a consequence, each enzyme works better under some conditions than under other conditions, because these *optimal conditions* favor the most active shape for the enzyme.

Temperature and pH are environmental factors important in the activity of an enzyme. Up to a point, the rate of an enzymatic reaction increases with increasing temperature, partly because substrates collide with active sites more frequently when the molecules move rapidly. Above that temperature, however, the speed of the enzymatic reaction drops sharply. The thermal agitation of the enzyme molecule disrupts the hydrogen bonds, ionic bonds, and other weak interactions that stabilize the active shape of the enzyme, and the protein molecule eventually denatures. Each enzyme has an optimal temperature at which its reaction rate is greatest. Without denaturing the enzyme, this temperature allows the greatest number of molecular collisions and the fastest conversion of the reactants to product molecules. Most human enzymes have optimal temperatures of about 35–40°C (close to human body temperature). The thermophilic bacteria that live in hot springs contain enzymes with optimal temperatures of 70°C or higher **(Figure 6.16a)**.

Just as each enzyme has an optimal temperature, it also has a pH at which it is most active. The optimal pH values for most enzymes fall in the range of pH 6–8, but there are exceptions. For example, pepsin, a digestive enzyme in the human stomach, works best at pH 2. Such an acidic environment denatures most enzymes, but pepsin is adapted to maintain its functional three-dimensional structure in the acidic environment of the stomach. In contrast, trypsin, a digestive enzyme residing in the alkaline environment of the human intestine, has an optimal pH of 8 and would be denatured in the stomach **(Figure 6.16b)**.

### Cofactors

Many enzymes require nonprotein helpers for catalytic activity. These adjuncts, called **cofactors**, may be bound tightly to the enzyme as permanent residents, or they may bind loosely and reversibly along with the substrate. The cofactors of some enzymes are inorganic, such as the metal atoms zinc, iron, and copper in ionic form. If the cofactor is an organic molecule, it is referred to, more specifically, as a **coenzyme**. Most vitamins

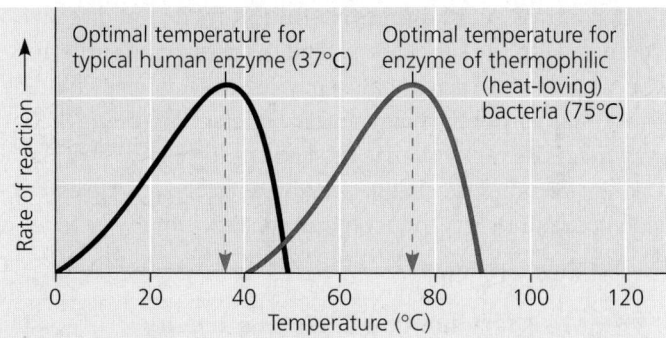

**(a)** The photo shows thermophilic cyanobacteria (green) thriving in the hot water of a Nevada geyser. The graph compares the optimal temperatures for an enzyme from the thermophilic bacterium *Thermus oshimai* (75°C) and human enzymes (body temperature, 37°C).

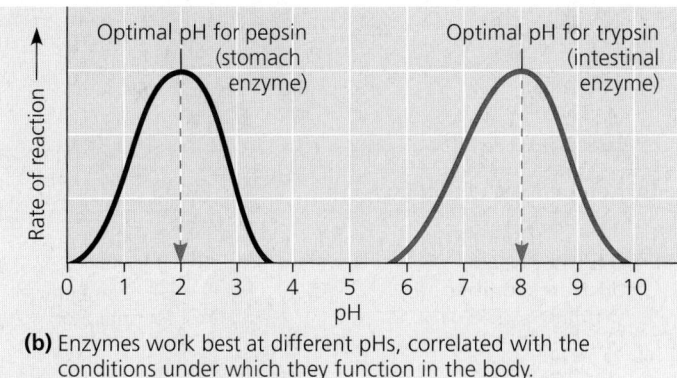

**(b)** Enzymes work best at different pHs, correlated with the conditions under which they function in the body.

▲ **Figure 6.16 Environmental factors affecting enzyme activity.** Each enzyme has an optimal **(a)** temperature and **(b)** pH that favor the most active shape of the protein molecule.

**DRAW IT** *Given that a mature lysosome has an internal pH of around 4.5, draw a curve in (b) showing what you would predict for a lysosomal enzyme, labeling its optimal pH.*

are important in nutrition because they act as coenzymes or raw materials from which coenzymes are made. Cofactors function in various ways, but in all cases where they are used, they perform a crucial chemical function in catalysis. You'll encounter examples of cofactors later in the book.

### Enzyme Inhibitors

Certain chemicals selectively inhibit the action of specific enzymes. Sometimes the inhibitor attaches to the enzyme by covalent bonds, in which case the inhibition is usually irreversible. Many enzyme inhibitors, however, bind to the enzyme by weak interactions, and when this occurs, the inhibition is reversible. Some reversible inhibitors resemble the normal substrate molecule and compete for admission into the active site **(Figure 6.17a** and **b)**. These mimics, called **competitive inhibitors**, reduce the productivity of enzymes by blocking substrates from entering active sites. This kind of inhibition can be overcome by increasing the concentration of substrate so that as active sites become available, more substrate molecules than inhibitor molecules are around to gain entry to the sites.

In contrast, **noncompetitive inhibitors** do not directly compete with the substrate to bind to the enzyme at the active site. Instead, they impede enzymatic reactions by binding to another part of the enzyme. This interaction causes the enzyme molecule to change its shape in such a way that the active site becomes much less effective at catalyzing the conversion of substrate to product **(Figure 6.17c)**.

Toxins and poisons are often irreversible enzyme inhibitors. An example is sarin, a nerve gas. Sarin was released by terrorists in the Tokyo subway in 1995, killing several people and injuring many others. This small molecule binds covalently to the R group on the amino acid serine, which is found in the active site of acetylcholinesterase, an enzyme important in the nervous system. Other examples include the pesticides DDT and parathion, inhibitors of key enzymes in the nervous system. Finally, many antibiotics are inhibitors of specific enzymes in bacteria. For instance, penicillin blocks the active site of an enzyme that many bacteria use to make cell walls.

Citing enzyme inhibitors that are metabolic poisons may give the impression that enzyme inhibition is generally abnormal and harmful. In fact, molecules naturally present in the cell often regulate enzyme activity by acting as inhibitors. Such regulation—selective inhibition—is essential to the control of cellular metabolism, as we will discuss in Concept 6.5.

### The Evolution of Enzymes

**EVOLUTION** Thus far, biochemists have identified more than 4,000 different enzymes in various species, most likely a very small fraction of all enzymes. How did this grand profusion of enzymes arise? Recall that most enzymes are proteins, and proteins are encoded by genes. A permanent change in a gene, known as a *mutation*, can result in a protein with one or more changed amino acids. In the case of an enzyme, if the changed amino acids are in the active site or some other crucial region, the altered enzyme might have a novel activity or might bind to a different substrate. Under environmental conditions where the new function benefits the organism, natural selection would tend to favor the mutated form of the gene, causing it to persist in the population. This simplified model is generally accepted as the main way in which the multitude of different enzymes arose over the past few billion years of life's history.

#### CONCEPT CHECK 6.4

1. Many spontaneous reactions occur very slowly. Why don't all spontaneous reactions occur instantly?
2. Why do enzymes act only on very specific substrates?
3. **WHAT IF?** Malonate is an inhibitor of the enzyme succinate dehydrogenase. How would you determine whether malonate is a competitive or noncompetitive inhibitor?

For suggested answers, see Appendix A.

**CONCEPT 6.5**

# Regulation of enzyme activity helps control metabolism

Chemical chaos would result if all of a cell's metabolic pathways were operating simultaneously. Intrinsic to life's processes is a cell's ability to tightly regulate its metabolic pathways by controlling when and where its various enzymes are active. It does this either by switching on and off the genes

▼ **Figure 6.17 Inhibition of enzyme activity.**

**(a) Normal binding**

A substrate can bind normally to the active site of an enzyme.

Substrate
Active site
Enzyme

**(b) Competitive inhibition**

A competitive inhibitor mimics the substrate, competing for the active site.

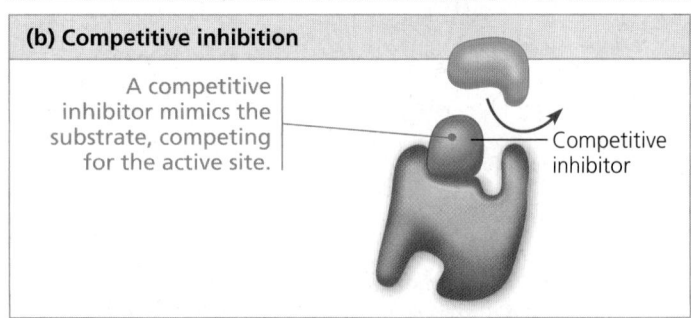

Competitive inhibitor

**(c) Noncompetitive inhibition**

A noncompetitive inhibitor binds to the enzyme away from the active site, altering the shape of the enzyme so that even if the substrate can bind, the active site functions less effectively, if at all.

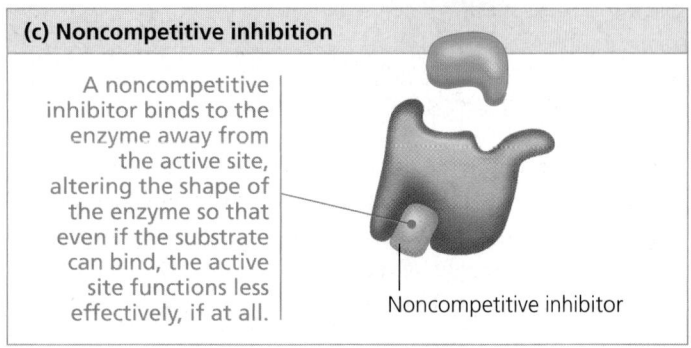

Noncompetitive inhibitor

that encode specific enzymes (as we will discuss in Unit Two) or, as we discuss here, by regulating the activity of enzymes once they are made.

## Allosteric Regulation of Enzymes

In many cases, the molecules that naturally regulate enzyme activity in a cell behave something like reversible noncompetitive inhibitors (see Figure 6.17c): These regulatory molecules change an enzyme's shape and the functioning of its active site by binding to a site elsewhere on the molecule, via non-covalent interactions. **Allosteric regulation** is the term used to describe any case in which a protein's function at one site is affected by the binding of a regulatory molecule to a separate site. It may result in either inhibition or stimulation of an enzyme's activity.

### Allosteric Activation and Inhibition

Most enzymes known to be allosterically regulated are constructed from two or more subunits, each composed of a polypeptide chain with its own active site. The entire complex oscillates between two different shapes, one catalytically active and the other inactive **(Figure 6.18a)**. In the simplest kind of allosteric regulation, an activating or inhibiting regulatory molecule binds to a regulatory site (sometimes called an allosteric site), often located where subunits join. The binding of an *activator* to a regulatory site stabilizes the shape that has functional active sites, whereas the binding of an *inhibitor* stabilizes the inactive form of the enzyme. The subunits of an allosteric enzyme fit together in such a way that a shape change in one subunit is transmitted to all others. Through this interaction of subunits, a single activator or inhibitor molecule that binds to one regulatory site will affect the active sites of all subunits.

Fluctuating concentrations of regulators can cause a sophisticated pattern of response in the activity of cellular enzymes. The products of ATP hydrolysis (ADP and $(P)_i$), for example, play a complex role in balancing the flow of traffic between anabolic and catabolic pathways by their effects on key enzymes. ATP binds to several catabolic enzymes allosterically, lowering their affinity for substrate and thus inhibiting their activity. ADP, however, functions as an activator of the same enzymes. This is logical because catabolism functions in regenerating ATP. If ATP production lags behind its use, ADP accumulates and activates the enzymes that speed up catabolism, producing more ATP. If the supply of ATP exceeds demand, then catabolism slows down as ATP molecules accumulate and bind to the same enzymes, inhibiting them. (You'll see examples of this type of regulation when you learn about cellular respiration in the next chapter.) ATP, ADP, and other related molecules also affect key enzymes in anabolic pathways. In this way, allosteric enzymes control the rates of important reactions in both sorts of metabolic pathways.

In another kind of allosteric activation, a *substrate* molecule binding to one active site in a multisubunit enzyme

▼ **Figure 6.18 Allosteric regulation of enzyme activity.**

**(a) Allosteric activators and inhibitors**

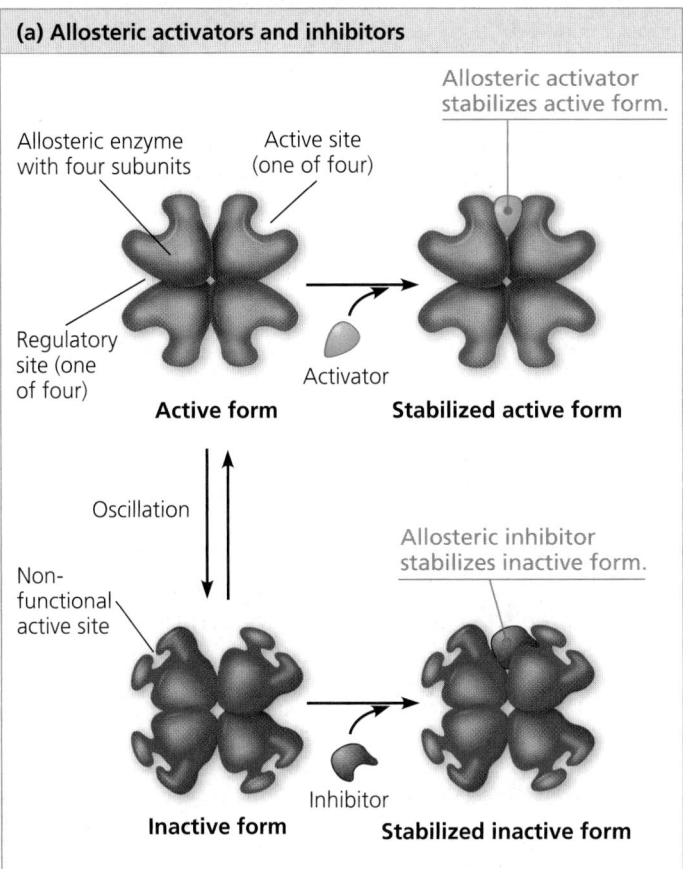

At low concentrations, activators and inhibitors dissociate from the enzyme. The enzyme can then oscillate again.

**(b) Cooperativity: another type of allosteric activation**

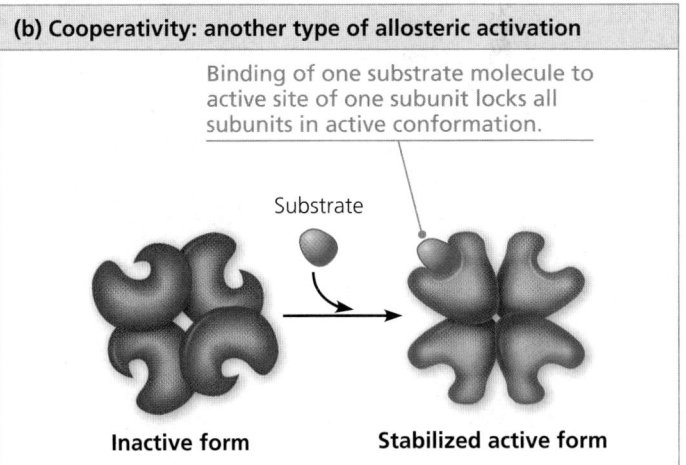

The inactive form shown on the left oscillates with the active form when the active form is not stabilized by substrate.

triggers a shape change in all the subunits, thereby increasing catalytic activity at the other active sites **(Figure 6.18b)**. Called **cooperativity**, this mechanism amplifies the response of enzymes to substrates: One substrate molecule primes an enzyme to act on additional substrate molecules more readily. Cooperativity is considered allosteric regulation because binding of the substrate to one active site affects catalysis in another active site.

Although hemoglobin is not an enzyme (it carries $O_2$ rather than catalyzing a reaction), classic studies of hemoglobin have elucidated the principle of cooperativity. Hemoglobin is made up of four subunits, each with an oxygen-binding site (see Figure 3.22). The binding of an oxygen molecule to one binding site increases the affinity for oxygen of the remaining binding sites. Thus, where oxygen is at high levels, such as in the lungs or gills, hemoglobin's affinity for oxygen increases as more binding sites are filled. In oxygen-deprived tissues, however, the release of each oxygen molecule decreases the oxygen affinity of the other binding sites, resulting in the release of oxygen where it is most needed. Cooperativity works similarly in multisubunit enzymes that have been studied.

### Feedback Inhibition

Earlier, we discussed the allosteric inhibition of an enzyme in an ATP-generating pathway by ATP itself. This is a common mode of metabolic control, called **feedback inhibition**, in which a metabolic pathway is halted by the inhibitory binding of its end product to an enzyme that acts early in the pathway. **Figure 6.19** shows an example of feedback inhibition operating on an anabolic pathway. Some cells use this five-step pathway to synthesize the amino acid isoleucine from threonine,

another amino acid. As isoleucine accumulates, it slows down its own synthesis by allosterically inhibiting the enzyme for the first step of the pathway. Feedback inhibition thereby prevents the cell from making more isoleucine than is necessary, and thus wasting chemical resources.

## Organization of Enzymes Within the Cell

The cell is not just a bag of chemicals with thousands of different kinds of enzymes and substrates in a random mix. The cell is compartmentalized, and cellular structures help bring order to metabolic pathways. In some cases, a team of enzymes for several steps of a metabolic pathway is assembled into a multienzyme complex. The arrangement facilitates the sequence of reactions, with the product from the first enzyme becoming the substrate for an adjacent enzyme in the complex, and so on, until the end product is released. Some enzymes and enzyme complexes have fixed locations within the cell and act as structural components of particular membranes. Others are in solution within particular membrane-enclosed eukaryotic organelles, each with its own internal chemical environment. For example, in eukaryotic cells, the enzymes for cellular respiration reside in specific locations within mitochondria **(Figure 6.20)**.

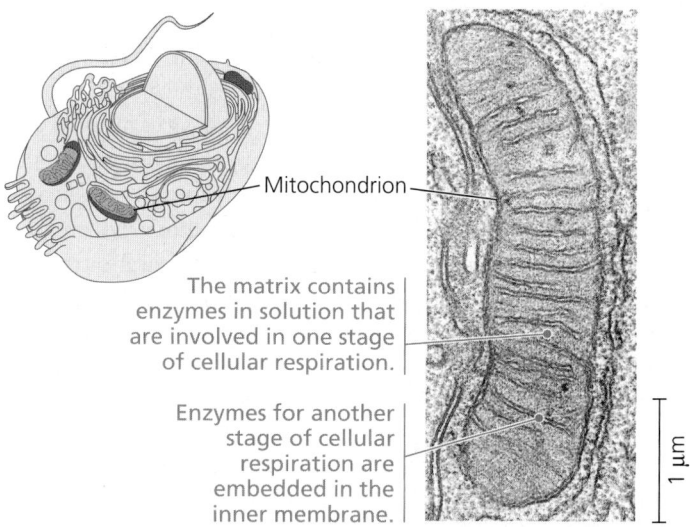

Mitochondrion

The matrix contains enzymes in solution that are involved in one stage of cellular respiration.

Enzymes for another stage of cellular respiration are embedded in the inner membrane.

1 μm

▲ **Figure 6.20 Organelles and structural order in metabolism.** Organelles such as the mitochondrion (TEM) contain enzymes that carry out specific functions, in this case cellular respiration.

In this chapter, you have learned that metabolism, the intersecting set of chemical pathways characteristic of life, is a choreographed interplay of thousands of different kinds of cellular molecules. In the next chapter, we'll explore cellular respiration, the major catabolic pathway that breaks down organic molecules, releasing energy that can be used for the crucial processes of life.

### CONCEPT CHECK 6.5

1. How do an activator and an inhibitor have different effects on an allosterically regulated enzyme?

   For suggested answers, see Appendix A.

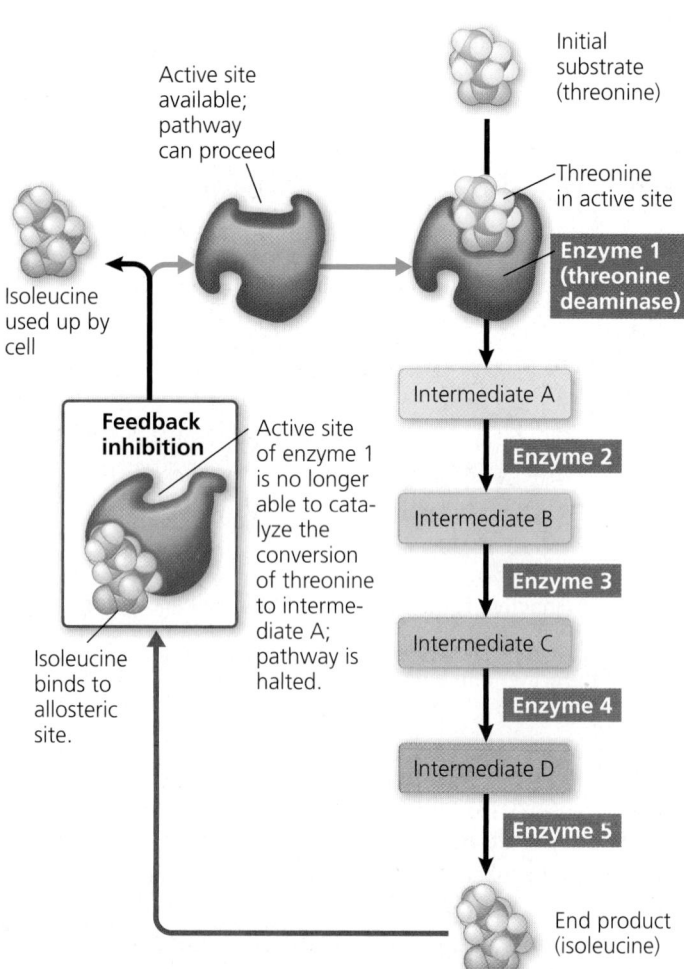

Active site available; pathway can proceed

Initial substrate (threonine)

Threonine in active site

**Enzyme 1 (threonine deaminase)**

Isoleucine used up by cell

**Feedback inhibition**

Active site of enzyme 1 is no longer able to catalyze the conversion of threonine to intermediate A; pathway is halted.

Isoleucine binds to allosteric site.

Intermediate A

**Enzyme 2**

Intermediate B

**Enzyme 3**

Intermediate C

**Enzyme 4**

Intermediate D

**Enzyme 5**

End product (isoleucine)

▲ **Figure 6.19 Feedback inhibition in isoleucine synthesis.**

Go to **MasteringBiology**® for Assignments, the eText, and the Study Area with Animations, Activities, Vocab Self-Quiz, and Practice Tests.

## SUMMARY OF KEY CONCEPTS

**VOCAB SELF-QUIZ**

goo.gl/gbai8v

### CONCEPT 6.1

**An organism's metabolism transforms matter and energy (pp. 122–125)**

- **Metabolism** is the collection of chemical reactions that occur in an organism. Enzymes catalyze reactions in intersecting **metabolic pathways**, which may be **catabolic** (breaking down molecules, releasing energy) or **anabolic** (building molecules, consuming energy).
- **Energy** is the capacity to cause change; some forms of energy do work by moving matter. **Kinetic energy** is associated with motion and includes **thermal energy**, associated with the random motion of atoms or molecules. **Heat** is thermal energy in transfer from one object to another. **Potential energy** is related to the location or structure of matter and includes **chemical energy** possessed by a molecule due to its structure.
- The **first law of thermodynamics**, conservation of energy, states that energy cannot be created or destroyed, only transferred or transformed. The **second law of thermodynamics** states that **spontaneous processes**, those requiring no outside input of energy, increase the **entropy** (disorder) of the universe.

**?** *Explain how the highly ordered structure of a cell does not conflict with the second law of thermodynamics.*

### CONCEPT 6.2

**The free-energy change of a reaction tells us whether or not the reaction occurs spontaneously (pp. 125–128)**

- A living system's **free energy** is energy that can do work under cellular conditions. Organisms live at the expense of free energy. The change in free energy ($\Delta G$) during a biological process tells us if the process is spontaneous. During a spontaneous process, free energy decreases and the stability of a system increases. At maximum stability, the system is at equilibrium and can do no work.
- In an **exergonic** (spontaneous) chemical reaction, the products have less free energy than the reactants ($-\Delta G$). **Endergonic** (nonspontaneous) reactions require an input of energy ($+\Delta G$). The addition of starting materials and the removal of end products prevent metabolism from reaching equilibrium.

**?** *Why are spontaneous reactions important in the metabolism of a cell?*

### CONCEPT 6.3

**ATP powers cellular work by coupling exergonic reactions to endergonic reactions (pp. 128–130)**

- **ATP** is the cell's energy shuttle. Hydrolysis of its terminal phosphate yields ADP and $\text{P}_i$ and releases free energy.
- Through **energy coupling**, the exergonic process of ATP hydrolysis drives endergonic reactions by transfer of a phosphate group to specific reactants, forming a **phosphorylated intermediate** that is more reactive. ATP hydrolysis (sometimes with protein phosphorylation) also causes changes in the shape and binding affinities of transport and motor proteins.
- Catabolic pathways drive regeneration of ATP from ADP + $\text{P}_i$.

**?** *Describe the ATP cycle: How is ATP used and regenerated in a cell?*

### CONCEPT 6.4

**Enzymes speed up metabolic reactions by lowering energy barriers (pp. 131–136)**

- In a chemical reaction, the energy necessary to break the bonds of the reactants is the **activation energy**, $E_A$.
- **Enzymes** lower the $E_A$ barrier:

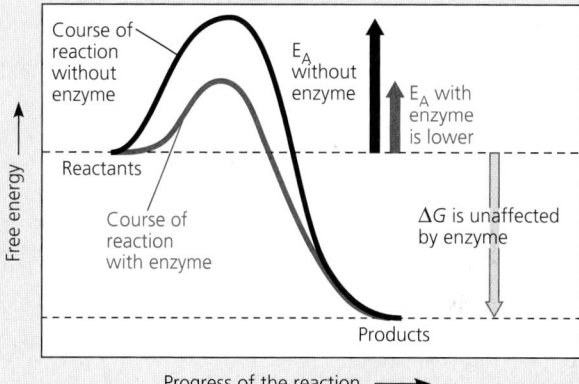

- Each type of enzyme has a unique **active site** that combines specifically with its **substrate(s)**, the reactant(s) on which it acts. It then changes shape, binding the substrate(s) more tightly (**induced fit**).
- The active site can lower an $E_A$ barrier by orienting substrates correctly, straining their bonds, providing a favorable microenvironment, or even covalently bonding with the substrate.
- Each enzyme has an optimal temperature and pH. Inhibitors reduce enzyme function. A **competitive inhibitor** binds to the active site, whereas a **noncompetitive inhibitor** binds to a different site on the enzyme.
- Natural selection, acting on organisms with variant enzymes, is responsible for the diversity of enzymes found in organisms.

**?** *How do both activation energy barriers and enzymes help maintain the structural and metabolic order of life?*

### CONCEPT 6.5

**Regulation of enzyme activity helps control metabolism (pp. 136–138)**

- Many enzymes are subject to **allosteric regulation**: Regulatory molecules, either activators or inhibitors, bind to specific regulatory sites, affecting the shape and function of the enzyme. In **cooperativity**, binding of one substrate molecule can stimulate binding or activity at other active sites. In **feedback inhibition**, the end product of a metabolic pathway allosterically inhibits the enzyme for a previous step in the pathway.
- Some enzymes are grouped into complexes, some are incorporated into membranes, and some are contained inside organelles, increasing the efficiency of metabolic processes.

**?** *What roles do allosteric regulation and feedback inhibition play in the metabolism of a cell?*

# TEST YOUR UNDERSTANDING

PRACTICE TEST

goo.gl/CRZjvS

## Level 1: Knowledge/Comprehension

**1.** Choose the pair of terms that correctly completes this sentence: Catabolism is to anabolism as _____ is to _____.
(A) exergonic; spontaneous
(B) exergonic; endergonic
(C) free energy; entropy
(D) work; energy

**2.** Most cells cannot harness heat to perform work because
(A) heat does not involve a transfer of energy.
(B) cells do not have much heat; they are relatively cool.
(C) temperature is usually uniform throughout a cell.
(D) heat can never be used to do work.

**3.** Which of the following metabolic processes can occur without a net influx of energy from some other process?
(A) $ADP + \text{(P)}_i \rightarrow ATP + H_2O$
(B) $C_6H_{12}O_6 + 6\,O_2 \rightarrow 6\,CO_2 + 6\,H_2O$
(C) $6\,CO_2 + 6\,H_2O \rightarrow C_6H_{12}O_6 + 6\,O_2$
(D) amino acids $\rightarrow$ protein

**4.** If an enzyme in solution is saturated with substrate, the most effective way to obtain a faster yield of products is to
(A) add more of the enzyme.
(B) heat the solution to 90°C.
(C) add more substrate.
(D) add an allosteric inhibitor.

**5.** Some bacteria are metabolically active in hot springs because
(A) they are able to maintain a lower internal temperature.
(B) high temperatures make catalysis unnecessary.
(C) their enzymes have high optimal temperatures.
(D) their enzymes are completely insensitive to temperature.

## Level 2: Application/Analysis

**6.** If an enzyme is added to a solution where its substrate and product are in equilibrium, what will occur?
(A) Additional product will be formed.
(B) The reaction will change from endergonic to exergonic.
(C) The free energy of the system will change.
(D) Nothing; the reaction will stay at equilibrium.

## Level 3: Synthesis/Evaluation

**7.** **DRAW IT** Using a series of arrows, draw the branched metabolic reaction pathway described by the following statements. Then answer the question at the end. Use red arrows and minus signs to indicate inhibition.

L can form either M or N.
M can form O.
O can form either P or R.
P can form Q.
R can form S.
O inhibits the reaction of L to form M.
Q inhibits the reaction of O to form P.
S inhibits the reaction of O to form R.

Which reaction would prevail if both Q and S were present in the cell at high concentrations?

(A) L → M
(B) M → O
(C) L → N
(D) O → P

**8. SCIENTIFIC INQUIRY**
**DRAW IT** A researcher has developed an assay to measure the activity of an important enzyme present in liver cells growing in culture. She adds the enzyme's substrate to a dish of cells and then measures the appearance of reaction products. The results are graphed as the amount of product on the $y$-axis versus time on the $x$-axis. The researcher notes four sections of the graph. For a short period of time, no products appear (section A). Then (section B) the reaction rate is quite high (the slope of the line is steep). Next, the reaction gradually slows down (section C). Finally, the graph line becomes flat (section D). Draw and label the graph, and propose a model to explain the molecular events occurring at each stage of this reaction profile.

**9. SCIENCE, TECHNOLOGY, AND SOCIETY**
Organophosphates (organic compounds containing phosphate groups) are commonly used as insecticides to improve crop yield. Organophosphates typically interfere with nerve signal transmission by inhibiting the enzymes that degrade transmitter molecules. They affect humans and other vertebrates as well as insects. Thus, the use of organophosphate pesticides poses some health risks. On the other hand, these molecules break down rapidly upon exposure to air and sunlight. As a consumer, what level of risk are you willing to accept in exchange for an abundant and affordable food supply? Explain your thinking.

**10. FOCUS ON EVOLUTION**
A recent revival of the antievolutionary "intelligent design" argument holds that biochemical pathways are too complex to have evolved, because all intermediate steps in a given pathway must be present to produce the final product. Critique this argument. How could you use the diversity of metabolic pathways that produce the same or similar products to support your case?

**11. FOCUS ON ENERGY AND MATTER**
Life requires energy. In a short essay (100–150 words), describe the basic principles of bioenergetics in an animal cell. How is the flow and transformation of energy different in a photosynthesizing cell? Include the role of ATP and enzymes in your discussion.

**12.** **SYNTHESIZE YOUR KNOWLEDGE**

Explain what is happening in this photo in terms of kinetic energy and potential energy. Include the energy conversions that occur when the penguins eat fish and climb back up on the glacier. Describe the role of ATP and enzymes in the underlying molecular processes, including what happens to the free energy of some of the molecules involved.

*For selected answers, see Appendix A.*

# 7 Cellular Respiration and Fermentation

## KEY CONCEPTS

**7.1** Catabolic pathways yield energy by oxidizing organic fuels

**7.2** Glycolysis harvests chemical energy by oxidizing glucose to pyruvate

**7.3** After pyruvate is oxidized, the citric acid cycle completes the energy-yielding oxidation of organic molecules

**7.4** During oxidative phosphorylation, chemiosmosis couples electron transport to ATP synthesis

**7.5** Fermentation and anaerobic respiration enable cells to produce ATP without the use of oxygen

**7.6** Glycolysis and the citric acid cycle connect to many other metabolic pathways

▲ **Figure 7.1** How do these leaves power the work of life for this giraffe?

## Life Is Work

Living cells require transfusions of energy from outside sources to perform their many tasks—for example, assembling polymers, pumping substances across membranes, moving, and reproducing. The giraffe in **Figure 7.1** obtains energy for its cells by eating plants; some animals feed on other organisms that eat plants. The energy stored in the organic molecules of food ultimately comes from the sun. Energy flows into an ecosystem as sunlight and exits as heat; in contrast, the chemical elements essential to life are recycled **(Figure 7.2)**. Photosynthesis generates oxygen and organic molecules that are used by the mitochondria of eukaryotes (including plants and algae) as fuel for cellular respiration. Respiration breaks down this fuel, generating ATP. The waste products of this type of respiration, carbon dioxide and water, are the raw materials for photosynthesis.

In this chapter, we'll consider how cells harvest the chemical energy stored in organic molecules and use it to generate ATP, the molecule that drives most cellular work. After presenting some basics about respiration, we'll focus on three key pathways of respiration: glycolysis, the citric acid cycle, and oxidative phosphorylation. We'll also consider fermentation, a somewhat simpler pathway coupled to glycolysis that has deep evolutionary roots.

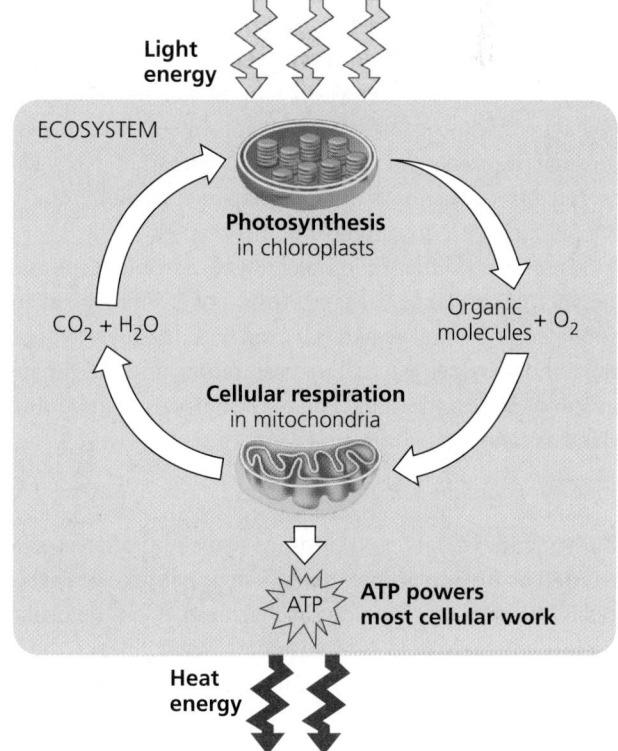

▲ **Figure 7.2 Energy flow and chemical recycling in ecosystems.** Energy flows into an ecosystem as sunlight and ultimately leaves as heat, while the chemical elements essential to life are recycled.

## CONCEPT 7.1

# Catabolic pathways yield energy by oxidizing organic fuels

Metabolic pathways that release stored energy by breaking down complex molecules are called catabolic pathways (see Concept 6.1). Electron transfer plays a major role in these pathways. In this section, we'll consider these processes, which are central to cellular respiration.

## Catabolic Pathways and Production of ATP

Organic compounds possess potential energy as a result of the arrangement of electrons in the bonds between their atoms. Compounds that can participate in exergonic reactions can act as fuels. Through the activity of enzymes, a cell systematically degrades complex organic molecules that are rich in potential energy to simpler waste products that have less energy. Some of the energy taken out of chemical storage can be used to do work; the rest is dissipated as heat.

One catabolic process, **fermentation**, is a partial degradation of sugars or other organic fuel that occurs without the use of oxygen. However, the most efficient catabolic pathway is **aerobic respiration**, in which oxygen is consumed as a reactant along with the organic fuel (*aerobic* is from the Greek *aer*, air, and *bios*, life). The cells of most eukaryotic and many prokaryotic organisms can carry out aerobic respiration. Some prokaryotes use substances other than oxygen as reactants in a similar process that harvests chemical energy without oxygen; this process is called *anaerobic respiration* (the prefix *an-* means "without"). Technically, the term **cellular respiration** includes both aerobic and anaerobic processes. However, it originated as a synonym for aerobic respiration because of the relationship of that process to organismal respiration, in which an animal breathes in oxygen. Thus, *cellular respiration* is often used to refer to the aerobic process, a practice we follow in most of this chapter.

Although very different in mechanism, aerobic respiration is in principle similar to the combustion of gasoline in an automobile engine after oxygen is mixed with the fuel (hydrocarbons). Food provides the fuel for respiration, and the exhaust is carbon dioxide and water. The overall process can be summarized as follows:

Organic compounds + Oxygen → Carbon dioxide + Water + Energy

Carbohydrates, fats, and proteins can all be processed and consumed as fuel. In animal diets, a major source of carbohydrates is starch, a storage polysaccharide that can be broken down into glucose ($C_6H_{12}O_6$) subunits. We will learn the steps of cellular respiration by tracking the degradation of the sugar glucose:

$$C_6H_{12}O_6 + 6\,O_2 \rightarrow 6\,CO_2 + 6\,H_2O + \text{Energy (ATP + heat)}$$

This breakdown of glucose is exergonic, having a free-energy change of −686 kcal (2,870 kJ) per mole of glucose

decomposed ($\Delta G = -686$ kcal/mol). Recall that a negative $\Delta G$ indicates that the products of the chemical process store less energy than the reactants and that the reaction can happen spontaneously—in other words, without an input of energy.

Catabolic pathways do not directly move flagella, pump solutes across membranes, polymerize monomers, or perform other cellular work. Catabolism is linked to work by a chemical drive shaft—ATP (which you learned about in Concepts 3.1 and 6.3). To keep working, the cell must regenerate its supply of ATP from ADP and $P_i$ (see Figure 6.11). To understand how cellular respiration accomplishes this, let's examine the fundamental chemical processes known as oxidation and reduction.

## Redox Reactions: Oxidation and Reduction

How do the catabolic pathways that decompose glucose and other organic fuels yield energy? The answer is based on the transfer of electrons during the chemical reactions. The relocation of electrons releases energy stored in organic molecules, and this energy ultimately is used to synthesize ATP.

### The Principle of Redox

In many chemical reactions, there is a transfer of one or more electrons ($e^-$) from one reactant to another. These electron transfers are called oxidation-reduction reactions, or **redox reactions** for short. In a redox reaction, the loss of electrons from one substance is called **oxidation**, and the addition of electrons to another substance is known as **reduction**. (Note that *adding* electrons is called *reduction*; adding negatively charged electrons to an atom *reduces* the amount of positive charge of that atom.) To take a simple, nonbiological example, consider the reaction between the elements sodium (Na) and chlorine (Cl) that forms table salt:

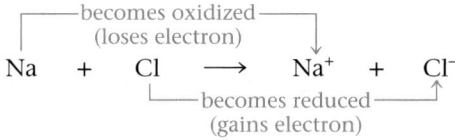

We could generalize a redox reaction this way:

$$\begin{array}{c}\overbrace{Xe^- + Y \longrightarrow X + Ye^-}\end{array}$$

In the generalized reaction, substance $Xe^-$, the electron donor, is called the **reducing agent**; it reduces Y, which accepts the donated electron. Substance Y, the electron acceptor, is the **oxidizing agent**; it oxidizes $Xe^-$ by removing its electron. Because an electron transfer requires both an electron donor and an acceptor, oxidation and reduction always go hand in hand.

Not all redox reactions involve the complete transfer of electrons from one substance to another; some change the degree of electron sharing in covalent bonds. Methane combustion,

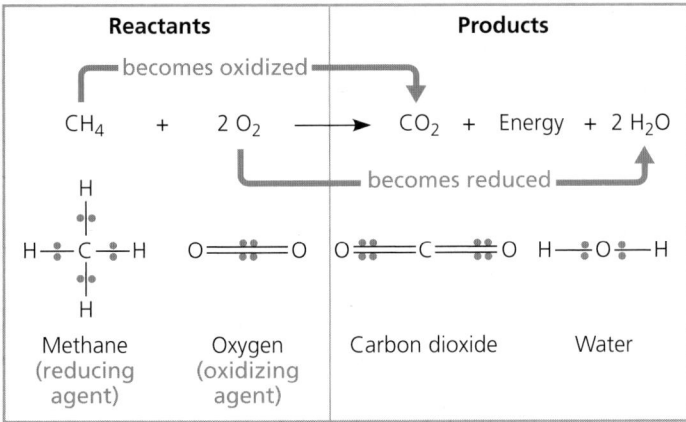

**▲ Figure 7.3 Methane combustion as an energy-yielding redox reaction.** The reaction releases energy to the surroundings because the electrons lose potential energy when they end up being shared unequally, spending more time near electronegative atoms such as oxygen.

shown in **Figure 7.3**, is an example. The covalent electrons in methane are shared nearly equally between the bonded atoms because carbon and hydrogen have about the same affinity for valence electrons; they are about equally electronegative. But when methane reacts with oxygen, forming carbon dioxide, electrons end up shared less equally between the carbon atom and its new covalent partners, the oxygen atoms, which are very electronegative. In effect, the carbon atom has partially "lost" its shared electrons; thus, methane has been oxidized.

Now let's examine the fate of the reactant $O_2$. The two atoms of the oxygen molecule ($O_2$) share their electrons equally. But when oxygen reacts with the hydrogen from methane, forming water, the electrons of the covalent bonds spend more time near the oxygen (see Figure 7.3). In effect, each oxygen atom has partially "gained" electrons, so the oxygen molecule has been reduced. Because oxygen is so electronegative, it is one of the most powerful of all oxidizing agents.

Energy must be added to pull an electron away from an atom, just as energy is required to push a ball uphill. The more electronegative the atom (the stronger its pull on electrons), the more energy is required to take an electron away from it. An electron loses potential energy when it shifts from a less electronegative atom toward a more electronegative one, just as a ball loses potential energy when it rolls downhill. A redox reaction that moves electrons closer to oxygen, such as the burning (oxidation) of methane, therefore releases chemical energy that can be put to work.

### Oxidation of Organic Fuel Molecules During Cellular Respiration

The oxidation of methane by oxygen is the main combustion reaction that occurs at the burner of a gas stove. The combustion of gasoline in an automobile engine is also a redox reaction; the energy released pushes the pistons. But the energy-yielding redox process of greatest interest to biologists

is respiration: the oxidation of glucose and other molecules in food. Examine again the summary equation for cellular respiration, but this time think of it as a redox process:

$$\overbrace{C_6H_{12}O_6 \ + \ 6\,O_2 \ \longrightarrow \ 6\,CO_2}^{\text{becomes oxidized}} \underbrace{\ + \ 6\,H_2O}_{\text{becomes reduced}} \ + \ \text{Energy}$$

As in the combustion of methane or gasoline, the fuel (glucose) is oxidized and oxygen is reduced. The electrons lose potential energy along the way, and energy is released.

In general, organic molecules that have an abundance of hydrogen are excellent fuels because their bonds are a source of "hilltop" electrons, whose energy may be released as these electrons "fall" down an energy gradient when they are transferred to oxygen. The summary equation for respiration indicates that hydrogen is transferred from glucose to oxygen. But the important point, not visible in the summary equation, is that the energy state of the electron changes as hydrogen (with its electron) is transferred to oxygen. In respiration, the oxidation of glucose transfers electrons to a lower energy state, liberating energy that becomes available for ATP synthesis.

The main energy-yielding foods, carbohydrates and fats, are reservoirs of electrons associated with hydrogen. Only the barrier of activation energy holds back the flood of electrons to a lower energy state (see Figure 6.12). Without this barrier, a food substance like glucose would combine almost instantaneously with $O_2$. If we supply the activation energy by igniting glucose, it burns in air, releasing 686 kcal (2,870 kJ) of heat per mole of glucose (about 180 g). Body temperature is not high enough to initiate burning, of course. Instead, if you swallow some glucose, enzymes in your cells will lower the barrier of activation energy, allowing the sugar to be oxidized in a series of steps.

### Stepwise Energy Harvest via NAD$^+$ and the Electron Transport Chain

If energy is released from a fuel all at once, it cannot be harnessed efficiently for constructive work. For example, if a gasoline tank explodes, it cannot drive a car very far. Cellular respiration does not oxidize glucose (or any other organic fuel) in a single explosive step, either. Rather, glucose is broken down in a series of steps, each one catalyzed by an enzyme. At key steps, electrons are stripped from the glucose. As is often the case in oxidation reactions, each electron travels with a proton—thus, as a hydrogen atom. The hydrogen atoms are not transferred directly to oxygen, but instead are usually passed first to an electron carrier, a coenzyme called **NAD$^+$** (nicotinamide adenine dinucleotide, a derivative of the vitamin niacin). NAD$^+$ is well suited as an electron carrier because it can cycle easily between oxidized (NAD$^+$) and reduced (NADH) states. As an electron acceptor, NAD$^+$ functions as an oxidizing agent during respiration.

How does NAD$^+$ trap electrons from glucose and other organic molecules in food? Enzymes called dehydrogenases remove a pair of hydrogen atoms (2 electrons and 2 protons) from the

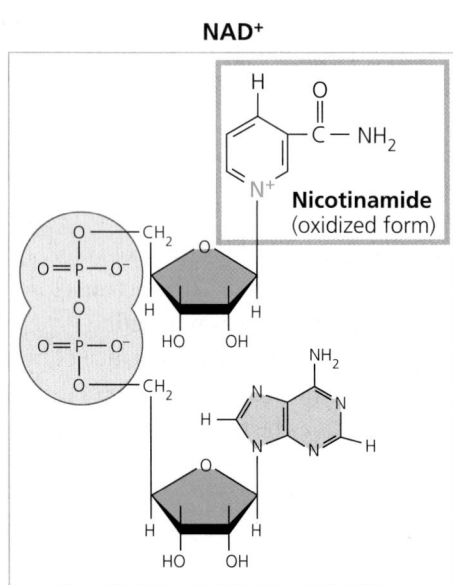

**NAD⁺**

H
O
$C - NH_2$

Nicotinamide
(oxidized form)

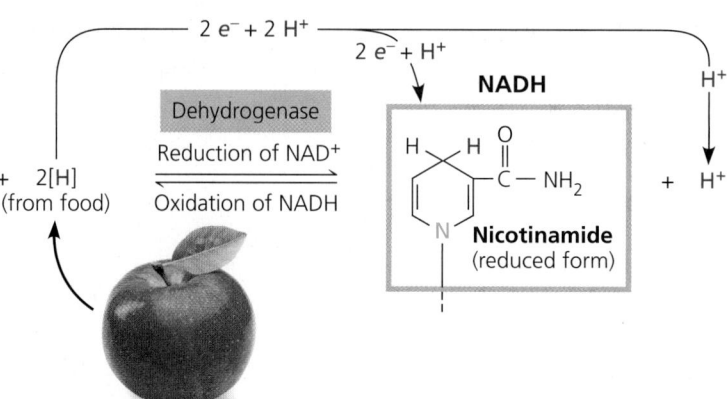

$2\,e^- + 2\,H^+$

$2\,e^- + H^+$

**Dehydrogenase**

Reduction of NAD⁺
⇌
Oxidation of NADH

$+ \quad 2[H]$
(from food)

**NADH**

$H^+$

H    H    O
$C - NH_2$

N

Nicotinamide
(reduced form)

$+ \quad H^+$

◄ **Figure 7.4 NAD⁺ as an electron shuttle.** The full name for NAD⁺, nicotinamide adenine dinucleotide, describes its structure: The molecule consists of two nucleotides joined together at their phosphate groups (shown in yellow). (Nicotinamide is a nitrogenous base, although not one that is present in DNA or RNA.) The enzymatic transfer of 2 electrons and 1 proton (H⁺) from an organic molecule in food to NAD⁺ reduces the NAD⁺ to NADH; the second proton (H⁺) is released. Most of the electrons removed from food are transferred initially to NAD⁺, forming NADH.

substrate (glucose, in this example), thereby oxidizing it. The enzyme delivers the 2 electrons along with 1 proton to its coenzyme, NAD⁺, forming NADH **(Figure 7.4)**. The other proton is released as a hydrogen ion (H⁺) into the surrounding solution:

$$H-\overset{|}{\underset{|}{C}}-OH + NAD^+ \xrightarrow{\text{Dehydrogenase}} \overset{|}{C}=O + NADH + H^+$$

By receiving 2 negatively charged electrons but only 1 positively charged proton, the nicotinamide portion of NAD⁺ has its charge neutralized when NAD⁺ is reduced to NADH. The name NADH shows the hydrogen that has been received in the reaction. NAD⁺ is the most versatile electron acceptor in cellular respiration and functions in several of the redox steps during the breakdown of glucose.

Electrons lose very little of their potential energy when they are transferred from glucose to NAD⁺. Each NADH molecule formed during respiration represents stored energy that can be tapped to make ATP when the electrons complete their "fall" down an energy gradient from NADH to oxygen.

How do electrons that are extracted from glucose and stored as potential energy in NADH finally reach oxygen? It will help to compare the redox chemistry of cellular respiration to a much simpler reaction: the reaction between hydrogen and oxygen to form water **(Figure 7.5a)**. Mix H₂ and O₂, provide a spark for activation energy, and the gases combine explosively. In fact, combustion of liquid

H₂ and O₂ is harnessed to help power the rocket engines that boost satellites into orbit and launch spacecraft. The explosion represents a release of energy as the electrons of hydrogen "fall" closer to the electronegative oxygen atoms. Cellular respiration also brings hydrogen and oxygen together to form water, but there are two important differences. First, in cellular respiration, the hydrogen that reacts with oxygen is derived from organic molecules rather than H₂. Second, instead of occurring in one explosive reaction, respiration uses an electron transport chain to break the fall of electrons

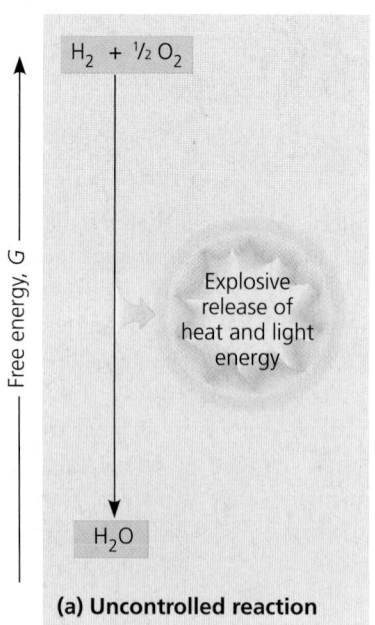

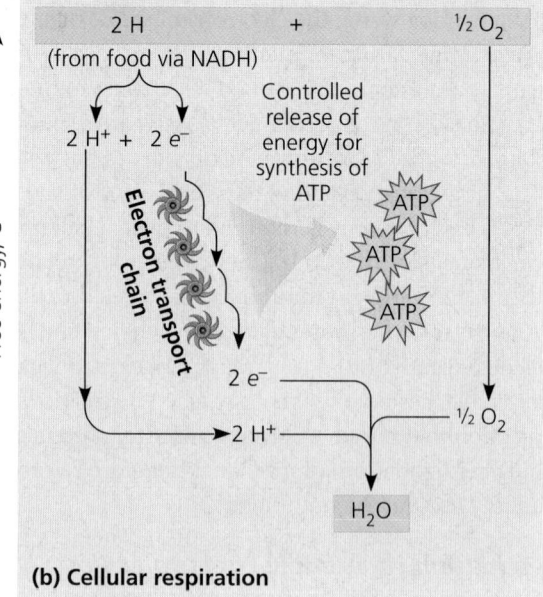

▲ **Figure 7.5 An introduction to electron transport chains. (a)** The one-step exergonic reaction of hydrogen with oxygen to form water releases a large amount of energy in the form of heat and light: an explosion. **(b)** In cellular respiration, the same reaction occurs in stages: An electron transport chain breaks the "fall" of electrons in this reaction into a series of smaller steps and stores some of the released energy in a form that can be used to make ATP. (The rest of the energy is released as heat.)

to oxygen into several energy-releasing steps **(Figure 7.5b)**. An **electron transport chain** consists of a number of molecules, mostly proteins, built into the inner membrane of the mitochondria of eukaryotic cells and the plasma membrane of aerobically respiring prokaryotes. Electrons removed from glucose are shuttled by NADH to the "top," higher-energy end of the chain. At the "bottom," lower-energy end, $O_2$ captures these electrons along with hydrogen nuclei ($H^+$), forming water.

Electron transfer from NADH to oxygen is an exergonic reaction with a free-energy change of −53 kcal/mol (−222 kJ/mol). Instead of this energy being released and wasted in a single explosive step, electrons cascade down the chain from one carrier molecule to the next in a series of redox reactions, losing a small amount of energy with each step until they finally reach oxygen, the terminal electron acceptor, which has a very great affinity for electrons. Each "downhill" carrier is more electronegative than, and thus capable of oxidizing, its "uphill" neighbor, with oxygen at the bottom of the chain. Therefore, the electrons transferred from glucose to $NAD^+$, forming NADH, will fall down an energy gradient in the electron transport chain to a far more stable location in the electronegative oxygen atom. Put another way, oxygen pulls electrons down the chain in an energy-yielding tumble analogous to gravity pulling objects downhill.

In summary, during cellular respiration, most electrons travel the following "downhill" route: glucose → NADH → electron transport chain → oxygen. Later in this chapter, you will learn more about how the cell uses the energy released from this exergonic electron fall to regenerate its supply of ATP. For now, having covered the basic redox mechanisms of cellular respiration, let's look at the entire process by which energy is harvested from organic fuels.

## The Stages of Cellular Respiration: *A Preview*

The harvesting of energy from glucose by cellular respiration is a cumulative function of three metabolic stages. We list them here along with a color-coding scheme that we will use throughout the chapter to help you keep track of the big picture.

1. GLYCOLYSIS (color-coded blue throughout the chapter)
2. PYRUVATE OXIDATION and the CITRIC ACID CYCLE (color-coded orange)
3. OXIDATIVE PHOSPHORYLATION: Electron transport and chemiosmosis (color-coded purple)

Biochemists usually reserve the term *cellular respiration* for stages 2 and 3 together. In this text, however, we include glycolysis as a part of cellular respiration because most respiring cells deriving energy from glucose use glycolysis to produce the starting material for the citric acid cycle.

As diagrammed in **Figure 7.6**, glycolysis and pyruvate oxidation followed by the citric acid cycle are the catabolic pathways that break down glucose and other organic fuels. **Glycolysis**, which occurs in the cytosol, begins the degradation process by breaking glucose into two molecules of a compound called pyruvate. In eukaryotes, pyruvate enters the mitochondrion and is oxidized to a compound called acetyl CoA, which enters the **citric acid cycle** (also called the Krebs cycle). There, the breakdown of glucose to carbon dioxide is completed. (In prokaryotes, these processes take place in the cytosol.) Thus, the carbon dioxide produced by respiration represents fragments of oxidized organic molecules.

Some of the steps of glycolysis and the citric acid cycle are redox reactions in which dehydrogenases transfer electrons from substrates to $NAD^+$, forming NADH. In the third stage

▶ **Figure 7.6 An overview of cellular respiration.** During glycolysis, each glucose molecule is broken down into two molecules of pyruvate. In eukaryotic cells, as shown here, the pyruvate enters the mitochondrion. There it is oxidized to acetyl CoA, which will be further oxidized to $CO_2$ in the citric acid cycle. The electron carriers NADH and $FADH_2$ transfer electrons derived from glucose to electron transport chains. During oxidative phosphorylation, electron transport chains convert the chemical energy to a form used for ATP synthesis in the process called chemiosmosis. (During earlier steps of cellular respiration, smaller amounts of ATP are synthesized in a process called substrate-level phosphorylation.)

 Visit the Study Area in **MasteringBiology** for the BioFlix® 3-D Animation on Cellular Respiration.

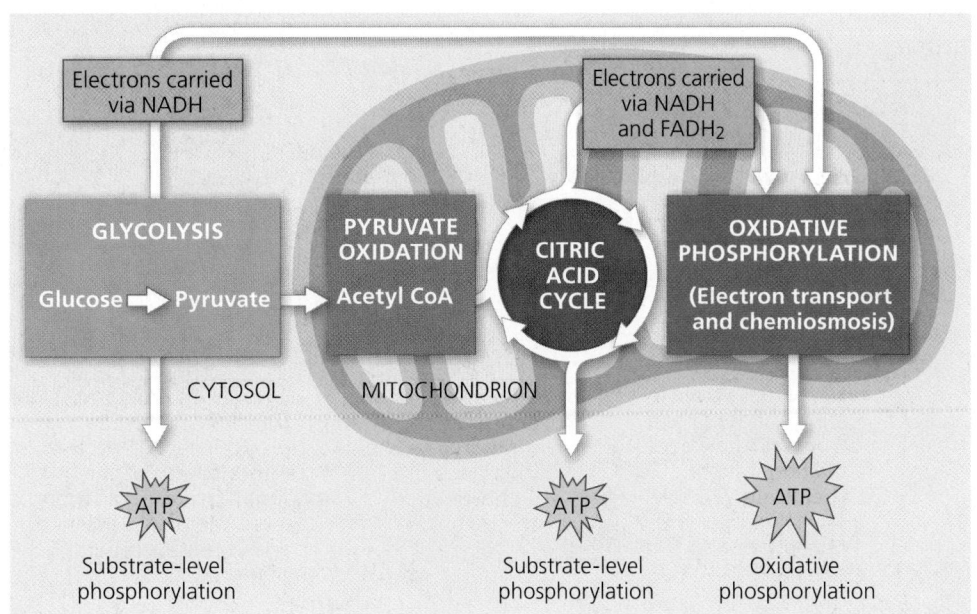

of respiration, the electron transport chain accepts electrons (most often via NADH) from the breakdown products of the first two stages and passes these electrons from one molecule to another. At the end of the chain, the electrons are combined with molecular oxygen and hydrogen ions ($H^+$), forming water (see Figure 7.5b). The energy released at each step of the chain is stored in a form the mitochondrion (or prokaryotic cell) can use to make ATP from ADP. This mode of ATP synthesis is called **oxidative phosphorylation** because it is powered by the redox reactions of the electron transport chain.

In eukaryotic cells, the inner membrane of the mitochondrion is the site of electron transport and chemiosmosis, the processes that together constitute oxidative phosphorylation. (In prokaryotes, these processes take place in the plasma membrane.) Oxidative phosphorylation accounts for almost 90% of the ATP generated by respiration. A smaller amount of ATP is formed directly in a few reactions of glycolysis and the citric acid cycle by a mechanism called **substrate-level phosphorylation (Figure 7.7)**. This mode of ATP synthesis occurs when an enzyme transfers a phosphate group from a substrate molecule to ADP, rather than adding an inorganic phosphate to ADP as in oxidative phosphorylation. "Substrate molecule" here refers to an organic molecule generated as an intermediate during the catabolism of glucose.

For each molecule of glucose degraded to carbon dioxide and water by respiration, the cell makes up to about 32 molecules of ATP, each with 7.3 kcal/mol of free energy. Respiration cashes in the large denomination of energy banked in a single molecule of glucose (686 kcal/mol) for the small change

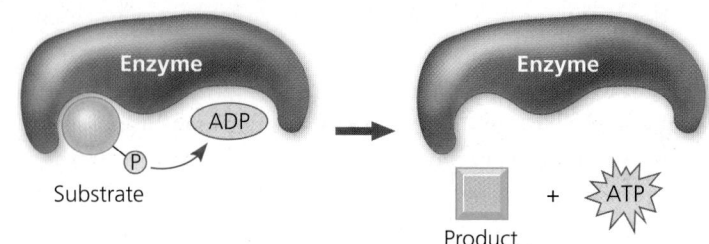

▲ **Figure 7.7 Substrate-level phosphorylation.** Some ATP is made by direct transfer of a phosphate group from an organic substrate to ADP by an enzyme. (For examples in glycolysis, see Figure 7.9, steps 7 and 10.)

**MAKE CONNECTIONS** *Review Figure 6.8. Do you think the potential energy is higher for the reactants or the products in the reaction shown above? Explain.*

of many molecules of ATP, which is more practical for the cell to spend on its work.

This preview has introduced you to how glycolysis, the citric acid cycle, and oxidative phosphorylation fit into the process of cellular respiration. We are now ready to take a closer look at each of these three stages of respiration.

**CONCEPT CHECK 7.1**

1. Compare and contrast aerobic and anaerobic respiration.
2. Name and describe the two ways in which ATP is made during cellular respiration. During what stage(s) in the process does each type occur?
3. **WHAT IF?** If the following redox reaction occurred, which compound would be oxidized? Which reduced?

$$C_4H_6O_5 + NAD^+ \rightarrow C_4H_4O_5 + NADH + H^+$$

**For suggested answers, see Appendix A.**

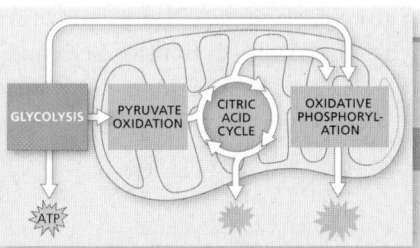

▼ **Figure 7.9 A closer look at glycolysis.** Note that glycolysis is a source of ATP and NADH.

**GLYCOLYSIS: Energy Investment Phase**

**WHAT IF?** *What would happen if you removed the dihydroxyacetone phosphate generated in step 4 as fast as it was produced?*

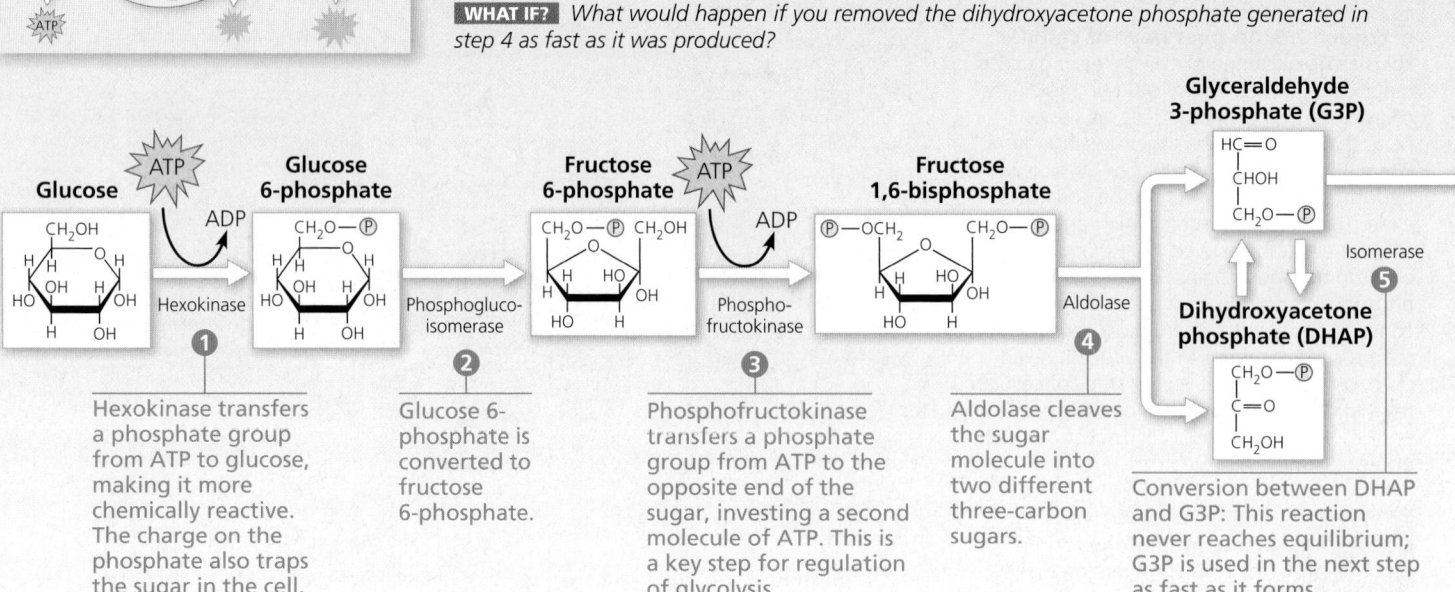

| | | | | |
|---|---|---|---|---|
| **①** Hexokinase transfers a phosphate group from ATP to glucose, making it more chemically reactive. The charge on the phosphate also traps the sugar in the cell. | **②** Glucose 6-phosphate is converted to fructose 6-phosphate. | **③** Phosphofructokinase transfers a phosphate group from ATP to the opposite end of the sugar, investing a second molecule of ATP. This is a key step for regulation of glycolysis. | **④** Aldolase cleaves the sugar molecule into two different three-carbon sugars. | **⑤** Conversion between DHAP and G3P: This reaction never reaches equilibrium; G3P is used in the next step as fast as it forms. |

## CONCEPT 7.2

# Glycolysis harvests chemical energy by oxidizing glucose to pyruvate

The word *glycolysis* means "sugar splitting," and that is exactly what happens during this pathway. Glucose, a six-carbon sugar, is split into two three-carbon sugars. These smaller sugars are then oxidized and their remaining atoms rearranged to form two molecules of pyruvate. (Pyruvate is the ionized form of pyruvic acid.)

As summarized in **Figure 7.8**, glycolysis can be divided into two phases: energy investment and energy payoff. During the energy investment phase, the cell actually spends ATP. This investment is repaid with interest during the energy payoff phase, when ATP is produced by substrate-level phosphorylation and $NAD^+$ is reduced to NADH by electrons released from the oxidation of glucose. The net energy yield from glycolysis, per glucose molecule, is 2 ATP plus 2 NADH.

Because glycolysis is a fundamental core process shared by bacteria, archaea, and eukaryotes alike, we will use it as an example of a biochemical pathway. The ten steps of the glycolytic pathway are shown in **Figure 7.9**.

All of the carbon originally present in glucose is accounted for in the two molecules of pyruvate; no carbon is released as $CO_2$ during glycolysis. Glycolysis occurs whether or not $O_2$ is present. However, if $O_2$ *is* present, the chemical energy stored in pyruvate and NADH can be extracted by pyruvate oxidation, the citric acid cycle, and oxidative phosphorylation.

▼ **Figure 7.8 The energy input and output of glycolysis.**

**Energy Investment Phase**

Glucose

$2$ ATP used $\longrightarrow$ $2$ ADP + $2$ $\textcircled{P}$

**Energy Payoff Phase**

$4$ ADP + $4$ $\textcircled{P}$ $\longrightarrow$ $4$ ATP formed

$2$ $NAD^+$ + $4$ $e^-$ + $4$ $H^+$ $\longrightarrow$ $2$ NADH + $2$ $H^+$

$\longrightarrow$ 2 Pyruvate + 2 $H_2O$

**Net**

Glucose $\longrightarrow$ 2 Pyruvate + 2 $H_2O$

4 ATP formed – 2 ATP used $\longrightarrow$ 2 ATP

$2$ $NAD^+$ + $4$ $e^-$ + $4$ $H^+$ $\longrightarrow$ 2 NADH + 2 $H^+$

## CONCEPT CHECK 7.2

1. During step 6 in Figure 7.9, which molecule acts as the oxidizing agent? The reducing agent?

   **For suggested answers, see Appendix A.**

---

The energy payoff phase occurs after glucose is split into two three-carbon sugars. Thus, the coefficient 2 precedes all molecules in this phase.

### GLYCOLYSIS: Energy Payoff Phase

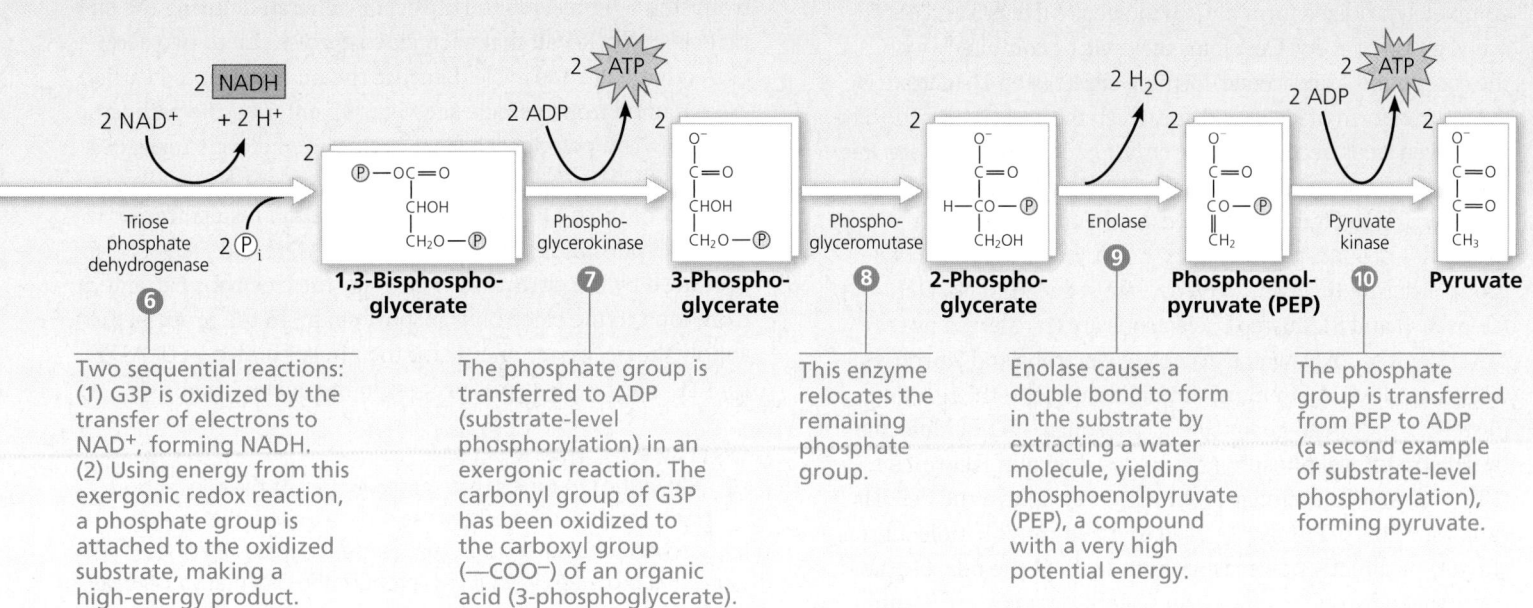

**6** Two sequential reactions: (1) G3P is oxidized by the transfer of electrons to $NAD^+$, forming NADH. (2) Using energy from this exergonic redox reaction, a phosphate group is attached to the oxidized substrate, making a high-energy product.

**7** The phosphate group is transferred to ADP (substrate-level phosphorylation) in an exergonic reaction. The carbonyl group of G3P has been oxidized to the carboxyl group (—COO⁻) of an organic acid (3-phosphoglycerate).

**8** This enzyme relocates the remaining phosphate group.

**9** Enolase causes a double bond to form in the substrate by extracting a water molecule, yielding phosphoenolpyruvate (PEP), a compound with a very high potential energy.

**10** The phosphate group is transferred from PEP to ADP (a second example of substrate-level phosphorylation), forming pyruvate.

# After pyruvate is oxidized, the citric acid cycle completes the energy-yielding oxidation of organic molecules

Glycolysis releases less than a quarter of the chemical energy in glucose that can be harvested by cells; most of the energy remains stockpiled in the two molecules of pyruvate. When $O_2$ is present, the pyruvate in eukaryotic cells enters a mitochondrion, where the oxidation of glucose is completed. (In aerobically respiring prokaryotic cells, this process occurs in the cytosol.)

Once inside the mitochondrion, pyruvate undergoes a series of enzymatic reactions that remove $CO_2$ and oxidizes the remaining fragment, forming NADH from $NAD^+$. The product is a highly reactive compound called acetyl coenzyme A, or **acetyl CoA**, which will feed its acetyl group into the citric acid cycle for further oxidation **(Figure 7.10)**.

The citric acid cycle (also known as the Krebs cycle) functions as a metabolic furnace that oxidizes organic fuel derived from pyruvate. Figure 7.10 summarizes the inputs and outputs as pyruvate is broken down to three $CO_2$ molecules, including the molecule of $CO_2$ released during the conversion of pyruvate to acetyl CoA. The cycle generates 1 ATP per turn by substrate-level phosphorylation, but most of the chemical energy is transferred to $NAD^+$ and a related electron carrier, the coenzyme FAD (flavin adenine dinucleotide, derived from riboflavin, a B vitamin), during the redox reactions. The reduced coenzymes, NADH and $FADH_2$, shuttle their cargo of high-energy electrons into the electron transport chain.

Now let's look at the citric acid cycle in more detail. The cycle has eight steps, each catalyzed by a specific enzyme. You can see in **Figure 7.11** that for each turn of the citric acid cycle, two carbons (red type) enter in the relatively reduced form of an acetyl group (step 1), and two different carbons (blue type) leave in the completely oxidized form of $CO_2$ molecules (steps 3 and 4). The acetyl group of acetyl CoA joins the cycle by combining with the compound oxaloacetate, forming citrate (step 1). (Citrate is the ionized form of citric acid, for which the cycle is named.) The next seven steps decompose the citrate back to oxaloacetate. It is this regeneration of oxaloacetate that makes the process a *cycle*.

Referring to Figure 7.11, we can tally the energy-rich molecules produced by the citric acid cycle. For each acetyl group entering the cycle, 3 $NAD^+$ are reduced to NADH (steps 3, 4, and 8). In step 6, electrons are transferred not to $NAD^+$, but to FAD, which accepts 2 electrons and 2 protons to become $FADH_2$. In many animal tissue cells, the reaction in step 5 produces a guanosine triphosphate (GTP) molecule by substrate-level phosphorylation, as shown in Figure 7.11. GTP is a molecule similar to ATP in its structure and cellular function. This GTP may be used to make an ATP molecule (as shown) or directly power work in the cell. In the cells of plants, bacteria, and some animal tissues, step 5 forms an ATP molecule directly by substrate-level phosphorylation. The output

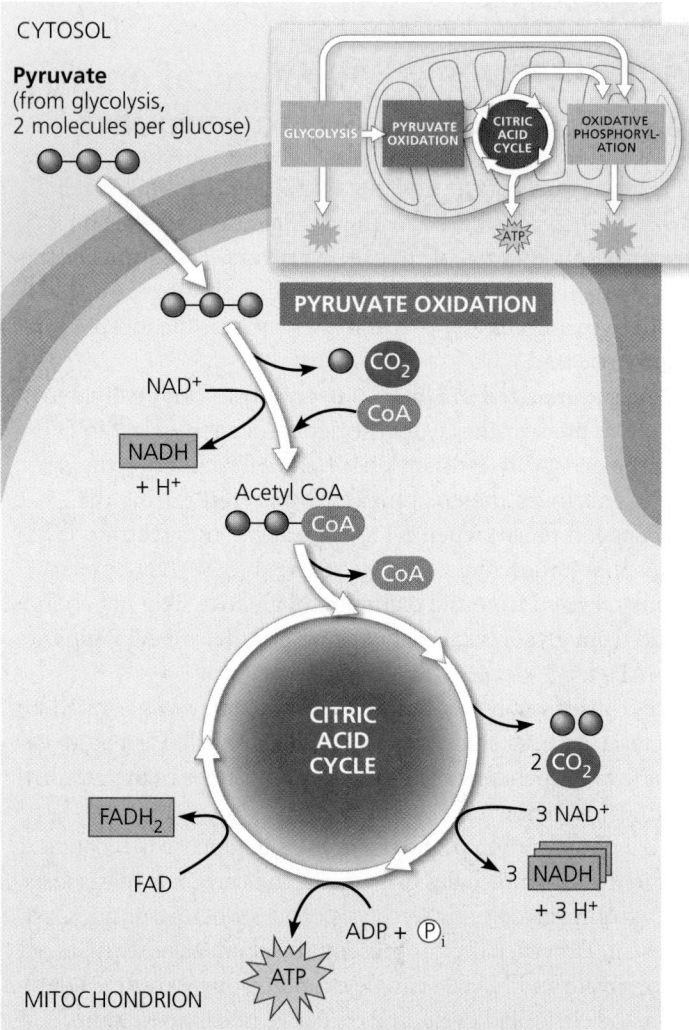

▲ **Figure 7.10 An overview of pyruvate oxidation and the citric acid cycle.** The inputs and outputs per pyruvate molecule are shown. To calculate on a per-glucose basis, multiply by 2, because each glucose molecule is split during glycolysis into two pyruvate molecules.

from step 5 represents the only ATP generated during the citric acid cycle. Recall that each glucose gives rise to two acetyl CoAs that enter the cycle. Because the numbers noted earlier are obtained from a single acetyl group entering the pathway, the total yield per glucose from the citric acid cycle turns out to be 6 NADHs, 2 $FADH_2$s, and the equivalent of 2 ATPs.

Most of the ATP produced by respiration results from oxidative phosphorylation, when the NADH and $FADH_2$ produced by the citric acid cycle relay the electrons extracted from food to the electron transport chain. In the process, they supply the necessary energy for the phosphorylation of ADP to ATP. We'll explore this process in the next section.

## CONCEPT CHECK 7.3

1. Name the molecules that conserve most of the energy from redox reactions of the citric acid cycle. How is this energy converted to a form that can be used to make ATP?
2. What processes in your cells produce the $CO_2$ that you exhale?

**For suggested answers, see Appendix A.**

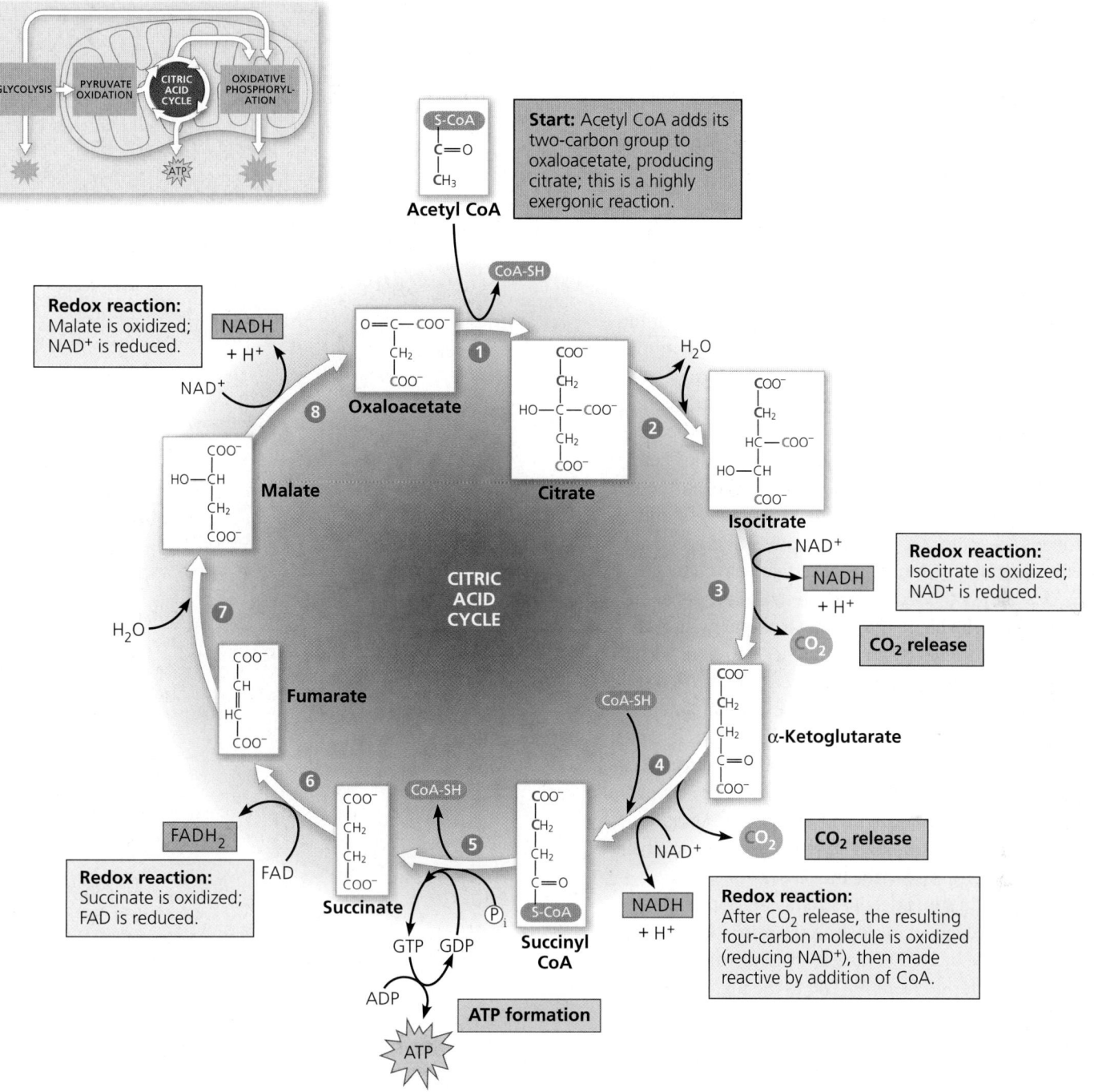

**Start:** Acetyl CoA adds its two-carbon group to oxaloacetate, producing citrate; this is a highly exergonic reaction.

Acetyl CoA

**Redox reaction:** Malate is oxidized; $NAD^+$ is reduced.

NADH + $H^+$

$NAD^+$

Oxaloacetate

Malate

**CITRIC ACID CYCLE**

Citrate

$H_2O$

Isocitrate

$NAD^+$

NADH + $H^+$

**Redox reaction:** Isocitrate is oxidized; $NAD^+$ is reduced.

$CO_2$

**$CO_2$ release**

α-Ketoglutarate

Fumarate

$H_2O$

FADH₂

FAD

**Redox reaction:** Succinate is oxidized; FAD is reduced.

Succinate

CoA-SH

$CO_2$

**$CO_2$ release**

$NAD^+$

NADH + $H^+$

**Redox reaction:** After $CO_2$ release, the resulting four-carbon molecule is oxidized (reducing $NAD^+$), then made reactive by addition of CoA.

Succinyl CoA

$P_i$

GTP GDP

ADP

**ATP formation**

ATP

▲ **Figure 7.11 A closer look at the citric acid cycle.** Key steps (redox reactions, $CO_2$ release, and ATP formation) are labeled. In the chemical structures, red type traces the fate of the two carbon atoms that enter the cycle via acetyl CoA (step 1), and blue type indicates the two carbons that exit the cycle as $CO_2$ in steps 3 and 4. (The red type goes only through step 5 because the succinate molecule is symmetrical; the two ends cannot be distinguished from each other.) Notice that the carbon atoms that enter the cycle from acetyl CoA do not leave the cycle in the same turn. They remain in the cycle, occupying a different location in the molecules on their next turn, after another acetyl group is added. Therefore, the oxaloacetate regenerated at step 8 is made up of different carbon atoms each time around.

# During oxidative phosphorylation, chemiosmosis couples electron transport to ATP synthesis

Our main objective in this chapter is to learn how cells harvest the energy of glucose and other nutrients in food to make ATP.

But the metabolic components of respiration we have dissected so far, glycolysis and the citric acid cycle, produce only 4 ATP molecules per glucose molecule, all by substrate-level phosphorylation: 2 net ATP from glycolysis and 2 ATP from the citric acid cycle. At this point, molecules of NADH (and FADH₂) account for most of the energy extracted from each glucose molecule. These electron escorts link glycolysis and the citric acid cycle to the machinery of oxidative phosphorylation, which uses energy released by the electron transport chain to power ATP synthesis. In this section, you will learn first how the electron transport chain works and then how electron flow down the chain is coupled to ATP synthesis.

## The Pathway of Electron Transport

The electron transport chain is a collection of molecules embedded in the inner membrane of the mitochondrion in eukaryotic cells. (In prokaryotes, these molecules reside in the plasma membrane.) The folding of the inner membrane to form cristae increases its surface area, providing space for thousands of copies of the electron transport chain in each mitochondrion. Once again, we see that structure fits function—the infolded membrane with its placement of electron carrier molecules in a row, one after the other, is well-suited for the series of sequential redox reactions that take place along the chain. Most components of the chain are proteins, which exist in multiprotein complexes numbered I through IV. Tightly bound to these proteins are *prosthetic groups*, nonprotein components essential for the catalytic functions of certain enzymes.

**Figure 7.12** shows the sequence of electron carriers in the electron transport chain and the drop in free energy as electrons travel down the chain. During this electron transport, electron carriers alternate between reduced and oxidized states as they accept and then donate electrons. Each component of the chain becomes reduced when it accepts electrons from its "uphill" neighbor, which has a lower affinity for electrons (is less electronegative). It then returns to its oxidized form as it passes electrons to its "downhill," more electronegative neighbor.

Now let's take a closer look at the electron transport chain in Figure 7.12. We'll first describe the passage of electrons through complex I in some detail as an illustration of the general principles involved in electron transport. Electrons acquired from glucose by $NAD^+$ during glycolysis and the citric acid cycle are transferred from NADH to the first molecule of the electron transport chain in complex I. This molecule is a flavoprotein, so named because it has a prosthetic group called flavin mononucleotide (FMN). In the next redox reaction, the flavoprotein returns to its oxidized form as it passes electrons to an iron-sulfur protein (Fe · S in complex I), one of a family of proteins with both iron and sulfur tightly bound. The iron-sulfur protein then passes the electrons to a compound called ubiquinone (Q in Figure 7.12). This electron carrier is a small hydrophobic molecule, the only member of the electron transport chain that is not a protein. Ubiquinone is individually mobile within the membrane rather than residing in a particular complex. (Another name for ubiquinone is coenzyme Q, or CoQ; you may have seen it sold as a nutritional supplement.)

Most of the remaining electron carriers between ubiquinone and oxygen are proteins called **cytochromes**. Their prosthetic group, called a heme group, has an iron atom that accepts and donates electrons. (The heme group in a cytochrome is similar to the heme group in hemoglobin, the protein of red

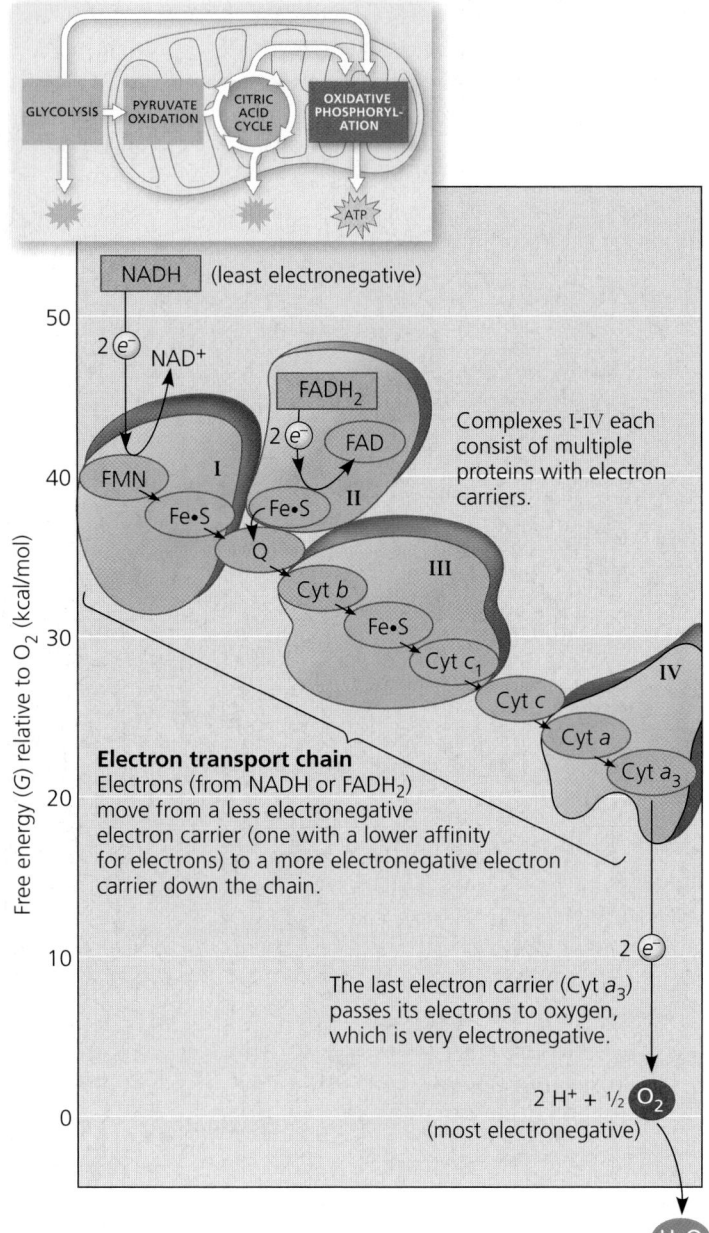

▲ **Figure 7.12 Free-energy change during electron transport.** The overall energy drop ($\Delta G$) for electrons traveling from NADH to oxygen is 53 kcal/mol, but this "fall" is broken up into a series of smaller steps by the electron transport chain. (An oxygen atom is represented here as $\frac{1}{2} O_2$ to emphasize that the electron transport chain reduces molecular oxygen, $O_2$, not individual oxygen atoms.)

blood cells, except that the iron in hemoglobin carries oxygen, not electrons.) The electron transport chain has several types of cytochromes, each a different protein with a slightly different electron-carrying heme group. The last cytochrome of the chain, Cyt $a_3$, passes its electrons to oxygen, which is *very* electronegative. Each oxygen atom also picks up a pair of hydrogen ions (protons) from the aqueous solution, neutralizing the $-2$ charge of the added electrons and forming water.

Another source of electrons for the transport chain is $FADH_2$, the other reduced product of the citric acid cycle.

Notice in Figure 7.12 that FADH$_2$ adds its electrons to the electron transport chain from within complex II, at a lower energy level than NADH does. Consequently, although NADH and FADH$_2$ each donate an equivalent number of electrons (2) for oxygen reduction, the electron transport chain provides about one-third less energy for ATP synthesis when the electron donor is FADH$_2$ rather than NADH. We'll see why in the next section.

The electron transport chain makes no ATP directly. Instead, it eases the fall of electrons from food to oxygen, breaking a large free-energy drop into a series of smaller steps that release energy in manageable amounts. How does the mitochondrion (or the plasma membrane in prokaryotes) couple this electron transport and energy release to ATP synthesis? The answer is a mechanism called chemiosmosis.

## Chemiosmosis: The Energy-Coupling Mechanism

Populating the inner membrane of the mitochondrion or the prokaryotic plasma membrane are many copies of a protein complex called **ATP synthase**, the enzyme that actually makes ATP from ADP and inorganic phosphate. ATP synthase works like an ion pump running in reverse. Ion pumps usually use ATP as an energy source to transport ions against their gradients. Enzymes can catalyze a reaction in either direction, depending on the $\Delta G$ for the reaction, which is affected by the local concentrations of reactants and products (see Concepts 6.2 and 6.3). Rather than hydrolyzing ATP to pump protons against their concentration gradient, under the conditions of cellular respiration ATP synthase uses the energy of an existing ion gradient to power ATP synthesis. The power source for ATP synthase is a difference in the concentration of H$^+$ on opposite sides of the inner mitochondrial membrane. (We can also think of this gradient as a difference in pH, since pH is a measure of H$^+$ concentration.) This process, in which energy stored in the form of a hydrogen ion gradient across a membrane is used to drive cellular work such as the synthesis of ATP, is called **chemiosmosis** (from the Greek *osmos*, push). We have previously used the word *osmosis* in discussing water transport, but here it refers to the flow of H$^+$ across a membrane.

From studying the structure of ATP synthase, scientists have learned how the flow of H$^+$ through this large enzyme powers ATP generation. ATP synthase is a multisubunit complex with four main parts, each made up of multiple polypeptides. Protons move one by one into binding sites on one of the parts (the rotor), causing it to spin in a way that catalyzes ATP production from ADP and inorganic phosphate **(Figure 7.13)**. The flow of protons thus behaves somewhat like a rushing stream that turns a waterwheel. ATP synthase is the smallest molecular rotary motor known in nature.

How does the inner mitochondrial membrane or the prokaryotic plasma membrane generate and maintain the H$^+$ gradient that drives ATP synthesis by the ATP synthase protein complex? Establishing the H$^+$ gradient across the inner

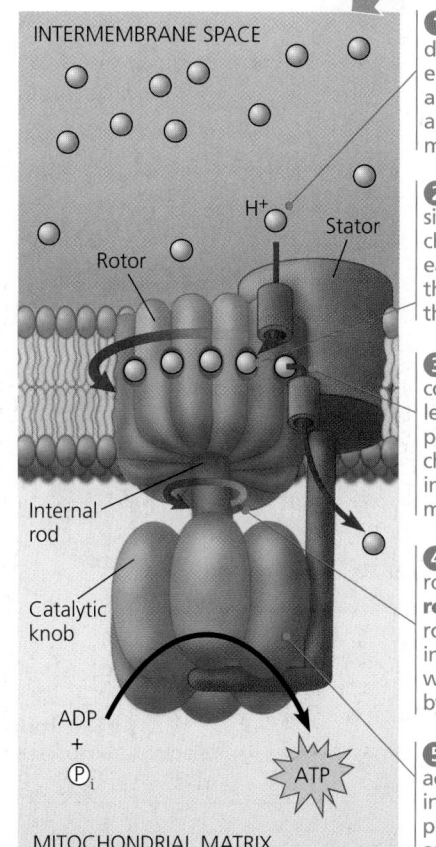

**1** H$^+$ ions flowing down their gradient enter a channel in a **stator**, which is anchored in the membrane.

**2** H$^+$ ions enter binding sites within a **rotor**, changing the shape of each subunit so that the rotor spins within the membrane.

**3** Each H$^+$ ion makes one complete turn before leaving the rotor and passing through a second channel in the stator into the mitochondrial matrix.

**4** Spinning of the rotor causes an internal **rod** to spin as well. This rod extends like a stalk into the **knob** below it, which is held stationary by part of the stator.

**5** Turning of the rod activates catalytic sites in the knob that produce ATP from ADP and P$_i$.

**(a)** The ATP synthase protein complex functions as a mill, powered by the flow of hydrogen ions.

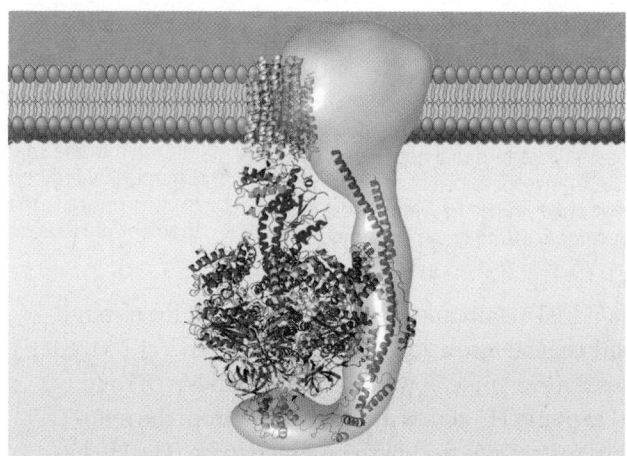

**(b)** This computer model shows the four parts of ATP synthase. Each part consists of a number of polypeptide subunits. The entire structure of the gray region has not yet been determined and is an area of active research.

▲ **Figure 7.13 ATP synthase, a molecular mill.** Multiple ATP synthases reside in eukaryotic mitochondrial and chloroplast membranes and in prokaryotic plasma membranes.

**DRAW IT** *Label the rotor, stator, internal rod, and catalytic knob in part (b), the computer model.*

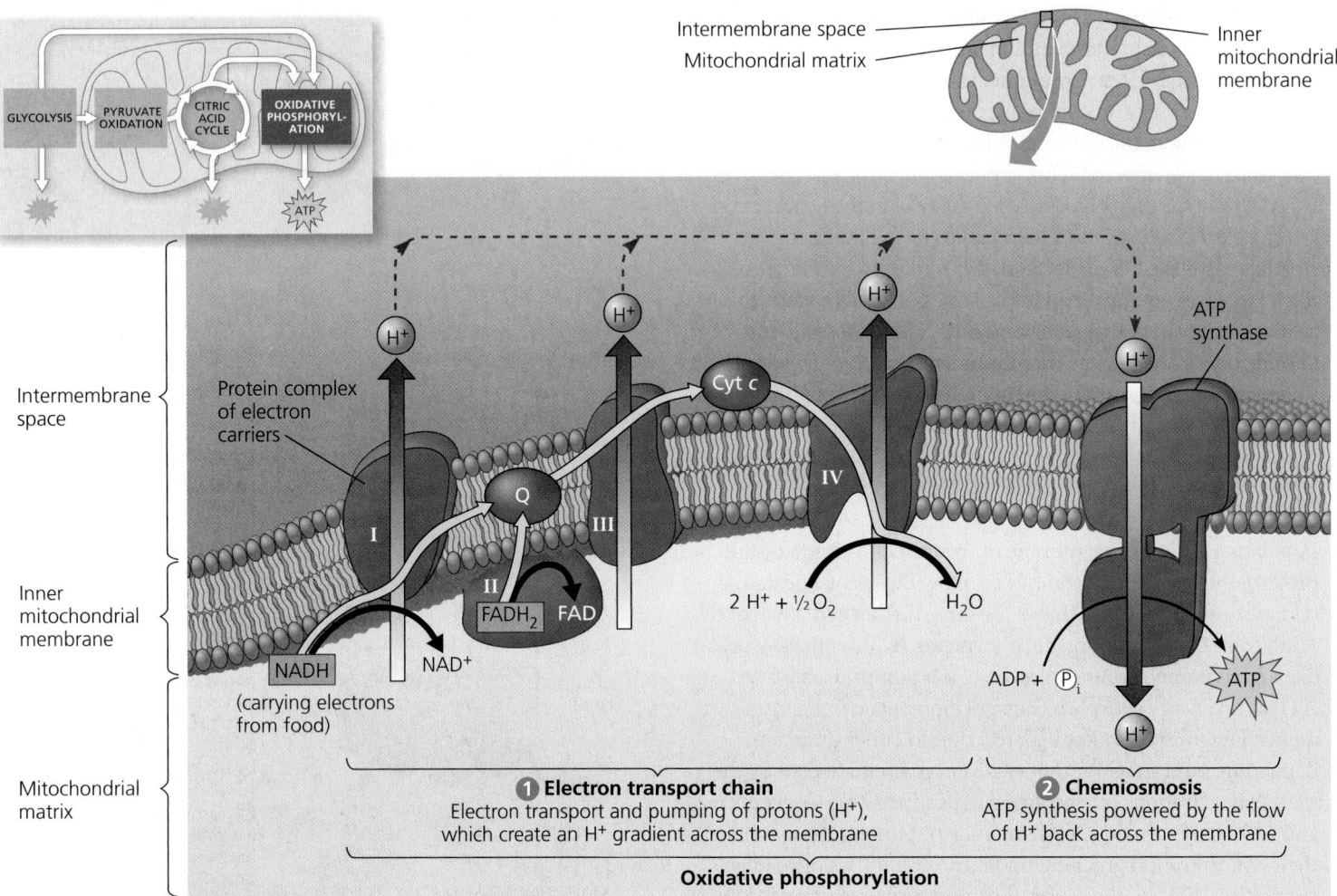

**Intermembrane space** — **Inner mitochondrial membrane**
**Mitochondrial matrix** —

GLYCOLYSIS → PYRUVATE OXIDATION → CITRIC ACID CYCLE → OXIDATIVE PHOSPHORYL-ATION

ATP

**Intermembrane space**

**Inner mitochondrial membrane**

**Mitochondrial matrix**

Protein complex of electron carriers

$H^+$

$H^+$

$H^+$

$H^+$

ATP synthase

Cyt c

I    II    III    IV

Q

FADH$_2$  FAD

$2 H^+ + \frac{1}{2} O_2$     $H_2O$

NADH    NAD$^+$

(carrying electrons from food)

ADP + Ⓟ$_i$     ATP

$H^+$

**1 Electron transport chain**
Electron transport and pumping of protons ($H^+$), which create an $H^+$ gradient across the membrane

**2 Chemiosmosis**
ATP synthesis powered by the flow of $H^+$ back across the membrane

**Oxidative phosphorylation**

▲ **Figure 7.14 Chemiosmosis couples the electron transport chain to ATP synthesis. 1** NADH and FADH$_2$ shuttle high-energy electrons extracted from food during glycolysis and the citric acid cycle into an electron transport chain built into the inner mitochondrial membrane. The gold arrows trace the transport of electrons, which are finally passed to a terminal acceptor (O$_2$, in the case of aerobic respiration) at the "downhill" end of the chain, forming water. Most of the electron carriers of the chain are grouped into four complexes (I–IV). Two mobile carriers, ubiquinone (Q) and cytochrome c (Cyt c), move rapidly, ferrying electrons between the large complexes. As the complexes shuttle electrons, they pump protons from the mitochondrial matrix into the intermembrane space. FADH$_2$ deposits its electrons via complex II and so results in fewer protons being pumped into the intermembrane space than occurs with NADH. Chemical energy originally harvested from food is transformed into a proton-motive force, a gradient of $H^+$ across the membrane.

**2** During chemiosmosis, the protons flow back down their gradient via ATP synthase, which is built into the membrane nearby. The ATP synthase harnesses the proton-motive force to phosphorylate ADP, forming ATP. Together, electron transport and chemiosmosis make up oxidative phosphorylation.

**WHAT IF?** *If complex IV were nonfunctional, could chemiosmosis produce any ATP, and if so, how would the rate of synthesis differ?*

mitochondrial membrane is a major function of the electron transport chain **(Figure 7.14)**. The chain is an energy converter that uses the exergonic flow of electrons from NADH and FADH$_2$ to pump $H^+$ across the membrane, from the mitochondrial matrix into the intermembrane space. The $H^+$ has a tendency to move back across the membrane, diffusing down its gradient. And the ATP synthases are the only sites that provide a route through the membrane for $H^+$. As we described previously, the passage of $H^+$ through ATP synthase uses the exergonic flow of $H^+$ to drive the phosphorylation of ADP. Thus, the energy stored in an $H^+$ gradient across a membrane couples the redox reactions of the electron transport chain to ATP synthesis.

At this point, you may be wondering how the electron transport chain pumps hydrogen ions. Researchers have found that certain members of the electron transport chain accept and release protons ($H^+$) along with electrons. (The aqueous solutions inside and surrounding the cell are a ready source of $H^+$.) At certain steps along the chain, electron transfers cause $H^+$ to be taken up and released into the surrounding solution. In eukaryotic cells, the electron carriers are spatially arranged in the inner mitochondrial membrane in such a way that $H^+$ is accepted from the mitochondrial matrix and deposited in the intermembrane space (see Figure 7.14). The $H^+$ gradient that results is referred to as a **proton-motive force**, emphasizing the capacity of the gradient to perform work. The force drives $H^+$ back across the membrane through the $H^+$ channels provided by ATP synthases.

In general terms, *chemiosmosis is an energy-coupling mechanism that uses energy stored in the form of an $H^+$ gradient* *across a membrane to drive cellular work.* In mitochondria, the energy for gradient formation comes from exergonic redox reactions, and ATP synthesis is the work performed. But chemiosmosis also occurs elsewhere and in other variations. Chloroplasts use chemiosmosis to generate ATP during photosynthesis; in these organelles, light (rather than chemical energy) drives both electron flow down an electron transport chain and the resulting $H^+$ gradient formation. Prokaryotes, as already mentioned, generate $H^+$ gradients across their plasma membranes. They then tap the proton-motive force not only to make ATP inside the cell but also to rotate their flagella and to pump nutrients and waste products across the membrane. Because of its central importance to energy conversions in prokaryotes and eukaryotes, chemiosmosis has helped unify the study of bioenergetics. Peter Mitchell was awarded the Nobel Prize in 1978 for originally proposing the chemiosmotic model.

## An Accounting of ATP Production by Cellular Respiration

In the last few sections, we have looked rather closely at the key processes of cellular respiration. Now let's take a step back and remind ourselves of its overall function: harvesting the energy of glucose for ATP synthesis.

During respiration, most energy flows in this sequence: glucose → NADH → electron transport chain → proton-motive

force → ATP. We can do some bookkeeping to calculate the ATP profit when cellular respiration oxidizes a molecule of glucose to six molecules of carbon dioxide. The three main departments of this metabolic enterprise are glycolysis, the citric acid cycle, and the electron transport chain, which drives oxidative phosphorylation. **Figure 7.15** gives a detailed accounting of the ATP yield per glucose molecule oxidized. The tally adds the 4 ATP produced directly by substrate-level phosphorylation during glycolysis and the citric acid cycle to the many more molecules of ATP generated by oxidative phosphorylation. Each NADH that transfers a pair of electrons from glucose to the electron transport chain contributes enough to the proton-motive force to generate a maximum of about 3 ATP.

Why are the numbers in Figure 7.15 inexact? There are three reasons we cannot state an exact number of ATP molecules generated by the breakdown of one molecule of glucose. First, phosphorylation and the redox reactions are not directly coupled to each other, so the ratio of the number of NADH molecules to the number of ATP molecules is not a whole number. We know that 1 NADH results in 10 $H^+$ being transported out across the inner mitochondrial membrane, but the exact number of $H^+$ that must reenter the mitochondrial matrix via ATP synthase to generate 1 ATP has long been debated. Based on experimental data, however, most biochemists now agree that the most accurate number is 4 $H^+$. Therefore, a single molecule of NADH generates enough proton-motive

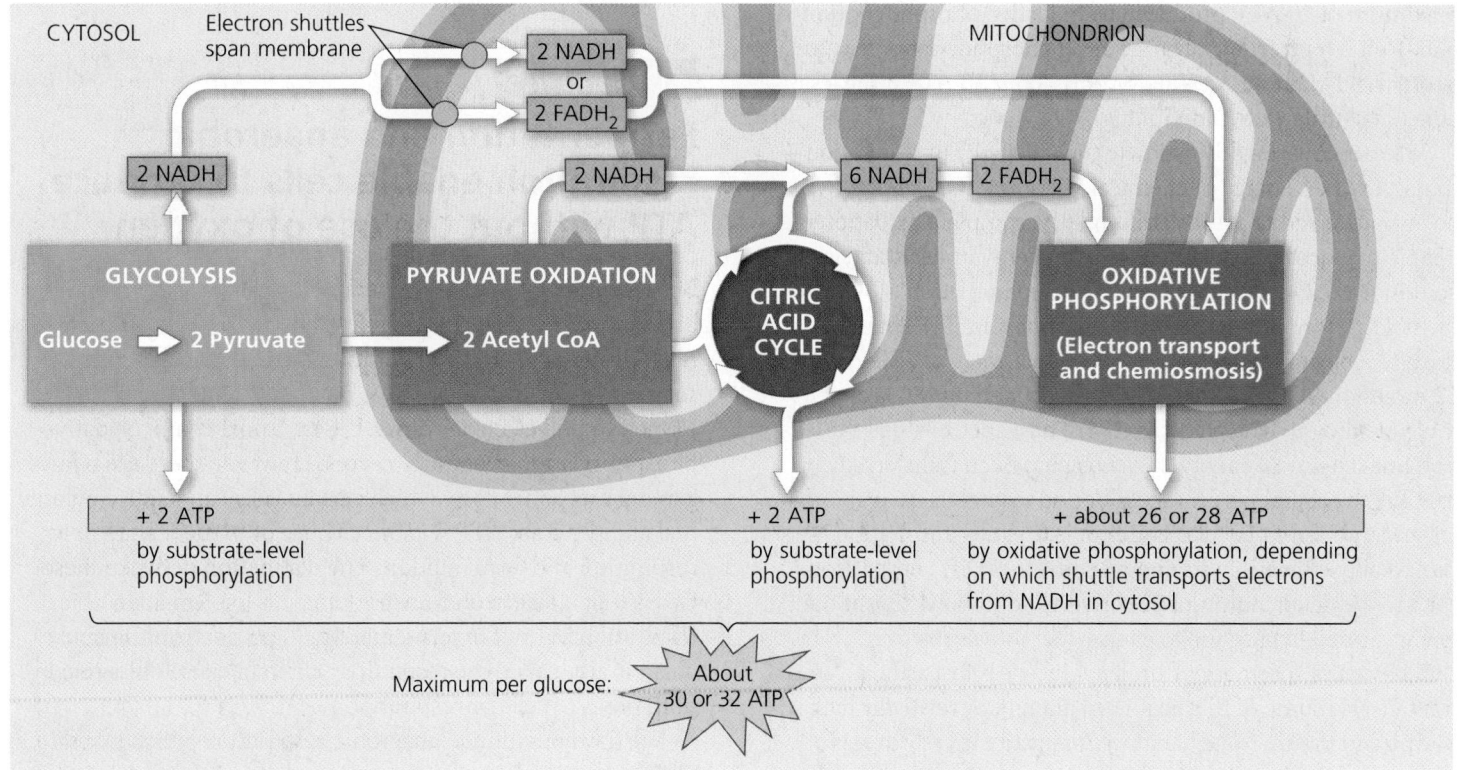

▲ **Figure 7.15 ATP yield per molecule of glucose at each stage of cellular respiration.**

**?** *Explain exactly how the total of 26 or 28 ATP (see the yellow bar) was calculated.*

force for the synthesis of 2.5 ATP. The citric acid cycle also supplies electrons to the electron transport chain via $FADH_2$, but since its electrons enter later in the chain, each molecule of this electron carrier is responsible for transport of only enough $H^+$ for the synthesis of 1.5 ATP. These numbers also take into account the slight energetic cost of moving the ATP formed in the mitochondrion out into the cytosol, where it will be used.

Second, the ATP yield varies slightly depending on the type of shuttle used to transport electrons from the cytosol into the mitochondrion. The mitochondrial inner membrane is impermeable to NADH, so NADH in the cytosol is segregated from the machinery of oxidative phosphorylation. The 2 electrons of NADH captured in glycolysis must be conveyed into the mitochondrion by one of several electron shuttle systems. Depending on the kind of shuttle in a particular cell type, the electrons are passed either to $NAD^+$ or to FAD in the mitochondrial matrix (see Figure 7.15). If the electrons are passed to FAD, as in brain cells, only about 1.5 ATP can result from each NADH that was originally generated in the cytosol. If the electrons are passed to mitochondrial $NAD^+$, as in liver cells and heart cells, the yield is about 2.5 ATP per NADH.

A third variable that reduces the yield of ATP is the use of the proton-motive force generated by the redox reactions of respiration to drive other kinds of work. For example, the proton-motive force powers the mitochondrion's uptake of pyruvate from the cytosol. However, if *all* the proton-motive force generated by the electron transport chain were used to drive ATP synthesis, one glucose molecule could generate a maximum of 28 ATP produced by oxidative phosphorylation plus 4 ATP (net) from substrate-level phosphorylation to give a total yield of about 32 ATP (or only about 30 ATP if the less efficient shuttle were functioning).

We can now roughly estimate the efficiency of respiration—that is, the percentage of chemical energy in glucose that has been transferred to ATP. Recall that the complete oxidation of a mole of glucose releases 686 kcal of energy under standard conditions ($\Delta G = -686$ kcal/mol). Phosphorylation of ADP to form ATP stores at least 7.3 kcal per mole of ATP. Therefore, the efficiency of respiration is 7.3 kcal per mole of ATP times 32 moles of ATP per mole of glucose divided by 686 kcal per mole of glucose, which equals 0.34. Thus, about 34% of the potential chemical energy in glucose has been transferred to ATP; the actual percentage is bound to vary as $\Delta G$ varies under different cellular conditions. Cellular respiration is remarkably efficient in its energy conversion. By comparison, the most efficient automobile converts only about 25% of the energy stored in gasoline to energy that moves the car.

The rest of the energy stored in glucose is lost as heat. We humans use some of this heat to maintain our relatively high body temperature (37°C), and we dissipate the rest through sweating and other cooling mechanisms.

Surprisingly, perhaps, it may be beneficial under certain conditions to reduce the efficiency of cellular respiration.

A remarkable adaptation is shown by hibernating mammals, which overwinter in a state of inactivity and lowered metabolism. Although their internal body temperature is lower than normal, it still must be kept significantly higher than the external air temperature. One type of tissue, called brown fat, is made up of cells packed full of mitochondria. The inner mitochondrial membrane contains a channel protein called the uncoupling protein, which allows protons to flow back down their concentration gradient without generating ATP. Activation of these proteins in hibernating mammals results in ongoing oxidation of stored fuel stores (fats), generating heat without any ATP production. In the absence of such an adaptation, the buildup of ATP would eventually cause cellular respiration to be shut down by regulatory mechanisms in the cell. In the **Scientific Skills Exercise**, you can work with data in a related but different case where a decrease in metabolic efficiency in cells is used to generate heat.

### CONCEPT CHECK 7.4

1. What effect would an absence of $O_2$ have on the process shown in Figure 7.14?
2. **WHAT IF?** In the absence of $O_2$, as in question 1, what do you think would happen if you decreased the pH of the intermembrane space of the mitochondrion? Explain your answer.
3. **MAKE CONNECTIONS** Membranes must be fluid to function properly (as you learned in Concept 5.1). How does the operation of the electron transport chain support that assertion?

For suggested answers, see Appendix A.

---

### CONCEPT 7.5

# Fermentation and anaerobic respiration enable cells to produce ATP without the use of oxygen

Because most of the ATP generated by cellular respiration is due to the work of oxidative phosphorylation, our estimate of ATP yield from aerobic respiration depends on an adequate supply of oxygen to the cell. Without the electronegative oxygen to pull electrons down the transport chain, oxidative phosphorylation eventually ceases. However, there are two general mechanisms by which certain cells can oxidize organic fuel and generate ATP *without* the use of oxygen: anaerobic respiration and fermentation. The distinction between these two is that an electron transport chain is used in anaerobic respiration but not in fermentation. (The electron transport chain is also called the respiratory chain because of its role in both types of cellular respiration.)

We have mentioned anaerobic respiration, which takes place in some prokaryotic organisms living in environments lacking oxygen. These organisms have an electron transport chain but do not use oxygen as a final electron acceptor at the

# Making a Bar Graph and Evaluating a Hypothesis

**Does Thyroid Hormone Level Affect Oxygen Consumption in Cells?** Some animals, such as mammals and birds, maintain a relatively constant body temperature, above that of their environment, by using heat produced as a by-product of metabolism. When the core temperature of these animals drops below an internal set point, their cells are triggered to reduce the efficiency of ATP production by the electron transport chains in mitochondria. At lower efficiency, extra fuel must be consumed to produce the same number of ATPs, generating additional heat. Because the response is moderated by the endocrine system, researchers hypothesized that thyroid hormone might trigger this cellular response. In this exercise, you will use a bar graph to visualize data from an experiment that compared the metabolic rates (by measuring oxygen consumption) in mitochondria of cells from animals with different levels of thyroid hormone.

**How the Experiment Was Done** Liver cells were isolated from sibling rats that had low, normal, or elevated thyroid hormone levels. The oxygen consumption rate due to activity of the mitochondrial electron transport chains of each type of cell was measured under controlled conditions.

### Data from the Experiment

| Thyroid Hormone Level | Oxygen Consumption Rate [nmol $O_2$/(min · mg cells)] |
|---|---|
| Low | 4.3 |
| Normal | 4.8 |
| Elevated | 8.7 |

**Data from** M. E. Harper and M. D. Brand, The quantitative contributions of mitochondrial proton leak and ATP turnover reactions to the changed respiration rates of hepatocytes from rats of different thyroid status, *Journal of Biological Chemistry* 268:14850–14860 (1993).

**INTERPRET THE DATA**

1. To visualize any differences in oxygen consumption between cell types, it will be useful to graph the data in a bar graph. First, set up the axes. (a) What is the independent variable (intentionally varied by the researchers), which goes on the x-axis? List the categories along the x-axis; because they are discrete rather than continuous, you can list them in any order. (b) What is the dependent variable (measured by the researchers), which goes on the y-axis? (c) What units (abbreviated) should go on the y-axis? Label the y-axis, including the units specified in the data table. Determine the range of values of the data that will need to go on the y-axis. What is the largest value? Draw evenly spaced tick marks and label them, starting with 0 at the bottom.

2. Graph the data for each sample. Match each x-value with its y-value and place a mark on the graph at that coordinate, then draw a bar from the x-axis up to the correct height for each sample. Why is a bar graph more appropriate than a scatter plot or line graph? (For additional information about graphs, see the Scientific Skills Review in Appendix F and in the Study Area in MasteringBiology.)

3. Examine your graph and look for a pattern in the data. (a) Which cell type had the highest rate of oxygen consumption, and which had the lowest? (b) Does this support the researchers' hypothesis? Explain. (c) Based on what you know about mitochondrial electron transport and heat production, predict which rats had the highest, and which had the lowest, body temperature.

(MB) A version of this Scientific Skills Exercise can be assigned in MasteringBiology.

---

end of the chain. Oxygen performs this function very well because it is extremely electronegative, but other, less electronegative substances can also serve as final electron acceptors. Some "sulfate-reducing" marine bacteria, for instance, use the sulfate ion ($SO_4^{2-}$) at the end of their respiratory chain. Operation of the chain builds up a proton-motive force used to produce ATP, but $H_2S$ (hydrogen sulfide) is made as a by-product rather than water. The rotten-egg odor you may have smelled while walking through a salt marsh or a mudflat signals the presence of sulfate-reducing bacteria.

Fermentation is a way of harvesting chemical energy without using either oxygen or any electron transport chain—in other words, without cellular respiration. How can food be oxidized without cellular respiration? Remember, oxidation simply refers to the loss of electrons to an electron acceptor, so it does not need to involve oxygen. Glycolysis oxidizes glucose to two molecules of pyruvate. The oxidizing agent of glycolysis is $NAD^+$, and neither oxygen nor any electron transfer chain is involved. Overall, glycolysis is exergonic, and some

of the energy made available is used to produce 2 ATP (net) by substrate-level phosphorylation. If oxygen *is* present, then additional ATP is made by oxidative phosphorylation when NADH passes electrons removed from glucose to the electron transport chain. But glycolysis generates 2 ATP whether oxygen is present or not—that is, whether conditions are aerobic or anaerobic.

As an alternative to respiratory oxidation of organic nutrients, fermentation is an extension of glycolysis that allows continuous generation of ATP by the substrate-level phosphorylation of glycolysis. For this to occur, there must be a sufficient supply of $NAD^+$ to accept electrons during the oxidation step of glycolysis. Without some mechanism to recycle $NAD^+$ from NADH, glycolysis would soon deplete the cell's pool of $NAD^+$ by reducing it all to NADH and would shut itself down for lack of an oxidizing agent. Under aerobic conditions, $NAD^+$ is recycled from NADH by the transfer of electrons to the electron transport chain. An anaerobic alternative is to transfer electrons from NADH to pyruvate, the end product of glycolysis.

## Types of Fermentation

Fermentation consists of glycolysis plus reactions that regenerate $NAD^+$ by transferring electrons from NADH to pyruvate or derivatives of pyruvate. The $NAD^+$ can then be reused to oxidize sugar by glycolysis, which nets two molecules of ATP by substrate-level phosphorylation. There are many types of fermentation, differing in the end products formed from pyruvate. Two common types are alcohol fermentation and lactic acid fermentation, and both are harnessed by humans in food production.

In **alcohol fermentation (Figure 7.16a)**, pyruvate is converted to ethanol (ethyl alcohol) in two steps. The first step releases carbon dioxide from the pyruvate, which is converted to the two-carbon compound acetaldehyde. In the second step, acetaldehyde is reduced by NADH to ethanol. This regenerates the supply of $NAD^+$ needed for the continuation of glycolysis. Many bacteria carry out alcohol fermentation under anaerobic conditions. Yeast (a fungus) also carries out alcohol fermentation. For thousands of years, humans have used yeast in brewing, winemaking, and baking. The $CO_2$ bubbles generated by baker's yeast during alcohol fermentation allow bread to rise.

During **lactic acid fermentation (Figure 7.16b)**, pyruvate is reduced directly by NADH to form lactate as an end product, with no release of $CO_2$. (Lactate is the ionized form of lactic acid.) Lactic acid fermentation by certain fungi and bacteria is used in the dairy industry to make cheese and yogurt.

Human muscle cells make ATP by lactic acid fermentation when oxygen is scarce. This occurs during strenuous exercise, when sugar catabolism for ATP production outpaces the muscle's supply of oxygen from the blood. Under these conditions, the cells switch from aerobic respiration to fermentation. The lactate that accumulates was previously thought to cause the muscle fatigue and pain that occurs a day or so after intense exercise. However, evidence shows that within an hour, blood carries the excess lactate from the muscles to the liver, where it is converted back to pyruvate by liver cells. Because oxygen is available, this pyruvate can then enter the mitochondria in liver cells and complete cellular respiration. Next-day muscle soreness is more likely caused by trauma to small muscle fibers, which leads to inflammation and pain.

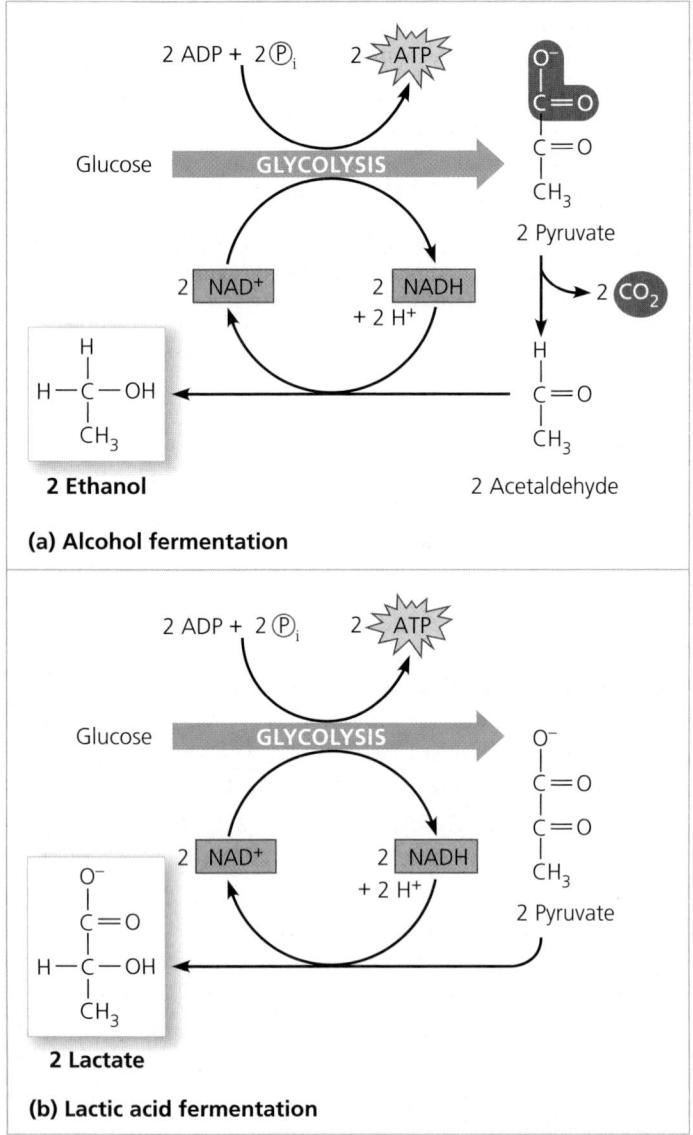

(a) Alcohol fermentation

(b) Lactic acid fermentation

▲ **Figure 7.16 Fermentation.** In the absence of oxygen, many cells use fermentation to produce ATP by substrate-level phosphorylation. Pyruvate, the end product of glycolysis, serves as an electron acceptor for oxidizing NADH back to $NAD^+$, which can then be reused in glycolysis. Two of the common end products formed from fermentation are **(a)** ethanol and **(b)** lactate, the ionized form of lactic acid.

## Comparing Fermentation with Anaerobic and Aerobic Respiration

Fermentation, anaerobic respiration, and aerobic respiration are three alternative cellular pathways for producing ATP by harvesting the chemical energy of food. All three use glycolysis to oxidize glucose and other organic fuels to pyruvate, with a net production of 2 ATP by substrate-level phosphorylation. And in all three pathways, $NAD^+$ is the oxidizing agent that accepts electrons from food during glycolysis.

A key difference is the contrasting mechanisms for oxidizing NADH back to $NAD^+$, which is required to sustain glycolysis. In fermentation, the final electron acceptor is an organic molecule such as pyruvate (lactic acid fermentation) or acetaldehyde (alcohol fermentation). In cellular respiration, by contrast, electrons carried by NADH are transferred to an electron transport chain, which regenerates the $NAD^+$ required for glycolysis.

Another major difference is the amount of ATP produced. Fermentation yields two molecules of ATP, produced by substrate-level phosphorylation. In the absence of an electron transport chain, the energy stored in pyruvate is unavailable. In cellular respiration, however, pyruvate is completely oxidized in the mitochondrion. Most of the chemical energy from this process is shuttled by NADH and FADH$_2$ in the form of electrons to the electron transport chain. There, the electrons move stepwise down a series of redox reactions to a final electron acceptor. (In aerobic respiration, the final electron acceptor is oxygen; in anaerobic respiration, the final acceptor is another molecule that is electronegative, although less so than oxygen.) Stepwise electron transport drives oxidative phosphorylation, yielding ATPs. Thus, cellular respiration harvests much more energy from each sugar molecule than fermentation can. In fact, aerobic respiration yields up to 32 molecules of ATP per glucose molecule—up to 16 times as much as does fermentation.

Some organisms, called **obligate anaerobes**, carry out only fermentation or anaerobic respiration. In fact, these organisms cannot survive in the presence of oxygen. A few cell types can carry out only aerobic oxidation of pyruvate, not fermentation. Other organisms, including yeasts and many bacteria, can make enough ATP to survive using either fermentation or respiration. Such species are called **facultative anaerobes**. On the cellular level, our muscle cells behave as facultative anaerobes. In such cells, pyruvate is a fork in the metabolic road that leads to two alternative catabolic routes **(Figure 7.17)**. Under aerobic conditions, pyruvate can be converted to acetyl CoA, which enters the citric acid cycle. Under anaerobic conditions, lactic acid fermentation occurs: Pyruvate is diverted from the citric acid cycle, serving instead as an electron acceptor to

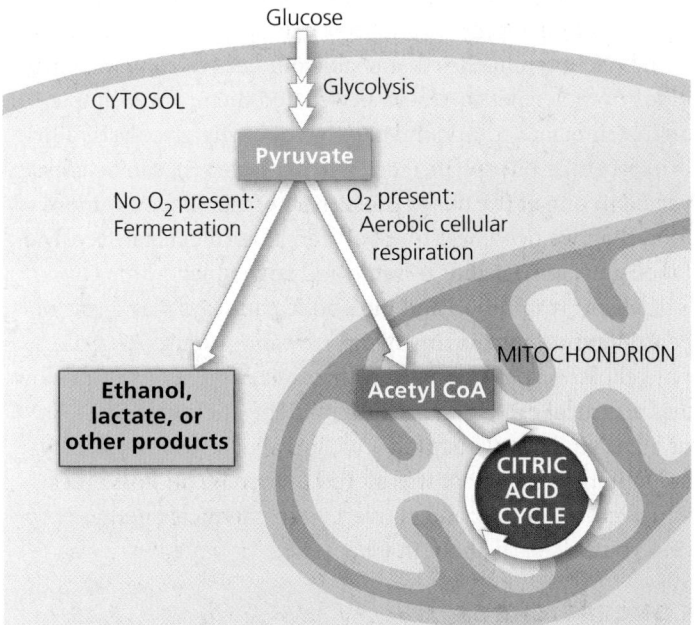

▲ **Figure 7.17 Pyruvate as a key juncture in catabolism.** Glycolysis is common to fermentation and cellular respiration. The end product of glycolysis, pyruvate, represents a fork in the catabolic pathways of glucose oxidation. In a facultative anaerobe or a muscle cell, which are capable of both aerobic cellular respiration and fermentation, pyruvate is committed to one of those two pathways, usually depending on whether or not oxygen is present.

recycle NAD$^+$. To make the same amount of ATP, a facultative anaerobe has to consume sugar at a much faster rate when fermenting than when respiring.

## The Evolutionary Significance of Glycolysis

**EVOLUTION** The role of glycolysis in both fermentation and respiration has an evolutionary basis. Ancient prokaryotes are thought to have used glycolysis to make ATP long before oxygen was present in Earth's atmosphere. The oldest known fossils of bacteria date back 3.5 billion years, but appreciable quantities of oxygen probably did not begin to accumulate in the atmosphere until about 2.7 billion years ago, produced by photosynthesizing cyanobacteria. Therefore, early prokaryotes may have generated ATP exclusively from glycolysis. The fact that glycolysis is today the most widespread metabolic pathway among Earth's organisms suggests that it evolved very early in the history of life. The cytosolic location of glycolysis also implies great antiquity; the pathway does not require any of the membrane-enclosed organelles of the eukaryotic cell, which evolved approximately 1 billion years after the prokaryotic cell. Glycolysis is a metabolic heirloom from early cells that continues to function in fermentation and as the first stage in the breakdown of organic molecules by respiration.

**CONCEPT CHECK 7.5**

1. Consider the NADH formed during glycolysis. What is the final acceptor for its electrons during fermentation? What is the final acceptor for its electrons during aerobic respiration?

2. **WHAT IF?** A glucose-fed yeast cell is moved from an aerobic environment to an anaerobic one. How would its rate of glucose consumption change if ATP were to be generated at the same rate?

For suggested answers, see Appendix A.

**CONCEPT 7.6**

# Glycolysis and the citric acid cycle connect to many other metabolic pathways

So far, we have treated the oxidative breakdown of glucose in isolation from the cell's overall metabolic economy. In this section, you will learn that glycolysis and the citric acid cycle are major intersections of the cell's catabolic (breakdown) and anabolic (biosynthetic) pathways.

## The Versatility of Catabolism

Throughout this chapter, we have used glucose as an example of a fuel for cellular respiration. But free glucose molecules are not common in the diets of humans and other animals. We obtain most of our calories in the form of fats, proteins, and carbohydrates such as sucrose and other disaccharides, and

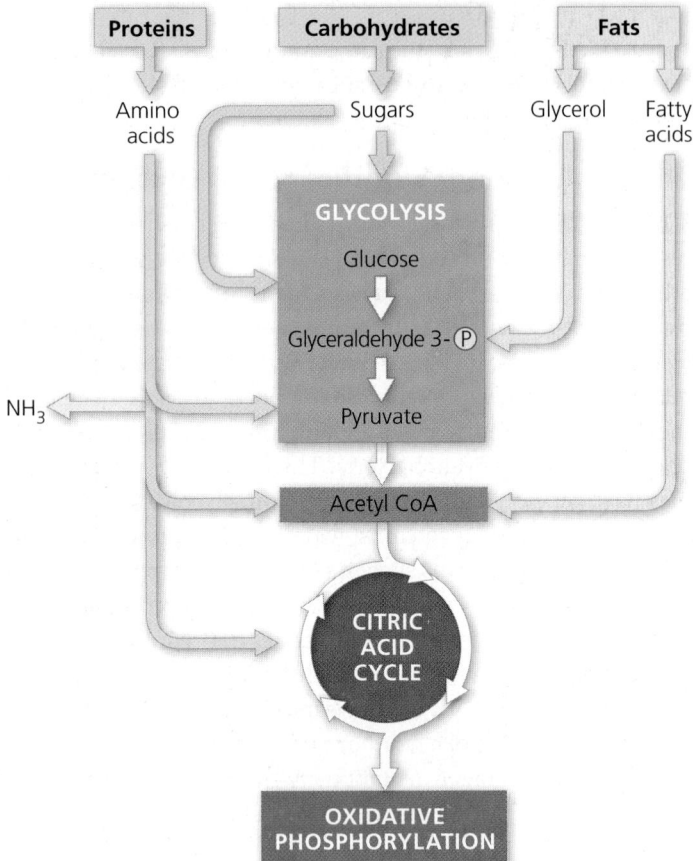

▲ **Figure 7.18 The catabolism of various molecules from food.** Carbohydrates, fats, and proteins can all be used as fuel for cellular respiration. Monomers of these molecules enter glycolysis or the citric acid cycle at various points. Glycolysis and the citric acid cycle are catabolic funnels through which electrons from all kinds of organic molecules flow on their exergonic fall to oxygen.

starch, a polysaccharide. All these organic molecules in food can be used by cellular respiration to make ATP (**Figure 7.18**).

Glycolysis can accept a wide range of carbohydrates for catabolism. In the digestive tract, starch is hydrolyzed to glucose, which is broken down in cells by glycolysis and the citric acid cycle. Similarly, glycogen, the polysaccharide that humans and many other animals store in their liver and muscle cells, can be hydrolyzed to glucose between meals as fuel for respiration. The digestion of disaccharides, including sucrose, provides glucose and other monosaccharides as fuel for respiration.

Proteins can also be used for fuel, but first they must be digested to their constituent amino acids. Many of the amino acids are used by the organism to build new proteins. Amino acids present in excess are converted by enzymes to intermediates of glycolysis and the citric acid cycle. Before amino acids can feed into glycolysis or the citric acid cycle, their amino groups must be removed, a process called *deamination*. The nitrogenous waste is excreted from the animal in the form of ammonia ($NH_3$), urea, or other waste products.

Catabolism can also harvest energy stored in fats obtained either from food or from fat cells. After fats are digested to glycerol and fatty acids, the glycerol is converted to glyceraldehyde 3-phosphate, an intermediate of glycolysis. Most

of the energy of a fat is stored in the fatty acids. A metabolic sequence called **beta oxidation** breaks the fatty acids down to two-carbon fragments, which enter the citric acid cycle as acetyl CoA. NADH and $FADH_2$ are also generated during beta oxidation, resulting in further ATP production. Fats make excellent fuels, largely due to their chemical structure and the high energy level of their electrons compared with those of carbohydrates. A gram of fat oxidized by respiration produces more than twice as much ATP as a gram of carbohydrate.

## Biosynthesis (Anabolic Pathways)

Cells need substance as well as energy. Not all the organic molecules of food are destined to be oxidized as fuel to make ATP. In addition to calories, food must also provide the carbon skeletons that cells require to make their own molecules. Some organic monomers obtained from digestion can be used directly. For example, as previously mentioned, amino acids from the hydrolysis of proteins in food can be incorporated into the organism's own proteins. Often, however, the body needs specific molecules that are not present as such in food. Compounds formed as intermediates of glycolysis and the citric acid cycle can be diverted into anabolic pathways as precursors from which the cell can synthesize the molecules it requires. For example, humans can make about half of the 20 amino acids in proteins by modifying compounds siphoned away from the citric acid cycle; the rest are "essential amino acids" that must be obtained in the diet. Also, glucose can be made from pyruvate, and fatty acids can be synthesized from acetyl CoA. Of course, these anabolic, or biosynthetic, pathways do not generate ATP, but instead consume it.

In addition, glycolysis and the citric acid cycle function as metabolic interchanges that enable our cells to convert some kinds of molecules to others as we need them. For example, an intermediate compound generated during glycolysis, dihydroxyacetone phosphate (see Figure 7.9, step 5), can be converted to one of the major precursors of fats. If we eat more food than we need, we store fat even if our diet is fat-free. Metabolism is remarkably versatile and adaptable.

Cellular respiration and metabolic pathways play a role of central importance in organisms. Examine Figure 7.2 again to put cellular respiration into the broader context of energy flow and chemical cycling in ecosystems. The energy that keeps us alive is *released*, not *produced*, by cellular respiration. We are tapping energy that was stored in food by photosynthesis, which captures light and converts it to chemical energy, a process you will learn about in Chapter 8.

### CONCEPT CHECK 7.6

1. **MAKE CONNECTIONS** Compare the structure of a fat (see Figure 3.13) with that of a carbohydrate (see Figure 3.8). What features of their structures make fat a much better fuel?
2. When might your body synthesize fat molecules?
3. **WHAT IF?** During intense exercise, can a muscle cell use fat as a concentrated source of chemical energy? Explain.

For suggested answers, see Appendix A.

7 Chapter Review

Go to **MasteringBiology**® for Assignments, the eText, and the Study Area with Animations, Activities, Vocab Self-Quiz, and Practice Tests.

## SUMMARY OF KEY CONCEPTS

VOCAB SELF-QUIZ
goo.gl/gbai8v

### CONCEPT 7.1

**Catabolic pathways yield energy by oxidizing organic fuels (pp. 142–146)**

- Cells break down glucose and other organic fuels to yield chemical energy in the form of ATP. **Fermentation** is a process that results in the partial degradation of glucose without the use of oxygen. **Cellular respiration** is a more complete breakdown of glucose; in **aerobic respiration**, oxygen is used as a reactant. The cell taps the energy stored in food molecules through **redox reactions**, in which one substance partially or totally shifts electrons to another. **Oxidation** is the loss of electrons from one substance, while **reduction** is the addition of electrons to the other.
- During aerobic respiration, glucose ($C_6H_{12}O_6$) is oxidized to $CO_2$, and $O_2$ is reduced to $H_2O$. Electrons lose potential energy during their transfer from glucose or other organic compounds to oxygen. Electrons are usually passed first to $NAD^+$, reducing it to NADH, and then from NADH to an **electron transport chain**, which conducts them to $O_2$ in energy-releasing steps. The energy is used to make ATP.
- Aerobic respiration occurs in three stages: (1) **glycolysis**, (2) pyruvate oxidation and the **citric acid cycle**, and (3) **oxidative phosphorylation** (electron transport and chemiosmosis).

? *Describe the difference between the two processes in cellular respiration that produce ATP: oxidative phosphorylation and substrate-level phosphorylation.*

### CONCEPT 7.2

**Glycolysis harvests chemical energy by oxidizing glucose to pyruvate (pp. 146–147)**

- Glycolysis ("splitting of sugar") is a series of reactions that breaks down glucose into two pyruvate molecules, which may go on to enter the citric acid cycle, and nets 2 ATP and 2 NADH per glucose molecule.

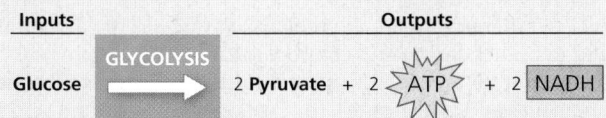

| Inputs | | Outputs |
|---|---|---|

Glucose → GLYCOLYSIS → 2 **Pyruvate** + 2 ATP + 2 NADH

? *Which reactions are the source of energy for the formation of ATP and NADH in glycolysis?*

### CONCEPT 7.3

**After pyruvate is oxidized, the citric acid cycle completes the energy-yielding oxidation of organic molecules (pp. 148–149)**

- In eukaryotic cells, pyruvate enters the mitochondrion and is oxidized to **acetyl CoA**, which is further oxidized in the citric acid cycle.

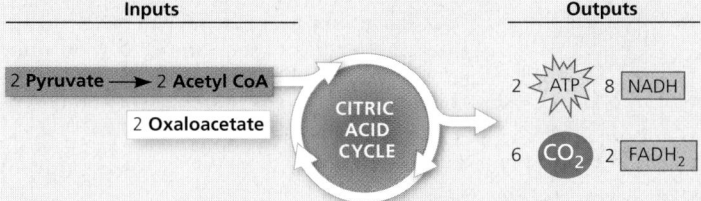

| Inputs | | | Outputs |
|---|---|---|---|

2 **Pyruvate** → 2 **Acetyl CoA**
2 **Oxaloacetate** → CITRIC ACID CYCLE → 2 ATP  8 NADH   6 $CO_2$  2 FADH$_2$

? *What molecular products indicate the complete oxidation of glucose during cellular respiration?*

### CONCEPT 7.4

**During oxidative phosphorylation, chemiosmosis couples electron transport to ATP synthesis (pp. 149–154)**

- NADH and FADH$_2$ transfer electrons to the electron transport chain. Electrons move down the chain, losing energy in several energy-releasing steps. Finally, electrons are passed to $O_2$, reducing it to $H_2O$.

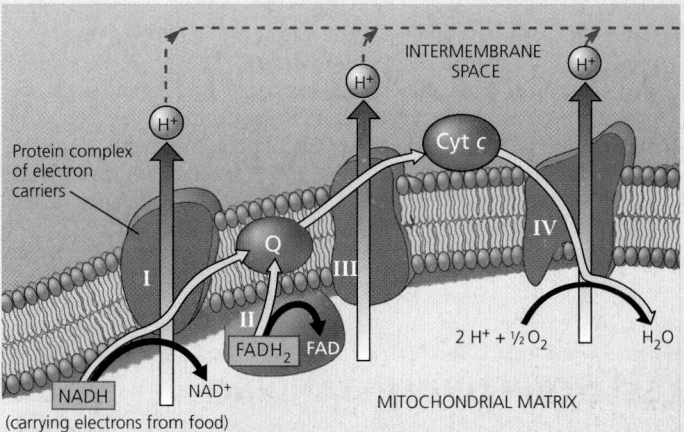

- Along the electron transport chain, electron transfer causes protein complexes to move $H^+$ from the mitochondrial matrix (in eukaryotes) to the intermembrane space, storing energy as a **proton-motive force** ($H^+$ gradient). As $H^+$ diffuses back into the matrix through **ATP synthase**, its passage drives the phosphorylation of ADP, an energy-coupling mechanism called **chemiosmosis**.
- About 34% of the energy stored in a glucose molecule is transferred to ATP during cellular respiration, producing a maximum of about 32 ATP.

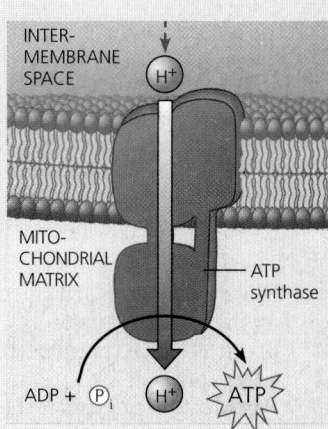

? *Briefly explain the mechanism by which ATP synthase produces ATP. List three locations in which ATP synthases are found.*

### CONCEPT 7.5

**Fermentation and anaerobic respiration enable cells to produce ATP without the use of oxygen (pp. 154–157)**

- Glycolysis nets 2 ATP by substrate-level phosphorylation, whether oxygen is present or not. Under anaerobic conditions, either anaerobic respiration or fermentation can take place. In anaerobic respiration, an electron transport chain is present with a final electron acceptor other than oxygen. In fermentation, the electrons from NADH are passed to pyruvate or a derivative of pyruvate, regenerating the $NAD^+$ required to oxidize more glucose. Two common types of fermentation are **alcohol fermentation** and **lactic acid fermentation**.

- Fermentation, anaerobic respiration, and aerobic respiration all use glycolysis to oxidize glucose, but they differ in their final electron acceptor and whether an electron transport chain is used (respiration) or not (fermentation). Respiration yields more ATP; aerobic respiration, with $O_2$ as the final electron acceptor, yields about 16 times as much ATP as does fermentation.
- Glycolysis occurs in nearly all organisms and is thought to have evolved in ancient prokaryotes before there was $O_2$ in the atmosphere.

**?** *Which process yields more ATP, fermentation or anaerobic respiration? Explain.*

## CONCEPT 7.6

### Glycolysis and the citric acid cycle connect to many other metabolic pathways (pp. 157–158)

- Catabolic pathways funnel electrons from many kinds of organic molecules into cellular respiration. Many carbohydrates can enter glycolysis, most often after conversion to glucose. Amino acids of proteins must be deaminated before being oxidized. The fatty acids of fats undergo **beta oxidation** to two-carbon fragments and then enter the citric acid cycle as acetyl CoA. Anabolic pathways can use small molecules from food directly or build other substances using intermediates of glycolysis or the citric acid cycle.

**?** *Describe how the catabolic pathways of glycolysis and the citric acid cycle intersect with anabolic pathways in the metabolism of a cell.*

## TEST YOUR UNDERSTANDING

### Level 1: Knowledge/Comprehension

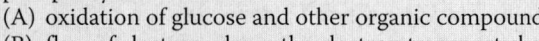

PRACTICE TEST

goo.gl/CRZjvS

1. The *immediate* energy source that drives ATP synthesis by ATP synthase during oxidative phosphorylation is the
   (A) oxidation of glucose and other organic compounds.
   (B) flow of electrons down the electron transport chain.
   (C) $H^+$ concentration gradient across the membrane holding ATP synthase.
   (D) transfer of phosphate to ADP.

2. Which metabolic pathway is common to both fermentation and cellular respiration of a glucose molecule?
   (A) the citric acid cycle
   (B) the electron transport chain
   (C) glycolysis
   (D) reduction of pyruvate to lactate

3. In mitochondria, exergonic redox reactions
   (A) are the source of energy driving prokaryotic ATP synthesis.
   (B) provide the energy that establishes the proton gradient.
   (C) reduce carbon atoms to carbon dioxide.
   (D) are coupled via phosphorylated intermediates to endergonic processes.

4. The final electron acceptor of the electron transport chain that functions in aerobic oxidative phosphorylation is
   (A) oxygen.                    (C) $NAD^+$.
   (B) water.                     (D) pyruvate.

### Level 2: Application/Analysis

5. What is the oxidizing agent in the following reaction?

   Pyruvate + NADH + $H^+$ → Lactate + $NAD^+$

   (A) oxygen                     (C) lactate
   (B) NADH                       (D) pyruvate

6. When electrons flow along the electron transport chains of mitochondria, which of the following changes occurs?
   (A) The pH of the matrix increases.
   (B) ATP synthase pumps protons by active transport.
   (C) The electrons gain free energy.
   (D) $NAD^+$ is oxidized.

7. Most $CO_2$ from catabolism is released during
   (A) glycolysis.               (C) lactate fermentation.
   (B) the citric acid cycle.    (D) electron transport.

### Level 3: Synthesis/Evaluation

8. **DRAW IT** The graph here shows the pH difference across the inner mitochondrial membrane over time in an actively respiring cell. At the time indicated by the vertical arrow, a metabolic poison is added that specifically and completely inhibits all function of mitochondrial ATP synthase. Draw what you would expect to see for the rest of the graphed line, and explain your graph.

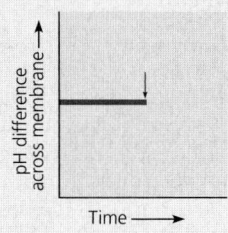

9. **INTERPRET THE DATA** Phosphofructokinase is an enzyme that acts on fructose 6-phosphate at an early step in glucose breakdown (step 3 in Figure 7.9). Negative regulation of this enzyme by ATP and positive regulation by AMP control whether the sugar will continue on in the glycolytic pathway. Considering this graph, under which condition is phosphofructokinase more active? Given this enzyme's role in glycolysis, explain why it makes sense that ATP and AMP have these effects.

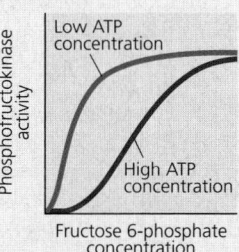

10. **SCIENTIFIC INQUIRY**
    In the 1930s, some physicians prescribed low doses of a compound called dinitrophenol (DNP) to help patients lose weight. This unsafe method was abandoned after some patients died. DNP uncouples the chemiosmotic machinery by making the lipid bilayer of the inner mitochondrial membrane leaky to $H^+$. Explain how this could cause weight loss and death.

11. **FOCUS ON EVOLUTION**
    ATP synthases are found in the prokaryotic plasma membrane and in mitochondria and chloroplasts. (a) Propose a hypothesis to account for an evolutionary relationship of these eukaryotic organelles and prokaryotes. (b) Explain how the amino acid sequences of the ATP synthases from the different sources might either support or fail to support your hypothesis.

12. **FOCUS ON ORGANIZATION**
    In a short essay (100–150 words), explain how oxidative phosphorylation—the production of ATP using energy derived from the redox reactions of a spatially organized electron transport chain followed by chemiosmosis—is an example of how new properties emerge at each level of the biological hierarchy.

13. **SYNTHESIZE YOUR KNOWLEDGE**

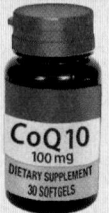

Coenzyme Q (CoQ) is sold as a nutritional supplement. One company uses this marketing slogan for CoQ: "Give your heart the fuel it craves most." Considering the role of coenzyme Q, how do you think this product might function as a nutritional supplement to benefit the heart? Is CoQ used as a "fuel" during cellular respiration?

*For selected answers, see Appendix A.*

# 8 Photosynthesis

▲ **Figure 8.1** How does sunlight help build the trunk, branches, and leaves of this broadleaf tree?

## The Process That Feeds the Biosphere

Life on Earth is solar powered. The chloroplasts in plants and other photosynthetic organisms capture light energy that has traveled 150 million km from the sun and convert it to chemical energy that is stored in sugar and other organic molecules. This conversion process is called **photosynthesis**. Let's begin by placing photosynthesis in its ecological context.

Photosynthesis nourishes almost the entire living world directly or indirectly. An organism acquires the organic compounds it uses for energy and carbon skeletons by one of two major modes: autotrophic nutrition or heterotrophic nutrition. **Autotrophs** are "self-feeders" (*auto-* means "self," and *trophos* means "feeder"); they sustain themselves without eating anything derived from other living beings. Autotrophs produce their organic molecules from $CO_2$ and other inorganic raw materials obtained from the environment. They are the ultimate sources of organic compounds for all nonautotrophic

organisms, and for this reason, biologists refer to autotrophs as the *producers* of the biosphere.

Almost all plants are autotrophs; the only nutrients they require are water and minerals from the soil and carbon dioxide from the air. Specifically, plants are *photo*autotrophs, organisms that use light as a source of energy to synthesize organic substances **(Figure 8.1)**. Photosynthesis also occurs in algae, certain other unicellular eukaryotes, and some prokaryotes.

**Heterotrophs** are unable to make their own food; they live on compounds produced by other organisms (*hetero-* means "other"). Heterotrophs are the biosphere's *consumers*. This "other-feeding" is most obvious when an animal eats plants or other animals, but heterotrophic nutrition may be more subtle. Some heterotrophs decompose and feed on the remains of dead organisms and organic litter such as feces and fallen leaves; these types of organisms are known as decomposers. Most fungi and many types of prokaryotes get their nourishment this way. Almost all heterotrophs, including humans, are completely dependent, either directly or indirectly, on photoautotrophs for food—and also for oxygen, a by-product of photosynthesis.

In this chapter, you'll learn how photosynthesis works. A variety of photosynthetic organisms are shown in **Figure 8.2**, including both eukaryotes and prokaryotes. Our discussion here will focus mainly on plants. (Variations in autotrophic nutrition that occur in prokaryotes and algae will be described in Concepts 24.2 and 25.4.) After discussing the general principles of photosynthesis, we'll consider the two stages of photosynthesis: the light reactions, which capture solar energy and transform it into chemical energy; and the Calvin cycle, which uses that chemical energy to make the organic molecules of food. Finally, we'll consider some aspects of photosynthesis from an evolutionary perspective.

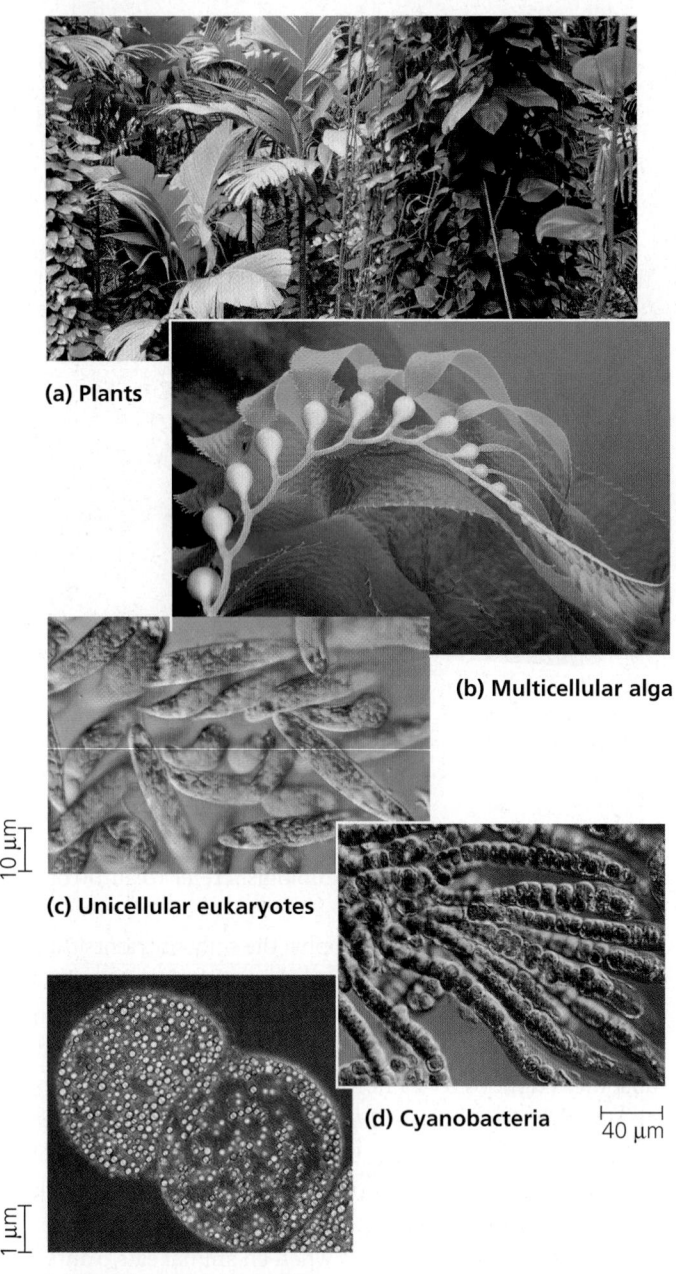

**(a) Plants**

**(b) Multicellular alga**

10 μm

**(c) Unicellular eukaryotes**

**(d) Cyanobacteria**    40 μm

1 μm

**(e) Purple sulfur bacteria**

▲ **Figure 8.2 Photoautotrophs.** These organisms use light energy to drive the synthesis of organic molecules from carbon dioxide and (in most cases) water. They feed themselves and the entire living world. **(a)** On land, plants are the predominant producers of food. In aquatic environments, photoautotrophs include unicellular and **(b)** multicellular algae, such as this kelp; **(c)** some non-algal unicellular eukaryotes, such as *Euglena*; **(d)** the prokaryotes called cyanobacteria; and **(e)** other photosynthetic prokaryotes, such as these purple sulfur bacteria, which produce sulfur (the yellow globules within the cells) (c–e, LMs).

# Photosynthesis converts light energy to the chemical energy of food

The remarkable ability of an organism to harness light energy and use it to drive the synthesis of organic compounds emerges from structural organization in the cell: Photosynthetic enzymes and other molecules are grouped together in a biological membrane, enabling the necessary series of chemical reactions to be carried out efficiently. The process of photosynthesis most likely originated in a group of bacteria that had infolded regions of the plasma membrane containing clusters of such molecules. In photosynthetic bacteria that exist today, infolded photosynthetic membranes function similarly to the internal membranes of the chloroplast, a eukaryotic organelle. According to the endosymbiont theory, the original chloroplast was a photosynthetic prokaryote that lived inside an ancestor of eukaryotic cells. (You learned about this theory in Concept 4.5, and it will be described more fully in Concept 25.1.) Chloroplasts are present in a variety of photosynthesizing organisms, but here we focus on chloroplasts in plants.

## Chloroplasts: The Sites of Photosynthesis in Plants

All green parts of a plant, including green stems and unripened fruit, have chloroplasts, but the leaves are the major sites of photosynthesis in most plants **(Figure 8.3)**. There are about half a million chloroplasts in a chunk of leaf with a top surface area of 1 mm². Chloroplasts are found mainly in the cells of the **mesophyll**, the tissue in the interior of the leaf. Carbon dioxide enters the leaf, and oxygen exits, by way of microscopic pores called **stomata** (singular, *stoma*; from the Greek, meaning "mouth"). Water absorbed by the roots is delivered to the leaves in veins. Leaves also use veins to export sugar to roots and other nonphotosynthetic parts of the plant.

A typical mesophyll cell has about 30–40 chloroplasts, each measuring about 2–4 μm by 4–7 μm. A chloroplast has an envelope of two membranes surrounding a dense fluid called the **stroma**. Suspended within the stroma is a third membrane system, made up of sacs called **thylakoids**, which segregates the stroma from the *thylakoid space* inside these sacs. In some places, thylakoid sacs are stacked in columns called *grana* (singular, *granum*). **Chlorophyll**, the green pigment that gives leaves their color, resides in the thylakoid membranes of the chloroplast. (The internal photosynthetic membranes of some prokaryotes are also called thylakoid membranes; see Figure 24.11b.) It is the light energy absorbed by chlorophyll that drives the synthesis of organic molecules in the chloroplast. Now that we have looked at the sites of photosynthesis in plants, we are ready to look more closely at the process of photosynthesis.

## Tracking Atoms Through Photosynthesis: *Scientific Inquiry*

Scientists have tried for centuries to piece together the process by which plants make food. Although some of the steps are still not completely understood, the overall photosynthetic equation has been known since the 1800s: In the presence of light, the green parts of plants produce organic compounds and oxygen from carbon dioxide and water. Using molecular formulas, we can summarize the complex series of chemical reactions in photosynthesis with this chemical equation:

$$6\ CO_2 + 12\ H_2O + \text{Light energy} \rightarrow C_6H_{12}O_6 + 6\ O_2 + 6\ H_2O$$

We use glucose ($C_6H_{12}O_6$) here to simplify the relationship between photosynthesis and respiration, but the direct product of photosynthesis is actually a three-carbon sugar that can be used to make glucose. Water appears on both sides of the equation because 12 molecules are consumed and 6 molecules are newly formed during photosynthesis. We can simplify the equation by indicating only the net consumption of water:

$$6\ CO_2 + 6\ H_2O + \text{Light energy} \rightarrow C_6H_{12}O_6 + 6\ O_2$$

Writing the equation in this form, we can see that the overall chemical change during photosynthesis is the reverse of the one that occurs during cellular respiration. Both of these metabolic processes occur in plant cells. However, as you will soon learn, chloroplasts do not synthesize sugars by simply reversing the steps of respiration.

Now let's divide the photosynthetic equation by 6 to put it in its simplest possible form:

$$CO_2 + H_2O \rightarrow [CH_2O] + O_2$$

Here, the brackets indicate that $CH_2O$ is not an actual sugar but represents the general formula for a carbohydrate. In other words, we are imagining the synthesis of a sugar molecule one carbon at a time. Let's now use this simplified formula to see how researchers tracked the elements C, H, and O from the reactants of photosynthesis to the products.

### The Splitting of Water

One of the first clues to the mechanism of photosynthesis came from the discovery that the $O_2$ given off by plants is

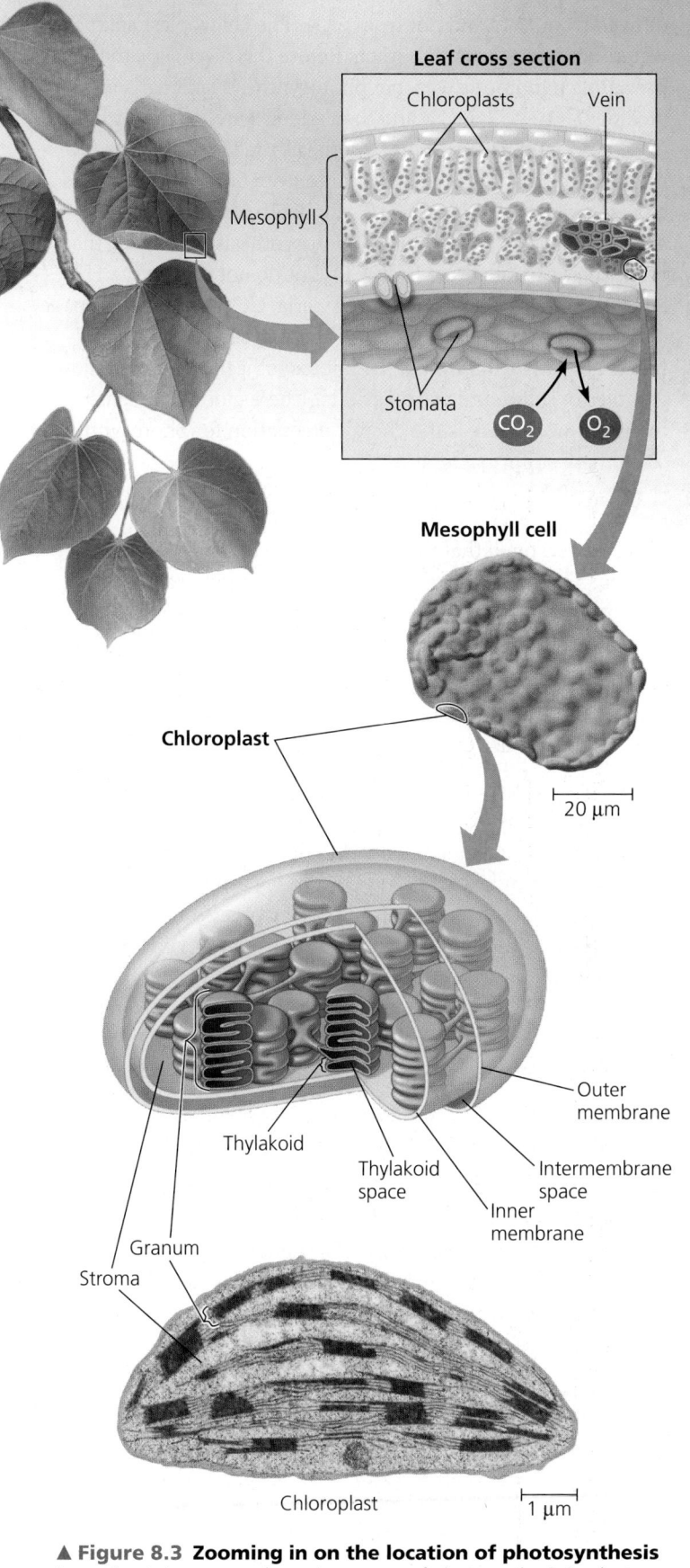

▲ **Figure 8.3 Zooming in on the location of photosynthesis in a plant.** Leaves are the major organs of photosynthesis in plants. These images take you into a leaf, then into a cell, and finally into a chloroplast, the organelle where photosynthesis occurs (middle, LM; bottom, TEM).

derived from $H_2O$ and not from $CO_2$. The chloroplast splits water into hydrogen and oxygen. Before this discovery, the prevailing hypothesis was that photosynthesis split carbon dioxide ($CO_2 \rightarrow C + O_2$) and then added water to the carbon ($C + H_2O \rightarrow [CH_2O]$). This hypothesis predicted that the $O_2$ released during photosynthesis came from $CO_2$. This idea was challenged in the 1930s by C. B. van Niel, of Stanford University. Van Niel was investigating photosynthesis in bacteria that make their carbohydrate from $CO_2$ but do not release $O_2$. He concluded that, at least in these bacteria, $CO_2$ is not split into carbon and oxygen. One group of bacteria used hydrogen sulfide ($H_2S$) rather than water for photosynthesis, forming yellow globules of sulfur as a waste product (these globules are visible in Figure 8.2e). Here is the chemical equation for photosynthesis in these sulfur bacteria:

$$CO_2 + 2\,H_2S \rightarrow [CH_2O] + H_2O + 2\,S$$

Van Niel reasoned that the bacteria split $H_2S$ and used the hydrogen atoms to make sugar. He then generalized that idea, proposing that all photosynthetic organisms require a hydrogen source but that the source varies:

Sulfur bacteria: $CO_2 + 2\,H_2S \rightarrow [CH_2O] + H_2O + 2\,S$
Plants: $CO_2 + 2\,H_2O \rightarrow [CH_2O] + H_2O + O_2$
General: $CO_2 + 2\,H_2X \rightarrow [CH_2O] + H_2O + 2\,X$

Thus, van Niel hypothesized that plants split $H_2O$ as a source of electrons from hydrogen atoms, releasing $O_2$ as a by-product.

Nearly 20 years later, scientists confirmed van Niel's hypothesis by using oxygen-18 ($^{18}O$), a heavy isotope, as a tracer to follow the fate of oxygen atoms during photosynthesis. The experiments showed that the $O_2$ from plants was labeled with $^{18}O$ *only* if water was the source of the tracer (experiment 1). If the $^{18}O$ was introduced to the plant in the form of $CO_2$, the label did not turn up in the released $O_2$ (experiment 2). In the following summary, red denotes labeled atoms of oxygen ($^{18}O$):

Experiment 1: $CO_2 + 2\,H_2\mathbf{O} \rightarrow [CH_2O] + H_2O + \mathbf{O_2}$
Experiment 2: $\mathbf{CO_2} + 2\,H_2O \rightarrow [CH_2\mathbf{O}] + H_2\mathbf{O} + O_2$

A significant result of the shuffling of atoms during photosynthesis is the extraction of hydrogen from water and its incorporation into sugar. The waste product of photosynthesis, $O_2$, is released to the atmosphere. **Figure 8.4** shows the fates of all atoms in photosynthesis.

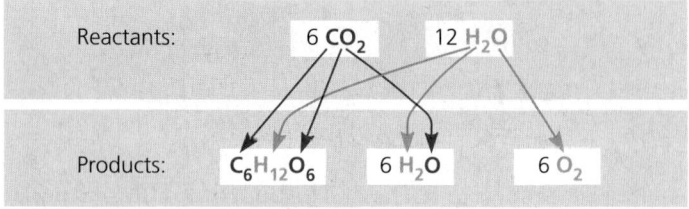

▲ **Figure 8.4 Tracking atoms through photosynthesis.** The atoms from $CO_2$ are shown in magenta, and the atoms from $H_2O$ are shown in blue.

### Photosynthesis as a Redox Process

Let's briefly compare photosynthesis with cellular respiration. Both processes involve redox reactions. During cellular respiration, energy is released from sugar when electrons associated with hydrogen are transported by carriers to oxygen, forming water as a by-product (see Concept 7.1). The electrons lose potential energy as they "fall" down the electron transport chain toward electronegative oxygen, and the mitochondrion harnesses that energy to synthesize ATP (see Figure 7.14). Photosynthesis reverses the direction of electron flow. Water is split, and electrons are transferred along with hydrogen ions from the water to carbon dioxide, reducing it to sugar.

$$\overbrace{\text{Energy} + 6\,CO_2 + 6\,H_2O \longrightarrow C_6H_{12}O_6}^{\text{becomes reduced}} + \underbrace{6\,O_2}_{\text{becomes oxidized}}$$

Because the electrons increase in potential energy as they move from water to sugar, this process requires energy—in other words, is endergonic. This energy boost is provided by light.

## The Two Stages of Photosynthesis: *A Preview*

The equation for photosynthesis is a deceptively simple summary of a very complex process. Actually, photosynthesis is not a single process, but two processes, each with multiple steps. These two stages of photosynthesis are known as the **light reactions** (the *photo* part of photosynthesis) and the **Calvin cycle** (the *synthesis* part) **(Figure 8.5)**.

The light reactions are the steps of photosynthesis that convert solar energy to chemical energy. Water is split, providing a source of electrons and protons (hydrogen ions, $H^+$) and giving off $O_2$ as a by-product. Light absorbed by chlorophyll drives a transfer of the electrons and hydrogen ions from water to an acceptor called **NADP$^+$** (nicotinamide adenine dinucleotide phosphate), where they are temporarily stored. The electron acceptor NADP$^+$ is first cousin to NAD$^+$, which functions as an electron carrier in cellular respiration; the two molecules differ only by the presence of an extra phosphate group in the NADP$^+$ molecule. The light reactions use solar energy to reduce NADP$^+$ to **NADPH** by adding a pair of electrons along with an $H^+$. The light reactions also generate ATP, using chemiosmosis to power the addition of a phosphate group to ADP, a process called **photophosphorylation**. Thus, light energy is initially converted to chemical energy in the form of two compounds: NADPH and ATP. NADPH, a source of electrons, acts as "reducing power" that can be passed along to an electron acceptor, reducing it, while ATP is the versatile energy currency of cells. Notice that the light reactions produce no sugar; that happens in the second stage of photosynthesis, the Calvin cycle.

The Calvin cycle is named for Melvin Calvin, who, along with his colleagues, began to elucidate its steps in the late

▶ **Figure 8.5 An overview of photosynthesis: cooperation of the light reactions and the Calvin cycle.** In the chloroplast, the thylakoid membranes (green) are the sites of the light reactions, whereas the Calvin cycle occurs in the stroma (gray). The light reactions use solar energy to make ATP and NADPH, which supply chemical energy and reducing power, respectively, to the Calvin cycle. The Calvin cycle incorporates $CO_2$ into organic molecules, which are converted to sugar. (Recall that most simple sugars have formulas that are some multiple of $CH_2O$.)

 **ANIMATION** Visit the Study Area in **MasteringBiology** for the BioFlix® 3-D Animation on Photosynthesis.

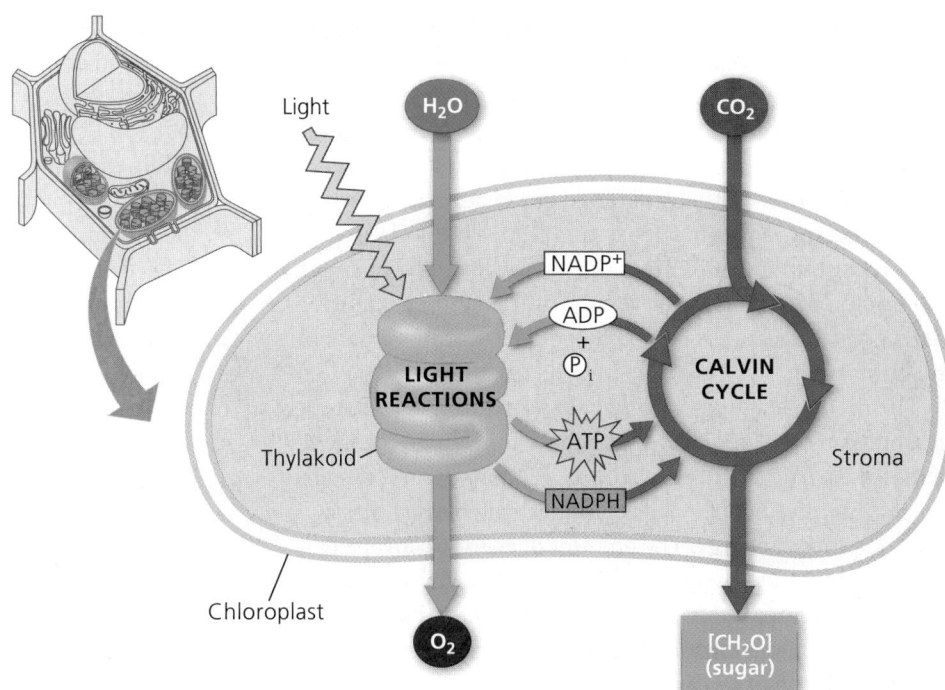

1940s. The cycle begins by incorporating $CO_2$ from the air into organic molecules already present in the chloroplast. This initial incorporation of carbon into organic compounds is known as **carbon fixation**. The Calvin cycle then reduces the fixed carbon to carbohydrate by the addition of electrons. The reducing power is provided by NADPH, which acquired its cargo of electrons in the light reactions. To convert $CO_2$ to carbohydrate, the Calvin cycle also requires chemical energy in the form of ATP, which is also generated by the light reactions. Thus, it is the Calvin cycle that makes sugar, but it can do so only with the help of the NADPH and ATP produced by the light reactions. The metabolic steps of the Calvin cycle are sometimes referred to as the dark reactions, or light-independent reactions, because none of the steps requires light *directly*. Nevertheless, the Calvin cycle in most plants occurs during daylight, for only then can the light reactions provide the NADPH and ATP that the Calvin cycle requires. In essence, the chloroplast uses light energy to make sugar by coordinating the two stages of photosynthesis.

As Figure 8.5 indicates, the thylakoids of the chloroplast are the sites of the light reactions, while the Calvin cycle occurs in the stroma. On the outside of the thylakoids, molecules of $NADP^+$ and ADP pick up electrons and phosphate, respectively, and NADPH and ATP are then released to the stroma, where they play crucial roles in the Calvin cycle. The two stages of photosynthesis are treated in this figure as metabolic modules that take in ingredients and crank out products. In the next two sections, we'll look more closely at how the two stages work, beginning with the light reactions.

**CONCEPT CHECK 8.1**

1. How do the reactant molecules of photosynthesis reach the chloroplasts in leaves?
2. How did the use of an oxygen isotope help elucidate the chemistry of photosynthesis?
3. **WHAT IF?** The Calvin cycle requires ATP and NADPH, products of the light reactions. If a classmate asserted that the light reactions don't depend on the Calvin cycle and, with continual light, could just keep on producing ATP and NADPH, how would you respond?

**For suggested answers, see Appendix A.**

**CONCEPT 8.2**

# The light reactions convert solar energy to the chemical energy of ATP and NADPH

Chloroplasts are chemical factories powered by the sun. Their thylakoids transform light energy into the chemical energy of ATP and NADPH. To understand this conversion better, we need to know about some important properties of light.

## The Nature of Sunlight

Light is a form of energy known as electromagnetic energy, also called electromagnetic radiation. Electromagnetic energy travels in rhythmic waves analogous to those created by dropping a pebble into a pond. Electromagnetic waves, however, are disturbances of electric and magnetic fields rather than disturbances of a material medium such as water.

The distance between the crests of electromagnetic waves is called the **wavelength**. Wavelengths range from less than a nanometer (for gamma rays) to more than a kilometer (for radio waves). This entire range of radiation is known as the **electromagnetic spectrum (Figure 8.6)**. The segment most important to life is the narrow band from about 380 nm to 750 nm in wavelength. This radiation is known as **visible light** because it can be detected as various colors by the human eye.

The model of light as waves explains many of light's properties, but in certain respects light behaves as though it consists of discrete particles, called **photons**. Photons are not tangible objects, but they act like objects in that each of them has a fixed quantity of energy. The amount of energy is inversely related to the wavelength of the light: The shorter the wavelength, the greater the energy of each photon of that light. Thus, a photon of violet light packs nearly twice as much energy as a photon of red light (see Figure 8.6).

Although the sun radiates the full spectrum of electromagnetic energy, the atmosphere acts like a selective window, allowing visible light to pass through while screening out a substantial fraction of other radiation. The part of the spectrum we can see—visible light—is also the radiation that drives photosynthesis.

## Photosynthetic Pigments: The Light Receptors

When light meets matter, it may be reflected, transmitted, or absorbed. Substances that absorb visible light are known as *pigments*. Different pigments absorb light of different wavelengths, and the wavelengths that are absorbed disappear. If a pigment is illuminated with white light, the color we see is the color most reflected or transmitted by the pigment. (If a pigment absorbs all wavelengths, it appears black.) We see green when we look at a leaf because chlorophyll absorbs violet-blue and red light while transmitting and reflecting green light **(Figure 8.7)**. The ability of a pigment to absorb various wavelengths of light can be measured with an instrument called a **spectrophotometer**. This machine directs beams of light of different wavelengths through a solution of the pigment and measures the fraction of the light transmitted at each wavelength. A graph plotting a pigment's light absorption versus wavelength is called an **absorption spectrum (Figure 8.8)**.

The absorption spectra of chloroplast pigments provide clues to the relative effectiveness of different wavelengths for driving photosynthesis, since light can perform work in chloroplasts only if it is absorbed. **Figure 8.9a** shows the absorption spectra of three types of pigments in chloroplasts: **chlorophyll *a***, the key light-capturing pigment that participates directly in the light reactions; the accessory pigment **chlorophyll *b***; and a separate group of accessory pigments called carotenoids. The spectrum of chlorophyll *a* suggests that violet-blue and red light work best for photosynthesis, since they are absorbed, while green is the least effective color. This is confirmed by an **action spectrum** for photosynthesis **(Figure 8.9b)**, which profiles the relative effectiveness of different wavelengths of radiation in driving the process. An action spectrum is prepared by illuminating chloroplasts with light of different colors and then plotting wavelength against some measure of photosynthetic rate, such as $CO_2$ consumption or

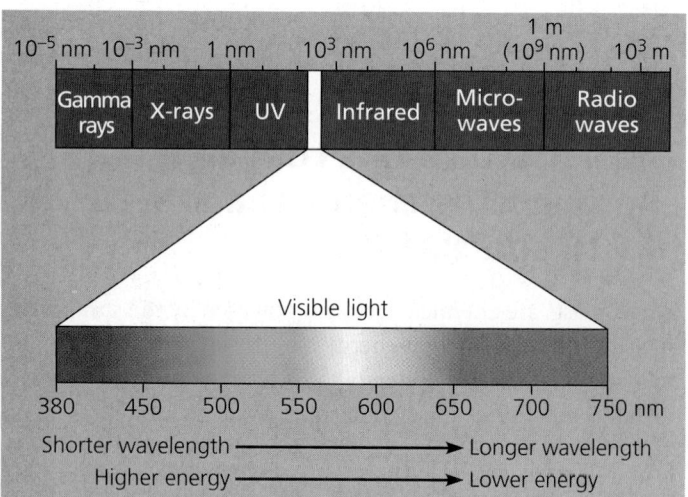

▲ **Figure 8.6 The electromagnetic spectrum.** White light is a mixture of all wavelengths of visible light. A prism can sort white light into its component colors by bending light of different wavelengths at different angles. (Droplets of water in the atmosphere can act as prisms, forming a rainbow.) Visible light drives photosynthesis.

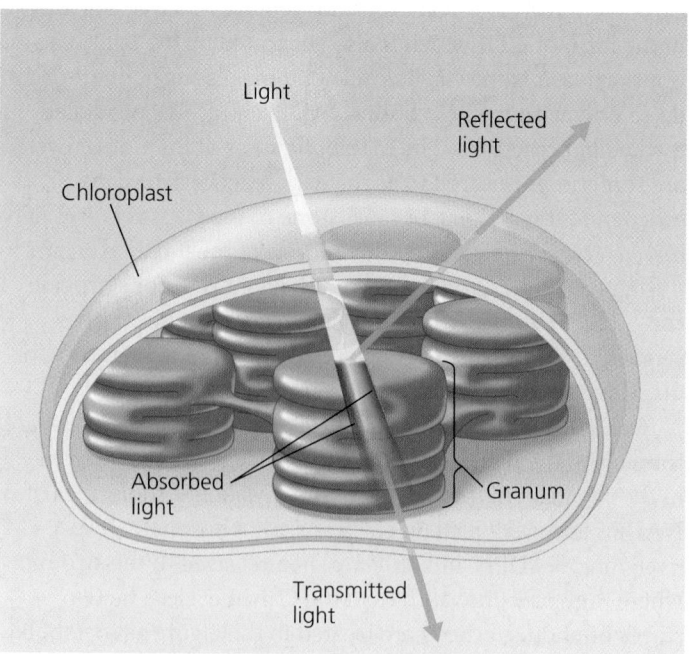

▲ **Figure 8.7 Why leaves are green: interaction of light with chloroplasts.** The chlorophyll molecules of chloroplasts absorb violet-blue and red light (the colors most effective in driving photosynthesis) and reflect or transmit green light. This is why leaves appear green.

## ▼ Figure 8.8 Research Method

### Determining an Absorption Spectrum

**Application** An absorption spectrum is a visual representation of how well a particular pigment absorbs different wavelengths of visible light. Absorption spectra of various chloroplast pigments help scientists decipher the role of each pigment in a plant.

**Technique** A spectrophotometer measures the relative amounts of light of different wavelengths absorbed and transmitted by a pigment solution.

**1** White light is separated into colors (wavelengths) by a prism.

**2** One by one, the different colors of light are passed through the sample (chlorophyll in this example). Green light and blue light are shown here.

**3** The transmitted light strikes a photoelectric tube, which converts the light energy to electricity.

**4** The electric current is measured by a galvanometer. The meter indicates the fraction of light transmitted through the sample, from which we can determine the amount of light absorbed.

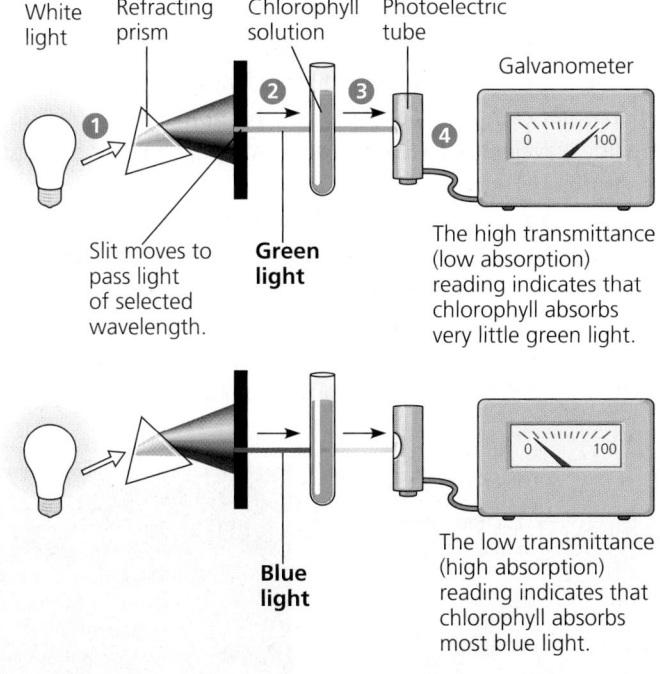

**Results** See Figure 8.9a for absorption spectra of three types of chloroplast pigments.

O₂ release. The action spectrum for photosynthesis was first demonstrated by Theodor W. Engelmann, a German botanist, in 1883. Before equipment for measuring O₂ levels had even been invented, Engelmann performed a clever experiment in which he used bacteria to measure rates of photosynthesis in filamentous algae **(Figure 8.9c)**. His results are a striking match to the modern action spectrum shown in Figure 8.9b.

Notice by comparing Figure 8.9a and 8.9b that the action spectrum for photosynthesis is much broader than the absorption spectrum of chlorophyll *a*. The absorption spectrum of chlorophyll *a* alone underestimates the effectiveness of

## ▼ Figure 8.9 Inquiry

### Which wavelengths of light are most effective in driving photosynthesis?

**Experiment** Absorption and action spectra, along with a classic experiment by Theodor W. Engelmann, reveal which wavelengths of light are photosynthetically important.

**Results**

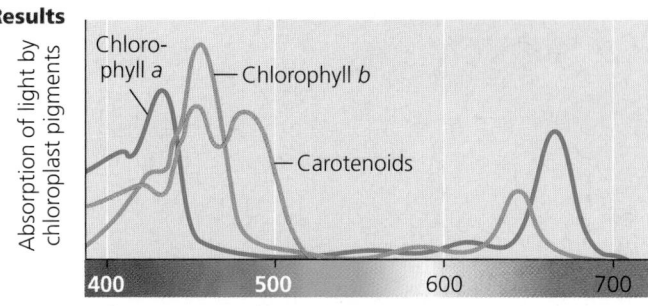

**(a) Absorption spectra.** The three curves show the wavelengths of light best absorbed by three types of chloroplast pigments.

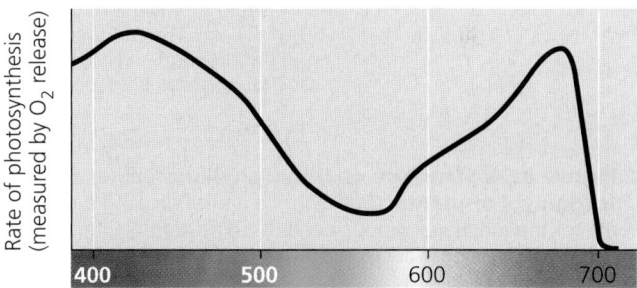

**(b) Action spectrum.** This graph plots the rate of photosynthesis versus wavelength. The resulting action spectrum resembles the absorption spectrum for chlorophyll *a* but does not match exactly (see part a). This is partly due to the absorption of light by accessory pigments such as chlorophyll *b* and carotenoids.

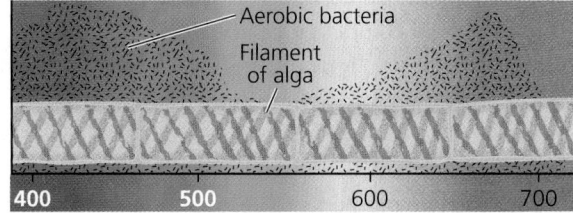

**(c) Engelmann's experiment.** In 1883, Theodor W. Engelmann illuminated a filamentous alga with light that had been passed through a prism, exposing different segments of the alga to different wavelengths. He used aerobic bacteria, which concentrate near an oxygen source, to determine which segments of the alga were releasing the most O₂ and thus photosynthesizing most. Bacteria congregated in greatest numbers around the parts of the alga illuminated with violet-blue or red light.

**Conclusion** Light in the violet-blue and red portions of the spectrum is most effective in driving photosynthesis.

**Data from** T. W. Engelmann, *Bacterium photometricum*. Ein Beitrag zur vergleichenden Physiologie des Licht-und Farbensinnes, *Archiv. für Physiologie* 30:95–124 (1883).

(MB) A related Experimental Inquiry Tutorial can be assigned in MasteringBiology.

**INTERPRET THE DATA** What wavelengths of light drive the highest rate of photosynthesis?

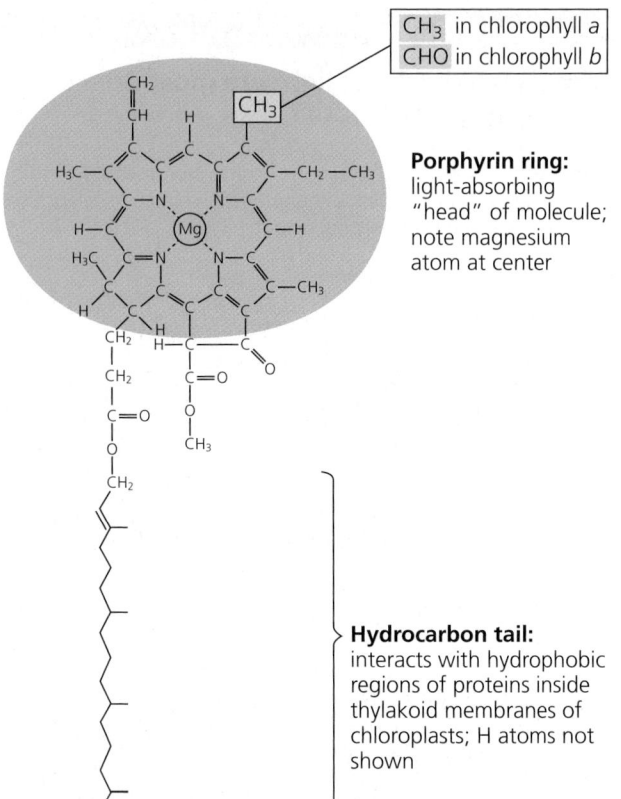

CH₃ in chlorophyll *a*
CHO in chlorophyll *b*

**Porphyrin ring:** light-absorbing "head" of molecule; note magnesium atom at center

**Hydrocarbon tail:** interacts with hydrophobic regions of proteins inside thylakoid membranes of chloroplasts; H atoms not shown

▲ **Figure 8.10 Structure of chlorophyll molecules in chloroplasts of plants.** Chlorophyll *a* and chlorophyll *b* differ only in one of the functional groups bonded to the porphyrin ring. (Also see the space-filling model of chlorophyll in Figure 1.3.)

certain wavelengths in driving photosynthesis. This is partly because accessory pigments with different absorption spectra also present in chloroplasts—including chlorophyll *b* and carotenoids—broaden the spectrum of colors that can be used for photosynthesis. **Figure 8.10** shows the structure of chlorophyll *a* compared with that of chlorophyll *b*. A slight structural difference between them is enough to cause the two pigments to absorb at slightly different wavelengths in the red and blue parts of the spectrum (see Figure 8.9a). As a result, chlorophyll *a* appears blue green and chlorophyll *b* olive green under visible light.

Other accessory pigments include **carotenoids**, hydrocarbons that are various shades of yellow and orange because they absorb violet and blue-green light (see Figure 8.9a). Carotenoids may broaden the spectrum of colors that can drive photosynthesis. However, a more important function of at least some carotenoids seems to be *photoprotection*: These compounds absorb and dissipate

excessive light energy that would otherwise damage chlorophyll or interact with oxygen, forming reactive oxidative molecules that are dangerous to the cell. Interestingly, carotenoids similar to the photoprotective ones in chloroplasts have a photoprotective role in the human eye. (Carrots, known for aiding night vision, are rich in carotenoids.)

## Excitation of Chlorophyll by Light

What exactly happens when chlorophyll and other pigments absorb light? The colors corresponding to the absorbed wavelengths disappear from the spectrum of the transmitted and reflected light, but energy cannot disappear. When a molecule absorbs a photon of light, one of the molecule's electrons is elevated to an electron shell where it has more potential energy (see Figure 2.5). When the electron is in its normal shell, the pigment molecule is said to be in its ground state. Absorption of a photon boosts an electron to a higher-energy electron shell, and the pigment molecule is then said to be in an excited state **(Figure 8.11a)**. The only photons absorbed are those whose energy is exactly equal to the energy difference between the ground state and an excited state, and this energy difference varies from one kind of molecule to another. Thus, a particular compound absorbs only photons corresponding to specific wavelengths, which is why each pigment has a unique absorption spectrum.

Once absorption of a photon raises an electron from the ground state to an excited state, the electron cannot stay there long. The excited state, like all high-energy states, is unstable. Generally, when isolated pigment molecules absorb light, their excited electrons drop back down to the ground-state electron shell in a billionth of a second, releasing their excess energy

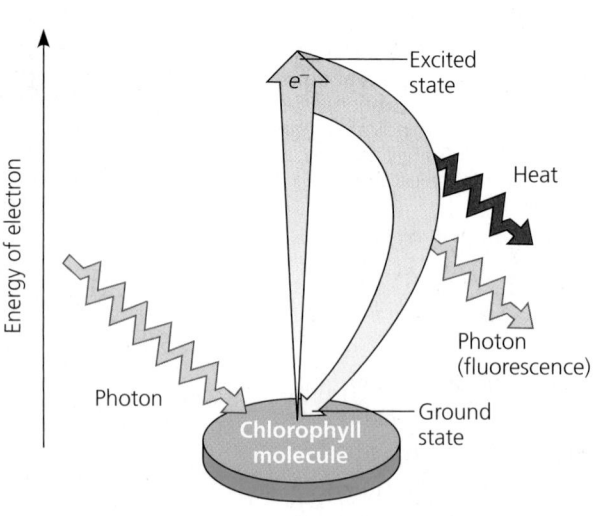

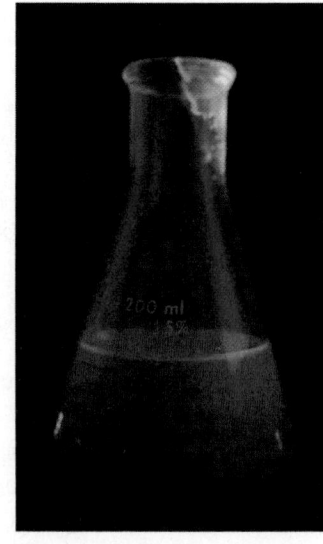

**(a) Excitation of isolated chlorophyll molecule**          **(b) Fluorescence**

▲ **Figure 8.11 Excitation of isolated chlorophyll by light. (a)** Absorption of a photon causes a transition of the chlorophyll molecule from its ground state to its excited state. The photon boosts an electron to an orbital where it has more potential energy. If the illuminated molecule exists in isolation, its excited electron immediately drops back down to the ground-state orbital, and its excess energy is given off as heat and fluorescence (light). **(b)** A chlorophyll solution excited with ultraviolet light fluoresces with a red-orange glow.

as heat. This conversion of light energy to heat is what makes the top of an automobile so hot on a sunny day. (White cars are coolest because their paint reflects all wavelengths of visible light.) In isolation, some pigments, including chlorophyll, emit light as well as heat after absorbing photons. As excited electrons fall back to the ground state, photons are given off, an afterglow called fluorescence. An illuminated solution of chlorophyll isolated from chloroplasts will fluoresce in the red-orange part of the spectrum and also give off heat. This is best seen by illuminating with ultraviolet light, which chlorophyll can also absorb **(Figure 8.11b)**. Viewed under visible light, the fluorescence would be hard to see against the green of the solution.

## A Photosystem: A Reaction-Center Complex Associated with Light-Harvesting Complexes

Chlorophyll molecules excited by the absorption of light energy produce very different results in an intact chloroplast than they do in isolation. In their native environment of the thylakoid membrane, chlorophyll molecules are organized along with other small organic molecules and proteins into complexes called photosystems.

A **photosystem** is composed of a **reaction-center complex** surrounded by several light-harvesting complexes **(Figure 8.12)**. The reaction-center complex is an organized association of proteins holding a special pair of chlorophyll *a* molecules. Each **light-harvesting complex** consists of various pigment molecules (which may include chlorophyll *a*, chlorophyll *b*, and multiple carotenoids) bound to proteins. The number and variety of pigment molecules enable a photosystem to harvest light over a larger surface area and a larger portion of the spectrum than could any single pigment molecule alone. Together, these light-harvesting complexes act as an antenna for the reaction-center complex. When a pigment molecule absorbs a photon, the energy is transferred from pigment molecule to pigment molecule within a light-harvesting complex, somewhat like a human "wave" at a sports arena, until it is passed into the reaction-center complex. The reaction-center complex also contains a molecule capable of accepting electrons and becoming reduced; this is called the **primary electron acceptor**. The pair of chlorophyll *a* molecules in the reaction-center complex are special because their molecular environment—their location and the other molecules with which they are associated—enables them to use the energy from light not only to boost one of their electrons to a higher energy level, but also to transfer it to a different molecule—the primary electron acceptor.

The solar-powered transfer of an electron from the reaction-center chlorophyll *a* pair to the primary electron acceptor is one of the first steps of the light reactions. As soon as the chlorophyll electron is excited to a higher energy level, the primary electron acceptor captures it; this is a redox reaction. In the flask shown in Figure 8.11b, isolated chlorophyll fluoresces because there is no electron acceptor, so electrons

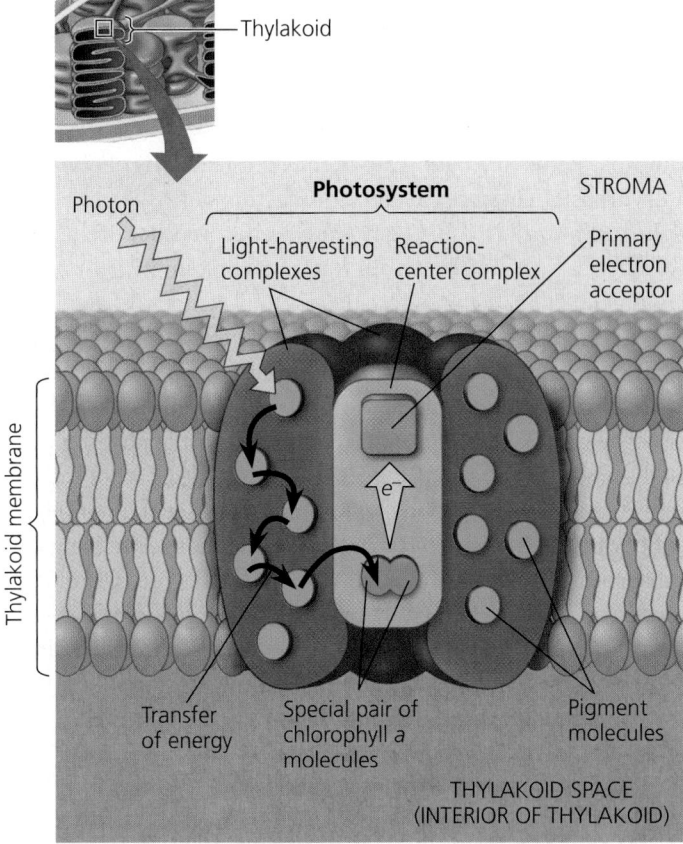

**(a) How a photosystem harvests light.** When a photon strikes a pigment molecule in a light-harvesting complex, the energy is passed from molecule to molecule until it reaches the reaction-center complex. Here, an excited electron from the special pair of chlorophyll *a* molecules is transferred to the primary electron acceptor.

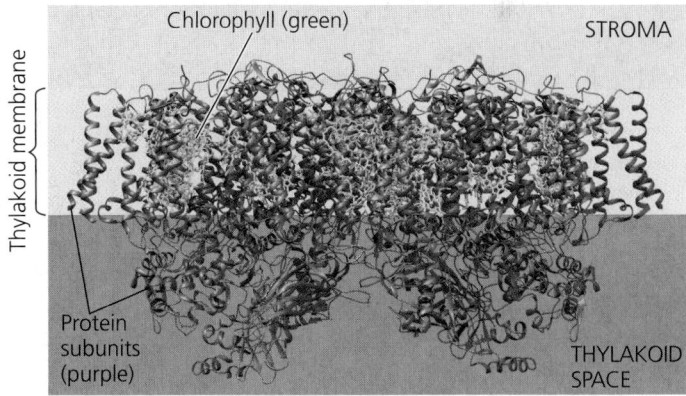

**(b) Structure of a photosystem.** This computer model, based on X-ray crystallography, shows two photosystem complexes side by side. Chlorophyll molecules (bright green ball-and-stick models within the membrane; the tails are not shown) are interspersed with protein subunits (purple ribbons; notice the many α helices spanning the membrane). For simplicity, a photosystem will be shown as a single complex in the rest of the chapter.

▲ **Figure 8.12 The structure and function of a photosystem.**

of photoexcited chlorophyll drop right back to the ground state. In the structured environment of a chloroplast, however, an electron acceptor is readily available, and the potential energy represented by the excited electron is not dissipated as light and heat. Thus, each photosystem—a reaction-center complex surrounded by light-harvesting complexes—functions in the chloroplast as a unit. It converts light energy to chemical energy, which will ultimately be used for the synthesis of sugar.

The thylakoid membrane is populated by two types of photosystems that cooperate in the light reactions of photosynthesis. They are called **photosystem II (PS II)** and **photosystem I (PS I)**. (They were named in order of their discovery, but photosystem II functions first in the light reactions.) Each has a characteristic reaction-center complex—a particular kind of primary electron acceptor next to a special pair of chlorophyll *a* molecules associated with specific proteins. The reaction-center chlorophyll *a* of photosystem II is known as P680 because this pigment is best at absorbing light having a wavelength of 680 nm (in the red part of the spectrum). The chlorophyll *a* at the reaction-center complex of photosystem I is called P700 because it most effectively absorbs light of wavelength 700 nm (in the far-red part of the spectrum). These two pigments, P680 and P700, are nearly identical chlorophyll *a* molecules. However, their association with different proteins in the thylakoid membrane affects the electron distribution in the two pigments and accounts for the slight differences in their light-absorbing properties. Now let's see how the two photosystems work together in using light energy to generate ATP and NADPH, the two main products of the light reactions.

## Linear Electron Flow

Light drives the synthesis of ATP and NADPH by energizing the two photosystems embedded in the thylakoid membranes of chloroplasts. The key to this energy transformation is a flow of electrons through the photosystems and other molecular components built into the thylakoid membrane. This is called **linear electron flow**, and it occurs during the light reactions of photosynthesis, as shown in **Figure 8.13**. The numbered steps in the text correspond to those in the figure.

1 A photon of light strikes one of the pigment molecules in a light-harvesting complex of PS II, boosting one of its electrons to a higher energy level. As this electron falls back to its ground state, an electron in a nearby pigment molecule is simultaneously raised to an excited state. The

▼ **Figure 8.13 How linear electron flow during the light reactions generates ATP and NADPH.** The gold arrows trace the current of light-driven electrons from water to NADPH.

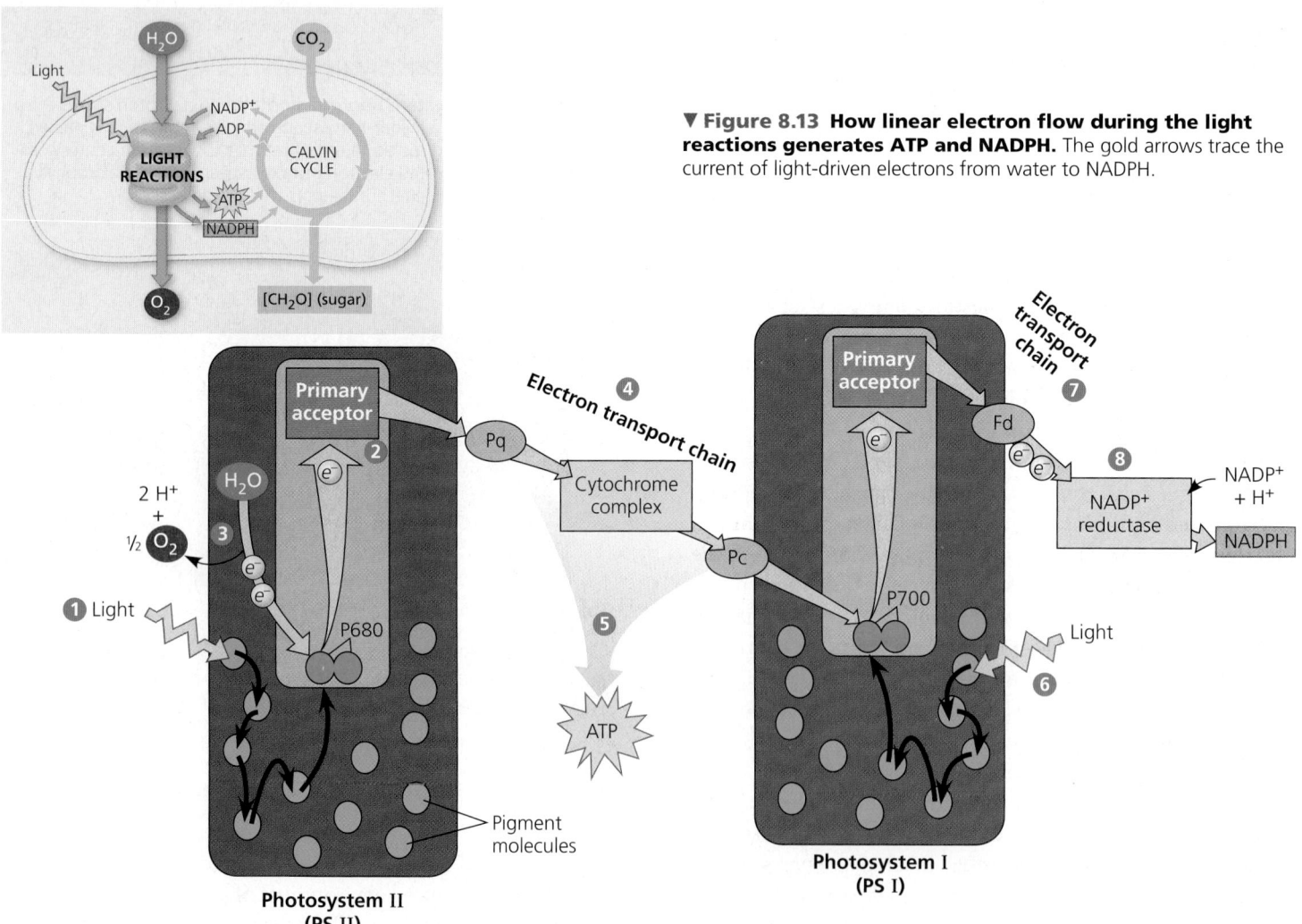

process continues, with the energy being relayed to other pigment molecules until it reaches the P680 pair of chlorophyll *a* molecules in the PS II reaction-center complex. It excites an electron in this pair of chlorophylls to a higher energy state.

**2** This electron is transferred from the excited P680 to the primary electron acceptor. We can refer to the resulting form of P680, missing an electron, as $P680^+$.

**3** An enzyme catalyzes the splitting of a water molecule into two electrons, two hydrogen ions ($H^+$), and an oxygen atom. The electrons are supplied one by one to the $P680^+$ pair, each electron replacing one transferred to the primary electron acceptor. ($P680^+$ is the strongest biological oxidizing agent known; its electron "hole" must be filled. This greatly facilitates the transfer of electrons from the split water molecule.) The $H^+$ are released into the thylakoid space. The oxygen atom immediately combines with an oxygen atom generated by the splitting of another water molecule, forming $O_2$.

**4** Each photoexcited electron passes from the primary electron acceptor of PS II to PS I via an electron transport chain, the components of which are similar to those of the electron transport chain that functions in cellular respiration. The electron transport chain between PS II and PS I is made up of the electron carrier plastoquinone (Pq), a cytochrome complex, and a protein called plastocyanin (Pc).

**5** The exergonic "fall" of electrons to a lower energy level provides energy for the synthesis of ATP. As electrons pass through the cytochrome complex, $H^+$ are pumped into the thylakoid space, contributing to the proton gradient that is then used in chemiosmosis, to be discussed shortly.

**6** Meanwhile, light energy has been transferred via light-harvesting complex pigments to the PS I reaction-center complex, exciting an electron of the P700 pair of chlorophyll *a* molecules located there. The photoexcited electron is then transferred to PS I's primary electron acceptor, creating an electron "hole" in the P700—which we now can call $P700^+$. In other words, $P700^+$ can now act as an electron acceptor, accepting an electron that reaches the bottom of the electron transport chain from PS II.

**7** Photoexcited electrons are passed in a series of redox reactions from the primary electron acceptor of PS I down a second electron transport chain through the protein ferredoxin (Fd). (This chain does not create a proton gradient and thus does not produce ATP.)

**8** The enzyme $NADP^+$ reductase catalyzes the transfer of electrons from Fd to $NADP^+$. Two electrons are required for its reduction to NADPH. This molecule is at a higher energy level than water, so its electrons are more readily available for the reactions of the Calvin cycle. This process also removes an $H^+$ from the stroma.

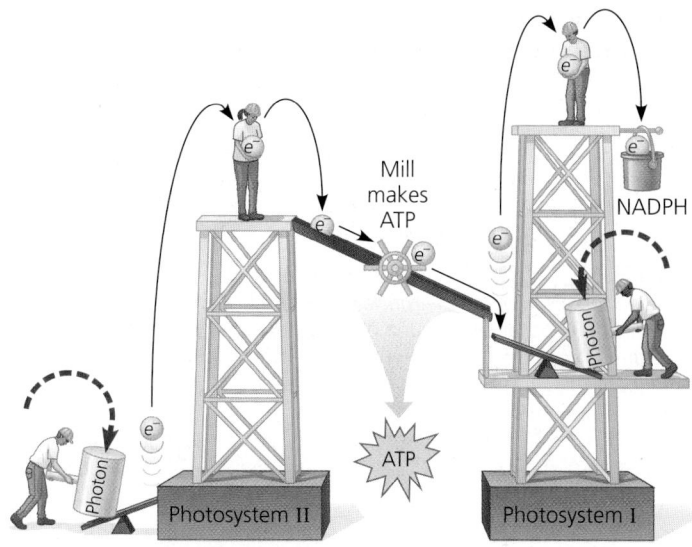

▲ **Figure 8.14 A mechanical analogy for linear electron flow during the light reactions.**

The energy changes of electrons during their linear flow through the light reactions are shown in a mechanical analogy in **Figure 8.14**. Although the scheme shown in Figures 8.13 and 8.14 may seem complicated, do not lose track of the big picture: The light reactions use solar power to generate ATP and NADPH, which provide chemical energy and reducing power, respectively, to the carbohydrate-synthesizing reactions of the Calvin cycle. Before we move on to the Calvin cycle, let's review chemiosmosis, the process that uses membranes to couple redox reactions to ATP production.

## A Comparison of Chemiosmosis in Chloroplasts and Mitochondria

Chloroplasts and mitochondria generate ATP by the same basic mechanism: chemiosmosis (see Figure 7.14). An electron transport chain assembled in a membrane pumps protons ($H^+$) across the membrane as electrons are passed through a series of carriers that are progressively more electronegative. Thus, electron transport chains transform redox energy to a proton-motive force, potential energy stored in the form of an $H^+$ gradient across a membrane. An ATP synthase complex in the same membrane couples the diffusion of hydrogen ions down their gradient to the phosphorylation of ADP, forming ATP. Some of the electron carriers, including the iron-containing proteins called cytochromes, are very similar in chloroplasts and mitochondria. The ATP synthase complexes of the two organelles are also quite similar. But there are noteworthy differences between photophosphorylation in chloroplasts and oxidative phosphorylation in mitochondria. Both work by way of chemiosmosis, but in chloroplasts, the high-energy electrons dropped down the transport chain come from water, whereas in mitochondria, they are extracted from organic molecules (which are thus oxidized). Chloroplasts do

▶ **Figure 8.15 Comparison of chemiosmosis in mitochondria and chloroplasts.** In both kinds of organelles, electron transport chains pump protons (H⁺) across a membrane from a region of low H⁺ concentration (light gray in this diagram) to one of high H⁺ concentration (dark gray). The protons then diffuse back across the membrane through ATP synthase, driving the synthesis of ATP.

**MAKE CONNECTIONS** *Describe how you would change the pH in order to artificially cause ATP synthesis (a) outside an isolated mitochondrion (assume H⁺ can freely cross the outer membrane) and (b) in the stroma of a chloroplast. Explain.*

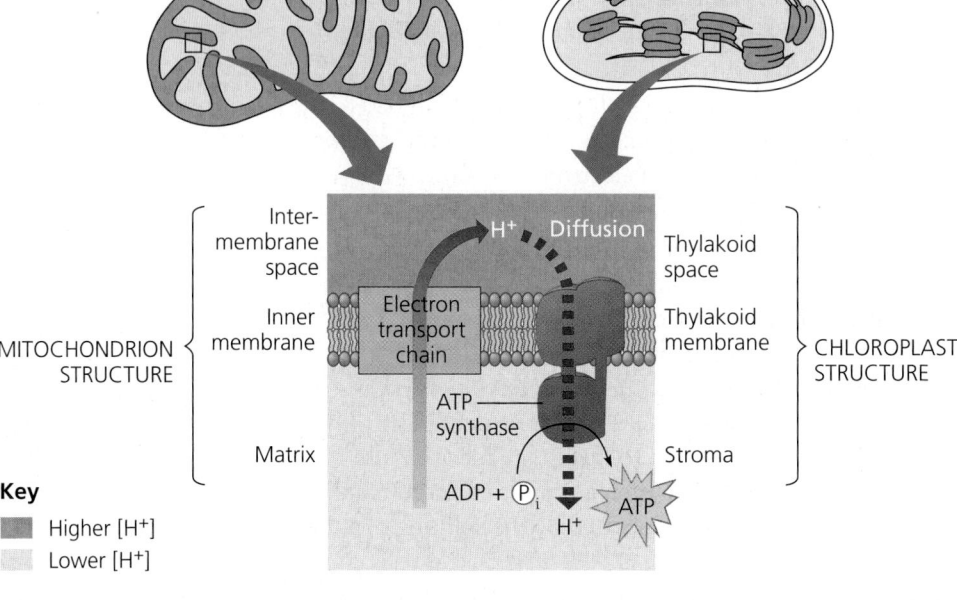

**Mitochondrion**

**Chloroplast**

MITOCHONDRION STRUCTURE

Intermembrane space

Inner membrane

Matrix

Electron transport chain

H⁺ Diffusion

ATP synthase

ADP + Ⓟ_i

H⁺

ATP

Thylakoid space

Thylakoid membrane

Stroma

CHLOROPLAST STRUCTURE

**Key**

■ Higher [H⁺]

■ Lower [H⁺]

not need molecules from food to make ATP; their photosystems capture light energy and use it to drive the electrons from water to the top of the transport chain. In other words, mitochondria use chemiosmosis to transfer chemical energy from food molecules to ATP, whereas chloroplasts use it to transform light energy into chemical energy in ATP.

Although the spatial organization of chemiosmosis differs slightly between chloroplasts and mitochondria, it is easy to see similarities in the two **(Figure 8.15)**. The inner membrane of the mitochondrion pumps protons from the mitochondrial matrix out to the intermembrane space, which then serves as a reservoir of hydrogen ions. The thylakoid membrane of the chloroplast pumps protons from the stroma into the thylakoid space (interior of the thylakoid), which functions as the H⁺ reservoir. If you imagine the cristae of mitochondria pinching off from the inner membrane, this may help you see how the thylakoid space and the intermembrane space are comparable spaces in the two organelles, while the mitochondrial matrix is analogous to the stroma of the chloroplast. In the mitochondrion, protons diffuse down their concentration gradient from the intermembrane space through ATP synthase to the matrix, driving ATP synthesis. In the chloroplast, ATP is synthesized as the hydrogen ions diffuse from the thylakoid space back to the stroma through ATP synthase complexes, whose catalytic knobs are on the stroma side of the membrane. Thus, ATP forms in the stroma, where it is used to help drive sugar synthesis during the Calvin cycle.

The proton (H⁺) gradient, or pH gradient, across the thylakoid membrane is substantial. When chloroplasts in an experimental setting are illuminated, the pH in the thylakoid space drops to about 5 (the H⁺ concentration increases), and the pH in the stroma increases to about 8 (the H⁺ concentration decreases). This gradient of three pH units corresponds

to a thousandfold difference in H⁺ concentration. If the lights are turned off, the pH gradient is abolished, but it can quickly be restored by turning the lights back on. Experiments such as this provided strong evidence in support of the chemiosmotic model.

Based on studies in several laboratories, **Figure 8.16** shows a current model for the organization of the light-reaction "machinery" within the thylakoid membrane. Each of the molecules and molecular complexes in the figure is present in numerous copies in each thylakoid. Notice that NADPH, like ATP, is produced on the side of the membrane facing the stroma, where the Calvin cycle reactions take place.

Let's summarize the light reactions. Electron flow pushes electrons from water, where they are at a low state of potential energy, ultimately to NADPH, where they are stored at a high state of potential energy. The light-driven electron flow also generates ATP. Thus, the equipment of the thylakoid membrane converts light energy to chemical energy stored in ATP and NADPH. (Oxygen is a by-product.) Let's now see how the Calvin cycle uses the products of the light reactions to synthesize sugar from $CO_2$.

**CONCEPT CHECK 8.2**

1. What color of light is *least* effective in driving photosynthesis? Explain.

2. In the light reactions, what is the initial electron donor? At the end of the light reactions, where are the electrons?

3. **WHAT IF?** In an experiment, isolated chloroplasts placed in an illuminated solution with the appropriate chemicals can carry out ATP synthesis. Predict what will happen to the rate of synthesis if a compound is added to the solution that makes membranes freely permeable to hydrogen ions.

For suggested answers, see Appendix A.

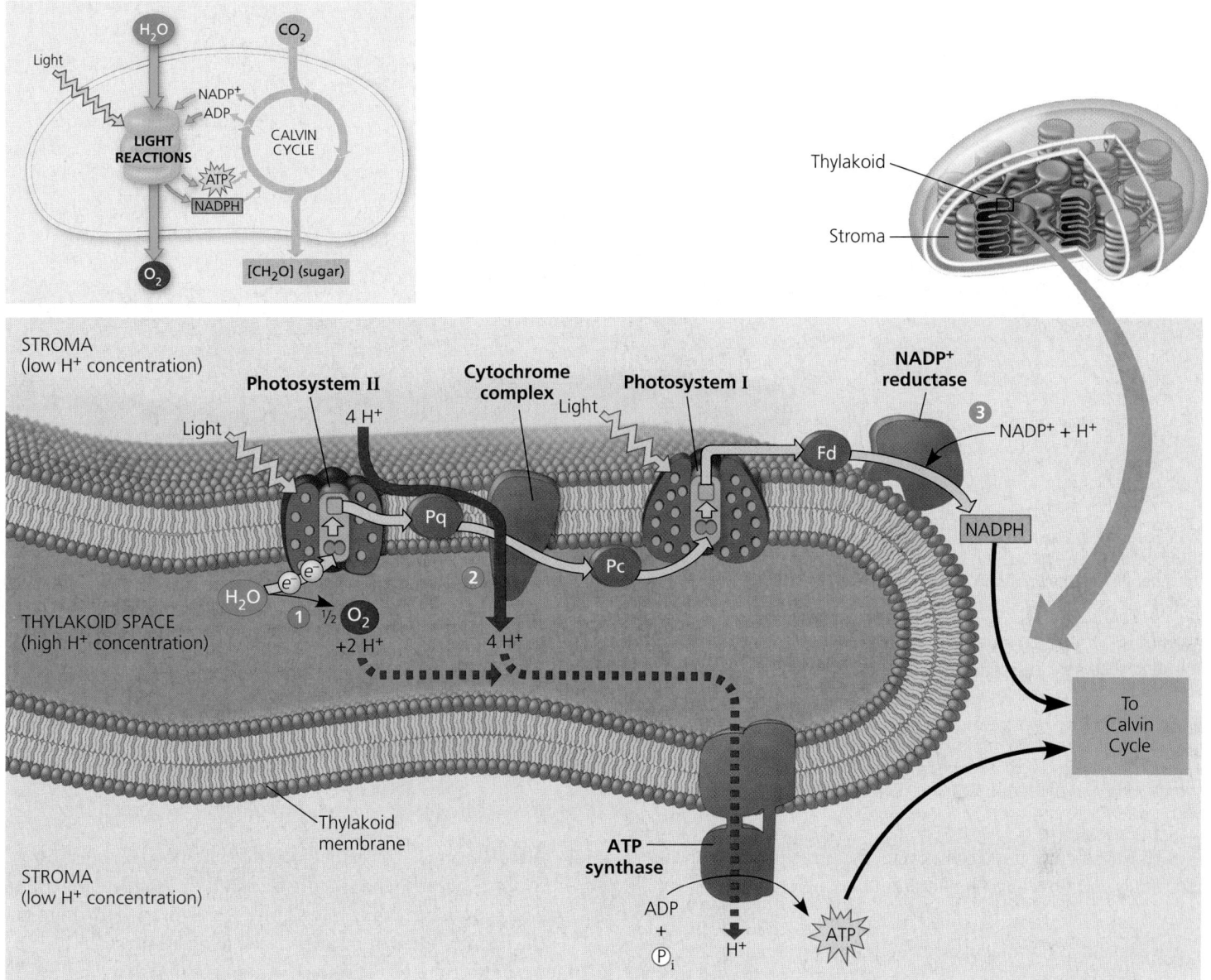

▲ **Figure 8.16 The light reactions and chemiosmosis: the current model of the organization of the thylakoid membrane.** The gold arrows track the linear electron flow outlined in Figure 8.13. At least three steps contribute to the H⁺ gradient across the thylakoid membrane: **①** Water is split by photosystem II on the side of the membrane facing the thylakoid space; **②** as plastoquinone (Pq) transfers electrons to the cytochrome complex, four protons are translocated across the membrane into the thylakoid space; and **③** a hydrogen ion is removed from the stroma when it is taken up by NADP⁺. Notice that in step 2, hydrogen ions are being pumped from the stroma into the thylakoid space, as in Figure 8.15. The diffusion of H⁺ from the thylakoid space back to the stroma (along the H⁺ concentration gradient) powers the ATP synthase.

CONCEPT 8.3

## The Calvin cycle uses the chemical energy of ATP and NADPH to reduce CO₂ to sugar

The Calvin cycle is similar to the citric acid cycle in that a starting material is regenerated after some molecules enter and others exit the cycle. However, the citric acid cycle is catabolic, oxidizing acetyl CoA and using the energy to synthesize ATP, while the Calvin cycle is anabolic, building carbohydrates from smaller molecules and consuming energy. Carbon enters the Calvin cycle in CO₂ and leaves in sugar. The cycle spends ATP as an energy source and consumes NADPH as reducing power for adding high-energy electrons to make sugar.

As mentioned in Concept 8.1, the carbohydrate produced directly from the Calvin cycle is not glucose. It is actually a three-carbon sugar named **glyceraldehyde 3-phosphate (G3P)**. For net synthesis of one molecule of G3P, the cycle must take place three times, fixing three molecules of CO₂—one per turn of the cycle. (Recall that the term *carbon fixation* refers to the initial incorporation of CO₂ into organic material.) As we trace the steps of the Calvin cycle, keep in mind that we are following three molecules of CO₂ through the reactions.

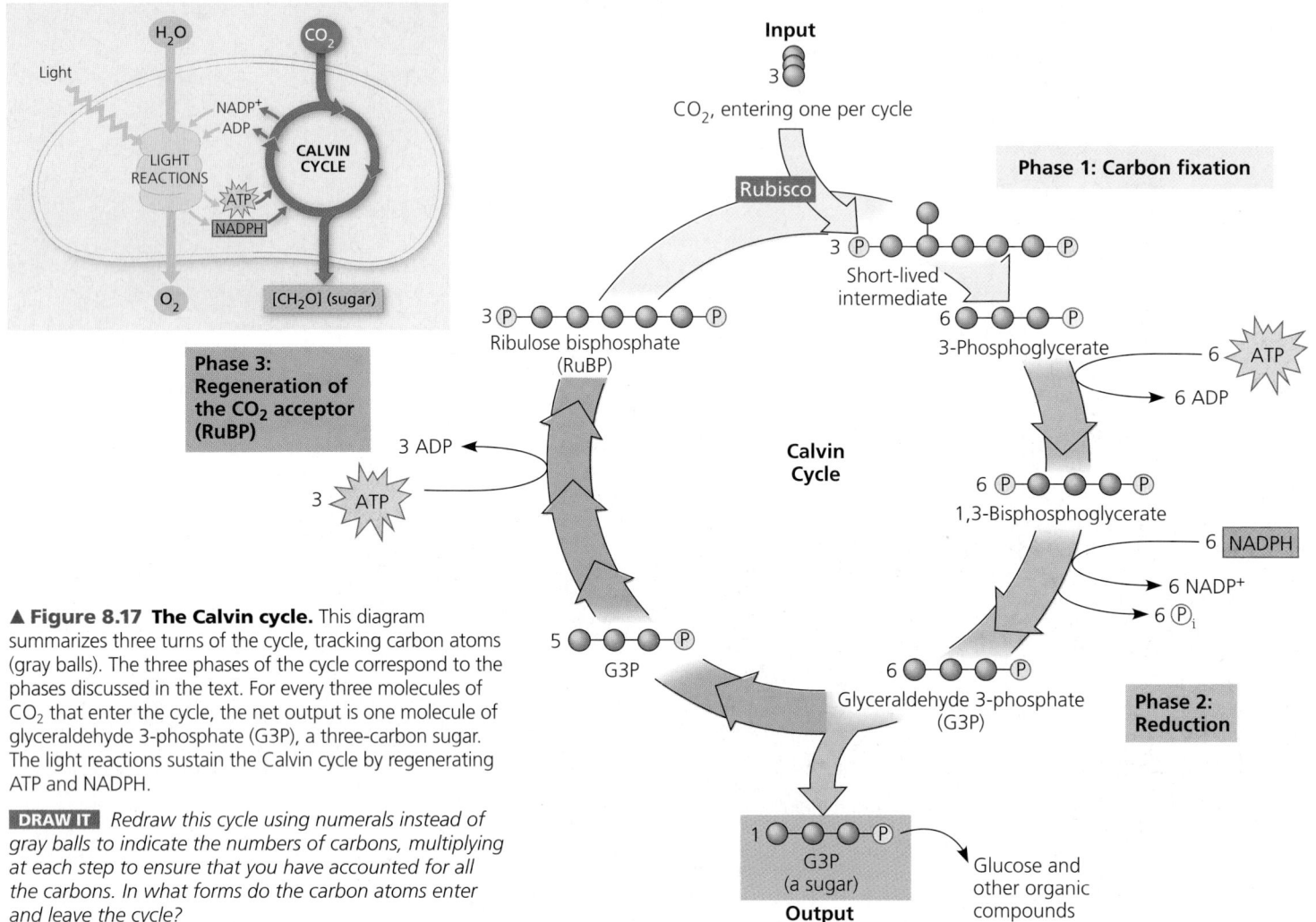

▲ **Figure 8.17 The Calvin cycle.** This diagram summarizes three turns of the cycle, tracking carbon atoms (gray balls). The three phases of the cycle correspond to the phases discussed in the text. For every three molecules of $CO_2$ that enter the cycle, the net output is one molecule of glyceraldehyde 3-phosphate (G3P), a three-carbon sugar. The light reactions sustain the Calvin cycle by regenerating ATP and NADPH.

**DRAW IT** *Redraw this cycle using numerals instead of gray balls to indicate the numbers of carbons, multiplying at each step to ensure that you have accounted for all the carbons. In what forms do the carbon atoms enter and leave the cycle?*

**Figure 8.17** divides the Calvin cycle into three phases: carbon fixation, reduction, and regeneration of the $CO_2$ acceptor.

**Phase 1: Carbon fixation.** The Calvin cycle incorporates each $CO_2$ molecule, one at a time, by attaching it to a five-carbon sugar named ribulose bisphosphate (abbreviated RuBP). The enzyme that catalyzes this first step is RuBP carboxylase/oxygenase, or **rubisco**. (This is the most abundant protein in chloroplasts and is also thought to be the most abundant protein on Earth.) The product of the reaction is a six-carbon intermediate so unstable that it immediately splits in half, forming two molecules of 3-phosphoglycerate (for each $CO_2$ fixed).

**Phase 2: Reduction.** Each molecule of 3-phosphoglycerate receives an additional phosphate group from ATP, becoming 1,3-bisphosphoglycerate. Next, a pair of electrons donated from NADPH reduces 1,3-bisphosphoglycerate, which also loses a phosphate group, becoming G3P. Specifically, the electrons from NADPH reduce a carboxyl group on 1,3-bisphosphoglycerate to the aldehyde group of G3P, which stores more potential energy. G3P is a sugar—the same three-carbon sugar formed in glycolysis by the splitting of glucose (see Figure 7.9). Notice in Figure 8.17 that for every *three* molecules of $CO_2$ that enter the cycle, there are *six* molecules of G3P formed. But only one molecule of this

three-carbon sugar can be counted as a net gain of carbohydrate, because the rest are required to complete the cycle. The cycle began with 15 carbons' worth of carbohydrate in the form of three molecules of the five-carbon sugar RuBP. Now there are 18 carbons' worth of carbohydrate in the form of six molecules of G3P. One molecule exits the cycle to be used by the plant cell, but the other five molecules must be recycled to regenerate the three molecules of RuBP.

**Phase 3: Regeneration of the $CO_2$ acceptor (RuBP).** In a complex series of reactions, the carbon skeletons of five molecules of G3P are rearranged by the last steps of the Calvin cycle into three molecules of RuBP. To accomplish this, the cycle spends three more ATPs. The RuBP is now prepared to receive $CO_2$ again, and the cycle continues.

For the net synthesis of one G3P molecule, the Calvin cycle consumes a total of nine molecules of ATP and six molecules of NADPH. The light reactions regenerate the ATP and NADPH. The G3P spun off from the Calvin cycle becomes the starting material for metabolic pathways that synthesize other organic compounds, including glucose (from two molecules of G3P) and other carbohydrates. Neither the light reactions nor the Calvin cycle alone can make sugar from $CO_2$. Photosynthesis is an emergent property of the intact chloroplast, which integrates the two stages of photosynthesis.

## Evolution of Alternative Mechanisms of Carbon Fixation in Hot, Arid Climates

**EVOLUTION** Ever since plants first moved onto land about 475 million years ago, they have been adapting to the problem of dehydration. The solutions often involve trade-offs. An example is the balance between photosynthesis and the prevention of excessive water loss from the plant. The $CO_2$ required for photosynthesis enters a leaf (and the resulting $O_2$ exits) via stomata, the pores on the leaf surface (see Figure 8.3). However, stomata are also the main avenues of the evaporative loss of water from leaves and may be partially or fully closed on hot, dry days. This prevents water loss, but it also reduces $CO_2$ levels.

In most plants, initial fixation of carbon occurs via rubisco, the Calvin cycle enzyme that adds $CO_2$ to ribulose bisphosphate. Such plants are called **$C_3$ plants** because the first organic product of carbon fixation is a three-carbon compound, 3-phosphoglycerate (see Figure 8.17). $C_3$ plants include important agricultural plants such as rice, wheat, and soybeans. When their stomata close on hot, dry days, $C_3$ plants produce less sugar because the declining level of $CO_2$ in the leaf starves the Calvin cycle. In addition, rubisco is capable of binding $O_2$ in place of $CO_2$. As $CO_2$ becomes scarce and $O_2$ builds up, rubisco adds $O_2$ to the Calvin cycle instead of $CO_2$. The product splits, forming a two-carbon compound that leaves the chloroplast and is broken down in the cell, releasing $CO_2$. The process is called **photorespiration** because it occurs in the light (*photo*) and consumes $O_2$ while producing $CO_2$ (*respiration*). However, unlike normal cellular respiration, photorespiration uses ATP rather than generating it. And unlike photosynthesis, photorespiration produces no sugar. In fact, photorespiration *decreases* photosynthetic output by siphoning organic material from the Calvin cycle and releasing $CO_2$ that would otherwise be fixed.

According to one hypothesis, photorespiration is evolutionary baggage—a metabolic relic from a much earlier time when the atmosphere had less $O_2$ and more $CO_2$ than it does today. In the ancient atmosphere that prevailed when rubisco first evolved, the ability of the enzyme's active site to bind $O_2$ would have made little difference. The hypothesis suggests that modern rubisco retains some of its chance affinity for $O_2$, which is now so concentrated in the atmosphere that a certain amount of photorespiration is inevitable. There is also some evidence that photorespiration may provide protection against damaging products of the light reactions that build up when the Calvin cycle slows due to low $CO_2$.

In some plant species, alternate modes of carbon fixation have evolved that minimize photorespiration and optimize the Calvin cycle—even in hot, arid climates. The two most important of these photosynthetic adaptations are $C_4$ photosynthesis and crassulacean acid metabolism (CAM).

### $C_4$ Plants

The **$C_4$ plants** are so named because they carry out a modified pathway for sugar synthesis that first fixes $CO_2$ into a four-carbon compound. When the weather is hot and dry, a $C_4$ plant partially closes its stomata, thus conserving water. Sugar continues to be made, however, through the function of two different types of photosynthetic cells: mesophyll cells and bundle-sheath cells **(Figure 8.18a)**. An enzyme in the mesophyll cells has a high affinity for $CO_2$ and can fix carbon even when the $CO_2$ concentration in the leaf is low. The resulting four-carbon compound then acts as a carbon shuttle; it moves into bundle-sheath cells, which are packed around the veins of the leaf, and releases $CO_2$. Thus, the $CO_2$ concentration in these cells remains high enough for the Calvin cycle to make sugars and avoid photorespiration. The $C_4$ pathway is believed to have evolved independently at least 45 times and is used by several thousand species in at least 19 plant families. Among the $C_4$ plants important to agriculture are sugarcane and corn (maize), members of the grass family. In the **Scientific Skills Exercise**, you will work with data to see how different concentrations of $CO_2$ affect growth in plants that use the $C_4$ pathway versus those that use the $C_3$ pathway.

### CAM Plants

A second photosynthetic adaptation to arid conditions has evolved in pineapples, many cacti, and other succulent (water-storing) plants, such as aloe and jade plants **(Figure 8.18b)**. These

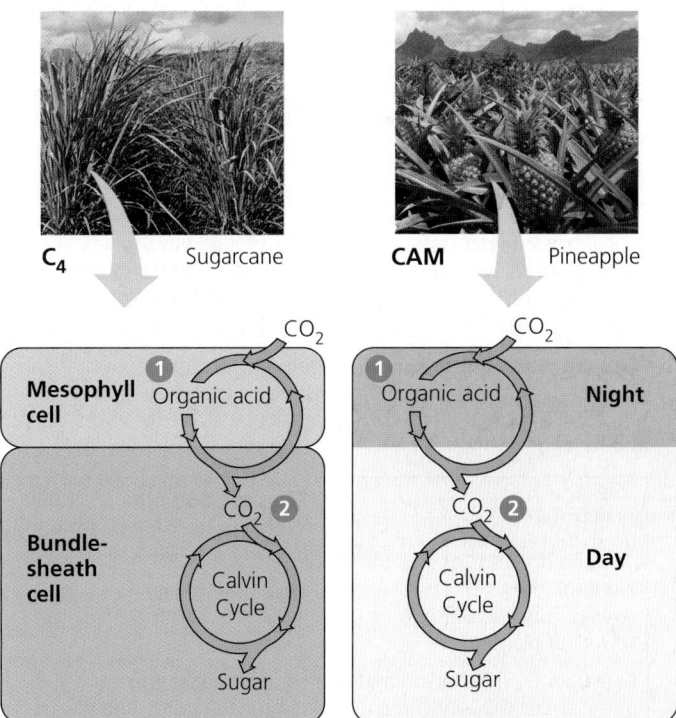

**(a) Spatial separation of steps.** In $C_4$ plants, carbon fixation and the Calvin cycle occur in different types of cells.

**(b) Temporal separation of steps.** In CAM plants, carbon fixation and the Calvin cycle occur in the same cell at different times.

▲ **Figure 8.18 $C_4$ and CAM photosynthesis compared.** Both adaptations are characterized by ❶ preliminary incorporation of $CO_2$ into organic acids, followed by ❷ transfer of $CO_2$ to the Calvin cycle. The $C_4$ and CAM pathways are two evolutionary solutions to the problem of maintaining photosynthesis with stomata partially or completely closed on hot, dry days.

plants open their stomata during the night and close them during the day, the reverse of how other plants behave. Closing stomata during the day helps desert plants conserve water, but it also prevents $CO_2$ from entering the leaves. During the night, when their stomata are open, these plants take up $CO_2$ and incorporate it into a variety of organic acids. This mode of carbon fixation is called **crassulacean acid metabolism (CAM)** after the plant family Crassulaceae, the succulents in which the process was first discovered. The mesophyll cells of **CAM plants** store the organic acids they make during the night in their vacuoles until morning, when the stomata close. During the day, when the light reactions can supply ATP and NADPH for the Calvin cycle, $CO_2$ is released from the organic acids made the night before to become incorporated into sugar in the chloroplasts.

Notice in Figure 8.18 that the CAM pathway is similar to the $C_4$ pathway in that carbon dioxide is first incorporated into organic intermediates before it enters the Calvin cycle. The difference is that in $C_4$ plants, the initial steps of carbon fixation are separated structurally from the Calvin cycle, whereas in CAM plants, the two steps occur within the same cell but at separate times. (Keep in mind that CAM, $C_4$, and $C_3$ plants all eventually use the Calvin cycle to make sugar from carbon dioxide.)

### CONCEPT CHECK 8.3

1. **MAKE CONNECTIONS** How are the large numbers of ATP and NADPH molecules used during the Calvin cycle consistent with the high value of glucose as an energy source? (Compare Figures 7.15 and 8.17.)
2. **WHAT IF?** Explain why a poison that inhibits an enzyme of the Calvin cycle will also inhibit the light reactions.
3. Describe how photorespiration lowers photosynthetic output.

For suggested answers, see Appendix A.

## The Importance of Photosynthesis: *A Review*

In this chapter, we have followed photosynthesis from photons to food. The light reactions capture solar energy and use it to make ATP and transfer electrons from water to $NADP^+$, forming NADPH. The Calvin cycle uses the ATP

---

## Scientific Skills Exercise

### Making Scatter Plots with Regression Lines

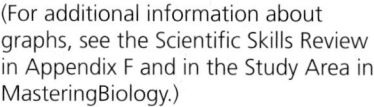

▶ Corn plant surrounded by invasive velvetleaf plants

**Does Atmospheric $CO_2$ Concentration Affect the Productivity of Agricultural Crops?** Atmospheric concentration of $CO_2$ has been rising globally, and scientists wondered whether this would affect $C_3$ and $C_4$ plants differently. In this exercise, you will make a scatter plot to examine the relationship between $CO_2$ concentration and growth of corn (maize), a $C_4$ crop plant, and velvetleaf, a $C_3$ weed found in cornfields.

**How the Experiment Was Done** Researchers grew corn and velvetleaf plants under controlled conditions for 45 days, where all plants received the same amounts of water and light. The plants were divided into three groups, and each was exposed to a different concentration of $CO_2$ in the air: 350, 600, or 1,000 ppm (parts per million).

**Data from the Experiment** The table shows the dry mass (in grams) of corn and velvetleaf plants grown at the three concentrations of $CO_2$. The dry mass values are averages of the leaves, stems, and roots of eight plants.

|  | 350 ppm $CO_2$ | 600 ppm $CO_2$ | 1,000 ppm $CO_2$ |
|---|---|---|---|
| Average dry mass of one corn plant (g) | 91 | 89 | 80 |
| Average dry mass of one velvetleaf plant (g) | 35 | 48 | 54 |

**Data from** D. T. Patterson and E. P. Flint, Potential effects of global atmospheric $CO_2$ enrichment on the growth and competitiveness of $C_3$ and $C_4$ weed and crop plants, *Weed Science* 28(1):71–75 (1980).

**INTERPRET THE DATA**

1. To explore the relationship between the two variables, it is useful to graph the data in a scatter plot and then draw a regression line. (a) First, place labels for the dependent and independent variables on the appropriate axes. Explain your choices. (b) Now plot the data points for corn and velvetleaf using different symbols for each set of data and add a key for the two symbols.

(For additional information about graphs, see the Scientific Skills Review in Appendix F and in the Study Area in MasteringBiology.)

2. Draw a "best-fit" line for each set of points. A best-fit line does not necessarily pass through all or even most points. It is a straight line that passes as close as possible to all data points from that set. Drawing a best-fit line is a matter of judgment, so two people may draw slightly different lines. The line that fits best, a regression line, can be identified by squaring the distances of all points to any candidate line, then selecting the line that minimizes the sum of the squares. (See the graph in the Scientific Skills Exercise in Chapter 2 for an example of a linear regression line.) Using a spreadsheet program (such as Excel) or a graphing calculator, enter the data points for each data set and have the program draw the regression lines. Compare them with the lines you drew.

3. Describe the trends shown by the regression lines. (a) Compare the relationship between increasing concentration of $CO_2$ and the dry mass of corn with that of velvetleaf. (b) Since velvetleaf is a weed invasive to cornfields, predict how increased $CO_2$ concentration may affect interactions between the two species.

4. Based on the data in the scatter plot, estimate the percentage change in dry mass of corn and velvetleaf plants if atmospheric $CO_2$ concentration increased from 390 ppm (current levels) to 800 ppm. (a) What is the estimated dry mass of corn and velvetleaf plants at 390 ppm? 800 ppm? (b) To calculate the percentage change in mass for each plant, subtract the mass at 390 ppm from the mass at 800 ppm (change in mass), divide by the mass at 390 ppm (initial mass), and multiply by 100. What is the estimated percentage change in dry mass for corn? For velvetleaf? (c) Do these results support the conclusion from other experiments that $C_3$ plants grow better than $C_4$ plants under increased $CO_2$ concentration? Why or why not?

(MB) A version of this Scientific Skills Exercise can be assigned in MasteringBiology.

and NADPH to produce sugar from carbon dioxide. The energy that enters the chloroplasts as sunlight becomes stored as chemical energy in organic compounds. The entire process is reviewed visually in **Figure 8.19**, where photosynthesis is also shown in its natural context.

As for the fates of photosynthetic products, enzymes in the chloroplast and cytosol convert the G3P made in the Calvin cycle to many other organic compounds. In fact, the sugar made in the chloroplasts supplies the entire plant with chemical energy and carbon skeletons for the synthesis of all the major organic molecules of plant cells. About 50% of the organic material made by photosynthesis is consumed as fuel for cellular respiration in plant cell mitochondria.

Green cells are the only autotrophic parts of the plant. Other cells depend on organic molecules exported from leaves via veins (see Figure 8.19, top). In most plants, carbohydrate is transported out of the leaves to the rest of the plant as sucrose, a disaccharide. After arriving at nonphotosynthetic cells, the sucrose provides raw material for cellular respiration and many anabolic pathways that synthesize proteins, lipids, and other products. A considerable amount of sugar in the form of glucose is linked together to make the polysaccharide cellulose (see Figure 3.11c), especially in plant cells that are still growing and maturing. Cellulose, the main ingredient of cell walls, is the most abundant organic molecule in the plant—and probably on the surface of the planet.

Most plants and other photosynthesizers manage to make more organic material each day than they need to use as respiratory fuel and precursors for biosynthesis. They stockpile the extra sugar by synthesizing starch, storing some in the chloroplasts themselves and some in storage cells of roots, tubers, seeds, and fruits. In accounting for the consumption of the food molecules produced by photosynthesis, let's not forget that most plants lose leaves, roots, stems, fruits, and sometimes their entire bodies to heterotrophs, including humans.

On a global scale, photosynthesis is responsible for the oxygen in our atmosphere. Furthermore, while each chloroplast is minuscule, their collective food production is prodigious: Photosynthesis makes an estimated 150 billion metric tons of carbohydrate per year (a metric ton is 1,000 kg, about 1.1 tons). That's organic matter equivalent in mass to a stack of about 60 trillion biology textbooks! Such a stack would reach 17 times the distance from Earth to the sun! No chemical process is more important than photosynthesis to the welfare of life on Earth.

In Chapters 3 through 8, you have learned about many activities of cells. **Figure 8.20** integrates these in the context of a working plant cell. As you study the figure, reflect on how each process fits into the big picture: As the most basic unit of living organisms, a cell performs all functions characteristic of life.

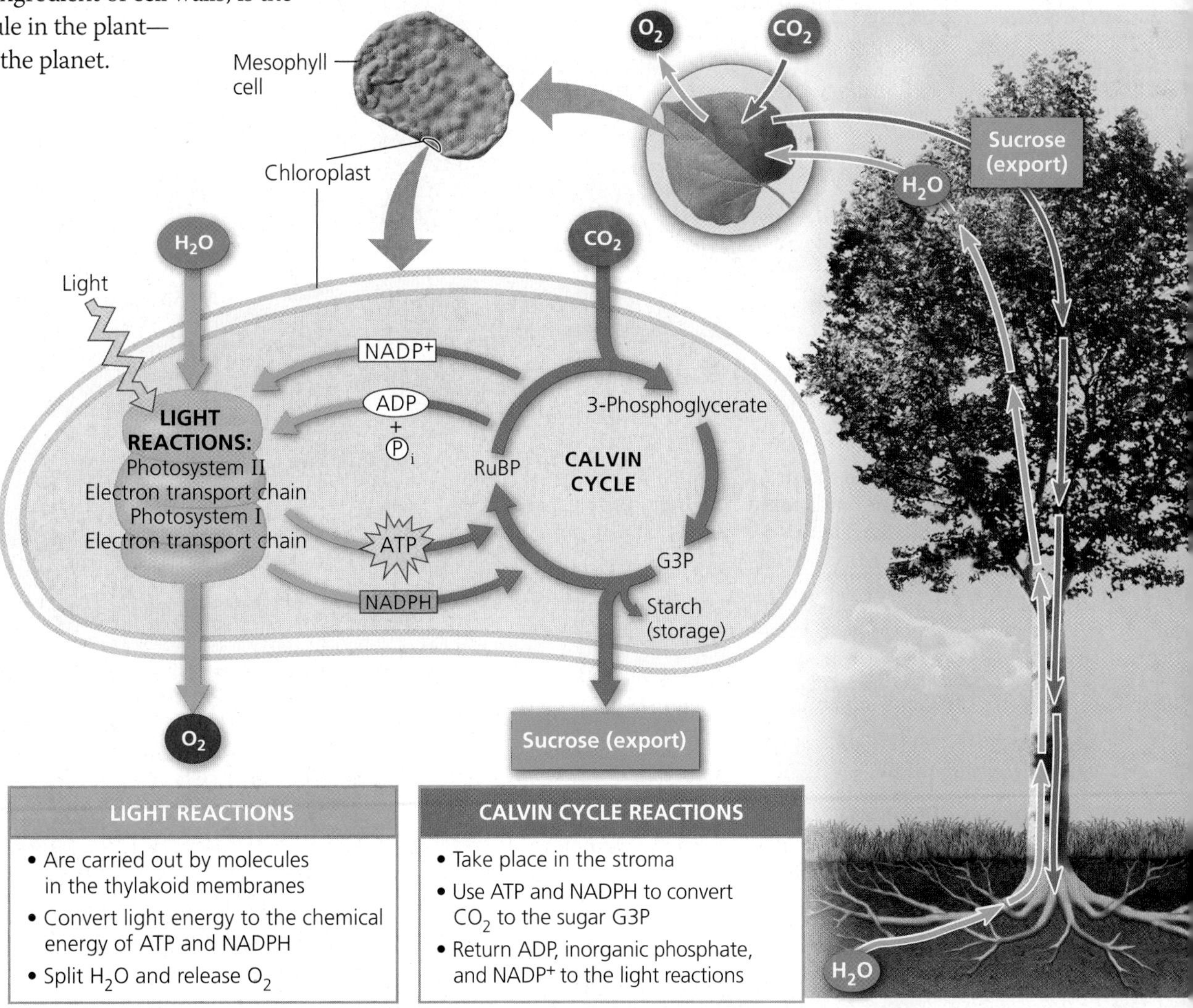

▶ **Figure 8.19 A review of photosynthesis.** This diagram shows the main reactants and products of photosynthesis as they move through the tissues of a tree (right) and a chloroplast (left).

**MAKE CONNECTIONS** *Can plants use the sugar they produce during photosynthesis to power the work of their cells? Explain. (See Figures 6.9, 6.10, and 7.6.)*

| LIGHT REACTIONS |
| --- |
| • Are carried out by molecules in the thylakoid membranes |
| • Convert light energy to the chemical energy of ATP and NADPH |
| • Split $H_2O$ and release $O_2$ |

| CALVIN CYCLE REACTIONS |
| --- |
| • Take place in the stroma |
| • Use ATP and NADPH to convert $CO_2$ to the sugar G3P |
| • Return ADP, inorganic phosphate, and $NADP^+$ to the light reactions |

This figure illustrates how a generalized plant cell functions, integrating the cellular activities you learned about in Chapters 3–8.

DNA

Nucleus

1

mRNA

Nuclear pore

2

Protein in vesicle

Rough endoplasmic reticulum (ER)

3

Protein

Ribosome    mRNA

4

Vesicle forming

Golgi apparatus

Protein

6

Plasma membrane

5

Cell wall

## Flow of Genetic Information in the Cell: DNA → RNA → Protein (Chapters 3–5)

1. In the nucleus, DNA serves as a template for the synthesis of mRNA, which moves to the cytoplasm. (See Figures 3.26 and 4.8.)

2. mRNA attaches to a ribosome, which remains free in the cytosol or binds to the rough ER. Proteins are synthesized. (See Figures 3.26 and 4.9.)

3. Proteins and membrane produced by the rough ER flow in vesicles to the Golgi apparatus, where they are processed. (See Figures 4.15 and 5.8.)

4. Transport vesicles carrying proteins pinch off from the Golgi apparatus. (See Figure 4.15.)

5. Some vesicles merge with the plasma membrane, releasing proteins by exocytosis. (See Figure 5.8.)

6. Proteins synthesized on free ribosomes stay in the cell and perform specific functions; examples include the enzymes that catalyze the reactions of cellular respiration and photosynthesis. (See Figures 7.7, 7.9, and 8.17.)

## Energy Transformations in the Cell: Photosynthesis and Cellular Respiration (Chapters 6–8)

**7** In chloroplasts, the process of photosynthesis uses the energy of light to convert $CO_2$ and $H_2O$ to organic molecules, with $O_2$ as a by-product. (See Figure 8.19.)

**8** In mitochondria, organic molecules are broken down by cellular respiration, capturing energy in molecules of ATP, which are used to power the work of the cell, such as protein synthesis and active transport. $CO_2$ and $H_2O$ are by-products. (See Figures 6.8–6.10, 7.2, and 7.15.)

## Movement Across Cell Membranes (Chapter 5)

**9** Water diffuses into and out of the cell directly through the plasma membrane and by facilitated diffusion through aquaporins. (See Figure 5.1.)

**10** By passive transport, the $CO_2$ used in photosynthesis diffuses into the cell and the $O_2$ formed as a by-product of photosynthesis diffuses out of the cell. Both solutes move down their concentration gradients. (See Figures 5.9 and 8.19.)

**11** In active transport, energy (usually supplied by ATP) is used to transport a solute against its concentration gradient. (See Figure 5.15.)

Exocytosis (shown in step 5) and endocytosis move larger materials out of and into the cell. (See Figures 5.8 and 5.18.)

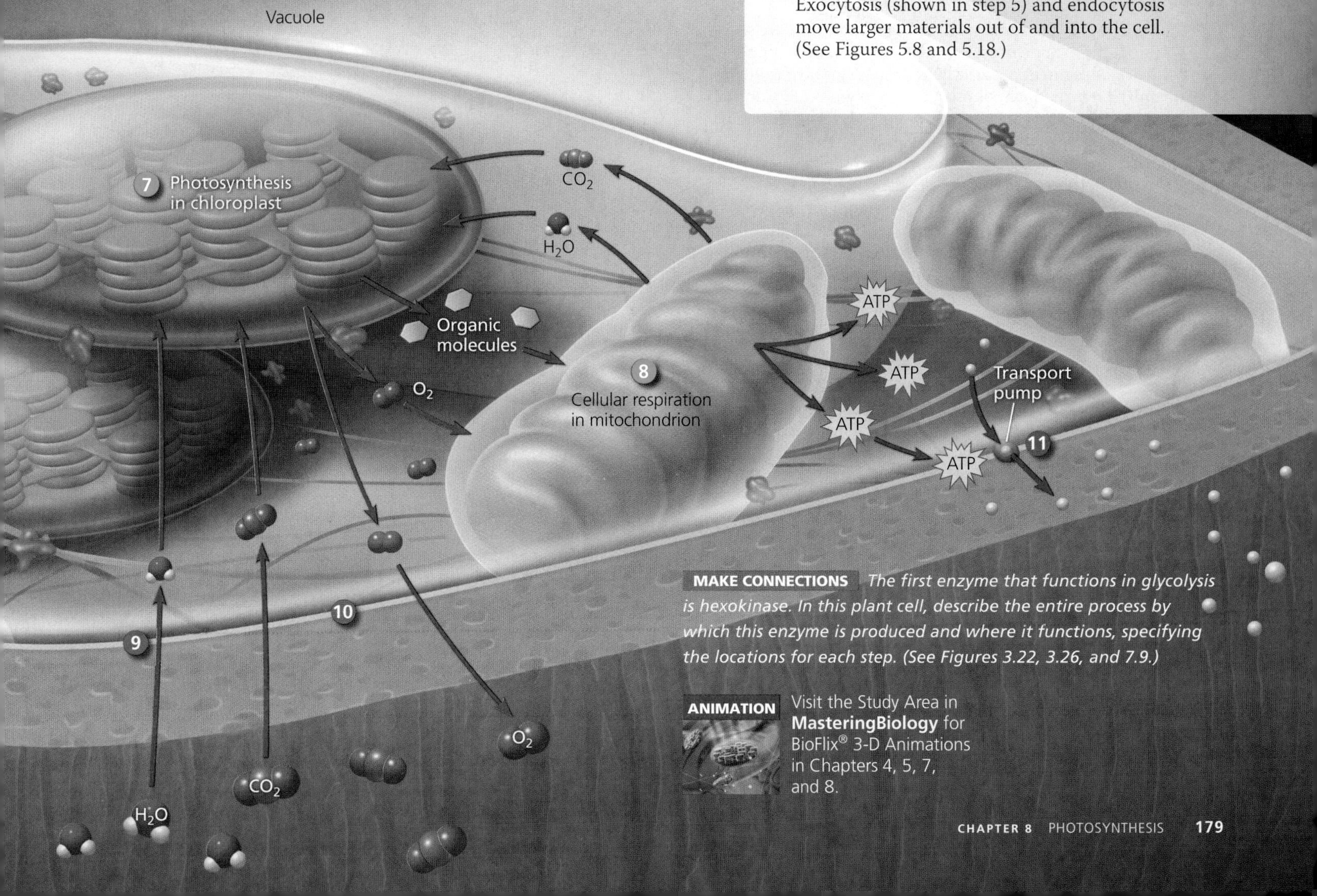

Vacuole

**7** Photosynthesis in chloroplast

$CO_2$

$H_2O$

Organic molecules

$O_2$

**8** Cellular respiration in mitochondrion

ATP

ATP

ATP

ATP

Transport pump

**11**

**10**

**9**

$O_2$

$CO_2$

$H_2O$

**MAKE CONNECTIONS** *The first enzyme that functions in glycolysis is hexokinase. In this plant cell, describe the entire process by which this enzyme is produced and where it functions, specifying the locations for each step. (See Figures 3.22, 3.26, and 7.9.)*

**ANIMATION** Visit the Study Area in **MasteringBiology** for BioFlix® 3-D Animations in Chapters 4, 5, 7, and 8.

# 8 Chapter Review

Go to **MasteringBiology**® for Assignments, the eText, and the Study Area with Animations, Activities, Vocab Self-Quiz, and Practice Tests.

## SUMMARY OF KEY CONCEPTS

**VOCAB SELF-QUIZ**

goo.gl/gbai8v

### CONCEPT 8.1

**Photosynthesis converts light energy to the chemical energy of food (pp. 162–165)**

- In eukaryotes that are **autotrophs**, photosynthesis occurs in **chloroplasts**, organelles containing **thylakoids**. Stacks of thylakoids form grana. **Photosynthesis** is summarized as

$$6\ CO_2 + 12\ H_2O + \text{Light energy} \rightarrow C_6H_{12}O_6 + 6\ O_2 + 6\ H_2O.$$

- Chloroplasts split water into hydrogen and oxygen, incorporating the electrons of hydrogen into sugar molecules. Photosynthesis is a redox process: $H_2O$ is oxidized, and $CO_2$ is reduced. The **light reactions** in the thylakoid membranes split water, releasing $O_2$, producing ATP, and forming **NADPH**. The **Calvin cycle** in the **stroma** forms sugars from $CO_2$, using ATP for energy and NADPH for reducing power.

> **?** *Compare the roles of $CO_2$ and $H_2O$ in respiration and photosynthesis.*

### CONCEPT 8.2

**The light reactions convert solar energy to the chemical energy of ATP and NADPH (pp. 165–173)**

- Light is a form of electromagnetic energy. The colors we see as **visible light** include those **wavelengths** that drive photosynthesis. A pigment absorbs light of specific wavelengths; **chlorophyll *a*** is the main photosynthetic pigment in plants. Other accessory pigments absorb different wavelengths of light and pass the energy on to chlorophyll *a*.
- A pigment goes from a ground state to an excited state when a **photon** of light boosts one of the pigment's electrons to a higher-energy electron shell. Electrons from isolated pigments tend to fall back to the ground state, giving off heat and/or light.
- A **photosystem** is composed of a **reaction-center complex** surrounded by **light-harvesting complexes** that funnel the energy of photons to the reaction-center complex. When a special pair of reaction-center chlorophyll *a* molecules absorbs energy, one of its electrons is boosted to a higher energy level and transferred to the **primary electron acceptor**. **Photosystem II** contains P680 chlorophyll *a* molecules in the reaction-center complex; **photosystem I** contains P700 molecules.
- **Linear electron flow** during the light reactions uses both photosystems and produces NADPH, ATP, and oxygen:

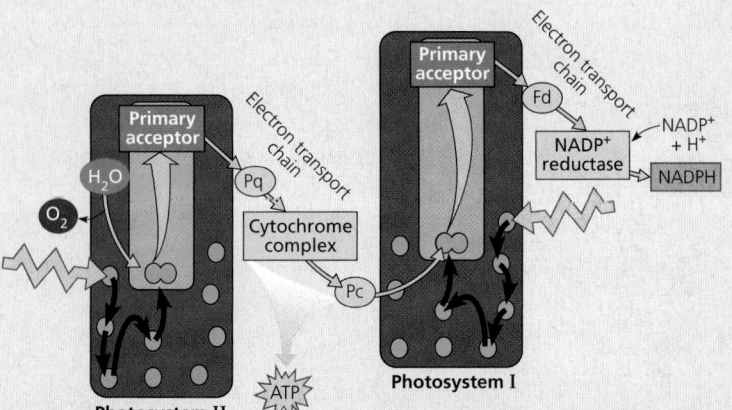

- During chemiosmosis in both mitochondria and chloroplasts, electron transport chains generate an $H^+$ (proton) gradient across a membrane. ATP synthase uses this proton-motive force to synthesize ATP.

> **?** *The absorption spectrum of chlorophyll* a *differs from the action spectrum of photosynthesis. Explain this observation.*

### CONCEPT 8.3

**The Calvin cycle uses the chemical energy of ATP and NADPH to reduce $CO_2$ to sugar (pp. 173–177)**

- The Calvin cycle occurs in the stroma, using electrons from NADPH and energy from ATP. One molecule of **G3P** exits the cycle per three $CO_2$ molecules fixed and is converted to glucose and other organic molecules.

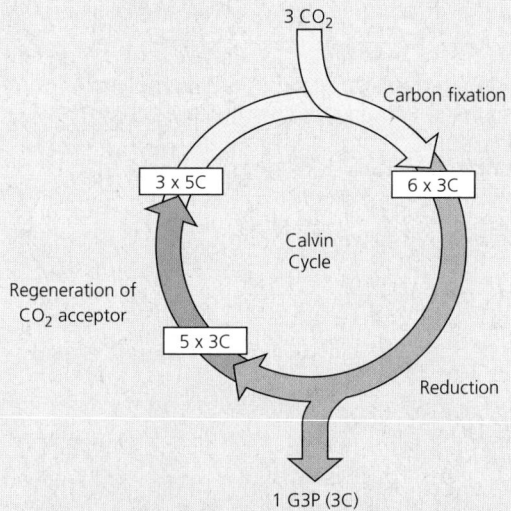

- On hot, dry days, **$C_3$ plants** close their stomata, conserving water but keeping $CO_2$ out and $O_2$ in. Under these conditions, **photorespiration** can occur: **Rubisco** binds $O_2$ instead of $CO_2$, leading to consumption of ATP and release of $CO_2$ without the production of sugar. Photorespiration may be an evolutionary relic and it may also play a protective role.
- **$C_4$ plants** are adapted to hot, dry climates. Even with their stomata partially or completely closed, they minimize the cost of photorespiration by incorporating $CO_2$ into four-carbon compounds in mesophyll cells. These compounds are exported to bundle-sheath cells, where they release carbon dioxide for use in the Calvin cycle.
- **CAM plants** are also adapted to hot, dry climates. They open their stomata at night, incorporating $CO_2$ into organic acids, which are stored in mesophyll cells. During the day, the stomata close, and the $CO_2$ is released from the organic acids for use in the Calvin cycle.
- Organic compounds produced by photosynthesis provide the energy and building material for Earth's ecosystems.

> **DRAW IT** *On the diagram above, draw where ATP and NADPH are used and where rubisco functions. Describe these steps.*

# TEST YOUR UNDERSTANDING

**PRACTICE TEST**

goo.gl/CRZjvS

## Level 1: Knowledge/Comprehension

1. The light reactions of photosynthesis supply the Calvin cycle with
   (A) light energy.
   (B) $CO_2$ and ATP.
   (C) $O_2$ and NADPH.
   (D) ATP and NADPH.

2. Which of the following sequences correctly represents the flow of electrons during photosynthesis?
   (A) NADPH → $O_2$ → $CO_2$
   (B) $H_2O$ → NADPH → Calvin cycle
   (C) $H_2O$ → photosystem I → photosystem II
   (D) NADPH → electron transport chain → $O_2$

3. How is photosynthesis similar in $C_4$ plants and CAM plants?
   (A) In both cases, electron transport is not used.
   (B) Both types of plants make sugar without the Calvin cycle.
   (C) In both cases, rubisco is not used to fix carbon initially.
   (D) Both types of plants make most of their sugar in the dark.

4. Which of the following statements is a correct distinction between autotrophs and heterotrophs?
   (A) Autotrophs, but not heterotrophs, can nourish themselves beginning with $CO_2$ and other nutrients that are inorganic.
   (B) Only heterotrophs require chemical compounds from the environment.
   (C) Cellular respiration is unique to heterotrophs.
   (D) Only heterotrophs have mitochondria.

5. Which of the following does *not* occur during the Calvin cycle?
   (A) carbon fixation
   (B) oxidation of NADPH
   (C) release of oxygen
   (D) regeneration of the $CO_2$ acceptor

## Level 2: Application/Analysis

6. In mechanism, photophosphorylation is most similar to
   (A) substrate-level phosphorylation in glycolysis.
   (B) oxidative phosphorylation in cellular respiration.
   (C) carbon fixation.
   (D) reduction of $NADP^+$.

7. Which process is most directly driven by light energy?
   (A) creation of a pH gradient by pumping protons across the thylakoid membrane
   (B) reduction of $NADP^+$ molecules
   (C) removal of electrons from chlorophyll molecules
   (D) ATP synthesis

8. To synthesize one glucose molecule, the Calvin cycle uses _____ molecules of $CO_2$, _____ molecules of ATP, and _____ molecules of NADPH.

## Level 3: Synthesis/Evaluation

9. **SCIENCE, TECHNOLOGY, AND SOCIETY**
   Scientific evidence indicates that the $CO_2$ added to the air by the burning of wood and fossil fuels is contributing to global warming, a rise in global temperature. Tropical rain forests are estimated to be responsible for approximately 20% of global photosynthesis, yet the consumption of large amounts of $CO_2$ by living trees is thought to make little or no *net* contribution to reduction of global warming. Explain why might this be the case. (*Hint*: What processes in both living and dead trees produce $CO_2$?)

10. **SCIENTIFIC INQUIRY**
    **DRAW IT** The following diagram represents an experiment with isolated thylakoids. The thylakoids were first made acidic by soaking them in a solution at pH 4. After the thylakoid space reached pH 4, the thylakoids were transferred to a basic solution at pH 8. The thylakoids then made ATP in the dark. (See Concept 2.5 to review pH.)

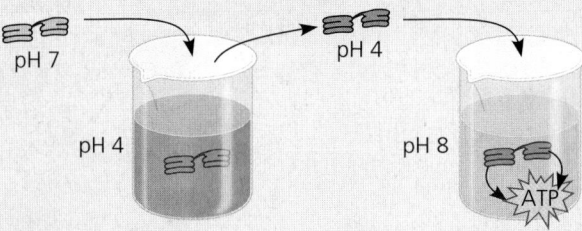

Draw an enlargement of part of the thylakoid membrane in the beaker with the solution at pH 8. Draw ATP synthase. Label the areas of high $H^+$ concentration and low $H^+$ concentration. Show the direction protons flow through the enzyme, and show the reaction where ATP is synthesized. Would ATP end up in the thylakoid or outside of it? Explain why the thylakoids in the experiment were able to make ATP in the dark.

11. **FOCUS ON EVOLUTION**
    Consider the endosymbiont theory (see Figure 4.16) and the fact that chloroplasts contain DNA molecules and ribosomes (see Figure 4.18). Chloroplasts, plant cell nuclei, and photosynthetic prokaryotes all have genes that code for ribosomal RNAs. Would you expect the DNA sequences of ribosomal RNA genes in chloroplasts to be more similar to those in plant cell nuclei or currently living photosynthetic prokaryotes? Explain. If sequencing studies show that your hypothesis is correct, what does this tell us about the evolution of photosynthesis?

12. **FOCUS ON EVOLUTION**
    Photorespiration can decrease soybeans' photosynthetic output by about 50%. Would you expect this figure to be higher or lower in wild relatives of soybeans? Explain.

13. **FOCUS ON ENERGY AND MATTER**
    Life is solar powered. Almost all the producers of the biosphere depend on energy from the sun to produce the organic molecules that supply the energy and carbon skeletons needed for life. In a short essay (100–150 words), describe how the process of photosynthesis in the chloroplasts of plants transforms the energy of sunlight into the chemical energy of sugar molecules.

14. **SYNTHESIZE YOUR KNOWLEDGE**

"Watermelon snow" in Antarctica is caused by a species of photosynthetic green algae that thrives in subzero temperatures (*Chlamydomonas nivalis*). These algae are also found in high-altitude year-round snowfields. In both locations, UV light levels tend to be high. Based on what you learned in this chapter, propose an explanation for why this photosynthetic alga appears reddish-pink.

*For selected answers, see Appendix A.*

# 9 The Cell Cycle

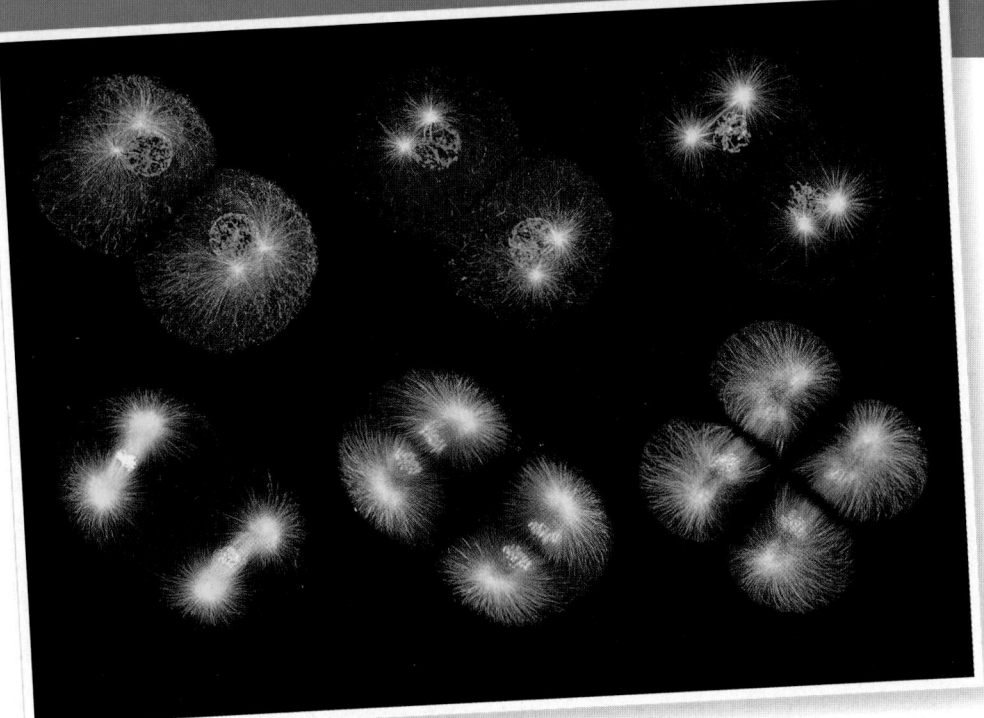

▲ **Figure 9.1** How do dividing cells distribute chromosomes to daughter cells?

## The Key Roles of Cell Division

The ability of organisms to produce more of their own kind is the one characteristic that best distinguishes living things from nonliving matter. This unique capacity to procreate, like all biological functions, has a cellular basis. In 1855, Rudolf Virchow, a German physician, put it this way: "Where a cell exists, there must have been a preexisting cell, just as the animal arises only from an animal and the plant only from a plant." He summarized this concept with the Latin axiom "*Omnis cellula e cellula*," meaning "Every cell from a cell." The continuity of life is based on the reproduction of cells, or **cell division**. The series of confocal fluorescence micrographs in **Figure 9.1**, starting at the upper left and reading across both rows left to right, follows the events of cell division as the cells of a two-celled embryo become four.

Cell division plays several important roles in life. The division of one prokaryotic cell reproduces an entire organism. The same is true of a unicellular eukaryote **(Figure 9.2a)**. Cell division also enables multicellular eukaryotes to develop from a single cell, like the fertilized egg that gave rise to the two-celled embryo in **Figure 9.2b**. And after such an organism is fully grown, cell division continues to function in renewal and repair, replacing cells that die from accidents or normal wear and tear. For example, dividing cells in your bone marrow continuously make new blood cells **(Figure 9.2c)**.

The cell division process is an integral part of the **cell cycle**, the life of a cell from the time it is first formed during division of a parent cell until its own division into two daughter cells. (Biologists use the words *daughter* or *sister* in relation to cells, but this is not meant to imply gender.) Passing identical genetic material to cellular offspring is a crucial function of cell division. In this chapter, you'll learn how this process occurs. After studying the mechanics of cell division in eukaryotes and bacteria, you'll learn about the molecular control system that regulates progress through the eukaryotic cell cycle and what happens when the control system malfunctions. Because a breakdown in cell cycle control plays a major role in cancer development, this aspect of cell biology is an active area of research.

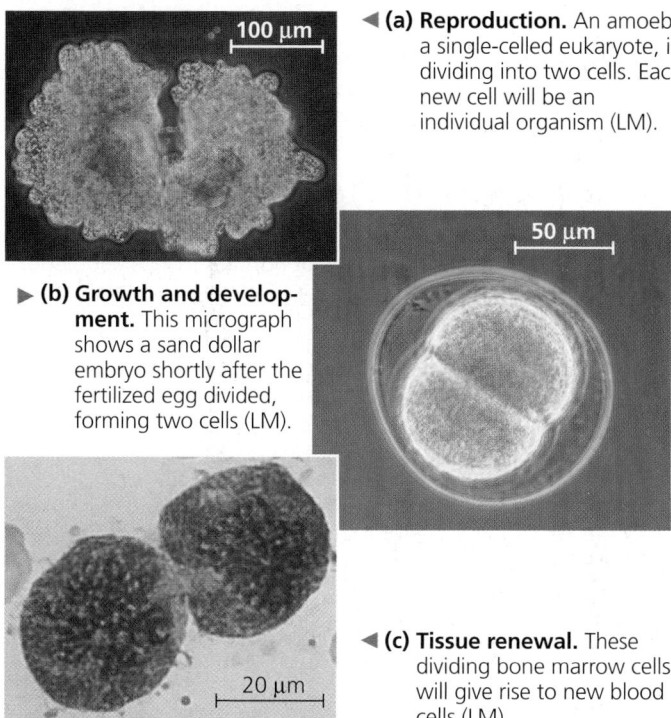

◀ **(a) Reproduction.** An amoeba, a single-celled eukaryote, is dividing into two cells. Each new cell will be an individual organism (LM).

100 μm

50 μm

▶ **(b) Growth and development.** This micrograph shows a sand dollar embryo shortly after the fertilized egg divided, forming two cells (LM).

◀ **(c) Tissue renewal.** These dividing bone marrow cells will give rise to new blood cells (LM).

20 μm

▲ **Figure 9.2 The functions of cell division.**

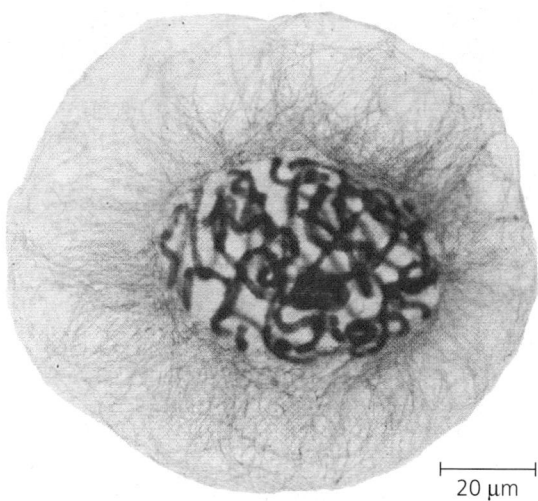

20 μm

▲ **Figure 9.3 Eukaryotic chromosomes.** Chromosomes (stained purple) are visible within the nucleus of this cell from an African blood lily. The thinner red threads in the surrounding cytoplasm are the cytoskeleton. The cell is preparing to divide (LM).

CONCEPT 9.1

# Most cell division results in genetically identical daughter cells

The reproduction of a cell, with all its complexity, cannot occur by a mere pinching in half; a cell is not like a soap bubble that simply enlarges and splits in two. In both prokaryotes and eukaryotes, most cell division involves the distribution of identical genetic material—DNA—to two daughter cells. (The exception is meiosis, the special type of eukaryotic cell division that can produce sperm and eggs.) What is most remarkable about cell division is the fidelity with which the DNA is passed from one generation of cells to the next. A dividing cell replicates its DNA, allocates the two copies to opposite ends of the cell, and only then splits into daughter cells.

## Cellular Organization of the Genetic Material

A cell's endowment of DNA, its genetic information, is called its **genome**. Although a prokaryotic genome is often a single DNA molecule, eukaryotic genomes usually consist of a number of DNA molecules. The overall length of DNA in a eukaryotic cell is enormous. A typical human cell, for example, has about 2 m of DNA—a length about 250,000 times greater than the cell's diameter. Before the cell can divide to form genetically identical daughter cells, all of this DNA must be copied, or replicated, and then the two copies must be separated so that each daughter cell ends up with a complete genome.

The replication and distribution of so much DNA is manageable because the DNA molecules are packaged into structures called **chromosomes** (from the Greek *chroma*, color, and *soma*, body), so named because they take up certain dyes used in microscopy **(Figure 9.3)**. Each eukaryotic chromosome consists of one very long, linear DNA molecule associated with many proteins (see Figure 4.8). The DNA molecule carries several hundred to a few thousand genes, the units of information that specify an organism's inherited traits. The associated proteins maintain the structure of the chromosome and help control the activity of the genes. Together, the entire complex of DNA and proteins that is the building material of chromosomes is referred to as **chromatin**. As you will soon see, the chromatin of a chromosome varies in its degree of condensation during the process of cell division.

Every eukaryotic species has a characteristic number of chromosomes in each cell's nucleus. For example, the nuclei of human **somatic cells** (all body cells except the reproductive cells) each contain 46 chromosomes, made up of two sets of 23, one set inherited from each parent. Reproductive cells, or **gametes**—sperm and eggs—have half as many chromosomes as somatic cells, or one set of 23 chromosomes in humans. The number of chromosomes in somatic cells varies widely among species: 18 in cabbage plants, 48 in chimpanzees, 56 in elephants, 90 in hedgehogs, and 148 in one species of alga. We'll now consider how these chromosomes behave during cell division.

## Distribution of Chromosomes During Eukaryotic Cell Division

When a cell is not dividing, and even as it replicates its DNA in preparation for cell division, each chromosome is in the form of a long, thin chromatin fiber. After DNA replication, however, the chromosomes condense as a part of cell division: Each chromatin fiber becomes densely coiled and folded,

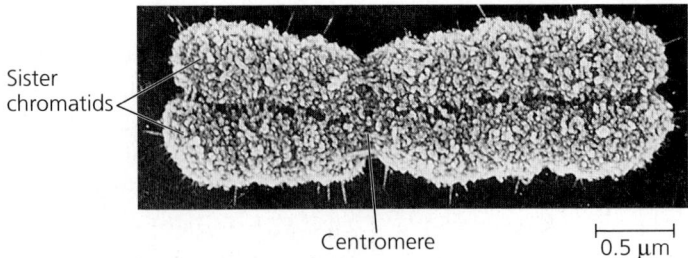

Sister
chromatids

Centromere

0.5 μm

▲ **Figure 9.4 A highly condensed, duplicated human chromosome (SEM).**

**DRAW IT** *Circle one sister chromatid of the chromosome in this micrograph.*

making the chromosomes much shorter and so thick that we can see them with a light microscope.

Each duplicated chromosome has two **sister chromatids**, which are joined copies of the original chromosome **(Figure 9.4)**. The two chromatids, each containing an identical DNA molecule, are initially attached all along their lengths by protein complexes called *cohesins*; this attachment is known as *sister chromatid cohesion*. Each sister chromatid has a **centromere**, a region of the chromosomal DNA where the chromatid is attached most closely to its sister chromatid. This attachment is mediated by proteins bound to the centromeric DNA sequences and gives the condensed, duplicated chromosome a narrow "waist." The part of a chromatid to either side of the centromere is referred to as an *arm* of the chromatid. (An uncondensed, unduplicated chromosome has a single centromere and two arms.)

Later in the cell division process, the two sister chromatids of each duplicated chromosome separate and move into two new nuclei, one forming at each end of the cell. Once the sister chromatids separate, they are no longer called sister chromatids but are considered individual chromosomes; this step essentially doubles the number of chromosomes in the cell. Thus, each new nucleus receives a collection of chromosomes identical to that of the parent cell **(Figure 9.5)**. **Mitosis**, the division of the genetic material in the nucleus, is usually followed immediately by **cytokinesis**, the division of the cytoplasm. One cell has become two, each the genetic equivalent of the parent cell.

From a fertilized egg, mitosis and cytokinesis produced the 200 trillion somatic cells that now make up your body, and the same processes continue to generate new cells to replace dead and damaged ones. In contrast,

you produce gametes—eggs or sperm—by a variation of cell division called *meiosis*, which yields daughter cells with only one set of chromosomes, half as many chromosomes as the parent cell. Meiosis in humans occurs only in special cells in the ovaries or testes (the gonads). Generating gametes, meiosis reduces the chromosome number from 46 (two sets) to 23 (one set). Fertilization fuses two gametes together and returns the chromosome number to 46 (two sets). Mitosis then conserves that number in every somatic cell nucleus of the new human individual. In Chapter 10, we'll examine the role of meiosis in reproduction and inheritance in more detail. In the remainder of this chapter, we focus on mitosis and the rest of the cell cycle in eukaryotes.

**CONCEPT CHECK 9.1**

1. How many chromosomes are drawn in each part of Figure 9.5? (Ignore the micrograph in part 2.)
2. **WHAT IF?** A chicken has 78 chromosomes in its somatic cells. How many chromosomes did the chicken inherit from each parent? How many chromosomes are in each of the chicken's gametes? How many chromosomes will be in each somatic cell of the chicken's offspring?

For suggested answers, see Appendix A.

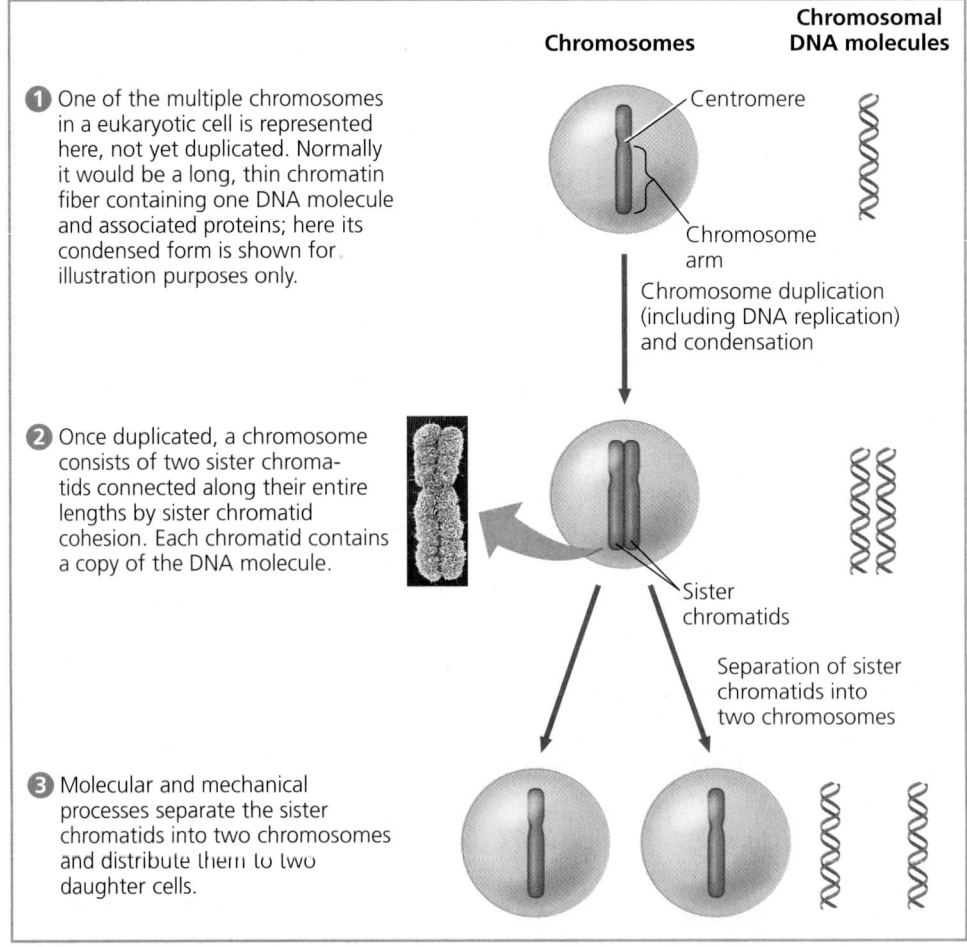

**①** One of the multiple chromosomes in a eukaryotic cell is represented here, not yet duplicated. Normally it would be a long, thin chromatin fiber containing one DNA molecule and associated proteins; here its condensed form is shown for illustration purposes only.

**②** Once duplicated, a chromosome consists of two sister chromatids connected along their entire lengths by sister chromatid cohesion. Each chromatid contains a copy of the DNA molecule.

**③** Molecular and mechanical processes separate the sister chromatids into two chromosomes and distribute them to two daughter cells.

Chromosomes

Chromosomal DNA molecules

Centromere

Chromosome arm

Chromosome duplication (including DNA replication) and condensation

Sister chromatids

Separation of sister chromatids into two chromosomes

▲ **Figure 9.5 Chromosome duplication and distribution during cell division.**

**?** *How many chromatid arms does the chromosome in step ② have? Identify the point in the figure where one chromosome becomes two.*

## CONCEPT 9.2

# The mitotic phase alternates with interphase in the cell cycle

In 1882, a German anatomist named Walther Flemming developed dyes that allowed him to observe, for the first time, the behavior of chromosomes during mitosis and cytokinesis. (In fact, Flemming coined the terms *mitosis* and *chromatin*.) It appeared to Flemming that during the period between one cell division and the next, the cell was simply growing larger. But we now know that many critical events occur during this stage in the life of a cell.

## Phases of the Cell Cycle

Mitosis is just one part of the cell cycle **(Figure 9.6)**. In fact, the **mitotic (M) phase**, which includes both mitosis and cytokinesis, is usually the shortest part of the cell cycle. The mitotic phase alternates with a much longer stage called **interphase**, which often accounts for about 90% of the cycle. Interphase can be divided into subphases: the $G_1$ **phase** ("first gap"), the **S phase** ("synthesis"), and the $G_2$ **phase** ("second gap"). The G phases were misnamed as "gaps" when they were first observed because the cells appeared inactive, but we now know that intense metabolic activity and growth occur throughout interphase. During all three subphases of interphase, in fact, a cell grows by producing proteins and cytoplasmic organelles such as mitochondria and endoplasmic reticulum. Duplication of the chromosomes, crucial for eventual division of the cell, occurs entirely during the S phase. (We will describe synthesis of DNA in Concept 13.2.) Thus, a cell grows ($G_1$), continues to grow as it copies its chromosomes (S), grows more as it completes preparations for cell division ($G_2$), and divides (M). The daughter cells may then repeat the cycle.

A particular human cell might undergo one division in 24 hours. Of this time, the M phase would occupy less than

1 hour, while the S phase might occupy about 10–12 hours, or about half the cycle. The rest of the time would be apportioned between the $G_1$ and $G_2$ phases. The $G_2$ phase usually takes 4–6 hours; in our example, $G_1$ would occupy about 5–6 hours. $G_1$ is the most variable in length in different types of cells. Some cells in a multicellular organism divide very infrequently or not at all. These cells spend their time in $G_1$ (or a related phase called $G_0$) doing their job in the organism—a nerve cell carries impulses, for example.

Mitosis is conventionally broken down into five stages: **prophase**, **prometaphase**, **metaphase**, **anaphase**, and **telophase**. Overlapping with the latter stages of mitosis, cytokinesis completes the mitotic phase. **Figure 9.7** describes these stages in an animal cell. Study this figure thoroughly before progressing to the next two sections, which examine mitosis and cytokinesis more closely.

## The Mitotic Spindle: *A Closer Look*

Many of the events of mitosis depend on the **mitotic spindle**, which begins to form in the cytoplasm during prophase. This structure consists of fibers made of microtubules and associated proteins. While the mitotic spindle assembles, the other microtubules of the cytoskeleton partially disassemble, providing the material used to construct the spindle. The spindle microtubules elongate (polymerize) by incorporating more subunits of the protein tubulin (see Table 4.1) and shorten (depolymerize) by losing subunits.

In animal cells, the assembly of spindle microtubules starts at the **centrosome**, a subcellular region containing material that functions throughout the cell cycle to organize the cell's microtubules. (It is also a type of *microtubule-organizing center*.) A pair of centrioles is located at the center of the centrosome, but they are not essential for cell division: If the centrioles are destroyed with a laser microbeam, a spindle nevertheless forms during mitosis. In fact, centrioles are not even present in plant cells, which do form mitotic spindles.

During interphase in animal cells, the single centrosome duplicates, forming two centrosomes, which remain together near the nucleus (see Figure 9.7). The two centrosomes move apart during prophase and prometaphase of mitosis as spindle microtubules grow out from them. By the end of prometaphase, the two centrosomes, one at each pole of the spindle, are at opposite ends of the cell. An **aster**, a radial array of short microtubules, extends from each centrosome. The spindle includes the centrosomes, the spindle microtubules, and the asters.

Each of the two sister chromatids of a duplicated chromosome has a **kinetochore**, a structure made up of proteins that have assembled on specific sections of chromosomal DNA at each centromere. The chromosome's two kinetochores face in opposite directions. During prometaphase, some of the spindle microtubules attach to the kinetochores; these are called kinetochore microtubules. (The number of microtubules attached to a kinetochore varies among species, from one microtubule in yeast cells to 40 or so in some mammalian cells.)

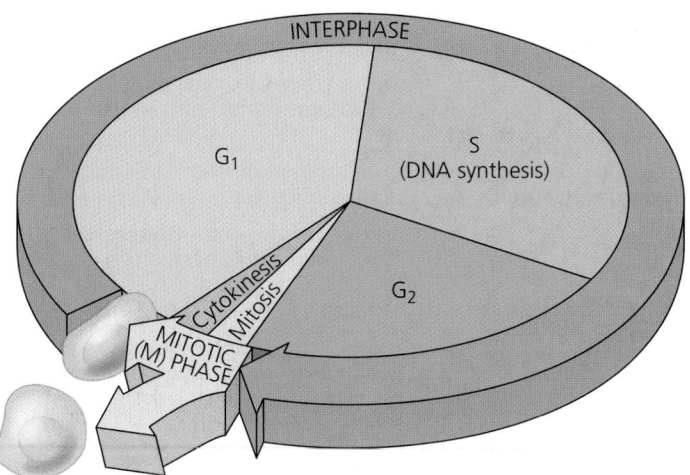

▲ **Figure 9.6 The cell cycle.** In a dividing cell, the mitotic (M) phase alternates with interphase, a growth period. The first part of interphase ($G_1$) is followed by the S phase, when the chromosomes duplicate; $G_2$ is the last part of interphase. In the M phase, mitosis distributes the daughter chromosomes to daughter nuclei, and cytokinesis divides the cytoplasm, producing two daughter cells.

**Exploring Mitosis in an Animal Cell**

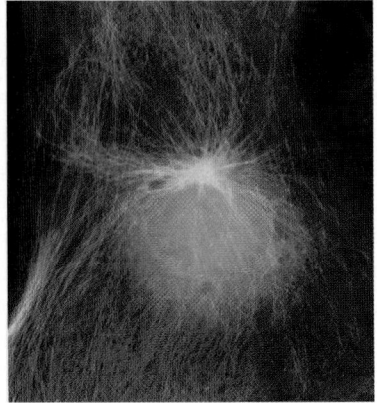

## G₂ of Interphase

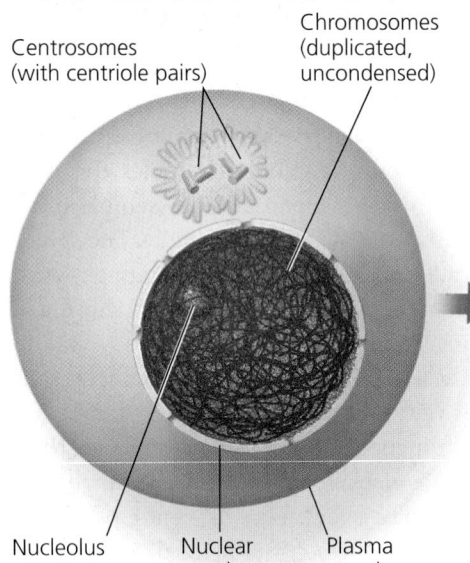

Centrosomes (with centriole pairs) — Chromosomes (duplicated, uncondensed)

Nucleolus — Nuclear envelope — Plasma membrane

### G₂ of Interphase

- A nuclear envelope encloses the nucleus.
- The nucleus contains one or more nucleoli (singular, *nucleolus*).
- Two centrosomes have formed by duplication of a single centrosome. Centrosomes are regions in animal cells that organize the microtubules of the spindle. Each centrosome contains two centrioles.
- Chromosomes, duplicated during S phase, cannot be seen individually because they have not yet condensed.

The fluorescence micrographs show dividing lung cells from a newt; this species has 22 chromosomes. Chromosomes appear blue, microtubules green, and intermediate filaments red. For simplicity, the drawings show only 6 chromosomes.

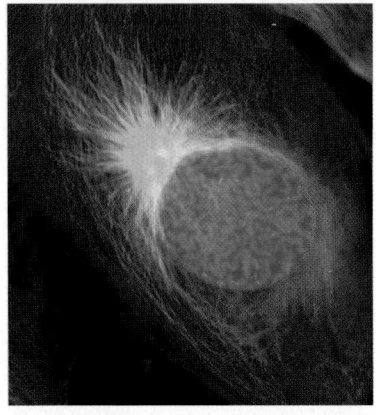

## Prophase

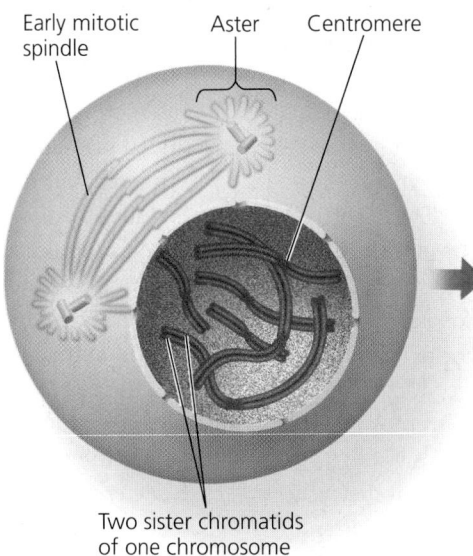

Early mitotic spindle — Aster — Centromere

Two sister chromatids of one chromosome

### Prophase

- The chromatin fibers become more tightly coiled, condensing into discrete chromosomes observable with a light microscope.
- The nucleoli disappear.
- Each duplicated chromosome appears as two identical sister chromatids joined at their centromeres and, in some species, all along their arms by cohesins (sister chromatid cohesion).
- The mitotic spindle (named for its shape) begins to form. It is composed of the centrosomes and the microtubules that extend from them. The radial arrays of shorter microtubules that extend from the centrosomes are called asters ("stars").
- The centrosomes move away from each other, propelled partly by the lengthening microtubules between them.

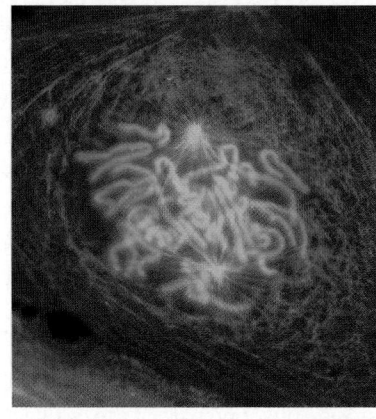

## Prometaphase

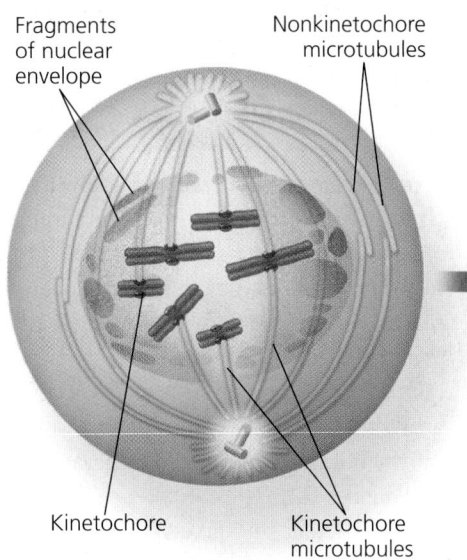

Fragments of nuclear envelope — Nonkinetochore microtubules

Kinetochore — Kinetochore microtubules

### Prometaphase

- The nuclear envelope fragments.
- The microtubules extending from each centrosome can now invade the nuclear area.
- The chromosomes have become even more condensed.
- Each of the two chromatids of each chromosome now has a kinetochore, a specialized protein structure at the centromere.
- Some of the microtubules attach to the kinetochores, becoming "kinetochore microtubules," which jerk the chromosomes back and forth.
- Nonkinetochore microtubules interact with those from the opposite pole of the spindle.

**?** *How many molecules of DNA are in the prometaphase drawing? How many molecules per chromosome? How many double helices are there per chromosome? Per chromatid?*

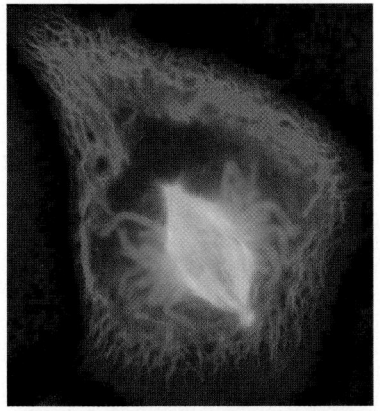

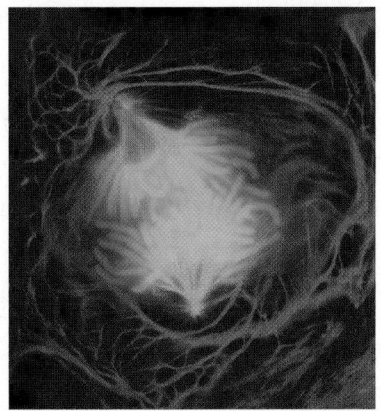

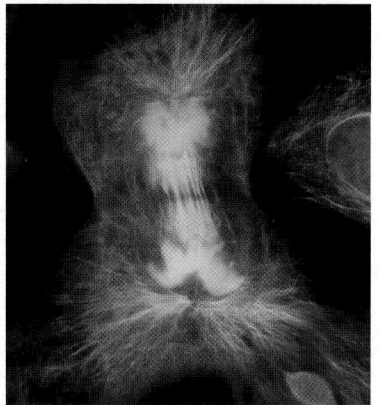

| Metaphase | Anaphase | Telophase and Cytokinesis |

10 μm

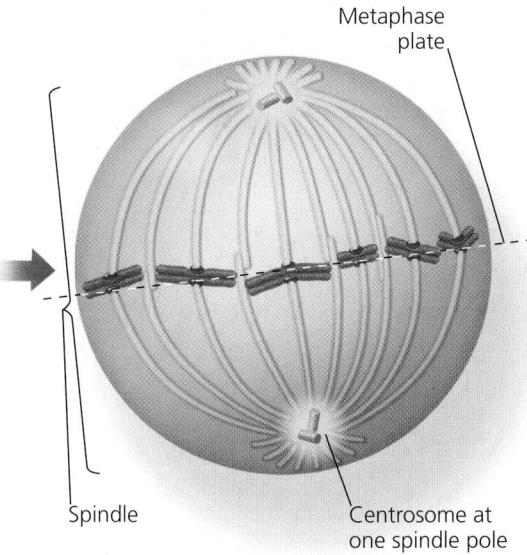

Metaphase plate

Spindle

Centrosome at one spindle pole

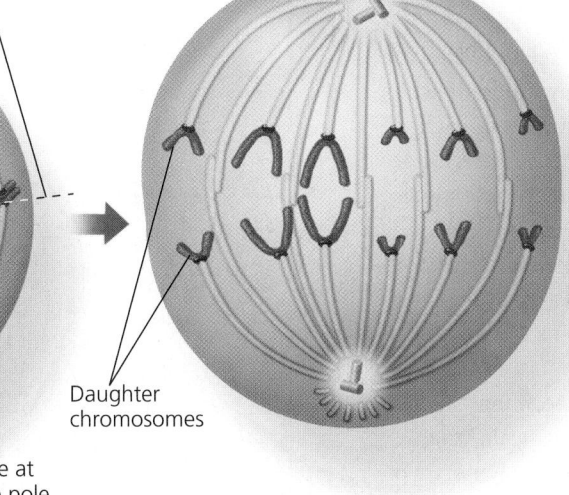

Daughter chromosomes

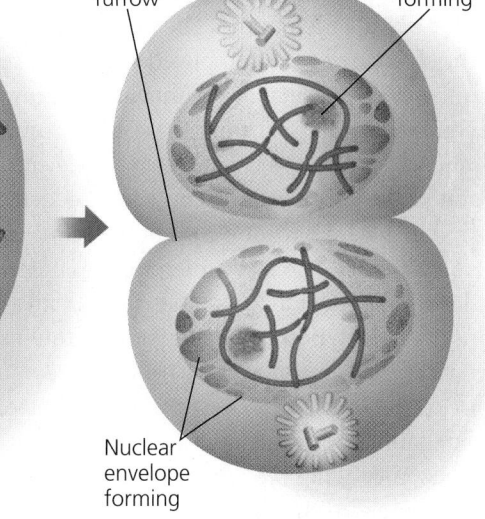

Cleavage furrow

Nucleolus forming

Nuclear envelope forming

## Metaphase

- The centrosomes are now at opposite poles of the cell.

- The chromosomes have all arrived at the *metaphase plate*, a plane that is equidistant between the spindle's two poles. The chromosomes' centromeres lie at the metaphase plate.

- For each chromosome, the kinetochores of the sister chromatids are attached to kinetochore microtubules coming from opposite poles.

## Anaphase

- Anaphase is the shortest stage of mitosis, often lasting only a few minutes.

- Anaphase begins when the cohesin proteins are cleaved. This allows the two sister chromatids of each pair to part suddenly. Each chromatid thus becomes a full-fledged chromosome.

- The two liberated daughter chromosomes begin moving toward opposite ends of the cell as their kinetochore microtubules shorten. Because these microtubules are attached at the centromere region, the chromosomes move centromere first (at about 1 μm/min).

- The cell elongates as the nonkinetochore microtubules lengthen.

- By the end of anaphase, the two ends of the cell have equivalent—and complete— collections of chromosomes.

## Telophase

- Two daughter nuclei form in the cell. Nuclear envelopes arise from the fragments of the parent cell's nuclear envelope and other portions of the endomembrane system.

- Nucleoli reappear.

- The chromosomes become less condensed.

- Any remaining spindle microtubules are depolymerized.

- Mitosis, the division of one nucleus into two genetically identical nuclei, is now complete.

### Cytokinesis

- The division of the cytoplasm is usually well under way by late telophase, so the two daughter cells appear shortly after the end of mitosis.

- In animal cells, cytokinesis involves the formation of a cleavage furrow, which pinches the cell in two.

**ANIMATION** Visit the Study Area in **MasteringBiology** for the BioFlix® 3-D Animation on Mitosis.

The kinetochore acts as a coupling device that attaches the motor of the spindle to the cargo that it moves—the chromosome. When one of a chromosome's kinetochores is "captured" by microtubules, the chromosome begins to move toward the pole from which those microtubules extend. However, this movement is checked as soon as microtubules from the opposite pole attach to the kinetochore on the other chromatid. What happens next is like a tug-of-war that ends in a draw. The chromosome moves first in one direction and then in the other, back and forth, finally settling midway between the two ends of the cell. At metaphase, the centromeres of all the duplicated chromosomes are on a plane midway between the spindle's two poles. This plane is called the **metaphase plate**, which is an imaginary plate rather than an actual cellular structure **(Figure 9.8)**. Meanwhile, microtubules that do not attach to kinetochores have been elongating, and by metaphase they overlap and interact with other nonkinetochore microtubules from the opposite pole of the spindle. By metaphase, the microtubules of the asters have also grown and are in contact with the plasma membrane. The spindle is now complete.

The structure of the completed spindle correlates well with its function during anaphase. Anaphase begins suddenly when the cohesins holding together the sister chromatids of each chromosome are cleaved by an enzyme called *separase*. Once separated, the chromatids become full-fledged chromosomes that move toward opposite ends of the cell.

How do the kinetochore microtubules function in this poleward movement of chromosomes? Apparently, two mechanisms are in play, both involving motor proteins. (To review how motor proteins move an object along a microtubule, see Figure 4.21.) Results of a cleverly designed experiment suggested that motor proteins on the kinetochores "walk" the chromosomes along the microtubules, which depolymerize at their kinetochore ends after the motor proteins have passed **(Figure 9.9)**. (This is referred to as the "Pac-man" mechanism because of its resemblance to the arcade game character that moves by eating all the dots in its path.) However, other researchers, working with different cell types or cells from other species, have shown that chromosomes are "reeled in" by motor proteins at the spindle poles and that the microtubules depolymerize after they pass by these motor proteins. The general consensus now is that both mechanisms are used and that their relative contributions vary among cell types.

In a dividing animal cell, the nonkinetochore microtubules are responsible for elongating the whole cell during anaphase. Nonkinetochore microtubules from opposite poles overlap each other extensively during metaphase (see Figure 9.8). During anaphase, the region of overlap is reduced as motor proteins attached to the microtubules walk them away from one another, using energy from ATP. As the microtubules push apart from each other, their spindle poles are pushed apart, elongating the cell. At the same time, the microtubules lengthen somewhat by the addition of tubulin subunits to their overlapping ends. As a result, the microtubules continue to overlap.

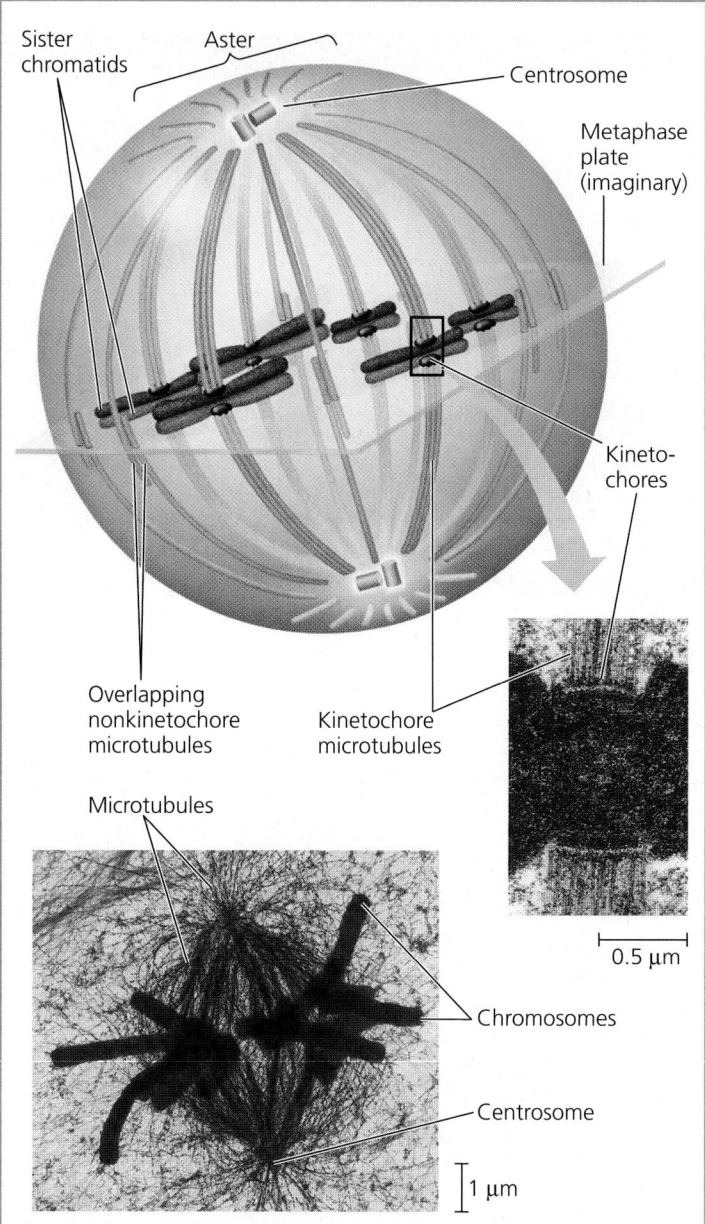

▲ **Figure 9.8 The mitotic spindle at metaphase.** The kinetochores of each chromosome's two sister chromatids face in opposite directions. Here, each kinetochore is attached to a cluster of kinetochore microtubules extending from the nearest centrosome. Nonkinetochore microtubules overlap at the metaphase plate (TEMs).

**DRAW IT** *On the lower micrograph, draw a line indicating the position of the metaphase plate. Circle an aster. Draw arrows indicating the directions of chromosome movement once anaphase begins.*

At the end of anaphase, duplicate groups of chromosomes have arrived at opposite ends of the elongated parent cell. Nuclei re-form during telophase. Cytokinesis generally begins during anaphase or telophase, and the spindle eventually disassembles by depolymerization of microtubules.

## Cytokinesis: *A Closer Look*

In animal cells, cytokinesis occurs by a process known as **cleavage**. The first sign of cleavage is the appearance of a **cleavage furrow**, a shallow groove in the cell surface near the old metaphase plate **(Figure 9.10a)**. On the cytoplasmic side

## ▼ Figure 9.9 Inquiry

### At which end do kinetochore microtubules shorten during anaphase?

**Experiment** Gary Borisy and colleagues at the University of Wisconsin wanted to determine whether kinetochore microtubules depolymerize at the kinetochore end or the pole end as chromosomes move toward the poles during mitosis. First they labeled the microtubules of a pig kidney cell in early anaphase with a yellow fluorescent dye. (Nonkinetochore microtubules are not shown.)

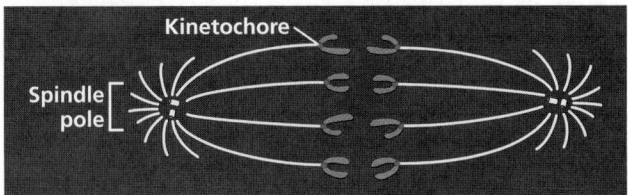

Then they marked a region of the kinetochore microtubules between one spindle pole and the chromosomes by using a laser to eliminate the fluorescence from that region, while leaving the microtubules intact (see below). As anaphase proceeded, they monitored the changes in microtubule length on either side of the mark.

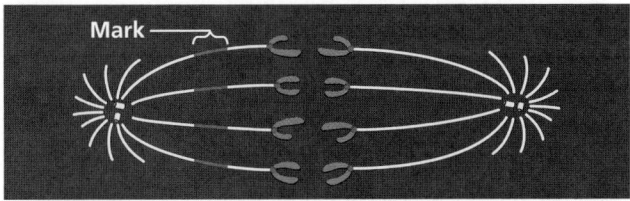

**Results** As the chromosomes moved poleward, the microtubule segments on the kinetochore side of the mark shortened, while those on the spindle pole side stayed the same length.

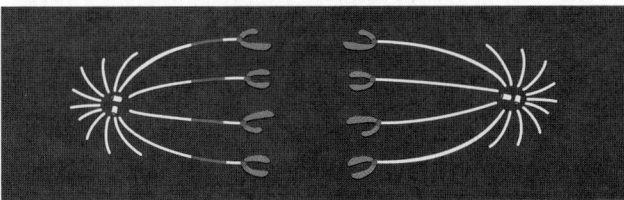

**Conclusion** During anaphase in this cell type, chromosome movement is correlated with kinetochore microtubules shortening at their kinetochore ends and not at their spindle pole ends. This experiment supports the hypothesis that during anaphase, a chromosome is walked along a microtubule as the microtubule depolymerizes at its kinetochore end, releasing tubulin subunits.

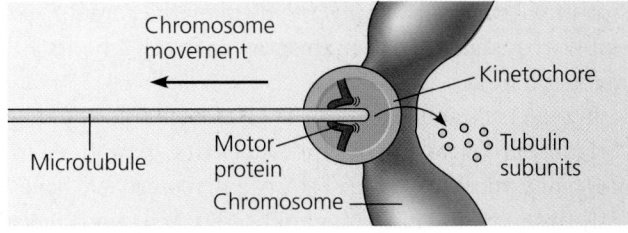

**Data from** G. J. Gorbsky, P. J. Sammak, and G. G. Borisy, Chromosomes move poleward in anaphase along stationary microtubules that coordinately disassemble from their kinetochore ends, *Journal of Cell Biology* 104:9–18 (1987).

**WHAT IF?** If this experiment had been done on a cell type in which "reeling in" at the poles was the main cause of chromosome movement, how would the mark have moved relative to the poles? How would the microtubule segment lengths have changed?

## ▼ Figure 9.10 Cytokinesis in animal and plant cells.

### (a) Cleavage of an animal cell (SEM)

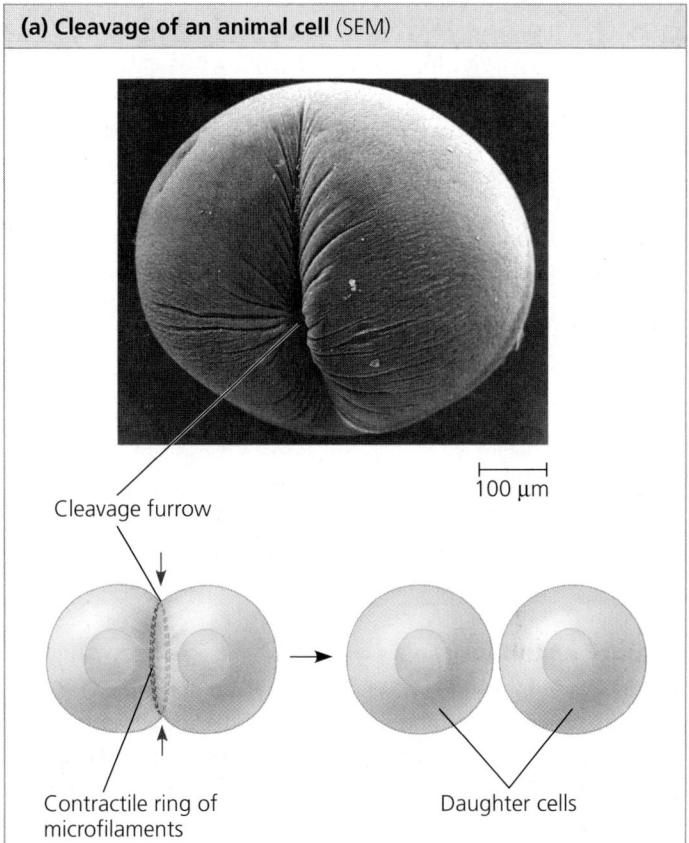

### (b) Cell plate formation in a plant cell (TEM)

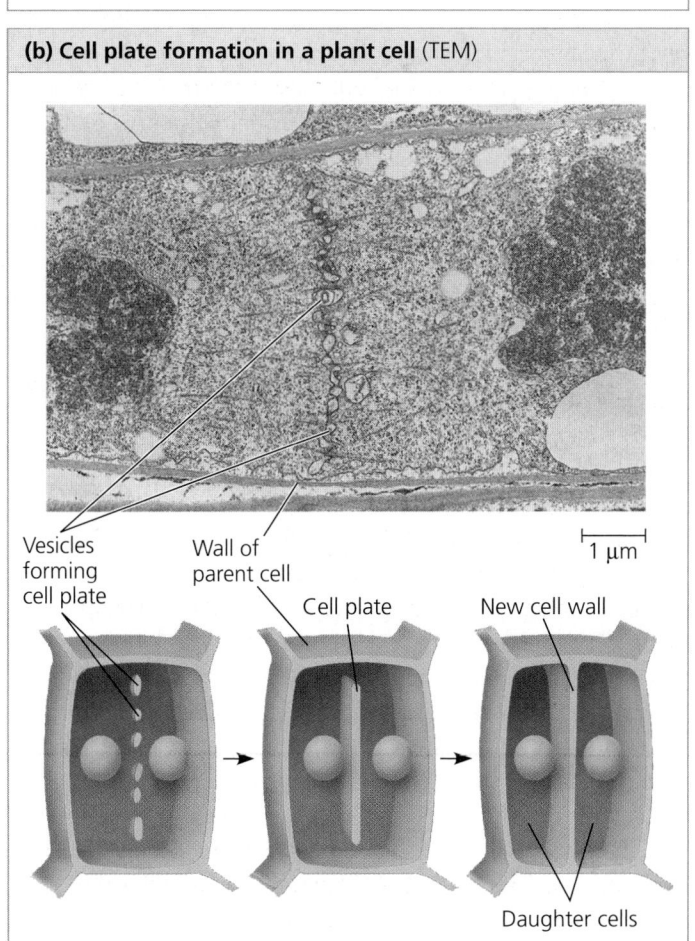

Nucleus
Nucleolus
Chromosomes condensing
Chromosomes
Cell plate
10 μm

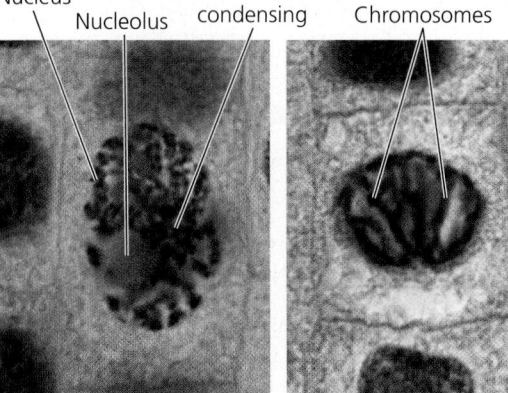

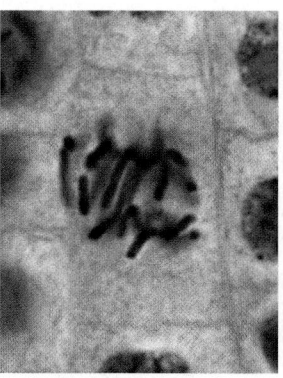
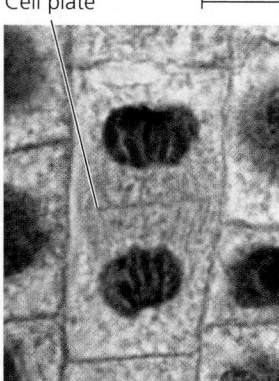

**❶ Prophase.** The chromosomes are condensing and the nucleolus is beginning to disappear. Although not yet visible in the micrograph, the mitotic spindle is starting to form.

**❷ Prometaphase.** Discrete chromosomes are now visible; each consists of two aligned, identical sister chromatids. Later in prometaphase, the nuclear envelope will fragment.

**❸ Metaphase.** The spindle is complete, and the chromosomes, attached to microtubules at their kinetochores, are all at the metaphase plate.

**❹ Anaphase.** The chromatids of each chromosome have separated, and the daughter chromosomes are moving to the ends of the cell as their kinetochore microtubules shorten.

**❺ Telophase.** Daughter nuclei are forming. Meanwhile, cytokinesis has started: The cell plate, which will divide the cytoplasm in two, is growing toward the perimeter of the parent cell.

▲ **Figure 9.11 Mitosis in a plant cell.** These light micrographs show mitosis in cells of an onion root.

of the furrow is a contractile ring of actin microfilaments associated with molecules of the protein myosin. The actin microfilaments interact with the myosin molecules, causing the ring to contract. The contraction of the dividing cell's ring of microfilaments is like the pulling of a drawstring. The cleavage furrow deepens until the parent cell is pinched in two, producing two completely separated cells, each with its own nucleus and its own share of cytosol, organelles, and other subcellular structures.

Cytokinesis in plant cells, which have cell walls, is markedly different. There is no cleavage furrow. Instead, during telophase, vesicles derived from the Golgi apparatus move along microtubules to the middle of the cell, where they coalesce, producing a **cell plate (Figure 9.10b)**. Cell wall materials carried in the vesicles collect inside the cell plate as it grows. The cell plate enlarges until its surrounding membrane fuses with the plasma membrane along the perimeter of the cell. Two daughter cells result, each with its own plasma membrane. Meanwhile, a new cell wall arising from the contents of the cell plate has formed between the daughter cells.

**Figure 9.11** is a series of micrographs of a dividing plant cell. Examining this figure will help you review mitosis and cytokinesis.

## Binary Fission in Bacteria

Prokaryotes (bacteria and archaea) undergo a type of reproduction in which the cell grows to roughly double its size and then divides to form two cells. The term **binary fission**, meaning "division in half," refers to this process and to the asexual reproduction of single-celled eukaryotes, such as the amoeba in Figure 9.2a. However, the process in eukaryotes involves mitosis, while that in prokaryotes does not.

In bacteria, most genes are carried on a single bacterial chromosome that consists of a circular DNA molecule and associated proteins. Although bacteria are smaller and simpler than eukaryotic cells, the challenge of replicating their genomes in an orderly fashion and distributing the copies equally to two daughter cells is still formidable. For example, when it is fully stretched out, the chromosome of the bacterium *Escherichia coli* is about 500 times as long as the cell. For such a long chromosome to fit within the cell, it must be highly coiled and folded.

In *E. coli*, the process of cell division is initiated when the DNA of the bacterial chromosome begins to replicate at a specific place on the chromosome called the **origin of replication**, producing two origins. As the chromosome continues to replicate, one origin moves rapidly toward the opposite end of the cell **(Figure 9.12)**. While the chromosome is replicating, the cell elongates. When replication is complete and the bacterium has reached about twice its initial size, its plasma membrane pinches inward, dividing the parent *E. coli* cell into two daughter cells. In this way, each cell inherits a complete genome.

Using the techniques of modern DNA technology to tag the origins of replication with molecules that glow green in fluorescence microscopy (see Figure 4.3), researchers have directly observed the movement of bacterial chromosomes. This movement is reminiscent of the poleward movements of the centromere regions of eukaryotic chromosomes during anaphase of mitosis, but bacteria don't have visible mitotic spindles or even microtubules. In most bacterial species studied, the two origins of replication end up at opposite ends of the cell or in some other very specific location, possibly anchored there by one or more proteins. How bacterial

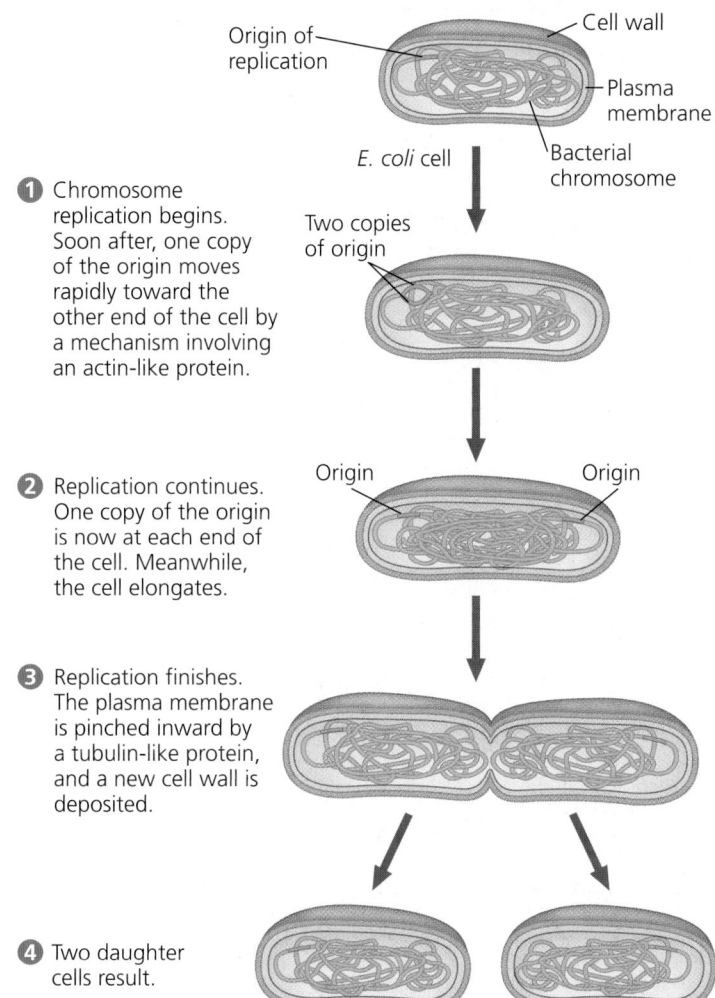

① **Chromosome replication begins.** Soon after, one copy of the origin moves rapidly toward the other end of the cell by a mechanism involving an actin-like protein.

Origin of replication
Cell wall
Plasma membrane
Bacterial chromosome
*E. coli* cell

② **Replication continues.** One copy of the origin is now at each end of the cell. Meanwhile, the cell elongates.

Two copies of origin

③ **Replication finishes.** The plasma membrane is pinched inward by a tubulin-like protein, and a new cell wall is deposited.

Origin          Origin

④ **Two daughter cells result.**

▲ **Figure 9.12 Bacterial cell division by binary fission.** The bacterium *E. coli*, shown here, has a single, circular chromosome.

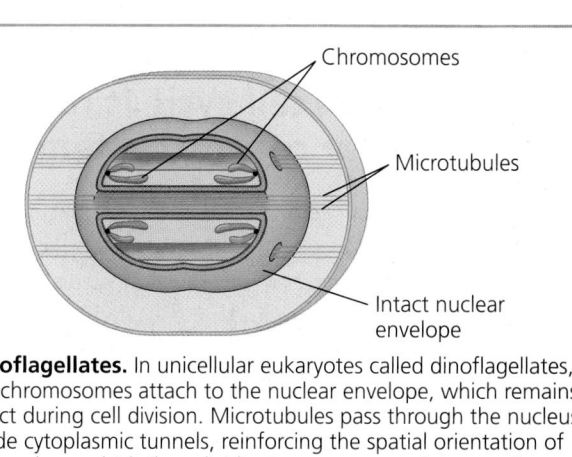

Chromosomes
Microtubules
Intact nuclear envelope

**(a) Dinoflagellates.** In unicellular eukaryotes called dinoflagellates, the chromosomes attach to the nuclear envelope, which remains intact during cell division. Microtubules pass through the nucleus inside cytoplasmic tunnels, reinforcing the spatial orientation of the nucleus, which then divides in a process reminiscent of bacterial binary fission.

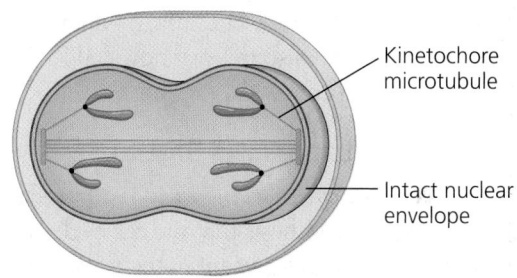

Kinetochore microtubule
Intact nuclear envelope

**(b) Diatoms and some yeasts.** In two other groups of unicellular eukaryotes, diatoms and some yeasts, the nuclear envelope also remains intact during cell division. In these organisms, the microtubules form a spindle *within* the nucleus. Microtubules separate the chromosomes, and the nucleus splits into two daughter nuclei.

▲ **Figure 9.13 Mechanisms of cell division.** Some unicellular eukaryotes existing today have mechanisms of cell division that may resemble intermediate steps in the evolution of mitosis.

chromosomes move and how their specific location is established and maintained are active areas of research. Several proteins have been identified that play important roles. Polymerization of one protein resembling eukaryotic actin apparently functions in bacterial chromosome movement during cell division, and another protein that is related to tubulin helps pinch the plasma membrane inward, separating the two bacterial daughter cells.

## The Evolution of Mitosis

**EVOLUTION** Given that prokaryotes preceded eukaryotes on Earth by more than a billion years, we might hypothesize that mitosis evolved from simpler prokaryotic mechanisms of cell reproduction. The fact that some of the proteins involved in bacterial binary fission are related to eukaryotic proteins that function in mitosis supports that hypothesis.

As eukaryotes with nuclear envelopes and larger genomes evolved, the ancestral process of binary fission, seen today in bacteria, somehow gave rise to mitosis. Variations on cell division exist in different groups. These variant processes may be similar to mechanisms used by ancestral species and thus may resemble steps in the evolution of mitosis from a binary fission-like process carried out by very early bacteria. Possible intermediate stages are suggested by two unusual types of nuclear division found today in certain unicellular eukaryotes—dinoflagellates, diatoms, and some yeasts **(Figure 9.13)**. These are thought to be cases where ancestral mechanisms have remained relatively unchanged over evolutionary time. In both types, the nuclear envelope remains intact, in contrast to what happens in most eukaryotic cells.

## CONCEPT CHECK 9.2

1. How many chromosomes are shown in the drawing in Figure 9.8? Are they duplicated? How many chromatids are shown?

2. Compare cytokinesis in animal cells and plant cells.

3. During which stages of the cell cycle does a chromosome consist of two identical chromatids?

4. Compare the roles of tubulin and actin during eukaryotic cell division with the roles of tubulin-like and actin-like proteins during bacterial binary fission.

For suggested answers, see Appendix A.

## CONCEPT 9.3

# The eukaryotic cell cycle is regulated by a molecular control system

The timing and rate of cell division in different parts of a plant or animal are crucial to normal growth, development, and maintenance. The frequency of cell division varies with the type of cell. For example, human skin cells divide frequently throughout life, whereas liver cells maintain the ability to divide but keep it in reserve until an appropriate need arises— say, to repair a wound. Some of the most specialized cells, such as fully formed nerve cells and muscle cells, do not divide at all in a mature human. These cell cycle differences result from regulation at the molecular level. The mechanisms of this regulation are of great interest, not only to understand the life cycles of normal cells but also to learn how cancer cells manage to escape the usual controls.

## Evidence for Cytoplasmic Signals

What controls the cell cycle? In the early 1970s, a variety of experiments led to the hypothesis that the cell cycle is driven by specific signaling molecules present in the cytoplasm. Some of the first strong evidence for this hypothesis came from experiments with mammalian cells grown in culture (**Figure 9.14**). In these experiments, two cells in different phases of the cell cycle were fused to form a single cell with two nuclei. If one of the original cells was in the S phase and the other was in $G_1$, the $G_1$ nucleus immediately entered the S phase, as though stimulated by signaling molecules present in the cytoplasm of the first cell. Similarly, if a cell undergoing mitosis (M phase) was fused with another cell in any stage of its cell cycle, even $G_1$, the second nucleus immediately entered mitosis, with condensation of the chromatin and formation of a mitotic spindle.

## Checkpoints of the Cell Cycle Control System

The experiment shown in Figure 9.14 and other experiments on animal cells and yeasts demonstrated that the sequential events of the cell cycle are directed by a distinct **cell cycle control system**, a cyclically operating set of molecules in the cell that both triggers and coordinates key events in the cell cycle. The cell cycle control system has been compared to the control device of a washing machine (**Figure 9.15**). Like the washer's timing device, the cell cycle control system proceeds on its own, according to a built-in clock. However, just as a washer's cycle is subject to both internal control (such as the sensor that detects when the tub is filled with water) and external adjustment (such as starting the machine), the cell cycle is regulated at certain checkpoints by both internal and external signals.

A **checkpoint** in the cell cycle is a control point where stop and go-ahead signals can regulate the cycle. (The signals are transmitted within the cell by the kinds of signal transduction pathways discussed in Concept 5.6.) Animal cells generally have built-in stop signals that halt the cell cycle at checkpoints

▼ **Figure 9.14**  Inquiry

### Do molecular signals in the cytoplasm regulate the cell cycle?

**Experiment**  Researchers at the University of Colorado wondered whether a cell's progression through the cell cycle is controlled by cytoplasmic molecules. They induced cultured mammalian cells at different phases of the cell cycle to fuse. Two experiments are shown.

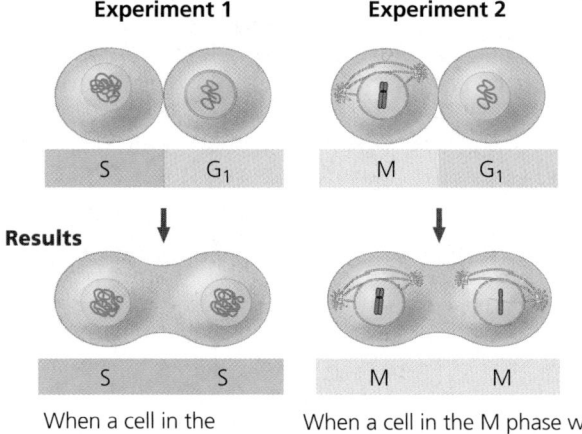

When a cell in the S phase was fused with a cell in $G_1$, the $G_1$ nucleus immediately entered the S phase—DNA was synthesized.

When a cell in the M phase was fused with a cell in $G_1$, the $G_1$ nucleus immediately began mitosis—a spindle formed and the chromosomes condensed, even though the chromosomes had not been duplicated.

**Conclusion**  The results of fusing a $G_1$ cell with a cell in the S or M phase of the cell cycle suggest that molecules present in the cytoplasm during the S or M phase control the progression to those phases.

**Data from**  R. T. Johnson and P. N. Rao, Mammalian cell fusion: Induction of premature chromosome condensation in interphase nuclei, *Nature* 226:717–722 (1970).

**WHAT IF?**  If the progression of phases did not depend on cytoplasmic molecules and each phase began when the previous one was complete, how would the results have differed?

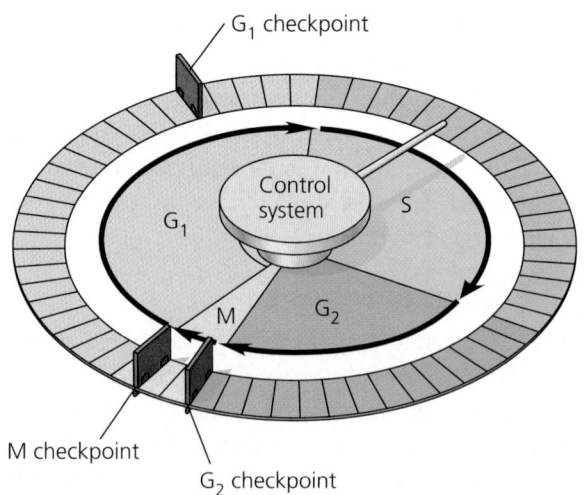

▲ **Figure 9.15  Mechanical analogy for the cell cycle control system.** In this diagram of the cell cycle, the flat "stepping stones" around the perimeter represent sequential events. Like the control device of an automatic washer, the cell cycle control system proceeds on its own, driven by a built-in clock. However, the system is subject to internal and external regulation at various checkpoints; three important checkpoints are shown (red).

until overridden by go-ahead signals. Many signals registered at checkpoints come from cellular surveillance mechanisms inside the cell. These signals report whether crucial cellular processes that should have occurred by that point have in fact been completed correctly and thus whether or not the cell cycle should proceed. Checkpoints also register signals from outside the cell, as we'll discuss later. Three important checkpoints are those in the $G_1$, $G_2$, and M phases (red gates in Figure 9.15).

For many cells, the $G_1$ checkpoint—dubbed the "restriction point" in mammalian cells—seems to be the most important. If a cell receives a go-ahead signal at the $G_1$ checkpoint, it will usually complete the $G_1$, S, $G_2$, and M phases and divide. If it does not receive a go-ahead signal at that point, it may exit the cycle, switching into a nondividing state called the **$G_0$ phase (Figure 9.16a)**. Most cells of the human body are actually in the $G_0$ phase. As mentioned earlier, mature nerve cells and muscle cells never divide. Other cells, such as liver cells, can be "called back" from the $G_0$ phase to the cell cycle by external cues, such as growth factors released during injury.

The cell cycle is regulated at the molecular level by a set of regulatory proteins and protein complexes, including proteins called *cyclins*, and other proteins interacting with cyclins that are kinases (enzymes that activate or inactivate other proteins by phosphorylating them; see Figure 5.24). To understand how a cell progresses through the cycle, let's consider the checkpoint signals that can make the cell cycle clock pause or continue.

Biologists are currently working out the pathways that link signals originating inside and outside the cell with the responses by cyclins, kinases, and other proteins. An example of an internal signal occurs at the third important checkpoint, the M phase checkpoint **(Figure 9.16b)**. Anaphase, the separation of sister chromatids, does not begin until all the chromosomes are properly attached to the spindle at the metaphase plate. Researchers have learned that as long as some kinetochores are unattached to spindle microtubules, the sister chromatids remain together, delaying anaphase. Only when the kinetochores of all the chromosomes are properly attached to the spindle does the appropriate regulatory protein complex become activated. Once activated, the complex sets off a chain

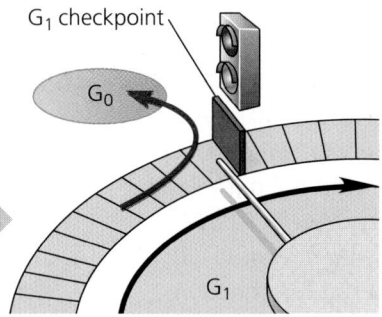 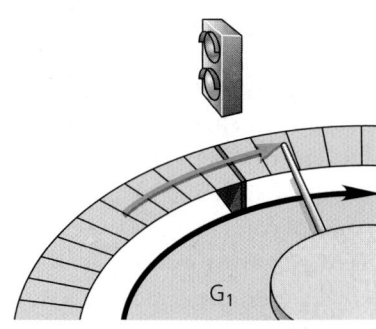

In the absence of a go-ahead signal, a cell exits the cell cycle and enters $G_0$, a nondividing state.

If a cell receives a go-ahead signal, the cell continues on in the cell cycle.

**(a) $G_1$ checkpoint**

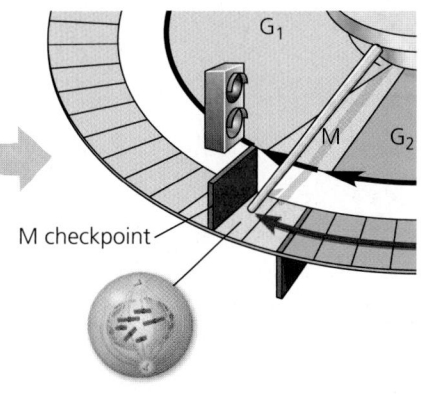

M checkpoint

Prometaphase

A cell in mitosis receives a stop signal when any of its chromosomes are not attached to spindle fibers.

Anaphase

$G_2$ checkpoint

Metaphase

When all chromosomes are attached to spindle fibers from both poles, a go-ahead signal allows the cell to proceed into anaphase.

**(b) M checkpoint**

▲ **Figure 9.16 Two important checkpoints.** At certain checkpoints in the cell cycle (red gates), cells do different things depending on the signals they receive. Events of the **(a)** $G_1$ and **(b)** M checkpoints are shown. In part (b), the $G_2$ checkpoint has already been passed by the cell.

**WHAT IF?** *What might be the result if the cell ignored either checkpoint and progressed through the cell cycle?*

of molecular events that activates the enzyme separase, which cleaves the cohesins, allowing the sister chromatids to separate. This mechanism ensures that daughter cells do not end up with missing or extra chromosomes.

Studies using animal cells in culture have led to the identification of many external factors, both chemical and physical, that can influence cell division. For example, cells fail to divide if an essential nutrient is lacking in the culture medium. (This is analogous to trying to run a washing machine without the water supply hooked up; an internal sensor won't allow the machine to continue past the point where water is needed.) And even if all other conditions are favorable, most types of mammalian cells divide in culture only if the growth medium includes specific growth factors. As mentioned in Concept 5.6, a **growth factor** is a protein released by certain cells that stimulates other cells to divide. Different cell types respond specifically to different growth factors or combinations of growth factors.

Consider, for example, *platelet-derived growth factor (PDGF)*, which is made by blood cell fragments called platelets. The experiment illustrated in **Figure 9.17** demonstrates that PDGF is required for the division of cultured fibroblasts, a type of connective tissue cell. Fibroblasts have PDGF receptors on their plasma membranes. The binding of PDGF molecules to these receptors triggers a signal transduction pathway that allows the cells to pass the $G_1$ checkpoint and divide. PDGF stimulates fibroblast division not only in the artificial conditions of cell culture, but also in an animal's

body. When an injury occurs, platelets release PDGF in the vicinity. The resulting proliferation of fibroblasts helps heal the wound.

The effect of an external physical factor on cell division is clearly seen in **density-dependent inhibition**, a phenomenon in which crowded cells stop dividing **(Figure 9.18a)**. As first observed many years ago, cultured cells normally divide until they form a single layer of cells on the inner surface of a culture flask, at which point the cells stop dividing. If some cells are removed, those bordering the open space begin dividing again and continue until the vacancy is filled. Follow-up studies revealed that the binding of a cell-surface protein to its counterpart on an adjoining cell sends a signal to both cells that inhibits cell division, preventing the cells from moving forward in the cell cycle. Growth factors also have a

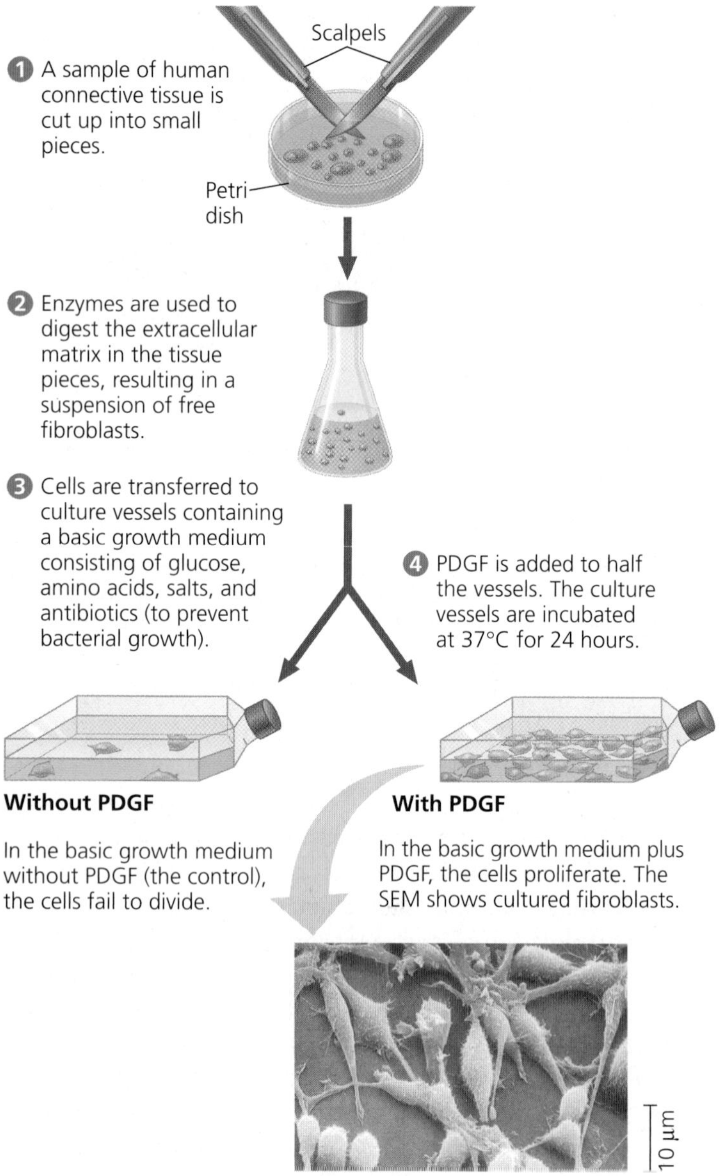

① A sample of human connective tissue is cut up into small pieces.

Scalpels

Petri dish

② Enzymes are used to digest the extracellular matrix in the tissue pieces, resulting in a suspension of free fibroblasts.

③ Cells are transferred to culture vessels containing a basic growth medium consisting of glucose, amino acids, salts, and antibiotics (to prevent bacterial growth).

④ PDGF is added to half the vessels. The culture vessels are incubated at 37°C for 24 hours.

**Without PDGF**

In the basic growth medium without PDGF (the control), the cells fail to divide.

**With PDGF**

In the basic growth medium plus PDGF, the cells proliferate. The SEM shows cultured fibroblasts.

10 μm

▲ **Figure 9.17 The effect of platelet-derived growth factor (PDGF) on cell division.**

**MAKE CONNECTIONS** *PDGF signals cells by binding to a cell-surface receptor that then becomes phosphorylated, activating it so that it sends a signal. If you added a chemical that blocked phosphorylation, how would the results differ? (See Figure 5.24.)*

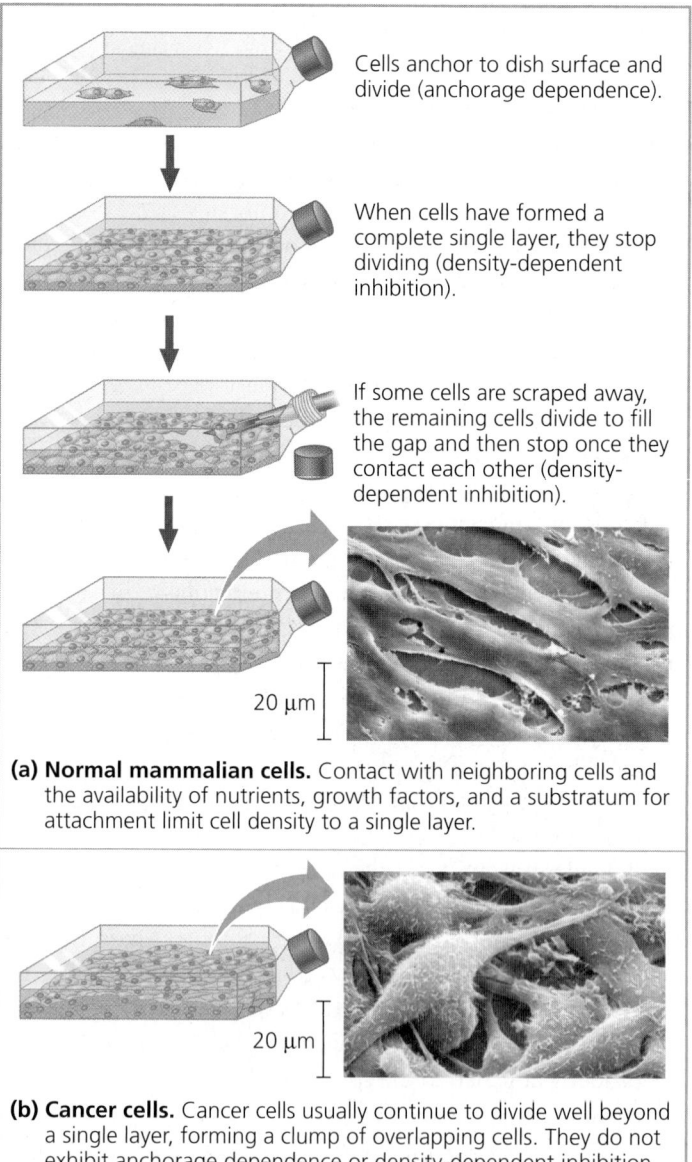

Cells anchor to dish surface and divide (anchorage dependence).

When cells have formed a complete single layer, they stop dividing (density-dependent inhibition).

If some cells are scraped away, the remaining cells divide to fill the gap and then stop once they contact each other (density-dependent inhibition).

20 μm

**(a) Normal mammalian cells.** Contact with neighboring cells and the availability of nutrients, growth factors, and a substratum for attachment limit cell density to a single layer.

20 μm

**(b) Cancer cells.** Cancer cells usually continue to divide well beyond a single layer, forming a clump of overlapping cells. They do not exhibit anchorage dependence or density-dependent inhibition.

▲ **Figure 9.18 Density-dependent inhibition and anchorage dependence of cell division.** Individual cells are shown disproportionately large in the drawings.

role in determining the density that cells attain before ceasing division.

Most animal cells also exhibit **anchorage dependence** (see Figure 9.18a). To divide, they must be attached to a substratum, such as the inside of a culture flask or the extracellular matrix of a tissue. Experiments suggest that like cell density, anchorage is signaled to the cell cycle control system via pathways involving plasma membrane proteins and elements of the cytoskeleton linked to them.

Density-dependent inhibition and anchorage dependence appear to function not only in cell culture but also in the body's tissues, checking the growth of cells at some optimal density and location during embryonic development and throughout an organism's life. Cancer cells, which we discuss next, exhibit neither density-dependent inhibition nor anchorage dependence **(Figure 9.18b)**.

## Loss of Cell Cycle Controls in Cancer Cells

Cancer cells do not heed the normal signals that regulate the cell cycle. In culture, they do not stop dividing when growth factors are depleted. A logical hypothesis is that cancer cells do not need growth factors in their culture medium to grow and divide. They may make a required growth factor themselves, or they may have an abnormality in the signaling pathway that conveys the growth factor's signal to the cell cycle control system even in the absence of that factor. Another possibility is an abnormal cell cycle control system. In these scenarios, the underlying basis of the abnormality is almost always a change in one or more genes (for example, a mutation) that alters the function of their protein products, resulting in faulty cell cycle control.

There are other important differences between normal cells and cancer cells that reflect derangements of the cell cycle. If and when they stop dividing, cancer cells do so at random points in the cycle, rather than at the normal checkpoints. Moreover, cancer cells can go on dividing indefinitely in culture if they are given a continual supply of nutrients; in essence, they are "immortal." A striking example is a cell line that has been reproducing in culture since 1951. Cells of this line are called HeLa cells because their original source was a tumor removed from a woman named *Henrietta Lacks*. Cells in culture that acquire the ability to divide indefinitely are said to have undergone a process called **transformation**, causing them to behave (in cell division, at least) like cancer cells. By contrast, nearly all normal, nontransformed mammalian cells growing in culture divide only about 20 to 50 times before they stop dividing, age, and die. Finally, cancer cells evade the normal controls that trigger a cell to undergo a type of programmed cell death called *apoptosis* when something is wrong—for example, when an irreparable mistake has occurred during DNA replication preceding mitosis.

The abnormal behavior of cancer cells can be catastrophic when it occurs in the body. The problem begins when a single cell in a tissue undergoes the first changes of the multistep process that converts a normal cell to a cancer cell. Such a cell often has altered proteins on its surface, and the body's immune system normally recognizes the cell as an insurgent—and destroys it. However, if the cell evades destruction, it may proliferate and form a tumor, a mass of abnormal cells within otherwise normal tissue. The abnormal cells may remain at the original site if they have too few genetic and cellular changes to survive at another site. In that case, the tumor is called a **benign tumor**. Most benign tumors do not cause serious problems and can be removed by surgery. In contrast, a **malignant tumor** includes cells whose genetic and cellular changes enable them to spread to new tissues and impair the functions of one or more organs; based on their ability to divide indefinitely in culture, these cells are also considered *transformed* cells. An individual with a malignant tumor is said to have cancer; **Figure 9.19** shows the development of breast cancer.

The changes that have occurred in cells of malignant tumors show up in many ways besides excessive proliferation. These

▼ **Figure 9.19 The growth and metastasis of a malignant breast tumor.** A series of genetic and cellular changes contribute to a tumor becoming malignant (cancerous). The cells of malignant tumors grow in an uncontrolled way and can spread to neighboring tissues and, via lymph and blood vessels, to other parts of the body (metastasis).

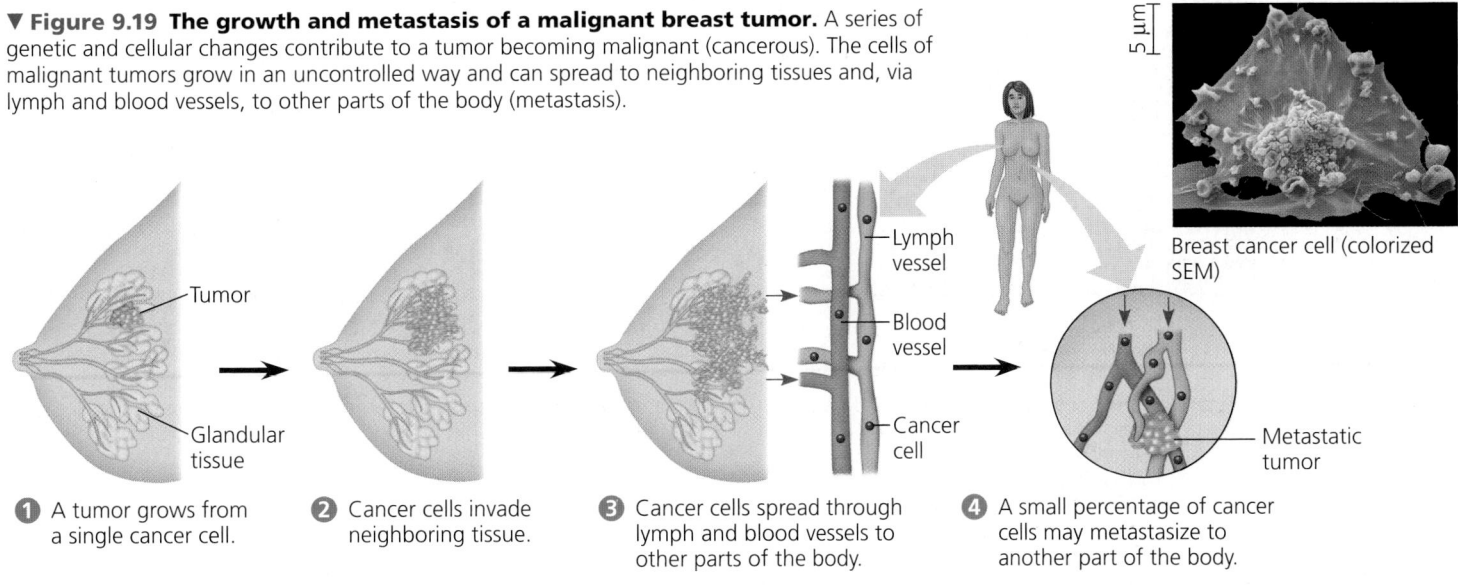

Breast cancer cell (colorized SEM)

❶ A tumor grows from a single cancer cell.

❷ Cancer cells invade neighboring tissue.

❸ Cancer cells spread through lymph and blood vessels to other parts of the body.

❹ A small percentage of cancer cells may metastasize to another part of the body.

cells may have unusual numbers of chromosomes, though whether this is a cause or an effect of transformation is an ongoing topic of debate. Their metabolism may be altered, and they may cease to function in any constructive way. Abnormal changes on the cell surface cause cancer cells to lose attachments to neighboring cells and the extracellular matrix, allowing them to spread into nearby tissues. Cancer cells may also secrete signaling molecules that cause blood vessels to grow toward the tumor. A few tumor cells may separate from the original tumor, enter blood vessels and lymph vessels, and travel to other parts of the body. There, they may proliferate and form a new tumor. This spread of cancer cells to locations distant from their original site is called **metastasis** (see Figure 9.19).

A tumor that appears to be localized may be treated with high-energy radiation, which damages DNA in cancer cells much more than DNA in normal cells, apparently because

the majority of cancer cells have lost the ability to repair such damage. To treat known or suspected metastatic tumors, chemotherapy is used, in which drugs that are toxic to actively dividing cells are administered through the circulatory system. As you might expect, chemotherapeutic drugs interfere with specific steps in the cell cycle. For example, the drug Taxol freezes the mitotic spindle by preventing microtubule depolymerization, which stops actively dividing cells from proceeding past metaphase and leads to their destruction. The side effects of chemotherapy are due to the drugs' effects on normal cells that divide often. For example, nausea results from chemotherapy's effects on intestinal cells, hair loss from effects on hair follicle cells, and susceptibility to infection from effects on immune system cells. You'll work with data from an experiment involving a potential chemotherapeutic agent in the **Scientific Skills Exercise**.

## Scientific Skills Exercise

### *Interpreting Histograms*

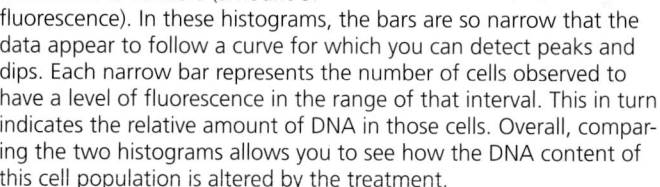

► Human glioblastoma cell

**At What Phase Is the Cell Cycle Arrested by an Inhibitor?** Many medical treatments are aimed at stopping cancer cell proliferation by blocking the cell cycle of cancerous tumor cells. One potential treatment is a cell cycle inhibitor derived from human umbilical cord stem cells. In this exercise, you will compare two histograms to determine where in the cell cycle the inhibitor blocks the division of cancer cells.

**How the Experiment Was Done** In the treated sample, human glioblastoma (brain cancer) cells were grown in tissue culture in the presence of the inhibitor, while control sample cells were grown in its absence. After 72 hours of growth, the two cell samples were harvested. To get a "snapshot" of the phase of the cell cycle each cell was in at that time, the samples were treated with a fluorescent chemical that binds to DNA and then run through a flow cytometer, an instrument that records the fluorescence level of each cell. Computer software then graphed the number of cells in each sample with a particular fluorescence level, as shown below.

**Data from the Experiment**

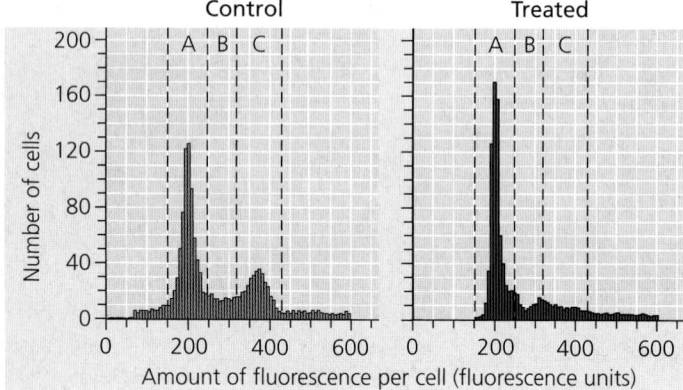

**Data from** K. K. Velpula et al., Regulation of glioblastoma progression by cord blood stem cells is mediated by downregulation of cyclin D1, *PLoS ONE* 6(3): e18017 (2011). doi:10.1371/journal.pone.0018017

The data are plotted in a type of graph called a histogram, which groups the values for a numeric variable on the *x*-axis into intervals.

A histogram allows you to see how all the experimental subjects (cells, in this case) are distributed along a continuous variable (amount of fluorescence). In these histograms, the bars are so narrow that the data appear to follow a curve for which you can detect peaks and dips. Each narrow bar represents the number of cells observed to have a level of fluorescence in the range of that interval. This in turn indicates the relative amount of DNA in those cells. Overall, comparing the two histograms allows you to see how the DNA content of this cell population is altered by the treatment.

#### INTERPRET THE DATA

1. Study the data in the histograms. (a) Which axis indirectly shows the relative amount of DNA per cell? Explain your answer. (b) In the control sample, compare the first peak in the histogram (in region A) to the second peak (in region C). Which peak shows the population of cells with the higher amount of DNA per cell? Explain. (For additional information about graphs, see the Scientific Skills Review in Appendix F and in the Study Area in MasteringBiology.)

2. (a) In the control sample histogram, identify the phase of the cell cycle ($G_1$, $S$, or $G_2$) of the population of cells in each region delineated by vertical lines. Label the histogram with these phases and explain your answer. (b) Does the S phase population of cells show a distinct peak in the histogram? Why or why not?

3. The histogram representing the treated sample shows the effect of growing the cancer cells alongside human umbilical cord stem cells that produce the potential inhibitor. (a) Label the histogram with the cell cycle phases. Which phase of the cell cycle has the greatest number of cells in the treated sample? Explain. (b) Compare the distribution of cells among $G_1$, $S$, and $G_2$ phases in the control and treated samples. What does this tell you about the cells in the treated sample? (c) Based on what you learned in Concept 9.3, propose a mechanism by which the stem cell–derived inhibitor might arrest the cancer cell cycle at this stage. (More than one answer is possible.)

(MB) A version of this Scientific Skills Exercise can be assigned in MasteringBiology.

Over the past several decades, researchers have produced a flood of valuable information about cell-signaling pathways and how their malfunction contributes to the development of cancer through effects on the cell cycle. Coupled with new molecular techniques, such as the ability to rapidly sequence the DNA of cells in a particular tumor, medical treatments for cancer are beginning to become more "personalized" to a particular patient's tumor (see Figure 16.21).

For example, the cells of roughly 20% of breast cancer tumors show abnormally high amounts of a cell-surface receptor tyrosine kinase called HER2, and many show an increase in the number of estrogen receptor (ER) molecules, intracellular receptors that can trigger cell division. Based on lab findings, a physician can prescribe chemotherapy with a molecule that blocks the function of the specific protein (Herceptin for HER2 and tamoxifen for ERs). Treatment using these agents, when appropriate, has led to increased survival rates and fewer cancer recurrences.

### CONCEPT CHECK 9.3

1. In Figure 9.14, why do the nuclei resulting from experiment 2 contain different amounts of DNA?
2. What phase are most of your body cells in?
3. Compare and contrast a benign tumor and a malignant tumor.
4. **WHAT IF?** What would happen if you performed the experiment in Figure 9.17 with cancer cells?

For suggested answers, see Appendix A.

# 9 | Chapter Review

Go to **MasteringBiology**® for Assignments, the eText, and the Study Area with Animations, Activities, Vocab Self-Quiz, and Practice Tests.

## SUMMARY OF KEY CONCEPTS

**VOCAB SELF-QUIZ**

goo.gl/gbai8v

- Unicellular organisms reproduce by **cell division**; multicellular organisms depend on cell division for their development from a fertilized egg and for growth and repair. Cell division is part of the **cell cycle**, an ordered sequence of events in the life of a cell.

### CONCEPT 9.1

**Most cell division results in genetically identical daughter cells (pp. 183–184)**

- The genetic material (DNA) of a cell—its **genome**—is partitioned among **chromosomes**. Each eukaryotic chromosome consists of one DNA molecule associated with many proteins. Together, the complex of DNA and associated proteins is called **chromatin**. The chromatin of a chromosome exists in different states of condensation at different times. In animals, **gametes** have one set of chromosomes, and **somatic cells** have two sets.
- Cells replicate their genetic material before they divide, each daughter cell receiving a copy of the DNA. Prior to cell division, chromosomes are duplicated. Each one then consists of two identical **sister chromatids** joined along their lengths by sister chromatid cohesion and held most tightly together at a constricted region at the **centromeres**. When this cohesion is broken, the chromatids separate during cell division, becoming the chromosomes of the daughter cells. Eukaryotic cell division consists of **mitosis** (division of the nucleus) and **cytokinesis** (division of the cytoplasm).

**?** *Differentiate between these terms: chromosome, chromatin, and chromatid.*

### CONCEPT 9.2

**The mitotic phase alternates with interphase in the cell cycle (pp. 185–191)**

- Between divisions, a cell is in **interphase**: the $G_1$, **S**, and $G_2$ **phases**. The cell grows throughout interphase, with DNA being replicated only during the synthesis (S) phase. Mitosis and cytokinesis make up the **mitotic (M) phase** of the cell cycle.

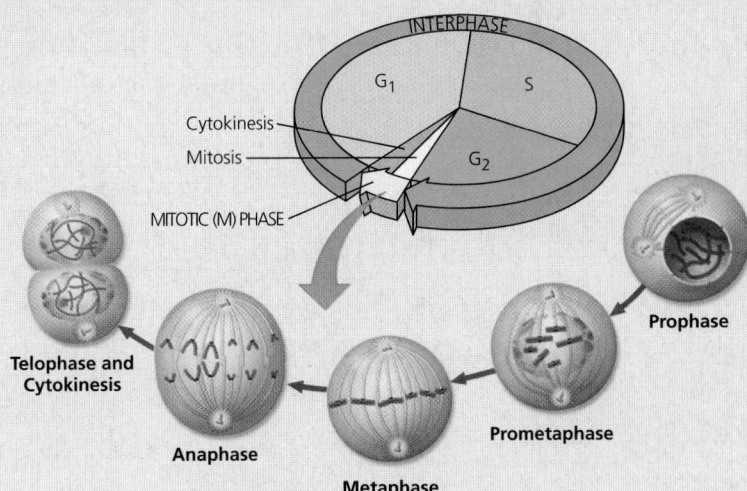

- The **mitotic spindle**, made up of microtubules, controls chromosome movement during mitosis. In animal cells, the mitotic spindle arises from the **centrosomes** and includes spindle microtubules and **asters**. Some spindle microtubules attach to the **kinetochores** of chromosomes and move the chromosomes to the **metaphase plate**. After sister chromatids separate, motor proteins move them along kinetochore microtubules toward opposite ends of the cell. The cell elongates when motor proteins push nonkinetochore microtubules from opposite poles away from each other.
- Mitosis is usually followed by cytokinesis. Animal cells carry out cytokinesis by **cleavage**, and plant cells form a **cell plate**.
- During **binary fission** in bacteria, the chromosome replicates and the daughter chromosomes actively move apart. Some of the proteins involved in bacterial binary fission are related to eukaryotic actin and tubulin. Since prokaryotes preceded eukaryotes by more than a billion years, it is likely that mitosis evolved from prokaryotic cell division.

**?** *In which of the three subphases of interphase and the stages of mitosis do chromosomes exist as single DNA molecules?*

## The eukaryotic cell cycle is regulated by a molecular control system (pp. 192–197)

- Signaling molecules present in the cytoplasm regulate progress through the cell cycle.
- The **cell cycle control system** is molecularly based; key regulatory proteins are cyclins and kinases. The cell cycle clock has specific **checkpoints** where the cell cycle stops until a go-ahead signal is received; important checkpoints occur in the $G_1$, $G_2$, and M phases. Cell culture has enabled researchers to study the molecular details of cell division. Both internal signals and external signals control the cell cycle checkpoints via signal transduction pathways. Most cells exhibit **density-dependent inhibition** of cell division as well as **anchorage dependence**.
- Cancer cells elude normal cell cycle regulation and divide unchecked, forming tumors. **Malignant tumors** invade nearby tissues and can undergo **metastasis**, exporting cancer cells to other sites, where they may form secondary tumors. Recent cell cycle and cell-signaling research, and new techniques for sequencing DNA, have led to improved cancer treatments.

**?** *Explain the significance of the $G_1$ and M checkpoints and the go-ahead signals involved in the cell cycle control system.*

# TEST YOUR UNDERSTANDING

**PRACTICE TEST**

goo.gl/CRZjvS

## Level 1: Knowledge/Comprehension

1. Through a microscope, you can see a cell plate beginning to develop across the middle of a cell and nuclei forming on either side of the cell plate. This cell is most likely
   - (A) an animal cell in the process of cytokinesis.
   - (B) a plant cell in the process of cytokinesis.
   - (C) a bacterial cell dividing.
   - (D) a plant cell in metaphase.

2. In the cells of some organisms, mitosis occurs without cytokinesis. This will result in
   - (A) cells with more than one nucleus.
   - (B) cells that are unusually small.
   - (C) cells lacking nuclei.
   - (D) cell cycles lacking an S phase.

3. Which of the following does *not* occur during mitosis?
   - (A) condensation of the chromosomes
   - (B) replication of the DNA
   - (C) separation of sister chromatids
   - (D) spindle formation

## Level 2: Application/Analysis

4. A particular cell has half as much DNA as some other cells in a mitotically active tissue. The cell in question is most likely in
   - (A) $G_1$.
   - (B) $G_2$.
   - (C) prophase.
   - (D) metaphase.

5. The drug cytochalasin B blocks the function of actin. Which of the following aspects of the animal cell cycle would be most disrupted by cytochalasin B?
   - (A) spindle formation
   - (B) spindle attachment to kinetochores
   - (C) cell elongation during anaphase
   - (D) cleavage furrow formation and cytokinesis

6. **DRAW IT** Draw one eukaryotic chromosome as it would appear during interphase, during each of the stages of mitosis, and during cytokinesis. Also draw and label the nuclear envelope and any microtubules attached to the chromosome(s).

7. The light micrograph shows dividing cells near the tip of an onion root. Identify a cell in each of the following stages: prophase, prometaphase, metaphase, anaphase, and telophase. Describe the major events occurring at each stage.

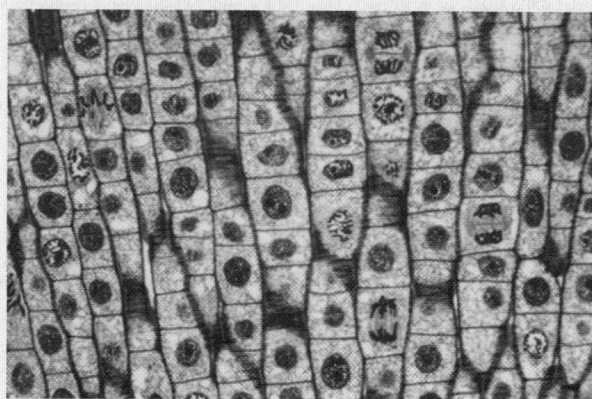

## Level 3: Synthesis/Evaluation

8. **SCIENTIFIC INQUIRY**
   Although both ends of a microtubule can gain or lose subunits, one end (called the plus end) polymerizes and depolymerizes at a higher rate than the other end (the minus end). For spindle microtubules, the plus ends are in the center of the spindle, and the minus ends are at the poles. Motor proteins that move along microtubules specialize in walking either toward the plus end or toward the minus end; the two types are called plus end directed and minus end directed motor proteins, respectively. Given what you know about chromosome movement and spindle changes during anaphase, predict which type of motor proteins would be present on (a) kinetochore microtubules and (b) nonkinetochore microtubules.

9. **FOCUS ON EVOLUTION**
   The result of mitosis is that the daughter cells end up with the same number of chromosomes that the parent cell had. Another way to maintain the number of chromosomes would be to carry out cell division first and then duplicate the chromosomes in each daughter cell. Assess whether this would be an equally good way of organizing the cell cycle. Explain why evolution has not led to this alternative.

10. **FOCUS ON INFORMATION**
    The continuity of life is based on heritable information in the form of DNA. In a short essay (100–150 words), explain how the process of mitosis faithfully parcels out exact copies of this heritable information in the production of genetically identical daughter cells.

11. **SYNTHESIZE YOUR KNOWLEDGE**

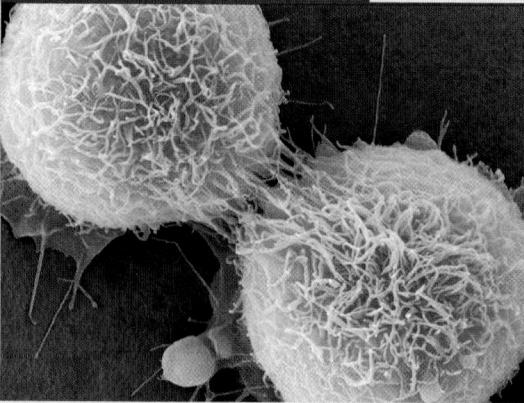

Shown here are two HeLa cancer cells that are just completing cytokinesis. Explain how the cell division of cancer cells like these is misregulated. What genetic and other changes might have caused these cells to escape normal cell cycle regulation?

*For selected answers, see Appendix A.*

# Unit 2 Genetics

## 10 Meiosis and Sexual Life Cycles

Sexually reproducing species alternate fertilization with **meiosis**, accurately passing on genetic information while generating genetic diversity.

## 11 Mendel and the Gene Idea

Although unaware of meiosis, Mendel did experiments that enabled him to describe the behavior of **genes**.

## 12 The Chromosomal Basis of Inheritance

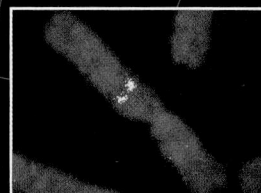

Genes are located on **chromosomes**, and chromosomal behavior underlies genetic inheritance.

## 13 The Molecular Basis of Inheritance

The nucleotide sequence of the DNA in chromosomes provides the **molecular basis for inheritance**.

## 14 Gene Expression: From Gene to Protein

An organism's characteristics emerge from **gene expression**, the process in which information in **genes** is transcribed into **RNAs** that can be translated into **proteins**.

## 15 Regulation of Gene Expression

An organism's different cell types and responses to its environment depend on **regulation of gene expression**.

## 16 Development, Stem Cells, and Cancer

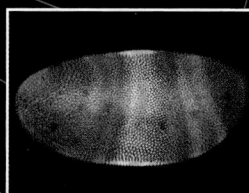

Coordinated gene regulation underlies **embryonic development**, while misregulation can contribute to **cancer**.

## 17 Viruses

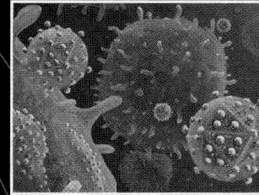

Our understanding of gene expression is informed by studying **viruses**, protein-coated packets of genetic information that hijack cellular resources and replicate themselves.

## 18 Genomes and Their Evolution

The **evolution of genomes** is the basis of life's diversity.

Chemistry and Cells

Ecology

Animals

Plants

History of Life

Evolution

Genetics

10 Meiosis and Sexual Life Cycles
11 Mendel and the Gene Idea
12 The Chromosomal Basis of Inheritance
13 The Molecular Basis of Inheritance
14 Gene Expression: From Gene to Protein
15 Regulation of Gene Expression
16 Development, Stem Cells, and Cancer
17 Viruses
18 Genomes and Their Evolution

1

# 10 Meiosis and Sexual Life Cycles

## KEY CONCEPTS

**10.1** Offspring acquire genes from parents by inheriting chromosomes

**10.2** Fertilization and meiosis alternate in sexual life cycles

**10.3** Meiosis reduces the number of chromosome sets from diploid to haploid

**10.4** Genetic variation produced in sexual life cycles contributes to evolution

▲ **Figure 10.1** What accounts for family resemblance?

## Variations on a Theme

Most people who send out birth announcements mention the sex of the baby, but they don't feel the need to specify that their offspring is a human being! One of the characteristics of life is the ability of organisms to reproduce their own kind—elephants produce little elephants, and oak trees generate oak saplings. Exceptions to this rule show up only as sensational but highly suspect stories in tabloid newspapers.

Another rule often taken for granted is that offspring resemble their parents more than they do unrelated individuals. If you examine the family members shown in **Figure 10.1**, you can pick out some similar features among them. The transmission of traits from one generation to the next is called inheritance, or **heredity** (from the Latin *heres*, heir). However, sons and daughters are not identical copies of either parent or of their siblings. Along with inherited similarity, there is also **variation**. Farmers have exploited the principles of heredity and variation for thousands of years, breeding plants and animals for desired traits. But what are the biological mechanisms leading to the hereditary similarity and variation that we call a "family resemblance"? A detailed answer to this question eluded biologists until the advance of genetics in the 20th century.

**Genetics** is the scientific study of heredity and inherited variation. In this unit, you'll learn about genetics at multiple levels, from organisms to cells to molecules. On the practical side, you'll see how genetics continues to revolutionize medicine, and you'll be asked to consider some social and ethical questions raised by our ability to manipulate DNA, the genetic material. At the end of the unit, you'll be able to stand back and consider the whole genome, an organism's entire complement of DNA. Rapid acquisition and analysis of the genome sequences of many species, including our own, have taught us a great deal about evolution on the molecular level—in other words, evolution of the genome itself. In fact, genetic methods and discoveries are catalyzing progress in all areas of biology, from cell biology to physiology, developmental biology, behavior, and even ecology.

We begin our study of genetics in this chapter by examining how chromosomes pass from parents to offspring in sexually reproducing organisms. The processes of meiosis (a special type of cell division) and fertilization (the fusion of sperm and egg) maintain a species' chromosome count during the sexual life cycle. We'll describe the cellular mechanics of meiosis and explain how this process differs from mitosis. Finally, we'll consider how both meiosis and fertilization contribute to genetic variation, such as the variation obvious in the family shown in Figure 10.1.

# Offspring acquire genes from parents by inheriting chromosomes

Family friends may tell you that you have your mother's nose or your father's eyes. Of course, parents do not, in any literal sense, give their children a nose, eyes, hair, or any other traits. What, then, *is* actually inherited?

## Inheritance of Genes

Parents endow their offspring with coded information in the form of hereditary units called **genes**. The genes we inherit from our mothers and fathers are our genetic link to our parents, and they account for family resemblances such as shared eye color or freckles. Our genes program specific traits that emerge as we develop from fertilized eggs into adults.

The genetic program is written in the language of DNA, the polymer of four different nucleotides (see Concepts 1.1 and 3.6). Inherited information is passed on in the form of each gene's specific sequence of DNA nucleotides, much as printed information is communicated in the form of meaningful sequences of letters. In both cases, the language is symbolic. Just as your brain translates the word *apple* into a mental image of the fruit, cells translate genes into freckles and other features. Most genes program cells to synthesize specific enzymes and other proteins, whose cumulative action produces an organism's inherited traits. The programming of these traits in the form of DNA is one of the unifying themes of biology.

The transmission of hereditary traits has its molecular basis in the replication of DNA, which produces copies of genes that can be passed from parents to offspring. In animals and plants, reproductive cells called **gametes** are the vehicles that transmit genes from one generation to the next. During fertilization, male and female gametes (sperm and eggs) unite, passing on genes of both parents to their offspring.

Except for small amounts of DNA in mitochondria and chloroplasts, the DNA of a eukaryotic cell is packaged into chromosomes within the nucleus. Every species has a characteristic number of chromosomes. For example, humans have 46 chromosomes in their **somatic cells**—all the cells of the body except the gametes and their precursors. Each chromosome consists of a single long DNA molecule elaborately coiled in association with various proteins. One chromosome includes several hundred to a few thousand genes, each of which is a specific sequence of nucleotides within the DNA molecule. A gene's specific location along the length of a chromosome is called the gene's **locus** (plural, *loci*; from the Latin, meaning "place"). Our genetic endowment (our genome) consists of the genes and other DNA that make up the chromosomes we inherited from our parents.

## Comparison of Asexual and Sexual Reproduction

Only organisms that reproduce asexually have offspring that are exact genetic copies of themselves. In **asexual reproduction**, a single individual is the sole parent and passes copies of all its genes to its offspring without the fusion of gametes. For example, single-celled eukaryotic organisms can reproduce asexually by mitotic cell division, in which DNA is copied and allocated equally to two daughter cells. The genomes of the offspring are virtually exact copies of the parent's genome. Some multicellular organisms are also capable of reproducing asexually **(Figure 10.2)**. Because the cells of the offspring arise via mitosis in the parent, the offspring is usually genetically identical to its parent. An individual that reproduces asexually gives rise to a **clone**, a group of genetically identical individuals. Genetic differences occasionally arise in asexually reproducing organisms as a result of changes in the DNA called mutations, which we will discuss in Chapter 14.

In **sexual reproduction**, two parents give rise to offspring that have unique combinations of genes inherited from the two parents. In contrast to a clone, offspring of sexual reproduction vary genetically from their siblings and both parents: They are variations on a common theme of family resemblance, not exact replicas. Genetic variation like that shown in Figure 10.1 is an important consequence of sexual reproduction. What mechanisms generate this genetic variation? The key is the behavior of chromosomes during the sexual life cycle.

0.5 mm

Parent

Bud

**(a) Hydra**　　　　　**(b) Redwoods**

▲ **Figure 10.2 Asexual reproduction in two multicellular organisms. (a)** This relatively simple animal, a hydra, reproduces by budding. The bud, a localized mass of mitotically dividing cells, develops into a small hydra, which detaches from the parent (LM). **(b)** All the trees in this circle of redwoods arose asexually from a single parent tree, whose stump is in the center of the circle.

1. **MAKE CONNECTIONS** Using what you know of gene expression in a cell, explain what causes traits of parents (such as hair color) to show up in their offspring. (See Concept 3.6.)
2. How does an asexually reproducing eukaryotic organism produce offspring that are genetically identical to each other and to their parent?
3. **WHAT IF?** A horticulturalist breeds orchids, trying to obtain a plant with a unique combination of desirable traits. After many years, she finally succeeds. To produce more plants like this one, should she crossbreed it with another plant or clone it? Why?

For suggested answers, see Appendix A.

## CONCEPT 10.2

# Fertilization and meiosis alternate in sexual life cycles

A **life cycle** is the generation-to-generation sequence of stages in the reproductive history of an organism, from conception to production of its own offspring. In this section, we use humans as an example to track the behavior of chromosomes through the sexual life cycle. We begin by considering the chromosome count in human somatic cells and gametes. We will then explore how the behavior of chromosomes relates to the human life cycle and other types of sexual life cycles.

## Sets of Chromosomes in Human Cells

In humans, each somatic cell has 46 chromosomes. During mitosis, the chromosomes become condensed enough to be visible under a light microscope. At this point, they can be distinguished from one another by their size, the position of their centromeres, and the pattern of colored bands produced by certain chromatin-binding stains.

Careful examination of a micrograph of the 46 human chromosomes from a single cell in mitosis reveals that there are two chromosomes of each of 23 types. This becomes clear when images of the chromosomes are arranged in pairs, starting with the longest chromosomes. The resulting ordered display is called a **karyotype (Figure 10.3)**. The two chromosomes of a pair have the same length, centromere position, and staining pattern: These are called **homologous chromosomes** (or **homologs**, for short) or a **homologous pair**. Both chromosomes of each pair carry genes controlling the same inherited characters. For example, if a gene for eye color is situated at a particular locus on a certain chromosome, its homologous chromosome will also have a version of the eye color gene at the equivalent locus.

The two chromosomes referred to as X and Y are an important exception to the general pattern of homologous chromosomes in human somatic cells. Human females have a homologous pair of X chromosomes (XX), but males have one X and one Y chromosome (XY; see Figure 10.3). Only

▼ **Figure 10.3** **Research Method**

## Preparing a Karyotype

**Application** A karyotype is a display of condensed chromosomes arranged in pairs. Karyotyping can be used to screen for defective chromosomes or abnormal numbers of chromosomes associated with certain congenital disorders, such as Down syndrome.

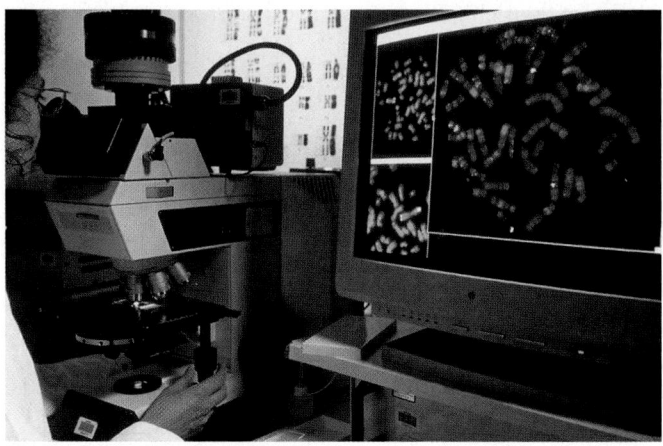

**Technique** Karyotypes are prepared from isolated somatic cells, which are treated with a drug to stimulate mitosis and then grown in culture for several days. Cells arrested when chromosomes are most highly condensed—in metaphase—are stained and then viewed with a microscope equipped with a digital camera. An image of the chromosomes is displayed on a computer monitor, and digital software is used to arrange them in pairs according to their appearance.

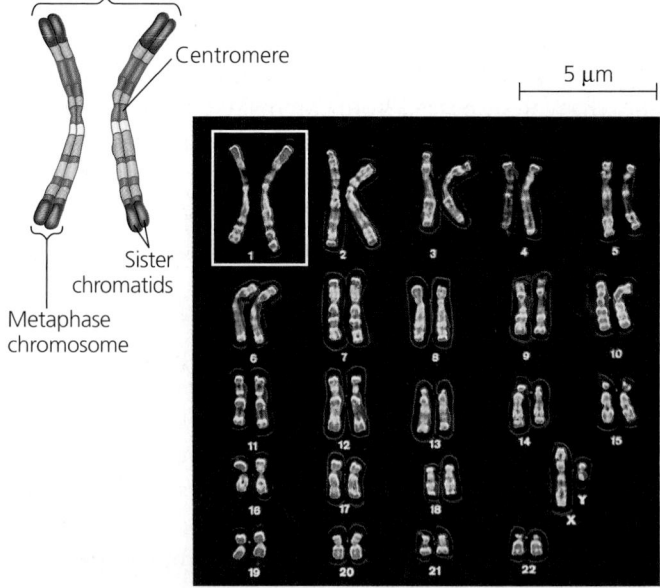

**Results** This karyotype shows the chromosomes from a human male, digitally colored to emphasize the chromosome banding patterns. The size of the chromosome, position of the centromere, and pattern of stained bands help identify specific chromosomes. Although difficult to discern in the karyotype, each metaphase chromosome consists of two closely attached sister chromatids (see the diagram of the boxed pair of homologous duplicated chromosomes).

small parts of the X and Y are homologous. Most of the genes carried on the X chromosome do not have counterparts on the tiny Y, and the Y chromosome has genes lacking on the X. Because they determine an individual's sex, the X and Y chromosomes are called **sex chromosomes**. The other chromosomes are called **autosomes**.

The occurrence of pairs of homologous chromosomes in each human somatic cell is a consequence of our sexual origins. We inherit one chromosome of each pair from each parent. Thus, the 46 chromosomes in our somatic cells are actually two sets of 23 chromosomes—a maternal set (from our mother) and a paternal set (from our father). The number of chromosomes in a single set is represented by $n$. Any cell with two chromosome sets is called a **diploid cell** and has a diploid number of chromosomes, abbreviated $2n$. For humans, the diploid number is 46 ($2n = 46$), the number of chromosomes in our somatic cells. In a cell in which DNA synthesis has occurred, all the chromosomes are duplicated, and therefore each consists of two identical sister chromatids, associated closely at the centromere and along the arms. (Even though the chromosomes are duplicated, we still say the cell is diploid, or $2n$, because it has only two sets of information.) **Figure 10.4** helps clarify the various terms that we use to describe duplicated chromosomes in a diploid cell.

Unlike somatic cells, gametes contain a single set of chromosomes. Such cells are called **haploid cells**, and each has a haploid number of chromosomes ($n$). For humans, the haploid number is 23 ($n = 23$). The set of 23 consists of the 22 autosomes plus a single sex chromosome. An unfertilized egg contains an X chromosome, but a sperm may contain an X or a Y chromosome.

Each sexually reproducing species has a characteristic diploid and haploid number. For example, the fruit fly *Drosophila melanogaster* has a diploid number ($2n$) of 8 and a haploid number ($n$) of 4, while for dogs, $2n$ is 78 and $n$ is 39.

Now let's consider chromosome behavior during sexual life cycles. We'll use the human life cycle as an example.

## Behavior of Chromosome Sets in the Human Life Cycle

The human life cycle begins when a haploid sperm from the father fuses with a haploid egg from the mother **(Figure 10.5)**. This union of gametes, culminating in fusion of their nuclei, is called **fertilization**. The resulting fertilized egg, or **zygote**, is diploid because it contains two haploid sets of chromosomes bearing genes representing the maternal and paternal family lines. As a human develops into a sexually mature adult, mitosis of the zygote and its descendant cells generates all the somatic cells of the body. Both chromosome sets in the zygote and all the genes they carry are passed with precision to the somatic cells.

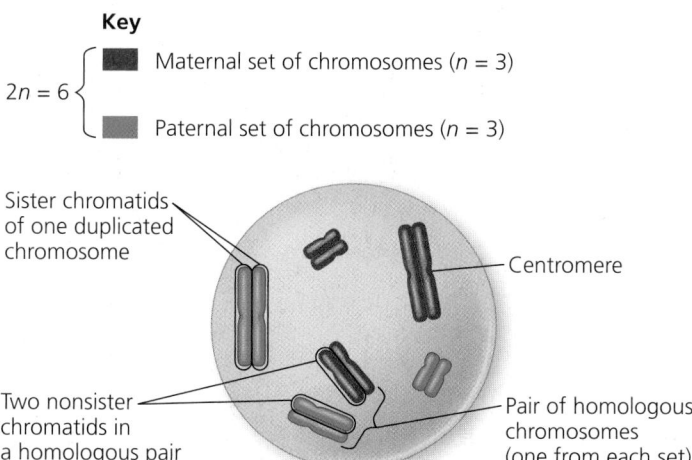

**Key**

⬛ Maternal set of chromosomes ($n = 3$)

$2n = 6$ {

▬ Paternal set of chromosomes ($n = 3$)

Sister chromatids of one duplicated chromosome

Centromere

Two nonsister chromatids in a homologous pair

Pair of homologous chromosomes (one from each set)

▲ **Figure 10.4 Describing chromosomes.** A cell from an organism with a diploid number of 6 ($2n = 6$) is depicted here following chromosome duplication and condensation. Each of the six duplicated chromosomes consists of two sister chromatids associated closely along their lengths. Each homologous pair is composed of one chromosome from the maternal set (red) and one from the paternal set (blue). Each set is made up of three chromosomes in this example (long, medium, and short). Together, one maternal and one paternal chromatid in a pair of homologous chromosomes are called nonsister chromatids.

**?** *How many sets of chromosomes are present in this diagram? How many pairs of homologs are present?*

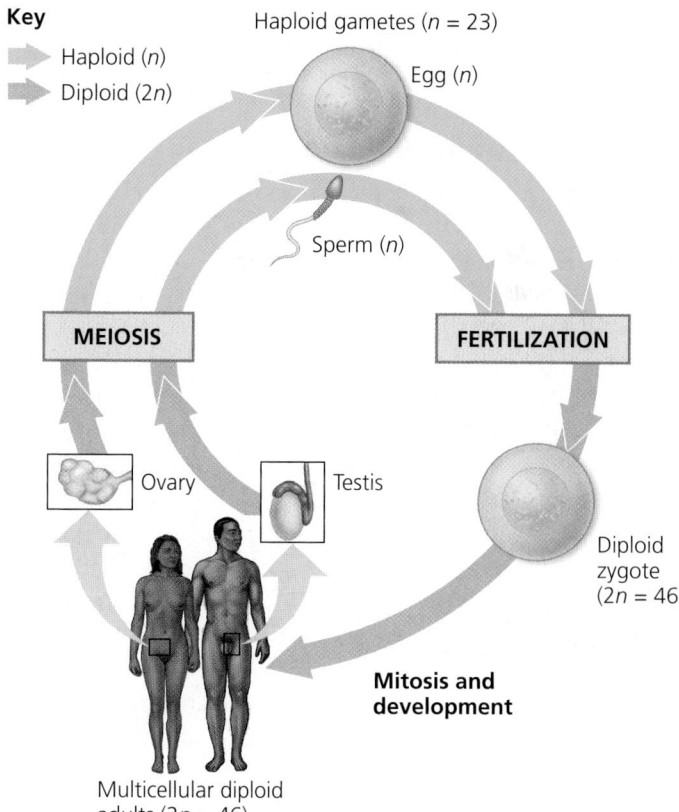

**Key**

➡ Haploid ($n$)
➡ Diploid ($2n$)

Haploid gametes ($n = 23$)

Egg ($n$)

Sperm ($n$)

**MEIOSIS**

**FERTILIZATION**

Ovary

Testis

Diploid zygote ($2n = 46$)

**Mitosis and development**

Multicellular diploid adults ($2n = 46$)

▲ **Figure 10.5 The human life cycle.** In each generation, the number of chromosome sets doubles at fertilization but is halved during meiosis. For humans, the number of chromosomes in a haploid cell is 23, consisting of one set ($n = 23$); the number of chromosomes in the diploid zygote and all somatic cells arising from it is 46, consisting of two sets ($2n = 46$).

This figure introduces a color code that will be used for other life cycles later in this book. The aqua arrows identify haploid stages of a life cycle, and the tan arrows identify diploid stages.

The only cells of the human body not produced by mitosis are the gametes, which develop from specialized cells called *germ cells* in the gonads—ovaries in females and testes in males (see Figure 10.5). Imagine what would happen if human gametes were made by mitosis: They would be diploid like the somatic cells. At the next round of fertilization, when two gametes fused, the normal chromosome number of 46 would double to 92, and each subsequent generation would double the number of chromosomes yet again. This does not happen, however, because in sexually reproducing organisms, gamete formation involves a type of cell division called **meiosis**. This type of cell division reduces the number of sets of chromosomes from two to one in the gametes, counterbalancing the doubling that occurs at fertilization. As a result of meiosis, each human sperm and egg is haploid ($n = 23$). Fertilization restores the diploid condition by combining two haploid sets of chromosomes, and the human life cycle is repeated, generation after generation (see Figure 10.5).

In general, the steps of the human life cycle are typical of many sexually reproducing animals. Indeed, the processes of fertilization and meiosis are the hallmarks of sexual reproduction in plants, fungi, and protists as well as in animals. Fertilization and meiosis alternate in sexual life cycles, maintaining a constant number of chromosomes in each species from one generation to the next.

## The Variety of Sexual Life Cycles

Although the alternation of meiosis and fertilization is common to all organisms that reproduce sexually, the timing of these two events in the life cycle varies, depending on the species. These variations can be grouped into three main types of life cycles. In the type that occurs in humans and most other animals, gametes are the only haploid cells **(Figure 10.6a)**. Meiosis occurs in germ cells during the production of gametes, which undergo no further cell division prior to fertilization. After fertilization, the diploid zygote divides by mitosis, producing a multicellular organism that is diploid.

Plants and some species of algae exhibit a second type of life cycle called **alternation of generations (Figure 10.6b)**. This type includes both diploid and haploid stages that are multicellular. The multicellular diploid stage is called the *sporophyte*. Meiosis in the sporophyte produces haploid cells called *spores*. Unlike a gamete, a haploid spore doesn't fuse with another cell but divides mitotically, generating a multicellular haploid stage called the *gametophyte*. Cells of the gametophyte give rise to gametes by mitosis. Fusion of two haploid gametes at fertilization results in a diploid zygote, which develops into the next sporophyte generation. Therefore, in this type of life cycle, the sporophyte generation produces a gametophyte as its offspring, and the gametophyte generation produces the next sporophyte generation. The term *alternation of generations* fits well as a name for this type of life cycle.

A third type of life cycle occurs in most fungi and some protists, including some algae **(Figure 10.6c)**. After gametes fuse and form a diploid zygote, meiosis occurs without a multicellular diploid offspring developing. Meiosis produces not gametes but haploid cells that then divide by mitosis and give rise to either unicellular descendants or a haploid multicellular adult organism. Subsequently, the haploid organism carries out further mitoses, producing the cells that develop into gametes. The only diploid stage found in these species is the single-celled zygote.

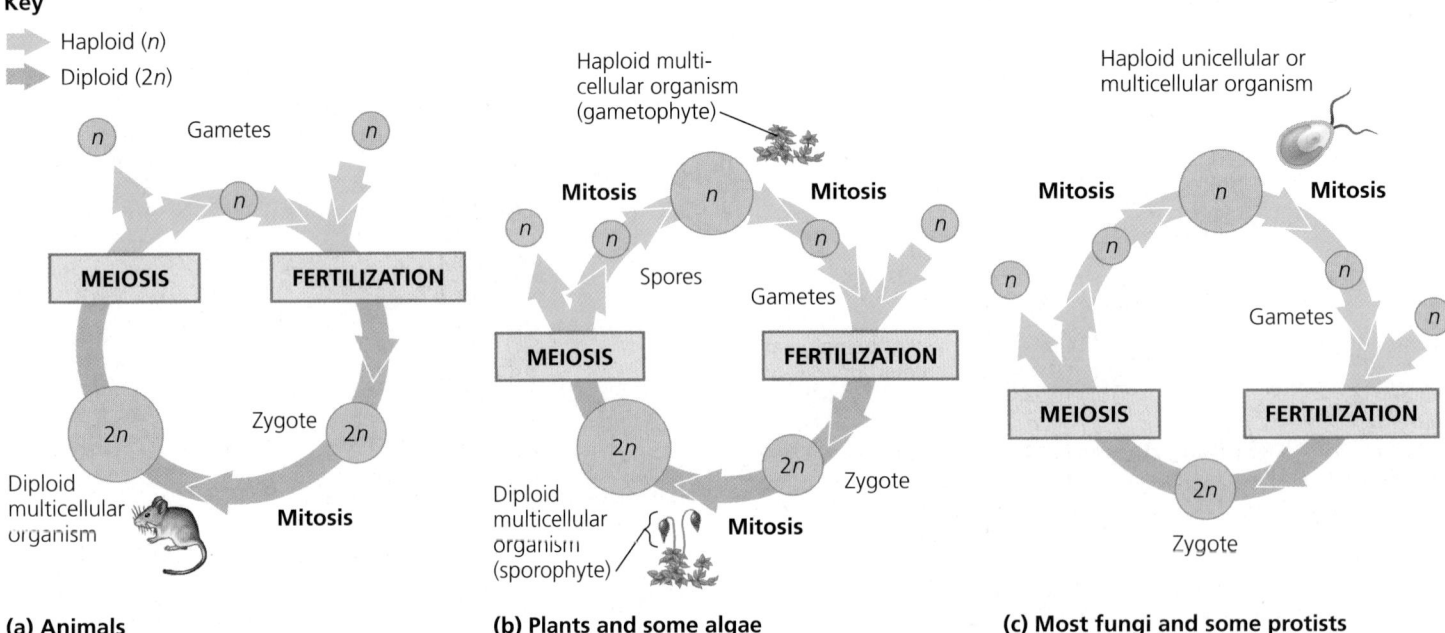

▲ **Figure 10.6 Three types of sexual life cycles.** The common feature of all three cycles is the alternation of meiosis and fertilization, key events that contribute to genetic variation among offspring. The cycles differ in the timing of these two key events. (Small circles are cells; large circles are organisms.)

Note that *either* haploid or diploid cells can divide by mitosis, depending on the type of life cycle. Only diploid cells, however, can undergo meiosis because haploid cells have a single set of chromosomes that cannot be further reduced. Though the three types of sexual life cycles differ in the timing of meiosis and fertilization, they share a fundamental result: genetic variation among offspring. A closer look at meiosis will reveal the sources of this variation.

## CONCEPT CHECK 10.2

1. **MAKE CONNECTIONS** In Figure 10.4, how many DNA molecules (double helices) are present (see Figure 9.5)? What is the haploid number of this cell? Is a set of chromosomes haploid or diploid?
2. In the karyotype shown in Figure 10.3, how many pairs of chromosomes are present? How many sets?
3. Each sperm of a pea plant contains seven chromosomes. What are the haploid and diploid numbers for this species?
4. **WHAT IF?** A certain eukaryote lives as a unicellular organism, but during environmental stress, it produces gametes. The gametes fuse, and the resulting zygote undergoes meiosis, generating new single cells. What type of organism could this be?

For suggested answers, see Appendix A.

## CONCEPT 10.3

# Meiosis reduces the number of chromosome sets from diploid to haploid

Several steps of meiosis closely resemble corresponding steps in mitosis. Meiosis, like mitosis, is preceded by the duplication of chromosomes. However, this single duplication is followed by not one but two consecutive cell divisions, called **meiosis I** and **meiosis II**. These two divisions result in four daughter cells (rather than the two daughter cells of mitosis), each with only half as many chromosomes as the parent cell—one set, rather than two.

## The Stages of Meiosis

The overview of meiosis in **Figure 10.7** shows, for a single pair of homologous chromosomes in a diploid cell, that both members of the pair are duplicated and the copies sorted into four haploid daughter cells. Recall that sister chromatids are two copies of *one* chromosome, closely associated all along their lengths; this association is called *sister chromatid cohesion*. Together, the sister chromatids make up one duplicated chromosome (see Figure 10.4). In contrast, the two chromosomes of a homologous pair are individual chromosomes that were inherited from different parents. Homologs appear alike in the microscope, but they may have different versions of genes at corresponding loci; each version is called an *allele* of that gene (see Figure 11.4). For example, one chromosome may have an allele for freckles, but the homologous chromosome may have

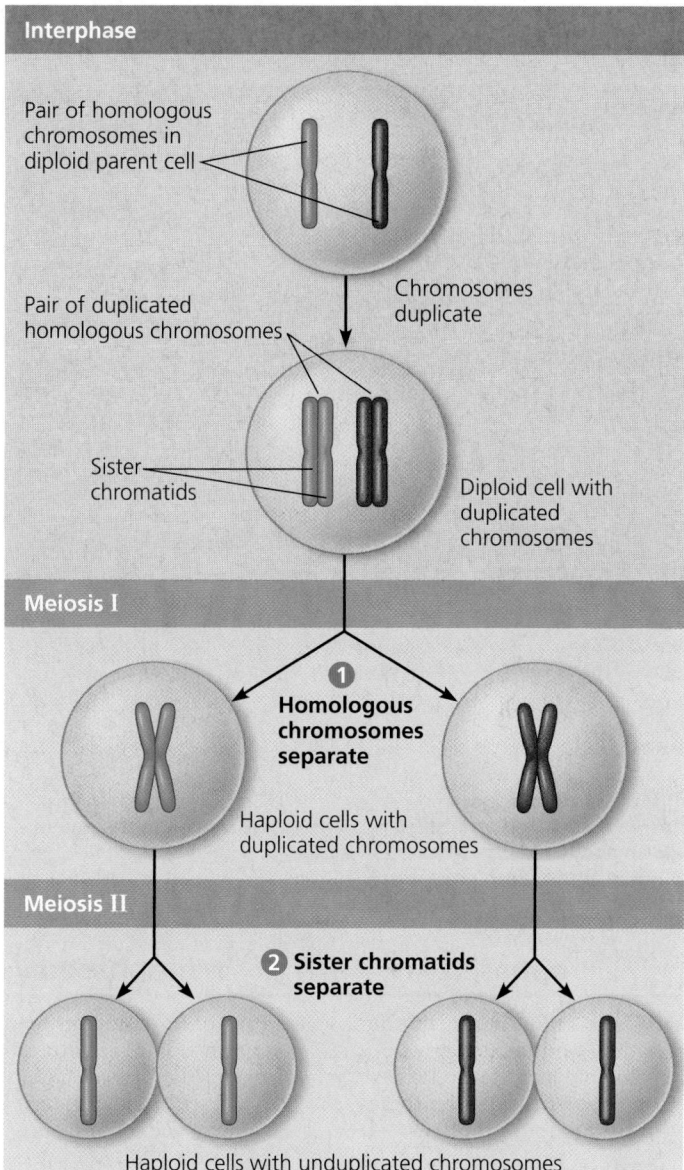

▲ **Figure 10.7 Overview of meiosis: how meiosis reduces chromosome number.** After the chromosomes duplicate in interphase, the diploid cell divides *twice*, yielding four haploid daughter cells. This overview tracks just one pair of homologous chromosomes, which for the sake of simplicity are drawn in the condensed state throughout. (They would not normally be condensed during interphase.) The red chromosome was inherited from the female parent, the blue chromosome from the male parent.

**DRAW IT** *Redraw the cells in this figure using a simple double helix to represent each DNA molecule.*

an allele for the absence of freckles at the same locus. Homologs are not associated with each other in any obvious way except during meiosis, as you will soon see.

**Figure 10.8** describes in detail the stages of the two divisions of meiosis for an animal cell whose diploid number is 6. Meiosis halves the total number of chromosomes in a very specific way, reducing the number of sets from two to one, with each daughter cell receiving one set of chromosomes. Study Figure 10.8 thoroughly before going on.

## MEIOSIS I: Separates homologous chromosomes

| Prophase I | Metaphase I | Anaphase I | Telophase I and Cytokinesis |
|---|---|---|---|

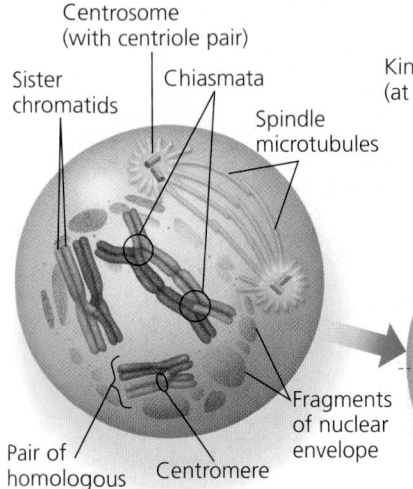

Duplicated homologous chromosomes (red and blue) pair up and exchange segments; $2n = 6$ in this example.

Chromosomes line up by homologous pairs.

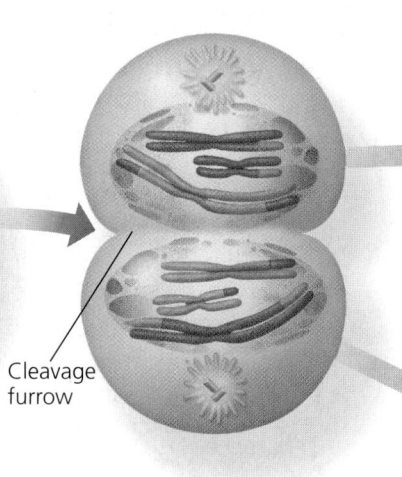

Each pair of homologous chromosomes separates.

Two haploid cells form; each chromosome still consists of two sister chromatids.

### Prophase I

- Centrosome movement, spindle formation, and nuclear envelope breakdown occur as in mitosis. Chromosomes condense progressively throughout prophase I.

- During early prophase I, before the stage shown above, each chromosome pairs with its homolog, aligned gene by gene, and **crossing over** occurs: The DNA molecules of nonsister chromatids are broken (by proteins) and are rejoined to each other.

- At the stage shown above, each homologous pair has one or more X-shaped regions called **chiasmata** (singular, *chiasma*), where crossovers have occurred.

- Later in prophase I, after the stage shown above, microtubules from one pole or the other attach to the kinetochores, one at the centromere of each homolog. (Each homolog acts as if it has a single kinetochore.) Microtubules move the homologous pairs toward the metaphase plate (see the metaphase I diagram).

### Metaphase I

- Pairs of homologous chromosomes are now arranged at the metaphase plate, with one chromosome of each pair facing each pole.

- Both chromatids of one homolog are attached to kinetochore microtubules from one pole; the chromatids of the other homolog are attached to microtubules from the opposite pole.

### Anaphase I

- Breakdown of proteins that are responsible for sister chromatid cohesion along chromatid arms allows homologs to separate.

- The homologs move toward opposite poles, guided by the spindle apparatus.

- Sister chromatid cohesion persists at the centromere, causing chromatids to move as a unit toward the same pole.

### Telophase I and Cytokinesis

- When telophase I begins, each half of the cell has a complete haploid set of duplicated chromosomes. Each chromosome is composed of two sister chromatids; one or both chromatids include regions of nonsister chromatid DNA.

- Cytokinesis (division of the cytoplasm) usually occurs simultaneously with telophase I, forming two haploid daughter cells.

- In animal cells like these, a cleavage furrow forms. (In plant cells, a cell plate forms.)

- In some species, chromosomes decondense and nuclear envelopes form.

- No chromosome duplication occurs between meiosis I and meiosis II.

## MEIOSIS II: Separates sister chromatids

| Prophase II | Metaphase II | Anaphase II | Telophase II and Cytokinesis |
|---|---|---|---|

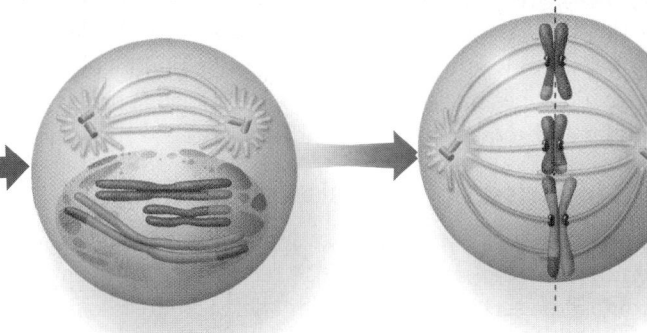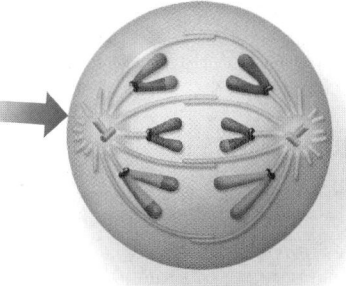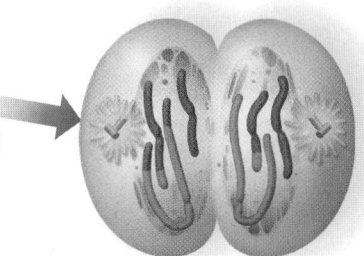

During another round of cell division, the sister chromatids finally separate;
four haploid daughter cells result, containing unduplicated chromosomes.

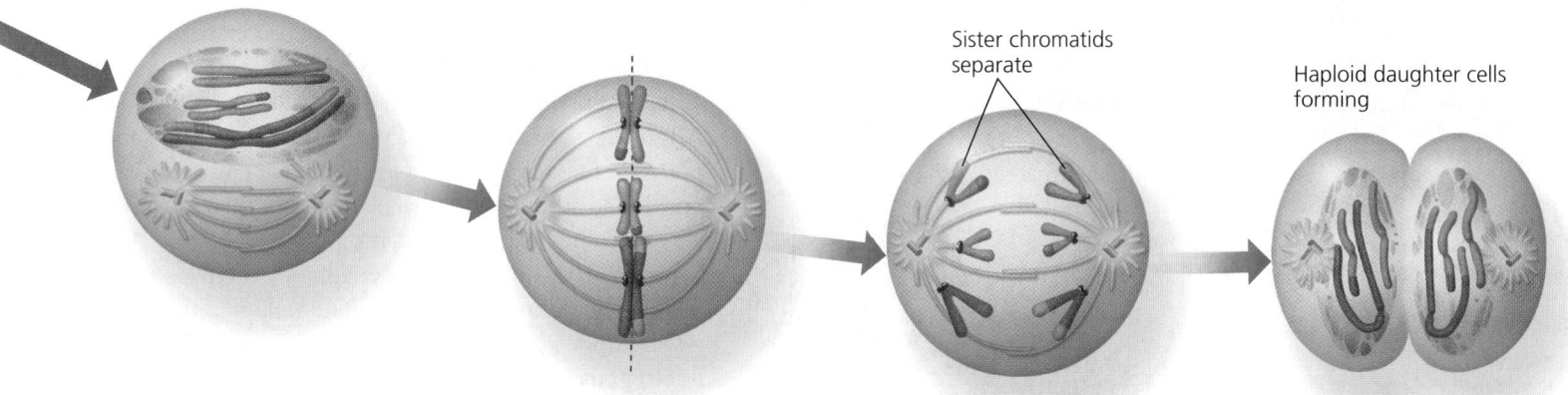

Sister chromatids
separate

Haploid daughter cells
forming

### Prophase II

• A spindle apparatus forms.

• In late prophase II (not shown here), chromosomes, each still composed of two chromatids associated at the centromere, are moved by microtubules toward the metaphase II plate.

### Metaphase II

• The chromosomes are positioned at the metaphase plate as in mitosis.

• Because of crossing over in meiosis I, the two sister chromatids of each chromosome are not genetically identical.

• The kinetochores of sister chromatids are attached to microtubules extending from opposite poles.

### Anaphase II

• Breakdown of proteins holding the sister chromatids together at the centromere allows the chromatids to separate. The chromatids move toward opposite poles as individual chromosomes.

### Telophase II and Cytokinesis

• Nuclei form, the chromosomes begin decondensing, and cytokinesis occurs.

• The meiotic division of one parent cell produces four daughter cells, each with a haploid set of (unduplicated) chromosomes.

• The four daughter cells are genetically distinct from one another and from the parent cell.

**MAKE CONNECTIONS** *Look at Figure 9.7 and imagine the two daughter cells undergoing another round of mitosis, yielding four cells. Compare the number of chromosomes in each of those four cells, after mitosis, with the number in each cell in Figure 10.8, after meiosis. Explain how the process of meiosis results in this difference, even though meiosis also includes two cell divisions.*

**ANIMATION** Visit the Study Area in **MasteringBiology** for the BioFlix® 3-D Animation on Meiosis.

## Crossing Over and Synapsis During Prophase I

Prophase I of meiosis is a very busy time. The prophase I cell shown in Figure 10.8 is at a point fairly late in prophase I, when pairing of homologous chromosomes, crossing over, and chromosome condensation have already taken place. The sequence of events leading up to that point is shown in more detail in **Figure 10.9**.

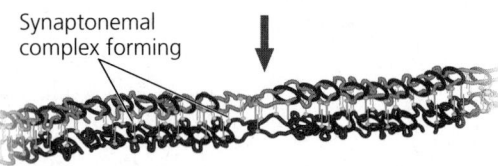

**DNA breaks    Centromere    DNA breaks**

**Cohesins**

Pair of homologous chromosomes:

Paternal sister chromatids

Maternal sister chromatids

**1** After interphase, the chromosomes have been duplicated, and sister chromatids are held together by proteins called cohesins (purple). Each pair of homologs associate along their length. The DNA molecules of two nonsister chromatids are broken at precisely corresponding points. The chromatin of the chromosomes starts to condense.

Synaptonemal complex forming

**2** A zipper-like protein complex, the synaptonemal complex (green), begins to form, attaching one homolog to the other. The chromatin continues to condense.

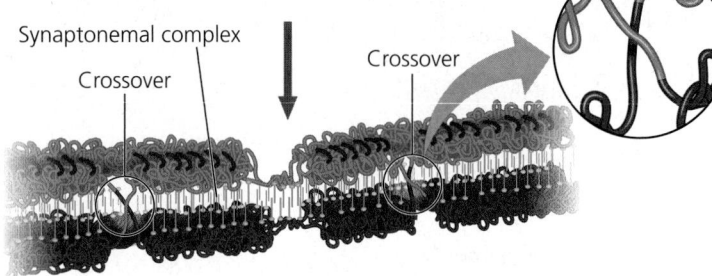

Synaptonemal complex

Crossover

Crossover

**3** The synaptonemal complex is fully formed; the two homologs are said to be in synapsis. During synapsis, the DNA breaks are closed up when each broken end is joined to the corresponding segment of the nonsister chromatid, producing crossovers.

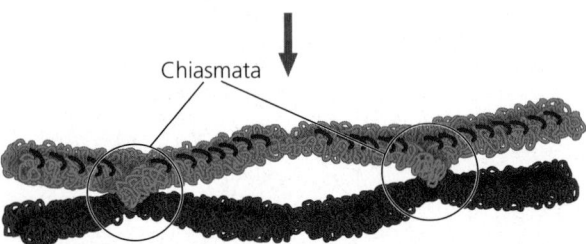

Chiasmata

**4** After the synaptonemal complex disassembles, the homologs move slightly apart from each other but remain attached because of sister chromatid cohesion, even though some of the DNA may no longer be attached to its original chromosome. The points of attachment where crossovers have occurred show up as chiasmata. The chromosomes continue to condense as they move toward the metaphase plate.

▲ **Figure 10.9 Crossing over and synapsis in prophase I.**

As shown in this figure, the two members of a homologous pair associate along their length, aligned allele by allele. The DNA molecules of a maternal and a paternal chromatid are broken at precisely matching points. A zipper-like structure called the **synaptonemal complex** forms, and during this attachment (**synapsis**), the DNA breaks are closed up so that a paternal chromatid is joined to a piece of maternal chromatid beyond the crossover point, and vice versa. At least one crossover per chromosome must occur, in conjunction with sister chromatid cohesion, in order for the homologous pair to stay together as it moves to the metaphase I plate.

## A Comparison of Mitosis and Meiosis

**Figure 10.10** summarizes the key differences between meiosis and mitosis in diploid cells. Meiosis reduces the number of chromosome sets from two to one, whereas mitosis conserves the number. Meiosis produces cells that differ genetically from the parent cell and from each other, whereas mitosis produces daughter cells that are genetically identical to the parent cell.

Three events unique to meiosis occur during meiosis I:

1. **Synapsis and crossing over.** During prophase I, duplicated homologs pair up, and crossing over occurs as described previously and in Figure 10.9.
2. **Alignment of homologous pairs at the metaphase plate.** At metaphase I of meiosis, pairs of homologs are positioned at the metaphase plate, rather than individual chromosomes as in metaphase of mitosis.
3. **Separation of homologs.** At anaphase I of meiosis, the duplicated chromosomes of each homologous pair move toward opposite poles, but the sister chromatids of each duplicated chromosome remain attached. In anaphase of mitosis, by contrast, sister chromatids separate.

Sister chromatids stay together due to sister chromatid cohesion, mediated by cohesion proteins. In mitosis, this attachment lasts until the end of metaphase, when enzymes cleave the cohesins, freeing the sister chromatids to move to opposite poles of the cell. In meiosis, sister chromatid cohesion is released in two steps, one at the start of anaphase I and one at anaphase II. In metaphase I, the two homologs of each pair are held together by cohesion between sister chromatid arms in regions beyond points of crossing over, where stretches of sister chromatids now belong to different chromosomes. The combination of crossing over and sister chromatid cohesion along the arms results in the formation of a chiasma. Chiasmata hold homologs together as the spindle forms for the first meiotic division. At the onset of anaphase I, the release of cohesion along sister chromatid arms allows homologs to separate. At anaphase II, the release of sister chromatid cohesion at the centromeres allows the sister chromatids to separate. Thus, sister chromatid cohesion and crossing over, acting together, play an essential role in the lining up of chromosomes by homologous pairs at metaphase I.

Meiosis I reduces the number of chromosome sets per cell from two (diploid) to one set (haploid). During the second

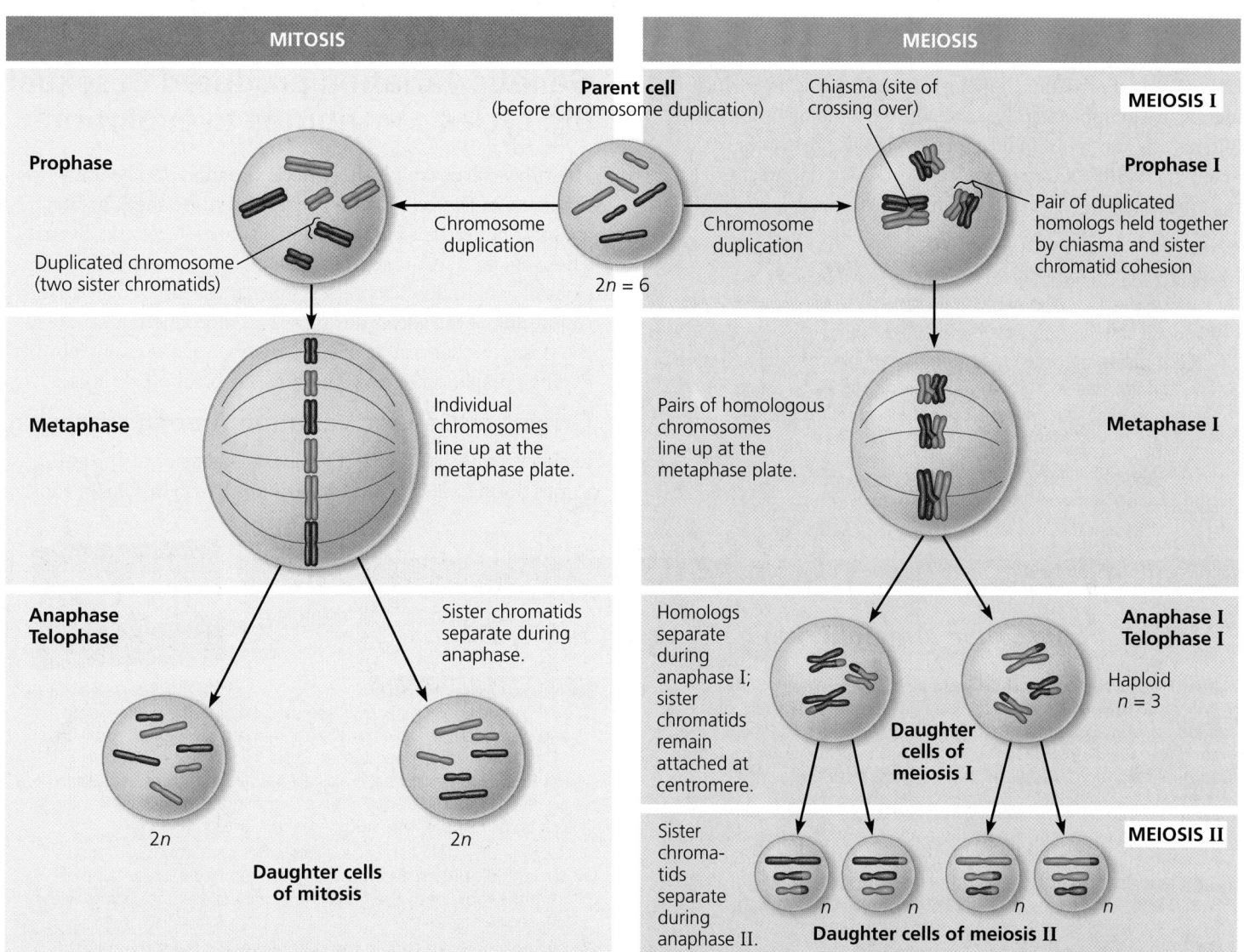

| | MITOSIS | | MEIOSIS | |
|---|---|---|---|---|

**Parent cell** (before chromosome duplication)

Chiasma (site of crossing over)

**MEIOSIS I**

**Prophase**

Duplicated chromosome (two sister chromatids)

Chromosome duplication

$2n = 6$

Chromosome duplication

**Prophase I**

Pair of duplicated homologs held together by chiasma and sister chromatid cohesion

**Metaphase**

Individual chromosomes line up at the metaphase plate.

Pairs of homologous chromosomes line up at the metaphase plate.

**Metaphase I**

**Anaphase Telophase**

Sister chromatids separate during anaphase.

$2n$ $2n$

**Daughter cells of mitosis**

Homologs separate during anaphase I; sister chromatids remain attached at centromere.

**Daughter cells of meiosis I**

**Anaphase I Telophase I**

Haploid $n = 3$

Sister chromatids separate during anaphase II.

**MEIOSIS II**

$n$ $n$ $n$ $n$

**Daughter cells of meiosis II**

## SUMMARY

| Property | Mitosis (occurs in both diploid and haploid cells) | Meiosis (can only occur in diploid cells) |
|---|---|---|
| DNA replication | Occurs during interphase before mitosis begins | Occurs during interphase before meiosis I begins |
| Number of divisions | One, including prophase, prometaphase, metaphase, anaphase, and telophase | Two, each including prophase, metaphase, anaphase, and telophase |
| Synapsis of homologous chromosomes | Does not occur | Occurs during prophase I along with crossing over between nonsister chromatids; resulting chiasmata hold pairs together due to sister chromatid cohesion |
| Number of daughter cells and genetic composition | Two, each genetically identical to the parent cell, with the same number of chromosomes | Four, each haploid ($n$); genetically different from the parent cell and from each other |
| Role in the animal or plant body | Enables multicellular animal or plant (gametophyte or sporophyte) to arise from a single cell; produces cells for growth, repair, and, in some species, asexual reproduction; produces gametes in the gametophyte plant | Produces gametes (in animals) or spores (in the sporophyte plant); reduces number of chromosome sets by half and introduces genetic variability among the gametes or spores |

▲ **Figure 10.10 A comparison of mitosis and meiosis.**

**DRAW IT** *Could any other combinations of chromosomes be generated during meiosis II from the specific cells shown in telophase I? Explain. (Hint: Draw the cells as they would appear in metaphase II.)*

meiotic division, meiosis II, the sister chromatids separate, producing haploid daughter cells. The mechanisms for separating sister chromatids in meiosis II and mitosis are virtually identical. The molecular basis of chromosome behavior during meiosis continues to be a focus of intense research. In the **Scientific Skills Exercise**, you can work with data that tracks the amount of DNA in cells during meiosis.

### CONCEPT CHECK 10.3

1. **MAKE CONNECTIONS** Compare the chromosomes in a cell at metaphase of mitosis with those in a cell at metaphase of meiosis II. (See Figures 9.7 and 10.8.)

2. **WHAT IF?** After the synaptonemal complex disappears, how would any pair of homologous chromosomes be associated if crossing over did not occur? What effect might this have on gamete formation?

For suggested answers, see Appendix A.

CONCEPT 10.4

# Genetic variation produced in sexual life cycles contributes to evolution

How do we account for the genetic variation of the family members in Figure 10.1? As you'll learn more about in later chapters, mutations are the original source of genetic diversity. These changes in an organism's DNA create the different versions of genes known as alleles. Once these differences arise, reshuffling of the alleles during sexual reproduction produces the variation that results in each member of a sexually reproducing population having a unique combination of traits.

## Origins of Genetic Variation Among Offspring

In species that reproduce sexually, the behavior of chromosomes during meiosis and fertilization is responsible for most

---

## Scientific Skills Exercise

### *Making a Line Graph and Converting Between Units of Data*

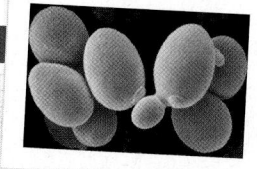
▶ Budding yeast

**How Does DNA Content Change as Budding Yeast Cells Proceed Through Meiosis?** When nutrients are low, cells of the budding yeast (*Saccharomyces cerevisiae*) exit the mitotic cell cycle and enter meiosis. In this exercise you will track the DNA content of a population of yeast cells as they progress through meiosis.

**How the Experiment Was Done** Researchers grew a culture of yeast cells in a nutrient-rich medium and then transferred the cells to a nutrient-poor medium to induce meiosis. At different times after induction, the DNA content per cell was measured in a sample of the cells, and the average DNA content per cell was recorded in femtograms (fg; 1 femtogram = $1 \times 10^{-15}$ gram).

**Data from the Experiment**

| Time After Induction (hours) | Average Amount of DNA per Cell (fg) |
|---|---|
| 0.0 | 24.0 |
| 1.0 | 24.0 |
| 2.0 | 40.0 |
| 3.0 | 47.0 |
| 4.0 | 47.5 |
| 5.0 | 48.0 |
| 6.0 | 48.0 |
| 7.0 | 47.5 |
| 7.5 | 25.0 |
| 8.0 | 24.0 |
| 9.0 | 23.5 |
| 9.5 | 14.0 |
| 10.0 | 13.0 |
| 11.0 | 12.5 |
| 12.0 | 12.0 |
| 13.0 | 12.5 |
| 14.0 | 12.0 |

**INTERPRET THE DATA**

1. First, set up your graph. (a) Place the labels for the independent and dependent variables on the appropriate axes, followed by units of measurement in parentheses. Explain your choices. (b) Add tick marks and values for each axis. Explain your choices. (For additional information about graphs, see the Scientific Skills Review in Appendix F and in the Study Area in MasteringBiology.)

2. Because the variable on the *x*-axis varies continuously, it makes sense to plot the data on a line graph. (a) Plot each data point from the table onto the graph. (b) Connect the data points with line segments.

3. Most of the yeast cells in the culture were in $G_1$ of the cell cycle before being moved to the nutrient-poor medium. (a) How many femtograms of DNA are there in each yeast cell in $G_1$? Estimate this value from the data in your graph. (b) How many femtograms of DNA should be present in each cell in $G_2$? (See Concept 9.2 and Figure 9.6.) At the end of meiosis I (MI)? At the end of meiosis II (MII)? (See Figure 10.7.) (c) Using these values as a guideline, distinguish the different phases by inserting vertical dashed lines in the graph between phases and label each phase ($G_1$, S, $G_2$, MI, MII). You can figure out where to put the dividing lines based on what you know about the DNA content of each phase (see Figure 10.7). (d) Think carefully about the point where the line at the highest value begins to slope downward. What specific point of meiosis does this "corner" represent? What stage(s) correspond to the downward sloping line?

4. Given the fact that 1 fg of DNA = $9.78 \times 10^5$ base pairs (on average), you can convert the amount of DNA per cell to the length of DNA in numbers of base pairs. (a) Calculate the number of base pairs of DNA in the haploid yeast genome. Express your answer in millions of base pairs (Mb), a standard unit for expressing genome size. Show your work. (b) How many base pairs per minute were synthesized during the S phase of these yeast cells?

**MB** A version of this Scientific Skills Exercise can be assigned in MasteringBiology.

**Further Reading** G. Simchen, Commitment to meiosis: What determines the mode of division in budding yeast? *BioEssays* 31:169–177 (2009). doi:10.1002/bies.200800124

of the variation that arises in each generation. Three mechanisms contribute to the genetic variation arising from sexual reproduction: independent assortment of chromosomes, crossing over, and random fertilization.

## Independent Assortment of Chromosomes

One aspect of sexual reproduction that generates genetic variation is the random orientation of pairs of homologous chromosomes at metaphase of meiosis I. At metaphase I, the homologous pairs, each consisting of one maternal and one paternal chromosome, are situated at the metaphase plate. (Note that the terms *maternal* and *paternal* refer, respectively, to whether the chromosome in question was contributed by the mother or the father of the individual whose cells are undergoing meiosis.) Each pair may orient with either its maternal or paternal homolog closer to a given pole—its orientation is as random as the flip of a coin. Thus, there is a 50% chance that a given daughter cell of meiosis I will get the maternal chromosome of a certain homologous pair and a 50% chance that it will get the paternal chromosome.

Because each pair of homologous chromosomes is positioned independently of the other pairs at metaphase I, the first meiotic division results in each pair sorting its maternal and paternal homologs into daughter cells independently of every other pair. This is called *independent assortment*. Each daughter cell represents one outcome of all possible combinations of maternal and paternal chromosomes. As shown in **Figure 10.11**, the number of combinations possible for daughter cells formed by meiosis of a diploid cell with two pairs of homologous chromosomes ($n = 2$) is four: two possible arrangements for the first pair times two possible arrangements for the second pair. Note that only two of the four combinations of daughter cells shown in the figure would result from meiosis of a *single* diploid cell, because a single parent cell would have one or the other possible chromosomal arrangement at metaphase I, but not both. However, the population of daughter cells resulting from meiosis of a large number of diploid cells contains all four types in approximately equal numbers. In the case of $n = 3$, eight combinations of chromosomes are possible for daughter cells. More generally, the number of possible combinations when chromosomes sort independently during meiosis is $2^n$, where $n$ is the haploid number of the organism.

In the case of humans ($n = 23$), the number of possible combinations of maternal and paternal chromosomes in the resulting gametes is $2^{23}$, or about 8.4 million. Each gamete that you produce in your lifetime contains one of roughly 8.4 million possible combinations of chromosomes.

## Crossing Over

As a consequence of the independent assortment of chromosomes during meiosis, each of us produces a collection of gametes differing greatly in their combinations of the chromosomes we inherited from our two parents. Figure 10.11 suggests that each chromosome in a gamete is exclusively maternal or paternal in origin. In fact, this is *not* the case because crossing over produces **recombinant chromosomes**, individual chromosomes that carry genes (DNA) from two different parents. In meiosis in humans, an average of one to three crossover events occur per chromosome pair, depending on the size of the chromosomes and the position of their centromeres. As you learned in Figure 10.9, crossing over produces chromosomes with new combinations of maternal and paternal alleles. At metaphase II, chromosomes that contain one or more recombinant chromatids can be oriented in two nonequivalent ways with respect to other chromosomes because their sister chromatids are no longer identical. The possible arrangements of nonidentical sister chromatids during meiosis II increase the number of genetic types of daughter cells that can result from meiosis.

▶ Figure 10.11 **The independent assortment of homologous chromosomes in meiosis.**

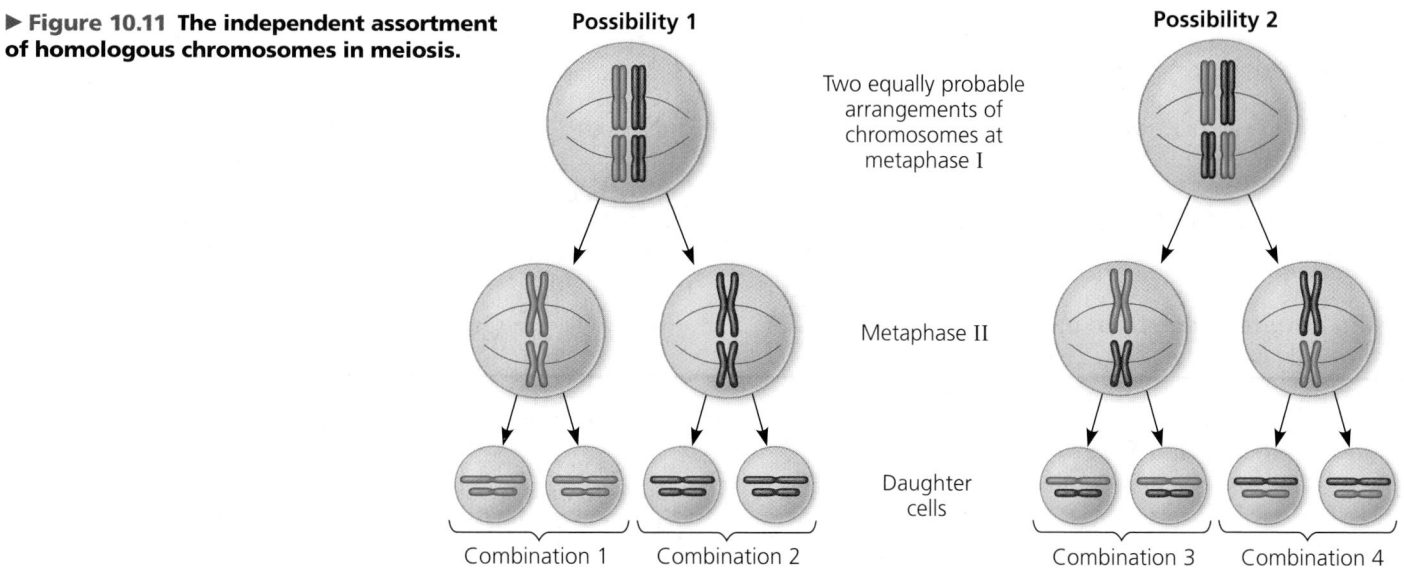

Possibility 1 — Possibility 2

Two equally probable arrangements of chromosomes at metaphase I

Metaphase II

Daughter cells

Combination 1    Combination 2    Combination 3    Combination 4

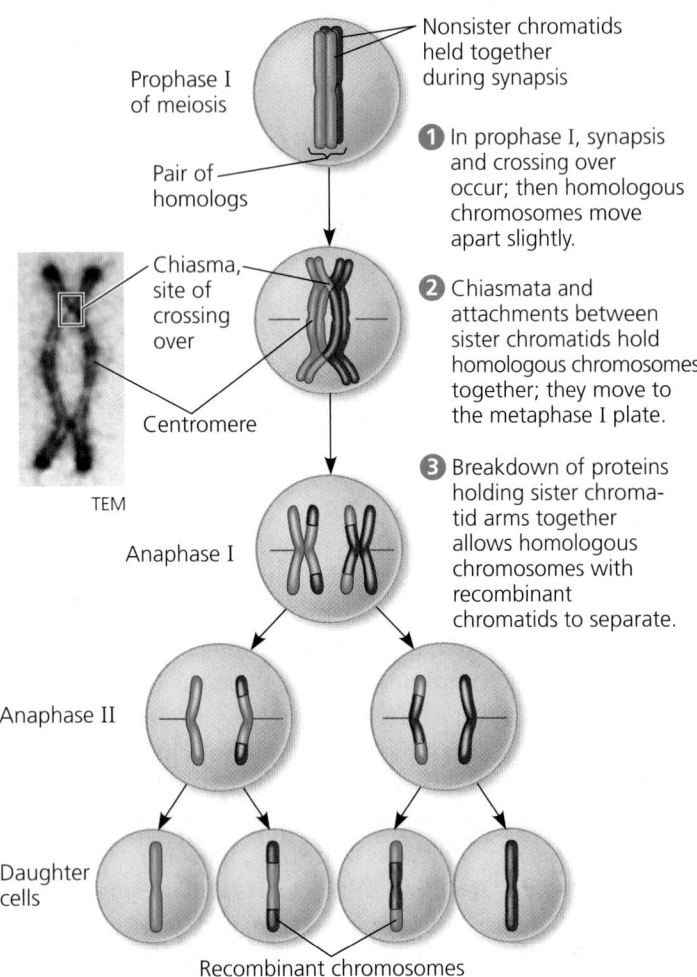

Prophase I of meiosis

Nonsister chromatids held together during synapsis

Pair of homologs

Chiasma, site of crossing over

Centromere

TEM

**1** In prophase I, synapsis and crossing over occur; then homologous chromosomes move apart slightly.

**2** Chiasmata and attachments between sister chromatids hold homologous chromosomes together; they move to the metaphase I plate.

**3** Breakdown of proteins holding sister chromatid arms together allows homologous chromosomes with recombinant chromatids to separate.

Anaphase I

Anaphase II

Daughter cells

Recombinant chromosomes

▲ **Figure 10.12 The results of crossing over during meiosis.**

**Figure 10.12** provides an overview of how crossing over increases genetic variation by producing recombinant chromosomes that include DNA from two parents. You'll learn more about crossing over in Chapter 12.

### Random Fertilization

The random nature of fertilization adds to the genetic variation arising from meiosis. In humans, each male and female gamete represents one of about 8.4 million ($2^{23}$) possible chromosome combinations due to independent assortment. The fusion of a male gamete with a female gamete during fertilization will produce a zygote with any of about 70 trillion ($2^{23} \times 2^{23}$) diploid combinations. With the variation from crossing over, the number of possibilities is astronomical. You really *are* unique.

## The Evolutionary Significance of Genetic Variation Within Populations

**EVOLUTION** How does the genetic variation in a population relate to evolution? A population evolves through differential reproductive success of its members. On average, those best suited to the environment leave more offspring, thereby transmitting their genes. Thus, natural selection leads to an increase in genetic variations favored by the environment. As the environment changes, the population may survive if some members can cope with the new conditions. Mutations, the original source of different alleles, are mixed and matched during meiosis. New combinations of alleles may work better than those that previously prevailed.

We have seen how sexual reproduction greatly increases the genetic variation present in a population. Although Darwin realized that heritable variation makes evolution possible, he could not explain why offspring resemble—but are not identical to—their parents. Ironically, Gregor Mendel, a contemporary of Darwin, published a theory of inheritance that helps explain genetic variation, but his discoveries had no impact on biologists until 1900, more than 15 years after Darwin (1809–1882) and Mendel (1822–1884) had died. In the next chapter, you'll learn how Mendel discovered the basic rules governing the inheritance of traits.

### CONCEPT CHECK 10.4

1. What is the source of variation among alleles of a gene?
2. **WHAT IF?** If maternal and paternal chromatids have the identical two alleles for every gene, will crossing over lead to genetic variation?

**For suggested answers, see Appendix A.**

---

Go to **Mastering**Biology® for Assignments, the eText, and the Study Area with Animations, Activities, Vocab Self-Quiz, and Practice Tests.

# 10 Chapter Review

## SUMMARY OF KEY CONCEPTS

VOCAB SELF-QUIZ
goo.gl/gbai8v

### CONCEPT 10.1

**Offspring acquire genes from parents by inheriting chromosomes (pp. 201–202)**

- Each **gene** in an organism's DNA exists at a specific **locus** on a certain chromosome.
- In **asexual reproduction**, a single parent produces genetically identical offspring by mitosis. **Sexual reproduction** combines genes from two parents, leading to genetically diverse offspring.

**?** *Explain why human offspring resemble their parents but are not identical to them.*

### CONCEPT 10.2

**Fertilization and meiosis alternate in sexual life cycles (pp. 202–205)**

- Normal human **somatic cells** are **diploid**. They have 46 chromosomes made up of two sets of 23 chromosomes, one set from each parent. Human diploid cells have 22 homologous pairs of **autosomes** and one pair of **sex chromosomes**; the latter determines whether the person is female (XX) or male (XY).
- In humans, ovaries and testes produce haploid gametes by **meiosis**, each gamete containing a single set of 23 chromosomes ($n = 23$). During **fertilization**, an egg and sperm unite, forming a diploid ($2n = 46$) single-celled **zygote**, which develops into a multicellular organism by mitosis.

- Sexual **life cycles** differ in the timing of meiosis relative to fertilization and in the point(s) of the cycle at which a multicellular organism is produced by mitosis.

> ? *Compare the life cycles of animals and plants, mentioning their similarities and differences.*

## CONCEPT 10.3

### Meiosis reduces the number of chromosome sets from diploid to haploid (pp. 205–210)

- **Meiosis I** and **meiosis II** produce four haploid daughter cells. The number of chromosome sets is reduced from two (diploid) to one (haploid) during meiosis I.
- Meiosis is distinguished from mitosis by three events of meiosis I:

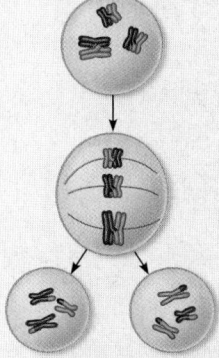

**Prophase I:** Each pair of homologous chromosomes undergoes **synapsis** and **crossing over** between nonsister chromatids with the subsequent appearance of **chiasmata**.

**Metaphase I:** Chromosomes line up as homologous pairs on the metaphase plate.

**Anaphase I:** Homologs separate from each other; sister chromatids remain joined at the centromere.

- Meiosis II separates the sister chromatids.
- Sister chromatid cohesion and crossing over allow chiasmata to hold homologs together until anaphase I. Cohesins are cleaved along the arms at anaphase I, allowing homologs to separate, and at the centromeres in anaphase II, releasing sister chromatids.

> ? *In prophase I, homologous chromosomes pair up and undergo crossing over. Can this also occur during prophase II? Explain.*

## CONCEPT 10.4

### Genetic variation produced in sexual life cycles contributes to evolution (pp. 210–212)

- Three events in sexual reproduction contribute to genetic variation in a population: independent assortment of chromosomes during meiosis, crossing over during meiosis I, and random fertilization of egg cells by sperm. During crossing over, DNA of nonsister chromatids in a homologous pair is broken and rejoined.
- Genetic variation is the raw material for evolution by natural selection. Mutations are the original source of this variation; recombination of variant genes generates additional diversity.

> ? *Explain how three processes unique to sexual reproduction generate a great deal of genetic variation.*

## TEST YOUR UNDERSTANDING

**PRACTICE TEST**

### Level 1: Knowledge/Comprehension

goo.gl/CRZjvS

1. A human cell containing 22 autosomes and a Y chromosome is
   (A) a sperm.
   (B) an egg.
   (C) a zygote.
   (D) a somatic cell of a male.

2. Homologous chromosomes move toward opposite poles of a dividing cell during
   (A) mitosis.
   (B) meiosis I.
   (C) meiosis II.
   (D) fertilization.

### Level 2: Application/Analysis

3. If the DNA content of a diploid cell in the $G_1$ phase of the cell cycle is represented by $x$, then the DNA content of the same cell at metaphase of meiosis I would be
   (A) $0.25x$.  (B) $0.5x$.  (C) $x$.  (D) $2x$.

4. If we continued to follow the cell lineage from question 3, the DNA content of a cell at metaphase of meiosis II would be
   (A) $0.25x$.  (B) $0.5x$.  (C) $x$.  (D) $2x$.

5. **DRAW IT** The diagram shows a cell in meiosis. (a) Label the appropriate structures with these terms: chromosome (label as duplicated or unduplicated), centromere, kinetochore, sister chromatids, nonsister chromatids, homologous pair (use a bracket when labeling), homolog (label each one), chiasma, sister chromatid cohesion, gene loci, alleles of the F gene, alleles of the H gene. (b) Identify the stage of meiosis shown. (c) Describe the makeup of a haploid set and a diploid set.

### Level 3: Synthesis/Evaluation

6. Explain how you can tell that the cell in question 5 is undergoing meiosis, not mitosis.

7. **SCIENTIFIC INQUIRY**
   The diagram in question 5 represents a meiotic cell. Assume the freckles gene is at the locus marked F, and the hair-color gene is located at the locus marked H, both on the long chromosome. The person from whom this cell was taken has inherited different alleles for each gene ("freckles" and "black hair" from one parent and "no freckles" and "blond hair" from the other). Predict allele combinations in the gametes resulting from this meiotic event. List other possible combinations of these alleles in this person's gametes.

8. **FOCUS ON EVOLUTION**
   Many species can reproduce either asexually or sexually. What might be the evolutionary significance of the switch from asexual to sexual reproduction that occurs in some organisms when the environment becomes unfavorable?

9. **FOCUS ON INFORMATION**
   The continuity of life is based on heritable information in the form of DNA. In a short essay (100–150 words), explain how chromosome behavior during sexual reproduction in animals ensures perpetuation of parental traits in offspring and, at the same time, genetic variation among offspring.

10. **SYNTHESIZE YOUR KNOWLEDGE**

The Cavendish banana, the world's most popular fruit, is threatened by extinction due to a fungus. This banana variety is "triploid" ($3n$, with three sets of chromosomes) and can only reproduce through cloning by cultivators. Explain how the triploid number accounts for its inability to form normal gametes. Discuss how the absence of sexual reproduction might make this species vulnerable to infection.

*For selected answers, see Appendix A.*

## KEY CONCEPTS

**11.1** Mendel used the scientific approach to identify two laws of inheritance

**11.2** Probability laws govern Mendelian inheritance

**11.3** Inheritance patterns are often more complex than predicted by simple Mendelian genetics

**11.4** Many human traits follow Mendelian patterns of inheritance

▲ **Figure 11.1** What principles of inheritance did Gregor Mendel discover by breeding pea plants?

## Drawing from the Deck of Genes

The crowd at a soccer match displays the marvelous variety and diversity of humankind. Brown, blue, or gray eyes; black, brown, or blond hair—these are just a few examples of heritable variations among individuals in a population. What principles account for the transmission of such traits from parents to offspring?

The explanation of heredity most widely in favor during the 1800s was the "blending" hypothesis: the idea that genetic material contributed by the two parents mixes, just as blue and yellow paints blend to make green. This hypothesis predicts that over many generations a freely mating population will give rise to a uniform population of individuals—something we don't see. The blending hypothesis also fails to explain how traits can reappear after skipping a generation.

An alternative to the blending model is a "particulate" hypothesis of inheritance: the gene idea. In this model, parents pass on discrete heritable units—genes—that retain their separate identities in offspring. An organism's collection of genes is more like a deck of cards than a bucket of paint. Like cards, genes can be shuffled and passed along, generation after generation, in undiluted form.

Modern genetics had its genesis in an abbey garden, where a monk named Gregor Mendel documented a particulate mechanism for inheritance using pea plants **(Figure 11.1)**. Mendel developed his theory of inheritance several decades before chromosomes were observed under the microscope and the significance of their behavior during mitosis or meiosis was understood. In this chapter, we'll step into Mendel's garden to re-create his experiments and explain how he arrived at his theory of inheritance. We'll also explore inheritance patterns more complex than those observed by Mendel in garden peas. Finally, we'll see how the Mendelian model applies to the inheritance of human variations, including hereditary disorders such as sickle-cell disease.

## CONCEPT 11.1

# Mendel used the scientific approach to identify two laws of inheritance

Mendel discovered the basic principles of heredity by breeding garden peas in carefully planned experiments. As we retrace his work, you'll recognize the key elements of the scientific process that were introduced in Chapter 1.

## Mendel's Experimental, Quantitative Approach

One reason Mendel probably chose to work with peas is that they are available in many varieties. For example, one variety has purple flowers, while another variety has white flowers. A heritable feature that varies among individuals, such as flower color, is called a **character**. Each variant for a character, such as purple or white color for flowers, is called a **trait**.

Mendel could strictly control mating between plants. Each pea flower has both pollen-producing organs (stamens) and an egg-bearing organ (carpel). In nature, pea plants usually self-fertilize: Pollen grains from the stamens land on the carpel of the same flower, and sperm released from the pollen grains fertilize eggs present in the carpel. To achieve cross-pollination of two plants, Mendel removed the immature stamens of a plant before they produced pollen and then dusted pollen from another plant onto the altered flowers **(Figure 11.2)**. Each resulting zygote then developed into a plant embryo encased in a seed (pea). His method allowed Mendel to always be sure of the parentage of new seeds.

Mendel chose to track only those characters that occurred in two distinct, alternative forms, such as purple or white flower color. He also made sure that he started his experiments with varieties that, over many generations of self-pollination, had produced only the same variety as the parent plant. Such plants are said to be **true-breeding**. For example, a plant with purple flowers is true-breeding if the seeds produced by self-pollination in successive generations all give rise to plants that also have purple flowers.

In a typical breeding experiment, Mendel cross-pollinated two contrasting, true-breeding pea varieties—for example, purple-flowered plants and white-flowered plants (see Figure 11.2). This mating, or *crossing*, of two true-breeding varieties is called **hybridization**. The true-breeding parents are referred to as the **P generation** (parental generation), and their hybrid offspring are the **F₁ generation** (first filial generation, the word *filial* from the Latin word for "son"). Allowing these F₁ hybrids to self-pollinate (or to cross-pollinate with other F₁ hybrids) produces an **F₂ generation** (second filial generation). Mendel usually followed traits for at least the P, F₁, and F₂ generations. Had Mendel stopped his experiments with the F₁ generation, the basic patterns of inheritance would have eluded him. Mendel's quantitative analysis of the F₂ plants from thousands of genetic crosses like these allowed him to

### Crossing Pea Plants

**Application**  By crossing (mating) two true-breeding varieties of an organism, scientists can study patterns of inheritance. In this example, Mendel crossed pea plants that varied in flower color.

**Technique**

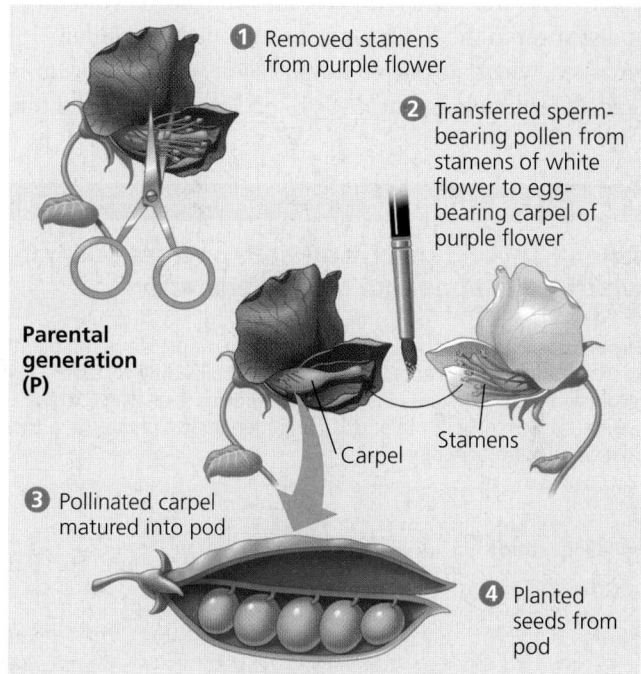

**Results**  When pollen from a white flower was transferred to a purple flower, the first-generation hybrids all had purple flowers. The result was the same for the reciprocal cross, which involved the transfer of pollen from purple flowers to white flowers.

deduce two fundamental principles of heredity, now called the law of segregation and the law of independent assortment.

## The Law of Segregation

If the blending model of inheritance were correct, the F₁ hybrids from a cross between purple-flowered and white-flowered pea plants would have pale purple flowers, a trait intermediate between those of the P generation. Notice in Figure 11.2 that the experiment produced a very different result: All the F₁ offspring had flowers of the same color as the purple-flowered parents. What happened to the white-flowered plants' genetic contribution to the hybrids? If it were lost, then the F₁ plants could produce only purple-flowered offspring in the $F_2$ generation. But when Mendel allowed the $F_1$ plants to self-pollinate and planted their seeds, the white-flower trait reappeared in the $F_2$ generation.

Mendel used very large sample sizes and kept accurate records of his results: 705 of the $F_2$ plants had purple flowers, and 224 had white flowers. These data fit a ratio of approximately three purple to one white **(Figure 11.3)**. Mendel reasoned that the heritable factor for white flowers did not disappear in the $F_1$ plants, but was somehow hidden, or masked, when the purple-flower factor was present. In Mendel's terminology, purple flower color is a *dominant* trait,

and white flower color is a *recessive* trait. The reappearance of white-flowered plants in the $F_2$ generation was evidence that the heritable factor causing white flowers had not been diluted or destroyed by coexisting with the purple-flower factor in the $F_1$ hybrids.

Mendel observed the same pattern of inheritance in six other characters, each represented by two distinctly different traits **(Table 11.1)**. For example, when Mendel crossed a true-breeding variety that produced smooth, round pea seeds with one that produced wrinkled seeds, all the $F_1$ hybrids produced round seeds; this is the dominant trait for seed shape. In the $F_2$ generation, approximately 75% of the seeds were round and 25% were wrinkled—a 3:1 ratio, as in Figure 11.3. Now let's see how Mendel deduced the law of segregation from his experimental results. In the discussion that follows, we will use modern terms instead of some of the terms used by Mendel. (For example, we'll use "gene" instead of Mendel's "heritable factor.")

---

**▼ Figure 11.3  Inquiry**

### When $F_1$ hybrid pea plants self- or cross-pollinate, which traits appear in the $F_2$ generation?

**Experiment** Mendel crossed true-breeding purple-flowered plants and white-flowered plants (crosses are symbolized by ×). The resulting $F_1$ hybrids were allowed to self-pollinate or were cross-pollinated with other $F_1$ hybrids. The $F_2$ generation plants were then observed for flower color.

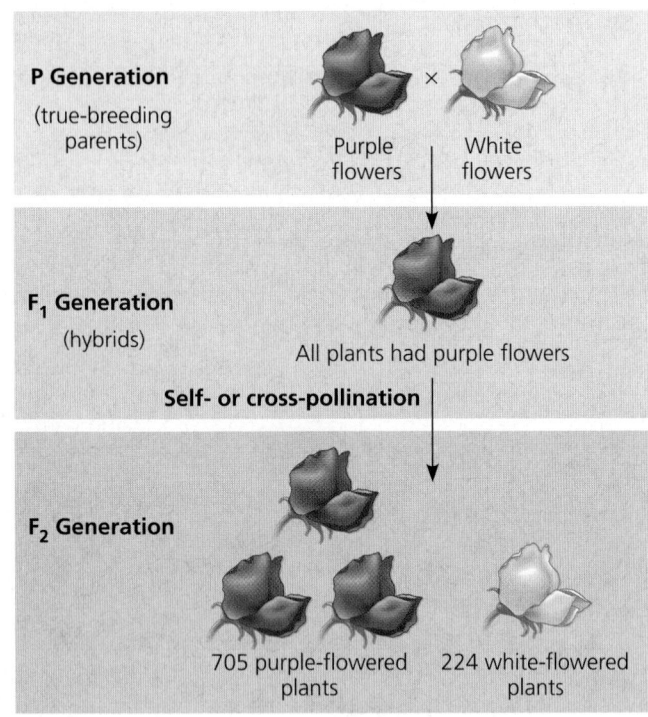

**Results** Both purple-flowered and white-flowered plants appeared in the $F_2$ generation, in a ratio of approximately 3:1.

**Conclusion** The "heritable factor" for the recessive trait (white flowers) had not been destroyed, deleted, or "blended" in the $F_1$ generation but was merely masked by the presence of the factor for purple flowers, which is the dominant trait.

**Data from** G. Mendel, Experiments in plant hybridization, *Proceedings of the Natural History Society of Brünn* 4:3–47 (1866).

**WHAT IF?** If you mated two purple-flowered plants from the P generation, what ratio of traits would you expect to observe in the offspring? Explain.

---

**Table 11.1 The Results of Mendel's $F_1$ Crosses for Seven Characters in Pea Plants**

| Character | Dominant Trait | × | Recessive Trait | $F_2$ Generation Dominant: Recessive | Ratio |
|---|---|---|---|---|---|
| Flower color | Purple | × | White | 705:224 | 3.15:1 |
| Seed color | Yellow | × | Green | 6,022:2,001 | 3.01:1 |
| Seed shape | Round | × | Wrinkled | 5,474:1,850 | 2.96:1 |
| Pod shape | Inflated | × | Constricted | 882:299 | 2.95:1 |
| Pod color | Green | × | Yellow | 428:152 | 2.82:1 |
| Flower position | Axial | × | Terminal | 651:207 | 3.14:1 |
| Stem length | Tall | × | Dwarf | 787:277 | 2.84:1 |

## Mendel's Model

Mendel developed a model to explain the 3:1 inheritance pattern that he consistently observed among the $F_2$ offspring in his pea experiments. We describe four related concepts making up this model, the fourth of which is the law of segregation.

First, *alternative versions of genes account for variations in inherited characters.* The gene for flower color in pea plants, for example, exists in two versions, one for purple flowers and the other for white flowers. These alternative versions of a gene are called **alleles**. Today, we can relate this concept to chromosomes and DNA. As shown in **Figure 11.4**, each gene is a sequence of nucleotides at a specific place, or locus, along a particular chromosome. The DNA at that locus, however, can vary slightly in its nucleotide sequence. This variation in information content can affect the function of the encoded protein and thus an inherited character of the organism. The purple-flower allele and the white-flower allele are two DNA sequence variations possible at the flower-color locus on one of a pea plant's chromosomes. The purple-flower allele sequence allows synthesis of purple pigment, and the white-flower allele sequence does not.

Second, *for each character, an organism inherits two copies (that is, two alleles) of a gene, one from each parent.* Remarkably, Mendel made this deduction without knowing about the role, or even the existence, of chromosomes. Each somatic cell in a diploid organism has two sets of chromosomes, one set inherited from each parent (see Figure 10.4). Thus, a genetic locus is actually represented twice in a diploid cell, once on each homolog of a specific pair of chromosomes. The two alleles at a particular locus may be identical, as in the true-breeding plants of Mendel's P generation, or the alleles may differ, as in the $F_1$ hybrids (see Figure 11.4).

Third, *if the two alleles at a locus differ, then one, the* **dominant allele**, *determines the organism's appearance; the other, the* **recessive allele**, *has no noticeable effect on the organism's appearance.* Accordingly, Mendel's $F_1$ plants had purple flowers because the allele for that trait is dominant and the allele for white flowers is recessive.

The fourth and final part of Mendel's model, the **law of segregation**, states that *the two alleles for a heritable character segregate (separate from each other) during gamete formation and end up in different gametes.* Thus, an egg or a sperm gets only one of the two alleles that are present in the somatic cells of the organism making the gamete. In terms of chromosomes, this segregation corresponds to the distribution of copies of the two members of a pair of homologous chromosomes to different gametes in meiosis (see Figure 10.7). Note that if an organism has identical alleles for a particular character—that is, the organism is true-breeding for that character—then that allele is present in all gametes. But if different alleles are present, as in the $F_1$ hybrids, then 50% of the gametes receive the dominant allele and 50% receive the recessive allele.

Does Mendel's segregation model account for the 3:1 ratio he observed in the $F_2$ generation of his numerous crosses? For the flower-color character, the model predicts that the two different alleles present in an $F_1$ individual will segregate into gametes such that half the gametes will have the purple-flower allele and half will have the white-flower allele. During self-pollination, gametes of each class unite randomly. An egg with a purple-flower allele has an equal chance of being fertilized by a sperm with a purple-flower allele or one with a white-flower allele. Since the same is true for an egg with a white-flower allele, there are four equally likely combinations of sperm

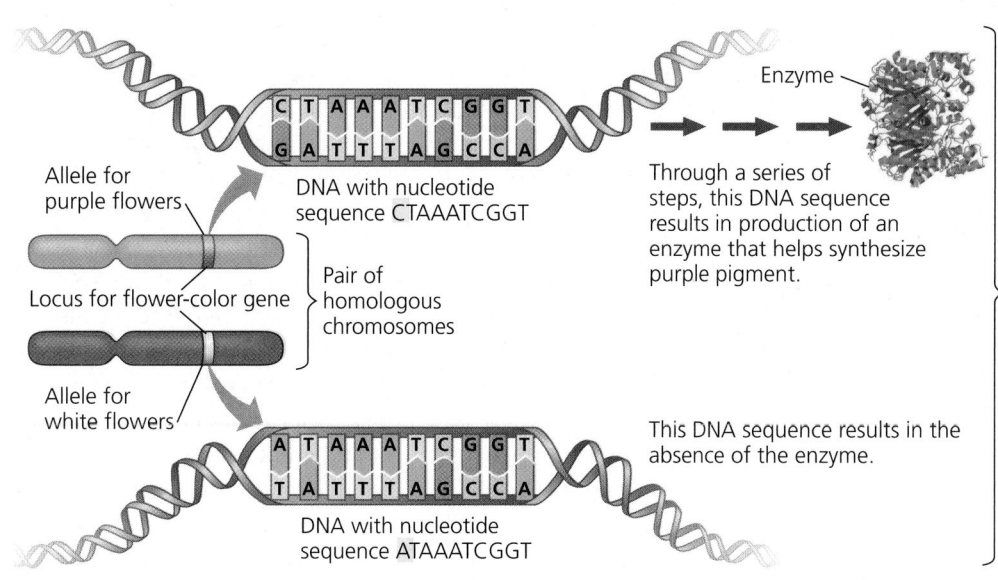

▶ **Figure 11.4 Alleles, alternative versions of a gene.** This diagram shows a pair of homologous chromosomes in an $F_1$ hybrid pea plant, with the actual DNA sequence from the flower color allele of each chromosome. The paternally inherited chromosome (blue) has an allele for purple flowers, which codes for a protein that indirectly controls synthesis of purple pigment. The maternally inherited chromosome (red) has an allele for white flowers, which results in no functional protein being made.

Allele for purple flowers

Locus for flower-color gene

Pair of homologous chromosomes

Allele for white flowers

CTAAATCGGT
GATTTAGCCA

DNA with nucleotide sequence CTAAATCGGT

ATAAATCGGT
TATTTAGCCA

DNA with nucleotide sequence ATAAATCGGT

Enzyme

Through a series of steps, this DNA sequence results in production of an enzyme that helps synthesize purple pigment.

This DNA sequence results in the absence of the enzyme.

One purple-flower allele results in sufficient pigment for purple flowers.

► **Figure 11.5 Mendel's law of segregation.** This diagram shows the genetic makeup of the generations in Figure 11.3. It illustrates Mendel's model for inheritance of the alleles of a single gene. Each plant has two alleles for the gene controlling flower color, one allele inherited from each of the plant's parents. To construct a Punnett square that predicts the F₂ generation offspring, we list all the possible gametes from one parent (here, the F₁ female) along the left side of the square and all the possible gametes from the other parent (here, the F₁ male) along the top. The boxes represent the offspring resulting from all the possible unions of male and female gametes.

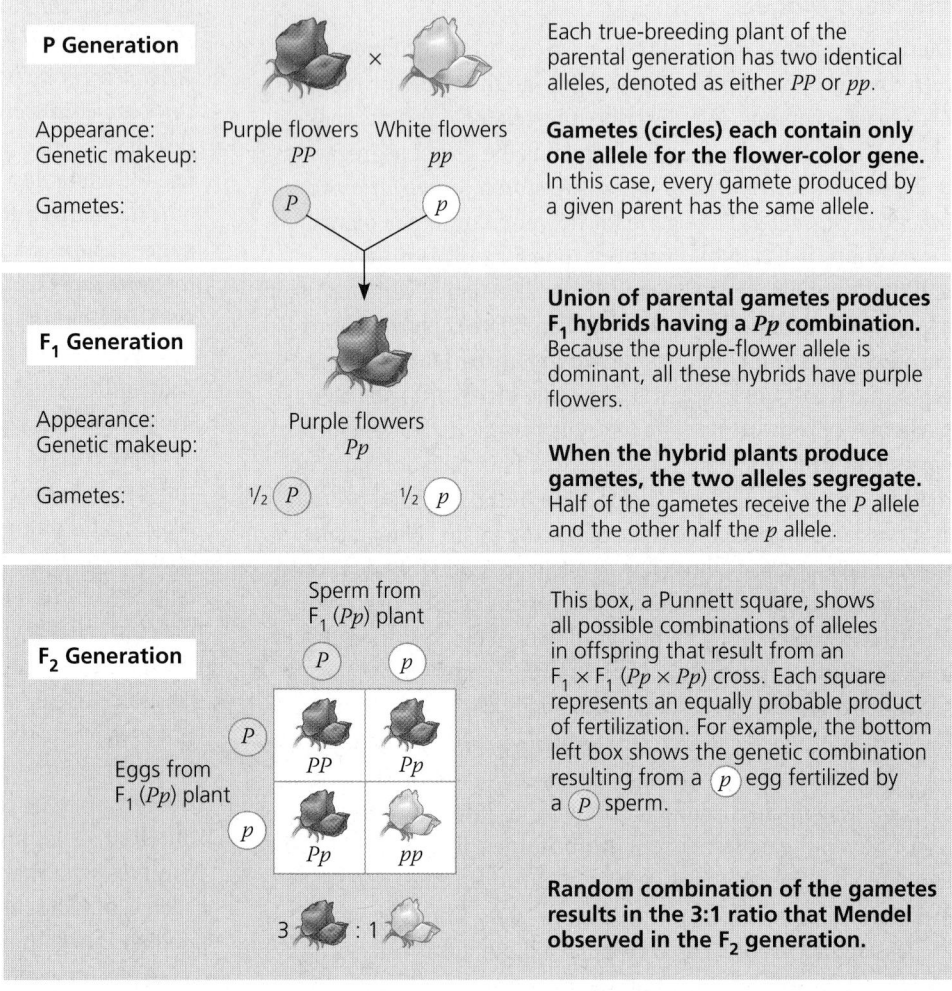

**P Generation**

Appearance: Purple flowers    White flowers
Genetic makeup:    *PP*      *pp*

Gametes:   *P*    *p*

Each true-breeding plant of the parental generation has two identical alleles, denoted as either *PP* or *pp*.

**Gametes (circles) each contain only one allele for the flower-color gene.** In this case, every gamete produced by a given parent has the same allele.

**F₁ Generation**

Appearance: Purple flowers
Genetic makeup: *Pp*

Gametes: ½ *P*    ½ *p*

**Union of parental gametes produces F₁ hybrids having a *Pp* combination.** Because the purple-flower allele is dominant, all these hybrids have purple flowers.

**When the hybrid plants produce gametes, the two alleles segregate.** Half of the gametes receive the *P* allele and the other half the *p* allele.

**F₂ Generation**

Sperm from F₁ (*Pp*) plant
*P*    *p*

Eggs from F₁ (*Pp*) plant
*P*    *p*

*PP*    *Pp*
*Pp*    *pp*

3 : 1

This box, a Punnett square, shows all possible combinations of alleles in offspring that result from an F₁ × F₁ (*Pp* × *Pp*) cross. Each square represents an equally probable product of fertilization. For example, the bottom left box shows the genetic combination resulting from a *p* egg fertilized by a *P* sperm.

**Random combination of the gametes results in the 3:1 ratio that Mendel observed in the F₂ generation.**

and egg. **Figure 11.5** illustrates these combinations using a **Punnett square**, a handy diagrammatic device for predicting the allele composition of all offspring resulting from a cross between individuals of known genetic makeup. Notice that we use a capital letter to symbolize a dominant allele and a lower-case letter for a recessive allele. In our example, *P* is the purple-flower allele, and *p* is the white-flower allele; the gene itself is sometimes referred to as the *P/p* gene.

In the F₂ offspring, what color will the flowers be? One-fourth of the plants have inherited two purple-flower alleles; these plants will have purple flowers. One-half of the F₂ offspring have inherited one purple-flower allele and one white-flower allele; these plants will also have purple flowers, the dominant trait. Finally, one-fourth of the F₂ plants have inherited two white-flower alleles and will express the recessive trait. Thus, Mendel's model accounts for the 3:1 ratio of traits that he observed in the F₂ generation.

### Useful Genetic Vocabulary

An organism that has a pair of identical alleles for a gene encoding a character is called a **homozygote** and is said to be **homozygous** for that gene. In the parental generation in Figure 11.5, the purple-flowered pea plant is homozygous for the dominant allele (*PP*), while the white-flowered plant is homozygous for the recessive allele (*pp*). Homozygous plants "breed true" because all of their gametes contain the same allele—either *P* or *p* in this example. If we cross dominant homozygotes with recessive homozygotes, every offspring will have two different alleles—*Pp* in the case of the F₁ hybrids of our flower-color experiment (see Figure 11.5). An organism that has two different alleles for a gene is called a **heterozygote** and is said to be **heterozygous** for that gene. Unlike homozygotes, heterozygotes produce gametes with different alleles, so they are not true-breeding. For example, *P*- and *p*-containing gametes are both produced by our F₁ hybrids. Self-pollination of the F₁ hybrids thus produces both purple-flowered and white-flowered offspring.

Because of the different effects of dominant and recessive alleles, an organism's traits do not always reveal its genetic composition. Therefore, we distinguish between an organism's appearance or observable traits, called its **phenotype**, and its genetic makeup, its **genotype**. In the case of flower color in pea plants, *PP* and *Pp* plants have the same phenotype (purple) but different genotypes. **Figure 11.6** reviews these terms. Note that the term *phenotype* refers to physiological traits as well as traits that relate directly to appearance. For example, one pea variety lacks the normal ability to self-pollinate, which is a phenotypic trait (non-self-pollination).

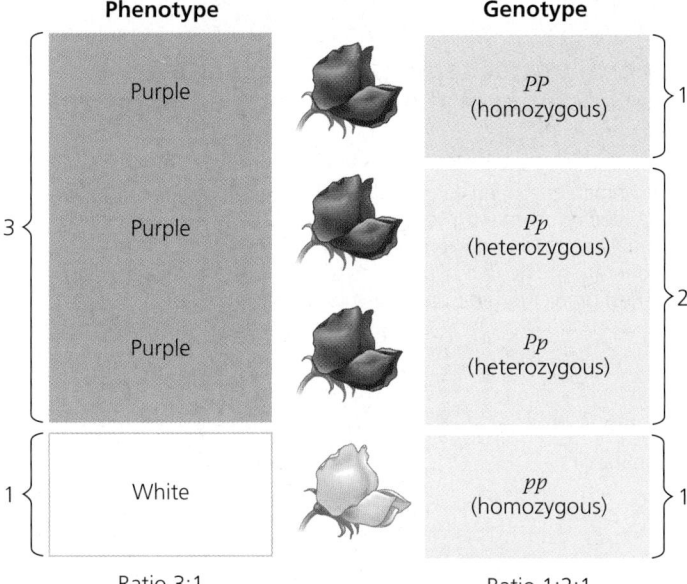

|  | Phenotype |  | Genotype |  |
|---|---|---|---|---|

3 {
- Purple — *PP* (homozygous) — 1
- Purple — *Pp* (heterozygous)
- Purple — *Pp* (heterozygous)

} 2

1 {
- White — *pp* (homozygous) — 1

Ratio 3:1          Ratio 1:2:1

▲ **Figure 11.6 Phenotype versus genotype.** Grouping F₂ offspring from a cross for flower color according to phenotype results in the typical 3:1 phenotypic ratio. In terms of genotype, however, there are actually two categories of purple-flowered plants, *PP* (homozygous) and *Pp* (heterozygous), giving a 1:2:1 genotypic ratio.

## The Testcross

Given a purple-flowered pea plant, we cannot tell if it is homozygous (*PP*) or heterozygous (*Pp*) because both genotypes result in the same purple phenotype. To determine the genotype, we can cross this plant with a white-flowered plant (*pp*), which will make only gametes with the recessive allele (*p*). The allele in the gamete contributed by the mystery plant will therefore determine the appearance of the offspring **(Figure 11.7)**. If all the offspring of the cross have purple flowers, then the purple-flowered mystery plant must be homozygous for the dominant allele, because a *PP* × *pp* cross produces all *Pp* offspring. But if both the purple and the white phenotypes appear among the offspring, then the purple-flowered parent must be heterozygous. The offspring of a *Pp* × *pp* cross will be expected to have a 1:1 phenotypic ratio. Breeding an organism of unknown genotype with a recessive homozygote is called a **testcross** because it can reveal the genotype of that organism. The testcross was devised by Mendel and continues to be used by geneticists.

## The Law of Independent Assortment

Mendel derived the law of segregation from experiments in which he followed only a *single* character, such as flower color. All the F₁ progeny produced in his crosses of true-breeding parents were **monohybrids**, meaning that they were heterozygous for the one particular character being followed in the cross. We refer to a cross between such heterozygotes as a **monohybrid cross**.

Mendel identified his second law of inheritance by following *two* characters at the same time, such as seed color and

### The Testcross

**Application** An organism that exhibits a dominant trait, such as purple flowers in pea plants, can be either homozygous for the dominant allele or heterozygous. To determine the organism's genotype, geneticists can perform a testcross.

**Technique** In a testcross, the individual with the unknown genotype is crossed with a homozygous individual expressing the recessive trait (white flowers in this example), and Punnett squares are used to predict the possible outcomes.

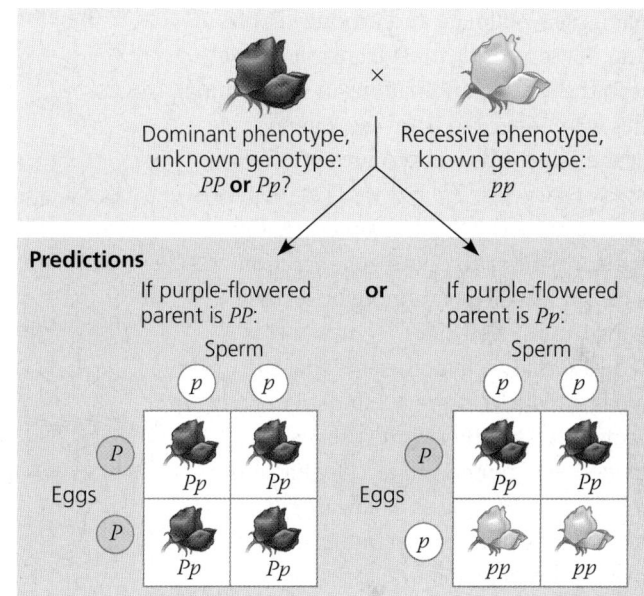

**Results** Matching the results to either prediction identifies the unknown parental genotype (either *PP* or *Pp* in this example). In this testcross, we transferred pollen from a white-flowered plant to the carpels of a purple-flowered plant; the opposite (reciprocal) cross would have led to the same results.

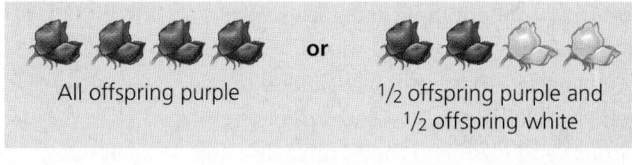

All offspring purple     or     ½ offspring purple and ½ offspring white

seed shape. Seeds (peas) may be either yellow or green. They also may be either round (smooth) or wrinkled. From single-character crosses, Mendel knew that the allele for yellow seeds (*Y*) is dominant and the allele for green seeds (*y*) is recessive. For the seed-shape character, the allele for round (*R*) is dominant, and the allele for wrinkled (*r*) is recessive.

Imagine crossing two true-breeding pea varieties that differ in *both* of these characters—a cross between a plant with yellow round seeds (*YYRR*) and a plant with green wrinkled seeds (*yyrr*). The F₁ plants will be **dihybrids**, individuals heterozygous for the two characters being followed in the cross (*YyRr*). But are these two characters transmitted from parents to offspring as a package? That is, will the *Y* and *R* alleles always stay together, generation after generation? Or are seed color and seed shape inherited independently? **Figure 11.8** shows how a **dihybrid cross**, a cross between $F_1$ dihybrids, can determine which of these two hypotheses is correct.

The $F_1$ plants, of genotype *YyRr*, exhibit both dominant phenotypes, yellow seeds with round shapes, no matter which hypothesis is correct. The key step in the experiment is to see what happens when $F_1$ plants self-pollinate and produce $F_2$ offspring. If the hybrids must transmit their alleles in the same combinations in which the alleles were inherited from the P generation, then the $F_1$ hybrids will produce only two classes of gametes: *YR* and *yr*. This "dependent assortment" hypothesis predicts that the phenotypic ratio of the $F_2$ generation will be 3:1, just as in a monohybrid cross (see Figure 11.8, left side).

The alternative hypothesis is that the two pairs of alleles segregate independently of each other. In other words, genes are packaged into gametes in all possible allelic combinations, as long as each gamete has one allele for each gene. In our example, an $F_1$ plant will produce four classes of gametes in equal quantities: *YR*, *Yr*, *yR*, and *yr*. If sperm of the four classes fertilize eggs of the four classes, there will be 16 (4 × 4) equally probable ways in which the alleles can combine in the $F_2$ generation, as shown in Figure 11.8, right side. These combinations result in four phenotypic categories with a ratio of 9:3:3:1 (nine yellow round to three green round to three yellow wrinkled to one green wrinkled). When Mendel did the experiment and classified the $F_2$ offspring, his results were close to the predicted 9:3:3:1 phenotypic ratio, supporting the hypothesis that the alleles for one gene—controlling seed color, for example—segregate into gametes independently of the alleles of any other gene, such as seed shape.

Mendel tested his seven pea characters in various dihybrid combinations and always observed a 9:3:3:1 phenotypic ratio in the $F_2$ generation. Is this consistent with the 3:1 phenotypic ratio seen for the monohybrid cross shown in Figure 11.5? If you calculate the ratio of yellow and green peas, ignoring shape, you will see that the color

### Do the alleles for one character segregate into gametes dependently or independently of the alleles for a different character?

**Experiment** To follow the characters of seed color and seed shape through the $F_2$ generation, Mendel crossed a true-breeding plant with yellow round seeds with a true-breeding plant with green wrinkled seeds, producing dihybrid $F_1$ plants. Self-pollination of the $F_1$ dihybrids produced the $F_2$ generation. The two hypotheses (dependent and independent "assortment" of the two genes) predict different phenotypic ratios.

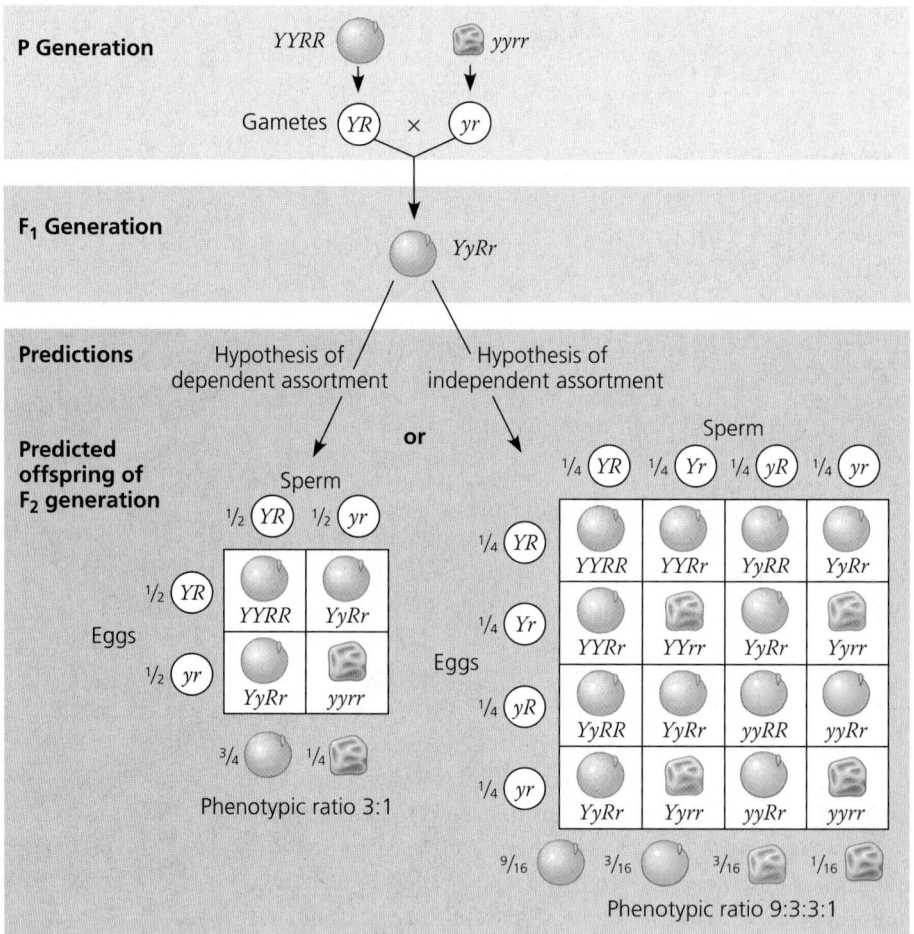

**Results**

315 ◯   108 ◯   101 ▨   32 ▨   Phenotypic ratio approximately 9:3:3:1

**Conclusion** Only the hypothesis of independent assortment predicts two of the observed phenotypes: green round seeds and yellow wrinkled seeds (see the right-hand Punnett square). The alleles for each gene segregate independently of those of the other, and the two genes are said to assort independently.

**Data from** G. Mendel, Experiments in plant hybridization, *Proceedings of the Natural History Society of Brünn* 4:3–47 (1866).

**WHAT IF?** Suppose Mendel had transferred pollen from an $F_1$ plant to the carpel of a plant that was homozygous recessive for both genes. Set up the cross and draw Punnett squares that predict the offspring for both hypotheses. Would this cross have supported the hypothesis of independent assortment equally well?

alleles segregate as if this were a monohybrid cross (3:1). The results of Mendel's dihybrid experiments are the basis for what we now call the **law of independent assortment**, which states that *two or more genes assort independently—that is, each pair of alleles segregates independently of any other pair during gamete formation.*

This law applies only to genes (allele pairs) located on different chromosomes—that is, on chromosomes that are not homologous—or very far apart on the same chromosome. (The latter case will be explained in Chapter 12, along with the more complex inheritance patterns of genes located near each other, which tend to be inherited together.) All the pea characters Mendel chose for analysis were controlled by genes on different chromosomes or far apart on the same chromosome; this greatly simplified the interpretation of his multicharacter pea crosses. All the examples we consider in the rest of this chapter involve genes located on different chromosomes.

### CONCEPT CHECK 11.1

1. **DRAW IT** Pea plants heterozygous for flower position and stem length (*AaTt*) are allowed to self-pollinate, and 400 of the resulting seeds are planted. Draw a Punnett square for this cross. How many offspring would be predicted to have terminal flowers and be dwarf? (See Table 11.1.)

2. List all gametes that could be made by a pea plant heterozygous for seed color, seed shape, and pod shape (*YyRrIi*; see Table 11.1). How large a Punnett square is needed to predict the offspring of a self-pollination of this "trihybrid"?

3. **MAKE CONNECTIONS** In some pea plant crosses, the plants are self-pollinated. Explain whether self-pollination is considered asexual or sexual reproduction (refer to Concept 10.1).

For suggested answers, see Appendix A.

---

### CONCEPT 11.2

# Probability laws govern Mendelian inheritance

Mendel's laws of segregation and independent assortment reflect the same rules of probability that apply to tossing coins, rolling dice, and drawing cards from a deck. The probability scale ranges from 0 to 1. An event that is certain to occur has a probability of 1, while an event that is certain *not* to occur has a probability of 0. With a coin that has heads on both sides, the probability of tossing heads is 1, and the probability of tossing tails is 0. With a normal coin, the chance of tossing heads is ½, and the chance of tossing tails is ½. The probability of drawing the ace of spades from a 52-card deck is $\frac{1}{52}$. The probabilities of all possible outcomes for an event must add up to 1. With a deck of cards, the chance of picking a card other than the ace of spades is $\frac{51}{52}$.

Tossing a coin illustrates an important lesson about probability. For every toss, the probability of heads is ½. The outcome of any particular toss is unaffected by what has happened on previous trials. We refer to phenomena such as coin tosses as independent events. Each toss of a coin, whether done sequentially with one coin or simultaneously with many, is independent of every other toss. And like two separate coin tosses, the alleles of one gene segregate into gametes independently of another gene's alleles (the law of independent assortment). We'll now look at two basic rules of probability that help us predict the outcome of the fusion of such gametes in simple monohybrid crosses and more complicated crosses as well.

## The Multiplication and Addition Rules Applied to Monohybrid Crosses

How do we determine the probability that two or more independent events will occur together in some specific combination? For example, what is the chance that two coins tossed simultaneously will both land heads up? The **multiplication rule** states that to determine this probability, we multiply the probability of one event (one coin coming up heads) by the probability of the other event (the other coin coming up heads). By the multiplication rule, then, the probability that both coins will land heads up is ½ × ½ = ¼.

We can apply the same reasoning to an $F_1$ monohybrid cross **(Figure 11.9)**. With seed shape in pea plants as the heritable character, the genotype of $F_1$ plants is *Rr*. Segregation in a heterozygous plant is like flipping a coin in terms of calculating

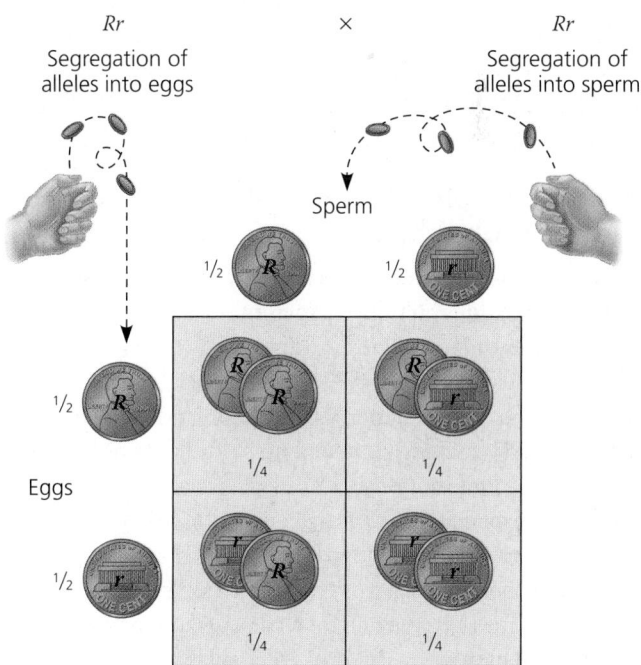

▲ **Figure 11.9 Segregation of alleles and fertilization as chance events.** When a heterozygote (*Rr*) forms gametes, whether a particular gamete ends up with an *R* or an *r* is like the toss of a coin. We can determine the probability for any genotype among the offspring of two heterozygotes by multiplying together the individual probabilities of an egg and sperm having a particular allele (*R* or *r* in this example).

the probability of each outcome: Each egg produced has a ½ chance of carrying the dominant allele (*R*) and a ½ chance of carrying the recessive allele (*r*). The same odds apply to each sperm cell produced. For a particular $F_2$ plant to have wrinkled seeds, the recessive trait, both the egg and the sperm that come together must carry the *r* allele. The probability that an *r* allele will be present in both gametes at fertilization is found by multiplying ½ (the probability that the egg will have an *r*) × ½ (the probability that the sperm will have an *r*). Thus, the multiplication rule tells us that the probability of an $F_2$ plant having wrinkled seeds (*rr*) is ¼ (see Figure 11.9). Likewise, the probability of an $F_2$ plant carrying both dominant alleles for seed shape (*RR*) is ¼.

To figure out the probability that an $F_2$ plant from a monohybrid cross will be heterozygous rather than homozygous, we need to invoke a second rule. Notice in Figure 11.9 that the dominant allele can come from the egg and the recessive allele from the sperm, or vice versa. That is, $F_1$ gametes can combine to produce *Rr* offspring in two *mutually exclusive* ways: For any particular heterozygous $F_2$ plant, the dominant allele can come from the egg *or* the sperm, but not from both. According to the **addition rule**, the probability that any one of two or more mutually exclusive events will occur is calculated by adding their individual probabilities. As we have just seen, the multiplication rule gives us the individual probabilities that we will now add together. The probability for one possible way of obtaining an $F_2$ heterozygote—the dominant allele from the egg and the recessive allele from the sperm—is ¼. The probability for the other possible way—the recessive allele from the egg and the dominant allele from the sperm—is also ¼ (see Figure 11.9). Using the rule of addition, then, we can calculate the probability of an $F_2$ heterozygote as ¼ + ¼ = ½.

## Solving Complex Genetics Problems with the Rules of Probability

We can also apply the rules of probability to predict the outcome of crosses involving multiple characters. Recall that each allelic pair segregates independently during gamete formation (the law of independent assortment). Thus, a dihybrid or other multi-character cross is equivalent to two or more independent monohybrid crosses occurring simultaneously. By applying what we have learned about monohybrid crosses, we can determine the probability of specific genotypes occurring in the $F_2$ generation without having to construct unwieldy Punnett squares.

Consider the dihybrid cross between *YyRr* heterozygotes shown in Figure 11.8. We will focus first on the seed-color character. For a monohybrid cross of *Yy* plants, we can use a simple Punnett square to determine that the probabilities of the offspring genotypes are ¼ for *YY*, ½ for *Yy*, and ¼ for *yy*. We can draw a second Punnett square to determine that the same probabilities apply to the offspring genotypes for seed shape: ¼ *RR*, ½ *Rr*, and ¼ *rr*. Knowing these probabilities, we can simply use the multiplication rule to determine the probability of each of the genotypes in the $F_2$ generation. To give two examples, the

calculations for finding the probabilities of two of the possible $F_2$ genotypes (*YYRR* and *YyRR*) are shown below:

| | |
|---|---|
| Probability of *YYRR* = ¼ (probability of *YY*) × ¼ (*RR*) = $\frac{1}{16}$ |
| Probability of *YyRR* = ½ (*Yy*) × ¼ (*RR*) = ⅛ |

The *YYRR* genotype corresponds to the upper left box in the larger Punnett square in Figure 11.8 (one box = $\frac{1}{16}$). Looking closely at the larger Punnett square in Figure 11.8, you will see that 2 of the 16 boxes (⅛) correspond to the *YyRR* genotype.

Now let's see how we can combine the multiplication and addition rules to solve even more complex problems in Mendelian genetics. Imagine a cross of two pea varieties in which we track the inheritance of three characters. Let's cross a trihybrid with purple flowers and yellow round seeds (heterozygous for all three genes) with a plant with purple flowers and green wrinkled seeds (heterozygous for flower color but homozygous recessive for the other two characters). Using Mendelian symbols, our cross is *PpYyRr* × *Ppyyrr*. What fraction of offspring from this cross is predicted to exhibit the recessive phenotypes for *at least two* of the three characters?

To answer this question, we can start by listing all genotypes we could get that fulfill this condition: *ppyyRr*, *ppYyrr*, *Ppyyrr*, *PPyyrr*, and *ppyyrr*. (Because the condition is *at least two* recessive traits, it includes the last genotype, which shows all three recessive traits.) Next, we calculate the probability for each of these genotypes resulting from our *PpYyRr* × *Ppyyrr* cross by multiplying together the individual probabilities for the allele pairs, just as we did in our dihybrid example. Note that in a cross involving heterozygous and homozygous allele pairs (for example, *Yy* × *yy*), the probability of heterozygous offspring is ½ and the probability of homozygous offspring is ½. Finally, we use the addition rule to add the probabilities for all the different genotypes that fulfill the condition of at least two recessive traits, as shown below:

| | | |
|---|---|---|
| *ppyyRr* | ¼ (probability of *pp*) × ½ (*yy*) × ½ (*Rr*) | = $\frac{1}{16}$ |
| *ppYyrr* | ¼ × ½ × ½ | = $\frac{1}{16}$ |
| *Ppyyrr* | ½ × ½ × ½ | = $\frac{2}{16}$ |
| *PPyyrr* | ¼ × ½ × ½ | = $\frac{1}{16}$ |
| *ppyyrr* | ¼ × ½ × ½ | = $\frac{1}{16}$ |
| Chance of *at least two* recessive traits | | = $\frac{6}{16}$ or ⅜ |

In time, you'll be able to solve genetics problems faster by using the rules of probability than by filling in Punnett squares.

We cannot predict with certainty the exact numbers of progeny of different genotypes resulting from a genetic cross. But the rules of probability give us the *likelihood* of various outcomes. Usually, the larger the sample size, the closer the results will conform to our predictions. The reason Mendel counted so many offspring from his crosses is that he understood this statistical feature of inheritance and had a keen sense of the rules of chance.

## CONCEPT 11.3

# Inheritance patterns are often more complex than predicted by simple Mendelian genetics

In the 20th century, geneticists extended Mendelian principles not only to diverse organisms but also to patterns of inheritance more complex than those described by Mendel. For the work that led to his two laws of inheritance, Mendel chose pea plant characters that turn out to have a relatively simple genetic basis: Each character is determined by one gene, for which there are only two alleles, one completely dominant and the other completely recessive. (There is one exception: Mendel's pod-shape character is actually determined by two genes.) Few heritable characters are determined so simply, and the relationship between genotype and phenotype is rarely so straightforward. Mendel himself realized that he could not explain the more complicated patterns he observed in crosses involving other pea characters or other plant species. This does not diminish the utility of Mendelian genetics, however, because the basic principles of segregation and independent assortment apply even to more complex patterns of inheritance. In this section, we'll extend Mendelian genetics to hereditary patterns that were not reported by Mendel.

## Extending Mendelian Genetics for a Single Gene

The inheritance of characters determined by a single gene deviates from simple Mendelian patterns when alleles are not completely dominant or recessive, when a particular gene has more than two alleles, or when a single gene produces multiple phenotypes. We'll describe examples of each of these situations in this section.

### Degrees of Dominance

Alleles can show different degrees of dominance and recessiveness in relation to each other. In Mendel's classic pea crosses,

the F$_1$ offspring always looked like one of the two parental varieties because one allele in a pair showed **complete dominance** over the other. In such situations, the phenotypes of the heterozygote and the dominant homozygote are indistinguishable.

For some genes, however, neither allele is completely dominant, and the F$_1$ hybrids have a phenotype somewhere between those of the two parental varieties. This phenomenon, called **incomplete dominance**, is seen when red snapdragons are crossed with white snapdragons: All the F$_1$ hybrids have pink flowers **(Figure 11.10)**. This third, intermediate phenotype results from flowers of the heterozygotes having less red pigment than the red homozygotes. (This is unlike the case of Mendel's pea plants, where the *Pp* heterozygotes make enough pigment for the flowers to be purple, indistinguishable from those of *PP* plants.)

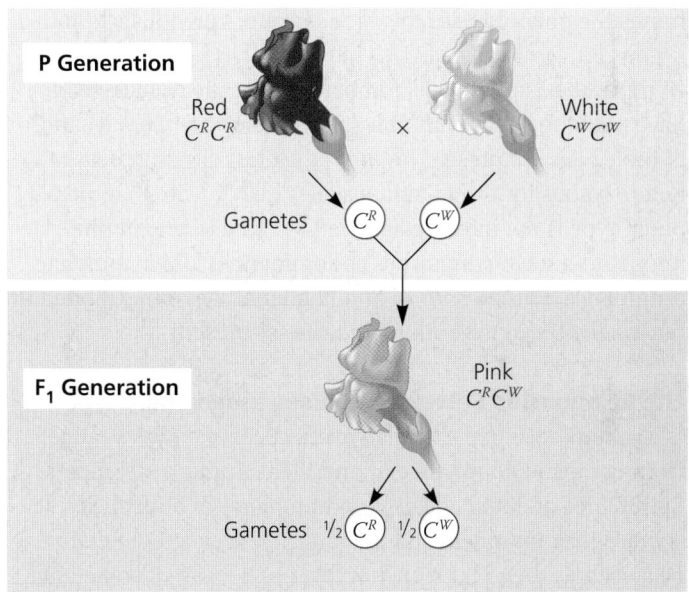

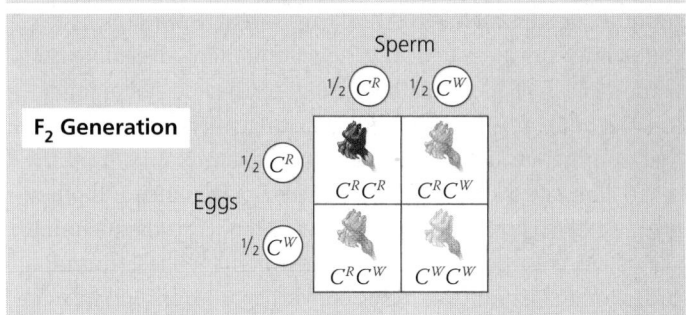

▲ **Figure 11.10 Incomplete dominance in snapdragon color.** When red snapdragons are crossed with white ones, the F$_1$ hybrids have pink flowers. Segregation of alleles into gametes of the F$_1$ plants results in an F$_2$ generation with a 1:2:1 ratio for both genotype and phenotype. Neither allele is dominant, so rather than using upper- and lowercase letters, we use the letter *C* with a superscript to indicate an allele for flower color: $C^R$ for red and $C^W$ for white.

**?** *Suppose a classmate argues that this figure supports the blending hypothesis for inheritance. What might your classmate say, and how would you respond?*

At first glance, incomplete dominance of either allele seems to provide evidence for the blending hypothesis of inheritance, which would predict that the red or white trait could never reappear among offspring of the pink hybrids. In fact, inter-breeding $F_1$ hybrids produces $F_2$ offspring with a phenotypic ratio of one red to two pink to one white. (Because heterozygotes have a separate phenotype, the genotypic and phenotypic ratios for the $F_2$ generation are the same, 1:2:1.) The segregation of the red-flower and white-flower alleles in the gametes produced by the pink-flowered plants confirms that the alleles for flower color are heritable factors that maintain their identity in the hybrids; that is, inheritance is particulate.

Another variation on dominance relationships between alleles is called **codominance**; in this variation, the two alleles each affect the phenotype in separate, distinguishable ways. For example, the human MN blood group is determined by codominant alleles for two specific molecules located on the surface of red blood cells, the M and N molecules. A single gene locus, at which two allelic variations are possible, determines the phenotype of this blood group. Individuals homozygous for the *M* allele (*MM*) have red blood cells with only M molecules; individuals homozygous for the *N* allele (*NN*) have red blood cells with only N molecules. But *both* M and N molecules are present on the red blood cells of individuals heterozygous for the *M* and *N* alleles (*MN*). Note that the MN phenotype is *not* intermediate between the M and N phenotypes, which distinguishes codominance from incomplete dominance. Rather, *both* M and N phenotypes are exhibited by heterozygotes, since both molecules are present.

### The Relationship Between Dominance and Phenotype -

We've now seen that the relative effects of two alleles range from complete dominance of one allele, through incomplete dominance of either allele, to codominance of both alleles. It is important to understand that an allele is called *dominant* because it is seen in the phenotype, not because it somehow subdues a recessive allele. Alleles are simply variations in a gene's nucleotide sequence. When a dominant allele coexists with a recessive allele in a heterozygote, they do not actually interact at all. It is in the pathway from genotype to phenotype that dominance and recessiveness come into play.

To illustrate the relationship between dominance and phenotype, we can use one of the characters Mendel studied—round versus wrinkled pea seed shape. The dominant allele (round) codes for an enzyme that helps convert an unbranched form of starch to a branched form in the seed. The recessive allele (wrinkled) codes for a defective form of this enzyme, leading to an accumulation of unbranched starch, which causes excess water to enter the seed by osmosis. Later, when the seed dries, it wrinkles. If a dominant allele is present, no excess water enters the seed and it does not wrinkle when it dries. One dominant allele results in enough of the enzyme to synthesize adequate amounts of branched starch, which means that dominant homozygotes and heterozygotes have the same phenotype: round seeds.

A closer look at the relationship between dominance and phenotype reveals an intriguing fact: For any character, the observed dominant/recessive relationship of alleles depends on the level at which we examine the phenotype. **Tay-Sachs disease**, an inherited disorder in humans, provides an example. The brain cells of a child with Tay-Sachs disease cannot metabolize certain lipids because a crucial enzyme does not work properly. As these lipids accumulate in brain cells, the child begins to suffer seizures, blindness, and degeneration of motor and mental performance and dies within a few years.

Only children who inherit two copies of the Tay-Sachs allele (homozygotes) have the disease. Thus, at the *organismal* level, the Tay-Sachs allele qualifies as recessive. However, the activity level of the lipid-metabolizing enzyme in heterozygotes is intermediate between the activity level in individuals homozygous for the normal allele and the activity level in individuals with Tay-Sachs disease. (The term *normal* is used in the genetic sense to refer to the allele coding for the enzyme that functions properly.) The intermediate phenotype observed at the *biochemical* level is characteristic of incomplete dominance of either allele. Fortunately, the heterozygote condition does not lead to disease symptoms, apparently because half the normal enzyme activity is sufficient to prevent lipid accumulation in the brain. Extending our analysis to yet another level, we find that heterozygous individuals produce equal numbers of normal and dysfunctional enzyme molecules. Thus, at the *molecular* level, the normal allele and the Tay-Sachs allele are codominant. As you can see, whether alleles appear to be completely dominant, incompletely dominant, or codominant depends on the level at which the phenotype is analyzed.

**Frequency of Dominant Alleles** While you might assume that the dominant allele for a particular character would be more common in a population than the recessive one, this is not always so. For example, about one baby out of 400 in the United States is born with extra digits (fingers or toes), a condition known as polydactyly. Some cases are caused by the presence of a dominant allele. The low frequency of polydactyly indicates that the recessive allele, which results in five digits per appendage when homozygous, is far more prevalent than the dominant allele in the population.

### *Multiple Alleles*

Only two alleles exist for each of the seven pea characters that Mendel studied, but most genes exist in more than two allelic forms. The ABO blood groups in humans, for instance, are determined by that person's two alleles of the blood group gene; there are three possible alleles: $I^A$, $I^B$, and $i$. A person's blood group may be one of four types: A, B, AB, or O. These letters refer to two carbohydrates—A and B—that are found on the surface of red blood cells. An individual's blood cells may have carbohydrate A (type A blood), carbohydrate B (type B), both (type AB), or neither (type O), as shown schematically in **Figure 11.11**. Matching compatible blood groups is critical for safe blood transfusions.

**(a) The three alleles for the ABO blood groups and their carbohydrates.** Each allele codes for an enzyme that may add a specific carbohydrate (designated by the superscript on the allele and shown as a triangle or circle) to red blood cells.

| Allele | $I^A$ | $I^B$ | $i$ |
|---|---|---|---|
| Carbohydrate | A △ | B ○ | none |

**(b) Blood group genotypes and phenotypes.** There are six possible genotypes, resulting in four different phenotypes.

| Genotype | $I^A I^A$ or $I^A i$ | $I^B I^B$ or $I^B i$ | $I^A I^B$ | $ii$ |
|---|---|---|---|---|
| Red blood cell appearance | 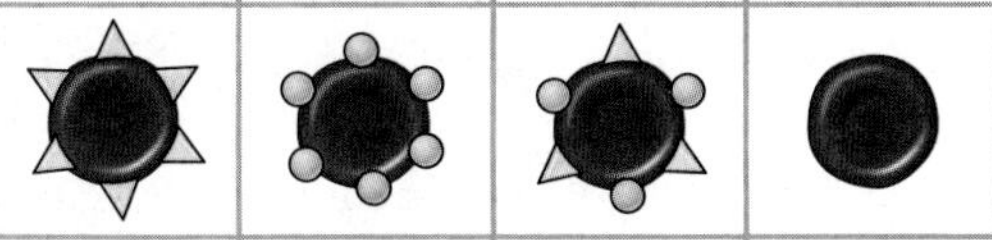 | | | |
| Phenotype (blood group) | A | B | AB | O |

▲ **Figure 11.11 Multiple alleles for the ABO blood groups.** The four blood groups result from different combinations of three alleles.

*Based on the surface carbohydrate phenotype in (b), what are the dominance relationships among the alleles?*

## Pleiotropy

So far, we have treated Mendelian inheritance as though each gene affects only one phenotypic character. Most genes, however, have multiple phenotypic effects, a property called **pleiotropy** (from the Greek *pleion*, more). In humans, for example, pleiotropic alleles are responsible for the multiple symptoms associated with certain hereditary diseases, such as cystic fibrosis and sickle-cell disease, discussed later in this chapter. In the garden pea, the gene that determines flower color also affects the color of the coating on the outer surface of the seed, which can be gray or white. Given the intricate molecular and cellular interactions responsible for an organism's development and physiology, it isn't surprising that a single gene can affect a number of characters.

## Extending Mendelian Genetics for Two or More Genes

Dominance relationships, multiple alleles, and pleiotropy all have to do with the effects of the alleles of a single gene. We now consider two situations in which two or more genes are involved in determining a particular phenotype. In the first case, one gene affects the phenotype of another because the two gene products interact, whereas in the second case, multiple genes independently affect a single trait.

## Epistasis

In **epistasis** (from the Greek for "standing upon"), the phenotypic expression of a gene at one locus alters that of a gene at a second locus. An example will help clarify this concept. In Labrador retrievers (commonly called Labs), black coat color

is dominant to brown. Let's designate *B* and *b* as the two alleles for this character. For a Lab to have brown fur, its genotype must be *bb*; these dogs are called chocolate Labs. But there is more to the story. A second gene determines whether or not pigment will be deposited in the hair. The dominant allele, symbolized by *E*, results in the deposition of either black or brown pigment, depending on the genotype at the first locus. But if the Lab is homozygous recessive for the second locus (*ee*), then the coat is yellow, regardless of the genotype at the black/brown locus. In this case, the gene for pigment deposition (*E/e*) is said to be epistatic to the gene that codes for black or brown pigment (*B/b*).

What happens if we mate black Labs that are heterozygous for both genes (*BbEe*)? Although the two genes affect the same phenotypic character (coat color), they follow the law of independent assortment. Thus, our breeding experiment represents an $F_1$ dihybrid cross, like those that produced a 9:3:3:1 ratio in Mendel's experiments. We can use a Punnett square to represent the genotypes of the $F_2$ offspring (**Figure 11.12**). As a result of epistasis, the phenotypic ratio among the $F_2$ offspring is 9 black to 3 chocolate to 4 yellow Labs. Other types of epistatic interactions produce different ratios, but all are modified versions of 9:3:3:1.

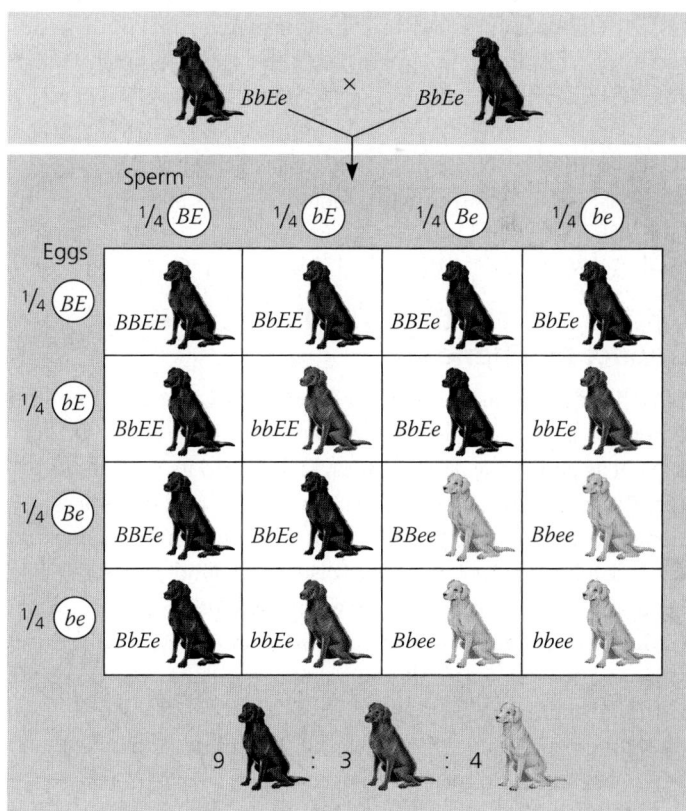

▲ **Figure 11.12 An example of epistasis.** This Punnett square illustrates the genotypes and phenotypes predicted for offspring of matings between two black Labrador retrievers of genotype *BbEe*. The *E/e* gene, which is epistatic to the *B/b* gene coding for hair pigment, controls whether or not pigment of any color will be deposited in the hair.

## Polygenic Inheritance

Mendel studied characters that could be classified on an either-or basis, such as purple versus white flower color. But many characters, such as human skin color and height, are not one of two discrete characters, but instead vary in the population in gradations along a continuum. These are called **quantitative characters**. Quantitative variation usually indicates **polygenic inheritance**, an additive effect of two or more genes on a single phenotypic character. (In a way, this is the converse of pleiotropy, where a single gene affects several phenotypic characters.) Height is a good example of polygenic inheritance: Genomic studies have identified at least 180 gene variations that affect height.

Skin pigmentation in humans is also controlled by many separately inherited genes. Here, we'll simplify the story in order to understand the concept of polygenic inheritance. Let's consider three genes, with a dark-skin allele for each gene (*A*, *B*, or *C*) contributing one "unit" of darkness (also a simplification) to the phenotype and being incompletely dominant to the other, light-skin allele (*a*, *b*, or *c*). In our model, an *AABBCC* person would be very dark, whereas an *aabbcc* individual would be very light. An *AaBbCc* person would have skin of an intermediate shade. Because the alleles have a cumulative effect, the genotypes *AaBbCc* and *AABbcc* would make the same genetic contribution (three units) to skin darkness. There are seven skin-color phenotypes that could result from a mating between *AaBbCc* heterozygotes, as shown in **Figure 11.13**. In a large number of such matings, the majority of offspring would be expected to have intermediate phenotypes (skin color in the middle range). You can graph the predictions from the Punnett square in the **Scientific Skills Exercise**. Environmental factors, such as exposure to the sun, also affect the skin-color phenotype.

## Nature and Nurture: The Environmental Impact on Phenotype

Another departure from simple Mendelian genetics arises when the phenotype for a character depends on environment as well as genotype. A single tree, locked into its inherited genotype, has leaves that vary in size, shape, and greenness, depending on their exposure to wind and sun. In humans, nutrition influences height, exercise alters build, sun-tanning darkens the skin, and experience improves performance on intelligence tests. Even identical twins, who are genetic equals, accumulate phenotypic differences as a result of their unique experiences.

Whether human characters are more influenced by genes or the environment—in everyday terms, nature versus nurture—is a debate that we will not attempt to settle here. We can say, however, that a genotype generally is not associated with a rigidly defined phenotype, but rather with a range of phenotypic possibilities due to environmental influences. For some characters, such as the ABO blood group system, the range is extremely narrow; that is, a given genotype mandates

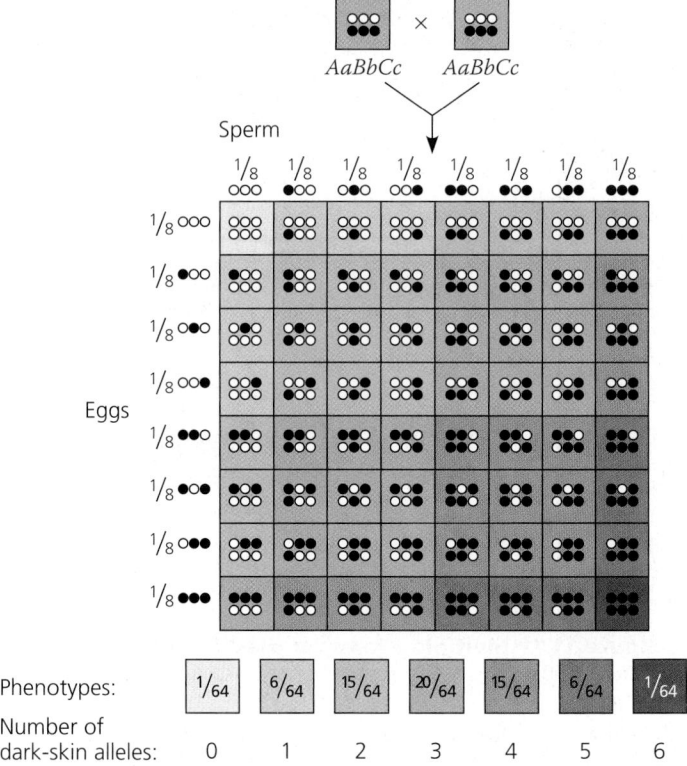

**▲ Figure 11.13 A simplified model for polygenic inheritance of skin color.** In this model, three separately inherited genes affect skin color. The heterozygous individuals (*AaBbCc*) represented by the two rectangles at the top of this figure each carry three dark-skin alleles (black circles, representing *A*, *B*, or *C*) and three light-skin alleles (white circles, representing *a*, *b*, or *c*). The Punnett square shows all the possible genetic combinations in gametes and offspring of many hypothetical matings between these heterozygotes. The results are summarized by the phenotypic frequencies (fractions) under the Punnett square. (The phenotypic ratio of the skin colors shown in the boxes is 1:6:15:20:15:6:1.)

a very specific phenotype. Other characters, such as a person's blood count of red and white cells, vary quite a bit, depending on such factors as the altitude, the customary level of physical activity, and the presence of infectious agents.

Generally, the phenotypic range is broadest for polygenic characters. Environment contributes to the quantitative nature of these characters, as we have seen in the continuous variation of skin color. Geneticists refer to such characters as **multifactorial**, meaning that many factors, both genetic and environmental, collectively influence phenotype.

## A Mendelian View of Heredity and Variation

We have now broadened our view of Mendelian inheritance by exploring degrees of dominance as well as multiple alleles, pleiotropy, epistasis, polygenic inheritance, and the phenotypic impact of the environment. How can we integrate these refinements into a comprehensive theory of Mendelian genetics? The key is to make the transition from the reductionist emphasis on single genes and phenotypic characters to the emergent properties of the organism as a whole, one of the themes of this book.

## Making a Histogram and Analyzing a Distribution Pattern

**What Is the Distribution of Phenotypes Among Offspring of Two Parents Who Are Both Heterozygous for Three Additive Genes?** Human skin color is a polygenic trait that is determined by the additive effects of many different genes. In this exercise, you will work with a simplified model of skin-color genetics where only three genes are assumed to affect the darkness of skin color and where each gene has two alleles—dark or light (see Figure 11.13). In this model, each dark allele contributes equally to the darkness of skin color, and each pair of alleles segregates independently of each other pair. Using a type of graph called a histogram, you will determine the distribution of phenotypes of offspring with different numbers of dark-skin alleles. (For additional information about graphs, see the Scientific Skills Review in Appendix F and in the Study Area in MasteringBiology.)

**How This Model Is Analyzed** To predict the phenotypes of the offspring of parents heterozygous for the three genes in our simplified model, we can use the Punnett square in Figure 11.13. The heterozygous individuals (*AaBbCc*) represented by the two rectangles at the top of that figure each carry three dark-skin alleles (black circles, which represent *A*, *B*, or *C*) and three light-skin alleles (white circles, which represent *a*, *b*, or *c*). The Punnett square shows all the possible genetic combinations in gametes and in offspring of a large number of hypothetical matings between these heterozygotes.

**Predictions from the Punnett Square** If we assume that each square in the Punnett square represents one offspring of the heterozygous *AaBbCc* parents, then the squares below show the possible phenotypes. Below the squares are the predicted phenotypic frequencies of individuals with the same number of dark-skin alleles.

Phenotypes:

| | 1/64 | 6/64 | 15/64 | 20/64 | 15/64 | 6/64 | 1/64 |
|---|---|---|---|---|---|---|---|

Number of dark-skin alleles:   0    1    2    3    4    5    6

**INTERPRET THE DATA**

1. A histogram is a bar graph that shows the distribution of numeric data (here, the number of dark skin alleles). To make a histogram of the allele distribution, put skin color (as the number of dark-skin alleles) along the *x*-axis and number of offspring (out of 64) with each phenotype on the *y*-axis. There are no gaps in our allele data, so draw the bars with no space between them.

2. You can see that the skin-color phenotypes are not distributed uniformly. (a) Which phenotype has the highest frequency? Draw a vertical dashed line through that bar. (b) Distributions of values like this one tend to show one of several common patterns. Sketch a rough curve that approximates the values and look at its shape. Is it symmetrically distributed around a central peak value (a "normal distribution," sometimes called a bell curve); is it skewed to one end of the *x*-axis or the other (a "skewed distribution"); or does it show two apparent groups of frequencies (a "bimodal distribution")? Explain the reason for the curve's shape. (It will help to read the text description that supports Figure 11.13.)

(MB) A version of this Scientific Skills Exercise can be assigned in MasteringBiology.

**Further Reading** R. A. Sturm, A golden age of human pigmentation genetics, *Trends in Genetics* 22:464–468 (2006).

---

The term *phenotype* can refer not only to specific characters, such as flower color and blood group, but also to an organism in its entirety—*all* aspects of its physical appearance, internal anatomy, physiology, and behavior. Similarly, the term *genotype* can refer to an organism's entire genetic makeup, not just its alleles for a single genetic locus. In most cases, a gene's impact on phenotype is affected by other genes and by the environment. In this integrated view of heredity and variation, an organism's phenotype reflects its overall genotype and unique environmental history.

Considering all that can occur in the pathway from genotype to phenotype, it is indeed impressive that Mendel could uncover the fundamental principles governing the transmission of individual genes from parents to offspring. Mendel's laws of segregation and independent assortment explain heritable variations in terms of alternative forms of genes (hereditary "particles," now known as the alleles of genes) that are passed along, generation after generation, according to simple rules of probability. This theory of inheritance is equally valid for peas, flies, fishes, birds, and human beings—indeed, for any organism with a sexual life cycle. Furthermore, by extending the principles of

segregation and independent assortment to help explain such hereditary patterns as epistasis and quantitative characters, we begin to see how broadly Mendelian genetics applies. From Mendel's abbey garden came a theory of particulate inheritance that anchors modern genetics. In the last section of this chapter, we'll apply Mendelian genetics to human inheritance, with emphasis on the transmission of hereditary diseases.

**CONCEPT CHECK 11.3**

1. *Incomplete dominance* and *epistasis* are both terms that define genetic relationships. What is the most basic distinction between these terms?

2. If a man with type AB blood marries a woman with type O, what blood types would you expect in their children? What fraction would you expect of each type?

3. **WHAT IF?** A rooster with gray feathers and a hen of the same phenotype produce 15 gray, 6 black, and 8 white chicks. What is the simplest explanation for the inheritance of these colors in chickens? What phenotypes would you expect in the offspring of a cross between a gray rooster and a black hen?

For suggested answers, see Appendix A.

# Many human traits follow Mendelian patterns of inheritance

Peas are convenient subjects for genetic research, but humans are not. The human generation span is long—about 20 years—and human parents produce many fewer offspring than peas and most other species. Even more important, it wouldn't be ethical to ask pairs of humans to breed so that the phenotypes of their offspring could be analyzed! In spite of these constraints, the study of human genetics continues, spurred on by our desire to understand our own inheritance. New molecular biological techniques have led to many breakthrough discoveries, but basic Mendelian genetics endures as the foundation of human genetics.

## Pedigree Analysis

Unable to manipulate the matings of people, geneticists instead analyze the results of matings that have already occurred. They do so by collecting information about a family's history for a particular trait and assembling this information into a family tree describing the traits of parents and children across the generations—a family **pedigree**.

**Figure 11.14a** shows a three-generation pedigree that traces the occurrence of a pointed contour of the hairline on the forehead. This trait, called a widow's peak, is due to a dominant allele, *W*. Because the widow's-peak allele is dominant, all individuals who lack a widow's peak must be homozygous recessive (*ww*). The two grandparents with widow's peaks must have the *Ww* genotype, since some of their offspring are homozygous recessive. The offspring in the second generation who *do* have widow's peaks must also be heterozygous, because they are the products of *Ww* × *ww* matings. The third generation in this pedigree consists of two sisters. The one who has a widow's peak could be either homozygous (*WW*) or heterozygous (*Ww*), given what we know about the genotypes of her parents (both *Ww*).

**Figure 11.14b** is a pedigree of the same family, but this time we focus on a recessive trait, the inability of individuals to taste a chemical called PTC (phenylthiocarbamide). Compounds similar to PTC are found in broccoli, brussels sprouts, and related vegetables and account for the bitter taste some people report when eating these foods. We'll use *t* for the recessive allele and *T* for the dominant allele, which results in the ability to taste PTC. As you work your way through the pedigree, notice once again that you can apply what you have learned about Mendelian inheritance to understand the genotypes shown for the family members.

An important application of a pedigree is to help us calculate the probability that a future child will have a particular genotype and phenotype. Suppose that the couple represented in the second generation of Figure 11.14 decides to have one

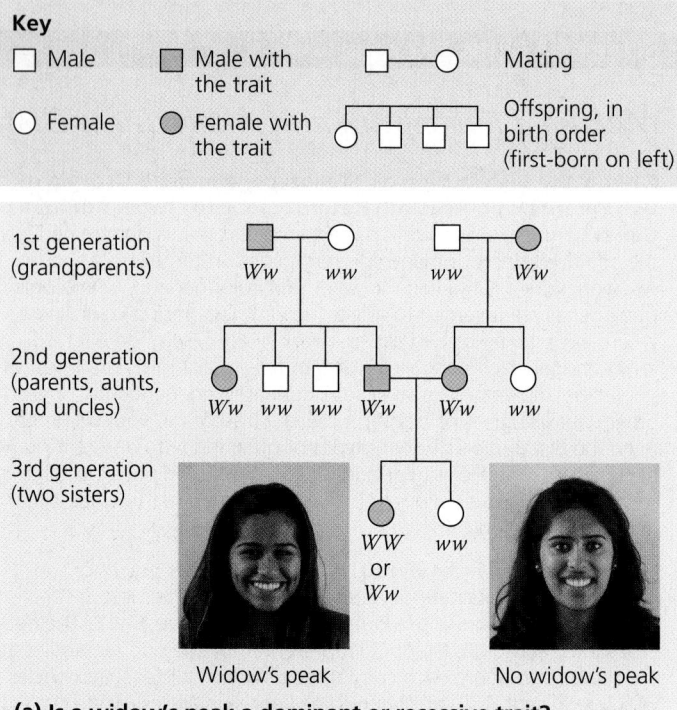

**(a) Is a widow's peak a dominant or recessive trait?**
**Tips for pedigree analysis:** Notice in the third generation that the second-born daughter lacks a widow's peak, although both of her parents had the trait. Such a pattern indicates that the trait is due to a dominant allele. If it were due to a *recessive* allele, and both parents had the recessive phenotype (straight hairline), *all* of their offspring would also have the recessive phenotype.

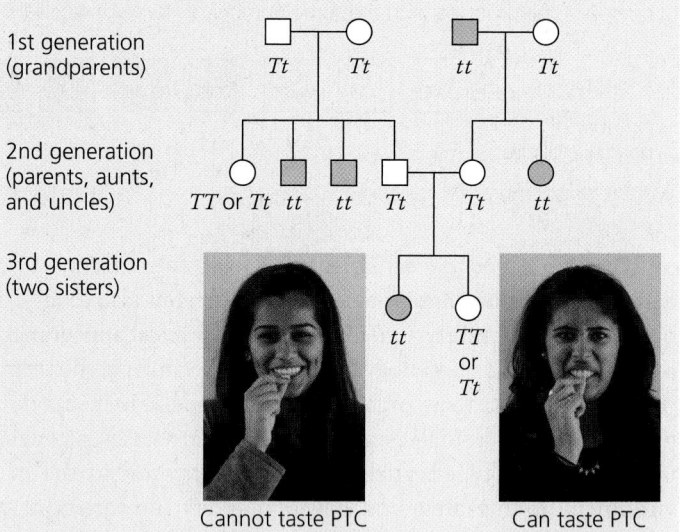

**(b) Is the inability to taste a chemical called PTC a dominant or recessive trait?**
**Tips for pedigree analysis:** Notice that the first-born daughter in the third generation has the trait (is unable to taste PTC), although both parents lack that trait (they *can* taste PTC). Such a pattern is explained if the non-taster phenotype is due to a recessive allele. (If it were due to a *dominant* allele, then at least one parent would also have had the trait.)

▲ **Figure 11.14 Pedigree analysis.** Each pedigree traces a trait through three generations of the same family. The two traits have different inheritance patterns. For the sake of understanding genetic principles, each trait is shown as being determined by two alleles of a single gene. This is a simplification because other genes may also affect each of these characters.

more child. What is the probability that the child will have a widow's peak? This is equivalent to a Mendelian $F_1$ monohybrid cross ($Ww \times Ww$), and therefore the probability that a child will inherit a dominant allele and have a widow's peak is ¾ (¼ $WW$ + ½ $Ww$). What is the probability that the child will be unable to taste PTC? We can also treat this as a monohybrid cross ($Tt \times Tt$), but this time we want to know the chance that the offspring will be homozygous recessive ($tt$). The probability is ¼. Finally, what is the chance the child will have a widow's peak *and* be unable to taste PTC? Assuming that the genes for these two characters are on different chromosomes, the two pairs of alleles will assort independently in this dihybrid cross ($WwTt \times WwTt$). Thus, we can use the multiplication rule: ¾ (chance of widow's peak) × ¼ (chance of inability to taste PTC) = ³⁄₁₆ (chance of widow's peak and inability to taste PTC).

Pedigrees are a more serious matter when the alleles in question cause disabling or deadly diseases instead of an innocuous human variation, such as hairline configuration or ability to taste an innocuous chemical. However, for disorders inherited as simple Mendelian traits, the same techniques of pedigree analysis apply.

## Recessively Inherited Disorders

Thousands of genetic disorders are known to be inherited as simple recessive traits. These disorders range in severity from relatively mild, such as albinism (lack of pigmentation, which results in susceptibility to skin cancers and vision problems), to life-threatening, such as cystic fibrosis.

### The Behavior of Recessive Alleles

How can we account for the behavior of alleles that cause recessively inherited disorders? Recall that genes code for proteins of specific function. An allele that causes a genetic disorder (let's call it allele *a*) codes for either a malfunctioning protein or no protein at all. In the case of disorders classified as recessive, heterozygotes ($Aa$) typically have the normal phenotype because one copy of the normal allele ($A$) produces a sufficient amount of the specific protein. Thus, a recessively inherited disorder shows up only in the homozygous individuals ($aa$) who inherit a recessive allele from each parent. Although phenotypically normal with regard to the disorder, heterozygotes may transmit the recessive allele to their offspring and thus are called **carriers**. **Figure 11.15** illustrates these ideas using albinism as an example.

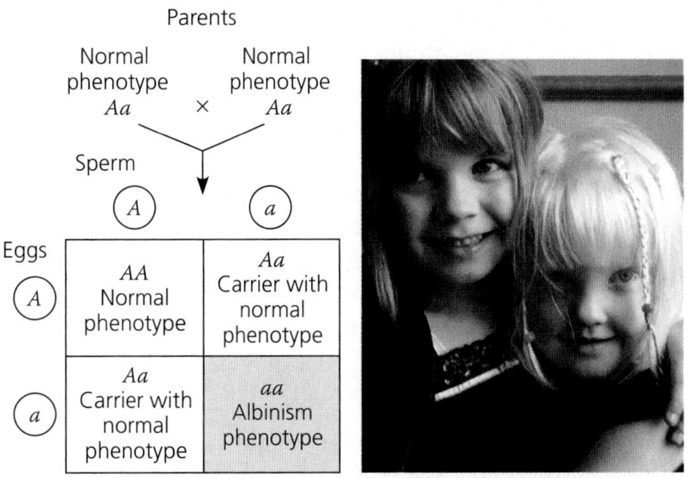

Most people who have recessive disorders are born to parents who are carriers of the disorder but have a normal phenotype, as is the case shown in the Punnett square in Figure 11.15. A mating between two carriers corresponds to a Mendelian $F_1$ monohybrid cross, so the predicted genotypic ratio for offspring is 1 $AA$ : 2 $Aa$ : 1 $aa$. Thus, each child has a ¼ chance of inheriting a double dose of the recessive allele; in the case of albinism, such a child will have albinism. From the genotypic ratio, we also can see that out of three offspring

▲ **Figure 11.15 Albinism: a recessive trait.** One of the two sisters shown here has normal coloration; the other has albinism. Most recessive homozygotes are born to parents who are carriers of the disorder but themselves have a normal phenotype, the case shown in the Punnett square.

**?** *What is the probability that the sister with normal coloration is a carrier of the albinism allele?*

with the normal phenotype (one $AA$ plus two $Aa$), two are predicted to be heterozygous carriers, a ⅔ chance. Recessive homozygotes could also result from $Aa \times aa$ and $aa \times aa$ matings, but if the disorder is lethal before reproductive age or results in sterility (neither of which is true for albinism), no $aa$ individuals will reproduce. Even if recessive homozygotes are able to reproduce, this will occur relatively rarely because such individuals account for a much smaller percentage of the population than heterozygous carriers.

In general, genetic disorders are not evenly distributed among all groups of people. For example, the incidence of Tay-Sachs disease, which we described earlier in this chapter, is disproportionately high among Ashkenazic Jews, Jewish people whose ancestors lived in central Europe. In that population, Tay-Sachs disease occurs in one out of 3,600 births, an incidence about 100 times greater than that among non-Jews or Mediterranean (Sephardic) Jews. This uneven distribution results from the different genetic histories of the world's peoples during less technological times, when populations were more geographically (and hence genetically) isolated.

When a disease-causing recessive allele is rare, it is relatively unlikely that two carriers of the same harmful allele will meet and mate. The probability of passing on recessive traits increases greatly, however, if the man and woman are close relatives (for example, siblings or first cousins). This is because people with recent common ancestors are more likely to carry the same recessive alleles than are unrelated people. Thus, these consanguineous ("same blood") matings, indicated in pedigrees by double lines, are more likely to produce offspring homozygous for recessive traits—including harmful ones. Such effects can be observed in many types of domesticated and zoo animals that have become inbred.

There is debate among geneticists about exactly how much human consanguinity increases the risk of inherited diseases. Many harmful alleles have such severe effects that a homozygous embryo spontaneously aborts long before birth. Still, most societies and cultures have laws or taboos forbidding marriages between close relatives. These rules may have evolved out of empirical observation that in most populations, stillbirths and birth defects are more common when parents are closely related. Social and economic factors have also influenced the development of customs and laws against consanguineous marriages.

### Cystic Fibrosis

The most common lethal genetic disease in the United States is **cystic fibrosis**, which strikes one out of every 2,500 people of European descent but is much rarer in other groups. Among people of European descent, one out of 25 (4%) are carriers of the cystic fibrosis allele. The normal allele for this gene codes for a membrane protein that functions in the transport of chloride ions between certain cells and the extracellular fluid. These chloride transport channels are defective or absent in the plasma membranes of children who inherit two recessive alleles for cystic fibrosis. The result is an abnormally high concentration of extracellular chloride, which causes the mucus that coats certain cells to become thicker and stickier than normal. The mucus builds up in the pancreas, lungs, digestive tract, and other organs, leading to multiple (pleiotropic) effects, including poor absorption of nutrients from the intestines, chronic bronchitis, and recurrent bacterial infections.

Untreated, cystic fibrosis can cause death by the age of 5. Daily doses of antibiotics to stop infection, gentle pounding on the chest to clear mucus from clogged airways, and other therapies can prolong life. In the United States, more than half of those with cystic fibrosis now survive into their 30s and beyond.

### Sickle-Cell Disease: A Genetic Disorder with Evolutionary Implications

**EVOLUTION** The most common inherited disorder among people of African descent is **sickle-cell disease**, which affects one out of 400 African-Americans. Sickle-cell disease is caused by the substitution of a single amino acid in the hemoglobin protein of red blood cells; in homozygous individuals, all hemoglobin is of the sickle-cell (abnormal) variety. When the oxygen content of an affected individual's blood is low (at high altitudes or under physical stress, for instance), the sickle-cell hemoglobin molecules aggregate into long rods that deform the red cells into a sickle shape (see Figure 3.23). Sickled cells may clump and clog small blood vessels, often leading to other symptoms throughout the body, including physical weakness, pain, organ damage, and even stroke and paralysis. Regular blood transfusions can ward off brain damage in children with sickle-cell disease, and new drugs can help prevent or treat other problems. There is currently no widely available cure, but the disease is the target of ongoing gene therapy research.

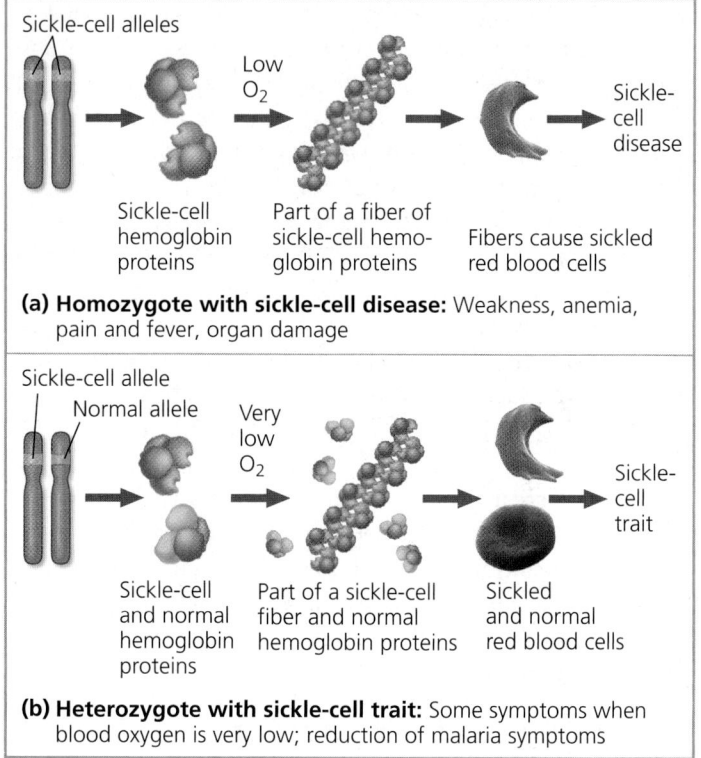

▼ **Figure 11.16 Sickle-cell disease and sickle-cell trait.**

Sickle-cell alleles → Sickle-cell hemoglobin proteins → Low O$_2$ → Part of a fiber of sickle-cell hemoglobin proteins → Fibers cause sickled red blood cells → Sickle-cell disease

**(a) Homozygote with sickle-cell disease:** Weakness, anemia, pain and fever, organ damage

Sickle-cell allele / Normal allele → Sickle-cell and normal hemoglobin proteins → Very low O$_2$ → Part of a sickle-cell fiber and normal hemoglobin proteins → Sickled and normal red blood cells → Sickle-cell trait

**(b) Heterozygote with sickle-cell trait:** Some symptoms when blood oxygen is very low; reduction of malaria symptoms

Although two sickle-cell alleles are necessary for an individual to manifest full-blown sickle-cell disease, the presence of one sickle-cell allele can affect the phenotype. Thus, at the organismal level, the normal allele is incompletely dominant to the sickle-cell allele **(Figure 11.16)**. At the molecular level, the two alleles are codominant; both normal and abnormal (sickle-cell) hemoglobins are made in heterozygotes (carriers), who are said to have *sickle-cell trait*. Heterozygotes are usually healthy but may suffer some symptoms during long periods of reduced blood oxygen.

About one out of ten African-Americans have sickle-cell trait, an unusually high frequency of heterozygotes for an allele with severe detrimental effects in homozygotes. Why haven't evolutionary processes resulted in the disappearance of the allele among this population? One explanation is that having a single copy of the sickle-cell allele reduces the frequency and severity of malaria attacks, especially among young children. The malaria parasite spends part of its life cycle in red blood cells (see Figure 25.26), and the presence of even heterozygous amounts of sickle-cell hemoglobin results in lower parasite densities and hence reduced malaria symptoms. Thus, in tropical Africa, where infection with the malaria parasite is common, the sickle-cell allele confers an advantage to heterozygotes even though it is harmful in the homozygous state. (The balance between these two effects will be discussed in Chapter 21; see Make Connections Figure 21.15.) The relatively high frequency of African-Americans with sickle-cell trait is a vestige of their African ancestry.

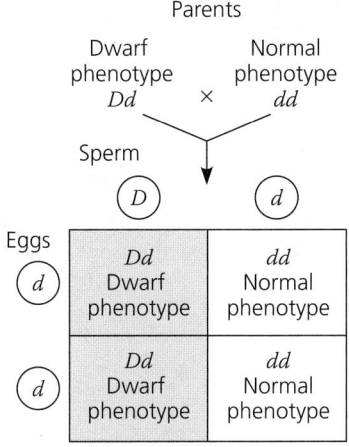

Parents

Dwarf phenotype
$Dd$
×
Normal phenotype
$dd$

Sperm

| | $D$ | $d$ |
|---|---|---|
| Eggs $d$ | $Dd$ Dwarf phenotype | $dd$ Normal phenotype |
| $d$ | $Dd$ Dwarf phenotype | $dd$ Normal phenotype |

▲ **Figure 11.17 Achondroplasia: a dominant trait.** Dr. Michael C. Ain has achondroplasia, a form of dwarfism caused by a dominant allele. This has inspired his work: He is a specialist in the repair of bone defects caused by achondroplasia and other disorders. The dominant allele ($D$) might have arisen as a mutation in the egg or sperm of a parent or could have been inherited from an affected parent, as shown for an affected father in the Punnett square.

## Dominantly Inherited Disorders

Although many harmful alleles are recessive, a number of human disorders are due to dominant alleles. One example is *achondroplasia*, a form of dwarfism that occurs in one of every 25,000 people. Heterozygous individuals have the dwarf phenotype **(Figure 11.17)**. Therefore, all people who do not have achondroplasia—99.99% of the population—are homozygous for the recessive allele. Like the presence of extra fingers or toes mentioned earlier, achondroplasia is a trait for which the recessive allele is much more prevalent than the corresponding dominant allele.

Unlike achondroplasia, which is relatively harmless, some dominant alleles cause lethal diseases. Those that do are much less common than recessive alleles that have lethal effects. A lethal recessive allele is only lethal when homozygous; it can be passed from one generation to the next by heterozygous carriers because the carriers themselves have normal phenotypes. A lethal dominant allele, however, often causes the death of afflicted individuals before they can mature and reproduce, and in this case the allele is not passed on to future generations.

A lethal dominant allele may be passed on, though, if the lethal disease symptoms first appear after reproductive age. In these cases, the individual may already have transmitted the allele to his or her children. For example, a degenerative disease of the nervous system called **Huntington's disease** is caused by a lethal dominant allele that has no obvious phenotypic effect until the individual is about 35 to 45 years old. Once the deterioration of the nervous system begins, it is irreversible and inevitably fatal. As with other dominant traits, a child born to a parent with the Huntington's disease allele has a 50% chance of inheriting the allele and the disorder (see the Punnett square in Figure 11.17). In the United States, this devastating disease afflicts about one in 10,000 people.

At one time, the onset of symptoms was the only way to know if a person had inherited the Huntington's allele, but this is no longer the case. By analyzing DNA samples from a large family with a high incidence of the disorder, geneticists tracked the Huntington's allele to a locus near the tip of chromosome 4, and the gene was sequenced in 1993. This information led to the development of a test that could detect the presence of the Huntington's allele in an individual's genome. The availability of this test poses an agonizing dilemma for those with a family history of Huntington's disease. Some individuals may want to be tested for this disease, whereas others may decide it would be too stressful to find out.

## Multifactorial Disorders

The hereditary diseases we have discussed so far are sometimes described as simple Mendelian disorders because they result from abnormality of one or both alleles at a single genetic locus. Many more people are susceptible to diseases that have a multifactorial basis—a genetic component plus a significant environmental influence. Heart disease, diabetes, cancer, alcoholism, certain mental illnesses such as schizophrenia and bipolar disorder, and many other diseases are multifactorial. In these cases, the hereditary component is polygenic. For example, many genes affect cardiovascular health, making some of us more prone than others to heart attacks and strokes. No matter what our genotype, however, our lifestyle has a tremendous effect on phenotype for cardiovascular health and other multifactorial characters. Exercise, a healthful diet, abstinence from smoking, and an ability to handle stressful situations all reduce our risk of heart disease and some types of cancer.

## Genetic Counseling Based on Mendelian Genetics

Avoiding simple Mendelian disorders is possible when the risk of a particular genetic disorder can be assessed before a child is conceived or during the early stages of the pregnancy. Many hospitals have genetic counselors who can provide information to prospective parents concerned about a family history for a specific disease.

Consider the case of a hypothetical couple, John and Carol. Each had a brother who died from the same recessively inherited lethal disease. Before conceiving their first child, John and Carol seek genetic counseling to determine the risk of having a child with the disease. From the information about their brothers, we know that both parents of John and both parents of Carol must have been carriers of the recessive allele. Thus, John and Carol are both products of $Aa \times Aa$ crosses, where $a$ symbolizes the allele that causes this particular disease. We also know that John and Carol are not homozygous recessive ($aa$), because they do not have the disease. Therefore, their genotypes are either $AA$ or $Aa$.

Given a genotypic ratio of 1 $AA$ : 2 $Aa$ : 1 $aa$ for offspring of an $Aa \times Aa$ cross, John and Carol each have a ⅔ chance of being carriers ($Aa$). According to the rule of multiplication, the overall probability of their firstborn having the disorder is ⅔ (the chance that John is a carrier) times ⅔ (the chance that Carol is a carrier) times ¼ (the chance of two carriers having a child with the disease), which equals ⅑. Suppose that Carol and John decide to have a child—after all, there is an ⁸⁄₉ chance that their baby will not have the disorder. If, despite these odds, their child is born with the disease, then we would know that *both* John and Carol are, in fact, carriers (*Aa* genotype). If both John and Carol are carriers, there is a ¼ chance that any subsequent child this couple has will have the disease. The probability is higher for subsequent children because the diagnosis of the disease in the first child established that both parents are carriers, not because the genotype of the first child affects in any way that of future children.

When we use Mendel's laws to predict possible outcomes of matings, it is important to remember that each child represents an independent event in the sense that its genotype is unaffected by the genotypes of older siblings. Suppose that John and Carol have three more children, and *all three* have the hypothetical hereditary disease. There is only one chance in 64 (¼ × ¼ × ¼) that such an outcome will occur. Despite this run of misfortune, the chance that still another child of this couple will have the disease remains ¼.

Genetic counseling like this relies on the Mendelian model of inheritance. We owe the "gene idea"—the concept

of heritable factors transmitted according to simple rules of chance—to the elegant quantitative experiments of Gregor Mendel. The importance of his discoveries was overlooked by most biologists until early in the 20th century, decades after he reported his findings. In the next chapter, you'll learn how Mendel's laws have their physical basis in the behavior of chromosomes during sexual life cycles and how the synthesis of Mendelian genetics and a chromosome theory of inheritance catalyzed progress in genetics.

## CONCEPT CHECK 11.4

1. Beth and Tom each have a sibling with cystic fibrosis, but neither Beth nor Tom nor any of their parents have the disease. Calculate the probability that if this couple has a child, the child will have cystic fibrosis. What would be the probability if a test revealed that Tom is a carrier but Beth is not? Explain your answers.

2. **MAKE CONNECTIONS** In Table 11.1, note the phenotypic ratio of the dominant to recessive trait in the F₂ generation for the monohybrid cross involving flower color. Then determine the phenotypic ratio for the offspring of the second-generation couple in Figure 11.14b. What accounts for the difference in the two ratios?

**For suggested answers, see Appendix A.**

---

# 11    Chapter Review

Go to **MasteringBiology**® for Assignments, the eText, and the Study Area with Animations, Activities, Vocab Self-Quiz, and Practice Tests.

## SUMMARY OF KEY CONCEPTS

**VOCAB SELF-QUIZ**

goo.gl/gbai8v

### CONCEPT 11.1

**Mendel used the scientific approach to identify two laws of inheritance (pp. 215–221)**

- Gregor Mendel formulated a theory of inheritance based on experiments with garden peas, proposing that parents pass on to their offspring discrete genes that retain their identity through generations. This theory includes two "laws."

- The **law of segregation** states that genes have alternative forms, or **alleles**. In a diploid organism, the two alleles of a gene segregate (separate) during meiosis and gamete formation; each sperm or egg carries only one allele of each pair. This law explains the 3:1 ratio of F₂ phenotypes observed when **monohybrids** self-pollinate. Each organism inherits one allele for each gene from each parent. In **heterozygotes**, the two alleles are different: expression of the **dominant allele** masks the phenotypic effect of the **recessive**

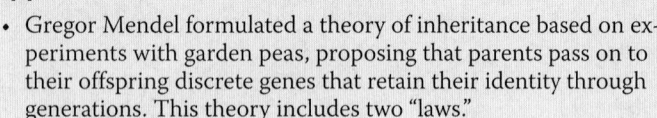

*PP* (homozygous)

*Pp* (heterozygous)

*Pp* (heterozygous)

*pp* (homozygous)

**allele**. **Homozygotes** have identical alleles of a given gene and are **true-breeding**.

- The **law of independent assortment** states that the pair of alleles for a given gene segregates into gametes independently of the pair of alleles for any other gene. In a cross between **dihybrids** (individuals heterozygous for two genes), the offspring have four phenotypes in a 9:3:3:1 ratio.

> ❓ *When Mendel did crosses of true-breeding purple- and white-flowered pea plants, the white-flowered trait disappeared from the F₁ generation but reappeared in the F₂ generation. Use genetic terms to explain why that happened.*

### CONCEPT 11.2

**Probability laws govern Mendelian inheritance (pp. 221–223)**

- The **multiplication rule** states that the probability of two or more events occurring together is equal to the product of the individual probabilities of the independent single events. The **addition rule** states that the probability of an event that can occur in two or more independent, mutually exclusive ways is the sum of the individual probabilities.

- The rules of probability can be used to solve complex genetics problems. A dihybrid or other multicharacter cross is equivalent to two or more independent monohybrid crosses occurring simultaneously. In calculating the chances of the various offspring genotypes from such crosses, each character is first considered separately and then the individual probabilities are multiplied.

**DRAW IT** *Redraw the Punnett square on the right side of Figure 11.8 as two smaller monohybrid Punnett squares, one for each gene. Below each square, list the fractions of each phenotype produced. Use the rule of multiplication to compute the overall fraction of each possible dihybrid phenotype. What is the phenotypic ratio?*

## CONCEPT 11.3

### Inheritance patterns are often more complex than predicted by simple Mendelian genetics (pp. 223–227)

- Extensions of Mendelian genetics for a single gene:

| Relationship among alleles of a single gene | Description | Example |
|---|---|---|
| Complete dominance of one allele | Heterozygous phenotype same as that of homozygous dominant | $PP$   $Pp$ |
| Incomplete dominance of either allele | Heterozygous phenotype intermediate between the two homozygous phenotypes | $C^R C^R$   $C^R C^W$   $C^W C^W$ |
| Codominance | Both phenotypes expressed in heterozygotes | $I^A I^B$ |
| Multiple alleles | In the population, some genes have more than two alleles | ABO blood group alleles $I^A, I^B, i$ |
| Pleiotropy | One gene affects multiple phenotypic characters | Sickle-cell disease |

- Extensions of Mendelian genetics for two or more genes:

| Relationship among two or more genes | Description | Example |
|---|---|---|
| Epistasis | The phenotypic expression of one gene affects the expression of another gene | $BbEe$ × $BbEe$ <br> 9 : 3 : 4 |
| Polygenic inheritance | A single phenotypic character is affected by two or more genes | $AaBbCc$ × $AaBbCc$ |

- The expression of a genotype can be affected by environmental influences. Polygenic characters that are also influenced by the environment are called **multifactorial** characters.

- An organism's overall phenotype reflects its overall genotype and unique environmental history. Even in more complex inheritance patterns, Mendel's fundamental laws still apply.

**?** *Which genetic relationships listed in the first column of the two tables are demonstrated by the inheritance pattern of the ABO blood group alleles? For each genetic relationship, explain why this inheritance pattern is or is not an example.*

## CONCEPT 11.4

### Many human traits follow Mendelian patterns of inheritance (pp. 228–232)

- Analysis of family **pedigrees** can be used to deduce the possible genotypes of individuals and make predictions about future offspring. Such predictions are statistical probabilities rather than certainties.

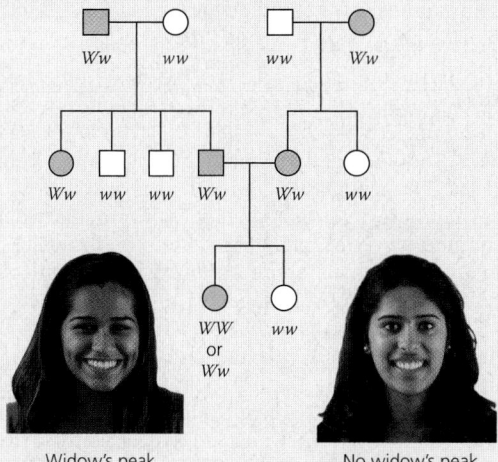

Widow's peak      No widow's peak

- Many genetic disorders are inherited as simple recessive traits. Most affected (homozygous recessive) individuals are children of phenotypically normal, heterozygous **carriers**.

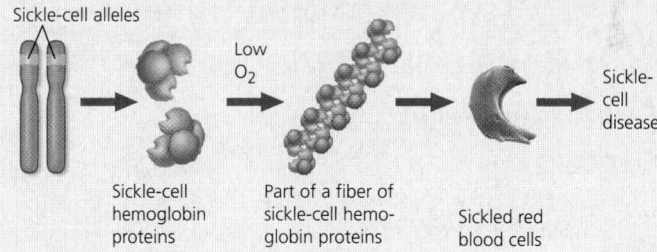

Sickle-cell alleles

Sickle-cell hemoglobin proteins

Part of a fiber of sickle-cell hemoglobin proteins

Low O₂

Sickled red blood cells

Sickle-cell disease

- The sickle-cell allele has probably persisted for evolutionary reasons: Heterozygotes have an advantage because one copy of the sickle-cell allele reduces both the frequency and severity of malaria attacks.
- Lethal dominant alleles are eliminated from the population if affected people die before reproducing. Nonlethal dominant alleles and lethal alleles that are expressed relatively late in life are inherited in a Mendelian way.
- Many human diseases are multifactorial—that is, they have both genetic and environmental components and do not follow simple Mendelian inheritance patterns.
- Using family histories, genetic counselors help couples determine the probability of their children having genetic disorders.

**?** *Both members of a couple know that they are carriers of the cystic fibrosis allele. None of their three children has cystic fibrosis, but any one of them might be a carrier. The couple would like to have a fourth child but are worried that he or she would very likely have the disease, since the first three do not. What would you tell the couple? Would it remove some more uncertainty in their prediction if they could find out whether the three children are carriers?*

# TIPS FOR GENETICS PROBLEMS

1. Write down symbols for the alleles. (These may be given in the problem.) When represented by single letters, the dominant allele is uppercase and the recessive allele is lowercase.

2. Write down the possible genotypes, as determined by the phenotype.
   a. If the phenotype is that of the dominant trait (for example, purple flowers), then the genotype is either homozygous dominant or heterozygous (*PP* or *Pp*, in this example).
   b. If the phenotype is that of the recessive trait, the genotype must be homozygous recessive (for example, *pp*).
   c. If the problem says "true-breeding," the genotype is homozygous.

3. Determine what the problem is asking. If asked to do a cross, write it out in the form [genotype] × [genotype], using the alleles you've decided on.

4. To figure out the outcome of a cross, set up a Punnett square.
   a. Put the gametes of one parent at the top and those of the other on the left. To determine the allele(s) in each gamete for a given genotype, set up a systematic way to list all the possibilities. (Remember, each gamete has one allele of each gene.) Note that there are $2^n$ possible types of gametes, where $n$ is the number of gene loci that are heterozygous. For example, an individual with genotype *AaBbCc* would produce $2^3 = 8$ types of gametes. Write the genotypes of the gametes in circles above the columns and to the left of the rows.
   b. Fill in the Punnett square as if each possible sperm were fertilizing each possible egg, making all of the possible offspring. In a cross of *AaBbCc* × *AaBbCc*, for example, the Punnett square would have 8 columns and 8 rows, so there are 64 different offspring; you would know the genotype of each and thus the phenotype. Count genotypes and phenotypes to obtain the genotypic and phenotypic ratios. Because the Punnett square is so large, this method is not the most efficient. Instead, see tip 5.

5. You can use the rules of probability if the Punnett square would be too big. (For example, see the question at the end of the summary for Concept 11.2 and question 7.) You can consider each gene separately (see the section Solving Complex Genetics Problems with the Rules of Probability in Concept 11.2).

6. If the problem gives you the phenotypic ratios of offspring but not the genotypes of the parents in a given cross, the phenotypes can help you deduce the parents' unknown genotypes.
   a. For example, if ½ of the offspring have the recessive phenotype and ½ the dominant, you know that the cross was between a heterozygote and a homozygous recessive.
   b. If the ratio is 3:1, the cross was between two heterozygotes.
   c. If two genes are involved and you see a 9:3:3:1 ratio in the offspring, you know that each parent is heterozygous for both genes. *Caution*: Don't assume that the reported numbers will exactly equal the predicted ratios. For example, if there are 13 offspring with the dominant trait and 11 with the recessive, assume that the ratio is one dominant to one recessive.

7. For pedigree problems, use the tips in Figure 11.14 and below to determine what kind of trait is involved.
   a. If parents without the trait have offspring with the trait, the trait must be recessive and both of the parents must be carriers.
   b. If the trait is seen in every generation, it is most likely dominant (see the next possibility, though).
   c. If both parents have the trait, then in order for it to be recessive, all offspring must show the trait.
   d. To determine the likely genotype of a certain individual in a pedigree, first label the genotypes of all the family members you can. Even if some of the genotypes are incomplete, label what you do know. For example, if an individual has the dominant phenotype, the genotype must be *AA* or *Aa*; you can write this as *A*–. Try different possibilities to see which fits the results. Use the rules of probability to calculate the probability of each possible genotype being the correct one.

# TEST YOUR UNDERSTANDING

## Level 1: Knowledge/Comprehension

1. **DRAW IT** Two pea plants heterozygous for the characters of pod color and pod shape are crossed. Draw a Punnett square to determine the phenotypic ratios of the offspring.

2. A man with type A blood marries a woman with type B blood. Their child has type O blood. What are the genotypes of these three individuals? What genotypes, and in what frequencies, would you expect in future offspring from this marriage?

3. A man has six fingers on each hand and six toes on each foot. His wife and their daughter have the normal number of digits. Remember that extra digits is a dominant trait. What fraction of this couple's children would be expected to have extra digits?

4. **DRAW IT** A pea plant heterozygous for inflated pods (*Ii*) is crossed with a plant homozygous for constricted pods (*ii*). Draw a Punnett square for this cross to predict genotypic and phenotypic ratios. Assume that pollen comes from the *ii* plant.

## Level 2: Application/Analysis

5. Flower position, stem length, and seed shape are three characters that Mendel studied. Each is controlled by an independently assorting gene and has dominant and recessive expression as indicated in Table 11.1. If a plant that is heterozygous for all three characters is allowed to self-fertilize, what proportion of the offspring would you expect to be as follows? (*Note*: Use the rules of probability instead of a huge Punnett square.)
   (a) homozygous for the three dominant traits
   (b) homozygous for the three recessive traits
   (c) heterozygous for all three characters
   (d) homozygous for axial and tall, heterozygous for seed shape

6. Hemochromatosis is an inherited disease caused by a recessive allele. If a woman and her husband, who are both carriers, have three children, what is the probability of each of the following?
   (a) All three children are of normal phenotype.
   (b) One or more of the three children have the disease.
   (c) All three children have the disease.
   (d) At least one child is phenotypically normal.

   (*Note*: It will help to remember that the probabilities of all possible outcomes always add up to 1.)

7. The genotype of $F_1$ individuals in a tetrahybrid cross is *AaBbCcDd*. Assuming independent assortment of these four genes, what are the probabilities that $F_2$ offspring will have the following genotypes?
   (a) *aabbccdd*
   (b) *AaBbCcDd*
   (c) *AABBCCDD*
   (d) *AaBBccDd*
   (e) *AaBBCCdd*

8. What is the probability that each of the following pairs of parents will produce the indicated offspring? (Assume independent assortment of all gene pairs.)
   (a) *AABBCC* × *aabbcc* → *AaBbCc*
   (b) *AABbCc* × *AaBbCc* → *AAbbCC*
   (c) *AaBbCc* × *AaBbCc* → *AaBbCc*
   (d) *aaBbCC* × *AABbcc* → *AaBbCc*

9. Karen and Steve each have a sibling with sickle-cell disease. Neither Karen nor Steve nor any of their parents have the disease, and none of them have been tested to see if they have the sickle-cell trait. Based on this incomplete information, calculate the probability that if this couple has a child, the child will have sickle-cell disease.

10. In 1981, a stray black cat with unusual rounded, curled-back ears was adopted by a family in California. Hundreds of descendants of the cat have since been born, and cat fanciers hope to develop the curl cat into a show breed. Suppose you owned the first curl cat and wanted to develop a true-breeding variety. How would you determine whether the curl allele is dominant or recessive? How would you obtain true-breeding curl cats? How could you be sure they are true-breeding?

11. In tigers, a recessive allele that is pleiotropic causes an absence of fur pigmentation (a white tiger) and a cross-eyed condition. If two phenotypically normal tigers that are heterozygous at this locus are mated, what percentage of their offspring will be cross-eyed? What percentage of cross-eyed tigers will be white?

12. In maize (corn) plants, a dominant allele *I* inhibits kernel color, while the recessive allele *i* permits color when homozygous. At a different locus, the dominant allele *P* causes purple kernel color, while the homozygous recessive genotype *pp* causes red kernels. If plants heterozygous at both loci are crossed, what will be the phenotypic ratio of the offspring?

13. The pedigree below traces the inheritance of alkaptonuria, a biochemical disorder. Affected individuals, indicated here by the colored circles and squares, are unable to metabolize a substance called alkapton, which colors the urine and stains body tissues. Does alkaptonuria appear to be caused by a dominant allele or by a recessive allele? Fill in the genotypes of the individuals whose genotypes can be deduced. What genotypes are possible for each of the other individuals?

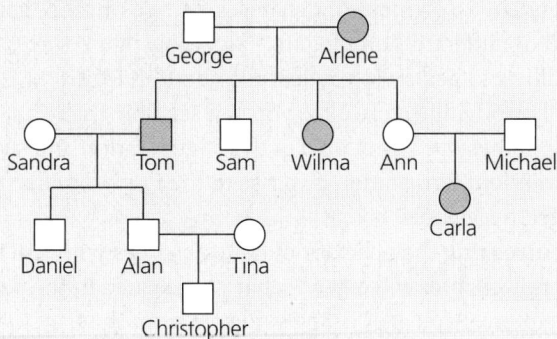

14. Imagine that you are a genetic counselor, and a couple planning to start a family comes to you for information. Charles was married once before, and he and his first wife had a child with cystic fibrosis. The brother of his current wife, Elaine, died of cystic fibrosis. What is the probability that Charles and Elaine will have a baby with cystic fibrosis? (Neither Charles, nor Elaine, nor their parents have cystic fibrosis.)

## Level 3: Synthesis/Evaluation

15. **FOCUS ON EVOLUTION**
    Over the past half century, there has been a trend in the United States and other developed countries for people to marry and start families later in life than did their parents and grandparents. Describe what effects this trend might have on the incidence (frequency) of late-acting dominant lethal alleles in the population.

16. **SCIENTIFIC INQUIRY**
    You are handed a mystery pea plant with tall stems and axial flowers and asked to determine its genotype as quickly as possible. You know that the allele for tall stems (*T*) is dominant to that for dwarf stems (*t*) and that the allele for axial flowers (*A*) is dominant to that for terminal flowers (*a*).
    (a) Identify ALL the possible genotypes for your mystery plant.
    (b) Describe the ONE cross you would do, out in your garden, to determine the exact genotype of your mystery plant.
    (c) While waiting for the results of your cross, you predict the results for each possible genotype listed in part a. Explain how you do this and why this is not called "performing a cross."
    (d) Explain how the results of your cross and your predictions will help you learn the genotype of your mystery plant.

17. **FOCUS ON INFORMATION**
    The continuity of life is based on heritable information in the form of DNA. In a short essay (100–150 words), explain how the passage of genes from parents to offspring, in the form of particular alleles, ensures perpetuation of parental traits in offspring and, at the same time, genetic variation among offspring. Use genetic terms in your explanation.

18. **SYNTHESIZE YOUR KNOWLEDGE**

Just for fun, imagine that "shirt-striping" is a phenotypic character caused by a single gene. Make up a genetic explanation for the appearance of the family in the above photograph, consistent with their "shirt phenotypes." Include in your answer the presumed allele combinations for "shirt-striping" in each family member. What is the inheritance pattern shown by the child?

*For selected answers, see Appendix A.*

# 12 The Chromosomal Basis of Inheritance

## KEY CONCEPTS

**12.1** Morgan showed that Mendelian inheritance has its physical basis in the behavior of chromosomes: *scientific inquiry*

**12.2** Sex-linked genes exhibit unique patterns of inheritance

**12.3** Linked genes tend to be inherited together because they are located near each other on the same chromosome

**12.4** Alterations of chromosome number or structure cause some genetic disorders

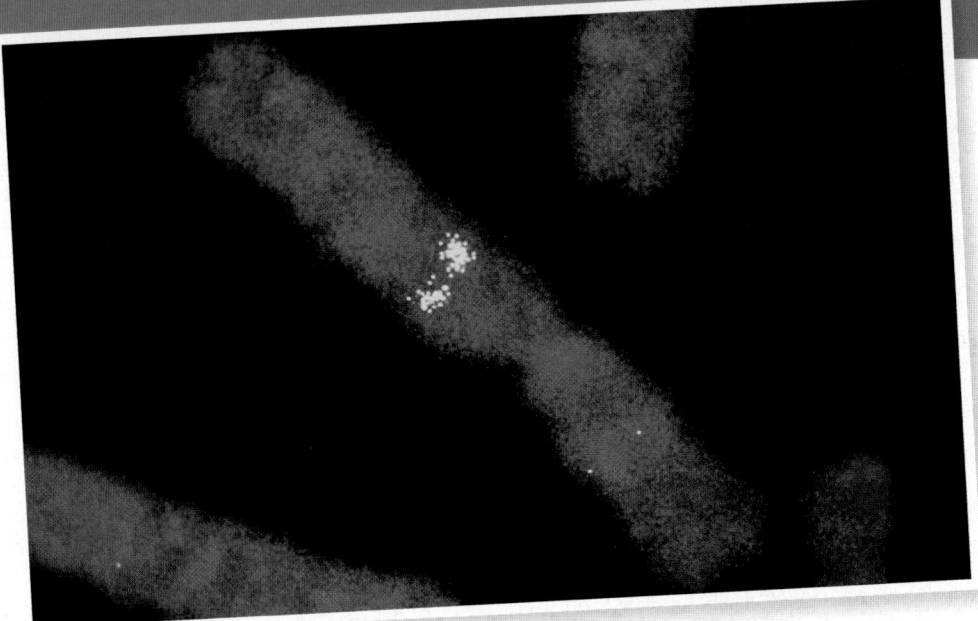

▲ **Figure 12.1** Where in the cell are Mendel's hereditary factors located?

## Locating Genes Along Chromosomes

Today, we know that genes—Mendel's "factors"—are segments of DNA located along chromosomes. We can see the location of a particular gene by tagging chromosomes with a fluorescent dye that highlights that gene. For example, the two yellow regions in **Figure 12.1** mark a specific gene on human chromosome 6. (The chromosome has duplicated, so the allele on that chromosome is present as two copies, one per sister chromatid.) However, Gregor Mendel's "hereditary factors" were purely an abstract concept when he proposed their existence in 1860. At that time, no cellular structures had been identified that could house these imaginary units, and most biologists were skeptical about Mendel's proposed laws of inheritance.

Using improved techniques of microscopy, cytologists worked out the process of mitosis in 1875 and meiosis in the 1890s. Cytology and genetics converged as biologists began to see parallels between the behavior of Mendel's proposed hereditary factors during sexual life cycles and the behavior of chromosomes. As shown in **Figure 12.2**, chromosomes and genes are both present in pairs in diploid cells; homologous chromosomes separate and alleles segregate during the process of meiosis. After meiosis, fertilization restores the paired condition for both chromosomes and genes. Around

1902, Walter S. Sutton, Theodor Boveri, and others noted these parallels and began to develop the **chromosome theory of inheritance**. According to this theory, Mendelian genes have specific loci (positions) along chromosomes, and it is the chromosomes that undergo segregation and independent assortment.

As you can see in Figure 12.2, the separation of homologs during anaphase I accounts for the segregation of the two alleles of a gene into separate gametes, and the random arrangement of chromosome pairs at metaphase I accounts for independent assortment of the alleles for two or more genes located on different homologous pairs. This figure traces the same dihybrid pea cross you learned about in Figure 11.8. By carefully studying Figure 12.2, you can see how the behavior of chromosomes during meiosis in the $F_1$ generation and subsequent random fertilization give rise to the $F_2$ phenotypic ratio observed by Mendel.

In correlating the behavior of chromosomes with that of genes, this chapter will extend what you learned in the past two chapters. After describing evidence from the fruit fly that strongly supported the chromosome theory, we'll explore the chromosomal basis for the transmission of genes from parents to offspring, including what happens when two genes are linked on the same chromosome. Finally, we'll discuss some important exceptions to the standard mode of inheritance.

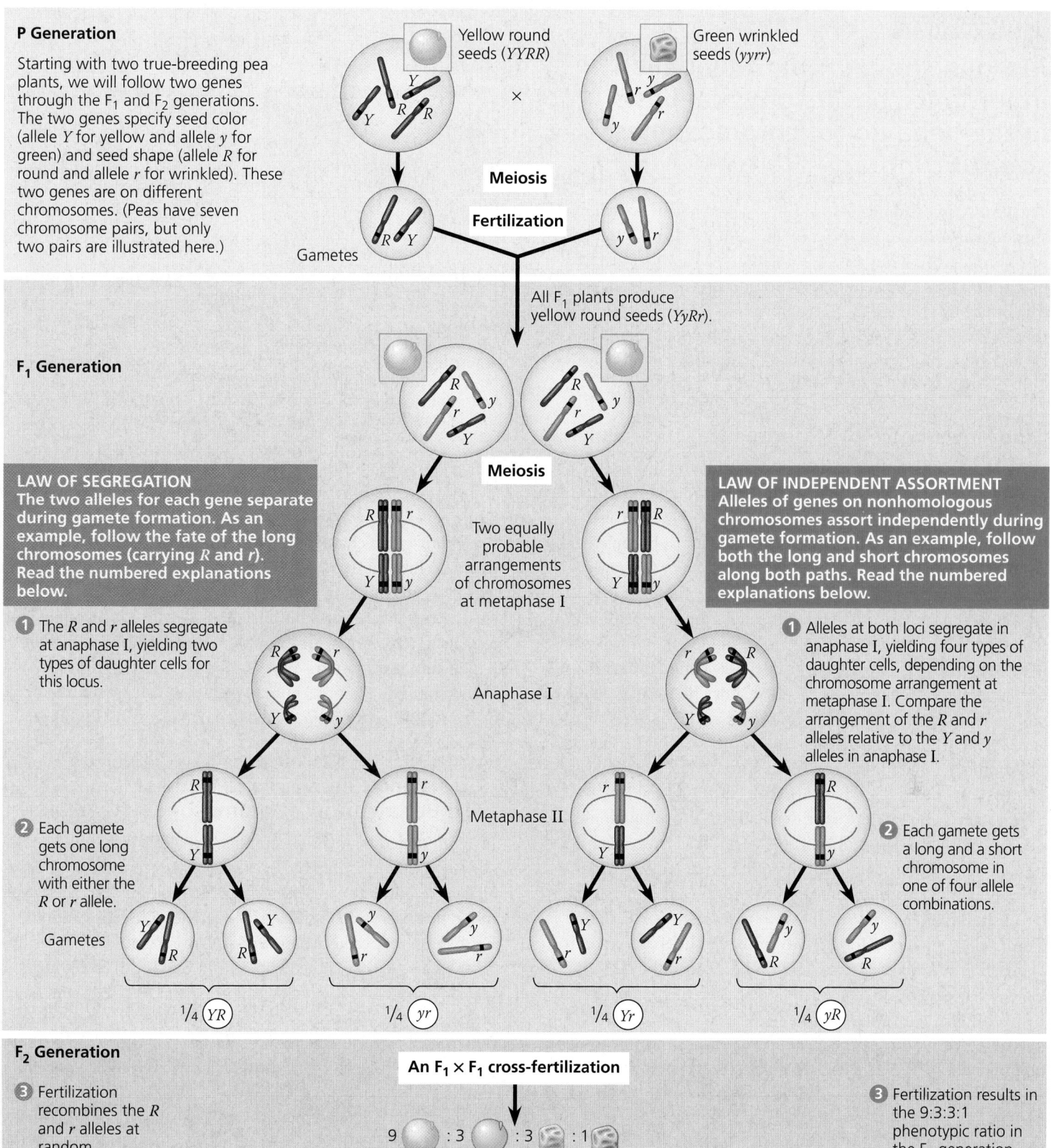

**P Generation**

Starting with two true-breeding pea plants, we will follow two genes through the $F_1$ and $F_2$ generations. The two genes specify seed color (allele $Y$ for yellow and allele $y$ for green) and seed shape (allele $R$ for round and allele $r$ for wrinkled). These two genes are on different chromosomes. (Peas have seven chromosome pairs, but only two pairs are illustrated here.)

Yellow round seeds (*YYRR*)

Green wrinkled seeds (*yyrr*)

**Meiosis**

**Fertilization**

Gametes

All $F_1$ plants produce yellow round seeds (*YyRr*).

**$F_1$ Generation**

**LAW OF SEGREGATION**
The two alleles for each gene separate during gamete formation. As an example, follow the fate of the long chromosomes (carrying $R$ and $r$). Read the numbered explanations below.

**Meiosis**

Two equally probable arrangements of chromosomes at metaphase I

**LAW OF INDEPENDENT ASSORTMENT**
Alleles of genes on nonhomologous chromosomes assort independently during gamete formation. As an example, follow both the long and short chromosomes along both paths. Read the numbered explanations below.

❶ The $R$ and $r$ alleles segregate at anaphase I, yielding two types of daughter cells for this locus.

Anaphase I

❶ Alleles at both loci segregate in anaphase I, yielding four types of daughter cells, depending on the chromosome arrangement at metaphase I. Compare the arrangement of the $R$ and $r$ alleles relative to the $Y$ and $y$ alleles in anaphase I.

Metaphase II

❷ Each gamete gets one long chromosome with either the $R$ or $r$ allele.

❷ Each gamete gets a long and a short chromosome in one of four allele combinations.

Gametes

¼ *YR*    ¼ *yr*    ¼ *Yr*    ¼ *yR*

**$F_2$ Generation**

❸ Fertilization recombines the $R$ and $r$ alleles at random.

**An $F_1 \times F_1$ cross-fertilization**

9 ◯ : 3 ◯ : 3 ▨ : 1 ▨

❸ Fertilization results in the 9:3:3:1 phenotypic ratio in the $F_2$ generation.

▲ **Figure 12.2 The chromosomal basis of Mendel's laws.** Here we correlate the results of one of Mendel's dihybrid crosses (see Figure 11.8) with the behavior of chromosomes during meiosis (see Figure 10.8). The arrangement of chromosomes at metaphase I of meiosis and their movement during anaphase I account for, respectively, the independent assortment and segregation of the alleles for seed color and shape. Each cell that undergoes meiosis in an $F_1$ plant produces two kinds of gametes. If we count the results for all cells, however, each $F_1$ plant produces equal numbers of all four kinds of gametes because the alternative chromosome arrangements at metaphase I are equally likely.

**?** *If you crossed an $F_1$ plant with a plant that was homozygous recessive for both genes* (yyrr), *how would the phenotypic ratio of the offspring compare with the 9:3:3:1 ratio seen here?*

## CONCEPT 12.1

# Morgan showed that Mendelian inheritance has its physical basis in the behavior of chromosomes: *scientific inquiry*

The first solid evidence associating a specific gene with a specific chromosome came early in the 1900s from the work of Thomas Hunt Morgan, an experimental embryologist at Columbia University. Although Morgan was initially skeptical about both Mendelian genetics and the chromosome theory, his early experiments provided convincing evidence that chromosomes are indeed the location of Mendel's heritable factors.

## Morgan's Choice of Experimental Organism

Many times in the history of biology, important discoveries have come to those insightful or lucky enough to choose an experimental organism suitable for the research problem being tackled. Mendel chose the garden pea because a number of distinct varieties were available. For his work, Morgan selected a species of fruit fly, *Drosophila melanogaster*, a common insect that feeds on the fungi growing on fruit. Fruit flies are prolific breeders; a single mating will produce hundreds of offspring, and a new generation can be bred every two weeks. Morgan's laboratory began using this convenient organism for genetic studies in 1907 and soon became known as "the fly room."

Another advantage of the fruit fly is that it has only four pairs of chromosomes, which are easily distinguishable with a light microscope. There are three pairs of autosomes and one pair of sex chromosomes. Female fruit flies have a pair of homologous X chromosomes, and males have one X chromosome and one Y chromosome.

While Mendel could readily obtain different pea varieties from seed suppliers, Morgan was probably the first person to want different varieties of the fruit fly. He faced the tedious task of carrying out many matings and then microscopically inspecting large numbers of offspring in search of naturally occurring variant individuals. After many months of this, he complained, "Two years' work wasted. I have been breeding those flies for all that time and I've got nothing out of it." Morgan persisted, however, and was finally rewarded with the discovery of a single male fly with white eyes instead of the usual red. The phenotype for a character most commonly observed in natural populations, such as red eyes in *Drosophila*, is called the **wild type** **(Figure 12.3)**. Traits that are alternatives to the wild type, such as white eyes in *Drosophila*, are called *mutant phenotypes* because they are due to alleles assumed to have originated as changes, or mutations, in the wild-type allele.

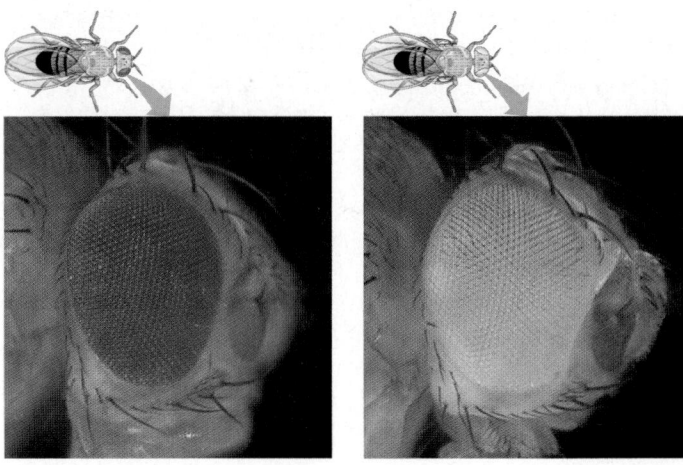

▲ **Figure 12.3 Morgan's first mutant.** Wild-type *Drosophila* flies have red eyes (left). Among his flies, Morgan discovered a mutant male with white eyes (right). This variation made it possible for Morgan to trace a gene for eye color to a specific chromosome (LMs).

Morgan and his students invented a notation for symbolizing alleles in *Drosophila* that is still widely used for fruit flies. For a given character in flies, the gene takes its symbol from the first mutant (non-wild type) discovered. Thus, the allele for white eyes in *Drosophila* is symbolized by *w*. The superscript + identifies the allele for the wild-type trait: $w^+$ for the allele for red eyes, for example. Over the years, a variety of gene notation systems have been developed for different organisms. For example, human genes are usually written in all capitals, such as *HTT* for the gene involved in Huntington's disease.

## Correlating Behavior of a Gene's Alleles with Behavior of a Chromosome Pair

Morgan mated his white-eyed male fly with a red-eyed female. All the $F_1$ offspring had red eyes, suggesting that the wild-type allele is dominant. When Morgan bred the $F_1$ flies to each other, he observed the classical 3:1 phenotypic ratio among the $F_2$ offspring. However, there was a surprising additional result: The white-eye trait showed up only in males. All the $F_2$ females had red eyes, while half the males had red eyes and half had white eyes. Therefore, Morgan concluded that somehow a fly's eye color was linked to its sex. (If the eye-color gene were unrelated to sex, half of the white-eyed flies would have been female.)

Recall that a female fly has two X chromosomes (XX), while a male fly has an X and a Y (XY). The correlation between the trait of white eye color and the male sex of the affected $F_2$ flies suggested to Morgan that the gene involved in his white-eyed mutant was located exclusively on the X chromosome, with no corresponding allele present on the Y chromosome. His reasoning can be followed in **Figure 12.4**. For a male, a single copy of the mutant allele would confer white eyes; since a male has only one

## ▼ Figure 12.4 Inquiry

### In a cross between a wild-type female fruit fly and a mutant white-eyed male, what color eyes will the F₁ and F₂ offspring have?

**Experiment** Thomas Hunt Morgan wanted to analyze the behavior of two alleles of a fruit fly eye-color gene. In crosses similar to those done by Mendel with pea plants, Morgan and his colleagues mated a wild-type (red-eyed) female with a mutant white-eyed male.

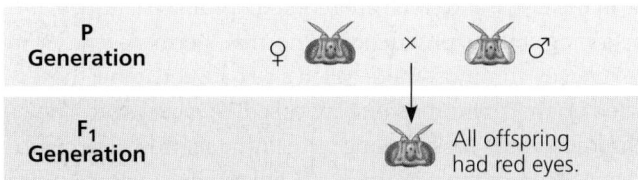

Morgan then bred an F₁ red-eyed female to an F₁ red-eyed male to produce the F₂ generation.

**Results** The F₂ generation showed a typical Mendelian ratio of 3 red-eyed flies to 1 white-eyed fly. However, all white-eyed flies were males; no females displayed the white-eye trait.

**Conclusion** All F₁ offspring had red eyes, so the mutant white-eye trait ($w$) must be recessive to the wild-type red-eye trait ($w^+$). Since the recessive trait—white eyes—was expressed only in males in the F₂ generation, Morgan deduced that this eye-color gene is located on the X chromosome and that there is no corresponding locus on the Y chromosome.

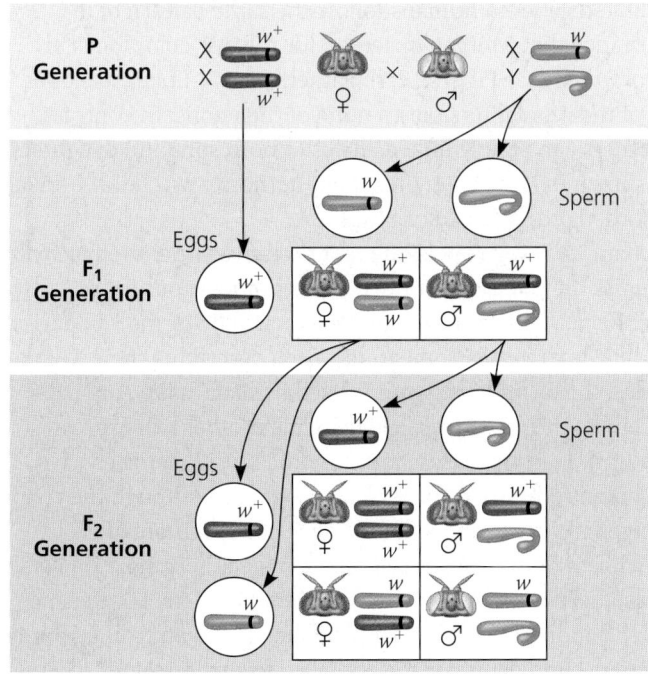

**Data from** T. H. Morgan, Sex-limited inheritance in *Drosophila, Science* 32:120–122 (1910).

(MB) A related Experimental Inquiry Tutorial can be assigned in MasteringBiology.

**WHAT IF?** Suppose this eye-color gene were located on an autosome. Predict the phenotypes (including gender) of the F₂ flies in this hypothetical cross. (*Hint*: Draw a Punnett square.)

---

X chromosome, there can be no wild-type allele ($w^+$) present to mask the recessive allele. However, a female could have white eyes only if both her X chromosomes carried the recessive mutant allele ($w$). This was impossible for the F₂ females in Morgan's experiment because all the F₁ fathers had red eyes, so each F₂ female received a $w^+$ allele on the X chromosome inherited from her father.

Morgan's finding of the correlation between a particular trait and an individual's sex provided support for the chromosome theory of inheritance, namely, that a specific gene is carried on a specific chromosome (in this case, an eye-color gene on the X chromosome). In addition, Morgan's work indicated that genes located on a sex chromosome exhibit unique inheritance patterns, which we'll discuss in the next section. Recognizing the importance of Morgan's early work, many bright students were attracted to his fly room.

### CONCEPT CHECK 12.1

1. Which one of Mendel's laws relates to the inheritance of alleles for a single character? Which law relates to the inheritance of alleles for two characters in a dihybrid cross?
2. **MAKE CONNECTIONS** Review the description of meiosis (see Figure 10.8) and Mendel's laws of segregation and independent assortment (see Concept 11.1). What is the physical basis for each of Mendel's laws?
3. **WHAT IF?** Propose a possible reason that the first naturally occurring mutant fruit fly Morgan saw involved a gene on a sex chromosome and was found in a male.

For suggested answers, see Appendix A.

---

## CONCEPT 12.2

# Sex-linked genes exhibit unique patterns of inheritance

As you just learned, Morgan's discovery of a trait (white eyes) that correlated with the sex of flies was a key episode in the development of the chromosome theory of inheritance. Because the identity of the sex chromosomes in an individual could be inferred by observing the sex of the fly, the behavior of the two members of the pair of sex chromosomes could be correlated with the behavior of the two alleles of the eye-color gene. In this section, we'll take a closer look at the role of sex chromosomes in inheritance.

## The Chromosomal Basis of Sex

Whether we are anatomically male or female may be one of our more obvious phenotypic characters. Although the anatomical and physiological differences between women and men are numerous, the chromosomal basis for determining sex is rather simple. Humans and other mammals have two types of sex chromosomes, designated X and Y. The Y chromosome is much smaller than the X chromosome

**(Figure 12.5).** A person who inherits two X chromosomes, one from each parent, usually develops as a female; a male inherits one X chromosome and one Y chromosome **(Figure 12.6)**. Short segments at either end of the Y chromosome are the only regions that are homologous with regions on the X. These homologous regions allow the X and Y chromosomes in males to pair and behave like homologous chromosomes during meiosis in the testes.

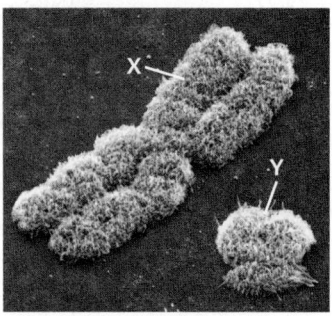

▲ **Figure 12.5 Human sex chromosomes.**

In mammalian testes and ovaries, the two sex chromosomes segregate during meiosis. Each egg receives one X chromosome. In contrast, sperm fall into two categories: Half the sperm cells a male produces receive an X chromosome, and half receive a Y chromosome. We can trace the sex of each offspring to the events of conception: If a sperm cell bearing an X chromosome fertilizes an egg, the zygote is XX, a female; if a sperm cell containing a Y chromosome fertilizes an egg, the zygote is XY, a male (see Figure 12.6). Thus, sex determination is a matter of chance—a fifty-fifty chance. In *Drosophila*, while males are also XY, sex depends not on the presence of the Y, but on the ratio between the number of X chromosomes and the number of autosome sets. There are other chromosomal systems as well, besides the X-Y system, for determining sex.

In humans, the anatomical signs of sex begin to emerge when the embryo is about 2 months old. Before then, the rudiments of the gonads are generic—they can develop into either testes or ovaries, depending on whether or not a Y chromosome is present. In 1990, a British research team identified a gene on the Y chromosome required for the development of testes. They named the gene *SRY*, for <u>s</u>ex-determining <u>r</u>egion

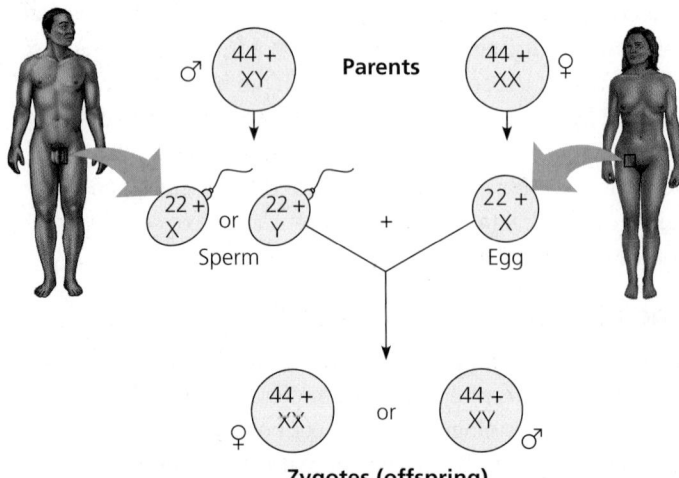

▲ **Figure 12.6 The mammalian X-Y chromosomal system of sex determination.** In mammals, the sex of an offspring depends on whether the sperm cell contains an X chromosome or a Y. (All eggs have an X.) Numerals indicate the number of autosomes.

of Y. In the absence of *SRY*, the gonads develop into ovaries. The biochemical, physiological, and anatomical features that distinguish males and females are complex, and many genes are involved in their development. In fact, *SRY* codes for a protein that regulates other genes.

Researchers have sequenced the human Y chromosome and have identified 78 genes that code for about 25 proteins (some genes are duplicates). About half of these genes are expressed only in the testis, and some are required for normal testicular functioning and the production of normal sperm. A gene located on either sex chromosome is called a **sex-linked gene**; those located on the Y chromosome are called *Y-linked genes*. The Y chromosome is passed along virtually intact from a father to all his sons. Because there are so few Y-linked genes, very few disorders are transferred from father to son on the Y chromosome. A rare example is that in the absence of certain Y-linked genes, an XY individual is male but does not produce normal sperm.

The human X chromosome contains approximately 1,100 genes, which are called **X-linked genes**. The fact that males and females inherit a different number of X chromosomes leads to a pattern of inheritance different from that produced by genes located on autosomes.

## Inheritance of X-Linked Genes

While most Y-linked genes help determine sex, the X chromosomes have genes for many characters unrelated to sex. X-linked genes in humans follow the same pattern of inheritance that Morgan observed for the eye-color locus in *Drosophila* (see Figure 12.4). Fathers pass X-linked alleles to all of their daughters but to none of their sons. In contrast, mothers can pass X-linked alleles to both sons and daughters, as shown in **Figure 12.7** for the inheritance of a mild X-linked disorder, red-green color blindness.

If an X-linked trait is due to a recessive allele, a female will express the phenotype only if she is homozygous for that allele. Because males have only one locus, the terms *homozygous* and *heterozygous* lack meaning when describing their X-linked genes; the term *hemizygous* is used in such cases. Any male receiving the recessive allele from his mother will express the trait. For this reason, far more males than females have X-linked recessive disorders. However, even though the chance of a female inheriting a double dose of the mutant allele is much less than the probability of a male inheriting a single dose, there *are* females with X-linked disorders. For instance, color blindness is almost always inherited as an X-linked trait. A color-blind daughter may be born to a color-blind father whose mate is a carrier (see Figure 12.7c). Because the X-linked allele for color blindness is relatively rare, however, the probability that such a man and woman will mate is low.

A number of human X-linked disorders are much more serious than color blindness, such as **Duchenne muscular dystrophy**, which affects about one out of 3,500 males born in the United States. The disease is characterized by a progressive weakening of the muscles and loss of coordination. Affected

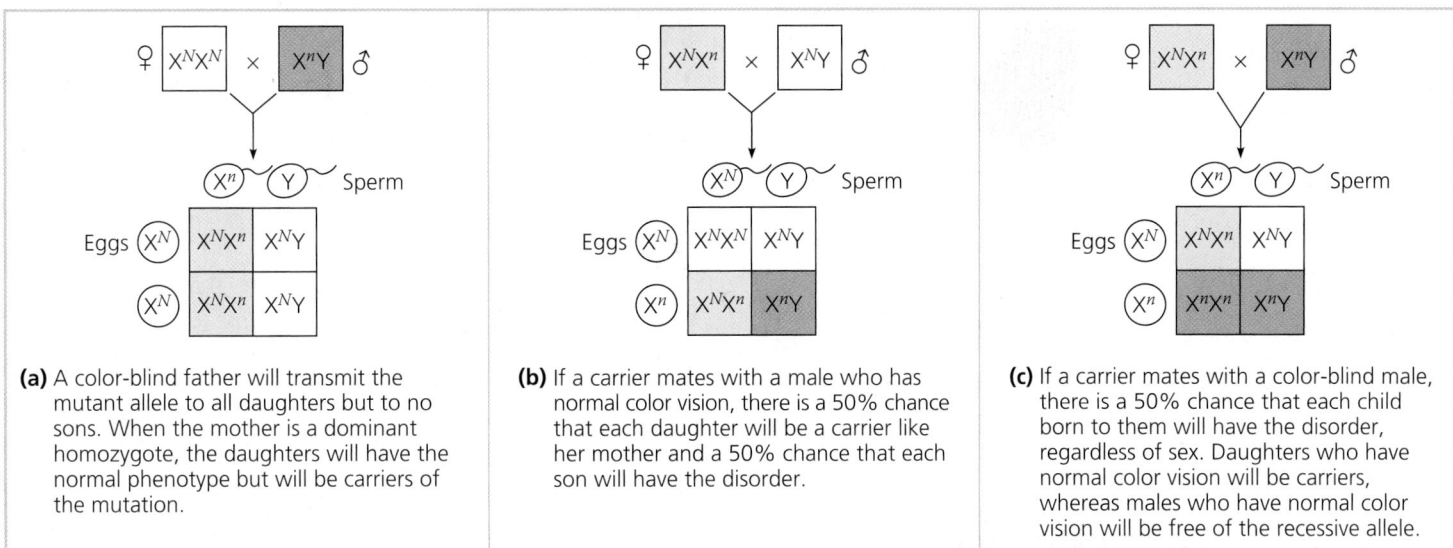

**(a)** A color-blind father will transmit the mutant allele to all daughters but to no sons. When the mother is a dominant homozygote, the daughters will have the normal phenotype but will be carriers of the mutation.

**(b)** If a carrier mates with a male who has normal color vision, there is a 50% chance that each daughter will be a carrier like her mother and a 50% chance that each son will have the disorder.

**(c)** If a carrier mates with a color-blind male, there is a 50% chance that each child born to them will have the disorder, regardless of sex. Daughters who have normal color vision will be carriers, whereas males who have normal color vision will be free of the recessive allele.

▲ **Figure 12.7 The transmission of X-linked recessive traits.** In this diagram, red-green color blindness is used as an example. The superscript $N$ represents the dominant allele for normal color vision carried on the X chromosome, while $n$ represents the recessive allele, which has a mutation causing color blindness. White boxes indicate unaffected individuals, light orange boxes indicate carriers, and dark orange boxes indicate color-blind individuals.

**?** *If a color-blind woman married a man who had normal color vision, what would be the probable phenotypes of their children?*

individuals rarely live past their early 20s. Researchers have traced the disorder to the absence of a key muscle protein called dystrophin and have mapped the gene for this protein to a specific locus on the X chromosome.

**Hemophilia** is an X-linked recessive disorder defined by the absence of one or more of the proteins required for blood clotting. When a person with hemophilia is injured, bleeding is prolonged because a firm clot is slow to form. Small cuts in the skin are usually not a problem, but bleeding in the muscles or joints can be painful and can lead to serious damage. In the 1800s, hemophilia was widespread among the royal families of Europe. Queen Victoria of England is known to have passed the allele to several of her descendants. Subsequent intermarriage with royal family members of other nations, such as Spain and Russia, further spread this X-linked trait, and its incidence is well documented in royal pedigrees. A few years ago, new genomic techniques allowed sequencing of DNA from tiny amounts isolated from the buried remains of royal family members. The genetic basis of the mutation, and how it resulted in a nonfunctional blood-clotting factor, is now understood. Today, people with hemophilia are treated as needed with intravenous injections of the protein that is missing.

## X Inactivation in Female Mammals

Female mammals, including human females, inherit two X chromosomes—twice the number inherited by males—so you may wonder if females make twice as much of the proteins encoded by X-linked genes. In fact, almost all of one X chromosome in each cell in female mammals becomes inactivated during early embryonic development. As a result, the cells of females and males have the same effective dose (one active

copy) of most X-linked genes. The inactive X in each cell of a female condenses into a compact object called a **Barr body** (discovered by Canadian anatomist Murray Barr), which lies along the inside of the nuclear envelope. Most of the genes of the X chromosome that forms the Barr body are not expressed. In the ovaries, however, Barr-body chromosomes are reactivated in the cells that give rise to eggs, resulting in every female gamete (egg) having an active X after meiosis.

British geneticist Mary Lyon demonstrated that the selection of which X chromosome will form the Barr body occurs randomly and independently in each embryonic cell present at the time of X inactivation. As a consequence, females consist of a *mosaic* of two types of cells: those with the active X derived from the father and those with the active X derived from the mother. After an X chromosome is inactivated in a particular cell, all mitotic descendants of that cell have the same inactive X. Thus, if a female is heterozygous for a sex-linked trait, about half of her cells will express one allele, while the others will express the alternate allele. **Figure 12.8** shows how this mosaicism results in the patchy coloration of a tortoiseshell cat. In humans, mosaicism can be observed in a recessive X-linked mutation that prevents the development of sweat glands. A woman who is heterozygous for this trait has patches of normal skin and patches of skin lacking sweat glands.

Inactivation of an X chromosome involves modification of the DNA and proteins bound to it called histones, including attachment of methyl groups ($-CH_3$) to DNA nucleotides. (The regulatory role of DNA methylation is discussed further in Concept 15.2.) A particular region of each X chromosome contains several genes involved in the inactivation process. The two regions, one on each X chromosome, associate briefly with

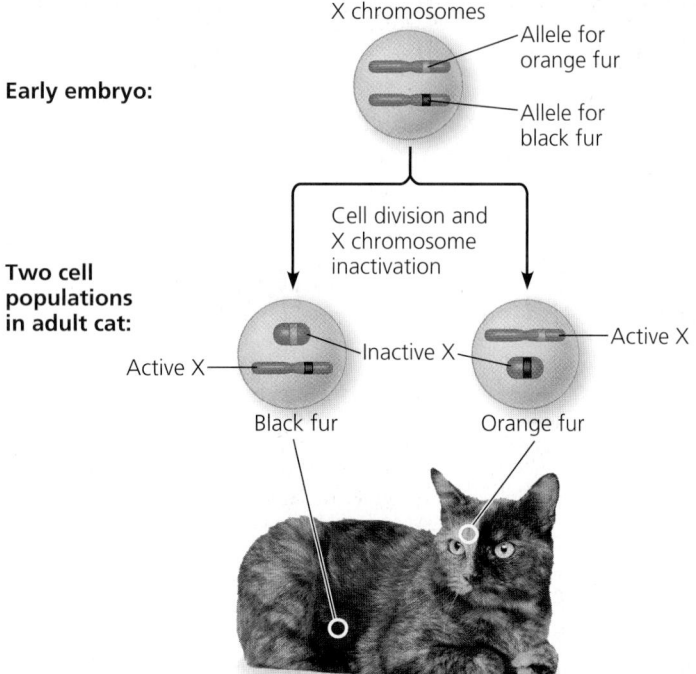

Early embryo:

X chromosomes

Allele for orange fur

Allele for black fur

Cell division and X chromosome inactivation

Two cell populations in adult cat:

Active X

Inactive X

Active X

Active X

Black fur

Orange fur

▲ **Figure 12.8 X inactivation and the tortoiseshell cat.** The tortoiseshell gene is on the X chromosome, and the tortoiseshell phenotype requires the presence of two different alleles, one for orange fur and one for black fur. Normally, only females can have both alleles, because only they have two X chromosomes. If a female cat is heterozygous for the tortoiseshell gene, she is tortoiseshell. Orange patches are formed by populations of cells in which the X chromosome with the orange allele is active; black patches have cells in which the X chromosome with the black allele is active. ("Calico" cats also have white areas, which are determined by yet another gene.)

each other in each cell at an early stage of embryonic development. Then one of the genes, called *XIST* (for X-inactive specific transcript), becomes active *only* on the chromosome that will become the Barr body. Multiple copies of the RNA product of this gene apparently attach to the X chromosome on which they are made, eventually almost covering it. Interaction of this RNA with the chromosome initiates X inactivation, and the RNA products of nearby genes help to regulate the process.

## CONCEPT CHECK 12.2

1. A white-eyed female *Drosophila* is mated with a red-eyed (wild-type) male, the reciprocal cross of the one shown in Figure 12.4. What phenotypes and genotypes do you predict for the offspring?

2. Neither Tim nor Rhoda has Duchenne muscular dystrophy, but their firstborn son does have it. What is the probability that a second child of this couple will have the disease? What is the probability if the second child is a boy? A girl?

3. **MAKE CONNECTIONS** Consider what you learned about dominant and recessive alleles in Concept 11.1. If a disorder were caused by a dominant X-linked allele, how would the inheritance pattern differ from what we see for recessive X-linked disorders?

For suggested answers, see Appendix A.

# Linked genes tend to be inherited together because they are located near each other on the same chromosome

The number of genes in a cell is far greater than the number of chromosomes; in fact, each chromosome (except the Y) has hundreds or thousands of genes. Genes located near each other on the same chromosome tend to be inherited together in genetic crosses; such genes are said to be genetically linked and are called **linked genes**. (Note the distinction between the terms *sex-linked gene*, referring to a single gene on a sex chromosome, and *linked genes*, referring to two or more genes on the same chromosome that tend to be inherited together.) When geneticists follow linked genes in breeding experiments, the results deviate from those expected from Mendel's law of independent assortment.

## How Linkage Affects Inheritance

To see how linkage between genes affects the inheritance of two different characters, let's examine another of Morgan's *Drosophila* experiments. In this case, the characters are body color and wing size, each with two different phenotypes. Wild-type flies have gray bodies and normal-sized wings. In addition to these flies, Morgan had managed to obtain, through breeding, doubly mutant flies with black bodies and wings much smaller than normal, called vestigial wings. The mutant alleles are recessive to the wild-type alleles, and neither gene is on a sex chromosome. In his investigation of these two genes, Morgan carried out the crosses shown in **Figure 12.9**. The first was a P generation cross to generate $F_1$ dihybrid flies, and the second was a testcross.

The resulting flies had a much higher proportion of the combinations of traits seen in the P generation flies (called parental phenotypes) than would be expected if the two genes assorted independently. Morgan thus concluded that body color and wing size are usually inherited together in specific (parental) combinations because the genes for these characters are near each other on the same chromosome:

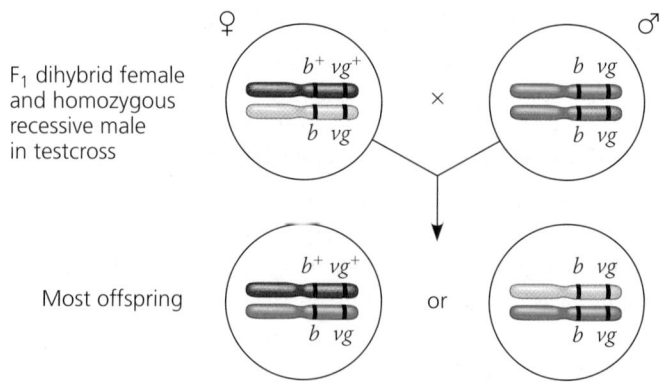

$F_1$ dihybrid female and homozygous recessive male in testcross

♀

$b^+$ $vg^+$

$b$ $vg$

♂

$b$ $vg$

$b$ $vg$

×

Most offspring

$b^+$ $vg^+$

$b$ $vg$

or

$b$ $vg$

$b$ $vg$

## ▼ Figure 12.9  Inquiry

### How does linkage between two genes affect inheritance of characters?

**Experiment** Morgan wanted to know whether the genes for body color and wing size are genetically linked and, if so, how this affects their inheritance. The alleles for body color are $b^+$ (gray) and $b$ (black), and those for wing size are $vg^+$ (normal) and $vg$ (vestigial).

Morgan mated true-breeding P (parental) generation flies—wild-type flies with black, vestigial-winged flies—to produce heterozygous F₁ dihybrids ($b^+ b \ vg^+ vg$), all of which are wild-type in appearance.

He then mated wild-type F₁ dihybrid females with homozygous recessive males. This testcross will reveal the genotype of the eggs made by the dihybrid female.

The male's sperm contributes only recessive alleles, so the phenotype of the offspring reflects the genotype of the female's eggs.

Note: Although only females (with pointed abdomens) are shown, half the offspring in each class would be males (with rounded abdomens).

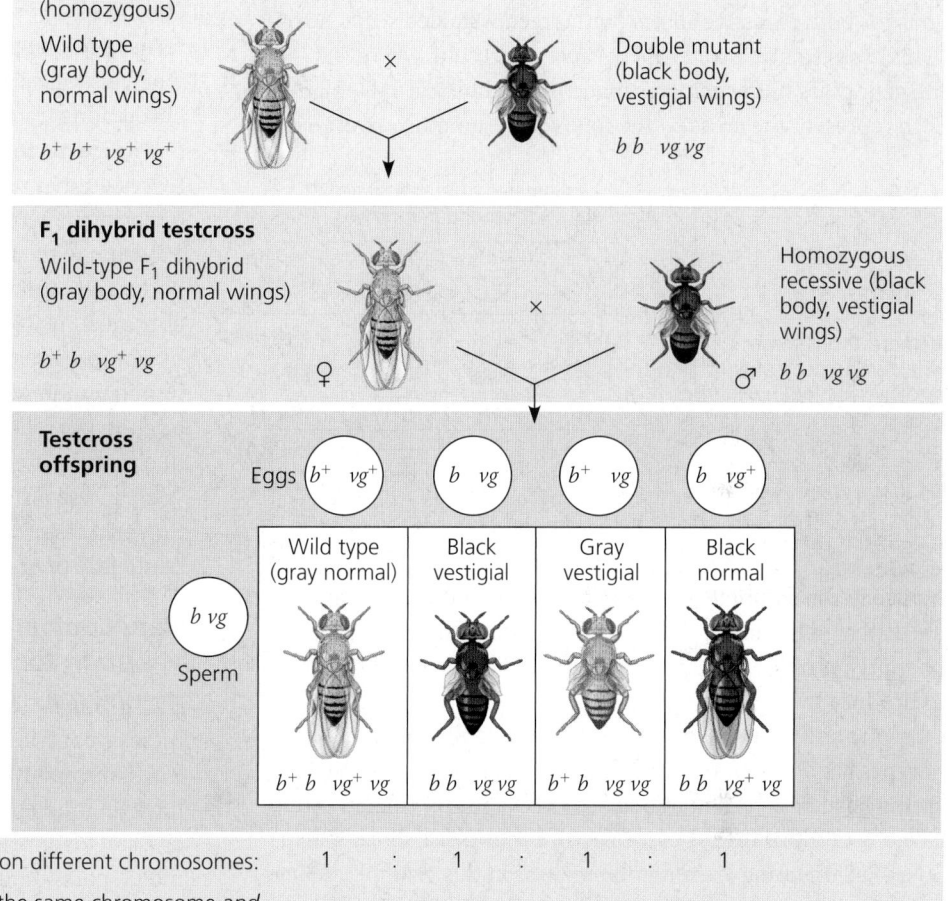

**PREDICTED RATIOS**

| | | | | | | | |
|---|---|---|---|---|---|---|---|
| Predicted ratio if genes are located on different chromosomes: | 1 | : | 1 | : | 1 | : | 1 |
| Predicted ratio if genes are located on the same chromosome *and* parental alleles are always inherited together: | 1 | : | 1 | : | 0 | : | 0 |
| **Results**   Data from Morgan's experiment: | 965 | : | 944 | : | 206 | : | 185 |

**Conclusion** Since most offspring had a parental (P generation) phenotype, Morgan concluded that the genes for body color and wing size are genetically linked on the same chromosome. However, the production of a relatively small number of offspring with nonparental phenotypes indicated that some mechanism occasionally breaks the linkage between specific alleles of genes on the same chromosome.

**Data from** T. H. Morgan and C. J. Lynch, The linkage of two factors in *Drosophila* that are not sex-linked, *Biological Bulletin* 23:174–182 (1912).

**WHAT IF?** If the parental (P generation) flies had been true-breeding for gray body with vestigial wings and black body with normal wings, which phenotypic class(es) would be largest among the testcross offspring?

---

However, as Figure 12.9 shows, both of the combinations of traits not seen in the P generation (called nonparental phenotypes) were also produced in Morgan's experiments, suggesting that the body-color and wing-size alleles are not always linked genetically. To understand this conclusion, we need to further explore **genetic recombination**, the production of offspring with combinations of traits that differ from those found in either P generation parent.

### Genetic Recombination and Linkage

Meiosis and random fertilization generate genetic variation among offspring of sexually reproducing organisms due to independent assortment of chromosomes and crossing over in meiosis I and the possibility of any sperm fertilizing any egg (see Concept 10.4). Here we'll examine the chromosomal basis of recombination of alleles in relation to the genetic findings of Mendel and Morgan.

## Recombination of Unlinked Genes: Independent Assortment of Chromosomes

Mendel learned from crosses in which he followed two characters that some offspring have combinations of traits that do not match those of either parent. For example, consider a cross of a dihybrid pea plant with yellow round seeds, heterozygous for both seed color and seed shape (*YyRr*), with a plant homozygous for both recessive alleles (with green wrinkled seeds, *yyrr*). (This cross acts as a testcross because the results will reveal the genotypes of the gametes made in the dihybrid *YyRr* plant.) Let's represent the cross by the following Punnett square:

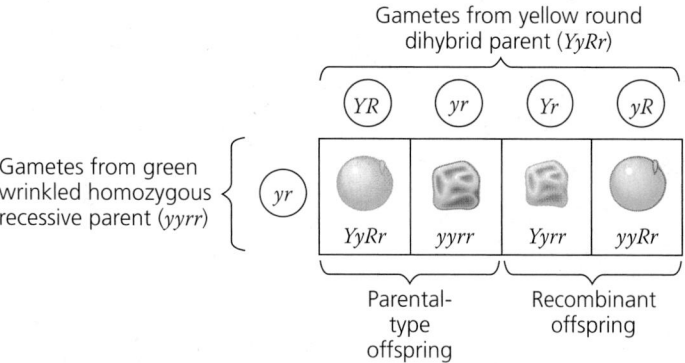

Gametes from yellow round dihybrid parent (*YyRr*)

YR    yr    Yr    yR

Gametes from green wrinkled homozygous recessive parent (*yyrr*)

yr

YyRr    yyrr    Yyrr    yyRr

Parental-type offspring          Recombinant offspring

Notice in this Punnett square that one-half of the offspring are expected to inherit a phenotype that matches either of the phenotypes of the P (parental) generation originally crossed to produce the F₁ dihybrid (see Figure 12.2). These matching offspring are called **parental types**. But two nonparental phenotypes are also found among the offspring. Because these offspring have new combinations of seed shape and color, they are called **recombinant types**, or **recombinants** for short. When 50% of all offspring are recombinants, as in this example, geneticists say that there is a 50% frequency of recombination. The predicted phenotypic ratios among the offspring are similar to what Mendel actually found in his *YyRr* × *yyrr* crosses.

A 50% frequency of recombination in such testcrosses is observed for any two genes that are located on different chromosomes and thus cannot be linked. The physical basis of recombination between unlinked genes is the random orientation of homologous chromosomes at metaphase I of meiosis, which leads to the independent assortment of the two unlinked genes (see Figure 10.11 and the question in the Figure 12.2 legend).

### Recombination of Linked Genes: Crossing Over

Now let's explain the results of the *Drosophila* testcross in Figure 12.9. Recall that most of the offspring from the testcross for body color and wing size had parental phenotypes. That suggested that the two genes were on the same chromosome, since the occurrence of parental types with a frequency greater than 50% indicates that the genes are linked. About 17% of offspring, however, were recombinants.

Seeing these results, Morgan proposed that some process must occasionally break the physical connection between specific alleles of genes on the same chromosome. Later experiments showed that this process, now called **crossing over**, accounts for the recombination of linked genes. In crossing over, which occurs while replicated homologous chromosomes are paired during prophase of meiosis I, a set of proteins orchestrates an exchange of corresponding segments of one maternal and one paternal chromatid (see Figure 10.12). In effect, when a single crossover occurs, end portions of two nonsister chromatids trade places.

**Figure 12.10** shows how crossing over in a dihybrid female fly resulted in recombinant eggs and ultimately recombinant offspring in Morgan's testcross. Most of the eggs had a chromosome with either the $b^+ vg^+$ or $b\ vg$ parental genotype for body color and wing size, but some eggs had a recombinant chromosome ($b^+ vg$ or $b\ vg^+$). Fertilization of all classes of eggs by homozygous recessive sperm ($b\ vg$) produced an offspring population in which 17% exhibited a nonparental, recombinant phenotype, reflecting combinations of alleles not seen before in either P generation parent. In the **Scientific Skills Exercise**, you can use a statistical test to analyze the results from an F₁ dihybrid testcross and see whether the two genes assort independently or are linked.

### New Combinations of Alleles: Variation for Natural Selection

**EVOLUTION** The physical behavior of chromosomes during meiosis contributes to the generation of variation in offspring (see Concept 10.4). Each pair of homologous chromosomes lines up independently of other pairs during metaphase I, and crossing over prior to that, during prophase I, can mix and match parts of maternal and paternal homologs. Mendel's elegant experiments show that the behavior of the abstract entities known as genes—or, more concretely, alleles of genes—also leads to variation in offspring (see Concept 11.1). Now, putting these different ideas together, you can see that the recombinant chromosomes resulting from crossing over may bring alleles together in new combinations, and the subsequent events of meiosis distribute to gametes the recombinant chromosomes in a multitude of combinations, such as the new variants seen in Figures 12.9 and 12.10. Random fertilization then increases even further the number of variant allele combinations that can be created.

This abundance of genetic variation provides the raw material on which natural selection works. If the traits conferred by particular combinations of alleles are better suited for a given environment, organisms possessing those genotypes will be expected to thrive and leave more offspring, ensuring the continuation of their genetic complement. In the next generation, of course, the alleles will be shuffled anew. Ultimately, the interplay between environment and genotype will determine which genetic combinations persist over time.

**▶ Figure 12.10 Chromosomal basis for recombination of linked genes.** In these diagrams recreating the testcross in Figure 12.9, we track chromosomes as well as genes. The maternal chromosomes (present in the wild-type F₁ dihybrid) are color-coded red and pink to distinguish one homolog from the other before any meiotic crossing over has occurred. Because crossing over between the $b^+/b$ and $vg^+/vg$ loci occurs in some, but not all, egg-producing cells, more eggs with parental-type chromosomes than with recombinant ones are produced in the mating females. Fertilization of the eggs by sperm of genotype $b\ vg$ gives rise to some recombinant offspring. The recombination frequency is the percentage of recombinant flies in the total pool of offspring.

**DRAW IT** *Suppose, as in the question at the bottom of Figure 12.9, the parental (P generation) flies were true-breeding for gray body with vestigial wings and black body with normal wings. Draw the chromosomes in each of the four possible kinds of eggs from an F₁ female, and label each chromosome as "parental" or "recombinant."*

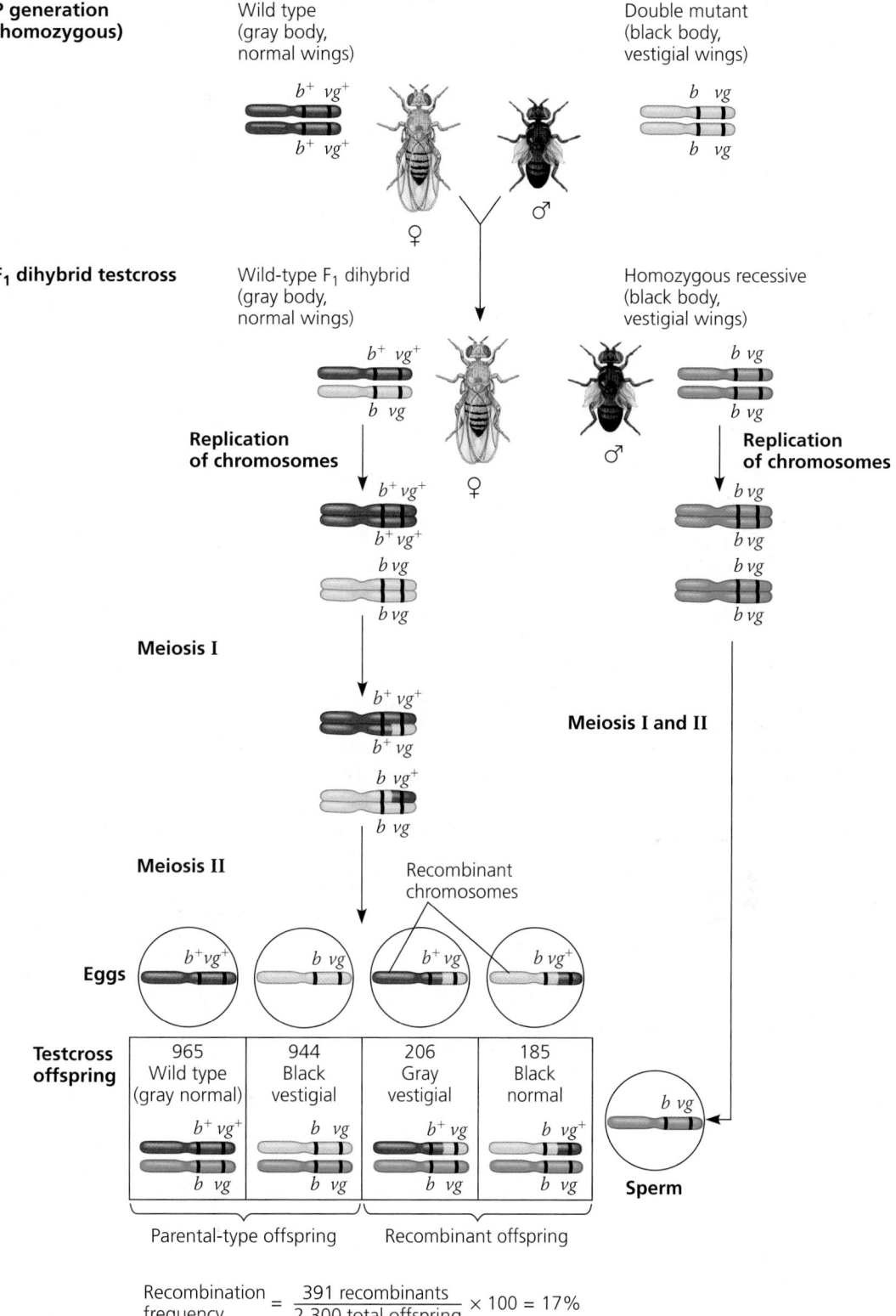

Recombination frequency $= \dfrac{391 \text{ recombinants}}{2{,}300 \text{ total offspring}} \times 100 = 17\%$

## Mapping the Distance Between Genes Using Recombination Data: *Scientific Inquiry*

The discovery of linked genes and recombination due to crossing over motivated one of Morgan's students, Alfred H. Sturtevant, to work out a method for constructing a **genetic map**, an ordered list of the genetic loci along a particular chromosome.

Sturtevant hypothesized that the percentage of recombinant offspring, the *recombination frequency*, calculated from experiments like the one in Figures 12.9 and 12.10, depends on the distance between genes on a chromosome. He assumed that crossing over is a random event, with the chance of crossing over approximately equal at all

# Using the Chi-Square ($\chi^2$) Test

**Are Two Genes Linked or Unlinked?** Genes that are in close proximity on the same chromosome will result in the linked alleles being inherited together more often than not. But how can you tell if certain alleles are inherited together due to linkage or whether they just happen to assort together? In this exercise, you will use a simple statistical test, the chi-square ($\chi^2$) test, to analyze phenotypes of $F_1$ testcross progeny in order to see whether two genes are linked or unlinked.

**How These Experiments Are Done** If genes are unlinked and assorting independently, the phenotypic ratio of offspring from an $F_1$ testcross is expected to be 1:1:1:1 (see Figure 12.9). If the two genes are linked, however, the observed phenotypic ratio of the offspring will not match that ratio. Given that random fluctuations in the data do occur, how much must the observed numbers deviate from the expected numbers for us to conclude that the genes are not assorting independently but may instead be linked? To answer this question, scientists use a statistical test. This test, called a chi-square ($\chi^2$) test, compares an observed data set with an expected data set predicted by a hypothesis (here, that the genes are unlinked) and measures the discrepancy between the two, thus determining the "goodness of fit." If the discrepancy between the observed and expected data sets is so large that it is unlikely to have occurred by random fluctuation, we say there is statistically significant evidence against the hypothesis (or, more specifically, evidence for the genes being linked). If the discrepancy is small, then our observations are well explained by random variation alone. In this case, we say that the observed data are consistent with our hypothesis, or that the discrepancy is statistically insignificant. Note, however, that consistency with our hypothesis is not the same as proof of our hypothesis. Also, the size of the experimental data set is important: With small data sets like this one, even if the genes are linked, discrepancies might be small by chance alone if the linkage is weak. (For simplicity, we overlook the effect of sample size here.)

**Data from the Simulated Experiment** In cosmos plants, purple stem (A) is dominant to green stem (a), and short petals (B) is dominant to long petals (b). In a simulated cross, AABB plants were crossed with aabb plants to generate $F_1$ dihybrids (AaBb), which were then testcrossed (AaBb × aabb). A total of 900 offspring plants were scored for stem color and flower petal length.

| Offspring from testcross of AaBb ($F_1$) × aabb | Purple stem/short petals (A–B–) | Green stem/short petals (aaB–) | Purple stem/long petals (A–bb) | Green stem/long petals (aabb) |
|---|---|---|---|---|
| Expected ratio if the genes are unlinked | 1 | 1 | 1 | 1 |
| Expected number of offspring (of 900) | | | | |
| Observed number of offspring (of 900) | 220 | 210 | 231 | 239 |

## INTERPRET THE DATA

1. The results in the data table are from a simulated $F_1$ dihybrid testcross. The hypothesis that the two genes are unlinked predicts that the offspring phenotypic ratio will be 1:1:1:1. Using this ratio, calculate the expected number of each phenotype out of the 900 total offspring, and enter the values in that data table.

2. The goodness of fit is measured by $\chi^2$. This statistic measures the amounts by which the observed values differ from their respective predictions to indicate how closely the two sets of values match. The formula for calculating this value is

$$\chi^2 = \Sigma \frac{(o - e)^2}{e}$$

where $o$ = observed and $e$ = expected. Calculate the $\chi^2$ value for the data using the table below. Fill out that table, carrying out the operations indicated in the top row. Then add up the entries in the last column to find the $\chi^2$ value.

| Testcross offspring | Expected (e) | Observed (o) | Deviation (o − e) | (o − e)² | (o − e)²/e |
|---|---|---|---|---|---|
| (A–B–) | | 220 | | | |
| (aaB–) | | 210 | | | |
| (A–bb) | | 231 | | | |
| (aabb) | | 239 | | | |
| | | | | $\chi^2$ = Sum | |

3. The $\chi^2$ value means nothing on its own—it is used to find the probability that, assuming the hypothesis is true, the observed data set could have resulted from random fluctuations. A low probability suggests that the observed data are not consistent with the hypothesis, and thus the hypothesis should be rejected. A standard cutoff point used by biologists is a probability of 0.05 (5%). If the probability corresponding to the $\chi^2$ value is 0.05 or less, the differences between observed and expected values are considered statistically significant and the hypothesis (that the genes are unlinked) should be rejected. If the probability is above 0.05, the results are not statistically significant; the observed data are consistent with the hypothesis. To find the probability, locate your $\chi^2$ value in the $\chi^2$ Distribution Table in Appendix F. The "degrees of freedom" (df) of your data set is the number of categories (here, 4 phenotypes) minus 1, so df = 3. (a) Determine which values on the df = 3 line of the table your calculated $\chi^2$ value lies between. (b) The column headings for these values show the probability range for your $\chi^2$ number. Based on whether there are nonsignificant ($p > 0.05$) or significant ($p \leq 0.05$) differences between the observed and expected values, are the data consistent with the hypothesis that the two genes are unlinked and assorting independently, or is there enough evidence to reject this hypothesis?

(MB) A version of this Scientific Skills Exercise can be assigned in MasteringBiology.

points along a chromosome. Based on these assumptions, Sturtevant predicted that *the farther apart two genes are, the higher the probability that a crossover will occur between them and therefore the higher the recombination frequency.*

His reasoning was simple: The greater the distance between two genes, the more points there are between them where crossing over can occur. Using recombination data from various fruit fly crosses, Sturtevant proceeded to assign relative

## Constructing a Linkage Map

**Application** A linkage map shows the relative locations of genes along a chromosome.

**Technique** A linkage map is based on the assumption that the probability of a crossover between two genetic loci is proportional to the distance separating the loci. The recombination frequencies used to construct a linkage map for a particular chromosome are obtained from experimental crosses, such as the cross depicted in Figures 12.9 and 12.10. The distances between genes are expressed as map units, with one map unit equivalent to a 1% recombination frequency. Genes are arranged on the chromosome in the order that best fits the data.

**Results** In this example, the observed recombination frequencies between three *Drosophila* gene pairs (*b* and *cn*, 9%; *cn* and *vg*, 9.5%; and *b* and *vg*, 17%) best fit a linear order in which *cn* is positioned about halfway between the other two genes:

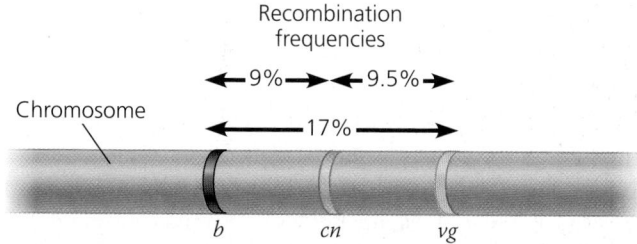

Recombination frequencies

←—9%—→ ←—9.5%—→

Chromosome

←——— 17% ———→

*b*      *cn*      *vg*

The *b-vg* recombination frequency (17%) is slightly less than the sum of the *b-cn* and *cn-vg* frequencies (9 + 9.5 = 18.5%) because of the few times that one crossover occurs between *b* and *cn* and another crossover occurs between *cn* and *vg*. The second crossover would "cancel out" the first, reducing the observed *b-vg* recombination frequency while contributing to the frequency between each of the closer pairs of genes. The value of 18.5% (18.5 map units) is closer to the actual distance between the genes. In practice, a geneticist would add the smaller distances in constructing a map.

---

positions to genes on the same chromosomes—that is, to *map* genes.

A genetic map based on recombination frequencies is called a **linkage map**. **Figure 12.11** shows Sturtevant's linkage map of three genes: the body-color (*b*) and wing-size (*vg*) genes depicted in Figure 12.10 and a third gene, called cinnabar (*cn*). Cinnabar is one of many *Drosophila* genes affecting eye color. Cinnabar eyes, a mutant phenotype, are a brighter red than the wild-type color. The recombination frequency between *cn* and *b* is 9%; that between *cn* and *vg*, 9.5%; and that between *b* and *vg*, 17%. In other words, crossovers between *cn* and *b* and between *cn* and *vg* are about half as frequent as crossovers between *b* and *vg*. Only a map that locates *cn* about midway between *b* and *vg* is consistent with these data, as you can prove to yourself by drawing alternative maps. Sturtevant expressed the distances between genes in **map units**, defining one map unit as equivalent to a 1% recombination frequency.

In practice, the interpretation of recombination data is more complicated than this example suggests. Some genes on a chromosome are so far from each other that a crossover between them is virtually certain. The observed frequency of recombination in crosses involving two such genes can have a maximum value of 50%, a result indistinguishable from that for genes on different chromosomes. In this case, the physical connection between genes on the same chromosome is not reflected in the results of genetic crosses. Despite being on the same chromosome and thus being *physically connected*, the genes are *genetically unlinked*; alleles of such genes assort independently, as if they were on different chromosomes. In fact, at least two of the genes for pea characters that Mendel studied are now known to be on the same chromosome, but the distance between them is so great that linkage is not observed in genetic crosses. Consequently, the two genes behaved as if they were on different chromosomes in Mendel's experiments. Genes located far apart on a chromosome are mapped by adding the recombination frequencies from crosses involving closer pairs of genes lying between the two distant genes.

Using recombination data, Sturtevant and his colleagues were able to map numerous *Drosophila* genes in linear arrays. They found that the genes clustered into four groups of linked genes (*linkage groups*). Light microscopy had revealed four pairs of chromosomes in *Drosophila*, so the linkage map provided additional evidence that genes are located on chromosomes. Each chromosome has a linear array of specific genes, each gene with its own locus **(Figure 12.12)**.

Because a linkage map is based strictly on recombination frequencies, it gives only an approximate picture of a chromosome. The frequency of crossing over is not actually uniform

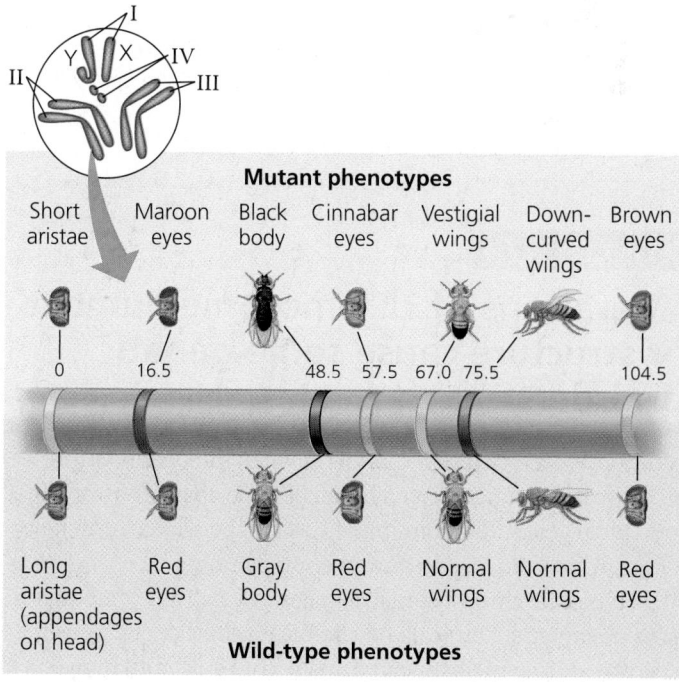

**Mutant phenotypes**

| Short aristae | Maroon eyes | Black body | Cinnabar eyes | Vestigial wings | Down-curved wings | Brown eyes |

0      16.5              48.5    57.5   67.0  75.5              104.5

| Long aristae (appendages on head) | Red eyes | Gray body | Red eyes | Normal wings | Normal wings | Red eyes |

**Wild-type phenotypes**

▲ **Figure 12.12  A partial genetic (linkage) map of a *Drosophila* chromosome.** This simplified map shows just seven of the genes that have been mapped on *Drosophila* chromosome 2. (DNA sequencing has revealed over 9,000 genes on that chromosome.) The number at each gene locus indicates the number of map units between that locus and the locus for arista length (left). Notice that more than one gene can affect a given phenotypic character, such as eye color.

over the length of a chromosome, as Sturtevant assumed, and therefore map units do not correspond to actual physical distances (in nanometers, for instance). A linkage map does portray the order of genes along a chromosome, but it does not accurately portray the precise locations of those genes. Other methods enable geneticists to construct **cytogenetic maps** of chromosomes, which locate genes with respect to chromosomal features, such as stained bands, that can be seen in the microscope. Technical advances over the last two decades have enormously increased the rate and affordability of DNA sequencing. Today, most researchers sequence whole genomes to map the locations of genes of a given species. The entire nucleotide sequence is the ultimate physical map of a chromosome, revealing the physical distances between gene loci in DNA nucleotides (see Concept 18.1). Comparing a linkage map with such a physical map or with a cytogenetic map of the same chromosome, we find that the linear order of genes is identical in all the maps, but the spacing between genes is not.

### CONCEPT CHECK 12.3

1. When two genes are located on the same chromosome, what is the physical basis for the production of recombinant offspring in a testcross between a dihybrid parent and a double-mutant (recessive) parent?

2. For each type of offspring of the testcross in Figure 12.9, explain the relationship between its phenotype and the alleles contributed by the female parent. (It will be useful to draw out the chromosomes of each fly and follow the alleles throughout the cross.)

3. **WHAT IF?** Genes *A*, *B*, and *C* are located on the same chromosome. Testcrosses show that the recombination frequency between *A* and *B* is 28% and that between *A* and *C* is 12%. Can you determine the linear order of these genes? Explain.

**For suggested answers, see Appendix A.**

---

### CONCEPT 12.4

## Alterations of chromosome number or structure cause some genetic disorders

As you have learned so far in this chapter, the phenotype of an organism can be affected by small-scale changes involving individual genes. Random mutations are the source of all new alleles, which can lead to new phenotypic traits.

Large-scale chromosomal changes can also affect an organism's phenotype. Physical and chemical disturbances, as well as errors during meiosis, can damage chromosomes in major ways or alter their number in a cell. Large-scale chromosomal alterations in humans and other mammals often lead to spontaneous abortion (miscarriage) of a fetus, and individuals born with these types of genetic defects commonly exhibit various developmental disorders. Plants appear to tolerate such genetic defects better than animals do.

## Abnormal Chromosome Number

Ideally, the meiotic spindle distributes chromosomes to daughter cells without error. But there is an occasional mishap, called a **nondisjunction**, in which the members of a pair of homologous chromosomes do not move apart properly during meiosis I or sister chromatids fail to separate during meiosis II **(Figure 12.13)**. In nondisjunction, one gamete receives two of the same type of chromosome and another gamete receives no copy. The other chromosomes are usually distributed normally.

If either of the aberrant gametes unites with a normal one at fertilization, the zygote will also have an abnormal number of a particular chromosome, a condition known as **aneuploidy**. Fertilization involving a gamete that has no copy of a particular chromosome will lead to a missing chromosome in the zygote (so that the cell has $2n - 1$ chromosomes); the aneuploid zygote is said to be **monosomic** for that chromosome. If a chromosome is present in triplicate in the zygote (so that the cell has $2n + 1$ chromosomes), the aneuploid cell is **trisomic** for that chromosome. Mitosis will subsequently transmit the anomaly to all embryonic cells. Monosomy and trisomy are estimated to occur in between 10 and 25% of human conceptions and are the main reason for pregnancy loss. If the organism survives, it usually has a set of traits caused by the abnormal dose of the genes associated with the extra or missing chromosome. Down syndrome is an example of trisomy in humans that will be discussed later. Nondisjunction can also occur during mitosis. If such an error takes place early

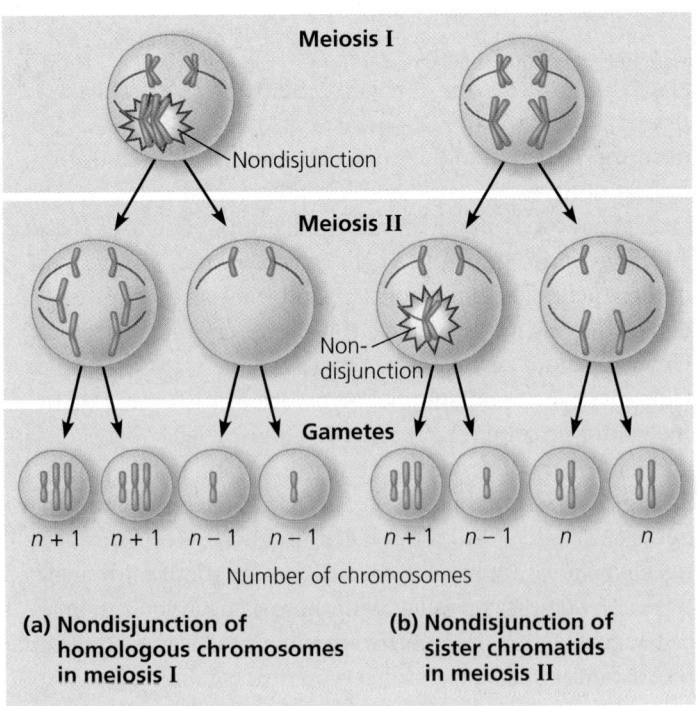

**▲ Figure 12.13 Meiotic nondisjunction.** Gametes with an abnormal chromosome number can arise by nondisjunction in either meiosis I or meiosis II. For simplicity, the figure does not show the spores formed by meiosis in plants. Ultimately, spores form gametes that have the defects shown. (See Figure 10.6.)

in embryonic development, then the aneuploid condition is passed along by mitosis to a large number of cells and is likely to have a substantial effect on the organism.

Some organisms have more than two complete chromosome sets in all somatic cells. The general term for this chromosomal alteration is **polyploidy**; the specific terms *triploidy* (3*n*) and *tetraploidy* (4*n*) indicate three and four chromosomal sets, respectively. One way a triploid cell may arise is by the fertilization of an abnormal diploid egg produced by nondisjunction of all its chromosomes. Tetraploidy could result from the failure of a 2*n* zygote to divide after replicating its chromosomes. Subsequent normal mitotic divisions would then produce a 4*n* embryo.

Polyploidy is fairly common in plants; the spontaneous origin of polyploid individuals plays an important role in the evolution of plants (see Figure 22.9 and the associated text). Many of the plant species we eat are polyploid; for example, bananas are triploid, wheat is hexaploid (6*n*), and strawberries are octoploid (8*n*).

## Alterations of Chromosome Structure

Errors in meiosis or damaging agents such as radiation can cause breakage of a chromosome, which can lead to four types of changes in chromosome structure **(Figure 12.14)**. A **deletion** occurs when a chromosomal fragment is lost. The affected chromosome is then missing certain genes. The "deleted" fragment may become attached as an extra segment to a sister chromatid, producing a **duplication** of a portion of that chromosome. Alternatively, a detached fragment could attach to a nonsister chromatid of a homologous chromosome. In that case, though, the "duplicated" segments might not be identical because the homologs could carry different alleles of certain genes. A chromosomal fragment may also reattach to the original chromosome but in the reverse orientation, producing an **inversion**. A fourth possible result of chromosomal breakage is for the fragment to join a nonhomologous chromosome, a rearrangement called a **translocation**.

Deletions and duplications are especially likely to occur during meiosis. In crossing over, nonsister chromatids sometimes exchange unequal-sized segments of DNA, so that one partner gives up more genes than it receives. The products of such an unequal crossover are one chromosome with a deletion and one chromosome with a duplication.

A diploid embryo that is homozygous for a large deletion (or has a single X chromosome with a large deletion, in a male) is usually missing a number of essential genes, a condition that is typically lethal. Duplications and translocations also tend to be harmful. In reciprocal translocations, in which segments are exchanged between nonhomologous chromosomes, and in inversions, the balance of genes is not abnormal—all genes are present in their normal doses. Nevertheless, translocations and inversions can alter phenotype because a gene's expression can be influenced by its location among neighboring genes, which can have devastating effects.

▼ **Figure 12.14 Alterations of chromosome structure.** Red arrows indicate breakage points. Dark purple highlights the chromosomal parts affected by the rearrangements.

**(a) Deletion**

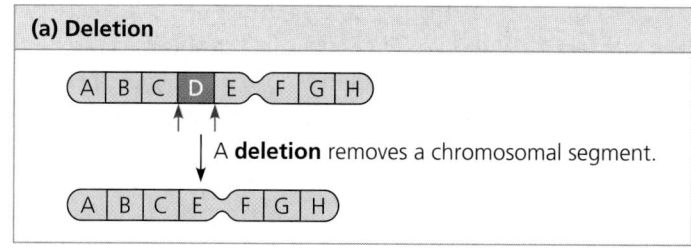

A **deletion** removes a chromosomal segment.

**(b) Duplication**

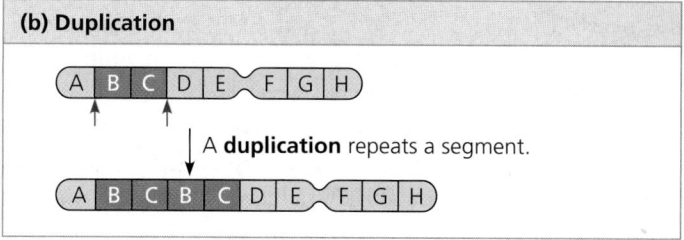

A **duplication** repeats a segment.

**(c) Inversion**

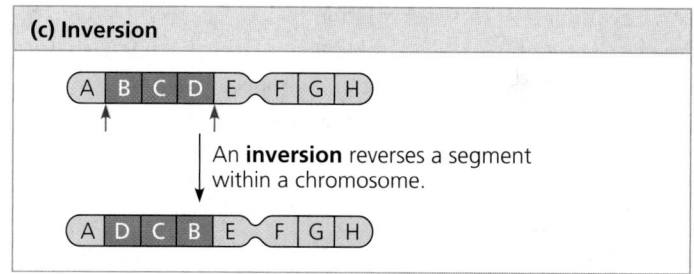

An **inversion** reverses a segment within a chromosome.

**(d) Translocation**

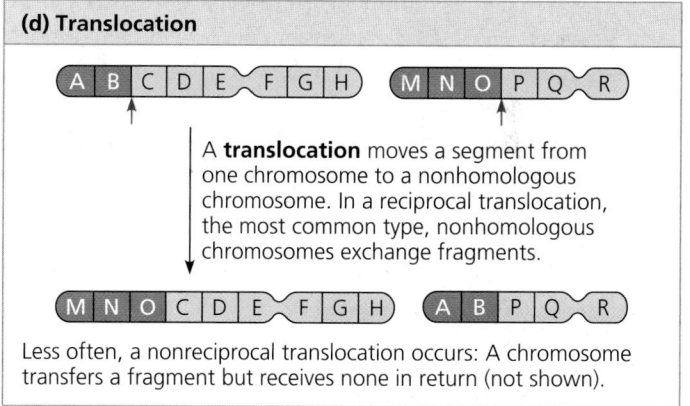

A **translocation** moves a segment from one chromosome to a nonhomologous chromosome. In a reciprocal translocation, the most common type, nonhomologous chromosomes exchange fragments.

Less often, a nonreciprocal translocation occurs: A chromosome transfers a fragment but receives none in return (not shown).

## Human Disorders Due to Chromosomal Alterations

Alterations of chromosome number and structure are associated with a number of serious human disorders. As described earlier, nondisjunction in meiosis results in aneuploidy in gametes and any resulting zygotes. Although the frequency of aneuploid zygotes may be quite high in humans, most of these chromosomal alterations are so disastrous to development that the affected embryos are spontaneously aborted long before birth. However, some types of aneuploidy appear to upset the genetic balance less than others, where individuals with certain aneuploid conditions can survive to birth and beyond. These individuals have a set of traits—a *syndrome*—characteristic of the type of aneuploidy. Genetic disorders caused by aneuploidy can be diagnosed before birth by genetic testing of the fetus.

### Down Syndrome (Trisomy 21)

One aneuploid condition, **Down syndrome**, affects approximately one out of every 830 children born in the United States **(Figure 12.15)**. Down syndrome is usually the result of an extra chromosome 21, so that each body cell has a total of 47 chromosomes. Because the cells are trisomic for chromosome 21, Down syndrome is often called *trisomy 21*. Down syndrome includes characteristic facial features, short stature, correctable heart defects, and developmental delays. Individuals with Down syndrome have an increased chance of developing leukemia and Alzheimer's disease but have a lower rate of high blood pressure, atherosclerosis (hardening of the arteries), stroke, and many types of solid tumors. Although people with Down syndrome, on average, have a life span shorter than normal, most, with proper medical treatment, live to middle age and beyond. Many live independently or at home with their families, are employed, and are valuable contributors to their communities. Almost all males and about half of females with Down syndrome are sexually underdeveloped and sterile.

The frequency of Down syndrome increases with the age of the mother. While the disorder occurs in just 0.04% of children born to women under age 30, the risk climbs to 0.92% for mothers at age 40 and is even higher for older mothers. The correlation of Down syndrome with maternal age has not yet been explained. Most cases result from nondisjunction during meiosis I, and some research points to an age-dependent abnormality in meiosis. Medical experts recommend that prenatal screening for trisomies in the embryo be offered to all pregnant women, due to its low risk and useful results. A law passed in 2008 stipulates that medical practitioners give accurate, up-to-date information about any prenatal or postnatal diagnosis received by parents and that they connect parents with appropriate support services.

### Aneuploidy of Sex Chromosomes

Aneuploid conditions involving sex chromosomes appear to upset the genetic balance less than those involving autosomes. This may be because the Y chromosome carries relatively few genes. Also, extra copies of the X chromosome simply become inactivated as Barr bodies in somatic cells.

An extra X chromosome in a male, producing an XXY genotype, occurs approximately once in every 500 to 1,000 live male births. People with this disorder, called *Klinefelter syndrome*, have male sex organs, but the testes are abnormally small and the man is sterile. Even though the extra X is inactivated, some breast enlargement and other female body characteristics are common. Affected individuals may have subnormal intelligence. About one of every 1,000 males is born with an extra Y chromosome (XYY). These males undergo normal sexual development and do not exhibit any well-defined syndrome.

Females with trisomy X (XXX), which occurs once in approximately 1,000 live female births, are healthy and have no unusual physical features other than being slightly taller than average. Triple-X females are at risk for learning disabilities but are fertile. Monosomy X, called *Turner syndrome*, occurs about once in every 2,500 female births and is the only known viable monosomy in humans. Although these X0 individuals are phenotypically female, they are sterile because their sex organs do not mature. When provided with estrogen replacement therapy, girls with Turner syndrome do develop secondary sex characteristics. Most have normal intelligence.

### Disorders Caused by Structurally Altered Chromosomes

Many deletions in human chromosomes, even in a heterozygous state, cause severe problems. One such syndrome, known as *cri du chat* ("cry of the cat"), results from a specific deletion in chromosome 5. A child born with this deletion is severely intellectually disabled, has a small head with unusual facial features, and has a cry that sounds like the mewing of a distressed cat. Such individuals usually die in infancy or early childhood.

Chromosomal translocations can also occur during mitosis; some have been implicated in certain cancers, including *chronic myelogenous leukemia (CML)*. This disease occurs when a reciprocal translocation happens during mitosis of pre-white blood cells. The exchange of a large portion of chromosome 22 with a small fragment from a tip of chromosome 9 produces a much shortened, easily recognized chromosome 22, called the *Philadelphia chromosome* **(Figure 12.16)**. Such an exchange causes cancer by creating a new "fused" gene that leads to uncontrolled cell cycle progression. (The mechanism of gene activation will be discussed in Concept 16.3.)

▲ **Figure 12.15 Down syndrome.** The karyotype shows trisomy 21, the most common cause of Down syndrome. The child exhibits the facial features characteristic of this disorder.

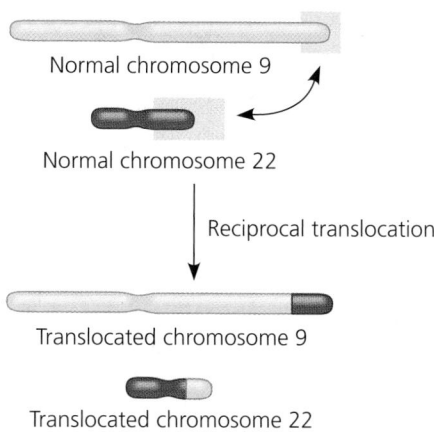

Normal chromosome 9

Normal chromosome 22

Reciprocal translocation

Translocated chromosome 9

Translocated chromosome 22
(Philadelphia chromosome)

▲ **Figure 12.16 Translocation associated with chronic myelogenous leukemia (CML).** The cancerous cells in nearly all CML patients contain an abnormally short chromosome 22, the so-called Philadelphia chromosome, and an abnormally long chromosome 9. These altered chromosomes result from the reciprocal translocation shown here, which presumably occurred in a single white blood cell precursor undergoing mitosis and was then passed along to all descendant cells.

## CONCEPT CHECK 12.4

1. About 5% of individuals with Down syndrome have a chromosomal translocation in which a third copy of chromosome 21 is attached to chromosome 14. If this translocation occurred in a parent's gonad, how could it lead to Down syndrome in a child?
2. **WHAT IF?** The ABO blood type locus has been mapped on chromosome 9. A father who has type AB blood and a mother who has type O blood have a child with trisomy 9 and type A blood. Using this information, can you tell in which parent the nondisjunction occurred? Explain your answer.
3. **MAKE CONNECTIONS** The gene that is activated on the Philadelphia chromosome codes for an intracellular kinase. Review the discussion of cell cycle control and cancer in Concept 9.3, and explain how the activation of this gene could contribute to the development of cancer.
4. Women born with an extra X chromosome (XXX) are generally healthy and indistinguishable in appearance from XX women. What is a likely explanation for this finding? How could you test this explanation?

For suggested answers, see Appendix A.

---

Go to **MasteringBiology**® for Assignments, the eText, and the Study Area with Animations, Activities, Vocab Self-Quiz, and Practice Tests.

# 12 Chapter Review

## SUMMARY OF KEY CONCEPTS

**VOCAB SELF-QUIZ**

goo.gl/gbai8v

### CONCEPT 12.1

**Morgan showed that Mendelian inheritance has its physical basis in the behavior of chromosomes:** *scientific inquiry* **(pp. 238–239)**

- Morgan's work with an eye-color gene in *Drosophila* led to the **chromosome theory of inheritance**, which states that genes are located on chromosomes and that the behavior of chromosomes during meiosis accounts for Mendel's laws.
- Morgan's discovery that transmission of the X chromosome in *Drosophila* correlates with inheritance of an eye-color trait was the first solid evidence indicating that a specific gene is associated with a specific chromosome.

> **?** *What characteristic of the sex chromosomes allowed Morgan to correlate their behavior with that of the alleles of the eye-color gene?*

### CONCEPT 12.2

**Sex-linked genes exhibit unique patterns of inheritance (pp. 239–242)**

- Sex is often chromosomally based. Humans and other mammals have an X-Y system in which sex is determined by whether a Y chromosome is present.
- The sex chromosomes carry **sex-linked genes**, virtually all of which are on the X chromosome (X-linked). Any male who inherits a recessive X-linked allele (from his mother) will express the trait, such as color blindness.
- In mammalian females, one of the two X chromosomes in each cell is randomly inactivated during early embryonic development, becoming highly condensed into a **Barr body**.

> **?** *Why are males affected by X-linked disorders much more often than females?*

### CONCEPT 12.3

**Linked genes tend to be inherited together because they are located near each other on the same chromosome (pp. 242–248)**

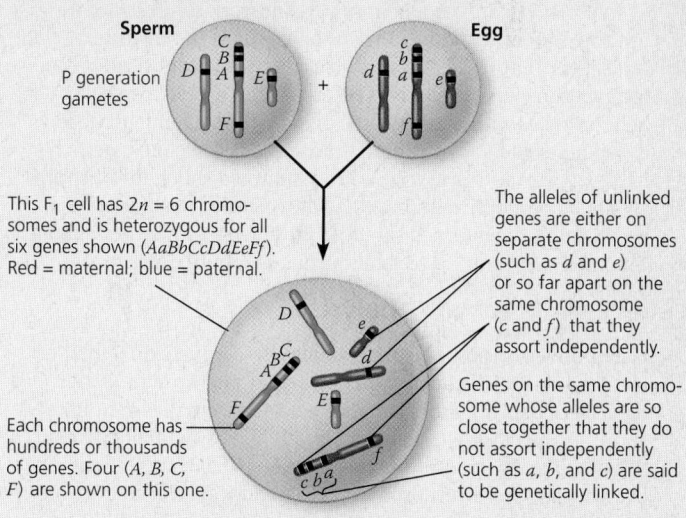

Sperm

Egg

P generation gametes

This F₁ cell has $2n = 6$ chromosomes and is heterozygous for all six genes shown (*AaBbCcDdEeFf*). Red = maternal; blue = paternal.

The alleles of unlinked genes are either on separate chromosomes (such as *d* and *e*) or so far apart on the same chromosome (*c* and *f*) that they assort independently.

Each chromosome has hundreds or thousands of genes. Four (*A, B, C, F*) are shown on this one.

Genes on the same chromosome whose alleles are so close together that they do not assort independently (such as *a, b,* and *c*) are said to be genetically linked.

- An F₁ testcross yields **parental types** with the same combination of traits as those in the P generation parents and **recombinant types** with new combinations of traits. Unlinked genes exhibit a 50% frequency of recombination in the gametes. For genetically **linked genes**, **crossing over** accounts for the observed recombinants, always less than 50%.
- Recombination frequencies observed in genetic crosses allow construction of a **linkage map** (a type of **genetic map**).

> **?** *Why are specific alleles of two distant genes more likely to show recombination than those of two closer genes?*

## Alterations of chromosome number or structure cause some genetic disorders (pp. 248–251)

- **Aneuploidy**, an abnormal chromosome number, results from **nondisjunction** during meiosis. When a normal gamete unites with one containing two copies or no copies of a particular chromosome, the resulting zygote and its descendant cells either have one extra copy of that chromosome (**trisomy**, $2n + 1$) or are missing a copy (**monosomy**, $2n - 1$). **Polyploidy** (extra sets of chromosomes) can result from complete nondisjunction.
- Chromosome breakage can result in alterations of chromosome structure: **deletions**, **duplications**, **inversions**, and **translocations**.

**?** *Why are inversions and reciprocal translocations less likely to be lethal than are aneuploidy, duplications, and deletions?*

# TEST YOUR UNDERSTANDING

**PRACTICE TEST**

goo.gl/CRZjvS

## Level 1: Knowledge/Comprehension

1. A man with hemophilia (a recessive, sex-linked condition) has a daughter without the condition, who marries a man who does not have hemophilia. What is the probability that their daughter will have the condition? Their son? If they have four sons, that all will be affected?

2. Pseudohypertrophic muscular dystrophy is an inherited disorder that causes gradual deterioration of the muscles. It is seen almost exclusively in boys born to apparently unaffected parents and usually results in death in the early teens. Is this disorder caused by a dominant or a recessive allele? Is its inheritance sex-linked or autosomal? How do you know? Explain why this disorder is almost never seen in girls.

3. A space probe discovers a planet inhabited by creatures that reproduce with the same hereditary patterns seen in humans. Three phenotypic characters are height ($T$ = tall, $t$ = dwarf), head appendages ($A$ = antennae, $a$ = no antennae), and nose morphology ($S$ = upturned snout, $s$ = downturned snout). Since the creatures are not "intelligent," Earth scientists are able to do some controlled breeding experiments using various heterozygotes in testcrosses. For tall heterozygotes with antennae, the offspring are tall-antennae, 46; dwarf-antennae, 7; dwarf-no antennae, 42; tall-no antennae, 5. For heterozygotes with antennae and an upturned snout, the offspring are antennae-upturned snout, 47; antennae-downturned snout, 2; no antennae-downturned snout, 48; no antennae-upturned snout, 3. Calculate the recombination frequencies for both experiments.

## Level 2: Application/Analysis

4. Using the information from problem 3, scientists do a further testcross using a heterozygote for height and nose morphology. The offspring are tall-upturned snout, 40; dwarf-upturned snout, 9; dwarf-downturned snout, 42; tall-downturned snout, 9. Calculate the recombination frequency from these data; then use your answer from problem 3 to determine the correct order of the three linked genes.

5. A man with red-green color blindness (a recessive, sex-linked condition) marries a woman with normal vision whose father was color-blind. What is the probability that they will have a color-blind daughter? That their first son will be color-blind? (Note the different wording in the two questions.)

6. You design *Drosophila* crosses to provide recombination data for gene $a$, which is located on the chromosome shown in Figure 12.12. Gene $a$ has recombination frequencies of 14% with the vestigial-wing locus and 26% with the brown-eye locus. Approximately where is gene $a$ located along the chromosome?

7. A wild-type fruit fly (heterozygous for gray body color and red eyes) is mated with a black fruit fly with purple eyes. The offspring are wild-type, 721; black-purple, 751; gray-purple, 49; black-red, 45. What is the recombination frequency between these genes for body color and eye color? Using information from Figure 12.9, what fruit flies (genotypes and phenotypes) would you mate to determine the sequence of the body-color, wing-size, and eye-color genes on the chromosome?

8. Assume that genes $A$ and $B$ are 50 map units apart on the same chromosome. An animal heterozygous at both loci is crossed with one that is homozygous recessive at both loci. What percentage of the offspring will show recombinant phenotypes? Without knowing that these genes are on the same chromosome, how would you interpret the results of this cross?

9. Two genes of a flower, one controlling blue ($B$) versus white ($b$) petals and the other controlling round ($R$) versus oval ($r$) stamens, are linked and are 10 map units apart. You cross a homozygous blue-oval plant with a homozygous white-round plant. The resulting $F_1$ progeny are crossed with homozygous white-oval plants, and 1,000 $F_2$ progeny are obtained. How many $F_2$ plants of each of the four phenotypes do you expect?

## Level 3: Synthesis/Evaluation

10. **SCIENTIFIC INQUIRY**

    **DRAW IT** Assume you are mapping genes $A$, $B$, $C$, and $D$ in *Drosophila*. You know that these genes are linked on the same chromosome, and you determine the recombination frequencies between each pair of genes to be as follows: $A$ and $B$, 8%; $A$ and $C$, 28%; $A$ and $D$, 25%; $B$ and $C$, 20%, $B$ and $D$, 33%.
    
    (a) Describe how you determined the recombination frequency for each pair of genes.
    
    (b) Draw a chromosome map based on your data.

11. **FOCUS ON EVOLUTION**

    Crossing over is thought to be evolutionarily advantageous because it continually shuffles genetic alleles into novel combinations. Until recently, it was thought that Y-linked genes might degenerate because they lack homologous genes on the X chromosome with which to recombine. However, when the Y chromosome was sequenced, eight large regions were found to be internally homologous to each other, and quite a few of the 78 genes represent duplicates. Explain how this might be beneficial.

12. **FOCUS ON INFORMATION**

    The continuity of life is based on heritable information in the form of DNA. In a short essay (100–150 words), relate the structure and behavior of chromosomes to inheritance in both asexually and sexually reproducing species.

13. **SYNTHESIZE YOUR KNOWLEDGE**

    Butterflies have an X-Y sex determination system that is different from that of flies or humans. Female butterflies may be either XY or XO, while butterflies with two or more X chromosomes are males. This photograph shows a tiger swallowtail *gynandromorph*, an individual that is half male (left side) and half female (right side). Given that the first division of the zygote divides the embryo into the future right and left halves of the butterfly, propose a hypothesis that explains how nondisjunction during the first mitosis might have produced this unusual-looking butterfly.

*For selected answers, see Appendix A.*

# 13 The Molecular Basis of Inheritance

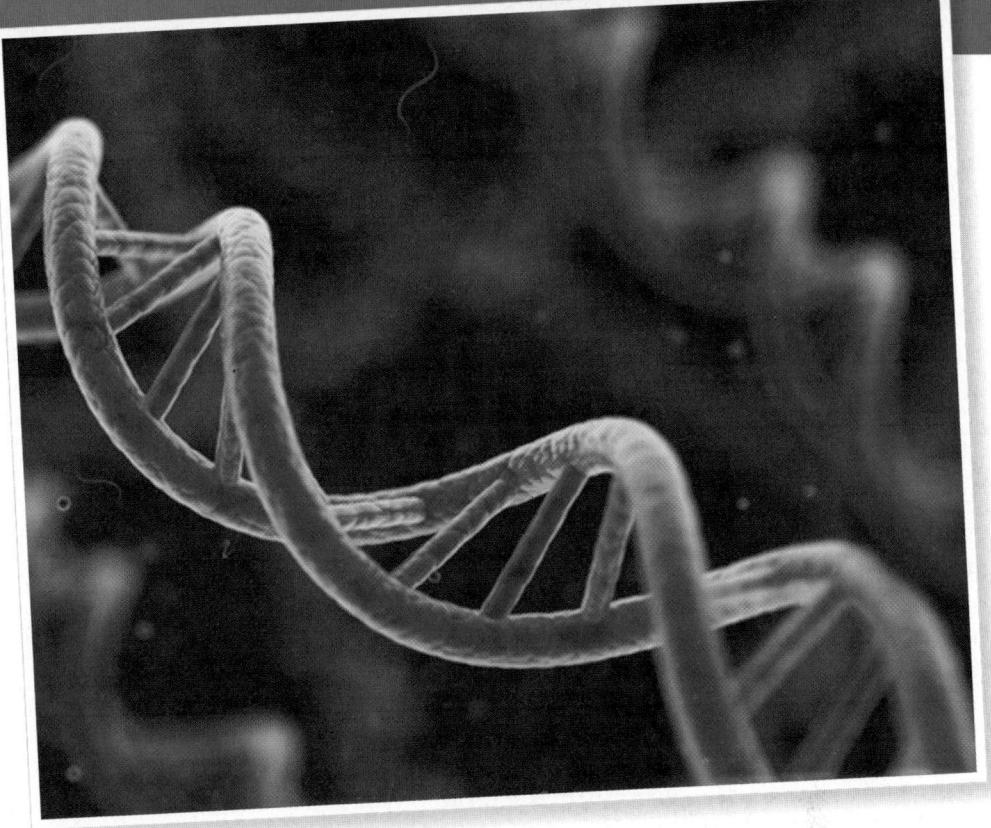

▲ **Figure 13.1** What is the structure of DNA?

## Life's Operating Instructions

The elegant double-helical structure of deoxyribonucleic acid, or DNA, has become an icon of modern biology **(Figure 13.1)**. James Watson and Francis Crick shook the scientific world in April 1953 with their DNA model, which they constructed from sheet metal and wire **(Figure 13.2)**. Gregor Mendel's heritable factors and Thomas Hunt Morgan's genes on chromosomes are, in fact, composed of DNA. Chemically speaking, your genetic endowment is the DNA you inherited from your parents. DNA, the substance of inheritance, is the most celebrated molecule of our time.

Of all nature's molecules, nucleic acids are unique in their ability to direct their own replication from monomers. Indeed, the resemblance of offspring to their parents has its basis in the accurate replication of DNA and its transmission from one generation to the next. Hereditary information in DNA directs the development of your biochemical, anatomical, physiological, and, to some extent, behavioral traits. In this chapter, you'll discover how biologists deduced that DNA is the genetic material and how Watson and Crick worked out its structure. You'll also learn how a molecule of DNA is copied during **DNA replication** and how cells repair their DNA. Next, you'll see how DNA is packaged together with proteins in a chromosome. Finally, you'll explore how an understanding of DNA-related processes has allowed scientists to directly manipulate genes for practical purposes.

◄ **Figure 13.2** James Watson (left) and Francis Crick with their DNA model.

# CONCEPT 13.1

## DNA is the genetic material

Today, even schoolchildren have heard of DNA, and scientists routinely manipulate DNA in the laboratory, often to change the heritable traits of cells in their experiments. Early in the 20th century, however, identifying the molecules of inheritance loomed as a major challenge to biologists.

### The Search for the Genetic Material: *Scientific Inquiry*

Once T. H. Morgan's group showed that genes exist as parts of chromosomes (described in Concept 12.1), the two chemical components of chromosomes—DNA and protein—emerged as the leading candidates for the genetic material. Until the 1940s, the case for proteins seemed stronger: Biochemists had identified proteins as a class of macromolecules with great heterogeneity and specificity of function, essential requirements for the hereditary material. Moreover, little was known about nucleic acids, whose physical and chemical properties seemed far too uniform to account for the multitude of specific inherited traits exhibited by every organism. This view gradually changed as the role of DNA in heredity was worked out in studies of bacteria and the viruses that infect them, systems far simpler than fruit flies or humans. Let's trace the search for the genetic material in some detail as a case study in scientific inquiry.

### Evidence That DNA Can Transform Bacteria

In 1928, a British medical officer named Frederick Griffith was trying to develop a vaccine against pneumonia. He was studying *Streptococcus pneumoniae*, a bacterium that causes pneumonia in mammals. Griffith had two strains (varieties) of the bacterium, one pathogenic (disease-causing) and one nonpathogenic (harmless). He was surprised to find that when he killed the pathogenic bacteria with heat and then mixed the cell remains with living bacteria of the nonpathogenic strain, some of the living cells became pathogenic (**Figure 13.3**). Furthermore, this newly acquired trait of pathogenicity was inherited by all the descendants of the transformed bacteria. Apparently, some chemical component of the dead pathogenic cells caused this heritable change, although the identity of the substance was not known. Griffith called the phenomenon **transformation**, now defined as a change in genotype and phenotype due to the assimilation of external DNA by a cell. Later work by Oswald Avery and others identified the transforming substance as DNA.

Scientists remained skeptical, however, since many still viewed proteins as better candidates for the genetic material. Also, many biologists were not convinced that bacterial genes would be similar in composition and function to those of more complex organisms. But the major reason for the continued doubt was that so little was known about DNA.

## Can a genetic trait be transferred between different bacterial strains?

**Experiment** Frederick Griffith studied two strains of the bacterium *Streptococcus pneumoniae*. The S (smooth) strain can cause pneumonia in mice; it is pathogenic because the cells have an outer capsule that protects them from an animal's immune system. Cells of the R (rough) strain lack a capsule and are nonpathogenic. To test for the trait of pathogenicity, Griffith injected mice with the two strains:

| Living S cells (pathogenic control) | Living R cells (nonpathogenic control) | Heat-killed S cells (nonpathogenic control) | Mixture of heat-killed S cells and living R cells |

**Results**

| Mouse dies | Mouse healthy | Mouse healthy | Mouse dies |

In blood sample, living S cells were found. They could reproduce, yielding more S cells.

**Conclusion** The living R bacteria had been transformed into pathogenic S bacteria by an unknown, heritable substance from the dead S cells that enabled the R cells to make capsules.

**Data from** F. Griffith, The significance of pneumococcal types, *Journal of Hygiene* 27:113–159 (1928).

**WHAT IF?** How did this experiment rule out the possibility that the R cells simply used the dead S cells' capsules to become pathogenic?

### Evidence That Viral DNA Can Program Cells

Additional evidence that DNA was the genetic material came from studies of viruses that infect bacteria. These viruses are called **bacteriophages** (meaning "bacteria-eaters"), or **phages** for short. Viruses are much simpler than cells. A **virus** is little more than DNA (or sometimes RNA) enclosed by a protective coat, which is often simply protein (**Figure 13.4**). To produce

▶ Figure 13.4 **A virus infecting a bacterial cell.** A phage called T2 attaches to a host cell and injects its genetic material through the plasma membrane while the head and tail parts remain on the outer bacterial surface (colorized TEM).

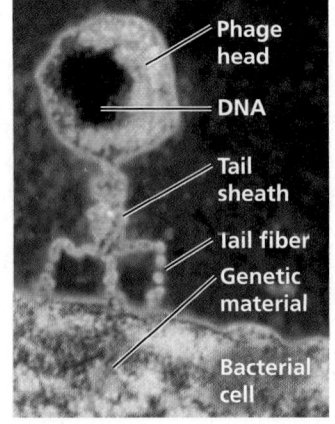

Phage head

DNA

Tail sheath

Tail fiber

Genetic material

Bacterial cell

100 nm

more viruses, a virus must infect a cell and take over the cell's metabolic machinery.

Phages have been widely used as tools by researchers in molecular genetics. In 1952, Alfred Hershey and Martha Chase performed experiments showing that DNA is the genetic material of a phage known as T2. This is one of many phages that infect *Escherichia coli* (*E. coli*), a bacterium that normally lives in the intestines of mammals and is a model organism for molecular biologists. At that time, biologists already knew that T2, like many other phages, was composed almost entirely of DNA and protein. They also knew that the T2 phage could quickly turn an *E. coli* cell into a T2-producing factory that released many copies of new phages when the cell ruptured. Somehow,

T2 could reprogram its host cell to produce viruses. But which viral component—protein or DNA—was responsible?

Hershey and Chase answered this question by devising an experiment showing that only one of the two components of T2 actually enters the *E. coli* cell during infection **(Figure 13.5)**. In their experiment, they used a radioactive isotope of sulfur to tag protein in one batch of T2 and a radioactive isotope of phosphorus to tag DNA in a second batch. Because protein, but not DNA, contains sulfur, radioactive sulfur atoms were incorporated only into the protein of the phage. In a similar way, the atoms of radioactive phosphorus labeled only the DNA, not the protein, because nearly all the phage's phosphorus is in its DNA. In the experiment, separate samples

## ▼ Figure 13.5 Inquiry

### Is protein or DNA the genetic material of phage T2?

**Experiment** Alfred Hershey and Martha Chase used radioactive sulfur and phosphorus to trace the fates of protein and DNA, respectively, of T2 phages that infected bacterial cells. They wanted to see which of these molecules entered the cells and could reprogram them to make more phages.

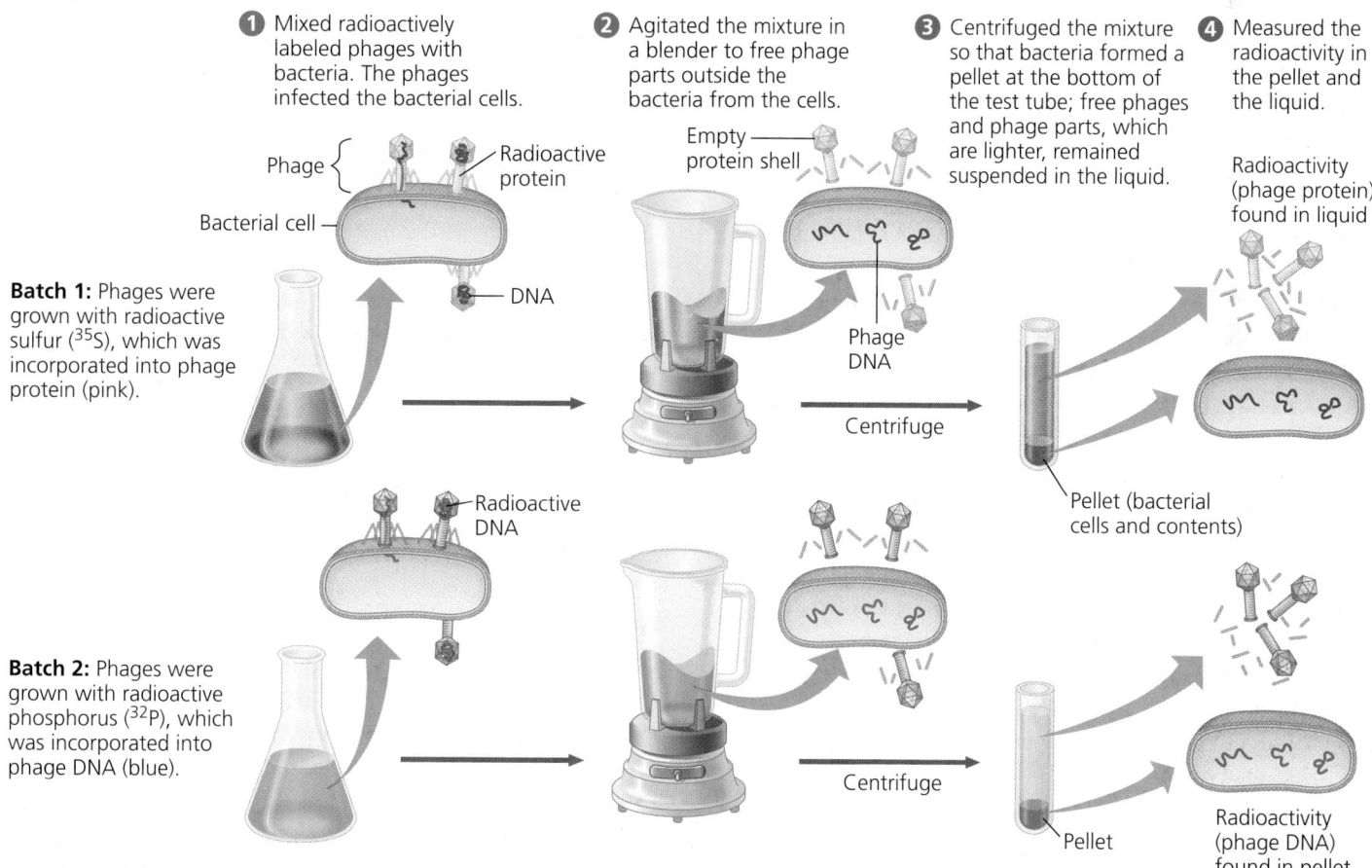

**Results** When proteins were labeled (batch 1), radioactivity remained outside the cells, but when DNA was labeled (batch 2), radioactivity was found inside the cells. Cells containing radioactive phage DNA released new phages with some radioactive phosphorus.

**Conclusion** Phage DNA entered bacterial cells, but phage proteins did not. Hershey and Chase concluded that DNA, not protein, functions as the genetic material of phage T2.

**Data from** A. D. Hershey and M. Chase, Independent functions of viral protein and nucleic acid in growth of bacteriophage, *Journal of General Physiology* 36:39–56 (1952).

**WHAT IF?** How would the results have differed if proteins carried the genetic information?

of nonradioactive *E. coli* cells were infected with the protein-labeled and DNA-labeled batches of T2. The researchers then tested the two samples shortly after the onset of infection to see which type of molecule—protein or DNA—had entered the bacterial cells and would therefore have been capable of reprogramming them.

Hershey and Chase found that the phage DNA entered the host cells, but the phage protein did not. Moreover, when these bacteria were returned to a culture medium and the infection ran its course, the *E. coli* released phages that contained some radioactive phosphorus. This result further showed that the DNA inside the cell played an ongoing role during the infection process.

Hershey and Chase concluded that the DNA injected by the phage must be the molecule carrying the genetic information that makes the cells produce new viral DNA and proteins. The Hershey-Chase experiment was a landmark study because it provided powerful evidence that nucleic acids, rather than proteins, are the hereditary material, at least for certain viruses.

### Additional Evidence That DNA Is the Genetic Material

Further evidence that DNA is the genetic material came from the laboratory of biochemist Erwin Chargaff. It was already known that DNA is a polymer of nucleotides, each consisting of three components: a nitrogenous (nitrogen-containing) base, a pentose sugar called deoxyribose, and a phosphate group **(Figure 13.6)**. The base can be adenine (A), thymine (T), guanine (G), or cytosine (C). Chargaff analyzed the base composition of DNA from a number of different organisms. In 1950, he reported that the base composition of DNA varies from one species to another. For example, 32.8% of sea urchin DNA nucleotides have the base A, whereas only 24.7% of the DNA nucleotides from the bacterium *E. coli* have an A. This evidence of molecular diversity among species, which had been presumed absent from DNA, made DNA a more credible candidate for the genetic material.

Chargaff also noticed a peculiar regularity in the ratios of nucleotide bases. In the DNA of each species he studied, the number of adenines approximately equaled the number of thymines, and the number of guanines approximately equaled the number of cytosines. In sea urchin DNA, for example, the four bases are present in these percentages: A = 32.8% and T = 32.1%; G = 17.7% and C = 17.3%. (The percentages are not exactly the same because of limitations in Chargaff's techniques.)

These two findings became known as *Chargaff's rules*: (1) DNA base composition varies between species, and (2) for each species, the percentages of A and T bases are roughly equal, as are those of G and C bases. In the **Scientific Skills Exercise**, you can use Chargaff's rules to predict unknown percentages of nucleotide bases. The basis for these rules remained unexplained until the discovery of the double helix.

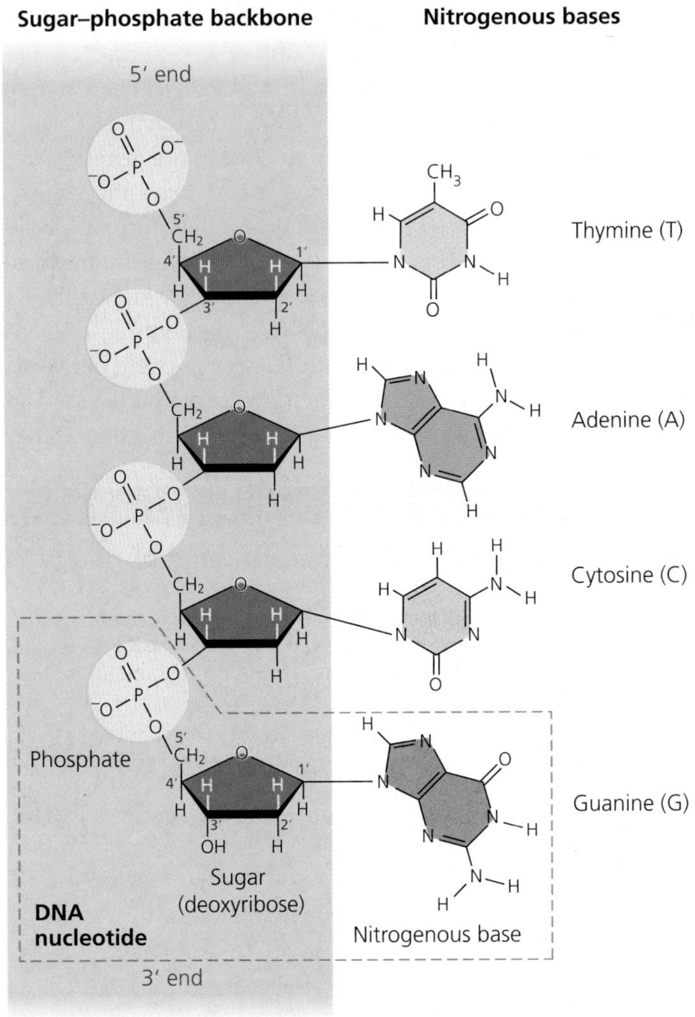

▲ **Figure 13.6 The structure of a DNA strand.** Each DNA nucleotide monomer consists of a nitrogenous base (T, A, C, or G), the sugar deoxyribose (blue), and a phosphate group (yellow). The phosphate group of one nucleotide is attached to the sugar of the next by a covalent bond, forming a "backbone" of alternating phosphates and sugars from which the bases project. The polynucleotide strand has directionality, from the 5' end (with the phosphate group) to the 3' end (with the —OH group of the sugar). 5' and 3' refer to the numbers assigned to the carbons in the sugar ring.

### Building a Structural Model of DNA: *Scientific Inquiry*

Once most biologists were convinced that DNA was the genetic material, the challenge was to determine how the structure of DNA could account for its role in inheritance. By the early 1950s, the arrangement of covalent bonds in a nucleic acid polymer was well established (see Figure 13.6), and researchers focused on discovering the three-dimensional structure of DNA. Among the scientists working on the problem were Linus Pauling, at the California Institute of Technology, and Maurice Wilkins and Rosalind Franklin, at King's College in London. First to come up with the complete answer, however, were two scientists who were relatively unknown at the time—the American James Watson and the Englishman Francis Crick.

## Working with Data in a Table

**Given the Percentage Composition of One Nucleotide in a Genome, Can We Predict the Percentages of the Other Three Nucleotides?** Even before the structure of DNA was elucidated, Erwin Chargaff and his coworkers noticed a pattern in the base composition of nucleotides from different species: The percentage of adenine (A) bases roughly equaled that of thymine (T) bases, and the percentage of cytosine (C) bases roughly equaled that of guanine (G) bases. Further, the percentage of each pair (A/T or C/G) varied from species to species. We now know that the 1:1 A/T and C/G ratios are due to complementary base pairing between A and T and between C and G in the DNA double helix, and interspecies differences are due to the unique sequences of bases along a DNA strand. In this exercise, you will apply Chargaff's rules to predict the composition of bases in a genome.

**How the Experiments Were Done** In Chargaff's experiments, DNA was extracted from the given organism, hydrolyzed to break apart the individual nucleotides, and then analyzed chemically. These experiments provided approximate values for each type of nucleotide. (Today, whole-genome sequencing allows base composition analysis to be done more precisely directly from the sequence data.)

**Data from the Experiments** Tables are useful for organizing sets of data representing a common set of values (here, percentages of A, G, C, and T) for a number of different samples (in this case, from different species). You can apply the patterns that you see in the known data to predict unknown values. In the table, complete base distribution data are given for sea urchin DNA and salmon DNA; you will use Chargaff's rules to fill in the rest of the table with predicted values.

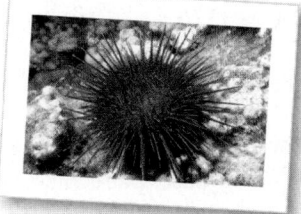

▶ Sea urchin

| Source of DNA | Base Percentage | | | |
|---|---|---|---|---|
| | Adenine | Guanine | Cytosine | Thymine |
| Sea urchin | 32.8 | 17.7 | 17.3 | 32.1 |
| Salmon | 29.7 | 20.8 | 20.4 | 29.1 |
| Wheat | 28.1 | 21.8 | 22.7 | |
| *E. coli* | 24.7 | 26.0 | | |
| Human | 30.4 | | | 30.1 |
| Ox | 29.0 | | | |
| **Average %** | | | | |

**Data from** several papers by Chargaff, for example, E. Chargaff et al., Composition of the desoxypentose nucleic acids of four genera of sea-urchin, *Journal of Biological Chemistry* 195:155–160 (1952).

**INTERPRET THE DATA**

1. Explain how the sea urchin and salmon data demonstrate both of Chargaff's rules.

2. Using Chargaff's rules, fill in the table with your predictions of the missing percentages of bases, starting with the wheat genome and proceeding through *E. coli*, human, and ox. Show how you arrived at your answers.

3. If Chargaff's rule—that the amount of A equals the amount of T and the amount of C equals the amount of G—is valid, then hypothetically we could extrapolate this to the combined DNA of all species on Earth (like one huge Earth genome). To see whether the data in the table support this hypothesis, calculate the average percentage for each base in your completed table by averaging the values in each column. Does Chargaff's equivalence rule still hold true?

(MB) A version of this Scientific Skills Exercise can be assigned in MasteringBiology.

---

The brief but celebrated partnership that solved the puzzle of DNA structure began soon after Watson journeyed to Cambridge University, where Crick was studying protein structure with a technique called X-ray crystallography (see Figure 3.25). While visiting the laboratory of Maurice Wilkins, Watson saw an X-ray diffraction image of DNA produced by Wilkins's accomplished colleague Rosalind Franklin **(Figure 13.7a)**. Images produced by X-ray crystallography are not actually pictures of molecules. The spots and smudges in **Figure 13.7b** were produced by X-rays that were diffracted (deflected) as they passed through aligned fibers of purified DNA. Watson was familiar with the type of X-ray diffraction pattern that helical molecules produce, and an examination of the photo that Wilkins showed him confirmed that DNA was helical in shape. The photo also augmented earlier data obtained by Franklin and others suggesting the width of the helix and the spacing of the nitrogenous bases along it. The pattern in this photo implied that the helix was made up of two strands, contrary to a three-stranded model that Linus Pauling had

**(a) Rosalind Franklin**

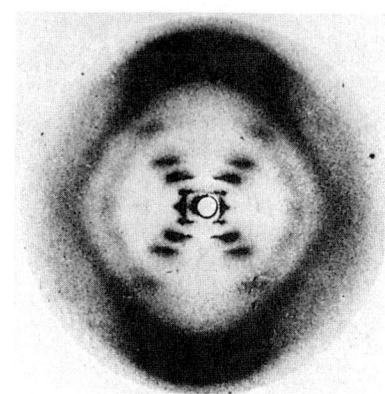

**(b) Franklin's X-ray diffraction photograph of DNA**

▲ **Figure 13.7 Rosalind Franklin and her X-ray diffraction photo of DNA.** Franklin, a very accomplished X-ray crystallographer, conducted critical experiments resulting in the photo that allowed Watson and Crick to deduce the double-helical structure of DNA.

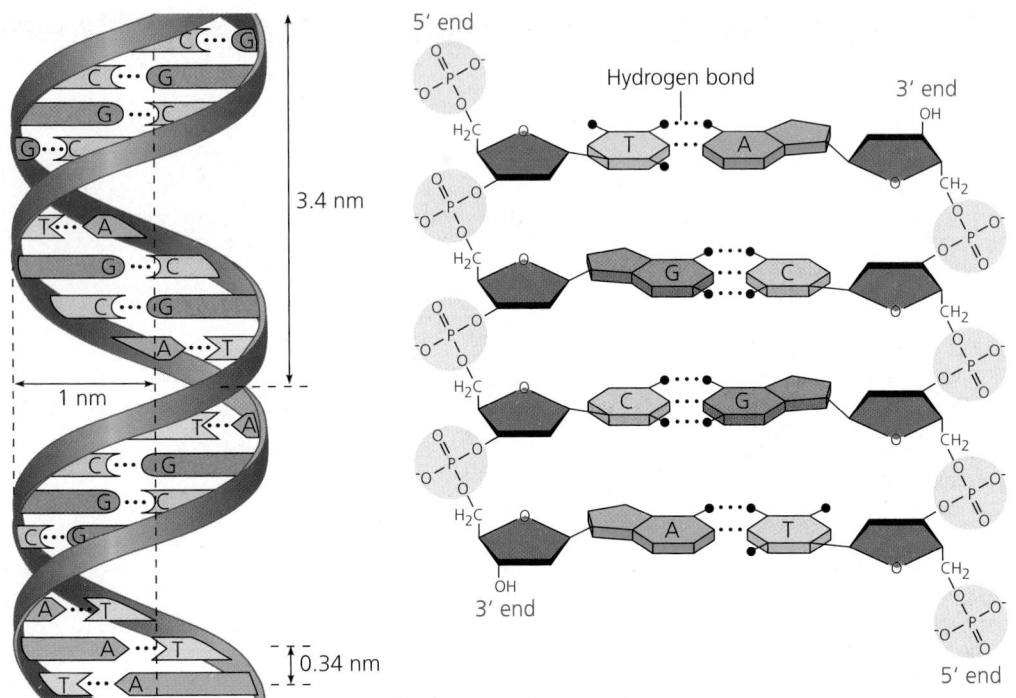

**(a) Key features of DNA structure.** The "ribbons" in this diagram represent the sugar-phosphate backbones of the two DNA strands. The helix is "right-handed," curving up to the right in the front. The two strands are held together by hydrogen bonds (dotted lines) between the nitrogenous bases, which are paired in the interior of the double helix.

**(b) Partial chemical structure.** For clarity, the two DNA strands are shown untwisted in this partial chemical structure. Strong covalent sugar-phosphate bonds link the nucleotides of each strand, while weaker hydrogen bonds between the bases hold one strand to the other. Notice that the strands are antiparallel, meaning that they are oriented in opposite directions, like the lanes of a divided street. From top to bottom, the left strand is oriented in the 5' to 3' direction, and the right strand in the 3' to 5' direction (see also Figure 13.6).

**(c) Space-filling model.** This space-filling model generated by a computer shows that the base pairs are tightly stacked. Van der Waals interactions between the stacked pairs play a major role in holding the molecule together.

▲ **Figure 13.8 The structure of the double helix.**

proposed a short time earlier. The presence of two strands accounts for the now-familiar term **double helix (Figure 13.8)**.

Watson and Crick began building models of a double helix that would conform to the X-ray measurements and what was then known about the chemistry of DNA, including Chargaff's rules. They knew that Franklin had concluded that the sugar-phosphate backbones were on the outside of the DNA molecule. This arrangement was appealing because it put the negatively charged phosphate groups facing the aqueous surroundings, while the relatively hydrophobic nitrogenous bases were hidden in the interior. Watson constructed such a model (see Figure 13.2). In this model, the two sugar-phosphate backbones are **antiparallel**—that is, their subunits run in opposite directions (see Figure 13.8b). You can imagine the overall arrangement as a rope ladder with rigid rungs. The side ropes represent the sugar-phosphate backbones, and the rungs represent pairs of nitrogenous bases. Now imagine twisting the ladder to form a helix. Franklin's X-ray data indicated that the helix makes one full turn every 3.4 nm along its length. With the bases stacked just 0.34 nm apart, there are ten "rungs" of base pairs in each full turn of the helix.

The nitrogenous bases of the double helix are paired in specific combinations: adenine (A) with thymine (T), and guanine (G) with cytosine (C). It was mainly by trial and error that

Watson and Crick arrived at this key feature of DNA. At first, Watson imagined that the bases paired like with like—for example, A with A and C with C. But this model did not fit the X-ray data, which suggested that the double helix had a uniform diameter. Why is this requirement inconsistent with like-with-like pairing of bases? Adenine and guanine are purines, nitrogenous bases with two organic rings, while cytosine and thymine are nitrogenous bases called pyrimidines, which have a single ring. Pairing a purine with a pyrimidine is the only combination that results in a uniform diameter for the double helix **(Figure 13.9)**.

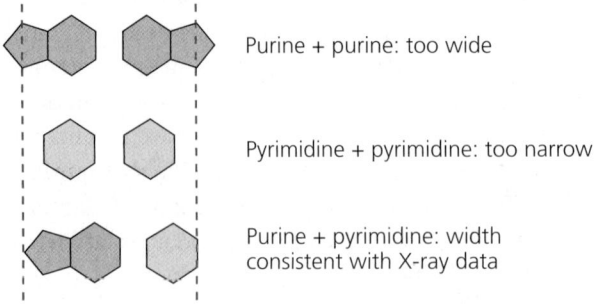

Purine + purine: too wide

Pyrimidine + pyrimidine: too narrow

Purine + pyrimidine: width consistent with X-ray data

▲ **Figure 13.9 Possible base pairings in the DNA double helix.** Purines (A and G) are about twice as wide as pyrimidines (C and T). A purine-purine pair is too wide and a pyrimidine-pyrimidine pair is too narrow to account for the 2-nm diameter of the double helix, while a purine-pyrimidine pair fits the data well.

► **Figure 13.10 Base pairing in DNA.** The pairs of nitrogenous bases in a DNA double helix are held together by hydrogen bonds, shown here as black dotted lines.

**Adenine (A)**　　**Thymine (T)**

**Guanine (G)**　　**Cytosine (C)**

Watson and Crick reasoned that there must be additional specificity of pairing dictated by the structure of the bases. Each base has chemical side groups that can form hydrogen bonds with its appropriate partner: Adenine can form two hydrogen bonds with thymine and only thymine; guanine forms three hydrogen bonds with cytosine and only cytosine. In shorthand, A pairs with T, and G pairs with C **(Figure 13.10)**.

The Watson-Crick model took into account Chargaff's ratios and ultimately explained them. Wherever one strand of a DNA molecule has an A, the partner strand has a T. Similarly, a G in one strand is always paired with a C in the complementary strand. Therefore, in the DNA of any organism, the amount of adenine equals the amount of thymine, and the amount of guanine equals the amount of cytosine. (Modern DNA sequencing techniques have confirmed that the amounts are exactly equal.) Although the base-pairing rules dictate the combinations of nitrogenous bases that form the "rungs" of the double helix, they do not restrict the sequence of nucleotides *along* each DNA strand. The linear sequence of the four bases

can be varied in countless ways, and each gene has a unique order, or base sequence.

In April 1953, Watson and Crick surprised the scientific world with a succinct, one-page paper that reported their molecular model for DNA: the double helix, which has since become the symbol of molecular biology. Watson and Crick, along with Maurice Wilkins, were awarded the Nobel Prize in 1962 for this work. (Sadly, Rosalind Franklin had died at the age of 38 in 1958 and was thus ineligible for the prize.) The beauty of the double helix model was that the structure of DNA suggested the basic mechanism of its replication.

### CONCEPT CHECK 13.1

1. Given a polynucleotide sequence such as GAATTC, explain what further information you would need in order to identify which end is the 5′ end. (See Figure 13.6.)
2. **WHAT IF?** Griffith did not expect transformation to occur in his experiment. What results was he expecting? Explain.

**For suggested answers, see Appendix A.**

## CONCEPT 13.2

# Many proteins work together in DNA replication and repair

The relationship between structure and function is manifest in the double helix. The idea that there is specific pairing of nitrogenous bases in DNA was the flash of inspiration that led Watson and Crick to the double helix. At the same time, they saw the functional significance of the base-pairing rules. They ended their classic paper with this wry statement: "It has not escaped our notice that the specific pairing we have postulated immediately suggests a possible copying mechanism for the genetic material." In this section, you'll learn about the basic principle of DNA replication **(Figure 13.11)**, as well as some important details of the process.

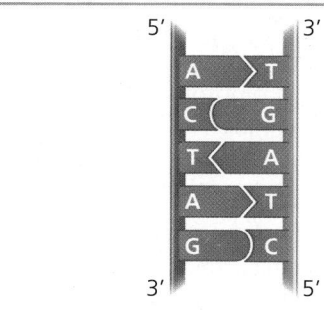

**(a)** The parental molecule has two complementary strands of DNA. Each base is paired by hydrogen bonding with its specific partner, A with T and G with C.

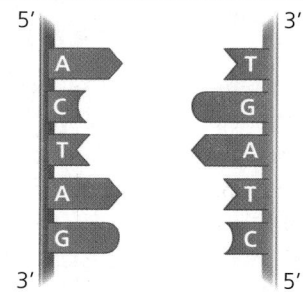

**(b)** First, the two DNA strands are separated. Each parental strand can now serve as a template for a new, complementary strand.

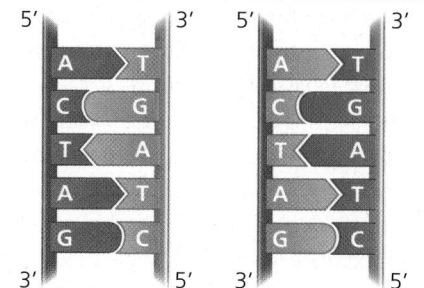

**(c)** Nucleotides complementary to the parental (dark blue) strand are connected to form the sugar-phosphate backbones of the new "daughter" (light blue) strands.

▲ **Figure 13.11 A model for DNA replication: the basic concept.** In this simplified illustration, a short segment of DNA has been untwisted. Simple shapes symbolize the four kinds of bases. Dark blue represents DNA strands present in the parental molecule; light blue represents newly synthesized DNA.

## The Basic Principle: Base Pairing to a Template Strand

In a second paper, Watson and Crick stated their hypothesis for how DNA replicates:

> Now our model for deoxyribonucleic acid is, in effect, a pair of templates, each of which is complementary to the other. We imagine that prior to duplication the hydrogen bonds are broken, and the two chains unwind and separate. Each chain then acts as a template for the formation on to itself of a new companion chain, so that eventually we shall have two pairs of chains, where we only had one before. Moreover, the sequence of the pairs of bases will have been duplicated exactly.*

Figure 13.11 illustrates Watson and Crick's basic idea. To make it easier to follow, we show only a short section of double helix, in untwisted form. Notice that if you cover one of the two DNA strands of Figure 13.11a, you can still determine its linear sequence of nucleotides by referring to the uncovered strand and applying the base-pairing rules. The two strands are complementary; each stores the information necessary to reconstruct the other. When a cell copies a DNA molecule, each strand serves as a template for ordering nucleotides into a new, complementary strand. Nucleotides line up along the template strand according to the base-pairing rules and are linked to form the new strands. Where there was one double-stranded DNA molecule at the beginning of the process, there are soon two, each an exact replica of the "parental" molecule.

This model of DNA replication remained untested for several years following publication of the DNA structure. The requisite experiments were simple in concept but difficult to perform. Watson and Crick's model predicts that when a double helix replicates, each of the two daughter molecules will have one old strand, from the parental molecule, and one newly made strand. This **semiconservative model** can be distinguished from a conservative model of replication, in which the two parental strands somehow come back together after the process (that is, the parental molecule is conserved). In yet a third model, called the dispersive model, all four strands of DNA following replication have a mixture of old and new DNA. These three models are shown in **Figure 13.12**. Although mechanisms for conservative or dispersive DNA replication are not easy to come up with, these models remained possibilities until they could be ruled out. After two years of preliminary work at the California Institute of Technology in the late 1950s, Matthew Meselson and Franklin Stahl devised a clever experiment that distinguished between the three models, described in detail in **Figure 13.13**. Their results supported the semiconservative model of DNA replication, predicted by Watson and Crick, and their experiment is widely recognized among biologists as a classic example of elegant design.

*J. D. Watson and F. H. C. Crick, Genetical implications of the structure of deoxyribonucleic acid, *Nature* 171:964–967 (1953).

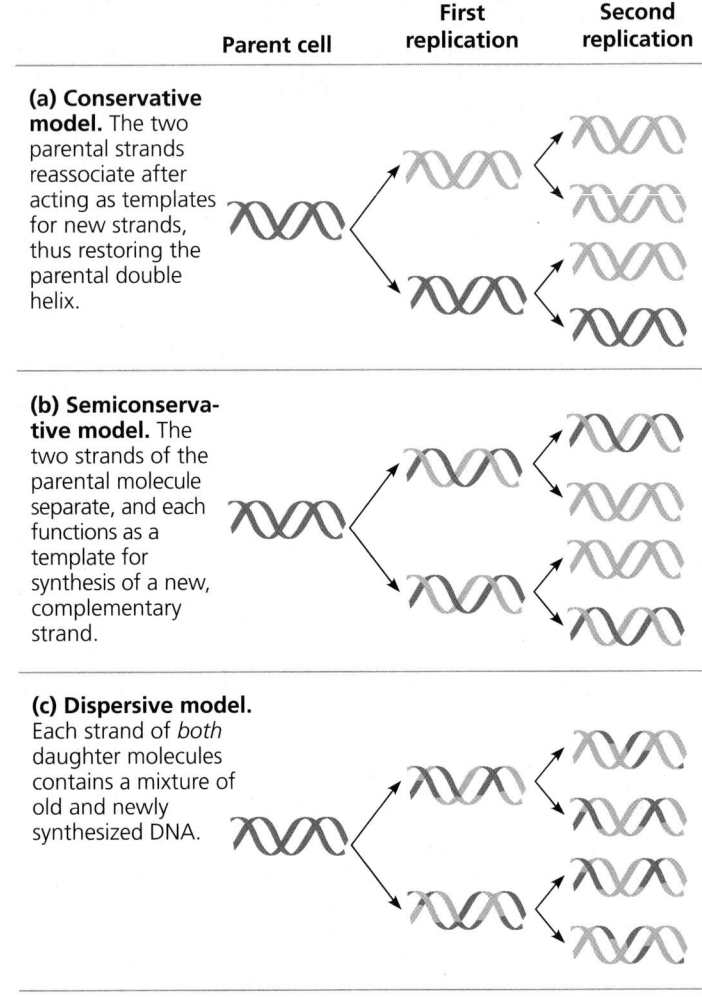

|  | Parent cell | First replication | Second replication |
|---|---|---|---|

**(a) Conservative model.** The two parental strands reassociate after acting as templates for new strands, thus restoring the parental double helix.

**(b) Semiconservative model.** The two strands of the parental molecule separate, and each functions as a template for synthesis of a new, complementary strand.

**(c) Dispersive model.** Each strand of *both* daughter molecules contains a mixture of old and newly synthesized DNA.

▲ **Figure 13.12 Three alternative models of DNA replication.** Each segment of double helix symbolizes the DNA within a cell. Beginning with a parent cell, we follow the DNA for two more generations of cells—two rounds of DNA replication. Parental DNA is dark blue; newly made DNA is light blue.

The basic principle of DNA replication is conceptually simple. However, the actual process involves some complicated biochemical gymnastics, as we will now see.

## DNA Replication: *A Closer Look*

The bacterium *E. coli* has a single chromosome of about 4.6 million nucleotide pairs. In a favorable environment, an *E. coli* cell can copy all this DNA and divide to form two genetically identical daughter cells in less than an hour. Each of *your* cells has 46 DNA molecules in its nucleus, one long double-helical molecule per chromosome. In all, that represents about 6 billion nucleotide pairs, or over a thousand times more DNA than is found in most bacterial cells. If we were to print the one-letter symbols for these bases (A, G, C, and T) the size of the type you are now reading, the 6 billion nucleotide pairs of information in a diploid human cell would fill about 1,400 biology textbooks. Yet it takes one of your cells just a few hours to copy all of this DNA. This replication of an enormous amount

## ▼ Figure 13.13 Inquiry

### Does DNA replication follow the conservative, semiconservative, or dispersive model?

**Experiment** Matthew Meselson and Franklin Stahl cultured *E. coli* for several generations in a medium containing nucleotide precursors labeled with a heavy isotope of nitrogen, $^{15}N$. They then transferred the bacteria to a medium with only $^{14}N$, a lighter isotope. They took one sample after the first DNA replication and another after the second replication. They extracted DNA from the bacteria in the samples and then centrifuged each DNA sample to separate DNA of different densities.

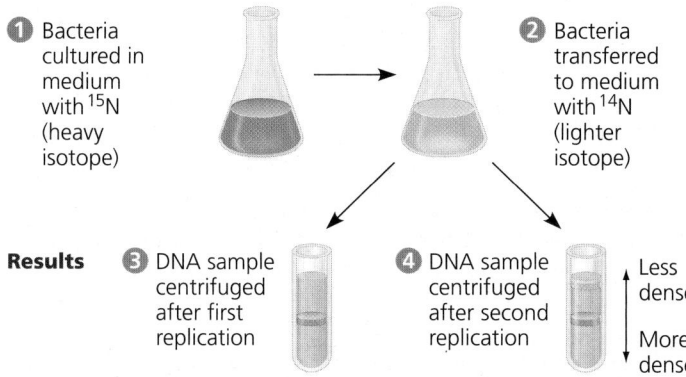

**①** Bacteria cultured in medium with $^{15}N$ (heavy isotope)

**②** Bacteria transferred to medium with $^{14}N$ (lighter isotope)

**Results** **③** DNA sample centrifuged after first replication

**④** DNA sample centrifuged after second replication

Less dense ↑

More dense ↓

**Conclusion** Meselson and Stahl compared their results to those predicted by each of the three models in Figure 13.12, as shown below. The first replication in the $^{14}N$ medium produced a band of hybrid ($^{15}N$ -$^{14}N$) DNA. This result eliminated the conservative model. The second replication produced both light and hybrid DNA, a result that refuted the dispersive model and supported the semiconservative model. They therefore concluded that DNA replication is semiconservative.

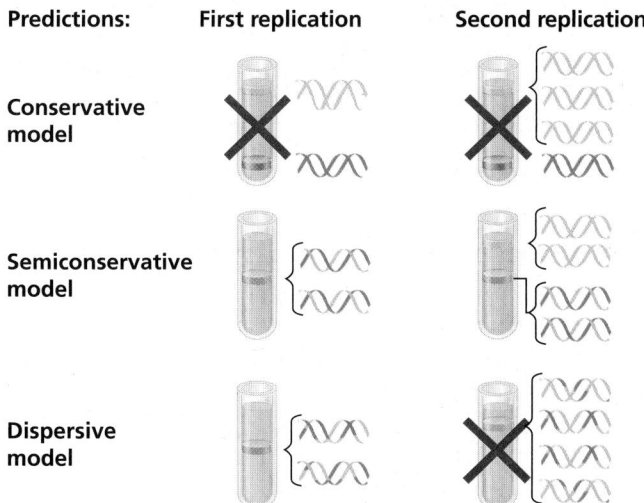

| Predictions: | First replication | Second replication |
|---|---|---|
| **Conservative model** | | |
| **Semiconservative model** | | |
| **Dispersive model** | | |

**Data from** M. Meselson and F. W. Stahl, The replication of DNA in *Escherichia coli*, *Proceedings of the National Academy of Sciences USA* 44:671–682 (1958).

**Inquiry in Action** Read and analyze the original paper in *Inquiry in Action: Interpreting Scientific Papers*.

**(MB)** A related Experimental Inquiry Tutorial can be assigned in MasteringBiology.

**WHAT IF?** If Meselson and Stahl had first grown the cells in $^{14}N$-containing medium and then moved them into $^{15}N$-containing medium before taking samples, what would have been the result?

---

of genetic information is achieved with very few errors—only about one per 10 billion nucleotides. The copying of DNA is remarkable in its speed and accuracy.

More than a dozen enzymes and other proteins participate in DNA replication. Much more is known about how this "replication machine" works in bacteria (such as *E. coli*) than in eukaryotes, and we will describe the basic steps of the process for *E. coli*, except where otherwise noted. What scientists have learned about eukaryotic DNA replication suggests, however, that most of the process is fundamentally similar for prokaryotes and eukaryotes.

### Getting Started

The replication of chromosomal DNA begins at particular sites called **origins of replication**, short stretches of DNA that have a specific sequence of nucleotides. Proteins that initiate DNA replication recognize this sequence and attach to the DNA, separating the two strands and opening up a replication "bubble." At each end of a bubble is a **replication fork**, a Y-shaped region where the parental strands of DNA are being unwound. Several kinds of proteins participate in the unwinding **(Figure 13.14)**. **Helicases** are enzymes that untwist the double helix at the replication forks, separating the two parental strands and making them available as template strands. After the parental strands separate, **single-strand binding proteins** bind to the unpaired DNA strands, keeping them from re-pairing. The untwisting of the double helix causes tighter twisting and strain ahead of the replication fork. **Topoisomerase** helps relieve this strain by breaking, swiveling, and rejoining DNA strands.

The *E. coli* chromosome, like many other bacterial chromosomes, is circular and has a single origin of replication, forming

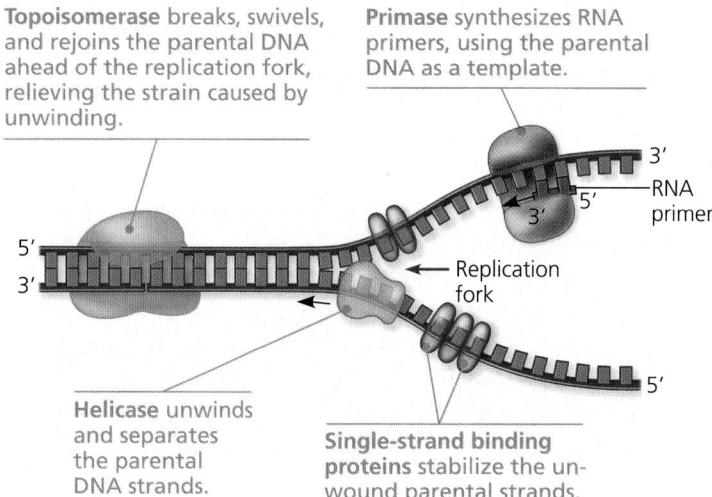

**Topoisomerase** breaks, swivels, and rejoins the parental DNA ahead of the replication fork, relieving the strain caused by unwinding.

**Primase** synthesizes RNA primers, using the parental DNA as a template.

RNA primer

Replication fork

**Helicase** unwinds and separates the parental DNA strands.

**Single-strand binding proteins** stabilize the unwound parental strands.

▲ **Figure 13.14 Some of the proteins involved in the initiation of DNA replication.** The same proteins function at both replication forks in a replication bubble. For simplicity, only the left-hand fork is shown, and the DNA bases are drawn much larger in relation to the proteins than they are in reality.

one replication bubble **(Figure 13.15a)**. Replication of DNA then proceeds in both directions until the entire molecule is copied. In contrast to a bacterial chromosome, a eukaryotic chromosome may have hundreds or even a few thousand replication origins. Multiple replication bubbles form and eventually fuse, thus speeding up the copying of the very long DNA molecules **(Figure 13.15b)**. As in bacteria, eukaryotic DNA replication proceeds in both directions from each origin.

### Synthesizing a New DNA Strand

Within a bubble, the unwound sections of parental DNA strands are available to serve as templates for the synthesis of new complementary DNA strands. However, the enzymes that synthesize DNA cannot *initiate* the synthesis of a polynucleotide; they can only add nucleotides to the end of an already existing chain that is base-paired with the template strand. The initial nucleotide chain that is produced during DNA synthesis is actually a short stretch of RNA, not DNA. This RNA chain

is called a **primer** and is synthesized by the enzyme **primase** (see Figure 13.14). Primase starts a complementary RNA chain with a single RNA nucleotide and adds RNA nucleotides one at a time, using the parental DNA strand as a template. The completed primer, generally 5–10 nucleotides long, is thus base-paired to the template strand. The new DNA strand will start from the 3′ end of the RNA primer.

Enzymes called **DNA polymerases** catalyze the synthesis of new DNA by adding nucleotides at the 3′ end of a preexisting chain. In *E. coli*, there are several DNA polymerases, but two appear to play the major roles in DNA replication: DNA polymerase III and DNA polymerase I. The situation in eukaryotes is more complicated, with at least 11 different DNA polymerases discovered so far, but the general principles are the same.

Most DNA polymerases require a primer and a DNA template strand along which complementary DNA nucleotides are lined up. In *E. coli*, DNA polymerase III (abbreviated DNA pol III) adds a DNA nucleotide to the RNA primer and then

▼ **Figure 13.15 Origins of replication in *E. coli* and eukaryotes.** The red arrows indicate the movement of the replication forks and thus the overall directions of DNA replication within each bubble.

**(a) Origin of replication in an *E. coli* cell**

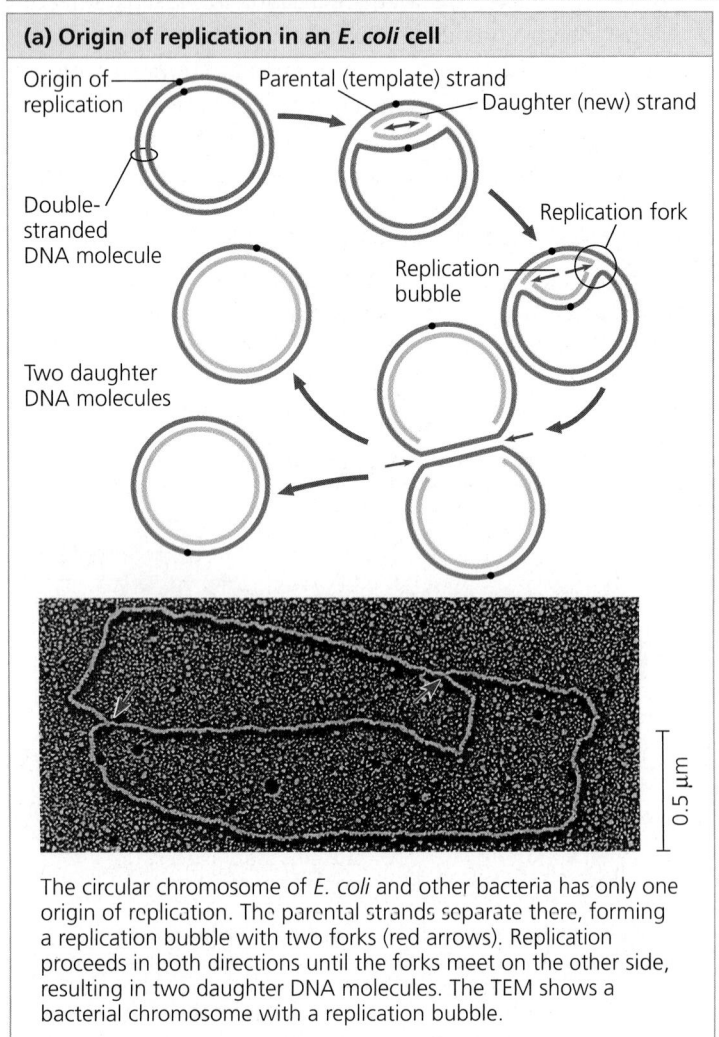

The circular chromosome of *E. coli* and other bacteria has only one origin of replication. The parental strands separate there, forming a replication bubble with two forks (red arrows). Replication proceeds in both directions until the forks meet on the other side, resulting in two daughter DNA molecules. The TEM shows a bacterial chromosome with a replication bubble.

**(b) Origins of replication in a eukaryotic cell**

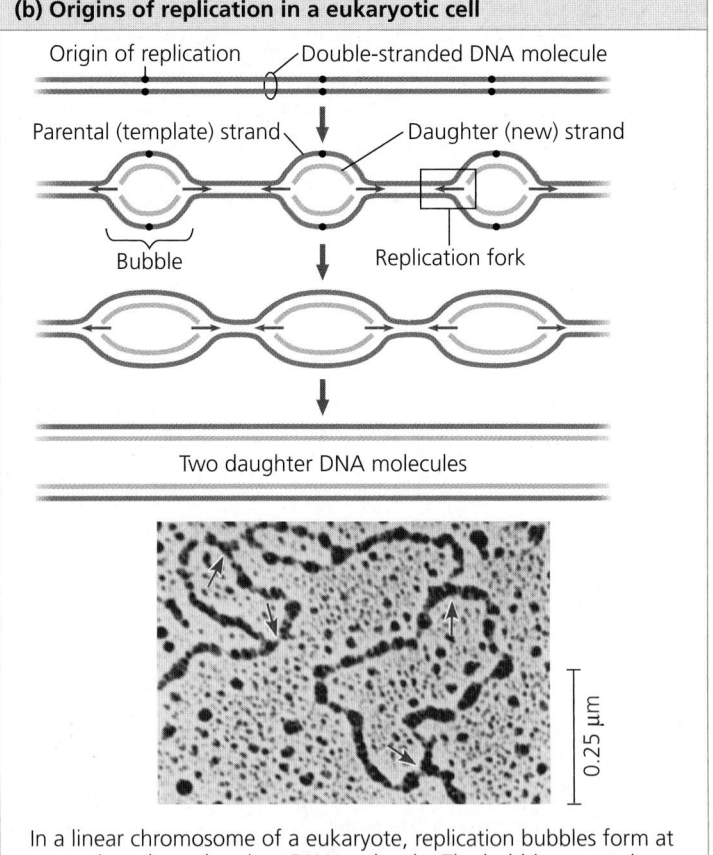

In a linear chromosome of a eukaryote, replication bubbles form at many sites along the giant DNA molecule. The bubbles expand as replication proceeds in both directions (red arrows). Eventually, the bubbles fuse and synthesis of the daughter strands is complete. The TEM shows three replication bubbles along the DNA of a cultured Chinese hamster cell.

**DRAW IT** *In the TEM above, add arrows for the third bubble.*

continues adding DNA nucleotides, complementary to the parental DNA template strand, to the growing end of the new DNA strand. The rate of elongation is about 500 nucleotides per second in bacteria and 50 per second in human cells.

Each nucleotide to be added to a growing DNA strand consists of a sugar attached to a base and to three phosphate groups. You have already encountered such a molecule—ATP (adenosine triphosphate; see Figure 6.8). The only difference between the ATP of energy metabolism and dATP, the adenine nucleotide used to make DNA, is the sugar component, which is deoxyribose in the building block of DNA but ribose in ATP. Like ATP, the nucleotides used for DNA synthesis are chemically reactive, partly because their triphosphate tails have an unstable cluster of negative charge. As each monomer joins the growing end of a DNA strand in a dehydration reaction (see Figure 3.7a) catalyzed by DNA polymerase, two phosphate groups are lost as a molecule of pyrophosphate ($P$—$P_i$). Subsequent hydrolysis of the pyrophosphate to two molecules of inorganic phosphate ($P_i$) is a coupled exergonic reaction that helps drive the polymerization reaction **(Figure 13.16)**.

### Antiparallel Elongation

As we have noted previously, the two ends of a DNA strand are different, giving each strand directionality, like a one-way street (see Figure 13.6). In addition, the two strands of DNA in a double helix are antiparallel, meaning that they are oriented in opposite directions to each other, like the two sides of a divided street (see Figure 13.16). Therefore, the two new strands formed during DNA replication must also be antiparallel to their template strands.

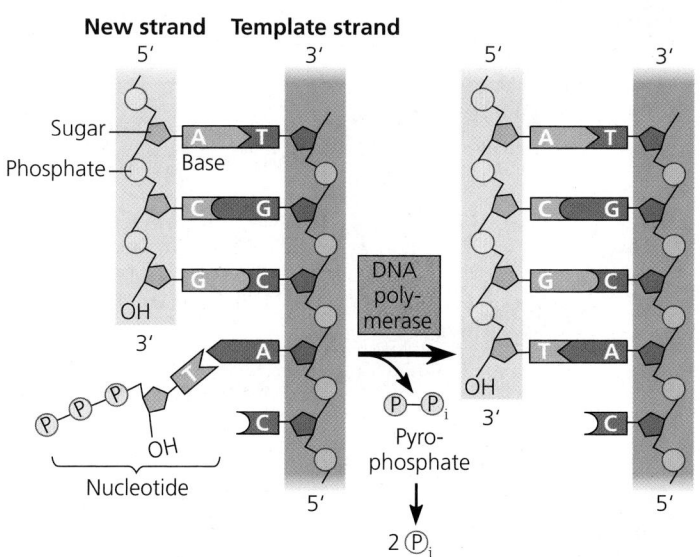

▲ **Figure 13.16 Addition of a nucleotide to a DNA strand.** DNA polymerase catalyzes the addition of a nucleotide to the 3′ end of a growing DNA strand, with the release of two phosphates.

**?** *Use this diagram to explain what we mean when we say that each DNA strand has directionality.*

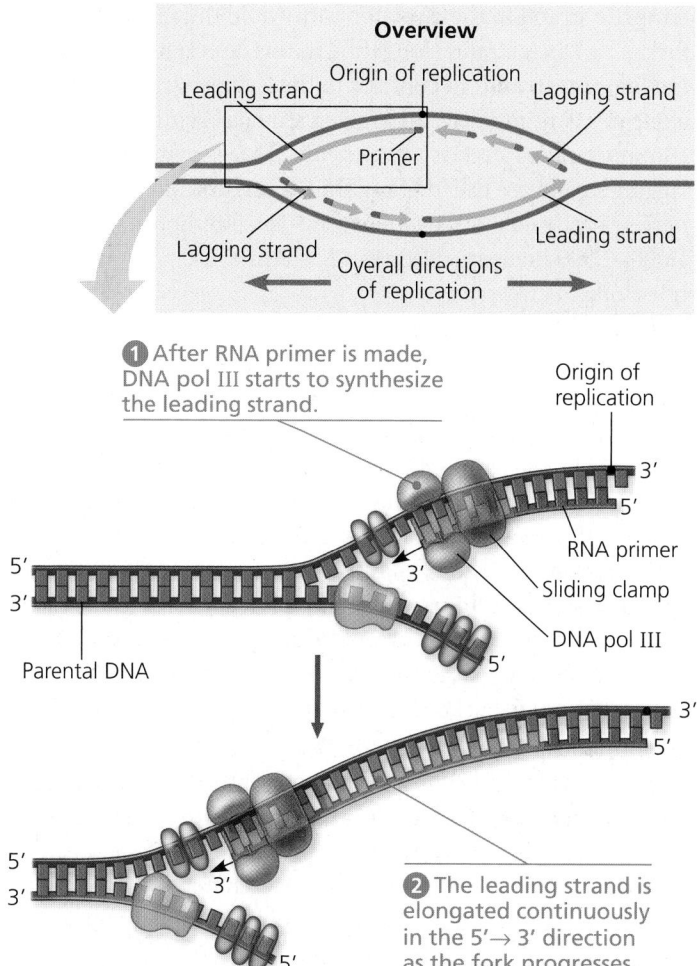

▲ **Figure 13.17 Synthesis of the leading strand during DNA replication.** This diagram focuses on the left replication fork shown in the overview box. DNA polymerase III (DNA pol III), shaped like a cupped hand, is shown closely associated with a protein called the "sliding clamp" that encircles the newly synthesized double helix like a doughnut. The sliding clamp moves DNA pol III along the DNA template strand.

How does the antiparallel arrangement of the double helix affect replication? Because of their structure, DNA polymerases can add nucleotides only to the free 3′ end of a primer or growing DNA strand, never to the 5′ end (see Figure 13.16). Thus, a new DNA strand can elongate only in the 5′ → 3′ direction. With this in mind, let's examine one of the two replication forks in a bubble **(Figure 13.17)**. Along one template strand, DNA polymerase III can synthesize a complementary strand continuously by elongating the new DNA in the mandatory 5′ → 3′ direction. DNA pol III remains in the replication fork on that template strand and continuously adds nucleotides to the new complementary strand as the fork progresses. The DNA strand made by this mechanism is called the **leading strand**. Only one primer is required for DNA pol III to synthesize the entire leading strand.

To elongate the other new strand of DNA in the mandatory 5′ → 3′ direction, DNA pol III must work along the other template strand in the direction *away from* the replication fork. The DNA strand elongating in this direction is called the **lagging strand**. In contrast to the leading strand, which elongates continuously, the lagging strand is synthesized discontinuously, as a series of segments. These segments of the lagging strand are called **Okazaki fragments**, after the Japanese scientist who discovered them. The fragments are about 1,000–2,000 nucleotides long in *E. coli* and 100–200 nucleotides long in eukaryotes.

**Figure 13.18** illustrates the steps in the synthesis of the lagging strand at one fork. Whereas only one primer is required on the leading strand, each Okazaki fragment on the lagging strand must be primed separately (steps **1** and **4**). After DNA pol III forms an Okazaki fragment (steps **2** to **4**), another DNA polymerase, DNA polymerase I (DNA pol I), replaces the RNA nucleotides of the adjacent primer with DNA nucleotides (step **5**). But DNA pol I cannot join the final nucleotide of this replacement DNA segment to the first DNA nucleotide of the adjacent Okazaki fragment. Another enzyme, **DNA ligase**, accomplishes this task, joining the sugar-phosphate backbones of all the Okazaki fragments into a continuous DNA strand (step **6**).

Synthesis of the leading strand and synthesis of the lagging strand occur concurrently and at the same rate. The lagging strand is so named because its synthesis is delayed slightly relative to synthesis of the leading strand; each new fragment of the lagging strand cannot be started until enough template has been exposed at the replication fork.

**Figure 13.19** summarizes DNA replication. Please study it carefully before proceeding.

### The DNA Replication Complex

It is traditional—and convenient—to represent DNA polymerase molecules as locomotives moving along a DNA railroad track, but such a model is inaccurate in two important ways. First, the various proteins that participate in DNA replication actually form a single large complex, a "DNA replication machine." Many protein-protein interactions facilitate the efficiency of this complex. For example, by interacting with other proteins at the fork, primase apparently acts as a molecular brake, slowing progress of the replication fork and coordinating the placement of primers and the rates of replication on the leading and lagging strands. Second, the DNA replication complex may not move along the DNA; rather, the DNA may move through the complex during the replication process. In eukaryotic cells, multiple copies of the complex, perhaps grouped into "factories," may be anchored to the nuclear matrix, a framework of fibers extending through the interior of the nucleus. Experimental evidence supports a model in which two DNA polymerase molecules, one on each template strand, "reel in" the parental DNA and extrude newly made daughter DNA molecules. In this so-called trombone model, the lagging strand is also looped back through the complex **(Figure 13.20)**.

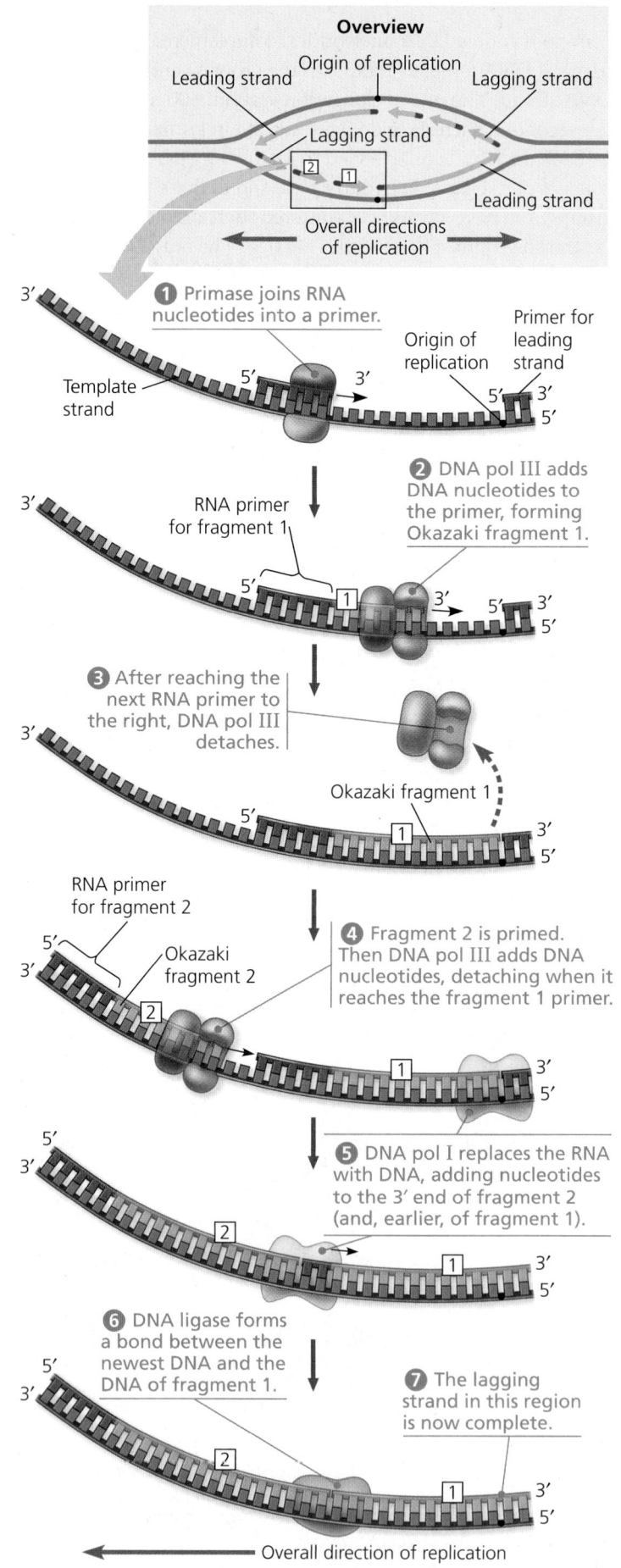

▲ Figure 13.18 **Synthesis of the lagging strand.**

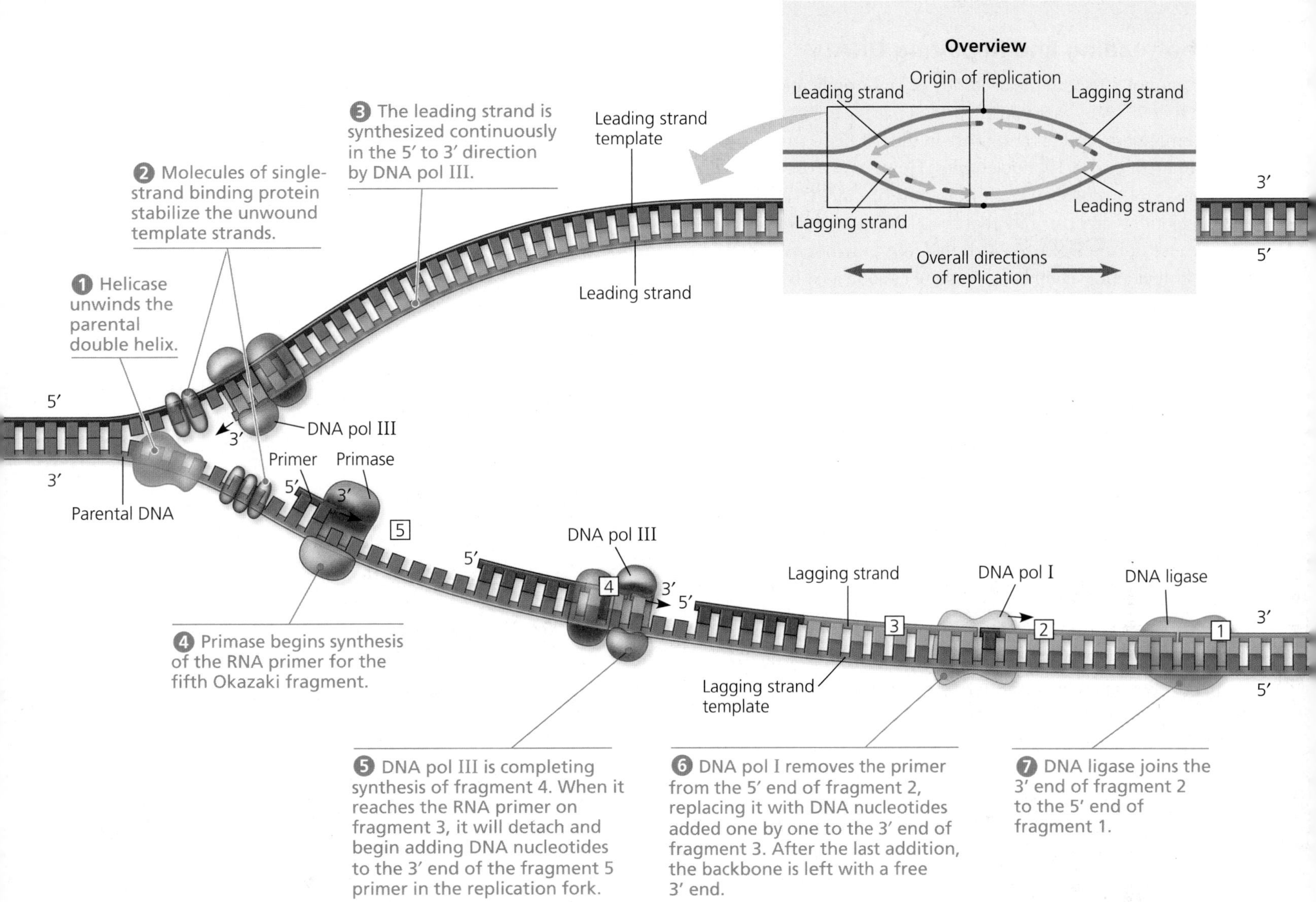

**Overview**

Origin of replication

Leading strand · Lagging strand

Leading strand · Lagging strand

Leading strand

3′

5′

Overall directions of replication

❸ The leading strand is synthesized continuously in the 5′ to 3′ direction by DNA pol III.

❷ Molecules of single-strand binding protein stabilize the unwound template strands.

❶ Helicase unwinds the parental double helix.

Leading strand template

Leading strand

5′

3′

3′

DNA pol III

Primer    Primase

5′    3′

Parental DNA

❹ Primase begins synthesis of the RNA primer for the fifth Okazaki fragment.

5

5′

DNA pol III

4

3′

5′

DNA pol III

Lagging strand

DNA pol I

DNA ligase

3

2

1

3′

5′

Lagging strand template

❺ DNA pol III is completing synthesis of fragment 4. When it reaches the RNA primer on fragment 3, it will detach and begin adding DNA nucleotides to the 3′ end of the fragment 5 primer in the replication fork.

❻ DNA pol I removes the primer from the 5′ end of fragment 2, replacing it with DNA nucleotides added one by one to the 3′ end of fragment 3. After the last addition, the backbone is left with a free 3′ end.

❼ DNA ligase joins the 3′ end of fragment 2 to the 5′ end of fragment 1.

▲ **Figure 13.19 A summary of bacterial DNA replication.** The detailed diagram shows the left-hand replication fork of the replication bubble in the overview (upper right). Viewing each daughter strand in its entirety in the overview, you can see that half of it is made continuously as the leading strand, while the other half (on the other side of the origin) is synthesized in fragments as the lagging strand.

**DRAW IT** *Draw a diagram showing the right-hand fork of the bubble in Figure 13.19. Number the Okazaki fragments and label all 5′ and 3′ ends.*

▶ **Figure 13.20 The "trombone" model of the DNA replication complex.** Two DNA polymerase III molecules work together in a complex, one on each template strand. The lagging strand template DNA loops through the complex, resembling the slide of a trombone.

**DRAW IT** *Draw a line tracing the lagging strand template along the entire stretch of DNA shown here.*

**ANIMATION** Visit the Study Area in **MasteringBiology** for the BioFlix® 3-D Animation on DNA Replication.

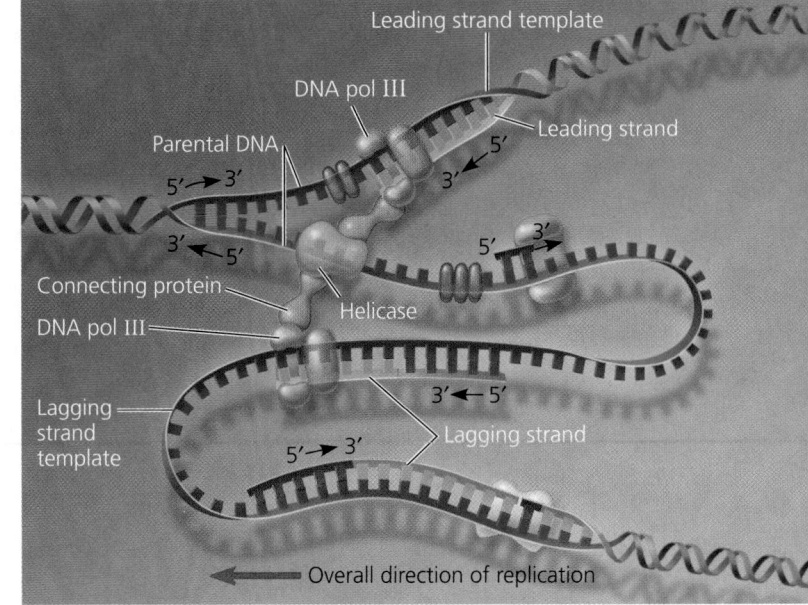

Leading strand template

DNA pol III

Leading strand

Parental DNA

5′→3′

3′

5′

3′

5′

3′←5′

Connecting protein

Helicase

DNA pol III

Lagging strand template

5′→3′

3′←5′

Lagging strand

Overall direction of replication

## Proofreading and Repairing DNA

We cannot attribute the accuracy of DNA replication solely to the specificity of base pairing. Initial pairing errors between incoming nucleotides and those in the template strand occur at a rate of one in $10^5$ nucleotides. However, errors in the completed DNA molecule amount to only one in $10^{10}$ (10 billion) nucleotides, an error rate that is 100,000 times lower. This is because during DNA replication, DNA polymerases proofread each nucleotide against its template as soon as it is added to the growing strand. Upon finding an incorrectly paired nucleotide, the polymerase removes the nucleotide and then resumes synthesis. (This action is similar to fixing a texting error by deleting the wrong letter and then entering the correct letter.)

Mismatched nucleotides sometimes do evade proofreading by a DNA polymerase. In **mismatch repair**, other enzymes remove and replace incorrectly paired nucleotides resulting from replication errors. Researchers highlighted the importance of such repair enzymes when they found that a hereditary defect in one of them is associated with a form of colon cancer. Apparently, this defect allows cancer-causing errors to accumulate in the DNA faster than normal.

Incorrectly paired or altered nucleotides can also arise after replication. In fact, maintenance of the genetic information encoded in DNA requires frequent repair of various kinds of damage to existing DNA. DNA molecules are constantly subjected to potentially harmful chemical and physical agents, such as cigarette smoke and X-rays (as we'll discuss in Concept 14.5). In addition, DNA bases often undergo spontaneous chemical changes under normal cellular conditions. However, these changes in DNA are usually corrected before they become permanent changes—*mutations*—perpetuated through successive replications. Each cell continuously monitors and repairs its genetic material. Because repair of damaged DNA is so important to the survival of an organism, it is no surprise that many different DNA repair enzymes have evolved. Almost 100 are known in *E. coli*, and about 130 have been identified so far in humans.

Most cellular systems for repairing incorrectly paired nucleotides, whether they are due to DNA damage or to replication errors, use a mechanism that takes advantage of the base-paired structure of DNA. In many cases, a segment of the strand containing the damage is cut out (excised) by a DNA-cutting enzyme—a **nuclease**—and the resulting gap is then filled in with nucleotides, using the undamaged strand as a template. The enzymes involved in filling the gap are a DNA polymerase and DNA ligase. One such DNA repair system is called **nucleotide excision repair (Figure 13.21)**.

An important function of DNA repair enzymes in our skin cells is to repair genetic damage caused by the ultraviolet rays of sunlight. One example of this damage is when adjacent thymine bases on a DNA strand become covalently linked. Such *thymine dimers* cause the DNA to buckle (see Figure 13.21), interfering with DNA replication. The importance of repairing

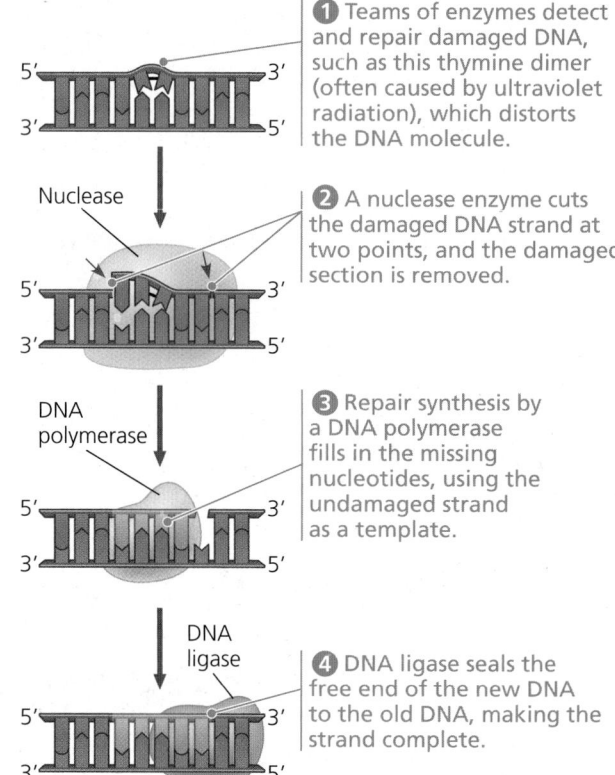

① Teams of enzymes detect and repair damaged DNA, such as this thymine dimer (often caused by ultraviolet radiation), which distorts the DNA molecule.

② A nuclease enzyme cuts the damaged DNA strand at two points, and the damaged section is removed.

③ Repair synthesis by a DNA polymerase fills in the missing nucleotides, using the undamaged strand as a template.

④ DNA ligase seals the free end of the new DNA to the old DNA, making the strand complete.

▲ **Figure 13.21 Nucleotide excision repair of DNA damage.**

this kind of damage is underscored by the disorder xeroderma pigmentosum (XP), which in most cases is caused by an inherited defect in a nucleotide excision repair enzyme. Individuals with XP are hypersensitive to sunlight; mutations in their skin cells caused by ultraviolet light are left uncorrected, often resulting in skin cancer. The effects are extreme: Without sun protection, children who have XP can develop skin cancer by age 10.

## Evolutionary Significance of Altered DNA Nucleotides

**EVOLUTION** Faithful replication of the genome and repair of DNA damage are important for the functioning of the organism and for passing on a complete, accurate genome to the next generation. The error rate after proofreading and repair is extremely low, but rare mistakes do slip through. Once a mismatched nucleotide pair is replicated, the sequence change is permanent in the daughter molecule that has the incorrect nucleotide as well as in any subsequent copies. As you know, a permanent change in the DNA sequence is called a mutation.

Mutations can change the phenotype of an organism (as you'll learn in Concept 14.5). And if they occur in germ cells, which give rise to gametes, mutations can be passed on from generation to generation. The vast majority of such changes are harmful, but a very small percentage can be beneficial. In either case, mutations are the source of the variation on which natural selection operates during evolution and are ultimately responsible for the appearance of new species. (You'll learn more about this process in Unit Three.) The balance between complete fidelity of DNA replication and repair and a low mutation rate has, over long periods of time, allowed the evolution of the rich diversity of species we see on Earth today.

## Replicating the Ends of DNA Molecules

For linear DNA, such as the DNA of eukaryotic chromosomes, the usual replication machinery cannot complete the 5′ ends of daughter DNA strands. (This is a consequence of the fact that a DNA polymerase can add nucleotides only to the 3′ ends of a pre-existing polynucleotide.) As a result, repeated rounds of replication produce shorter and shorter DNA molecules with uneven ends.

What protects the genes near the ends of eukaryotic chromosomes from being eroded away during successive replications? Eukaryotic chromosomal DNA molecules have special nucleotide sequences called **telomeres** at their ends **(Figure 13.22)**. Telomeres do not contain genes; instead, the DNA typically consists of multiple repetitions of one short nucleotide sequence. In each human telomere, for example, the sequence TTAGGG is repeated 100 to 1,000 times. Telomeric DNA acts as a buffer zone that protects the organism's genes.

Telomeres do not prevent the erosion of genes near ends of chromosomes; they merely postpone it. As you would expect, telomeres tend to be shorter in cultured cells that have divided many times and in dividing somatic cells of older individuals. Shortening of telomeres is proposed to play a role in the aging process of some tissues and even of the organism as a whole.

If the chromosomes of germ cells became shorter in every cell cycle, essential genes would eventually be missing from the gametes they produce. However, this does not occur: An enzyme called *telomerase* catalyzes the lengthening of telomeres in eukaryotic germ cells, thus restoring their original length and compensating for the shortening that occurs during DNA replication. Telomerase is not active in most human somatic cells, but shows inappropriate activity in some cancer cells that may remove limits to a cell's normal life span. Thus, telomerase is under study as a target for cancer therapies.

**CONCEPT CHECK 13.2**

1. What role does base pairing play in the replication of DNA?
2. Make a table listing the functions of seven proteins involved in DNA replication in *E. coli*.
3. **MAKE CONNECTIONS** What is the relationship between DNA replication and the S phase of the cell cycle? See Figure 9.6.

For suggested answers, see Appendix A.

# A chromosome consists of a DNA molecule packed together with proteins

Now that you have learned about the structure and replication of DNA, let's take a step back and examine how DNA is packaged into chromosomes, the structures that carry genetic information. The main component of the genome in most bacteria is one double-stranded, circular DNA molecule that is associated with a small amount of protein. Although we refer to this structure as a bacterial chromosome, it is very different from a eukaryotic chromosome, which consists of one linear DNA molecule associated with a large amount of protein. In *E. coli*, the chromosomal DNA consists of about 4.6 million nucleotide pairs, representing about 4,400 genes. This is 100 times more DNA than is found in a typical virus, but only about one-thousandth as much DNA as in a human somatic cell. Still, that is a lot of DNA to be packaged in such a small container.

Stretched out, the DNA of an *E. coli* cell would measure about a millimeter in length, 500 times longer than the cell. Within a bacterium, however, certain proteins cause the chromosome to coil and "supercoil," densely packing it so that it fills only part of the cell. Unlike the nucleus of a eukaryotic cell, this dense region of DNA in a bacterium, called the **nucleoid**, is not surrounded by membrane (see Figure 4.4).

Each eukaryotic chromosome contains a single linear DNA double helix that, in humans, averages about $1.5 \times 10^8$ nucleotide pairs. This is an enormous amount of DNA relative to a chromosome's condensed length. If completely stretched out, such a DNA molecule would be about 4 cm long, thousands of times the diameter of a cell nucleus—and that's not even considering the DNA of the other 45 human chromosomes!

In the cell, eukaryotic DNA is precisely combined with a large amount of protein. Together, this complex of DNA and protein, called **chromatin**, fits into the nucleus through an elaborate, multilevel system of packing.

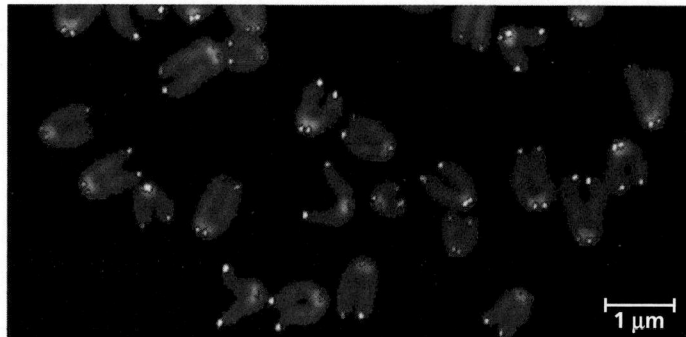

▲ **Figure 13.22 Telomeres.** Eukaryotes have repetitive, noncoding sequences called telomeres at the ends of their DNA. Telomeres are stained orange in these mouse chromosomes (LM).

Chromatin undergoes striking changes in its degree of packing during the course of the cell cycle (see Figure 9.7). In interphase cells stained for light microscopy, the chromatin usually appears as a diffuse mass within the nucleus, suggesting that the chromatin is highly extended. As a cell prepares for mitosis, its chromatin coils and folds up (condenses), eventually forming a characteristic number of short, thick metaphase chromosomes that are distinguishable from each other with the light microscope. Our current view of the successive levels of DNA packing in a chromosome is outlined in **Figure 13.23**.

Though interphase chromatin is generally much less condensed than the chromatin of mitotic chromosomes, it shows several of the same levels of higher-order packing. Some of the chromatin comprising a chromosome seems to be present as a 10-nm fiber, but much is compacted into a 30-nm fiber, which in some regions is further folded into looped domains. Even during interphase, the centromeres of chromosomes, as well as other chromosomal regions in some cells, exist in a highly condensed state similar to that seen in a metaphase chromosome. This type of interphase chromatin, visible as irregular clumps with a light

---

**▼ Figure 13.23**    **Exploring Chromatin Packing in a Eukaryotic Chromosome**

This illustration, accompanied by transmission electron micrographs, depicts a current model for the progressive levels of DNA coiling and folding. The illustration zooms out from a single molecule of DNA to a metaphase chromosome, which is large enough to be seen with a light microscope.

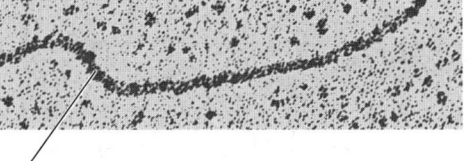

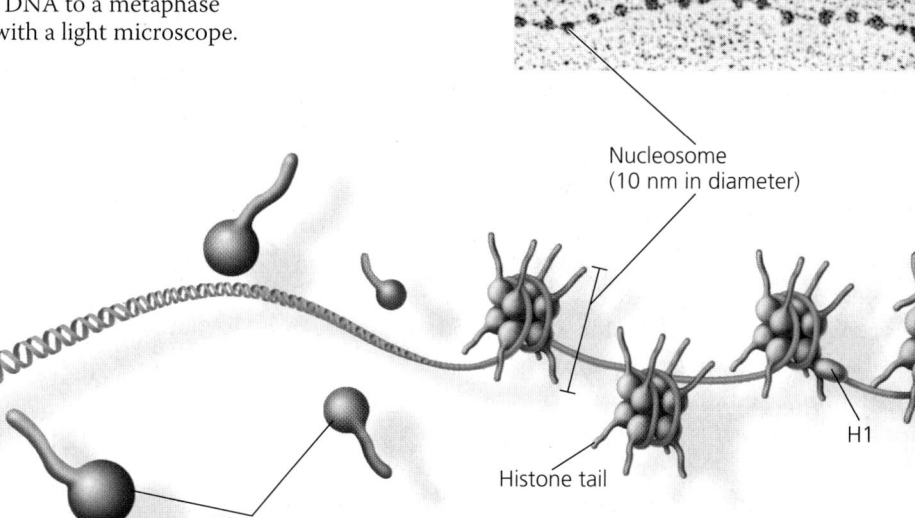

Nucleosome
(10 nm in diameter)

DNA
double helix
(2 nm in diameter)

H1

Histone tail

Histones

## DNA, the double helix

Shown above is a ribbon model of DNA, with each ribbon representing one of the polynucleotide strands. Recall that the phosphate groups along the backbone contribute a negative charge along the outside of each strand. The TEM shows a molecule of naked (protein-free) DNA; the double helix alone is 2 nm across.

## Histones

Proteins called **histones** are responsible for the first level of DNA packing in chromatin. Although each histone is small—containing only about 100 amino acids—the total mass of histone in chromatin roughly equals the mass of DNA. More than a fifth of a histone's amino acids are positively charged (lysine or arginine) and therefore bind tightly to the negatively charged DNA.

Four types of histones are most common in chromatin: H2A, H2B, H3, and H4. The histones are very similar among eukaryotes; for example, all but two of the amino acids in cow H4 are identical to those in pea H4. The apparent conservation of histone genes during evolution probably reflects the important role of histones in organizing DNA within cells.

These four types of histones are critical to the next level of DNA packing. (A fifth type of histone, called H1, is involved in a further stage of packing.)

## Nucleosomes, or "beads on a string" (10-nm fiber)

In electron micrographs, unfolded chromatin is 10 nm in diameter (the *10-nm fiber*). Such chromatin resembles beads on a string (see the TEM). Each "bead" is a **nucleosome**, the basic unit of DNA packing; the "string" between beads is called *linker* DNA.

A nucleosome consists of DNA wound twice around a protein core of eight histones, two each of the main histone types (H2A, H2B, H3, and H4). The amino end (N-terminus) of each histone (the *histone tail*) extends outward from the nucleosome.

In the cell cycle, the histones leave the DNA only briefly during DNA replication. Generally, they do the same during transcription, another process that requires access to the DNA by the cell's molecular machinery. Nucleosomes, and in particular their histone tails, are involved in the regulation of gene expression.

microscope, is called **heterochromatin**, to distinguish it from the less compacted, more dispersed **euchromatin** ("true chromatin"). Because of its compaction, heterochromatic DNA is largely inaccessible to the machinery in the cell responsible for transcribing the genetic information coded in the DNA, a crucial early step in gene expression. In contrast, the looser packing of euchromatin makes its DNA accessible to this machinery, so the genes present in euchromatin can be transcribed.

The chromosome is a dynamic structure that is condensed, loosened, modified, and remodeled as necessary for various cell processes, including mitosis, meiosis, and gene activity. Certain chemical modifications of histones affect the state of chromatin condensation and also have multiple effects on gene activity (as you'll see in Concept 15.2).

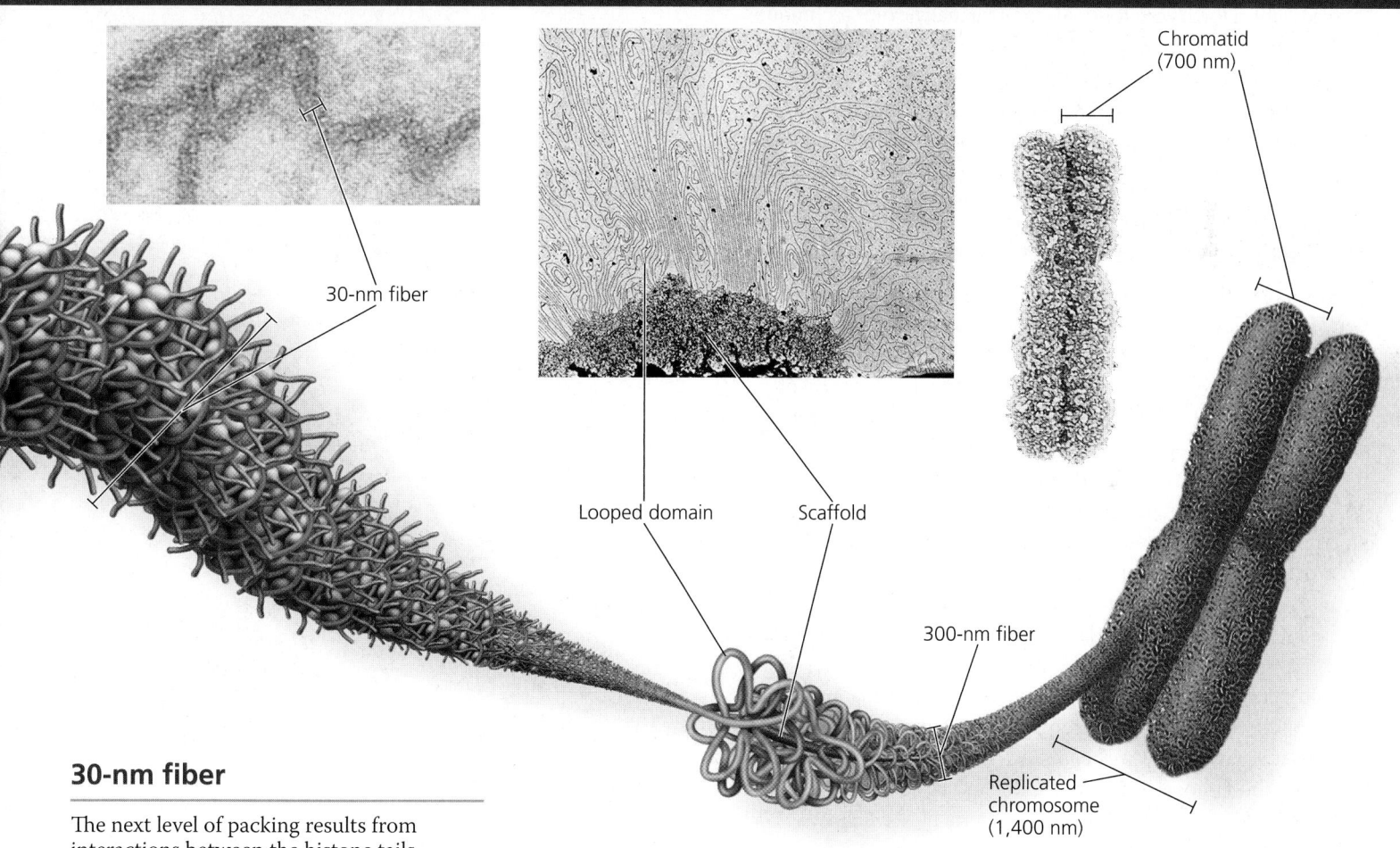

Chromatid (700 nm)

30-nm fiber

Looped domain

Scaffold

300-nm fiber

Replicated chromosome (1,400 nm)

## 30-nm fiber

The next level of packing results from interactions between the histone tails of one nucleosome and the linker DNA and nucleosomes on either side. The fifth histone, H1, is involved at this level. These interactions cause the extended 10-nm fiber to coil or fold, forming a chromatin fiber roughly 30 nm in thickness, the *30-nm fiber*. Although the 30-nm fiber is quite prevalent in the interphase nucleus, the packing arrangement of nucleosomes in this form of chromatin is still a matter of some debate.

## Looped domains (300-nm fiber)

The 30-nm fiber, in turn, forms loops called *looped* domains attached to a chromosome scaffold composed of proteins, thus making up a *300-nm fiber*. The scaffold is rich in one type of topoisomerase, and H1 molecules also appear to be present.

## Metaphase chromosome

In a mitotic chromosome, the looped domains themselves coil and fold in a manner not yet fully understood, further compacting all the chromatin to produce the characteristic metaphase chromosome (also shown in the micrograph above). The width of one chromatid is 700 nm. Particular genes always end up located at the same places in metaphase chromosomes, indicating that the packing steps are highly specific and precise.

# Understanding DNA structure and replication makes genetic engineering possible

The discovery of the structure of DNA marked a milestone in biology and changed the course of biological research. Most notable was the realization that the two strands of a DNA molecule are complementary to each other. This fundamental structural property of DNA is the basis for **nucleic acid hybridization**, the base pairing of one strand of a nucleic acid to a complementary sequence on another strand. Nucleic acid hybridization forms the foundation of virtually every technique used in **genetic engineering**, the direct manipulation of genes for practical purposes. Genetic engineering has launched a revolution in fields as varied as agriculture, criminal law, and medical and basic biological research. In this section, we'll describe several of the most important techniques and their uses.

## DNA Cloning: Making Multiple Copies of a Gene or Other DNA Segment

A molecular biologist studying a particular gene faces a challenge. Naturally occurring DNA molecules are very long, and a single molecule usually carries many genes. Moreover, in many eukaryotic genomes, genes occupy only a small proportion of the chromosomal DNA, the rest being noncoding nucleotide sequences. A single human gene, for example, might constitute only 1/100,000 of a chromosomal DNA molecule. As a further complication, the distinctions between a gene and the surrounding DNA are subtle, consisting only of differences in nucleotide sequence. To work directly with specific genes, scientists have developed methods for preparing well-defined segments of DNA in multiple identical copies, a process called *DNA cloning*.

Most methods for cloning pieces of DNA in the laboratory share certain general features. One common approach uses bacteria, most often *E. coli*. Recall from Figure 13.15 that the *E. coli* chromosome is a large circular molecule of DNA. In addition, *E. coli* and many other bacteria have **plasmids**, small circular DNA molecules that are replicated separately. A plasmid has only a small number of genes; these genes may be useful when the bacterium is in a particular environment but may not be required for survival or reproduction under most conditions.

To clone pieces of DNA using bacteria, researchers first obtain a plasmid (originally isolated from a bacterial cell and genetically engineered for efficient cloning) and insert DNA from another source ("foreign" DNA) into it **(Figure 13.24)**. The resulting plasmid is now a **recombinant DNA molecule**, a molecule containing DNA from two different sources, very often different species. The plasmid is then returned to a bacterial cell, producing a *recombinant bacterium*. This single cell

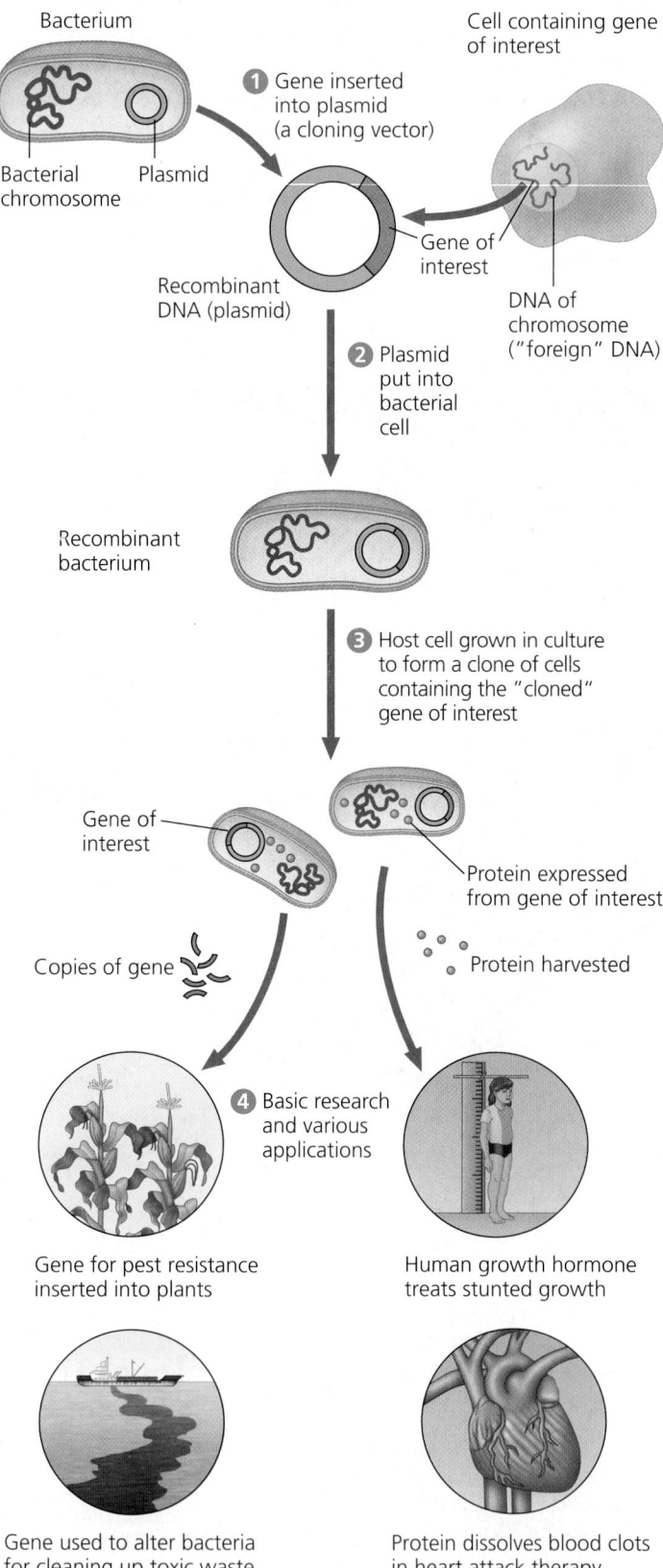

▲ **Figure 13.24 An overview of gene cloning and some uses of cloned genes.** In this simplified diagram of gene cloning, we start with a plasmid (originally isolated from a bacterial cell) and a gene of interest from another organism. Only one plasmid and one copy of the gene of interest are shown at the top of the figure, but the starting materials would include many of each.

reproduces through repeated cell divisions to form a clone of cells, a population of genetically identical cells. Because the dividing bacteria replicate the recombinant plasmid and pass it on to their descendants, the foreign DNA and any genes it carries are cloned at the same time. The production of multiple copies of a single gene is a type of DNA cloning called **gene cloning**. In our example in Figure 13.24, the plasmid acts as a **cloning vector**, a DNA molecule that can carry foreign DNA into a cell and be replicated there. The foreign DNA segment could be a gene from a eukaryotic cell; we will describe in more detail how it was obtained later in this section.

Gene cloning is useful for two basic purposes: to make many copies of, or *amplify*, a particular gene and to produce a protein product from it. Researchers can isolate copies of a cloned gene from bacteria for use in basic research or to endow another organism with a new metabolic trait, such as pest resistance. For example, a resistance gene present in one crop species might be cloned and transferred into plants of another species. (Such organisms are called Genetically Modified Organisms, or GMOs; see Concept 30.3.) Alternatively, a protein with medical uses, such as human growth hormone, can be harvested in large quantities from cultures of bacteria carrying a cloned gene for the protein. Since a single gene is usually a very small part of the total DNA in a cell, the ability to amplify such rare DNA fragments is crucial for any application involving a single gene.

## Using Restriction Enzymes to Make a Recombinant DNA Plasmid

Gene cloning and genetic engineering generally rely on the use of enzymes that cut DNA molecules at a limited number of specific locations. These enzymes, called restriction endonucleases, or **restriction enzymes**, were discovered in the late 1960s by biologists doing basic research on bacteria. Restriction enzymes protect the bacterial cell by cutting up foreign DNA from other organisms or phages.

Hundreds of different restriction enzymes have been identified and isolated. Each restriction enzyme is very specific, recognizing a particular short DNA sequence, or **restriction site**, and cutting both DNA strands at precise points within this restriction site. The DNA of a bacterial cell is protected from the cell's own restriction enzymes by the addition of methyl groups ($-CH_3$) to adenines or cytosines within the sequences recognized by the enzymes.

**Figure 13.25** shows how restriction enzymes are used to clone a foreign DNA fragment into a bacterial plasmid. At the top is a plasmid (like the one in Figure 13.24) that has a single restriction site recognized by a particular restriction enzyme from *E. coli*. (Various such plasmids are available from commercial suppliers.) As shown in this example, most restriction sites are symmetrical. That is, the sequence of nucleotides is the same on both strands when read in the 5′ → 3′ direction. The most commonly used restriction enzymes recognize sequences containing four to eight nucleotide pairs. Because any sequence this short

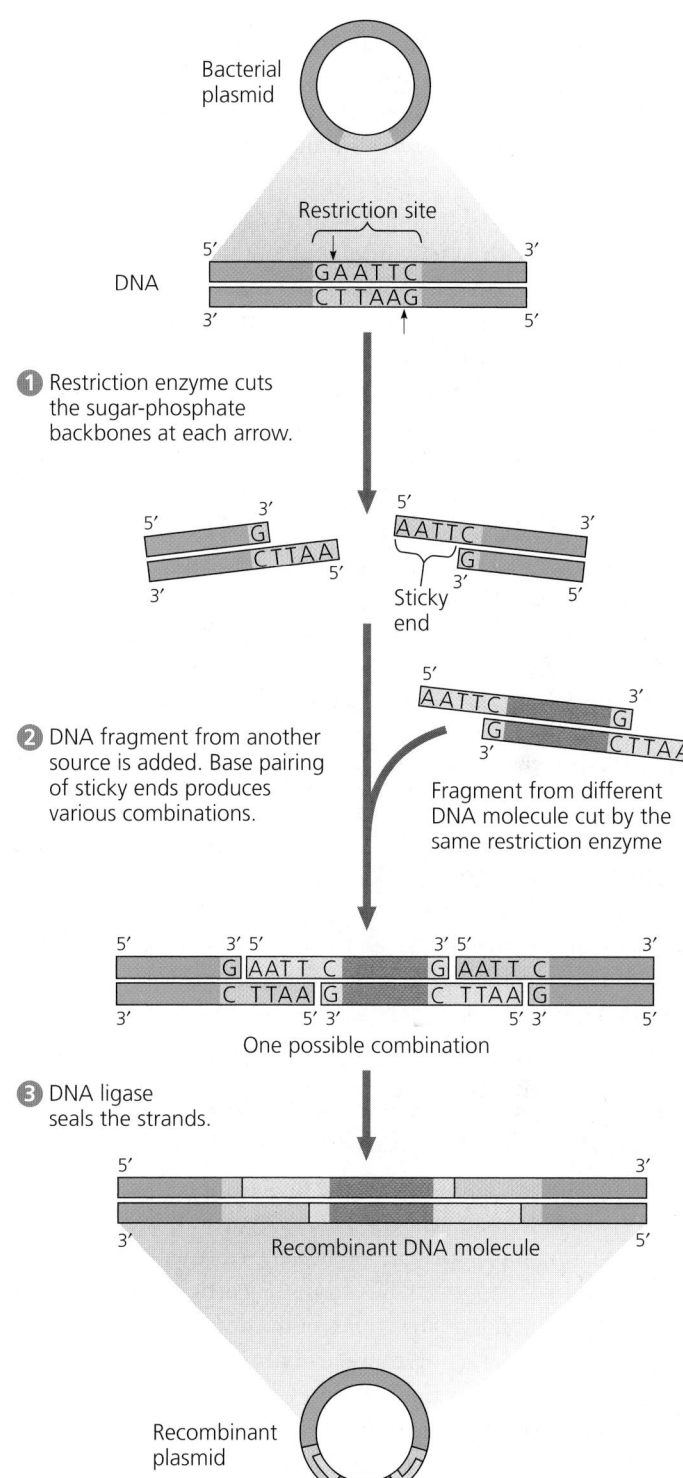

▲ **Figure 13.25 Using a restriction enzyme and DNA ligase to make a recombinant DNA plasmid.** The restriction enzyme in this example (called *Eco*RI) recognizes a single six-base-pair restriction site present in the plasmid. It makes staggered cuts in the sugar-phosphate backbones, producing fragments with "sticky ends." Foreign DNA fragments with complementary sticky ends can base-pair with the plasmid ends; the ligated product is a recombinant plasmid. (If the two plasmid sticky ends base-pair, the original non-recombinant plasmid would re-form.)

**DRAW IT** *The restriction enzyme Hind III recognizes the sequence 5′-AAGCTT-3′, cutting between the two A's. Draw the double-stranded sequence before and after the enzyme cuts.*

usually occurs (by chance) many times in a long DNA molecule, a restriction enzyme will make many cuts in such a DNA molecule, yielding a set of **restriction fragments**. All copies of a given DNA molecule always yield the same set of restriction fragments when exposed to the same restriction enzyme.

The most useful restriction enzymes cleave the sugar-phosphate backbones in the two DNA strands in a staggered manner, as indicated in Figure 13.25. The resulting double-stranded restriction fragments have at least one single-stranded end, called a **sticky end**. These short extensions can form hydrogen-bonded base pairs (hybridize) with complementary sticky ends on any other DNA molecules cut with the same enzyme, such as the inserted DNA in Figure 13.25. The associations formed in this way are only temporary but can be made permanent by DNA ligase, which catalyzes the formation of covalent bonds that close up the sugar-phosphate backbones of DNA strands (see Figure 13.18). You can see at the bottom of Figure 13.25 that the ligase-catalyzed joining of DNA from the plasmid and the foreign DNA produces a stable recombinant DNA molecule, in this example a recombinant plasmid.

To check the recombinant plasmids after they have been copied many times in host cells, researchers might cut the products again using the same restriction enzyme, expecting two DNA fragments, one the size of the plasmid and one the size of the inserted DNA. To separate and visualize the fragments, they would next carry out a technique called **gel electrophoresis**, which uses a gel made of a polymer as a sieve to separate a mixture of nucleic acid fragments by length **(Figure 13.26)**. Gel electrophoresis is used in conjunction with many different techniques in molecular biology.

## Amplifying DNA: The Polymerase Chain Reaction (PCR) and Its Use in Cloning

Today, most researchers have some information about the sequence of the foreign gene or DNA fragment they want to clone. Using this information, they can start with the entire collection of genomic DNA from the particular species of interest and obtain many copies of the desired gene by using a technique called the **polymerase chain reaction**, or **PCR**. **Figure 13.27** illustrates the steps in PCR. Within a few hours, this technique can make billions of copies of a specific target DNA segment in a sample, even if that segment makes up less than 0.001% of the total DNA in the sample.

In the PCR procedure, a three-step cycle brings about a chain reaction that produces an exponentially growing population of identical DNA molecules. During each cycle, the reaction mixture is heated to denature (separate) the DNA strands and then cooled to allow annealing (hybridization) of short, single-stranded DNA primers complementary to sequences on opposite strands at each end of the target segment; finally, a DNA polymerase extends the primers in the $5' \rightarrow 3'$ direction. If a standard DNA polymerase were used, the protein would be denatured along with the DNA during the first heating step and would have to be replaced after each cycle.

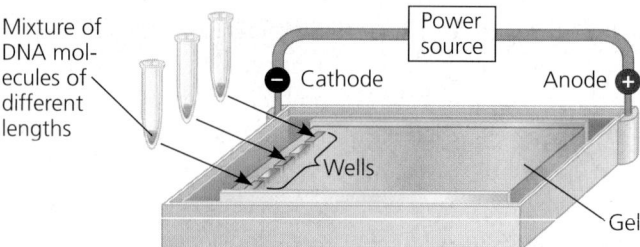

**(a)** Each sample, a mixture of different DNA molecules, is placed in a separate well near one end of a thin slab of agarose gel. The gel is set into a small plastic support and immersed in an aqueous, buffered solution in a tray with electrodes at each end. The current is then turned on, causing the negatively charged DNA molecules to move toward the positive electrode.

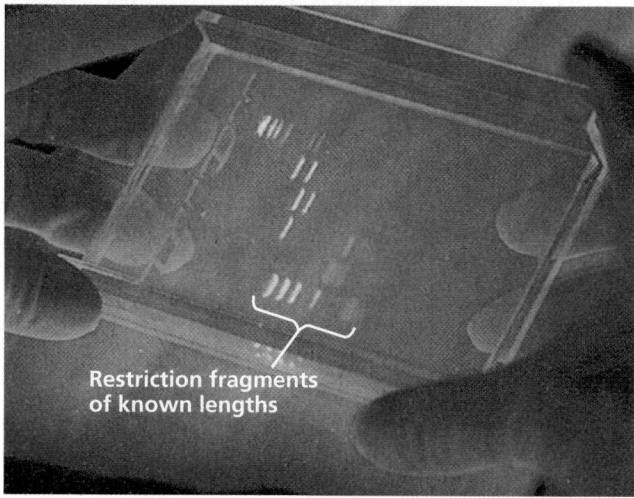

**(b)** Shorter molecules are slowed down less than longer ones, so they move faster through the gel. After the current is turned off, a DNA-binding dye is added that fluoresces pink in UV light. Each pink band corresponds to many thousands of DNA molecules of the same length. The horizontal ladder of bands at the bottom of the gel is a set of restriction fragments of known lengths for comparison with samples of unknown length.

▲ **Figure 13.26 Gel electrophoresis.** A gel made of a polymer acts as a molecular sieve to separate nucleic acids or proteins differing in size, electrical charge, or other physical properties as they move in an electric field. In the example shown here, DNA molecules are separated by length in a gel made of a polysaccharide called agarose.

The key to automating PCR was the discovery of an unusually heat-stable DNA polymerase called *Taq* polymerase, named after the bacterial species from which it was first isolated. This bacterial species, *Thermus aquaticus*, lives in hot springs, and the stability of its DNA polymerase at high temperatures is an evolutionary adaptation that enables the enzyme to function at temperatures up to 95°C. Today, researchers also use a DNA polymerase from the archaean species *Pyrococcus furiosus*. This enzyme, called *Pfu* polymerase, is more accurate, more stable, and costlier than *Taq* polymerase.

PCR is speedy and very specific. Only minuscule amounts of DNA need be present in the starting material, and this DNA can be partially degraded, as long as a few molecules contain the complete target segment. The key to this high specificity is the primers, the sequences of which are chosen so they

## The Polymerase Chain Reaction (PCR)

**Application** With PCR, any specific segment—the so-called target sequence—in a DNA sample can be copied many times (amplified) within a test tube.

**Technique** PCR requires double-stranded DNA containing the target sequence, a heat-resistant DNA polymerase, all four nucleotides, and two 15- to 20-nucleotide DNA strands that serve as primers. One primer is complementary to one end of the target sequence on one strand; the second primer is complementary to the other end of the sequence on the other strand.

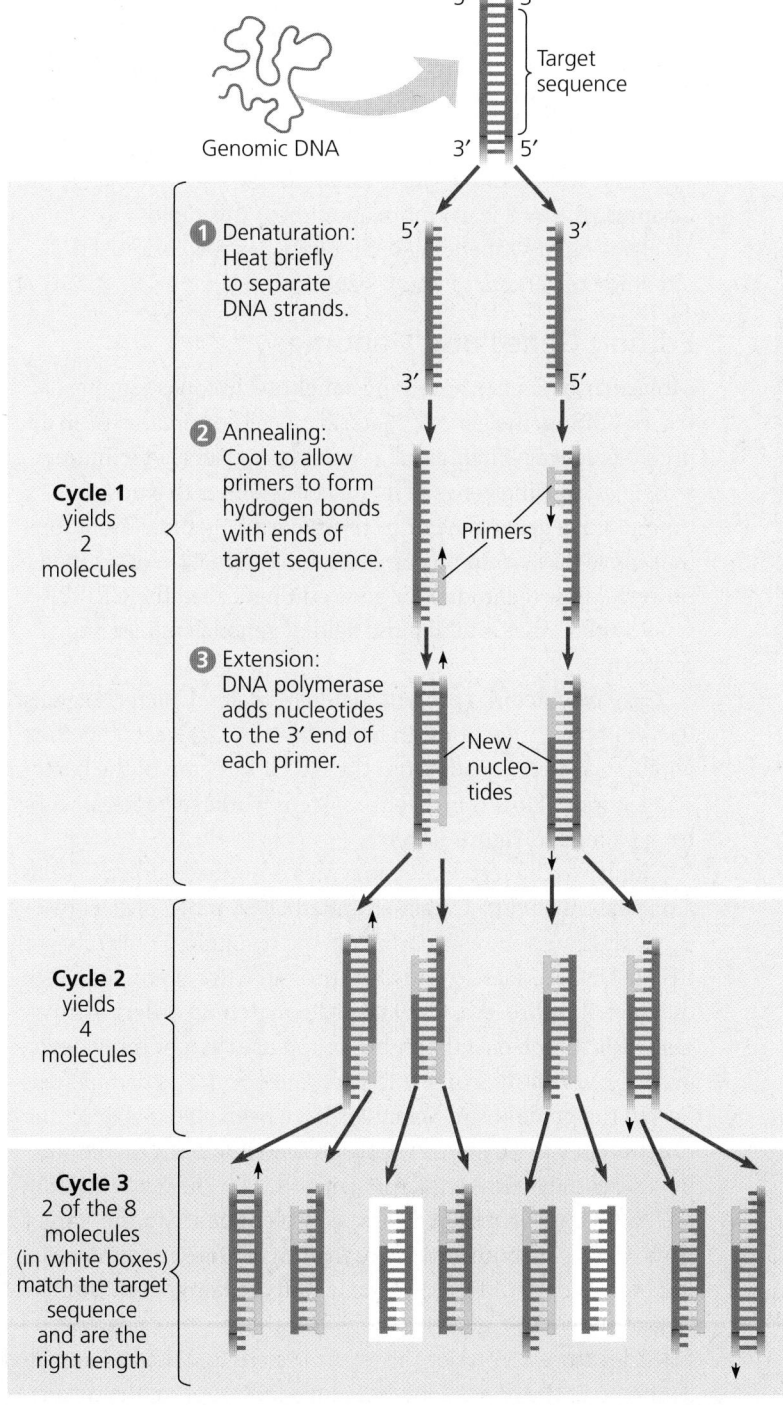

**Results** After three cycles, two molecules match the target sequence exactly. After 30 more cycles, over 1 billion ($10^9$) molecules match the target sequence.

hybridize *only* with complementary sequences at opposite ends of the target segment. (For high specificity, the primers must be at least 15 or so nucleotides long.) With each successive cycle, the number of target segment molecules of the correct length doubles, so the number of molecules equals $2n$, where $n$ is the number of cycles. After 30 or so cycles, about a billion copies of the target sequence are present!

Despite its speed and specificity, PCR amplification alone cannot substitute for gene cloning in cells to make large amounts of a gene. This is because occasional errors during PCR replication limit the number of good copies and the length of DNA fragments that can be copied. Instead, PCR is used to provide the specific DNA fragment for cloning. PCR primers are synthesized to include a restriction site at each end of the DNA fragment that matches the site in the cloning vector, and the fragment and vector are cut and ligated together **(Figure 13.28)**. The resulting plasmids are sequenced so that those with error-free inserts can be selected.

Devised in 1985, PCR has had a major impact on biological research and genetic engineering. PCR has been used to amplify DNA from a wide variety of sources: a 40,000-year-old frozen woolly mammoth; fingerprints or tiny amounts of blood, tissue, or semen found at crime scenes; single embryonic cells for rapid prenatal diagnosis of genetic disorders; and cells infected with viruses that are difficult to detect, such as HIV (in the latter case, viral genes are amplified).

## DNA Sequencing

Once a gene is cloned, researchers can exploit the principle of complementary base pairing to determine the gene's complete nucleotide sequence, a process called **DNA sequencing**. In the last 15 years, "next-generation" sequencing techniques have been developed that are rapid and inexpensive. In machines that carry out next-generation

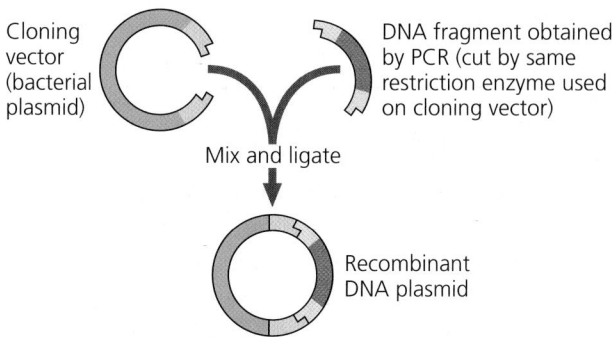

▲ **Figure 13.28 Use of a restriction enzyme and PCR in gene cloning.** In a closer look at the process shown at the top of Figure 13.24, PCR is used to produce the DNA fragment or gene of interest that will be ligated into a cloning vector, in this case a bacterial plasmid. Both the plasmid and the DNA fragments are cut with the same restriction enzyme, combined so the sticky ends can hybridize, and ligated together. The resulting plasmids will then be introduced into bacterial cells.

sequencing **(Figure 13.29a)**, a single template strand is immobilized, and DNA polymerase and other reagents are added that allow so-called *sequencing by synthesis* of the complementary strand, one nucleotide at a time. A specialized chemical technique enables electronic monitors to identify which of the four nucleotides is being added, allowing determination of the sequence **(Figure 13.29b)**.

Next-generation sequencing is rapidly being complemented or even replaced by "third-generation" sequencing techniques, with each new technique being faster and less expensive than the previous one. In some of these new methods, the DNA is neither cut into fragments nor amplified. Instead, a single, very long DNA molecule is sequenced on its own. Several groups have been working on moving a single strand of a DNA molecule through a very small pore (a *nanopore*) in a membrane. This is one of the many approaches to further increase the

**(a) Next-generation sequencing machines**

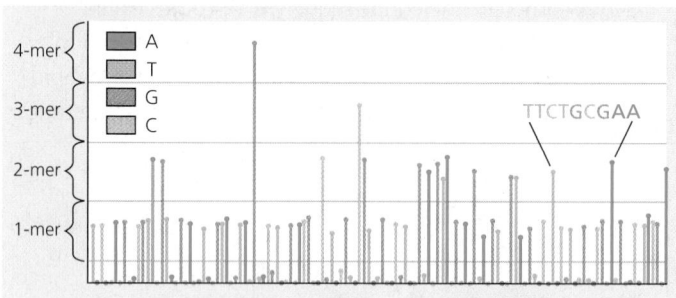

**(b) A "flow-gram" from a next-generation sequencing machine**

▲ **Figure 13.29 Next-generation sequencing. (a)** Next-generation sequencing machines use "sequencing by synthesis" to sequence many 300-nucleotide fragments in parallel. In this way, one machine can sequence about 2 billion nucleotides in 24 hours. **(b)** The results for one fragment are displayed as a "flow-gram," where the nucleotides are identified by color, one by one. The sequences of the entire set of fragments are analyzed using computer software, which "stitches" them together into a whole sequence—often, an entire genome.

**INTERPRET THE DATA** *If the template strand has two or more identical nucleotides in a row, the complementary nucleotides on the synthesized strand will be added, one after the other, in the same step. How are two or more of the same nucleotide in a row displayed in the flow-gram? (See example sequence on the right.) Write out the sequence of the first 25 nucleotides in the flow-gram, starting from the left as the 5' end. (Ignore the very short lines.)*

▶ **Figure 13.30 An example of a third-generation sequencing technique.** In this approach, a single strand of an uncut DNA molecule would be passed, nucleotide by nucleotide, through a nanopore in a membrane. Here, the pore is a protein channel embedded in a lipid membrane. The nucleotides are identified by slight differences in the amount of time they interrupt an electrical current across the opening.

rate and reduce the cost of sequencing; one model of this approach is shown in **Figure 13.30**. In Chapter 18, you'll learn more about how this rapid acceleration of sequencing technology has revolutionized our study of genes and whole genomes.

## Editing Genes and Genomes

Molecular biologists have long sought techniques for altering, or editing, the genetic material of cells or organisms in a predictable way. Their aim has been to use such a technique to change specific genes in living cells—either to study the function of a given gene or to try to correct genetic mutations that cause disease. In recent years, biologists have developed a powerful new technique for gene editing, called the CRISPR-Cas9 system, that is taking the field of genetic engineering by storm.

Cas9 is a bacterial protein that helps defend bacteria against bacteriophage infections. In bacterial cells, Cas9 acts together with "guide RNA" made from the CRISPR region of the bacterial genome. (How this defense system works in bacteria will be explained in Figure 17.6.)

Similar to the restriction enzymes described earlier, Cas9 is a nuclease that cuts double-stranded DNA molecules. However, while a given restriction enzyme recognizes only one particular DNA sequence, the Cas9 protein will cut any sequence to which it is directed. Cas9 takes its marching orders from a guide RNA molecule that it binds and uses as a homing device, cutting both strands of any DNA sequence that is complementary to the guide RNA. Scientists have been able to exploit the function of Cas9 by introducing a Cas9–guide RNA complex into a cell they wish to alter **(Figure 13.31)**. The guide RNA in the complex is engineered to be complementary to the "target" gene. Cas9 cuts both strands of the target DNA, and the resulting broken ends of DNA trigger a DNA repair system (similar to that shown in Figure 13.21). When there is no undamaged DNA for the enzymes of the repair system to use as a template, as shown at the bottom left of Figure 13.31, part a, the repair enzymes introduce or remove random nucleotides while rejoining the ends. Generally, this alters the DNA sequence so that the gene no longer works properly. This technique is a

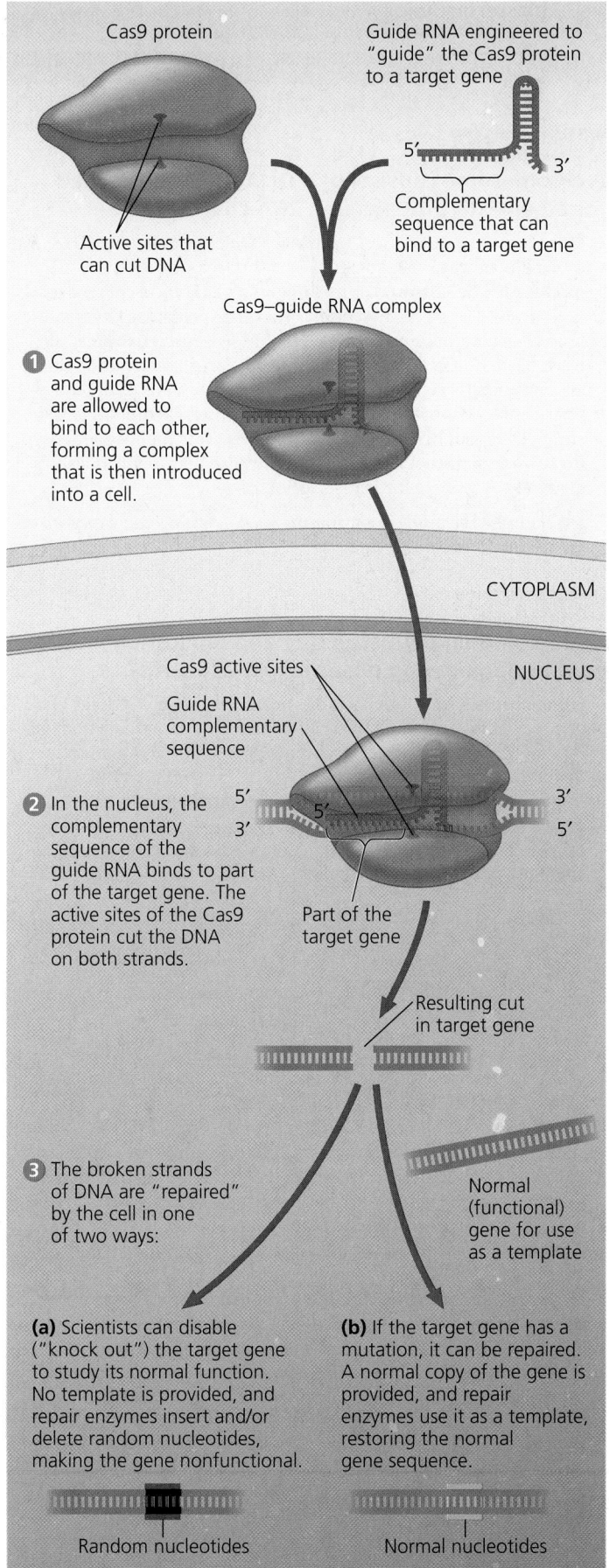

① **Cas9 protein and guide RNA are allowed to bind to each other, forming a complex that is then introduced into a cell.**

Cas9 protein

Guide RNA engineered to "guide" the Cas9 protein to a target gene

Active sites that can cut DNA

5′ 3′

Complementary sequence that can bind to a target gene

Cas9–guide RNA complex

CYTOPLASM

NUCLEUS

Cas9 active sites

Guide RNA complementary sequence

② **In the nucleus, the complementary sequence of the guide RNA binds to part of the target gene. The active sites of the Cas9 protein cut the DNA on both strands.**

5′ 3′
3′ 5′
5′

Part of the target gene

Resulting cut in target gene

③ **The broken strands of DNA are "repaired" by the cell in one of two ways:**

Normal (functional) gene for use as a template

**(a)** Scientists can disable ("knock out") the target gene to study its normal function. No template is provided, and repair enzymes insert and/or delete random nucleotides, making the gene nonfunctional.

**(b)** If the target gene has a mutation, it can be repaired. A normal copy of the gene is provided, and repair enzymes use it as a template, restoring the normal gene sequence.

Random nucleotides

Normal nucleotides

▲ **Figure 13.31 Gene editing using the CRISPR-Cas9 system.**

highly successful way for researchers to "knock out" (disable) a given gene to study what that gene does in an organism.

But what about using this system to help treat genetic diseases? Researchers have modified the technique so that the CRISPR-Cas9 system can be used to repair a gene that has a mutation. They introduce a segment from the normal (functional) gene along with the CRISPR-Cas9 system. After Cas9 cuts the target DNA, repair enzymes can use the normal DNA as a template to repair the target DNA at the break point. In this way, the CRISPR-Cas9 system edits the defective gene so that it is corrected (see the bottom right of Figure 13.31, part b).

In 2014, a group of researchers reported success in correcting a genetic defect in mice using CRISPR technology. The lab mice had been genetically engineered to have a mutation in a gene encoding a liver enzyme that metabolizes the amino acid tyrosine, mimicking a fatal genetic disorder in humans called tyrosinemia. A guide RNA molecule complementary to the mutated region of the gene was introduced into the mouse along with the Cas9 protein and a segment of DNA from the same region of the normal gene for use as a template. Subsequent analysis indicated that the faulty gene had been corrected in enough of the liver cells that the amount of functional enzyme made was sufficient to alleviate the disease symptoms. There are still many hurdles to overcome before this approach can be tried in humans, but the CRISPR technology is sparking widespread excitement among researchers and physicians alike.

In this section, you have learned how understanding the elegant structure of DNA has led to powerful techniques for genetic engineering. Earlier in the chapter, you also saw how DNA molecules are arranged in chromosomes and how DNA replication provides the copies of genes that parents pass to offspring. However, it is not enough that genes be copied and transmitted; the information they carry must be used by the cell. In other words, genes must also be "expressed." In the next few chapters, we'll examine how the cell expresses the genetic information encoded in DNA. We'll also return to the subject of genetic engineering by exploring a few techniques for analyzing gene expression.

### CONCEPT CHECK 13.4

1. **MAKE CONNECTIONS** The restriction site for an enzyme called *Pvu*I is the following sequence:

   5′-C G A T C G-3′
   3′-G C T A G C-5′

   Staggered cuts are made between the T and C on each strand. What type of bonds are being cleaved? (See Concept 3.6.)

2. **DRAW IT** One strand of a DNA molecule has the following sequence: 5′-CCTTGACGATCGTTACCG-3′. Draw the other strand. Will *Pvu*I (see question 1) cut this molecule? If so, draw the products.

3. Describe the role of complementary base pairing during cloning, PCR, DNA sequencing, and gene editing using the CRISPR-Cas9 system.

For suggested answers, see Appendix A.

# 13 | Chapter Review

Go to **MasteringBiology®** for Assignments, the eText, and the Study Area with Animations, Activities, Vocab Self-Quiz, and Practice Tests.

## SUMMARY OF KEY CONCEPTS

**VOCAB SELF-QUIZ**

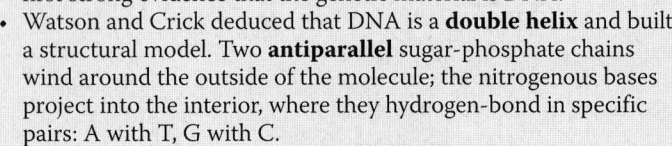

goo.gl/gbai8v

### CONCEPT 13.1

**DNA is the genetic material (pp. 254–259)**

- Experiments with bacteria and **phages** provided the first strong evidence that the genetic material is DNA.
- Watson and Crick deduced that DNA is a **double helix** and built a structural model. Two **antiparallel** sugar-phosphate chains wind around the outside of the molecule; the nitrogenous bases project into the interior, where they hydrogen-bond in specific pairs: A with T, G with C.

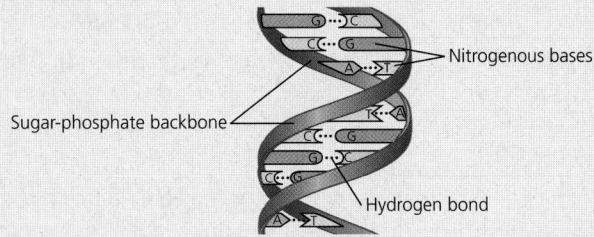

Nitrogenous bases

Sugar-phosphate backbone

Hydrogen bond

? *What does it mean when we say that the two DNA strands in the double helix are antiparallel? What would an end of the double helix look like if the strands were parallel?*

### CONCEPT 13.2

**Many proteins work together in DNA replication and repair (pp. 259–267)**

- The Meselson-Stahl experiment showed that **DNA replication** is **semiconservative**: The parental molecule unwinds, and each strand then serves as a template for the synthesis of a new strand according to base-pairing rules.
- DNA replication at one **replication fork** is summarized here:

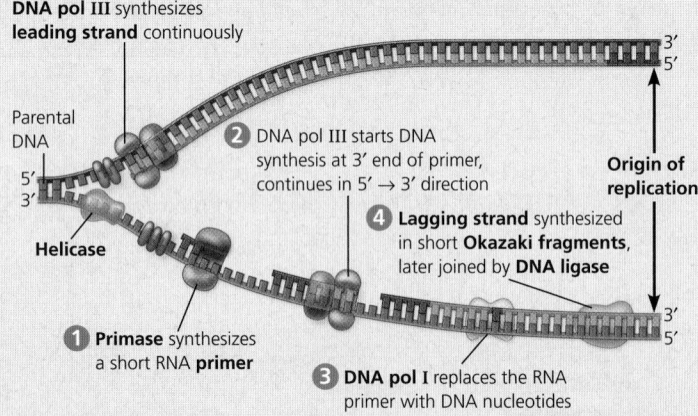

**DNA pol III** synthesizes **leading strand** continuously

Parental DNA

**Helicase**

**1 Primase** synthesizes a short RNA **primer**

**2** DNA pol III starts DNA synthesis at 3′ end of primer, continues in 5′ → 3′ direction

**3** DNA pol I replaces the RNA primer with DNA nucleotides

**4 Lagging strand** synthesized in short **Okazaki fragments**, later joined by **DNA ligase**

Origin of replication

- **DNA polymerases** proofread new DNA, replacing incorrect nucleotides. In **mismatch repair**, enzymes correct errors that persist. **Nucleotide excision repair** is a general process by which **nucleases** cut out and replace damaged stretches of DNA.

? *Compare DNA replication on the leading and lagging strands, including both similarities and differences.*

### CONCEPT 13.3

**A chromosome consists of a DNA molecule packed together with proteins (pp. 267–269)**

- The chromosome of most bacterial species is a circular DNA molecule with some associated proteins, making up the **nucleoid**. The **chromatin** making up a eukaryotic chromosome is composed of DNA, **histones**, and other proteins. The histones bind to each other and to the DNA to form **nucleosomes**, the most basic units of DNA packing. Additional coiling and folding lead ultimately to the highly condensed chromatin of the metaphase chromosome. In interphase cells, most chromatin is less compacted (**euchromatin**), but some remains highly condensed (**heterochromatin**). Euchromatin, but not heterochromatin, is generally accessible for transcription of genes.

? *Describe the levels of chromatin packing you would expect to see in an interphase nucleus.*

### CONCEPT 13.4

**Understanding DNA structure and replication makes genetic engineering possible (pp. 270–275)**

- **Gene cloning** (or DNA cloning) produces multiple copies of a gene (or DNA segment) that can be used to manipulate and analyze DNA and to produce useful new products or organisms with beneficial traits.
- In **genetic engineering**, bacterial **restriction enzymes** are used to cut DNA molecules within short, specific nucleotide sequences (**restriction sites**), yielding a set of double-stranded **restriction fragments** with single-stranded **sticky ends**.

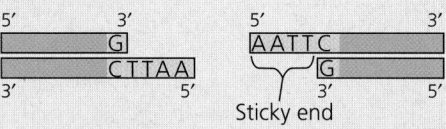

Sticky end

- DNA fragments of different lengths can be separated and their lengths assessed by **gel electrophoresis**.
- The sticky ends on restriction fragments from one DNA source—such as a bacterial **plasmid** or other **cloning vector**—can base-pair with complementary sticky ends on fragments from other DNA molecules; sealing the base-paired fragments with DNA ligase produces **recombinant DNA molecules**.
- The **polymerase chain reaction (PCR)** can amplify (produce many copies of) a specific target segment of DNA for use as a DNA fragment for cloning. PCR uses primers that bracket the desired segment and requires a heat-resistant DNA polymerase.
- The rapid development of fast, inexpensive techniques for **DNA sequencing** is based on *sequencing by synthesis*: DNA polymerase is used to replicate a stretch of DNA from a single-stranded template, and the order in which nucleotides are added reveals the sequence.
- The CRISPR-Cas9 system allows researchers to edit genes in a specific, desired way. This may ultimately be used for treatment of genetic diseases.

? *Describe how the process of gene cloning results in a cell clone containing a recombinant plasmid.*

# TEST YOUR UNDERSTANDING

**PRACTICE TEST**

goo.gl/CRZjvS

## Level 1: Knowledge/Comprehension

1. In his work with pneumonia-causing bacteria and mice, Griffith found that
   (A) the protein coat from pathogenic cells was able to transform nonpathogenic cells.
   (B) heat-killed pathogenic cells caused pneumonia.
   (C) some substance from pathogenic cells was transferred to nonpathogenic cells, making them pathogenic.
   (D) the polysaccharide coat of bacteria caused pneumonia.

2. What is the basis for the difference in how the leading and lagging strands of DNA molecules are synthesized?
   (A) DNA polymerase can join new nucleotides only to the 3′ end of a preexisting strand.
   (B) Helicases and single-strand binding proteins work at the 5′ end.
   (C) The origins of replication occur only at the 5′ end.
   (D) DNA ligase works only in the 3′ → 5′ direction.

3. In analyzing the number of different bases in a DNA sample, which result would be consistent with the base-pairing rules?
   (A) A = G          (C) A + T = G + C
   (B) A + G = C + T  (D) A = C

4. The elongation of the leading strand during DNA synthesis
   (A) progresses away from the replication fork.
   (B) occurs in the 3′ → 5′ direction.
   (C) produces Okazaki fragments.
   (D) depends on the action of DNA polymerase.

5. In a nucleosome, the DNA is wrapped around
   (A) polymerase molecules.   (C) histones.
   (B) ribosomes.              (D) a thymine dimer.

6. Which of the following sequences in double-stranded DNA is most likely to be recognized as a cutting site for a restriction enzyme?
   (A) AAGG   (B) GGCC   (C) ACCA   (D) AAAA
      TTCC        CCGG        TGGT        TTTT

## Level 2: Application/Analysis

7. *E. coli* cells grown on $^{15}$N medium are transferred to $^{14}$N medium and allowed to grow for two more generations (two rounds of DNA replication). DNA extracted from these cells is centrifuged. What density distribution of DNA would you expect in this experiment?
   (A) one high-density and one low-density band
   (B) one intermediate-density band
   (C) one high-density and one intermediate-density band
   (D) one low-density and one intermediate-density band

8. A student isolates, purifies, and combines in a test tube a variety of molecules needed for DNA replication. After adding some DNA to the mixture, replication occurs, but each DNA molecule consists of a normal strand paired with numerous segments of DNA a few hundred nucleotides long. What has the student probably left out of the mixture?
   (A) DNA polymerase    (C) Okazaki fragments
   (B) DNA ligase        (D) primase

9. The spontaneous loss of amino groups from adenine in DNA results in hypoxanthine, an uncommon base, opposite thymine. What combination of proteins could repair such damage?
   (A) nuclease, DNA polymerase, DNA ligase
   (B) topoisomerase, primase, DNA polymerase
   (C) topoisomerase, helicase, single-strand binding protein
   (D) DNA ligase, replication fork proteins, adenylyl cyclase

10. **MAKE CONNECTIONS** Although the proteins that cause the *E. coli* chromosome to coil are not histones, identify a property you would expect them to share with histones, given their ability to bind to DNA (see Figure 3.18).

## Level 3: Synthesis/Evaluation

11. **SCIENTIFIC INQUIRY**

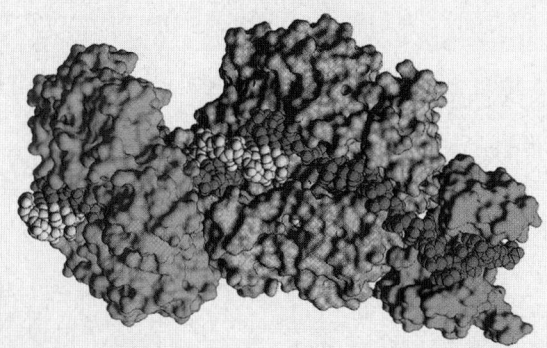

**DRAW IT** Model building can be an important part of the scientific process. The illustration shown here is a computer-generated model of a DNA replication complex. The parental and newly synthesized DNA strands are color-coded differently, as are each of the following three proteins: DNA pol III, the sliding clamp, and single-strand binding protein. (a) Using what you've learned in this chapter to clarify this model, label each DNA strand and each protein. (b) Draw an arrow to indicate the overall direction of DNA replication.

12. **FOCUS ON EVOLUTION**
    Some bacteria may be able to respond to environmental stress by increasing the rate at which mutations occur during cell division. How might this be accomplished? Might there be an evolutionary advantage of this ability? Explain.

13. **FOCUS ON ORGANIZATION**
    The continuity of life is based on heritable information in the form of DNA, and structure and function are correlated at all levels of biological organization. In a short essay (100–150 words), describe how the structure of DNA is correlated with its role as the molecular basis of inheritance.

14. **SYNTHESIZE YOUR KNOWLEDGE**

This image shows DNA interacting with a computer-generated model of a TAL protein, one of a family of proteins found only in a species of the bacterium *Xanthomonas*. The bacterium uses proteins like this one to find particular gene sequences in cells of the organisms it infects, such as tomatoes, rice, and citrus fruits. Researchers are excited about working with this family of proteins. Their goal is to generate modified versions that can home in on specific gene sequences. Such proteins could then be used in an approach called gene therapy to "fix" mutated genes in individuals with genetic diseases. Given what you know about DNA structure and considering the image shown here, discuss what the TAL protein's structure suggests about how it functions.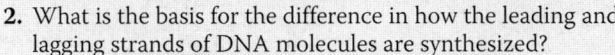

*For selected answers, see Appendix A.*

# 14

# Gene Expression: From Gene to Protein

## KEY CONCEPTS

**14.1** Genes specify proteins via transcription and translation

**14.2** Transcription is the DNA-directed synthesis of RNA: *a closer look*

**14.3** Eukaryotic cells modify RNA after transcription

**14.4** Translation is the RNA-directed synthesis of a polypeptide: *a closer look*

**14.5** Mutations of one or a few nucleotides can affect protein structure and function

▲ **Figure 14.1** How does a single faulty gene result in the dramatic appearance of an albino donkey?

## The Flow of Genetic Information

The island of Asinara lies off the coast of Sardinia, an Italian island. The name Asinara probably originated from the Latin word *sinuaria*, which means "sinus-shaped." A second meaning of Asinara is "donkey-inhabited," which is particularly appropriate because Asinara is home to a wild population of albino donkeys **(Figure 14.1)**. The donkeys were brought to Asinara in the early 1800s and abandoned there in 1885 when the 500 residents were forced to leave the island so it could be used as a penal colony. What is responsible for the phenotype of the albino donkey, strikingly different from its pigmented relative?

Inherited traits are determined by genes, and the trait of albinism is caused by a recessive allele of a pigmentation gene (see Concept 11.4). The information content of genes is in the form of specific sequences of nucleotides along strands of DNA, the genetic material. But how does this information determine an organism's traits? Put another way, what does a gene actually say? And how is its message

translated by cells into a specific trait, such as brown hair, type A blood, or, in the case of an albino donkey, a total lack of pigment? The albino donkey has a faulty version of a key protein, an enzyme required for pigment synthesis, and this protein is faulty because the gene that codes for it contains incorrect information.

This example illustrates the main point of this chapter: The DNA inherited by an organism leads to specific traits by dictating the synthesis of proteins and of RNA molecules involved in protein synthesis. In other words, proteins are the link between genotype and phenotype. **Gene expression** is the process by which DNA directs the synthesis of proteins (or, in some cases, just RNAs). The expression of genes that code for proteins includes two stages: transcription and translation. This chapter describes the flow of information from gene to protein and explains how genetic mutations affect organisms through their proteins. Understanding the processes of gene expression, which are similar in all three domains of life, will allow us to revisit the concept of the gene in more detail at the end of the chapter.

# CONCEPT 14.1

# Genes specify proteins via transcription and translation

Before going into the details of how genes direct protein synthesis, let's step back and examine how the fundamental relationship between genes and proteins was discovered.

## Evidence from the Study of Metabolic Defects

In 1902, British physician Archibald Garrod was the first to suggest that genes dictate phenotypes through enzymes, proteins that catalyze specific chemical reactions in the cell. He postulated that the symptoms of an inherited disease reflect an inability to make a particular enzyme. For example, people with a disease called alkaptonuria have black urine because it contains a chemical called alkapton. Garrod reasoned that these people cannot make an enzyme that breaks down alkapton, so the chemical is expelled in their urine.

Several decades later, research supported Garrod's hypothesis that a gene dictates the production of a specific enzyme. Biochemists learned that cells synthesize and degrade most organic molecules via metabolic pathways, in which each chemical reaction in a sequence is catalyzed by a specific enzyme (see Concept 6.1). Such metabolic pathways lead, for instance, to the synthesis of the pigments that give the brown donkey in Figure 14.1 its fur color or fruit flies (*Drosophila*) their eye color (see Figure 12.3). In the 1930s, the American geneticist George Beadle and his French colleague Boris Ephrussi speculated that in *Drosophila* each mutation affecting eye color blocks pigment synthesis at a specific step by preventing production of the enzyme that catalyzes that step. But neither the chemical reactions nor the enzymes that catalyze them were known.

## *Nutritional Mutants:* Scientific Inquiry

A breakthrough in demonstrating the relationship between genes and enzymes came a few years later at Stanford University, where Beadle and Edward Tatum began experimenting with the bread mold *Neurospora crassa*. Unlike the diploid organisms studied by Mendel (peas) and T. H. Morgan (fruit flies), *Neurospora* is a haploid species. In order to observe a change in a mutant's phenotype, Beadle and Tatum needed to disable just one allele (rather than two) of a protein-coding gene required for a specific metabolic activity. By cleverly electing to work with *Neurospora*, they could cause a mutation in one allele in a cell and directly deduce the function of the wild-type gene.

Another advantage was that *Neurospora* has modest food requirements. For this species, the *minimal medium*, the culture containing minimal nutrients for growth of wild-type cells, is a simple solution of inorganic salts, glucose, and the vitamin biotin. On a gel-like substance called agar, saturated with minimal medium, single wild-type cells can synthesize all the nutrients they need for growth, dividing repeatedly and forming visible colonies of genetically identical cells.

Beadle and Tatum designed an experiment in which they generated different "nutritional mutants" of *Neurospora* cells, each of which was unable to synthesize a particular essential nutrient. Such cells could not grow on minimal medium but could grow on *complete medium*, which contains all nutrients needed for growth, including any that a mutant cell can't synthesize. For *Neurospora*, the complete medium consists of the minimal medium supplemented with all 20 amino acids and a few other nutrients. Beadle and Tatum hypothesized that in each nutritional mutant, the gene for the enzyme that synthesizes a particular nutrient had been disabled. **Figure 14.2** summarizes their experimental approach, in which they tested each type of mutant to determine which nutrient it was unable to synthesize.

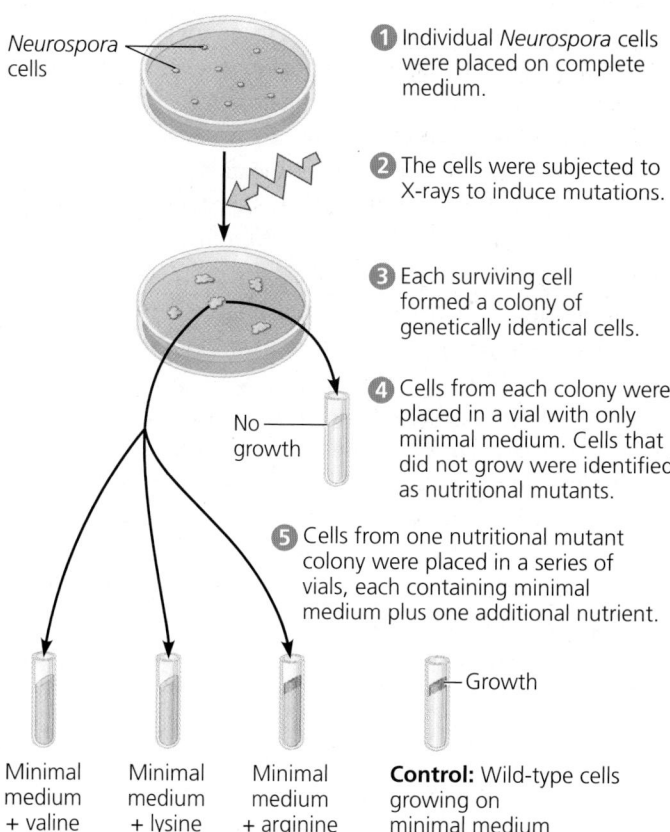

*Neurospora* cells

1. Individual *Neurospora* cells were placed on complete medium.

2. The cells were subjected to X-rays to induce mutations.

3. Each surviving cell formed a colony of genetically identical cells.

4. Cells from each colony were placed in a vial with only minimal medium. Cells that did not grow were identified as nutritional mutants.

No growth

5. Cells from one nutritional mutant colony were placed in a series of vials, each containing minimal medium plus one additional nutrient.

Growth

| Minimal medium + valine | Minimal medium + lysine | Minimal medium + arginine | **Control:** Wild-type cells growing on minimal medium |

6. The vials were observed for growth. In this example, the mutant cells grew only on minimal medium + arginine, indicating that this mutant was missing the enzyme for the synthesis of arginine.

▲ **Figure 14.2 The experimental approach of Beadle and Tatum.** To obtain nutritional mutants, Beadle and Tatum exposed *Neurospora* cells to X-rays to induce mutations. They then screened mutants with new nutritional requirements, such as arginine, as shown here.

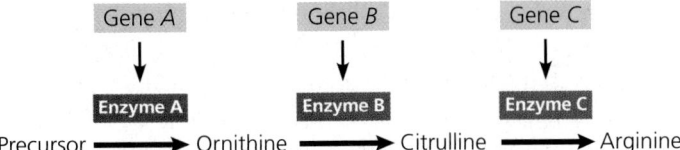

Precursor ——→ Ornithine ——→ Citrulline ——→ Arginine

▲ **Figure 14.3 The one gene–one protein hypothesis.** Based on results from work in their lab on nutritional mutants, Beadle and Tatum proposed that the function of a specific gene is to dictate production of a specific enzyme that catalyzes a particular reaction. The model shown here for the arginine-synthesizing pathway illustrates their hypothesis.

Beadle and Tatum amassed a valuable collection of mutant strains of *Neurospora*, catalogued not only by their defects in a particular metabolic pathway but also at different steps in that pathway. For example, a series of experiments on mutants requiring the amino acid arginine revealed that they could be grouped into distinct classes, each corresponding to a particular blocked step in the biochemical pathway for arginine synthesis. Presumably, the enzyme that normally catalyzes that step was unable to do so as a result of a mutation in its encoding gene **(Figure 14.3)**.

Beadle and Tatum concluded that, taken together, the collected results provided strong support for their so-called *one gene–one enzyme hypothesis*: that the function of a gene is to dictate the production of a specific enzyme. Further support for this hypothesis came from experiments that identified the specific enzymes lacking in the mutants. Beadle and Tatum shared a Nobel Prize in 1958 for "their discovery that genes act by regulating definite chemical events," in the words of the Nobel committee. Their experimental approach of disabling genes and observing the results plays a central role in genetic research today.

Today, we know of countless examples in which a mutation in a gene causes a faulty enzyme that in turn leads to an identifiable condition. The albino donkey in Figure 14.1 lacks a key enzyme called tyrosinase in the metabolic pathway that produces melanin, a dark pigment. The absence of melanin causes white fur and other effects throughout the donkey's body. Its nose, ears, and hooves, as well as its eyes, are pink because no melanin is present to mask the reddish color of the blood vessels that run through those structures.

### The Products of Gene Expression: A Developing Story

As researchers learned more about proteins, they made revisions to the one gene–one enzyme hypothesis. First of all, not all proteins are enzymes. Keratin, the structural protein of animal hair, and the hormone insulin are two examples of nonenzyme proteins. Because proteins that are not enzymes are nevertheless gene products, molecular biologists began to think in terms of one gene–one protein. However, many proteins are constructed from two or more different polypeptide

chains, and each polypeptide is specified by its own gene. For example, hemoglobin, the oxygen-transporting protein of vertebrate red blood cells, contains two kinds of polypeptides, and thus two genes code for this protein (Figure 3.22). Beadle and Tatum's idea was therefore restated as the *one gene–one polypeptide hypothesis*. Even this description is not entirely accurate, though. First, in many cases, a eukaryotic gene can code for a set of closely related polypeptides via a process called alternative splicing, which you will learn about later in this chapter. Second, quite a few genes code for RNA molecules that have important functions in cells even though they are never translated into protein. For now, we will focus on genes that do code for polypeptides. (Note that it is common to refer to these gene products as proteins—a practice you'll encounter in this text—rather than more precisely as polypeptides.)

### Basic Principles of Transcription and Translation

Genes provide the instructions for making specific proteins. But a gene does not build a protein directly. The bridge between DNA and protein synthesis is the nucleic acid RNA. RNA is chemically similar to DNA except that it contains ribose instead of deoxyribose as its sugar and has the nitrogenous base uracil rather than thymine (see Concept 3.6). Thus, each nucleotide along a DNA strand has A, G, C, or T as its base, and each nucleotide along an RNA strand has A, G, C, or U as its base. An RNA molecule usually consists of a single strand.

It is customary to describe the flow of information from gene to protein in linguistic terms. Just as specific sequences of letters communicate information in a language like English, both nucleic acids and proteins are polymers with specific sequences of monomers that convey information. In DNA or RNA, the monomers are the four types of nucleotides, which differ in their nitrogenous bases. Genes are typically hundreds or thousands of nucleotides long, each gene having a specific sequence of nucleotides. Each polypeptide of a protein also has monomers arranged in a particular linear order (the protein's primary structure; see Figure 3.22), but its monomers are amino acids. Thus, nucleic acids and proteins contain information written in two different chemical languages. Getting from DNA to protein requires two major stages: transcription and translation.

**Transcription** is the synthesis of RNA using information in the DNA. The two nucleic acids are written in different forms of the same language, and the information is simply transcribed, or "rewritten," from DNA to RNA. Just as a DNA strand provides a template for making a new complementary strand during DNA replication, it also can serve as a template for assembling a complementary sequence of RNA nucleotides. For a protein-coding gene, the resulting RNA molecule is a faithful transcript of the gene's protein-building instructions.

This type of RNA molecule is called **messenger RNA (mRNA)** because it carries a genetic message from the DNA to the protein-synthesizing machinery of the cell. (Transcription is the general term for the synthesis of *any* kind of RNA on a DNA template. Later, you'll learn about some other types of RNA produced by transcription.)

**Translation** is the synthesis of a polypeptide using the information in the mRNA. During this stage, there is a change in language: The cell must translate the nucleotide sequence of an mRNA molecule into the amino acid sequence of a polypeptide. The sites of translation are **ribosomes**, molecular complexes that facilitate the orderly linking of amino acids into polypeptide chains.

Transcription and translation occur in all organisms. Because most studies of transcription and translation have used bacteria and eukaryotic cells, they are our main focus in this chapter. While our understanding of transcription and translation in archaea lags behind, we do know that archaeal cells share some features of gene expression with bacteria and others with eukaryotes.

The basic mechanics of transcription and translation are similar for bacteria and eukaryotes, but there is an important difference in the flow of genetic information within the cells. Bacteria do not have nuclei. Therefore, nuclear membranes do not separate bacterial DNA and mRNA from ribosomes and the other protein-synthesizing equipment **(Figure 14.4a)**. This lack of compartmentalization allows translation of an mRNA to begin while its transcription is still in progress, as you'll see later. By contrast, eukaryotic cells have nuclei. The presence of a nuclear envelope separates transcription from translation in space and time **(Figure 14.4b)**. Transcription occurs in the nucleus, but the mRNA must be transported to the cytoplasm for translation. Before eukaryotic RNA transcripts from protein-coding genes can leave the nucleus, they are modified in various ways to produce the final, functional mRNA. The transcription of a protein-coding eukaryotic gene results in *pre-mRNA*, and further processing yields the finished mRNA. The initial RNA transcript from any gene, including those specifying RNA that is not translated into protein, is more generally called a **primary transcript**.

To summarize: Genes program protein synthesis via genetic messages in the form of messenger RNA. Put another way, cells are governed by a molecular chain of command with a directional flow of genetic information, shown here by arrows:

This concept was dubbed the *central dogma* by Francis Crick in 1956. How has the concept held up over time? In the 1970s, scientists were surprised to discover that some enzymes

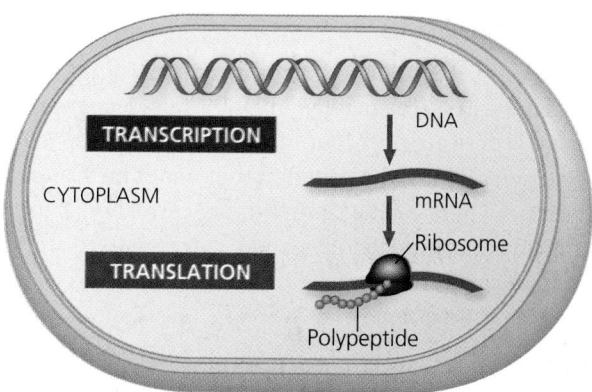

**(a) Bacterial cell.** In a bacterial cell, which lacks a nucleus, mRNA produced by transcription is immediately translated without additional processing.

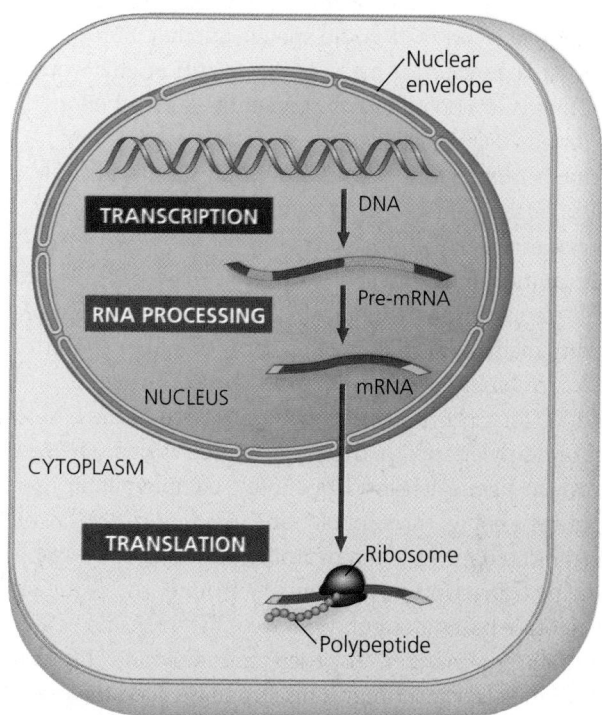

**(b) Eukaryotic cell.** The nucleus provides a separate compartment for transcription. The original RNA transcript, called pre-mRNA, is processed in various ways before leaving the nucleus as mRNA.

▲ **Figure 14.4 Overview: the roles of transcription and translation in the flow of genetic information.** In a cell, inherited information flows from DNA to RNA to protein. The two main stages of information flow are transcription and translation. A miniature version of part (a) or (b) accompanies several figures later in the chapter as an orientation diagram to help you see where a particular figure fits into the overall scheme of gene expression.

exist that use RNA molecules as templates for DNA synthesis (a process you'll read about in Chapter 17). However, these exceptions do not invalidate the idea that, in general, genetic information flows from DNA to RNA to protein. Now let's discuss how the instructions for assembling amino acids into a specific order are encoded in nucleic acids.

# The Genetic Code

When biologists began to suspect that the instructions for protein synthesis were encoded in DNA, they recognized a problem: There are only four nucleotide bases to specify 20 amino acids. Thus, the genetic code cannot be a language like Chinese, where each written symbol corresponds to a word. How many nucleotides, then, correspond to an amino acid?

## Codons: Triplets of Nucleotides

If each kind of nucleotide base were translated into an amino acid, only four amino acids could be specified, one per nucleotide base. Would a language of two-letter code words suffice? The two-nucleotide sequence AG, for example, could specify one amino acid, and GT could specify another. Since there are four possible nucleotide bases in each position, this would give us 16 (that is, $4^2$) possible arrangements—still not enough to code for all 20 amino acids.

Triplets of nucleotide bases are the smallest units of uniform length that can code for all the amino acids. If each arrangement of three consecutive nucleotide bases specifies an amino acid, there can be 64 (that is, $4^3$) possible code words—more than enough to specify all the amino acids. Experiments have verified that the flow of information from gene to protein is based on a **triplet code**: The genetic instructions for a polypeptide chain are written in the DNA as a series of nonoverlapping, three-nucleotide words. The series of words in a gene is transcribed into a complementary series of nonoverlapping, three-nucleotide words in mRNA, which is then translated into a chain of amino acids **(Figure 14.5)**.

During transcription, the gene determines the sequence of nucleotide bases along the length of the RNA molecule that is being synthesized. For each gene, only one of the two DNA strands is transcribed. This strand is called the **template strand** because it provides the pattern, or template, for the sequence of nucleotides in an RNA transcript. For any given gene, the same strand is used as the template every time the gene is transcribed. However, on the same DNA molecule farther along the chromosome, the opposite strand may be the one that functions as the template for another gene.

An mRNA molecule is complementary rather than identical to its DNA template because RNA nucleotides are assembled on the template according to base-pairing rules (see Figure 14.5). The pairs are similar to those that form during DNA replication, except that U (the RNA substitute for T) pairs with A and the mRNA nucleotides contain ribose instead of deoxyribose. Like a new strand of DNA, the RNA molecule is synthesized in an antiparallel direction to the template strand of DNA. (To review what is meant by "antiparallel" and the 5′ and 3′ ends of a nucleic acid chain, see Figure 13.8.) In the example in Figure 14.5, the nucleotide

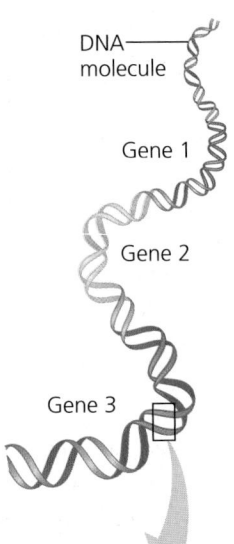

▶ **Figure 14.5 The triplet code.** For each gene, one DNA strand functions as a template for transcription of RNAs, such as mRNA. The base-pairing rules for DNA synthesis also guide transcription, except that uracil (U) takes the place of thymine (T) in RNA. During translation, the mRNA is read as a sequence of nucleotide triplets, called codons. Each codon specifies an amino acid to be added to the growing polypeptide chain. The mRNA is read in the 5′ → 3′ direction.

**?** *Compare the sequence of the mRNA to that of the nontemplate DNA strand, in both cases reading from 5′ → 3′.*

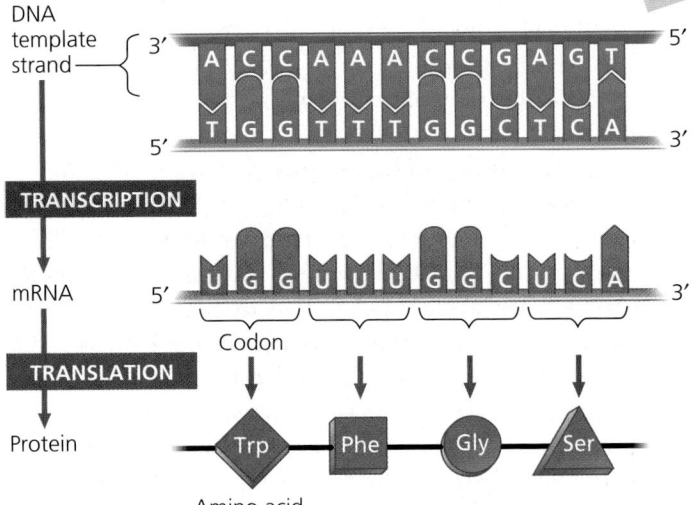

triplet ACC along the DNA template strand (written as 3′-ACC-5′) provides a template for 5′-UGG-3′ in the mRNA molecule. The mRNA nucleotide triplets are called **codons**, and they are customarily written in the 5′ → 3′ direction. In our example, UGG is the codon for the amino acid tryptophan (abbreviated Trp). The term *codon* is also used for the DNA nucleotide triplets along the *nontemplate* strand. These codons are complementary to the template strand and thus identical in sequence to the mRNA, except that they have T wherever there is a U in the mRNA. (For this reason, the nontemplate DNA strand is often called the *coding strand*.)

During translation, the sequence of codons along an mRNA molecule is decoded, or translated, into a sequence of amino acids making up a polypeptide chain. The codons are read by the translation machinery in the 5′ → 3′ direction along the mRNA. Each codon specifies which one of the 20 amino acids will be incorporated at the corresponding position along a polypeptide. Because codons are nucleotide triplets, the number of nucleotides making up a genetic message must be three times the number of amino acids in the protein product. For example, it takes 300 nucleotides along an mRNA strand to code for the amino acids in a polypeptide that is 100 amino acids long.

### Cracking the Code

Molecular biologists cracked the genetic code of life in the early 1960s when a series of elegant experiments disclosed the amino acid translations of each of the RNA codons. The first codon was deciphered in 1961 by Marshall Nirenberg, of the National Institutes of Health, along with his colleagues. Nirenberg synthesized an artificial mRNA by linking together many identical RNA nucleotides containing uracil as their base. No matter where the genetic message started or stopped, it could contain only one codon (UUU) over and over. Nirenberg added this "poly-U" polynucleotide to a test-tube mixture that contained all 20 amino acids, ribosomes, and the other components required for protein synthesis. His artificial system translated the poly-U mRNA into a polypeptide containing many units of the amino acid phenylalanine (Phe), strung together as a polyphenylalanine chain. Thus, Nirenberg determined that the mRNA codon UUU specifies the amino acid phenylalanine. Soon, the amino acids specified by the codons AAA, GGG, and CCC were determined in the same way.

Although more elaborate techniques were required to decode mixed triplets such as AUA and CGA, all 64 codons were deciphered by the mid-1960s. As **Figure 14.6** shows, 61 of the 64 triplets code for amino acids. The three codons that do not designate amino acids are "stop" signals, or termination codons, marking the end of translation. Notice that the codon AUG has a dual function: It codes for the amino acid methionine (Met) and also functions as a "start" signal, or initiation codon. Genetic messages usually begin with the mRNA codon AUG, which signals the protein-synthesizing machinery to begin translating the mRNA at that location. (Because AUG also stands for methionine, polypeptide chains begin with methionine when they are synthesized. However, an enzyme may subsequently remove this starter amino acid from the chain.)

Notice in Figure 14.6 that there is redundancy in the genetic code, but no ambiguity. For example, although codons GAA and GAG both specify glutamic acid (redundancy), neither of them ever specifies any other amino acid (no ambiguity). The redundancy in the code is not altogether random. In many cases, codons that are synonyms for a particular amino acid differ only in the third nucleotide base of the triplet. We'll consider the significance of this redundancy later in the chapter.

Our ability to extract the intended message from a written language depends on reading the symbols in the correct groupings—that is, in the correct **reading frame**. Consider this statement: "The red dog ate the bug." Group the letters

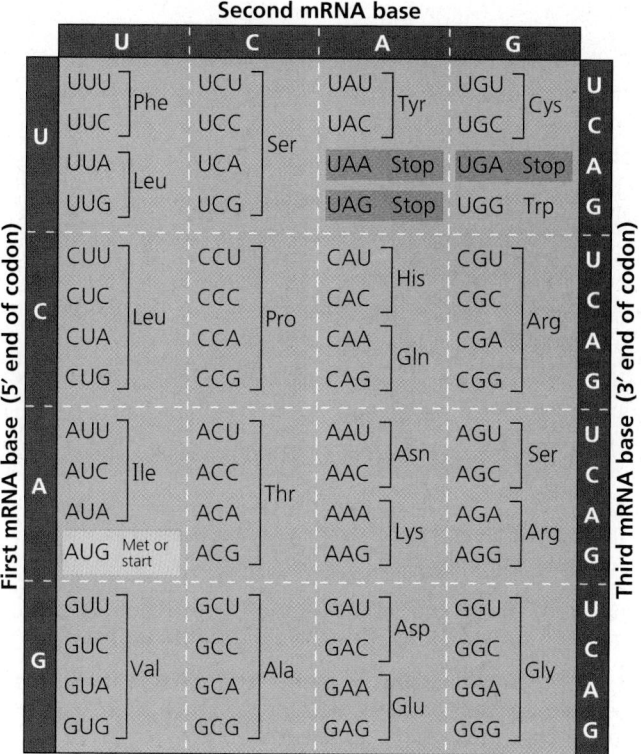

**Second mRNA base**

▲ **Figure 14.6 The codon table for mRNA.** The three nucleotide bases of an mRNA codon are designated here as the first, second, and third bases, reading in the 5′ → 3′ direction along the mRNA. (Practice using this table by finding the codons in Figure 14.5.) The codon AUG not only stands for the amino acid methionine (Met) but also functions as a "start" signal for ribosomes to begin translating the mRNA at that point. Three of the 64 codons function as "stop" signals, marking where ribosomes end translation. See Figure 3.18 for a list of the full names of all the amino acids.

incorrectly by starting at the wrong point, and the result will probably be gibberish: for example, "her edd oga tet heb ug." The reading frame is also important in the molecular language of cells. The short stretch of polypeptide shown in Figure 14.5, for instance, will be made correctly only if the mRNA nucleotides are read from left to right (5′ → 3′) in the groups of three shown in the figure: UGG UUU GGC UCA. Although a genetic message is written with no spaces between the codons, the cell's protein-synthesizing machinery reads the message as a series of nonoverlapping three-letter words. The message is *not* read as a series of overlapping words—UGGUUU, and so on—which would convey a very different message.

### Evolution of the Genetic Code

**EVOLUTION**   The genetic code is nearly universal, shared by organisms from the simplest bacteria to the most complex plants and animals. The mRNA codon CCG, for instance, is translated as the amino acid proline in all organisms whose genetic code has been examined. In laboratory experiments, genes can be transcribed and translated after being transplanted from one species to another, sometimes with quite

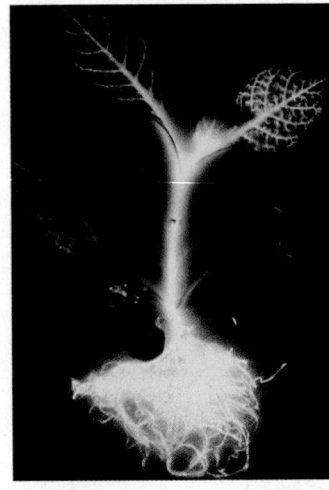

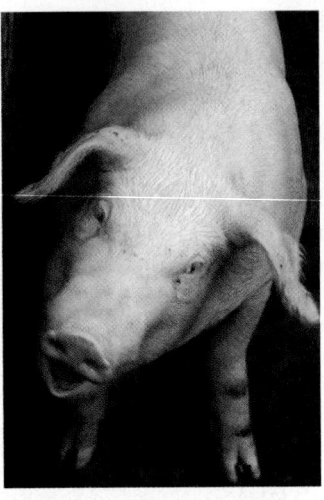

**(a) Tobacco plant expressing a firefly gene.** The yellow glow is produced by a chemical reaction catalyzed by the protein product of the firefly gene.

**(b) Pig expressing a jellyfish gene.** Researchers injected the gene for a fluorescent protein into fertilized pig eggs. One of the eggs developed into this fluorescent pig.

▲ **Figure 14.7 Expression of genes from different species.** Because diverse forms of life share a common genetic code, one species can be programmed to produce proteins characteristic of a second species by introducing DNA from the second species into the first.

striking results, as shown in **Figure 14.7**. Bacteria can be programmed by the insertion of human genes to synthesize certain human proteins for medical use, such as insulin. Such applications have produced many exciting developments in the area of genetic engineering (see Concept 13.4).

Despite a small number of exceptions in which a few codons differ from the standard ones, the evolutionary significance of the code's *near* universality is clear. A language shared by all living things must have been operating very early in the history of life—early enough to be present in the common ancestor of all present-day organisms. A shared genetic vocabulary is a reminder of the kinship that bonds all life on Earth.

### CONCEPT CHECK 14.1

1. **MAKE CONNECTIONS** In a research article about alkaptonuria published in 1902, Garrod suggested that humans inherit two "characters" (alleles) for a particular enzyme and that both parents must contribute a faulty version for the offspring to have the disorder. Today, would this disorder be called dominant or recessive? (See Concept 11.4.)

2. What polypeptide product would you expect from a poly-G mRNA that is 30 nucleotides long?

3. **DRAW IT** The template strand of a gene contains the sequence 3'-TTCAGTCGT-5'. Suppose that the nontemplate sequence could be transcribed instead of the template sequence. Draw the nontemplate sequence in 3' to 5' order. Then draw the mRNA sequence and translate it using Figure 14.6. (Be sure to pay attention to the 5' and 3' ends, remembering that the mRNA is antiparallel to the DNA strand.) Predict how well the protein synthesized from the nontemplate strand would function, if at all.

*For suggested answers, see Appendix A.*

# Transcription is the DNA-directed synthesis of RNA: *a closer look*

Now that we have considered the linguistic logic and evolutionary significance of the genetic code, we are ready to reexamine transcription, the first stage of gene expression, in more detail.

## Molecular Components of Transcription

Messenger RNA, the carrier of information from DNA to the cell's protein-synthesizing machinery, is transcribed from the template strand of a gene. An enzyme called an **RNA polymerase** pries the two strands of DNA apart and joins together RNA nucleotides complementary to the DNA template strand, thus elongating the RNA polynucleotide **(Figure 14.8)**. Like the DNA polymerases that function in DNA replication, RNA polymerases can assemble a polynucleotide only in its 5' → 3' direction. Unlike DNA polymerases, however, RNA polymerases are able to start a chain from scratch; they don't need a primer.

Specific sequences of nucleotides along the DNA mark where transcription of a gene begins and ends. The DNA sequence where RNA polymerase attaches and initiates transcription is known as the **promoter**; in bacteria, the sequence that signals the end of transcription is called the **terminator**. (The termination mechanism is different in eukaryotes; we'll describe it later.) Molecular biologists refer to the direction of transcription as "downstream" and the other direction as "upstream." These terms are also used to describe the positions of nucleotide sequences within the DNA or RNA. Thus, the promoter sequence in DNA is said to be upstream from the terminator. The stretch of DNA downstream from the promoter that is transcribed into an RNA molecule is called a **transcription unit**.

Bacteria have a single type of RNA polymerase that synthesizes not only mRNA but also other types of RNA that function in protein synthesis, such as ribosomal RNA. In contrast, eukaryotes have at least three types of RNA polymerase in their nuclei; the one used for pre-mRNA synthesis is called RNA polymerase II. In the discussion that follows, we start with the features of mRNA synthesis common to both bacteria and eukaryotes and then describe some key differences.

## Synthesis of an RNA Transcript

The three stages of transcription, as shown in Figure 14.8 and described next, are initiation, elongation, and termination of the RNA chain. Study Figure 14.8 to familiarize yourself with the stages and the terms used to describe them.

### RNA Polymerase Binding and Initiation of Transcription

The promoter of a gene includes within it the transcription **start point** (the nucleotide where RNA synthesis actually begins) and typically extends several dozen or more nucleotide pairs upstream from the start point. RNA polymerase binds in a precise location and orientation on the promoter, thereby determining where transcription starts and which of the two strands of the DNA helix is used as the template.

Certain sections of a promoter are especially important for binding RNA polymerase. In bacteria, part of the RNA polymerase itself specifically recognizes and binds to the promoter.

In eukaryotes, a collection of proteins called **transcription factors** mediate the binding of RNA polymerase and the initiation of transcription. Only after transcription factors are attached to the promoter does RNA polymerase II bind to it. The whole complex of transcription factors and RNA polymerase II bound to the promoter is called a **transcription initiation complex**. **Figure 14.9** shows the role of transcription factors

---

**Promoter**

Transcription unit

5'
3'

DNA

Start point

RNA polymerase

**1 Initiation.** After RNA polymerase binds to the promoter, the DNA strands unwind, and the polymerase initiates RNA synthesis at the start point on the template strand.

Nontemplate strand of DNA

5'
3'

3'
5'

Unwound DNA

RNA transcript

Template strand of DNA

**2 Elongation.** The polymerase moves downstream, unwinding the DNA and elongating the RNA transcript 5' → 3'. In the wake of transcription, the DNA strands re-form a double helix.

Rewound DNA

5'
3'

3'
5'

5'

3'

RNA transcript

**3 Termination.** Eventually, the RNA transcript is released, and the polymerase detaches from the DNA.

5'
3'

3'
5'

5'

3'

Completed RNA transcript

Direction of transcription ("downstream")

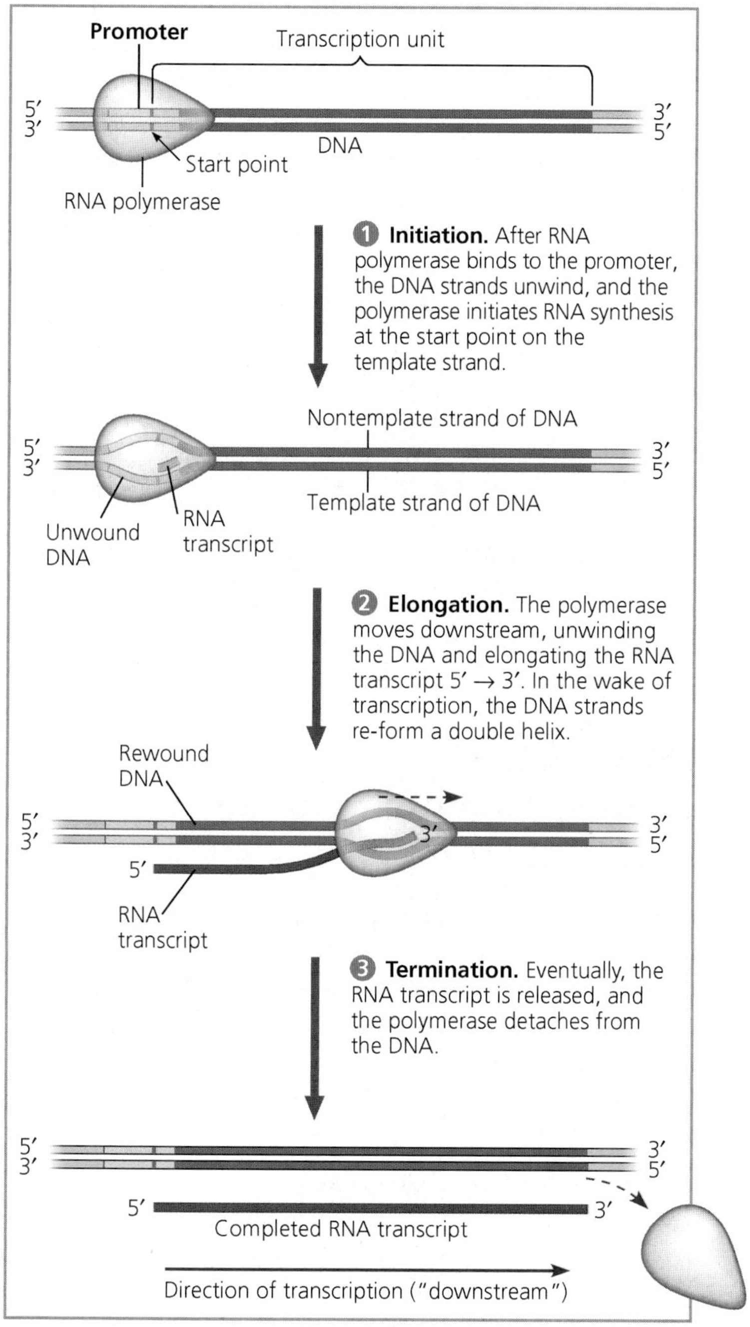

▲ **Figure 14.8 The stages of transcription: initiation, elongation, and termination.** This general depiction of transcription applies to both bacteria and eukaryotes, but the details of termination differ, as described in the text. Also, in a bacterium, the RNA transcript is immediately usable as mRNA; in a eukaryote, the RNA transcript must first undergo processing.

**MAKE CONNECTIONS** *Compare the use of a template strand during transcription and replication. See Figure 13.19.*

---

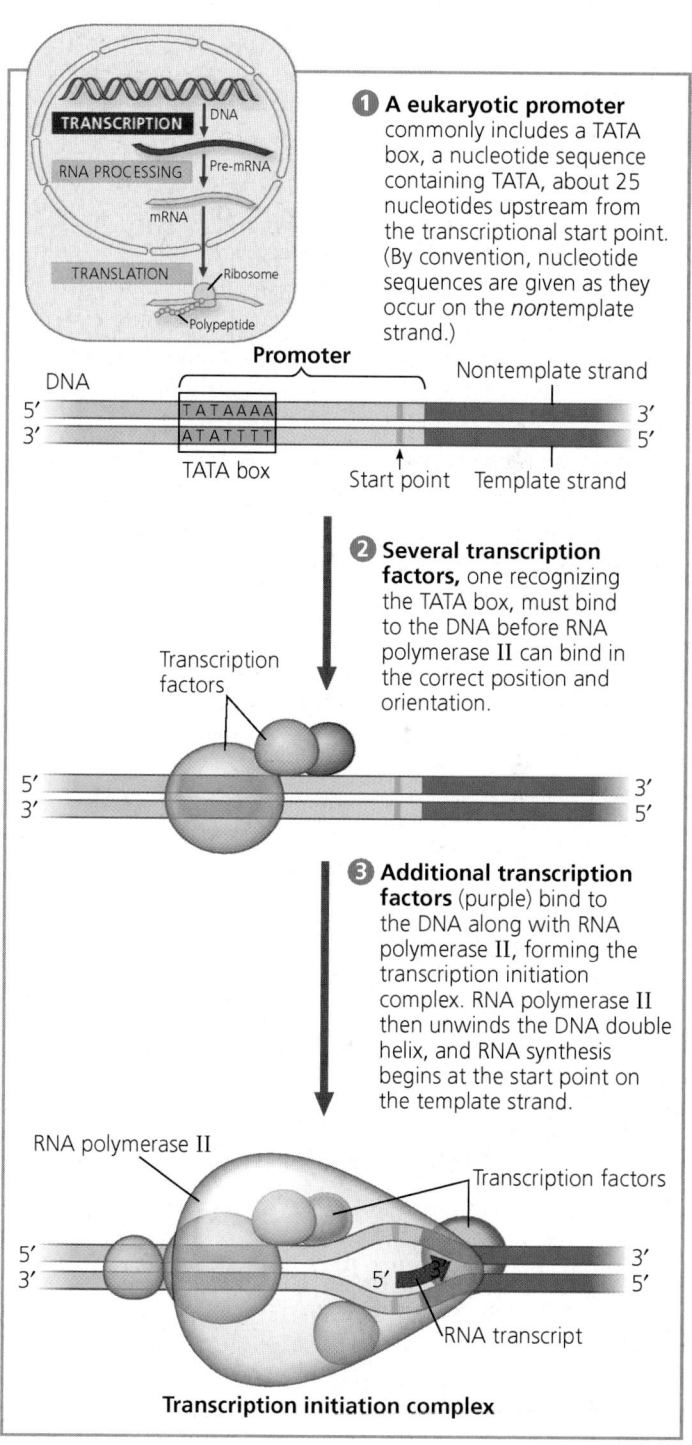

TRANSCRIPTION — DNA

RNA PROCESSING — Pre-mRNA

mRNA

TRANSLATION — Ribosome

Polypeptide

**1 A eukaryotic promoter** commonly includes a TATA box, a nucleotide sequence containing TATA, about 25 nucleotides upstream from the transcriptional start point. (By convention, nucleotide sequences are given as they occur on the *non*template strand.)

DNA

**Promoter**

Nontemplate strand

5'
3'

T A T A A A A
A T A T T T T

3'
5'

TATA box

Start point   Template strand

**2 Several transcription factors,** one recognizing the TATA box, must bind to the DNA before RNA polymerase II can bind in the correct position and orientation.

Transcription factors

5'
3'

3'
5'

**3 Additional transcription factors** (purple) bind to the DNA along with RNA polymerase II, forming the transcription initiation complex. RNA polymerase II then unwinds the DNA double helix, and RNA synthesis begins at the start point on the template strand.

RNA polymerase II

Transcription factors

5'
3'

3'
5'

5'   3'

RNA transcript

**Transcription initiation complex**

▲ **Figure 14.9 The initiation of transcription at a eukaryotic promoter.** In eukaryotic cells, proteins called transcription factors mediate the initiation of transcription by RNA polymerase II.

**?** *Explain how the interaction of RNA polymerase with the promoter would differ if the figure showed transcription initiation for bacteria.*

and a crucial promoter DNA sequence called the **TATA box** in forming the initiation complex at a eukaryotic promoter.

The interaction between eukaryotic RNA polymerase II and transcription factors is an example of the importance of protein-protein interactions in controlling eukaryotic transcription. Once the appropriate transcription factors are firmly attached to the promoter DNA and the polymerase is bound in the correct orientation, the enzyme unwinds the two DNA strands and begins transcribing the template strand at the start point.

### Elongation of the RNA Strand

As RNA polymerase moves along the DNA, it untwists the double helix, exposing about 10–20 DNA nucleotides at a time for pairing with RNA nucleotides **(Figure 14.10)**. The enzyme adds nucleotides to the 3′ end of the growing RNA molecule as it continues along the double helix. In the wake of this advancing wave of RNA synthesis, the new RNA molecule peels away from its DNA template, and the DNA double helix re-forms. Transcription progresses at a rate of about 40 nucleotides per second in eukaryotes.

A single gene can be transcribed simultaneously by several molecules of RNA polymerase following each other like trucks in a convoy. A growing strand of RNA trails off from each polymerase, with the length of each new strand reflecting how far along the template the enzyme has traveled from the start point (see the mRNA molecules in Figure 14.23). The congregation of many polymerase molecules simultaneously transcribing a single gene increases the amount of mRNA transcribed from it, which helps the cell make the encoded protein in large amounts.

### Termination of Transcription

The mechanism of termination differs between bacteria and eukaryotes. In bacteria, transcription proceeds through a terminator sequence in the DNA. The transcribed terminator (an RNA sequence) functions as the termination signal, causing the polymerase to detach from the DNA and release the transcript, which requires no further modification before translation. In eukaryotes, RNA polymerase II transcribes a sequence on the DNA called the polyadenylation signal sequence, which specifies a polyadenylation signal (AAUAAA) in the pre-mRNA. This is called a "signal" because when this stretch of six RNA nucleotides appears, it is immediately bound by certain proteins in the nucleus. Then, at a point about 10–35 nucleotides downstream from the AAUAAA signal, these proteins cut the RNA transcript free from the polymerase, releasing the pre-mRNA. The pre-mRNA then undergoes processing, the topic of the next section.

**CONCEPT 14.3**

# Eukaryotic cells modify RNA after transcription

Enzymes in the eukaryotic nucleus modify pre-mRNA in specific ways before the genetic message is dispatched to the cytoplasm. During this **RNA processing**, both ends of the primary transcript are altered. Also, in most cases, certain interior sections of the RNA molecule are cut out and the remaining parts spliced together. These modifications produce an mRNA molecule ready for translation.

## Alteration of mRNA Ends

Each end of a pre-mRNA molecule is modified in a particular way **(Figure 14.11)**. The 5′ end, which is synthesized first, receives a **5′ cap**, a modified form of a guanine (G) nucleotide added onto the 5′ end after transcription of the first 20–40 nucleotides. The 3′ end of the pre-mRNA molecule is also modified before the mRNA exits the nucleus. Recall that the pre-mRNA is released soon after the polyadenylation signal, AAUAAA, is transcribed. At the 3′ end, an enzyme then adds 50–250 more adenine (A) nucleotides, forming a **poly-A tail**. The 5′ cap and poly-A tail share several important functions. First, they seem to facilitate the export of the mature

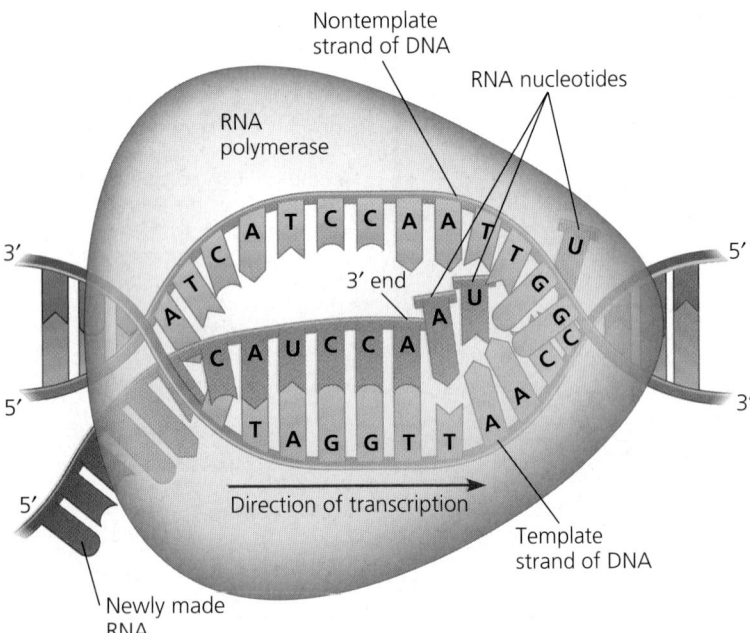

▲ **Figure 14.10 Transcription elongation.** RNA polymerase moves along the DNA template strand, joining complementary RNA nucleotides to the 3′ end of the growing RNA transcript. Behind the polymerase, the new RNA peels away from the template strand, which re-forms a double helix with the nontemplate strand.

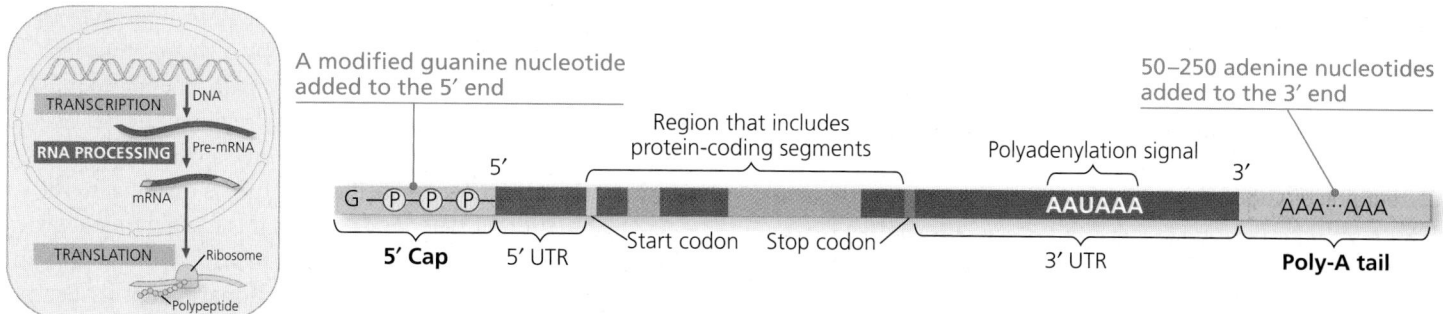

**▲ Figure 14.11 RNA processing: Addition of the 5′ cap and poly-A tail.** Enzymes modify the two ends of a eukaryotic pre-mRNA molecule. The modified ends may promote the export of mRNA from the nucleus, and they help protect the mRNA from degradation. When the mRNA reaches the cytoplasm, the modified ends, in conjunction with certain cytoplasmic proteins, facilitate ribosome attachment. The 5′ cap and poly-A tail are not translated into protein, nor are the regions called the 5′ untranslated region (5′ UTR) and 3′ untranslated region (3′ UTR). The pink segments will be described shortly.

mRNA from the nucleus. Second, they help protect the mRNA from degradation by hydrolytic enzymes. And third, they help ribosomes attach to the 5′ end of the mRNA once the mRNA reaches the cytoplasm. Figure 14.11 shows a diagram of a eukaryotic mRNA molecule with cap and tail. The figure also shows the untranslated regions (UTRs) at the 5′ and 3′ ends of the mRNA (referred to as the 5′ UTR and 3′ UTR). The UTRs are parts of the mRNA that will not be translated into protein, but they have other functions, such as ribosome binding.

## Split Genes and RNA Splicing

A remarkable stage of RNA processing in the eukaryotic nucleus is the removal of large portions of the RNA molecule and reconnection of the remaining portions. This cut-and-paste job, called **RNA splicing**, is similar to editing a video. The average length of a transcription unit along a human DNA molecule is about 27,000 nucleotide pairs, so the primary RNA transcript is also that long. However, the average-sized protein of 400 amino acids requires only 1,200 nucleotides in RNA to code for it. (Remember, each amino acid is encoded by a *triplet* of nucleotides.) This is because most eukaryotic genes and their RNA transcripts

have long noncoding stretches of nucleotides, regions that are not translated. Even more surprising is that most of these noncoding sequences are interspersed between coding segments of the gene and thus between coding segments of the pre-mRNA. In other words, the sequence of DNA nucleotides that codes for a eukaryotic polypeptide is usually not continuous; it is split into segments. The noncoding segments of nucleic acid that lie between coding regions are called *in*tervening sequences, or **introns**. The other regions are called **exons**, because they are eventually *ex*pressed, usually by being translated into amino acid sequences. (Exceptions include the UTRs of the exons at the ends of the RNA, which make up part of the mRNA but are not translated into protein. Because of these exceptions, you may prefer to think of exons as sequences of RNA that *exit* the nucleus.) The terms *intron* and *exon* are used for both RNA sequences and the DNA sequences that specify them.

In making a primary transcript from a gene, RNA polymerase II transcribes both introns and exons from the DNA, but the mRNA molecule that enters the cytoplasm is an abridged version **(Figure 14.12)**. The introns are cut out from the molecule and the exons joined together, forming an mRNA

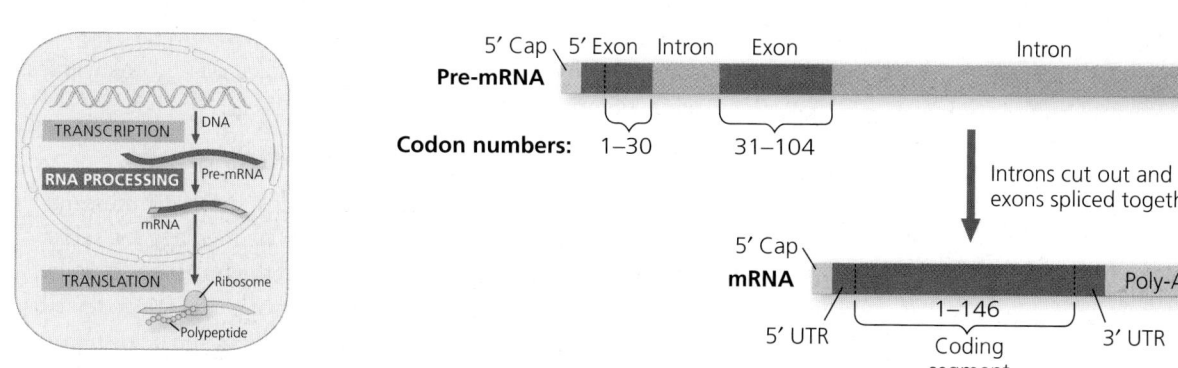

**▲ Figure 14.12 RNA processing: RNA splicing.** The RNA molecule shown here codes for β-globin, one of the polypeptides of hemoglobin. The numbers under the RNA refer to codons; β-globin is 146 amino acids long. The β-globin gene and its pre-mRNA transcript have three exons, corresponding to sequences that will leave the nucleus as mRNA. (The 5′ UTR and 3′ UTR are parts of exons because they are included in the mRNA; however, they do not code for protein.) During RNA processing, the introns are cut out and the exons spliced together. In many genes, the introns are much larger than the exons.

**?** *On the mRNA, indicate the sites of the start and stop codons.*

molecule with a continuous coding sequence. This is the process of RNA splicing.

One important consequence of the presence of introns in genes is that a single gene can encode more than one kind of polypeptide. Many genes are known to give rise to two or more different polypeptides, depending on which segments are treated as exons during RNA processing; this is called **alternative RNA splicing** (see Figure 15.12). Because of alternative splicing, the number of different protein products an organism produces can be much greater than its number of genes.

How is pre-mRNA splicing carried out? The removal of introns is accomplished by a large complex made of proteins and small RNAs called a **spliceosome**. This complex binds to several short nucleotide sequences along the intron, including key sequences at each end **(Figure 14.13)**. The intron is then released (and rapidly degraded), and the spliceosome joins together the two exons that flanked the intron. It turns out that the small RNAs in the spliceosome not only participate in the assembly of the spliceosome and recognition of the splice site but also catalyze the splicing process.

### Ribozymes

The idea of a catalytic role for the RNAs in the spliceosome arose from the discovery of **ribozymes**, RNA molecules that function as enzymes. In some organisms, RNA splicing can occur without proteins or even additional RNA molecules: The intron RNA functions as a ribozyme and catalyzes its own excision! For example, in the ciliate protist *Tetrahymena*, self-splicing occurs in the production of ribosomal RNA (rRNA), a component of the organism's ribosomes. The pre-rRNA actually removes its own introns. The discovery of ribozymes rendered obsolete the idea that all biological catalysts are proteins.

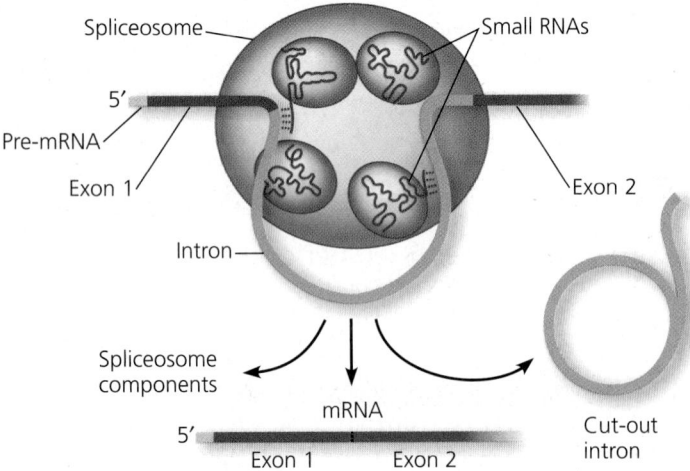

▲ **Figure 14.13 A spliceosome splicing a pre-mRNA.** The diagram shows a portion of a pre-mRNA transcript, with an intron (pink) flanked by two exons (red). Small RNAs within the spliceosome base-pair with nucleotides at specific sites along the intron. Next, small spliceosome RNAs also catalyze cutting of the pre-mRNA and the splicing together of the exons, releasing the intron for rapid degradation. (These RNAs are *ribozymes*, discussed in the next section.)

Three properties of RNA enable some RNA molecules to function as enzymes. First, because RNA is single-stranded, a region of an RNA molecule may base-pair, in an antiparallel arrangement, with a complementary region elsewhere in the same molecule; this gives the molecule a particular three-dimensional structure. A specific structure is essential to the catalytic function of ribozymes, just as it is for enzymatic proteins. Second, like certain amino acids in an enzymatic protein, some of the bases in RNA contain functional groups that can participate in catalysis. Third, the ability of RNA to hydrogen-bond with other nucleic acid molecules (either RNA or DNA) adds specificity to its catalytic activity. For example, complementary base pairing between the RNA of the spliceosome and the RNA of a primary RNA transcript precisely locates the region where the ribozyme catalyzes splicing. Later in this chapter, you'll see how these properties of RNA also allow it to perform important noncatalytic roles in the cell, such as recognition of the three-nucleotide codons on mRNA.

**CONCEPT CHECK 14.3**

1. Given that there are about 20,000 human genes, how can human cells make 75,000–100,000 different proteins?
2. How is RNA splicing similar to how you would watch a TV show recorded earlier using a DVR? In what ways is it different? What would introns correspond to in this analogy?
3. **WHAT IF?** What would be the effect of treating cells with an agent that removed the cap from mRNAs?

For suggested answers, see Appendix A.

---

**CONCEPT 14.4**

# Translation is the RNA-directed synthesis of a polypeptide: *a closer look*

We will now examine in greater detail how genetic information flows from mRNA to protein—the process of translation. As we did for transcription, we'll concentrate on the basic steps of translation that occur in both bacteria and eukaryotes, while pointing out key differences.

## Molecular Components of Translation

In the process of translation, a cell "reads" a genetic message and builds a polypeptide accordingly. The message is a series of codons along an mRNA molecule, and the translator is called a **transfer RNA (tRNA)**. The function of a tRNA is to transfer amino acids from the cytoplasmic pool of amino acids to a growing polypeptide in a ribosome. A cell keeps its cytoplasm stocked with all 20 amino acids, either by synthesizing them from other compounds or by taking them up from the surrounding solution. The ribosome, a structure made of proteins and RNAs, adds each amino acid brought to it by a tRNA to the growing end of a polypeptide chain **(Figure 14.14)**.

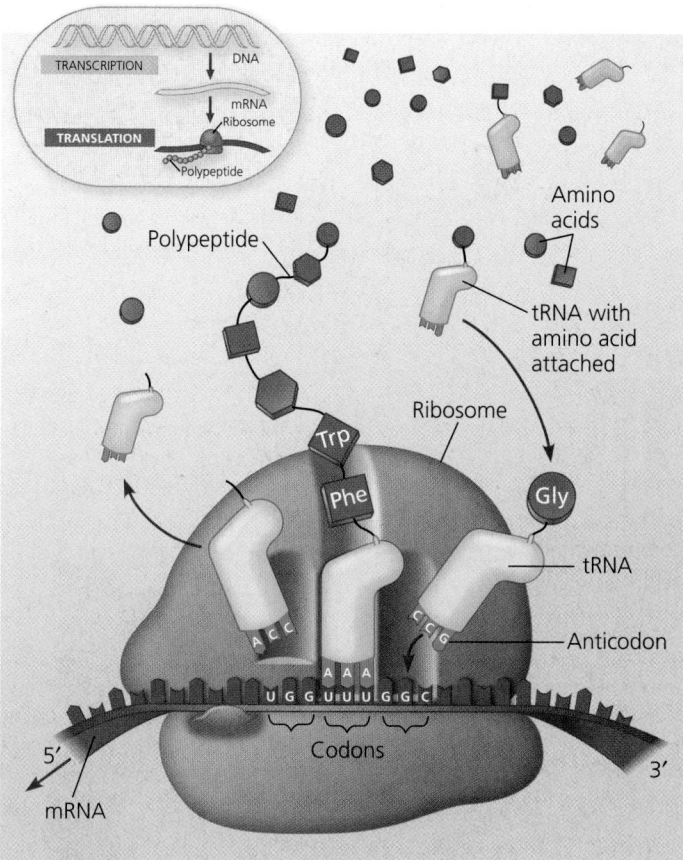

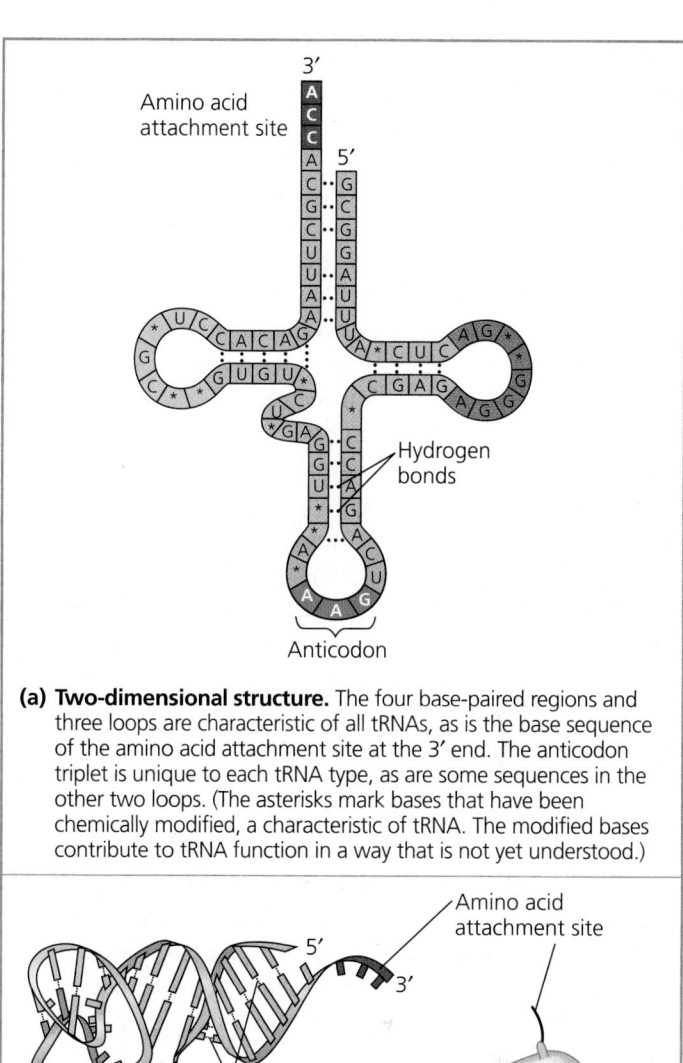

**(a) Two-dimensional structure.** The four base-paired regions and three loops are characteristic of all tRNAs, as is the base sequence of the amino acid attachment site at the 3′ end. The anticodon triplet is unique to each tRNA type, as are some sequences in the other two loops. (The asterisks mark bases that have been chemically modified, a characteristic of tRNA. The modified bases contribute to tRNA function in a way that is not yet understood.)

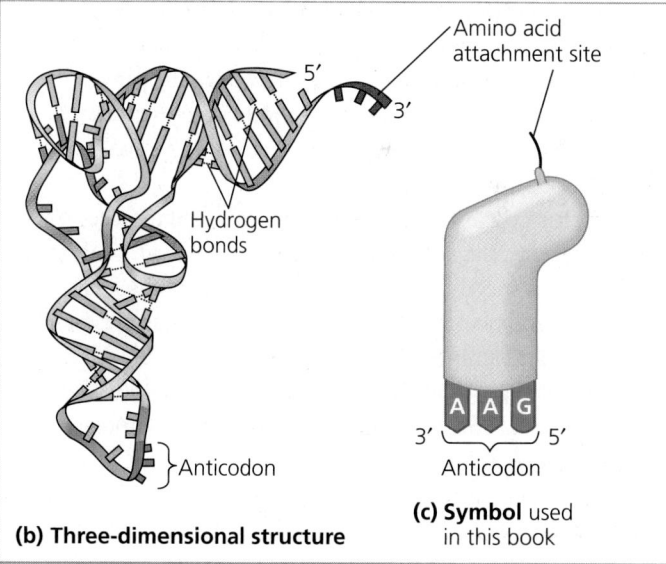

**(b) Three-dimensional structure**

**(c) Symbol** used in this book

▲ **Figure 14.14 Translation: the basic concept.** As a molecule of mRNA is moved through a ribosome, codons are translated into amino acids, one by one. The translators, or interpreters, are tRNA molecules, each type with a specific nucleotide triplet called an anticodon at one end and a corresponding amino acid at the other end. A tRNA adds its amino acid cargo to a growing polypeptide chain after the anticodon hydrogen-bonds to the complementary codon on the mRNA. The figures that follow show some of the details of translation in a bacterial cell.

 **ANIMATION** Visit the Study Area in **MasteringBiology** for the BioFlix® 3-D Animation on Protein Synthesis.

Translation is simple in principle but complex in its biochemistry and mechanics. In dissecting translation, we'll focus on the slightly less complicated version of the process that occurs in bacteria. We'll first look at the major players in this cellular process.

### The Structure and Function of Transfer RNA

The key to translating a genetic message into a specific amino acid sequence is the fact that each tRNA enables translation of a given mRNA codon into a certain amino acid. This is possible because a tRNA bears a specific amino acid at one end of its three-dimensional structure, while at the other end is a nucleotide triplet that can base-pair with the complementary codon on mRNA.

A tRNA molecule consists of a single RNA strand that is only about 80 nucleotides long (compared to hundreds of

▲ **Figure 14.15 The structure of transfer RNA (tRNA).**

nucleotides for most mRNA molecules). Because of the presence of complementary stretches of nucleotide bases that can hydrogen-bond to each other, this single strand can fold back on itself and form a molecule with a three-dimensional structure. Flattened into one plane to clarify this base pairing, a tRNA molecule looks like a cloverleaf **(Figure 14.15a)**. The tRNA actually twists and folds into a compact three-dimensional structure that is roughly L-shaped, with the 5′ and 3′ ends of the linear tRNA both located near one end of the structure **(Figure 14.15b)**. The protruding 3′ end acts as the attachment site for an amino acid. The loop extending from the other end of the L includes the **anticodon**, the particular nucleotide triplet that base-pairs to a specific mRNA codon. Thus, the structure of a tRNA molecule fits its function.

Anticodons are conventionally written $3' \rightarrow 5'$ to align properly with codons written $5' \rightarrow 3'$ (see Figure 14.14). (For base pairing, RNA strands must be antiparallel, like DNA.) As an example of how tRNAs work, consider the mRNA codon 5'-GGC-3', which is translated as the amino acid glycine. The tRNA that base-pairs with this codon by hydrogen bonding has 3'-CCG-5' as its anticodon and carries glycine at its other end (see the incoming tRNA approaching the ribosome in Figure 14.14). As an mRNA molecule is moved through a ribosome, glycine will be added to the polypeptide chain whenever the codon 5'-GGC-3' is presented for translation. Codon by codon, the genetic message is translated as tRNAs position each amino acid, in the order prescribed, and the ribosome adds that amino acid onto the growing polypeptide chain. The tRNA molecule is a translator in the sense that, in the context of the ribosome, it can read a nucleic acid word (the mRNA codon) and interpret it as a protein word (the amino acid).

Like mRNA and other types of cellular RNA, transfer RNA molecules are transcribed from DNA templates. In a eukaryotic cell, tRNA, like mRNA, is made in the nucleus and then travels from the nucleus to the cytoplasm, where translation occurs. In both bacterial and eukaryotic cells, each tRNA molecule is used repeatedly, picking up its designated amino acid in the cytosol, depositing this cargo onto a polypeptide chain at the ribosome, and then leaving the ribosome, ready to pick up another of the same amino acid.

The accurate translation of a genetic message requires two instances of molecular recognition. First, a tRNA that binds to an mRNA codon specifying a particular amino acid must carry that amino acid, and no other, to the ribosome. The correct matching up of tRNA and amino acid is carried out by a family of related enzymes called **aminoacyl-tRNA synthetases (Figure 14.16)**. The active site of each type of aminoacyl-tRNA synthetase fits only a specific combination of amino acid and tRNA. (Regions of both the amino acid attachment end and the anticodon end of the tRNA are instrumental in ensuring the specific fit.) There are 20 different synthetases, one that joins each amino acid to the right tRNA; each synthetase is able to bind to all the different tRNAs that code for its particular amino acid. The synthetase catalyzes the covalent attachment of the amino acid to its tRNA in a process driven by the hydrolysis of ATP. The resulting aminoacyl tRNA, also called a charged tRNA, is released from the enzyme and is then available to deliver its amino acid to a growing polypeptide chain on a ribosome.

The second instance of molecular recognition is the pairing of the tRNA anticodon with the appropriate mRNA codon. If one tRNA variety existed for each mRNA codon specifying an amino acid, there would be 61 tRNAs (see Figure 14.6). In fact, there are only about 45, signifying that some tRNAs must be able to bind to more than one codon. Such versatility

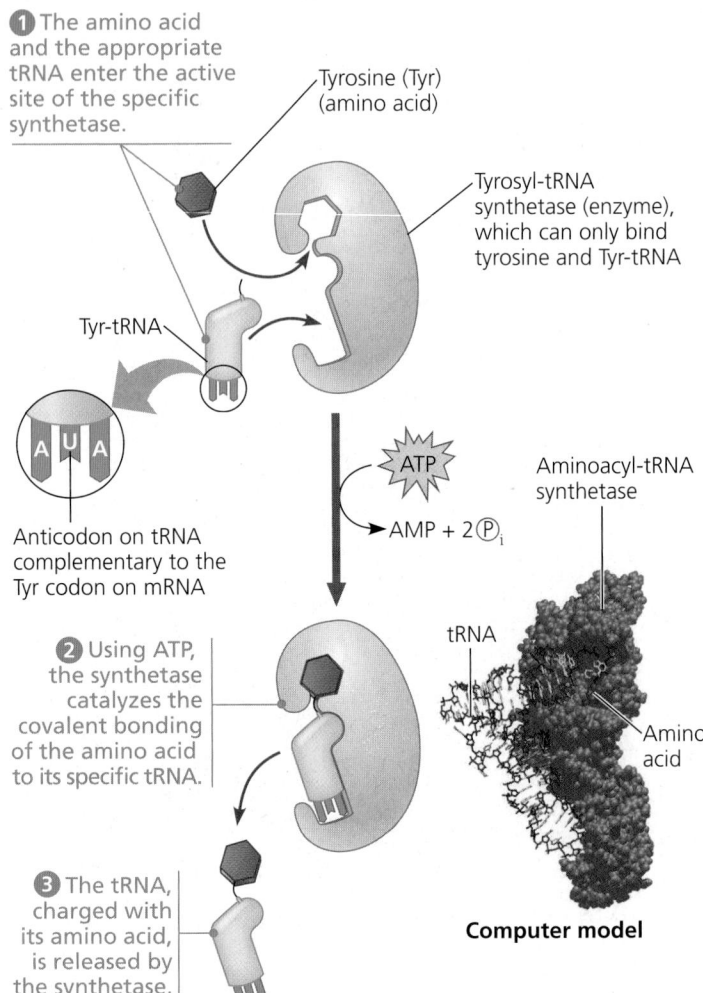

**1** The amino acid and the appropriate tRNA enter the active site of the specific synthetase.

Tyrosine (Tyr) (amino acid)

Tyrosyl-tRNA synthetase (enzyme), which can only bind tyrosine and Tyr-tRNA

Tyr-tRNA

A U A

Anticodon on tRNA complementary to the Tyr codon on mRNA

ATP

AMP + 2 (P)$_i$

Aminoacyl-tRNA synthetase

**2** Using ATP, the synthetase catalyzes the covalent bonding of the amino acid to its specific tRNA.

tRNA

Amino acid

**Computer model**

**3** The tRNA, charged with its amino acid, is released by the synthetase.

▲ **Figure 14.16 Aminoacyl-tRNA synthetases provide specificity in joining amino acids to their tRNAs.** Linkage of a tRNA to its amino acid is an endergonic process that occurs at the expense of ATP (which loses two phosphate groups, becoming AMP).

is possible because the rules for base pairing between the third nucleotide base of a codon and the corresponding base of a tRNA anticodon are relaxed compared to those at other codon positions. For example, the nucleotide base U at the 5' end of a tRNA anticodon can pair with A or G in the third position (at the 3' end) of an mRNA codon. The flexible base pairing at this codon position is called **wobble**. Wobble explains why synonymous codons for an amino acid most often differ in their third nucleotide base but not in the other bases. One such case is that a tRNA with the anticodon 3'-UCU-5' can base-pair with either the mRNA codon 5'-AGA-3' or 5'-AGG-3', both of which code for arginine (see Figure 14.6).

### Ribosomes

Ribosomes facilitate the specific coupling of tRNA anticodons with mRNA codons during protein synthesis. A ribosome consists of a large subunit and a small subunit, each made up of proteins and one or more **ribosomal RNAs (rRNAs) (Figure 14.17)**. In eukaryotes, the subunits are made in the nucleolus. Ribosomal RNA genes are transcribed, and the RNA

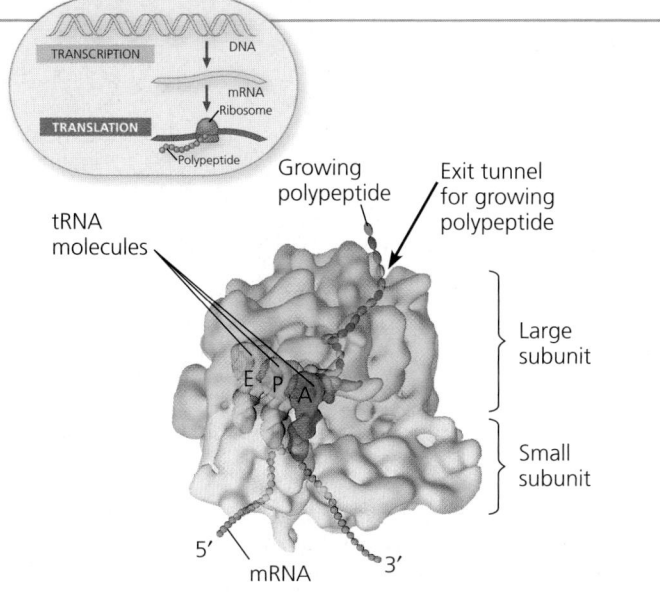

(a) **Computer model of functioning ribosome.** This is a model of a bacterial ribosome, showing its overall shape. The eukaryotic ribosome is roughly similar. A ribosomal subunit is a complex of ribosomal RNA molecules and proteins.

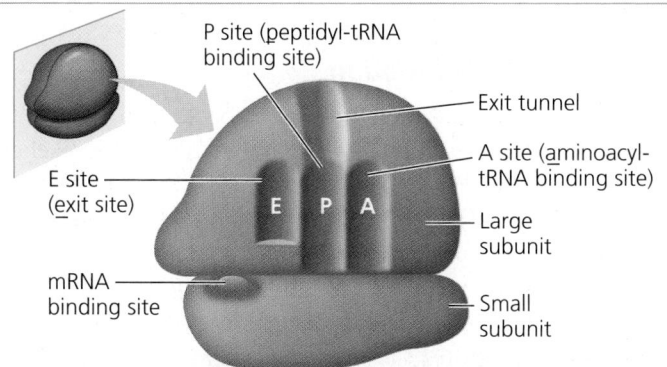

(b) **Schematic model showing binding sites.** A ribosome has an mRNA binding site and three tRNA binding sites, known as the A, P, and E sites. This schematic ribosome will appear in later diagrams.

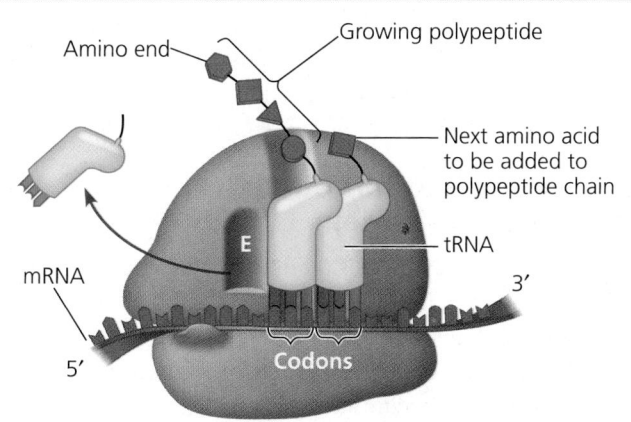

(c) **Schematic model with mRNA and tRNA.** A tRNA fits into a binding site when its anticodon base-pairs with an mRNA codon. The P site holds the tRNA attached to the growing polypeptide. The A site holds the tRNA carrying the next amino acid to be added to the polypeptide chain. Discharged tRNAs leave from the E site. The polypeptide grows at its carboxyl end.

▲ **Figure 14.17 The anatomy of a functioning ribosome.**

is processed and assembled with proteins imported from the cytoplasm. Completed ribosomal subunits are then exported via nuclear pores to the cytoplasm. In both bacteria and eukaryotes, a large and a small subunit join to form a functional ribosome only when attached to an mRNA molecule. About one-third of the mass of a ribosome is made up of proteins; the rest consists of rRNAs, either three molecules (in bacteria) or four (in eukaryotes). Because most cells contain thousands of ribosomes, rRNA is the most abundant type of cellular RNA.

Although the ribosomes of bacteria and eukaryotes are very similar in structure and function, eukaryotic ribosomes are slightly larger, as well as differing somewhat from bacterial ribosomes in their molecular composition. The differences are medically significant. Certain antibiotic drugs can inactivate bacterial ribosomes without affecting eukaryotic ribosomes to make proteins. These drugs, including tetracycline and streptomycin, are used to combat bacterial infections.

The structure of a ribosome reflects its function of bringing mRNA together with tRNAs carrying amino acids. In addition to a binding site for mRNA, each ribosome has three binding sites for tRNA, as described in Figure 14.17. The **P site** (peptidyl-tRNA binding site) holds the tRNA carrying the growing polypeptide chain, while the **A site** (aminoacyl-tRNA binding site) holds the tRNA carrying the next amino acid to be added to the chain. Discharged tRNAs leave the ribosome from the **E site** (exit site). The ribosome holds the tRNA and mRNA in close proximity and positions the new amino acid so that it can be added to the carboxyl end of the growing polypeptide. It then catalyzes the formation of the peptide bond. As the polypeptide becomes longer, it passes through an *exit tunnel* in the ribosome's large subunit. When the polypeptide is complete, it is released through the exit tunnel.

The widely accepted model is that rRNA, not the proteins, is primarily responsible for both the structure and the function of the ribosome. The proteins, which are largely on the exterior, support the shape changes of the rRNA molecules as they carry out catalysis during translation. Ribosomal RNA is the main constituent of the A and P sites and of the interface between the two ribosomal subunits; it also acts as the catalyst of peptide bond formation. Thus, a ribosome could actually be considered one colossal ribozyme!

## Building a Polypeptide

We can divide translation, the synthesis of a polypeptide chain, into three stages: initiation, elongation, and termination. All three stages require protein "factors" that aid in the translation process. For certain aspects of chain initiation and elongation, energy is also required. It is provided by the hydrolysis of guanosine triphosphate (GTP).

### Ribosome Association and Initiation of Translation

The initiation stage of translation brings together mRNA, a tRNA bearing the first amino acid of the polypeptide, and the two subunits of a ribosome. First, a small ribosomal subunit binds to both mRNA and a specific initiator tRNA, which carries the amino acid methionine. In bacteria, the small subunit can bind these two in either order; it binds the mRNA at a specific RNA sequence, just upstream of the start codon, AUG. In eukaryotes, the small subunit, with the initiator tRNA already bound, binds to the 5′ cap of the mRNA and then moves, or *scans*, downstream along the mRNA until it reaches the AUG start codon, where the initiator tRNA hydrogen-bonds, as shown in step 1 of **Figure 14.18**. In either case, the start codon signals the start of translation; this is important because it establishes the codon reading frame for the mRNA. In the **Scientific Skills Exercise**, you can work with DNA sequences encoding the ribosomal binding sites on the mRNAs of a group of *Escherichia coli* (*E. coli*) genes.

The union of mRNA, initiator tRNA, and a small ribosomal subunit is followed

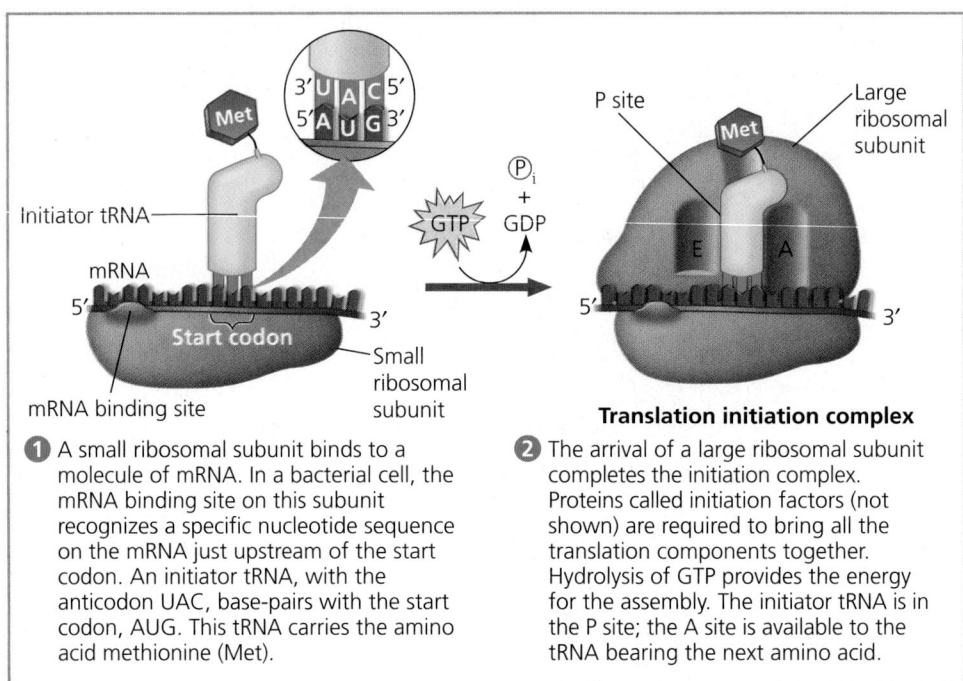

**① A** small ribosomal subunit binds to a molecule of mRNA. In a bacterial cell, the mRNA binding site on this subunit recognizes a specific nucleotide sequence on the mRNA just upstream of the start codon. An initiator tRNA, with the anticodon UAC, base-pairs with the start codon, AUG. This tRNA carries the amino acid methionine (Met).

**② The** arrival of a large ribosomal subunit completes the initiation complex. Proteins called initiation factors (not shown) are required to bring all the translation components together. Hydrolysis of GTP provides the energy for the assembly. The initiator tRNA is in the P site; the A site is available to the tRNA bearing the next amino acid.

▲ **Figure 14.18 The initiation of translation.**

by the attachment of a large ribosomal subunit, completing the *translation initiation complex*. Proteins called *initiation factors* are required to bring all these components together. The cell also expends energy obtained by hydrolysis of a GTP molecule to form the initiation complex. At the completion of the initiation process, the initiator tRNA sits in the P site of the ribosome, and the vacant A site is ready for the next aminoacyl tRNA. Note that a polypeptide is always synthesized in one direction, from the initial methionine at the amino end, also called the N-terminus, toward the final amino acid at the carboxyl end, also called the C-terminus (see Figure 3.19).

### Elongation of the Polypeptide Chain

In the elongation stage of translation, amino acids are added one by one to the previous amino acid at the C-terminus of the growing chain. Each addition involves the participation of several proteins called *elongation factors* and occurs in a three-step cycle described in **Figure 14.19**. Energy expenditure occurs in the first and third steps. Codon recognition requires hydrolysis of one molecule of GTP, which increases the accuracy and efficiency of this step. One more GTP is hydrolyzed to provide energy for the translocation step.

The mRNA is moved through the ribosome in one direction only, 5′ end first; this is equivalent to the ribosome moving 5′ → 3′ on the mRNA. The important point is that the ribosome and the mRNA move relative to each other, unidirectionally, codon by codon. The elongation cycle takes less than a tenth of a second in bacteria and is repeated as each amino acid is added to the chain until the polypeptide is completed. The empty tRNAs that are released from the E site return to the cytoplasm, where they will be reloaded with the proper amino acid (see Figure 14.16).

### Termination of Translation

The final stage of translation is termination **(Figure 14.20)**. Elongation continues until a stop codon in the mRNA reaches the A site. The nucleotide base triplets UAG, UAA, and UGA (all written 5′ → 3′) do not code for amino acids but instead act as signals to stop translation. A *release factor*, a protein shaped like an aminoacyl tRNA, binds directly to the stop codon in the A site. The release factor causes the addition of a water molecule instead of an amino acid to the polypeptide chain. (Water molecules are abundant in the aqueous cellular environment.) This reaction breaks (hydrolyzes) the bond between the completed polypeptide and the tRNA in the P site, releasing the polypeptide through the exit tunnel of the ribosome's large subunit. The remainder of the translation assembly then comes apart in a multistep process, aided by other protein factors. Breakdown of the translation assembly requires the hydrolysis of two more GTP molecules.

## Completing and Targeting the Functional Protein

The process of translation is often not sufficient to make a functional protein. In this section, you'll learn about modifications that polypeptide chains undergo after the translation process as well as some of the mechanisms used to target completed proteins to specific sites in the cell.

### Protein Folding and Post-Translational Modifications

During its synthesis, a polypeptide chain begins to coil and fold spontaneously as a consequence of its amino acid sequence (primary structure), forming a protein with a specific shape: a three-dimensional molecule with secondary and tertiary

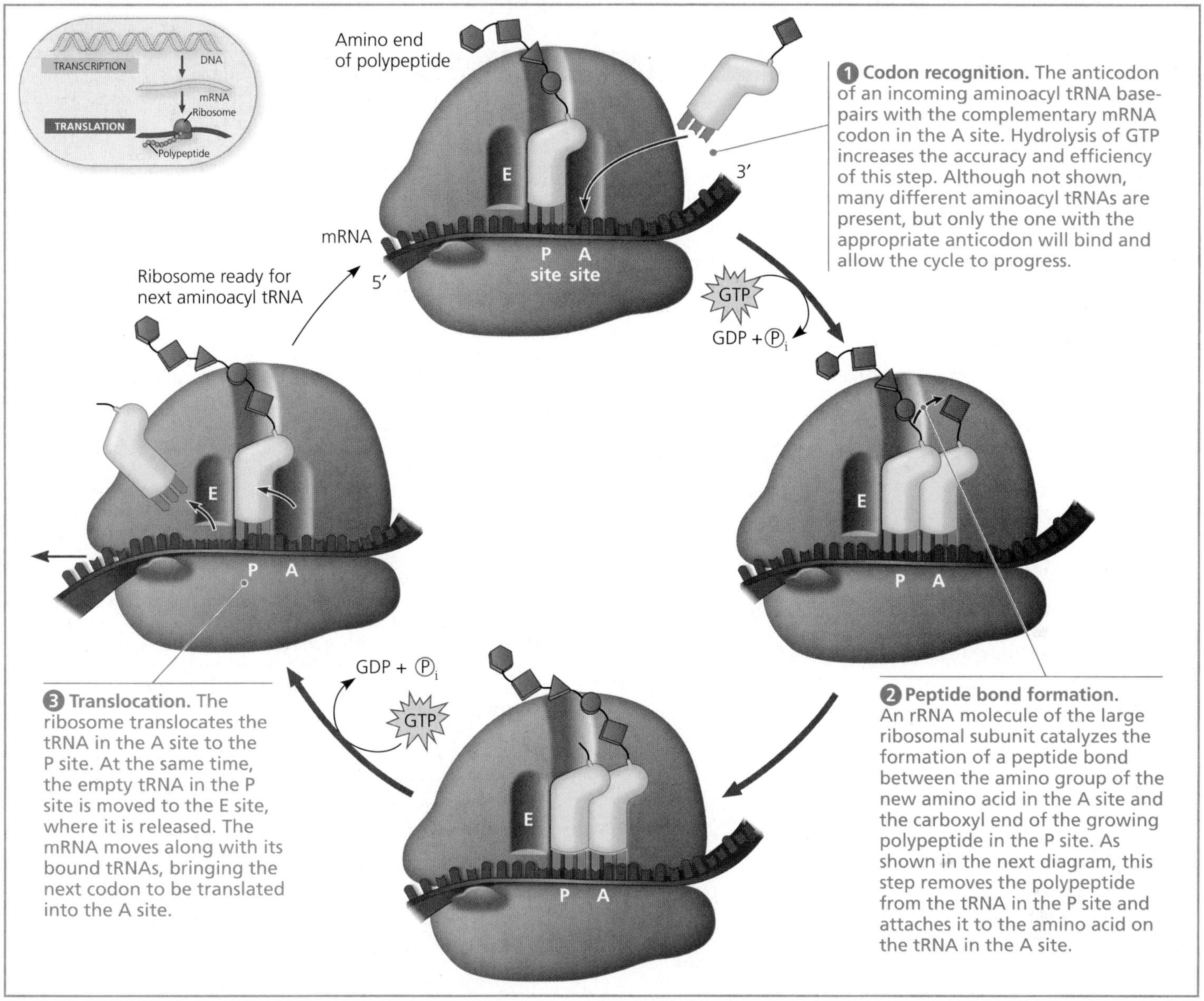

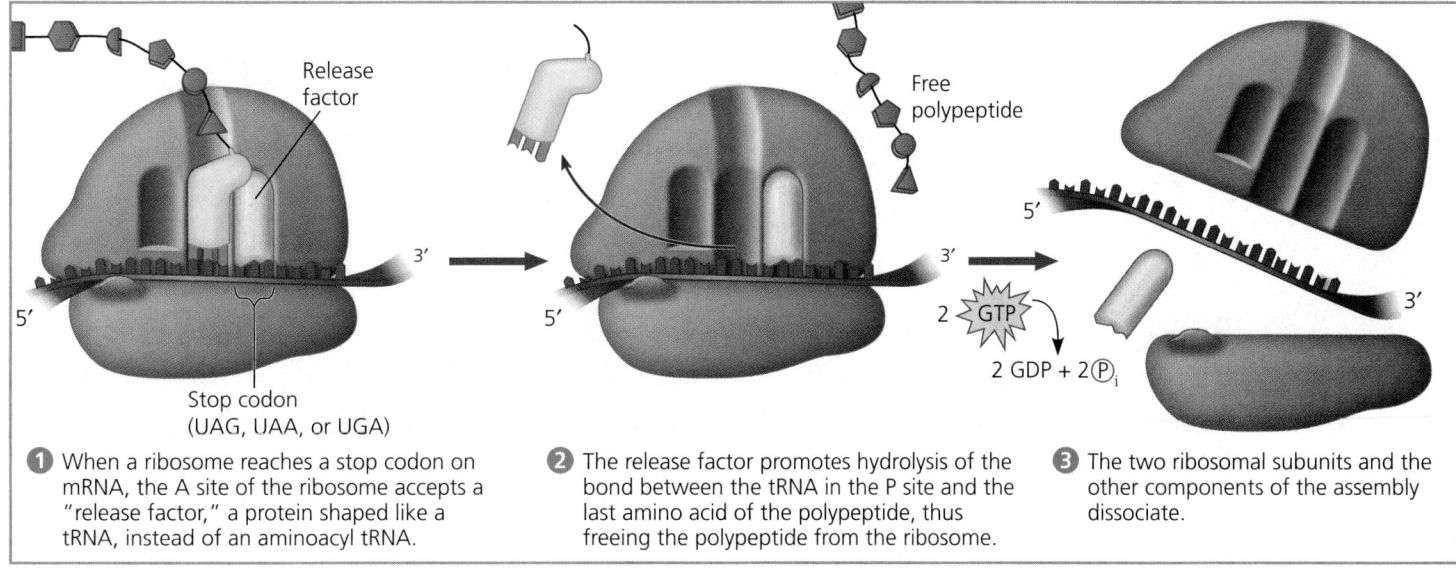

**Amino end of polypeptide**

**mRNA**

**Ribosome ready for next aminoacyl tRNA**

**❶ Codon recognition.** The anticodon of an incoming aminoacyl tRNA base-pairs with the complementary mRNA codon in the A site. Hydrolysis of GTP increases the accuracy and efficiency of this step. Although not shown, many different aminoacyl tRNAs are present, but only the one with the appropriate anticodon will bind and allow the cycle to progress.

GTP
GDP + Ⓟᵢ

**❸ Translocation.** The ribosome translocates the tRNA in the A site to the P site. At the same time, the empty tRNA in the P site is moved to the E site, where it is released. The mRNA moves along with its bound tRNAs, bringing the next codon to be translated into the A site.

GDP + Ⓟᵢ
GTP

**❷ Peptide bond formation.** An rRNA molecule of the large ribosomal subunit catalyzes the formation of a peptide bond between the amino group of the new amino acid in the A site and the carboxyl end of the growing polypeptide in the P site. As shown in the next diagram, this step removes the polypeptide from the tRNA in the P site and attaches it to the amino acid on the tRNA in the A site.

▲ **Figure 14.19 The elongation cycle of translation.** The hydrolysis of GTP plays an important role in the elongation process. Not shown are the proteins called elongation factors.

**Release factor**

**Free polypeptide**

**Stop codon (UAG, UAA, or UGA)**

2 GTP
2 GDP + 2 Ⓟᵢ

❶ When a ribosome reaches a stop codon on mRNA, the A site of the ribosome accepts a "release factor," a protein shaped like a tRNA, instead of an aminoacyl tRNA.

❷ The release factor promotes hydrolysis of the bond between the tRNA in the P site and the last amino acid of the polypeptide, thus freeing the polypeptide from the ribosome.

❸ The two ribosomal subunits and the other components of the assembly dissociate.

▲ **Figure 14.20 The termination of translation.** Like elongation, termination requires GTP hydrolysis as well as additional protein factors, which are not shown here.

## Interpreting a Sequence Logo

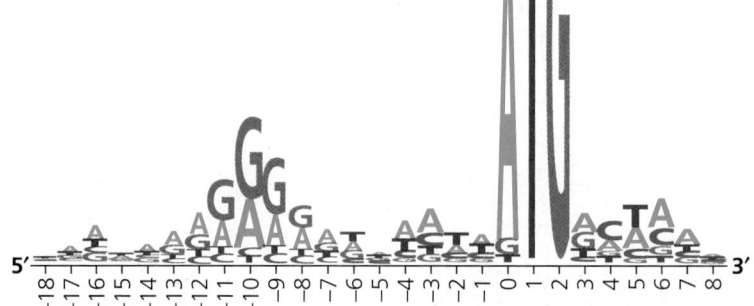

Ribosome binding site on mRNA

**How Can a Sequence Logo Be Used to Identify Ribosome Binding Sites?** When initiating translation, ribosomes bind to an mRNA at a ribosome binding site upstream of the 5'-AUG-3' start codon. Because mRNAs from different genes all bind to a ribosome, the genes encoding these mRNAs are likely to have a similar base sequence where the ribosomes bind. Therefore, candidate ribosome binding sites on mRNA can be identified by comparing DNA sequences (and thus the mRNA sequences) of multiple genes in a species, searching the region upstream of the start codon for shared ("conserved") stretches of bases. In this exercise you will analyze DNA sequences from multiple such genes, represented by a visual graphic called a sequence logo.

**How the Experiment Was Done** The DNA sequences of 149 genes from the *E. coli* genome were aligned and analyzed using computer software. The aim was to identify similar base sequences—at the appropriate location in each gene—as potential ribosome binding sites. Rather than presenting the data as a series of 149 sequences aligned in a column (a sequence alignment), the researchers used a sequence logo.

**Data from the Experiment** To show how sequence logos are made, the potential ribosome binding regions from 10 of the *E. coli* genes are shown in a sequence alignment, followed by the sequence logo derived from the aligned sequences. Note that the DNA shown is the nontemplate (coding) strand, which is how DNA sequences are typically presented.

```
thrA   G G T A A C G A G G T A A C A A C C A T G C G A G T G
lacA   C A T A A C G G A G T G A T C G C A T T G A A C A T G
lacY   C G C G T A A G G A A A T C C A T T A T G T A C T A T
lacZ   T T C A C A C A G G A A A C A G C T A T G A C C A T G
lacI   C A A T T C A G G G T G G T G A A T G T G A A A C C A
recA   G G C A T G A C A G G A G T A A A A A T G G C T A T C
galR   A C C C A C T A A G G T A T T T T C A T G G C G A C C
metJ   A A G A G G A T T A A G T A T C T C A T G G C T G A A
lexA   A T A C A C C C A G G G G G G C G G A A T G A A A G C G
trpR   T A A C A A T G G C G A C A T A T T A T G G C C A A
```

5'  —18 —17 —16 —15 —14 —13 —12 —11 —10 —9 —8 —7 —6 —5 —4 —3 —2 —1 0 1 2 3 4 5 6 7 8  3'

▲ Sequence alignment

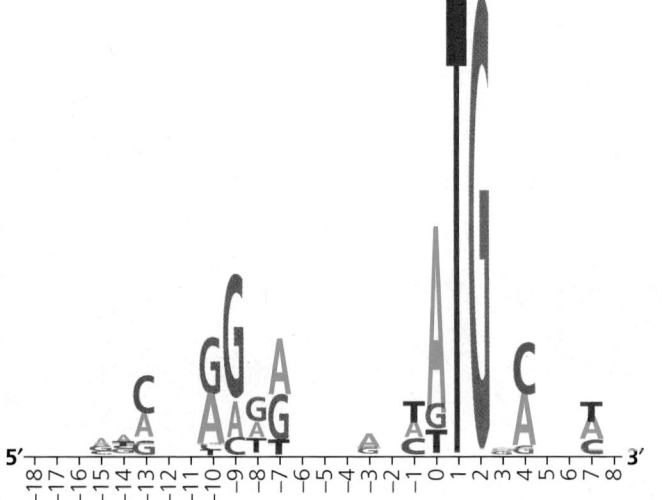

5'  —18 —17 —16 —15 —14 —13 —12 —11 —10 —9 —8 —7 —6 —5 —4 —3 —2 —1 0 1 2 3 4 5 6 7 8  3'

▲ Sequence logo

**Further Reading** T. D. Schneider and R. M. Stephens, Sequence logos: A new way to display consensus sequences, *Nucleic Acids Research* 18:6097–6100 (1990).

**INTERPRET THE DATA**

1. In the sequence logo (bottom, left), the horizontal axis shows the primary sequence of the DNA by nucleotide position. Letters for each base are stacked on top of each other according to their relative frequency at that position among the aligned sequences, with the most common base as the largest letter at the top of the stack. The height of each letter represents the relative frequency of that base *at that position*. (a) In the sequence alignment, count the number of each base at position −9 and order them from most to least frequent. Compare this to the size and placement of each base at −9 in the logo. (b) Do the same for positions 0 and 1.

2. The height of a stack of letters in a logo indicates the predictive power of that stack (determined statistically). If the stack is tall, we can be more confident in predicting what base will be in that position if a new sequence is added to the logo. For example, at position 2, all 10 sequences have a G; the probability of finding a G there in a new sequence is very high, as is the stack. For short stacks, the bases all have about the same frequency, and so it's hard to predict a base at those positions. (a) Which two positions have the most predictable bases? What bases do you predict would be at those positions in a newly sequenced gene? (b) Which 12 positions have the least predictable bases? How do you know? How does this reflect the relative frequencies of the bases shown in the 10 sequences? Use the two left-most positions of the 12 as examples in your answer.

3. In the actual experiment, the researchers used 149 sequences to build their sequence logo, which is shown below. There is a stack at each position, even if short, because the sequence logo includes more data. (a) Which three positions in this sequence logo have the most predictable bases? Name the most frequent base at each. (b) Which positions have the least predictable bases? How can you tell?

All data from Thomas D. Schneider.

4. A consensus sequence identifies the base occurring most often at each position in the set of sequences. (a) Write out the consensus sequence of this (the nontemplate) strand. In any position where the base can't be determined, put a dash. (b) Which provides more information—the consensus sequence or the sequence logo? What is lost in the less informative method?

5. (a) Based on the logo, what five adjacent base positions in the 5' UTR region are most likely to be involved in ribosome binding? Explain. (b) What is represented by the bases in positions 0–2?

**MB** A version of this Scientific Skills Exercise can be assigned in MasteringBiology.

structure (see Figure 3.22). Thus, a gene determines primary structure, and primary structure in turn determines shape.

Additional steps—*post-translational modifications*—may be required before the protein can begin doing its particular job in the cell. Certain amino acids may be chemically modified by the attachment of sugars, lipids, phosphate groups, or other additions. Enzymes may remove one or more amino acids from the leading (amino) end of the polypeptide chain. In some cases, a polypeptide chain may be enzymatically cleaved into two or more pieces. For example, the protein insulin is first synthesized as a single polypeptide chain but becomes active only after an enzyme cuts out a central part of the chain, leaving a protein made up of two shorter polypeptide chains connected by disulfide bridges. In other cases, two or more polypeptides that are synthesized separately may come together, becoming the subunits of a protein that has quaternary structure. A familiar example is hemoglobin (see Figure 3.22).

### Targeting Polypeptides to Specific Locations

In electron micrographs of eukaryotic cells active in protein synthesis, two populations of ribosomes are evident: free and bound (see Figure 4.9). Free ribosomes are suspended in the cytosol and mostly synthesize proteins that stay in the cytosol and function there. In contrast, bound ribosomes are attached to the cytosolic side of the endoplasmic reticulum (ER) or to the nuclear envelope. Bound ribosomes make proteins of the endomembrane system (the nuclear envelope, ER, Golgi apparatus, lysosomes, vacuoles, and plasma membrane) as well as proteins secreted from the cell, such as insulin. It is important to note that the ribosomes themselves are identical and can alternate between being free ribosomes one time they are used and being bound ribosomes the next time.

What determines whether a ribosome is free in the cytosol or bound to rough ER? Polypeptide synthesis always begins in the cytosol as a free ribosome starts to translate an mRNA molecule. There the process continues to completion—*unless* the growing polypeptide itself cues the ribosome to attach to the ER. The polypeptides of proteins destined for the endomembrane system or for secretion are marked by a **signal peptide**, which targets the protein to the ER **(Figure 14.21)**. The signal peptide, a sequence of about 20 amino acids at or near the leading end (N-terminus) of the polypeptide, is recognized as it emerges from the ribosome by a protein-RNA complex called a **signal-recognition particle (SRP)**. This particle functions as an escort that brings the ribosome to a receptor protein built into the ER membrane. The receptor is part of a multiprotein translocation complex. Polypeptide synthesis continues there, and the growing polypeptide snakes across the membrane into the ER lumen via a protein pore. The signal peptide is usually removed by an enzyme. The rest of the completed polypeptide, if it is to be secreted from the cell, is released into solution within the ER lumen (as in Figure 14.21). Alternatively, if the polypeptide is to be a membrane protein, it remains partially embedded in the ER membrane. In either

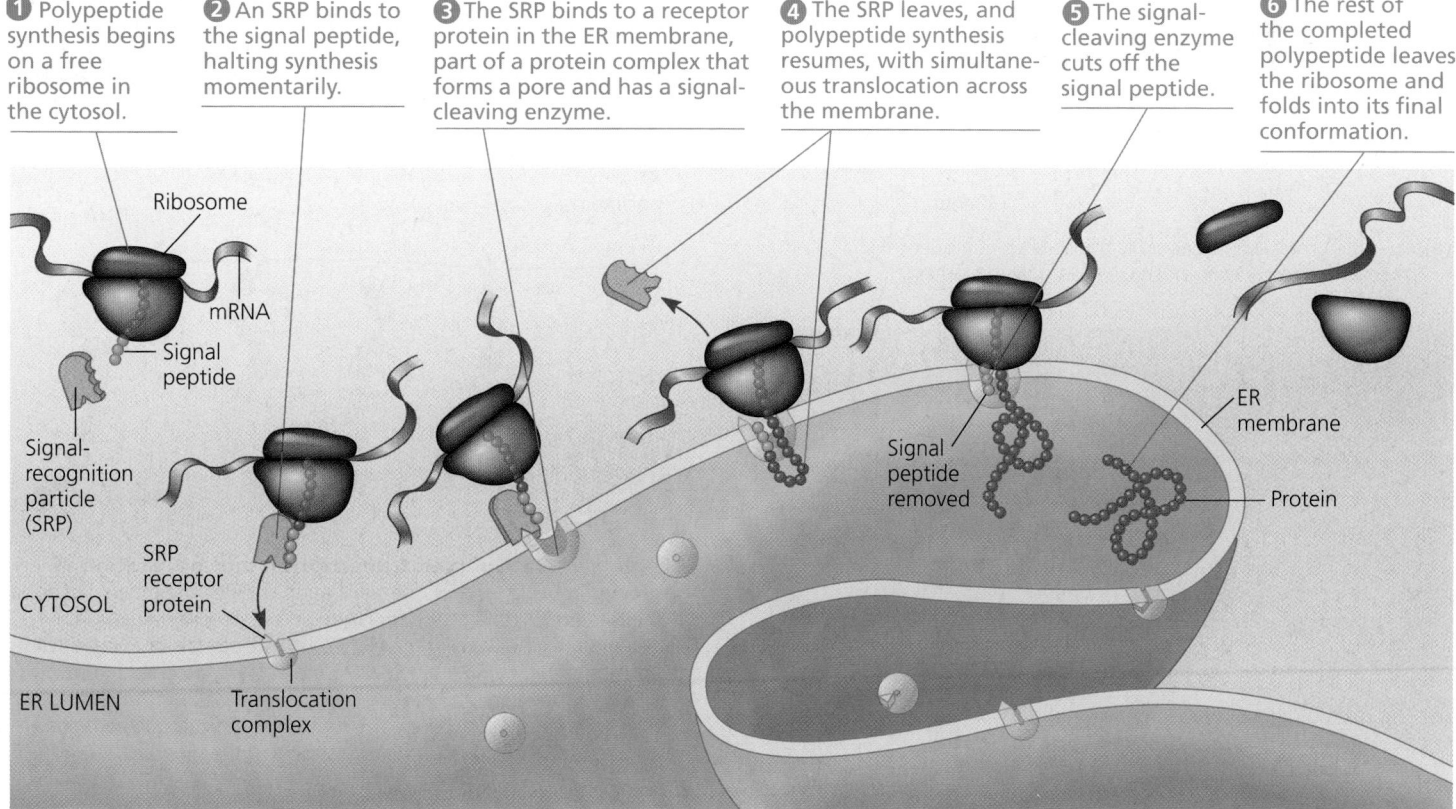

**1** Polypeptide synthesis begins on a free ribosome in the cytosol.

**2** An SRP binds to the signal peptide, halting synthesis momentarily.

**3** The SRP binds to a receptor protein in the ER membrane, part of a protein complex that forms a pore and has a signal-cleaving enzyme.

**4** The SRP leaves, and polypeptide synthesis resumes, with simultaneous translocation across the membrane.

**5** The signal-cleaving enzyme cuts off the signal peptide.

**6** The rest of the completed polypeptide leaves the ribosome and folds into its final conformation.

Ribosome

mRNA

Signal peptide

Signal-recognition particle (SRP)

SRP receptor protein

CYTOSOL

Translocation complex

ER LUMEN

Signal peptide removed

ER membrane

Protein

▲ **Figure 14.21 The signal mechanism for targeting proteins to the ER.**

case, it travels in a transport vesicle to the plasma membrane (see Figure 5.8).

Other kinds of signal peptides are used to target polypeptides to mitochondria, chloroplasts, the interior of the nucleus, and other organelles that are not part of the endomembrane system. The critical difference in these cases is that translation is completed in the cytosol before the polypeptide is imported into the organelle. Translocation mechanisms also vary, but in all cases studied to date, the "postal zip codes" that address proteins for secretion or to cellular locations are signal peptides of some sort. Bacteria also employ signal peptides to target proteins to the plasma membrane for secretion.

## Making Multiple Polypeptides in Bacteria and Eukaryotes

In previous sections, you have learned how a single polypeptide is synthesized using the information encoded in an mRNA molecule. When a polypeptide is required in a cell, though, the need is for many copies, not just one.

In both bacteria and eukaryotes, multiple ribosomes translate an mRNA at the same time **(Figure 14.22)**; that is, a single mRNA is used to make many copies of a polypeptide simultaneously. Once a ribosome is far enough past the start codon, a second ribosome can attach to the mRNA, eventually resulting in a number of ribosomes trailing along the mRNA. Such strings of ribosomes, called polyribosomes (or polysomes), can be seen with an electron microscope (see Figure 14.22); they can be either free or bound. They enable a cell to make many copies of a polypeptide very quickly.

Another way both bacteria and eukaryotes augment the number of copies of a polypeptide is by transcribing multiple mRNAs from the same gene, as we mentioned earlier. However, the coordination of the two processes—transcription and translation—differs in the two groups. The most important differences between bacteria and eukaryotes arise from the bacterial cell's lack of compartmental organization. Like a one-room workshop, a bacterial cell ensures a streamlined operation by coupling the two processes. In the absence of a nucleus, it can simultaneously transcribe and translate the same gene **(Figure 14.23)**, and the newly made protein can quickly diffuse to its site of function.

In contrast, the eukaryotic cell's nuclear envelope segregates transcription from translation and provides a compartment for extensive RNA processing. This processing stage includes additional steps whose regulation can help coordinate the eukaryotic cell's elaborate activities (see Concept 15.2). **Figure 14.24** summarizes the path from gene to polypeptide in a eukaryotic cell.

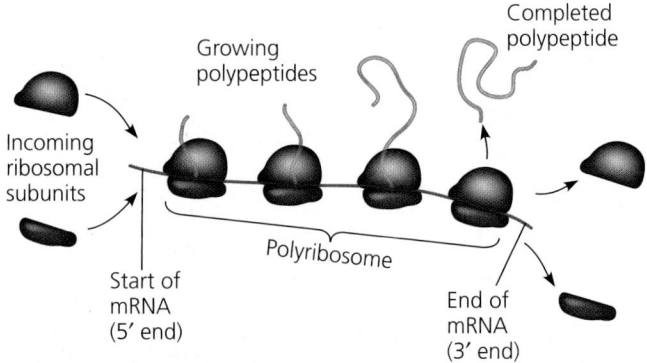

**(a)** An mRNA molecule is generally translated simultaneously by several ribosomes in clusters called polyribosomes.

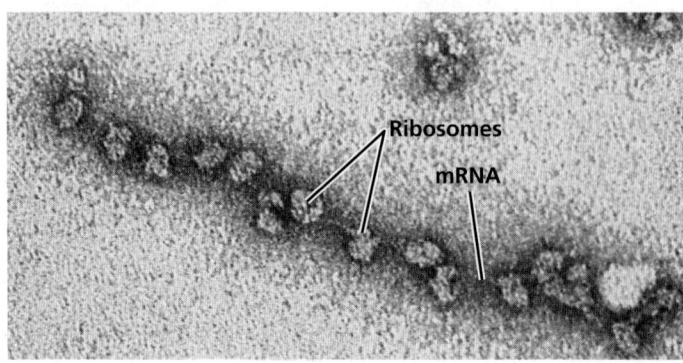

**(b)** This micrograph shows a large polyribosome in a bacterial cell. Growing polypeptides are not visible here (TEM). 0.1 μm

▲ **Figure 14.22 Polyribosomes.**

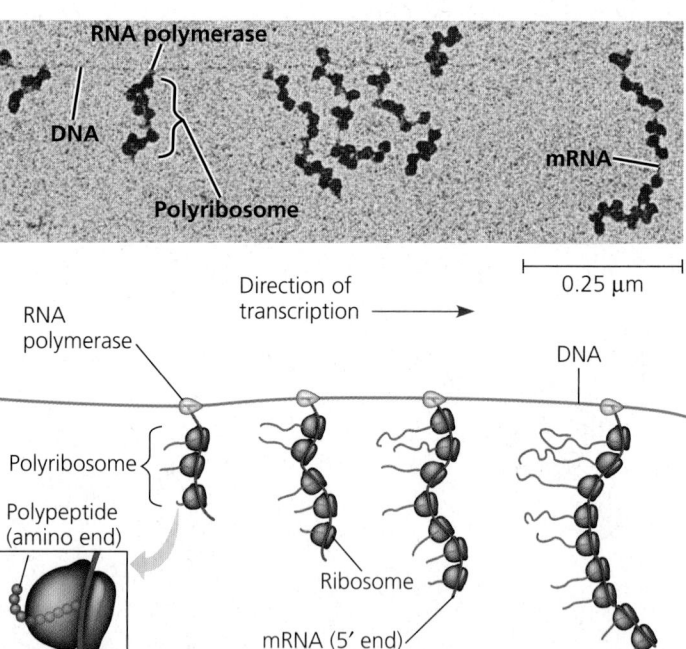

▲ **Figure 14.23 Coupled transcription and translation in bacteria.** In bacterial cells, the translation of mRNA can begin as soon as the leading (5′) end of the mRNA molecule peels away from the DNA template. The micrograph (TEM) shows a stretch of *E. coli* DNA being transcribed by RNA polymerase molecules. Attached to each RNA polymerase molecule is a growing strand of mRNA, which is already being translated by ribosomes. The newly synthesized polypeptides are not visible in the micrograph but are shown in the diagram.

? *Which one of the mRNA molecules started being transcribed first? On that mRNA, which ribosome started translating first?*

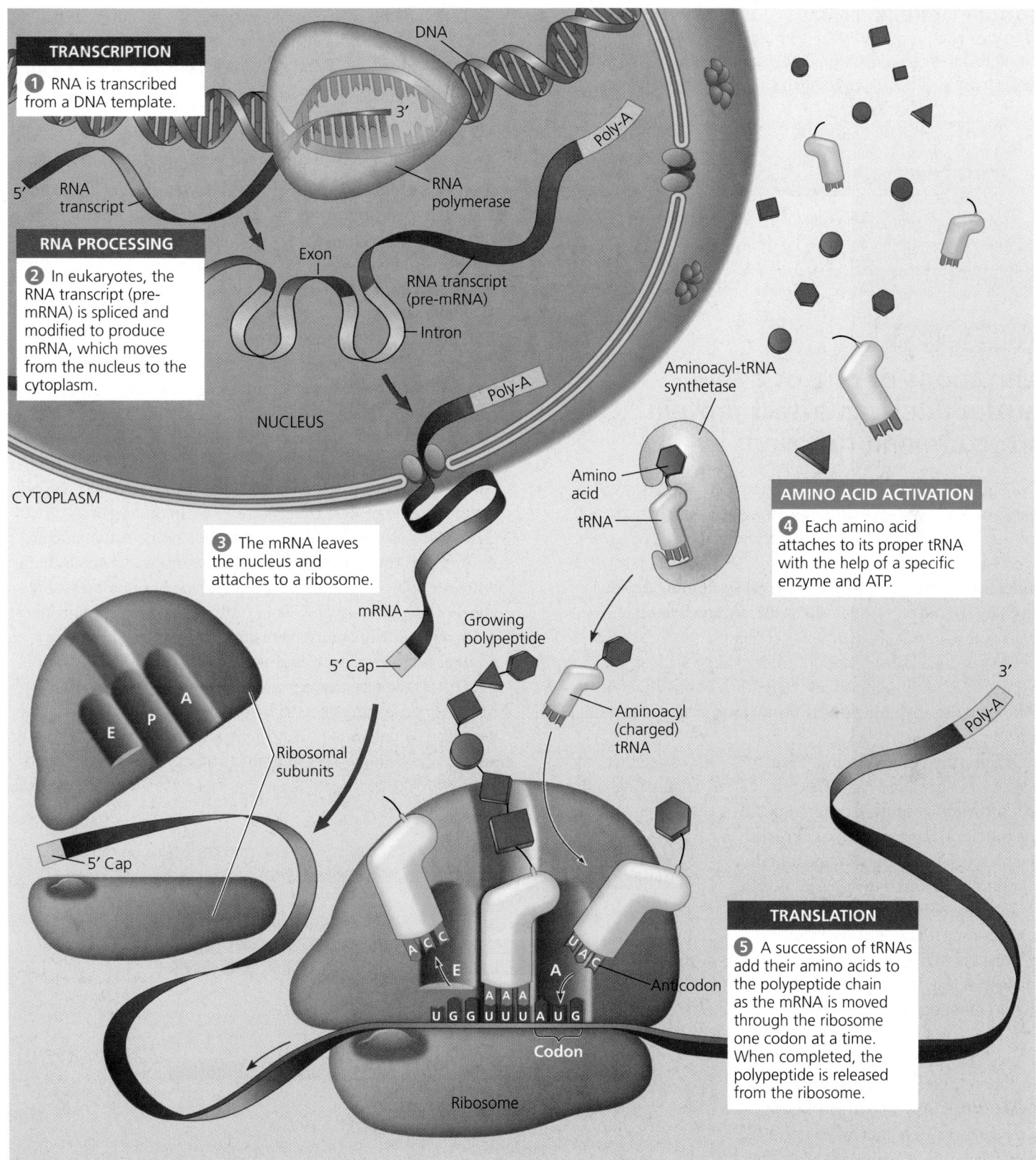

**TRANSCRIPTION**

**1** RNA is transcribed from a DNA template.

DNA

3'

RNA transcript

5'

RNA polymerase

Poly-A

**RNA PROCESSING**

**2** In eukaryotes, the RNA transcript (pre-mRNA) is spliced and modified to produce mRNA, which moves from the nucleus to the cytoplasm.

Exon

RNA transcript (pre-mRNA)

Intron

Poly-A

NUCLEUS

CYTOPLASM

**3** The mRNA leaves the nucleus and attaches to a ribosome.

mRNA

5' Cap

Growing polypeptide

Aminoacyl-tRNA synthetase

Amino acid

tRNA

**AMINO ACID ACTIVATION**

**4** Each amino acid attaches to its proper tRNA with the help of a specific enzyme and ATP.

Aminoacyl (charged) tRNA

3'

Poly-A

E P A

E

Ribosomal subunits

5' Cap

A C C

E

A

U A C

Anticodon

A A A

U G G U U U A U G

Codon

Ribosome

**TRANSLATION**

**5** A succession of tRNAs add their amino acids to the polypeptide chain as the mRNA is moved through the ribosome one codon at a time. When completed, the polypeptide is released from the ribosome.

▲ **Figure 14.24 A summary of transcription and translation in a eukaryotic cell.** This diagram shows the path from one gene to one polypeptide. Keep in mind that each gene in the DNA can be transcribed repeatedly into many identical RNA molecules and that each mRNA can be translated repeatedly to yield many identical polypeptide molecules. (Also, remember that the final products of some genes are not polypeptides but RNA molecules that don't get translated, including tRNA and rRNA.) In general, the steps of transcription and translation are similar in bacterial, archaeal, and eukaryotic cells. The major difference is the occurrence of RNA processing in the eukaryotic nucleus. Other significant differences are found in the initiation stages of both transcription and translation and in the termination of transcription.

1. What two processes ensure that the correct amino acid is added to a growing polypeptide chain?
2. Discuss the ways in which rRNA structure likely contributes to ribosomal function.
3. Describe how a polypeptide to be secreted is transported to the endomembrane system.
4. **WHAT IF?** **DRAW IT** Draw a tRNA with the anticodon 3′-CGU-5′. What two different codons could it bind to? Draw each codon on an mRNA, labeling all 5′ and 3′ ends, the tRNA, and the amino acid it carries.

For suggested answers, see Appendix A.

## CONCEPT 14.5

# Mutations of one or a few nucleotides can affect protein structure and function

Now that you have explored the process of gene expression, you are ready to understand the effects of changes to the genetic information of a cell (or virus). These changes, called **mutations**, are responsible for the huge diversity of genes found among organisms because mutations are the ultimate source of new genes. Earlier, we considered chromosomal rearrangements that affect long segments of DNA (see Figure 12.14); these are considered large-scale mutations. Here we'll examine small-scale mutations of one or a few nucleotide pairs, including **point mutations**, changes in a single nucleotide pair of a gene.

If a point mutation occurs in a gamete or in a cell that gives rise to gametes, it may be transmitted to offspring and to a succession of future generations. If the mutation has an adverse effect on the phenotype of a person, the mutant condition is referred to as a genetic disorder or hereditary disease. For example, we can trace the genetic basis of sickle-cell disease to the mutation of a single nucleotide pair in the gene that encodes the β-globin polypeptide of hemoglobin. The change of a single nucleotide in the DNA's template strand leads to an altered mRNA and the production of an abnormal protein **(Figure 14.25**; also see Figure 3.23). In individuals who are homozygous for the mutant allele, the sickling of red blood cells caused by the altered hemoglobin produces the multiple symptoms associated with sickle-cell disease (see Concept 11.4 and Figure 21.15). Another disorder caused by a point mutation is a heart condition called familial

cardiomyopathy, which is responsible for some incidents of sudden death in young athletes. Point mutations in several genes encoding muscle proteins have been identified, any of which can lead to this disorder.

## Types of Small-Scale Mutations

Let's now consider how small-scale mutations affect proteins. Small-scale mutations within a gene can be divided into two general categories: (1) single nucleotide-pair substitutions and (2) nucleotide-pair insertions or deletions. Insertions and deletions can involve one or more nucleotide pairs.

### Substitutions

A **nucleotide-pair substitution** is the replacement of one nucleotide and its partner with another pair of nucleotides **(Figure 14.26a)**. Some substitutions have no effect on the encoded protein, owing to the redundancy of the genetic code. For example, if 3′-CCG-5′ on the template strand mutated to 3′-CCA-5′, the mRNA codon that used to be GGC would become GGU, but a glycine would still be inserted at the proper location in the protein (see Figure 14.6). In other words, a change in a nucleotide pair may transform one codon into another that is translated into the same amino acid. Such a change is an example of a **silent mutation**, which has no observable effect on the phenotype. (Silent mutations can occur outside genes as well.) Substitutions that change one amino acid to another one are called **missense mutations**. Such a mutation may have little effect on the protein: The new amino acid may have properties similar to those of the amino acid it replaces, or it may be in a region of the protein where the exact sequence of amino acids is not essential to the protein's function.

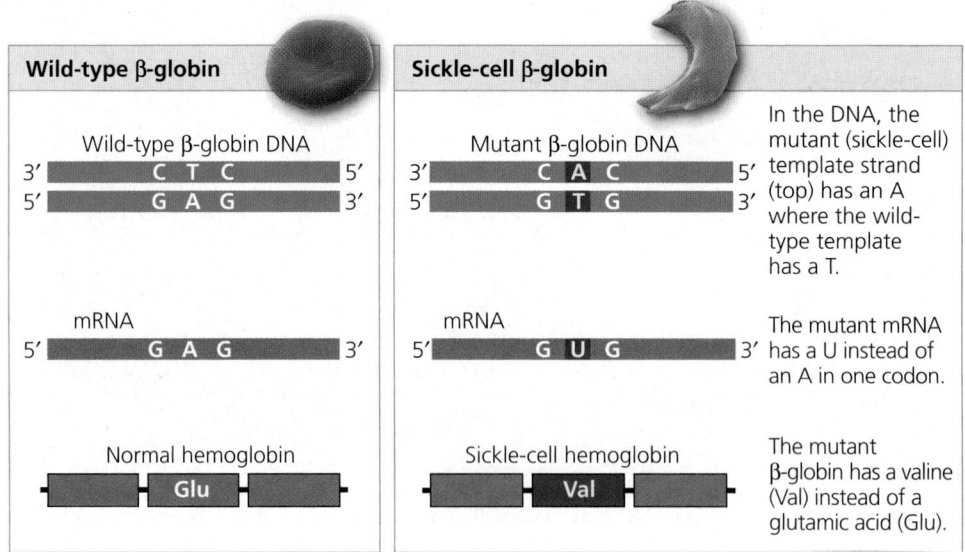

| Wild-type β-globin | Sickle-cell β-globin | |
|---|---|---|
| Wild-type β-globin DNA | Mutant β-globin DNA | In the DNA, the mutant (sickle-cell) template strand (top) has an A where the wild-type template has a T. |
| 3′ **C T C** 5′<br>5′ **G A G** 3′ | 3′ **C A C** 5′<br>5′ **G T G** 3′ | |
| mRNA | mRNA | The mutant mRNA has a U instead of an A in one codon. |
| 5′ **G A G** 3′ | 5′ **G U G** 3′ | |
| Normal hemoglobin | Sickle-cell hemoglobin | The mutant β-globin has a valine (Val) instead of a glutamic acid (Glu). |
| **Glu** | **Val** | |

▲ **Figure 14.25 The molecular basis of sickle-cell disease: a point mutation.** The allele that causes sickle-cell disease differs from the wild-type (normal) allele by a single DNA nucleotide pair. The micrographs are SEMs of a normal red blood cell (on the left) and a sickled red blood cell (right) from individuals homozygous for wild-type and mutant alleles, respectively.

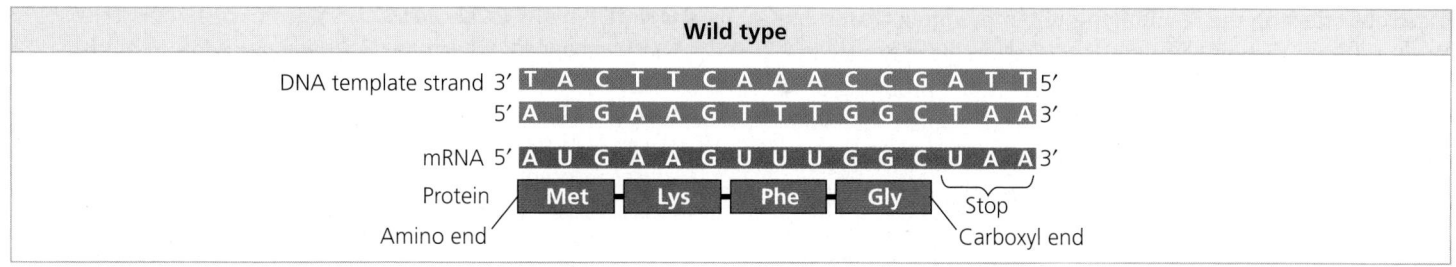

**Wild type**

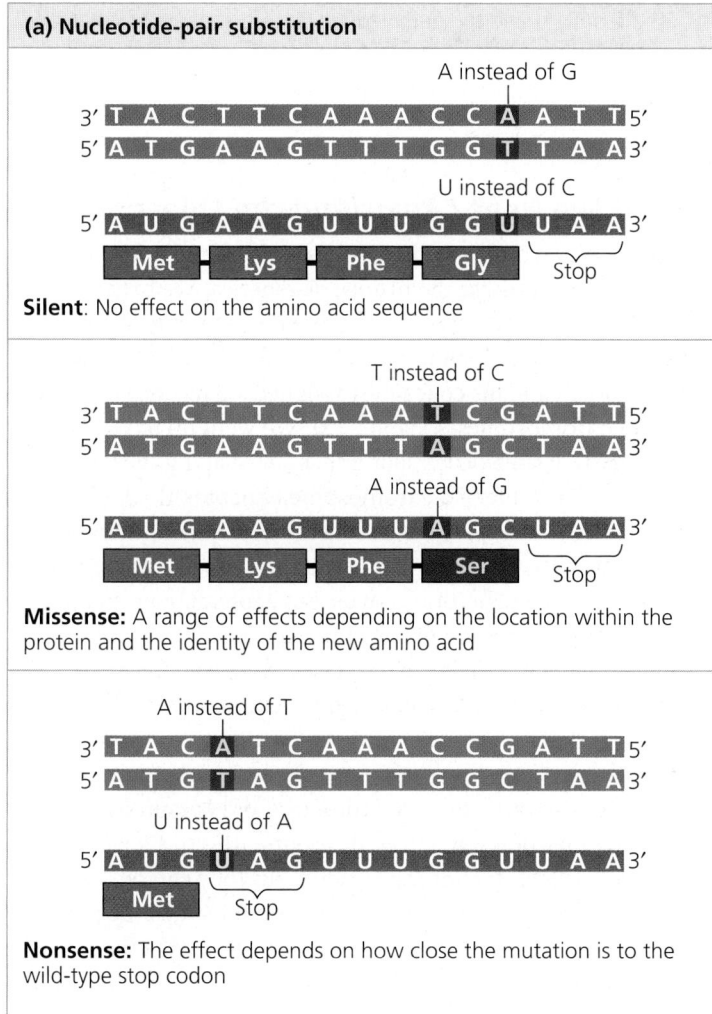

**(a) Nucleotide-pair substitution**

**Silent:** No effect on the amino acid sequence

**Missense:** A range of effects depending on the location within the protein and the identity of the new amino acid

**Nonsense:** The effect depends on how close the mutation is to the wild-type stop codon

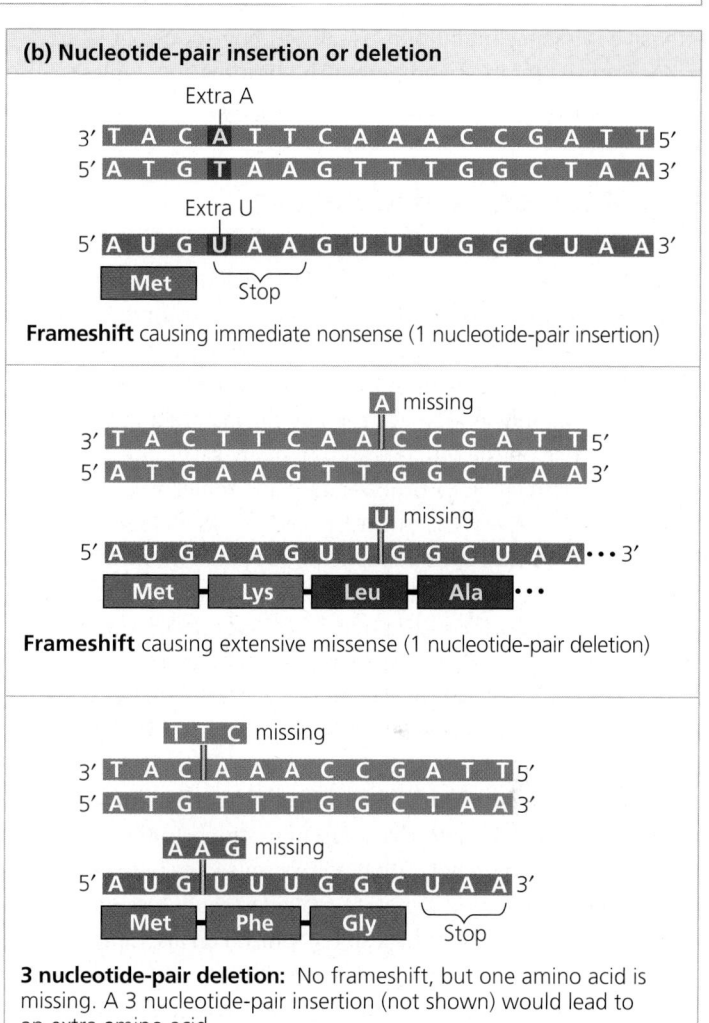

**(b) Nucleotide-pair insertion or deletion**

**Frameshift** causing immediate nonsense (1 nucleotide-pair insertion)

**Frameshift** causing extensive missense (1 nucleotide-pair deletion)

**3 nucleotide-pair deletion:** No frameshift, but one amino acid is missing. A 3 nucleotide-pair insertion (not shown) would lead to an extra amino acid.

▲ **Figure 14.26 Types of small-scale mutations that affect mRNA sequence.** All but one of the types shown here also affect the amino acid sequence of the encoded polypeptide.

However, the nucleotide-pair substitutions of greatest interest are those that cause a major change in a protein. The alteration of a single amino acid in a crucial area of a protein—such as in the part of the β-globin subunit of hemoglobin shown in Figure 14.25 or in the active site of an enzyme—can significantly alter protein activity. Occasionally, such a mutation leads to an improved protein or one with novel capabilities, but much more often such mutations are neutral or detrimental, leading to a useless or less active protein that impairs cellular function.

Substitution mutations are usually missense mutations; that is, the altered codon still codes for an amino acid and thus makes sense, although not necessarily the *right* sense. But a point mutation can also change a codon for an amino acid into a stop codon. This is called a **nonsense mutation**, and it causes translation to be terminated prematurely; the resulting polypeptide will be shorter than the polypeptide encoded by the normal gene. Nearly all nonsense mutations lead to nonfunctional proteins.

### Insertions and Deletions

**Insertions** and **deletions** are additions or losses of nucleotide pairs in a gene **(Figure 14.26b)**. These mutations have a disastrous effect on the resulting protein more often than substitutions do. Insertion or deletion of nucleotides may alter the reading frame of the genetic message, the triplet grouping of nucleotides on the mRNA that is read during translation. Such a mutation, called a **frameshift mutation**, occurs whenever the number of nucleotides inserted or deleted is not a multiple of three. All nucleotides downstream of the deletion or insertion will be improperly grouped into codons; the result will be extensive missense mutations, usually ending sooner or later in a nonsense mutation and premature termination. Unless the frameshift is very near the end of the gene, the protein is almost certain to be nonfunctional.

## New Mutations and Mutagens

Mutations can arise in a number of ways. Errors during DNA replication or recombination can lead to nucleotide-pair substitutions, insertions, or deletions, as well as to mutations affecting longer stretches of DNA. If an incorrect nucleotide is added to a growing chain during replication, for example, the base on that nucleotide will then be mismatched with the nucleotide base on the other strand. In many cases, the error will be corrected by DNA proofreading and repair systems (see Concept 13.2). Otherwise, the incorrect base will be used as a template in the next round of replication, resulting in a mutation. Such mutations are called *spontaneous mutations*. It is difficult to calculate the rate at which such mutations occur. Rough estimates have been made of the rate of mutation during DNA replication for both *E. coli* and eukaryotes, and the numbers are similar: About one nucleotide in every $10^{10}$ is altered, and the change is passed on to the next generation of cells.

A number of physical and chemical agents, called **mutagens**, interact with DNA in ways that cause mutations. In the 1920s, Hermann Muller discovered that X-rays caused genetic changes in fruit flies, and he used X-rays to make *Drosophila* mutants for his genetic studies. But he also recognized an alarming implication of his discovery: X-rays and other forms of high-energy radiation pose hazards to the genetic material of people as well as laboratory organisms. Mutagenic radiation, a physical mutagen, includes ultraviolet (UV) light, which can cause disruptive thymine dimers in DNA (see Figure 13.21).

Chemical mutagens fall into several categories. Nucleotide analogs are chemicals that are similar to normal DNA nucleotides but that pair incorrectly during DNA replication. Some other chemical mutagens interfere with correct DNA replication by inserting themselves into the DNA and distorting the double helix. Still other mutagens cause chemical changes in bases that change their pairing properties.

Researchers have developed a variety of methods to test the mutagenic activity of chemicals. A major application of these tests is the preliminary screening of chemicals to identify those that may cause cancer. This approach makes sense because most carcinogens (cancer-causing chemicals) are mutagenic, and conversely, most mutagens are carcinogenic.

**CONCEPT CHECK 14.5**

1. What happens when one nucleotide pair is lost from the middle of the coding sequence of a gene?
2. **MAKE CONNECTIONS** Individuals heterozygous for the sickle-cell allele show effects of the allele under some circumstances (see Concept 11.4). Explain in terms of gene expression.
3. **WHAT IF?** **DRAW IT** The template strand of a gene includes this sequence: 3'-TACTTGTCCGATATC-5'. It is mutated to 3'-TACTTGTCCAATATC-5'. For both versions, draw the DNA, the mRNA, and the encoded amino acid sequence. What is the effect on the amino acid sequence?

For suggested answers, see Appendix A.

## What Is a Gene? *Revisiting the Question*

Our definition of a gene has evolved over the past few chapters, as it has through the history of genetics. We began with the Mendelian concept of a gene as a discrete unit of inheritance that affects a phenotypic character (Chapter 11). We saw that Morgan and his colleagues assigned such genes to specific loci on chromosomes (Chapter 12). We went on to view a gene as a region of specific nucleotide sequence along the length of the DNA molecule of a chromosome (Chapter 13). Finally, in this chapter, we have considered a functional definition of a gene as a DNA sequence that codes for a specific polypeptide chain. All these definitions are useful, depending on the context in which genes are being studied.

We now realize that saying a gene codes for a polypeptide is overly simplistic. Most eukaryotic genes contain noncoding segments (such as introns), so large portions of these genes have no corresponding segments in polypeptides. Molecular biologists also often include promoters and certain other regulatory regions of DNA within the boundaries of a gene. These DNA sequences are not transcribed, but they can be considered part of the functional gene because they must be present for transcription to occur. Our definition of a gene must also be broad enough to include the DNA that is transcribed into rRNA, tRNA, and other RNAs that are not translated. These genes have no polypeptide products but play crucial roles in the cell. Thus, we arrive at the following definition: *A gene is a region of DNA that can be expressed to produce a final functional product that is either a polypeptide or an RNA molecule.*

When considering phenotypes, however, it is useful to focus on genes that code for polypeptides. In this chapter, you have learned how a typical gene is expressed—by transcription into RNA and then translation into a polypeptide that forms a protein of specific structure and function. Proteins, in turn, bring about an organism's observable phenotype.

A given type of cell expresses only a subset of its genes. This is an essential feature in multicellular organisms: Gene expression is precisely regulated. We'll explore gene regulation in the next chapter, beginning with the simpler case of bacteria and continuing with eukaryotes.

Go to **MasteringBiology**® for Assignments, the eText, and the Study Area with Animations, Activities, Vocab Self-Quiz, and Practice Tests.

## SUMMARY OF KEY CONCEPTS

**VOCAB SELF-QUIZ**

goo.gl/gbai8v

### CONCEPT 14.1

**Genes specify proteins via transcription and translation (pp. 279–284)**

- Beadle and Tatum's studies of mutant strains of *Neurospora* led to the one gene–one polypeptide hypothesis. During **gene expression**, the information encoded in genes is used to make specific polypeptide chains (enzymes and other proteins) or RNA molecules.
- **Transcription** is the synthesis of RNA complementary to a **template strand** of DNA. **Translation** is the synthesis of a polypeptide whose amino acid sequence is specified by the nucleotide sequence in **mRNA**.
- Genetic information is encoded as a sequence of nonoverlapping nucleotide triplets, or **codons**. A codon in messenger RNA (mRNA) either is translated into an amino acid (61 of the 64 codons) or serves as a stop signal (3 codons). Codons must be read in the correct **reading frame**.

? *Describe the process of gene expression, by which a gene affects the phenotype of an organism.*

### CONCEPT 14.2

**Transcription is the DNA-directed synthesis of RNA: *a closer look* (pp. 284–286)**

- RNA synthesis is catalyzed by **RNA polymerase**, which links together RNA nucleotides complementary to a DNA template strand. This process follows the same base-pairing rules as DNA replication, except that in RNA, uracil substitutes for thymine.

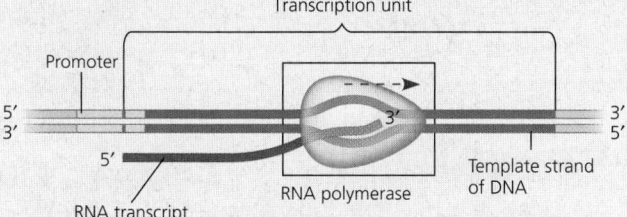

- The three stages of transcription are initiation, elongation, and termination. A **promoter**, often including a **TATA box** in eukaryotes, establishes where RNA synthesis is initiated. **Transcription factors** help eukaryotic RNA polymerase recognize promoter sequences, forming a **transcription initiation complex**. Termination differs in bacteria and eukaryotes.

? *What are the similarities and differences in the initiation of gene transcription in bacteria and eukaryotes?*

### CONCEPT 14.3

**Eukaryotic cells modify RNA after transcription (pp. 286–288)**

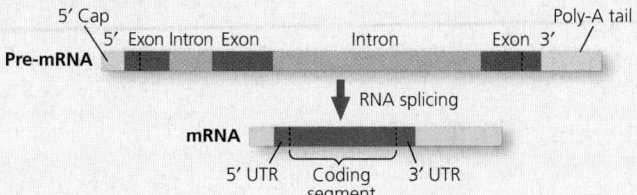

- Eukaryotic pre-mRNAs undergo **RNA processing**, which includes RNA splicing, the addition of a modified nucleotide **5′ cap** to the 5′ end, and the addition of a **poly-A tail** to the 3′ end. The processed mRNA includes an untranslated region (5′ UTR or 3′ UTR) at each end of the coding segment.
- Most eukaryotic genes are split into segments: They have **introns** interspersed among the **exons** (regions included in the mRNA). In **RNA splicing**, introns are removed and exons joined. RNA splicing is typically carried out by **spliceosomes**, but in some cases, RNA alone catalyzes its own splicing. The catalytic ability of some RNA molecules, called **ribozymes**, derives from the properties of RNA. The presence of introns allows for **alternative RNA splicing**.

? *What function do the 5′ cap and the poly-A tail serve on a eukaryotic mRNA?*

### CONCEPT 14.4

**Translation is the RNA-directed synthesis of a polypeptide: *a closer look* (pp. 288–298)**

- A cell translates an mRNA message into protein using **transfer RNAs (tRNAs)**. After being bound to a specific amino acid by an **aminoacyl-tRNA synthetase**, a tRNA lines up via its **anticodon** at the complementary codon on mRNA. A ribosome, made up of **ribosomal RNAs (rRNAs)** and proteins, facilitates this coupling with binding sites for mRNA and tRNA.
- Ribosomes coordinate the three stages of translation: initiation, elongation, and termination. The formation of peptide bonds between amino acids is catalyzed by ribosomal RNAs as tRNAs move through the **A** and **P sites** and exit through the **E site**.
- After translation, modifications to proteins can affect their shape. Free ribosomes in the cytosol initiate synthesis of all proteins, but proteins with a **signal peptide** are synthesized on the ER.
- A gene can be transcribed by multiple RNA polymerases simultaneously. Also, a single mRNA molecule can be translated simultaneously by a number of ribosomes, forming a polyribosome. In bacteria, these processes are coupled, but in eukaryotes they are separated in time and space by the nuclear membrane.

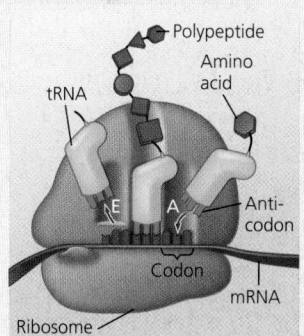

? *What function do tRNAs serve in the process of translation?*

### CONCEPT 14.5

**Mutations of one or a few nucleotides can affect protein structure and function (pp. 298–300)**

- Small-scale **mutations** include **point mutations**, changes in one DNA nucleotide pair, which may lead to production of nonfunctional proteins. **Nucleotide-pair substitutions** can cause **missense** or **nonsense mutations**. Nucleotide-pair **insertions** or **deletions** may produce **frameshift mutations**.
- Spontaneous mutations can occur during DNA replication, recombination, or repair. Chemical and physical **mutagens** cause DNA damage that can alter genes.

## TEST YOUR UNDERSTANDING

**PRACTICE TEST**

goo.gl/CRZjvS

### Level 1: Knowledge/Comprehension

1. In eukaryotic cells, transcription cannot begin until
   (A) the two DNA strands have completely separated and exposed the promoter.
   (B) several transcription factors have bound to the promoter.
   (C) the 5′ caps are removed from the mRNA.
   (D) the DNA introns are removed from the template.

2. Which of the following is *not* true of a codon?
   (A) It may code for the same amino acid as another codon.
   (B) It never codes for more than one amino acid.
   (C) It extends from one end of a tRNA molecule.
   (D) It is the basic unit of the genetic code.

3. The anticodon of a particular tRNA molecule is
   (A) complementary to the corresponding mRNA codon.
   (B) complementary to the corresponding triplet in rRNA.
   (C) the part of tRNA that bonds with a specific amino acid.
   (D) catalytic, making the tRNA a ribozyme.

4. Which of the following is *not* true of RNA processing?
   (A) Exons are cut out before mRNA leaves the nucleus.
   (B) Nucleotides may be added at both ends of the RNA.
   (C) Ribozymes may function in RNA splicing.
   (D) RNA splicing can be catalyzed by spliceosomes.

5. Which component is *not* directly involved in translation?
   (A) GTP        (C) tRNA
   (B) DNA        (D) ribosomes

### Level 2: Application/Analysis

6. Using Figure 14.6, identify a 5′ → 3′ sequence of nucleotides in the DNA template strand for an mRNA coding for the polypeptide sequence Phe-Pro-Lys.
   (A) 5′-UUUCCCAAA-3′
   (B) 5′-GAACCCCTT-3′
   (C) 5′-CTTCGGGAA-3′
   (D) 5′-AAACCCUUU-3′

7. Which of the following mutations would be *most* likely to have a harmful effect on an organism?
   (A) a deletion of three nucleotides near the middle of a gene
   (B) a single nucleotide deletion in the middle of an intron
   (C) a single nucleotide deletion near the end of the coding sequence
   (D) a single nucleotide insertion downstream of, and close to, the start of the coding sequence

8. Would the coupling of the processes shown in Figure 14.23 be found in a eukaryotic cell? Explain why or why not.

9. Fill in the following table:

| Type of RNA | Functions |
|---|---|
| Messenger RNA (mRNA) | |
| Transfer RNA (tRNA) | |
| | Plays catalytic (ribozyme) roles and structural roles in ribosomes |
| Primary transcript | |
| Small RNAs in the spliceosome | |

### Level 3: Synthesis/Evaluation

10. **SCIENTIFIC INQUIRY**
    Knowing that the genetic code is almost universal, a scientist uses molecular biological methods to insert the human β-globin gene (shown in Figure 14.12) into bacterial cells, hoping the cells will express it and synthesize functional β-globin protein. Instead, the protein produced is nonfunctional and is found to contain many fewer amino acids than does β-globin made by a eukaryotic cell. Explain why.

11. **FOCUS ON EVOLUTION**
    Most amino acids are coded for by a set of similar codons (see Figure 14.6). What evolutionary explanation can you give for this pattern?

12. **FOCUS ON INFORMATION**
    Evolution accounts for the unity and diversity of life, and the continuity of life is based on heritable information in the form of DNA. In a short essay (100–150 words), discuss how the fidelity with which DNA is inherited is related to the processes of evolution. (Review the discussion of proofreading and DNA repair in Concept 13.2.)

13. **SYNTHESIZE YOUR KNOWLEDGE**

Some mutations result in proteins that function well at one temperature but are nonfunctional at a different (usually higher) temperature. Siamese cats have such a "temperature-sensitive" mutation in a gene encoding an enzyme that makes dark pigment in the fur. The mutation results in the breed's distinctive point markings and lighter body color (see the photo). Using this information and what you learned in the chapter, explain the pattern of the cat's fur pigmentation.

*For selected answers, see Appendix A.*

▲ **Figure 15.1** How can this fish's eyes see equally well in both air and water?

## Beauty in the Eye of the Beholder

The fish shown in **Figure 15.1** is keeping an eye out for predators—or, more precisely, both halves of each eye! *Anableps anableps*, commonly known as "cuatro ojos" ("four eyes"), glides through freshwater lakes and ponds in Central and South America with the upper half of each eye protruding from the water. The eye's upper half is particularly well-suited for aerial vision and the lower half for aquatic vision. The molecular basis of this specialization has recently been revealed: The cells of the two parts of the eye express a slightly different set of genes involved in vision, even though these two groups of cells are quite similar and contain identical genomes.

A hallmark of prokaryotic and eukaryotic cells alike—from a bacterium to the cells of a fish—is their intricate and precise regulation of gene expression. Both prokaryotes and eukaryotes must alter their patterns of gene expression in response to changes in environmental conditions. Multicellular eukaryotes must also develop and maintain multiple cell types, each expressing a different subset of genes. This is a significant challenge in gene regulation.

In this chapter, we'll first explore how bacteria regulate expression of their genes in response to different environmental conditions. We'll then examine the general mechanisms by which eukaryotes regulate gene expression. Next, we'll look at the many roles played by RNA molecules in regulating gene expression in eukaryotes. Finally, we'll describe a few genetic engineering techniques that have been developed to investigate gene expression. Elucidating how gene expression is regulated in different cells is crucial to our understanding of living systems.

### CONCEPT 15.1

## Bacteria often respond to environmental change by regulating transcription

Bacterial cells that can conserve resources and energy have a selective advantage over cells that are unable to do so. Thus, natural selection has favored bacteria that express only the genes whose products are needed by the cell.

Consider, for instance, an individual *Escherichia coli* (*E. coli*) cell living in the erratic environment of a human colon, dependent for its nutrients on the whimsical eating habits of its host. If the environment is lacking in the amino acid tryptophan, which the bacterium needs to survive, the cell responds by activating a metabolic pathway that makes tryptophan from another compound. If the human host later eats a tryptophan-rich meal, the bacterial cell stops producing tryptophan, thus avoiding wasting resources to produce a substance that is readily available from the surrounding solution.

A metabolic pathway can be controlled on two levels, as shown for the synthesis of tryptophan in **Figure 15.2**. First, cells can adjust the activity of enzymes already present. This is a fairly rapid response, which relies on the sensitivity of many enzymes to chemical cues that increase or decrease their catalytic activity (see Concept 6.5). The activity of the first enzyme in the tryptophan synthesis pathway is inhibited by tryptophan, the pathway's end product **(Figure 15.2a)**. Thus, if tryptophan accumulates in a cell, it shuts down the synthesis of more tryptophan by inhibiting enzyme activity. Such *feedback inhibition*, typical of anabolic (biosynthetic) pathways, allows a cell to adapt to short-term fluctuations in the supply of a substance it needs.

Second, cells can adjust the production of certain enzymes; that is, they can regulate the expression of the genes encoding the enzymes. If, in our example, the environment provides all the tryptophan the cell needs, the cell stops making the enzymes that catalyze the synthesis of tryptophan **(Figure 15.2b)**. In this case, the control of enzyme production occurs at the level of transcription, the synthesis of messenger RNA from the genes that code for these enzymes.

Regulation of the tryptophan synthesis pathway is just one example of how bacteria tune their metabolism to changing environments. Many genes of the bacterial genome are switched on or off by changes in the metabolic status of the cell. One basic mechanism for this control of gene expression in bacteria, described as the *operon model*, was discovered in 1961 by François Jacob and Jacques Monod at the Pasteur Institute in Paris. Let's see what an operon is and how it works, using the control of tryptophan synthesis as our first example.

## Operons: The Basic Concept

*E. coli* synthesizes the amino acid tryptophan from a precursor molecule in the three-step pathway shown in Figure 15.2. Each reaction in the pathway is catalyzed by a specific enzyme, and the five genes that code for the subunits of these enzymes are clustered together on the bacterial chromosome. A single promoter serves all five genes, which together constitute a transcription unit. (Recall that a promoter is a site where RNA polymerase can bind to DNA and begin transcription; see Figure 14.8.) Thus, transcription gives rise to one long mRNA molecule that codes for the five polypeptides making up the enzymes in the tryptophan pathway **(Figure 15.3a)**. The cell can translate this one mRNA into five separate polypeptides because the mRNA is punctuated with start and stop codons that signal where the coding sequence for each polypeptide begins and ends.

A key advantage of grouping genes of related function into one transcription unit is that a single "on-off switch" can control the whole cluster of functionally related genes; in other words, these genes are *coordinately controlled*. When an *E. coli* cell must make tryptophan for itself because its surrounding environment lacks this amino acid, all the enzymes for the metabolic pathway are synthesized at one time. The "on-off switch" is a segment of DNA called an **operator**. Both its location and name suit its function: Positioned within the promoter or, in some cases, between the promoter and the enzyme-coding genes, the operator controls the access of RNA polymerase to the genes. All together, the operator, the promoter, and the genes they control—the entire stretch of DNA required for enzyme production for the tryptophan pathway—constitute an **operon**. The *trp* operon (*trp* for tryptophan) is one of many operons in the *E. coli* genome (see Figure 15.3a).

If the operator is the operon's switch for controlling transcription, how does this switch work? By itself, the *trp* operon is turned on; that is, RNA polymerase can bind to the

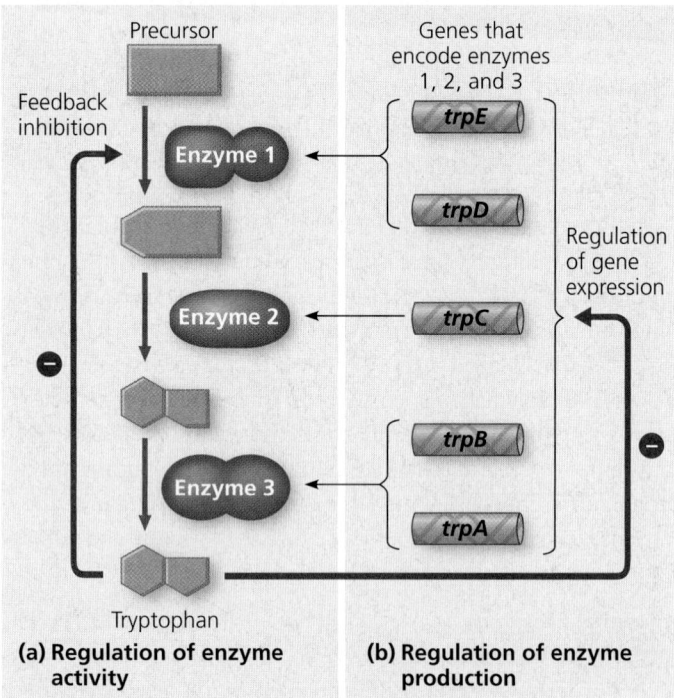

▲ **Figure 15.2 Regulation of a metabolic pathway.** In the pathway for tryptophan synthesis, an abundance of tryptophan can both **(a)** inhibit the activity of the first enzyme in the pathway (feedback inhibition), a rapid response, and **(b)** repress expression of the genes encoding all subunits of the enzymes in the pathway, a longer-term response. Genes *trpE* and *trpD* encode the two subunits of enzyme 1, and genes *trpB* and *trpA* encode the two subunits of enzyme 3. (The genes were named before the order in which they functioned in the pathway was determined.) The ⊖ symbol stands for inhibition.

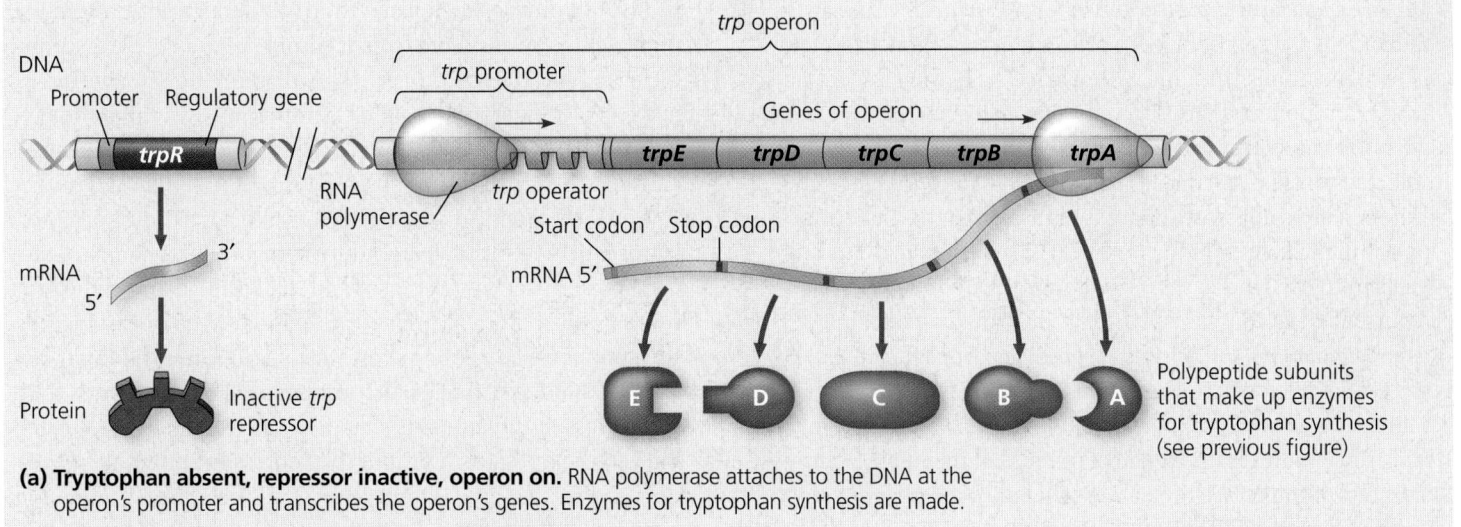

**(a) Tryptophan absent, repressor inactive, operon on.** RNA polymerase attaches to the DNA at the operon's promoter and transcribes the operon's genes. Enzymes for tryptophan synthesis are made.

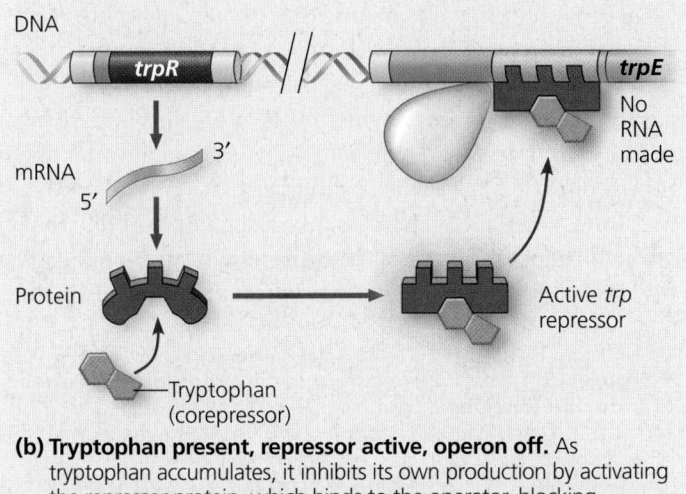

**(b) Tryptophan present, repressor active, operon off.** As tryptophan accumulates, it inhibits its own production by activating the repressor protein, which binds to the operator, blocking transcription. Enzymes for tryptophan synthesis are not made.

▲ **Figure 15.3 The *trp* operon in *E. coli*: regulated synthesis of repressible enzymes.** Tryptophan is an amino acid, produced by an anabolic pathway catalyzed by three enzymes (see Figure 15.2). **(a)** The five genes encoding the polypeptide subunits of the enzymes in this pathway are grouped, along with a promoter, into the *trp* operon. The *trp* operator (the repressor binding site) is located within the *trp* promoter (the RNA polymerase binding site). **(b)** Accumulation of tryptophan, the end product of the pathway, represses transcription of the *trp* operon, thus blocking synthesis of all the enzymes in the pathway and shutting down tryptophan production.

**?** *Describe what happens to the* trp *operon as the cell uses up its store of tryptophan.*

promoter and transcribe the genes of the operon. The operon can be switched off by a protein called the *trp* repressor. A **repressor** binds to the operator and blocks the attachment of RNA polymerase to the promoter, preventing transcription of the genes. A repressor protein is specific for the operator of a particular operon. For example, the *trp* repressor, which switches off the *trp* operon by binding to the *trp* operator, has no effect on other operons in the *E. coli* genome.

The *trp* repressor is the protein product of a **regulatory gene** called *trpR*, which is located some distance from the *trp* operon and has its own promoter. Regulatory genes are expressed continuously, although at a low rate, and a few *trp* repressor molecules are always present in *E. coli* cells. Why, then, is the *trp* operon not switched off permanently? First, the binding of repressors to operators is reversible. An operator alternates between two states: one with the repressor bound and one without. The relative duration of the repressor-bound state is higher when more active repressor molecules are present. Second, the *trp* repressor, like most regulatory proteins, is an allosteric protein, with two alternative shapes: active and inactive

(see Figure 6.18). The *trp* repressor is synthesized in the inactive form, which has little affinity for the *trp* operator. Only when tryptophan binds to the *trp* repressor at an allosteric site does the repressor protein change to the active form that can attach to the operator, turning the operon off **(Figure 15.3b)**.

Tryptophan functions in this system as a **corepressor**, a small molecule that cooperates with a repressor protein to switch an operon off. As tryptophan accumulates, more tryptophan molecules associate with *trp* repressor molecules, which can then bind to the *trp* operator and shut down production of the tryptophan pathway enzymes. If the cell's tryptophan level drops, transcription of the operon's genes resumes. The *trp* operon is one example of how gene expression can respond to changes in the cell's internal and external environment.

## Repressible and Inducible Operons: Two Types of Negative Gene Regulation

The *trp* operon is said to be a *repressible operon* because its transcription is usually on but can be inhibited (repressed) when a specific small molecule (in this case, tryptophan) binds allosterically to a regulatory protein. In contrast, an *inducible operon* is usually off but can be stimulated (induced) when a specific small molecule interacts with a regulatory protein. The classic example of an inducible operon is the *lac* operon (*lac* for lactose).

The disaccharide lactose (milk sugar) is available to *E. coli* in the human colon if the host drinks or eats a dairy product. Lactose metabolism begins with hydrolysis of the disaccharide into its component monosaccharides (glucose and galactose), a reaction catalyzed by the enzyme β-galactosidase. Only a few molecules of this enzyme are present in an *E. coli* cell growing in the absence of lactose. If lactose is added to the bacterium's environment, however, the number of β-galactosidase molecules in the cell increases 1,000-fold within about 15 minutes. How can a cell ramp up enzyme production this quickly?

The gene for β-galactosidase (*lacZ*) is part of the *lac* operon, which includes two other genes coding for enzymes that function in the use of lactose (**Figure 15.4**). The entire transcription unit is under the command of one main operator and promoter. The regulatory gene, *lacI*, located outside the operon, codes for an allosteric repressor protein that can switch off the *lac* operon by binding to the operator. So far, this sounds just like regulation of the *trp* operon, but there is one

important difference. Recall that the *trp* repressor protein is inactive by itself and requires tryptophan as a corepressor in order to bind to the operator. The *lac* repressor, in contrast, is active by itself, binding to the operator and switching the *lac* operon off. In this case, a specific small molecule, called an **inducer**, *inactivates* the repressor.

For the *lac* operon, the inducer is allolactose, an isomer of lactose formed in small amounts from lactose that enters the cell. In the absence of lactose (and hence allolactose), the *lac* repressor is in its active shape and binds to the operator; thus, the genes of the *lac* operon are silenced (**Figure 15.4a**). If lactose is added to the cell's surroundings, allolactose binds to the *lac* repressor and alters its shape; the repressor can no longer bind to the operator. Without the repressor bound, the *lac* operon is transcribed into mRNA, and the enzymes for using lactose are made (**Figure 15.4b**).

In the context of gene regulation, the enzymes of the lactose pathway are referred to as *inducible enzymes* because their synthesis is induced by a chemical signal (allolactose, in this case). Analogously, the enzymes for tryptophan synthesis are said to be repressible. *Repressible enzymes* generally function in anabolic pathways, which synthesize essential end products from raw materials (precursors). By suspending production of an end product when it is already present in sufficient quantity, the cell can allocate its organic precursors and energy for

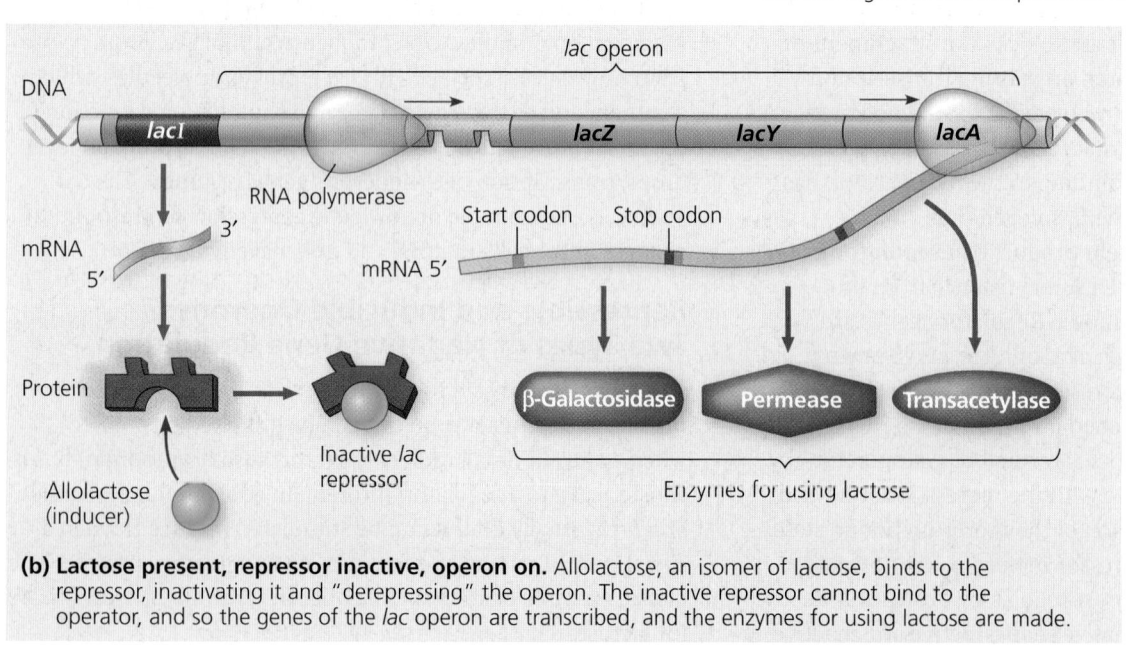

(a) **Lactose absent, repressor active, operon off.** The *lac* repressor is innately active, and in the absence of lactose it switches off the operon by binding to the operator. The enzymes for using lactose are not made.

▼ **Figure 15.4 The *lac* operon in *E. coli*: regulated synthesis of inducible enzymes.** *E. coli* uses three enzymes to take up and metabolize lactose, the genes for which are clustered in the *lac* operon. The first gene, *lacZ*, codes for β-galactosidase, which hydrolyzes lactose to glucose and galactose. The second gene, *lacY*, codes for a permease, a membrane protein that transports lactose into the cell. The third gene, *lacA*, codes for an enzyme that detoxifies other molecules entering the cell via the permease. Unusually, the gene for the *lac* repressor, *lacI*, is adjacent to the *lac* operon. The function of the teal region within the promoter will be revealed in Figure 15.5.

(b) **Lactose present, repressor inactive, operon on.** Allolactose, an isomer of lactose, binds to the repressor, inactivating it and "derepressing" the operon. The inactive repressor cannot bind to the operator, and so the genes of the *lac* operon are transcribed, and the enzymes for using lactose are made.

other uses. In contrast, inducible enzymes usually function in catabolic pathways, which break down a nutrient to simpler molecules. By producing the appropriate enzymes only when the nutrient is available, the cell avoids wasting energy and precursors making proteins that are not needed.

Regulation of both the *trp* and *lac* operons involves the *negative* control of genes, because the operons are switched off by the active form of the repressor protein (see Figures 15.3b and 15.4a). It may be easier to see this for the *trp* operon, but it is also true for the *lac* operon. In the case of the *lac* operon, allolactose induces enzyme synthesis not by directly activating the *lac* operon, but by freeing it from the negative effect of the repressor (see Figure 15.4b). Gene regulation is said to be *positive* only when a regulatory protein interacts directly with the genome to switch transcription on. Let's look at an example of the positive control of genes, again involving the *lac* operon.

## Positive Gene Regulation

When glucose and lactose are both present in its environment, *E. coli* preferentially uses glucose. The enzymes for glucose breakdown in glycolysis (see Figure 7.9) are continually present. Only when lactose is present *and* glucose is in short supply does *E. coli* use lactose as an energy source, and only then does it synthesize appreciable quantities of the enzymes for lactose breakdown.

How does the *E. coli* cell sense the glucose concentration and relay this information to the genome? Again, the mechanism depends on the interaction of an allosteric regulatory protein with a small organic molecule, in this case **cyclic AMP (cAMP)**, which accumulates when glucose is scarce. The regulatory protein, called *cAMP receptor protein (CRP)*, is an **activator**, a protein that binds to DNA and stimulates transcription of a gene. When cAMP binds to this regulatory protein, CRP assumes its active shape and can attach to a specific site at the upstream end of the *lac* promoter **(Figure 15.5a)**. This attachment increases the affinity of RNA polymerase for the *lac* promoter, which is actually rather low even when no *lac* repressor is bound to the operator. By facilitating the binding of RNA polymerase to the promoter and thereby increasing the rate of transcription of the *lac* operon, the attachment of CRP to the promoter directly stimulates gene expression. Therefore, this mechanism qualifies as positive regulation.

If the amount of glucose in the cell increases, the cAMP concentration falls, and without cAMP, CRP detaches from the operon. Because CRP is inactive, RNA polymerase binds less efficiently to the promoter, and transcription of the *lac* operon proceeds at only a low level, even in the presence of lactose **(Figure 15.5b)**. Thus, the *lac* operon is under dual control: negative control by the *lac* repressor and positive control by CRP. The state of the *lac* repressor (with or without bound allolactose) determines whether or not transcription of the *lac* operon's genes occurs at all; the state of CRP (with or without bound cAMP) controls the *rate* of transcription if the operon

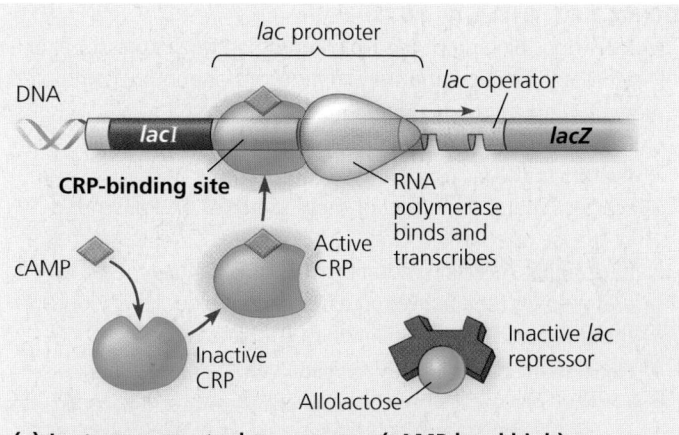

**(a) Lactose present, glucose scarce (cAMP level high): abundant *lac* mRNA synthesized.** If glucose is scarce, the high level of cAMP activates CRP, which binds to the promoter and increases RNA polymerase binding there. The *lac* operon produces large amounts of mRNA coding for the enzymes that the cell needs for use of lactose.

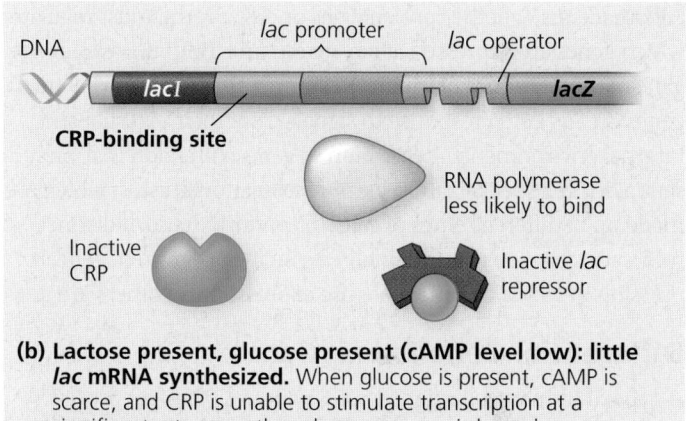

**(b) Lactose present, glucose present (cAMP level low): little *lac* mRNA synthesized.** When glucose is present, cAMP is scarce, and CRP is unable to stimulate transcription at a significant rate, even though no repressor is bound.

▲ **Figure 15.5 Positive control of the *lac* operon by cAMP receptor protein (CRP).** RNA polymerase has high affinity for the *lac* promoter only when CRP is bound to a DNA site at the upstream end of the promoter. CRP, in turn, attaches to its DNA site only when associated with cyclic AMP (cAMP), whose concentration in the cell rises when the glucose concentration falls. Thus, when glucose is present, even if lactose also is available, the cell preferentially catabolizes glucose and makes very little of the enzymes for using lactose.

is repressor-free. It is as though the operon has both an on-off switch and a volume control.

In addition to regulating the *lac* operon, CRP helps regulate other operons that encode enzymes used in catabolic pathways. All told, it may affect the expression of more than 100 genes in *E. coli*. When glucose is plentiful and CRP is inactive, the synthesis of enzymes that catabolize compounds other than glucose generally slows down. The ability to catabolize other compounds, such as lactose, enables a cell deprived of glucose to survive. The compounds present in the cell at any given moment determine which operons are switched on—the result of simple interactions of activator and repressor proteins with the promoters of the genes in question.

## CONCEPT 15.2

# Eukaryotic gene expression is regulated at many stages

All organisms, whether prokaryotes or eukaryotes, must regulate which genes are expressed at any given time. Both unicellular organisms and the cells of multicellular organisms continually turn genes on and off in response to signals from their external and internal environments. Regulation of gene expression is also essential for cell specialization in multicellular organisms, which are made up of different types of cells. To perform its own distinct role, each cell type must maintain a specific program of gene expression in which certain genes are expressed and others are not.

## Differential Gene Expression

A typical human cell might express about 20% of its protein-coding genes at any given time. Highly differentiated cells, such as muscle or nerve cells, express an even smaller fraction of their genes. Almost all the cells in a multicellular organism contain an identical genome. (Cells of the immune system are one exception, as you will see in Figure 35.10.) A subset of genes is expressed in each cell type; some of those are "housekeeping" genes, expressed by many cell types, while others are unique to that cell type. The uniquely expressed genes allow these cells to carry out their specific function. The differences between cell types, therefore, are due not to different genes being present, but to **differential gene expression**, the expression of different genes by cells with the same genome.

The function of any cell, whether a single-celled eukaryote or a particular cell type in a multicellular organism, depends on the appropriate set of genes being expressed. The transcription factors of a cell must locate the right genes at the right time, a task on a par with finding a needle in a haystack. When gene expression proceeds abnormally, serious imbalances and diseases, including cancer, can arise.

**Figure 15.6** summarizes the process of gene expression in a eukaryotic cell, highlighting key stages in the expression of a protein-coding gene. Each stage depicted in Figure 15.6 is a potential control point at which gene expression can be turned on or off, accelerated, or slowed down.

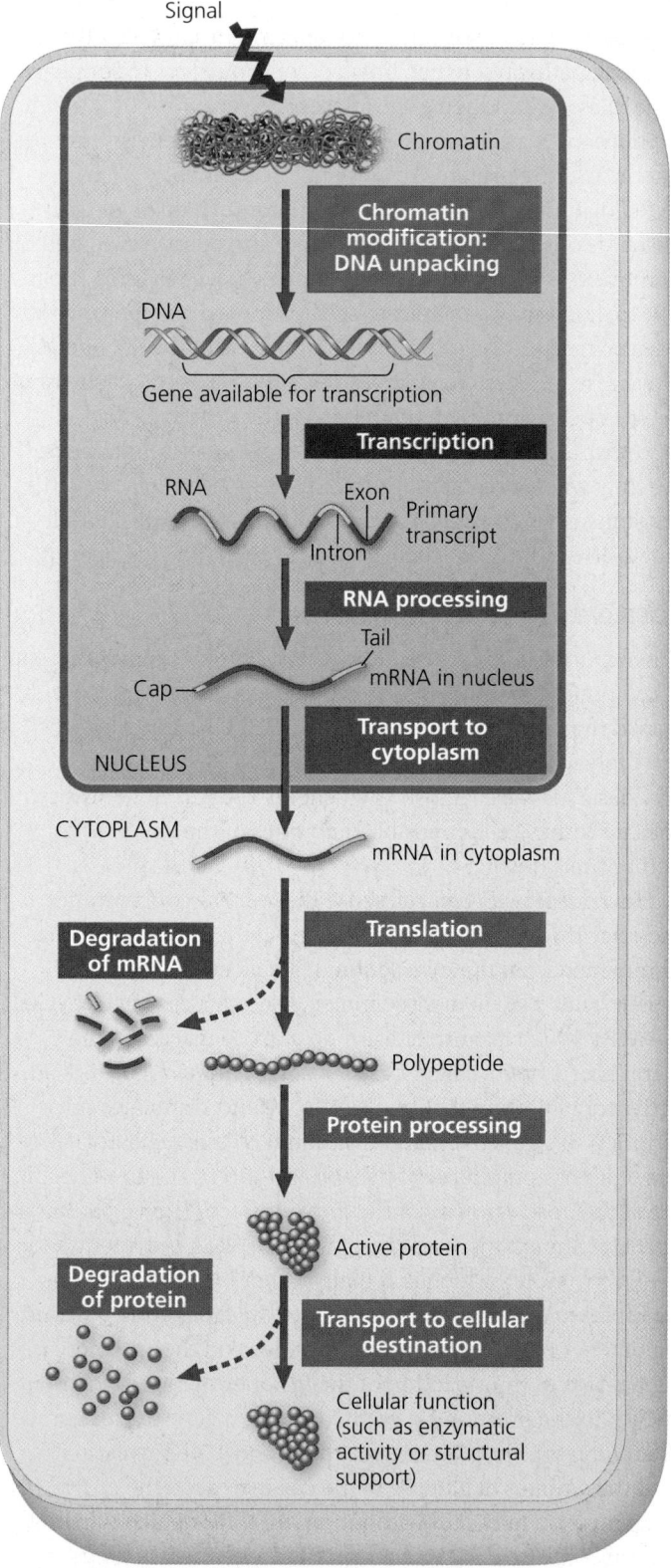

▲ **Figure 15.6 Stages in gene expression that can be regulated in eukaryotic cells.** In this diagram, the colored boxes indicate the processes most often regulated; each color indicates the type of molecule that is affected (blue = DNA, red/orange = RNA, purple = protein). The nuclear envelope separating transcription from translation in eukaryotic cells offers an opportunity for post-transcriptional control in the form of RNA processing that is absent in prokaryotes. In addition, eukaryotes have a greater variety of control mechanisms operating before transcription and after translation. A miniature version of this figure accompanies several figures later in the chapter as an orientation diagram.

Fifty years ago, an understanding of the mechanisms that control gene expression in eukaryotes seemed almost hopelessly out of reach. Since then, new research methods, notably advances in DNA technology that we will discuss in Concept 15.4, have enabled molecular biologists to uncover many details of eukaryotic gene regulation. In all organisms, gene expression is commonly controlled at transcription; regulation at this stage often occurs in response to signals coming from outside the cell, such as hormones or other signaling molecules. For this reason, the term *gene expression* is often equated with transcription, for both bacteria and eukaryotes. While this may often be the case for bacteria, the greater complexity of eukaryotic cell structure and function provides opportunities for regulating gene expression at many additional stages (see Figure 15.6). In the remainder of this section, we'll examine some of the important control points of eukaryotic gene expression more closely.

## Regulation of Chromatin Structure

Recall that the DNA of eukaryotic cells is packaged with proteins in an elaborate complex known as chromatin, the basic unit of which is the nucleosome. A nucleosome consists of a cluster of eight histone proteins around which the DNA double helix is wrapped (see Figure 13.23). The structural organization of chromatin not only packs a cell's DNA into a compact form that fits inside the nucleus, but also helps regulate gene expression in several ways. Whether or not a gene is transcribed is affected by the location of nucleosomes along a gene's promoter and also the sites where the promoter DNA attaches to the protein scaffolding of the chromosome (see Figure 13.23). In addition, genes within highly condensed chromatin (heterochromatin) are usually not expressed. Lastly, certain chemical modifications to the histone proteins and to the DNA of chromatin can influence both chromatin structure and gene expression. Here we examine the effects of these modifications, which are catalyzed by specific enzymes.

### Histone Modifications and DNA Methylation

Chemical modifications to histones play a direct role in the regulation of gene transcription. The N-terminus of each histone protein in a nucleosome protrudes outward from the nucleosome. These so-called *histone tails* are accessible to various modifying enzymes that catalyze the addition or removal of specific chemical groups, such as acetyl ($-COCH_3$), methyl, and phosphate groups. Generally, **histone acetylation** appears to promote transcription by opening up the chromatin structure **(Figure 15.7)**, while addition of methyl groups can lead to the condensation of chromatin and reduced transcription.

Rather than modifying proteins associated with DNA, a different set of enzymes can methylate the DNA itself on certain bases, usually cytosine. Such **DNA methylation** occurs in most plants, animals, and fungi. Long stretches of inactive DNA, such as that of inactivated mammalian X chromosomes (see Figure 12.8), are generally more methylated than regions of actively transcribed DNA (although there are exceptions).

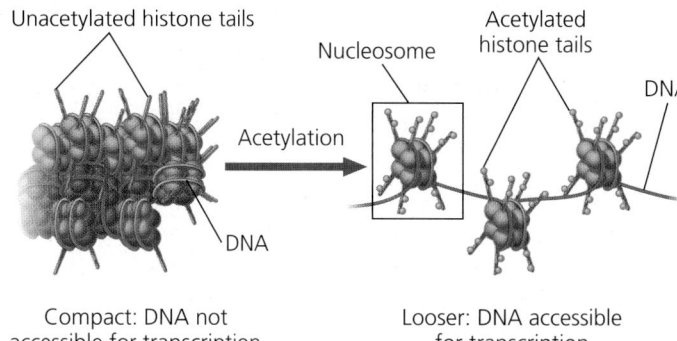

▲ **Figure 15.7 A simple model of the effect of histone acetylation.** Amino acids in histone tails may be chemically modified. For example, acetyl groups (green balls) may be added, causing the chromatin to be less compact and making the DNA accessible for transcription.

On a smaller scale, the DNA of individual genes is usually more heavily methylated in cells in which those genes are not expressed. Once methylated, genes usually stay that way through successive cell divisions in a given individual. At DNA sites where one strand is already methylated, enzymes methylate the correct daughter strand after each round of DNA replication. In this way, methylation patterns can be inherited.

### Epigenetic Inheritance

The chromatin modifications we just discussed do not change the DNA sequence, yet they still may be passed along to future generations of cells. Inheritance of traits transmitted by mechanisms not involving the nucleotide sequence itself is called **epigenetic inheritance**. Whereas mutations in DNA are permanent, modifications to the chromatin can be reversed. For example, DNA methylation patterns are largely erased and reestablished during gamete formation.

Researchers are amassing more and more evidence for the importance of epigenetic information in the regulation of gene expression. Epigenetic variations might help explain why one identical twin acquires a genetically based disease, such as schizophrenia, but the other does not, despite their identical genomes. Alterations in normal patterns of DNA methylation are seen in some cancers, where they are associated with inappropriate gene expression. Evidently, enzymes that modify chromatin structure are integral parts of the eukaryotic cell's machinery for regulating transcription.

## Regulation of Transcription Initiation

Chromatin-modifying enzymes provide initial control of gene expression by making a region of DNA either more or less able to bind the transcription machinery. Once the chromatin of a gene is optimally modified for expression, the initiation of transcription is the next major step at which gene expression is regulated. As in bacteria, the regulation of transcription initiation in eukaryotes involves proteins that bind to DNA and either facilitate or inhibit binding of RNA polymerase. The process is more complicated in eukaryotes, however. Before looking at how eukaryotic cells control their transcription, let's review the structure of a typical eukaryotic gene and its transcript.

## Organization of a Typical Eukaryotic Gene and Its Transcript

A eukaryotic gene and the DNA elements (segments) that control it are typically organized as shown in **Figure 15.8**, which extends what you learned about eukaryotic genes in Chapter 14. Recall that a cluster of proteins called a *transcription initiation complex* assembles on the promoter sequence at the "upstream" end of the gene (see Figure 14.9). One of these proteins, RNA polymerase II, then proceeds to transcribe the gene, synthesizing a primary RNA transcript (pre-mRNA). RNA processing includes enzymatic addition of a 5′ cap and a poly-A tail, as well as splicing out of introns, to yield a mature mRNA. Associated with most eukaryotic genes are multiple **control elements**, segments of noncoding DNA having particular nucleotide sequences that serve as binding sites for the proteins called transcription factors. Control elements on the DNA and the transcription factors that bind to them are critical to the precise regulation of gene expression seen in different cell types.

### The Roles of General and Specific Transcription Factors

Transcription factors are of two types: General transcription factors act at the promoter of all genes, while some genes require specific transcription factors that bind to control elements close to or further away from the promoter.

**General Transcription Factors at the Promoter** To initiate transcription, eukaryotic RNA polymerase requires the assistance of transcription factors. Some transcription factors

(such as those illustrated in Figure 14.9) are essential for the transcription of *all* protein-coding genes; therefore, they are often called *general transcription factors*. A few general transcription factors bind to a DNA sequence such as the TATA box within the promoter, but many bind to proteins, including other transcription factors and RNA polymerase II. Protein-protein interactions are crucial to the initiation of eukaryotic transcription. Only when the complete initiation complex has assembled can the polymerase begin to move along the DNA template strand, producing a complementary strand of RNA.

The interaction of general transcription factors and RNA polymerase II with a promoter usually leads to only a low rate of initiation and production of few RNA transcripts from genes that are not expressed all the time, but, instead, are regulated. In eukaryotes, high levels of transcription of these particular genes at the appropriate time and place depend on the interaction of control elements with another set of proteins, which can be thought of as *specific transcription factors*.

**Enhancers and Specific Transcription Factors** As you can see in Figure 15.8, some control elements, named *proximal control elements*, are located close to the promoter. (Although some biologists consider proximal control elements part of the promoter, in this text we do not.) The more distant *distal control elements*, groupings of which are called **enhancers**, may be thousands of nucleotides upstream or downstream of a gene or even within an intron. A given gene may have multiple enhancers, each active at a different time or in a different cell

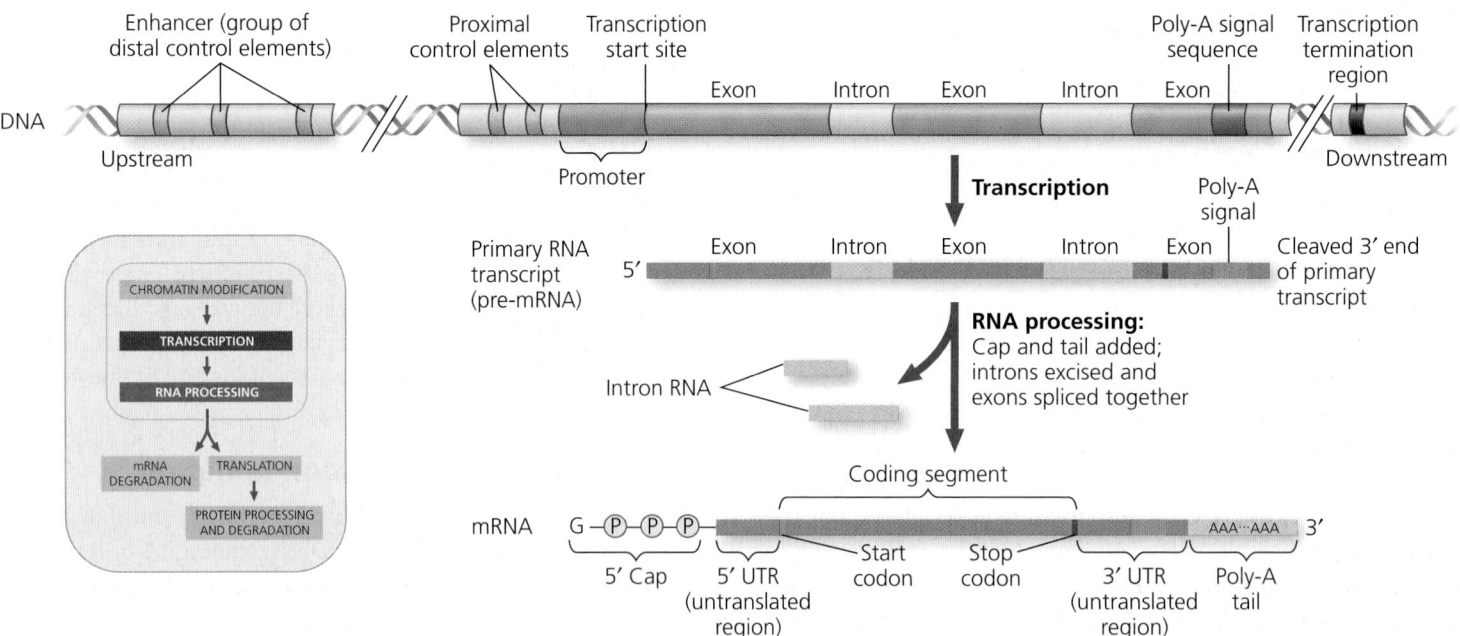

▲ **Figure 15.8 A eukaryotic gene and its transcript.** Each eukaryotic gene has a promoter, a DNA sequence where RNA polymerase binds and starts transcription, proceeding "downstream." A number of control elements (gold) are involved in regulating the initiation of transcription; these are DNA sequences located near (proximal to) or far from

(distal to) the promoter. Distal control elements can be grouped together as enhancers, one of which is shown for this gene. At the other end of the gene, a polyadenylation (poly-A) signal sequence in the last exon of the gene is transcribed into an RNA sequence that signals where the transcript is cleaved and the poly-A tail added. Transcription may continue for

hundreds of nucleotides beyond the poly-A signal before terminating. RNA processing of the primary transcript into a functional mRNA involves three steps: addition of the 5′ cap, addition of the poly-A tail, and splicing. In the cell, the 5′ cap is added soon after transcription is initiated, and splicing occurs while transcription is still under way (see Figure 14.12).

type or location in the organism. Each enhancer, however, is generally associated with only that gene and no other.

In eukaryotes, the rate of gene expression can be strongly increased or decreased by the binding of specific transcription factors, either activators or repressors, to the control elements of enhancers. Hundreds of transcription activators have been discovered in eukaryotes; the structure of one example is shown in **Figure 15.9**. Researchers have identified two types of structural domains that are commonly found in a large number of activator proteins: a *DNA-binding domain*—a part of the protein's three-dimensional structure that binds to DNA—and one or more *activation domains*. Activation domains bind other regulatory proteins or components of the transcription machinery, facilitating a series of protein-protein interactions that result in enhanced transcription of a given gene.

**Figure 15.10** shows the currently accepted model for how binding of activators to an enhancer located far from the promoter can influence transcription. Protein-mediated bending of the DNA brings the bound activators into contact with a

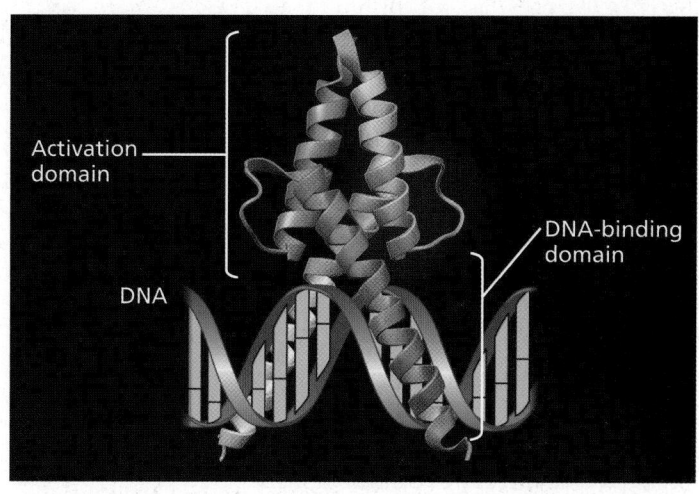

▲ **Figure 15.9 MyoD, a transcriptional activator.** The MyoD protein is made up of two subunits (purple and salmon) with extensive regions of α helix. Each subunit has one DNA-binding domain and one activation domain. The latter includes binding sites for the other subunit and for other proteins. MyoD is involved in muscle development in vertebrate embryos (see Concept 16.1).

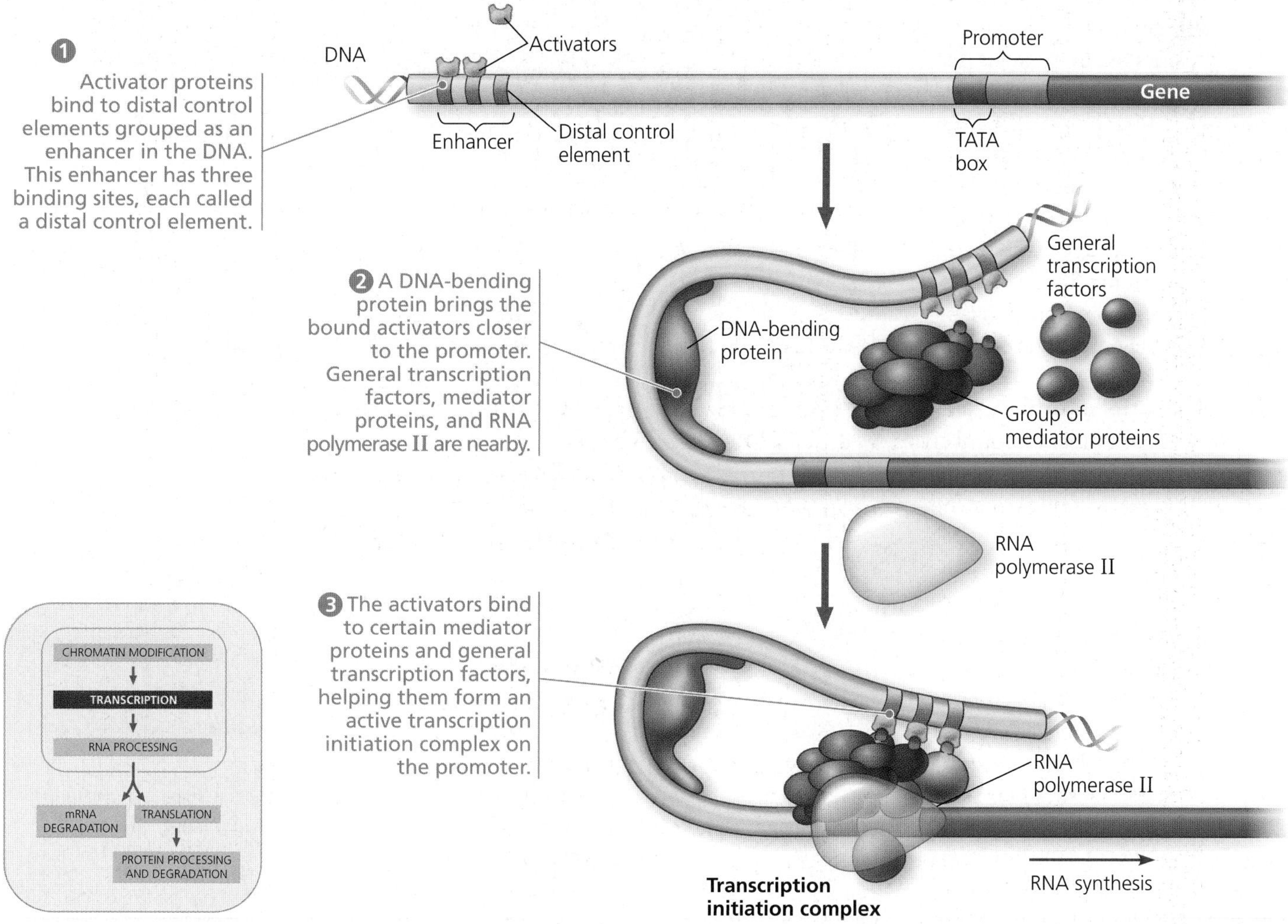

**❶** Activator proteins bind to distal control elements grouped as an enhancer in the DNA. This enhancer has three binding sites, each called a distal control element.

DNA

Activators

Promoter

Gene

Enhancer

Distal control element

TATA box

**❷** A DNA-bending protein brings the bound activators closer to the promoter. General transcription factors, mediator proteins, and RNA polymerase II are nearby.

DNA-bending protein

General transcription factors

Group of mediator proteins

RNA polymerase II

**❸** The activators bind to certain mediator proteins and general transcription factors, helping them form an active transcription initiation complex on the promoter.

RNA polymerase II

**Transcription initiation complex**

RNA synthesis

CHROMATIN MODIFICATION

TRANSCRIPTION

RNA PROCESSING

mRNA DEGRADATION

TRANSLATION

PROTEIN PROCESSING AND DEGRADATION

▲ **Figure 15.10 A model for the action of enhancers and transcription activators.** Bending of the DNA by a protein enables enhancers to influence a promoter hundreds or even thousands of nucleotides away. Specific transcription factors called activators bind to the enhancer DNA sequences and then to a group of mediator proteins, which in turn bind to general transcription factors and then RNA polymerase II, assembling the transcription initiation complex. These protein-protein interactions lead to correct positioning of the complex on the promoter and the initiation of RNA synthesis. Only one enhancer (with three gold control elements) is shown here, but a gene may have several enhancers that act at different times or in different cell types.

group of *mediator proteins*, which interact with proteins at the promoter. These interactions help assemble and position the initiation complex on the promoter. One of the studies supporting this model shows that proteins regulating a mouse globin gene contact both the gene's promoter and an enhancer about 50,000 nucleotides upstream. Protein interactions allow these two DNA regions to come together in a very specific fashion, in spite of the many nucleotide pairs between them. In the **Scientific Skills Exercise**, you can work with data from an experiment that identified the control elements in the enhancer of a particular human gene.

Specific transcription factors that function as repressors can inhibit gene expression in several different ways. Some repressors bind directly to control element DNA, blocking activator binding. Other repressors interfere with the activator itself so it can't bind the DNA.

In addition to influencing transcription directly, some activators and repressors indirectly affect chromatin structure. Studies using yeast and mammalian cells show that some activators recruit proteins that acetylate histones near promoters of specific genes, promoting transcription (see Figure 15.7). Some repressors recruit proteins that remove acetyl groups from histones, leading to reduced transcription, a phenomenon called *silencing*. The recruitment of proteins that modify chromatin seems to be the most common mechanism of repression in eukaryotic cells.

**Combinatorial Control of Gene Activation** In eukaryotes, the precise control of transcription depends largely on the binding of activators to DNA control elements. Considering that many genes must be regulated in a typical animal or plant cell, the number of different nucleotide sequences in control elements is surprisingly small. A dozen or so short nucleotide sequences appear again and again in the control elements for different genes. On average, each enhancer is composed of about ten control elements, each binding only one or two specific transcription factors. It is the particular *combination* of control elements in an enhancer associated with a gene, rather than a unique control element, that is important in regulating transcription of the gene.

Even with only a dozen control element sequences available, many combinations are possible. Each combination can activate transcription only when the appropriate activator proteins are present, which may occur at a precise time during development or in a particular cell type. **Figure 15.11** shows how different combinations of a few control elements can allow differential regulation of transcription in two cell types—liver and lens cells. This can occur because each cell type contains a different group of activator proteins. (Concept 16.1 will explore how these groups came to differ during embryonic development.)

### Coordinately Controlled Genes in Eukaryotes

How does the eukaryotic cell deal with a group of genes of related function that need to be turned on or off at the same time? You have learned that in bacteria such *coordinately*

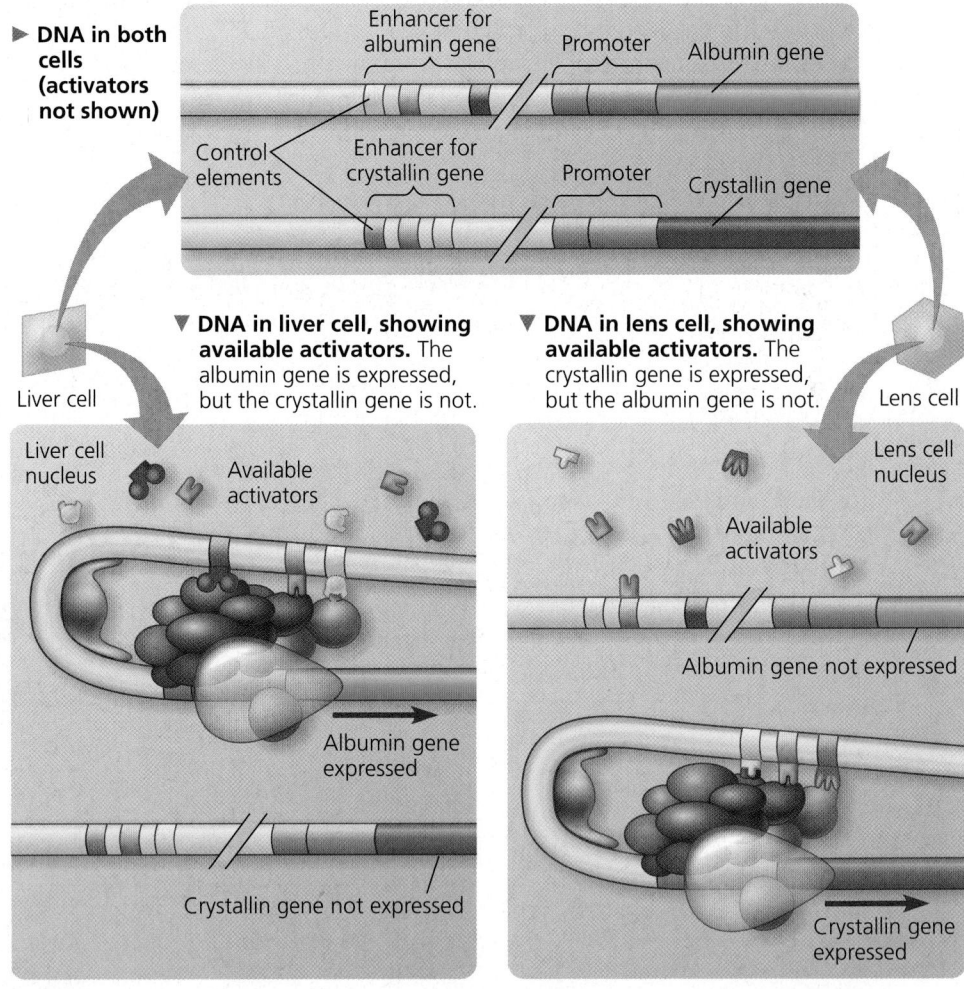

▲ **Figure 15.11 Cell type–specific transcription.** Both liver cells and lens cells have the genes for making the proteins albumin and crystallin, but only liver cells make albumin (a blood protein) and only lens cells make crystallin (the main protein of the lens of the eye). The specific transcription factors made in a cell determine which genes are expressed. In this example, the genes for albumin and crystallin are shown at the top, each with an enhancer made up of three different control elements. Although the enhancers for the two genes both have a gray control element, each enhancer has a unique *combination* of elements. All the activator proteins required for high-level expression of the albumin gene are present in liver cells only **(left)**, whereas the activators needed for expression of the crystallin gene are present in lens cells only **(right)**. For simplicity, we consider only the role of activators here, although the presence or absence of repressors may also influence transcription in certain cell types.

**?** *Describe the enhancer for the albumin gene in each type of cell. How would the nucleotide sequence of this enhancer in the liver cell compare with that in the lens cell?*

# Analyzing DNA Deletion Experiments

**What Control Elements Regulate Expression of the *mPGES-1* Gene?** The promoter of a gene includes the DNA immediately upstream of the transcription start site, but the control elements regulating the level of transcription of the gene (grouped in an enhancer) may be thousands of base pairs upstream of the promoter. Because the distance and spacing of control elements make them difficult to identify, scientists begin by deleting possible control elements and measuring the effect on gene expression. In this exercise, you will analyze data obtained from DNA deletion experiments that tested possible control elements for the human gene *mPGES-1*. This gene codes for an enzyme that synthesizes a type of prostaglandin, a chemical made during tissue inflammation.

**How the Experiment Was Done** The researchers hypothesized that there were three possible control elements in an enhancer region located 8–9 kilobases upstream of the *mPGES-1* gene. Control elements regulate whatever gene is in the appropriate downstream location. Thus, to test the activity of the possible elements, researchers first synthesized molecules of DNA ("constructs") that had the intact enhancer region upstream of a "reporter gene," a gene whose mRNA product could be easily measured experimentally. Next, they made three more DNA constructs with one of the three proposed control elements deleted in each (see left side of figure). The researchers then introduced each DNA construct into a separate human cell culture, where the cells took up the DNA constructs. After 48 hours, the amount of reporter gene mRNA made by the cells was measured. Comparing these amounts allowed researchers to determine if any of the deletions had an effect on expression of the reporter gene, mimicking the effect of the deletions on *mPGES-1* gene expression. (The *mPGES-1* gene itself couldn't be used to measure expression levels because the cells express their own *mPGES-1* gene. The mRNA from that gene would confuse the results.)

**Data from the Experiment** The diagrams on the left side of the figure show the intact DNA sequence (top) and the three experimental DNA constructs. A red X is located on the possible control element (1, 2, or 3) that was deleted in each experimental DNA construct. The area between the slashes represents the approximately 8 kilobases of DNA located between the promoter and the enhancer region. The horizontal bar graph on the right shows the amount of reporter gene mRNA that was present in each cell culture after 48 hours relative to the amount that was in the culture containing the intact enhancer region (top bar = 100%).

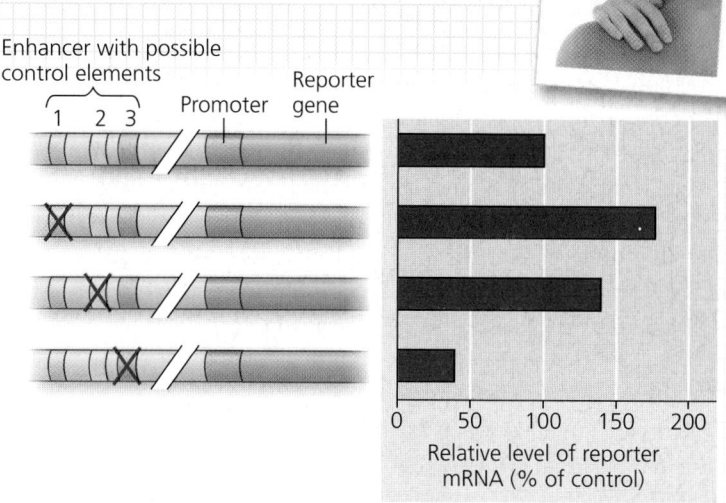

Enhancer with possible control elements

1 2 3    Promoter    Reporter gene

Relative level of reporter mRNA (% of control)

**Data from** J. N. Walters et al., Regulation of human microsomal prostaglandin E synthase-1 by IL-1β requires a distal enhancer element with a unique role for C/EBPβ, *Biochemical Journal* 443:561–571 (2012).

**INTERPRET THE DATA**

1. (a) What is the independent variable in the graph? (b) What is the dependent variable? (c) What was the control treatment in this experiment? Label it on the diagram.

2. Do the data suggest that any of these possible control elements are actual control elements? Explain.

3. (a) Did deletion of any of the possible control elements cause a *reduction* in reporter gene expression? If so, which one(s), and how can you tell? (b) If loss of a control element causes a reduction in gene expression, what must be the normal role of that control element? Provide a biological explanation for how the loss of such a control element could lead to a reduction in gene expression.

4. (a) Did deletion of any of the possible control elements cause an *increase* in reporter gene expression relative to the control? If so, which one(s), and how can you tell? (b) If loss of a control element causes an increase in gene expression, what must be the normal role of that control element? Propose a biological explanation for how the loss of such a control element could lead to an increase in gene expression.

**MB** A version of this Scientific Skills Exercise can be assigned in MasteringBiology.

*controlled* genes are often clustered into an operon regulated by a promoter and transcribed into an mRNA molecule. Thus, the genes are expressed together, and the encoded proteins are produced concurrently. With few exceptions, operons that work in this way have *not* been found in eukaryotic cells.

Eukaryotic genes that are co-expressed are typically scattered over different chromosomes. Thus, coordinate gene expression depends on every gene of a dispersed group having a specific combination of control elements. Activator proteins bind to the control elements, promoting simultaneous transcription of the genes, no matter where they are in the genome.

Coordinate control of dispersed genes in a eukaryotic cell often occurs in response to chemical signals from outside the cell. A steroid hormone, for example, enters a cell and binds to a receptor protein, forming a hormone-receptor complex that acts as a transcription activator (see Figure 5.23). Every gene whose transcription is stimulated by a given steroid hormone, on any chromosome, has a control element recognized by that hormone-receptor complex. This is how estrogen activates a group of genes that stimulate cell division in uterine cells, preparing the uterus for pregnancy.

Many signaling molecules, such as nonsteroid hormones and growth factors, bind to receptors on a cell's surface and never enter the cell. Such molecules can control gene expression indirectly by triggering signal transduction pathways that activate particular transcription factors (see Figure 5.26). Coordinate regulation in such pathways is the same as for steroid hormones: Genes with the same sets of control elements are activated by the same chemical signals. Because this system for coordinating gene regulation is so widespread, biologists think that it probably arose early in evolutionary history.

## Mechanisms of Post-transcriptional Regulation

Transcription alone does not constitute gene expression. The expression of a protein-coding gene is measured by the amount of functional protein a cell makes, and much happens between synthesis of the RNA transcript and the activity of the protein in the cell. Many regulatory mechanisms operate at various stages after transcription (see Figure 15.6). These mechanisms allow a cell to fine-tune gene expression rapidly in response to environmental changes without altering its transcription patterns. Here we discuss how cells regulate gene expression.

### RNA Processing

RNA processing in the nucleus and the export of mature RNA to the cytoplasm provide opportunities for regulating gene expression that are not available in prokaryotes. One example of regulation at the RNA-processing level is **alternative RNA splicing**, in which different mRNA molecules are produced from the same primary transcript, depending on which RNA segments are treated as exons and which as introns. Regulatory proteins specific to a cell type control intron-exon choices by binding to RNA sequences within the primary transcript.

A simple example of alternative RNA splicing is shown in **Figure 15.12** for the troponin T gene, which encodes two different (though related) proteins. Other genes code for many more possible products. For instance, researchers have found a gene in *Drosophila* with enough alternatively spliced exons to generate about 19,000 membrane proteins with different extracellular domains. At least 17,500 (94%) of the alternative

mRNAs are actually synthesized. Each developing nerve cell in the fly appears to synthesize a different form of the protein, which acts as a unique identifier on the cell surface.

It is clear that alternative RNA splicing can significantly expand the repertoire of a eukaryotic genome. In fact, alternative splicing was proposed as one explanation for the surprisingly low number of human genes counted when the human genome was sequenced. The number of human genes was found to be similar to that of a soil worm (nematode), mustard plant, or sea anemone. Scientists wondered what, if not the total number of genes, accounts for the more complex morphology (external form) of humans. It turns out that more than 90% of human protein-coding genes probably undergo alternative splicing. Thus, the extent of alternative splicing greatly multiplies the number of possible human proteins, which may be better correlated with complexity of form than the number of genes.

### Initiation of Translation and mRNA Degradation

Translation is another occasion when gene expression is regulated, most commonly at the initiation stage (see Figure 14.18). For some mRNAs, the initiation of translation can be blocked by regulatory proteins that bind to specific sequences or structures within the untranslated region (UTR) at the 5′ or 3′ end, preventing the attachment of ribosomes. (Recall from Concept 14.3 that both the 5′ cap and the poly-A tail of an mRNA molecule are important for ribosome binding.)

Alternatively, translation of *all* the mRNAs in a cell may be regulated simultaneously. In a eukaryotic cell, such "global" control usually involves activation or inactivation of one or more protein factors required to initiate translation. This mechanism plays a role in starting translation of mRNAs stored in eggs. Just after fertilization, translation is triggered by sudden activation of translation initiation factors. The response is a burst of synthesis of proteins encoded by the stored mRNAs. Some plants and algae store mRNAs during periods of darkness; light triggers reactivation of the translational apparatus.

The life span of mRNA molecules in the cytoplasm is important in determining the pattern of protein synthesis in a cell. Bacterial mRNA molecules typically are degraded by enzymes within a few minutes. This short life span of mRNAs is one reason bacteria can change their patterns of protein synthesis so quickly in response to environmental changes. In contrast, mRNAs in multicellular eukaryotes typically survive for hours, days, or even weeks. For instance, the mRNAs for the hemoglobin polypeptides (α-globin and β-globin) in developing red blood cells are unusually stable, and these mRNAs are translated repeatedly in these cells. Nucleotide sequences that affect how long an mRNA remains intact are often found in the untranslated region (UTR) at the 3′ end of the molecule (see Figure 15.8).

Other mechanisms that degrade or block expression of mRNA molecules have come to light. They involve a group of newly discovered RNA molecules that regulate gene expression at several levels, as we'll discuss shortly.

▼ **Figure 15.12 Alternative RNA splicing of the troponin T gene.** The primary transcript of this gene can be spliced in more than one way, generating different mRNA molecules. Notice that one mRNA molecule has ended up with exon 3 (green) and the other with exon 4 (purple). These two mRNAs are translated into different but related muscle proteins.

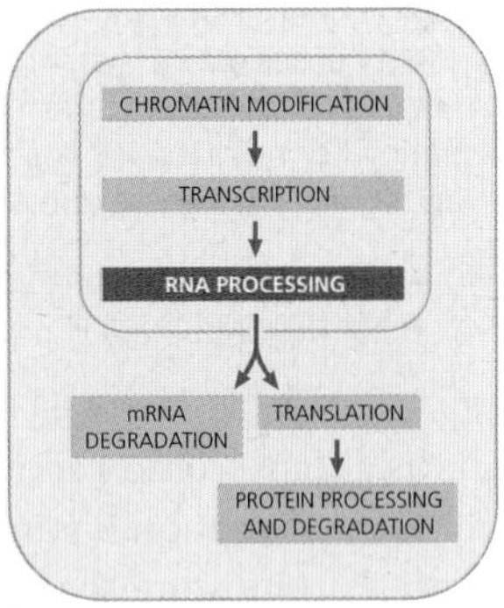

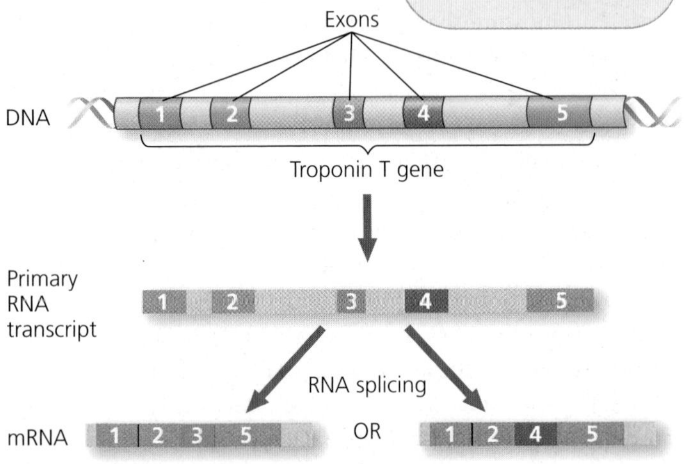

### Protein Processing and Degradation

The final opportunities for controlling gene expression occur after translation. Often, eukaryotic polypeptides must be processed to yield functional protein molecules. For instance, cleavage of the initial insulin polypeptide (pro-insulin) forms the active hormone. In addition, many proteins undergo chemical modifications that make them functional. Regulatory proteins are commonly activated or inactivated by the reversible addition of phosphate groups, and proteins destined for the surface of animal cells acquire sugars. Cell-surface proteins and many others must also be transported to target destinations in the cell in order to function. Regulation might occur at any of the steps involved in modifying or transporting a protein.

Finally, the length of time each protein functions in the cell is strictly regulated by selective degradation. Many proteins, such as the cyclins involved in regulating the cell cycle, must be short-lived if the cell is to function appropriately. To mark a protein for destruction, the cell commonly attaches molecules of a small protein called ubiquitin to the protein, which triggers its destruction by protein complexes.

#### CONCEPT CHECK 15.2

1. In general, what are the effects of histone acetylation and DNA methylation on gene expression?
2. Compare the roles of general and specific transcription factors in regulating gene expression.
3. **WHAT IF?** Suppose you compared the nucleotide sequences of the distal control elements in the enhancers of three genes that are expressed only in muscle cells. What would you expect to find? Why?

For suggested answers, see Appendix A.

---

### CONCEPT 15.3

# Noncoding RNAs play multiple roles in controlling gene expression

Genome sequencing has revealed that protein-coding DNA accounts for only 1.5% of the human genome and a similarly small percentage of the genomes of many other multicellular eukaryotes. A very small fraction of the non-protein-coding DNA consists of genes for RNAs such as ribosomal RNA and transfer RNA. Scientists assumed until recently that most of the remaining DNA was not transcribed, thinking that since it didn't specify proteins or the few known types of RNA, such DNA didn't contain meaningful genetic information—in fact, it was called "junk DNA." However, a flood of recent data has contradicted this idea. For example, a massive study of the entire human genome showed that roughly 75% of the genome is transcribed at some point in any given cell. Introns account for only a fraction of this transcribed RNA, most of which is untranslated. These and other results suggest that a significant amount of the genome may be transcribed into non-protein-coding RNAs—also called *noncoding RNAs*, including a variety

of small RNAs. Researchers are uncovering more evidence of the biological roles of these RNAs every day.

These discoveries have revealed a large, diverse population of RNA molecules in the cell that play crucial roles in regulating gene expression—and that have gone largely unnoticed until recently. Our long-standing view, that because mRNAs code for proteins they are the most important RNAs functioning in the cell, demands revision. This represents a major shift in the thinking of biologists, one that you are witnessing as students entering this field of study.

## Effects on mRNAs by MicroRNAs and Small Interfering RNAs

Regulation by both small and large noncoding RNAs occurs at several points in the pathway of gene expression, including mRNA translation and chromatin modification. We'll examine two types of small noncoding RNAs, the importance of which was acknowledged when their discovery was the focus of the 2006 Nobel Prize in Physiology or Medicine.

Since 1993, a number of research studies have uncovered small, single-stranded RNA molecules, called **microRNAs (miRNAs)**, that are capable of binding to complementary sequences in mRNA molecules. A longer RNA precursor is processed by cellular enzymes into an miRNA, a single-stranded RNA of about 22 nucleotides that forms a complex with one or more proteins **(Figure 15.13)**. The miRNA allows

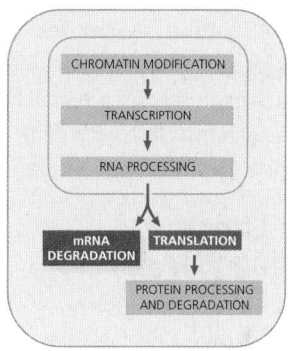

▼ **Figure 15.13 Regulation of gene expression by microRNAs (miRNAs).** An miRNA of about 22 nucleotides, formed by enzymatic processing of an RNA precursor, associates with one or more proteins in a complex that can degrade or block translation of target mRNAs.

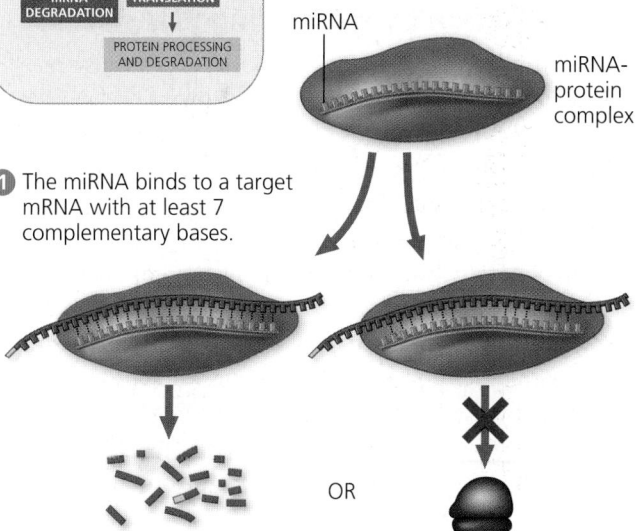

1 The miRNA binds to a target mRNA with at least 7 complementary bases.

OR

mRNA degraded          Translation blocked

2 If miRNA and mRNA bases are complementary all along their length, the mRNA is degraded (left); if the match is less complete, translation is blocked (right).

the complex to bind to any mRNA molecule with at least 7 or 8 nucleotides of complementary sequence. The miRNA-protein complex then either degrades the target mRNA or blocks its translation. Biologists estimate that expression of at least one-half of all human genes may be regulated by miRNAs, a remarkable figure given that the existence of miRNAs was unknown 25 years ago.

Another class of small RNAs are called **small interfering RNAs (siRNAs)**. These are similar in size and function to miRNAs—both can associate with the same proteins, producing similar results. In fact, if siRNA precursor RNA molecules are injected into a cell, the cell's machinery can process them into siRNAs that turn off expression of genes with related sequences, similarly to how miRNAs function. The distinction between miRNAs and siRNAs is based on subtle differences in the structure of their precursors, which in both cases are RNA molecules that are mostly double-stranded. The blocking of gene expression by siRNAs is called **RNA interference (RNAi)**, and it is used in the laboratory as a means of disabling specific genes to investigate their function.

**EVOLUTION** How did the RNAi pathway evolve? As you will learn in Chapter 17, some viruses have double-stranded RNA genomes. Given that the cellular RNAi pathway can process double-stranded RNAs into homing devices that lead to the destruction of related RNAs, some researchers hypothesize that this pathway may have evolved as a natural defense against infection by such viruses. However, the fact that RNAi can also affect the expression of nonviral cellular genes may reflect a different evolutionary origin for the RNAi pathway. Moreover, many species, including mammals, apparently produce their own long, double-stranded RNA precursors to small RNAs such as siRNAs. Once produced, these RNAs can interfere with gene expression at stages other than translation, as we'll discuss next.

## Chromatin Remodeling and Effects on Transcription by Noncoding RNAs

In addition to affecting mRNAs, small RNAs can cause remodeling of chromatin structure. In the S phase of the cell cycle, the centromeric regions of DNA must be loosened for chromosomal replication and then recondensed into heterochromatin in preparation for mitosis. In some yeasts, siRNAs produced by the yeast cells themselves are required for this formation of heterochromatin at the centromeres of chromosomes. Exactly how the process occurs is still under debate, but biologists agree on the general idea: The siRNA system interacts with other noncoding RNAs and with chromatin-modifying enzymes to condense the centromere chromatin into heterochromatin. In most mammalian cells, siRNAs have not been found, and the mechanism for centromere condensation is not yet understood. However, it may turn out to involve other small noncoding RNAs.

A newly discovered class of small noncoding RNAs called *piwi-associated RNAs* (*piRNAs*) also induces formation of

heterochromatin, blocking expression of some parasitic DNA elements in the genome known as transposons. (Transposons are discussed in Concepts 18.4 and 18.5.) Usually 24–31 nucleotides in length, piRNAs are probably processed from a longer, single-stranded RNA precursor. They play an indispensable role in the germ cells of many animal species, where they appear to help reestablish appropriate methylation patterns in the genome during gamete formation.

The role of noncoding RNAs in regulation of gene expression adds yet another layer to the complex and intricate process described in the previous section. As more is learned about the multiple interacting ways a cell can fine-tune expression of its genes, the goal is to understand how a specific set of genes is expressed in a particular cell. In the next section, we'll describe a few methods that researchers use to monitor expression of specific genes, such as in different cell types.

### CONCEPT CHECK 15.3

1. **WHAT IF?** Suppose the mRNA being degraded in Figure 15.13 coded for a protein that promotes cell division in a multicellular organism. What would happen if a mutation disabled the gene for the miRNA that triggers this degradation?
2. **MAKE CONNECTIONS** Inactivation of one of the X chromosomes in female mammals results in a Barr body (see Concept 12.2). Suggest a model for how the noncoding RNA described in Concept 12.2 (*XIST* RNA) functions to cause Barr body formation.

For suggested answers, see Appendix A.

---

**CONCEPT 15.4**

# Researchers can monitor expression of specific genes

The diverse mechanisms of regulating gene expression discussed in this chapter underlie one basic generality: Cells of a given multicellular organism differ from each other because they express different genes from an identical genome. Biologists driven to understand the assorted cell types of a multicellular organism, cancer cells, or the developing tissues of an embryo first try to discover which genes are expressed by the cells of interest. The most straightforward way to do this is usually to identify the mRNAs being made. Techniques related to those developed for genetic engineering (see Concept 13.4) are widely used to track expression of mRNAs. In this section we'll first examine techniques that look for patterns of expression of specific individual genes. Next, we'll explore techniques that characterize groups of genes being expressed by cells or tissues of interest. As you will see, all of these techniques depend in some way on base pairing between complementary nucleotide sequences.

## Studying the Expression of Single Genes

Suppose we have cloned a gene that may play an important role in the embryonic development of *Drosophila* (the fruit fly). The first thing we might want to know is which embryonic cells express the gene—in other words, where in the embryo is the corresponding mRNA found? We can detect the mRNA using the technique of **nucleic acid hybridization**, the base pairing of one strand of a nucleic acid to the complementary sequence on another strand. The complementary molecule, a short, single-stranded nucleic acid that can be either RNA or DNA, is called a **nucleic acid probe**. Using our cloned gene as a template, we can synthesize a probe complementary to the mRNA. For example, if part of the sequence on the mRNA were

5'  ···CUCAUCACCGGC···  3'

then we would synthesize this single-stranded DNA probe:

3'  GAGTAGTGGCCG  5'

Each probe molecule is labeled with a fluorescent tag so we can follow it. A solution with the probe is applied to *Drosophila* embryos, allowing probe molecules to hybridize to any complementary sequences on the many mRNAs in embryonic cells in which the gene is being transcribed **(Figure 15.14)**.

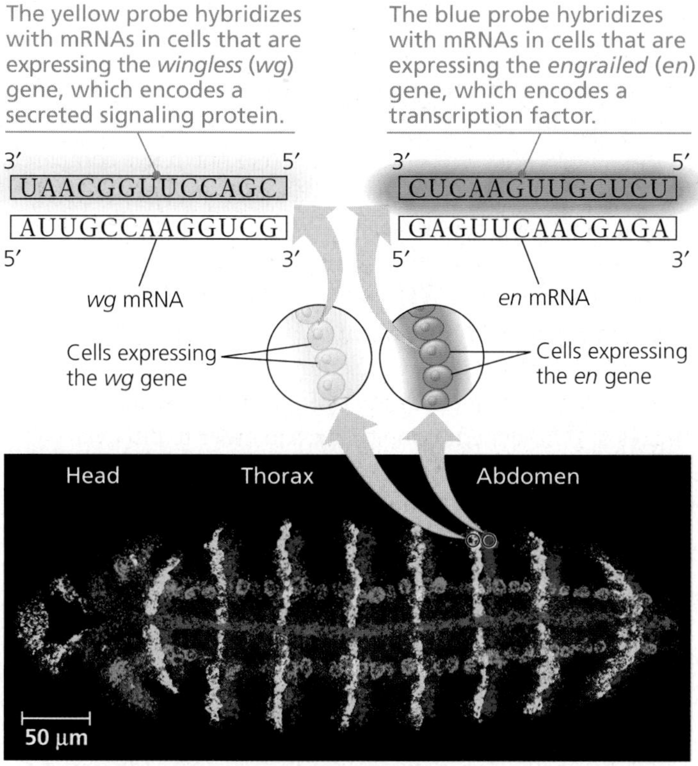

Because this technique allows us to see the mRNA in place (or *in situ*, in Latin) in the intact organism, it is called ***in situ* hybridization**. Different probes can be labeled with different fluorescent dyes, sometimes with strikingly beautiful results. (see Figures 15.14 and 16.1).

Other mRNA detection techniques may be preferable for comparing the amounts of a specific mRNA in several samples at the same time—for example, in different cell types or in embryos of different stages. One method that is widely used is called the **reverse transcriptase–polymerase chain reaction (RT-PCR)**. RT-PCR begins by turning sample sets of mRNAs into double-stranded DNAs with the corresponding sequences. This feat is accomplished by an enzyme called *reverse transcriptase*, isolated in the late 1980s from a type of virus called a retrovirus. (You'll learn more about retroviruses, including HIV, in Concept 17.2.) Reverse transcriptase is able to synthesize a complementary DNA copy of an mRNA, thus making a *reverse transcript* **(Figure 15.15)**. Following enzymatic degradation of the mRNA, a second DNA strand, complementary to the first, is synthesized by DNA polymerase. The resulting double-stranded DNA is called **complementary DNA (cDNA)**, and the reverse transcription step accounts for the "RT" in the name RT-PCR. To analyze the timing of expression of the *Drosophila* gene of interest, for example,

The yellow probe hybridizes with mRNAs in cells that are expressing the *wingless* (*wg*) gene, which encodes a secreted signaling protein.

The blue probe hybridizes with mRNAs in cells that are expressing the *engrailed* (*en*) gene, which encodes a transcription factor.

3'  UAACGGUUCCAGC  5'
5'  AUUGCCAAGGUCG  3'
*wg* mRNA

3'  CUCAAGUUGCUCU  5'
5'  GAGUUCAACGAGA  3'
*en* mRNA

Cells expressing the *wg* gene

Cells expressing the *en* gene

Head  Thorax  Abdomen

50 µm

▲ **Figure 15.14 Determining where genes are expressed by *in situ* hybridization analysis.** This *Drosophila* embryo was incubated in a solution containing probes for five different mRNAs, each probe labeled with a different fluorescently colored tag. The embryo was then viewed using fluorescence microscopy. Each color marks cells in which a specific gene is expressed as mRNA. The arrows from the groups of yellow and blue cells above the micrograph point to a magnified view of nucleic acid hybridization of the appropriately colored probe to the mRNA.

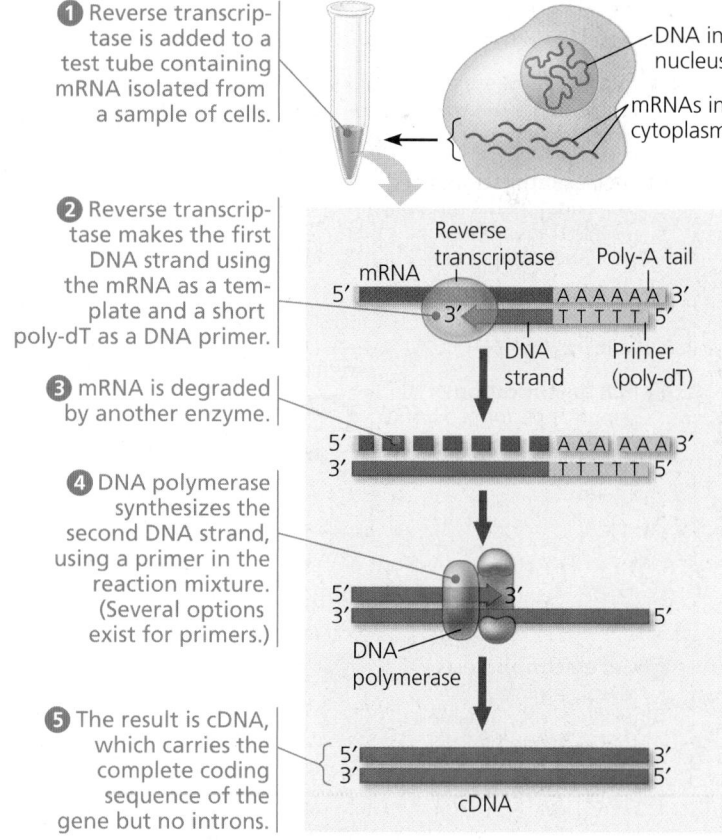

1 Reverse transcriptase is added to a test tube containing mRNA isolated from a sample of cells.

DNA in nucleus

mRNAs in cytoplasm

2 Reverse transcriptase makes the first DNA strand using the mRNA as a template and a short poly-dT as a DNA primer.

mRNA
Reverse transcriptase
Poly-A tail
AAAAAA 3'
TTTTT 5'
DNA strand
Primer (poly-dT)

3 mRNA is degraded by another enzyme.

5' □□□□□□□□□□ AAA AAA 3'
3' TTTTT 5'

4 DNA polymerase synthesizes the second DNA strand, using a primer in the reaction mixture. (Several options exist for primers.)

5'
3'
5'
DNA polymerase

5 The result is cDNA, which carries the complete coding sequence of the gene but no introns.

5' 3'
3' 5'
cDNA

▲ **Figure 15.15 Making complementary DNA (cDNA) from eukaryotic genes.** Complementary DNA is DNA made in a test tube using mRNA as a template for the first strand. Only one mRNA is shown here, but the final collection of cDNAs would reflect all the mRNAs that were present in the cell.

we would first isolate all the mRNAs from different stages of *Drosophila* embryos and then make cDNA from each stage (**Figure 15.16**).

Next in RT-PCR is the PCR step (see Figure 13.27). As you may recall, PCR is a way of rapidly making many copies of one specific stretch of double-stranded DNA, using primers that hybridize to the opposite ends of the region of interest. In our case, we would add primers corresponding to a region of our *Drosophila* gene, using the cDNA from each sample as a template for PCR amplification. When the products are run on a gel, copies of the amplified region will be observed as bands only in samples that originally contained mRNA from the gene we're focusing on. A recent enhancement involves using a fluorescent dye that fluoresces only when bound to a double-stranded PCR product. The newer PCR machines can detect the light and measure the PCR product, thus avoiding the need for electrophoresis while also providing quantitative data, a distinct advantage. RT-PCR can also be carried out with

---

▼ Figure 15.16 **Research Method**

## RT-PCR Analysis of the Expression of Single Genes

**Application** RT-PCR uses the enzyme reverse transcriptase (RT) in combination with PCR and gel electrophoresis. RT-PCR can be used to compare gene expression in different embryonic stages, in different tissues, or in the same type of cell under different conditions.

**Technique** In this example, samples containing mRNAs from six embryonic stages of *Drosophila* were processed as shown below. (The mRNAs from only one stage are shown here.)

❶ **cDNA synthesis** is carried out by incubating the mRNAs with reverse transcriptase and other necessary components.

mRNAs

cDNAs

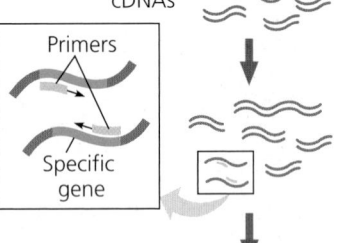

Primers

❷ **PCR amplification** of the sample is performed using primers specific to the *Drosophila* gene of interest.

Specific gene

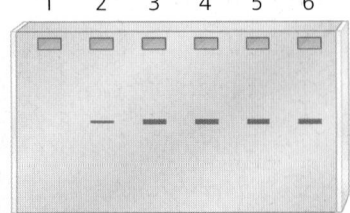

Embryonic stages
1  2  3  4  5  6

❸ **Gel electrophoresis** will reveal amplified DNA products only in samples that contained mRNA transcribed from the specific *Drosophila* gene.

**Results** The mRNA for this gene is expressed from stage 2 through stage 6. The size of the amplified fragment (shown by its position on the gel) depends on the distance between the primers that were used.

---

mRNAs collected from different tissues at one time to discover which tissue is producing a specific mRNA.

## Studying the Expression of Groups of Genes

A major goal of biologists is to learn how genes act together to produce and maintain a functioning organism. Now that the genomes of a number of species have been sequenced, it is possible to study the expression of large groups of genes—the so-called *systems approach*. Researchers use what is known about the whole genome to investigate which groups of genes are transcribed in different tissues or at different stages of development. One aim is to identify networks of gene expression across an entire genome.

Genome-wide expression studies can be carried out using **DNA microarray assays**. A DNA microarray consists of tiny amounts of a large number of single-stranded DNA fragments representing different genes fixed to a glass slide in a tightly spaced array, or grid. (The microarray is also called a *DNA chip* by analogy to a computer chip.) Ideally, these fragments represent all the genes in the genome of an organism.

The basic strategy in such studies is to isolate the mRNAs made in a cell of interest and use these mRNAs as templates for making the corresponding cDNAs by reverse transcription. In microarray assays, these cDNAs are labeled with fluorescent molecules and then allowed to hybridize to a DNA microarray. Most often, the cDNAs from two samples (for example, two tissues) are labeled with molecules that emit different colors and tested on the same microarray. **Figure 15.17** shows the result of such an experiment, identifying the subsets of genes in the genome that are being expressed in one tissue compared with another. DNA technology makes such studies possible; with automation, they are easily performed on a large scale. Scientists can now measure the expression of thousands of genes at one time.

Alternatively, with the advent of rapid, inexpensive DNA-sequencing methods (see Concept 13.4), researchers can now afford to simply sequence the cDNA samples from different tissues or different embryonic stages in order to discover which genes are expressed. This straightforward method is called *RNA sequencing* or *RNA-seq*, even though it is the cDNA that is actually sequenced. As the price of sequencing plummets, this method is growing more widespread.

By characterizing sets of genes that are expressed together in some tissues but not others, genome-wide gene expression studies may contribute to a better understanding of diseases and suggest new diagnostic techniques or therapies. For instance, comparing patterns of gene expression in breast cancer tumors and noncancerous breast tissue has already resulted in more informed and effective treatment protocols (see Make Connections Figure 16.21). Ultimately, information from these methods should provide more of a big-picture view of how ensembles of genes interact to form an organism. The genetic basis of embryonic development and disease will be considered in the next chapter.

Each dot is a well containing identical copies of DNA fragments that carry a specific gene.

The genes in the red wells are expressed in one tissue and bind the red cDNAs.

The genes in the green wells are expressed in the other tissue and bind the green cDNAs.

The genes in the yellow wells are expressed in both tissues and bind both red and green cDNAs, appearing yellow.

The genes in the black wells are not expressed in either tissue and do not bind either cDNA.

◀ DNA microarray (actual size)

▲ **Figure 15.17 DNA microarray assay of gene expression levels.** Researchers synthesized two sets of cDNAs, fluorescently labeled red or green, from mRNAs from two different human tissues. These cDNAs were hybridized with a microarray containing 5,760 human genes (about 25% of human genes), resulting in the pattern shown here. The intensity of fluorescence at each spot measures the relative expression in the two samples of the gene represented by that spot: Red indicates expression in one sample, green in the other, yellow in both, and black in neither.

**CONCEPT CHECK 15.4**

1. Describe the role of complementary base pairing during RT-PCR and DNA microarray analysis.
2. **WHAT IF?** Study the microarray in Figure 15.17. If a sample from normal tissue is labeled with a green fluorescent dye, and a sample from cancerous tissue is labeled red, what color spots would represent genes that you would be interested in if you were studying cancer? Explain.

For suggested answers, see Appendix A.

---

Go to **MasteringBiology**® for Assignments, the eText, and the Study Area with Animations, Activities, Vocab Self-Quiz, and Practice Tests.

# 15 Chapter Review

## SUMMARY OF KEY CONCEPTS

**VOCAB SELF-QUIZ**

goo.gl/gbai8v

### CONCEPT 15.1

**Bacteria often respond to environmental change by regulating transcription (pp. 303–308)**

- In bacteria, genes are often clustered into an operon with a single promoter. An **operator** site on the DNA switches the **operon** on or off, resulting in coordinate regulation of the genes.
- Both repressible and inducible operons are examples of negative gene regulation: Binding of a specific **repressor** protein to the operator shuts off transcription. (The repressor is encoded by a separate **regulatory gene**.) In a repressible operon (usually encoding anabolic enzymes), the repressor is active when bound to a **corepressor**:

**Repressible operon:**

**Genes expressed**

Promoter

Genes

Operator

Inactive repressor: no corepressor present

**Genes not expressed**

Active repressor: corepressor bound

Corepressor

In an inducible operon (usually encoding catabolic enzymes), binding of an **inducer** to an innately active repressor inactivates the repressor and turns on transcription.
- Some operons have positive gene regulation. A stimulatory **activator** protein (such as CRP, when activated by **cyclic AMP**), binds to a site within the promoter and stimulates transcription.

**?** *Compare and contrast the roles of a corepressor and an inducer in negative regulation of an operon.*

### CONCEPT 15.2

**Eukaryotic gene expression is regulated at many stages (pp. 308–315)**

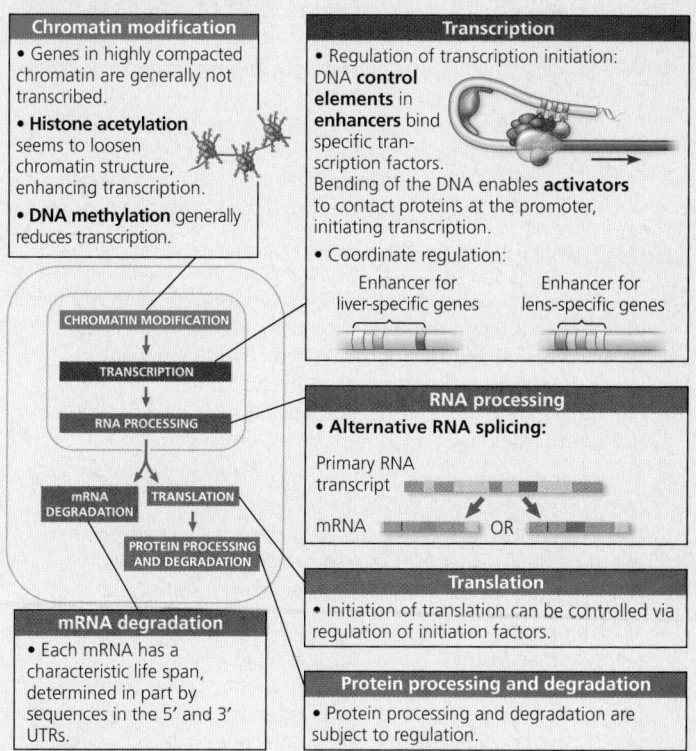

**Chromatin modification**
- Genes in highly compacted chromatin are generally not transcribed.
- **Histone acetylation** seems to loosen chromatin structure, enhancing transcription.
- **DNA methylation** generally reduces transcription.

CHROMATIN MODIFICATION
↓
TRANSCRIPTION
↓
RNA PROCESSING
↓
mRNA DEGRADATION / TRANSLATION
↓
PROTEIN PROCESSING AND DEGRADATION

**Transcription**
- Regulation of transcription initiation: DNA **control elements** in **enhancers** bind specific transcription factors. Bending of the DNA enables **activators** to contact proteins at the promoter, initiating transcription.
- Coordinate regulation:

Enhancer for liver-specific genes

Enhancer for lens-specific genes

**RNA processing**
- **Alternative RNA splicing:**

Primary RNA transcript

mRNA ⬛⬛⬛⬛⬛ OR ⬛⬛⬛⬛⬛

**mRNA degradation**
- Each mRNA has a characteristic life span, determined in part by sequences in the 5′ and 3′ UTRs.

**Translation**
- Initiation of translation can be controlled via regulation of initiation factors.

**Protein processing and degradation**
- Protein processing and degradation are subject to regulation.

**?** *Describe what must happen in a cell before a gene specific to that type of cell can be transcribed.*

## CONCEPT 15.3

### Noncoding RNAs play multiple roles in controlling gene expression (pp. 315–316)

- Noncoding RNAs (for example, **miRNAs** and **siRNAs**) can block translation or cause degradation of mRNAs.

? *Why are miRNAs called noncoding RNAs? Explain how they participate in gene regulation.*

## CONCEPT 15.4

### Researchers can monitor expression of specific genes (pp. 316–319)

- In **nucleic acid hybridization**, a **nucleic acid probe** is used to detect the presence of a specific mRNA.
- *In situ* **hybridization** and **RT-PCR** can detect the presence of a given mRNA in a tissue or an RNA sample, respectively.
- **DNA microarrays** are used to identify sets of genes co-expressed by a group of cells. Their **cDNAs** can also be sequenced (RNA-seq).

? *What does detecting expression of specific genes tell a researcher?*

# TEST YOUR UNDERSTANDING

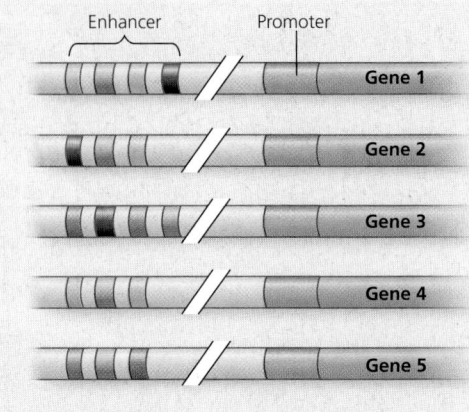

**PRACTICE TEST**
goo.gl/CRZjvS

## Level 1: Knowledge/Comprehension

1. If a particular operon encodes enzymes for making an essential amino acid and is regulated like the *trp* operon, then
   (A) the amino acid inactivates the repressor.
   (B) the repressor is active in the absence of the amino acid.
   (C) the amino acid acts as a corepressor.
   (D) the amino acid turns on transcription of the operon.

2. The functioning of enhancers is an example of
   (A) a eukaryotic equivalent of prokaryotic promoter function.
   (B) transcriptional control of gene expression.
   (C) the stimulation of translation by initiation factors.
   (D) post-translational control that activates certain proteins.

3. Which of the following is an example of post-transcriptional control of gene expression?
   (A) the addition of methyl groups to cytosine bases of DNA
   (B) the binding of transcription factors to a promoter
   (C) the removal of introns and alternative splicing of exons
   (D) the binding of RNA polymerase to transcription factors

## Level 2: Application/Analysis

4. What would occur if the repressor of an inducible operon were mutated so it could not bind the operator?
   (A) irreversible binding of the repressor to the promoter
   (B) reduced transcription of the operon's genes
   (C) buildup of a substrate for the pathway controlled by the operon
   (D) continuous transcription of the operon's genes

5. Which statement about DNA in one of your brain cells is true?
   (A) Most of the DNA codes for protein.
   (B) The majority of genes are likely to be transcribed.
   (C) It is the same as the DNA in one of your liver cells.
   (D) Many genes are grouped into operon-like clusters.

6. Which of the following would *not* be true of cDNA produced using human brain tissue as the starting material?
   (A) It could be amplified by the polymerase chain reaction.
   (B) It was produced from pre-mRNA using reverse transcriptase.
   (C) It could be labeled and used as a probe to detect genes expressed in the brain.
   (D) It lacks the introns of the pre-mRNA.

## Level 3: Synthesis/Evaluation

7. **DRAW IT** The diagram shows five genes, including their enhancers, from the genome of a certain species. Imagine that orange, blue, green, black, red, and purple activator proteins exist that can bind to the appropriately color-coded control elements in the enhancers of these genes. (a) Draw an X above enhancer elements (of all the genes) that would have activators bound in a cell in which only gene 5 is transcribed. Identify the colored activators that would have to be present. (b) Draw a dot above all enhancer elements that would have activators bound in a cell in which the green, blue, and orange activators are present. Identify the gene(s) that would be transcribed. (c) Imagine that genes 1, 2, and 4 code for nerve-specific proteins and genes 3 and 5 are skin-specific. Identify the activators that would ensure transcription of the appropriate genes.

Enhancer    Promoter
Gene 1
Gene 2
Gene 3
Gene 4
Gene 5

8. **SCIENTIFIC INQUIRY**
   Imagine you want to study one of the mouse crystallins, proteins present in the lens of the eye. Assuming that the gene has been cloned, describe two ways to study its expression in the embryo.

9. **FOCUS ON EVOLUTION**
   DNA sequences can act as "tape measures of evolution" (see Concept 3.7). Some highly conserved regions of the human genome (similar to comparable regions in other species) don't code for proteins. Propose a possible explanation.

10. **FOCUS ON INTERACTIONS**
    In a short essay (100–150 words), discuss how the processes shown in Figure 15.2 are examples of feedback mechanisms regulating biological systems in bacterial cells.

11. **SYNTHESIZE YOUR KNOWLEDGE**

    The flashlight fish has an organ under its eye that emits light, which startles predators, attracts prey, and allows the fish to communicate with other fish. Some species can rotate the organ, so the light appears to flash on and off. The light is actually emitted by bacteria that live in the organ in a mutualistic relationship with the fish. (The bacteria receive nutrients from the fish.) The bacteria must multiply until they reach a certain density in the organ, at which point they all begin emitting light at the same time. There is a group of six or so genes, called *lux* genes, whose gene products are necessary for light formation. Given that these bacterial genes are regulated together, propose a hypothesis for how the genes are organized and regulated.

*For selected answers, see Appendix A.*

# 16 Development, Stem Cells, and Cancer

## KEY CONCEPTS

**16.1** A program of differential gene expression leads to the different cell types in a multicellular organism

**16.2** Cloning of organisms showed that differentiated cells could be "reprogrammed" and ultimately led to the production of stem cells

**16.3** Abnormal regulation of genes that affect the cell cycle can lead to cancer

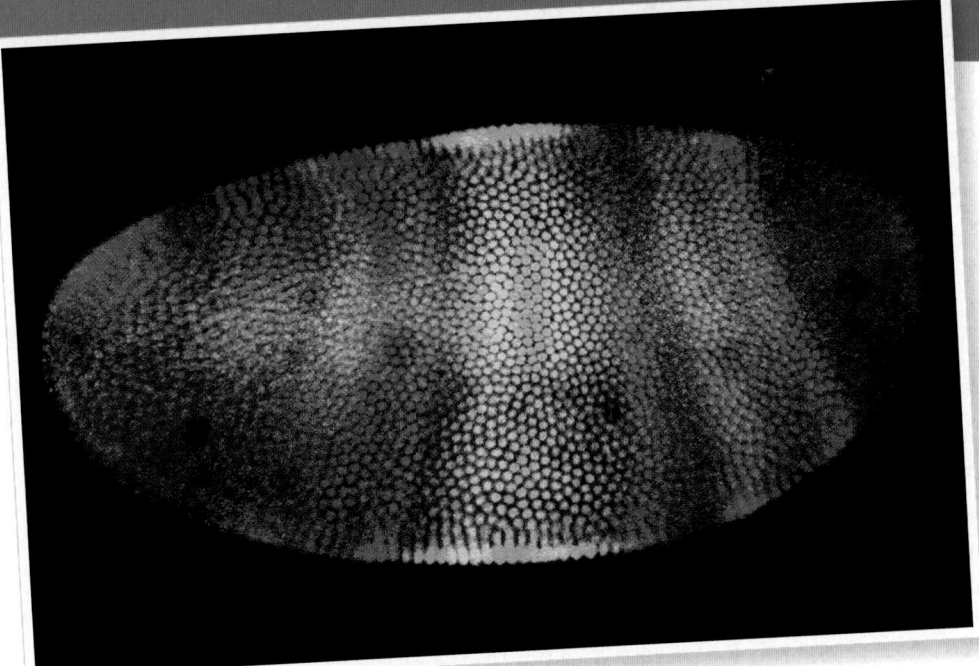

▲ **Figure 16.1** What regulates the precise pattern of gene expression in the developing fly embryo?

## Orchestrating Life's Processes

The development of the fertilized egg, a single cell, into an embryo and later an adult is an astounding transformation that requires a precisely regulated program of gene expression. All of the levels of eukaryotic gene regulation you learned about in the previous chapter come into play during embryonic development. The elaborate sequence of genes being turned on and off in different cells is the ultimate example of regulation of gene expression.

Understanding the genetic underpinnings of development has progressed mainly by studying the process in **model organisms**, species that are easy to raise in the lab and use in experiments. A prime example is the fruit fly *Drosophila melanogaster*. An adult fruit fly develops from a fertilized egg into an embryo then a wormlike stage called a larva that metamorphoses into the adult fly. At every stage, gene expression is carefully regulated, ensuring that the right genes are expressed only at the correct time and place. The embryo shown in **Figure 16.1** has been analyzed by *in situ* hybridization (see Figure 15.14) to reveal the mRNA for three genes—labeled red, blue, and green.

(Red and green together appear yellow; red and blue together appear purple.) The intricate pattern of expression for each gene is the same from embryo to embryo at this stage, and it provides a graphic display of the precision of gene regulation. But what is the molecular basis for this pattern? Why is one particular gene expressed only in the few hundred cells that appear blue or purple in this image and not in the other cells?

Part of the answer involves the transcription factors and other regulatory molecules you learned about in the previous chapter. But how do they come to be different in distinct cell types? In this chapter, we'll first explain the mechanisms that send cells down diverging genetic pathways to adopt different fates. Then we'll take a closer look at *Drosophila* development. Next, we'll describe the discovery of stem cells, a powerful cell type that is key to the developmental process. Understanding these cells offers hope for medical treatments as well. Finally, after having explored embryonic development and stem cells, we will underscore the crucial role played by regulation of gene expression by investigating how cancer can result when this regulation goes awry. Orchestrating proper gene expression by all cells is crucial to the functions of life.

# A program of differential gene expression leads to the different cell types in a multicellular organism

In the embryonic development of multicellular organisms, a fertilized egg (a zygote) gives rise to cells of many different types, each with a different structure and corresponding function. Typically, cells are organized into tissues, tissues into organs, organs into organ systems, and organ systems into the whole organism. Thus, any developmental program must produce cells of different types that form higher-level structures arranged in a particular way in three dimensions. Here, we'll focus on the program of regulation of gene expression that orchestrates development using a few animal species as examples.

## A Genetic Program for Embryonic Development

The photos in **Figure 16.2** illustrate the dramatic difference between a frog zygote (fertilized egg) and the tadpole it becomes. This remarkable transformation results from three interrelated processes: cell division, cell differentiation, and morphogenesis. Through a succession of mitotic cell divisions, the zygote gives rise to a large number of cells. Cell division alone, however, would merely produce a great ball of identical cells, nothing like a tadpole. During embryonic development, cells not only increase in number but also undergo cell **differentiation**, the process by which cells become specialized in structure and function. Moreover, the different kinds of cells are not randomly distributed but are organized into tissues and organs in a particular three-dimensional arrangement. The physical processes that give an organism its shape constitute **morphogenesis**, the development of the form of an organism and its structures.

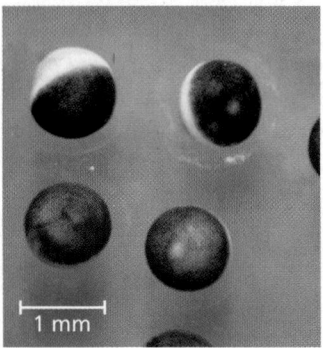
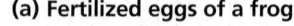
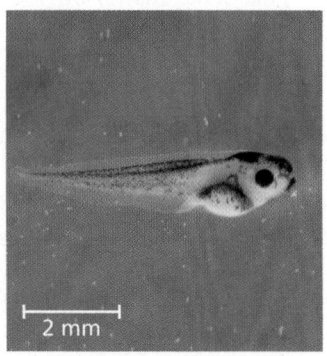

(a) Fertilized eggs of a frog  (b) Newly hatched tadpole

▲ **Figure 16.2 From fertilized egg to animal: What a difference four days makes.** It takes just four days for cell division, differentiation, and morphogenesis to transform each of the fertilized frog eggs shown in **(a)** into a tadpole like the one in **(b)**.

All three processes are rooted in cellular behavior. Even morphogenesis, the shaping of the organism, can be traced to changes in the motility, shape, and other characteristics of the cells that make up the regions of the embryo. As you know, the activities of a cell depend on the genes it expresses and the proteins it produces. Almost all cells in an organism have the same genome; therefore, differential gene expression results from the genes being regulated differently in each cell type.

In Figure 15.11, you saw a simplified view of how differential gene expression occurs in two cell types, a liver cell and a lens cell. Each of these fully differentiated cells has a particular mix of specific activators that turn on the collection of genes whose products are required in the cell. The fact that both cells arose through a series of mitoses from a common fertilized egg inevitably leads to a question: How do different sets of activators come to be present in the two cells?

It turns out that materials placed into the egg by the mother set up a sequential program of gene regulation that is carried out as cells divide, and this program coordinates cell differentiation during embryonic development. To understand how this works, we'll consider two basic developmental processes: First, we'll explore how cells that arise from early embryonic mitoses develop the differences that start each cell along its own differentiation pathway. Second, we'll see how cellular differentiation leads to one particular cell type, using muscle development as an example.

## Cytoplasmic Determinants and Inductive Signals

What generates the first differences among cells in an early embryo? And what controls the differentiation of all the various cell types as development proceeds? You can probably deduce the answer: The specific genes expressed in any particular cell of a developing organism determine its path. Two sources of information, used to varying extents in different species, "tell" a cell which genes to express at any given time during embryonic development.

One important source of information early in development is the egg's cytoplasm, which contains both RNA and proteins encoded by the mother's DNA. The cytoplasm of an unfertilized egg is not homogeneous. Messenger RNA, proteins, other substances, and organelles are distributed unevenly in the unfertilized egg, and this unevenness has a profound impact on the development of the future embryo in many species. Maternal substances in the egg that influence the course of early development are called **cytoplasmic determinants** **(Figure 16.3a)**. After fertilization, early mitotic divisions distribute the zygote's cytoplasm into separate cells. The nuclei of these cells may thus be exposed to different cytoplasmic determinants, depending on which portions of the zygotic cytoplasm a cell received. The combination of cytoplasmic determinants in a cell helps determine its developmental fate by regulating gene expression during cell differentiation.

## ▼ Figure 16.3 Sources of developmental information for the early embryo.

### (a) Cytoplasmic determinants in the egg

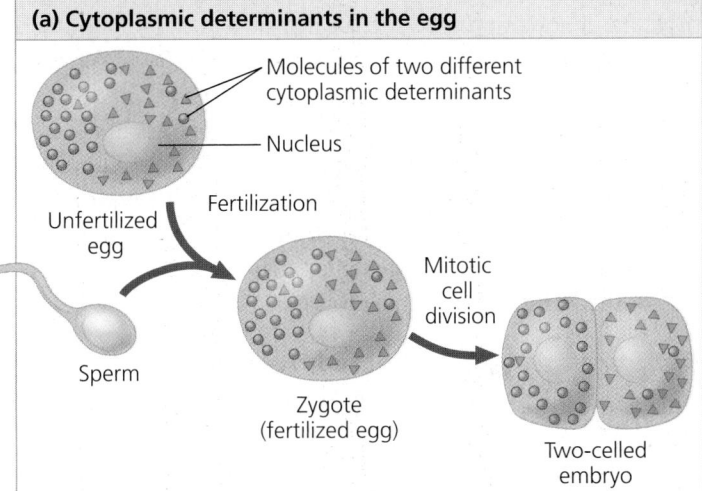

The unfertilized egg has molecules in its cytoplasm, encoded by the mother's genes, that influence development. Many of these cytoplasmic determinants, like the two shown here, are unevenly distributed in the egg. After fertilization and mitotic division, the cell nuclei of the embryo are exposed to different sets of cytoplasmic determinants and, as a result, express different genes.

### (b) Induction by nearby cells

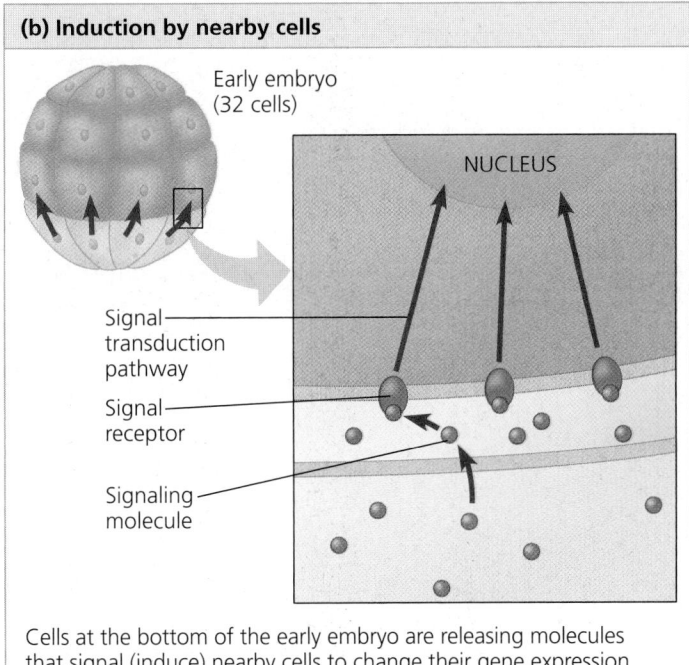

Cells at the bottom of the early embryo are releasing molecules that signal (induce) nearby cells to change their gene expression.

The other major source of developmental information, which becomes increasingly important as the number of embryonic cells increases, is the environment around a particular cell. Most influential are the signals conveyed to an embryonic cell from other embryonic cells in the vicinity, including contact with cell-surface molecules on neighboring cells and the binding of growth factors secreted by neighboring cells (see Concept 5.6). Such signals cause changes in the target cells, a process called **induction (Figure 16.3b)**. The molecules passing along these signals within the target cell are cell-surface receptors and other signaling pathway proteins. In general, the signaling molecules send a cell down a specific developmental path by causing changes in its gene expression that eventually result in observable cellular changes. Thus, interactions between embryonic cells help induce differentiation into the many specialized cell types making up a new organism.

## Sequential Regulation of Gene Expression during Cellular Differentiation

The earliest changes that set a cell on a path to specialization are subtle ones, showing up only at the molecular level. Before biologists knew much about the molecular changes occurring in embryos, they coined the term **determination** to refer to the point at which an embryonic cell is irreversibly committed to becoming a particular cell type. Once an embryonic cell has undergone determination, it can be experimentally placed at another site in the embryo and will still differentiate into the cell type that is its normal fate. Differentiation, then, is the process by which a cell attains its determined fate. As the tissues and organs of an embryo develop, their cells differentiate, becoming more noticeably different in structure and function.

### Differentiation of Cell Types

Today we understand determination in terms of molecular changes. The outcome of determination, observable cell differentiation, is marked by the expression of genes for *tissue-specific proteins*. These are found only in a specific cell type and give the cell its characteristic structure and function. The first sign of differentiation is the appearance of mRNAs for these proteins. Eventually, differentiation is observable with a microscope as changes in cellular structure. On the molecular level, different sets of genes are sequentially expressed in a regulated manner as new cells arise from division of their precursors. Some steps in gene expression may be regulated during differentiation, transcription being the most common. In the fully differentiated cell, transcription is the principal regulatory point for maintaining appropriate gene expression.

Differentiated cells are specialists at making tissue-specific proteins. For example, as a result of transcriptional regulation, liver cells specialize in making albumin, and lens cells specialize in making crystallin (see Figure 15.11). Skeletal muscle cells in vertebrates are another instructive example. Each of these cells is a long fiber containing many nuclei within a single plasma membrane. Skeletal muscle cells have high concentrations of muscle-specific versions of the contractile proteins myosin and actin, as well as membrane receptor proteins that detect signals from nerve cells.

Muscle cells develop from embryonic precursor cells that have the potential to become a number of cell types, including cartilage cells and fat cells, but particular conditions commit them to becoming muscle cells. Although the committed cells appear unchanged under the microscope, determination has occurred, and they are now *myoblasts*. Eventually, myoblasts start to churn out large amounts of muscle-specific proteins and fuse to form mature, elongated, multinucleate skeletal muscle cells.

Researchers have worked out what happens at the molecular level during muscle cell determination. To do so, they grew embryonic precursor cells in culture and analyzed them using molecular techniques like those described in Concepts 13.4 and 15.4 . In a series of experiments, they isolated different genes, caused each to be expressed in a separate precursor cell, and then looked for differentiation into myoblasts and muscle cells. In this way, they identified several so-called "master regulatory genes" whose protein products commit the cells to becoming skeletal muscle. Thus, in the case of muscle cells, the molecular basis of determination is the expression of one or more of these master regulatory genes.

To understand more about how determination occurs in muscle cell differentiation, let's focus on the master regulatory gene called *myoD* **(Figure 16.4)**. This gene encodes MyoD protein, a transcription factor that binds to specific control elements in the enhancers of various target genes and stimulates their expression (see Figure 15.9). Some target genes for MyoD encode still other muscle-specific transcription factors. MyoD also stimulates expression of the *myoD* gene itself, an example of positive feedback that perpetuates MyoD's effect in maintaining the cell's

differentiated state. Since all the genes activated by MyoD have enhancer control elements recognized by MyoD, they are coordinately controlled. Finally, the secondary transcription factors activate the genes for proteins such as myosin and actin that confer the unique properties of skeletal muscle cells. The MyoD protein deserves its designation as a master regulatory gene.

The regulation of genes that play important roles in development of embryonic tissues and structures is often complex. In the **Scientific Skills Exercise,** you'll work with data from an experiment that tested how different regulatory regions in DNA affect expression of a gene that helps establish the pattern of the different digits in a mouse's paw.

### Apoptosis: A Type of Programmed Cell Death

During the time when most cells are differentiating, some cells in the developing organism are genetically programmed to die. The best-understood type of "programmed cell death" is **apoptosis** (from the Greek, meaning "falling off," and used in a classic Greek poem to refer to leaves falling from a tree). Apoptosis also occurs in cells of the mature organism that are infected, damaged, or have reached the end of their functional life span. During this process, cellular agents chop up the DNA and fragment the organelles and other cytoplasmic

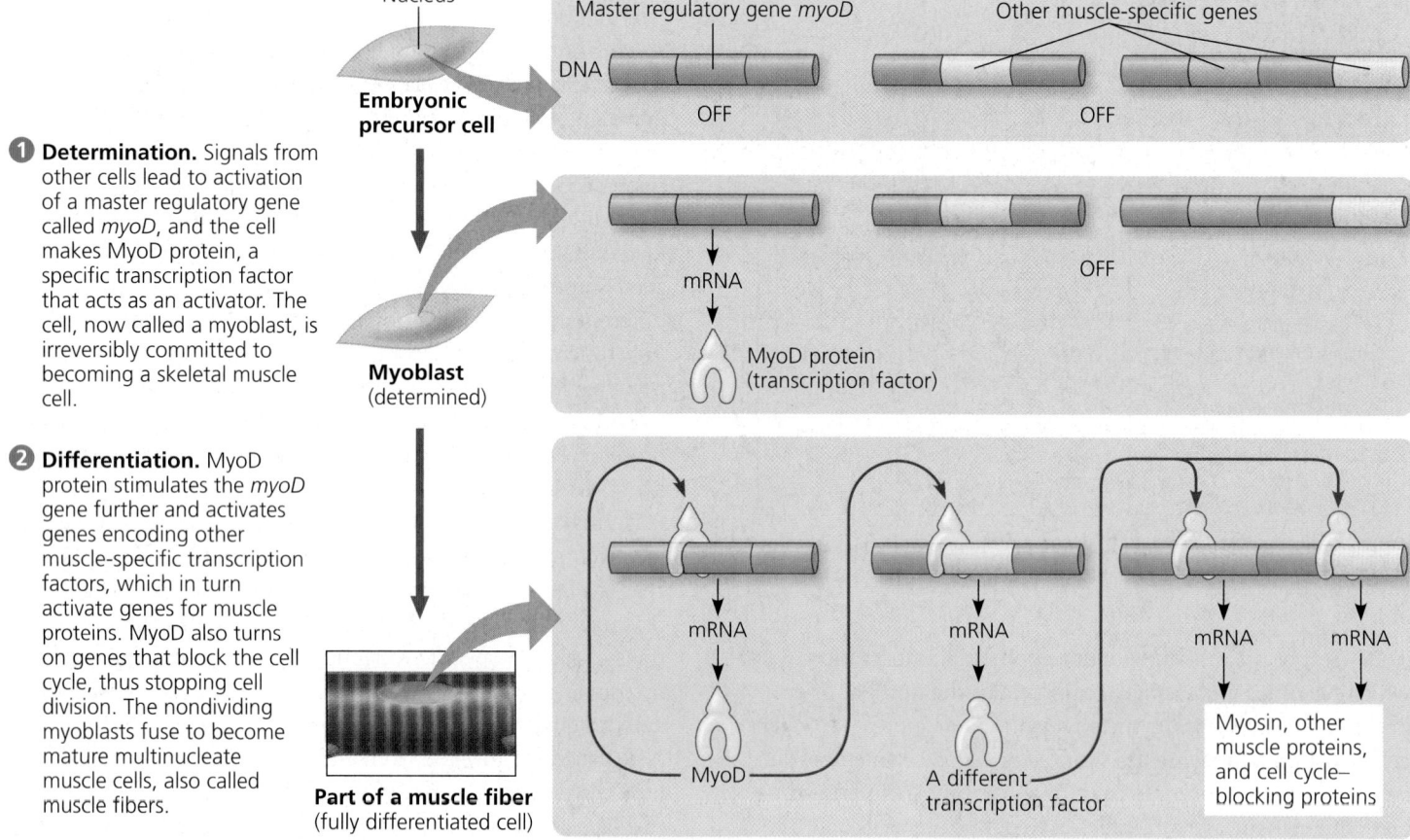

**① Determination.** Signals from other cells lead to activation of a master regulatory gene called *myoD*, and the cell makes MyoD protein, a specific transcription factor that acts as an activator. The cell, now called a myoblast, is irreversibly committed to becoming a skeletal muscle cell.

**② Differentiation.** MyoD protein stimulates the *myoD* gene further and activates genes encoding other muscle-specific transcription factors, which in turn activate genes for muscle proteins. MyoD also turns on genes that block the cell cycle, thus stopping cell division. The nondividing myoblasts fuse to become mature multinucleate muscle cells, also called muscle fibers.

Nucleus

**Embryonic precursor cell**

**Myoblast** (determined)

**Part of a muscle fiber** (fully differentiated cell)

Master regulatory gene *myoD*    Other muscle-specific genes

DNA    OFF    OFF

mRNA

MyoD protein (transcription factor)

OFF

mRNA    mRNA    mRNA    mRNA

MyoD    A different transcription factor    Myosin, other muscle proteins, and cell cycle–blocking proteins

▲ **Figure 16.4 Determination and differentiation of muscle cells.** Skeletal muscle cells arise from embryonic cells as a result of changes in gene expression. (In this depiction, the process of gene activation is greatly simplified.)

**WHAT IF?** *What would happen if a mutation in the* myoD *gene resulted in a MyoD protein that could not activate the* myoD *gene?*

# Analyzing Quantitative and Spatial Gene Expression Data

### How Is a Particular *Hox* Gene Regulated During Paw Development?

*Hox* genes code for transcription factor proteins, which in turn control sets of genes important for animal development (see Concept 18.6 for more information on *Hox* genes). One group of *Hox* genes, the *Hoxd* genes, plays a role in establishing the pattern of the different digits (fingers and toes) at the end of a limb. Unlike the *mPGES-1* gene mentioned in the Chapter 15 Scientific Skills Exercise, *Hox* genes have very large, complicated regulatory regions, including control elements that may be hundreds of kilobases (kb; thousands of nucleotides) away from the gene.

In cases like this, how do biologists locate the DNA segments that contain important elements? They begin by removing (deleting) large segments of DNA and studying the effect on gene expression. In this exercise, you'll compare data from two different but complementary approaches that look at the expression of a specific *Hoxd* gene (*Hoxd13*). One approach quantifies overall expression; the other approach is less quantitative but gives important spatial localization information.

### How the Experiment Was Done

Researchers interested in the regulation of *Hoxd13* gene expression genetically engineered a set of mice that had different segments of DNA deleted upstream of the gene. They then compared levels and patterns of *Hoxd13* gene expression in developing paws of 12.5-day-old mouse embryos that had the DNA deletions with gene expression seen in wild-type mouse embryos of the same age.

They used two different approaches: In some mice, they extracted the mRNA from the embryonic paws and quantified the overall level of *Hoxd13* mRNA in the whole paw. In another set of the same mice, they used *in situ* hybridization (see Figure 15.14) to pinpoint exactly where in the paws the *Hoxd13* gene was expressed as mRNA. The particular technique that was used causes the *Hoxd13* mRNA to appear blue or black, which indicates the highest mRNA levels.

### Data from the Experiment

The topmost diagram depicts the very large regulatory region upstream of the *Hoxd13* gene. The area between the slashes represents the long stretch of DNA located between the promoter and the regulatory region.

The diagrams to the left of the bar graph show, first, the intact DNA (830 kb) and, next, the three altered DNA sequences. (Each is called a "deletion" because a particular section of DNA has been deleted from it.) A red X indicates the segment (A, B, and/or C) that was deleted in each experimental treatment.

The horizontal bar graph shows the amount of *Hoxd13* mRNA that was present in the digit formation zone of each 12.5-day-old embryo with a deletion, relative to the amount that was in the digit formation zone of the mouse with the intact regulatory region (top bar = 100%).

The images to the right of the graph are fluorescent micrographs of the embryo paws showing the location of the *Hoxd13* mRNA (the stain appears blue or black). The white triangles show the location where the thumb will form.

### INTERPRET THE DATA

1. The researchers hypothesized that all three regulatory segments (A, B, and C) were required for full expression of the *Hoxd13*

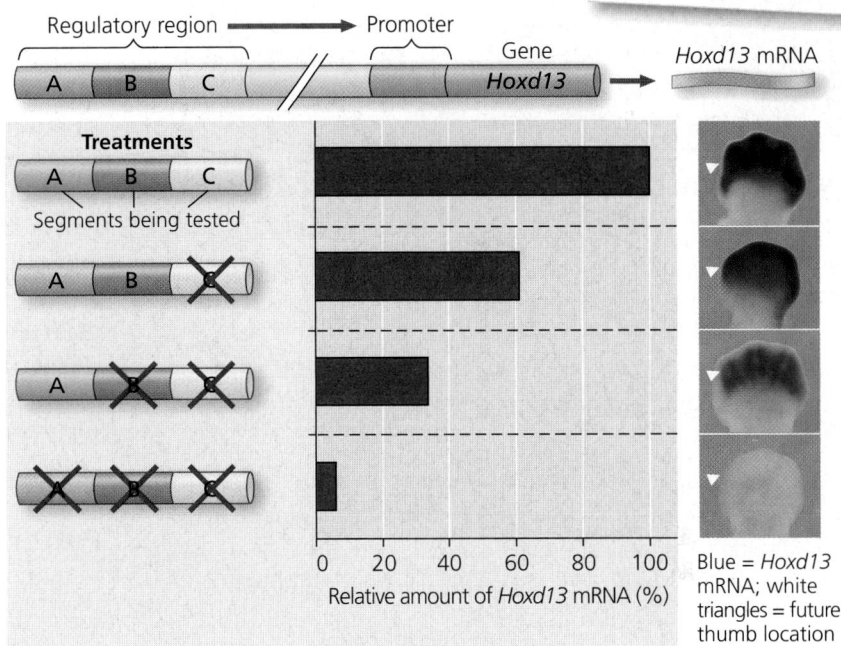

Blue = *Hoxd13* mRNA; white triangles = future thumb location

**Data from** T. Montavon et al., A regulatory archipelago controls *Hox* genes transcription in digits, *Cell* 147:1132–1145 (2011). doi:10.1016/j.cell.2011.10.023

gene. By measuring the amount of *Hoxd13* mRNA in the embryo paw zones where digits develop, they could measure the effect of the regulatory segments individually and in combination. Refer to the graph to answer these questions, noting that the segments being tested are shown on the vertical axis and the relative amount of *Hoxd13* mRNA is shown on the horizontal axis. (a) Which of the four treatments was used as a control for the experiment? (b) The hypothesis is that all three segments together are required for highest expression of the *Hoxd13* gene. Is this supported by the results? Explain your answer.

2. (a) What is the effect on the amount of *Hoxd13* mRNA when segments B and C are both deleted, compared with the control? (b) Is this effect visible in the blue-stained regions of the *in situ* hybridizations? How would you describe the spatial pattern of gene expression in the embryo paws that lack segments B and C? (You'll need to look carefully at different regions of each paw and how they differ.)

3. (a) What is the effect on the amount of *Hoxd13* mRNA when just segment C is deleted, compared with the control? (b) Is this effect visible in the *in situ* hybridizations? How would you describe the spatial pattern of gene expression in embryo paws that lack just segment C, compared with the control and with the paws that lack segments B and C?

4. If the researchers had only measured the amount of *Hoxd13* mRNA and not done the *in situ* hybridizations, what important information about the role of the regulatory segments in *Hoxd13* gene expression during paw development would have been missed? Conversely, if the researchers had only done the *in situ* hybridizations, what information would have been inaccessible?

(MB) A version of this Scientific Skills Exercise can be assigned in MasteringBiology.

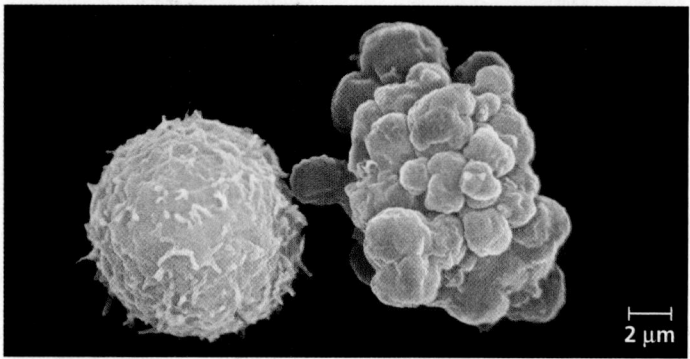

▲ **Figure 16.5 Apoptosis of a human white blood cell.** We can compare a normal white blood cell (left) with a white blood cell undergoing apoptosis (right) (colorized SEM). The apoptotic cell is shrinking and forming lobes ("blebs"), which eventually are shed as membrane-enclosed cell fragments.

components. The cell becomes multilobed, a change called "blebbing" **(Figure 16.5)**, and the cell's parts are packaged in vesicles. These "blebs" are engulfed by scavenger cells, leaving no trace. Apoptosis protects neighboring cells from damage that they would suffer if a dying cell merely leaked out all its contents, including its digestive enzymes.

Apoptosis plays a crucial role in the developing embryo. The molecular mechanisms underlying apoptosis were worked out by studying embryonic development of a small soil worm, a nematode called *Caenorhabditis elegans* that has now become a popular model organism for genetic studies. Because the adult worm has only about a thousand cells, researchers worked out the complete ancestry of each cell. The timely suicide of cells occurs exactly 131 times during normal development of *C. elegans*, at precisely the same points in the cell lineage of each worm. In worms and other species, apoptosis is triggered by signal transduction pathways (see Figure 5.20). These activate a cascade of apoptotic "suicide" proteins in the cells destined to die, including the enzymes that break down and package cellular molecules in the "blebs."

Apoptosis is essential to development and maintenance in all animals. There are similarities in genes encoding apoptotic proteins in nematodes and mammals, and apoptosis is known to occur as well in multicellular fungi and single-celled yeasts, evidence that the basic mechanism evolved early among eukaryotes. In vertebrates, apoptosis is essential for normal development of the nervous system and for normal morphogenesis of hands and feet in humans and paws in other mammals **(Figure 16.6)**. The level of apoptosis between the developing digits is lower in the webbed feet of ducks and other water birds than in the nonwebbed feet of land birds, such as chickens. In the case of humans, the failure of appropriate apoptosis can result in webbed fingers and toes.

We have seen how different programs of gene expression that are activated in the fertilized egg can result in differentiated cells and tissues as well as the death of some cells. But for tissues to function effectively in the organism as a whole, the organism's *body plan*—its overall three-dimensional arrangement—must be established and superimposed on the differentiation process. Next we'll look at the molecular basis for establishing the body plan, using the well-studied fruit fly *Drosophila* as an example.

## Pattern Formation: Setting Up the Body Plan

Cytoplasmic determinants and inductive signals both contribute to the development of a spatial organization in which the tissues and organs of an organism are all in their characteristic places. This process is called **pattern formation**.

Just as the locations of the front, back, and sides of a new building are determined before construction begins, pattern formation in animals begins in the early embryo, when the major axes of an animal are established. In a bilaterally symmetrical animal, the relative positions of head and tail, right and left sides, and back and front—the three major body axes—are set up before the organs appear. The molecular cues that control pattern formation, collectively called **positional information**, are provided by cytoplasmic determinants and inductive signals (see Figure 16.3). These cues tell a cell its location relative to the body axes and to neighboring cells and determine how the cell and its descendants will respond to future molecular signals.

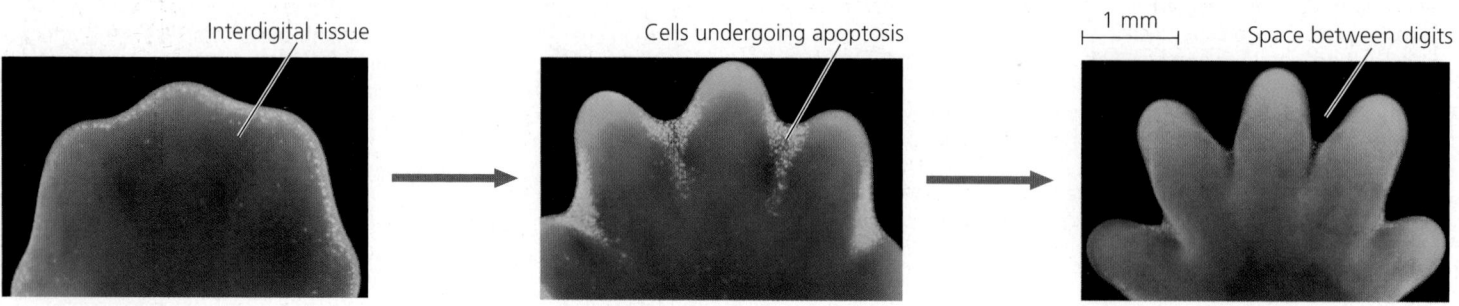

▲ **Figure 16.6 Effect of apoptosis during paw development in the mouse.** In mice, humans, other mammals, and land birds, the embryonic region that develops into feet or hands initially has a solid, platelike structure. Apoptosis eliminates the cells in the interdigital regions, thus forming the digits. The embryonic mouse paws shown in these fluorescence light micrographs are stained so that cells undergoing apoptosis appear bright yellow. Apoptosis of cells begins at the margin of each interdigital region (left), peaks as the tissue in these regions is reduced (middle), and is no longer visible when the interdigital tissue has been eliminated (right). (Note that the Scientific Skills Exercise shows a different genetic process involved in mouse paw development.)

During the first half of the 20th century, classical embryologists made detailed anatomical observations of embryonic development in a number of species and performed experiments in which they manipulated embryonic tissues. This research laid the groundwork for understanding the mechanisms of development, but it did not reveal the specific molecules that guide development or determine how patterns are established.

In the 1940s, scientists began using the genetic approach— the study of mutants—to investigate *Drosophila* development. That approach has had spectacular success. Genetic studies have established that genes control development and have led to an understanding of the key roles that specific molecules play in defining position and directing differentiation. By combining anatomical, genetic, and biochemical approaches to the study of *Drosophila* development, researchers have discovered developmental principles common to many other species, including humans.

### The Life Cycle of Drosophila

Fruit flies and other arthropods have a modular construction, an ordered series of segments. These segments make up the body's three major parts: the head, the thorax (the mid-body, from which the wings and legs extend), and the abdomen **(Figure 16.7a)**. Like other bilaterally symmetrical animals, *Drosophila* has an anterior-posterior (head-to-tail) axis, a dorsal-ventral (back-to-belly) axis, and a right-left axis. In *Drosophila*, cytoplasmic determinants that are localized in the unfertilized egg provide positional information for the placement of anterior-posterior and dorsal-ventral axes even before fertilization. We'll focus here on the molecules involved in establishing the anterior-posterior axis as a case in point.

The *Drosophila* egg develops in one of the female's ovaries, surrounded by ovarian cells called nurse cells and follicle cells **(Figure 16.7b**, top). These support cells supply the egg with nutrients, mRNAs, and other substances needed for development, and make the egg shell. After fertilization and laying of the egg, embryonic development results in the formation of a segmented larva, which goes through three larval stages. Then, in a process much like that by which a caterpillar becomes a butterfly, the larva forms a pupa in which it metamorphoses into the adult fly pictured in Figure 16.7a.

## Genetic Analysis of Early Development: *Scientific Inquiry*

Edward B. Lewis was a visionary American biologist who, in the 1940s, first showed the value of the genetic approach to studying embryonic development in *Drosophila*. Lewis studied bizarre mutant flies with developmental defects that led to extra wings or legs in the wrong place **(Figure 16.8)**. He located the mutations on the fly's genetic map, connecting the abnormalities to specific genes. This research supplied the first evidence that genes direct the developmental processes studied by embryologists. The genes Lewis discovered, called

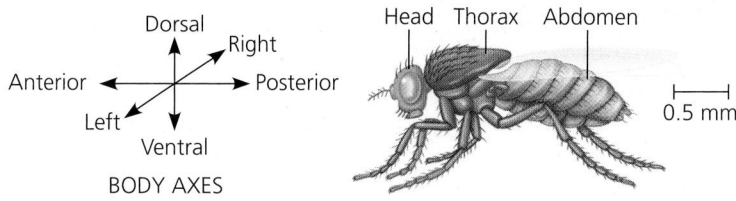

BODY AXES

**(a) Adult.** The adult fly is segmented, and multiple segments make up each of the three main body parts—head, thorax, and abdomen. The body axes are shown by arrows.

**1** **Developing egg** within one ovarian follicle (among many in an ovary). The egg (yellow) is surrounded by support cells (follicle cells).

**2** **Mature, unfertilized egg.** The developing egg enlarges as nutrients and mRNAs are supplied to it by other support cells (nurse cells), which shrink. Eventually, the mature egg fills the egg shell that is secreted by the follicle cells.

**3** **Fertilized egg.** The egg is fertilized within the mother and then laid.

**4** **Segmented embryo.** The egg develops into a segmented embryo.

**5** **Larva.** The embryo develops into a larva, which has three stages. The third stage forms a pupa (not shown), within which the larva metamorphoses into the adult shown in (a).

Follicle cell — Nucleus — Egg — Nurse cell — Egg shell — Depleted nurse cells — Fertilization — Laying of egg — Embryonic development — Body segments — 0.1 mm — Hatching

**(b) Development from egg to larva.**

▲ **Figure 16.7 Key events in *Drosophila* development.**

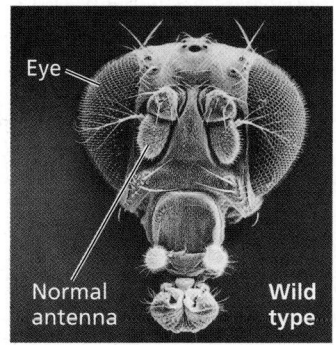
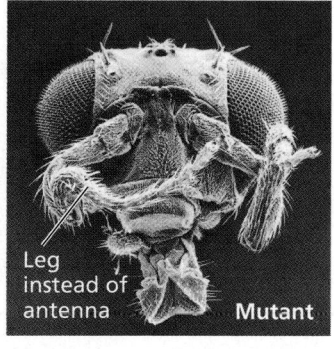

Eye — Normal antenna — **Wild type** — Leg instead of antenna — **Mutant**

▲ **Figure 16.8 Abnormal pattern formation in *Drosophila*.** Mutations in homeotic genes cause misplacement of structures. These scanning electron micrographs contrast the head of a wild-type fly with that of a homeotic mutant (a fly with a mutation in a homeotic gene) bearing a pair of legs in place of antennae.

**homeotic genes**, are regulatory genes that control pattern formation in the late embryo, larva, and adult.

Further insight into pattern formation during early embryonic development did not come for another 30 years, when two researchers in Germany, Christiane Nüsslein-Volhard and Eric Wieschaus, set out to identify *all* the genes that affect segment formation in *Drosophila*. The project was daunting for three reasons. The first was the sheer number of *Drosophila* protein-coding genes, now known to total about 14,000. The genes affecting segmentation might be just a few needles in a haystack or might be so numerous and varied that the scientists would be unable to make sense of them. Second, mutations affecting a process as fundamental as segmentation would surely be **embryonic lethals**, mutations with phenotypes causing death at the embryonic or larval stage. Since organisms with embryonic lethal mutations never reproduce, they cannot be bred for study. The researchers dealt with this problem by looking for recessive mutations, which can be propagated in heterozygous flies that act as genetic carriers. Third, cytoplasmic determinants in the egg were known to play a role in axis formation, so the researchers knew they would have to study the mother's genes as well as those of the embryo. It is the mother's genes that we will discuss further as we focus on how the anterior-posterior body axis is set up in the developing egg.

Nüsslein-Volhard and Wieschaus began their search for segmentation genes by exposing flies to a mutagenic chemical that affected the flies' gametes. They mated the mutagenized flies and then scanned their descendants for dead embryos or larvae with abnormal segmentation or other defects. For example, to find genes that might set up the anterior-posterior axis, they looked for embryos or larvae with abnormal ends, such as two heads or two tails, predicting that such abnormalities would arise from mutations in maternal genes required for correctly setting up the offspring's head or tail end.

Using this approach, Nüsslein-Volhard and Wieschaus eventually identified about 1,200 genes essential for pattern formation during embryonic development. Of these, about 120 were essential for normal segmentation. Over several years, the researchers were able to group these segmentation genes by general function, to map them, and to clone many of them for further study in the lab. The result was a detailed molecular understanding of the early steps in pattern formation in *Drosophila*.

When the results of Nüsslein-Volhard and Wieschaus were combined with Lewis's earlier work, a coherent picture of *Drosophila* development emerged. In recognition of their discoveries, the three researchers were awarded a Nobel Prize in 1995. Next, let's consider a specific example of the genes that Nüsslein-Volhard, Wieschaus, and co-workers found.

### Axis Establishment

As we mentioned earlier, cytoplasmic determinants in the egg are the substances that initially establish the axes of the *Drosophila* body. These substances are encoded by genes of the mother, fittingly called maternal effect genes. A **maternal effect gene** is a gene that, when mutant in the mother, results in a mutant phenotype in the offspring, regardless of the offspring's own genotype. In fruit fly development, the mRNA or protein products of maternal effect genes are placed in the egg while it is still in the mother's ovary. When the mother has a mutation in such a gene, she makes a defective gene product (or none at all), and her eggs are defective; when these eggs are fertilized, they fail to develop properly.

Because maternal effect genes control the orientation (polarity) of the egg and consequently of the fly, they are also called *egg-polarity genes*. One group of these genes sets up the anterior-posterior axis of the embryo, while a second group establishes the dorsal-ventral axis. Like mutations in segmentation genes, mutations in maternal effect genes are generally embryonic lethals.

**Bicoid: A Morphogen That Determines Head Structures**
To see how maternal effect genes determine the body axes of the offspring, we'll focus on one such gene, called **bicoid**, a term meaning "two-tailed." An embryo whose mother has two mutant *bicoid* alleles lacks the front half of its body and has posterior structures at both ends **(Figure 16.9)**. This phenotype suggested to Nüsslein-Volhard and her colleagues that the product of the mother's *bicoid* gene is essential for setting up the anterior end of the fly and might be concentrated at the future anterior end of the embryo. This hypothesis is an example of the *morphogen gradient hypothesis* first proposed by embryologists a century ago, in which gradients of substances called **morphogens** establish an embryo's axes and other features of its form.

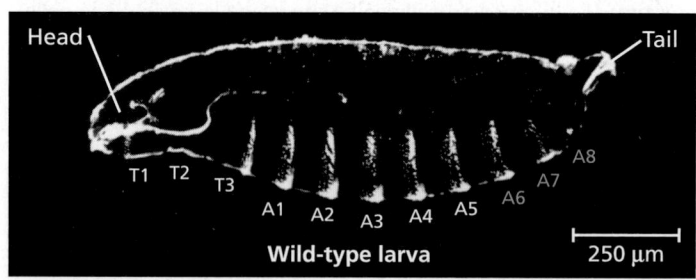

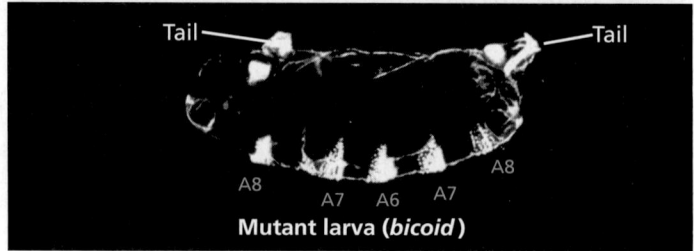

▲ **Figure 16.9 Effect of the *bicoid* gene on *Drosophila* development.** A wild-type fruit fly larva has a head, three thoracic (T) segments, eight abdominal (A) segments, and a tail. A larva whose mother has two mutant alleles of the *bicoid* gene has two tails and lacks all anterior structures (LMs).

DNA technology and other modern biochemical methods enabled the researchers to test whether the *bicoid* product, a protein called Bicoid, is in fact a morphogen that determines the anterior end of the fly. The first question they asked was whether the mRNA and protein products of this gene are located in the egg in a position consistent with the hypothesis. They found that *bicoid* mRNA is highly concentrated at the extreme anterior end of the mature egg, as predicted by the hypothesis (**Figure 16.10**). After the egg is fertilized, the mRNA is translated into protein. The Bicoid protein then diffuses from the anterior end toward the posterior, resulting in a gradient of protein within the early embryo, with the highest concentration at the anterior end. These results are consistent with the hypothesis that Bicoid protein specifies the fly's anterior end. To test the hypothesis more specifically, scientists injected pure *bicoid* mRNA into various regions of early embryos. The protein that resulted from its translation caused anterior structures to form at the injection sites.

The *bicoid* research was groundbreaking for several reasons. First, it led to the identification of a specific protein required for some of the earliest steps in pattern formation. It thus helped us understand how different regions of the egg can give rise to cells that go down different developmental pathways. Second, it increased our understanding of the mother's critical role in the initial phases of embryonic development. Finally, the principle that a gradient of morphogens can determine polarity and position has proved to be a key developmental concept for a number of species, just as early embryologists had hypothesized.

Maternal mRNAs are crucial during development of many species. In *Drosophila*, gradients of specific proteins encoded by maternal mRNAs not only determine the posterior and anterior ends, but also establish the dorsal-ventral axis. As the fly embryo grows, it reaches a point when the embryonic program of gene expression takes over, and the maternal mRNAs must be destroyed. (This process involves miRNAs in *Drosophila* and other species.) Later, positional information encoded by the embryo's genes, operating on an ever finer scale, establishes a specific number of correctly oriented segments and triggers the formation of each segment's characteristic structures. When the genes operating in this final step are abnormal, the pattern of the adult is abnormal, as you saw in Figure 16.8.

**EVOLUTION** The fly with legs emerging from its head in Figure 16.8 is the result of a single mutation in one gene. The gene does not encode any antenna protein, however. Instead, it encodes a transcription factor that regulates other genes, and its malfunction leads to misplaced structures like legs instead of antennae. The observation that a change in gene regulation during development could lead to such a fantastic change in body form prompted some scientists to consider whether these types of mutations could contribute to evolution by generating novel body shapes. Ultimately this line of inquiry gave rise to the field of evolutionary developmental biology, so-called "evo-devo," which will be discussed further in Concept 18.6.

---

▼ **Figure 16.10** **Inquiry**

## Could Bicoid be a morphogen that determines the anterior end of a fruit fly?

**Experiment** Using a genetic approach to study *Drosophila* development, Christiane Nüsslein-Volhard and colleagues at two research institutions in Germany analyzed expression of the *bicoid* gene. The researchers hypothesized that *bicoid* normally codes for a morphogen that specifies the head (anterior) end of the embryo. To begin to test this hypothesis, they used molecular techniques to determine whether the mRNA and protein encoded by this gene were found in the anterior end of the fertilized egg and early embryo of wild-type flies.

**Results** *Bicoid* mRNA (dark blue in the light micrographs and drawings) was confined to the anterior end of the unfertilized egg. Later in development, Bicoid protein (dark orange) was seen to be concentrated in cells at the anterior end of the embryo.

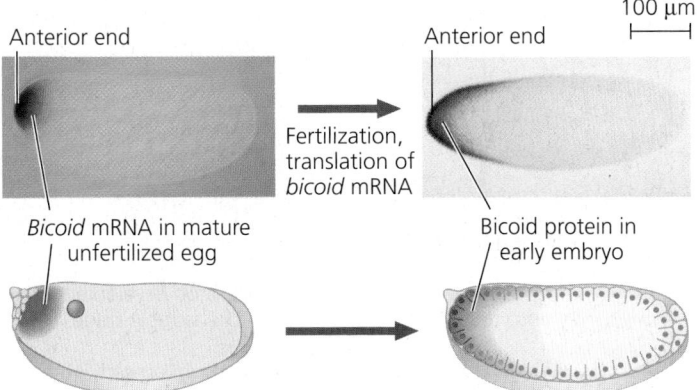

**Conclusion** The location of *bicoid* mRNA and the diffuse gradient of Bicoid protein seen later are consistent with the hypothesis that Bicoid protein is a morphogen specifying formation of head-specific structures.

**Further Reading** C. Nüsslein-Volhard et al., Determination of anteroposterior polarity in *Drosophila*, *Science* 238:1675–1681 (1987); W. Driever and C. Nüsslein-Volhard, A gradient of Bicoid protein in *Drosophila* embryos, *Cell* 54:83–93 (1988); T. Berleth et al., The role of localization of *bicoid* RNA in organizing the anterior pattern of the *Drosophila* embryo, *EMBO Journal* 7:1749–1756 (1988).

**WHAT IF?** The researchers needed further evidence, so they injected *bicoid* mRNA into the anterior end of an egg from a female with a mutation disabling the *bicoid* gene. Given that the hypothesis was supported, what must their results have been?

---

**CONCEPT CHECK 16.1**

1. **MAKE CONNECTIONS** As you learned in Chapter 9, mitosis gives rise to two daughter cells that are genetically identical to the parent cell. Yet you, the product of many mitotic divisions, are not composed of identical cells. Why?

2. **MAKE CONNECTIONS** Explain how the signaling molecules released by an embryonic cell can induce changes in a neighboring cell without entering the cell. (See Figure 5.20.)

3. How do fruit fly maternal effect genes determine the polarity of the egg and the embryo?

For suggested answers, see Appendix A.

## CONCEPT 16.2

# Cloning of organisms showed that differentiated cells could be "reprogrammed" and ultimately led to the production of stem cells

When the field of developmental biology (then called embryology) was first taking shape in the early 1900s, a major question was whether all the cells of an organism have the same genes or whether cells lose genes during the process of differentiation. Today, we know that genes are not lost—but the question is whether each cell can express all of its genes.

One way to answer this question is to see whether a differentiated cell has the potential to generate a whole organism. Because an organism generated in this way develops from a single cell without either meiosis or fertilization, this is known as "cloning." In this context, cloning produces one or more organisms genetically identical to the "parent" that donated the single cell. This is often called *organismal cloning* to differentiate it from gene cloning and, more significantly, from cell cloning—the division of an asexually reproducing cell such as a bacterium into a group of genetically identical cells. (The common theme for all types of cloning is that the product is genetically identical to the parent. In fact, the word *clone* comes from the Greek *klon*, meaning "twig.")

The current interest in organismal cloning arises primarily from its potential to generate **stem cells**, relatively unspecialized cells that can both reproduce themselves and, under the right conditions, differentiate into many different tissues. We'll return to stem cells later. For now, we'll discuss a series of experiments that provides a conceptual framework for thinking about what determines the potential of a cell—in other words, what genes that cell can express at any given time.

## Cloning Plants: Single-Cell Cultures

The successful cloning of whole plants from single differentiated cells was accomplished during the 1950s by F. C. Steward and his students at Cornell University, who worked with carrot plants. They found that single differentiated cells taken from the root (the carrot) and incubated in culture medium could grow into normal adult plants, each genetically identical to the parent plant. These results showed that differentiation does not necessarily involve irreversible changes in the DNA. In plants, at least, mature cells can "dedifferentiate" and then give rise to all the specialized cell types of the organism. Any cell with this potential is said to be **totipotent**.

## Cloning Animals: Nuclear Transplantation

Differentiated cells from animals generally do not divide in culture, much less develop into the multiple cell types of a new organism. Therefore, early researchers had to use a different approach to the question of whether differentiated animal cells are

totipotent. Their approach was to remove the nucleus of an egg (creating an *enucleated* egg) and replace it with the nucleus of a differentiated cell, a procedure called *nuclear transplantation*. If the nucleus from the differentiated donor cell retains its full genetic capability, then it should be able to direct development of the recipient cell into all the tissues and organs of an organism.

Such experiments were conducted on one species of frog (*Rana pipiens*) by Robert Briggs and Thomas King in the 1950s and on another frog species (*Xenopus laevis*) by John Gurdon in the 1970s (**Figure 16.11**). These researchers transplanted a

▼ **Figure 16.11** Inquiry

### Can the nucleus from a differentiated animal cell direct development of an organism?

**Experiment** John Gurdon and colleagues at Oxford University, in England, destroyed the nuclei of frog (*Xenopus laevis*) eggs by exposing the eggs to ultraviolet light. They then transplanted nuclei from cells of frog embryos and tadpoles into the enucleated eggs.

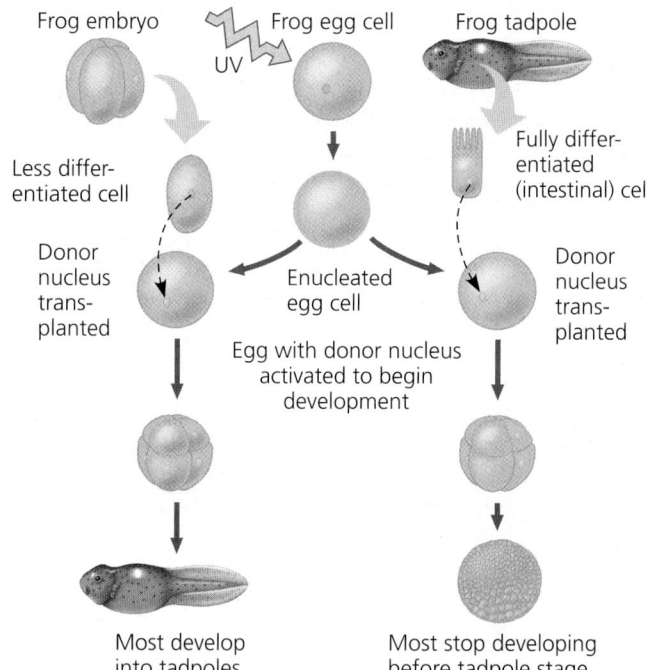

**Results** When the transplanted nuclei came from an early embryo, the cells of which are relatively undifferentiated, most of the recipient eggs developed into tadpoles. But when the nuclei came from the fully differentiated intestinal cells of a tadpole, fewer than 2% of the eggs developed into normal tadpoles, and most of the embryos stopped developing at a much earlier stage.

**Conclusion** The nucleus from a differentiated frog cell can direct development of a tadpole. However, its ability to do so decreases as the donor cell becomes more differentiated, presumably because of changes in the nucleus.

**Data from** J. B. Gurdon et al., The developmental capacity of nuclei transplanted from keratinized cells of adult frogs, *Journal of Embryology and Experimental Morphology* 34:93–112 (1975).

**WHAT IF?** If each cell in a four-cell embryo were already so specialized that it was not totipotent, what results would you predict for the experiment on the left side of the figure?

nucleus from an embryonic or tadpole cell into an enucleated egg of the same species. In Gurdon's experiments, the transplanted nucleus was often able to support normal development of the egg into a tadpole. However, he found that the potential of a transplanted nucleus to direct normal development was inversely related to the age of the donor: The older the donor nucleus, the lower the percentage of normal tadpoles.

From these results, Gurdon concluded that something in the nucleus *does* change as animal cells differentiate. In frogs and most other animals, nuclear potential tends to be restricted more and more as embryonic development and cell differentiation progress. These were foundational experiments that ultimately led to stem cell technology, and Gurdon received the 2012 Nobel Prize in Medicine for this work.

### Reproductive Cloning of Mammals

In addition to cloning frogs, researchers had long been able to clone mammals using early embryonic cells as a source of donor nuclei. Until about 20 years ago, though, it was not known whether a nucleus from a fully differentiated cell could be reprogrammed successfully to act as a donor nucleus. In 1997, researchers in Scotland announced the birth of Dolly, a lamb cloned from an adult sheep by nuclear transplantation from a differentiated mammary gland cell **(Figure 16.12)**. Using a technique related to that shown in Figure 16.11, the researchers implanted early embryos into surrogate mothers. Out of several hundred embryos, one successfully completed normal development, and Dolly was born, a genetic clone of the nucleus donor. At the age of 6, Dolly suffered complications from a lung disease usually seen only in much older sheep and was euthanized. Dolly's premature death, as well as an arthritic condition, led to speculation that her cells were in some way not quite as healthy as those of a normal sheep, possibly reflecting incomplete reprogramming of the original transplanted nucleus.

Since that time, researchers have cloned numerous other mammals, including mice, cats, cows, horses, pigs, dogs, and monkeys. In most cases, their goal has been the production of new individuals; this is known as *reproductive cloning*. We have already learned a lot from such experiments. For example, cloned animals of the same species do *not* always look or behave identically. In a herd of cows cloned from the same line

▶ **Figure 16.12 Dolly, the first mammal cloned by nuclear transplantation.** Dolly, shown here as a lamb, has a very different appearance from her surrogate mother, standing behind her.

© Photo courtesy of the Roslin Institute, The University of Edinburgh

▶ **Figure 16.13 CC ("Carbon Copy"), the first cloned cat (right), and her single parent.** Rainbow (left) donated the nucleus in a cloning procedure that resulted in CC. However, the two cats are not identical: Rainbow has orange patches on her fur, but CC does not.

of cultured cells, certain cows are dominant in behavior and others are more submissive. Another example of nonidentity in clones is the first cloned cat, named CC for Carbon Copy **(Figure 16.13)**. She has a calico coat, like her single female parent, but the color and pattern are different because of random X chromosome inactivation, which is a normal occurrence during embryonic development (see Figure 12.8). And identical human twins, which are naturally occurring "clones," are always slightly different. Evidently, environmental influences and random phenomena can play a significant role during development.

### Faulty Gene Regulation in Cloned Animals

In most nuclear transplantation studies thus far, only a small percentage of cloned embryos develop normally to birth. And like Dolly, many cloned animals exhibit defects. Cloned mice, for instance, are prone to obesity, pneumonia, liver failure, and premature death. Scientists assert that even cloned animals that appear normal are likely to have subtle defects.

Researchers have uncovered some reasons for the low efficiency of cloning and the high incidence of abnormalities. In the nuclei of fully differentiated cells, a small subset of genes is turned on and expression of the rest of the genes is repressed. This regulation often is the result of epigenetic changes in chromatin, such as acetylation of histones or methylation of DNA (see Figure 15.7). During the nuclear transfer procedure, many of these changes must be reversed in the later-stage nucleus from a donor animal for genes to be expressed or repressed appropriately in early stages of development. Researchers have found that the DNA in cells from cloned embryos, like that of differentiated cells, often has more methyl groups than does the DNA in equivalent cells from normal embryos of the same species. This finding suggests that the reprogramming of donor nuclei requires more accurate and complete chromatin restructuring than occurs during cloning procedures. Because DNA methylation helps regulate gene expression, misplaced or extra methyl groups in the DNA of donor nuclei may interfere with the pattern of gene expression necessary for normal embryonic development. In fact, the success of a cloning attempt may depend in large part on whether or not the chromatin in the donor nucleus can be artificially "rejuvenated" to resemble that of a newly fertilized egg.

## Stem Cells of Animals

Progress in cloning mammalian embryos, including primates, has heightened speculation about the cloning of humans, which has not yet been achieved. The main reason researchers are trying to clone human embryos is not for reproduction, but for the production of stem cells to treat human diseases. Recall that a stem cell is a relatively unspecialized cell that can both reproduce itself indefinitely and, under appropriate conditions, differentiate into specialized cells of one or more types **(Figure 16.14)**. Thus, stem cells are able to both replenish their own population and generate cells that travel down specific differentiation pathways.

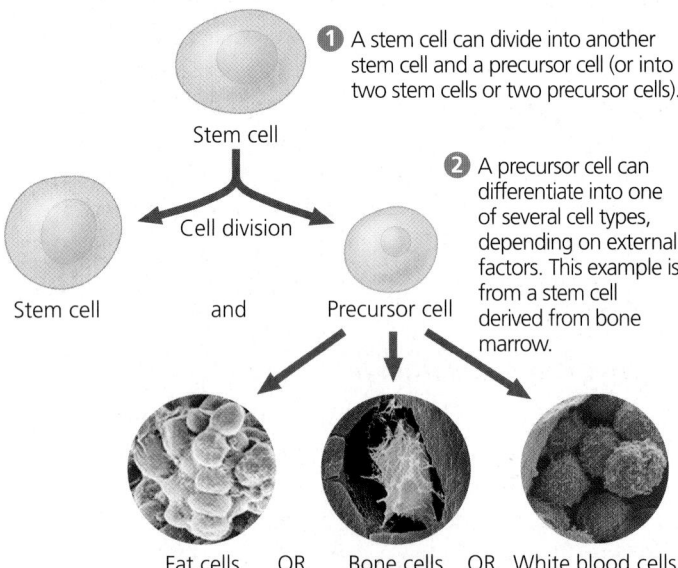

**1** A stem cell can divide into another stem cell and a precursor cell (or into two stem cells or two precursor cells).

Stem cell

Cell division

Stem cell    and    Precursor cell

**2** A precursor cell can differentiate into one of several cell types, depending on external factors. This example is from a stem cell derived from bone marrow.

Fat cells    OR    Bone cells    OR    White blood cells

▲ **Figure 16.14 How stem cells maintain their own population and generate differentiated cells.**

### Embryonic and Adult Stem Cells

Many early animal embryos contain stem cells capable of giving rise to differentiated embryonic cells of any type. Stem cells can be isolated from early embryos at a stage called the blastula stage or its human equivalent, the blastocyst stage. In culture, these *embryonic stem (ES) cells* reproduce indefinitely, and depending on culture conditions, they can be made to differentiate into a wide variety of specialized cells **(Figure 16.15)**, including even eggs and sperm.

The adult body also has stem cells, which serve to replace nonreproducing specialized cells as needed. In contrast to ES cells, *adult stem cells* are not able to give rise to all cell types in the organism, though they can generate multiple types. For example, one of the several types of stem cells in bone marrow can generate all the different kinds of blood cells (see Figure 16.15), and another type of bone marrow stem cell can differentiate into bone, cartilage, fat, muscle, and the linings of blood vessels. To the surprise of many, the adult brain has been found to contain stem cells that continue to produce certain kinds of nerve cells there. Researchers have also reported

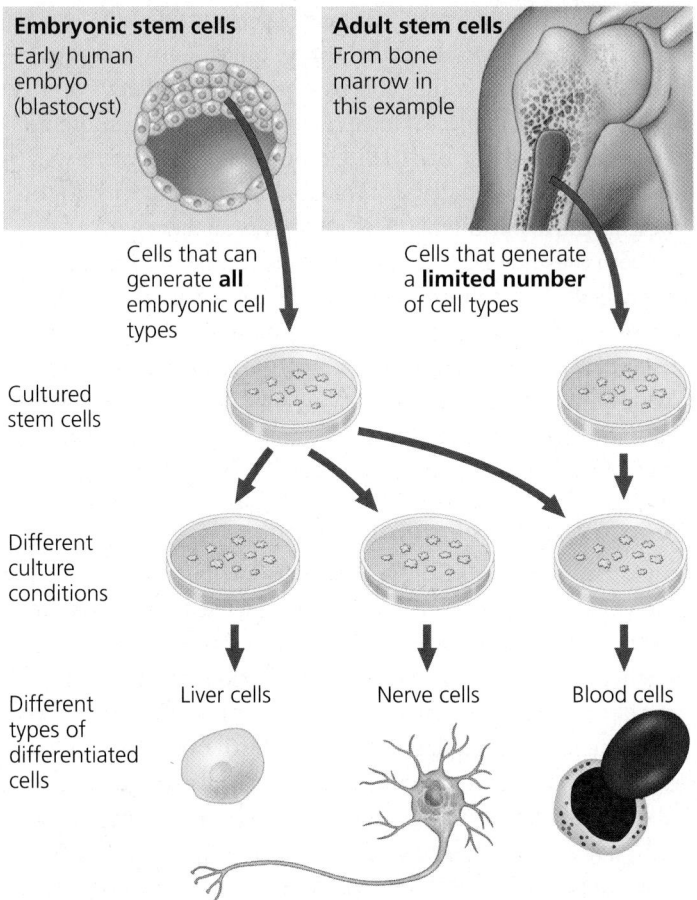

**Embryonic stem cells** Early human embryo (blastocyst)

**Adult stem cells** From bone marrow in this example

Cells that can generate **all** embryonic cell types

Cells that generate a **limited number** of cell types

Cultured stem cells

Different culture conditions

Different types of differentiated cells

Liver cells    Nerve cells    Blood cells

▲ **Figure 16.15 Working with stem cells.** Animal stem cells, which can be isolated from early embryos or adult tissues and grown in culture, are self-perpetuating, relatively undifferentiated cells. Embryonic stem cells are easier to grow than adult stem cells and can theoretically give rise to *all* types of cells in an organism. The range of cell types that can arise from adult stem cells is not yet fully understood.

finding stem cells in skin, hair, eyes, and dental pulp. Although adult animals have only tiny numbers of stem cells, scientists are learning to identify and isolate these cells from various tissues and, in some cases, to grow them in culture. With the right culture conditions (for instance, the addition of specific growth factors), cultured stem cells from adult animals have been made to differentiate into multiple types of specialized cells, although none are as versatile as ES cells.

Research with embryonic or adult stem cells is a source of valuable data about differentiation and has enormous potential for medical applications. The ultimate aim is to supply cells for the repair of damaged or diseased organs: for example, insulin-producing pancreatic cells for people with type 1 diabetes or certain kinds of brain cells for people with Parkinson's disease or Huntington's disease. Adult stem cells from bone marrow have long been used as a source of immune system cells in patients whose own immune systems are nonfunctional because of genetic disorders or radiation treatments for cancer.

The developmental potential of adult stem cells is limited to certain tissues. ES cells hold more promise than adult stem cells for most medical applications because ES cells are

**pluripotent**, capable of differentiating into many different cell types. Until recently, ES cells were obtained only from embryos donated (with informed consent) by patients undergoing infertility treatment or from long-term cell cultures originally established with cells isolated from donated embryos. However, in 2013, researchers in Oregon were able to obtain human ES cells from cloned human blastocysts and to establish several ES cell lines from them. When the main aim of cloning is to produce ES cells to treat disease, the process is called *therapeutic cloning*. Although most people believe that reproductive cloning of humans is unethical, opinions vary about the morality of therapeutic cloning.

### Induced Pluripotent Stem (iPS) Cells

Resolving the debate about therapeutic cloning now seems less urgent because researchers have learned to turn back the clock in fully differentiated cells, reprogramming them to act like ES cells. This major feat was announced in 2007, first by labs using mouse skin cells and then by additional groups using cells from human skin and other organs or tissues. In all these cases, researchers transformed the differentiated cells into a type of ES cell by using viruses called retroviruses to introduce extra cloned copies of four "stem cell" master regulatory genes. The "deprogrammed" cells are known as *induced pluripotent stem (iPS) cells* because, in using this fairly simple laboratory technique to return them to their undifferentiated state, pluripotency has been restored. The experiments that first transformed human differentiated cells into iPS cells are described in **Figure 16.16**. Shinya Yamanaka received the 2012 Nobel Prize in Medicine for this work, shared with John Gurdon (see Figure 16.11).

By many criteria, iPS cells can perform most of the functions of ES cells, but there are some differences in gene expression and other cellular functions, such as cell division. At least until these differences are fully understood, the study of ES cells will continue to make important contributions to the development of stem cell therapies. In the meantime, work is proceeding using iPS cells that have been experimentally produced.

There are two major potential uses for human iPS cells. First, cells from patients suffering from diseases can be reprogrammed to become iPS cells, which can act as model cells for studying the disease and potential treatments. Human iPS cell lines have been developed from people with type 1 diabetes, Parkinson's disease, and at least a dozen other diseases. Second, in the field of regenerative medicine, a patient's cells could be reprogrammed into iPS cells and then used to replace nonfunctional tissues, such as insulin-producing cells of the pancreas or cells of the retina in the eye. In fact, Dr. Masayo Takahashi, of the RIKEN Center for Developmental Biology in Japan, has been leading a study on patients with age-related macular degeneration, a disease of the retina. She obtains skin cells from each patient and turns them into iPS cells, and then into retinal cells, ultimately transplanting them into the patient's eye. While this technique is currently too expensive to be used as a standard treatment for the disorder, it holds promise for the future.

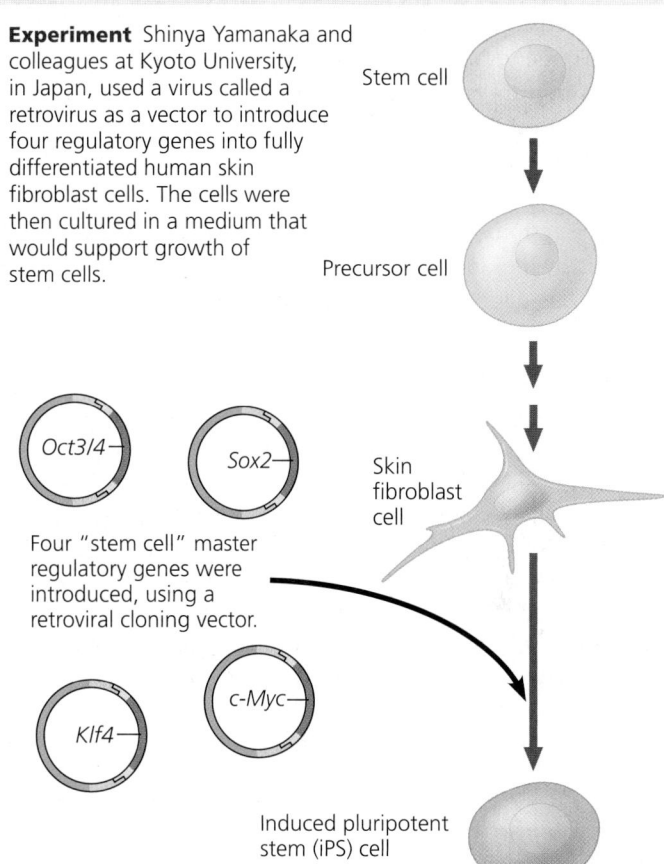

▼ **Figure 16.16** **Inquiry**

## Can a fully differentiated human cell be "deprogrammed" to become a stem cell?

**Experiment** Shinya Yamanaka and colleagues at Kyoto University, in Japan, used a virus called a retrovirus as a vector to introduce four regulatory genes into fully differentiated human skin fibroblast cells. The cells were then cultured in a medium that would support growth of stem cells.

Stem cell

Precursor cell

Oct3/4

Sox2

Skin fibroblast cell

Four "stem cell" master regulatory genes were introduced, using a retroviral cloning vector.

Klf4

c-Myc

Induced pluripotent stem (iPS) cell

**Results** Two weeks later, the cells resembled embryonic stem cells in appearance and were actively dividing. Their gene expression patterns, gene methylation patterns, and other characteristics were also consistent with those of embryonic stem cells. The iPS cells were able to differentiate into heart muscle cells, as well as other cell types.

**Conclusion** The four regulatory genes induced differentiated skin cells to become pluripotent stem cells, with characteristics of embryonic skin cells.

**Data from** K. Takahashi et al., Induction of pluripotent stem cells from adult human fibroblasts by defined factors, *Cell* 131:861–872 (2007).

**WHAT IF?** When organs are transplanted from a donor to a diseased recipient, the recipient's immune system may reject the transplant, a condition with serious and often fatal consequences. Would the same risk be faced if the patients' own cells were used in an iPS-based procedure? Explain.

In another surprising recent development, researchers have identified genes that can reprogram a differentiated cell into another type of differentiated cell without passing through a pluripotent state. In the first reported example, one type of cell in the pancreas was transformed into another type. However, the two types of cells do not need to be very closely related: Another research group has been able to directly reprogram a skin fibroblast into a nerve cell. Developing techniques that direct iPS cells or even differentiated cells to become cell types for regenerative medicine is an area of intense research that has seen some success. The iPS cells created in this way could eventually provide tailor-made "replacement" cells for patients without using human eggs or embryos, thus avoiding most ethical objections.

The research described in this and the preceding section on stem cells and cell differentiation has underscored the key role of gene regulation in embryonic development. The genetic program is carefully balanced between turning on the genes for differentiation in the right place and turning off other genes. Even when an organism is fully developed, gene expression is regulated in a similarly fine-tuned manner. In the final section of the chapter, we'll consider how fine this tuning is by looking at how specific changes in expression of one or a few genes can lead to the development of cancer.

**CONCEPT CHECK 16.2**

1. Based on current knowledge, how would you explain the difference in the percentage of tadpoles that developed from the two kinds of donor nuclei in Figure 16.11?
2. Dolly's egg donor and surrogate mother were "Scottish blackface" sheep, while the donor of the nucleus was a white-faced sheep. Explain why Dolly has a white face (see Figure 16.12).
3. **WHAT IF?** If you were a doctor who wanted to use iPS cells to treat a patient with severe type 1 diabetes, what new technique would have to be developed?

For suggested answers, see Appendix A.

## CONCEPT 16.3

# Abnormal regulation of genes that affect the cell cycle can lead to cancer

In Concept 9.3, we considered cancer as a type of disease in which cells escape from the control mechanisms that normally limit their growth. Now that we have discussed the molecular basis of gene expression and its regulation, we are ready to look at cancer more closely. The gene regulation systems that go wrong during cancer turn out to be the very same systems that play important roles in embryonic development, the maintenance of stem cell populations, and many other biological processes. Thus, research into the molecular basis of cancer has both benefited from and informed many other fields of biology.

## Types of Genes Associated with Cancer

The genes that normally regulate cell growth and division during the cell cycle include genes for growth factors, their receptors, and the intracellular molecules of signaling pathways. (To review regulation of the cell cycle, see Concept 9.3; for cell signaling, see Concept 5.6.) Mutations that alter any of these genes in somatic cells can lead to cancer. The agent of such change can be random spontaneous mutation. However, it is also likely that many cancer-causing mutations result from environmental influences, such as chemical carcinogens, X-rays and other high-energy radiation, and some viruses.

Cancer research led to the discovery of cancer-causing genes called **oncogenes** (from the Greek *onco*, tumor) in certain types of viruses. Subsequently, close counterparts of viral oncogenes were found in the genomes of humans and other animals. The normal versions of the cellular genes, called **proto-oncogenes**, code for proteins that stimulate normal cell growth and division.

How might a proto-oncogene—a gene that has an essential function in normal cells—become an oncogene, a cancer-causing gene? In general, an oncogene arises from a genetic change that leads to an increase either in the amount of the proto-oncogene's protein product or in the intrinsic activity of each protein molecule. The genetic changes that convert proto-oncogenes to oncogenes fall into three main categories: movement of DNA within the genome, amplification of a proto-oncogene, and point mutations in a control element or in the proto-oncogene itself **(Figure 16.17)**.

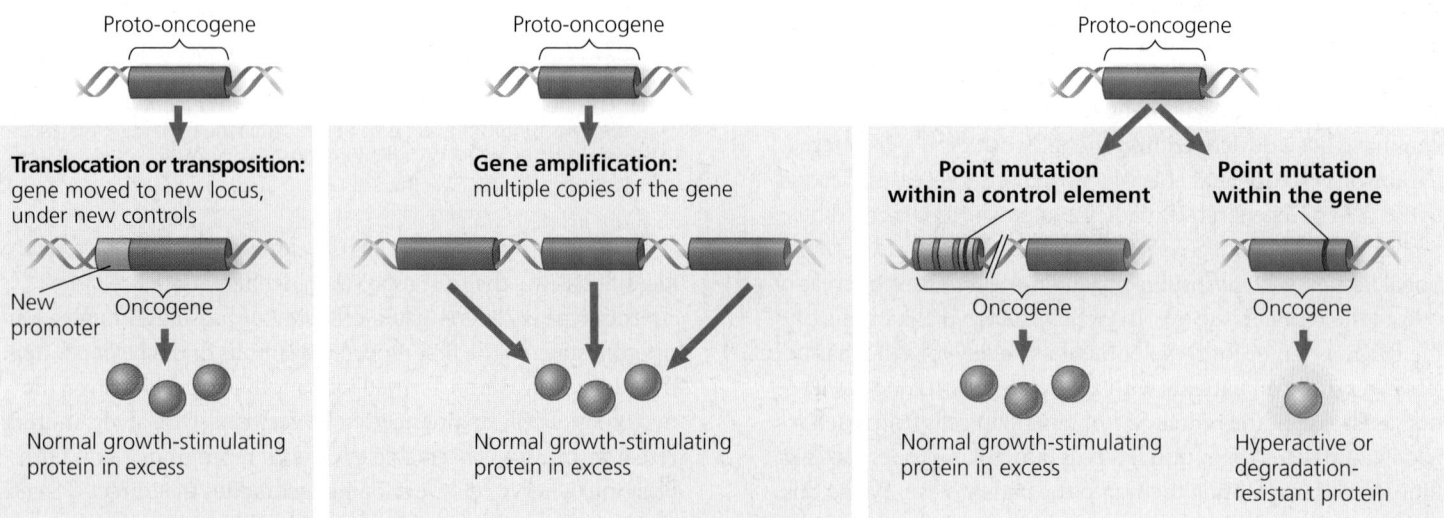

▲ **Figure 16.17 Genetic changes that can turn proto-oncogenes into oncogenes.**

Cancer cells are frequently found to contain chromosomes that have broken and rejoined incorrectly, translocating fragments from one chromosome to another (see Figure 12.14). Having learned how gene expression is regulated, you can now see the possible consequences of such translocations. If a translocated proto-oncogene ends up near an especially active promoter (or other control element), its transcription may increase, making it an oncogene. The second main type of genetic change, amplification, increases the number of copies of the proto-oncogene in the cell through repeated gene duplication (discussed in Concept 18.5). The third possibility is a point mutation either in the promoter or an enhancer that controls a proto-oncogene, causing an increase in its expression, or in the coding sequence of the proto-oncogene, changing the gene's product to a protein that is more active or more resistant to degradation than the normal protein. These mechanisms can lead to abnormal stimulation of the cell cycle and put the cell on the path to becoming a cancer cell.

In addition to genes whose products normally promote cell division, cells contain genes whose normal products *inhibit* cell division. Such genes are called **tumor-suppressor genes**, since the proteins they encode help prevent uncontrolled cell growth. Any mutation that decreases the normal activity of a tumor-suppressor protein may contribute to the onset of cancer, in effect stimulating growth through the absence of suppression.

Proteins produced from tumor-suppressor genes have various functions. Some repair damaged DNA, a function that prevents the cell from accumulating cancer-causing mutations.

Other tumor-suppressor proteins control the adhesion of cells to each other or to the extracellular matrix; proper cell anchorage is crucial in normal tissues—and is often absent in cancers. Still other tumor-suppressor proteins are components of cell-signaling pathways that inhibit the cell cycle.

## Interference with Cell-Signaling Pathways

The proteins encoded by many proto-oncogenes and tumor-suppressor genes are components of cell-signaling pathways. Let's take a closer look at how such proteins function in normal cells and what goes wrong with their function in cancer cells. We'll focus on the products of two key genes, the *ras* proto-oncogene and the *p53* tumor-suppressor gene. Mutations in *ras* occur in about 30% of human cancers, and mutations in *p53* in more than 50%.

The Ras protein, encoded by the ***ras* gene** (named for r̲a̲t sarcoma, a connective tissue cancer), is a G protein that relays a signal from a growth factor receptor on the plasma membrane to a cascade of protein kinases (see Figures 5.21 and 5.24). The cellular response at the end of the pathway is the synthesis of a protein that stimulates the cell cycle **(Figure 16.18a)**. Normally, such a pathway will not operate unless triggered by the appropriate growth factor. But certain mutations in the *ras* gene can lead to production of a hyperactive Ras protein that triggers the kinase cascade even in the absence of growth factor, resulting in increased cell division **(Figure 16.18b)**. In fact, hyperactive versions or excess amounts of any of the pathway's components can have the same outcome: excessive cell division.

▶ **Figure 16.18 Normal and mutant cell cycle–stimulating pathway.** **(a)** The normal pathway is triggered by ❶ a growth factor that binds to ❷ its receptor in the plasma membrane. The signal is relayed to ❸ a G protein called Ras. Like all G proteins, Ras is active when GTP is bound to it. Ras passes the signal to ❹ a series of protein kinases. The last kinase activates ❺ a transcription factor (activator) that turns on one or more genes for ❻ a protein that stimulates the cell cycle. **(b)** If a mutation makes Ras or any other pathway component abnormally active, excessive cell division and cancer may result.

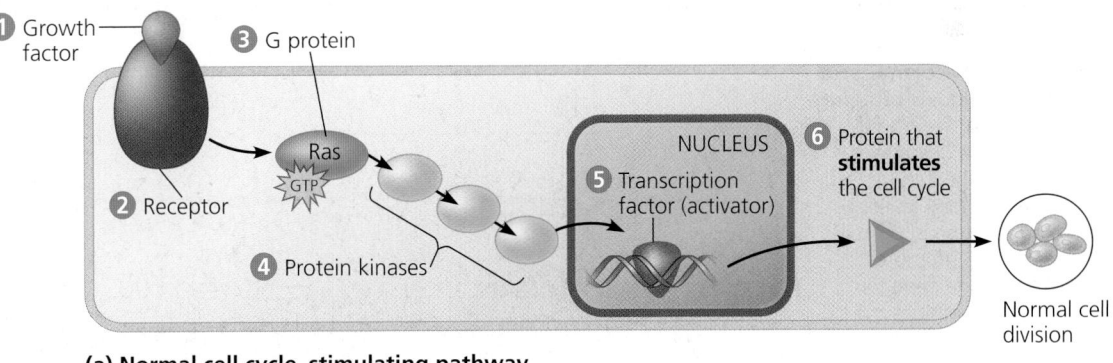

**(a) Normal cell cycle–stimulating pathway**

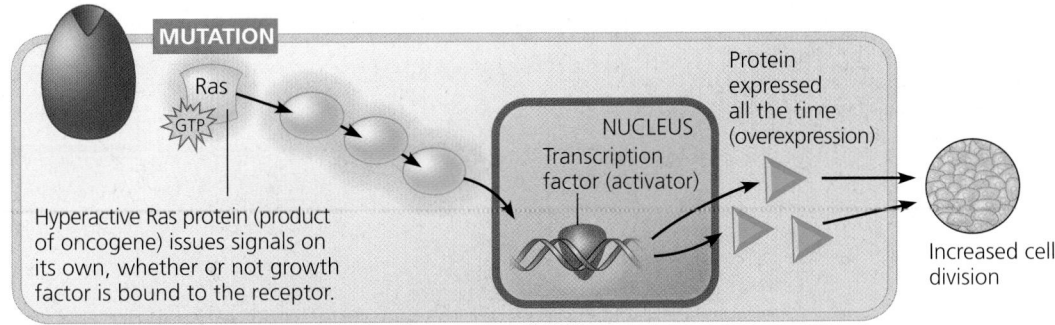

**(b) Mutant cell cycle–stimulating pathway**

**Figure 16.19a** shows a pathway in which an intracellular signal leads to the synthesis of a protein that suppresses the cell cycle. In this case, the signal is damage to the cell's DNA, perhaps as the result of exposure to ultraviolet light. Operation of this signaling pathway blocks the cell cycle until the damage has been repaired. Otherwise, the damage might contribute to tumor formation by causing mutations or chromosomal abnormalities. Thus, the genes for the components of the pathway act as tumor-suppressor genes. The *p53* **gene**, named for the 53,000-dalton molecular weight of its protein product, is a tumor-suppressor gene. The protein it encodes is a specific transcription factor that promotes the synthesis of cell cycle–inhibiting proteins. That is why a mutation that knocks out the *p53* gene, like a mutation that leads to a hyperactive Ras protein, can lead to excessive cell growth and cancer **(Figure 16.19b)**.

The *p53* gene has been called the "guardian angel of the genome." Once the gene is activated—for example, by DNA damage—the p53 protein functions as an activator for several other genes. Often it activates a gene called *p21*, whose product halts the cell cycle by binding to cyclin-dependent kinases, allowing time for the cell to repair the DNA. Researchers recently showed that p53 also activates expression of a group of miRNAs, which in turn inhibit the cell cycle. In addition, the p53 protein can turn on genes directly involved in DNA repair. Finally, when DNA damage is irreparable, p53 activates "suicide" genes, whose protein products bring about apoptosis, as described in the first section of this chapter. Thus, p53 acts in several ways to prevent a cell from passing on mutations due to DNA damage. If mutations do accumulate and the cell survives through many divisions—as is more likely if the *p53* tumor-suppressor gene is defective or missing—cancer may ensue.

The many functions of p53 suggest a complex picture of regulation in normal cells, one that we do not yet fully understand.

For the present, the diagrams in Figures 16.18 and 16.19 are an accurate view of how mutations contribute to cancer, but we don't know exactly how a cell becomes a cancer cell. As we discover previously unknown aspects of gene regulation, it is informative to study their role in the onset of cancer. Such studies have shown, for instance, that DNA methylation and histone modification patterns differ in normal and cancer cells and that miRNAs probably participate in cancer development. While we've learned a lot about cancer by studying cell-signaling pathways, there is still a lot to learn.

## The Multistep Model of Cancer Development

More than one somatic mutation is generally needed to produce all the changes characteristic of a full-fledged cancer cell. This may help explain why the incidence of cancer increases greatly with age. If cancer results from an accumulation of mutations and if mutations occur throughout life, then the longer we live, the more likely we are to develop cancer.

The model of a multistep path to cancer is well supported by studies of one of the best-understood types of human cancer, colorectal cancer, which affects the colon and/or rectum. About 140,000 new cases of colorectal cancer are diagnosed each year in the United States, and the disease causes 50,000 deaths each year. Like most cancers, colorectal cancer develops gradually **(Figure 16.20)**. The first sign is often a polyp, a small, benign growth in the colon lining. The cells of the polyp look normal, although they divide unusually frequently. The tumor grows and may eventually become malignant, invading other tissues. The development of a malignant

▶ **Figure 16.19 Normal and mutant cell cycle–inhibiting pathway. (a)** In the normal pathway, ❶ DNA damage is an intracellular signal that is passed via ❷ protein kinases, leading to activation of ❸ p53. ❹ Activated p53 promotes transcription of the gene for ❺ a protein that inhibits the cell cycle. The resulting suppression of cell division ensures that the damaged DNA is not replicated. If the DNA damage is irreparable, then the p53 signal leads to apoptosis. **(b)** Mutations causing deficiencies in any pathway component can contribute to the development of cancer.

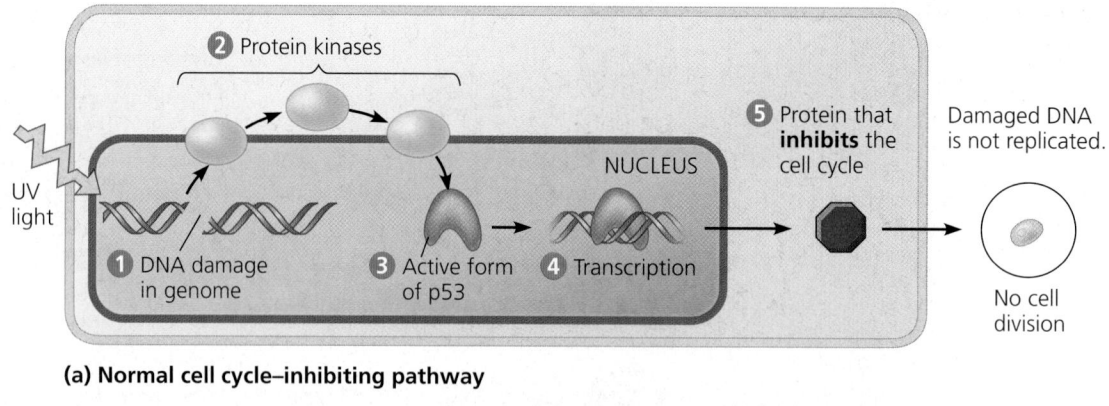

**(a) Normal cell cycle–inhibiting pathway**

**(b) Mutant cell cycle–inhibiting pathway**

tumor is paralleled by a gradual accumulation of mutations that convert proto-oncogenes to oncogenes and knock out tumor-suppressor genes. A *ras* oncogene and a mutated *p53* tumor-suppressor gene are often involved.

About half a dozen changes must occur at the DNA level for a cell to become fully cancerous. These changes usually include the appearance of at least one active oncogene and the mutation or loss of several tumor-suppressor genes. Furthermore, since mutant tumor-suppressor alleles are usually recessive, in most cases mutations must knock out *both* alleles in a cell's genome to block tumor suppression. (Most oncogenes, on the other hand, behave as dominant alleles.)

Since we understand the progression of this type of cancer, routine screenings (colonoscopies, for example) are recommended to identify and remove any suspicious polyps. The colorectal cancer rate has been declining for the past 20 years, due to increased screening and improved treatments. Treatments for other cancers have improved as well. Advances in the sequencing of DNA and mRNA allow medical researchers to compare the genes expressed by different types of tumors and by the same type in different people. These comparisons have led to personalized treatments based on the molecular characteristics of a person's tumor.

Breast cancer is the second most common form of cancer in the United States, and the most common form of cancer among women. Each year, this cancer strikes over 230,000 women (and some men) in the United States and kills 40,000 (450,000 worldwide). A major problem with understanding breast cancer is its heterogeneity: Tumors differ in significant ways. Identifying differences between types of breast cancer is expected to improve treatment and decrease the mortality rate. In 2012, The Cancer Genome Atlas Network, sponsored by the National Institutes of Health, published the results of a genomics approach to profile subtypes of breast cancer based on their molecular signatures. Four major types of breast cancer were identified **(Figure 16.21)**.

## Inherited Predisposition and Other Factors Contributing to Cancer

The fact that multiple genetic changes are required to produce a cancer cell helps explain the observation that cancers can run in families. An individual inheriting an oncogene or a mutant allele of a tumor-suppressor gene is one step closer to accumulating the necessary mutations for cancer to develop than is an individual without any such mutations.

Geneticists are identifying inherited cancer alleles so that predisposition to certain cancers can be detected early in life. About 15% of colorectal cancers, for example, involve inherited mutations. Many affect the tumor-suppressor gene called *adenomatous polyposis coli*, or *APC* (see Figure 16.20). This gene has multiple functions in the cell, including regulation of cell migration and adhesion. Even in patients with no family history of the disease, the *APC* gene is mutated in 60% of colorectal cancers. In these people, new mutations must occur in both *APC* alleles before the gene's function is lost. Since only 15% of colorectal cancers are associated with known inherited mutations, researchers are trying to identify "markers" that could predict the risk of developing this type of cancer.

Given the prevalence and significance of breast cancer, it is not surprising that it was one of the first cancers for which the role of inheritance was investigated. It turns out that for 5–10% of patients with breast cancer, there is evidence of a strong inherited predisposition. Geneticist Mary-Claire King began working on this problem in the mid-1970s. After 16 years of research, she demonstrated that mutations in one gene—*BRCA1*—were associated with increased susceptibility to breast cancer, a finding that flew in the face of medical opinion at the time. (*BRCA* stands for breast cancer.) Mutations in that gene or a gene called *BRCA2* are found in at least half of inherited breast cancers, and tests using DNA sequencing can detect these mutations. A woman who inherits one mutant *BRCA1* allele has a 60% probability of developing breast cancer before the age of 50, compared with only a 2% probability for an individual homozygous for the normal allele.

Both *BRCA1* and *BRCA2* are considered tumor-suppressor genes because their wild-type alleles protect against breast cancer and their mutant alleles are recessive. (Note that mutations in *BRCA1* are commonly found in the genomes of cells from

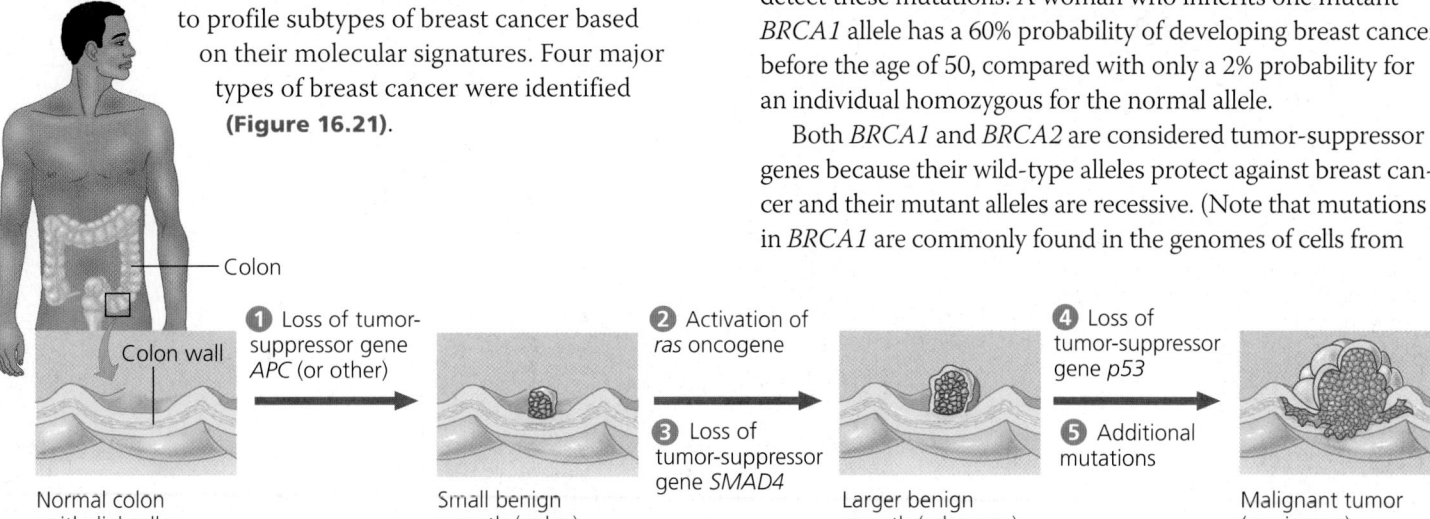

▲ **Figure 16.20 A multistep model for the development of colorectal cancer.** This type of cancer is one of the best understood. Changes in a tumor parallel a series of genetic changes, including mutations affecting several tumor-suppressor genes (such as *p53*) and the *ras* proto-oncogene. Mutations of tumor-suppressor genes often entail loss (deletion) of the gene. *APC* stands for "adenomatous polyposis coli," and *SMAD4* is a gene involved in signaling that results in apoptosis.

# Genomics, Cell Signaling, and Cancer

Modern medicine that melds genome-wide molecular studies with cell-signaling research is transforming the treatment of many diseases, such as breast cancer. Using microarray analysis (see Figure 15.17) and other techniques, researchers measured the relative levels of mRNA transcripts for every gene in hundreds of breast cancer tumor samples. They identified four major subtypes of breast cancer that differ in their expression of three signal receptors involved in regulating cell growth and division:

- Estrogen receptor alpha (ERα)
- Progesterone receptor (PR)
- HER2, a type of receptor called a receptor tyrosine kinase

(ERα and PR are steroid receptors; see Figure 5.23.) The absence or excess expression of these receptors can cause aberrant cell signaling, leading in some cases to inappropriate cell division, which may contribute to cancer (see Figure 16.18).

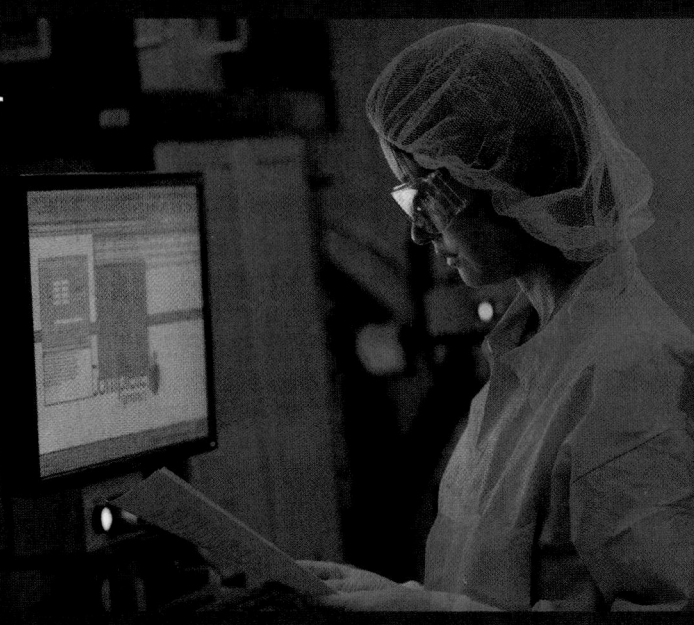

▲ A research scientist examines DNA sequencing data from breast cancer samples.

Milk duct

Mammary gland lobule

Epithelial milk-secreting cell

Support cell

## Normal Breast Cells in a Milk Duct

In a normal breast cell, the three signal receptors are at normal levels (indicated by +):
- ERα+
- PR+
- HER2+

Duct interior

Estrogen receptor alpha (ERα)

Progesterone receptor (PR)

HER2 receptor

Extracellular matrix

# Breast Cancer Subtypes

Each breast cancer subtype is characterized by the overexpression (indicated by ++ or +++) or absence (−) of three signal receptors: ERα, PR, and HER2. Breast cancer treatments are becoming more effective because they can be tailored to the specific cancer subtype.

## Luminal A | Luminal B

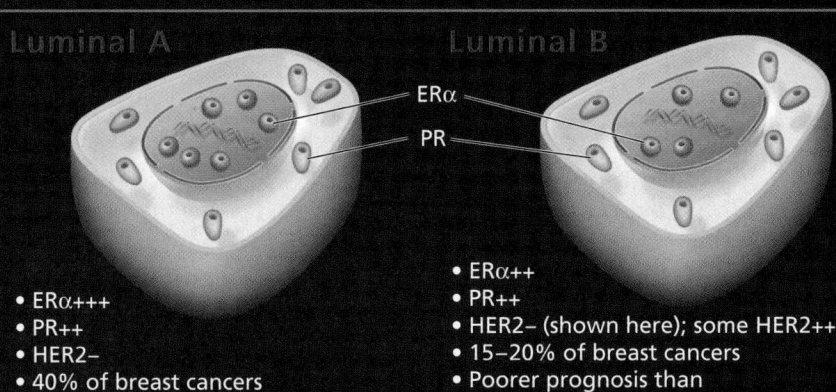
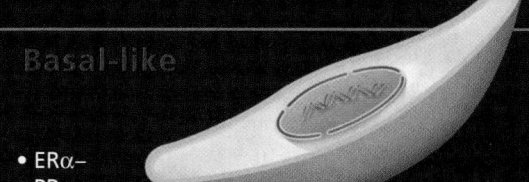

ERα
PR

**Luminal A**
- ERα+++
- PR++
- HER2−
- 40% of breast cancers
- Best prognosis

**Luminal B**
- ERα++
- PR++
- HER2− (shown here); some HER2++
- 15–20% of breast cancers
- Poorer prognosis than luminal A subtype

Both luminal subtypes overexpress ERα (luminal A more than luminal B) and PR, and usually lack expression of HER2. Both can be treated with drugs that target ERα and inactivate it, the most well-known drug being tamoxifen. These subtypes can also be treated with drugs that inhibit estrogen synthesis.

## Basal-like

- ERα−
- PR−
- HER2−
- 15–20% of breast cancers
- More aggressive; poorer prognosis than other subtypes

The basal-like subtype is "triple negative"—it does not express ERα, PR, or HER2. It often has a mutation in the tumor-suppressor gene *BRCA1* (see Concept 16.3). Treatments that target ER, PR, or HER2 are not effective, but new treatments are being developed. Currently, patients are treated with cytotoxic chemo-therapy, which selectively kills fast-growing cells.

## HER2

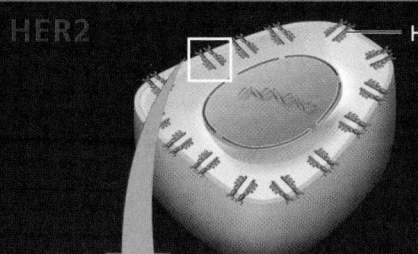

HER2

- ERα−
- PR−
- HER2++
- 10–15% of breast cancers
- Poorer prognosis than luminal A subtype

The HER2 subtype overexpresses HER2. Because it does not express either ERα or PR at normal levels, the cells are unresponsive to therapies that target those two receptors. However, patients with the HER2 subtype can be treated with Herceptin, an antibody protein that inactivates HER2 (see Concept 9.3).

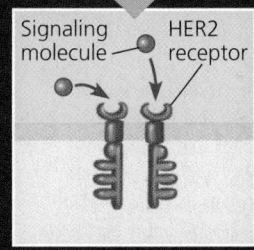
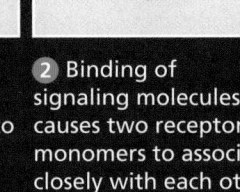

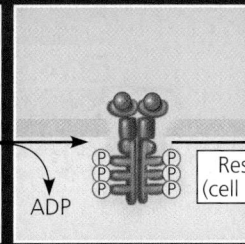

Signaling molecule — HER2 receptor

Dimer

ATP → ADP

Response (cell division)

**1** Signaling molecules (such as a growth factor) bind to HER2 receptor monomers (single receptor proteins).

**2** Binding of signaling molecules causes two receptor monomers to associate closely with each other, forming a dimer.

**3** Formation of a dimer activates each monomer.

**4** Each monomer adds phosphate from ATP to the other monomer, triggering a signal transduction pathway.

**5** The signal is transduced through the cell, which leads to a cellular response—in this case, turning on genes that trigger cell division. HER2 cells have up to 100 times as many HER2 receptors as normal cells, so they undergo uncontrolled cell division.

## Treatment with Herceptin for the HER2 subtype

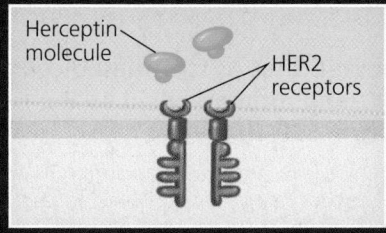
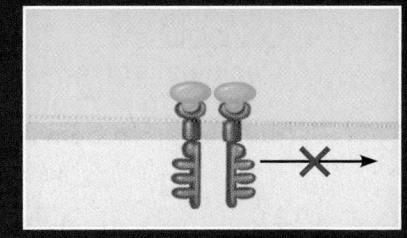

Herceptin molecule

HER2 receptors

**1** The drug Herceptin binds to HER2 receptors in place of the usual signaling molecules, preventing signaling.

**2** In certain patients with the HER2 subtype, signaling is blocked and excessive cell division does not occur.

**MAKE CONNECTIONS** *When researchers compared gene expression in normal breast cells and cells from breast cancers, they found that the genes showing the most significant differences in expression encoded signal receptors, as shown here. Given what you learned in Chapters 5, 9, and this chapter, explain why this result is not surprising.*

basal-like breast cancers; see Figure 16.21.) The BRCA1 and BRCA2 proteins appear to function in the cell's DNA damage repair pathway. More is known about BRCA2. Along with another protein, BRCA2 helps repair breaks that occur in both strands of DNA, crucial for maintaining undamaged DNA.

Because DNA breakage can contribute to cancer, it makes sense that the risk of cancer can be lowered by minimizing exposure to DNA-damaging agents, such as the ultraviolet radiation in sunlight and chemicals found in cigarette smoke. Novel genomics-based analyses of specific cancers, such as the approach described in Figure 16.21, are contributing to both early diagnosis and development of treatments that interfere with expression of key genes in tumors. Ultimately, such approaches are expected to lower the death rate from cancer.

The study of genes associated with cancer, inherited or not, increases our basic understanding of how disruption of normal gene regulation can result in this disease. In addition to the mutations and other genetic alterations described in this section, a number of *tumor viruses* can cause cancer in various animals, including humans. In fact, one of the earliest breakthroughs in understanding cancer came in 1911, when Peyton Rous, an American pathologist, discovered a virus that causes cancer in chickens. The Epstein-Barr virus, which causes infectious mononucleosis, has been linked to several types of cancer in humans, notably Burkitt's lymphoma. Papillomaviruses

are associated with cancer of the cervix, and a virus called HTLV-1 causes a type of adult leukemia. Viruses play a role in about 15% of the cases of human cancer.

Viruses may at first seem very different from mutations as a cause of cancer. However, we now know that viruses can interfere with gene regulation in several ways if they integrate their genetic material into the DNA of a cell. Viral integration may donate an oncogene to the cell, disrupt a tumor-suppressor gene, or convert a proto-oncogene to an oncogene. Some viruses produce proteins that inactivate p53 and other tumor-suppressor proteins, making the cell more prone to becoming cancerous. Viruses are powerful biological agents. You'll learn more about them in Chapter 17.

### CONCEPT CHECK 16.3

1. The p53 protein can activate genes involved in apoptosis. Review Concept 16.1 and discuss how mutations in genes coding for proteins that function in apoptosis could contribute to cancer.
2. Under what circumstances is cancer considered to have a hereditary component?
3. **WHAT IF?** Cancer-promoting mutations are likely to have different effects on the activities of proteins encoded by proto-oncogenes compared with proteins encoded by tumor-suppressor genes. Explain.

   **For suggested answers, see Appendix A.**

Go to **MasteringBiology®** for Assignments, the eText, and the Study Area with Animations, Activities, Vocab Self-Quiz, and Practice Tests.

# 16 Chapter Review

## SUMMARY OF KEY CONCEPTS

**VOCAB SELF-QUIZ**
goo.gl/gbai8v

### CONCEPT 16.1

**A program of differential gene expression leads to the different cell types in a multicellular organism (pp. 322–329)**

- Embryonic cells undergo **differentiation**, becoming specialized in structure and function. **Morphogenesis** encompasses the processes that give shape to the organism and its various structures. Cells differ in structure and function not because they contain different genomes but because they express different genes.
- Localized **cytoplasmic determinants** in the unfertilized egg are distributed differentially to daughter cells, where they regulate the expression of genes that affect those cells' developmental fates. In the process called **induction**, signaling molecules from embryonic cells cause transcriptional changes in nearby target cells.

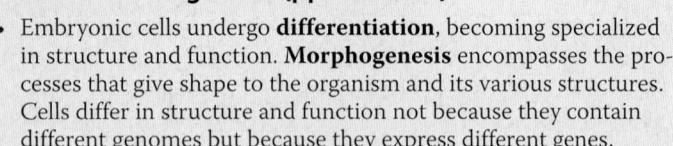

Cytoplasmic determinants          Induction

- Differentiation is heralded by the appearance of tissue-specific proteins, which enable differentiated cells to carry out their roles.

- **Apoptosis** is a type of programmed cell death in which cell components are disposed of in an orderly fashion, without damage to neighboring cells. Studies of the soil worm *Caenorhabditis elegans* showed that apoptosis occurs at defined times during embryonic development. Related apoptotic signaling pathways exist in the cells of humans and other mammals, as well as yeasts.
- In animals, **pattern formation**, the development of a spatial organization of tissues and organs, begins in the early embryo. **Positional information**, the molecular cues that control pattern formation, tells a cell its location relative to the body's axes and to other cells. In *Drosophila melanogaster*, gradients of **morphogens** encoded by **maternal effect genes** determine the body axes. For example, the gradient of **Bicoid** protein determines the anterior-posterior axis.

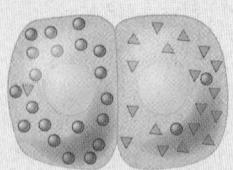 *Describe the two main processes that cause embryonic cells to head down different pathways to their final fates.*

### CONCEPT 16.2

**Cloning of organisms showed that differentiated cells could be "reprogrammed" and ultimately led to the production of stem cells (pp. 330–334)**

- Studies showing genomic equivalence (that an organism's cells all have the same genome) are the first cases of organismal cloning.
- Single differentiated cells from plants are often **totipotent**: capable of generating all the tissues of a complete new plant.
- Transplantation of the nucleus from a differentiated animal cell into an enucleated egg can sometimes give rise to a new animal.

- Certain embryonic **stem cells** (ES cells) from animal embryos or adult stem cells from adult tissues can reproduce and differentiate in a test tube as well as in the organism, offering the potential for medical use. ES cells are **pluripotent** but difficult to acquire. Induced pluripotent stem (iPS) cells resemble ES cells in their capacity to differentiate; they can be generated by reprogramming differentiated cells. iPS cells hold promise for medical research.

> **?** *Describe how a researcher could carry out organismal cloning, production of ES cells, and generation of iPS cells, focusing on how the cells are reprogrammed and using mice as an example. (The procedures are basically the same in humans and mice.)*

## CONCEPT 16.3

### Abnormal regulation of genes that affect the cell cycle can lead to cancer (pp. 334–340)

- The products of **proto-oncogenes** and **tumor-suppressor genes** control cell division. A DNA change that makes a proto-oncogene excessively active converts it to an **oncogene**, which then may promote excessive cell division and cancer. A tumor-suppressor gene encodes a protein that inhibits abnormal cell division. A mutation that reduces activity of its protein product may lead to excessive cell division and cancer.
- Many proto-oncogenes and tumor-suppressor genes encode components of growth-stimulating and growth-inhibiting signaling pathways, respectively, and mutations in these genes can interfere with normal cell-signaling pathways. A hyperactive version of a protein in a stimulatory pathway, such as **Ras** (a G protein), functions as an oncogene protein. A defective version of a protein in an inhibitory pathway, such as **p53** (a transcription activator), fails to function as a tumor suppressor.

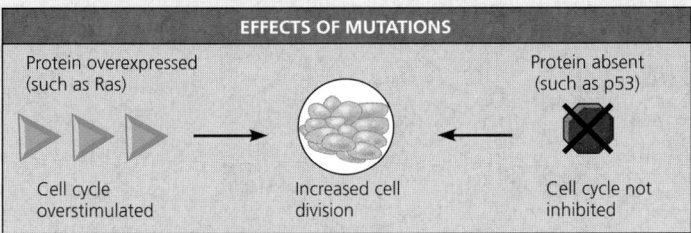

**EFFECTS OF MUTATIONS**

Protein overexpressed (such as Ras) → Increased cell division ← Protein absent (such as p53)

Cell cycle overstimulated | Increased cell division | Cell cycle not inhibited

- In the multistep model of cancer development, normal cells are converted to cancer cells by the accumulation of mutations affecting proto-oncogenes and tumor-suppressor genes. Technical advances in DNA and mRNA sequencing are enabling cancer treatments that are more individually based
- Genomics studies have resulted in four proposed subtypes of breast cancer, based on expression of genes by tumor cells.
- Individuals who inherit a mutant allele of a proto-oncogene or tumor-suppressor gene have a predisposition to develop a particular cancer. Certain viruses promote cancer by integration of viral DNA into a cell's genome.

> **?** *Compare the usual functions of proteins encoded by proto-oncogenes with the functions of proteins encoded by tumor-suppressor genes.*

## TEST YOUR UNDERSTANDING

### Level 1: Knowledge/Comprehension

1. Muscle cells differ from nerve cells mainly because they
   (A) express different genes.
   (B) contain different genes.
   (C) use different genetic codes.
   (D) have unique ribosomes.

**PRACTICE TEST** goo.gl/CRZjvS

2. Cell differentiation always involves
   (A) the transcription of the *myoD* gene.
   (B) the movement of cells.
   (C) the production of tissue-specific proteins.
   (D) the selective loss of certain genes from the genome.

### Level 2: Application/Analysis

3. Apoptosis involves all but which of the following?
   (A) fragmentation of the DNA
   (B) cell-signaling pathways
   (C) lysis of the cell
   (D) digestion of cellular contents by scavenger cells

4. Absence of *bicoid* mRNA from a *Drosophila* egg leads to the absence of anterior larval body parts and mirror-image duplication of posterior parts. This is evidence that the product of the *bicoid* gene
   (A) normally leads to formation of head structures.
   (B) normally leads to formation of tail structures.
   (C) is transcribed in the early embryo.
   (D) is a protein present in all head structures.

5. Proto-oncogenes can change into oncogenes that cause cancer. Which of the following best explains the presence of these potential time bombs in eukaryotic cells?
   (A) Proto-oncogenes first arose from viral infections.
   (B) Proto-oncogenes are mutant versions of normal genes.
   (C) Proto-oncogenes are genetic "junk."
   (D) Proto-oncogenes normally help regulate cell division.

### Level 3: Synthesis/Evaluation

6. **SCIENTIFIC INQUIRY**
   Prostate cells usually require testosterone and other androgens to survive. But some prostate cancer cells thrive despite treatments that eliminate androgens. One hypothesis is that estrogen, often considered a female hormone, may be activating genes normally controlled by an androgen in these cancer cells. Describe one or more experiments to test this hypothesis. (See Figure 5.23 to review the action of these steroid hormones.)

7. **FOCUS ON EVOLUTION**
   Cancer cells can be considered a population that undergoes evolutionary processes such as random mutation and natural selection. Apply what you learned about evolution in Chapter 1 and about cancer in this chapter to discuss this concept.

8. **FOCUS ON ORGANIZATION**
   The property of life emerges at the biological level of the cell. The highly regulated process of apoptosis is not simply the destruction of a cell; it is also an emergent property. In a short essay (about 100–150 words), briefly explain the role of apoptosis in the development and proper functioning of an animal and describe how this form of programmed cell death is a process that emerges from the orderly integration of signaling pathways.

9. **SYNTHESIZE YOUR KNOWLEDGE**

Recently, new anti-aging skin creams have been developed that claim to harness the power of stem cells. One such skin cream contains stem cells from red grape plants, and the manufacturer claims that it will "restore our skin's stem cells." Do you think this cream will fulfill the manufacturer's promise? Explain.

*For selected answers, see Appendix A.*

# 17 Viruses

## KEY CONCEPTS

**17.1** A virus consists of a nucleic acid surrounded by a protein coat

**17.2** Viruses replicate only in host cells

**17.3** Viruses and prions are formidable pathogens in animals and plants

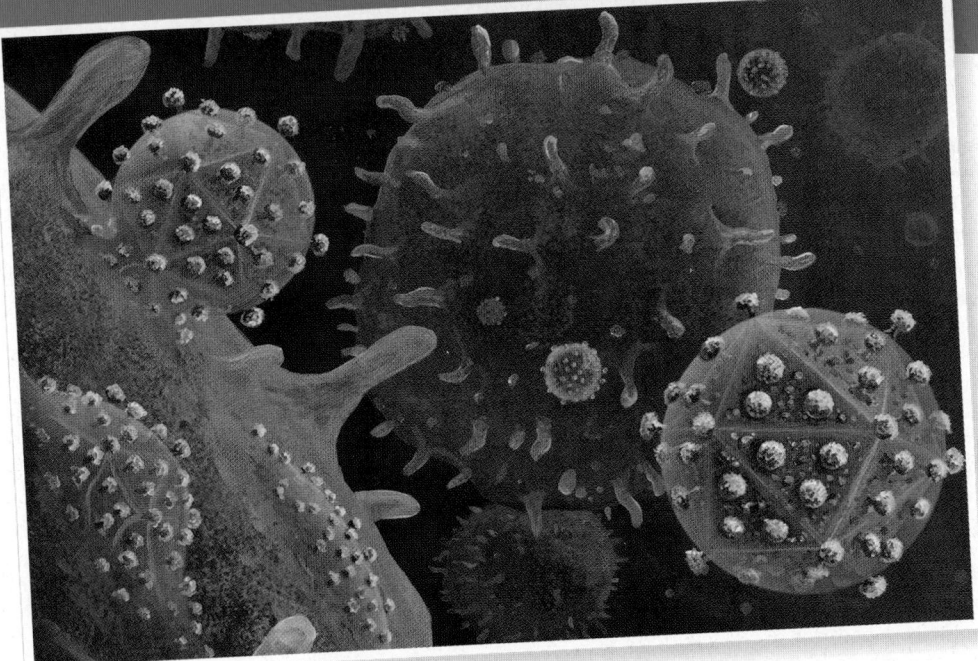

▲ **Figure 17.1** Are the viruses (red) budding from these cells alive?

## A Borrowed Life

The illustration in **Figure 17.1** shows a remarkable event: Human immune cells (purple) infected by human immunodeficiency viruses (HIV) are releasing more HIV viruses. These viruses (red, surrounded by purple membrane from the immune cell) will infect other cells. By injecting its genetic information into a cell, a virus hijacks the cell, recruiting cellular machinery to manufacture many new viruses and promote further infection. Left untreated, HIV causes acquired immunodeficiency syndrome (AIDS) by destroying vital immune system cells.

Compared with eukaryotic and even prokaryotic cells, viruses are much smaller and simpler in structure. Lacking the structures and metabolic machinery found in a cell, a **virus** is an infectious particle consisting of little more than genes packaged in a protein coat.

Are viruses living or nonliving? Because viruses are capable of causing a wide variety of diseases, researchers in the late 1800s saw a parallel with bacteria and proposed that viruses were the simplest of living forms. However, viruses cannot reproduce or carry out metabolic activities outside of a host cell. Most biologists studying viruses today would probably agree that they are not alive but exist in a shady area between life-forms and chemicals. The simple phrase used recently by two researchers describes them aptly enough: Viruses lead "a kind of borrowed life."

In this chapter, we'll explore the biology of viruses, beginning with their structure and then describing how they replicate. We'll next look at the role of viruses as disease-causing agents, or pathogens, of plants and animals. Finally, we'll consider some even simpler infectious agents called prions.

### CONCEPT 17.1

## A virus consists of a nucleic acid surrounded by a protein coat

The tiniest viruses are only 20 nm in diameter—smaller than a ribosome. Millions could easily fit on a pinhead. Even the largest known virus, which has a diameter of several hundred nanometers, is barely visible under the light microscope. An early discovery that some viruses could be crystallized was exciting and puzzling news. Not even the simplest of cells can aggregate into regular crystals. But if viruses are not cells, then what are they? Examining the structure of a virus more closely reveals that it is an infectious particle consisting of nucleic acid enclosed in a protein coat and, for some viruses, surrounded by a membranous envelope.

## Viral Genomes

We usually think of genes as being made of double-stranded DNA, but many viruses defy this convention. Their genomes may consist of double-stranded DNA, single-stranded DNA, double-stranded RNA, or single-stranded RNA, depending on the type of virus. A virus is called a DNA virus or an RNA virus, based on the kind of nucleic acid that makes up its genome. In either case, the genome is usually organized as a single linear or circular molecule of nucleic acid, although the genomes of some viruses consist of multiple molecules of nucleic acid. The smallest viruses known have only three genes in their genome, while the largest have several hundred to a thousand. For comparison, bacterial genomes contain about 200 to a few thousand genes.

## Capsids and Envelopes

The protein shell enclosing the viral genome is called a **capsid**. Depending on the type of virus, the capsid may be rod-shaped, polyhedral, or more complex in shape. Capsids are built from a large number of protein subunits called *capsomeres*, but the number of different *kinds* of proteins in a capsid is usually small. Tobacco mosaic virus (TMV), for example, has a rigid, rod-shaped capsid made from over a thousand molecules of a single type of protein arranged in a helix; rod-shaped viruses are commonly called *helical viruses* for this reason **(Figure 17.2a)**. Adenoviruses, which infect the respiratory tracts of animals, have 252 identical protein molecules arranged in a polyhedral capsid with 20 triangular facets—an icosahedron; thus, these and other similarly shaped viruses are referred to as *icosahedral viruses* **(Figure 17.2b)**.

Some viruses have accessory structures that help them infect their hosts. For instance, a membranous envelope surrounds the capsids of influenza viruses and many other viruses found in animals **(Figure 17.2c)**. These **viral envelopes**, which are derived from the membranes of the host cell, contain host cell phospholipids and membrane proteins. They also contain proteins and glycoproteins of

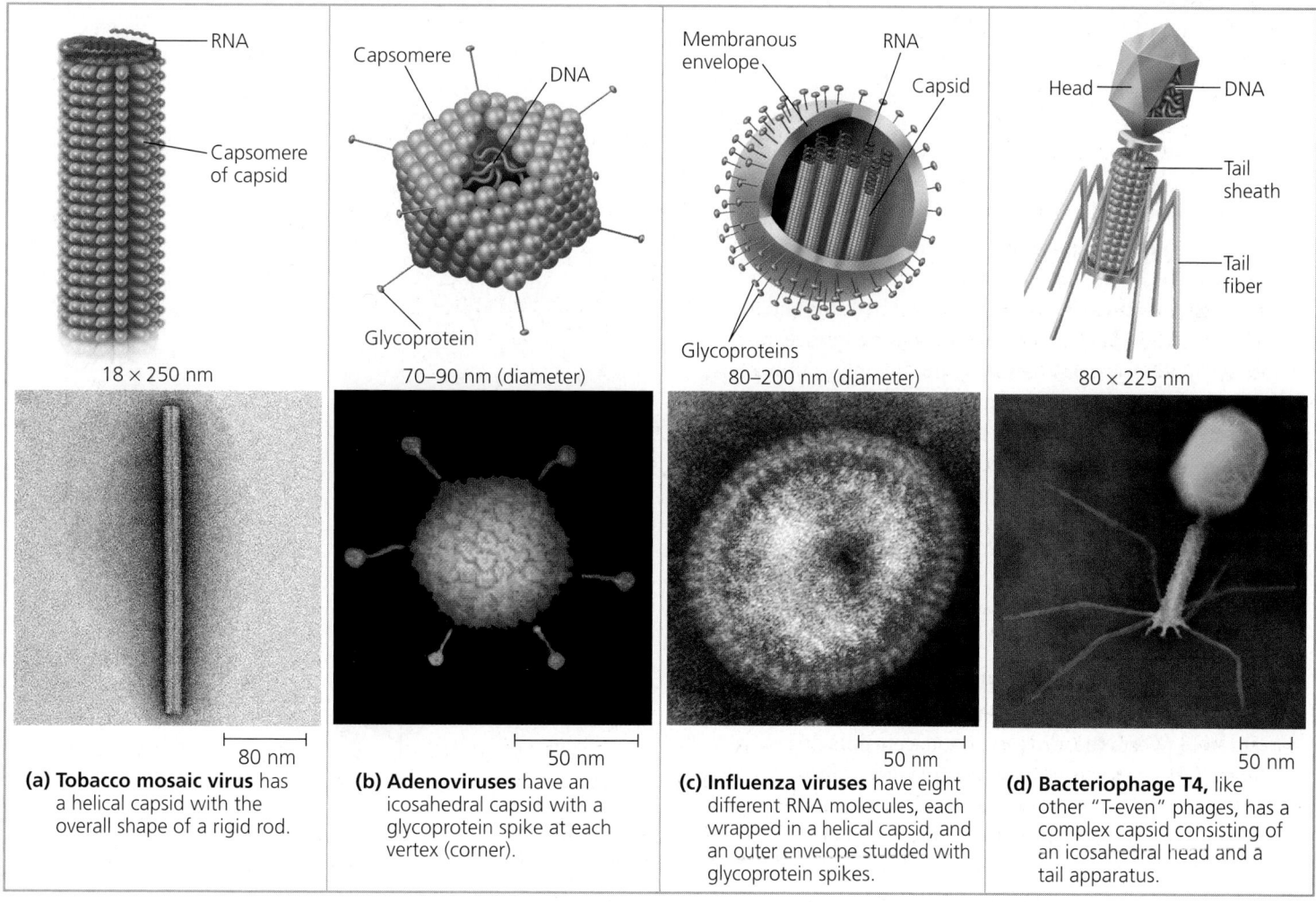

18 × 250 nm

70–90 nm (diameter)

80–200 nm (diameter)

80 × 225 nm

80 nm

50 nm

50 nm

50 nm

**(a) Tobacco mosaic virus** has a helical capsid with the overall shape of a rigid rod.

**(b) Adenoviruses** have an icosahedral capsid with a glycoprotein spike at each vertex (corner).

**(c) Influenza viruses** have eight different RNA molecules, each wrapped in a helical capsid, and an outer envelope studded with glycoprotein spikes.

**(d) Bacteriophage T4,** like other "T-even" phages, has a complex capsid consisting of an icosahedral head and a tail apparatus.

▲ **Figure 17.2 Viral structure.** Viruses are made up of nucleic acid (DNA or RNA) enclosed in a protein coat (the capsid) and sometimes further wrapped in a membranous envelope. The individual protein subunits making up the capsid are called capsomeres. Although diverse in size and shape, viruses have many common structural features. (All micrographs are colorized TEMs.)

viral origin. (Glycoproteins are proteins with carbohydrates covalently attached.) Some viruses carry a few viral enzyme molecules within their capsids.

Many of the most complex capsids are found among the viruses that infect bacteria, called **bacteriophages**, or simply **phages**. The first phages studied included seven that infect *Escherichia coli*. These seven phages were named type 1 (T1), type 2 (T2), and so forth, in the order of their discovery. The three "T-even" phages (T2, T4, and T6) turned out to be very similar in structure. Their capsids have elongated icosahedral heads enclosing their DNA. Attached to the head is a protein tail piece with fibers by which the phages attach to a bacterium **(Figure 17.2d)**. In the next section, we'll examine how these few viral parts function together with cellular components to produce large numbers of viral progeny.

### CONCEPT CHECK 17.1

1. Compare the structures of tobacco mosaic virus and influenza virus (see Figure 17.2).
2. **MAKE CONNECTIONS** Bacteriophages were used to provide evidence that DNA carries genetic information (see Figure 13.5). Briefly describe the experiment carried out by Hershey and Chase, including in your description why the researchers chose to use phages.

For suggested answers, see Appendix A.

---

### CONCEPT 17.2

# Viruses replicate only in host cells

Viruses lack metabolic enzymes and equipment for making proteins, such as ribosomes. They are obligate intracellular parasites; in other words, they can replicate only within a host cell. It is fair to say that viruses in isolation are merely packaged sets of genes in transit from one host cell to another.

Each particular virus can infect cells of only a limited number of host species, called the **host range** of the virus. This host specificity results from the evolution of recognition systems by the virus. Viruses usually identify host cells by a "lock-and-key" fit between viral surface proteins and specific receptor molecules on the outside of cells. According to one model, such receptor molecules originally carried out functions that benefited the host cell but were co-opted later by viruses as portals of entry. Some viruses have broad host ranges. For example, West Nile virus and equine encephalitis virus are distinctly different viruses that can each infect mosquitoes, birds, horses, and humans. Other viruses have host ranges so narrow that they infect only a single species. Measles virus, for instance, can infect only humans. Furthermore, viral infection of multicellular eukaryotes is usually limited to particular tissues. Human cold viruses infect only the cells lining the upper respiratory tract, and the HIV seen in Figure 17.1 binds to receptors present only on certain types of immune cells.

## General Features of Viral Replicative Cycles

A viral infection begins when a virus binds to a host cell and the viral genome makes its way inside **(Figure 17.3)**. The mechanism of genome entry depends on the type of virus and the type of host cell. For example, T-even phages use their elaborate tail apparatus to inject DNA into a bacterium (see Figure 17.2d). Other viruses are taken up by endocytosis or, in the case of enveloped viruses, by fusion of the viral envelope with the host's plasma membrane. Once the viral genome is inside, the proteins it encodes can commandeer the host, reprogramming the cell to copy the viral genome and manufacture viral proteins. The host provides the nucleotides for making viral nucleic acids, as well as enzymes, ribosomes, tRNAs,

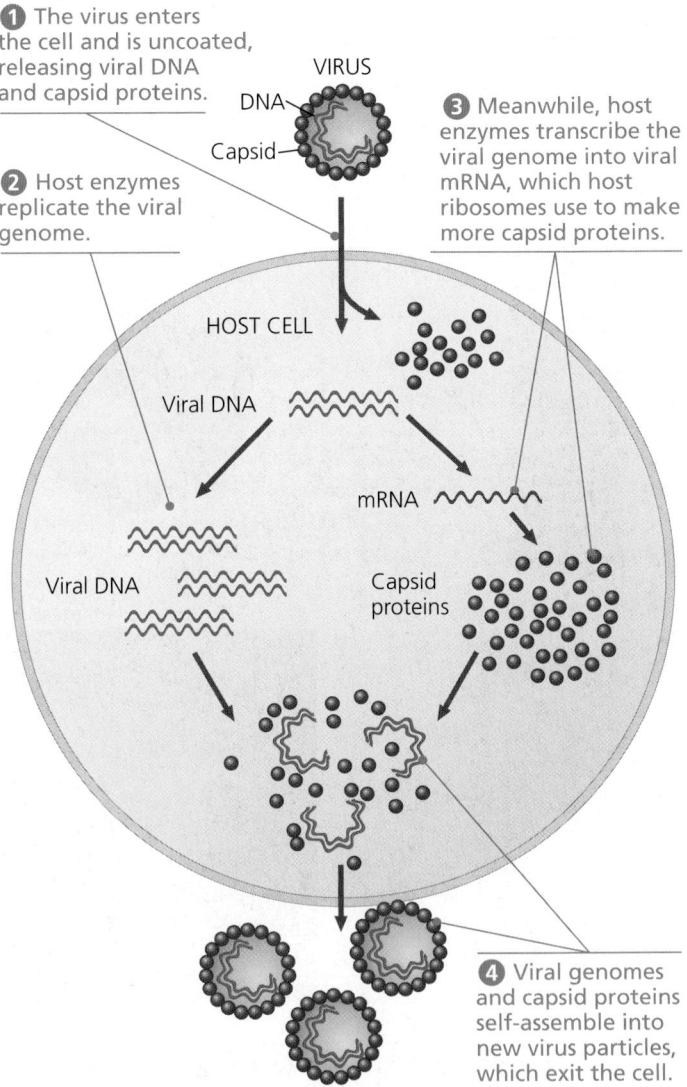

1 The virus enters the cell and is uncoated, releasing viral DNA and capsid proteins.

2 Host enzymes replicate the viral genome.

3 Meanwhile, host enzymes transcribe the viral genome into viral mRNA, which host ribosomes use to make more capsid proteins.

4 Viral genomes and capsid proteins self-assemble into new virus particles, which exit the cell.

VIRUS
DNA
Capsid
HOST CELL
Viral DNA
mRNA
Viral DNA
Capsid proteins

▲ **Figure 17.3 A simplified viral replicative cycle.** A virus is an intracellular parasite that uses the equipment and small molecules of its host cell to replicate. In this simplest of viral cycles, the parasite is a DNA virus with a capsid consisting of a single type of protein.

**MAKE CONNECTIONS** *Label each of the straight gray arrows with one word representing the name of the process that is occurring. (Review Figure 14.24.)*

amino acids, ATP, and other components needed for making the viral proteins. Many DNA viruses use the DNA polymerases of the host cell to synthesize new genomes along the templates provided by the viral DNA. In contrast, to replicate their genomes, RNA viruses use virally encoded RNA polymerases that can use RNA as a template. (Uninfected cells generally make no enzymes for carrying out this process.)

After the viral nucleic acid molecules and capsomeres are produced, they spontaneously self-assemble into new viruses. In fact, researchers can separate the RNA and capsomeres of TMV and then reassemble complete viruses simply by mixing the components together under the right conditions. The simplest type of viral replicative cycle ends with the exit of hundreds or thousands of viruses from the infected host cell, a process that often damages or destroys the cell. Such cellular damage and death, as well as the body's responses to this destruction, cause many of the symptoms associated with viral infections. The viral progeny that exit a cell have the potential to infect additional cells, spreading the viral infection.

There are many variations on the simplified viral replicative cycle we have just described. We'll now take a look at some of these variations in bacterial viruses (phages) and animal viruses; later in the chapter, we'll consider plant viruses.

## Replicative Cycles of Phages

Phages are the best understood of all viruses, although some of them are also among the most complex. Research on phages led to the discovery that some double-stranded DNA viruses can replicate by two alternative mechanisms: the lytic cycle and the lysogenic cycle.

### The Lytic Cycle

A phage replicative cycle that culminates in death of the host cell is known as a **lytic cycle**. The term refers to the last stage of infection, during which the bacterium lyses (breaks open) and releases the phages that were produced within the cell. Each of these phages can then infect a healthy cell, and a few successive lytic cycles can destroy an entire bacterial population in just a few hours. A phage that replicates only by a lytic cycle is called a **virulent phage**. **Figure 17.4** illustrates the major steps in the lytic cycle of T4, a typical virulent phage. Study this figure before proceeding.

### The Lysogenic Cycle

Instead of lysing their host cells, many phages coexist with them in a state called lysogeny. In contrast to the lytic cycle,

▶ **Figure 17.4 The lytic cycle of phage T4, a virulent phage.** Phage T4 has almost 300 genes, which are transcribed and translated using the host cell's machinery. One of the first phage genes translated after the viral DNA enters the host cell codes for an enzyme that degrades the host cell's DNA (step ❷); the phage DNA is protected from breakdown because it contains a modified form of cytosine that is not recognized by the phage enzyme. The entire lytic cycle, from the phage's first contact with the cell surface to cell lysis, takes only 20–30 minutes at 37°C.

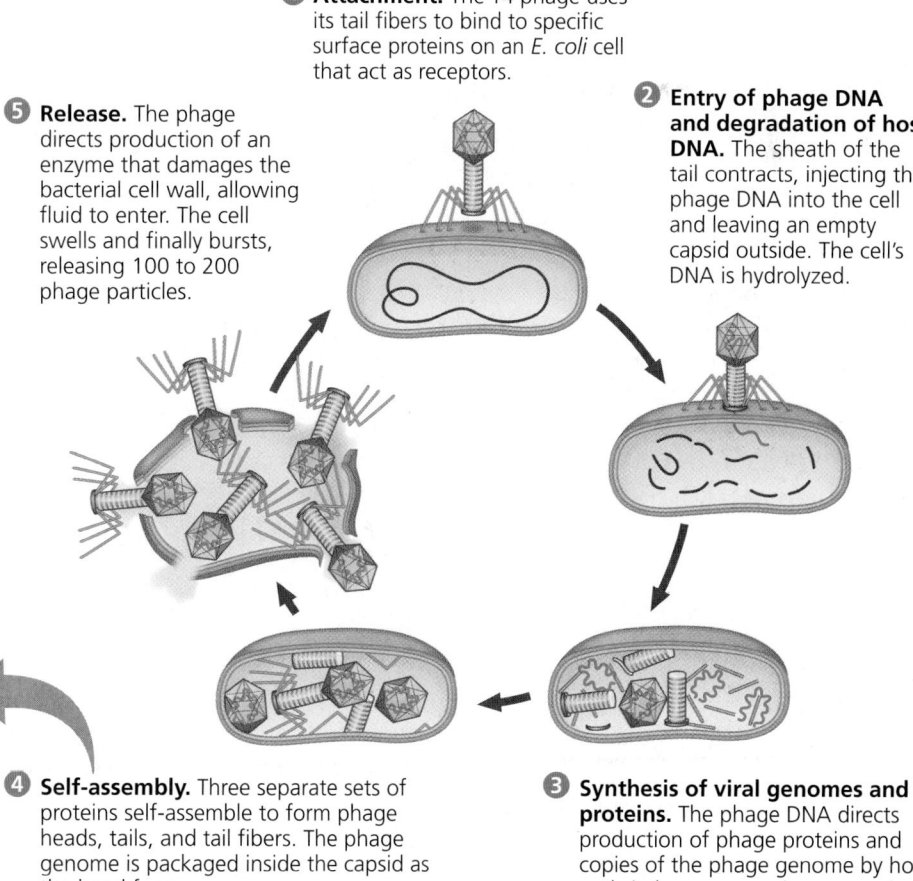

**Phage assembly**

Head    Tail    Tail fibers

❶ **Attachment.** The T4 phage uses its tail fibers to bind to specific surface proteins on an *E. coli* cell that act as receptors.

❷ **Entry of phage DNA and degradation of host DNA.** The sheath of the tail contracts, injecting the phage DNA into the cell and leaving an empty capsid outside. The cell's DNA is hydrolyzed.

❸ **Synthesis of viral genomes and proteins.** The phage DNA directs production of phage proteins and copies of the phage genome by host and viral enzymes, using components within the cell.

❹ **Self-assembly.** Three separate sets of proteins self-assemble to form phage heads, tails, and tail fibers. The phage genome is packaged inside the capsid as the head forms.

❺ **Release.** The phage directs production of an enzyme that damages the bacterial cell wall, allowing fluid to enter. The cell swells and finally bursts, releasing 100 to 200 phage particles.

which kills the host cell, the **lysogenic cycle** allows replication of the phage genome without destroying the host. Phages capable of using both modes of replicating within a bacterium are called **temperate phages**. A temperate phage called lambda, written with the Greek letter λ, has been widely used in biological research. Phage λ resembles T4, but its tail has only one short tail fiber.

Infection of an *E. coli* cell by phage λ begins when the phage binds to the surface of the cell and injects its linear DNA genome **(Figure 17.5)**. Within the host, the λ DNA molecule forms a circle. What happens next depends on the replicative mode: lytic cycle or lysogenic cycle. During a lytic cycle, the viral genes immediately turn the host cell into a λ-producing factory, and the cell soon lyses and releases its virus progeny. During a lysogenic cycle, however, the λ DNA molecule is incorporated into a specific site on the *E. coli* chromosome by viral proteins that break both circular DNA molecules and join them to each other. When integrated into the bacterial chromosome in this way, the viral DNA is known as a **prophage**. One prophage gene codes for a protein that prevents transcription of most of the other prophage genes. Thus, the phage genome is mostly silent within the bacterium. Every time the *E. coli* cell prepares to divide, it replicates the phage DNA along with its own chromosome

such that each daughter cell inherits a prophage. A single infected cell can quickly give rise to a large population of bacteria carrying the virus in prophage form. This mechanism enables viruses to propagate without killing the host cells on which they depend.

The term *lysogenic* signifies that prophages are capable of generating active phages that lyse their host cells. This occurs when the λ genome (or that of another temperate phage) is induced to exit the bacterial chromosome and initiate a lytic cycle. An environmental signal, such as a certain chemical or high-energy radiation, usually triggers the switchover from the lysogenic to the lytic mode.

In addition to the gene for the viral protein that prevents transcription, a few other prophage genes may be expressed during lysogeny. Expression of these genes may alter the host's phenotype, a phenomenon that can have important medical significance. For example, the three species of bacteria that cause the human diseases diphtheria, botulism, and scarlet fever would not be so harmful to humans without certain prophage genes that cause the host bacteria to make toxins. And the difference between the *E. coli* strain in our intestines and the O157:H7 strain that has caused several deaths by food poisoning appears to be the presence of toxin genes of prophages in the O157:H7 strain.

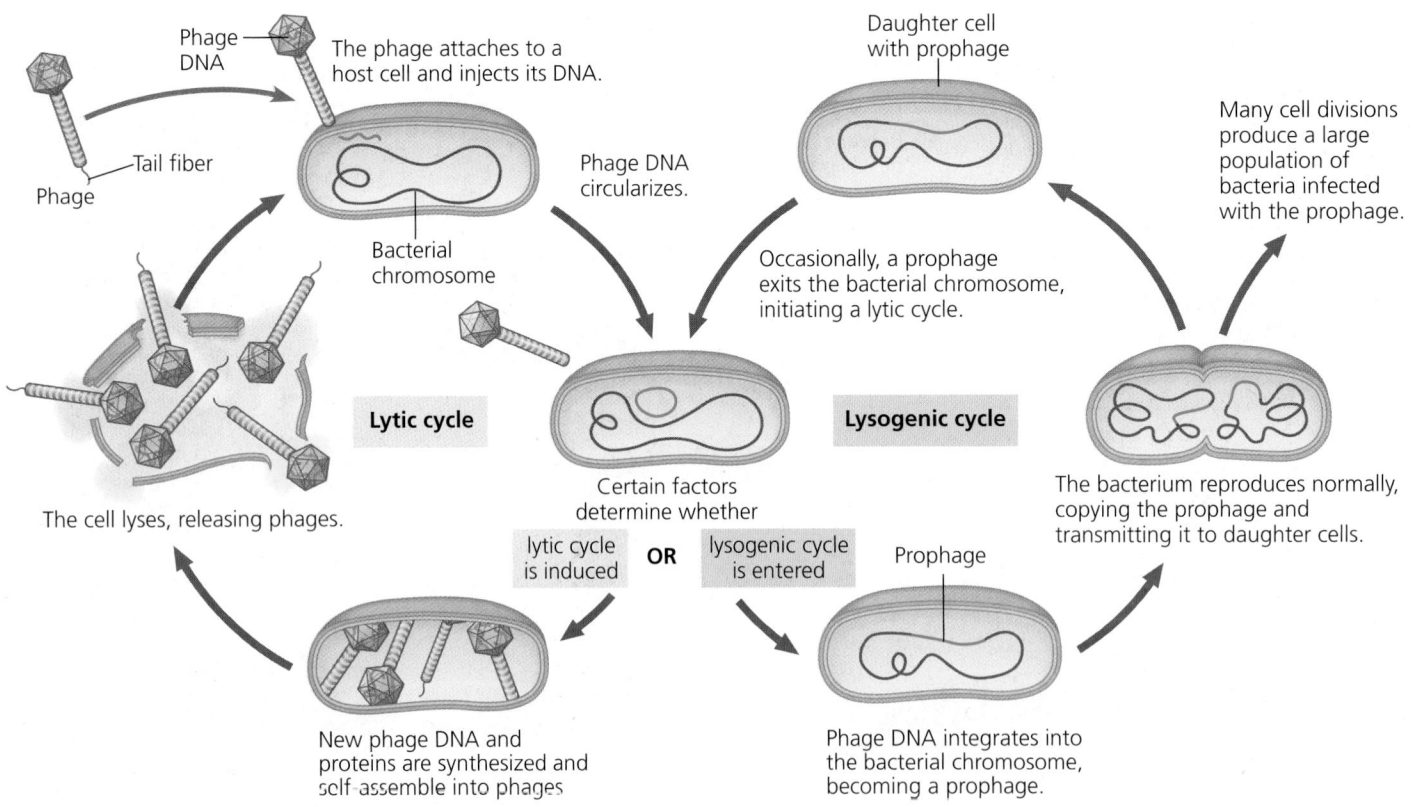

▲ **Figure 17.5 The lytic and lysogenic cycles of phage λ, a temperate phage.** After entering the bacterial cell and circularizing, the λ DNA can immediately initiate the production of a large number of progeny phages (lytic cycle) or integrate into the bacterial chromosome (lysogenic cycle). In most cases, phage λ follows the lytic pathway, which is similar to that detailed in Figure 17.4. However, once a lysogenic cycle begins, the prophage may be carried in the host cell's chromosome for many generations. Phage λ has one main tail fiber, which is short.

## Bacterial Defenses Against Phages

After reading about the lytic cycle, you may have wondered why phages haven't exterminated all bacteria. Lysogeny is one major reason why bacteria have been spared from extinction caused by phages. Bacteria also have their own defenses against phages. First, natural selection favors bacterial mutants with surface proteins that are no longer recognized as receptors by a particular type of phage. Second, when phage DNA does enter a bacterium, the DNA often is identified as foreign and cut up by **restriction enzymes**, so named because they *restrict* a phage's ability to replicate within the bacterium. (Restriction enzymes are used in molecular biology and DNA cloning techniques; see Concept 13.4.) The bacterium's own DNA is methylated in a way that prevents attack by its own restriction enzymes. A third defense is the CRISPR-Cas system, which exists in both bacteria and archaea. In Concept 13.4, you learned how this system is used to alter genes in other cells. Here we'll focus on how it works within bacterial cells.

The CRISPR-Cas system was discovered during a study of repetitive DNA sequences present in the genomes of many prokaryotes. These sequences, which puzzled scientists, were named c̲lustered r̲egularly i̲nterspaced s̲hort p̲alindromic r̲epeats (CRISPRs) because each sequence read the same forward and backward (a palindrome), with different stretches of "spacer DNA" in between the repeats. At first, scientists assumed the spacer DNA sequences were random and meaningless, but analysis by several research groups showed that each spacer sequence corresponded to DNA from a particular phage that had infected the cell. Further studies revealed that particular nuclease proteins interact with the CRISPR region. These nucleases, called Cas (C̲RISPR-a̲ssociated) proteins, can identify and cut phage DNA, thereby defending the bacterium against phage infection.

When a phage infects a bacterial cell that has the CRISPR-Cas system, the DNA of the invading phage is integrated into the genome between two repeat sequences. If the cell survives the infection, any further attempt by the same type of phage to infect this cell (or its offspring) triggers transcription of the CRISPR region into RNA molecules **(Figure 17.6)**. These RNAs are cut into pieces and then bound by Cas proteins, such as the Cas9 protein (see Figure 13.31). A Cas protein uses a portion of the phage-related RNA as a homing device to identify the invading phage DNA and cut it, leading to its destruction.

Just as natural selection favors bacteria that have receptors altered by mutation or that have enzymes that cut phage DNA, it also favors phage mutants that can bind to altered receptors or that are resistant to enzymes. Thus, the bacterium-phage relationship is in constant evolutionary flux.

## Replicative Cycles of Animal Viruses

Everyone has suffered from viral infections, whether cold sores, influenza, or the common cold. Like all viruses, those

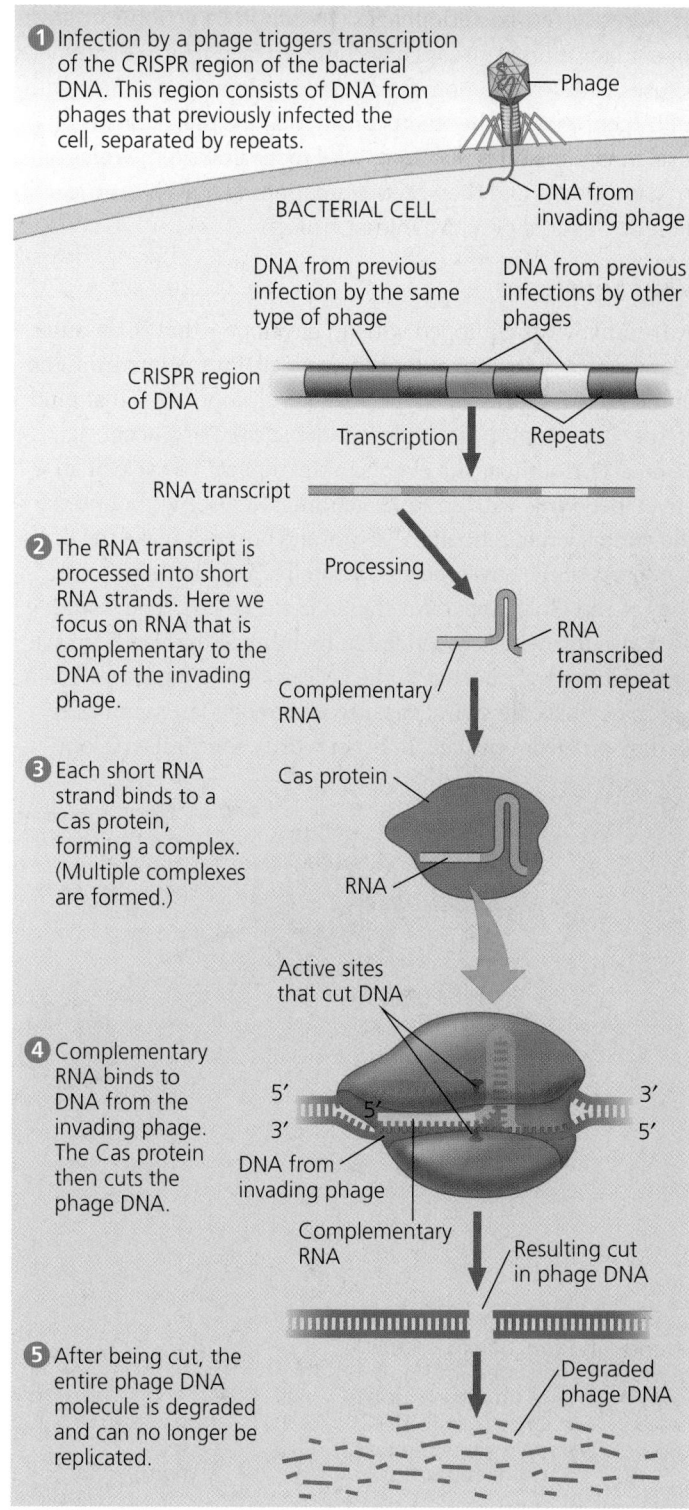

**① Infection** by a phage triggers transcription of the CRISPR region of the bacterial DNA. This region consists of DNA from phages that previously infected the cell, separated by repeats.

Phage

BACTERIAL CELL

DNA from invading phage

DNA from previous infection by the same type of phage

DNA from previous infections by other phages

CRISPR region of DNA

Transcription     Repeats

RNA transcript

**②** The RNA transcript is processed into short RNA strands. Here we focus on RNA that is complementary to the DNA of the invading phage.

Processing

RNA transcribed from repeat

Complementary RNA

**③** Each short RNA strand binds to a Cas protein, forming a complex. (Multiple complexes are formed.)

Cas protein

RNA

Active sites that cut DNA

**④** Complementary RNA binds to DNA from the invading phage. The Cas protein then cuts the phage DNA.

5′     3′

3′     5′     5′

DNA from invading phage

Complementary RNA

Resulting cut in phage DNA

**⑤** After being cut, the entire phage DNA molecule is degraded and can no longer be replicated.

Degraded phage DNA

▲ **Figure 17.6 The CRISPR system: a type of bacterial immune system.**

that cause illness in humans and other animals can replicate only inside host cells. Many variations on the basic scheme of viral infection and replication are represented among the animal viruses. Key variables are the nature of the viral genome (double- or single-stranded DNA or RNA) and the presence or absence of an envelope.

Whereas few bacteriophages have an RNA genome or envelope, many animal viruses have both. In fact, nearly all animal viruses with RNA genomes have an envelope, as do some with DNA genomes. Rather than consider all the mechanisms of viral infection and replication, we'll focus first on the roles of viral envelopes and then on the functioning of RNA as the genetic material of many animal viruses.

## Viral Envelopes

An animal virus equipped with an envelope—that is, an outer membrane—uses it to enter the host cell. Protruding from the outer surface of this envelope are viral glycoproteins that bind to specific receptor molecules on the surface of a host cell. **Figure 17.7** outlines the events in the replicative cycle of an enveloped virus with an RNA genome. Ribosomes bound to the endoplasmic reticulum (ER) of the host cell make the protein parts of the envelope glycoproteins; cellular enzymes in the ER and Golgi apparatus then add the sugars. The resulting viral glycoproteins, embedded in membrane derived from the host cell, are transported to the cell surface. In a process much like exocytosis, new viral capsids are wrapped in membrane as they bud from the cell. In other words, the viral envelope is usually derived from the host cell's plasma membrane, although all or most of the molecules of this membrane are specified by viral genes. The enveloped viruses are now free to infect other cells. This replicative cycle does not necessarily kill the host cell, in contrast to the lytic cycles of phages.

Some viruses have envelopes that are not derived from plasma membrane. Herpesviruses, for example, are temporarily cloaked in membrane derived from the nuclear envelope of the host; they then shed this membrane in the cytoplasm and acquire a new envelope made from membrane of the Golgi apparatus. These viruses have a double-stranded DNA genome and replicate within the host cell nucleus, using a combination of viral and cellular enzymes to replicate and transcribe their DNA. In the case of herpesviruses, copies of the viral DNA can remain behind as mini-chromosomes in the nuclei of certain nerve cells. There they remain latent until some sort of physical or emotional stress triggers a new round of active virus production. The infection of other cells by these new viruses causes the blisters characteristic of herpes, such as cold sores or genital sores. Once someone acquires a herpesvirus infection, flare-ups may recur throughout the person's life.

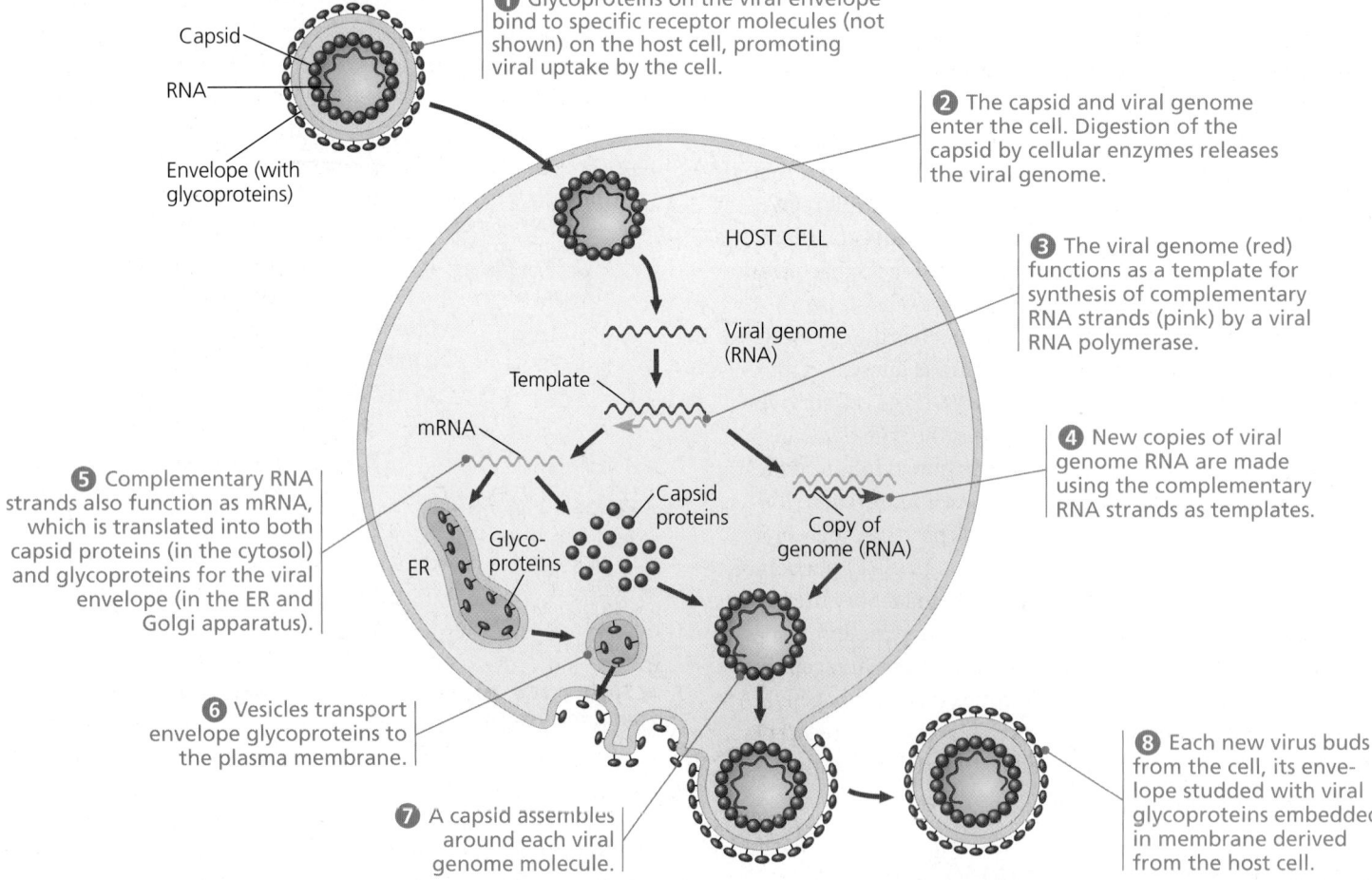

**1** Glycoproteins on the viral envelope bind to specific receptor molecules (not shown) on the host cell, promoting viral uptake by the cell.

**2** The capsid and viral genome enter the cell. Digestion of the capsid by cellular enzymes releases the viral genome.

**3** The viral genome (red) functions as a template for synthesis of complementary RNA strands (pink) by a viral RNA polymerase.

**4** New copies of viral genome RNA are made using the complementary RNA strands as templates.

**5** Complementary RNA strands also function as mRNA, which is translated into both capsid proteins (in the cytosol) and glycoproteins for the viral envelope (in the ER and Golgi apparatus).

**6** Vesicles transport envelope glycoproteins to the plasma membrane.

**7** A capsid assembles around each viral genome molecule.

**8** Each new virus buds from the cell, its envelope studded with viral glycoproteins embedded in membrane derived from the host cell.

Capsid
RNA
Envelope (with glycoproteins)
HOST CELL
Viral genome (RNA)
Template
mRNA
Capsid proteins
Copy of genome (RNA)
Glyco-proteins
ER

▲ **Figure 17.7 The replicative cycle of an enveloped RNA virus.** Shown here is a virus with a single-stranded RNA genome that functions as a template for synthesis of mRNA. Some enveloped viruses enter the host cell by fusion of the envelope with the cell's plasma membrane; others enter by endocytosis. For all enveloped RNA viruses, formation of new envelopes for progeny viruses occurs by the mechanism depicted in this figure.

### RNA as Viral Genetic Material

Although some phages and most plant viruses are RNA viruses, the broadest variety of RNA genomes is found among the viruses that infect animals. There are three types of single-stranded RNA genomes found in animal viruses. In the first type, the viral genome can directly serve as mRNA and thus can be translated into viral protein immediately after infection. In a second type, the RNA genome serves instead as a *template* for mRNA synthesis. The RNA genome is transcribed into complementary RNA strands, which function both as mRNA and as templates for the synthesis of additional copies of genomic RNA. All viruses that use an RNA genome as a template for mRNA transcription require RNA → RNA synthesis. These viruses use a viral enzyme capable of carrying out this process; there are no such enzymes in most cells. The viral enzyme used in this process is packaged during viral self-assembly with the genome inside the viral capsid.

The RNA animal viruses with the most complicated replicative cycles are the third type, the **retroviruses**. These viruses are equipped with an enzyme called **reverse transcriptase**, which transcribes an RNA template into DNA, providing an RNA → DNA information flow, the opposite of the usual direction. (Reverse transcriptase is the enzyme used in the technique called RT-PCR, described in Concept 15.4.) This unusual phenomenon is the source of the name retroviruses (*retro* means "backward"). Of particular medical importance is **HIV (human immunodeficiency virus)**, the retrovirus that causes **AIDS (acquired immunodeficiency syndrome)**. HIV and other retroviruses are enveloped viruses that contain two identical molecules of single-stranded RNA and two molecules of reverse transcriptase.

The HIV replicative cycle (traced in **Figure 17.8**) is typical of a retrovirus. After HIV enters a host cell, its reverse transcriptase molecules are released into the cytoplasm, where they catalyze synthesis of viral DNA. The newly made viral DNA then enters the cell's nucleus and integrates into the DNA of a chromosome. The integrated viral DNA, called a **provirus**, never leaves the host's genome, remaining a permanent resident of the cell. (Recall that a prophage, in contrast, leaves the host's genome at the start of a lytic cycle.) The host's RNA polymerase transcribes the proviral DNA into RNA molecules, which can function both as mRNA for the synthesis of viral proteins and as genomes for the new viruses that will be assembled and released from the cell. In Concept 35.3, we'll describe how HIV causes the deterioration of the immune system that occurs in AIDS.

## Evolution of Viruses

EVOLUTION We began this chapter by asking whether or not viruses are alive. Viruses do not really fit our definition of living organisms. An isolated virus is biologically inert, unable to replicate its genes or regenerate its own ATP. Yet it has a genetic program written in the universal language of life. Do we think of viruses as nature's most complex associations of molecules or as the simplest forms of life? Either way, we must bend our usual definitions. Although viruses cannot replicate or carry out metabolic activities independently, their use of the genetic code makes it hard to deny their evolutionary connection to the living world.

How did viruses originate? Viruses have been found that infect every form of life—not just bacteria, animals, and plants, but also archaea, fungi, and algae and other protists. Because they depend on cells for their own propagation, it seems likely that viruses are not the descendants of precellular forms of life, but evolved—possibly multiple times—*after* the first cells appeared. Most molecular biologists favor the hypothesis that viruses originated from naked bits of cellular nucleic acids that moved from one cell to another, perhaps via injured cell surfaces. The evolution of genes coding for capsid proteins may have allowed viruses to bind cell membranes, thus facilitating the infection of uninjured cells.

Candidates for the original sources of viral genomes include plasmids and transposons. *Plasmids* are small, circular DNA molecules found in bacteria and in the unicellular eukaryotes called yeasts. Plasmids exist apart from and can replicate independently of the genome and are occasionally transferred between cells. (Use of plasmids in gene cloning was discussed in Concept 13.4.) *Transposons* are DNA segments that can move from one location to another within a cell's genome. Thus, plasmids, transposons, and viruses all share an important feature: They are *mobile genetic elements*. (We'll discuss plasmids in Concept 24.3 and transposons in Concept 18.4.) Consistent with this notion of pieces of DNA shuttling from cell to cell is the observation that a viral genome can have more in common with the genome of its host than with the genomes of viruses that infect other hosts.

Debate about the origin of viruses was revived in 2003 when an extremely large virus was discovered: Mimivirus is a double-stranded DNA (dsDNA) virus with an icosahedral capsid that is 400 nm in diameter, the size of a small bacterium. Its genome has 1.2 million bases (Mb)—approximately 100 times as many as influenza virus—and an estimated 1,000 genes. Surprisingly, its genome included genes previously found only in cellular genomes. In 2013 an even larger virus was discovered, unrelated to any known virus. This virus is 1 μm (1,000 nm) in diameter, with a dsDNA genome of around 2–2.5 Mb, larger than that of some small eukaryotes. What's more, over 90% of its 2,000 or so genes are not related to cellular genes, so it was fittingly named pandoravirus. How these and all other viruses fit in the tree of life is an intriguing question.

The ongoing evolutionary relationship between viruses and the genomes of their host cells is an association that makes viruses very useful experimental systems in molecular biology. Knowledge about viruses also allows many practical applications, since viruses have a tremendous impact on all organisms through their ability to cause disease.

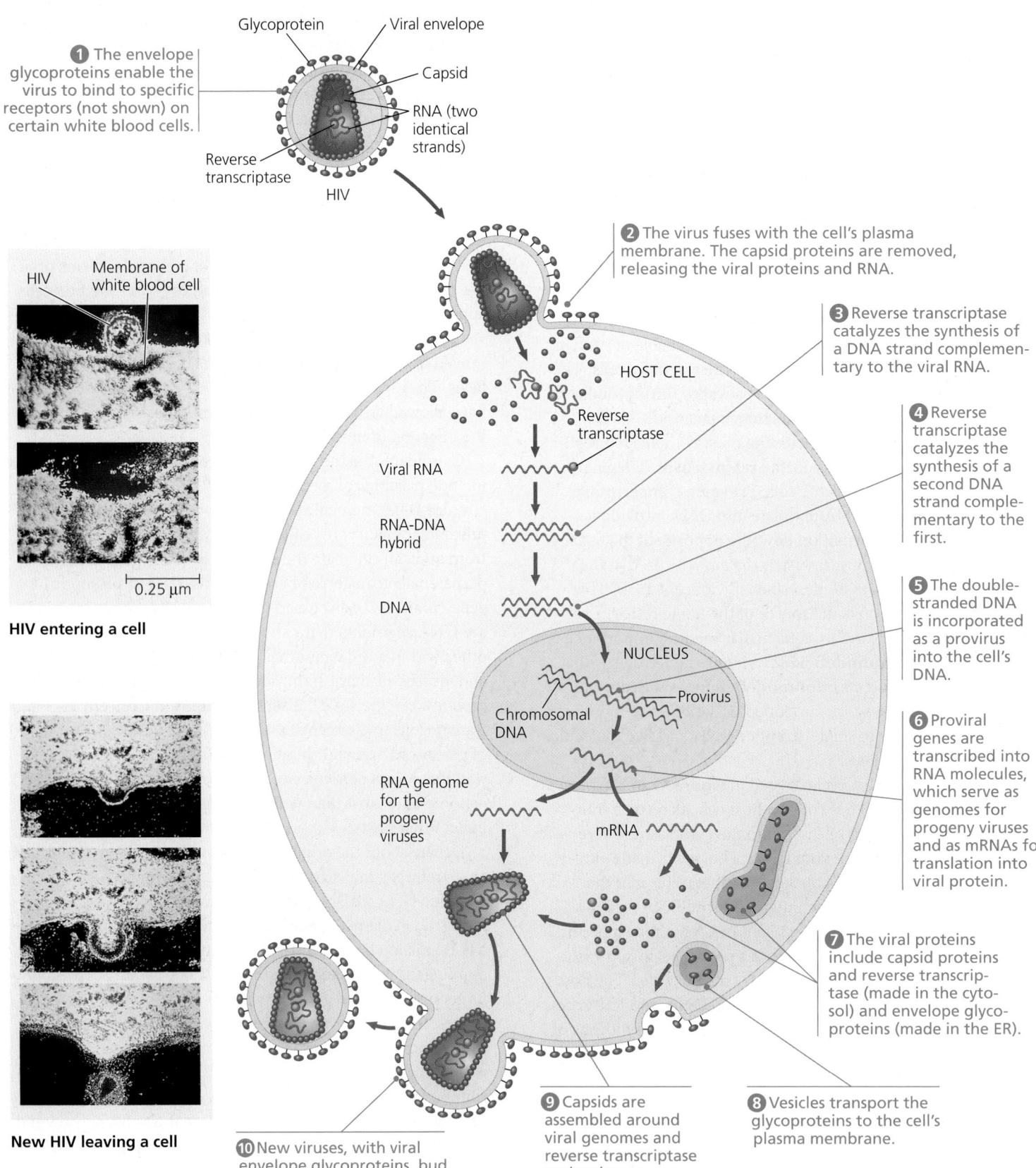

**①** The envelope glycoproteins enable the virus to bind to specific receptors (not shown) on certain white blood cells.

Glycoprotein

Viral envelope

Capsid

RNA (two identical strands)

Reverse transcriptase

HIV

HIV

Membrane of white blood cell

0.25 µm

**HIV entering a cell**

**New HIV leaving a cell**

**②** The virus fuses with the cell's plasma membrane. The capsid proteins are removed, releasing the viral proteins and RNA.

**③** Reverse transcriptase catalyzes the synthesis of a DNA strand complementary to the viral RNA.

**④** Reverse transcriptase catalyzes the synthesis of a second DNA strand complementary to the first.

**⑤** The double-stranded DNA is incorporated as a provirus into the cell's DNA.

**⑥** Proviral genes are transcribed into RNA molecules, which serve as genomes for progeny viruses and as mRNAs for translation into viral protein.

**⑦** The viral proteins include capsid proteins and reverse transcriptase (made in the cytosol) and envelope glycoproteins (made in the ER).

**⑧** Vesicles transport the glycoproteins to the cell's plasma membrane.

**⑨** Capsids are assembled around viral genomes and reverse transcriptase molecules.

**⑩** New viruses, with viral envelope glycoproteins, bud from the host cell.

HOST CELL

Reverse transcriptase

Viral RNA

RNA-DNA hybrid

DNA

NUCLEUS

Provirus

Chromosomal DNA

RNA genome for the progeny viruses

mRNA

▲ **Figure 17.8 The replicative cycle of HIV, the retrovirus that causes AIDS.** Note in step ⑤ that DNA synthesized from the viral RNA genome is integrated as a provirus into the host cell chromosomal DNA, a characteristic unique to retroviruses. For simplicity, the cell-surface proteins that act as receptors for HIV are not shown (see Figure 17.3). The photos on the left (artificially colored TEMs) show HIV entering and leaving a human white blood cell.

**WHAT IF?** *If you were a researcher trying to combat HIV infection, what molecular processes could you attempt to block?*

1. Compare the effect on the host cell of a lytic (virulent) phage and a lysogenic (temperate) phage.
2. **MAKE CONNECTIONS** The RNA virus in Figure 17.7 has a viral RNA polymerase that functions in step 3 of the virus's replicative cycle. Compare this with a cellular RNA polymerase in terms of template and overall function (see Figure 14.10).
3. Why is HIV called a retrovirus?
4. **MAKE CONNECTIONS** Compare the CRISPR system to the miRNAs discussed in Concept 15.3, including their mechanisms and their functions.

For suggested answers, see Appendix A.

## CONCEPT 17.3

# Viruses and prions are formidable pathogens in animals and plants

Diseases caused by viral infections afflict humans, crops, and livestock worldwide. Other smaller, less complex entities known as prions also cause disease in animals. We'll first discuss animal viruses.

## Viral Diseases in Animals

A viral infection can produce symptoms by a number of different routes. Viruses may damage or kill cells by causing the release of hydrolytic enzymes from lysosomes. Some viruses cause infected cells to produce toxins that lead to disease symptoms, and some have molecular components that are toxic, such as envelope proteins. How much damage a virus causes depends partly on the ability of the infected tissue to regenerate by cell division. People usually recover completely from colds because the epithelium of the respiratory tract, which the viruses infect, can efficiently repair itself. In contrast, damage inflicted by poliovirus to mature nerve cells is permanent because these cells do not divide and usually cannot be replaced. Many of the temporary symptoms associated with viral infections, such as fever and body aches, actually result from the body's own efforts to defend itself against infection rather than from cell death caused by the virus.

The immune system is a complex and critical part of the body's natural defenses (see Chapter 35). It is also the basis for the major medical tool for preventing viral infections—vaccines. A **vaccine** is a harmless variant or derivative of a pathogen that stimulates the immune system to mount defenses against the harmful pathogen. Smallpox, a viral disease that was at one time a devastating scourge in many parts of the world, was eradicated by a vaccination program carried out by the World Health Organization (WHO). The very narrow host range of the smallpox virus—it infects only humans—was a critical factor in the success of this program. Similar worldwide vaccination campaigns are currently under way to eradicate polio and measles. Effective vaccines are also available to protect against rubella, mumps, hepatitis A and B, and a number of other viral diseases.

Although vaccines can prevent certain viral illnesses, medical technology can do little, at present, to cure most viral infections once they occur. The antibiotics that help us recover from bacterial infections are powerless against viruses. Antibiotics kill bacteria by inhibiting enzymes specific to bacteria but have no effect on eukaryotic or virally encoded enzymes. However, the few enzymes that are encoded only by viruses have provided targets for other drugs. Most antiviral drugs resemble nucleosides and as a result interfere with viral nucleic acid synthesis. One such drug is acyclovir, which impedes herpesvirus replication by inhibiting the viral polymerase that synthesizes viral DNA but not the eukaryotic one. Similarly, azidothymidine (AZT) curbs HIV replication by interfering with the synthesis of DNA by reverse transcriptase. In the past two decades, much effort has gone into developing drugs to treat HIV. Currently, multidrug treatments, sometimes called "cocktails," have been found to be most effective. Such treatments commonly include a combination of two nucleoside mimics and a protease inhibitor, which interferes with an enzyme required for assembly of the viruses.

Receptor proteins on the surface of cells are important factors to consider in the development of ways to prevent or treat viral infection. For example, medical researchers knew that a protein called CD4 helped HIV infect immune cells. They also noticed, however, that despite multiple exposures to HIV, a small number of people failed to develop AIDS and showed no evidence of harboring HIV-infected cells. Comparing their genes with the genes of infected individuals, the researchers learned that resistant people have an unusual form of a gene that codes for an immune cell-surface protein called CCR5. Further work showed that although CD4 is the main HIV receptor, HIV must also bind to CCR5 as a "co-receptor" to infect most cells **(Figure 17.9a)**. An absence of CCR5 on the cells of resistant individuals, due to the gene alteration, prevents the virus from entering the cells **(Figure 17.9b)**.

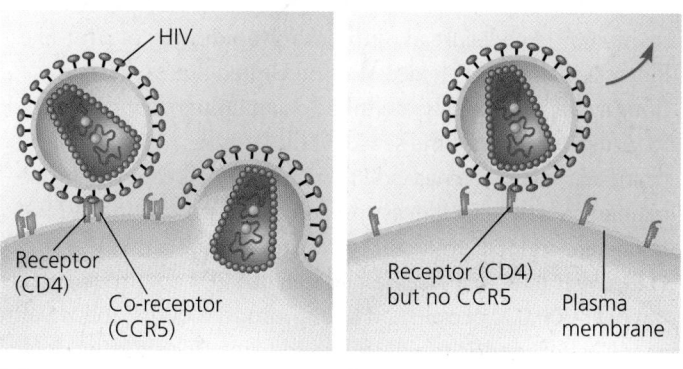

**(a)** **(b)**

▲ **Figure 17.9 HIV resistance due to differences in cell-surface proteins. (a)** HIV can infect a cell with CCR5 on its surface, as in most people. **(b)** HIV cannot infect a cell lacking CCR5 on its surface, as in resistant individuals.

**MAKE CONNECTIONS** *Study Figures 2.14 and 3.21, both of which show pairs of molecules binding to each other. What would you predict about CCR5 that would allow HIV to bind to it? How could a drug molecule interfere with this binding?*

This information has been key to developing a treatment for HIV infection. Interfering with CD4 could cause dangerous side effects because CD4 performs many vital functions in the cell (and was probably co-opted later for use as an HIV entry gate). Discovery of the CCR5 co-receptor provided a safer target for development of drugs that mask this protein and block HIV entry. One such drug, maraviroc (brand name Selzentry), was approved for treatment of HIV in 2007 and is still being used today. A clinical trial began in 2012 to test whether this drug might also work to prevent HIV infection in uninfected, at-risk patients.

## Emerging Viruses

Viruses that suddenly become apparent are often referred to as *emerging viruses*. HIV, the AIDS virus, is a classic example: This virus appeared in San Francisco in the early 1980s, seemingly out of nowhere, although later studies uncovered a case in the Belgian Congo in 1959. The deadly Ebola virus **(Figure 17.10)**, recognized initially in 1976 in central Africa, is one of several emerging viruses that cause *hemorrhagic fever*, an often fatal syndrome (set of symptoms) characterized by fever, vomiting, massive bleeding, and circulatory system collapse. In 2014, a widespread outbreak of Ebola virus caused the World Health Organization to declare an

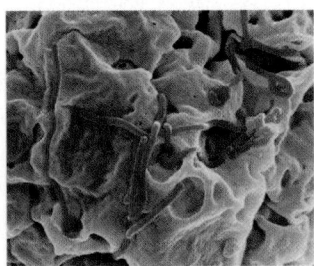

▲ Figure 17.10
**Ebola viruses infecting a monkey cell.**

international health emergency. A number of other dangerous emerging viruses cause encephalitis, inflammation of the brain. One example is the West Nile virus, which appeared in North America for the first time in 1999 and has now spread to all 48 contiguous states in the United States. By 2013, there had been about 40,000 cases total, with over 1,600 deaths.

In 2009, a widespread outbreak, or **epidemic**, of a flu-like illness appeared in Mexico and the United States. The infectious agent was quickly identified as an influenza virus related to viruses that cause the seasonal flu **(Figure 17.11a)**. This particular virus was named H1N1 for reasons that will be explained shortly. The illness spread rapidly, prompting WHO to declare a global epidemic, or **pandemic**, shortly thereafter. Half a year later, the disease had reached 207 countries, infecting over 600,000 people and killing almost 8,000. Public health agencies responded rapidly with guidelines for shutting down schools and other public places, and vaccine development and screening efforts were accelerated **(Figure 17.11b)**.

How do such viruses burst on the human scene, giving rise to harmful diseases that were previously rare or even unknown? Three processes contribute to the emergence of viral diseases. The first, and perhaps most important, is the mutation of existing viruses. RNA viruses tend to have an unusually high rate of mutation because viral RNA polymerases

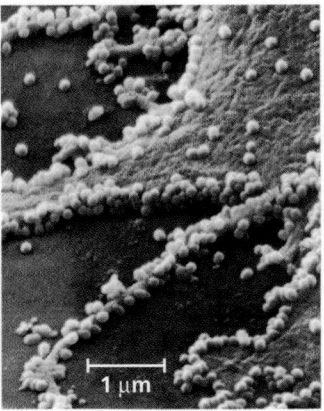
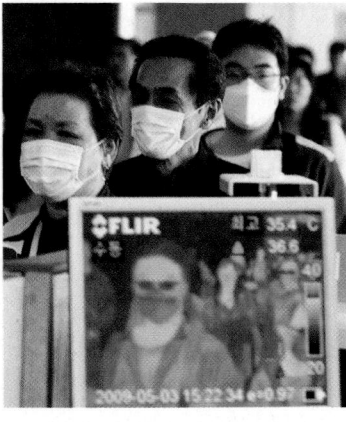

**(a) 2009 pandemic H1N1 influenza A virus.** Viruses (blue) are seen on an infected cell (green) in this colorized SEM.

**(b) 2009 pandemic screening.** At a South Korean airport, thermal scans were used to detect passengers with a fever who might have the H1N1 flu.

▲ **Figure 17.11 Influenza in humans.**

do not proofread and correct errors in replicating their RNA genomes. Some mutations change existing viruses into new genetic varieties (strains) that can cause disease, even in individuals who are immune to the ancestral virus. For instance, seasonal flu epidemics are caused by new strains of influenza virus genetically different enough from earlier strains that people have little immunity to them. You'll see an example of this process in the **Scientific Skills Exercise**, where you'll analyze genetic changes in variants of the H1N1 flu virus and correlate them with spread of the disease.

A second process that can lead to the emergence of viral diseases is the dissemination of a viral disease from a small, isolated human population. For instance, AIDS went unnamed and virtually unnoticed for decades before it began to spread around the world. In this case, technological and social factors, including affordable international travel, blood transfusions, sexual promiscuity, and the abuse of intravenous drugs, allowed a previously rare human disease to become a global scourge.

A third source of new viral diseases in humans is the spread of existing viruses from other animals. Scientists estimate that about three-quarters of new human diseases originate in this way. Animals that harbor and can transmit a particular virus but are generally unaffected by it are said to act as a natural reservoir for that virus. For example, the H1N1 virus that caused the 2009 flu pandemic mentioned earlier was likely passed to humans from pigs; for this reason, the disease it caused was originally called "swine flu."

In general, flu epidemics provide an instructive example of the effects of viruses moving between species. There are three types of influenza virus: types B and C, which infect only humans and have never caused an epidemic, and type A, which infects a wide range of animals, including birds, pigs, horses, and humans. Influenza A strains have caused four major flu epidemics among humans in the last 100 years. The worst was the first one, the "Spanish flu" pandemic of 1918–1919, which killed 40–50 million people, including many World War I soldiers.

# Analyzing a Sequence-Based Phylogenetic Tree to Understand Viral Evolution

▶ H1N1 flu vaccination

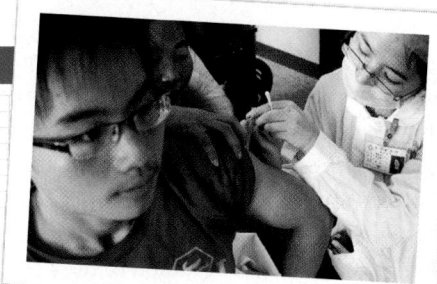

## How Can Sequence Data Be Used to Track Flu Virus Evolution?

In 2009, an influenza A H1N1 virus caused a pandemic, and the virus has continued to resurface in outbreaks across the world. Researchers in Taiwan were curious about why the virus kept appearing despite widespread flu vaccine initiatives. They hypothesized that newly evolved variant strains of the H1N1 virus were able to evade human immune system defenses. To test this hypothesis, they needed to determine if each wave of flu infection was caused by a different H1N1 variant strain.

**How the Experiment Was Done** Scientists obtained the genome sequences for 4,703 virus isolates collected from patients with H1N1 flu in Taiwan. They compared the sequences in different strains for the viral hemagglutinin (HA) gene, and based on mutations that had occurred, arranged the isolates into a phylogenetic tree (see Figure 20.5 for information on how to read phylogenetic trees).

**Data from the Experiment** In the phylogenetic tree, each branch tip is one variant strain of the H1N1 virus with a unique HA gene sequence. The tree is a way to visualize a working hypothesis about the evolutionary relationships between H1N1 variants.

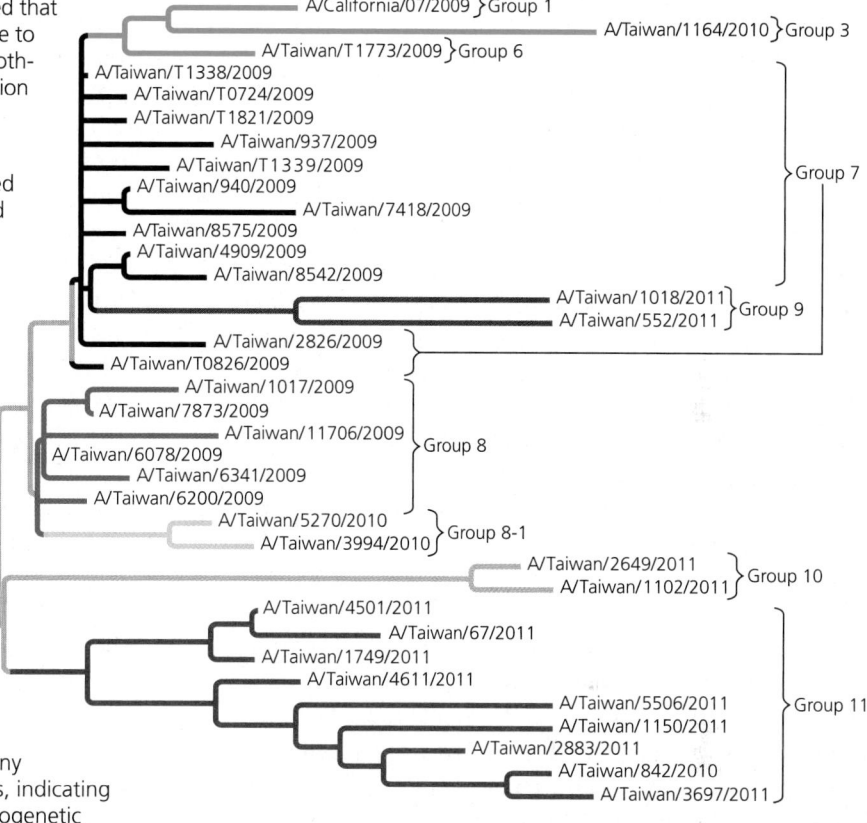

### INTERPRET THE DATA

1. The phylogenetic tree shows the hypothesized evolutionary relationship between the variant strains of H1N1 virus. The more closely connected two variants are, the more alike they are in terms of HA gene sequence. Each fork in a branch, called a node, shows where two lineages separate due to different accumulated mutations. The length of the branches is a measure of how many sequence differences there are between the variants, indicating how distantly related they are. Referring to the phylogenetic tree, which variants are more closely related to each other: A/Taiwan/1018/2011 and A/Taiwan/552/2011 or A/Taiwan/1018/2011 and A/Taiwan/8542/2009? Explain your answer.

2. The scientists arranged the branches into groups made up of one ancestral variant and all of its descendant, mutated variants. They are color-coded in the figure. Using Group 11 as an example, trace the lineage of its variants. (a) Do all of the nodes have the same number of branches or branch tips? (b) Are all of the branches in the group the same length? (c) What do these results indicate?

3. The graph shows the number of isolates collected (each from an ill patient) on the *y*-axis and the month and year that the isolates were collected on the *x*-axis. Each group of variants is plotted separately with a line color that matches the tree diagram. (a) Which group of variants was the earliest to cause the first wave of H1N1 flu in over 100 patients in Taiwan? (b) After a group of variants had a peak number of infections, did members of that same group cause another (later) wave of infection? (c) One variant in Group 1 (green, uppermost branch) was used to make a vaccine that was distributed very early in the pandemic. Based on the graphed data, does it look like the vaccine was effective?

4. Groups 9, 10, and 11 all had H1N1 variants that caused a large number of infections at the same time in Taiwan. Does this mean that the scientists' hypothesis, that new variants cause new waves of infection, was incorrect? Explain your answer.

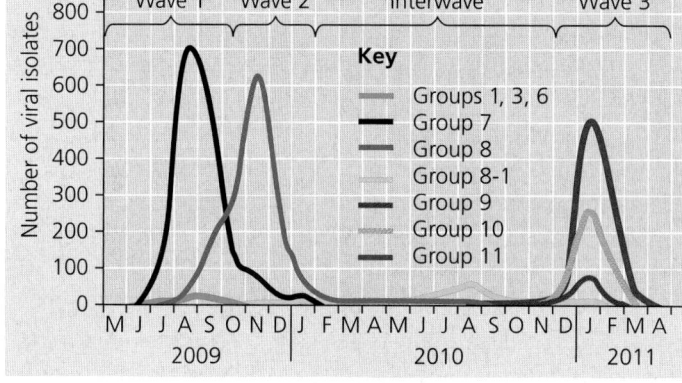

▲ Scientists graphed the number of isolates by the month and year of isolate collection to show the period in which each viral variant was actively causing illness in people.

**Data from** J.-R. Yang et al , New variants and age shift to high fatality groups contribute to severe successive waves in the 2009 influenza pandemic in Taiwan, *PLoS ONE* 6(11):e28288 (2011).

(MB) A version of this Scientific Skills Exercise can be assigned in MasteringBiology.

Different strains of influenza A are given standardized names; for example, both the strain that caused the 1918 flu and the one that caused the 2009 pandemic flu are called H1N1. The name identifies which forms of two viral surface proteins are present: hemagglutinin (H) and neuraminidase (N). There are 16 different types of hemagglutinin, a protein that helps the flu virus attach to host cells, and 9 types of neuraminidase, an enzyme that helps release new virus particles from infected cells. Waterbirds have been found that carry viruses with all possible combinations of H and N.

A likely scenario for the 1918 pandemic and others is that the virus mutated as it passed from one host species to another. When an animal like a pig or a bird is infected with more than one strain of flu virus, the different strains can undergo genetic recombination if the RNA molecules making up their genomes mix and match during viral assembly. Pigs were probably the main hosts for recombination that led to the 2009 flu virus, which turned out to contain sequences from bird, pig, and human flu viruses. Coupled with mutation, these reassortments can lead to the emergence of a viral strain capable of infecting human cells. People who have never been exposed to that particular strain before will lack immunity, and the recombinant virus has the potential to be highly pathogenic. If such a flu virus recombines with viruses that circulate widely among humans, it may acquire the ability to spread easily from person to person, dramatically increasing the potential for a major human outbreak.

One potential long-term threat is the avian flu caused by an H5N1 virus carried by wild and domestic birds. The first documented transmission from birds to humans occurred in Hong Kong in 1997. Since then, the overall mortality rate due to H5N1 has been greater than 50% of those infected—an alarming number. Also, the host range of H5N1 is expanding, which provides increasing chances for reassortment between different strains. If the H5N1 avian flu virus evolves so that it can spread easily from person to person, it could represent a major global health threat akin to that of the 1918 pandemic.

How easily could this happen? Recently, scientists working with ferrets, small mammals that are animal models for human flu, found that only a few mutations of the avian flu virus would allow infection of cells in the human nasal cavity and windpipe. Furthermore, when the scientists transferred nasal swabs serially from ferret to ferret, the virus became transmissible through the air. Reports of this startling discovery at a scientific conference in 2011 ignited a firestorm of debate about whether to publish the results. Ultimately, the scientific community decided the benefits of potentially understanding how to prevent pandemics would outweigh the risks of the information being used for harmful purposes, and the work was published in 2012.

As we have seen, emerging viruses are generally not new; rather, they are existing viruses that mutate, disseminate more widely in the current host species, or spread to new host species. Changes in host behavior or environmental changes can increase the viral traffic responsible for emerging diseases. For instance, new roads built through remote areas can allow viruses to spread between previously isolated human populations. Also, the destruction of forests to expand cropland can bring humans into contact with other animals that may host viruses capable of infecting humans.

## Viral Diseases in Plants

More than 2,000 types of viral diseases of plants are known, and together they account for an estimated annual loss of $15 billion worldwide due to their destruction of agricultural and horticultural crops. Common signs of viral infection include bleached or brown spots on leaves and fruits **(Figure 17.12)**, stunted growth, and damaged flowers or roots, all tending to diminish the yield and quality of crops.

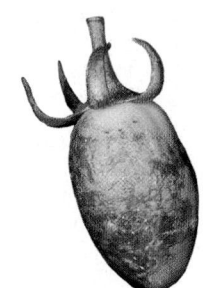

▲ **Figure 17.12 Virus-infected tomato.**

Plant viruses have the same basic structure and mode of replication as animal viruses. Most plant viruses discovered thus far, including tobacco mosaic virus (TMV), have an RNA genome. Many have a helical capsid, like TMV, while others have an icosahedral capsid (see Figure 17.2b).

Viral diseases of plants spread by two major routes. In the first route, called *horizontal transmission*, a plant is infected from an external source of the virus. Because the invading virus must get past the plant's outer protective layer of cells (the epidermis), a plant becomes more susceptible to viral infections if it has been damaged by wind, injury, or herbivores. Herbivores, especially insects, pose a double threat because they can also act as carriers of viruses, transmitting disease from plant to plant. Moreover, farmers and gardeners may transmit plant viruses inadvertently on pruning shears and other tools. The other route of viral infection is *vertical transmission*, in which a plant inherits a viral infection from a parent. Vertical transmission can occur in asexual propagation (for example, through cuttings) or in sexual reproduction via infected seeds.

Once a virus enters a plant cell and begins replicating, viral genomes and associated proteins can spread throughout the plant by means of plasmodesmata, the cytoplasmic connections that penetrate the walls between adjacent plant cells (see Figure 4.25). The passage of viral macromolecules from cell to cell is facilitated by virally encoded proteins that cause enlargement of plasmodesmata. Scientists have not yet devised cures for most viral plant diseases. Consequently, research efforts are focused largely on reducing the transmission of such diseases and on breeding resistant varieties of crop plants.

Earlier in this chapter, we mentioned the ongoing evolutionary relationship between viruses and the genomes of their host cells. In fact, the original source of viral genetic material may have been transposons, mobile genetic elements that are present in multiple copies in many genomes. In the next chapter, we'll discuss the structure of genomes and how they evolve.

## Prions: Proteins as Infectious Agents

The viruses we have discussed in this chapter are infectious agents that spread diseases, and their genetic material is composed of nucleic acids, whose ability to be replicated is well known. Surprisingly, there are also *proteins* that are known to be infectious. Proteins called **prions** appear to cause a number of degenerative brain diseases in various animal species. These diseases include scrapie in sheep; mad cow disease, which has plagued the European beef industry in recent years; and Creutzfeldt-Jakob disease in humans, which has caused the death of some 150 people in Great Britain. Prions can be transmitted in food, as may occur when people eat prion-laden beef from cattle with mad cow disease. Kuru, another human disease caused by prions, was identified in the early 1900s among the South Fore natives of New Guinea. A kuru epidemic peaked there in the 1960s, puzzling scientists, who at first thought the disease had a genetic basis. Eventually, however, anthropological investigations ferreted out how the disease was spread: ritual cannibalism, a widespread practice among South Fore natives at that time.

Two characteristics of prions are especially alarming. First, prions act very slowly, with an incubation period of at least ten years before symptoms develop. The lengthy incubation period prevents sources of infection from being identified until long after the first cases appear, allowing many more infections to occur. Second, prions are virtually indestructible; they are not destroyed or deactivated by heating to normal cooking temperatures. To date, there is no known cure for prion diseases, and the only hope for developing effective treatments lies in understanding the process of infection.

How can a protein, which cannot replicate itself, be a transmissible pathogen? According to the leading model, a prion is a misfolded form of a protein normally present in brain cells. When the prion gets into a cell containing the normal form of the protein, the prion somehow converts normal protein molecules to the misfolded prion versions. Several prions then aggregate into a complex that can convert other normal proteins to prions, which join the chain (**Figure 17.13**). Prion aggregation interferes with normal cellular functions and causes disease symptoms. This model was greeted with much skepticism when it was first proposed by Stanley Prusiner in the early 1980s, but it is now widely accepted. Prusiner was awarded the Nobel Prize in 1997 for his work on prions. He has recently proposed that prions are also involved in neurodegenerative disorders such as Alzheimer's and Parkinson's disease. There are many outstanding questions about these small infectious agents.

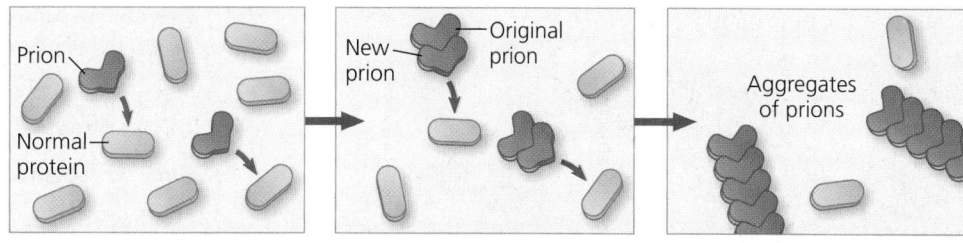

▲ **Figure 17.13 Model for how prions propagate.** Prions are misfolded versions of normal brain proteins. When a prion contacts a normally folded version of the same protein, it may induce the normal protein to assume the abnormal shape. The resulting chain reaction may continue until high levels of prion aggregation cause cellular malfunction and eventual degeneration of the brain.

### CONCEPT CHECK 17.3

1. Describe two ways in which a preexisting virus can become an emerging virus.
2. Contrast horizontal and vertical transmission of viruses in plants.
3. **WHAT IF?** TMV has been isolated from virtually all commercial tobacco products. Why, then, is TMV infection not an additional hazard for smokers?

   **For suggested answers, see Appendix A.**

---

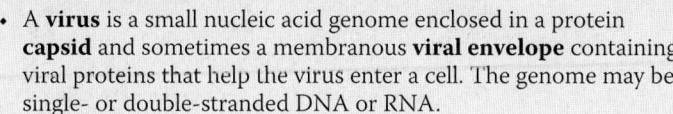

# 17 Chapter Review

Go to **MasteringBiology**® for Assignments, the eText, and the Study Area with Animations, Activities, Vocab Self-Quiz, and Practice Tests.

## SUMMARY OF KEY CONCEPTS

**VOCAB SELF-QUIZ**

goo.gl/gbai8v

### CONCEPT 17.1

**A virus consists of a nucleic acid surrounded by a protein coat (pp. 342–344)**

- A **virus** is a small nucleic acid genome enclosed in a protein **capsid** and sometimes a membranous **viral envelope** containing viral proteins that help the virus enter a cell. The genome may be single- or double-stranded DNA or RNA.

  **?** *Are viruses generally considered living or nonliving? Explain.*

### CONCEPT 17.2

**Viruses replicate only in host cells (pp. 344–351)**

- Viruses use enzymes, ribosomes, and small molecules of host cells to synthesize progeny viruses during replication. Each type of virus has a characteristic **host range**, affected by whether cell-surface proteins that viral surface proteins can bind to are present.
- Phages (viruses that infect bacteria) can replicate by two alternative mechanisms: the **lytic cycle** and the **lysogenic cycle**.

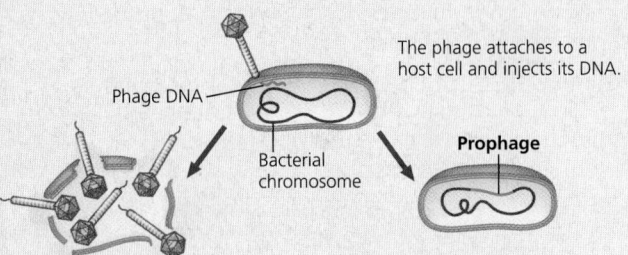

The phage attaches to a host cell and injects its DNA.

Phage DNA

Bacterial chromosome

**Prophage**

**Lytic cycle**
• **Virulent** or **temperate phage**
• Destruction of host DNA
• Production of new phages
• Lysis of host cell causes release of progeny phages

**Lysogenic cycle**
• **Temperate phage** only
• Genome integrates into bacterial chromosome as **prophage**, which
  (1) is replicated and passed on to daughter cells and
  (2) can be induced to leave the chromosome and initiate a lytic cycle

• Bacteria have various ways of defending themselves against phage infections, including the CRISPR system.
• Many animal viruses have an envelope. **Retroviruses** (such as **HIV**) use the enzyme **reverse transcriptase** to copy their RNA genome into DNA, which can be integrated into the host genome as a **provirus**.
• Since viruses can replicate only within cells, they probably evolved after the first cells appeared, perhaps as packaged fragments of cellular nucleic acid.

**?** *Describe enzymes that are not found in most cells but are necessary for the replication of certain types of viruses.*

### CONCEPT 17.3

### Viruses and prions are formidable pathogens in animals and plants (pp. 351–355)

• Symptoms of viral diseases may be caused by direct viral harm to cells or by the body's immune response. **Vaccines** stimulate the immune system to defend the host against specific viruses.
• Outbreaks of emerging viral diseases in humans are usually caused by existing viruses that expand their host territory. The H1N1 2009 flu virus was a new combination of pig, human, and avian viral genes that caused a **pandemic**.
• Viruses enter plant cells through damaged cell walls (horizontal transmission) or are inherited from a parent (vertical transmission).
• **Prions** are slow-acting, virtually indestructible infectious proteins that cause brain diseases in mammals.

**?** *What aspect of an RNA virus makes it more likely than a DNA virus to become an emerging virus?*

## TEST YOUR UNDERSTANDING

### Level 1: Knowledge/Comprehension

**PRACTICE TEST**

goo.gl/CRZjvS

1. Which of the following characteristics, structures, or processes is common to both bacteria and viruses?
   (A) metabolism
   (B) ribosomes
   (C) genetic material composed of nucleic acid
   (D) cell division

2. Emerging viruses arise by
   (A) mutation of existing viruses.
   (B) the spread of existing viruses to new host species.
   (C) the spread of existing viruses more widely within their host species.
   (D) all of the above.

3. A human pandemic is
   (A) a viral disease that infects all humans.
   (B) a flu that kills more than 1 million people.
   (C) an epidemic that extends around the world.
   (D) a virus that increases in mortality rate as it spreads.

### Level 2: Application/Analysis

4. A bacterium is infected with an experimentally constructed bacteriophage composed of the T2 phage protein coat and T4 phage DNA. The new phages produced would have
   (A) T2 protein and T4 DNA.
   (B) T4 protein and T2 DNA.
   (C) T2 protein and T2 DNA.
   (D) T4 protein and T4 DNA.

5. RNA viruses require their own supply of certain enzymes because
   (A) host cells rapidly destroy the viruses.
   (B) host cells lack enzymes that can replicate the viral genome.
   (C) these enzymes translate viral mRNA into proteins.
   (D) these enzymes penetrate host cell membranes.

6. **DRAW IT** Redraw Figure 17.8 to show the replicative cycle of a virus with a single-stranded genome that can function as mRNA.

### Level 3: Synthesis/Evaluation

7. **SCIENTIFIC INQUIRY**
   When bacteria infect an animal, the number of bacteria in the body increases in an exponential fashion (graph A). After infection by a virulent animal virus with a lytic replicative cycle, there is no evidence of infection for a while. Then the number of viruses rises suddenly and subsequently increases in a series of steps (graph B). Explain the difference in the curves.

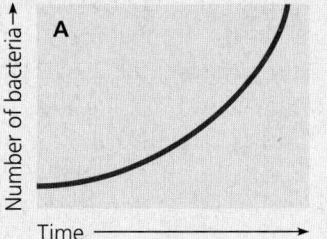
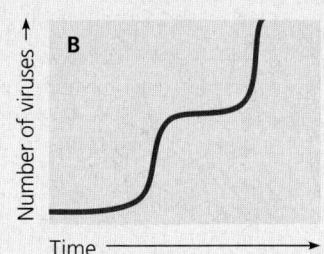

8. **FOCUS ON EVOLUTION**
   The success of some viruses lies in their ability to evolve rapidly within the host. Such viruses evade the host's defenses by mutating and producing many altered progeny viruses before the body can mount an attack. Thus, the viruses present late in infection differ from those that initially infected the body. Discuss this as an example of evolution in microcosm. Which viral lineages tend to predominate?

9. **FOCUS ON ORGANIZATION**
   While viruses are considered by most scientists to be nonliving, they do show some characteristics of life, including the correlation of structure and function. In a short essay (100–150 words), discuss how the structure of a virus correlates with its function.

10. **SYNTHESIZE YOUR KNOWLEDGE**

Oseltamivir (Tamiflu), an antiviral drug prescribed for the flu, inhibits the enzyme neuraminidase. Explain how this drug could prevent infection in someone who is exposed to the flu or could shorten the course of flu in an infected patient (the reasons for which it is prescribed).

*For selected answers, see Appendix A.*

# 18 Genomes and Their Evolution

▲ **Figure 18.1** What genomic information distinguishes a human from a chimpanzee?

## Reading Leaves from the Tree of Life

The chimpanzee (*Pan troglodytes*) is our closest living relative on the evolutionary tree of life. The boy in **Figure 18.1** and his chimpanzee companion are intently studying the same leaf, but only one of them is able to talk about what he sees. What accounts for this difference between two primates that share so much of their evolutionary history? With the advent of recent techniques for rapidly sequencing complete genomes, we have now started to address the genetic basis of intriguing questions like this.

The chimpanzee genome was sequenced two years after sequencing of the human genome was largely completed. Now that we can compare our genome, base by base, with that of the chimpanzee, we can tackle the more general issue of what differences in genetic information account for the distinct characteristics of these two species of primates.

In addition to determining the sequences of the human and chimpanzee genomes, researchers have obtained complete genome sequences for *Escherichia coli* and other prokaryotes, as well as eukaryotes such as *Zea mays* (corn), *Mus musculus* (house mouse), and *Pongo pygmaeus* (orangutan). Even the genome of an extinct species closely related to present-day humans, *Homo neanderthalensis,* has been sequenced. These

whole and partial genomes provide important insights into evolution. Comparing human and chimpanzee genomes with those of other primates should reveal the sets of genes that control group-defining characteristics. Further comparisons with the genomes of bacteria, fungi, protists, and plants will enlighten us about the long evolutionary history of shared ancient genes.

With the genomes of many species fully sequenced, scientists can study whole sets of genes and their interactions, an approach called **genomics**. The sequencing efforts that feed this approach have generated, and continue to generate, enormous volumes of data. The need to deal with this ever-increasing flood of information has spawned the field of **bioinformatics**, the application of computational methods to the storage and analysis of biological data.

We'll begin this chapter by discussing genome sequencing and some of the advances in bioinformatics and its applications. We'll then summarize what has been learned from the genomes that have been sequenced thus far. Next, we'll describe the composition of the human genome as a representative genome of a complex multicellular eukaryote. Finally, we'll explore current ideas about how genomes evolve and about how the evolution of developmental mechanisms could have generated the great diversity of life on Earth today.

# The Human Genome Project fostered development of faster, less expensive sequencing techniques

Sequencing of the human genome, an ambitious undertaking, officially began as the **Human Genome Project** in 1990. Organized by an international, publicly funded consortium of scientists at universities and research institutes, the project involved 20 large sequencing centers in six countries plus a host of other labs working on smaller parts of the project.

After the human genome sequence was largely completed in 2003, the sequence of each chromosome was analyzed and described in a series of papers, the last of which covered chromosome 1 and was published in 2006. With this refinement, researchers declared the sequencing "virtually complete."

The ultimate goal in mapping any genome is to determine the complete nucleotide sequence of each chromosome. For the human genome, this was accomplished by sequencing machines. Even with automation, though, the sequencing of all 3 billion base pairs in a haploid set of human chromosomes presented a formidable challenge. In fact, a major thrust of the Human Genome Project was the development of technology for faster sequencing, as described in Concept 13.4. Improvements over the years chipped away at each time-consuming step, enabling the rate of sequencing to accelerate impressively: Whereas a productive lab could typically sequence 1,000 base pairs a day in the 1980s, by the year 2000 each research center working on the Human Genome Project was sequencing 1,000 base pairs *per second*, 24 hours a day, seven days a week—and this rate has only continued to improve. Methods that can analyze biological materials very rapidly and produce enormous volumes of data are said to be "high-throughput." Sequencing machines are an example of high-throughput devices.

Two complementary approaches were used to obtain the complete sequence. The initial approach was a methodical one that built on an earlier storehouse of human genetic information. In 1998, however, molecular biologist J. Craig Venter set up a company (Celera Genomics) and declared his intention to sequence the entire human genome using an alternative strategy. The **whole-genome shotgun approach** starts with the cloning and sequencing of DNA fragments from randomly cut DNA. Powerful computer programs then assemble the resulting very large number of overlapping short sequences into a single continuous sequence **(Figure 18.2)**.

Today, the whole-genome shotgun approach is widely used. Also, the development of newer sequencing techniques, generally called *sequencing by synthesis* (see Concept 13.4), has resulted in massive increases in speed and decreases in the cost of sequencing entire genomes. In these new techniques, many very small DNA fragments (each about 300 base pairs long) are sequenced at the same time, and computer software rapidly

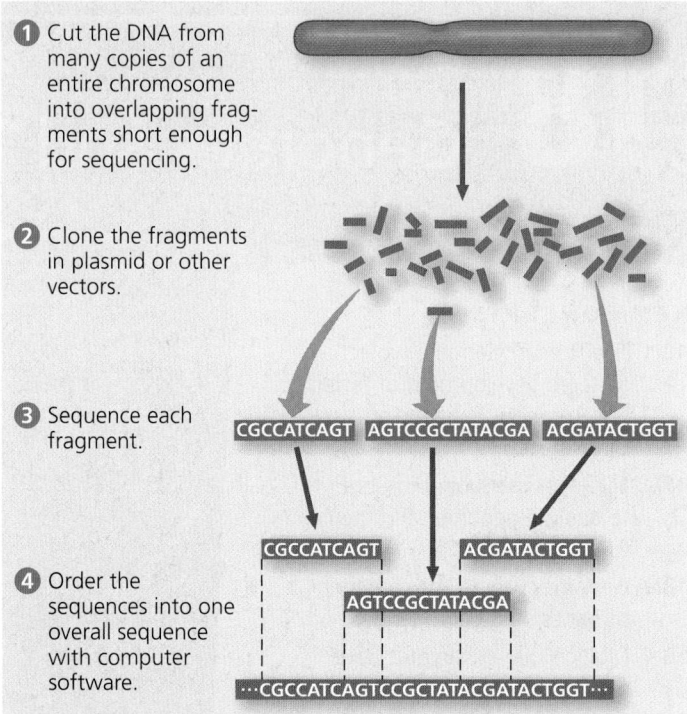

1. Cut the DNA from many copies of an entire chromosome into overlapping fragments short enough for sequencing.

2. Clone the fragments in plasmid or other vectors.

3. Sequence each fragment.

CGCCATCAGT  AGTCCGCTATACGA  ACGATACTGGT

4. Order the sequences into one overall sequence with computer software.

CGCCATCAGT    ACGATACTGGT
AGTCCGCTATACGA
···CGCCATCAGTCCGCTATACGATACTGGT···

▲ **Figure 18.2  Whole-genome shotgun approach to sequencing.** In this approach, developed by J. Craig Venter and colleagues at Celera Genomics, random DNA fragments are cloned (see Figure 13.24), sequenced, and then ordered relative to each other.

**?** *The fragments in stage 2 of this figure are depicted as scattered, rather than arranged in an ordered array. How does this depiction accurately reflect the approach?*

assembles the complete sequence. Because of the sensitivity of these techniques, the fragments can be sequenced directly; the cloning step (stage 2 in Figure 18.2) is unnecessary. By 2010, the worldwide output was astronomical: close to 100 *billion* bases per day, with the rate estimated to double every 9 months. Whereas sequencing the first human genome took 13 years and cost $100 million, biologist James Watson's genome was sequenced during 4 months in 2007 for about $1 million, and at least one technology company claims their machine can sequence an individual's genome in hours for less than $1,000!

These technological advances have also facilitated an approach called **metagenomics** (from the Greek *meta*, beyond), in which DNA from an entire group of species (a *metagenome*) is collected from an environmental sample and sequenced. Again, computer software sorts out the partial sequences and assembles them into the individual specific genomes. So far, this approach has been applied to microbial communities found in environments as diverse as the Sargasso Sea and the human intestine. A 2012 study characterized the astounding diversity of the human "microbiome"—the many species of bacteria that coexist within and upon our bodies and contribute to our survival. The ability to sequence the DNA of mixed populations eliminates the need to culture each species separately in the lab, a difficulty that has limited the study of many microbial species.

At first glance, genome sequences of humans and other organisms are simply dry lists of nucleotide bases—millions of A's, T's, C's, and G's in mind-numbing succession. Crucial to making sense of this massive amount of data have been new analytical approaches, which we discuss next.

**CONCEPT CHECK 18.1**

1. Describe the whole-genome shotgun approach.

For suggested answers, see Appendix A.

---

## CONCEPT 18.2

# Scientists use bioinformatics to analyze genomes and their functions

Each of the 20 or so sequencing centers around the world working on the Human Genome Project in the 1990s churned out voluminous amounts of DNA sequence day after day. As the data began to accumulate, the need to coordinate efforts to keep track of all the sequences became clear. Thanks to the foresight of research scientists and government officials involved in the Human Genome Project, its goals included establishing databases and refining analytical software. These databases and software programs are now centralized and accessible on the Internet. Accomplishing this aim has accelerated progress in DNA sequence analysis by making bioinformatics resources available to researchers worldwide and speeding up dissemination of information.

## Centralized Resources for Analyzing Genome Sequences

Government-funded agencies carried out their mandate to establish databases and provide software with which scientists could analyze the sequence data. For example, in the United States, a joint endeavor between the National Library of Medicine and the National Institutes of Health (NIH) created the National Center for Biotechnology Information (NCBI), which maintains a website (www.ncbi.nlm.nih.gov) with extensive bioinformatics resources. On this site are links to databases, software, and a wealth of information about genomics and related topics. Similar websites have been established by three genome centers with which the NCBI collaborates: the European Molecular Biology Laboratory, the DNA Data Bank of Japan, and BGI (formerly known as the Beijing Genome Institute) in Shenzhen, China. These large, comprehensive websites are complemented by others maintained by individual or small groups of laboratories. Smaller websites often provide databases and software designed for a narrower purpose, such as studying genetic and genomic changes in one particular type of cancer.

The NCBI database of sequences is called GenBank. As of June 2015, it included the sequences of 182 million fragments of genomic DNA, totaling 190 billion base pairs! GenBank is constantly updated, and the amount of data it contains is estimated to double approximately every 18 months. Any

sequence in the database can be retrieved and analyzed using software from the NCBI website or elsewhere.

One software program available on the NCBI website, called BLAST, allows the user to compare a DNA sequence with every sequence in GenBank, base by base. A researcher might search for similar regions in other genes of the same species, or among the genes of other species. Another program allows comparison of predicted protein sequences. A third can search any protein sequence for common stretches of amino acids (domains) for which a function is known or suspected, and it can show a three-dimensional model of the domain alongside other relevant information **(Figure 18.3)**. There is even a software program that can compare a collection of sequences, either nucleic acids or polypeptides, and diagram them in the form of an evolutionary tree based on the sequence relationships. (One such diagram is shown in Figure 18.15.)

Two research institutions, Rutgers University and the University of California, San Diego, also maintain a worldwide Protein Data Bank, a database of all three-dimensional protein structures that have been determined. (The database is accessible at www.wwpdb.org.) These structures can be rotated by the viewer to show all sides of the protein. Throughout this book, you'll find images of protein structures that have been obtained from the Protein Data Bank.

There is a vast array of resources available for researchers anywhere in the world to use free of charge. Now let's consider the types of questions scientists can address using these resources.

## Understanding the Functions of Protein-Coding Genes

The identities of about half of the human genes were known before the Human Genome Project began. But what about the others, the previously unknown genes revealed by analysis of DNA sequences? Clues about their identities and functions come from using software to compare sequences that might be genes with those of known genes from other organisms. Due to redundancy in the genetic code, the DNA sequence itself may vary more among species than the protein sequence does. Thus, scientists interested in proteins often compare the predicted amino acid sequence of a protein with that of other proteins.

Sometimes a newly identified sequence will match, at least partially, the sequence of a gene or protein in another species whose function is well known. For example, a plant researcher working on signaling pathways in the muskmelon would be excited to see that a partial amino acid sequence from a gene she had identified matched sequences in other species encoding a so-called "WD40 domain" (see Figure 18.3). These WD40 domains are present in many eukaryotes and are known to function in signal transduction pathways. Alternatively, a new gene sequence might be similar to a previously encountered sequence whose function is still unknown. Another possibility is that the sequence is entirely unlike anything ever seen before. This was true for about a third of the genes of *E. coli* when its genome was sequenced. In such a case, protein function is

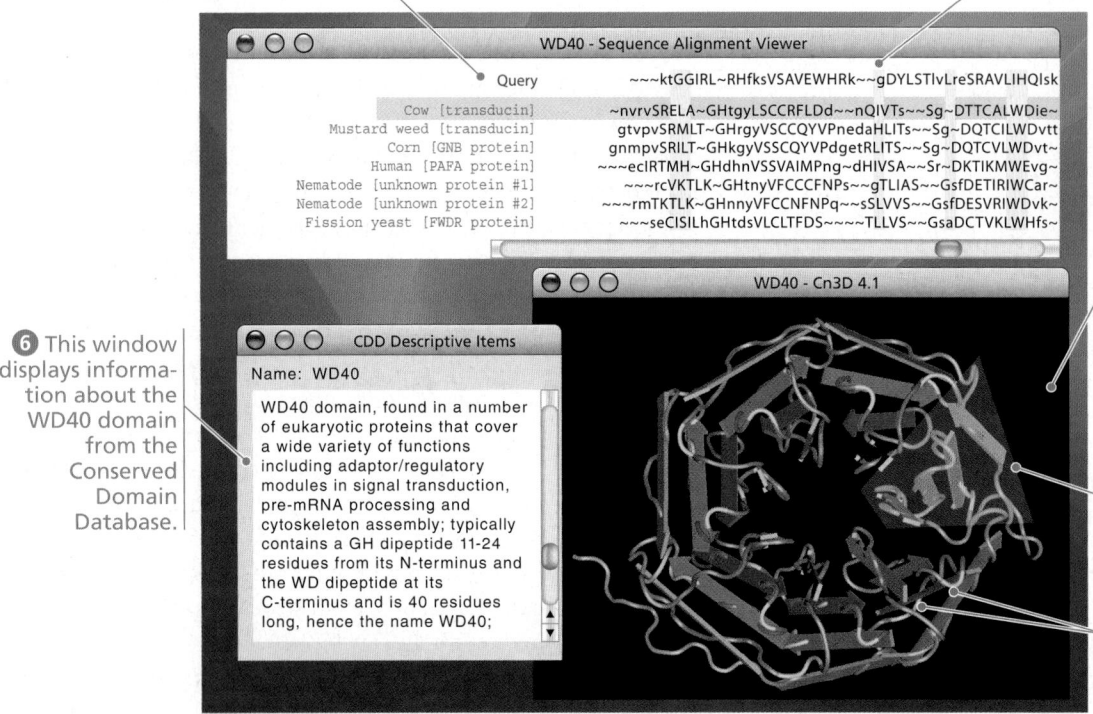

**①** In this window, a partial amino acid sequence from an unknown muskmelon protein ("Query") is aligned with sequences from other proteins that the computer program found to be similar. Each sequence represents a domain called WD40.

**②** Four hallmarks of the WD40 domain are highlighted in yellow. (Sequence similarity is based on chemical aspects of the amino acids, so the amino acids in each hallmark region are not always identical.)

**③** The Cn3D program displays a three-dimensional ribbon model of cow transducin (the protein highlighted in purple in the Sequence Alignment Viewer). This protein is the only one of those shown for which a structure has been determined. The sequence similarity of the other proteins to cow transducin suggests that their structures are likely to be similar.

**④** Cow transducin contains seven WD40 domains, one of which is highlighted here in gray.

**⑤** The yellow segments correspond to the WD40 hallmarks highlighted in yellow in the window above.

**⑥** This window displays information about the WD40 domain from the Conserved Domain Database.

WD40 - Sequence Alignment Viewer

| | |
|---|---|
| Query | ~~~ktGGIRL~RHfksVSAVEWHRk~~gDYLSTlvLreSRAVLIHQlsk |
| Cow [transducin] | ~nvrvSRELA~GHtgyLSCCRFLDd~~nQIVTs~~Sg~DTTCALWDie~ |
| Mustard weed [transducin] | gtvpvSRMLT~GHrgyVSCCQYVPnedaHLITs~~Sg~DQTCILWDvtt |
| Corn [GNB protein] | gnmpvSRILT~GHkgyVSSCQYVPdgetRLITS~~Sg~DQTCVLWDvt~ |
| Human [PAFA protein] | ~~~eclRTMH~GHdhnVSSVAIMPng~dHIVSA~~Sr~DKTIKMWEvg~ |
| Nematode [unknown protein #1] | ~~~rcVKTLK~GHtnyVFCCCFNPs~~gTLIAS~~GsfDETIRIWCar~ |
| Nematode [unknown protein #2] | ~~~rmTKTLK~GHnnyVFCCNFNPq~~sSLVVS~~GsfDESVRIWDvk~ |
| Fission yeast [FWDR protein] | ~~~seCISILhGHtdsVLCLTFDS~~~~TLLVS~~GsaDCTVKLWHfs~ |

WD40 - Cn3D 4.1

CDD Descriptive Items

Name: WD40

WD40 domain, found in a number of eukaryotic proteins that cover a wide variety of functions including adaptor/regulatory modules in signal transduction, pre-mRNA processing and cytoskeleton assembly; typically contains a GH dipeptide 11-24 residues from its N-terminus and the WD dipeptide at its C-terminus and is 40 residues long, hence the name WD40;

▲ **Figure 18.3 Bioinformatics tools that are available on the Internet.** A website maintained by the National Center for Biotechnology Information (NCBI) allows scientists and the public to access DNA and protein sequences and other stored data. The site includes a link to a protein structure database (Conserved Domain Database, CDD) that can find and describe similar domains in related proteins, as well as software (Cn3D, "See in 3-D") that displays models of domains. Some results are shown from a search for regions of proteins similar to an amino acid sequence in a muskmelon protein. The WD40 domain is very common in proteins encoded by eukaryotic genomes. It often plays a key role in molecular interactions during signal transduction.

usually deduced through a combination of biochemical and functional studies. The biochemical approach aims to determine the three-dimensional structure of the protein as well as other attributes, such as potential binding sites for other molecules. Functional studies usually involve blocking or disabling the gene in an organism to see how the phenotype is affected.

## Understanding Genes and Gene Expression at the Systems Level

The impressive computational power provided by the tools of bioinformatics allows the study of whole sets of genes and their interactions, as well as the comparison of genomes from different species. Genomics is a rich source of new insights into fundamental questions about genome organization, regulation of gene expression, embryonic development, and evolution.

One informative approach has been taken by an ongoing research project called ENCODE (Encyclopedia of DNA Elements), which began in 2003. The aim of the project is to learn everything possible about the functionally important elements in the human genome using multiple experimental techniques. Investigators have sought to identify protein-coding genes and genes for noncoding RNAs, along with sequences that regulate gene expression, such as enhancers and promoters.

In addition, they have extensively characterized DNA and histone modifications and chromatin structure. The second phase of the ENCODE project, involving more than 440 scientists in 32 research groups, culminated with the simultaneous publication of 30 papers in 2012, describing over 1,600 large data sets. The power of ENCODE is that it provides the opportunity to compare the results from different projects, yielding a much richer picture of the whole genome.

Perhaps the most striking finding is that about 75% of the genome is transcribed at some point in at least one of the cell types studied, even though less than 2% codes for proteins. Furthermore, biochemical functions have been assigned to DNA elements making up at least 80% of the genome. To learn more about the different types of functional elements, parallel projects are analyzing in a similar way the genomes of two model organisms, the soil nematode *Caenorhabditis elegans* (typically referred to as *C. elegans*) and the fruit fly *Drosophila melanogaster* (usually called *D. melanogaster* or simply *Drosophila*). Because genetic and biochemical experiments using DNA technology can be performed on these species, testing the activities of potentially functional DNA elements in their genomes is expected to illuminate the workings of the human genome.

### Systems Biology

The scientific progress resulting from sequencing genomes and studying large sets of genes has encouraged scientists to attempt similar systematic studies of sets of proteins and their properties (such as their abundance, chemical modifications, and interactions), an approach called **proteomics**. (A **proteome** is the entire set of proteins expressed by a cell or group of cells.) Proteins, not the genes that encode them, carry out most of the activities of the cell. Therefore, we must study when and where proteins are produced in an organism, as well as how they interact in networks, if we are to understand the functioning of cells and organisms.

Genomics and proteomics are enabling molecular biologists to approach the study of life from an increasingly global perspective. Using the tools we have described, biologists have begun to compile catalogs of genes and proteins—listings of all the "parts" that contribute to the operation of cells, tissues, and organisms. With such catalogs in hand, researchers have shifted their attention from the individual parts to their functional integration in biological systems. As you may recall, Concept 1.1 discussed this approach, called **systems biology**, which aims to model the dynamic behavior of whole biological systems based on the study of the interactions among the system's parts. Because of the vast amounts of data generated in these types of studies, advances in computer technology and bioinformatics have been crucial in making systems biology possible.

### Application of Systems Biology to Medicine

The Cancer Genome Atlas is an example of systems biology in which a large group of interacting genes and gene products is analyzed together. This project, under the joint leadership of the National Cancer Institute and NIH, aims to determine how changes in biological systems lead to cancer. A three-year pilot project ending in 2010 set out to find all the common mutations in three types of cancer—lung cancer, ovarian cancer, and glioblastoma of the brain—by comparing gene sequences and patterns of gene expression in cancer cells with those in normal cells. Work on glioblastoma confirmed the role of several suspected genes and identified a few previously unknown ones, suggesting possible new targets for therapies. The approach proved so fruitful for these three types of cancer that it has been extended to ten other types, chosen because they are common and often lethal in humans.

As high-throughput techniques become more rapid and less expensive, they are being increasingly applied to the problem of cancer. Rather than sequencing only protein-coding genes, sequencing the whole genomes of many tumors of a particular type allows scientists to uncover common chromosomal abnormalities, as well as any other consistent changes in these aberrant genomes.

In addition to whole-genome sequencing, silicon and glass "chips" that hold a microarray of most of the known human genes are now used to analyze gene expression patterns in patients suffering from various cancers and other diseases **(Figure 18.4)**. Analyzing which genes are over- or underexpressed in a particular cancer may allow physicians to tailor patients' treatment to their unique genetic makeup and the specifics of their cancers. This approach has been used to characterize subsets of particular cancers, enabling more refined treatments. Breast cancer is one example (see Figure 16.21).

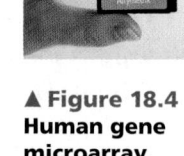

▲ **Figure 18.4 Human gene microarray chip.**

Ultimately, medical records may include an individual's DNA sequence, a sort of genetic bar code, with regions highlighted that predispose the person to specific diseases. The use of such sequences for personalized medicine—disease prevention and treatment—has great potential.

Systems biology is a very efficient way to study emergent properties at the molecular level. Novel properties emerge at each level of biological complexity as a result of the arrangement of building blocks at the underlying level (see Concept 1.1). The more we can learn about the arrangement and interactions of the components of genetic systems, the deeper will be our understanding of whole organisms. The rest of this chapter surveys what we've learned from genomic studies.

### CONCEPT CHECK 18.2

1. What role does the Internet play in current genomics and proteomics research?
2. Explain the advantage of the systems biology approach to studying cancer versus studying a single gene at a time.
3. **MAKE CONNECTIONS** The ENCODE pilot project found that at least 75% of the genome is transcribed into RNAs, far more than could be accounted for by protein-coding genes. Suggest some roles that these RNAs might play. (Review Concepts 14.3 and 15.3.)

For suggested answers, see Appendix A.

---

### CONCEPT 18.3

## Genomes vary in size, number of genes, and gene density

The sequences of thousands of genomes have been completed, with tens of thousands of genomes either in progress or considered permanent drafts (because they require more work than it would be worth to complete them). Among sequences in progress are 550 metagenomes. In the completely sequenced group, about 3,000 are genomes of bacteria, and 180 are genomes of archaea. There are 60 completed eukaryotic species and 875 permanent drafts. Among these are vertebrates, invertebrates, protists, fungi, and plants. Next, we'll discuss what we've learned about genome size, number of genes, and gene density, focusing on general trends.

## Genome Size

Comparing the three domains (Bacteria, Archaea, and Eukarya), we find a general difference in genome size between prokaryotes and eukaryotes (Table 18.1). While there are some exceptions, most bacterial genomes have between 1 and 6 million base pairs (Mb); the genome of *E. coli*, for instance, has 4.6 Mb. Genomes of archaea are, for the most part, within the size range of bacterial genomes. (Keep in mind, however, that many fewer genomes of archaea have been completely sequenced, so this picture may change.) Eukaryotic genomes tend to be larger: The genome of the single-celled yeast *Saccharomyces cerevisiae* (a fungus) has about 12 Mb, while most animals and plants, which are multicellular, have genomes of at least 100 Mb. There are 165 Mb in the fruit fly genome, while humans have 3,000 Mb, about 500 to 3,000 times as many as a typical bacterium.

Aside from this general difference between prokaryotes and eukaryotes, a comparison of genome sizes among eukaryotes fails to reveal any systematic relationship between genome size and the organism's phenotype. For instance, the genome of *Paris japonica*, the Japanese canopy plant, contains 149 billion base pairs (149,000 Mb), about 50 times the size of the human genome. On a finer scale, comparing two insect species, the cricket (*Anabrus simplex*) genome turns out to have 11 times as many base pairs as the *Drosophila melanogaster* genome. There is a wide range of genome sizes within the groups of unicellular eukaryotes, insects, amphibians, and plants and less of a range within mammals and reptiles.

## Number of Genes

The number of genes also varies between prokaryotes and eukaryotes: Bacteria and archaea, in general, have fewer genes than eukaryotes. Free-living bacteria and archaea have from 1,500 to 7,500 genes, while the number of genes in eukaryotes ranges from about 5,000 for unicellular fungi (yeasts) to at least 40,000 for some multicellular eukaryotes.

Within the eukaryotes, the number of genes in a species is often lower than expected from considering simply the size of its genome. Looking at Table 18.1, you can see that the genome of the nematode *C. elegans* is 100 Mb in size and contains 20,100 genes. The *Drosophila* genome, in comparison, is much bigger (165 Mb) but has about two-thirds the number of genes—14,000 genes.

Considering an example closer to home, we noted that the human genome contains 3,000 Mb, well over ten times the size of either the *D. melanogaster* or *C. elegans* genome. At the outset of the Human Genome Project, biologists expected somewhere between 50,000 and 100,000 genes to be identified in the completed sequence, based on the number of known human proteins. As the project progressed, the estimate was revised downward several times, and the ENCODE project has established the number to be fewer than 21,000. This relatively low number, similar to the number of genes in the nematode

**Table 18.1 Genome Sizes and Estimated Numbers of Genes\***

| Organism | Haploid Genome Size (Mb)[†] | Number of Genes | Genes per Mb[†] |
|---|---|---|---|
| **Bacteria** | | | |
| *Haemophilus influenzae* | 1.8 | 1,700 | 940 |
| *Escherichia coli* | 4.6 | 4,400 | 950 |
| **Archaea** | | | |
| *Archaeoglobus fulgidus* | 2.2 | 2,500 | 1,130 |
| *Methanosarcina barkeri* | 4.8 | 3,600 | 750 |
| **Eukaryotes** | | | |
| *Saccharomyces cerevisiae* (yeast, a fungus) | 12 | 6,300 | 525 |
| *Caenorhabditis elegans* (nematode) | 100 | 20,100 | 200 |
| *Arabidopsis thaliana* (mustard family plant) | 120 | 27,000 | 225 |
| *Drosophila melanogaster* (fruit fly) | 165 | 14,000 | 85 |
| *Oryza sativa* (rice) | 430 | 42,000 | 95 |
| *Zea mays* (corn) | 2,300 | 32,000 | 14 |
| *Ailuropoda melanoleuca* (giant panda) | 2,400 | 21,000 | 9 |
| *Homo sapiens* (human) | 3,000 | <21,000 | 7 |
| *Paris japonica* (Japanese canopy plant) | 149,000 | ND[‡] | ND[‡] |

\*Some values given here are likely to be revised as genome analysis continues.
[†]Mb = million base pairs. [‡]ND = not determined.

*C. elegans*, has surprised biologists, who had been expecting many more human genes.

What genetic attributes allow humans (and other vertebrates) to get by with no more genes than nematodes? An important factor is that vertebrate genomes "get more bang for the buck" from their coding sequences because of extensive alternative splicing of RNA transcripts. Recall that this process generates more than one polypeptide from a single gene (see Figure 15.12). A typical human gene contains about ten exons, and an estimated 90% or more of these multi-exon genes are spliced in at least two different ways. Some genes are expressed in hundreds of alternatively spliced forms, others in just two. Scientists have not yet catalogued all of the different forms, but it is clear that the number of different polypeptides encoded in the human genome far exceeds the proposed number of genes.

Additional polypeptide diversity could result from varying post-translational modifications in different cell types or at different developmental stages. Finally, the discovery of miRNAs and other small RNAs that play regulatory roles has added a new variable to the mix (see Concept 15.3). Some scientists think that this added level of regulation may contribute to greater organismal complexity for some genes.

## Gene Density and Noncoding DNA

We can take both genome size and number of genes into account by comparing gene density in different species. In other words, we can ask how many genes are in a given length of DNA. When we compare the genomes of bacteria, archaea, and eukaryotes, we see that eukaryotes generally have larger genomes but fewer genes in a given number of base pairs. Humans have hundreds or thousands of times as many base pairs in their genome as most bacteria, as we already noted, but only 5 to 15 times as many genes; thus, gene density is lower in humans (see Table 18.1). Even unicellular eukaryotes, such as yeasts, have fewer genes per million base pairs than bacteria and archaea. Among the genomes that have been sequenced completely, humans and other mammals have the lowest gene density.

In all bacterial genomes studied so far, most of the DNA consists of genes for protein, tRNA, or rRNA; the small amount remaining consists mainly of nontranscribed regulatory sequences, such as promoters. The sequence of nucleotides along a bacterial protein-coding gene is not interrupted by noncoding sequences (introns). In eukaryotic genomes, by contrast, most of the DNA neither encodes proteins nor is transcribed into RNA molecules of known function, and the DNA includes more complex regulatory sequences. In fact, humans have 10,000 times as much noncoding DNA as bacteria. Some of this DNA in multicellular eukaryotes is present as introns within genes. Indeed, introns account for most of the difference in average length between human genes (27,000 base pairs) and bacterial genes (1,000 base pairs).

In addition to introns, multicellular eukaryotes have a vast amount of non-protein-coding DNA between genes. Next, we'll discuss these stretches of DNA in the human genome.

### CONCEPT CHECK 18.3

1. The best estimate is that the human genome contains fewer than 21,000 genes. However, there is evidence that human cells produce many more than 21,000 different polypeptides. What processes might account for this discrepancy?
2. The Genomes Online Database (GOLD) website of the Joint Genome Institute has information about genome sequencing projects. Go to https://gold.jgi-psf.org/statistics, scroll through the page, and describe the information you find there. What percent of bacterial genome projects have medical relevance?
3. **WHAT IF?** What evolutionary processes might account for prokaryotes having smaller genomes than eukaryotes?

For suggested answers, see Appendix A.

---

## CONCEPT 18.4

# Multicellular eukaryotes have much noncoding DNA and many multigene families

We have spent most of this chapter, and indeed this unit, focusing on genes that code for proteins. Yet the coding regions of these genes and the genes for noncoding RNA products such as rRNA, tRNA, and miRNA make up a small portion of the genomes of most multicellular eukaryotes. For example, the sequencing of the human genome revealed that only a tiny part—about 1.5%—codes for proteins or is transcribed into rRNAs or tRNAs. **Figure 18.5** shows what is known about the remaining 98.5% of the genome.

Gene-related regulatory sequences and introns account, respectively, for 5% and about 20% of the human genome. The rest, located between functional genes, includes some unique (single-copy) noncoding DNA, such as gene fragments and **pseudogenes**, former genes that have accumulated mutations over a long time and no longer produce functional proteins. (The genes that produce small noncoding RNAs are a tiny percentage of the genome, distributed between the 20% introns and the 15% unique noncoding DNA.) Most of the DNA between functional genes, however, is **repetitive DNA**, which consists of sequences that are present in multiple copies in the genome.

The bulk of many eukaryotic genomes likewise consists of DNA sequences that neither code for proteins nor are transcribed to produce RNAs with known functions; this

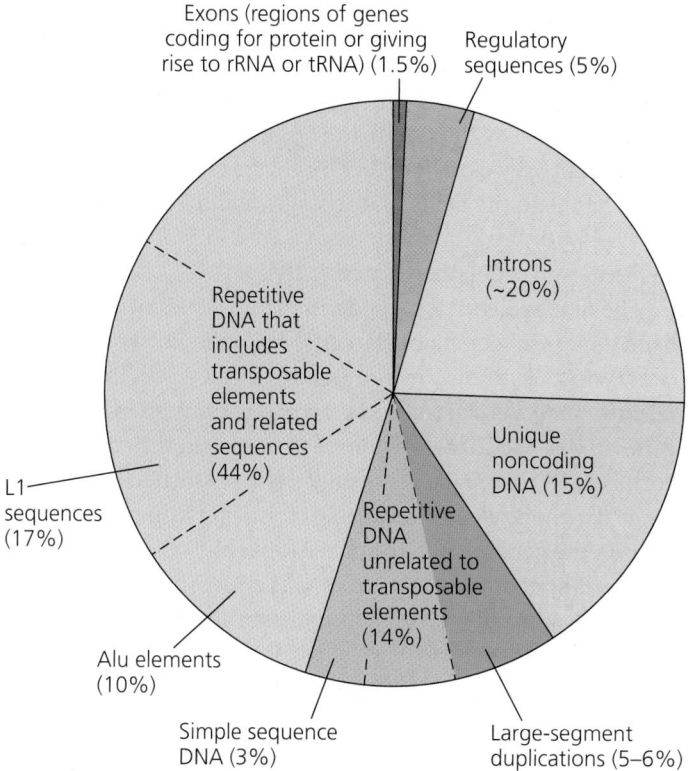

▲ **Figure 18.5 Types of DNA sequences in the human genome.** The gene sequences that code for proteins or are transcribed into rRNA or tRNA molecules make up only about 1.5% of the human genome (dark purple in the pie chart), while introns and regulatory sequences associated with genes (light purple) make up about a quarter. The vast majority of the human genome does not code for proteins or give rise to known RNAs, and much of it is repetitive DNA (dark and light green and teal). Because repetitive DNA is the most difficult to sequence and analyze, classification of some portions is tentative, and the percentages given here may shift slightly as genome analysis proceeds.

noncoding DNA was often described in the past as "junk DNA." However, genome comparisons over the past 10 years have revealed the persistence of this DNA in diverse genomes over many hundreds of generations. For example, the genomes of humans, rats, and mice contain almost 500 regions of non-coding DNA that are *identical* in sequence in all three species. This is a higher level of sequence conservation than is seen for protein-coding regions in these species, strongly suggesting that the noncoding regions have important functions. The results of the ENCODE project have underscored the key roles played by much of this noncoding DNA. Next, we'll examine how genes and noncoding DNA sequences are organized within genomes of multicellular eukaryotes, using the human genome as our main example. Genome organization tells us a lot about how genomes have evolved and continue to evolve, as we'll discuss in Concept 18.5.

## Transposable Elements and Related Sequences

Both prokaryotes and eukaryotes have stretches of DNA that can move from one location to another within the genome. These stretches are known as *transposable genetic elements*, or simply **transposable elements**. During the process called *transposition*, a transposable element moves from one site in a cell's DNA to a different target site by a type of recombination process. Transposable elements are sometimes called "jumping genes," but actually they never completely detach from the cell's DNA. Instead, the original and new DNA sites are brought very close together by enzymes and other proteins that bend the DNA. Surprisingly, about 75% of repetitive DNA (44% of the human genome) is made up of transposable elements and sequences related to them.

The first evidence for wandering DNA segments came from American geneticist Barbara McClintock's breeding experiments with Indian corn (maize) in the 1940s and 1950s **(Figure 18.6)**. As she tracked corn plants through multiple generations, McClintock identified changes in the color of corn kernels that made sense only if she postulated the existence of genetic elements capable of moving from other locations in the genome into the genes for kernel color, disrupting the genes so that the kernel color was changed. McClintock's discovery was met with great skepticism and virtually discounted at the time. Her careful work and insightful ideas were finally validated many years later when transposable elements were found in bacteria. In 1983, at the age of 81, McClintock received the Nobel Prize for her pioneering research.

### Movement of Transposons and Retrotransposons

Eukaryotic transposable elements are of two types. The first type, **transposons**, can move within a genome by means of a DNA intermediate. Transposons can move by a "cut-and-paste" mechanism, which removes the element from the original site, or by a "copy-and-paste" mechanism, which leaves a copy behind **(Figure 18.7)**. Both mechanisms require an enzyme called *transposase*, which is generally encoded by the transposon.

◄ **Figure 18.6 The effect of transposable elements on corn kernel color.** Barbara McClintock first proposed the idea of mobile genetic elements after observing variegations in the color of the kernels on a corn cob.

Most transposable elements in eukaryotic genomes are of the second type, **retrotransposons**, which move by means of an RNA intermediate that is a transcript of the retrotransposon DNA. Thus, retrotransposons always leave a copy at the original site during transposition **(Figure 18.8)**. To insert at another site, the RNA intermediate is first converted back to DNA by reverse transcriptase, an enzyme encoded by the retrotransposon. (Reverse transcriptase is also encoded by retroviruses, as you learned in Concept 17.2. In fact, retroviruses may have evolved from retrotransposons.) Another cellular enzyme catalyzes insertion of the reverse-transcribed DNA at a new site.

### Sequences Related to Transposable Elements

Multiple copies of transposable elements and sequences related to them are scattered throughout eukaryotic genomes. A single unit is usually hundreds to thousands of base pairs long, and the dispersed copies are similar but usually not identical to each other. Some of these are transposable elements that can move; the enzymes required for this movement may be encoded by any transposable element, including the one that is moving. Others are related sequences that have lost the ability to move altogether. Transposable elements and related sequences make up 25–50% of most mammalian genomes (see Figure 18.5) and even higher percentages in amphibians and

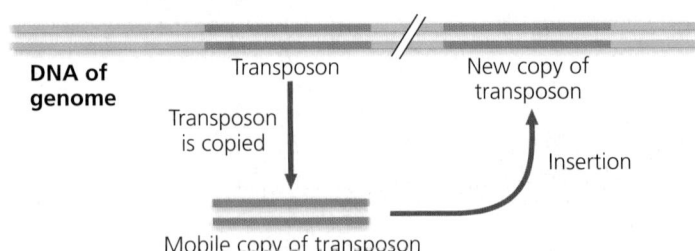

**DNA of genome**

Transposon

New copy of transposon

Transposon is copied

Insertion

Mobile copy of transposon

▲ **Figure 18.7 Transposon movement.** Movement of transposons by either the copy-and-paste mechanism (shown here) or the cut-and-paste mechanism involves a double-stranded DNA intermediate that is inserted into the genome.

**?** *How would this figure differ if it showed the cut-and-paste mechanism?*

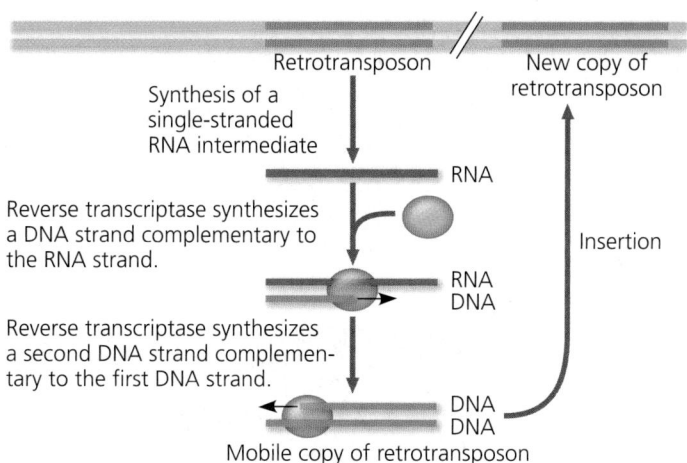

▲ **Figure 18.8 Retrotransposon movement.** Movement begins with synthesis of a single-stranded RNA intermediate. The remaining steps are essentially identical to part of the retrovirus replicative cycle (see Figure 17.8).

Synthesis of a single-stranded RNA intermediate

Reverse transcriptase synthesizes a DNA strand complementary to the RNA strand.

Reverse transcriptase synthesizes a second DNA strand complementary to the first DNA strand.

Retrotransposon

New copy of retrotransposon

RNA

RNA DNA

DNA DNA

Insertion

Mobile copy of retrotransposon

many plants. In fact, the very large size of some plant genomes is accounted for by extra transposable elements rather than by extra genes. For example, transposable elements make up 85% of the corn genome!

In humans and other primates, a large portion of transposable element–related DNA consists of a family of similar sequences called *Alu elements*. These sequences alone account for approximately 10% of the human genome. Alu elements are about 300 nucleotides long, much shorter than most functional transposable elements, and they do not code for any protein. However, many Alu elements are transcribed into RNA, and at least some of these RNAs are thought to help regulate gene expression.

An even larger percentage (17%) of the human genome is made up of a type of retrotransposon called *LINE-1*, or *L1*. These sequences are much longer than Alu elements—about 6,500 base pairs—and typically have a very low rate of transposition. However, researchers working with rats have found L1 retrotransposons to be more active in cells of the developing brain. They have proposed that different effects on gene expression of L1 retrotransposition in developing neurons may contribute to the great diversity of neuronal cell types (see Concept 37.1).

Although many transposable elements encode proteins, these proteins do not carry out normal cellular functions. Therefore, transposable elements are usually included in the noncoding DNA category, along with other repetitive sequences.

## Other Repetitive DNA, Including Simple Sequence DNA

Repetitive DNA that is not related to transposable elements has probably arisen from mistakes during DNA replication or recombination. Such DNA accounts for about 14% of the human genome (see Figure 18.5). About a third of this (5–6% of the human genome) consists of duplications of long stretches of DNA, with each segment ranging from 10,000 to 300,000 base pairs. These long segments seem to have been copied from one

chromosomal location to another site on the same or a different chromosome and probably include some functional genes.

In contrast to scattered copies of long sequences, **simple sequence DNA** contains many copies of tandemly repeated short sequences, as in the following example (showing one DNA strand only):

. . . GTTACGTTACGTTACGTTACGTTACGTTAC . . .

In this case, the repeated unit (GTTAC) consists of 5 nucleotides. Repeated units may contain as many as 500 nucleotides, but often contain fewer than 15 nucleotides, as in this example. When the unit contains 2–5 nucleotides, the series of repeats is called a **short tandem repeat (STR)**. The number of copies of the repeated unit can vary from site to site within a given genome. There could be as many as several hundred thousand repetitions of the GTTAC unit at one site, but only half that number at another.

The repeat number also varies from person to person, and since humans are diploid, each person has two alleles per site, which can differ. This diversity produces variation that can be analyzed by PCR (see Figure 13.27) and used to identify a unique set of genetic markers for each individual, his or her **genetic profile**. Forensic scientists can use STR analysis on DNA extracted from samples of tissues or body fluids to identify victims of a crime scene or natural disaster. In such an application, STR analysis is performed on STR sites selected because they have relatively few repeats and are easily sequenced. This technique has also been used by the Innocence Project, a nonprofit organization, to free more than 325 wrongly convicted people from prison.

Altogether, simple sequence DNA makes up 3% of the human genome. Much of a genome's simple sequence DNA is located at chromosomal telomeres and centromeres, suggesting that this DNA plays a structural role for chromosomes. The DNA at centromeres is essential for the separation of chromatids in cell division (see Concept 9.2). Centromeric DNA, along with simple sequence DNA located elsewhere, may also help organize the chromatin within the interphase nucleus. The simple sequence DNA located at telomeres, at the tips of chromosomes, binds proteins that protect the ends of a chromosome from degradation and from joining to other chromosomes.

Short repetitive sequences provide a challenge for wholegenome shotgun sequencing because they can hinder accurate reassembly of fragment sequences by computers. Regions of simple sequence DNA account for much of the uncertainty present in estimates of whole-genome sizes, and are the reason some genome sequences are considered "permanent drafts."

## Genes and Multigene Families

We finish our discussion of the various types of DNA sequences in eukaryotic genomes with a closer look at genes. Recall that DNA sequences that code for proteins or give rise to tRNA or rRNA compose a mere 1.5% of the human genome (see Figure 18.5). If we include introns and regulatory sequences associated with genes, the total amount of DNA that is gene related—coding and noncoding—constitutes about 25% of the human genome. Put another way, only about 6% (1.5% out of 25%) of the length of the average gene is represented in the final gene product.

Many eukaryotic genes are present as unique sequences, with only one copy per haploid set of chromosomes. But in the human genome and the genomes of many other animals and plants, such unique genes make up less than half of the total gene-related DNA. The rest occur in **multigene families**, collections of two or more identical or very similar genes.

In multigene families consisting of *identical* DNA sequences, those sequences are usually clustered tandemly and, with the notable exception of the genes for histone proteins, have RNAs as their final products. An example is the family of identical DNA sequences that are the genes for the three largest rRNA molecules **(Figure 18.9a)**. These rRNA molecules are transcribed from a single transcription unit that is repeated tandemly hundreds to thousands of times in one or several clusters in the genome of a multicellular eukaryote. The many copies of this rRNA transcription unit help cells to quickly make the millions of ribosomes needed for active protein synthesis. The primary transcript is cleaved to yield the three rRNA molecules, which combine with proteins and one other kind of rRNA (5S rRNA) to form ribosomal subunits.

The classic examples of multigene families of *nonidentical* genes are two related families of genes that encode globins, a group of proteins that include the α and β polypeptide subunits of hemoglobin. One family, located on chromosome 16 in humans, encodes various forms of α-globin; the other, on chromosome 11, encodes forms of β-globin **(Figure 18.9b)**. The different forms of each globin subunit are expressed at different times in development, allowing hemoglobin to function effectively in the changing environment of the developing animal. In humans, for example, the embryonic and fetal forms of hemoglobin have a higher affinity for oxygen than the adult forms, ensuring the efficient transfer of oxygen from mother to fetus. Also found in the globin gene family clusters are several pseudogenes.

In Concept 18.5, we'll consider the evolution of these two globin gene families, as we explore how arrangements of genes provide insight into the evolution of genomes. We'll also examine some processes that have shaped the genomes of different species over evolutionary time.

## CONCEPT CHECK 18.4

1. Discuss the characteristics of mammalian genomes that make them larger than prokaryotic genomes.
2. Which of the three mechanisms described in Figures 18.7 and 18.8 result(s) in a copy remaining at the original site as well as a copy appearing in a new location?
3. Contrast the organizations of the rRNA gene family and the globin gene families. For each, explain how the existence of a family of genes benefits the organism.

For suggested answers, see Appendix A.

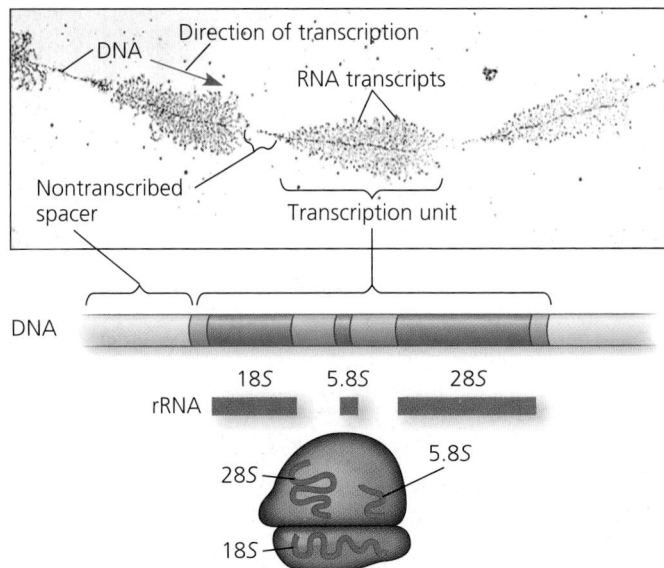

**(a) Part of the ribosomal RNA gene family.** The TEM at the top shows three of the hundreds of copies of rRNA transcription units in the rRNA gene family of a salamander genome. Each "feather" corresponds to a single unit being transcribed by about 100 molecules of RNA polymerase (dark dots along the DNA), moving left to right (red arrow). The growing RNA transcripts extend from the DNA, accounting for the feather-like appearance. In the diagram of a transcription unit below the TEM, the genes for three types of rRNA (darker blue) are adjacent to regions that are transcribed but later removed (medium blue). A single transcript is processed to yield one of each of the three rRNAs (red), key components of the ribosome.

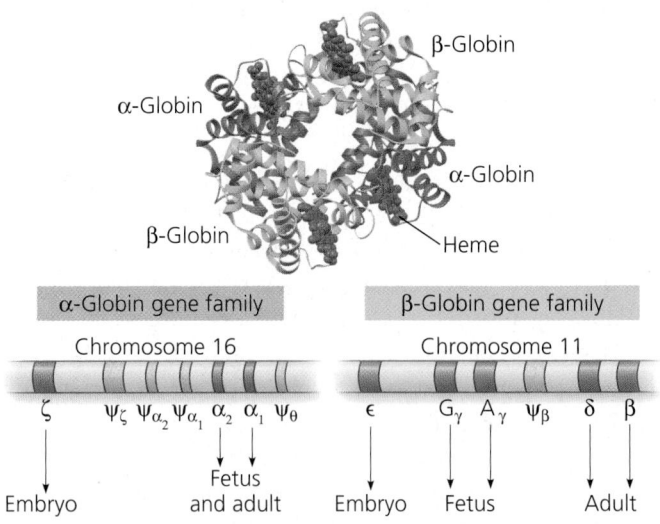

**(b) The human α-globin and β-globin gene families.** Adult hemoglobin is composed of two α-globin and two β-globin polypeptide subunits, as shown in the molecular model. The genes (darker blue) encoding α- and β-globins are found in two families, organized as shown here. The noncoding DNA (light blue) separating the functional genes within each family includes pseudogenes (ψ; gold), versions of the functional genes that no longer encode functional polypeptides. Genes and pseudogenes are named with Greek letters, as you have seen previously for the α- and β-globins. Some genes are expressed only in the embryo or fetus.

▲ **Figure 18.9  Gene families.**

**?** *In the TEM at the top of part (a), how could you determine the direction of transcription if it weren't indicated by the red arrow?*

## CONCEPT 18.5

# Duplication, rearrangement, and mutation of DNA contribute to genome evolution

**EVOLUTION**   Now that we have explored the makeup of the human genome, let's see what its composition reveals about how the genome evolved. The basis of change at the genomic level is mutation, which underlies much of genome evolution. It seems likely that the earliest forms of life had a minimal number of genes—those necessary for survival and reproduction. If this were indeed the case, one aspect of evolution must have been an increase in the size of the genome, with the extra genetic material providing the raw material for gene diversification. In this section, we'll first describe how extra copies of all or part of a genome can arise and then consider subsequent processes that can lead to the evolution of proteins (or RNA products) with slightly different or entirely new functions.

## Duplication of Entire Chromosome Sets

An accident in meiosis can result in one or more extra sets of chromosomes, a condition known as polyploidy. Although such accidents would most often be lethal, in rare cases they could facilitate the evolution of genes. In a polyploid organism, one set of genes can provide essential functions for the organism. The genes in the one or more extra sets can diverge by accumulating mutations; these variations may persist if the organism carrying them survives and reproduces. In this way, genes with novel functions can evolve. As long as one copy of an essential gene is expressed, the divergence of another copy can lead to its encoded protein acting in a novel way, thereby changing the organism's phenotype. The outcome of this accumulation of mutations may eventually be the branching off of a new species. While polyploidy is rare in animals, it is relatively common in plants, especially flowering plants (see Concept 22.2).

## Alterations of Chromosome Structure

With the recent explosion in genomic sequence information, we can now compare the chromosomal organizations of many species in detail. This information allows us to make inferences about the evolutionary processes that shape chromosomes and may drive speciation. For example, scientists have long known that sometime in the last 6 million years, when the ancestors of humans and chimpanzees diverged as species, the fusion of two ancestral chromosomes in the human line led to different haploid numbers for humans ($n = 23$) and chimpanzees ($n = 24$). The banding patterns in stained chromosomes suggested that the ancestral versions of current chimpanzee chromosomes 12 and 13 fused end to end, forming chromosome 2 in an ancestor of the human lineage. Sequencing and analysis of human chromosome 2 during the Human Genome Project provided very strong supporting evidence for the model we have just described **(Figure 18.10)**.

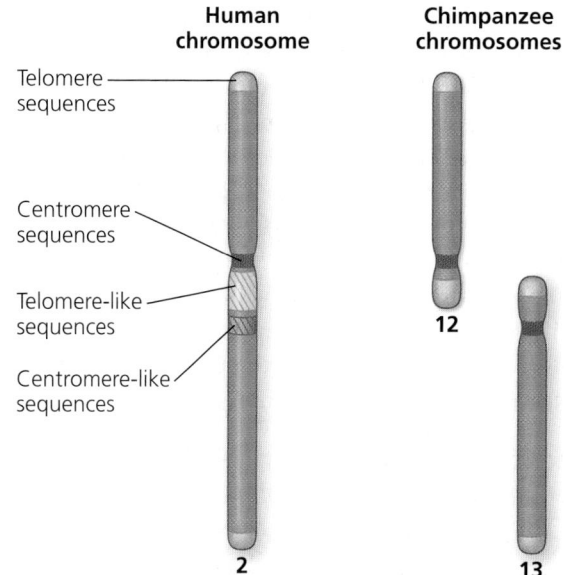

▲ **Figure 18.10  Human and chimpanzee chromosomes.** The positions of telomere-like and centromere-like sequences on human chromosome 2 (left) match those of telomeres on chimpanzee chromosomes 12 and 13 and the centromere on chimpanzee chromosome 13 (right). This suggests that chromosomes 12 and 13 in a human ancestor fused end to end to form human chromosome 2. The centromere from ancestral chromosome 12 remained functional on human chromosome 2, while the one from ancestral chromosome 13 did not.

In another study of broader scope, researchers compared the DNA sequence of each human chromosome with the whole-genome sequence of the mouse. One part of their study showed that large blocks of genes on human chromosome 16 are found on four mouse chromosomes, indicating that the genes in each block stayed together in both the mouse and human lineages during their divergent evolution from a common ancestor **(Figure 18.11)**.

Performing the same comparison of chromosomes of humans and six other mammalian species allowed the researchers to reconstruct the evolutionary history of chromosomal rearrangements in these eight species. They found many duplications and inversions of large portions of chromosomes, the result of errors during meiotic recombination in which the DNA broke and was rejoined incorrectly. The rate of these events seems to have begun accelerating about 100 million

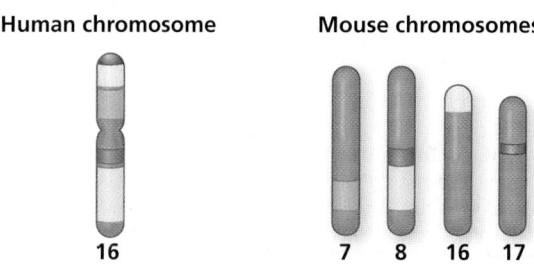

▲ **Figure 18.11  Human and mouse chromosomes.** Here, we can see that DNA sequences very similar to large blocks of human chromosome 16 (colored areas in this diagram) are found on mouse chromosomes 7, 8, 16, and 17. This finding suggests that the DNA sequence in each block has stayed together in the mouse and human lineages since the time they diverged from a common ancestor.

years ago, around 35 million years before large dinosaurs became extinct and the number of mammalian species began rapidly increasing. The apparent coincidence is interesting because chromosomal rearrangements are thought to contribute to the generation of new species. Although two individuals with different arrangements could still mate and produce offspring, the offspring would have two nonequivalent sets of chromosomes, making meiosis inefficient or even impossible. Thus, chromosomal rearrangements would reduce the success of matings between members of the two populations, a step on the way to the populations becoming two separate species. (You'll learn more about this in Concept 22.2.)

## Duplication and Divergence of Gene-Sized Regions of DNA

Errors during meiosis can also lead to the duplication of chromosomal regions that are smaller than the ones we've just discussed, including segments the length of individual genes. Unequal crossing over during prophase I of meiosis, for instance, can result in one chromosome with a deletion and another with a duplication of a particular gene. Transposable elements can provide homologous sites where nonsister chromatids can cross over, even when other chromatid regions are not correctly aligned **(Figure 18.12)**.

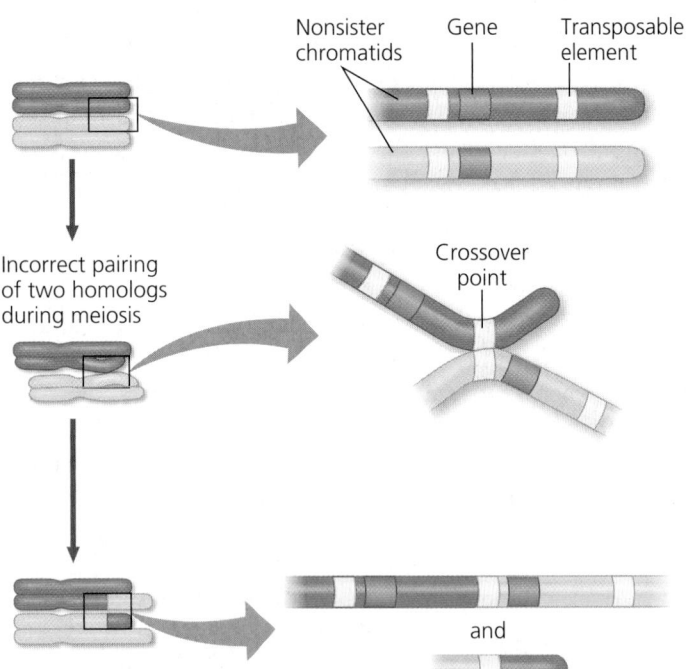

▲ **Figure 18.12 Gene duplication due to unequal crossing over.** One mechanism by which a gene (or other DNA segment) can be duplicated is recombination during meiosis between copies of a transposable element flanking the gene. Such recombination between misaligned nonsister chromatids of homologous chromosomes produces one chromatid with two copies of the gene and one chromatid with no copy.

**MAKE CONNECTIONS** *Examine how crossing over occurs in Figure 10.12. In the middle panel above, draw a line through the portions that result in the upper chromatid in the bottom panel. Use a different color to do the same for the other chromatid.*

Also, slippage can occur during DNA replication, such that the template shifts with respect to the new complementary strand, and a part of the template strand is either skipped by the replication machinery or used twice as a template. As a result, a segment of DNA is deleted or duplicated. It is easy to imagine how such errors could occur in regions of repeats. The variable number of repeated units of simple sequence DNA at a given site, used for STR analysis, is probably due to errors like these. Evidence that unequal crossing over and template slippage during DNA replication lead to duplication of genes is found in the existence of multigene families, such as the globin family.

### *Evolution of Genes with Related Functions: The Human Globin Genes*

In Figure 18.9b, you saw the organization of the α-globin and β-globin gene families as they exist in the human genome today. Now, let's consider how events such as duplications can lead to the evolution of genes with related functions like the globin genes. A comparison of gene sequences within a multigene family can suggest the order in which the genes arose. Re-creating the evolutionary history of the globin genes using this approach indicates that they all evolved from one common ancestral globin gene that underwent duplication and divergence into the α-globin and β-globin ancestral genes about 450–500 million years ago **(Figure 18.13)**. Each of these genes was later duplicated several times, and the copies then diverged from each other in sequence, yielding the current family members. In fact, the common ancestral globin gene also gave rise to the oxygen-binding muscle protein myoglobin and to the plant protein leghemoglobin. The latter two proteins function as monomers, and their genes are included in a "globin superfamily."

After the duplication events, the differences between the genes in the globin families undoubtedly arose from mutations that accumulated in the gene copies over many generations. The current model is that the necessary function provided by an α-globin protein, for example, was fulfilled by one gene, while other copies of the α-globin gene accumulated random mutations. Many mutations may have had an adverse effect on the organism, and others may have had no effect. However, a few mutations must have altered the function of the protein product in a way that helped the organism at a particular life stage without substantially changing the protein's oxygen-carrying function. Presumably, natural selection acted on these altered genes, maintaining them in the population.

In the **Scientific Skills Exercise**, you can compare amino acid sequences of the globin family members and see how such comparisons were used to generate the model for globin gene evolution shown in Figure 18.13. The existence of several pseudogenes among the functional globin genes provides additional evidence for this model (see Figure 18.9b): Random mutations in these "genes" over evolutionary time have destroyed their function.

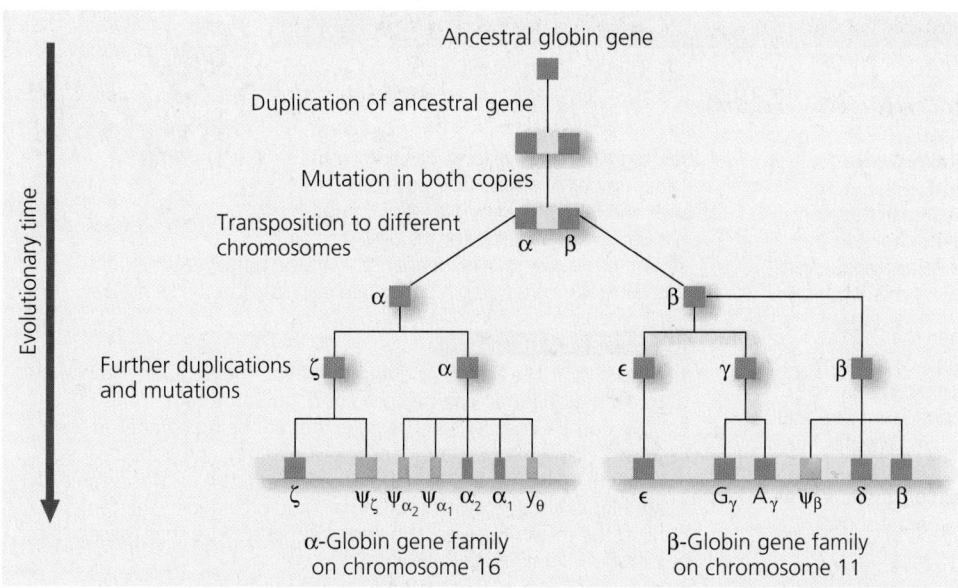

▲ **Figure 18.13** A proposed model for the sequence of events in the evolution of the human α-globin and β-globin gene families from a single ancestral globin gene.

**?** *The gold elements are pseudogenes. Explain how they could have arisen after gene duplication.*

### Evolution of Genes with Novel Functions

In the evolution of the globin gene families, gene duplication and subsequent divergence produced family members whose protein products performed similar functions (oxygen transport). Alternatively, one copy of a duplicated gene can undergo alterations that lead to a completely new function for the protein product. The genes for lysozyme and α-lactalbumin are good examples.

Lysozyme is an enzyme that helps protect animals against bacterial infection by hydrolyzing bacterial cell walls; α-lactalbumin is a nonenzymatic protein that plays a role in milk production in mammals. The two proteins are quite similar in their amino acid sequences and three-dimensional structures:

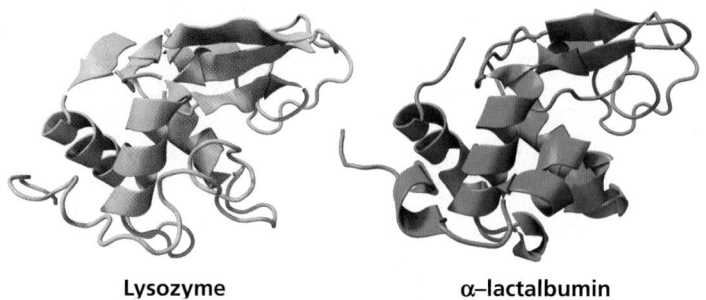

**Lysozyme**          **α–lactalbumin**

Both the lysozyme and the α-lactalbumin genes are found in mammals, but only the lysozyme gene is present in birds. These findings suggest that at some time after the lineages leading to mammals and birds had separated, the lysozyme gene was duplicated in the mammalian lineage but not in the avian lineage. Subsequently, one copy of the duplicated lysozyme gene evolved into a gene encoding α-lactalbumin, a protein with a completely different function affecting a key characteristic of mammals.

Besides the duplication and divergence of whole genes, rearrangement of existing DNA sequences within genes has also contributed to genome evolution. The presence of introns may have promoted the evolution of new proteins by facilitating the duplication or shuffling of exons, as we'll discuss next.

### Rearrangements of Parts of Genes: Exon Duplication and Exon Shuffling

Proteins often have a modular architecture consisting of discrete structural and functional regions called **domains**. One domain of an enzyme, for example, might include the active site, while another might allow the enzyme to bind to a cellular membrane. In quite a few cases, different exons code for the different domains of a protein.

We've already seen that unequal crossing over during meiosis can lead to duplication of a gene on one chromosome and its loss from the homologous chromosome (see Figure 18.12). By a similar process, a particular exon within a gene could be duplicated on one chromosome and deleted from the other. The gene with the duplicated exon would code for a protein containing a second copy of the encoded domain. This change in the protein's structure might augment its function by increasing its stability, enhancing its ability to bind a particular ligand, or altering some other property. Quite a few protein-coding genes have multiple copies of related exons, which presumably arose by duplication and then diverged. The gene encoding the extracellular matrix protein collagen is a good example. Collagen is a structural protein with a highly repetitive amino acid sequence, which reflects the repetitive pattern of exons in the collagen gene.

Alternatively, we can imagine the occasional mixing and matching of different exons either within a gene or between two different (nonallelic) genes owing to errors in meiotic recombination. This process, termed *exon shuffling*, could lead to new proteins with novel combinations of functions. As an example, let's consider the gene for tissue plasminogen activator (TPA). The TPA protein is an extracellular protein that helps control blood clotting. It has four domains of three types, each encoded by an exon; one exon is present in two copies. Because each type of exon is also found in other proteins, the current version of the gene for TPA is thought to have arisen by several

# Reading an Amino Acid Sequence Identity Table

▶ Hemoglobin

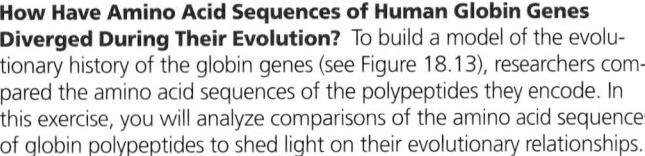

**How Have Amino Acid Sequences of Human Globin Genes Diverged During Their Evolution?** To build a model of the evolutionary history of the globin genes (see Figure 18.13), researchers compared the amino acid sequences of the polypeptides they encode. In this exercise, you will analyze comparisons of the amino acid sequences of globin polypeptides to shed light on their evolutionary relationships.

**How the Experiment Was Done** Scientists obtained the DNA sequences for each of the eight globin genes and "translated" them into amino acid sequences. They then used a computer program to align the sequences (with dashes indicating gaps in one sequence) and calculate a percent identity value for each pair of globins. The percent identity reflects the number of positions with identical amino acids relative to the total number of amino acids in a globin polypeptide. The data were displayed in a table to show the pairwise comparisons.

**Data from the Experiment** The following table shows an example of a pairwise alignment—that of the $\alpha_1$-globin (alpha-1 globin) and $\zeta$-globin (zeta globin) amino acid sequences—using the standard single-letter symbols for amino acids. To the left of each line of amino acid sequence is the number of the first amino acid in that line. The percent identity value for the $\alpha_1$- and $\zeta$-globin amino acid sequences was calculated by counting the number of matching amino acids

(86, highlighted in yellow), dividing by the total number of amino acid positions (143), and then multiplying by 100. This resulted in a 60% identity value for the $\alpha_1$-$\zeta$ pair, as shown in the amino acid identity table at the bottom of the page. The values for other globin pairs were calculated in the same way.

### INTERPRET THE DATA

1. Note that in the amino acid identity table, the data are arranged so each globin pair can be compared. (a) Some cells in the table have dashed lines. Given the pairs that are being compared for these cells, what percent identity value is implied by the dashed lines? (b) Notice that the cells in the lower left half of the table are blank. Using the information already provided in the table, fill in the missing values. Why does it make sense that these cells were left blank in the table?

2. The earlier that two genes arose from a duplicated gene, the more their nucleotide sequences can have diverged, which may result in amino acid differences in the protein products. (a) Based on that premise, identify which two genes are most divergent from each other. What is the percent amino acid identity between their polypeptides? (b) Using the same approach, identify which two globin genes are the most recently duplicated. What is the percent identity between them?

3. The model of globin gene evolution shown in Figure 18.13 suggests that an ancestral gene duplicated and mutated to become $\alpha$- and $\beta$-globin genes, and then each one was further duplicated and mutated. What features of the data set support the model?

4. Make a list of all the percent identity values from the table, starting with 100% at the top. Next to each number write the globin pair(s) with that percent identity value. Use one color for the globins from the $\alpha$ family and a different color for the globins from the $\beta$ family. (a) Compare the order of pairs on your list with their positions in the model shown in Figure 18.13. Does the order of pairs describe the same relative "closeness" of globin family members seen in the model? (b) Compare the percent identity values for pairs within the $\alpha$ or $\beta$ group to the values for between-group pairs.

Ⓜ️ A version of this Scientific Skills Exercise can be assigned in MasteringBiology.

**Further Reading** R. C. Hardison, Globin genes on the move, *Journal of Biology* 7:35.1–35.5 (2008). doi:10.1186/jbiol92

| Globin | Alignment of Globin Amino Acid Sequences |
|--------|------------------------------------------|
| $\alpha_1$ | 1 MVLSPADKTNVKAAWGKVGAHAGEYGAEAL |
| $\zeta$ | 1 MSLTKTERTIIVSMWAKISTQADTIGTETL |
| | |
| $\alpha_1$ | 31 ERMFLSFPTTKTYFPHFDLSH-GSAQVKGH |
| $\zeta$ | 31 ERLFLSHPQTKTYFPHFDL-HPGSAQLRAH |
| | |
| $\alpha_1$ | 61 GKKVADALTNAVAHVDDMPNALSALSDLHA |
| $\zeta$ | 61 GSKVVAAVGDAVKSIDDIGGALSKLSELHA |
| | |
| $\alpha_1$ | 91 HKLRVDPVNFKLLSHCLLVTLAAHLPAEFT |
| $\zeta$ | 91 YILRVDPVNFKLLSHCLLVTLAARFPADFT |
| | |
| $\alpha_1$ | 121 PAVHASLDKFLASVSTVLTSKYR |
| $\zeta$ | 121 AEAHAAWDKFLSVVSSVLTEKYR |

| Amino Acid Identity Table | | | | | | | | |
|---|---|---|---|---|---|---|---|---|
| | | **α Family** | | | **β Family** | | | |
| | | $\alpha_1$ (alpha 1) | $\alpha_2$ (alpha 2) | $\zeta$ (zeta) | $\beta$ (beta) | $\delta$ (delta) | $\epsilon$ (epsilon) | $A_\gamma$ (gamma A) | $G_\gamma$ (gamma G) |
| **α Family** | $\alpha_1$ | ----- | 100 | 60 | 45 | 44 | 39 | 42 | 42 |
| | $\alpha_2$ | | ----- | 60 | 45 | 44 | 39 | 42 | 42 |
| | $\zeta$ | | | ----- | 38 | 40 | 41 | 41 | 41 |
| **β Family** | $\beta$ | | | | ----- | 93 | 76 | 73 | 73 |
| | $\delta$ | | | | | | 73 | 71 | 72 |
| | $\epsilon$ | | | | | | ----- | 80 | 80 |
| | $A_\gamma$ | | | | | | | ----- | 99 |
| | $G_\gamma$ | | | | | | | | ----- |

Compiled using data from the National Center for Biotechnology Information (NCBI).

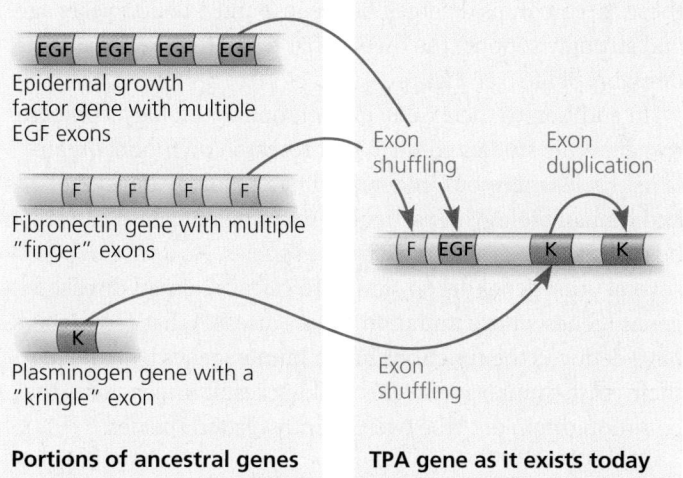

| Portions of ancestral genes | TPA gene as it exists today |
|---|---|

EGF EGF EGF EGF
Epidermal growth
factor gene with multiple
EGF exons

F F F F
Fibronectin gene with multiple
"finger" exons

K
Plasminogen gene with a
"kringle" exon

Exon shuffling
Exon duplication
Exon shuffling

F EGF K K

▲ **Figure 18.14 Evolution of a new gene by exon shuffling.**
Meiotic errors could have moved exons, each encoding a particular
domain, from ancestral forms of the genes for epidermal growth
factor, fibronectin, and plasminogen (left) into the evolving gene for
tissue plasminogen activator, TPA (right). Duplication of the "kringle"
exon (K) from the plasminogen gene after its movement could account
for the two copies of this exon in the TPA gene existing today.

**?** *How could the presence of transposable elements within introns
have facilitated the exon shuffling shown here?*

instances of exon shuffling during errors in meiotic recombination and subsequent duplication **(Figure 18.14)**.

## How Transposable Elements Contribute to Genome Evolution

The persistence of transposable elements as a large fraction of
some eukaryotic genomes is consistent with the idea that they
play an important role in shaping a genome over evolutionary
time. These elements can contribute to the evolution of the
genome in several ways. They can promote recombination,
disrupt cellular genes or control elements, and carry entire
genes or individual exons to new locations.

Transposable elements of similar sequence scattered
throughout the genome facilitate recombination between different chromosomes by providing homologous regions for
crossing over (see Figure 18.12). Most of these events are probably detrimental, causing chromosomal translocations and
other changes in the genome that may be lethal. But over the
course of evolutionary time, an occasional recombination event
of this sort may help the organism. (For the change to be heritable, it must happen in a cell that will give rise to a gamete.)

The movement of a transposable element can have a variety
of consequences. For instance, a transposable element that
"jumps" into a protein-coding sequence will prevent production of a normal transcript of the gene. If it inserts within a
regulatory sequence, the transposition may lead to increased
or decreased production of one or more proteins. Transposition caused both types of effects on the genes coding for
pigment-synthesizing enzymes in McClintock's corn kernels.
Again, while such changes are usually harmful, in the long run
some may provide a survival advantage.

During transposition, a transposable element may carry
along a gene or even a group of genes to a new position in the
genome. This mechanism probably accounts for the location
of the α-globin and β-globin gene families on different human
chromosomes, as well as the dispersion of the genes of certain
other gene families. By a similar tag-along process, an exon
from one gene may be inserted into another gene in a mechanism similar to that of exon shuffling during recombination.
For example, an exon may be inserted by transposition into the
intron of a protein-coding gene. If the inserted exon is retained
in the RNA transcript during RNA splicing, the protein that is
synthesized will have an additional domain, which may confer
a new function on the protein.

Most often, the processes discussed in this section produce
harmful effects, which may be lethal, or no effect. In a few
cases, however, small heritable changes may occur that are beneficial. Over many generations, the resulting genetic diversity
provides valuable raw material for natural selection. Diversification of genes and their products is an important factor in the
evolution of new species. Thus, the accumulation of changes in
the genome of each species provides a record of its evolutionary history. To read this record, we must be able to identify
genomic changes. Comparing the genomes of different species
allows us to do that, increasing our understanding of how genomes evolve. You'll learn more about these topics next.

**CONCEPT CHECK 18.5**

1. Describe three examples of errors in cellular processes that
lead to DNA duplications.
2. Explain how multiple exons might have arisen in the ancestral
EGF and fibronectin genes shown in Figure 18.14 (left).
3. What are three ways that transposable elements are thought
to contribute to genome evolution?

For suggested answers, see Appendix A.

**CONCEPT 18.6**

# Comparing genome sequences provides clues to evolution and development

**EVOLUTION** One researcher has likened the current state
of biology to the Age of Exploration in the 1400s, which occurred soon after major improvements in navigation and ship
design. In the last 25 years, we have seen rapid advances in
genome sequencing and data collection, new techniques for
assessing gene activity across the whole genome, and refined
approaches for understanding how genes and their products
work together in complex systems. We are truly poised on the
brink of a new world.

Comparisons of genome sequences from different species
reveal a lot about the evolutionary history of life, from very
ancient to more recent. Similarly, comparative studies of the

genetic programs that direct embryonic development in different species are beginning to clarify the mechanisms that generated the great diversity of life-forms present today. In this final section of the chapter, we'll discuss what has been learned from these two approaches.

## Comparing Genomes

The more similar in sequence the genes and genomes of two species are, the more closely related those species are in their evolutionary history. Comparing genomes of closely related species sheds light on more recent evolutionary events, whereas comparing genomes of very distantly related species helps us understand ancient evolutionary history. In either case, learning about characteristics that are shared or divergent between groups enhances our picture of the evolution of organisms and biological processes. Evolutionary relationships between species can be represented by a diagram in the form of a tree (often turned sideways), where each branch point marks the divergence of two lineages (see Figure 1.17). **Figure 18.15** shows the evolutionary relationships of some groups and species we'll now discuss.

### Comparing Distantly Related Species

Determining which genes have remained similar—that is, are *highly conserved*—in distantly related species can help clarify evolutionary relationships among species that diverged from each other long ago. Indeed, comparisons of specific gene sequences of bacteria, archaea, and eukaryotes indicate that

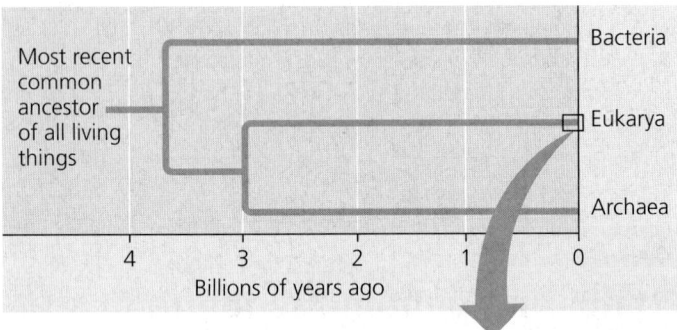

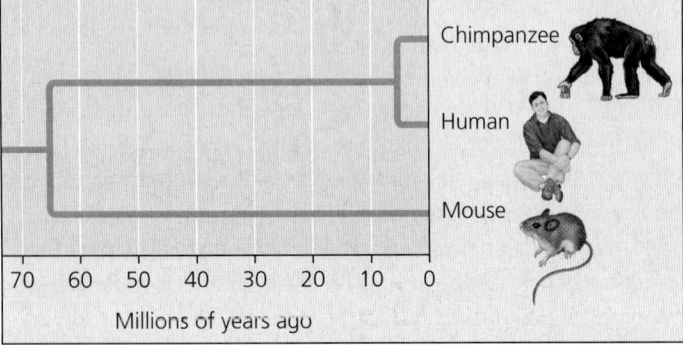

▲ **Figure 18.15 Evolutionary relationships of the three domains of life.** The tree diagram at the top shows the ancient divergence of bacteria, archaea, and eukaryotes. A portion of the eukaryote lineage is expanded in the inset to show the more recent divergence of three mammalian species discussed in this chapter.

these three groups diverged between 2 and 4 billion years ago and strongly support the theory that they are the fundamental domains of life (see Figure 18.15).

In addition to their value in evolutionary biology, comparative genomic studies confirm that research on model organisms is relevant to our understanding of biology in general and human biology in particular. Very ancient genes can still be surprisingly similar in disparate species. As a case in point, several yeast genes are so similar to certain human disease genes (genes whose mutation causes disease) that researchers have deduced the functions of the human genes by studying their yeast counterparts. This striking result underscores the common origin of these two distantly related species.

### Comparing Closely Related Species

The genomes of two closely related species are likely to be organized similarly because of their relatively recent divergence. In the past, this kind of similarity allowed the fully sequenced genome of one species to be used as a scaffold for assembling the genomic sequences of a closely related species, accelerating mapping of the second genome. For instance, using the human genome sequence as a guide, researchers were able to quickly sequence the entire chimpanzee genome. With the advent of new and faster sequencing techniques, most genomes are assembled individually, as has been done recently for the bonobo and gorilla genomes. (Along with chimpanzees, bonobos are the other African ape species that are the closest living relatives to humans.)

The recent divergence of two closely related species also underlies the small number of gene differences that are found when their genomes are compared. The particular genetic differences can therefore be more easily correlated with phenotypic differences between the two species. An exciting application of this type of analysis is seen as researchers compare the human genome with the genomes of the chimpanzee, mouse, rat, and other mammals. Identifying the genes shared by all of these species but not by nonmammals will give us clues about what it takes to make a mammal, while finding the genes shared by chimpanzees and humans but not by rodents tells us something about primates. And, of course, comparing the human genome with that of the chimpanzee helps us answer the tantalizing question we asked at the beginning of the chapter: What genomic information makes a human or a chimpanzee?

An analysis of the overall composition of the human and chimpanzee genomes, which are thought to have diverged only about 6 million years ago (see Figure 18.15), reveals some general differences. Considering single nucleotide substitutions, the two genomes differ by only 1.2%. When researchers looked at longer stretches of DNA, however, they were surprised to find a further 2.7% difference due to insertions or deletions of larger regions in the genome of one or the other species; many of the insertions were duplications or other repetitive DNA. In fact, a third of the human duplications are not present in the chimpanzee genome, and some of these duplications contain regions associated with human diseases. There are more Alu elements in the human genome than in the chimpanzee genome, and the latter contains many copies of a retroviral provirus not present in humans. All of these observations provide clues to the forces that might have swept the two genomes along different paths, but we don't have a complete picture yet.

The sequencing of the bonobo genome, completed in 2012, revealed that in some regions, human sequences were more closely related to either chimpanzee or bonobo sequences than chimpanzee or bonobo sequences were to each other. Such a fine-grained comparison of three closely related species allows even more detail to be worked out in reconstructing their related evolutionary history.

We don't know how the genetic differences revealed by genome sequencing might account for the distinct characteristics of each species. To discover the basis for the phenotypic differences between chimpanzees and humans, biologists are studying specific genes and types of genes that differ between the two species and comparing them with their counterparts in other mammals. This approach has revealed a number of genes that are apparently changing (evolving) faster in the human than in either the chimpanzee or the mouse. Among them are genes involved in defense against malaria and tuberculosis as well as at least one gene that regulates brain size. When genes are classified by function, the genes that seem to be evolving the fastest are those that code for transcription factors. This discovery makes sense because transcription factors regulate gene expression and thus play a key role in orchestrating the overall genetic program.

One transcription factor whose gene shows evidence of rapid change in the human lineage is called *FOXP2*. Several lines of evidence suggest that the *FOXP2* gene functions in vocalization in vertebrates. For one thing, mutations in this gene can produce severe speech and language impairment in humans. Moreover, the *FOXP2* gene is expressed in the brains of zebra finches and canaries at the time when these songbirds are learning their songs. But perhaps the strongest evidence comes from a "knockout" experiment in which researchers disrupted the *FOXP2* gene in mice and analyzed the resulting phenotype. Normal mice produce ultrasonic squeaks (whistles) to communicate stress, but mice that were homozygous for a mutated form of *FOXP2* had malformed brains and failed to vocalize normally **(Figure 18.16)**. Heterozygous mice, with one faulty copy of the gene, also showed vocalization defects. These results augmented the evidence from birds and humans, supporting the idea that the *FOXP2* gene product turns on genes involved in vocalization.

In 2010, the Neanderthal genome was sequenced from a very small amount of preserved genomic DNA. Neanderthals (*Homo neanderthalensis*) are members of the same genus to which humans (*Homo sapiens*) belong (see Concept 27.6). A reconstruction of their evolutionary history based on genomic comparisons between the two species suggests that some groups of humans and Neanderthals coexisted and interbred

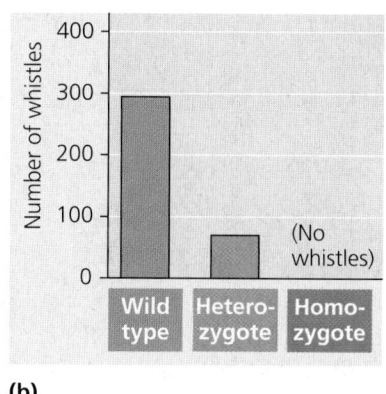

**(a)**            **(b)**

▲ **Figure 18.16 The function of *FOXP2*, a gene that is rapidly evolving in the human lineage. (a)** Wild-type mice emit ultrasonic squeaks (whistles) to communicate stress. **(b)** Researchers used genetic engineering to produce mice in which one or both copies of *FOXP2* were disrupted, separated each newborn pup from its mother, and recorded the number of ultrasonic whistles produced by the pup. No vocalization was observed in homozygous mutants, and the effect on heterozygotes was also extreme.

for a period of time before Neanderthals went extinct about 30,000 years ago. While Neanderthals have sometimes been portrayed as primitive beings that could only grunt, their *FOXP2* gene sequence encodes an identical protein to that of humans. This suggests that Neanderthals may have been capable of speech and, along with other observed genetic similarities, forces us to reevaluate our image of our recent extinct relatives.

The *FOXP2* story is an excellent example of how different approaches can complement each other in uncovering biological phenomena of widespread importance. The *FOXP2* experiments used mice as a model for humans because it would be unethical (as well as impractical) to carry out such experiments in humans. Mice and humans, which diverged about 65.5 million years ago (see Figure 18.15), share about 85% of their genes. This genetic similarity can be exploited in studying human genetic disorders. If researchers know the organ or tissue that is affected by a particular genetic disorder, they can look for genes that are expressed in these locations in mice.

Further research efforts are under way to extend genomic studies to many more species, including neglected species from diverse branches of the tree of life. These studies will advance our understanding of evolution, as well as all other aspects of biology, from human health to ecology.

### Comparing Genomes Within a Species

Another exciting consequence of our ability to analyze genomes is our growing understanding of the spectrum of genetic variation in humans. Because the history of the human species is so short—probably about 200,000 years—the amount of DNA variation among humans is small compared with that of many other species. Much of our diversity seems to be in the form of **single nucleotide polymorphisms** (**SNPs**, pronounced "snips"), defined as single base-pair sites where variation is found in at least 1% of the population. Usually detected by sequencing of DNA, SNPs occur on average about once in 100–300 base pairs in the human genome. Scientists have already identified the location of several million human SNP sites and continue to find more.

In the course of this search, they have also found other variations—including chromosomal regions with inversions, deletions, and duplications. The most surprising discovery has been the widespread occurrence of *copy-number variants* (*CNVs*), loci where some individuals have one or multiple copies of a particular gene or genetic region, rather than the standard two copies (one on each homolog). CNVs result from regions of the genome being duplicated or deleted inconsistently within the population. One study of 40 people found more than 8,000 CNVs involving 13% of the genes in the genome, and these CNVs probably represent just a small subset of the total. Since these variants encompass much longer stretches of DNA than the single nucleotides of SNPs, CNVs are more likely to have phenotypic consequences and to play a role in complex diseases and disorders. At the very least, the high incidence of copy-number variation blurs the meaning of the phrase "a normal human genome."

Copy-number variants, SNPs, and variations in repetitive DNA such as short tandem repeats (STRs) are useful genetic markers for studying human evolution. In one study, the genomes of two Africans from different communities were sequenced: Archbishop Desmond Tutu, the South African civil rights advocate and a member of the Bantu tribe, the majority population in southern Africa, and !Gubi, a hunter-gatherer from the Khoisan community in Namibia, a minority African population that is probably the human group with the oldest known lineage. The comparison revealed many differences, as you might expect. The analysis was then broadened to compare the protein-coding regions of !Gubi's genome with those of three other Khoisan community members (self-identified Bushmen) living nearby. Remarkably, the four African genomes differed more from each other than a European would from an Asian. These data highlight the extensive diversity among African genomes. Extending this approach will help us answer questions about the differences between human populations and the migratory routes of human populations.

## Widespread Conservation of Developmental Genes Among Animals

Biologists in the field of evolutionary developmental biology, or **evo-devo** as it is often called, compare developmental processes of different multicellular organisms. Their aim is to understand how these processes have evolved and how changes in them can modify existing organismal features or lead to new ones. With the advent of molecular techniques and the recent flood of genomic information, we are beginning to realize that the genomes of related species with strikingly different forms may have only minor differences in gene sequence or, perhaps more importantly, in gene regulation. Discovering the molecular basis of these differences helps us understand the origins of diverse forms, thus informing our study of evolution.

In Concept 16.1, you learned about the homeotic genes in *Drosophila melanogaster*, which encode transcription factors that regulate gene expression and specify the identity of body segments in the fruit fly (see Figure 16.8). Molecular analysis of the homeotic genes in *Drosophila* has shown that they all include a 180-nucleotide sequence called a **homeobox**, which codes for a 60-amino-acid *homeodomain* in the encoded proteins. An identical or very similar nucleotide sequence has been discovered in the homeotic genes of many invertebrates and vertebrates. As shown in **Figure 18.17**, the resemblance even extends to the organization of these genes: The vertebrate genes homologous to the homeotic genes of fruit flies have

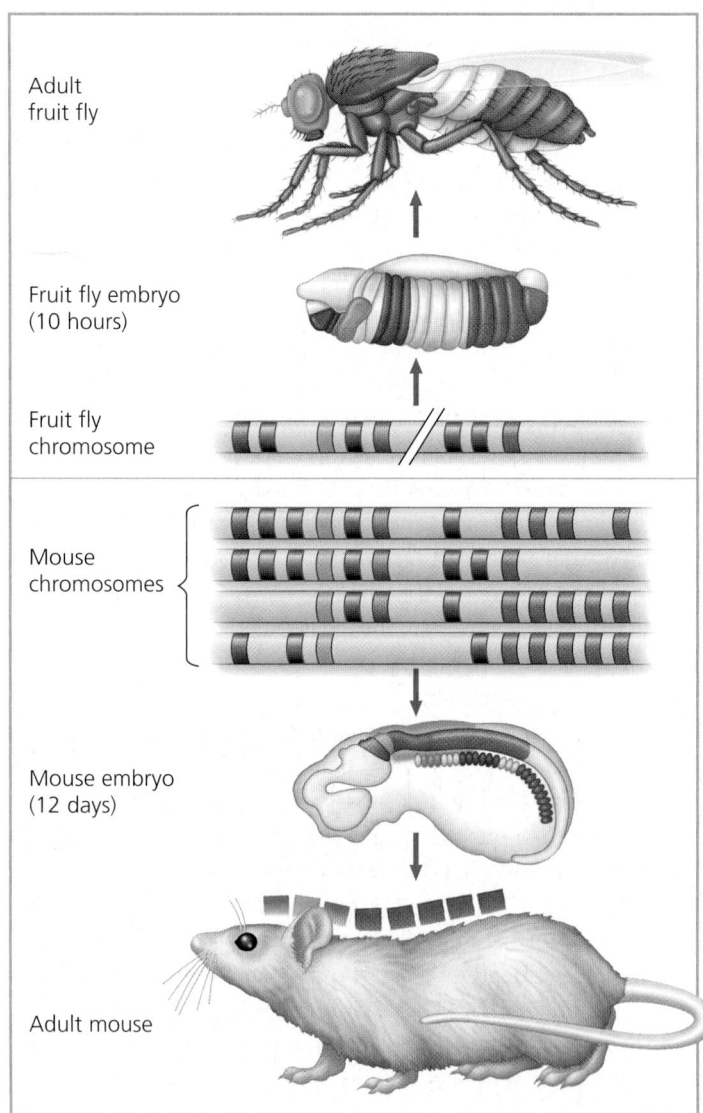

▲ **Figure 18.17 Conservation of homeotic genes in a fruit fly and a mouse.** Homeotic genes that control the form of anterior and posterior structures of the body occur in the same linear sequence on chromosomes in *Drosophila* and mice. Each colored band on the chromosomes shown here represents a homeotic gene. In fruit flies, all homeotic genes are found on one chromosome. The mouse and other mammals have the same or similar sets of genes on four chromosomes. The color code indicates the parts of the embryos in which these genes are expressed and the adult body regions that result. All of these genes are essentially identical in flies and mice, except for those represented by black bands, which are less similar in the two animals.

kept the same chromosomal arrangement. In fact, the nucleotide sequences are so similar between humans and fruit flies that one researcher has whimsically referred to flies as "little people with wings." Homeobox-containing sequences have also been found in regulatory genes of much more distantly related eukaryotes, including plants and yeasts. From these similarities, we can deduce that the homeobox DNA sequence evolved very early and was sufficiently beneficial to organisms to have been conserved in animals and plants virtually unchanged for hundreds of millions of years.

Homeotic genes in animals were named *Hox* genes, short for *homeobox*-containing genes, because homeotic genes were the first genes found to have this sequence. Other homeobox-containing genes were later found that do not act as homeotic genes; that is, they do not directly control the identity of body parts. However, most of these genes, in animals at least, are associated with development, suggesting their ancient and fundamental importance in that process. In *Drosophila*, for example, homeoboxes are present not only in the homeotic genes but also in the egg-polarity gene *bicoid* (see Figures 16.9 and 16.10), in several of the segmentation genes, and in a master regulatory gene for eye development.

Researchers have discovered that the homeobox-encoded homeodomain is the part of a protein that binds to DNA when the protein functions as a transcription factor. Elsewhere in the protein, domains that are more variable interact with other transcription factors, allowing the homeodomain-containing protein to recognize specific enhancers and regulate the associated genes. Proteins with homeodomains probably regulate development by coordinating the transcription of batteries of developmental genes, switching them on or off. In embryos of *Drosophila* and other animal species, different combinations of homeobox genes are active in different parts of the embryo. This selective expression of regulatory genes, varying over time and space, is central to pattern formation.

Developmental biologists have found that in addition to homeotic genes, many other genes involved in development are highly conserved from species to species. These include numerous genes encoding components of signaling pathways. The extraordinary similarity among some developmental genes in different animal species raises a question: How can the same genes be involved in the development of animals whose forms are so very different from each other?

Ongoing studies are suggesting answers to this question. In some cases, small changes in regulatory sequences of particular genes cause changes in gene expression patterns that can lead to major changes in body form. For example, the differing patterns of expression of the *Hox* genes along the body axis in a crustacean and an insect can explain the variation in the number of leg-bearing segments among these closely related animals **(Figure 18.18)**. In other cases, similar genes direct different developmental processes in various organisms, resulting in diverse body shapes. Several *Hox* genes, for instance, are expressed in the embryonic and larval stages of the sea urchin,

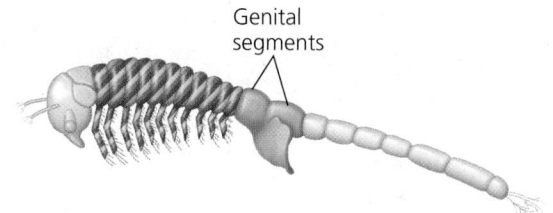

**(a) Expression of four *Hox* genes in the brine shrimp *Artemia***

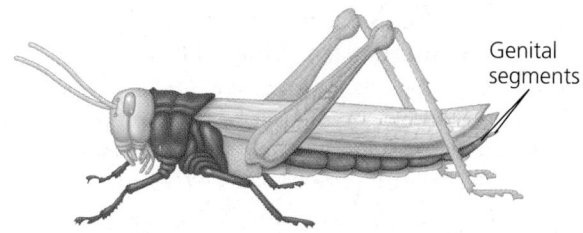

**(b) Expression of the grasshopper versions of the same four *Hox* genes**

© 1995 The Royal Society

▲ **Figure 18.18 Effect of differences in *Hox* gene expression in a crustacean and an insect.** Changes in the expression patterns of *Hox* genes have occurred over evolutionary time since insects diverged from a crustacean ancestor. These changes account in part for the different body plans of the brine shrimp *Artemia*, a crustacean (top), and the grasshopper, an insect. Shown here are regions of the adult body color-coded for expression of four *Hox* genes that determine the formation of particular body parts during embryonic development. Each color represents a specific *Hox* gene. Colored stripes on *Artemia* indicate co-expression of three *Hox* genes.

a nonsegmented animal that has a body plan quite different from those of insects and mice. Sea urchin adults make the pincushion-shaped shells you may have seen on the beach. Sea urchins are among the organisms long used in classical embryological studies (see Concept 36.4).

In this final chapter of the genetics unit, you have learned how studying genomic composition and comparing genomes of different species can illuminate the process by which genomes evolve. Furthermore, comparing developmental programs, we can see that the unity of life is reflected in the similarity of molecular and cellular mechanisms used to establish body pattern, although the genes directing development may differ among organisms. The similarities between genomes reflect the common ancestry of life on Earth. But the differences are also crucial, for they have created the huge diversity of organisms that have evolved. In the remaining chapters, we expand our perspective beyond the level of molecules, cells, and genes to explore this diversity on the organismal level.

**CONCEPT CHECK 18.6**

1. Would you expect the genome of the macaque (a monkey) to be more similar to the mouse genome or to the human genome? Explain.
2. DNA sequences called homeoboxes help homeotic genes in animals direct development. Given that they are common to flies and mice, explain why these animals are so different.

For suggested answers, see Appendix A.

# 18 Chapter Review

Go to **MasteringBiology**® for Assignments, the eText, and the Study Area with Animations, Activities, Vocab Self-Quiz, and Practice Tests.

## SUMMARY OF KEY CONCEPTS

**VOCAB SELF-QUIZ**

goo.gl/gbai8v

### CONCEPT 18.1

**The Human Genome Project fostered development of faster, less expensive sequencing techniques (pp. 358–359)**

- The **Human Genome Project** was largely completed in 2003, aided by major advances in sequencing technology.
- In the **whole-genome shotgun approach**, the whole genome is cut into many small, overlapping fragments that are sequenced; computer software then assembles the genome sequence.

**?** *How did the Human Genome Project result in more rapid, less expensive DNA sequencing technology?*

### CONCEPT 18.2

**Scientists use bioinformatics to analyze genomes and their functions (pp. 359–361)**

- Computer analysis of genome sequences aids the identification of protein-coding sequences. Methods to determine gene function include comparing the sequences of newly discovered genes with those of known genes in other species, and also observing the phenotypic effects of experimentally inactivating the genes.
- In **systems biology**, researchers aim to model the dynamic behavior of whole biological systems based on the study of the interactions among the system's parts. For example, scientists use the computer-based tools of bioinformatics to compare genomes and to study sets of genes and proteins as whole systems (**genomics** and **proteomics**, respectively). Studies include large-scale analyses of functional DNA elements and of genes contributing to medical conditions.

**?** *What has been the most significant finding of the ENCODE pilot project? Why was the project expanded to include nonhuman species?*

### CONCEPT 18.3

**Genomes vary in size, number of genes, and gene density (pp. 361–363)**

|  | Bacteria | Archaea | Eukarya |
|---|---|---|---|
| **Genome size** | Most are 1–6 Mb | | Most are 10–4,000 Mb, but a few are much larger |
| **Number of genes** | 1,500–7,500 | | 5,000–40,000 |
| **Gene density** | Higher than in eukaryotes | | Lower than in prokaryotes (Within eukaryotes, lower density is correlated with larger genomes.) |
| **Introns** | None in protein-coding genes | Present in some genes | Present in most genes of multicellular eukaryotes, but only in some genes of unicellular eukaryotes |
| **Other noncoding DNA** | Very little | | Can exist in large amounts; generally more repetitive noncoding DNA in multicellular eukaryotes |

**?** *Compare genome size, gene number, and gene density (a) in the three domains and (b) among eukaryotes.*

### CONCEPT 18.4

**Multicellular eukaryotes have much noncoding DNA and many multigene families (pp. 363–366)**

- Only 1.5% of the human genome codes for proteins or gives rise to rRNAs or tRNAs; the rest is noncoding DNA, including **pseudogenes** and **repetitive DNA** of unknown function.
- The most abundant type of repetitive DNA in multicellular eukaryotes consists of **transposable elements** and related sequences. In eukaryotes, there are two types of transposable elements: **transposons**, which move via a DNA intermediate, and **retrotransposons**, which are more prevalent and move via an RNA intermediate.
- Other repetitive DNA includes short noncoding sequences that are tandemly repeated thousands of times (**simple sequence DNA**, which includes **STRs**); these sequences are especially prominent in centromeres and telomeres, where they probably play structural roles in the chromosome.
- Though many eukaryotic genes are present in one copy per haploid chromosome set, others are members of a gene family, such as the human globin gene families:

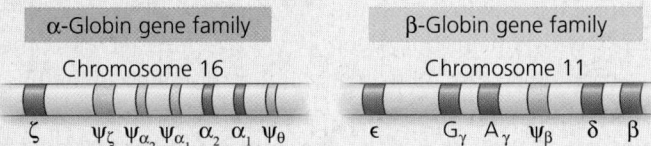

**?** *Explain how the function of transposable elements might account for their prevalence in human noncoding DNA.*

### CONCEPT 18.5

**Duplication, rearrangement, and mutation of DNA contribute to genome evolution (pp. 367–371)**

- Errors in cell division can lead to extra copies of all or part of entire chromosome sets, which may then diverge if one set accumulates sequence changes. Polyploidy occurs more often in plants than animals and contributes to speciation.
- The chromosomal organization of genomes can be compared among species, providing information about evolutionary relationships. Within a given species, rearrangements of chromosomes are thought to contribute to the emergence of new species.
- The genes encoding the various related but different globin proteins evolved from one common ancestral globin gene, which duplicated and diverged into α-globin and β-globin ancestral genes. Subsequent duplication and random mutation gave rise to the present globin genes, all of which code for oxygen-binding proteins. The copies of some duplicated genes have diverged so much that the functions of their encoded proteins (such as lysozyme and α-lactalbumin) are now substantially different.
- Each exon may code for a **domain**, a discrete structural and functional region of a protein. Rearrangement of exons within and between genes during evolution has led to genes containing multiple copies of similar exons and/or several different exons derived from other genes.
- Movement of transposable elements or recombination between copies of the same element can generate new sequence combinations

that are beneficial to the organism. These may alter the functions of genes or their patterns of expression and regulation.

> **?** *How could chromosomal rearrangements lead to the emergence of new species?*

CONCEPT 18.6

## Comparing genome sequences provides clues to evolution and development (pp. 371–375)

- Comparisons of genomes from widely divergent and closely related species provide valuable information about ancient and more recent evolutionary history, respectively. Analysis of **single nucleotide polymorphisms (SNPs)** and copy-number variants (CNVs) within a species can also shed light on the evolution of that species.
- Evolutionary developmental (**evo-devo**) biologists have shown that homeotic genes and some other genes associated with animal development contain a **homeobox** region whose sequence is highly conserved among diverse species. Related sequences are present in the genes of plants and yeasts.

> **?** *What type of information can be obtained by comparing the genomes of closely related species? Of very distantly related species?*

# TEST YOUR UNDERSTANDING

**PRACTICE TEST**

goo.gl/CRZjvS

## Level 1: Knowledge/Comprehension

1. Bioinformatics includes all of the following except
   - (A) using computer programs to align DNA sequences.
   - (B) using DNA technology to combine DNA from two different sources in a test tube.
   - (C) developing computer-based tools for genome analysis.
   - (D) using mathematical tools to analyze biological systems.

2. Homeotic genes
   - (A) encode transcription factors that control the expression of genes responsible for specific anatomical structures.
   - (B) are found only in *Drosophila* and other arthropods.
   - (C) are the only genes that contain the homeobox domain.
   - (D) encode proteins that form anatomical structures in the fly.

## Level 2: Application/Analysis

3. Two eukaryotic proteins have one domain in common but are otherwise very different. Which of the following processes is most likely to have contributed to this similarity?
   - (A) gene duplication
   - (B) alternative splicing
   - (C) exon shuffling
   - (D) random point mutations

4. **DRAW IT** Below are the amino acid sequences (using the single-letter code; see Figure 3.18) of four short segments of the *FOXP2* protein from six species: chimpanzee (C), orangutan (O), gorilla (G), rhesus macaque (R), mouse (M), and human (H). These segments contain all of the amino acid differences between the *FOXP2* proteins of these species.

   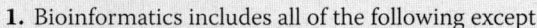

   1. ATETI...PKSSD...TSSTT...NARRD
   2. ATETI...PKSSE...TSSTT...NARRD
   3. ATETI...PKSSD...TSSTT...NARRD
   4. ATETI...PKSSD...TSSNT...SARRD
   5. ATETI...PKSSD...TSSTT...NARRD
   6. VTETI...PKSSD...TSSTT...NARRD

Use a highlighter to color any amino acid that varies among the species. (Color that amino acid in all sequences.)
   - (a) The C, G, R sequences are identical. Identify which lines correspond to those sequences.
   - (b) The H sequence differs from that of the C, G, R species at two amino acids. Underline the two differences in the H sequence.
   - (c) The O sequence differs from the C, G, R sequences at one amino acid (having V instead of A) and from the H sequence at three amino acids. Identify the O sequence.
   - (d) In the M sequence, circle the amino acid(s) that differ from the C, G, R sequences, and draw a square around those that differ from the H sequence. Describe these differences.
   - (e) Primates and rodents diverged between 60 and 100 million years ago, and chimpanzees and humans diverged about 6 million years ago. Compare the amino acid differences between the mouse and the C, G, R species with those between the human and the C, G, R species. What can you conclude?

## Level 3: Synthesis/Evaluation

5. **SCIENTIFIC INQUIRY**
   The scientists mapping human SNPs noticed that groups of SNPs tended to be inherited together, in blocks known as "haplotypes," ranging from 5,000 to 200,000 base pairs. There are only four or five commonly occurring combinations of SNPs per haplotype. Integrating what you've learned throughout this chapter and this unit, propose an explanation for this observation.

6. **FOCUS ON EVOLUTION**
   Genes important in the embryonic development of animals, such as homeobox-containing genes, have been relatively well conserved during evolution; that is, they are more similar among different species than are many other genes. Explain why.

7. **FOCUS ON INFORMATION**
   The continuity of life is based on heritable information in the form of DNA. In a short essay (100–150 words), explain how mutations in protein-coding genes and regulatory DNA contribute to evolution.

8. **SYNTHESIZE YOUR KNOWLEDGE**

Insects have three thoracic (trunk) segments. While researchers have found fossils with wings on all segments, modern insects have wings or related structures on only the second and third segments. *Hox* gene products inhibit wing formation on the first segment. The treehopper is an exception. Its first segment has an ornate helmet that is a modified fused pair of "wings," which provide camouflage in branches, reducing risk of predation. Explain how changes in gene regulation could have led to this evolution.

*For selected answers, see Appendix A.*

# Unit 3 Evolution

## 19 Descent with Modification

Darwin proposed that the diversity of life and the match between organisms and their environments arose through **natural selection** over time, as species adapted to their environments.

## 20 Phylogeny

As organisms adapt to their environments over time, they become increasingly different from their ancestors. To reconstruct an organism's evolutionary history, or **phylogeny**, biologists use data ranging from fossils to molecules.

## 21 The Evolution of Populations

The evolutionary impact of natural selection appears in the genetic changes of a **population** of organisms over time.

## 22 The Origin of Species

Evolutionary changes in a population ultimately can result in **speciation**, a process in which one species gives rise to two or more species.

## 23 Broad Patterns of Evolution

As speciation occurs again and again, new groups of organisms arise while others disappear. These changes make up the **broad patterns of evolutionary change** documented in the fossil record.

Genetics

Chemistry and Cells

Ecology

Animals

Plants

History of Life

Evolution

19 Descent with Modification

20 Phylogeny

21 The Evolution of Populations

22 The Origin of Species

23 Broad Patterns of Evolution

## KEY CONCEPTS

**19.1** The Darwinian revolution challenged traditional views of a young Earth inhabited by unchanging species

**19.2** Descent with modification by natural selection explains the adaptations of organisms and the unity and diversity of life

**19.3** Evolution is supported by an overwhelming amount of scientific evidence

▲ **Figure 19.1** How is its resemblance to a fallen leaf helpful to this moth?

## Endless Forms Most Beautiful

A hungry bird in the Peruvian rain forest would have to look very closely to spot a "dead-leaf moth" (*Oxytenis modestia*), which blends in well with its forest floor habitat **(Figure 19.1)**. This distinctive moth is a member of a diverse group, the more than 120,000 species of lepidopteran insects (moths and butterflies). All lepidopterans have a juvenile stage characterized by a well-developed head and many chewing mouthparts: the ravenous, efficient feeding machines we call caterpillars. As adults, all lepidopterans share other features, such as three pairs of legs and two pairs of wings covered with small scales. But the many lepidopterans also differ from one another, in both their caterpillar and adult forms. How did there come to be so many different moths and butterflies, and what causes their similarities and differences?

This moth and its many close relatives illustrate three key observations about life:

- the striking ways in which organisms are suited for life in their environments*

- the many shared characteristics (unity) of life
- the rich diversity of life

More than a century and a half ago, Charles Darwin was inspired to develop a scientific explanation for these three broad observations. When he published his hypothesis in his book *The Origin of Species*, Darwin ushered in a scientific revolution—the era of evolutionary biology.

For now, we will define **evolution** as *descent with modification*, a phrase Darwin used in proposing that Earth's many species are descendants of ancestral species that were different from the present-day species. Evolution can also be defined more narrowly as a change in the genetic composition of a population from generation to generation (as discussed further in Chapter 21).

Whether it is defined broadly or narrowly, we can view evolution in two related but different ways: as a pattern and as a process. The *pattern* of evolutionary change is revealed by data from many scientific disciplines, including biology, geology, physics, and chemistry. These data are facts—they are observations about the natural world. The *process* of evolution

---

*Here and throughout this book, the term *environment* refers to other organisms as well as to the physical aspects of an organism's surroundings.

consists of the mechanisms that produce the observed pattern of change. These mechanisms represent natural causes of the natural phenomena we observe. Indeed, the power of evolution as a unifying theory is its ability to explain and connect a vast array of observations about the living world.

As with all general theories in science, we continue to test our understanding of evolution by examining whether it can account for new observations and experimental results. In this and the following chapters, we'll examine how ongoing discoveries shape what we know about the pattern and process of evolution. To set the stage, we'll first retrace Darwin's quest to explain the adaptations, unity, and diversity of what he called life's "endless forms most beautiful."

CONCEPT 19.1

## The Darwinian revolution challenged traditional views of a young Earth inhabited by unchanging species

What impelled Darwin to challenge the prevailing views about Earth and its life? Darwin developed his revolutionary proposal over time, influenced by the work of others and by his travels **(Figure 19.2)**. As we'll see, his ideas also had deep historical roots.

### *Scala Naturae* and Classification of Species

Long before Darwin was born, several Greek philosophers suggested that life might have changed gradually over time. But one philosopher who greatly influenced early Western science, Aristotle (384–322 BCE), viewed species as fixed (unchanging). Through his observations of nature, Aristotle recognized certain "affinities" among organisms. He concluded that life-forms could be arranged on a ladder, or scale, of increasing complexity, later called the *scala naturae* ("scale

of nature"). Each form of life, perfect and permanent, had its allotted rung on this ladder.

These ideas were generally consistent with the Old Testament account of creation, which holds that species were individually designed by God and therefore perfect. In the 1700s, many scientists interpreted the often remarkable match of organisms to their environment as evidence that the Creator had designed each species for a particular purpose.

One such scientist was Carolus Linnaeus (1707–1778), a Swedish physician and botanist who sought to classify life's diversity, in his words, "for the greater glory of God." Linnaeus developed the two-part, or *binomial*, format for naming species (such as *Homo sapiens* for humans) that is still used today. In contrast to the linear hierarchy of the *scala naturae*, Linnaeus adopted a nested classification system, grouping similar species into increasingly general categories. For example, similar species are grouped in the same genus, similar genera (plural of genus) are grouped in the same family, and so on (see Figure 20.3).

Linnaeus did not ascribe the resemblances among species to evolutionary kinship, but rather to the pattern of their creation. A century later, however, Darwin argued that classification should be based on evolutionary relationships. He also noted that scientists using the Linnaean system often grouped organisms in ways that reflected those relationships.

### Ideas About Change over Time

Among other sources of information, Darwin drew from the work of scientists studying **fossils**, the remains or traces of organisms from the past. Many fossils are found in sedimentary rocks formed from the sand and mud that settle to the bottom of seas, lakes, and swamps **(Figure 19.3)**. New layers

▲ **Figure 19.2 Unusual species inspired novel ideas.** Darwin observed this species of marine iguana and many other unique animals when he visited the Galápagos Islands in 1835.

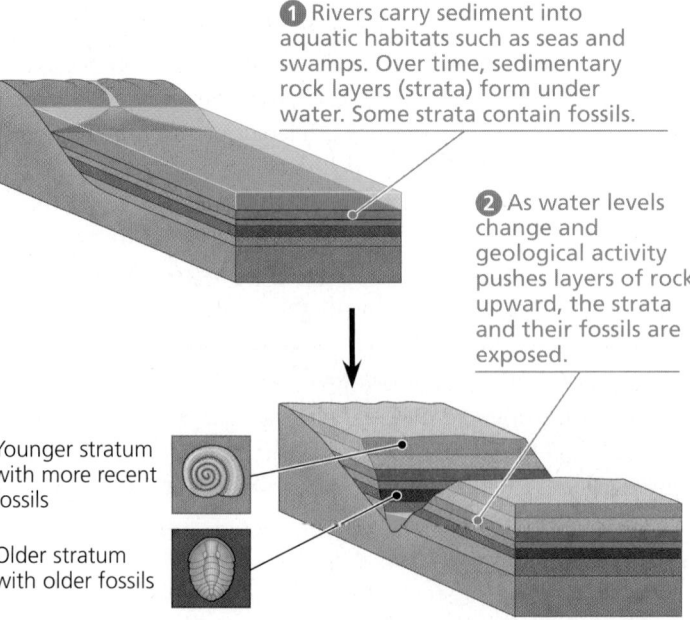

❶ Rivers carry sediment into aquatic habitats such as seas and swamps. Over time, sedimentary rock layers (strata) form under water. Some strata contain fossils.

❷ As water levels change and geological activity pushes layers of rock upward, the strata and their fossils are exposed.

Younger stratum with more recent fossils

Older stratum with older fossils

▲ **Figure 19.3 Formation of sedimentary strata with fossils.**

of sediment cover older ones and compress them into layers of rock called **strata** (singular, *stratum*). The fossils in a particular stratum provide a glimpse of some of the organisms that populated Earth at the time that layer formed. Later, erosion may carve through upper (younger) strata, revealing deeper (older) strata that had been buried.

**Paleontology**, the study of fossils, was developed in large part by French scientist Georges Cuvier (1769–1832). In examining strata near Paris, Cuvier noted that the older the stratum, the more dissimilar its fossils were to current life-forms. He also observed that from one layer to the next, some new species appeared while others disappeared. He inferred that extinctions must have been a common occurrence, but he staunchly opposed the idea of evolution. Cuvier speculated that each boundary between strata represented a sudden catastrophic event, such as a flood, that had destroyed many of the species living in that area. Such regions, he reasoned, were later repopulated by different species immigrating from other areas.

In contrast to Cuvier's emphasis on sudden events, other scientists suggested that profound change could take place through the cumulative effect of slow but continuous processes. In 1795, Scottish geologist James Hutton (1726–1797) proposed that Earth's geologic features could be explained by gradual mechanisms, such as valleys being formed by rivers wearing through rocks. The leading geologist of Darwin's time, Charles Lyell (1797–1875), incorporated Hutton's thinking into his proposal that the same geologic processes are operating today as in the past, and at the same rate.

Hutton and Lyell's ideas strongly influenced Darwin's thinking. Darwin agreed that if geologic change results from slow, continuous actions rather than from sudden events, then Earth must be much older than the widely accepted age of a few thousand years. It would, for example, take a very long time for a river to carve a canyon by erosion. He later reasoned that perhaps similarly slow and subtle processes could produce substantial biological change. However, Darwin was not the first to apply the idea of gradual change to biological evolution.

## Lamarck's Hypothesis of Evolution

Although some 18th-century naturalists suggested that life evolves as environments change, only one proposed a mechanism for *how* life changes over time: French biologist Jean-Baptiste de Lamarck (1744–1829). Alas, Lamarck is primarily remembered today *not* for his visionary recognition that evolutionary change explains patterns in fossils and the match of organisms to their environments, but for the incorrect mechanism he proposed.

Lamarck published his hypothesis in 1809, the year Darwin was born. By comparing living species with fossil forms, Lamarck had found what appeared to be several lines of

◀ **Figure 19.4 Acquired traits cannot be inherited.** This bonsai tree was "trained" to grow as a dwarf by pruning and shaping. However, seeds from this tree would produce offspring of normal size.

descent, each a chronological series of older to younger fossils leading to a living species. He explained his findings using two principles that were widely accepted at the time. The first was *use and disuse*, the idea that parts of the body that are used extensively become larger and stronger, while those that are not used deteriorate. Among many examples, he cited a giraffe stretching its neck to reach leaves on high branches. The second principle, *inheritance of acquired characteristics*, stated that an organism could pass these modifications to its offspring. Lamarck reasoned that the long, muscular neck of the living giraffe had evolved over many generations as giraffes stretched their necks ever higher.

Lamarck also thought that evolution happens because organisms have an innate drive to become more complex. Darwin rejected this idea, but he, too, thought that variation was introduced into the evolutionary process in part through inheritance of acquired characteristics. Today, however, our understanding of genetics refutes this mechanism: Experiments show that traits acquired by use during an individual's life are not inherited in the way proposed by Lamarck **(Figure 19.4)**.

Lamarck was vilified in his own time, especially by Cuvier, who denied that species ever evolve. In retrospect, however, Lamarck did recognize that the fact that organisms are well suited for life in their environments can be explained by gradual evolutionary change, and he did propose a testable explanation for how this change occurs.

### CONCEPT CHECK 19.1

1. How did Hutton's and Lyell's ideas influence Darwin's thinking about evolution?
2. **MAKE CONNECTIONS** Scientific hypotheses must be testable (see Concept 1.3). Applying this criterion, are Cuvier's explanation of the fossil record and Lamarck's hypothesis of evolution scientific? Explain your answer in each case.

For suggested answers, see Appendix A.

**CONCEPT 19.2**

# Descent with modification by natural selection explains the adaptations of organisms and the unity and diversity of life

As the 19th century dawned, it was generally thought that species had remained unchanged since their creation. A few clouds of doubt about the permanence of species were beginning to gather, but no one could have forecast the thundering storm just beyond the horizon. How did Charles Darwin become the lightning rod for a revolutionary view of life?

## Darwin's Research

Charles Darwin (1809–1882) was born in Shrewsbury, England. He had a consuming interest in nature—reading nature books, fishing, hunting, and collecting insects. Darwin's father, a physician, could see no future for his son as a naturalist and sent him to medical school in Edinburgh. But Charles found medicine boring and surgery before the days of anesthesia horrifying. He quit medical school and enrolled at Cambridge University, intending to become a clergyman. (At that time many scholars of science belonged to the clergy.)

At Cambridge, Darwin became the protégé of John Henslow, a botany professor. Soon after Darwin graduated, Henslow recommended him to Captain Robert FitzRoy, who was preparing the survey ship HMS *Beagle* for a long voyage around the world. FitzRoy, who was himself an accomplished scientist, accepted Darwin because he was a skilled naturalist and because they were of similar age and social class.

### The Voyage of the Beagle

Darwin embarked from England on the *Beagle* in December 1831. The primary mission of the voyage was to chart poorly known stretches of the South American coastline. Darwin, however, spent most of his time on shore, observing and collecting thousands of plants and animals. He described features of organisms that made them well suited to such diverse environments as Brazil's humid jungles, Argentina's broad grasslands, and the Andes' towering peaks.

Darwin observed that the plants and animals in temperate regions of South America more closely resembled species living in the South American tropics than species living in temperate regions of Europe. Furthermore, the fossils he found, though clearly different from living species, distinctly resembled the living organisms of South America.

Darwin also read Lyell's *Principles of Geology* during the voyage. He experienced geologic change firsthand when a violent earthquake shook the coast of Chile, and he observed afterward that rocks along the coast had been thrust upward by several meters. Finding fossils of ocean organisms high in the Andes, Darwin inferred that the rocks containing the fossils must have been raised there by many similar earthquakes. These observations reinforced what he had learned from Lyell: Physical evidence did not support the traditional view that Earth was only a few thousand years old.

Darwin's interest in the species (or fossils) found in an area was further stimulated by the *Beagle*'s stop at the Galápagos, a group of volcanic islands located near the equator about 900 km west of South America **(Figure 19.5)**. Darwin was fascinated by the unusual organisms there. The birds he collected included several kinds of mockingbirds. These mockingbirds, though similar to each other, seemed to be different species. Some were unique

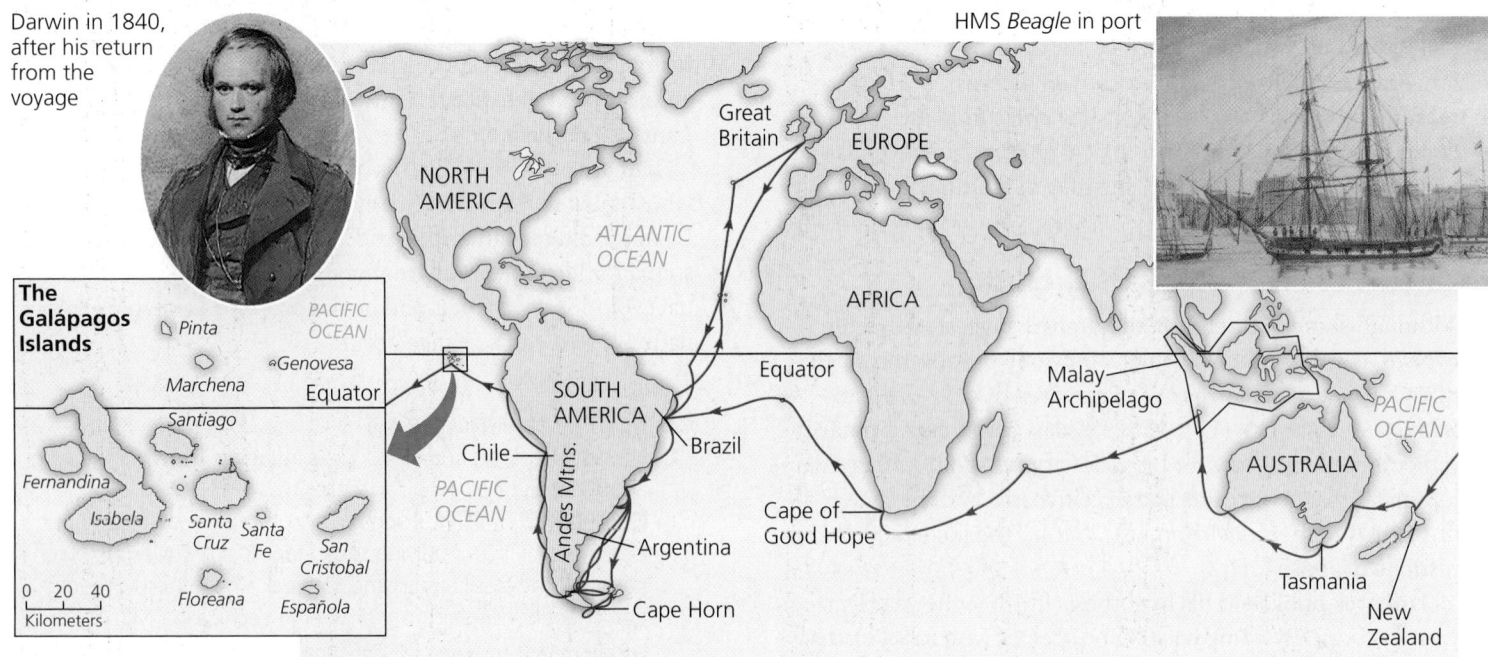

▲ Figure 19.5 **The voyage of HMS *Beagle* (December 1831–October 1836).**

to individual islands, while others lived on two or more adjacent islands. Furthermore, although the animals on the Galápagos resembled species living on the South American mainland, most of the Galápagos species were not known from anywhere else in the world. Darwin hypothesized that the Galápagos had been colonized by organisms that had strayed from South America and then diversified, giving rise to new species on the various islands.

### Darwin's Focus on Adaptation

During the voyage of the *Beagle*, Darwin observed many examples of **adaptations**, inherited characteristics of organisms that enhance their survival and reproduction in specific environments. Later, as he reassessed his observations, he began to perceive adaptation to the environment and the origin of new species as closely related processes. Could a new species arise from an ancestral form by the gradual accumulation of adaptations to a different environment? From studies made years after Darwin's voyage, biologists have concluded that this is indeed what happened to a diverse group of finches found on the Galápagos Islands (see Figure 1.17). The finches' various beaks and behaviors are adapted to the specific foods available on their home islands **(Figure 19.6)**. Darwin realized that explaining such adaptations was essential to understanding evolution. His explanation of how adaptations arise centered on **natural selection**, a process in which individuals that have certain inherited traits tend to survive and reproduce at higher rates than do other individuals *because of* those traits.

By the early 1840s, Darwin had worked out the major features of his hypothesis. He set these ideas on paper in 1844, when he wrote a long essay on descent with modification and its underlying mechanism, natural selection. Yet he was still reluctant to publish his ideas, in part because he anticipated the uproar they would cause. During this time, Darwin continued to compile evidence in support of his hypothesis. By the mid-1850s, he had described his ideas to Lyell and a few others. Lyell, who was not yet convinced of evolution, nevertheless urged Darwin to publish on the subject before someone else came to the same conclusions and published first.

In June 1858, Lyell's prediction came true. Darwin received a manuscript from Alfred Russel Wallace (1823–1913), a British naturalist working in the South Pacific islands of the Malay Archipelago **(Figure 19.7)**. Wallace had developed a hypothesis of natural selection nearly identical to Darwin's. He asked Darwin to evaluate his paper and forward it to Lyell if it merited publication. Darwin complied, writing to Lyell: "Your words have come true with a vengeance. . . . I never saw

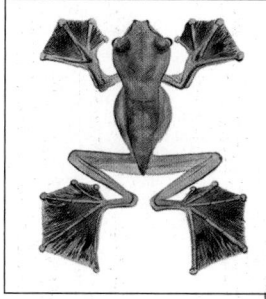

**(a) Cactus-eater.** The long, sharp beak of the common cactus finch (*Geospiza scandens*) helps it tear and eat cactus flowers and pulp.

**(c) Insect-eater.** The green warbler finch (*Certhidea olivacea*) uses its narrow, pointed beak to grasp insects.

**(b) Seed-eater.** The large ground finch (*Geospiza magnirostris*) has a large beak adapted for cracking seeds on the ground.

▲ **Figure 19.6 Three examples of beak variation in Galápagos finches.** The Galápagos Islands are home to more than a dozen species of closely related finches, some found only on a single island. A striking difference among them is their beaks, which are adapted for specific diets.

**MAKE CONNECTIONS** *Review Figure 1.17. To which of the other two species shown above is the common cactus finch more closely related (that is, with which species does the common cactus finch share a more recent common ancestor)?*

▶ **Figure 19.7 Alfred Russel Wallace.** The inset is a painting Wallace made of a flying tree frog from the Malay Archipelago.

a more striking coincidence . . . so all my originality, whatever it may amount to, will be smashed." On July 1, 1858, Lyell and a colleague presented Wallace's paper, along with extracts from Darwin's unpublished 1844 essay, to the Linnean Society of London. Darwin quickly finished his book, titled *On the Origin of Species by Means of Natural Selection* (commonly referred to as *The Origin of Species*), and published it the next year. Although Wallace had submitted his ideas for publication first, he admired Darwin and thought that Darwin had developed and tested the idea of natural selection so extensively that he should be known as its main architect.

Within a decade, Darwin's book and its proponents had convinced most scientists that life's diversity is the product of evolution. Darwin succeeded where previous evolutionists had failed, mainly by presenting a plausible scientific mechanism with immaculate logic and an avalanche of supporting evidence.

## Ideas from *The Origin of Species*

In his book, Darwin amassed evidence that descent with modification by natural selection explains three broad observations about nature—the unity of life, the diversity of life, and the striking ways in which organisms are suited for life in their environments.

### Descent with Modification

In the first edition of *The Origin of Species*, Darwin never used the word *evolution* (although the final word of the book is "evolved"). Rather, he discussed *descent with modification*, a phrase that summarized his view of life. Organisms share many characteristics, leading Darwin to perceive unity in life. He attributed the unity of life to the descent of all organisms from an ancestor that lived in the remote past. He also thought that as the descendants of that ancestral organism lived in various habitats, they gradually accumulated diverse modifications, or adaptations, that fit them to specific ways of life. Darwin reasoned that over a long period of time, descent with modification eventually led to the rich diversity of life we see today.

Darwin viewed the history of life as a tree, with multiple branchings from a common trunk out to the tips of the youngest twigs **(Figure 19.8)**. In his diagram, the tips of the twigs that are labeled A through D represent several groups of organisms living in the present day, while the unlabeled branches represent groups that are extinct. Each fork of the tree represents the most recent common ancestor of all the lines of evolution that subsequently branch from that point. Darwin reasoned that such a branching process, along with past extinction events, could explain the large morphological gaps (differences in form) that sometimes exist between related groups of organisms.

As an example, consider the three living species of elephants: the Asian elephant (*Elephas maximus*) and two species of African elephants (*Loxodonta africana* and *L. cyclotis*). These closely related species are very similar because they shared the same line of descent until a relatively recent split from their common ancestor, as shown in the tree diagram in **Figure 19.9**.

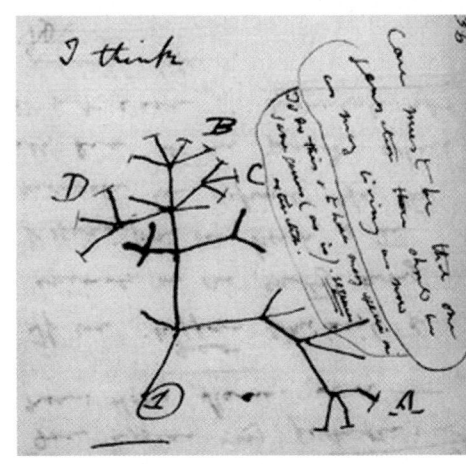

► Figure 19.8 **"I think . . ."** In this 1837 sketch, Darwin envisioned the branching pattern of evolution. Branches that end in twigs labeled A–D represent particular groups of living organisms; all other branches represent extinct groups.

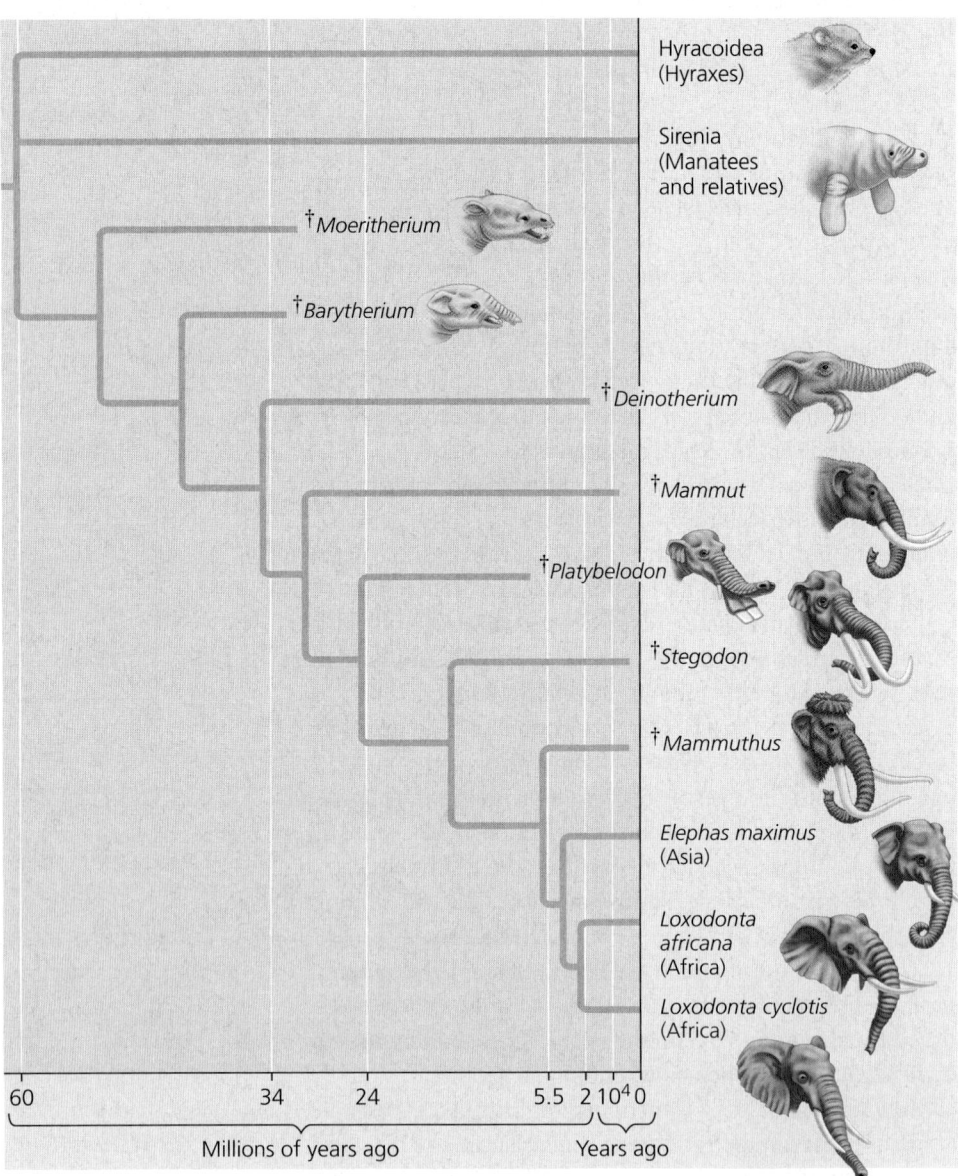

▲ **Figure 19.9 Descent with modification.** This evolutionary tree of elephants and their relatives is based mainly on fossils—their anatomy, order of appearance in strata, and geographic distribution. Note that most branches of descent ended in extinction (denoted by the dagger symbol †). (Time line not to scale.)

**?** *Based on the tree shown here, approximately when did the most recent ancestor shared by Mammuthus (woolly mammoths), Asian elephants, and African elephants live?*

Note that seven lineages related to elephants have become extinct over the past 32 million years. As a result, there are no living species that fill the morphological gap between the elephants and their nearest relatives today, the hyraxes and manatees. Such extinctions are not uncommon. In fact, many evolutionary branches, even some major ones, are dead ends: Scientists estimate that over 99% of all species that have ever lived are now extinct. As in Figure 19.9, fossils of extinct species can document the divergence of present-day groups by "filling in" gaps between them.

## Artificial Selection, Natural Selection, and Adaptation

Darwin proposed the mechanism of natural selection to explain the observable patterns of evolution. He crafted his argument carefully, hoping to persuade even the most skeptical readers. First he discussed familiar examples of selective breeding of domesticated plants and animals. Humans have modified other species over many generations by selecting and breeding individuals that possess desired traits, a process called **artificial selection (Figure 19.10)**. As a result of artificial selection, crops, livestock animals, and pets often bear little resemblance to their wild ancestors.

Darwin then argued that a similar process occurs in nature. He based his argument on two observations, from which he drew two inferences.

**Observation #1:** Members of a population often vary in their inherited traits **(Figure 19.11)**.

**Observation #2:** All species can produce more offspring than their environment can support **(Figure 19.12)**, and many of these offspring fail to survive and reproduce.

▲ **Figure 19.10 Artificial selection.** These different vegetables have all been selected from one species of wild mustard. By selecting variations in different parts of the plant, breeders have obtained these divergent results.

**Inference #1:** Individuals whose inherited traits give them a higher probability of surviving and reproducing in a given environment tend to leave more offspring than do other individuals.

**Inference #2:** This unequal ability of individuals to survive and reproduce will lead to the accumulation of favorable traits in the population over generations.

As these two inferences suggest, Darwin saw an important connection between natural selection and the capacity of organisms to "overreproduce." He began to make this connection after reading an essay by economist Thomas Malthus, who contended that much of human suffering—disease, famine, and war—resulted from the human population's potential to

▲ **Figure 19.11 Variation in a population.** Individuals in this population of Asian ladybird beetles vary in color and spot pattern. Natural selection may act on these variations only if (1) they are heritable and (2) they affect the beetles' ability to survive and reproduce.

◀ **Figure 19.12 Overproduction of offspring.** A single puffball fungus can produce billions of spores that give rise to offspring. If all of these offspring and their descendants survived to maturity, they would carpet the surrounding land.

increase faster than food supplies and other resources. Similarly, Darwin realized that the capacity to overreproduce was characteristic of all species. Of the many eggs laid, young born, and seeds spread, only a tiny fraction complete their development and leave offspring of their own. The rest are eaten, starved, diseased, unmated, or unable to tolerate physical conditions of the environment such as salinity or temperature.

An organism's heritable traits can influence not only its own performance, but also how well its offspring cope with environmental challenges. For example, an organism might have a trait that gives its offspring an advantage in escaping predators, obtaining food, or tolerating physical conditions. When such advantages increase the number of offspring that survive and reproduce, the traits that are favored will likely appear at a greater frequency in the next generation. Thus, over time, natural selection resulting from factors such as predators, lack of food, or adverse physical conditions can lead to an increase in the proportion of favorable traits in a population.

How rapidly do such changes occur? Darwin reasoned that if artificial selection can bring about dramatic change in a relatively short period of time, then natural selection should be capable of substantial modification of species over many hundreds of generations. Even if the advantages of some heritable traits over others are slight, the advantageous variations will gradually accumulate in the population, and less favorable variations will diminish. Over time, this process will increase the frequency of individuals with favorable adaptations, hence increasing the degree to which organisms are well suited for life in their environment.

### Key Features of Natural Selection

Let's now recap the main ideas of natural selection:

- Natural selection is a process in which individuals that have certain heritable traits survive and reproduce at a higher rate than do other individuals because of those traits.
- Over time, natural selection can increase the frequency of adaptations that are favorable in a given environment **(Figure 19.13)**.
- If an environment changes, or if individuals move to a new environment, natural selection may result in adaptation to these new conditions, sometimes giving rise to new species.

One subtle but important point is that although natural selection occurs through interactions between individual organisms and their environment, *individuals do not evolve*. Rather, it is the population that evolves over time.

A second key point is that natural selection can amplify or diminish only those heritable traits that differ among the individuals in a population. Thus, even if a trait is heritable, if all the individuals in a population are genetically identical for that trait, evolution by natural selection cannot occur.

Third, remember that environmental factors vary from place to place and over time. A trait that is favorable in one place or time may be useless—or even detrimental—in other places or times. Natural selection is always operating, but

▲ **Figure 19.13 Camouflage as an example of evolutionary adaptation.** Related species of the insects called mantises have diverse shapes and colors that evolved in different environments, as seen in this Malaysian orchid mantis (top) and South African flower-eyed mantis (bottom).

**?** *Explain how these mantises demonstrate the three key observations about life introduced at the beginning of this chapter: how organisms are well suited for life in their environments, unity, and diversity.*

which traits are favored depends on the context in which a species lives and mates.

Next, we'll survey the wide range of observations that support a Darwinian view of evolution by natural selection.

### CONCEPT CHECK 19.2

1. How does the concept of descent with modification explain both the unity and diversity of life?
2. **WHAT IF?** Predict whether a fossil of an extinct mammal that lived high in the Andes would more closely resemble present-day mammals that live in South American jungles or present-day mammals that live high in Asian mountains. Explain.
3. **MAKE CONNECTIONS** Review the relationship between genotype and phenotype (see Figure 11.6). Suppose that in a particular pea population, flowers with the white phenotype are favored by natural selection. Predict what would happen over time to the frequency of the $p$ allele in the population, and explain your reasoning.

For suggested answers, see Appendix A.

# Evolution is supported by an overwhelming amount of scientific evidence

In *The Origin of Species*, Darwin marshaled a broad range of evidence to support the concept of descent with modification. Still—as he readily acknowledged—there were instances in which key evidence was lacking. For example, Darwin referred to the origin of flowering plants as an "abominable mystery," and he lamented the lack of fossils showing how earlier groups of organisms gave rise to new groups.

In the last 150 years, new discoveries have filled many of the gaps that Darwin identified. The origin of flowering plants, for example, is much better understood (see Concept 26.4), and many fossils have been discovered that signify the origin of new groups of organisms (see Concept 23.1). In this section, we'll consider four types of data that document the pattern of evolution and illuminate how it occurs: direct observations, homology, the fossil record, and biogeography.

## Direct Observations of Evolutionary Change

Biologists have documented evolutionary change in thousands of scientific studies. We'll examine many such studies throughout this unit, but let's look at two examples here.

### Natural Selection in Response to Introduced Species

Animals that eat plants, called herbivores, often have adaptations that help them feed efficiently on their primary food sources. What happens when herbivores switch to a new food source with different characteristics?

An opportunity to study this question in nature is provided by soapberry bugs, which use their "beak," a hollow, needlelike mouthpart, to feed on seeds located within the fruits of various plants. In southern Florida, the soapberry bug (*Jadera haematoloma*) feeds on the seeds of a native plant, the balloon vine (*Cardiospermum corindum*). In central Florida, however, balloon vines have become rare. Instead, soapberry bugs in that region now feed on seeds of the goldenrain tree (*Koelreuteria elegans*), a species recently introduced from Asia.

Soapberry bugs feed most effectively when the length of their beak is similar to the depth at which seeds are found within the fruit. Goldenrain tree fruit consists of three flat lobes, and its seeds are much closer to the fruit surface than are the seeds of the plump, round fruit of the native balloon vine. These differences led researchers to predict that in populations that feed on goldenrain tree, natural selection would result in beaks that are *shorter* than those in populations that feed on balloon vine **(Figure 19.14)**. Indeed, beak lengths are shorter in the populations that feed on goldenrain tree.

Researchers have also studied beak length evolution in soapberry bug populations that feed on plants introduced to Louisiana, Oklahoma, and Australia. In each of these locations,

## ▼ Figure 19.14 Inquiry

### Can a change in a population's food source result in evolution by natural selection?

**Field Study** Soapberry bugs feed most effectively when the length of their "beak" closely matches the depth of the seeds within the fruit. Scott Carroll and his colleagues measured beak lengths in soapberry bug populations feeding on the native balloon vine. They also measured beak lengths in populations feeding on the introduced goldenrain tree. The researchers then compared the measurements with those of museum specimens collected in the two areas before the goldenrain tree was introduced.

Soapberry bug with beak inserted in balloon vine fruit

**Results** Beak lengths were shorter in populations feeding on the introduced species than in populations feeding on the native species, in which the seeds are buried more deeply. The average beak length in museum specimens from each population (indicated by red arrows) was similar to beak lengths in populations feeding on native species.

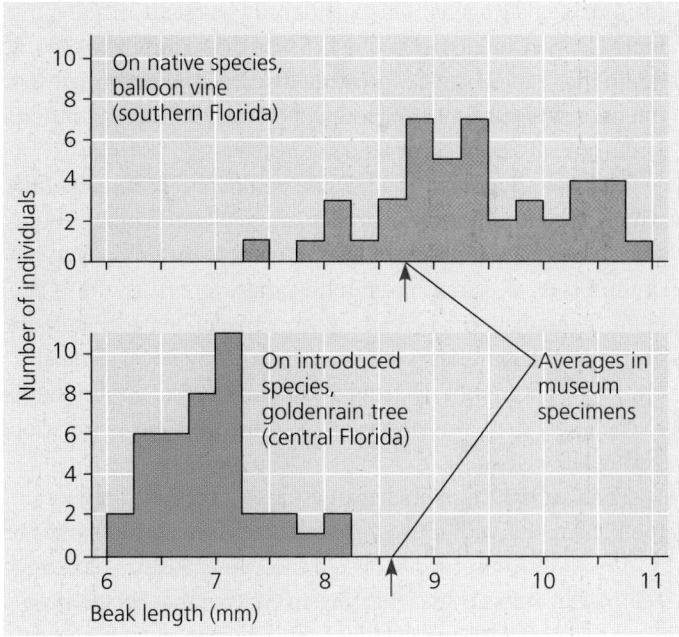

**Conclusion** Museum specimens and contemporary data suggest that a change in the size of the soapberry bug's food source can result in evolution by natural selection for matching beak size.

**Data from** S. P. Carroll and C. Boyd, Host race radiation in the soapberry bug: natural history with the history, *Evolution* 46:1052–1069 (1992).

**WHAT IF?** Data from additional studies showed that when soapberry bug eggs from a population fed on balloon vine fruits were reared on goldenrain tree fruits (or vice versa), the beak lengths of the adult insects matched those in the population from which the eggs were obtained. Interpret these results.

the fruit of the introduced plants is larger than the fruit of the native plant. Thus, in populations feeding on introduced species in these regions, researchers predicted that natural selection would result in the evolution of *longer* beaks. Again, data collected in field studies upheld this prediction.

The observed changes in beak lengths had important consequences. In Australia, for example, the increase in beak length nearly doubled the success with which soapberry bugs could eat the seeds of the introduced species. Furthermore, since historical data show that the goldenrain tree reached central Florida just 35 years before the scientific studies were initiated, the results demonstrate that natural selection can cause rapid evolution in a wild population.

### The Evolution of Drug-Resistant Bacteria

An example of ongoing natural selection that dramatically affects humans is the evolution of drug-resistant pathogens (disease-causing organisms and viruses). This is a particular problem with bacteria and viruses because resistant strains of these pathogens can proliferate very quickly.

Consider the evolution of drug resistance in the bacterium *Staphylococcus aureus*. About one in three people harbor this species on their skin or in their nasal passages with no negative effects. However, certain genetic varieties (strains) of this species, known as methicillin-resistant *S. aureus* (MRSA), are formidable pathogens. The past decade has seen an alarming increase in virulent forms of MRSA such as clone USA300, a strain that can cause "flesh-eating disease" and potentially fatal infections **(Figure 19.15)**. How did clone USA300 and other strains of MRSA become so dangerous?

The story begins in 1943, when penicillin became the first widely used antibiotic. Since then, penicillin and other antibiotics have saved millions of lives. However, by 1945, over 20% of the *S. aureus* strains seen in hospitals were resistant to penicillin. These bacteria had an enzyme, penicillinase, that could destroy penicillin. Researchers responded by developing antibiotics that were not destroyed by penicillinase, but resistance to each new drug was observed in some *S. aureus* populations within a few years.

Then, in 1959, doctors began using the powerful antibiotic methicillin. But within two years, methicillin-resistant strains of *S. aureus* were observed. How did these resistant strains emerge? Methicillin works by deactivating an enzyme that bacteria use to synthesize their cell walls. However, different *S. aureus* populations exhibited variations in how strongly their members were affected by the drug. In particular, some individuals were able to synthesize their cell walls using a different enzyme that was not affected by methicillin. These individuals survived the methicillin treatments and reproduced at higher rates than did other individuals. Over time, these resistant individuals became increasingly common, leading to the spread of MRSA.

Initially, MRSA could be controlled by antibiotics that work differently from the way methicillin works. But this has become less effective because some MRSA strains are resistant to

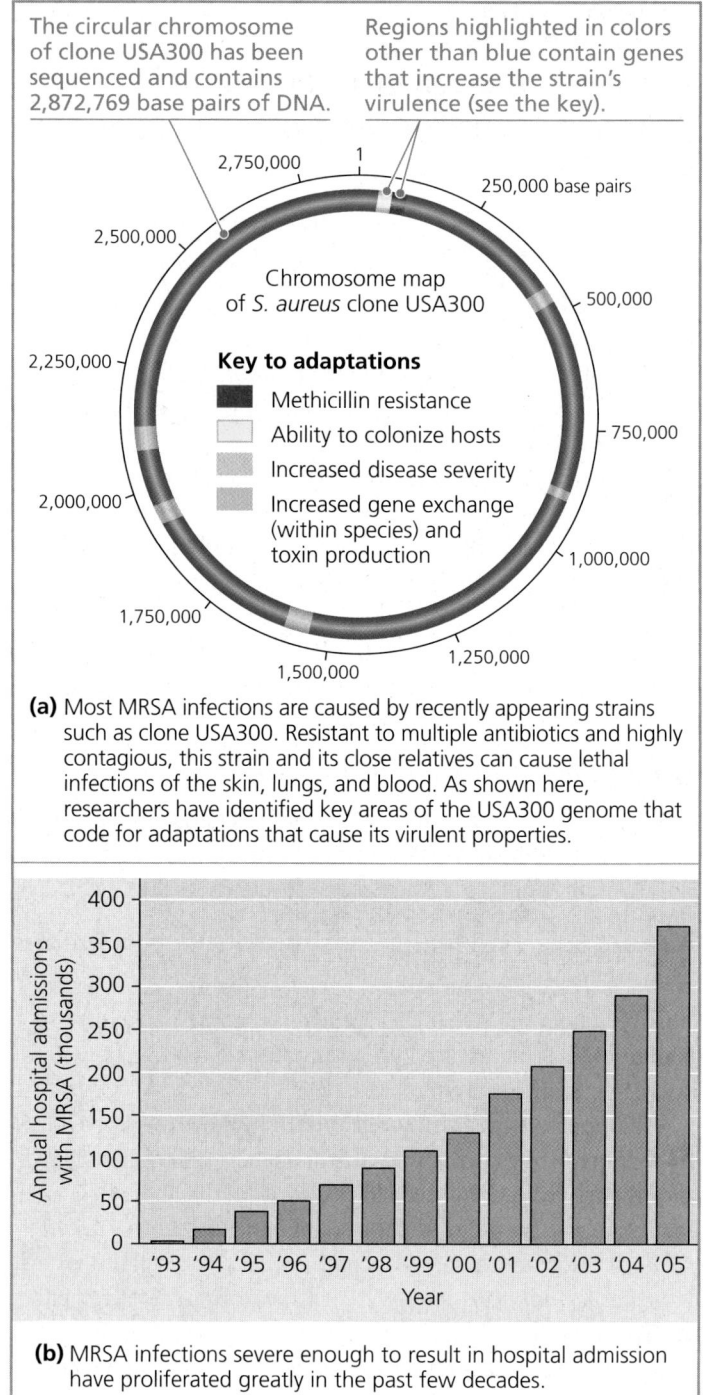

**(a)** Most MRSA infections are caused by recently appearing strains such as clone USA300. Resistant to multiple antibiotics and highly contagious, this strain and its close relatives can cause lethal infections of the skin, lungs, and blood. As shown here, researchers have identified key areas of the USA300 genome that code for adaptations that cause its virulent properties.

**(b)** MRSA infections severe enough to result in hospital admission have proliferated greatly in the past few decades.

▲ **Figure 19.15 The rise of methicillin-resistant *Staphylococcus aureus* (MRSA).**

multiple antibiotics—probably because bacteria can exchange genes with members of their own and other species (see Figure 24.17). Thus, the multidrug-resistant strains of today may have emerged over time as MRSA strains that were resistant to different antibiotics exchanged genes.

The *S. aureus* and soapberry bug examples highlight two key points about natural selection. First, natural selection is a process of editing, not a creative mechanism. A drug does not *create* resistant pathogens; it *selects for* resistant individuals that are already present in the population. Second, natural selection depends on time and place. It favors those characteristics in a genetically variable population that provide an advantage in the current, local environment. What is beneficial in one situation may be useless or even harmful in another. Beak lengths suitable for the size of the typical fruit eaten by members of a particular soapberry bug population are favored by natural selection. However, a beak length suitable for fruit of one size can be disadvantageous when the bug is feeding on fruit of another size.

## Homology

A second type of evidence for evolution comes from analyzing similarities among different organisms. As we've discussed, evolution is a process of descent with modification: Characteristics present in an ancestral organism are altered (by natural selection) in its descendants over time as they face different environmental conditions. As a result, related species can have characteristics that have an underlying similarity yet function differently. Similarity resulting from common ancestry is known as **homology**. As we'll describe in this section, an understanding of homology can be used to make testable predictions and explain observations that are otherwise puzzling.

### *Anatomical and Molecular Homologies*

The view of evolution as a remodeling process leads to the prediction that closely related species should share similar features—and they do. Of course, closely related species share the features used to determine their relationship, but they also share many other features. Some of these shared features make little sense except in the context of evolution. For example, the forelimbs of all mammals, including humans, cats, whales, and bats, show the same arrangement of bones from the shoulder to the tips of the digits, even though the appendages have very different functions: lifting, walking, swimming, and flying **(Figure 19.16)**. Such striking anatomical resemblances would be highly unlikely if these structures had arisen anew in each

species. Rather, the underlying skeletons of the arms, forelegs, flippers, and wings of different mammals are **homologous structures** that represent variations on a structural theme that was present in their common ancestor.

Comparing early stages of development in different animal species reveals additional anatomical homologies not visible in adult organisms. For example, at some point in their development, all vertebrate embryos have a tail located posterior to (behind) the anus, as well as structures called pharyngeal (throat) arches **(Figure 19.17)**. These homologous throat arches ultimately develop into structures with very different functions, such as gills in fishes and parts of the ears and throat in humans and other mammals.

Some of the most intriguing homologies concern "leftover" structures of marginal, if any, importance to the organism. These **vestigial structures** are remnants of features that served a function in the organism's ancestors. For instance, the skeletons of some snakes retain vestiges of the pelvis and leg bones of walking ancestors. Another example is provided by

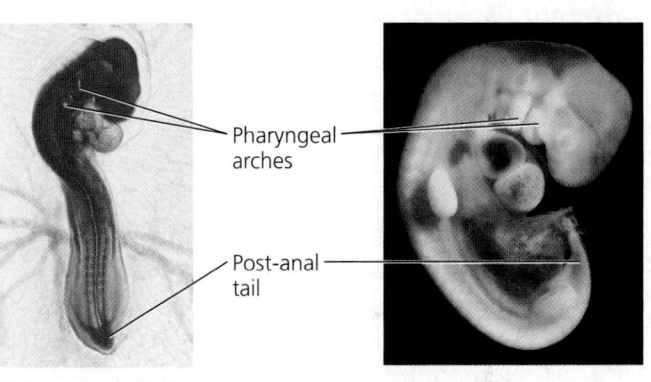

Pharyngeal arches

Post-anal tail

Chick embryo (LM)     Human embryo

▲ **Figure 19.17 Anatomical similarities in vertebrate embryos.** At some stage in their embryonic development, all vertebrates have a tail located posterior to the anus (referred to as a post-anal tail), as well as pharyngeal (throat) arches. Descent from a common ancestor can explain such similarities.

▶ **Figure 19.16 Mammalian forelimbs: homologous structures.** Even though they have become adapted for different functions, the forelimbs of all mammals are constructed from the same basic skeletal elements: one large bone (purple), attached to two smaller bones (orange and tan), attached to several small bones (gold), attached to several metacarpals (green), attached to approximately five digits, each of which is composed of multiple phalanges (blue).

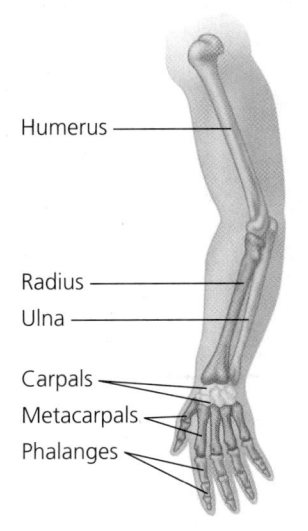
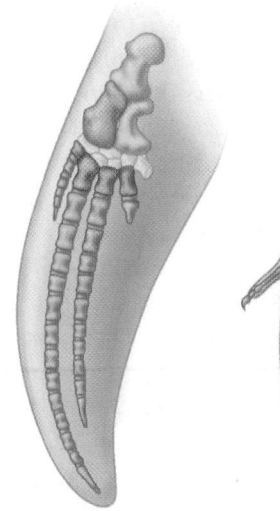
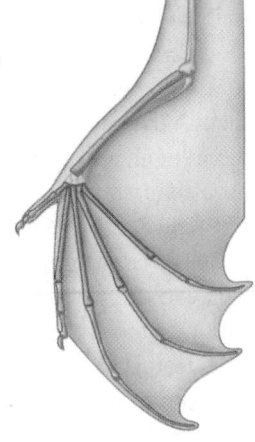

Humerus

Radius
Ulna

Carpals
Metacarpals
Phalanges

Human    Cat    Whale    Bat

eye remnants that are buried under scales in blind species of cave fishes. We would not expect to see these vestigial structures if snakes and blind cave fishes had origins separate from those of other vertebrate animals.

Biologists also observe similarities among organisms at the molecular level. All forms of life use essentially the same genetic code, suggesting that all species descended from common ancestors that used this code. But molecular homologies go beyond a shared code. For example, organisms as dissimilar as humans and bacteria share genes inherited from a very distant common ancestor. Some of these homologous genes have acquired new functions, while others, such as those coding for the ribosomal subunits used in protein synthesis (see Figure 14.17), have retained their original functions. It is also common for organisms to have genes that have lost their function, even though the homologous genes in related species may be fully functional. Like vestigial structures, it appears that such inactive "pseudogenes" may be present simply because a common ancestor had them.

### A Different Cause of Resemblance: Convergent Evolution

Although organisms that are closely related share characteristics because of common descent, distantly related organisms can resemble one another for a different reason: **convergent evolution**, the independent evolution of similar features in different lineages. Consider marsupial mammals, many of which live in Australia. Marsupials are distinct from another group of mammals—the eutherians—few of which live in Australia. (Eutherians complete their embryonic development in the uterus, whereas marsupials are born as embryos and complete their development in an external pouch.) Some Australian marsupials have eutherian look-alikes with superficially similar adaptations.

For instance, as shown in **Figure 19.18**, the sugar glider, a forest-dwelling Australian marsupial, looks very similar to flying squirrels, gliding eutherians that live in North American forests. But the sugar glider has many other characteristics that make it a marsupial, much more closely related to kangaroos and other Australian marsupials than to flying squirrels or other eutherians. Once again, our understanding of evolution can explain these observations. Although they evolved independently from different ancestors, these two mammals have adapted to similar environments in similar ways. In such examples in which species share features because of convergent evolution, the resemblance is said to be **analogous**, not homologous. Analogous features share similar function, but not common ancestry, while homologous features share common ancestry, but not necessarily similar function.

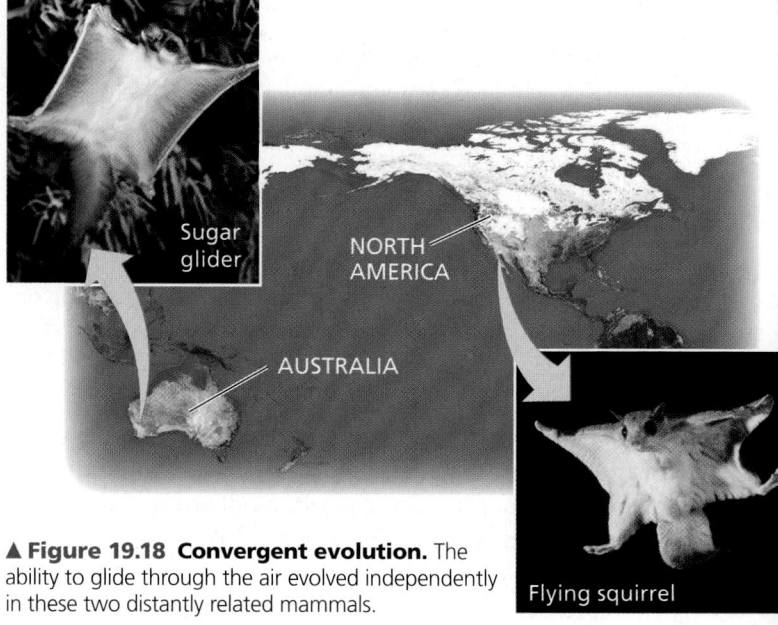

▲ **Figure 19.18 Convergent evolution.** The ability to glide through the air evolved independently in these two distantly related mammals.

## The Fossil Record

A third type of evidence for evolution comes from fossils. The fossil record documents the pattern of evolution, showing that past organisms differed from present-day organisms and that many species have become extinct. Fossils also show the evolutionary changes that have occurred in various groups of organisms. To give one of hundreds of possible examples, researchers found that the pelvic bone in fossil stickleback fish became greatly reduced in size over time in a number of different lakes. The consistent nature of this change suggests that the reduction in the size of the pelvic bone may have been driven by natural selection.

Fossils can also shed light on the origins of new groups of organisms. An example is the fossil record of cetaceans, the mammalian order that includes whales, dolphins, and porpoises. As shown in **Figure 19.19**, some of these fossils provided strong support for a hypothesis based on DNA sequence data: that cetaceans are closely related to even-toed ungulates, a group that includes deer, pigs, camels, and cows.

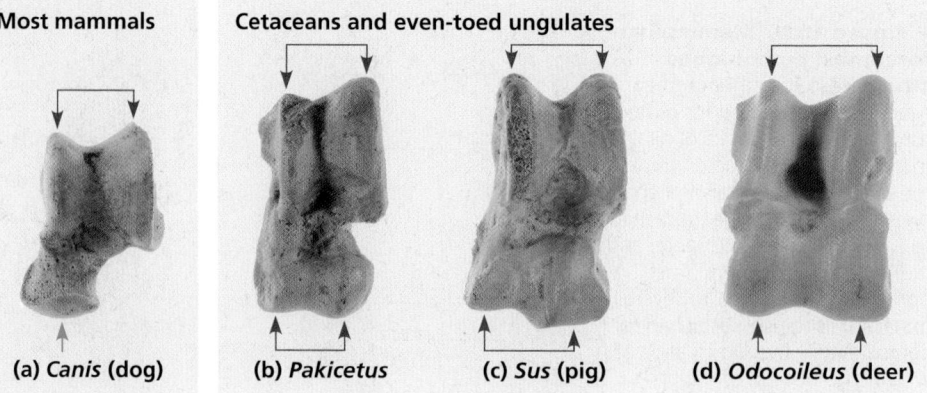

▲ **Figure 19.19 Ankle bones: one piece of the puzzle.** Comparing fossils and present-day examples of the astragalus (a type of ankle bone) indicates that cetaceans are closely related to even-toed ungulates. **(a)** In most mammals, the astragalus is shaped like that of a dog, with a double hump on one end (indicated by the red arrows) but not at the opposite end (blue arrow). **(b)** Fossils show that the early cetacean *Pakicetus* had an astragalus with double humps at both ends, a shape otherwise found only in **(c)** pigs, **(d)** deer, and all other even-toed ungulates.

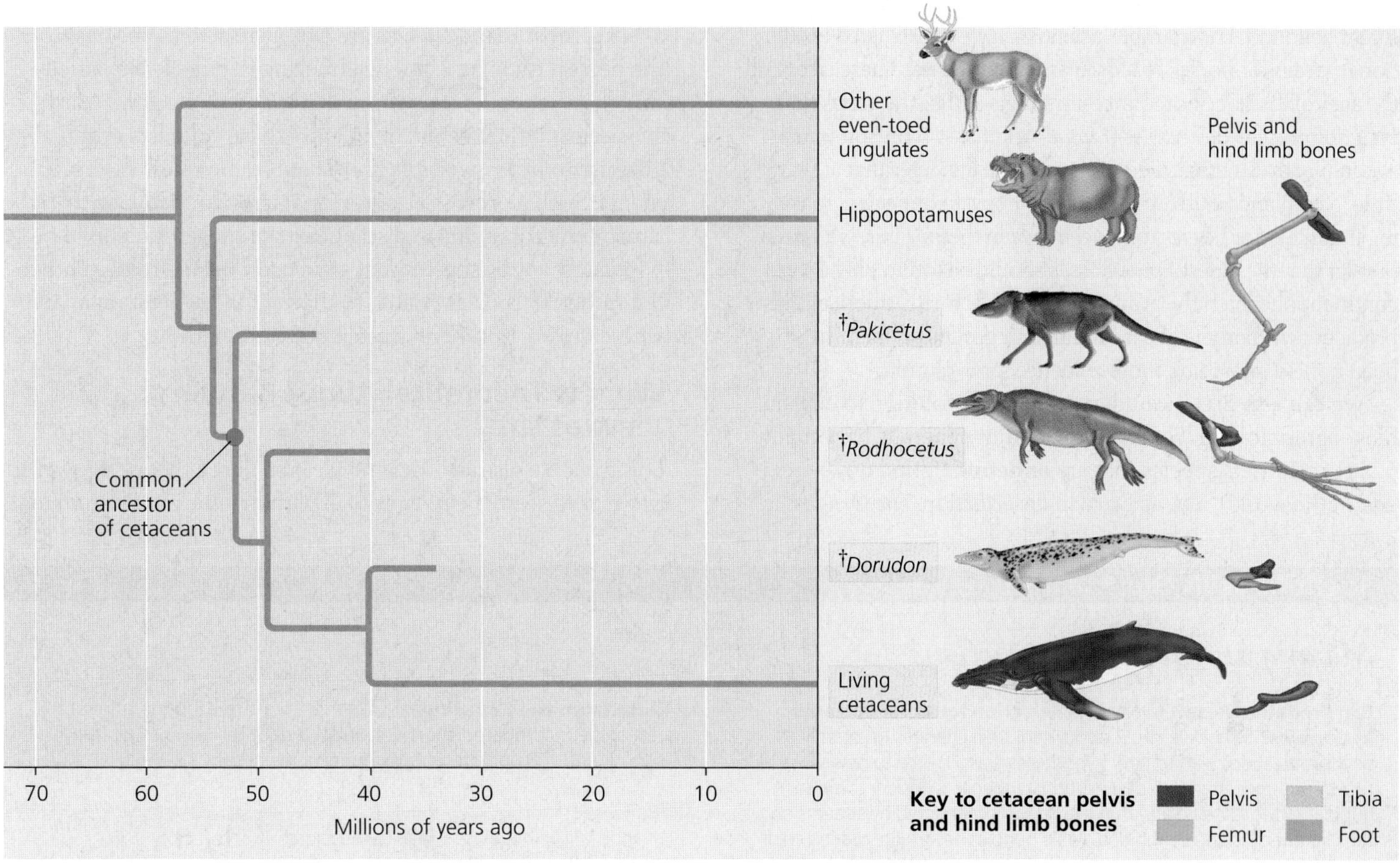

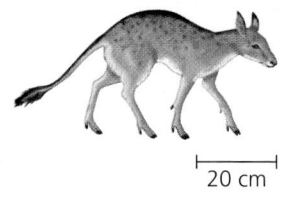

**▲ Figure 19.20 The transition to life in the sea.** Multiple lines of evidence support the hypothesis that cetaceans (highlighted in yellow) evolved from terrestrial mammals. Fossils document the reduction over time in the pelvis and hind limb bones of extinct (†) cetacean ancestors, including *Pakicetus*, *Rodhocetus*, and *Dorudon*. DNA sequence data support the hypothesis that cetaceans are most closely related to hippopotamuses.

**?** *Which happened first during the evolution of cetaceans: changes in hind limb structure or the origin of tail flukes? Explain.*

What else can fossils tell us about cetacean origins? The earliest cetaceans lived 50–60 million years ago. The fossil record indicates that prior to that time, most mammals were terrestrial. Although scientists had long realized that whales and other cetaceans originated from land mammals, few fossils had been found that revealed how cetacean limb structure had changed over time, leading eventually to the loss of hind limbs and the development of flukes (the lobes on a whale's tail) and flippers. In the past few decades, however, a series of remarkable fossils have been discovered in Pakistan, Egypt, and North America. These fossils document steps in the transition from life on land to life in the sea, filling in some of the gaps between ancestral and living cetaceans **(Figure 19.20)**.

Collectively, the recent fossil discoveries document the origin of a new group of mammals, the cetaceans. These discoveries also show that cetaceans and their close living relatives (hippopotamuses, deer, and other even-toed ungulates) are much more different from each other than were *Pakicetus* and early even-toed ungulates, such as *Diacodexis* **(Figure 19.21)**. Similar patterns are seen in fossils documenting the origins of other new groups of organisms, including mammals (see Figure 23.5), flowering plants (see Concept 26.4), and

tetrapods (see Figure 27.23). In each of these cases, the fossil record shows that over time, descent with modification produced increasingly large differences among related groups of organisms, ultimately resulting in the diversity of life we see today.

**▲ Figure 19.21 *Diacodexis*.**

## Biogeography

A fourth type of evidence for evolution has to do with **biogeography**, the scientific study of the geographic distributions of species. The geographic distributions of organisms are influenced by many factors, including *continental drift*, the slow movement of Earth's continents over time. About 250 million years ago, these movements united all of Earth's landmasses into a single large continent called **Pangaea** (see Figure 23.9). Roughly 200 million years ago, Pangaea began to break apart; by 20 million years ago, the continents we know today were within a few hundred kilometers of their present locations.

We can use our understanding of evolution and continental drift to predict where fossils of different groups of organisms might be found. For example, scientists have constructed evolutionary trees for horses based on anatomical data. These trees and the ages of fossils of horse ancestors suggest that the genus that includes present-day horses (*Equus*) originated 5 million years ago in North America. Geologic evidence indicates that at that time, North and South America were not yet connected, making it difficult for horses to travel between them. Thus, we would predict that the oldest *Equus* fossils should be found only on the continent on which the group originated—North America. This prediction and others like it for different groups of organisms have been upheld, providing more evidence for evolution.

We can also use our understanding of evolution to explain biogeographic data. For example, islands generally have many plant and animal species that are **endemic** (found nowhere else in the world). Yet, as Darwin described in *The Origin of*

*Species*, most island species are closely related to species from the nearest mainland or a neighboring island. He explained this observation by suggesting that islands are colonized by species from the nearest mainland. These colonists eventually give rise to new species as they adapt to their new environments. Such a process also explains why two islands with similar environments in distant parts of the world tend to be populated not by species that are closely related to each other, but rather by species related to those of the nearest mainland, where the environment is often quite different.

## What Is Theoretical About Darwin's View of Life?

Some people dismiss Darwin's ideas as "just a theory." However, as we have seen, the *pattern* of evolution—the observation that

---

### Scientific Skills Exercise

# *Making and Testing Predictions*

**Can Predation Result in Natural Selection for Color Patterns in Guppies?** What we know about evolution changes constantly as new observations lead to new hypotheses—and hence to new ways to test our understanding of evolutionary theory. Consider the wild guppies (*Poecilia reticulata*) that live in pools connected by streams on the Caribbean island of Trinidad. Male guppies have highly varied color patterns that are controlled by genes that are only expressed in adult males. Female guppies choose males with bright color patterns as mates more often than they choose males with drab coloring. But the bright colors that attract females also can make the males more conspicuous to predators. Researchers observed that in pools with few predator species, the benefits of bright colors appear to "win out," and males are more brightly colored than in pools where predation is more intense.

One guppy predator, the killifish, preys on juvenile guppies that have not yet displayed their adult coloration. Researchers predicted that if adult guppies with drab colors were transferred to a pool with only killifish, eventually the descendants of these guppies would be more brightly colored (because of the female preference for brightly colored males).

**How the Experiment Was Done** Researchers transplanted 200 guppies from pools containing pike-cichlid fish, intense predators of adult guppies, to pools containing killifish, less active predators that prey mainly on juvenile guppies. They tracked the number of bright-colored spots and the total area of those spots on male guppies in each generation.

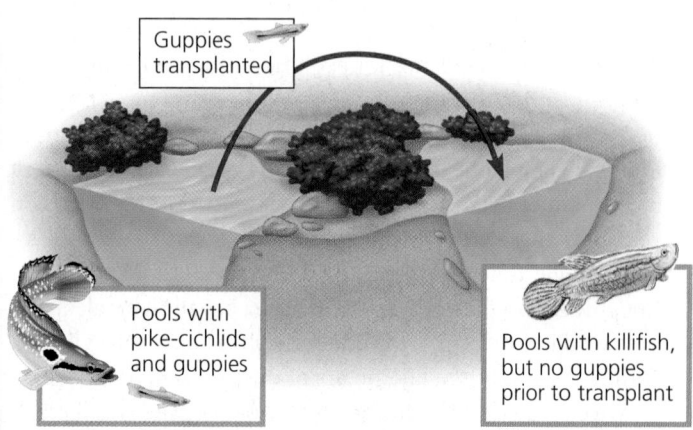

Guppies transplanted

Pools with pike-cichlids and guppies

Pools with killifish, but no guppies prior to transplant

**Data from the Experiment** After 22 months (15 generations), researchers compared the color pattern data for guppies from the source and transplanted populations.

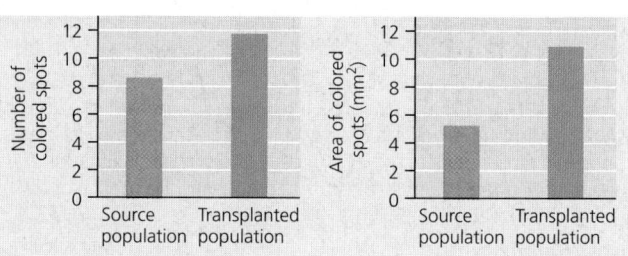

**Data from** J. A. Endler, Natural selection on color patterns in *Poecilia reticulata*, *Evolution* 34:76–91 (1980).

#### INTERPRET THE DATA

1. Identify the following elements of hypothesis-based science in this example: (a) question, (b) hypothesis, (c) prediction, (d) control group, and (e) experimental group. (For additional information about hypothesis-based science, see Chapter 1 and the Scientific Skills Review in Appendix F and the Study Area of MasteringBiology.)

2. Explain how the types of data the researchers chose to collect enabled them to test their prediction.

3. What conclusion do you draw from the data presented above?

4. Predict what would happen if, after 22 months, guppies from the transplanted population were returned to the source pool. Describe an experiment to test your prediction.

(MB) A related version of this Scientific Skills Exercise can be assigned in MasteringBiology.

life has evolved over time—has been documented directly and is supported by a great deal of evidence. In addition, Darwin's explanation of the *process* of evolution—that natural selection is the primary cause of the observed pattern of evolutionary change—makes sense of massive amounts of data. The effects of natural selection also can be observed and tested in nature.

What, then, is theoretical about evolution? Keep in mind that the scientific meaning of the term *theory* is very different from its meaning in everyday use. The colloquial use of the word *theory* comes close to what scientists mean by a hypothesis. In science, a theory is more comprehensive than a hypothesis. A theory, such as the theory of evolution by natural selection, accounts for many observations and explains and integrates a great variety of phenomena. Such a unifying theory does not become widely accepted unless its predictions stand up to thorough and continual testing by experiment and additional observation (see Concept 1.3). As the rest of this unit demonstrates, this has certainly been the case with the theory of evolution by natural selection.

The skepticism of scientists as they continue to test theories prevents these ideas from becoming dogma. For example, although Darwin thought that evolution was a very slow process, we now know that this isn't always true. New species can form in relatively short periods of time—a few thousand years or less. Furthermore, evolutionary biologists now recognize that natural selection is not the only mechanism responsible

for evolution. Indeed, the study of evolution today is livelier than ever as scientists use a wide range of experimental approaches and genetic analyses to test predictions based on natural selection and other evolutionary mechanisms. In the **Scientific Skills Exercise**, you'll work with data from an experiment on natural selection in wild guppies.

Although Darwin's theory attributes the diversity of life to natural processes, the diverse products of evolution nevertheless remain elegant and inspiring. As Darwin wrote in the final sentence of *The Origin of Species*, "There is grandeur in this view of life . . . [in which] endless forms most beautiful and most wonderful have been, and are being, evolved."

## CONCEPT CHECK 19.3

1. Explain how the following statement is inaccurate: "Antibiotics have created drug resistance in MRSA."
2. How does evolution account for (a) the similar mammalian forelimbs with different functions shown in Figure 19.16 and (b) the similar forms of the two distantly related mammals shown in Figure 19.18?
3. **WHAT IF?** Fossils show that dinosaurs originated between 200 and 250 million years ago. Would you expect the geographic distribution of early dinosaur fossils to be broad (on many continents) or narrow (on one or a few continents only)? Explain.

For suggested answers, see Appendix A.

---

# 19 Chapter Review

Go to **MasteringBiology®** for Assignments, the eText, and the Study Area with Animations, Activities, Vocab Self-Quiz, and Practice Tests.

## SUMMARY OF KEY CONCEPTS

VOCAB SELF-QUIZ

goo.gl/gbai8v

### CONCEPT 19.1

**The Darwinian revolution challenged traditional views of a young Earth inhabited by unchanging species (pp. 380–381)**

- Darwin proposed that life's diversity arose from ancestral species through natural selection, a departure from prevailing views.
- Cuvier studied **fossils** but denied that evolution occurs; he proposed that sudden catastrophic events in the past caused species to disappear from an area. Hutton and Lyell thought that geologic change could result from gradual, continuous mechanisms. Lamarck hypothesized that species evolve, but the underlying mechanisms he proposed are not supported by evidence.

**?** *Why was the age of Earth important for Darwin's ideas about evolution?*

### CONCEPT 19.2

**Descent with modification by natural selection explains the adaptations of organisms and the unity and diversity of life (pp. 382–386)**

- Darwin's voyage on the *Beagle* gave rise to his idea that species originate from ancestral forms through the accumulation of

**adaptations**. He refined his theory for many years and finally published it in 1859 after learning that Wallace had come to the same idea.
- In *The Origin of Species*, Darwin proposed that over long periods of time, descent with modification produced the rich diversity of life through the mechanism of **natural selection**.

**Observations**

| Individuals in a population vary in their heritable characteristics. | Organisms produce more offspring than the environment can support. |

**Inferences**

Individuals that are well suited to their environment tend to leave more offspring than other individuals.

and

Over time, favorable traits accumulate in the population.

**?** *Describe how overreproduction and heritable variation relate to evolution by natural selection.*

## Evolution is supported by an overwhelming amount of scientific evidence (pp. 387–393)

- Researchers have directly observed natural selection leading to adaptive evolution in many studies, including research on soapberry bug populations and on MRSA.
- Organisms share characteristics because of common descent (**homology**) or because natural selection affects independently evolving species in similar environments in similar ways (**convergent evolution**).
- Fossils show that past organisms differed from living organisms, that many species have become extinct, and that species have evolved over long periods of time; fossils also document the origin of major new groups of organisms.
- Evolutionary theory can explain some biogeographic patterns.

**?** *Summarize the different lines of evidence supporting the hypothesis that cetaceans descended from land mammals and are closely related to even-toed ungulates.*

# TEST YOUR UNDERSTANDING

**PRACTICE TEST**

goo.gl/CRZjvS

## Level 1: Knowledge/Comprehension

1. Which of the following is *not* an observation or inference on which natural selection is based?
   (A) There is heritable variation among individuals.
   (B) Poorly adapted individuals never produce offspring.
   (C) Species produce more offspring than the environment can support.
   (D) Only a fraction of an individual's offspring may survive.

2. Which of the following observations helped Darwin shape his concept of descent with modification?
   (A) Species diversity declines farther from the equator.
   (B) Fewer species live on islands than on the nearest continents.
   (C) Birds live on islands located farther from the mainland than the birds' maximum nonstop flight distance.
   (D) South American temperate plants are more similar to the tropical plants of South America than to the temperate plants of Europe.

## Level 2: Application/Analysis

3. Within six months of effectively using methicillin to treat *S. aureus* infections in a community, all new *S. aureus* infections were caused by MRSA. How can this best be explained?
   (A) A patient must have become infected with MRSA from another community.
   (B) In response to the drug, *S. aureus* began making drug-resistant versions of the protein targeted by the drug.
   (C) Some drug-resistant bacteria were present at the start of treatment, and natural selection increased their frequency.
   (D) *S. aureus* evolved to resist vaccines.

4. The upper forelimbs of humans and bats have fairly similar skeletal structures, whereas the corresponding bones in whales have very different shapes and proportions. However, genetic data suggest that all three kinds of organisms diverged from a common ancestor at about the same time. Which of the following is the most likely explanation for these data?
   (A) Forelimb evolution was adaptive in people and bats, but not in whales.
   (B) Natural selection in an aquatic environment resulted in significant changes to whale forelimb anatomy.
   (C) Genes mutate faster in whales than in humans or bats.
   (D) Whales are not properly classified as mammals.

5. DNA sequences in many human genes are very similar to the sequences of corresponding genes in chimpanzees. The most likely explanation for this result is that
   (A) humans and chimpanzees share a relatively recent common ancestor.
   (B) humans evolved from chimpanzees.
   (C) chimpanzees evolved from humans.
   (D) convergent evolution led to the DNA similarities.

## Level 3: Synthesis/Evaluation

6. **SCIENTIFIC INQUIRY**
   **DRAW IT** Mosquitoes resistant to the pesticide DDT first appeared in India in 1959, but now are found throughout the world. (a) Graph the data in the table below. (b) Examining the graph, hypothesize why the percentage of mosquitoes resistant to DDT rose rapidly. (c) Suggest an explanation for the global spread of DDT resistance.

| Month | 0 | 8 | 12 |
|---|---|---|---|
| Mosquitoes Resistant* to DDT | 4% | 45% | 77% |

Data from C. F. Curtis et al., Selection for and against insecticide resistance and possible methods of inhibiting the evolution of resistance in mosquitoes, *Ecological Entomology* 3:273–287 (1978).

*Mosquitoes were considered resistant if they were not killed within 1 hour of receiving a dose of 4% DDT.

7. **FOCUS ON EVOLUTION**
   Explain why anatomical and molecular features often fit a similar nested pattern. In addition, describe a process that can cause this not to be the case.

8. **FOCUS ON INTERACTIONS**
   Write a short essay (about 100–150 words) evaluating whether changes to an organism's physical environment are likely to result in evolutionary change. Use an example to support your reasoning.

9. **SYNTHESIZE YOUR KNOWLEDGE**

This honeypot ant (genus *Myrmecocystus*) can store liquid food inside its expandable abdomen. Consider other ants you are familiar with, and explain how a honeypot ant exemplifies three key features of life: adaptation, unity, and diversity.

*For selected answers, see Appendix A.*

# 20 Phylogeny

## KEY CONCEPTS

**20.1** Phylogenies show evolutionary relationships

**20.2** Phylogenies are inferred from morphological and molecular data

**20.3** Shared characters are used to construct phylogenetic trees

**20.4** Molecular clocks help track evolutionary time

**20.5** New information continues to revise our understanding of evolutionary history

▲ **Figure 20.1** What kind of organism is this?

## Investigating the Evolutionary History of Life

Look closely at the organism in **Figure 20.1**. Although it resembles a snake, this animal is actually a legless lizard known as the eastern glass lizard (*Ophisaurus ventralis*). Why isn't this glass lizard considered a snake? More generally, how do biologists distinguish and categorize the millions of species on Earth?

An understanding of evolutionary relationships suggests one way to address these questions: We can decide in which category to place a species by comparing its traits with those of potential close relatives. For example, the eastern glass lizard does not have a highly mobile jaw, a large number of vertebrae, or a short tail located behind the anus, three traits shared by all snakes. These and other characteristics suggest that despite a superficial resemblance, the glass lizard is not a snake.

Snakes and lizards are part of the continuum of life extending from the earliest organisms to the great variety of species alive today. To help make sense of that diversity, biologists trace **phylogeny**, the evolutionary history of a species or group of species. A phylogeny of lizards and snakes, for example, indicates that both the eastern glass lizard and snakes evolved

from lizards with legs—but they evolved from different lineages of legged lizards **(Figure 20.2)**. Thus, it appears that their legless conditions evolved independently.

In fact, a broader survey of the lizards reveals that a snakelike body form has evolved in many different groups of lizards. Most lizards with such a body form are burrowers or live in grasslands. The repeated evolution of a snakelike body form in a consistent set of environments suggests that this change has

▼ **Figure 20.2 Convergent evolution of limbless bodies.**
A phylogeny based on DNA sequence data reveals that a legless body form evolved independently from legged ancestors in the lineages leading to the eastern glass lizard and to snakes.

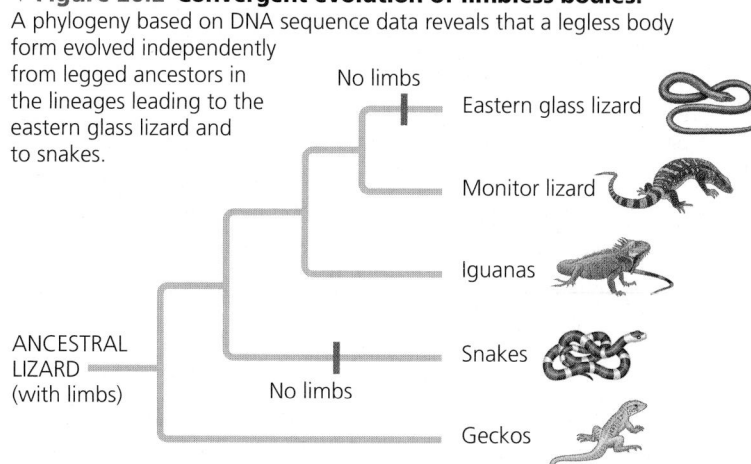

No limbs — Eastern glass lizard

Monitor lizard

Iguanas

ANCESTRAL LIZARD (with limbs)

No limbs — Snakes

Geckos

been driven by natural selection: The legs of these organisms became reduced in size, or even disappeared, over generations as the species adapted to their environments.

In this chapter, we'll examine how biologists reconstruct and interpret phylogenies using **systematics**, a discipline focused on classifying organisms and determining their evolutionary relationships.

CONCEPT 20.1

# Phylogenies show evolutionary relationships

Organisms share many characteristics because of common ancestry (see Concept 19.3). As a result, we can learn a great deal about a species if we know its evolutionary history. For example, an organism is likely to share many of its genes, metabolic pathways, and structural proteins with its close relatives. We'll consider practical applications of such information later in this section, but first we'll examine how organisms are named and classified, the scientific discipline of **taxonomy**. We'll also look at how we can interpret and use diagrams that represent evolutionary history.

## Binomial Nomenclature

Common names for organisms—such as monkey, finch, and lilac—convey meaning in casual usage, but they can also cause confusion. Each of these names, for example, refers to more than one species. Moreover, some common names do not accurately reflect the kind of organism they signify. Consider these three "fishes": jellyfish (a cnidarian), crayfish (a small lobsterlike crustacean), and silverfish (an insect). And of course, a given organism has different names in different languages.

To avoid ambiguity when communicating about their research, biologists refer to organisms by Latin scientific names. The two-part format of the scientific name, commonly called a **binomial**, was instituted in the 18th century by Carolus Linnaeus (see Concept 19.1). The first part of a binomial is the name of the **genus** (plural, *genera*) to which the species belongs. The second part, called the specific epithet, is unique for each species within the genus. An example of a binomial is *Panthera pardus*, the scientific name for the large cat commonly called the leopard. Notice that the first letter of the genus is capitalized and the entire binomial is italicized. (Newly created scientific names are also "latinized": You can name an insect you discover after a friend, but you must add a Latin ending.) Many of the more than 11,000 binomials assigned by Linnaeus are still used today, including the optimistic name he gave our own species—*Homo sapiens*, meaning "wise man."

## Hierarchical Classification

In addition to naming species, Linnaeus also grouped them into a hierarchy of increasingly inclusive categories. The first grouping is built into the binomial: Species that appear to be closely related are grouped into the same genus. For example, the leopard (*Panthera pardus*) belongs to a genus that also includes the African lion (*Panthera leo*), the tiger (*Panthera tigris*), and the jaguar (*Panthera onca*). Beyond genera, taxonomists employ progressively more comprehensive categories of classification. The taxonomic system named after Linnaeus, the Linnaean system, places related genera into the same **family**, families into **orders**, orders into **classes**, classes into **phyla** (singular, *phylum*), phyla into **kingdoms**, and, more recently, kingdoms into **domains (Figure 20.3)**. The resulting biological classification of a particular organism is somewhat like a postal address identifying a person in a particular apartment, in a building with many apartments, on a street with many apartment buildings, in a city with many streets, and so on.

The named taxonomic unit at any level of the hierarchy is called a **taxon** (plural, *taxa*). In the leopard example, *Panthera* is a taxon at the genus level, and Mammalia is a taxon at the

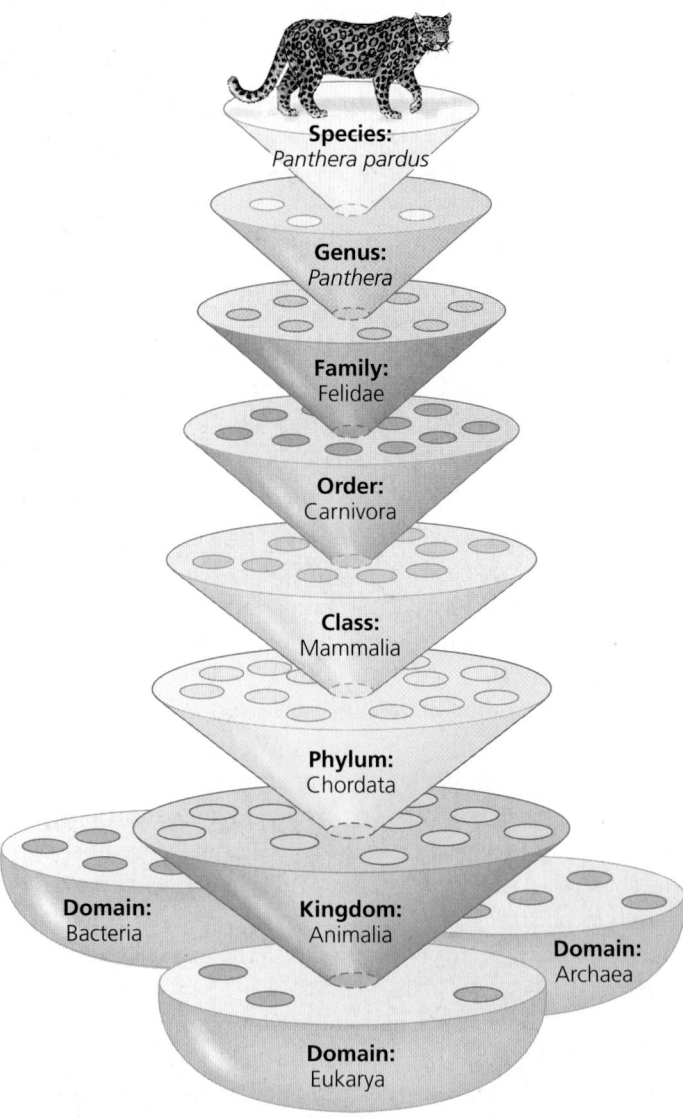

**Species:**
*Panthera pardus*

**Genus:**
*Panthera*

**Family:**
Felidae

**Order:**
Carnivora

**Class:**
Mammalia

**Phylum:**
Chordata

**Domain:**
Bacteria

**Kingdom:**
Animalia

**Domain:**
Archaea

**Domain:**
Eukarya

▲ **Figure 20.3 Linnaean classification.** At each level, or "rank," species are placed in groups within more inclusive groups.

class level that includes all the many orders of mammals. Note that in the Linnaean system, taxa broader than the genus are not italicized, though they are capitalized.

Classifying species is a way to structure our human view of the world. We lump together various species of trees to which we give the common name of pines and distinguish them from other trees that we call firs. Taxonomists have decided that pines and firs are different enough to be placed in separate genera, yet similar enough to be grouped into the same family, Pinaceae. As with pines and firs, higher levels of classification are usually defined by particular characters chosen by taxonomists. However, characters that are useful for classifying one group of organisms may not be appropriate for other organisms. For this reason, the larger categories often are not comparable between lineages; that is, an order of snails does not exhibit the same degree of morphological or genetic diversity as an order of mammals. Furthermore, as we'll see, the placement of species into orders, classes, and so on does not necessarily reflect evolutionary history.

## Linking Classification and Phylogeny

The evolutionary history of a group of organisms can be represented in a branching diagram called a **phylogenetic tree**. As in **Figure 20.4**, the branching pattern often matches how

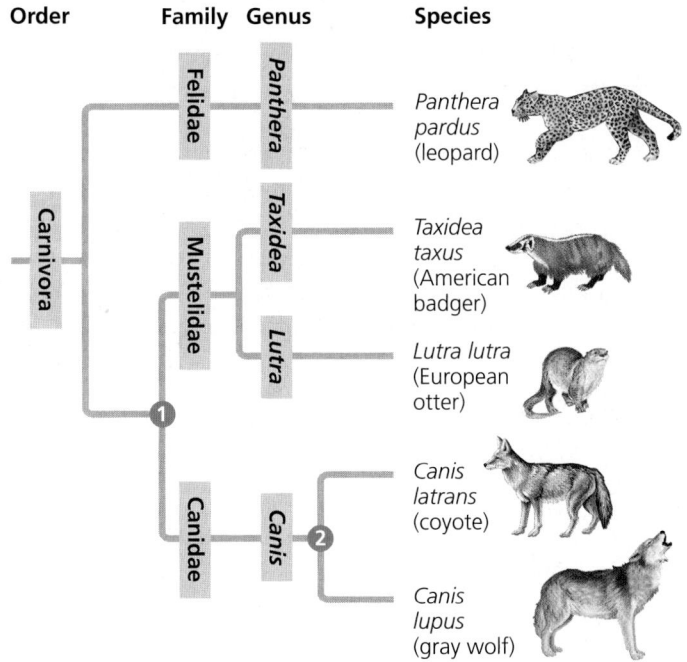

▲ **Figure 20.4 The connection between classification and phylogeny.** Hierarchical classification can reflect the branching patterns of phylogenetic trees. This tree traces possible evolutionary relationships between some of the taxa within order Carnivora, itself a branch of class Mammalia. The branch point ❶ represents the most recent common ancestor of all members of the weasel (Mustelidae) and dog (Canidae) families. The branch point ❷ represents the most recent common ancestor of coyotes and gray wolves.

❓ *What does this phylogenetic tree indicate about the evolutionary relationships between the leopard, badger, and wolf?*

taxonomists have classified groups of organisms nested within more inclusive groups. Sometimes, however, taxonomists have placed a species within a genus (or other group) to which it is *not* most closely related. One reason for such a mistake might be that over the course of evolution, a species has lost a key feature shared by its close relatives. If DNA or other new evidence indicates that an organism has been misclassified, the organism may be reclassified to accurately reflect its evolutionary history. Another issue is that while the Linnaean system may distinguish groups, such as amphibians, mammals, reptiles, and other classes of vertebrates, it tells us nothing about these groups' evolutionary relationships to one another. Such difficulties in aligning Linnaean classification with phylogeny have led many systematists to propose that classification be based entirely on evolutionary relationships.

Regardless of how groups are named, a phylogenetic tree represents a hypothesis about evolutionary relationships. These relationships often are depicted as a series of dichotomies, or two-way **branch points**. Each branch point represents the divergence of two evolutionary lineages from a common ancestor. In **Figure 20.5**, for example, branch point ❸ represents the common ancestor of taxa A, B, and C. The position of branch point ❹ to the right of ❸ indicates that taxa B and C diverged after their shared lineage split from the lineage leading to taxon A. Note also that tree branches can be rotated around a branch point without changing their evolutionary relationships.

In Figure 20.5, taxa B and C are **sister taxa**, groups of organisms that share an immediate common ancestor (branch point ❹) and hence are each other's closest relatives. In addition, this tree, like most of the phylogenetic trees in this book, is **rooted**, which means that a branch point within the tree (often drawn farthest to the left) represents the most recent

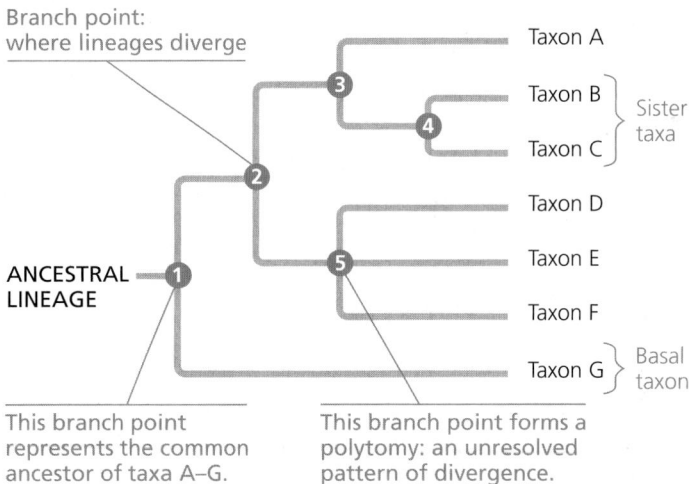

▲ **Figure 20.5 How to read a phylogenetic tree.**

**DRAW IT** *Redraw this tree, rotating the branches around branch points ❷ and ❹. Does your new version tell a different story about the evolutionary relationships between the taxa? Explain.*

common ancestor of all taxa in the tree. The term **basal taxon** refers to a lineage that diverges from all other lineages in its group early in the history of the group. Hence, like taxon G in Figure 20.5, a basal taxon lies on a branch that originates near the common ancestor of the group. Finally, on the lineage leading to taxa D–F, ❺ represents a **polytomy**, a branch point from which more than two descendant groups emerge. A polytomy signifies that evolutionary relationships among the taxa are not yet clear.

## What We Can and Cannot Learn from Phylogenetic Trees

Let's summarize three key points about phylogenetic trees. First, they are intended to show patterns of descent, not phenotypic similarity. Although closely related organisms often resemble one another due to their common ancestry, they may not if their lineages have evolved at different rates or faced very different environmental conditions. For example, even though crocodiles are more closely related to birds than to lizards (see Figure 20.16), they look more like lizards because morphology has changed dramatically in the bird lineage.

Second, the sequence of branching in a tree does not necessarily indicate the actual (absolute) ages of the particular species. For example, the tree in Figure 20.4 does not indicate that the wolf evolved more recently than the European otter; rather, the tree shows only that the most recent common ancestor of the wolf and otter (branch point ❶) lived before the most recent common ancestor of the wolf and coyote (❷). To indicate when wolves and otters evolved, the tree would need to include additional divergences in each evolutionary lineage, as well as the dates when those splits occurred. Generally, unless given specific information about what the branch lengths in a phylogenetic tree mean—for example, that they are proportional to time—we should interpret the diagram solely in terms of patterns of descent. No assumptions should be made about when particular species evolved or how much change occurred in each lineage.

Third, we should not assume that a taxon on a phylogenetic tree evolved from the taxon next to it. Figure 20.4 does not indicate that wolves evolved from coyotes or vice versa. We can infer only that the lineage leading to wolves and the lineage leading to coyotes both evolved from the common ancestor ❷. That ancestor, which is now extinct, was neither a wolf nor a coyote. However, its descendants include the two *extant* (living) species shown here, wolves and coyotes.

## Applying Phylogenies

Understanding phylogeny can have practical applications. Consider maize (corn), which originated in the Americas and is now an important food crop worldwide. From a phylogeny of maize based on DNA data, researchers have been able to identify two species of wild grasses that may be maize's closest living relatives. These two close relatives may be useful as "reservoirs" of beneficial alleles that can be transferred to cultivated maize by cross-breeding or genetic engineering (see Concept 13.4).

A different use of phylogenetic trees is to infer species identities by analyzing the relatedness of DNA sequences from different organisms. Researchers have used this approach to investigate whether "whale meat" had been harvested illegally from whale species protected under international law rather than from species that can be harvested legally, such as Minke whales caught in the Southern Hemisphere (**Figure 20.6**).

How do researchers construct trees like those we've considered here? In the next section, we'll begin to answer that question by examining the data that are used to infer phylogenies.

▼ **Figure 20.6**   **Inquiry**

### What is the species identity of food being sold as whale meat?

**Experiment**   C. S. Baker and S. R. Palumbi purchased 13 samples of "whale meat" from Japanese fish markets. They sequenced part of the mitochondrial DNA (mtDNA) from each sample and compared their results with the comparable mtDNA sequence from known whale species. To infer the species identity of each sample, the team constructed a *gene tree*, a phylogenetic tree that shows patterns of relatedness among DNA sequences rather than among taxa.

**Results**   The analysis yielded the following gene tree:

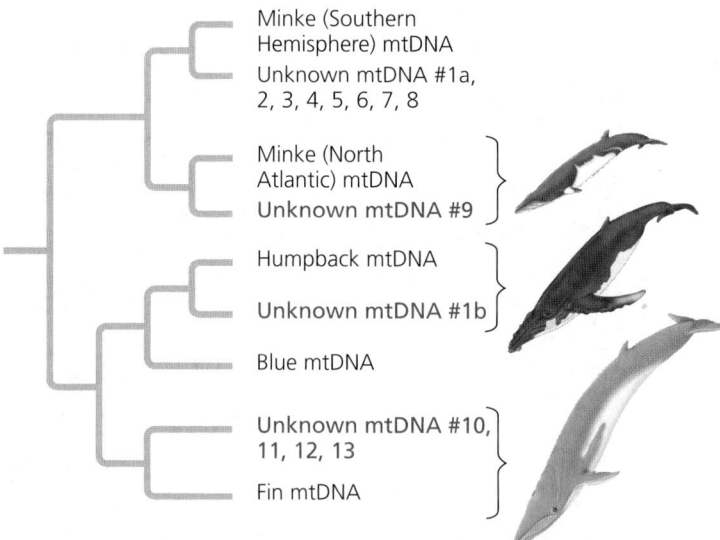

Minke (Southern Hemisphere) mtDNA
Unknown mtDNA #1a, 2, 3, 4, 5, 6, 7, 8

Minke (North Atlantic) mtDNA
Unknown mtDNA #9

Humpback mtDNA
Unknown mtDNA #1b

Blue mtDNA

Unknown mtDNA #10, 11, 12, 13

Fin mtDNA

**Conclusion**   The mtDNA sequences of six of the unknown samples (in red) were most similar to mtDNA sequences of whales that are not legal to harvest, indicating that the unknown samples were from illegally harvested whales.

**Data from**   C. S. Baker and S. R. Palumbi, Which whales are hunted? A molecular genetic approach to monitoring whaling, *Science* 265:1538–1539 (1994).

**WHAT IF?**   What different results would have indicated that *none* of the whale meat had been harvested illegally?

## CONCEPT CHECK 20.1

1. Which levels of the classification in Figure 20.3 do humans share with leopards?

2. Which of the trees shown here depicts an evolutionary history different from the other two? Explain.

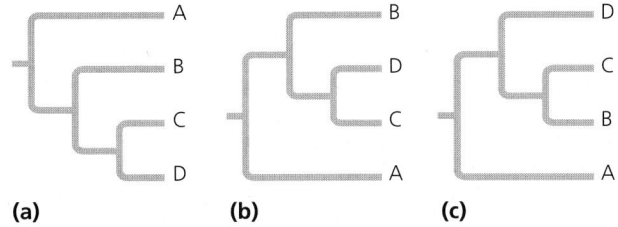

(a)      (b)      (c)

3. **WHAT IF?** Suppose new evidence indicates that taxon E in Figure 20.5 is the sister taxon of a group consisting of taxa D and F. Redraw the tree accordingly.

For suggested answers, see Appendix A.

## CONCEPT 20.2

# Phylogenies are inferred from morphological and molecular data

To infer phylogeny, systematists must gather as much information as possible about the morphology, genes, and biochemistry of the relevant organisms. It is important to focus on features that result from common ancestry, because only such features reflect evolutionary relationships.

## Morphological and Molecular Homologies

Recall that phenotypic and genetic similarities due to shared ancestry are called *homologies*. For example, the similarity in the number and arrangement of bones in the forelimbs of mammals is due to their descent from a common ancestor with the same bone structure; this is an example of a morphological homology (see Figure 19.16). In the same way, genes or other DNA sequences are homologous if they are descended from sequences carried by a common ancestor.

In general, organisms that share very similar morphologies or similar DNA sequences are likely to be more closely related than organisms with vastly different structures or sequences. In some cases, however, the morphological divergence between related species can be great and their genetic divergence small (or vice versa). Consider the Hawaiian silversword plants: Some of these species are tall, twiggy trees, while others are dense, ground-hugging shrubs (see Figure 23.16). But despite these striking phenotypic differences, the silverswords' genes are very similar. Based on these small molecular divergences, scientists estimate that the silversword group began to diverge 5 million years ago, which is also about the time when the oldest of the current Hawaiian islands formed. We'll discuss how scientists use molecular data to estimate such divergence times later in this chapter.

## Sorting Homology from Analogy

A potential source of confusion in constructing a phylogeny is similarity between organisms that is due to convergent evolution—called **analogy**—rather than to shared ancestry (homology). Convergent evolution occurs when similar environmental pressures and natural selection produce similar (analogous) adaptations in organisms from different evolutionary lineages. For example, the two mole-like animals shown in **Figure 20.7** are similar in their external appearance. However, their internal anatomy, physiology, and reproductive systems are very dissimilar. Indeed, genetic and fossil evidence indicates that the common ancestor of these animals lived 140 million years ago. This common ancestor and most of its descendants were not mole-like, but analogous characteristics evolved independently in these two lineages as they became adapted to similar lifestyles.

Australian marsupial mole

North American eutherian mole

▲ **Figure 20.7 Convergent evolution in burrowers.** A long body, large front paws, small eyes, and a pad of thick skin that protects the nose all evolved independently in these species.

Distinguishing between homology and analogy is critical in reconstructing phylogenies. To see why, consider bats and birds, both of which have adaptations that enable flight. This superficial resemblance might imply that bats are more closely related to birds than they are to cats, which cannot fly. But a closer examination reveals that a bat's wing is more similar to the forelimbs of cats and other mammals than to a bird's wing. Bats and birds descended from a common tetrapod ancestor that lived about 320 million years ago. This common ancestor could not fly. Thus, although the underlying skeletal systems of bats and birds are homologous, their *wings* are not. Flight is enabled in different ways—stretched membranes in the bat wing versus feathers in the bird wing. Fossil evidence also documents that bat wings and bird wings arose independently from the forelimbs of different tetrapod ancestors. Thus, with respect to flight, a bat's wing is *analogous*, not homologous, to a bird's wing. Analogous structures that arose independently are also called **homoplasies** (from the Greek, meaning "to mold in the same way").

Besides corroborative similarities and fossil evidence, another clue to distinguishing between homology and analogy is the complexity of the characters being compared. The more elements that are similar in two complex structures, the more likely it is that the structures evolved from a common ancestor. For instance, the skulls of an adult human and an adult chimpanzee both consist of many bones fused together. The compositions of the skulls match almost perfectly, bone for bone. It is highly improbable that such complex structures, matching in so many details, have separate origins. More likely, the genes involved in the development of both skulls were inherited from a common ancestor. The same argument applies to comparisons at the gene level. Genes are sequences of thousands of nucleotides, each of which represents an inherited character in the form of one of the four DNA bases: A (adenine), G (guanine), C (cytosine), or T (thymine). If genes in two organisms share many portions of their nucleotide sequences, it is likely that the genes are homologous.

## Evaluating Molecular Homologies

Comparing DNA molecules often poses challenges for researchers. The first step after sequencing the DNA is to align comparable sequences from the species being studied. If the species are very closely related, the sequences probably differ at only one or a few sites. In contrast, comparable nucleic acid sequences in distantly related species usually have different bases at many sites and may have different lengths. This is because insertions and deletions accumulate over long periods of time.

Suppose, for example, that certain noncoding DNA sequences near a particular gene are very similar in two species, except that the first base of the sequence has been deleted in one of the species. The effect is that the remaining sequence shifts back one notch. A comparison of the two sequences that does not take this deletion into account would overlook what in fact is a very good match. As described in **Figure 20.8**,

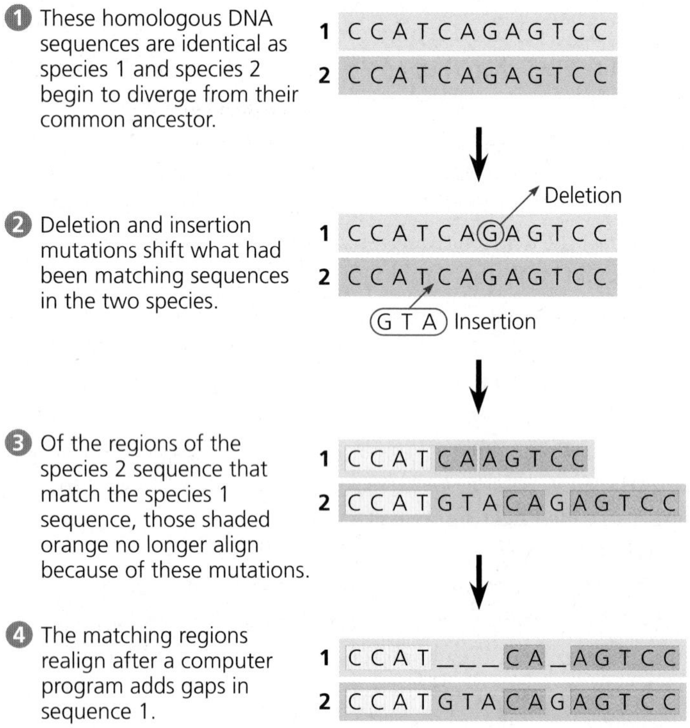

**①** These homologous DNA sequences are identical as species 1 and species 2 begin to diverge from their common ancestor.

1 CCATCAGAGTCC
2 CCATCAGAGTCC

**②** Deletion and insertion mutations shift what had been matching sequences in the two species.

1 CCATCA(G)AGTCC — Deletion
2 CCAT̲CAGAGTCC
(G T A) Insertion

**③** Of the regions of the species 2 sequence that match the species 1 sequence, those shaded orange no longer align because of these mutations.

1 CCATCAAGTCC
2 CCATGTACAGAGTCC

**④** The matching regions realign after a computer program adds gaps in sequence 1.

1 CCAT___CA_AGTCC
2 CCATGTACAGAGTCC

▲ **Figure 20.8 Aligning segments of DNA.** Systematists search for similar sequences along DNA segments from two species (only one DNA strand is shown for each species). In this example, 11 of the original 12 bases have not changed since the species diverged. Hence, those portions still align once the length is adjusted.

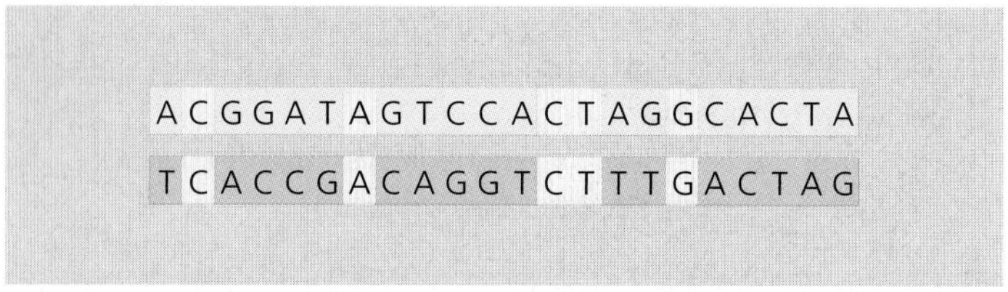

ACGGATAGTCCACTAGGCACTA
TCACCGACAGGTCTTTGACTAG

▲ **Figure 20.9 A molecular homoplasy.** These two DNA sequences from organisms that are not closely related coincidentally share 23% of their bases. Statistical tools have been developed to determine whether DNA sequences that share more than 25% of their bases do so because they are homologous.

**?** *Why might you expect organisms that are not closely related to nevertheless share roughly 25% of their bases?*

computer programs can help researchers identify such matches by testing possible alignments for comparable DNA segments of differing lengths.

Such molecular comparisons reveal that many base substitutions and other differences have accumulated in the comparable genes of an Australian mole and a North American mole. The many differences indicate that their lineages have diverged greatly since their common ancestor; thus, we say that the living species are not closely related. In contrast, the high degree of gene sequence similarity among the silversword plants indicates that they are all very closely related, in spite of their considerable morphological differences.

Just as with morphological characters, it is necessary to distinguish homology from analogy in evaluating molecular similarities for evolutionary studies. Two sequences that resemble each other at many points along their length most likely are homologous (see Figure 20.8). But in organisms that do not appear to be closely related, the bases that their otherwise very different sequences happen to share may simply be coincidental matches, called molecular homoplasies **(Figure 20.9)**. Scientists have developed statistical tools that can help distinguish "distant" homologies from such coincidental matches in extremely divergent sequences.

## CONCEPT CHECK 20.2

1. Decide whether each of the following pairs of structures more likely represents analogy or homology, and explain your reasoning: (a) a porcupine's quills and a cactus's spines; (b) a cat's paw and a human's hand; (c) an owl's wing and a hornet's wing.

2. **WHAT IF?** Suppose that two species, A and B, have similar appearances but very divergent gene sequences, while species B and C have very different appearances but similar gene sequences. Which pair of species is more likely to be closely related: A and B, or B and C? Explain.

For suggested answers, see Appendix A.

# Shared characters are used to construct phylogenetic trees

As we've discussed, a key step in reconstructing phylogenies is to distinguish homologous features from analogous ones (since only homology reflects evolutionary history). We must also choose a method of inferring phylogeny from these homologous characters. A widely used set of methods is known as cladistics.

## Cladistics

In the approach to systematics called **cladistics**, common ancestry is the primary criterion used to classify organisms. Using this methodology, biologists attempt to place species into groups called **clades**, each of which includes an ancestral species and all of its descendants **(Figure 20.10a)**. Clades, like taxonomic categories of the Linnaean system, are nested within larger clades. In Figure 20.4, for example, the cat group (Felidae) represents a clade within a larger clade (Carnivora) that also includes the dog group (Canidae).

However, a taxon is equivalent to a clade only if it is **monophyletic** (from the Greek, meaning "single tribe"), signifying that it consists of an ancestral species and all of its descendants (see Figure 20.10a). Contrast this with a **paraphyletic** ("beside the tribe") group, which consists of an ancestral species and some, but not all, of its descendants **(Figure 20.10b)**, or a **polyphyletic** ("many tribes") group, which includes distantly related species but does not include their most recent common ancestor **(Figure 20.10c)**.

Note that in a paraphyletic group, the most recent common ancestor of all members of the group *is* part of the group, whereas in a polyphyletic group the most recent common ancestor *is not* part of the group. For example, a group consisting of even-toed ungulates (hippopotamuses, deer, and their relatives) and their common ancestor is paraphyletic because it does not include cetaceans (whales, dolphins, and porpoises), which descended from that ancestor **(Figure 20.11)**.

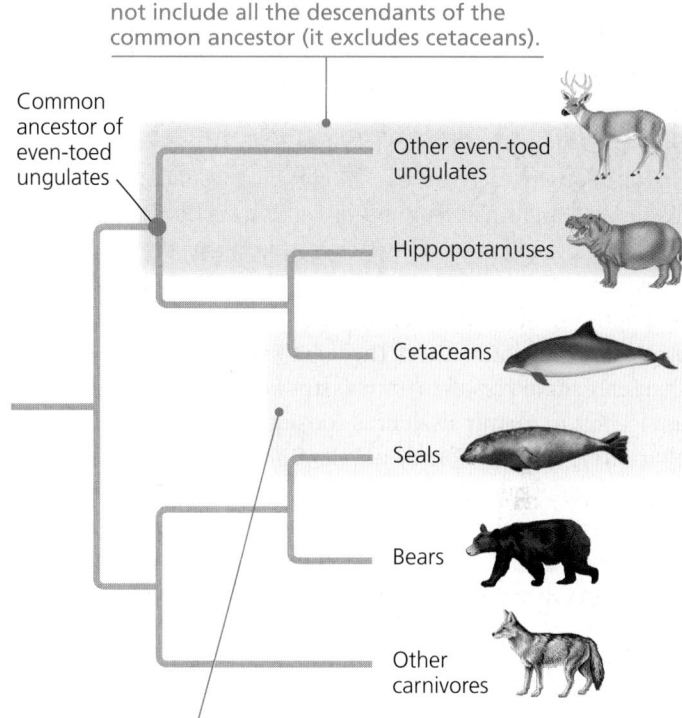

This group is paraphyletic because it does not include all the descendants of the common ancestor (it excludes cetaceans).

Common ancestor of even-toed ungulates

Other even-toed ungulates

Hippopotamuses

Cetaceans

Seals

Bears

Other carnivores

This group is polyphyletic because it does not include the most recent common ancestor of its members.

▲ **Figure 20.11 Paraphyletic vs. polyphyletic groups.**

**?** *Circle the branch point that represents the most recent common ancestor of cetaceans and seals. Explain why that ancestor would not be part of a cetacean-seal group defined by their similar body forms.*

▼ **Figure 20.10 Monophyletic, paraphyletic, and polyphyletic groups.**

**(a) Monophyletic group (clade)**

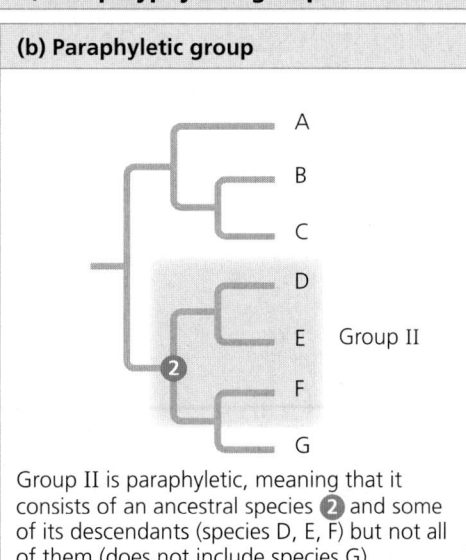

A
B   Group I
C
D
E
F
G

Group I, consisting of three species (A, B, C) and their common ancestor ❶, is a monophyletic group (clade), meaning that it consists of an ancestral species and *all* of its descendants.

**(b) Paraphyletic group**

A
B
C
D
E   Group II
F
G

Group II is paraphyletic, meaning that it consists of an ancestral species ❷ and some of its descendants (species D, E, F) but not all of them (does not include species G).

**(c) Polyphyletic group**

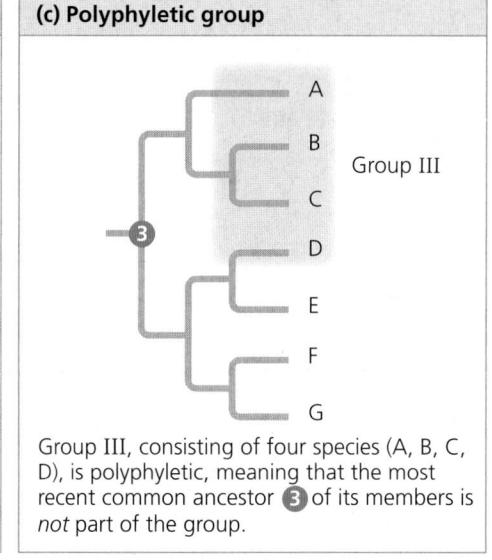

A
B   Group III
C
D
E
F
G

Group III, consisting of four species (A, B, C, D), is polyphyletic, meaning that the most recent common ancestor ❸ of its members is *not* part of the group.

In contrast, a group consisting of seals and cetaceans (based on their similar body forms) would be polyphyletic because it does not include the common ancestor of seals and cetaceans. Biologists avoid defining such polyphyletic groups; if new evidence indicates that an existing group is polyphyletic, organisms in that group are reclassified.

### Shared Ancestral and Shared Derived Characters

As a result of descent with modification, organisms have characteristics they share with their ancestors, and they also have characteristics that differ from those of their ancestors. For example, all mammals have backbones, but a backbone does not distinguish mammals from other vertebrates because *all* vertebrates have backbones. The backbone predates the branching of mammals from other vertebrates. Thus, for mammals, the backbone is a **shared ancestral character**, a character that originated in an ancestor of the taxon. In contrast, hair is a character shared by all mammals but *not* found in their ancestors. Thus, in mammals, hair is considered a **shared derived character**, an evolutionary novelty unique to a clade.

Note that it is a relative matter whether a character is considered ancestral or derived. A backbone can also qualify as a shared derived character, but only at a deeper branch point that distinguishes all vertebrates from other animals.

### Inferring Phylogenies Using Derived Characters

Shared derived characters are unique to particular clades. Because all features of organisms arose at some point in the history of life, it should be possible to determine the clade in which each shared derived character first appeared and to use that information to infer evolutionary relationships.

To see an example of this approach, consider the set of characters shown in **Figure 20.12a** for each of five vertebrates—a leopard, turtle, frog, bass, and lamprey (a jawless aquatic vertebrate). As a basis of comparison, we need to select an outgroup. An **outgroup** is a closely related species or group of species from a lineage that is known to have diverged before the lineage that includes the species we are studying (the **ingroup**). A suitable outgroup can be determined based on evidence from morphology, paleontology, embryonic development, and gene sequences. An appropriate outgroup for our example is the lancelet, a small animal that lives in mudflats and (like vertebrates) is a member of the more inclusive group called the chordates. Unlike the vertebrates, however, the lancelet does not have a backbone.

By comparing members of the ingroup with each other and with the outgroup, we can determine which characters were derived at the various branch points of vertebrate evolution. For example, *all* of the vertebrates in the ingroup have backbones: This character was present in the ancestral vertebrate, but not in the outgroup. Now note that hinged jaws are a character absent in lampreys but present in other members of the ingroup; this character helps us to identify an early branch point in the vertebrate clade. Proceeding in this way, we can translate the data in our table of characters into a phylogenetic tree that places all the ingroup taxa into a hierarchy based on their derived characters **(Figure 20.12b)**.

### Phylogenetic Trees with Proportional Branch Lengths

In the phylogenetic trees we have presented so far, the lengths of the tree's branches do not indicate the degree of

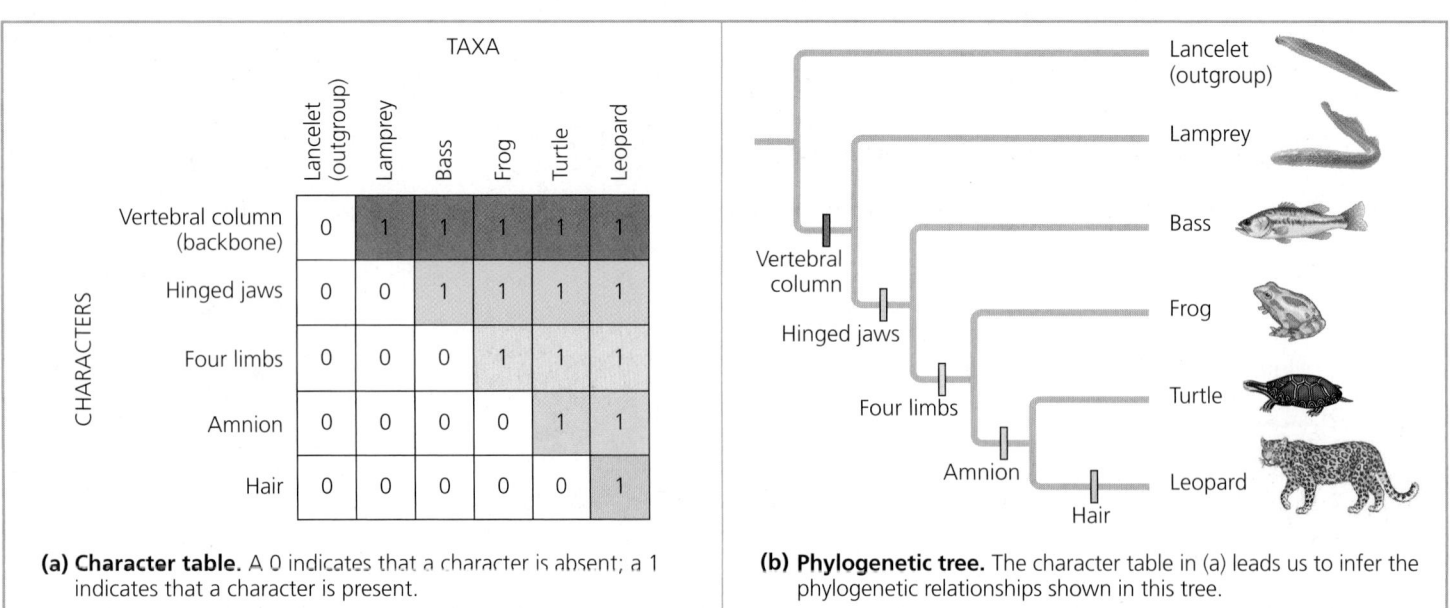

**(a) Character table.** A 0 indicates that a character is absent; a 1 indicates that a character is present.

**(b) Phylogenetic tree.** The character table in (a) leads us to infer the phylogenetic relationships shown in this tree.

▲ **Figure 20.12 Using derived characters to infer phylogeny.** The derived characters used here include the amnion, a membrane that encloses the embryo inside a fluid-filled sac (see Figure 27.28). Note that a different set of characters could lead us to infer a different phylogenetic tree.

**DRAW IT** *In (b), circle the most inclusive clade for which a hinged jaw is a shared ancestral character.*

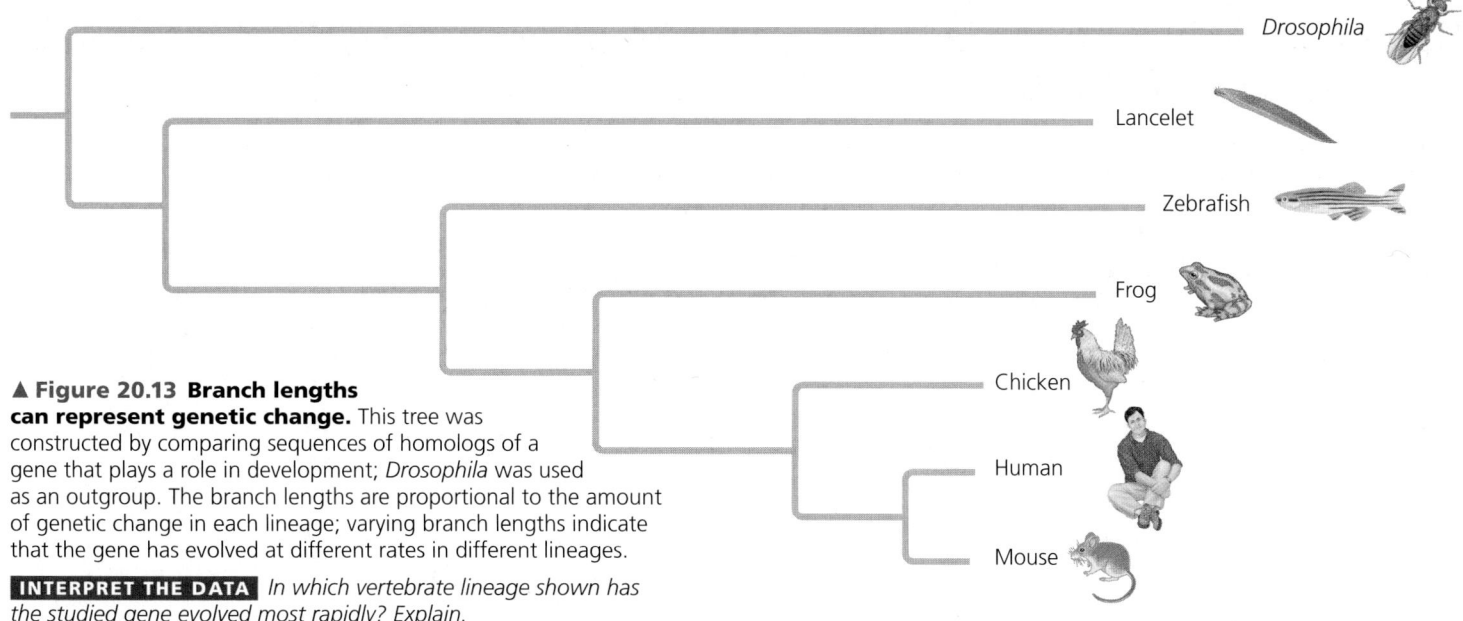

▲ **Figure 20.13 Branch lengths can represent genetic change.** This tree was constructed by comparing sequences of homologs of a gene that plays a role in development; *Drosophila* was used as an outgroup. The branch lengths are proportional to the amount of genetic change in each lineage; varying branch lengths indicate that the gene has evolved at different rates in different lineages.

**INTERPRET THE DATA** *In which vertebrate lineage shown has the studied gene evolved most rapidly? Explain.*

evolutionary change in each lineage. Furthermore, the chronology represented by the branching pattern of the tree is relative (earlier versus later) rather than absolute (how many millions of years ago). But in some tree diagrams, branch lengths are proportional to the amount of evolutionary change or to the length of time since particular events occurred.

In **Figure 20.13**, for example, the branch length of the phylogenetic tree reflects the number of changes that have taken place in a particular DNA sequence in that lineage. Note that the total length of the horizontal lines from the base of the tree to the mouse is less than that of the line leading to the outgroup species, the fruit fly *Drosophila*. This implies that in the time since the mouse and fly lineages diverged from their common ancestor, more genetic changes have occurred in the *Drosophila* lineage than in the mouse lineage.

Even though the branches of a phylogenetic tree may have different lengths, among organisms alive today, all the different lineages that descend from a common ancestor have survived for the same number of years. To take an extreme example, humans and bacteria had a common ancestor that lived over 3 billion years ago. Fossils and genetic evidence indicate that this ancestor was a single-celled prokaryote. Even though bacteria have apparently changed little in their morphology since that common ancestor, there have nonetheless been 3 billion years of evolution in the bacterial lineage, just as there have been 3 billion years of evolution in the lineage that ultimately gave rise to humans.

These equal spans of chronological time can be represented in a phylogenetic tree whose branch lengths are proportional to time **(Figure 20.14)**. Such a tree draws on fossil data to place branch points in the context of geologic time. Additionally, it is possible to combine these two types of trees by labeling branch points with information about rates of genetic change or dates of divergence.

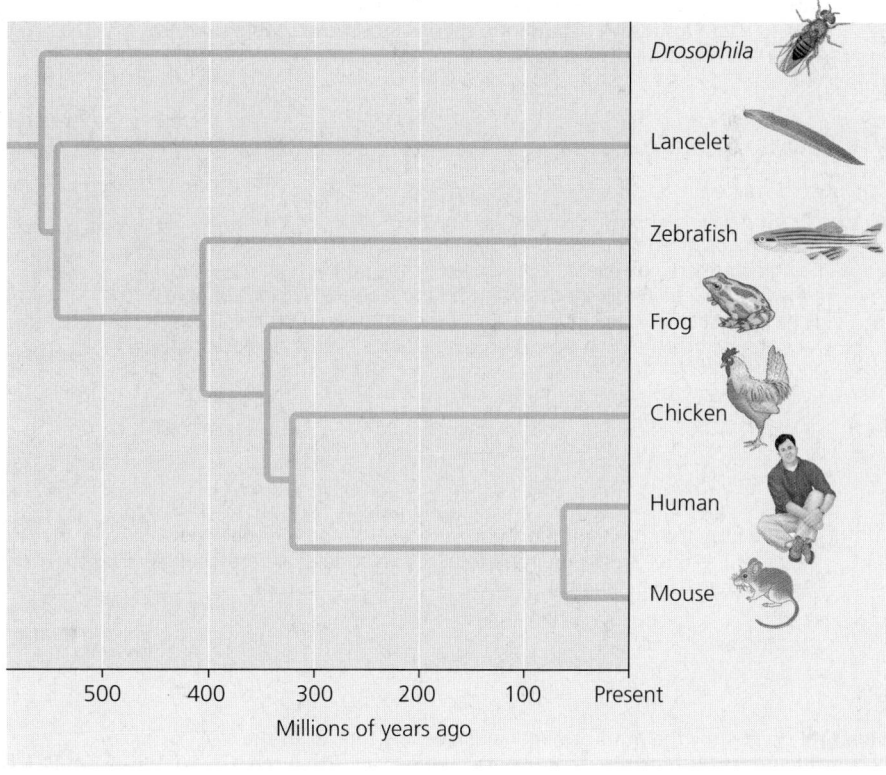

▲ **Figure 20.14 Branch lengths can indicate time.** This tree is based on the same molecular data as the tree in Figure 20.13, but here the branch points are mapped to dates based on fossil evidence. Thus, the branch lengths are proportional to time. Each lineage has the same total length from the base of the tree to the branch tip, indicating that all the lineages have diverged from the common ancestor for equal amounts of time.

## Applying Parsimony to a Problem in Molecular Systematics

**Application**  In considering possible phylogenies for a group of species, systematists compare molecular data for the species. An efficient way to begin is by identifying the most parsimonious hypothesis—the one that requires the fewest evolutionary events (molecular changes) to have occurred.

**Technique**  Follow the numbered steps as we apply the principle of parsimony to a hypothetical phylogenetic problem involving three closely related bird species.

Species I          Species II          Species III

❶ First, draw the three possible phylogenies for the species. (Although only 3 trees are possible when ordering 3 species, the number of possible trees increases rapidly with the number of species: There are 15 trees for 4 species and 34,459,425 trees for 10 species.)

Three phylogenetic hypotheses:

❷ Tabulate the molecular data for the species. In this simplified example, the data represent a DNA sequence consisting of just four nucleotide bases. Data from several outgroup species (not shown) were used to infer the ancestral DNA sequence.

|  | Site | | | |
|---|---|---|---|---|
|  | 1 | 2 | 3 | 4 |
| Species I | C | T | A | T |
| Species II | C | T | T | C |
| Species III | A | G | A | C |
| Ancestral sequence | A | G | T | T |

❸ Now focus on site 1 in the DNA sequence. In the tree on the left, a single base-change event, represented by the purple hatchmark on the branch leading to species I and II (and labeled 1/C, indicating a change at site 1 to nucleotide C), is sufficient to account for the site 1 data. In the other two trees, two base-change events are necessary.

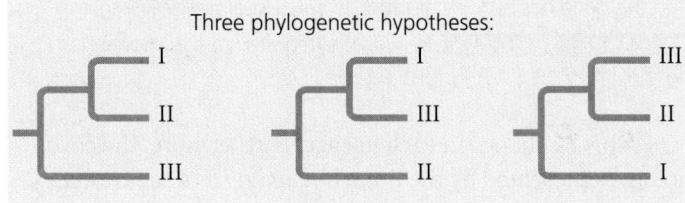

❹ Continuing the comparison of bases at site 2, 3, and 4 reveals that each of the three trees requires a total of five additional base-change events (purple hatchmarks).

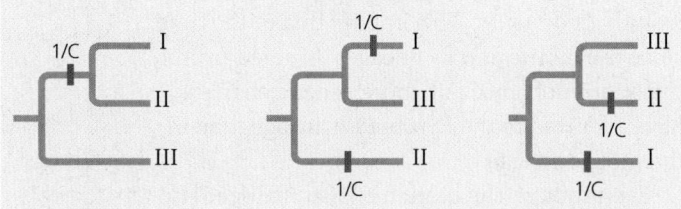

**Results**  To identify the most parsimonious tree, we total all of the base-change events noted in steps 3 and 4. We conclude that the first tree is the most parsimonious of the three possible phylogenies. (In a real example, many more sites would be analyzed. Hence, the trees would often differ by more than one base-change event.)

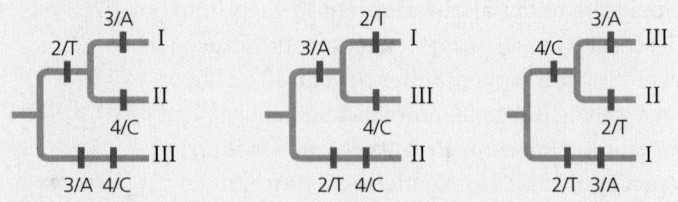

6 events          7 events          7 events

## Maximum Parsimony

As the database of DNA sequences that enables us to study more species grows, the difficulty of building the phylogenetic tree that best describes their evolutionary history also grows. What if you are analyzing data for 50 species? There are $3 \times 10^{76}$ different ways to arrange 50 species into a tree! And which tree in this huge forest reflects the true phylogeny? Systematists can never be sure of finding the most accurate tree in such a large data set, but they can narrow the possibilities by applying the principle of maximum parsimony.

According to the principle of **maximum parsimony**, we should first investigate the simplest explanation that is consistent with the facts. (The parsimony principle is also called "Occam's razor" after William of Occam, a 14th-century English philosopher who advocated this minimalist problem-solving approach of "shaving away" unnecessary complications.) In the case of trees based on morphology, the most parsimonious tree requires the fewest evolutionary events, as measured by the origin of shared derived morphological characters. For phylogenies based on DNA, the most parsimonious tree requires the fewest base changes.

Scientists have developed many computer programs to search for trees that are parsimonious. When a large amount of accurate data is available, the methods used in these programs usually yield similar trees. As an example of one method, **Figure 20.15** walks you through the process of identifying the most parsimonious molecular tree for a three-species problem. Computer programs use the principle of parsimony to estimate phylogenies in a similar way: They examine large numbers of possible trees and identify those that require the fewest evolutionary changes.

## Phylogenetic Trees as Hypotheses

This is a good place to reiterate that any phylogenetic tree represents a hypothesis about how the organisms in the tree are related to one another. The best hypothesis is the one that best fits all the available data. A phylogenetic hypothesis may be modified when new evidence compels systematists to revise their trees. Indeed, while many older phylogenetic hypotheses have been supported by new morphological and molecular data, others have been changed or rejected.

Thinking of phylogenies as hypotheses also allows us to use them in a powerful way: We can make and test predictions based on the assumption that a particular phylogeny—our hypothesis—is correct. For example, in an approach known as *phylogenetic bracketing*, we can predict (by parsimony) that features shared by two groups of closely related organisms are present in their common ancestor and all of its descendants unless independent data indicate otherwise. (Note that "prediction" can refer to unknown past events as well as to evolutionary changes yet to occur.)

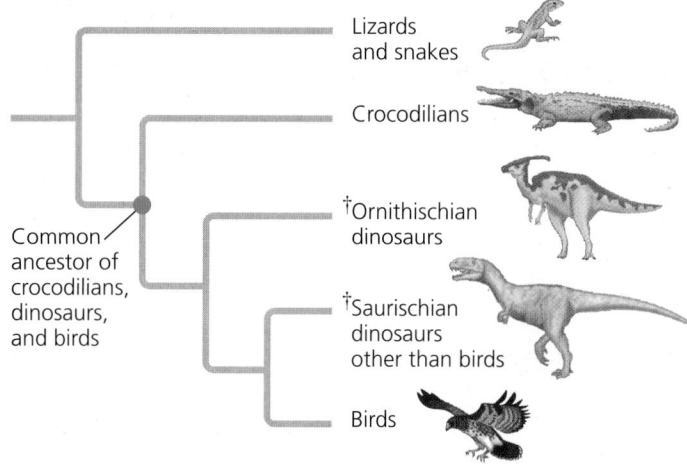

▲ **Figure 20.16  A phylogenetic tree of birds and their close relatives.** († indicates extinct lineages.)

**?** *What is the most basal taxon represented in this tree?*

This approach has been used to make novel predictions about dinosaurs. For example, there is evidence that birds descended from the theropods, a group of bipedal saurischian dinosaurs. As seen in **Figure 20.16**, the closest living relatives of birds are crocodiles. Birds and crocodiles share numerous features: They have four-chambered hearts, they "sing" to defend territories and attract mates (although a crocodile's "song" is more like a bellow), and they build nests (**Figure 20.17**). Both birds and crocodiles also care for their eggs by *brooding*, a behavior in which a parent warms the eggs with its body. Birds brood by sitting on their eggs, whereas crocodiles cover their eggs with their neck. Reasoning that any feature shared by birds and crocodiles is likely to have been present in their common ancestor (denoted by the blue dot in Figure 20.16) and *all* of its descendants, biologists predicted that dinosaurs had four-chambered hearts, sang, built nests, and exhibited brooding.

▲ **Figure 20.17  A crocodile guards its nest.** After building its nest mound, this female African dwarf crocodile will care for the eggs until they hatch.

Internal organs, such as the heart, rarely fossilize, and it is, of course, difficult to test whether dinosaurs sang to defend territories and attract mates. However, fossilized dinosaur eggs and nests have provided evidence supporting the prediction of brooding in dinosaurs. First, a fossil embryo of an *Oviraptor* dinosaur was found, still inside its egg. This egg was identical to those found in another fossil, one that showed an *Oviraptor* crouching over a group of eggs in a posture similar to that seen in brooding birds today **(Figure 20.18)**. Researchers suggested that the *Oviraptor* dinosaur preserved in this second fossil died while incubating or protecting its eggs. The broader conclusion that emerged from this work—that dinosaurs built nests and exhibited brooding—has since been strengthened by additional fossil discoveries that show that other species of dinosaurs built nests and sat on their eggs. Finally, fossil discoveries of nests and brooding in dinosaurs support predictions based on the phylogenetic hypothesis shown in Figure 20.16. These fossils thus provide independent data which suggest that the hypothesis is correct.

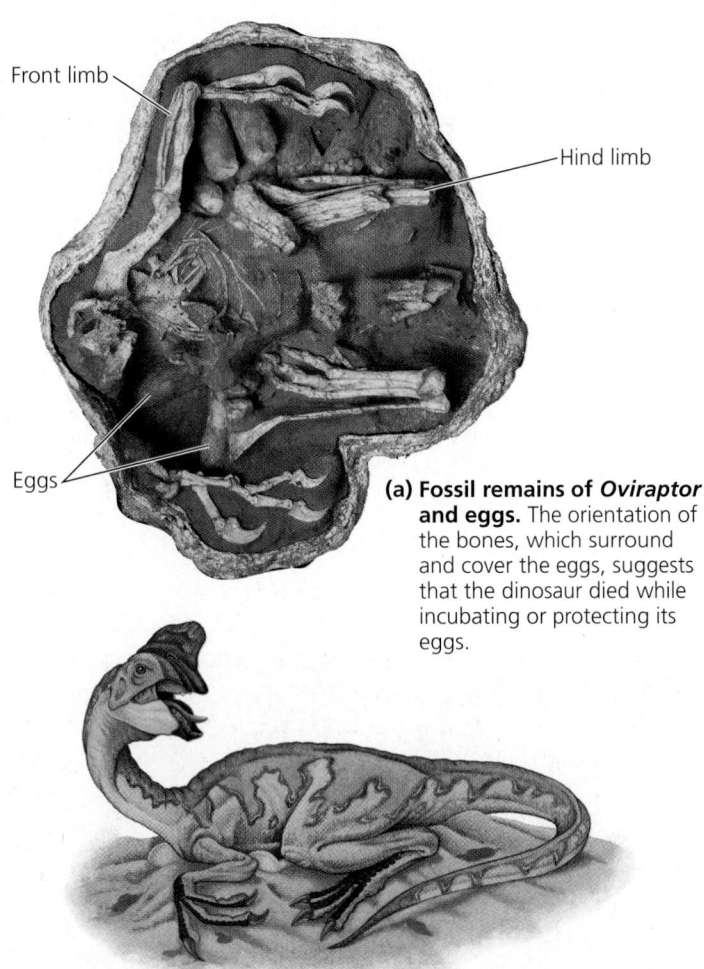

(a) **Fossil remains of *Oviraptor* and eggs.** The orientation of the bones, which surround and cover the eggs, suggests that the dinosaur died while incubating or protecting its eggs.

(b) **Artist's reconstruction of the dinosaur's posture based on the fossil findings.**

▲ **Figure 20.18 Fossil support for a phylogenetic prediction: Dinosaurs built nests and brooded their eggs.**

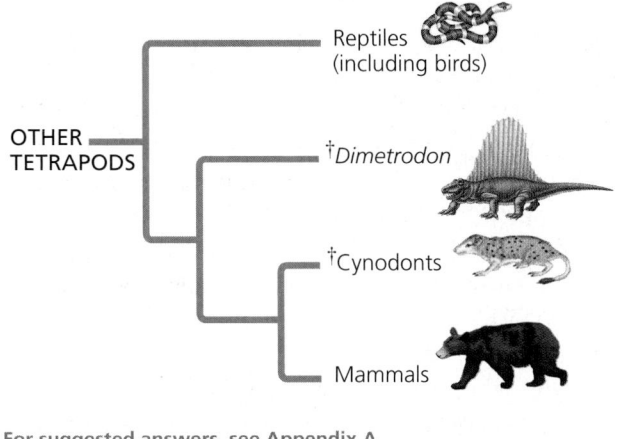
## CONCEPT 20.4

# Molecular clocks help track evolutionary time

One goal of evolutionary biology is to understand the relationships among all organisms, including those for which there is no fossil record. However, if we attempt to determine the timing of phylogenies that extend beyond the fossil record, we must rely on an important assumption about how change occurs at the molecular level.

## Molecular Clocks

We stated earlier that researchers have estimated that the common ancestor of Hawaiian silversword plants lived about 5 million years ago. How did they make this estimate? They relied on the concept of a **molecular clock**, an approach for measuring the absolute time of evolutionary change based on the observation that some genes and other regions of genomes appear to evolve at constant rates. An assumption underlying the molecular clock is that the number of nucleotide substitutions in related genes is proportional to the time that has elapsed since the genes branched from their common ancestor (divergence time).

We can calibrate the molecular clock of a gene that has a reliable average rate of evolution by graphing the number of

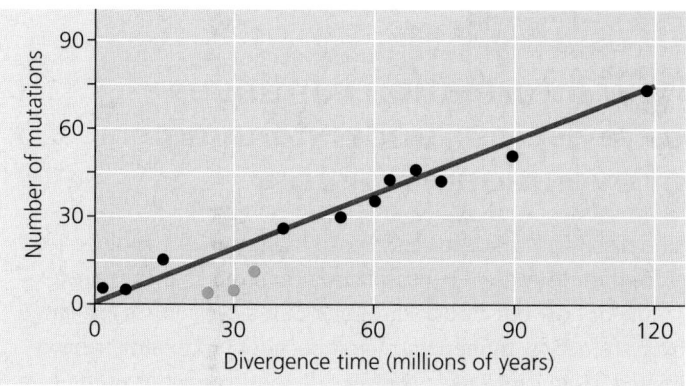

▲ **Figure 20.19 A molecular clock for mammals.** The number of accumulated mutations in seven proteins has increased over time in a consistent manner for most mammal species. The three green data points represent primate species, whose proteins appear to have evolved more slowly than those of other mammals. The divergence time for each data point was based on fossil evidence.

**INTERPRET THE DATA** *Use the graph to estimate the divergence time for a mammal with a total of 30 mutations in the seven proteins.*

genetic differences—for example, nucleotide, codon, or amino acid differences—against the dates of evolutionary branch points that are known from the fossil record **(Figure 20.19)**. The average rates of genetic change inferred from such graphs can then be used to estimate the dates of events that cannot be discerned from the fossil record, such as the origin of the silverswords discussed earlier.

Of course, no gene marks time with complete precision. In fact, some portions of the genome appear to have evolved in irregular bursts that are not at all clocklike. And even those genes that seem to act as reliable molecular clocks are accurate only in the statistical sense of showing a fairly smooth *average* rate of change. Over time, there may still be deviations from that average rate. Furthermore, the same gene may evolve at different rates in different groups of organisms. Finally, when comparing genes that are clocklike, the rate of the clock may vary greatly from one gene to another; some genes evolve a million times faster than others.

### Differences in Clock Speed

What causes such differences in the speed at which clocklike genes evolve? The answer stems from the fact that some mutations are selectively neutral—neither beneficial nor detrimental. Of course, many new mutations are harmful and are removed quickly by selection. But if most of the rest are neutral and have little or no effect on survival and reproduction, then the rate of evolution of those neutral mutations should indeed be regular, like a clock. Differences in the clock rate for different genes are a function of how important a gene is. If the exact sequence of amino acids that a gene specifies is essential to survival, most of the mutational changes will be harmful and only a few will be neutral. As a result,

such genes change only slowly. But if the exact sequence of amino acids is less critical, fewer of the new mutations will be harmful and more will be neutral. Such genes change more quickly.

### Potential Problems with Molecular Clocks

In fact, molecular clocks do not run as smoothly as would be expected if the underlying mutations were selectively neutral. Many irregularities are likely to be the result of natural selection in which certain DNA changes are favored over others. Indeed, evidence suggests that almost half the amino acid differences in proteins of two *Drosophila* species, *D. simulans* and *D. yakuba*, are not neutral but have resulted from natural selection. But because the direction of natural selection may change repeatedly over long periods of time (and hence may average out), some genes experiencing selection can nevertheless serve as approximate markers of elapsed time.

Another question arises when researchers attempt to extend molecular clocks beyond the time span documented by the fossil record. Although some fossils are more than 3 billion years old, these are very rare. An abundant fossil record extends back only about 550 million years, but molecular clocks have been used to date evolutionary divergences that occurred a billion or more years ago. These estimates assume that the clocks have been constant for all that time. Such estimates are highly uncertain.

In some cases, problems may be avoided by calibrating molecular clocks with data on the rates at which genes have evolved in different taxa. In other cases, problems may be avoided by using many genes rather just using one or a few genes. By using many genes, fluctuations in evolutionary rate due to natural selection or other factors that vary over time may average out. For example, one group of researchers constructed molecular clocks of vertebrate evolution from published sequence data for 658 nuclear genes. Despite the broad period of time covered (nearly 600 million years) and the fact that natural selection probably affected some of these genes, their estimates of divergence times agreed closely with fossil-based estimates. As this example suggests, if used with care, molecular clocks can aid our understanding of evolutionary relationships.

### Applying a Molecular Clock: Dating the Origin of HIV

Researchers have used a molecular clock to date the origin of HIV infection in humans. Phylogenetic analysis shows that HIV, the virus that causes AIDS, is descended from viruses that infect chimpanzees and other primates. (Most of these viruses do not cause AIDS-like diseases in their native hosts.) When did HIV jump to humans? There is no simple answer, because the virus has spread to humans more than once. The multiple origins of HIV are reflected in the variety of strains

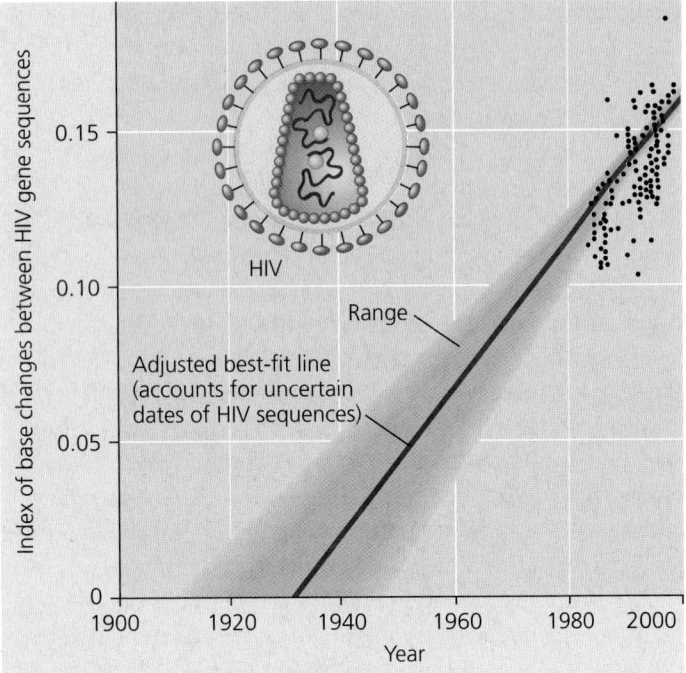

**▲ Figure 20.20 Dating the origin of HIV-1 M.** The black data points are based on DNA sequences of an HIV gene in patients' blood samples. (The dates when these individual HIV gene sequences arose are not certain because a person can harbor the virus for years before symptoms occur.) Projecting the gene's rate of change backward in time by this method suggests that the virus originated in the 1930s.

(genetic types) of the virus. HIV's genetic material is made of RNA, and like other RNA viruses, it evolves quickly.

The most widespread strain in humans is HIV-1 M. To pinpoint the earliest HIV-1 M infection, researchers compared samples of the virus from various times during the epidemic, including a sample from 1959. A comparison of gene sequences showed that the virus has evolved in a clock-like fashion **(Figure 20.20)**. Extrapolating backward in time using the molecular clock indicates that the HIV-1 M strain originated around 1930. A later study, which dated the origin of HIV using a more complex molecular clock approach than that covered in this book, estimated that the HIV-1 M strain originated about 1910.

## CONCEPT CHECK 20.4

1. What is a molecular clock? What assumption underlies the use of a molecular clock?
2. **MAKE CONNECTIONS** Review Concept 14.5. Then explain how numerous base changes could occur in an organism's DNA yet have no effect on its survival and reproduction.
3. **WHAT IF?** Suppose a molecular clock dates the divergence of two taxa at 80 million years ago, but new fossil evidence shows that the taxa diverged at least 120 million years ago. Explain how this could happen.

**For suggested answers, see Appendix A.**

# New information continues to revise our understanding of evolutionary history

The discovery that the glass lizard in Figure 20.1 evolved from a different lineage of legless lizards than did snakes is one example of how our understanding of life's diversity is informed by systematics. Indeed, in recent decades, systematists have gained insight into even the very deepest branches of the tree of life by analyzing DNA sequence data.

## From Two Kingdoms to Three Domains

Taxonomists once classified all known species into two kingdoms: plants and animals. Classification schemes with more than two kingdoms gained broad acceptance in the late 1960s, when many biologists recognized five kingdoms: Monera (prokaryotes), Protista (a diverse kingdom consisting mostly of unicellular organisms), Plantae, Fungi, and Animalia. This system highlighted the two fundamentally different types of cells, prokaryotic and eukaryotic, and set the prokaryotes apart from all eukaryotes by placing them in their own kingdom, Monera.

However, phylogenies based on genetic data soon began to reveal a problem with this system: Some prokaryotes differ as much from each other as they do from eukaryotes. Such difficulties have led biologists to adopt a three-domain system **(Figure 20.21)**. The three domains—Bacteria, Archaea, and Eukarya—are a taxonomic level higher than the kingdom level. The validity of these domains is supported by many studies, including a recent study that analyzed nearly 100 completely sequenced genomes.

The domain Bacteria contains most of the currently known prokaryotes, while the domain Archaea consists of a diverse group of prokaryotic organisms that inhabit a wide variety of environments. The domain Eukarya consists of all the organisms that have cells containing true nuclei. This domain includes many groups of single-celled organisms as well as multicellular plants, fungi, and animals. Figure 20.21 represents one possible phylogenetic tree for the three domains and some of the many lineages they encompass.

The three-domain system highlights the fact that much of the history of life has been about single-celled organisms. The two prokaryotic domains consist entirely of single-celled organisms, and even in Eukarya, only the branches labeled in red type (plants, fungi, and animals) are dominated by multicellular organisms. Of the five kingdoms previously recognized by taxonomists, most biologists continue to recognize Plantae, Fungi, and Animalia, but not Monera and Protista. The kingdom Monera is obsolete because it would have members in two different domains. The kingdom Protista has also

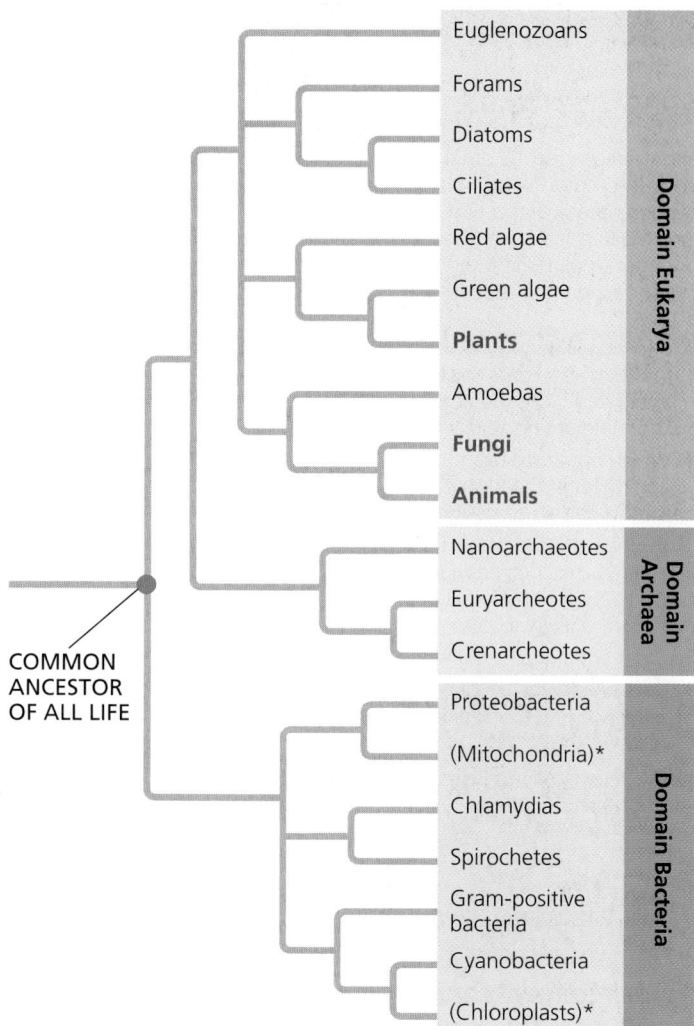

▲ **Figure 20.21 The three domains of life.** This phylogenetic tree is based on sequence data for rRNA and other genes. For simplicity, only some of the major branches in each domain are shown. Lineages within Eukarya that are dominated by multicellular organisms (plants, fungi, and animals) are in bold red type, while the two lineages denoted by an asterisk are based on DNA from cellular organelles. All other lineages consist solely or mainly of single-celled organisms.

**MAKE CONNECTIONS** *After reviewing endosymbiont theory (see Figure 4.16), explain the specific positions of the mitochondrion and chloroplast lineages on this tree.*

crumbled because it includes members that are more closely related to plants, fungi, or animals than to other protists (see Concept 25.3).

## The Important Role of Horizontal Gene Transfer

In the phylogeny shown in Figure 20.21, the first major split in the history of life occurred when bacteria diverged from other organisms. If this tree is correct, eukaryotes and archaea are more closely related to each other than either is to bacteria.

This reconstruction of the tree of life is based in part on sequence comparisons of rRNA genes, which code for the RNA components of ribosomes. However, some other genes

reveal a different set of relationships. For example, researchers have found that many of the genes that influence metabolism in yeast (a unicellular eukaryote) are more similar to genes in the domain Bacteria than they are to genes in the domain Archaea—a finding that suggests that the eukaryotes may share a more recent common ancestor with bacteria than with archaea.

What causes trees based on data from different genes to yield such different results? Comparisons of complete genomes from the three domains show that there have been substantial movements of genes between organisms in the different domains. These took place through **horizontal gene transfer**, a process in which genes are transferred from one genome to another through mechanisms such as exchange of transposable elements and plasmids (see Concept 24.3), viral infection, and perhaps fusions of organisms (as when a host and its endosymbiont become a single organism). Because phylogenetic trees are based on the assumption that genes are passed vertically from one generation to the next, the occurrence of such horizontal transfer events helps to explain why trees built using different genes can give inconsistent results.

Recent research further highlights the importance of horizontal gene transfer. For example, a 2008 study showed that on average, 80% of the genes in 181 prokaryotic genomes had moved between species at some point during the course of evolution. Such findings have led some biologists to hypothesize that horizontal gene transfer was so common in the early history of life that this history should be represented as a tangled network of connected branches **(Figure 20.22)** rather than as a dichotomously branching tree like that in

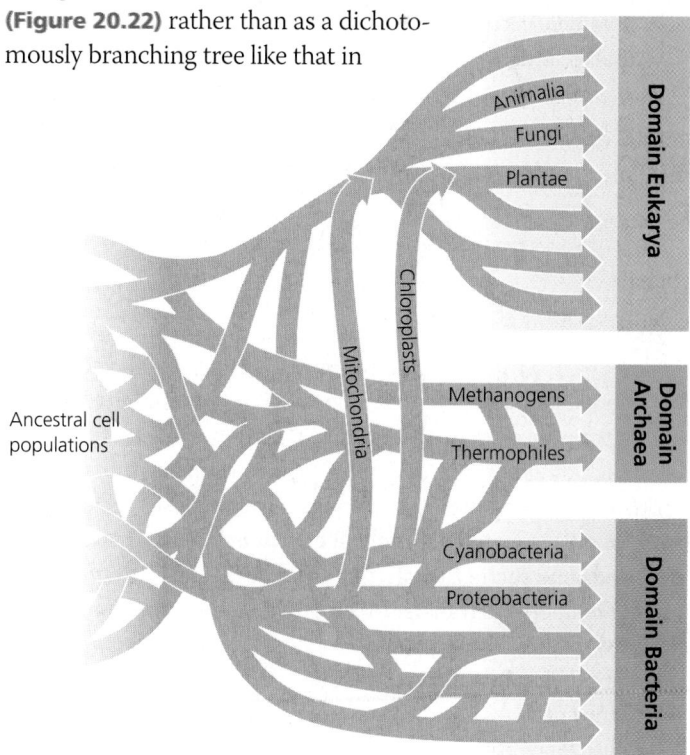

▲ **Figure 20.22 A tangled web of life.** Horizontal gene transfer may have been so common in the early history of life that the base of a "tree of life" might be more accurately portrayed as a tangled web.

# Using Protein Sequence Data to Test an Evolutionary Hypothesis

**Did Aphids Acquire Their Ability to Make Carotenoids Through Horizontal Gene Transfer?** Carotenoids are colored molecules that have diverse functions in many organisms, such as photosynthesis in plants and light detection in animals. Plants and many microorganisms can synthesize carotenoids from scratch, but animals generally cannot (they must obtain carotenoids from their diet). One exception is the pea aphid *Acyrthosiphon pisum*, a small plant-dwelling insect whose genome includes a full set of genes for the enzymes needed to make carotenoids. Because other animals lack these genes, it is unlikely that aphids inherited them from a single-celled common ancestor shared with microorganisms and plants. So where did they come from? Evolutionary biologists hypothesize that an aphid ancestor acquired these genes by horizontal gene transfer from distantly related organisms.

**How the Experiment Was Done** Scientists obtained the DNA sequences for the carotenoid-biosynthesis genes from several species, including aphids, fungi, bacteria, and plants. A computer "translated" these sequences into amino acid sequences of the encoded polypeptides and aligned the amino acid sequences. This allowed the team to compare the corresponding polypeptides in the different organisms.

**Data from the Experiment** The sequences below show the first 60 amino acids of one polypeptide of the carotenoid-biosynthesis enzymes in the plant *Arabidopsis thaliana* (bottom) and the corresponding amino acids in five nonplant species, using the one-letter

abbreviations for the amino acids (see Figure 3.18). A hyphen (-) indicates a gap inserted in a sequence to optimize its alignment with the corresponding sequence in *Arabidopsis*.

**INTERPRET THE DATA**

1. In the rows of data for the organisms being compared with the aphid, highlight the amino acids that are identical to the corresponding amino acids in the aphid.

2. Which organism has the most amino acids in common with the aphid? Rank the partial polypeptides from the other four organisms in degree of similarity to that of the aphid.

3. Do these data support the hypothesis that aphids acquired the gene for this polypeptide by horizontal gene transfer? Why or why not? If horizontal gene transfer did occur, what type of organism is likely to have been the source?

4. What additional sequence data would support your hypothesis?

5. How would you account for the similarities between the aphid sequence and the sequences for the bacteria and plant?

(MB) A version of this Scientific Skills Exercise can be assigned in MasteringBiology.

| Organism | Alignment of Amino Acid Sequences |
|---|---|
| *Acyrthosiphon* (aphid) | IKIIIIGSGV GGTAAAARLS KKGFQVEVYE KNSYNGGRCS IIR-HNGHRF DQGPSL--YL |
| *Pantoea* (bacterium) | KRTFVIGAGF GGLALAIRLQ AAGIATTVLE QHDKPGGRAY VWQ-DQGFTF DAGPTV--IT |
| *Staphylococcus* (bacterium) | MKIAVIGAGV TGLAAAARIA SQGHEVTIFE KNNNVGGRMN QLK-KDGFTF DMGPTI--VM |
| *Ustilago* (fungus) | KKVVIIGAGA GGTALAARLG RRGYSVTVLE KNSFGGGRCS LIH-HDGHRW DQGPSL--YL |
| *Gibberella* (fungus) | KSVIVIGAGV GGVSTAARLA KAGFKVTILE KNDFTGGRCS LIH-NDGHRF DQGPSL--LL |
| *Arabidopsis* (plant) | WDAVVIGGGH NGLTAAAYLA RGGLSVAVLE RRHVIGGAAV TEEIVPGFKF SRCSYLQGLL |

**Data from** Nancy A. Moran, Yale University. See N. A. Moran and T. Jarvik, Lateral transfer of genes from fungi underlies carotenoid production in aphids, *Science* 328:624–627 (2010).

Figure 20.21. Moreover, horizontal gene transfer can also occur between eukaryotes. For example, over 200 cases of the horizontal transfer of transposons have been reported in eukaryotes, including humans and other primates, plants, birds, and reptiles. Nuclear genes have also been transferred horizontally from one eukaryote to another. The **Scientific Skills Exercise** describes one such example, giving you the opportunity to interpret data on the transfer of a pigment gene to an aphid from another species.

Overall, horizontal gene transfer has played a key role throughout the evolutionary history of life and it continues to occur today. Although scientists continue to debate whether early steps in the history of life are best represented as a tree or a tangled web, in recent decades there have been many exciting discoveries about evolutionary events that occurred over

time. We'll explore the mechanisms that underlie such events in the rest of this unit's chapters, beginning with factors that cause genetic change in populations.

## CONCEPT CHECK 20.5

1. Why is the kingdom Monera no longer considered a valid taxon?

2. Explain why phylogenies based on different genes can yield different branching patterns for the tree of all life.

3. **WHAT IF?** Draw the three possible dichotomously branching trees showing evolutionary relationships for the domains Bacteria, Archaea, and Eukarya. Two of these trees have been supported by genetic data. Is it likely that the third tree might also receive such support? Explain your answer.

For suggested answers, see Appendix A.

## SUMMARY OF KEY CONCEPTS

**VOCAB SELF-QUIZ**

goo.gl/gbai8v

### CONCEPT 20.1

**Phylogenies show evolutionary relationships (pp. 396–399)**

- Linnaeus's **binomial** classification system gives organisms two-part names: a **genus** plus a specific epithet.
- In the Linnaean system, species are grouped in increasingly broad taxa: Related genera are placed in the same **family**, families in **orders**, orders in **classes**, classes in **phyla**, phyla in **kingdoms**, and (more recently) kingdoms in **domains**.
- Systematists depict evolutionary relationships as branching **phylogenetic trees**. Many systematists propose that classification be based entirely on evolutionary relationships.

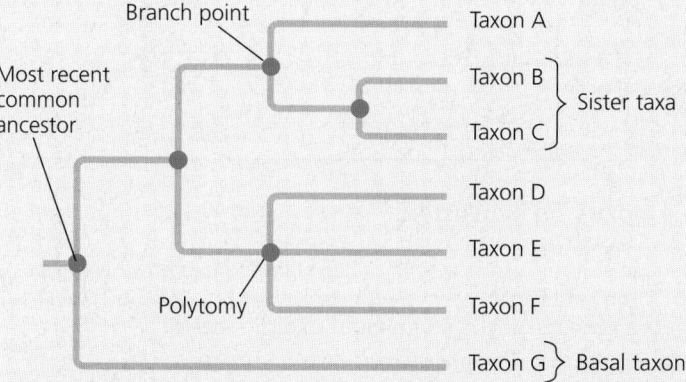

- Unless branch lengths are proportional to time or genetic change, a phylogenetic tree indicates only patterns of descent.
- Much information can be learned about a species from its evolutionary history; hence, phylogenies are useful in a wide range of applications.

? *Humans and chimpanzees are sister species. Explain what that means.*

### CONCEPT 20.2

**Phylogenies are inferred from morphological and molecular data (pp. 399–400)**

- Organisms with similar morphologies or DNA sequences are likely to be more closely related than organisms with very different structures and genetic sequences.
- To infer phylogeny, homology (similarity due to shared ancestry) must be distinguished from **analogy** (similarity due to convergent evolution).
- Computer programs are used to align comparable DNA sequences and to distinguish molecular homologies from coincidental matches between taxa that diverged long ago.

? *Why is it necessary to distinguish homology from analogy to infer phylogeny?*

### CONCEPT 20.3

**Shared characters are used to construct phylogenetic trees (pp. 401–406)**

- A **clade** is a **monophyletic** group that includes an ancestral species and all of its descendants.

- Clades can be distinguished by their **shared derived characters**.

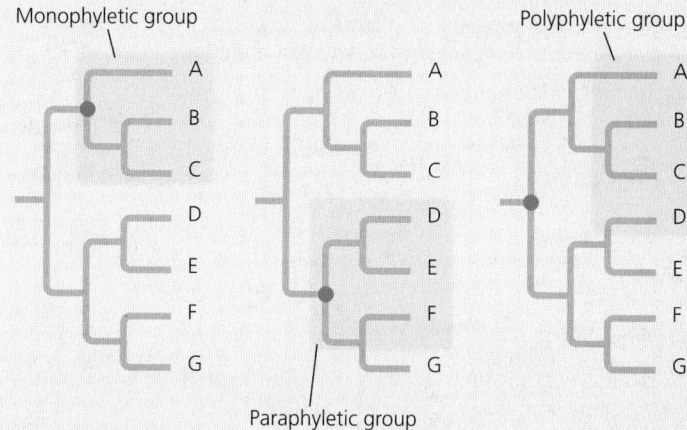

- Branch lengths can be proportional to amount of evolutionary change or time.
- Among phylogenies, the most parsimonious tree is the one that requires the fewest evolutionary changes.
- Well-supported phylogenetic hypotheses are consistent with a wide range of data.

? *Explain the logic of using shared derived characters to infer phylogeny.*

### CONCEPT 20.4

**Molecular clocks help track evolutionary time (pp. 406–408)**

- Some regions of DNA change at a rate consistent enough to serve as a **molecular clock**, in which the amount of genetic change is used to estimate the date of past evolutionary events. Other DNA regions change in a less predictable way.
- Molecular clock analyses suggest that the most common strain of HIV jumped from primates to humans in the early 1900s.

? *Describe some assumptions and limitations of molecular clocks.*

### CONCEPT 20.5

**New information continues to revise our understanding of evolutionary history (pp. 408–410)**

- Past classification systems have given way to the current view of the tree of life, which consists of three great domains: Bacteria, Archaea, and Eukarya.
- Phylogenies based in part on rRNA genes suggest that eukaryotes are most closely related to archaea, while data from some other genes suggest a closer relationship to bacteria.
- Genetic analyses indicate that extensive **horizontal gene transfer** has occurred throughout the evolutionary history of life.

? *Why was the five-kingdom system abandoned for a three-domain system?*

# TEST YOUR UNDERSTANDING

**PRACTICE TEST**

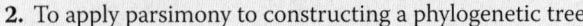

goo.gl/CRZjvS

## Level 1: Knowledge/Comprehension

1. In a comparison of birds and mammals, the condition of having four limbs is
   (A) a shared ancestral character.
   (B) a shared derived character.
   (C) a character useful for distinguishing birds from mammals.
   (D) an example of analogy rather than homology.

2. To apply parsimony to constructing a phylogenetic tree,
   (A) choose the tree that assumes all evolutionary changes are equally probable.
   (B) choose the tree in which the branch points are based on as many shared derived characters as possible.
   (C) choose the tree that represents the fewest evolutionary changes, in either DNA sequences or morphology.
   (D) choose the tree with the fewest branch points.

## Level 2: Application/Analysis

3. In Figure 20.4, which similarly inclusive taxon descended from the same common ancestor as Canidae?
   (A) Felidae
   (B) Mustelidae
   (C) Carnivora
   (D) *Lutra*

4. Three living species X, Y, and Z share a common ancestor T, as do extinct species U and V. A grouping that consists of species T, X, Y, and Z (but not U or V) makes up
   (A) a monophyletic taxon.
   (B) an ingroup, with species U as the outgroup.
   (C) a polyphyletic group.
   (D) a paraphyletic group.

5. Based on the tree below, which statement is *not* correct?

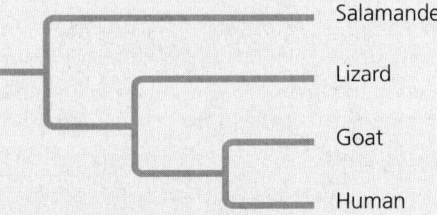

   (A) The salamander lineage is a basal taxon.
   (B) Salamanders are a sister group to the group containing lizards, goats, and humans.
   (C) Salamanders are as closely related to goats as to humans.
   (D) Lizards are more closely related to salamanders than to humans.

6. If you were using cladistics to build a phylogenetic tree of cats, which of the following would be the best outgroup?
   (A) wolf
   (B) domestic cat
   (C) frog
   (D) leopard

7. The relative lengths of the frog and mouse branches in the phylogenetic tree in Figure 20.13 indicate that
   (A) frogs evolved before mice.
   (B) mice evolved before frogs.
   (C) the homolog has evolved more rapidly in mice.
   (D) the homolog has evolved more slowly in mice.

## Level 3: Synthesis/Evaluation

8. **SCIENTIFIC INQUIRY**
   **DRAW IT** (a) Draw a phylogenetic tree based on characters 1–5 in the table below. Place hatch marks on the tree to indicate the origin(s) of characters 1–6. (b) Assume that tuna and dolphins are sister species, and redraw the phylogenetic tree accordingly. Use hatch marks to indicate the origin(s) of characters 1–6. (c) How many evolutionary changes are required in each tree? Which tree is most parsimonious?

| Character | Lancelet (outgroup) | Lamprey | Tuna | Salamander | Turtle | Leopard | Dolphin |
|---|---|---|---|---|---|---|---|
| 1. Backbone | 0 | 1 | 1 | 1 | 1 | 1 | 1 |
| 2. Hinged jaw | 0 | 0 | 1 | 1 | 1 | 1 | 1 |
| 3. Four limbs | 0 | 0 | 0 | 1 | 1 | 1 | 1* |
| 4. Amnion | 0 | 0 | 0 | 0 | 1 | 1 | 1 |
| 5. Milk | 0 | 0 | 0 | 0 | 0 | 1 | 1 |
| 6. Dorsal fin | 0 | 0 | 1 | 0 | 0 | 0 | 1 |

*Although adult dolphins have only two obvious limbs (their flippers), as embryos they have two hind-limb buds, for a total of four limbs.

9. **FOCUS ON EVOLUTION**
   Darwin suggested looking at a species' close relatives to learn what its ancestors may have been like. Explain how his suggestion anticipates recent methods, such as phylogenetic bracketing and the use of outgroups in cladistic analysis.

10. **FOCUS ON INFORMATION**
    In a short essay (100–150 words), explain how genetic information—along with an understanding of the process of descent with modification—enables scientists to reconstruct phylogenies that extend hundreds of millions of years back in time.

11. **SYNTHESIZE YOUR KNOWLEDGE**

This West Indian manatee (*Trichechus manatus*) is an aquatic mammal. Like amphibians and reptiles, mammals are tetrapods (vertebrates with four limbs). Explain why manatees are considered tetrapods even though they lack hind limbs, and suggest traits that manatees likely share with leopards and other mammals (see Figure 20.12b). How might early members of the manatee lineage have differed from today's manatees?

*For selected answers, see Appendix A.*

**▲ Figure 21.1** Is this finch evolving?

**KEY CONCEPTS**

**21.1** Genetic variation makes evolution possible

**21.2** The Hardy-Weinberg equation can be used to test whether a population is evolving

**21.3** Natural selection, genetic drift, and gene flow can alter allele frequencies in a population

**21.4** Natural selection is the only mechanism that consistently causes adaptive evolution

## The Smallest Unit of Evolution

One common misconception about evolution is that individual organisms evolve. It is true that natural selection acts on individuals: Each organism's traits affect its survival and reproductive success compared with that of other individuals. But the evolutionary impact of natural selection is only apparent in how a *population* of organisms changes over time.

Consider the medium ground finch (*Geospiza fortis*), a seed-eating bird that inhabits the Galápagos Islands **(Figure 21.1)**. In 1977, the *G. fortis* population on the island of Daphne Major was decimated by a long period of drought: Of some 1,200 birds, only 180 survived. Researchers Peter and Rosemary Grant observed that during the drought, small, soft seeds were in short supply. The finches mostly fed on large, hard seeds that were more plentiful. Birds with larger, deeper beaks were better able to crack and eat these larger seeds, and they survived at a higher rate than finches with smaller beaks. Since beak depth is an inherited trait in these birds, the offspring of surviving birds also tended to have deeper beaks. As a result, the average beak depth in the next generation of *G. fortis* was greater than it had been in the pre-drought

population **(Figure 21.2)**. The finch population had evolved by natural selection. However, the *individual* finches did not evolve. Each bird had a beak of a particular size, which did not grow larger during the drought. Rather, the proportion of large beaks in the population increased from generation to generation: The population evolved, not its individual members.

Focusing on evolutionary change in populations, we can define evolution on its smallest scale, called **microevolution**, as a change in allele frequencies in a population over generations. Let's apply this definition to the changes that occurred in

**► Figure 21.2 Evidence of selection by food source.** The data represent adult beak depth measurements of medium ground finches hatched in the generations before and after the 1977 drought. In a single generation, evolution by natural selection resulted in a larger average beak size in the population.

**MB** A related Experimental Inquiry Tutorial can be assigned in MasteringBiology.

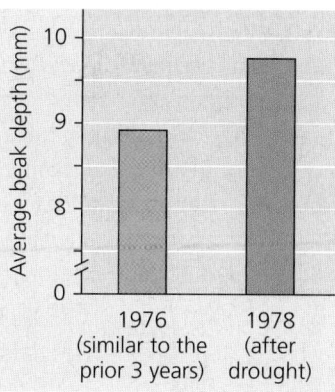

the *G. fortis* population on Daphne Major. As the proportion of individuals with large beaks increased from one generation to the next, so too did the frequency of alleles that encoded large beaks.

As the *G. fortis* example suggests, natural selection can cause allele frequencies to change over time. Moreover, as you will see in this chapter, natural selection is not the only cause of microevolution. In fact, there are three main mechanisms that can cause allele frequency change: natural selection, genetic drift (chance events that alter allele frequencies), and gene flow (the transfer of alleles between populations). Each of these mechanisms has distinctive effects on the genetic composition of populations. However, only natural selection consistently improves the degree to which organisms are well suited for life in their environment (adaptation). Before we examine natural selection and adaptation more closely, let's revisit a prerequisite for these processes in a population: genetic variation.

▲ **Figure 21.3 Phenotypic variation in horses.** In horses, coat color varies along a continuum and is influenced by multiple genes.

CONCEPT 21.1

# Genetic variation makes evolution possible

In *The Origin of Species*, Darwin provided abundant evidence that life on Earth has evolved over time, and he proposed natural selection as the primary mechanism for that change. He observed that individuals differ in their inherited traits and that selection acts on such differences, leading to evolutionary change. Although Darwin realized that variation in heritable traits is a prerequisite for evolution, he did not know precisely how organisms pass heritable traits to their offspring.

Just a few years after Darwin published *The Origin of Species*, Gregor Mendel wrote a groundbreaking paper on inheritance in pea plants (see Concept 11.1). In that paper, Mendel proposed a model of inheritance in which organisms transmit discrete heritable units (now called genes) to their offspring. Although Darwin did not know about genes, Mendel's paper set the stage for understanding the genetic differences on which evolution is based. Here we'll examine such genetic differences and how they are produced.

## Genetic Variation

Individuals within all species vary in their phenotypic traits. Among humans, for example, you can easily observe phenotypic variation in facial features, height, and voice. And though you cannot identify a person's blood group (A, B, AB, or O) from his or her appearance, this and many other molecular traits also vary extensively among individuals.

Such phenotypic variations often reflect **genetic variation**, differences among individuals in the composition of their genes or other DNA sequences. Some heritable phenotypic differences occur on an "either-or" basis, such as the flower colors of Mendel's pea plants: Each plant had flowers that were either purple or white (see Figure 11.3). Characters that vary in this way are typically determined by a single gene locus, with different alleles producing distinct phenotypes. In contrast, other phenotypic differences vary in gradations along a continuum. Such variation usually results from the influence of two or more genes on a single phenotypic character. In fact, many phenotypic characters are influenced by multiple genes, including coat color in horses **(Figure 21.3)**, seed number in maize (corn), and height in humans.

How much do genes and other DNA sequences vary from one individual to another? Genetic variation at the whole-gene level (*gene variability*) can be quantified as the average percentage of loci that are heterozygous. (Recall that a heterozygous individual has two different alleles for a given locus, whereas a homozygous individual has two identical alleles for that locus.) As an example, on average the fruit fly *Drosophila melanogaster* is heterozygous for about 1,920 of its 13,700 loci (14%) and homozygous for all the rest.

Considerable genetic variation can also be measured at the molecular level of DNA (*nucleotide variability*). But little of this variation results in phenotypic variation. Why? Many of the nucleotide variations occur within *introns*, noncoding segments of DNA lying between *exons*, the regions retained in mRNA after RNA processing (see Figure 14.12). And of the variations that occur within exons, most do not cause a change in the amino acid sequence of the protein encoded by the gene. For example, in the sequence comparison shown in **Figure 21.4**, there are 43 nucleotide sites with variable base pairs (where substitutions have occurred), as well as several sites where insertions or deletions have occurred. Although 18 variable sites occur within the four exons of the *Adh* gene, only one of these variations (at site 1,490) results in an amino acid change. Note, however, that this single variable site is enough to cause genetic variation at the level of the gene—and hence two different forms of the Adh enzyme are produced.

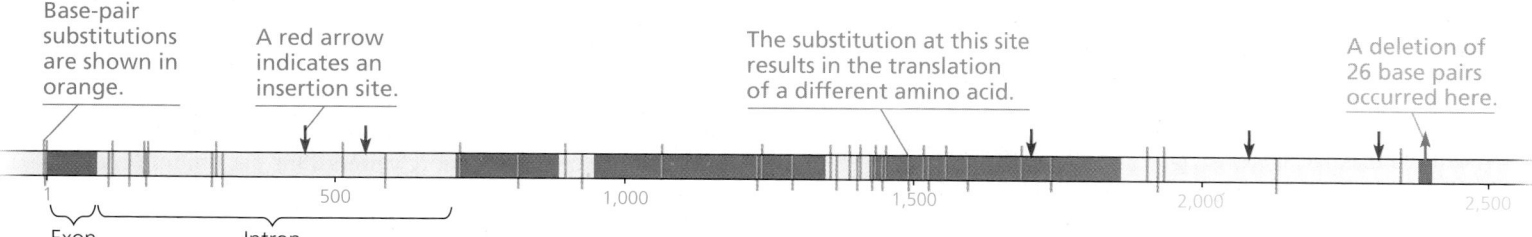

Base-pair substitutions are shown in orange.

A red arrow indicates an insertion site.

The substitution at this site results in the translation of a different amino acid.

A deletion of 26 base pairs occurred here.

1          500          1,000          1,500          2,000          2,500

Exon          Intron

▲ **Figure 21.4 Extensive genetic variation at the molecular level.** This diagram summarizes data from a study comparing the DNA sequence of the alcohol dehydrogenase (*Adh*) gene in several fruit flies (*Drosophila melanogaster*). The *Adh* gene has four exons (dark blue) separated by introns (light blue); the exons include the coding regions that are ultimately translated into the amino acids of the Adh enzyme. Only one substitution has a phenotypic effect, producing a different form of the Adh enzyme.

**MAKE CONNECTIONS** *Review Figures 14.6 and 14.12. Explain how a base-pair substitution that alters a coding region of the* Adh *locus could have no effect on amino acid sequence. Then explain how an insertion in an exon could have no effect on the protein produced.*

It is important to bear in mind that some phenotypic variation does not result from genetic differences among individuals (**Figure 21.5** shows a striking example in a caterpillar of the southwestern United States). Phenotype is the product of an inherited genotype and many environmental influences (see Concept 11.3). In a human example, bodybuilders alter their phenotypes dramatically but do not pass their huge muscles on to the next generation. In general, only the genetically determined part of phenotypic variation can have evolutionary consequences. As such, genetic variation provides the raw material for evolutionary change: Without genetic variation, evolution cannot occur.

## Sources of Genetic Variation

The genetic variation on which evolution depends originates when mutation, gene duplication, or other processes produce new alleles and new genes. Genetic variants can be produced rapidly in organisms with short generation times. Sexual reproduction can also result in genetic variation as existing genes are arranged in new ways.

### Formation of New Alleles

New alleles can arise by *mutation*, a change in the nucleotide sequence of an organism's DNA. A mutation is like a shot in the dark—we cannot predict accurately which segments of DNA will be altered or in what way. In multicellular organisms, only mutations in cell lines that produce gametes can be passed to offspring. In plants and fungi, this is not as limiting as it may sound, since many different cell lines can produce gametes. But in most animals, the majority of mutations occur in somatic cells and are not passed to offspring.

A change of as little as one base in a gene—a "point mutation"—can have a significant impact on phenotype, as in sickle-cell disease (see Figure 14.25). Organisms reflect many generations of past selection, and hence their phenotypes tend to be well-suited for life in their environments. As a result,

▲ **Figure 21.5 Nonheritable variation.** These caterpillars of the moth *Nemoria arizonaria* owe their different appearances to chemicals in their diets, not to differences in their genotypes. **(a)** Caterpillars raised on a diet of oak flowers resemble the flowers, whereas **(b)** their siblings raised on oak leaves resemble oak twigs.

most new mutations that alter a phenotype are at least slightly harmful. In some cases, natural selection quickly removes such harmful alleles. In diploid organisms, however, harmful alleles that are recessive can be hidden from selection. Indeed, a harmful recessive allele can persist for generations by propagation in heterozygous individuals (where its harmful effects are masked by the more favorable dominant allele). Such "heterozygote protection" maintains a huge pool of alleles that might not be favored under present conditions but that could be beneficial if the environment changes.

While many mutations are harmful, many others are not. Recall that much of the DNA in eukaryotic genomes does not encode proteins (see Figure 18.5). Point mutations in these noncoding regions generally result in **neutral variation**, differences in DNA sequence that do not confer a selective advantage or disadvantage. The redundancy in the genetic code is another source of neutral variation: Even a point mutation in a gene that encodes a protein will have no effect on the protein's function if the amino acid composition is not changed. And even where there is a change in the amino acid, it may not affect the protein's shape and function. Finally, as you will see later in this chapter, a mutant allele may on rare occasions actually make its bearer better suited to the environment, enhancing reproductive success.

### Altering Gene Number or Position

Chromosomal changes that delete, disrupt, or rearrange many loci are usually harmful. However, when such large-scale changes leave genes intact, they may not affect the organisms' phenotypes. In rare cases, chromosomal rearrangements may even be beneficial. For example, the translocation of part of one chromosome to a different chromosome could link genes in a way that produces a positive effect.

A key potential source of variation is the duplication of genes due to errors in meiosis (such as unequal crossing over), slippage during DNA replication, or the activities of transposable elements (see Concept 18.4). Duplications of large chromosome segments, like other chromosomal aberrations, are often harmful, but the duplication of smaller pieces of DNA may not be. Gene duplications that do not have severe effects can persist over generations, allowing mutations to accumulate. The result is an expanded genome with new genes that may take on new functions.

Such increases in gene number appear to have played a major role in evolution. For example, the remote ancestors of mammals had a single gene for detecting odors that has since been duplicated many times. As a result, humans today have about 380 functional olfactory receptor genes, and mice have 1,200. This proliferation of olfactory genes has probably helped mammals over the course of evolution, enabling them to detect faint odors in their environment and to distinguish among many different smells.

### Rapid Reproduction

Mutation rates tend to be low in plants and animals, averaging about one mutation in every 100,000 genes per generation, and they are often even lower in prokaryotes. But prokaryotes have many more generations per unit of time, so mutations can quickly generate genetic variation in their populations. The same is true of viruses. For instance, HIV has a generation time of about two days (that is, it takes two days for a newly formed virus to produce the next generation of viruses). HIV also has an RNA genome, which has a much higher mutation rate than a typical DNA genome because of the lack of RNA repair mechanisms in host cells. For this reason, single-drug treatments are unlikely to be effective against HIV: Mutant forms of the virus that are resistant to a particular drug would tend to proliferate in relatively short order. The most effective AIDS treatments to date have been drug "cocktails" that combine several medications. This approach has worked well because it is less likely that a set of mutations that together confer resistance to *all* the drugs will occur in a short time period.

### Sexual Reproduction

In organisms that reproduce sexually, most of the genetic variation in a population results from the unique combination of alleles that each individual receives from its parents. Of course, at the nucleotide level, all the differences among these alleles have originated from past mutations. Sexual reproduction then shuffles existing alleles and deals them at random to produce individual genotypes.

Three mechanisms contribute to this shuffling: crossing over, independent assortment of chromosomes, and fertilization (see Concept 10.4). During meiosis, homologous chromosomes, one inherited from each parent, trade some of their alleles by crossing over. These homologous chromosomes and the alleles they carry are then distributed at random into gametes. Then, because myriad possible mating combinations exist in a population, fertilization brings together gametes that are likely to have different genetic backgrounds. The combined effects of these three mechanisms ensure that sexual reproduction rearranges existing alleles into fresh combinations each generation, providing much of the genetic variation that makes evolution possible.

#### CONCEPT CHECK 21.1

1. Explain why genetic variation within a population is a prerequisite for evolution.
2. Of all the mutations that occur in a population, why do only a small fraction become widespread?
3. **MAKE CONNECTIONS** If a population stopped reproducing sexually (but still reproduced asexually), how would its genetic variation be affected over time? Explain. (See Concept 10.4.)

For suggested answers, see Appendix A.

# The Hardy-Weinberg equation can be used to test whether a population is evolving

Although the individuals in a population must differ genetically for evolution to occur, the presence of genetic variation does not guarantee that a population will evolve. For that to happen, one or more factors that cause evolution must be at work. In this section, we'll explore one way to test whether evolution is occurring in a population. First, let's clarify what we mean by a population.

## Gene Pools and Allele Frequencies

A **population** is a group of individuals of the same species that live in the same area and interbreed, producing fertile offspring. Different populations of a species may be isolated geographically from one another, exchanging genetic material only rarely. Such isolation is common for species that live on widely separated islands or in different lakes. But not all populations are isolated **(Figure 21.6)**. Still, members of a population typically breed with one another and thus on average are more closely related to each other than to members of other populations.

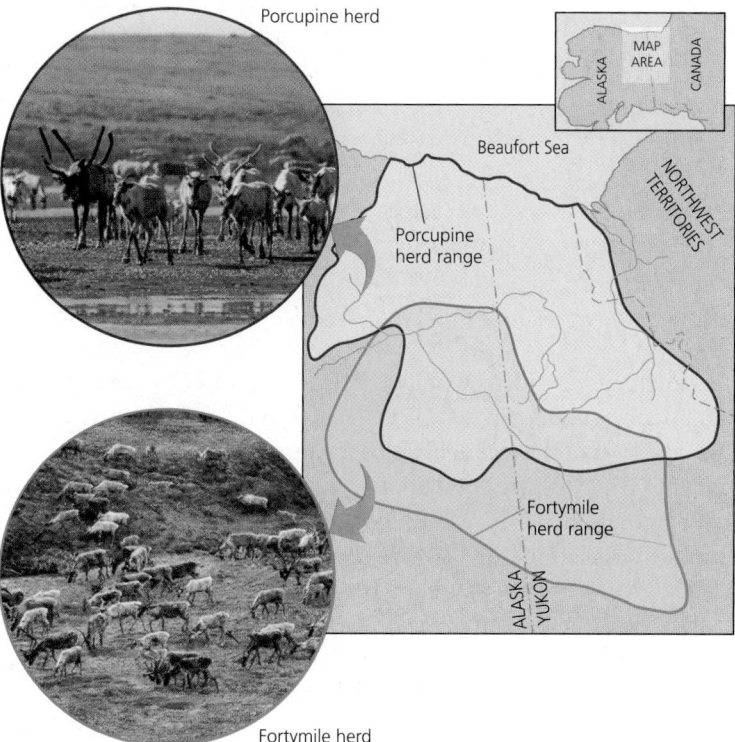

▲ **Figure 21.6 One species, two populations.** These two caribou populations in the Yukon are not totally isolated; they sometimes share the same area. Still, members of either population are most likely to breed within their own population.

We can characterize a population's genetic makeup by describing its **gene pool**, which consists of all copies of every type of allele at every locus in all members of the population. If only one allele exists for a particular locus in a population, that allele is said to be *fixed* in the gene pool, and all individuals are homozygous for that allele. But if there are two or more alleles for a particular locus in a population, individuals may be either homozygous or heterozygous.

For example, imagine a population of 500 wildflower plants with two alleles, $C^R$ and $C^W$, for a locus that codes for flower pigment. These alleles show incomplete dominance; thus, each genotype has a distinct phenotype. Plants homozygous for the $C^R$ allele ($C^R C^R$) produce red pigment and have red flowers; plants homozygous for the $C^W$ allele ($C^W C^W$) produce no red pigment and have white flowers; and heterozygotes ($C^R C^W$) produce some red pigment and have pink flowers.

Each allele has a frequency (proportion) in the population. For example, suppose our population has 320 plants with red flowers, 160 with pink flowers, and 20 with white flowers. Because these are diploid organisms, these 500 individuals have a total of 1,000 copies of the gene for flower color. The $C^R$ allele accounts for 800 of these copies ($320 \times 2 = 640$ for $C^R C^R$ plants, plus $160 \times 1 = 160$ for $C^R C^W$ plants). Thus, the frequency of the $C^R$ allele is $800/1,000 = 0.8$ (80%).

When studying a locus with two alleles, the convention is to use $p$ to represent the frequency of one allele and $q$ to represent the frequency of the other allele. Thus, $p$, the frequency of the $C^R$ allele in the gene pool of this population, is 0.8 (80%). And because there are only two alleles for this gene, the frequency of the $C^W$ allele, represented by $q$, must be $1 - p = 0.2$ (20%). For loci that have more than two alleles, the sum of all allele frequencies must still equal 1 (100%).

Next we'll see how allele and genotype frequencies can be used to test whether evolution is occurring in a population.

## The Hardy-Weinberg Equation

One way to assess whether natural selection or other factors are causing evolution at a particular locus is to determine what the genetic makeup of a population would be if it were *not* evolving at that locus. We can then compare that scenario with the data we actually observed for the population. If there are no differences, we can conclude that the population is not evolving. If there are differences, this suggests that the population may be evolving—and then we can try to figure out why.

### Hardy-Weinberg Equilibrium

In a population that is not evolving, allele and genotype frequencies will remain constant from generation to generation, provided that only Mendelian segregation and recombination of alleles are at work. Such a population is said to be in **Hardy-Weinberg equilibrium**, named for the British mathematician and German physician, respectively, who independently developed this idea in 1908.

To determine whether a population is in Hardy-Weinberg equilibrium, it is helpful to think about genetic crosses in a new way. Previously, we used Punnett squares to determine the genotypes of offspring in a genetic cross (see Figure 11.5). Here, instead of considering the possible allele combinations from one cross, we'll consider the combination of alleles in *all* of the crosses in a population.

Imagine that all the alleles for a given locus from all the individuals in a population are placed in a large bin **(Figure 21.7)**. We can think of this bin as holding the population's gene pool for that locus. "Reproduction" occurs by selecting alleles at random from the bin; somewhat similar events occur in nature when fish release sperm and eggs into the water or when pollen (containing plant sperm) is blown about by the wind. By viewing reproduction as a process of randomly selecting and combining alleles from the bin (the gene pool), we are in effect assuming that mating occurs at random—that is, that all male-female matings are equally likely.

Let's apply the bin analogy to the hypothetical wildflower population discussed earlier. In that population of 500 flowers,

the frequency of the allele for red flowers ($C^R$) is $p = 0.8$, and the frequency of the allele for white flowers ($C^W$) is $q = 0.2$. In other words, a bin holding all 1,000 copies of the flower-color gene in the population would contain 800 $C^R$ alleles and 200 $C^W$ alleles. Assuming that gametes are formed by selecting alleles at random from the bin, the probability that an egg or sperm contains a $C^R$ or $C^W$ allele is equal to the frequency of these alleles in the bin. Thus, as shown in Figure 21.7, each egg has an 80% chance of containing a $C^R$ allele and a 20% chance of containing a $C^W$ allele; the same is true for each sperm.

Using the rule of multiplication (see Figure 11.9), we can now calculate the frequencies of the three possible genotypes, assuming random unions of sperm and eggs. The probability that two $C^R$ alleles will come together is $p \times p = p^2 = 0.8 \times 0.8 = 0.64$. Thus, about 64% of the plants in the next generation will have the genotype $C^R C^R$. The frequency of $C^W C^W$ individuals is expected to be about $q \times q = q^2 = 0.2 \times 0.2 = 0.04$, or 4%. $C^R C^W$ heterozygotes can arise in two different ways. If the sperm provides the $C^R$ allele and the egg provides the $C^W$ allele, the resulting heterozygotes will be $p \times q = 0.8 \times 0.2 = 0.16$, or 16% of the total. If the sperm provides the $C^W$ allele and the egg the $C^R$ allele, the heterozygous offspring will make up $q \times p = 0.2 \times 0.8 = 0.16$, or 16%. The frequency of heterozygotes is thus the sum of these possibilities: $pq + qp = 2pq = 0.16 + 0.16 = 0.32$, or 32%.

As shown in **Figure 21.8**, the genotype frequencies in the next generation must add up to 1 (100%). Thus, the equation for Hardy-Weinberg equilibrium states that at a locus with two alleles, the three genotypes will appear in the following proportions:

$$
\underset{\substack{\text{Expected}\\\text{frequency}\\\text{of genotype}\\C^R C^R}}{p^2} + \underset{\substack{\text{Expected}\\\text{frequency}\\\text{of genotype}\\C^R C^W}}{2pq} + \underset{\substack{\text{Expected}\\\text{frequency}\\\text{of genotype}\\C^W C^W}}{q^2} = 1
$$

Note that for a locus with two alleles, only three genotypes are possible (in this case, $C^R C^R$, $C^R C^W$, and $C^W C^W$). As a result, the sum of the frequencies of the three genotypes must equal 1 (100%) in *any* population—regardless of whether the population is in Hardy-Weinberg equilibrium. The key point is that a population is in Hardy-Weinberg equilibrium only if the observed genotype frequency of one homozygote is $p^2$, the observed frequency of the other homozygote is $q^2$, and the observed frequency of heterozygotes is $2pq$. Finally, as suggested by Figure 21.8, if a population such as our wildflowers is in Hardy-Weinberg equilibrium and its members continue to mate randomly generation after generation, allele and genotype frequencies will remain constant. The system operates somewhat like a deck of cards: No matter how many times the deck is reshuffled to deal out new hands, the deck itself remains the same. Aces do not grow more numerous than jacks.

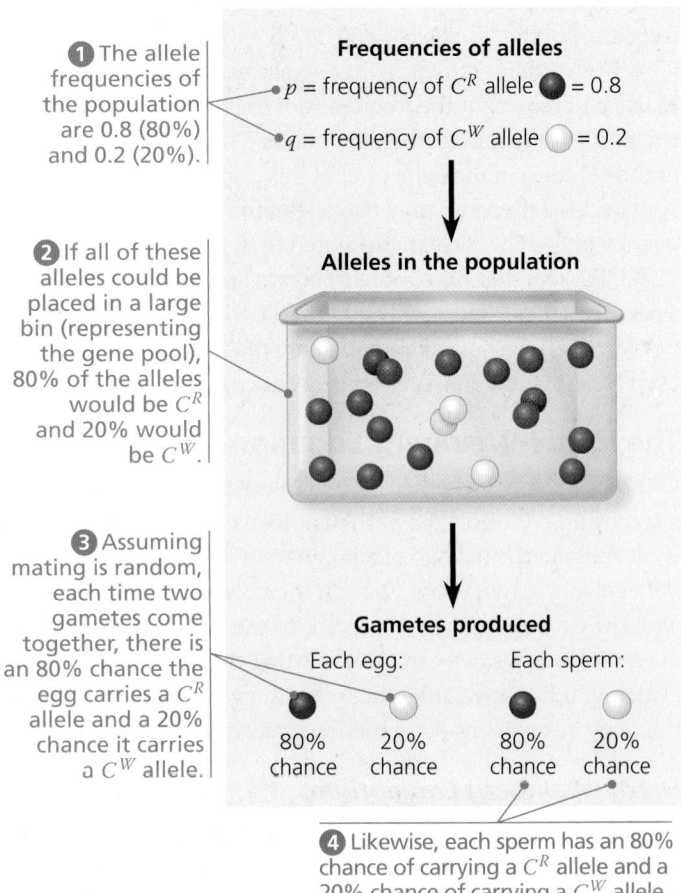

① The allele frequencies of the population are 0.8 (80%) and 0.2 (20%).

**Frequencies of alleles**

$p$ = frequency of $C^R$ allele ● = 0.8

$q$ = frequency of $C^W$ allele ○ = 0.2

② If all of these alleles could be placed in a large bin (representing the gene pool), 80% of the alleles would be $C^R$ and 20% would be $C^W$.

**Alleles in the population**

③ Assuming mating is random, each time two gametes come together, there is an 80% chance the egg carries a $C^R$ allele and a 20% chance it carries a $C^W$ allele.

**Gametes produced**

Each egg:

80% chance   20% chance

Each sperm:

80% chance   20% chance

④ Likewise, each sperm has an 80% chance of carrying a $C^R$ allele and a 20% chance of carrying a $C^W$ allele.

▲ **Figure 21.7 Selecting alleles at random from a gene pool.**

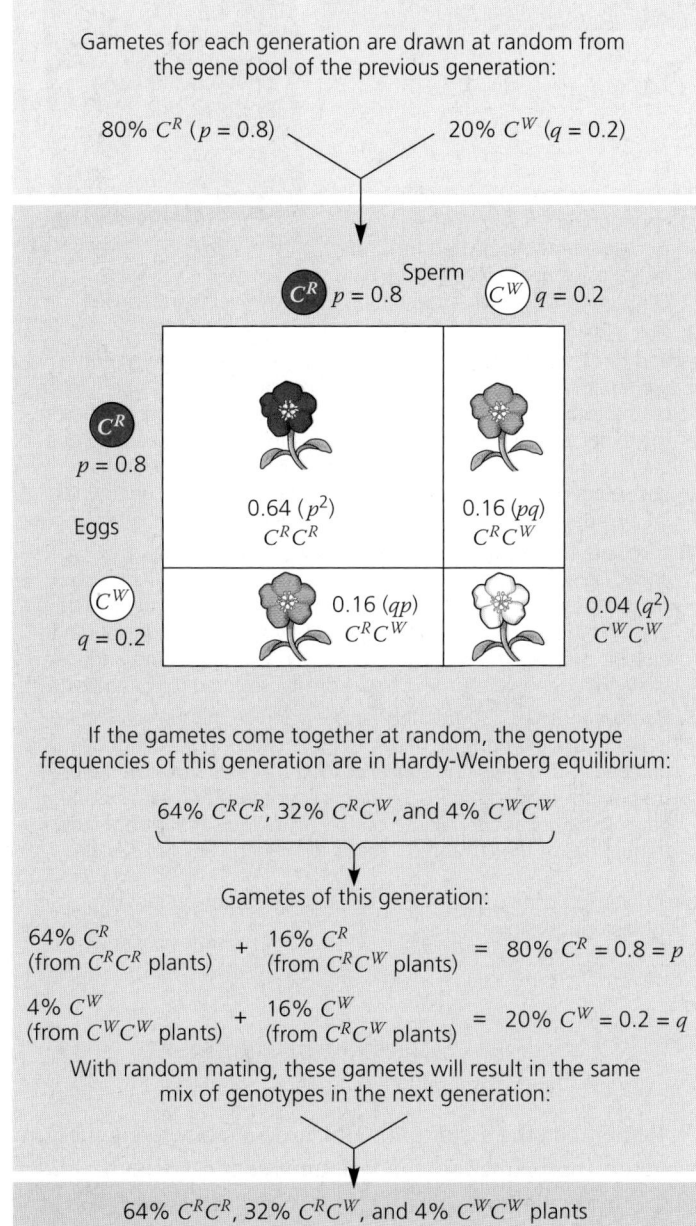

Gametes for each generation are drawn at random from the gene pool of the previous generation:

80% $C^R$ ($p = 0.8$)     20% $C^W$ ($q = 0.2$)

Sperm
$C^R$ $p = 0.8$     $C^W$ $q = 0.2$

$C^R$
$p = 0.8$

Eggs

0.64 ($p^2$)
$C^R C^R$

0.16 ($pq$)
$C^R C^W$

$C^W$
$q = 0.2$

0.16 ($qp$)
$C^R C^W$

0.04 ($q^2$)
$C^W C^W$

If the gametes come together at random, the genotype frequencies of this generation are in Hardy-Weinberg equilibrium:

64% $C^R C^R$, 32% $C^R C^W$, and 4% $C^W C^W$

Gametes of this generation:

64% $C^R$
(from $C^R C^R$ plants) + 16% $C^R$
(from $C^R C^W$ plants) = 80% $C^R = 0.8 = p$

4% $C^W$
(from $C^W C^W$ plants) + 16% $C^W$
(from $C^R C^W$ plants) = 20% $C^W = 0.2 = q$

With random mating, these gametes will result in the same mix of genotypes in the next generation:

64% $C^R C^R$, 32% $C^R C^W$, and 4% $C^W C^W$ plants

▲ **Figure 21.8 Hardy-Weinberg equilibrium.** In our wildflower population, the gene pool remains constant from one generation to the next. Mendelian processes alone do not alter frequencies of alleles or genotypes.

**?** *If the frequency of the $C^R$ allele is 0.6, predict the frequencies of the $C^R C^R$, $C^R C^W$, and $C^W C^W$ genotypes.*

And the repeated shuffling of a population's gene pool over the generations cannot, in itself, change the frequency of one allele relative to another.

## Conditions for Hardy-Weinberg Equilibrium

The Hardy-Weinberg equation describes a hypothetical population that is not evolving. But in real populations, the allele and genotype frequencies often *do* change over time. Such changes can occur when at least one of the following five conditions of Hardy-Weinberg equilibrium is not met.

1. **No mutations.** The gene pool is modified if mutations alter alleles or if entire genes are deleted or duplicated.
2. **Random mating.** If individuals tend to mate within a subset of the population, such as their near neighbors or close relatives (inbreeding), random mixing of gametes does not occur, and genotype frequencies change.
3. **No natural selection.** Allele frequencies can change when individuals with different genotypes differ in their survival or reproductive success.
4. **Extremely large population size.** The smaller the population, the more likely it is that allele frequencies will fluctuate by chance from one generation to the next (a process called genetic drift).
5. **No gene flow.** By moving alleles into or out of populations, gene flow can alter allele frequencies.

Departure from these conditions usually results in evolutionary change, which, as we've already described, is common in natural populations. But it is also common for natural populations to be in Hardy-Weinberg equilibrium for specific genes. One way this can occur is if selection alters allele frequencies at some loci but not others. In addition, some populations evolve so slowly that the changes in their allele and genotype frequencies are difficult to distinguish from those predicted for a non-evolving population.

### Applying the Hardy-Weinberg Equation

The Hardy-Weinberg equation is often used as an initial test of whether evolution is occurring in a population (Concept Check 21.2, question 3 is an example). The equation also has medical applications, such as estimating the percentage of a population carrying the allele for an inherited disease. For example, consider phenylketonuria (PKU), a metabolic disorder that results from homozygosity for a recessive allele. This disorder occurs in about one out of every 10,000 babies born in the United States. Left untreated, PKU results in mental disability and other problems. (Newborns are now tested for PKU, and symptoms can be largely avoided with a diet very low in phenylalanine. For this reason, products that contain phenylalanine, such as diet colas, carry warning labels.)

To apply the Hardy-Weinberg equation, we must assume that no new PKU mutations are being introduced into the population (condition 1) and that people neither choose their mates on the basis of whether or not they carry this gene nor generally mate with close relatives (condition 2). We must also ignore any effects of differential survival and reproductive success among PKU genotypes (condition 3) and assume that there are no effects of genetic drift (condition 4) or of gene flow from

# Using the Hardy-Weinberg Equation to Interpret Data and Make Predictions

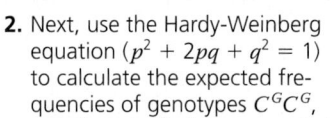

**Is Evolution Occurring in a Soybean Population?** One way to test whether evolution is occurring in a population is to compare the observed genotype frequencies at a locus with those expected for a non-evolving population based on the Hardy-Weinberg equation. In this exercise, you'll test whether a soybean population is evolving at a locus with two alleles, $C^G$ and $C^Y$, that affect chlorophyll production and hence leaf color.

**How the Experiment Was Done** Students planted soybean seeds and then counted the number of seedlings of each genotype at day 7 and again at day 21. Seedlings of each genotype could be distinguished visually because the $C^G$ and $C^Y$ alleles show incomplete dominance: $C^GC^G$ seedlings have green leaves, $C^GC^Y$ seedlings have green-yellow leaves, and $C^YC^Y$ seedlings have yellow leaves.

**Data from the Experiment**

| Time (days) | Number of Seedlings | | | Total |
| | Green ($C^GC^G$) | Green-yellow ($C^GC^Y$) | Yellow ($C^YC^Y$) | |
|---|---|---|---|---|
| 7 | 49 | 111 | 56 | 216 |
| 21 | 47 | 106 | 20 | 173 |

**INTERPRET THE DATA**

1. Use the observed genotype frequencies from the day 7 data to calculate the frequencies of the $C^G$ allele ($p$) and the $C^Y$ allele ($q$). (Remember that the frequency of an allele in a gene pool is the number of copies of that allele divided by the total number of copies of all alleles at that locus.)

2. Next, use the Hardy-Weinberg equation ($p^2 + 2pq + q^2 = 1$) to calculate the expected frequencies of genotypes $C^GC^G$, $C^GC^Y$, and $C^YC^Y$ for a population in Hardy-Weinberg equilibrium.

3. Calculate the observed frequencies of genotypes $C^GC^G$, $C^GC^Y$, and $C^YC^Y$ at day 7. (The observed frequency of a genotype in a gene pool is the number of individuals with that genotype divided by the total number of individuals.) Compare these frequencies to the expected frequencies calculated in step 2. Is the seedling population in Hardy-Weinberg equilibrium at day 7, or is evolution occurring? Explain your reasoning and identify which genotypes, if any, appear to be selected for or against.

4. Calculate the observed frequencies of genotypes $C^GC^G$, $C^GC^Y$, and $C^YC^Y$ at day 21. Compare these frequencies to the expected frequencies calculated in step 2 and the observed frequencies at day 7. Is the seedling population in Hardy-Weinberg equilibrium at day 21, or is evolution occurring? Explain your reasoning and identify which genotypes, if any, appear to be selected for or against.

5. Homozygous $C^YC^Y$ individuals cannot produce chlorophyll. The ability to photosynthesize becomes more critical as seedlings age and begin to exhaust the supply of food that was stored in the seed from which they emerged. Develop a hypothesis that explains the data for days 7 and 21. Based on this hypothesis, predict how the frequencies of the $C^G$ and $C^Y$ alleles will change beyond day 21.

(MB) A version of this Scientific Skills Exercise can be assigned in MasteringBiology.

---

other populations into the United States (condition 5). These assumptions are reasonable: The mutation rate for the PKU gene is low, inbreeding and other forms of nonrandom mating are not common in the United States, selection occurs only against the rare homozygotes (and then only if dietary restrictions are not followed), the U.S. population is very large, and populations outside the country have PKU allele frequencies similar to those seen in the United States.

If all these assumptions hold, then the frequency of individuals in the population born with PKU will correspond to $q^2$ in the Hardy-Weinberg equation ($q^2$ = frequency of homozygotes). Because the allele is recessive, we must estimate the number of heterozygotes rather than counting them directly as we did with the pink flowers. Since we know there is one PKU occurrence per 10,000 births ($q^2 = 0.0001$), the frequency ($q$) of the recessive allele for PKU is

$$q = \sqrt{0.0001} = 0.01$$

and the frequency of the dominant allele is

$$p = 1 - q = 1 - 0.01 = 0.99$$

The frequency of carriers, heterozygous people who do not have PKU but may pass the PKU allele to offspring, is

$$2pq = 2 \times 0.99 \times 0.01 = 0.0198$$
(approximately 2% of the U.S. population)

Remember, the assumption of Hardy-Weinberg equilibrium yields an approximation; the real number of carriers may differ. Still, our calculations suggest that harmful recessive alleles at this and other loci can be concealed in a population because they are carried by healthy heterozygotes. The **Scientific Skills Exercise** provides another opportunity for you to apply the Hardy-Weinberg equation to allele data.

## CONCEPT CHECK 21.2

1. A population has 700 individuals, 85 of genotype *AA*, 320 of genotype *Aa*, and 295 of genotype *aa*. What are the frequencies of alleles *A* and *a*?

2. The frequency of allele *a* is 0.45 for a population in Hardy-Weinberg equilibrium. What are the expected frequencies of genotypes *AA*, *Aa*, and *aa*?

3. **WHAT IF?** A locus that affects susceptibility to a degenerative brain disease has two alleles, *V* and *v*. In a population, 16 people have genotype *VV*, 92 have genotype *Vv*, and 12 have genotype *vv*. Is this population evolving? Explain.

For suggested answers, see Appendix A.

# Natural selection, genetic drift, and gene flow can alter allele frequencies in a population

Note again the five conditions required for a population to be in Hardy-Weinberg equilibrium. A deviation from any of these conditions is a potential cause of evolution. New mutations (violation of condition 1) can alter allele frequencies, but because mutations are rare, the change from one generation to the next is likely to be very small. Nonrandom mating (violation of condition 2) can affect the frequencies of homozygous and heterozygous genotypes but by itself has no effect on allele frequencies in the gene pool. (Allele frequencies can change if individuals with certain inherited traits are more likely than other individuals to obtain mates. However, such a situation not only causes a deviation from random mating, but also violates condition 3, no natural selection.)

For the rest of this section, we will focus on the three mechanisms that alter allele frequencies directly and cause most evolutionary change: natural selection, genetic drift, and gene flow (violations of conditions 3–5).

## Natural Selection

The concept of natural selection is based on differential success in survival and reproduction: Individuals in a population exhibit variations in their heritable traits, and those with traits that are better suited to their environment tend to produce more offspring than those with traits that are not as well suited.

In genetic terms, selection results in alleles being passed to the next generation in proportions that differ from those in the present generation. For example, the fruit fly *D. melanogaster* has an allele that confers resistance to several insecticides, including DDT. This allele has a frequency of 0% in laboratory strains of *D. melanogaster* established from flies collected in the wild in the early 1930s, prior to DDT use. However, in strains established from flies collected after 1960 (following 20 or more years of DDT use), the allele frequency is 37%. We can infer that this allele either arose by mutation between 1930 and 1960 or was present in 1930, but very rare. In any case, the rise in frequency of this allele most likely occurred because DDT is a powerful poison that is a strong selective force in exposed fly populations.

As the *D. melanogaster* example suggests, an allele that confers resistance to an insecticide will increase in frequency in a population exposed to that insecticide. Such changes are not coincidental. By consistently favoring some alleles over others, natural selection can cause *adaptive evolution* (a process in which traits that enhance survival or reproduction tend to increase in frequency over time). We'll explore this process in more detail later in this chapter.

## Genetic Drift

If you flip a coin 1,000 times, a result of 700 heads and 300 tails might make you suspicious about that coin. But if you flip a coin only 10 times, an outcome of 7 heads and 3 tails would not be surprising. The smaller the number of coin flips, the more likely it is that chance alone will cause a deviation from the predicted result. (In this case, the prediction is an equal number of heads and tails.) Chance events can also cause allele frequencies to fluctuate unpredictably from one generation to the next, especially in small populations—a process called **genetic drift**.

**Figure 21.9** models how genetic drift might affect a small population of our wildflowers. In this example, drift leads to the

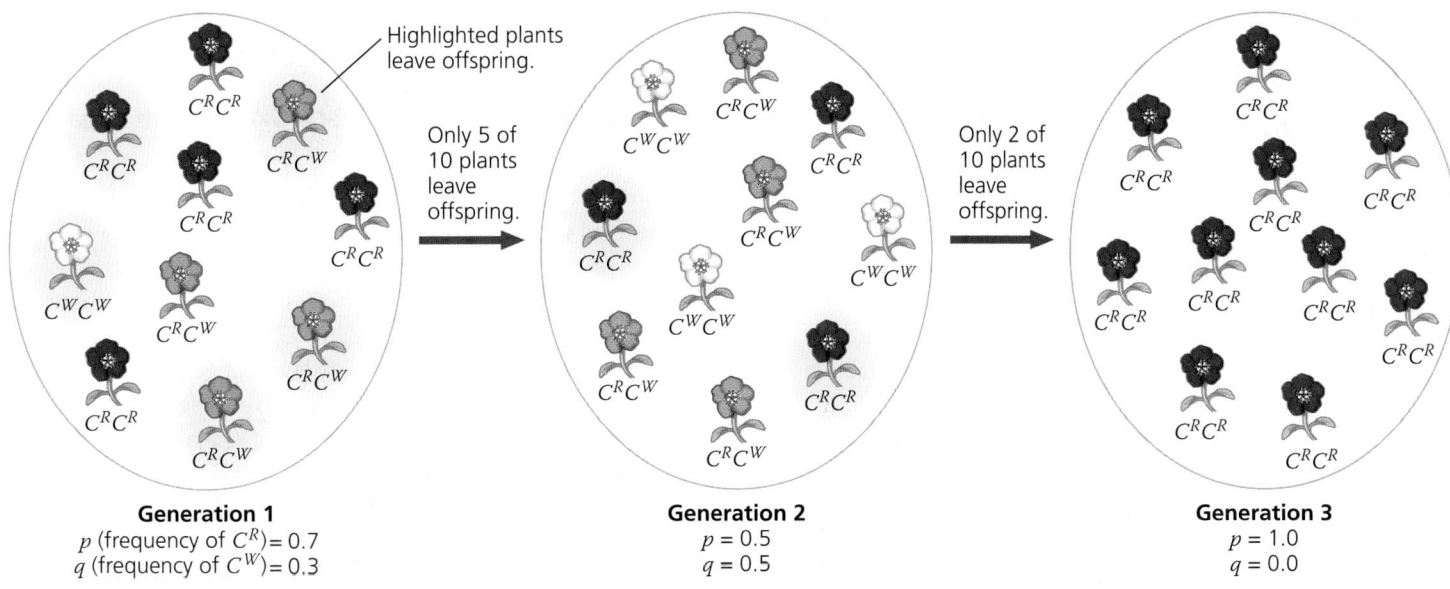

**Generation 1**
$p$ (frequency of $C^R$) = 0.7
$q$ (frequency of $C^W$) = 0.3

**Generation 2**
$p = 0.5$
$q = 0.5$

**Generation 3**
$p = 1.0$
$q = 0.0$

▲ **Figure 21.9 Genetic drift.** This small wildflower population has a stable size of ten plants. Suppose that by chance only five plants of generation 1 (those highlighted in yellow) produce fertile offspring. (This could occur, for example, if only those plants happened to grow in a location that provided enough nutrients to support the production of offspring.) Again by chance, only two plants of generation 2 leave fertile offspring. As a result, by chance the frequency of the $C^W$ allele first increases in generation 2 and then falls to zero in generation 3.

 **ANIMATION** Visit the Study Area in **MasteringBiology** for the BioFix® 3-D Animation on Mechanisms of Evolution.

loss of an allele from the gene pool, but it is a matter of chance that the $C^W$ allele is lost and not the $C^R$ allele. Such unpredictable changes in allele frequencies can be caused by chance events associated with survival and reproduction. Perhaps a large animal such as a moose stepped on the three $C^W C^W$ individuals in generation 2, killing them and increasing the chance that only the $C^R$ allele would be passed to the next generation. Allele frequencies can also be affected by chance events that occur during fertilization. For example, suppose two individuals of genotype $C^R C^W$ had a small number of offspring. By chance alone, every egg and sperm pair that generated offspring could happen to have carried the $C^R$ allele and not the $C^W$ allele.

Certain circumstances can result in genetic drift having a significant impact on a population. Two examples are the founder effect and the bottleneck effect.

### The Founder Effect

When a few individuals become isolated from a larger population, this smaller group may establish a new population whose gene pool differs from the source population; this is called the **founder effect**. The founder effect might occur, for example, when a few members of a population are blown by a storm to a new island. Genetic drift, in which chance events alter allele frequencies, can occur in such a case because the storm indiscriminately transports some individuals (and their alleles), but not others, from the source population.

The founder effect probably accounts for the relatively high frequency of certain inherited disorders among isolated human populations. For example, in 1814, 15 British colonists founded a settlement on Tristan da Cunha, a group of small islands in the Atlantic Ocean midway between Africa and South America. Apparently, one of the colonists carried a recessive allele for retinitis pigmentosa, a progressive form of blindness that afflicts homozygous individuals. Of the founding colonists' 240 descendants on the island in the late 1960s, 4 had retinitis pigmentosa. The frequency of the allele that causes this disease is ten times higher on Tristan da Cunha than in the populations from which the founders came.

### The Bottleneck Effect

A sudden change in the environment, such as a fire or flood, may drastically reduce the size of a population. A severe drop in population size can cause the **bottleneck effect**, so named because the population has passed through a "bottleneck" that reduces its size **(Figure 21.10)**. By chance alone, certain alleles may be overrepresented among the survivors, others may be underrepresented, and some may be absent altogether. Ongoing genetic drift is likely to have substantial effects on the gene pool until the population becomes large enough that chance events have less impact. But even if a population that has passed through a bottleneck ultimately recovers in size, it may have low levels of genetic variation for a long period of time—a legacy of the genetic drift that occurred when the population was small.

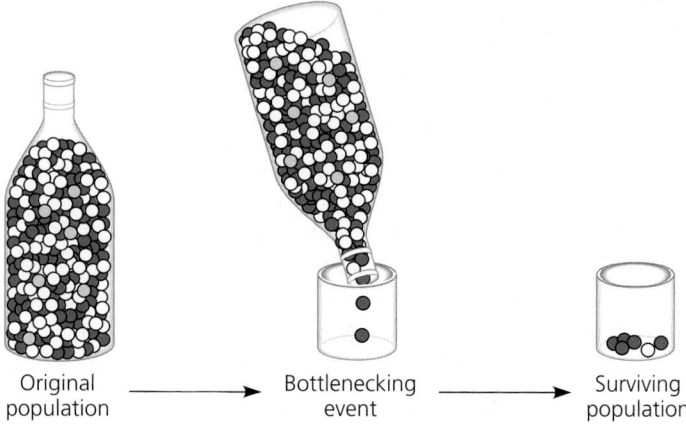

**(a)** Shaking just a few marbles through the narrow neck of a bottle is analogous to a drastic reduction in the size of a population. By chance, blue marbles are overrepresented in the surviving population, and gold marbles are absent.

**(b)** Similarly, bottlenecking a wild population tends to reduce genetic variation, as in the Florida panther (*Puma concolor coryi*), a subspecies in danger of extinction.

▲ **Figure 21.10  The bottleneck effect.**

One reason it is important to understand the bottleneck effect is that human actions sometimes create severe bottlenecks for other species, as the following example shows.

### Case Study: *Impact of Genetic Drift on the Greater Prairie Chicken*

Millions of greater prairie chickens (*Tympanuchus cupido*) once lived on the prairies of Illinois. As these prairies were converted to farmland and other uses during the 19th and 20th centuries, the number of greater prairie chickens plummeted **(Figure 21.11a)**. By 1993, fewer than 50 birds remained. These few surviving birds had low levels of genetic variation, and less than 50% of their eggs hatched, compared with much higher hatching rates of the larger populations in Kansas and Nebraska **(Figure 21.11b)**.

These data suggest that genetic drift during the bottleneck may have led to a loss of genetic variation and an increase in the frequency of harmful alleles. To investigate this hypothesis, researchers extracted DNA from 15 museum specimens of Illinois greater prairie chickens. Of the 15 birds, 10 had been collected in the 1930s, when there were 25,000 greater prairie chickens in Illinois, and 5 had been collected in the 1960s,

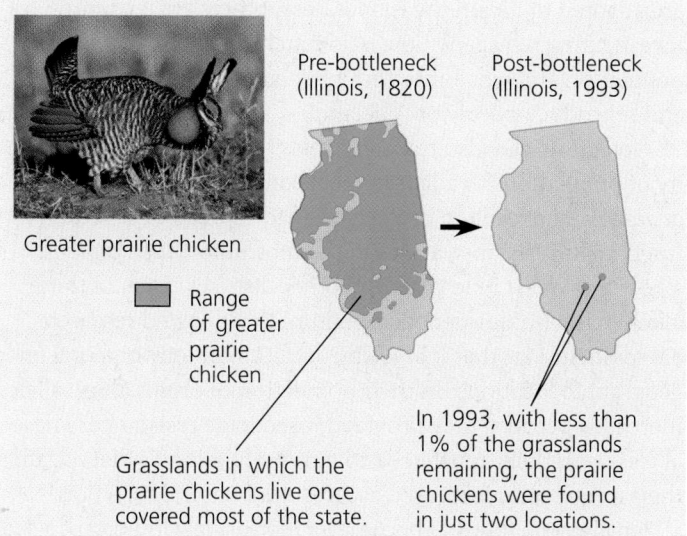

Greater prairie chicken

Range of greater prairie chicken

Pre-bottleneck (Illinois, 1820)    Post-bottleneck (Illinois, 1993)

Grasslands in which the prairie chickens live once covered most of the state.

In 1993, with less than 1% of the grasslands remaining, the prairie chickens were found in just two locations.

**(a)** The Illinois population of greater prairie chickens dropped from millions of birds in the 1800s to fewer than 50 birds in 1993.

| Location | Population size | Number of alleles per locus | Percentage of eggs hatched |
|---|---|---|---|
| Illinois | | | |
| 1930–1960s | 1,000–25,000 | 5.2 | 93 |
| 1993 | <50 | 3.7 | <50 |
| Kansas, 1998 (no bottleneck) | 750,000 | 5.8 | 99 |
| Nebraska, 1998 (no bottleneck) | 75,000–200,000 | 5.8 | 96 |

**(b)** As a consequence of the drastic reduction in the size of the Illinois population, genetic drift resulted in a drop in the number of alleles per locus (averaged across six loci studied) and a decrease in the percentage of eggs that hatched.

▲ **Figure 21.11 Genetic drift and loss of genetic variation.**

when there were 1,000 greater prairie chickens in Illinois. By studying the DNA of these specimens, the researchers were able to obtain a minimum, baseline estimate of how much genetic variation was present in the Illinois population *before* the population shrank to extremely low numbers. This baseline estimate is a key piece of information that is not usually available in cases of population bottlenecks.

The researchers surveyed six loci and found that the 1993 population had fewer alleles per locus than the pre-bottleneck Illinois or the current Kansas and Nebraska populations (see Figure 21.11b). Thus, as predicted, drift had reduced the genetic variation of the small 1993 population. Drift may also have increased the frequency of harmful alleles, leading to the low egg-hatching rate. To counteract these negative effects,

271 birds from neighboring states were added to the Illinois population over four years. This strategy succeeded: New alleles entered the population, and the egg-hatching rate improved to over 90%. Overall, studies on the Illinois greater prairie chicken illustrate the powerful effects of genetic drift in small populations and provide hope that in at least some populations, these effects can be reversed.

### *Effects of Genetic Drift:* A Summary

The examples we've described highlight four key points:

1. **Genetic drift is significant in small populations.** Chance events can cause an allele to be disproportionately over- or underrepresented in the next generation. Although chance events occur in populations of all sizes, they tend to alter allele frequencies substantially only in small populations.

2. **Genetic drift can cause allele frequencies to change at random.** Because of genetic drift, an allele may increase in frequency one year and then decrease the next; the change from year to year is not predictable. Thus, unlike natural selection, which in a given environment consistently favors some alleles over others, genetic drift causes allele frequencies to change at random over time.

3. **Genetic drift can lead to a loss of genetic variation within populations.** By causing allele frequencies to fluctuate randomly over time, genetic drift can eliminate alleles from a population. Because evolution depends on genetic variation, such losses can influence how effectively a population can adapt to a change in the environment.

4. **Genetic drift can cause harmful alleles to become fixed.** Alleles that are neither harmful nor beneficial can be lost or become fixed by chance through genetic drift. In very small populations, genetic drift can also cause alleles that are slightly harmful to become fixed. When this occurs, the population's survival can be threatened (as in the case of the greater prairie chicken).

## Gene Flow

Natural selection and genetic drift are not the only phenomena affecting allele frequencies. Allele frequencies can also change by **gene flow**, the transfer of alleles into or out of a population due to the movement of fertile individuals or their gametes. For example, suppose that near our original hypothetical wildflower population there is another population consisting primarily of white-flowered individuals ($C^W C^W$). Insects carrying pollen from these plants may fly to and pollinate plants in our original population. The introduced $C^W$ alleles would modify our original population's allele frequencies in the next generation. Because alleles are transferred between populations, gene flow tends to reduce the genetic differences between populations. In fact, if it is extensive enough, gene flow can result in two populations combining into a single population with a common gene pool.

Alleles transferred by gene flow can also affect how well populations are adapted to local environmental conditions. Researchers studying the songbird *Parus major* (great tit) on the small Dutch island of Vlieland noted survival differences between two populations on the island. The survival rate of females born in the eastern population is twice that of females born in the central population, regardless of where the females eventually settle and raise offspring (**Figure 21.12**). This finding suggests that females born in the eastern population are better adapted to life on the island than females born in the central population. But field studies also showed that the two populations are connected by high levels of gene flow (mating), which should reduce genetic differences between them. So how can the eastern population be better adapted to life on Vlieland than the central population?

The answer lies in the unequal amounts of gene flow from the mainland. In any given year, 43% of the first-time breeders in the central population are immigrants from the mainland, compared with only 13% in the eastern population. Birds with mainland genotypes survive and reproduce poorly on Vlieland, and in the eastern population, selection reduces the frequency of these genotypes. In the central population, however, gene flow from the mainland is so high that it overwhelms the effects of selection. As a result, females born in the central population have many immigrant genes, reducing the degree to which members of that population

are adapted to life on the island. Researchers are currently investigating why gene flow is so much higher in the central population and why birds with mainland genotypes survive and reproduce poorly on Vlieland.

Gene flow can also transfer alleles that improve the ability of populations to adapt to local conditions. For example, gene flow has resulted in the worldwide spread of several insecticide-resistance alleles in the mosquito *Culex pipiens*, a vector of West Nile virus and other diseases. Each of these alleles has a unique genetic signature that allowed researchers to document that it arose by mutation in only one or a few geographic locations. In their population of origin, these alleles increased because they provided insecticide resistance. These alleles were then transferred to new populations, where again, their frequencies increased as a result of natural selection.

Finally, gene flow has become an increasingly important agent of evolutionary change in human populations. Humans today move much more freely about the world than in the past. As a result, mating is more common between members of populations that previously had very little contact, leading to an exchange of alleles and fewer genetic differences between those populations.

### CONCEPT CHECK 21.3

1. In what sense is natural selection more "predictable" than genetic drift?
2. Distinguish genetic drift from gene flow in terms of (a) how they occur and (b) their implications for future genetic variation in a population.
3. **WHAT IF?** Suppose two plant populations exchange pollen and seeds. In one population, individuals of genotype *AA* are most common (9,000 *AA*, 900 *Aa*, 100 *aa*), while the opposite is true in the other population (100 *AA*, 900 *Aa*, 9,000 *aa*). If neither allele has a selective advantage, what will happen over time to the allele and genotype frequencies of these populations?

For suggested answers, see Appendix A.

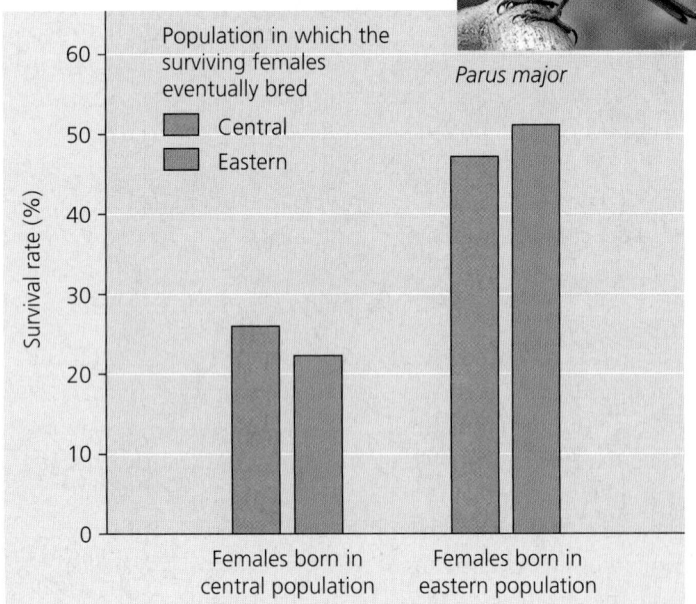

▲ **Figure 21.12 Gene flow and local adaptation.** In *Parus major* populations on Vlieland, the yearly survival rate of females born in the central population is lower than that of females born in the eastern population. Gene flow from the mainland is much higher to the central population than it is to the eastern population, and birds from the mainland are selected against in both populations. These data suggest that gene flow from the mainland has prevented the central population from adapting fully to its local conditions.

### CONCEPT 21.4

# Natural selection is the only mechanism that consistently causes adaptive evolution

Evolution by natural selection is a blend of chance and "sorting": chance in the creation of new genetic variations (as in mutation) and sorting as natural selection favors some alleles over others. Because of this favoring process, the outcome of natural selection is *not* random. Instead, natural selection consistently increases the frequencies of alleles that provide reproductive advantage, thus leading to **adaptive evolution**.

## Natural Selection: *A Closer Look*

In examining how natural selection brings about adaptive evolution, we'll begin with the concept of relative fitness and the different ways that an organism's phenotype is subject to natural selection.

## Relative Fitness

The phrases "struggle for existence" and "survival of the fittest" are commonly used to describe natural selection, but these expressions are misleading if taken to mean direct competitive contests among individuals. There *are* animal species in which individuals, usually the males, lock horns or otherwise do combat to determine mating privilege. But reproductive success is generally more subtle and depends on many factors besides outright battle. For example, a barnacle that is more efficient at collecting food than its neighbors may have greater stores of energy and hence be able to produce a larger number of eggs. A moth may have more offspring than other moths in the same population because its body colors more effectively conceal it from predators, improving its chance of surviving long enough to produce more offspring. These examples illustrate how in a given environment, certain traits can lead to greater **relative fitness**: the contribution an individual makes to the gene pool of the next generation *relative to* the contributions of other individuals.

Although we often refer to the relative fitness of a genotype, remember that the entity that is subjected to natural selection is the whole organism, not the underlying genotype. Thus, selection acts more directly on the phenotype than on the genotype; it acts on the genotype indirectly, via how the genotype affects the phenotype.

## Directional, Disruptive, and Stabilizing Selection

Natural selection can occur in three ways, depending on which phenotypes in a population are favored. These three modes of selection are called directional selection, disruptive selection, and stabilizing selection.

**Directional selection** occurs when conditions favor individuals exhibiting one extreme of a phenotypic range, thereby shifting a population's frequency curve for the phenotypic character in one direction or the other **(Figure 21.13a)**. Directional selection is common when a population's environment changes or when members of a population migrate to a new (and different) habitat. For instance, an increase in the relative abundance of large seeds over small seeds led to an increase in beak depth in a population of Galápagos finches (see Figure 21.2).

▼ **Figure 21.13 Modes of selection.** These cases describe three ways in which a hypothetical deer mouse population with heritable variation in fur coloration from light to dark might evolve. The graphs show how the frequencies of individuals with different fur colors change over time. The large white arrows symbolize selective pressures against certain phenotypes.

**MAKE CONNECTIONS** *Review Figure 19.14. Which mode of selection has occurred in soapberry bug populations that feed on the introduced goldenrain tree? Explain.*

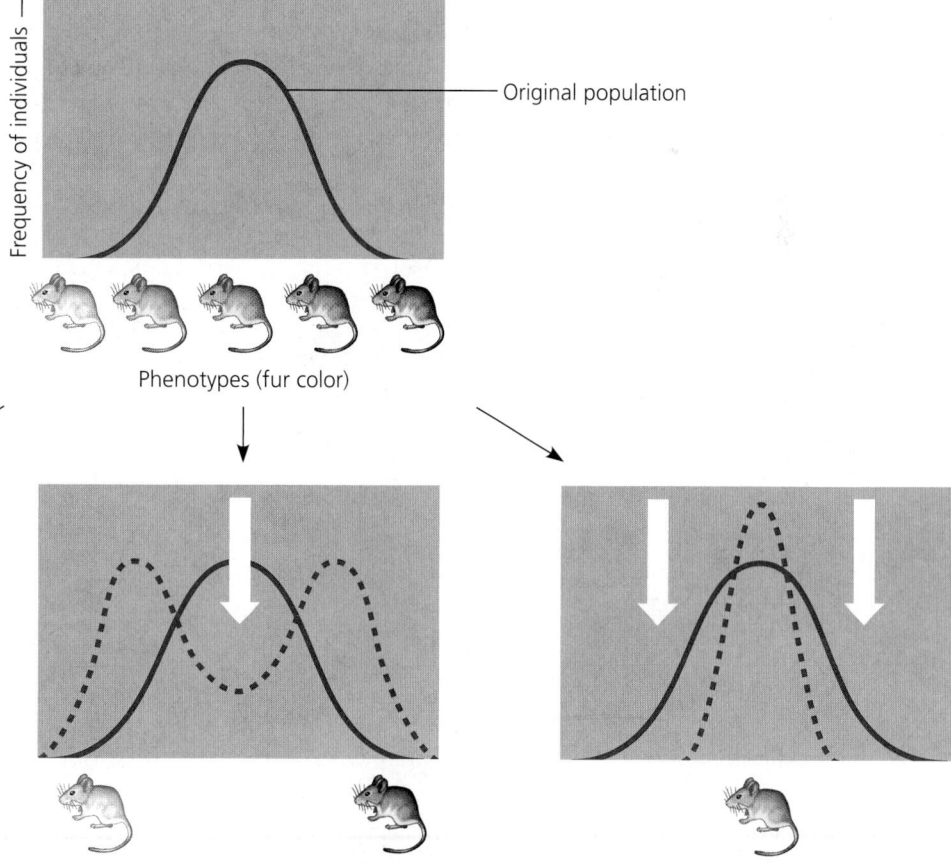

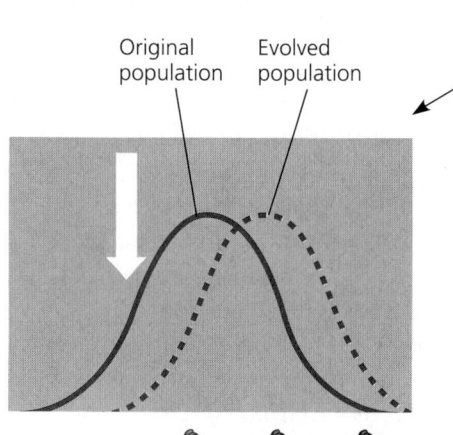

**(a) Directional selection** shifts the overall makeup of the population by favoring variants that are at one extreme of the distribution. In this case, lighter mice are selected against because they live among dark rocks, making it harder for them to hide from predators.

**(b) Disruptive selection** favors variants at both ends of the distribution. These mice have colonized a patchy habitat made up of light and dark rocks, with the result that mice of an intermediate color are selected against.

**(c) Stabilizing selection** removes extreme variants from the population and preserves intermediate types. If the environment consists of rocks of an intermediate color, both light and dark mice will be selected against.

**Disruptive selection (Figure 21.13b)** occurs when conditions favor individuals at both extremes of a phenotypic range over individuals with intermediate phenotypes. One example is a population of black-bellied seedcracker finches in Cameroon whose members display two distinctly different beak sizes. Small-billed birds feed mainly on soft seeds, whereas large-billed birds specialize in cracking hard seeds. It appears that birds with intermediate-sized bills are relatively inefficient at cracking both types of seeds and thus have lower relative fitness.

**Stabilizing selection (Figure 21.13c)** acts against both extreme phenotypes and favors intermediate variants. This mode of selection reduces variation and tends to maintain the status quo for a particular phenotypic character. For example, the birth weights of most human babies lie in the range of 3–4 kg (6.6–8.8 pounds); babies who are either much smaller or much larger suffer higher rates of mortality.

Regardless of the mode of selection, however, the basic mechanism remains the same. Selection favors individuals whose heritable phenotypic traits provide higher reproductive success than do the traits of other individuals.

## The Key Role of Natural Selection in Adaptive Evolution

The adaptations of organisms include many striking examples. Certain octopuses, for example, have the ability to change color rapidly, enabling them to blend into different backgrounds. Another example is the remarkable jaws of snakes **(Figure 21.14)**, which allow them to swallow prey much larger

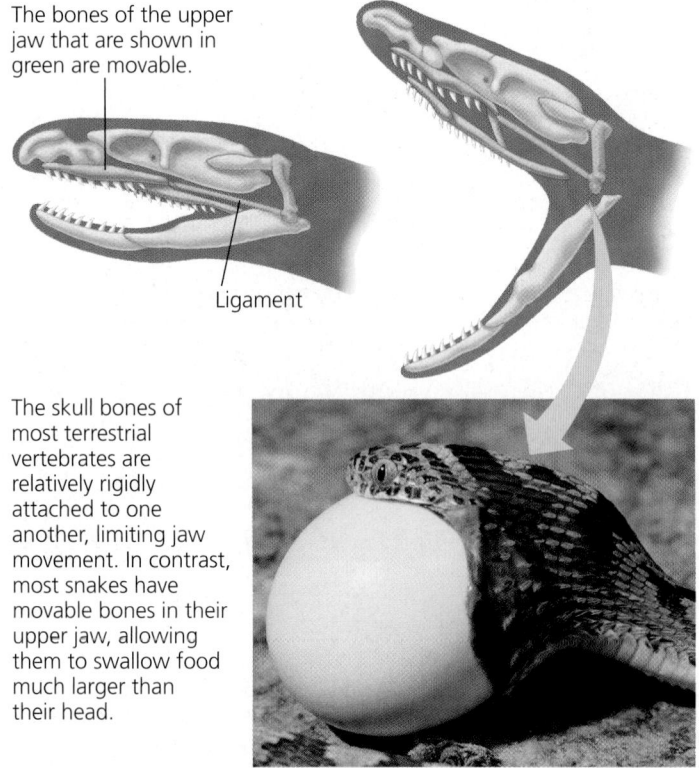

The bones of the upper jaw that are shown in green are movable.

Ligament

The skull bones of most terrestrial vertebrates are relatively rigidly attached to one another, limiting jaw movement. In contrast, most snakes have movable bones in their upper jaw, allowing them to swallow food much larger than their head.

▲ **Figure 21.14 Movable jaw bones in snakes.**

than their own head (a feat analogous to a person swallowing a whole watermelon). Other adaptations, such as a version of an enzyme that shows improved function in cold environments, may be less visually dramatic but just as important for survival and reproduction.

Such adaptations can arise gradually over time as natural selection increases the frequencies of alleles that enhance survival or reproduction. As the proportion of individuals that have favorable traits increases, the degree to which a species is well suited for life in its environment improves; that is, adaptive evolution occurs. However, the physical and biological components of an organism's environment may change over time. As a result, what constitutes a "good match" between an organism and its environment can be a moving target, making adaptive evolution a continuous, dynamic process.

And what about genetic drift and gene flow? Both can, in fact, increase the frequencies of alleles that enhance survival or reproduction, but neither does so consistently. Genetic drift can cause the frequency of a slightly beneficial allele to increase, but it also can cause the frequency of such an allele to decrease. Similarly, gene flow may introduce alleles that are advantageous or ones that are disadvantageous. Natural selection is the only evolutionary mechanism that consistently leads to adaptive evolution.

## Balancing Selection

As we've seen, genetic variation is often found at loci affected by selection. What prevents natural selection from reducing the variation at those loci by culling all unfavorable alleles? As mentioned earlier, in diploid organisms, many unfavorable recessive alleles persist because they are hidden from selection when in heterozygous individuals. In addition, selection itself may preserve variation at some loci, thus maintaining two or more phenotypic forms in a population. Known as **balancing selection**, this type of selection includes heterozygote advantage and frequency-dependent selection.

### Heterozygote Advantage

If individuals who are heterozygous at a particular locus have greater fitness than do both kinds of homozygotes, they exhibit **heterozygote advantage**. In such a case, natural selection tends to maintain two or more alleles at that locus. Note that heterozygote advantage is defined in terms of *genotype*, not phenotype. Thus, whether heterozygote advantage represents stabilizing or directional selection depends on the relationship between the genotype and the phenotype. For example, if the phenotype of a heterozygote is intermediate to the phenotypes of both homozygotes, heterozygote advantage is a form of stabilizing selection.

An example of heterozygote advantage occurs at the locus in humans that codes for the β polypeptide subunit of hemoglobin, the oxygen-carrying protein of red blood cells. In homozygous individuals, a recessive allele at that locus causes sickle-cell disease. The red blood cells of people with sickle-cell disease become distorted in shape, or *sickled*, under low-oxygen conditions (see Figure 3.23). These sickled cells can clump together and block the flow of blood, damaging organs such as the kidney, heart, and brain. Although some red blood cells become sickled in heterozygotes, not enough become sickled to cause sickle-cell disease.

Heterozygotes for the sickle-cell allele are protected against the most severe effects of malaria, a disease caused by a parasite that infects red blood cells (see Figure 25.26). One reason for this partial protection is that the body destroys sickled red blood cells rapidly, killing the parasites they harbor. Malaria is a major killer in some tropical regions. In such areas, selection favors heterozygotes over homozygous dominant individuals, who are more vulnerable to the effects of malaria, and also over homozygous recessive individuals, who develop sickle-cell disease. As we explore further in **Figure 21.15** on the next two pages, these selective pressures have enabled the frequency of the sickle-cell allele to exist at relatively high levels in areas where the malaria parasite is common.

### Frequency-Dependent Selection

In **frequency-dependent selection**, the fitness of a phenotype depends on how common it is in the population. Consider the scale-eating fish (*Perissodus microlepis*) of Lake Tanganyika, in Africa. These fish attack other fish from behind, darting in to remove a few scales from the flank of their prey. Of interest here is a peculiar feature of the scale-eating fish: Some are "left-mouthed" and some are "right-mouthed." Simple Mendelian inheritance determines these phenotypes, with the right-mouthed allele being dominant to the left-mouthed allele.

Because their mouth twists to the left, left-mouthed fish always attack their prey's right flank **(Figure 21.16)**. (To see why, twist your lower jaw and lips to the left and imagine trying to take a bite from the left side of a fish, approaching it from behind.) Similarly, right-mouthed fish always attack from the left. Prey species guard against attack from whatever phenotype of scale-eating fish is most common in the lake. Thus, from year to year, selection favors whichever mouth phenotype is least common. As a result, the frequency of left- and right-mouthed fish oscillates over time, and balancing selection (due to frequency dependence) keeps the frequency of each phenotype close to 50%.

## Sexual Selection

Charles Darwin was the first to explore the implications of **sexual selection**, a process in which individuals with certain inherited characteristics are more likely than other individuals of the same sex to obtain mates. Sexual selection can result in **sexual dimorphism**, a difference in secondary sexual characteristics between males and females of the same species **(Figure 21.17)**. These distinctions include differences in size, color, ornamentation, and behavior.

How does sexual selection operate? There are several ways. In *intrasexual selection*, meaning selection within the same sex, individuals of one sex compete directly for mates of the

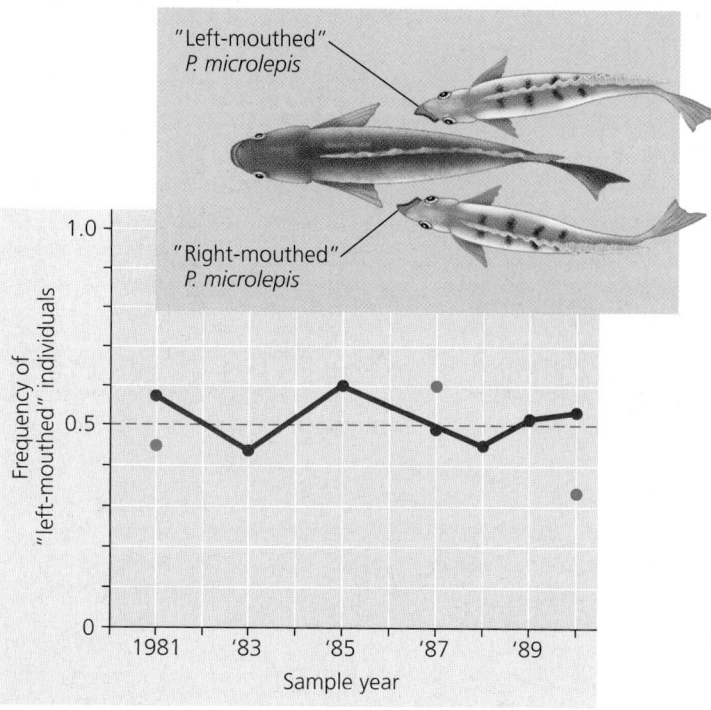

© 1993 AAAS

▲ **Figure 21.16 Frequency-dependent selection.** In a population of the scale-eating fish *Perissodus microlepis*, the frequency of left-mouthed individuals (red data points) rises and falls in a regular manner. The frequency of left-mouthed individuals among adults that reproduced was also recorded in three sample years (green data points).

**INTERPRET THE DATA** *For 1981, 1987, and 1990, compare the frequency of left-mouthed individuals among breeding adults to the frequency of left-mouthed individuals in the entire population. What do the data indicate about when natural selection favors left-mouthed individuals over right-mouthed individuals (or vice versa)? Explain.*

▲ **Figure 21.17 Sexual dimorphism and sexual selection.** Peacocks (left) and peahens (right) show extreme sexual dimorphism. Intrasexual selection between competing males is followed by intersexual selection when the females choose among the showiest males.

opposite sex. In many species, intrasexual selection occurs among males. For example, a single male may patrol a group of females and prevent other males from mating with them. The patrolling male may defend his status by defeating smaller, weaker, or less fierce males in combat. More often, this male is

# The Sickle-Cell Allele

This child has sickle-cell disease, a genetic disorder that strikes individuals who have two copies of the sickle-cell allele. This allele causes an abnormality in the structure and function of hemoglobin, the oxygen-carrying protein in red blood cells. Although sickle-cell disease is lethal if not treated, in some regions the sickle-cell allele can reach frequencies as high as 15–20%. How can such a harmful allele be so common?

## Events at the Molecular Level

- Due to a point mutation, the sickle-cell allele differs from the wild-type allele by a single nucleotide. (See Figure 14.25.)
- The resulting change in one amino acid leads to hydrophobic interactions between the sickle-cell hemoglobin proteins under low-oxygen conditions.
- As a result, the sickle-cell proteins bind to each other in chains that together form a fiber.

## Consequences for Cells

- The abnormal hemoglobin fibers distort the red blood cell into a sickle shape under low-oxygen conditions, such as those found in small blood vessels returning to the heart.

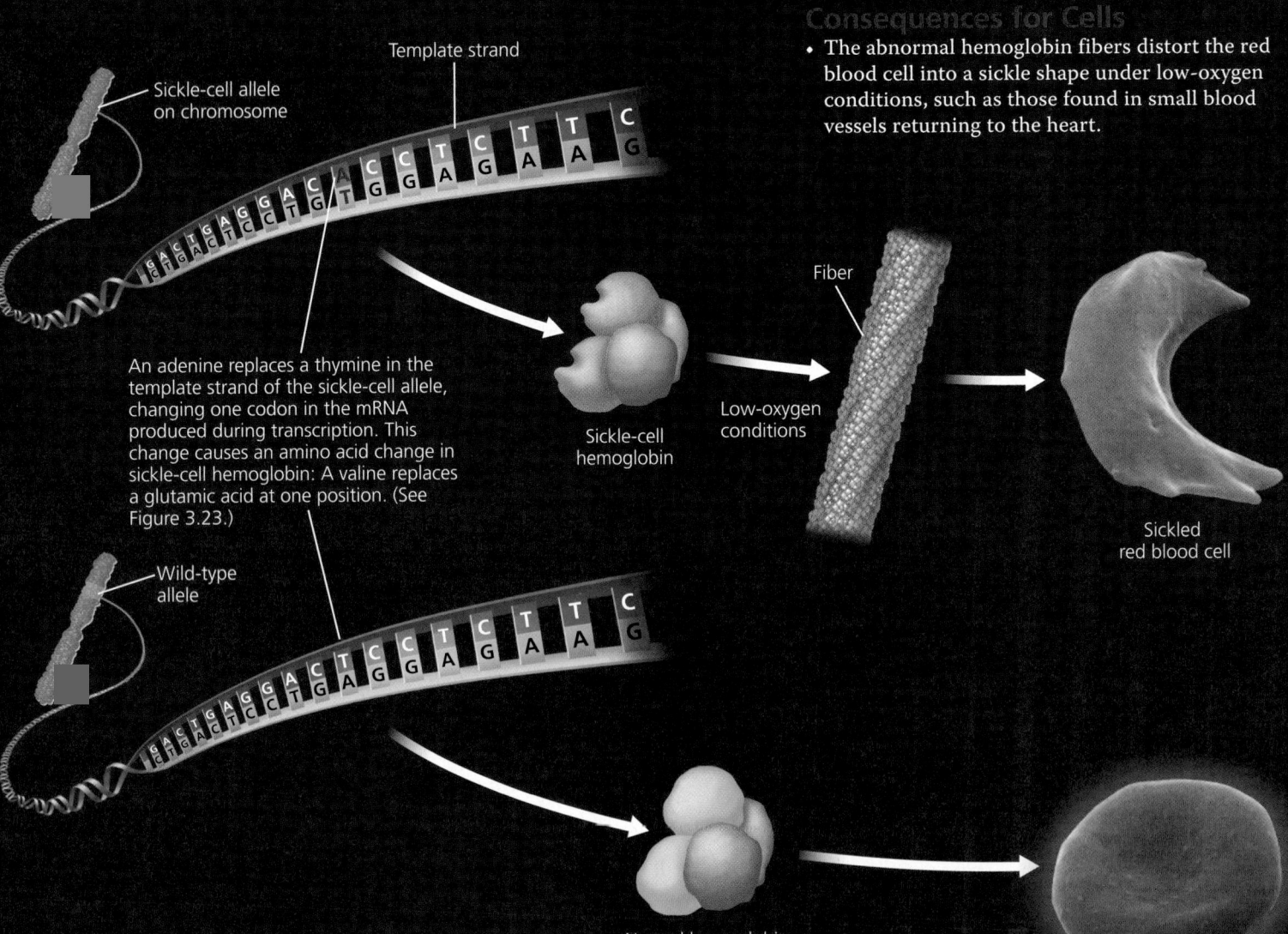

Sickle-cell allele on chromosome

Template strand

An adenine replaces a thymine in the template strand of the sickle-cell allele, changing one codon in the mRNA produced during transcription. This change causes an amino acid change in sickle-cell hemoglobin: A valine replaces a glutamic acid at one position. (See Figure 3.23.)

Sickle-cell hemoglobin

Low-oxygen conditions

Fiber

Sickled red blood cell

Wild-type allele

Normal hemoglobin (does not aggregate into fibers)

Normal red blood cell

Infected mosquitoes spread malaria when they bite people. (See Figure 25.26.)

## Evolution in Populations

- Homozygotes with two sickle-cell alleles are strongly selected against because of mortality caused by sickle-cell disease. In contrast, heterozygotes experience few harmful effects from sickling yet are more likely to survive malaria than are homozygotes.
- In regions where malaria is common, the net effect of these opposing selective forces is heterozygote advantage. This has caused evolutionary change in populations—the products of which are the areas of relatively high frequencies of the sickle-cell allele shown in the map.

## Effects on Individual Organisms

- The formation of sickled red blood cells causes homozygotes with two copies of the sickle-cell allele to have sickle-cell disease.
- Some sickling also occurs in heterozygotes, but not enough to cause the disease; they have sickle-cell trait.

The sickled blood cells of a homozygote block small blood vessels, causing great pain and damage to organs such as the heart, kidney, and brain.

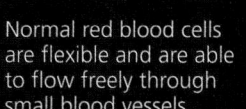

Normal red blood cells are flexible and are able to flow freely through small blood vessels.

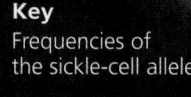

**Key**
Frequencies of the sickle-cell allele

3.0–6.0%

6.0–9.0%

9.0–12.0%

12.0–15.0%

>15.0%

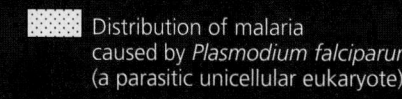
Distribution of malaria caused by *Plasmodium falciparum* (a parasitic unicellular eukaryote)

**MAKE CONNECTIONS** *In a region free of malaria, would individuals who are heterozygous for the sickle-cell allele be selected for or selected against? Explain.*

the psychological victor in ritualized displays that discourage would-be competitors but do not risk injury that would reduce his own fitness (see Figure 39.23). Intrasexual selection also occurs among females in a variety of species, including ring-tailed lemurs and broad-nosed pipefish.

In *intersexual selection*, also called *mate choice*, individuals of one sex (usually the females) are choosy in selecting their mates from the other sex. In many cases, the female's choice depends on the showiness of the male's appearance or behavior (see Figure 21.17). What intrigued Darwin about mate choice is that male showiness may not seem adaptive in any other way and may in fact pose some risk. For example, bright plumage may make male birds more visible to predators. But if such characteristics help a male gain a mate, and if this benefit outweighs the risk from predation, then both the bright plumage and the female preference for it will be reinforced because they enhance overall reproductive success.

How do female preferences for certain male characteristics evolve in the first place? One hypothesis is that females prefer male traits that are correlated with "good genes." If the trait preferred by females is indicative of a male's overall genetic quality, both the male trait and female preference for it should increase in frequency. **Figure 21.18** describes one experiment testing this hypothesis in gray tree frogs.

Other researchers have shown that in several bird species, the traits preferred by females are related to overall male health. Here, too, female preference appears to be based on traits that reflect "good genes," in this case alleles indicative of a robust immune system.

## Why Natural Selection Cannot Fashion Perfect Organisms

Though natural selection leads to adaptation, nature abounds with examples of organisms that are less than ideally suited for their lifestyles. There are several reasons why.

1. **Selection can act only on existing variations.** Natural selection favors only the fittest phenotypes among those currently in the population, which may not be the ideal traits. New advantageous alleles do not arise on demand.

2. **Evolution is limited by historical constraints.** Each species has a legacy of descent with modification from ancestral forms. Evolution does not scrap the ancestral anatomy and build each new complex structure from scratch; rather, evolution co-opts existing structures and adapts them to new situations. We could imagine that if a terrestrial animal were to adapt to an environment in which flight would be advantageous, it might be best just to grow an extra pair of limbs that would serve as wings. However, evolution does not work this way; instead, it operates on the traits an organism already has. Thus, in birds and bats, an existing pair of limbs took on new functions for flight as these organisms evolved from nonflying ancestors.

---

### ▼ Figure 21.18   Inquiry

### Do females select mates based on traits indicative of "good genes"?

**Experiment** Female gray tree frogs (*Hyla versicolor*) prefer to mate with males that give long mating calls. Allison Welch and colleagues, at the University of Missouri, tested whether the genetic makeup of long-calling (LC) males is superior to that of short-calling (SC) males. The researchers fertilized half the eggs of each female with sperm from an LC male and fertilized the remaining eggs with sperm from an SC male. In two separate experiments (one in 1995, the other in 1996), the resulting half-sibling offspring were raised in a common environment and their survival and growth were monitored.

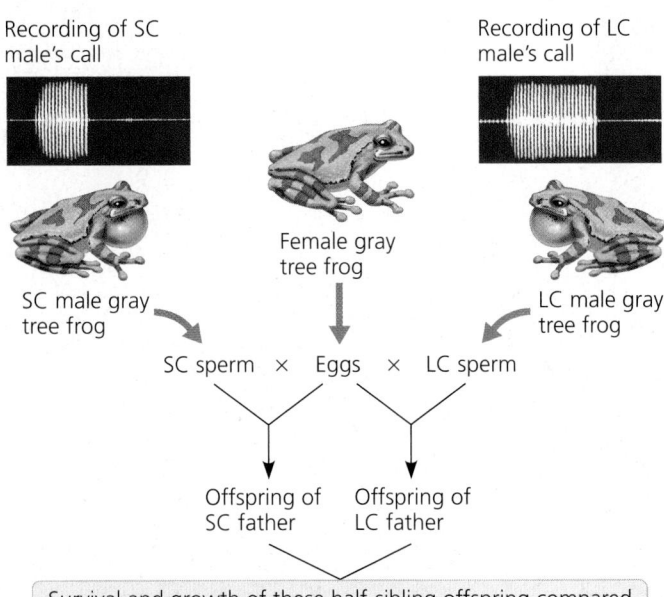

**Results**

| Offspring Performance | 1995 | 1996 |
|---|---|---|
| Larval survival | LC better | NSD |
| Larval growth | NSD | LC better |
| Time to metamorphosis | LC better (shorter) | LC better (shorter) |

NSD = no significant difference; LC better = offspring of LC males superior to offspring of SC males.

**Conclusion** Because offspring fathered by an LC male outperformed their half-siblings fathered by an SC male, the team concluded that the duration of a male's mating call is indicative of the male's overall genetic quality. This result supports the hypothesis that female mate choice can be based on a trait that indicates whether the male has "good genes."

**Data from** A. M. Welch et al., Call duration as an indicator of genetic quality in male gray tree frogs, *Science* 280:1928–1930 (1998).

**Inquiry in Action** Read and analyze the original paper in *Inquiry in Action: Interpreting Scientific Papers*.

**WHAT IF?** Why did the researchers split each female frog's eggs into two batches for fertilization by different males? Why didn't they mate each female with a single male frog?

3. **Adaptations are often compromises.** Each organism must do many different things. A seal spends part of its time on rocks; it could probably walk better if it had legs instead of flippers, but then it would not swim nearly as well. We humans owe much of our versatility and athleticism to our prehensile hands and flexible limbs, but these also make us prone to sprains, torn ligaments, and dislocations: Structural reinforcement has been compromised for agility. **Figure 21.19** depicts another example of evolutionary compromise.

4. **Chance, natural selection, and the environment interact.** Chance events can affect the subsequent evolutionary history of populations. For instance, when a storm blows insects or birds hundreds of kilometers over an ocean to an island, the wind does not necessarily transport those individuals that are best suited to the new environment. Thus, not all alleles present in the founding population's gene pool are better suited to the new environment than the alleles that are "left behind." In addition, the environment at a particular location may change unpredictably from year to year, again limiting the extent to which adaptive evolution results in organisms being well suited to current environmental conditions.

With these four constraints, evolution does not tend to craft perfect organisms. Natural selection operates on a "better than" basis. We can, in fact, see evidence for evolution in the many imperfections of the organisms it produces.

▲ **Figure 21.19 Evolutionary compromise.** The loud call that enables a Túngara frog to attract mates also attracts more dangerous characters in the neighborhood—in this case, a bat about to seize a meal.

## CONCEPT CHECK 21.4

1. What is the relative fitness of a sterile mule? Explain.
2. Explain why natural selection is the only evolutionary mechanism that consistently leads to adaptive evolution.
3. **WHAT IF?** Consider a population in which heterozygotes at a certain locus have an extreme phenotype (such as being larger than homozygotes) that confers a selective advantage. Does such a situation represent directional, disruptive, or stabilizing selection? Explain your answer.

For suggested answers, see Appendix A.

---

Go to **MasteringBiology**® for Assignments, the eText, and the Study Area with Animations, Activities, Vocab Self-Quiz, and Practice Tests.

# 21 | Chapter Review

## SUMMARY OF KEY CONCEPTS

**VOCAB SELF-QUIZ**

goo.gl/gbai8v

### CONCEPT 21.1

**Genetic variation makes evolution possible (pp. 414–416)**

- **Genetic variation** refers to genetic differences among individuals within a population.
- The nucleotide differences that provide the basis of genetic variation originate when mutation and gene duplication produce new alleles and new genes.
- New genetic variants are produced rapidly in organisms with short generation times. In sexually reproducing organisms, most of the genetic differences among individuals result from crossing over, the independent assortment of chromosomes, and fertilization.

 *Typically, most of the nucleotide variability that occurs within a genetic locus does not affect the phenotype. Explain why.*

### CONCEPT 21.2

**The Hardy-Weinberg equation can be used to test whether a population is evolving (pp. 417–420)**

- A **population**, a localized group of organisms belonging to one species, is united by its **gene pool**, the aggregate of all the alleles in the population.

- For a population in **Hardy-Weinberg equilibrium**, the allele and genotype frequencies will remain constant if the population is large, mating is random, mutation is negligible, there is no gene flow, and there is no natural selection. For such a population, if $p$ and $q$ represent the frequencies of the only two possible alleles at a particular locus, then $p^2$ is the frequency of one kind of homozygote, $q^2$ is the frequency of the other kind of homozygote, and $2pq$ is the frequency of the heterozygous genotype.

? *Is it circular reasoning to calculate* p *and* q *from observed genotype frequencies and then use those values of* p *and* q *to test whether the population is in Hardy-Weinberg equilibrium? Explain your answer. (Hint: Consider a specific case, such as a population with 195 individuals of genotype* AA*, 10 of genotype* Aa*, and 195 of genotype* aa*.)*

### CONCEPT 21.3

**Natural selection, genetic drift, and gene flow can alter allele frequencies in a population (pp. 421–424)**

- In natural selection, individuals that have certain inherited traits tend to survive and reproduce at higher rates than other individuals *because of* those traits.
- In **genetic drift**, chance fluctuations in allele frequencies over generations tend to reduce genetic variation.

- **Gene flow**, the transfer of alleles between populations, tends to reduce genetic differences between populations over time.

> **?** *Would two small, geographically isolated populations in very different environments be likely to evolve in similar ways? Explain.*

### CONCEPT 21.4

## Natural selection is the only mechanism that consistently causes adaptive evolution (pp. 424–431)

- One organism has greater **relative fitness** than another if it leaves more fertile descendants. The modes of natural selection differ in their effect on phenotype (the white arrows in the summary diagram below represent selective pressure on a population).

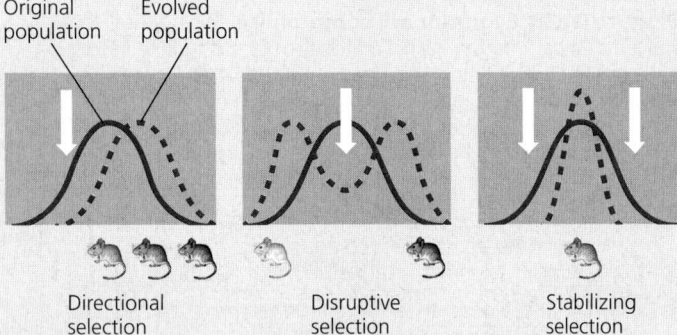

- Unlike genetic drift and gene flow, natural selection consistently increases the frequencies of alleles that enhance survival and reproduction, thus improving the degree to which organisms are well suited for life in their environment.
- **Sexual selection** can result in secondary sex characteristics that can give individuals advantages in mating.
- **Balancing selection** occurs when natural selection maintains two or more forms in a population.
- There are constraints to evolution: Natural selection can act only on available variation; structures result from modified ancestral anatomy; adaptations are often compromises; and chance, natural selection, and the environment interact.

> **?** *How might secondary sex characteristics differ between males and females in a species in which females compete for mates?*

## TEST YOUR UNDERSTANDING

### Level 1: Knowledge/Comprehension

**PRACTICE TEST**
goo.gl/CRZjvS

1. Natural selection changes allele frequencies because some _____ survive and reproduce better than others.
   - (A) alleles
   - (B) loci
   - (C) species
   - (D) individuals

2. No two people are genetically identical, except for identical twins. The main source of genetic variation among humans is
   - (A) new mutations that occurred in the preceding generation.
   - (B) genetic drift.
   - (C) the reshuffling of alleles in sexual reproduction.
   - (D) environmental effects.

3. Sparrows with average-sized wings survive severe storms better than those with longer or shorter wings, illustrating
   - (A) the bottleneck effect.
   - (B) disruptive selection.
   - (C) frequency-dependent selection.
   - (D) stabilizing selection.

### Level 2: Application/Analysis

4. If the nucleotide variability of a locus equals 0%, what is the gene variability and number of alleles at that locus?
   - (A) gene variability = 0%; number of alleles = 0
   - (B) gene variability = 0%; number of alleles = 1
   - (C) gene variability = 0%; number of alleles = 2
   - (D) gene variability > 0%; number of alleles = 2

5. There are 25 individuals in population 1, all with genotype *AA*, and there are 40 individuals in population 2, all with genotype *aa*. Assume that these populations are located far from each other and that their environmental conditions are very similar. Based on the information given here, the observed genetic variation most likely resulted from
   - (A) genetic drift.
   - (B) gene flow.
   - (C) nonrandom mating.
   - (D) directional selection.

6. A fruit fly population has a gene with two alleles, *A1* and *A2*. Tests show that 70% of the gametes produced in the population contain the *A1* allele. If the population is in Hardy-Weinberg equilibrium, what proportion of the flies carry both *A1* and *A2*?
   - (A) 0.7
   - (B) 0.49
   - (C) 0.42
   - (D) 0.21

### Level 3: Synthesis/Evaluation

7. **SCIENTIFIC INQUIRY**

   **INTERPRET THE DATA** Researchers studied genetic variation in the marine mussel *Mytilus edulis* around Long Island, New York. They measured the frequency of a particular allele (*lap⁹⁴*) for an enzyme involved in regulating the mussel's internal saltwater balance. The researchers presented their data as a series of pie charts linked to sampling sites within Long Island Sound, where the salinity is highly variable, and along the coast of the open ocean, where salinity is constant:

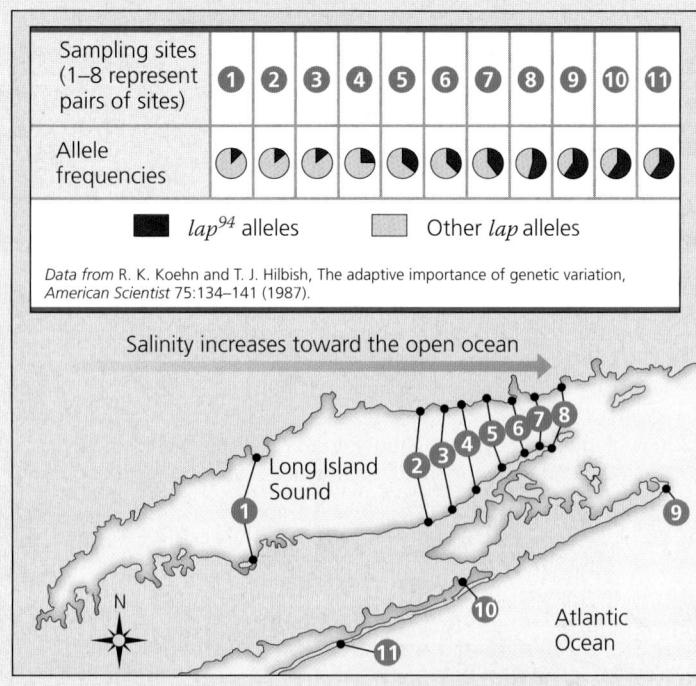

(a) Create a data table for the 11 sampling sites by estimating the frequency of *lap⁹⁴* from the pie charts. (*Hint:* Think of each pie chart as a clock face to help you estimate the proportion of the shaded area.) (b) Graph the frequencies for sites 1–8 to show how the frequency of this allele changes with increasing salinity in Long Island Sound (from southwest to northeast). Evaluate how the data from sites 9–11 compare with the data from the sites within the Sound. (c) Construct a hypothesis that explains the patterns you observe in the data and that accounts for the following observations: (1) The $lap^{94}$ allele helps mussels maintain osmotic balance in water with a high salt concentration but is costly to use in less salty water; and (2) mussels produce larvae that can disperse long distances before they settle on rocks and grow into adults.

### 8. FOCUS ON EVOLUTION

Using at least TWO examples, explain how the process of evolution is revealed by the imperfections of living organisms.

### 9. FOCUS ON ORGANIZATION

Heterozygotes at the sickle-cell locus produce both normal and abnormal (sickle-cell) hemoglobin (see Concept 11.4). When hemoglobin molecules are packed into a heterozygote's red blood cells, some cells receive relatively large quantities of abnormal hemoglobin, making these cells prone to sickling. In a short essay (approximately 100–150 words), explain how these molecular and cellular events lead to emergent properties at the individual and population levels of biological organization.

### 10. SYNTHESIZE YOUR KNOWLEDGE

This kettle lake formed 14,000 years ago when a glacier that covered the surrounding area melted. Initially devoid of animal life, over time the lake was colonized by invertebrates and other animals. Hypothesize how mutation, natural selection, genetic drift, and gene flow may have affected populations that colonized the lake.

*For selected answers, see Appendix A.*

## KEY CONCEPTS

**22.1** The biological species concept emphasizes reproductive isolation

**22.2** Speciation can take place with or without geographic separation

**22.3** Hybrid zones reveal factors that cause reproductive isolation

**22.4** Speciation can occur rapidly or slowly and can result from changes in few or many genes

▲ **Figure 22.1** How did this flightless bird come to live on the isolated Galápagos Islands?

## That "Mystery of Mysteries"

When Darwin came to the Galápagos Islands, he noted that these volcanic islands were teeming with plants and animals found nowhere else in the world **(Figure 22.1)**. Later he realized that these species had formed relatively recently. He wrote in his diary: "Both in space and time, we seem to be brought somewhat near to that great fact—that mystery of mysteries—the first appearance of new beings on this Earth."

The "mystery of mysteries" that captivated Darwin is **speciation**, the process by which one species splits into two or more species. Speciation fascinated Darwin (and many biologists since) because it has produced the tremendous diversity of life, repeatedly yielding new species that differ from existing ones. Speciation also helps to explain the many features that organisms share (the unity of life). When one species splits into two, the species that result share many characteristics because they are descended from this common ancestor. At the DNA sequence level, for example, such similarities indicate that the flightless cormorant (*Phalacrocorax harrisi*) in Figure 22.1 is closely related to flying cormorants found in the Americas. This suggests that the flightless cormorant originated from an ancestral cormorant species that flew from the mainland to the Galápagos.

Speciation also forms a conceptual bridge between **microevolution**, changes over time in allele frequencies in a population, and **macroevolution**, the broad pattern of evolution above the species level. An example of macroevolutionary change is the origin of new groups of organisms, such as mammals or flowering plants, through a series of speciation events. We examined microevolutionary mechanisms in Chapter 21, and we'll turn to macroevolution in Chapter 23. In this chapter, we'll explore the "bridge" between microevolution and macroevolution—the mechanisms by which new species originate from existing ones. First, let's establish what we actually mean by a "species."

## CONCEPT 22.1

# The biological species concept emphasizes reproductive isolation

The word *species* is Latin for "kind" or "appearance." In daily life, we commonly distinguish between various "kinds" of organisms—dogs and cats, for instance—from differences in their appearance. But are organisms truly divided into the discrete units we call species, or is this classification an arbitrary attempt to impose order on the natural world? To answer this question, biologists compare not only the morphology (body form) of different groups of organisms but also less obvious differences in physiology, biochemistry, and DNA sequences. The results generally confirm that morphologically distinct species are indeed discrete groups, differing in many ways besides their body forms.

## The Biological Species Concept

The primary definition of species used in this textbook is the **biological species concept**. According to this concept, a **species** is a group of populations whose members have the potential to interbreed in nature and produce viable, fertile offspring—but do not produce viable, fertile offspring with members of other such groups **(Figure 22.2)**. Thus, the members of a biological species are united by being reproductively compatible, at least potentially. All human beings, for example, belong to the same species. A businesswoman in Manhattan may be unlikely to meet a dairy farmer in Mongolia, but if the two should happen to meet and mate, they could have viable babies who develop into fertile adults. In contrast, humans and chimpanzees remain distinct biological species even where they live in the same region, because many factors keep them from interbreeding and producing fertile offspring.

What holds the gene pool of a species together, causing its members to resemble each other more than they resemble members of other species? To answer this question, recall the evolutionary mechanism of *gene flow*, the transfer of alleles between populations (see Concept 21.3). Typically, gene flow occurs between the different populations of a species. This ongoing exchange of alleles tends to hold the populations together genetically. But as we'll explore in this chapter, a reduction or lack of gene flow can play a key role in the formation of new species.

### Reproductive Isolation

Because biological species are defined in terms of reproductive compatibility, the formation of a new species hinges on **reproductive isolation**—the existence of biological factors (barriers) that impede members of two species from interbreeding and producing viable, fertile offspring. Such barriers block gene flow between the species and limit the formation of **hybrids**, offspring that result from an interspecific mating. Although a single barrier may not prevent all gene flow, a combination of several barriers can effectively isolate a species' gene pool.

Clearly, a fly cannot mate with a frog or a fern, but the reproductive barriers between more closely related species are not so obvious. These barriers can be classified according to whether they contribute to reproductive isolation before or after fertilization. **Prezygotic barriers** ("before the zygote") block fertilization from occurring. Such barriers typically act in

**(a) Similarity between different species.** The eastern meadowlark (*Sturnella magna*, left) and the western meadowlark (*Sturnella neglecta*, right) have similar body shapes and colorations. Nevertheless, they are distinct biological species because their songs and other behaviors are different enough to prevent interbreeding should they meet in the wild.

**(b) Diversity within a species.** As diverse as we may be in appearance, all humans belong to a single biological species (*Homo sapiens*), defined by our capacity to interbreed successfully.

▲ **Figure 22.2 The biological species concept is based on the potential to interbreed, not on physical similarity.**

one of three ways: by impeding members of different species from attempting to mate, by preventing an attempted mating from being completed successfully, or by hindering fertilization if mating is completed successfully. If a sperm cell from one species overcomes prezygotic barriers and fertilizes an ovum from another species, a variety of **postzygotic barriers** ("after the zygote") may contribute to reproductive isolation after the hybrid zygote is formed. For example, developmental errors may reduce survival among hybrid embryos. Or problems after birth may cause hybrids to be infertile or decrease their chance of surviving long enough to reproduce. **Figure 22.3** describes prezygotic and postzygotic barriers in more detail.

## ▼ Figure 22.3    Exploring Reproductive Barriers

**Prezygotic barriers impede mating or hinder fertilization if mating does occur**

| Habitat Isolation | Temporal Isolation | Behavioral Isolation | Mechanical Isolation |
|---|---|---|---|

Individuals of different species → 🚫 | 🚫 | 🚫 | **MATING ATTEMPT** | 🚫

**Habitat Isolation**

Two species that occupy different habitats within the same area may encounter each other rarely, if at all, even though they are not isolated by obvious physical barriers, such as mountain ranges.

**Example:** These two fly species in the genus *Rhagoletis* occur in the same geographic areas, but the apple maggot fly (*Rhagoletis pomonella*) feeds and mates on hawthorns and apples **(a)** while its close relative, the blueberry maggot fly (*R. mendax*), mates and lays its eggs only on blueberries **(b)**.

**Temporal Isolation**

Species that breed during different times of the day, different seasons, or different years cannot mix their gametes.

**Example:** In North America, the geographic ranges of the western spotted skunk (*Spilogale gracilis*) **(c)** and the eastern spotted skunk (*Spilogale putorius*) **(d)** overlap, but *S. gracilis* mates in late summer and *S. putorius* mates in late winter.

**Behavioral Isolation**

Courtship rituals that attract mates and other behaviors unique to a species are effective reproductive barriers, even between closely related species. Such behavioral rituals enable *mate recognition*—a way to identify potential mates of the same species.

**Example:** Blue-footed boobies, inhabitants of the Galápagos, mate only after a courtship display unique to their species. Part of the "script" calls for the male to high-step **(e)**, a behavior that calls the female's attention to his bright blue feet.

**Mechanical Isolation**

Mating is attempted, but morphological differences prevent its successful completion.

**Example:** The shells of two species of snails in the genus *Bradybaena* spiral in different directions: Moving inward to the center, one spirals in a counterclockwise direction (**f**, left), the other in a clockwise direction (**f**, right). As a result, the snails' genital openings (indicated by arrows) are not aligned, and mating cannot be completed.

(a)

(c)

(e)

(f)

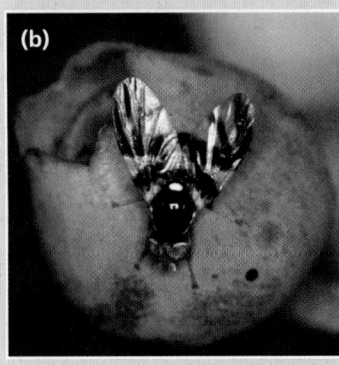

(b)

(d)

| Gametic Isolation | Reduced Hybrid Viability | Reduced Hybrid Fertility | Hybrid Breakdown |

FERTILIZATION

VIABLE, FERTILE OFFSPRING

Sperm of one species may not be able to fertilize the eggs of another species. For instance, sperm may not be able to survive in the reproductive tract of females of the other species, or biochemical mechanisms may prevent the sperm from penetrating the membrane surrounding the other species' eggs.

**Example:** Gametic isolation separates certain closely related species of aquatic animals, such as sea urchins **(g)**. Sea urchins release their sperm and eggs into the surrounding water, where they fuse and form zygotes. It is difficult for gametes of different species, such as the red and purple urchins shown here, to fuse because proteins on the surfaces of the eggs and sperm bind very poorly to each other.

The genes of different parent species may interact in ways that impair the hybrid's development or survival in its environment.

**Example:** Some salamander subspecies of the genus *Ensatina* live in the same regions and habitats, where they may occasionally hybridize. But most of the hybrids do not complete development, and those that do are frail **(h)**.

Even if hybrids are vigorous, they may be sterile. If the chromosomes of the two parent species differ in number or structure, meiosis in the hybrids may fail to produce normal gametes. Since the infertile hybrids cannot produce offspring when they mate with either parent species, genes cannot flow freely between the species.

**Example:** The hybrid offspring of a male donkey **(i)** and a female horse **(j)** is a mule **(k)**, which is robust but sterile. A "hinny" (not shown), the offspring of a female donkey and a male horse, is also sterile.

Some first-generation hybrids are viable and fertile, but when they mate with one another or with either parent species, offspring of the next generation are feeble or sterile.

**Example:** Strains of cultivated rice have accumulated different mutant recessive alleles at two loci in the course of their divergence from a common ancestor. Hybrids between them are vigorous and fertile (**l**, left and right), but plants in the next generation that carry too many of these recessive alleles are small and sterile (**l**, center). Although these rice strains are not yet considered different species, they have begun to be separated by postzygotic barriers.

### Limitations of the Biological Species Concept

One strength of the biological species concept is that it directs our attention to a way by which speciation can occur: by the evolution of reproductive isolation. However, the number of species to which this concept can be usefully applied is limited. There is, for example, no way to evaluate the reproductive isolation of fossils. The biological species concept also does not apply to organisms that reproduce asexually all or most of the time, such as prokaryotes. (Many prokaryotes do transfer genes among themselves, as we will discuss in Concept 24.3, but this is not part of their reproductive process.) Furthermore, in the biological species concept, species are designated by the *absence* of gene flow. But there are many pairs of species that are morphologically and ecologically distinct, and yet gene flow occurs between them. An example is the grizzly bear (*Ursus arctos*) and polar bear (*Ursus maritimus*), whose hybrid offspring have been dubbed "grolar bears" **(Figure 22.4)**. As we'll discuss, natural selection can cause such species to remain distinct even though some gene flow occurs between them. Because of the limitations to the biological species concept, alternative species concepts are useful in certain situations.

◀ Grizzly bear (*U. arctos*)

▼ Polar bear (*U. maritimus*)

▲ Hybrid "grolar bear"

▲ **Figure 22.4 Hybridization between two species of bears in the genus *Ursus*.**

### Other Definitions of Species

While the biological species concept emphasizes the *separateness* of different species due to reproductive barriers, several other definitions emphasize the *unity within* a species. For example, the **morphological species concept** distinguishes a species by body shape and other structural features. The morphological species concept can be applied to asexual and sexual organisms, and it can be useful even without information on the extent of gene flow. In practice, scientists often distinguish species using morphological criteria. A disadvantage of this approach, however, is that it relies on subjective criteria; researchers may disagree on which structural features distinguish a species.

The **ecological species concept** defines a species in terms of its ecological niche, the sum of how members of the species interact with the nonliving and living parts of their environment. For example, two species of oak trees might differ in their size or in their ability to tolerate dry conditions, yet still occasionally interbreed. Because they occupy different ecological niches, these oaks would be considered separate species even though they are connected by some gene flow. Unlike the biological species concept, the ecological species concept can accommodate asexual as well as sexual species. It also emphasizes the role of disruptive natural selection as organisms adapt to different environments.

The **phylogenetic species concept** defines a species as the smallest group of individuals that share a common ancestor, forming one branch on the tree of life. Biologists trace the phylogenetic history of a species by comparing its characteristics, such as morphology or molecular sequences, with those of other organisms. Such analyses can distinguish groups of individuals that are sufficiently different to be considered separate species. Of course, the difficulty with this species concept is determining the degree of difference required to indicate separate species.

In addition to those discussed here, more than 20 other species definitions have been proposed. The usefulness of each definition depends on the situation and the research questions being asked. For our purposes of studying how species originate, the biological species concept, with its focus on reproductive barriers, is particularly helpful.

#### CONCEPT CHECK 22.1

1. (a) Which species concept(s) could you apply to both asexual and sexual species? (b) Which would be most useful for identifying species in the field? Explain.
2. **WHAT IF?** Suppose you are studying two bird species that live in a forest and are not known to interbreed. One species feeds and mates in the treetops and the other on the ground. But in captivity, the birds can interbreed and produce viable, fertile offspring. What type of reproductive barrier most likely keeps these species separate in nature? Explain.

For suggested answers, see Appendix A.

# Speciation can take place with or without geographic separation

Having discussed what constitutes a species, let's return to the process by which new species arise from existing species. We'll focus on the geographic setting in which gene flow is interrupted between populations of the existing species—in allopatric speciation the populations are geographically isolated, while in sympatric speciation they are not **(Figure 22.5)**.

## Allopatric ("Other Country") Speciation

In **allopatric speciation** (from the Greek *allos*, other, and *patra*, homeland), gene flow is interrupted when a population is divided into geographically isolated subpopulations. For example, the water level in a lake may subside, resulting in two or more smaller lakes that are now home to separated populations (see Figure 22.5a). Or a river may change course and divide a population of animals that cannot cross it. Allopatric speciation can also occur without geologic remodeling, such as when individuals colonize a remote area and their descendants become geographically isolated from the parent population. The flightless cormorant shown in Figure 22.1 probably originated in this way from an ancestral flying species that reached the Galápagos Islands.

### The Process of Allopatric Speciation

How formidable must a geographic barrier be to promote allopatric speciation? The answer depends on the ability of the organisms to move about. Birds, mountain lions, and coyotes can cross rivers and canyons—as can the windblown pollen of pine trees and the seeds of many flowering plants. In contrast, small rodents may find a wide river or deep canyon a formidable barrier.

Once geographic isolation has occurred, the separated gene pools may diverge. Different mutations arise, and natural selection and genetic drift may alter allele frequencies in different ways in the separated populations. Reproductive isolation may then evolve as a by-product of these genetic changes.

Let's consider an example. On Andros Island, in the Bahamas, populations of the mosquitofish *Gambusia hubbsi* colonized a series of ponds that later became isolated from one another. Genetic analyses indicate that little or no gene flow currently occurs between the ponds. The environments of these ponds are very similar except that some contain predatory fishes, while others do not. In the ponds with predatory fishes, selection has favored the evolution of a mosquitofish body shape that enables rapid bursts of speed **(Figure 22.6)**. In ponds without predatory fishes, selection has favored a different body shape, one that improves the ability to swim for long periods of time. How have these different selective pressures affected the evolution of reproductive barriers? Researchers studied this question by bringing together mosquitofish from the two types of ponds. They found that female mosquitofish prefer to mate with males whose body shape is similar to their own. This preference establishes a behavioral barrier to reproduction between mosquitofish from ponds with predators and those from ponds without predators. Thus, as a by-product of selection for avoiding predators, reproductive barriers have formed in these allopatric populations.

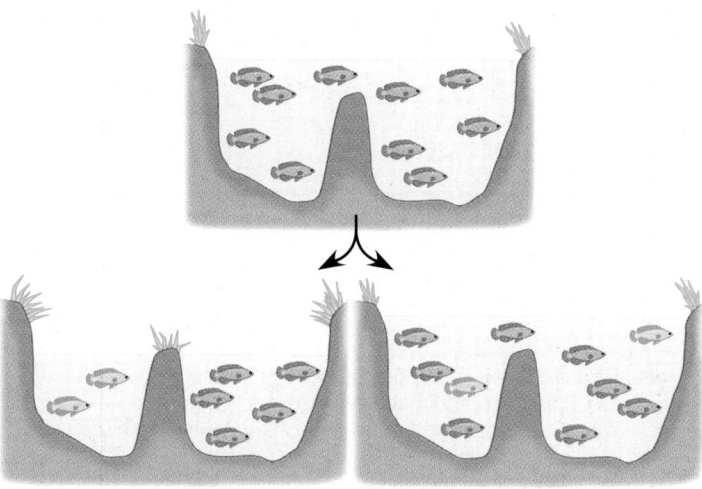

(a) **Allopatric speciation.** A population forms a new species while geographically isolated from its parent population.

(b) **Sympatric speciation.** A subset of a population forms a new species without geographic separation.

▲ **Figure 22.5 The geography of speciation.**

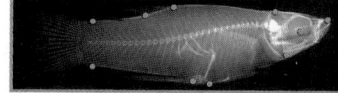 

In ponds with predatory fishes, the mosquitofish's head is streamlined and the tail is powerful, enabling rapid bursts of speed.

In ponds without predatory fishes, mosquitofish have a different body shape that favors long, steady swimming.

**(a) Differences in body shape**

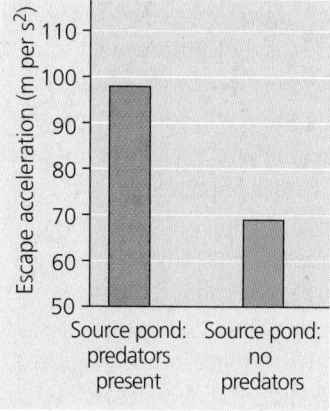 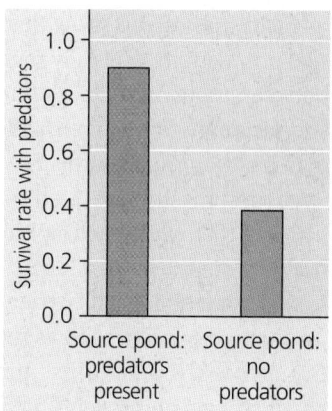

**(b) Differences in escape acceleration and survival**

▲ **Figure 22.6 Evolution in mosquitofish populations.** Different body shapes have evolved in mosquitofish populations from ponds with and without predators. These differences affect how quickly the fish can accelerate to escape and their survival rate when they are exposed to predators.

## Evidence of Allopatric Speciation

Many studies provide evidence that speciation can occur in allopatric populations. For example, laboratory studies show that reproductive barriers can develop when populations are isolated experimentally and subjected to different environmental conditions **(Figure 22.7)**.

Field studies indicate that allopatric speciation also can occur in nature. Consider the 30 species of snapping shrimp in the genus *Alpheus* that live off the Isthmus of Panama, the land bridge that connects South and North America. Fifteen of these species live on the Atlantic side of the isthmus, while the other 15 live on the Pacific side. Before the isthmus formed, gene flow could occur between the Atlantic and Pacific populations of snapping shrimp. Did the species on different sides of the isthmus originate by allopatric speciation? Morphological and genetic data group these shrimp into 15 pairs of *sister species*, pairs whose member species are each other's closest relative (see Figure 20.5). In each of these 15 pairs, one of the sister species lives on the Atlantic side of the isthmus, while the other lives on the Pacific side **(Figure 22.8)**. This fact strongly suggests that in each case, the two species arose as a consequence of geographic separation. Furthermore, genetic analyses indicate that the *Alpheus* species originated from 9 to 3 million years ago, with the sister species that live in the deepest water diverging first. These divergence times are consistent with geologic evidence that the isthmus formed gradually, starting 10 million years ago and closing completely about 3 million years ago.

The importance of allopatric speciation is also suggested by the fact that regions that are isolated or highly subdivided by barriers typically have more species than do otherwise similar regions that lack such features. For example, many unique plants and animals are found on the geographically isolated Hawaiian Islands (we'll return to the origin of Hawaiian species in Concept 23.2). Similarly, unusually high numbers of butterfly species are found in regions of South America that are subdivided by many rivers.

Field studies also show that reproductive isolation between two populations generally increases as the geographic distance between them increases, a finding consistent with allopatric speciation. In the **Scientific Skills Exercise**, you will analyze data from one such study that examined reproductive isolation in geographically separated salamander populations.

Note that while geographic isolation prevents interbreeding between members of allopatric populations, physical separation is not a biological barrier to reproduction. Biological reproductive barriers such as those described in Figure 22.3 are intrinsic to the organisms themselves. Hence, it is biological barriers that can prevent interbreeding when members of different populations come into contact with one another.

## Sympatric ("Same Country") Speciation

In **sympatric speciation** (from the Greek *syn*, together), speciation occurs in populations that live in the same geographic area (see Figure 22.5b). How can reproductive barriers form between sympatric populations while their members remain in contact

▼ **Figure 22.7** Inquiry

## Can divergence of allopatric populations lead to reproductive isolation?

**Experiment** A researcher divided a laboratory population of the fruit fly *Drosophila pseudoobscura*, raising some flies on a starch medium and others on a maltose medium. After one year (about 40 generations), natural selection resulted in divergent evolution: Populations raised on starch digested starch more efficiently, while those raised on maltose digested maltose more efficiently. The researcher then put flies from the same or different populations in mating cages and measured mating frequencies. All flies used in the mating preference tests were reared for one generation on a standard cornmeal medium.

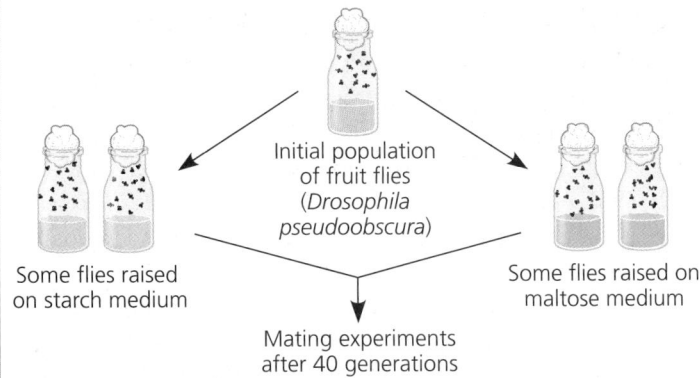

Some flies raised on starch medium

Initial population of fruit flies (*Drosophila pseudoobscura*)

Some flies raised on maltose medium

Mating experiments after 40 generations

**Results** Mating patterns among populations of flies raised on different media are shown below. When flies from "starch populations" were mixed with flies from "maltose populations," the flies tended to mate with like partners. But in the control group (shown on the right), flies from different populations adapted to starch were about as likely to mate with each other as with flies from their own population; similar results were obtained for control groups adapted to maltose.

| | | Female | |
|---|---|---|---|
| | | Starch | Maltose |
| **Male** Starch | | 22 | 9 |
| **Male** Maltose | | 8 | 20 |

Number of matings in experimental group

| | | Female | |
|---|---|---|---|
| | | Starch population 1 | Starch population 2 |
| **Male** Starch population 1 | | 18 | 15 |
| **Male** Starch population 2 | | 12 | 15 |

Number of matings in control group

**Conclusion** In the experimental group, the strong preference of "starch flies" and "maltose flies" to mate with like-adapted flies indicates that a reproductive barrier was forming between these fly populations. Although this barrier was not absolute (some mating between starch flies and maltose flies did occur), after 40 generations reproductive isolation appeared to be increasing. This barrier may have been caused by differences in courtship behavior that arose as an incidental by-product of differing selective pressures as these allopatric populations adapted to different sources of food.

**Data from** D. M. B. Dodd, Reproductive isolation as a consequence of adaptive divergence in *Drosophila pseudoobscura*, *Evolution* 43:1308–1311 (1989).

**WHAT IF?** Why were all flies used in the mating preference tests reared on a standard medium (rather than on starch or maltose)?

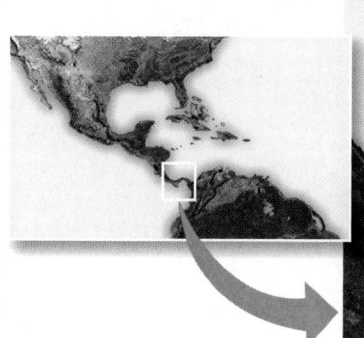

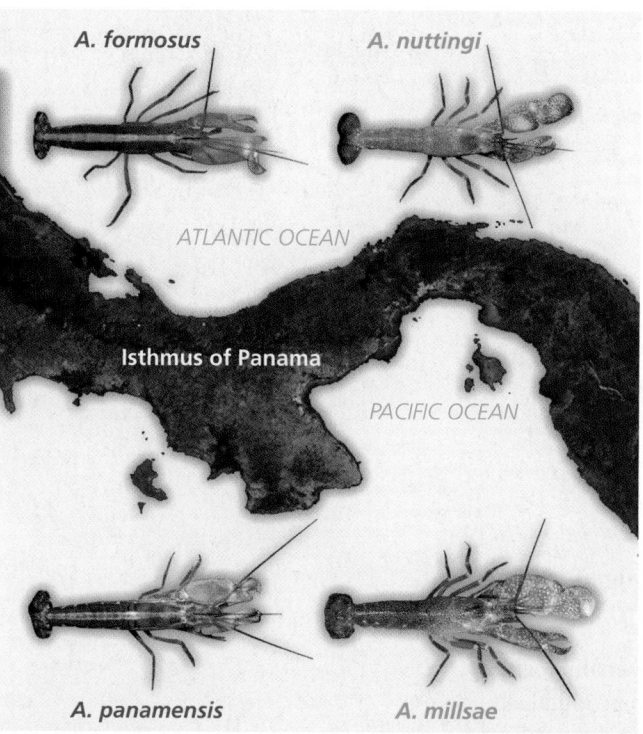

A. formosus    A. nuttingi

ATLANTIC OCEAN

Isthmus of Panama

PACIFIC OCEAN

A. panamensis    A. millsae

▶ **Figure 22.8**
**Allopatric speciation in snapping shrimp (*Alpheus*).** The shrimps pictured are just 2 of the 15 pairs of sister species that arose as populations were divided by the formation of the Isthmus of Panama. The color-coded type indicates the sister species.

with each other? Although such contact (and the ongoing gene flow that results) makes sympatric speciation less common than allopatric speciation, sympatric speciation can occur if gene flow is reduced by such factors as polyploidy, habitat differentiation, and sexual selection. (Note that these factors can also promote allopatric speciation.)

### Polyploidy

A species may originate from an accident during cell division that results in extra sets of chromosomes, a condition called **polyploidy**. Polyploid speciation occasionally occurs in animals; for example, the gray tree frog *Hyla versicolor* (see Figure 21.18) is thought to have originated in this way. However, polyploidy is far more common in plants. Botanists estimate that more than 80% of the plant species alive today are descended from

## Scientific Skills Exercise

### Identifying Independent and Dependent Variables, Making a Scatter Plot, and Interpreting Data

**Does Distance Between Salamander Populations Increase Their Reproductive Isolation?** Allopatric speciation begins when populations become geographically isolated, preventing mating between individuals in different populations and thus stopping gene flow. It is logical that as distance between populations increases, so will their degree of reproductive isolation. To test this hypothesis, researchers studied populations of dusky salamanders (*Desmognathus ochrophaeus*) living on different mountains in the southern Appalachians.

**How the Experiment Was Done** The researchers tested the reproductive isolation of pairs of salamander populations by leaving one male and one female together and later checking the females for the presence of sperm. Four mating combinations were tested for each pair of populations (A and B)—two *within* the same population (female A with male A and female B with male B) and two *between* populations (female A with male B and female B with male A).

**Data from the Experiment** The researchers used an index of reproductive isolation that ranged from a value of 0 (no isolation) to a value of 2 (full isolation). The proportion of successful matings for each mating combination was measured, with 100% success = 1 and no success = 0. The reproductive isolation value for two populations is the sum of the proportion of successful matings of each type within populations (AA + BB) minus the sum of the proportion of successful matings of each type between populations (AB + BA). The

table provides distance and reproductive isolation data for 27 pairs of dusky salamander populations.

**INTERPRET THE DATA**

1. State the researchers' hypothesis, and identify the independent and dependent variables in this study. Explain why the researchers used four mating combinations for each pair of populations.

2. Calculate the value of the reproductive isolation index if (a) *all* of the matings within a population were successful, but *none* of the matings between populations were successful; (b) salamanders are equally successful in mating with members of their own population and members of another population.

3. Make a scatter plot to help you visualize any patterns that might indicate a relationship between the variables. Plot the independent variable on the *x*-axis and the dependent variable on the *y*-axis. (For additional information about graphs, see the Scientific Skills Review in Appendix F and the Study Area of MasteringBiology.)

4. Interpret your graph by (a) explaining in words any pattern indicating a possible relationship between the variables and (b) hypothesizing the possible cause of such a relationship.

(MB) A version of this Scientific Skills Exercise can be assigned in MasteringBiology.

| Geographic Distance (km) | 15 | 32 | 40 | 47 | 42 | 62 | 63 | 81 | 86 | 107 | 107 | 115 | 137 | 147 |
|---|---|---|---|---|---|---|---|---|---|---|---|---|---|---|
| Reproductive Isolation Value | 0.32 | 0.54 | 0.50 | 0.50 | 0.82 | 0.37 | 0.67 | 0.53 | 1.15 | 0.73 | 0.82 | 0.81 | 0.87 | 0.87 |
| Distance (continued) | 137 | 150 | 165 | 189 | 219 | 239 | 247 | 53 | 55 | 62 | 105 | 179 | 169 | |
| Isolation (continued) | 0.50 | 0.57 | 0.91 | 0.93 | 1.50 | 1.22 | 0.82 | 0.99 | 0.21 | 0.56 | 0.41 | 0.72 | 1.15 | |

**Data from** S. G. Tilley et al., Correspondence between sexual isolation and allozyme differentiation: A test in the salamander *Desmognathus ochrophaeus*, *Proceedings of the National Academy of Sciences USA* 87:2715–2719 (1990).

ancestors that formed by polyploid speciation.

Two distinct forms of polyploidy have been observed in plant (and a few animal) populations. An **autopolyploid** (from the Greek *autos*, self) is an individual that has more than two chromosome sets that are all derived from a single species. In plants, for example, a failure of cell division could double a cell's chromosome number from the original number (2*n*) to a tetraploid number (4*n*) (**Figure 22.9**).

A tetraploid can produce fertile tetraploid offspring by self-pollinating or by mating with other tetraploids. In addition, the tetraploids are reproductively isolated from 2*n* plants of the original population, because the triploid (3*n*) offspring of such unions have reduced fertility. Thus, in just one generation, autopolyploidy can generate reproductive isolation without any geographic separation.

A second form of polyploidy can occur when two different species interbreed and produce hybrid offspring. Most such hybrids are sterile because the set of chromosomes from one species cannot pair during meiosis with the set of chromosomes from the other species. However, an infertile hybrid may be able to propagate itself asexually (as many plants can do). In subsequent generations, various mechanisms can change a sterile hybrid into a fertile polyploid called an **allopolyploid** (**Figure 22.10**). The allopolyploids are fertile when mating with each other but cannot interbreed with either parent species; thus, they represent a new biological species.

Although it can be challenging to study speciation in the field, scientists have documented at least five new plant species that have originated by polyploidy speciation since 1850. One of these examples involves the origin of a new species of goatsbeard plant (genus *Tragopogon*) in the Pacific Northwest. *Tragopogon* first arrived in the region when humans introduced three European species in the early 1900s: *T. pratensis*, *T. dubius*, and *T. porrifolius*. These three species are now common weeds in abandoned parking lots and other urban sites. In 1950, a new *Tragopogon* species was discovered near the Idaho-Washington border, a region where all three European species also were found. Genetic analyses revealed that this new species, *Tragopogon miscellus*, is a hybrid of two of the European species (**Figure 22.11**). Although the *T. miscellus* population grows mainly by reproduction of its own members, additional episodes of hybridization between the parent species continue to add new members to the *T. miscellus* population. Later, scientists discovered

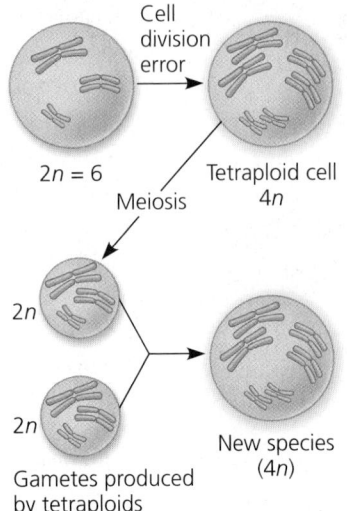

▲ Figure 22.9 **Sympatric speciation by autopolyploidy.**

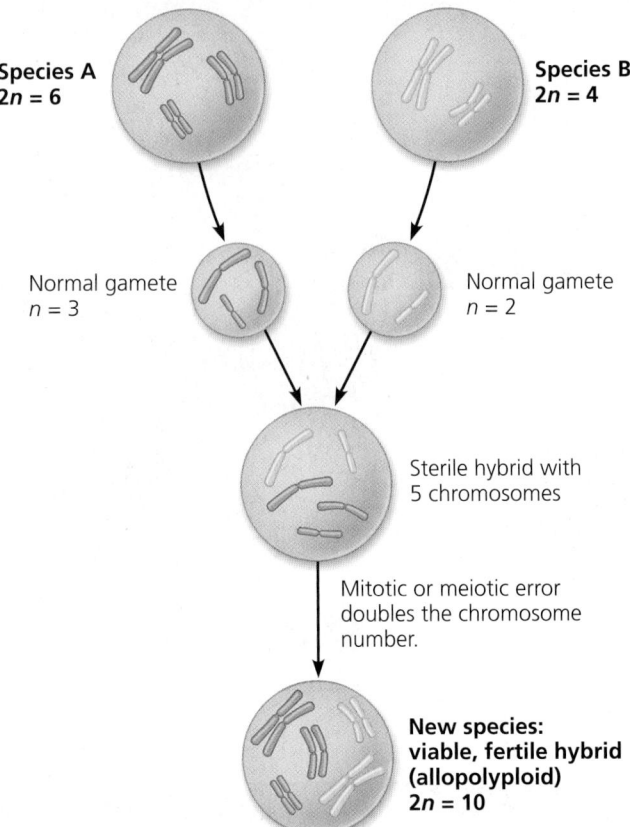

▲ Figure 22.10 **One mechanism for allopolyploid speciation in plants.** Most hybrids are sterile because their chromosomes are not homologous and cannot pair during meiosis. However, such a hybrid may be able to reproduce asexually. This diagram traces one mechanism that can produce fertile hybrids (allopolyploids) as new species. The new species has a diploid chromosome number equal to the sum of the diploid chromosome numbers of the two parent species.

another new *Tragopogon* species, *T. mirus*—this one a hybrid of *T. dubius* and *T. porrifolius* (see Figure 22.11). The *Tragopogon* story is just one of several well-studied examples in which scientists have observed speciation in progress.

Many important agricultural crops—such as oats, cotton, potatoes, tobacco, and wheat—are polyploids. The wheat used for bread, *Triticum aestivum*, is an allohexaploid (six sets of chromosomes, two sets from each of three different species). The first of the polyploidy events that eventually led to modern wheat probably occurred about 8,000 years ago in the Middle East as a spontaneous hybrid of an early cultivated wheat species and a wild grass. Today, plant geneticists generate new polyploids in the laboratory by using chemicals that induce meiotic and mitotic errors. By harnessing the evolutionary process, researchers can produce new hybrid species with desired qualities, such as a hybrid that combines the high yield of wheat with the hardiness of rye.

### Habitat Differentiation

Sympatric speciation can also occur when a subpopulation exploits a habitat or resource not used by the parent population. Consider the North American apple maggot fly (*Rhagoletis pomonella*), a pest of apples. The fly's original habitat was the

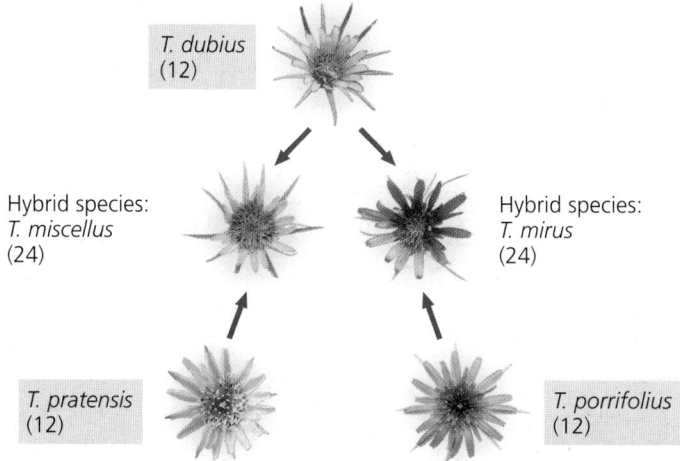

▲ Figure 22.11 **Allopolyploid speciation in *Tragopogon*.** The gray boxes indicate the three parent species. The diploid chromosome number of each species is shown in parentheses.

native hawthorn tree, but about 200 years ago, some populations colonized apple trees that had been introduced by European settlers. Apple maggot flies usually mate on or near their host plant. This results in a prezygotic barrier (habitat isolation) between populations that feed on apples and populations that feed on hawthorns (such as the fly shown in Figure 22.3a). Furthermore, because apples mature more quickly than hawthorn fruit, natural selection has favored apple-feeding flies with rapid development. These apple-feeding populations now show temporal isolation from the hawthorn-feeding *R. pomonella*, providing a second prezygotic restriction to gene flow between the two populations. Researchers have also identified alleles that benefit the flies that use one host plant but harm the flies that use the other host plant. Natural selection operating on these alleles has provided a postzygotic barrier to reproduction, further limiting gene flow. Altogether, although the two populations are still classified as subspecies rather than separate species, sympatric speciation appears to be well under way.

## Sexual Selection

There is evidence that sympatric speciation can also be driven by sexual selection. Clues to how this can occur have been found in cichlid fish from one of Earth's hot spots of animal speciation, East Africa's Lake Victoria. This lake was once home to as many as 600 species of cichlids. Genetic data indicate that these species originated within the last 100,000 years from a small number of colonizing species that arrived from other lakes and rivers. How did so many species—more than double the number of freshwater fish species known in all of Europe—originate within a single lake?

One hypothesis is that subgroups of the original cichlid populations adapted to different food sources and the resulting genetic divergence contributed to speciation in Lake Victoria. But sexual selection, in which (typically) females select males based on their appearance (see Concept 21.4), may also have been a factor. Researchers have studied two closely related sympatric species of cichlids that differ mainly

▼ **Figure 22.12** **Inquiry**

### Does sexual selection in cichlids result in reproductive isolation?

**Experiment** Researchers placed males and females of *Pundamilia pundamilia* and *P. nyererei* together in two aquarium tanks, one with natural light and one with a monochromatic orange lamp. Under normal light, the two species are noticeably different in male breeding coloration; under monochromatic orange light, the two species are very similar in color. The researchers then observed the mate choices of the females in each tank.

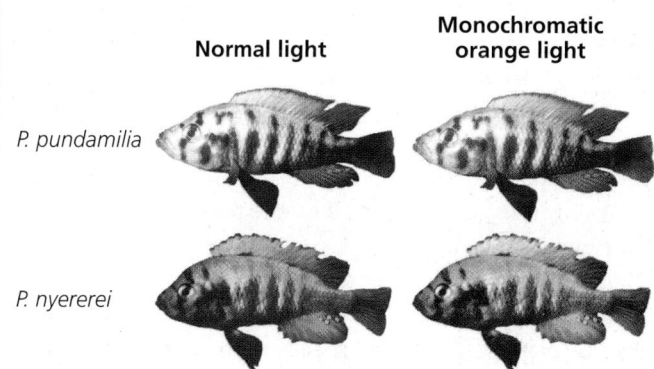

**Results** Under normal light, females of each species strongly preferred males of their own species. But under orange light, females of each species responded indiscriminately to males of both species. The resulting hybrids were viable and fertile.

**Conclusion** The researchers concluded that mate choice by females based on male breeding coloration can act as a reproductive barrier that keeps the gene pools of these two species separate. Since the species can still interbreed when this prezygotic behavioral barrier is breached in the laboratory, the genetic divergence between the species is likely to be small. This suggests that speciation in nature has occurred relatively recently.

**Data from** O. Seehausen and J. J. M. van Alphen, The effect of male coloration on female mate choice in closely related Lake Victoria cichlids (*Haplochromis nyererei* complex), *Behavioral Ecology and Sociobiology* 42:1–8 (1998).

**WHAT IF?** Suppose that female cichlids living in the murky waters of a polluted lake could not distinguish colors well. In such waters, how might the gene pools of these species change over time?

in the coloration of breeding males: Breeding *Pundamilia pundamilia* males have a blue-tinged back, whereas breeding *Pundamilia nyererei* males have a red-tinged back **(Figure 22.12)**. Their results suggest that mate choice based on male breeding coloration can act as a reproductive barrier that keeps the gene pools of these two species separate.

## Allopatric and Sympatric Speciation: *A Review*

Now let's recap the processes by which new species form. In allopatric speciation, a new species forms in geographic isolation from its parent population. Geographic isolation severely restricts gene flow. Intrinsic barriers to reproduction with the parent population may then arise as a by-product of genetic changes that occur within the isolated population.

Many different processes can produce such genetic changes, including natural selection under different environmental conditions, genetic drift, and sexual selection. Once formed, reproductive barriers that arise in allopatric populations can prevent interbreeding with the parent population even if the populations come back into contact.

Sympatric speciation, in contrast, requires the emergence of a reproductive barrier that isolates a subset of a population from the remainder of the population in the same area. Though rarer than allopatric speciation, sympatric speciation can occur when gene flow to and from the isolated subpopulation is blocked. This can occur as a result of polyploidy, a condition in which an organism has extra sets of chromosomes. Sympatric speciation also can occur when a subset of a population becomes reproductively isolated because of natural selection that results from a switch to a habitat or food source not used by the parent population. Finally, sympatric speciation can result from sexual selection.

Having reviewed the geographic context in which species originate, we'll next explore in more detail what can happen when new or partially formed species come into contact.

**CONCEPT CHECK 22.2**

1. Contrast allopatric and sympatric speciation. Which type of speciation is more common, and why?
2. **WHAT IF?** Is allopatric speciation more likely to occur on an island close to a mainland or on a more isolated island of the same size? Explain your prediction.
3. **MAKE CONNECTIONS** Review the process of meiosis in Figure 10.8. Describe how an error during meiosis could lead to polyploidy.

For suggested answers, see Appendix A.

**CONCEPT 22.3**

# Hybrid zones reveal factors that cause reproductive isolation

What happens if species with incomplete reproductive barriers come into contact with one another? One possible outcome is the formation of a **hybrid zone**, a region in which members of different species meet and mate, producing at least some offspring of mixed ancestry **(Figure 22.13)**. In this section, we'll

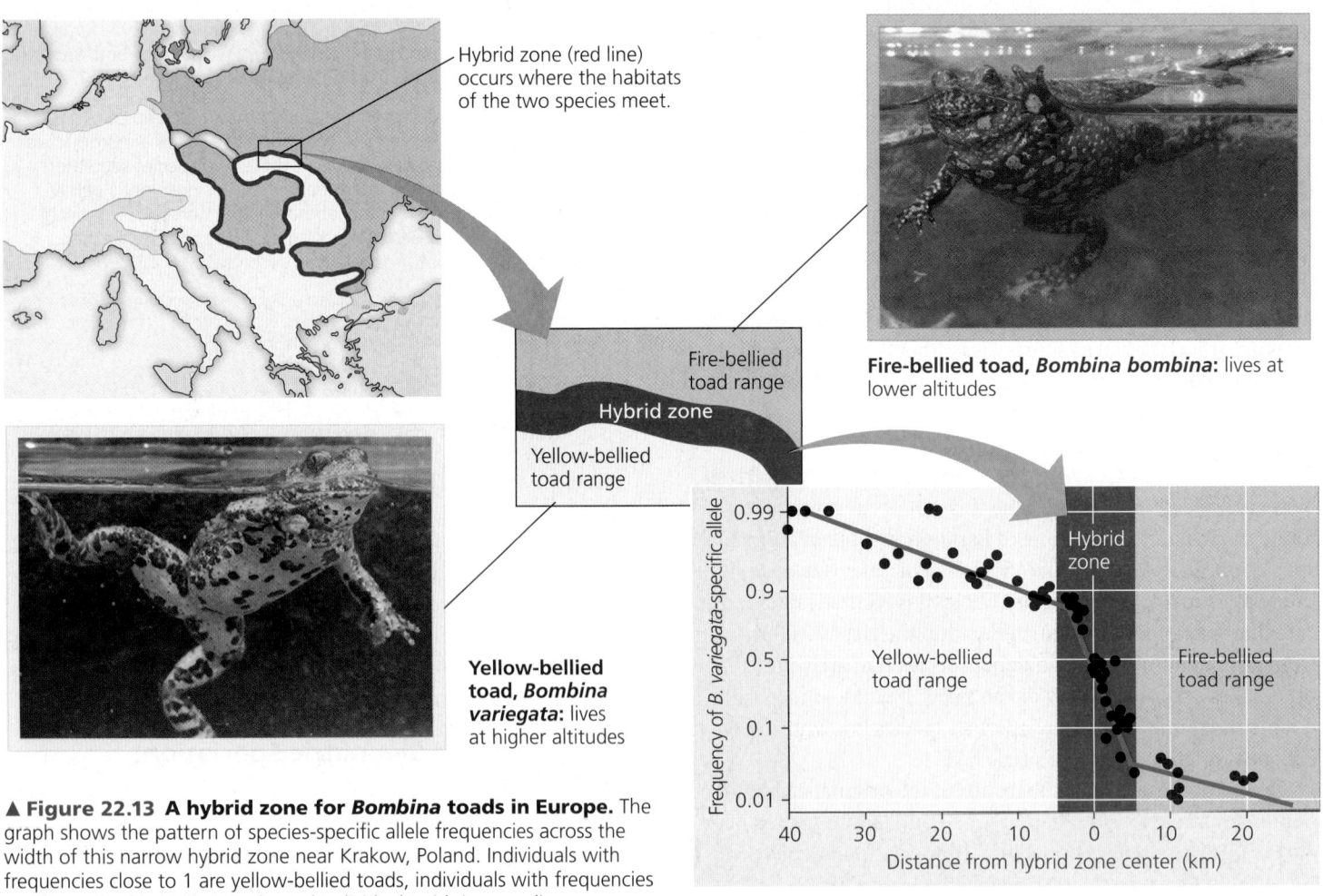

Hybrid zone (red line) occurs where the habitats of the two species meet.

Fire-bellied toad range

Hybrid zone

Yellow-bellied toad range

**Fire-bellied toad, *Bombina bombina*:** lives at lower altitudes

**Yellow-bellied toad, *Bombina variegata*:** lives at higher altitudes

▲ **Figure 22.13 A hybrid zone for *Bombina* toads in Europe.** The graph shows the pattern of species-specific allele frequencies across the width of this narrow hybrid zone near Krakow, Poland. Individuals with frequencies close to 1 are yellow-bellied toads, individuals with frequencies close to 0 are fire-bellied toads, and individuals with intermediate frequencies are considered hybrids.

**?** *Does the graph indicate that gene flow is spreading fire-bellied toad alleles into the range of the yellow-bellied toad? Explain.*

explore hybrid zones and what they reveal about factors that cause the evolution of reproductive isolation.

## Patterns Within Hybrid Zones

Some hybrid zones form as narrow bands, such as the one depicted in Figure 22.13 for the yellow-bellied toad (*Bombina variegata*) and its close relative, the fire-bellied toad (*B. bombina*). This hybrid zone, represented by the red line on the map, extends for 4,000 km but is less than 10 km wide in most places. The hybrid zone occurs where the higher-altitude habitat of the yellow-bellied toad meets the lowland habitat of the fire-bellied toad. Across a given "slice" of the zone, the frequency of alleles specific to yellow-bellied toads typically decreases from close to 100% at the edge where only yellow-bellied toads are found, to 50% in the central portion of the zone, to 0% at the edge where only fire-bellied toads are found.

What causes such a pattern of allele frequencies across a hybrid zone? We can infer that there is an obstacle to gene flow—otherwise, alleles from one parent species would also be common in the gene pool of the other parent species. Are geographic barriers reducing gene flow? Not in this case, since the toads can move throughout the hybrid zone. A more important factor is that hybrid toads have increased rates of embryonic mortality and a variety of morphological abnormalities, including ribs that are fused to the spine and malformed tadpole mouthparts. Because the hybrids have poor survival and reproduction, they produce few viable offspring with members of the parent species. As a result, hybrid individuals rarely serve as a stepping-stone from which alleles are passed from one species to the other. Outside the hybrid zone, additional obstacles to gene flow may be provided by natural selection in the different environments in which the parent species live.

Hybrid zones typically are located wherever the habitats of the interbreeding species meet. Those regions often resemble a group of isolated patches scattered across the landscape—more like the complex pattern of spots on a Dalmatian than the continuous band shown in Figure 22.13. But regardless of whether they have complex or simple spatial patterns, hybrid zones form when two species lacking complete barriers to reproduction come into contact. Once formed, how does a hybrid zone change over time?

## Hybrid Zones over Time

Studying a hybrid zone is like observing a naturally occurring experiment on speciation. Will the hybrids become reproductively isolated from their parents and form a new species, as occurred by polyploidy in the goatsbeard plant of the Pacific Northwest? If not, there are three possible outcomes for the hybrid zone over time: reinforcement of barriers, fusion of species, or stability **(Figure 22.14)**. We'll discuss each of these outcomes in turn.

- **Reinforcement:** Hybrids often are less fit than members of their parent species. In such cases, natural selection should strengthen prezygotic barriers to reproduction, reducing the formation of unfit hybrids. Because this process involves *reinforcing* reproductive barriers, it is called reinforcement. If reinforcement is occurring, a logical prediction is that barriers to reproduction between species should be stronger

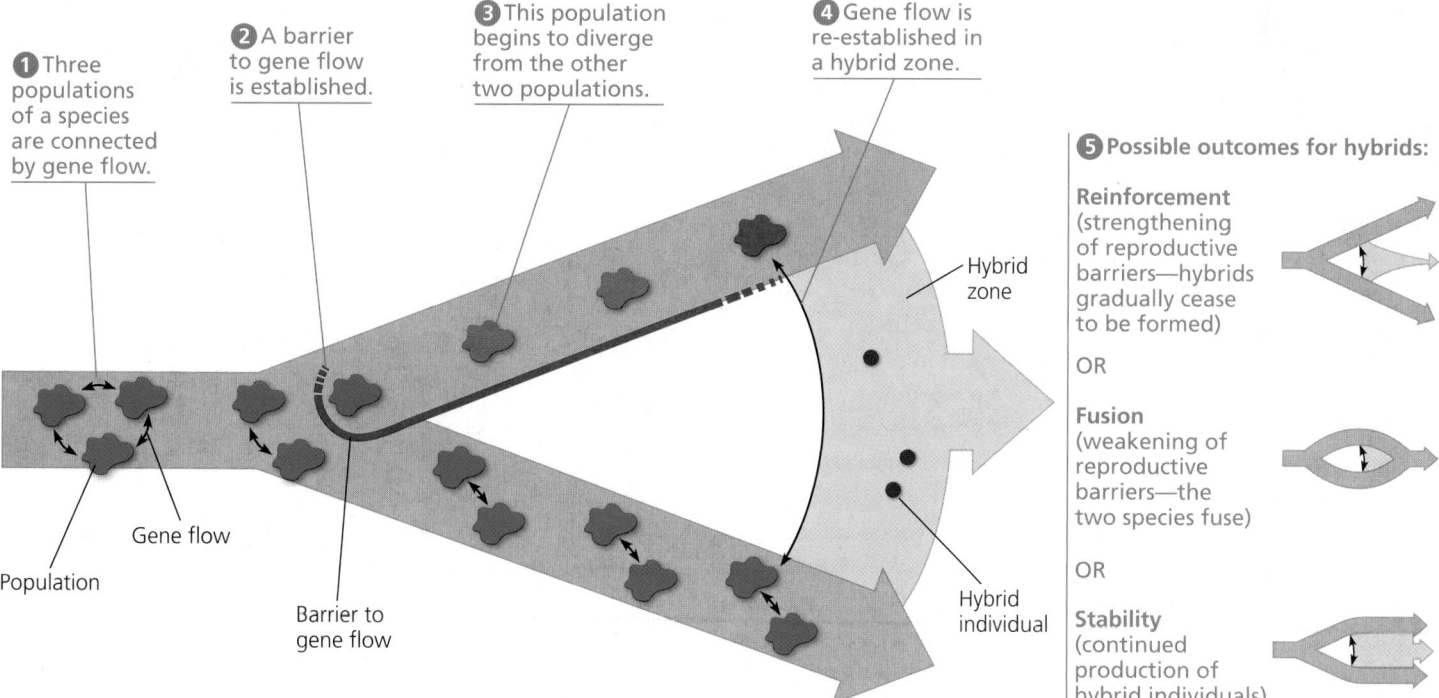

**1** Three populations of a species are connected by gene flow.

**2** A barrier to gene flow is established.

**3** This population begins to diverge from the other two populations.

**4** Gene flow is re-established in a hybrid zone.

**5** Possible outcomes for hybrids:

**Reinforcement** (strengthening of reproductive barriers—hybrids gradually cease to be formed)

OR

**Fusion** (weakening of reproductive barriers—the two species fuse)

OR

**Stability** (continued production of hybrid individuals)

Gene flow

Population

Barrier to gene flow

Hybrid zone

Hybrid individual

▲ **Figure 22.14 Formation of a hybrid zone and possible outcomes for hybrids over time.** The thick colored arrows represent the passage of time.

**WHAT IF?** *Predict what might happen if gene flow were re-established at step 3 in this process.*

*Pundamilia nyererei*

*Pundamilia pundamilia*

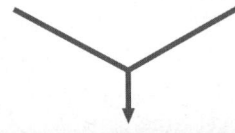

*Pundamilia "turbid water,"*
hybrid offspring from a location
with turbid water

▲ **Figure 22.15 Fusion: The breakdown of reproductive barriers.** Increasingly cloudy water in Lake Victoria over the past several decades may have weakened reproductive barriers between *P. nyererei* and *P. pundamilia*. In areas of cloudy water, the two species have hybridized extensively, causing their gene pools to fuse.

for sympatric populations than for allopatric populations. Evidence in support of this prediction has been observed in birds, fishes, insects, plants, and other organisms.

• **Fusion:** Barriers to reproduction may be weak when two species meet in a hybrid zone. Indeed, so much gene flow may occur that reproductive barriers weaken further and the gene pools of the two species become increasingly alike. In effect, the speciation process reverses, eventually causing the two hybridizing species to fuse into a single species. For example, genetic and morphological evidence indicate that the recent loss of the large tree finch from the Galápagos island of Floreana resulted from extensive hybridization with another finch species on that island. Such a situation also may be occurring among Lake Victoria cichlids. Many pairs of ecologically similar cichlid species are reproductively isolated because the females of one species prefer to mate with males of one color, while females of the other species prefer to mate with males of a different color (see Figure 22.12). Murky waters caused by pollution may have reduced the ability of females to use color to distinguish males of their own species from males of closely related species. In some polluted waters, many hybrids have been produced, leading to fusion of the parent species' gene pools and a loss of species **(Figure 22.15)**.

• **Stability:** Many hybrid zones are stable in the sense that hybrids continue to be produced. In some cases, this occurs

because the hybrids survive or reproduce better than members of either parent species, at least in certain habitats or years. But stable hybrid zones have also been observed in cases where the hybrids are selected *against*—an unexpected result. For example, hybrids continue to form in the *Bombina* hybrid zone even though they are strongly selected against. What could explain this finding? One possibility relates to the narrowness of the *Bombina* hybrid zone (see Figure 22.13). Evidence suggests that members of both parent species migrate into the zone from the parent populations located outside the zone, thus leading to the continued production of hybrids. If the hybrid zone were wider, this would be less likely to occur, since the center of the zone would receive little gene flow from distant parent populations located outside the hybrid zone.

Sometimes the outcomes in hybrid zones match our predictions (cichlid fishes), and sometimes they don't (*Bombina*). But whether our predictions are upheld or not, events in hybrid zones can shed light on how barriers to reproduction between closely related species change over time. In the next section, we'll examine how interactions between hybridizing species can also provide a glimpse into the speed and genetic control of speciation.

**CONCEPT CHECK 22.3**

1. What are hybrid zones, and why can they be viewed as "natural laboratories" in which to study speciation?

2. **WHAT IF?** Consider two species that diverged while geographically separated but resumed contact before reproductive isolation was complete. Predict what would happen over time if the two species mated indiscriminately and (a) hybrid offspring survived and reproduced more poorly than offspring from intraspecific matings or (b) hybrid offspring survived and reproduced as well as offspring from intraspecific matings.

For suggested answers, see Appendix A.

**CONCEPT 22.4**

# Speciation can occur rapidly or slowly and can result from changes in few or many genes

Darwin faced many questions when he began to ponder that "mystery of mysteries"—speciation. He found answers to some of those questions when he realized that evolution by natural selection helps explain both the diversity of life and the adaptations of organisms. But biologists since Darwin have continued to ask fundamental questions about speciation. For example, how long does it take for new species to form? And how many genes change when one species splits into two? Answers to these questions are also emerging.

**(a)** In a punctuated model, new species change most as they branch from a parent species and then change little for the rest of their existence.

Time ⟶

**(b)** In a gradual model, species diverge from one another more slowly and steadily over time.

▲ **Figure 22.16 Two models for the tempo of speciation.**

## The Time Course of Speciation

We can gather information about how long it takes new species to form from broad patterns in the fossil record and from studies that use morphological data (including fossils) or molecular data to assess the time interval between speciation events in particular groups of organisms.

### Patterns in the Fossil Record

The fossil record includes many episodes in which new species appear suddenly in a geologic stratum, persist essentially unchanged through several strata, and then disappear. For example, there are dozens of species of marine invertebrates that make their debut in the fossil record with novel morphologies, but then change little for millions of years before becoming extinct. Paleontologists Niles Eldredge and Stephen Jay Gould coined the term **punctuated equilibria** to describe these periods of apparent stasis punctuated by sudden change **(Figure 22.16a)**. Other species do not show a punctuated pattern; instead, they appear to have changed more gradually over long periods of time **(Figure 22.16b)**.

What might punctuated and gradual patterns tell us about how long it takes new species to form? Suppose that a species survived for 5 million years, but most of the morphological changes that caused it to be designated a new species occurred during the first 50,000 years of its existence—just 1% of its total lifetime. Time periods this short (in geologic terms) often cannot be distinguished in fossil strata, in part because the rate of sediment accumulation may be too slow to separate layers this close in time. Thus, based on its fossils, the species would seem to have appeared suddenly and then lingered with little or no change before becoming extinct. Even though such a species may have originated more slowly than its fossils suggest (in this case taking up to 50,000 years), a punctuated pattern indicates that speciation occurred relatively rapidly. For species whose fossils changed much more gradually, we also cannot tell exactly when a new biological

species formed, since information about reproductive isolation does not fossilize. However, it is likely that speciation in such groups occurred relatively slowly, perhaps taking millions of years.

### Speciation Rates

The existence of fossils that display a punctuated pattern suggests that once the process of speciation begins, it can be completed relatively rapidly—a suggestion supported by a growing number of studies. For example, rapid speciation appears to have produced the wild sunflower *Helianthus anomalus*. Genetic evidence indicates that this species originated by the hybridization of two other sunflower species, *H. annuus* and *H. petiolaris*. The hybrid species *H. anomalus* is ecologically distinct and reproductively isolated from both parent species **(Figure 22.17)**.

▲ **Figure 22.17 A hybrid sunflower species and its dry sand dune habitat.** The wild sunflower *Helianthus anomalus* shown here originated via the hybridization of two other sunflowers, *H. annuus* and *H. petiolaris*, which live in nearby but moister environments.

Unlike the outcome of allopolyploid speciation, in which there is a change in chromosome number after hybridization, in these sunflowers the two parent species and the hybrid all have the same number of chromosomes ($2n = 34$). How, then, did speciation occur? To study this question, researchers performed an experiment designed to mimic events in nature: They crossed the two parent species and followed the fate of the hybrid offspring over several generations (**Figure 22.18**). Their results indicated that natural selection could produce extensive genetic changes in hybrid populations over short periods of time. These changes appear to have caused the hybrids to diverge reproductively from their parents and form a new species, *H. anomalus*.

The sunflower example, along with the apple maggot fly, Lake Victoria cichlid, and fruit fly examples discussed earlier, suggests that new species can arise rapidly *once divergence begins*. But what is the total length of time between speciation events? This interval consists of the time that elapses before populations of a newly formed species start to diverge from one another plus the time it takes for speciation to be complete once divergence begins. It turns out that the total time between speciation events varies considerably. In a survey of data from 84 groups of plants and animals, speciation intervals ranged from 4,000 years (in cichlids of Lake Nabugabo, Uganda) to 40 million years (in some beetles). Overall, the time between speciation events averaged 6.5 million years and was rarely less than 500,000 years.

These data suggest that on average, millions of years may pass before a newly formed plant or animal species will itself give rise to another new species. As we'll see in Concept 23.2, this finding has implications for how long it takes life on Earth to recover from mass extinction events. Moreover, the extreme variability in the time it takes new species to form indicates that organisms do not have an internal "speciation clock" that causes them to produce new species at regular intervals. Instead, speciation begins only after gene flow between populations is interrupted, perhaps by changing environmental conditions or by unpredictable events, such as a storm that transports a few individuals to a new area. Furthermore, once gene flow is interrupted, the populations must diverge genetically to such an extent that they become reproductively isolated—all before other events cause gene flow to resume, possibly reversing the speciation process (see Figure 22.14).

## Studying the Genetics of Speciation

Studies of ongoing speciation (as in hybrid zones) can reveal traits that cause reproductive isolation. By identifying the genes that control those traits, scientists can explore a fundamental question of evolutionary biology: How many genes influence the formation of new species?

In some cases, the evolution of reproductive isolation results from the effects of a single gene. For example, in Japanese snails of the genus *Euhadra*, a change in a single gene results in a mechanical barrier to reproduction. This gene controls the direction in which the shells spiral. When their shells spiral in different directions, the snails' genitalia are

▼ **Figure 22.18** **Inquiry**

## How does hybridization lead to speciation in sunflowers?

**Experiment** Researchers crossed the two parent sunflower species, *H. annuus* and *H. petiolaris*, to produce experimental hybrids in the laboratory (for each gamete, only two of the $n = 17$ chromosomes are shown).

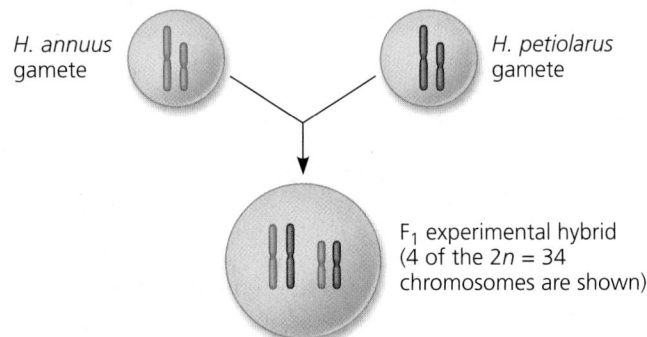

*H. annuus* gamete

*H. petiolarus* gamete

$F_1$ experimental hybrid (4 of the $2n = 34$ chromosomes are shown)

Note that in the first ($F_1$) generation, each chromosome of the experimental hybrids consisted entirely of DNA from one or the other parent species. The researchers then tested whether the $F_1$ and subsequent generations of experimental hybrids were fertile. They also used species-specific genetic markers to compare the chromosomes in the experimental hybrids with the chromosomes in the naturally occurring hybrid *H. anomalus*.

**Results** Although only 5% of the $F_1$ experimental hybrids were fertile, after just four more generations the hybrid fertility rose to more than 90%. The chromosomes of individuals from this fifth hybrid generation differed from those in the $F_1$ generation (see above) but were similar to those in *H. anomalus* individuals from natural populations:

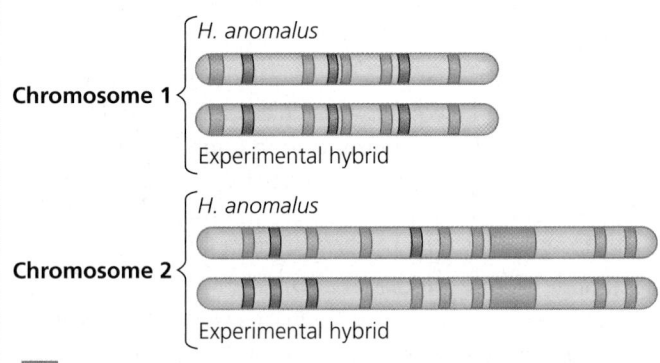

**Chromosome 1**
*H. anomalus*
Experimental hybrid

**Chromosome 2**
*H. anomalus*
Experimental hybrid

Comparison region containing *H. annuus*–specific marker

Comparison region containing *H. petiolarus*–specific marker

**Conclusion** Over time, the chromosomes in the population of experimental hybrids became similar to the chromosomes of *H. anomalus* individuals from natural populations. This suggests that the observed rise in the fertility of the experimental hybrids may have occurred as selection eliminated regions of DNA from the parent species that were not compatible with one another. Overall, it appeared that the initial steps of the speciation process occurred rapidly and could be mimicked in a laboratory experiment.

**Data from** L. H. Rieseberg et al., Role of gene interactions in hybrid speciation: Evidence from ancient and experimental hybrids, *Science* 272:741–745 (1996).

**WHAT IF?** The increased fertility of the experimental hybrids could have resulted from natural selection for thriving under laboratory conditions. Evaluate this alternative explanation for the result.

oriented in a manner that prevents mating (Figure 22.3f shows a similar example). Recent genetic analyses have uncovered other single genes that cause reproductive isolation in fruit flies or mice.

A major barrier to reproduction between two closely related species of monkey flower, *Mimulus cardinalis* and *M. lewisii*, also appears to be influenced by a relatively small number of genes. These two species are isolated by several prezygotic and postzygotic barriers. Of these, one prezygotic barrier, pollinator choice, accounts for most of the isolation: In a hybrid zone between *M. cardinalis* and *M. lewisii*, nearly 98% of pollinator visits were restricted to one species or the other.

The two monkey flower species are visited by different pollinators: Hummingbirds prefer the red-flowered *M. cardinalis*, and bumblebees prefer the pink-flowered *M. lewisii*.

**(a) Typical *Mimulus lewisii***

**(b) *M. lewisii* with an *M. cardinalis* flower-color allele**

**(c) Typical *Mimulus cardinalis***

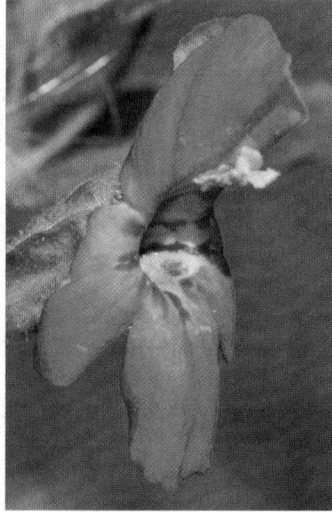

**(d) *M. cardinalis* with an *M. lewisii* flower-color allele**

▲ **Figure 22.19 A locus that influences pollinator choice.** Pollinator preferences provide a strong barrier to reproduction between *Mimulus lewisii* and *M. cardinalis*. After transferring the *M. lewisii* allele for a flower-color locus into *M. cardinalis* and vice versa, researchers observed a shift in some pollinators' preferences.

**WHAT IF?** *If* M. cardinalis *individuals that had the* M. lewisii yup *allele were planted in an area that housed both monkey flower species, how might the production of hybrid offspring be affected?*

Pollinator choice is affected by at least two loci in the monkey flowers, one of which, the "yellow upper," or *yup*, locus, influences flower color **(Figure 22.19)**. By crossing the two parent species to produce F$_1$ hybrids and then performing repeated backcrosses of these F$_1$ hybrids to each parent species, researchers succeeded in transferring the *M. cardinalis* allele at this locus into *M. lewisii*, and vice versa. In a field experiment, *M. lewisii* plants with the *M. cardinalis yup* allele received 68-fold more visits from hummingbirds than did wild-type *M. lewisii*. Similarly, *M. cardinalis* plants with the *M. lewisii yup* allele received 74-fold more visits from bumblebees than did wild-type *M. cardinalis*. Thus, a mutation at a single locus can influence pollinator preference and hence contribute to reproductive isolation in monkey flowers.

In other organisms, the speciation process is influenced by larger numbers of genes and gene interactions. For example, hybrid sterility between two subspecies of the fruit fly *Drosophila pseudoobscura* results from gene interactions among at least four loci, and postzygotic isolation in the sunflower hybrid zone discussed earlier is influenced by at least 26 chromosome segments (and an unknown number of genes). Overall, studies suggest that few or many genes can influence the evolution of reproductive isolation and hence the emergence of a new species.

## From Speciation to Macroevolution

As you've seen, speciation may begin with differences as small as the color on a cichlid's back. However, as speciation occurs again and again, such differences can accumulate and become more pronounced, eventually leading to the formation of new groups of organisms that differ greatly from their ancestors (as in the origin of whales from terrestrial mammals; see Figure 19.20). Moreover, as one group of organisms increases in size by producing many new species, another group of organisms may shrink, losing species to extinction. The cumulative effects of many such speciation and extinction events have helped shape the sweeping evolutionary changes that are documented in the fossil record. In the next chapter, we turn to such large-scale evolutionary changes as we begin our study of macroevolution.

**CONCEPT CHECK 22.4**

1. Speciation can occur rapidly between diverging populations, yet the time between speciation events is often more than a million years. Explain this apparent contradiction.

2. Summarize evidence that the *yup* flower-color locus acts as a prezygotic barrier to reproduction in two species of monkey flowers. Do these results demonstrate that the *yup* locus alone controls barriers to reproduction between these species? Explain.

3. **MAKE CONNECTIONS** Compare Figure 10.11 with Figure 22.18. What cellular process could cause the hybrid chromosomes in Figure 22.18 to contain DNA from both parent species? Explain.

For suggested answers, see Appendix A.

# 22 Chapter Review

Go to **MasteringBiology®** for Assignments, the eText, and the Study Area with Animations, Activities, Vocab Self-Quiz, and Practice Tests.

## SUMMARY OF KEY CONCEPTS

### CONCEPT 22.1

**The biological species concept emphasizes reproductive isolation (pp. 434–438)**

- A biological **species** is a group of populations whose individuals may interbreed and produce viable, fertile offspring with each other but not with members of other species. The **biological species concept** emphasizes reproductive isolation through prezygotic and postzygotic barriers that separate gene pools.
- Although helpful in thinking about how speciation occurs, the biological species concept has limitations. For instance, it cannot be applied to organisms known only as fossils or to organisms that reproduce only asexually. Thus, scientists use other species concepts, such as the **morphological species concept**, in certain circumstances.

❓ *Explain the role of gene flow in the biological species concept.*

### CONCEPT 22.2

**Speciation can take place with or without geographic separation (pp. 439–444)**

- In **allopatric speciation**, gene flow is reduced when two populations of one species become geographically separated from each other. One or both populations may undergo evolutionary change during the period of separation, resulting in the establishment of prezygotic or postzygotic barriers to reproduction.
- In **sympatric speciation**, a new species originates while remaining in the same geographic area as the parent species. Plant species (and, more rarely, animal species) have evolved sympatrically through **polyploidy**. Sympatric speciation can also result from habitat shifts and sexual selection.

Original population

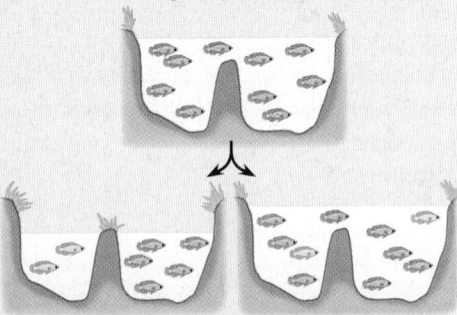

Allopatric speciation      Sympatric speciation

❓ *Can factors that cause sympatric speciation also cause allopatric speciation? Explain.*

### CONCEPT 22.3

**Hybrid zones reveal factors that cause reproductive isolation (pp. 444–446)**

- Many groups of organisms form **hybrid zones** in which members of different species meet and mate, producing at least some offspring of mixed ancestry.

- Many hybrid zones exhibit **stability** in that hybrid offspring continue to be produced over time. In others, **reinforcement** strengthens prezygotic barriers to reproduction, thus decreasing the formation of unfit hybrids. In still other hybrid zones, barriers to reproduction may weaken over time, resulting in the **fusion** of the species' gene pools (reversing the speciation process).

❓ *What factors can support the long-term stability of a hybrid zone if the parent species live in different environments?*

### CONCEPT 22.4

**Speciation can occur rapidly or slowly and can result from changes in few or many genes (pp. 446–449)**

- New species can form rapidly once divergence begins—but it can take millions of years for that to happen. The time interval between speciation events varies considerably, from a few thousand years to tens of millions of years.
- New developments in genetics have enabled researchers to identify specific genes involved in some cases of speciation. Results show that speciation can be driven by few or many genes.

❓ *Is speciation something that happened only in the distant past, or are new species continuing to arise today? Explain.*

## TEST YOUR UNDERSTANDING

### Level 1: Knowledge/Comprehension

1. The *largest* unit within which gene flow can readily occur is a
   (A) population.          (C) genus.
   (B) species.             (D) hybrid.

2. Males of different species of the fruit fly *Drosophila* that live in the same parts of the Hawaiian Islands have different elaborate courtship rituals. These rituals involve fighting other males and making stylized movements that attract females. What type of reproductive isolation does this represent?
   (A) habitat isolation    (C) behavioral isolation
   (B) temporal isolation   (D) gametic isolation

3. According to the punctuated equilibria model,
   (A) natural selection is unimportant as a mechanism of evolution.
   (B) given enough time, most existing species will branch gradually into new species.
   (C) most evolution occurs in sympatric populations.
   (D) most new species accumulate their unique features relatively rapidly as they come into existence, then change little for the rest of their duration as a species.

### Level 2: Application/Analysis

4. Bird guides once listed the myrtle warbler and Audubon's warbler as distinct species. Recently, these birds were reclassified as eastern and western forms of a single species, the yellow-rumped warbler. Which of the following pieces of evidence, if true, would be cause for this reclassification?
   (A) The two forms interbreed often in nature, and their offspring survive and reproduce well.
   (B) The two forms live in similar habitats and have similar food requirements.
   (C) The two forms have many genes in common.
   (D) The two forms are very similar in appearance.

**5.** Which of the following factors would *not* contribute to allopatric speciation?

(A) A population becomes geographically isolated from the parent population.

(B) The separated population is small, and genetic drift occurs.

(C) The isolated population is exposed to different selection pressures than the ancestral population.

(D) Gene flow between the two populations is extensive.

**6.** Plant species A has a diploid number of 12. Plant species B has a diploid number of 16. A new species, C, arises as an allopolyploid from A and B. The diploid number for species C would probably be

(A) 14.　　　(B) 16.　　　(C) 28.　　　(D) 56.

## Level 3: Synthesis/Evaluation

**7. SCIENTIFIC INQUIRY**

**DRAW IT** In this chapter, you read that bread wheat (*Triticum aestivum*) is an allohexaploid, containing two sets of chromosomes from each of three different parent species. Genetic analysis suggests that the three species pictured following this question each contributed chromosome sets to *T. aestivum*. (The capital letters here represent sets of chromosomes rather than individual genes.) Evidence also indicates that the first polyploidy event was a spontaneous hybridization of the early cultivated wheat species *T. monococcum* and a wild *Triticum* grass species. Based on this information, draw a diagram of one possible chain of events that could have produced the allohexaploid *T. aestivum*.

**Ancestral species:**

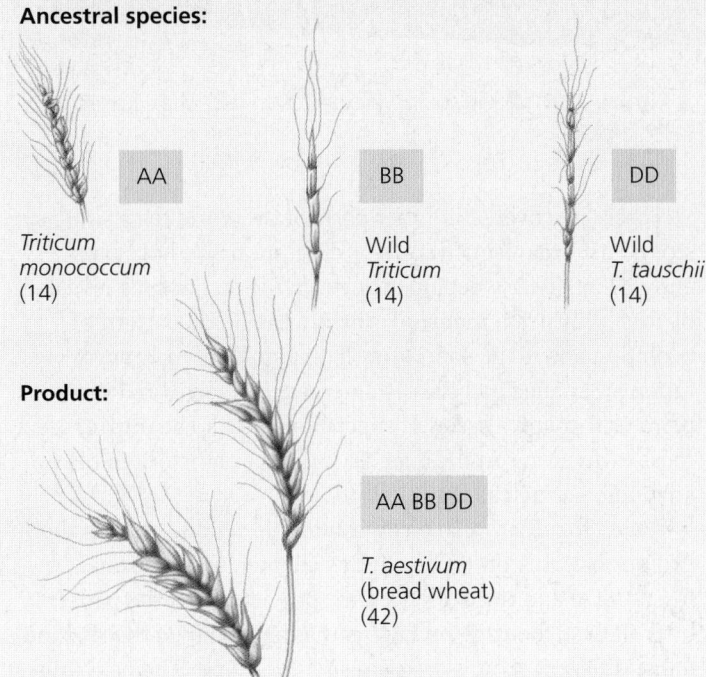

Triticum
monococcum
(14)　　　AA

Wild
Triticum
(14)　　　BB

Wild
T. tauschii
(14)　　　DD

**Product:**

AA BB DD

T. aestivum
(bread wheat)
(42)

**8. SCIENCE, TECHNOLOGY, AND SOCIETY**

In the United States, the rare red wolf (*Canis lupus*) has been known to hybridize with coyotes (*Canis latrans*), which are much more numerous. Although red wolves and coyotes differ in terms of morphology, DNA, and behavior, genetic evidence suggests that living red wolf individuals are actually hybrids. Red wolves are designated as an endangered species and hence receive legal protection under the Endangered Species Act. Some people think that their endangered status should be withdrawn because the remaining red wolves are hybrids, not members of a "pure" species. Do you agree? Why or why not?

**9. FOCUS ON EVOLUTION**

Explain the biological basis for assigning all human populations to a single species. Can you think of a scenario by which a second human species could originate in the future?

**10. FOCUS ON INFORMATION**

In sexually reproducing species, each individual begins life with DNA inherited from both parent organisms. In a short essay (100–150 words), apply this idea to what occurs when organisms of two species that have homologous chromosomes mate and produce ($F_1$) hybrid offspring. What percentage of the DNA in the $F_1$ hybrids' chromosomes comes from each parent species? As the hybrids mate and produce $F_2$ and later-generation hybrid offspring, describe how recombination and natural selection may affect whether the DNA in hybrid chromosomes is derived from one parent species or the other.

**11.** **SYNTHESIZE YOUR KNOWLEDGE**

Suppose that females of one population of strawberry poison dart frogs (*Dendrobates pumilio*) prefer to mate with males that have a bright red and black coloration. In a different population, the females prefer males with yellow skin. Propose a hypothesis to explain how such differences could have arisen in allopatric versus sympatric populations.

*For selected answers, see Appendix A.*

# 23 Broad Patterns of Evolution

## KEY CONCEPTS

**23.1** The fossil record documents life's history

**23.2** The rise and fall of groups of organisms reflect differences in speciation and extinction rates

**23.3** Major changes in body form can result from changes in the sequences and regulation of developmental genes

**23.4** Evolution is not goal oriented

▲ **Figure 23.1** Would you have expected to find whale bones buried here?

## A Surprise in the Desert

With its dry, wind-sculpted sands and searing heat, the Sahara Desert seems an unlikely place to discover the bones of whales. But starting in the 1870s, researchers uncovered fossils of ancient whales at several locations near the scene shown in **Figure 23.1**. For example, a nearly complete skeleton of *Dorudon atrox*, an extinct whale that lived 35 million years ago **(Figure 23.2)**, was discovered in a region that came to be called Wadi Hitan, the "Valley of Whales." Collectively, the whale fossils found in the Sahara were spectacular not only for where they were found, but also for documenting early steps in the transition from life on land to life in the sea.

▼ **Figure 23.2 Fossil of *Dorudon atrox*, an ancient whale.**

Fossils discovered in other parts of the world tell a similar story: Past organisms were very different from those presently living. The sweeping changes in life on Earth as revealed by fossils illustrate **macroevolution**, the broad pattern of evolution above the species level. Examples of macroevolutionary change include the emergence of terrestrial vertebrates through a series of speciation events, the impact of mass extinctions on the diversity of life, and the origin of key adaptations, such as flight in birds.

Taken together, such changes provide a grand view of the evolutionary history of life. In this chapter, we'll examine how fossils form and the evidence they provide about the pattern of evolution, focusing on factors that have shaped the rise and fall of different groups of organisms over time. The next unit (Chapters 24–27) will explore major steps in the history of life.

CONCEPT 23.1

## The fossil record documents life's history

Starting with the earliest traces of life, the fossil record opens a window into the world of long ago and provides glimpses of the evolution of life over billions of years **(Figure 23.3)**. In this

▼ **Figure 23.3 Documenting the history of life.** These fossils illustrate representative organisms from different points in time. Although prokaryotes and unicellular eukaryotes are shown only at the base of the diagram, these organisms continue to thrive today. In fact, most organisms on Earth are unicellular.

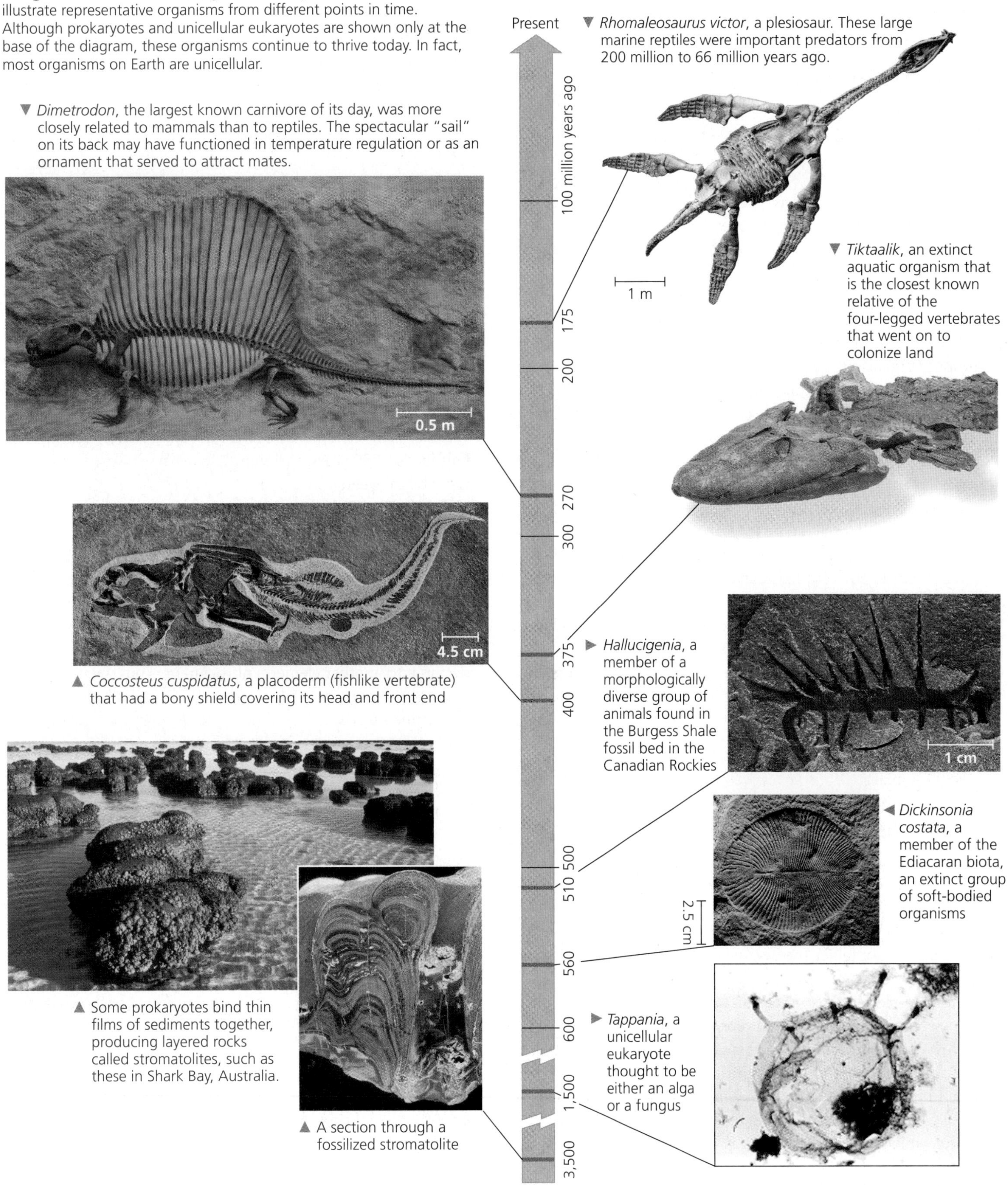

▼ *Dimetrodon*, the largest known carnivore of its day, was more closely related to mammals than to reptiles. The spectacular "sail" on its back may have functioned in temperature regulation or as an ornament that served to attract mates.

0.5 m

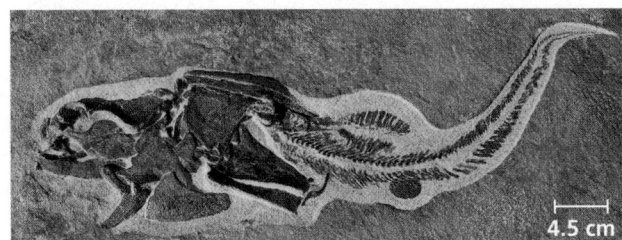

4.5 cm

▲ *Coccosteus cuspidatus*, a placoderm (fishlike vertebrate) that had a bony shield covering its head and front end

▲ Some prokaryotes bind thin films of sediments together, producing layered rocks called stromatolites, such as these in Shark Bay, Australia.

▲ A section through a fossilized stromatolite

Present

100 million years ago
175
200
270
300
375
400
500
510
560
600
1,500
3,500

▼ *Rhomaleosaurus victor*, a plesiosaur. These large marine reptiles were important predators from 200 million to 66 million years ago.

1 m

▼ *Tiktaalik*, an extinct aquatic organism that is the closest known relative of the four-legged vertebrates that went on to colonize land

▶ *Hallucigenia*, a member of a morphologically diverse group of animals found in the Burgess Shale fossil bed in the Canadian Rockies

1 cm

◀ *Dickinsonia costata*, a member of the Ediacaran biota, an extinct group of soft-bodied organisms

2.5 cm

▶ *Tappania*, a unicellular eukaryote thought to be either an alga or a fungus

section, we'll examine fossils as a form of scientific evidence: how fossils form, how scientists date and interpret them, and what they can and cannot tell us about changes in the history of life.

## The Fossil Record

Sedimentary rocks are the richest source of fossils. As a result, the fossil record is based primarily on the sequence in which fossils have accumulated in sedimentary rock layers, called *strata* (see Figure 19.3). Useful information is also provided by other types of fossils, such as insects preserved in amber (fossilized tree sap) and mammals frozen in ice.

The fossil record shows that there have been great changes in the kinds of organisms on Earth at different points in time. Many past organisms were unlike organisms living today, and many organisms that once were common are now extinct. As we'll see later in this section, fossils also document how new groups of organisms arose from previously existing ones.

As substantial and significant as the fossil record is, keep in mind that it is an incomplete chronicle of evolutionary change. Many of Earth's organisms did not die in the right place and time to be preserved as fossils. Of those fossils that were formed, many were destroyed by later geologic processes, and only a fraction of the others have been discovered. As a result, the known fossil record is biased in favor of species that existed for a long time, were abundant and widespread in certain kinds of environments, and had hard shells, skeletons, or other parts that facilitated their fossilization. Even with its limitations, however, the fossil record is a remarkably detailed account of biological change over the vast scale of geologic time. Furthermore, as shown by the recently unearthed fossils of whale ancestors with hind limbs (see Figure 23.1 and Concept 19.3), gaps in the fossil record continue to be filled by new discoveries. Although some of these discoveries are fortuitous, others illustrate the predictive nature of paleontology (see Figure 27.25).

## How Rocks and Fossils Are Dated

Fossils are valuable data for reconstructing the history of life, but only if we can determine where they fit in that unfolding story. While the order of fossils in rock strata tells us the sequence in which the fossils were laid down—their relative ages—it does not tell us their actual (absolute) ages. Examining the relative positions of fossils is like peeling off layers of wallpaper in an old house. You can infer the sequence in which the layers were applied, but not the year each layer was added.

How can we determine the absolute age of a fossil? (Note that "absolute" does not mean errorless, but that an age is given in years rather than relative terms such as *before* and *after*.) One of the most common techniques is **radiometric dating**, which is based on the decay of radioactive isotopes (see Concept 2.2). In this process, a radioactive "parent" isotope decays to a "daughter" isotope at a characteristic rate. The rate of decay is expressed by the **half-life**, the time required for 50% of the parent isotope to decay **(Figure 23.4)**. Each type

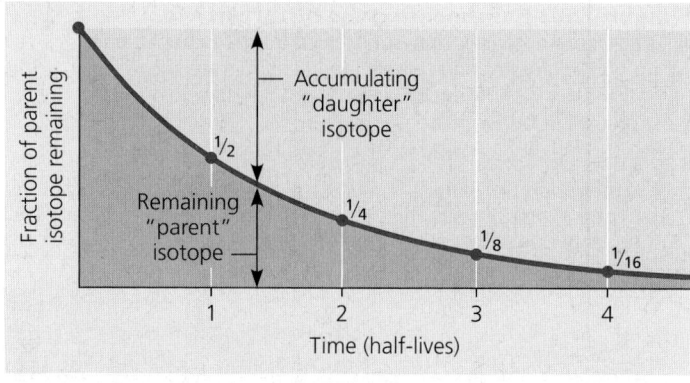

▲ **Figure 23.4 Radiometric dating.** In this diagram, each unit of time represents one half-life of a radioactive isotope.

**DRAW IT** *Relabel the x-axis of this graph in years to illustrate the radioactive decay of uranium-238 (half-life = 4.5 billion years).*

of radioactive isotope has a characteristic half-life, which is not affected by temperature, pressure, or other environmental variables. For example, carbon-14 decays relatively quickly; its half-life is 5,730 years. Uranium-238 decays slowly; its half-life is 4.5 billion years.

Fossils contain isotopes of elements that accumulated in the organisms when they were alive. For example, a living organism contains the most common carbon isotope, carbon-12, as well as a radioactive isotope, carbon-14. When the organism dies, it stops accumulating carbon, and the amount of carbon-12 in its tissues does not change over time. However, the carbon-14 that it contains at the time of death slowly decays into another element, nitrogen-14. Thus, by measuring the ratio of carbon-14 to carbon-12 in a fossil, we can determine the fossil's age. This method works for fossils up to about 75,000 years old; fossils older than that contain too little carbon-14 to be detected with current techniques. Radioactive isotopes with longer half-lives are used to date older fossils.

Determining the age of these older fossils in sedimentary rocks is challenging. Organisms do not use radioisotopes with long half-lives, such as uranium-238, to build their bones or shells. In addition, the sedimentary rocks themselves tend to consist of sediments of differing ages. So while we may not be able to date these older fossils directly, an indirect method can be used to infer the age of fossils that are sandwiched between two layers of volcanic rock. As lava cools into volcanic rock, radioisotopes from the surrounding environment become trapped in the newly formed rock. Some of the trapped radioisotopes have long half-lives, allowing geologists to estimate the ages of ancient volcanic rocks. If two volcanic layers surrounding fossils are determined to be 525 million and 535 million years old, for example, then the fossils are roughly 530 million years old.

## Fossils Frame the Geologic Record

The study of fossils has helped geologists establish a **geologic record**, a standard time scale that divides Earth's history into four eons and further subdivisions **(Table 23.1)**. The first three

**Table 23.1** **The Geologic Record**

| Relative Duration of Eons | Era | Period | Epoch | Age (Millions of Years Ago) | Some Important Events in the History of Life |
|---|---|---|---|---|---|
| Phanerozoic | Cenozoic | Quaternary | Holocene | | Historical time |
| | | | | 0.01 | |
| | | | Pleistocene | | Ice ages; origin of genus *Homo* |
| | | | | 2.6 | |
| | | Neogene | Pliocene | | Appearance of bipedal human ancestors |
| | | | | 5.3 | |
| | | | Miocene | | Continued radiation of mammals and angiosperms; earliest direct human ancestors |
| | | | | 23 | |
| | | Paleogene | Oligocene | | Origins of many primate groups |
| | | | | 33.9 | |
| | | | Eocene | | Angiosperm dominance increases; continued radiation of most present-day mammalian orders |
| | | | | 56 | |
| | | | Paleocene | | Major radiation of mammals, birds, and pollinating insects |
| | | | | 66 | |
| Proterozoic | Mesozoic | Cretaceous | | | Flowering plants (angiosperms) appear and diversify; many groups of organisms, including most dinosaurs, become extinct at end of period |
| | | | | 145 | |
| | | Jurassic | | | Gymnosperms continue as dominant plants; dinosaurs abundant and diverse |
| | | | | 201 | |
| | | Triassic | | | Cone-bearing plants (gymnosperms) dominate landscape; dinosaurs evolve and radiate; origin of mammals |
| | | | | 252 | |
| | Paleozoic | Permian | | | Radiation of reptiles; origin of most present-day groups of insects; extinction of many marine and terrestrial organisms at end of period |
| | | | | 299 | |
| | | Carboniferous | | | Extensive forests of vascular plants form; first seed plants appear; origin of reptiles; amphibians dominant |
| | | | | 359 | |
| | | Devonian | | | Diversification of bony fishes; first tetrapods and insects appear |
| | | | | 419 | |
| | | Silurian | | | Diversification of early vascular plants |
| | | | | 443 | |
| | | Ordovician | | | Marine algae abundant; colonization of land by diverse fungi, plants, and animals |
| | | | | 485 | |
| | | Cambrian | | | Sudden increase in diversity of many animal phyla (Cambrian explosion) |
| | | | | 541 | |
| Archaean | | Ediacaran | | | Diverse algae and soft-bodied invertebrate animals appear |
| | | | | 635 | |
| | | | | 1,800 | Oldest fossils of eukaryotic cells appear |
| | | | | 2,500 | |
| | | | | 2,700 | Concentration of atmospheric oxygen begins to increase |
| | | | | 3,500 | Oldest fossils of cells (prokaryotes) appear |
| Hadean | | | | 4,000 | Oldest known rocks on Earth's surface |
| | | | | Approx. 4,600 | Origin of Earth |

eons—the Hadean, Archaean, and Proterozoic—together lasted about 4 billion years. The Phanerozoic eon, roughly the last half billion years, encompasses most of the time that animals have existed on Earth. It is divided into three eras: the Paleozoic, Mesozoic, and Cenozoic. Each era represents a distinct age in the history of Earth and its life. For example, the Mesozoic era is sometimes called the "age of reptiles" because of its abundance of reptilian fossils, including those of dinosaurs. The boundaries between the eras correspond to major extinction events seen in the fossil record, when many forms of life disappeared and were replaced by forms that evolved from the survivors.

The earliest direct evidence of life comes from the Archaean eon, based on 3.5 billion-year-old fossils of stromatolites (see Figure 23.3). **Stromatolites** are layered rocks that form when certain prokaryotes bind thin films of sediment together. These and other early prokaryotes were Earth's sole inhabitants for about 1.5 billion years. Early prokaryotes included species that transformed life on our planet by releasing oxygen to the atmosphere during the water-splitting step of photosynthesis (see Concept 24.1).

The ensuing increase in atmospheric oxygen—a process that began about 2.4 billion years ago—led to the extinction of some organisms and the proliferation of others. One group that flourished was the eukaryotes, which originated about 1.8 billion years ago. The rise of the eukaryotes was associated with a series of other key events in the history of life, including the origin of multicellular organisms and the colonization of land (see Concepts 25.2 and 25.3). Fossil evidence and molecular clock estimates based on DNA sequence data suggest that simple multicellular organisms emerged about 1.5 billion years ago. Later, more complex multicellular organisms arose independently in several groups of eukaryotes, including those that eventually moved onto land: plants, fungi, and animals (see Chapters 26 and 27).

### The Origin of New Groups of Organisms

Some fossils provide a detailed look at the origin of new groups of organisms. Such fossils are central to our understanding of evolution; they illustrate how new features arise and how long it takes for such changes to occur. We'll examine one such case here: the origin of mammals.

Along with amphibians and reptiles, mammals belong to the group of animals called *tetrapods* (from the Greek *tetra*, four, and *pod*, foot), named for having four limbs. Mammals have a number of unique anatomical features that fossilize readily, allowing scientists to trace their origin. For example, the lower jaw is composed of one bone (the dentary) in mammals but several bones in other tetrapods. In addition, the lower and upper jaws hinge between a different set of bones in mammals than in other tetrapods. Mammals also have a unique set of three bones that transmit sound in the middle ear (the hammer, anvil, and stirrup), whereas other tetrapods

have only one such bone (the stirrup). Finally, the teeth of mammals are differentiated into incisors (for tearing), canines (for piercing), and the multi-pointed premolars and molars (for crushing and grinding). In contrast, the teeth of other tetrapods usually consist of a row of undifferentiated, single-pointed teeth.

As detailed in **Figure 23.5**, the fossil record shows that the unique features of mammalian jaws and teeth evolved gradually over time, in a series of steps. As you study Figure 23.5, bear in mind that it includes just a few examples of the fossil skulls that document the origin of mammals. If all the known fossils in the sequence were arranged by shape and placed side by side, their features would blend smoothly from one group to the next. Some of these fossils would reflect how the features of a group that dominates life today, the mammals, gradually arose in a previously existing group, the cynodonts. Others would reveal side branches on the tree of life—groups of organisms that thrived for millions of years but ultimately left no descendants that survive today.

#### CONCEPT CHECK 23.1

1. Your measurements indicate that a fossilized skull you unearthed has a carbon-14/carbon-12 ratio about 1/16 that of the skulls of present-day animals. What is the approximate age of the fossilized skull?
2. Describe an example from the fossil record that shows how life has changed over time.
3. **WHAT IF?** What might a fossil record of life today look like?
4. **WHAT IF?** Suppose researchers discover a fossil of an organism that lived 300 million years ago but had mammalian teeth and a mammalian jaw hinge. What inferences might you draw from this fossil about the origin of mammals and the evolution of novel skeletal structures? Explain.

For suggested answers, see Appendix A.

---

### CONCEPT 23.2

## The rise and fall of groups of organisms reflect differences in speciation and extinction rates

From its beginnings, life on Earth has been marked by the rise and fall of groups of organisms. Anaerobic prokaryotes originated, flourished, and then declined as the oxygen content of the atmosphere rose. Billions of years later, the first tetrapods emerged from the sea, giving rise to several major new groups of organisms. One of these, the amphibians, went on to dominate life on land for 100 million years, until other tetrapods (including dinosaurs and, later, mammals) replaced them as the dominant terrestrial vertebrates.

The rise and fall of these and other major groups of organisms have shaped the history of life. Narrowing our focus, we can also see that the rise or fall of any particular group is

Over the course of 120 million years, mammals originated gradually from a group of tetrapods called synapsids. Shown here are a few of the many fossil organisms whose morphological features represent intermediate steps between living mammals and their early synapsid ancestors. The evolutionary context of the origin of mammals is shown in the tree diagram at right (the dagger symbol † indicates extinct lineages).

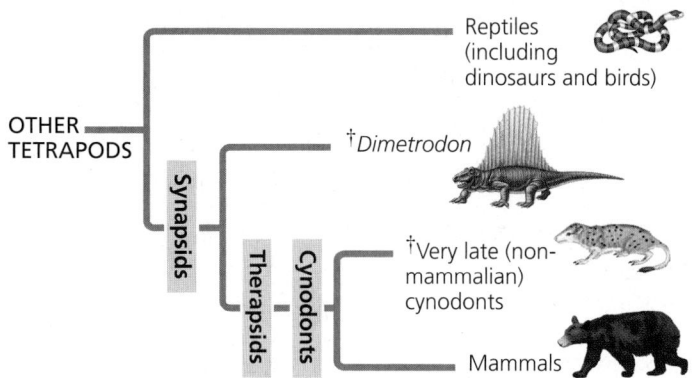

### Key to skull bones

- Articular
- Quadrate
- Dentary
- Squamosal

## Synapsid (300 mya)

Early synapsids had multiple bones in the lower jaw and single-pointed teeth. The jaw hinge was formed by the articular and quadrate bones. Early synapsids also had an opening called the *temporal fenestra* behind the eye socket. Powerful cheek muscles for closing the jaws probably passed through the temporal fenestra. Over time, this opening enlarged and moved in front of the hinge between the lower and upper jaws, thereby increasing the power and precision with which the jaws could be closed (much as moving a doorknob away from the hinge makes a door easier to close).

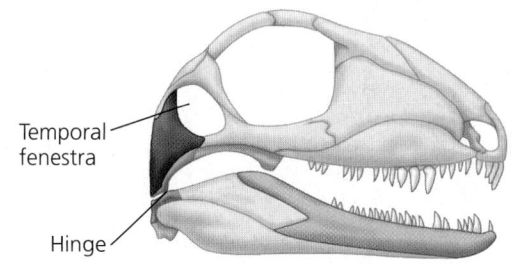

## Therapsid (280 mya)

Later, a group of synapsids called therapsids appeared. Therapsids had large dentary bones, long faces, and the first examples of specialized teeth, large canines. These trends continued in a group of therapsids called cynodonts.

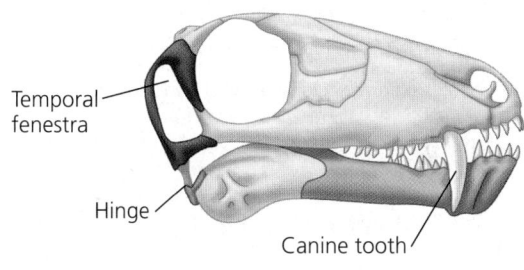

## Early cynodont (260 mya)

In early cynodont therapsids, the dentary was the largest bone in the lower jaw, the temporal fenestra was large and positioned forward of the jaw hinge, and teeth with several cusps first appeared (not visible in the diagram). As in earlier synapsids, the jaw had an articular-quadrate hinge.

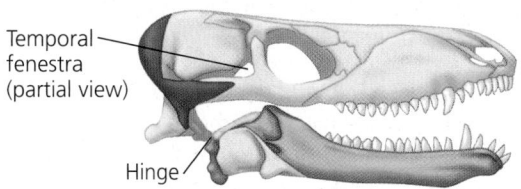

## Later cynodont (220 mya)

Later cynodonts had teeth with complex cusp patterns, and their lower and upper jaws hinged in two locations: They retained the original articular-quadrate hinge and formed a new, second hinge between the dentary and squamosal bones. (The temporal fenestra is not visible in this or the below cynodont skull at the angles shown.)

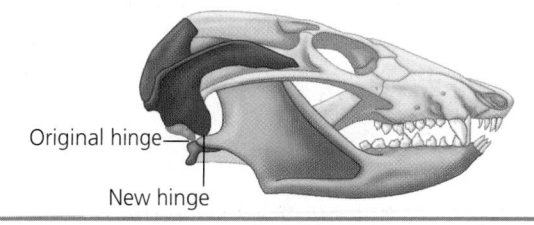

## Very late cynodont (195 mya)

In some very late (nonmammalian) cynodonts and early mammals, the original articular-quadrate hinge was lost, leaving the dentary-squamosal hinge as the only hinge between the lower and upper jaws, as in living mammals. The articular and quadrate bones migrated into the ear region (not shown), where they functioned in transmitting sound. In the mammal lineage, these two bones later evolved into the familiar hammer (malleus) and anvil (incus) bones of the ear.

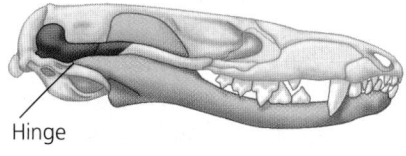

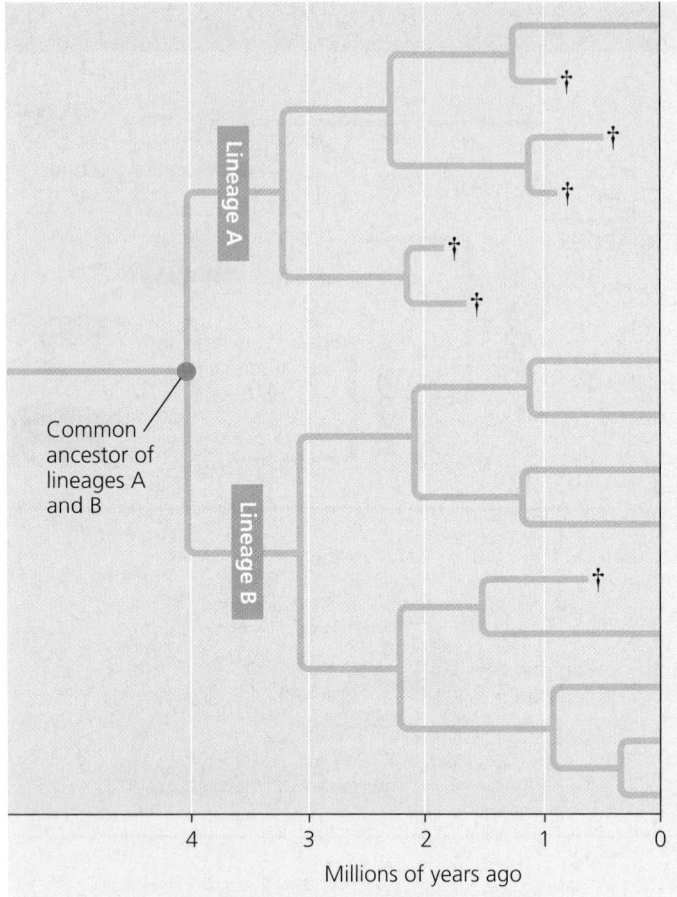

▲ **Figure 23.6 How speciation and extinction affect diversity.** The species diversity of an evolutionary lineage will increase when more new member species originate than are lost to extinction. In this hypothetical example, by 2 million years ago both lineage A and lineage B have given rise to four species, and no species have become extinct (denoted by a dagger symbol †). By time 0, however, lineage A contains only one species, while lineage B contains eight species.

**INTERPRET THE DATA** *Consider the period from 2 million years ago to time 0. For each lineage, determine how many speciation and extinction events occurred during that time.*

related to the speciation and extinction rates of its member species **(Figure 23.6)**. Just as a population increases in size when there are more births than deaths, the rise of a group of organisms occurs when more new species are produced than are lost to extinction. The reverse occurs when a group is in decline. In the **Scientific Skills Exercise**, you will interpret data from the fossil record about changes in a group of snail species in the early Paleogene period. Such changes in the fates of groups of organisms have been influenced by large-scale processes such as plate tectonics, mass extinctions, and adaptive radiations.

## Plate Tectonics

If photographs of Earth were taken from space every 10,000 years and spliced together to make a movie, it would show something many of us find hard to imagine: The seemingly "rock solid" continents we live on move over time.

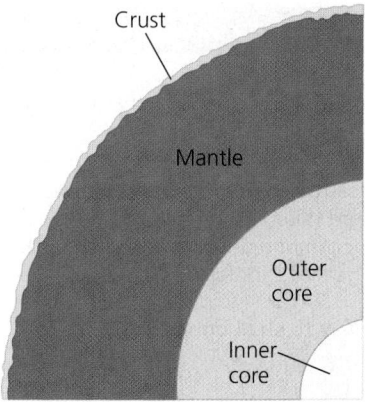

▶ **Figure 23.7 Cutaway view of Earth.** The thickness of the crust is exaggerated here.

Over the past billion years, there have been three occasions (1 billion, 600 million, and 250 million years ago) when most of the landmasses of Earth came together to form a supercontinent, then later broke apart. Each time, this breakup yielded a different configuration of continents. Based on the directions in which the continents are moving today, some geologists have estimated that a new supercontinent will form roughly 250 million years from now.

According to the theory of **plate tectonics**, the continents are part of great plates of Earth's crust that essentially float on the hot, underlying portion of the mantle **(Figure 23.7)**. Movements in the mantle cause the plates to move over time in a process called *continental drift*. Geologists can measure the rate at which the plates are moving now, usually only a few centimeters per year. They can also infer the past locations of the continents using the magnetic signal recorded in rocks at the time of their formation. This method works because as a continent shifts its position over time, the direction of magnetic north recorded in its newly formed rocks also changes.

Earth's major tectonic plates are shown in **Figure 23.8**. Many important geologic processes, including the formation of mountains and islands, occur at plate boundaries. In some cases, two plates are moving away from each other, as are the North American and Eurasian plates, which are currently drifting apart at a rate of about 2 cm per year. In other cases, two plates slide past each other, forming regions where earthquakes are common. California's infamous San Andreas Fault is part of a border where two plates are sliding past each other. In still other cases, two plates collide, producing violent upheavals and forming new mountains along the plate boundaries. One spectacular example of this occurred 45 million years ago, when the Indian plate crashed into the Eurasian plate, starting the formation of the Himalayan mountains.

### Consequences of Continental Drift

Plate movements rearrange geography slowly, but their cumulative effects are dramatic. In addition to reshaping the physical features of our planet, continental drift also has a major impact on life on Earth.

► **Figure 23.8 Earth's major tectonic plates.** The arrows indicate direction of movement. The reddish orange dots represent zones of violent tectonic activity.

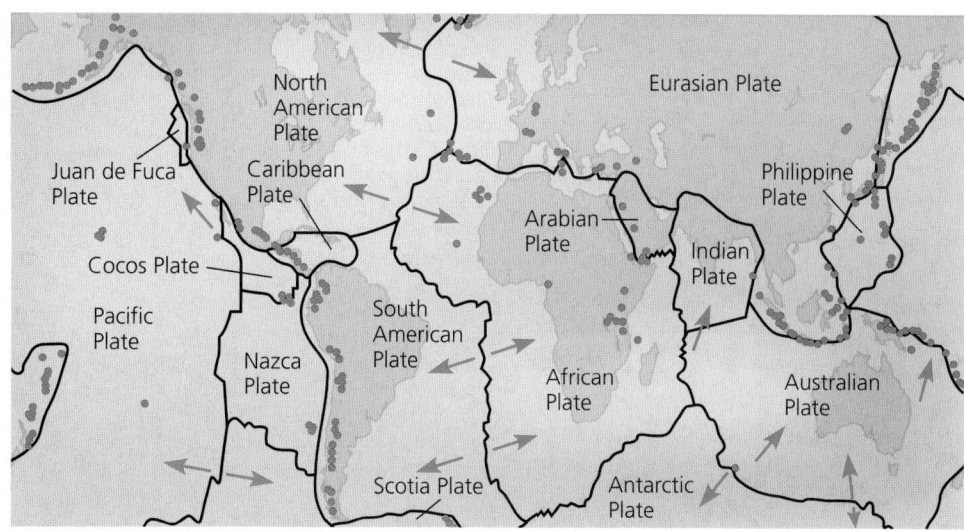

## Scientific Skills Exercise

# Estimating Quantitative Data from a Graph and Developing Hypotheses

**Do Ecological Factors Affect Evolutionary Rates?** Researchers studied the fossil record to investigate whether differing modes of larval dispersal might explain species longevity within a family of marine snails, the Volutidae. Some of the snail species had nonplanktonic larvae: They developed directly into adults without a swimming stage. Other species had planktonic larvae: They had a swimming stage and could disperse very long distances. The adults of these planktonic species tended to have broad geographic distributions, whereas nonplanktonic species tended to be more isolated.

**How the Research Was Done** The researchers studied the stratigraphic distribution of volutes in outcrops of sedimentary rocks located along North America's Gulf coast. These rocks, which formed from 66 to 37 million years ago, early in the Paleogene period, are an excellent source of well-preserved snail fossils. The researchers were able to classify each fossil species of volute snail as having planktonic or nonplanktonic larvae based on features of the earliest formed whorls of the snail's shell. Each bar in the graph shows how long one species of snail persisted in the fossil record.

**INTERPRET THE DATA**

1. You can estimate quantitative data (fairly precisely) from a graph. The first step is to obtain a conversion factor by measuring along an axis that has a scale. In this case, 25 million years (my; from 60 to 35 million years ago [mya] on the *x*-axis) is represented by a distance of 7.0 cm. This yields a conversion factor (a ratio) of 25 my/7.0 cm = 3.6 my/cm. To estimate the time period represented by a horizontal bar on this graph, measure the length of that bar in centimeters and multiply that measurement by the conversion factor, 3.6 my/cm. For example, a bar that measures 1.1 cm on the graph represents a persistence time of 1.1 cm × 3.6 my/cm = 4 million years.

2. Calculate the mean (average) persistence times for species with planktonic larvae and species with nonplanktonic larvae.

3. Count the number of new species that form in each group beginning at 60 mya (the first three species in each group were present around 64 mya, the first time period sampled, so we don't know when those species first appear in the fossil record).

4. Propose a hypothesis to explain the differences in longevity of snail species with planktonic and nonplanktonic larvae.

(MB) A version of this Scientific Skills Exercise can be assigned in MasteringBiology.

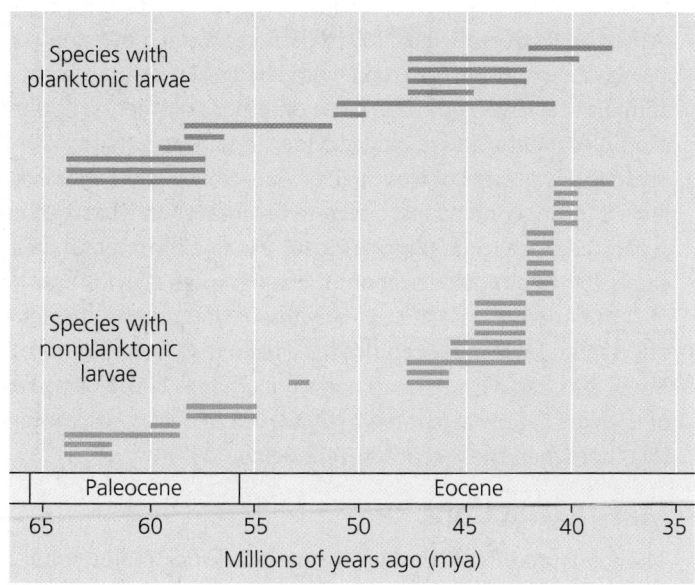

**Data from** T. Hansen, Larval dispersal and species longevity in Lower Tertiary gastropods, *Science* 199:885–887 (1978).

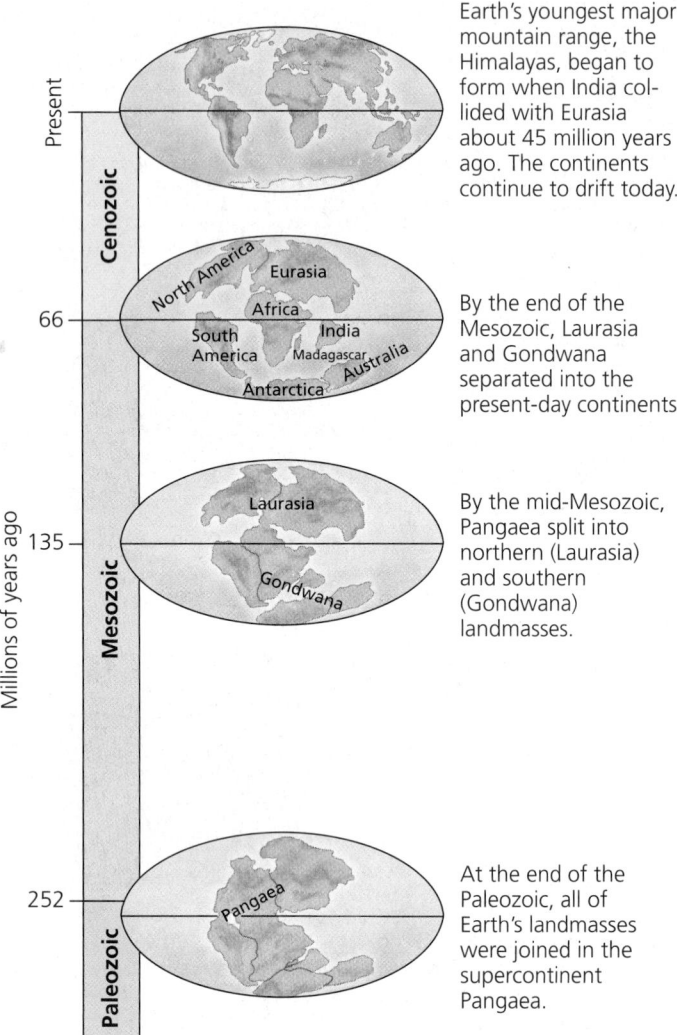

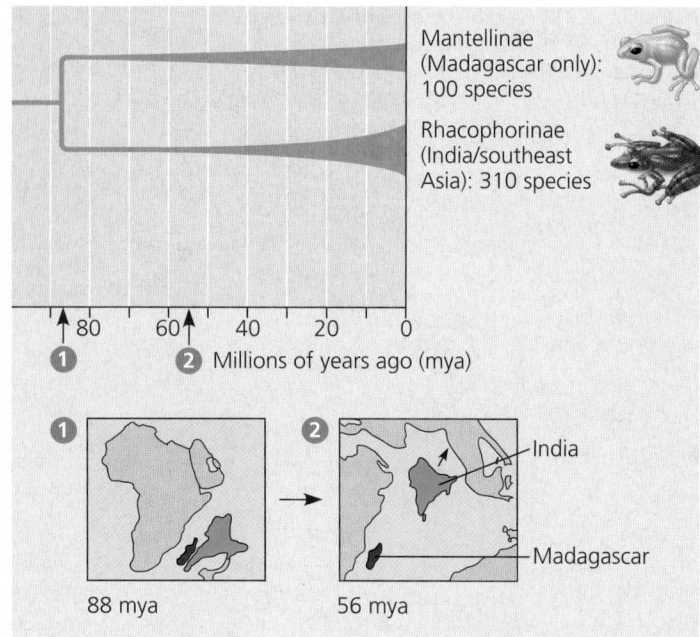

▲ **Figure 23.10 Speciation in frogs as a result of continental drift.** When present-day Madagascar began to separate from India 88 million years ago (**1**), the frog subfamilies Mantellinae and Rhacophorinae started to diverge, ultimately forming hundreds of new species in each location. The maps show the movement of Madagascar (red) and India (green) over time.

Figure 23.9 labels and descriptions:

Present — Earth's youngest major mountain range, the Himalayas, began to form when India collided with Eurasia about 45 million years ago. The continents continue to drift today.

66 — By the end of the Mesozoic, Laurasia and Gondwana separated into the present-day continents.

135 — By the mid-Mesozoic, Pangaea split into northern (Laurasia) and southern (Gondwana) landmasses.

252 — At the end of the Paleozoic, all of Earth's landmasses were joined in the supercontinent Pangaea.

▲ **Figure 23.9 The history of continental drift during the Phanerozoic eon.**

**?** *Is the Australian plate's current direction of movement (see Figure 23.8) similar to the direction it traveled over the past 66 million years?*

One reason for this is that continental drift alters the habitats in which organisms live. Consider the changes shown in **Figure 23.9**. About 250 million years ago, plate movements brought previously separated landmasses together into a supercontinent named **Pangaea**. Ocean basins became deeper, which lowered sea levels and drained shallow coastal seas. At that time, as now, most marine species inhabited shallow waters, and the formation of Pangaea destroyed much of that habitat. Pangaea's interior was cold and dry, probably an even more severe environment than that of central Asia today. Overall, the formation of Pangaea greatly altered the physical environment and climate, which drove some species to extinction and provided new opportunities for groups of organisms that survived the crisis.

Organisms are also affected by the climate change that results when a continent shifts its location. The southern tip of Labrador, Canada, for example, once was located in the tropics but has moved 40° to the north over the last 200 million years.

When faced with the changes in climate that such shifts in position entail, organisms adapt, move to a new location, or become extinct (this last outcome occurred for many organisms stranded on Antarctica, which separated from Australia 40 million years ago).

Continental drift also promotes allopatric speciation on a grand scale. When supercontinents break apart, regions that once were connected become isolated. As the continents drifted apart over the last 200 million years, each became a separate evolutionary arena, with lineages of plants and animals that diverged from those on other continents. For example, genetic and geologic evidence indicates that two present-day groups of frog species, the subfamilies Mantellinae and Rhacophorinae, began to diverge when Madagascar separated from India **(Figure 23.10)**. Finally, continental drift can help explain puzzles about the geographic distribution of extinct organisms, such as why fossils of the same species of Permian freshwater reptiles have been discovered in both Brazil and the West African nation of Ghana. These two parts of the world, now separated by 3,000 km of ocean, were joined together when these reptiles were living.

## Mass Extinctions

The fossil record shows that the overwhelming majority of species that ever lived are now extinct. A species may become extinct for many reasons. Its habitat may have been destroyed, or its environment may have changed in a manner unfavorable to the species. For example, if ocean temperatures fall by

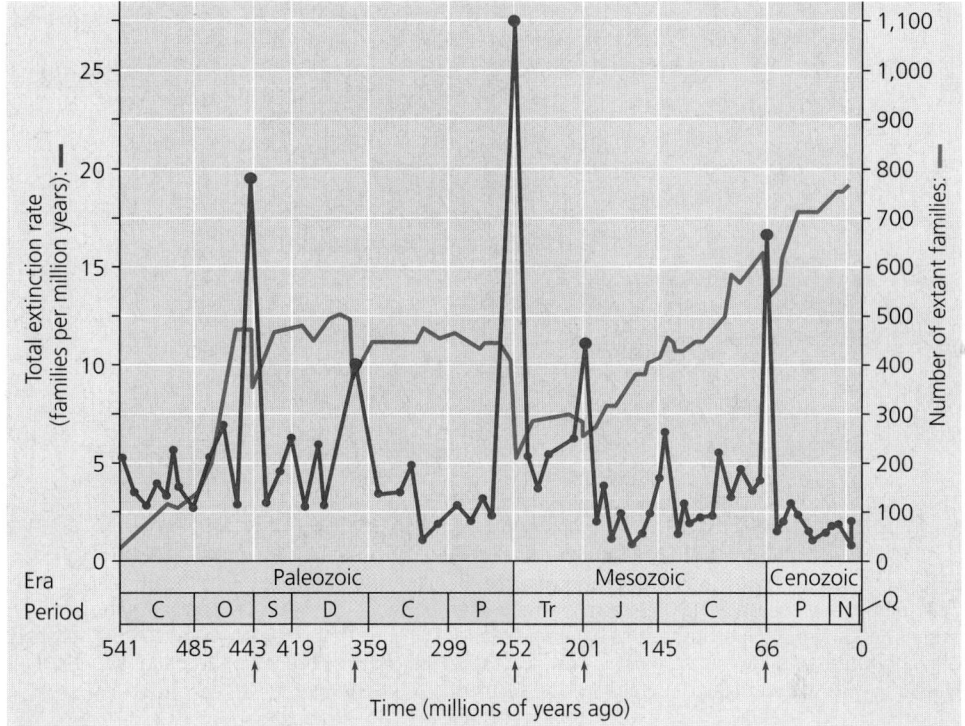

► **Figure 23.11 Mass extinction and the diversity of life.** The five generally recognized mass extinction events, indicated by red arrows, represent peaks in the extinction rate of marine animal families (red line and left vertical axis). These mass extinctions interrupted the overall increase over time in the number of extant families of marine animals (blue line and right vertical axis).

**INTERPRET THE DATA** *As mentioned in the text, 96% of marine animal species became extinct in the Permian mass extinction. Explain why the blue curve shows only a 50% drop at that time.*

even a few degrees, species that are otherwise well adapted may perish. Even if physical factors in the environment remain stable, biological factors may change—the origin of one species can spell doom for another.

Although extinction occurs regularly, at certain times disruptive changes to the global environment have caused the rate of extinction to increase dramatically. The result is a **mass extinction**, in which large numbers of species become extinct worldwide.

### The "Big Five" Mass Extinction Events

Five mass extinctions are documented in the fossil record over the past 500 million years **(Figure 23.11)**. These events are particularly well documented for the decimation of hard-bodied animals that lived in shallow seas, the organisms for which the fossil record is most complete. In each mass extinction, 50% or more of marine species became extinct.

Two mass extinctions—the Permian and the Cretaceous—have received the most attention. The Permian mass extinction, which defines the boundary between the Paleozoic and Mesozoic eras (252 million years ago), claimed about 96% of marine animal species and drastically altered life in the ocean. Terrestrial life was also affected. For example, 8 out of 27 known orders of insects were wiped out. This mass extinction occurred in less than 500,000 years, possibly in just a few thousand years—an instant in the context of geologic time.

The Permian mass extinction occurred during the most extreme episode of volcanism in the past 500 million years. Geologic data indicate that 1.6 million km² (roughly half the size of western Europe) in Siberia was covered with lava hundreds of meters thick. The eruptions are thought to have produced enough carbon dioxide to warm the global climate

by an estimated 6°C, harming many temperature-sensitive species. The rise in atmospheric $CO_2$ levels would also have led to ocean acidification (see Figure 2.24), thereby reducing the availability of calcium carbonate ($CaCO_3$), which is required by reef-building corals and many shell-building species. The explosions would also have added nutrients such as phosphorus to marine ecosystems, stimulating the growth of microorganisms. Upon their deaths, these microorganisms would have provided food for bacterial decomposers. Bacteria use oxygen as they decompose the bodies of dead organisms, thus causing oxygen concentrations to drop. This would have harmed oxygen-breathers and promoted the growth of anaerobic bacteria that emit a poisonous metabolic by-product, hydrogen sulfide ($H_2S$) gas. Overall, the volcanic eruptions may have triggered a series of catastrophic events that together resulted in the Permian mass extinction.

The Cretaceous mass extinction occurred 66 million years ago and marks the boundary between the Mesozoic and Cenozoic eras. This event extinguished more than half of all marine species and eliminated many families of terrestrial plants and animals, including all dinosaurs (except birds, which are members of the same group; see Concept 27.4). One clue to a possible cause of the Cretaceous mass extinction is a thin layer of clay enriched in iridium that dates to the time of the mass extinction. Iridium is an element that is very rare on Earth but common in many of the meteorites and other extraterrestrial objects that occasionally fall to Earth. As a result, researchers proposed that this clay is fallout from a huge cloud of debris that billowed into the atmosphere when an asteroid or large comet collided with Earth. This cloud would have blocked sunlight and severely disturbed the global climate for several months.

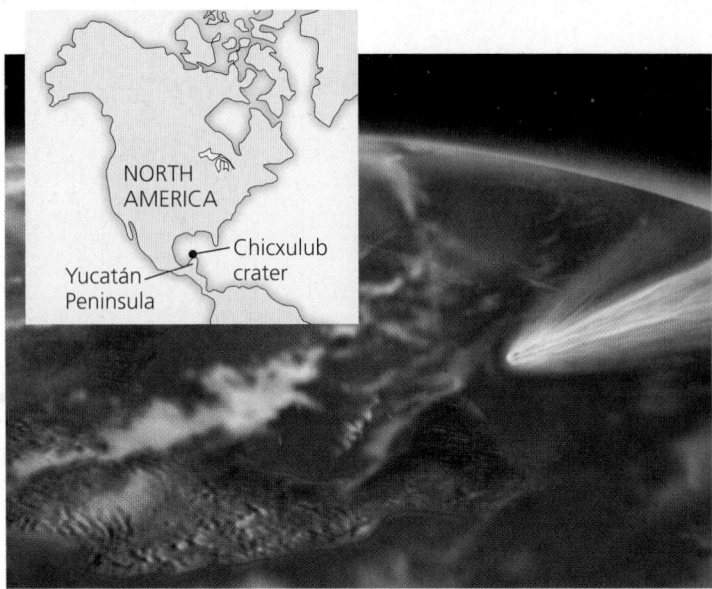

▲ **Figure 23.12 Trauma for Cretaceous life.** Beneath the Caribbean Sea, the 66-million-year-old Chicxulub crater measures 180 km across. The horseshoe shape of the crater and the pattern of debris in sedimentary rocks indicate that an asteroid or comet struck at a low angle from the southeast. This drawing represents the impact and its immediate effect: a cloud of hot vapor and debris that could have killed many of the plants and animals in North America within hours.

Is there evidence of such an asteroid or comet? Research has focused on the Chicxulub crater, a 66-million-year-old scar beneath sediments off the Yucatán coast of Mexico **(Figure 23.12)**. The crater is the right size to have been caused by an object with a diameter of 10 km. Critical evaluation of this and other hypotheses for mass extinctions continues.

### Is a Sixth Mass Extinction Under Way?

As you will read further in Concept 43.1, human actions, such as habitat destruction, are modifying the global environment to such an extent that many species are threatened with extinction. More than a thousand species have become extinct in the last 400 years. Scientists estimate that this rate is 100 to 1,000 times the typical background rate seen in the fossil record. Is a sixth mass extinction now in progress?

This question is difficult to answer, in part because it is hard to document the total number of extinctions occurring today. Tropical rain forests, for example, harbor many undiscovered species. As a result, destroying tropical forest may drive species to extinction before we even learn of their existence. Such uncertainties make it hard to assess the full extent of the current extinction crisis. Even so, it is clear that losses to date have not reached those of the "big five" mass extinctions, in which large percentages of Earth's species became extinct. This does not in any way discount the seriousness of today's situation. Monitoring programs show that many species are declining at an alarming rate due to habitat loss, introduced species, overharvesting, and other factors. Recent studies on a variety of organisms, including lizards, pine trees, and polar bears, suggest that climate change may hasten some of these declines. The fossil record also highlights the potential importance of climate change: Over the last 500 million years, extinction rates

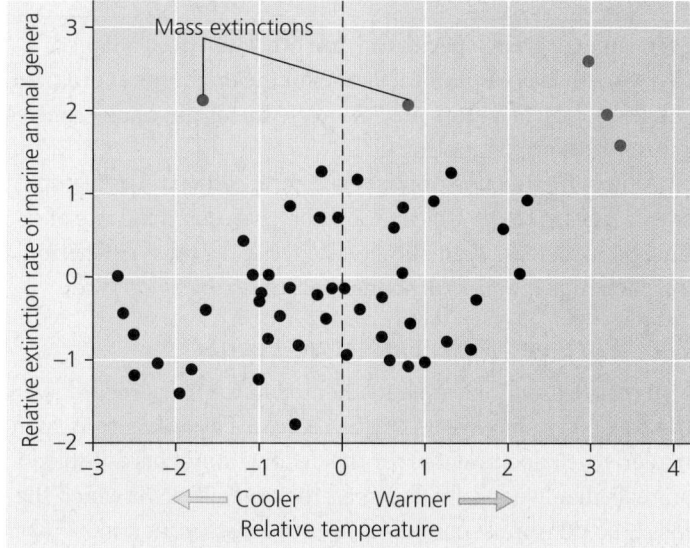

▲ **Figure 23.13 Fossil extinctions and temperature.** Extinction rates increased when global temperatures were high. Temperatures were estimated using ratios of oxygen isotopes and converted to an index in which 0 is the overall average temperature.

have tended to increase when global temperatures were high **(Figure 23.13)**. Overall, the evidence suggests that unless dramatic actions are taken, a sixth, human-caused mass extinction is likely to occur within the next few centuries.

### Consequences of Mass Extinctions

Mass extinctions have significant and long-term effects. By eliminating large numbers of species, a mass extinction can reduce a thriving and complex ecological community to a pale shadow of its former self. And once an evolutionary lineage disappears, it cannot reappear. The course of evolution is

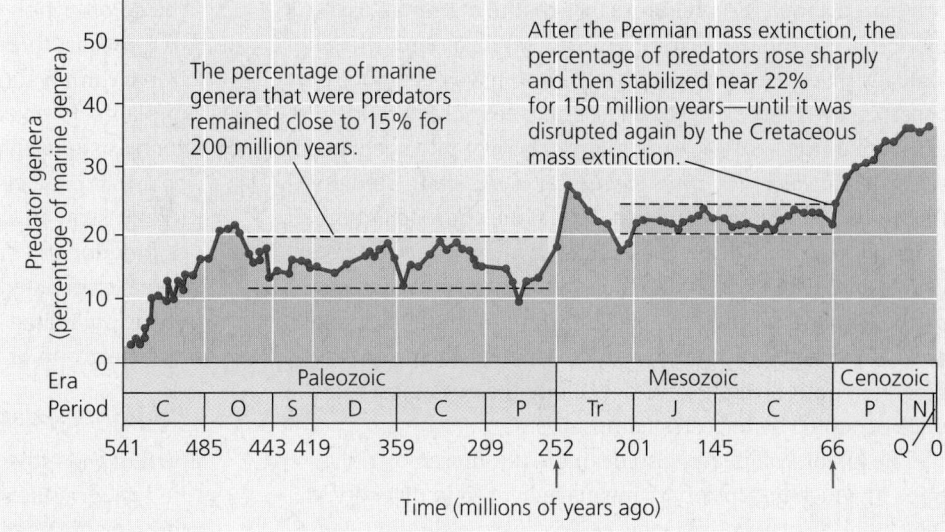

► **Figure 23.14 Mass extinctions and ecology.** The Permian and Cretaceous mass extinctions (indicated by red arrows) altered the ecology of the oceans by increasing the percentage of marine genera that were predators.

The percentage of marine genera that were predators remained close to 15% for 200 million years.

After the Permian mass extinction, the percentage of predators rose sharply and then stabilized near 22% for 150 million years—until it was disrupted again by the Cretaceous mass extinction.

changed forever. Consider what would have happened if the early primates living 66 million years ago had died out in the Cretaceous mass extinction. Humans would not exist, and life on Earth would differ greatly from what it is today.

The fossil record shows that it typically takes 5–10 million years for the diversity of life to recover to previous levels after a mass extinction. In some cases, it has taken much longer than that: It took about 100 million years for the number of marine families to recover after the Permian mass extinction (see Figure 23.11). These data have sobering implications. If current trends continue and a sixth mass extinction occurs, it will take millions of years for life on Earth to recover.

Mass extinctions can also alter ecological communities by changing the types of organisms residing there. For example, after the Permian and Cretaceous mass extinctions, the percentage of marine organisms that were predators increased substantially **(Figure 23.14)**. A rise in the number of predators can increase both the risks faced by prey and the competition among predators for food. In addition, mass extinctions can curtail lineages with novel and advantageous features. For example, in the late Triassic period a group of gastropods (snails and their relatives) arose that could drill through the shells of bivalves (such as clams) and feed on the animals inside. Although shell drilling provided access to a new and abundant source of food, this newly formed group was wiped out during the mass extinction at the end of the Triassic period (about 200 million years ago). Another 120 million years passed before another group of gastropods (the oyster drills) exhibited the ability to drill through shells. As their predecessors might have done if they had not originated at an unfortunate time, oyster drills have since diversified into many new species. Finally, by eliminating so many species, mass extinctions can pave the

way for adaptive radiations, in which new groups of organisms proliferate.

## Adaptive Radiations

The fossil record shows that the diversity of life has increased over the past 250 million years (see blue line in Figure 23.11). This increase has been fueled by **adaptive radiations**, periods of evolutionary change in which groups of organisms form many new species whose adaptations allow them to fill different ecological roles, or niches, in their communities. Large-scale adaptive radiations occurred after each of the big five mass extinctions, when survivors became adapted to the many vacant ecological niches. Adaptive radiations have also occurred in groups of organisms that possessed major evolutionary innovations, such as seeds or armored body coverings, or that colonized regions in which they faced little competition from other species.

### Worldwide Adaptive Radiations

Fossil evidence indicates that mammals underwent a dramatic adaptive radiation after the extinction of terrestrial dinosaurs 66 million years ago **(Figure 23.15)**. Although mammals

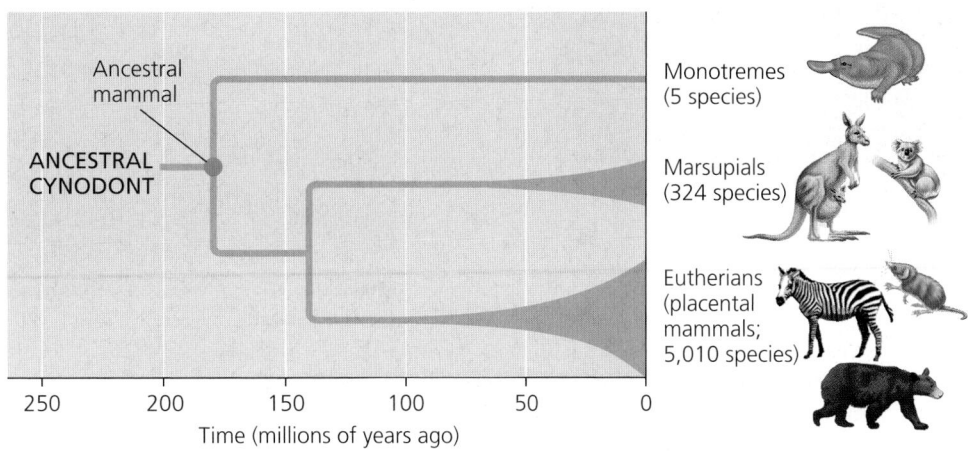

▲ **Figure 23.15 Adaptive radiation of mammals.**

originated about 180 million years ago, the mammal fossils older than 66 million years are mostly small and not morphologically diverse. Many species appear to have been nocturnal based on their large eye sockets, similar to those in living nocturnal mammals. A few early mammals were intermediate in size, such as *Repenomamus giganticus*, a 1-m-long predator that lived 130 million years ago—but none approached the size of many dinosaurs. Early mammals may have been restricted in size and diversity because they were eaten or outcompeted by the larger and more diverse dinosaurs. With the disappearance of the dinosaurs (except for birds), mammals expanded greatly in both diversity and size, filling the ecological roles once occupied by terrestrial dinosaurs.

The history of life has also been greatly altered by radiations in which groups of organisms increased in diversity as they came to play entirely new ecological roles in their communities. As we'll explore in later chapters, examples include the rise of photosynthetic prokaryotes, the evolution of large predators in the early Cambrian, and the radiations following the colonization of land by plants, insects, and tetrapods. Each of these last three radiations was associated with major evolutionary innovations that facilitated life on land. The radiation of plants, for example, was associated with key adaptations, such as stems that support plants against gravity and a waxy coat that protects leaves from water loss. Finally, organisms that arise in an adaptive radiation can serve as a new source of food for still other organisms. In fact, the diversification of plants stimulated a series of adaptive radiations in insects that ate or pollinated plants, one reason that insects are the most diverse group of animals on Earth today.

### Regional Adaptive Radiations

Striking adaptive radiations have also occurred over more limited geographic areas. Such radiations can be initiated when a few organisms make their way to a new, often distant location in which they face relatively little competition from other organisms. The Hawaiian archipelago is one of the world's great showcases of this type of adaptive radiation **(Figure 23.16)**.

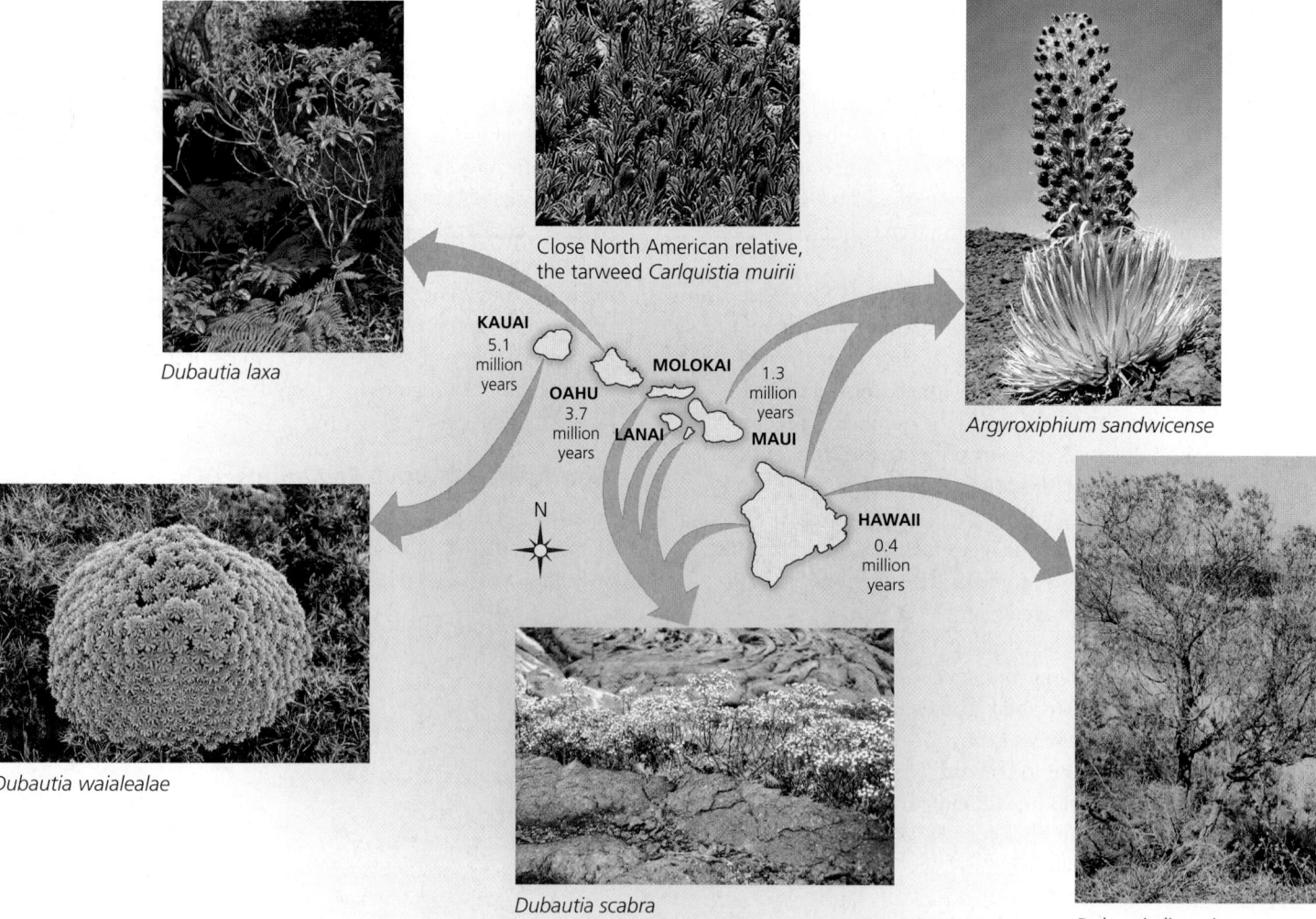

Close North American relative, the tarweed *Carlquistia muirii*

*Dubautia laxa*

*Argyroxiphium sandwicense*

*Dubautia waialealae*

*Dubautia scabra*

*Dubautia linearis*

KAUAI
5.1 million years

MOLOKAI

OAHU
3.7 million years

1.3 million years

LANAI

MAUI

HAWAII
0.4 million years

N

▲ **Figure 23.16 Adaptive radiation on the Hawaiian Islands.** Molecular analysis indicates that these remarkably varied Hawaiian plants, known collectively as the "silversword alliance," are all descended from an ancestral tarweed that arrived on the islands from North America about 5 million years ago. Members of the silversword alliance have since spread into different habitats and formed new species with strikingly different adaptations.

Located about 3,500 km from the nearest continent, the volcanic islands are progressively older as one follows the chain toward the northwest; the youngest island, Hawaii, is less than a million years old and still has active volcanoes. Each island was born "naked" and was gradually populated by stray organisms that rode the ocean currents and winds either from far-distant land areas or from older islands of the archipelago.

The physical diversity of each island in the Hawaiian archipelago, including immense variation in soil conditions, elevation, and rainfall, provides many opportunities for evolutionary divergence by natural selection. Multiple invasions followed by speciation events have ignited an explosion of adaptive radiation in Hawaii. As a result, thousands of species that inhabit the islands are found nowhere else on Earth. Among plants, for example, about 1,100 species are unique to the Hawaiian islands. Unfortunately, many of these species are now facing an elevated risk of extinction due to human actions such as habitat destruction and the introduction of non-native plant species.

**CONCEPT CHECK 23.2**

1. Explain the consequences of plate tectonics for life on Earth.
2. Summarize how mass extinctions affect the evolutionary history of life.
3. What factors promote adaptive radiations?
4. **WHAT IF?** Suppose that an invertebrate species was lost in a mass extinction caused by a sudden catastrophic event. Would the last appearance of this species in the fossil record necessarily be close to when the extinction actually occurred? Would the answer to this question differ depending on whether the species was common (abundant and widespread) or rare? Explain.

For suggested answers, see Appendix A.

**CONCEPT 23.3**

# Major changes in body form can result from changes in the sequences and regulation of developmental genes

The fossil record tells us what the great changes in the history of life have been and when they occurred. Moreover, an understanding of plate tectonics, mass extinction, and adaptive radiation provides a picture of how those changes came about. But we can also seek to understand the intrinsic biological mechanisms that underlie changes seen in the fossil record. For this, we turn to genetic mechanisms of change, paying particular attention to genes that influence development.

## Effects of Developmental Genes

As you read in Concept 18.6, evo-devo—research at the interface between evolutionary biology and developmental

biology—is illuminating how slight genetic differences can produce major morphological differences between species. In particular, large morphological differences can result from genes that alter the rate, timing, and spatial pattern of change in an organism's physical form as it develops from a zygote into an adult.

### Changes in Rate and Timing

Many striking evolutionary transformations are the result of **heterochrony** (from the Greek *hetero*, different, and *chronos*, time), an evolutionary change in the rate or timing of developmental events. For example, an organism's shape depends in part on the relative growth rates of different body parts during development. Changes to these rates can alter the adult form substantially, as seen in the contrasting shapes of human and chimpanzee skulls **(Figure 23.17)**.

Other examples of the dramatic evolutionary effects of heterochrony include how increased growth rates of finger bones

Chimpanzee infant                    Chimpanzee adult

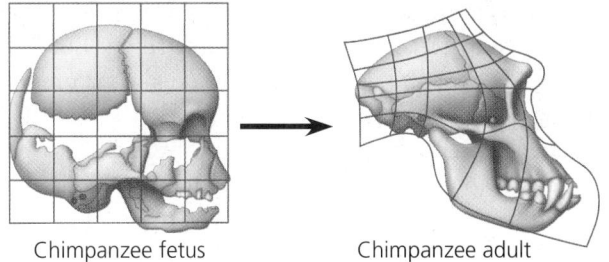

Chimpanzee fetus          Chimpanzee adult

Human fetus          Human adult

▲ **Figure 23.17 Relative skull growth rates.** In the human evolutionary lineage, mutations slowed the growth of the jaw relative to other parts of the skull. As a result, in humans the skull of an adult is more similar to the skull of an infant than is the case for chimpanzees.

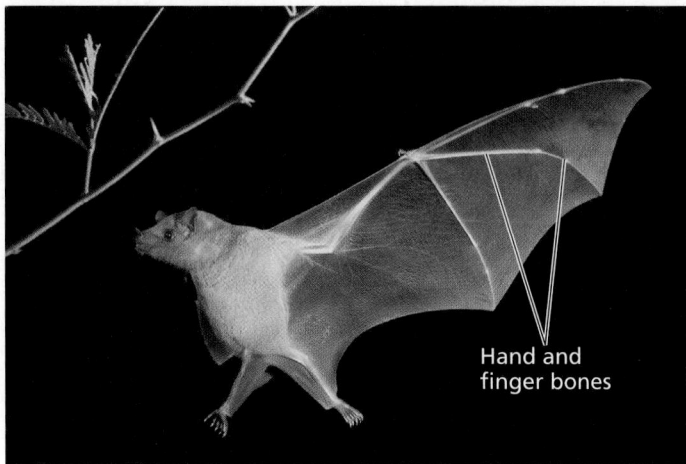

▲ **Figure 23.18 Elongated hand and finger bones in a bat wing.** Heterochrony is responsible for the increased total length of hand and finger bones in a bat compared to that of other mammals.

MAKE CONNECTIONS *Locate the bat's wrist and elbow joints (see Figure 19.16). Calculate the ratio of the length of the bat's longest set of hand and finger bones to the length of its radius. Compare this ratio to the ratio of the bones in your own hand and arm.*

▲ **Figure 23.19  Paedomorphosis.** The adults of some species retain features that were juvenile in ancestors. This salamander is an axolotl, an aquatic species that becomes a sexually mature adult while retaining certain larval (tadpole) characteristics, including gills.

yielded the skeletal structure of wings in bats **(Figure 23.18)** and how slowed growth of leg and pelvic bones led to the reduction and eventual loss of hind limbs in whales (see Figure 19.20).

Heterochrony can also alter the timing of reproductive development relative to the development of nonreproductive organs. If the development of reproductive organs accelerates compared to that of other organs, the sexually mature stage of a species may retain body features that were juvenile structures in an ancestral species, a condition called **paedomorphosis** (from the Greek *paedos*, of a child, and *morphosis*, formation). For example, most salamander species have aquatic larvae that undergo metamorphosis in becoming adults. But some species grow to adult size and become sexually mature while retaining gills and other larval features **(Figure 23.19)**. Such an evolutionary alteration of developmental timing can produce animals that appear very different from their ancestors, even though the overall genetic change may be small. Indeed, recent evidence indicates that a change at a single locus was probably sufficient to bring about paedomorphosis in the axolotl salamander, although other genes may have contributed as well.

### Changes in Spatial Pattern

Substantial evolutionary changes can also result from alterations in genes that control the spatial organization of body parts. For example, master regulatory genes called **homeotic genes** (described in Concept 16.1) determine such basic features as where a pair of wings and legs will develop on a bird or how a plant's flower parts are arranged.

The products of one class of homeotic genes, the *Hox* genes, provide positional information in an animal embryo. This information prompts cells to develop into structures appropriate for a particular location. Changes in *Hox* genes or in how they are expressed can have a profound impact on morphology. For example, among crustaceans, a change in the location where two *Hox* genes (*Ubx* and *Scr*) are expressed correlates with the conversion of a swimming appendage to a feeding appendage. Similarly, when comparing plant species, changes to the expression of homeotic genes known as *MADS-box* genes can produce flowers that differ dramatically in form.

### The Evolution of Development

Large members of most animal phyla appear suddenly in fossils formed 535–525 million years ago. This rapid diversification of animals is referred to as the *Cambrian explosion* (see Concept 27.2). Yet the discovery of 560-million-year-old fossils of Ediacaran animals (see Figure 23.3) suggests that a set of genes sufficient to produce complex animals existed at least 25 million years before the Cambrian explosion. If such genes have existed for so long, how can we explain the astonishing increases in diversity seen during and since the Cambrian explosion?

Adaptive evolution by natural selection provides one answer to this question. As we've seen throughout this unit, by sorting among differences in the sequences of protein-encoding genes, selection can improve adaptations rapidly. In addition, new genes (created by gene duplication events) can take on new metabolic and structural functions, as can existing genes that are regulated in new ways.

Examples in the previous section suggest that developmental genes may have been particularly important. Thus, we'll turn next to how new morphological forms can arise from changes in the nucleotide sequences or regulation of developmental genes.

## Changes in Gene Sequence

New developmental genes arising after gene duplication events probably facilitated the origin of novel morphological forms. But since other genetic changes also may have occurred at such times, it can be difficult to establish causal links between genetic and morphological changes that occurred in the past.

This difficulty was sidestepped in a study of developmental changes associated with the divergence of six-legged insects from crustacean ancestors that had more than six legs. (Insects arose from within a subgroup of the crustaceans, the traditional name for organisms such as shrimp, crabs, and lobsters.) Researchers noted differences between crustaceans and insects in the expression and the effects of the *Hox* gene *Ubx*: in particular, in insects, *Ubx* suppresses leg formation where it is expressed (**Figure 23.20**).

To examine the workings of this gene, researchers cloned the *Ubx* gene from an insect, the fruit fly *Drosophila*, and from a crustacean, the brine shrimp *Artemia*. Next, they genetically engineered fruit fly embryos to express either the *Drosophila Ubx* gene or the *Artemia Ubx* gene throughout their bodies. The *Drosophila* gene suppressed 100% of the limbs in the embryos, as expected, whereas the *Artemia* gene suppressed only 15%.

The researchers then sought to uncover key steps involved in the evolutionary transition from an ancestral *Ubx* gene to an insect *Ubx* gene. Their approach was to identify mutations that would cause the *Artemia Ubx* gene to suppress leg formation, thus making its gene act more like an insect *Ubx* gene. To do this, they constructed a series of "hybrid" *Ubx* genes, each of which contained known segments of the *Drosophila Ubx* gene and known segments of the *Artemia Ubx* gene. By inserting these hybrid genes into fruit fly embryos (one hybrid gene per embryo) and observing their effects on leg development, the researchers were able to pinpoint the exact amino acid changes responsible for the suppression of additional limbs in insects. In so doing, this study provided evidence that particular changes in the nucleotide sequence of a developmental gene contributed to a major evolutionary change: the origin of the six-legged insect body plan.

## Changes in Gene Regulation

While a change in the nucleotide sequence of a gene may affect its function wherever the gene is expressed, changes in the regulation of gene expression can be limited to one cell type (see Concept 16.1). Thus, a change in the regulation of a developmental gene may have fewer harmful side effects than a change to the sequence of the gene. This reasoning has prompted researchers to suggest that changes in the form of organisms may often be caused by mutations that affect the regulation of developmental genes—not their sequences.

This idea is supported by studies of a variety of species, including threespine stickleback fish. These fish live in the open ocean and in shallow, coastal waters. In western Canada, they also live in lakes formed when the coastline receded during the past 12,000 years. Marine stickleback fish have a pair of spines on their ventral (lower) surface, which deter some predators. These spines are often reduced or absent in stickleback fish living in lakes that lack predatory fishes and that are also low in calcium. Spines may have been lost in such lakes because they are not advantageous in the absence of predators, and the limited calcium is needed for purposes other than constructing spines.

At the genetic level, the developmental gene *Pitx1* was known to influence whether stickleback fish have ventral spines. Was the reduction of spines in some lake populations due to changes in the *Pitx1* gene or to changes in how the gene is expressed (**Figure 23.21**)? Data from an experiment testing this question indicate that the regulation of gene expression has changed, not the DNA sequence. Moreover, lake stickleback fish do express the *Pitx1* gene in tissues not related to the production of spines (such as the mouth), illustrating how morphological change can be caused by altering the expression of a developmental gene in some parts of the body but not others. In a 2010 follow-up study, researchers showed that changes to the *Pel* enhancer, a noncoding DNA region that affects expression of the *Pitx1* gene, resulted in the reduction of ventral spines in lake sticklebacks. Overall, results from studies on stickleback fish provide a clear and detailed example of how changes in gene regulation can alter the form of individual organisms and ultimately lead to evolutionary change in populations.

▶ **Figure 23.20 Effects of the *Hox* gene *Ubx* on the insect body plan.** In crustaceans, the *Hox* gene *Ubx* is expressed in the region shaded green, the body segments between the head and genital segments. In insects, *Ubx* is expressed in only a subset (shaded pink) of the homologous body segments, where it suppresses leg formation.

Hox gene 7

Ubx

Genital segments

Crustacean ancestor: *Ubx* expressed in the body segments shaded green; does not suppress legs

Changes over time in the expression and effect of the *Ubx* gene

Hox gene 7

Ubx

Genital segments

Present-day insect: *Ubx* expressed in the body segments shaded pink; suppresses legs

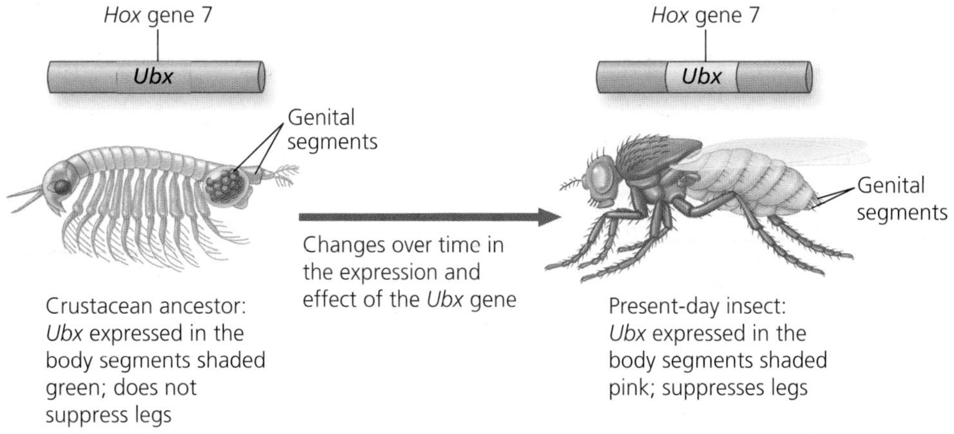

## What causes the loss of spines in lake stickleback fish?

**Experiment** Marine populations of the threespine stickleback fish (*Gasterosteus aculeatus*) have a set of protective spines on their lower (ventral) surface; however, these spines have been lost or reduced in some lake populations of this fish. Working at Stanford University, Michael Shapiro, David Kingsley, and colleagues performed genetic crosses and found that most of the reduction in spine size resulted from the effects of a single developmental gene, *Pitx1*. The researchers then tested two hypotheses about how *Pitx1* causes this morphological change.

**Hypothesis A:** A change in the DNA sequence of *Pitx1* had caused spine reduction in lake populations. To test this idea, the team used DNA sequencing to compare the coding sequence of the *Pitx1* gene between marine and lake stickleback populations.

**Hypothesis B:** A change in the regulation of the expression of *Pitx1* had caused spine reduction. To test this idea, the researchers monitored where in the developing embryo the *Pitx1* gene was expressed. They conducted whole-body *in situ* hybridization experiments (see Concept 15.4) using *Pitx1* DNA as a probe to detect *Pitx1* mRNA in the fish.

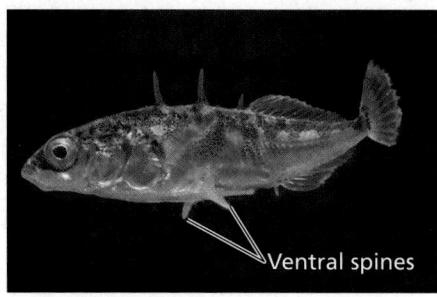

Ventral spines

▲ **Threespine stickleback (*Gasterosteus aculeatus*)**

**Results**

**Test of Hypothesis A:** Are there differences in the coding sequence of the *Pitx1* gene in marine and lake stickleback fish?

**Result: No** →

The 283 amino acids of the Pitx1 protein are identical in marine and lake stickleback populations.

**Test of Hypothesis B:** Are there any differences in the regulation of expression of *Pitx1*?

**Result: Yes** →

Red arrows (→) indicate regions of *Pitx1* gene expression in the photographs below. *Pitx1* is expressed in the ventral spine and mouth regions of developing marine stickleback fish but only in the mouth region of developing lake stickleback fish.

**Marine stickleback embryo**

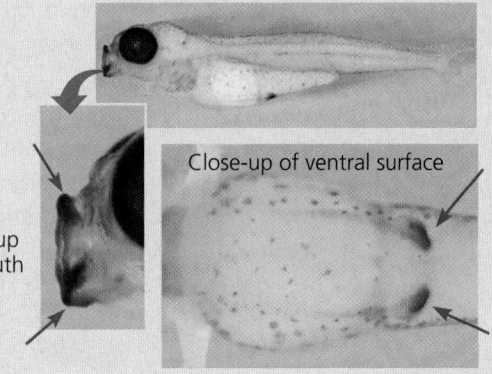

Close-up of ventral surface

Close-up of mouth

**Lake stickleback embryo**

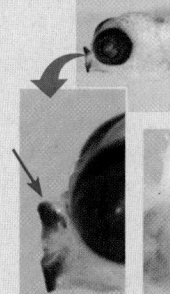

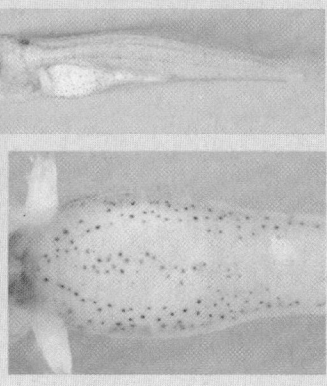

**Conclusion** The loss or reduction of ventral spines in lake populations of threespine stickleback fish appears to have resulted primarily from a change in the regulation of *Pitx1* gene expression, not from a change in the gene's sequence.

**Data from** M. D. Shapiro et al., Genetic and developmental basis of evolutionary pelvic reduction in three-spine sticklebacks, *Nature* 428:717–723 (2004).

**WHAT IF?** Describe the set of results that would have led researchers to the conclusion that a change in the coding sequence of the *Pitx1* gene was more important than a change in regulation of gene expression.

---

## CONCEPT CHECK 23.3

1. Explain how new body forms can originate by heterochrony.
2. Why is it likely that *Hox* genes have played a major role in the evolution of novel morphological forms?
3. **MAKE CONNECTIONS** Given that changes in morphology are often caused by changes in the regulation of gene expression, predict whether noncoding DNA is likely to be affected by natural selection. (Review Concepts 15.2 and 15.3.)

For suggested answers, see Appendix A.

**CONCEPT 23.4**

# Evolution is not goal oriented

What does our study of macroevolution tell us about how evolution works? One lesson is that throughout the history of life, the origin of new species has been affected by both the small-scale factors described in Concept 21.3 (such as natural selection operating in populations) and the large-scale factors described in this chapter (such as continental drift promoting bursts of speciation throughout the globe). Moreover, to paraphrase the Nobel Prize–winning geneticist François Jacob, evolution is like tinkering—a process in which new forms arise by the modification of existing structures or existing developmental genes. Over time, such tinkering has led to the three key features of the natural world described on the opening page of Chapter 19: the striking ways in which organisms are suited for life in their environments, the many shared characteristics of life, and the rich diversity of life.

## Evolutionary Novelties

François Jacob's view of evolution harkens back to Darwin's concept of descent with modification. As new species form, novel and complex structures can arise as gradual modifications of ancestral structures. In many cases, complex structures have evolved in increments from simpler versions that performed the same basic function. For example, consider the human eye, an intricate organ constructed from numerous parts that work together in forming an image and transmitting it to the brain. How could the human eye have evolved in gradual increments? Some argue that if the eye needs all of its components to function, a partial eye could not have been of use to our ancestors.

The flaw in this argument, as Darwin himself noted, lies in the assumption that only complicated eyes are useful. In fact, many animals depend on eyes that are far less complex than our own **(Figure 23.22)**. The simplest eyes that we know of are patches of light-sensitive photoreceptor cells. These simple eyes appear to have had a single evolutionary origin and are now found in a variety of animals, including small molluscs called limpets. Such eyes have no equipment for focusing images, but they do enable the animal to distinguish light from dark. Limpets cling more tightly to their rock when a shadow falls on them, a behavioral adaptation that reduces the risk of being eaten. Limpets have had a long evolutionary history, demonstrating that their "simple" eyes are quite adequate to support their survival and reproduction.

In the animal kingdom, complex eyes have evolved independently from such basic structures many times. Some molluscs, such as squids and octopuses, have eyes as complex as those of humans and other vertebrates (see Figure 23.22). Although complex mollusc eyes

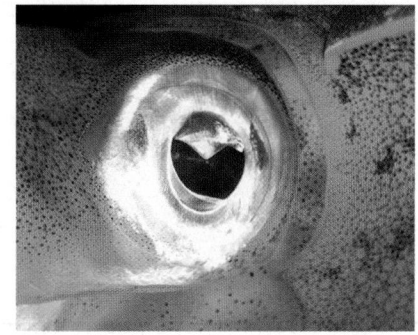

▲ **The complex eye of a bigfin reef squid (*Sepioteuthis lessoniana*).**

evolved independently of vertebrate eyes, both evolved from a simple cluster of photoreceptor cells present in a common ancestor. In each case, the complex eye evolved through a series of steps that benefited the eyes' owners at every stage.

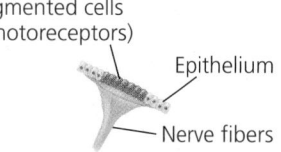

▼ **Figure 23.22 A range of eye complexity among molluscs.**

**(a) Patch of pigmented cells**

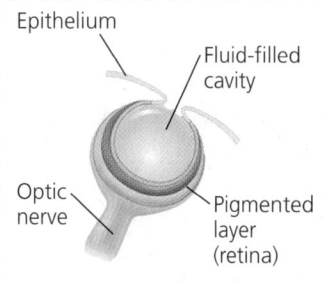

Pigmented cells (photoreceptors)
Epithelium
Nerve fibers

The limpet *Patella* has a simple patch of photoreceptors.

**(b) Eyecup**

Pigmented cells
Nerve fibers

The slit shell mollusc *Pleurotomaria* has an eyecup.

**(c) Pinhole camera-type eye**

Epithelium
Fluid-filled cavity
Optic nerve
Pigmented layer (retina)

The *Nautilus* eye functions like a pinhole camera (an early type of camera lacking a lens).

**(d) Eye with primitive lens**

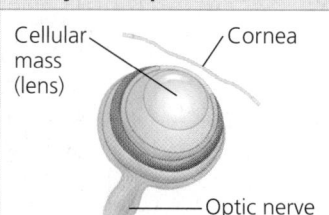

Cellular mass (lens)
Cornea
Optic nerve

The marine snail *Murex* has a primitive lens consisting of a mass of crystal-like cells. The cornea is a transparent region of tissue that protects the eye and helps focus light.

**(e) Complex camera lens-type eye**

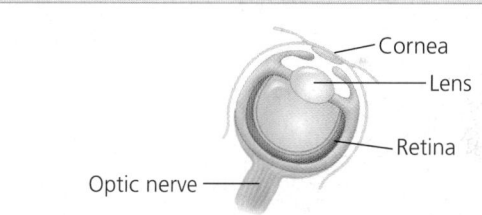

Cornea
Lens
Retina
Optic nerve

The squid *Loligo* has a complex eye with features (cornea, lens, and retina) similar to those of vertebrate eyes. However, the squid eye evolved independently from vertebrate eyes.

Evidence of their independent evolution can be found in their structure: Vertebrate eyes detect light at the back layer of the retina and conduct nerve impulses toward the front, while complex mollusc eyes do the reverse.

Throughout their evolutionary history, eyes retained their basic function of vision. But evolutionary novelties can also arise when structures that originally played one role gradually acquire a different one. For example, as cynodonts gave rise to early mammals, bones that formerly comprised the jaw hinge (the articular and quadrate; see Figure 23.5) were incorporated into the ear region of mammals, where they eventually took on a new function: the transmission of sound. Structures that evolve in one context but become co-opted for another function are sometimes called *exaptations* to distinguish them from the adaptive origin of the original structure. Note that the concept of exaptation does not imply that a structure somehow evolves in anticipation of future use. Natural selection cannot predict the future; it can only improve a structure in the context of its *current* utility. Novel features, such as the new jaw hinge and ear bones of early mammals, can arise gradually via a series of intermediate stages, each of which has some function in the organism's current context.

## Evolutionary Trends

What else can we learn from patterns of macroevolution? Consider evolutionary "trends" observed in the fossil record. For instance, some evolutionary lineages exhibit a trend toward larger or smaller body size. An example is the evolution of the present-day horse (genus *Equus*), a descendant of the 55-million-year-old *Hyracotherium* (**Figure 23.23**). About the

size of a large dog, *Hyracotherium* had four toes on its front feet, three toes on its hind feet, and teeth adapted for browsing on bushes and trees. In comparison, present-day horses are larger, have only one toe on each foot, and possess teeth modified for grazing on grasses.

Extracting a single evolutionary progression from the fossil record can be misleading, however; it is like describing a bush as growing toward a single point by tracing only the branches that lead to that twig. For example, by selecting certain species from the available fossils, it is possible to arrange a succession of animals intermediate between *Hyracotherium* and living horses that shows a trend toward large, single-toed species

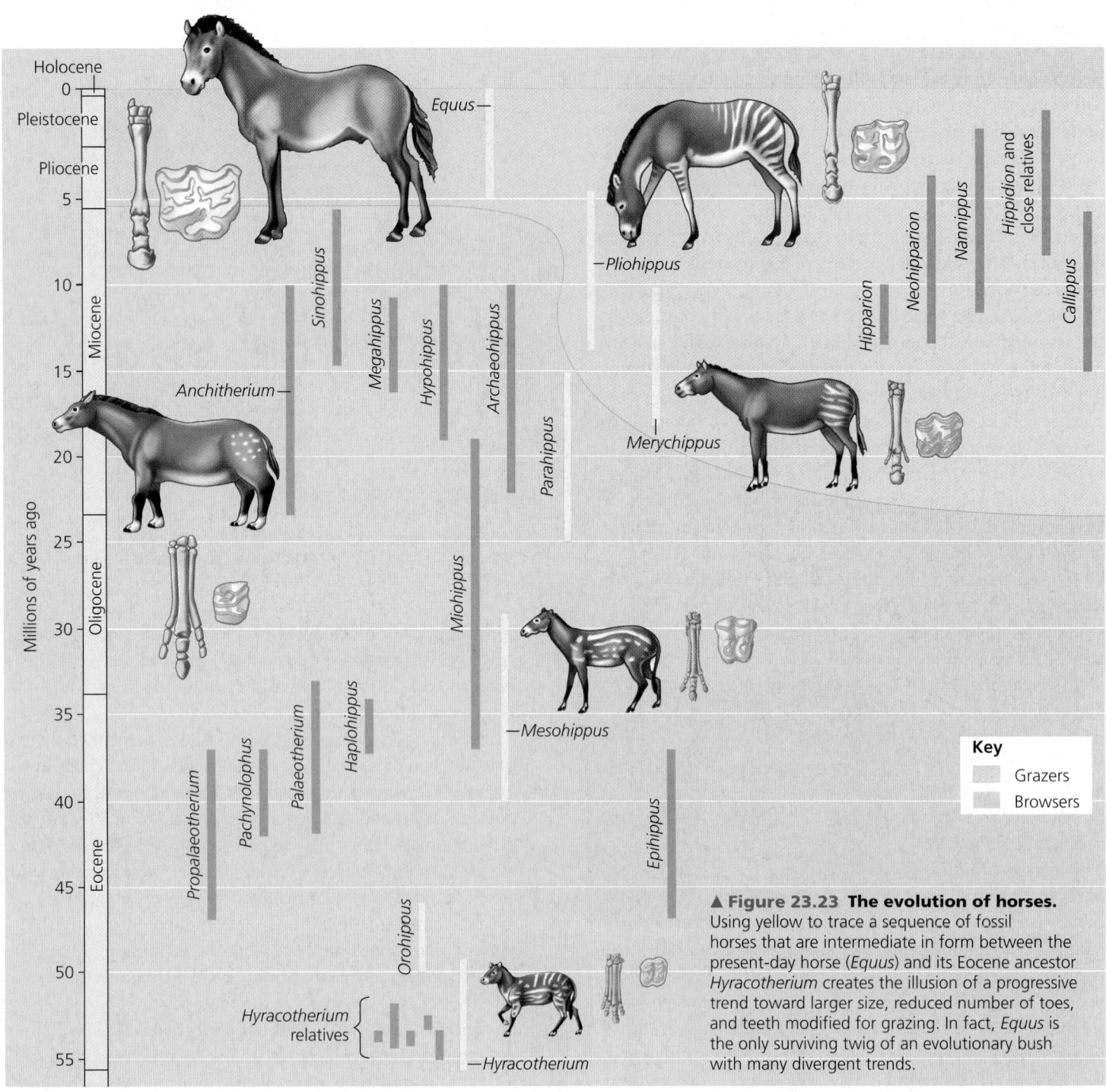

▲ **Figure 23.23  The evolution of horses.** Using yellow to trace a sequence of fossil horses that are intermediate in form between the present-day horse (*Equus*) and its Eocene ancestor *Hyracotherium* creates the illusion of a progressive trend toward larger size, reduced number of toes, and teeth modified for grazing. In fact, *Equus* is the only surviving twig of an evolutionary bush with many divergent trends.

**Key**
Grazers
Browsers

(follow the yellow highlighting in Figure 23.23). However, if we consider *all* fossil horses known today, this apparent trend vanishes. The genus *Equus* did not evolve in a straight line; it is the only surviving twig of an evolutionary tree that is so branched that it is more like a bush. *Equus* actually descended through a series of speciation episodes that included several adaptive radiations, not all of which led to large, one-toed, grazing horses. In fact, phylogenetic analyses suggest that all lineages that include grazers are closely related to *Parahippus*; the many other horse lineages, all of which are now extinct, remained multi-toed browsers for 35 million years.

Branching evolution *can* result in a real evolutionary trend even if some species counter the trend. One model of long-term trends views species as analogous to individuals: Speciation is their birth, extinction is their death, and new species that diverge from them are their offspring. In this model, just as populations of individual organisms undergo natural selection, species undergo *species selection*. The species that endure the longest and generate the most new offspring species determine the direction of major evolutionary trends. The species selection model suggests that "differential speciation success" plays a role in macroevolution similar to the role of differential reproductive success in microevolution. Evolutionary trends can also result directly from natural selection. For example, when horse ancestors invaded the grasslands that spread during the mid-Cenozoic, there was strong selection for grazers that could escape predators by running faster. This trend would not have occurred without open grasslands.

Whatever its cause, an evolutionary trend does not imply that there is some intrinsic drive toward a particular phenotype. Evolution is the result of the interactions between organisms and their current environments; if environmental conditions change, an evolutionary trend may cease or even reverse itself. The cumulative effect of these ongoing interactions between organisms and their environments is enormous: It is through them that the staggering diversity of life—Darwin's "endless forms most beautiful"—has arisen.

## CONCEPT CHECK 23.4

1. How can the Darwinian concept of descent with modification explain the evolution of such complex structures as the vertebrate eye?
2. **WHAT IF?** The myxoma virus kills up to 99.8% of infected European rabbits in populations with no previous exposure to the virus. The virus is transmitted between living rabbits by mosquitoes. Describe an evolutionary trend (in either the rabbit or virus) that might occur after a rabbit population first encounters the virus.

For suggested answers, see Appendix A.

# 23 Chapter Review

Go to **MasteringBiology®** for Assignments, the eText, and the Study Area with Animations, Activities, Vocab Self-Quiz, and Practice Tests.

## SUMMARY OF KEY CONCEPTS

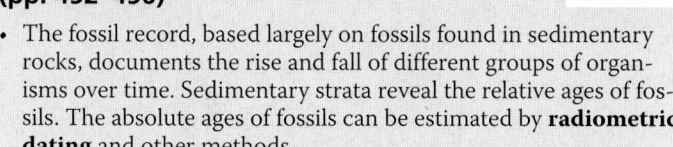

VOCAB SELF-QUIZ

goo.gl/gbai8v

### CONCEPT 23.1

**The fossil record documents life's history (pp. 452–456)**

- The fossil record, based largely on fossils found in sedimentary rocks, documents the rise and fall of different groups of organisms over time. Sedimentary strata reveal the relative ages of fossils. The absolute ages of fossils can be estimated by **radiometric dating** and other methods.
- The study of fossils has helped geologists establish a **geologic record** of Earth's history.
- The fossil record shows how new groups of organisms can arise via the gradual modification of preexisting organisms.

**?** *What are the challenges of estimating the absolute ages of old fossils? Explain how these challenges may be overcome in some circumstances.*

### CONCEPT 23.2

**The rise and fall of groups of organisms reflect differences in speciation and extinction rates (pp. 456–465)**

- In **plate tectonics**, continental plates move gradually over time, altering the physical geography and climate of Earth. These changes lead to extinctions in some groups of organisms and bursts of speciation in others.
- Evolutionary history has been punctuated by five **mass extinctions** that radically altered the history of life. Some of these extinctions may have been caused by changes in continent positions, volcanic activity, or impacts from meteorites or comets.
- Large increases in the diversity of life have resulted from **adaptive radiations** that followed mass extinctions. Adaptive radiations have also occurred in groups of organisms that possessed major evolutionary innovations or that colonized new regions in which there was little competition from other organisms.

**?** *Explain how the broad evolutionary changes seen in the fossil record are the cumulative result of speciation and extinction events.*

### CONCEPT 23.3

**Major changes in body form can result from changes in the sequences and regulation of developmental genes (pp. 465–468)**

- Developmental genes affect morphological differences between species by influencing the rate, timing, and spatial patterns of change in an organism's form as it develops into an adult.
- The evolution of new forms can be caused by changes in the nucleotide sequences or regulation of developmental genes.

**?** *How could changes in a single gene or DNA region ultimately lead to the origin of a new group of organisms?*

CONCEPT 23.4

## Evolution is not goal oriented (pp. 468–471)

- Novel and complex biological structures can evolve through a series of incremental modifications, each of which benefits the organism that possesses it.
- Evolutionary trends can be caused by factors such as natural selection in a changing environment or species selection. Like all aspects of evolution, evolutionary trends result from interactions between organisms and their current environments.

**?** *Explain the reasoning behind the statement "Evolution is not goal oriented."*

# TEST YOUR UNDERSTANDING

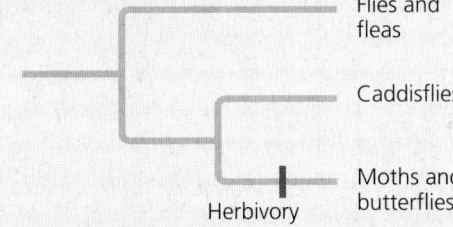

PRACTICE TEST

goo.gl/CRZjvS

## Level 1: Knowledge/Comprehension

1. Which factor most likely caused animals and plants in India to differ greatly from species in nearby southeast Asia?
   - (A) The species became separated by convergent evolution.
   - (B) The climates of the two regions are similar.
   - (C) India is in the process of separating from the rest of Asia.
   - (D) India was a separate continent until 45 million years ago.

2. Adaptive radiations can be a direct consequence of three of the following four factors. Select the exception.
   - (A) vacant ecological niches
   - (B) genetic drift
   - (C) colonization of an isolated region that contains suitable habitat and few competitor species
   - (D) evolutionary innovation

3. A researcher discovers a fossil of what appears to be one of the oldest-known multicellular organisms. The researcher could estimate the age of this fossil based on
   - (A) the amount of uranium-238 in the fossil.
   - (B) the amount of carbon-14 in the fossil.
   - (C) the amount of uranium-238 in volcanic layers surrounding the fossil.
   - (D) the amount of carbon-14 in volcanic layers surrounding the fossil.

## Level 2: Application/Analysis

4. A genetic change that caused a certain *Hox* gene to be expressed along the tip of a vertebrate limb bud instead of farther back helped make possible the evolution of the tetrapod limb. This type of change is illustrative of
   - (A) the influence of environment on development.
   - (B) paedomorphosis.
   - (C) a change in a developmental gene or in its regulation that altered the spatial organization of body parts.
   - (D) heterochrony.

5. A swim bladder is a gas-filled sac that helps fish maintain buoyancy. The evolution of the swim bladder from the air-breathing organ (a simple lung) of an ancestral fish is an example of
   - (A) exaptation.
   - (B) changes in *Hox* gene expression.
   - (C) paedomorphosis.
   - (D) adaptive radiation.

## Level 3: Synthesis/Evaluation

6. **SCIENTIFIC INQUIRY**
   Herbivory (plant eating) has evolved repeatedly in insects, typically from meat-eating or detritus-feeding ancestors (detritus

is dead organic matter). Moths and butterflies, for example, eat plants, whereas their "sister group" (the insect group to which they are most closely related), the caddisflies, feed on animals, fungi, or detritus. As illustrated in this phylogenetic tree, the combined moth/butterfly and caddisfly group shares a common ancestor with flies and fleas. Like caddisflies, flies and fleas are thought to have evolved from ancestors that did not eat plants.

Flies and fleas

Caddisflies

Herbivory

Moths and butterflies

There are 140,000 species of moths and butterflies and 7,000 species of caddisflies. State a hypothesis about the impact of herbivory on adaptive radiations in insects. How could this hypothesis be tested?

7. **FOCUS ON EVOLUTION**
   Describe how gene flow, genetic drift, and natural selection all can influence macroevolution.

8. **FOCUS ON ORGANIZATION**
   You have seen many examples of how form fits function at all levels of the biological hierarchy. However, we can imagine forms that would function better than some forms actually found in nature. For example, if the wings of a bird were not formed from its forelimbs, such a hypothetical bird could fly yet also hold objects with its forelimbs. In a short essay (100–150 words), use the concept of "evolution as tinkering" to explain why there are limits to the functionality of forms in nature.

9. **SYNTHESIZE YOUR KNOWLEDGE**

In 2010, the Soufrière Hills volcano on the Caribbean island of Montserrat erupted violently, spewing huge clouds of ash and gases into the sky. Explain how the volcanic eruptions at the end of the Permian period and the formation of Pangaea, both of which occurred about 252 million years ago, set in motion events that altered evolutionary history.

*For selected answers, see Appendix A.*

# Unit 4 The Evolutionary History of Life

## 24 Early Life and the Diversification of Prokaryotes

Life on Earth began 3.5 billion years ago with the origin of single-celled **prokaryotes**. Over this long history, a wide range of metabolic adaptations have evolved in prokaryotes, enabling them to thrive throughout the biosphere.

## 25 The Origin and Diversification of Eukaryotes

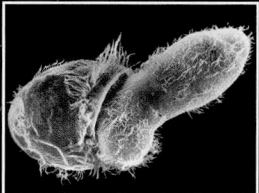

Following the metabolic diversification of prokaryotes, the origin of **eukaryotes** 1.8 billion years ago led to the evolution of a vast array of structurally complex organisms—the protists, plants, fungi, and animals that fill our world today.

## 26 The Colonization of Land

The **colonization of land** by plants and fungi 500 million years ago transformed terrestrial environments from a "green slime" consisting of bacteria and single-celled eukaryotes to lush forests and other plant communities.

## 27 The Rise of Animal Diversity

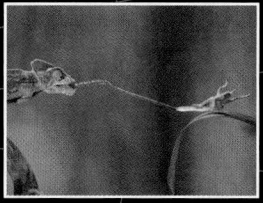

The earliest animals were microscopic and lived in marine environments. By 530 million years ago, the origin of larger, mobile animals with complex nervous and digestive systems led to an explosive **radiation of animals**, transforming the microorganism-only world to a world inhabited also by predators and other large eukaryotes.

The Evolutionary History of Life

24 Early Life and the Diversification of Prokaryotes
25 The Origin and Diversification of Eukaryotes
26 The Colonization of Land
27 The Rise of Animal Diversity

Evolution
Genetics
Chemistry and Cells
Ecology
Animals
Plants

# 24 Early Life and the Diversification of Prokaryotes

## KEY CONCEPTS

**24.1** Conditions on early Earth made the origin of life possible

**24.2** Diverse structural and metabolic adaptations have evolved in prokaryotes

**24.3** Rapid reproduction, mutation, and genetic recombination promote genetic diversity in prokaryotes

**24.4** Prokaryotes have radiated into a diverse set of lineages

**24.5** Prokaryotes play crucial roles in the biosphere

▲ **Figure 24.1** What organisms lived on early Earth?

## The First Cells

Our planet formed 4.6 billion years ago, condensing from a vast cloud of dust and rocks that surrounded the young sun. For its first few hundred million years, Earth was bombarded by huge chunks of rock and ice left over from the birth of the solar system. The collisions generated so much heat that all of the available water was vaporized, preventing the formation of seas and lakes. As a result, life probably could not have originated or survived during this time.

This massive bombardment ended about 4 billion years ago, setting the stage for the origin of life on our young planet. While chemical signatures of life date back to 3.8 billion years ago, the earliest direct evidence comes from fossils that are 3.5 billion years old. These fossils are of **prokaryotes**, an informal term for single-celled organisms in domains Bacteria and Archaea (see Figure 20.21). Some of the earliest prokaryotic cells lived in dense mats similar to those that resemble stepping stones in **Figure 24.1**; others lived as free-floating, individual cells. These early prokaryotes were Earth's first organisms, and their prokaryotic descendants had the planet to themselves for about 1.5 billion years—until eukaryotes first appeared about 1.8 billion years ago (see Concept 25.1).

Over their long evolutionary history, descendants of Earth's first cells have given rise to the vast diversity of prokaryotes living today. This diversity includes "extreme" species such as *Deinococcus radiodurans*, which can survive 3 million rads of radiation (3,000 times the dose fatal to humans). Other prokaryotes live in environments that are too cold or hot or salty for most other organisms, and some have even been found living in rocks 3.2 km (2 miles) below Earth's surface.

But prokaryotic species also thrive in more "normal" habitats—the lands and waters in which most other species are found. And within these lands and waters, prokaryotes have colonized the bodies of other organisms that live there, including humans **(Figure 24.2)**. Their ability to live in a broad range of habitats helps explain why prokaryotes are the most abundant organisms on Earth—indeed, the number of prokaryotes

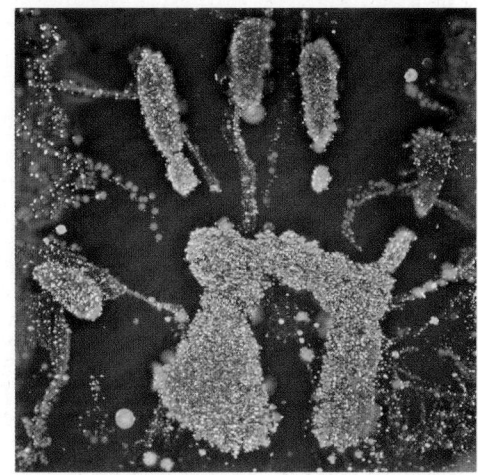

▶ **Figure 24.2**
**Bacteria that inhabit the human body.**
Touching an agar gel led to the handprint-shaped growth of *Staphylococcus epidermidis*, just one of more than 1,000 species of bacteria that live on or in the human body.

in a handful of fertile soil is greater than the number of people who have ever lived. In this chapter, we'll examine the origin, adaptations, diversity, and enormous ecological impact of these remarkable organisms.

## CONCEPT 24.1

# Conditions on early Earth made the origin of life possible

The earliest fossils are of prokaryotes that lived 3.5 billion years ago. But how did the first living cells appear? Observations and experiments in chemistry, geology, and physics have led scientists to propose one scenario that we'll examine here. They hypothesize that chemical and physical processes could have produced simple cells through a sequence of four main stages:

1. The abiotic (nonliving) synthesis of small organic molecules, such as amino acids and nitrogenous bases
2. The joining of these small molecules into macromolecules, such as proteins and nucleic acids
3. The packaging of these molecules into **protocells**, droplets with membranes that maintained an internal chemistry different from that of their surroundings
4. The origin of self-replicating molecules that eventually made inheritance possible

Though speculative, this scenario leads to predictions that can be tested in the laboratory. In this section, we'll examine some of the evidence for each stage.

## Synthesis of Organic Compounds on Early Earth

As the bombardment of early Earth ended, the first atmosphere had little oxygen and was probably thick with water vapor, along with various compounds released by volcanic eruptions, including nitrogen and its oxides, carbon dioxide, methane, ammonia, and hydrogen. As Earth cooled, the water vapor condensed into oceans, and much of the hydrogen escaped into space.

During the 1920s, Russian chemist A. I. Oparin and British scientist J. B. S. Haldane independently hypothesized that

Earth's early atmosphere was a reducing (electron-adding) environment, in which organic compounds could have formed from simpler molecules. The energy for this synthesis could have come from lightning and UV radiation. Haldane suggested that the early oceans were a solution of organic molecules, a "primitive soup" from which life arose.

In 1953, Stanley Miller, working with Harold Urey at the University of Chicago, tested the Oparin-Haldane hypothesis by creating laboratory conditions comparable to those that scientists at the time thought existed on early Earth. His apparatus yielded a variety of amino acids found in organisms today, along with other organic compounds. Many laboratories have since repeated Miller's classic experiment using different recipes for the atmosphere, some of which also produced organic compounds.

However, some evidence suggests that the early atmosphere was made up primarily of nitrogen and carbon dioxide and was neither reducing nor oxidizing (electron removing). Recent Miller-Urey–type experiments using such "neutral" atmospheres have also produced organic molecules. In addition, small pockets of the early atmosphere, such as those near the openings of volcanoes, may have been reducing. Perhaps the first organic compounds formed near volcanoes or deep-sea vents, where hot water and minerals gush into the ocean from Earth's interior. In a 2008 test of the volcanic-atmosphere hypothesis, researchers used modern equipment to reanalyze molecules that Miller had saved from one of his experiments. The 2008 study found that numerous amino acids had formed under conditions that simulated a volcanic eruption **(Figure 24.3)**.

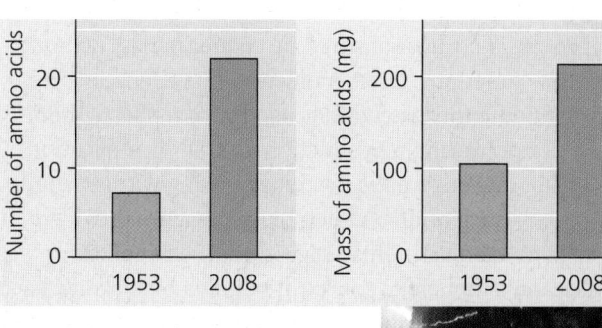

▲ **Figure 24.3 Amino acid synthesis in a simulated volcanic eruption.** In addition to his classic 1953 study, Miller also conducted an experiment simulating a volcanic eruption. In a 2008 reanalysis of those results, researchers found that far more amino acids were produced under simulated volcanic conditions than were produced in the conditions of the original 1953 experiment.

**MAKE CONNECTIONS** *Explain how more than 20 amino acids could have been produced in the 2008 experiment. (See Concept 3.5.)*

Miller-Urey–type experiments show that the abiotic synthesis of organic molecules is possible under various conditions. Another source of organic molecules may have been meteorites. For example, fragments of the Murchison meteorite, a 4.5-billion-year-old rock that landed in Australia in 1969, contain more than 80 amino acids, some in large amounts. These amino acids cannot be contaminants from Earth because they consist of an equal mix of two different structural forms—only one of which is typically produced or used by organisms on our planet. Recent studies have shown that the Murchison meteorite also contained other key organic molecules, including lipids, simple sugars, and nitrogenous bases such as uracil.

## Abiotic Synthesis of Macromolecules

The presence of small organic molecules, such as amino acids and nitrogenous bases, is not sufficient for the emergence of life as we know it. Every cell has many types of macromolecules, including enzymes and other proteins and the nucleic acids needed for self-replication. Could such macromolecules have formed on early Earth? A 2009 study demonstrated that one key step, the abiotic synthesis of RNA monomers, can occur spontaneously from simple precursor molecules. In addition, by dripping solutions of amino acids or RNA nucleotides onto hot sand, clay, or rock, researchers have produced polymers of these molecules. The polymers formed spontaneously, without the help of enzymes or ribosomes. Unlike proteins, the amino acid polymers are a complex mix of linked and cross-linked amino acids. Still, it is possible that such polymers acted as weak catalysts for a variety of chemical reactions on early Earth.

## Protocells

All organisms must be able to carry out both reproduction and energy processing (metabolism). Life cannot persist without both of these functions. DNA molecules carry genetic information, including the instructions needed to replicate themselves accurately during reproduction. But DNA replication requires elaborate enzymatic machinery, along with an abundant supply of nucleotide building blocks that are provided by the cell's metabolism. This suggests that self-replicating molecules and a metabolic source of building blocks may have appeared together in early protocells. How did that happen?

The necessary conditions may have been met in *vesicles*, fluid-filled compartments enclosed by a membrane-like structure. Recent experiments show that abiotically produced vesicles can exhibit certain properties of life, including simple reproduction and metabolism, as well as the maintenance of an internal chemical environment different from that of their surroundings.

For example, vesicles can form spontaneously when lipids or other organic molecules are added to water. When this occurs, molecules that have both a hydrophobic region and a hydrophilic region can organize into a bilayer similar to the lipid bilayer of a plasma membrane. Adding substances

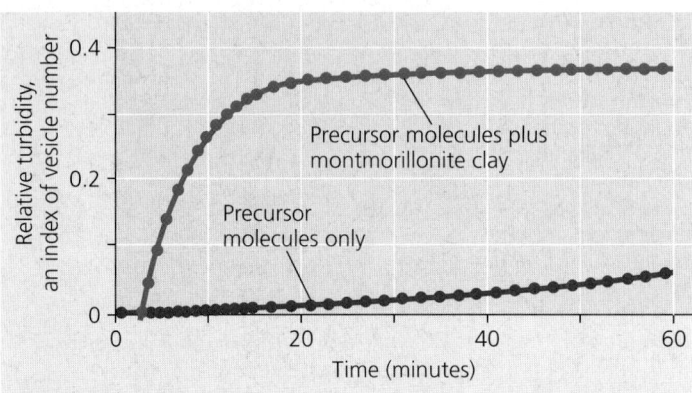

**(a) Self-assembly.** The presence of montmorillonite clay greatly increases the rate of vesicle self-assembly.

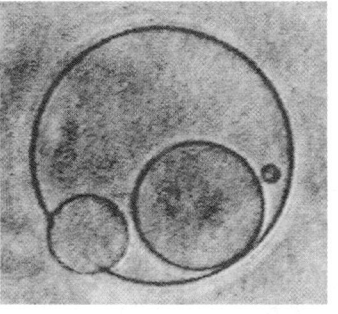

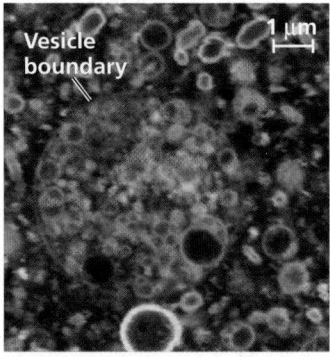

**(b) Reproduction.** Vesicles can divide on their own, as in this vesicle "giving birth" to smaller vesicles (LM).

**(c) Absorption of RNA.** This vesicle has incorporated montmorillonite clay particles coated with RNA (orange).

▲ **Figure 24.4 Features of abiotically produced vesicles.**

**MAKE CONNECTIONS** *Explain how molecules with both a hydrophobic region and a hydrophilic region can self-assemble into a bilayer when in water. (See Concept 3.4.)*

such as *montmorillonite*, a soft mineral clay produced by the weathering of volcanic ash, greatly increases the rate of vesicle self-assembly **(Figure 24.4a)**. This clay, which is thought to have been common on early Earth, provides surfaces on which organic molecules become concentrated, increasing the likelihood that the molecules will react with each other and form vesicles. Abiotically produced vesicles can "reproduce" on their own **(Figure 24.4b)**, and they can increase in size ("grow") without dilution of their contents. Vesicles also can absorb montmorillonite particles, including those on which RNA and other organic molecules have become attached **(Figure 24.4c)**. Finally, experiments have shown that some vesicles have a selectively permeable bilayer and can perform metabolic reactions using an external source of reagents—another important prerequisite for life.

## Self-Replicating RNA

The first genetic material was most likely RNA, not DNA. RNA plays a central role in protein synthesis, but it can also function as an enzyme-like catalyst (see Figure 14.13). Such

RNA catalysts are called **ribozymes**. Some ribozymes can make complementary copies of short pieces of RNA, provided that they are supplied with nucleotide building blocks.

Natural selection on the molecular level has produced ribozymes capable of self-replication in the laboratory. How does this occur? Unlike double-stranded DNA, which takes the form of a uniform helix, single-stranded RNA molecules assume a variety of specific three-dimensional shapes mandated by their nucleotide sequences. In a given environment, RNA molecules with certain nucleotide sequences may have shapes that enable them to replicate faster and with fewer errors than other sequences. The RNA molecule with the greatest ability to replicate itself will leave the most descendant molecules. Occasionally, a copying error will result in a molecule with a shape that is even more adept at self-replication than the ancestral sequence. Similar selection events may have occurred on early Earth. Thus, life as we know it may have been preceded by an "RNA world," in which small RNA molecules were able to replicate and to store genetic information about the vesicles that carried them.

A vesicle with self-replicating, catalytic RNA would differ from its many neighbors that lacked such molecules. If that vesicle could grow, split, and pass its RNA molecules to its daughters, the daughters would be protocells that had some of the properties of their parent. Although the first such protocells likely carried only limited amounts of genetic information, specifying only a few properties, their inherited characteristics could have been acted on by natural selection. The most successful of the early protocells would have increased in number because they could exploit their resources effectively and pass their abilities on to subsequent generations.

Once RNA sequences that carried genetic information appeared in protocells, many additional changes would have been possible. For example, RNA could have provided the template on which DNA nucleotides were assembled. Double-stranded DNA is a more chemically stable repository for genetic information than is the more fragile RNA. DNA also can be replicated more accurately. Accurate replication was advantageous as genomes grew larger through gene duplication and other processes and as more properties of the protocells became coded in genetic information. Once DNA appeared, the stage was set for a blossoming of new forms of life—a change we see documented in the fossil record.

## Fossil Evidence of Early Life

Many of the oldest known fossils are of *stromatolites*, layered rocks that form from the activities of certain prokaryotes **(Figure 24.5)**. The earliest stromatolites date to 3.5 billion years ago. For several hundred million years, all such fossils were similar in overall structure and all were from shallow marine bays; stromatolites are still found in such bays today (see Figure 23.3). By 3.1 billion years ago, stromatolites with two distinctly different morphologies had appeared, and by 2.8 billion years ago, stromatolites occurred in salty lakes as well as marine environments. Thus, early fossil stromatolites show signs of ecological and evolutionary change over time.

Ancient fossils of individual prokaryotic cells have also been discovered that are nearly as old as the oldest stromatolites (see Figure 24.5). For example, a 2011 study found fossilized prokaryotic cells in 3.4-billion-year-old rocks from Australia. In South Africa, other researchers have found 3.4-billion-year-old fossils of prokaryotes that resemble cyanobacteria, a group of photosynthetic bacteria living today. Some scientists question whether the South African fossils really were cyanobacteria, but by 2.5 billion years ago, diverse communities of cyanobacteria lived in the oceans. Cyanobacteria remained the main photosynthetic organisms for over a billion years, and they continue to be one of the most important groups of photosynthetic organisms alive today.

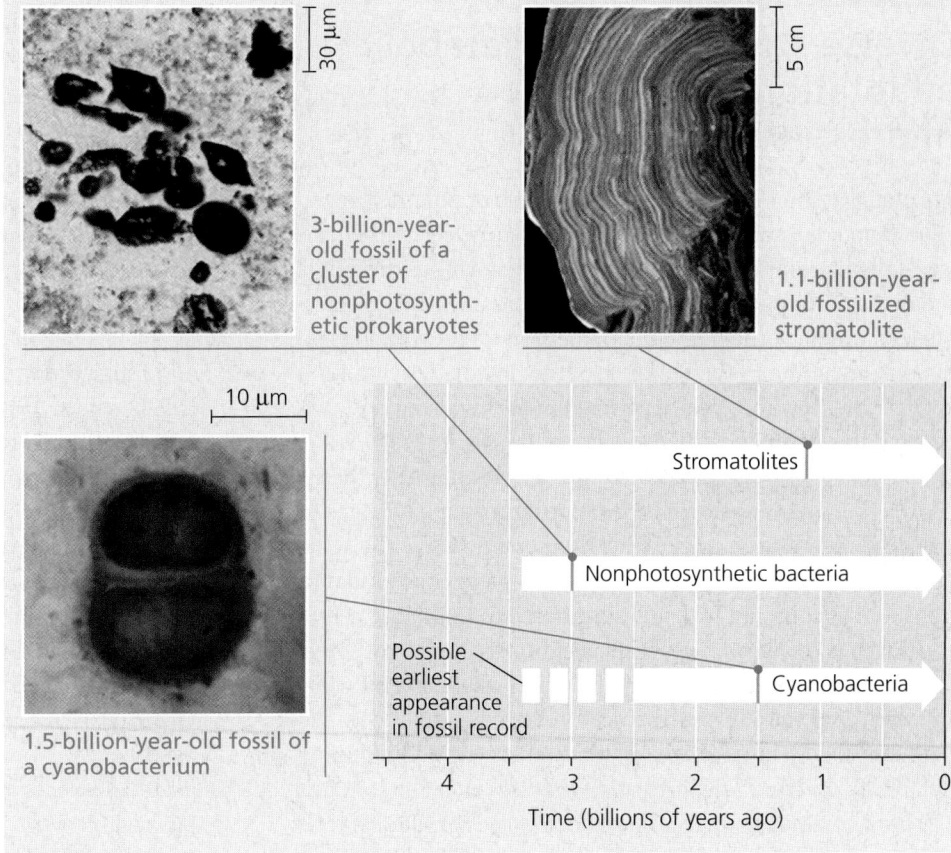

3-billion-year-old fossil of a cluster of nonphotosynthetic prokaryotes

1.1-billion-year-old fossilized stromatolite

10 μm

1.5-billion-year-old fossil of a cyanobacterium

Stromatolites

Nonphotosynthetic bacteria

Possible earliest appearance in fossil record

Cyanobacteria

Time (billions of years ago)

▲ **Figure 24.5 Appearance in the fossil record of early prokaryote groups.**

Early cyanobacteria began what is arguably the greatest impact organisms have ever had on our planet: the release of oxygen to Earth's atmosphere during the water-splitting step of photosynthesis. In some of its chemical forms, oxygen attacks chemical bonds and can inhibit enzymes and damage cells. As a result, the rising concentration of atmospheric $O_2$ probably doomed many prokaryotic groups. Some species survived in habitats that remained anaerobic, where we find their descendants living today. As we'll see, among other survivors, diverse adaptations to the changing atmosphere evolved, including cellular respiration, which uses $O_2$ in the process of harvesting the energy stored in organic molecules.

## CONCEPT CHECK 24.1

1. What hypothesis did Miller test in his classic experiment?
2. How would the appearance of protocells have represented a key step in the origin of life?
3. Summarize fossil evidence of early prokaryotes. Describe how these organisms altered Earth's atmosphere.
4. **MAKE CONNECTIONS** In changing from an "RNA world" to today's "DNA world," genetic information must have flowed from RNA to DNA. After reviewing Figures 14.4 and 17.7, suggest how this could have occurred. Does such a flow occur today?

For suggested answers, see Appendix A.

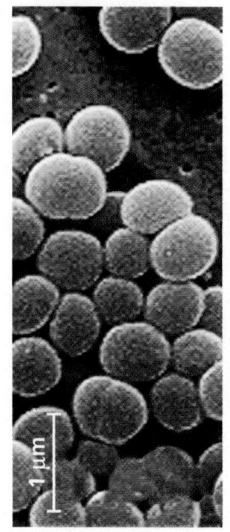 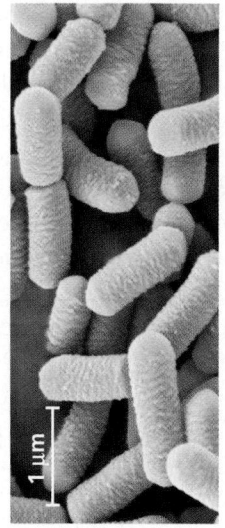 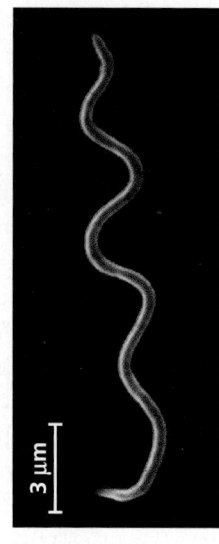

**(a) Spherical**    **(b) Rod-shaped**    **(c) Spiral**

▲ **Figure 24.6  The most common shapes of prokaryotes.**
**(a)** Cocci (singular, *coccus*) are spherical prokaryotes. They occur singly, in pairs (diplococci), in chains of many cells (streptococci), and in clusters resembling bunches of grapes (staphylococci). **(b)** Bacilli (singular, *bacillus*) are rod-shaped prokaryotes. They are usually solitary, but in some forms the rods are arranged in chains (streptobacilli). **(c)** Spiral prokaryotes include spirilla, which range from comma-like shapes to loose coils, and spirochetes (shown here), which are corkscrew-shaped (colorized SEMs).

---

## CONCEPT 24.2

# Diverse structural and metabolic adaptations have evolved in prokaryotes

Throughout their long history, prokaryotic populations have been (and continue to be) subjected to natural selection in all kinds of environments, resulting in their enormous diversity today. As described in Concept 24.1, fossils of early prokaryotes document some of the major steps in their evolutionary history, including the appearance of the first photosynthetic organisms. However, prokaryotic populations have also evolved in ways that cannot be seen in the fossil record, including changes in the type and efficiency of their enzymes. Although we cannot trace the time course of such changes in the fossil record, we can examine their end results—the adaptations found in prokaryotes today. We'll survey those adaptations here, beginning with a description of prokaryotic cells.

Most prokaryotes are unicellular, although the cells of some species remain attached to each other after cell division. Prokaryotic cells typically have diameters of 0.5–5 µm, much smaller than the 10–100 µm diameter of many eukaryotic cells. (One notable exception, *Thiomargarita namibiensis*, can be as large as 750 µm in diameter—bigger than the dot on this i.) Prokaryotic cells have a variety of shapes **(Figure 24.6)**. Finally, although they are unicellular and small, prokaryotes are well organized, achieving all of an organism's life functions within a single cell.

## Cell-Surface Structures

A key feature of nearly all prokaryotic cells is the cell wall, which maintains cell shape, protects the cell, and prevents it from bursting in a hypotonic environment (see Concept 5.3). In a hypertonic environment, most prokaryotes lose water and shrink away from their wall (plasmolyze). Such water losses can inhibit cell reproduction. Thus, salt can be used to preserve foods because it causes food-spoiling prokaryotes to lose water, preventing them from rapidly multiplying.

The cell walls of prokaryotes differ in structure from those of eukaryotes. In eukaryotes that have cell walls, such as plants and fungi, the walls are usually made of cellulose or chitin (see Concept 3.3). In contrast, most bacterial cell walls contain **peptidoglycan**, a polymer composed of modified sugars cross-linked by short polypeptides. This molecular fabric encloses the entire bacterium and anchors other molecules that extend from its surface. Archaeal cell walls contain a variety of polysaccharides and proteins but lack peptidoglycan.

Using a staining technique developed by the Dutch scientist Hans Christian Gram, biologists can categorize many bacterial species according to cell wall composition **(Figure 24.7)**. **Gram-positive** bacteria have relatively simple walls composed of a thick layer of peptidoglycan. The walls of **gram-negative** bacteria have less peptidoglycan and are structurally more complex, with an outer membrane that contains lipopolysaccharides (carbohydrates bonded to lipids). These differences in cell wall composition can have medical implications. The lipid portions of the lipopolysaccharides in the walls of many gram-negative

▼ **Figure 24.7 Gram staining.**

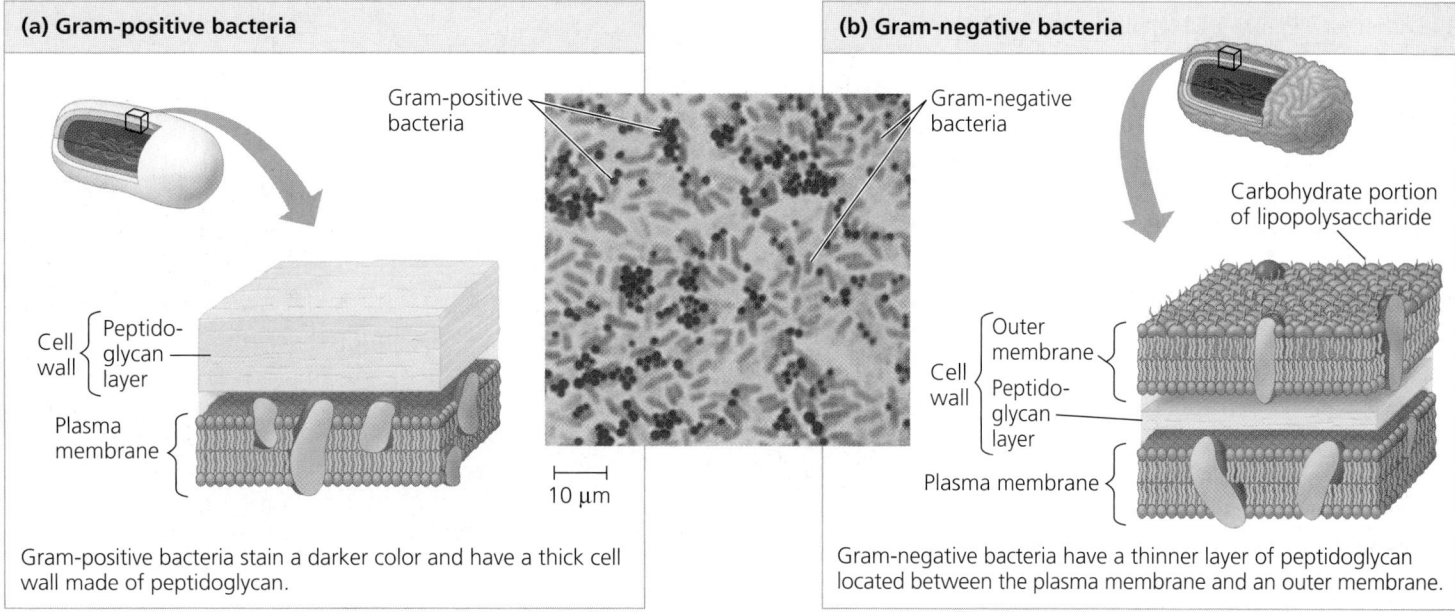

**(a) Gram-positive bacteria**

Gram-positive bacteria

Cell wall — Peptido-glycan layer

Plasma membrane

10 µm

Gram-positive bacteria stain a darker color and have a thick cell wall made of peptidoglycan.

**(b) Gram-negative bacteria**

Gram-negative bacteria

Carbohydrate portion of lipopolysaccharide

Cell wall — Outer membrane — Peptido-glycan layer

Plasma membrane

Gram-negative bacteria have a thinner layer of peptidoglycan located between the plasma membrane and an outer membrane.

bacteria are toxic, causing fever or shock. Furthermore, the outer membrane of a gram-negative bacterium helps protect it from the body's defenses. Gram-negative bacteria also tend to be more resistant than gram-positive species to antibiotics because the outer membrane impedes entry of the drugs. However, some gram-positive species have virulent strains that are resistant to one or more antibiotics. (Figure 19.15 discusses one example: methicillin-resistant *Staphylococcus aureus*, or MRSA, which can cause lethal skin infections.)

The cell wall of many prokaryotes is surrounded by a sticky layer of polysaccharide or protein. This layer is called a **capsule** if it is dense and well defined **(Figure 24.8)** or a *slime layer* if it is not as well organized. Both kinds of sticky outer layers enable prokaryotes to adhere to their substrate or to other individuals in a colony. Some capsules and slime layers protect against dehydration, and some shield pathogenic prokaryotes from attacks by their host's immune system.

Other bacteria develop resistant cells called **endospores** when they lack water or essential nutrients. The original cell produces a copy of its chromosome and surrounds that copy with a multilayered structure, forming the endospore. Water is removed from the endospore, and its metabolism halts. The original cell then lyses, releasing the endospore. Most endospores are so durable that they can survive in boiling water; killing them requires heating lab equipment to 121°C under high pressure. In less hostile environments, endospores can remain dormant but viable for centuries, able to rehydrate and resume metabolism when their environment improves.

Finally, some prokaryotes stick to their substrate or to one another by means of hairlike appendages called **fimbriae** (singular, *fimbria*) **(Figure 24.9)**. For example, the bacterium that causes gonorrhea, *Neisseria gonorrhoeae*, uses fimbriae to fasten itself to the mucous membranes of its host. Fimbriae are usually shorter and more numerous than **pili** (singular, *pilus*),

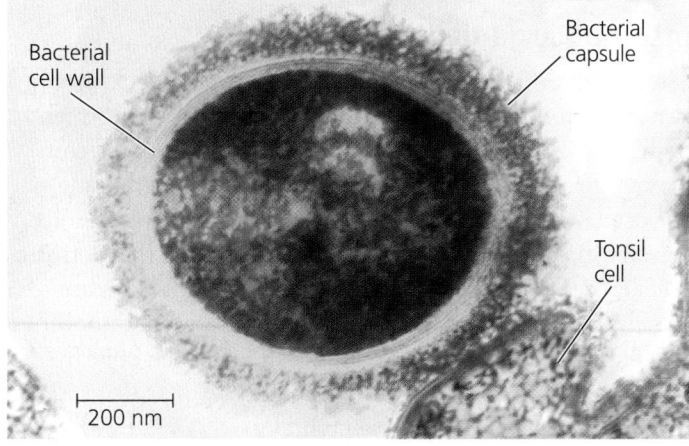

Bacterial cell wall

Bacterial capsule

Tonsil cell

200 nm

▲ **Figure 24.8 Capsule.** The polysaccharide capsule around this *Streptococcus* bacterium enables the prokaryote to attach to cells in the respiratory tract—in this colorized TEM, a tonsil cell.

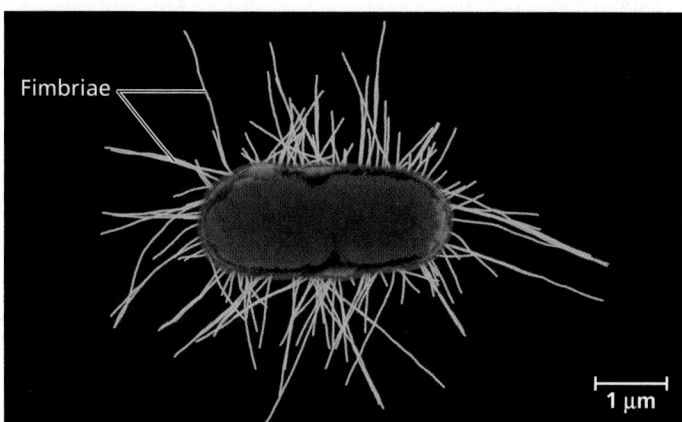

Fimbriae

1 µm

▲ **Figure 24.9 Fimbriae.** These numerous protein-containing appendages enable some prokaryotes to attach to surfaces or to other cells (colorized TEM).

appendages that pull two cells together prior to DNA transfer from one cell to the other (see Figure 24.16); pili are sometimes referred to as *sex pili*.

## Motility

About half of all prokaryotes are capable of **taxis**, a directed movement toward or away from a stimulus (from the Greek *taxis*, to arrange). For example, prokaryotes that exhibit *chemotaxis* change their movement pattern in response to chemicals. They may move *toward* nutrients or oxygen (positive chemotaxis) or *away from* a toxic substance (negative chemotaxis). Some species can move at velocities exceeding 50 μm/sec—up to 50 times their body length per second. For perspective, consider that a person 1.7 m tall moving that fast would be running 306 km (190 miles) per hour!

Of the various structures that enable prokaryotes to move, the most common are flagella **(Figure 24.10)**. Flagella (singular, *flagellum*) may be scattered over the entire surface of the cell or concentrated at one or both ends. Prokaryotic flagella differ greatly from eukaryotic flagella: They are one-tenth the width and typically are not covered by an extension of the plasma membrane (see Figure 4.23). The flagella of prokaryotes and eukaryotes also differ in their molecular composition and their mechanism of propulsion. Among prokaryotes, bacterial and archaeal flagella are similar in size and rotational mechanism, but they are composed of entirely different and unrelated proteins. Overall, these structural and molecular comparisons indicate that the flagella of bacteria, archaea, and eukaryotes arose independently. Since current evidence shows that the flagella of organisms in the three domains perform similar functions but are not related by common descent, they are described as analogous, not homologous, structures (see Concept 20.2).

### *Evolutionary Origins of Bacterial Flagella*

The bacterial flagellum shown in Figure 24.10 has three main parts (the motor, hook, and filament) that are themselves composed of 42 different kinds of proteins. How could such a complex structure evolve? In fact, much evidence indicates that bacterial flagella originated as simpler structures that were modified in a stepwise fashion over time. As in the case of the human eye (see Concept 23.4), biologists asked whether a less complex version of the flagellum could still benefit its owner. Analyses of hundreds of bacterial genomes indicate that only half of the flagellum's protein components appear to be necessary for it to function; the others are inessential or not encoded in the genomes of some species. Of the 21 proteins required

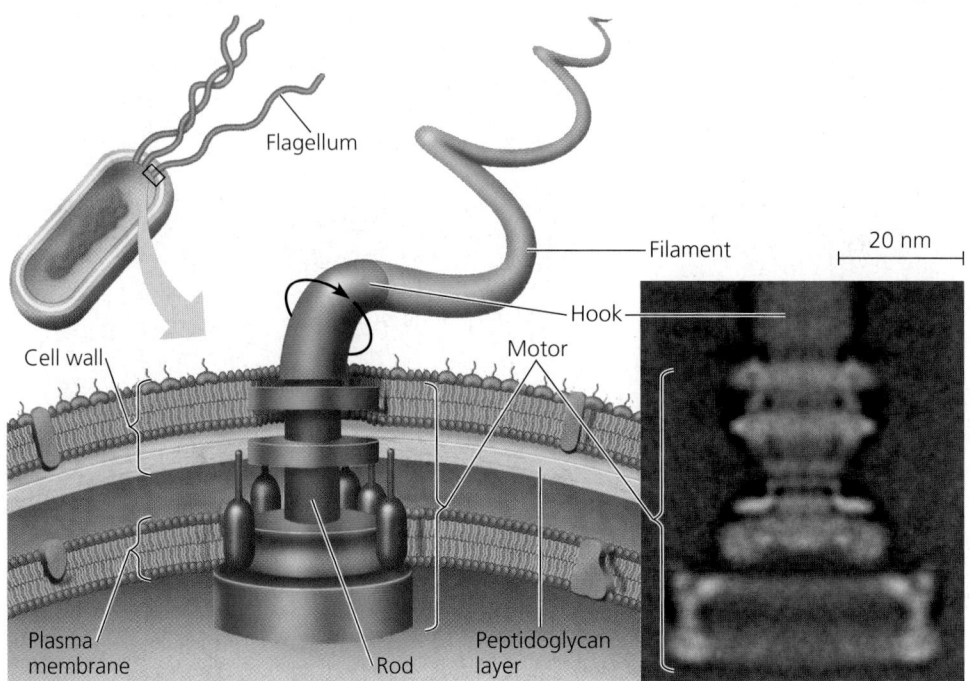

▲ **Figure 24.10  A prokaryotic flagellum.** The motor of a prokaryotic flagellum consists of a system of rings embedded in the cell wall and plasma membrane (TEM). The electron transport chain pumps protons out of the cell. The diffusion of protons back into the cell provides the force that turns a curved hook and thereby causes the attached filament to rotate and propel the cell. (This diagram shows flagellar structures characteristic of gram-negative bacteria.)

by all species studied to date, 19 are modified versions of proteins that perform other tasks in bacteria. For example, a set of 10 proteins in the motor is homologous to 10 similar proteins in a secretory system found in bacteria. (A secretory system is a protein complex that enables a cell to produce and release certain macromolecules.) Two other proteins in the motor are homologous to proteins that function in ion transport. The proteins that comprise the rod, hook, and filament are all related to each other and are descended from an ancestral protein that formed a pilus-like tube. These findings suggest that the bacterial flagellum evolved as other proteins were added to an ancestral secretory system. This is an example of *exaptation*, the process in which structures originally adapted for one purpose take on new functions through descent with modification.

## Internal Organization and DNA

The cells of prokaryotes are simpler than those of eukaryotes in both their internal structure and the physical arrangement of their DNA (see Figure 4.4). Prokaryotic cells lack the complex compartmentalization associated with the membrane-enclosed organelles found in eukaryotic cells. However, some prokaryotic cells do have specialized membranes that perform metabolic functions **(Figure 24.11)**. These membranes are usually infoldings of the plasma membrane. Recent discoveries also indicate that some prokaryotes can store metabolic by-products in simple compartments composed of proteins; these compartments do not have a membrane.

The genome of a prokaryote is structurally different from a eukaryotic genome and in most cases has considerably less

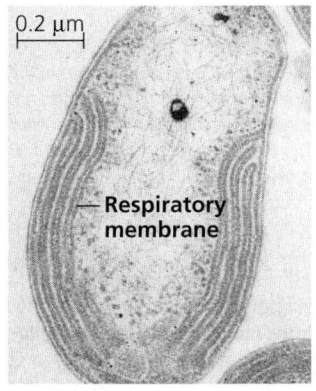

**(a) Aerobic prokaryote**

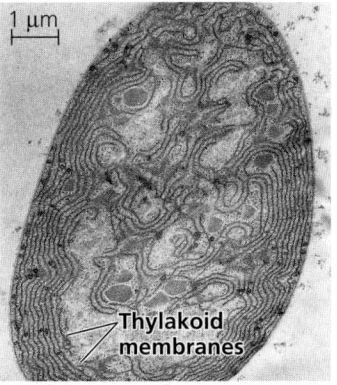

**(b) Photosynthetic prokaryote**

▲ **Figure 24.11 Specialized membranes of prokaryotes.**
**(a)** Infoldings of the plasma membrane, reminiscent of the cristae of mitochondria, function in cellular respiration in some aerobic prokaryotes (TEM). **(b)** Photosynthetic prokaryotes called cyanobacteria have thylakoid membranes, much like those in chloroplasts (TEM).

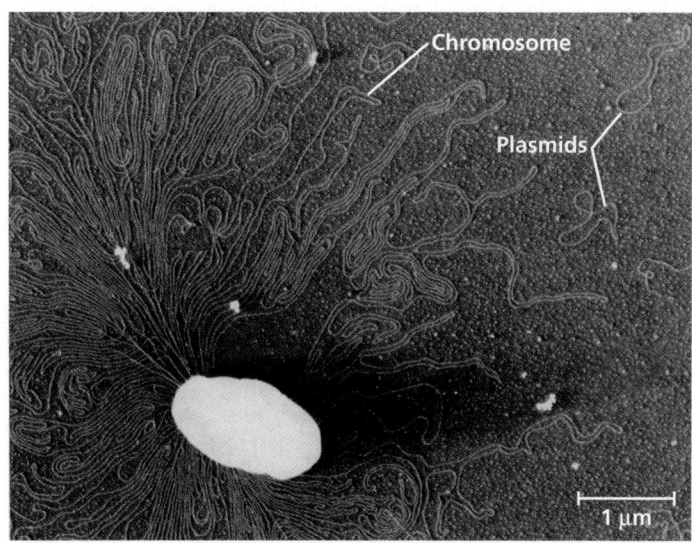

▲ **Figure 24.12 A prokaryotic chromosome and plasmids.**
The thin, tangled loops surrounding this ruptured *E. coli* cell are parts of the cell's large, circular chromosome (colorized TEM). Three of the cell's plasmids, the much smaller rings of DNA, are also shown.

DNA. Prokaryotes generally have circular chromosomes **(Figure 24.12)**, whereas eukaryotes have linear chromosomes. In addition, in prokaryotes the chromosome is associated with many fewer proteins than are the chromosomes of eukaryotes. Also unlike eukaryotes, prokaryotes lack a nucleus; their chromosome is located in the **nucleoid**, a region of cytoplasm that is not enclosed by a membrane. In addition to its single chromosome, a typical prokaryotic cell may also have much smaller rings of independently replicating DNA molecules called **plasmids** (see Figure 24.12), most carrying only a few genes.

Although DNA replication, transcription, and translation are fundamentally similar processes in prokaryotes and eukaryotes, some of the details are different (see Chapter 14). For example, prokaryotic ribosomes are slightly smaller than eukaryotic ribosomes and differ in their protein and RNA content. These differences allow certain antibiotics, such as erythromycin and tetracycline, to bind to ribosomes and block protein synthesis in prokaryotes but not in eukaryotes. As a result, people can use these antibiotics to kill or inhibit the growth of bacteria without harming themselves.

## Nutritional and Metabolic Adaptations

Like all organisms, prokaryotes can be categorized by how they obtain energy and the carbon used in building organic molecules. Every type of nutrition observed in eukaryotes is represented among prokaryotes, along with some nutritional modes unique to prokaryotes. In fact, prokaryotes have an astounding range of metabolic adaptations, much broader than that found in eukaryotes.

Organisms that obtain energy from light are called *phototrophs*, and those that obtain

energy from chemicals are called *chemotrophs*. Organisms that need only $CO_2$ or related compounds as a carbon source are called *autotrophs*. In contrast, *heterotrophs* require at least one organic nutrient, such as glucose, to make other organic compounds. Combining possible energy sources and carbon sources results in four major modes of nutrition, summarized in **Table 24.1**.

### The Role of Oxygen in Metabolism

Prokaryotic metabolism also varies with respect to oxygen ($O_2$). *Obligate aerobes* must use $O_2$ for cellular respiration and cannot grow without it. *Obligate anaerobes,* on the other hand, are poisoned by $O_2$. Some obligate anaerobes live exclusively by fermentation; others extract chemical energy by

| Table 24.1 Major Nutritional Modes | | | |
|---|---|---|---|
| **Mode** | **Energy Source** | **Carbon Source** | **Types of Organisms** |
| **Autotroph** | | | |
| **Photoautotroph** | Light | $CO_2$, $HCO_3^-$, or related compound | Photosynthetic prokaryotes (for example, cyanobacteria); plants; certain protists (for example, algae) |
| **Chemoautotroph** | Inorganic chemicals (such as $H_2S$, $NH_3$, or $Fe^{2+}$) | $CO_2$, $HCO_3^-$, or related compound | Unique to certain prokaryotes (for example, *Sulfolobus*) |
| **Heterotroph** | | | |
| **Photoheterotroph** | Light | Organic compounds | Unique to certain aquatic and salt-loving prokaryotes (for example, *Rhodobacter*, *Chloroflexus*) |
| **Chemoheterotroph** | Organic compounds | Organic compounds | Many prokaryotes (for example, *Clostridium*) and protists; fungi; animals; some plants |

**anaerobic respiration**, in which substances other than $O_2$, such as nitrate ions ($NO_3^-$) or sulfate ions ($SO_4^{2-}$), accept electrons at the "downhill" end of electron transport chains. *Facultative anaerobes* use $O_2$ if it is present but can also carry out fermentation or anaerobic respiration in an anaerobic environment.

### Nitrogen Metabolism

Nitrogen is essential for the production of amino acids and nucleic acids in all organisms. Whereas eukaryotes can obtain nitrogen from only a limited group of nitrogen compounds, prokaryotes can metabolize nitrogen in a wide variety of forms. For example, some cyanobacteria and some methanogens (a group of archaea) convert atmospheric nitrogen ($N_2$) to ammonia ($NH_3$), a process called **nitrogen fixation**. The cells can then incorporate this "fixed" nitrogen into amino acids and other organic molecules. In terms of their nutrition, nitrogen-fixing cyanobacteria are some of the most self-sufficient organisms, since they need only light, $CO_2$, $N_2$, water, and some minerals to grow.

Nitrogen fixation by prokaryotes has a large impact on other organisms. For example, it can increase the nitrogen available to plants, which cannot use atmospheric nitrogen but can use the nitrogen compounds that the prokaryotes produce from ammonia. Concept 42.4 discusses this and other essential roles of prokaryotes in the nitrogen cycles of ecosystems.

### Metabolic Cooperation

Cooperation between prokaryotic cells allows them to use environmental resources they could not use as individual cells. In some cases, this cooperation takes place between specialized cells of a filament. For instance, the cyanobacterium *Anabaena* has genes that encode proteins for photosynthesis and for nitrogen fixation, but a single cell cannot carry out both processes at the same time. The reason is that photosynthesis produces $O_2$, which inactivates the enzymes involved in nitrogen fixation. Instead of living as isolated cells, *Anabaena* forms filamentous chains **(Figure 24.13)**. Most cells in a filament carry out only photosynthesis, while a few specialized cells called **heterocysts** (sometimes called *heterocytes*) carry out only nitrogen fixation. Each heterocyst is surrounded by a thickened cell wall that

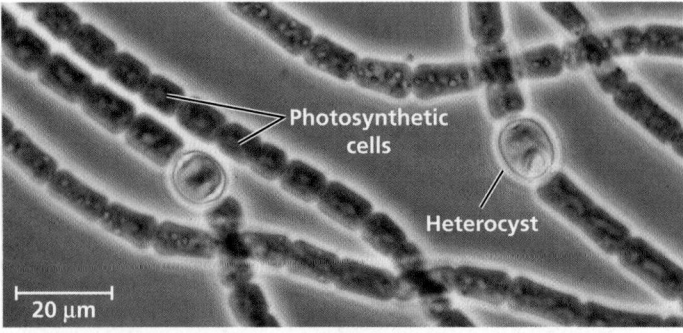

▲ **Figure 24.13 Metabolic cooperation in a prokaryote.** In the filamentous cyanobacterium *Anabaena*, cells called heterocysts fix nitrogen, while the other cells carry out photosynthesis (LM). *Anabaena* is found in many freshwater lakes.

restricts entry of $O_2$ produced by neighboring photosynthetic cells. Intercellular connections allow heterocysts to transport fixed nitrogen to neighboring cells and to receive carbohydrates.

Metabolic cooperation between different prokaryotic species often occurs in surface-coating colonies known as **biofilms**. Cells in a biofilm secrete signaling molecules that recruit nearby cells, causing the colonies to grow. The cells also produce polysaccharides and proteins that stick the cells to the substrate and to one another. Channels in the biofilm allow nutrients to reach cells in the interior and wastes to be expelled. Biofilms are common in nature, but they can cause problems by contaminating industrial products and medical equipment and contributing to tooth decay and more serious health problems. Altogether, damage caused by biofilms costs billions of dollars annually.

## Reproduction

Many prokaryotes can reproduce quickly in favorable environments. By *binary fission* (see Figure 9.12), a single prokaryotic cell divides into 2 cells, which then divide into 4, 8, 16, and so on. Under optimal conditions, many prokaryotes can divide every 1–3 hours; some species can produce a new generation in only 20 minutes. At this rate, a single prokaryotic cell could give rise to a colony outweighing Earth in only two days!

In reality, of course, this does not occur. The cells eventually exhaust their nutrient supply, poison themselves with metabolic wastes, face competition from other microorganisms, or are consumed by other organisms. Still, the fact that many prokaryotic species can divide after short periods of time draws attention to three key features of their biology: *They are small, they reproduce by binary fission, and they often have short generation times.* As a result, prokaryotic populations can consist of many trillions of individuals—far more than populations of multicellular eukaryotes, such as plants or animals.

## Adaptations of Prokaryotes: *A Summary*

Let's step back and examine the big picture of the adaptations that have arisen in prokaryotic populations. We've described some of their key structural features, such as cell walls, endospores, fimbriae, and flagella. But prokaryotic cells are simpler structurally than are eukaryotic cells—they do not vary as much in shape or size, and they lack the complex compartmentalization associated with the membrane-enclosed organelles of eukaryotic cells. Indeed, the ongoing success of prokaryotes is not primarily a story of structural diversification; rather, their success is an extraordinary example of physiological and metabolic diversification. As we've seen, prokaryotes thrive under a wide variety of physical and chemical conditions, and they have an astonishing range of metabolic adaptations that allow them to obtain energy and carbon in these environments.

Overall, the metabolic diversification of prokaryotes can be viewed as a first great wave of adaptive radiation in the evolutionary history of life. Bearing that broad perspective in mind, we turn now to the genetic diversity that has enabled the adaptations found in prokaryotic populations.

1. Contrast the cellular and DNA structures of prokaryotes and eukaryotes.
2. Distinguish between the four major modes of nutrition, noting which are unique to prokaryotes.
3. **MAKE CONNECTIONS** Suggest a hypothesis to explain why the thylakoid membranes of chloroplasts resemble those of cyanobacteria. Refer to Figures 4.16 and 20.21.
4. **WHAT IF?** Describe what you might eat for a typical meal if humans, like cyanobacteria, could fix nitrogen.

For suggested answers, see Appendix A.

## CONCEPT 24.3

# Rapid reproduction, mutation, and genetic recombination promote genetic diversity in prokaryotes

As we saw in Unit Three, evolution cannot occur without genetic variation. The evolutionary changes seen in the prokaryotic fossil record and the diverse adaptations found in prokaryotes living today suggest that their populations must have considerable genetic variation—and they do. In this section, we'll examine three factors that give rise to high levels of genetic diversity in prokaryotes: rapid reproduction, mutation, and genetic recombination.

## Rapid Reproduction and Mutation

In eukaryotes, sexual reproduction is a source of genetic variation, but prokaryotes do not reproduce sexually. Moreover, when a prokaryotic cell divides (by binary fission) to form two daughter cells, the generation of a novel allele by a new mutation is rare for any particular gene. How, then, does the extensive genetic variation of prokaryotes arise? In many species, the key is rapid reproduction combined with mutation, even at a low rate.

Consider the bacterium *Escherichia coli* as it reproduces by binary fission in a human intestine, one of its natural environments. After repeated rounds of division, most of the offspring cells are genetically identical to the original parent cell. However, if errors occur during DNA replication, some of the offspring cells may differ genetically. The probability of such a mutation occurring in a given *E. coli* gene is about one in 10 million $(1 \times 10^{-7})$ per cell division. But among the $2 \times 10^{10}$ new *E. coli* cells that arise each day in a person's intestine, there will be approximately $(2 \times 10^{10}) \times (1 \times 10^{-7}) = 2,000$ bacteria that have a mutation in that gene. Thus, the total number of new mutations when all 4,300 *E. coli* genes are considered is about $4,300 \times 2,000$—more than 8 million per day per human host.

The key point is that new mutations, though rare on a per gene basis, can increase genetic diversity quickly in species with short generation times and large populations. This diversity, in turn, can lead to rapid evolution: Individuals that

are genetically better equipped for their environment tend to survive and reproduce at higher rates than other individuals **(Figure 24.14)**. The ability of prokaryotes to adapt rapidly to new conditions highlights the point that although the structure

## ▼ Figure 24.14 Inquiry

### Can prokaryotes evolve rapidly in response to environmental change?

**Experiment** Vaughn Cooper and Richard Lenski tested the ability of *E. coli* populations to adapt to a new environment. They established 12 populations, each founded by a single cell from an *E. coli* strain, and followed these populations for 20,000 generations (3,000 days). To maintain a continual supply of resources, each day the researchers performed a *serial transfer*: They transferred 0.1 mL of each population to a new tube containing 9.9 mL of fresh growth medium. The growth medium used throughout the experiment provided a challenging environment that contained only low levels of glucose and other resources needed for growth.

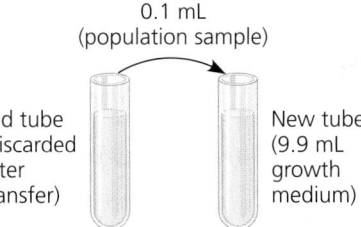

**Daily serial transfer**

0.1 mL
(population sample)

Old tube
(discarded after transfer)

New tube
(9.9 mL growth medium)

Samples were periodically removed from the 12 populations and grown in competition with the common ancestral strain in the experimental (low-glucose) environment.

**Results** The fitness of the experimental populations, as measured by the growth rate of each population, increased rapidly for the first 5,000 generations (2 years) and more slowly for the next 15,000 generations. The graph shows the averages for the 12 populations.

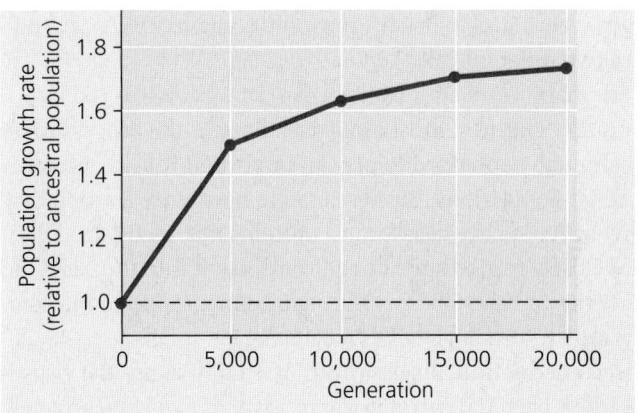

**Conclusion** Populations of *E. coli* continued to accumulate beneficial mutations for 20,000 generations, allowing rapid evolution of increased population growth rates in their new environment.

**Data from** V. S. Cooper and R. E. Lenski, The population genetics of ecological specialization in evolving *Escherichia coli* populations, *Nature* 407:736–739 (2000).

**WHAT IF?** Suggest possible functions of the genes whose sequence or expression was altered as the experimental populations evolved in the low-glucose environment.

of their cells is simpler than that of eukaryotic cells, prokaryotes are not "primitive" or "inferior" in an evolutionary sense. They are, in fact, highly evolved: For 3.5 billion years, prokaryotic populations have responded successfully to many types of environmental challenges.

## Genetic Recombination

Although new mutations are a major source of variation in prokaryotic populations, additional diversity arises from *genetic recombination*, the combining of DNA from two sources. In eukaryotes, the sexual processes of meiosis and fertilization combine DNA from two individuals in a single zygote. But meiosis and fertilization do not occur in prokaryotes. Instead, three other mechanisms—transformation, transduction, and conjugation—can bring together prokaryotic DNA from different individuals (that is, different cells). When the individuals are members of different species, this movement of genes from one organism to another is called *horizontal gene transfer*. Although scientists have found evidence that each of these mechanisms can transfer DNA within and between species in both domain Bacteria and domain Archaea, to date most of our knowledge comes from research on bacteria.

### Transformation and Transduction

In **transformation**, the genotype and possibly phenotype of a prokaryotic cell are altered by the uptake of foreign DNA from its surroundings. For example, a harmless strain of *Streptococcus pneumoniae* can be transformed into pneumonia-causing cells if the cells are exposed to DNA from a pathogenic strain (see Concept 13.1). This transformation occurs when a nonpathogenic cell takes up a piece of DNA carrying the allele for pathogenicity and replaces its own allele with the foreign allele, an exchange of homologous DNA segments. The cell is now a recombinant: Its chromosome contains DNA derived from two different cells.

For many years after transformation was discovered in laboratory cultures, most biologists thought the process to be too rare and haphazard to play an important role in natural bacterial populations. But researchers have since learned that many bacteria have cell-surface proteins that recognize DNA from closely related species and transport it into the cell. Once inside the cell, the foreign DNA can be incorporated into the genome by homologous DNA exchange.

In **transduction**, phages (from "bacteriophages," the viruses that infect bacteria) carry prokaryotic genes from one host cell to another. In most cases, transduction results from accidents that occur during the phage replicative cycle (**Figure 24.15**). A virus that carries prokaryotic DNA may not be able to replicate because it lacks some or all of its own genetic material. However, the virus can attach to another prokaryotic cell (a recipient) and inject prokaryotic DNA acquired from the first cell (the donor). If some of this DNA is then incorporated into the recipient cell's chromosome by crossing over, a recombinant cell is formed.

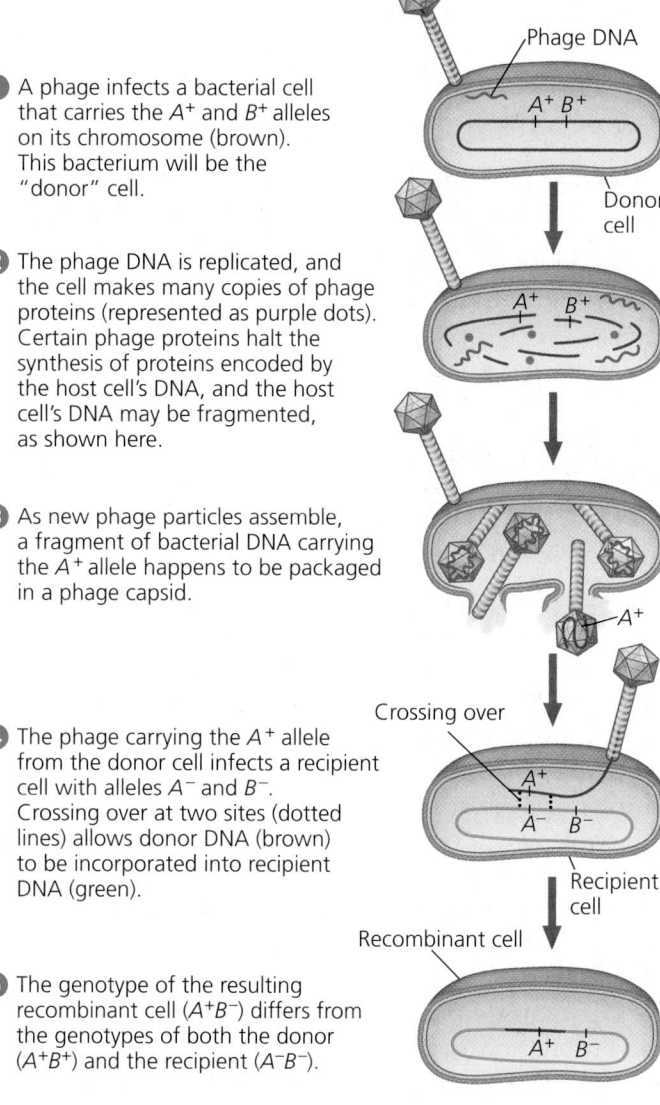

❶ A phage infects a bacterial cell that carries the $A^+$ and $B^+$ alleles on its chromosome (brown). This bacterium will be the "donor" cell.

❷ The phage DNA is replicated, and the cell makes many copies of phage proteins (represented as purple dots). Certain phage proteins halt the synthesis of proteins encoded by the host cell's DNA, and the host cell's DNA may be fragmented, as shown here.

❸ As new phage particles assemble, a fragment of bacterial DNA carrying the $A^+$ allele happens to be packaged in a phage capsid.

❹ The phage carrying the $A^+$ allele from the donor cell infects a recipient cell with alleles $A^-$ and $B^-$. Crossing over at two sites (dotted lines) allows donor DNA (brown) to be incorporated into recipient DNA (green).

❺ The genotype of the resulting recombinant cell ($A^+B^-$) differs from the genotypes of both the donor ($A^+B^+$) and the recipient ($A^-B^-$).

▲ **Figure 24.15 Transduction.** Phages may carry pieces of a bacterial chromosome from one cell (the donor) to another (the recipient). If crossing over occurs after the transfer, genes from the donor may be incorporated into the recipient's genome.

**?** *Under what circumstances would a transduction event result in horizontal gene transfer?*

### Conjugation and Plasmids

In a process called **conjugation**, DNA is transferred between two prokaryotic cells (usually of the same species) that are temporarily joined. In bacteria, the DNA transfer is always one-way: One cell donates the DNA, and the other receives it. We'll focus here on the mechanism used by *E. coli*.

First, a pilus of the donor cell attaches to the recipient (**Figure 24.16**). The pilus then retracts, pulling the two cells together, like a grappling hook. The next step is thought to be the formation of a temporary structure between the two cells, a "mating bridge" through which the donor may transfer DNA to the recipient. However, the mechanism by which DNA transfer occurs is unclear; indeed, recent evidence indicates that DNA may pass directly through the hollow pilus.

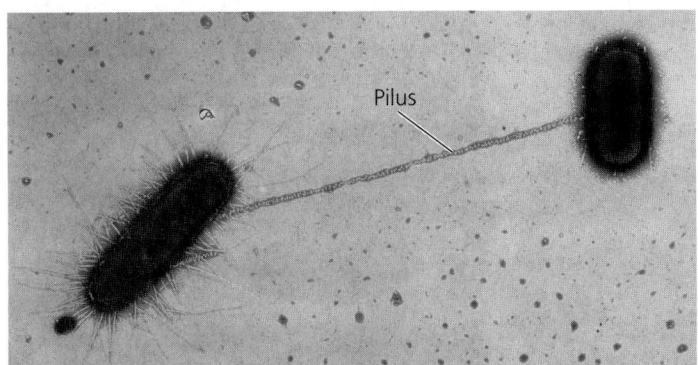

▲ **Figure 24.16 Bacterial conjugation.** The *E. coli* donor cell (left) extends a pilus that attaches to a recipient cell, a key first step in the transfer of DNA. The pilus is a flexible tube of protein subunits (TEM).

**The F Factor** The ability to form pili and donate DNA during conjugation results from the presence of a particular piece of DNA called the **F factor** (F for fertility). The F factor of *E. coli* consists of about 25 genes, most required for the production of pili. The F factor can exist either as a plasmid or as a segment of DNA within the bacterial chromosome.

The F factor in its plasmid form is called the **F plasmid**. Cells containing the F plasmid, designated F⁺ cells, function as DNA donors during conjugation. Cells lacking the F factor, designated F⁻, function as DNA recipients during conjugation. The F⁺ condition is transferable in the sense that an F⁺ cell converts an F⁻ cell to F⁺ if a copy of the entire F plasmid is transferred **(Figure 24.17)**. Even if this does not occur, as long as some of the F plasmid's DNA is transferred successfully to the recipient cell, that cell is now a recombinant cell.

A donor cell's F factor can also be integrated into the chromosome. In this case, chromosomal genes can be transferred to a recipient cell during conjugation. When this occurs, homologous regions of the donor and recipient chromosomes may align, allowing segments of their DNA to be exchanged. As a result, the recipient cell becomes a recombinant bacterium that has genes derived from the chromosomes of two different cells—a new genetic variant on which evolution can act.

**R Plasmids and Antibiotic Resistance** During the 1950s in Japan, physicians started noticing that some hospital patients with bacterial dysentery, which produces severe diarrhea, did not respond to antibiotics that had been effective in the past. Apparently, resistance to these antibiotics had evolved in some strains of *Shigella*, the bacterium that causes the disease.

Eventually, researchers began to identify the specific genes that confer antibiotic resistance in *Shigella* and other pathogenic bacteria. Sometimes, mutation in a chromosomal gene of the pathogen can confer resistance. For example, a mutation in one gene may make it less likely that the pathogen will transport a particular antibiotic into its cell. Mutation in a different gene may alter the intracellular target protein for an antibiotic molecule, reducing its inhibitory effect. In other cases, bacteria have "resistance genes," which code for enzymes that specifically destroy or otherwise hinder the effectiveness of certain antibiotics, such as tetracycline or ampicillin. Such resistance genes are often carried by plasmids known as **R plasmids** (R for resistance).

Exposing a bacterial population to a specific antibiotic will kill antibiotic-sensitive bacteria but not those that happen to have R plasmids with genes that counter the antibiotic. Under these circumstances, we would predict that natural selection would cause the fraction of the bacterial population carrying genes for antibiotic resistance to increase, and that is exactly

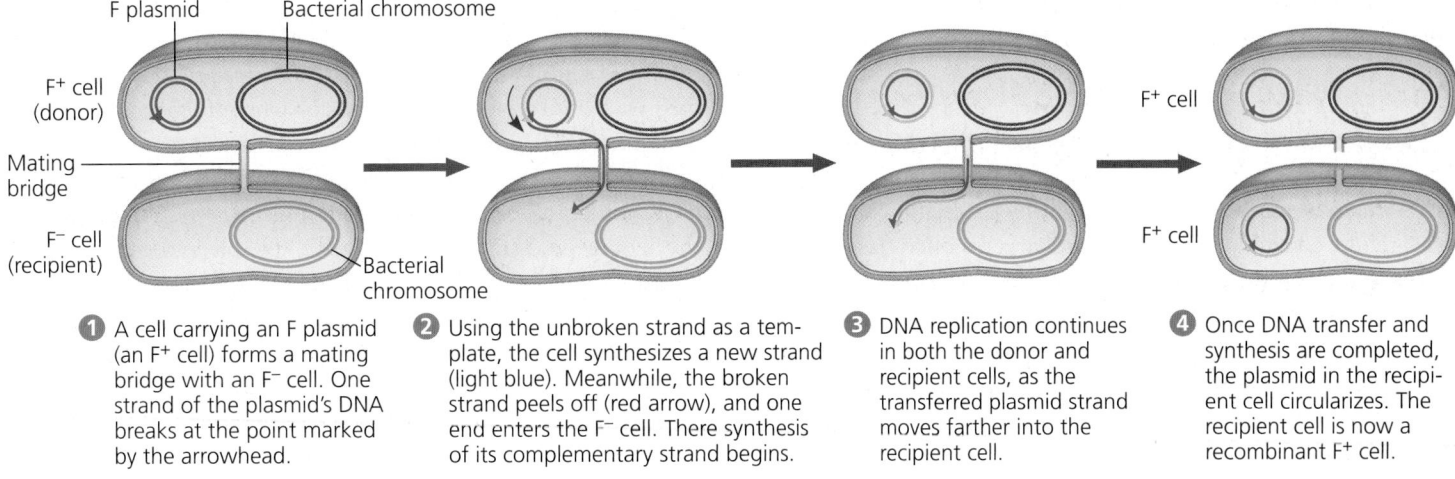

① A cell carrying an F plasmid (an F⁺ cell) forms a mating bridge with an F⁻ cell. One strand of the plasmid's DNA breaks at the point marked by the arrowhead.

② Using the unbroken strand as a template, the cell synthesizes a new strand (light blue). Meanwhile, the broken strand peels off (red arrow), and one end enters the F⁻ cell. There synthesis of its complementary strand begins.

③ DNA replication continues in both the donor and recipient cells, as the transferred plasmid strand moves farther into the recipient cell.

④ Once DNA transfer and synthesis are completed, the plasmid in the recipient cell circularizes. The recipient cell is now a recombinant F⁺ cell.

▲ **Figure 24.17 Conjugation and transfer of an F plasmid, resulting in recombination.** The DNA replication that accompanies the transfer of an F plasmid is called *rolling circle replication*. In effect, the intact circular DNA strand from the donor cell's F plasmid "rolls" as its other strand peels off and a new complementary strand is synthesized.

what happens. The medical consequences are also predictable: Resistant strains of pathogens are becoming more common, making the treatment of certain bacterial infections more difficult. The problem is compounded by the fact that many R plasmids, like F plasmids, have genes that encode pili and enable DNA transfer from one bacterial cell to another by conjugation. Making the problem still worse, some R plasmids carry genes for resistance to as many as ten antibiotics.

**CONCEPT CHECK 24.3**

1. Although rare on a per gene basis, new mutations can add considerable genetic variation to prokaryotic populations in each generation. Explain how this occurs.
2. Distinguish between the three mechanisms by which bacteria can transfer DNA from one bacterial cell to another.
3. In a rapidly changing environment, which bacterial population would likely be more successful, one that includes individuals capable of conjugation or one that does not? Explain.
4. **WHAT IF?** If a nonpathogenic bacterium were to acquire resistance to antibiotics, could this strain pose a health risk to people? In general, how does DNA transfer among bacteria affect the spread of resistance genes?

For suggested answers, see Appendix A.

**CONCEPT 24.4**

# Prokaryotes have radiated into a diverse set of lineages

Since their origin 3.5 billion years ago, prokaryotic populations have radiated extensively as they acquired diverse structural and metabolic adaptations. Collectively, these adaptations have enabled prokaryotes to inhabit every environment known to support life; if there are organisms in a particular place, some of those organisms are prokaryotes. Yet despite their obvious success, it is only in recent decades that advances in genomics have begun to reveal the full extent of prokaryotic diversity.

## An Overview of Prokaryotic Diversity

Microbiologists began comparing the sequences of prokaryotic genes in the 1970s. For example, using small-subunit ribosomal RNA as a marker for evolutionary relationships, researchers concluded that many prokaryotes once classified as bacteria are actually more closely related to eukaryotes and belong in a domain of their own: Archaea. Microbiologists have since analyzed larger amounts of genetic data—including more than 1,700 entire genomes—and have concluded that a few traditional taxonomic groups, such as cyanobacteria, are monophyletic. However, other traditional groups, such as gram-negative bacteria, are scattered throughout several lineages. **Figure 24.18** shows one phylogenetic hypothesis for some of the major taxa of prokaryotes based on molecular systematics.

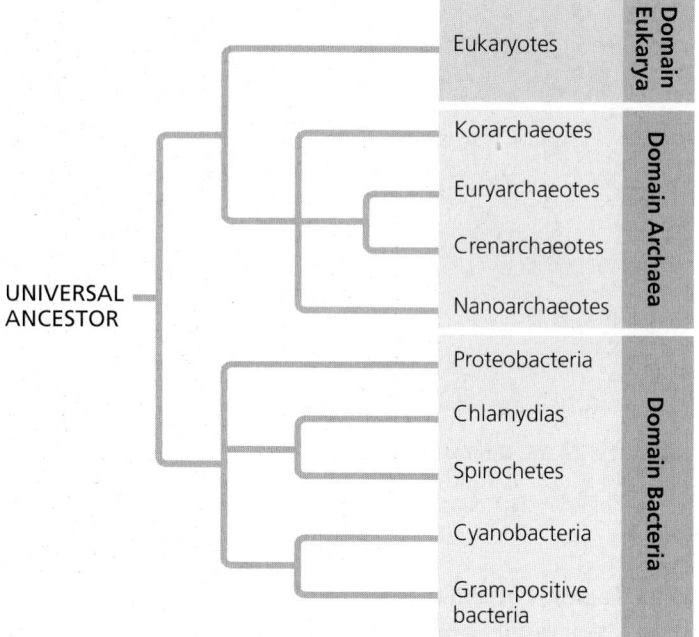

▲ **Figure 24.18  A simplified phylogeny of prokaryotes.** This phylogenetic tree based on molecular data shows one of several debated hypotheses of the relationships between the major prokaryotic groups discussed in this chapter. Within Archaea, the placement of the korarchaeotes and nanoarchaeotes remains unclear.

**?** *Which domain is the sister group of Archaea?*

One lesson from studying prokaryotic phylogeny is that the genetic diversity of prokaryotes is immense. When researchers began to sequence the genes of prokaryotes, they could investigate only the small fraction of species that could be cultured in the laboratory. In the 1980s, researchers began using the polymerase chain reaction (PCR; see Figure 13.27) to analyze the genes of prokaryotes collected from the environment (such as from soil or water samples). Such "genetic prospecting" is now widely used; in fact, today entire prokaryotic genomes can be obtained from environmental samples using *metagenomics* (see Concept 18.1). Each year, these techniques add new branches to the tree of life. While only about 10,600 prokaryotic species have been assigned scientific names, a single handful of soil could contain 10,000 prokaryotic species by some estimates. Taking full stock of this diversity will require many years of research.

Another important lesson from molecular systematics is that horizontal gene transfer has played a key role in the evolution of prokaryotes. Over hundreds of millions of years, prokaryotes have acquired genes from even distantly related species, and they continue to do so today. As a result, significant portions of the genomes of many prokaryotes are actually mosaics of genes imported from other species. For example, a 2011 study of 329 sequenced bacterial genomes found that an average of 75% of the genes in each genome had been transferred horizontally at some point in their evolutionary history. As we saw in Concept 20.5, such gene transfers can make it difficult to determine phylogenetic relationships. Still, it is clear that for billions of years, the prokaryotes have evolved in two separate lineages: the bacteria and the archaea (see Figure 24.18).

## Bacteria

As surveyed in **Figure 24.19**, on the next two pages, bacteria include the vast majority of prokaryotic species familiar to most people, from the pathogenic species that cause strep throat and tuberculosis to the beneficial species used to make Swiss cheese and yogurt. Every major mode of nutrition and metabolism is represented among bacteria, and even a small taxonomic group of bacteria may contain species exhibiting many different nutritional modes. The diverse nutritional and metabolic capabilities of bacteria—and archaea—are behind the great impact these organisms have on Earth and its life.

## Archaea

Archaea share certain traits with bacteria and other traits with eukaryotes **(Table 24.2)**. However, archaea also have many unique characteristics, as we would expect in a taxon that has followed a separate evolutionary path for so long.

The first prokaryotes assigned to domain Archaea live in environments so extreme that few other organisms can survive

there. Such organisms are called **extremophiles**, meaning "lovers" of extreme conditions (from the Greek *philos*, lover), and include extreme halophiles and extreme thermophiles.

**Extreme halophiles** (from the Greek *halo*, salt) live in highly saline environments, such as the Great Salt Lake and the Dead Sea. Some species merely tolerate salinity, while others require an environment that is several times saltier than seawater (which has a salinity of 3.5%). For example, the proteins and cell walls of archaea in the genus *Halobacterium* have unusual features that improve function in extremely salty environments but render these organisms incapable of survival if the salinity drops below 9%.

**Extreme thermophiles** (from the Greek *thermos*, hot) thrive in very hot environments **(Figure 24.20)**. For example, archaea in the genus *Sulfolobus* live in sulfur-rich volcanic springs as hot as 90°C. At temperatures this high, the cells of most organisms die because their DNA does not remain in a double helix and many of their proteins denature. *Sulfolobus* and other extreme thermophiles avoid this fate because they have structural and biochemical adaptations that make their DNA and proteins stable at high temperatures. One extreme thermophile that lives near deep-sea hot springs called *hydrothermal vents* is informally known as "strain 121," since it can reproduce even at 121°C. Another extreme thermophile, *Pyrococcus furiosus*, is used in biotechnology as a source of DNA polymerase for the PCR technique (see Figure 13.27).

Many other archaea live in more moderate environments. Consider the **methanogens**, archaea that release methane as

**Table 24.2 A Comparison of the Three Domains of Life**

| CHARACTERISTIC | DOMAIN | | |
|---|---|---|---|
| | Bacteria | Archaea | Eukarya |
| Nuclear envelope | Absent | Absent | Present |
| Membrane-enclosed organelles | Absent | Absent | Present |
| Peptidoglycan in cell wall | Present | Absent | Absent |
| Membrane lipids | Unbranched hydrocarbons | Some branched hydrocarbons | Unbranched hydrocarbons |
| RNA polymerase | One kind | Several kinds | Several kinds |
| Initiator amino acid for protein synthesis | Formyl-methionine | Methionine | Methionine |
| Introns in genes | Very rare | Present in some genes | Present in many genes |
| Response to the antibiotics streptomycin and chloramphenicol | Growth usually inhibited | Growth not inhibited | Growth not inhibited |
| Histones associated with DNA | Absent | Present in some species | Present |
| Circular chromosome | Present | Present | Absent |
| Growth at temperatures > 100°C | No | Some species | No |

▲ **Figure 24.20 Extreme thermophiles.** Orange and yellow colonies of thermophilic prokaryotes grow in the hot water of Yellowstone National Park's Grand Prismatic Spring.

**MAKE CONNECTIONS** *How might the enzymes of thermophiles differ from those of other organisms? (Review enzymes in Concept 6.4.)*

## Proteobacteria

This large and diverse clade of gram-negative bacteria includes photo-autotrophs, chemoautotrophs, and heterotrophs. Some are anaerobic, while others are aerobic. Molecular systematists currently recognize five subgroups of proteobacteria; the phylogenetic tree at right shows their relationships based on molecular data.

Alpha
Beta
Gamma  Proteobacteria
Delta
Epsilon

### Subgroup: Alpha Proteobacteria

Many of the species in this subgroup are closely associated with eukaryotic hosts. For example, *Rhizobium* species live in nodules within the roots of legumes (plants of the pea/bean family), where the bacteria convert atmospheric $N_2$ to compounds the host plant can use to make proteins. Species in the genus *Agrobacterium* produce tumors in plants; genetic engineers use these bacteria to carry foreign DNA into the genomes of crop plants. Scientists hypothesize that mitochondria evolved from aerobic alpha proteobacteria through endosymbiosis.

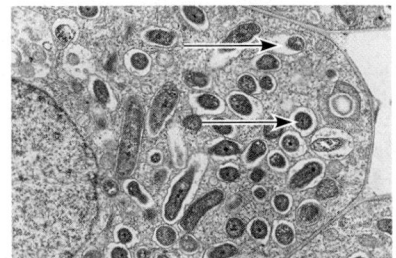

*Rhizobium* (arrows) inside a root cell of a legume (TEM)

2.5 μm

### Subgroup: Beta Proteobacteria

This nutritionally diverse subgroup includes *Nitrosomonas*, a genus of soil bacteria that play an important role in nitrogen recycling by oxidizing ammonium ($NH_4^+$), producing nitrite ($NO_2^-$) as a waste product. Other members of this subgroup include a wide range of aquatic species, such as the photoheterotroph *Rubrivivax*, along with pathogens such as the species that causes the sexually transmitted disease gonorrhea, *Neisseria gonorrhoeae*.

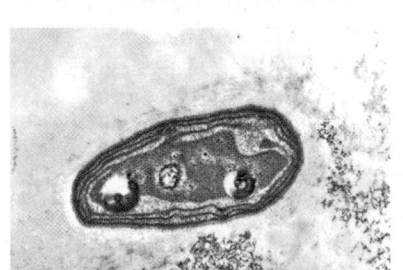
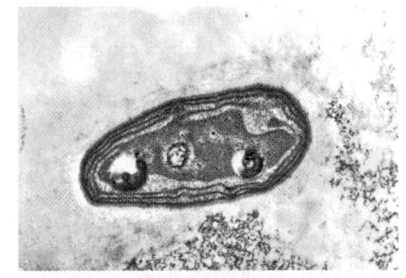

*Nitrosomonas* (colorized TEM)

1 μm

### Subgroup: Gamma Proteobacteria

This subgroup's autotrophic members include sulfur bacteria, such as *Thiomargarita namibiensis*, which obtain energy by oxidizing $H_2S$, producing sulfur as a waste product (the small globules in the photograph at right). Some heterotrophic gamma proteobacteria are pathogens; for example, *Legionella* causes Legionnaires' disease, *Salmonella* is responsible for some cases of food poisoning, and *Vibrio cholerae* causes cholera. *Escherichia coli*, a common resident of the intestines of humans and other mammals, normally is not pathogenic.

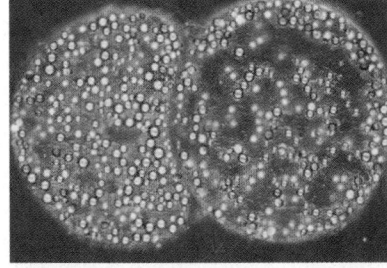

*Thiomargarita namibiensis* containing sulfur wastes (LM)

200 μm

### Subgroup: Delta Proteobacteria

This subgroup includes the slime-secreting myxobacteria. When the soil dries out or food is scarce, the cells congregate into a fruiting body that releases resistant "myxospores." These cells found new colonies in favorable environments. Another group of delta proteobacteria, the bdellovibrios, attack other bacteria, charging at up to 100 μm/sec (comparable to a human running 240 km/hr). The attack begins when a bdellovibrio attaches to specific molecules found on the outer covering of some bacterial species. The bdellovibrio then drills into its prey by using digestive enzymes and spinning at 100 revolutions per second.

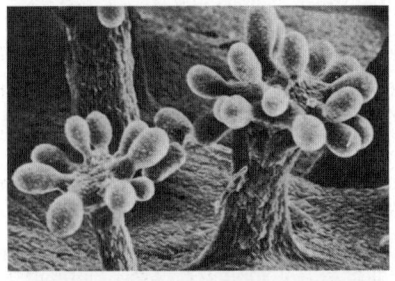

Fruiting bodies of *Chondromyces crocatus*, a myxobacterium (SEM)

300 μm

### Subgroup: Epsilon Proteobacteria

Most species in this subgroup are pathogenic to humans or other animals. Epsilon proteobacteria include *Campylobacter*, which causes blood poisoning and intestinal inflammation, and *Helicobacter pylori*, which causes stomach ulcers.

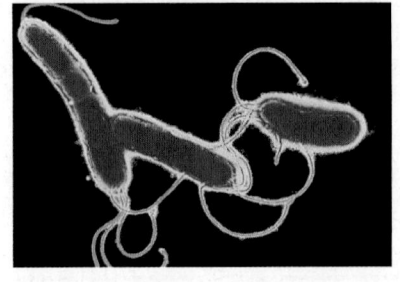

*Helicobacter pylori* (colorized TEM)

2 μm

## Chlamydias

These parasites can survive only within animal cells, depending on their hosts for resources as basic as ATP. The gram-negative walls of chlamydias are unusual in that they lack peptidoglycan. One species, *Chlamydia trachomatis*, is the most common cause of blindness in the world and also causes nongonococcal urethritis, the most common sexually transmitted disease in the United States.

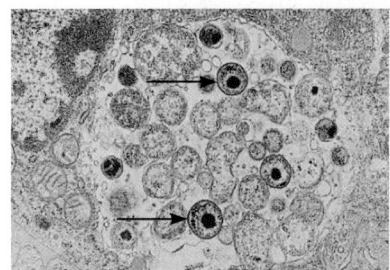

*Chlamydia* (arrows) inside an animal cell (colorized TEM)

## Spirochetes

These helical gram-negative heterotrophs spiral through their environment by means of rotating, internal, flagellum-like filaments. Many spirochetes are free-living, but others are notorious pathogenic parasites: *Treponema pallidum* causes syphilis, and *Borrelia burgdorferi* causes Lyme disease.

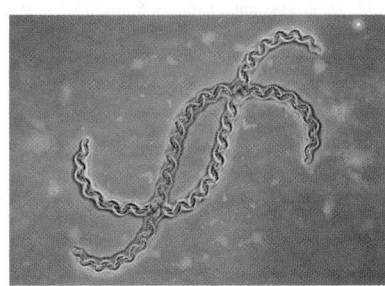

*Leptospira,* a spirochete (colorized TEM)

## Cyanobacteria

These gram-negative photoautotrophs are the only prokaryotes with plantlike, oxygen-generating photosynthesis. (In fact, chloroplasts are thought to have evolved from an endosymbiotic cyanobacterium.) Both solitary and filamentous cyanobacteria are abundant components of freshwater and marine *phytoplankton*, the collection of photosynthetic organisms that drift near the water's surface. Some filaments have cells specialized for nitrogen fixation, the process that incorporates atmospheric $N_2$ into inorganic compounds that can be used in the synthesis of amino acids and other organic molecules.

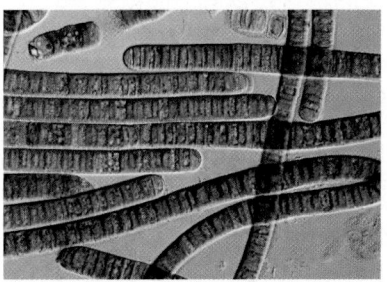

*Oscillatoria,* a filamentous cyanobacterium

## Gram-Positive Bacteria

Gram-positive bacteria rival the proteobacteria in diversity. Species in one subgroup, the actinomycetes (from the Greek *mykes,* fungus, for which these bacteria were once mistaken), form colonies containing branched chains of cells. Two species of actinomycetes cause tuberculosis and leprosy. However, most actinomycetes are free-living species that help decompose the organic matter in soil; their secretions are partly responsible for the "earthy" odor of rich soil. Soil-dwelling species in the genus *Streptomyces* (top) are cultured by pharmaceutical companies as a source of many antibiotics, including streptomycin.

Gram-positive bacteria include many solitary species, such as *Bacillus anthracis,* which causes anthrax, and *Clostridium botulinum,* which causes botulism. The various species of *Staphylococcus* and *Streptococcus* are also gram-positive bacteria.

Mycoplasmas (bottom) are the only bacteria known to lack cell walls. They are also the tiniest known cells, with diameters as small as 0.1 μm, only about five times as large as a ribosome. Mycoplasmas have small genomes—*Mycoplasma genitalium* has only 517 genes, for example. Many mycoplasmas are free-living soil bacteria, but others are pathogens.

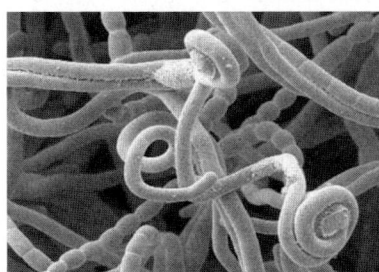

*Streptomyces,* the source of many antibiotics (SEM)

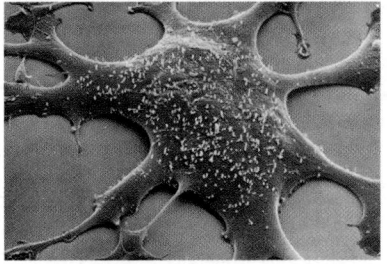

Hundreds of mycoplasmas covering a human fibroblast cell (colorized SEM)

▲ **Figure 24.21 A highly thermophilic methanogen.** The archaean *Methanopyrus kandleri* (inset) lives in the extreme heat of "black smoker" hydrothermal vents on the ocean floor.

a by-product of their unique ways of obtaining energy. Many methanogens use $CO_2$ to oxidize $H_2$, a process that produces both energy and methane waste. Among the strictest of anaerobes, methanogens are poisoned by $O_2$. Although some methanogens live in extreme environments, such as around deep-sea hydrothermal vents **(Figure 24.21)**, others live in swamps and marshes where other microorganisms have consumed all the $O_2$. The "marsh gas" found in such environments is the methane released by these archaea. Other species inhabit the anaerobic guts of cattle, termites, and other herbivores, playing an essential role in the nutrition of these animals. Methanogens are also useful to humans as decomposers in sewage treatment facilities.

Many extreme halophiles and all known methanogens are archaea in the clade Euryarchaeota (from the Greek *eurys*, broad, a reference to their wide habitat range). The euryarchaeotes also include some extreme thermophiles, though most thermophilic species belong to a second clade, Crenarchaeota (*cren* means "spring," such as a hydrothermal spring). Recent metagenomic studies have identified many species of euryarchaeotes and crenarchaeotes that are not extremophiles. These archaea exist in habitats ranging from farm soils to lake sediments to the surface waters of the open ocean.

New findings continue to inform our understanding of archaeal phylogeny. In 1996, researchers sampling a hot spring in Yellowstone National Park discovered archaea that do not appear to belong to either Euryarchaeota or Crenarchaeota. They placed these archaea in a new clade, Korarchaeota (from the Greek *koron*, young man). In 2002, researchers exploring hydrothermal vents off the coast of Iceland discovered archaeal cells only 0.4 μm in diameter attached to a much larger crenarchaeote. The genome of the smaller archaean is one of the smallest known of any organism, containing only 500,000 base pairs. Genetic analysis indicates that this prokaryote belongs to a fourth archaeal clade, Nanoarchaeota (from the Greek *nanos*, dwarf). Within a year after this clade was named, three other DNA sequences from nanoarchaeote species were isolated: one

from Yellowstone's hot springs, one from hot springs in Siberia, and one from a hydrothermal vent in the Pacific. As metagenomic prospecting continues, the tree in Figure 24.18 may well undergo further changes.

**CONCEPT CHECK 24.4**

1. Explain how molecular systematics and metagenomics have contributed to our understanding of the phylogeny and evolution of prokaryotes.
2. **WHAT IF?** What would the discovery of a bacterial species that is a methanogen imply about the evolution of the methane-producing pathway?

   **For suggested answers, see Appendix A.**

**CONCEPT 24.5**

# Prokaryotes play crucial roles in the biosphere

If people were to disappear from the planet tomorrow, life on Earth would change for many species, but few would be driven to extinction. In contrast, prokaryotes are so important to the biosphere that if they were to disappear, the prospects of survival for many other species would be dim.

## Chemical Recycling

The atoms that make up the organic molecules in all living things were at one time part of inorganic substances in the soil, air, and water. Sooner or later, those atoms will return to the nonliving environment. Ecosystems depend on the continual recycling of chemical elements between the living and nonliving components of the environment, and prokaryotes play a major role in this process. For example, some chemoheterotrophic prokaryotes function as **decomposers**, breaking down dead organisms as well as waste products and thereby unlocking supplies of carbon, nitrogen, and other elements. Without the actions of prokaryotes and other decomposers such as fungi, life as we know it would cease. (See Concept 42.4 for a detailed discussion of chemical cycles.)

Prokaryotes also convert some molecules to forms that can be taken up by other organisms. Cyanobacteria and other autotrophic prokaryotes use $CO_2$ to make organic compounds such as sugars, which are then passed up through food chains. Cyanobacteria also produce atmospheric $O_2$, and a variety of prokaryotes fix atmospheric nitrogen ($N_2$) into forms that other organisms can use to make the building blocks of proteins and nucleic acids. Under some conditions, prokaryotes can increase the availability of nutrients that plants require for growth, such as nitrogen, phosphorus, and potassium **(Figure 24.22)**. Prokaryotes can also *decrease* the availability of key plant nutrients; this occurs when prokaryotes "immobilize" nutrients by using them to synthesize molecules that remain within their cells. Thus, prokaryotes can have complex effects

► **Figure 24.22 Impact of bacteria on soil nutrient availability.** Pine seedlings grown in sterile soils to which one of three strains of the bacterium *Burkholderia glathei* had been added absorbed more potassium (K⁺) than did seedlings grown in soil without any bacteria. Other results (not shown) demonstrated that strain 3 increased the amount of K⁺ released from mineral crystals to the soil.

**WHAT IF?** *Estimate the average uptake of K⁺ for seedlings in soils with bacteria. What would you expect this average to be if bacteria had no effect on nutrient availability?*

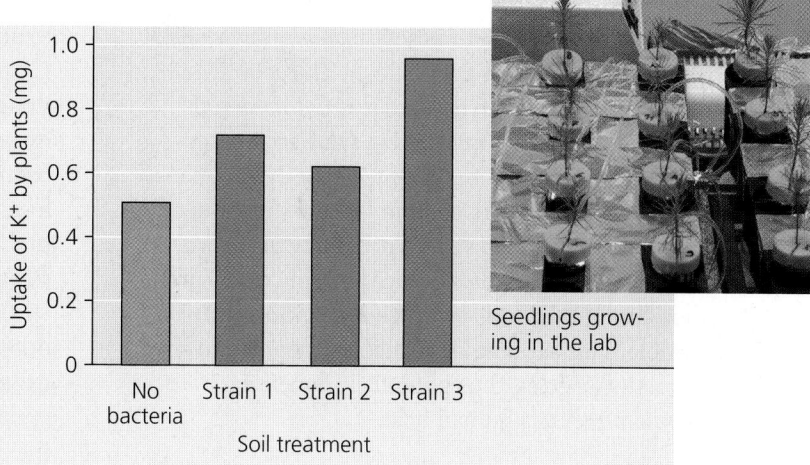

Seedlings growing in the lab

on soil nutrient concentrations. In marine environments, an archaean from the clade Crenarchaeota can perform nitrification, a key step in the nitrogen cycle (see Figure 42.13). Crenarchaeotes dominate the oceans by numbers, comprising an estimated $10^{28}$ cells. The sheer abundance of these organisms suggests that they may have a large impact on the global nitrogen cycle; scientists are investigating this possibility.

## Ecological Interactions

Prokaryotes play a central role in many ecological interactions. Consider **symbiosis** (from a Greek word meaning "living together"), an ecological relationship in which two species live in close contact with each other. Prokaryotes often form symbiotic associations with much larger organisms. In general, the larger organism in a symbiotic relationship is known as the **host**, and the smaller is known as the **symbiont**. There are many cases in which a prokaryote and its host participate in **mutualism**, an ecological interaction between two species in which both benefit **(Figure 24.23)**. Other interactions take the form of **commensalism**, an ecological relationship in which one species benefits while the other is not harmed or helped in any significant way. For example, more than 150 bacterial species live on the surface of your body, covering portions of your skin with up to 10 million cells per square centimeter. Some

of these species are commensalists: You provide them with food, such as the oils that exude from your pores, and a place to live, while they neither harm nor benefit you. Finally, some prokaryotes engage in **parasitism**, an interaction in which a **parasite** eats the cell contents, tissues, or body fluids of its host. As a group, parasites harm but usually do not kill their host, at least not immediately (unlike a predator). Parasites that cause disease are known as **pathogens**, many of which are prokaryotic. (We'll discuss mutualism, commensalism, and parasitism in greater detail in Concept 41.1.)

The very existence of an ecosystem can depend on prokaryotes. For example, consider the diverse ecological communities found at hydrothermal vents. These communities are densely populated by many different kinds of animals, including worms, clams, crabs, and fishes. But since sunlight does not penetrate to the deep ocean floor, the community does not include photosynthetic organisms. Instead, the energy that supports the community is derived from the metabolic activities of chemoautotrophic bacteria. These bacteria harvest chemical energy from compounds such as hydrogen sulfide ($H_2S$) that are released from the vent. An active hydrothermal vent may support hundreds of eukaryotic species, but when the vent stops releasing chemicals, the chemoautotrophic bacteria cannot survive. As a result, the entire vent community collapses.

## Impact on Humans

Although the best-known prokaryotes tend to be the bacteria that cause human illness, these pathogens represent only a small fraction of prokaryotic species. Many other prokaryotes have positive interactions with people, and some play essential roles in agriculture and industry.

### Mutualistic Bacteria

As is true for many other eukaryotes, human well-being can depend on mutualistic prokaryotes. For example, our intestines are home to an estimated 500–1,000 species of bacteria; collectively, their cells outnumber all human cells in the body by a factor of ten. Different species live in different portions of the intestines, and they vary in their ability to process different

▲ **Figure 24.23 Mutualism: bacterial "headlights."** The glowing oval below the eye of the flashlight fish (*Photoblepharon palpebratus*) is an organ harboring bioluminescent bacteria. The fish uses the light to attract prey and to signal potential mates. The bacteria receive nutrients from the fish.

foods. Many of these species are mutualists, digesting food that our own intestines cannot break down. The genome of one of these gut mutualists, *Bacteroides thetaiotaomicron*, includes a large array of genes involved in synthesizing carbohydrates, vitamins, and other nutrients needed by humans. Signals from the bacterium activate human genes that build the network of intestinal blood vessels necessary to absorb nutrient molecules. Other signals induce human cells to produce antimicrobial compounds to which *B. thetaiotaomicron* is not susceptible. This action may reduce the population sizes of other, competing species, thus potentially benefiting both *B. thetaiotaomicron* and its human host.

### Pathogenic Bacteria

All the pathogenic prokaryotes known to date are bacteria, and they deserve their negative reputation. Bacteria cause about half of all human diseases. For example, more than 1 million people die each year of the lung disease tuberculosis, caused by *Mycobacterium tuberculosis*. And another 2 million people die each year from diarrheal diseases caused by various bacteria.

Some bacterial diseases are transmitted by other species, such as fleas or ticks. In the United States, the most widespread pest-carried disease is Lyme disease, which infects 15,000 to 20,000 people each year **(Figure 24.24)**. Caused by a bacterium carried by ticks that live on deer and field mice, Lyme disease can result in debilitating arthritis, heart disease, nervous disorders, and death if untreated.

Pathogenic prokaryotes usually cause illness by producing poisons, which are classified as exotoxins or endotoxins. **Exotoxins** are proteins secreted by certain bacteria and other organisms. Cholera, a dangerous diarrheal disease, is caused by an exotoxin secreted by the proteobacterium *Vibrio cholerae*. The exotoxin stimulates intestinal cells to release chloride ions into the gut, and water follows by osmosis. In another example, the potentially fatal disease botulism is caused by botulinum

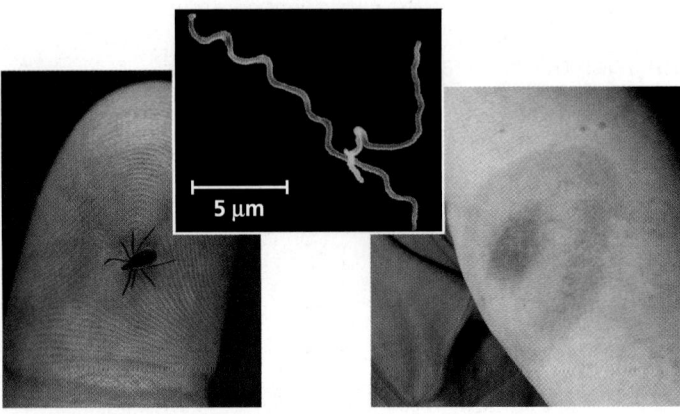

▲ **Figure 24.24 Lyme disease.** Ticks in the genus *Ixodes* spread the disease by transmitting the spirochete *Borrelia burgdorferi* (colorized SEM). A rash may develop at the site of the tick's bite; the rash may be large and ring-shaped (as shown) or much less distinctive.

toxin, an exotoxin secreted by the gram-positive bacterium *Clostridium botulinum* as it ferments various foods, including improperly canned meat, seafood, and vegetables. Like other exotoxins, the botulinum toxin can produce disease even if the bacteria that manufacture it are no longer present when the food is eaten. Another species in the same genus, *C. difficile*, produces exotoxins that cause severe diarrhea, resulting in more than 12,000 deaths per year in the United States alone.

**Endotoxins** are lipopolysaccharide components of the outer membrane of gram-negative bacteria. In contrast to exotoxins, endotoxins are released only when the bacteria die and their cell walls break down. Endotoxin-producing bacteria include species in the genus *Salmonella*, such as *Salmonella typhi*, which causes typhoid fever. You might have heard of food poisoning caused by other *Salmonella* species that can be found in poultry and some fruits and vegetables.

Since the 19th century, improved sanitation systems in the industrialized world have greatly reduced the threat of pathogenic bacteria. Antibiotics have saved a great many lives and reduced the incidence of disease. However, resistance to antibiotics is currently evolving in many bacterial strains. As you read earlier, the rapid reproduction of bacteria enables cells carrying resistance genes to quickly give rise to large populations as a result of natural selection, and these genes can also spread to other species by horizontal gene transfer.

Horizontal gene transfer can also spread genes associated with virulence, turning normally harmless bacteria into potent pathogens. *E. coli*, for instance, is ordinarily a harmless symbiont in the human intestines, but pathogenic strains that cause bloody diarrhea have emerged. One of the most dangerous strains, O157:H7, is a global threat; in the United States alone, there are 75,000 cases of O157:H7 infection per year, often from contaminated beef or produce. Scientists have sequenced the genome of O157:H7 and compared it with the genome of a harmless strain of *E. coli* called K-12. They discovered that 1,387 out of the 5,416 genes in O157:H7 have no counterpart in K-12. Many of these 1,387 genes are found in chromosomal regions that include phage DNA. This suggests that at least some of the 1,387 genes were incorporated into the genome of O157:H7 through phage-mediated horizontal gene transfer (transduction). Some of the genes found only in O157:H7 are associated with virulence, including genes that code for adhesive fimbriae that enable O157:H7 to attach itself to the intestinal wall and extract nutrients.

### Prokaryotes in Research and Technology

On a positive note, we reap many benefits from the metabolic capabilities of both bacteria and archaea. For example, people have long used bacteria to convert milk to cheese and yogurt. Bacteria are also used in the production of beer and wine, pepperoni, fermented cabbage (sauerkraut), and soy sauce. In recent years, our greater understanding of prokaryotes has led to an explosion of new applications in biotechnology. Examples include the use of *E. coli* in gene cloning (see Figure 13.24) and the use of DNA polymerase from *Pyrococcus furiosus* in the PCR technique (see Figure 13.27). Through genetic engineering, we can now modify bacteria to produce vitamins, antibiotics, hormones, and other products (see Concept 13.4). In addition, naturally occurring soil bacteria may have potential for combating diseases that affect crop plants, as you can explore in the **Scientific Skills Exercise**.

Recently, the prokaryotic CRISPR-Cas system, which helps bacteria and archaea defend against attack by viruses (see Figure 17.6), has been developed into a powerful new tool for altering genes in virtually any organism. The genomes of many prokaryotes contain short DNA repeats, called CRISPRs, that interact with the so-called Cas (CRISPR-associated) proteins. Cas proteins, acting together with "guide RNA" made from the CRISPR region, can cut any DNA sequence to which they are directed. Scientists have been able to exploit this system by introducing a Cas protein (Cas9) and guide RNA into cells whose DNA they want to alter (see Figure 13.31). Among other applications, the CRISPR-Cas9 system has already opened new lines of research on HIV, the virus that causes AIDS **(Figure 24.25)**. While the CRISPR-Cas9 system can potentially be used in many different ways, care must be taken to guard against the unintended consequences that could arise when applying such a new and powerful technology.

Another valuable application of bacteria is to reduce our use of petroleum. Consider the plastics industry. Globally, each year about 350 billion pounds of plastic are produced from petroleum and used to make toys, storage containers, soft drink bottles, and many other items. These products degrade slowly, creating environmental problems. Bacteria can produce natural plastics. For example, some bacteria produce a type of organic polymer known as PHA (polyhydroxyalkanoate), which they use to store chemical energy

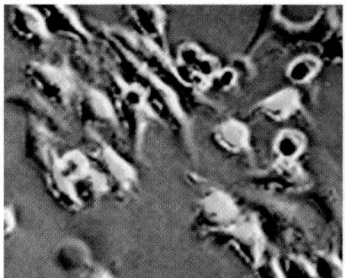 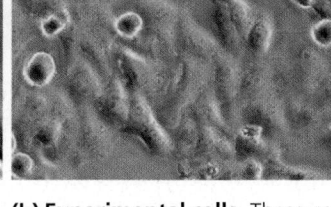

**(a) Control cells.** The green color indicates infection by HIV.

**(b) Experimental cells.** These cells were treated with a CRISPR-Cas9 system that targets HIV.

▲ **Figure 24.25 CRISPR: Opening new avenues of research on treating HIV infection. (a)** In laboratory experiments, untreated (control) human cells were susceptible to infection by HIV, the virus that causes AIDS. **(b)** In contrast, cells treated with a CRISPR-Cas9 system that targets HIV were resistant to viral infection. The CRISPR-Cas9 system was also able to remove HIV proviruses (see Figure 17.8) that had become incorporated into the DNA of human cells.

## Making a Bar Graph and Interpreting Data

**Do Soil Microorganisms Protect Against Crop Disease?** The soil layer surrounding plant roots, called the *rhizosphere*, is a complex community in which archaea, bacteria, fungi, and plants interact with one another. When crop plants are attacked by fungal or bacterial pathogens, in some cases soil from the rhizosphere protects plants from future attacks. Such protective soil is called disease-suppressive soil. Plants grown in disease-suppressive soils appear to be less vulnerable to pathogen attack. In this exercise, you'll interpret data from an experiment studying whether microorganisms were responsible for the protective effects of disease-suppressive soils.

**How the Experiment Was Done** The researchers obtained disease-suppressive soil from 25 random sites in an agricultural field in the Netherlands in which sugar beet crops had previously been attacked by *Rhizoctonia solani*, a fungal pathogen that also afflicts potatoes and rice. The researchers collected other soil samples from the grassy margins of the field where sugar beets had not been grown. The researchers predicted that these soil samples from the margins would not offer protection against pathogens.

The researchers then planted and raised sugar beets in greenhouses, using five different soil treatments. Each soil treatment was applied to four pots, and each pot contained eight plants. The pots were inoculated with *R. solani*. After 20 days, the percentage of infected sugar beet seedlings was determined for each plot and then averaged for each soil treatment.

**Data from the Experiment**

| Soil Treatment | Average % of Seedlings Afflicted with Fungal Disease |
|---|---|
| Disease-suppressive soil | 3 |
| Soil from margin of field | 62 |
| Soil from margin of field + 10% disease-suppressive soil | 39 |
| Disease-suppressive soil heated to 50°C for 1 hour | 31 |
| Disease-suppressive soil heated to 80°C for 1 hour | 70 |

**Data from** R. Mendes et al., Deciphering the rhizosphere for disease-suppressive bacteria, *Science* 332:1097–1100 (2011).

**INTERPRET THE DATA**

1. What hypothesis were the researchers testing in this study? What is the independent variable in this study? What is the dependent variable?

2. What is the total number of pots used in this experiment, and how many plants received each soil treatment? Explain why multiple pots and plants were used for each treatment.

3. Use the data in the table to create a bar graph. Then, in words, describe and compare the results for the five soil treatments.

4. The researchers stated, "Collectively, these results indicated that disease suppressiveness [of soil] toward *Rhizoctonia solani* was microbiological in nature." Is this statement supported by the results shown in the graph? Explain.

**(MB)** A version of this Scientific Skills Exercise can be assigned in MasteringBiology.

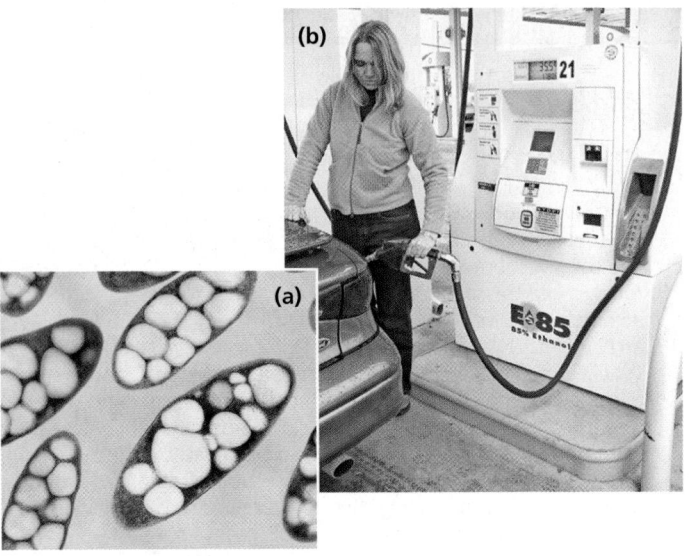

▲ **Figure 24.26 Products from prokaryotes. (a)** These bacteria synthesize and store PHA, which can be extracted and used to make biodegradable plastic products. **(b)** Researchers are developing bacteria that produce ethanol (E-85) fuel from renewable plant products.

**(Figure 24.26a)**. The PHA can be extracted, formed into pellets, and used to make durable, yet biodegradable, plastics. Researchers are also seeking to reduce use of petroleum and other fossil fuels by engineering bacteria that can produce ethanol from various forms of biomass, including agricultural waste, switchgrass, municipal waste (such as paper products that are not recycled), and corn **(Figure 24.26b)**.

Yet another way to harness prokaryotes is in **bioremediation**, the use of organisms to remove pollutants from soil, air, or water. For example, anaerobic bacteria and archaea decompose the organic matter in sewage, converting it to material that can be used as landfill or fertilizer after chemical sterilization. Other applications of bioremediation include cleaning up oil spills **(Figure 24.27)** and precipitating radioactive material (such as uranium) out of groundwater.

▲ **Figure 24.27 Bioremediation of an oil spill.** Spraying fertilizer stimulates the growth of native bacteria that metabolize oil, increasing the breakdown process up to fivefold.

The usefulness of prokaryotes largely derives from their diverse forms of nutrition and metabolism. All this metabolic versatility evolved prior to the appearance of the structural novelties that heralded the evolution of eukaryotic organisms, to which we devote the remainder of this unit.

**CONCEPT CHECK 24.5**

1. Explain how prokaryotes, though small, can be considered giants in their collective impact on Earth and its life.
2. A pathogenic bacterium's toxin causes symptoms that increase the bacterium's chance of spreading from host to host. Does this information indicate whether the poison is an exotoxin or endotoxin? Explain.
3. **MAKE CONNECTIONS** Review photosynthesis in Figure 8.5. Then summarize the main steps by which cyanobacteria produce $O_2$ and use $CO_2$ to make organic compounds.
4. **WHAT IF?** How might a sudden, dramatic change in your diet affect the diversity of prokaryotic species in your gut?

For suggested answers, see Appendix A.

---

# 24 Chapter Review

Go to **MasteringBiology®** for Assignments, the eText, and the Study Area with Animations, Activities, Vocab Self-Quiz, and Practice Tests.

## SUMMARY OF KEY CONCEPTS

**VOCAB SELF-QUIZ**

goo.gl/gbai8v

### CONCEPT 24.1

**Conditions on early Earth made the origin of life possible (pp. 475–478)**

- Experiments simulating possible early atmospheres have produced organic molecules from inorganic precursors. Amino acids, lipids, sugars, and nitrogenous bases have also been found in meteorites.
- Amino acids and RNA nucleotides polymerize when dripped onto hot sand, clay, or rock. Organic compounds can spontaneously assemble into **protocells**, membrane-bounded droplets that have some properties of cells.
- The first genetic material may have been self-replicating, catalytic RNA. Early protocells containing such RNA would have increased in abundance through natural selection.
- Fossil evidence of early prokaryotes dates to 3.5 billion years ago. By 2.8 billion years ago, prokaryotes included stromatolites that differed in morphology and habitat. Early prokaryotes also included cyanobacteria that released oxygen as a by-product of photosynthesis, thereby changing Earth's atmosphere and altering the course of evolution.

**?** *Describe the roles that montmorillonite clay and vesicles may have played in the origin of life.*

## Diverse structural and metabolic adaptations have evolved in prokaryotes (pp. 478–483)

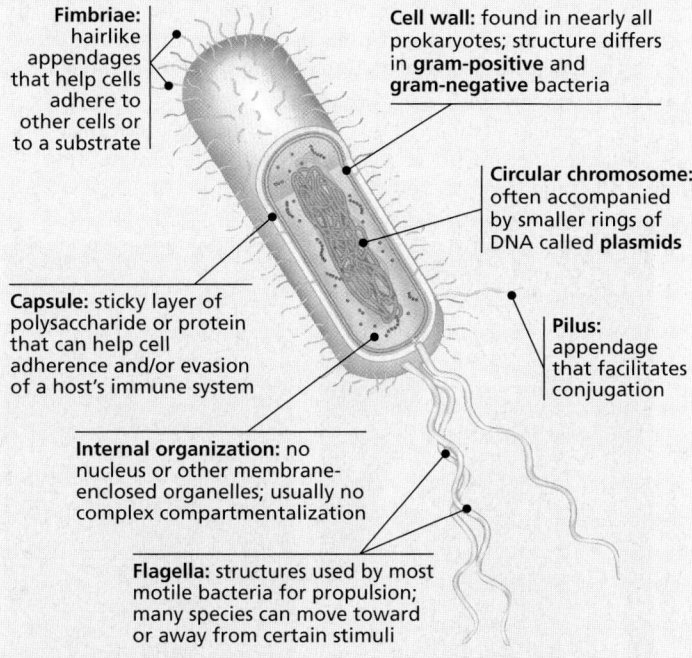

**Fimbriae:** hairlike appendages that help cells adhere to other cells or to a substrate

**Cell wall:** found in nearly all prokaryotes; structure differs in **gram-positive** and **gram-negative** bacteria

**Circular chromosome:** often accompanied by smaller rings of DNA called **plasmids**

**Capsule:** sticky layer of polysaccharide or protein that can help cell adherence and/or evasion of a host's immune system

**Pilus:** appendage that facilitates conjugation

**Internal organization:** no nucleus or other membrane-enclosed organelles; usually no complex compartmentalization

**Flagella:** structures used by most motile bacteria for propulsion; many species can move toward or away from certain stimuli

- Nutritional diversity is much greater in prokaryotes than in eukaryotes. As a group, prokaryotes perform all four modes of nutrition: photoautotrophy, chemoautotrophy, photoheterotrophy, and chemoheterotrophy.
- Among prokaryotes, obligate aerobes require $O_2$, obligate anaerobes are poisoned by $O_2$, and facultative anaerobes can survive with or without $O_2$.
- Unlike eukaryotes, prokaryotes can metabolize nitrogen in many different forms. Some can convert atmospheric nitrogen to ammonia, a process called **nitrogen fixation**.
- Prokaryotic cells and even species may cooperate metabolically. In *Anabaena*, photosynthetic cells and nitrogen-fixing cells exchange metabolic products. Metabolic cooperation also occurs in surface-coating **biofilms** that include different species.
- Prokaryotes can reproduce quickly by binary fission, leading to the formation of populations containing enormous numbers of individuals. Some form **endospores**, which can remain viable in harsh conditions for centuries.

**?** *Describe features of prokaryotes that enable them to thrive in a wide range of different environments.*

## Rapid reproduction, mutation, and genetic recombination promote genetic diversity in prokaryotes (pp. 483–486)

- Because prokaryotes can often proliferate rapidly, mutations can quickly increase a population's genetic variation. As a result, prokaryotic populations often can evolve in short periods of time in response to changing conditions.
- Genetic diversity in prokaryotes also can arise by recombination of the DNA from two different cells (via **transformation**, **transduction**, or **conjugation**). By transferring advantageous alleles, such as ones for antibiotic resistance, genetic

recombination can promote adaptive evolution in prokaryotic populations.

**?** *Mutations are rare and prokaryotes reproduce asexually, yet their populations can have high genetic diversity. Explain how this can occur.*

## Prokaryotes have radiated into a diverse set of lineages (pp. 486–490)

- Molecular systematics is helping biologists classify prokaryotes and identify major new clades.
- Diverse nutritional types are scattered among the major groups of bacteria. The two largest groups are the proteobacteria and gram-positive bacteria.
- Some archaea, such as **extreme thermophiles** and **extreme halophiles**, live in extreme environments. Other archaea live in moderate environments, such as soils and lakes.

**?** *How have molecular data informed prokaryotic phylogeny?*

## Prokaryotes play crucial roles in the biosphere (pp. 490–494)

- Decomposition by heterotrophic prokaryotes and the synthetic activities of autotrophic and nitrogen-fixing prokaryotes contribute to the recycling of elements in ecosystems.
- Many prokaryotes have a symbiotic relationship with a **host**; the relationships between prokaryotes and their hosts range from **mutualism** to **commensalism** to **parasitism**.
- People depend on mutualistic prokaryotes, including hundreds of species that live in our intestines and help digest food.
- Pathogenic bacteria typically cause disease by releasing **exotoxins** or **endotoxins**. Horizontal gene transfer can spread genes associated with virulence to harmless species or strains.
- Prokaryotes can be used in **bioremediation**, production of biodegradable plastics, and the synthesis of vitamins, antibiotics, and other products.

**?** *In what ways are prokaryotes key to the survival of many species?*

# TEST YOUR UNDERSTANDING

**PRACTICE TEST**

goo.gl/CRZjvS

## Level 1: Knowledge/Comprehension

1. Which of the following steps has *not* yet been accomplished by scientists studying the origin of life?
   (A) synthesis of small RNA polymers by ribozymes
   (B) abiotic synthesis of polypeptides
   (C) formation of molecular aggregates with selectively permeable membranes
   (D) formation of protocells that use DNA to direct the polymerization of amino acids

2. Fossilized stromatolites
   (A) more than 2.8 billion years old have not been discovered.
   (B) formed around deep-sea vents.
   (C) resemble structures formed by bacterial communities that are found today in some shallow marine bays.
   (D) provide evidence that photosynthesis was occurring in the oceans by 2.5 billion years ago.

**3.** Genetic variation in bacterial populations cannot result from
  (A) meiosis.
  (B) transformation.
  (C) transduction.
  (D) mutation.

**4.** Photoautotrophs use
  (A) light as an energy source and methane as a carbon source.
  (B) light as an energy source and $CO_2$ as a carbon source.
  (C) $H_2S$ as an energy source and $CO_2$ as a carbon source.
  (D) $CO_2$ as both an energy source and a carbon source.

**5.** Which of the following statements is *not* true?
  (A) Archaea and bacteria have different membrane lipids.
  (B) The cell walls of archaea lack peptidoglycan.
  (C) Only bacteria have histones associated with DNA.
  (D) Only some archaea use $CO_2$ to oxidize $H_2$, releasing methane.

**6.** Bacteria participate in many ecological interactions. Which of the following bacterial roles typically does *not* involve symbiosis?
  (A) living as commensalists on the skin of humans
  (B) providing bioluminescence in fish
  (C) digesting food as mutualists in animal intestines
  (D) decomposing dead organisms and organic wastes

**7.** Plantlike photosynthesis that releases $O_2$ occurs in
  (A) cyanobacteria.
  (B) chlamydias.
  (C) archaea.
  (D) chemoautotrophic bacteria.

## Level 2: Application/Analysis

**8. SCIENTIFIC INQUIRY**
  **INTERPRET THE DATA** The nitrogen-fixing bacterium *Rhizobium* infects the roots of some plant species, forming a mutualism in which the bacterium provides nitrogen and the plant provides carbohydrates. Scientists measured the 12-week growth of one such plant species (*Acacia irrorata*) when infected by six different *Rhizobium* strains. (a) Graph the data. (b) Interpret your graph.

| *Rhizobium* strain | 1 | 2 | 3 | 4 | 5 | 6 |
|---|---|---|---|---|---|---|
| Plant mass (g) | 0.91 | 0.06 | 1.56 | 1.72 | 0.14 | 1.03 |

Data from J. J. Burdon et al., Variation in the effectiveness of symbiotic associations between native rhizobia and temperate Australian *Acacia*: Within-species interactions, *Journal of Applied Ecology* 36:398–408 (1999).

Note: Without *Rhizobium*, after 12 weeks, *Acacia* plants have a mass of about 0.1 g.

## Level 3: Synthesis/Evaluation

**9. FOCUS ON EVOLUTION**
  In patients infected with nonresistant strains of the tuberculosis bacterium, antibiotics can relieve symptoms in a few weeks. However, it takes much longer to halt the infection, and patients may discontinue treatment while bacteria are still present. How might this result in the evolution of drug-resistant pathogens?

**10. FOCUS ON ENERGY AND MATTER**
  In a short essay (about 100–150 words), discuss how prokaryotes and other members of hydrothermal vent communities transfer and transform energy.

**11.** SYNTHESIZE YOUR KNOWLEDGE

  Explain how the small size and rapid reproduction rate of bacteria (such as the population shown here on the tip of a pin) contribute to their large population sizes and high genetic variation.

*For selected answers, see Appendix A.*

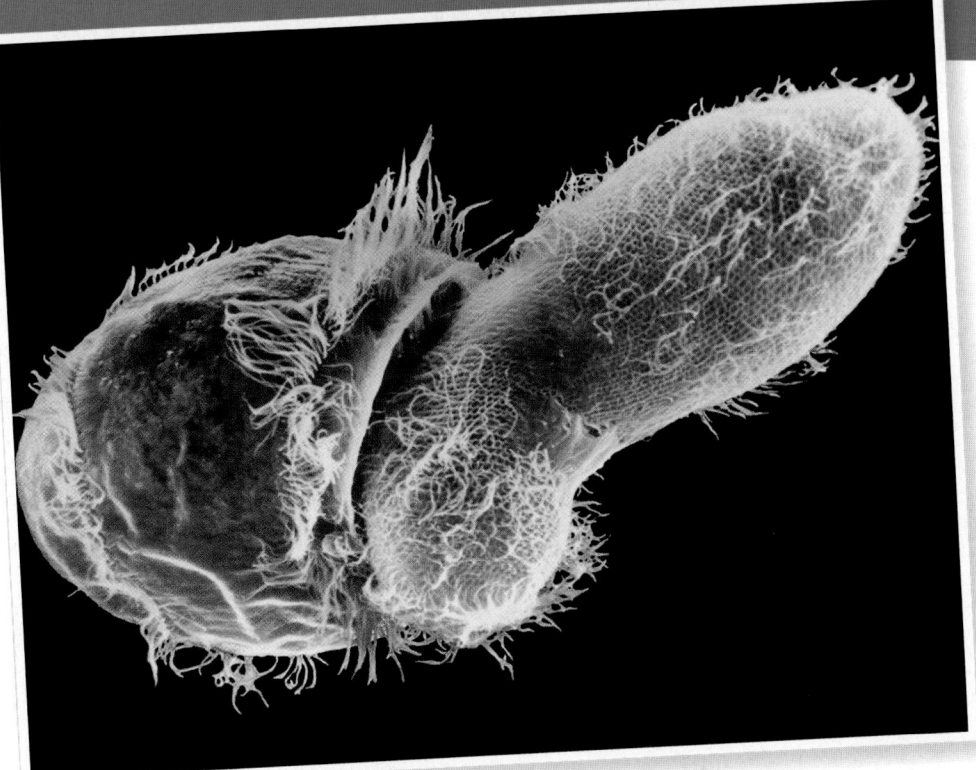

▲ **Figure 25.1** What enables the cell on the left to engulf its prey?

## Shape Changers

The organisms in **Figure 25.1** are ciliates, a diverse group of single-celled eukaryotes named after the small appendages—cilia—that cover much of their bodies and enable them to move. The ciliate on the left, *Didinium*, has begun a seemingly impossible task: it will completely engulf the *Paramecium* (right), even though the *Paramecium* is as large as it is.

Reflect for a moment on the magnitude of this feat. If we humans could do this, in a single swallow we could ingest more food than we would typically eat in a month. Even prokaryotes cannot engulf food items their own size—although prokaryotes can metabolize an astonishing range of compounds, they can only absorb small particles of food. What enables *Didinium* to tackle food items that could easily evade a hungry prokaryote?

One key to *Didinium*'s success lies within its cells—it has a complex set of cytoskeletal proteins that enable the cell to change in shape dramatically as it feeds. *Didinium* also has small structures similar to miniature harpoons that it can eject to help ensnare its prey. These two features illustrate the structural complexity that characterizes the cells of *Didinium* and the other diverse, mostly unicellular groups of eukaryotes informally known as **protists**.

As we'll see, some protists change their form as they creep along using blob-like appendages, others are shaped like tiny trumpets, and still others resemble miniature jewelry. In this chapter, we'll explore how these shape-changing, structurally complex eukaryotic cells arose from their morphologically simpler prokaryotic ancestors. We'll also examine another major step in the evolutionary history of life: the origin of multicellular eukaryotes such as plants, fungi, and animals. Finally, we'll consider how single-celled eukaryotes affect ecosystems and human health.

### CONCEPT 25.1

## Eukaryotes arose by endosymbiosis more than 1.8 billion years ago

As we discussed in Concept 24.1, all organisms were unicellular early in the history of life. The evolution of eukaryotes did not immediately change this, but it did involve fundamental changes in the structure of these individual cells. For example, unlike the cells of prokaryotes, the cells of all eukaryotes have a nucleus and other membrane-enclosed organelles, such as mitochondria and the Golgi apparatus. Such organelles provide specific locations where particular cellular functions are accomplished, making the structure and organization of eukaryotic cells more complex than that of prokaryotic cells.

Another key eukaryote characteristic is a well-developed cytoskeleton that extends throughout the cell (see Figure 4.20).

The cytoskeleton provides structural support that enables eukaryotic cells to have asymmetric (irregular) forms, as well as to change shape as they feed, move, or grow. Although some prokaryotes have proteins related to eukaryotic cytoskeletal proteins, their rigid cell walls and lack of a well-developed cytoskeleton limit the extent to which their cells can maintain asymmetric forms or change shape over time.

The fossil record indicates that prokaryotes were inhabiting Earth at least 3.5 billion years ago. At what point did irregular

## ▼ Figure 25.2    Exploring the Early Evolution of Eukaryotes

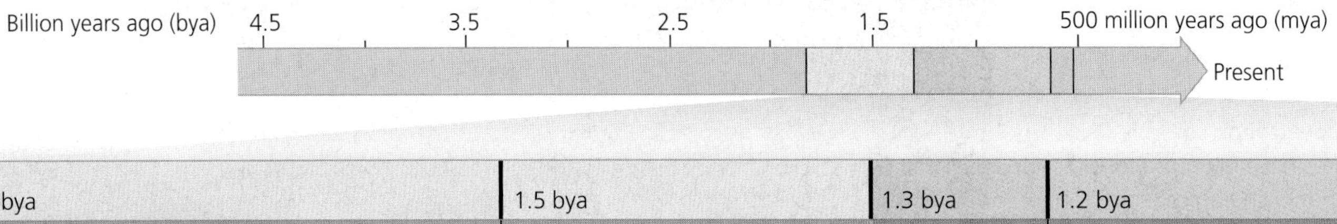

**(a) A 1.8-billion-year-old fossil eukaryote.** This ancient organism had a cell wall structure similar to that of some living algae, but different from that of prokaryotes—indicating this fossil organism was an early eukaryote.

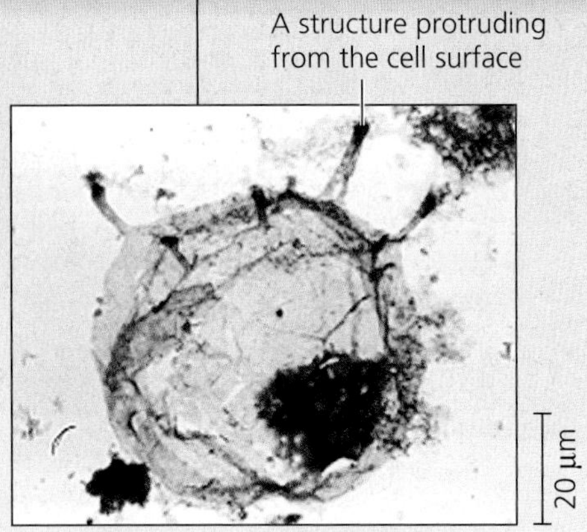

20 μm

**(b) *Tappania*, a 1.5-billion-year-old fossil that may represent an early alga or fungus.** Many fossils of this organism have been found; the number, size, and location of the structures protruding from the cell's outer surface vary from one individual to another.

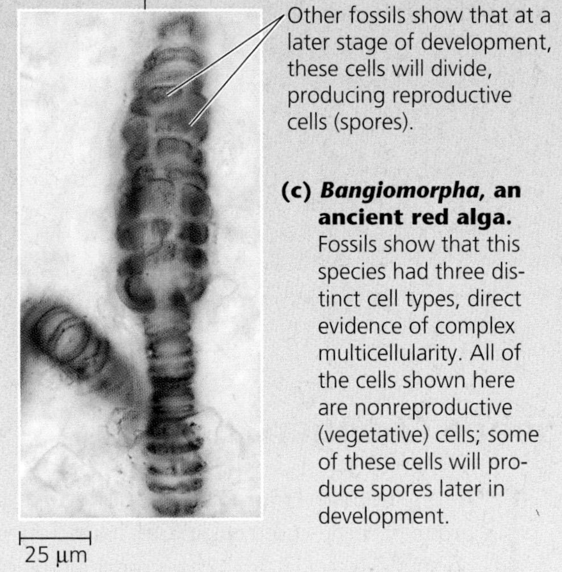

A structure protruding from the cell surface

20 μm

Other fossils show that at a later stage of development, these cells will divide, producing reproductive cells (spores).

**(c) *Bangiomorpha*, an ancient red alga.** Fossils show that this species had three distinct cell types, direct evidence of complex multicellularity. All of the cells shown here are nonreproductive (vegetative) cells; some of these cells will produce spores later in development.

25 μm

### Initial Diversification

The fossil record shows that a moderate diversity of single-celled eukaryotes was already present 1.8 billion years ago, including the fossil specimen in **(a)**. Researchers have discovered other fossil eukaryotes of the same age that vary in size and had shapes ranging from spherical to elliptical (some that are spindle-shaped, or tapering at each end). This variety suggests that the eukaryotes may have originated well before 1.8 billion years ago.

Early members of domain Eukarya had a nucleus, a flexible membrane, and a cytoskeleton capable of supporting an irregular cell shape, like that seen in **(b)**. Other fossils from this time period include several different types of simple filaments thought to be small, multicellular eukaryotes. Although fossil eukaryotes from this time range are moderately diverse, none of these organisms can be assigned with confidence to an extant group of eukaryotes.

(*Time line not to scale.*)

### Appearance of Novel Features

The oldest known eukaryotic fossils that can be resolved taxonomically are of small red algae that lived 1.2 billion years ago **(c)**; note that algae is a general term that includes all groups of photosynthetic protists. Other fossils from this period include green algae **(d)** and certain types of amoebas that lived within vase-shaped structures **(e)**, as well as a variety of colonial and multicellular protists of unknown taxonomic affinity.

The taxonomic diversification seen in these fossils was accompanied by a suite of novel biological features, including the origins of complex multicellularity (a term that applies to multicellular organisms with differentiated cell types), sexual life cycles, and eukaryotic forms and other novel features of eukaryotes appear, signifying the origin of the group? Fossils and molecular data provide clues to when and how eukaryotes arose from their prokaryotic ancestors.

## The Fossil Record of Early Eukaryotes

Complex lipids that are synthesized by eukaryotes (but not by prokaryotes) have been found in rocks dated to 2.7 billion years ago. Although such chemical evidence is consistent with eukaryotes having lived at that time, the oldest widely accepted fossils of eukaryotic organisms are 1.8 billion years old. Over time, the descendants of these organisms gave rise to the rich diversity of protists and other eukaryotes alive today.

**Figure 25.2** surveys how that diversity arose, focusing on three stages documented by the fossil record: an initial diversification (1.8–1.3 billion years ago), the origin of multicellularity and other novel features (1.3 billion–635 million years ago), and the emergence of large eukaryotes (635–535 million years ago).

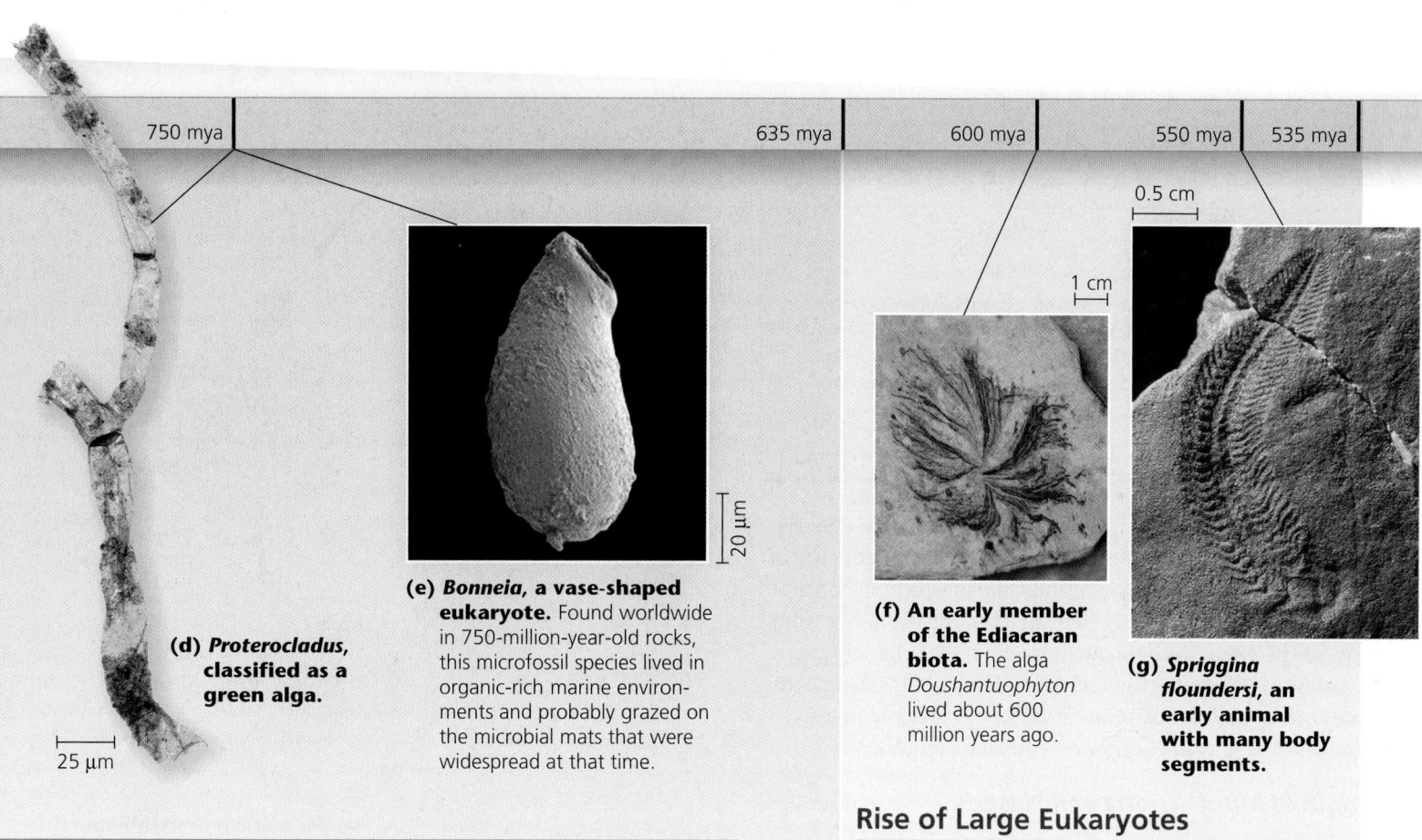

750 mya          635 mya     600 mya     550 mya   535 mya

**(d) Proterocladus, classified as a green alga.**

25 μm

**(e) Bonneia, a vase-shaped eukaryote.** Found worldwide in 750-million-year-old rocks, this microfossil species lived in organic-rich marine environments and probably grazed on the microbial mats that were widespread at that time.

20 μm

1 cm

0.5 cm

**(f) An early member of the Ediacaran biota.** The alga *Doushantuophyton* lived about 600 million years ago.

**(g) Spriggina floundersi, an early animal with many body segments.**

photosynthesis. Some of these features can be observed directly, while others are inferred from living members of the group to which a fossil belongs. For example, fossils show that *Bangiomorpha* (c) had "holdfasts" similar to those that anchor living red algae to their substrate; it also had a pattern of cell division that is only known to occur in a particular group of extant red algae. Hence, *Bangiomorpha* is classified as a member of this group of red algae. Living members of this group have sexual life cycles and are photosynthetic, so biologists infer that *Bangiomorpha* also had these features.

A range of other fossils show that by 800–750 million years ago, increasingly complex communities of eukaryotes had emerged, with algae at the bottom of the food chain and species that ate algae (or each other) above. These organisms remained small, however, and the entire community began to decline with the onset of a series of severe ice ages and other environmental changes.

### Rise of Large Eukaryotes

For nearly 3 billion years, life on Earth was a world of microscopic forms. Larger multicellular eukaryotes do not appear in the fossil record until the Ediacaran period, 635–541 million years ago (**f, g**). These fossils, referred to as the Ediacaran biota, were of soft-bodied organisms, some over 1 m long.

More generally, the fossil record from 635 to 535 million years ago documents changes in the history of life: maximum body size, taxonomic diversity, and the extent of morphological differences all increased dramatically. In addition, the average time that species persisted in the fossil record dropped considerably. Indeed, the entire Ediacaran biota declined 535 million years ago with the onset of another great wave of evolutionary diversification—the so-called "Cambrian explosion."

As discussed in Figure 25.2, large, multicellular eukaryotes did not appear until about 600 million years ago. Prior to that time, Earth was a microbial world: Its only inhabitants were single-celled prokaryotes and eukaryotes, along with an assortment of microscopic, multicellular eukaryotes. We'll return to the rise of large, multicellular eukaryotes in Chapters 26 and 27.

## Endosymbiosis in Eukaryotic Evolution

The fossil record documents when early eukaryotes lived and when key eukaryotic traits, such as a well-developed cytoskeleton and sexual life cycles, first appeared. Additional insights into the origin of eukaryotes have come from molecular studies. In particular, DNA sequence data suggest that eukaryotes are "combination" organisms, with some of their genes and cellular characteristics being derived from archaea, and others from bacteria **(Table 25.1)**.

**Table 25.1 Inferred Origins of Key Eukaryotic Features**

| Feature | Original Source |
|---|---|
| DNA replication enzymes | Archaeal |
| Transcription enzymes | Archaeal |
| Translation enzymes | Mostly archaeal |
| Cell division apparatus | Mostly archaeal |
| Endoplasmic reticulum | Archaeal and bacterial |
| Mitochondrion | Bacterial |
| Metabolic genes | Mostly bacterial |

How did eukaryotes come to have both archaeal and bacterial features? This mixture of features may be a consequence of **endosymbiosis**, a symbiotic relationship in which one organism lives inside the body or cell of another organism. According to this hypothesis, the defining moment in the origin of eukaryotes occurred when an archaeal cell (or a cell with archaeal ancestors) engulfed a bacterium that would later become an organelle found in all eukaryotes—the mitochondrion.

### Origin of Mitochondria and Plastids

The idea that eukaryotes are "combination" organisms is related to the **endosymbiont theory**, which holds that mitochondria and plastids (a general term for chloroplasts and related organelles) were formerly small prokaryotes that began living within larger cells **(Figure 25.3)**. The term *endosymbiont* refers to a cell that lives within another cell, called the *host cell*. The prokaryotic ancestors of mitochondria and plastids probably gained entry to the host cell as undigested prey or internal parasites. Though such a process may seem unlikely, scientists have directly observed cases in which endosymbionts that began as prey or parasites came to have a mutually beneficial relationship with the host in as little as five years.

By whatever means the relationship began, we can hypothesize how the symbiosis could have become mutually beneficial. For example, in a world that was gradually becoming

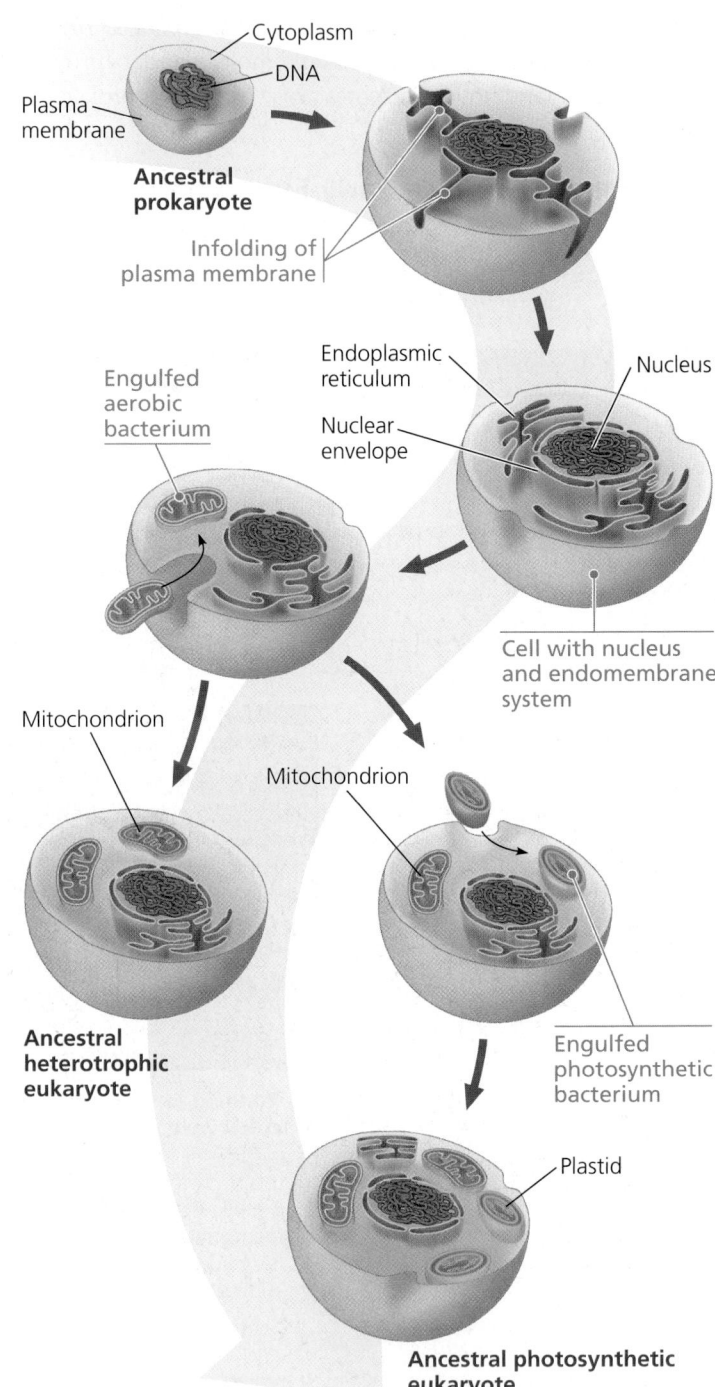

▲ **Figure 25.3 A hypothesis for the origin of eukaryotes through endosymbiosis.** The proposed host was an archaean or a cell descended from archaeal ancestors. The proposed ancestors of mitochondria were aerobic, heterotrophic prokaryotes, whereas those of plastids were photosynthetic prokaryotes. In this figure, the arrows represent change over evolutionary time.

more aerobic, a host that was itself an anaerobe would have benefited from endosymbionts that could make use of the oxygen. Over time, the host and endosymbionts would have become a single organism, its parts inseparable. Although all eukaryotes have mitochondria or remnants of these organelles, they do not all have plastids. Thus, the **serial endosymbiosis** hypothesis supposes that mitochondria evolved before plastids through a sequence of endosymbiotic events (see Figure 25.3).

A great deal of evidence supports the endosymbiotic origin of mitochondria and plastids:

- The inner membranes of both organelles have enzymes and transport systems that are homologous to those found in the plasma membranes of living prokaryotes.
- Mitochondria and plastids replicate by a splitting process that is similar to that of certain prokaryotes. Mitochondria and plastids both contain circular DNA molecules that, like the chromosomes of bacteria, are not associated with histones or large amounts of other proteins.
- As might be expected of organelles descended from free-living organisms, mitochondria and plastids also have the cellular machinery (including ribosomes) needed to transcribe and translate their DNA into proteins.
- Finally, in terms of size, RNA sequences, and sensitivity to certain antibiotics, the ribosomes of mitochondria and plastids are more similar to prokaryotic ribosomes than they are to the cytoplasmic ribosomes of eukaryotic cells.

Which prokaryotic lineages gave rise to mitochondria? To answer this question, researchers have compared the DNA sequences of mitochondrial genes (mtDNA) with those found in major clades of bacteria and archaea. In the **Scientific Skills Exercise**, you will interpret one such set of DNA sequence comparisons. Collectively, such studies indicate that mitochondria arose from an alpha proteobacterium (see Figure 24.19). Researchers have also compared genome sequences of various alpha proteobacteria with the entire mtDNA sequences of animals, plants, fungi, and protists. Such studies indicate that eukaryotic mitochondria descended from a single common ancestor, suggesting that mitochondria arose only once over the course of evolution. Similar analyses provide evidence that plastids descended from a single common

---

## Scientific Skills Exercise

### Interpreting Comparisons of Genetic Sequences

► Wheat, used as the source of mitochondrial RNA

**Which Prokaryotes Are Most Closely Related to Mitochondria?**
The first eukaryotes acquired mitochondria by endosymbiosis: A host cell engulfed an aerobic prokaryote that persisted within the cytoplasm to the mutual benefit of both cells. In studying which living prokaryotes might be most closely related to mitochondria, researchers compared ribosomal RNA (rRNA) sequences. Ribosomes perform critical cell functions. Hence, rRNA sequences are under strong selection and change slowly over time, making them suitable for comparing even distantly related species. In this exercise, you'll interpret some of their results to draw conclusions about the phylogeny of mitochondria.

**How the Research Was Done** Researchers isolated and cloned nucleotide sequences from the gene that codes for the small-subunit rRNA molecule for wheat (a eukaryote) and five bacterial species:

- Wheat, used as the source of mitochondrial rRNA genes
- *Agrobacterium tumefaciens*, an alpha proteobacterium that lives within plant tissue and produces tumors in the host
- *Comamonas testosteroni*, a beta proteobacterium
- *Escherichia coli*, a well-studied gamma proteobacterium that inhabits human intestines
- *Mycoplasma capricolum*, a gram-positive mycoplasma, which is the only group of bacteria lacking cell walls
- *Anacystis nidulans*, a cyanobacterium

**Data from the Research** Cloned rRNA gene sequences for the six organisms were aligned and compared. The data table below, called a *comparison matrix*, summarizes the comparison of 617 nucleotide positions from the gene sequences. Each value in the table is the percentage of the 617 nucleotide positions for which the pair of organisms have the same base. Any positions that were identical across the rRNA genes of all six organisms were omitted from this comparison matrix.

**INTERPRET THE DATA**

1. First, make sure you understand how to read the comparison matrix. Find the cell that represents the comparison of *C. testosteroni* and *E. coli*. What value is given in this cell? What does that value signify about the comparable rRNA gene sequences in those two organisms? Explain why some cells have a dash rather than a value. Why are some cells shaded gray, with no value?
2. Why did the researchers choose one plant mitochondrion and five bacterial species to include in the comparison matrix?
3. Which bacterium has an rRNA gene that is most similar to that of the wheat mitochondrion? What is the significance of this similarity?

**Data from** D. Yang et al., Mitochondrial origins, *Proceedings of the National Academy of Sciences USA* 82:4443–4447 (1985).

(MB) A version of this Scientific Skills Exercise can be assigned in MasteringBiology.

| | Wheat mitochondrion | A. tumefaciens | C. testosteroni | E. coli | M. capricolum | A. nidulans |
|---|---|---|---|---|---|---|
| **Wheat mitochondrion** | – | 48 | 38 | 35 | 34 | 34 |
| **A. tumefaciens** | | – | 55 | 57 | 52 | 53 |
| **C. testosteroni** | | | – | 61 | 52 | 52 |
| **E. coli** | | | | – | 48 | 52 |
| **M. capricolum** | | | | | – | 50 |
| **A. nidulans** | | | | | | – |

ancestor—a cyanobacterium that was engulfed by a eukaryotic host cell.

As we've seen, the prokaryotic lineages that gave rise to plastids and mitochondria have been identified. However, questions remain about the identity of the host cell that engulfed the alpha proteobacterium that gave rise to the mitochondrion—an event that ultimately led to the origin of the eukaryotes. According to recent genomic studies, this host cell came from an archaeal lineage, but which lineage remains undetermined. Alternatively, the host could have been a member of a lineage that was related to, but had diverged from, its archaeal ancestors. In this case, the host may have been a "protoeukaryote" in which certain features of eukaryotic cells had evolved, such as an endomembrane system and a cytoskeleton that enabled it to change shape (and thereby engulf the alpha proteobacterium).

### Plastid Evolution: A Closer Look

As you've seen, current evidence indicates that mitochondria are descended from a bacterium that was engulfed by a cell from an archaeal lineage. This event gave rise to the eukaryotes. There is also much evidence that later in eukaryotic history, a lineage of heterotrophic eukaryotes acquired an additional endosymbiont—a photosynthetic cyanobacterium—that then evolved into plastids. According to the hypothesis illustrated

in **Figure 25.4**, this plastid-bearing lineage gave rise to two lineages of photosynthetic protists, red algae and green algae.

Let's examine some of the steps shown in Figure 25.4 in more detail. First, recall that cyanobacteria are gram-negative and that gram-negative bacteria have two cell membranes, an inner plasma membrane and an outer membrane that is part of the cell wall (see Figure 24.7). Plastids in red algae and green algae are also surrounded by two membranes. Transport proteins in these membranes are homologous to proteins in the inner and outer membranes of cyanobacteria, providing further support for the hypothesis that plastids originated from a cyanobacterial endosymbiont.

On several occasions during eukaryotic evolution, red algae and green algae underwent **secondary endosymbiosis**; that is, the algal cells were ingested in the food vacuoles of heterotrophic eukaryotes and became endosymbionts themselves. For example, as shown in Figure 25.4, protists known as chlorarachniophytes likely evolved when a heterotrophic eukaryote engulfed a green alga. Evidence for this process can be found within the engulfed cell, which contains a tiny vestigial nucleus, called a *nucleomorph*. Genes from the nucleomorph are still transcribed, and their DNA sequences indicate that the engulfed cell was a green alga. Also consistent with the hypothesis that chlorarachniophytes evolved from a eukaryote that

▼ **Figure 25.4 Diversity of plastids produced by endosymbiosis.** Studies of plastid-bearing eukaryotes suggest that plastids evolved from a cyanobacterium that was engulfed by an ancestral heterotrophic eukaryote (primary endosymbiosis). That ancestor then diversified into red algae and green algae, some of which were subsequently engulfed by other eukaryotes (secondary endosymbiosis).

**MAKE CONNECTIONS** *How many distinct genomes does a chlorarachniophyte cell contain? Explain. (See Figure 4.16.)*

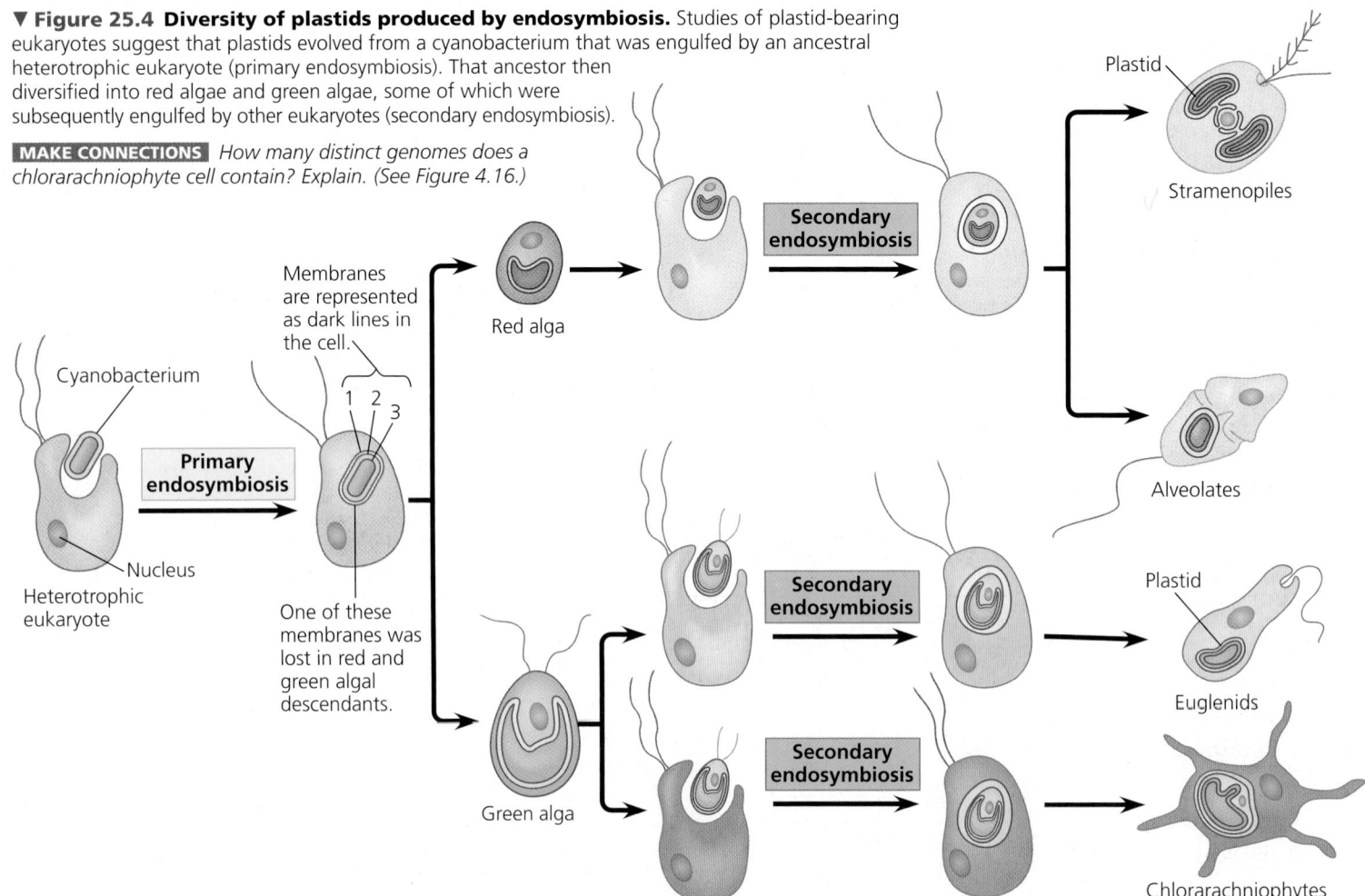

engulfed another eukaryote, their plastids are surrounded by *four* membranes. The two inner membranes originated as the inner and outer membranes of the ancient cyanobacterium, the third membrane is derived from the engulfed alga's plasma membrane, and the outermost membrane is derived from the membrane of the heterotrophic eukaryote's food vacuole.

**CONCEPT CHECK 25.1**

1. Describe major events in the evolution of early eukaryotes that are documented in the fossil record.
2. Explain why eukaryotes are said to be "combination" organisms, and summarize the role of endosymbiosis in eukaryotic evolution.
3. **WHAT IF?** Suppose the photosynthetic organelle of a protist is discovered to be most closely related to a different cyanobacterium than the one that gave rise to plastids. What would this result suggest about the origin of eukaryotic photosynthesis?

For suggested answers, see Appendix A.

CONCEPT 25.2

# Multicellularity has originated several times in eukaryotes

An orchestra can play a greater variety of musical compositions than a violin soloist can; the increased complexity of the orchestra makes more variations possible. Likewise, the origin of structurally complex eukaryotic cells sparked the evolution of greater morphological diversity than was possible for the simpler prokaryotic cells. This burst of evolutionary change resulted in the immense variety of unicellular protists that continue to flourish today. Another wave of diversification also occurred: Some single-celled eukaryotes gave rise to multicellular forms, whose descendants include a variety of algae, plants, fungi, and animals.

## Multicellular Colonies

The first multicellular forms were *colonies*, collections of cells that are connected to one another but show little or no cellular differentiation. Multicellular colonies consisting of simple filaments, balls, or cell sheets occur early and often in the eukaryotic fossil record, and they remain common today **(Figure 25.5)**. Such simple colonies are often found in eukaryotic lineages whose members have rigid cell walls. In such organisms, a colony may take shape as the cells divide and remain attached to one another by their shared cell walls.

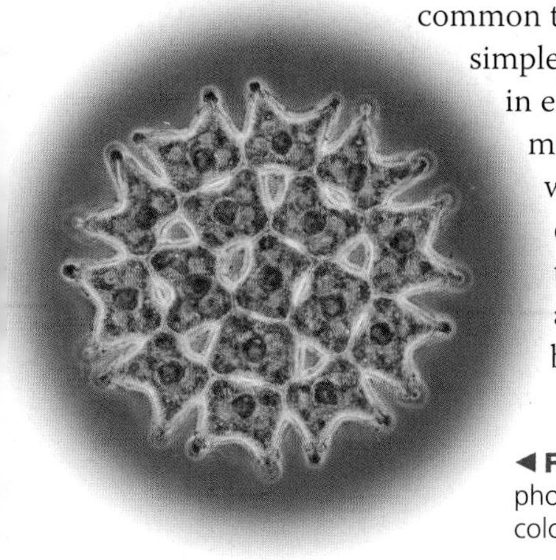

◀ **Figure 25.5 *Pediastrum.*** This photosynthetic eukaryote forms flat colonies (LM).

Simple colonies are also found in eukaryotes that lack rigid cell walls, but in this case a colony may form when dividing cells are held together by proteins that physically connect adjacent cells to one another.

Some simple colonies have features that are intermediate between those of single-celled eukaryotes and those of more complex multicellular forms, such as plants, fungi, and animals. As you'll see in the following sections, such differences between unicellular, colonial, and multicellular eukaryotes can reveal clues to the origin of multicellularity.

## Independent Origins of Complex Multicellularity

Although they occur in fewer lineages than do simple colonies, multicellular organisms with differentiated cells originated multiple times over the course of eukaryotic evolution. Examples include lineages of red, green, and brown algae, as well as plants, fungi, and animals. Genetic and morphological data indicate that these different lineages of complex multicellular eukaryotes arose independently of one another. For example, although both fungi and animals arose from single-celled ancestors, they arose from *different* single-celled ancestors.

The fact that complex multicellularity has originated multiple times allows us to examine the similarities and differences in how these independent groups arose. One example is the evolutionary lineage that includes the multicellular green alga *Volvox* **(Figure 25.6)**. (*Volvox* has two types of differentiated

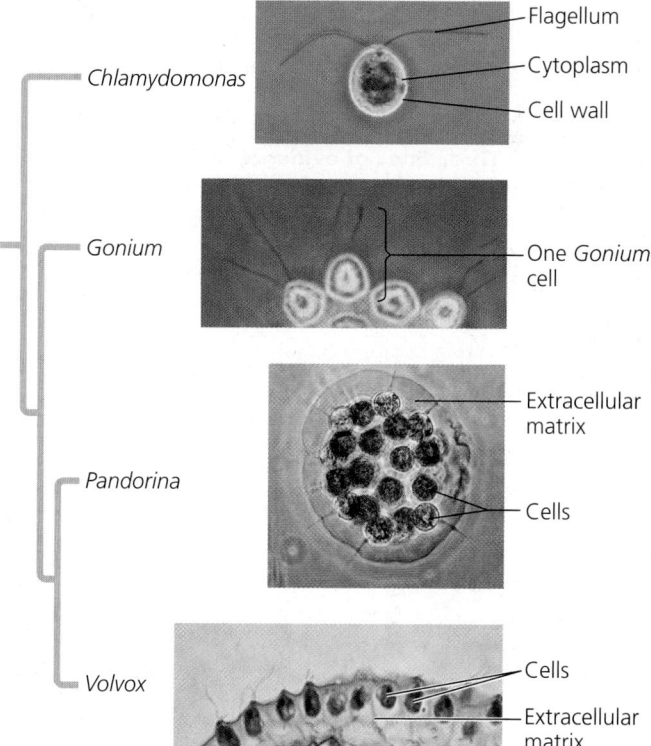

▲ **Figure 25.6 Morphological change in the *Volvox* lineage.** Each *Gonium* cell resembles a *Chlamydomonas* cell, and the structures that attach *Gonium* cells to one another contain proteins homologous to those in the *Chlamydomonas* cell wall. In *Pandorina* and *Volvox*, the cells are embedded in an extracellular matrix containing proteins also homologous to those found in the *Chlamydomonas* cell wall.

cells, and for this reason most researchers refer to this species as "multicellular," as opposed to "colonial.") DNA evidence indicates that *Volvox* forms a monophyletic group with a single-celled alga (*Chlamydomonas*) and several colonial species (see Figure 25.6). *Volvox* cells are embedded in an extracellular matrix composed of proteins homologous to those in the *Chlamydomonas* cell wall; the same is true for the colonial species that branch between *Chlamydomonas* and *Volvox*. This suggests that multicellularity in *Volvox* may have originated as descendants of a single-celled common ancestor gave rise to a series of larger and more complex colonial forms.

A 2010 comparison of the *Chlamydomonas* and *Volvox* genomes yielded a further surprising result: *Volvox* has few novel genes that could account for the differences in morphology between these species. This suggests that the transition to multicellularity may not require the origin of many new genes. Instead, this transition may result primarily from changes in how existing genes are used—a conclusion also supported by recent studies on the origin of multicellularity in animals.

## Steps in the Origin of Multicellular Animals

Although the origin of animals was a pivotal moment in the history of life, until recently little was known about the genetic toolkit that facilitated the emergence of multicellular animals from their single-celled ancestors. One way to gather information about this toolkit is to identify protist groups that are closely related to animals. As shown in **Figure 25.7**, a combination of morphological and molecular evidence points to choanoflagellates as the closest living relatives of animals.

Based on such evidence, researchers have hypothesized that the common ancestor of choanoflagellates and living animals may have been a unicellular suspension feeder that resembled present-day choanoflagellates.

Note that the origin of multicellularity in animals required the evolution of new ways for cells to adhere (attach) and signal (communicate) to each other. In an effort to learn more about such mechanisms, researchers compared the genome of the unicellular choanoflagellate *Monosiga brevicollis* with those of representative animals.

This analysis uncovered 78 protein domains in *M. brevicollis* that were otherwise only known to occur in animals. (A *domain* is a key structural or functional region of a protein.) In animals, many of these shared protein domains function in cell adherence or cell signaling. To give just two examples, *M. brevicollis* has genes that encode domains of certain proteins (known as cadherins) that play key roles in how animal cells attach to one another, as well as genes that encode protein domains that animals (and only animals) use in cell-signaling pathways.

New research has also enabled us to take a closer look at specific proteins that played important roles in the origin of multicellularity in animals. Were these proteins composed mostly of domains found in ancestral choanoflagellate proteins? Or did they have a more novel structure? Consider the cadherin attachment proteins mentioned earlier. DNA sequence analyses show that animal cadherin proteins are composed primarily of domains that are also found in a cadherin-like protein of choanoflagellates **(Figure 25.8)**. However, animal cadherin proteins that attach cells to one another also

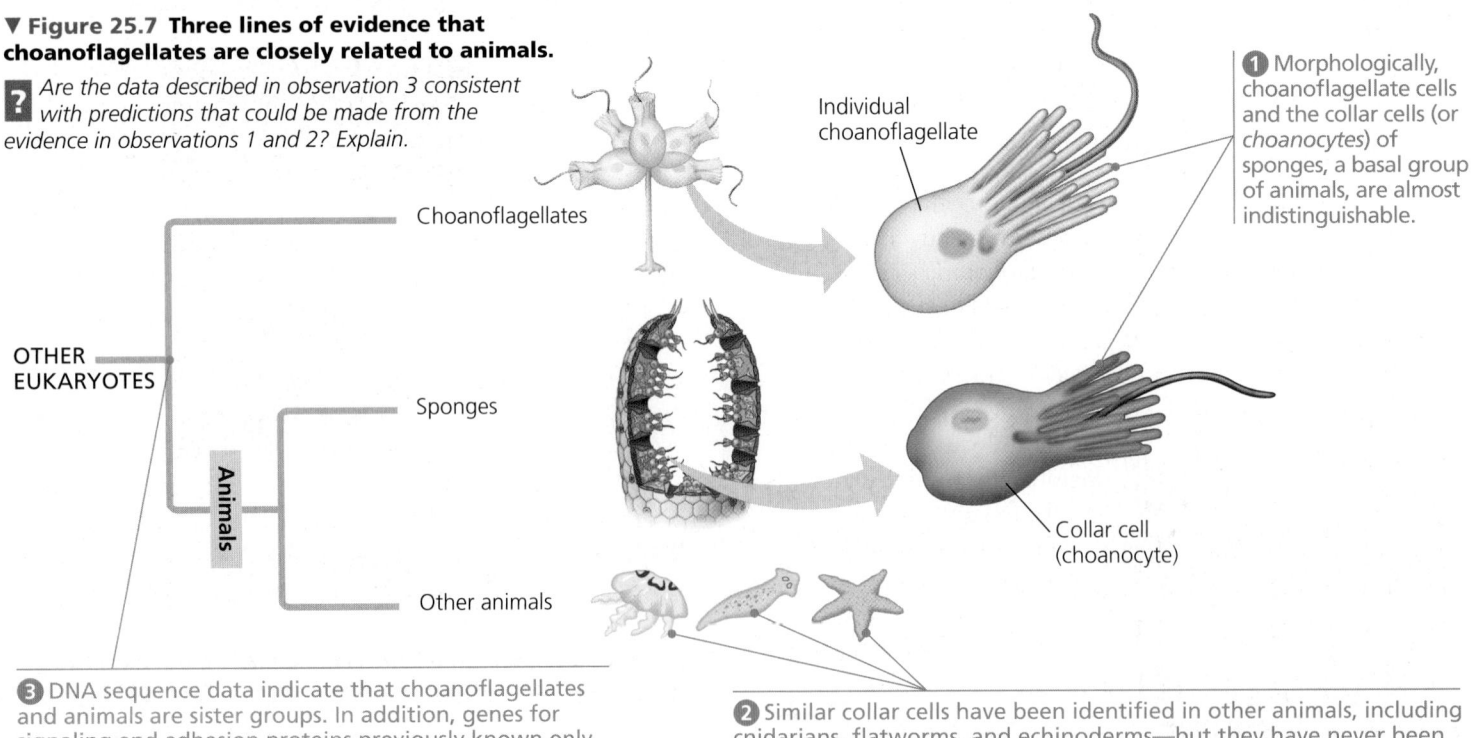

▼ **Figure 25.7 Three lines of evidence that choanoflagellates are closely related to animals.**

**?** *Are the data described in observation 3 consistent with predictions that could be made from the evidence in observations 1 and 2? Explain.*

Choanoflagellates

OTHER EUKARYOTES

Sponges

Animals

Other animals

Individual choanoflagellate

**❶** Morphologically, choanoflagellate cells and the collar cells (or *choanocytes*) of sponges, a basal group of animals, are almost indistinguishable.

Collar cell (choanocyte)

**❸** DNA sequence data indicate that choanoflagellates and animals are sister groups. In addition, genes for signaling and adhesion proteins previously known only from animals have been discovered in choanoflagellates.

**❷** Similar collar cells have been identified in other animals, including cnidarians, flatworms, and echinoderms—but they have never been observed in non-choanoflagellate protists or in plants or fungi.

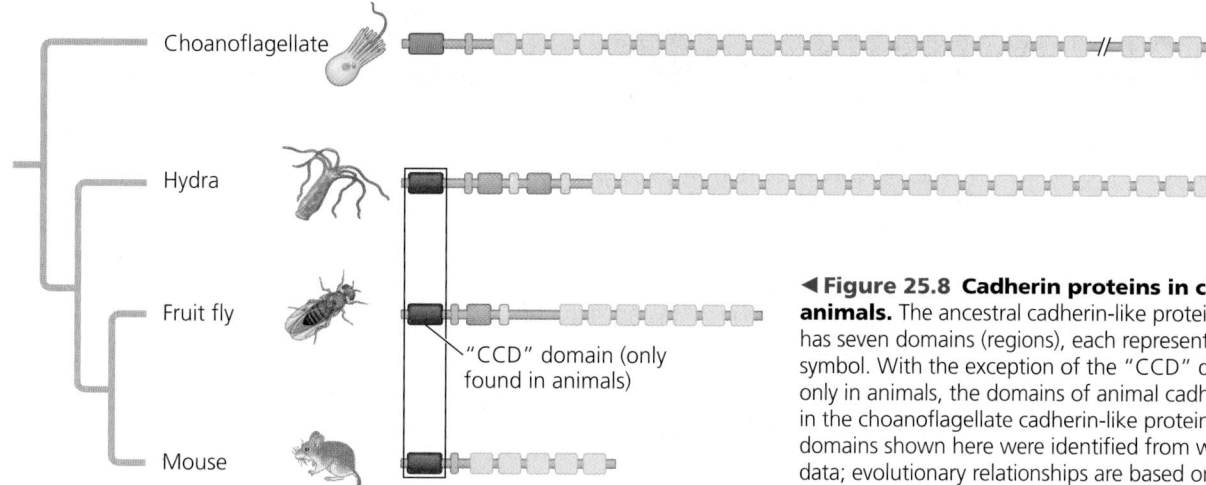

◀ **Figure 25.8 Cadherin proteins in choanoflagellates and animals.** The ancestral cadherin-like protein of choanoflagellates has seven domains (regions), each represented here by a particular symbol. With the exception of the "CCD" domain, which is found only in animals, the domains of animal cadherin proteins are present in the choanoflagellate cadherin-like protein. The cadherin protein domains shown here were identified from whole-genome sequence data; evolutionary relationships are based on morphological and DNA sequence data.

contain a highly conserved region not found in the choano-flagellate protein (the "CCD" domain shown in Figure 25.8). These results suggest that the origin of the cadherin attachment protein occurred by the rearrangement of protein domains found in choanoflagellates—along with the incorporation of a novel domain, the conserved CCD region.

Overall, comparisons of choanoflagellate and animal genome sequences tell us that key steps in the transition to multicellularity in animals involved new ways of using proteins or parts of proteins that were encoded by genes found in choanoflagellates. Thus, as we also saw for the origin of multicellularity in *Volvox*, the origin of multicellularity in animals may have resulted mostly from the co-opting of genes used for other purposes in choanoflagellates—not from the evolution of a genetic toolkit composed of many novel genes.

**CONCEPT CHECK 25.2**

1. Summarize the evidence that choanoflagellates are the sister group of animals.
2. **MAKE CONNECTIONS** Describe how the origin of multicellularity in animals illustrates Darwin's concept of descent with modification (see Concept 19.2).
3. **WHAT IF?** Cells in *Volvox*, plants, and fungi are similar in being bounded by a cell wall. Predict whether the cell-to-cell attachments of these organisms form using similar or different molecules. Explain.

For suggested answers, see Appendix A.

**CONCEPT 25.3**

# Four "supergroups" of eukaryotes have been proposed based on morphological and molecular data

How have events described so far in this chapter influenced the diversity of eukaryotes living today? First, by their very nature, eukaryotes are "combination" organisms. Having

originated by endosymbiosis, they had archaeal and bacterial genes and they possessed endosymbionts with novel metabolic capabilities. These features promoted the diversification of unicellular protists seen in the fossil record and still evident today in a drop of pond water. The independent origins of complex multicellularity in several eukaryotic lineages also had a major influence. Each of these independent groups evolved different solutions to the various challenges that all organisms face, thus contributing to the rich diversity of eukaryotes alive today. We'll survey that diversity here, beginning with an overview of the big picture: the four eukaryotic "supergroups."

## Four Supergroups of Eukaryotes

Our understanding of the evolutionary history of eukaryotes has been in a state of flux in recent years. Genetic and morphological studies have shown that some protists are more closely related to plants, fungi, or animals than they are to other protists. As a result, the kingdom in which all protists once were classified, Protista, has been abandoned. Other hypotheses have been discarded as well. For example, biologists once thought that the most basal eukaryotic lineage consisted of the *amitochondriate protists*, organisms without conventional mitochondria. But recent structural and DNA data have undermined this hypothesis. Many of the so-called amitochondriate protists have been shown to have mitochondria—though reduced ones—and some of these organisms are now classified in distantly related groups.

The ongoing changes in our understanding of the phylogeny of eukaryotes pose challenges to students and instructors alike. Hypotheses about these relationships are a focus of scientific activity, changing rapidly as new data cause previous ideas to be modified or discarded. We'll focus here on one current hypothesis: the four supergroups of eukaryotes shown in **Figure 25.9**. Because the root of the eukaryotic tree is not known, all four supergroups are shown as diverging simultaneously from a common ancestor. We know that this is not correct, but we do not know which supergroup was the first to diverge from the other three. In addition, while some of the

The tree below represents a phylogenetic hypothesis for the relationships among eukaryotes on Earth today. The eukaryotic groups at the branch tips are related in larger "supergroups," labeled vertically at the far right of the tree. Groups that were formerly classified in the kingdom Protista are highlighted in yellow. Dotted lines indicate evolutionary relationships that are uncertain and proposed clades that are under active debate. For clarity, this tree only includes representative clades from each supergroup. In addition, the recent discoveries of many new groups of eukaryotes indicate that eukaryotic diversity is actually much greater than shown here.

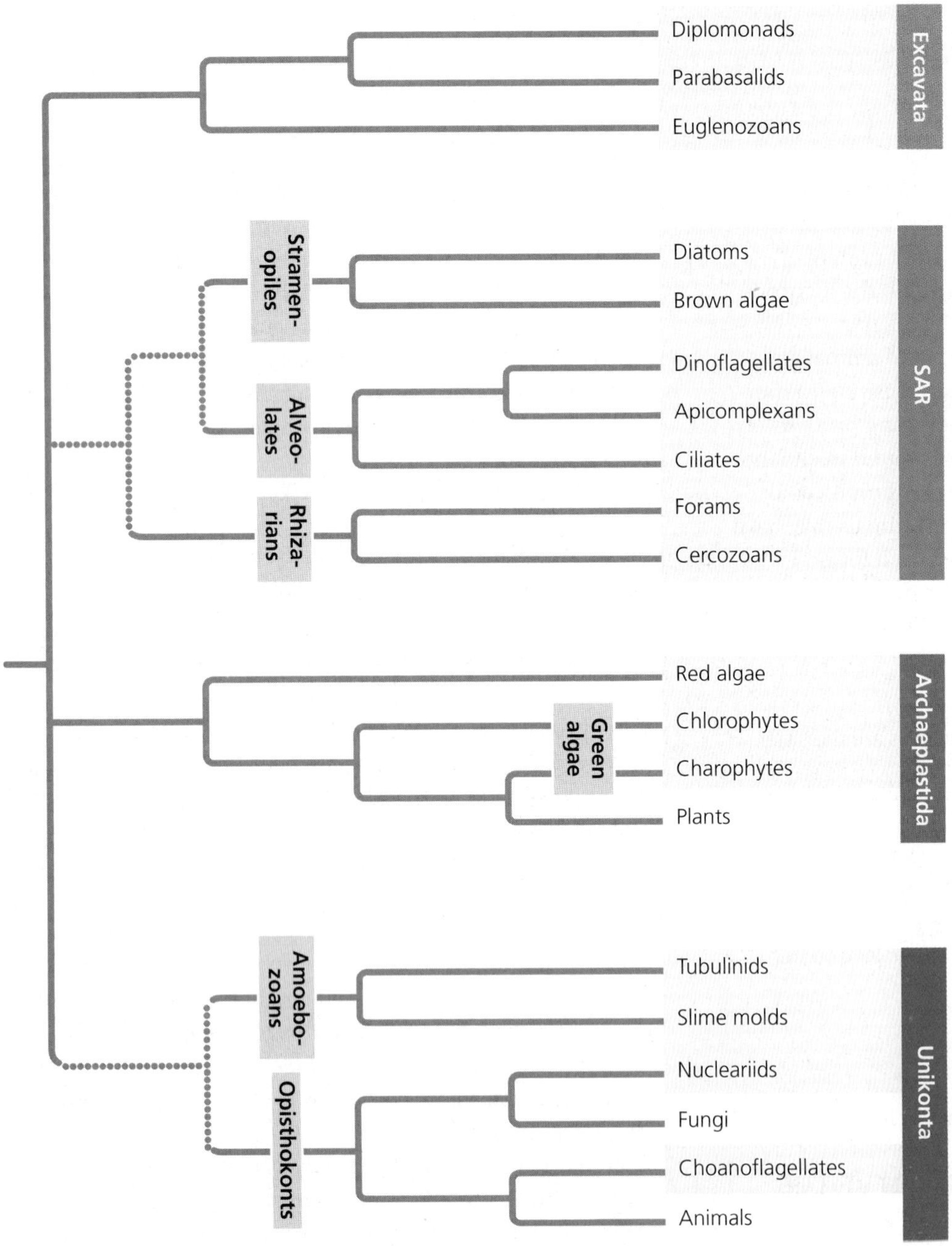

## Phylogenetic Tree

**Excavata**
- Diplomonads
- Parabasalids
- Euglenozoans

**SAR**
- Stramenopiles
  - Diatoms
  - Brown algae
- Alveolates
  - Dinoflagellates
  - Apicomplexans
  - Ciliates
- Rhizarians
  - Forams
  - Cercozoans

**Archaeplastida**
- Red algae
- Green algae
  - Chlorophytes
  - Charophytes
- Plants

**Unikonta**
- Amoebozoans
  - Tubulinids
  - Slime molds
- Opisthokonts
  - Nucleariids
  - Fungi
  - Choanoflagellates
  - Animals

? *Based on the fossil record of early eukaryotes and the tree shown here, by what date had the supergroups begun to diverge from one another? Explain.*

## ■ Excavata

Some members of this supergroup have an "excavated" groove on one side of the cell body. Two major clades (the parabasalids and diplomonads) have highly reduced mitochondria; members of a third clade (the euglenozoans) have flagella that differ in structure from those of other organisms. Excavates include parasites such as *Giardia*, as well as many predatory and photosynthetic species.

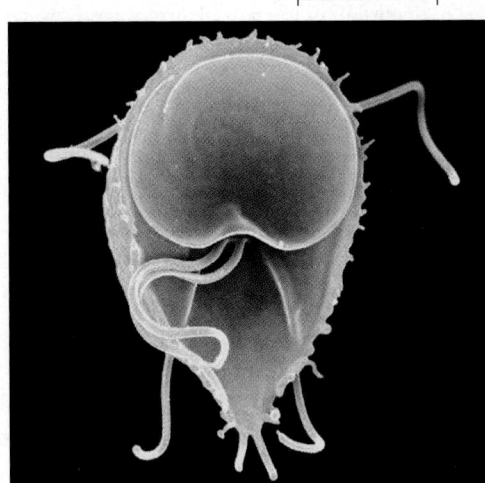

5 μm

***Giardia intestinalis*, a diplomonad parasite.** This diplomonad (colorized SEM), which lacks the characteristic surface groove of the Excavata, inhabits the intestines of mammals. It can infect people when they drink water contaminated with feces containing *Giardia* cysts. Drinking such water—even from a seemingly pristine stream—can cause severe diarrhea. Boiling the water kills the parasite.

## ■ SAR

This supergroup contains (and is named after) three large and very diverse clades: Stramenopila, Alveolata, and Rhizaria. Stramenopiles include some of the most important photosynthetic organisms on Earth, such as the diatoms shown here. Alveolates also include photosynthetic species, as well as important pathogens, such as *Plasmodium*, which causes malaria. According to one current hypothesis, stramenopiles and alveolates originated by secondary endosymbiosis when a heterotrophic protist engulfed a red alga.

50 μm

**Diatom diversity.** These beautiful single-celled protists are important photosynthetic organisms in aquatic communities (LM).

The rhizarian subgroup of SAR includes many species of amoebas, most of which have pseudopodia that are threadlike in shape. Pseudopodia are extensions that can bulge from any portion of the cell; they are used in movement and in the capture of prey.

100 μm

**Globigerina, a rhizarian in the SAR supergroup.** This species is a foram, a group whose members have threadlike pseudopodia that extend through pores in the shell, or test (LM). The inset shows a foram test, which is hardened by calcium carbonate.

## ■ Archaeplastida

This supergroup of eukaryotes includes red algae and green algae, along with plants. Red algae and green algae include unicellular species, colonial species, and multicellular species (including the green alga *Volvox*). Many of the large algae known informally as "seaweeds" are multicellular red or green algae. Protists in Archaeplastida include key photosynthetic species that form the base of the food web in many aquatic communities.

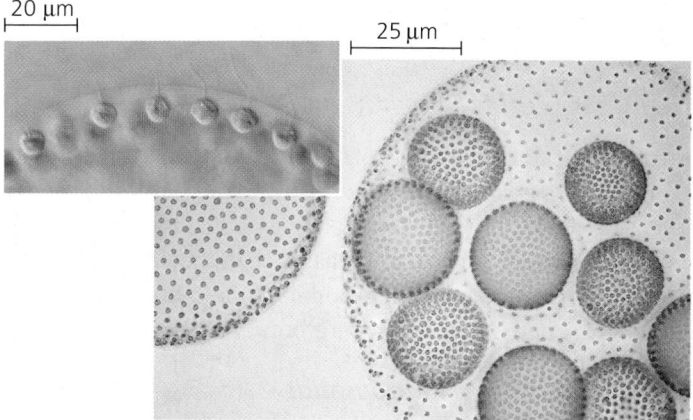

20 μm

25 μm

***Volvox*, a multicellular freshwater green alga.** This alga resembles a hollow ball whose wall is composed of hundreds of biflagellated cells (see inset LM) embedded in a gelatinous extracellular matrix. The cells in the wall are usually connected by cytoplasmic strands; if isolated, these cells cannot reproduce. However, the alga also contains cells that are specialized for either sexual or asexual reproduction. The large algae shown here will eventually release the small "daughter" algae that can be seen within them (LM).

## ■ Unikonta

This supergroup of eukaryotes includes amoebas that have lobe- or tube-shaped pseudopodia, as well as animals, fungi, and non-amoeba protists that are closely related to animals or fungi. According to one current hypothesis, the unikonts were the first eukaryotic supergroup to diverge from all other eukaryotes; however, this hypothesis has yet to be widely accepted.

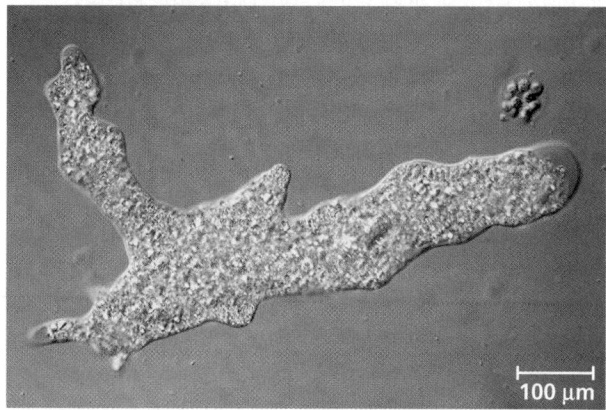

100 μm

**A unikont amoeba.** This amoeba, the tubulinid *Amoeba proteus*, is using its pseudopodia to move.

groups in Figure 25.9 are well supported by morphological and DNA data, others are more controversial.

We'll now examine some representative members of the four supergroups. As you read about these groups, it may be helpful to focus less on the specific names of their members and more on why these organisms are important and how ongoing research is elucidating their evolutionary relationships.

## Excavates

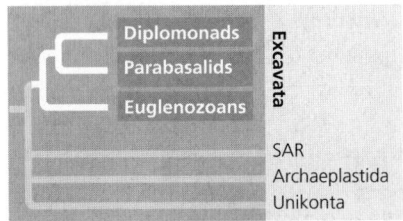

The clade **Excavata** (the excavates) was originally proposed based on morphological studies of the cytoskeleton. The name derives from the fact that some members of this diverse group have an "excavated" feeding groove on one side of the cell body. The excavates include the diplomonads, parabasalids, and euglenozoans. Molecular data indicate that each of these three groups is monophyletic, and recent genomic studies support the monophyly of the excavate supergroup.

### Diplomonads and Parabasalids

The protists in these two groups lack plastids and have highly reduced mitochondria (until recently, they were thought to lack mitochondria altogether). Most diplomonads and parabasalids are found in anaerobic environments.

**Diplomonads** have reduced mitochondria called *mitosomes*. These organelles lack functional electron transport chains and hence cannot use oxygen to help extract energy from carbohydrates and other organic molecules. Instead, diplomonads get the energy they need from anaerobic biochemical pathways. Many diplomonads are parasites, including the infamous *Giardia intestinalis* (see Figure 25.9), which inhabits the intestines of mammals. *Giardia* and other diplomonads propel themselves using multiple flagella.

**Parabasalids** also have reduced mitochondria; called *hydrogenosomes*, these organelles generate some energy anaerobically, releasing hydrogen gas as a by-product. The best-known parabasalid is *Trichomonas vaginalis*, a sexually transmitted parasite that infects some 5 million people each year. *T. vaginalis* travels along the mucus-coated lining of the human reproductive and urinary tracts by moving its flagella and by undulating part of its plasma membrane **(Figure 25.10)**.

### Euglenozoans

Protists called **euglenozoans** belong to a diverse clade that includes predatory heterotrophs, photosynthetic autotrophs, and parasites. The main morphological feature that distinguishes protists in this clade is the presence of a rod with either a

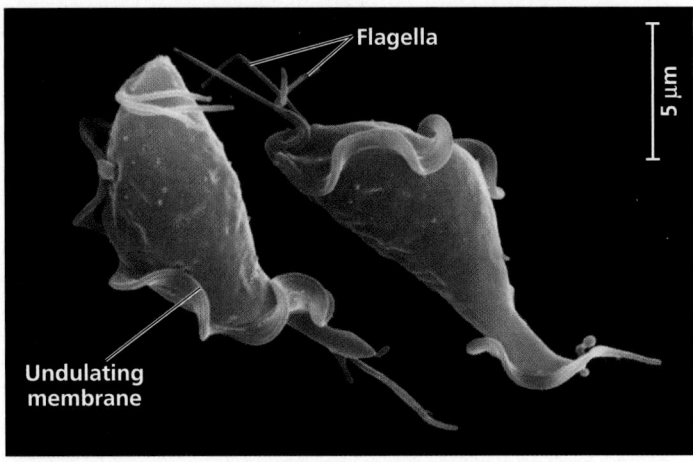

▲ **Figure 25.10 The parabasalid parasite *Trichomonas vaginalis*** (colorized SEM).

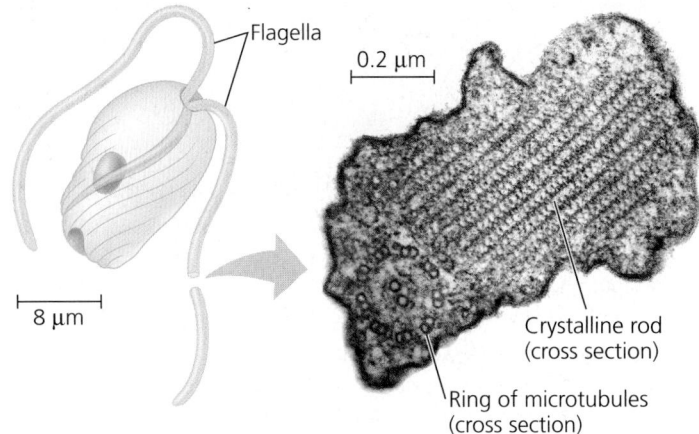

▲ **Figure 25.11 Euglenozoan flagellum.** Most euglenozoans have a crystalline rod inside one of their flagella (the TEM is a flagellum shown in cross section).

spiral or a crystalline structure inside each of their flagella **(Figure 25.11)**. The two best-studied groups of euglenozoans are the euglenids and the kinetoplastids.

A *euglenid* has a pocket at one end of the cell from which one or two flagella emerge (see the drawing in Figure 25.11). Some euglenids perform photosynthesis when sunlight is available; when sunlight is not available, they can become heterotrophic, absorbing organic nutrients from their environment. Many other euglenids engulf prey by phagocytosis.

A *kinetoplastid* has a single, large mitochondrion that contains an organized mass of DNA called a kinetoplast. These protists include species that feed on prokaryotes in aquatic ecosystems, as well as species that parasitize animals, plants, and other protists. For example, kinetoplastids in the genus *Trypanosoma* infect humans and cause sleeping sickness, a neurological disease that is invariably fatal if not treated **(Figure 25.12)**.

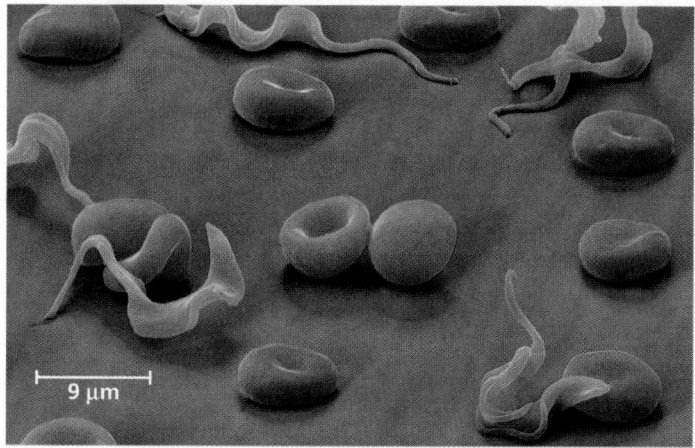

▲ **Figure 25.12** *Trypanosoma*, **the kinetoplastid that causes sleeping sickness.** The purple, ribbon-shaped cells among these red blood cells are the trypanosomes (colorized SEM).

## SAR: Stramenopiles, Alveolates, and Rhizarians

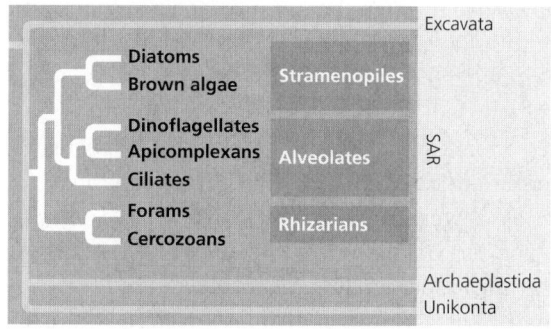

Recent genomic studies have led researchers to propose that three major clades of protists—the stramenopiles, alveolates, and rhizarians—form a monophyletic supergroup referred to as **SAR**, after the first letters of its member clades.

### Stramenopiles

One major subgroup of SAR, the **stramenopiles**, is thought to have arisen by secondary endosymbiosis (along with their sister group, the alveolates; see Figure 25.4). The stramenopiles include some of the most important photosynthetic organisms on the planet. Here we'll focus on two clades of stramenopiles: diatoms and brown algae.

**Diatoms** A key group of photosynthetic protists, **diatoms** are unicellular algae that have a unique glass-like wall made of silicon dioxide embedded in an organic matrix **(Figure 25.13)**. The wall consists of two parts that overlap like a shoe box and its lid. These walls provide effective protection from the crushing jaws of predators: Live diatoms can withstand pressures as great as 1.4 million kg/m², equal to the pressure under each leg of a table supporting an elephant!

With an estimated 100,000 living species, diatoms are a highly diverse group of protists. They are among the most abundant photosynthetic organisms both in the ocean and in lakes: One

▶ **Figure 25.13 The diatom** *Triceratium morlandii* (colorized SEM).

bucket of water scooped from the surface of the sea may contain millions of these microscopic algae. As we'll discuss later in the chapter, the photosynthetic activity of these widespread and abundant algae can affect global carbon dioxide ($CO_2$) levels.

**Brown Algae** The largest and most complex algae are **brown algae**. All are multicellular, and most are marine. Brown algae are especially common along temperate coasts that have cold-water currents. They owe their characteristic brown or olive color to the carotenoids in their plastids.

Many of the species commonly called "seaweeds" are brown algae. Some brown algal seaweeds have specialized structures that resemble organs in plants, such as a rootlike **holdfast**, which anchors the alga, and a stemlike **stipe**, which supports the leaflike **blades (Figure 25.14)**. Unlike plants, however, brown algae lack true tissues and organs. Moreover, morphological and DNA evidence shows that these similarities

▲ **Figure 25.14 Seaweeds: adapted to life at the ocean's margins.** The sea palm (*Postelsia*) lives on rocks along the coast of the northwestern United States and western Canada. This brown alga is well adapted to maintaining a firm foothold despite the crashing surf.

evolved independently in the algal and plant lineages and are thus analogous, not homologous. In addition, while plants have adaptations (such as rigid stems) that provide support against gravity, brown algae have adaptations that enable their main photosynthetic surfaces (the leaflike blades) to be near the water surface. Some brown algae accomplish this task with gas-filled, bubble-shaped floats. Giant brown algae known as kelps that live in deep waters have such floats in their blades, which are attached to stipes that can rise as much as 60 m from the seafloor—more than half the length of a football field.

### Alveolates

Members of the next subgroup of SAR, the **alveolates**, have membrane-enclosed sacs (alveoli) just under the plasma membrane **(Figure 25.15)**; like stramenopiles, alveolates may have originated by secondary endosymbiosis. Alveolates are abundant in many habitats and include a wide range of photosynthetic and heterotrophic protists. We'll discuss two alveolate clades here, a group of flagellates (the dinoflagellates) and a group of protists that move using cilia (the ciliates); we'll discuss a third clade (the apicomplexans) that parasitizes animals in Concept 25.4.

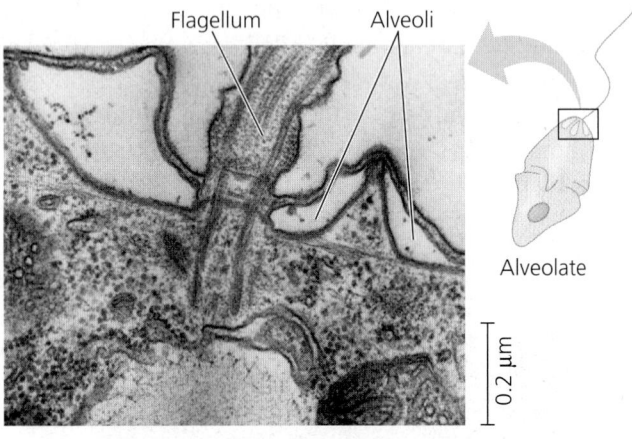

**Flagellum**     **Alveoli**

**Alveolate**

0.2 μm

▲ **Figure 25.15 Alveoli.** These sacs under the plasma membrane are a characteristic that distinguishes alveolates from other eukaryotes (TEM).

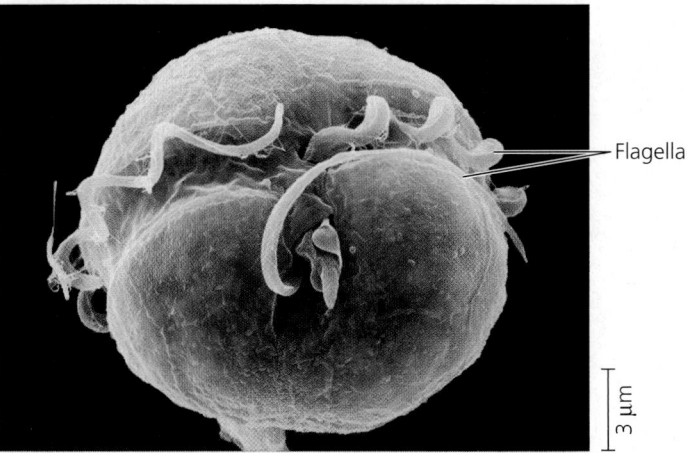

Flagella

3 μm

▲ **Figure 25.16 *Pfiesteria shumwayae*, a dinoflagellate.** Beating of the spiral flagellum, which lies in a groove that encircles the cell, makes this alveolate spin (colorized SEM).

**Dinoflagellates** The cells of many **dinoflagellates** are reinforced by cellulose plates. Two flagella located in grooves in this "armor" make dinoflagellates (from the Greek *dinos*, whirling) spin as they move through the waters of their marine and freshwater communities **(Figure 25.16)**. Although their ancestors may have originated by secondary endosymbiosis (see Figure 25.4), roughly half of all dinoflagellates are now purely heterotrophic. Others are important photosynthetic species, while still others are **mixotrophs**, organisms that combine photosynthesis *and* heterotrophic nutrition.

Periods of explosive population growth, or *blooms*, in dinoflagellates sometimes cause a phenomenon called "red tide." The blooms make coastal waters appear brownish red or pink because of the presence of carotenoids, the most common pigments in dinoflagellate plastids. Toxins produced by certain dinoflagellates have caused massive kills of invertebrates and fishes. Humans who eat molluscs that have accumulated the toxins are affected as well, sometimes fatally.

**Ciliates** The **ciliates** are a large and varied group of protists named for their use of cilia to move and feed **(Figure 25.17)**.

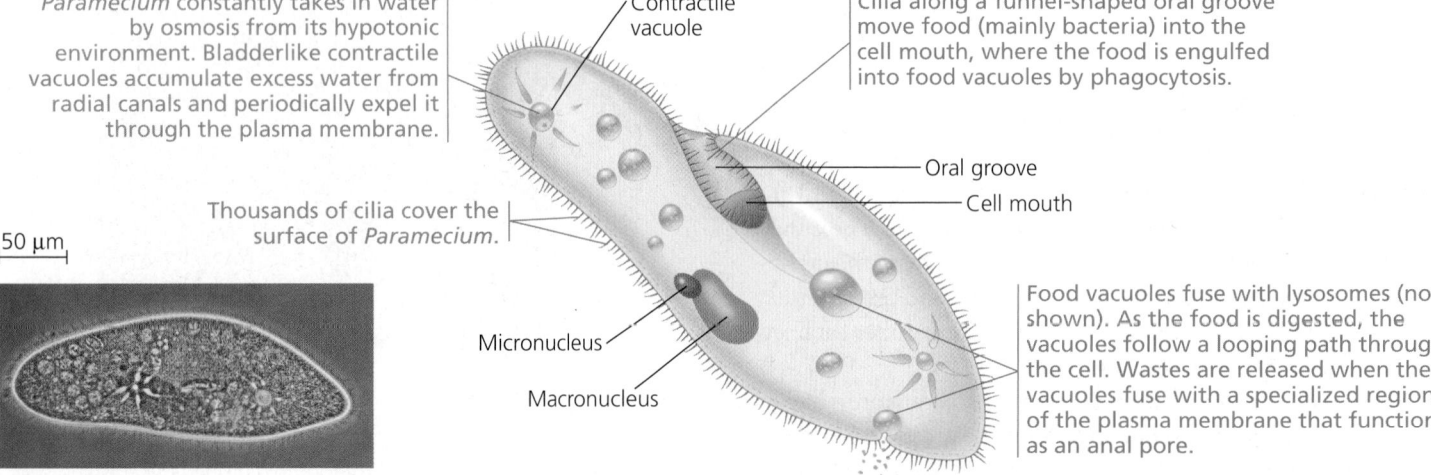

*Paramecium* constantly takes in water by osmosis from its hypotonic environment. Bladderlike contractile vacuoles accumulate excess water from radial canals and periodically expel it through the plasma membrane.

Contractile vacuole

Cilia along a funnel-shaped oral groove move food (mainly bacteria) into the cell mouth, where the food is engulfed into food vacuoles by phagocytosis.

Oral groove

Cell mouth

Thousands of cilia cover the surface of *Paramecium*.

50 μm

Micronucleus

Macronucleus

Food vacuoles fuse with lysosomes (not shown). As the food is digested, the vacuoles follow a looping path through the cell. Wastes are released when the vacuoles fuse with a specialized region of the plasma membrane that functions as an anal pore.

▲ **Figure 25.17 Structure and function in the ciliate *Paramecium caudatum*.**

Most ciliates are predators, typically of bacteria or of other protists (see Figure 25.1). Their cilia may completely cover the cell surface or may be clustered in a few rows or tufts. In certain species, rows of tightly packed cilia function collectively in locomotion. Other ciliates scurry about on leg-like structures constructed from many cilia bonded together.

### Rhizarians

Our next subgroup of SAR is the **rhizarians**. Many species in this group are **amoebas**, protists that move and feed by means of **pseudopodia**, extensions that may bulge from almost anywhere on the cell surface. An amoeba moves by extending a pseudopodium and anchoring the tip; more cytoplasm then streams into the pseudopodium. Amoebas do not constitute a monophyletic group; instead, they are dispersed across many distantly related eukaryotic taxa. Most amoebas that are rhizarians differ morphologically from other amoebas by having threadlike pseudopodia. Rhizarians also include flagellated (non-amoeboid) protists that feed using threadlike pseudopodia.

We'll examine two groups of rhizarians here: the forams and the cercozoans.

**Forams** The protists called **foraminiferans** (from the Latin *foramen*, little hole, and *ferre*, to bear), or **forams**, are named for their porous shells, called **tests** (see Figure 25.9). Foram tests consist of a single piece of organic material that typically is hardened with calcium carbonate. The pseudopodia that extend through the pores function in swimming, test formation, and feeding. Many forams also derive nourishment from the photosynthesis of symbiotic algae that live within the tests. Found in both lakes and oceans, most forams live in sand or attach themselves to rocks or algae, but some drift in currents near the water's surface. The largest forams, though single-celled, have tests measuring several centimeters in diameter.

**Cercozoans** First identified in molecular phylogenies, the **cercozoans** are a large group of amoeboid and flagellated protists that feed using threadlike pseudopodia. Common in marine, freshwater, and soil ecosystems, many cercozoans are parasites of plants, animals, or other protists; many others are predators that feed on bacteria, fungi, and other protists. One small group of cercozoans, the chlorarachniophytes (mentioned earlier in the discussion of secondary endosymbiosis), are mixotrophic: These organisms ingest smaller protists and bacteria as well as perform photosynthesis. At least one other cercozoan, *Paulinella chromatophora*, is an autotroph, deriving its energy from light and its carbon from $CO_2$. As described in **Figure 25.18**, *Paulinella* appears to represent an intriguing additional evolutionary example of a eukaryotic lineage that obtained its photosynthetic apparatus directly from a cyanobacterium.

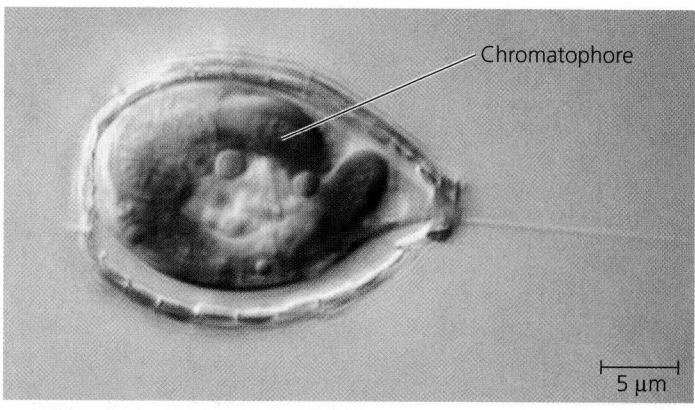

▲ **Figure 25.18 A second case of primary endosymbiosis?**
The cercozoan *Paulinella* conducts photosynthesis in a unique sausage-shaped structure called a chromatophore (LM). Chromatophores are surrounded by a membrane with a peptidoglycan layer, suggesting that they are derived from a bacterium. DNA evidence indicates that chromatophores are derived from a different cyanobacterium than that from which other plastids are derived.

## Archaeplastids

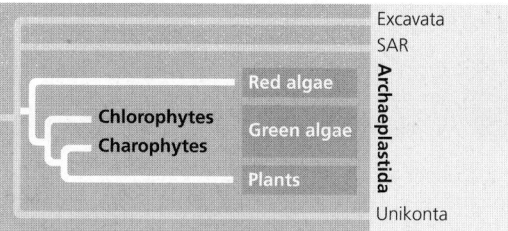

As described earlier, morphological and molecular evidence indicates that plastids arose when a heterotrophic protist acquired a cyanobacterial endosymbiont. Later, photosynthetic descendants of this ancient protist evolved into red algae and green algae (see Figure 25.4), and the lineage that produced green algae then gave rise to plants. Together, red algae, green algae, and plants make up our third eukaryotic supergroup, which is called **Archaeplastida**. We will examine plants and the colonization of land in Chapter 26; here we will look at the diversity of their closest algal relatives, red algae and green algae.

### Red Algae

Many of the 6,000 known species of **red algae** (rhodophytes, from the Greek *rhodos*, red) are reddish, owing to a photosynthetic pigment called phycoerythrin, which masks the green of chlorophyll. However, other species (those adapted to more shallow water) have less phycoerythrin. As a result, red algal species may be greenish red in very shallow water, bright red at moderate depths, and almost black in deep water. Some species lack pigmentation altogether and function heterotrophically as parasites on other red algae.

Red algae are abundant in the warm coastal waters of tropical oceans. Some of their photosynthetic pigments, including phycoerythrin, allow them to absorb blue and green

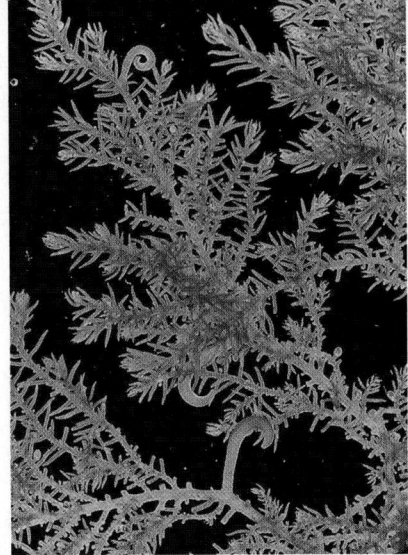

► **Bonnemaisonia hamifera.** This red alga has a filamentous form.

▼ **Nori.** The red alga *Porphyra* has a leafy form and is the source of a traditional Japanese food.

|— 8 mm —|

The seaweed is grown on nets in shallow coastal waters.

After being dried, the paper-thin, glossy sheets of nori make a mineral-rich wrap for rice, seafood, and vegetables in sushi.

▲ **Figure 25.19 Red algae.**

(a) *Ulva,* **or sea lettuce.** This edible chlorophyte has leaflike blades and a holdfast that anchors the alga.

(b) *Caulerpa,* **an intertidal chlorophyte.** The filaments lack cross-walls and thus are multinucleate. In effect, the algal body is one huge "supercell."

▲ **Figure 25.20 Multicellular chlorophytes.**

light, which penetrate relatively far into the water. A species of red alga has been discovered near the Bahamas at a depth of more than 260 m. There are also a small number of freshwater and terrestrial species. Most red algae are multicellular **(Figure 25.19)**. Although none are as big as the giant brown kelps, the largest multicellular red algae are included in the informal designation "seaweeds." You may have eaten one of these multicellular red algae, *Porphyra* (Japanese "nori"), as crispy sheets or as a wrap for sushi. Red algae reproduce sexually. However, unlike other algae, red algae do not have flagellated gametes, so they depend on water currents to bring gametes together for fertilization.

### Green Algae

The grass-green chloroplasts of **green algae** have a structure and pigment composition much like the chloroplasts of plants. Molecular systematics and cellular morphology leave little doubt that green algae and plants are closely related. In fact, some systematists now advocate including green algae in an expanded "plant" kingdom, Viridiplantae (from the Latin *viridis*, green). Phylogenetically, this change makes sense, since otherwise the green algae are a paraphyletic group.

Green algae can be divided into two main groups, the charophytes and the chlorophytes. The charophytes include the algae most closely related to plants, and we will discuss them along with plants in Chapter 26.

The second group, the chlorophytes (from the Greek *chloros*, green), includes more than 7,000 species. Most live in fresh water, but there are also many marine and some terrestrial species. Nearly all species of chlorophytes reproduce sexually by means of biflagellated gametes that have cup-shaped chloroplasts. The simplest chlorophytes are unicellular organisms such as *Chlamydomonas* (see Figure 25.6), which resemble gametes of more complex chlorophytes. Some unicellular chlorophytes live independently in aquatic habitats, while others live symbiotically within other eukaryotes, contributing part of their photosynthetic output to the food supply of their hosts. Larger size and greater complexity are found in various multicellular chlorophytes, including *Volvox* (see Figure 25.9) and *Ulva* **(Figure 25.20)**.

### Unikonts

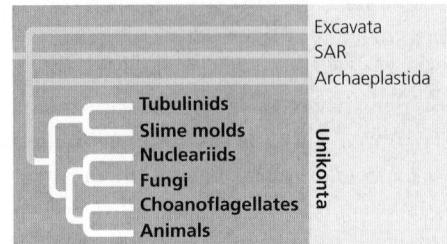

The fourth supergroup, **Unikonta**, is an extremely diverse group that includes animals, fungi, and some protists. There are two major clades of unikonts, the amoebozoans and the opisthokonts (animals, fungi, and closely related protist groups). Each of these two major clades is strongly supported by molecular systematics. The close relationship between amoebozoans and opisthokonts is more controversial. Support for this close relationship is provided by comparisons of myosin proteins and by some (but not all) studies based on multiple genes or whole genomes.

Another controversy involving the unikonts concerns the root of the eukaryotic tree. Recall that the root of a phylogenetic tree anchors the tree in time: Branch points close to the root are the oldest. At present, the root of the eukaryotic tree is uncertain; hence, we do not know which supergroup of eukaryotes was the first to diverge from all other eukaryotes. Some hypotheses, such as the amitochondriate hypothesis described earlier, have been abandoned, but researchers have yet to agree on an alternative. If the root of the eukaryotic tree were known, it would help scientists infer characteristics of the common ancestor of all eukaryotes.

In trying to determine the root of the eukaryotic tree, researchers have based their phylogenies on different sets of genes, some of which have produced conflicting results. Researchers have also tried a different approach based on tracing the occurrence of a rare evolutionary event **(Figure 25.21)**. Results from this "rare event" approach indicate that Excavata, SAR, and Archaeplastida share a more recent common ancestor than any of them does with Unikonta. This suggests that the root of the tree is located between the unikonts and all other eukaryotes, which implies that the unikonts were the first eukaryotic supergroup to diverge from all other eukaryotes. This idea remains controversial and will require more supporting evidence to be widely accepted.

### Amoebozoans

The **amoebozoan** clade includes many species of amoebas that have lobe- or tube-shaped pseudopodia, rather than the thread-like pseudopodia found in rhizarians. One major group of unicellular amoebozoans, the tubulinids, are ubiquitous in soil as well as freshwater and marine environments. Most are heterotrophs that actively seek and consume bacteria and other protists; one such tubulinid, *Amoeba proteus*, is shown in Figure 25.9. Some tubulinids also feed on detritus (nonliving organic matter).

Amoebozoans also include the multicellular slime molds. Slime molds once were thought to be fungi because, like fungi, they produce fruiting bodies that aid in spore dispersal. However, the resemblance between slime molds and fungi appears to be another case of evolutionary convergence. DNA sequence analyses indicate that slime molds descended from different unicellular ancestors than did fungi, making them another example of the independent evolution of multicellularity in eukaryotes (see Concept 25.2).

The life cycle of some slime molds can prompt us to question what it means to be an individual organism. Consider the cellular slime mold *Dictyostelium*. The feeding stage of this organism consists of solitary cells that function individually; but when food is depleted, the cells form an aggregate

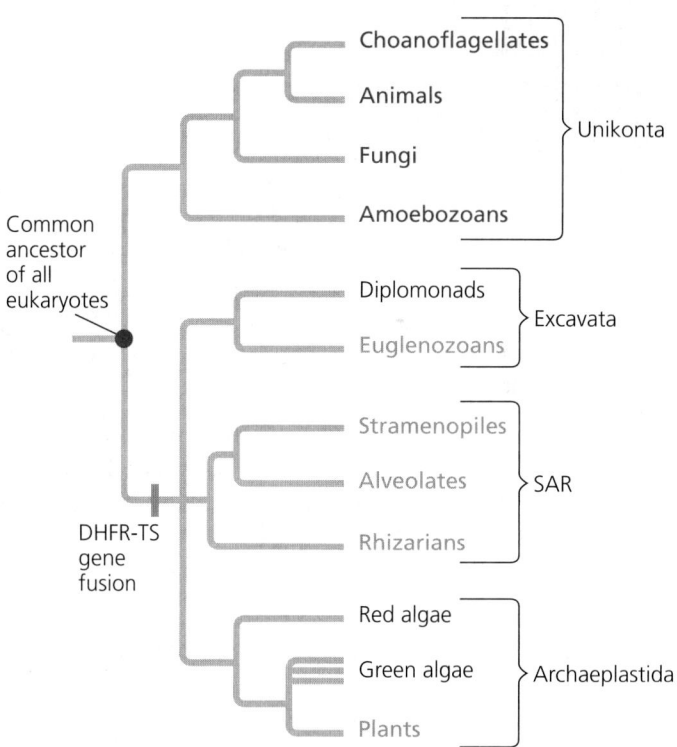

**▼ Figure 25.21  Inquiry**

### What is the root of the eukaryotic tree?

**Experiment**  Responding to the difficulty in determining the root of the eukaryotic phylogenetic tree, Alexandra Stechmann and Thomas Cavalier-Smith proposed a new approach. They studied two genes, one coding for the enzyme dihydrofolate reductase (DHFR), and the other coding for the enzyme thymidylate synthase (TS). The scientists' approach took advantage of a rare evolutionary event: In some organisms, the genes for DHFR and TS have fused, leading to the production of a single protein with both enzyme activities. Stechmann and Cavalier-Smith amplified (using PCR; see Figure 13.27) and sequenced the genes for DHFR and TS in nine species (one choanoflagellate, two amoebozoans, one euglenozoan, one stramenopile, one alveolate, and three rhizarians). They combined their data with previously published data for species of bacteria, animals, plants, and fungi.

**Results**  The bacteria studied all have separate genes coding for DHFR and TS, suggesting that this is the ancestral condition (red dot on the tree below). Other taxa with separate genes are denoted by red type. Fused genes are a derived character, found in certain members (blue type) of the supergroups Excavata, SAR, and Archaeplastida:

**Conclusion**  The results show that Excavata, SAR, and Archaeplastida form a clade, which supports the hypothesis that the root of the tree is located between the unikonts and all other eukaryotes. Because support for this hypothesis is based on only one trait—the fusion of the genes for DHFR and TS—more data are needed to evaluate its validity.

**Data from**  A. Stechmann and T. Cavalier-Smith, Rooting the eukaryote tree by using a derived gene fusion, *Science* 297:89–91 (2002).

**WHAT IF?**  Stechmann and Cavalier-Smith wrote that their conclusions are "valid only if the genes fused just once and were never secondarily split." Why is this assumption critical to their approach?

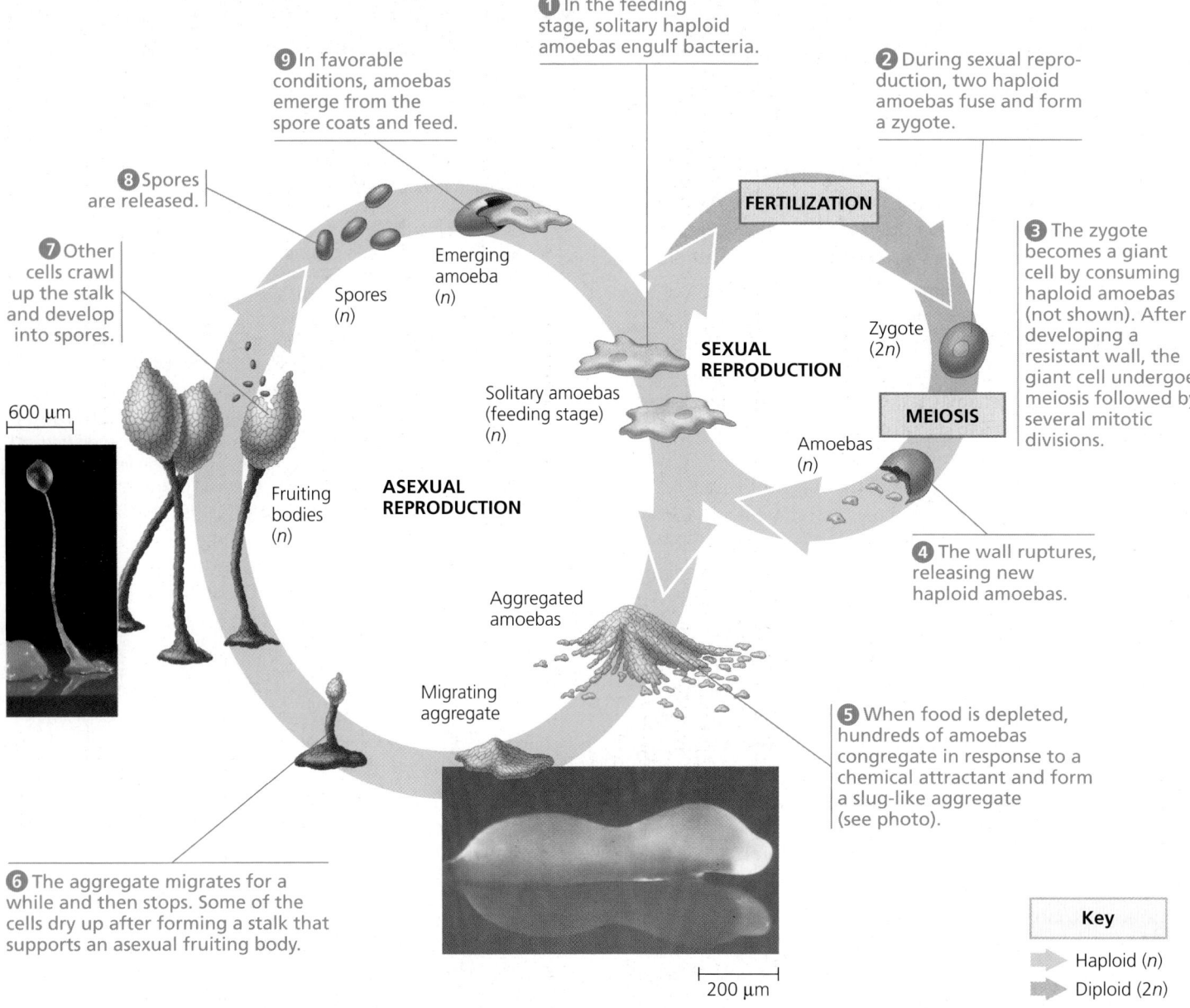

**1** In the feeding stage, solitary haploid amoebas engulf bacteria.

**2** During sexual reproduction, two haploid amoebas fuse and form a zygote.

**9** In favorable conditions, amoebas emerge from the spore coats and feed.

**8** Spores are released.

**7** Other cells crawl up the stalk and develop into spores.

**FERTILIZATION**

**3** The zygote becomes a giant cell by consuming haploid amoebas (not shown). After developing a resistant wall, the giant cell undergoes meiosis followed by several mitotic divisions.

Emerging amoeba (*n*)

Spores (*n*)

Zygote (2*n*)

**SEXUAL REPRODUCTION**

Solitary amoebas (feeding stage) (*n*)

**MEIOSIS**

Amoebas (*n*)

600 μm

Fruiting bodies (*n*)

**ASEXUAL REPRODUCTION**

**4** The wall ruptures, releasing new haploid amoebas.

Aggregated amoebas

Migrating aggregate

**5** When food is depleted, hundreds of amoebas congregate in response to a chemical attractant and form a slug-like aggregate (see photo).

**6** The aggregate migrates for a while and then stops. Some of the cells dry up after forming a stalk that supports an asexual fruiting body.

200 μm

**Key**

Haploid (*n*)
Diploid (2*n*)

▲ **Figure 25.22 The life cycle of *Dictyostelium*, a cellular slime mold.**

that functions as a unit **(Figure 25.22)**. These aggregated cells eventually form the slime mold's fruiting body stage. During this stage, the cells that form the stalk die as they dry out, while the spore cells at the top survive and have the potential to disperse and later reproduce.

### Opisthokonts

**Opisthokonts** are an extremely diverse group of eukaryotes that includes animals, fungi, and several groups of protists. We will discuss the colonization of land and the evolutionary history of fungi and animals in Chapters 26 and 27. Of the opisthokont protists, we will discuss the nucleariids in Chapter 26 because they are more closely related to fungi than they are to other protists. And as we discussed earlier in this chapter, the choanoflagellates are more closely related to animals than they are to other protists. The nucleariids and choanoflagellates illustrate why scientists have abandoned the former kingdom

Protista: A monophyletic group that includes these single-celled eukaryotes would also have to include the multicellular animals and fungi that are closely related to them.

### CONCEPT CHECK 25.3

1. Briefly describe the organisms found in each of the four eukaryotic supergroups.
2. **MAKE CONNECTIONS** Review Figures 7.2 and 8.5. Summarize how $CO_2$ and $O_2$ are both used and produced by aerobic algae.
3. **WHAT IF?** DNA sequence data for a diplomonad, a euglenozoan, a plant, and an unidentified protist suggest that the unidentified species is most closely related to the diplomonad. Further studies reveal that the unknown species has fully functional mitochondria. Based on these data, at what point on the phylogenetic tree in Figure 25.9 did the mystery protist's lineage probably diverge from other eukaryotic lineages? Explain.

For suggested answers, see Appendix A.

# Single-celled eukaryotes play key roles in ecological communities and affect human health

As our survey of the four eukaryotic supergroups suggests, the large, multicellular organisms that we know best—plants, animals, and fungi—are the tips of just a few branches on the eukaryotic tree of life. All the other branches are lineages of protists, and these protists exhibit an impressive range of structural and functional diversity, as we'll discuss. We'll then examine the effects of protists on ecological communities and human societies. (We focus on protists here, but we'll address similar topics for plants, fungi, and animals in Chapters 26 and 27.)

## Structural and Functional Diversity in Protists

Most protists are unicellular, although there are some colonial and multicellular species. Single-celled protists are justifiably considered the simplest eukaryotes, but at the cellular level, many protists are very complex—the most elaborate of all cells. In multicellular organisms, essential biological functions are carried out by organs. Unicellular protists carry out the same functions, but they do so using subcellular organelles, not multicellular organs: the nucleus, endoplasmic reticulum, Golgi apparatus, and lysosomes.

Most protists are aquatic, and they are found almost anywhere there is water, including moist terrestrial habitats such as damp soil and leaf litter. In oceans and lakes, many protists attach to the bottom or creep through the sand and silt, while others float near the water's surface. The protists living in these varied habitats also show a wide range of nutritional diversity. As we've seen, many protists are photoautotrophs and contain chloroplasts. Many others are heterotrophs, absorbing organic molecules or ingesting larger food particles; such heterotrophic protists include important mutualistic and parasitic species. Still other protists are mixotrophs that combine photosynthesis *and* heterotrophic nutrition. Photoautotrophy, heterotrophy, and mixotrophy have all arisen independently in many different protist lineages. In part as a result of this nutritional and taxonomic diversity, protist producers and symbionts are abundant in natural communities and have large ecological effects.

## Photosynthetic Protists

Many protists are important **producers**, organisms that use energy from light (or in some prokaryotes, inorganic chemicals) to convert $CO_2$ to organic compounds. Producers form the base of ecological food webs. In aquatic communities, the main producers are photosynthetic protists and prokaryotes. All other organisms in the community depend on them for food, either directly (by eating them) or indirectly (by eating an organism that ate a producer; **Figure 25.23**). Scientists estimate that roughly 30% of the world's photosynthesis is performed by diatoms, dinoflagellates, multicellular algae, and other aquatic

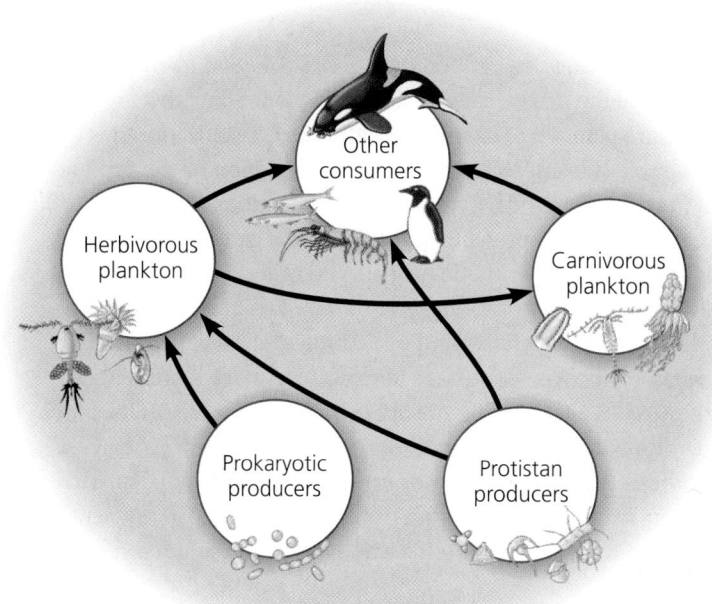

▲ **Figure 25.23 Protists: key producers in aquatic communities.** Arrows in this simplified food web lead from food sources to the organisms that eat them.

protists. Photosynthetic prokaryotes contribute another 20%, and plants are responsible for the remaining 50%.

Because producers form the foundation of food webs, factors that affect producers can affect their entire community. In aquatic environments, photosynthetic protists are often held in check by low concentrations of nitrogen, phosphorus, or iron. Various human actions can increase the concentrations of these elements in aquatic communities. For example, some of the fertilizer applied to a field may be washed by rain into a river that drains into a lake or ocean. When people add nutrients to aquatic communities in this or other ways, the abundance of photosynthetic protists can increase spectacularly.

Such increases can have major ecological consequences, including the formation of large "dead zones" in marine ecosystems (see Figure 43.21). As another example, earlier in the chapter we mentioned that diatoms can affect global $CO_2$ levels. This effect can result from a chain of events that occurs when ample nutrients produce a rapid increase (a bloom) in diatom abundance. Typically, diatoms are eaten by a variety of protists and invertebrates, but during a bloom, many escape this fate. When these uneaten diatoms die, their bodies sink to the ocean floor. It takes decades, or even centuries, for diatoms that sink to the ocean floor to be broken down by bacteria and other decomposers. As a result, the carbon in their bodies remains there long-term, rather than being released immediately as $CO_2$ as the decomposers respire. The overall effect of these events is that $CO_2$ absorbed by diatoms during photosynthesis is transported, or "pumped," to the ocean floor. With an eye toward reducing global warming by lowering atmospheric $CO_2$ levels, some scientists advocate promoting diatom blooms by fertilizing the ocean with essential nutrients such as iron. In a 2012 study, researchers found that $CO_2$ was indeed pumped to the ocean floor after iron was added to a small region of the ocean, causing a diatom bloom. Further tests are planned to examine whether iron fertilization has undesirable side effects (such as oxygen depletion or the production of nitrous oxide, a more potent greenhouse gas than $CO_2$).

A related and pressing question is how global warming will affect photosynthetic protists and other producers. Satellite data and historical observations show that the growth of photosynthetic protists and prokaryotes has declined in many ocean regions as sea surface temperatures have increased **(Figure 25.24)**. By what mechanism do rising sea surface temperatures reduce the growth of marine producers? One hypothesis relates to the rise, or upwelling, of cold, nutrient-rich waters from below. Many marine producers rely on nutrients brought to the surface in this way. However, rising sea surface temperatures can cause the formation of a layer of light, warm water that acts as a barrier to nutrient upwelling—thus reducing the growth of marine producers. If sustained, these changes would likely have far-reaching effects on marine ecosystems, fishery yields, and the global carbon cycle (see Concept 42.4).

## Symbiotic Protists

Many protists form symbiotic associations with other species. For example, photosynthetic dinoflagellates are food-providing symbiotic partners of the coral polyps that build coral reefs. Coral reefs are highly diverse ecological communities. That diversity ultimately depends on corals—and on the mutualistic protist symbionts that nourish them. Corals support reef diversity by providing food to some species and habitat to many others.

Another example is the wood-digesting protists that inhabit the gut of many termite species **(Figure 25.25)**. Unaided, termites cannot digest wood, and they rely on protistan or prokaryotic symbionts to do so. Termites cause over $3.5 billion in damage annually to wooden homes in the United States.

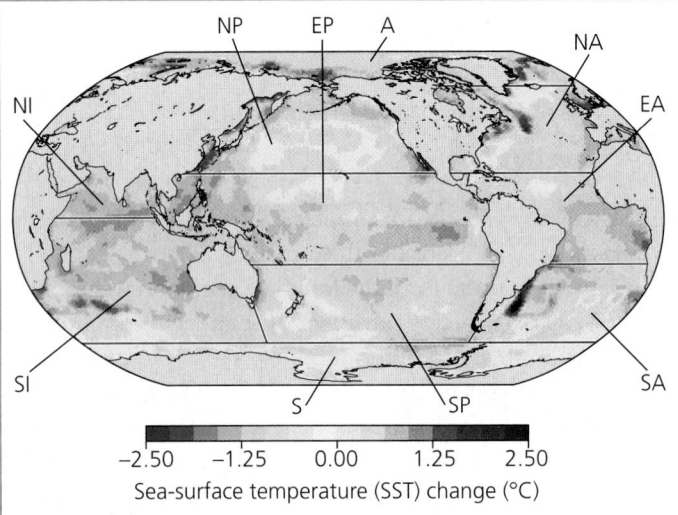

**(a)** Researchers studied 10 ocean regions, identified with letters on the map (see (b) for the corresponding names). SSTs have increased since 1950 in most areas of these regions.

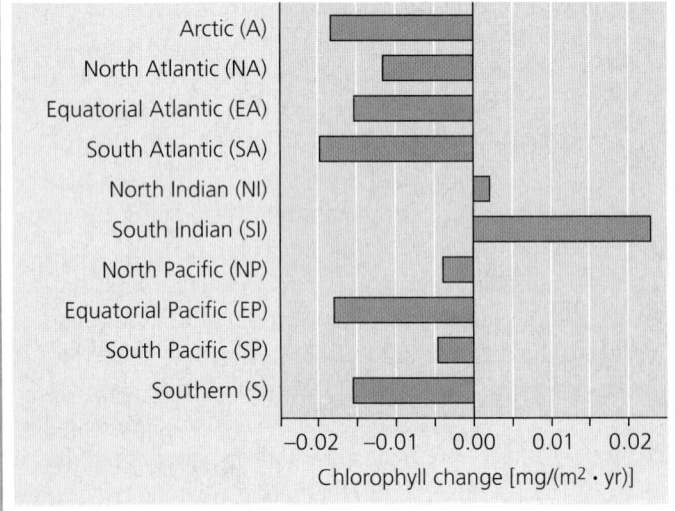

**(b)** The concentration of chlorophyll, an index for the biomass and growth of marine producers, has decreased over the same time period in most ocean regions.

▲ **Figure 25.24 Effects of climate change on marine producers.**

▶ **Figure 25.25 A symbiotic protist.** This organism is a hypermastigote, a member of a group of parabasalids that live in the gut of termites and certain cockroaches and enable the hosts to digest wood (SEM).

Symbiotic protists also include parasites that feed on the tissues of plants or animals. Among the species that parasitize plants, the stramenopile *Phytophthora ramorum* has emerged as a major new forest pathogen. This species causes sudden oak death (SOD), a disease that has killed millions of oaks and other trees in California and Oregon (see Concept 41.5). A closely related species, *P. infestans*, causes potato late blight, which turns the stalk and stem of potato plants to black slime. Late blight contributed to the devastating Irish famine of the 19th century, in which a million people died and at least that many were forced to leave Ireland. The disease remains a major problem today, destroying as much as 70% of the crop in some areas.

We'll close the chapter by taking a closer look at the parasitic protists that cause disease in humans.

## Effects on Human Health

Our bodies are home to many symbiotic species, including some protists that can cause disease. While bacteria and viruses may be the pathogens that most readily come to mind, protists that cause infectious disease can pose major challenges, both to our immune systems and to public health.

Consider *Trypanosoma*, the excavate that causes sleeping sickness (see Figure 25.12). This disease is fatal if not treated. Trypanosomes use an effective "bait-and-switch" defense that enables them to evade the immune responses of their human hosts. The surface of a trypanosome is coated with millions of copies of a single protein. However, before the host's immune system can recognize the protein and mount an attack, new generations of the parasite switch to another surface protein with a different molecular structure. Frequent changes in the surface protein prevent the host from developing immunity. About a third of *Trypanosoma*'s genome is dedicated to producing these surface proteins.

A group of alveolates, the **apicomplexans**, includes protists that cause serious human diseases such as malaria. Nearly all apicomplexans are parasites of animals—and virtually all animal species examined so far are attacked by these parasites. Although apicomplexans are not photosynthetic, they retain a modified plastid (*apicoplast*), most likely of red algal origin (see Figure 25.4). Apicomplexans typically have intricate life cycles with both sexual and asexual stages. Those life cycles often require two or more host species for completion. For example, *Plasmodium*, the parasite that causes malaria, lives in both mosquitoes and humans (**Figure 25.26**).

Historically, malaria has rivaled tuberculosis (which is caused by a bacterium) as the leading cause of human death by infectious disease. The incidence of malaria was diminished in the 1960s by insecticides that reduced carrier populations of

▶ **Figure 25.26 The two-host life cycle of *Plasmodium*, the apicomplexan that causes malaria.** The parasite enters its human host as tiny infectious cells called sporozoites.

❓ *Studies indicate that the merozoite apicoplast has only one essential function: It synthesizes a chemical the parasite requires and cannot otherwise make. Why would drugs that target the metabolic pathway by which this chemical is made probably not harm humans?*

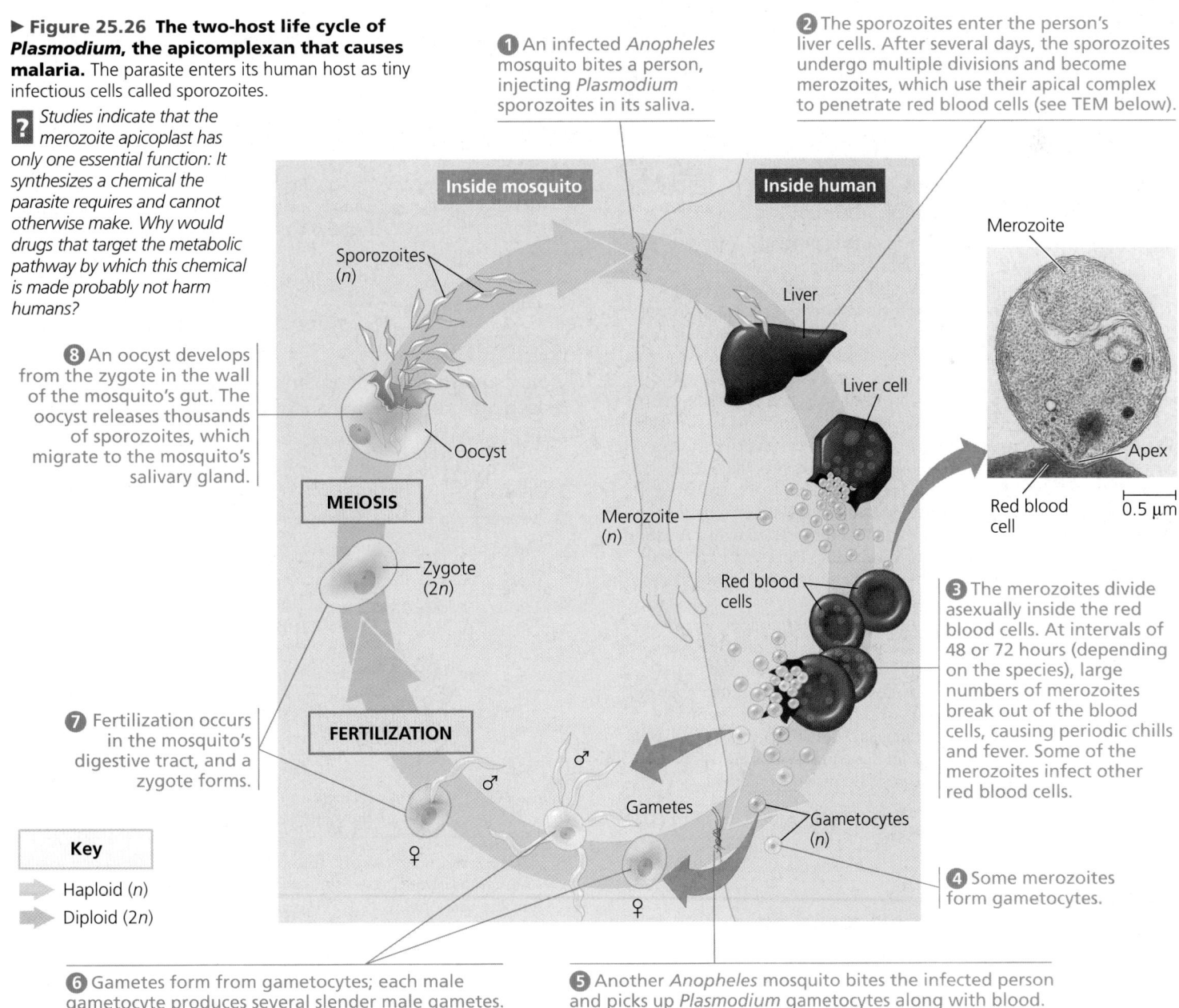

❶ An infected *Anopheles* mosquito bites a person, injecting *Plasmodium* sporozoites in its saliva.

❷ The sporozoites enter the person's liver cells. After several days, the sporozoites undergo multiple divisions and become merozoites, which use their apical complex to penetrate red blood cells (see TEM below).

Inside mosquito

Inside human

Sporozoites (*n*)

Liver

Liver cell

Merozoite

Merozoite (*n*)

Red blood cell

0.5 µm

Apex

❽ An oocyst develops from the zygote in the wall of the mosquito's gut. The oocyst releases thousands of sporozoites, which migrate to the mosquito's salivary gland.

Oocyst

**MEIOSIS**

Zygote (2*n*)

Red blood cells

❸ The merozoites divide asexually inside the red blood cells. At intervals of 48 or 72 hours (depending on the species), large numbers of merozoites break out of the blood cells, causing periodic chills and fever. Some of the merozoites infect other red blood cells.

❼ Fertilization occurs in the mosquito's digestive tract, and a zygote forms.

**FERTILIZATION**

♂

♀

Gametes

Gametocytes (*n*)

❹ Some merozoites form gametocytes.

**Key**

Haploid (*n*)

Diploid (2*n*)

♂

♀

❻ Gametes form from gametocytes; each male gametocyte produces several slender male gametes.

❺ Another *Anopheles* mosquito bites the infected person and picks up *Plasmodium* gametocytes along with blood.

*Anopheles* mosquitoes and by drugs that killed *Plasmodium* in humans. But the emergence of resistant varieties of both *Anopheles* and *Plasmodium* has led to a resurgence of malaria. About 250 million people in the tropics are currently infected, and 900,000 die each year. Efforts are under way to develop new methods of treatment, including drugs that target the apicoplast. This approach may be effective because the apicoplast is a modified plastid; as such, it descended from a cyanobacterium and hence has different metabolic pathways from those in humans.

As we've seen in this chapter, the origin of eukaryotes had an enormous impact on the history of life, leading to a great increase in the structural diversity of cells and ultimately to the rise of large, multicellular organisms. These changes set the stage for the events we'll describe in the next two chapters: the colonization of land by plants and fungi (Chapter 26) and the ecological and evolutionary effects resulting from the origin of animals (Chapter 27).

## CONCEPT CHECK 25.4

1. Justify the claim that photosynthetic protists are among the biosphere's most important organisms.
2. Describe three symbioses that include protists.
3. **WHAT IF?** High water temperatures and pollution can cause corals to expel their dinoflagellate symbionts. Predict how such "coral bleaching" would affect corals and other species in the community.

For suggested answers, see Appendix A.

# 25 Chapter Review

Go to **MasteringBiology®** for Assignments, the eText, and the Study Area with Animations, Activities, Vocab Self-Quiz, and Practice Tests.

## SUMMARY OF KEY CONCEPTS

**VOCAB SELF-QUIZ**

goo.gl/gbai8v

### CONCEPT 25.1

**Eukaryotes arose by endosymbiosis more than 1.8 billion years ago (pp. 497–503)**

- Domain Eukarya contains many groups of **protists**, along with plants, animals, and fungi. Eukaryotic cells have a nucleus and other membrane-enclosed organelles, unlike the cells of prokaryotes. These membrane-enclosed organelles make the cells of eukaryotes more complex than the cells of prokaryotes. Eukaryotic cells also have a well-developed cytoskeleton that enables them to have asymmetric forms and to change in shape as they move, feed, or grow.
- The oldest fossils of eukaryotes are of single-celled organisms that lived 1.8 billion years ago. By 1.5 billion years ago, some fossil eukaryotes had asymmetric forms, indicating a well-developed cytoskeleton. Other biological innovations, such as complex multicellularity and sexual life cycles, were in place by 1.2 billion years ago. Larger eukaryotes appeared in the fossil record about 600 million years ago.
- DNA sequence analyses indicate that eukaryotes contain a mixture of archaeal and bacterial genes and cellular characteristics. According to **endosymbiont theory**, this mixture of features likely resulted because eukaryotes originated when an archaeal host (or a host with archaeal ancestors) engulfed a bacterium that would later become an organelle found in all eukaryotes, the mitochondrion.
- In addition to mitochondria, plastids are also thought to be descendants of bacteria that were engulfed by an early eukaryote and became endosymbionts. The plastid-bearing lineage eventually evolved into red algae and green algae. Other groups of photosynthetic protists evolved from secondary endosymbiotic events in which red algae or green algae were themselves engulfed.

**?** *What evidence indicates that mitochondria arose before plastids in eukaryotic evolution?*

### CONCEPT 25.2

**Multicellularity has originated several times in eukaryotes (pp. 503–505)**

- The first multicellular eukaryotes were colonies, collections of cells that are connected to one another but show little or no cellular differentiation.

- Complex multicellular forms—those with differentiated cell types—arose independently in a variety of eukaryotic groups, including plants, fungi, animals, and several lineages of algae.
- Genomic analyses suggest that a transition from unicellularity to multicellularity does not require the origin of large numbers of novel genes; instead, such transitions can result primarily from changes in how existing genes are used.

**?** *Describe an example that illustrates the role of co-opting genes in the origin of complex multicellular eukaryotes from their unicellular ancestors.*

### CONCEPT 25.3

**Four "supergroups" of eukaryotes have been proposed based on morphological and molecular data (pp. 505–514)**

- In one hypothesis, eukaryotes are grouped into four supergroups, each a monophyletic clade. Each eukaryotic supergroup contains a great diversity of organisms, most of which are unicellular.

| Supergroup | Major Clades | Specific Example | |
|---|---|---|---|
| Excavata | Diplomonads, parabasalids, euglenozoans | *Euglena* | |
| SAR | Stramenopiles, alveolates, rhizarians | *Plasmodium* | |
| Archaeplastida | Red algae, green algae, plants | *Chlamydomonas* | |
| Unikonta | Amoebozoans, opisthokonts | *Amoeba* | |

- The root of the eukaryotic tree is not known. An approach based on tracing the occurrence of a rare evolutionary event suggests that the unikonts were the first eukaryotic supergroup to diverge from other eukaryotes. This hypothesis will require more supporting evidence before it is widely accepted.

**?** *Summarize recent changes in our understanding of the evolutionary history of eukaryotes, beginning with an explanation for why kingdom Protista has been abandoned.*

## Single-celled eukaryotes play key roles in ecological communities and affect human health (pp. 515–518)

- Unicellular protists use subcellular organelles to accomplish the essential biological functions that multicellular organisms perform with organs. Protists include many lineages of photoautotrophic, heterotrophic, and mixotrophic species.
- Photosynthetic protists are among the most important **producers** in aquatic communities.
- Protists form a wide range of mutualistic and parasitic relationships that affect their symbiotic partners and many other members of the community.

**?** *Describe several protists that are ecologically important.*

# TEST YOUR UNDERSTANDING

**PRACTICE TEST**

goo.gl/CRZjvS

## Level 1: Knowledge/Comprehension

1. The oldest fossil eukaryote that can be resolved taxonomically is of
   (A) a red alga that lived 1.2 billion years ago.
   (B) a green alga that lived 635 million years ago.
   (C) a fungus that lived 2 billion years ago.
   (D) an Ediacaran that lived 550 million years ago.

2. The evolution of complex multicellularity in eukaryotes
   (A) occurred only once, in the common ancestor of all eukaryotes.
   (B) occurred only once, in the common ancestor of all multicellular eukaryotes.
   (C) occurred only once, in the animal lineage.
   (D) occurred independently in several eukaryotic lineages.

3. Plastids that are surrounded by more than two membranes are evidence of
   (A) evolution from mitochondria.
   (B) fusion of plastids.
   (C) origin of the plastids from archaea.
   (D) secondary endosymbiosis.

4. Biologists think that endosymbiosis gave rise to mitochondria before plastids partly because
   (A) the products of photosynthesis could not be metabolized without mitochondrial enzymes.
   (B) all eukaryotes have mitochondria (or their remnants), whereas many eukaryotes do not have plastids.
   (C) mitochondrial DNA is less similar to prokaryotic DNA than is plastid DNA.
   (D) without mitochondrial $CO_2$ production, photosynthesis could not occur.

5. Which group is *incorrectly* paired with its description?
   (A) diatoms—important producers in aquatic communities
   (B) red algae—acquired plastids by secondary endosymbiosis
   (C) apicomplexans—parasites with intricate life cycles
   (D) diplomonads—protists with modified mitochondria

## Level 2: Application/Analysis

6. Based on the phylogenetic tree in Figure 25.9, which of the following statements is correct?
   (A) The most recent common ancestor of Excavata is older than that of SAR.
   (B) The most recent common ancestor of red algae and plants is older than that of nucleariids and fungi.
   (C) The most basal (first to diverge) eukaryotic supergroup cannot be determined.
   (D) Excavata is the most basal eukaryotic supergroup.

## Level 3: Synthesis/Evaluation

7. **MAKE CONNECTIONS** The bacterium *Wolbachia* is a symbiont that lives in mosquito cells and spreads rapidly through mosquito populations. *Wolbachia* can make mosquitoes resistant to infection by *Plasmodium*; researchers are seeking a strain that confers resistance and does not harm mosquitoes. Compare evolutionary changes that could occur if malaria control is attempted using such a *Wolbachia* strain versus using insecticides to kill mosquitoes. (Review Figure 25.26 and Concept 21.3.)

8. **FOCUS ON EVOLUTION**
   **DRAW IT** Medical researchers seek to develop drugs that can kill or restrict the growth of human pathogens yet have few harmful effects on patients. These drugs often work by disrupting the metabolism of the pathogen or by targeting its structural features.

   Draw and label a phylogenetic tree that includes an ancestral prokaryote and the following groups of organisms: Excavata, SAR, Archaeplastida, Unikonta, and, within Unikonta, amoebozoans, animals, choanoflagellates, fungi, and nucleariids. Based on this tree, hypothesize whether it would be most difficult to develop drugs to combat human pathogens that are prokaryotes, protists, animals, or fungi. (You do not need to consider the evolution of drug resistance by the pathogen.)

9. **FOCUS ON INTERACTIONS**
   Organisms interact with each other and the physical environment. In a short essay (100–150 words), explain how the response of diatom populations to a drop in nutrient availability can affect both other organisms and aspects of the physical environment (such as carbon dioxide concentrations).

10. **SYNTHESIZE YOUR KNOWLEDGE**

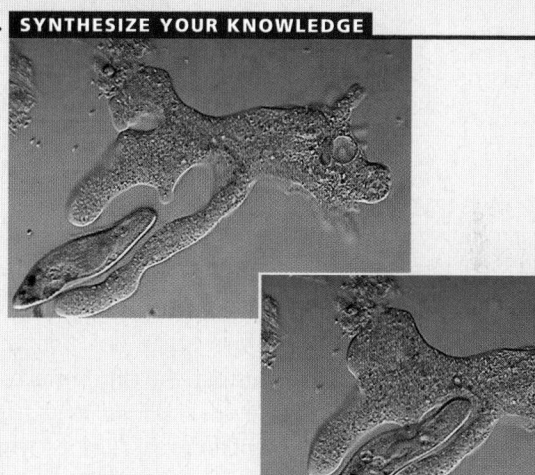

These micrographs show one single-celled eukaryote, a tubulinid amoeba, engulfing another, a ciliate. Describe a key feature of eukaryotes suggested by these images, and summarize the role of endosymbiosis in the evolutionary history of the eukaryotes. Are tubulinid amoebas more closely related to all other protists than they are to plants, fungi, or animals? Explain.

*For selected answers, see Appendix A.*

# 26 The Colonization of Land

## KEY CONCEPTS

**26.1** Fossils show that plants colonized land more than 470 million years ago

**26.2** Though not closely related to plants, fungi played a key role in the colonization of land

**26.3** Early plants radiated into a diverse set of lineages

**26.4** Seeds and pollen grains are key adaptations for life on land

**26.5** Plants and fungi fundamentally changed chemical cycling and biotic interactions

▲ **Figure 26.1** How did the colonization of land change the world?

## The Greening of Earth

Looking at a lush landscape, such as the forest scene in **Figure 26.1**, it is difficult to imagine the terrestrial environment without plants or other organisms. Yet for more than 2 billion years of Earth's history, the land surface was largely lifeless. Geochemical analyses and fossil evidence suggest that this had changed by 1.2 billion years ago, by which time thin coatings of cyanobacteria and protists existed on land. But it was only within the last 500 million years that fungi as well as small plants and animals joined them ashore. Finally, by about 385 million years ago, tall plants appeared, leading to the formation of the first forests (which consisted of very different species than those in Figure 26.1).

Although a few plant species returned to aquatic habitats during their evolution, most present-day plants live on land. In this text, we distinguish plants from algae, which are photosynthetic protists.

In this chapter, we'll examine the colonization of land by plants and fungi; we'll turn to animals in Chapter 27. Although plants and fungi are not closely related **(Figure 26.2)**, we discuss them together in this chapter in part because fossil evidence suggests that they both arrived on land before animals, which depend on them to survive. For example, plants supply oxygen and are a key source of food for terrestrial animals, while fungi break down organic material and recycle nutrients, allowing animals and other organisms to assimilate essential chemical elements.

Another reason for discussing plants and fungi together is that fossil evidence suggests that plants colonized land in partnership with fungi. As we'll see, this partnership and the diversification of plants and fungi that occurred in terrestrial environments fundamentally changed biotic interactions and chemical cycling. We'll begin this story with the origin of plants, an event that occurred over millions of years as the algal ancestors of early plants adapted to life in a new environment—land.

▶ **Figure 26.2 Fungi and plants are not closely related.** The lineage leading to fungi diverged from that leading to plants more than 1.2 billion years ago. In contrast, the fungal and animal lineages diverged more recently.

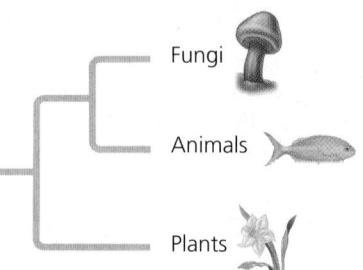

Fungi

Animals

Plants

## CONCEPT 26.1

# Fossils show that plants colonized land more than 470 million years ago

Evidence in the form of fossils documents key steps in the origin of plants from their algal ancestors. As you read in Concept 25.3, researchers have identified green algae called charophytes as the closest living relatives of plants. After discussing evidence for this relationship, we'll describe the terrestrial adaptations and fossil record of early plants.

## Evidence of Algal Ancestry

Many key traits of plants also appear in some algae. For example, plants are multicellular, eukaryotic, photosynthetic autotrophs, as are brown, red, and certain green algae. Plants have cell walls made of cellulose, and so do green algae, dinoflagellates, and brown algae. And chloroplasts with chlorophylls *a* and *b* are present in green algae, euglenids, and a few dinoflagellates, as well as in plants.

However, the charophytes are the only present-day algae that share certain distinctive traits with plants, a fact that suggests that they are the closest living relatives of plants. For example, the cells of both plants and charophytes have distinctive circular rings of proteins embedded in the plasma membrane

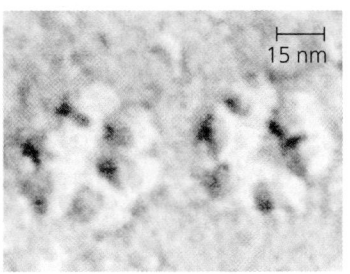

▲ **Figure 26.3 Rings of cellulose-synthesizing proteins** (SEM).

**(Figure 26.3)**; these protein rings synthesize the cellulose found in the cell wall. In contrast, noncharophyte algae have linear sets of proteins that synthesize cellulose. Likewise, in species of plants that have flagellated sperm, the structure of the sperm closely resembles that of charophyte sperm.

Biochemical studies and analyses of nuclear, chloroplast, and mitochondrial DNA from a wide range of plant and algal species suggest more specifically that certain groups of charophytes—such as *Zygnema* and *Coleochaete*—are the closest living relatives of plants **(Figure 26.4)**. While this finding indicates that plants arose from within a group of charophyte algae, it does not mean that plants are descended from these living algae; even so, present-day charophytes may tell us something about what the algal ancestors of plants were like.

## Adaptations Enabling the Move to Land

Many species of charophyte algae inhabit shallow waters around the edges of ponds and lakes, where they are subject to occasional drying. In such environments, natural selection favors individual algae that can survive periods when they are not submerged in water. In charophytes, a layer of a durable

◄ *Zygnema* sp., a common filamentous pond alga featuring two star-shaped chloroplasts in each cell (LM)

▼ *Coleochaete orbicularis,* a disk-shaped charophyte that also lives in ponds (LM)

▲ **Figure 26.4 Examples of charophytes, the closest algal relatives of plants.**

polymer called **sporopollenin** prevents exposed zygotes from drying out. A similar chemical adaptation is found in the tough sporopollenin walls that encase the spores of plants.

The accumulation of such traits by at least one population of (now extinct) charophyte algae probably enabled their descendants—the first plants—to live permanently above the waterline. This ability opened a new frontier: a terrestrial habitat that offered enormous benefits. The bright sunlight was unfiltered by water and plankton; the atmosphere offered more plentiful carbon dioxide ($CO_2$) than did water; and the soil by the water's edge was rich in some mineral nutrients. But these benefits were accompanied by challenges: a relative scarcity of water and a lack of structural support against gravity. (To appreciate why such support is important, picture how the soft body of a jellyfish sags when taken out of water.) Plants diversified as new adaptations arose that enabled them to thrive despite these challenges.

Today, what adaptations are unique to plants? The answer depends on where you draw the boundary dividing plants from algae **(Figure 26.5)**. Since the placement of this boundary is

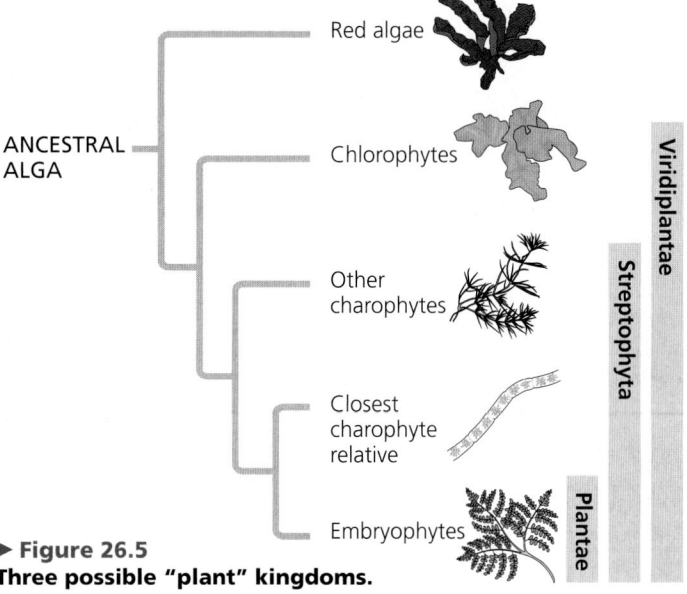

Red algae

ANCESTRAL ALGA

Chlorophytes

Other charophytes

Closest charophyte relative

Embryophytes

Viridiplantae

Streptophyta

Plantae

► **Figure 26.5 Three possible "plant" kingdoms.**

Charophyte algae lack the key traits of plants described in this figure: alternation of generations and the associated trait of multicellular, dependent embryos. As described in the main text, charophyte algae also lack walled spores produced in sporangia and apical meristems. This suggests that these four traits were absent in the ancestor common to plants and charophytes but instead evolved as derived traits of plants.

## Alternation of Generations

The life cycles of all plants alternate between two generations of distinct multicellular organisms: gametophytes and sporophytes. As shown in the diagram below (using a fern as an example), each generation gives rise to the other, a process that is called **alternation of generations**. This type of reproductive cycle evolved in various groups of algae but does not occur in the charophytes, the algae most closely related to plants. Take care not to confuse the alternation of generations in plants with the haploid and diploid stages in the life cycles of other sexually reproducing organisms. Alternation of generations is distinguished by the fact that the life cycle includes both multicellular haploid organisms and multicellular diploid organisms. The multicellular haploid **gametophyte** ("gamete-producing plant") is named for its production by mitosis of haploid gametes—eggs and sperm—that fuse during fertilization, forming diploid zygotes. Mitotic division of the zygote produces a multicellular diploid **sporophyte** ("spore-producing plant"). Meiosis in a mature sporophyte produces haploid **spores**, reproductive cells that can develop into a new haploid organism without fusing with another cell. Mitotic division of the spore cell produces a new multicellular gametophyte, and the cycle begins again.

**Alternation of generations: five generalized steps**

① The gametophyte produces haploid gametes by mitosis.

② Two gametes unite (fertilization) and form a diploid zygote.

③ The zygote develops into a multicellular diploid sporophyte.

④ The sporophyte produces unicellular haploid spores by meiosis.

⑤ The spores develop into multicellular haploid gametophytes.

Gametophyte (n)

Gamete from another plant

Gamete

Mitosis

Mitosis

MEIOSIS

FERTILIZATION

Spore

Zygote

Mitosis

Sporophyte (2n)

**Key**

→ Haploid (n)

→ Diploid (2n)

## Multicellular, Dependent Embryos

As part of a life cycle with alternation of generations, multicellular plant embryos develop from zygotes that are retained within the tissues of the female parent (a gametophyte). The parental tissues protect the developing embryo from harsh environmental conditions and provide nutrients such as sugars and amino acids. The embryo has specialized *placental transfer* cells that enhance the transfer of nutrients to the embryo through elaborate ingrowths of the wall surface (plasma membrane and cell wall). The multicellular, dependent embryo of plants is such a significant derived trait that plants are also known as **embryophytes**.

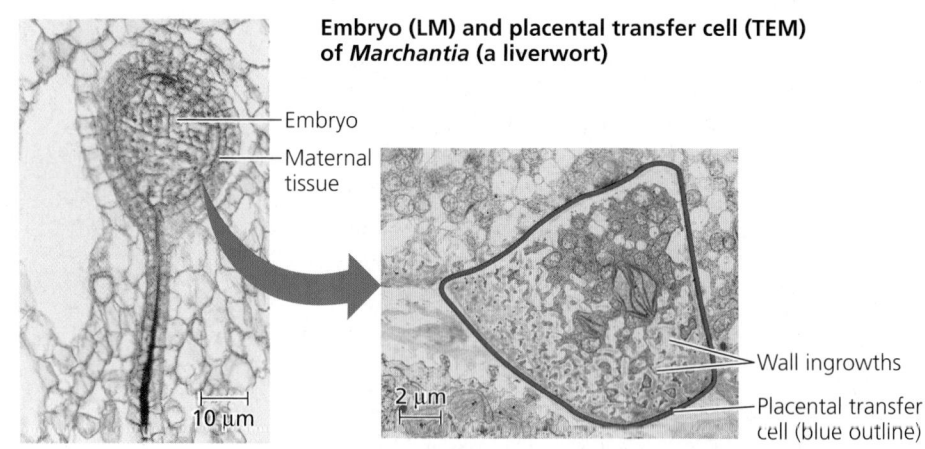

**Embryo (LM) and placental transfer cell (TEM) of *Marchantia* (a liverwort)**

Embryo

Maternal tissue

10 µm

2 µm

Wall ingrowths

Placental transfer cell (blue outline)

**MAKE CONNECTIONS** *Review sexual life cycles in Figure 10.6. Identify which type of sexual life cycle has alternation of generations, and summarize how it differs from other life cycles.*

the subject of ongoing debate, this text uses a traditional definition that equates the kingdom Plantae with embryophytes (plants with embryos). In this context, let's now examine the derived traits that separate plants from their closest algal relatives.

## Derived Traits of Plants

Several adaptations that facilitate survival and reproduction on dry land emerged after plants diverged from their algal relatives. Examples of such traits that are found in plants but not in the charophyte algae include the following:

- **Alternation of generations.** This type of life cycle, consisting of multicellular forms that give rise to each other in turn, is described in **Figure 26.6**.
- **Walled spores produced in sporangia.** The sporophyte stage of the plant life cycle has multicellular organs called **sporangia** (singular, *sporangium*) that produce spores **(Figure 26.7)**. The polymer sporopollenin makes the walls of these spores resistant to harsh environments, enabling plant spores to be dispersed through dry air without harm.
- **Apical meristems.** Plants also differ from their algal relatives in having **apical meristems**, localized regions of cell division at the tips of roots and shoots (see Figure 28.16). Apical meristem cells can divide throughout the plant's life, enabling its roots and shoots to elongate, thus increasing the plant's exposure to environmental resources.

Additional derived traits that relate to terrestrial life have evolved in many plant species. For example, the epidermis in many species has a covering, the **cuticle**, that consists of wax and other polymers. Permanently exposed to the air, plants run a far greater risk of desiccation (drying out) than do their algal relatives. The cuticle acts as waterproofing, helping prevent excessive water loss from the aboveground plant organs, while also providing some protection from microbial attack. Most plants also have specialized pores called **stomata** (singular,

*stoma*), which support photosynthesis by allowing the exchange of $CO_2$ and $O_2$ between the outside air and the plant (see Figure 8.3). Stomata are also the main avenues by which water evaporates from the plant; in hot, dry conditions, the stomata close, minimizing water loss. As we describe in the next section, fossil evidence documents the appearance of stomata and other novel traits in early plants.

## Early Plants

The algae most closely related to plants include many unicellular and small, colonial species. Since it is likely that the first plants were similarly small, the search for the earliest fossils of plants has focused on the microscopic world. As mentioned earlier, microorganisms colonized land as early as 1.2 billion years ago. But the microscopic fossils that document life on land changed dramatically 470 million years ago with the appearance of spores from early plants.

What distinguishes these spores from those of algae or fungi? One clue comes from their chemical composition, which matches that found in plant spores but differs from that in the spores of other organisms. In addition, the walls of these ancient spores have structural features that today are found only in the spores of certain plants (liverworts). And in rocks dating to 450 million years ago, researchers have discovered similar spores embedded in plant cuticle material that resembles spore-bearing tissue in living plants **(Figure 26.8)**.

It is not surprising that spores provide the earliest fossil evidence of plants. For one thing, plants produce large numbers of widely dispersed spores. In addition, recall that plant spores contain sporopollenin, a compound that makes the spores extremely durable and thus likely to be well represented in the fossil record.

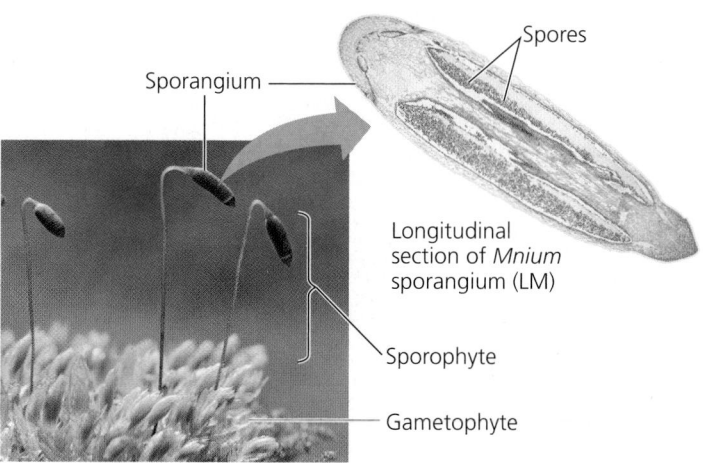

Sporangium

Spores

Longitudinal section of *Mnium* sporangium (LM)

Sporophyte

Gametophyte

▲ **Figure 26.7 Sporophytes and sporangia of a moss in the genus *Mnium*.** Each of the many spores produced by a sporangium is encased by a durable, sporopollenin-enriched wall.

**(a) Fossilized spores.** The chemical composition and the physical structure of the walls of these 450 million-year-old spores match those found in certain plants today.

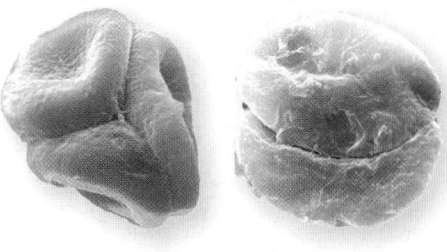

**(b) Fossilized sporophyte tissue.** The spores in (a) were embedded in tissue that appears to be from plants.

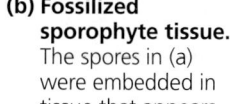

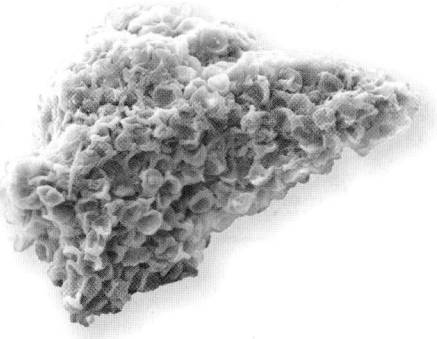

▲ **Figure 26.8 Ancient plant spores and tissue** (colorized SEMs).

► **Figure 26.9 Fossil of a *Cooksonia* sporangium (425 million years old).**

0.3 mm

Larger plant structures, such as the spore-producing structure (sporangium) from *Cooksonia* (**Figure 26.9**), first appear in the fossil record dating to 425 million years ago. By 400 million years ago, a diverse assemblage of plants lived on land. Fossil evidence shows that some of these early plants had key traits not found in their algal ancestors, including specialized tissues for water transport, cuticles, stomata, and branched sporophytes (**Figure 26.10**). Although these early plants were less than 20 cm tall, their branching enabled their bodies to become more complex. Over time, as plant bodies increased in complexity, competition for space and sunlight probably also increased. That competition may have stimulated still more evolution in later plant lineages—eventually leading to the formation of the first forests.

Overall, the fossil record shows that by 400 million years ago, early plants had a variety of novel features that facilitated life on land, significantly distinguishing them from their algal ancestors. These early plants also formed a key symbiotic association with the group we turn to next, the fungi.

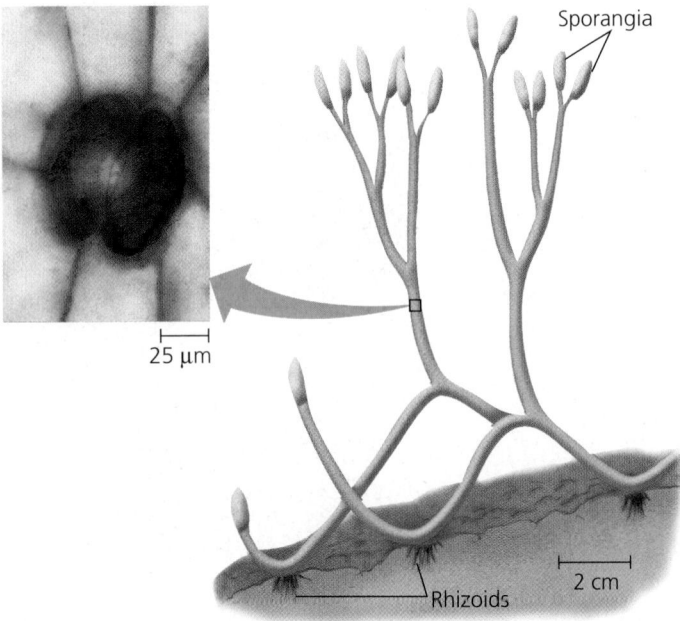

Sporangia

25 μm

Rhizoids

2 cm

▲ **Figure 26.10 *Aglaophyton major,* an early plant.** This reconstruction from 405-million-year-old fossils exhibits dichotomous (Y-shaped) branching with sporangia at the ends of the branches. *Aglaophyton* had structures called rhizoids that anchored it to the ground. The inset shows a fossilized stoma of *A. major* (colorized LM).

**CONCEPT 26.2**

# Though not closely related to plants, fungi played a key role in the colonization of land

Although the earliest plants had some adaptations for life on land, they lacked true roots and leaves. Without roots, how did these plants absorb nutrients from the soil? Fossils reveal evidence of an adaptation that may have facilitated nutrient uptake by these early plants: They formed symbiotic associations with fungi (as do most plant species living today).

How do close associations with fungi help plants take up nutrients? The answer relates to the nutritional mode and morphology of fungi. Like animals, fungi are heterotrophs: They cannot make their own food as plants and algae can. But unlike animals, fungi do not ingest (eat) their food. Instead, a fungus absorbs nutrients from the environment outside of its body. Furthermore, the morphology of many fungi enhances their capacity for absorption: Their bodies consist of a network of filaments called **hyphae** (singular, *hypha*) that provide a large surface area across which absorption can occur. In the plant-fungal symbiotic associations called *mycorrhizae*, fungal hyphae transfer nutrients absorbed from soil to their plant partner (**Figure 26.11**). This benefit may have helped plants without roots to colonize land.

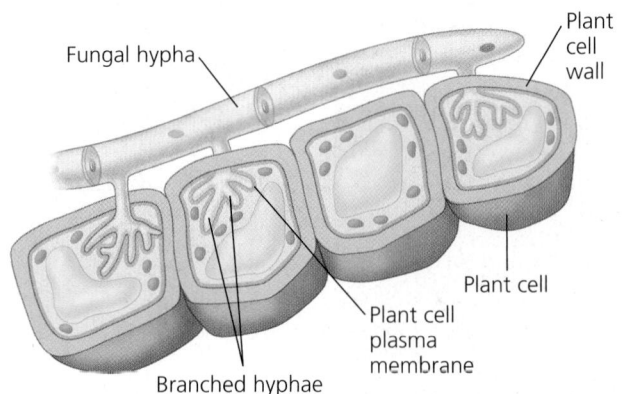

Fungal hypha

Plant cell wall

Plant cell

Plant cell plasma membrane

Branched hyphae

▲ **Figure 26.11 Mycorrhizae: Plant-fungal symbioses.** Some mycorrhizal fungi grow branched extensions of their hyphae that exchange nutrients with living plant cells. The branched hyphae remain separated from a plant cell's cytoplasm by the plasma membrane of the plant cell (orange).

We'll discuss mycorrhizae in more detail later in this section, when we examine fungal adaptations for life on land. First, we'll take a step back to consider the origin of fungi and how they are related to other groups of eukaryotes.

## The Origin of Fungi

Molecular data show that fungi and animals are more closely related to each other than either group is to plants or to most other eukaryotes (see Figures 26.2 and 25.9). DNA sequence data also indicate that fungi are more closely related to several groups of single-celled protists than they are to animals, suggesting that the ancestor of fungi was unicellular. One such group of unicellular protists, the **nucleariids**, consists of amoebas that feed on algae and bacteria. As we discussed in Concept 25.2, animals are more closely related to a *different* group of protists (the choanoflagellates) than they are to either fungi or nucleariids. Together, these results suggest that multicellularity evolved in animals and fungi independently, from different single-celled ancestors.

Based on molecular clock analyses, scientists have estimated that the ancestors of animals and fungi diverged into separate lineages more than a billion years ago. Fossils of certain unicellular, marine eukaryotes that lived as early as 1.5 billion years ago have been interpreted as fungi, but those claims remain controversial. Furthermore, although most biologists think that fungi originated in aquatic environments, the oldest fossils that are widely accepted as fungi are of terrestrial species that lived about 460 million years ago **(Figure 26.12)**. Overall, additional fossil discoveries are needed to clarify when fungi originated and what features were present in their earliest lineages.

## Fungal Adaptations for Life on Land

Some researchers have hypothesized that life on land before the arrival of plants was a "green slime" that consisted of cyanobacteria, algae, and a variety of small, heterotrophic species, including fungi. As we'll explore in this section, fungi have adaptations that would have made them well suited for feeding on other early terrestrial organisms (or their remains).

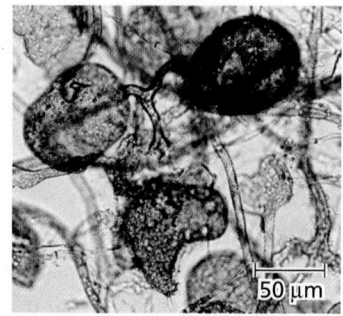

▲ **Figure 26.12 Fossil fungi.** These fossils of the filaments and spores of terrestrial fungi are 460 million years old (LM).

### Morphological Adaptations

As we mentioned earlier, fungi are heterotrophs that feed by absorption. Many fungi accomplish this task by secreting hydrolytic enzymes into their surroundings. These enzymes break down complex molecules to smaller organic compounds that the fungi can absorb into their bodies and use. Collectively, fungi can digest compounds from a wide range of sources, living or dead.

What fungal traits facilitate feeding by absorption? One such trait is a cell wall strengthened by **chitin**, a strong but flexible polysaccharide. As fungi absorb nutrients from their environment, the concentration of those nutrients in their cells increases; that, in turn, causes water to move into fungal cells by osmosis. The movement of water into fungal cells creates pressure that would cause their cells to burst if they were not surrounded by a rigid cell wall.

In multicellular fungi, another trait that supports efficient absorption is their morphology. As it grows, a multicellular fungus forms a network of hyphae consisting of tubular cell walls that surround the plasma membrane and cytoplasm of the cells **(Figure 26.13)**. The hyphae form an interwoven mass called a **mycelium** (plural, *mycelia*) that grows into and absorbs nutrients from the material on which the fungus feeds. (Note that some fungi can grow as single cells—**yeasts**—either some or all of the time. Yeasts often inhabit moist environments, including plant sap and animal tissues, where there is a ready supply of soluble nutrients, such as sugars and amino acids.)

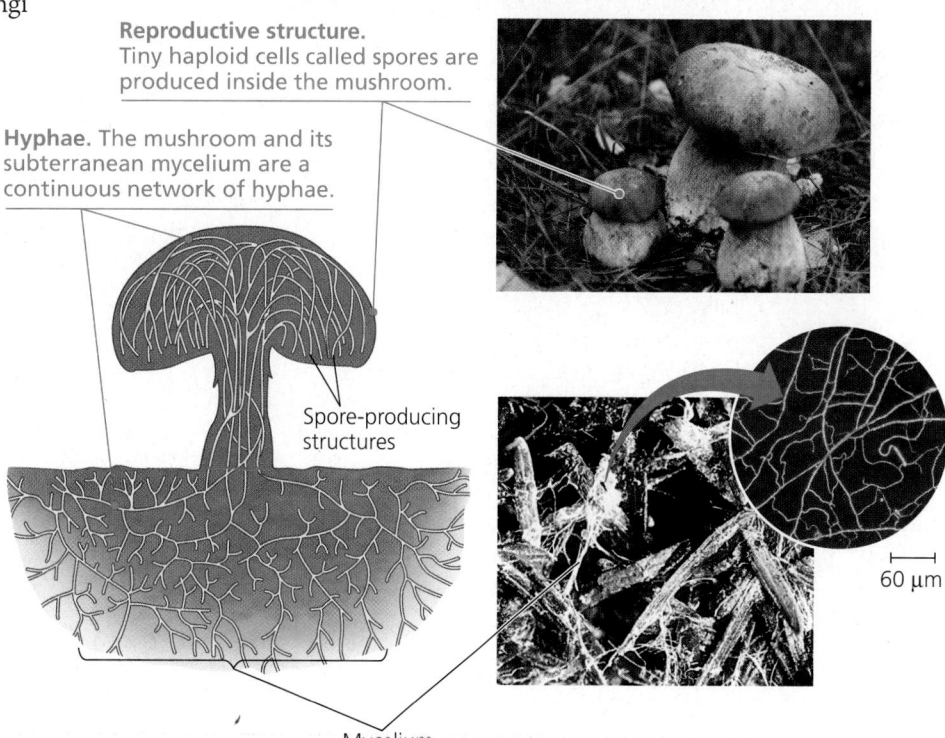

**Reproductive structure.** Tiny haploid cells called spores are produced inside the mushroom.

**Hyphae.** The mushroom and its subterranean mycelium are a continuous network of hyphae.

Spore-producing structures

Mycelium

▲ **Figure 26.13 Structure of a multicellular fungus.** The top photograph shows the sexual structures, in this case called mushrooms, of the penny bun fungus (*Boletus edulis*). The bottom photograph shows a mycelium growing on fallen conifer needles. The inset SEM shows hyphae.

**?** *Although the mushrooms in the top photograph appear to be different individuals, could their DNA be identical? Explain.*

## MAKE CONNECTIONS

# Maximizing Surface Area

In general, the amount of metabolic or chemical activity a cell or organism can carry out is proportional to its mass or volume. Maximizing metabolic rate, however, requires the efficient uptake of energy and raw materials, such as nutrients and oxygen, as well as the effective disposal of waste products. For large cells, plants, and animals, these exchange processes have the potential to be limiting due to simple geometry. When a cell or organism grows without changing shape, its volume increases more rapidly than its surface area (see Figure 4.6). As a result, there is proportionately less surface area available to support chemical activity. The challenge posed by the relationship of surface area and volume occurs in diverse contexts and organisms, but the evolutionary adaptations that meet this challenge are similar. Structures that maximize surface area through branching, flattening, folding, and projections have an essential role in biological systems.

These diagrams compare surface area (SA) for two different shapes with the same volume (V). Note which shape has the greater surface area.

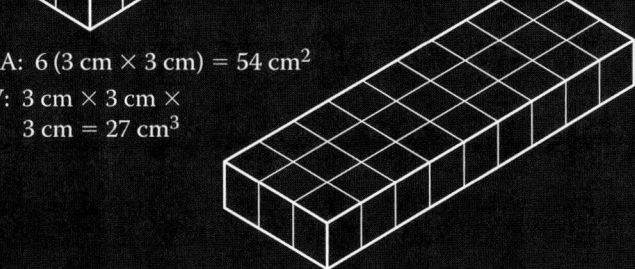

SA: $6 (3 \text{ cm} \times 3 \text{ cm}) = 54 \text{ cm}^2$
V: $3 \text{ cm} \times 3 \text{ cm} \times 3 \text{ cm} = 27 \text{ cm}^3$

SA: $2 (3 \text{ cm} \times 1 \text{ cm}) + 2 (9 \text{ cm} \times 1 \text{ cm}) + 2 (3 \text{ cm} \times 9 \text{ cm}) = 78 \text{ cm}^2$
V: $1 \text{ cm} \times 3 \text{ cm} \times 9 \text{ cm} = 27 \text{ cm}^3$

## Branching

Water uptake relies on passive diffusion. The highly branched filaments of a fungal mycelium increase the surface area across which water and minerals can be absorbed from the environment. (See Figure 26.13.)

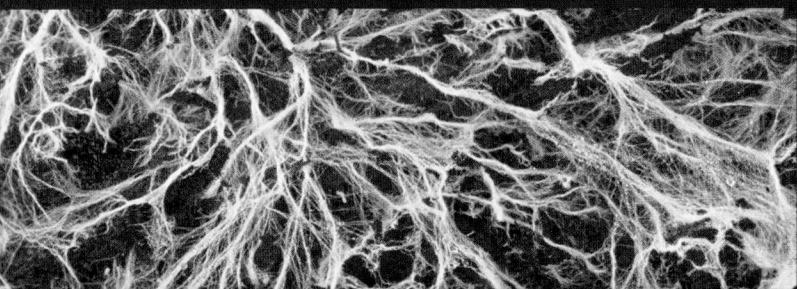

## Folding

This TEM shows portions of two chloroplasts in a plant leaf. Photosynthesis occurs in chloroplasts, which have a flattened and interconnected set of internal membranes called thylakoid membranes. The foldings of the thylakoid membranes increase their surface area, enhancing the exposure to light and thus increasing the rate of photosynthesis. (See Figure 8.3.)

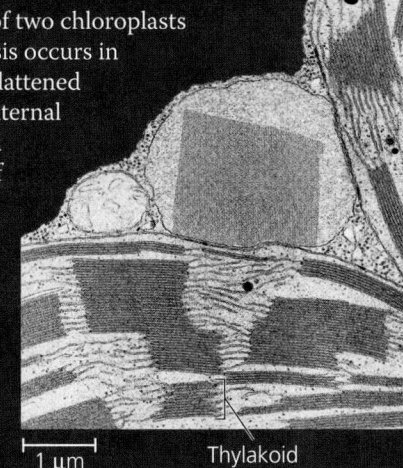

1 μm          Thylakoid

## Flattening

By having a body that is only a few cells thick, an organism such as this flatworm can use its entire body surface for exchange. (See Figure 34.2.)

## Projections

In vertebrates, the small intestine is lined with finger-like projections called villi that absorb nutrients released by the digestion of food. Each of the villi shown here is covered with large numbers of microscopic projections called microvilli, resulting in a total surface area of about 300 m² in humans, as large as a tennis court. (See Figure 33.10.)

**MAKE CONNECTIONS** *Find other examples of flattening, folding, branching, and projections (see Chapters 4, 7, 28, and 34). How is maximizing surface area important to the structure's function in each example?*

The structure of a mycelium maximizes its surface-to-volume ratio, making feeding very efficient. Just 1 cm³ of rich soil may contain as much as 1 km of hyphae with a total surface area of 300 cm² in contact with the soil. As shown in **Figure 26.14**, the branching of a fungal mycelium is one of several structural features that maximize surface area and have arisen (by convergent evolution) in different organisms.

### Mycorrhizae: A Key Adaptation to Life on Land

In addition to forming mycelia, some fungi also have specialized hyphae through which they can exchange nutrients with the root cells of a plant. Such mutually beneficial relationships between fungi and plant roots are called **mycorrhizae** (the term means "fungus roots").

There are two main types of mycorrhizal fungi (fungi that form mycorrhizae). **Arbuscular mycorrhizal fungi** (from the Latin *arbor*, tree) extend branching hyphae through the root cell wall and into tubes formed by invagination (pushing inward) of the root cell plasma membrane (see Figures 26.11 and 29.14b). **Ectomycorrhizal fungi** (from the Greek *ektos*, out) form unique sheaths of hyphae over the surface of a root and typically grow into the extracellular spaces of the root cortex (see Figure 29.14a). In both types of mycorrhizae, the specialized hyphae that contact or penetrate plant roots cells are connected to a fungal mycelium that grows into the soil.

What benefits do plants and fungi receive from mycorrhizal associations? Such associations can improve delivery of phosphate ions and other minerals to plants because the vast mycelial networks of the fungi are more efficient than the plants' roots at acquiring these minerals from the soil. In exchange, the plants supply the fungi with organic nutrients such as carbohydrates. A similar, mutually beneficial exchange of minerals for carbohydrates may have occurred in early plant-fungus symbioses. Fossil evidence shows that once on land, some fungi formed mycorrhizal associations with early plants. For example, 405-million-year-old fossils of the early plant *Aglaophyton* (see Figure 26.10) contain evidence of branching hyphae that have penetrated within the plant cells. These branching hyphae formed structures called *arbuscules* resembling those formed today by arbuscular mycorrhizae **(Figure 26.15)**. Similar structures have been found in a variety of other early plants, suggesting that plants probably existed in beneficial relationships with fungi from the earliest periods of colonization of land.

Support for the antiquity of mycorrhizal associations has also come from recent molecular studies. For a mycorrhizal fungus and its plant partner to establish a symbiotic relationship, certain genes must be expressed by the fungus, and other genes must be expressed by the plant. Researchers focused on three plant genes (called *sym* genes) whose expression is required for the formation of mycorrhizae in flowering plants. They found that these genes were present in all major plant lineages, including liverworts (see Figure 26.18), a lineage that

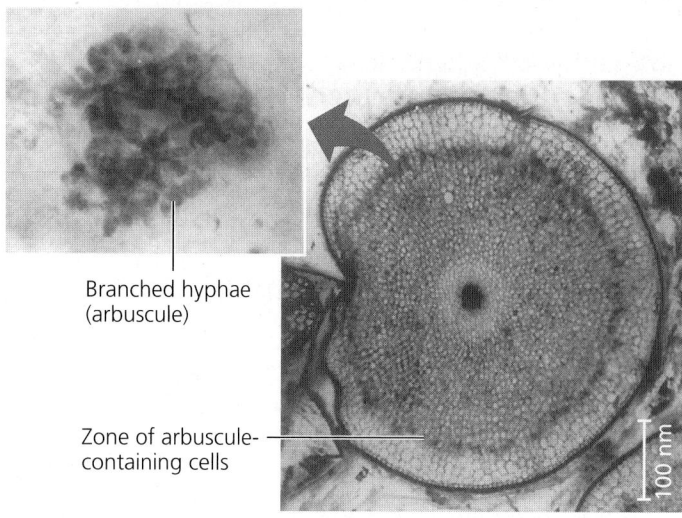

▲ **Figure 26.15 An ancient symbiosis.** This 405-million-year-old fossil stem (cross section) documents mycorrhizae in the early plant *Aglaophyton major*. The inset shows an enlarged view of a cell containing an *arbuscule*, a branched hyphal structure that forms after a mycorrhizal fungus penetrates a plant cell; the fossil arbuscule resembles those seen today in arbuscular mycorrhizae (see Figure 29.14b).

*Labels in figure:* Branched hyphae (arbuscule); Zone of arbuscule-containing cells; 100 nm

diverged from all other plants more than 400 million years ago. Furthermore, after the researchers transferred a liverwort *sym* gene to a flowering plant mutant that could not form mycorrhizae, the mutant recovered its ability to form mycorrhizae. These results suggest that mycorrhizal *sym* genes were present in the common ancestor of plants—and that the function of these genes has been conserved for hundreds of millions of years as plants continued to adapt to life on land.

Given the importance and long history of mycorrhizal associations, scientists have sought to uncover genes that affect how mycorrhizal fungi interact with their plant partners. In the **Scientific Skills Exercise**, you'll examine results from a study that compares genomic data from fungi that form mycorrhizae and fungi that do not.

## Diversification of Fungi

Once on land, fungi diversified into a wide range of living species. In the past decade, molecular analyses have helped clarify the evolutionary relationships between fungal groups, although there are still areas of uncertainty. **Figure 26.16** presents a simplified version of one current hypothesis.

The groups shown in Figure 26.16 may represent only a small fraction of the diversity of extant fungal groups. (Extant lineages are those that have surviving members.) While there are roughly 100,000 known species of fungi, there may actually be close to 1.5 million species. Two metagenomic studies published in 2011 support such higher estimates: Entirely new groups of unicellular fungi were discovered, and the genetic variation in some of these groups is as large as that displayed across all of the groups in Figure 26.16.

Along with their extensive genetic variation, fungi also show considerable variation in their life cycles and means of reproduction. Most fungi propagate by producing vast numbers

The phylogeny of fungi is currently the subject of much research. Most mycologists recognize five major groups of fungi, although the chytrids and zygomycetes are probably paraphyletic (as indicated by the parallel lines).

Hyphae                                              25 µm

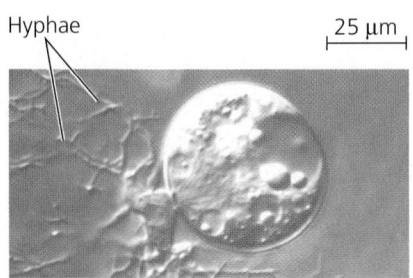

## Chytrids (1,000 species)

In chytrids such as *Chytriomyces*, the globular fruiting body forms multicellular, branched hyphae (LM); other species are single-celled. Ubiquitous in lakes and soil, chytrids have flagellated spores and are thought to include some of the earliest fungal groups to diverge from other fungi.

## Zygomycetes (1,000 species)

The hyphae of some zygomycetes, including this mold in the genus *Mucor* (LM), grow rapidly in foods such as fruits and bread. As such, the fungi may act as decomposers (if the food is not alive) or parasites; other species live as neutral (commensal) symbionts.

## Glomeromycetes (160 species)

The glomeromycetes form arbuscular mycorrhizae with plant roots, supplying minerals and other nutrients to the roots; about 80% of all plant species have such mutualistic partnerships with glomeromycetes. This SEM shows the branched hyphae—an arbuscule—of *Glomus mosseae* bulging into a plant root cell (the root has been treated to remove the cytoplasm).

2.5 µm

## Ascomycetes (65,000 species)

Also called sac fungi, members of this diverse group are common to many marine, freshwater, and terrestrial habitats. The cup-shaped ascocarp (fruiting body) of the ascomycete shown here (*Aleuria aurantia*) gives this species its common name: orange peel fungus.

## Basidiomycetes (30,000 species)

Often important as decomposers and ectomycorrhizal fungi, basidiomycetes, or club fungi, are unusual in having a long-lived, heterokaryotic stage in which each cell has two nuclei (one from each parent). The fruiting bodies—commonly called mushrooms—of this fly agaric (*Amanita muscaria*) are a familiar sight in coniferous forests of the Northern Hemisphere.

## Interpreting Genomic Data and Generating Hypotheses

**What Can Genomic Analysis of a Mycorrhizal Fungus Reveal About Mycorrhizal Interactions?** The first genome of a mycorrhizal fungus to be sequenced was that of the basidiomycete *Laccaria bicolor* (see photo). In nature, *L. bicolor* is a common ectomycorrhizal fungus of trees such as poplar and fir, as well as a free-living soil organism. In forest nurseries, it is used in large-scale inoculation programs to enhance seedling growth. The fungus can easily be grown alone in culture and can establish mycorrhizae with tree roots in the laboratory. Researchers hope that studying the genome of *Laccaria* will yield clues to the processes by which it interacts with its mycorrhizal partners—and by extension, to mycorrhizal interactions involving other fungi.

**How the Study Was Done** Using the whole-genome shotgun method (see Figure 18.2), researchers sequenced the *L. bicolor* genome and compared it with the genomes of some nonmycorrhizal basidiomycete fungi. The team used microarrays to compare gene expression levels for different protein-coding genes and for the same genes in a mycorrhizal mycelium and a free-living mycelium. They could thus identify the genes for fungal proteins made specifically in mycorrhizae.

**Data from the Study**

**Table 1. Numbers of Genes in *L. bicolor* and Four Nonmycorrhizal Fungal Species**

|  | *L. bicolor* | 1 | 2 | 3 | 4 |
|---|---|---|---|---|---|
| Protein-coding genes | 20,614 | 13,544 | 10,048 | 7,302 | 6,522 |
| Genes for membrane transporters | 505 | 412 | 471 | 457 | 386 |
| Genes for small secreted proteins (SSPs) | 2,191 | 838 | 163 | 313 | 58 |

**Table 2. *L. bicolor* Genes Most Highly Upregulated in Ectomycorrhizal Mycelium (ECM) of Douglas Fir or Poplar versus Free-Living Mycelium (FLM)**

| Protein ID | Protein Feature or Function | Douglas Fir ECM/FLM Ratio | Poplar ECM/FLM Ratio |
|---|---|---|---|
| 298599 | SSP | 22,877 | 12,913 |
| 293826 | Enzyme inhibitor | 14,750 | 17,069 |
| 333839 | SSP | 7,844 | 1,931 |
| 316764 | Enzyme | 2,760 | 1,478 |

**Data from** F. Martin et al., The genome of *Laccaria bicolor* provides insights into mycorrhizal symbiosis, *Nature* 452:88–93 (2008).

### INTERPRET THE DATA

1. (a) From the data in Table 1, which fungal species has the most genes encoding membrane transporters (membrane transport proteins; see Concept 5.2)? (b) Why might these genes be of particular importance to *L. bicolor*?

2. The phrase "small secreted proteins" (SSPs) refers to proteins fewer than 100 amino acids in length that the fungi secrete; their function is not yet known. (a) Describe the Table 1 data on SSPs. (b) The researchers found that the SSP genes shared a common feature that indicated the encoded proteins were destined for secretion. Based on Figure 14.21, predict what this common feature of the SSP genes was. (c) Suggest a hypothesis for the roles of SSPs in mycorrhizae.

3. Table 2 shows data from gene expression studies for the four *L. bicolor* genes whose transcription was most increased ("upregulated") in mycorrhizae. (a) For the gene encoding the first protein listed, what does the number 22,877 indicate? (b) Do the data in Table 2 support your hypothesis in 2(c)? Explain. (c) Compare the data for poplar mycorrhizae with those for Douglas fir and hypothesize what might account for any differences.

(MB) A version of this Scientific Skills Exercise can be assigned in MasteringBiology.

---

of spores, either sexually or asexually. Spores can be carried long distances by wind or water. If a spore lands in a moist place where there is food, it germinates, producing a new mycelium. **Figure 26.17** generalizes the many different life cycles that can produce fungal spores. Many fungi reproduce both sexually and asexually, as shown in this figure; others, however, reproduce only sexually or asexually. The sexual portion of a fungal life cycle typically occurs in two stages. First, the cytoplasms of two parent mycelia fuse, an event known as **plasmogamy**. Hours, days, or (in some fungi) even centuries may pass between plasmogamy and the next stage in the sexual cycle,

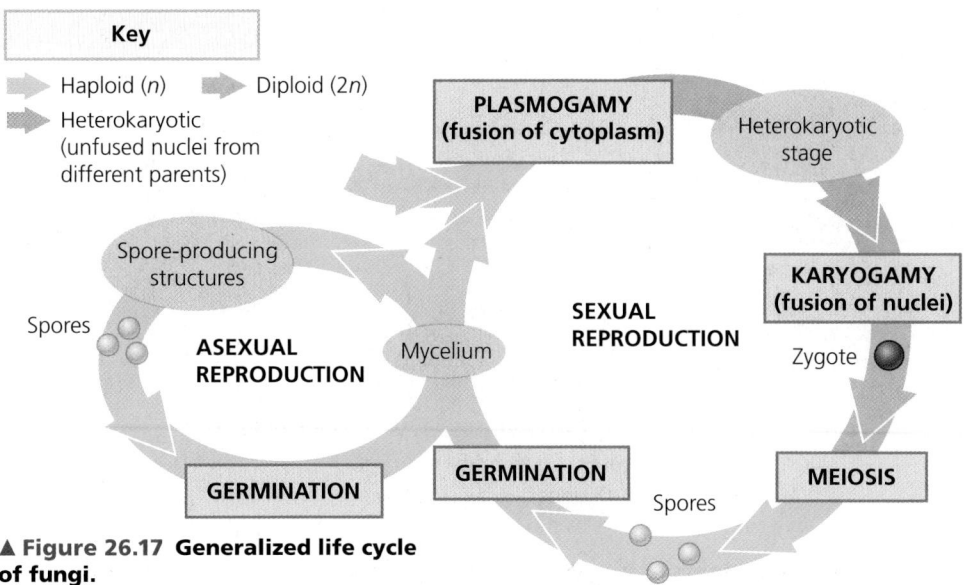

▲ **Figure 26.17 Generalized life cycle of fungi.**

karyogamy. During karyogamy, the haploid nuclei contributed by the two parents fuse, producing diploid cells. Zygotes and other structures formed during karyogamy are the only diploid stage in most fungi. Meiosis then restores the haploid condition, ultimately leading to the production of genetically diverse spores. The sexual processes of karyogamy and meiosis generate extensive genetic variation, a prerequisite for natural selection. Many fungi then reproduce asexually by growing as filamentous fungi that produce (haploid) spores by mitosis; such species are informally referred to as *molds* if they form visible mycelia. Other species reproduce asexually as single-celled yeasts that divide to produce genetically identical daughter cells.

Their diverse life cycles and the phylogenetic data show that fungi radiated extensively after they colonized land. So, too, did the plants that fungi helped ashore, as we'll discuss next.

## CONCEPT CHECK 26.2

1. Compare and contrast the nutritional mode of a fungus with your own nutritional mode.
2. Describe the importance of mycorrhizae, both today and in the colonization of land. What evidence supports the antiquity of mycorrhizal associations?
3. **MAKE CONNECTIONS** Review Figures 8.3 and 8.5. If a plant has mycorrhizae, where might carbon that enters the plant's stomata as $CO_2$ eventually be deposited: in the plant, the fungus, or both? Explain.

For suggested answers, see Appendix A.

# Early plants radiated into a diverse set of lineages

As early plants adapted to terrestrial environments, they gave rise to lineages that ultimately produced the vast diversity of present-day plants. An overview of that diversity is provided by **Figure 26.18**, which summarizes the evolutionary history of extant plant groups.

One way to distinguish plant groups is whether they have an extensive system of **vascular tissue**, cells joined into tubes that transport water and nutrients throughout the plant body. Most present-day plants have a complex vascular tissue system and are therefore called **vascular plants**. We'll return to vascular plants later in this section, but first we'll discuss the nonvascular plants, or **bryophytes** (from the Greek *bryon*, moss, and *phyton*, plant), an informal name for plants that lack an extensive transport system.

## Bryophytes: A Collection of Basal Plant Lineages

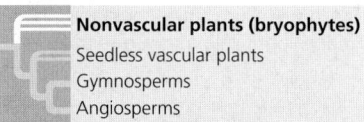

**Nonvascular plants (bryophytes)**
Seedless vascular plants
Gymnosperms
Angiosperms

The nonvascular plants (bryophytes) are represented today by three clades of small herbaceous (nonwoody) plants: *liverworts, mosses,* and *hornworts* (**Figure 26.19**). Current evidence indicates that these

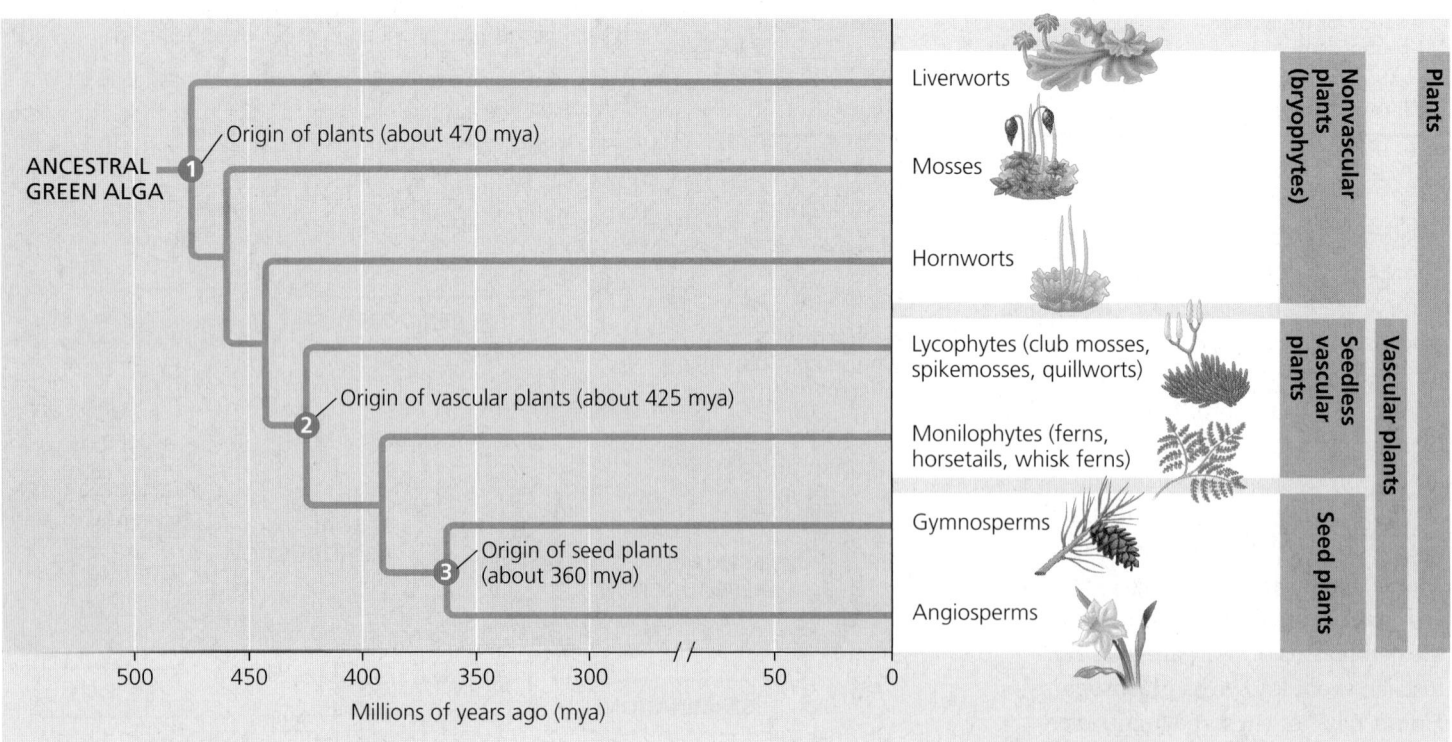

▲ **Figure 26.18 Highlights of plant evolution.** The phylogeny shown here illustrates a leading hypothesis about the relationships between plant groups.

**(a) *Plagiochila deltoidea*, a liverwort.** The name "liverwort" refers to the shape of this clade's gametophytes. In medieval times, their shape was thought to be a sign that the plants could help treat liver diseases.

Capsule
Seta
Sporophyte (a sturdy plant that takes months to grow)
Gametophyte

**(b) *Polytrichum commune*, a moss.** Moss gametophytes are less than 15 cm tall in most species but can reach up to 2 m. The familiar carpet of moss you observe consists mainly of gametophytes.

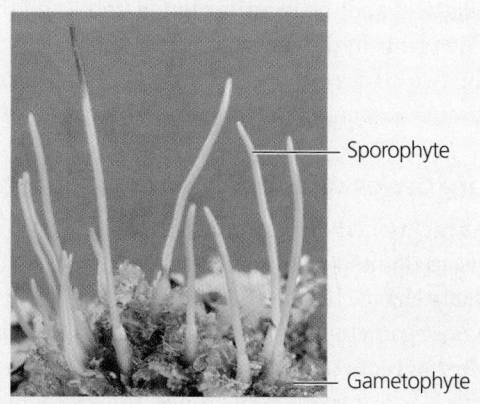

Sporophyte
Gametophyte

**(c) *Anthoceros* sp., a hornwort.** This group's name refers to the long, tapered shape of the sporophyte, which can grow to about 5 cm high.

▲ **Figure 26.19 Bryophytes (nonvascular plants).** Molecular and morphological data indicate that bryophytes are paraphyletic; they do not form a single clade.

three clades are basal taxa, meaning they diverged from other plant lineages early in the history of plant evolution (see Figure 26.18). Indeed, some of the oldest known fossils of plants have features similar to those found in present-day bryophytes. For example, as mentioned earlier, the earliest spores of plants (dating to 450–470 million years ago) have structural features that today are found only in the spores of liverworts. By 430 million years ago, spores similar to those of mosses and hornworts also occur in the fossil record.

In addition, as in some early plants, the bryophytes of today are anchored to the ground by **rhizoids**, structures that lack specialized conducting cells and do not play a primary role in water and mineral absorption. Living bryophytes are typically found in moist habitats—as you might expect, since they have flagellated sperm that must swim through a film of water to fertilize an egg. Unlike most plants today, in bryophytes the haploid gametophytes are the dominant stage of the life cycle: The gametophytes are usually larger and longer-living than the sporophytes (see Figure 26.19). The gametophytes of mosses and other bryophytes typically form ground-hugging carpets, partly because their body parts are too thin to support a tall plant. A second constraint on the height of many bryophytes is the absence of vascular tissue, which would enable long-distance transport of water and nutrients. In contrast, the group we turn to next, the vascular plants, does not share these constraints.

## Seedless Vascular Plants: The First Plants to Grow Tall

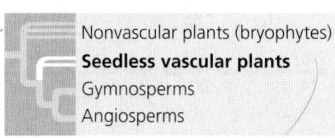

Nonvascular plants (bryophytes)
**Seedless vascular plants**
Gymnosperms
Angiosperms

During the first 100 million years of plant evolution, bryophytes were prominent members of the vegetation. But it is vascular plants that dominate most landscapes today. The earliest fossils of vascular plants date to

425 million years ago. These plants lacked seeds but had well-developed vascular systems, an evolutionary novelty that set the stage for plants to grow tall.

The rise of vascular plants was accompanied by other evolutionary changes as well, resulting in life cycles with dominant sporophytes and the origin of well-developed roots and leaves. Our focus here will be on the two clades of vascular plants shown in **Figure 26.20**, the **lycophytes** (club mosses and their

2.5 cm
2.5 cm

Strobili (cone-like structures in which spores are produced)

**(a) *Diphasiastrum tristachyum*, a lycophyte.** The spores of lycophytes such as this club moss are released in clouds and are so rich in oil that magicians and photographers once ignited them to create smoke or flashes of light.

**(b) *Matteuccia struthiopteris* (ostrich fern), a monilophyte.** The sporophytes of ferns typically have horizontal stems that give rise to large leaves called fronds, which grow as their coiled tips unfurl.

▲ **Figure 26.20 Lycophytes and monilophytes (seedless vascular plants).** Although lycophytes and monilophytes each form a monophyletic group, seedless vascular plants are paraphyletic. These photographs are of sporophytes; the small, free-living gametophytes are not shown.

relatives) and the **monilophytes** (ferns and their relatives). The plants in these clades lack seeds, which is why collectively the two clades are often called **seedless vascular plants**. We'll discuss vascular plants that have seeds in Concept 26.4.

## Life Cycles with Dominant Sporophytes

As mentioned earlier, mosses and other bryophytes have life cycles dominated by gametophytes. Fossil evidence suggests that a change began to occur in relatives of vascular plants, whose gametophytes and sporophytes were about equal in size. Further reductions in gametophyte size occurred among extant vascular plants; in these groups, the sporophyte generation is the larger and more complex plant form in the alternation of generations **(Figure 26.21)**. In ferns, for example, the familiar leafy plants are the sporophytes. You would have to get down on your hands and knees and search the ground carefully to find fern gametophytes, which are tiny structures that often grow on or just below the soil surface. As in nonvascular plants, the sperm of ferns and all other seedless vascular plants are flagellated and must swim through a film of water to reach eggs.

## Transport in Xylem and Phloem

Vascular plants have two types of vascular tissue: xylem and phloem (see Figure 28.9). **Xylem** conducts most of the water and minerals. The xylem of most vascular plants includes **tracheids**, tube-shaped cells that carry water and minerals up from the roots. The water-conducting cells in vascular plants are *lignified*; that is, their cell walls are strengthened by the polymer **lignin**. The tissue called **phloem** has cells arranged into tubes that distribute sugars, amino acids, and other organic products.

Lignified vascular tissue helped enable vascular plants to grow tall. Their stems became strong enough to provide support against gravity, and their vascular tissue could transport water and mineral nutrients high above the ground. Tall plants could also outcompete short plants for access to the sunlight needed for photosynthesis. In addition, the spores of tall plants could disperse farther than those of short plants, enabling tall species to colonize new environments rapidly. Overall, the ability to grow tall gave vascular plants a competitive edge over nonvascular plants, which rarely grow above 20 cm in height. Competition

| | PLANT GROUP | | |
|---|---|---|---|
| | **Mosses and other nonvascular plants** | **Ferns and other seedless vascular plants** | **Seed plants (gymnosperms and angiosperms)** |
| **Gametophyte** | Dominant | Reduced, independent (photosynthetic and free-living) | Reduced (usually microscopic), dependent on surrounding sporophyte tissue for nutrition |
| **Sporophyte** | Reduced, dependent on gametophyte for nutrition | Dominant | Dominant |
| **Example** | 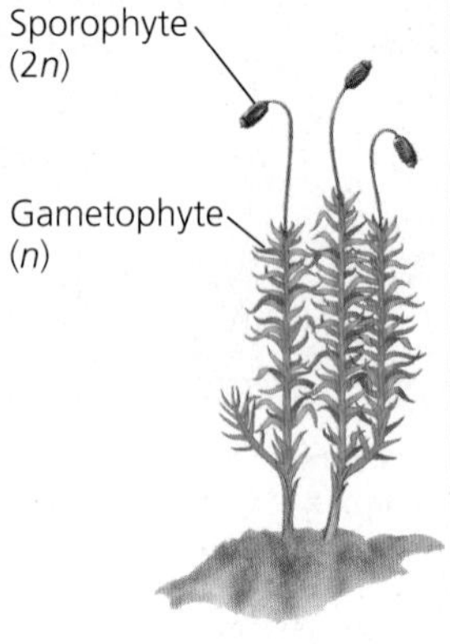 |  | 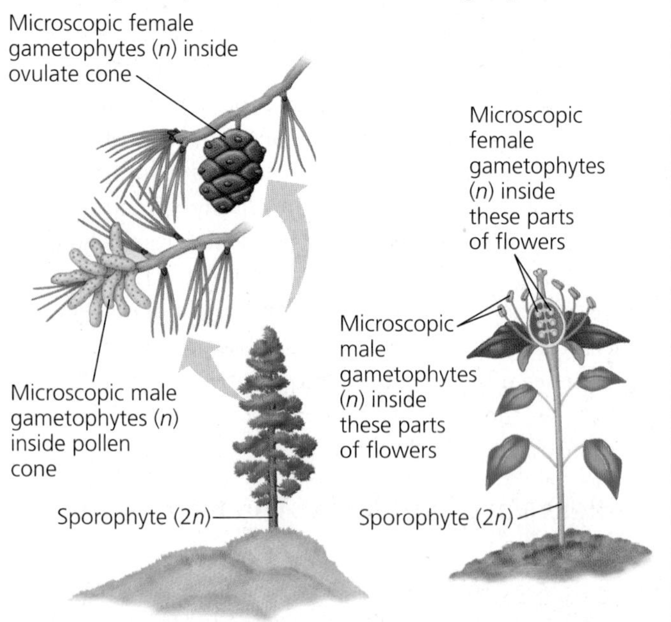 |

▲ **Figure 26.21 Gametophyte-sporophyte relationships in different plant groups.**

among vascular plants also would have increased, leading to selection for taller growth forms—a process that eventually gave rise to the trees that formed the first forests about 385 million years ago.

### Evolution of Roots and Leaves

Vascular tissue also provides benefits below ground. In contrast to the rhizoids of bryophytes, roots with vascular tissue evolved in the sporophytes of almost all vascular plants. **Roots** are organs that absorb water and nutrients from the soil; roots also anchor vascular plants.

**Leaves** increase the surface area of the plant body and serve as the primary photosynthetic organ of vascular plants. In terms of size and complexity, leaves can be classified as either microphylls or megaphylls **(Figure 26.22)**. All of the lycophytes—and only the lycophytes—have **microphylls**, small, usually spine-shaped leaves supported by a single strand of vascular tissue. Almost all other vascular plants have **megaphylls**, leaves with a highly branched vascular system; a few species have reduced leaves that appear to have evolved from megaphylls. Megaphylls are typically larger and support greater photosynthetic productivity than microphylls. Microphylls first appear in the fossil record 410 million years ago, but megaphylls do not emerge until about 370 million years ago.

Seedless vascular plants were abundant in the swampy forests and other moist ecosystems of the Carboniferous period (359–299 million years ago). Growing along with these seedless

▼ Figure 26.22 Microphyll and megaphyll leaves.

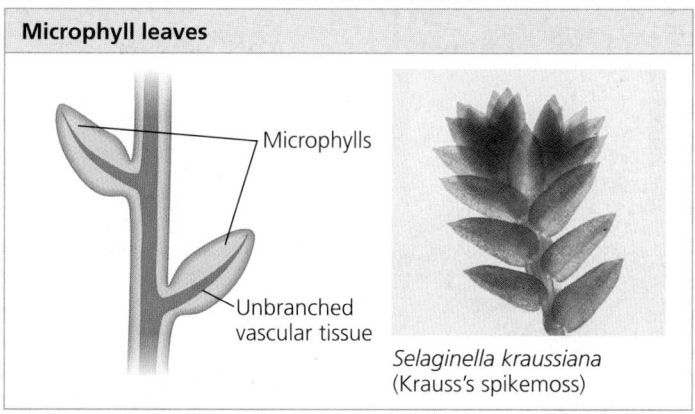

**Microphyll leaves**

Microphylls

Unbranched vascular tissue

*Selaginella kraussiana* (Krauss's spikemoss)

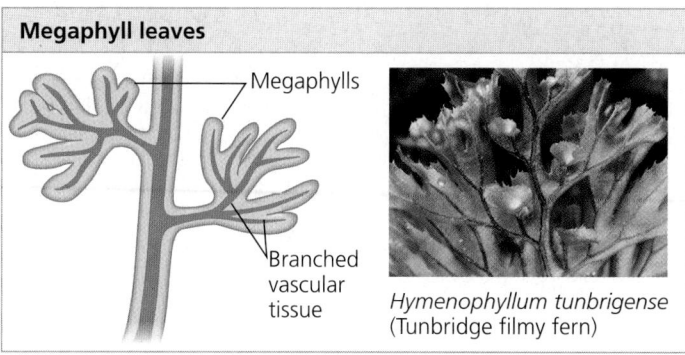

**Megaphyll leaves**

Megaphylls

Branched vascular tissue

*Hymenophyllum tunbrigense* (Tunbridge filmy fern)

plants were early seed plants. Though seed plants were not dominant at that time, they rose to prominence after the climate became drier at the end of the Carboniferous period. In Concept 26.4, we'll trace the origin and diversification of seed plants, continuing the story of adaptation to life on land.

#### CONCEPT CHECK 26.3

1. How do the main similarities and differences between seedless vascular plants and nonvascular plants influence function in these plants?
2. **MAKE CONNECTIONS** Figure 26.18 identifies lineages as plants, nonvascular plants, vascular plants, seedless vascular plants, and seed plants. Which of these categories are monophyletic, and which are paraphyletic (see Figure 20.10)? Explain.
3. **MAKE CONNECTIONS** Monilophytes and seed plants both have megaphylls, as well as other traits not found in lycophytes. Explain this observation using Figure 26.18 and the concept of descent with modification (see Concept 19.2).

**For suggested answers, see Appendix A.**

---

CONCEPT 26.4

## Seeds and pollen grains are key adaptations for life on land

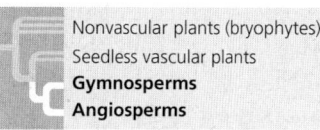

Nonvascular plants (bryophytes)
Seedless vascular plants
**Gymnosperms**
**Angiosperms**

Seed plants originated about 360 million years ago. As this new group of plants became established, they expanded into a broad range of terrestrial environments, dramatically altering the course of plant evolution. This large impact was due in part to the innovation for which this group of plants is named: the seed. A **seed** consists of an embryo and its food supply, surrounded by a protective coat. When mature, seeds are dispersed from their parent by wind or other means.

Extant seed plants can be divided into two major clades, gymnosperms (pines and their relatives) and angiosperms (flowering plants). **Gymnosperms** (from the Greek *gymnos*, naked, and *sperm*, seed) are grouped together as "naked seed" plants because their seeds are not enclosed in chambers. In contrast, the seeds of **angiosperms** (from the Greek *angion*, container) develop inside chambers called ovaries. We'll begin our discussion of seed plants with an overview of their adaptations for life on land. Then we'll turn to their origin and evolutionary history.

### Terrestrial Adaptations in Seed Plants

In addition to seeds, all seed plants have highly reduced gametophytes, as well as ovules and pollen. These adaptations provided new ways for seed plants to cope with terrestrial conditions such as drought and exposure to the ultraviolet (UV) radiation in sunlight. These adaptations also freed seed plants from requiring water for fertilization, enabling reproduction to occur under a broader range of conditions than in seedless plants.

### Reduced Gametophytes

Unlike mosses and other bryophytes, ferns and other seedless vascular plants have sporophyte-dominated life cycles. This evolutionary trend of gametophyte reduction continued further in the vascular plant lineage that led to seed plants. While the gametophytes of seedless vascular plants are visible to the naked eye, the gametophytes of most seed plants are microscopic (see Figure 26.21).

This miniaturization allowed for an important evolutionary innovation in seed plants: Their tiny gametophytes can develop from spores retained within the parental sporophyte. The moist reproductive tissues of the sporophyte shield the gametophytes from UV radiation and protect them from drying out. This relationship also enables the developing gametophytes to obtain nutrients from the parental sporophyte. In contrast, the free-living gametophytes of seedless vascular plants must fend for themselves.

### Ovules and Pollen

Seed plants are unique in retaining the structures that develop into a female gametophyte within the parent sporophyte. Early in this process, a layer of sporophyte tissue called **integument** envelops and protects the tissues that will eventually give rise to the female gametophyte. The integument and the tissues it encloses together make up an **ovule (Figure 26.23a)**. Inside each ovule, an egg-producing female gametophyte develops

from a haploid spore. Spores that produce female gametophytes are called *megaspores* because they are larger than spores that produce male gametophytes (*microspores*).

A microspore develops into a **pollen grain** that consists of a male gametophyte enclosed within the pollen wall. The pollen wall, which contains sporopollenin, protects the gametophyte as it is transported from the parent plant by wind or by hitchhiking on the body of an animal. The transfer of pollen to the part of a seed plant that contains the ovules is called **pollination**. If a pollen grain germinates (begins growing), it gives rise to a pollen tube that discharges sperm into the female gametophyte within the ovule, as shown in **Figure 26.23b**.

Recall that in nonvascular plants and seedless vascular plants such as ferns, free-living gametophytes release flagellated sperm that swim through a film of water to reach eggs. Given this requirement, it is not surprising that many of these species are found in moist habitats. But in seed plants, a sperm-producing male gametophyte inside a pollen grain can be carried long distances by wind or animals, eliminating the dependence on water for sperm transport. The ability of seed plants to transfer sperm without water likely contributed to their successful colonization of dry habitats.

### The Evolutionary Advantage of Seeds

If a sperm fertilizes an egg of a seed plant, the zygote grows into a sporophyte embryo. As shown in **Figure 26.23c**, the whole ovule develops into a seed: the embryo, along with a food supply, packaged within a protective coat derived from the integument.

Until the advent of seeds, the spore was the only protective stage in any plant life cycle. What advantages do seeds provide

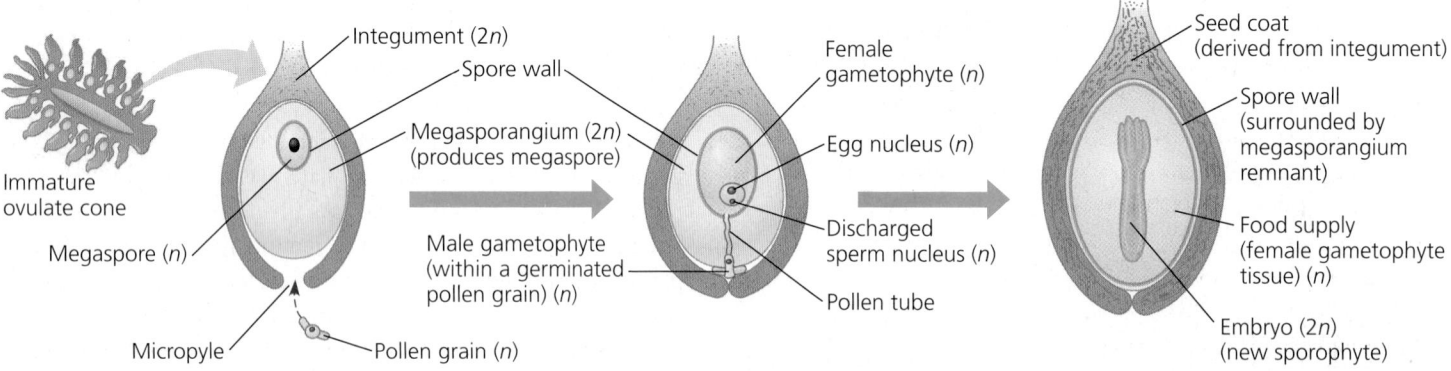

**(a) Unfertilized ovule.** In this longitudinal section through the ovule of a pine (a gymnosperm), a fleshy megasporangium is surrounded by a protective layer of tissue called an integument. The micropyle, the only opening through the integument, allows entry of a pollen grain.

**(b) Fertilized ovule.** A megaspore develops into a female gametophyte, which produces an egg. The pollen grain, which had entered through the micropyle, contains a male gametophyte. The male gametophyte develops a pollen tube that discharges sperm, thereby fertilizing the egg.

**(c) Gymnosperm seed.** Fertilization initiates the transformation of the ovule into a seed, which consists of a sporophyte embryo, a food supply, and a protective seed coat derived from the integument. The megasporangium dries out and collapses.

▲ **Figure 26.23 From ovule to seed in a gymnosperm.**

**?** *A gymnosperm seed contains cells from how many different plant generations? Identify the cells and whether each is haploid or diploid.*

over spores? Spores are usually single-celled, whereas seeds are multicellular, consisting of an embryo protected by a layer of tissue, the seed coat. A seed can remain dormant for days, months, or even years after being released from the parent plant, whereas most spores have shorter lifetimes. Also, unlike spores, seeds have a supply of stored food. Under favorable conditions, the seed can emerge from dormancy and germinate, with its stored food providing critical support for growth as the sporophyte embryo emerges as a seedling.

## Early Seed Plants and the Rise of Gymnosperms

Recall from Figure 26.18 that extant seed plants form two sister clades: gymnosperms and angiosperms. How did these two groups arise?

Fossils reveal that by the late Devonian period (about 380 million years ago), some plants had acquired features found in seed plants, such as the megaspores and microspores mentioned earlier. But these plants did not bear seeds and hence are not classified as seed plants. The first seed plants to appear in the fossil record date from around 360 million years ago, 55 million years before the first fossils classified as gymnosperms and more than 200 million years before the first fossils of angiosperms. These early seed plants became extinct, and it remains uncertain which of these extinct lineages ultimately gave rise to the gymnosperms.

The earliest fossils of species belonging to an extant lineage of gymnosperms are about 305 million years old. These early gymnosperms lived in moist Carboniferous ecosystems that were dominated by lycophytes, ferns, and other seedless vascular plants. As the Carboniferous period gave way to the Permian (299 million years ago), the climate became much drier. As a result, the lycophytes and ferns that dominated moist Carboniferous swamps were largely replaced by gymnosperms, which were better suited to the drier climate.

Gymnosperms thrived as the climate dried in part because they have the key terrestrial adaptations found in all seed plants, such as seeds and pollen. In addition, some gymnosperms were particularly well suited to arid conditions because of the thick cuticles and relatively small surface areas of their needle-shaped leaves.

Gymnosperms dominated terrestrial ecosystems through much of the Mesozoic era, which lasted from 252 to 66 million years ago. In addition to serving as the food supply for giant herbivorous dinosaurs, these gymnosperms were involved in many other interactions with animals. Recent fossil discoveries, for example, show that some gymnosperms were pollinated by insects more than 100 million years ago—the earliest evidence of insect pollination in any plant group.

Late in the Mesozoic, gymnosperms began to be replaced by angiosperms in some ecosystems. Even so, today gymnosperms remain an important part of Earth's flora **(Figure 26.24)**. For example, vast regions in northern

**(a) Sago palm (*Cycas revoluta*).** This "palm" is actually a cycad, the next largest group of gymnosperms after the conifers (true palms are flowering plants). Cycads have large cones and palmlike leaves.

**(b) Douglas fir (*Pseudotsuga menziesii*).** This conifer dominates large forested regions and provides more timber than any other North American tree species.

**(c) Creeping juniper (*Juniperus horizontalis*).** The "berries" of this low-growing conifer are actually ovule-producing cones consisting of fleshy sporophylls.

▲ **Figure 26.24 Examples of gymnosperms.**

latitudes are covered by forests of cone-bearing gymnosperms called **conifers**, which include spruce, pine, fir, and redwood. Yet despite the ongoing importance of gymnosperms, most terrestrial ecosystems are now dominated by the group we turn to next, the angiosperms.

## The Origin and Diversification of Angiosperms

Commonly known as flowering plants, angiosperms are seed plants that produce the reproductive structures called flowers and fruits. Today, angiosperms are the most diverse and widespread of all plants, with more than 250,000 species (about 90% of all plant species). Before considering the evolution of angiosperms, we'll examine their two key adaptations—flowers and fruits.

### Flowers and Fruits

The **flower** is a unique angiosperm structure that is specialized for sexual reproduction. In many angiosperm species, insects or other animals transfer pollen from one flower to the sex organs on another flower, which makes pollination more directed than the wind-dependent pollination of most species of gymnosperms.

A flower is a specialized shoot that can have up to four types of modified leaves called floral organs: sepals, petals, stamens, and carpels **(Figure 26.25)**. Starting at the base of the flower are the **sepals**, which are usually green and enclose the flower before it opens (think of a rosebud). Interior to the sepals are the **petals**, which are brightly colored in most flowers and can aid in attracting pollinators. Flowers that are wind-pollinated, however, generally lack brightly colored parts. In all angiosperms, the sepals and petals are sterile floral organs, meaning that they do not produce sperm or eggs. Within the petals are two whorls of fertile floral organs, the stamens and carpels. **Stamens** produce pollen grains containing male gametophytes. A stamen consists of a stalk called the filament and a terminal sac, the anther, where pollen is produced. **Carpels** make ovules. The carpel is the "container" mentioned earlier in which seeds are enclosed; as such, it is a key structure that distinguishes angiosperms

from gymnosperms. At the tip of the carpel is a sticky stigma that receives pollen. A style leads from the stigma to a structure at the base of the carpel, the **ovary**; the ovary contains one or more ovules. As in gymnosperms, each angiosperm ovule contains a female gametophyte. If fertilized, an ovule develops into a seed.

A flower may have one or more carpels. In many species, mutliple carpels are fused into one structure. The term **pistil** is sometimes used to refer to a single carpel (a simple pistil) or two or more fused carpels (a compound pistil).

As seeds develop from ovules after fertilization, the ovary wall thickens and the ovary matures into a **fruit**. A pea pod is an example of a fruit, with seeds (mature ovules, the peas) encased in the ripened ovary (the pod). Fruits protect seeds and aid in their dispersal (see Figure 30.11). For example, the seeds of some flowering plants, such as dandelions and maples, are contained within fruits that function like parachutes or propellers, adaptations that enhance dispersal by wind. Many other angiosperms rely on animals to carry seeds. Some of these plants have fruits modified as burrs that cling to animal fur (or the clothes of humans). Other angiosperms produce edible fruits, which are usually nutritious, sweet tasting, and vividly colored, advertising their ripeness. When an animal eats the fruit, it digests the fruit's fleshy part, but the tough seeds usually pass unharmed through the animal's digestive tract. When the animal defecates, it may deposit the seeds, along with a supply of natural fertilizer, many kilometers from where the fruit was eaten.

### Angiosperm Evolution

Charles Darwin once referred to the origin of angiosperms as an "abominable mystery." He was particularly troubled by the relatively sudden and geographically widespread appearance of angiosperms in the fossil record (about 100 million years ago based on fossils known to Darwin). Recent fossil discoveries and phylogenetic analyses have led to progress in solving Darwin's mystery, but we still do not fully understand how angiosperms arose from earlier seed plants.

**Fossil Evidence** Angiosperms are thought to have originated in the early Cretaceous period, about 140 million years ago. By the mid-Cretaceous (100 million years ago), angiosperms began to dominate some terrestrial ecosystems. Landscapes changed dramatically as conifers and other gymnosperms gave way to flowering plants in many parts of the world. The Cretaceous ended 66 million years ago with mass extinctions of dinosaurs and many other animal groups, as well as further increases in the diversity and importance of angiosperms.

What evidence suggests that angiosperms arose 140 million years ago? First, although pollen grains are common in rocks from the Jurassic period (201–145 million years ago), none of these pollen fossils have features diagnostic of angiosperms, suggesting that angiosperms may have originated after the

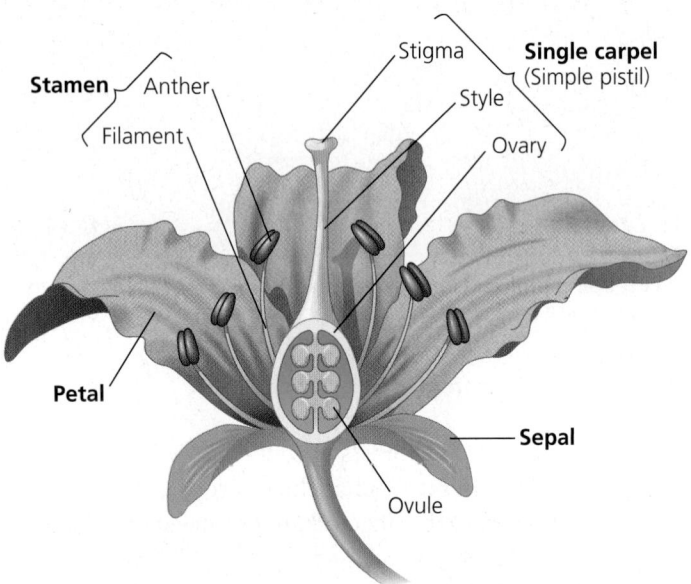

▲ **Figure 26.25 The structure of an idealized flower.**

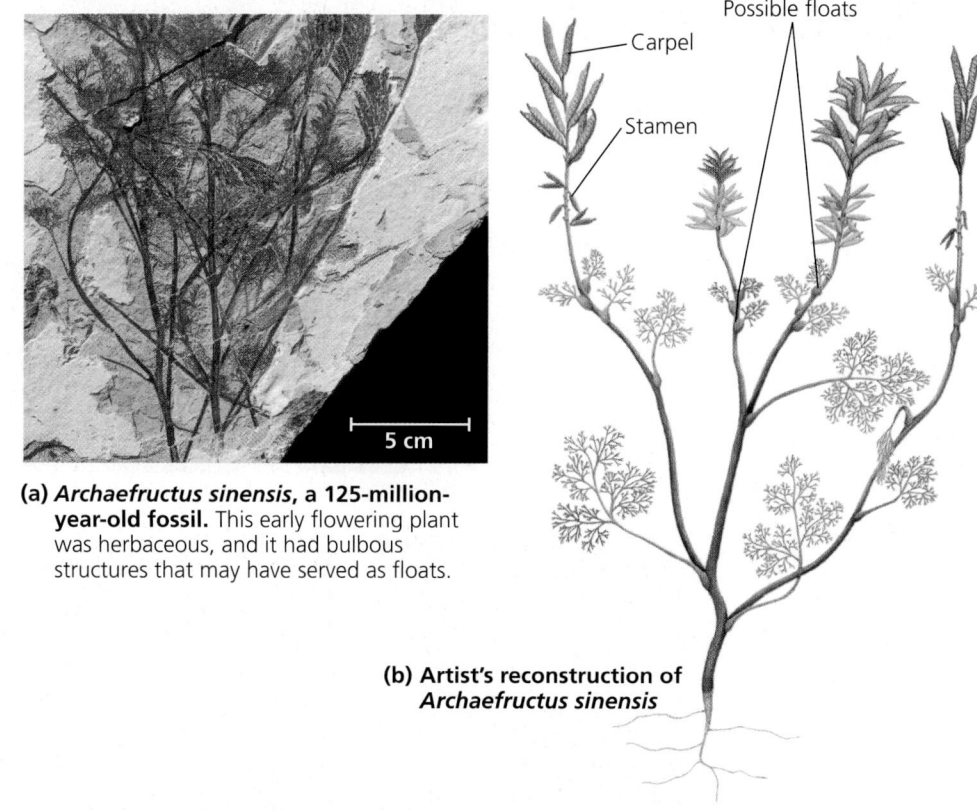

**(a) *Archaefructus sinensis*, a 125-million-year-old fossil.** This early flowering plant was herbaceous, and it had bulbous structures that may have served as floats.

Possible floats

Carpel

Stamen

**(b) Artist's reconstruction of *Archaefructus sinensis***

▲ **Figure 26.26 An early flowering plant.**

Jurassic. Indeed, the earliest fossils with distinctive angiosperm features are 130-million-year-old pollen grains discovered in China, Israel, and England. Early fossils of larger flowering plant structures include those of *Archaefructus* **(Figure 26.26)** and *Leefructus*, both of which were discovered in China in rocks that are about 125 million years old. Overall, early angiosperm fossils indicate that the group arose and began to diversify over a 20- to 30-million-year period—a less sudden event than was suggested by the fossils that were known during Darwin's lifetime.

Can we infer traits of the common ancestor of angiosperms from traits found in early fossil angiosperms? *Archaefructus*, for example, was herbaceous and had bulbous structures that may have served as floats, suggesting it was aquatic. But investigating whether the common ancestor of angiosperms was herbaceous and aquatic also requires examining fossils of other seed plants thought to have been closely related to angiosperms. All of those plants were woody, indicating that the common ancestor was probably woody. As we'll see, this conclusion has been supported by some recent phylogenetic analyses.

**Angiosperm Phylogeny** Molecular and morphological evidence suggests that the lineages leading to angiosperms and gymnosperms diverged from each other about 305 million years ago. Indeed, extant angiosperms are thought to be more closely related to several extinct lineages

of woody seed plants than they are to living gymnosperms. One such lineage is the Bennettitales **(Figure 26.27)**, a group with flowerlike structures that may have been pollinated by insects. However, the Bennettitales and other similar extinct lineages of woody seed plants did not have carpels or flowers and hence are not classified as angiosperms.

Making sense of the origin of angiosperms also depends on working out the order in which angiosperm clades diverged from one another. Here, dramatic progress has been made in recent years. Molecular and morphological evidence suggests that a small South Pacific shrub called *Amborella trichopoda* and water lilies are living representatives of two lineages that diverged from other angiosperms early in the history of the group **(Figure 26.28)**. *Amborella* is woody, supporting the conclusion mentioned earlier that the angiosperm common ancestor was likely woody. Among the other lineages shown in Figure 26.28, the monocots and eudicots in particular have radiated extensively: There are now 70,000 species of monocots and 170,000 species of eudicots. Monocots are named for the single cotyledon (seed leaf) that emerges when their seeds germinate, while eudicots typically have two cotyledons (see also Figure 28.2).

From their humble beginnings in the Cretaceous period, angiosperms have diversified into more than 250,000 species, making them by far the largest group of living plants. This large group, along with the nonflowering plants and the fungi, has enormous ecological and evolutionary effects on other species.

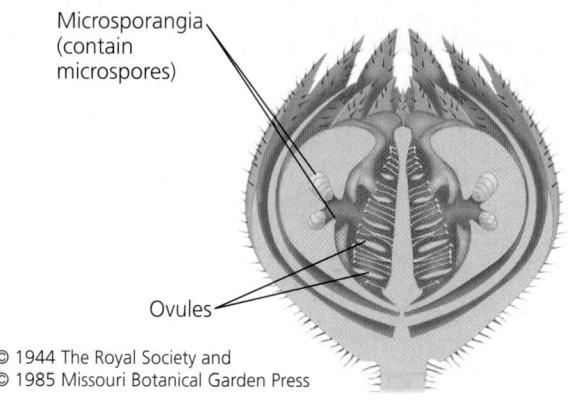

Microsporangia (contain microspores)

Ovules

© 1944 The Royal Society and
© 1985 Missouri Botanical Garden Press

▲ **Figure 26.27 A close relative of the angiosperms?** This reconstruction shows a longitudinal section through the flowerlike structures found in Bennettitales, an extinct group of seed plants hypothesized to be more closely related to extant angiosperms than to living gymnosperms.

The phylogenetic tree below represents one current hypothesis of angiosperm evolutionary relationships, based on morphological and molecular evidence.

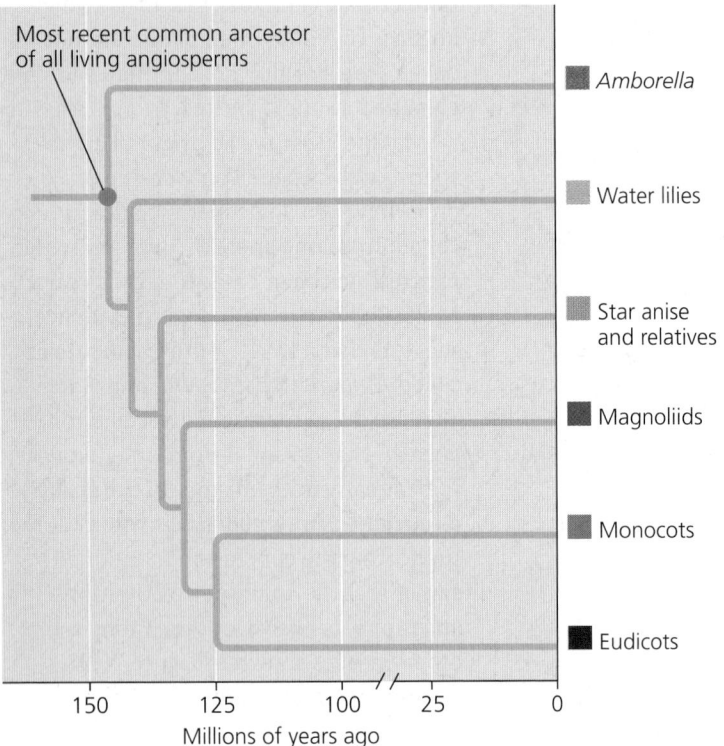

Most recent common ancestor of all living angiosperms

Amborella

Water lilies

Star anise and relatives

Magnoliids

Monocots

Eudicots

150   125   100   25   0

Millions of years ago

**Amborella.** This small shrub (*Amborella trichopoda*), found only on the South Pacific island of New Caledonia, may be the sole survivor of a branch at the base of the angiosperm tree. *Amborella* lacks vessels, efficient water-conducting cells found in most angiosperm lineages.

**Water lilies.** Species of water lilies (genus *Nymphaea*) are found in aquatic habitats throughout the world. Like *Amborella*, water lilies are a basal group that lack vessels (efficient water-conducting cells).

**Star anise.** Some of the shrubs and small trees in this genus (*Illicium*) are native to southeast Asia, others to the southeastern United States. Living species in the genus probably descended from ancestors whose populations were separated by continental drift.

**Magnoliids.** This clade consists of about 8,000 woody and herbaceous species, including such familiar and economically important plants as magnolias, laurels, avocado, cinnamon, and black pepper. The variety of southern magnolia shown here (*Magnolia grandiflora*, also called "Goliath"), has flowers that can measure a third of a meter across.

**Monocots.** Over 25% of extant angiosperms are monocots. This large clade includes the most important crop plants in the world today: grains such as maize, rice, and wheat. Other monocots are widely used as ornamental plants, such as this pygmy date palm (*Phoenix roebelenii*). The monocots also include plants such as orchids, grasses, irises, and onions.

**Eudicots.** Nearly 70% of living flowering plants are eudicots. One example, zucchini, a subspecies of *Cucurbita pepo*, is an important crop, as are acorn squash, pumpkin, and other *C. pepo* subspecies. The eudicots also include sunflowers, roses, cacti, clovers, oaks, and a wide range of other species.

## CONCEPT CHECK 26.4

1. What features not present in seedless plants have contributed to the enormous success of seed plants on land?
2. Explain why Darwin called the origin of angiosperms an "abominable mystery," and describe what has been learned from fossil evidence and phylogenetic analyses.
3. **MAKE CONNECTIONS** Suppose the Bennettitales and extant angiosperms are sister taxa. Draw a phylogenetic tree that includes Bennettitales, angiosperms, gymnosperms, monilophytes, and lycophytes. Identify the common ancestor and circle the basal taxon (see Figure 20.5).

For suggested answers, see Appendix A.

## CONCEPT 26.5

# Plants and fungi fundamentally changed chemical cycling and biotic interactions

Throughout Unit Four, we are highlighting major steps in the evolutionary history of life. We have focused on great waves of adaptive radiation, such as the metabolic diversification of prokaryotes (Chapter 24) and the rise in structural diversity that followed the origin of eukaryotes (Chapter 25). In this chapter, we've examined another major step in the history of life: the colonization of land by plants and fungi. Let's now explore how the colonization of land has altered the physical environment and the organisms that live there.

## Physical Environment and Chemical Cycling

Fungi and plants have profound effects on the physical environment. Consider a **lichen**, a symbiotic association between a fungus and a photosynthetic microorganism (an alga or cyanobacterium). Lichens are important pioneers on cleared rock and soil surfaces, such as volcanic flows and burned forests; they are also common in tundra and other nutrient-poor environments. As pioneers on cleared rock or soil, lichens break down the surface by physically penetrating and chemically altering it, and they trap windblown soil. These processes affect the formation of soil and make it possible for plants to grow. **Figure 26.29** shows two examples of the diverse forms of lichens along with the structure of a lichen composed of a fungus and a green alga. Fossils show that lichens were on land 420 million years ago. These early lichens may have modified rocks and soil much as they do today, helping pave the way for plants.

The colonization of land by plants also has resulted in great changes to the physical environment. By their very presence, plants such as the trees in a forest or the grasses in a grassland physically create the habitats of animals and many other organisms. Moreover, like lichens, plants affect the formation of soil: Their roots hold the soil in place, and leaf litter and other decaying plant parts add nutrients to

▼ Crustose (encrusting) lichens    ▼ A foliose (leaflike) lichen

**(a) Two common lichen growth forms**

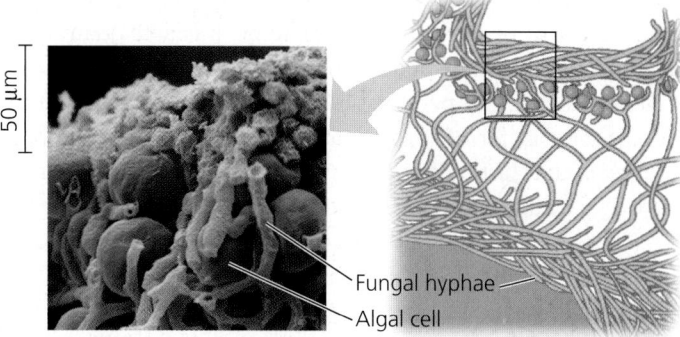

50 μm

Fungal hyphae
Algal cell

**(b) Anatomy of a lichen involving an ascomycete fungus and an alga**

▲ **Figure 26.29 Lichens.** Often found growing on rocks or rotting logs, lichens are a symbiotic association between a fungus and a photosynthetic microorganism (a green alga or a cyanobacterium).

the soil. Plants also have altered the composition of Earth's atmosphere, perhaps most importantly by releasing oxygen to the air as a by-product of photosynthesis—thereby helping to replenish the supply of oxygen that animals and many other organisms need for cellular respiration.

Plants and fungi also have profound effects on the cycling of chemicals in ecosystems (see Figure 1.9). This process begins when plants absorb nutrients from the physical environment. Next, those nutrients pass to organisms that eat plants. Decomposers then break down the bodies of dead organisms, thereby returning nutrients to the physical environment and completing the cycle. Fungi are well adapted as decomposers of organic material. In fact, almost any carbon-containing substrate—even jet fuel and house paint—can be consumed by at least some fungi. (The same is true of bacteria.) As a result, fungi and bacteria play a central role in keeping ecosystems stocked with the inorganic nutrients essential for plant growth. Without these decomposers, carbon, nitrogen, and other elements would remain tied up in organic matter. If that were to happen, plants and the animals that eat them could not exist because elements taken from the soil would not be returned. As such, life as we know it would cease.

Let's take a closer look at how plants affect carbon recycling. Carbon forms the basis of the organic compounds that are essential for life. During photosynthesis, plants remove large quantities of $CO_2$ from the atmosphere—an action that can influence the global climate. A dramatic example occurred when seedless vascular plants formed the first forests during the Devonian and early Carboniferous **(Figure 26.30)**. These forests contributed to a large drop in atmospheric $CO_2$ levels during the Carboniferous (359–299 million years ago), causing global cooling and widespread glacier formation. What caused $CO_2$ levels to fall?

One contributing factor was that with the evolution of leaves and other vascular tissue, plants accelerated their rate of photosynthesis, thus increasing the removal of $CO_2$ from the atmosphere. In the stagnant waters of Carboniferous swamps, the dead bodies of trees did not decay completely. Eventually, their bodies were converted to coal, removing large quantities of $CO_2$ from the atmosphere. Coal was crucial to the Industrial Revolution, and people still burn 6 billion tons a year. It is ironic that coal, formed from plants that contributed to global cooling in the Carboniferous, now contributes to global warming by returning carbon to the atmosphere (see Figure 43.26).

The trees of early forests also contributed to the drop in $CO_2$ levels during the Carboniferous by the actions of their roots. The roots of vascular plants secrete acids that break down rocks, releasing calcium and magnesium into the soil. These chemicals react with carbon dioxide dissolved in rain water, forming compounds that ultimately wash into the oceans and are incorporated into rocks. The net effect of these processes—which were set in motion by plants—is that $CO_2$ is removed from the air and stored in marine rocks. Today, plants continue to affect carbon cycling and hence both the global climate and the extent of climate change (see Concepts 42.4 and 43.4).

## Biotic Interactions

The colonization of land by plants and fungi also had a dramatic effect on interactions between members of different species. Such biotic interactions include those in which both species benefit (mutualism) and those in which one species benefits while the other is harmed (as when a parasite feeds on its host).

Plants and fungi had such large effects on biotic interactions because their presence on land increased the availability of energy and nutrients for other organisms. For example, during photosynthesis, plants convert light energy to the chemical energy of food. That chemical energy supports all life on land, either directly (as when an insect eats a plant leaf) or indirectly (as when a bird eats an insect that ate a plant). Likewise, nitrogen and other nutrients are first absorbed by plants and then passed to organisms that eat plants; ultimately, these nutrients are returned to the environment by the actions of fungi and other decomposers. If plants and fungi had not colonized land, biotic interactions would still result in the transfer of energy and nutrients, but those transfers would likely occur on a much smaller scale, such as that of the "green slime" mentioned earlier in the chapter.

The previous paragraphs describe the big picture of how plants and fungi have affected biotic interactions. We'll close the chapter with several specific examples.

Fern    Lycophyte trees    Horsetail    Tree trunk covered with small leaves    Lycophyte tree reproductive structures

▲ **Figure 26.30 Artist's conception of a Carboniferous forest based on fossil evidence.** In addition to plants, animals, including giant dragonflies like the one in the foreground, also thrived in Carboniferous forests.

### Fungi as Mutualists and Pathogens

The different enzymes found in various fungal species can digest compounds from a wide range of sources, living or dead. This diversity of food sources corresponds to the varied roles of fungi in ecological communities, with different species living as decomposers, mutualists, or parasites. Having already described the importance of fungi as decomposers, we'll focus here on mutualism and parasitism.

Mutualistic fungi absorb nutrients from a host organism, but they reciprocate with actions that benefit the host—as in the case of the important mycorrhizal associations that fungi form with most vascular plants. In addition, all plant species studied to date appear to harbor symbiotic **endophytes**, fungi (or bacteria) that live inside leaves or other plant parts without causing harm.

## Do endophytes benefit a woody plant?

**Experiment** Endophytes are symbiotic fungi found within the bodies of all plants examined to date. Researchers tested whether endophytes benefit the cacao tree (*Theobroma cacao*). This tree, whose name means "food of the gods" in Greek, is the source of the beans used to make chocolate, and it is cultivated throughout the tropics. Endophytes were added to the leaves of some cacao seedlings, but not others. (In cacao, endophytes colonize leaves after the seedling germinates.) The seedlings were then inoculated with a virulent pathogen, the protist *Phytophthora*.

**Results** Fewer leaves were killed by the pathogen in seedlings with endophytes than in seedlings without endophytes. Among leaves that survived, pathogens damaged less of the leaf surface area in seedlings with endophytes than in seedlings without endophytes.

Endophyte not present; pathogen present (E–P+)

Both endophyte and pathogen present (E+P+)

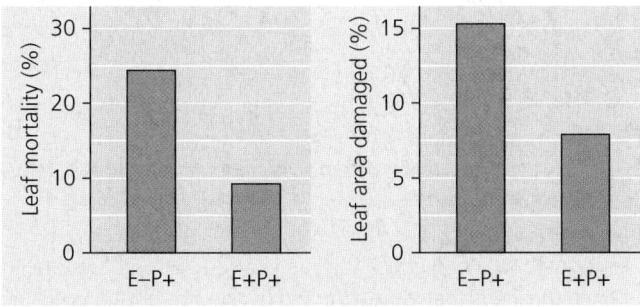

**Conclusion** Endophytes appear to benefit cacao trees by reducing the leaf mortality and damage caused by *Phytophthora*.

**Data from** A. E. Arnold et al., Fungal endophytes limit pathogen damage in a tropical tree, *Proceedings of the National Academy of Sciences USA* 100: 15649–15654 (2003).

**WHAT IF?** The researchers also performed control treatments. Suggest two controls they might have used, and explain how each would be helpful in interpreting the results described here.

Fungal endophytes have been shown to benefit certain grasses by making toxins that deter herbivores or by increasing host plant tolerance of heat, drought, or heavy metals. As described in **Figure 26.31**, researchers studying how fungal endophytes affect a woody plant tested whether leaf endophytes benefit seedlings of the cacao tree, *Theobroma cacao*. Their findings show that the fungal endophytes of woody flowering plants can play an important role in defending against pathogens.

Parasitic fungi also absorb nutrients from the cells of living hosts, but they provide no benefits in return **(Figure 26.32)**. Some parasitic fungi are pathogenic, including many species that cause diseases in plants. For example, *Cryphonectria parasitica*, the ascomycete fungus that causes chestnut blight, dramatically changed the landscape of the northeastern United States. Accidentally introduced via trees imported from Asia in the early 1900s, spores of the fungus enter cracks in the chestnut bark and produce hyphae, killing the tree. The once-common chestnuts now survive mainly as sprouts from the stumps of former trees.

**(b) Tar spot fungus on maple leaves**

**(a) Corn smut on corn**

▲ **Figure 26.32 Examples of fungal diseases of plants.** About 30% of the 100,000 known species of fungi make a living as parasites or pathogens, mostly of plants.

**(c) Ergots on rye**

## Plant-Animal Interactions

Plants and animals have interacted for hundreds of millions of years, and those interactions have led to evolutionary change. For example, herbivores can reduce a plant's reproductive success by eating its roots, leaves, or seeds. As a result, if an effective defense against herbivores originates in a group of plants, those plants may be favored by natural selection—as will any herbivores that can overcome this new defense.

Interactions between plants and animals also may have affected the rates at which new species form. Consider the impact of flower shape, which can be symmetric in one direction only (*bilateral symmetry*) or symmetric in all directions (*radial symmetry*). On a flower with bilateral symmetry, an insect pollinator may only be able to enter the flower from a certain direction. This constraint can make it more likely that as an insect moves from flower to flower, pollen is placed on a part of the insect's body that will come into contact with the stigma of a flower of the same species.

**Bilateral symmetry**

**Radial symmetry**

Such specificity of pollen transfer tends to reduce gene flow between diverging populations and hence could lead to increased rates of plant speciation in plants with bilateral symmetry. This hypothesis can be tested using the approach illustrated in this diagram:

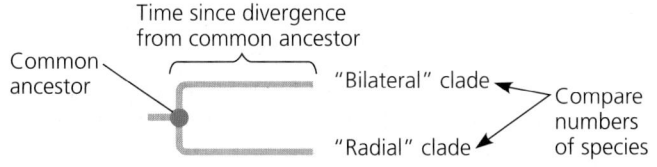

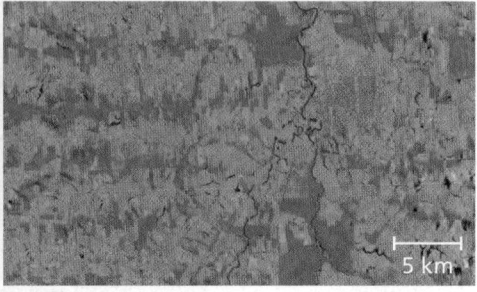

▲ **Figure 26.33 Clear-cutting of tropical forests.** Over the past several hundred years, nearly half of Earth's tropical forests have been cut down and converted to farmland and other uses. A satellite image from 1975 (left) shows a dense forest in Brazil. By 2012, much of this forest had been cut down. Deforested and urban areas are shown as light purple.

A key step is to identify cases in which a clade with bilaterally symmetric flowers shares an immediate common ancestor with a clade whose members have radially symmetric flowers. One recent study identified 19 such pairs of closely related "bilateral" and "radial" clades. On average, the clade with bilaterally symmetric flowers had nearly 2,400 more species than did its closely related clade with radially symmetric flowers. This result suggests that flower shape can affect the rate at which new species form—perhaps because of how flower shape affects the behavior of insect pollinators.

Angiosperms have been dominant members of Earth's ecological communities for over 100 million years. Although angiosperms remain dominant today, they and other plant groups are threatened by the exploding human population and its demand for space and resources. The problem is especially severe in the tropics, where more than two-thirds of the human population live and where population growth is fastest. About 63,000 km² (15 million acres) of tropical rain forest are cleared each year **(Figure 26.33)**, a rate that would completely eliminate the remaining 11 million km² of tropical forests in 175 years. As forests disappear, so do large numbers of plant species. Of course, once a species becomes extinct, it can never return.

The loss of plant species is often accompanied by the loss of insects and other rain forest animals. Scientists estimate that if current rates of loss in the tropics and elsewhere continue, 50% or more of Earth's species will become extinct within the next few centuries. Such losses would constitute a global mass extinction, rivaling the Permian and Cretaceous mass extinctions and changing the evolutionary history of life—including that of the animals, the group we'll turn to in Chapter 27.

### CONCEPT CHECK 26.5

1. Describe how terrestrial fungi and plants have affected the physical environment.
2. Discuss the importance of fungi as mutualists and parasites.
3. **MAKE CONNECTIONS** Figure 1.9 illustrates the transfer of energy and matter in ecosystems. Draw a simple diagram of energy flow and chemical cycling in a terrestrial ecosystem; circle the steps that were affected by the colonization of land by plants and fungi.
4. **WHAT IF?** Explain why researchers testing whether flower shape (bilateral versus radial) affected speciation rates only analyzed cases in which a bilateral clade shared an immediate common ancestor with a radial clade.

For suggested answers, see Appendix A.

---

# 26 | Chapter Review

Go to **MasteringBiology®** for Assignments, the eText, and the Study Area with Animations, Activities, Vocab Self-Quiz, and Practice Tests.

## SUMMARY OF KEY CONCEPTS

**VOCAB SELF-QUIZ**
goo.gl/gbai8v

### CONCEPT 26.1

**Fossils show that plants colonized land more than 470 million years ago (pp. 521–524)**

- Morphological and biochemical traits, as well as similarities in nuclear and chloroplast gene sequences, indicate that plants arose from a lineage of charophyte green algae.
- A protective layer of **sporopollenin** and other traits allow charophytes to tolerate occasional drying along the edges of ponds and lakes. Such traits may have enabled the algal ancestors of plants to survive in terrestrial conditions, opening the way to the colonization of dry land.
- Derived traits that distinguish plants from charophytes, their closest algal relatives, include **apical meristems**, **cuticles**, **stomata**, and the two shown here:

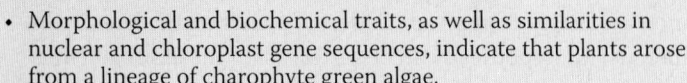

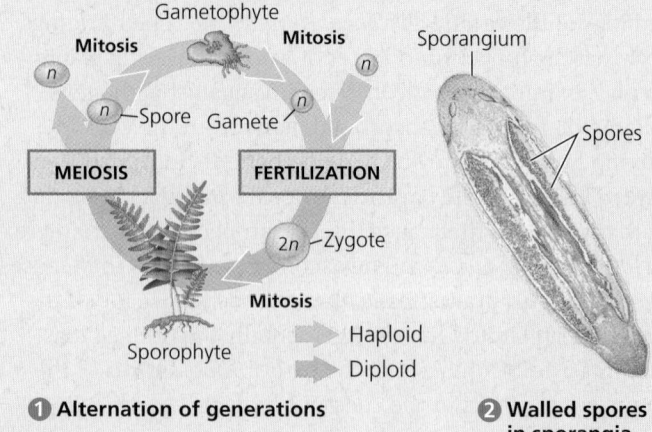

❶ **Alternation of generations**

❷ **Walled spores in sporangia**

- Fossil evidence indicates that plants were inhabiting land by 470 million years ago. By 400 million years ago, a diverse assemblage of fossil plant species lived on land, some of which had specialized tissues for water transport, cuticles, stomata, and branched sporophytes.

> **?** *The oldest fossil representing a large structure from a plant is 425 million years old, yet scientists think that plants colonized land 470 million years ago. What evidence supports this idea?*

CONCEPT 26.2

### Though not closely related to plants, fungi played a key role in the colonization of land (pp. 524–530)

- All fungi are heterotrophs that acquire nutrients by absorption. Many fungi secrete enzymes that break down complex molecules to smaller molecules that can be absorbed.
- Molecular data show that fungi arose from a single-celled protist and are more closely related to animals than to plants or most other eukaryotes.
- The cell walls of fungi are strengthened by **chitin**, a strong but flexible polysaccharide; these strong cell walls enable the cell to absorb nutrients and water without bursting.
- Most fungi grow as thin, multicellular filaments called **hyphae**; some species can grow as single-celled **yeasts**. In their multicellular form, fungi consist of **mycelia**, networks of branched hyphae adapted for absorption. Mycorrhizal fungi have specialized hyphae that enable them to form a mutually beneficial relationship with plants.
- Fungi typically propagate themselves by producing spores, either sexually or asexually. Spores can be transported by wind or water; if they are deposited in a moist place that has food, they germinate, producing new mycelia.
- Although fungi likely colonized land before plants, the earliest fossils of fungi date to 460 million years ago. Once on land, some fungi formed mycorrhizal associations with early plants—a symbiosis that probably helped plants without roots to colonize land.
- Since colonizing land, fungi have radiated into a diverse set of lineages.

> **?** *Explain how the morphology of multicellular fungi affects the efficiency of nutrient absorption and may have played a role in the colonization of land by plants.*

CONCEPT 26.3

### Early plants radiated into a diverse set of lineages (pp. 530–533)

- The three extant phyla of nonvascular plants, or **bryophytes**—liverworts, mosses, and hornworts—are basal plant lineages.
- In bryophytes, the dominant generation consists of haploid gametophytes, such as those that make up a carpet of moss. The flagellated sperm require a film of water to travel to the eggs.
- Fossils of the forerunners of today's **vascular plants** date back 420–425 million years and show that these small plants lacked seeds but had independent, branching sporophytes and a well-developed vascular system.
- Over time, other derived traits of living vascular plants arose, such as a life cycle with dominant sporophytes, lignified vascular tissue, and well-developed **roots** and **leaves**.
- **Seedless vascular plants** formed the first forests about 385 million years ago. Today, seedless vascular plants include

the **lycophytes** (club mosses and their relatives) and the **monilophytes** (ferns and their relatives).

> **?** *What trait(s) allowed vascular plants to grow tall, and why might increased height have been advantageous?*

CONCEPT 26.4

### Seeds and pollen grains are key adaptations for life on land (pp. 533–539)

- Derived traits of seed plants include **seeds** (which survive better than spores), highly reduced gametophytes (which are nourished and protected by the sporophyte), **ovules** (which house female gametophytes), and pollen (which eliminates dependency on water for fertilization).
- Seed plants originated 360 million years ago. Living seed plants can be divided into two monophyletic groups: **gymnosperms** and **angiosperms**. Gymnosperms appear early in the seed plant fossil record and dominated many terrestrial ecosystems until angiosperms (flowering plants) began to replace them 100 million years ago.
- **Flowers** typically have four whorls of modified leaves: **sepals, petals, stamens**, and **carpels**. **Ovaries** ripen into **fruits**, which often carry seeds by wind, water, or animals to new locations.

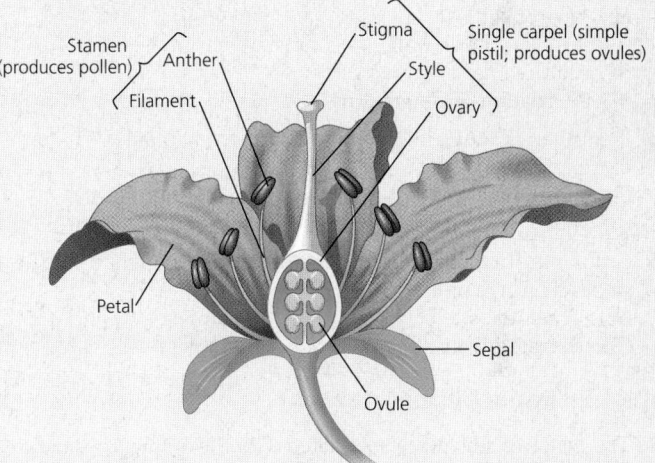

- Angiosperms arose and diversified greatly during the Cretaceous period. Fossils and phylogenetic analyses offer insights into the origin of flowering plants, which today are by far the largest group of extant plants. The two most diverse angiosperm clades are monocots and eudicots.

> **?** *Summarize fossil and phylogenetic evidence that suggests that the angiosperm common ancestor was likely woody.*

CONCEPT 26.5

### Plants and fungi fundamentally changed chemical cycling and biotic interactions (pp. 539–542)

- **Lichens** and plants affect soil formation. Plants also alter the composition of Earth's atmosphere by releasing oxygen to the air as a by-product of photosynthesis.
- Plants play a central role in chemical cycling by absorbing nutrients from the physical environment; those nutrients then pass to organisms that eat plants. Fungal decomposers break down the bodies of dead organisms; this returns nutrients to the physical environment, completing the cycle.

- Since colonizing land, the activities of plants and fungi have altered biotic interactions by increasing the availability of energy and nutrients for other organisms.
- Fungi play key ecological roles as decomposers, mutualists (such as **endophytes** that help protect plants from herbivores and pathogens), and parasites.
- Interactions between plants and animals have led to natural selection in plant and animal populations and may have affected speciation rates. Destruction of habitat threatens the extinction of many plant species and the animal species they support.

> **?** *Summarize how plants and fungi have increased the availability of energy and nutrients for other organisms, and explain how this affects biotic interactions.*

# TEST YOUR UNDERSTANDING

**PRACTICE TEST**

goo.gl/CRZjvS

## Level 1: Knowledge/Comprehension

1. *All* fungi are
   (A) symbiotic.
   (B) heterotrophic.
   (C) flagellated.
   (D) decomposers.

2. Which of the following characteristics of plants is absent in their closest relatives, the charophyte algae?
   (A) chlorophyll *b*
   (B) cellulose in cell walls
   (C) multicellularity
   (D) alternation of generations

3. Identify each of the following structures as haploid or diploid.
   (a) sporophyte      (c) gametophyte
   (b) spore           (d) zygote

4. Among the organisms listed here, which are thought to be the closest relatives of fungi?
   (A) slime molds     (C) animals
   (B) vascular plants  (D) mosses

## Level 2: Application/Analysis

5. The adaptive advantage associated with the filamentous nature of fungal mycelia is primarily related to
   (A) the ability to parasitize other organisms.
   (B) avoiding sexual reproduction until the environment changes.
   (C) the potential to inhabit almost all terrestrial habitats.
   (D) an extensive surface area well suited for invasive growth and absorptive nutrition.

6. **DRAW IT** Use the letters a–d to label where on the phylogenetic tree each of the following derived characters appears.
   (a) flowers         (c) seeds
   (b) embryos         (d) vascular tissue

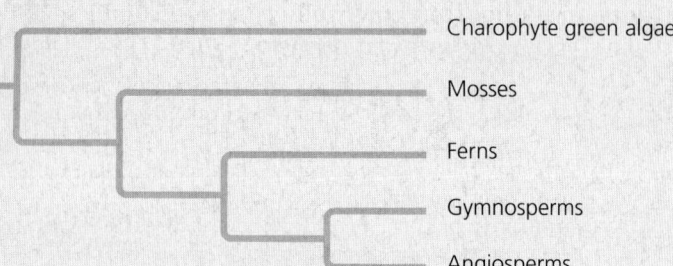

Charophyte green algae

Mosses

Ferns

Gymnosperms

Angiosperms

7. **SCIENTIFIC INQUIRY**
   **INTERPRET THE DATA** The grass *Dichanthelium lanuginosum* lives in hot soils and houses fungi of the genus *Curvularia* as endophytes. Researchers tested the impact of *Curvularia* on the heat tolerance of this grass. They grew plants without (E−) and with (E+) *Curvularia* endophytes at different temperatures and measured plant mass and the number of new shoots the plants produced. Draw a bar graph for plant mass versus temperature and interpret it.

| Soil Temp. | *Curvularia* + or − | Plant Mass (g) | No. of New Shoots |
|---|---|---|---|
| 30°C | E− | 16.2 | 32 |
|  | E+ | 22.8 | 60 |
| 35°C | E− | 21.7 | 43 |
|  | E+ | 28.4 | 60 |
| 40°C | E− | 8.8 | 10 |
|  | E+ | 22.2 | 37 |
| 45°C | E− | 0 | 0 |
|  | E+ | 15.1 | 24 |

**Data from** R. S. Redman et al., Thermotolerance generated by plant/fungal symbiosis, *Science* 298:1581 (2002).

## Level 3: Synthesis/Evaluation

8. **FOCUS ON EVOLUTION**
   The history of life has been punctuated by several mass extinctions. For example, the impact of a meteorite may have wiped out most of the dinosaurs and many forms of marine life at the end of the Cretaceous period (see Figure 23.12). Fossils indicate that plants were less severely affected by this mass extinction. What adaptations may have enabled plants to withstand this disaster better than animals?

9. **FOCUS ON INTERACTIONS**
   Giant lycophyte trees of Earth's early forests (see Figure 26.30) had microphylls, whereas ferns and seed plants have megaphylls. Write a short essay (100–150 words) describing how a forest of lycophyte trees may have differed from a forest of large ferns or seed plants. In your answer, consider how the type of forest in which they grew may have affected interactions among small plants growing beneath the tall ones.

10. **SYNTHESIZE YOUR KNOWLEDGE**

These stomata are from the leaf of a common horsetail. Describe how stomata and other adaptations facilitated life on land and ultimately led to the formation of the first forests.

*For selected answers, see Appendix A.*

# 27 The Rise of Animal Diversity

**KEY CONCEPTS**

27.1 Animals originated more than 700 million years ago

27.2 The diversity of large animals increased dramatically during the "Cambrian explosion"

27.3 Diverse animal groups radiated in aquatic environments

27.4 Vertebrates have been the ocean's dominant predators for more than 400 million years

27.5 Several animal groups had features facilitating their colonization of land

27.6 Amniotes have key adaptations for life in a wide range of terrestrial environments

27.7 Animals have transformed ecosystems and altered the course of evolution

▲ **Figure 27.1 What adaptations make a chameleon a fearsome predator?**

## Life Becomes Dangerous

Although slow-moving on its feet, the chameleon in **Figure 27.1** can wield its long, sticky tongue with blinding speed to capture unsuspecting prey. Other animals overwhelm their prey using their strength, speed, or toxins, while still others build traps or blend into their surroundings, enabling them to capture unwary prey. And hunting animals are not the only ones that pose a threat to other organisms. Herbivorous animals can strip the plants they eat bare of leaves or seeds, while parasitic animals weaken their hosts by consuming their tissues or body fluids.

As these examples suggest, animals can make life dangerous for the organisms around them. Most animals are mobile and can detect, capture, and eat other organisms—including those that are themselves mobile and can flee from attack. Indeed, all but the simplest animals have specialized muscle and nerve cells that allow them to move and respond rapidly to changing environmental conditions. Most animals also have a complete digestive tract, which has a mouth at one end and an anus at the other. This type of digestive system enables animals to efficiently break down most organic compounds. Together, their mobility, nervous system, and digestive tract, accompanied by

often complex behaviors, make animals highly effective eating machines.

Animals are so integral to life today that it is difficult to imagine what Earth would be like without them. The fossil record, however, paints an intriguing picture. Large eukaryotes were once soft-bodied and lived in a relatively safe world—until the appearance of animals changed everything. In this chapter, we'll examine how animals have evolved over time and influenced the world around them.

## CONCEPT 27.1

## Animals originated more than 700 million years ago

Current evidence indicates that animals evolved from single-celled eukaryotes similar to present-day choanoflagellates (see Figure 25.7). The descendants of these early animals include a vast diversity of living animal species: To date, biologists have named more than 1.3 million species, and estimates of the actual number run far higher—nearly 8 million species according to one recent study. When did this diverse group originate?

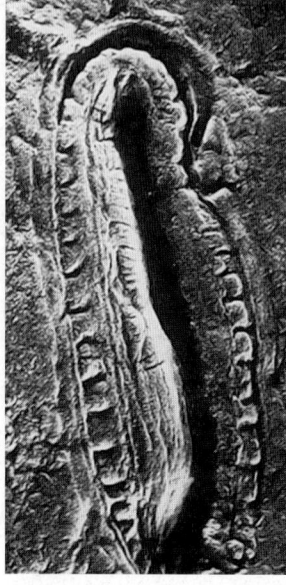

**(a)** *Dickinsonia costata* (taxonomic affiliation unknown)

⊢———⊣ 2.5 cm

▲ **Figure 27.2 Ediacaran fossils.** Fossils dating to about 560 million years ago include the earliest macroscopic fossils of animals, including these two species. Earlier members of the Ediacaran biota include the alga *Doushantuophyton* (see Figure 25.2).

**(b)** *Kimberella*, a mollusc

⊢———⊣ 1 cm

## Fossil and Molecular Evidence

Researchers have unearthed 710-million-year-old sediments containing chemical evidence of steroids that today are primarily produced by a particular group of sponges. Since sponges are animals, these "fossil steroids" suggest that animals had arisen by 710 million years ago.

DNA analyses generally agree with this fossil biochemical evidence; for example, one recent molecular clock study estimated that sponges originated about 700 million years ago. These findings are also consistent with molecular analyses

suggesting that the common ancestor of all extant (living) animal species lived about 770 million years ago.

Despite the data from molecular clocks and fossil steroids indicating an earlier origin, the first generally accepted macroscopic fossils of animals date from about 560 million years ago **(Figure 27.2)**. These fossils are members of an early group of soft-bodied multicellular eukaryotes known collectively as the **Ediacaran biota**. The name comes from the Ediacara Hills of Australia, where fossils of these organisms were first discovered. Similar fossils have since been found on other continents. Among the Ediacaran fossils that resemble animals, some may be sponges, while others are thought to be cnidarians (sea anemones and their relatives) or molluscs (snails and their relatives). Still others are difficult to classify, as they do not seem to be closely related to any living animals or algae.

## Early-Diverging Animal Groups

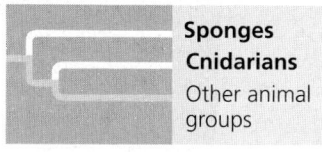

Sponges
Cnidarians
Other animal groups

Based on fossil evidence and molecular analyses, scientists have identified certain animal groups as having diverged from all other animals early in animal evolution. Two such groups are sponges and cnidarians.

### Sponges

Animals in the phylum Porifera are known informally as sponges **(Figure 27.3)**. (Recent phylogenomic studies indicate that sponges are monophyletic, and that is the phylogeny we present here; this remains under debate, however, as some

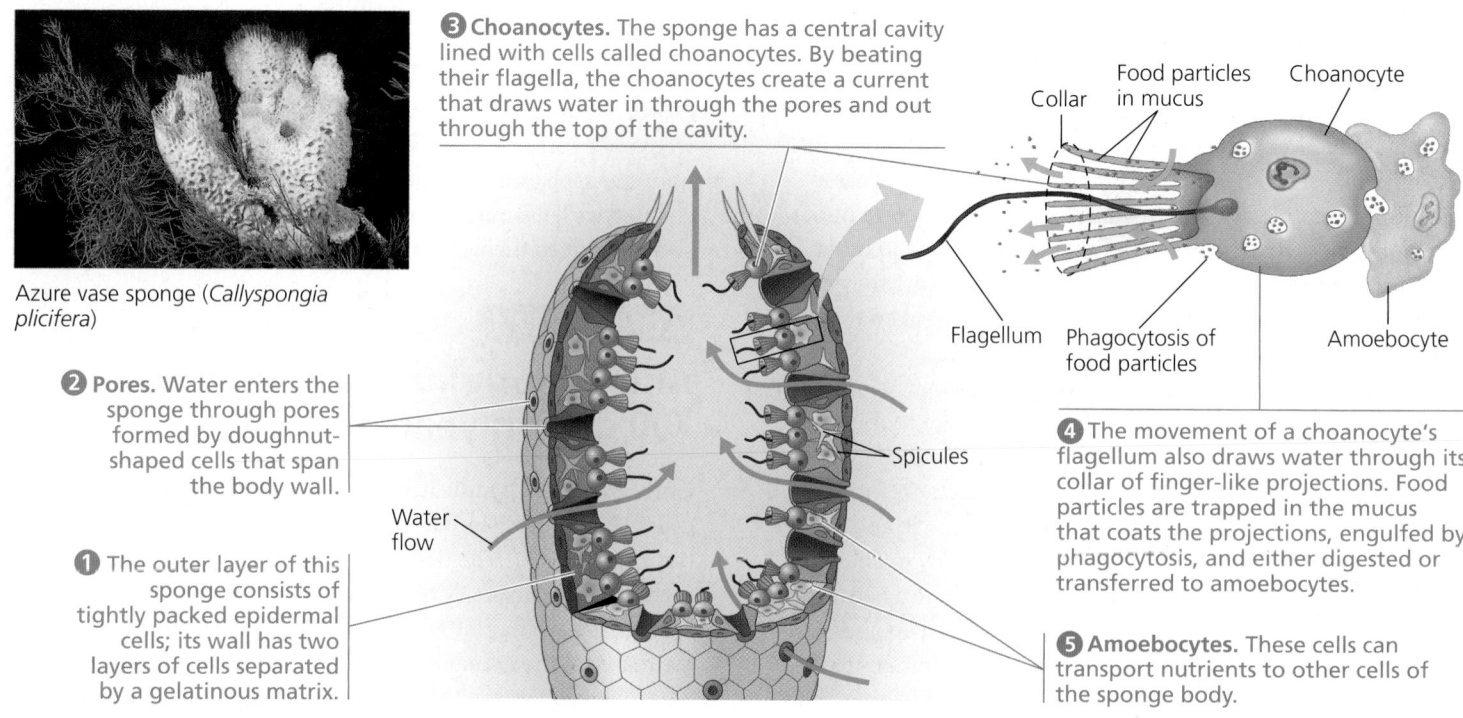

❸ **Choanocytes.** The sponge has a central cavity lined with cells called choanocytes. By beating their flagella, the choanocytes create a current that draws water in through the pores and out through the top of the cavity.

Azure vase sponge (*Callyspongia plicifera*)

❷ **Pores.** Water enters the sponge through pores formed by doughnut-shaped cells that span the body wall.

Water flow

❶ The outer layer of this sponge consists of tightly packed epidermal cells; its wall has two layers of cells separated by a gelatinous matrix.

Spicules

Collar
Food particles in mucus
Choanocyte

Flagellum
Phagocytosis of food particles
Amoebocyte

❹ The movement of a choanocyte's flagellum also draws water through its collar of finger-like projections. Food particles are trapped in the mucus that coats the projections, engulfed by phagocytosis, and either digested or transferred to amoebocytes.

❺ **Amoebocytes.** These cells can transport nutrients to other cells of the sponge body.

▲ **Figure 27.3 Anatomy of a sponge.**

studies suggest that sponges are paraphyletic.) Among the simplest of animals, sponges are sedentary and were mistaken for plants by the ancient Greeks. Most species are marine, and they range in size from a few millimeters to a few meters. Sponges are **filter feeders**: They filter out food particles suspended in the surrounding water as they draw the water through their body (see Figure 27.3).

Sponges represent a lineage that originates near the root of the phylogenetic tree of animals; thus, they are said to be *basal animals*. Unlike nearly all other animals, sponges lack **tissues**, groups of similar cells that act as a functional unit. However, the sponge body does contain several different cell types. For example, the interior of the body is lined with flagellated **choanocytes**, or collar cells (named for the finger-like projections that form a "collar" around the flagellum). These cells engulf bacteria and other food particles by phagocytosis. Choanocytes resemble the cells of choanoflagellates, a finding that is consistent with the similarities between the DNA sequences of sponges and those of choanoflagellates. Together, these results suggest that animals may have evolved from a choanoflagellate-like ancestor (see Figure 25.7). Sponges also have mobile cells called **amoebocytes**, named for their use of pseudopodia. As these cells move through the sponge body, they take up food from the surrounding water and from choanocytes, digest it, and carry nutrients to other cells.

### Cnidarians

All animals except sponges and a few other groups are *eumetazoans* ("true animals"), members of a clade of animals that have tissues. One of the first groups to have diverged from all other eumetazoans is the phylum Cnidaria, which originated about 680 million years ago according to molecular clock analyses. Cnidarians have diversified into a wide range of sessile and motile forms, including hydrozoans, jellies, and sea anemones **(Figure 27.4)**.

The basic morphology of a cnidarian is a sac with a central digestive compartment, the **gastrovascular cavity**. A single opening to this cavity functions as both mouth and anus. Cnidarians are carnivores that typically use tentacles arranged in a ring around their mouth to capture prey and pass the food into their gastrovascular cavity. Enzymes are then secreted into the cavity, breaking down the prey into a nutrient-rich broth. Cells lining the cavity then absorb these nutrients and complete the digestive process; any undigested remains are expelled through the cnidarian's mouth/anus.

Muscles and nerves occur in their simplest forms in cnidarians. Movements are coordinated by a noncentralized nerve net. Cnidarians have no brain, and the nerve net is associated with sensory structures distributed around the body. Thus, the animal can detect and respond to stimuli from all directions.

#### CONCEPT CHECK 27.1

1. Summarize fossil and DNA evidence documenting the origin and early diversification of animals.
2. **MAKE CONNECTIONS** Unlike nearly all other animals, sponges lack tissues. Are sponges therefore less successful or less "highly evolved" than other animal groups? Explain. (See Concept 23.4.)

For suggested answers, see Appendix A.

### CONCEPT 27.2

# The diversity of large animals increased dramatically during the "Cambrian explosion"

As we've seen, the oldest fossils of large animals date to 560 million years ago and include members of just a few extant groups—sponges, cnidarians, and molluscs. In fossils formed in the early Cambrian period (between 535 and 525 million years ago), large forms of many other present-day animal phyla suddenly appear, a phenomenon referred to as the **Cambrian explosion**. What factors may have spurred this rapid (in geologic terms) increase in the diversity of large animals?

### Evolutionary Change in the Cambrian Explosion

Strata formed during the Cambrian explosion contain the oldest fossils of about half of all extant animal phyla, including the first arthropods, chordates, and echinoderms

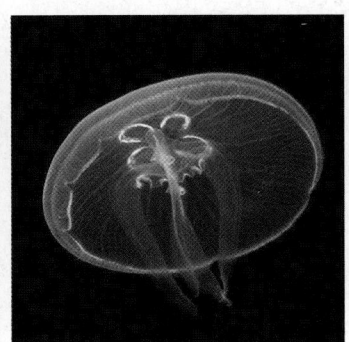

**(a) Hydrozoa.** Some species, such as this one, live as colonial polyps.

**(b) Scyphozoa.** Many jellies (commonly called jellyfish) are bioluminescent. Some species stun their prey with specialized stinging cells called nematocysts located on their tentacles.

**(c) Anthozoa.** Sea anemones and other anthozoans exist only as polyps. Many anthozoans form symbiotic relationships with photosynthetic algae.

▲ **Figure 27.4 Major groups of cnidarians.**

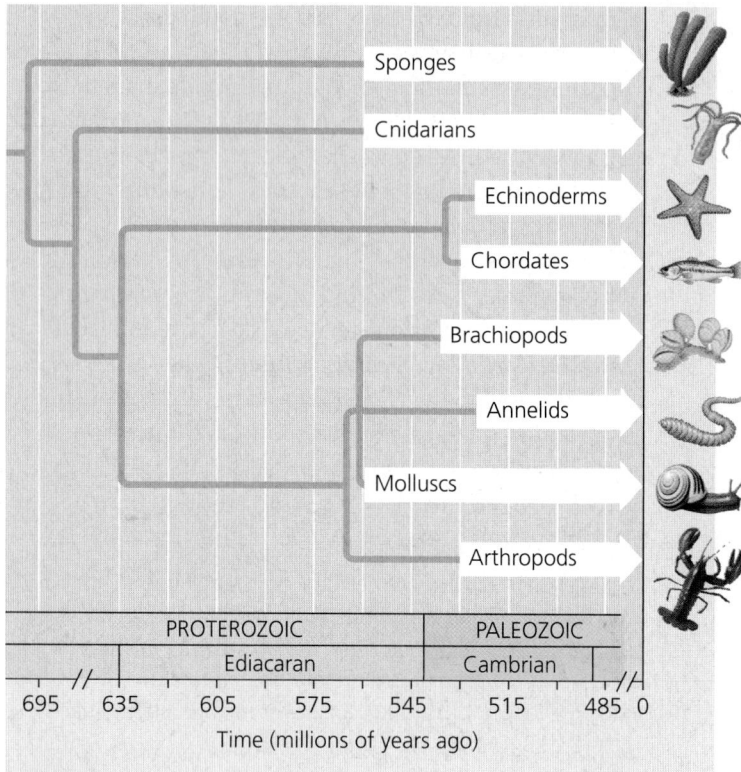

**▲ Figure 27.5 Appearance of selected animal groups.** The white bars indicate earliest appearances of these animal groups in the fossil record.

**DRAW IT** *Circle the branch point that represents the most recent common ancestor of chordates and annelids. What is a minimum estimate of that ancestor's age?*

**(Figure 27.5).** Many of these fossils, which include the first animals with hard, mineralized skeletons, look very different from most living animals **(Figure 27.6)**. Even so, paleontologists have established that these Cambrian fossils are members of extant animal phyla, or at least are close relatives. In particular, most of the fossils from the Cambrian explosion are of **bilaterians**, an enormous clade whose members (unlike sponges and cnidarians) typically have a complete digestive tract and a two-sided (bilaterally symmetric) form. As we'll discuss later in the chapter, bilaterians include molluscs, arthropods, chordates, and most other living animal phyla.

Entirely new sorts of animals made their debut during the Cambrian explosion. Previously, virtually all large animals were soft-bodied. In addition, the fossils of large pre-Cambrian animals reveal little evidence of predation. Instead, these animals seem to have been grazers (feeding on mats of algae and bacteria), filter feeders, or scavengers, not hunters.

In a relatively short period of time (10 million years), predators emerged that were over 1 m long and had claws and other features for capturing prey; simultaneously, new defensive adaptations, such as sharp spines and heavy body armor, appeared in their prey (see Figure 27.6). These and other changes set the stage for many of the key events in the history of life over the last 500 million years.

The increase in the diversity of large animals during the Cambrian explosion was accompanied by a decline in the diversity of Ediacaran life-forms. What caused these trends? Fossil evidence suggests that during the Cambrian period, predators acquired novel adaptations, such as forms of locomotion that helped them catch prey, while prey species acquired new defenses, such as protective shells. As new predator-prey relationships emerged, natural selection may have led to the decline of soft-bodied Ediacaran species and to the rise of various bilaterian phyla. Another hypothesis focuses on an increase in atmospheric oxygen that preceded the Cambrian explosion. More plentiful oxygen would have enabled animals with higher metabolic rates and larger body sizes to thrive, while potentially harming other species. A third hypothesis proposes that the origin of *Hox* genes (see Concept 23.3) and other genetic changes affecting the regulation of developmental genes facilitated the evolution of new body forms. These hypotheses are not mutually exclusive, however; predator-prey relationships, atmospheric changes, and changes in the regulation of development may each have played a role.

## Dating the Origin of Bilaterians

Although the radiation of bilaterians during the Cambrian explosion had an enormous impact on life on Earth, it is possible

530-million-year-old fossil of *Hallucigenia*

**▲ Figure 27.6 A Cambrian seascape.** This artist's reconstruction depicts a diverse array of organisms found in fossils from the Burgess Shale site in British Columbia, Canada. The animals include *Pikaia* (eel-like chordate at top left), *Marella* (arthropod swimming at left), *Anomalocaris* (large animal with anterior grasping limbs and a circular mouth), and *Hallucigenia* (animals with toothpick-like spikes on the seafloor and in inset).

that many animal phyla originated long before that time. As we've seen, molecular clock analyses suggest that sponges and cnidarians had evolved by 680–700 million years ago. Molecular estimates also suggest that bilaterians had evolved by 670 million years ago, which was 135 million years *before* the Cambrian explosion.

Turning to the fossil record, fossil steroids corroborate the molecular dates for the origin of sponges. However, no fossil bilaterians are close in age to the molecular clock estimates for when this group originated. The oldest fossil bilaterian is the mollusc *Kimberella* (see Figure 27.2), which lived 560 million years ago. Thus, the fossil evidence differs from molecular clock estimates by more than 100 million years.

Seeking to resolve this discrepancy, researchers have taken a closer look at the fossil record from the Ediacaran period (635–541 million years ago). Prior to the Ediacaran, eukaryotes were microscopic and smooth-walled, and such species often persisted in the fossil record for hundreds of millions of years **(Figure 27.7a)**. Then eukaryotic life changed dramatically. Some eukaryotic lineages gave rise to large organisms, such as the 600-million-year-old alga shown in Figure 25.2. Organisms in other eukaryotic lineages remained relatively small, but defensive structures such as spines began to appear on their outer surfaces **(Figure 27.7b)**. Additional fossil evidence shows that new species of these well-defended eukaryotes originated more frequently but persisted in the fossil record for shorter periods of time than did their smooth-walled, pre-Ediacaran counterparts.

What triggered these dramatic changes? Recall from the beginning of this chapter that present-day animals are dangerous feeding machines because of their mobility, nervous system, and efficient digestive tract. Nearly all bilaterians have these features, suggesting that early bilaterians did as well. With mobility, a nervous system, and an efficient digestive tract, early bilaterians would have decimated populations of the small, soft-bodied organisms on which they fed. Thus, the feeding activities of early bilaterians may have resulted in

natural selection for increased size or new defensive structures in the organisms that they ate—exactly the change seen in the fossil record during the Ediacaran period.

Overall, the fossil record and molecular clock data suggest that bilaterians arose sometime between 635 and 670 million years ago. Possibly aided by a later rise in the atmospheric concentration of oxygen, these early bilaterians then diversified explosively during the Cambrian and beyond.

**CONCEPT CHECK 27.2**

1. What is the "Cambrian explosion"? Why is it significant?
2. **WHAT IF?** Suppose a well-defended prey species arose that was difficult for predators to catch or eat. How might this affect ongoing evolutionary changes in predator and prey populations?

For suggested answers, see Appendix A.

---

**CONCEPT 27.3**

# Diverse animal groups radiated in aquatic environments

By the end of the Cambrian explosion, many of the big steps in animal evolution were well under way. Animals with legs or leg-like appendages walked on the ocean floor, and worms burrowed through the sediments. Swimming in the waters above were predators that used sharp claws and mandibles to capture and break apart their prey. Other animals had protective spikes or armor, as well as modified mouthparts that enabled their bearers to filter food from the water.

As these examples suggest, the animals in early Cambrian oceans were very diverse in morphology, way of life, and taxonomic affiliation. We'll examine that diversity here, beginning with an overview of how to categorize the morphological variation found in different animal groups.

## Animal Body Plans

The diversity in form of the animals that emerged from the Cambrian explosion consists of relatively few major "body plans." A **body plan** is a particular set of morphological and developmental traits, integrated into a functional whole—the living animal. Note that the term *plan* here does not imply that animal forms are the result of conscious planning or invention. But body plans do provide a succinct way to compare and contrast key animal features. We'll focus on three aspects of animal body plans: symmetry, tissues, and body cavities.

### Symmetry

A basic feature of animal bodies is their type of symmetry— or absence of symmetry. (Many sponges, for example, lack symmetry altogether.) Some animals exhibit **radial symmetry**, the type of symmetry found in a flowerpot

**(a)** *Valeria* (800 mya): roughly spherical, no structural defenses, soft-bodied

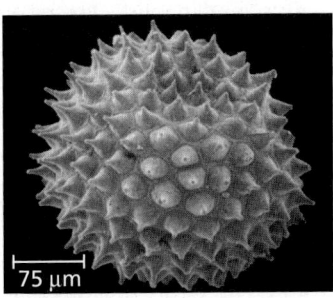

**(b)** Spiny acritarch (575 mya): about five times larger than *Valeria* and covered in hard spines

▲ **Figure 27.7 Indirect evidence of bilaterians?** The rise in the fossil record of larger, well-defended eukaryotes during the Ediacaran period (635–541 million years ago) suggests that bilaterian animals with a complete digestive tract may have originated by that time.

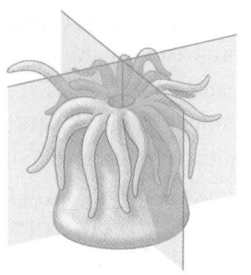
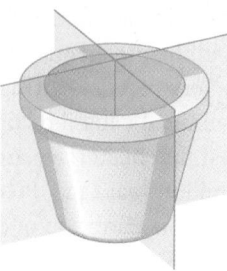

**(a) Radial symmetry.** A radial animal, such as a sea anemone (phylum Cnidaria), does not have a left side and a right side. Any imaginary slice through the central axis divides the animal into mirror images.

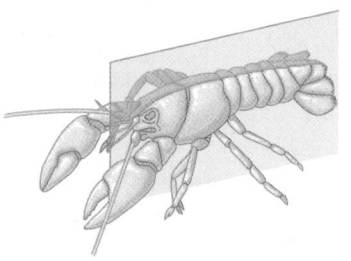
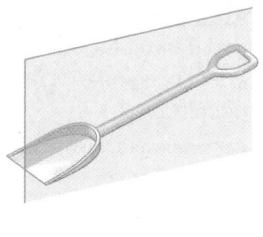

**(b) Bilateral symmetry.** A bilateral animal, such as a lobster (phylum Arthropoda), has a left side and a right side. Only one imaginary cut divides the animal into mirror-image halves.

▲ **Figure 27.8 Body symmetry.** The flowerpot and shovel are included to help you remember the radial-bilateral distinction.

**(Figure 27.8a).** Sea anemones, for example, have a top side (where the mouth is located) and a bottom side. But they have no front and back ends and no left and right sides.

By contrast, the two-sided symmetry of a shovel is an example of **bilateral symmetry (Figure 27.8b).** A bilateral animal has two axes of orientation: front to back and top to bottom. Such animals have a **dorsal** (top) side and a **ventral** (bottom) side, a left side and a right side, and an **anterior** (front) end and a **posterior** (back) end. Nearly all animals with a bilaterally symmetric body plan (such as arthropods and mammals) have sensory equipment concentrated at their anterior end, including a central nervous system ("brain") in the head.

The symmetry of an animal generally fits its lifestyle. Many radial animals are sessile (living attached to a substrate) or planktonic (drifting or weakly swimming, such as jellies). Their symmetry equips them to meet the environment equally well from all sides. In contrast, bilateral animals typically move actively from place to place. Most bilateral animals have a central nervous system that enables them to coordinate the complex movements involved in crawling, burrowing, flying, or swimming.

## Tissues

Animal body plans also vary with regard to tissue organization. Recall that tissues are collections of specialized cells that act as a functional unit; in animals, true tissues are isolated from other tissues by membranous layers. Sponges and a few other groups lack true tissues. In all other animals, the embryo becomes layered during development; these layers, called *germ layers*, form the various tissues and organs of the body **(Figure 27.9).**

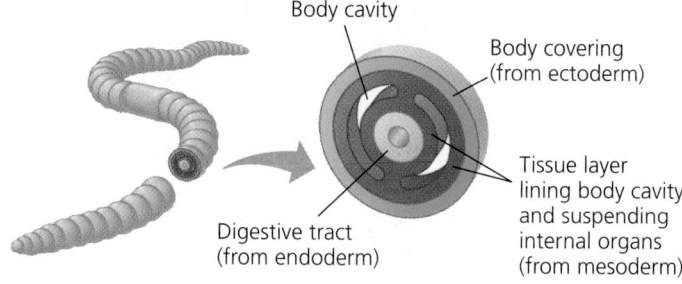

▲ **Figure 27.9 Tissue layers in bilaterians.** The organ systems of a bilaterally symmetric animal develop from the three germ layers that form in the embryo. Blue represents tissue derived from ectoderm, red from mesoderm, and yellow from endoderm. The internal organs of most bilaterians are suspended in a "body cavity," a fluid- or air-filled space that helps protect the organs from injury.

**Ectoderm,** the germ layer covering the surface of the embryo, gives rise to the outer covering of the animal and, in some phyla, to the central nervous system. **Endoderm,** the innermost germ layer, gives rise to the lining of the digestive tract (or cavity) and organs such as the liver and lungs of vertebrates.

Cnidarians and a few other animal groups have only these two germ layers. In contrast, all bilaterally symmetric animals have a third germ layer, called the **mesoderm,** which fills much of the space between the ectoderm and endoderm. In bilaterally symmetric animals, the mesoderm forms the muscles and most other organs between the digestive tract and the outer covering of the animal.

### Body Cavities

Most bilaterians have a **body cavity,** a fluid- or air-filled space located between the digestive tract and the outer body wall (see Figure 27.9). This body cavity is also called a *coelom*. The inner and outer layers of tissue that surround the cavity connect and form structures that suspend the internal organs.

A body cavity has many functions. Its fluid cushions the suspended organs, helping to prevent internal injury. In soft-bodied bilaterians, such as earthworms, the coelom contains noncompressible fluid that acts like a skeleton against which muscles can work. The cavity also enables the internal organs to grow and move independently of the outer body wall. If it were not for your coelom, every beat of your heart or ripple of your intestine would warp your body's surface.

## The Diversification of Animals

As animals with different body plans radiated in the early Cambrian, some lineages arose, thrived for a period of time, and then became extinct, leaving no descendants. However, by 500 million years ago, most animal phyla with members alive today were established.

Evolutionary relationships among living animals provide a helpful framework for studying the rise of animals. These relationships have been estimated using ribosomal RNA (rRNA) genes, *Hox* genes, and hundreds of protein-coding nuclear genes, as well as mitochondrial genes and morphological

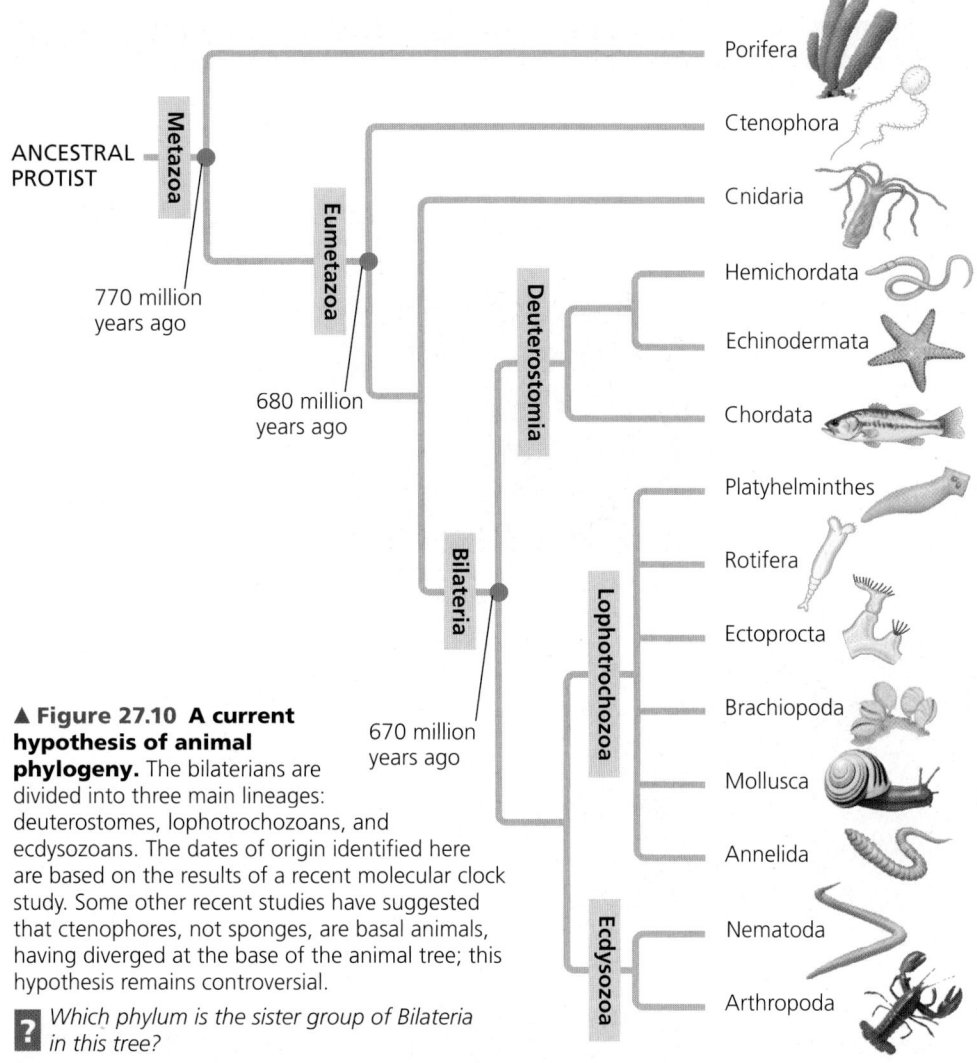

**▲ Figure 27.10 A current hypothesis of animal phylogeny.** The bilaterians are divided into three main lineages: deuterostomes, lophotrochozoans, and ecdysozoans. The dates of origin identified here are based on the results of a recent molecular clock study. Some other recent studies have suggested that ctenophores, not sponges, are basal animals, having diverged at the base of the animal tree; this hypothesis remains controversial.

**?** *Which phylum is the sister group of Bilateria in this tree?*

5. **Most animals are invertebrates.** The members of most animal phyla are **invertebrates**, animals that lack a backbone. Only one animal phylum, Chordata, includes **vertebrates**, animals with a backbone.

With the phylogeny in Figure 27.10 providing the overall context for the rise of animals, let's examine the diversification of bilaterians during the Cambrian in more detail. We'll begin here with the radiation of bilaterian invertebrates; we'll turn to the radiation of aquatic vertebrates in Concept 27.4.

## Bilaterian Radiation I: Diverse Invertebrates

As shown in Figure 27.10, bilaterian animals have diversified into three major clades: Lophotrochozoa, Ecdysozoa, and Deuterostomia. The species in these clades dominated life in the Cambrian oceans—and initially, at least, all of these species were invertebrates.

### An Overview of Invertebrate Diversity

Bilaterian invertebrates account for 95% of known animal species. They occupy almost every habitat on Earth, from the scalding water released by deep-sea hydrothermal vents to the frozen ground of Antarctica. Evolution in these varied environments has produced an immense diversity of forms, ranging from tiny worms with a flat body shape to species with features such as silk-spinning glands, pivoting spines, and tentacles covered with suction cups. Bilaterian invertebrates also show enormous variation in size, from microscopic organisms to organisms that can grow to 18 m long (1.5 times the length of a school bus).

The morphological diversity found in invertebrate animals is mirrored by their taxonomic diversity: There are literally millions of species of invertebrates. The vast majority of these species are members of two of the bilaterian clades that emerged from the Cambrian explosion: Lophotrochozoa and Ecdysozoa **(Figure 27.11)**. The third major bilaterian clade, Deuterostomia, also includes some invertebrates.

We can illustrate the radiation of invertebrates further using the example of Lophotrochozoa, the bilaterian clade whose members show the greatest range in body form. This diversity of form is reflected in the number of phyla classified in the group: Lophotrochozoa includes 18 animal phyla, more than twice the number in any other clade of bilaterians. In addition,

traits. Zoologists currently recognize about three dozen animal phyla, 14 of which are shown in **Figure 27.10**. Notice how the following points are reflected in this phylogeny.

1. **All animals share a common ancestor.** Current evidence indicates that animals are monophyletic, forming a clade called Metazoa: All extant and extinct animal lineages have descended from a common ancestor.

2. **Sponges are basal animals.** Among the extant taxa, sponges (phylum Porifera) branch from the base of the animal tree.

3. **Eumetazoa is a clade of animals with true tissues.** All animals except for sponges and a few others belong to a clade of **eumetazoans** ("true animals"). True tissues evolved in the common ancestor of living eumetazoans. Some eumetazoans, such as ctenophores and cnidarians, have two germ layers and generally have radial symmetry.

4. **Most animal phyla belong to the clade Bilateria.** Bilateral symmetry and the presence of three germ layers are shared derived characters that help define the clade Bilateria. This clade contains the majority of animal phyla, and its members are known as bilaterians. The Cambrian explosion was primarily a rapid diversification of bilaterians.

**Exploring the Diversity of Invertebrate Bilaterians**

The overwhelming majority of the 1.3 million known animal species are bilaterians, and most bilaterians are invertebrates. This figure highlights just a few of the invertebrate groups within the three major clades of bilaterians: Lophotrochozoa, Ecdysozoa, and Deuterostomia.

## Lophotrochozoa

Ectoprocts

**Ectoprocta (4,500 species)** Ectoprocts (also known as bryozoans) live as sessile colonies. Most species have a hard exoskeleton studded with pores; ciliated tentacles extend through the pores and trap food particles from the surrounding water.

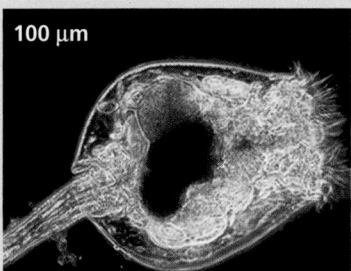

100 μm

A rotifer (LM)

**Rotifera (1,800 species)** Despite their microscopic size, rotifers have specialized organ systems, including a complete digestive tract with both a mouth and an anus. They feed on microorganisms.

**Annelida (16,500 species)** Annelids, or segmented worms, are distinguished from other worms by their body segmentation. Earthworms are the most familiar annelids, but the phylum consists primarily of marine and freshwater species.

A fireworm, a marine annelid

## Ecdysozoa

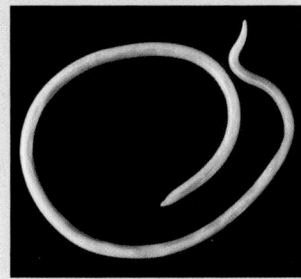

A roundworm

**Nematoda (25,000 species)** Also called roundworms, nematodes are enormously abundant and diverse in the soil and in aquatic habitats; many species parasitize plants and animals. Their most distinctive feature is a tough cuticle that coats their body.

**Arthropoda (1,000,000 species)** The vast majority of known animal species, including insects, millipedes, crabs, and arachnids, are arthropods. All arthropods have a segmented exoskeleton and jointed appendages.

A web-building spider (an arachnid)

## Deuterostomia

An acorn worm

**Hemichordata (85 species)** Hemichordates share some traits with chordates, such as gill slits and a dorsal nerve cord. The largest group of hemichordates is the acorn worms, marine animals that may grow to more than 2 m long.

**Echinodermata (7,000 species)** Echinoderms, such as sea stars, sea urchins, and sand dollars, are marine animals that are bilaterally symmetric as larvae but not as adults. They move and feed using unique "tube feet" whose gripping action results from the secretion of adhesive chemicals.

Sea urchins and a sea star

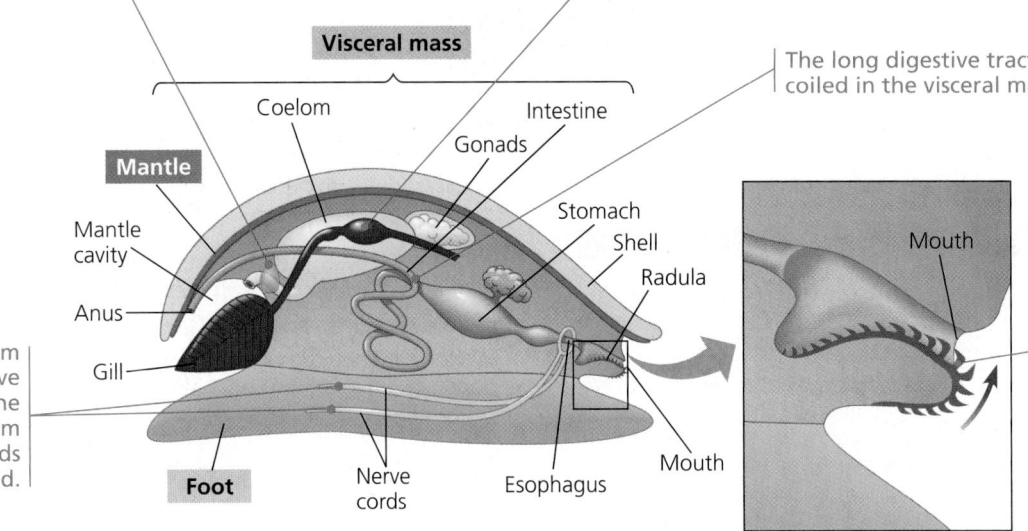

**Metanephridium.** Excretory organs called metanephridia remove metabolic wastes from the hemolymph.

**Heart.** Most molluscs have an open circulatory system. The dorsally located heart pumps circulatory fluid called hemolymph through arteries into sinuses (body spaces). The organs of the mollusc are thus continually bathed in hemolymph.

Visceral mass

Coelom

Intestine

The long digestive tract is coiled in the visceral mass.

Gonads

Mantle

Stomach

Mantle cavity

Shell

Radula

Mouth

**Radula.** The mouth region in many mollusc species contains a rasp-like feeding organ called a radula. This belt of backward-curved teeth repeatedly thrusts outward and then retracts into the mouth, scraping and scooping like a backhoe.

Anus

The nervous system consists of a nerve ring around the esophagus, from which nerve cords extend.

Gill

Foot

Nerve cords

Esophagus

Mouth

▲ **Figure 27.12 The body plan of a mollusc.**

the species within many of these 18 phyla vary greatly in body form. Consider Mollusca (the molluscs), a lophotrochozoan phylum that includes 100,000 different species. The body of a mollusc has three main parts **(Figure 27.12)**: a muscular **foot**, usually used for movement; a **visceral mass** containing most of the internal organs; and a **mantle**, a fold of tissue that drapes over the visceral mass and secretes a shell (if one is present). Although all molluscs have a foot, visceral mass, and mantle, they vary considerably in size and body form **(Figure 27.13)**.

The molluscs, along with the seven phyla shown in Figure 27.11, serve here as representatives of the great diversity of invertebrate bilaterians. Next, we'll examine the origin of one of these phyla, Arthropoda, the most species-rich (by far) of all animal groups. We focus on this group because its members were among the first animals to colonize land (see Concept 27.5).

### Arthropod Origins

Zoologists estimate that there are about a billion billion ($10^{18}$) arthropods living on Earth. More than 1 million arthropod species have been described, most of which are insects. In fact, two out of every three known species are arthropods, and members of this group can be found in nearly all habitats of the biosphere. By the criteria of species diversity, distribution, and sheer numbers, arthropods must be regarded as the most successful of all animal phyla.

Biologists hypothesize that the diversity and success of **arthropods** are related to their body plan—their segmented body, hard exoskeleton, and jointed appendages (*arthropod* means "jointed feet"). How did this body plan arise and what advantages did it provide?

The earliest fossils of arthropods are from the Cambrian explosion (535–525 million years ago), indicating that the arthropods are at least that old. The fossil record of the

◄ **Nudibranchs**, or sea slugs, lost their shell over the course of their evolution.

► **Bivalves** get their name from their two-part, hinged shells. Each half of this scallop's shell is lined with many dark blue eyes.

◄ **Cephalopods** include octopi, which are considered among the most intelligent invertebrates.

▲ **Figure 27.13 Examples of morphological diversity in molluscs.** The molluscs include more than 100,000 species—more than any other phylum of animals except the arthropods.

Cambrian explosion also contains many species of *lobopods*, a group from which arthropods may have evolved. Lobopods such as *Hallucigenia* (see Figure 27.6) had segmented

bodies, but most of their body segments were identical to one another. Early arthropods, such as the trilobites, also showed little variation from segment to segment (Figure 27.14). As arthropods continued to evolve, the segments tended to fuse and become fewer, and the appendages became specialized for a variety of functions. These evolutionary changes resulted not only

▶ Figure 27.14
**A fossil of a trilobite.**

in great diversification but also in an efficient body plan that permits the division of labor among different body regions.

What genetic changes led to the increasing complexity of the arthropod body plan? Arthropods today have two unusual *Hox* genes, both of which influence segmentation. To test whether these genes could have driven the evolution of increased body segment diversity in arthropods, researchers studied *Hox* genes in onychophorans, close relatives of arthropods (Figure 27.15). Their results indicate that the diversity of arthropod body plans did *not* arise from the acquisition of new *Hox* genes. Instead, the evolution of body segment diversity in arthropods was probably driven by changes in the sequence or regulation of existing *Hox* genes (see Concept 23.3).

We'll return to the arthropods later in the chapter, where we'll discuss how features of their body plan facilitated the colonization of land by a key group of animals, the insects.

---

▼ **Figure 27.15**   Inquiry

### Did the arthropod body plan result from new *Hox* genes?

**Experiment**   One hypothesis suggests that the arthropod body plan resulted from the origin (by gene duplication and subsequent mutations) of two unusual *Hox* genes found in arthropods: *Ultrabithorax* (*Ubx*) and *abdominal-A* (*abd-A*). Researchers tested this hypothesis using onychophorans, a group of invertebrates closely related to arthropods. Unlike many living arthropods, onychophorans have a body plan in which most body segments are identical to one another. If the origin of the *Ubx* and *abd-A Hox* genes drove the evolution of body segment diversity in arthropods, these genes probably arose on the arthropod branch of the evolutionary tree:

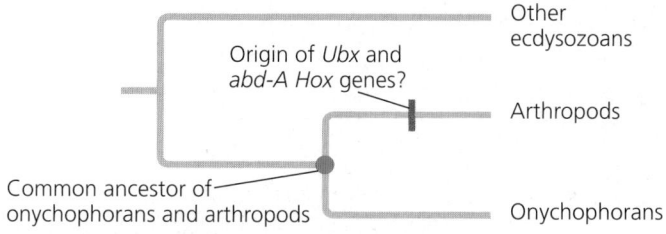

Origin of *Ubx* and *abd-A Hox* genes?

Other ecdysozoans

Arthropods

Common ancestor of onychophorans and arthropods

Onychophorans

According to this hypothesis, *Ubx* and *abd-A* would not have been present in the common ancestor of arthropods and onychophorans; hence, onychophorans should not have these genes. The team examined the *Hox* genes of the onychophoran *Acanthokara kaputensis*.

**Results**   The onychophoran *A. kaputensis* has all arthropod *Hox* genes, including *Ubx* and *abd-A*.

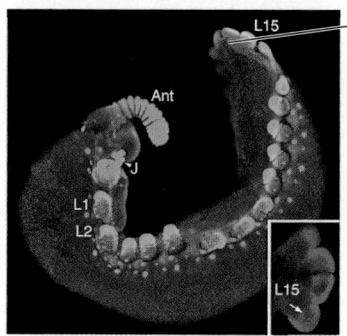

L15
Ant
J
L1
L2
L15

Red indicates the body regions of this onychophoran embryo in which *Ubx* or *abd-A* genes were expressed. (The inset shows this area enlarged.)

**Ant** = antenna
**J** = jaws
**L1–L15** = body segments

**Conclusion**   The evolution of increased body segment diversity in arthropods was not related to the origin of new *Hox* genes.

**Data from**   J. K. Grenier et al., Evolution of the entire arthropod *Hox* gene set predated the origin and radiation of the onychophoran/arthropod clade, *Current Biology* 7:547–553 (1997).

**WHAT IF?**   Suppose *A. kaputensis* did *not* have the *Ubx* and *abd-A Hox* genes. How would the conclusions of this study have been affected? Explain.

---

**CONCEPT CHECK 27.3**

1. Explain what is meant by "body plan" and describe three key features of animal body plans.

2. **WHAT IF?**   Would it be accurate to describe the Cambrian explosion as consisting of three "explosions," not one? Explain your reasoning and summarize the major steps in animal evolution shown in Figure 27.10.

For suggested answers, see Appendix A.

---

CONCEPT 27.4

# Vertebrates have been the ocean's dominant predators for more than 400 million years

As we discussed earlier, an explosive radiation of bilaterian invertebrates began about 530 million years ago. During that time of dramatic change, an inconspicuous, 3-cm-long creature could also be found gliding through the water: *Myllokunmingia fengjiaoa* (Figure 27.16). Although lacking armor and appendages, this ancient species was closely related to one of the most successful groups of animals ever to swim, walk, slither, or fly: the vertebrates, which derive their name from vertebrae, the series of bones that make up the backbone. We turn now to the origin and initial radiation of this group, focusing on adaptations that have enabled vertebrates to dominate life in the oceans for more than 400 million years.

## Bilaterian Radiation II: Aquatic Vertebrates

Vertebrates are members of the phylum Chordata. As seen in Figure 27.10, **chordates** are bilaterian animals that belong to the animal clade Deuterostomia. Among the deuterostomes that

▲ **Figure 27.16 A 530-million-year-old fossil of the chordate** *Myllokunmingia fengjiaoa.*

radiated during the Cambrian, we will focus on the chordates and how early members of this group gave rise to vertebrates.

### Early Chordate Evolution

All chordates share a set of derived characters, though many species possess some of these traits only during embryonic development. **Figure 27.17** illustrates four key characters of chordates: a **notochord**; a dorsal, hollow nerve cord; **pharyngeal slits** (or **pharyngeal clefts**); and a muscular, post-anal tail.

Among extant chordates, a group of blade-shaped animals called *lancelets* **(Figure 27.18a)** closely resemble the idealized chordate shown in Figure 27.17. The lancelet lineage diverged from other chordates at the base of the chordate phylogenetic tree. Like lancelets, *tunicates* **(Figure 27.18b)** diverged from other chordates early in chordate evolution. Also like lancelets, tunicates display key chordate traits, but only as larvae (adult tunicates have a highly modified body plan). These findings suggest that the ancestral chordate may have looked something like a lancelet—that is, it had an anterior end with a mouth; a

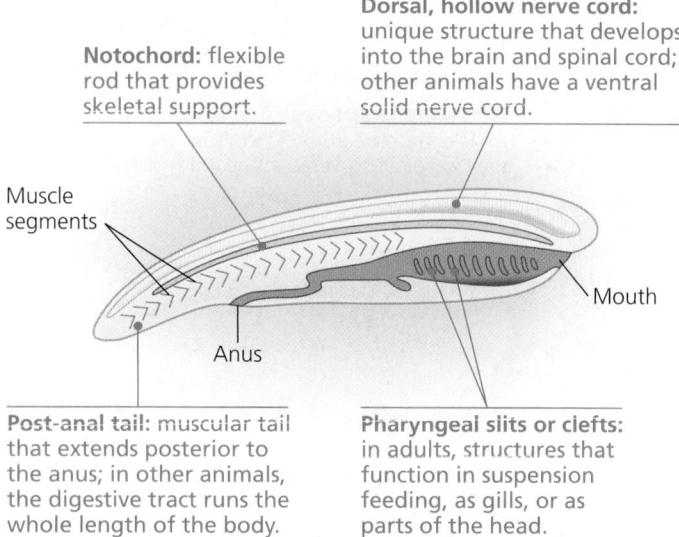

**Notochord:** flexible rod that provides skeletal support.

**Dorsal, hollow nerve cord:** unique structure that develops into the brain and spinal cord; other animals have a ventral solid nerve cord.

Muscle segments

Mouth

Anus

**Post-anal tail:** muscular tail that extends posterior to the anus; in other animals, the digestive tract runs the whole length of the body.

**Pharyngeal slits or clefts:** in adults, structures that function in suspension feeding, as gills, or as parts of the head.

▲ **Figure 27.17 Chordate characteristics.** All chordates possess the four highlighted structural trademarks at some point during their development.

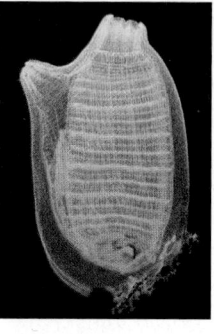

**(a) Lancelet**

**(b) Tunicate**

▲ **Figure 27.18 Lancelets and tunicates, chordates that lack a backbone.**

notochord; a dorsal, hollow nerve cord; pharyngeal slits; and a post-anal tail.

After the evolution of the basic chordate body plan, another major step in early chordate evolution was the origin of vertebrates. Unlike lancelets and tunicates, vertebrates have a backbone, a skull, and a well-defined head with a brain as well as eyes and other sensory organs.

Some of the fossils that formed during the Cambrian explosion 530 million years ago appear to straddle the transition to vertebrates. For example, some of these fossil chordates resembled lancelets, yet (unlike lancelets) they had a brain and eyes. *Myllokunmingia* (see Figure 27.16) not only had a brain and eyes, but also had parts of a skull surrounding its eyes and ears, making it one of the earliest chordates with a well-defined head. (The earliest "ears" were organs for maintaining balance, a function still performed by the ears of humans and other living vertebrates.) Although *Myllokunmingia* had a head, it lacked vertebrae and hence is not classified as a vertebrate.

### The Rise of Vertebrates

Vertebrates originated about 500 million years ago. With a more complex nervous system and a more elaborate skeleton than those of their ancestors, vertebrates became more efficient at two essential tasks: capturing food and avoiding being eaten.

Some of the earliest fossil vertebrates are of *conodonts*, soft-bodied, jawless vertebrates that hunted by impaling prey on a set of barbed hooks in their mouth. Other early vertebrates had paired fins and an inner ear with two semicircular canals that provided a sense of balance. Like conodonts, these vertebrates lacked jaws, but they had a muscular pharynx, which they may have used to suck in bottom-dwelling organisms or detritus. They were also armored with mineralized bone, which covered varying amounts of their body and may have offered protection from predators.

Only two lineages of jawless vertebrates (**cyclostomes**) survive today: the *hagfishes* and *lampreys* **(Figure 27.19)**. Living jawless vertebrates are far outnumbered by jawed vertebrates, known as **gnathostomes**. Gnathostomes appeared in the fossil

## ▼ Figure 27.19  Exploring Vertebrate Diversity

This phylogenetic hypothesis shows the relationships among major clades of vertebrates. Derived characters are listed for some clades; for example, only gnathostomes have a jaw. In some lineages, derived traits have been lost over time or occur in reduced form; for example, hagfishes and lampreys are vertebrates with highly reduced vertebrae.

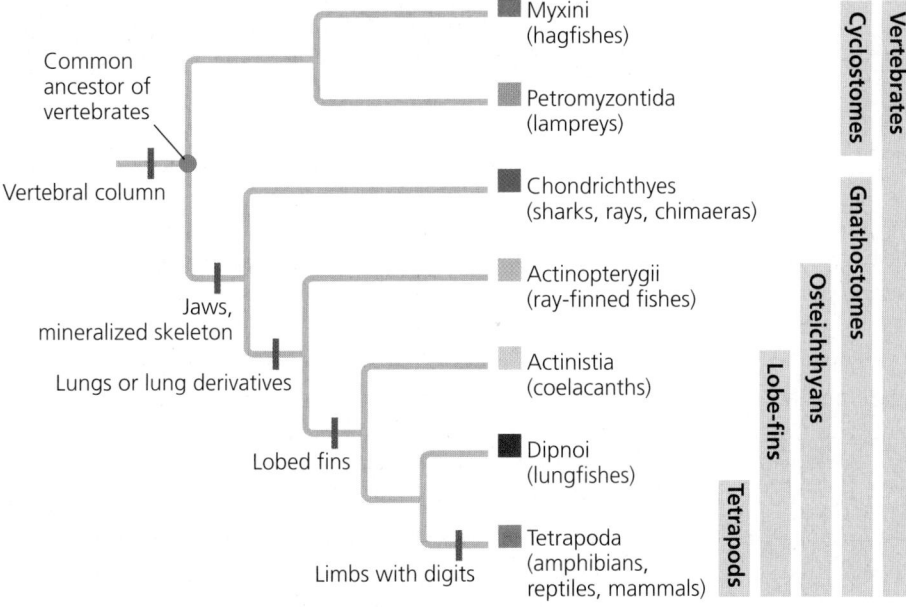

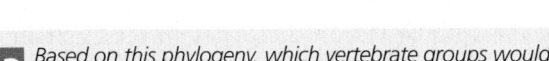

? Based on this phylogeny, which vertebrate groups would you expect to have lungs or lung derivatives? Explain.

**Myxini (30 species).** Hagfishes are scavengers that live and feed on the seafloor. They have slime-secreting glands that function in defense.

**Petromyzontida (35 species).** Most species of lampreys are parasites that use their mouth (inset) and tongue to bore a hole in the side of a fish. The lamprey then ingests the blood and other tissues of its host.

**Chondrichthyes (1,000 species).** Chondrichthyans such as this reef shark have skeletons made primarily of cartilage; the group also includes rays and chimaeras.

**Actinopterygii (over 27,000 species).** The diverse species of ray-finned fishes include this lionfish, which can inject venom through its spines.

**Actinistia (2 species).** Coelacanths were thought to have become extinct 75 million years ago until they were rediscovered in the Indian Ocean in 1938.

**Dipnoi (6 species).** Lungfishes have both gills and lungs and can gulp air into their lungs.

**Tetrapoda (over 29,000 species).** Tetrapods have limbs with digits; this group includes a diverse collection of amphibians, reptiles, and mammals (such as this giraffe).

record about 440 million years ago and steadily became more diverse. Their success probably resulted from a combination of anatomical features: Their paired fins and tail allowed them to swim efficiently after prey, and their jaws facilitated feeding by enabling them to grab prey whole or simply bite off chunks of flesh.

The earliest gnathostomes include extinct lineages of armored vertebrates known collectively as *placoderms*, which means "plate-skinned." Most placoderms were less than a meter long, though some giants measured more than 10 m (Figure 27.20). By 420 million years ago, gnathostomes had diverged further to include the three lineages of jawed vertebrates that survive today: chondrichthyans, ray-finned fishes, and lobe-fins. (Despite its name, this last group includes humans and other terrestrial animals with legs.)

0.5 m

▲ **Figure 27.20 Fossil of an early gnathostome.** A formidable predator, *Dunkleosteus* grew up to 10 m in length. Its jaw structure indicates that *Dunkleosteus* could exert a force of 560 kg/cm² (8,000 pounds per square inch) at the tip of its jaws.

**Chondrichthyans** Sharks, rays, and their relatives include some of the biggest and most successful vertebrate predators in the oceans today (see Figure 27.19). They belong to the clade Chondrichthyes, which means "cartilage fish." As their name indicates, the **chondrichthyans** have a skeleton composed predominantly of cartilage, though often impregnated with calcium. There are about 1,000 species of living chondrichthyans, many of which are threatened by overfishing.

**Ray-Finned Fishes** The vast majority of vertebrates belong to the clade of gnathostomes called Osteichthyes. Unlike sharks and their relatives, living **osteichthyans** typically have an ossified (bony) endoskeleton; they also have lungs or lung derivatives. Osteichthyans have diversified into two lineages, the ray-finned fishes and the lobe-fins. Nearly all the aquatic

osteichthyans familiar to us are among the **ray-finned fishes** (see Figure 27.19), named for the bony rays that support their fins. Today, there are more than 27,000 species of ray-finned fishes—almost as many species as in all other vertebrate groups combined.

**Lobe-Fins** Along with the ray-finned fishes, the other major lineage of osteichthyans is the **lobe-fins** (see Figure 27.19). A key derived character of lobe-fins is the presence of rod-shaped bones surrounded by a thick layer of muscle in their pectoral and pelvic fins. During the Devonian (419–359 million years ago), many lobe-fins lived in brackish waters, such as in coastal wetlands.

Today, however, only three lineages of lobe-fins survive. Two of these lineages are the coelacanths and the lungfishes (see Figure 27.19), but the third surviving lineage of lobe-fins is far more diverse. As you'll see in Concept 27.5, these organisms adapted to life on land and gave rise to the **tetrapods**, vertebrates with limbs and digits.

## *Summary:* Effects of Bilaterian Radiations I and II

Let's pause to review some of the events and consequences of bilaterian radiations I and II: the diversification of marine invertebrates and the rise of aquatic vertebrates. Prior to these radiations, the ocean's large eukaryotes were slow-moving and soft-bodied. With the radiation of marine invertebrates that began during the early Cambrian (535 million years ago), large invertebrate predators with sharp claws and other means of dismembering prey appeared, as did a rich diversity of large, well-defended prey.

Ocean life became still more dangerous with the rise of aquatic vertebrates, which excelled at capturing food and avoiding being eaten. In particular, the gnathostomes (jawed vertebrates), which originated 440 million years ago, have been the dominant predators in the oceans for more than 400 million years. Their ongoing success has resulted from their paired fins and tail (adaptations for swimming) and their jaws (an adaptation for feeding). Overall, the major bilaterian radiations greatly changed life in the oceans: The safe, soft-bodied world that existed before these events was gone forever.

### CONCEPT CHECK 27.4

1. Identify four derived characters that all chordates have at some point during their life.
2. Describe two key adaptations of aquatic gnathostomes.
3. **MAKE CONNECTIONS** The radiation of bilaterians in marine environments from 535 to 400 million years ago demonstrates that evolution is not goal oriented—it is not, for example, directed toward the origin of terrestrial vertebrates. Explain. (Review Concept 23.4.)

For suggested answers, see Appendix A.

# CONCEPT 27.5

## Several animal groups had features facilitating their colonization of land

Following the Cambrian explosion and its transformation of marine communities, some bilaterian animals colonized land, leading to profound changes there as well.

### Early Land Animals

Life on land provided opportunities for early animal colonists. The atmosphere had higher concentrations of oxygen than did aquatic environments, and there were new sources of food (such as early plants) and few competitors. But there were challenges as well: Water was scarcer, temperatures fluctuated more greatly than in aquatic environments, and there was no support against gravity.

Such challenges are potentially lethal. The soft body of a jelly, for example, provides no support against gravity; hence, a jelly cannot move or survive for long when stranded on land. But despite the challenges, members of many animal groups made the transition to terrestrial life. Arthropods, for example, invaded land multiple times, including a relatively recent event (4 million years ago) in which a crab lineage colonized the island

of Jamaica. The same is true of other animal groups, such as marine snails that have given rise to terrestrial species repeatedly over the course of evolution. When did animals begin to colonize land, and what enabled successful colonists to do so?

Fossils suggest that arthropods were among the first animals to colonize land, roughly 450 million years ago. The evidence includes fragments of arthropod remains, as well as possible millipede burrows. Well-preserved fossils from several continents indicate that by 410 million years ago, millipedes, centipedes, spiders, and a variety of wingless insects all had colonized land. Vertebrates colonized land 365 million years ago, by which time early forests had also formed. By 360 million years ago, terrestrial animal communities were broadly similar to those of today and included predators, detritivores (animals that eat decaying organic matter, such as plant debris), and herbivores.

Land animals often bear a striking resemblance to their close aquatic relatives (**Figure 27.21**). In some cases, the resemblance is so strong that it appears as if the land animals simply walked or crawled ashore, as in terrestrial snails and crabs that live mostly on land (land crabs return to water to reproduce). In other cases, more extensive changes took place, as we'll describe shortly for the colonization of land by vertebrates. But even in vertebrates, the evolutionary changes involved in the transition to terrestrial life were less extensive

| | | GREEN ALGA | MARINE CRUSTACEAN | AQUATIC LOBE-FIN |
|---|---|---|---|---|
| **CLOSE AQUATIC RELATIVE** | | 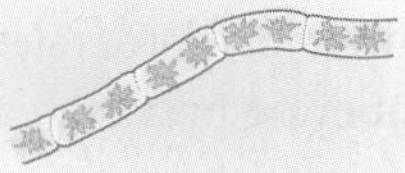 | 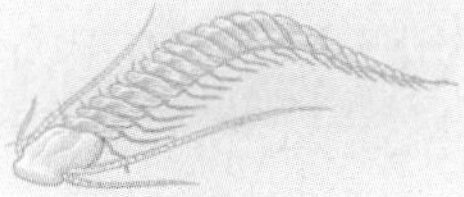 | 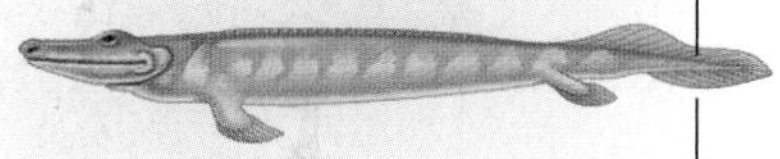 |
| **CHARACTER** | Anchoring structure | Derived (roots) | N/A | N/A |
| | Support structure | Derived (lignin/stems) | Ancestral | Ancestral (skeletal system) Derived (limbs) |
| | Internal transport | Derived (vascular system) | Ancestral | Ancestral |
| | Muscle/nerve cells | N/A | Ancestral | Ancestral |
| | Protection against desiccation | Derived (cuticle) | Ancestral | Derived (amniotic egg/scales) |
| | Gas exchange | Derived (stomata) | Derived (tracheal system) | Ancestral |
| **TERRESTRIAL ORGANISM** | | PLANTS | INSECTS | TERRESTRIAL VERTEBRATES |

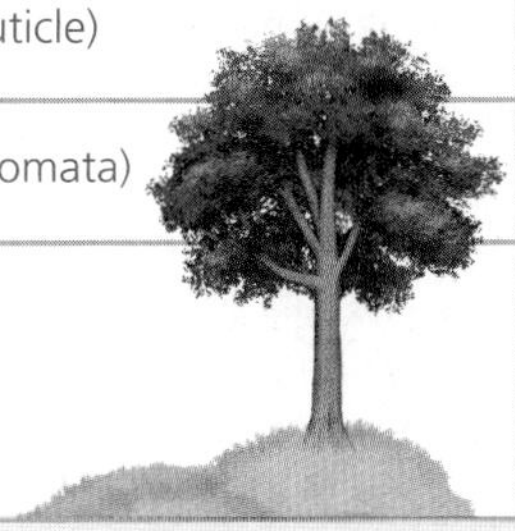

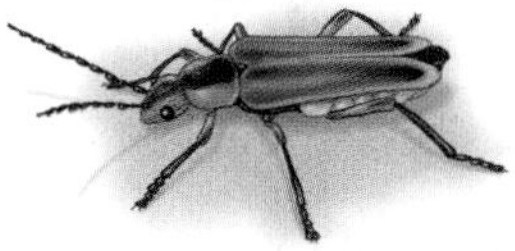

▲ **Figure 27.21 Descent with modification during the colonization of land.** This chart identifies some key characters that enable three major groups of terrestrial organisms—plants, insects, and terrestrial vertebrates—to live on land. Red type indicates adaptations that have evolved since the lineages diverged from their aquatic relatives. In plants, most terrestrial adaptations evolved after the split. In contrast, two large clades of terrestrial animals—the insects and the vertebrates—display many ancestral characters that facilitated their transition to life on land.

than those that occurred in plants. For example, the aquatic relatives of animals that colonized land typically had well-developed skeletal, muscular, digestive, circulatory, respiratory, and nervous systems—all of which facilitated the colonization of land. Plants, in contrast, arose from a small green alga whose structure bore little resemblance to those of its descendants, the plants that now cover Earth (see Figure 27.21).

## Colonization of Land by Arthropods

As mentioned earlier, terrestrial lineages have arisen in several different arthropod groups, including millipedes, spiders, crabs, and insects. After describing general features of arthropods, we'll focus on the colonization of land by what is now their largest clade, the insects.

### General Characteristics of Arthropods

Over the course of evolution, the appendages of some arthropods have become modified, specializing in functions such as walking, feeding, sensory reception, reproduction, and defense. Like the appendages from which they were derived, these modified structures are jointed and come in pairs. **Figure 27.22** shows the diverse appendages of a representative arthropod.

The body of an arthropod is completely covered by the **cuticle**, an exoskeleton constructed from layers of protein and the polysaccharide chitin. As you know if you've ever eaten a crab or lobster, the cuticle can be thick and hard over some parts of the body and thin and flexible over others, such as the joints. The rigid exoskeleton protects the animal and provides points of attachment for the muscles that move the appendages. Later, the exoskeleton enabled some arthropods to live on land. The exoskeleton's relative impermeability to water helped prevent desiccation, and its strength provided support when arthropods left the buoyancy of water.

A variety of specialized gas exchange organs have evolved in arthropods. Most aquatic species have gills with thin, feathery extensions that place an extensive surface area in contact with the surrounding water. Terrestrial arthropods generally have internal surfaces specialized for gas exchange. Most insects, for instance, have tracheal systems, branched air ducts leading into the interior from pores in the cuticle. These ducts infiltrate the body, carrying oxygen directly to cells.

### Insects

One arthropod group that colonized land, the insects **(Figure 27.23)**, is now found in almost every terrestrial habitat. Insects also live in fresh water, and flying insects fill the air. Insects are rare, though not absent, in marine habitats.

The oldest insect fossils date to about 416 million years ago. Later, an explosion in insect diversity took place when insect flight evolved during the Carboniferous and Permian periods (359–252 million years ago). An animal that can fly can escape predators, find food and mates, and disperse to new habitats more effectively than an animal that must crawl about on the ground. Many insects have one or two pairs of wings that

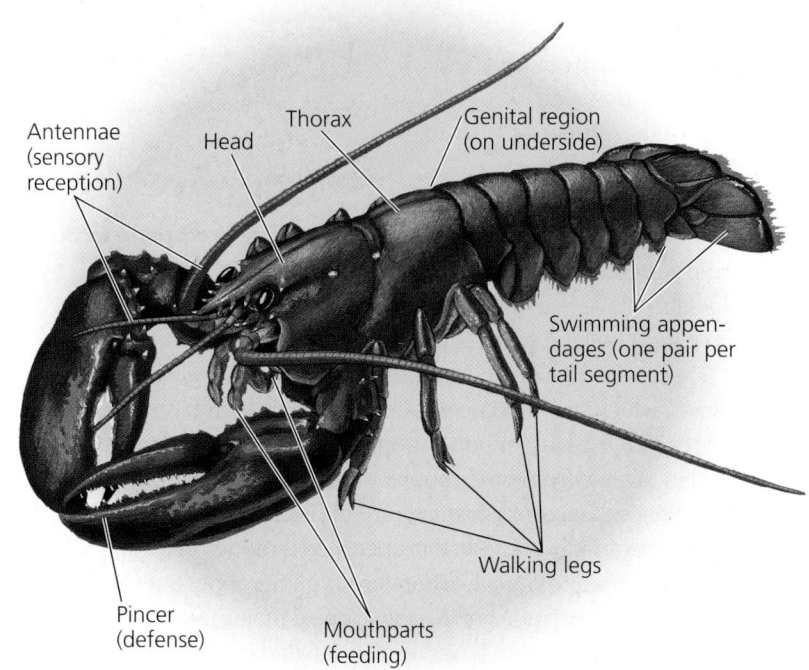

Antennae (sensory reception)  Head  Thorax  Genital region (on underside)  Swimming appendages (one pair per tail segment)  Walking legs  Pincer (defense)  Mouthparts (feeding)

▲ **Figure 27.22 External anatomy of an arthropod.** Many of the distinctive features of arthropods are apparent in this dorsal view of a lobster. The body is segmented, but this character is obvious only in the post-genital (behind the genitals) region, or "tail." The appendages (including antennae, pincers, mouthparts, walking legs, and swimming appendages) are jointed. The head bears a pair of compound (multilens) eyes. The whole body, including appendages, is covered by an exoskeleton.

▶ **Lepidopterans**—moths and butterflies—undergo complete metamorphosis: The larval stage (called a caterpillar), which is specialized for eating and growing, looks completely different from the adult stage, which is specialized for dispersal and reproduction.

◀ **Hymenopterans** include ants, bees, and wasps. They undergo complete metamorphosis, and most are highly social insects.

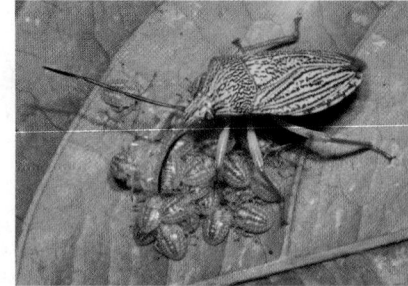

▶ **Hemipterans** include stink bugs, bed bugs, and other so-called "true bugs." They have piercing mouthparts and undergo incomplete metamorphosis: The young (nymphs) resemble the adults but are smaller and lack wings.

▲ **Figure 27.23 Insect diversity.** Insects have more described species than all other eukaryotic groups combined; here, we show just three of the many insect clades.

emerge from the dorsal side of the thorax **(Figure 27.24)**. Because the wings are extensions of the cuticle, insects can fly without sacrificing any walking legs. By contrast, the flying vertebrates—birds and bats—have one of their two pairs of walking legs modified into wings, making some of these species clumsy on the ground.

▲ **Figure 27.24 Ladybird beetle in flight.**

Insects also radiated in response to the origin of new plant species, which provided new sources of food. As you read in Concept 22.2, an insect population feeding on a new plant species can diverge from other populations, eventually forming a new species of insect. A fossil record of diverse insect mouthparts, for example, suggests that specialized modes of feeding on gymnosperms and other Carboniferous plants contributed to early adaptive radiations of insects. Later, a major increase in insect diversity appears to have been stimulated by the evolutionary expansion of flowering plants during the mid-Cretaceous period (about 100 million years ago). Although insect and plant diversity decreased during the Cretaceous mass extinction, both groups have rebounded over the past 66 million years.

## Terrestrial Vertebrates

Another key event in the colonization of land by animals took place 365 million years ago, when the fins of a lineage of lobe-fins gradually evolved into the limbs and feet of tetrapods. Until then, all vertebrates had shared the same basic fishlike anatomy. After the colonization of land, early tetrapods gave rise to many new forms, from leaping frogs to flying eagles to bipedal humans.

The most significant character of tetrapods gives the group its name, which means "four feet" in Greek. In place of pectoral and pelvic fins, tetrapods have limbs with digits. Limbs support a tetrapod's weight on land, while feet with digits efficiently transmit muscle-generated forces to the ground when the tetrapod walks. Tetrapods also differ from earlier vertebrates in having a neck, which consists of vertebrae that separate the head from the rest of the body; the neck allows the head to move up and down and from side to side, independently of the rest of the body. In addition, the bones of the pelvic girdle, to which the hind legs are attached, are fused to the backbone, permitting forces generated by the hind legs against the ground to be transferred to the rest of the body.

### The Origin of Tetrapods

During the Devonian (419–359 million years ago), many lobe-fins lived in coastal wetlands. Those that entered shallow, oxygen-poor water could have used their lungs to breathe air. Some species probably used their stout fins to swim and "walk" underwater across the bottom (moving their fins in an alternating gait, as do some living lobe-fins). This suggests that the tetrapod body plan did not evolve "out of nowhere" but was simply a modification of a preexisting body plan.

The recent discovery of a fossil called *Tiktaalik* provided new details on how this process occurred **(Figure 27.25)**. Like

Shoulder bones

Ribs

Scales

Neck

Head

Eyes on top of skull

Flat skull

Humerus

Ulna

"Wrist"

Elbow

Radius

Fin skeleton

Fin

▲ **Figure 27.25 Discovery of a "fishapod": *Tiktaalik*.** Paleontologists were on the hunt for fossils that could shed light on the evolutionary origin of tetrapods. Based on the ages of previously discovered fossils, researchers were looking for a dig site with rocks about 365–385 million years old. Ellesmere Island, in the Canadian Arctic, was one of the few such sites that was also likely to contain fossils, because it was once a river. The search at this site was rewarded by the discovery of fossils of a 375-million-year-old lobe-fin, named *Tiktaalik*. As shown in the chart and photographs, *Tiktaalik* exhibits both fish and tetrapod characters.

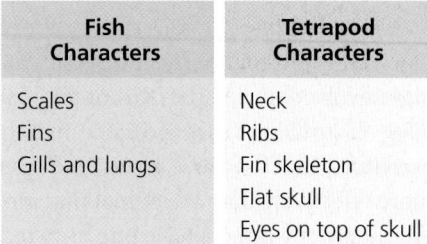

| Fish Characters | Tetrapod Characters |
|---|---|
| Scales | Neck |
| Fins | Ribs |
| Gills and lungs | Fin skeleton |
| | Flat skull |
| | Eyes on top of skull |

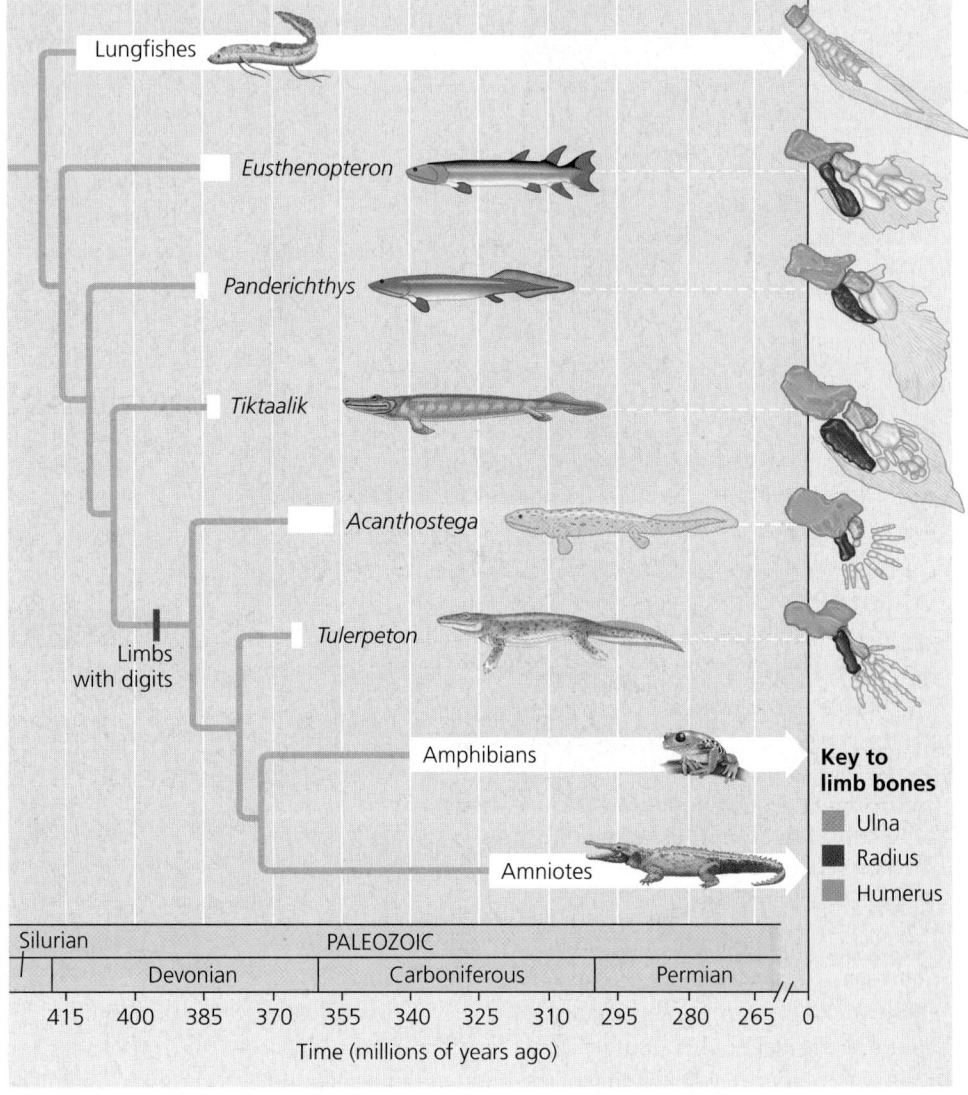

► **Figure 27.26 Steps in the origin of limbs with digits.** The white bars on the branches of this diagram place known fossils in time; arrowheads indicate lineages that extend to today. The drawings of extinct organisms are based on fossilized skeletons, but the colors are fanciful.

**WHAT IF?** *If the most recent common ancestor of* Tulerpeton *and living tetrapods originated 370 million years ago, what range of dates would include the origin of amphibians?*

a fish, this species had fins, gills, and lungs, and its body was covered in scales. But unlike a fish, *Tiktaalik* had a full set of ribs that would have helped it breathe air and support its body. Also unlike a fish, *Tiktaalik* had a neck and shoulders, allowing it to move its head about. Finally, the bones of *Tiktaalik's* front fin had the same basic pattern found in all limbed animals: one bone (the humerus), followed by two bones (the radius and ulna), followed by a group of small bones that comprise the wrist. Although it is unlikely that *Tiktaalik* could walk on land, its front fin skeleton suggests that it could prop itself up in water on its fins. Since *Tiktaalik* predates the oldest known tetrapod, its features suggest that key "tetrapod" traits, such as a wrist, ribs, and a neck, were in fact ancestral to the tetrapod lineage.

*Tiktaalik* and other extraordinary fossil discoveries have allowed paleontologists to reconstruct how fins became progressively more limb-like over time, culminating in the appearance in the fossil record of the first tetrapods 365 million years ago **(Figure 27.26)**. Over the next 60 million years, a great diversity

of tetrapods arose. Some of these species retained functional gills and had weak limbs, while others had lost their gills and had stronger limbs that facilitated walking on land. Judging from the morphology and locations of their fossils, most of these early tetrapods probably remained tied to water, a characteristic they share with some members of the most basal group of living tetrapods, the amphibians.

## Amphibians

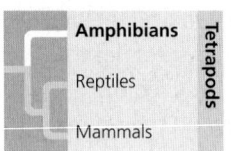

The **amphibians** are represented today by about 6,150 species in three clades: salamanders (clade Urodela, "tailed ones"), frogs (clade Anura, "tailless ones"), and caecilians (clade Apoda, "legless ones"). Many amphibians live first in water during a larval stage, and then on land as adults. Such species typically undergo metamorphosis, as when a gill-breathing tadpole that lives in water transforms into a lung-breathing frog that lives on land (see Figure 32.10).

**(a)** Salamanders retain their tails as adults.

**(b)** Frogs and toads lack tails as adults.

**(c)** Caecilians have no legs and are mainly burrowing animals.

▲ **Figure 27.27 Amphibian diversity.**

However, other amphibians do not live a dual—that is, both aquatic and terrestrial—life. For example, some salamanders **(Figure 27.27a)** are entirely aquatic, but others live on land as adults or throughout life. Most salamanders that live on land walk with a side-to-side bending of the body, a trait also found in early terrestrial tetrapods.

Frogs **(Figure 27.27b)** are better suited than salamanders for moving on land. Adult frogs use their powerful hind legs to hop along the terrain. Although often distinctive in appearance, the animals known as "toads" are simply frogs that have leathery skin or other adaptations for life on land.

Finally, the caecilians **(Figure 27.27c)** are legless and nearly blind. Their lack of legs is a secondary adaptation, as they evolved from a legged ancestor. Caecilians inhabit tropical areas, where most species burrow in moist forest soil.

Most amphibians are found in damp habitats such as swamps and rain forests. Even those adapted to drier habitats spend much of their time in burrows or under moist leaves, where humidity is high. One reason amphibians require relatively wet habitats is that they rely heavily on their moist skin for gas exchange—if their skin dries out, they cannot get enough oxygen. In addition, amphibians typically lay their eggs in water or in moist environments on land; their eggs lack a shell and dehydrate quickly in dry air.

Over the past 30 years, zoologists have documented a rapid and alarming decline in amphibian populations in locations throughout the world. There appear to be several causes, including the spread of a disease-causing chytrid fungus, habitat loss, climate change, and pollution. In some cases, declines have become extinctions. Recent studies indicate that at least 9 amphibian species have become extinct within the last four decades; more than 100 other species have not been observed in that time and are considered possibly extinct.

### CONCEPT CHECK 27.5

1. Describe two adaptations that have enabled insects to thrive on land.
2. **MAKE CONNECTIONS** Compare and contrast how the colonization of land by plants and by vertebrates exemplifies descent with modification. (Review Concepts 19.2 and 26.1.)

For suggested answers, see Appendix A.

---

### CONCEPT 27.6

## Amniotes have key adaptations for life in a wide range of terrestrial environments

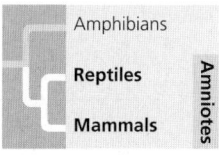

Compared to the amphibians, a more extensive colonization of dry habitats occurred in the **amniotes**, a group of tetrapods whose extant members are the reptiles (including birds, as we'll discuss shortly) and mammals. We'll begin by considering terrestrial adaptations in amniotes, after which we'll turn to their origin and diversity today.

### Terrestrial Adaptations in Amniotes

Amniotes are named for the major derived character of the clade, the **amniotic egg**, which contains four specialized membranes: the amnion, the chorion, the yolk sac, and the allantois **(Figure 27.28)**. The amniotic egg was a key evolutionary innovation for terrestrial life: It allowed the embryo to develop on land in its own private "pond," reducing the dependence of tetrapods on an aqueous environment for reproduction.

In contrast to the shell-less eggs of amphibians, the amniotic eggs of most reptiles and some mammals have a shell. A shell slows dehydration of the egg in air, an adaptation that helped amniotes to occupy a wider range of terrestrial habitats than amphibians, their closest living relatives. (Seeds played a similar role in the evolution of plants, as discussed in Concept 26.4.) Most mammals have lost the eggshell over the course of their evolution, and the embryo avoids desiccation by developing within the amnion inside the mother's body.

Another trait that enables amniotes to thrive on land is their use of the rib cage to ventilate their lungs. Rib cage ventilation is more efficient than throat-based ventilation, which

► **Figure 27.28 The amniotic egg.** The embryos of reptiles and mammals form four extraembryonic membranes: the amnion, chorion, yolk sac, and allantois. This diagram shows these membranes in the shelled egg of a reptile.

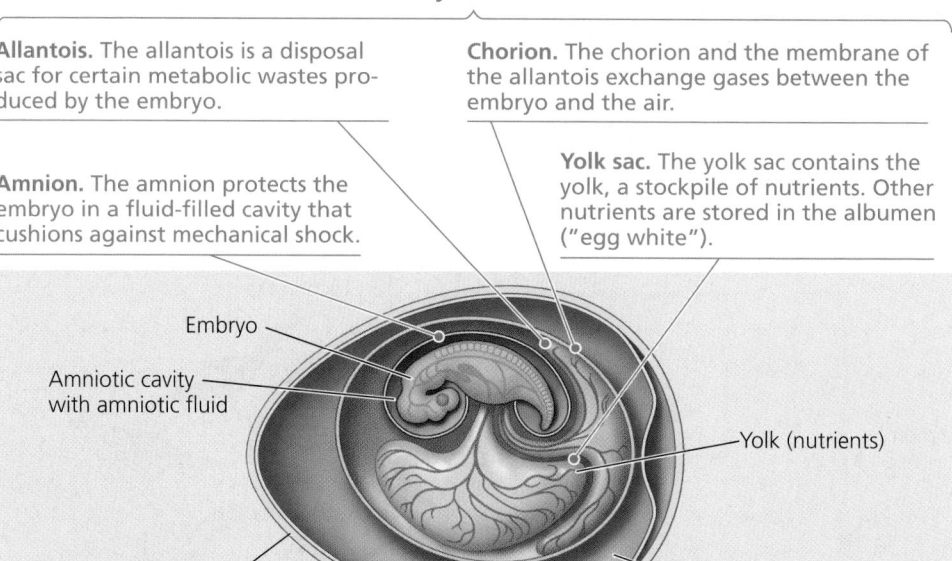

**Extraembryonic membranes**

**Allantois.** The allantois is a disposal sac for certain metabolic wastes produced by the embryo.

**Chorion.** The chorion and the membrane of the allantois exchange gases between the embryo and the air.

**Amnion.** The amnion protects the embryo in a fluid-filled cavity that cushions against mechanical shock.

**Yolk sac.** The yolk sac contains the yolk, a stockpile of nutrients. Other nutrients are stored in the albumen ("egg white").

Embryo

Amniotic cavity with amniotic fluid

Yolk (nutrients)

Shell

Albumen

amphibians use as a supplement to breathing through their skin. The increased efficiency of rib cage ventilation may have allowed amniotes to abandon breathing through their skin and develop less permeable skin, thereby conserving water.

## The Origin and Radiation of Amniotes

The most recent common ancestor of living amphibians and amniotes lived about 350 million years ago. Based on where their fossils have been found, the earliest amniotes appear to have lived in warm, moist environments, as did earlier tetrapods. Over time, however, early amniotes expanded into a wide range of new environments, including dry and high-latitude regions. Fossil evidence shows that the earliest amniotes resembled small lizards with sharp teeth, a sign that they were predators. Later groups of amniotes also included herbivores, as evidenced by their grinding teeth and other features.

Amniotes today include two large clades of terrestrial vertebrates: reptiles and mammals. Next, let's examine the diversity of these groups, paying particular attention to adaptations that facilitate life in dry environments.

### Reptiles

Living members of the **reptile** clade include tuataras, lizards and snakes, turtles, crocodilians, and birds **(Figure 27.29)**. There are about 18,300 species of reptiles, the majority of which are squamates (lizards and snakes; 7,900 species) or birds (10,000 species). Notice in Figure 27.29 that dinosaurs are reptiles and that birds originated from saurischian dinosaurs (a group that includes *Tyrannosaurus rex*); as a result, birds are also considered reptiles.

Fossil evidence indicates that the earliest reptiles lived about 310 million years ago and resembled lizards. Reptiles have

diverged greatly since then, but as a group they share several derived characters that distinguish them from other tetrapods. For example, unlike amphibians, reptiles have scales that contain the protein keratin (as does a human nail). Scales help protect the animal's skin from desiccation and abrasion. In addition, most reptiles lay their shelled eggs on land; the shell protects the egg from drying out. Fertilization must occur internally, before the eggshell is secreted.

Reptiles such as lizards and snakes are sometimes described as "cold-blooded" because they do not use their metabolism extensively to control their body temperature. However, they do regulate their body temperature through behavioral adaptations. For example, many lizards bask in the sun when the air is cool and seek shade when the air is warm. A more accurate description of these reptiles is to say that they are **ectothermic**, which means that they absorb external heat as their main source of body heat. However, the reptile clade is not entirely ectothermic; birds are **endothermic**, capable of maintaining body temperature through metabolic activity.

Extant reptiles comprise two large clades, one that includes tuataras and squamates (lizards and snakes) and one that includes turtles, crocodilians, and birds (see Figure 27.29). We'll focus here on birds, a diverse group of flying reptiles that arose about 160 million years ago.

**Birds** Many of the characters of birds are adaptations that facilitate flight, including weight-saving modifications that make flying more efficient. For example, birds lack a urinary bladder, and the females of most species have only one ovary. The gonads of both females and males are usually small, except during the breeding season, when they increase in size. Living birds are also toothless, an adaptation that trims the weight of the head.

The reptile clade consists of five groups with living members, shown below, along with extinct groups (denoted by the dagger symbol †) such as the pterosaurs and nonflying dinosaurs.

**?** *Are pterosaurs dinosaurs? Are birds? Explain.*

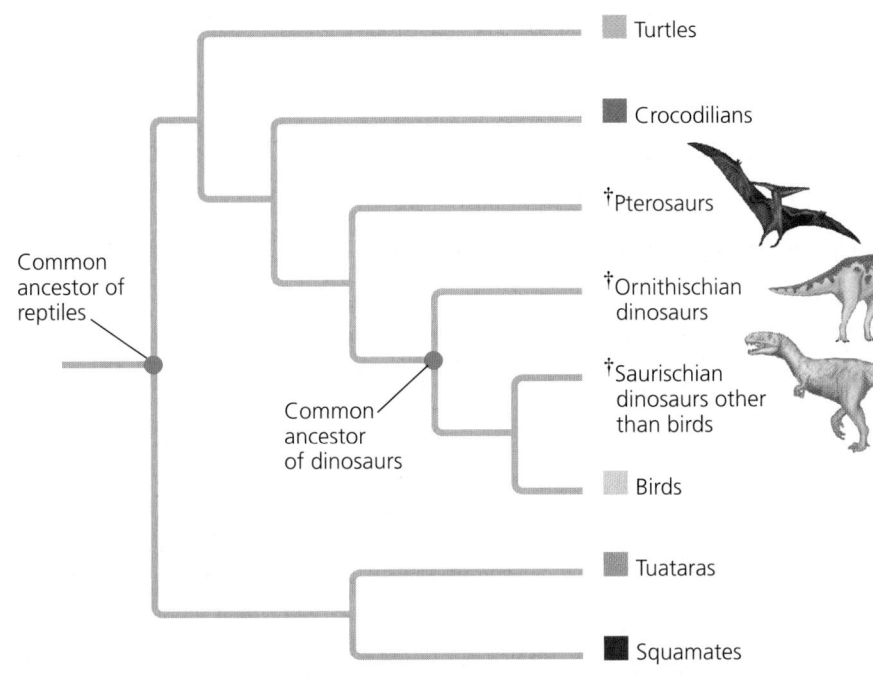

**Crocodilians (23 species).** Crocodiles and alligators (collectively called crocodilians) belong to a lineage whose earliest members lived on land more than 200 million years ago. Later, some species adapted to life in water, breathing air through their upturned nostrils.

**Birds (10,000 species).** The anatomy of birds includes many adaptations that facilitate flight. A hummingbird has an even more specialized skeleton that allows its wings to rotate in all directions, enabling it to hover and fly backward. Its beak is also specialized to feed on the nectar of its particular food source.

**Turtles (307 species).** The species in this group have a boxlike shell fused to their skeletons. Some turtles live on land, while others live in freshwater or marine habitats, but all are air-breathing.

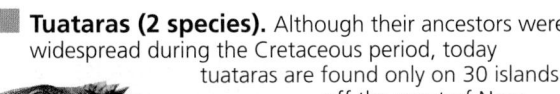

**Tuataras (2 species).** Although their ancestors were widespread during the Cretaceous period, today tuataras are found only on 30 islands off the coast of New Zealand.

**Squamates (7,910 species).** Snakes, together with lizards, make up the squamate lineage of reptiles. Snakes are carnivorous, and despite their lack of legs, have adaptations that make them effective predators, including the ability of various species to detect heat, chemicals, or vibrations that signal the presence of prey.

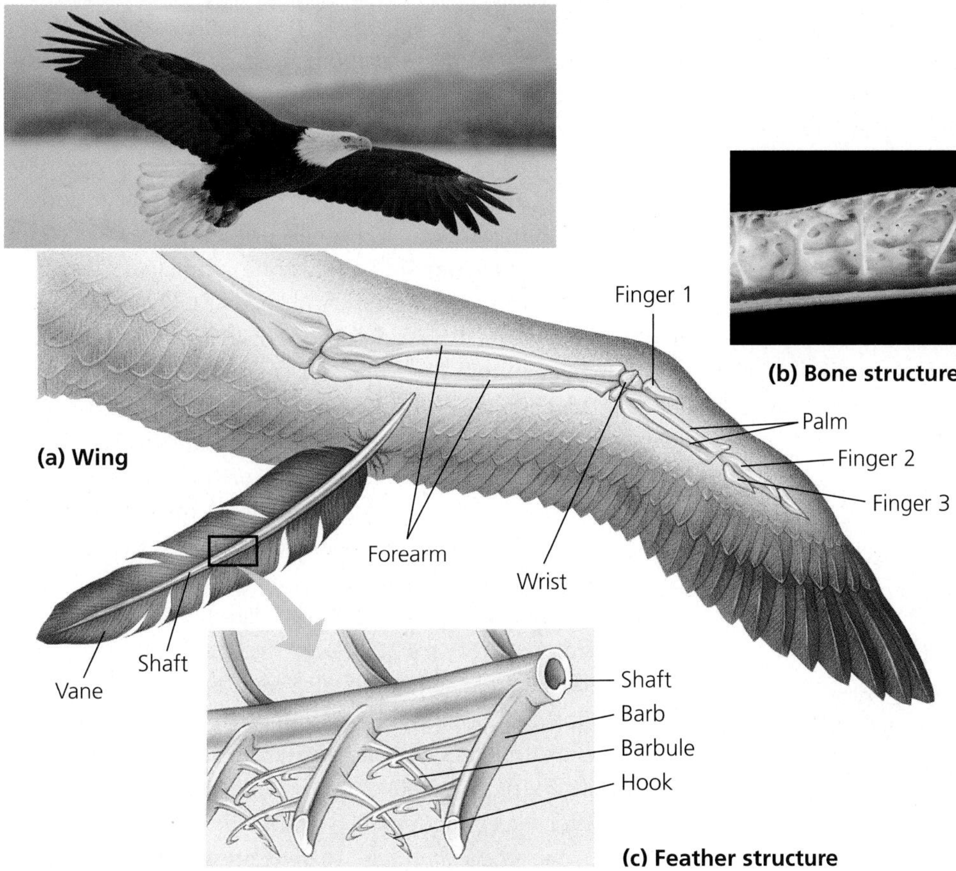

## (a) Wing

Finger 1

Palm

Finger 2

Finger 3

Forearm

Wrist

Shaft

Vane

## (b) Bone structure

Shaft

Barb

Barbule

Hook

## (c) Feather structure

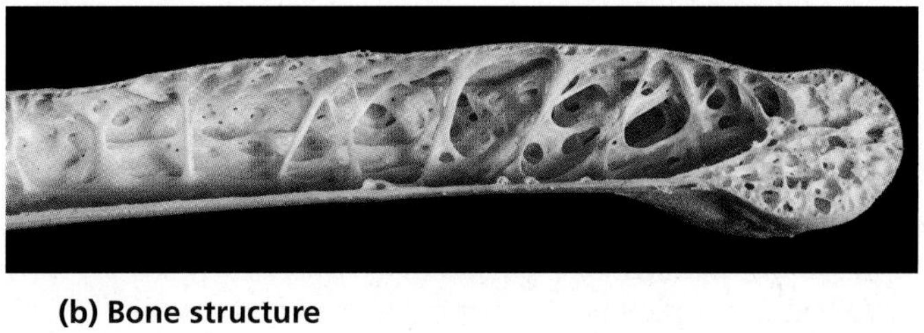

◀ **Figure 27.30 Form fits function: the avian wing and feather. (a)** A wing is a remodeled version of the tetrapod forelimb. **(b)** The bones of many birds have a honeycombed internal structure and are filled with air. **(c)** A feather consists of a central air-filled shaft, from which radiate the vanes. Birds have contour feathers and downy feathers. Contour feathers are stiff and contribute to the aerodynamic shapes of the wings and body. When a bird preens, it runs the length of each contour feather through its beak, engaging the hooks on the barbs, and uniting the barbs into a precise shape. Downy feathers lack hooks, and the free-form arrangement of their barbs produces a fluffiness that provides insulation by trapping air.

A bird's most obvious adaptations for flight are its wings and feathers **(Figure 27.30)**. The shape and arrangement of the feathers form the wings into airfoils, and they illustrate some of the same principles of aerodynamics as the wings of an airplane. Flapping the wings is powered by contractions of large pectoral (breast) muscles anchored to the sternum (breastbone).

Flight provides numerous benefits. It enables escape from earthbound predators and enhances scavenging and hunting. Flight also enables some birds to migrate great distances and gain access to different food resources and seasonal breeding areas.

Flying requires a great expenditure of energy from an active metabolism. The lungs of birds have tiny tubes leading to and from elastic air sacs that improve airflow and oxygen uptake. This efficient respiratory system and a circulatory system with a four-chambered heart keep tissues well supplied with oxygen and nutrients, supporting a high rate of metabolism.

Flight also requires both acute vision and fine muscle control. Birds have color vision and excellent eyesight. The visual and motor areas of the brain are well developed, and the brain is proportionately larger than those of amphibians and nonbird reptiles.

### Mammals

Besides the reptiles, the other living lineage of amniotes is our own, the **mammals**, named for their distinctive mammary glands, which produce milk for offspring. Hair, another mammalian character, and a fat layer under the skin provide insulation that can conserve water and protect the body against the extremes of heat or cold found in many terrestrial

environments. Another mammalian adaptation for life on land is the kidney (see Figure 32.21), which is efficient at conserving water when removing wastes from the body. Some mammals, such as kangaroo rats, are so adept at conserving water that they can survive in arid environments while drinking little or no water at all **(Figure 27.31)**.

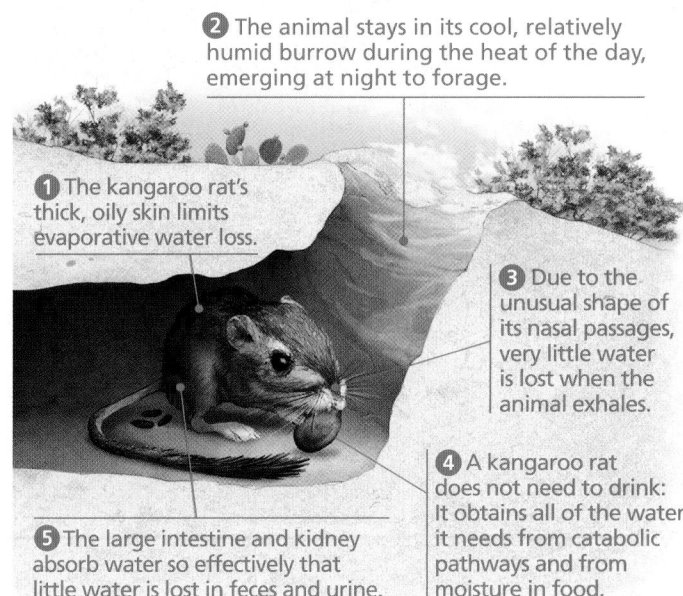

❷ The animal stays in its cool, relatively humid burrow during the heat of the day, emerging at night to forage.

❶ The kangaroo rat's thick, oily skin limits evaporative water loss.

❸ Due to the unusual shape of its nasal passages, very little water is lost when the animal exhales.

❹ A kangaroo rat does not need to drink: It obtains all of the water it needs from catabolic pathways and from moisture in food.

❺ The large intestine and kidney absorb water so effectively that little water is lost in feces and urine.

▲ **Figure 27.31 Some adaptations of the kangaroo rat to its extremely dry habitat.**

**MAKE CONNECTIONS** *Explain how the catabolic pathways mentioned in ❹ could provide a kangaroo rat with water. (See Concept 7.1.)*

Like birds, mammals are endothermic, and most have a high metabolic rate. Efficient respiratory and circulatory systems (including a four-chambered heart) support a mammal's metabolism. Also as in birds, mammals generally have a larger brain than other vertebrates of equivalent size, and many species are capable learners. In addition, whereas the teeth of reptiles are generally uniform in size and shape, the jaws of mammals bear a variety of teeth with sizes and shapes adapted for chewing many kinds of foods. Humans, like most mammals, have teeth modified for shearing (incisors and canine

teeth) and for crushing and grinding (premolars and molars; see Figure 33.13).

Among extant vertebrate taxa, mammals and reptiles are sister groups. Even so, mammals and reptiles have evolved independently for a long time—their lineages diverged more than 300 million years ago. Mammals originated from a group of amniotes called **synapsids**. Early nonmammalian synapsids lacked hair, had a sprawling gait, and laid eggs. Over the course of 120 million years, these ancestors gave rise to a series of increasingly mammal-like synapsids (see Figure 23.5). Finally, about 180 million years ago, the first true mammals arose. A diverse set of mammals coexisted with dinosaurs from 180 to 66 million years ago, but these species were not abundant, and most measured less than 1 m. One possible explanation for their small size is that dinosaurs already occupied the ecological niches of large-bodied animals.

By 140 million years ago, the three major lineages of mammals had emerged **(Figure 27.32)**: the lineages leading to **monotremes** (egg-laying mammals), **marsupials** (mammals whose young are born as embryos and complete their development in an external pouch), and **eutherians** (placental mammals whose young complete embryonic development in the uterus). After the extinction of terrestrial dinosaurs (66 million years ago), mammals continued to diversify, ultimately resulting in the more than 5,300 species living today.

### Primates: A Closer Look

Among extant mammals, we'll focus here on the primates, a group of eutherians that includes lemurs, tarsiers, monkeys, and apes **(Figure 27.33)**. Humans are members of the ape group.

The earliest known primates were tree-dwellers, and many derived characters of primates are adaptations to the demands of living in the trees. For example, grasping hands and feet allow primates to hang onto tree branches. All primates also have a thumb that is relatively movable and separate from the fingers. In addition, monkeys and apes have a fully **opposable thumb**; that is, they can touch the ventral surface (fingerprint side) of the tip of all four fingers with the ventral surface of the thumb of the same hand. Relative to other mammals, primates also have a large brain and short jaws, giving them a flat face. Their forward-looking eyes are close together on the front of the face. The overlapping visual fields of their forward-facing eyes enhance depth perception—an obvious advantage when swinging from branch to branch in trees.

Today, nonhuman apes are found exclusively in tropical regions of Africa and Asia. All living apes have relatively long arms, short legs, and no tail. Although all nonhuman apes spend time in trees, only gibbons and orangutans are primarily arboreal. Social organization varies among the apes; gorillas and chimpanzees are highly social. Finally, compared to other primates, apes have a larger brain in proportion to their body size, and their behavior is more flexible. These two characteristics are especially prominent in the next group that we'll consider, the hominins.

---

▼ **Figure 27.32 The major mammalian lineages.**

**Monotremes**

The platypus and four species of spiny anteaters comprise the five extant species of monotremes, which are found only in Australia and New Guinea. Monotremes have hair and produce milk, but they lack nipples. They are the only mammals that lay eggs (inset).

**Marsupials**

Kangaroos, opossums, and koalas are examples of marsupials (324 species). Like eutherians, they have nipples that provide milk and they give birth to live young. Offspring are born early in development; they finish their growth while nursing from a nipple (in their mother's pouch in most species).

**Eutherians**

Most mammals are eutherians, a clade of 5,010 species that include primates, whales, rodents, and many other mammal groups. Eutherians have a longer pregnancy than marsupials, and they have a more complex **placenta** (a structure in which nutrients diffuse into the embryo from the mother's blood).

**?** *Monotremes are basal mammals. Draw a phylogenetic tree showing evolutionary relationships among the three lineages.*

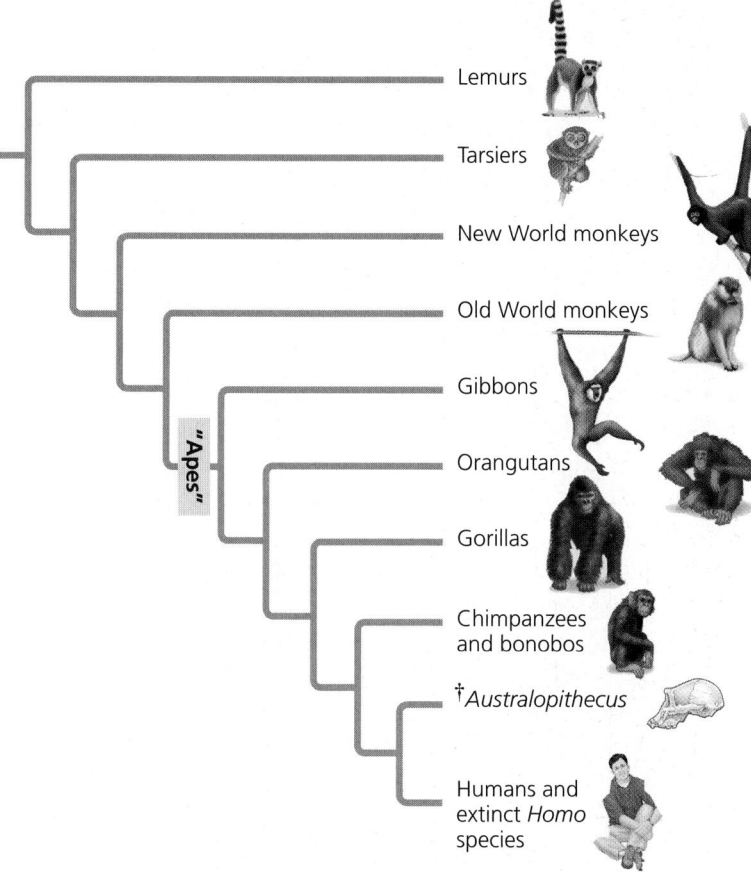

**▲ Figure 27.33 The primate evolutionary tree.**
The fossil record shows that monkeys and apes began diverging from other primates about 50 million years ago. The lineage leading to humans and *Australopithecus* branched off from other apes between 6 and 7 million years ago.

## Human Evolution

Humans (*Homo sapiens*) are primates, nested within the group informally called apes (see Figure 27.33). Unlike other apes, humans stand upright and are bipedal (walk on two legs). Humans also have a larger brain and are capable of language, symbolic thought, artistic expression, and the use of complex tools. How did these distinctive traits arise?

### Early Human Ancestors

Molecular data show that among living apes, humans and chimpanzees are each other's closest relatives—in fact, the human and chimpanzee genomes are 99% identical. But as Figure 27.33 shows, we are even more closely related to *Australopithecus* and other extinct human ancestors. Fossil evidence and genetic analyses indicate that the evolutionary lineage leading to humans diverged from other apes between 6 and 7 million years ago. The species in this lineage are known as **hominins**, a group consisting of humans and the extinct species that are more closely related to us than to chimpanzees.

The oldest hominin fossils are of *Sahelanthropus tchadensis*, which lived about 6.5 million years ago.

*Sahelanthropus* and other early hominins shared some of the derived characters of humans. For example, they had reduced canine teeth, and some fossils suggest that they had relatively flat faces. They also show signs of having been more upright and bipedal than other apes.

Fossil evidence suggests that early hominins stood upright and were bipedal long before their brains increased in size. Consider *Ardipithecus ramidus* **(Figure 27.34)**. This species showed signs of bipedalism, yet its brain (325 cm$^3$ in volume) was much smaller than that of *H. sapiens* (1,300 cm$^3$). Fossils show that by 2.5 million years ago, our extinct relatives walked upright and used tools—yet they still had a brain the size of a softball.

The earliest fossils placed in our genus, *Homo*, include those of *Homo habilis*, which lived 2.4 to 1.6 million years ago. Compared to earlier hominins, *H. habilis* had a shorter jaw and a larger brain volume, about 675 cm$^3$. Brain size, body size, and tool use continued to increase over time in various fossil *Homo* species, some of which may have lived as recently as from 40,000 to 18,000 years ago. One such species, *H. neanderthalensis* (commonly called a Neanderthal), lived in Europe by 350,000 years ago. Neanderthals had a brain larger than that of present-day humans, buried their dead, and made hunting tools from stone and wood. But despite their adaptations and culture, Neanderthals became extinct at some point between 28,000 and 40,000 years ago—by which time our own species had reached Europe and many other regions around the globe.

### Homo sapiens

Fossil evidence indicates that the ancestors of humans originated in Africa. Older species gave rise to later species, ultimately including our own species, *H. sapiens*. The oldest known fossils of our species have been found at two different sites in Ethiopia and include specimens that are

**▲ Figure 27.34 The skeleton of "Ardi," a 4.4-million-year-old fossil of *Ardipithecus ramidus*.**

195,000 and 160,000 years old **(Figure 27.35)**. Recent DNA analyses have also shed light on human origins, revealing, for example, that all living humans are more closely related to one another than to Neanderthals, and that all living humans have ancestors that originated as *H. sapiens* in Africa. (Such studies are possible because scientists have been able to extract DNA from Neanderthal fossils and sequence the Neanderthal genome.)

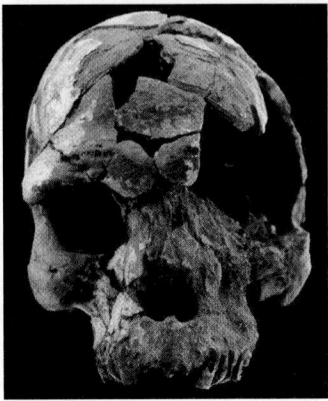

▲ **Figure 27.35 A 160,000-year-old fossil of *Homo sapiens*.**

The oldest fossils of *H. sapiens* outside Africa are from the Middle East and date back about 115,000 years. Fossil evidence and genetic analyses suggest that humans spread beyond Africa in one or more waves, first into Asia and then to Europe and Australia. The date of the first arrival of humans in the New World is uncertain, although the oldest generally accepted evidence puts that date at about 15,000 years ago.

As humans spread from Africa to the rest of the world, they entered regions inhabited by Neanderthals and other (now-extinct) hominins. A long-standing question is whether humans mated with any of these species, leading to interspecific gene flow. Researchers have argued that evidence of such gene flow can be found in fossils that show a mixture of human and Neanderthal characteristics. Others have disputed this conclusion. In 2015, the most extensive evidence yet of such gene flow was reported: Analysis of a human jawbone excavated from a Romanian cave showed that it contained long stretches of Neanderthal DNA **(Figure 27.36)**. In fact, the amount of Neanderthal DNA indicated that this individual's great-great-great-grandparent was a Neanderthal.

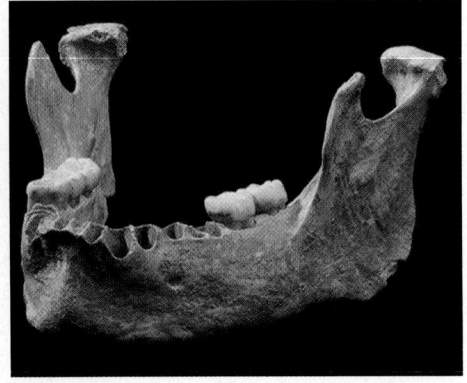

► **Figure 27.36 Fossil evidence of human-Neanderthal interbreeding.** This jawbone belonged to a human who lived about 40,000 years ago and had a relatively recent Neanderthal ancestor.

## CONCEPT CHECK 27.6

1. Describe three key amniote adaptations for life on land.
2. Why is it incorrect to say that humans evolved from chimps?
3. **WHAT IF?** Which came first, the chicken or the egg? Explain, basing your answer on evolutionary principles.

For suggested answers, see Appendix A.

# Animals have transformed ecosystems and altered the course of evolution

The rise of animals coincided with one of the most monumental changes in the history of life: the transformation of a microbe-only world to a world filled with large producers, predators, and prey. This change affected all aspects of ecological communities, in the sea and on land.

## Ecological Effects of Animals

Until 600 million years ago, life in the oceans was almost entirely microscopic. Among other differences from life today, early marine communities had no large *suspension feeders* (which remove food particles suspended in water). As a result, researchers think that ocean waters were cloudy, thick with microorganisms and suspended organic matter **(Figure 27.37a)**. Geologic and fossil evidence suggests that these turbid waters also had low oxygen levels and were dominated by cyanobacteria. Marine ecosystems remained in this condition for over a billion years, despite the fact that algae and a variety of heterotrophic eukaryotes were present for most of that time. What changes did the rise of animals bring?

### Marine Ecosystems

Fossil biochemical evidence suggests that the abundance of cyanobacteria decreased in the early Cambrian. This decrease may have been caused by the activities of crustaceans and other animals with suspension-feeding mouthparts. Such suspension feeders can process an enormous amount of water: Every 20 days, animals filter a volume of ocean water that is estimated to be equal to that in which most organisms live (the top 500 m). As early suspension-feeding animals removed

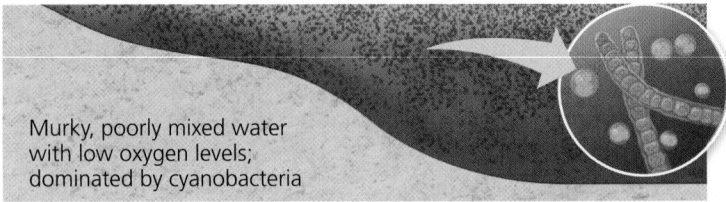

Murky, poorly mixed water with low oxygen levels; dominated by cyanobacteria

**(a) Ocean conditions before 600 mya: before origin of suspension feeders**

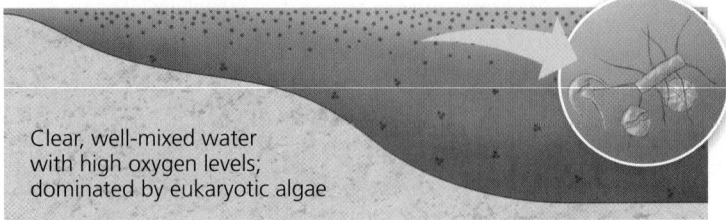

Clear, well-mixed water with high oxygen levels; dominated by eukaryotic algae

**(b) Changes to ocean conditions by 530 mya: after origin of suspension feeders**

▲ **Figure 27.37 A sea change for Earth's oceans.**

cyanobacteria and other suspended matter from the water, the ocean waters would have become clearer. As a result, algae, which require more light for photosynthesis than do cyanobacteria, increased in abundance and moved to deeper waters **(Figure 27.37b)**.

Along with changes in water clarity and a shift to algae as the dominant producers, a different set of feeding relationships also emerged. A host of small animals that ate marine producers and detritus evolved. Those small animals, in turn, were eaten by larger animals—which were themselves eaten by still larger animals. Overall, the explosion of animal diversity in the early Cambrian marked the end of the microbial world and the beginning of ocean life as we know it today—a world filled with predators, grazers, suspension feeders, and scavengers of all shapes and sizes.

### Terrestrial Ecosystems

Before animals joined plants and fungi onshore, terrestrial ecosystems had a simple structure: Producers (early plants) harnessed energy from the sun and drew essential nutrients from the soil, while decomposers (fungi and bacteria) returned nutrients to the soil. By 410 million years ago, animals had transformed these ecosystems. Plants and decomposers continued to be important, of course, but new biotic interactions were also in place: Plants were being consumed by herbivorous animals, and they, in turn, were being eaten by predators. Still other animals (detritivores) consumed organic debris, making for a complex network of ecological interactions—much of it driven by animals.

The lesser snow goose (*Chen caerulescens*), a migratory bird that breeds in marsh lands bordering Canada's Hudson Bay, illustrates the impact that animals can have on terrestrial communities. These birds feed on grasses and other marsh plants. At low population numbers, lesser snow geese improve the growth of marsh plants. This positive effect may be due to the fact that the birds defecate every few minutes as they feed, thereby adding nitrogen (which plants need to grow) to the soil. At high numbers, however, the feeding activities of the birds can destroy a marsh, converting it to a mudflat **(Figure 27.38)**.

Predators can have equally large effects. For example, the introduction of the arctic fox (*Vulpes lagopus*) to islands between Alaska and Russia transformed habitats on these islands from grasslands to tundra. Field observations and experimental results indicate that this change occurred because the foxes fed upon seabirds, drastically reducing their numbers—and that in turn meant less bird guano, the primary source of nutrients for plants on these islands. The reduction in nutrients favored slow-growing tundra shrubs instead of grasses, causing the grasslands to be replaced.

## Evolutionary Effects of Animals

The rise of animals also set in motion a series of profound evolutionary changes. As we've seen, many of these changes resulted from the fact that animals can make life dangerous: The origin of mobile, heterotrophic animals with a complete digestive tract drove some species to extinction and initiated ongoing "arms races" between bilaterian predators and prey.

In this section, we'll consider the related topic of whether increases in animal diversity have led to other evolutionary radiations. Then we'll examine the ongoing evolutionary effects of one particular animal species—humans.

### Evolutionary Radiations

Two species that interact can exert selective pressures on one another. A plant (or any other species) that interacts with an animal may evolve in response to selection imposed by the animal—and the animal, in turn, may evolve in response to evolutionary changes in the plant **(Figure 27.39)**. In the

▼ **Figure 27.39 Results of reciprocal selection.** The Madagascar orchid *Angraecum sesquipedale* secretes a sugary nectar solution to the base of its unusually long floral tube. Based on the tube's length, Charles Darwin predicted the existence of a pollinating moth with a 28-cm-long proboscis—long enough to reach the bottom of the tube. Such a moth (*Xanthopan morganii*, shown here) was discovered two decades after Darwin's death.

Proboscis

▶ **Figure 27.38 Effects of herbivory.** The area inside the fence, which the geese could not access, shows the original state of the marsh.

## Understanding Experimental Design and Interpreting Data

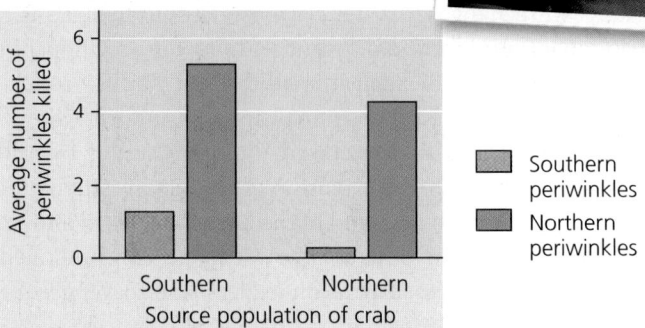

**Is There Evidence of Selection for Defensive Adaptations in Mollusc Populations Exposed to Predators?** The fossil record shows that historically, increased risk to prey species from predators is often accompanied by increased incidence and expression of prey defenses. Researchers tested whether populations of the predatory European green crab (*Carcinus maenas*) have exerted similar selective pressures on its prey, the flat periwinkle (*Littorina obtusata*). Periwinkles from southern sites in the Gulf of Maine have experienced predation by European green crabs for over 100 generations, at about one generation per year. Periwinkles from northern sites in the Gulf have been interacting with the invasive green crabs for relatively few generations, as the invasive crabs spread to the northern Gulf comparatively recently.

Previous research shows that (1) flat periwinkle shells recently collected from the Gulf are thicker than those collected in the late 1800s, and (2) periwinkle populations from southern sites in the Gulf have thicker shells than periwinkle populations from northern sites. In this exercise, you'll interpret the design and results of the researchers' experiment studying the rates of predation by European green crabs on periwinkles from northern and southern populations.

**How the Experiment Was Done** The researchers collected periwinkles and crabs from sites in the northern and southern Gulf of Maine, separated by 450 km of coastline. A single crab was placed in a cage with eight periwinkles of different sizes. After three days, researchers assessed the fate of the eight periwinkles. Four different treatments were set up, with crabs from northern or southern populations offered periwinkles from northern and southern populations. All crabs were of similar size and included equal numbers of males and females. Each experimental treatment was tested 12 to 14 times.

In a second part of the experiment, the bodies of periwinkles from northern and southern populations were removed from their shells and presented to crabs from northern and southern populations.

**Data from the Experiment**

Data from R. Rochette et al., Interaction between an invasive decapod and a native gastropod: Predator foraging tactics and prey architectural defenses, *Marine Ecology Progress Series* 330:179–188 (2007).

When the researchers presented the crabs with unshelled periwinkles, all the unshelled periwinkles were consumed in less than an hour.

**INTERPRET THE DATA**

1. What hypothesis were the researchers testing in this study? What are the independent variables in this study? What are the dependent variables in this study?
2. Why did the research team set up four different treatments?
3. Why did researchers present unshelled periwinkles to the crabs? What do the results of this part of the experiment indicate?
4. Summarize the results of the experiment in words. Do these results support the hypothesis you identified in question 1? Explain.
5. Suggest how natural selection may have affected populations of flat periwinkles in the southern Gulf of Maine over the last 100 years.

(MB) A version of this Scientific Skills Exercise can be assigned in MasteringBiology.

---

**Scientific Skills Exercise,** you can interpret data from a study of selection occurring in a predator-prey interaction over time. Such reciprocal selective pressures also occur when the origin of new species in one group of organisms stimulates further radiations in other organisms, especially those that can eat, escape from, or compete effectively with the new group.

As animal groups have diversified, they have often had this effect. For example, the origin of a new group of animals provides new sources of food for *parasites*, organisms that feed on the tissues of another organism (the *host*). Many parasites feed on a single host species. As a result, the ongoing diversification of animals has led to evolutionary radiations in many groups of parasites—the animals, fungi, protists, and bacteria that can feed on newly evolved animal hosts.

### Human Impacts on Evolution

As can be seen from satellite photographs or the window of an airplane, humans have dramatically altered the global

environment. By making large changes to the environment, we have also altered the selective pressures faced by many species. This suggests that we are likely causing evolutionary change—and we are. For example, by using antibiotics to kill bacteria, we have (inadvertently) caused the evolution of resistance in bacterial populations (see Concept 19.3). We have also caused evolutionary change in species that we hunt for sport or food. For example, in cod fisheries, commercial fishing operations target older and larger fish. This has led to a reduction in the age and size at which individuals reach sexual maturity **(Figure 27.40)**. Natural selection has favored fish that mature at a younger age and smaller size because such individuals are more likely to reproduce before they are caught than are individuals that mature when they are older and larger.

In addition to causing evolution by natural selection, human actions can also drive species to extinction, thereby altering the future course of evolution. Species extinction rates have increased greatly in the last 400 years, raising concern

▲ **Figure 27.40 Reproducing at a younger age.** Age at sexual maturity has dropped over time in heavily fished populations of northern cod (*Gadus morhua*). Size at maturity has also dropped (not shown).

**?** *Fish that reproduce when they are younger and smaller typically have fewer offspring than fish that reproduce when they are older and larger. Predict how evolution in response to fishing will affect the ability of cod populations to recover from overfishing.*

that unless dramatic preventative measures are taken, a sixth, human-caused mass extinction may occur (see Concept 23.2). Among the many taxa under threat, molluscs have the dubious distinction of being the animal group with the largest number of documented extinctions **(Figure 27.41a)**. For example, pearl mussels, a group of freshwater molluscs that can make natural pearls, are among the world's most endangered animals. Thirty of the pearl mussel species that once lived in North America have become extinct in the last 100 years, and nearly 200 of the 270 that remain are threatened by extinction.

Threats faced by pearl mussels and other molluscs include habitat loss, pollution, and competition or predation by non-native species, and overharvesting **(Figure 27.41b)**. Is it too late to protect these molluscs? In some locations, reducing water pollution and changing how water is released from dams have led to dramatic rebounds in pearl mussel populations. Such results provide hope that with corrective measures, other endangered species can be revived.

Our discussion of how humans affect evolution brings this unit on the history of life to an end. But this organization isn't meant to imply that life consists of a ladder leading from lowly microorganisms to lofty humanity. The history of life shows that biological diversity is the product of branching phylogeny, not ladderlike "progress," however we choose to measure it. The fact that there are almost as many species of ray-finned fishes alive today as in all other vertebrate groups combined is a clear indication that our finned relatives are not outmoded underachievers that failed to leave the water. Similarly, the ubiquity of diverse prokaryotes throughout

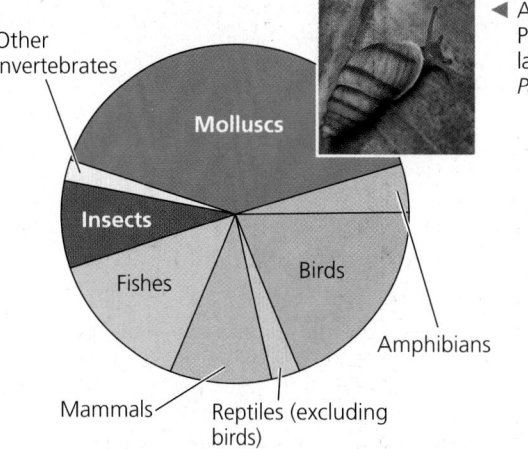

◀ An endangered Pacific island land snail, *Partula suturalis*

**(a)** Recorded extinctions of animal species

**(b)** Workers on a mound of pearl mussels killed to make buttons (ca. 1919)

▲ **Figure 27.41 The silent extinction. (a)** Molluscs account for a largely unheralded but sobering 40% of all documented extinctions of animal species. These extinctions have resulted from habitat loss, pollution, introduced species, overharvesting, and other human actions. Land snails such as the species pictured above are highly vulnerable to these threats; they are among the world's most imperiled animal groups. **(b)** Many pearl mussel populations were driven to extinction by overharvesting for their shells, which were used to make buttons and other goods.

the biosphere today is a reminder of the enduring ability of these relatively simple organisms to keep up with the times through adaptive evolution. Biology exalts all of life's diversity, past and present.

**CONCEPT CHECK 27.7**

1. Describe how ocean communities changed in the early Cambrian period, and explain how animals may have influenced those changes.
2. How did the colonization of land by animals affect terrestrial communities?
3. **MAKE CONNECTIONS** Human actions often break large areas of forest or grassland into small remnant parcels that support fewer individuals and are far apart from one another. Predict how gene flow, genetic drift, and extinction risk would differ between the original and the remnant populations. (Review Concept 21.3.)

For suggested answers, see Appendix A.

# 27 | Chapter Review

Go to **MasteringBiology**® for Assignments, the eText, and the Study Area with Animations, Activities, Vocab Self-Quiz, and Practice Tests.

## SUMMARY OF KEY CONCEPTS

**VOCAB SELF-QUIZ**

goo.gl/gbai8v

### CONCEPT 27.1

**Animals originated more than 700 million years ago (pp. 545–547)**

- The earliest evidence of animal life comes from fossil steroids indicative of sponges that date to 710 million years ago.
- The first fossils of large animals date to 560 million years ago and include sponges as well as fossil organisms that resemble living cnidarians and molluscs.
- Unlike nearly all other animals, sponges lack true tissues. Cnidarians are an early-diverging group of **eumetazoans**, an animal clade whose members have tissues.

**?** *What features are shared by sponges and choanoflagellates? Interpret these observations.*

### CONCEPT 27.2

**The diversity of large animals increased dramatically during the "Cambrian explosion" (pp. 547–549)**

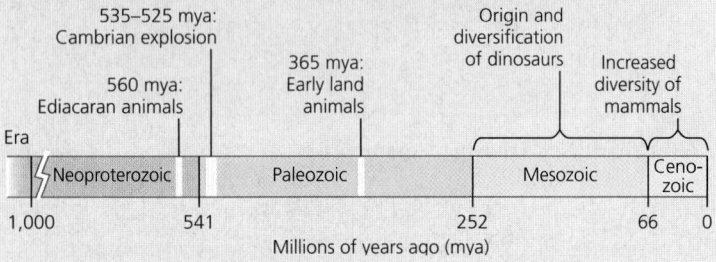

- Prior to the **Cambrian explosion** (535–525 million years ago), virtually all large animals were soft-bodied and poorly defended. Fossils dating to the Cambrian explosion include the oldest known members of many living animal phyla.
- Fossil and molecular evidence suggests that **bilaterians** had emerged by 635 million years ago.

**?** *What caused the Cambrian explosion? Describe current hypotheses.*

### CONCEPT 27.3

**Diverse animal groups radiated in aquatic environments (pp. 549–554)**

- The diverse animals that emerged from the Cambrian explosion can be categorized by their **body plan**, the morphological and developmental traits that are integrated into a functional whole.
- Most living animals are bilaterians, bilaterally symmetric animals with three tissue layers and a complete digestive tract.
- Bilaterally symmetric animals have diverged into three major clades: Lophotrochozoa, Ecdysozoa, and Deuterostomia. These clades include diverse invertebrate phyla whose early members dominated life in the Cambrian oceans.

**DRAW IT** *Draw a phylogenetic tree showing relationships among Lophotrochozoa, Cnidaria, Ecdysozoa, Ctenophora, Porifera, and Deuterostomia. Label the common ancestor of animals and the origin of three germ layers, true tissues, and bilateral symmetry.*

### CONCEPT 27.4

**Vertebrates have been the ocean's dominant predators for more than 400 million years (pp. 554–557)**

- Vertebrates originated 500 million years ago. The earliest vertebrates lacked jaws.
- **Gnathostomes** (vertebrates with jaws) arose 440 million years ago. Three lineages of jawed vertebrates survive today: **chondrichthyans**, **ray-finned fishes**, and **lobe-fins**.

**?** *How would the origin of organisms with jaws have altered ecological interactions? Provide supporting evidence.*

### CONCEPT 27.5

**Several animal groups had features facilitating their colonization of land (pp. 558–562)**

- Unlike plants, whose ancestors colonized land only once, many animal groups have made the transition to terrestrial life.
- Animals that colonized land were "pre-adapted" for their new environment in that they typically had a complete digestive tract and well-developed skeletal, muscle, and nervous systems.
- Arthropods were the first animals to colonize land, about 450 million years ago. The insects radiated explosively and now comprise more known species than all other eukaryotic groups combined.
- Vertebrates colonized land 365 million years ago when early tetrapods arose from aquatic lobe-fins. Early tetrapods remained tied to water, a characteristic they share with most amphibians.

**?** *Describe the origin of tetrapods and describe some of their key derived traits.*

### CONCEPT 27.6

**Amniotes have key adaptations for life in a wide range of terrestrial environments (pp. 562–568)**

- A more extensive colonization of dry habitats occurred in **amniotes**, a group of tetrapods that originated 350 million years ago and whose living members are reptiles and mammals.
- Amniote adaptations for life on land include the **amniotic egg**, rib cage ventilation, and skin (covered by scales or hair) that limits evaporative water loss.
- Extant **reptiles** include tuataras, lizards and snakes, turtles, crocodilians, and birds. Birds are the most diverse reptile group; many traits in birds are adaptations facilitating flight.
- Unique traits of **mammals** include mammary glands and hair. Mammals have diversified into the egg-laying **monotremes**, the pouched **marsupials**, and the placental **eutherians**.
- **Hominins**—humans and extinct species more closely related to humans than to chimpanzees—originated in Africa 6–7 million years ago. Early hominins walked upright but had a small brain. *Homo sapiens* originated in Africa 195,000 years ago and spread from there to other continents.

**?** *Describe the amniotic egg and evaluate its significance.*

### CONCEPT 27.7

**Animals have transformed ecosystems and altered the course of evolution (pp. 568–571)**

- The rise of animals coincided with the change from a microbe-only world to a world filled with large producers, scavengers, predators, and prey.

- The origin of animals with suspension-feeding mouthparts may have caused sweeping changes in early oceans, such as an increase in water clarity and a shift from cyanobacteria to algae as the dominant producers.
- The diversification of bilaterians in the sea and on land has changed biotic interactions and stimulated evolutionary radiations in other groups of organisms.
- Human actions have caused evolution by natural selection and have the potential to cause a mass extinction.

**?** *Explain how the activities of animals (including humans) can lead to evolutionary change, and provide an example.*

## TEST YOUR UNDERSTANDING

### Level 1: Knowledge/Comprehension

**1.** Which of the following clades contains the greatest number of animal species?
(A) the vertebrates     (C) the deuterostomes
(B) the bilaterians     (D) the insects

**2.** Fossil steroid and molecular clock evidence suggests that animals originated
(A) between 710 and 770 million years ago.
(B) more than 100 million years before the oldest known fossils of large animals.
(C) during the Cambrian explosion.
(D) both A and B

**3.** Which of the following was probably the *least* important factor in bringing about the Cambrian explosion?
(A) the emergence of predator-prey relationships among animals
(B) the accumulation of sufficient atmospheric oxygen to support the more active metabolism of mobile animals
(C) the movement of animals onto land
(D) the origin of *Hox* genes and other genetic changes affecting the regulation of developmental genes

**4.** Which of the following could be considered the most recent common ancestor of living tetrapods?
(A) a sturdy-finned, shallow-water lobe-fin whose appendages had skeletal supports similar to those of terrestrial vertebrates
(B) an armored gnathostome with two pairs of appendages
(C) an early ray-finned fish that developed bony skeletal supports in its paired fins
(D) a salamander that had legs supported by a bony skeleton but moved with the side-to-side bending typical of fishes

### Level 2: Application/Analysis

**5.** Which clade does *not* include humans?
(A) synapsids     (C) lophotrochozoans
(B) lobe-fins     (D) tetrapods

**6.** In Figure 27.10, the clade Deuterostomia is most closely related to which two main clades?
(A) Ctenophora and Cnidaria
(B) Lophotrochozoa and Ecdysozoa
(C) Cnidaria and Platyhelminthes
(D) Echinodermata and Hemichordata

### Level 3: Synthesis/Evaluation

**7. SCIENTIFIC INQUIRY**

**DRAW IT** As a consequence of size alone, organisms that are large tend to have larger brains than organisms that are small. However, some organisms have brains that are considerably larger than expected for an animal of their size. There are high energy costs associated with the development and maintenance of brains that are large relative to body size.

(a) The fossil record documents trends in which brains that are large relative to body size evolved in certain lineages, including ancestors of humans. In such lineages, what can you infer about the relative costs and benefits of large brains?
(b) Hypothesize how natural selection might favor the evolution of large brains despite their high maintenance costs.
(c) Data for 14 bird species are listed below. Graph the data, placing deviation from expected brain size on the *x*-axis and mortality rate on the *y*-axis. What can you conclude about the relationship between brain size and mortality?

| Deviation from Expected Brain Size* | –2.4 | –2.1 | –2.0 | –1.8 | –1.0 | 0.0 | 0.3 | 0.7 | 1.2 | 1.3 | 2.0 | 2.3 | 3.0 | 3.2 |
|---|---|---|---|---|---|---|---|---|---|---|---|---|---|---|
| Mortality Rate | 0.9 | 0.7 | 0.5 | 0.9 | 0.4 | 0.7 | 0.8 | 0.4 | 0.8 | 0.3 | 0.6 | 0.6 | 0.3 | 0.6 |

**Data from** D. Sol et al., Big-brained birds survive better in nature, *Proceedings of the Royal Society B* 274:763–769 (2007).

*Values < 0 indicate brain sizes smaller than expected; values > 0 indicate sizes larger than expected.

**8. FOCUS ON EVOLUTION**
In Figure 27.29, circle the smallest monophyletic group that includes dinosaurs. Explain your answer and list the taxa that are in this clade. Knowing that birds are endothermic and crocodiles are ectothermic, can you use phylogenetic bracketing (see Concept 20.3) to predict whether dinosaurs other than birds are ectothermic or endothermic? Explain.

**9. FOCUS ON ORGANIZATION**
Early tetrapods had a sprawling gait, like that of a lizard: As the right front foot moved forward, the body twisted to the left and the left rib cage and lung were compressed; the reverse occurred with the next step. Normal breathing, in which both lungs expand equally with each breath, was hindered during walking and prevented during running. In a short essay (100–150 words), explain how the origin of dinosaurs, whose gait allowed them to move without compressing their lungs, could have led to emergent properties in biological communities.

**10. SYNTHESIZE YOUR KNOWLEDGE**

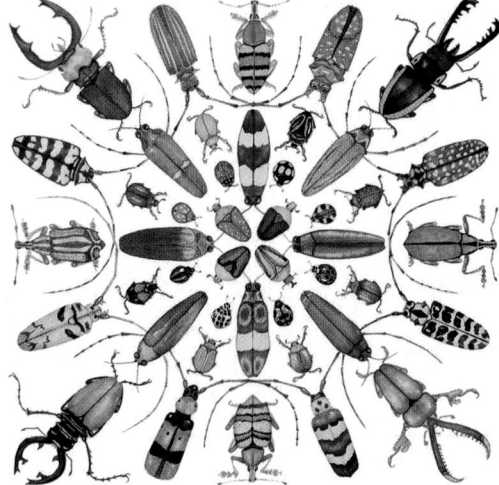

Collectively, do these beetles and all other invertebrate species combined form a monophyletic group? Explain your answer and give an overview of invertebrate evolutionary history.

*For selected answers, see Appendix A.*

# Unit 5 Plant Form and Function

## 28 Plant Structure and Growth

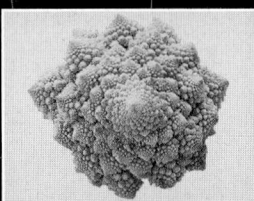

Plants make their living by harvesting the energy of light by photosynthesis. The ability of plants to photosynthesize efficiently depends on the **structure and function** of roots, stems, and leaves and their **growth**.

## 29 Resource Acquisition, Nutrition, and Transport in Vascular Plants

Land plants inhabit two worlds: above ground where they gather carbon dioxide and sunlight for photosynthesis, and below ground where they acquire water and minerals. In vascular plants, these **resources are transported** throughout the plant by the vascular system.

## 30 Reproduction and Domestication of Flowering Plants

**Reproduction** is a fundamental feature of all life, including plants. In addition to sexual reproduction, many plants have the ability to reproduce asexually. Plant reproduction is central to human survival because it serves as the basis for most agriculture.

## 31 Plant Responses to Internal and External Signals

Plants are exquisitely sensitive organisms; for the most part, they respond to challenges and opportunities in their environments not by movement as do animals but by growth. Many chemical regulators are involved in coordinating **plant responses to the environment.**

### Plant Form and Function

- 28 Plant Structure and Growth
- 29 Resource Acquistion, Nutrition, and Transport in Vascular Plants
- 30 Reproduction and Domestication of Flowering Plants
- 31 Plant Responses to Internal and External Signals

History of Life
Evolution
Genetics
Chemistry and Cells
Ecology
Animals

# 28 Plant Structure and Growth

## KEY CONCEPTS

**28.1** Plants have a hierarchical organization consisting of organs, tissues, and cells

**28.2** Different meristems generate new cells for primary and secondary growth

**28.3** Primary growth lengthens roots and shoots

**28.4** Secondary growth increases the diameter of stems and roots in woody plants

▲ **Figure 28.1** Computer art?

## Are Plants Computers?

The object in **Figure 28.1** is not the creation of a computer genius with a flair for the artistic. It is a head of romanesco, an edible relative of broccoli. Romanesco's mesmerizing beauty is attributable to the fact that each of its smaller buds resembles in miniature the entire vegetable. (Mathematicians refer to such repetitive patterns as *fractals*.) If romanesco looks as if it were generated by a computer, it's because its growth pattern follows a repetitive sequence of instructions. As in most plants, the growing shoot tips lay down a pattern of stem . . . leaf . . . bud, over and over again. These repetitive developmental patterns are genetically determined and subject to natural selection. For example, a mutation that shortens the stem segments between leaves will generate a bushier plant. If this altered architecture enhances the plant's ability to access resources such as light and, by doing so, to produce more offspring, then this trait will occur more frequently in later generations—the population will have evolved.

Romanesco is unusual in adhering so rigidly to its basic body organization. Most plants show much greater diversity in their individual forms because the growth of most plants, much more than in animals, is affected by local environmental conditions. All adult lions, for example, have four legs and are of roughly the same size, but oak trees vary in the number and arrangement of their branches. This is because plants respond to challenges and opportunities in their local environment by altering their growth. (In contrast, animals typically respond by movement.) Illumination of a plant from the side, for example, creates asymmetries in its basic body plan. Branches grow more quickly from the illuminated side of a shoot than from the shaded side, an architectural change of obvious benefit for photosynthesis. Changes in growth and development are critical in facilitating the plant's acquisition of resources from the local environment.

Chapter 26 described the evolution of nonvascular and vascular plants. In this chapter we focus on vascular plants, particularly angiosperms (flowering plants) because they are the primary producers in many terrestrial ecosystems and are of great agricultural importance. Taxonomists split the angiosperms into two major clades: *Monocots* typically have a single cotyledon (seed leaf), whereas *eudicots* typically have two. Monocots and eudicots have several other structural differences as well **(Figure 28.2)**. We'll explore the structure and growth of both of these types of flowering plants.

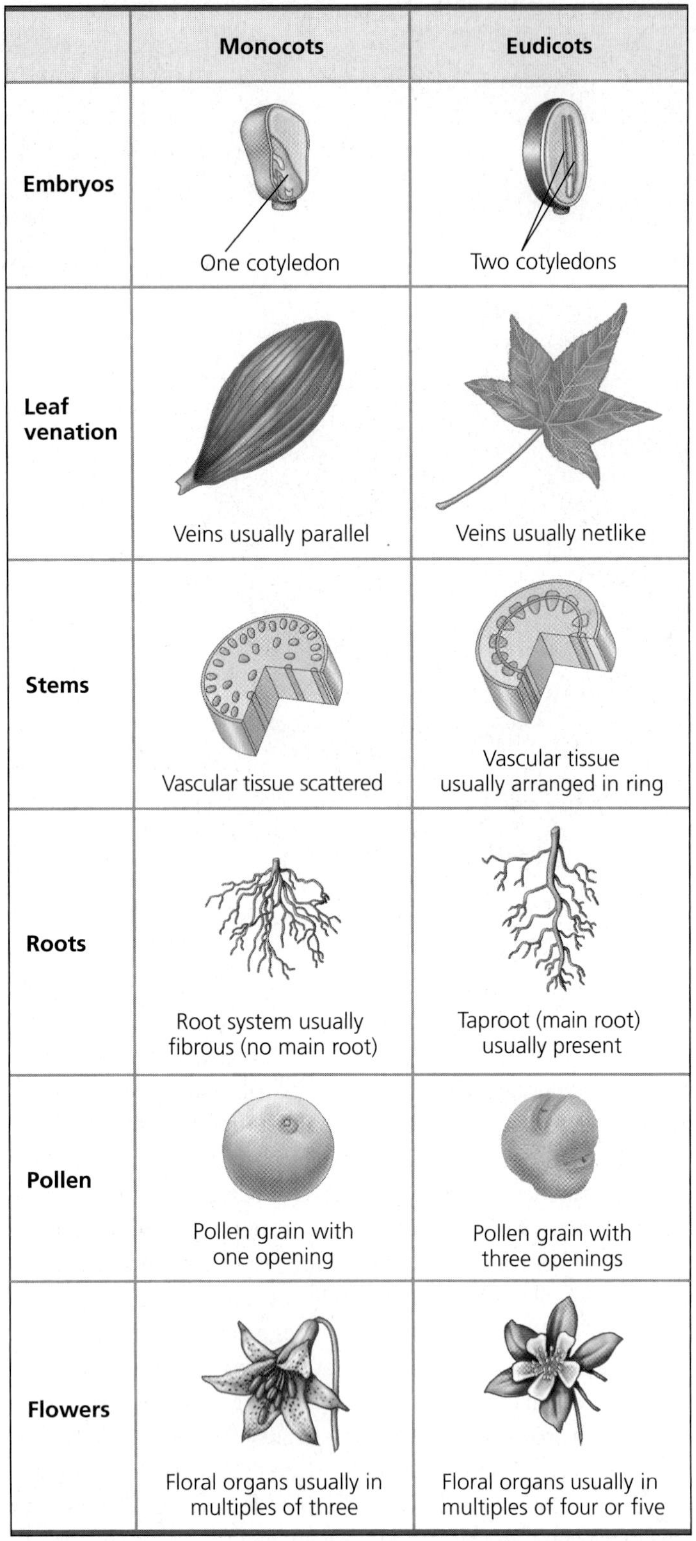

| | Monocots | Eudicots |
|---|---|---|
| **Embryos** | One cotyledon | Two cotyledons |
| **Leaf venation** | Veins usually parallel | Veins usually netlike |
| **Stems** | Vascular tissue scattered | Vascular tissue usually arranged in ring |
| **Roots** | Root system usually fibrous (no main root) | Taproot (main root) usually present |
| **Pollen** | Pollen grain with one opening | Pollen grain with three openings |
| **Flowers** | Floral organs usually in multiples of three | Floral organs usually in multiples of four or five |

▲ **Figure 28.2 A comparison of monocots and eudicots.** These classes of angiosperms are named for the number of cotyledons (seed leaves) they possess. Monocots typically have one cotyledon. Eudicots typically have two cotyledons. Monocots include orchids, bamboos, palms, and lilies, as well as grasses, such as wheat, maize, and rice. A few examples of eudicots are beans, sunflowers, maples, and oaks.

# Plants have a hierarchical organization consisting of organs, tissues, and cells

Plants, like most animals, are composed of organs, tissues, and cells. An **organ** consists of several types of tissues that together carry out particular functions. A **tissue** is a group of cells consisting of one or more cell types that together perform a specialized function. In considering this hierarchy of structures, we begin with plant organs because they are most familiar. Vegetative growth—the production of leaves, stems, and roots—is only one stage in a plant's life. Most plants also undergo growth relating to sexual reproduction. In angiosperms, reproductive growth is associated with the production of flowers, a topic we'll visit in Concept 30.1.

## The Three Basic Plant Organs: Roots, Stems, and Leaves

**EVOLUTION** The basic morphology of vascular plants reflects their evolutionary history as terrestrial organisms that inhabit and draw resources from two very different environments—below the ground and above the ground. A typical plant must absorb water and minerals from below the ground surface and $CO_2$ and light from above. The ability to acquire these resources efficiently is traceable to the evolution of three basic organs—roots, stems, and leaves. These organs form a **root system** and a **shoot system**, the latter consisting of stems and leaves **(Figure 28.3)**. Vascular plants rely on both systems for survival. Roots are almost never photosynthetic; they starve unless *photosynthates*, the sugars and other carbohydrates produced during photosynthesis, are imported from the shoot system. Conversely, the shoot system depends on the water and minerals that roots absorb from the soil.

### Roots

A **root** is an organ that anchors a vascular plant in the soil, absorbs minerals and water, and often stores carbohydrates. Tall, erect plants with large shoot masses generally have a *taproot system*, consisting of one main vertical root, the **taproot**, which penetrates the soil deeply and helps prevent the plant from toppling. In taproot systems, the role of absorption is restricted largely to tips of **lateral roots**, which branch off from the taproot (see Figure 28.3). A taproot, although energetically expensive to make, allows the plant to be taller, thereby giving it access to more favorable light conditions and, in some cases, providing an advantage for pollen and seed dispersal.

Small plants or those that have a trailing growth habit are particularly susceptible to grazing animals that can potentially uproot the plant and kill it. Such plants are most efficiently anchored by a *fibrous root system*, a mat of thin roots spreading out below the soil surface (see Figure 28.2). In plants that have

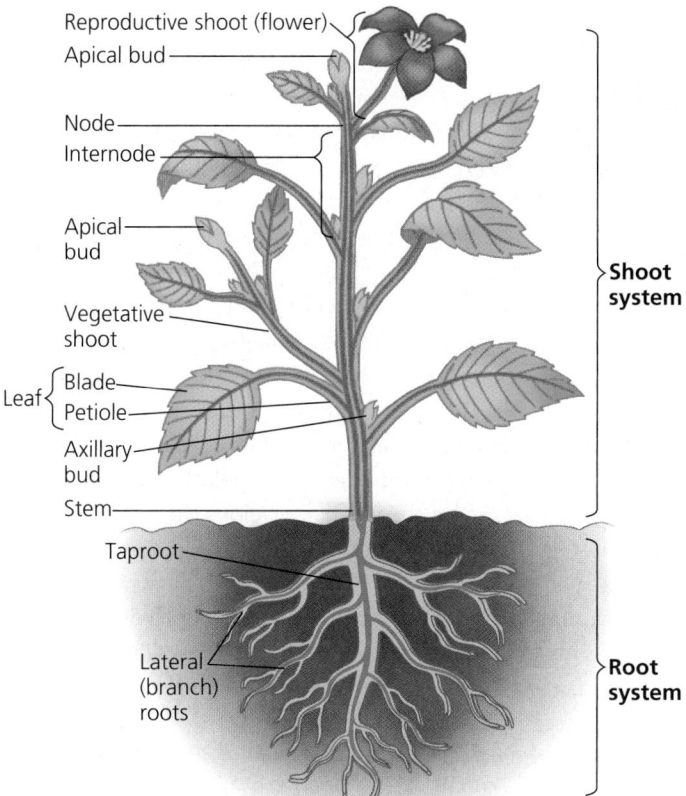

Reproductive shoot (flower)
Apical bud
Node
Internode
Apical bud
Vegetative shoot
Leaf { Blade, Petiole }
Axillary bud
Stem
Shoot system

Taproot
Lateral (branch) roots
Root system

▲ **Figure 28.3 An overview of a flowering plant.** The plant body is divided into a root system and a shoot system, connected by vascular tissue (purple strands in this diagram) that is continuous throughout the plant. The plant shown is an idealized eudicot.

fibrous root systems, including most monocots, the primary root dies early on and does not form a taproot. Instead, many small roots emerge from the stem. Such roots are said to be *adventitious*, a term describing a plant organ that grows from an unusual source, such as roots arising from stems or leaves. Each root forms its own lateral roots, which in turn form their own lateral roots, thereby creating a thick mat of slender roots. Because this mat of roots holds the topsoil in place, plants such as grasses that have dense fibrous root systems are especially good at preventing soil erosion.

In most plants, the absorption of water and minerals occurs primarily near the tips of elongating roots, where vast numbers of **root hairs**, thin, finger-like extensions of root epidermal cells, emerge and increase the surface area of the root enormously **(Figure 28.4)**. Most terrestrial plant root systems

◀ **Figure 28.4 Root hairs of a radish seedling.** Root hairs grow by the thousands near the tip of each root. By increasing the root's surface area, they greatly enhance the absorption of water and minerals from the soil.

also form *mycorrhizal associations*, symbiotic interactions with soil fungi that increase a plant's ability to absorb minerals (see Figure 29.14). Many plants have root adaptations with specialized functions **(Figure 28.5)**. Some of these arise from the roots, and others are adventitious, developing from stems

▲ **Pneumatophores.** Also known as air roots, pneumatophores are produced by trees such as mangroves that inhabit tidal swamps. By projecting above the water's surface at low tide, they enable the root system to obtain oxygen, which is lacking in the thick, waterlogged mud.

▲ **Storage roots.** Many plants, such as the common beet, store food and water in their roots.

▶ **"Strangling" aerial roots.** Strangler fig seeds germinate in crevices of tall trees. Aerial roots grow to the ground, wrapping around the host tree and objects such as this Cambodian temple. Shoots grow upward and shade out the host tree, killing it.

▲ **Figure 28.5 Evolutionary adaptations of roots.**

or, in rare cases, leaves. Some modified roots add support and anchorage. Others store water and nutrients or absorb oxygen from the air.

### Stems

A **stem** is a plant organ bearing leaves and buds. Its chief function is to elongate and orient the shoot in a way that maximizes photosynthesis by the leaves. Another function of stems is to elevate reproductive structures, thereby facilitating the dispersal of pollen and fruit. Green stems may also perform a limited amount of photosynthesis. Each stem consists of an alternating system of **nodes**, the points at which leaves are attached, and **internodes**, the stem segments between nodes (see Figure 28.3). Most of the growth of a young shoot is concentrated near the growing shoot tip, or **apical bud**. Apical buds are not the only types of buds found in shoots. In the upper angle (axil) formed by each leaf and the stem is an **axillary bud**, which can potentially form a lateral branch or, in some cases, a thorn or flower.

Some plants have stems with alternative functions, such as food storage or asexual reproduction. Many of these modified stems, including rhizomes, bulbs, stolons, and tubers, are often mistaken for roots **(Figure 28.6)**.

◀ **Rhizomes.** The base of this iris plant is an example of a rhizome, a horizontal shoot that grows just below the surface. Vertical shoots emerge from axillary buds on the rhizome.

▶ **Stolons.** Shown here on a strawberry plant, stolons are horizontal shoots that grow along the surface. These "runners" enable a plant to reproduce asexually, as plantlets form at nodes along each runner.

◀ **Tubers.** Tubers, such as these potatoes, are enlarged ends of rhizomes or stolons specialized for storing food. The "eyes" of a potato are clusters of axillary buds that mark the nodes.

▲ **Figure 28.6 Evolutionary adaptations of stems.**

### Leaves

In most vascular plants, the **leaf** is the main photosynthetic organ. In addition to intercepting light, leaves exchange gases with the atmosphere, dissipate heat, and defend themselves from herbivores and pathogens. These functions may have conflicting physiological, anatomical, or morphological requirements. For example, a dense covering of hairs may help repel herbivorous insects but may also trap air near the leaf surface, thereby reducing gas exchange and, consequently, photosynthesis. Because of these conflicting demands and trade-offs, leaves vary extensively in form. In general, however, a leaf consists of a flattened **blade** and a stalk, the **petiole**, which joins the leaf to the stem at a node (see Figure 28.3). Grasses and many other monocots lack petioles; instead, the base of the leaf forms a sheath that envelops the stem.

Monocots and eudicots differ in the arrangement of **veins**, the vascular tissue of leaves. Most monocots have parallel major veins of equal diameter that run the length of the blade. Eudicots generally have a branched network of veins arising from a major vein (the midrib) that runs down the center of the blade (see Figure 28.2).

The morphological features of leaves are often products of genetic programs that are tweaked to varying extents by environmental influences. Interpret the data in the **Scientific Skills Exercise** to explore the roles of genetics and the environment in determining the leaf morphology in red maple trees.

Almost all leaves are specialized for photosynthesis. However, some species have leaves with adaptations that enable them to perform additional functions, such as support, protection, storage, or reproduction **(Figure 28.7)**.

## Dermal, Vascular, and Ground Tissue Systems

All plant organs—roots, stems, and leaves—are composed of three fundamental tissue systems: dermal, vascular, and ground tissue. These tissue systems are continuous throughout the plant, but their specific characteristics and spatial relationships to one another vary in different organs **(Figure 28.8)**.

The **dermal tissue system** is the plant's outer protective covering. Like our skin, it forms the first line of defense against physical damage and pathogens. In nonwoody plants, it consists of a single tissue called the **epidermis**, a layer of tightly packed cells. In leaves and most stems, the **cuticle**, a waxy coating on the epidermal surface, helps prevent water loss. In woody plants, protective tissues called **periderm** replace the epidermis in older regions of stems and roots. In addition to protecting the plant from water loss and disease, the epidermis has specialized characteristics in each organ. For example, a root hair is an extension of an epidermal cell near the tip of a root. *Trichomes* are hairlike outgrowths of the shoot epidermis. In some desert species, trichomes reduce water loss and reflect excess light, but their most common function is to defend against herbivores and pathogens by forming a mechanical barrier or secreting chemicals.

► **Tendrils.** The tendrils by which this pea plant clings to a support are modified leaves. After it has "lassoed" a support, a tendril forms a coil that brings the plant closer to the support. Tendrils are typically modified leaves, but some tendrils are modified stems, as in grapevines.

◄ **Spines.** The spines of cacti, such as this prickly pear, are actually leaves; photosynthesis is carried out by the fleshy green stems.

◄ **Storage leaves.** Bulbs, such as this cut onion, have a short underground stem and modified leaves that store food.

Plantlet

Storage leaves

Stem

◄ **Reproductive leaves.** The leaves of some succulents, such as *Kalanchoë daigremontiana*, produce adventitious plantlets, which fall off the leaf and take root in the soil.

▲ **Figure 28.7 Evolutionary adaptations of leaves.**

The chief functions of the **vascular tissue system** are to facilitate the transport of materials through the plant and to provide mechanical support. The two types of vascular tissues are xylem and phloem. **Xylem** conducts water and dissolved minerals upward from roots into the shoots. **Phloem** transports sugars, the products of photosynthesis, from where they are made (usually the leaves) to where they are needed or stored—usually roots and sites of growth, such as developing leaves and fruits. The vascular tissue of a root or stem is collectively called the **stele** (the Greek word for "pillar"). The arrangement of the stele varies, depending on the species and organ. In angiosperms, for example, the root stele is a solid central *vascular cylinder* of xylem and phloem, whereas the stele of stems and leaves consists of *vascular bundles*, separate strands containing xylem and phloem (see Figure 28.8). Both xylem and phloem are composed of a variety of cell types, including cells that are highly specialized for transport or support.

Tissues that are neither dermal nor vascular are part of the **ground tissue system**. Ground tissue that is internal to the vascular tissue is known as **pith**, and ground tissue that is external to the vascular tissue is called **cortex**. The ground tissue system is not just filler. It may include cells specialized for functions such as storage, photosynthesis, support, or short-distance transport.

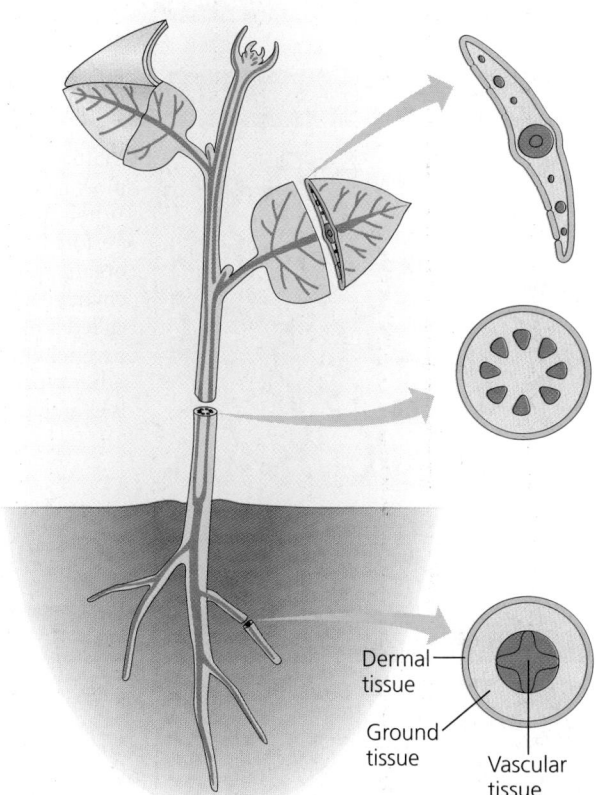

Dermal tissue

Ground tissue

Vascular tissue

▲ **Figure 28.8 The three tissue systems.** The dermal tissue system (blue) provides a protective cover for the entire body of a plant. The vascular tissue system (purple), which transports materials between the root and shoot systems, is also continuous throughout the plant but is arranged differently in each organ. The ground tissue system (yellow), which is responsible for most of the plant's metabolic functions, is located between the dermal tissue and the vascular tissue in each organ.

## Common Types of Plant Cells

**Figure 28.9** focuses on the major types of plant cells: parenchyma cells, collenchyma cells, sclerenchyma cells, the water-conducting cells of the xylem, and the sugar-conducting cells of the phloem. Notice the structural adaptations in the different cells that make their specific functions possible. You may also wish to review basic plant cell structure (see Figures 4.7 and 4.25).

**CONCEPT CHECK 28.1**

1. How does the vascular tissue system enable leaves and roots to function together in supporting growth and development of the whole plant?
2. **WHAT IF?** If humans were photoautotrophs, making food by capturing light energy for photosynthesis, how might our anatomy be different?
3. **MAKE CONNECTIONS** Explain how central vacuoles and cellulose cell walls contribute to plant growth (see Concepts 4.4 and 4.7).

For suggested answers, see Appendix A.

## Parenchyma Cells

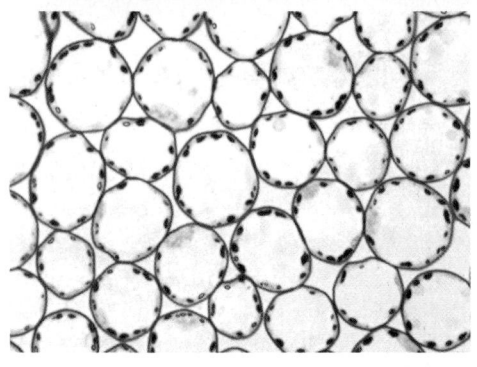

Mature **parenchyma cells** have primary walls that are relatively thin and flexible, and most lack secondary walls. (See Figure 4.25 to review primary and secondary cell walls.) When mature, parenchyma cells generally have a large central vacuole. Parenchyma cells perform most of the metabolic functions of the plant, synthesizing and storing various organic products. For example, photosynthesis occurs within the chloroplasts of parenchyma cells in the leaf. Some parenchyma cells in stems and roots have colorless plastids called amyloplasts that store starch. The fleshy tissue of many fruits is composed mainly of parenchyma cells. Most parenchyma cells retain the ability to divide and differentiate into other types of plant cells under particular conditions—during wound repair, for example. It is even possible to grow an entire plant from a single parenchyma cell.

**Parenchyma cells** in a privet (*Ligustrum*) leaf (LM)    25 μm

## Collenchyma Cells

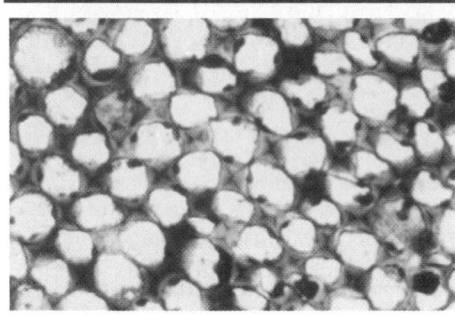

Grouped in strands, **collenchyma cells** (seen here in cross section) help support young parts of the plant shoot. Collenchyma cells are generally elongated cells that have thicker primary walls than parenchyma cells, though the walls are unevenly thickened. Young stems and petioles often have strands of collenchyma cells just below their epidermis. Collenchyma cells provide flexible support without restraining growth. At maturity, these cells are living and flexible, elongating with the stems and leaves they support—unlike sclerenchyma cells, which we discuss next.

**Collenchyma cells** (in *Helianthus* stem) (LM)    5 μm

## Sclerenchyma Cells

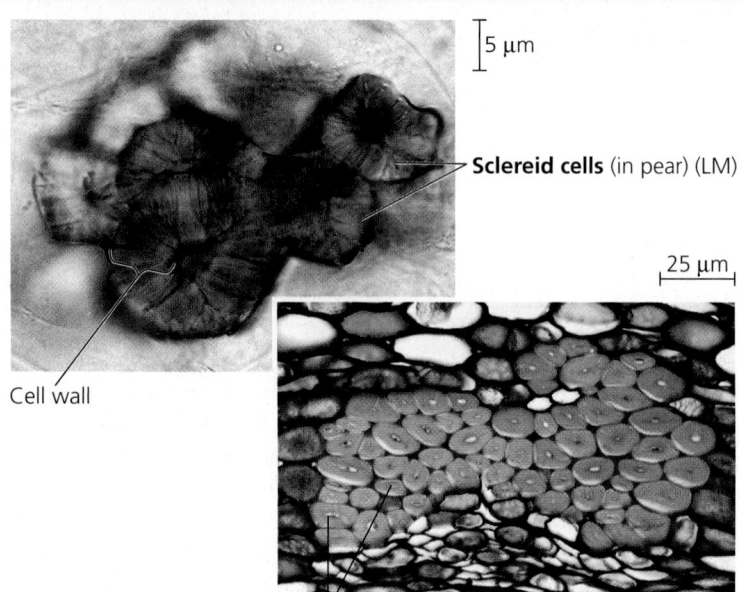

5 μm

**Sclereid cells** (in pear) (LM)

Cell wall

25 μm

**Fiber cells** (cross section from ash tree) (LM)

**Sclerenchyma cells** also function as supporting elements in the plant but are much more rigid than collenchyma cells. In sclerenchyma cells, the secondary cell wall, produced after cell elongation has ceased, is thick and contains large amounts of **lignin**, a relatively indigestible strengthening polymer that accounts for more than a quarter of the dry mass of wood. Lignin is present in all vascular plants but not in bryophytes. Mature sclerenchyma cells cannot elongate, and they occur in regions of the plant that have stopped growing in length. Sclerenchyma cells are so specialized for support that many are dead at functional maturity, but they produce secondary walls before the protoplast (the living part of the cell) dies. The rigid walls remain as a "skeleton" that supports the plant, in some cases for hundreds of years.

Two types of sclerenchyma cells, known as **sclereids** and **fibers**, are specialized entirely for support and strengthening. Sclereids, which are boxier than fibers and irregular in shape, have very thick, lignified secondary walls. Sclereids impart the hardness to nutshells and seed coats and the gritty texture to pear fruits. Fibers, which are usually grouped in strands, are long, slender, and tapered. Some are used commercially, such as hemp fibers for making rope and flax fibers for weaving into linen.

## Water-Conducting Cells of the Xylem

The two types of water-conducting cells, **tracheids** and **vessel elements**, are tubular, elongated cells that are dead and lignified at functional maturity. Tracheids occur in the xylem of all vascular plants. In addition to tracheids, most angiosperms, as well as a few gymnosperms and a few seedless vascular plants, have vessel elements. When the living cellular contents of a tracheid or vessel element disintegrate, the cell's thickened walls remain behind, forming a nonliving conduit through which water can flow. The secondary walls of tracheids and vessel elements are often interrupted by pits, thinner regions where only primary walls are present (see Figure 4.25 to review primary and secondary walls). Water can migrate laterally between neighboring cells through pits.

Tracheids are long, thin cells with tapered ends. Water moves from cell to cell mainly through the pits, where it does not have to cross thick secondary walls.

Vessel elements are generally wider, shorter, thinner walled, and less tapered than the tracheids. They are aligned end to end, forming long pipes known as vessels that in some cases are visible with the naked eye. The end walls of vessel elements have perforation plates that enable water to flow freely through the vessels.

The secondary walls of tracheids and vessel elements are hardened with lignin. This hardening provides support and prevents collapse under the tension of water transport.

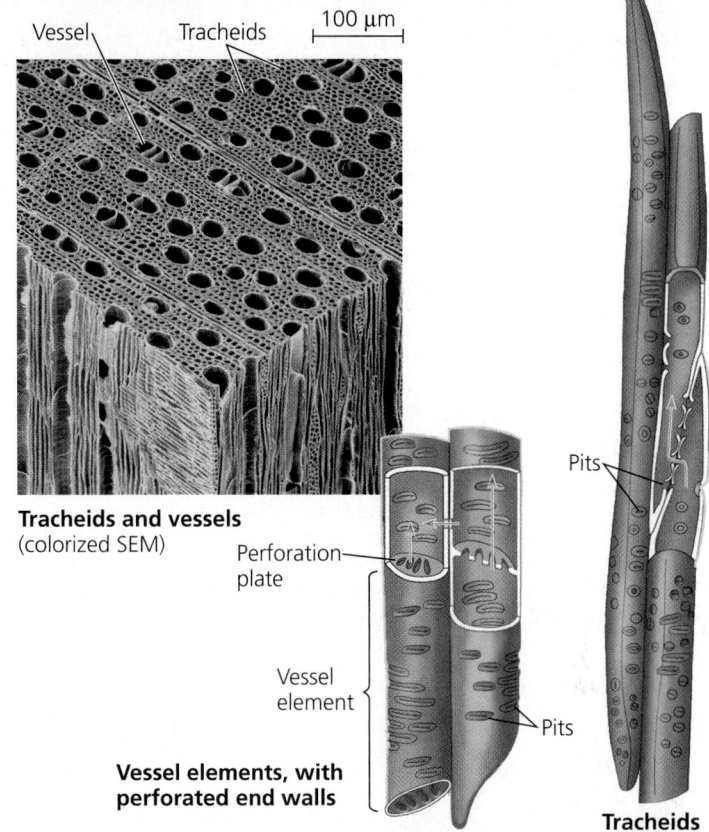

**Tracheids and vessels** (colorized SEM)

Vessel · Tracheids · 100 µm · Pits · Perforation plate · Vessel element · Pits

**Vessel elements, with perforated end walls**

**Tracheids**

## Sugar-Conducting Cells of the Phloem

Unlike the water-conducting cells of the xylem, the sugar-conducting cells of the phloem are alive at functional maturity. In seedless vascular plants and gymnosperms, sugars and other organic nutrients are transported through long, narrow cells called sieve cells. In the phloem of angiosperms, these nutrients are transported through sieve tubes, which consist of chains of cells that are called **sieve-tube** elements, or sieve-tube members.

Though alive, sieve-tube elements lack a nucleus, ribosomes, a distinct vacuole, and cytoskeletal elements. This reduction in cell contents enables nutrients to pass more easily through the cell. The end walls between sieve-tube elements, called **sieve plates**, have pores that facilitate the flow of fluid from cell to cell along the sieve tube. Alongside each sieve-tube element is a nonconducting cell called a **companion cell**, which is connected to the sieve-tube element by numerous plasmodesmata (see Figure 4.25). The nucleus and ribosomes of the companion cell serve not only that cell itself but also the adjacent sieve-tube element. In some plants, the companion cells in leaves also help load sugars into the sieve-tube elements, which then transport the sugars to other parts of the plant.

**ANIMATION**

Visit the Study Area in **MasteringBiology** for the BioFlix® 3-D Animation Tour of a Plant Cell.

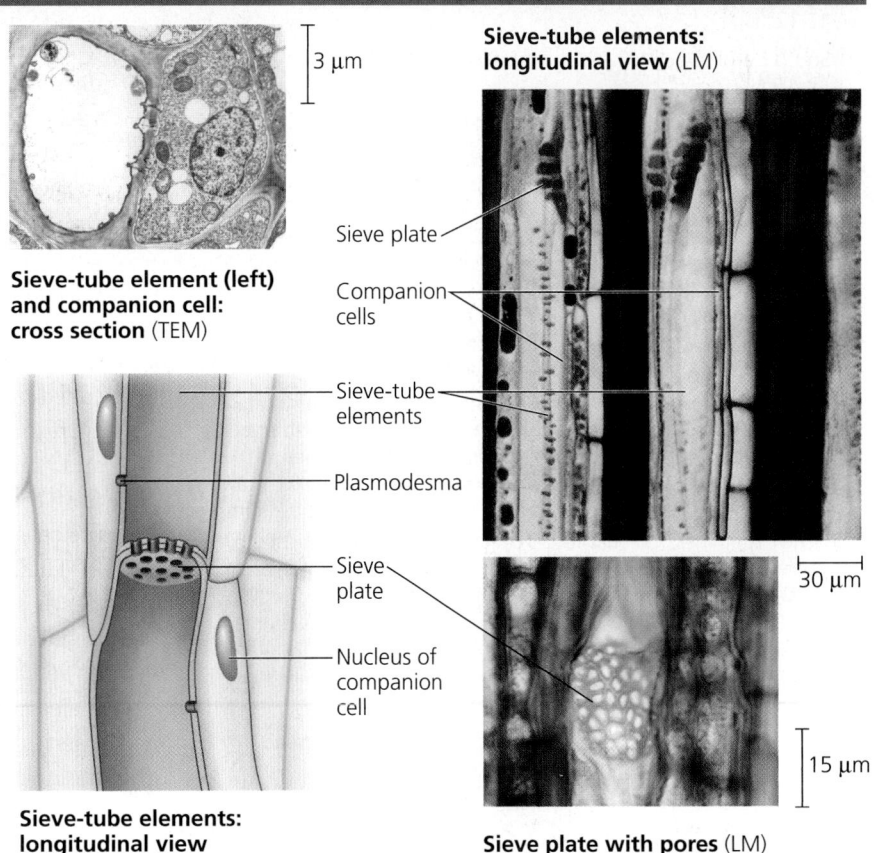

3 µm

**Sieve-tube element (left) and companion cell: cross section** (TEM)

**Sieve-tube elements: longitudinal view**

Sieve plate · Companion cells · Sieve-tube elements · Plasmodesma · Sieve plate · Nucleus of companion cell

**Sieve-tube elements: longitudinal view (LM)**

30 µm

15 µm

**Sieve plate with pores** (LM)

## Scientific Skills Exercise

### Using Bar Graphs to Interpret Data

**Nature versus Nurture: Why Are Leaves from Northern Red Maples "Toothier" Than Leaves from Southern Red Maples?** Not all leaves of red maple (*Acer rubrum*) are the same. The "teeth" along the margins of leaves growing in northern locations differ in size and number compared with their southern counterparts. (The leaf seen here has an intermediate appearance.) Are these morphological differences due to genetic differences between northern and southern *A. rubrum* populations, or do they arise from environmental differences between northern and southern locations, such as average temperature, that affect gene expression?

**How the Experiment Was Done** Seeds of *Acer rubrum* were collected from four latitudinally distinct sites: Ontario (Canada), Pennsylvania, South Carolina, and Florida. The seeds collected from the four locations were then grown in a northern location (Rhode Island) and a southern location (Florida). After a few years of growth, leaves were harvested from the four sets of plants growing in the two locations. The average area of single teeth and the average number of teeth per leaf area were determined.

**Data from the Experiment**

| Seed Collection Site | Average Area of a Single Tooth (cm²) | | Number of Teeth per cm² of Leaf Area | |
|---|---|---|---|---|
| | Grown in Rhode Island | Grown in Florida | Grown in Rhode Island | Grown in Florida |
| Ontario 43.32°N | 0.017 | 0.017 | 3.9 | 3.2 |
| Pennsylvania 42.12°N | 0.020 | 0.014 | 3.0 | 3.5 |
| South Carolina 33.45°N | 0.024 | 0.028 | 2.3 | 1.9 |
| Florida 30.65°N | 0.027 | 0.047 | 2.1 | 0.9 |

**Data from** D. L. Royer et al., Phenotypic plasticity of leaf shape along a temperature gradient in *Acer rubrum*, *PLoS ONE* 4(10):e7653 (2009).

#### INTERPRET THE DATA

1. Make a bar graph for tooth size and a bar graph for number of teeth. (For additional information about bar graphs, see the Scientific Skills Review in Appendix F and in the Study Area in MasteringBiology.) From north to south, what is the general trend in tooth size and number of teeth in leaves of *Acer rubrum*?

2. Based on the data above, would you estimate that leaf tooth traits in red maple are largely determined by genetic heritage (genotype) or the capacity for responding to environmental change within a single genotype (phenotypic plasticity) or both? Make specific reference to the data in answering the question.

3. The "toothiness" of leaf fossils of known age has been used by paleoclimatologists to estimate past temperatures in a region. If a 10,000-year-old fossilized red maple leaf from South Carolina had an average of 4.2 teeth per square centimeter of leaf area, what could you infer about the relative temperature of South Carolina 10,000 years ago compared with today? Explain your reasoning.

(MB) A version of this Scientific Skills Exercise can be assigned in MasteringBiology.

## CONCEPT 28.2

# Different meristems generate new cells for primary and secondary growth

How do plant organs develop? A major difference between plants and most animals is that plant growth is not limited to an embryonic or juvenile period. Instead, growth occurs throughout the plant's life, a process known as **indeterminate growth**. Plants can keep growing because they have perpetually dividing, unspecialized tissues called **meristems** that divide when conditions permit, leading to new cells that elongate and become specialized. At any given time, a typical plant has embryonic, developing, and mature organs. Except for dormant periods, most plants grow continuously. In contrast, most animals and some plant organs—such as leaves, thorns, and flowers—undergo **determinate growth**; that is, they stop growing after reaching a certain size.

There are two main types of meristems: apical meristems and lateral meristems **(Figure 28.10)**. **Apical meristems**, located at the tips of roots and shoots and in axillary buds of shoots, provide additional cells that enable growth in length, a process known as **primary growth**. Primary growth allows roots to extend throughout the soil and shoots to increase their exposure to light. In herbaceous (nonwoody) plants, primary growth produces all, or almost all, of the plant body. Woody plants, however, also grow in circumference in the parts of stems and roots that no longer grow in length. This growth in thickness, known as **secondary growth**, is caused by **lateral meristems** called the vascular cambium and cork cambium. These cylinders of dividing cells extend along the length of roots and stems. The **vascular cambium** adds layers of vascular tissue called secondary xylem (wood) and secondary phloem. The **cork cambium** replaces the epidermis with the thicker, tougher periderm.

The cells within meristems divide relatively frequently, generating additional cells. Some new cells remain in the meristem and produce more cells, while others differentiate and are incorporated into tissues and organs of the growing plant. Cells that remain as sources of new cells have traditionally been called *initials* but are increasingly being called *stem cells* to correspond to animal stem cells, which also perpetually divide and remain functionally unspecialized. The new cells displaced from the meristem, which are known as *derivatives*, divide until the cells they produce become specialized in mature tissues.

The relationship between primary and secondary growth is clearly seen in the winter twig of a deciduous tree. At the shoot tip is the dormant apical bud, enclosed by scales that protect its apical meristem **(Figure 28.11)**. In spring, the bud sheds its scales and begins a new spurt of primary growth, producing a series of nodes and internodes. On each growth segment, the nodes are marked by scars that were left when leaves fell.

**582** UNIT FIVE PLANT FORM AND FUNCTION

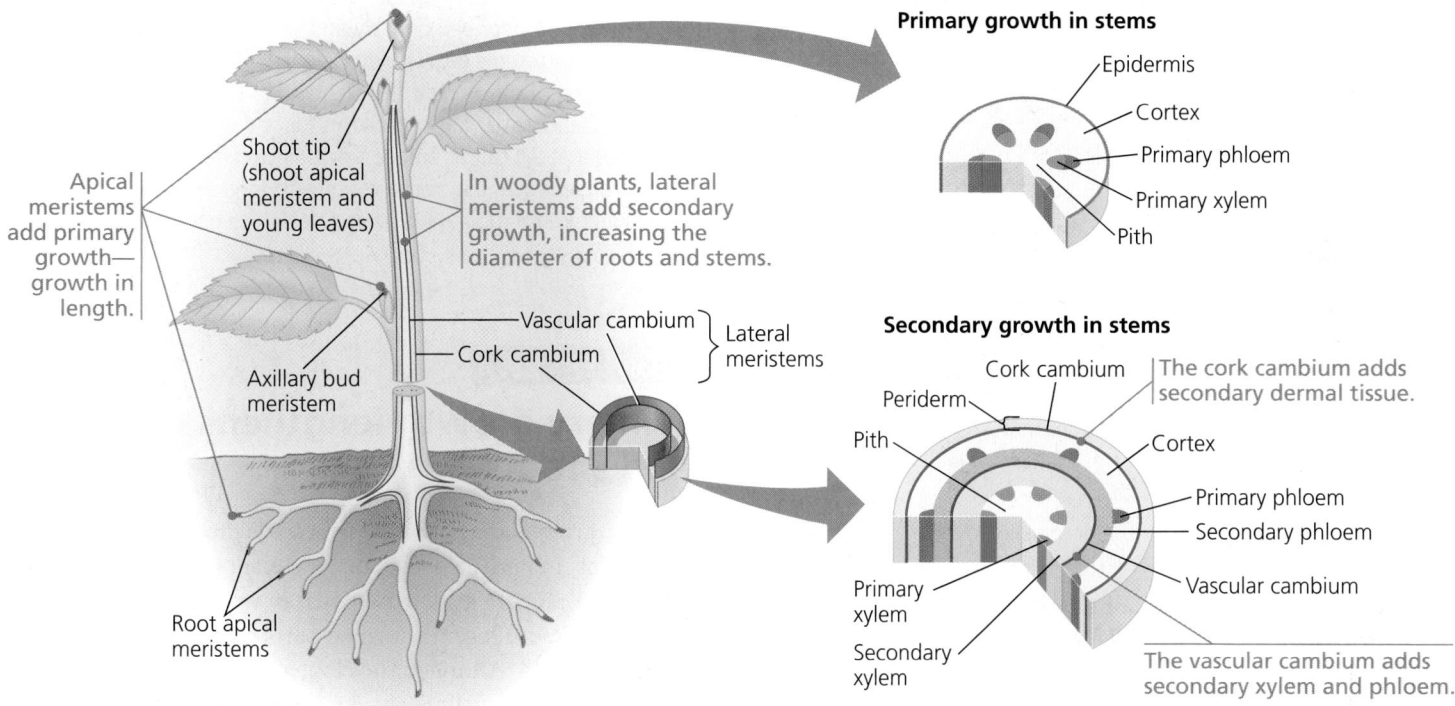

▲ **Figure 28.10  An overview of primary and secondary growth.**

Primary growth in stems
- Epidermis
- Cortex
- Primary phloem
- Primary xylem
- Pith

Secondary growth in stems
- Cork cambium
- Periderm | The cork cambium adds secondary dermal tissue.
- Pith
- Cortex
- Primary phloem
- Secondary phloem
- Vascular cambium
- Primary xylem
- Secondary xylem | The vascular cambium adds secondary xylem and phloem.

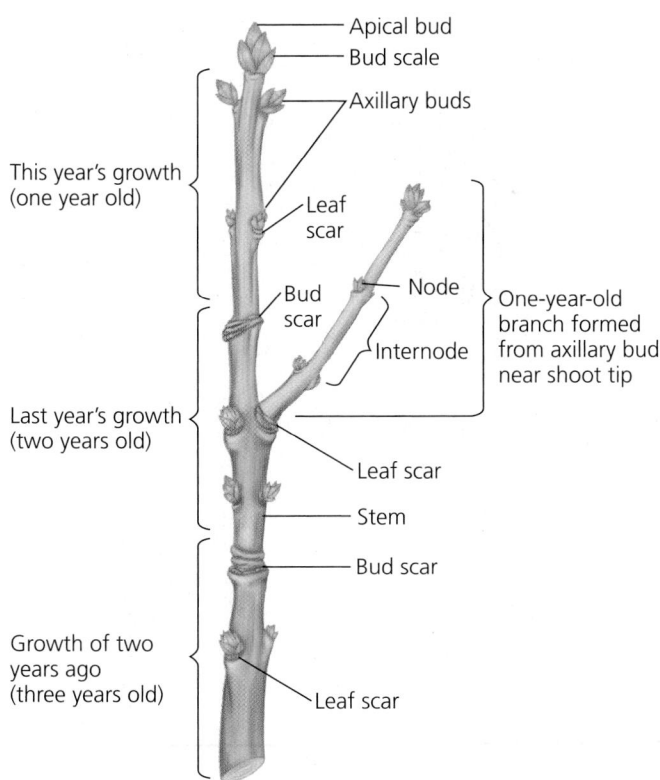

▲ **Figure 28.11  Three years' growth in a winter twig.**

Above each leaf scar is an axillary bud or a branch formed by an axillary bud. Farther down the twig are bud scars from the whorls of scales that enclosed the apical bud during the previous winter. During each growing season, primary growth extends the shoots, and secondary growth increases the diameter of the parts that formed in previous years.

## Gene Expression and Control of Cell Differentiation

Derivative cells can diverge in structure and function even though they share a common genome. Such cell differentiation depends, to a large degree, on the control of gene expression—the regulation of transcription and translation, resulting in the production of specific proteins. Although cell differentiation depends on the control of gene expression, the fate of a plant cell is determined by its final position in the developing organ.

Evidence suggests that the activation or inactivation of specific genes involved in cell differentiation depends largely on cell-to-cell communication. For example, two cell types arise in the root epidermis of the model plant *Arabidopsis thaliana*: root hair cells and hairless epidermal cells. Cell fate is associated with the position of the epidermal cells. The immature epidermal cells that are in contact with two underlying cells of the root cortex differentiate into root hair cells, whereas the immature epidermal cells in contact with only one cortical cell differentiate into mature hairless cells. Differential expression of a gene called *GLABRA-2* (from the Latin *glaber*, bald) is required for the appropriate root hair distribution.

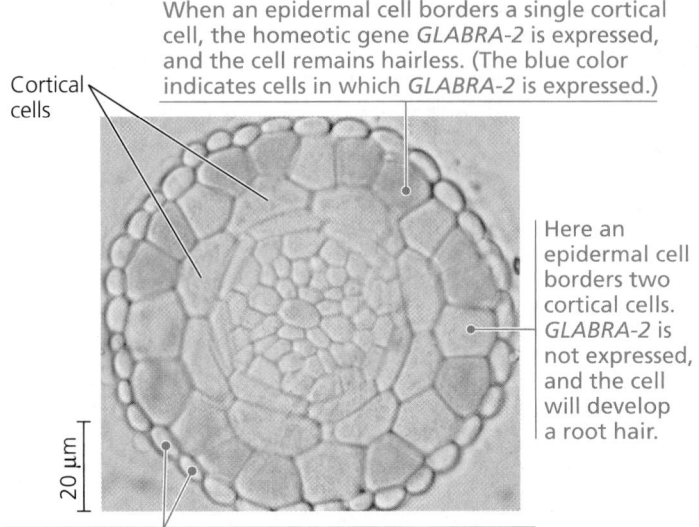

Cortical cells

When an epidermal cell borders a single cortical cell, the homeotic gene *GLABRA-2* is expressed, and the cell remains hairless. (The blue color indicates cells in which *GLABRA-2* is expressed.)

Here an epidermal cell borders two cortical cells. *GLABRA-2* is not expressed, and the cell will develop a root hair.

20 μm

The root cap cells external to the epidermal layer will be sloughed off before root hairs emerge.

▲ **Figure 28.12 Control of root hair differentiation by a master regulatory gene (LM).**

**WHAT IF?** *What would the roots look like if* GLABRA-2 *were rendered dysfunctional by a mutation?*

Researchers have demonstrated this requirement by coupling the *GLABRA-2* gene to a "reporter gene" that causes every cell expressing *GLABRA-2* in the root to turn pale blue following a certain treatment. The *GLABRA-2* gene is normally expressed only in epidermal cells that will not develop root hairs **(Figure 28.12)**.

## Meristematic Control of the Transition to Flowering and the Life Spans of Plants

*Vegetative growth*—the production of leaves, stems, and roots (as well as asexual, or vegetative, reproduction)—is only one phase in a plant's life. Most angiosperms at some point in their life direct some or all of their shoot apical meristems to undergo a transition from vegetative growth to *reproductive growth*, the production of flowers, fruits, and seeds. This transition is triggered by a combination of environmental cues, such as day length, and internal signals, such as hormones. (You will learn more about the roles of these signals in flowering in Concept 31.2.) Unlike vegetative growth, which is indeterminate, reproductive growth is determinate: The production of a flower by a shoot apical meristem stops the primary growth of that shoot. Some plants may also go through a juvenile phase during which they are incapable of reproductive growth.

Based on the timing and completeness of a plant species' switch from vegetative to reproductive growth, flowering plants can be categorized as annuals, biennials, or perennials. *Annuals* complete their life cycle—from germination to flowering to seed production to death—in a single year or less. Many wildflowers are annuals, as are most staple food crops, including legumes and cereal grains such as wheat and rice. *Biennials*, such as turnips, generally require two growing seasons to complete their life cycle, flowering and fruiting only in their second year. *Perennials* live many years and include trees, shrubs, and some grasses.

**CONCEPT 28.3**

# Primary growth lengthens roots and shoots

Primary growth arises directly from cells produced by apical meristems. In herbaceous plants, the plant is produced almost entirely by primary growth, whereas in woody plants only the nonwoody, more recently formed parts of the plant represent primary growth. Although the elongation of both roots and shoots arises from cells derived from apical meristems, the primary growth of roots and shoots differs in many ways.

## Primary Growth of Roots

The tip of a root is covered by a thimble-like **root cap** **(Figure 28.13)**, which protects the delicate apical meristem

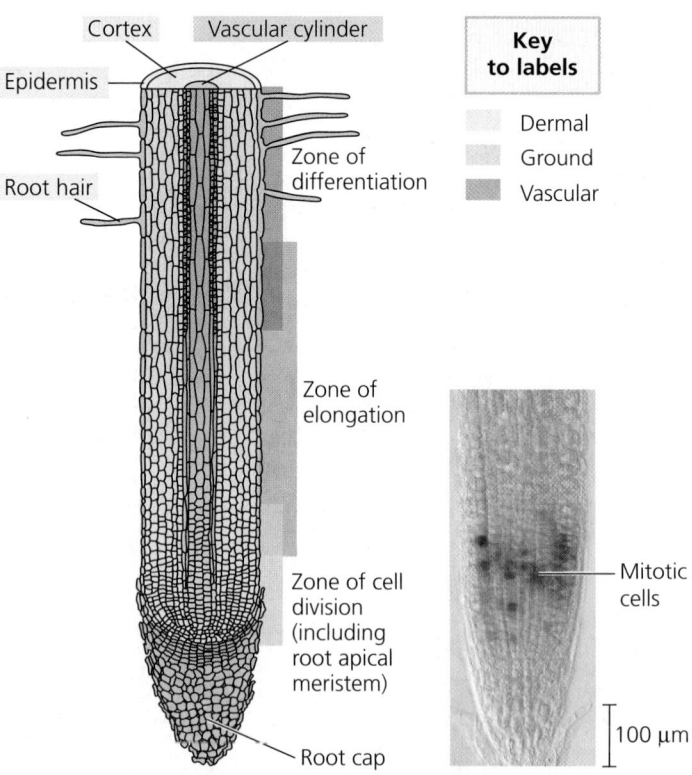

Cortex Vascular cylinder

Epidermis

Root hair

Zone of differentiation

**Key to labels**

Dermal
Ground
Vascular

Zone of elongation

Zone of cell division (including root apical meristem)

Mitotic cells

100 μm

Root cap

▲ **Figure 28.13 Primary growth of a typical eudicot root.** The root apical meristem produces all the cells of the root and the root cap. Most lengthening of the root occurs in the zone of elongation. In the micrograph, cells undergoing mitosis in the apical meristem are revealed by staining for cyclin, a protein that plays an important role in cell division (LM).

as the root pushes through the abrasive soil during primary growth. The cells of the root cap also secrete a polysaccharide slime that lubricates the soil around the tip of the root. Growth occurs just behind the tip in three overlapping zones of cells at successive stages of primary growth. These are the zones of cell division, elongation, and differentiation (see Figure 28.13).

The *zone of cell division* includes the root apical meristem and its derivatives. New root cells are produced in this region, including cells of the root cap. Typically a few millimeters behind the tip of the root is the *zone of elongation*, where most of the growth occurs as root cells elongate—sometimes to more than ten times their original length. Cell elongation in this zone pushes the tip farther into the soil. Meanwhile, the root apical meristem keeps adding cells to the younger end of the zone of elongation. Even before the root cells finish lengthening, many begin specializing in structure and function; for example, roots hairs start to form. In the *zone of differentiation*, or zone of maturation, cells complete their differentiation and become distinct cell types.

The primary growth of a root produces its epidermis, ground tissue, and vascular tissue. Root hairs, although typically only living a few weeks, are one of the more prominent features of the root epidermis. Together, they can make up 70–90% of the total root surface area. A 4-month-old rye plant has an estimated 14 billion root hairs. Laid end-to-end, they would cover 10,000 km, one-quarter the length of the equator. The ground tissue of roots, consisting mostly of parenchyma cells, is found in the cortex, the region between the vascular cylinder and epidermis. In addition to storing carbohydrates, cortical cells transport water and salts from the root hairs to the center of the root. The cortex, because of its large intercellular spaces, also allows for the *extracellular* diffusion of water, minerals, and oxygen from the root hairs inward. The innermost layer of the cortex is called the **endodermis**, a cylinder one cell thick that forms the boundary with the vascular cylinder. The endodermis is a selective barrier that regulates passage of substances from the soil into the vascular cylinder (see Figure 29.16). In angiosperm roots, the stele is a vascular cylinder, consisting of a solid core of xylem and phloem tissues. In most eudicot roots, the xylem has a starlike appearance in cross section, and the phloem occupies the indentations between the arms of the xylem "star" **(Figure 28.14a)**. In many monocot roots, the vascular tissue consists of a central core of unspecialized parenchyma cells surrounded by a ring of alternating xylem and phloem tissues **(Figure 28.14b)**.

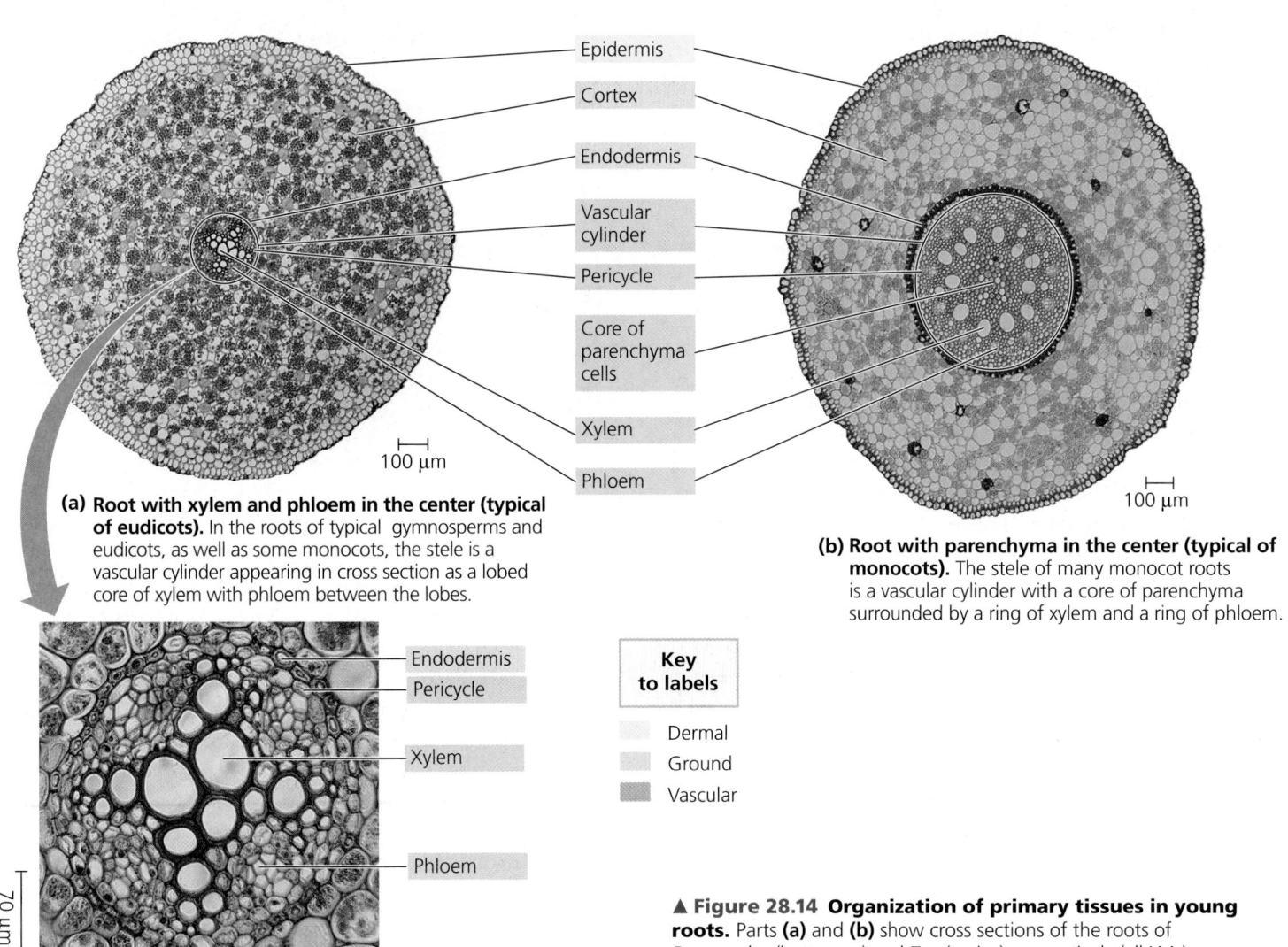

Epidermis
Cortex
Endodermis
Vascular cylinder
Pericycle
Core of parenchyma cells
Xylem
Phloem

100 μm

**(a) Root with xylem and phloem in the center (typical of eudicots).** In the roots of typical gymnosperms and eudicots, as well as some monocots, the stele is a vascular cylinder appearing in cross section as a lobed core of xylem with phloem between the lobes.

100 μm

**(b) Root with parenchyma in the center (typical of monocots).** The stele of many monocot roots is a vascular cylinder with a core of parenchyma surrounded by a ring of xylem and a ring of phloem.

Endodermis
Pericycle
Xylem
Phloem

70 μm

**Key to labels**

Dermal
Ground
Vascular

▲ **Figure 28.14 Organization of primary tissues in young roots.** Parts **(a)** and **(b)** show cross sections of the roots of *Ranunculus* (buttercup) and *Zea* (maize), respectively (all LMs).

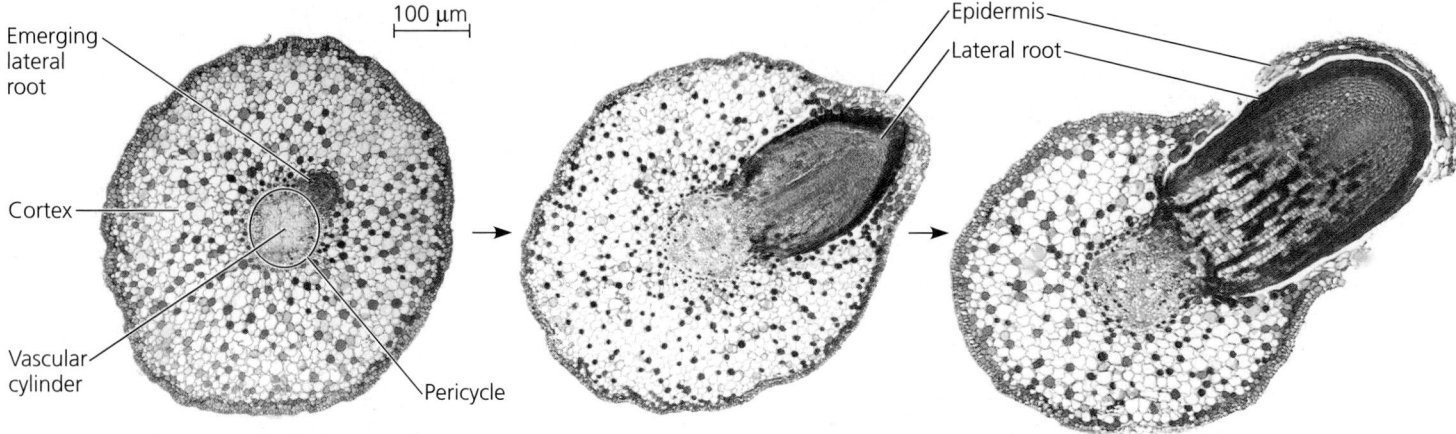

▲ **Figure 28.15 The formation of a lateral root.** A lateral root originates in the pericycle, the outermost layer of the vascular cylinder of a root, and grows out through the cortex and epidermis. In this series of light micrographs, the view of the original root is a cross section, while the view of the lateral root is a longitudinal section.

Lateral roots arise from meristematically active regions of the **pericycle**, the outermost cell layer in the vascular cylinder, which is adjacent to and just inside the endodermis (see Figure 28.14). The lateral roots destructively push through the cortex and epidermis until they emerge from the established root **(Figure 28.15)**.

## Primary Growth of Shoots

A shoot apical meristem is a dome-shaped mass of dividing cells at the shoot tip **(Figure 28.16)**. Leaves develop from **leaf primordia** (singular, *primordium*), projections shaped like a cow's horns that emerge along the sides of the shoot apical meristem. Within a bud, young leaves are spaced close together because the internodes are very short. Shoot elongation is due to the lengthening of internode cells below the shoot tip.

The branching of shoots, which is also part of primary growth, arises from the activation of axillary buds, each of which has its own shoot apical meristem. Because of chemical communication by plant hormones, the closer an axillary bud is to an active apical bud, the more inhibited it is, a phenomenon called **apical dominance**. (The specific hormonal changes underlying apical dominance are discussed in Concept 31.1.) If an animal eats the end of the shoot or if shading results in the light being more intense on the side of the shoot, the chemical communication underlying apical dominance is disrupted. As a result, the axillary buds break dormancy and start to grow. Released from dormancy, an axillary bud eventually gives rise to a lateral shoot, complete with its own apical bud, leaves, and axillary buds. When gardeners prune shrubs and pinch back houseplants, they are reducing the number of apical buds a plant has, thereby allowing branches to elongate and giving the plants a fuller, bushier appearance.

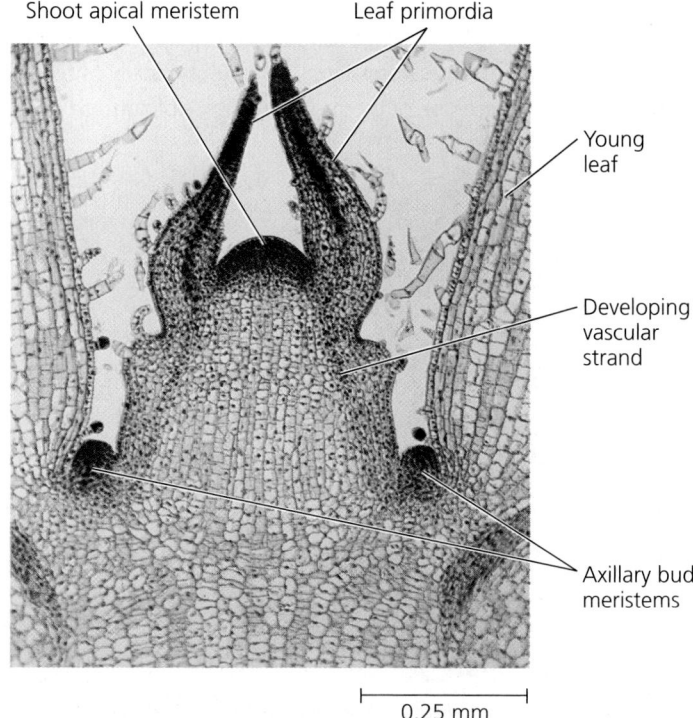

▲ **Figure 28.16 The shoot tip.** Leaf primordia arise from the flanks of the dome of the apical meristem. This is a longitudinal section of the shoot tip of *Coleus* (LM).

In some monocots, particularly grasses, meristematic activity occurs at the bases of stems and leaves. These areas, called *intercalary meristems*, allow damaged leaves to rapidly regrow, which accounts for the ability of lawns to grow following mowing. The ability of grasses to regrow leaves by intercalary meristems enables the plant to recover more effectively from damage incurred from grazing herbivores.

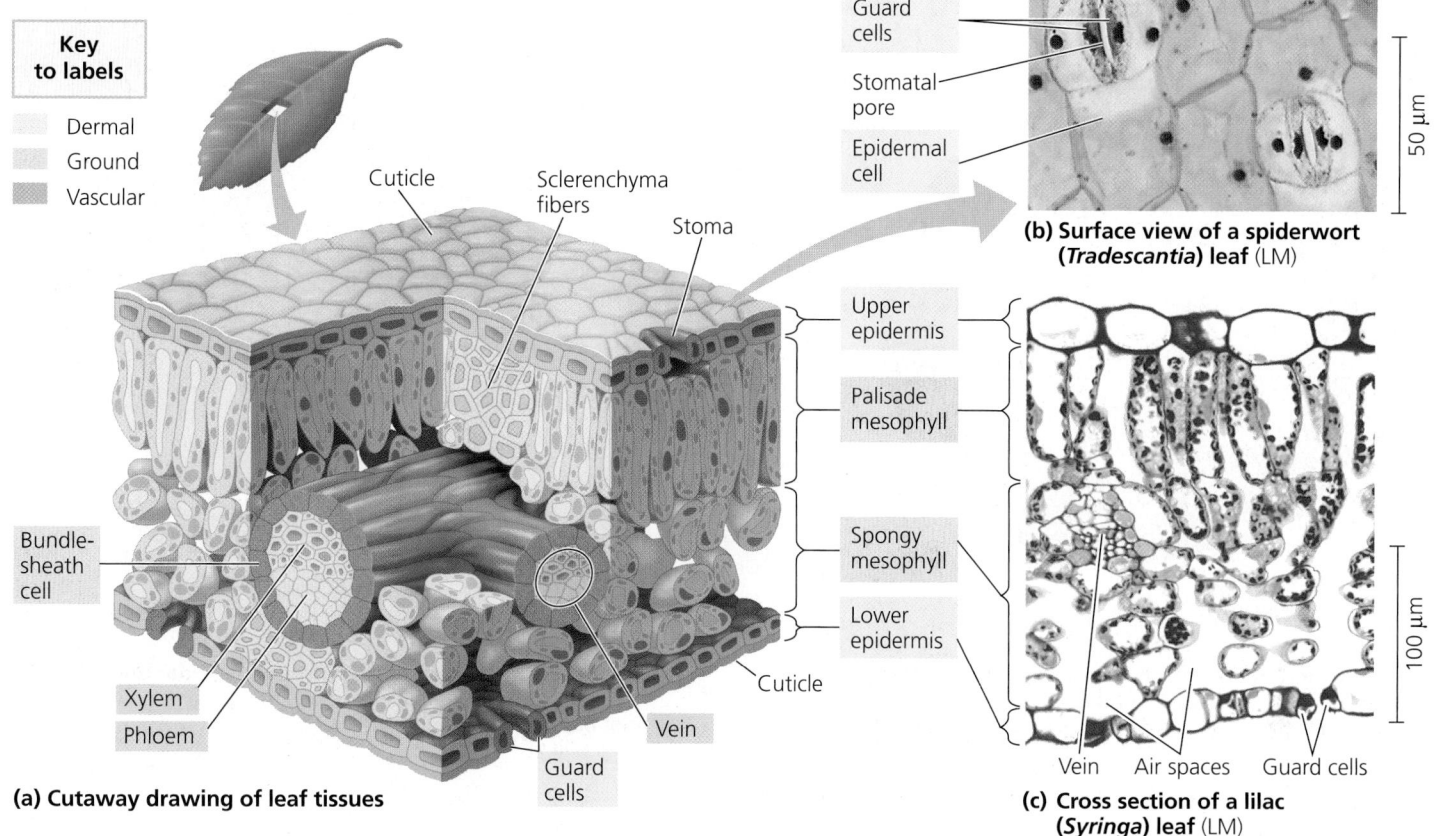

Key
to labels
- Dermal
- Ground
- Vascular

Cuticle

Sclerenchyma
fibers

Stoma

Guard
cells

Stomatal
pore

Epidermal
cell

**(b) Surface view of a spiderwort
(*Tradescantia*) leaf** (LM)

50 μm

Upper
epidermis

Palisade
mesophyll

Spongy
mesophyll

Lower
epidermis

Bundle-
sheath
cell

Xylem

Phloem

Cuticle

Vein

Guard
cells

**(a) Cutaway drawing of leaf tissues**

Vein     Air spaces    Guard cells

100 μm

**(c) Cross section of a lilac
(*Syringa*) leaf** (LM)

▲ **Figure 28.17 Leaf anatomy.**

## Tissue Organization of Leaves

**Figure 28.17** provides an overview of leaf structure. The epidermis of leaves is interrupted by pores called **stomata** (singular, *stoma*), which allow exchange of $CO_2$ and $O_2$ between the surrounding air and the photosynthetic cells inside the leaf. In addition to regulating $CO_2$ uptake for photosynthesis, stomata are major avenues for the evaporative loss of water. The term *stoma* can refer to the stomatal pore or to the entire stomatal complex consisting of a pore flanked by two specialized epidermal cells called **guard cells**, which regulate the opening and closing of the pore. (In Concept 29.6, we will discuss stomata in detail.)

The leaf's ground tissue, called the **mesophyll** (from the Greek *mesos*, middle, and *phyll*, leaf), is sandwiched between the upper and lower epidermal layers. Mesophyll consists mainly of parenchyma cells specialized for photosynthesis. The mesophyll in many eudicot leaves has two distinct layers: palisade and spongy. *Palisade mesophyll* consists of one or more layers of elongated parenchyma cells on the upper part of the leaf. *Spongy mesophyll* is below the palisade mesophyll. These parenchyma cells are more loosely arranged, with a labyrinth of air spaces through which $CO_2$ and $O_2$ circulate around the cells and up to the palisade region. The air spaces are

particularly large in the vicinity of stomata, where $CO_2$ is taken up from the outside air and $O_2$ is released.

The vascular tissue of each leaf is continuous with the vascular tissue of the stem. Veins subdivide repeatedly and branch throughout the mesophyll. This network brings xylem and phloem into close contact with the photosynthetic tissue, which obtains water and minerals from the xylem and loads its sugars and other organic products into the phloem for transport to other parts of the plant. The vascular structure also functions as a framework that reinforces the leaf. Each vein is enclosed by a protective *bundle sheath*, a layer of cells that regulates the movement of substances between the vascular tissue and the mesophyll. Bundle-sheath cells are particularly prominent in leaves of plant species that carry out $C_4$ photosynthesis (see Concept 8.3).

## Tissue Organization of Stems

The epidermis covers stems as part of the continuous dermal tissue system. Vascular tissue runs the length of a stem in vascular bundles. Unlike lateral roots, which arise from vascular tissue deep within a root and disrupt the vascular cylinder, cortex, and epidermis as they emerge (see Figure 28.15), lateral shoots develop from axillary bud meristems on the stem's surface and disrupt no other tissues (see Figure 28.16). Near the

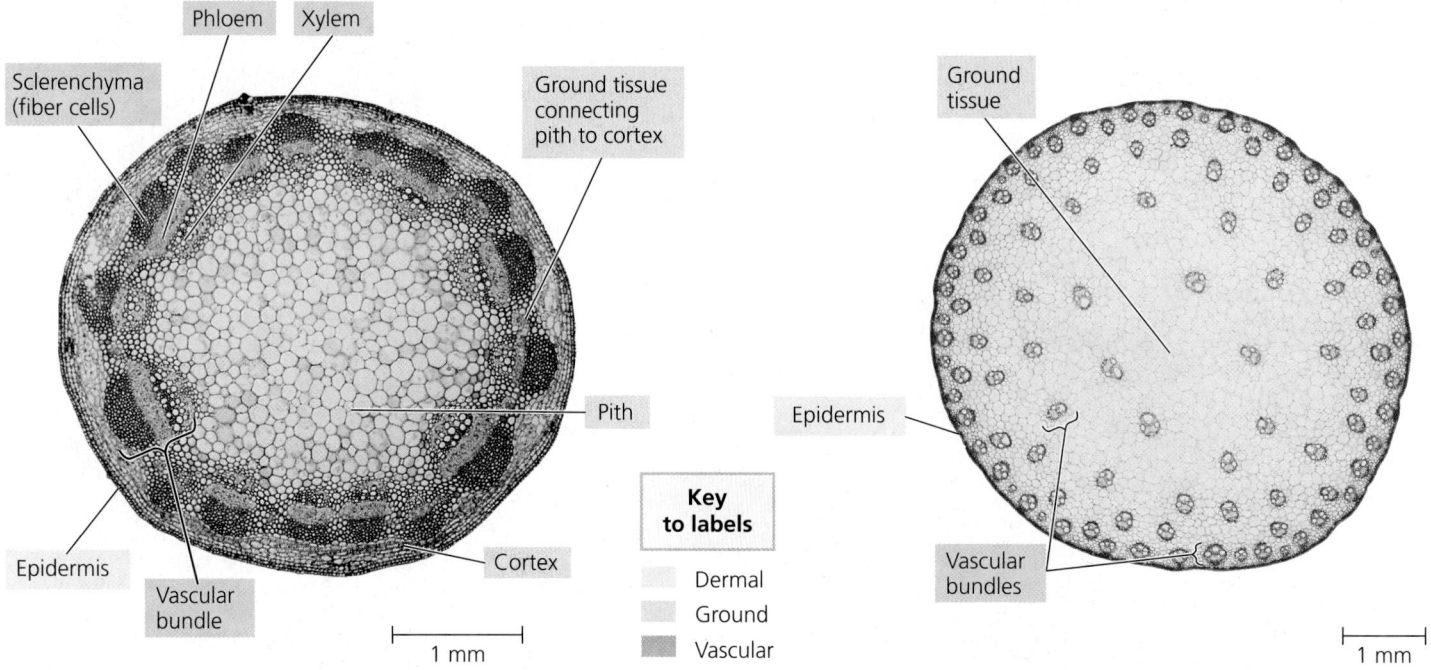

**Phloem**   **Xylem**

**Sclerenchyma (fiber cells)**

**Ground tissue connecting pith to cortex**

**Ground tissue**

**Pith**

**Epidermis**

**Epidermis**

**Cortex**

**Vascular bundle**

**Vascular bundles**

**Key to labels**

Dermal
Ground
Vascular

1 mm

1 mm

**(a) Cross section of stem with vascular bundles forming a ring (typical of eudicots).** Ground tissue toward the inside is called pith, and ground tissue toward the outside is called cortex (LM).

**(b) Cross section of stem with scattered vascular bundles (typical of monocots).** In such an arrangement, ground tissue is not partitioned into pith and cortex (LM).

▲ **Figure 28.18 Organization of primary tissues in young stems.**

**?** *Why aren't the terms* pith *and* cortex *used to describe the ground tissue of monocot stems?*

soil surface, in the transition zone between shoot and root, the bundled vascular arrangement of the stem converges with the solid vascular cylinder of the root.

In most eudicot species, the vascular tissue of stems consists of vascular bundles arranged in a ring (**Figure 28.18a**). The xylem in each vascular bundle is adjacent to the pith, and the phloem in each bundle is adjacent to the cortex. In most monocot stems, the vascular bundles are scattered throughout the ground tissue rather than forming a ring (**Figure 28.18b**). In the stems of both monocots and eudicots, the ground tissue consists mostly of parenchyma cells. However, collenchyma cells just beneath the epidermis strengthen many stems during primary growth. Sclerenchyma cells, especially fiber cells, also provide support in those parts of the stems that are no longer elongating.

**CONCEPT CHECK 28.3**

1. Contrast primary growth in roots and shoots.
2. **WHAT IF?** If a plant species has vertically oriented leaves, would you expect its mesophyll to be divided into spongy and palisade layers? Explain.
3. **MAKE CONNECTIONS** How are root hairs and microvilli analogous structures? (See Figure 4.7 and the discussion of analogy in Concept 20.2.)

For suggested answers, see Appendix A.

---

**CONCEPT 28.4**

# Secondary growth increases the diameter of stems and roots in woody plants

Many land plants display secondary growth, the growth in thickness produced by lateral meristems. The advent of secondary growth during plant evolution allowed the production of novel plant forms ranging from massive forest trees to woody vines. All gymnosperm species and many eudicot species undergo secondary growth, but it is rare in monocots. It occurs in the stems and roots of woody plants, but rarely and only to a limited extent in leaves.

Secondary growth consists of the tissues produced by the vascular cambium and cork cambium. The vascular cambium adds secondary xylem (wood) and secondary phloem, thereby increasing vascular flow and support for the shoots. The cork cambium produces a tough, thick covering of waxy cells that protect the stem from water loss and from invasion by insects, bacteria, and fungi.

In woody plants, primary growth and secondary growth occur simultaneously. As primary growth adds leaves and lengthens stems and roots in the younger regions of a plant, secondary growth increases the diameter of stems and roots in older regions where primary growth has ceased. The process is similar in shoots and roots. **Figure 28.19** provides an overview of growth in a woody stem.

**(a) Primary and secondary growth in a two-year-old woody stem**

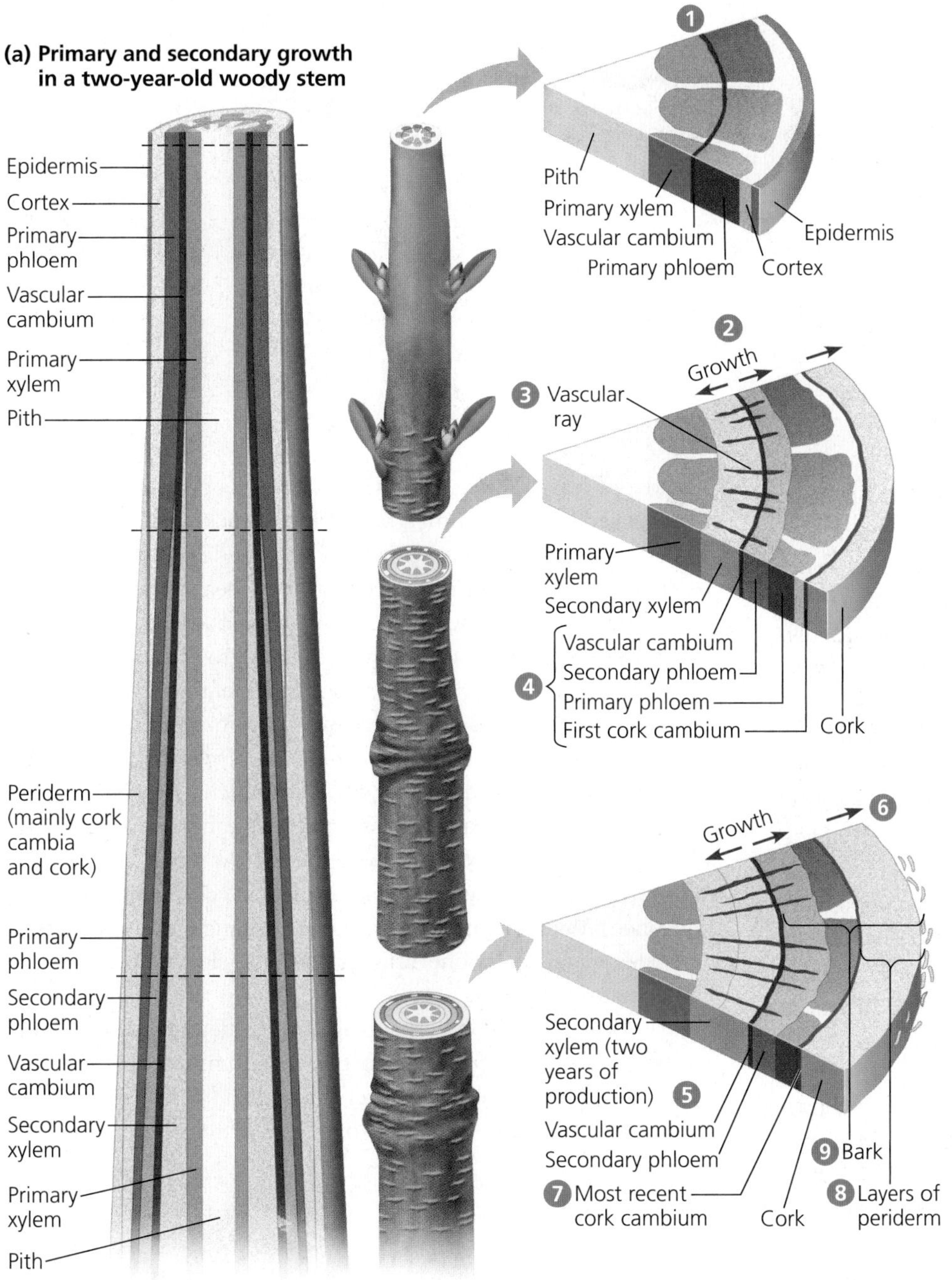

Epidermis
Cortex
Primary phloem
Vascular cambium
Primary xylem
Pith

Periderm (mainly cork cambia and cork)

Primary phloem
Secondary phloem
Vascular cambium
Secondary xylem
Primary xylem
Pith

**1**
Pith
Primary xylem
Vascular cambium
Primary phloem
Epidermis
Cortex

**2**
Growth

**3** Vascular ray
Primary xylem
Secondary xylem
Vascular cambium
Secondary phloem
Primary phloem
First cork cambium
Cork
**4**

Growth
**6**

Secondary xylem (two years of production) **5**
Vascular cambium
Secondary phloem
**7** Most recent cork cambium
Cork
**9** Bark
**8** Layers of periderm

**1** Primary growth from the activity of the apical meristem is nearing completion. The vascular cambium has just formed.

**2** Although primary growth continues in the apical bud, only secondary growth occurs in this region. The stem thickens as the vascular cambium forms secondary xylem to the inside and secondary phloem to the outside.

**3** Some initials of the vascular cambium give rise to vascular rays.

**4** As the vascular cambium's diameter increases, the secondary phloem and other tissues external to the cambium can't keep pace because their cells no longer divide. As a result, these tissues, including the epidermis, will eventually rupture. A second lateral meristem, the cork cambium, develops from parenchyma cells in the cortex. The cork cambium produces cork cells, which replace the epidermis.

**5** In year 2 of secondary growth, the vascular cambium produces more secondary xylem and phloem, and the cork cambium produces more cork.

**6** As the stem's diameter increases, the outermost tissues exterior to the cork cambium rupture and are sloughed off.

**7** In many cases, the cork cambium re-forms deeper in the cortex. When none of the cortex is left, the cambium develops from phloem parenchyma cells.

**8** Each cork cambium and the tissues it produces form a layer of periderm.

**9** Bark consists of all tissues exterior to the vascular cambium.

▲ **Figure 28.19 Primary and secondary growth of a woody stem.** The progress of secondary growth can be tracked by examining the sections through sequentially older parts of the stem.

**?** *How does the vascular cambium cause some tissues to rupture?*

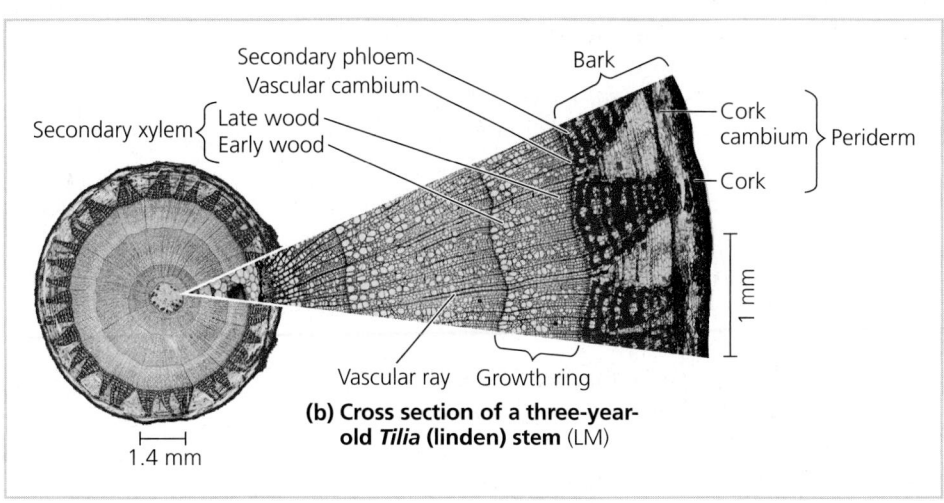

Secondary phloem
Vascular cambium
Bark
Secondary xylem { Late wood / Early wood }
Cork cambium } Periderm
Cork
Vascular ray    Growth ring
1 mm
1.4 mm

**(b) Cross section of a three-year-old *Tilia* (linden) stem** (LM)

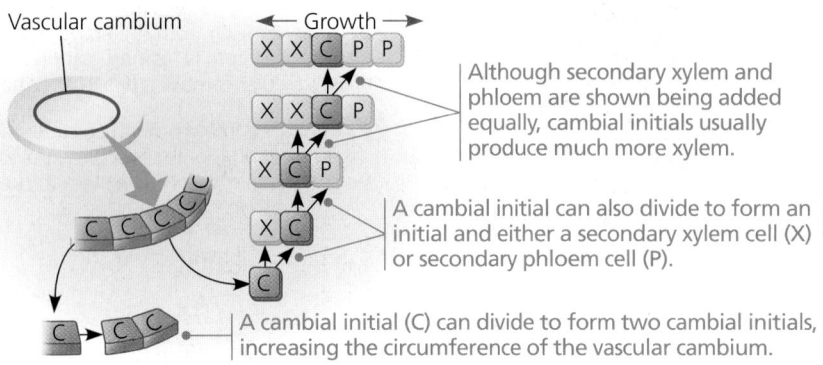

Although secondary xylem and phloem are shown being added equally, cambial initials usually produce much more xylem.

A cambial initial can also divide to form an initial and either a secondary xylem cell (X) or secondary phloem cell (P).

A cambial initial (C) can divide to form two cambial initials, increasing the circumference of the vascular cambium.

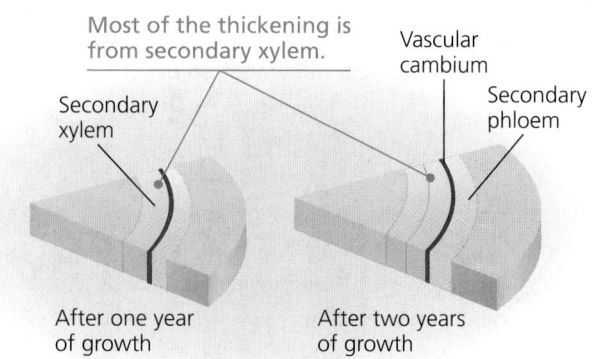

Most of the thickening is from secondary xylem.

After one year of growth

After two years of growth

▲ Figure 28.20 **Secondary growth produced by the vascular cambium.**

## The Vascular Cambium and Secondary Vascular Tissue

The vascular cambium, a cylinder of meristematic cells only one cell thick, is wholly responsible for the production of secondary vascular tissue. In a typical woody stem, the vascular cambium is located outside the pith and primary xylem and to the inside of the primary phloem and the cortex. In a typical woody root, the vascular cambium forms exterior to the primary xylem and interior to the primary phloem and pericycle.

In cross section, the vascular cambium appears as a ring of meristematic cells (see Figure 28.19). As these cells divide, they increase the cambium's circumference and add secondary xylem to the inside and secondary phloem to the outside **(Figure 28.20)**. Each ring is larger than the previous ring, increasing the diameter of roots and stems.

Some of the initials produced by the vascular cambium are elongated and are oriented with their long axis parallel to the axis of the stem or root. They produce cells such as the tracheids, vessel elements, and fibers of the xylem, as well as the sieve-tube elements, companion cells, axially oriented parenchyma, and fibers of the phloem. The other initials are shorter and are oriented perpendicular to the axis of the stem or root. They produce *vascular rays*—radial files of mostly parenchyma cells that connect the secondary xylem and phloem (see Figure 28.19b). These cells move water and nutrients between the secondary xylem and phloem, store carbohydrates and other reserves, and aid in wound repair.

As secondary growth continues, layers of secondary xylem (wood) accumulate, consisting mainly of tracheids, vessel elements, and fibers (see Figure 28.9). Tracheids are the only kind of water-conducting cell found in the xylem tissue of most gymnosperms, whereas both tracheids and vessel elements are found in most angiosperms. The walls of secondary xylem cells are heavily lignified and account for the hardness and strength of wood.

In temperate regions, wood that develops early in the spring, known as early (or spring) wood, usually has secondary xylem cells with large diameters and thin cell walls (see Figure 28.19b). This structure maximizes delivery of water

to new leaves. Wood produced later in the growing season is called late (or summer) wood. It has thick-walled cells that do not transport as much water but provide more support. Because there is a marked contrast between the large cells of the new early wood and the smaller cells of the late wood of the previous growing season, a year's growth appears as a distinct *growth ring* in the cross sections of most tree trunks and roots. Therefore, researchers can estimate a tree's age by counting its growth rings. *Dendrochronology* (from the Greek *dendron*, trees, and *chronos*, time) is the science of analyzing tree growth ring patterns. Growth rings can vary in thickness, depending on seasonal growth. Trees grow well in wet and warm years but may grow hardly at all in cold or dry years. Since a thick ring indicates a warm year and a thin ring indicates a cold or dry one, scientists can use ring patterns to study climate changes.

As a tree or woody shrub ages, the older layers of secondary xylem no longer transport water and minerals (a solution called xylem sap). These layers are called *heartwood* because they are closer to the center of a stem or root **(Figure 28.21)**.

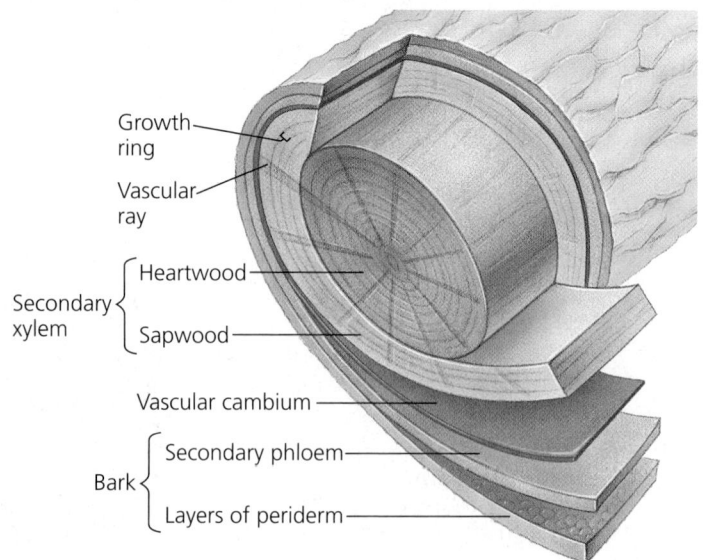

Growth ring

Vascular ray

Secondary xylem {
Heartwood
Sapwood

Vascular cambium

Bark {
Secondary phloem
Layers of periderm

▲ Figure 28.21 **Anatomy of a tree trunk.**

The newest, outer layers of secondary xylem still transport xylem sap and are therefore known as *sapwood*. Sapwood allows a large tree to survive even if the center of its trunk is hollow. Because each new layer of secondary xylem has a larger circumference, secondary growth enables the xylem to transport more sap each year, supplying an increasing number of leaves. Heartwood is generally darker than sapwood because of resins and other compounds that permeate the cell cavities and help protect the core of the tree from fungi and wood-boring insects.

Only the youngest secondary phloem, closest to the vascular cambium, functions in sugar transport. As a stem or root increases in circumference, the older secondary phloem is sloughed off, which is one reason secondary phloem does not accumulate as extensively as secondary xylem.

## The Cork Cambium and the Production of Periderm

During the early stages of secondary growth, the epidermis is pushed outward, causing it to split, dry, and fall off the stem or root. It is replaced by tissues produced by the first cork cambium, a cylinder of dividing cells that arises in the outer cortex of stems (see Figure 28.19a) and in the outer layer of the pericycle in roots. The cork cambium gives rise to *cork cells* that accumulate to the exterior of the cork cambium. As cork cells mature, they deposit a waxy, hydrophobic material called *suberin* in their walls and then die. Because cork cells have suberin and are usually compacted together, most of the periderm is impermeable to water and gases, unlike the epidermis. Cork thus functions as a barrier that helps protect the stem or root from water loss, physical damage, and pathogens. "Cork" is commonly and incorrectly referred to as "bark." In plant biology, **bark** includes all tissues external to the vascular

cambium. Its main components are the secondary phloem (produced by the vascular cambium) and, external to that, the most recent periderm and all the older layers of periderm (see Figure 28.21).

How can living cells in the interior tissues of woody organs absorb oxygen and respire if they are surrounded by a waxy periderm? Dotting the periderm are small, raised areas called **lenticels**, in which there is more space between cork cells, enabling living cells within a woody stem or root to exchange gases with the outside air. Lenticels often appear as horizontal slits, as shown on the stem in Figure 28.19a.

In examining the parts of plants in a dissected fashion as we have done in this chapter, it is important not to lose sight of the fact that the whole plant functions as an integrated organism. In the following chapters, you'll learn more about how materials are absorbed and transported by vascular plants (Chapter 29), how flowering plants reproduce (Chapter 30), and how plant functions are coordinated (Chapter 31). When thinking about plants, bear in mind that plant structures largely reflect evolutionary adaptations to the challenges of a photoautotrophic existence on land.

**CONCEPT CHECK 28.4**

1. A sign is hammered into a tree 2 m from the tree's base. If the tree is 10 m tall and elongates 1 m each year, how high will the sign be after 10 years?
2. Would you expect a tropical tree to have distinct growth rings? Why or why not?
3. **WHAT IF?** If a complete ring of bark is removed around a tree trunk (a process called girdling), would the tree die slowly (in weeks) or quickly (in days)? Explain why.

For suggested answers, see Appendix A.

Go to **MasteringBiology®** for Assignments, the eText, and the Study Area with Animations, Activities, Vocab Self-Quiz, and Practice Tests.

# 28 Chapter Review

## SUMMARY OF KEY CONCEPTS

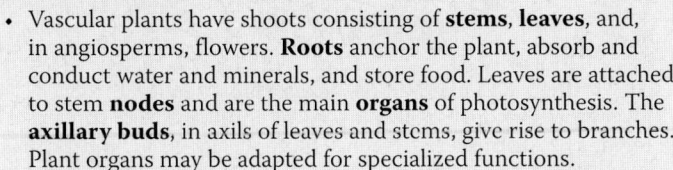

**VOCAB SELF-QUIZ**

goo.gl/gbai8v

### CONCEPT 28.1

**Plants have a hierarchical organization consisting of organs, tissues, and cells (pp. 576–582)**

- Vascular plants have shoots consisting of **stems**, **leaves**, and, in angiosperms, flowers. **Roots** anchor the plant, absorb and conduct water and minerals, and store food. Leaves are attached to stem **nodes** and are the main **organs** of photosynthesis. The **axillary buds**, in axils of leaves and stems, give rise to branches. Plant organs may be adapted for specialized functions.
- Vascular plants have three tissue systems—dermal, vascular, and ground—which are continuous throughout the plant. The **dermal tissue system** protects against pathogens, herbivores, and drought and aids in the absorption of water, minerals, and carbon dioxide. The **vascular tissue system** (**xylem** and **phloem**

tissues) facilitates the long-distance transport of substances. The **ground tissue system** functions in storage, metabolism, and regeneration.
- **Parenchyma cells** are relatively unspecialized and thin-walled cells that retain the ability to divide; they perform most of the plant's metabolic functions of synthesis and storage. **Collenchyma cells** have unevenly thickened walls; they support young, growing parts of the plant. **Sclerenchyma cells**—**sclereids** and **fibers**—have thick, lignified walls that help support mature, nongrowing parts of the plant. **Tracheids** and **vessel elements**, the water-conducting cells of xylem, have thick walls and are dead at functional maturity. **Sieve-tube elements** are living but highly modified cells that are largely devoid of internal organelles; they function in the transport of sugars through the phloem of angiosperms.

? *Describe at least three specializations in plant organs and plant cells that are adaptations to life on land.*

## CONCEPT 28.2

**Different meristems generate new cells for primary and secondary growth (pp. 582–584)**

**?** *Which plant organs originate from the activity of meristems?*

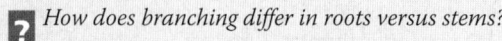

Shoot tip (shoot apical meristem and young leaves)

Vascular cambium

Cork cambium

Lateral meristems

Axillary bud meristem

Root apical meristems

## CONCEPT 28.3

**Primary growth lengthens roots and shoots (pp. 584–588)**

- The root **apical meristem** is located near the tip of the root, where it generates cells for the growing root axis and the **root cap**.
- The **primary growth** of shoots arises from shoot apical meristems found in **apical** and **axillary buds**. The closer an axillary bud is to an active apical bud, the more inhibited it is, a phenomenon called **apical dominance**.
- **Mesophyll** cells are adapted for photosynthesis. **Stomata**, the epidermal pores formed by pairs of **guard cells**, allow for gaseous exchange and are major avenues for water loss.
- Eudicot stems have vascular bundles in a ring, whereas monocot stems have scattered vascular bundles.

**?** *How does branching differ in roots versus stems?*

## CONCEPT 28.4

**Secondary growth increases the diameter of stems and roots in woody plants (pp. 588–591)**

- The **vascular cambium** is a meristematic cylinder that produces secondary xylem and secondary phloem during **secondary growth**. Older layers of secondary xylem (heartwood) become inactive, whereas younger layers (sapwood) still conduct water.
- The **cork cambium** gives rise to a thick protective covering called the **periderm**, which consists of the cork cambium plus the layers of cork cells it produces.

**?** *What advantages did plants gain from the evolution of secondary growth?*

# TEST YOUR UNDERSTANDING

**PRACTICE TEST**

goo.gl/CRZjvS

## Level 1: Knowledge/Comprehension

1. Most of the growth of a plant body is the result of
   - (A) cell differentiation.
   - (B) cell division.
   - (C) morphogenesis.
   - (D) cell elongation.

2. The innermost layer of the root cortex is the
   - (A) core.
   - (B) pericycle.
   - (C) endodermis.
   - (D) pith.

3. Heartwood and sapwood consist of
   - (A) bark.
   - (B) periderm.
   - (C) secondary xylem.
   - (D) secondary phloem.

## Level 2: Application/Analysis

4. Which of the following arise, directly or indirectly, from meristematic activity?
   - (A) secondary xylem
   - (B) dermal tissue
   - (C) leaves
   - (D) all of the above

5. Which of the following would not be seen in a cross section through the woody part of a root?
   - (A) sclerenchyma cells
   - (B) sieve-tube elements
   - (C) parenchyma cells
   - (D) root hairs

6. **DRAW IT** On this cross section from a woody eudicot, label a growth ring, late wood, early wood, and a vessel element. Then draw an arrow in the pith-to-cork direction.

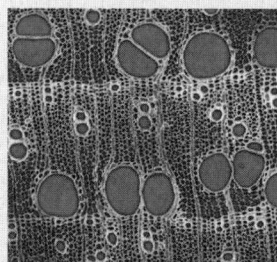

## Level 3: Synthesis/Evaluation

7. **SCIENTIFIC INQUIRY**
   Grasslands typically do not flourish when large herbivores are removed. In fact, they are soon replaced by broad-leaved herbaceous eudicots, shrubs, and trees. Based on your knowledge of the structure and growth habits of monocots versus eudicots, suggest a reason why.

8. **FOCUS ON EVOLUTION**
   Evolutionary biologists have coined the term *exaptation* to describe a common occurrence in the evolution of life: A limb or organ evolves in a particular context but over time takes on a new function (see Concept 23.4). What are some examples of exaptations in plant organs?

9. **FOCUS ON ORGANIZATION**
   In a short essay (100–150 words), explain how the evolution of lignin affected vascular plant structure and function.

10. **SYNTHESIZE YOUR KNOWLEDGE**

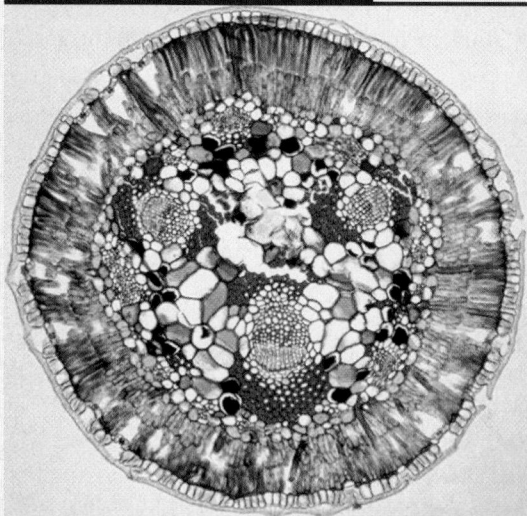

This stained light micrograph shows a cross section through a plant organ from *Hakea purpurea*, a shrub native to some arid regions of Australia. **(a)** Review Figures 28.14, 28.17, and 28.18 to identify whether this is a root, leaf, or stem. Explain your reasoning. **(b)** How might this organ be an adaptation to dry conditions?

*For selected answers, see Appendix A.*

# 29 Resource Acquisition, Nutrition, and Transport in Vascular Plants

## KEY CONCEPTS

**29.1** Adaptations for acquiring resources were key steps in the evolution of vascular plants

**29.2** Different mechanisms transport substances over short or long distances

**29.3** Plant roots absorb essential elements from the soil

**29.4** Plant nutrition often involves relationships with other organisms

**29.5** Transpiration drives the transport of water and minerals from roots to shoots via the xylem

**29.6** The rate of transpiration is regulated by stomata

**29.7** Sugars are transported from sources to sinks via the phloem

▲ **Figure 29.1** Why do aspens quake?

## A Whole Lot of Shaking Going On

If you walk amidst an aspen (*Populus tremuloides*) forest on a clear day, you will be treated to a fantastic light display **(Figure 29.1)**. Even on a day with little wind, the trembling of aspen leaves causes shafts of brilliant sunlight to dapple the forest floor with ever-changing flecks of radiance. The mechanism underlying these passive leaf movements is not difficult to discern: The petiole of each leaf is flattened along its sides, permitting the leaf to flop only in the horizontal plane. Perhaps more curious is why this peculiar adaptation has evolved in *Populus*.

Many hypotheses have been put forward to explain how leaf quaking benefits *Populus*. Old ideas that leaf trembling helps replace the $CO_2$-depleted air near the leaf surface, or deters herbivores, have not been supported by experiments. The leading hypothesis is that leaf trembling increases the photosynthetic productivity of the whole plant by allowing more light to reach the lower leaves of the tree. If not for the shafts of transient sunlight provided by leaf trembling, the lower leaves would be too shaded to photosynthesize sufficiently.

In this chapter, we'll examine various adaptations that help plants acquire water, minerals, $CO_2$, and light more efficiently. We'll look at what nutrients plants require and how plant nutrition often involves other organisms. The acquisition of resources, however, is just the beginning of the story. Resources must be transported to where they are needed. Thus, we will also examine how water, minerals, and sugars are transported through the plant.

# Adaptations for acquiring resources were key steps in the evolution of vascular plants

**EVOLUTION**  Plants typically inhabit two worlds—above ground, where shoots acquire sunlight and $CO_2$, and below ground, where roots acquire water and minerals. Without adaptations that allow the acquisition of these resources, plants could not have colonized land.

The algal ancestors of plants absorbed water, minerals, and $CO_2$ directly from the water in which they lived. Transport in these algae was simple because every cell was close to the source of these substances. The earliest land plants were nonvascular plants that grew photosynthetic shoots above the shallow fresh water in which they lived. These leafless shoots typically had waxy cuticles and few stomata, which allowed them to avoid excessive water loss while still permitting some exchange of $CO_2$ and $O_2$ for photosynthesis. The anchoring and absorbing functions of these early plants were assumed by the base of the stem or by threadlike rhizoids.

As plants evolved and increased in number, the competition for light, water, and nutrients intensified. Taller plants with broad, flat appendages had an advantage in absorbing light. This increase in surface area, however, resulted in more evaporation and therefore a greater need for water. Larger shoots also required stronger anchorage. These needs favored the production of multicellular, branching roots. Meanwhile, as greater shoot heights further separated the photosynthesizing leaves from the nonphotosynthetic parts below ground, natural selection favored plants capable of efficient long-distance transport of water, minerals, and products of photosynthesis.

The evolution of vascular tissue consisting of xylem and phloem made possible the development of extensive root and shoot systems that carry out long-distance transport (see Figure 28.9). The **xylem** transports water and minerals from roots to shoots. The **phloem** transports products of photosynthesis from where they are made or stored to where they are needed. **Figure 29.2** provides an overview of resource acquisition and transport in a vascular plant during the day.

## Shoot Architecture and Light Capture

Because plant success is generally related to photosynthesis, natural selection has resulted in many structural adaptations in shoots that allow for the more efficient acquisition of light from the sun and $CO_2$ from the air. Water use efficiency is another important driving factor that has affected the evolution of shoot architecture.

The enormous variety of shoot architectures seen in nature, including differences in the dimensions, shapes, orientations,

▼ **Figure 29.2  An overview of resource acquisition and transport in a vascular plant during the day.**

**MAKE CONNECTIONS** *When photosynthesis stops at night, cellular respiration continues. Explain how this affects gas exchange in photosynthetic cells at night.* (See Figure 8.20 to review gas exchange between chloroplasts and mitochondria.)

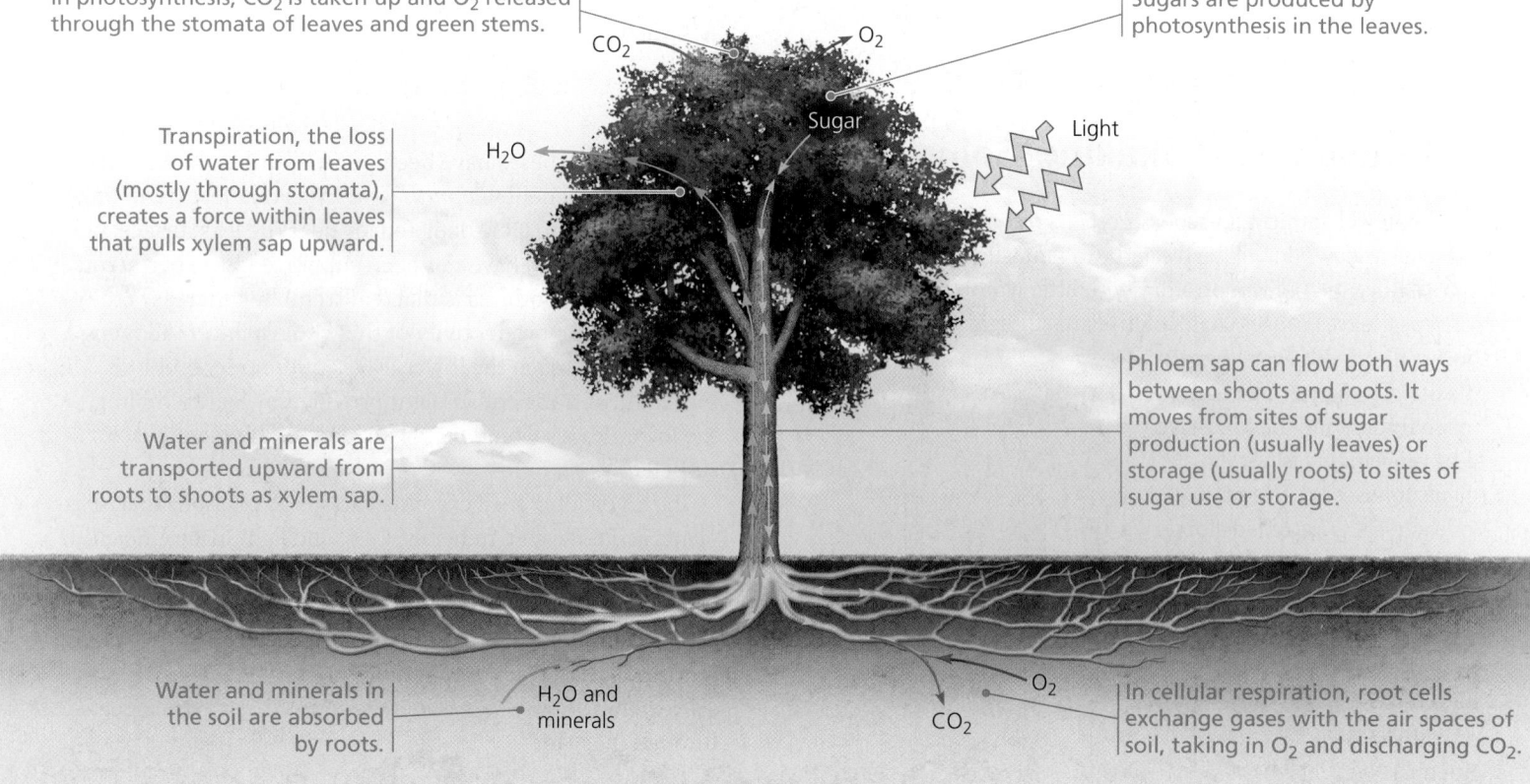

In photosynthesis, $CO_2$ is taken up and $O_2$ released through the stomata of leaves and green stems.

Sugars are produced by photosynthesis in the leaves.

Transpiration, the loss of water from leaves (mostly through stomata), creates a force within leaves that pulls xylem sap upward.

Phloem sap can flow both ways between shoots and roots. It moves from sites of sugar production (usually leaves) or storage (usually roots) to sites of sugar use or storage.

Water and minerals are transported upward from roots to shoots as xylem sap.

Water and minerals in the soil are absorbed by roots.

$H_2O$ and minerals

In cellular respiration, root cells exchange gases with the air spaces of soil, taking in $O_2$ and discharging $CO_2$.

and other attributes of stems and leaves, are largely adaptations for the efficient absorption of light in the ecological niche each species occupies.

Stem lengths, widths, and branching patterns are three architectural features affecting light capture. Plants that grow tall avoid shading from neighboring plants. Most tall plants require thick stems, which enable greater vascular flow to and from the leaves and stronger mechanical support for them. Vines are an exception, relying on other objects (usually other plants) to support their stems. In woody plants, stems become thicker through secondary growth (see Figure 28.10). Branching generally enables plants to harvest sunlight for photosynthesis more effectively. However, some species, such as the coconut palm, do not branch at all. Why is there so much variation in branching patterns? Plants have only a finite amount of energy to devote to shoot growth. If most of that energy goes into branching, there is less available for growing tall, and the risk of being shaded by taller plants increases. Conversely, if most of the energy goes into growing tall, the plants are not optimally harvesting sunlight.

The broad surface of most leaves favors light capture, while stomatal pores allow for diffusion of $CO_2$ into the photosynthetic tissues. Open stomatal pores, however, also promote evaporation of water. Consequently, the adaptations of plants represent compromises between enhancing photosynthesis and minimizing water loss. Later, we'll discuss how plants enhance $CO_2$ uptake and minimize water loss by regulating the opening of stomatal pores.

The arrangement of leaves on a stem, known as **phyllotaxy**, is another feature important in light capture. Phyllotaxy is determined by the shoot apical meristem (see Figure 28.16) and is specific to each species **(Figure 29.3)**. A species may have one leaf per node (alternate, or spiral, phyllotaxy), two leaves per node (opposite phyllotaxy), or more (whorled phyllotaxy). Most angiosperms have alternate phyllotaxy, with leaves arranged in an ascending spiral around the stem, each successive leaf emerging 137.5° from the site of the previous one. Why 137.5°? One hypothesis is that this angle minimizes shading of the lower leaves. In environments where intense sunlight can harm leaves, the greater shading provided by oppositely arranged leaves may be advantageous.

The total area of the leafy portions of all the plants in a community affects each plant's productivity. When there are many layers of vegetation, the shading of the lower leaves is so great that they photosynthesize less than they respire. When this happens, the nonproductive leaves or branches undergo programmed cell death and are eventually shed, a process called *self-pruning*.

Another factor affecting light capture is leaf orientation. Some plants have horizontally oriented leaves; others, such as grasses, have leaves that are vertically oriented. In low-light conditions, horizontal leaves capture sunlight much more effectively than vertical leaves. In grasslands or other sunny regions, however, horizontal orientation may expose upper leaves to overly intense light, injuring leaves and reducing photosynthesis. But if a plant's leaves are nearly vertical, light rays are essentially

▲ **Figure 29.3 Emerging phyllotaxy of Norway spruce.** This SEM, taken from above a shoot tip, shows the pattern of emergence of leaves. The leaves are numbered, with 1 being the youngest. (Some numbered leaves are not visible in the close-up.)

**?** *With your finger, trace the progression of leaf emergence, moving from leaf number 29 to 28 and so on. What is the pattern?*

parallel to the leaf surfaces, so no leaf receives too much light, and light penetrates more deeply to the lower leaves.

## Root Architecture and Acquisition of Water and Minerals

Just as $CO_2$ and sunlight are resources exploited by the shoot system, soil contains resources mined by the root system. Plants can rapidly adjust the architecture and physiology of their roots to exploit patches of available nutrients in the soil. The roots of many plants, for example, respond to pockets of low nitrate availability in soils by extending straight through the pockets instead of branching within them. Conversely, when encountering a pocket rich in nitrate, a root will often branch extensively there. Root cells also respond to high soil nitrate levels by synthesizing more proteins involved in nitrate transport and assimilation. Thus, the plant devotes more mass to exploiting a nitrate-rich patch, and the cells absorb nitrate more efficiently.

The efficient absorption of limited nutrients is also enhanced by reduced competition within the root system. For example, cuttings from stolons of buffalo grass (*Buchloe dactyloides*) develop fewer and shorter roots in the presence of cuttings from the same plant than they do in the presence of cuttings from another buffalo grass plant. Researchers are trying to uncover how the plant distinguishes self from nonself.

Plant roots also form mutually beneficial relationships with microorganisms that enable them to exploit soil resources more efficiently. For example, the evolution of mutualistic associations between roots and fungi called mycorrhizae was a critical step in the successful colonization of land by plants. The role of mycorrhizae in plant nutrition will be examined in Concept 29.4.

Once acquired, resources must be transported to other parts of the plant. Next, we'll examine the processes and pathways that move water, minerals, and sugars throughout the plant.

## CONCEPT CHECK 29.1

1. Why is long-distance transport important for vascular plants?
2. What architectural features influence self-shading?
3. **WHAT IF?** Some plants can detect increased levels of light reflected from leaves of encroaching neighbors. This detection elicits stem elongation, production of erect leaves, and less branching. How do these responses help the plant compete?

For suggested answers, see Appendix A.

## CONCEPT 29.2

# Different mechanisms transport substances over short or long distances

Given the diversity of substances that move through plants and the great range of distances and barriers over which such substances must be transported, it is not surprising that plants employ a variety of transport processes. Before examining these processes, however, let's consider the two major pathways of transport: the apoplast and the symplast.

## The Apoplast and Symplast: Transport Continuums

Plant tissues have two major compartments—the apoplast and the symplast. The **apoplast** consists of everything external to the plasma membranes of living cells and includes cell walls, extracellular spaces, and the interior of dead cells such as vessel elements and tracheids (see Figure 28.9). The **symplast** consists of the entire mass of cytosol of all the living cells in a plant, as well as the plasmodesmata, the cytoplasmic channels that interconnect them.

The compartmental structure of plants provides three routes for transport within a plant tissue or organ: the apoplastic, symplastic, and transmembrane routes (**Figure 29.4**). In the *apoplastic route*, water and solutes (dissolved chemicals) move along the continuum of cell walls and extracellular spaces. In the *symplastic route*, water and solutes move along the continuum of cytosol. This route requires substances to cross a plasma membrane once, when they first enter the plant. After entering one cell, substances can move from cell to cell via plasmodesmata. In the *transmembrane route*, water and solutes move out of one cell, across the cell wall, and into the neighboring cell, which may pass them to the next cell in the same way. The transmembrane route requires repeated crossings of plasma membranes as substances exit one cell and enter the next. These three routes are not mutually exclusive, and some substances may use more than one route to varying degrees.

## Short-Distance Transport of Solutes Across Plasma Membranes

In plants, as in any organism, the selective permeability of the plasma membrane controls the short-distance movement of substances into and out of cells (see Concept 5.2). Both active and passive transport mechanisms occur in plants, and plant cell membranes are equipped with the same *general* types of pumps and transport proteins (channel proteins, carrier proteins, and cotransporters) that function in other cells. There are, however, *specific* differences between the membrane transport processes of plant and animal cells. Unlike in animal cells, hydrogen ions ($H^+$) rather than sodium ions ($Na^+$) play the primary role in basic transport processes in plant cells. For example, in plant cells the membrane potential (the voltage across the membrane) is established mainly through the pumping of $H^+$ by proton pumps (see Figure 5.16), rather than the pumping of $Na^+$ by sodium-potassium pumps. Also, $H^+$ is most often cotransported in plants, whereas $Na^+$ is typically cotransported in animals. During cotransport, plant cells use the energy in the $H^+$ gradient and membrane potential to drive the active transport of many different solutes. For instance, cotransport with $H^+$ is responsible for absorption of neutral solutes, such as the sugar sucrose, by phloem cells and other plant cells. An $H^+$/sucrose cotransporter couples movement of sucrose against its concentration gradient with movement of $H^+$ down its electrochemical gradient (see Figure 5.17). Cotransport with $H^+$ also facilitates movement of ions across plant cell membranes.

The membranes of plant cells also have ion channels that allow only certain ions to pass (see Figure 5.13). As in animal cells, most channels are gated, opening or closing in response to stimuli such as chemicals, pressure, or voltage. Later in this chapter, we'll discuss how potassium ($K^+$) ion channels in guard cells function in opening and closing stomata. Ion channels are also involved in producing electrical signals analogous to the action potentials of animals (see Concept 37.3). For example, the phloem conducts nerve-like electrical signals that help integrate whole-plant function. However, these signals are 1,000 times slower than in animals and employ calcium ($Ca^{2+}$)-activated anion channels rather than the sodium ($Na^+$) ion channels used by animal cells.

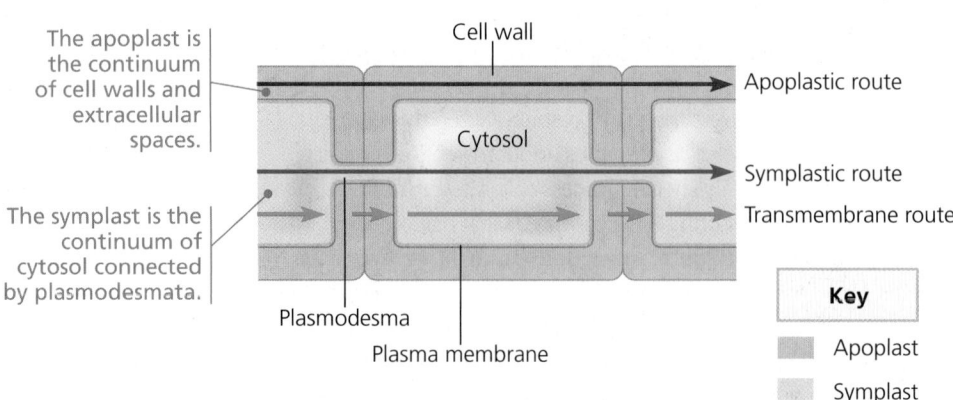

The apoplast is the continuum of cell walls and extracellular spaces.

The symplast is the continuum of cytosol connected by plasmodesmata.

Cell wall

Cytosol

Apoplastic route

Symplastic route

Transmembrane route

Plasmodesma

Plasma membrane

**Key**

Apoplast

Symplast

▲ **Figure 29.4 Cell compartments and routes for short-distance transport.** Some substances may use more than one transport route.

## Short-Distance Transport of Water Across Plasma Membranes

The absorption or loss of water by a cell occurs by **osmosis**, the diffusion of free water—water that is not bound to solutes or surfaces—across a membrane (see Figure 5.10). The physical property that predicts the direction in which water will flow is called **water potential**, a quantity that includes the effects of solute concentration and physical pressure. Free water moves from regions of higher water potential to regions of lower water potential if there is no barrier to its flow. For example, if a plant cell or seed is immersed in a solution that has a higher water potential than the cell, water will move into the cell or seed, causing it to expand. The expansion of plant cells and seeds can be a powerful force: Growing tree roots, for example, can break sidewalks, and the swelling of wet grain within the holds of damaged ships has led to complete hull failure resulting in sinking of the ships. Given the strong forces generated, it is interesting to consider whether water uptake by seeds is an active process, a question examined in the **Scientific Skills Exercise**, which explores the effect of temperature on this process.

Water potential is abbreviated by the Greek letter $\psi$ (psi, pronounced "sigh"). Plant biologists measure $\psi$ in a unit of pressure called a **megapascal (MPa)**. By definition, the $\psi$ of pure water in a container open to the atmosphere under standard conditions (at sea level and at room temperature) is 0 MPa.

### How Solutes and Pressure Affect Water Potential

Solute concentration and physical pressure are the major determinants of water potential in hydrated plants, as expressed in the *water potential equation*:

$$\psi = \psi_S + \psi_P$$

where $\psi$ is the water potential, $\psi_S$ is the solute potential (osmotic potential), and $\psi_P$ is the pressure potential. The **solute potential ($\psi_S$)** of a solution is directly proportional to its molarity. Solute potential is also called *osmotic potential* because solutes affect the direction of osmosis. The solutes in plants are typically ions and sugars. By definition, the $\psi_S$ of pure water is 0. When solutes are added, they bind water molecules. As a result, there are fewer free water molecules, reducing the capacity of the water to move and do work. In this way, an increase in solute concentration has a negative effect on water potential, which is why the $\psi_S$ of a solution is always expressed as a negative number. For example, a 0.1 $M$ solution of a sugar has a $\psi_S$ of $-0.23$ MPa. As the solute concentration increases, $\psi_S$ becomes more negative.

---

### Scientific Skills Exercise

## Calculating and Interpreting Temperature Coefficients

**Does the Initial Uptake of Water by Seeds Depend on Temperature?** One way to answer this question is to soak seeds in water at different temperatures and measure the rate of water uptake at each temperature. The collected data can be used to calculate the temperature coefficient, $Q_{10}$, the factor by which a physiological process or reaction rate increases when the temperature is raised by 10°C:

$$Q_{10} = \left(\frac{k_2}{k_1}\right)^{\frac{10}{t_2 - t_1}}$$

where $t_2$ = higher temperature (°C), $t_1$ = lower temperature, $k_2$ = reaction rate at $t_2$, and $k_1$ = reaction rate at $t_1$. (Note that if $t_2 - t_1 = 10$, as in this exercise, the math is simplified.)

$Q_{10}$ values may be used to make inferences about the physiological process under investigation. Chemical (metabolic) processes involving large-scale protein shape changes are highly dependent on temperature and have higher $Q_{10}$ values, closer to 2 or 3. In contrast, many, but not all, physical parameters are relatively independent of temperature and have $Q_{10}$ values closer to 1. For example, the $Q_{10}$ of the change in the viscosity of water is 1.2–1.3. In this exercise, you will calculate $Q_{10}$ using data for radish seeds (*Raphanus sativus*) to assess whether the initial uptake of water by seeds is more likely to be a physical or a chemical process.

**How the Experiment Was Done** Samples of radish seeds were weighed and placed in water at four different temperatures. After 30 minutes, the seeds were removed, blotted dry, and reweighed. The researchers then calculated the percent increase in mass due to water uptake for each sample.

**Data from the Experiment** The table shows the percent increases.

| Temperature | % Increase in Mass Due to Water Uptake After 30 Minutes |
|---|---|
| 5°C | 18.5% |
| 15°C | 26.0% |
| 25°C | 31.0% |
| 35°C | 36.2% |

**Data from** J. D. Murphy and D. L. Noland, Temperature effects on seed imbibition and leakage mediated by viscosity and membranes, *Plant Physiology* 69:428–431 (1982).

**INTERPRET THE DATA**

1. Based on the data, does the initial uptake of water by radish seeds vary with temperature? What is the relationship between temperature and water uptake?

2. (a) Using the data for 35°C and 25°C, calculate $Q_{10}$ for water uptake by radish seeds. Repeat the calculation using the data for 25°C and 15°C and the data for 15°C and 5°C. (b) What is the average $Q_{10}$? (c) Do your results imply that the uptake of water by radish seeds is mainly a physical process or a chemical (metabolic) process? (d) Given that the $Q_{10}$ for the change in the viscosity of water is 1.2–1.3, could the slight temperature dependence of water uptake by seeds be a reflection of the slight temperature dependence of the viscosity of water?

3. Besides temperature, what other independent variables could you alter to test whether radish seed swelling is essentially a physical process or a chemical process?

4. Would you expect plant growth to have a $Q_{10}$ closer to 1 or 3? Why?

(MB) A version of this Scientific Skills Exercise can be assigned in MasteringBiology.

**Pressure potential ($\psi_P$)** is the physical pressure on a solution. Unlike $\psi_S$, $\psi_P$ can be positive or negative relative to atmospheric pressure. For example, when a solution is being withdrawn by a syringe, it is under negative pressure; when it is being expelled from a syringe, it is under positive pressure. The water in living cells is usually under positive pressure due to the osmotic uptake of water. Specifically, the **protoplast** (the living part of the cell, which includes the plasma membrane) presses against the cell wall, creating what is called **turgor pressure**. This pushing effect of internal pressure, much like the air in an inflated tire, is critical for plant function because it helps maintain the stiffness of plant tissues and also serves as the driving force for cell elongation. Conversely, the water in the hollow, nonliving xylem cells (tracheids and vessel elements) of a plant is often under a negative pressure potential (tension) of less than $-2$ MPa.

As you learn to apply the water potential equation, keep in mind the key point: *Water moves from regions of higher water potential to regions of lower water potential.*

### Water Movement Across Plant Cell Membranes

Now let's consider how water potential affects absorption and loss of water by a living plant cell. First, imagine a cell that is **flaccid** (limp) as a result of losing water. The cell has a $\psi_P$ of 0 MPa. Suppose this flaccid cell is bathed in a solution of higher solute concentration (more negative solute potential) than the cell itself **(Figure 29.5a)**. Since the external solution has the lower (more negative) water potential, water diffuses out of the cell. The cell's protoplast undergoes **plasmolysis**—that is, it shrinks and pulls away from the cell wall. If we place the same flaccid cell in pure water ($\psi = 0$ MPa) **(Figure 29.5b)**, the cell, because it contains solutes, has a lower water potential than the water, and water enters the cell by osmosis. The contents of

▲ **Figure 29.6 A moderately wilted plant can regain its turgor when watered.**

the cell begin to swell and press the plasma membrane against the cell wall. The partially elastic wall, exerting turgor pressure, confines the pressurized protoplast. When this pressure is enough to offset the tendency for water to enter because of the solutes in the cell, then $\psi_P$ and $\psi_S$ are equal, and $\psi = 0$. This matches the water potential of the extracellular environment—in this example, 0 MPa. A dynamic equilibrium has been reached, and there is no further net movement of water.

In contrast to a flaccid cell, a walled cell with a greater solute concentration than its surroundings is **turgid**, or very firm. When turgid cells push against each other, the tissue stiffens. The effects of turgor loss are seen during **wilting**, when leaves and stems droop as a result of cells losing water **(Figure 29.6)**.

### Aquaporins: Facilitating Diffusion of Water

A difference in water potential determines the *direction* of water movement across membranes, but how do water molecules actually cross the membranes? The movement of water molecules across biological membranes is too rapid to

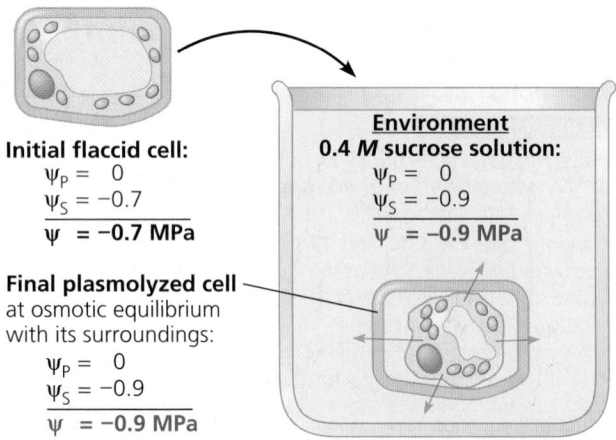

**(a) Initial conditions: cellular $\psi$ > environmental $\psi$.** The protoplast loses water, and the cell plasmolyzes. After plasmolysis is complete, the water potentials of the cell and its surroundings are the same.

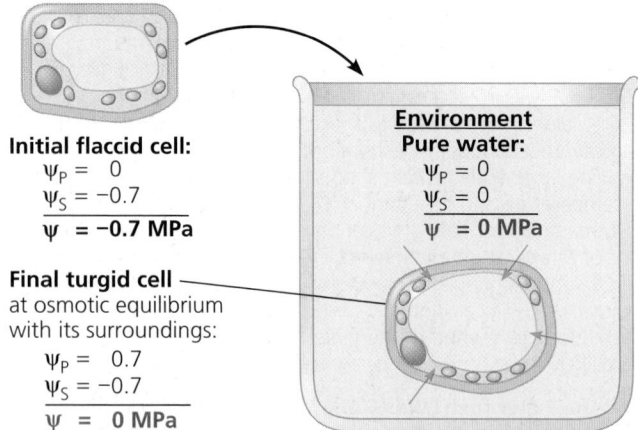

**(b) Initial conditions: cellular $\psi$ < environmental $\psi$.** There is a net uptake of water by osmosis, causing the cell to become turgid. When this tendency for water to enter is offset by the back pressure of the elastic wall, water potentials are equal for the cell and its surroundings. (The volume change of the cell is exaggerated in this diagram.)

▲ **Figure 29.5 Water relations in plant cells.** In these experiments, flaccid cells (cells in which the protoplast contacts the cell wall but lacks turgor pressure) are placed in two environments. The blue arrows indicate initial net water movement.

be explained by unaided diffusion. Transport proteins called **aquaporins** (see Figure 5.1) facilitate the transport of water molecules across plant cell plasma membranes. Aquaporin channels, which can open and close, affect the *rate* at which water moves osmotically across the membrane.

## Long-Distance Transport: The Role of Bulk Flow

Long-distance transport in plants occurs through **bulk flow**, the movement of liquid in response to a pressure gradient. The bulk flow of material always occurs from higher to lower pressure. Unlike osmosis, bulk flow is independent of solute concentration. Long-distance bulk flow occurs within the tracheids and vessel elements of the xylem and within the sieve-tube elements of the phloem. The structures of these conducting cells facilitate bulk flow. Mature tracheids and vessel elements are dead cells (see Figure 28.9) and therefore have no cytoplasm, and the cytoplasm of sieve-tube elements is almost devoid of internal organelles. If you have ever dealt with a partially clogged drain, you know that the volume of flow depends on the pipe's diameter. Clogs reduce the effective diameter of the drainpipe. Such experiences help us understand how the structures of plant cells specialized for bulk flow fit their function. Like the unclogging of a kitchen drain, the absence or reduction of cytoplasm in a plant's "plumbing" facilitates bulk flow through the xylem and phloem.

Diffusion, active transport, and bulk flow act in concert to transport resources throughout the whole plant. In the following sections, we examine in more detail the transport of water and minerals from roots to shoots, the control of evaporation, and the transport of sugars.

### CONCEPT CHECK 29.2

1. If a plant cell immersed in distilled water has a $\psi_S$ of $-0.7$ MPa and a $\psi$ of 0 MPa, what is the cell's $\psi_P$? If you put it in an open beaker of solution that has a $\psi$ of $-0.4$ MPa, what would be its $\psi_P$ at equilibrium?
2. How would a reduction in the number of aquaporin channels affect a plant cell's ability to adjust to new osmotic conditions?
3. **WHAT IF?** What would happen if you put plant protoplasts in pure water? Explain.

For suggested answers, see Appendix A.

---

## CONCEPT 29.3

# Plant roots absorb essential elements from the soil

Water, air, and soil minerals all contribute to plant growth. A plant's water content can be measured by comparing the mass before and after drying. Typically, 80–90% of a plant's fresh mass is water. Some 96% of the remaining dry mass consists of carbohydrates such as cellulose that are produced by photosynthesis. Thus, the components of carbohydrates— carbon, oxygen, and hydrogen—are the most abundant elements in dried plant residue. Inorganic substances from the soil, although essential for plant survival, generally account for only about 4% of a plant's dry mass.

## Macronutrients and Micronutrients

The inorganic substances in plants consist of more than 50 chemical elements. In studying the chemical composition of plants, we must distinguish elements that are essential from those that are merely present in the plant. A chemical element is considered an **essential element** only if it is required for a plant to complete its life cycle and produce another generation.

To determine which chemical elements are essential, researchers often use **hydroponic culture**, in which plants are grown in mineral solutions instead of soil **(Figure 29.7)**. Such studies have helped identify 17 essential elements needed by all plants. Hydroponic culture is also used on a small scale to grow some greenhouse crops.

Nine of the essential elements are called **macronutrients** because plants require them in relatively large amounts. Six of these are the major components of organic compounds forming a plant's structure: carbon, oxygen, hydrogen, nitrogen, phosphorus, and sulfur. The other three macronutrients are potassium, calcium, and magnesium. Of all the mineral nutrients, nitrogen contributes the most to plant growth and crop yields. Plants require nitrogen as a component of proteins, nucleic

---

### ▼ Figure 29.7   Research Method

#### Hydroponic Culture

**Application**  In hydroponic culture, plants are grown in mineral solutions without soil. One use of hydroponic culture is to identify essential elements in plants.

**Technique**  Plant roots are bathed in aerated solutions of known mineral composition. Aerating the water provides the roots with oxygen for cellular respiration. (*Note*: The flasks would normally be opaque to prevent algal growth.) A mineral, such as iron, can be omitted to test whether it is essential.

**Control:** Solution containing all minerals

**Experimental:** Solution without iron

**Results**  If the omitted mineral is essential, mineral deficiency symptoms occur, such as stunted growth and discolored leaves. By definition, the plant would not be able to complete its life cycle. Deficiencies of different elements may have different symptoms, which can aid in diagnosing mineral deficiencies in soil.

acids, chlorophyll, and other important organic molecules. **Table 29.1** summarizes the functions of the macronutrients.

The other essential elements are called **micronutrients** because plants need them in only tiny quantities. They are chlorine, iron, manganese, boron, zinc, copper, nickel, and molybdenum. Sodium is a ninth essential micronutrient for plants that use the CAM or $C_4$ pathway of photosynthesis.

Micronutrients function in plants mainly as cofactors, non-protein helpers in enzymatic reactions (see Concept 6.4). Iron, for example, is a metallic component of cytochromes, the proteins in the electron transport chains of chloroplasts and mitochondria. It is because micronutrients generally play catalytic roles that plants need only tiny quantities. The requirement for molybdenum, for instance, is so modest that there is only one atom of this rare element for every 60 million atoms of hydrogen in dried plant material. Yet a deficiency of molybdenum or any other micronutrient can weaken or kill a plant.

## Symptoms of Mineral Deficiency

The symptoms of a deficiency depend partly on the mineral's function as a nutrient. For example, a deficiency of magnesium, a component of chlorophyll, causes *chlorosis*, yellowing of the leaves. In some cases, the relationship between a mineral deficiency and its symptoms is less direct. For instance, iron deficiency can cause chlorosis even though chlorophyll contains no iron, because iron ions are required as a cofactor in an enzymatic step of chlorophyll synthesis.

Mineral deficiency symptoms depend not only on the role of the nutrient but also on its mobility within the plant. If a nutrient moves about freely, symptoms appear first in older organs, because young, growing tissues are a greater sink for nutrients that are in short supply. For example, magnesium is relatively mobile and is shunted preferentially to young leaves. Therefore, a plant deficient in magnesium first shows signs of chlorosis in its older leaves. In contrast, a deficiency of a mineral that is relatively immobile affects young parts of the plant first. Older tissues may have adequate amounts that they retain during

periods of short supply. For example, iron does not move freely within a plant, and an iron deficiency causes yellowing of young leaves before any effect on older leaves is visible. The mineral requirements of a plant also change with the age of the plant. Young seedlings, for example, rarely show mineral deficiency symptoms because their mineral needs are met largely by the mineral reserves stored in the seeds themselves.

The symptoms of a mineral deficiency in a given plant species are often distinctive enough to aid in diagnosis **(Figure 29.8)**. Deficiencies of phosphorus, potassium, and especially nitrogen are most common. Micronutrient shortages are less common and reflect local differences in soil composition. The amount of a micronutrient needed to correct a deficiency is usually quite small. For example, a zinc deficiency in fruit trees can usually be

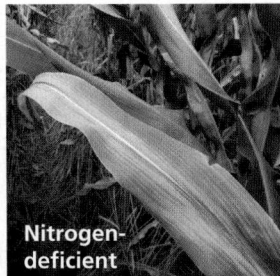

▲ **Figure 29.8 The most common mineral deficiencies, as seen in maize leaves.** Mineral deficiency symptoms may vary in different species. In maize, nitrogen deficiency is evident in a yellowing that starts at the tip and moves along the center (midrib) of older leaves. Phosphorus-deficient maize plants have reddish purple margins, particularly in young leaves. Potassium-deficient maize plants exhibit "firing," or drying, along tips and margins of older leaves.

| Table 29.1 Macronutrients in Plants | | | |
|---|---|---|---|
| Element | Form Primarily Absorbed by Plants | % Mass in Dry Tissue | Major Functions |
| Carbon | $CO_2$ | 45% | Major component of plant's organic compounds |
| Oxygen | $CO_2$ | 45% | Major component of plant's organic compounds |
| Hydrogen | $H_2O$ | 6% | Major component of plant's organic compounds |
| Nitrogen | $NO_3^-$, $NH_4^+$ | 1.5% | Component of nucleic acids, proteins, and chlorophyll |
| Potassium | $K^+$ | 1.0% | Cofactor of many enzymes; major solute functioning in water balance; operation of stomata |
| Calcium | $Ca^{2+}$ | 0.5% | Important component of middle lamella and cell walls; maintains membrane function; signal transduction |
| Magnesium | $Mg^{2+}$ | 0.2% | Component of chlorophyll; cofactor of many enzymes |
| Phosphorus | $H_2PO_4^-$, $HPO_4^{2-}$ | 0.2% | Component of nucleic acids, phospholipids, ATP |
| Sulfur | $SO_4^{2-}$ | 0.1% | Component of proteins |

cured by hammering a few zinc nails into each tree trunk. Moderation is important because overdoses of many nutrients can be detrimental or toxic to plants. Too much nitrogen, for example, can lead to excessive vine growth in tomato plants at the expense of good fruit production.

## Soil Management

Ancient farmers recognized that yields on a particular plot of land decreased over the years. Moving to uncultivated areas, they observed the same pattern of reduced yields over time. Eventually, they realized that fertilization could make soil a renewable resource that enabled crops to be cultivated season after season at a fixed location. This sedentary agriculture facilitated a new way of life. People began to build permanent dwellings—the first villages. They also stored food for use between harvests, and food surpluses enabled some people to specialize in nonfarming occupations. In short, the early discovery of soil fertilization helped prepare the way for modern societies.

### Fertilization

In natural ecosystems, mineral nutrients are usually recycled by the excretion of animal wastes and the decomposition of **humus**, the remains of dead organisms and other organic matter. Over many harvests, agricultural fields eventually become depleted of nutrients, a major cause of global soil degradation. Farmers must reverse nutrient depletion by means of fertilization, the addition of mineral nutrients to the soil.

Today, most farmers in industrialized nations use fertilizers containing minerals that are either mined or prepared by energy-intensive processes. These fertilizers are usually enriched in nitrogen (N), phosphorus (P), and potassium (K)—the nutrients most commonly deficient in depleted soils. You may have seen fertilizers labeled with a three-number code, called the N–P–K ratio. A fertilizer marked "15–10–5," for instance, is 15% N (as ammonium or nitrate), 10% P (as phosphate), and 5% K (as the mineral potash).

Manure, fishmeal, and compost are called "organic" fertilizers because they are of biological origin and contain decomposing organic material. Before plants can use organic material, however, it must be decomposed into the inorganic nutrients that roots can absorb. Whether from organic fertilizer or a chemical factory, the minerals a plant extracts are in the same form. However, organic fertilizers release them gradually, whereas minerals in commercial fertilizers are immediately available but may not be retained by the soil for long. A drawback of modern fertilization practices is that minerals not absorbed by roots are often leached from the soil by rainwater or irrigation. To make matters worse, mineral runoff into lakes may lead to explosions in algal populations that can deplete oxygen levels and decimate fish populations.

### Adjusting Soil pH

Soil pH is an important factor that influences mineral availability. Depending on the soil pH, a particular mineral may be bound too tightly to soil particles or may be in a chemical form that the plant cannot absorb. Most plants prefer slightly acidic soil because the high $H^+$ concentrations can displace positively charged minerals from soil particles, making them more available for absorption. Adjusting soil pH for optimal crop growth is tricky because a change in $H^+$ concentration may make one mineral more available but another less available. At pH 8, for instance, plants can absorb calcium, but iron is almost unavailable. The soil pH should be matched to a crop's mineral needs. If the soil is too alkaline, adding sulfate will lower the pH. Soil that is too acidic can be adjusted by adding pulverized limestone.

When the soil pH dips to 5 or lower, toxic aluminum ions ($Al^{3+}$) become more soluble and are absorbed by roots, stunting root growth. Some plants cope with high $Al^{3+}$ levels by secreting organic anions that bind $Al^{3+}$ and render it harmless. However, low soil pH and $Al^{3+}$ toxicity continue to pose problems, especially in tropical regions, where the pressure of producing food for a growing population is often most acute.

Soil mismanagement is a major problem facing the world. More than 30% of the world's farmland has reduced productivity stemming from poor soil conditions, such as chemical contamination, mineral deficiencies, acidity, salinity, and poor drainage. As the world's population continues to grow, the demand for food increases. Because soil quality is a major determinant of crop yield, the need to manage soil resources prudently has never been greater.

## The Living, Complex Ecosystem of Soil

The successful cultivation of plants in soil-free hydroponic systems demonstrates that plants do not need soil to complete their life cycles. Still, most terrestrial plants do grow in soil, and it is from the topsoil that they usually acquire mineral nutrients. Thus, an understanding of the properties of soil is important for understanding plants and their growth. We begin by discussing the basic physical properties of soil: its texture and composition.

### Soil Texture

The texture of soil depends on the size of its particles. Soil particles can range from coarse sand (0.02–2 mm in diameter) to silt (0.002–0.02 mm) to microscopic clay particles (less than 0.002 mm). These different-sized particles arise ultimately from the weathering of rock. Water freezing in the crevices of rocks causes mechanical fracturing, and weak acids in the soil break rocks down chemically. When organisms penetrate the rock, they accelerate breakdown by chemical and mechanical means. Roots, for example, secrete acids that dissolve the rock, and their growth in fissures leads to mechanical fracturing. Mineral particles released by weathering become mixed with living organisms and humus, forming topsoil.

In the topsoil, plants are nourished by the soil solution, the water and dissolved minerals in the pores between soil particles. The pores also contain air pockets. After a heavy rainfall, water drains away from the larger spaces in the soil, but smaller spaces retain water because water molecules are attracted to the negatively charged surfaces of clay and other soil particles.

The most fertile topsoils are **loams**, which are composed of roughly equal amounts of sand, silt, and clay. Loamy soils have enough small silt and clay particles to provide ample surface area for the adhesion and retention of minerals and water. Meanwhile, the large spaces between sand particles enable efficient diffusion of oxygen to the roots. Sandy soils generally don't retain enough water to support vigorous plant growth, and clayey soils tend to retain too much water. When soil does not drain adequately, the air is replaced by water, and the roots suffocate from lack of oxygen. Typically, the most fertile topsoils have pores that are about half water and half air, providing a good balance between aeration, drainage, and water storage capacity. The physical properties of soils can be adjusted by adding soil amendments, such as peat moss, compost, manure, or sand.

### Topsoil Composition

A soil's composition encompasses its inorganic (mineral) and organic chemical components. The organic components include the many life-forms that inhabit the soil.

**Inorganic Components** The surface charges of soil particles determine their ability to bind many nutrients. Most soil particles are negatively charged, so positively charged ions (cations), such as potassium ($K^+$), bind strongly to them. Negatively charged ions (anions), such as nitrate ($NO_3^-$), do not bind to these soil particles and are therefore easily lost by leaching, the percolation of water through the soil.

Roots do not absorb mineral cations directly from soil particles; they absorb them from the soil solution. Mineral cations enter the soil solution by **cation exchange**, a process in which cations are displaced from soil particles by other cations, particularly $H^+$ **(Figure 29.9)**. A soil's capacity to exchange cations is determined by the number of cation adhesion sites and by the soil's pH. Soils with more small clay particles and organic matter have higher cation exchange capacity.

**Organic Components** The major organic component of topsoil is humus, which consists of organic material produced by the decomposition of fallen leaves, dead organisms, feces, and other organic matter by bacteria and fungi. Humus prevents clay particles from packing together and forms a crumbly soil that retains water but is still porous enough to aerate roots adequately. Humus also increases the soil's capacity to exchange cations and serves as a reservoir of mineral nutrients that return gradually to the soil as microorganisms decompose the organic matter.

Topsoil is home to an astonishing number and variety of organisms. A teaspoon of topsoil has about 5 billion bacteria, which cohabit with fungi, algae and other protists, insects, earthworms, nematodes, and plant roots. The activities of all these organisms affect the soil's physical and chemical properties. Earthworms, for example, consume organic matter and derive their nutrition from the bacteria and fungi growing on this material. They excrete wastes and move large amounts of material to the soil surface. In addition, they move organic matter into deeper layers of the soil. Earthworms mix and

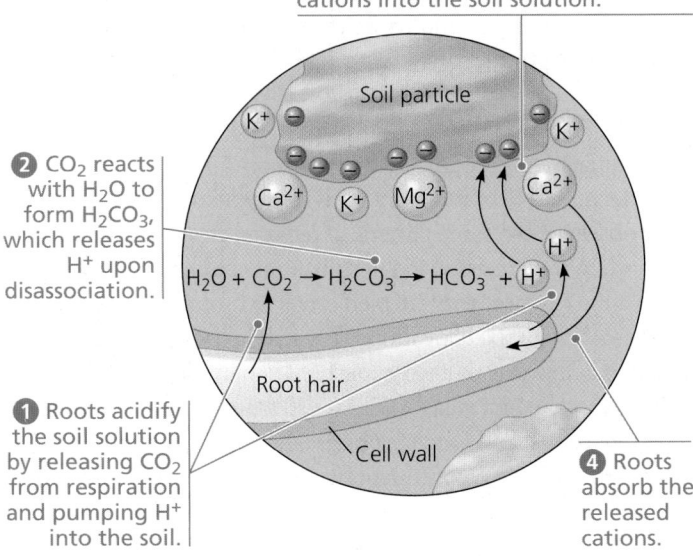

❸ $H^+$ ions in the soil solution neutralize the negative charge of soil particles, causing release of mineral cations into the soil solution.

❷ $CO_2$ reacts with $H_2O$ to form $H_2CO_3$, which releases $H^+$ upon disassociation.

$$H_2O + CO_2 \rightarrow H_2CO_3 \rightarrow HCO_3^- + H^+$$

❶ Roots acidify the soil solution by releasing $CO_2$ from respiration and pumping $H^+$ into the soil.

❹ Roots absorb the released cations.

Soil particle

Root hair

Cell wall

▲ **Figure 29.9 Cation exchange in soil.**

**?** *Which are more likely to be leached from the soil by heavy rains—cations or anions? Explain.*

clump the soil particles, allowing for better gaseous diffusion and retention of water. Roots also affect soil texture and composition. For example, by binding the soil, they reduce erosion, and by excreting acids, they lower soil pH.

**CONCEPT CHECK 29.3**

1. Are some essential elements more important than others? Explain.
2. **WHAT IF?** If an element increases the growth rate of a plant, can it be defined as an essential element?
3. **MAKE CONNECTIONS** Explain why ethanol accumulates in plant roots subjected to waterlogging (see Figure 7.17).

For suggested answers, see Appendix A.

**CONCEPT 29.4**

# Plant nutrition often involves relationships with other organisms

To this point, we have portrayed plants as exploiters of soil resources. But plants and soil have a two-way relationship. Dead plants provide much of the energy needed by soil microorganisms, while sugar-rich secretions from living roots support a wide variety of soil microbes in the near-root environment. Plants derive benefits from their associations with many of these microbes. Although such mutually beneficial relationships across kingdoms are not rare in nature, they are of particular importance to plants **(Figure 29.10)**. In this Concept, we'll explore some important *mutualisms* between plants and soil organisms as well as some unusual, nonmutualistic forms of plant nutrition.

MAKE CONNECTIONS

# Mutualism Across Kingdoms and Domains

Some toxic species of fish don't make their own poison. How is that possible? Some species of ants chew leaves but don't eat them. Why? The answers lie in some amazing mutualisms, relationships between different species in which each species provides a substance or service that benefits the other (see Concept 41.1). Sometimes mutualisms occur within the same kingdom, such as between two species of animals. Many mutualisms, however, involve species from different kingdoms or domains, as in these examples.

## Fungus–Bacterium

A lichen is a mutualistic association between a fungus and a photosynthetic partner. In the lichen *Peltigera*, the photosynthetic partner is a species of cyanobacterium. The cyanobacterium supplies carbohydrates, while the fungus provides anchorage, protection, minerals, and water. (See Figure 26.29.)

The lichen *Peltigera*

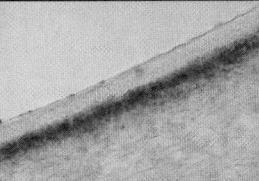

A longitudinal section of the lichen *Peltigera* showing green photosynthetic bacteria sandwiched between layers of fungus

## Animal–Bacterium

Fugu is the Japanese name for puffer fish and the delicacy made from it, which can be deadly. Most species of puffer fish contain lethal amounts of the nerve toxin tetrodotoxin in their organs, especially the liver, ovaries, and intestines. Therefore, a specially trained chef must remove the poisonous parts. The tetrodotoxin is synthesized by mutualistic bacteria (various *Vibrio* species) associated with the fish. The fish gains a potent chemical defense, while the bacteria live in a high-nutrient, low-competition environment.

Puffer fish (fugu)

## Plant–Bacterium

The floating fern *Azolla* provides carbohydrates for a nitrogen-fixing cyanobacterium that resides in the air spaces of the leaves. In return, the fern receives nitrogen from the cyanobacterium. (See Concept 24.5.)

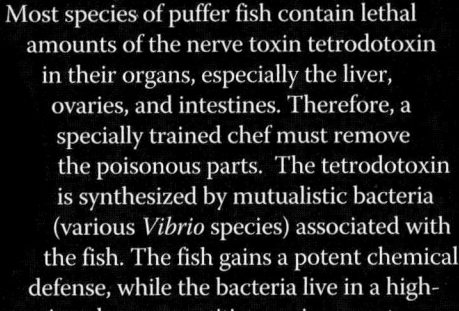

The floating fern *Azolla*

## Animal–Fungus

Leaf-cutter ants harvest leaves that they carry back to their nest, but the ants do not eat the leaves. Instead, a fungus grows by absorbing nutrients from the leaves, and the ants eat part of the fungus that they have cultivated.

Leaf-cutter ants bringing leaves to a nest

Ants tending a fungal garden in a nest

## Plant–Fungus

Most plant species have mycorrhizae, mutualistic associations between roots and fungi. The fungus absorbs carbohydrates from the roots. In return, the fungus's mycelium, a dense network of filaments called hyphae, increases the surface area for the uptake of water and minerals by the roots. (See Figure 26.11.)

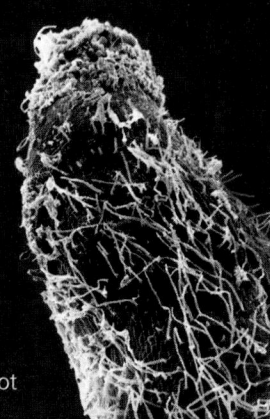

A fungus growing on the root of a sorghum plant (SEM)

## Plant–Animal

Some species of *Acacia* plants are aggressively defended from predators and competitors by ants that live within the plant's hollow thorns. The plant provides nourishment for the ants in the forms of protein-rich structures at the bases of leaves and carbohydrate-rich nectar. (See Figure 41.7.)

Protective ants harvesting protein-rich structures from an *Acacia* plant

**MAKE CONNECTIONS** *Describe four more examples of mutualisms. (See Figure 1.10, Figure 24.23, Concept 30.1, and Concept 33.4.)*

## Bacteria and Plant Nutrition

Some soil bacteria engage in mutually beneficial chemical exchanges with plant roots. Others enhance the decomposition of organic materials and increase nutrient availability. Some even live inside roots and convert nitrogen from the air.

### Rhizobacteria

A variety of mutualistic bacteria play roles in plant nutrition. **Rhizobacteria** live in the **rhizosphere**, the soil closely surrounding the plant's roots. **Endophytes** are nonpathogenic bacteria (or fungi) that live between cells within the plant itself but do not form deep, intimate associations with the cells or alter their morphology. Both endophytic bacteria and rhizobacteria depend on nutrients such as sugars, amino acids, and organic acids that are secreted by plant cells. In the case of the rhizosphere, up to 20% of a plant's photosynthetic production

fuels the organisms in this miniature ecosystem. In turn, endophytic bacteria and rhizobacteria enhance plant growth by a variety of mechanisms. Some produce chemicals that stimulate plant growth. Others produce antibiotics that protect roots from disease. Still others absorb toxic metals or make nutrients more available to roots. Inoculation of seeds with plant-growth-promoting rhizobacteria can increase crop yield and reduce the need for fertilizers and pesticides.

Both the intercellular spaces occupied by endophytic bacteria and the rhizosphere associated with each plant root system contain a unique and complex cocktail of root secretions and microbial products that differ from those of the surrounding soil. A recent metagenomics study has revealed that the compositions of bacterial communities living endophytically and in the rhizosphere are not identical **(Figure 29.11)**. A better understanding of the bacteria within and around roots could have profound agricultural benefits.

---

▼ **Figure 29.11**  Inquiry

## How variable are the compositions of bacterial communities inside and outside of roots?

**Experiment** The bacterial communities found within and immediately outside of root systems are known to improve plant growth. In order to devise agricultural strategies to increase the benefits of these bacterial communities, it is necessary to determine how complex they are and what factors affect their composition. A problem inherent in studying these bacterial communities is that a handful of soil contains as many as 10,000 types of bacteria, more than all the bacterial species that have been described. One cannot simply culture each species and use a taxonomic key to identify them; a molecular approach is needed.

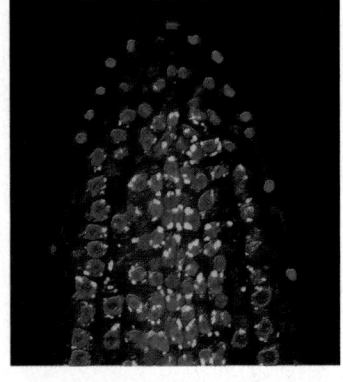

Bacteria (green) on surface of root (fluorescent LM)

Researchers estimated the number of bacterial "species" in various samples using a technique called *metagenomics* (see Concept 18.1). The bacterial community samples they studied differed in location (endophytic, rhizospheric, or outside the rhizosphere), soil type (clayey or porous), and developmental stage of the root system with which they were associated (old or young). The DNA from each sample was purified, and the polymerase chain reaction (PCR) was used to amplify the DNA that codes for the 16S ribosomal RNA subunits. Many thousands of DNA sequence variations were found in each sample. The researchers then lumped the sequences that were more than 97% identical into "taxonomic units" or "species." (The word *species* is in quotation marks because "two organisms having a single gene that is more than 97% identical" is not explicit in any definition of species.) Having established the types of "species" in each community, the researchers constructed a tree diagram showing the percent of bacterial "species" that were found in common in each community.

**Results** This tree diagram breaks down the relatedness of bacterial communities into finer and finer levels of detail. The two explanatory labels give examples of how to interpret the diagram.

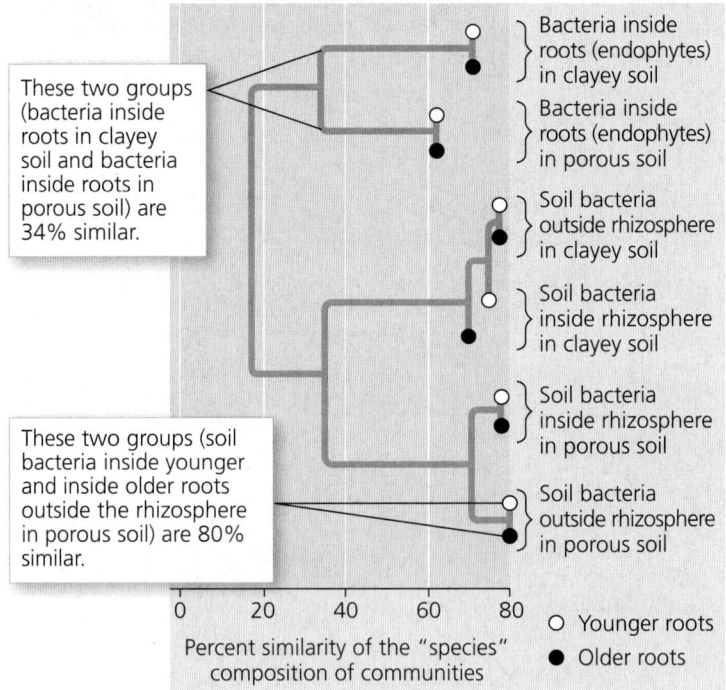

These two groups (bacteria inside roots in clayey soil and bacteria inside roots in porous soil) are 34% similar.

These two groups (soil bacteria inside younger and inside older roots outside the rhizosphere in porous soil) are 80% similar.

- Bacteria inside roots (endophytes) in clayey soil
- Bacteria inside roots (endophytes) in porous soil
- Soil bacteria outside rhizosphere in clayey soil
- Soil bacteria inside rhizosphere in clayey soil
- Soil bacteria inside rhizosphere in porous soil
- Soil bacteria outside rhizosphere in porous soil

Percent similarity of the "species" composition of communities

○ Younger roots
● Older roots

**Conclusion** The "species" composition of the bacterial communities varied markedly according to the location inside the root versus outside the root and according to soil type.

**INTERPRET THE DATA** **(a)** Which of the three community locations was least like the other two? **(b)** Rank the three variables (community location, developmental stage of roots, and soil type) in terms of how strongly they affect the "species" composition of the bacterial communities.

**Data from** D. S. Lundberg et al., Defining the core *Arabidopsis thaliana* root microbiome, *Nature* 488:86–94 (2012).

## Bacteria in the Nitrogen Cycle

Plants have mutualistic relationships with several groups of bacteria that help make nitrogen more available. From a global perspective, no mineral nutrient is more limiting to plant growth than nitrogen, which is required in large amounts for synthesizing proteins and nucleic acids. Here we focus on processes leading directly to nitrogen assimilation by plants.

Ammonium ions ($NH_4^+$) and nitrate ions ($NO_3^-$) are the forms of nitrogen that plants can use. Some soil nitrogen derives from the weathering of rocks. Lightning also produces small amounts of $NO_3^-$ that get carried to the soil in rain. Most of the nitrogen available to plants, however, comes from the activity of bacteria **(Figure 29.12)**.

The **nitrogen cycle** describes transformations of nitrogen and nitrogenous compounds in nature (see Figure 42.13). When an organism dies, or an animal expels waste, the initial form of nitrogen is organic. Decomposers convert the organic nitrogen within the remains back into ammonium ($NH_4^+$), a process called ammonification. Other sources of soil $NH_4^+$ are *nitrogen-fixing bacteria* that convert gaseous nitrogen ($N_2$) to $NH_3$, which then picks up another $H^+$ in the soil solution, forming $NH_4^+$.

In addition to $NH_4^+$, plants can also acquire nitrogen in the form of nitrate ($NO_3^-$). Soil $NO_3^-$ is largely formed by a two-step process called *nitrification*, which consists of the oxidation of ammonia ($NH_3$) to nitrite ($NO_2^-$), followed by the oxidation of $NO_2^-$ to $NO_3^-$. Different types of *nitrifying bacteria* mediate each step. After the roots absorb $NO_3^-$, a plant enzyme reduces it back to $NH_4^+$, which other enzymes incorporate into amino acids and other organic compounds. Most plant species export nitrogen from roots to shoots via the xylem as $NO_3^-$ or organic compounds synthesized in the roots. Some soil nitrogen is lost, particularly in anaerobic soils, when denitrifying bacteria convert $NO_3^-$ to $N_2$, which diffuses into the atmosphere.

### Nitrogen-Fixing Bacteria: A Closer Look

Although Earth's atmosphere is 79% nitrogen ($N_2$), plants cannot use gaseous nitrogen ($N_2$) directly because there is a triple bond between the two nitrogen atoms, making the molecule almost inert. For $N_2$ to be of use to plants, it must be reduced to $NH_3$ by a process called **nitrogen fixation**. All nitrogen-fixing organisms are bacteria. Some nitrogen-fixing bacteria are free-living in the soil (see Figure 29.12), whereas others are endophytic. Still others, particularly members of the genus *Rhizobium*, form efficient and intimate associations with the roots of legumes (such as peas, soybeans, alfalfa, and peanuts), altering the structure of the hosts' roots markedly, as will be discussed shortly.

The multistep conversion of $N_2$ to $NH_3$ by nitrogen fixation can be summarized as follows:

$$N_2 + 8\,e^- + 8\,H^+ + 16\,ATP \rightarrow 2\,NH_3 + H_2 + 16\,ADP + 16\,\text{\textcircled{P}}_i$$

The reaction is driven by the enzyme complex *nitrogenase*. Because the process of $N_2$ fixation requires 16 ATP molecules for every 2 $NH_3$ synthesized, nitrogen-fixing bacteria require a rich supply of carbohydrates from decaying material, root secretions, or (in the case of the *Rhizobium* bacteria) the vascular tissue of roots.

The mutualism between *Rhizobium* ("root living") bacteria and legume roots involves dramatic changes in root structure. Along a legume's roots are swellings called **nodules**, composed

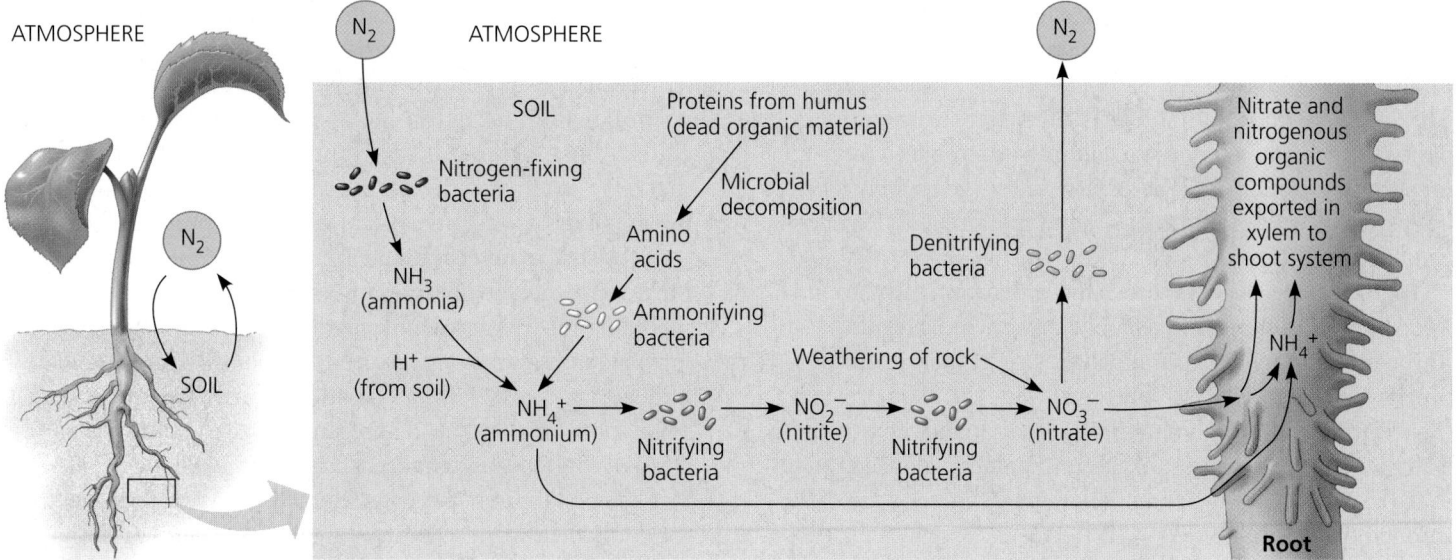

▲ **Figure 29.12 The roles of soil bacteria in the nitrogen nutrition of plants.** Ammonium is made available to plants by two types of soil bacteria: those that fix atmospheric $N_2$ (nitrogen-fixing bacteria) and those that decompose organic material (ammonifying bacteria). Although plants absorb some ammonium from the soil, they absorb mainly nitrate, which is produced from ammonium by nitrifying bacteria as well as by the weathering of rocks. Inside the plant, nitrate is reduced to ammonium before assimilation into organic compounds.

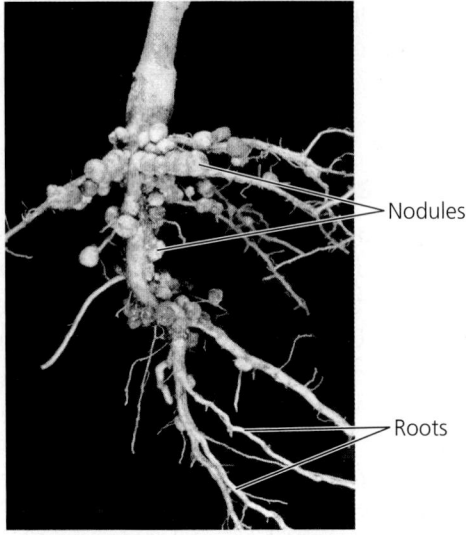

▲ **Figure 29.13 Soybean root nodules.** The spherical structures along this soybean root system are nodules containing *Rhizobium* bacteria. The bacteria fix nitrogen and obtain photosynthetic products made by the plant.

**?** *How is the relationship between legume plants and* Rhizobium *bacteria mutualistic?*

of plant cells that have been "infected" by *Rhizobium* **(Figure 29.13)**. Inside each nodule, *Rhizobium* bacteria assume a form called **bacteroids**, which are contained within vesicles formed in the root cells. Legume-*Rhizobium* relationships generate more usable nitrogen for plants than all industrial fertilizers used today—and at virtually no cost to the farmer.

The location of the bacteroids inside living, nonphotosynthetic cells facilitates nitrogen fixation, which requires an anaerobic environment. Lignified external layers of root nodules also limit gas exchange. Some root nodules appear reddish because of a molecule called leghemoglobin (*leg-* for "legume"), an iron-containing protein that binds reversibly to oxygen (similar to the hemoglobin in human red blood cells). This protein is an oxygen "buffer," reducing the concentration of free oxygen and thereby providing an anaerobic environment for $N_2$ fixation while regulating the oxygen supply for the intense cellular respiration required to produce ATP for nitrogen fixation.

Each legume species is associated with a particular strain of *Rhizobium*. The symbiotic relationship between a legume and nitrogen-fixing bacteria is mutualistic in that the bacteria supply the host plant with fixed nitrogen while the plant provides the bacteria with carbohydrates and other organic compounds. The root nodules use most of the ammonium produced to make amino acids, which are then transported up to the shoot through the xylem.

## Fungi and Plant Nutrition

Certain species of soil fungi also form mutualistic relationships with roots and play a major role in plant nutrition.

Some of these fungi are endophytic, but the most important relationships are **mycorrhizae** ("fungus roots"), the intimate mutualistic associations of roots and fungi **(Figure 29.14)**. The host plant provides the fungus with a steady supply of sugar. Meanwhile, the fungus increases the surface area for water uptake and also supplies the plant with phosphorus and other minerals absorbed from the soil. The fungi of mycorrhizae also secrete growth factors that stimulate roots to grow and branch, as well as antibiotics that help protect the plant from soil pathogens.

### Mycorrhizae and Plant Evolution

**EVOLUTION** Mycorrhizae are not oddities: most plants form them. Fossil evidence suggests that mycorrhizae were an early evolutionary adaptation that helped plants colonize the land (see Concept 26.2). When the earliest plants, which evolved from green algae, invaded the land 400 to 500 million years ago, they encountered harsh conditions. Since the soil lacked organic matter, rain probably leached away much of its soluble minerals. The barren land, however, was a place of opportunity because there was little competition, and light and carbon dioxide were plentiful. Alone, neither the early land plants nor early land fungi were fully equipped to exploit the terrestrial environment. The early plants lacked the ability to extract essential nutrients from the soil, while the fungi were unable to manufacture carbohydrates. By forming mycorrhizal associations, both groups of organisms were able to succeed.

### The Two Main Types of Mycorrhizae

One type of mycorrhiza—the **ectomycorrhizae**—forms a dense sheath, or mantle, of mycelia (mass of branching hyphae) over the surface of the root **(Figure 29.14a)**. Fungal hyphae extend from the mantle into the soil, greatly increasing the surface area for water and mineral absorption. Hyphae also grow into the root cortex, forming an apoplastic network within the extracellular spaces that facilitates nutrient exchange. Of the 10% of plant families that have species that form ectomycorrhizae, the majority are woody, including members of the pine, birch, and eucalyptus families.

**Arbuscular mycorrhizae** are more common than ectomycorrhizae and are found in over 85% of plant species, including most crops. Unlike ectomycorrhizae, they do not form a dense mantle ensheathing the root **(Figure 29.14b)**. Arbuscular mycorrhizal associations start when microscopic soil hyphae respond to the presence of a root by growing toward it, establishing contact, and growing along its surface. Hyphae penetrate between epidermal cells and then enter the root cortex. These hyphae digest small patches of the cortical cell walls, but they do not actually pierce the plasma membrane and enter the cytoplasm. Instead, a hypha grows into a tube formed by invagination of the root cell's membrane. The process is analogous to poking a finger gently

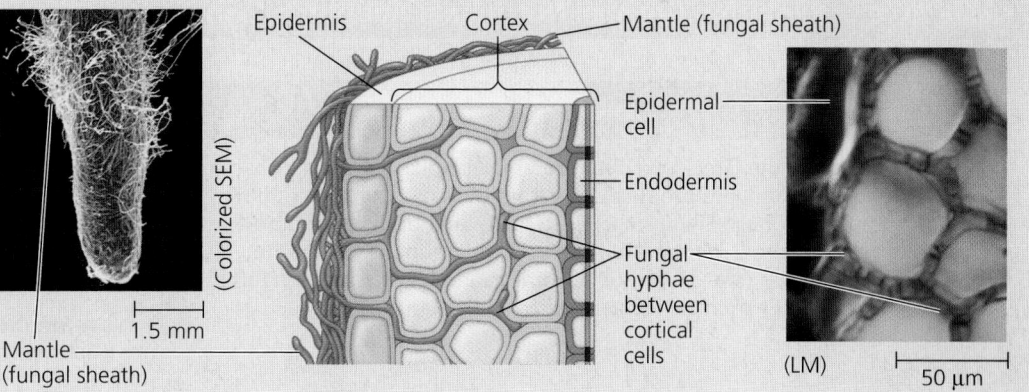

**(a) Ectomycorrhizae.** The mantle of the fungal mycelium ensheathes the root. Fungal hyphae extend from the mantle into the soil, absorbing water and minerals, especially phosphorus. Hyphae also extend into the extracellular spaces of the root cortex, providing extensive surface area for nutrient exchange between the fungus and its host plant.

(Colorized SEM)

Epidermis    Cortex    Mantle (fungal sheath)

Epidermal cell

Endodermis

Fungal hyphae between cortical cells

Mantle (fungal sheath)

1.5 mm

(LM)    50 µm

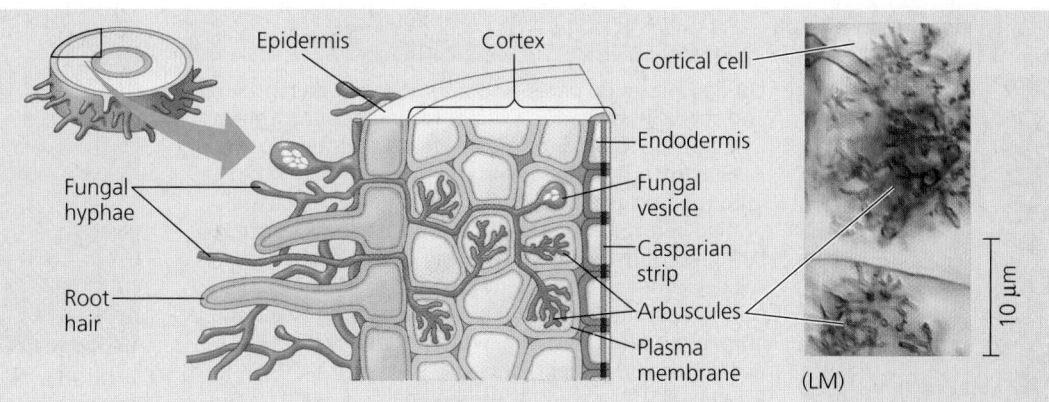

**(b) Arbuscular mycorrhizae (endomycorrhizae).** No mantle forms around the root, but microscopic fungal hyphae extend into the root. Within the root cortex, the fungus makes extensive contact with the plant through branching of hyphae that form arbuscules, providing an enormous surface area for nutrient swapping. The hyphae penetrate the cell walls, but not the plasma membranes, of cells within the cortex.

Epidermis    Cortex

Cortical cell

Fungal hyphae

Root hair

Endodermis

Fungal vesicle

Casparian strip

Arbuscules

Plasma membrane

(LM)    10 µm

▲ **Figure 29.14  Mycorrhizae.**

into a balloon without popping it; your finger is like the fungal hypha, and the balloon skin is like the root cell's membrane. After the fungal hyphae have penetrated in this way, some branch densely, forming structures called arbuscules ("little trees"), which are important sites of nutrient transfer between the fungus and the plant. Within the hyphae themselves, oval vesicles may form, possibly serving as food storage sites for the fungus. To the unaided eye, arbuscular mycorrhizae look like "normal" roots with root hairs, but a microscope reveals the enormous extent of the mutualistic relationship.

### Agricultural and Ecological Importance of Mycorrhizae

Roots can form mycorrhizal symbioses only if exposed to the appropriate species of fungus. In most ecosystems, these fungi are present in the soil, but if seeds are collected in one environment and planted in foreign soil, the plants may show signs of malnutrition (particularly phosphorus deficiency), resulting from the absence of fungal partners. Treating seeds with mycorrhizal fungal spores can help seedlings form mycorrhizae and improve crop yield.

Mycorrhizal associations are also important in understanding ecological relationships. Invasive exotic plants sometimes colonize areas by disrupting interactions between native organisms. For example, garlic mustard (*Alliaria petiolata*),

a native of Europe, has invaded woodlands throughout the eastern and middle United States, suppressing tree seedlings and other native plants. Recent evidence suggests that its invasive properties may stem from its ability to slow the growth of other plant species by preventing the growth of arbuscular mycorrhizal fungi. Mustards themselves are unusual in not forming mycorrhizal associations.

### Epiphytes, Parasitic Plants, and Carnivorous Plants

Almost all plant species have mutualistic symbiotic relationships with soil fungi or bacteria or both. Though rarer, there are also plant species with nutritional adaptations that use other organisms in nonmutualistic ways. **Figure 29.15** provides an overview of three unusual adaptations: epiphytes, parasitic plants, and carnivorous plants.

**CONCEPT CHECK 29.4**

1. Why is the study of the rhizosphere critical to understanding plant nutrition?
2. How do soil bacteria and mycorrhizae contribute to plant nutrition?
3. **WHAT IF?**  A soybean farmer finds that the older leaves of his plant are turning yellow following a long period of wet weather. Suggest a reason why.

For suggested answers, see Appendix A.

## Epiphytes

An **epiphyte** (from the Greek *epi*, upon, and *phyton*, plant) is a plant that grows on another plant. Epiphytes produce and gather their own nutrients; they do not tap into their hosts for sustenance. Usually anchored to the branches or trunks of living trees, epiphytes absorb water and minerals from rain, mostly through leaves rather than roots. Some examples are staghorn ferns, bromeliads, and many orchids, including the vanilla plant.

▶ **Staghorn fern,** an epiphyte

## Parasitic Plants

Unlike epiphytes, parasitic plants absorb water, minerals, and sometimes products of photosynthesis from their living hosts. Many species have roots that function as haustoria, nutrient-absorbing projections that tap into the host plant. Some parasitic species, such as orange-colored, spaghetti-like dodder (genus *Cuscuta*), lack chlorophyll entirely, whereas others, such as mistletoe (genus Phoradendron), are photosynthetic. Still others, such as Indian pipe (*Monotropa uniflora*), absorb nutrients from the hyphae of mycorrhizae associated with other plants.

◀ **Mistletoe**, a photosynthetic parasite

▲ **Dodder**, a nonphotosynthetic parasite (orange)

▲ **Indian pipe**, a nonphotosynthetic parasite of mycorrhizae

## Carnivorous Plants

Carnivorous plants are photosynthetic but supplement their mineral diet by capturing insects and other small animals. They live in acid bogs and other habitats where soils are poor in nitrogen and other minerals. Pitcher plants such as *Nepenthes* and *Sarracenia* have water-filled funnels into which prey slip and drown, eventually to be digested by enzymes. Sundews (genus *Drosera*) exude a sticky fluid from tentacle-like glands on highly modified leaves. Stalked glands secrete sweet mucilage that attracts and ensnares insects, and they also release digestive enzymes. Other glands then absorb the nutrient "soup." The highly modified leaves of Venus flytrap (*Dionaea muscipula*) close quickly but partially when a prey hits two trigger hairs in rapid enough succession. Smaller insects can escape, but larger ones are trapped by the teeth lining the margins of the lobes. Excitation by the prey causes the trap to narrow more and digestive enzymes to be released.

▲ **Sundew**

◀ **Pitcher plants**

▼ **Venus flytraps**

# Transpiration drives the transport of water and minerals from roots to shoots via the xylem

An average-sized tree, despite having neither heart nor muscle, transports nearly 800 L of water on a warm, sunny day. How do trees accomplish this feat? To answer this question, we'll follow each step in the journey of water and minerals from the tips of roots to leaves.

## Absorption of Water and Minerals by Root Cells

Although all living plant cells absorb nutrients across their plasma membranes, the cells near the tips of roots are particularly important because most of the absorption of water and minerals occurs there. In this region, the epidermal cells are permeable to water, and many are differentiated into root hairs, modified cells that account for much of the absorption of water by roots (see Figure 28.4). The root hairs absorb the

soil solution, which consists of water molecules and dissolved mineral ions that are not bound tightly to soil particles. The soil solution is drawn into the hydrophilic walls of epidermal cells and passes freely along the cell walls and the extracellular spaces into the root cortex. This flow enhances the exposure of the cells of the cortex to the soil solution, providing a much greater membrane surface area for absorption than the surface area of the epidermis alone. Although the soil solution usually has a low mineral concentration, active transport enables roots to accumulate essential minerals to concentrations hundreds of times greater than in the soil.

## Transport of Water and Minerals into the Xylem

Water and minerals that pass from the soil into the root cortex cannot be transported to the rest of the plant until they enter the xylem of the vascular cylinder, or stele. The **endodermis**, the innermost layer of the cortex, functions as a last checkpoint for the selective passage of minerals into the vascular cylinder **(Figure 29.16)**. Minerals already in the symplast when they reach the endodermis continue through

▼ **Figure 29.16 Transport of water and minerals from root hairs to the xylem.**

**?** *How does the Casparian strip force water and minerals to pass through the plasma membranes of endodermal cells?*

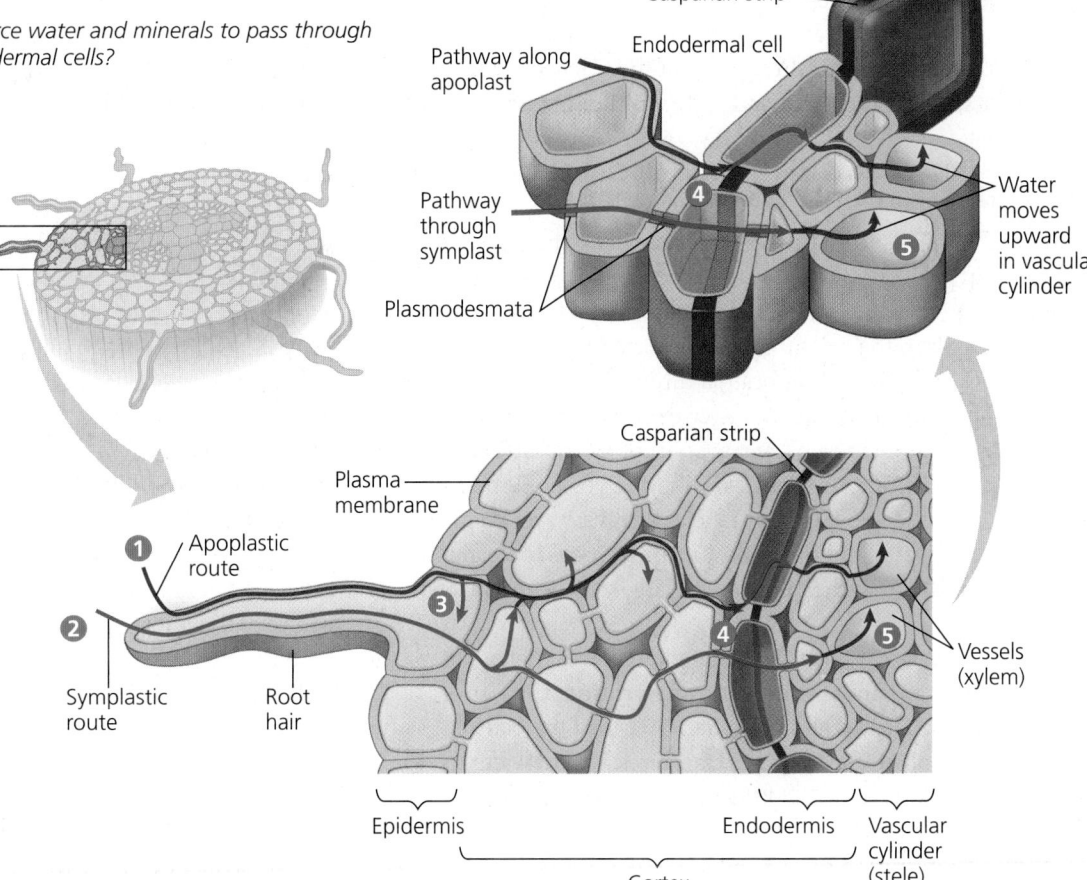

❶ **Apoplastic route.** Uptake of soil solution by the hydrophilic walls of root hairs provides access to the apoplast. Water and minerals can then diffuse into the cortex along this matrix of walls and extracellular spaces.

❷ **Symplastic route.** Minerals and water that cross the plasma membranes of root hairs can enter the symplast.

❸ **Transmembrane route.** As soil solution moves along the apoplast, some water and minerals are transported into the protoplasts of cells of the epidermis and cortex and then move inward via the symplast.

❹ **The endodermis: controlled entry to the vascular cylinder (stele).** Within the transverse and radial walls of each endodermal cell is the Casparian strip, a belt of waxy material (purple band) that blocks the passage of water and dissolved minerals. Only minerals already in the symplast or entering that pathway by crossing the plasma membrane of an endodermal cell can detour around the Casparian strip and pass into the vascular cylinder (stele).

❺ **Transport in the xylem.** Endodermal cells and also living cells within the vascular cylinder discharge water and minerals into their walls (apoplast). The xylem vessels then transport the water and minerals by bulk flow upward into the shoot system.

the plasmodesmata of endodermal cells and pass into the vascular cylinder. These minerals were already screened by the plasma membrane they had to cross to enter the symplast in the epidermis or cortex. Those minerals that reach the endodermis via the apoplast encounter a dead end that blocks their passage into the vascular cylinder. This barrier, located in the transverse and radial walls of each endodermal cell, is the **Casparian strip**, a belt made of suberin, a waxy material impervious to water and dissolved minerals (see Figure 29.16). The Casparian strip forces water and minerals that are passively moving through the apoplast to cross the plasma membrane of an endodermal cell before they can enter the vascular cylinder. The endodermis, with its Casparian strip, ensures that no minerals can reach the vascular tissue of the root without crossing a selectively permeable plasma membrane. The endodermis also prevents solutes that have accumulated in the xylem from leaking back into the soil solution.

The last segment in the soil-to-xylem pathway is the passage of water and minerals into the tracheids and vessel elements of the xylem. These water-conducting cells lack protoplasts when mature and are therefore parts of the apoplast. Endodermal cells, as well as living cells within the vascular cylinder, discharge minerals from their protoplasts into their own cell walls. Both diffusion and active transport are involved in this transfer of solutes from symplast to apoplast, and the water and minerals are now free to enter the tracheids and vessel elements, where they are transported to the shoot system by bulk flow.

## Bulk Flow Transport via the Xylem

Water and minerals from the soil enter the plant through the epidermis of roots, cross the root cortex, and pass into the vascular cylinder. From there, the **xylem sap**, the water and dissolved minerals in the xylem, gets transported long distances by bulk flow to the veins that branch throughout each leaf. As noted earlier, bulk flow is much faster than diffusion or active transport. Peak velocities in the transport of xylem sap can range from 15 to 45 m/hr for trees with wide vessel elements. Stems and leaves depend on this efficient delivery system for their supply of water and minerals. Xylem sap rises to heights of more than 120 m in the tallest trees, largely by being pulled upward.

The process of transporting xylem sap involves the loss of an astonishing amount of water by **transpiration**, the loss of water vapor from leaves and other aerial parts of the plant. A single maize plant, for example, transpires 60 L of water (the equivalent of 170 12-ounce bottles) during a growing season. A maize crop growing at a typical density of 60,000 plants per hectare transpires almost 4 million L of water per hectare every growing season (about 400,000 gallons of water per acre per growing season). Unless the transpired water is replaced by water transported up from the roots, the leaves will wilt, and the plants may eventually die.

## Pulling Xylem Sap: The Cohesion-Tension Hypothesis

The xylem sap that rises through a tree does not require living xylem cells to do so. Early researchers demonstrated that leafy stems with their lower end immersed in toxic solutions of copper sulfate or acid readily draw these poisons up if the stem is cut below the surface of the liquid. As the toxic solutions ascend, they kill all living cells in their path, eventually arriving in the transpiring leaves and killing the leaf cells as well. Nevertheless, the uptake of the toxic solutions and the loss of water from the dead leaves can continue for weeks.

The **cohesion-tension hypothesis** is almost universally accepted by plant biologists as the mechanism underlying the ascent of xylem sap. According to this hypothesis, transpiration provides the pull for the ascent of xylem sap, and the cohesion of water molecules transmits this pull along the entire length of the xylem from shoots to roots. Hence, xylem sap is normally under negative pressure, or tension. Since transpiration is a "pulling" process, our exploration of the rise of xylem sap by the cohesion-tension mechanism begins with the leaves, where the driving force for transpirational pull begins.

**Transpirational Pull** Stomata on a leaf's surface lead to a maze of internal air spaces that expose the mesophyll cells to the $CO_2$ required for photosynthesis. The air in these spaces is saturated with water vapor. On most days, the air outside the leaf is drier; that is, it has lower water potential than the air inside the leaf. Therefore, water vapor in the air spaces of a leaf diffuses down its water potential gradient and exits the leaf via the stomata. Transpiration refers to this loss of water vapor from plants by diffusion and evaporation.

But how does loss of water vapor from the leaf translate into a pulling force for upward movement of water through a plant? The negative pressure potential that causes water to move up through the xylem develops at the surface of mesophyll cell walls in the leaf **(Figure 29.17)**. The cell wall acts like a very thin capillary network. Water adheres to the cellulose microfibrils and other hydrophilic components of the cell wall. As water evaporates from the water film that covers the cell walls of mesophyll cells, the air-water interface retreats farther into the cell wall. Because of the high surface tension of water, the curvature of the interface induces a tension, or negative pressure potential, in the water. As more water evaporates from the cell wall, the curvature of the air-water interface increases and the pressure of the water becomes more negative. Water molecules from the more hydrated parts of the leaf are then pulled toward this area, reducing the tension. These pulling forces are transferred to the xylem because each water molecule is cohesively bound to the next by hydrogen bonds. Thus, transpirational pull depends on several of the properties of water discussed in Chapter 2: adhesion, cohesion, and surface tension.

The role of negative pressure potential in transpiration is consistent with the water potential equation because negative

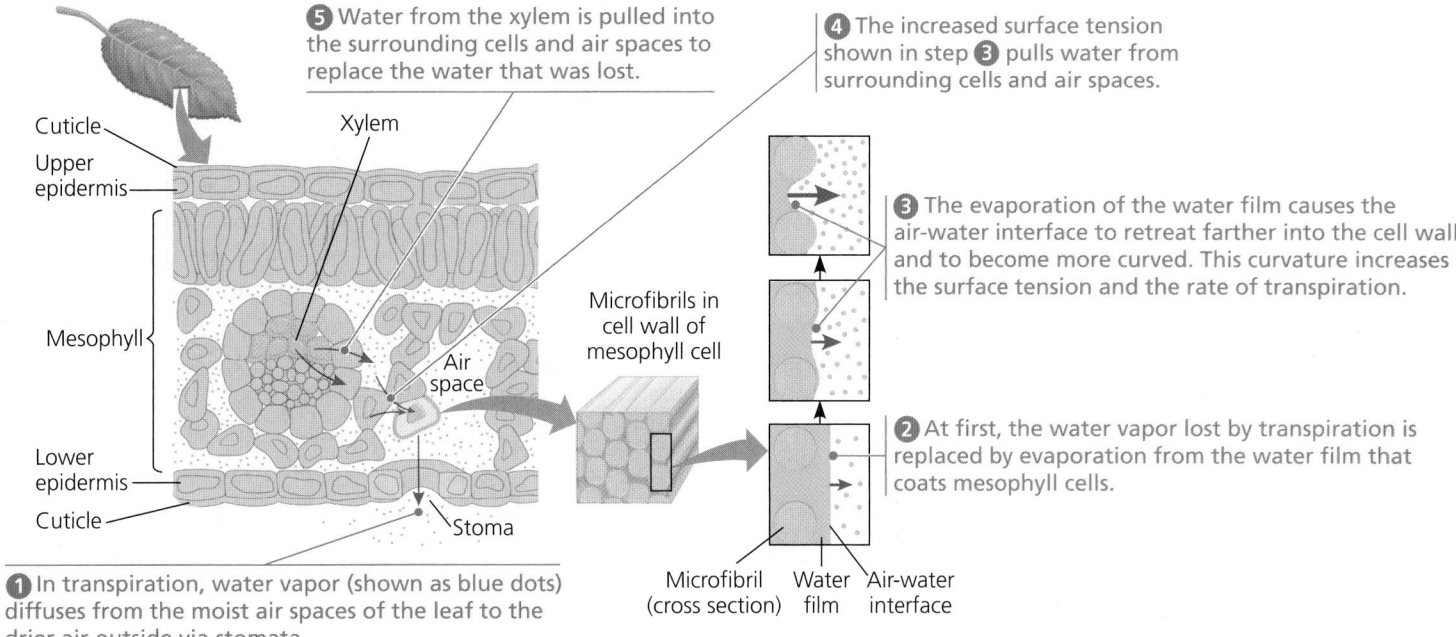

**5** Water from the xylem is pulled into the surrounding cells and air spaces to replace the water that was lost.

**4** The increased surface tension shown in step **3** pulls water from surrounding cells and air spaces.

**3** The evaporation of the water film causes the air-water interface to retreat farther into the cell wall and to become more curved. This curvature increases the surface tension and the rate of transpiration.

Microfibrils in cell wall of mesophyll cell

**2** At first, the water vapor lost by transpiration is replaced by evaporation from the water film that coats mesophyll cells.

Cuticle
Upper epidermis
Xylem
Mesophyll
Air space
Lower epidermis
Cuticle
Stoma

Microfibril (cross section)    Water film    Air-water interface

**1** In transpiration, water vapor (shown as blue dots) diffuses from the moist air spaces of the leaf to the drier air outside via stomata.

▲ **Figure 29.17 Generation of transpirational pull.** Negative pressure (tension) at the air-water interface in the leaf is the basis of transpirational pull, which draws water out of the xylem.

pressure potential (tension) *lowers* water potential (see Figure 29.5). Because water moves from areas of higher water potential to areas of lower water potential, the more negative pressure potential at the air-water interface causes water in xylem cells to be "pulled" into mesophyll cells, which lose water to the air spaces, with the water eventually diffusing out through stomata. In this way, the negative water potential of leaves provides the "pull" in transpirational pull. The transpirational pull on xylem sap is transmitted all the way from the leaves to the root tips and even into the soil solution **(Figure 29.18)**.

▶ **Figure 29.18 Ascent of xylem sap.**
Hydrogen bonding forms an unbroken chain of water molecules extending from leaves to the soil. The force driving the ascent of xylem sap is a gradient of water potential ($\psi$). For bulk flow over long distance, the $\psi$ gradient is due mainly to a gradient of the pressure potential ($\psi_P$). Transpiration results in the $\psi_P$ at the leaf end of the xylem being lower than the $\psi_P$ at the root end. The $\psi$ values shown on the left side of the figure are a "snapshot." They may vary during daylight, but the direction of the $\psi$ gradient remains the same.

 **ANIMATION** Visit the Study Area in **MasteringBiology** for the BioFlix® 3-D Animation on Water Transport in Plants.

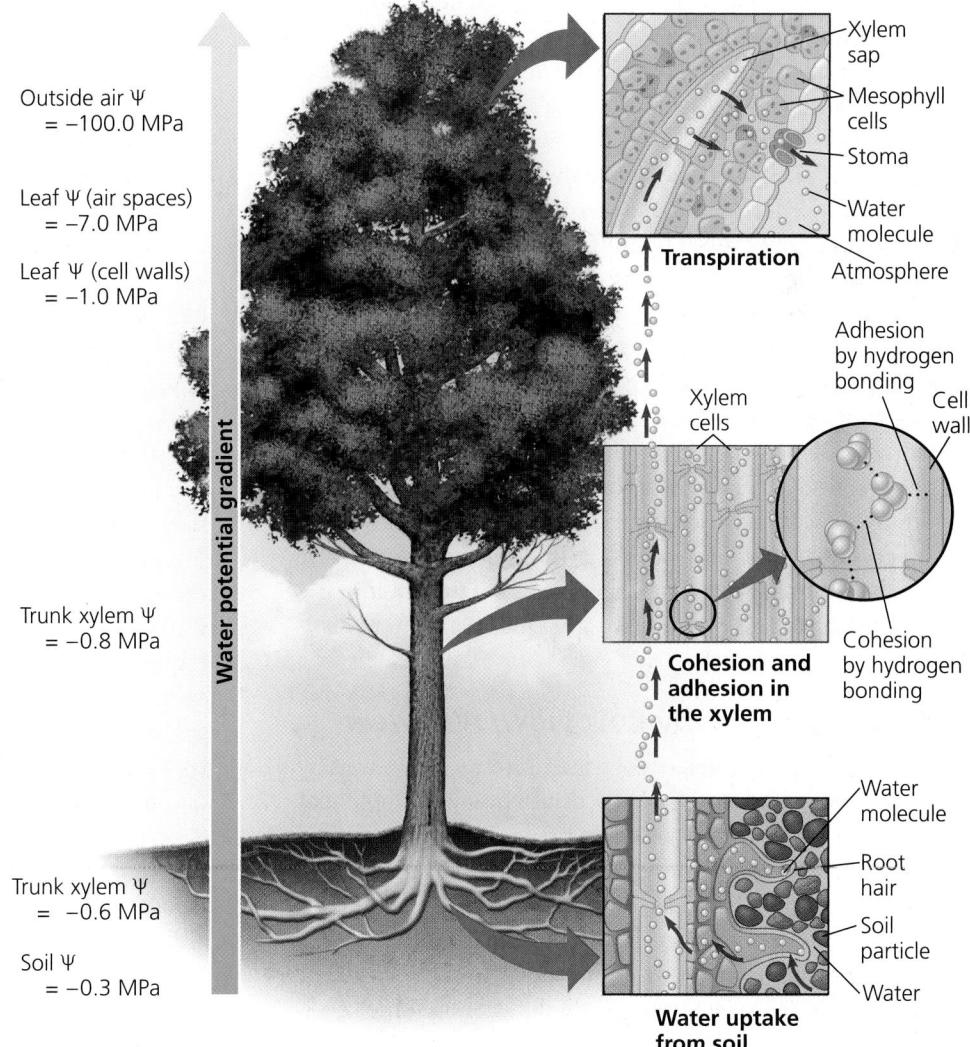

Outside air $\psi$ = −100.0 MPa

Leaf $\psi$ (air spaces) = −7.0 MPa

Leaf $\psi$ (cell walls) = −1.0 MPa

Trunk xylem $\psi$ = −0.8 MPa

Trunk xylem $\psi$ = −0.6 MPa

Soil $\psi$ = −0.3 MPa

Water potential gradient

Xylem sap
Mesophyll cells
Stoma
Water molecule
Atmosphere
**Transpiration**

Adhesion by hydrogen bonding
Cell wall
Xylem cells
Cohesion by hydrogen bonding

**Cohesion and adhesion in the xylem**

Water molecule
Root hair
Soil particle
Water

**Water uptake from soil**

## Cohesion and Adhesion in the Ascent of Xylem Sap

Cohesion and adhesion facilitate the transport of water by bulk flow. Cohesion is the attractive force between molecules of the same substance. Water has an unusually high cohesive force due to the hydrogen bonds each water molecule can potentially make with other water molecules. Water's cohesive force within the xylem gives it a tensile strength equivalent to that of a steel wire of similar diameter. The cohesion of water makes it possible to pull a column of xylem sap from above without the water molecules separating. Water molecules exiting the xylem in the leaf tug on adjacent water molecules, and this pull is relayed, molecule by molecule, down the entire column of water in the xylem. Meanwhile, the strong adhesion of water molecules (again by hydrogen bonds) to the hydrophilic walls of xylem cells helps offset the downward force of gravity.

The upward pull on the sap creates tension within the vessel elements and tracheids, which are like elastic pipes. Positive pressure causes an elastic pipe to swell, whereas tension pulls the walls of the pipe inward. On a warm day, a decrease in the diameter of a tree trunk can even be measured. As transpirational pull puts the vessel elements and tracheids under tension, their thick secondary walls prevent them from collapsing, much as wire rings maintain the shape of a vacuum cleaner hose. The tension produced by transpirational pull lowers water potential in the root xylem to such an extent that water flows passively from the soil, across the root cortex, and into the vascular cylinder.

Transpirational pull can extend down to the roots only through an unbroken chain of water molecules. Cavitation, the formation of a water vapor pocket, breaks the chain. It is more common in wide vessel elements than in tracheids and can occur during drought stress or when xylem sap freezes in winter. The air bubbles resulting from cavitation expand and block the water channels of the xylem. The interruption of xylem sap transport by cavitation is not always permanent. The chain of water molecules can detour around the air bubbles through pits between adjacent tracheids or vessel elements (see Figure 28.9). Moreover, secondary growth adds a layer of new xylem each year, and only the youngest, outermost secondary xylem layers transport water. Finally, an active though minor force called *root pressure* enables some small plants to refill blocked vessel elements.

## Xylem Sap Ascent by Bulk Flow: *A Review*

In the long-distance transport of water from roots to leaves by bulk flow, the movement of fluid is driven by a water potential difference at opposite ends of xylem tissue. The water potential difference is created at the leaf end of the xylem by the evaporation of water from leaf cells. Evaporation lowers the water potential at the air-water interface, thereby generating the negative pressure (tension) that pulls water through the xylem.

Bulk flow in the xylem differs from diffusion in some key ways. First, it is driven by differences in pressure potential ($\psi_P$);

solute potential ($\psi_S$) is not a factor. Therefore, the water potential gradient within the xylem is essentially a pressure gradient. Also, the flow does not occur across plasma membranes of living cells, but instead within hollow, dead cells. Furthermore, it moves the entire solution together—not just water or solutes—and at much greater speed than diffusion.

The plant expends no energy to lift xylem sap by bulk flow. Instead, the absorption of sunlight drives most of transpiration by causing water to evaporate from the moist walls of mesophyll cells and by lowering the water potential in the air spaces within a leaf. Thus, the ascent of xylem sap, like the process of photosynthesis, is ultimately solar powered.

### CONCEPT CHECK 29.5

1. A scientist adds a water-soluble inhibitor of photosynthesis to roots of a transpiring plant, but photosynthesis is not reduced. Why?
2. **WHAT IF?** Suppose an *Arabidopsis* mutant lacking functional aquaporin proteins has a root mass three times greater than that of wild-type plants. Suggest an explanation.
3. **MAKE CONNECTIONS** How are the Casparian strip and tight junctions similar? See Figure 4.27.

For suggested answers, see Appendix A.

---

**CONCEPT 29.6**

# The rate of transpiration is regulated by stomata

Leaves generally have large surface areas and high surface-to-volume ratios. The large surface area enhances light absorption for photosynthesis. The high surface-to-volume ratio aids in $CO_2$ absorption during photosynthesis as well as in the release of $O_2$, a by-product of photosynthesis. Upon diffusing through the stomata, $CO_2$ enters a honeycomb of air spaces formed by the spongy mesophyll cells (see Figure 28.17). Because of the irregular shapes of these cells, the leaf's internal surface area may be 10 to 30 times greater than the external surface area.

Although large surface areas and high surface-to-volume ratios increase the rate of photosynthesis, they also increase water loss by way of the stomata. Thus, a plant's tremendous requirement for water is largely a consequence of the shoot system's need for ample exchange of $CO_2$ and $O_2$ for photosynthesis. By opening and closing the stomata, guard cells help balance the plant's requirement to conserve water with its requirement for photosynthesis.

## Stomata: Major Pathways for Water Loss

About 95% of the water a plant loses escapes through stomata, although these pores account for only 1–2% of the external leaf surface. The waxy cuticle limits water loss through the remaining surface of the leaf. Each stoma is flanked by a pair of guard cells. Guard cells control the diameter of the stoma by changing shape, thereby widening or narrowing the gap between the two guard cells. Under the same environmental conditions, the amount of water lost by a leaf depends largely on the number of stomata and the average size of their pores.

The stomatal density of a leaf is under both genetic and environmental control. For example, desert plants are genetically programmed to have lower stomatal densities than do marsh plants. Stomatal density, however, is a developmentally plastic feature of many plants. High light exposures and low $CO_2$ levels during leaf development lead to increased density in many species. By measuring the stomatal density of dried herbarium specimens, scientists have gained insight into the levels of atmospheric $CO_2$ in the past. A recent British survey found that stomatal density of many woodland species has decreased since 1927, when a similar survey was made. This observation is consistent with other findings that atmospheric $CO_2$ levels increased dramatically during the late 20th century.

## Mechanisms of Stomatal Opening and Closing

When guard cells absorb water from neighboring cells, they become more turgid. In most angiosperm species, the cell walls of guard cells are uneven in thickness, and the cellulose microfibrils are oriented in a direction that causes the guard cells to bow outward when turgid (**Figure 29.19a**). This bowing outward increases the size of the pore between the guard cells. When the cells lose water and become flaccid, they become less bowed, and the pore closes.

The changes in turgor pressure in guard cells result primarily from the reversible absorption and loss of potassium ions ($K^+$). Stomata open when guard cells actively accumulate $K^+$ from neighboring epidermal cells (**Figure 29.19b**). The flow of $K^+$ across the plasma membrane of the guard cell is coupled to the generation of a membrane potential by proton pumps. Stomatal opening correlates with active transport of $H^+$ out of the guard cell. The resulting voltage (membrane potential) drives $K^+$ into the cell through specific membrane channels. The absorption of $K^+$ causes the water potential to become more negative within the guard cells, and the cells become more turgid as water enters by osmosis. Stomatal closing results from a loss of $K^+$ from guard cells to neighboring cells, which leads to an osmotic loss of water. Aquaporins also help regulate the osmotic swelling and shrinking of guard cells.

## Stimuli for Stomatal Opening and Closing

In general, stomata are open during the day and mostly closed at night, preventing the plant from losing water under conditions when photosynthesis cannot occur. At least three cues contribute to stomatal opening at dawn: light, $CO_2$ depletion, and an internal "clock" in guard cells.

The light stimulates guard cells to accumulate $K^+$ and become turgid. This response is triggered by illumination of blue-light receptors in the plasma membrane of guard cells.

**Guard cells turgid/Stoma open**     **Guard cells flaccid/Stoma closed**

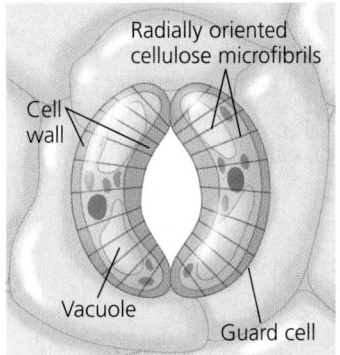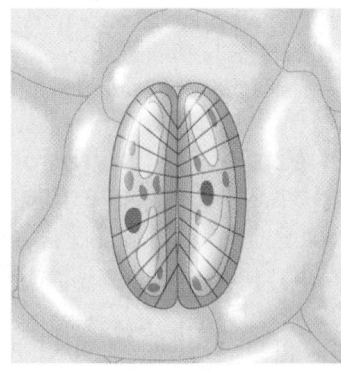

**(a) Changes in guard cell shape and stomatal opening and closing (surface view).** Guard cells of a typical angiosperm are illustrated in their turgid (stoma open) and flaccid (stoma closed) states. The radial orientation of cellulose microfibrils in the cell walls causes the guard cells to increase more in length than width when turgor increases. Since the two guard cells are tightly joined at their tips, they bow outward when turgid, causing the stomatal pore to open.

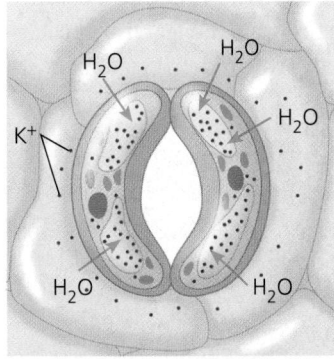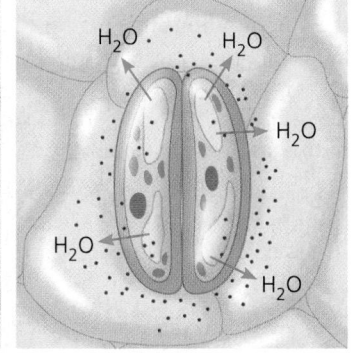

**(b) Role of potassium ions ($K^+$) in stomatal opening and closing.** The transport of $K^+$ (symbolized here as red dots) across the plasma membrane and vacuolar membrane causes the turgor changes of guard cells. The uptake of anions, such as malate and chloride ions (not shown), also contributes to guard cell swelling.

▲ **Figure 29.19 Mechanisms of stomatal opening and closing.**

Activation of these receptors stimulates the activity of proton pumps in the plasma membrane of the guard cells, in turn promoting absorption of $K^+$.

The stomata also open in response to depletion of $CO_2$ within the leaf's air spaces as a result of photosynthesis. As $CO_2$ concentrations decrease during the day, the stomata progressively open if sufficient water is supplied to the leaf.

The internal "clock" in the guard cells ensures that stomata continue their daily rhythm of opening and closing. This rhythm occurs even if a plant is kept in a dark location. All eukaryotic organisms have internal clocks that regulate cyclic processes. Cycles with intervals of approximately 24 hours are called **circadian rhythms** (see Concept 31.2).

Drought stress can also cause stomata to close. A hormone called **abscisic acid (ABA)** is produced in roots and leaves in response to water deficiency and signals guard cells to close stomata. This response reduces wilting but also restricts $CO_2$ absorption, thereby slowing photosynthesis. ABA also directly inhibits photosynthesis. Water availability is so tied to plant productivity not because water is needed as a substrate in photosynthesis but because freely available water allows plants to keep stomata open and take up more $CO_2$.

## Effects of Transpiration on Wilting and Leaf Temperature

As long as most stomata remain open, transpiration is greatest on days that are sunny, warm, dry, and windy because these environmental factors increase evaporation. If transpiration cannot pull sufficient water to the leaves, the shoot becomes slightly wilted as cells lose turgor pressure. Although plants respond to such mild drought stress by rapidly closing stomata, some evaporative water loss still occurs through the cuticle. Under prolonged drought conditions, the leaves can become irreversibly injured.

Transpiration also results in evaporative cooling, which can lower a leaf's temperature by as much as 10°C compared with the surrounding air. This cooling prevents the leaf from reaching temperatures that could lead to protein denaturation.

## Adaptations That Reduce Evaporative Water Loss

Many species of desert plants avoid drying out by completing their short life cycles during the brief rainy seasons. Rain comes infrequently in deserts, but when it arrives, the vegetation is transformed as dormant seeds of annual species quickly germinate and bloom, completing their life cycle before dry conditions return. Longer-lived species have unusual physiological or morphological adaptations that enable them to withstand the harsh desert conditions. Plants adapted to arid environments are called **xerophytes** (from the Greek *xero*, dry). **Figure 29.20** shows other examples. Many xerophytes, such as cacti, have highly reduced leaves that resist excessive water loss; they carry out photosynthesis mainly in their stems. The stems of many xerophytes are fleshy because they store water for use during long dry periods.

Another adaptation to arid habitats is **crassulacean acid metabolism (CAM)**, a specialized form of photosynthesis found in succulents of the family Crassulaceae and several other families. Because the leaves of CAM plants take in $CO_2$

▶ Ocotillo (*Fouquieria splendens*) is common in the southwestern region of the United States and northern Mexico. It is leafless during most of the year, thereby avoiding excessive water loss (right). Immediately after a heavy rainfall, it produces small leaves (below and inset). As the soil dries, the leaves quickly shrivel and die.

▼ Oleander (*Nerium oleander*), shown in the inset, is commonly found in arid climates. Its leaves have a thick cuticle and multiple-layered epidermal tissue that reduce water loss. Stomata are recessed in cavities called "crypts," an adaptation that reduces the rate of transpiration by protecting the stomata from hot, dry wind. Trichomes help minimize transpiration by breaking up the flow of air, allowing the chamber of the crypt to have a higher humidity than the surrounding atmosphere (LM).

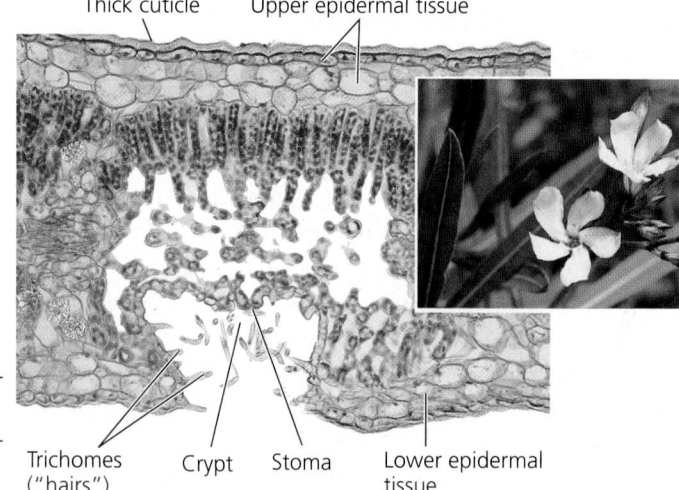

Thick cuticle | Upper epidermal tissue

100 μm

Trichomes ("hairs") | Crypt | Stoma | Lower epidermal tissue

▶ The long, white hairlike bristles along the stem of the old man cactus (*Cephalocereus senilis*) help reflect the intense sunlight of the Mexican desert.

▲ **Figure 29.20 Some xerophytic adaptations.**

at night, the stomata can remain closed during the day, when evaporative stresses are greater.

## CONCEPT CHECK 29.6

1. The pathogenic fungus *Fusicoccum amygdali* secretes a toxin called fusicoccin that activates the plasma membrane proton pumps of plant cells and leads to uncontrolled water loss. Suggest a mechanism by which the activation of proton pumps could lead to severe wilting.
2. **WHAT IF?** If you buy cut flowers, why might the florist recommend cutting the stems underwater and then transferring the flowers to a vase while the cut ends are still wet?
3. **MAKE CONNECTIONS** Explain why the evaporation of water from leaves lowers their temperature. See Concept 2.5.

For suggested answers, see Appendix A.

## CONCEPT 29.7

# Sugars are transported from sources to sinks via the phloem

The unidirectional flow of water and minerals from soil to roots to leaves through the xylem is largely in an upward direction. In contrast, the movement of photosynthates often runs in the opposite direction, transporting sugars from mature leaves to lower parts of the plant, such as root tips that require large amounts of sugars for energy and growth. The transport of the products of photosynthesis, known as **translocation**, is carried out by another tissue, the phloem.

## Movement from Sugar Sources to Sugar Sinks

Sieve-tube elements are specialized cells in angiosperms that serve as conduits for translocation. Arranged end to end, they form long sieve tubes (see Figure 28.9). Between these cells are sieve plates, structures that allow the flow of sap along the sieve tube. **Phloem sap**, the solution that flows through sieve tubes, differs markedly from the xylem sap that is transported by tracheids and vessel elements. By far the most prevalent solute in phloem sap is sugar, typically sucrose in most species. The sucrose concentration may be as high as 30% by weight, giving the sap a syrupy thickness. Phloem sap may also contain amino acids, hormones, and minerals.

In contrast to the unidirectional transport of xylem sap from roots to leaves, phloem sap moves from sites of sugar production to sites of sugar use or storage (see Figure 29.2). A **sugar source** is a plant organ that is a net producer of sugar, by photosynthesis or by breakdown of starch. A **sugar sink** is an organ that is a net consumer or depository of sugar. Growing roots, buds, stems, and fruits are sugar sinks. Although expanding leaves are sugar sinks, mature leaves, if well illuminated, are sugar sources. A storage organ, such as a tuber or a bulb, may be a source or a sink, depending on the season. When stockpiling carbohydrates in the summer, it is a sugar sink. After breaking dormancy in the spring, it becomes a sugar source because its starch is broken down to sugar, which is carried to the growing shoot tips.

Sinks usually receive sugar from the nearest sugar sources. For each sieve tube, the direction of transport depends on the locations of the sugar source and sugar sink that are connected by that tube. Therefore, neighboring sieve tubes may carry sap in opposite directions if they originate and end in different locations.

Sugar must be transported, or loaded, into sieve-tube elements before being exported to sugar sinks. In some species, it moves from mesophyll cells to sieve-tube elements via the symplast, passing through plasmodesmata. In other species, it moves by symplastic and apoplastic pathways. During apoplastic loading, sugar is accumulated by nearby sieve-tube elements, either directly or through the companion cells **(Figure 29.21a)**. In some plants, the walls of the companion

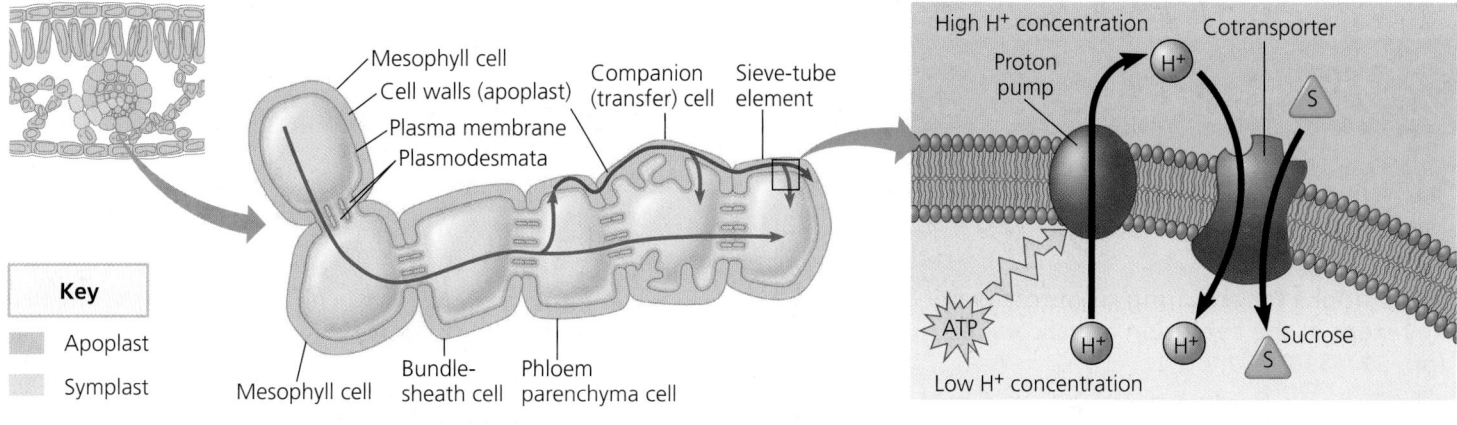

**(a)** Sucrose manufactured in mesophyll cells can travel via the symplast (blue arrows) to sieve-tube elements. In some species, sucrose exits the symplast near sieve tubes and travels through the apoplast (red arrow). It is then actively accumulated from the apoplast by sieve-tube elements and their companion cells.

**(b)** A chemiosmotic mechanism is responsible for the active transport of sucrose into companion cells and sieve-tube elements. Proton pumps generate an H+ gradient, which drives sucrose accumulation with the help of a cotransport protein that couples sucrose transport to the diffusion of H+ back into the cell.

▲ **Figure 29.21 Loading of sucrose into phloem.**

cells feature many ingrowths, enhancing solute transfer between apoplast and symplast.

In most plants, sugar movement into the phloem requires active transport because sucrose is more concentrated in sieve-tube elements and companion cells than in mesophyll. Proton pumping and $H^+$/sucrose cotransport enable sucrose to move from mesophyll cells to sieve-tube elements or companion cells **(Figure 29.21b)**.

Sucrose is unloaded at the sink end of a sieve tube. The process varies by species and organ. However, the concentration of free sugar in the sink is always lower than in the sieve tube because the unloaded sugar is consumed during growth and metabolism of the cells of the sink or converted to insoluble polymers such as starch. As a result of this sugar concentration gradient, sugar molecules diffuse from the phloem into the sink tissues, and water follows by osmosis.

## Bulk Flow by Positive Pressure: The Mechanism of Translocation in Angiosperms

Phloem sap flows from source to sink at rates as great as 1 m/hr, much faster than diffusion or cytoplasmic streaming. The translocation of phloem sap through sieve tubes by bulk flow is driven by positive pressure, or *pressure flow* **(Figure 29.22)**. The building of pressure at the source and reduction of that pressure at the sink cause sap to flow from source to sink. Sinks vary in energy demands and capacity to unload sugars. Sometimes there are more sinks than can be supported by sources. In such cases, a plant might abort some flowers, seeds, or fruits—a phenomenon called *self-thinning*. Removing sinks can also be a horticulturally useful practice. For example, since large apples command a much better price than small ones, growers sometimes remove flowers or young fruits so that their trees produce fewer but larger apples.

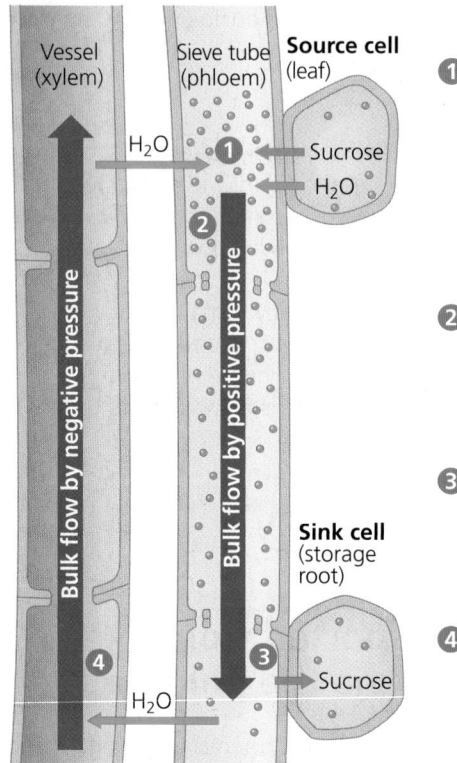

① Loading of sugar (green dots) into the sieve tube at the source reduces water potential inside the sieve-tube elements. This causes the tube to take up water by osmosis.

② This uptake of water generates a positive pressure that forces the sap to flow along the tube.

③ The pressure is relieved by the unloading of sugar and the consequent loss of water at the sink.

④ In leaf-to-root translocation, xylem recycles water from sink to source.

▲ **Figure 29.22 Bulk flow by positive pressure (pressure flow) in a sieve tube.**

### CONCEPT CHECK 29.7

1. Identify plant organs that are sugar sources, organs that are sugar sinks, and organs that might be either. Explain.
2. Why can xylem transport water and minerals using dead cells, whereas phloem requires living cells?
3. **WHAT IF?** Apple growers in Japan sometimes make a nonlethal spiral slash around the bark of trees that will be removed after the growing season. This makes the apples sweeter. Why?

   **For suggested answers, see Appendix A.**

Go to **MasteringBiology**® for Assignments, the eText, and the Study Area with Animations, Activities, Vocab Self-Quiz, and Practice Tests.

# 29 Chapter Review

## SUMMARY OF KEY CONCEPTS

**VOCAB SELF-QUIZ**

goo.gl/gbai8v

### CONCEPT 29.1

**Adaptations for acquiring resources were key steps in the evolution of vascular plants (pp. 594–596)**

- Leaves typically function in gathering sunlight and $CO_2$. Stems serve as supporting structures for leaves and as conduits for the long-distance transport of water and nutrients. Roots mine the soil for water and minerals and anchor the whole plant.
- Natural selection has produced plant architectures that fine-tune resource acquisition in the ecological niche in which the plant species naturally exists.

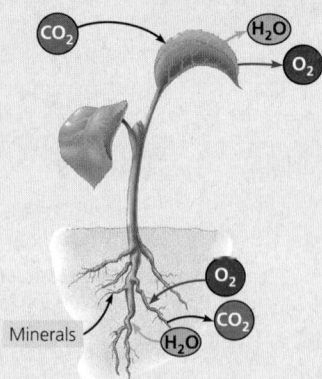

**?** *How did the evolution of xylem and phloem contribute to the successful colonization of land by vascular plants?*

## CONCEPT 29.2

### Different mechanisms transport substances over short or long distances (pp. 596–599)

- The selective permeability of the plasma membrane controls the movement of substances into and out of cells. Both active and passive transport mechanisms occur in plants.
- Plant tissues have two major compartments: the **apoplast** (everything outside the cells' plasma membranes) and the **symplast** (the cytosol and connecting plasmodesmata).
- The direction of water movement depends on the **water potential**, a quantity incorporating solute concentration and physical pressure. The osmotic uptake of water by plant cells and the resulting internal pressure that builds up make plant cells **turgid**.
- Long-distance transport occurs through **bulk flow**, the movement of liquid in response to a pressure gradient. Bulk flow occurs within the tracheids and vessel elements of the xylem and within the sieve-tube elements of the phloem.

**?** *Is xylem sap usually pulled or pushed up the plant?*

## CONCEPT 29.3

### Plant roots absorb essential elements from the soil (pp. 599–602)

- **Macronutrients**, elements required in relatively large amounts, include carbon, oxygen, hydrogen, nitrogen, and other major ingredients of organic compounds. **Micronutrients**, elements required in very small amounts, typically have catalytic functions as cofactors of enzymes.
- Deficiency of a mobile nutrient usually affects older organs more than younger ones; the reverse is true for nutrients that are less mobile within a plant. Deficiencies of nitrogen, phosphorus, and potassium are most common.
- Soil particles of various sizes derived from the breakdown of rock are found in soil. Soil particle size affects the availability of water, oxygen, and minerals in the soil.
- A soil's composition refers to its inorganic and organic components. Topsoil is a complex ecosystem teeming with bacteria, fungi, protists, animals, and the roots of plants.

**?** *Do plants need soil to grow? Explain.*

## CONCEPT 29.4

### Plant nutrition often involves relationships with other organisms (pp. 602–608)

- **Rhizobacteria** derive their energy from the **rhizosphere**, a microbe-enriched ecosystem intimately associated with roots. Plant secretions support the energy needs of the rhizosphere. Some rhizobacteria produce antibiotics, whereas others make nutrients more available for plants. Most are free-living, but some live inside plants. Plants satisfy most of their huge needs for nitrogen from the decomposition of humus and the fixation of gaseous $N_2$ by bacteria.
- $N_2$-fixing bacteria convert atmospheric $N_2$ to nitrogenous minerals that plants can absorb as a nitrogen source for organic synthesis. The most efficient mutualism between plants and nitrogen-fixing bacteria occurs in the **nodules** formed by *Rhizobium* bacteria growing in the roots of legumes. These bacteria obtain sugar from the plant and supply the plant with fixed nitrogen.

- **Mycorrhizae** are mutualistic associations formed between roots and certain soil fungi that aid in the absorption of minerals and water.
- **Epiphytes** grow on the surfaces of other plants but acquire water and minerals from rain. Parasitic plants absorb nutrients from host plants. Carnivorous plants supplement their mineral nutrition by digesting animals.

**?** *Do all plants gain their energy directly from photosynthesis? Explain.*

## CONCEPT 29.5

### Transpiration drives the transport of water and minerals from roots to shoots via the xylem (pp. 609–612)

- Water and minerals from the soil enter the plant through the epidermis of roots, cross the root cortex, and then pass into the vascular cylinder by way of the selectively permeable cells of the **endodermis**. From the vascular cylinder, the **xylem sap** is transported long distances by bulk flow to the veins that branch throughout each leaf.
- The **cohesion-tension hypothesis** proposes that the movement of xylem sap is driven by a water potential difference created at the leaf end of the xylem by the evaporation of water from leaf cells. Evaporation lowers the water potential at the air-water interface, thereby generating the negative pressure that pulls water through the xylem.

**?** *Why is the ability of water molecules to form hydrogen bonds important for the movement of xylem sap?*

## CONCEPT 29.6

### The rate of transpiration is regulated by stomata (pp. 612–615)

- **Transpiration** is the loss of water vapor from plants. **Wilting** occurs when the water lost by transpiration is not replaced by absorption from roots.
- Stomata are the major pathway for water loss from plants. Guard cells widen or narrow the stomatal pores. When guard cells take up $K^+$, the pore widens. The opening and closing of stomata are controlled by light, $CO_2$, the drought hormone **abscisic acid**, and a **circadian rhythm**.
- Reduced leaves and **CAM** photosynthesis are examples of adaptations to arid environments.

**?** *Why are stomata necessary?*

## CONCEPT 29.7

### Sugars are transported from sources to sinks via the phloem (pp. 615–616)

- Mature leaves are the main **sugar sources**, although storage organs can be seasonal sources. Growing organs such as roots, stems, and fruits are the main **sugar sinks**.
- Phloem loading depends on the active transport of sucrose. Sucrose is cotransported with $H^+$, which diffuses down a gradient generated by proton pumps. Loading of sugar at the source and unloading at the sink maintain a pressure difference that keeps sap flowing through a sieve tube.

**?** *Why is phloem transport considered an active process?*

CHAPTER 29 RESOURCE ACQUISITION, NUTRITION, AND TRANSPORT IN VASCULAR PLANTS **617**

# TEST YOUR UNDERSTANDING

PRACTICE TEST

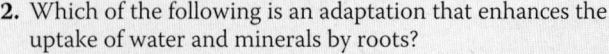

goo.gl/CRZjvS

## Level 1: Knowledge/Comprehension

**1.** Which structure or compartment is part of the symplast?
(A) the interior of a vessel element
(B) the interior of a sieve tube
(C) the cell wall of a mesophyll cell
(D) an extracellular air space

**2.** Which of the following is an adaptation that enhances the uptake of water and minerals by roots?
(A) mycorrhizae
(B) cavitation
(C) active uptake by vessel elements
(D) rhythmic contractions by cortical cells

**3.** Movement of xylem sap from roots to leaves
(A) occurs through the apoplast of sieve-tube elements.
(B) usually depends on tension, or negative pressure potential.
(C) depends on active transport.
(D) depends on the pumping of water through aquaporins.

## Level 2: Application/Analysis

**4.** What would enhance water uptake by a plant cell?
(A) decreasing the $\psi$ of the surrounding solution
(B) positive pressure on the surrounding solution
(C) the loss of solutes from the cell
(D) increasing the $\psi$ of the cytoplasm

**5.** A plant cell with a $\psi_S$ of $-0.65$ MPa maintains a constant volume when bathed in a solution that has a $\psi_S$ of $-0.30$ MPa and is in an open container. The cell has a
(A) $\psi_P$ of $+0.65$ MPa.          (C) $\psi_P$ of $+0.35$ MPa.
(B) $\psi$ of $-0.65$ MPa.          (D) $\psi$ of 0 MPa.

**6.** Compared with a cell with few aquaporin proteins in its membrane, a cell containing many aquaporin proteins will have a
(A) faster rate of osmosis.      (C) higher water potential.
(B) lower water potential.      (D) faster rate of active transport.

**7.** Two groups of tomatoes were grown in the laboratory, one with humus added to the soil and the other a control without humus. The leaves of the plants grown without humus were yellowish (less green) compared with those of the plants grown in humus-enriched soil. The best explanation for this difference is that
(A) the healthy plants used carbohydrates in the decomposing leaves of the humus for energy to make chlorophyll.
(B) the humus made the soil more loosely packed, so water penetrated more easily to the roots.
(C) the humus contained minerals such as magnesium and iron, needed for the synthesis of chlorophyll.
(D) the heat released by the decomposing leaves of the humus caused more rapid growth and chlorophyll synthesis.

**8.** **DRAW IT** Trace the uptake of water and minerals from root hairs to the endodermis in a root, following a symplastic route and an apoplastic route. Label the routes on the diagram below.

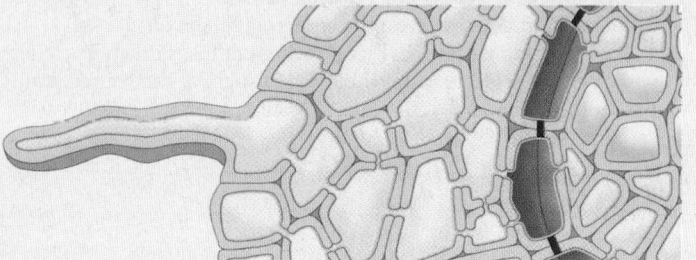

## Level 3: Synthesis/Evaluation

**9. SCIENTIFIC INQUIRY**
Acid precipitation has an abnormally high concentration of hydrogen ions ($H^+$). One effect of acid precipitation is to deplete the soil of nutrients such as calcium ($Ca^{2+}$), potassium ($K^+$), and magnesium ($Mg^{2+}$). Devise a hypothesis to explain how acid precipitation washes the nutrients from the soil. How might you test your hypothesis?

**10. SCIENTIFIC INQUIRY**
A Minnesota gardener notes that the plants immediately bordering a walkway are stunted compared with those farther away. Suspecting that the soil near the walkway may be contaminated from salt added to the walkway in winter, the gardener tests the soil. The composition of the soil near the walkway is identical to that farther away except that it contains an additional 50 m$M$ NaCl. Assuming that the NaCl is completely ionized, calculate how much it will lower the solute potential of the soil at 20°C using the *solute potential equation*:

$$\psi_S = -iCRT$$

where $i$ = the ionization constant (2 for NaCl), $C$ = the molar concentration (in moles/liter), $R$ = the pressure constant [$R$ = 0.00831 (liter MPa)/(mole K)], and $T$ = temperature in Kelvin (273 + °C). Describe how this change in the solute potential of the soil would affect the water potential of the soil. Explain how the change in the water potential of the soil would affect the movement of water in or out of the roots.

**11. FOCUS ON EVOLUTION**
Large brown algae called kelps can grow as tall as 25 m. Kelps consist of a holdfast anchored to the ocean floor, blades that float at the surface and collect light, and a long stalk connecting the blades to the holdfast (see Figure 25.14). Specialized cells in the stalk, although nonvascular, can transport sugar. Suggest a reason why these structures analogous to sieve-tube elements might have evolved in kelps. Explain your thinking.

**12. FOCUS ON INTERACTIONS**
The soil in which plants grow teems with organisms from every taxonomic kingdom. In a short essay (100–150 words), discuss examples of how the mutualistic interactions of plants with bacteria, fungi, and animals improve plant nutrition.

**13. SYNTHESIZE YOUR KNOWLEDGE**

Imagine yourself as a water molecule in the soil solution of a forest. In a short essay (100–150 words), explain what pathways and what forces would be necessary to carry you to the leaves of these trees.

*For selected answers, see Appendix A.*

# 30 Reproduction and Domestication of Flowering Plants

## KEY CONCEPTS

**30.1** Flowers, double fertilization, and fruits are unique features of the angiosperm life cycle

**30.2** Flowering plants reproduce sexually, asexually, or both

**30.3** People modify crops through breeding and genetic engineering

▲ **Figure 30.1** Why have blowflies laid their eggs on this flower?

## Flowers of Deceit

The only visible part of *Rhizanthes lowii*, a denizen of the rain forests of southeast Asia, is its large flesh-colored flower **(Figure 30.1)**. It derives all its energy from tapping into a species of tropical vine that it parasitizes. Its thieving ways also extend to its mode of pollination. Upon opening, a *Rhizanthes* flower emits a foul odor reminiscent of a decaying corpse. Female blowflies, insects that normally deposit their eggs on carrion, find the scent irresistible and lay their eggs on the flower. There, the blowflies are peppered with sticky pollen grains that adhere to their bodies. Eventually, the pollen-coated blowflies fly away, hopefully, from the plant's perspective, to another *Rhizanthes* flower.

An unusual aspect of the *Rhizanthes* example is that the insect does not profit from interacting with the flower. In fact, the blowfly maggots that emerge on *Rhizanthes* flowers find no carrion to eat and quickly perish. More typically, a plant lures an animal pollinator to its flowers not with offers of false carrion but with rewards of energy-rich nectar or pollen. Thus, both plant and pollinator benefit. Participating in such mutually beneficial relationships with other organisms is common in the plant kingdom. In fact, in recent evolutionary

times, some flowering plants have formed relationships with an animal that not only disperses their seeds but also provides the plants with water and mineral nutrients and vigorously protects them from encroaching competitors, pathogens, and predators. In return for these favors, the animal typically gets to eat a fraction of some part of the plants, such as their seeds or fruits. The plants involved in these relationships are called crops; the animals are humans.

Since the origins of crop domestication over 10,000 years ago, plant breeders have genetically manipulated the traits of a few hundred wild angiosperm species by artificial selection, transforming them into the crops we grow today. Genetic engineering has dramatically increased the variety of ways and the speed with which we can now modify plants.

In Chapter 26, we approached plant reproduction from an evolutionary perspective, tracing the descent of land plants from algal ancestors. Because angiosperms are the most important group of plants in agriculture and in most terrestrial ecosystems, we'll explore their reproductive biology in detail in this chapter. After discussing the sexual and asexual reproduction of angiosperms, we'll examine the role that people have played in domesticating crop species, as well as the controversies surrounding modern plant biotechnology.

# Flowers, double fertilization, and fruits are unique features of the angiosperm life cycle

The life cycles of plants are characterized by an alternation of generations, in which multicellular haploid (*n*) and multicellular diploid (*2n*) generations alternately produce each other (see Figures 10.6b and 26.6). The diploid plant, the *sporophyte*, produces haploid spores by meiosis. These spores divide by mitosis, giving rise to multicellular *gametophytes*, the male and female haploid plants that produce gametes (sperm and eggs). **Fertilization**, the fusion of gametes, results in a diploid zygote, which divides by mitosis and forms a new sporophyte. In angiosperms, the sporophyte is the dominant generation: It is larger, more conspicuous, and longer-lived than the gametophyte. Over the course of seed plant evolution, gametophytes became reduced in size and wholly dependent on the sporophyte for nutrients. Angiosperm gametophytes are the most reduced of all plants, consisting of only a few cells. The key traits of the angiosperm life cycle can be remembered as the "three Fs"—*f*lowers, double *f*ertilization, and *f*ruits.

## Flower Structure and Function

Flowers, the reproductive shoots of angiosperm sporophytes, are typically composed of four types of floral organs: **carpels**, **stamens**, **petals**, and **sepals (Figure 30.2)**. When viewed from above, these organs appear as concentric whorls. Carpels form the first (innermost) whorl, stamens form the second, petals form the third, and sepals form the fourth (outermost) whorl. All four types of floral organs are attached to a part of the stem called the **receptacle**. Unlike vegetative shoots, flowers are determinate shoots; they cease growing after the flower and fruit are formed.

Carpels and stamens are reproductive organs; sepals and petals are sterile. A carpel has an **ovary** at its base and a long, slender neck called the **style**. At the top of the style is a sticky structure called the **stigma** that captures pollen. Within the ovary are one or more **ovules**; the number depends on the species. The flower in Figure 30.2 has a single carpel, but many species have multiple carpels. In most species, two or more carpels are fused into one structure; the result is an ovary with two or more chambers, each containing one or more ovules. The term **pistil** is sometimes used to refer to a single carpel (a simple pistil) or two or more fused carpels (a compound pistil). A stamen consists of a stalk called the *filament* and a terminal structure called the **anther**; within the anther are chambers called microsporangia (pollen sacs) that produce pollen. Petals are typically more brightly colored than sepals and advertise the flower to insects and other pollinators. Sepals, which enclose and protect unopened floral buds, usually resemble leaves more than the other floral organs do.

**Complete flowers** have all four basic floral organs (see Figure 30.2). Some species have **incomplete flowers**, lacking

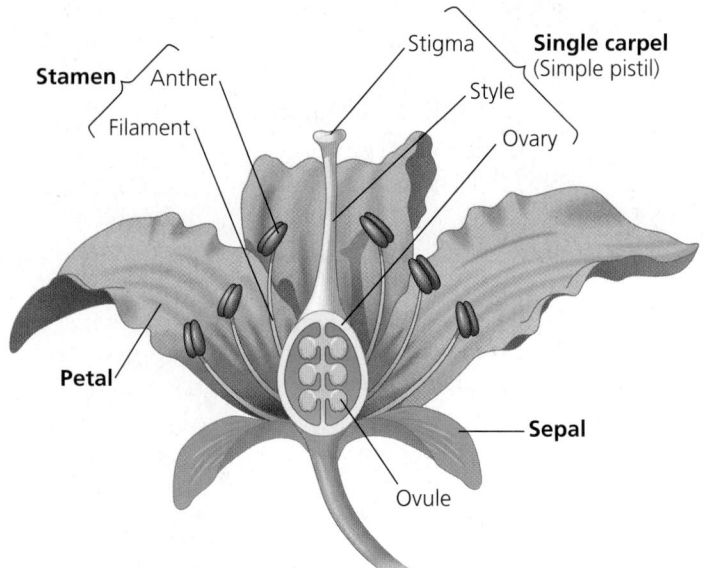

▲ **Figure 30.2 The structure of an idealized flower.**

sepals, petals, stamens, or carpels. For example, most grass flowers lack petals. Some incomplete flowers are sterile, lacking functional stamens and carpels; others are *unisexual*, lacking either stamens or carpels. Flowers also vary in size, shape, color, odor, organ arrangement, and time of opening. Some are borne singly, while others are arranged in showy clusters called **inflorescences**. For example, a daisy is actually an inflorescence consisting of a central disk composed of hundreds of tiny complete flowers surrounded by sterile, incomplete flowers that look like white petals. Much of floral diversity represents adaptations to specific pollinators.

## Flower Formation

The flowers of a given plant species typically appear suddenly at a specific time of year. Such synchrony promotes outbreeding, the main advantage of sexual reproduction. Flower formation involves a developmental switch in the shoot apical meristem from a vegetative to a reproductive growth mode. This transition into a *floral meristem* is triggered by a combination of environmental cues, such as day length, and internal signals. (You'll learn more about the roles of these signals in flowering in Chapter 31.) Once the transition to flowering has begun, the order of each organ's emergence from the floral meristem determines whether it will develop into a sepal, petal, stamen, or carpel. Several organ identity genes have been identified that encode transcription factors that regulate the development of this floral organization. A mutation in one of these genes can cause abnormal floral development, such as petals growing in place of stamens **(Figure 30.3)**.

By studying flower mutants, researchers have developed a model called the **ABC hypothesis** to explain how three floral organ identity genes direct the formation of the four types of floral organs. According to the slightly simplified version of the ABC hypothesis presented in **Figure 30.4a**, each class of organ identity genes is switched on in two specific whorls of the floral meristem. Normally, *A* genes are switched on in the two outer whorls (sepals and petals), *B* genes are switched on in the two middle whorls (petals and stamens), and *C* genes are switched on in the two inner whorls (stamens and carpels). Sepals arise from those parts of the floral meristems in which only *A* genes are active, petals where *A* and *B* genes are active, stamens where *B* and *C* genes are active, and carpels where only *C* genes are active. The ABC hypothesis can account for the phenotypes of mutants lacking *A*, *B*, or *C* gene activity, with one addition: Where gene *A* activity is present, it inhibits *C*, and vice versa. If either *A* or *C* is missing, the other takes its place. **Figure 30.4b** shows the floral patterns of mutants lacking each of the three classes of organ identity genes and depicts how the hypothesis accounts for their floral phenotypes.

▲ **Normal *Arabidopsis* flower.** *Arabidopsis* normally has four whorls of flower parts: sepals (Se), petals (Pe), stamens (St), and carpels (Ca).

▶ **Abnormal *Arabidopsis* flower.** Researchers have identified several mutations of organ identity genes that cause abnormal flowers to develop. This flower has an extra set of petals in place of stamens and an internal flower where normal plants have carpels.

▲ **Figure 30.3 Organ identity genes and pattern formation in flower development.**

**MAKE CONNECTIONS** *Provide another example of a homeotic gene mutation that leads to organs being produced in the wrong place (see Concept 16.1).*

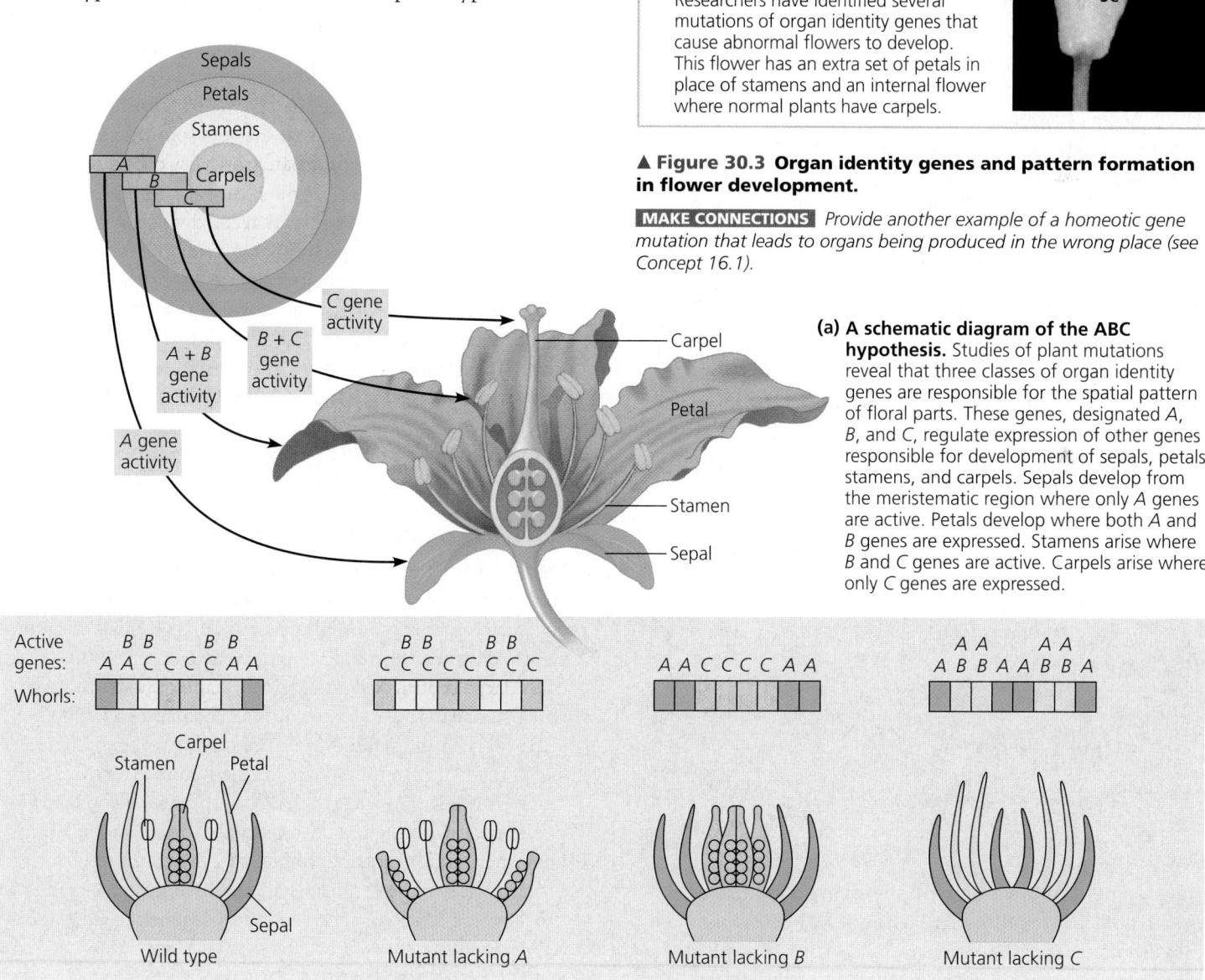

**(a) A schematic diagram of the ABC hypothesis.** Studies of plant mutations reveal that three classes of organ identity genes are responsible for the spatial pattern of floral parts. These genes, designated *A*, *B*, and *C*, regulate expression of other genes responsible for development of sepals, petals, stamens, and carpels. Sepals develop from the meristematic region where only *A* genes are active. Petals develop where both *A* and *B* genes are expressed. Stamens arise where *B* and *C* genes are active. Carpels arise where only *C* genes are expressed.

**(b) Side view of flowers with organ identity mutations.** The phenotype of mutants lacking a functional *A*, *B*, or *C* organ identity gene can be explained by combining the model in part (a) with the rule that if *A* or *C* activity is missing, the other activity occurs through all four whorls.

▲ **Figure 30.4 The ABC hypothesis for the functioning of organ identity genes in flower development.**

**WHAT IF?** *What would a flower look like if the* A *genes and* B *genes were inactivated?*

## The Angiosperm Life Cycle: An Overview

**Figure 30.5** shows the angiosperm life cycle, including gametophyte development, pollination, double fertilization, and seed development. We'll begin by examining the development of gametophytes.

### Gametophyte Development

Over the course of seed plant evolution, gametophytes became reduced in size and wholly dependent on the sporophyte for nutrients (see Figure 26.19). The gametophytes of angiosperms are the most reduced of all plants, consisting of only a few cells: They are microscopic, and their development is obscured by protective tissues.

### Development of Female Gametophytes (Embryo Sacs)

As a carpel develops, one or more ovules form deep within its ovary, its swollen base. A female gametophyte, also known as an **embryo sac**, develops inside each ovule. The process of embryo sac formation occurs in a tissue called the megasporangium ❶ within each ovule. Two *integuments* (layers of protective sporophytic tissue that will develop into the seed coat) surround each megasporangium, except at a gap called the *micropyle*. Female gametophyte development begins when one cell in the megasporangium of each ovule, the *megasporocyte* (or megaspore mother cell), enlarges and undergoes meiosis, producing four haploid **megaspores**. Only one megaspore survives; the others degenerate.

The nucleus of the surviving megaspore divides by mitosis three times without cytokinesis, resulting in one large cell with eight haploid nuclei. The multinucleate mass is then divided by membranes to form the embryo sac. The cell fates of the nuclei are determined by a gradient of the hormone auxin originating near the micropyle. At the micropylar end of the embryo sac, two cells called synergids flank the egg and help attract and guide the pollen tube to the embryo sac. At the opposite end of the embryo sac are three antipodal cells of unknown function. The other two nuclei, called polar nuclei, are not partitioned into separate cells but share the cytoplasm of the large central cell of the embryo sac. The mature embryo sac thus consists of eight nuclei contained within seven cells. The ovule, which will become a seed if fertilized, now consists of the embryo sac, enclosed by the megasporangium (which eventually withers) and two surrounding integuments.

### Development of Male Gametophytes in Pollen Grains

As the stamens are produced, each anther ❷ develops four microsporangia, also called pollen sacs. Within the microsporangia are many diploid cells called *microsporocytes*, or microspore mother cells. Each microsporocyte undergoes meiosis, forming four haploid **microspores**, ❸ each of which eventually gives rise to a haploid male gametophyte. Each microspore then undergoes mitosis, producing a haploid male

gametophyte consisting of only two cells: the *generative cell* and the *tube cell*. Together, these two cells *and* the spore wall constitute a **pollen grain**. The spore wall, which consists of material produced by both the microspore and the anther, usually exhibits an elaborate pattern unique to the species. During maturation of the male gametophyte, the generative cell passes into the tube cell: The tube cell now has a completely freestanding cell inside it.

### Pollination

After the microsporangium breaks open and releases the pollen, a pollen grain may be transferred to a receptive surface of a stigma—the act of **pollination**. Here we'll focus on how a pollen grain delivers sperm after pollination. Later we'll look at the various ways that a pollen grain can be transported from an anther to a stigma.

At the time of pollination, the pollen grain typically consists of only the tube cell and the generative cell. It then absorbs water and germinates by producing a **pollen tube**, a long cellular protuberance that delivers sperm to the female gametophyte. A pollen tube can grow very quickly, at a rate of 1 cm/hr or more. As the pollen tube elongates through the style, the nucleus of the generative cell divides by mitosis and produces two sperm, which remain inside the tube cell. The tube nucleus then leads the two sperm as the tip of the pollen tube grows toward the micropyle in response to chemical attractants produced by the synergids. The arrival of the pollen tube initiates the death of one of the two synergids, thereby providing a passageway into the embryo sac. The tube nucleus then degenerates, and the two sperm are discharged from the pollen tube ❹ in the vicinity of the female gametophyte.

### Double Fertilization

**Fertilization**, the fusion of gametes, occurs after the two sperm reach the female gametophyte. One sperm fertilizes the egg, forming the zygote. The other sperm combines with the two polar nuclei, forming a triploid ($3n$) nucleus in the center of the large central cell of the female gametophyte. This cell will give rise to the **endosperm**, a food-storing tissue of the seed. ❺ The union of the two sperm cells with different nuclei of the female gametophyte is called **double fertilization**. Double fertilization ensures that endosperm develops only in ovules where the egg has been fertilized, thereby preventing angiosperms from squandering nutrients on infertile ovules. Near the time of double fertilization, the tube nucleus, the other synergid, and the antipodal cells degenerate.

### Seed Development

❻ After double fertilization, each ovule develops into a seed. Meanwhile, the ovary develops into a fruit, which encloses the seeds and aids in their dispersal by wind or animals. As the sporophyte embryo develops from the zygote, the seed stockpiles proteins, oils, and starch to varying degrees, depending on the species. This is why seeds are such a major nutrient drain. Initially, carbohydrates and other nutrients are stored in the seed's endosperm, but later, depending on the species, the swelling cotyledons (seed leaves) of the embryo may take over this function. When a seed germinates, ❼ the embryo develops into a new sporophyte. The mature sporophyte produces its own flowers and fruits.

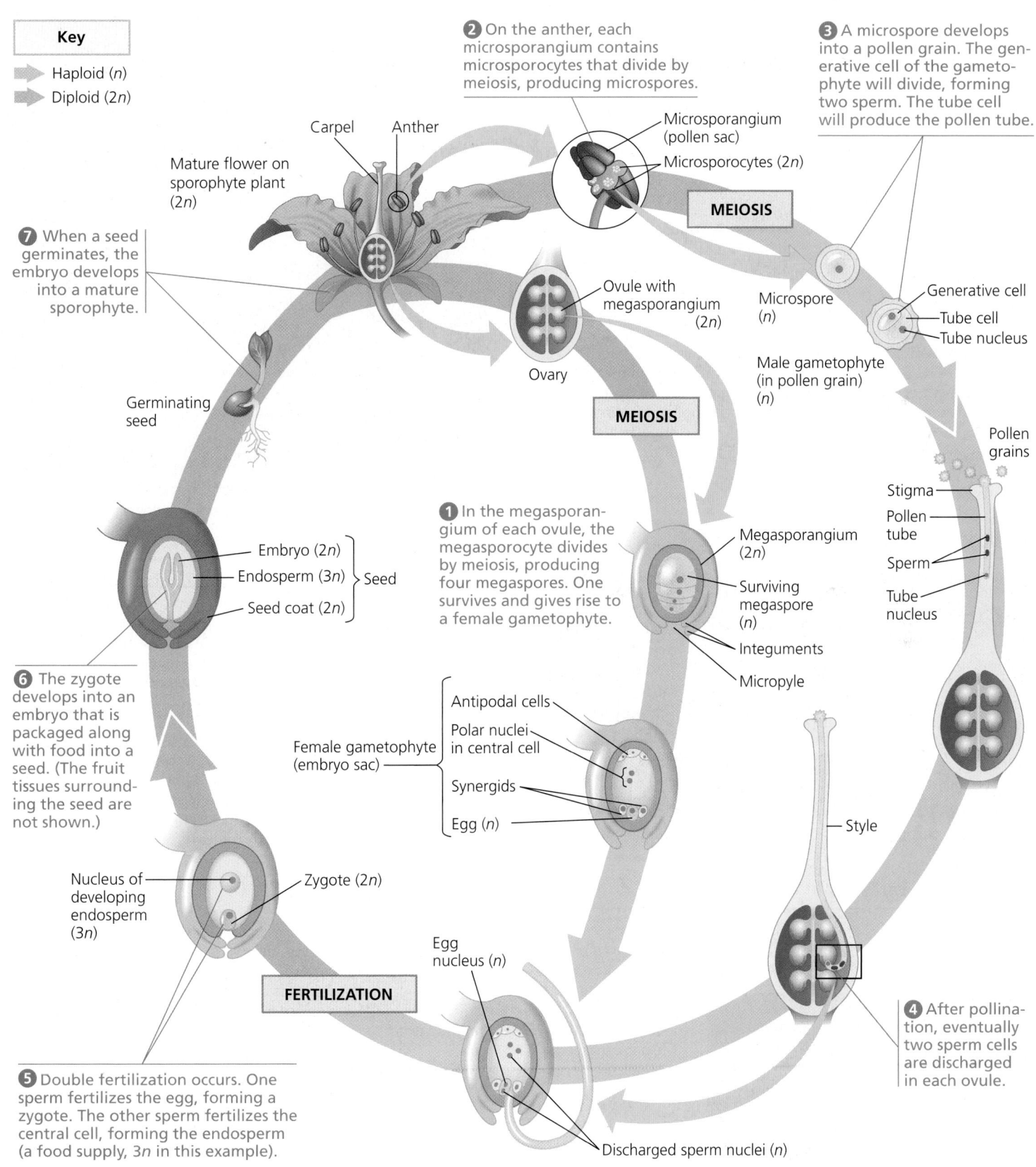

▲ Figure 30.5 **The life cycle of an angiosperm.**

## Pollination: *A Closer Look*

One detail unexplored in the preceding discussion of the angiosperm life cycle is the question of how pollination, the transfer of pollen, is accomplished. The answer is by wind, water, or animals **(Figure 30.6)**. At certain times of the year, the air is loaded with pollen grains, as anyone plagued with pollen allergies can attest. Some aquatic plants rely on water to disperse pollen. Most angiosperm species, however, depend on insects, birds, or other animal pollinators to transfer pollen from one flower to another.

---

### ▼ Figure 30.6    Exploring Flower Pollination

Most angiosperm species rely on a living (biotic) or nonliving (abiotic) pollinating agent that can move pollen from the anther of a flower on one plant to the stigma of a flower on another plant. Approximately 80% of all angiosperm pollination is biotic, employing animal go-betweens. Among abiotically pollinated species, 98% rely on wind and 2% on water.

## Abiotic Pollination by Wind

Since the reproductive success of wind-pollinated angiosperms does not depend on attracting pollinators, there has been no selective pressure favoring colorful or scented flowers. Accordingly, the flowers of wind-pollinated species are often small, green, and inconspicuous, and they produce neither nectar nor scent. Most temperate trees and grasses are wind-pollinated. The flowers of hazel (*Corylus avellana*) and many other temperate, wind-pollinated trees appear in early spring, when leaves are not present to interfere with pollen movement. The relative inefficiency of wind pollination is compensated for by production of enormous numbers of pollen grains.

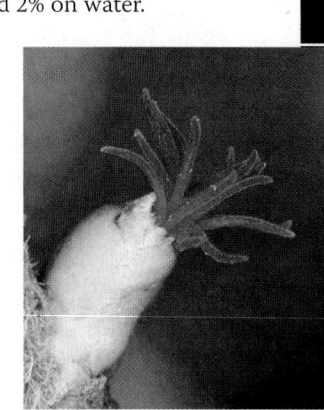

▲ Hazel carpellate flower (carpels only)

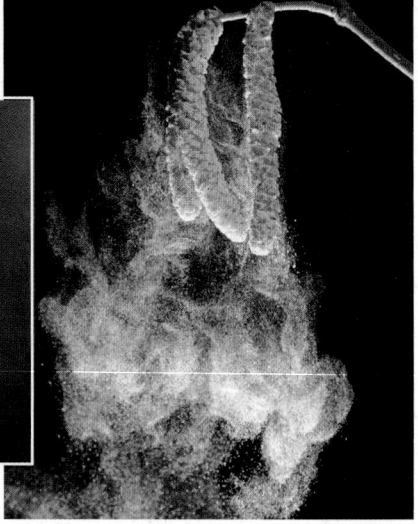

▲ Hazel staminate flowers (stamens only) releasing clouds of pollen

## Pollination by Insects

▲ Common dandelion under normal light

▲ Common dandelion under ultraviolet light

About 65% of all flowering plants require insects for pollination; the percentage is even greater for major crops. Pollinating insects include bees, moths, butterflies, flies, and beetles. Bees are the most important, and there is great concern that honeybee populations are in decline. Pollinating bees depend on pollen and the sugary solution called nectar for food. The main function of nectar, which is produced by glands called nectaries at the base of many flowers, is to "reward" the pollinator. Typically, bee-pollinated flowers have a delicate, sweet fragrance. Bees are attracted to bright colors, primarily yellow and blue. Red appears dull to them, but they can see ultraviolet radiation. Many bee-pollinated flowers, such as the common dandelion (*Taraxacum officinale*), have markings called "nectar guides" that help insects locate the nectaries; some of these markings are visible to human eyes only under ultraviolet light.

## Pollination by Bats

Bat-pollinated flowers are light-colored and aromatic, attracting their nocturnal pollinators. The lesser long-nosed bat (*Leptonycteris curasoae yerbabuenae*) pollinates agave and cactus flowers in the southwestern United States and Mexico as it feeds on their nectar and pollen.

◄ Long-nosed bat feeding on agave flowers at night

## Pollination by Birds

Bird-pollinated flowers, such as columbine flowers, are usually large and bright red or yellow, but they have little odor since many birds do not have a well-developed sense of smell. Nectar helps meet the high energy demands of pollinating birds. The petals of bird-pollinated flowers are often fused, forming a bent floral tube that fits the curved beak of the bird.

▶ Hummingbird drinking nectar of columbine flower

**EVOLUTION** Many species of flowering plants have evolved with specific pollinators. The joint evolution of two interacting species, each in response to selection imposed by the other, is called **coevolution**. Natural selection favors individual plants or insects having slight deviations of structure that enhance the flower-pollinator mutualism. For example, some species have flower petals fused together, forming long, tubelike structures bearing nectaries tucked deep inside. Charles Darwin suggested that natural selection might lead to correspondences between the lengths of a floral tube and an insect's proboscis, a straw-like mouthpart. Imagine an insect with a proboscis long enough to drink the nectar of flowers without picking up pollen on its body. The resulting failure of these plants to fertilize others would render them less evolutionarily fit. Natural selection would favor flowers with longer tubes. At the same time, an insect with a proboscis too short for the tube couldn't reach the nectar and therefore would be at a selective disadvantage. As a result, flower shapes and sizes often closely correspond to the pollen-adhering parts of pollinators. In fact, based on the length of a long, tubular flower in Madagascar, Darwin predicted the existence of a moth with a 28-cm-long proboscis. Such a moth was discovered decades after Darwin's death (see Figure 27.39).

## Seed Development and Structure

After successful pollination and double fertilization, a seed begins to form. During this process both the endosperm and the embryo develop. When mature, a **seed** consists of a dormant embryo surrounded by stored food and protective layers.

### Endosperm Development

Endosperm usually develops before the embryo does. After double fertilization, the triploid nucleus of the ovule's central cell divides, forming a multinucleate "supercell" that has a milky consistency. This liquid mass, the endosperm, becomes multicellular when cytokinesis partitions the cytoplasm by forming membranes between the nuclei. Eventually, these "naked" cells produce cell walls, and the endosperm becomes solid. Coconut "milk" and "meat" are examples of liquid and solid endosperm, respectively. The white fluffy part of popcorn is also endosperm.

In grains and most other species of monocots, as well as many eudicots, the endosperm stores nutrients that can be used by the seedling after germination. In other eudicot seeds, the food reserves of the endosperm are completely exported to the cotyledons before the seed completes its development; consequently, the mature seed lacks endosperm.

### Embryo Development

The first mitotic division of the zygote splits the fertilized egg into a basal cell and a terminal cell **(Figure 30.7)**. The terminal cell eventually gives rise to most of the embryo. The basal cell continues to divide, producing a thread of cells called the suspensor, which anchors the embryo to the parent plant. The suspensor helps in transferring nutrients to the embryo from the parent plant and, in some species, from the endosperm. As the

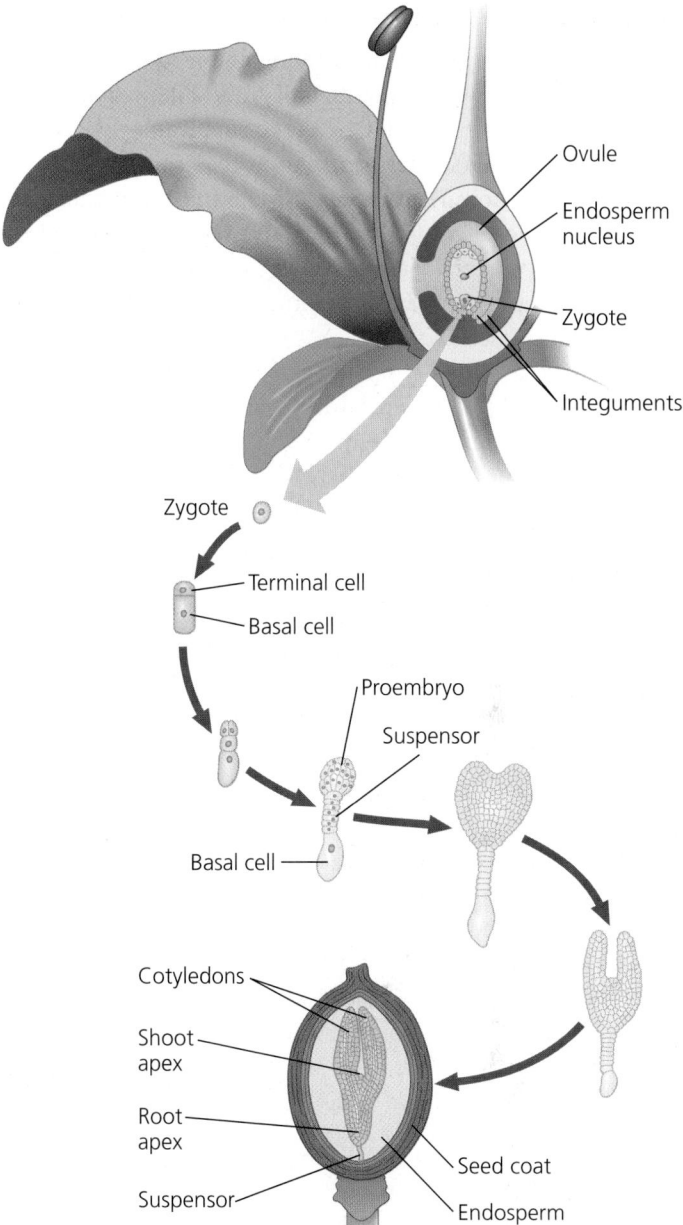

▲ **Figure 30.7 The development of a eudicot plant embryo.** By the time the ovule becomes a mature seed and the integuments harden and thicken into the seed coat, the zygote has given rise to an embryonic plant with rudimentary organs.

suspensor elongates, it pushes the embryo deeper into the nutritive and protective tissues. Meanwhile, the terminal cell divides several times and forms a spherical proembryo (early embryo) attached to the suspensor. The cotyledons begin to form as bumps on the proembryo. A eudicot, with its two cotyledons, is heart-shaped at this stage. Only one cotyledon develops in monocots.

Soon after the rudimentary cotyledons appear, the embryo elongates. Cradled between the two cotyledons is the embryonic shoot apex. At the opposite end of the embryo's axis, where the suspensor attaches, an embryonic root apex forms. After the seed germinates—indeed, for the rest of the plant's life—the apical meristems at the apices of shoots and roots sustain primary growth (see Figure 28.10).

## Structure of the Mature Seed

During the last stages of its maturation, the seed dehydrates until its water content is only about 5–15% of its weight. The embryo, which is surrounded by a food supply (cotyledons, endosperm, or both), enters **dormancy**; that is, it stops growing and its metabolism nearly ceases. The embryo and its food supply are enclosed by a hard, protective **seed coat** formed from the integuments of the ovule. In some species, dormancy is imposed by the presence of an intact seed coat rather than by the embryo itself.

You can look closely at one type of eudicot seed by splitting open the seed of a common garden bean. The embryo consists of an elongate structure, the embryonic axis, attached to fleshy cotyledons **(Figure 30.8a)**. Below where the two cotyledons are attached, the embryonic axis is called the **hypocotyl** (from the Greek *hypo*, under). The hypocotyl terminates in the **radicle**, or embryonic root. The portion of the embryonic axis above where the cotyledons are attached and below the first pair of miniature leaves is the **epicotyl** (from the Greek *epi*, on, over). The epicotyl, young leaves, and shoot apical meristem are collectively called the *plumule*.

The cotyledons of the common garden bean are packed with starch before the seed germinates because they absorbed carbohydrates from the endosperm when the seed was developing. However, the seeds of some eudicot species, such as castor beans (*Ricinus communis*), retain their food supply in the endosperm and have very thin cotyledons. The cotyledons absorb nutrients from the endosperm and transfer them to the rest of the embryo when the seed germinates.

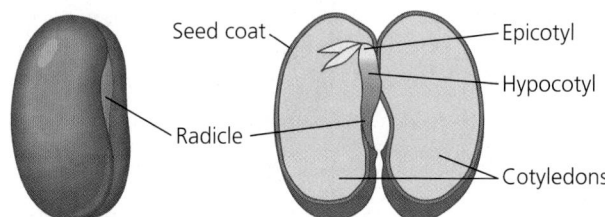

**(a) Common garden bean, a eudicot with thick cotyledons.** The fleshy cotyledons store food absorbed from the endosperm before the seed germinates.

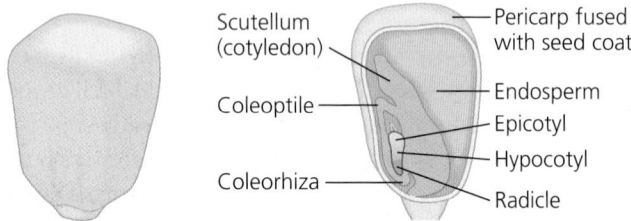

**(b) Maize, a monocot.** Like all monocots, maize has only one cotyledon. Maize and other grasses have a large cotyledon called a scutellum. The rudimentary shoot is sheathed in a structure called the coleoptile, and the coleorhiza covers the young root.

▲ **Figure 30.8 Seed structure.**

MAKE CONNECTIONS *In addition to cotyledon number, what are some other ways that the structures of monocots and eudicots differ? (See Figure 28.2.)*

The embryos of monocots possess only a single cotyledon **(Figure 30.8b)**. Grasses, including maize and wheat, have a specialized cotyledon called a *scutellum* (from the Latin *scutella*, small shield, a reference to its shape). The scutellum, which has a large surface area, is pressed against the endosperm, from which it absorbs nutrients during germination. The embryo of a grass seed is enclosed within two protective sheaths: a **coleoptile**, which covers the young shoot, and a **coleorhiza**, which covers the young root. Both structures aid in soil penetration after germination.

Seed weights range from less than 1 µg for some orchids to 20 kg for Coco-de-Mer palms. Orchid seeds have almost no food reserves and must bond symbiotically with mycorrhizae prior to germination. Large, endosperm-rich palm seeds are an adaptation for seedling establishment on nutrient-poor beaches.

## Seed Dormancy: An Adaptation for Tough Times

Environmental conditions required to break seed dormancy vary among species. Seeds of some species germinate as soon as they are in a suitable environment. Others remain dormant, even if sown in a favorable place, until a specific environmental cue causes them to break dormancy.

The requirement for specific cues to break seed dormancy increases the chances that germination will occur at a time and place most advantageous to the seedling. Seeds of many desert plants, for instance, germinate only after a substantial rainfall. If they were to germinate after a mild drizzle, the soil might soon become too dry to support the seedlings. Where natural fires are common, many seeds require intense heat or smoke to break dormancy; seedlings are therefore most abundant after fire has cleared away competing vegetation. Where winters are harsh, seeds may require extended exposure to cold before they germinate; seeds sown during summer or fall will therefore not germinate until the following spring, ensuring a long growth season before the next winter. Certain small seeds, such as those of some lettuce varieties, require light for germination and will break dormancy only if buried shallow enough for the seedlings to poke through the soil surface. Some seeds have coats that must be weakened by chemical attack as they pass through an animal's digestive tract and thus are usually carried a long distance before germinating from feces.

The length of time a dormant seed remains viable and capable of germinating varies from a few days to decades or even longer, depending on the plant species and environmental conditions. The oldest carbon-14–dated seed that has grown into a viable plant was a 2,000-year-old date palm seed recovered from excavations of Herod's palace in Israel. Most seeds are durable enough to last a year or two until conditions are favorable for germinating. Thus, the soil has a bank of ungerminated seeds that may have accumulated for several years. This is one reason vegetation reappears so rapidly after an environmental disruption such as fire.

# Germination, Growth, and Flowering

When environmental conditions become ripe for growth, seed dormancy is broken and seed germination and vegetative growth begin.

## Seed Germination

Seed germination is initiated by **imbibition**, the uptake of water due to the low water potential of the dry seed. Imbibition causes the seed to expand and rupture its coat and triggers changes in the embryo that enable it to resume growth. Following hydration, enzymes digest the storage materials of the endosperm or cotyledons, and the nutrients are transferred to the growing regions of the embryo.

The first organ to emerge from the germinating seed is the radicle, the embryonic root. The development of a root system anchors the seedling in the soil and supplies it with water necessary for cell expansion. A ready supply of water is a prerequisite for the next step, the emergence of the shoot tip into the drier conditions encountered above ground. In garden beans and many other eudicots, a hook forms in the hypocotyl, and growth pushes the hook above ground **(Figure 30.9a)**. In

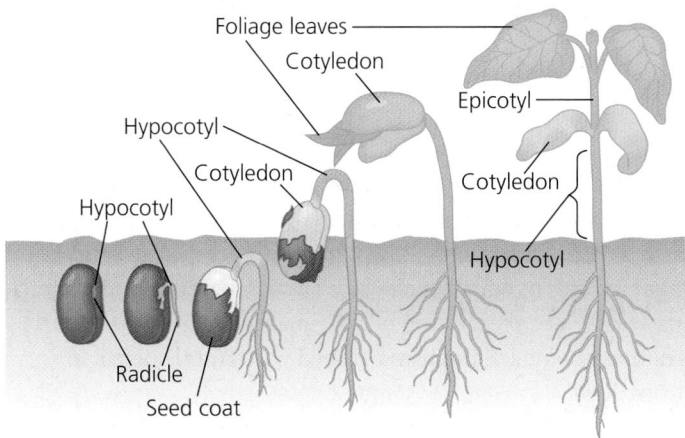

**(a) Common garden bean.** In common garden beans, straightening of a hook in the hypocotyl pulls the cotyledons from the soil.

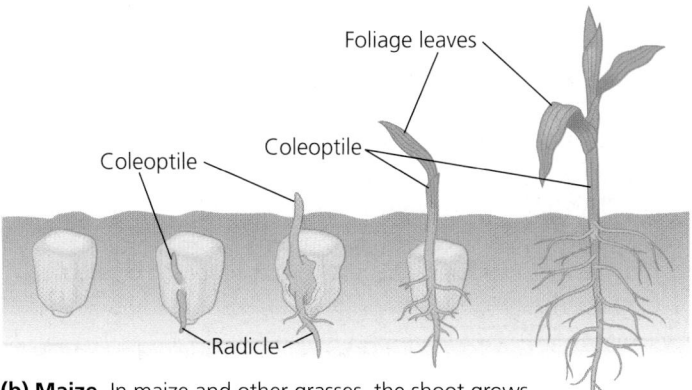

**(b) Maize.** In maize and other grasses, the shoot grows straight up through the tube of the coleoptile.

▲ **Figure 30.9 Two common types of seed germination.**

? *How do bean and maize seedlings protect their shoot systems as they push through the soil?*

response to light, the hypocotyl straightens, the cotyledons separate, and the delicate epicotyl, now exposed, spreads its first true leaves (as distinct from the cotyledons, or seed leaves). These leaves expand, become green, and begin making food by photosynthesis. The cotyledons shrivel and fall away, their food reserves having been exhausted by the germinating embryo.

Some monocots, such as maize and other grasses, use a different method for breaking ground when they germinate **(Figure 30.9b)**. The coleoptile pushes up through the soil and into the air. The shoot tip grows through the tunnel provided by the coleoptile and breaks through the coleoptile's tip.

## Vegetative Growth and Flowering

Once a seed has germinated and started to photosynthesize, most of the plant's resources are devoted to vegetative growth. Vegetative growth, including both primary and secondary growth, arises from the activity of meristematic cells (see Concept 28.2). During this stage, usually the best strategy is to photosynthesize and grow as much as possible before the reproductive phase (flowering).

The flowers of a given plant species typically appear suddenly and simultaneously at a specific time of year. Such synchrony promotes outbreeding, the main advantage of sexual reproduction. Flower formation involves a developmental switch in the shoot apical meristem from a vegetative to a reproductive growth mode. This transition into a *floral meristem* is triggered by a combination of environmental cues (such as day length) and internal signals, as you'll learn in Concept 31.2. Once the transition to flowering has begun, the order of each organ's emergence from the floral meristem determines whether it will develop into a sepal, petal, stamen, or carpel (see Figure 30.4).

## Fruit Structure and Function

Before a seed can germinate and develop into a mature plant, it must be deposited in suitable soil. Fruits play a key role in this process. A **fruit** is the mature ovary of a flower. It protects the enclosed seeds and, when mature, aids in their dispersal.

Fertilization triggers hormonal changes that cause the ovary to begin its transformation into a fruit. If a flower has not been pollinated, fruit typically does not develop, and the flower usually withers and falls away. During fruit development, the ovary wall becomes the *pericarp*, the thickened wall of the fruit. In some fruits, such as soybean pods, the ovary wall dries out completely at maturity, whereas in other fruits, such as grapes, it remains fleshy. In still others, such as peaches, the inner part of the ovary becomes stony (the pit) while the outer parts stay fleshy. As the ovary grows, the other parts of the flower usually wither and are shed.

Fruits are classified into several types, depending on their developmental origin. Most fruits are derived from a single

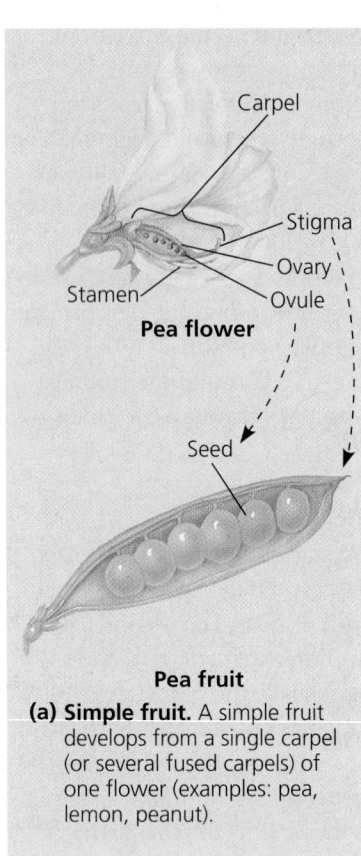

Carpel
Stigma
Ovary
Stamen
Ovule

**Pea flower**

Seed

**Pea fruit**

**(a) Simple fruit.** A simple fruit develops from a single carpel (or several fused carpels) of one flower (examples: pea, lemon, peanut).

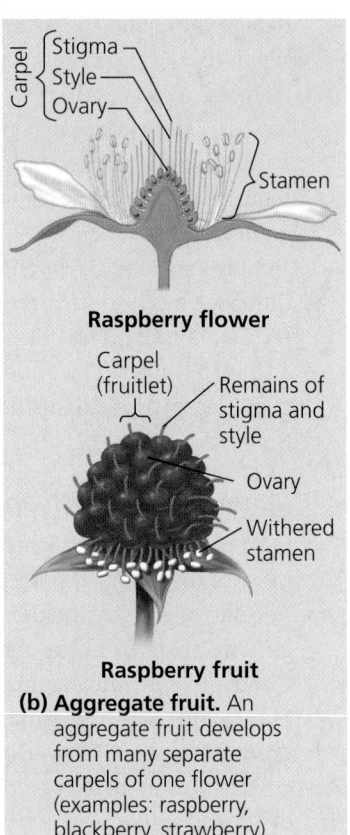

Carpel
Stigma
Style
Ovary
Stamen

**Raspberry flower**

Carpel (fruitlet)
Remains of stigma and style
Ovary
Withered stamen

**Raspberry fruit**

**(b) Aggregate fruit.** An aggregate fruit develops from many separate carpels of one flower (examples: raspberry, blackberry, strawberry).

Flowers

**Pineapple inflorescence**

Each segment develops from the carpel of one flower

**Pineapple fruit**

**(c) Multiple fruit.** A multiple fruit develops from many carpels of the many flowers that form an inflorescence (examples: pineapple, fig).

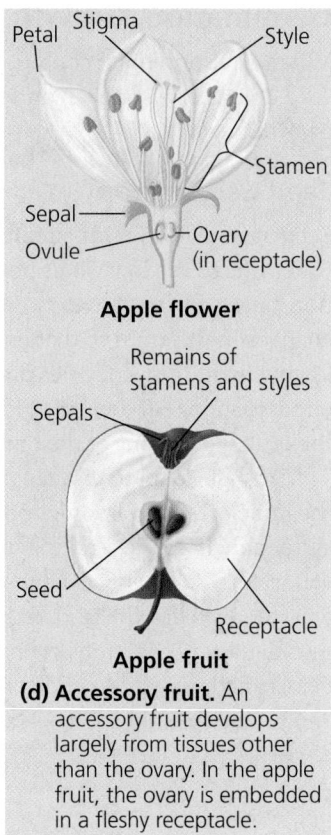

Petal
Stigma
Style
Stamen
Sepal
Ovule
Ovary (in receptacle)

**Apple flower**

Remains of stamens and styles
Sepals
Seed
Receptacle

**Apple fruit**

**(d) Accessory fruit.** An accessory fruit develops largely from tissues other than the ovary. In the apple fruit, the ovary is embedded in a fleshy receptacle.

▲ **Figure 30.10 Developmental origins of fruits.**

carpel or several fused carpels and are called **simple fruits** (**Figure 30.10a**). An **aggregate fruit** results from a single flower that has more than one separate carpel, each forming a small fruit (**Figure 30.10b**). These "fruitlets" are clustered together on a single receptacle, as in a raspberry. A **multiple fruit** develops from an inflorescence, a group of flowers tightly clustered together. When the walls of the many ovaries start to thicken, they fuse together and become incorporated into one fruit, as in a pineapple (**Figure 30.10c**).

In some angiosperms, other floral parts contribute to what we commonly call the fruit. Such fruits are called **accessory fruits**. In apple flowers, the ovary is embedded in the receptacle, and the fleshy part of this simple fruit is derived mainly from the enlarged receptacle; only the apple core develops from the ovary (**Figure 30.10d**). Another example is the strawberry, an aggregate fruit consisting of an enlarged receptacle studded with tiny, partially embedded fruits, each bearing a single seed.

A fruit usually ripens about the same time that its seeds complete their development. Whereas the ripening of a dry fruit, such as a soybean pod, involves the aging and drying out of fruit tissues, the process in a fleshy fruit is more elaborate. Complex interactions of hormones result in an edible fruit that entices animals that disperse the seeds. The fruit's "pulp"

becomes softer as enzymes digest components of cell walls. The color usually changes from green to a more overt color, such as red, orange, or yellow. The fruit becomes sweeter as organic acids or starch molecules are converted to sugar, which may reach a concentration of 20% in a ripe fruit. **Figure 30.11** examines some mechanisms of seed and fruit dispersal in more detail.

In this section, you have learned about the key features of sexual reproduction in angiosperms—flowers, double fertilization, and fruits. Next, we'll examine asexual reproduction.

**CONCEPT CHECK 30.1**

1. **WHAT IF?** If flowers had shorter styles, pollen tubes would more easily reach the embryo sac. Suggest an explanation for why very long styles have evolved in most flowering plants.

2. **WHAT IF?** In some species, sepals look like petals, and both are collectively called "tepals." Suggest a possible extension to the ABC hypothesis that could account for the origin of tepals.

3. **MAKE CONNECTIONS** Does the life cycle of animals have any structures analogous to plant gametophytes? Explain your answer (see Figure 10.6).

For suggested answers, see Appendix A.

A plant's life depends on finding fertile ground. But a seed that falls and sprouts beneath the parent plant will stand little chance of competing successfully for nutrients. To prosper, seeds must be widely dispersed. Plants use biotic dispersal agents as well as abiotic agents such as water and wind.

## Dispersal by Water

▶ Some buoyant seeds and fruits can survive months or years at sea. In coconut, the seed embryo and fleshy white "meat" (endosperm) are within a hard layer (endocarp) surrounded by a thick and buoyant fibrous husk.

## Dispersal by Wind

▶ With a wingspan of 12 cm, the giant seed of the tropical Asian climbing gourd *Alsomitra macrocarpa* glides through the air of the rain forest in wide circles when released.

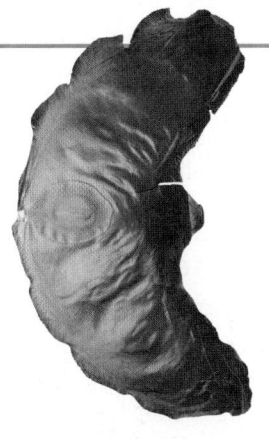

▼ The winged fruit of a maple spins like a helicopter blade, slowing descent and increasing the chance of being carried farther by horizontal winds.

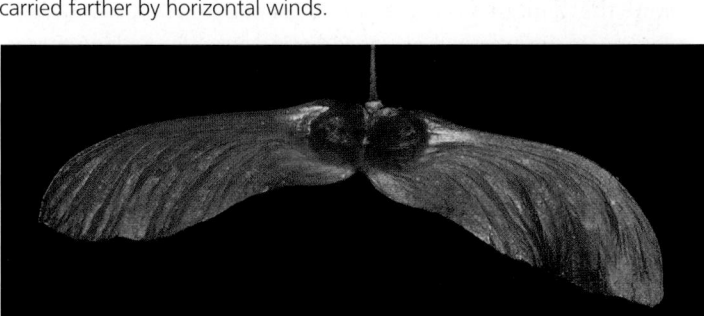

Dandelion fruit

▶ Tumbleweeds break off at the ground and tumble across the terrain, scattering their seeds.

▲ Some seeds and fruits are attached to umbrella-like "parachutes" that are made of intricately branched hairs and often produced in puffy clusters. These dandelion "seeds" (actually one-seeded fruits) are carried aloft by the slightest gust of wind.

## Dispersal by Animals

◀ The sharp, tack-like spines on the fruits of puncture vine (*Tribulus terrestris*) can pierce bicycle tires and injure animals, including humans. When these painful "tacks" are removed and discarded, the seeds are dispersed.

◀ Some animals, such as squirrels, hoard seeds or fruits in underground caches. If the animal dies or forgets the cache's location, the buried seeds are well positioned to germinate.

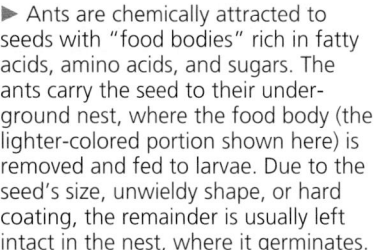

▶ Seeds in edible fruits are often dispersed in feces, such as the black bear feces shown here. Such dispersal may carry seeds far from the parent plant.

▶ Ants are chemically attracted to seeds with "food bodies" rich in fatty acids, amino acids, and sugars. The ants carry the seed to their underground nest, where the food body (the lighter-colored portion shown here) is removed and fed to larvae. Due to the seed's size, unwieldy shape, or hard coating, the remainder is usually left intact in the nest, where it germinates.

# Flowering plants reproduce sexually, asexually, or both

Imagine chopping off your finger and watching it develop into an exact copy of you. If this could actually occur, it would be an example of **asexual reproduction**, in which offspring are derived from a single parent without fusion of egg and sperm. The result would be a clone, an asexually produced, genetically identical organism. Asexual reproduction is common in angiosperms, as well as in other plants, and for some plant species it is the predominant mode of reproduction.

## Mechanisms of Asexual Reproduction

Asexual reproduction in plants is typically an extension of the capacity for indeterminate growth. Plant growth can be sustained or renewed indefinitely by meristems, regions of undifferentiated, dividing cells (see Concept 28.2). In addition, parenchyma cells throughout the plant can divide and differentiate into more specialized types of cells, enabling plants to regenerate lost parts. Detached vegetative fragments of some plants can develop into whole offspring; for example, pieces of a potato with an "eye" (vegetative bud) can each regenerate a whole plant. Such **fragmentation**, the separation of a parent plant into parts that develop into whole plants, is one of the most common modes of asexual reproduction. The adventitious plantlets on *Kalanchoë* leaves exemplify an unusual type of fragmentation (see Figure 28.7). In other cases, the root system of a single parent, such as an aspen tree, can give rise to many adventitious shoots that become separate shoot systems **(Figure 30.12)**. One aspen clone in Utah has been estimated to be composed of 47,000 stems of genetically identical trees. Although it is likely that some of the root system connections have been severed, making some of the trees isolated from the rest of the clone, each tree still shares a common genome.

An entirely different mechanism of asexual reproduction has evolved in dandelions and some other plants. These plants can sometimes produce seeds without pollination or fertilization. This asexual production of seeds is called **apomixis** (from the Greek words meaning "away from the act of mixing") because there is no joining or, indeed, production of sperm and egg. Instead, a diploid cell in the ovule gives rise to the embryo, and the ovules mature into seeds, which in the dandelion are dispersed by windblown fruits. Thus, these plants clone themselves by an asexual process but have the advantage of seed dispersal, usually associated with sexual reproduction. Introducing apomixis into hybrid crops is of great interest to plant breeders because apomixis would allow hybrid plants to pass on their desirable genomes intact to their offspring.

## Advantages and Disadvantages of Asexual Versus Sexual Reproduction

**EVOLUTION** An advantage of asexual reproduction is that there is no need for a pollinator. This may be beneficial if plants of the same species are sparsely distributed and unlikely to be visited by the same pollinator. Asexual reproduction also allows the plant to pass on all of its genetic legacy intact to its progeny. In contrast, when reproducing sexually, a plant passes on only half of its alleles. If a plant is superbly suited to its environment, asexual reproduction can be advantageous. A vigorous plant can potentially clone many copies of itself, and if the environmental circumstances remain stable, these offspring will also be genetically well adapted to the same environmental conditions under which the parent flourished.

Generally, the progeny produced by asexual reproduction are stronger than seedlings produced by sexual reproduction. The offspring usually arise from mature vegetative fragments from the parent plant, which is why asexual reproduction in plants is also known as **vegetative reproduction**. In contrast, seed germination is a precarious stage in a plant's life. The tough seed gives rise to a fragile seedling that may face exposure to predation, competition, and other hazards. In the wild, only a small fraction of seedlings survive to become parents themselves. Production of enormous numbers of seeds compensates for the odds against individual survival and gives natural selection ample genetic variations to screen. However, this is an expensive means of reproduction in terms of the resources consumed in flowering and fruiting.

Because sexual reproduction generates variation in offspring and populations, it can be advantageous in unstable environments where evolving pathogens and other fluctuating conditions affect survival and reproductive success. In contrast, the genotypic uniformity of asexually produced plants puts them at great risk of local extinction if there is a catastrophic environmental change, such as a new strain of disease. Moreover, seeds (which are almost always produced

▲ **Figure 30.12 Asexual reproduction in aspen trees.** Some aspen groves, such as those shown here, consist of thousands of trees that are the result of asexual reproduction. Each grove of trees derives from the root system of one parent. Thus, the grove is a clone. Notice that genetic differences between groves descended from different parents result in different timing for the development of fall color.

sexually) facilitate the dispersal of offspring to more distant locations. Finally, seed dormancy allows growth to be suspended until environmental conditions become more favorable. In the **Scientific Skills Exercise**, you can use data to determine which species of monkey flower are mainly asexual reproducers and which are mainly sexual reproducers.

Although sexual reproduction involving two genetically different plants has the benefit of producing the most genetically diverse offspring, some plants, such as garden peas, usually self-fertilize. This process, called "selfing," is a desirable attribute in some crop plants because it ensures that every ovule will develop into a seed. In many angiosperm species, however, mechanisms have evolved that make it difficult or impossible for a flower to fertilize itself, as we'll discuss next.

## Mechanisms That Prevent Self-Fertilization

The various mechanisms that prevent self-fertilization contribute to genetic variety by ensuring that the sperm and egg come from different parents. In the case of **dioecious** species, plants cannot self-fertilize because different individuals have either staminate flowers (lacking carpels) or carpellate flowers (lacking stamens) **(Figure 30.13a)**. Other plants have flowers with functional stamens and carpels that mature at different times or are structurally arranged in such a way that it is unlikely that an animal pollinator could transfer pollen from an anther to a stigma of the same flower **(Figure 30.13b)**. However, the most common anti-selfing mechanism in flowering plants is **self-incompatibility**, the ability of a plant to reject its own pollen and the pollen of closely related individuals. If a pollen grain lands on a stigma of a flower of the same or a closely related individual, a biochemical block prevents the pollen from completing its development and fertilizing an egg.

Researchers are unraveling the molecular mechanisms involved in self-incompatibility. Recognition of "self" pollen is based on genes for self-incompatibility, called $S$-genes. In the gene pool of a plant population, there can be dozens of alleles of an $S$-gene. If a pollen grain has an allele that matches an allele of the stigma on which it lands, either the pollen fails to germinate or it germinates but its tube fails to grow through the style to the ovary.

There are two types of self-incompatibility—gametophytic and sporophytic. In gametophytic self-incompatibility, the $S$-allele in the pollen genome governs the blocking of fertilization. For example, an $S_1$ pollen grain from an $S_1S_2$ parental sporophyte cannot fertilize eggs of an $S_1S_2$ flower but can fertilize an $S_2S_3$ flower. An $S_2$ pollen grain cannot fertilize either flower. In some plant families, self-incompatibility of this kind involves the enzymatic destruction of RNA within a pollen tube. RNA-hydrolyzing enzymes are produced by the style and enter the pollen tube. If the pollen tube is a "self" type, these enzymes destroy its RNA.

In sporophytic self-incompatibility, fertilization is blocked by $S$-allele gene products in tissues of the parental sporophyte

**(a)** Some species, such as *Sagittaria latifolia* (common arrowhead), are dioecious, having plants that produce only staminate flowers (left) or carpellate flowers (right).

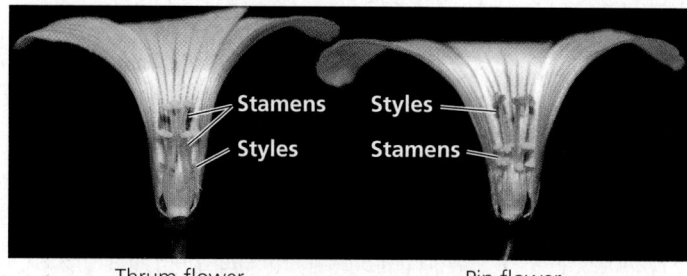

Thrum flower          Pin flower

**(b)** Some species, such as *Oxalis alpina* (alpine woodsorrel), produce two types of flowers on different individuals: "thrums," which have short styles and long stamens, and "pins," which have long styles and short stamens. An insect foraging for nectar would collect pollen on different parts of its body; thrum pollen would be deposited on pin stigmas, and vice versa.

▲ **Figure 30.13  Some floral adaptations that prevent self-fertilization.**

that adhere to the pollen grain wall. For example, neither an $S_1$ nor $S_2$ pollen grain from an $S_1S_2$ parental sporophyte can fertilize eggs of an $S_1S_2$ flower or $S_2S_3$ flower due to the $S_1S_2$ parental tissue attached to the pollen wall. Sporophytic incompatibility involves a signal transduction pathway in epidermal cells of the stigma that prevents germination of the pollen grain.

Plant breeders frequently hybridize different varieties of a crop plant to combine the best traits of the varieties and counter the loss of vigor that can often result from excessive inbreeding. To obtain hybrid seeds, plant breeders prevent self-fertilization either by laboriously removing the anthers from the parent plants that provide the seeds (as Mendel did) or by developing male-sterile plants. The latter option is increasingly common. Eventually, it may also be possible to impose self-incompatibility genetically on crop species that are normally self-compatible. Basic research on the mechanisms of self-incompatibility may thus have agricultural applications.

## Totipotency, Vegetative Reproduction, and Tissue Culture

In a multicellular organism, any cell that can divide and asexually generate a clone of the original organism is **totipotent**. Totipotency is found to a high degree in many plants and is generally associated with meristematic tissues. In some plants,

# Using Positive and Negative Correlations to Interpret Data

**Do Monkey Flower Species Differ in Allocating Energy to Sexual Versus Asexual Reproduction?** Over the course of its life span, a plant captures only a finite amount of resources and energy, which must be allocated to best meet the plant's individual requirements for maintenance, growth, defense, and reproduction. Researchers examined how five species of monkey flower (genus *Mimulus*) use their resources for sexual and asexual reproduction.

**How the Experiment Was Done** After growing specimens of each species in separate pots in the open, the researchers determined averages for nectar volume, nectar concentration, seeds produced per flower, and the number of times the plants were visited by broad-tailed

hummingbirds (*Selasphorus platycercus*). Using greenhouse-grown specimens, they determined the average number of rooted branches per gram fresh shoot weight for each species. The phrase *rooted branches* refers to asexual reproduction through horizontal shoots that develop roots.

**INTERPRET THE DATA**

1. A correlation is a way to describe the relationship between two variables. In a positive correlation, as the values of one of the variables increase, the values of the second variable also increase. In a negative correlation, as the values of one of the variables increase, the values of the second variable decrease. Or there may be no correlation between two variables. If researchers know how two variables are correlated, they can make a prediction about one variable based on what they know about the other variable. (a) Which variable(s) is (are) positively correlated with the volume of nectar production in this genus? (b) Which is (are) negatively correlated? (c) Which show(s) no clear relationship?

2. (a) Which *Mimulus* species would you categorize as mainly asexual reproducers? Why? (b) Which species would you categorize as mainly sexual reproducers? Why?

3. (a) Which species would probably fare better in response to a pathogen that infects all *Mimulus* species? (b) Which species would fare better if a pathogen were to cause hummingbird populations to dwindle? Explain.

(MB) A version of this Scientific Skills Exercise can be assigned in MasteringBiology.

**Data from the Experiment**

| Species | Nectar Volume (µL) | Nectar Concentration (% wt of sucrose/ total wt) | Seeds per Flower | Visits per Flower | Rooted Branches per Gram Shoot Weight |
|---|---|---|---|---|---|
| *M. rupestris* | 4.93 | 16.6 | 2.2 | 0.22 | 0.673 |
| *M. eastwoodiae* | 4.94 | 19.8 | 25.0 | 0.74 | 0.488 |
| *M. nelsonii* | 20.25 | 17.1 | 102.5 | 1.08 | 0.139 |
| *M. verbenaceus* | 38.96 | 16.9 | 155.1 | 1.26 | 0.091 |
| *M. cardinalis* | 50.00 | 19.9 | 283.7 | 1.75 | 0.069 |

**Data from** S. Sutherland and R. K. Vickery, Jr., Trade-offs between sexual and asexual reproduction in the genus *Mimulus*, *Oecologia* 76:330–335 (1988).

however, even differentiated cells can dedifferentiate and become meristematic.

## Vegetative Propagation and Grafting

Vegetative reproduction occurs naturally in many plants, but it can often be facilitated or induced by humans, in which case it is called **vegetative propagation**. Most houseplants, woody ornamentals, and orchard trees, for example, are asexually reproduced from plant fragments called cuttings. In most cases, shoot cuttings are used. At the cut end of the shoot, a mass of dividing, undifferentiated cells called a **callus** forms, and adventitious roots then develop from the callus. If the shoot fragment includes a node, then adventitious roots form without a callus stage.

In grafting, a twig or bud from one plant is joined to a plant of a closely related species or a different variety of the same species. Grafting makes it possible to combine the best qualities of different species or varieties into a single plant. The plant that provides the root system is called the **stock**; the twig grafted onto the stock is referred to as the **scion**. For example, scions from French varieties of vines that produce superior wine grapes are grafted onto rootstocks of American varieties

that produce inferior grapes but are more resistant to certain soil pathogens. The genes of the scion determine the quality of the fruit. During grafting, a callus first forms between the adjoining cut ends of the scion and stock; cell differentiation then completes the functional unification of the grafted individuals.

## Test-Tube Cloning and Related Techniques

Plant biologists have adopted *in vitro* methods to clone plants for research or horticulture. Whole plants can be obtained by culturing small pieces of tissue from the parent plant on an artificial medium containing nutrients and hormones. The cells or tissues can come from any part of a plant, but growth may vary depending on the plant part, species, and artificial medium. In some media, the cultured cells divide and form a callus of undifferentiated cells **(Figure 30.14a)**. When the concentrations of hormones and nutrients are manipulated appropriately, a callus can sprout shoots and roots with fully differentiated cells **(Figure 30.14b and c)**. If desired, the plantlets can then be transferred to soil, where they continue their growth. A single plant can be cloned into thousands of genetically identical copies by dividing calluses as they grow.

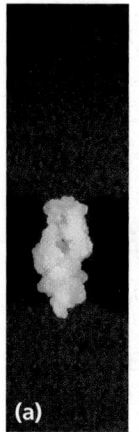

**(a)** **(b)** **(c)** Developing root

▲ **Figure 30.14 Laboratory cloning of a garlic plant.**
**(a)** A root from a garlic clove gave rise to this callus culture, a mass of undifferentiated cells. **(b** and **c)** The differentiation of a callus into a plantlet depends on the nutrient levels and hormone concentrations in the artificial medium, as can be seen in these cultures grown for different lengths of time.

Plant tissue culture is important in eliminating weakly pathogenic viruses from vegetatively propagated varieties. Although the presence of weak viruses may not be obvious, yield or quality may be substantially reduced as a result of infection. Strawberry plants, for example, are susceptible to more than 60 viruses, and typically the plants must be replaced each year because of viral infection. However, the distribution of viruses in a plant is not uniform, and the apical meristems are sometimes virus-free. Therefore, apical meristems can be excised and used to produce virus-free material for tissue culture.

Plant tissue culture also facilitates genetic engineering. Most techniques for the introduction of foreign genes into plants require small pieces of plant tissue or single plant cells as the starting material. Test-tube culture makes it possible to regenerate genetically modified (GM) plants from a single plant cell into which the foreign DNA has been incorporated. In the next section, we'll take a closer look at some of the promises and challenges surrounding the use of GM plants in agriculture.

**CONCEPT CHECK 30.2**

1. What are three ways that flowering plants avoid self-fertilization?
2. The seedless banana, the world's most popular fruit, is losing the battle against two fungal epidemics. Why do such epidemics generally pose a greater risk to asexually propagated crops?
3. Self-fertilization, or selfing, seems to have obvious disadvantages as a reproductive "strategy" in nature, and it has even been called an "evolutionary dead end." So it is surprising that about 20% of angiosperm species primarily rely on selfing. Suggest a reason why selfing might be advantageous and still be an evolutionary dead end.

For suggested answers, see Appendix A.

CONCEPT 30.3

# People modify crops through breeding and genetic engineering

People have intervened in the reproduction and genetic makeup of plants since the dawn of agriculture. Maize, for example, owes its existence to humans. Left on its own in nature, maize would soon become extinct for the simple reason that it cannot spread its seeds. Maize kernels are not only permanently attached to the central axis (the "cob") but also permanently protected by tough, overlapping leaf sheaths (the "husk") **(Figure 30.15)**. These attributes arose by artificial selection by humans. (See Concept 19.2 to review the basic concept of artificial selection.) Despite having no understanding of the scientific principles underlying plant breeding, early farmers domesticated most of our crop species over a relatively short period about 10,000 years ago. But genetic modification began long before people started altering crops by artificial selection. For example, the wheat species we rely on for much of our food evolved by the natural hybridization between different species of grasses. Such hybridization is common in plants and has long been exploited by breeders to introduce genetic variation for artificial selection and crop improvement.

## Plant Breeding

Plant breeding is the art and science of changing the traits of plants in order to produce desired characteristics. Breeders scrutinize their fields carefully and travel far and wide searching for domesticated varieties or wild relatives with desirable traits. Such traits occasionally arise spontaneously through mutation, but the natural rate of mutation is too slow and unreliable to produce all the mutations that breeders would like to study. Breeders sometimes hasten mutations by treating large batches of seeds or seedlings with radiation or chemicals.

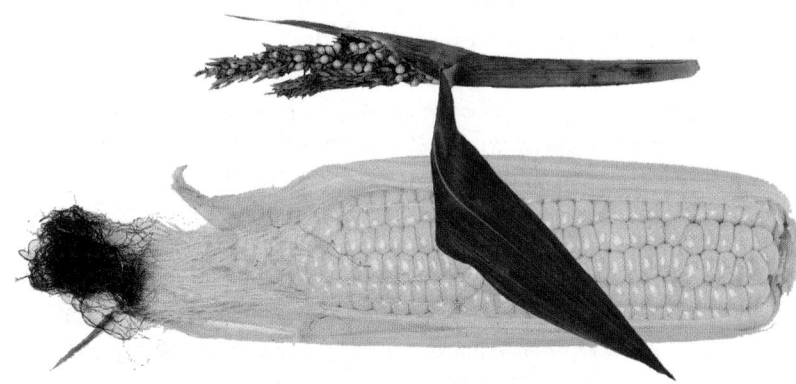

▲ **Figure 30.15 Maize: a product of artificial selection.**
Modern maize (bottom) was derived from teosinte (top). Teosinte kernels are tiny, and each row has a husk that must be removed to get at the kernel. The seeds are loose at maturity, allowing dispersal, which probably made harvesting difficult for early farmers. Ancient farmers selected seeds from plants with larger cob and kernel size as well as the permanent attachment of seeds to the cob and the encasing of the entire cob by a tough husk.

In traditional plant breeding, when a desirable trait is identified in a wild species, the wild species is crossed with a domesticated variety. Generally, those progeny that have inherited the desirable trait from the wild parent have also inherited many traits that are not desirable for agriculture, such as small fruits or low yields. The progeny that express the desired trait are again crossed with members of the domesticated species and their progeny examined for the desired trait. This process is continued until the progeny with the desired wild trait resemble the original domesticated parent in their other agricultural attributes.

While most breeders cross-pollinate plants of a single species, some breeding methods rely on hybridization between two distant species of the same genus. Such crosses sometimes result in the abortion of the hybrid seed during development. Often in these cases, the embryo begins to develop, but the endosperm does not. Hybrid embryos are sometimes rescued by surgically removing them from the ovule and culturing them *in vitro*.

## Plant Biotechnology and Genetic Engineering

Plant biotechnology has two meanings. In the general sense, it refers to innovations in the use of plants (or substances obtained from plants) to make products of use to people—an endeavor that began in prehistory. In a more specific sense, biotechnology refers to the use of GM organisms in agriculture and industry. Indeed, in the last two decades, genetic engineering has become such a powerful force that the terms *genetic engineering* and *biotechnology* have become synonymous in the media.

Unlike traditional plant breeders, modern plant biotechnologists, using techniques of genetic engineering, are not limited to the transfer of genes between closely related species or genera. For example, traditional breeding techniques could not be used to insert a desired gene from daffodil into rice because the many intermediate species between rice and daffodil and their common ancestor are extinct. In theory, if breeders had the intermediate species, over the course of several centuries they could probably introduce a daffodil gene into rice by traditional hybridization and breeding methods. With genetic engineering, however, such gene transfers can be done more quickly, more specifically, and without the need for intermediate species. The term **transgenic** is used to describe organisms that have been engineered to express a gene from another species.

In the remainder of this chapter, we examine the prospects and controversies surrounding the use of GM crops. The advocates of plant biotechnology contend that the genetic engineering of crop plants is the key to overcoming some of the most pressing problems of the 21st century, including world hunger and fossil fuel dependency.

### *Reducing World Hunger and Malnutrition*

Currently, 800 million people suffer from nutritional deficiencies, with 40,000 dying each day of malnutrition, half of them children. There is much disagreement about the causes of such hunger. Some argue that food shortages arise from inequities in distribution and that the dire poor simply cannot afford food. Others regard food shortages as evidence that the world is overpopulated—that the human species has exceeded the carrying capacity of the planet (see Concept 40.5). Whatever the social and demographic causes of malnutrition, increasing food production is a humane objective. Because land and water are the most limiting resources, the best option is to increase yields on already existing farmland. Indeed, there is very little "extra" land that can be farmed, especially if the few remaining pockets of wilderness are to be preserved. Based on conservative estimates of population growth, farmers will have to produce 40% more grain per hectare to feed the human population in 2030. Plant biotechnology can help make these crop yields possible.

The commercial use of transgenic crops has been one of the most dramatic examples of rapid technology adoption in the history of agriculture. These crops include varieties and hybrids of cotton, maize, and potatoes that contain genes from the bacterium *Bacillus thuringiensis*. These "transgenes" encode a protein (*Bt* toxin) that is toxic to insect pests. The use of such plant varieties greatly reduces the need for chemical insecticides. The *Bt* toxin used in crops is produced in the plant as a harmless protoxin that only becomes toxic if activated by alkaline conditions, such as occur in the guts of insects. Because vertebrates have highly acidic stomachs, protoxin consumed by humans or farm animals is rendered harmless by denaturation.

Considerable progress has also been made in developing transgenic crops that tolerate certain herbicides. The cultivation of these plants may reduce production costs by enabling farmers to "weed" crops with herbicides that do not damage the transgenic crop plants, instead of using heavy tillage, which can cause soil erosion. Researchers are also engineering plants with enhanced resistance to disease. In one case, a transgenic papaya that is resistant to a ring spot virus was introduced into Hawaii, thereby saving its papaya industry.

The nutritional quality of plants is also being improved. For example, some 250,000 to 500,000 children go blind each year because of vitamin A deficiencies. More than half of these children die within a year of becoming blind. In response to this crisis, genetic engineers have created "Golden Rice," a transgenic variety supplemented with transgenes that enable it to produce grain with increased levels of beta-carotene, a precursor of vitamin A. This genetically modified rice, named for its yellow color from the beta-carotene, is still undergoing testing. Another target for improvement by genetic engineering is cassava, a staple for 800 million of the poorest people on our planet **(Figure 30.16)**.

### *Reducing Fossil Fuel Dependency*

Global sources of inexpensive fossil fuels, particularly oil, are rapidly being depleted. Moreover, most climatologists attribute global warming mainly to the rampant burning of fossil fuels,

► **Figure 30.16 Fighting world hunger with transgenic cassava (*Manihot esculenta*).** This starchy root crop is the primary food for 800 million of the world's poor, but it does not provide a balanced diet. Moreover, it must be processed to remove chemicals that release cyanide, a toxin. Transgenic cassava plants have been developed with greatly increased levels of iron and beta-carotene (a vitamin A precursor). Researchers have also created cassava plants with root masses twice the normal size that contain almost no cyanide-producing chemicals.

such as coal and oil, and the resulting release of the greenhouse gas $CO_2$. How can the world meet its energy demands in the 21st century in an economical and nonpolluting way? In certain localities, wind or solar power may become economically viable, but such alternative energy sources are unlikely to fill the global energy demands completely. Many scientists predict that **biofuels**—fuels derived from living biomass—could produce a sizable fraction of the world's energy needs in the not-too-distant future. **Biomass** is the total mass of organic matter in a group of organisms in a particular habitat. The use of biofuels from plant biomass would reduce the net emission of $CO_2$. Whereas burning fossil fuels increases atmospheric $CO_2$ concentrations, biofuel crops reabsorb by photosynthesis the $CO_2$ emitted when biofuels are burned, creating a cycle that is carbon neutral.

In working to create biofuel crops from wild precursors, scientists are focusing their domestication efforts on fast-growing plants, such as switchgrass (*Panicum virgatum*) and poplar (*Populus trichocarpa*) that can grow on soil that is too poor for food production. Scientists do not envisage the plant biomass being burned directly. Instead, the polymers in cell walls, such as cellulose and hemicellulose, which constitute the most abundant organic compounds on Earth, would be broken down into sugars by enzymatic reactions. These sugars, in turn, would be fermented into alcohol, which would be distilled to yield biofuels. In addition to increasing plant polysaccharide content and overall biomass, researchers are trying to genetically engineer plants with cell wall properties, such as reduced lignin content, that will lower the costs of biofuel production.

## The Debate over Plant Biotechnology

Much of the debate about GM organisms (GMOs) in agriculture is political, social, economic, or ethical and therefore outside the scope of this book. But we *should* consider the biological concerns about GM crops. Some biologists, particularly ecologists, are concerned about the unknown risks associated

with the release of GMOs into the environment. The debate centers on the extent to which GMOs could harm the environment or human health. Those who want to proceed more slowly with agricultural biotechnology (or end it) are concerned about the unstoppable nature of the "experiment." If a drug trial produces unanticipated harmful results, the trial is stopped. But we may not be able to stop the "trial" of introducing novel organisms into the biosphere. Here, we examine some of the proposed negative consequences of using GM crops, including their effects on human health and nontarget organisms and the potential for transgene escape.

### Issues of Human Health

Many GMO opponents worry that genetic engineering may inadvertently transfer allergens, molecules to which some people are allergic, from a species that produces an allergen to a plant used for food. However, biotechnologists are already engaged in removing genes that encode allergenic proteins from soybeans and other crops. So far, there is no credible evidence that GM plants specifically designed for human consumption have adverse effects on human health. In fact, some GM foods are potentially healthier than non-GM foods. For example, *Bt* maize (the transgenic variety with the *Bt* toxin) contains 90% less of a fungal toxin that causes cancer and birth defects than non-*Bt* maize. Called fumonisin, this toxin is highly resistant to degradation and has been found in alarmingly high concentrations in some batches of processed maize products, ranging from cornflakes to beer. Fumonisin is produced by a fungus (*Fusarium*) that infects insect-damaged maize. Because *Bt* maize generally suffers less insect damage than non-GM maize, it contains much less fumonisin.

Nevertheless, because of health concerns, GMO opponents lobby for the clear labeling of all foods containing products of GMOs. Some also argue for strict regulations against the mixing of GM foods with non-GM foods during food transport, storage, and processing. Biotechnology advocates, however, note that similar demands were not made when "transgenic" crops produced by traditional plant-breeding techniques were put on the market. There are, for example, some commercially grown varieties of wheat derived by traditional plant-breeding techniques that contain entire chromosomes (and thousands of genes) from rye.

### Possible Effects on Nontarget Organisms

Many ecologists are concerned that the growing of GM crops might have unforeseen effects on nontarget organisms. One laboratory study indicated that the larvae (caterpillars) of monarch butterflies responded adversely and even died after eating milkweed leaves (their preferred food) heavily dusted with pollen from transgenic *Bt* maize. This study has since been discredited, affording a good example of the self-correcting nature of science. As it turns out, when the original researcher shook the male maize inflorescences onto the milkweed leaves in the laboratory, the filaments of stamens, opened microsporangia, and other floral parts also rained onto the leaves. Subsequent research found that it was these other floral parts, *not* the pollen, that contained *Bt* toxin in high concentrations. Unlike pollen, these floral parts would not be carried by the wind to neighboring milkweed plants when shed under natural field conditions. Only one *Bt* maize line, accounting for less than 2% of commercial *Bt* maize production (and now discontinued), produced pollen with high *Bt* toxin concentrations.

In considering the negative effects of *Bt* pollen on monarch butterflies, we must also weigh the effects of an alternative to the cultivation of *Bt* maize—the spraying of non-*Bt* maize with chemical pesticides. Subsequent studies have revealed that chemical spraying is much more harmful to nearby monarch populations than is *Bt* maize production. Although the effects of *Bt* maize pollen on monarch butterfly larvae appear to be minor, the controversy has emphasized the need for accurate field testing of all GM crops and the importance of targeting gene expression to specific tissues to improve safety.

### Addressing the Problem of Transgene Escape

Perhaps the most serious concern raised about GM crops is the possibility of the introduced genes escaping from a transgenic crop into related weeds through crop-to-weed hybridization. The fear is that the spontaneous hybridization between a crop engineered for herbicide resistance and a wild relative might give rise to a "superweed" that would have a selective advantage over other weeds in the wild and would be much more difficult to control in the field. GMO advocates point out that the likelihood of transgene escape depends on the ability of the crop and weed to hybridize and on how the transgenes affect the overall fitness of the hybrids. A desirable crop trait—a dwarf phenotype, for example—might be disadvantageous to a weed growing in the wild. In other instances, there are no weedy relatives nearby with which to hybridize; soybean, for example, has no wild relatives in the United States. However, canola, sorghum, and many other crops do hybridize readily with weeds, and crop-to-weed transgene escape has occurred. In 2003 a transgenic variety of creeping bentgrass (*Agrostis stolonifera*) genetically engineered to resist the herbicide glyphosate escaped from an experimental plot in Oregon following a windstorm. Despite efforts to eradicate the escapee, 62% of the *Agrostis* plants found in the vicinity three years later were glyphosate resistant. So far, the ecological impact of this event appears to be minor, but that not may be the case with future transgenic escapes.

Many different strategies are being pursued with the goal of preventing transgene escape. For example, if male sterility could be engineered into plants, these plants would still produce seeds and fruit if pollinated by nearby nontransgenic plants, but they would produce no viable pollen. A second approach involves genetically engineering apomixis into transgenic crops. When a seed is produced by apomixis, the embryo and endosperm develop without fertilization. The transfer of this trait to transgenic crops would therefore minimize the possibility of transgene escape via pollen because plants could be male-sterile without compromising seed or fruit production. A third approach is to engineer the transgene into the chloroplast DNA of the crop. Chloroplast DNA in many plant species is inherited strictly from the egg, so transgenes in the chloroplast cannot be transferred by pollen. A fourth approach for preventing transgene escape is to genetically engineer flowers that develop normally but fail to open. Consequently, self-pollination would occur, but pollen would be unlikely to escape from the flower. This solution would require modifications to flower design. Several floral genes have been identified that could be manipulated to this end.

The continuing debate about GMOs in agriculture exemplifies one of this textbook's recurring ideas: the relationship of science and technology to society. Technological advances almost always involve some risk of unintended outcomes. In plant biotechnology, zero risk is probably unattainable. Therefore, scientists and the public must assess on a case-by-case basis the possible benefits of transgenic products versus the risks that society is willing to take. The best scenario is for these discussions and decisions to be based on sound scientific information and rigorous testing rather than on reflexive fear or blind optimism.

---

**CONCEPT CHECK 30.3**

1. Compare traditional plant-breeding methods with genetic engineering.
2. Why does *Bt* maize have less fumonisin than non-GM maize?
3. **WHAT IF?** In a few species, chloroplast genes are inherited only from sperm. How might this influence efforts to prevent transgene escape?

For suggested answers, see Appendix A.

# 30 Chapter Review

Go to **MasteringBiology**® for Assignments, the eText, and the Study Area with Animations, Activities, Vocab Self-Quiz, and Practice Tests.

## SUMMARY OF KEY CONCEPTS

**VOCAB SELF-QUIZ**

goo.gl/gbai8v

### CONCEPT 30.1

**Flowers, double fertilization, and fruits are unique features of the angiosperm life cycle (pp. 620–629)**

- Angiosperm reproduction involves an alternation of generations between a multicellular diploid sporophyte generation and multicellular haploid gametophyte generation. Flowers, produced by the sporophyte, function in sexual reproduction.
- The four floral organs are sepals, petals, stamens, and carpels. **Sepals** protect the floral bud. **Petals** help attract pollinators. **Stamens** bear **anthers** in which haploid **microspores** develop into **pollen grains** containing a male gametophyte. **Carpels** contain **ovules** (immature seeds) in their swollen bases. Within the ovules, **embryo sacs** (female gametophytes) develop from **megaspores**.
- **Pollination**, which precedes **fertilization**, is the placing of pollen on the **stigma** of a carpel. After pollination, the **pollen tube** discharges two sperm into the female gametophyte. Two sperm are needed for **double fertilization**, a process in which one sperm fertilizes the egg, forming a zygote and eventually an embryo, while the other sperm combines with the polar nuclei, giving rise to food-storing **endosperm**.

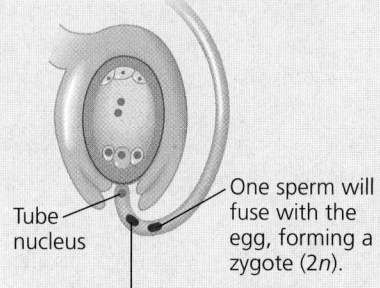

Tube nucleus

One sperm will fuse with the egg, forming a zygote (2n).

One sperm cell will fuse with the two polar nuclei, forming an endosperm nucleus (3n).

- The **seed coat** encloses the embryo along with a food supply stocked in either the endosperm or the **cotyledons**. Seed **dormancy** ensures that seeds germinate only when conditions for seedling survival are optimal. The breaking of dormancy often requires environmental cues, such as temperature or lighting changes.
- The **fruit** protects the enclosed seeds and aids in wind dispersal or in the attraction of seed-dispersing animals.

**?** *What changes occur to the four types of floral parts as a flower changes into a fruit?*

### CONCEPT 30.2

**Flowering plants reproduce sexually, asexually, or both (pp. 630–633)**

- **Asexual reproduction** enables successful plants to proliferate quickly. Sexual reproduction generates most of the genetic variation that makes evolutionary adaptation possible.
- Plants have evolved many mechanisms to avoid self-fertilization, including having male and female flowers on different individuals (**dioecious** species), asynchronous production of male and female parts within a single flower, and **self-incompatibility**, in which pollen grains that bear an allele identical to one in the female are rejected.

- Plants can be cloned from single cells, which can be genetically manipulated before being allowed to develop into a plant.

**?** *What are the advantages of asexual reproduction in plants? What are the advantages of sexual reproduction?*

### CONCEPT 30.3

**People modify crops through breeding and genetic engineering (pp. 633–636)**

- Hybridization of different varieties and even species of plants is common in nature and has been used by breeders, ancient and modern, to introduce new genes into crops. After two plants are successfully hybridized, plant breeders select those progeny that have the desired traits.
- In genetic engineering, genes from unrelated organisms are incorporated into plants. Genetically modified (GM) plants have the potential of increasing the quality and quantity of food worldwide and may also become increasingly important as biofuels.
- GM crops include Golden Rice, which provides more vitamin A, and *Bt* maize, which is insect resistant.
- There are concerns about the unknown risks of releasing GM organisms into the environment, but the potential benefits of **transgenic** crops need to be considered.

**?** *Give three specific examples of genetic engineering strategies to improve food quality or agricultural productivity.*

## TEST YOUR UNDERSTANDING

### Level 1: Knowledge/Comprehension

**PRACTICE TEST**

goo.gl/CRZjvS

1. A seed develops from an
   (A) ovum.
   (B) embryo.
   (C) ovule.
   (D) ovary.

2. A fruit is a
   (A) mature ovary.
   (B) mature ovule.
   (C) seed plus its integuments.
   (D) fused carpel.

3. Double fertilization means that
   (A) flowers must be pollinated twice to yield fruits and seeds.
   (B) every egg must receive two sperm to produce an embryo.
   (C) one sperm is needed to fertilize the egg, and a second sperm is needed to fertilize the polar nuclei.
   (D) every sperm has two nuclei.

4. "Golden Rice"
   (A) is resistant to various herbicides, making it practical to weed rice fields with those herbicides.
   (B) is resistant to a virus that commonly attacks rice fields.
   (C) includes bacterial genes that produce a toxin that reduces damage from insect pests.
   (D) has a modified genome that includes transgenes that lead to an increase vitamin A content.

5. Which statement concerning grafting is correct?
   (A) Stocks and scions refer to twigs of different species.
   (B) Stocks and scions must come from unrelated species.
   (C) Stocks provide root systems for grafting.
   (D) Grafting creates new species.

## Level 2: Application/Analysis

6. Some dioecious species have the XY genotype for male and XX for female. After double fertilization, what would be the genotypes of the embryos and endosperm nuclei?
   (A) embryo X/endosperm XX or embryo Y/endosperm XY
   (B) embryo XX/endosperm XX or embryo XY/endosperm XY
   (C) embryo XY/endosperm XXX or embryo XX/endosperm XXY
   (D) embryo XX/endosperm XXX or embryo XY/endosperm XXY

7. A small flower with green petals is most likely
   (A) bee-pollinated.              (C) bat-pollinated.
   (B) bird-pollinated.             (D) wind-pollinated.

8. The pollen produced by wind-pollinated plants is often smaller than the pollen produced by animal-pollinated plants. A reason for this might be that
   (A) wind-pollinated plants, in general, are smaller than animal-pollinated plants.
   (B) wind-pollinated flowers don't need large pollen grains because they don't have to attract animal pollinators.
   (C) small pollen grains can be carried farther by the wind.
   (D) animal pollinators are more facile at picking up large pollen grains.

9. The black dots that cover strawberries are actually individual fruits from a flower with multiple carpels. The fleshy and tasty portion of a strawberry derives from the receptacle of the flower. Therefore, a strawberry is both
   (A) a multiple fruit and an aggregate fruit.
   (B) a multiple fruit and an accessory fruit.
   (C) a simple fruit and an aggregate fruit.
   (D) an aggregate fruit and an accessory fruit.

10. **DRAW IT** Draw a flower and label the parts.

## Level 3: Synthesis/Evaluation

11. **SCIENTIFIC INQUIRY**
    Critics of GM foods have argued that foreign genes may disturb normal cellular functioning, causing unexpected and potentially harmful substances to appear inside cells. Toxic intermediary substances that normally occur in very small amounts may arise in larger amounts, or new substances may appear. The disruption may also lead to loss of substances that help maintain normal metabolism. If you were your nation's chief scientific advisor, how would you respond to these criticisms?

12. **SCIENCE, TECHNOLOGY, AND SOCIETY**
    People have engaged in genetic manipulation for millennia, producing plant and animal varieties through selective breeding and hybridization processes that significantly modify the genomes of organisms. Why do you think modern genetic engineering, which often entails introducing or modifying only one or a few genes, has met with so much public opposition? Should some forms of genetic engineering be of greater concern than others? Explain.

13. **FOCUS ON EVOLUTION**
    With respect to sexual reproduction, some plant species are fully self-fertile, others are fully self-incompatible, and some exhibit a "mixed strategy" with partial self-incompatibility. These reproductive strategies differ in their implications for evolutionary potential. How might these three strategies fare in a small founder population (see Concept 21.3)?

14. **FOCUS ON ORGANIZATION**
    In a short essay (100–150 words), discuss how the ability of a flower to reproduce with other flowers of the same species is an emergent property that arises from its floral parts and their organization.

15. **SYNTHESIZE YOUR KNOWLEDGE**

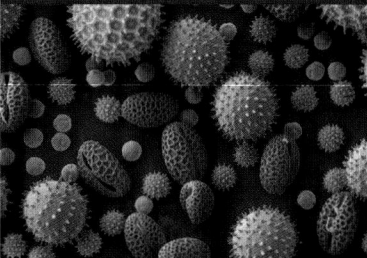

This colorized SEM shows pollen grains from six plant species. Explain how pollen develops, what its function is, and how it contributed to the dominance of angiosperms and other seed plants.

*For selected answers, see Appendix A.*

## KEY CONCEPTS

**31.1** Plant hormones help coordinate growth, development, and responses to stimuli

**31.2** Responses to light are critical for plant success

**31.3** Plants respond to a wide variety of stimuli other than light

**31.4** Plants respond to attacks by herbivores and pathogens

▲ **Figure 31.1** Can a plant camouflage itself?

## A Chameleon Vine

The Chilean vine *Boquila trifoliata* has the amazing ability to mimic the leaves of the trees supporting it in up to six different traits, including size, shape, color, orientation, petiole length, and tip shape. **Figure 31.1**, for example, shows the leaves formed by *Boquila* (at center, with three

*Boquila trifoliata    Ranunculus repens*

leaflets) when growing on *Rhaphithamnus spinosus*, while the other photo shows an impressive attempt by *Boquila* to mimic the creeping buttercup (*Ranunculus repens*). If *Boquila* happens to switch hosts in the course of climbing through the canopy, its new leaves mimic those of the new host. The leaves of unsupported *Boquila* vines do not differ from those of vines climbing on leafless trunks, suggesting that *Boquila* somehow senses the leaves of the host plant and modifies its appearance accordingly. The apparent advantage of this behavior is that leaves that have morphed to mimic their hosts are less susceptible to damage by feeding animals. The mechanism by which *Boquila* performs this feat of mimicry is a mystery but may involve the sensing of volatile emissions from the host plant leaves.

The story of *Boquila* reminds us that plants do not exist in isolation: They interact with a wide range of organisms as well as the environment. All of these biotic and abiotic interactions and all of the internal chemical changes they initiate involve signal transduction pathways that, in the big picture, are not too far removed from some of the pathways that you use to sense your environment. In effect, at the levels of signal reception and signal transduction, your cells are not all that different from those of a plant—certainly the similarities far outweigh the differences. As an animal, however, your responses to environmental stimuli are generally quite different from those of plants. Animals commonly respond to environmental challenges and opportunities by movement; plants respond by altering their growth and development.

Architectural modifications due to altered growth and development lead to changes in spatial orientation, but plants also adjust to changes in time. The passage of seasons is an example of a temporal variable that plants need to measure to compete successfully. In this chapter, first we'll discuss the internal chemicals (hormones) that regulate plant growth and development; then we'll explore how plants perceive and respond to light and other environmental signals.

# Plant hormones help coordinate growth, development, and responses to stimuli

A **hormone**, in the original meaning of the term, is a signaling molecule that is produced in tiny amounts by one part of an organism's body and transported to other parts, where it binds to a specific receptor and triggers responses in target cells and tissues. In animals, hormones are usually transported through the circulatory system, a criterion often included in definitions of the term. Some have argued that the hormone concept, which originated from studies of animals, is too limiting to describe plant physiological processes. For example, plants don't have circulating blood to transport hormone-like signaling molecules. Moreover, some signaling molecules that are considered plant hormones act only locally. Finally, some signaling molecules in plants, such as sucrose, typically occur at concentrations hundreds of thousands times greater than that of a typical hormone. Nevertheless, they are transported through plants and activate signal transduction pathways that greatly alter the functioning of plants in a manner similar to a hormone. Thus, many plant biologists prefer the broader term *plant growth regulator* to describe organic compounds, whether natural or synthetic, that modify or control one or more specific physiological processes within a plant. Currently, the terms *plant hormone* and *plant growth regulator* are used about equally, but for historical continuity we will use the term *plant hormone* and adhere to the criterion that plant hormones are active at very low concentrations.

Plant hormones are produced in very low concentrations, but a tiny amount of hormone can have a profound effect on plant growth and development. Virtually every aspect of plant growth and development is under hormonal control to some degree. Each hormone has multiple effects, depending on its site of action, its concentration, and the developmental stage of the plant. Conversely, multiple hormones can influence a single process. Response to a hormone often depends not so much on the amount of that hormone as on its relative concentration compared with other hormones. It is often the interactions between different hormones, rather than hormones acting in isolation, that control growth and development. These interactions will become apparent in the following survey of hormone function.

## The Discovery of Plant Hormones

The idea that chemical messengers exist in plants emerged from a series of experiments on how stems respond to light. Plant shoots typically grow toward light. Any growth response that results in plant organs curving toward or away from stimuli is called a **tropism** (from the Greek *tropos*, turn). The growth of a plant organ toward light or away from it is called **phototropism**; shoots generally exhibit positive phototropism, whereas roots exhibit negative phototropism. In nature, positive phototropism directs shoot growth toward the sunlight that powers photosynthesis. This response involves cells on the darker side elongating faster than the cells on the brighter side.

Charles Darwin and his son Francis conducted some of the earliest experiments on phototropism **(Figure 31.2)**. They observed that a grass seedling ensheathed in its coleoptile (see Figure 30.9b) could bend toward light only if the tip of the coleoptile was present. If the tip was removed, the coleoptile did not curve. The seedling also failed to grow toward light if the tip was covered with an opaque cap, but neither a transparent cap over the tip nor an opaque shield placed below the coleoptile tip prevented the phototropic response. It was the tip of the coleoptile, the Darwins concluded, that was responsible for sensing light. However, they noted that the differential growth response that led to curvature of the coleoptile occurred some distance below the tip. The Darwins postulated that some signal was transmitted downward from the tip to the elongating region of the coleoptile. A few decades later, the Danish scientist Peter Boysen-Jensen demonstrated that the signal was a mobile chemical substance. He separated the tip from the remainder of the coleoptile by a cube of gelatin, which prevented cellular contact but allowed chemicals to pass through. These seedlings responded normally, bending toward light. However, if the tip was experimentally separated from the lower coleoptile by an impermeable barrier, such as the mineral mica, no phototropic response occurred.

In 1926, Frits Went extracted the chemical messenger for phototropism by modifying the experiments of Boysen-Jensen **(Figure 31.3)**. Went removed the coleoptile tip and placed it on a cube of agar, a gelatinous material. The chemical messenger moving from the tip, Went reasoned, should diffuse into the agar, and the agar block should then be able to substitute for the coleoptile tip. Went placed the agar blocks on decapitated coleoptiles that were kept in the dark. A block that was centered on top of the coleoptile caused the stem to grow straight upward. However, when the block was placed off center, the coleoptile began to bend away from the side with the agar block, as though growing toward light. Went concluded that the agar block contained a chemical produced in the coleoptile tip, that this chemical stimulated growth as it passed down the coleoptile, and that a coleoptile curved toward light because of a higher concentration of the growth-promoting chemical on the darker side of the coleoptile. For this chemical messenger, or hormone, Went chose the name auxin (from the Greek *auxein*, to increase). The major type of auxin was later purified, and its chemical structure was then determined to be indoleacetic acid (IAA). These classic experiments support the idea that an asymmetric distribution of auxin moving down from the coleoptile tip causes cells on the darker side to elongate faster than cells on the brighter side.

## ▼ Figure 31.2 Inquiry

### What part of a grass coleoptile senses light, and how is the signal transmitted?

**Experiment** In 1880, Charles and Francis Darwin removed and covered parts of grass coleoptiles to determine what part senses light. In 1913, Peter Boysen-Jensen separated coleoptiles with different materials to determine how the signal for phototropism is transmitted.

**Results**

**Control**

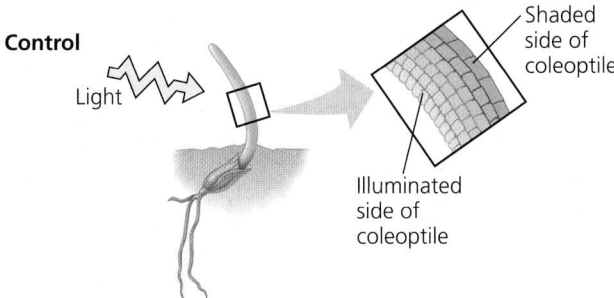

Light

Shaded side of coleoptile

Illuminated side of coleoptile

**Darwin and Darwin: Phototropism occurs only when the tip is illuminated.**

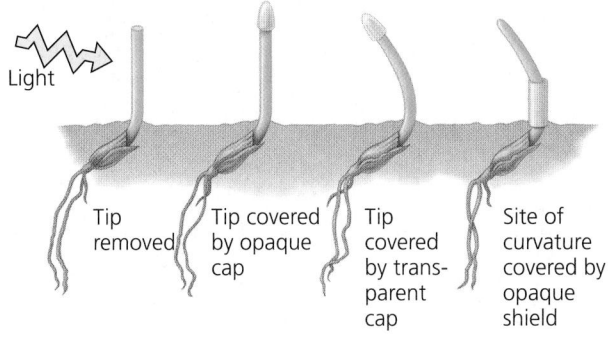

Light

Tip removed

Tip covered by opaque cap

Tip covered by transparent cap

Site of curvature covered by opaque shield

**Boysen-Jensen: Phototropism occurs when the tip is separated by a permeable barrier but not an impermeable barrier.**

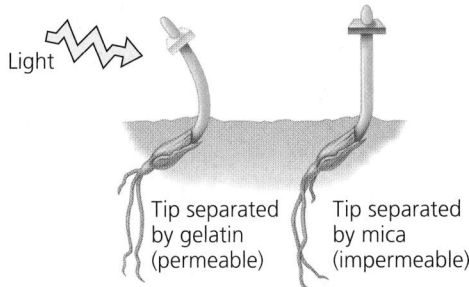

Light

Tip separated by gelatin (permeable)

Tip separated by mica (impermeable)

**Conclusion** The Darwins' experiment suggested that only the tip of the coleoptile senses light. The phototropic bending, however, occurred at a distance from the site of light perception (the tip). Boysen-Jensen's results suggested that the signal for the bending is a light-activated mobile chemical.

**Data from** C. R. Darwin, *The Power of Movement in Plants*, John Murray, London (1880). P. Boysen-Jensen, Concerning the performance of phototropic stimuli on the *Avena* coleoptile, *Berichte der Deutschen Botanischen Gesellschaft* (*Reports of the German Botanical Society*) 31:559–566 (1913).

**WHAT IF?** How could you experimentally determine which colors of light cause the most phototropic bending?

## ▼ Figure 31.3 Inquiry

### Does asymmetric distribution of a growth-promoting chemical cause a coleoptile to bend?

**Experiment** In 1926, Frits Went's experiment identified how a growth-promoting chemical causes a coleoptile to grow toward light. He placed coleoptiles in the dark and removed their tips, putting some tips on agar cubes that he predicted would absorb the growth-promoting chemical. On a control coleoptile, he placed a cube that lacked the chemical. On others, he placed cubes containing the chemical, either centered on top of the coleoptile to distribute the chemical evenly or offset to increase the concentration on one side.

**Results** The coleoptile grew straight if the growth-promoting chemical was distributed evenly. If the chemical was distributed unevenly, the coleoptile curved away from the side with the cube, as if growing toward light, even though it was grown in the dark.

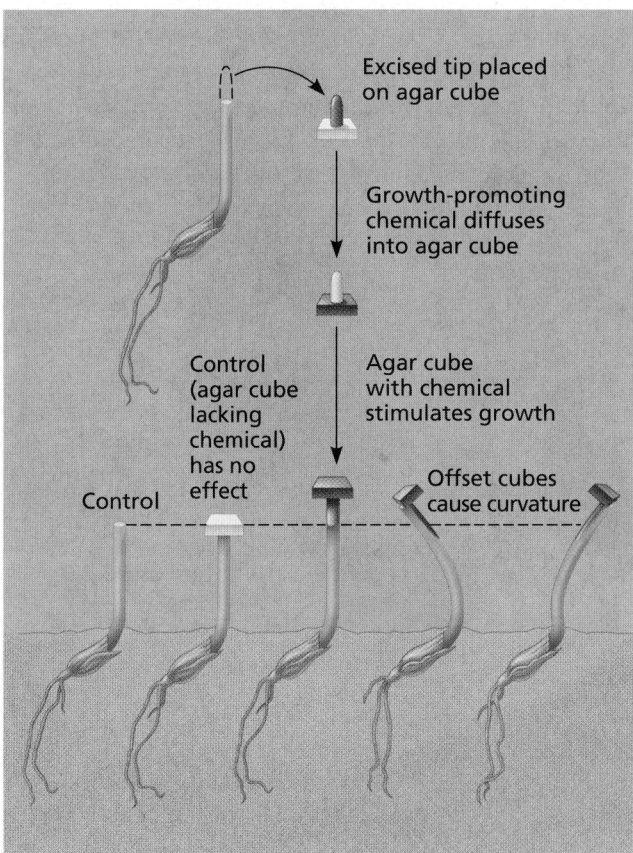

Excised tip placed on agar cube

Growth-promoting chemical diffuses into agar cube

Control (agar cube lacking chemical) has no effect

Agar cube with chemical stimulates growth

Control

Offset cubes cause curvature

**Conclusion** Went concluded that a coleoptile curves toward light because its dark side has a higher concentration of the growth-promoting chemical, which he named auxin.

**Data from** F. Went, A growth substance and growth, *Recueils des Travaux Botaniques Néerlandais* (*Collections of Dutch Botanical Works*) 25:1–116 (1928).

**(MB)** A related Experimental Inquiry Tutorial can be assigned in MasteringBiology.

**WHAT IF?** Triiodobenzoic acid (TIBA) inhibits auxin transport. If a tiny agar bead containing TIBA were placed off center on the tip of an intact coleoptile, which way would the coleoptile bend: toward the side with the bead or away from it? Explain.

# A Survey of Plant Hormones

The discovery of auxin stimulated the search for other plant hormones. **Table 31.1** previews some major classes of plant hormones: auxin, cytokinins, gibberellins, brassinosteroids, abscisic acid, and ethylene (this list is by no means exhaustive).

## Auxin

The term **auxin** is used for any chemical substance that promotes elongation of coleoptiles, although auxins have multiple functions in flowering plants. The major natural auxin in plants is indoleacetic acid (IAA), although several other compounds, including some synthetic ones, have auxin activity.

Auxin is produced predominantly in shoot tips and is transported from cell to cell down the stem at a rate of about 1 cm/hr. It moves only from tip to base, not in the reverse direction. This unidirectional transport of auxin is called *polar transport*. Polar transport is unrelated to gravity; experiments have shown that auxin travels upward when a stem or coleoptile segment is placed upside down. Rather, the polarity of auxin movement is attributable to the polar distribution of auxin transport protein in the cells. Concentrated at the basal end of a cell, the auxin transporters move the hormone out of the cell. The auxin can then enter the apical end of the neighboring cell **(Figure 31.4)**. Auxin has a variety of effects, including stimulating cell elongation and regulating plant architecture.

**The Role of Auxin in Cell Elongation** Auxin stimulates cell elongation within young developing shoots. As auxin from the shoot tip (see Figure 28.16) moves down to the region of cell elongation, the hormone stimulates cell growth, probably by binding to a receptor in the plasma membrane. Auxin stimulates growth only over a certain concentration range, from about $10^{-8}$ to $10^{-4}M$. At higher concentrations, auxin may inhibit cell elongation by inducing the production of ethylene, a hormone that generally hinders growth. We'll return to this hormonal interaction when we discuss ethylene.

According to a model called the *acid growth hypothesis*, proton pumps play a major role in the growth response of cells to auxin. In a shoot's zone of elongation, auxin stimulates the plasma membrane's proton ($H^+$) pumps. This pumping of $H^+$ increases the voltage difference across the membrane (membrane potential) and lowers the pH in the cell wall within

| Table 31.1 Overview of Plant Hormones | | |
|---|---|---|
| **Hormone** | **Where Produced or Found in Plant** | **Major Functions** |
| **Auxin** | Shoot apical meristems and young leaves are the primary sites of auxin synthesis. Root apical meristems also produce auxin, although the root depends on the shoot for much of its auxin. Developing seeds and fruits contain high levels of auxin, but it is unclear whether it is newly synthesized or transported from maternal tissues. | Stimulates stem elongation (low concentration only); promotes the formation of lateral and adventitious roots; regulates development of fruit; enhances apical dominance; functions in phototropism and gravitropism; promotes vascular differentiation; retards leaf abscission. |
| **Cytokinins** | These are synthesized primarily in roots and transported to other organs, although there are many minor sites of production as well. | Regulate cell division in shoots and roots; modify apical dominance and promote lateral bud growth; promote movement of nutrients into sink tissues; stimulate seed germination; delay leaf senescence. |
| **Gibberellins** | Meristems of apical buds and roots, young leaves, and developing seeds are the primary sites of production. | Stimulate stem elongation, pollen development, pollen tube growth, fruit growth, and seed development and germination; regulate sex determination and the transition from juvenile to adult phases. |
| **Brassinosteroids** | These compounds are present in all plant tissues, although different intermediates predominate in different organs. Internally produced brassinosteroids act near the site of synthesis. | Promote cell expansion and cell division in shoots; promote root growth at low concentrations; inhibit root growth at high concentrations; promote xylem differentiation and inhibit phloem differentiation; promote seed germination and pollen tube elongation. |
| **Abscisic acid (ABA)** | Almost all plant cells have the ability to synthesize abscisic acid, and its presence has been detected in every major organ and living tissue; may be transported in the phloem or xylem. | Inhibits growth; promotes stomatal closure during drought stress; promotes seed dormancy and inhibits early germination; promotes leaf senescence; promotes desiccation tolerance. |
| **Ethylene** | This gaseous hormone can be produced by most parts of the plant. It is produced in high concentrations during senescence, leaf abscission, and the ripening of some types of fruit. Synthesis is also stimulated by wounding and stress. | Promotes ripening of many types of fruit, leaf abscission, and the triple response in seedlings (inhibition of stem elongation, promotion of lateral expansion, and horizontal growth); enhances the rate of senescence; promotes root and root hair formation; promotes flowering in the pineapple family. |

## What causes polar movement of auxin from shoot tip to base?

**Experiment** To investigate how auxin is transported unidirectionally, Leo Gälweiler and colleagues designed an experiment to identify the location of the auxin transport protein. They used a greenish yellow fluorescent molecule to label antibodies that bind to the auxin transport protein. Then they applied the antibodies to longitudinally sectioned *Arabidopsis* stems.

**Results** The light micrograph on the left shows that auxin transport proteins are not found in all stem tissues, but only in the xylem parenchyma. In the light micrograph on the right, a higher magnification reveals that these proteins are primarily localized at the basal ends of the cells.

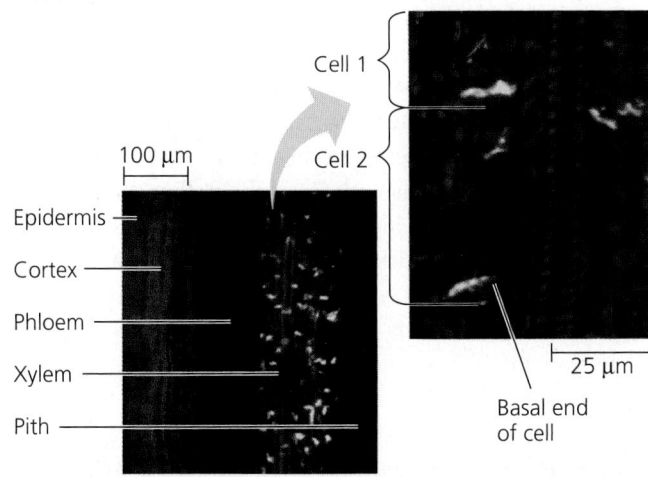

**Conclusion** The results support the hypothesis that concentration of the auxin transport protein at the basal ends of cells mediates the polar transport of auxin.

**Data from** L. Gälweiler et al., Regulation of polar auxin transport by AtPIN1 in *Arabidopsis* vascular tissue, *Science* 282:2226–2230 (1998).

**WHAT IF?** If auxin transport proteins were equally distributed at both ends of the cells, would polar auxin still be possible? Explain.

---

minutes **(Figure 31.5)**. Acidification of the wall activates enzymes called **expansins** that break the cross-links (hydrogen bonds) between cellulose microfibrils and other cell wall constituents, loosening the wall's fabric. Increasing the voltage difference across the membrane enhances ion uptake into the cell, which causes osmotic uptake of water and increased turgor. Increased turgor and increased cell wall plasticity enable the cell to elongate.

Auxin also rapidly alters gene expression, causing cells in the zone of elongation to produce new proteins within minutes. Some of these proteins are short-lived transcription factors that regulate gene expression. For sustained growth after this initial spurt, cells must make more cytoplasm and wall material. Auxin also stimulates this sustained growth response.

**Auxin's Role in Plant Development** The polar transport of auxin is a central element controlling the spatial organization, or *pattern formation*, of the developing plant. Auxin is synthesized in shoot tips, and it carries integrated information about the development, size, and environment of individual branches. This flow of information controls branching patterns. A reduced flow of auxin from a branch, for example, indicates that the branch is not being sufficiently productive: New branches are needed elsewhere. Thus, lateral buds below the branch are released from dormancy and begin to grow.

The transport of auxin also plays a key role in establishing *phyllotaxy* (see Figure 29.3), the arrangement of leaves on the stem. A leading model proposes that polar auxin transport in the shoot tip generates local peaks in auxin concentration that determine the site of leaf primordium formation and thereby the different phyllotaxies found in nature.

Auxin, cytokinins, and newly discovered plant hormones called strigolactones interact in the control of apical

---

❸ Wedge-shaped expansins (red), activated by low pH, separate cellulose microfibrils (brown) from cross-linking polysaccharides (green). The exposed cross-linking polysaccharides are now more accessible to cell wall–loosening enzymes (purple).

❷ The cell wall becomes more acidic.

❶ Auxin increases the activity of proton pumps.

CELL WALL

❹ Cell wall-loosening enzymes (purple) cleave cross-linking polysaccharides (green), allowing cellulose microfibrils to slide. The extensibility of the cell wall is increased. Turgor causes the cell to expand.

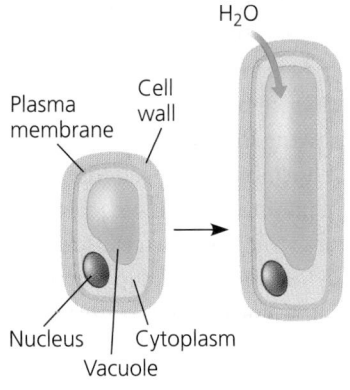

❺ With the cellulose loosened, the cell can elongate.

▲ **Figure 31.5 Cell elongation in response to auxin: the acid growth hypothesis.**

dominance, the ability of the apical bud to suppress the development of axillary buds. If the apical bud, the primary source of auxin, is removed, the inhibition of axillary buds is removed and the plant becomes bushier. Applying auxin to the cut surface of the decapitated shoot resuppresses the growth of the lateral buds.

**Practical Uses for Auxins** Auxins, both natural and synthetic, have many commercial applications. One example is spraying auxins on greenhouse tomato plants to induce normal fruit development. When tomatoes are grown outdoors, each fruit usually develops normally because enough auxin is produced by its seeds. However, greenhouse-grown tomatoes have fewer seeds and therefore are often malformed unless more auxin is supplied by spraying. A second example is the use of the natural auxin indolebutyric acid (IBA) in the vegetative propagation of plants by cuttings. Treating a detached leaf or stem with powder containing IBA often causes adventitious roots to form near the cut surface. Another example is the use of the synthetic auxin 2,4-dichlorophenoxyacetic acid (2,4-D) as a herbicide. Spraying it eliminates eudicot (broadleaf) weeds. Monocots, such as maize and turfgrass, can rapidly inactivate 2,4-D, but eudicots die from a "hormonal overdose."

## Cytokinins

Trial-and-error attempts to find chemical additives that would enhance the growth and development of plant cells in tissue culture led to the discovery of **cytokinins**. Researchers found that they could induce cultured tobacco cells to divide by adding degraded DNA samples. The active ingredients turned out to be modified forms of adenine, a component of nucleic acids. These growth regulators were named cytokinins because they stimulate cytokinesis, or cell division. The most common natural cytokinin is zeatin, so named because it was discovered first in maize. Cytokinins influence cell division and differentiation, apical dominance, and aging.

**Control of Cell Division and Differentiation** Cytokinins are produced in actively growing tissues, particularly in roots, embryos, and fruits. Cytokinins produced in roots reach their target tissues by moving up the plant in the xylem sap. Acting in concert with auxin, cytokinins stimulate cell division and influence organ formation. The effects of cytokinins on cells growing in tissue culture provide clues about how this class of hormones may function in an intact plant. When a piece of parenchyma tissue from a stem is cultured in the absence of cytokinins, the cells grow very large but do not divide. But if cytokinins are added along with auxin, the cells divide. Cytokinins alone have no effect. The ratio of cytokinins to auxin controls cell differentiation. When the concentrations of these two hormones are at certain levels, the mass of cells continues to grow, but it remains a cluster of undifferentiated cells

called a callus (see Figure 30.14). If cytokinin levels increase, shoot buds develop from the callus. If auxin levels increase, roots form.

**Anti-aging Effects** Cytokinins slow the aging of certain plant organs by inhibiting protein breakdown, stimulating RNA and protein synthesis, and mobilizing nutrients from surrounding tissues. If leaves removed from a plant are dipped in a cytokinin solution, they stay green much longer than otherwise.

## Gibberellins

In the early 1900s, farmers in Asia noticed that some rice seedlings in their paddies grew so tall and spindly that they toppled over before they could mature. In 1926, it was discovered that a fungus of the genus *Gibberella* causes this "foolish seedling disease." By the 1930s, it was found that the fungus causes hyperelongation of rice stems by secreting a chemical, which was given the name **gibberellin**. In the 1950s, researchers determined that plants also produce gibberellins (GAs). Since that time, scientists have identified more than 100 types of gibberellins that can occur naturally in plants, although a much smaller number occur in each plant species. Gibberellins have a variety of effects, such as stem elongation, fruit growth, and seed germination. The rapidly elongating "foolish seedlings" apparently suffer from too much gibberellin.

**Stem Elongation** The major sites of gibberellin production are young roots and leaves. Gibberellins are best known for stimulating stem and leaf growth by enhancing cell elongation *and* cell division. One hypothesis proposes that they activate enzymes that loosen cell walls, facilitating entry of expansin proteins. Thus, gibberellins act in concert with auxin to promote stem elongation.

The effects of gibberellins in enhancing stem elongation are evident when certain dwarf (mutant) varieties of plants are treated with gibberellins. For instance, some dwarf pea plants grow tall if treated with gibberellins. But there is often no response if the gibberellins are applied to wild-type plants. Apparently, these plants already produce an optimal dose of the hormone. The most dramatic example of gibberellin-induced stem elongation is *bolting*, rapid growth of the floral stalk **(Figure 31.6a)**.

**Fruit Growth** In many plants, both auxin and gibberellins must be present for fruit to develop. The most important commercial application of gibberellins is in the spraying of Thompson seedless grapes **(Figure 31.6b)**. The hormone makes the individual grapes grow larger, a trait valued by the consumer. The gibberellin sprays also make the internodes of the grape bunch elongate, allowing more space for the individual grapes. By enhancing air circulation between the grapes, this increase in space also makes it harder for yeasts and other microorganisms to infect the fruit.

**(b)** The Thompson seedless grape bunch on the left is from an untreated control vine. The bunch on the right is growing from a vine that was sprayed with gibberellin during fruit development.

**(a)** Some plants develop in a rosette form, low to the ground with very short internodes, as in the *Arabidopsis* plant shown at the left. As the plant switches to reproductive growth, a surge of gibberellins induces bolting: Internodes elongate rapidly, elevating floral buds that develop at stem tips (right).

▲ **Figure 31.6 Effects of gibberellins on stem elongation and fruit growth.**

**Germination** The embryo of a seed is a rich source of gibberellins. After water is imbibed, the release of gibberellins from the embryo signals the seed to break dormancy and germinate. Some seeds that normally require particular environmental conditions to germinate, such as exposure to light or low temperatures, break dormancy if they are treated with gibberellins. Gibberellins support the growth of cereal seedlings by stimulating the synthesis of digestive enzymes such as α-amylase that mobilize stored nutrients (**Figure 31.7**).

## Brassinosteroids

**Brassinosteroids** are steroids similar to cholesterol and the sex hormones of animals. They induce cell elongation and division in stem segments and seedlings at concentrations as low as $10^{-12}M$. They also slow leaf abscission (leaf drop) and promote xylem differentiation. These effects are so qualitatively similar to those of auxin that it took years for plant physiologists to determine that brassinosteroids were not types of auxins.

The identification of brassinosteroids as plant hormones arose from studies of an *Arabidopsis* mutant that exhibited morphological features similar to those of light-grown plants even when grown in the dark. The researchers discovered that the mutation affects a gene that normally codes for an enzyme similar to one involved in steroid synthesis in mammals. They also found that this brassinosteroid-deficient mutant could be restored to the wild-type phenotype by applying brassinosteroids.

## Abscisic Acid (ABA)

In the 1960s, one research group studying the chemical changes that precede bud dormancy and leaf abscission in deciduous trees and another team investigating chemical changes preceding abscission of cotton fruits isolated the same compound, **abscisic acid (ABA)**. Ironically, ABA is no longer thought to play a major role in bud dormancy or leaf abscission, but it is important in other functions. Unlike the growth-stimulating hormones we have discussed so far—auxin, cytokinins, gibberellins, and brassinosteroids—ABA *slows* growth. ABA often antagonizes the actions of growth hormones, and the ratio of ABA to one or more growth hormones determines the final outcome. Promoting seed dormancy and drought tolerance are two of ABA's many effects.

**Seed Dormancy** Seed dormancy increases the likelihood that seeds will germinate only when there are sufficient amounts of light, temperature, and moisture for the seedlings to survive (see Concept 30.1). What prevents seeds dispersed in autumn from germinating immediately, only to die in winter? What

**①** After a seed imbibes water, the embryo releases gibberellin (GA), which sends a signal to the aleurone, the thin outer layer of the endosperm.

**②** The aleurone responds to GA by synthesizing and secreting digestive enzymes that hydrolyze nutrients stored in the endosperm. One example is α-amylase, which hydrolyzes starch.

**③** Sugars and other nutrients absorbed from the endosperm by the scutellum (cotyledon) are consumed during growth of the embryo into a seedling.

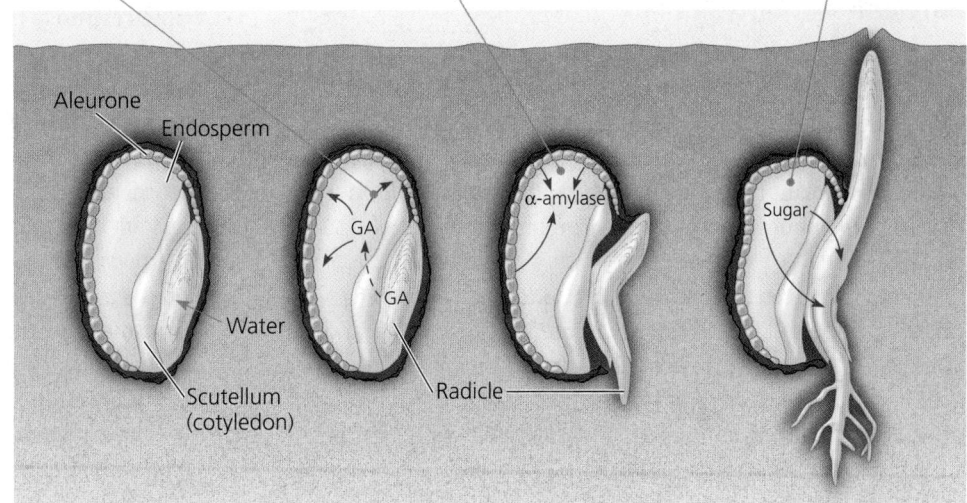

▲ **Figure 31.7 Mobilization of nutrients by gibberellins during the germination of grain seeds such as barley.**

mechanisms ensure that such seeds do not germinate until spring? For that matter, what prevents seeds from germinating in the dark, moist interior of the fruit? The answer to these questions is ABA. The levels of ABA may increase 100-fold during seed maturation. The high levels of ABA in maturing seeds inhibit germination and induce the production of proteins that help the seeds withstand the extreme dehydration that accompanies maturation.

Many types of dormant seeds germinate when ABA is removed or inactivated. The seeds of some desert plants break dormancy only when heavy rains wash ABA out of them. Other seeds require light or prolonged exposure to cold to inactivate ABA. Often, the ratio of ABA to gibberellins determines whether seeds remain dormant or germinate, and adding ABA to seeds that are primed to germinate makes them dormant again. Low levels of ABA or ABA sensitivity can lead to precocious (early) germination **(Figure 31.8)**. Precocious germination of red mangrove seeds, due to low ABA levels, is actually an adaptation that helps the young seedlings to plant themselves like darts in the soft mud below the parent tree.

**Drought Tolerance** ABA plays a major role in drought signaling. When a plant begins to wilt, ABA accumulates in the leaves and causes stomata to close rapidly, reducing transpiration and preventing further water loss. By affecting second messengers such as calcium, ABA causes potassium channels in the plasma membrane of guard cells to open, leading to a massive loss of potassium ions from the cells. The accompanying osmotic loss of water reduces guard cell turgor and leads to closing of the stomatal pores (see Figure 29.19). In some cases, water shortage stresses the root system before the shoot system, and ABA transported from roots to leaves may function as an "early warning system."

### Ethylene

During the 1800s, when coal gas was used as fuel for streetlights, leakage from gas pipes caused nearby trees to drop leaves prematurely. In 1901, the gas **ethylene** was demonstrated to be the active factor in coal gas. But the idea that ethylene is a plant hormone was not widely accepted until the advent of a technique called gas chromatography simplified its identification.

Plants produce ethylene in response to stresses such as drought, flooding, mechanical pressure, insect damage, and infection. Ethylene is also produced during fruit ripening and programmed cell death and in response to high concentrations of externally applied auxin. Indeed, many effects previously ascribed to auxin, such as inhibition of root elongation, may be due to auxin-induced ethylene production. We will focus here on four of ethylene's many effects: response to mechanical stress, senescence, leaf abscission, and fruit ripening.

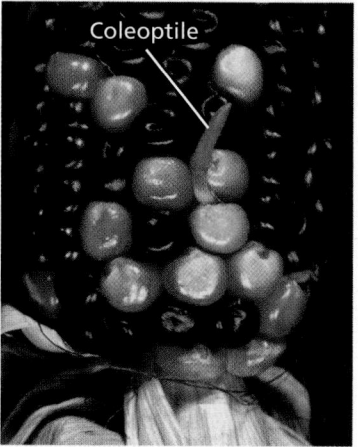

◀ Red mangrove (*Rhizophora mangle*) seeds produce only low levels of ABA, and their seeds germinate while still on the tree. In this case, early germination is a useful adaptation. When released, the radicle of the dart-like seedling deeply penetrates the soft mudflats in which the mangroves grow.

Coleoptile

▲ Precocious germination in this maize mutant is caused by lack of a functional transcription factor required for ABA action.

▲ **Figure 31.8 Precocious germination of wild-type mangrove and mutant maize seeds.**

**The Triple Response to Mechanical Stress** Imagine a pea seedling pushing upward through the soil, only to come up against a stone. As it pushes against the obstacle, the stress in its delicate tip induces the seedling to produce ethylene. The hormone then instigates a growth maneuver known as the **triple response** that enables the shoot to avoid the obstacle. The three parts of this response are a slowing of stem elongation, a thickening of the stem (which makes it stronger), and a curvature that causes the stem to start growing horizontally. As the effects of the initial ethylene pulse lessen, the stem resumes vertical growth. If it again contacts a barrier, another burst of ethylene is released, and horizontal growth resumes. However, if the upward touch detects no solid object, then ethylene production decreases, and the stem, now clear of the obstacle, resumes its normal upward growth. It is ethylene that induces the stem to grow horizontally rather than the physical obstruction itself; when ethylene is applied to normal seedlings growing free of physical impediments, they still undergo the triple response.

Studies of *Arabidopsis* mutants with abnormal triple responses are an example of how biologists identify a signal transduction pathway. Scientists isolated ethylene-insensitive (*ein*) mutants, which fail to undergo the triple response after

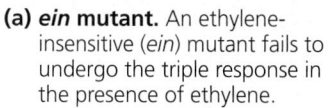

*ein* mutant

*ctr* mutant

**(a)** *ein* **mutant.** An ethylene-insensitive (*ein*) mutant fails to undergo the triple response in the presence of ethylene.

**(b)** *ctr* **mutant.** A constitutive triple-response (*ctr*) mutant undergoes the triple response even in the absence of ethylene.

▲ **Figure 31.9 Triple-response mutants in *Arabidopsis*.**

exposure to ethylene **(Figure 31.9a)**. Some types of *ein* mutants are insensitive to ethylene because they lack a functional ethylene receptor. Mutants of a different sort undergo the triple response even out of soil, in the air, where there are no physical obstacles. Some of these mutants have a regulatory defect that causes them to produce ethylene at 20 times the normal rate. The phenotype of such ethylene-overproducing (*eto*) mutants can be restored to wild-type by treating the seedlings with inhibitors of ethylene synthesis. Other mutants, called constitutive triple-response (*ctr*) mutants, undergo the triple response in air but do not respond to inhibitors of ethylene synthesis **(Figure 31.9b)**. (Constitutive genes are genes that are continually expressed in all cells of an organism.) In *ctr* mutants, ethylene signal transduction is permanently turned on, even though ethylene is not present.

The affected gene in *ctr* mutants codes for a protein kinase. The fact that this mutation *activates* the ethylene response suggests that the normal kinase product of the wild-type allele is a *negative* regulator of ethylene signal transduction. Thus, binding of the hormone ethylene to the ethylene receptor normally leads to inactivation of the kinase, and the inactivation of this negative regulator allows synthesis of the proteins required for the triple response.

**Senescence** Consider the shedding of a leaf in autumn or the death of an annual after flowering. Or think about the final step in differentiation of a vessel element, when its living contents are destroyed, leaving a hollow tube behind. Such events involve **senescence**—the programmed death of certain

cells or organs or the entire plant. Cells, organs, and plants genetically programmed to die on a schedule do not simply shut down cellular machinery and await death. Instead, at the molecular level, the onset of the programmed cell death called *apoptosis* is a busy time in a cell's life, requiring new gene expression. During apoptosis, newly formed enzymes break down many chemical components, including chlorophyll, DNA, RNA, proteins, and membrane lipids. The plant salvages many of the breakdown products. A burst of ethylene is almost always associated with the apoptosis of cells during senescence.

**Leaf Abscission** The loss of leaves from deciduous trees helps prevent desiccation during seasonal periods when the availability of water to the roots is severely limited. Before dying leaves abscise, many essential elements are salvaged from them and stored in stem parenchyma cells. These nutrients are recycled back to developing leaves the next growth season. Autumn leaf color is due to newly made red pigments as well as yellow and orange carotenoids (see Concept 8.2) that were already present in the leaf and are rendered visible by the breakdown of the dark-green chlorophyll in autumn.

When an autumn leaf falls, it detaches from the stem at an abscission layer that develops near the base of the petiole **(Figure 31.10)**. The small parenchyma cells of this layer have very thin walls, and there are no fiber cells around the vascular tissue. The abscission layer is further weakened when enzymes hydrolyze cell wall polysaccharides. Finally, the weight of the leaf, with the help of the wind, causes a separation within the abscission layer. Even before the leaf falls, a layer of cork forms a protective scar on the twig side of the abscission layer, preventing pathogens from invading the plant.

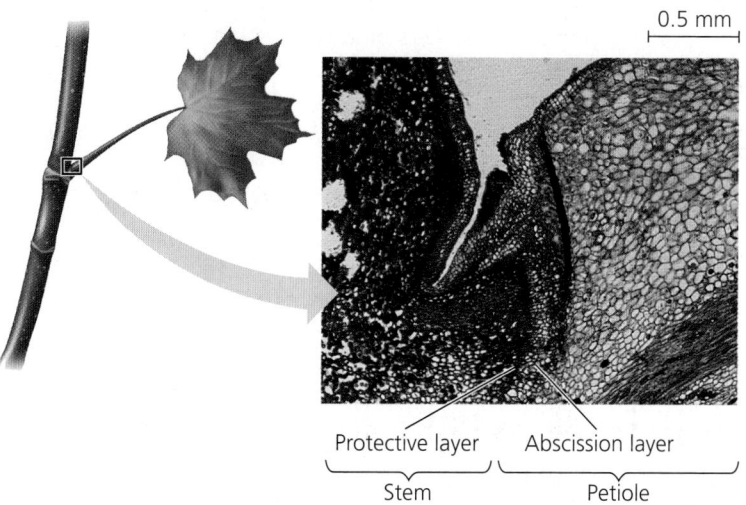

0.5 mm

Protective layer          Abscission layer

Stem          Petiole

▲ **Figure 31.10 Abscission of a maple leaf.** Abscission is controlled by a change in the ratio of ethylene to auxin. The abscission layer is seen in this longitudinal section as a vertical band at the base of the petiole. After the leaf falls, a protective layer of cork becomes the leaf scar that helps prevent pathogens from invading the plant (LM).

A change in the ratio of ethylene to auxin controls abscission. An aging leaf produces less and less auxin, rendering the cells of the abscission layer more sensitive to ethylene. As the influence of ethylene on the abscission layer prevails, the cells produce enzymes that digest the cellulose and other components of cell walls.

**Fruit Ripening** Immature fleshy fruits are generally tart, hard, and green—features that help protect the developing seeds from herbivores. After ripening, the mature fruits help *attract* animals that disperse the seeds (see Figure 30.11). In many cases, a burst of ethylene production in the fruit triggers the ripening process. The enzymatic breakdown of cell wall components softens the fruit, and the conversion of starches and acids to sugars makes the fruit sweet. The production of new scents and colors helps advertise ripeness to animals, which eat the fruits and disperse the seeds.

A chain reaction occurs during ripening: Ethylene triggers ripening, and ripening triggers more ethylene production. The result is a huge burst in ethylene production. Because ethylene is a gas, the signal to ripen spreads from fruit to fruit. If you pick or buy green fruit, you may be able to speed ripening by storing the fruit in a paper bag, allowing ethylene to accumulate. On a commercial scale, many kinds of fruits are ripened in huge storage containers in which ethylene levels are enhanced. In other cases, fruit producers take measures to slow ripening caused by natural ethylene. Apples, for instance, are stored in bins flushed with carbon dioxide. Circulating the air prevents ethylene from accumulating, and carbon dioxide inhibits synthesis of new ethylene. Stored in this way, apples picked in autumn can still be shipped to grocery stores the following summer.

### CONCEPT CHECK 31.1

1. Fusicoccin is a fungal toxin that stimulates the plasma membrane H⁺ pumps of plant cells. How would it affect the growth of isolated stem sections?
2. **WHAT IF?** If a plant has the double mutation *ctr* and *ein*, what is its triple-response phenotype? Explain your answer.
3. **MAKE CONNECTIONS** What type of feedback process is exemplified by production of ethylene during fruit ripening? Explain. (See Concept 32.2.)

For suggested answers, see Appendix A.

---

### CONCEPT 31.2

# Responses to light are critical for plant success

Light is an especially important environmental factor in the lives of plants. In addition to being required for photosynthesis, light triggers many key events in plant growth and development, collectively known as **photomorphogenesis**. Light reception also allows plants to measure the passage of days and seasons.

## Photomorphogenesis

As an example of photomorphogenesis, consider a sprouting potato **(Figure 31.11a)**. This modified underground stem, or tuber, has sprouted shoots from its "eyes" (axillary buds). These shoots, however, scarcely resemble those of a typical plant. Instead of sturdy stems and broad green leaves, this plant has pale stems and unexpanded leaves, as well as short, stubby roots. These morphological adaptations for growing in darkness, collectively referred to as **etiolation**, make sense if we consider that a young potato plant in nature usually encounters continuous darkness when sprouting underground. Under these circumstances, expanded leaves would be a hindrance to soil penetration and would be damaged as the shoots pushed through the soil. Because the leaves are unexpanded and underground, there is little evaporative loss of water and little requirement for an extensive root system to replace the water lost by transpiration. Moreover, the energy expended in producing green chlorophyll would be wasted because there is no light for photosynthesis. Instead, a potato plant growing in the dark allocates as much energy as possible to elongating its stems. This adaptation enables the shoots to break ground before the nutrient reserves in the tuber are exhausted.

When a shoot reaches light, the plant undergoes profound changes, collectively called **de-etiolation** (informally known as greening). Stem elongation slows; leaves expand; roots elongate; and the shoot produces chlorophyll. In short, it begins to resemble a typical plant **(Figure 31.11b)**. How do light signals initiate this remarkable change in form? As in all signal transduction processes, the signal must first be detected by a receptor protein.

Plants detect not only the presence of light signals but also their direction, intensity, and wavelength (color). A graph called an **action spectrum** depicts the relative effectiveness

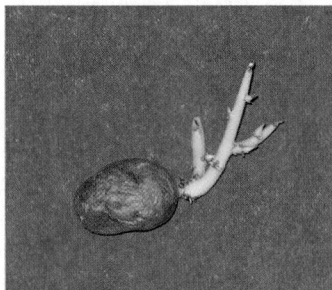 

**(a) Before exposure to light.** A dark-grown potato has tall, spindly stems and nonexpanded leaves—morphological adaptations that enable the shoots to penetrate the soil. The roots are short, but there is little need for water absorption because little water is lost by the shoots.

**(b) After a week's exposure to natural daylight.** The potato plant begins to resemble a typical plant with broad green leaves, short sturdy stems, and long roots. This transformation begins with the reception of light by a specific pigment, phytochrome.

▲ **Figure 31.11 Light-induced de-etiolation (greening) of dark-grown potatoes.**

of different wavelengths of radiation in driving a particular process, such as photosynthesis (see Figure 8.9b). Action spectra are useful in studying *any* process that depends on light, including phototropism **(Figure 31.12)**. By comparing action spectra of various plant responses, researchers determine which responses are mediated by the same photoreceptor (pigment). They also compare action spectra with absorption spectra of pigments; a close correspondence for a given pigment suggests that the pigment is the photoreceptor mediating the response. In the case of photomorphogenesis, action

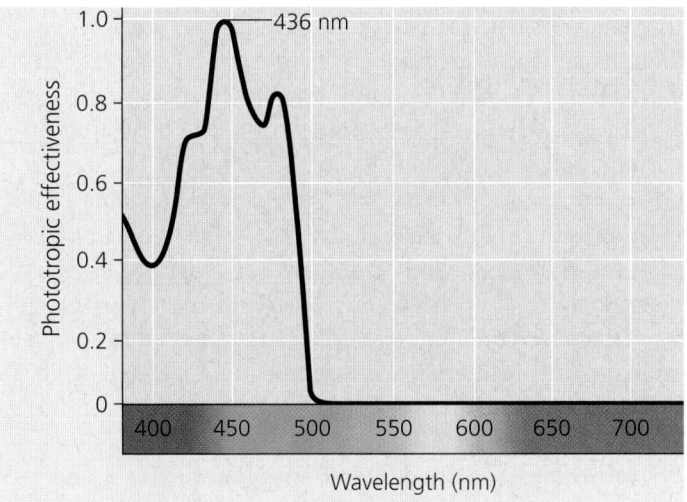

**(a)** This action spectrum illustrates that only light wavelengths below 500 nm (blue and violet light) induce curvature.

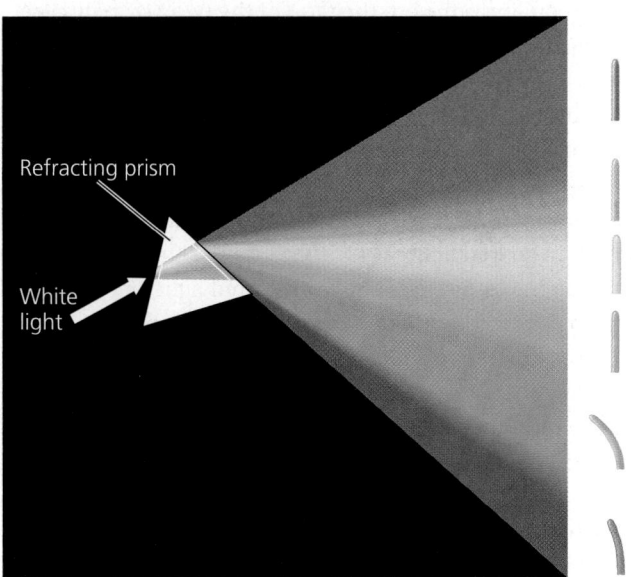

**(b)** When coleoptiles are exposed to light of various wavelengths as shown here, violet light induces slight curvature toward the light and blue light induces the most curvature. The other colors do not induce any curvature.

▲ **Figure 31.12 Action spectrum for blue-light-stimulated phototropism in maize coleoptiles.** Phototropic bending toward light is controlled by phototropin, a photoreceptor sensitive to blue and violet light, particularly blue light.

spectra reveal that red and blue light are the most important colors. In fact, there are two major classes of photoreceptors in plants: **blue-light photoreceptors** and **phytochromes**, photoreceptors that absorb mostly red light.

### Blue-Light Photoreceptors

Blue light initiates a variety of responses in plants, including phototropism, the light-induced opening of stomata (see Figure 29.19), and the light-induced slowing of hypocotyl elongation that occurs when a seedling breaks ground. The biochemical identity of the blue-light photoreceptor was so elusive that in the 1970s, plant physiologists began to call this receptor "cryptochrome" (from the Greek *kryptos*, hidden, and *chrom*, pigment). In the 1990s, molecular biologists analyzing *Arabidopsis* mutants found that plants use different types of pigments to detect blue light. *Cryptochromes*, molecular relatives of DNA repair enzymes, are involved in the blue-light-induced inhibition of stem elongation that occurs, for example, when a seedling first emerges from the soil. *Phototropin* is a protein kinase involved in mediating blue-light-mediated stomatal opening, chloroplast movements in response to light, and phototropic curvatures, such as those studied by the Darwins.

### Phytochrome Photoreceptors

Phytochromes, another class of photoreceptors, regulate many plant responses to light, including seed germination and shade avoidance.

**Phytochromes and Seed Germination** Studies of seed germination led to the discovery of phytochromes. Because of limited nutrient reserves, many types of seeds, especially small ones, germinate only when the light environment and other conditions are near optimal. Such seeds often remain dormant for years until light conditions change. For example, the death of a shading tree or the plowing of a field may create a favorable light environment.

In the 1930s, scientists determined the action spectrum for light-induced germination of lettuce seeds. They exposed water-swollen seeds to a few minutes of single-colored light of various wavelengths and then stored the seeds in the dark. After two days, the researchers counted the number of seeds that had germinated under each light regimen. They found that red light of wavelength 660 nm increased the germination percentage of lettuce seeds maximally, whereas far-red light—that is, light of wavelengths near the upper edge of human visibility (730 nm)—*inhibited* germination compared with dark controls **(Figure 31.13)**. What happens when the lettuce seeds are subjected to a flash of red light followed by a flash of far-red light or, conversely, to far-red light followed by red light? The *last* flash of light determines the seeds' response: The effects of red and far-red light are reversible.

The photoreceptors responsible for the opposing effects of red and far-red light are phytochromes. So far, researchers have

### How does the order of red and far-red illumination affect seed germination?

**Experiment** Scientists at the U.S. Department of Agriculture briefly exposed batches of lettuce seeds to red light or far-red light to test the effects on germination. After the light exposure, the seeds were placed in the dark, and the results were compared with control seeds that were not exposed to light.

**Results** The bar below each photo indicates the sequence of red light exposure, far-red light exposure, and darkness. The germination rate increased greatly in groups of seeds that were last exposed to red light (left). Germination was inhibited in groups of seeds that were last exposed to far-red light (right).

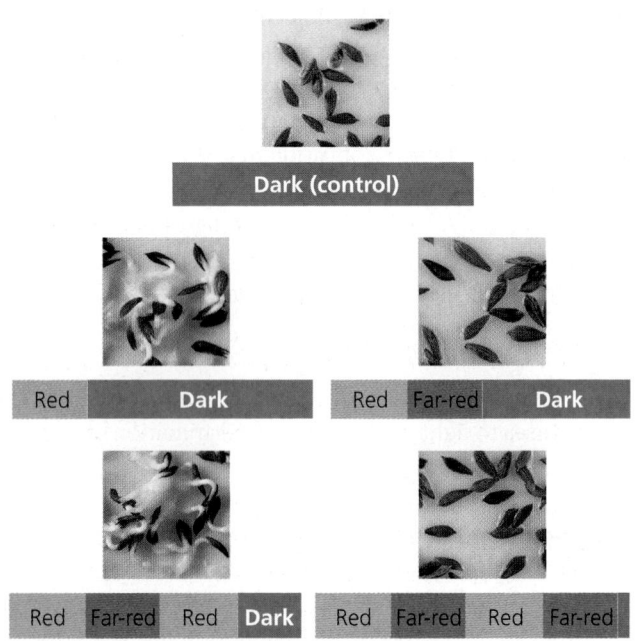

**Conclusion** Red light stimulates germination, and far-red light inhibits germination. The final light exposure is the determining factor. The effects of red and far-red light are reversible.

**Data from** H. Borthwick et al., A reversible photoreaction controlling seed germination, *Proceedings of the National Academy of Sciences, USA* 38:662–666 (1952).

**WHAT IF?** Phytochrome responds faster to red light than to far-red light. If the seeds had been placed in white light instead of the dark after their red and far-red light treatments, would the results have been different? Explain.

identified five phytochromes in *Arabidopsis*, each with a slightly different polypeptide component. In most phytochromes, the light-absorbing portion is photoreversible, converting back and forth between two forms, depending on the color of light it is exposed to. In its $P_r$ form, a phytochrome absorbs red (r) light maximally and is converted to the $P_{fr}$ form; in its $P_{fr}$ form, it absorbs far-red (fr) light and is converted to its $P_r$ form **(Figure 31.14)**. This $P_r \leftrightarrow P_{fr}$ interconversion is a switching mechanism that controls various light-induced events in the life of the plant. $P_{fr}$ is the form of phytochrome that triggers many of a plant's developmental responses to light. For example, $P_r$ in lettuce seeds exposed to red light is converted to $P_{fr}$, stimulating

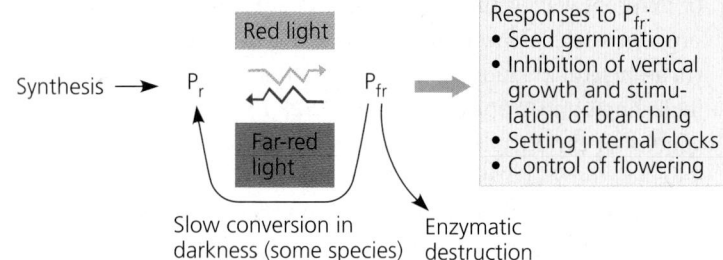

▲ **Figure 31.14 Phytochrome: a molecular switching mechanism.** Absorption of red light causes $P_r$ to change to $P_{fr}$. Far-red light reverses this conversion.

the cellular responses that lead to germination. When red-illuminated seeds are then exposed to far-red light, $P_{fr}$ is converted back to $P_r$, inhibiting the germination response.

How does phytochrome switching explain light-induced germination in nature? Plants synthesize phytochrome as $P_r$, and if seeds are kept in the dark, the pigment remains almost entirely in the $P_r$ form (see Figure 31.14). Sunlight contains both red light and far-red light, but the conversion to $P_{fr}$ is faster than the conversion to $P_r$. Therefore, the ratio of $P_{fr}$ to $P_r$ increases in sunlight. When seeds are exposed to adequate sunlight, the production and accumulation of $P_{fr}$ will trigger their germination.

**Phytochromes and Shade Avoidance** The phytochrome system also provides the plant with information about the *quality* of light. Because sunlight includes both red and far-red radiation, during the day the $P_r \leftrightarrow P_{fr}$ interconversion reaches a dynamic equilibrium, with the ratio of the two phytochrome forms indicating the relative amounts of red and far-red light. This sensing mechanism enables plants to adapt to changes in light conditions. Consider, for example, the "shade avoidance" response of a tree that requires relatively high light intensity. If other trees in a forest shade this tree, the phytochrome ratio shifts in favor of $P_r$ because the forest canopy screens out more red light than far-red light. This is because the chlorophyll pigments in the leaves of the canopy absorb red light and allow far-red light to pass through. The shift in the ratio of red to far-red light induces the tree to allocate more of its resources to growing taller. In contrast, direct sunlight increases the proportion of $P_{fr}$, which stimulates branching and inhibits vertical growth.

In addition to helping plants detect light, phytochrome helps a plant keep track of the passage of days and seasons. To understand phytochrome's role in these timekeeping processes, we must first examine the nature of the plant's internal clock.

### Biological Clocks and Circadian Rhythms

Many plant processes, such as transpiration and the synthesis of certain enzymes, undergo a daily oscillation. Some of these cyclic variations are responses to the changes in light levels and temperature that accompany the 24-hour cycle of day and night. We can control these external factors by growing plants in growth chambers under rigidly maintained conditions of

Noon                    10:00 PM

▲ **Figure 31.15 Sleep movements of a bean plant
(*Phaseolus vulgaris*).** The movements are caused by reversible
changes in the turgor pressure of cells on opposing sides of the pulvini,
motor organs of the leaf.

light and temperature. But even under artificially constant
conditions, many physiological processes in plants, such as the
opening and closing of stomata and the production of photosynthetic enzymes, continue to oscillate with a frequency of
about 24 hours. For example, many legumes lower their leaves
in the evening and raise them in the morning (**Figure 31.15**).
A bean plant continues these "sleep movements" even if kept
in constant light or constant darkness; the leaves are not
simply responding to sunrise and sunset. Such cycles, with a
frequency of about 24 hours and not directly controlled by any
known environmental variable, are called **circadian rhythms**
(from the Latin *circa*, approximately, and *dies*, day).

Recent research supports the idea that the molecular "gears"
of the circadian clock really are internal and not a daily response to some subtle but pervasive environmental cycle, such
as geomagnetism or cosmic radiation. Organisms, including
plants and people, continue their rhythms even after being
placed in deep mine shafts or orbiting in spacecraft, conditions
that alter these subtle geophysical periodicities. However, daily
signals from the environment can entrain (set) the circadian
clock to a period of precisely 24 hours.

If an organism is kept in a constant environment, its circadian rhythms deviate from a 24-hour period (a period is
the duration of one cycle). These free-running periods, as
they are called, vary from about 21 to 27 hours, depending
on the particular rhythmic response. The sleep movements
of bean plants, for instance, have a period of 26 hours when
the plants are kept in the free-running condition of constant
darkness. Deviation of the free-running period from exactly
24 hours does not mean that biological clocks drift erratically.
Free-running clocks are still keeping perfect time, but they are
not synchronized with the outside world. To understand the
mechanisms underlying circadian rhythms, we must distinguish between the clock and the rhythmic processes it controls. For example, the leaves of the bean plant in Figure 31.15
are the clock's "hands" but are not the essence of the clock
itself. If bean leaves are restrained for several hours and then
released, they will reestablish the position appropriate for the

time of day. We can interfere with a biological rhythm, but the
underlying clockwork continues to tick.

At the heart of the molecular mechanisms underlying circadian rhythms are oscillations in the transcription of certain
genes. Mathematical models propose that the 24-hour period
arises from negative-feedback loops involving the transcription of a few central "clock genes." Some clock genes may encode transcription factors that inhibit, after a time delay, the
transcription of the gene that encodes the transcription factor
itself. Such negative-feedback loops, together with a time delay,
are enough to produce oscillations.

### The Effect of Light on the Biological Clock

As we have discussed, the free-running period of the circadian
rhythm of bean leaf movements is 26 hours. Consider a bean
plant placed at dawn in a dark cabinet for 72 hours: Its leaves
would not rise again until 2 hours after natural dawn on the
second day, 4 hours after natural dawn on the third day, and
so on. Shut off from environmental cues, the plant becomes
desynchronized. Desynchronization happens to humans when
we fly across time zones; when we reach our destination, the
clocks on the wall are not synchronized with our internal
clocks. Most organisms are probably prone to jet lag.

The factor that entrains the biological clock to precisely
24 hours every day is light. Both phytochromes and blue-light
photoreceptors can entrain circadian rhythms in plants, but
our understanding of how phytochromes do this is more complete. The mechanism involves turning cellular responses on
and off by means of the $P_r \leftrightarrow P_{fr}$ switch.

Consider again the photoreversible system in Figure 31.14. In
darkness, the phytochrome ratio shifts gradually in favor of the
$P_r$ form, partly as a result of turnover in the overall phytochrome
pool. The pigment is synthesized in the $P_r$ form, and enzymes
destroy more $P_{fr}$ than $P_r$. In some plant species, $P_{fr}$ present at
sundown slowly converts to $P_r$. In darkness, there is no means for
the $P_r$ to be reconverted to $P_{fr}$, but upon illumination, the $P_{fr}$ level
suddenly increases again as $P_r$ is rapidly converted. This increase
in $P_{fr}$ each day at dawn resets the biological clock: Bean leaves
reach their most extreme night position 16 hours after dawn.

In nature, interactions between phytochrome and the biological clock enable plants to measure the passage of night and
day. The relative lengths of night and day, however, change over
the course of the year (except at the equator). Plants use this
change to adjust their activities in synchrony with the seasons.

## Photoperiodism and Responses to Seasons

Imagine the consequences if a plant produced flowers when
pollinators were not present or if a deciduous tree produced
leaves in the middle of winter. Seasonal events are of critical
importance in the life cycles of most plants. Seed germination,
flowering, and the onset and breaking of bud dormancy are
all stages that usually occur at specific times of the year. The
environmental stimulus that plants use most often to detect
the time of year is the photoperiod, the relative lengths of night

and day. A physiological response to photoperiod, such as flowering, is called **photoperiodism**.

### Photoperiodism and Control of Flowering

An early clue to how plants detect seasons came from a mutant variety of tobacco, Maryland Mammoth, that grew tall but failed to flower during summer. It finally bloomed in a greenhouse in December. Researchers determined that the shortening days of winter stimulated this variety to flower. If the plants were kept in light-tight boxes so that lamps could manipulate "day" and "night," flowering occurred only if the day length was 14 hours or shorter. The plants did not flower during summer because at Maryland's latitude, the summer days were too long.

The researchers called Maryland Mammoth a **short-day plant** because it apparently required a light period *shorter* than a critical length to flower. Chrysanthemums, poinsettias, and some soybean varieties are also short-day plants, which generally flower in late summer, fall, or winter. Another group of plants flower only when the light period is *longer* than a certain number of hours. These **long-day plants** generally flower in late spring or early summer. Spinach, for example, flowers when days are 14 hours or longer. Radishes, lettuce, irises, and many cereal varieties are also long-day plants. **Day-neutral plants**, such as tomatoes, rice, and dandelions, are unaffected by photoperiod and flower when they reach a certain stage of maturity, regardless of day length.

**Critical Night Length**  In the 1940s, researchers learned that flowering and other responses to photoperiod are actually controlled by night length, not day length. Many of these scientists worked with cocklebur (*Xanthium strumarium*), a short-day plant that flowers only when days are 16 hours or shorter (and nights are at least 8 hours long). These researchers found that if the light portion of the photoperiod is broken by a brief exposure to darkness, flowering proceeds. However, if the dark part of the photoperiod is interrupted by even a few minutes of dim light, cocklebur will not flower, and this turned out to be true for other short-day plants **(Figure 31.16a)**. Cocklebur is unresponsive to day length, but it requires at least 8 hours of continuous darkness to flower. Short-day plants are really long-night plants, but the older term is embedded firmly in the lexicon of plant physiology. Similarly, long-day plants are actually short-night plants. A long-day plant grown on photoperiods of long nights that would not normally induce flowering will flower if the period of continuous darkness is interrupted by a few minutes of light **(Figure 31.16b)**. Notice that we distinguish long-day from short-day plants *not* by an absolute night length but by whether the critical night length sets a maximum (long-day plants) or minimum (short-day plants) number of hours of darkness required for flowering. In both cases, the actual number of hours in the critical night length is specific to each species of plant.

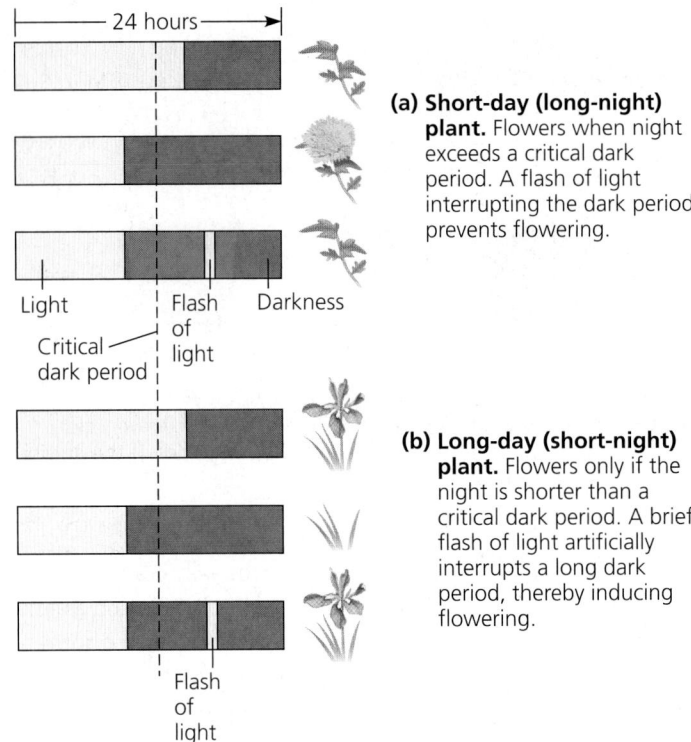

**(a) Short-day (long-night) plant.** Flowers when night exceeds a critical dark period. A flash of light interrupting the dark period prevents flowering.

**(b) Long-day (short-night) plant.** Flowers only if the night is shorter than a critical dark period. A brief flash of light artificially interrupts a long dark period, thereby inducing flowering.

▲ **Figure 31.16 Photoperiodic control of flowering.**

Red light is the most effective color in interrupting the dark portion of the photoperiod **(Figure 31.17)**. For example, if a flash of red (R) light during the dark period is followed by a flash of far-red (FR) light, then the plant detects no interruption of night length. As in the case of phytochrome-mediated seed germination, red/far-red photoreversibility, a hallmark of phytochrome responses, is evident in the night interruption.

Plants detect night length very precisely; some short-day plants will not flower if night is even 1 minute shorter than the critical length. Some plant species always flower on the same day each year. It appears that plants use their biological clock, entrained by night length with the help of phytochrome, to tell the season of the year. The floriculture (flower-growing) industry applies this knowledge to produce flowers out of season. Chrysanthemums, for instance, are short-day plants that normally bloom in fall, but their blooming can be stalled until Mother's Day in May by punctuating each long night with a flash of light, thus turning one long night into two short nights.

Some plants bloom after a single exposure to the photoperiod required for flowering. Other species need several successive days of the appropriate photoperiod. Still others respond to a photoperiod only if they have been previously exposed to some other environmental stimulus, such as a period of cold. Winter wheat, for example, will not flower unless it has been exposed to several weeks of temperatures below 10°C. The use of pretreatment with cold to induce flowering is called

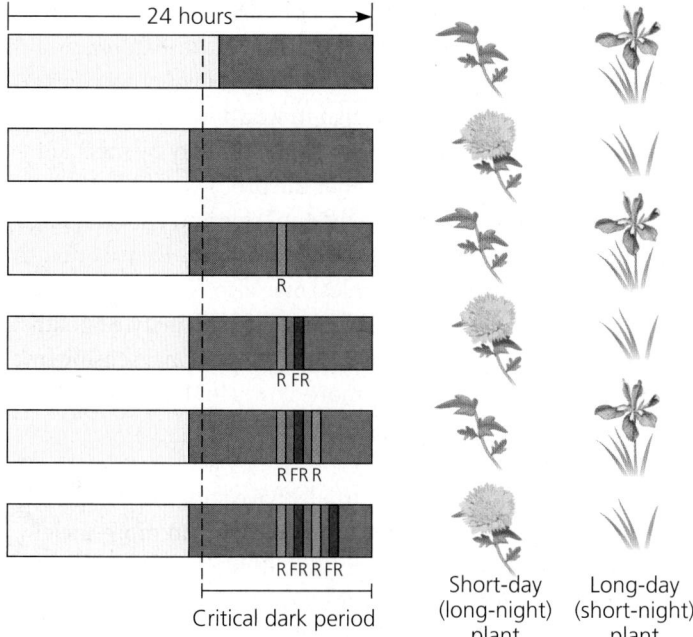

▲ **Figure 31.17 Reversible effects of red and far-red light on photoperiodic response.** A flash of red (R) light shortens the dark period. A subsequent flash of far-red (FR) light cancels the red flash's effect.

**?** *How would a single flash of full-spectrum light affect each plant?*

**vernalization** (from the Latin for "spring"). Several weeks after winter wheat is vernalized, a photoperiod with long days (short nights) induces flowering.

### A Flowering Hormone?

Although flowers form from apical or axillary bud meristems, it is leaves that detect changes in photoperiod and produce signaling molecules that cue buds to develop as flowers. In many short-day and long-day plants, exposing just one leaf to the appropriate photoperiod is enough to induce flowering. Indeed, as long as one leaf is left on the plant, photoperiod is detected and floral buds are induced. If all leaves are removed, the plant is insensitive to photoperiod.

Classic experiments revealed that the floral stimulus could move across a graft from an induced plant to a noninduced plant and trigger flowering in the latter. Moreover, the flowering stimulus appears to be the same for short-day and long-day plants, despite the different photoperiodic conditions required for leaves to send this signal **(Figure 31.18)**. The hypothetical signaling molecule for flowering, called **florigen**, remained unidentified for over 70 years as scientists focused on small hormone-like molecules. However, large macromolecules, including proteins, can move by the symplastic route via plasmodesmata and regulate plant development. It now appears that florigen is a protein. A gene called *FLOWERING LOCUS T* (*FT*) is activated in leaf cells during conditions favoring flowering, and the FT protein travels through the

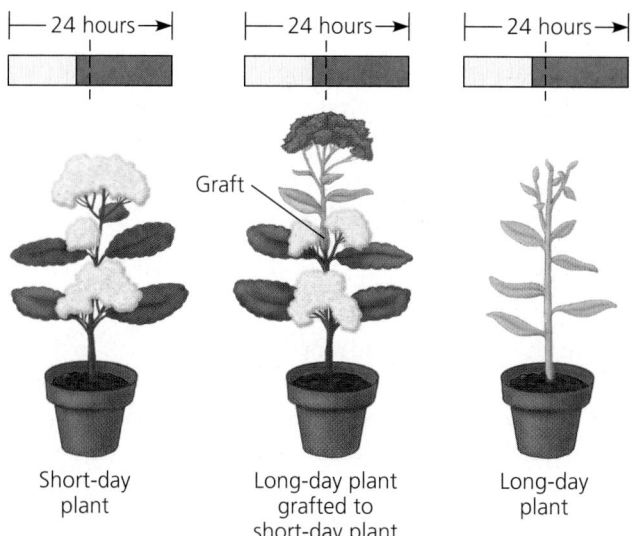

Graft

Short-day plant | Long-day plant grafted to short-day plant | Long-day plant

▲ **Figure 31.18 Experimental evidence for a flowering hormone.** If grown individually under short-day conditions, a short-day plant will flower and a long-day plant will not. However, both will flower if grafted together and exposed to short days. This result indicates that a flower-inducing substance (florigen) is transmitted across grafts and induces flowering in both short-day and long-day plants.

**WHAT IF?** *If flowering were inhibited in both parts of the grafted plants, what would you conclude?*

symplasm to the shoot apical meristem and initiates the transition of a bud's meristem from a vegetative to a flowering state.

### CONCEPT CHECK 31.2

1. If an enzyme in field-grown soybean leaves is most active at noon and least active at midnight, is its activity under circadian regulation?
2. **WHAT IF?** If a plant flowers in a controlled chamber with a daily cycle of 10 hours of light and 14 hours of darkness, is it a short-day plant? Explain.
3. **MAKE CONNECTIONS** Plants detect the quality of their light environment by using blue-light photoreceptors and red-light-absorbing phytochromes. Suggest a reason why plants are so sensitive to these colors of light. (See Figure 8.9.)

For suggested answers, see Appendix A.

---

**CONCEPT 31.3**

# Plants respond to a wide variety of stimuli other than light

Plants are immobile, but adaptations have evolved by natural selection that enable them to adjust to a wide range of environmental circumstances by developmental or physiological means. Light is so important in the life of a plant that we devoted the entire previous section to a plant's reception of and response to this one environmental factor. In this section, we examine responses to some of the other environmental stimuli that a plant commonly encounters.

## Gravity

Because plants are photoautotrophs, it is not surprising that mechanisms for growing toward sunlight have evolved. But what environmental cue does the shoot of a young seedling use to grow upward when it is completely underground and there is no light for it to detect? Similarly, what environmental factor prompts the young root to grow downward? The answer to both questions is gravity.

Place a plant on its side, and it adjusts its growth so that the shoot bends upward and the root curves downward. In their responses to gravity, or **gravitropism**, roots display positive gravitropism **(Figure 31.19a)** and shoots exhibit negative gravitropism. Gravitropism occurs as soon as a seed germinates, ensuring that the root grows into the soil and the shoot grows toward sunlight, regardless of how the seed is oriented when it lands.

Plants may detect gravity by means of **statoliths**, dense cytoplasmic components that settle under the influence of gravity to the lower portions of the cell. The statoliths of vascular plants are specialized plastids containing dense starch grains **(Figure 31.19b)**. In roots, statoliths are located in certain cells of the root cap. According to one hypothesis, the aggregation of statoliths at the low points of these cells triggers a redistribution of calcium, which causes lateral transport of auxin within the root. The calcium and auxin accumulate on the lower side of the root's zone of elongation. At high concentration, auxin inhibits cell elongation, an effect that slows growth on the root's lower side. The more rapid elongation of cells on the upper side eventually causes the root to grow straight downward.

Falling statoliths, however, may not be indispensable for gravitropism. For example, there are mutants of *Arabidopsis* and tobacco that lack statoliths but are still capable of gravitropism, though the response is slower than in wild-type plants. It could be that the entire cell helps the root sense gravity by mechanically pulling on proteins that tether the protoplast to the cell wall, stretching the proteins on the "up" side and compressing the proteins on the "down" side of the root cells. Dense organelles, in addition to starch granules, may also contribute by distorting the cytoskeleton as they are pulled by gravity. Statoliths, because of their density, may enhance gravitational sensing by a mechanism that simply works more slowly in their absence.

## Mechanical Stimuli

Trees in windy environments usually have shorter, stockier trunks than trees of the same species growing in more sheltered locations. The advantage of this stunted morphology is that it enables the plant to resist strong gusts of wind. The term **thigmomorphogenesis** (from the Greek *thigma*, touch) refers to the changes in form that result from mechanical perturbation. Plants are very sensitive to mechanical stress: Even the act of measuring the length of a leaf with a ruler alters its subsequent growth. Rubbing the stems of a young plant a couple of times daily results in plants that are shorter than controls **(Figure 31.20)**.

Some plant species have become, over the course of their evolution, "touch specialists." Acute responsiveness to mechanical stimuli is an integral part of these plants' "life strategies." Most vines and other climbing plants have tendrils that coil rapidly around supports (see Figure 28.7). These grasping organs usually grow straight until they touch something; the contact stimulates a coiling response caused by differential growth of cells on opposite sides of the tendril. This directional

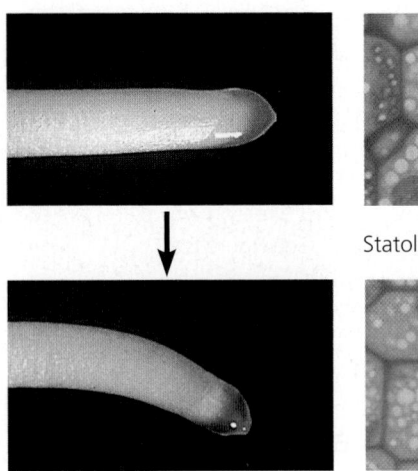
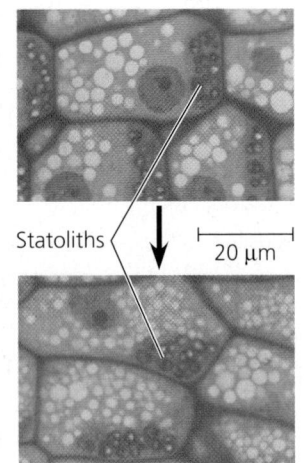

**(a)** Over the course of hours, a horizontally oriented primary root of maize bends gravitropically until its growing tip becomes vertically oriented (LMs).

**(b)** Within minutes after the root is placed horizontally, plastids called statoliths begin settling to the lowest sides of root cap cells. This settling may be the gravity-sensing mechanism that leads to redistribution of auxin and differing rates of elongation by cells on opposite sides of the root (LMs).

Statoliths

20 μm

▲ **Figure 31.19 Positive gravitropism in roots: the statolith hypothesis.**

▲ **Figure 31.20 Thigmorphogenesis in *Arabidopsis*.** The shorter plant on the left was rubbed twice a day. The untouched plant (right) grew much taller.

growth in response to touch is called **thigmotropism**, and it allows the vine to take advantage of whatever mechanical supports it comes across as it climbs upward toward a forest canopy.

Other examples of touch specialists are plants that undergo rapid leaf movements in response to mechanical stimulation. For example, when the compound leaf of the sensitive plant *Mimosa pudica* is gently touched, its leaflets fold together **(Figure 31.21)**. This response, which takes only a second or two, results from a rapid loss of turgor in specialized motor cells located at the base of each leaflet. The

**(a) Unstimulated state (leaflets spread apart)**    **(b) Stimulated state (leaflets folded)**

▲ **Figure 31.21 Rapid turgor movements by the sensitive plant (*Mimosa pudica*).**

motor cells suddenly become flaccid after stimulation because they lose potassium ions, causing water to leave the cells by osmosis. It takes about 10 minutes for the cells to regain their turgor and restore the "unstimulated" form of the leaf. The function of the sensitive plant's behavior invites speculation. Perhaps by folding its leaves and reducing its surface area when jostled, the plant appears less leafy and appetizing to herbivores.

A remarkable feature of rapid leaf movements is the mode of transmission of the stimulus through the plant. If one leaflet on a sensitive plant is touched, first that leaflet responds, then the adjacent leaflet responds, and so on, until all the leaflet pairs have folded together. From the point of stimulation, the signal that produces this response travels at a speed of about 1 cm/sec. An electrical impulse traveling at the same rate can be detected when electrodes are attached to the leaf. These impulses, called **action potentials**, resemble nerve impulses in animals, though the action potentials of plants are thousands of times slower. Action potentials have been discovered in many species of algae and plants and may be used as a form of internal communication. For example, in the Venus flytrap (*Dionaea muscipula*), action potentials are transmitted from sensory hairs in the trap to the cells that respond by closing the trap (see Figure 29.15). In the case of *Mimosa pudica*, more violent stimuli, such as touching a leaf with a hot needle, causes *all* the leaves and leaflets on a plant to droop, but this whole-plant response involves the spread of signaling molecules released from the injured area to other parts of the shoot.

## Environmental Stresses

Certain factors in the environment may change severely enough to have a potentially adverse effect on a plant's survival, growth, and reproduction. Environmental stresses, such as flooding, drought, or extreme temperatures, can have a devastating impact on crop yields in agriculture. In natural ecosystems, plants that cannot tolerate an environmental stress will either succumb or be outcompeted by other plants. Thus, environmental stresses are an important factor in determining the geographic ranges of plants. Here we'll consider some of the more common **abiotic** (nonliving) stresses

that plants encounter. In the last section of this chapter, we'll examine the defensive responses of plants to common **biotic** (living) stresses, such as herbivores and pathogens.

### Drought

On a dry, sunny day, a plant may wilt because its water loss by transpiration exceeds water absorption from the soil. Prolonged drought, of course, will kill a plant, but plants have control systems that enable them to cope with less extreme water deficits.

Many of a plant's responses to water deficit help the plant conserve water by reducing the rate of transpiration. Water deficit in a leaf causes stomata to close, thereby slowing transpiration dramatically (see Figure 29.19). Water deficit stimulates increased synthesis and release of ABA in the leaves and roots; this hormone helps keep stomata closed by acting on guard cell membranes. Leaves respond to water deficit in several other ways. For example, when the leaves of grasses wilt, they roll into a tubelike shape that reduces transpiration by exposing less leaf surface to dry air and wind. Other plants, such as ocotillo (see Figure 29.20), shed their leaves in response to seasonal drought. Although these leaf responses conserve water, they also reduce photosynthesis, which is one reason why a drought diminishes crop yield. Plants can even take advantage of "early warnings" in the form of chemical signals from wilting neighbors and prime themselves to respond more readily and intensely to impending drought stress—a phenomenon you'll explore in the **Scientific Skills Exercise**.

### Flooding

Too much water is also problematic for plants. An overwatered houseplant may suffocate because the soil lacks the air spaces that provide oxygen for cellular respiration in the roots. Some plants are structurally adapted to very wet habitats. For example, the submerged roots of mangroves, which inhabit coastal marshes, are continuous with aerial roots exposed to oxygen (see Figure 28.5). But how do less specialized plants cope with oxygen deprivation in waterlogged soils? Oxygen deprivation stimulates the production of ethylene, which causes

## *Interpreting Experimental Results from a Bar Graph*

**Do Drought-Stressed Plants Communicate Their Condition to Their Neighbors?** Researchers wanted to learn if plants can communicate drought-induced stress to neighboring plants and, if so, whether they use aboveground or belowground signals.

**How the Experiment Was Done** Eleven potted pea plants (*Pisum sativum*) were placed equidistantly in a row. The root systems of plants 6–11 were connected to their immediate neighbors by tubes, which allowed chemicals to move from the roots of one plant to the roots of the next plant without moving through the soil. The root systems of plants 1–6 were not connected. Osmotic shock was inflicted on plant 6 using a highly concentrated solution of mannitol, a natural sugar commonly used to mimic drought stress in vascular plants.

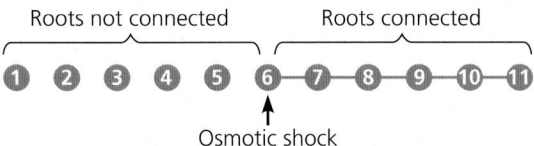

Fifteen minutes following the osmotic shock to plant 6, researchers measured the width of stomatal openings in leaves from all the plants. A control experiment was also done in which water was added to plant 6 instead of mannitol.

### INTERPRET THE DATA

1. How do the widths of the stomatal openings of plants 6–8 and plants 9 and 10 compare with those of the other plants in the experiment? What does this indicate about the state of plants 6–8 and 9 and 10? (For information about reading graphs, see the Scientific Skills Review in Appendix F and in the Study Area in MasteringBiology.)

2. Do the data support the idea that plants can communicate their drought-stressed condition to their neighbors? If so, is the communication via the shoot system or the root system? Refer specifically to the data in answering both questions.

3. Why was it necessary to make sure that chemicals could not move through the soil from one plant to the next?

▶ Pea plant (*Pisum sativum*)

**Data from the Experiment** The graph below summarizes the stomatal measurements.

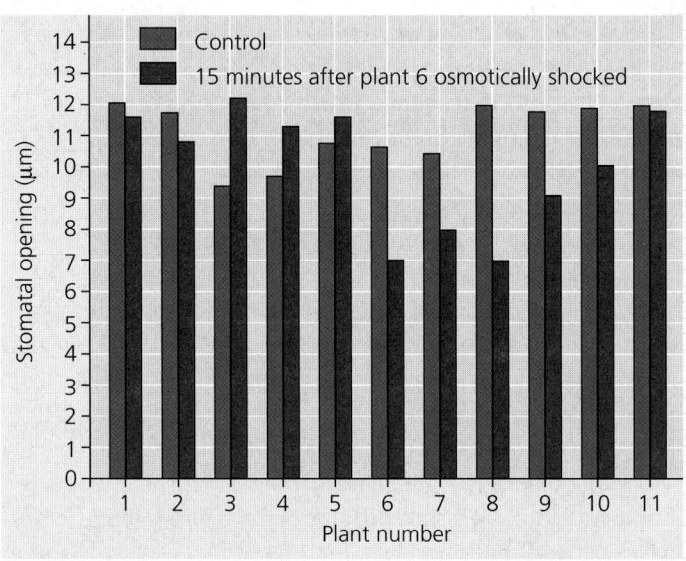

**Data from** O. Falik et al., Rumor has it...: Relay communication of stress cues in plants, *PLoS ONE* 6(11): e23625 (2011).

4. When the experiment was run for 1 hour rather than 15 minutes, the results were about the same except that the stomatal openings of plants 9–11 were comparable to those of plants 6–8. Suggest a reason why.

5. Why was water added to plant 6 instead of mannitol in the control experiment? What do the results of this experiment indicate?

(MB) A version of this Scientific Skills Exercise can be assigned in MasteringBiology.

---

some cells in the root cortex to die. The destruction of these cells creates air tubes that function as "snorkels," providing oxygen to the submerged roots **(Figure 31.22)**.

### Salt Stress

An excess of sodium chloride or other salts in the soil threatens plants for two reasons. First, by lowering the water potential of the soil solution, salt can cause a water deficit in plants even though the soil has plenty of water. As the water potential of the soil solution becomes more negative, the water potential gradient from soil to roots is lowered, thereby reducing water uptake and growth (see Figure 29.18). Another problem with saline soil is that sodium and certain other ions are toxic to plants when their concentrations are too high. Many plants can respond to moderate soil salinity by producing solutes that are well tolerated at high concentrations: These mostly organic compounds keep the water potential of cells more

negative than that of the soil solution without admitting toxic quantities of salt. However, most plants cannot survive salt stress for long. The exceptions are *halophytes*, salt-tolerant plants with adaptations such as salt glands that pump salts out across the leaf epidermis.

### Heat Stress

Excessive heat may harm and even kill a plant by denaturing its enzymes. Transpiration helps cool leaves by evaporative cooling. On a warm day, for example, the temperature of a leaf may be 3–10°C below the ambient air temperature. Hot, dry weather also tends to dehydrate many plants; the closing of stomata in response to this stress conserves water but then sacrifices evaporative cooling. This dilemma is one reason why very hot, dry days take a toll on most plants. Most plants have a backup response that enables them to survive heat stress. Above a certain temperature—about 40°C for most

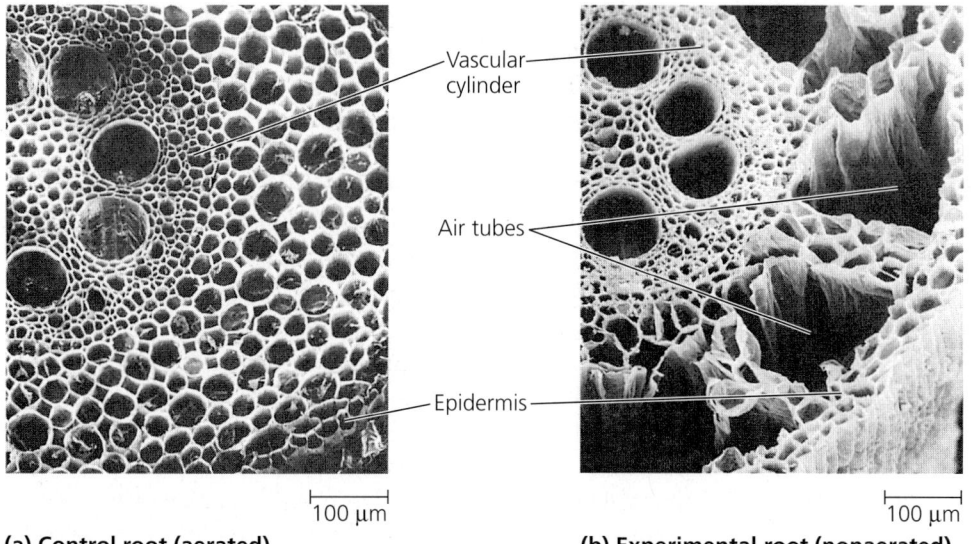

Vascular cylinder

Air tubes

Epidermis

100 μm

**(a) Control root (aerated)**

100 μm

**(b) Experimental root (nonaerated)**

▲ **Figure 31.22 A developmental response of maize roots to flooding and oxygen deprivation. (a)** A cross section of a control root grown in an aerated hydroponic medium. **(b)** A root grown in a nonaerated hydroponic medium. Ethylene-stimulated apoptosis (programmed cell death) creates the air tubes (SEMs).

plants in temperate regions—plant cells begin synthesizing **heat-shock proteins**, which help protect other proteins from denaturing.

### Cold Stress

One problem plants face when the temperature of the environment falls is a change in the fluidity of cell membranes. When a membrane cools below a critical point, membranes lose their fluidity as the lipids become locked into crystalline structures. This alters solute transport across the membrane and also adversely affects the functions of membrane proteins. Plants respond to cold stress by altering the lipid composition of their membranes. For example, membrane lipids increase in their proportion of unsaturated fatty acids, which have shapes that help keep membranes more fluid at low temperatures.

Freezing is another type of cold stress. At subfreezing temperatures, ice forms in the cell walls and intercellular spaces of most plants. The cytosol generally does not freeze at the cooling rates encountered in nature because it contains more solutes than the very dilute solution found in the cell wall, and solutes lower the freezing point of a solution. The reduction in liquid water in the cell wall caused by ice formation lowers the extracellular water potential, causing water to leave the cytoplasm. The resulting increase in the concentration of ions in the cytoplasm is harmful and can lead to cell death. Whether the cell survives depends largely on how well it resists dehydration. In regions with cold winters, native plants are adapted to cope with freezing stress. For example, before the onset of winter, the cells of many frost-tolerant species increase cytoplasmic levels of specific solutes, such as sugars, that are well tolerated at high concentrations and that help reduce the loss of water from the cell during

extracellular freezing. The unsaturation of membrane lipids also increases, thereby maintaining proper levels of membrane fluidity.

**EVOLUTION** Many organisms, including certain vertebrates, fungi, bacteria, and many species of plants, have special proteins that hinder ice crystals from growing, helping the organism escape freezing damage. First described in Arctic fish in the 1950s, these *antifreeze proteins* permit survival at temperatures below 0°C. Antifreeze proteins bind to small ice crystals and inhibit their growth or, in the case of plants, prevent the crystallization of ice. The five major classes of antifreeze proteins differ markedly in their amino acid sequences but have a similar three-dimensional structure, suggesting convergent evolution. Progress is being made in increasing the freezing tolerance of crop plants by genetically engineering antifreeze protein genes into their genomes.

### CONCEPT CHECK 31.3

1. Thermal images are photographs of the heat emitted by an object. Researchers have used thermal imaging of plants to isolate mutants that overproduce abscisic acid. Suggest a reason why these mutants are warmer than wild-type plants under conditions that are normally nonstressful.

2. A greenhouse worker finds that potted chrysanthemums nearest to the aisles are often shorter than those in the middle of the bench. Explain this "edge effect," a common problem in horticulture.

3. **WHAT IF?** If you removed the root cap from a root, would the root still respond to gravity? Explain.

For suggested answers, see Appendix A.

### CONCEPT 31.4

# Plants respond to attacks by herbivores and pathogens

**EVOLUTION** Through natural selection, plants have evolved many types of interactions with other species in their communities. Some interspecific interactions are mutually beneficial, such as the associations of plants with mycorrhizal fungi (see Figure 29.14) or with pollinators (see Figure 30.6). Many plant interactions with other organisms, however, do not benefit the plant. As primary producers, plants are at the base of most food webs and are subject to attack by a wide range of plant-eating (herbivorous) animals. A plant is also subject to infection by diverse pathogens that can damage tissues or even kill the plant. Plants counter these threats by means of defense systems that deter herbivores or protect against pathogens.

## Defenses Against Herbivores

**Herbivory**, animals eating plants, is a stress that plants face in any ecosystem. The mechanical damage caused by herbivores reduces the size of plants, hindering their ability to acquire resources. It can also restrict growth because many species divert some of their energy to defend against herbivores. Furthermore, it opens up portals for infection by viruses, bacteria, and fungi. Plants prevent excessive herbivory through a multitude of mechanisms, including physical defenses such as thorns, trichomes, and spines **(Figure 31.23a)**, chemical defenses such as the production of distasteful or toxic compounds **(Figure 31.23b)**, and behavioral defenses such as the recruitment of predatory animals that help defend the plant against specific herbivores **(Figure 31.23c)**.

The volatile molecules a plant releases in response to herbivore damage can also function as an early warning system for nearby plants of the same species. For example, lima bean plants infested with spider mites release a cocktail of volatile chemicals that signal "news" of the attack to neighboring, noninfested lima bean plants. In response to these volatile compounds, the neighbors instigate biochemical changes that make themselves less susceptible, including the release of volatile chemicals that attract another predatory mite species that feeds on spider mites. Researchers have even transgenically engineered *Arabidopsis* plants to produce volatile chemicals that attract carnivorous predatory mites in other plants. The mites become attracted to the genetically modified *Arabidopsis*, a finding that could have implications for the genetic engineering of insect resistance in crop plants.

## Defenses Against Pathogens

A plant's first line of defense against infection is the physical barrier presented by the epidermis and periderm of the plant body (see Figure 28.19). This line of defense, however, is not impenetrable. The mechanical wounding of leaves by herbivores, for example, opens up portals for invasion by pathogens. Even when plant tissues are intact, viruses, bacteria, and the spores and hyphae of fungi can still enter the plant through natural openings in the epidermis, such as stomata. Once the physical lines of defense are breached, a plant's next lines of defense are two types of immune responses: PAMP-triggered immunity and effector-triggered immunity.

### PAMP-Triggered Immunity

Once successfully invaded by a pathogen, the plant mounts the first of two lines of immune defense, which results in a chemical attack that isolates the pathogen and prevents its spread from the site of infection. This first line of immune defense, called *PAMP-triggered immunity*, depends on the plant's ability to recognize **pathogen-associated molecular patterns (PAMPs)**, molecular sequences that are specific to certain pathogens. For example, bacterial *flagellin*, a major protein found in bacterial flagella, is a PAMP. Some pathogenic soil bacteria can be splashed onto the shoots of plants by raindrops. If these bacteria penetrate the plant, a specific amino acid sequence within flagellin is detected by a Toll-like receptor, a type of receptor also found in animals, where it plays a key role in the innate immune system (see Concept 35.1). The innate immune system is an evolutionarily old defense and is the dominant immune system in plants, fungi, insects, and primitive multicellular organisms. Unlike vertebrates, plants do not have an adaptive immune system: Plants neither generate antibody or T cell responses nor possess mobile cells that detect and attack pathogens.

PAMP recognition in plants triggers a chain of signaling events that lead to the local production of broad-spectrum antimicrobial chemicals called *phytoalexins*, which are compounds having fungicidal and bactericidal properties. The plant cell wall is also toughened, hindering further progress of the pathogen. Similar but even stronger defenses are initiated by the second plant immune response, effector-triggered immunity.

### Effector-Triggered Immunity

**EVOLUTION** Over the course of evolution, plants and pathogens have engaged in an arms race. PAMP-triggered immunity can be overcome by the evolution of pathogens that evade detection by the plant. These pathogens deliver **effectors**, pathogen-encoded proteins that cripple the plant's innate immune system, directly into plant cells. For example, some bacteria deliver effectors that block the perception of flagellin.

**(a)** Physical defenses include bristles on the spines of some cacti, which have fearsome barbs that tear flesh during removal.

◄ Bristles on cactus spines

**(b)** Chemical defenses include narcotics in fruits of the opium poppy (*Papaver somniferum*). When any herbivore damages the fruit, chemicals that intoxicate the animal are released in a milky sap.

▲ Opium poppy fruit

**(c)** Behavioral defenses include the release of volatile chemicals by a leaf being eaten by a caterpillar. The chemicals attract parasitoid wasps, which inject their eggs into the caterpillar. When the eggs hatch, the larvae eat their way through the caterpillar from the inside out. Then they form cocoons on the caterpillar before emerging as adults.

◄ Wasp cocoons on a caterpillar

▼ Adult wasp emerging from a cocoon

▲ **Figure 31.23 Some defense responses against herbivores.**

These effectors allow the pathogen to redirect the host's metabolism to the pathogen's advantage.

The suppression of PAMP-triggered immunity by pathogen effectors led to the evolution of *effector-triggered immunity*. Because there are thousands of effectors, this plant defense is typically made up of hundreds of disease resistance (*R*) genes. Each *R* gene codes for an R protein that can be activated by a specific effector. Signal transduction pathways then lead to an arsenal of defense responses, including a local defense called the *hypersensitive response* and a general defense called *systemic acquired resistance*. Local and systemic responses to pathogens require extensive genetic reprogramming and commitment of cellular resources. Therefore, a plant activates these defenses only after detecting an invading pathogen.

**The Hypersensitive Response** The **hypersensitive response** refers to the local tissue death that occurs at and near the infection site. In many cases, the hypersensitive response restricts the spread of a pathogen. As indicated in **Figure 31.24**, the hypersensitive response is initiated as part of effector-triggered immunity. The hypersensitive response is part of a complex defense response that involves the production of enzymes and chemicals

that impair the pathogen's cell wall integrity, metabolism, or reproduction. Effector-triggered immunity also stimulates the formation of lignin and the cross-linking of molecules within the plant cell wall, responses that hinder the spread of the pathogen to other parts of the plant. As shown in the upper right of the figure, the hypersensitive response results in localized lesions on a leaf. As "sick" as such a leaf appears, it will still survive, and its defensive response will help protect the rest of the leaf.

**Systemic Acquired Resistance** The hypersensitive response is localized and specific. However, as noted previously, pathogen invasions can also produce signaling molecules that "sound the alarm" of infection to the whole plant. The resulting **systemic acquired resistance** arises from the plant-wide expression of defense genes. It is nonspecific, providing protection against a diversity of pathogens that can last for days. A signaling molecule called methylsalicylic acid is produced around the infection site, carried by the phloem throughout the plant, and then converted to **salicylic acid** in areas remote from the sites of infection. Salicylic acid activates a signal transduction pathway that readies the defense system to respond rapidly to another infection (see Figure 31.24).

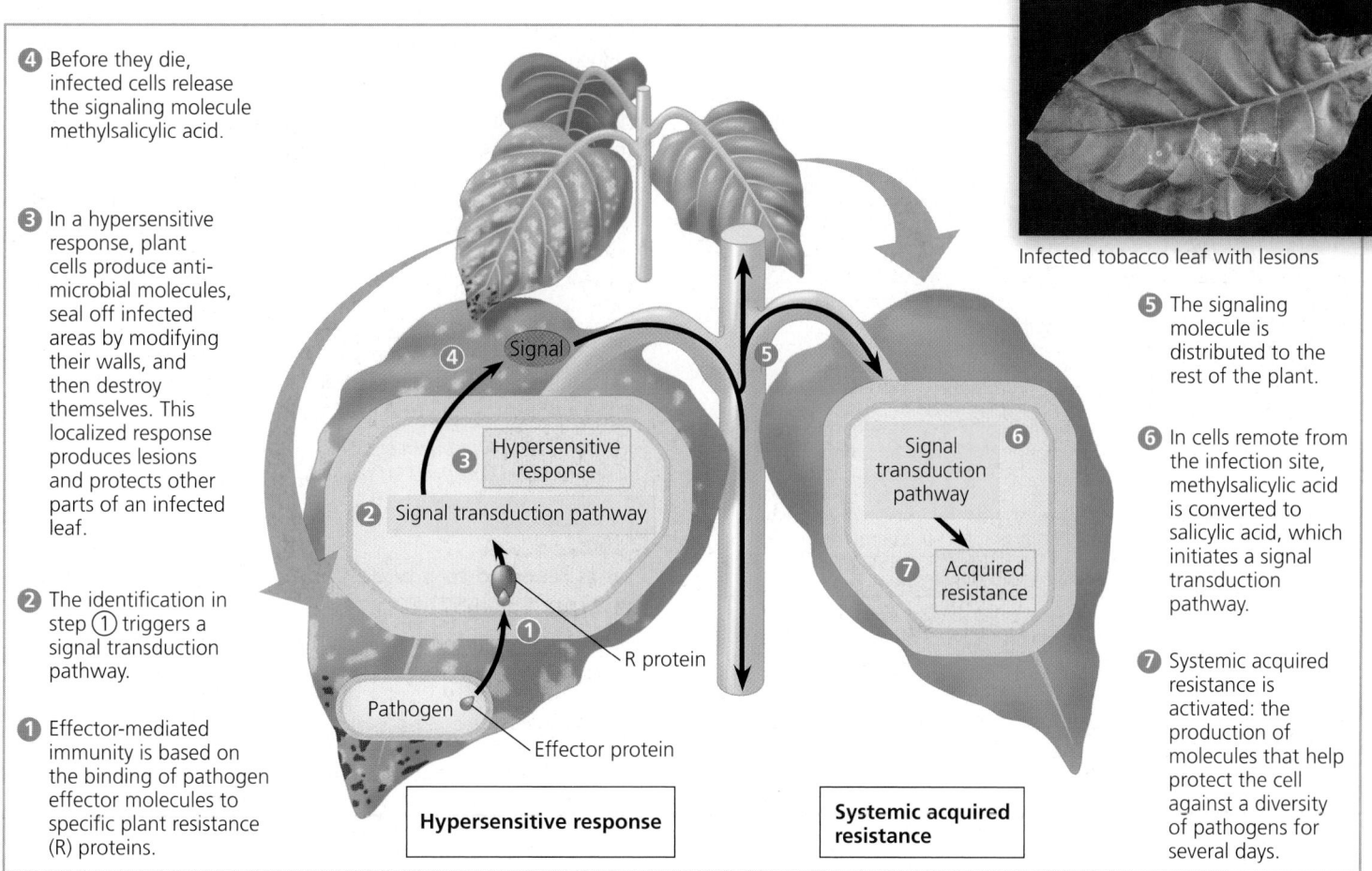

❹ Before they die, infected cells release the signaling molecule methylsalicylic acid.

❸ In a hypersensitive response, plant cells produce antimicrobial molecules, seal off infected areas by modifying their walls, and then destroy themselves. This localized response produces lesions and protects other parts of an infected leaf.

❷ The identification in step ① triggers a signal transduction pathway.

❶ Effector-mediated immunity is based on the binding of pathogen effector molecules to specific plant resistance (R) proteins.

Infected tobacco leaf with lesions

Signal

❹

Hypersensitive response
❸

❷ Signal transduction pathway

R protein

Pathogen

Effector protein

**Hypersensitive response**

❺

Signal transduction pathway ❻

Acquired resistance ❼

**Systemic acquired resistance**

❺ The signaling molecule is distributed to the rest of the plant.

❻ In cells remote from the infection site, methylsalicylic acid is converted to salicylic acid, which initiates a signal transduction pathway.

❼ Systemic acquired resistance is activated: the production of molecules that help protect the cell against a diversity of pathogens for several days.

▲ **Figure 31.24 Effector-triggered defense responses against pathogens.** Plants can often prevent the systemic spread of infection by instigating a hypersensitive response. This response helps isolate the pathogen by producing lesions that form "rings of death" around the sites of infection.

Plant disease epidemics, such as the potato late blight (see Concept 25.4) that caused the Irish potato famine of the 1840s, can lead to incalculable human misery. Other diseases, such as chestnut blight (see Concept 26.5) and sudden oak death (see Concept 41.5), can dramatically alter community structures. Plant epidemics are often the result of infected plants or timber being inadvertently transported around the world. As global commerce increases, such epidemics will become increasingly more common. To prepare for such outbreaks, plant biologists are stockpiling the seeds of wild relatives of crop plants in special storage facilities. Scientists hope that undomesticated relatives may have genes that will be able to curb the next plant epidemic.

## CONCEPT CHECK 31.4

1. What are some drawbacks of spraying fields with general-purpose insecticides?
2. Chewing insects mechanically damage plants and lessen the surface area of leaves for photosynthesis. In addition, these insects make plants more vulnerable to pathogen attack. Suggest a reason why.
3. **WHAT IF?** Suppose a scientist finds that a population of plants growing in a breezy location is more prone to herbivory by insects than a population of the same species growing in a sheltered area. Suggest a hypothesis to account for this observation.

For suggested answers, see Appendix A.

---

# 31 Chapter Review

Go to **MasteringBiology®** for Assignments, the eText, and the Study Area with Animations, Activities, Vocab Self-Quiz, and Practice Tests.

## SUMMARY OF KEY CONCEPTS

**VOCAB SELF-QUIZ**

goo.gl/gbai8v

### CONCEPT 31.1

**Plant hormones help coordinate growth, development, and responses to stimuli (pp. 640–648)**

- **Hormones** control plant growth and development by affecting the division, elongation, and differentiation of cells. Some hormones also mediate the responses of plants to environmental stimuli.

| Plant Hormone | Major Responses |
|---|---|
| Auxin | Stimulates cell elongation; regulates branching and organ bending |
| Cytokinins | Stimulate plant cell division; promote later bud growth; slow organ death |
| Gibberellins | Promote stem elongation; help seeds break dormancy and use stored reserves |
| Brassinosteroids | Chemically similar to the sex hormones of animals; induce cell elongation and division |
| Abscisic acid (ABA) | Promotes stomatal closure in response to drought; promotes seed dormancy |
| Ethylene | Mediates fruit ripening |

**?** *Is there any truth to the old adage "One bad apple spoils the whole bunch"? Explain.*

### CONCEPT 31.2

**Responses to light are critical for plant success (pp. 648–653)**

- **Blue-light photoreceptors** control hypocotyl elongation, stomatal opening, and phototropism.

- **Phytochromes** act like molecular "on-off" switches. Red light turns phytochrome "on," and far-red light turns it "off":

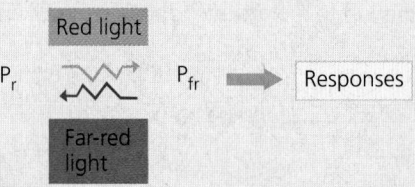

Phytochromes regulate shade avoidance and the germination of many seed types.
- Many daily rhythms in plant behavior are controlled by an internal circadian clock. Free-running **circadian rhythms** are approximately 24 hours long but are entrained to exactly 24 hours by dawn and dusk effects on phytochrome form.
- Phytochrome conversion also provides information about the relative lengths of day and night (photoperiod) and hence the time of year. **Photoperiodism** regulates the time of flowering in many species. **Short-day plants** require a night longer than a critical length to flower. **Long-day plants** need a night length shorter than a critical period to flower.

**?** *Why did plant physiologists propose the existence of a mobile molecule (florigen) that triggers flowering?*

### CONCEPT 31.3

**Plants respond to a wide variety of stimuli other than light (pp. 653–657)**

- **Gravitropism** is the bending of an organ in response to gravity. Roots show positive gravitropism, and stems show negative gravitropism. **Statoliths**, starch-filled plastids, enable plant roots to detect gravity.
- Plants are highly sensitive to touch. **Thigmotropism** is a growth response to touch. Rapid leaf movements involve transmission of electrical impulses called **action potentials**.
- Plants are sensitive to environmental stresses, including drought, flooding, high salinity, and extremes of temperature.

| Environmental Stress | Major Response |
|---|---|
| Drought | ABA production, reducing water loss by closing stomata |
| Flooding | Formation of air tubes that help roots survive oxygen deprivation |
| Salt | Avoiding osmotic water loss by producing solutes tolerated at high concentrations |
| Heat | Synthesis of heat-shock proteins, which reduce protein denaturation at high temperatures |
| Cold | Adjusting membrane fluidity; avoiding osmotic water loss; producing antifreeze proteins |

**?** *Plants that have acclimated to drought stress are often more resistant to freezing stress as well. Suggest a reason why.*

## CONCEPT 31.4

### Plants respond to attacks by herbivores and pathogens (pp. 657–660)

- In addition to physical defenses such as thorns and trichomes, plants produce distasteful or toxic chemicals, as well as attractants that recruit animals that destroy herbivores.
- In PAMP-triggered immunity, plants recognize **pathogen-associated molecular patterns** (**PAMPs**), molecular sequences that are specific to certain pathogens. PAMP recognition leads to the local production of broad-spectrum antimicrobial chemicals and toughening of the plant cell wall.
- Effector-triggered immunity includes the **hypersensitive response**, which seals off an infection and destroys both pathogen and host cells in the region, and **systemic acquired resistance**, a generalized defense response in organs distant from the infection site.

**?** *What are three ways herbivory negatively impacts plant productivity?*

## TEST YOUR UNDERSTANDING

**PRACTICE TEST**

### Level 1: Knowledge/Comprehension

goo.gl/CRZjvS

1. The hormone that helps plants respond to drought is
   (A) auxin.
   (B) abscisic acid.
   (C) cytokinin.
   (D) ethylene.

2. Auxin enhances cell elongation in all of these ways *except*
   (A) increased uptake of solutes.
   (B) gene activation.
   (C) acid-induced denaturation of cell wall proteins.
   (D) increased activity of plasma membrane proton pumps.

3. Charles and Francis Darwin discovered that
   (A) red light is most effective in shoot phototropism.
   (B) auxin can pass through agar.
   (C) light destroys auxin.
   (D) light is perceived by the tips of coleoptiles.

### Level 2: Application/Analysis

4. The signaling molecule for flowering might be released earlier than usual in a long-day plant exposed to flashes of
   (A) far-red light during the night.
   (B) red light during the night.
   (C) far-red light during the day.
   (D) red light during the day.

5. If a long-day plant has a critical night length of 9 hours, which 24-hour cycle would prevent flowering?
   (A) 16 hours light/8 hours dark
   (B) 14 hours light/10 hours dark
   (C) 4 hours light/8 hours dark/4 hours light/8 hours dark
   (D) 8 hours light/8 hours dark/light flash/8 hours dark

6. A plant mutant that showed normal gravitropic bending but did not store starch in its plastids would require a reevaluation of the role of _____ in gravitropism.
   (A) auxin
   (B) statoliths
   (C) light
   (D) differential growth

7. **DRAW IT** Indicate the response of *Arabidopsis* to each condition by drawing either a straight seedling or one undergoing the triple response.

|  | Control | Ethylene added | Ethylene synthesis inhibitor |
|---|---|---|---|
| Wild-type |  |  |  |
| Ethylene insensitive (*ein*) |  |  |  |
| Ethylene overproducing (*eto*) |  |  |  |
| Constitutive triple response (*ctr*) |  |  |  |

### Level 3: Synthesis/Evaluation

8. **SCIENTIFIC INQUIRY**
   A field biologist notes that a caterpillar quits feeding on a certain plant after a while, rejects nearby uneaten plants, and begins feeding anew on plants some distance away from the original. The biologist hypothesizes that insect-damaged leaves emit volatile "alarm" chemicals that signal nearby plants to initiate plant defense responses. Devise a test of this hypothesis.

9. **FOCUS ON EVOLUTION**
   As a general rule, light-sensitive germination is more pronounced in small seeds than in large seeds. Suggest a reason why. Explain your thinking.

10. **FOCUS ON INTERACTIONS**
    In a short essay (100–150 words), summarize phytochrome's role in altering shoot growth for the enhancement of light capture.

11. **SYNTHESIZE YOUR KNOWLEDGE**

This mule deer is grazing on the shoot tips of a shrub. Describe how this event will alter the physiology, biochemistry, structure, and health of the plant, and identify which hormones are involved in making these changes.

*For selected answers, see Appendix A.*

# Unit 6 **Animal Form and Function**

### 32 The Internal Environment of Animals: Organization and Regulation

The activities of the cells, tissues, and organs that make up the animal body are controlled and coordinated by hormones, which are signals from the **endocrine** system, and by the nervous system. One major result is **homeostasis**, the maintenance of a balanced internal environment.

### 33 Animal Nutrition

Animals meet their **nutritional** needs by the stepwise digestion of ingested food and the efficient absorption of released nutrients.

### 34 Circulation and Gas Exchange

Cells and tissues throughout a multilayered animal body rely on **circulation** and **gas exchange** systems to carry out an interchange of oxygen, nutrients, and wastes with the external environment.

### 35 The Immune System

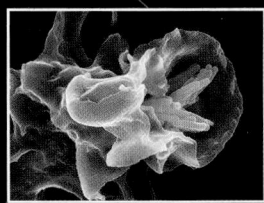

An **immune system** provides barriers to infection and distinguishes self from nonself in initiating defense against foreign cells and viruses.

### 36 Reproduction and Development

Sexual reproduction involves the fertilization of egg by sperm. In animals, the process of embryonic **development** involves cell division, specialization, and movement.

### 37 Neurons, Synapses, and Signaling

**Neurons** receive, retrieve, and transmit information by **signaling** along cellular extensions and across specialized cell junctions called **synapses**.

### 38 Nervous and Sensory Systems

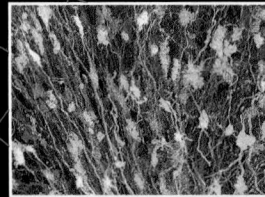

**Sensory** receptors tuned to chemicals, light, vibrations, and other stimuli transfer information to the **nervous system** for processing and integration.

### 39 Motor Mechanisms and Behavior

Animals respond to their environment through **motor mechanisms**, including muscle contractions, that bring about **behavior**.

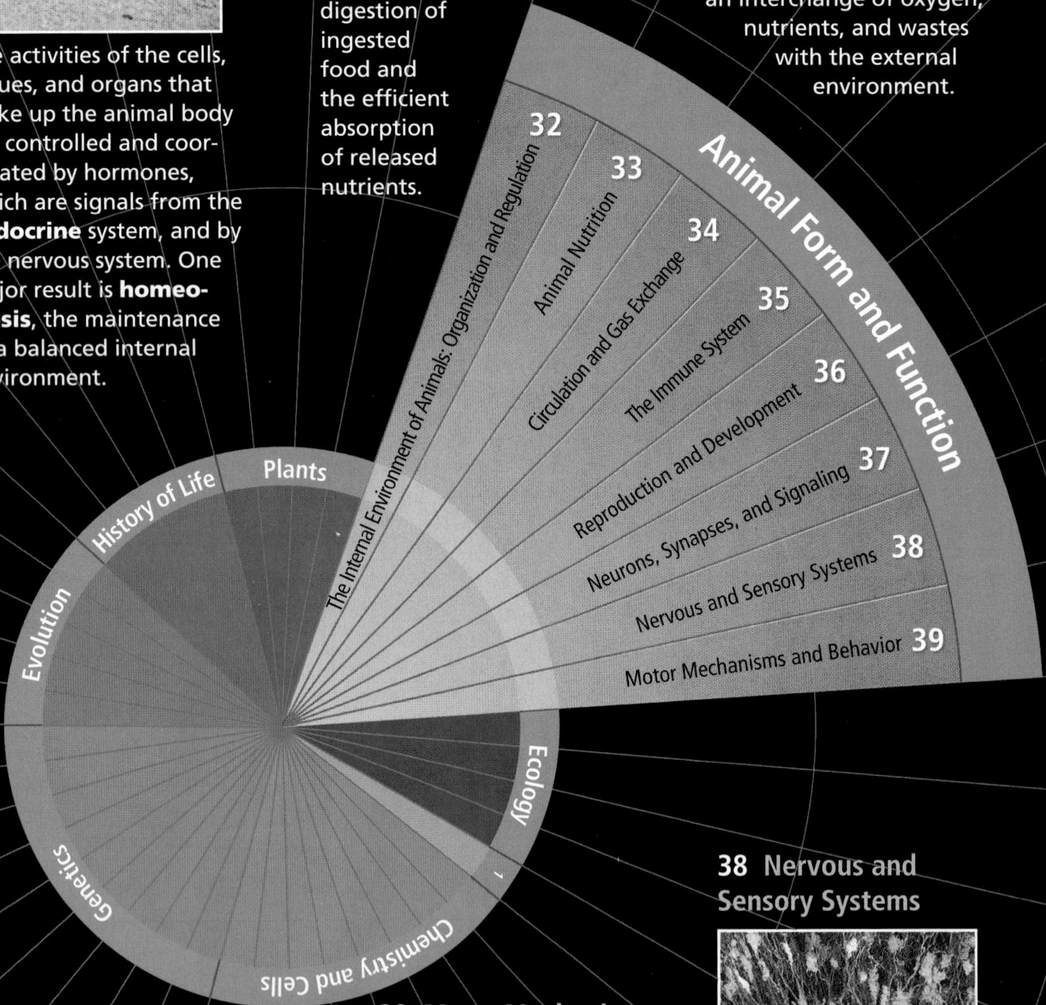

*Circular diagram labels:*

History of Life
Plants
Evolution
Genetics
Chemistry and Cells
Ecology

Animal Form and Function

32 The Internal Environment of Animals: Organization and Regulation
33 Animal Nutrition
34 Circulation and Gas Exchange
35 The Immune System
36 Reproduction and Development
37 Neurons, Synapses, and Signaling
38 Nervous and Sensory Systems
39 Motor Mechanisms and Behavior

# 32 The Internal Environment of Animals: Organization and Regulation

## KEY CONCEPTS

**32.1** Animal form and function are correlated at all levels of organization

**32.2** The endocrine and nervous systems act individually and together in regulating animal physiology

**32.3** Feedback control maintains the internal environment in many animals

**32.4** A shared system mediates osmoregulation and excretion in many animals

**32.5** The mammalian kidney's ability to conserve water is a key terrestrial adaptation

▲ **Figure 32.1** How do long legs help this scavenger survive in the scorching desert heat?

## Diverse Forms, Common Challenges

The desert ant (*Cataglyphis*) in **Figure 32.1** scavenges insects that have succumbed to the daytime heat of the Sahara Desert. To gather corpses for feeding, the ant must forage when surface temperatures on the sunbaked sand exceed 60°C (140°F), well above the thermal limit for virtually all animals. How does the desert ant survive these conditions? To address this question, we need to consider the relationship of **anatomy**, or biological form, to species survival.

Over the course of its life, an ant faces the same fundamental challenges as any other animal, whether hydra, hawk, or human. All animals must obtain nutrients and oxygen, fight off infection, and produce offspring. Given that they share these and other basic requirements, why do species vary so enormously in makeup, complexity, organization, and appearance? The answer is adaptation: Natural selection favors those variations in a population that increase relative fitness (see Concept 21.4). The evolutionary adaptations that enable survival vary among environments and species, but they frequently result in a close match of form to function.

Because form and function are correlated, examining anatomy often provides clues to **physiology**—biological function. In the case of the desert ant, researchers noted that its stilt-like legs are disproportionately long, elevating the rest of the ant 4 mm above the sand. At this height, the ant's body is exposed to a temperature 6°C lower than that at ground level. The ant's long legs also facilitate rapid locomotion: Researchers have found that desert ants can run as fast as 1 m/sec, close to the top speed recorded for a running arthropod. Speedy sprinting minimizes the time that the ant is exposed to the sun. Thus, the long legs of the desert ant are adaptations that allow it to be active during the heat of the day, when competition for food and the risk of predation are lowest.

We will begin our study of animal form and function by examining the organization of cells and tissues in the animal body, the systems for coordinating the activities of different body parts, and the general means by which animals control their internal environment. We then apply these ideas to two challenges of particular relevance for desert animals: regulating body temperature and maintaining proper balance of body salts and water.

# Animal form and function are correlated at all levels of organization

For animals, as for other multicellular organisms, having many cells facilitates specialization. For example, a hard outer covering helps protect against predators, and large muscles facilitate rapid escape. In a multicellular body, the immediate environment of most cells is the internal body fluid. Control systems that regulate the composition of this solution allow the animal to maintain a relatively stable internal environment, even if the external environment is variable. To understand how these control systems operate, we first need to explore the layers of organization that characterize animal bodies.

Cells form a working animal body through their emergent properties, which arise from successive levels of structural and functional organization. Cells are organized into **tissues**, groups of cells with a similar appearance and a common function. Different types of tissues are further organized into functional units called **organs**. (The simplest animals, such as sponges, lack organs or even true tissues.) Groups of organs that work together, providing an additional level of organization and coordination, make up an **organ system** (Table 32.1). Thus, for example, the skin is an organ of the *integumentary system*, which protects against infection and helps regulate body temperature.

Many organs have more than one physiological role. If the roles are distinct enough, we consider the organ to belong to more than one organ system. The pancreas, for instance, produces enzymes critical to the function of the digestive system, but also regulates the level of sugar in the blood as a vital part of the endocrine system.

Just as viewing the body's organization from the "bottom up" (from cells to organ systems) reveals emergent properties, a "top-down" view of the hierarchy reveals the multilayered basis of specialization. Consider the human digestive system—each organ has specific roles. In the case of the stomach, one role is to initiate protein breakdown. This process requires a churning motion powered by stomach muscles, as well as digestive juices secreted by the stomach lining. Producing digestive juices, in turn, requires highly specialized cell types: One cell type secretes a protein-digesting enzyme, a second generates concentrated hydrochloric acid, and a third produces mucus, which protects the stomach lining.

The specialized and complex organ systems of animals are built from a limited set of cell and tissue types. For example, lungs and blood vessels have different functions but are lined by tissues that are of the same basic type and that therefore share many properties. Animal tissues are commonly grouped into four main types: epithelial, connective, muscle, and nervous (**Figure 32.2**).

As you read in Unit Five, plants also have a hierarchical organization. Although plant anatomy and animal anatomy differ, they are adapted to a shared set of challenges, as shown in **Figure 32.3**.

## CONCEPT CHECK 32.1

1. What properties do all types of epithelia share?
2. Suggest why a disease that damages connective tissue is likely to affect most of the body's organs.

For suggested answers, see Appendix A.

---

**Table 32.1  Organ Systems in Mammals**

| Organ System | Main Components | Main Functions |
|---|---|---|
| Digestive | Mouth, pharynx, esophagus, stomach, intestines, liver, pancreas, anus | Food processing (ingestion, digestion, absorption, elimination) |
| Circulatory | Heart, blood vessels, blood | Internal distribution of materials |
| Respiratory | Lungs, trachea, other breathing tubes | Gas exchange (uptake of oxygen; disposal of carbon dioxide) |
| Immune and lymphatic | Bone marrow, lymph nodes, thymus, spleen, lymph vessels, white blood cells | Body defense (fighting infections and virally induced cancers) |
| Excretory | Kidneys, ureters, urinary bladder, urethra | Disposal of metabolic wastes; regulation of osmotic balance of blood |
| Endocrine | Pituitary, thyroid, pancreas, adrenal, and other hormone-secreting glands | Coordination of body activities (such as digestion and metabolism) |
| Reproductive | Ovaries or testes and associated organs | Reproduction |
| Nervous | Brain, spinal cord, nerves, sensory organs | Coordination of body activities; detection of stimuli and formulation of responses to them |
| Integumentary | Skin and its derivatives (such as hair, claws, sweat glands) | Protection against mechanical injury, infection, dehydration; thermoregulation |
| Skeletal | Skeleton (bones, tendons, ligaments, cartilage) | Body support, protection of internal organs, movement |
| Muscular | Skeletal muscles | Locomotion and other movement |

## Epithelial Tissue

Occurring as sheets of closely packed cells, **epithelial tissue** covers the outside of the body and lines organs and cavities. Epithelial tissue functions as a barrier against mechanical injury, pathogens, and fluid loss. It also forms active interfaces with the environment. For example, the **epithelium** (plural, epithelia) that lines the intestines secretes digestive juices and absorbs nutrients.

All epithelia are polarized, meaning that they have two different sides. The *apical* surface faces the lumen (cavity) or outside of the organ and is therefore exposed to fluid or air. The basal surface is attached to a *basal lamina*, a dense mat of extracellular matrix that separates the epithelium from the underlying tissue.

Lumen — 10 µm

Apical surface

Epithelial tissue

Basal surface

**Epithelial tissue lining small intestine**

## Nervous Tissue

**Nervous tissue** functions in the receipt, processing, and transmission of information. Specialized cells called **neurons** are the basic units of the nervous system. A neuron receives nerve impulses from other neurons via its cell body and multiple extensions called dendrites. Neurons transmit impulses to neurons, muscles, or other cells via extensions called axons, which are often bundled together into nerves. Nervous tissue also contains support cells called **glial cells**, or simply **glia**. The various types of glia help nourish, insulate, and replenish neurons and in some cases modulate neuron function. In many animals, a concentration of nervous tissue forms a brain, an information-processing center.

**Nervous tissue in brain**

Axons of neurons

Blood vessel

(Confocal LM)

Glia

20 µm

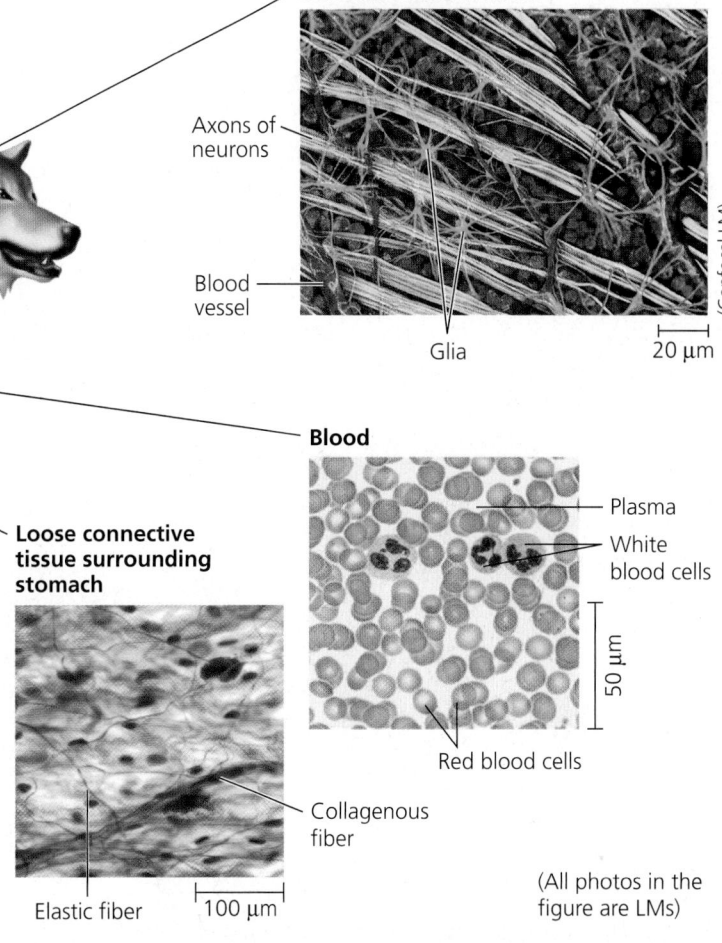

**Blood**

Plasma

White blood cells

50 µm

Red blood cells

**Loose connective tissue surrounding stomach**

Collagenous fiber

Elastic fiber     100 µm

(All photos in the figure are LMs)

**Skeletal muscle tissue**

Nuclei

Muscle cell

100 µm

## Muscle Tissue

Vertebrates have three types of **muscle tissue**: skeletal, cardiac, and smooth. All muscle cells consist of filaments containing the proteins actin and myosin, which together enable muscles to contract. Attached to bones by tendons, **skeletal muscle**, or striated muscle, is responsible for voluntary movements. The arrangement of contractile units along the cells gives them a striped (striated) appearance. Cardiac muscle, which is also striated, forms the contractile wall of the heart. Smooth muscle, which lacks striations and has spindle-shaped cells, is found in the walls of many internal organs. Smooth muscles are responsible for involuntary activities, such as churning of the stomach and constriction of arteries.

## Connective Tissue

**Connective tissue** consists of cells scattered through an extracellular matrix, often forming a web of fibers embedded in a liquid, jellylike, or solid foundation. Within the matrix are cells called *fibroblasts*, which secrete collagen and other matrix proteins, and *macrophages*, which engulf foreign particles and cell debris.

In vertebrates, the many forms of connective tissue include loose connective tissue, which holds skin and other organs in place; fibrous connective tissue, found in tendons and ligaments; adipose tissue, which stores fat; blood, which consists of cells and cell fragments suspended in a liquid called plasma; cartilage, which provides flexible support in the spine and elsewhere; and bone, a hard mineral of calcium, magnesium, and phosphate ions in a matrix of collagen.

# Life Challenges and Solutions in Plants and Animals

Multicellular organisms face a common set of challenges. Comparing the solutions that have evolved in plants and animals reveals both unity (shared elements) and diversity (distinct features) across these two lineages.

## Nutritional Mode

All living things must obtain energy and carbon from the environment to grow, survive, and reproduce. Plants are autotrophs, obtaining their energy through photosynthesis and their carbon from inorganic sources, whereas animals are heterotrophs, obtaining their energy and carbon from food. Evolutionary adaptations in plants and animals support these different nutritional modes. The broad surface of many leaves (left) enhances light capture for photosynthesis. When hunting, a bobcat relies on stealth, speed, and sharp claws (right). (See Figure 29.2 and Figure 33.14.)

## Growth and Regulation

The growth and physiology of both plants and animals are regulated by hormones. In plants, hormones may act in a local area or be transported in the body. They control growth patterns, flowering, fruit development, and more (left). In animals, hormones circulate throughout the body and act in specific target tissues, controlling homeostatic processes and developmental events such as molting (below). (See Table 31.1 and Figure 33.19.)

## Environmental Response

All forms of life must detect and respond appropriately to conditions in their environment. Specialized organs sense environmental signals. For example, the floral head of a sunflower (left) and an insect's eyes (right) both contain photoreceptors that detect light. Environmental signals activate specific receptor proteins, triggering signal transduction pathways that initiate cellular responses coordinated by chemical and electrical communication. (See Figure 31.12 and Figure 38.26.)

## Transport

All but the simplest
multicellular organisms
must transport nutrients and
waste products between locations in the body. A system of
tubelike vessels is the common evolutionary solution, while the
mechanism of circulation varies. Plants harness solar energy
to transport water, minerals, and sugars through specialized
tubes (left). In animals, a pump (heart) moves circulatory fluid
through vessels (right). (See Figure 28.9 and Figure 34.3.)

## Reproduction

In sexual reproduction,
specialized tissues and
structures produce and
exchange gametes. Offspring
are generally supplied with
nutritional stores that facilitate
rapid growth and development.
For example, seeds (left) have
stored food reserves that supply energy to the young seedling,
while milk provides sustenance for juvenile mammals (right).
(See Figure 30.8 and Figure 32.7.)

## Gas Exchange

The exchange of certain
gases with the environment
is essential for life.
Respiration by plants and
animals requires taking up
oxygen ($O_2$) and releasing carbon dioxide ($CO_2$). In photosynthesis,
net exchange occurs in the opposite direction: $CO_2$ uptake and $O_2$
release. In both plants and animals, highly convoluted surfaces that
increase the area available for gas exchange have evolved, such as
the spongy mesophyll of leaves (left) and the alveoli of lungs (right).
(See Figure 28.17 and Figure 34.20.)

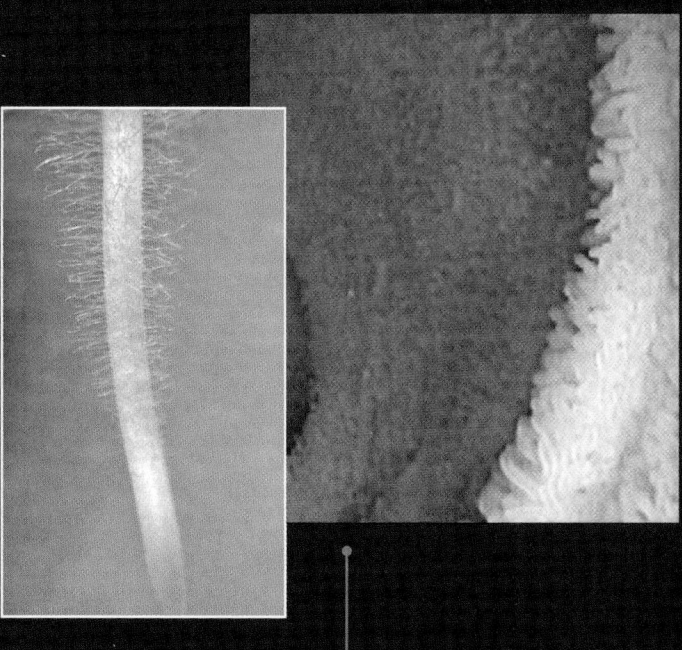

## Absorption

Organisms need to absorb nutrients. The root hairs
of plants (left) and the villi (projections) that line
the intestines of vertebrates (right) increase the
surface area available for absorption. (See Figure 28.4
and Figure 33.10.)

**MAKE CONNECTIONS** *Compare the adaptations that enable plants
and animals to respond to the challenges of living in hot and cold
environments. See Concepts 31.3 and 32.3.*

**ANIMATION** Visit the Study Area in **MasteringBiology** for the BioFlix®
3-D Animations on Water Transport in Plants (Chapter 29),
Homeostasis: Regulating Blood Sugar (Chapter 33), and
Gas Exchange (Chapter 34).

# The endocrine and nervous systems act individually and together in regulating animal physiology

For an animal's tissues and organ systems to perform their specialized functions effectively, they must act in concert with one another. For example, when the wolf shown in Figure 32.2 is hunting, it regulates blood flow in its circulatory system to bring adequate nutrients and gases to its leg muscles, which in turn are controlled by the brain in response to cues detected by the nose. What signals are used to coordinate activity? How do the signals move within the body?

## An Overview of Coordination and Control

Animals have two major systems for coordinating and controlling responses to stimuli: the endocrine system and the nervous system **(Figure 32.4)**. In the **endocrine system**, signaling molecules released into the bloodstream by endocrine cells are carried to all locations in the body. In the **nervous system**, neurons transmit signals along dedicated routes connecting specific locations in the body.

The signaling molecules broadcast throughout the body by the endocrine system are called **hormones** (from the Greek *horman*, to excite). Different hormones cause distinct effects, and only cells that have receptors for a particular hormone respond (see Figure 32.4a). Depending on which cells have receptors for that hormone, the hormone may have an effect in just a single location or in sites throughout the body. It takes many seconds for hormones to be released into the bloodstream and carried throughout the body. The effects are often long-lasting, however, because hormones can remain in the bloodstream for minutes or even hours.

In the nervous system, signals called nerve impulses travel to specific target cells along communication lines consisting mainly of extensions called axons (see Figure 32.4b). Nerve impulses can act on other neurons, on muscle cells, and on cells and glands that produce secretions. Unlike the endocrine system, the nervous system conveys information by the particular *pathway* the signal takes. For example, a person can distinguish different musical notes because within the ear each note's frequency activates neurons that connect to slightly different regions of the brain.

Communication in the nervous system usually involves more than one type of signal. Nerve impulses travel along axons, sometimes over long distances, as changes in voltage. In contrast, passing information from one neuron to another often involves very short-range chemical signals. Overall, transmission in the nervous system is extremely fast: Nerve impulses take only a fraction of a second to reach the target and last only a fraction of a second.

▼ **Figure 32.4 Signaling in the endocrine and nervous systems.**

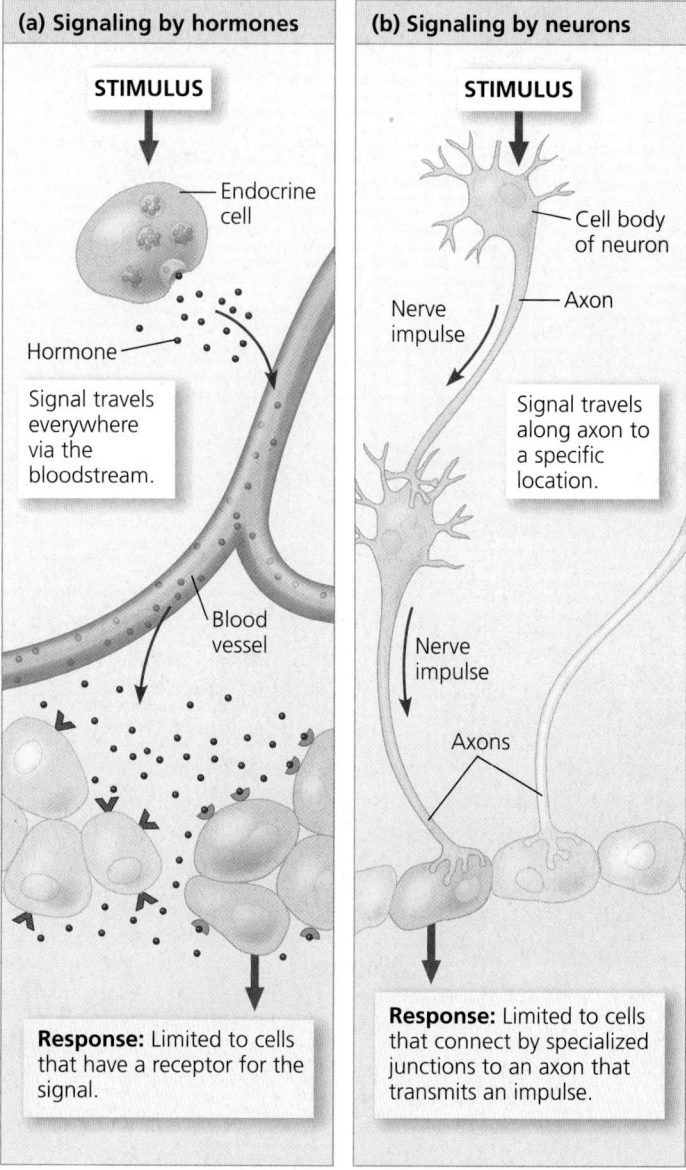

**(a) Signaling by hormones**

STIMULUS

Endocrine cell

Hormone

Signal travels everywhere via the bloodstream.

Blood vessel

**Response:** Limited to cells that have a receptor for the signal.

**(b) Signaling by neurons**

STIMULUS

Cell body of neuron

Axon

Nerve impulse

Signal travels along axon to a specific location.

Nerve impulse

Axons

**Response:** Limited to cells that connect by specialized junctions to an axon that transmits an impulse.

The two major communication systems of the body differ in signal type, transmission, speed, and duration. All these differences reflect adaptation to different functions. The endocrine system is especially well adapted for coordinating gradual changes that affect the entire body, such as growth, development, reproduction, metabolic processes, and digestion. The nervous system is well suited for directing immediate and rapid responses to the environment, such as reflexes and other rapid movements. Although the general functions of the endocrine and nervous systems are distinct, the two systems often work in close coordination, as we will explore shortly.

We'll investigate nervous system organization and function in detail in Chapters 37 and 38. Here we'll focus on the components of the endocrine system and the organization of pathways for endocrine signaling.

## Endocrine Glands and Hormones

Within the body, endocrine cells are often grouped in ductless organs called **endocrine glands**. The major endocrine glands of humans and the hormones that they produce are illustrated in **Figure 32.5**. Some endocrine cells are instead found in organs that are part of other organ systems. For example, the stomach contains isolated endocrine cells that regulate digestion by secreting the hormone gastrin in response to ingested food.

Endocrine cells and glands secrete hormones directly into the surrounding fluid. From there, the hormones enter the circulatory system. In contrast, *exocrine glands*, such as salivary glands, have ducts that carry enzymes or other secreted substances into body cavities or onto body surfaces.

## Regulation of Endocrine Signaling

The stimuli that cause endocrine cells and glands to release hormones are varied. In some cases, organic molecules or ions trigger the endocrine response. For example, high levels of glucose stimulate the pancreas to secrete insulin, a hormone that causes a decrease in the level of glucose in the blood. In other cases, the nervous system provides the stimulus for hormone release, a type of control called *neuroendocrine signaling.*

The **hypothalamus**, an almond-sized region of the brain (see Figure 32.5), controls most neuroendocrine signaling in mammals. Finally, some hormones are secreted in response to other hormones. Hormones that regulate other hormones have essential roles in growth, metabolism, and reproduction, as you will learn in this and later chapters.

Regulation of a signaling process involves not only its initiation but also its termination. How is an endocrine pathway turned off? Typically, the control process involves **negative feedback**, also called feedback inhibition, a control circuit or loop that reduces, or "damps," the stimulus. In the case of insulin, for example, the secreted hormone triggers a reduction in blood glucose levels, which in turn eliminates the stimulus for further insulin release. Because negative feedback prevents excessive pathway activity, this type of control circuit is common in endocrine pathways that keep physiological systems within normal limits.

A few hormone pathways are controlled by **positive feedback**, a control mechanism in which the response reinforces the stimulus, leading to an even greater response. Whereas a negative feedback loop prevents excessive pathway activity, a positive feedback loop helps drive a process to completion. Positive feedback plays a central role in several processes associated with reproduction, including the uterine contractions of childbirth.

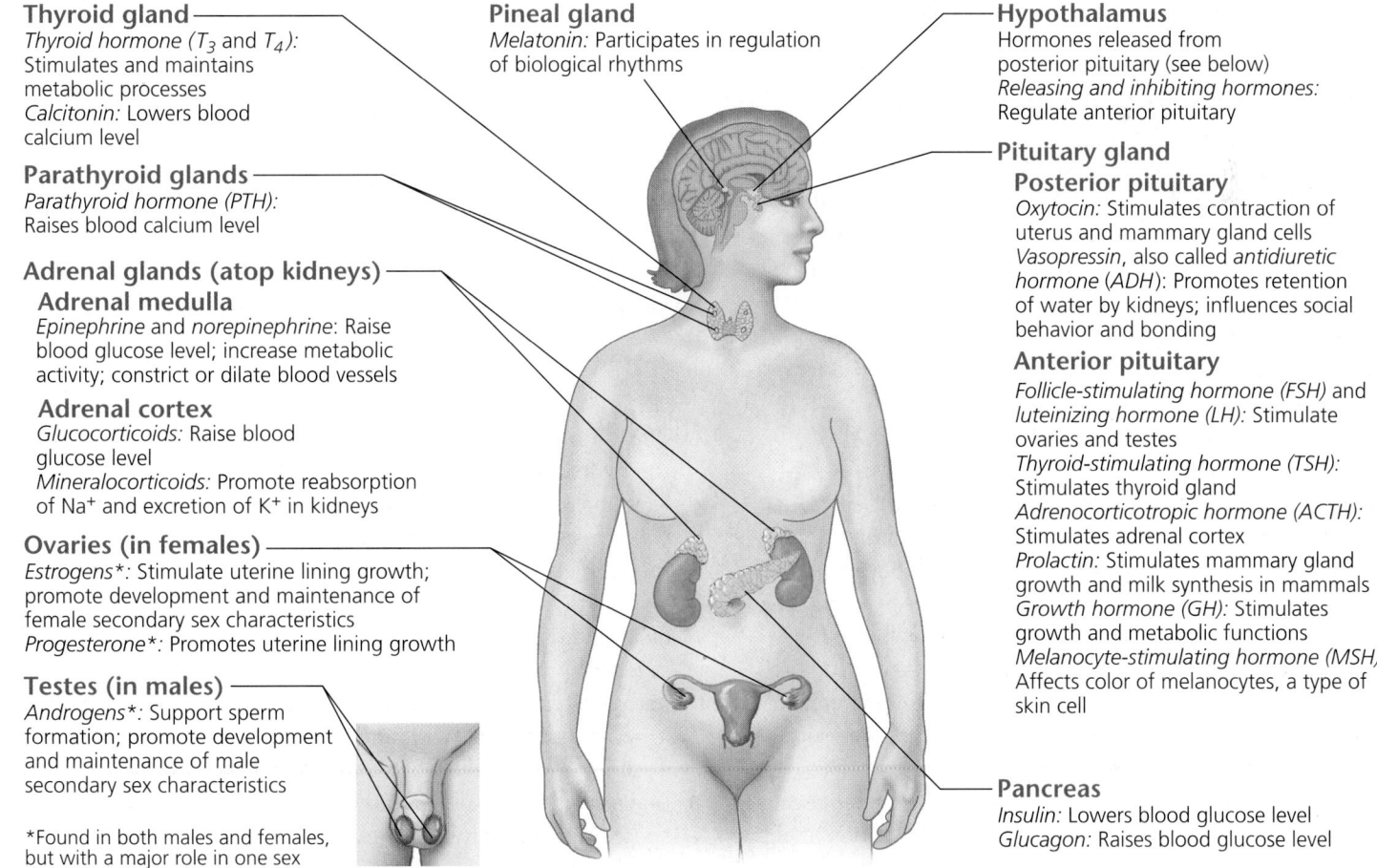

**Thyroid gland**
*Thyroid hormone ($T_3$ and $T_4$):* Stimulates and maintains metabolic processes
*Calcitonin:* Lowers blood calcium level

**Parathyroid glands**
*Parathyroid hormone (PTH):* Raises blood calcium level

**Adrenal glands (atop kidneys)**
 **Adrenal medulla**
 *Epinephrine* and *norepinephrine*: Raise blood glucose level; increase metabolic activity; constrict or dilate blood vessels
 **Adrenal cortex**
 *Glucocorticoids:* Raise blood glucose level
 *Mineralocorticoids:* Promote reabsorption of $Na^+$ and excretion of $K^+$ in kidneys

**Ovaries (in females)**
*Estrogens\*:* Stimulate uterine lining growth; promote development and maintenance of female secondary sex characteristics
*Progesterone\*:* Promotes uterine lining growth

**Testes (in males)**
*Androgens\*:* Support sperm formation; promote development and maintenance of male secondary sex characteristics

\*Found in both males and females, but with a major role in one sex

**Pineal gland**
*Melatonin:* Participates in regulation of biological rhythms

**Hypothalamus**
Hormones released from posterior pituitary (see below)
*Releasing and inhibiting hormones:* Regulate anterior pituitary

**Pituitary gland**
 **Posterior pituitary**
 *Oxytocin:* Stimulates contraction of uterus and mammary gland cells
 *Vasopressin*, also called *antidiuretic hormone* (ADH): Promotes retention of water by kidneys; influences social behavior and bonding
 **Anterior pituitary**
 *Follicle-stimulating hormone (FSH)* and *luteinizing hormone (LH):* Stimulate ovaries and testes
 *Thyroid-stimulating hormone (TSH):* Stimulates thyroid gland
 *Adrenocorticotropic hormone (ACTH):* Stimulates adrenal cortex
 *Prolactin:* Stimulates mammary gland growth and milk synthesis in mammals
 *Growth hormone (GH):* Stimulates growth and metabolic functions
 *Melanocyte-stimulating hormone (MSH):* Affects color of melanocytes, a type of skin cell

**Pancreas**
*Insulin:* Lowers blood glucose level
*Glucagon:* Raises blood glucose level

▲ **Figure 32.5  The human endocrine system.**

## Simple Endocrine Pathways

In the simplest endocrine pathways, endocrine cells respond directly to an internal or environmental stimulus by secreting a particular hormone. The hormone travels in the bloodstream to target cells, where it interacts with its specific receptors. Signal transduction within target cells brings about a response.

As an example of a simple endocrine pathway, we will again turn to the human digestive system. After processing of food in the stomach, partially digested material passes to the *duodenum*, the first part of the small intestine. The digestive juices of the stomach are extremely acidic and must be neutralized before further steps of digestion can occur. To learn how neutral pH is restored, we will consider the pathway outlined in **Figure 32.6**.

Endocrine cells called S cells, found in the lining of the duodenum, recognize the low pH of the partially digested food arriving from the stomach. The S cells respond by secreting the hormone secretin into the bloodstream. Secretin in turn triggers release of bicarbonate from the **pancreas**, a gland located behind the stomach. The bicarbonate travels along ducts leading to the duodenum. There, the bicarbonate neutralizes the acidic contents, raising pH and allowing digestion to proceed.

Note that the pH increase that results from secretin signaling eliminates the stimulus and thus shuts off the endocrine pathway. Secretin signaling is thus an example of a pathway under negative feedback control.

## Neuroendocrine Signaling

For many hormones, secretion is triggered when the nervous system detects and processes a stimulus. In vertebrates, such neuroendocrine signaling involves the hypothalamus and the **pituitary gland**, found at its base. The pituitary is actually two glands fused together. One is the **posterior pituitary**, which is an extension of the hypothalamus. The posterior pituitary stores and secretes hormones synthesized in the hypothalamus. In contrast, the **anterior pituitary** is an endocrine gland that both synthesizes and secretes hormones.

### Posterior Pituitary Pathways

As an example of a neuroendocrine pathway, we will consider the regulation of milk release during nursing in mammals. When an infant suckles, it stimulates sensory neurons in the nipples, generating nerve impulses that reach the hypothalamus. This input triggers the release of the hormone **oxytocin** from the posterior pituitary **(Figure 32.7)**. Oxytocin in turn stimulates the mammary glands, which respond by secreting milk.

Milk released in response to oxytocin leads to more suckling and therefore more stimulation. Pathway activation continues until the baby stops suckling. Thus, oxytocin regulation of milk release involves positive feedback control. Other functions of oxytocin, such as stimulating contractions of the uterus during birthing, also exhibit positive feedback.

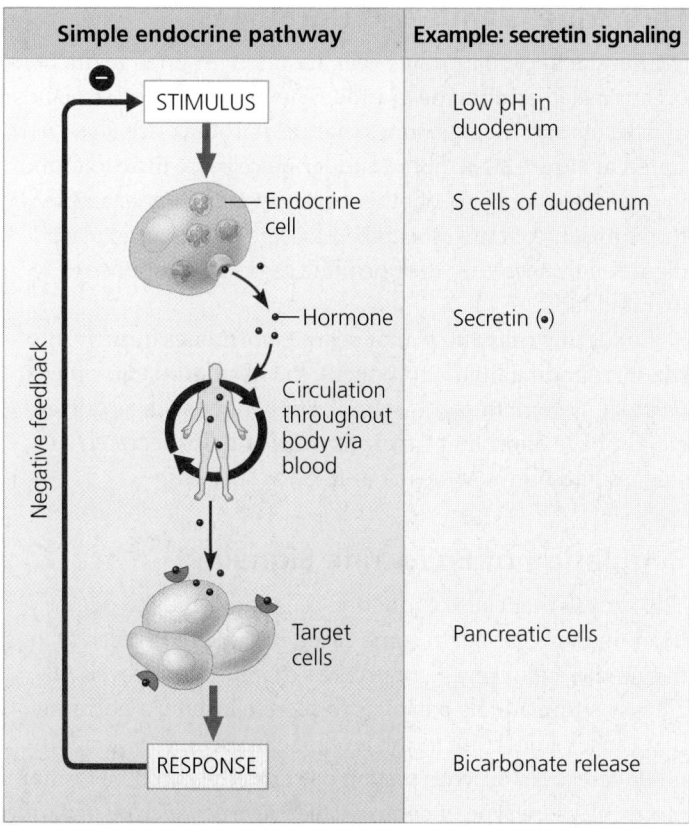

| Simple endocrine pathway | Example: secretin signaling |
|---|---|
| STIMULUS | Low pH in duodenum |
| Endocrine cell | S cells of duodenum |
| Hormone | Secretin (•) |
| Circulation throughout body via blood | |
| Target cells | Pancreatic cells |
| RESPONSE | Bicarbonate release |

▲ **Figure 32.6 A simple endocrine pathway.** Endocrine cells respond to a change in some internal or external variable—the stimulus—by secreting a particular hormone that triggers a specific response by target cells. In the case of secretin signaling, the simple endocrine pathway is self-limiting because the response to secretin (bicarbonate release) reduces the stimulus (low pH) through negative feedback.

Oxytocin is one of just two posterior pituitary hormones. The other, **antidiuretic hormone (ADH)**, also called **vasopressin**, will be discussed later in this chapter.

### Anterior Pituitary Pathways

The anterior pituitary synthesizes and secretes a diverse set of hormones, each regulated by one or more hormones from the hypothalamus. Anterior pituitary hormones range in function from reproduction and growth to metabolism and stress responses (see Figure 32.5). Some, such as growth hormone, regulate cells outside the endocrine system. Others, such as thyroid-stimulating hormone (TSH), regulate endocrine glands.

Anterior pituitary hormones that target endocrine tissues often form part of a *hormone cascade*. Such a cascade is kicked off when the nervous system conveys a stimulus to the hypothalamus. In response, the hypothalamus secretes a factor that regulates release of a specific anterior pituitary hormone. This hormone in turn stimulates an endocrine organ to secrete yet another hormone, which exerts effects on specific target tissues.

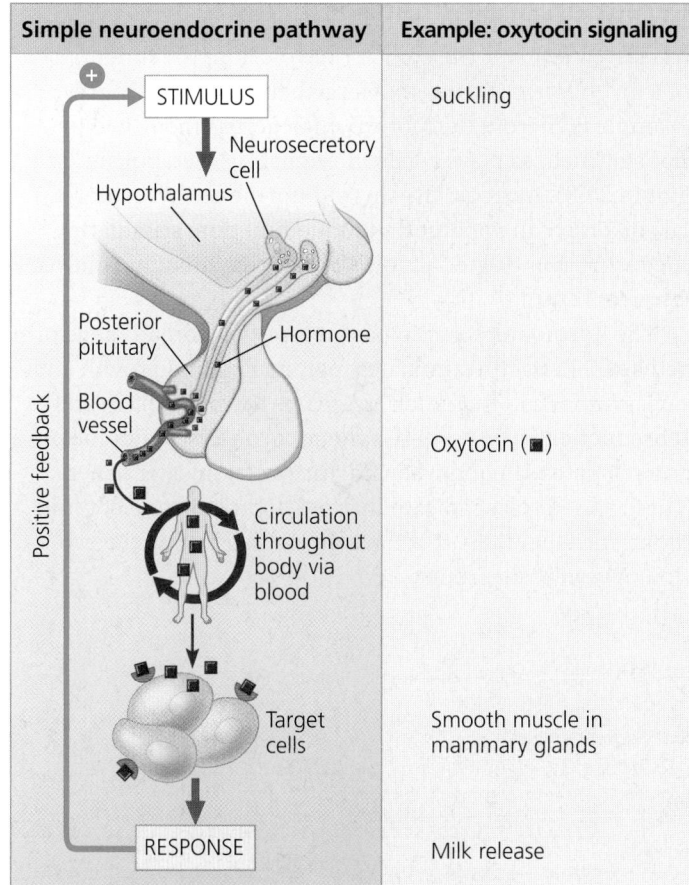

**Simple neuroendocrine pathway** | **Example: oxytocin signaling**

STIMULUS — Suckling
Neurosecretory cell
Hypothalamus
Posterior pituitary
Hormone
Blood vessel
Oxytocin (■)
Positive feedback
Circulation throughout body via blood
Target cells — Smooth muscle in mammary glands
RESPONSE — Milk release

▲ **Figure 32.7 A simple neuroendocrine pathway.** Sensory neurons respond to a stimulus by sending nerve impulses to a neurosecretory cell, triggering hormone secretion. Upon reaching its target cells, the hormone binds to its receptor, causing a specific response. In oxytocin signaling, the response increases the stimulus, forming a positive-feedback loop that amplifies signaling.

**Figure 32.8** outlines such a hormone cascade pathway, using the specific example of thyroid regulation.

Because hormone cascade pathways in a sense redirect signals from the hypothalamus to other endocrine glands, the anterior pituitary hormones in these pathways are sometimes called *tropic* hormones, or tropins, from the Greek word for bending or turning. Thus, for example, TSH is also known as thyrotropin.

Feedback regulation often occurs at multiple levels in hormone cascade pathways. For example, thyroid hormone exerts negative feedback on the hypothalamus and on the anterior pituitary, in each case blocking release of the hormone that promotes its production (see Figure 32.8).

## Hormone Solubility

Many hormones, including secretin, ADH, and oxytocin, are soluble in water but not in lipids. For this reason, they are unable to pass through the plasma membranes of target cells. Instead, they bind to cell-surface receptors, triggering events at the plasma membrane that result in a cellular response. The series of changes in cellular proteins that converts the

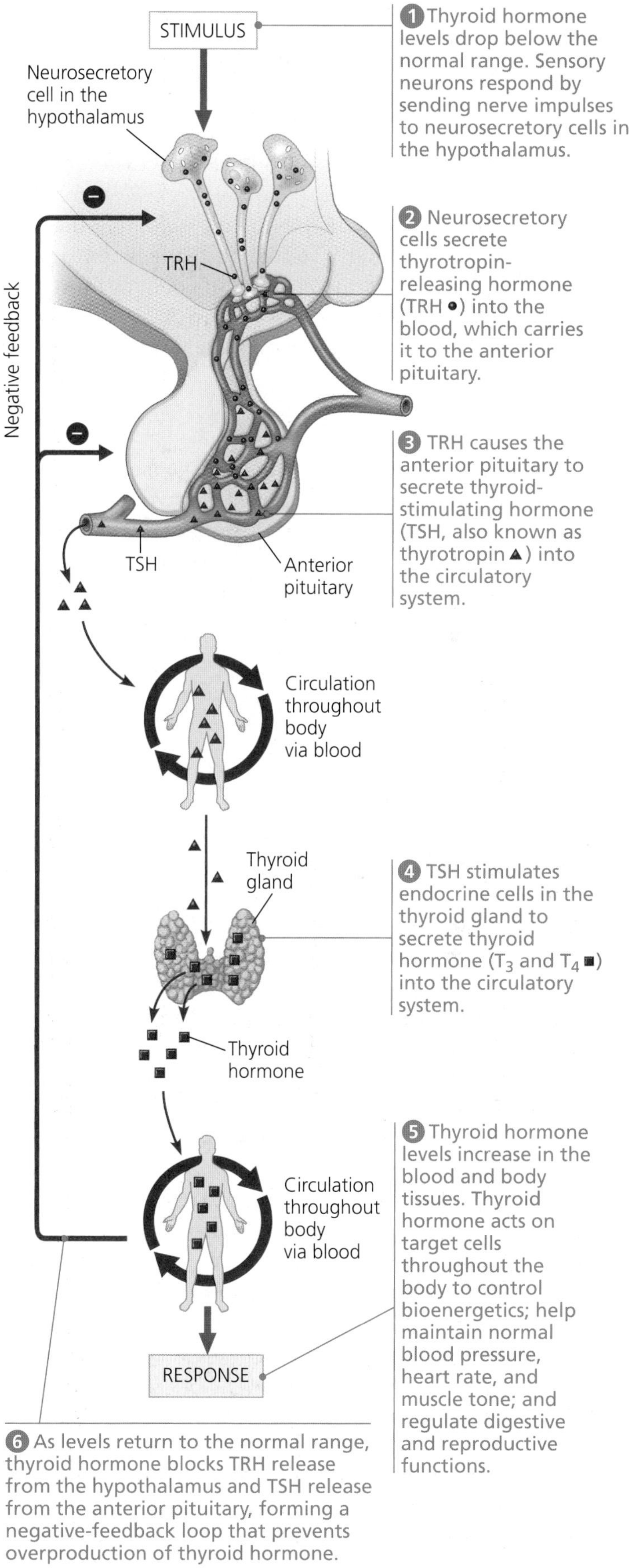

STIMULUS

Neurosecretory cell in the hypothalamus

Negative feedback

TRH

TSH

Anterior pituitary

Circulation throughout body via blood

Thyroid gland

Thyroid hormone

Circulation throughout body via blood

RESPONSE

❶ Thyroid hormone levels drop below the normal range. Sensory neurons respond by sending nerve impulses to neurosecretory cells in the hypothalamus.

❷ Neurosecretory cells secrete thyrotropin-releasing hormone (TRH ●) into the blood, which carries it to the anterior pituitary.

❸ TRH causes the anterior pituitary to secrete thyroid-stimulating hormone (TSH, also known as thyrotropin ▲) into the circulatory system.

❹ TSH stimulates endocrine cells in the thyroid gland to secrete thyroid hormone ($T_3$ and $T_4$ ■) into the circulatory system.

❺ Thyroid hormone levels increase in the blood and body tissues. Thyroid hormone acts on target cells throughout the body to control bioenergetics; help maintain normal blood pressure, heart rate, and muscle tone; and regulate digestive and reproductive functions.

❻ As levels return to the normal range, thyroid hormone blocks TRH release from the hypothalamus and TSH release from the anterior pituitary, forming a negative-feedback loop that prevents overproduction of thyroid hormone.

▲ **Figure 32.8 A hormone cascade pathway.**

extracellular signal to a specific intracellular response is called *signal transduction*. A signal transduction pathway typically has multiple steps, each involving specific molecular interactions (see Concept 5.6).

There are also hormones that are lipid-soluble, including the sex hormones estradiol and testosterone (as well as thyroid hormone). The major receptors for steroid hormones are located in the cytosol rather than on the cell surface. When a steroid hormone binds to its cytosolic receptor, a hormone-receptor complex forms, which moves into the nucleus (see Figure 5.23). There, the receptor portion of the complex alters transcription of particular genes.

## Multiple Effects of Hormones

Many hormones elicit more than one response. Consider, for example, epinephrine. Also called *adrenaline*, **epinephrine** is secreted by the *adrenal glands*, which lie atop the kidneys (see Figure 32.5). When you are in a stressful situation, perhaps running to catch a bus, the release of epinephrine rapidly triggers responses that help you chase the departing bus: raising blood glucose levels, increasing blood flow to muscles, and decreasing blood flow to the digestive system.

How can one hormone have such different effects? Target cells can vary in their response if they differ in their receptor type or in the molecules that produce the response. In the liver, epinephrine binds to a β-type epinephrine receptor in the plasma membrane of target cells. This receptor activates the enzyme protein kinase A, which regulates enzymes of glycogen metabolism, causing release of glucose into the bloodstream **(Figure 32.9a)**. In blood vessels supplying skeletal muscle, the same kinase activated by the same receptor inactivates a muscle-specific enzyme. The result is smooth muscle relaxation, vasodilation, and hence increased blood flow **(Figure 32.9b)**. In contrast, intestinal blood vessels have an α-type epinephrine receptor **(Figure 32.9c)**. Rather than activating protein kinase A, the α receptor triggers a distinct signaling pathway involving different enzymes. The result is smooth muscle contraction, vasoconstriction, and restricted blood flow to the intestines.

## Evolution of Hormone Function

**EVOLUTION** Over the course of evolution, the functions of a given hormone often diverge between species. An example is thyroid hormone, which across many evolutionary lineages plays a role in regulating metabolism (see Figure 32.8). In frogs, the thyroid hormone thyroxine ($T_4$) has taken on an apparently unique function: stimulating the resorption of the tadpole's tail during metamorphosis **(Figure 32.10)**.

The hormone prolactin has an especially broad range of activities. Prolactin stimulates mammary gland growth and milk synthesis in mammals, regulates fat metabolism and reproduction in birds, delays metamorphosis in amphibians, and regulates salt and water balance in freshwater fishes. These varied roles indicate that prolactin is an ancient hormone with functions that have diversified during the evolution of vertebrate groups.

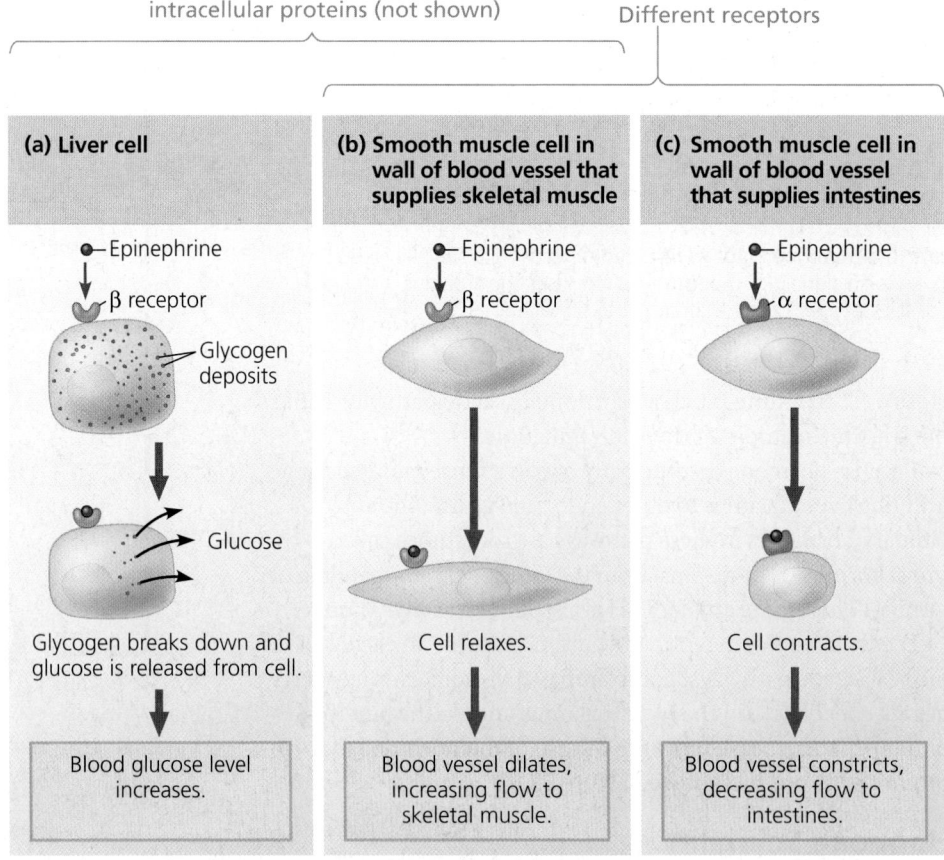

▲ **Figure 32.9 One hormone, different effects.** Epinephrine, the primary "fight-or-flight" hormone, produces different responses in different target cells. As shown by the brackets at the top of the figure, two types of differences between cells can alter the response to a hormone. Target cells with the same receptor exhibit different responses if they have different signal transduction pathways and/or effector proteins; compare **(a)** with **(b)**. Target cells with different receptors for the hormone often exhibit different responses; compare **(b)** with **(c)**.

▲ Tadpole

▲ Adult frog

▲ **Figure 32.10 Specialized role of a hormone in frog metamorphosis.** The hormone thyroxine is responsible for the resorption of the tadpole's tail as the frog develops into its adult form.

---

**CONCEPT CHECK 32.2**

1. Can cells differ in their response to a hormone if they have the same receptor for that hormone? Explain.
2. **WHAT IF?** If a hormone pathway provides a transient response to a stimulus, how would shortening the stimulus duration affect the need for negative feedback?
3. **MAKE CONNECTIONS** What parallels can you identify in the properties and effects of epinephrine and the plant hormone auxin (see Concept 31.1)?

For suggested answers, see Appendix A.

---

## Regulating and Conforming

An animal is a **regulator** for an environmental variable if it uses internal mechanisms to control internal change in the face of external fluctuation. The river otter in **Figure 32.11** is a regulator for temperature, keeping its body at a temperature that is largely independent of that of the water in which it swims. In contrast, an animal is a **conformer** for an environmental variable if it allows its internal condition to change in accordance with external changes. The bass in Figure 32.11 conforms to the temperature of the lake it inhabits. As the water warms or cools, so does the bass's body.

Note that an animal may regulate some internal conditions while allowing others to conform to the environment. For example, even though the bass conforms to the temperature of the surrounding water, it regulates the solute concentration in its blood and **interstitial fluid**, the fluid that surrounds body cells.

## Homeostasis

The steady body temperature of a river otter and the stable concentration of solutes in a freshwater bass are examples of **homeostasis**, which means "steady state," referring to the maintenance of internal balance. In achieving homeostasis, animals maintain a relatively constant internal environment even when the external environment changes significantly.

Many animals exhibit homeostasis for a range of physical and chemical properties. For example, humans maintain a fairly constant body temperature of about 37°C (98.6°F), a blood pH within 0.1 pH unit of 7.4, and a blood glucose concentration that is predominantly in the range of 70–110 mg per 100 mL of blood.

---

**CONCEPT 32.3**

# Feedback control maintains the internal environment in many animals

Managing an animal's internal environment can present a major challenge. Imagine if your body temperature soared every time you took a hot shower or drank a freshly brewed cup of coffee. Faced with environmental fluctuations, animals manage their internal environment by either regulating or conforming.

▶ **Figure 32.11 Regulating and conforming.** The river otter regulates its body temperature, keeping it stable across a wide range of environmental temperatures. The largemouth bass allows its internal environment to conform to the water temperature.

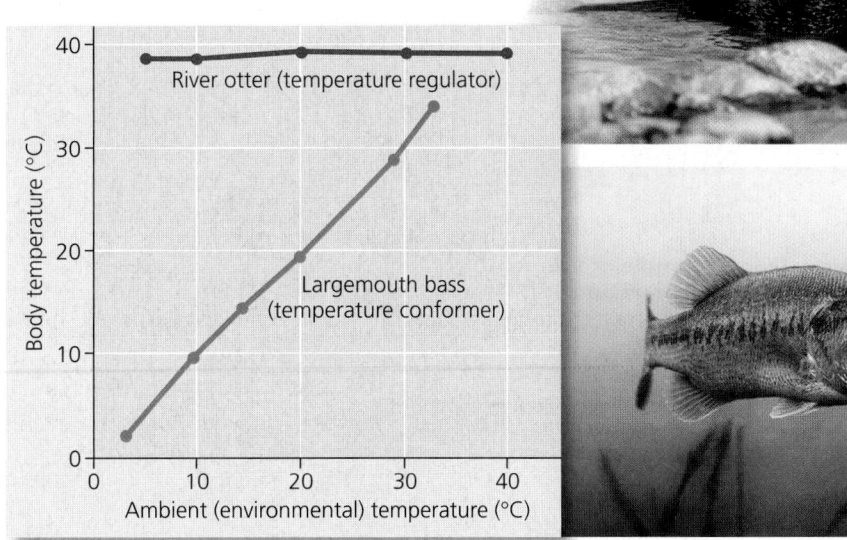

*Graph: Body temperature (°C) vs. Ambient (environmental) temperature (°C). River otter (temperature regulator) remains near 38–39°C. Largemouth bass (temperature conformer) rises linearly with ambient temperature.*

Before exploring homeostasis in animals, let's first consider a nonliving example: the regulation of room temperature **(Figure 32.12)**. Let's assume you want to keep a room at 20°C (68°F), a comfortable temperature for normal activity. You set a control device—the thermostat—to 20°C. A thermometer in the thermostat monitors the room temperature. If the temperature falls below 20°C, the thermostat responds by turning on a radiator, furnace, or other heater. Once the room temperature reaches 20°C, the thermostat switches off the heater. If the temperature then drifts below 20°C, the thermostat activates another heating cycle.

Like a home heating system, an animal achieves homeostasis by maintaining a variable, such as body temperature or solute concentration, at or near a particular value, or **set point**. A fluctuation in the variable above or below the set point serves as the **stimulus** detected by a **sensor**. Upon receiving a signal from the sensor, a *control center* generates output that triggers a **response**, a physiological activity that helps return the variable to the set point.

Like the circuit shown in Figure 32.12, homeostasis in animals relies largely on negative feedback. For example, when

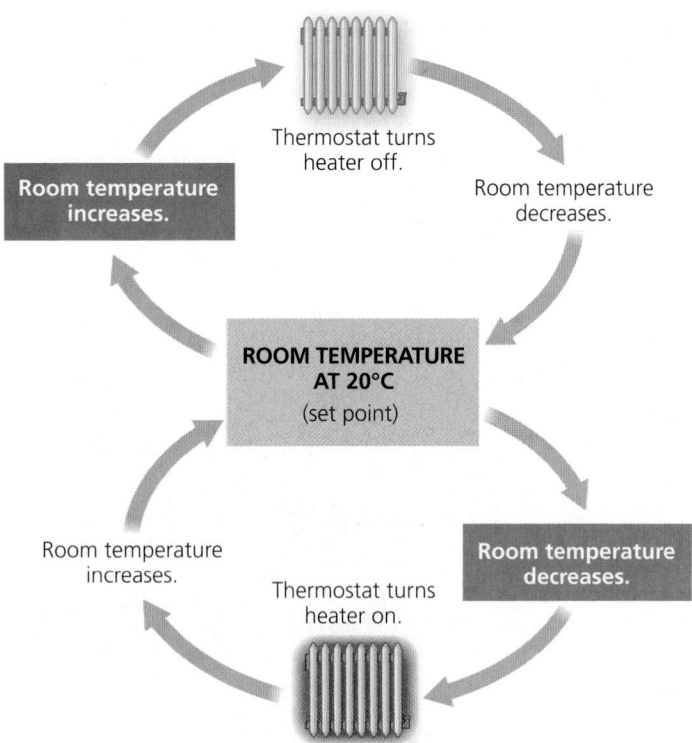

▲ **Figure 32.12 A nonliving example of temperature regulation: control of room temperature.** Regulating room temperature depends on a control center (a thermostat) that detects temperature change and activates mechanisms that reverse that change.

**WHAT IF?** *Label at least one stimulus, response, and sensor/control center in the above figure. How would adding an air conditioner to the system contribute to homeostasis?*

you exercise vigorously, you produce heat, which increases your body temperature. Your nervous system detects this increase and triggers sweating. As you sweat, the evaporation of moisture from your skin cools your body, helping return your body temperature to its set point.

Homeostasis moderates but doesn't eliminate changes in the internal environment. Fluctuation is greater if a variable has a *normal range*—an upper and lower limit—rather than a set point. This is equivalent to a heating system that begins producing heat when the temperature drops to 19°C (66°F) and stops heating when the temperature reaches 21°C (70°F).

Although the set points and normal ranges for homeostasis are usually stable, certain regulated changes in the internal environment are essential. Some of these changes are associated with a particular stage in life, such as the radical shift in hormone balance during puberty. Others are cyclic, such as the monthly variation in hormone levels responsible for a woman's menstrual cycle (see Figure 36.12).

## Thermoregulation: *A Closer Look*

As a physiological example of homeostasis, we'll examine **thermoregulation**, the process by which animals maintain their body temperature within a normal range. Body temperatures outside this range can reduce the efficiency of enzymatic reactions, alter the fluidity of cellular membranes, and affect other temperature-sensitive biochemical processes, potentially with fatal results.

### Endothermy and Ectothermy

Heat for thermoregulation can come from either internal metabolism or the external environment. Humans and other mammals, as well as birds, are **endothermic**, meaning that they are warmed mostly by heat generated by metabolism. In contrast, amphibians, many fishes and nonavian reptiles, and most invertebrates are **ectothermic**, meaning that they gain most of their heat from external sources. However, endothermy and ectothermy are not mutually exclusive. For example, although a bird is mainly endothermic, it may warm itself by basking in the sun on a cold morning, much as an ectothermic lizard does.

Endotherms can maintain a stable body temperature even in the face of large fluctuations in the environmental temperature. In a cold environment, an endotherm generates enough heat to keep its body substantially warmer than its surroundings **(Figure 32.13a)**. In a hot environment, endothermic vertebrates have mechanisms for cooling their bodies, enabling them to withstand heat loads that are intolerable for most ectotherms.

Many ectotherms can adjust their body temperature by behavioral means, such as seeking out shade or basking in the sun **(Figure 32.13b)**. Because their heat source is largely

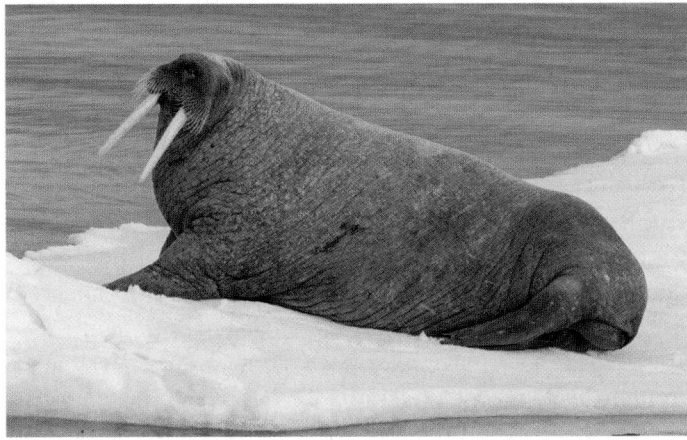

**(a) A walrus, an endotherm**

**(b) A lizard, an ectotherm**

▲ **Figure 32.13 Endothermy and ectothermy.** Endotherms obtain heat from their internal metabolism, whereas ectotherms rely on heat from their external environment.

**Radiation** is the emission of electromagnetic waves by all objects warmer than absolute zero. Here, a lizard absorbs heat radiating from the distant sun and radiates a smaller amount of energy to the surrounding air.

**Evaporation** is the removal of heat from the surface of a liquid that is losing some of its molecules as gas. Evaporation of water from a lizard's moist surfaces that are exposed to the environment has a strong cooling effect.

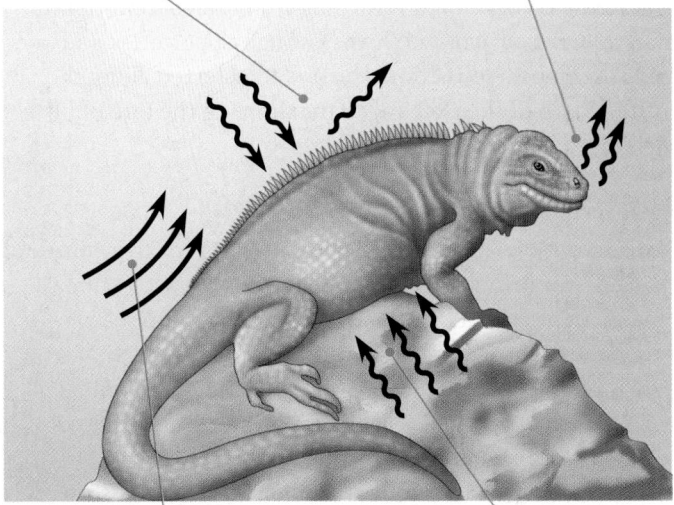

**Convection** is the transfer of heat by the movement of air or liquid past a surface, as when a breeze contributes to heat loss from a lizard's dry skin or when blood moves heat from the body core to the extremities.

**Conduction** is the direct transfer of thermal motion (heat) between molecules of objects in contact with each other, as when a lizard sits on a hot rock.

▲ **Figure 32.14 Heat exchange between an organism and its environment.**

**?** *Which type or types of heat exchange occur when you fan yourself on a hot day?*

environmental, ectotherms generally need to consume much less food than endotherms of equivalent size—an advantage if food supplies are limited. Overall, ectothermy is an effective and successful strategy in most environments, as shown by the abundance and diversity of insects and other ectotherms.

### Balancing Heat Loss and Gain

Thermoregulation depends on an animal's ability to control the exchange of heat with its environment. An organism, like any object, exchanges heat by radiation, evaporation, convection, and conduction **(Figure 32.14)**. Note that heat is always transferred from an object of higher temperature to one of lower temperature.

Numerous adaptations that enhance thermoregulation have evolved in animals. Mammals and birds, for instance, have insulation that reduces the flow of heat between an animal's body and its environment. Such insulation may include hair or feathers as well as layers of fat formed by adipose tissue, such as a whale's thick blubber.

### Circulatory Adaptations for Thermoregulation

Circulatory systems provide a major route for heat flow between the interior and exterior of the body. Adaptations that regulate the extent of blood flow near the body surface or that trap heat within the body core play a significant role in thermoregulation.

In response to changes in the temperature of their surroundings, many animals alter the amount of blood (and hence heat) flowing between their body core and surface. Nerve signals that relax the muscles of the vessel walls result in *vasodilation*, a widening of superficial blood vessels (those near the body surface). As a consequence, blood flow in the outer layer of the body increases. In endotherms, vasodilation usually warms the skin and increases the transfer of body heat to the environment. The reverse process, *vasoconstriction*, reduces blood flow and heat transfer by decreasing the diameter of superficial vessels.

In many birds and mammals, reducing heat loss from the body relies on **countercurrent exchange**, the transfer of heat (or solutes) between fluids that are flowing in opposite directions. In a countercurrent heat exchanger, arteries and veins are located adjacent to each other **(Figure 32.15)**. As warm blood leaves the body core in the arteries, it transfers heat to the colder blood returning from the extremities in the veins. Because blood flows through the arteries and veins in opposite directions, heat is transferred along the entire length of the exchanger, maximizing the rate of heat exchange.

### Acclimatization in Thermoregulation

Acclimatization—a physiological adjustment to environmental changes—contributes to thermoregulation in many animal species. In birds and mammals, acclimatization to seasonal temperature changes often includes adjusting insulation—growing a thicker coat of fur in the winter and shedding it in the summer, for example. These changes help endotherms keep a near constant body temperature year-round.

Acclimatization in ectotherms often includes adjustments at the cellular level. Cells may produce variants of enzymes that have the same function but different optimal temperatures. Also, the proportions of saturated and unsaturated lipids in membranes may change; unsaturated lipids help keep membranes fluid at lower temperatures (see Figure 5.5). Some ectotherms that experience subzero body temperatures produce antifreeze proteins that prevent ice formation in their cells. These compounds enable certain fishes to survive in Arctic or Antarctic water as cold as −2°C (28°F).

### Physiological Thermostats

The regulation of body temperature in humans and other mammals is based on feedback mechanisms. The sensors for thermoregulation are concentrated in the hypothalamus region of the brain. Within the hypothalamus, a group of nerve cells functions as a thermostat, responding to body temperatures outside a normal range by activating mechanisms that promote heat loss or gain **(Figure 32.16)**.

At body temperatures below the normal range, the thermostat inhibits heat loss mechanisms while activating mechanisms that either save heat, including vasoconstriction of vessels in the skin, or generate heat, such as shivering. In response to elevated body temperature, the thermostat shuts down heat retention mechanisms and promotes cooling by vasodilation of vessels in the skin, sweating, or panting.

In the course of certain bacterial and viral infections, mammals and birds develop *fever*, an elevated body temperature. Experiments have shown that fever reflects an increase in the biological thermostat's set point. Indeed, one hypothesis is that fever enhances the body's ability to fight infection, although how fever is beneficial remains a subject of debate.

▶ **Figure 32.15 A countercurrent heat exchanger.** A countercurrent heat exchange system traps heat in the body core, thus reducing heat loss from the extremities, particularly when they are immersed in cold water or in contact with ice or snow. In essence, heat in the arterial blood emerging from the body core is transferred directly to the returning venous blood instead of being lost to the environment.

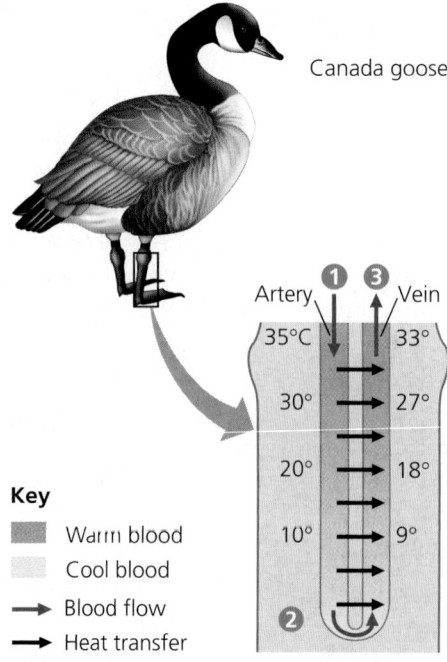

Canada goose

**1** Arteries carrying warm blood to the animal's extremities are in close contact with veins conveying cool blood in the opposite direction, back toward the trunk of the body. This arrangement facilitates heat transfer from arteries to veins along the entire length of the blood vessels.

**2** Near the end of the leg, where arterial blood has been cooled to far below the animal's core temperature, the artery can still transfer heat to the even colder blood in an adjacent vein. The blood in the veins continues to absorb heat as it passes warmer and warmer blood traveling in the opposite direction in the arteries.

**3** As the blood in the veins approaches the center of the body, it is almost as warm as the body core, minimizing the heat loss that results from supplying blood to body parts immersed in cold water.

Artery **1** **3** Vein
35°C  33°
30°  27°
20°  18°
10°  9°

**Key**

Warm blood
Cool blood
→ Blood flow
→ Heat transfer

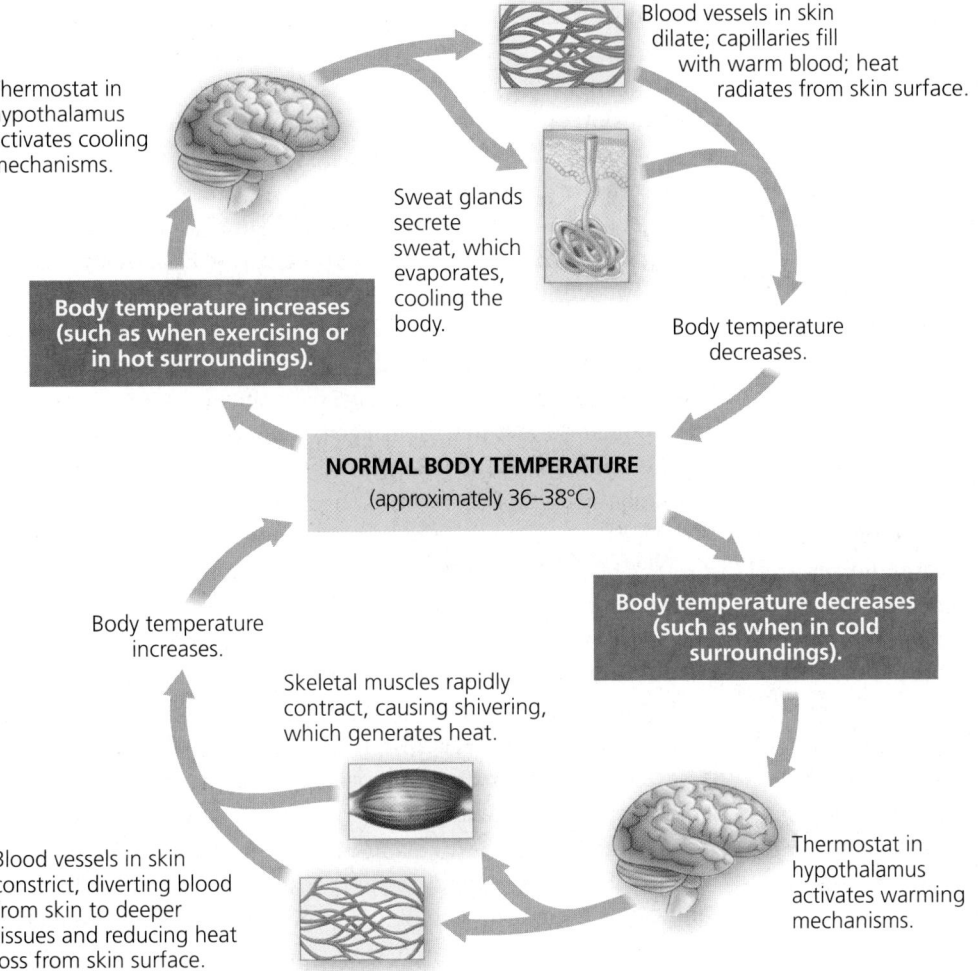

Thermostat in hypothalamus activates cooling mechanisms.

Blood vessels in skin dilate; capillaries fill with warm blood; heat radiates from skin surface.

Sweat glands secrete sweat, which evaporates, cooling the body.

**Body temperature increases (such as when exercising or in hot surroundings).**

Body temperature decreases.

**NORMAL BODY TEMPERATURE** (approximately 36–38°C)

Body temperature increases.

**Body temperature decreases (such as when in cold surroundings).**

Skeletal muscles rapidly contract, causing shivering, which generates heat.

Blood vessels in skin constrict, diverting blood from skin to deeper tissues and reducing heat loss from skin surface.

Thermostat in hypothalamus activates warming mechanisms.

---

## CONCEPT CHECK 32.3

1. Is it accurate to define homeostasis as a constant internal environment? Explain.

2. **MAKE CONNECTIONS** How does negative feedback in thermoregulation differ from feedback inhibition in an enzyme-catalyzed biosynthetic process (see Figure 6.19)?

3. **WHAT IF?** Suppose at the end of a hard run on a hot day you find that there are no drinks left in the cooler. If, out of desperation, you dunk your head into the cooler, how might the ice-cold water affect the rate at which your body temperature returns to normal?

**For suggested answers, see Appendix A.**

---

## CONCEPT 32.4

# A shared system mediates osmoregulation and excretion in many animals

Now that we've considered thermoregulation as an example of homeostasis, we'll turn to another example, the maintenance of salt and water balance in body fluids. Maintaining the fluid environment of animal tissues requires that the relative concentrations of water and solutes be kept within fairly narrow limits. In addition, ions such as sodium and calcium must be maintained at concentrations that permit normal activity of muscles, neurons, and other body cells. Homeostasis thus requires **osmoregulation**, the general term for the processes by which animals control solute concentrations in the interstitial fluid and balance water gain and loss.

In safeguarding their internal fluid environment, animals must deal with a hazardous metabolite produced by the dismantling of proteins and nucleic acids. Breakdown of *nitrogenous* (nitrogen-containing) molecules releases ammonia, a very toxic compound. Several different mechanisms have evolved for **excretion**, the process that rids the body of nitrogenous metabolites and other metabolic waste products. Because systems for excretion and osmoregulation are structurally and functionally linked in many animals, we will consider both of these processes here.

## Osmosis and Osmolarity

All animals—regardless of their habitat and the type of waste they produce—need to balance water uptake and loss. If animal cells take up too much water, the cells swell and burst; if the cells lose too much water, they shrivel and die (see Figure 5.11). Water enters and leaves cells by osmosis, which occurs whenever two solutions separated by a membrane differ in osmotic pressure, or **osmolarity** (total solute concentration expressed as molarity, that is, moles of solute per liter of solution).

If two solutions separated by a selectively permeable membrane have the same osmolarity, they are said to be *isoosmotic*. When two solutions differ in osmolarity, the one with the greater concentration of solutes is said to be *hyperosmotic*, and the more dilute solution is said to be *hypoosmotic*. Water flows by osmosis from a hypoosmotic solution to a hyperosmotic one.

## Osmoregulatory Challenges and Mechanisms

An animal can maintain water balance in two ways. One is to be an **osmoconformer**: to be isoosmotic with its surroundings. All osmoconformers are marine animals. The second way to maintain water balance is to be an **osmoregulator**: to control internal osmolarity independent of the environment. Osmoregulators are found in a wide range of environments, including fresh water and terrestrial habitats that are uninhabitable for osmoconformers.

The opposite osmoregulatory challenges of marine and freshwater environments are illustrated in **Figure 32.17**. For the marine cod (see Figure 32.17a), the ocean is a strongly dehydrating environment. Constantly losing water by osmosis, such fishes balance the water loss by drinking large amounts of seawater. In ridding themselves of salts, they make use of both their gills and kidneys. In the gills, specialized *chloride cells* actively transport chloride ions ($Cl^-$) out and allow sodium ions ($Na^+$) to follow passively. In the kidneys, excess calcium, magnesium, and sulfate ions are excreted with the loss of only small amounts of water.

The freshwater perch (see Figure 32.17b), lives in an environment with a very low osmolarity. As a result, it faces the problem of gaining water by osmosis and losing salts by diffusion. Like many freshwater animals, the perch solves this problem by drinking almost no water and excreting large amounts of very dilute urine. At the same time, salts lost by diffusion and in the urine are replenished by eating. Freshwater fishes such as the perch also replenish salts by uptake across the gills.

For land animals, the threat of dehydration is a major regulatory problem. Although most terrestrial animals have body coverings that help prevent dehydration, they lose water through many routes: in urine and feces, across their skin, and from the surfaces of gas exchange organs. Land animals maintain water balance by drinking and eating moist foods and by producing water metabolically through cellular respiration. In the **Scientific Skills Exercise**, you can examine water balance in one species of desert-dwelling mammal.

## Nitrogenous Wastes

Because most metabolic wastes must be dissolved in water to be excreted from the body, the type and quantity of an animal's waste products may have a large impact on osmoregulation. In this regard, some of the most significant waste products are the nitrogenous breakdown products of proteins and nucleic acids **(Figure 32.18)**. When proteins and nucleic acids are broken apart for energy or converted to carbohydrates or fats, enzymes remove nitrogen in the form of **ammonia** ($NH_3$). Ammonia is very toxic, in part because its ion, ammonium ($NH_4^+$), interferes with oxidative phosphorylation. Although some animals excrete ammonia directly, many species expend energy to convert it to a less toxic compound, either urea or uric acid, prior to excretion.

Animals that excrete nitrogenous wastes as ammonia need access to lots of water because ammonia can be tolerated only at very low concentrations. Therefore, ammonia excretion is most common in aquatic species. The highly soluble ammonia molecules, which interconvert between $NH_3$ and $NH_4^+$, easily

▼ **Figure 32.17 Osmoregulation in marine and freshwater bony fishes: a comparison.**

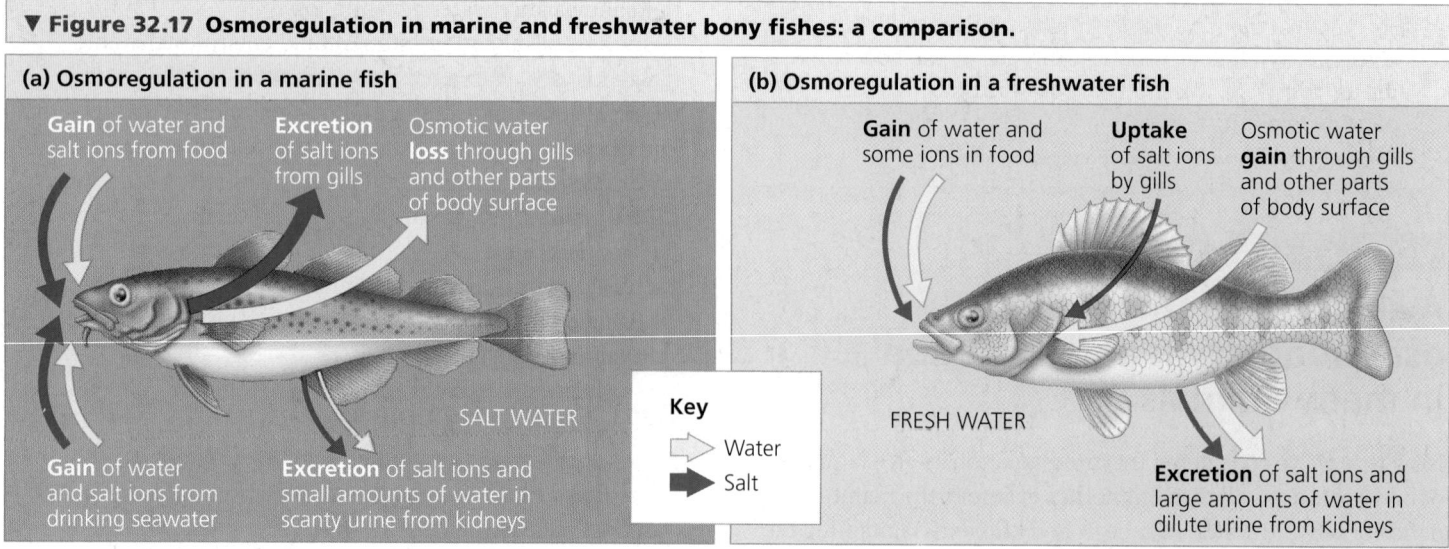

**(a) Osmoregulation in a marine fish**

**Gain** of water and salt ions from food

**Excretion** of salt ions from gills

Osmotic water **loss** through gills and other parts of body surface

SALT WATER

**Gain** of water and salt ions from drinking seawater

**Excretion** of salt ions and small amounts of water in scanty urine from kidneys

**(b) Osmoregulation in a freshwater fish**

**Gain** of water and some ions in food

**Uptake** of salt ions by gills

Osmotic water **gain** through gills and other parts of body surface

FRESH WATER

**Excretion** of salt ions and large amounts of water in dilute urine from kidneys

**Key**
▷ Water
▶ Salt

## Describing and Interpreting Quantitative Data

**How Do Desert Mice Maintain Osmotic Homeostasis?** The sandy inland mouse, recently reclassified as *Pseudomys hermannsburgensis*, is an Australian desert mammal that can survive indefinitely on a diet of dried seeds without drinking water. To study this species' adaptations to its arid environment, researchers conducted a laboratory experiment in which they controlled access to water. In this exercise, you will analyze some of the data from the experiment.

**How the Experiment Was Done** Nine captured mice were kept in an environmentally controlled room and given birdseed (10% water by weight) to eat. In Part A of the study, the mice had unlimited access to tap water for drinking; in Part B of the study, the mice were not given any additional water for 35 days, similar to conditions in their natural habitat. At the end of parts A and B, the researchers measured the osmolarity and urea concentration of the urine and blood of each mouse. The mice were also weighed three times a week.

### Data from the Experiment

| Access to Water | Mean Osmolarity (mOsm/L) | | Mean Urea Concentration (m*M*) | |
|---|---|---|---|---|
| | Urine | Blood | Urine | Blood |
| Part A: Unlimited | 490 | 350 | 330 | 7.6 |
| Part B: None | 4,700 | 320 | 2,700 | 11 |

In part A, the mice drank about 33% of their body weight each day. The change in body weight during the study was negligible for all mice.

**Data from** R. E. MacMillen et al., Water economy and energy metabolism of the sandy inland mouse, *Leggadina hermannsburgensis, Journal of Mammalogy* 53:529–539 (1972).

**INTERPRET THE DATA**

1. (a–d) In words, describe how the data differ between the unlimited-water and no-water conditions with regard to (a) urine osmolarity, (b) blood osmolarity, (c) urea concentration in urine, and (d) urea concentration in blood. (e) Does this data set provide evidence of homeostatic regulation? Explain.

2. (a) Calculate the ratio of urine osmolarity to blood osmolarity for mice with unlimited access to water. (b) Calculate this ratio for mice with no access to water. (c) What conclusion would you draw from these ratios?

3. If the amount of urine produced were different in the two conditions, how would that affect your calculation? Explain.

**MB** A version of this Scientific Skills Exercise can be assigned in MasteringBiology.

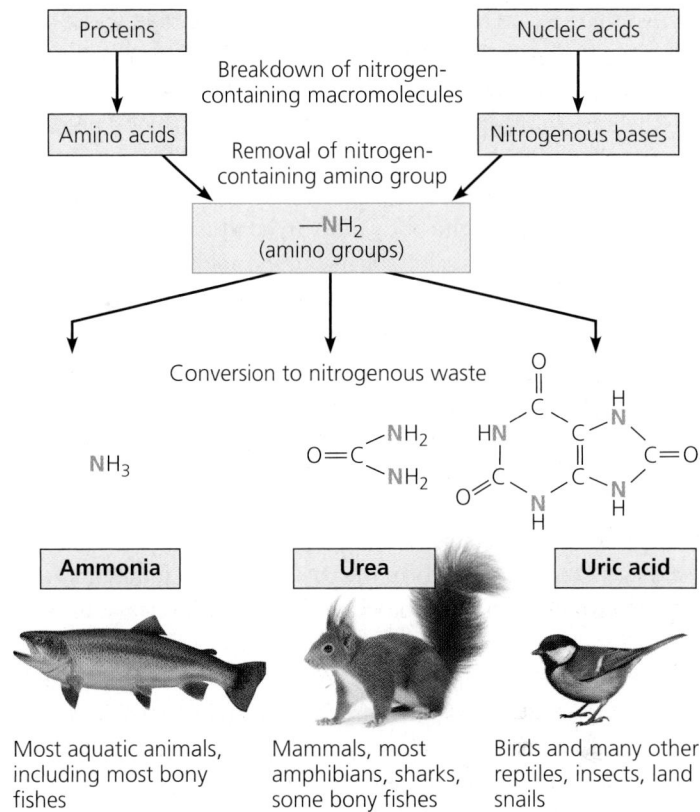

▲ **Figure 32.18 Forms of nitrogenous waste.**

Ammonia — Most aquatic animals, including most bony fishes

Urea — Mammals, most amphibians, sharks, some bony fishes

Uric acid — Birds and many other reptiles, insects, land snails

pass through membranes and are readily lost by diffusion to the surrounding water.

Most terrestrial animals and many marine species cannot afford to lose the amount of water necessary to routinely excrete ammonia. Instead, they mainly excrete a different nitrogenous waste, **urea**. In vertebrates, urea is the product of an energy-consuming metabolic cycle that combines ammonia with carbon dioxide in the liver. The main advantage of urea for nitrogenous waste excretion is its very low toxicity.

Insects, land snails, and many reptiles, including birds, excrete **uric acid** as their primary nitrogenous waste. Uric acid is relatively nontoxic and does not readily dissolve in water. It therefore can be excreted as a semisolid paste with very little water loss. However, uric acid is even more energetically expensive to produce than urea.

## Excretory Processes

In most animals, both osmoregulation and metabolic waste disposal rely on **transport epithelia**, one or more layers of epithelial cells specialized for moving particular solutes in controlled amounts in specific directions. Transport epithelia are typically arranged in complex tubular networks with extensive surface areas. Some transport epithelia face the outside environment directly, while others line channels connected to the outside by an opening on the body surface.

Animals across a range of species produce a fluid waste by the process outlined in **Figure 32.19**. First, blood, coelomic fluid, or hemolymph is brought in contact with a transport epithelium. In most cases, hydrostatic pressure (blood pressure in many animals) drives **filtration**. Cells, as well as proteins and other large molecules, cannot cross the epithelial membrane and remain in the body fluid. In contrast, water and small solutes, such as salts, sugars, amino acids, and nitrogenous wastes, cross the membrane, forming a solution called the **filtrate**. Selective **reabsorption** returns useful molecules and water from the filtrate to the body fluids. Valuable solutes such as glucose, vitamins, and amino acids are reabsorbed by active transport. Nonessential solutes and wastes are left in the filtrate or undergo selective **secretion** into the filtrate by active transport. Finally, the processed filtrate is released from the body as urine during excretion.

The systems that perform the basic excretory functions vary. We'll examine examples from invertebrates and vertebrates.

## Invertebrates

Flatworms (phylum Platyhelminthes), which lack a coelom or body cavity, have excretory systems called *protonephridia* **(Figure 32.20)**. Tubules connected to external openings branch throughout the body. Cellular units called *flame bulbs* cap each branch. Consisting of a tubule cell and a cap cell, each flame bulb has a tuft of cilia projecting into the tubule. During filtration, the beating cilia draw the interstitial fluid through the flame bulb, releasing filtrate into the tubule network. (The moving cilia resemble a flickering flame, hence the name *flame bulb*.) The processed filtrate is then emptied as urine into the external environment. Because the urine excreted by freshwater flatworms has a low solute concentration, its production helps to balance the osmotic uptake of water from the environment.

Natural selection has adapted protonephridia to different tasks in different environments. In the freshwater flatworms, protonephridia serve chiefly in osmoregulation. However, in some parasitic flatworms, which are isoosmotic to the surrounding fluids of their host organisms, the main function of protonephridia is the disposal of nitrogenous wastes.

In insects and other terrestrial arthropods, the filtration step is absent. Instead, the transport epithelium of organs called *Malpighian tubules* secretes certain solutes and wastes into the lumen of the tubule. The filtrate passes to the digestive tract, where most solutes are pumped back into the hemolymph, and water is reabsorbed by osmosis. Nitrogenous wastes are eliminated as nearly dry matter along with the feces, conserving water. Indeed, this excretory system was a very important adaptation for arthropod colonization of land.

## Vertebrates

In vertebrates and some other chordates, a compact organ called the **kidney (Figure 32.21)** functions in both osmoregulation and excretion. Like the excretory organs of most animal

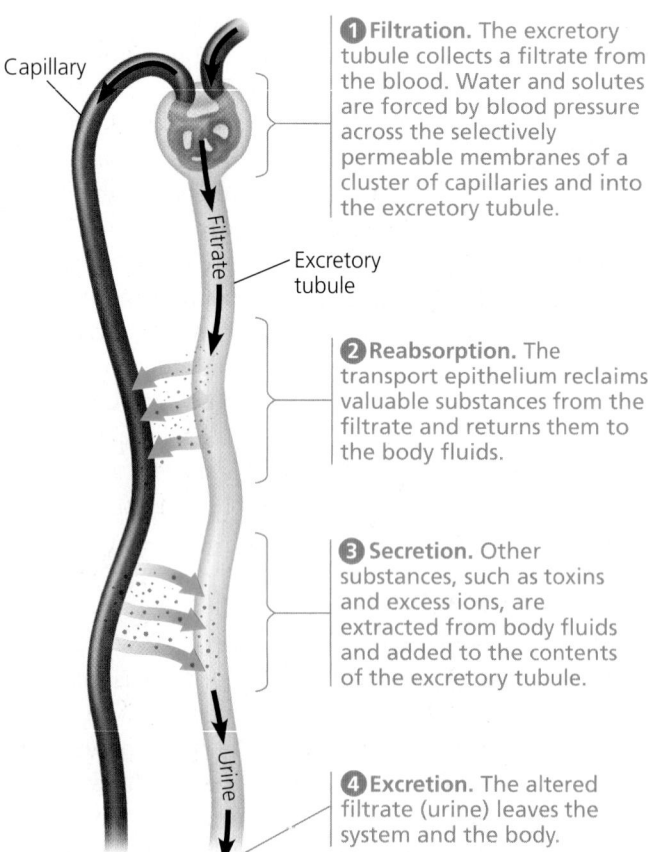

**Capillary**

**Filtrate**

**Excretory tubule**

**Urine**

**❶ Filtration.** The excretory tubule collects a filtrate from the blood. Water and solutes are forced by blood pressure across the selectively permeable membranes of a cluster of capillaries and into the excretory tubule.

**❷ Reabsorption.** The transport epithelium reclaims valuable substances from the filtrate and returns them to the body fluids.

**❸ Secretion.** Other substances, such as toxins and excess ions, are extracted from body fluids and added to the contents of the excretory tubule.

**❹ Excretion.** The altered filtrate (urine) leaves the system and the body.

▲ **Figure 32.19 Key steps of excretory system function: an overview.** Most excretory systems produce a filtrate by pressure-filtering body fluids and then modify the filtrate's contents. This diagram is modeled after the vertebrate excretory system.

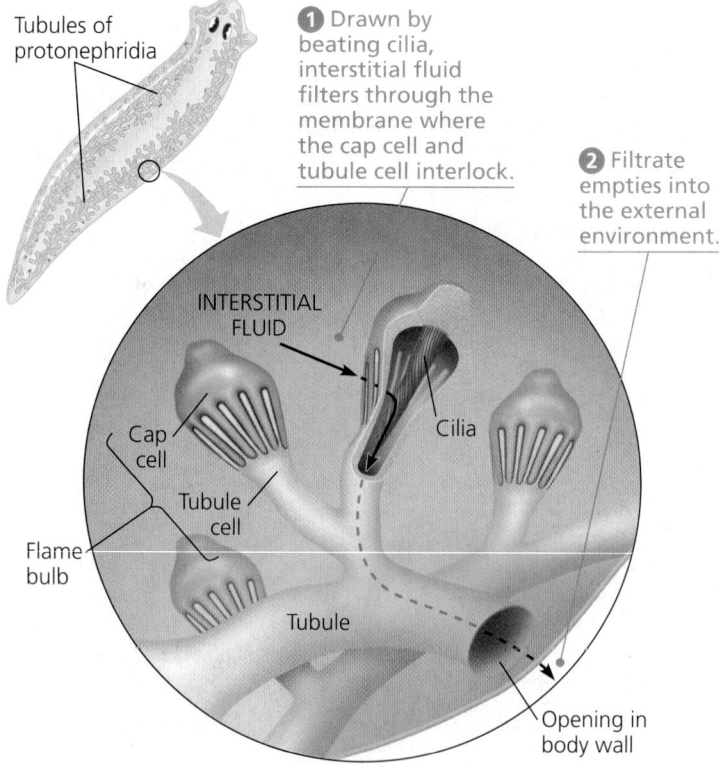

Tubules of protonephridia

**❶** Drawn by beating cilia, interstitial fluid filters through the membrane where the cap cell and tubule cell interlock.

**❷** Filtrate empties into the external environment.

INTERSTITIAL FLUID

Cap cell

Cilia

Tubule cell

Flame bulb

Tubule

Opening in body wall

▲ **Figure 32.20 Protonephridia in a planarian.**

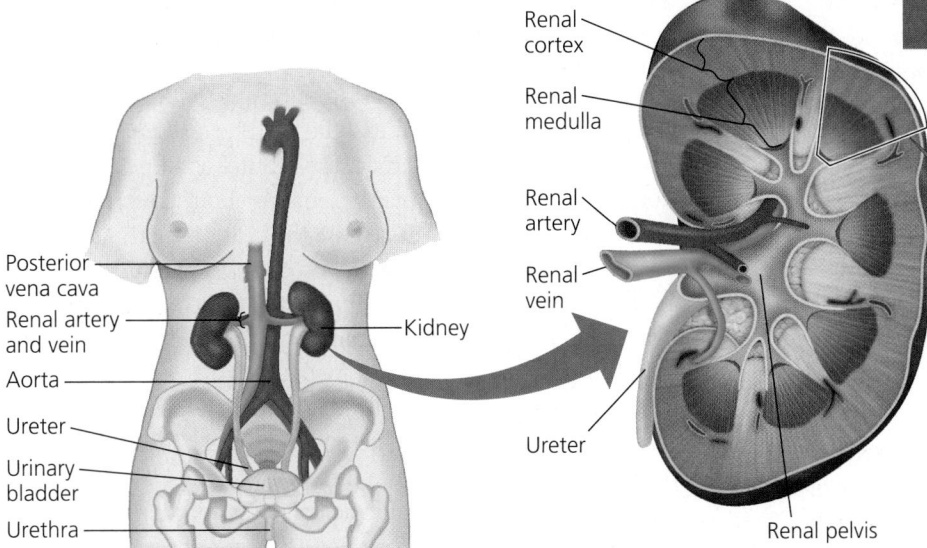

In humans, the excretory system consists of a pair of **kidneys**, bean-shaped organs about 10 cm in length, as well as organs for transporting and storing urine. Urine produced by each kidney exits through a duct called the ureter; the two ureters drain into a common sac called the **urinary bladder**. During urination, urine is expelled from the bladder through a tube called the **urethra**, which empties to the outside near the vagina in females and through the penis in males. Sphincter muscles near the junction of the urethra and bladder regulate urination.

## Kidney Structure

Each kidney has an outer **renal cortex** and an inner **renal medulla**. Both regions are supplied with blood by a renal artery and drained by a renal vein. Within the cortex and medulla lie tightly packed excretory tubules and associated blood vessels. The inner **renal pelvis** collects urine from the excretory tubules and passes it to the urinary bladder.

## Nephron Types

Weaving back and forth across the renal cortex and medulla are the **nephrons**, the functional units of the vertebrate kidney. Most of these are cortical nephrons, which reach only a short distance into the medulla. The remainder, the **juxtamedullary nephrons**, extend deep into the medulla. Juxtamedullary nephrons are essential for production of urine that is hyperosmotic to body fluids.

## Nephron Organization

Each nephron consists of a single long tubule and a ball of capillaries called the **glomerulus**. The blind end of the tubule forms a cup-shaped swelling, called **Bowman's capsule**, which surrounds the glomerulus. Filtrate is formed when blood pressure forces fluid from the blood in the glomerulus into the lumen of Bowman's capsule. Processing occurs as the filtrate passes through three major regions of the nephron: the **proximal tubule**, the **loop of Henle** (a hairpin turn with a descending limb and an ascending limb), and the **distal tubule**. A **collecting duct** receives processed filtrate from many nephrons and transports it to the renal pelvis. Each nephron is supplied with blood by an *afferent arteriole*, an offshoot of the renal artery that branches and forms the capillaries of the glomerulus. The capillaries converge as they leave the glomerulus, forming an *efferent arteriole*. Branches of this vessel form the **peritubular capillaries**, which surround the proximal and distal tubules. Other branches extend downward and form the **vasa recta**, hairpin-shaped capillaries that serve the renal medulla and surround the loop of Henle.

phyla, kidneys consist of tubules. The tubules of these organs are arranged in a highly organized manner and are closely associated with a network of capillaries. The vertebrate excretory system also includes ducts and other structures that carry urine from the tubules out of the kidney and, eventually, the body. Familiarizing yourself with the terms and diagrams in Figure 32.21 will provide you with a solid foundation for learning about filtrate processing in the kidney, the focus of the last section of this chapter.

### CONCEPT CHECK 32.4

1. What is the function of the filtration step in excretory systems?
2. What advantage does uric acid offer as a nitrogenous waste in arid environments?
3. **WHAT IF?** A camel standing in the sun requires much more water when its fur is shaved off, although its body temperature remains the same. What can you conclude about the relationship between osmoregulation and the insulation provided by fur?

For suggested answers, see Appendix A.

## CONCEPT 32.5

# The mammalian kidney's ability to conserve water is a key terrestrial adaptation

As illustrated in Figure 32.21, the basic unit of the mammalian kidney is the nephron. Here, we will consider the role of the nephron in forming the filtrate in the mammalian kidney. We will then focus on how tubules, capillaries, and the surrounding tissue of the nephron function together in processing that filtrate.

In the human kidney, filtrate forms when fluid passes from the bloodstream to the lumen of Bowman's capsule in each nephron. The glomerular capillaries retain blood cells and large molecules such as plasma proteins but are permeable to water and small solutes. Consequently, the filtrate contains salts, glucose, amino acids, vitamins, nitrogenous wastes, and other small molecules. Because such molecules pass freely between glomerular capillaries and Bowman's capsule, the concentrations of these substances in the initial filtrate are the same as those in blood plasma.

Under normal conditions, roughly 1,600 L of blood flows through a pair of human kidneys each day, yielding about 180 L of initial filtrate. Of this, about 99% of the water and nearly all of the sugars, amino acids, vitamins, and other organic nutrients are reabsorbed into the blood, leaving only about 1.5 L of urine to be transported to the bladder.

### From Blood Filtrate to Urine: *A Closer Look*

To explore how filtrate is processed into urine, we'll follow the filtrate along its path through a nephron (**Figure 32.22**). Each

circled number refers to the processing in transport epithelia as the filtrate moves through the cortex and medulla of the kidney.

**❶ Proximal tubule.** Reabsorption in the proximal tubule is critical for the recapture of ions, water, and valuable nutrients from the huge volume of initial filtrate. Because a large amount of water *and* salt is reabsorbed, the filtrate's volume decreases substantially, but its osmolarity remains about the same.

NaCl (salt) in the filtrate enters the cells of the transport epithelium by facilitated diffusion and cotransport mechanisms and then is transferred to the interstitial fluid by active transport (see Concept 5.4). This transfer of positive charge out of the tubule drives the passive transport of $Cl^-$.

As salt moves from the filtrate to the interstitial fluid, water follows by osmosis. The salt and water then diffuse from the interstitial fluid into the peritubular capillaries (see Figure 32.21). Glucose, amino acids, potassium ions ($K^+$), and other essential substances are also actively or passively transported from the filtrate to the interstitial fluid and then into the peritubular capillaries. In contrast, some toxic materials, such as drugs and toxins that have been processed in the liver, are actively secreted into filtrate by the transport epithelium.

**❷ Descending limb of the loop of Henle.** Upon leaving the proximal tubule, filtrate enters the loop of Henle, which further reduces filtrate volume via distinct stages of water and salt movement. In the first portion of the loop, the descending limb, numerous water channels formed by **aquaporin** proteins make the transport epithelium freely permeable to water. In contrast, there are almost no channels for salt and other small solutes, resulting in very low permeability for these substances.

For water to move out of the tubule by osmosis, the interstitial fluid bathing the tubule must be hyperosmotic to the filtrate. This condition is met along the entire length of the descending limb because the osmolarity of the interstitial fluid increases progressively from the cortex to the inner medulla of the kidney. Consequently, the filtrate loses water and increases in solute concentration all along its journey down the descending limb. The highest osmolarity (about 1,200 mOsm/L) occurs at the elbow of the loop of Henle.

**❸ Ascending limb of the loop of Henle.** Once the filtrate reaches the tip of the loop of Henle, it returns to the cortex within the ascending limb. Unlike the descending portion of the loop, the ascending limb has a transport epithelium that lacks water channels. As a result, in this region the epithelial membrane that faces the filtrate is impermeable to water.

The ascending limb has two specialized regions: a thin segment near the loop tip and a thick segment adjacent to the distal tubule. As filtrate ascends in the thin segment, NaCl, which became highly concentrated in the descending limb, diffuses out of the permeable tubule into the interstitial fluid. This movement of NaCl out of the tubule helps maintain the osmolarity of the interstitial fluid in the medulla.

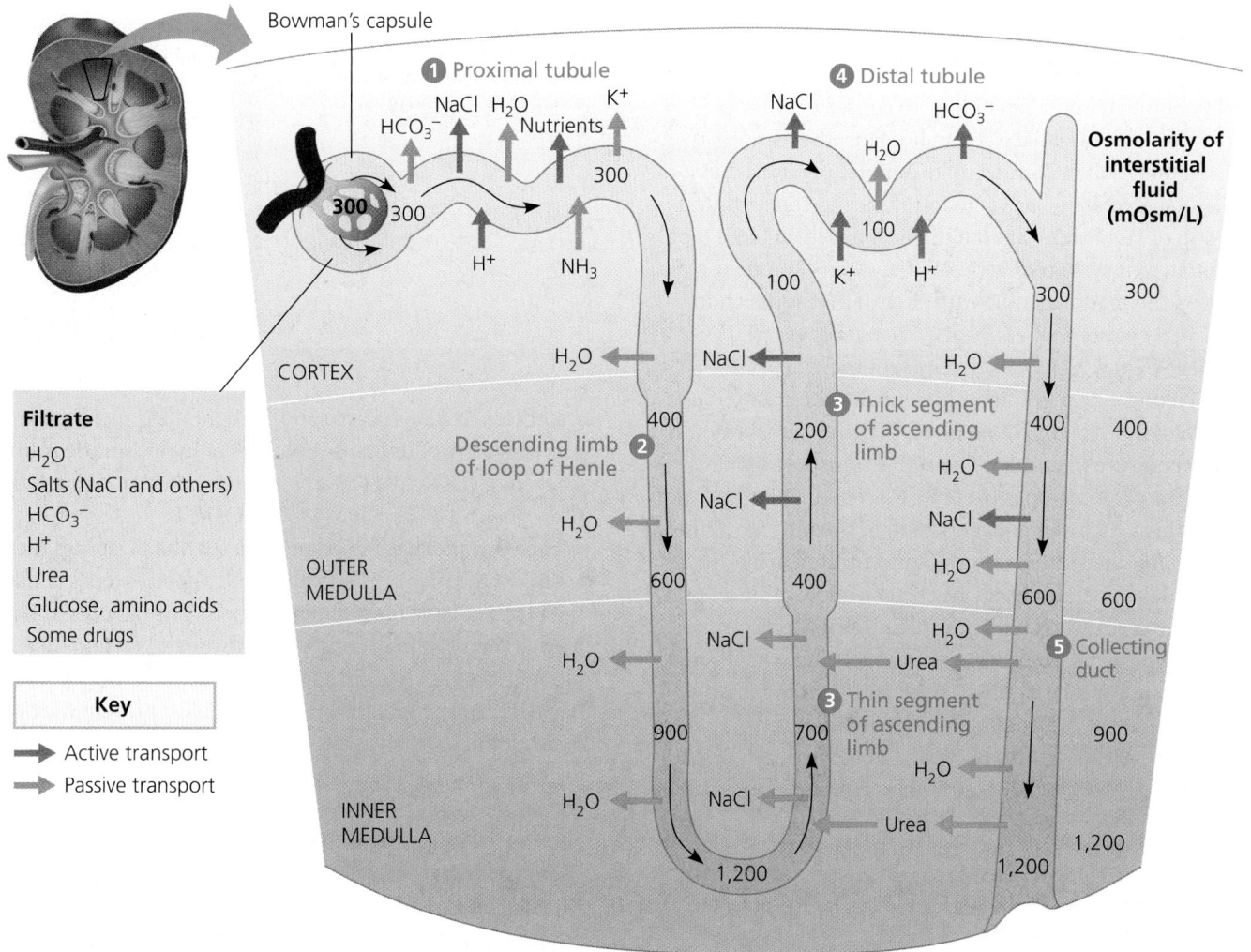

▲ **Figure 32.22 How the human kidney concentrates urine: regional functions of the transport epithelium.**
The numbered regions in this diagram are keyed to the circled numbers in the text discussion of kidney function.

**WHAT IF?** *The drug furosemide blocks the cotransporters for Na⁺and Cl⁻in the ascending limb of the loop of Henle. What effect would you expect this drug to have on urine volume?*

In the thick segment of the ascending limb, the movement of NaCl out of the filtrate continues. Here, however, the epithelium actively transports NaCl into the interstitial fluid. As a result of losing salt but not water, the filtrate becomes progressively more dilute as it moves up to the cortex in the ascending limb of the loop.

**4** **Distal tubule.** The distal tubule plays a key role in regulating the $K^+$ and NaCl concentrations of body fluids. This regulation involves variation in the amount of $K^+$ secreted into the filtrate as well as the amount of NaCl reabsorbed from the filtrate. The distal tubule also contributes to pH regulation by the controlled secretion of $H^+$ and reabsorption of $HCO_3^-$.

**5** **Collecting duct.** The collecting duct processes the filtrate into urine, which it carries to the renal pelvis (see Figure 32.21). As filtrate passes along the transport epithelium of the collecting duct, regulation of permeability and transport across the epithelium determines the extent to which the urine becomes concentrated.

When the kidneys are conserving water, aquaporin channels in the collecting duct allow water molecules to cross the epithelium. The filtrate becomes increasingly concentrated, losing more and more water by osmosis to the hyperosmotic interstitial fluid. In the inner medulla, the duct becomes permeable to urea. Because of the high urea concentration in the filtrate at this point, some urea diffuses out of the duct and into the interstitial fluid. The net result is urine that is hyperosmotic to the general body fluids.

When maintaining salt and water balance requires the production of dilute rather than concentrated urine, the collecting duct actively transports NaCl out of the filtrate and into the surrounding medulla. At the same time, aquaporin channels are removed from the collecting duct epithelium, with the result that water cannot follow the salts by osmosis.

As we will see, the state of the collecting duct epithelium is controlled by hormones that together maintain homeostasis for osmolarity, blood pressure, and blood volume.

## Concentrating Urine in the Mammalian Kidney

The ability of the mammalian kidney to conserve water is a key adaptation for terrestrial habitats. In humans, the osmolarity of blood is about 300 mOsm/L (milliOsmoles per liter), but the kidney can excrete urine up to four times as concentrated.

The loop of Henle and surrounding capillaries act as a type of countercurrent system to generate the steep osmotic gradient between the medulla and cortex. Recall that some endotherms have a countercurrent heat exchanger that reduces heat loss (see Figure 32.15). In that system there is passive movement along a heat gradient. In contrast, the countercurrent system of the loop of Henle involves active transport and thus an expenditure of energy. The active transport of NaCl from the filtrate in the upper part of the ascending limb of the loop maintains a high salt concentration in the interior of the kidney, enabling the kidney to form concentrated urine. Such a system, which expends energy to create a concentration gradient, is called a **countercurrent multiplier system**.

When the human kidney concentrates urine maximally, the urine reaches an osmolarity of 1,200 mOsm/L. Some mammals can do even better: Australian hopping mice, small marsupials that live in dry desert regions, can produce urine with an osmolarity of 9,300 mOsm/L, 25 times as concentrated as the animal's blood.

## Adaptations of the Vertebrate Kidney to Diverse Environments

**EVOLUTION** Vertebrates occupy habitats ranging from rain forests to deserts and from some of the saltiest bodies of water to the nearly pure waters of high mountain lakes. Variations in nephron structure and function equip the kidneys of different vertebrates for osmoregulation in their various habitats. These adaptations are made apparent by comparing species that inhabit a range of environments or by comparing the responses of different vertebrates to similar conditions.

Mammals that excrete the most hyperosmotic urine, such as hopping mice, kangaroo rats, and other desert mammals, have loops of Henle that extend deep into the medulla. Long loops maintain steep osmotic gradients in the kidney, resulting in urine becoming very concentrated as it passes from cortex to medulla in the collecting ducts.

Birds have loops of Henle that extend less far into the medulla than those of mammals. Thus, bird kidneys cannot concentrate urine to the high osmolarities achieved by mammalian kidneys. Although birds can produce hyperosmotic urine, their main water conservation adaptation is excreting their nitrogenous waste in the form of uric acid.

In mammals, both the volume and osmolarity of urine are adjusted according to an animal's water and salt balance. In situations of high salt intake and low water availability, a mammal can excrete small volumes of hyperosmotic urine with minimal

▶ **Figure 32.23 A vampire bat (*Desmodus rotundus*), a mammal with unique excretory challenges.**

water loss. If salt is scarce and fluid intake is high, the kidney can instead produce large volumes of hypoosmotic urine, getting rid of the excess water with little salt loss. At such times, the urine can be as dilute as 70 mOsm/L.

The vampire bat shown in **Figure 32.23** illustrates the versatility of the mammalian kidney. This species feeds at night on the blood of large birds and mammals. The bat uses its sharp teeth to make a small incision in the prey's skin and then laps up blood from the wound (the prey animal is typically not seriously harmed). Anticoagulants in the bat's saliva prevent the blood from clotting. Because a vampire bat may fly long distances to locate a suitable victim, when it does find prey it benefits from consuming as much blood as possible—often more than half its body mass. By itself, this blood intake would make the bat too heavy to fly. As the bat feeds, however, its kidneys enable it to excrete large volumes of dilute urine, up to 24% of body mass per hour. Having lost enough weight to take off, the bat can fly back to its roost in a cave or hollow tree, where it spends the day.

In the roost, the vampire bat faces a different regulatory problem. Most of the nutrition it derives from blood comes in the form of protein. Digesting proteins generates large quantities of urea, but roosting bats lack access to the drinking water necessary to dilute it. Instead, their kidneys shift to producing small quantities of highly concentrated urine (up to 4,600 mOsm/L), an adjustment that disposes of the urea load while conserving as much water as possible. The bat's ability to alternate rapidly between large amounts of dilute urine and small amounts of very hyperosmotic urine is an essential part of its adaptation to an unusual food source.

## Homeostatic Regulation of the Kidney

A combination of nervous and hormonal inputs regulates the osmoregulatory function of the mammalian kidney. Through their effect on the amount and osmolarity of urine, these inputs contribute to homeostasis for both blood pressure and blood volume.

### Antidiuretic Hormone

One key hormone in the regulatory circuitry of the kidney is antidiuretic hormone (ADH), also called vasopressin **(Figure 32.24)**. Osmoreceptor cells in the hypothalamus monitor

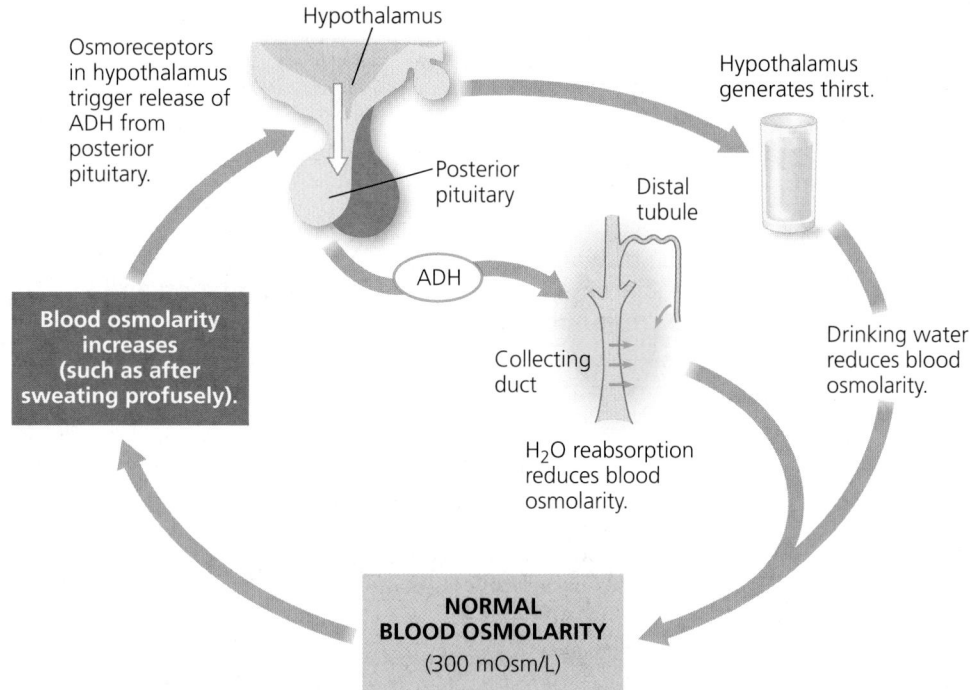

Osmoreceptors in hypothalamus trigger release of ADH from posterior pituitary.

Hypothalamus

Posterior pituitary

ADH

Hypothalamus generates thirst.

Distal tubule

Blood osmolarity increases (such as after sweating profusely).

Collecting duct

Drinking water reduces blood osmolarity.

$H_2O$ reabsorption reduces blood osmolarity.

**NORMAL BLOOD OSMOLARITY**
(300 mOsm/L)

▲ **Figure 32.24 Regulation of fluid retention in the kidney by antidiuretic hormone (ADH).** Osmoreceptors in the hypothalamus monitor blood osmolarity via its effect on the net diffusion of water into or out of the receptor cells. When blood osmolarity increases, signals from the osmoreceptors trigger a release of ADH from the posterior pituitary, as well as thirst. Drinking water and increased water reabsorption reduce blood osmolarity, inhibiting further ADH secretion and thereby completing the feedback circuit.

the osmolarity of blood and regulate release of ADH from the posterior pituitary. ADH binds to receptor molecules on epithelial cells in the collecting duct, leading to a temporary increase in the number of aquaporin proteins in the plasma membrane. Because aquaporin proteins form water channels, the net effect is an increased permeability of the epithelium to water.

To see how the response to ADH in the kidney contributes to osmoregulation, let's consider first what occurs when blood osmolarity rises, such as after eating salty food or losing water through sweating. When osmolarity rises above the set point (300 mOsm/L), ADH release into the bloodstream is increased. The collecting duct's permeability to water rises, resulting in water reabsorption, which concentrates urine, reduces urine volume, and lowers blood osmolarity back toward the set point. (Only the gain of additional water in food or drink can fully restore osmolarity to 300 mOsm/L.) As the osmolarity of the blood falls, a negative-feedback mechanism reduces the activity of osmoreceptor cells in the hypothalamus, and ADH secretion is reduced.

What happens if, instead of ingesting salt or sweating profusely, you drink a large amount of water? The resulting reduction in blood osmolarity below the set point causes a drop in ADH secretion to a very low level. The number of aquaporin channels decreases, lowering permeability of the collecting ducts. Water reabsorption is reduced, resulting in discharge of large volumes of dilute urine. (A high level of urine production

is called diuresis; ADH opposes this state and is therefore called antidiuretic hormone.)

### Coordination of Kidney Regulation

The release of ADH is a response to an increase in blood osmolarity, as when the body is dehydrated from excessive water loss or inadequate water intake. However, an excessive loss of both salt and body fluids—caused, for example, by a major wound or severe diarrhea—will reduce blood volume *without* increasing osmolarity. Given that this will not affect ADH release, how does the body respond? It turns out that an endocrine circuit called the *renin-angiotensin-aldosterone system* (*RAAS*) also regulates kidney function. The RAAS responds to the drop in blood volume and pressure by increasing water and $Na^+$ reabsorption. Thus, ADH and the RAAS are partners in homeostasis.

One product of the RAAS is a peptide called *angiotensin II*. Functioning as a hormone, angiotensin II triggers vasoconstriction, increasing blood pressure and decreasing blood flow to capillaries in the kidney (and elsewhere). Angiotensin II also triggers events that cause nephrons to increase $Na^+$ and water reabsorption, thus increasing blood volume and pressure. Because angiotensin II increases blood pressure, drugs that block angiotensin II production are widely used to treat hypertension (chronic high blood pressure). Many of these drugs are specific inhibitors of angiotensin converting enzyme (ACE), which catalyzes one of the steps in the production of angiotensin II.

In all animals, some of the intricate physiological machines we call organs work continuously in maintaining solute and water balance and excreting nitrogenous wastes. The details that we have reviewed in this chapter only hint at the great complexity of the neural and hormonal mechanisms involved in regulating these homeostatic processes.

### CONCEPT CHECK 32.5

1. Why could it be dangerous to drink a very large amount of water in a short period of time?
2. Many medications make the epithelium of the collecting duct less permeable to water. How would taking such a drug affect kidney output?
3. **WHAT IF?** If blood pressure in the afferent arteriole leading to a glomerulus decreased, how would the rate of blood filtration within Bowman's capsule be affected? Explain.

For suggested answers, see Appendix A.

# 32 Chapter Review

Go to **MasteringBiology**® for Assignments, the eText, and the Study Area with Animations, Activities, Vocab Self-Quiz, and Practice Tests.

## SUMMARY OF KEY CONCEPTS

**VOCAB SELF-QUIZ**

goo.gl/gbai8v

### CONCEPT 32.1

**Animal form and function are correlated at all levels of organization (pp. 664–667)**

- Animal bodies are based on a hierarchy of cells, **tissues**, **organs**, and **organ systems**. **Epithelial tissue** forms active interfaces on external and internal surfaces; **connective tissue** binds and supports other tissues; **muscle tissue** contracts, moving body parts; and **nervous tissue** transmits nerve impulses throughout the body.
- Animals and plants exhibit both shared and diverse adaptations to common life challenges.

**?** *Describe how the epithelial tissue that lines the stomach lumen is well suited to its function.*

### CONCEPT 32.2

**The endocrine and nervous systems act individually and together in regulating animal physiology (pp. 668–673)**

- In communicating between different locations in the body, the **endocrine system** broadcasts signaling molecules called **hormones** everywhere via the bloodstream. Only certain cells are responsive to each hormone. The **nervous system** uses dedicated cellular circuits involving electrical and chemical signals to send information to specific locations. Hormone pathways may be regulated by **negative feedback**, which damps the stimulus, or **positive feedback**, which amplifies the stimulus and drives the response to completion.

**?** *Why would a water-soluble hormone likely have no effect if injected directly into the cytosol of a target cell?*

### CONCEPT 32.3

**Feedback control maintains the internal environment in many animals (pp. 673–677)**

- Animals *regulate* certain internal variables while allowing other internal variables to *conform* to external changes. **Homeostasis** is the maintenance of a steady state despite internal and external changes.

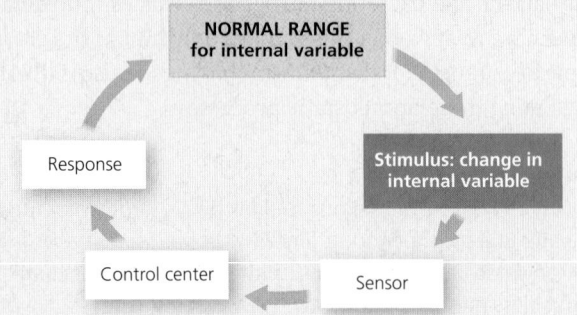

- An animal maintains its internal temperature within a tolerable range by **thermoregulation**. **Endotherms** are warmed mostly by heat generated by metabolism. **Ectotherms** get most of their heat from external sources. The **hypothalamus** acts as the thermostat in mammalian regulation of body temperature.

**?** *Given that humans thermoregulate, explain why your skin is cooler than your body core.*

### CONCEPT 32.4

**A shared system mediates osmoregulation and excretion in many animals (pp. 677–682)**

- Cells balance water gain and loss through **osmoregulation**, a process based on the controlled movement of solutes between internal fluids and the external environment and on the movement of water, which follows by osmosis.
- Protein and nucleic acid metabolism generates **ammonia**, which in many animals is converted to **urea** or **uric acid** for **excretion**. Most excretory systems carry out **filtration**, **reabsorption**, **secretion**, and **excretion**. Excretory tubules (consisting of **nephrons** and **collecting ducts**) and blood vessels pack the mammalian **kidney**.

**DRAW IT** *Construct a table summarizing the three major types of nitrogenous wastes and their relative toxicity, energy content, and associated water loss during excretion.*

### CONCEPT 32.5

**The mammalian kidney's ability to conserve water is a key terrestrial adaptation (pp. 682–685)**

- Within the nephron, selective secretion and reabsorption in the **proximal tubule** alter **filtrate** volume and composition. The *descending limb* of the **loop of Henle** is permeable to water but not salt, whereas the *ascending limb* is permeable to salt but not water. The **distal tubule** and collecting duct regulate $K^+$ and NaCl levels in body fluids.
- In a mammalian kidney, **a countercurrent multiplier system** involving the loop of Henle maintains the gradient of salt concentration in the kidney interior. In response to hormonal signals, urine can be concentrated in the collecting duct.
- Natural selection has shaped the form and function of nephrons in vertebrates to the challenges of the animals' habitats. For example, desert mammals, which excrete the most hyperosmotic urine, have loops of Henle that extend deep into the **renal medulla**.
- When blood **osmolarity** rises, the posterior pituitary releases **antidiuretic hormone (ADH)**, which increases permeability to water in collecting ducts by increasing the number of water channels. A second endocrine pathway, the renin-angiotensin-aldosterone system (RAAS), regulates blood pressure and volume, functions that partially overlap with those of ADH.

**?** *How do the nephrons of kangaroo rats and birds differ in structure and in their ability to concentrate urine?*

## TEST YOUR UNDERSTANDING

**PRACTICE TEST**

goo.gl/CRZjvS

### Level 1: Knowledge/Comprehension

1. The body tissue that consists largely of material located outside of cells is
   (A) epithelial tissue.
   (B) connective tissue.
   (C) muscle.
   (D) nervous tissue.

2. Which of the following would increase the rate of heat exchange between an animal and its environment?
(A) feathers or fur
(B) vasoconstriction
(C) wind blowing across the body surface
(D) countercurrent heat exchanger

3. Which process in the nephron is *least* selective?
(A) filtration
(B) reabsorption
(C) active transport
(D) secretion

## Level 2: Application/Analysis

4. Homeostasis typically relies on negative feedback because positive feedback
(A) requires a response but not a stimulus.
(B) drives processes to completion rather than to a balance point.
(C) acts within, but not beyond, a normal range.
(D) can decrease but not increase a variable.

5. Which of the following is an accurate statement about thermoregulation?
(A) Endotherms are regulators and ectotherms are conformers.
(B) Endotherms maintain a constant body temperature and ectotherms do not.
(C) Endotherms and ectotherms differ in their primary source of heat for thermoregulation.
(D) Endothermy has a lower energy cost than ectothermy.

6. In which of the following species should natural selection favor the highest proportion of nephrons with loops of Henle that extend deep into the renal medulla?
(A) a river otter
(B) a mouse species living in a temperate broadleaf forest
(C) a mouse species living in a desert
(D) a beaver

7. African lungfish, which are often found in small, stagnant pools of fresh water, produce urea as a nitrogenous waste. What is the advantage of this adaptation?
(A) Urea takes less energy to synthesize than ammonia.
(B) Small, stagnant pools do not provide enough water to dilute the toxic ammonia.
(C) Urea forms an insoluble precipitate.
(D) Urea makes lungfish tissue hypoosmotic to the pool.

## Level 3: Synthesis/Evaluation

8. **DRAW IT** Draw a model of the control circuit(s) required for driving an automobile at a fairly constant speed over a hilly road. Label each feature that represents a sensor, stimulus, or response.

9. **INTERPRET THE DATA** Use the data below to draw four pie charts for water gain and loss in a kangaroo rat and a human.

| | Kangaroo Rat | Human |
|---|---|---|
| **Water Gain (mL)** | | |
| Ingested in food | 0.2 | 750 |
| Ingested in liquid | 0 | 1,500 |
| Derived from metabolism | 1.8 | 250 |
| **Water Loss (mL)** | | |
| Urine | 0.45 | 1,500 |
| Feces | 0.09 | 100 |
| Evaporation | 1.46 | 900 |

Which routes of water gain and loss make up a much larger share of the total in a kangaroo rat than in a human?

10. **FOCUS ON EVOLUTION**
Merriam's kangaroo rats (*Dipodomys merriami*) live in North American habitats ranging from moist, cool woodlands to hot deserts. Assuming that natural selection has resulted in differences in water conservation between *D. merriami* populations, devise a hypothesis concerning the relative rates of evaporative water loss by populations that live in moist versus dry environments. Describe how you could test your hypothesis using a humidity sensor to detect evaporative water loss by kangaroo rats.

11. **FOCUS ON ORGANIZATION**
In a short essay (100–150 words), compare how membrane structures in the loop of Henle and collecting duct of the mammalian kidney enable water to be recovered from filtrate in the process of osmoregulation.

12. **SYNTHESIZE YOUR KNOWLEDGE**

These macaques (*Macaca fuscata*) are partially immersed in a hot spring in a snowy region of Japan. What are some ways that form, function, and behavior contribute to homeostasis for these animals?

*For selected answers, see Appendix A.*

# 33 Animal Nutrition

▲ **Figure 33.1** How does a crab help an otter make fur?

## The Need to Feed

Dinnertime has arrived for the sea otter in **Figure 33.1** (and for the crab, though in quite a different sense). The muscles and other tissues of the crab will be chewed into pieces, broken down by acid and enzymes in the otter's digestive system, and finally absorbed as small molecules into the body of the otter. Such a process is what is meant by animal **nutrition**: food being taken in, taken apart, and taken up.

Although dining on fish, crabs, urchins, and abalone is the sea otter's specialty, all animals eat other organisms—dead or alive, piecemeal or whole. Unlike plants, animals must consume food for both energy and the organic molecules used to assemble new molecules, cells, and tissues. Despite this shared need, animals have diverse diets. **Herbivores**, such as cattle, sea slugs, and caterpillars, dine mainly on plants or algae. **Carnivores**, such as sea otters, hawks, and spiders, mostly eat other animals. Rats and other **omnivores** (from the Latin *omnis*, all) don't in fact eat

everything, but they do regularly consume animals as well as plants or algae. We humans are typically omnivores, as are cockroaches and crows.

The terms *herbivore*, *carnivore*, and *omnivore* represent the kinds of food an animal usually eats. Keep in mind, however, that most animals are opportunistic feeders, eating foods outside their standard diet when their usual foods aren't available. For example, deer are herbivores, but in addition to feeding on grass and other plants, they occasionally eat insects, worms, or bird eggs. Note as well that microorganisms are an unavoidable "supplement" in every animal's diet.

In order to survive and reproduce, animals must balance their consumption, storage, and use of food. Sea otters, for example, support a high rate of metabolism by eating up to 25% of their body mass each day. Eating too little food, too much food, or the wrong mixture of foods can endanger an animal's health. In this chapter, we'll examine the nutritional requirements of animals, explore diverse evolutionary adaptations for obtaining and processing food, and investigate the regulation of energy intake and expenditure.

# An animal's diet must supply chemical energy, organic building blocks, and essential nutrients

Overall, an adequate diet must satisfy three nutritional needs: chemical energy for cellular processes, building blocks for large organic molecules, and essential nutrients.

The activities of cells, tissues, organs, and whole animals depend on sources of chemical energy in the diet. This energy is used to produce ATP, which powers processes ranging from DNA replication and cell division to vision and flight (see Concept 6.3). To meet the need for ATP, animals ingest and digest nutrients, including carbohydrates, proteins, and lipids, for use in cellular respiration and energy storage.

In addition to fuel for ATP production, an animal requires raw materials needed for biosynthesis (see Concepts 3.3 through 3.5). To build the complex molecules it needs to grow, maintain itself, and reproduce, an animal must obtain two types of organic molecules from its food: a source of organic carbon (such as sugar) and a source of organic nitrogen (such as protein). Starting with these materials, animal cells can construct a great variety of organic molecules.

An animal's diet must also provide **essential nutrients**, substances that an animal requires but cannot assemble from simple organic molecules.

## Essential Nutrients

Essential nutrients in the diet include essential amino acids, essential fatty acids, vitamins, and minerals. Essential nutrients have key functions in cells, including serving as substrates, coenzymes, and cofactors in biosynthetic reactions **(Figure 33.2)**.

In general, an animal can obtain all the essential amino acids and fatty acids, as well as vitamins and minerals, by feeding on plants or other animals. Needs for particular nutrients vary among species. For instance, some animals (including humans) must get ascorbic acid (vitamin C) from their diet, whereas many other animals can synthesize it from other nutrients.

### Essential Amino Acids and Fatty Acids

All organisms require a standard set of 20 amino acids to make a complete set of proteins (see Figure 3.18). Microorganisms and plants normally can produce all 20. Most animal species have the enzymes to synthesize about half of these amino acids, as long as their diet includes sulfur and organic nitrogen. The remaining amino acids must be obtained from food in prefabricated form and are therefore called **essential amino acids**. In many adult animals, including humans, 8 of the 20 amino acids are required in the diet.

Animal cells use fatty acids to produce cellular components that include membrane phospholipids, signaling molecules, and storage fats (see Concept 3.4). The **essential fatty acids**, which animals cannot synthesize (but plants can), contain one or more double bonds; an example is linoleic acid (see Figure 33.2). Animals typically obtain ample quantities of essential fatty acids from seeds, grains, and other plant matter in their diets.

### Vitamins

As Albert Szent-Györgyi, the discoverer of vitamin C, once quipped, "A vitamin is a substance that makes you ill if you *don't* eat it." **Vitamins** are organic molecules that are required in the diet in very small amounts (0.01–100 mg per day, depending on the vitamin). The 13 known human vitamins vary in both chemical properties and function. Some are water-soluble, including the eight B vitamins, which generally act as coenzymes (see Figure 33.2). Vitamin C, which is required

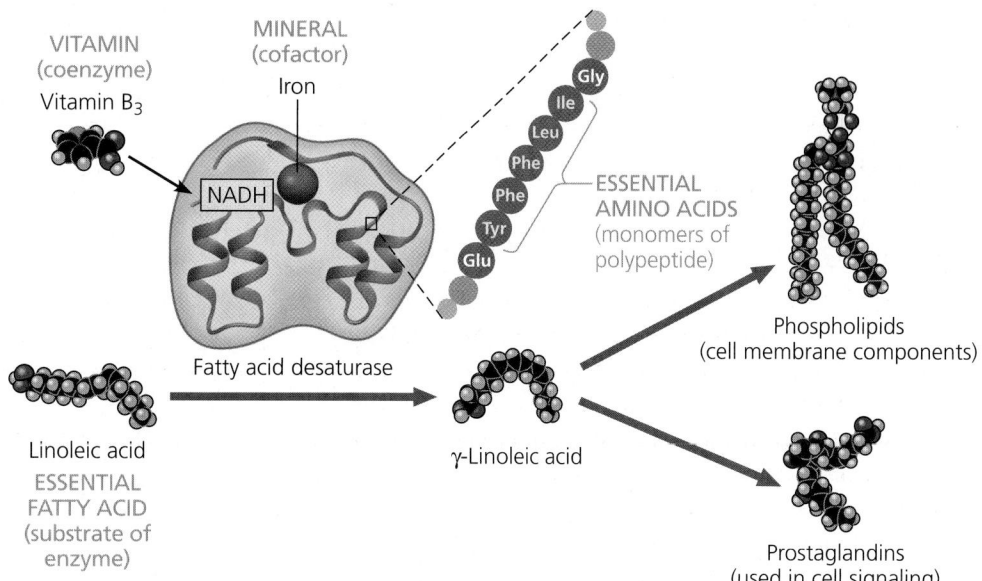

▶ **Figure 33.2 Roles of essential nutrients.** This example of a biosynthetic reaction illustrates some common functions for essential nutrients. When the enzyme fatty acid desaturase converts linoleic acid to γ-linoleic acid, all four classes of essential nutrients participate, as labeled in blue. Note that almost all enzymes and other proteins in animals contain some essential amino acids, as indicated in the partial sequence shown for fatty acid desaturase.

VITAMIN (coenzyme)
Vitamin B₃

MINERAL (cofactor)
Iron

NADH

Gly
Ile
Leu
Phe
Phe
Tyr
Glu

ESSENTIAL AMINO ACIDS (monomers of polypeptide)

Fatty acid desaturase

Linoleic acid
ESSENTIAL FATTY ACID (substrate of enzyme)

γ-Linoleic acid

Phospholipids (cell membrane components)

Prostaglandins (used in cell signaling)

for the production of connective tissue, is also water-soluble. Vitamins A, D, E, and K are fat-soluble. Vitamin K is important in blood clotting. Vitamin E is an antioxidant and helps prevent damage to cell membranes. Vitamin A is incorporated into visual pigments of the eye, and vitamin D aids in calcium absorption and bone formation. Our dietary requirement for vitamin D, unlike other vitamins, turns out to be variable. Why? During those times when our skin is exposed to sunlight, our bodies synthesize vitamin D, reducing our dietary need.

For people with imbalanced diets, taking supplements that provide vitamins at recommended daily levels is reasonable. It is far less clear whether massive doses of vitamins confer any health benefits or are even safe. In particular, excesses of fat-soluble vitamins are deposited in body fat, so overconsumption may cause them to accumulate to toxic levels.

## Minerals

Dietary **minerals** are inorganic nutrients, such as iron and sulfur, that are usually required in small amounts—from less than 1 mg to about 2,500 mg per day. Minerals have diverse functions in animal physiology. Some are assembled into the structure of proteins; iron, for example, is incorporated into the oxygen carrier protein hemoglobin as well as some enzymes (see Figure 33.2). In contrast, sodium, potassium, and chloride are important in the functioning of nerves and muscles and in maintaining osmotic balance between cells and the surrounding body fluid. In vertebrates, the mineral iodine is incorporated into thyroid hormone, which regulates metabolic rate. Vertebrates also require relatively large quantities of calcium and phosphorus for building and maintaining bone.

Ingesting large amounts of some minerals can impair health. For example, excess salt (sodium chloride) can contribute to high blood pressure. This is a particular problem in the United States, where the typical person consumes enough salt to provide about 20 times the required amount of sodium.

## Dietary Deficiencies

A diet that lacks one or more essential nutrients or consistently supplies less chemical energy than the body requires results in *malnutrition*, a failure to obtain adequate nutrition. Malnutrition affects one out of four children worldwide and has negative impacts on health and survival.

### Deficiencies in Essential Nutrients

Insufficient intake of essential nutrients can cause deformities, disease, and even death. For example, bone fragility can occur in deer or other herbivores that feed on plants where the soil lacks phosphorus. In such environments, some grazing animals obtain missing nutrients by consuming concentrated sources of salt or other minerals **(Figure 33.3)**. Similarly, some birds supplement their diet with snail shells.

Among humans, the most common type of malnutrition is protein deficiency, which results from insufficient intake of

▶ **Figure 33.3 Obtaining essential nutrients from an unusual source.** A juvenile chamois (*Rupicapra rupicapra*), an herbivore, licks salts from exposed rocks in its alpine habitat. This behavior is common among herbivores that live where soils and plants provide insufficient amounts of minerals, such as sodium, calcium, phosphorus, and iron.

essential amino acids. In children, protein deficiency may arise if their diet shifts from breast milk to foods that contain relatively little protein, such as rice. Such children, if they survive infancy, often have impaired development.

In populations subsisting on simple rice diets, individuals are often deficient in vitamin A, which can result in blindness or death. In response to this problem, scientists have engineered "Golden Rice," a strain of rice that synthesizes the orange-colored pigment beta-carotene, which the body converts to vitamin A. Golden Rice, which is undergoing field testing, is just one example of efforts to use plant biotechnology to address malnutrition (see Concept 30.3).

### Undernourishment

As mentioned earlier, malnutrition can also arise if a diet fails to provide adequate sources of chemical energy. When this occurs, the body first uses up stored carbohydrates and fat. It then begins breaking down its own proteins for fuel; muscles begin to decrease in size; and the brain may become protein-deficient. If energy intake remains less than energy expenditures, the animal will eventually die. Even if a seriously undernourished animal survives, some of the damage may be irreversible.

Inadequate nourishment among humans is most common when drought, war, or another crisis severely disrupts the food supply. In sub-Saharan Africa, where the AIDS epidemic has crippled both rural and urban communities, approximately 200 million children and adults cannot obtain enough food.

Sometimes undernourishment occurs within well-fed human populations as a result of eating disorders. For example, people with anorexia nervosa restrict food intake to the point where they lose more weight than is healthy for their age and height.

### CONCEPT CHECK 33.1

1. An animal requires 20 amino acids to make proteins. Why aren't all 20 essential to animal diets?
2. **MAKE CONNECTIONS** Considering how enzymes function (see Concept 6.4), explain why vitamins are required in very small amounts.
3. **WHAT IF?** If a zoo animal eating ample food shows signs of malnutrition, how might a researcher determine which nutrient is lacking in its diet?

For suggested answers, see Appendix A.

# Food processing involves ingestion, digestion, absorption, and elimination

In this section, we turn from nutritional requirements to the mechanisms by which animals process food. We will consider food processing in four stages: ingestion, digestion, absorption, and elimination.

Food processing begins with **ingestion**, the act of eating or feeding. We can divide the ways animals feed into four major groups. Blue whales, flamingos, and other *suspension feeders* filter, capture, or trap food particles from the surrounding medium. *Substrate feeders*, such as caterpillars, live in or on their food source. Mosquitoes, hummingbirds, and other *fluid feeders* suck nutrient-rich fluid from a living host. Most animals are *bulk feeders*, which eat relatively large pieces of food. We are bulk feeders, as are pythons **(Figure 33.4)**.

During **digestion**, the second stage of food processing, food is broken down into molecules small enough for the body to absorb. Typically both mechanical and chemical processes are required. The mechanical breakdown of food into smaller pieces by chewing or grinding increases surface area. The exposed food substances then undergo chemical digestion, which cleaves large molecules into component parts.

Chemical digestion is necessary because animals cannot directly use the proteins, carbohydrates, nucleic acids, fats, and phospholipids in food. These molecules are too large to pass through cell membranes and also are not all identical to those the animal needs for its particular tissues and functions. When large molecules in food are broken down, however, the animal can use the smaller products of digestion to assemble the large molecules it needs. For example, although an otter and a snake have very different diets, both convert proteins in their food to the same 20 amino acids from which they assemble all of the specific proteins in their bodies.

A cell makes a fat or macromolecule by linking together smaller components; it does so by removing a molecule of water for each new covalent bond formed. Chemical digestion by enzymes reverses this process by breaking bonds through the addition of water (see Figure 3.7). This splitting process is catalyzed by digestive enzymes and is called *enzymatic hydrolysis*. Polysaccharides and disaccharides are split into simple sugars; proteins are broken down into small peptides and amino acids; and nucleic acids are cleaved into nucleotides and their components. Enzymatic hydrolysis also releases fatty acids and other components from fats and phospholipids. In many animals, bacteria living in the digestive system carry out some chemical digestion.

The last two stages of food processing occur after the food is digested. In the third stage, **absorption**, the animal's cells take up (absorb) small molecules such as amino acids and simple sugars. **Elimination**, in which undigested material passes out of the digestive system, completes the process.

## Digestive Compartments

In our overview of food processing, we have seen that digestive enzymes hydrolyze the same biological materials (such as proteins, fats, and carbohydrates) that make up the bodies of the animals themselves. How, then, are animals able to digest food without digesting their own cells and tissues? The evolutionary adaptation found across a wide range of animal species is the processing of food within specialized intracellular or extracellular compartments.

▲ **Figure 33.4 Bulk feeding by a python.** In this amazing scene, a rock python is beginning to ingest a gazelle it has captured and killed. After swallowing its prey, the python will spend two weeks or longer digesting its meal.

## Intracellular Digestion

Food vacuoles—cellular organelles in which hydrolytic enzymes break down food—are the simplest digestive compartments. The hydrolysis of food inside vacuoles, called intracellular digestion, begins after a cell engulfs solid food by phagocytosis or liquid food by pinocytosis (see Figure 5.18). Newly formed food vacuoles fuse with lysosomes, organelles containing hydrolytic enzymes. This fusion of organelles brings food in contact with these enzymes, allowing digestion to occur safely within a compartment enclosed by a protective membrane. A few animals, such as sponges, digest their food entirely by this intracellular mechanism (see Figure 27.3).

## Extracellular Digestion

In most animal species, hydrolysis occurs largely by extracellular digestion, the breakdown of food in compartments that are continuous with the outside of the animal's body. Having one or more extracellular compartments for digestion enables an animal to devour much larger pieces of food than can be ingested by phagocytosis.

Animals with relatively simple body plans typically have a digestive compartment with a single opening **(Figure 33.5)**. This pouch, called a **gastrovascular cavity**, functions in digestion as well as in the distribution of nutrients throughout the body (hence the *vascular* part of the term). Small freshwater cnidarians called hydras provide a good example of how a gastrovascular cavity works. A carnivore, the hydra uses its tentacles to stuff captured prey through its mouth into its gastrovascular cavity. Specialized gland cells of the hydra's gastrodermis, the tissue layer that lines the cavity, then secrete digestive enzymes that break the soft tissues of the prey into tiny pieces. Other cells of the gastrodermis engulf these food particles, and most of the hydrolysis of macromolecules occurs intracellularly, as in

sponges. After the hydra has digested its meal, undigested materials that remain in its gastrovascular cavity, such as exoskeletons of small crustaceans, are eliminated through its mouth. Many flatworms also have a gastrovascular cavity.

Rather than a gastrovascular cavity, animals with complex body plans have a digestive tube with two openings: a mouth and an anus. Such a tube is called a *complete digestive tract* or, more commonly, an **alimentary canal (Figure 33.6)**. Because food moves along the alimentary canal in a single direction, the tube can be organized into specialized compartments that carry out digestion and nutrient absorption in a stepwise fashion.

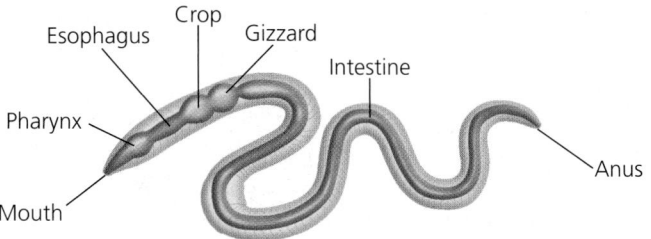

**(a) Earthworm.** The alimentary canal of an earthworm includes a muscular pharynx that sucks food in through the mouth. Food passes through the esophagus and is stored and moistened in the crop. Mechanical digestion occurs in the muscular gizzard, which pulverizes food with the aid of small bits of sand and gravel. Further digestion and absorption occur in the intestine.

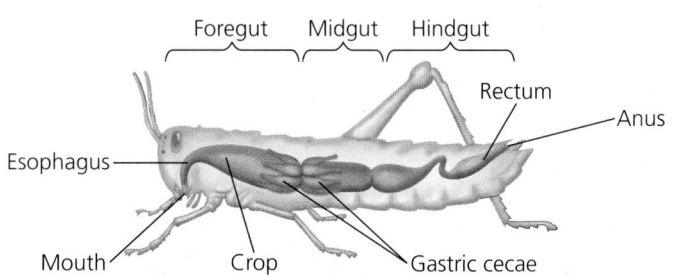

**(b) Grasshopper.** A grasshopper has several digestive chambers grouped into three main regions: a foregut, with an esophagus and crop; a midgut; and a hindgut. Food is moistened and stored in the crop, but most digestion occurs in the midgut. Pouches called gastric cecae (singular, *ceca*) extend from the beginning of the midgut and function in digestion and absorption.

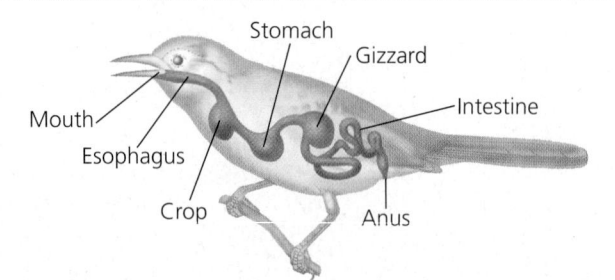

**(c) Bird.** Many birds have a crop for storing food and a stomach and gizzard for mechanically digesting it. Chemical digestion and absorption of nutrients occur in the intestine.

▲ **Figure 33.6 Alimentary canals.** These examples illustrate variation in the organization and structure of compartments that carry out nutrient digestion, storage, and absorption in different animals.

**①** Digestive enzymes are released from a gland cell.

**②** Enzymes break food down into small particles.

**③** Food particles are engulfed and digested in food vacuoles.

▲ **Figure 33.5 Digestion in a hydra.** Digestion begins in the gastrovascular cavity and is completed intracellularly after small food particles are engulfed by specialized cells of the gastrodermis.

An animal with an alimentary canal can ingest food while earlier meals are still being digested, a feat that is likely to be difficult or inefficient for an animal with a gastrovascular cavity.

CONCEPT CHECK 33.2

1. Distinguish the overall structure of a gastrovascular cavity from that of an alimentary canal.
2. In what sense are nutrients from a recently ingested meal not really "inside" your body prior to the absorption stage of food processing?
3. **WHAT IF?** Thinking in broad terms, what similarities can you identify between digestion in an animal body and the breakdown of gasoline in an automobile? (You don't have to know about auto mechanics.)

For suggested answers, see Appendix A.

## CONCEPT 33.3

# Organs specialized for sequential stages of food processing form the mammalian digestive system

In mammals, a number of accessory glands support food processing by secreting digestive juices through ducts into the alimentary canal. There are three pairs of salivary glands, as well as three individual glands: the pancreas, liver, and gallbladder **(Figure 33.7)**. To explore the coordinated function of the accessory glands and alimentary canal, we'll consider the steps in food processing as a meal travels along the canal in a human.

## The Oral Cavity, Pharynx, and Esophagus

As soon as you place a bite of food into your mouth, or **oral cavity**, food processing begins. Teeth of various shapes cut, mash, and grind, breaking the food into smaller pieces. This mechanical breakdown not only increases the surface area available for chemical breakdown but also facilitates swallowing. Meanwhile, the anticipation or arrival of food in the oral cavity triggers the release of saliva by the **salivary glands**.

Saliva is a complex mixture of materials with a number of vital functions. One major component is **mucus**, a viscous mixture of water, salts, cells, and slippery glycoproteins (carbohydrate-protein complexes). Mucus lubricates food for easier swallowing, protects the gums against abrasion, and facilitates taste and smell. Saliva also contains buffers, which help prevent tooth decay by neutralizing acid, and antimicrobial agents (such as lysozyme; see Figure 3.20), which protect against bacteria that enter the mouth with food.

Scientists have long been puzzled by the fact that saliva contains a large amount of the enzyme **amylase**, which breaks down starch (a glucose polymer from plants) and glycogen (a glucose polymer from animals). Most chemical digestion occurs not in the mouth but in the small intestine, where amylase is also present. Why, then, is there salivary amylase? Current thinking is that amylase in saliva releases food particles stuck

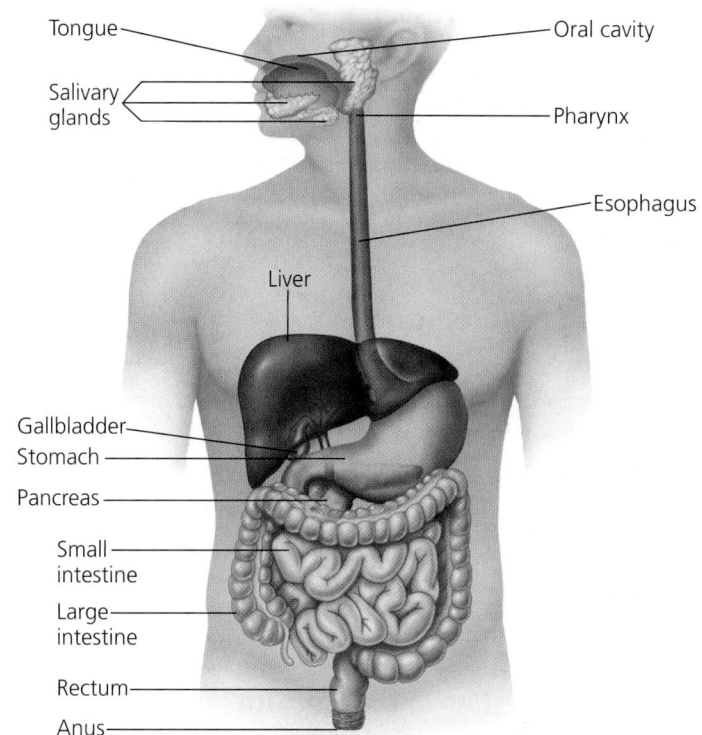

▲ **Figure 33.7 The human digestive system.** After food is chewed and swallowed, it takes 5–10 seconds for it to pass down the esophagus and into the stomach, where it spends 2–6 hours being partially digested. Final digestion and nutrient absorption occur in the small intestine over a period of 5–6 hours. Within 12–24 hours, any undigested material passes through the large intestine, and feces are expelled through the anus.

to teeth, thereby reducing the nutrients available to microorganisms living in the mouth.

The tongue also has important roles in food processing. Much as a doorman screens and assists people entering a fancy hotel, the tongue aids digestive processes by evaluating ingested material, distinguishing which foods should be processed, and then enabling their further passage. (See Concept 38.4 for a discussion of the sense of taste.) After food is deemed acceptable and chewing commences, tongue movements manipulate the mixture of saliva and food, helping shape it into a ball called a **bolus**. During swallowing, the tongue provides further assistance, pushing the bolus to the back of the oral cavity and into the throat.

Each bolus of food is received by the **pharynx**, or throat region, which leads to two passageways: the esophagus and the trachea. The **esophagus** is a muscular tube that connects to the stomach, whereas the trachea (windpipe) leads to the lungs. Swallowing must therefore be carefully choreographed to keep food and liquids from entering the airway. Each time you swallow, a flap of cartilage covers your vocal cords and the opening between them. Guided by the *larynx*, or voice box, these movements direct each bolus into the esophagus. Failure of this swallowing reflex can cause choking, a blockage of the trachea. The resulting lack of airflow into the lungs can be fatal if the material is not dislodged by vigorous coughing, a series of back slaps, or a forced upward thrust of the diaphragm (the Heimlich maneuver).

Within the esophagus, food is pushed along by **peristalsis**, alternating waves of smooth muscle contraction and relaxation. Upon reaching the end of the esophagus, the bolus encounters a **sphincter**, a ringlike valve of muscle. Acting like a drawstring, the sphincter regulates passage of the ingested food into the next compartment, the stomach.

## Digestion in the Stomach

The **stomach**, which is located just below the diaphragm, plays two major roles in digestion. The first is storage. With accordion-like folds and a very elastic wall, the stomach can stretch to accommodate about 2 L of food and fluid. The second major function is to process food into a liquid suspension. As shown in **Figure 33.8**, the stomach secretes the components of a digestive fluid called **gastric juice**. It then mixes these secretions with the food through a churning action, forming a mixture of ingested food and digestive juice called **chyme**.

### Chemical Digestion in the Stomach

Two components of gastric juice help liquefy food in the stomach. First, hydrochloric acid (HCl) disrupts the extracellular matrix that binds cells together in meat and plant material. The concentration of HCl is so high that the pH of gastric juice is about 2, acidic enough to dissolve iron nails (and to kill most bacteria). This low pH denatures (unfolds) proteins in food, increasing exposure of their peptide bonds. The exposed bonds are then attacked by the second component of gastric juice—a **protease**, or protein-digesting enzyme, called **pepsin**. Unlike most enzymes, pepsin works best in a very acidic environment. By breaking peptide bonds, it cleaves proteins into smaller polypeptides and further exposes the contents of ingested tissues.

What prevents gastric juice from destroying the stomach cells that make it? The answer is that the ingredients of gastric juice are kept inactive until they are released into the lumen (cavity) of the stomach. Cells in the gastric glands of the stomach produce the components of gastric juice. As detailed in Figure 33.8, *parietal cells* and *chief cells* function together to produce HCl and pepsin in the lumen of the stomach, not within the cells themselves.

Why don't HCl and pepsin eat through the lining of the stomach? For one thing, mucus secreted by cells in gastric glands protects against self-digestion. In addition, cell division adds a new epithelial layer every three days, replacing cells before the lining is fully eroded by digestive juices. Under certain circumstances, however, damaged areas of the stomach lining called gastric ulcers can appear. It had been thought that they were caused by psychological stress and resulting excess acid secretion. However, Australian researchers Barry Marshall and Robin Warren discovered that infection by the acid-tolerant bacterium *Helicobacter pylori* causes ulcers. They also demonstrated that an antibiotic treatment could cure most gastric ulcers. For these findings, they were awarded the Nobel Prize in 2005.

### Stomach Dynamics

Breakdown of food by gastric juices is enhanced by muscular activity of the stomach. The coordinated series of muscle contractions and relaxations that we call churning mixes the stomach contents about every 20 seconds. This churning facilitates chemical digestion by bringing all of the food into contact with the gastric juices secreted by the lining of the stomach. As a result, what began as a recently swallowed meal becomes the acidic, nutrient-rich broth known as chyme.

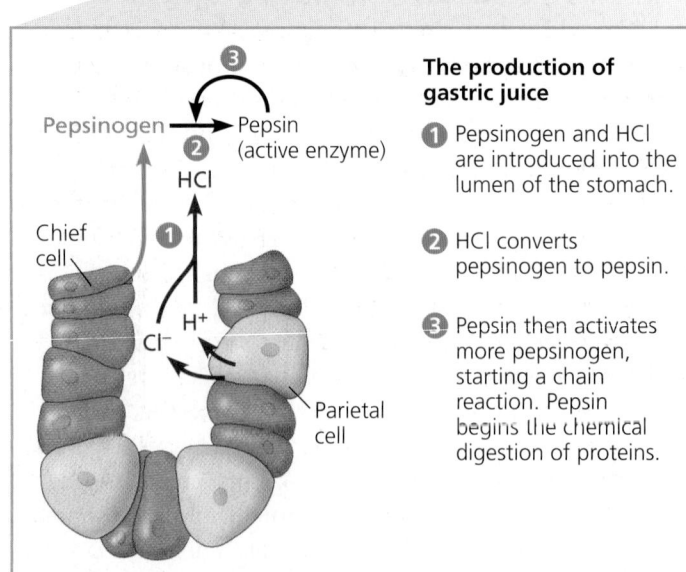

Stomach

**Interior surface of stomach.** The interior surface of the stomach wall is highly folded and dotted with pits leading into tubular gastric glands.

Epithelium

**Gastric gland.** The gastric glands have three types of cells that secrete different components of the gastric juice: mucous cells, chief cells, and parietal cells.

**Mucous cells** secrete mucus, which lubricates and protects the cells lining the stomach.

**Chief cells** secrete pepsinogen, an inactive form of the digestive enzyme pepsin.

**Parietal cells** produce the components of hydrochloric acid (HCl).

**The production of gastric juice**

1. Pepsinogen and HCl are introduced into the lumen of the stomach.

2. HCl converts pepsinogen to pepsin.

3. Pepsin then activates more pepsinogen, starting a chain reaction. Pepsin begins the chemical digestion of proteins.

Pepsinogen → Pepsin (active enzyme)

HCl

Chief cell

$H^+$

$Cl^-$

Parietal cell

▲ **Figure 33.8 The stomach and its secretions.**

Muscular activity of the stomach also promotes passage through the alimentary canal. In particular, peristaltic contractions typically move the contents of the stomach into the small intestine within 2–6 hours after a meal. The sphincter located where the stomach opens to the small intestine helps regulate passage into the small intestine, allowing only one squirt of chyme at a time.

Occasionally, the sphincter at the top of the stomach allows a backflow of chyme, that is, a movement or flux of chyme from the stomach into the lower end of the esophagus. The painful irritation of the esophagus that results from this process of acid reflux is what we commonly call "heartburn."

## Digestion in the Small Intestine

Although some chemical digestion occurs in the oral cavity and stomach, most enzymatic hydrolysis of macromolecules from food occurs in the **small intestine (Figure 33.9)**. The small intestine is the alimentary canal's longest compartment—over 6 m (20 feet) long in humans! Its name refers to its small diameter, compared with that of the large intestine. The

first 25 cm (10 inches) or so of the small intestine forms the **duodenum**. It is here that chyme from the stomach mixes with digestive juices from the pancreas, liver, and gallbladder, as well as from gland cells of the intestinal wall itself.

The arrival of chyme in the duodenum triggers release of the hormone secretin, which stimulates the **pancreas** to secrete bicarbonate (see Figure 32.6). Bicarbonate neutralizes the acidity of chyme and acts as a buffer for chemical digestion in the small intestine. The pancreas also secretes numerous digestive enzymes into the small intestine (see Figure 33.9). These include the proteases trypsin and chymotrypsin, which are produced in inactive forms. In a chain reaction similar to that for pepsinogen (see Figure 33.8), they are activated when safely located in the lumen of the duodenum.

The epithelial lining of the duodenum is the source of additional digestive enzymes (see Figure 33.9). Some are secreted into the lumen of the duodenum, whereas others are bound to the surface of epithelial cells. Together with the enzymes from the pancreas, they complete most digestion in the duodenum.

▼ **Figure 33.9 Chemical digestion in the human digestive system.** Most chemical digestion takes place in the duodenum, the first 25 cm of the small intestine. The timing and location of chemical breakdown are specific to each class of nutrients.

**?** *Pepsin is resistant to the denaturing effect of the low pH environment of the stomach. Thinking about the different digestive processes that occur in the small intestine, describe an adaptation shared by the digestive enzymes in that compartment.*

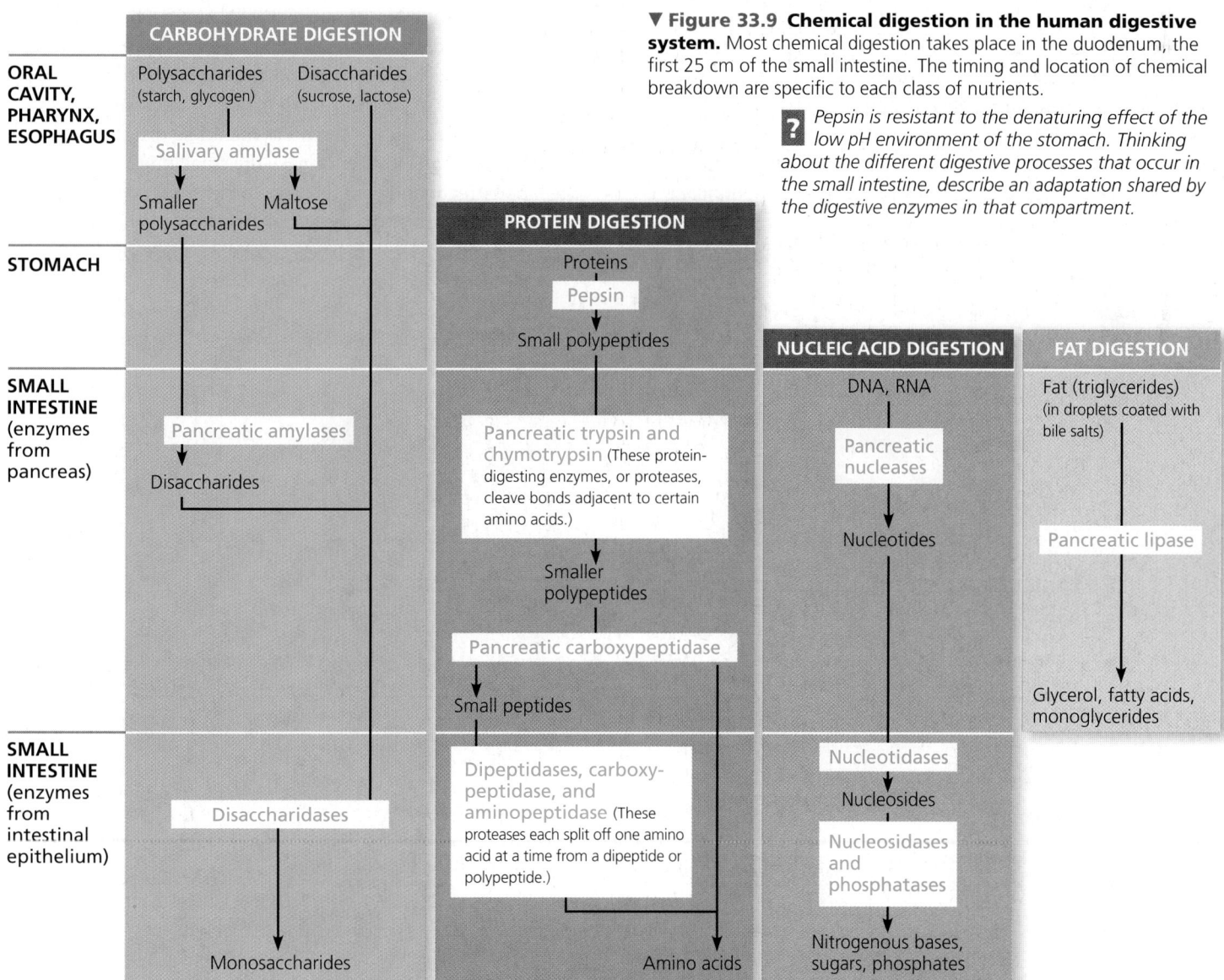

Fats present a special challenge for digestion. Insoluble in water, they form large globules that cannot be attacked efficiently by digestive enzymes. The evolutionary adaptation that enables fat digestion is the production of bile salts, which act as emulsifiers (detergents) that break apart fat and other lipid globules. Bile salts are a major component of **bile**, a secretion of the **liver** that is stored and concentrated in the **gallbladder**.

Bile production is integral to one of the other vital functions of the liver: the destruction of red blood cells that are no longer fully functional. In producing bile, the liver incorporates some pigments that are by-products of red blood cell disassembly. These bile pigments are then eliminated from the body with the feces. In some liver or blood disorders, bile pigments accumulate in the skin, resulting in a characteristic yellowing condition called jaundice.

## Absorption in the Small Intestine

With digestion largely complete, the contents of the duodenum move by peristalsis to the *jejunum* and *ileum*, the remaining regions of the small intestine. There, nutrient absorption occurs across the highly folded surface **(Figure 33.10)**. Large folds in the lining are studded with finger-like projections called **villi**. Within the villi, each epithelial cell has many microscopic projections, or **microvilli**, that are exposed to the intestinal lumen. The microvilli give cells of the intestinal epithelium a brush-like appearance that is reflected in the name *brush border*. Together, the folds, villi, and microvilli of the small intestine have a surface area of 200–300 m$^2$, roughly the size of a tennis court. This enormous surface area is an evolutionary adaptation that greatly increases the rate of nutrient absorption.

Depending on the nutrient, transport across the epithelial cells can be passive or active (see Concepts 5.3 and 5.4). The sugar fructose, for example, moves by facilitated diffusion down its concentration gradient from the lumen of the small intestine into the epithelial cells. From there, fructose exits the basal surface and is absorbed into microscopic blood vessels, or capillaries, at the core of each villus. Other nutrients, including amino acids, small peptides, vitamins, and most glucose molecules, are pumped against concentration gradients by the epithelial cells of the villus.

The capillaries and veins that carry nutrient-rich blood away from the villi converge into the **hepatic portal vein**, a blood vessel that leads directly to the liver. From the liver, blood travels to the heart and then to other tissues and organs. This arrangement serves two major functions. First, it allows the liver to regulate the distribution of nutrients to the rest of the body. Because the liver converts many nutrients to different forms for their use elsewhere, blood that leaves the liver may have a very different nutrient balance than the blood that entered. Second, the arrangement allows the liver to remove toxic substances before the blood circulates broadly. The liver is the primary site for the detoxification of many organic molecules that are foreign to the body, including drugs.

Although many nutrients leave the small intestine through the bloodstream and pass through the liver for processing, some products of fat (triglyceride) digestion take a different path **(Figure 33.11)**. Hydrolysis of a fat by lipase generates

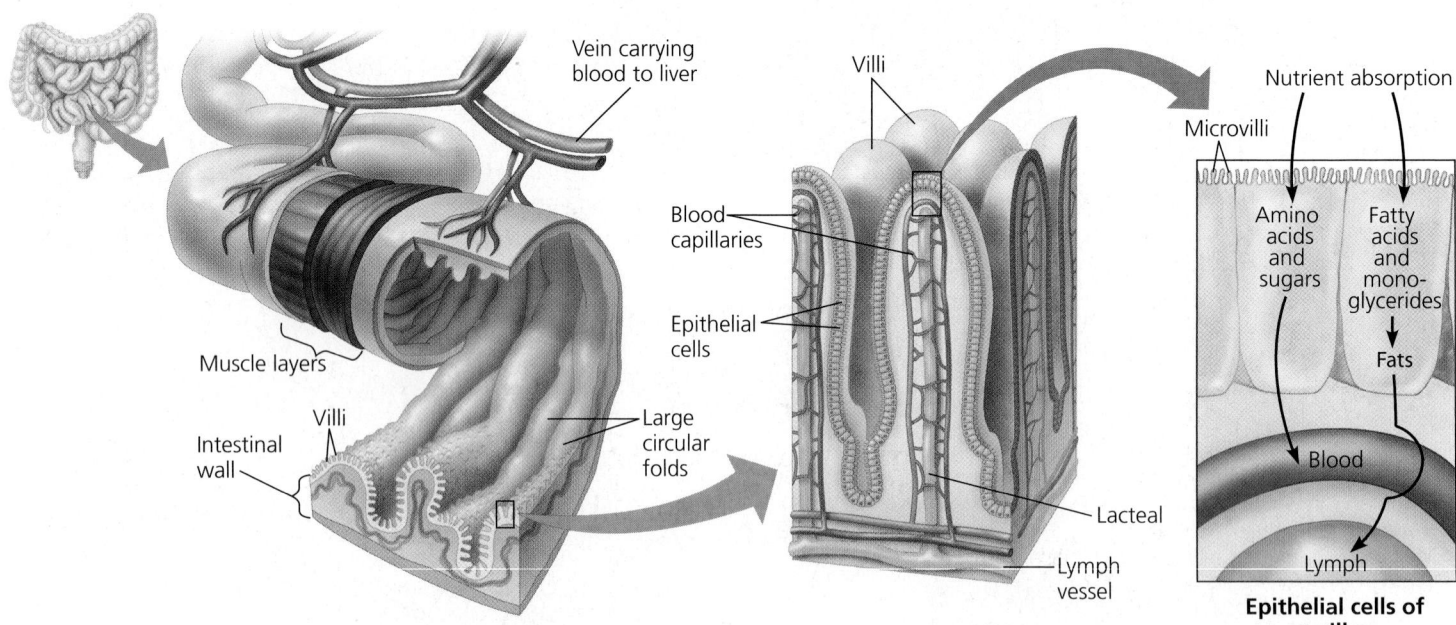

▲ **Figure 33.10 Nutrient absorption in the small intestine.** Water-soluble nutrients, such as amino acids and sugars, enter the bloodstream, whereas fats enter the lymphatic system.

**?** *Tapeworms sometimes infect the human alimentary canal, anchoring themselves to the wall of the small intestine. Based on how digestion is compartmentalized along the mammalian alimentary canal, what digestive functions would you expect these parasites to have?*

**ANIMATION** Visit the Study Area in **MasteringBiology** for the BioFlix® 3-D Animation on Membrane Transport.

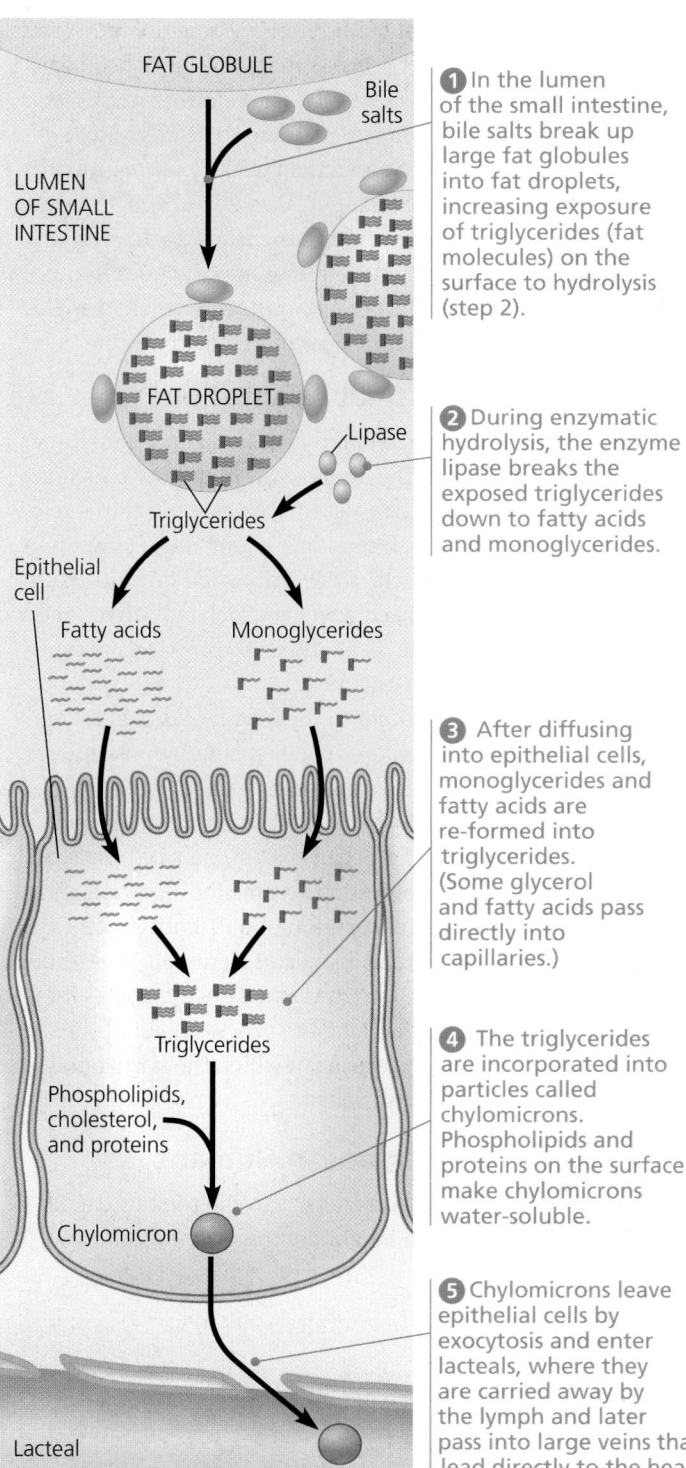

**FAT GLOBULE**

Bile salts

LUMEN OF SMALL INTESTINE

**1** In the lumen of the small intestine, bile salts break up large fat globules into fat droplets, increasing exposure of triglycerides (fat molecules) on the surface to hydrolysis (step 2).

**FAT DROPLET**

Lipase

Triglycerides

**2** During enzymatic hydrolysis, the enzyme lipase breaks the exposed triglycerides down to fatty acids and monoglycerides.

Epithelial cell

Fatty acids          Monoglycerides

**3** After diffusing into epithelial cells, monoglycerides and fatty acids are re-formed into triglycerides. (Some glycerol and fatty acids pass directly into capillaries.)

Triglycerides

Phospholipids, cholesterol, and proteins

**4** The triglycerides are incorporated into particles called chylomicrons. Phospholipids and proteins on the surface make chylomicrons water-soluble.

Chylomicron

**5** Chylomicrons leave epithelial cells by exocytosis and enter lacteals, where they are carried away by the lymph and later pass into large veins that lead directly to the heart.

Lacteal

▲ **Figure 33.11 Digestion and absorption of fats.** Fats, which are insoluble in water, are broken down in the lumen of the small intestine, reassembled in epithelial cells, and then transported in water-soluble globules called chylomicrons. The chylomicrons are carried by the lymph to large veins leading to the heart.

fatty acids and a monoglyceride (glycerol joined to a fatty acid). These products are absorbed by epithelial cells and re-combined into triglycerides. They are then coated with phospholipids, cholesterol, and proteins, forming water-soluble globules called **chylomicrons**.

In exiting the small intestine, chylomicrons first enter a **lacteal**, a vessel at the core of each villus (see Figures 33.10 and 33.11). Lacteals are part of the lymphatic system, which is a network of vessels filled with a clear fluid called lymph. Starting at the lacteals, lymph containing the chylomicrons passes into the larger vessels of the lymphatic system and eventually into large veins that return the blood directly to the heart.

In addition to absorbing nutrients, the small intestine recovers water and ions. Each day we consume about 2 L of water and secrete another 7 L in digestive juices. Typically all but 0.1 L of the water is reabsorbed in the intestines, with most of the recovery occurring in the small intestine. There is no mechanism for active transport of water. Instead, water is reabsorbed by osmosis when sodium and other ions are pumped out of the lumen of the small intestine.

## Processing in the Large Intestine

The alimentary canal ends with the **large intestine**, which includes the colon, cecum, and rectum. The small intestine connects to the large intestine at a T-shaped junction **(Figure 33.12)**. One arm of the T is the 1.5-m-long **colon**, which leads to the rectum and anus. The other arm is a pouch called the **cecum**. The cecum is important for fermenting ingested material, especially in animals that eat large amounts of plant material. Compared with many other mammals, humans have a small cecum. The **appendix**, a finger-like extension of the human cecum, has a minor and dispensable role in immunity.

The colon completes the recovery of water that began in the small intestine. What remains are the **feces**, the wastes of the digestive system, which become increasingly solid as they are moved along the colon by peristalsis. It takes approximately 12–24 hours for material to travel the length of the colon. If the lining of the colon is irritated—by a viral or bacterial infection, for instance—less water than normal may be reabsorbed, resulting in diarrhea. The opposite problem, constipation, occurs when the feces move along the colon too slowly. Too much water is reabsorbed and the feces become compacted.

The undigested material in feces includes cellulose fiber. Although fiber provides no caloric value (energy) to humans, it helps move food along the alimentary canal.

A rich community of mostly harmless bacteria lives on unabsorbed organic material in the human colon, contributing approximately one-third of the dry weight of feces. As by-products of their metabolism, many colon bacteria generate gases, including methane

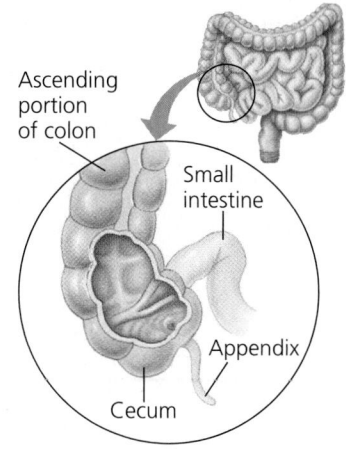

Ascending portion of colon

Small intestine

Appendix

Cecum

▲ **Figure 33.12 Junction of the small and large intestines.**

and hydrogen sulfide, the latter of which has an offensive odor. These gases and ingested air are expelled through the anus.

The terminal portion of the large intestine is the **rectum**, where the feces are stored until they can be eliminated. Between the rectum and the anus are two sphincters; the inner one is involuntary and the outer one is voluntary. Periodically, strong contractions of the colon create an urge to defecate. Because filling of the stomach triggers a reflex that increases the rate of contractions in the colon, the urge to defecate often follows a meal.

We have followed a meal from one opening (the mouth) of the alimentary canal to the other (the anus). Next we'll look at some adaptations of this general digestive plan in different animals.

**CONCEPT CHECK 33.3**

1. How does swallowed food reach the stomach of a weightless astronaut in orbit?
2. Explain why a proton pump inhibitor, such as the drug Prilosec, relieves the symptoms of acid reflux.
3. **WHAT IF?** If you mixed gastric juice with crushed food in a test tube, what would happen?

For suggested answers, see Appendix A.

## CONCEPT 33.4

# Evolutionary adaptations of vertebrate digestive systems correlate with diet

**EVOLUTION** The digestive systems of mammals and other vertebrates are variations on a common plan, but there are many intriguing adaptations, often associated with the animal's diet. To highlight how form fits function, we'll examine a few of them.

## Dental Adaptations

Dentition, an animal's assortment of teeth, is one example of structural variation reflecting diet **(Figure 33.13)**. The evolutionary adaptation of teeth for processing different kinds of food is one of the major reasons mammals have been so successful. For example, the sea otter in Figure 33.1 uses its sharp canine teeth to tear apart prey such as crabs and its slightly rounded molars to crush their shells. Nonmammalian vertebrates generally have less specialized dentition, but there are interesting exceptions. Venomous snakes, such as rattlesnakes, have fangs, modified teeth that inject venom into prey. Some fangs are hollow, like syringes, whereas others drip the toxin along grooves on the surfaces of the teeth.

## Stomach and Intestinal Adaptations

Evolutionary adaptations to differences in diet are sometimes apparent as variations in the dimensions of digestive organs. For example, large, expandable stomachs are common in carnivorous vertebrates, which may wait a long time between meals and must eat as much as they can when they do catch prey. An expandable stomach enables a rock python to ingest a whole gazelle (see Figure 33.4) and a 200-kg African lion to consume 40 kg of meat in one meal!

Adaptation is also apparent in the length of the digestive system in different vertebrates. In general, herbivores and omnivores have longer alimentary canals relative to their body size than do carnivores **(Figure 33.14)**. Plant matter is more difficult to digest than meat because it contains cell walls. A longer digestive tract furnishes more time for digestion and more surface area for the absorption of nutrients. As an example, consider the coyote and koala in Figure 33.14. Although these two mammals are about the same size, the koala's intestines are much longer, enhancing the processing of fibrous, protein-poor eucalyptus leaves from which the koala obtains nearly all of its nutrients and water.

## Mutualistic Adaptations in Humans

An estimated 10–100 trillion bacteria live in the human digestive system. One bacterial inhabitant, *Escherichia coli*, is so

---

▼ **Figure 33.13 Dentition and diet.**

| Carnivore | Herbivore | Omnivore |
|---|---|---|
| 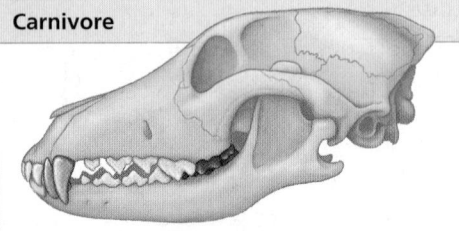 | 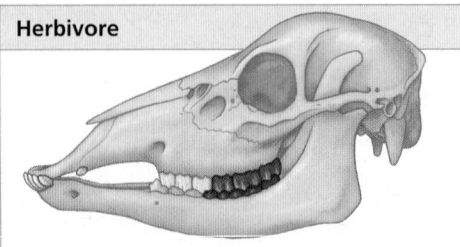 | 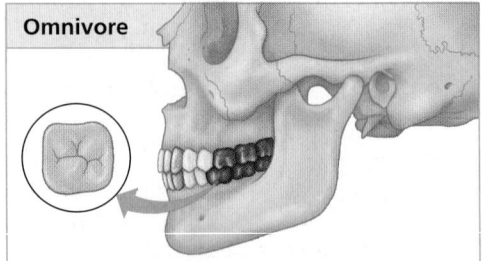 |
| Carnivores, such as members of the dog and cat families, generally have large, pointed incisors and canines that can be used to kill prey and rip or cut away pieces of flesh. The jagged premolars and molars crush and shred food. | Herbivores, such as horses and deer, usually have premolars and molars with broad, ridged surfaces that grind tough plant material. The incisors and canines are generally modified for biting off pieces of vegetation. In some herbivores, canines are absent. | As omnivores, humans are adapted to eating both plants and meat. Adults have 32 teeth. From front to back along either side of the mouth are four bladelike incisors for biting, a pair of pointed canines for tearing, four premolars for grinding, and six molars for crushing (see inset, top view). |

**Key**  Incisors  Canines ▢ Premolars ▪ Molars

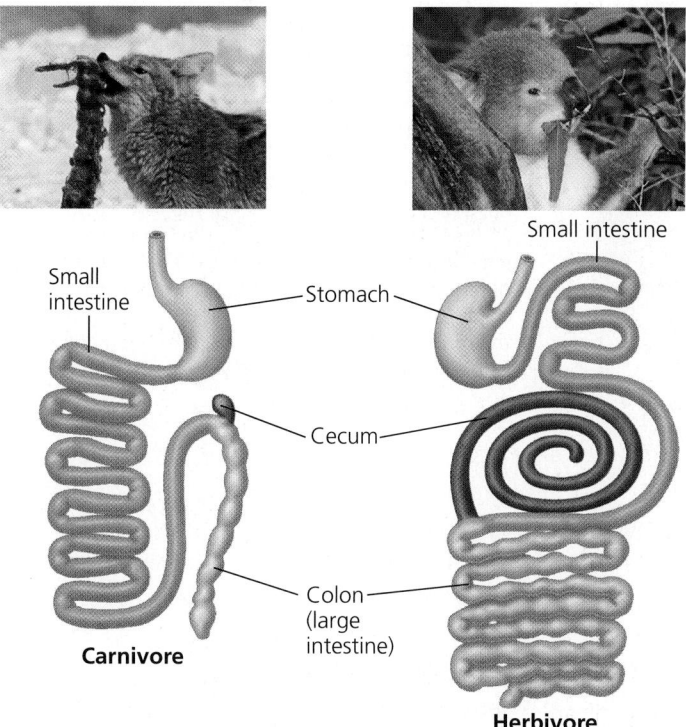

**▲ Figure 33.14 The alimentary canals of a carnivore (coyote) and herbivore (koala).** The relatively short digestive tract of the coyote is sufficient for digesting meat and absorbing its nutrients. In contrast, the koala's long alimentary canal is specialized for digesting eucalyptus leaves. Extensive chewing chops the leaves into tiny pieces, increasing exposure to digestive juices. In the long cecum and the upper portion of the colon, symbiotic bacteria further digest the shredded leaves, releasing nutrients that the koala can absorb.

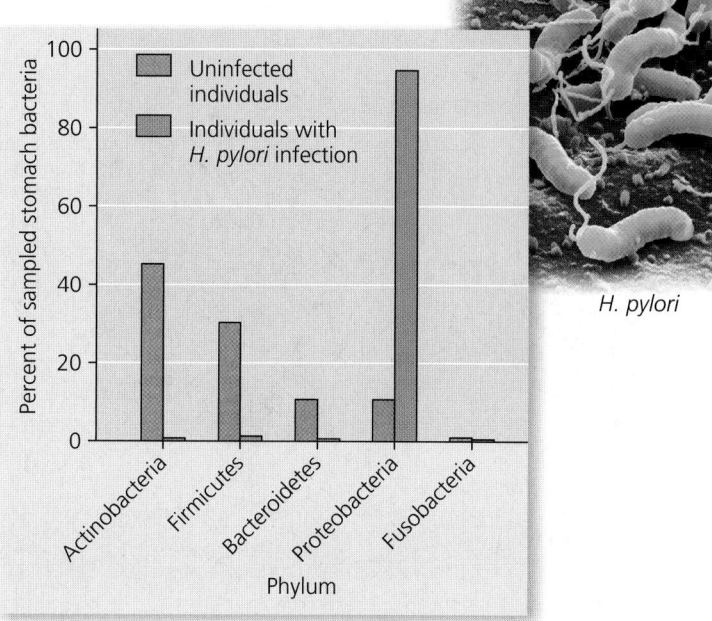

*H. pylori*

**▲ Figure 33.15 The stomach microbiome.** By copying and sequencing bacterial DNA in samples obtained from human stomachs, researchers characterized the bacterial community that makes up the stomach microbiome. In samples from individuals infected with *Helicobacter pylori*, more than 95% of the sequences were from that species, which belongs to the phylum Proteobacteria. The stomach microbiome in uninfected individuals was much more diverse.

common in the human digestive system that its presence in lakes and streams is a useful indicator of contamination by untreated sewage.

The coexistence of humans and many of these bacteria involves mutualistic symbiosis, an interaction between two species that benefits both (see Concept 41.1). For example, some intestinal bacteria produce vitamins, such as vitamin K, biotin, and folic acid, which supplement our dietary intake when absorbed into the blood. Intestinal bacteria also regulate the development of the intestinal epithelium and the function of the innate immune system. The bacteria in turn obtain a steady supply of nutrients and a stable host environment.

Recently, we have greatly expanded our knowledge of the collection of bacteria, called the *microbiome,* in the human digestive system. To identify these bacteria, both beneficial and harmful, scientists are using a DNA-sequencing approach based on the polymerase chain reaction (see Concept 13.4). They have found more than 400 bacterial species in the human digestive tract, a far greater number than had been identified through approaches relying on laboratory culture and characterization.

One recent microbiome study provided an important clue as to why the bacterium *H. pylori* can disrupt stomach health, leading to ulcers. After collecting stomach tissue from uninfected and *H. pylori*–infected adults, researchers identified

all the bacterial species in each sample. What they found was remarkable: *H. pylori* infection led to a near complete elimination from the stomach of all other bacterial species **(Figure 33.15)**. Such studies on differences in the microbiome associated with particular diseases hold promise for the development of new and more effective therapies.

## Mutualistic Adaptations in Herbivores

Mutualistic symbiosis is particularly important in herbivores. Much of the chemical energy in herbivore diets comes from the cellulose of plant cell walls, but animals do not produce enzymes that hydrolyze cellulose. Instead, many vertebrates (as well as termites, whose wooden diets consist largely of cellulose) host mutualistic bacteria and protists in fermentation chambers in their alimentary canals. These microorganisms have enzymes that can digest cellulose to simple sugars and other compounds that the animal can absorb. In many cases, the microorganisms also use the sugars from digested cellulose in the production of a variety of nutrients essential to the animal, such as vitamins and amino acids.

In horses, koalas, and elephants, symbiotic microorganisms are housed in a large cecum. In rabbits and some rodents, mutualistic bacteria live in the large intestine as well as the cecum. Since most nutrients are absorbed in the small intestine, nourishing by-products of fermentation by bacteria in the large intestine are initially lost with the feces. Rabbits and rodents recover these nutrients by *coprophagy* (from the Greek, meaning "dung eating"), feeding on some of their feces and then passing the food through the alimentary canal a second time.

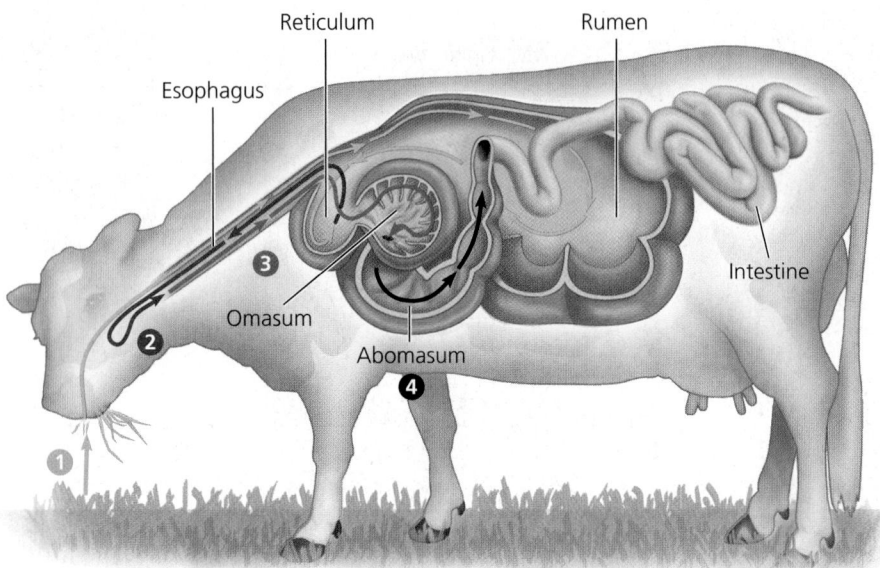

Reticulum          Rumen

Esophagus

③

Omasum

Abomasum

②

①

④

Intestine

◀ **Figure 33.16 Ruminant digestion.** The stomach of a cow, a ruminant, has four chambers. ① Chewed food first enters the rumen and reticulum, where mutualistic microorganisms digest cellulose in the plant material. ② Periodically, the cow regurgitates and rechews "cud" from the reticulum, further breaking down fibers and thereby enhancing microbial action. ③ The reswallowed cud passes to the omasum, where some water is removed. ④ It then passes to the abomasum for digestion by the cow's enzymes. In this way, the cow obtains significant nutrients from both the grass and the mutualistic microorganisms, which maintain a stable population in the rumen.

The familiar rabbit "pellets," which are not reingested, are the feces eliminated after food has passed through the digestive tract twice.

The most elaborate adaptations for a herbivorous diet have evolved in the animals called **ruminants**, the cud-chewing animals that include deer, sheep, and cattle **(Figure 33.16)**.

Having examined how animals optimize their extraction of nutrients from food, we will next turn to the challenge of balancing the use of these nutrients.

**CONCEPT CHECK 33.4**

1. What are two advantages of a longer alimentary canal for processing plant material that is difficult to digest?

2. What features of a mammal's digestive system make it an attractive habitat for mutualistic microorganisms?

3. **WHAT IF?** "Lactose-intolerant" people have a shortage of lactase, the enzyme that breaks down lactose in milk. As a result, they sometimes develop cramps, bloating, or diarrhea after consuming dairy products. Suppose such a person ate yogurt that contains bacteria that produce lactase. Why would eating yogurt likely provide at best only temporary relief of the symptoms?

For suggested answers, see Appendix A.

---

**CONCEPT 33.5**

# Feedback circuits regulate digestion, energy allocation, and appetite

In completing our consideration of animal nutrition, we'll explore the ways that obtaining and using nutrients are matched to an animal's circumstances and need for energy.

## Regulation of Digestion

For many animals, there are long gaps between meals. Under such circumstances, there is no need for their digestive systems to be active continuously. Instead, processing is activated stepwise. As food reaches each new compartment, it triggers the secretion of digestive juices for the next stage of processing. In addition, muscular contractions begin that move the contents farther along the canal. For example, you learned earlier that nervous reflexes stimulate the release of saliva when food enters the oral cavity and orchestrate swallowing when a bolus of food reaches the pharynx. Similarly, the arrival of food in the stomach triggers churning and the release of gastric juices. A branch of the nervous system called the *enteric division*, which is dedicated to the digestive organs, regulates these events as well as peristalsis in the small and large intestines.

The endocrine system also plays a critical role in controlling digestion. As described in **Figure 33.17**, a series of hormones released by the stomach and duodenum help ensure that digestive secretions are present only when needed. Like all hormones, they are transported through the bloodstream. This is true even for the hormone gastrin, which is secreted by the stomach and targets that same organ.

## Energy Allocation

Digested food provides animals with chemical energy to fuel metabolism and activity. In turn, the flow and transformation of energy in an animal—its **bioenergetics**—determine nutritional needs **(Figure 33.18)**. Energy extracted from nutrients is converted to ATP by cellular respiration and fermentation. Stores of ATP enable cells, organs, and organ systems to perform the functions that keep an animal alive. ATP is also used in biosynthesis, which is needed for body growth and repair, for energy storage, and for reproduction. The production and use of ATP generate heat, which the animal eventually gives off to its surroundings.

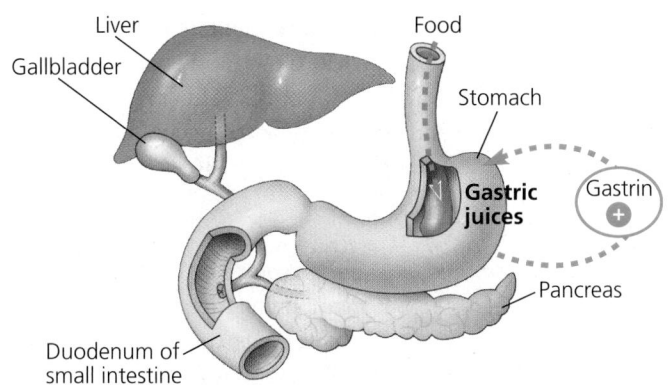

**1** As food arrives at the stomach, it stretches the stomach walls, triggering release of the hormone *gastrin*. Gastrin circulates via the bloodstream back to the stomach, where it stimulates production of gastric juices.

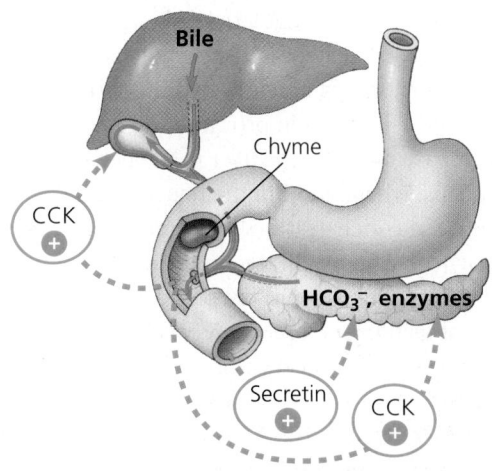

**2** Chyme—an acidic mixture of partially digested food—eventually passes from the stomach to the duodenum. The duodenum responds by releasing the digestive hormones cholecystokinin and secretin. *Cholecystokinin (CCK)* stimulates the release of digestive enzymes from the pancreas and of bile from the gallbladder. *Secretin* stimulates the pancreas to release bicarbonate ($HCO_3^-$), which neutralizes chyme.

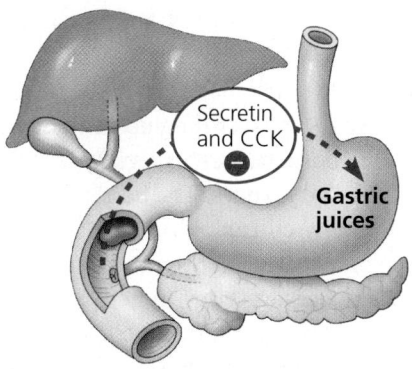

**3** If the chyme is rich in fats, the high levels of secretin and CCK released act on the stomach to inhibit peristalsis and secretion of gastric juices, thereby slowing digestion.

**Key** ⊕ Stimulation ⊖ Inhibition

▲ **Figure 33.17 Hormonal control of digestion.**

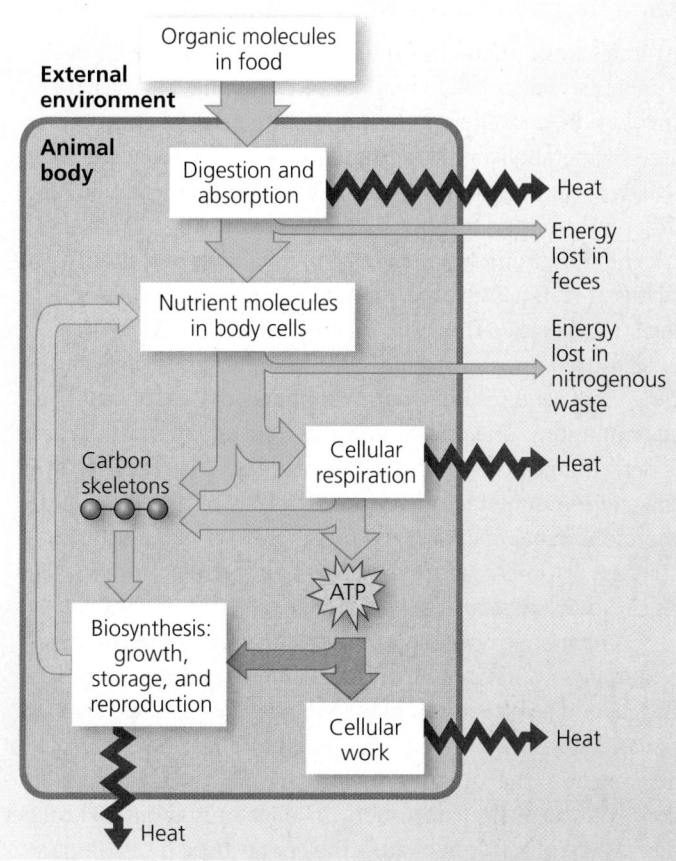

▲ **Figure 33.18 Bioenergetics of an animal: an overview.**

**MAKE CONNECTIONS** *Review the idea of energy coupling (see Concept 6.3). Then use that idea to explain why heat is produced in the absorption of nutrients, in cellular respiration, and in the synthesis of large biological molecules.*

How much of the total energy obtained from food does an animal need just to stay alive? How much energy must be expended to walk, run, swim, or fly from one place to another? What fraction of the energy intake is used for reproduction? Physiologists answer such questions by measuring the rate at which an animal uses chemical energy and how this rate changes in different circumstances.

The sum of all the energy an animal uses in a given time is called its **metabolic rate**. Energy is measured in joules or in calories and kilocalories; 1 kcal = 1,000 cal = 4,184 J. (The unit Calorie, with a capital C, as used by many nutritionists, is actually a kilocalorie.) Because nearly all chemical energy used eventually appears as heat, metabolic rate can be measured by monitoring an animal's rate of heat loss. For this approach, researchers use a calorimeter, which is a recording device in a closed, insulated chamber. Metabolic rate can also be determined from the amount of oxygen consumed or carbon dioxide produced. To calculate metabolic rate over longer periods, researchers record the rate of food consumption, the energy content of the food, and the chemical energy lost in waste products (see Figure 33.18).

## Minimum Metabolic Rate

Animals must maintain a minimum metabolic rate for basic functions such as cell maintenance, breathing, and heartbeat. In addition, endotherms, but not ectotherms, use heat generated by metabolism for thermoregulation (see Concept 32.3). Researchers therefore measure the minimum metabolic rate differently for endotherms and ectotherms.

The minimum metabolic rate of a nongrowing endotherm that is at rest, has an empty stomach, and is not experiencing stress is called the *basal metabolic rate* (*BMR*). BMR is measured under a "comfortable" temperature range—a range that requires no generation or shedding of heat above the minimum. The minimum metabolic rate of ectotherms is determined at a specific temperature because changes in the environmental temperature alter body temperature and therefore metabolic rate. The metabolic rate of a fasting, non-stressed ectotherm at rest at a particular temperature is called its *standard metabolic rate* (*SMR*).

Comparisons of minimum metabolic rates confirm that endothermy and ectothermy have different energy costs. The BMR per day for adult humans averages 1,600–1,800 kcal for males and 1,300–1,500 kcal for females. These BMRs are about equivalent to the rate of energy use by a 75-watt lightbulb. In contrast, the SMR of an American alligator is about 60 kcal per day at 20°C (68°F), less than $\frac{1}{20}$ the energy used by a comparably sized adult human.

For both ectotherms and endotherms, activity greatly affects metabolic rate. Even a person reading quietly at a desk or an insect twitching its wings consumes energy beyond the BMR or SMR. Maximum metabolic rates (the highest rates of ATP use) occur during peak activity, such as lifting heavy weights, sprinting, or high-speed swimming. In general, the maximum metabolic rate an animal can sustain is inversely related to the duration of activity.

For most terrestrial animals, the average daily rate of energy consumption is 2 to 4 times BMR (for endotherms) or SMR (for ectotherms). People in most developed countries have an unusually low average daily metabolic rate of about 1.5 times BMR—an indication of their relatively sedentary lifestyles.

## Regulation of Energy Storage

When an animal takes in more energy-rich molecules than it needs for metabolism and activity, it stores the excess energy. In humans, the first sites used for energy storage are liver and muscle cells. In these cells, excess energy from the diet is stored in glycogen, a polymer made up of many glucose units (see Figure 3.11). Once glycogen depots are full, any additional excess energy is usually stored in fat in adipose cells.

When fewer calories are taken in than are expended—perhaps because of sustained heavy exercise or lack of food—the human body generally expends liver glycogen first and then draws on muscle glycogen and fat. Fats are especially rich in energy; oxidizing a gram of fat liberates about twice the energy liberated from a gram of carbohydrate or protein. For this reason, adipose tissue provides the most space-efficient way for the body to store large amounts of energy. Most healthy people have enough stored fat to sustain them through several weeks without food.

## Glucose Homeostasis

The synthesis and breakdown of glycogen are central not only to energy storage but also to maintaining metabolic balance through glucose homeostasis. In humans, the normal range for the concentration of glucose in the blood is 70–110 mg/100 mL. Because glucose is a major fuel for cellular respiration and a key source of carbon skeletons for biosynthesis, maintaining blood glucose concentrations near this normal range is critical.

The pancreatic hormones **insulin** and **glucagon** maintain glucose homeostasis by tightly regulating the synthesis and breakdown of glycogen. The liver is a key site of action for both hormones **(Figure 33.19)**. After a carbohydrate-rich meal, for example, the rising level of insulin promotes biosynthesis of glycogen from glucose entering the liver in the hepatic portal vein. Between meals, when blood in the hepatic portal vein has a much lower glucose concentration, glucagon stimulates the liver to break down glycogen, convert amino acids and glycerol to glucose, and release glucose into the blood. Together, these opposing effects of insulin and glucagon ensure that blood exiting the liver has a glucose concentration in the normal range at nearly all times.

Glucagon and insulin are produced in the pancreas. Much of the pancreas is dedicated to producing and secreting bicarbonate ions and the digestive enzymes active in the small intestine. However, clusters of endocrine cells called pancreatic islets are scattered throughout this organ. Each pancreatic islet has *alpha cells*, which make glucagon, and *beta cells*, which make insulin. Like all hormones, insulin and glucagon are secreted into the interstitial fluid and from there enter the circulatory system.

## Diabetes Mellitus

A number of disorders can disrupt glucose homeostasis with potentially serious consequences, especially for the heart, blood vessels, eyes, and kidneys. The best known and most prevalent is **diabetes mellitus**, a disease caused by a deficiency of insulin or a decreased response to insulin in target cells. The blood glucose level rises, but cells are unable to take up enough glucose to meet metabolic needs. Instead, fat becomes the main substrate for cellular respiration. In severe cases, acidic metabolites formed during fat breakdown accumulate in the blood, threatening life by lowering blood pH and depleting sodium and potassium ions from the body.

In people with diabetes mellitus, the level of glucose in the blood may exceed the capacity of the kidneys to reabsorb this nutrient. Glucose that remains in the kidney filtrate is excreted. For this reason, the presence of sugar in urine is one

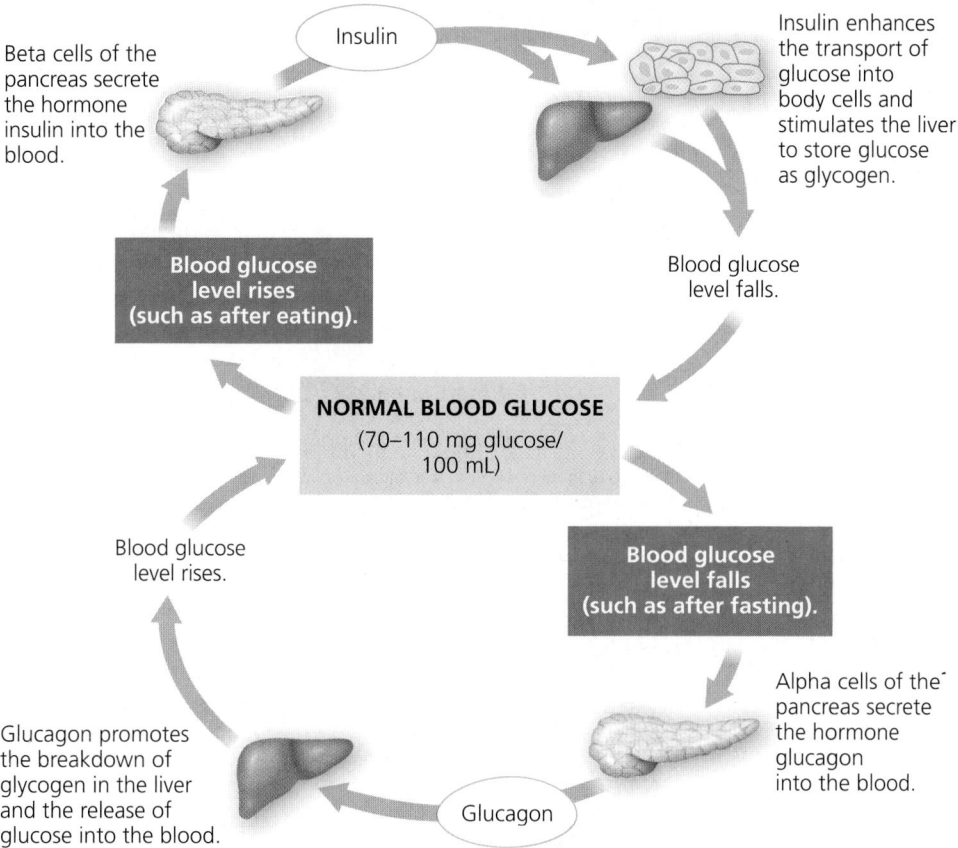

▲ **Figure 33.19 Homeostatic regulation of cellular fuel.** After a meal is digested, glucose and other monomers are absorbed into the blood from the digestive tract. The human body regulates the use and storage of glucose, a major cellular fuel.

**MAKE CONNECTIONS** *What form of feedback control does each of these regulatory circuits reflect (see Concept 32.2)?*

**ANIMATION** Visit the Study Area in **MasteringBiology** for the BioFlix® 3-D Animation on Homeostasis: Regulating Blood Sugar.

test for this disorder. As glucose is concentrated in the urine, more water is excreted along with it, resulting in excessive volumes of urine. *Diabetes* (from the Greek *diabainein*, to pass through) refers to this copious urination, and *mellitus* (from the Greek *meli*, honey) refers to the presence of sugar in urine.

There are two main types of diabetes mellitus. *Type 1 diabetes*, or insulin-dependent diabetes, is an autoimmune disorder in which the immune system destroys the insulin-producing beta cells of the pancreas. Treatment consists of insulin, typically injected several times daily. *Type 2 diabetes*, or non-insulin-dependent diabetes, is characterized by a failure of target cells to respond normally to insulin. Insulin is produced, but target cells fail to take up glucose from the blood, and blood glucose levels remain elevated. Although heredity can play a role in type 2 diabetes, excess body weight and lack of exercise significantly increase the risk of developing this disorder. Type 2 diabetes is the seventh most common cause of death in the United States.

## Regulation of Appetite and Consumption

Consuming more calories than the body needs for normal metabolism, or *overnourishment*, can lead to obesity, the excessive accumulation of fat. Obesity, in turn, contributes to a number of health problems, including type 2 diabetes, cancer of the colon and breast, and cardiovascular disease that can result in heart attacks and strokes. It is estimated that obesity is a factor in 300,000 deaths per year in the United States alone.

Researchers have discovered several homeostatic mechanisms that operate as feedback circuits, controlling the storage and metabolism of fat. A network of neurons relays and integrates information from the digestive system to control secretion of hormones that regulate long-term and short-term appetite. The target for these hormones is a "satiety center" in the brain. For example, *ghrelin*, a hormone secreted by the stomach wall, triggers feelings of hunger before meals. In contrast, both insulin and *PYY*, a hormone secreted by the small intestine after meals, suppress appetite. *Leptin*, a hormone produced by adipose (fat) tissue, also suppresses appetite and appears to play a major role in regulating body fat levels. In the **Scientific Skills Exercise**, you'll interpret data from an experiment studying genes that affect leptin production and function in mice.

Obtaining food, digesting it, and absorbing and storing nutrients are part of the larger story of how animals fuel their activities. Provisioning the body also involves distributing nutrients (circulation), and using nutrients for metabolism requires exchanging respiratory gases with the environment. These processes and the adaptations that facilitate them are the focus of Chapter 34.

### CONCEPT CHECK 33.5

1. Explain how people can become obese even if their intake of dietary fat is relatively low compared with carbohydrate intake.
2. The energy required to maintain each gram of body mass is much greater for a mouse than for an elephant. What can you conclude about metabolic rates for the mouse and the elephant?
3. **WHAT IF?** An insulinoma is a cancerous mass of pancreatic beta cells that secrete insulin but do not respond to feedback mechanisms. How would you expect an insulinoma to affect blood glucose level and liver activity?

For suggested answers, see Appendix A.

# Interpreting Data from an Experiment with Genetic Mutants

**What Are the Roles of the *ob* and *db* Genes in Appetite Regulation?** A mutation that disrupts a physiological process is often used to study the normal function of the mutated gene. Ideally, researchers use a standard set of conditions and compare animals that differ genetically only in whether a particular gene is mutant (nonfunctional) or wild-type (normal). In this way, a difference in phenotype, the physiological property being measured, can be attributed to a difference in genotype, the presence or absence of the mutation. To study the role of specific genes in regulating appetite, researchers used laboratory animals with known mutations in those genes.

Mice in which recessive mutations inactivate both copies of either the *ob* gene or the *db* gene eat voraciously and grow much more massive than wild-type mice. In the photograph, the mouse on the right is wild-type, whereas the obese mouse on the left has an inactivating mutation in both copies of the *ob* gene.

One hypothesis for the normal role of the *ob* and *db* genes is that they participate in a hormone pathway that suppresses appetite when caloric intake is sufficient. Before setting out to isolate the potential hormone, researchers explored this hypothesis genetically.

**How the Experiment Was Done** The researchers measured the mass of young subject mice of various genotypes and surgically linked the circulatory system of each one to that of another mouse. This procedure ensured that any factor circulating in the bloodstream of either mouse would be transferred to the other in the pair. After eight weeks, they again measured the mass of each subject mouse.

**INTERPRET THE DATA**

1. First, practice reading the genotype information given in the data table. For example, pairing (a) joined two mice that each had the wild-type version of both genes. Describe the two mice in pairing (b), pairing (c), and pairing (d). Explain how each pairing contributed to the experimental design.

## Data from the Experiment

| Genotype Pairing (red type indicates mutant genes) | | Average Change in Body Mass of Subject (g) |
|---|---|---|
| **Subject** | **Paired with** | |
| (a) $ob^+/ob^+$, $db^+/db^+$ | $ob^+/ob^+$, $db^+/db^+$ | 8.3 |
| (b) *ob/ob*, $db^+/db^+$ | *ob/ob*, $db^+/db^+$ | 38.7 |
| (c) *ob/ob*, $db^+/db^+$ | $ob^+/ob^+$, $db^+/db^+$ | 8.2 |
| (d) *ob/ob*, $db^+/db^+$ | $ob^+/ob^+$, *db/db* | −14.9* |

*Due to pronounced weight loss and weakening, subjects in this pairing were remeasured after less than eight weeks.

**Data from** D. L. Coleman, Effects of parabiosis of obese mice with diabetes and normal mice, *Diabetologia* 9:294–298 (1973).

2. Compare the results observed for pairing (a) and pairing (b) in terms of phenotype. If the results had been identical for these two pairings, what would that outcome have implied about the experimental design?

3. Compare the results observed for pairing (c) to those observed for pairing (b). Based on these results, does the $ob^+$ gene product appear to promote or suppress appetite? Explain your answer.

4. Describe the results observed for pairing (d). Note how these results differ from those for pairing (b). Suggest a hypothesis to explain this difference. How could you test your hypothesis using the kinds of mice in this study?

(MB) A version of this Scientific Skills Exercise, as well as a related Experimental Inquiry Tutorial, can be assigned in MasteringBiology.

---

# 33 Chapter Review

Go to **MasteringBiology®** for Assignments, the eText, and the Study Area with Animations, Activities, Vocab Self-Quiz, and Practice Tests.

## SUMMARY OF KEY CONCEPTS

**VOCAB SELF-QUIZ**

goo.gl/gbai8v

- Animals have diverse diets. **Herbivores** mainly eat plants; **carnivores** mainly eat other animals; and **omnivores** eat both. Animals must balance consumption, storage, and use of food.

### CONCEPT 33.1

**An animal's diet must supply chemical energy, organic building blocks, and essential nutrients (pp. 689–690)**

- Food provides animals with energy for ATP production, carbon skeletons for biosynthesis, and **essential nutrients**—nutrients that must be supplied in preassembled form. Essential nutrients include certain amino acids and fatty acids that animals cannot synthesize; **vitamins**, which are organic molecules; and **minerals**, which are inorganic substances.
- Malnutrition results from an inadequate intake of essential nutrients or a deficiency in sources of chemical energy. Studies of genetic defects and of disease at the population level help researchers determine human dietary requirements.

**?** *How can an enzyme cofactor needed for an essential process be an essential nutrient for only some animals?*

### CONCEPT 33.2

**Food processing involves ingestion, digestion, absorption, and elimination (pp. 691–693)**

- Food processing in animals involves **ingestion** (eating), **digestion** (enzymatic breakdown of large molecules), **absorption** (uptake of nutrients by cells), and **elimination** (passage of undigested materials out of the body in feces).
- Animals differ in the ways they obtain and ingest food. Many animals are **bulk feeders**, eating large pieces of food. Other strategies include filter feeding, suspension feeding, and fluid feeding.
- Compartmentalization is necessary to avoid self-digestion. In intracellular digestion, food particles are engulfed by endocytosis and digested within food vacuoles that have fused with lysosomes. In extracellular digestion, which is used by most animals, enzymatic hydrolysis occurs outside cells in a **gastrovascular cavity** or **alimentary canal**.

**?** *Propose an artificial diet that would eliminate the need for one of the first three steps in food processing.*

## CONCEPT 33.3

### Organs specialized for sequential stages of food processing form the mammalian digestive system (pp. 693–698)

• The mammalian digestive system has a tubular alimentary canal, into which accessory glands secrete digestive juices. **Peristalsis** pushes food along the alimentary canal. Digestion and absorption occur in specialized portions of the canal.

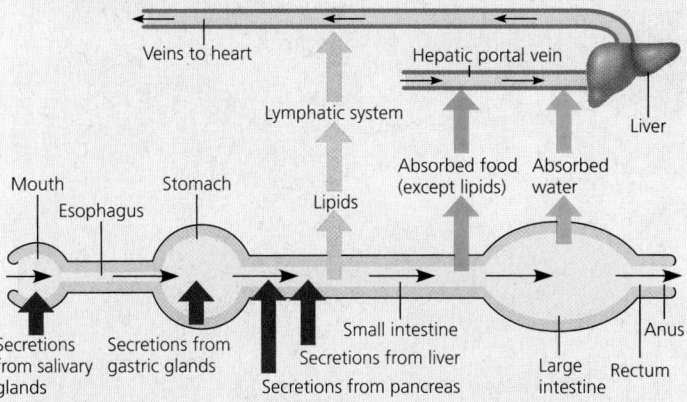

? *What structural feature of the small intestine makes it better suited for absorption of nutrients than the stomach?*

## CONCEPT 33.4

### Evolutionary adaptations of vertebrate digestive systems correlate with diet (pp. 698–700)

• Vertebrate digestive systems display evolutionary adaptations associated with diet. For example the assortment of teeth (dentition) generally correlates with diet. Also, many herbivores have fermentation chambers where mutualistic microorganisms digest cellulose. In addition, herbivores usually have longer alimentary canals than carnivores, reflecting the longer time needed to digest vegetation.

? *How does human anatomy indicate that our ancestors were not strict vegetarians?*

## CONCEPT 33.5

### Feedback circuits regulate digestion, energy allocation, and appetite (pp. 700–703)

• Food intake triggers nervous and hormonal responses that trigger secretion of digestive juices and promote movement of ingested material through the canal. The hormones insulin and glucagon control the synthesis and breakdown of glycogen, thereby regulating glucose availability.

• Animals obtain chemical energy from food. The total amount used in a unit of time defines an animal's metabolic rate. Animals allocate energy for basal (or standard) metabolism, activity, growth, and reproduction.

• Vertebrates store excess calories in glycogen (in liver and muscle cells) and in fat (in adipose cells). These energy stores can be tapped when an animal expends more calories than it consumes. If, however, an animal consumes more calories than it needs for normal metabolism, the resulting overnourishment can cause obesity.

• Several hormones, including leptin and insulin, regulate appetite by affecting the brain's satiety center.

? *Explain why your stomach might make growling noises when you skip a meal.*

## TEST YOUR UNDERSTANDING

### Level 1: Knowledge/Comprehension

1. The mammalian trachea and esophagus both connect to the
   (A) large intestine.       (C) pharynx.
   (B) stomach.               (D) rectum.

2. Which organ is *incorrectly* paired with its function?
   (A) stomach—protein digestion
   (B) large intestine—bile production
   (C) small intestine—nutrient absorption
   (D) pancreas—enzyme production

3. Which of the following is *not* a major activity of the stomach?
   (A) storage                (C) nutrient absorption
   (B) HCl production         (D) enzyme secretion

4. Fat digestion yields fatty acids and glycerol. Protein digestion yields amino acids. Both digestive processes
   (A) occur inside cells in most animals.
   (B) add a water molecule to break bonds.
   (C) require a low pH resulting from HCl production.
   (D) consume ATP.

### Level 2: Application/Analysis

5. After surgical removal of the gallbladder, a person might need to limit their dietary intake of
   (A) starch.     (B) protein.     (C) sugar.     (D) fat.

6. If you were to jog 1 km a few hours after lunch, which stored fuel would you probably tap?
   (A) muscle proteins            (C) fat in the liver
   (B) muscle and liver glycogen  (D) fat in adipose tissue

### Level 3: Synthesis/Evaluation

7. **DRAW IT** In a flowchart, summarize the events that occur after partially digested food leaves the stomach. Use the following terms: bicarbonate secretion, circulation, decrease in acidity, increase in acidity, secretin secretion, signal detection. Next to each term, indicate the compartment(s) involved. You may use terms more than once.

8. **SCIENTIFIC INQUIRY**
   In human populations of northern European origin, the disorder called hemochromatosis causes excess iron uptake from food and affects one in 200 adults. Men are ten times as likely as women to suffer from iron overload. Taking into account the existence of a menstrual cycle in humans, devise a hypothesis that explains this difference.

9. **FOCUS ON EVOLUTION**
   The human esophagus and trachea share a passage leading from the mouth and nasal passages, which can cause problems. After reviewing vertebrate evolution (see Concept 21.4), explain how descent with modification explains this "imperfect" anatomy.

10. **FOCUS ON ORGANIZATION**
    Hair is largely made up of the protein keratin. In a short essay (100–150 words), explain why a shampoo containing protein cannot replace the protein in damaged hair.

11. **SYNTHESIZE YOUR KNOWLEDGE**

Hummingbirds are well adapted to obtain sugary nectar from flowers, but they use some of the energy obtained from nectar when they forage for insects and spiders. Explain why this foraging is necessary.

*For selected answers, see Appendix A.*

# 34 Circulation and Gas Exchange

## KEY CONCEPTS

**34.1** Circulatory systems link exchange surfaces with cells throughout the body

**34.2** Coordinated cycles of heart contraction drive double circulation in mammals

**34.3** Patterns of blood pressure and flow reflect the structure and arrangement of blood vessels

**34.4** Blood components function in exchange, transport, and defense

**34.5** Gas exchange occurs across specialized respiratory surfaces

**34.6** Breathing ventilates the lungs

**34.7** Adaptations for gas exchange include pigments that bind and transport gases

▲ **Figure 34.1** How does a feathery fringe help this animal survive?

## Trading Places

The animal in **Figure 34.1** may look like a creature from a science fiction film, but it's actually an axolotl, a salamander native to shallow ponds in central Mexico. The feathery red appendages jutting out from the head of this albino adult are gills. Although external gills are uncommon in adult animals, they help the axolotl carry out a process common to all organisms—the exchange of substances between body cells and the environment.

The exchange of substances between an axolotl, or any other animal, and its surroundings ultimately occurs at the cellular level. The resources that an animal cell requires, such as nutrients and oxygen ($O_2$), enter the cytoplasm by crossing the plasma membrane. Metabolic by-products, such as carbon dioxide ($CO_2$), cross the same membrane in exiting the cell. In unicellular organisms, exchange occurs directly with the external environment. For most multicellular organisms, however, direct transfer of materials between every cell and the environment is not possible. Instead, these organisms rely on

specialized respiratory systems that carry out exchange with the environment and circulatory systems that transport materials between sites of exchange and the rest of the body.

The reddish color and the branching structure of the axolotl's gills reflect the intimate association between exchange and transport. Tiny blood vessels lie close to the surface of each filament in the gills. Across this surface, there is a net movement of $O_2$ from the surrounding water into the blood and of $CO_2$ from the blood into the water. The axolotl's heart pumps the $O_2$-rich blood from the gill filaments to all other tissues of the body. There, more short-range exchange occurs, involving nutrients and $O_2$ as well as $CO_2$ and other wastes.

Because internal transport and gas exchange are functionally related in most animals, not just axolotls, we'll discuss circulatory and respiratory systems together in this chapter. By considering examples of these systems from a range of species, we'll explore the common elements as well as the remarkable variation in form and organization. We'll also highlight the roles of circulatory and respiratory systems in maintaining homeostasis.

## CONCEPT 34.1

# Circulatory systems link exchange surfaces with cells throughout the body

The molecular trade that an animal carries out with its environment involves gaining $O_2$ and nutrients while shedding $CO_2$ and other waste products. How does this exchange take place? Small molecules, including $O_2$ and $CO_2$, undergo random thermal motion—**diffusion** (see Concept 5.3). When there is a difference in concentration, such as between a cell and its immediate surroundings, diffusion can result in net movement. But such movement is very slow for distances of more than a few millimeters. That's because the time it takes for a substance to diffuse from one place to another is proportional to the *square* of the distance. For example, a quantity of glucose that takes 1 second to diffuse 100 μm will take 100 seconds to diffuse 1 mm and almost 3 hours to diffuse 1 cm.

Given that net movement by diffusion is rapid only over very small distances, how does each cell of an animal participate in exchange? Natural selection has resulted in two basic adaptations that permit effective exchange for all of an animal's cells.

One adaptation for efficient exchange is a body that is only one or two cells thick. Each cell can thus exchange gases directly with the surrounding medium. Such an arrangement is characteristic of certain invertebrates, including cnidarians, such as hydras and jellies, and flatworms **(Figure 34.2)**. In these animals, a central **gastrovascular cavity** functions in digestion and in distributing throughout the body the nutrients released from food by digestion.

Animals that lack a simple body plan display an alternative adaptation for efficient exchange: a circulatory system, which moves fluid between each cell's immediate surroundings and the body tissues. As a result, exchange with the environment and exchange with body tissues both occur over very short distances.

## Open and Closed Circulatory Systems

A circulatory system has three basic components: a circulatory fluid, a set of interconnecting vessels, and a muscular pump,

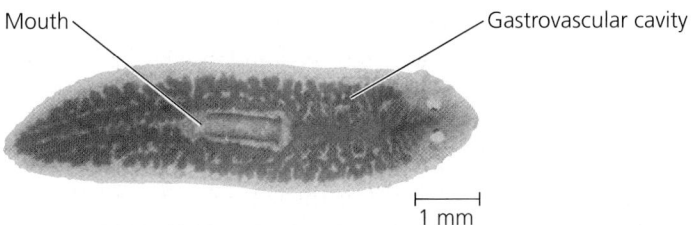

▲ **Figure 34.2 The flat body plan of the planarian *Dugesia* (LM).** The planarian's flat shape enables gas exchange to occur across the entire body surface. A gastrovascular cavity (here stained red) that branches throughout the body extracts and distributes nutrients from food that enters the mouth.

the **heart**. The heart powers circulation by using metabolic energy to elevate the circulatory fluid's hydrostatic pressure, the pressure the fluid exerts on surrounding vessels. The fluid then flows through the vessels and back to the heart.

Circulatory systems are either open or closed. In an **open circulatory system**, the circulatory fluid, called **hemolymph**, is also the *interstitial fluid* that bathes body cells. Arthropods, such as grasshoppers, and some molluscs, such as clams, have an open circulatory system. Contraction of the heart pumps the hemolymph through the circulatory vessels into interconnected sinuses, spaces surrounding the organs **(Figure 34.3a)**. There, chemical exchange occurs between the hemolymph and body cells. Relaxation of the heart draws hemolymph back in through pores, which have valves that close when the heart

▼ **Figure 34.3 Open and closed circulatory systems.**

**(a) An open circulatory system**

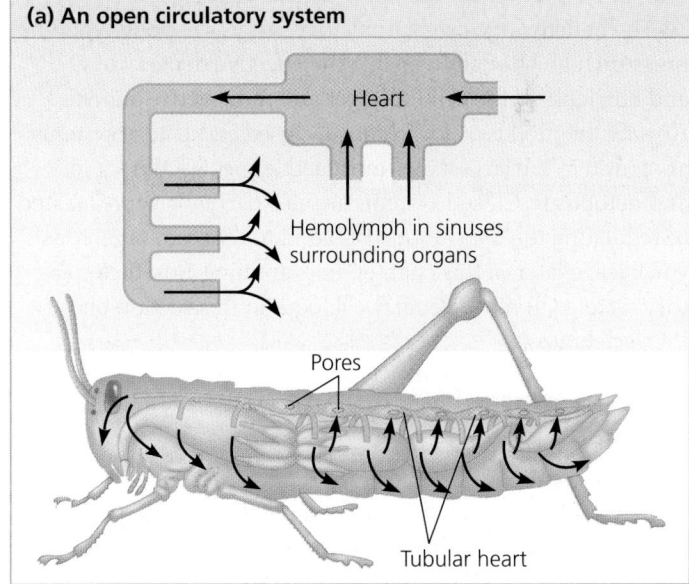

**(b) A closed circulatory system**

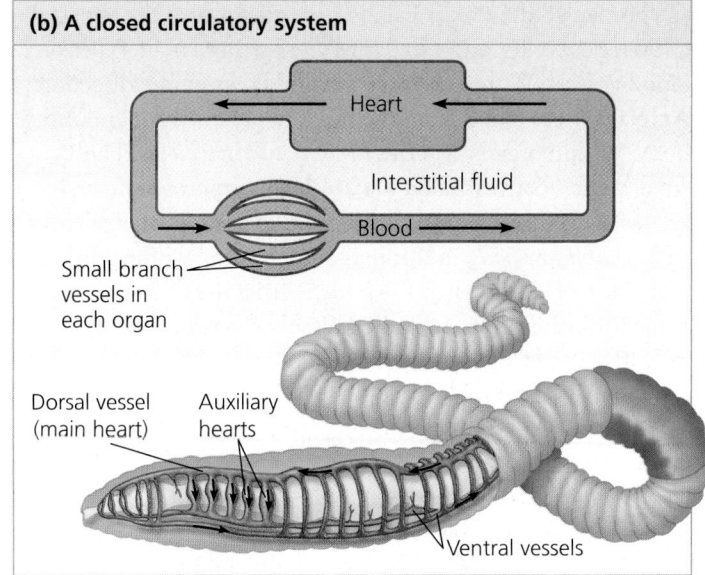

contracts. Body movements periodically squeeze the sinuses, helping circulate the hemolymph.

In a **closed circulatory system**, a circulatory fluid called **blood** is confined to vessels and is distinct from the interstitial fluid **(Figure 34.3b)**. This type of circulatory system is found in annelids (including earthworms), many molluscs, and all vertebrates. In closed circulatory systems, one or more hearts pump blood into large vessels that branch into smaller ones that infiltrate the tissues and organs. Chemical exchange occurs between the blood and the interstitial fluid, as well as between the interstitial fluid and body cells.

The fact that both open and closed circulatory systems are widespread among animals suggests that each system offers evolutionary advantages. The lower hydrostatic pressures typically associated with open circulatory systems make them less costly than closed systems in terms of energy expenditure. In some invertebrates, open circulatory systems serve additional functions. For example, spiders use the hydrostatic pressure of their open circulatory system to extend their legs.

The benefits of closed circulatory systems include blood pressure high enough to enable the effective delivery of $O_2$ and nutrients to the cells of larger and more active animals. Among the molluscs, for instance, closed circulatory systems are found in the largest and most active species, the squids and octopuses. Closed systems are also particularly well suited to regulating the distribution of blood to different organs, as you'll learn later in this chapter. In examining closed circulatory systems in more detail, we'll focus in this section on the vertebrates.

## Organization of Vertebrate Circulatory Systems

The term **cardiovascular system** is often used to refer to the heart and blood vessels of humans and other vertebrates. Blood circulates to and from the heart through an amazingly extensive network of vessels: The total length of blood vessels in an average human adult is twice Earth's circumference at the equator!

Arteries, veins, and capillaries are the three main types of blood vessels. Within each type, blood flows in one direction. **Arteries** carry blood from the heart to organs throughout the body. Within organs, arteries branch into *arterioles*. These small vessels convey blood to **capillaries**, microscopic vessels with very thin, porous walls. Local networks of capillaries, called *capillary beds*, infiltrate tissues, passing within a few cell diameters of every cell in the body. Across the thin walls of capillaries, chemicals, including dissolved gases, are exchanged by diffusion between the blood and the interstitial fluid around the tissue cells. At their "downstream" end, capillaries converge into *venules*, and venules converge into **veins**, the vessels that carry blood back to the heart.

Note that arteries and veins are distinguished by the *direction* in which they carry blood, not by the $O_2$ content or other characteristics of the blood they contain. Arteries carry

blood *away* from the heart toward capillaries, and veins return blood *toward* the heart from capillaries.

The hearts of all vertebrates contain two or more muscular chambers. The chambers that receive blood entering the heart are called **atria** (singular, *atrium*). The chambers responsible for pumping blood out of the heart are called **ventricles**. The number of chambers and the extent to which they are separated from one another differ among groups of vertebrates, as we'll discuss next. These important differences reflect the close fit of form to function that arises from natural selection.

### Single Circulation

In sharks, rays, and bony fishes, blood travels through the body and returns to its starting point in a single circuit (loop), an arrangement called **single circulation (Figure 34.4a)**. These animals have a heart that consists of two chambers: an atrium and a ventricle. Blood entering the heart collects in the atrium before transfer to the ventricle. Contraction of the ventricle pumps blood to a capillary bed in the gills, where there is a net diffusion of $O_2$ into the blood and of $CO_2$ out of the blood. As blood leaves the gills, the capillaries converge into a vessel that carries oxygen-rich blood to capillary beds throughout the body. Blood then returns to the heart.

In single circulation, blood that leaves the heart passes through two capillary beds (in the gills and elsewhere in the body) before returning to the heart. When blood flows through a capillary bed, blood pressure drops substantially, for reasons we'll explain shortly. The drop in blood pressure in the gills limits the rate of blood flow in the rest of the animal's body. As the animal swims, however, the contraction and relaxation of its muscles help accelerate the pace of circulation.

### Double Circulation

The circulatory systems of amphibians, reptiles, and mammals have two circuits of blood flow, an arrangement called **double circulation (Figure 34.4b** and **c)**. In animals with double circulation, the pumps for the two circuits are combined into a single organ, the heart. Having both pumps within a single heart simplifies coordination of the pumping cycles. One pump, the right side of the heart, delivers oxygen-poor blood to the capillary beds of the gas exchange tissues, where there is a net movement of $O_2$ into the blood and of $CO_2$ out of the blood. This part of the circulation is called a *pulmocutaneous circuit* if it includes capillaries in both the lungs and the skin, as in many amphibians. It is called a *pulmonary circuit* if the capillary beds involved are all in the lungs, as in reptiles and mammals.

After the oxygen-enriched blood leaves the gas exchange tissues, it enters the other pump, the left side of the heart. Heart contraction propels this blood to capillary beds in organs and tissues throughout the body. Following the exchange of $O_2$ and $CO_2$, as well as nutrients and waste products, the now oxygen-poor blood returns to the heart, completing the **systemic circuit**.

▼ **Figure 34.4 Examples of vertebrate circulatory system organization.**

| (a) Single circulation: fish | (b) Double circulation: amphibian | (c) Double circulation: mammal |
|---|---|---|
| 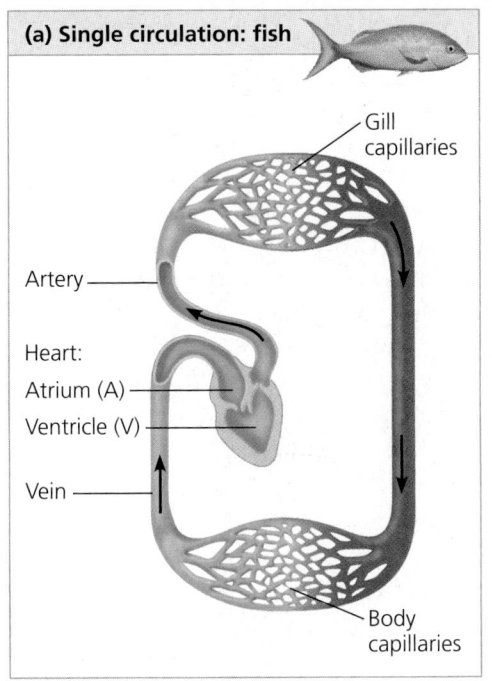 | 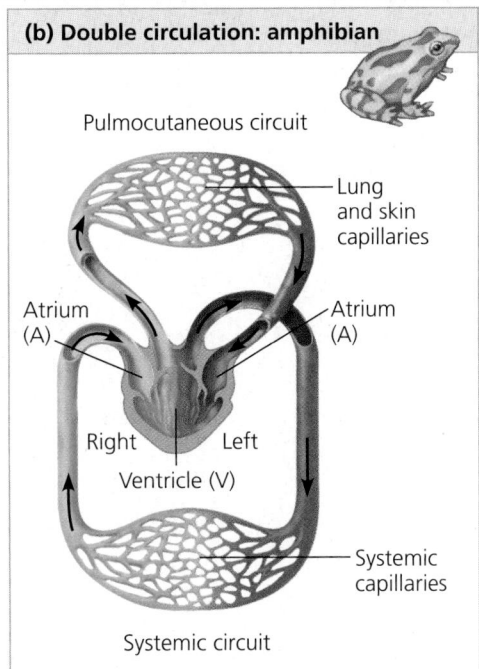 | 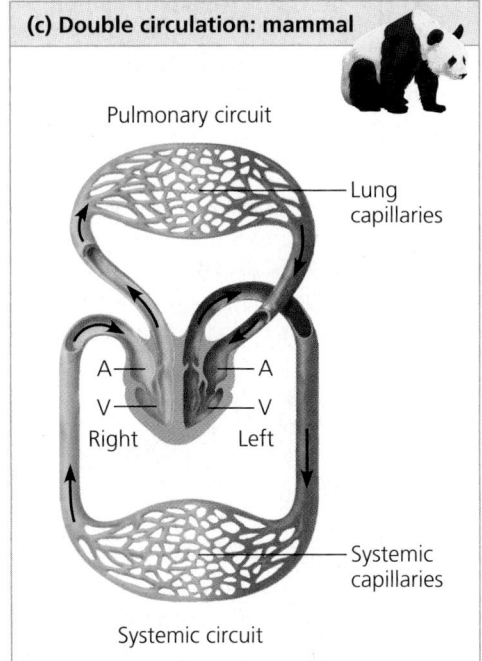 |

**Key** Oxygen-rich blood
Oxygen-poor blood

(Note that circulatory systems are shown as if the body were facing you: The right side of the heart is shown on the left, and vice versa.)

Double circulation provides a vigorous flow of blood to the brain, muscles, and other organs because the heart repressurizes the blood destined for these tissues after it passes through the capillary beds of the lungs or skin. Indeed, blood pressure is often much higher in the systemic circuit than in the gas exchange circuit. By contrast, in single circulation the blood flows under reduced pressure directly from the gas exchange organs to other organs.

## Evolutionary Variation in Double Circulation

**EVOLUTION** Some vertebrates with double circulation are intermittent breathers. For example, amphibians and many reptiles fill their lungs with air periodically, passing long periods either without gas exchange or by relying on another gas exchange tissue, typically the skin. A variety of adaptations found among intermittent breathers enable their circulatory systems to temporarily bypass the lungs in part or in whole:

- Frogs and other amphibians have a heart with three chambers—two atria and one ventricle (see Figure 34.4b). A ridge within the ventricle diverts most (about 90%) of the oxygen-rich blood from the left atrium into the systemic circuit and most of the oxygen-poor blood from the right atrium into the pulmocutaneous circuit. When a frog is underwater, the incomplete division of the ventricle allows the frog to adjust its circulation, largely shutting off blood flow to its temporarily ineffective lungs. Blood flow continues to the skin, which acts as the sole site of gas exchange while the frog is submerged.

- In the three-chambered heart of turtles, snakes, and lizards, an incomplete septum partially divides the single ventricle into right and left chambers. Two major arteries, called aortas, lead to the systemic circulation. As with amphibians, the circulatory system enables control of the relative amount of blood flowing to the lungs and the rest of the body.

- In alligators, caimans, and other crocodilians, the ventricles are divided by a complete septum, but the pulmonary and systemic circuits connect where the arteries exit the heart. This connection allows arterial valves to shunt blood flow away from the lungs temporarily, such as when the animal is underwater.

Double circulation in birds and mammals, which for the most part breathe continuously, differs from double circulation in other vertebrates. As shown for a panda in Figure 34.4c, the heart has two atria and two completely divided ventricles. The left side of the heart receives and pumps only oxygen-rich blood, while the right side receives and pumps only oxygen-poor blood. Unlike amphibians and many reptiles, birds and mammals cannot vary blood flow to the lungs without varying blood flow throughout the body in parallel.

How has natural selection shaped the double circulation of birds and mammals? As endotherms, they use about ten times as much energy as equal-sized ectotherms (see Concept 33.5). Their circulatory systems therefore need to deliver about ten times as much fuel and $O_2$ to their tissues and remove ten times as much $CO_2$ and other wastes. This large traffic of substances is made possible by the separate and independently powered systemic and pulmonary circuits and by large hearts. A powerful four-chambered heart arose independently in the distinct ancestors of birds and mammals and thus reflects convergent evolution.

In the next section, we'll restrict our focus to circulation in mammals and to the anatomy and physiology of the key circulatory organ—the heart.

**CONCEPT CHECK 34.1**

1. How is the flow of hemolymph through an open circulatory system similar to the flow of water through an outdoor fountain?
2. Three-chambered hearts with incomplete septa were once viewed as being less adapted to circulatory function than mammalian hearts. What advantage of such hearts did this viewpoint overlook?
3. **WHAT IF?** The heart of a normally developing human fetus has a hole between the left and right atria. In some cases, this hole does not close completely before birth. If the hole weren't surgically corrected, how would it affect the $O_2$ content of the blood entering the systemic circuit?

For suggested answers, see Appendix A.

CONCEPT 34.2

# Coordinated cycles of heart contraction drive double circulation in mammals

The timely delivery of $O_2$ to the body's organs is critical: Some brain cells, for example, die if their $O_2$ supply is interrupted for even a few minutes. How does the mammalian cardiovascular system meet the body's continuous (although variable) demand for $O_2$? To answer this question, we must consider how the parts of the system are arranged and how each part functions.

## Mammalian Circulation

Let's first examine the overall organization of the mammalian cardiovascular system, beginning with the pulmonary circuit. (The circled numbers refer to corresponding locations in **Figure 34.5**.) Contraction of ❶ the right ventricle pumps blood to the lungs via ❷ the pulmonary arteries. As the blood flows through ❸ capillary beds in the left and right lungs, it loads $O_2$ and unloads $CO_2$. Oxygen-rich blood returns from the lungs via the pulmonary veins to ❹ the left atrium of the heart. Next, the oxygen-rich blood flows into ❺ the heart's left ventricle, which pumps the oxygen-rich blood out to body tissues through the systemic circuit.

Blood leaves the left ventricle of the heart via ❻ the aorta, which conveys blood to arteries leading throughout the body. The first branches leading from the aorta are the coronary

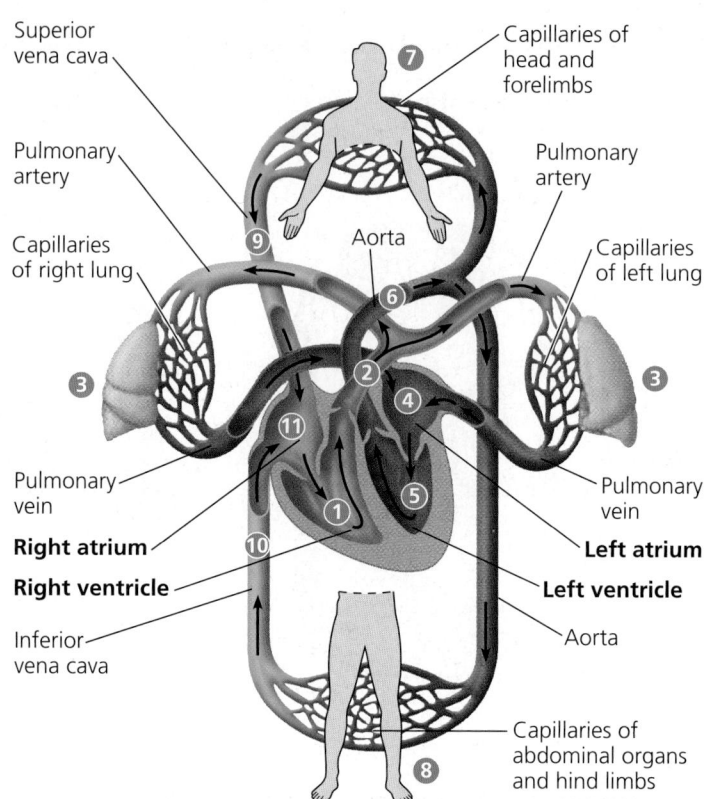

▲ **Figure 34.5 The mammalian cardiovascular system: an overview.** Note that the dual circuits operate simultaneously, not in the serial fashion that the numbering in the diagram suggests. The two ventricles contract almost in unison and pump the same volume of blood. However, the systemic circuit has a much greater total volume than the pulmonary circuit.

arteries (not shown), which supply blood to the heart muscle itself. Then branches lead to ❼ capillary beds in the head and arms (forelimbs). The aorta then descends into the abdomen, supplying oxygen-rich blood to arteries leading to ❽ capillary beds in the abdominal organs and legs (hind limbs). Within the capillary beds, there is a net diffusion of $O_2$ from the blood to the tissues and of $CO_2$ (produced by cellular respiration) into the blood. Capillaries rejoin, forming venules, which convey blood to veins.

As it moves back toward the heart, oxygen-poor blood from the head, neck, and forelimbs is channeled into a large vein, ❾ the superior vena cava. Another large vein, ❿ the inferior vena cava, drains blood from the trunk and hind limbs. The two venae cavae empty their blood into ⓫ the right atrium, from which the oxygen-poor blood flows into the right ventricle.

## The Mammalian Heart: *A Closer Look*

Located behind the sternum (breastbone), the human heart is about the size of a clenched fist and consists mostly of cardiac muscle. The two atria have relatively thin walls and serve as collection chambers for blood returning to the heart from the lungs or other body tissues **(Figure 34.6)**. Much of the blood that enters the atria flows into the ventricles while all

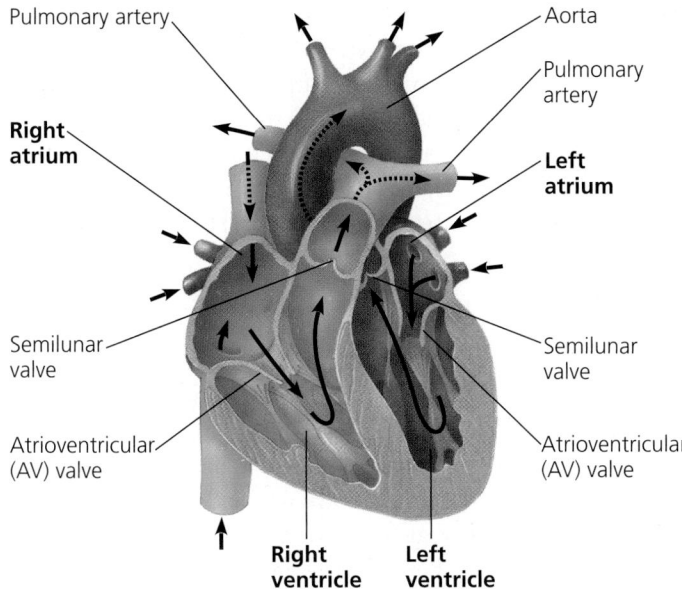

Pulmonary artery

Aorta

Pulmonary artery

**Right atrium**

**Left atrium**

Semilunar valve

Semilunar valve

Atrioventricular (AV) valve

Atrioventricular (AV) valve

**Right ventricle**

**Left ventricle**

▲ **Figure 34.6 The mammalian heart: a closer look.** Notice the locations of the valves, which prevent backflow of blood within the heart. Also notice how the atria and left and right ventricles differ in the thickness of their muscular walls.

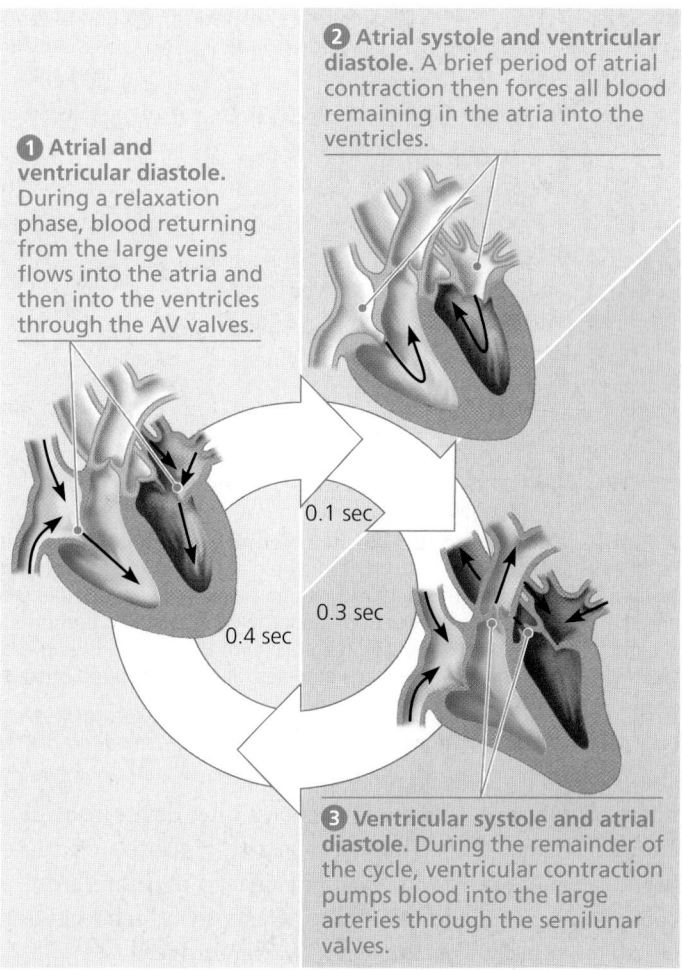

❶ **Atrial and ventricular diastole.** During a relaxation phase, blood returning from the large veins flows into the atria and then into the ventricles through the AV valves.

❷ **Atrial systole and ventricular diastole.** A brief period of atrial contraction then forces all blood remaining in the atria into the ventricles.

0.1 sec

0.3 sec

0.4 sec

❸ **Ventricular systole and atrial diastole.** During the remainder of the cycle, ventricular contraction pumps blood into the large arteries through the semilunar valves.

▲ **Figure 34.7 The cardiac cycle.** Note that during all but 0.1 second of the cardiac cycle, the atria are relaxed and are filling with blood returning via the veins.

heart chambers are relaxed. The remainder is transferred by contraction of the atria before the ventricles begin to contract. Compared to the atria, the ventricles have thicker walls and contract much more forcefully—especially the left ventricle, which pumps blood throughout the body via the systemic circuit. Although the left ventricle contracts with greater force than the right ventricle, it pumps the same volume of blood as the right ventricle during each contraction.

The heart contracts and relaxes in a rhythmic cycle. When it contracts, it pumps blood; when it relaxes, its chambers fill with blood. One complete sequence of pumping and filling is referred to as the **cardiac cycle (Figure 34.7)**. The contraction phase of the cycle is called **systole** (pronounced sis′-tō-lē), and the relaxation phase is called **diastole** (dī-as′-tō-lē).

The volume of blood each ventricle pumps per minute is the cardiac output. Two factors determine cardiac output: the rate of contraction, or heart rate (beats per minute), and the *stroke volume*, the amount of blood pumped by a ventricle in a single contraction. The average stroke volume in humans is about 70 mL. Multiplying this stroke volume by a resting heart rate of 72 beats per minute yields a cardiac output of 5 L/min—about equal to the total volume of blood in the human body. During heavy exercise, when $O_2$ demand is especially high, cardiac output increases as much as fivefold.

Four valves in the heart prevent backflow and keep blood moving in the correct direction (see Figures 34.6 and 34.7). Made of flaps of connective tissue, the valves open when pushed from one side and close when pushed from the other. An **atrioventricular (AV) valve** lies between each atrium and ventricle. Pressure generated by the powerful contraction of

the ventricles closes the AV valves, keeping blood from flowing back into the atria. **Semilunar valves** are located at the two exits of the heart: where the pulmonary artery leaves the right ventricle and where the aorta leaves the left ventricle. These valves are pushed open by the pressure generated during contraction of the ventricles. When the ventricles relax, blood pressure built up in the pulmonary artery and aorta closes the semilunar valves and prevents significant backflow.

You can follow the closing of the two sets of heart valves either with a stethoscope or by pressing your ear tightly against the chest of a friend (or a friendly dog). The sound pattern is "lub-dup, lub-dup, lub-dup." The first heart sound ("lub") is created by the recoil of blood against the closed AV valves. The second sound ("dup") is due to the vibrations caused by closing of the semilunar valves.

If blood squirts backward through a defective valve, it may produce an abnormal sound called a heart murmur. Some people are born with heart murmurs; in others, the valves may be damaged by infection (for instance, from rheumatic fever, an inflammation of the heart or other tissues triggered by

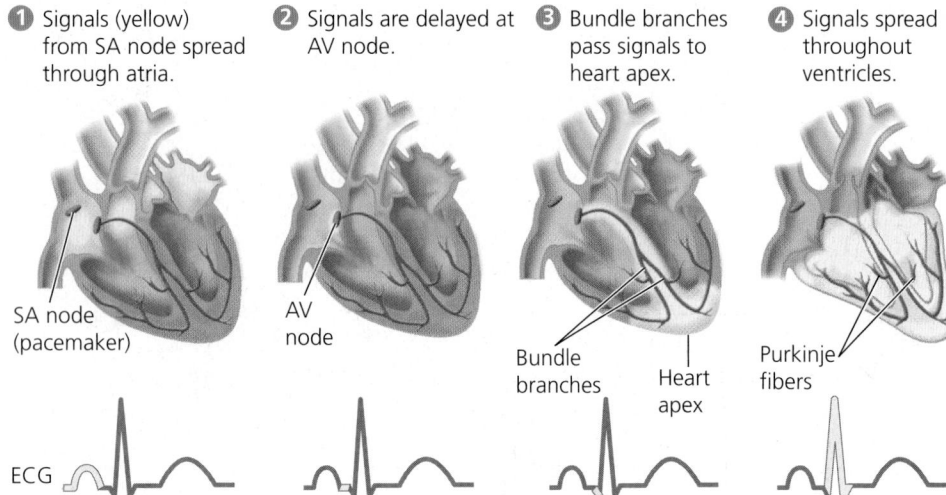

**❶** Signals (yellow) from SA node spread through atria.

**❷** Signals are delayed at AV node.

**❸** Bundle branches pass signals to heart apex.

**❹** Signals spread throughout ventricles.

SA node (pacemaker)

AV node

Bundle branches

Heart apex

Purkinje fibers

ECG

▲ **Figure 34.8 The control of heart rhythm.** Electrical signals follow a set path through the heart in establishing the heart rhythm. The diagrams at the top trace the movement of electrical signals (yellow) during the cardiac cycle; specialized cells involved in electrical control of the rhythm are indicated in orange. Under each step, the corresponding portion of an electrocardiogram (ECG) is highlighted in yellow. In step 4, the portion of the ECG to the right of the "spike" represents electrical activity that reprimes the ventricles for the next round of contraction.

**WHAT IF?** *If your doctor gave you a copy of your ECG recording over many cardiac cycles, how could you determine what your heart rate had been during the test?*

infection with certain bacteria). When a valve defect is severe enough to endanger health, surgeons may implant a mechanical replacement valve. However, not all heart murmurs are caused by a defect, and most valve defects do not reduce the efficiency of blood flow enough to warrant surgery.

### Maintaining the Heart's Rhythmic Beat

In vertebrates, the heartbeat originates in the heart itself. Some cardiac muscle cells are autorhythmic, meaning they can contract and relax repeatedly without any signal from the nervous system. A group of such cells forms the **sinoatrial (SA) node**, or *pacemaker*, which sets the rate and timing at which all cardiac muscle cells contract. (In contrast, some arthropods have pacemakers located in the nervous system.)

The SA node produces electrical impulses much like those produced by nerve cells. Because cardiac muscle cells are electrically coupled through gap junctions (see Figure 4.27), impulses from the SA node spread rapidly within heart tissue. In addition, these impulses generate currents that are conducted to the skin via body fluids. In an **electrocardiogram** (**ECG** or, often, **EKG**, from the German spelling), these currents are recorded by electrodes placed on the skin. The resulting graph of current against time has a characteristic shape that represents the stages in the cardiac cycle **(Figure 34.8)**.

Impulses from the SA node first spread rapidly through the walls of the atria, causing both atria to contract in unison. During atrial contraction, the impulses reach a relay point called the **atrioventricular (AV) node**, located in the wall between the left and right atria. Here the impulses are delayed for about 0.1 second before spreading to the heart apex. This delay allows the atria to empty completely before the ventricles contract. Then the signals from the AV node are conducted to the heart apex and throughout the ventricular walls.

Physiological cues alter heart tempo by regulating the pacemaker function of the SA node. For example, when you stand up and start walking, the nervous system speeds up your pacemaker. The resulting increase in heart rate provides the additional $O_2$ needed by the muscles that are powering your activity. If you then sit down and relax, the nervous system slows down your pacemaker, decreasing your heart rate and thus conserving energy. Hormones and temperature also influence the pacemaker. For instance, epinephrine, the "fight-or-flight" hormone secreted by the adrenal glands, causes the heart rate to increase, as does an increase in body temperature.

Having examined the operation of the circulatory pump, we turn in the next section to the forces and structures that influence blood flow in the vessels of each circuit.

### CONCEPT CHECK 34.2

1. Explain why blood in the pulmonary veins has a higher $O_2$ concentration than in the venae cavae, which are also veins.

2. Why is it important that the AV node delay the electrical impulse moving from the SA node and the atria to the ventricles?

3. **WHAT IF?** Suppose that after you exercise regularly for several months, your resting heart rate decreases, but your cardiac output at rest is unchanged. What other change in the function of your heart at rest must have occurred?

For suggested answers, see Appendix A.

---

**CONCEPT 34.3**

# Patterns of blood pressure and flow reflect the structure and arrangement of blood vessels

The vertebrate circulatory system enables blood to deliver oxygen and nutrients and remove wastes throughout the body. In doing so, the circulatory system relies on blood vessels that exhibit a close match of structure and function.

## Blood Vessel Structure and Function

All blood vessels contain a central lumen (cavity) lined with an **endothelium**, a single layer of flattened epithelial cells. Like the polished surface of a copper pipe, the smooth endothelial layer minimizes resistance to fluid flow. Surrounding the endothelium are tissue layers that differ among capillaries, arteries, and veins, reflecting distinct adaptations to the particular functions of these vessels in circulation.

Capillaries are the smallest blood vessels, having a diameter only slightly greater than that of a red blood cell (**Figure 34.9**). Capillaries also have very thin walls, which consist of just an endothelium and a surrounding extracellular layer called the basal lamina. The exchange of substances between the blood and interstitial fluid occurs only in capillaries because only there are the vessel walls thin enough to permit this exchange.

In contrast to capillaries, both arteries and veins have thick walls that consist of two layers of tissue surrounding the endothelium. The outer layer is formed by connective tissue that contains elastic fibers, which allow the vessel to stretch and recoil, and collagen, which provides strength. The layer next to the endothelium contains smooth muscle and more elastic fibers.

The walls of arteries are thick, strong, and elastic. They can thus accommodate blood pumped at high pressure by the heart, bulging outward as blood enters and recoiling when the heart relaxes between contractions. As we'll discuss shortly, this behavior of arterial walls has an essential role in maintaining blood pressure and flow to capillaries.

The smooth muscles in the walls of arteries and arterioles also help regulate the path of blood flow. Signals from the nervous system and circulating hormones act on the smooth muscle of these vessels, causing dilation or constriction that modulates blood flow to different parts of the body.

Because veins convey blood back to the heart at a lower pressure, they do not require thick walls. For a given blood vessel diameter, a vein has a wall only about a third as thick as that of an artery. Unlike arteries, veins contain valves, which maintain a unidirectional flow of blood despite the low blood pressure in these vessels.

We consider next how blood vessel diameter, vessel number, and blood pressure influence the speed at which blood flows in different locations within the body.

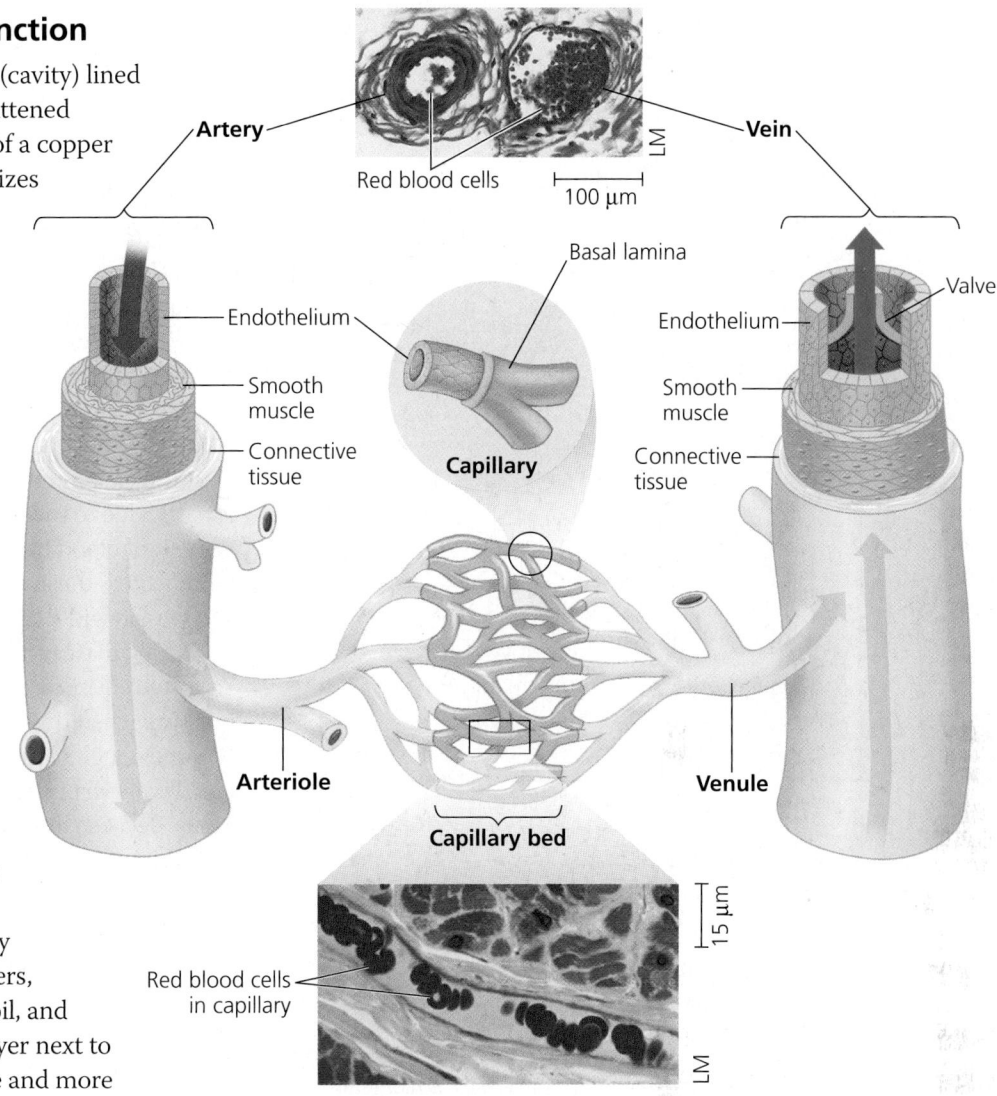

▲ Figure 34.9 **The structure of blood vessels.**

## Blood Flow Velocity

To understand how blood vessel diameter influences blood flow, consider how water flows through a thick hose connected to a faucet. When the faucet is turned on, water flows at the same velocity at each point along the hose. What happens when a narrow nozzle is attached to the hose? Because water doesn't compress under pressure, the volume of water moving through the nozzle in a given time must be the same as the volume moving through the rest of the hose. The cross-sectional area of the nozzle is smaller than that of the hose, so the water speeds up, exiting the nozzle at high velocity.

An analogous situation exists in the circulatory system, but blood *slows* as it moves from arteries to arterioles to the much narrower capillaries. Why? The reason is that the number of capillaries is enormous, roughly 7 billion in a human body. Each artery conveys blood to so many capillaries that the *total* cross-sectional area is much greater in capillary

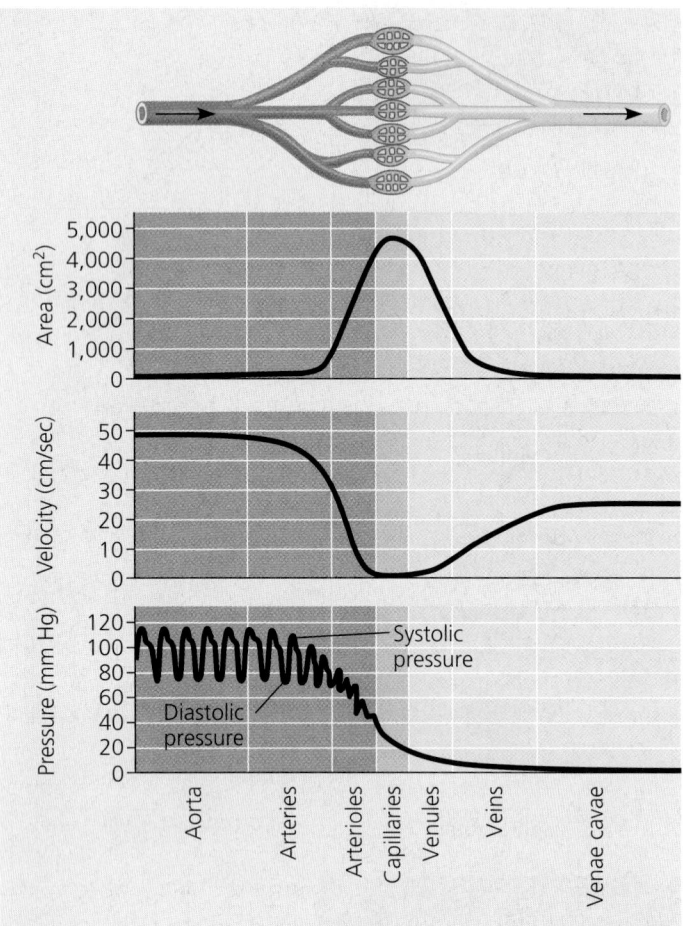

▲ **Figure 34.10 The interrelationship of cross-sectional area of blood vessels, blood flow velocity, and blood pressure.** Because total cross-sectional area increases in the arterioles and in the capillaries, blood flow velocity decreases markedly. Blood pressure, the main force driving blood from the heart to the capillaries, is highest in the aorta and other arteries.

beds than in the arteries or any other part of the circulatory system **(Figure 34.10)**. The result is a dramatic decrease in velocity from the arteries to the capillaries: Blood travels 500 times more slowly in the capillaries (about 0.1 cm/sec) than in the aorta (about 48 cm/sec). After passing through the capillaries, the blood speeds up as it enters the venules and veins, which have smaller *total* cross-sectional areas than the capillaries.

## Blood Pressure

Blood, like all fluids, flows from areas of higher pressure to areas of lower pressure. Contraction of a heart ventricle generates blood pressure, which exerts a force in all directions. The part of the force directed lengthwise in an artery causes the blood to flow away from the heart, the site of highest pressure. The part of the force exerted sideways stretches the wall of the artery. Following ventricular contraction, the recoil of the elastic arterial walls plays a critical role in maintaining blood pressure, and hence blood flow, throughout the cardiac cycle. Once the blood enters the millions of tiny arterioles and capillaries,

the narrow diameter of these vessels generates substantial resistance to flow. By the time the blood enters the veins, this resistance has dissipated much of the pressure generated by the pumping heart (see Figure 34.10).

### Changes in Blood Pressure During the Cardiac Cycle

Arterial blood pressure is highest when the heart contracts during ventricular systole. The pressure at this time is called *systolic pressure* (see Figure 34.10). With each ventricular contraction the resulting spike in blood pressure stretches the walls of the arteries. You can feel this **pulse**—the rhythmic bulging of the artery walls with each heartbeat—by placing your fingers on the inside of your wrist. The pressure surge is partly due to the narrow openings of arterioles impeding the exit of blood from the arteries. When the heart contracts, blood enters the arteries faster than it can leave, and the vessels stretch wider from the rise in pressure.

During diastole, the elastic walls of the arteries snap back. As a consequence, there is a lower but still substantial blood pressure when the ventricles are relaxed (*diastolic pressure*). Before enough blood has flowed into the arterioles to completely relieve pressure in the arteries, the heart contracts again. Because the arteries remain pressurized throughout the cardiac cycle (see Figure 34.10), blood continuously flows into arterioles and capillaries.

To measure blood pressure, doctors or nurses often use an inflatable cuff attached to a pressure gauge. The cuff is wrapped around the upper arm and inflated until the pressure closes the artery. Next, the cuff is deflated gradually. When the cuff pressure drops just below that in the artery, blood begins to pulse past the cuff, making sounds that can be heard with a stethoscope. The pressure measured at this point equals the systolic pressure. As deflation continues, the cuff pressure at some point no longer constricts blood movement. The reading on the gauge when the blood begins to flow freely and silently equals the diastolic pressure. For a healthy 20-year-old human at rest, arterial blood pressure in the systemic circuit is typically about 120 millimeters of mercury (mm Hg) at systole and 70 mm Hg at diastole, expressed as 120/70. (Arterial blood pressure in the pulmonary circuit is six to ten times lower.)

### Maintenance of Blood Pressure

Homeostatic mechanisms regulate arterial blood pressure by altering the diameter of arterioles. As the smooth muscles in arteriole walls contract, the arterioles narrow, a process called **vasoconstriction**. Vasoconstriction increases blood pressure upstream in the arteries. When the smooth muscles relax, the arterioles undergo **vasodilation**, an increase in diameter that causes blood pressure in the arteries to fall.

Researchers have identified nitric oxide (NO), a gas, as a major inducer of vasodilation and endothelin, a peptide, as the

# Blood components function in exchange, transport, and defense

As you read in Concept 34.1, the fluid transported by an open circulatory system is continuous with the fluid that surrounds all of the body cells and therefore has the same composition. In contrast, the fluid in a closed circulatory system can be more highly specialized, as is the case for the blood of vertebrates.

## Blood Composition and Function

Vertebrate blood is a connective tissue consisting of cells suspended in a liquid matrix called **plasma**. Separating the components of blood using a centrifuge reveals that cellular elements (cells and cell fragments) occupy about 45% of the volume of blood **(Figure 34.13)**. The remainder is plasma. Dissolved in the plasma are ions and proteins that, together with the blood cells, function in osmotic regulation, transport, and defense.

### Plasma

Among the many solutes in plasma are inorganic salts in the form of dissolved ions, sometimes referred to as blood electrolytes. The dissolved ions are an essential component of the blood. Some buffer the blood, which in humans normally has a pH of 7.4. Ions are also important in maintaining the osmotic balance of the blood. In addition, the concentration of ions in plasma directly affects the composition of the interstitial fluid, where many ions have a vital role in muscle and nerve activity. Serving all of these functions necessitates keeping plasma electrolytes within narrow concentration ranges via homeostatic mechanisms (see Concept 32.3).

Plasma proteins, including albumins, buffer against pH changes and help maintain the osmotic balance between blood and interstitial fluid. Certain plasma proteins have additional functions. Immunoglobulins, or antibodies, combat viruses and other foreign agents that invade the body (see Concept 35.3). Apolipoproteins escort lipids, which are insoluble in water and can travel in blood only when bound to proteins. Plasma also contains fibrinogens, which are clotting factors that help plug leaks when blood vessels are injured. (The term *serum* refers to plasma from which these clotting factors have been removed.)

Plasma also contains a wide variety of other substances in transit from one part of the body to another, including nutrients, metabolic wastes, respiratory gases, and hormones.

| Plasma 55% | |
|---|---|
| **Constituent** | **Major functions** |
| **Water** | Solvent |
| **Ions (blood electrolytes)**<br>Sodium<br>Potassium<br>Calcium<br>Magnesium<br>Chloride<br>Bicarbonate | Osmotic balance, pH buffering, and regulation of membrane permeability |
| **Plasma proteins**<br>Albumin | Osmotic balance, pH buffering |
| Immunoglobulins (antibodies) | Defense |
| Apolipoproteins | Lipid transport |
| Fibrinogen | Clotting |
| **Substances transported by blood** | |
| Nutrients (such as glucose, fatty acids, vitamins)<br>Waste products of metabolism<br>Respiratory gases ($O_2$ and $CO_2$)<br>Hormones | |

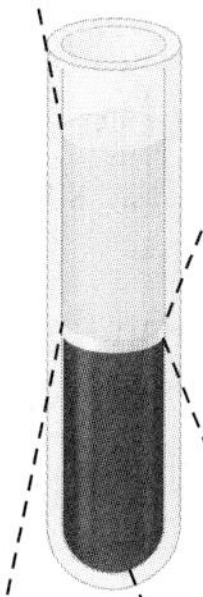

Separated blood elements

| Cellular elements 45% | | |
|---|---|---|
| **Cell type** | **Number** per µL ($mm^3$) of blood | **Functions** |
| **Leukocytes (white blood cells)** 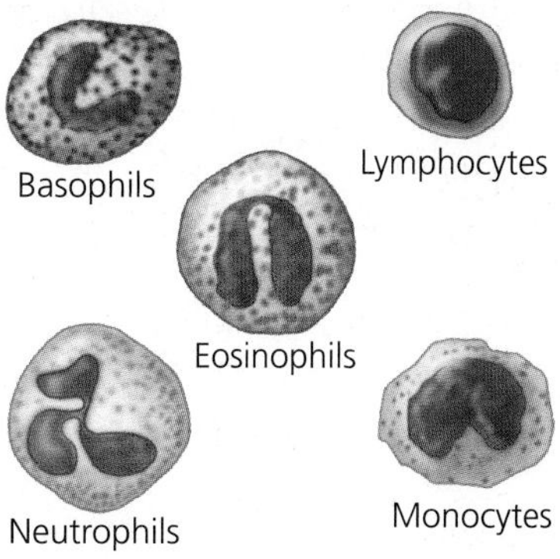<br>Basophils, Lymphocytes, Eosinophils, Neutrophils, Monocytes | 5,000–10,000 | Defense and immunity |
| **Platelets**  | 250,000–400,000 | Blood clotting |
| **Erythrocytes (red blood cells)** 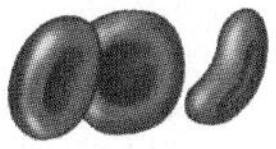 | 5,000,000–6,000,000 | Transport of $O_2$ and some $CO_2$ |

▲ **Figure 34.13 The composition of mammalian blood.**

## Cellular Elements

Blood contains two classes of cells: red blood cells, which transport $O_2$, and white blood cells, which function in defense. Also suspended in blood plasma are **platelets**, cell fragments that are involved in the clotting process.

All cellular elements of blood develop from **stem cells** located in the red marrow inside bones, particularly the ribs, vertebrae, sternum, and pelvis. As they divide and self-renew, these multipotent stem cells produce two sets of progenitor cells with a more limited capacity for self-renewal **(Figure 34.14)**. One set, the lymphoid progenitors, produces lymphocytes. The other set, the myeloid progenitors, produces all other white blood cells, red blood cells, and platelets.

**Erythrocytes** Red blood cells, or **erythrocytes**, are by far the most numerous blood cells. Each microliter (μL, or mm³) of human blood contains 5–6 million red cells, and there are about 25 trillion of these cells in the body's 5 L of blood. Their main function is $O_2$ transport, and their structure is closely related to this function. Human erythrocytes are small disks (7–8 μm in diameter) that are biconcave—thinner in the center than at the edges. This shape increases surface area, enhancing the rate of diffusion of $O_2$ across the plasma membrane. Mature mammalian erythrocytes lack nuclei. This unusual characteristic leaves more space in these tiny cells for **hemoglobin**, the iron-containing protein that transports $O_2$ (see Figure 3.22).

Despite its small size, an erythrocyte contains about 250 million molecules of hemoglobin (Hb). Because each molecule of hemoglobin binds up to four molecules of $O_2$, one erythrocyte can transport about 1 billion $O_2$ molecules. As erythrocytes pass through the capillary beds of lungs, gills, or other respiratory organs, $O_2$ diffuses into the erythrocytes and binds to hemoglobin. In the systemic capillaries, $O_2$ dissociates from hemoglobin and diffuses into body cells.

In **sickle-cell disease**, an abnormal form of hemoglobin (Hb$^S$) polymerizes into aggregates. Because the concentration of hemoglobin in erythrocytes is so high, these aggregates are large enough to distort the erythrocyte into an elongated, curved shape that resembles a sickle. This abnormality results from an alteration in the amino acid sequence of hemoglobin at a single position (see Figure 3.23).

Throughout a person's life, stem cells replace the worn-out cellular elements of blood. Erythrocytes are the shortest lived, circulating for only 120 days on average before being replaced. A feedback mechanism sensitive to $O_2$ levels in the blood controls erythrocyte production. If the tissues do not receive enough $O_2$, the kidneys synthesize and secrete *erythropoietin* (*EPO*), a hormone that stimulates erythrocyte production. Today, EPO produced by recombinant DNA technology is used to treat health problems such as *anemia*, a condition of lower-than-normal erythrocyte or hemoglobin levels. Some athletes inject themselves with EPO to increase their erythrocyte levels, although this practice has been outlawed by major sports organizations.

**Leukocytes** The blood contains five major types of white blood cells, or **leukocytes**. Their function is to fight infections. Some are phagocytic, engulfing and digesting microorganisms and debris from the body's own dead cells. Other leukocytes, called lymphocytes, develop into B and T cells that mount immune responses against foreign substances (as we'll discuss in Concept 35.2). Normally, 1 μL of human blood contains about 5,000–10,000 leukocytes; their numbers increase temporarily whenever the body is fighting an infection. Unlike erythrocytes, leukocytes are also found outside the circulatory system, patrolling interstitial fluid and the lymphatic system.

**Platelets** Platelets are pinched-off cytoplasmic fragments of specialized bone marrow cells. They are about 2–3 μm in diameter and have no nuclei. Platelets serve both structural and molecular functions in blood clotting.

## Blood Clotting

When blood vessels are broken by an injury such as a cut or scrape, a chain of events ensues that quickly seals the break, halting blood loss and exposure to infection. The key mechanical event in this response is coagulation, the conversion of liquid components of blood to a solid—a blood clot.

In the absence of injury, the coagulant, or sealant, circulates in an inactive form called fibrinogen. Blood clotting begins when injury causes blood to contact the proteins in a broken blood vessel wall. The exposed proteins attract platelets, which

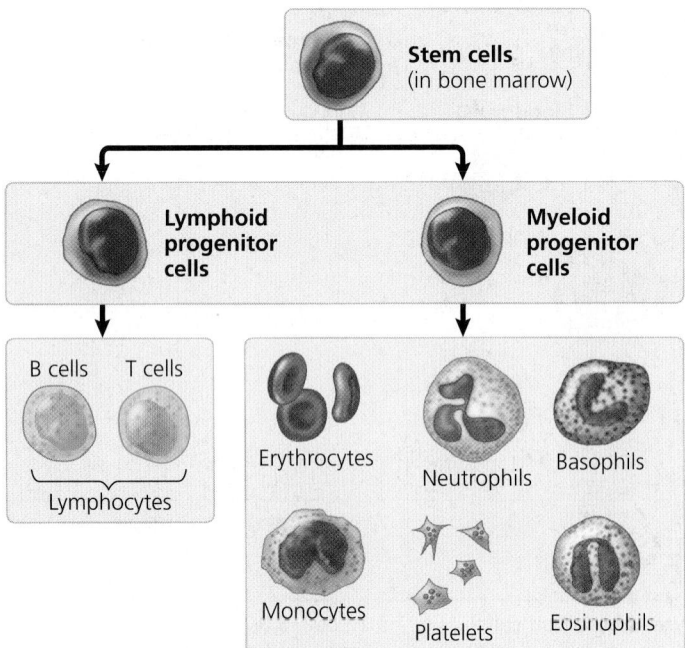

▲ **Figure 34.14 Differentiation of blood cells.** Cell divisions of multipotent stem cells in bone marrow give rise to two specialized sets of cells. The lymphoid progenitor cells in turn give rise to immune cells called lymphocytes, primarily B and T cells. The myeloid progenitor cells give rise to other immune cells, red blood cells (erythrocytes), and cell fragments called platelets.

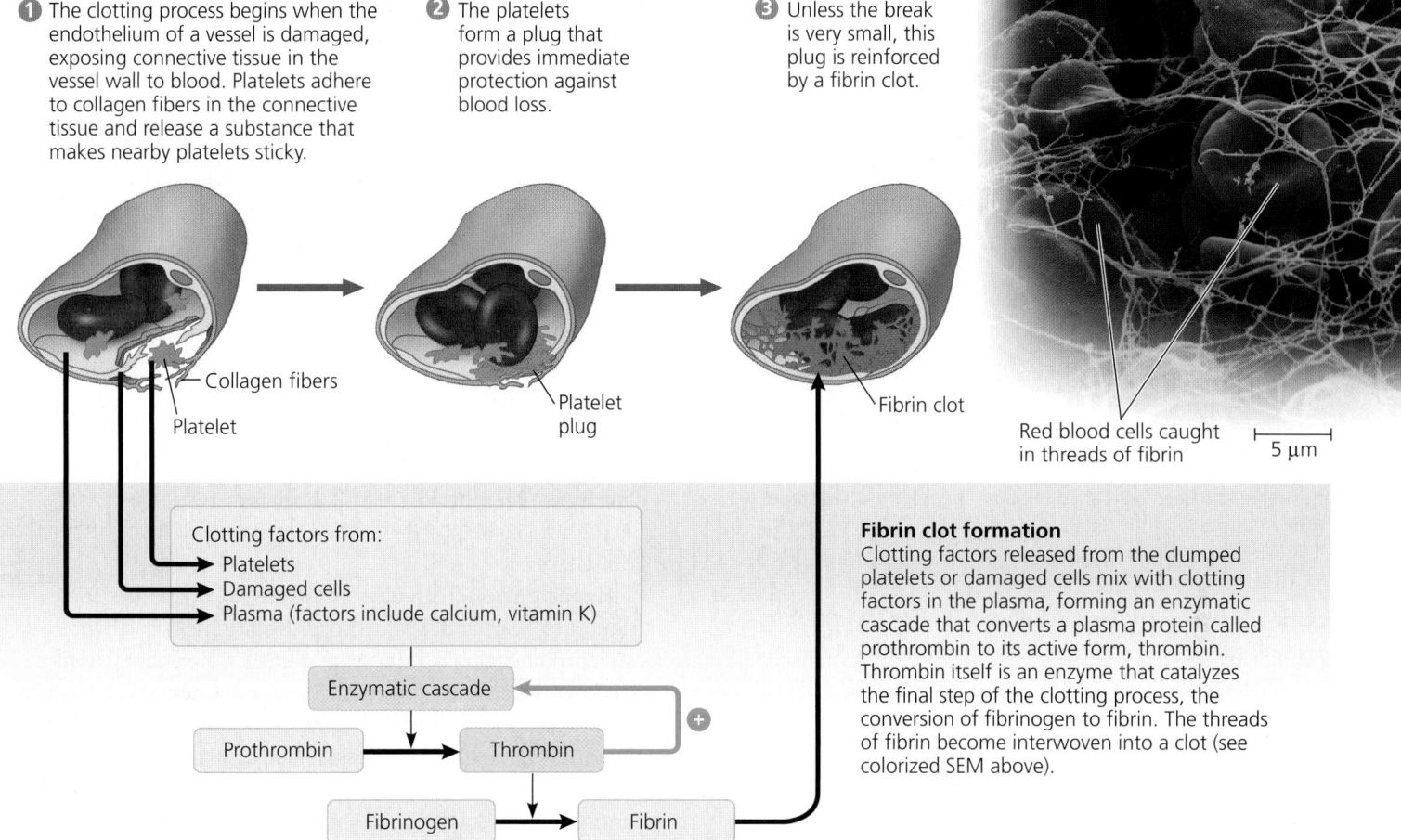

① The clotting process begins when the endothelium of a vessel is damaged, exposing connective tissue in the vessel wall to blood. Platelets adhere to collagen fibers in the connective tissue and release a substance that makes nearby platelets sticky.

② The platelets form a plug that provides immediate protection against blood loss.

③ Unless the break is very small, this plug is reinforced by a fibrin clot.

Collagen fibers

Platelet

Platelet plug

Fibrin clot

Red blood cells caught in threads of fibrin

5 μm

Clotting factors from:
Platelets
Damaged cells
Plasma (factors include calcium, vitamin K)

Enzymatic cascade

Prothrombin → Thrombin ⊕

Fibrinogen → Fibrin

**Fibrin clot formation**
Clotting factors released from the clumped platelets or damaged cells mix with clotting factors in the plasma, forming an enzymatic cascade that converts a plasma protein called prothrombin to its active form, thrombin. Thrombin itself is an enzyme that catalyzes the final step of the clotting process, the conversion of fibrinogen to fibrin. The threads of fibrin become interwoven into a clot (see colorized SEM above).

▲ **Figure 34.15 Blood clotting.**

gather at the site of injury and release clotting factors. These clotting factors trigger a cascade of reactions leading to the formation of an active enzyme, *thrombin*, from an inactive form, prothrombin **(Figure 34.15)**. Thrombin in turn converts fibrinogen to fibrin, which aggregates into threads that form the framework of the clot. Any mutation that disrupts these events can cause hemophilia, a disease characterized by excessive bleeding and bruising from even minor cuts and bumps.

As shown in Figure 34.15, clotting involves a positive feedback loop (see Concept 32.2). Initially, the clotting reactions convert only some of the prothrombin at the clot site to thrombin. However, thrombin itself stimulates the enzymatic cascade, leading to more conversion of prothrombin to thrombin and thus driving clotting to completion.

Anticlotting factors in the blood normally prevent spontaneous clotting in the absence of injury. Sometimes, however, clots form within a blood vessel, blocking the flow of blood. Such a clot is called a **thrombus**. We'll explore how a thrombus forms and the danger that it poses shortly.

## Cardiovascular Disease

Each year, cardiovascular diseases—disorders of the heart and blood vessels—kill more than 750,000 people in the United States. These diseases range from minor disturbances of vein

or heart valve function to a life-threatening disruptions of blood flow to the heart or brain.

### *Atherosclerosis, Heart Attacks, and Stroke*

Healthy arteries have a smooth inner lining that reduces resistance to blood flow. However, damage or infection can roughen the lining and lead to **atherosclerosis**, the hardening of the arteries by accumulation of fatty deposits. A key player in the development of atherosclerosis is cholesterol, a steroid that is important for maintaining normal membrane fluidity in animal cells (see Figure 5.5). Cholesterol travels in plasma in particles that consist of thousands of cholesterol molecules and other lipids bound to a protein. One type of particle— **low-density lipoprotein (LDL)**—delivers cholesterol to cells for membrane production. Another type—**high-density lipoprotein (HDL)**—scavenges excess cholesterol for return to the liver. Individuals with a high ratio of LDL to HDL are at substantially increased risk for atherosclerosis.

In atherosclerosis, damage to the arterial lining results in *inflammation*, the body's reaction to injury. Leukocytes are attracted to the inflamed area and begin to take up lipids, including cholesterol. A fatty deposit, called a plaque, grows steadily, incorporating fibrous connective tissue and additional cholesterol. As the plaque grows, the walls of the artery

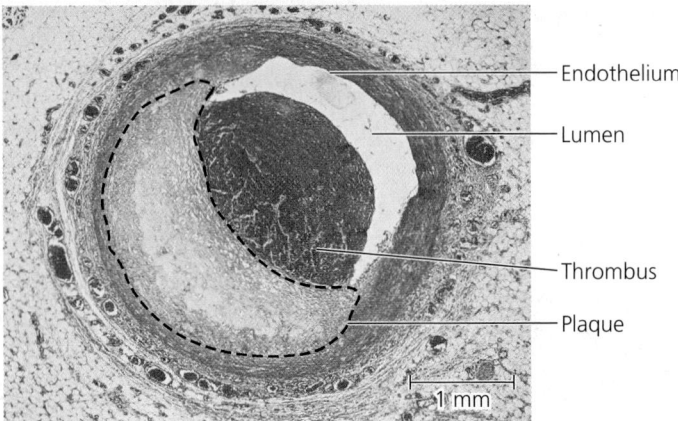

Endothelium

Lumen

Thrombus

Plaque

1 mm

▲ **Figure 34.16 Atherosclerosis.** In atherosclerosis, thickening of an arterial wall by plaque formation can restrict blood flow through the artery. Fragments of ruptured plaque can travel via the bloodstream and become lodged in other arteries. If the blockage is in an artery that supplies the heart or brain, the result could be a heart attack or stroke, respectively.

become thick and stiff, and the obstruction of the artery increases. If the plaque ruptures, a thrombus can form in the artery **(Figure 34.16)**, potentially triggering a heart attack or a stroke.

A **heart attack**, also called a *myocardial infarction*, is the damage or death of cardiac muscle tissue resulting from blockage of one or more coronary arteries, which supply oxygen-rich blood to the heart muscle. The coronary arteries are small in diameter and therefore especially vulnerable to obstruction by atherosclerotic plaques or thrombi. Such blockage can destroy cardiac muscle quickly because the constantly beating heart muscle requires a steady supply of $O_2$. If a large enough portion of the heart is affected, the heart will stop beating. Such cardiac arrest causes death if a heartbeat is not restored within a few minutes by cardiopulmonary resuscitation (CPR) or some other emergency procedure.

A **stroke** is the death of nervous tissue in the brain due to a lack of $O_2$. Strokes usually result from rupture or blockage of arteries in the neck or head. The effects of a stroke and the individual's chance of survival depend on the extent and location of the damaged brain tissue. If a stroke results from an arterial blockage by a thrombus, rapid administration of a clot-dissolving drug may help limit the damage.

Although atherosclerosis often isn't detected until critical blood flow is disrupted, there can be warning signs. Partial blockage of the coronary arteries may cause occasional chest pain, a condition known as angina pectoris, or more commonly angina. The pain is most likely to be felt when the heart is laboring under stress, and it signals that part of the heart is not receiving enough $O_2$. An obstructed coronary artery may be treated surgically, either by inserting a metal mesh tube called a stent to expand the artery or by transplanting a healthy blood vessel from the chest or a limb to bypass the blockage.

### Risk Factors and Treatment of Cardiovascular Disease

Although the tendency to develop particular cardiovascular diseases is inherited, it is also strongly influenced by lifestyle. For example, exercise decreases the LDL/HDL ratio, reducing the risk of cardiovascular disease. In contrast, consuming certain processed vegetable oils called *trans fats* and smoking increase LDL/HDL ratio. For many individuals at high risk, treatment with drugs called statins can lower LDL levels and thereby reduce the likelihood of heart attacks. In the **Scientific Skills Exercise**, you can interpret the effect of a genetic mutation on blood LDL levels.

The recognition that inflammation plays a central role in atherosclerosis and thrombus formation is also influencing the treatment of cardiovascular disease. For example, aspirin, which inhibits the inflammatory response, has been found to help prevent the recurrence of heart attacks and stroke.

**Hypertension** (high blood pressure) is yet another contributor to heart attack and stroke. According to one hypothesis, chronic high blood pressure damages the endothelium that lines the arteries, promoting plaque formation. The usual definition of hypertension in adults is a systolic pressure above 140 mm Hg or a diastolic pressure above 90 mm Hg. Fortunately, hypertension is simple to diagnose and can usually be controlled by dietary changes, exercise, medication, or a combination of these approaches.

#### CONCEPT CHECK 34.4

1. Explain why a physician might order a white blood cell count for a patient with symptoms of an infection.
2. Clots in arteries can cause heart attacks and strokes. Why, then, does it make sense to treat people with hemophilia by introducing clotting factors into their blood?
3. **WHAT IF?** Nitroglycerin (the key ingredient in dynamite) is sometimes prescribed for heart disease patients. Within the body, nitroglycerin is converted to nitric oxide (see Concept 34.3). Why would you expect nitroglycerin to increase blood flow to the heart and thus relieve chest pain?
4. **MAKE CONNECTIONS** How do stem cells from the bone marrow of an adult differ from embryonic stem cells (see Concept 16.2)?

For suggested answers, see Appendix A.

---

**CONCEPT 34.5**

# Gas exchange occurs across specialized respiratory surfaces

In the remainder of this chapter, we will focus on the process of **gas exchange**. Although this process is often called respiratory exchange or respiration, it should not be confused with the energy transformations of cellular respiration. Gas

# Interpreting Data in Histograms

**Does Inactivating the PCSK9 Enzyme Lower LDL Levels in Humans?** Researchers interested in genetic factors affecting susceptibility to cardiovascular disease examined the DNA of 15,000 individuals. They found that 3% of the individuals had a mutation that inactivates one copy of the gene for PCSK9, a liver enzyme. Because mutations that *increase* the activity of PCSK9 are known to *increase* levels of LDL cholesterol in the blood, the researchers hypothesized that *inactivating* mutations in this gene would *lower* LDL levels. In this exercise, you will interpret the results of an experiment they carried out to test this hypothesis.

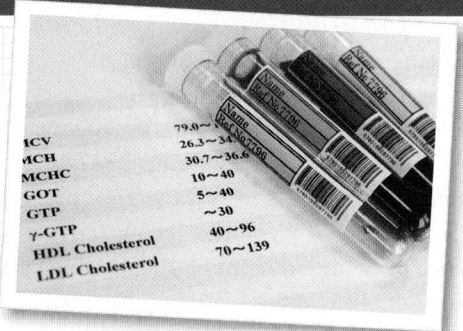

**How the Experiment Was Done** The researchers measured LDL cholesterol levels in blood plasma from 85 individuals with one copy of the *PCSK9* gene inactivated (the study group) and from 3,278 individuals with two functional copies of the gene (the control group).

**Data from the Experiment**

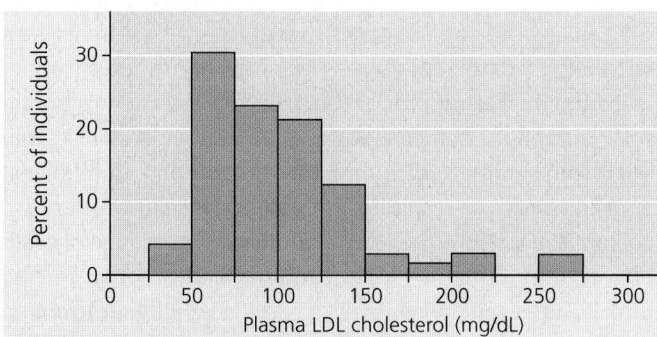

**Individuals with an inactivating mutation in one copy of *PCSK9* gene (study group)**

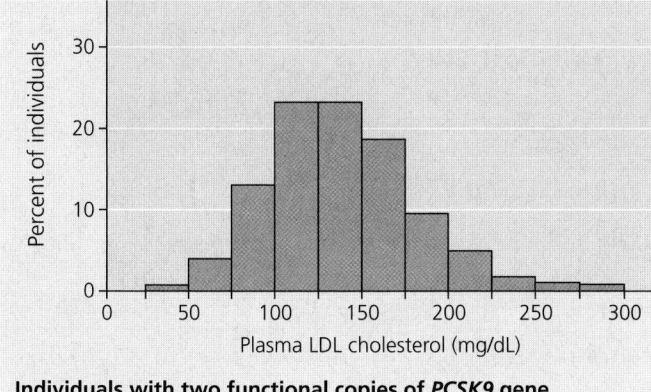

**Individuals with two functional copies of *PCSK9* gene (control group)**

**Data from** J. C. Cohen et al., Sequence variations in PCSK9, low LDL, and protection against coronary heart disease, *New England Journal of Medicine* 354:1264–1272 (2006).

**INTERPRET THE DATA**

1. The results are presented using a variant of a bar graph called a *histogram*. In a histogram, the variable on the x-axis is grouped into ranges. The height of each bar in this histogram reflects the percentage of samples that fall into the range specified on the x-axis for that bar. For example, in the top histogram, about 4% of individuals studied had plasma LDL cholesterol levels in the 25–50 mg/dL (milligrams per deciliter) range. Add the percentages for the relevant bars to calculate the percentage of individuals in the study and control groups that had an LDL level of 100 mg/dL or less. (For additional information about histograms, see the Scientific Skills Review in Appendix F and in the Study Area in MasteringBiology.)

2. Comparing the two histograms, do you find support for the researchers' hypothesis? Explain.

3. What if instead of graphing the data the researchers had compared the range of concentrations for plasma LDL cholesterol (low to high) in the control and study groups? How would their conclusions have differed?

4. What does the fact that the two histograms overlap as much as they do indicate about the extent to which PCSK9 determines plasma LDL cholesterol levels?

5. Comparing these two histograms allowed researchers to draw a conclusion regarding the effect of PCSK9 mutations on LDL cholesterol levels in blood. Consider two individuals with a plasma LDL level cholesterol of 160 mg/dL, one from the study group and one from the control group. What do you predict regarding their relative risk of developing cardiovascular disease? Explain how you arrived at your prediction. What role did the histograms play in making your prediction?

**MB** A version of this Scientific Skills Exercise can be assigned in MasteringBiology.

---

exchange is the uptake of molecular $O_2$ from the environment and the discharge of $CO_2$ to the environment. Exchange occurs across a membrane at specialized respiratory surfaces and relies on partial pressure gradients that drive net diffusion across the membrane. The respiratory medium—the source of oxygen in the environment—is either water or air.

## Partial Pressure Gradients in Gas Exchange

To understand the driving forces for gas exchange, we must consider **partial pressure**, which is the pressure exerted by a particular gas in a mixture of gases. Determining partial pressures enables us to predict the net movement of a gas at an exchange surface: A gas always undergoes net diffusion from a region of higher partial pressure to a region of lower partial pressure. To calculate partial pressures, we need to know the pressure that a gas mixture exerts and the fraction of the mixture represented by a particular gas. Let's consider $O_2$ as an example. At sea level, the atmosphere exerts a downward force equal to that of a column of mercury (Hg) 760 mm high. Atmospheric pressure at sea level is thus 760 mm Hg. Since the atmosphere is 21% $O_2$ by volume, the partial pressure of $O_2$ is $0.21 \times 760$, or about 160 mm Hg. This value is called the *partial pressure* of $O_2$ (abbreviated $P_{O_2}$) because it is the part of atmospheric pressure contributed by $O_2$. The partial pressure of $CO_2$ (abbreviated $P_{CO_2}$) is much, much less—only 0.29 mm Hg at sea level.

Partial pressures also apply to gases dissolved in a liquid, such as water. When water is exposed to air, an equilibrium state is reached in which the partial pressure of each gas in the water equals the partial pressure of that gas in the air. Thus, water exposed to air at sea level has a $P_{O_2}$ of 160 mm Hg, the same as in the atmosphere. However, the *concentrations* of $O_2$ in the air and water differ substantially because $O_2$ is much less soluble in water than in air. Furthermore, the warmer and saltier the water is, the less dissolved $O_2$ it can hold.

## Respiratory Media

The conditions for gas exchange vary considerably, depending on whether the respiratory medium—the source of $O_2$—is air or water. As already noted, $O_2$ is plentiful in air, making up about 21% of Earth's atmosphere by volume. Compared to water, air is much less dense and less viscous, so it is easier to move and to force through small passageways. As a result, breathing air is relatively easy and need not be particularly efficient. Humans, for example, extract only about 25% of the $O_2$ in inhaled air.

Water is a much more demanding gas exchange medium than air. The amount of $O_2$ dissolved in a given volume of water varies but is always less than in an equivalent volume of air: Water in many marine and freshwater habitats contains only about 7 mL of dissolved $O_2$ per liter, a concentration roughly 30 times less than in air. Water's lower $O_2$ content, greater density, and greater viscosity mean that aquatic animals such as fishes and lobsters must expend considerable energy to carry out gas exchange. In the context of these challenges, adaptations have evolved that enable most aquatic animals to be very efficient in gas exchange. Many of these adaptations involve the organization of the surfaces dedicated to exchange, as illustrated in a marine worm and a sea star **(Figure 34.17)**.

## Respiratory Surfaces

Specialization for gas exchange is apparent in the structure of the respiratory surface, the part of an animal's body where gas exchange occurs. Like all living cells, the cells that carry out gas exchange have a plasma membrane that must be in contact with an aqueous solution. Respiratory surfaces are therefore always moist.

The movement of $O_2$ and $CO_2$ across respiratory surfaces takes place by diffusion. The rate of diffusion is proportional to the surface area across which it occurs and inversely proportional to the square of the distance through which molecules must move. In other words, gas exchange is fast when the area for diffusion is large and the path for diffusion is short. As a result, respiratory surfaces tend to be large and thin.

In some relatively simple animals, such as sponges, cnidarians, and flatworms, every cell in the body is close enough to the external environment that gases can diffuse quickly between any cell and the environment. In many animals, however, the bulk of the body's cells lack immediate access to the environment. The respiratory surface in these animals is a thin, moist epithelium that constitutes a respiratory organ.

For some animals, including earthworms and amphibians, the skin serves as a respiratory organ. A dense network of capillaries just below the skin facilitates the exchange of gases between the circulatory system and the environment. For most animals, however, the general body surface lacks sufficient area

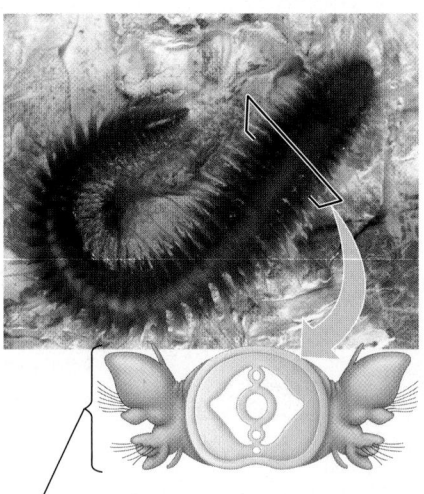

Parapodium (functions as gill)

**(a) Marine worm.** Many polychaetes (marine worms of the phylum Annelida) have a pair of flattened appendages called parapodia (singular, *parapodium*) on each body segment. The parapodia serve as gills and also function in crawling and swimming.

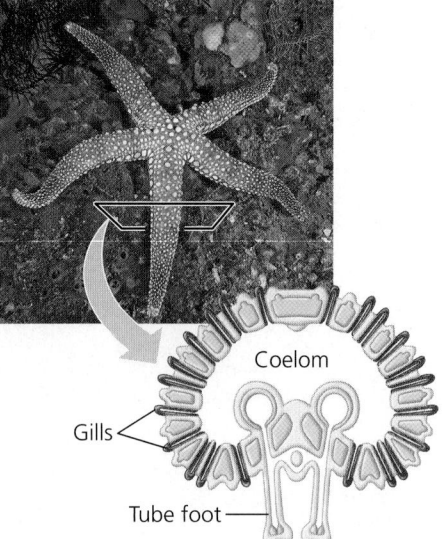

Coelom

Gills

Tube foot

**(b) Sea star.** The gills of a sea star are simple tubular projections of the skin. The hollow core of each gill is an extension of the coelom (body cavity). Gas exchange occurs by diffusion across the gill surfaces, and fluid in the coelom circulates in and out of the gills, aiding gas transport. The surfaces of a sea star's tube feet also function in gas exchange.

▲ **Figure 34.17 Diversity in the structure of gills, external body surfaces that function in gas exchange.**

to exchange gases for the whole organism. The evolutionary solution to this limitation is a respiratory organ that is extensively folded or branched, thereby enlarging the available surface area for gas exchange. Gills, tracheae, and lungs are three such organs.

## Gills in Aquatic Animals

Gills are outfoldings of the body surface that are suspended in the water. As illustrated in Figures 34.1 and 34.17, the distribution of gills over the body can vary considerably. Regardless of their distribution, gills often have a total surface area much greater than that of the rest of the body's exterior.

Movement of the respiratory medium over the respiratory surface, a process called ventilation, maintains the partial pressure gradients of $O_2$ and $CO_2$ across the gill that are necessary for gas exchange. To promote ventilation, most gill-bearing animals either move their gills through the water or move water over their gills. For example, crayfish and lobsters have paddle-like appendages that drive a current of water over the gills, whereas mussels and clams move water with cilia. Octopuses and squids ventilate their gills by taking in and ejecting water, with the significant side benefit of getting about by jet propulsion. Fishes use the motion of swimming or coordinated movements of the mouth and gill covers to ventilate their gills. In both cases, a current of water enters the mouth of the fish, passes through slits in the pharynx, flows over the gills, and then exits the body **(Figure 34.18)**.

In fishes, the efficiency of gas exchange is maximized by **countercurrent exchange**, the exchange of a substance or heat between two fluids flowing in opposite directions. In a fish gill, the two fluids are blood and water. Because blood flows in the direction opposite to that of water passing over the gills, at each point in its travel blood is less saturated with $O_2$ than the water it meets (see Figure 34.18). As blood enters a gill capillary, it encounters water that is completing its passage through the gill. Depleted of much of its dissolved $O_2$, this water nevertheless has a higher $P_{O_2}$ than the incoming blood, and $O_2$ transfer takes place. As the blood continues its passage, its $P_{O_2}$ steadily increases, but so does that of the water it encounters, since each successive position in the blood's travel corresponds to an earlier position in the water's passage over the gills. The result is a partial pressure gradient that favors the diffusion of $O_2$ from water to blood along the entire length of the capillary.

Countercurrent exchange mechanisms are remarkably efficient. In the fish gill, more than 80% of the $O_2$ dissolved in the water is removed as the water passes over the respiratory surface. In other settings, countercurrent mechanisms contribute to temperature regulation and to the functioning of the mammalian kidney (see Concepts 32.1 and 32.4).

## Tracheal Systems in Insects

In most terrestrial animals, respiratory surfaces are enclosed within the body, exposed to the atmosphere only through

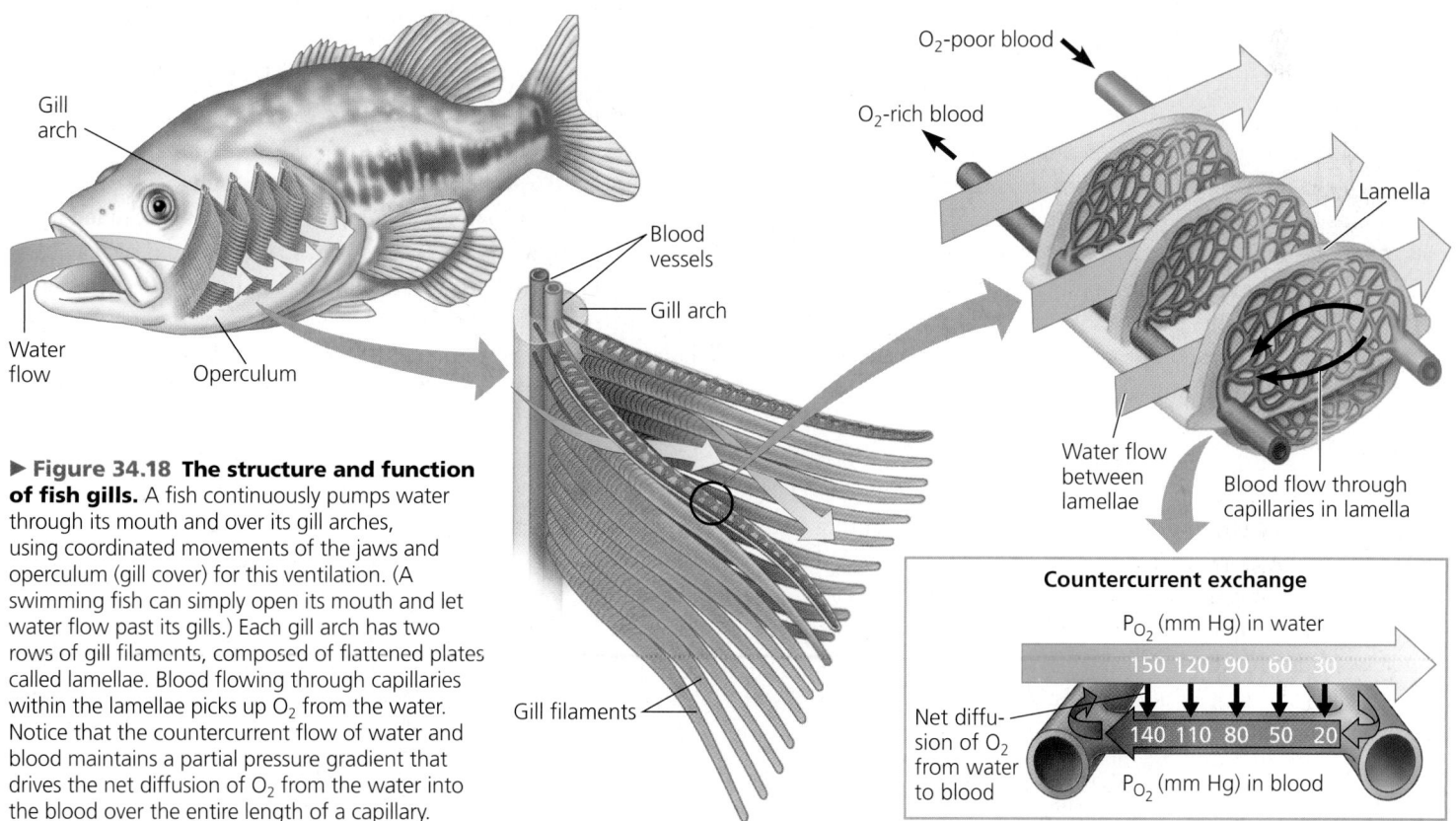

▶ **Figure 34.18 The structure and function of fish gills.** A fish continuously pumps water through its mouth and over its gill arches, using coordinated movements of the jaws and operculum (gill cover) for this ventilation. (A swimming fish can simply open its mouth and let water flow past its gills.) Each gill arch has two rows of gill filaments, composed of flattened plates called lamellae. Blood flowing through capillaries within the lamellae picks up $O_2$ from the water. Notice that the countercurrent flow of water and blood maintains a partial pressure gradient that drives the net diffusion of $O_2$ from the water into the blood over the entire length of a capillary.

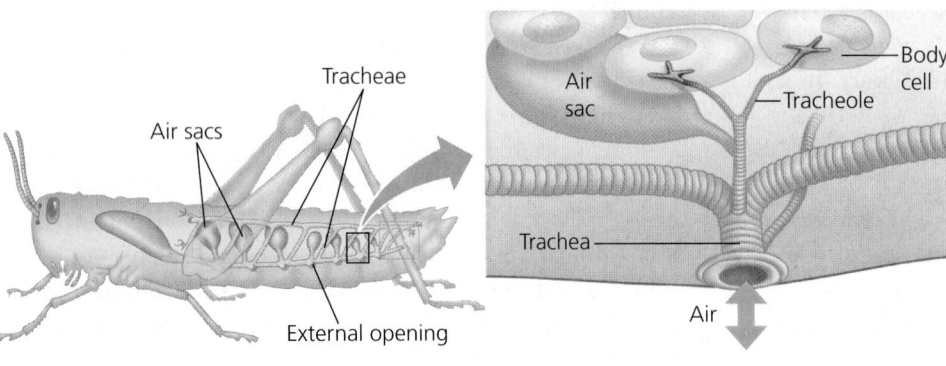

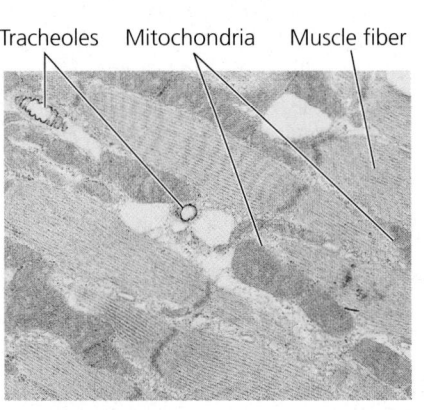

**(a)** The respiratory system of an insect consists of branched internal tubes. The largest tubes, called tracheae, connect to external openings spaced along the insect's body surface. Air sacs formed from enlarged portions of the tracheae are found near organs that require a large supply of oxygen.

**(b)** Rings of chitin keep the tracheae open, allowing air to enter and pass into smaller tubes called tracheoles. The branched tracheoles deliver air directly to cells throughout the body. Tracheoles have closed ends filled with fluid (blue-gray). When the animal is active and using more $O_2$, most of the fluid is withdrawn into the body. This increases the surface area of air-filled tracheoles in contact with cells.

**(c)** The TEM above shows cross sections of tracheoles in a tiny piece of insect flight muscle. Each of the numerous mitochondria in the muscle cells lies within about 5 µm of a tracheole.

▲ **Figure 34.19 The insect tracheal system.**

narrow tubes. Although the most familiar example of such an arrangement is the lung, the most common is the insect **tracheal system**, a network of air tubes that branch throughout the body **(Figure 34.19a)**. The largest tubes, called tracheae, open to the outside **(Figure 34.19b)**. At the tips of the finest branches, a moist epithelial lining enables gas exchange by diffusion. Because the tracheal system brings air within a very short distance of virtually every body cell in an insect, the efficient transport of $O_2$ and $CO_2$ does not require participation of the animal's open circulatory system.

Tracheal systems often exhibit adaptations directly related to bioenergetics. Consider, for example, a flying insect, which consumes 10 to 200 times more $O_2$ when in flight than it does at rest. In many flying insects, cycles of flight muscle contraction and relaxation pump air rapidly through the tracheal system. This pumping improves ventilation, bringing ample $O_2$ to the densely packed mitochondria that support the high metabolic rate of flight muscle **(Figure 34.19c)**.

## Lungs

Unlike tracheal systems, which branch throughout the insect body, **lungs** are localized respiratory organs. Representing an infolding of the body surface, they are typically subdivided into numerous pockets. Because the respiratory surface of a lung is not in direct contact with all other parts of the body, the gap must be bridged by the circulatory system, which transports gases between the lungs and the rest of the body. Lungs have evolved in organisms with open circulatory systems, such as spiders and land snails, as well as in vertebrates.

Among vertebrates that lack gills, the use of lungs for gas exchange varies. Amphibians rely heavily on diffusion across

external body surfaces, such as skin, to carry out gas exchange; lungs, if present, are relatively small. In contrast, most reptiles (including all birds) and all mammals depend entirely on lungs for gas exchange. Lungs and air breathing have evolved in a few aquatic vertebrates as adaptations to living in oxygen-poor water or to spending part of their time exposed to air (for instance, when the water level in a pond recedes).

### *Mammalian Respiratory Systems:* A Closer Look

In mammals, branching ducts convey air to the lungs, which are located in the *thoracic cavity*, enclosed by the ribs and diaphragm **(Figure 34.20)**. Air enters through the nostrils and is filtered by hairs, warmed, humidified, and sampled for odors as it flows through a maze of spaces in the nasal cavity. The nasal cavity leads to the pharynx, an intersection where the paths for air and food cross. When food is swallowed, the **larynx** (the upper part of the respiratory tract) moves upward and tips a flap of cartilage over the opening of the **trachea**, or windpipe. This allows food to go down the esophagus to the stomach. The rest of the time, the airway is open.

From the larynx, air passes into the trachea. The cartilage that reinforces the walls of both the larynx and the trachea keeps this part of the airway open. Within the larynx of most mammals, the exhaled air rushes by a pair of elastic bands of muscle called vocal folds, or, in humans, vocal cords. Sounds are produced when muscles in the larynx are tensed, stretching the cords so that they vibrate. High-pitched sounds result from tightly stretched cords vibrating rapidly; low-pitched sounds come from looser cords vibrating slowly.

The trachea branches into two **bronchi** (singular, *bronchus*), one leading to each lung. Within the lung, the bronchi branch

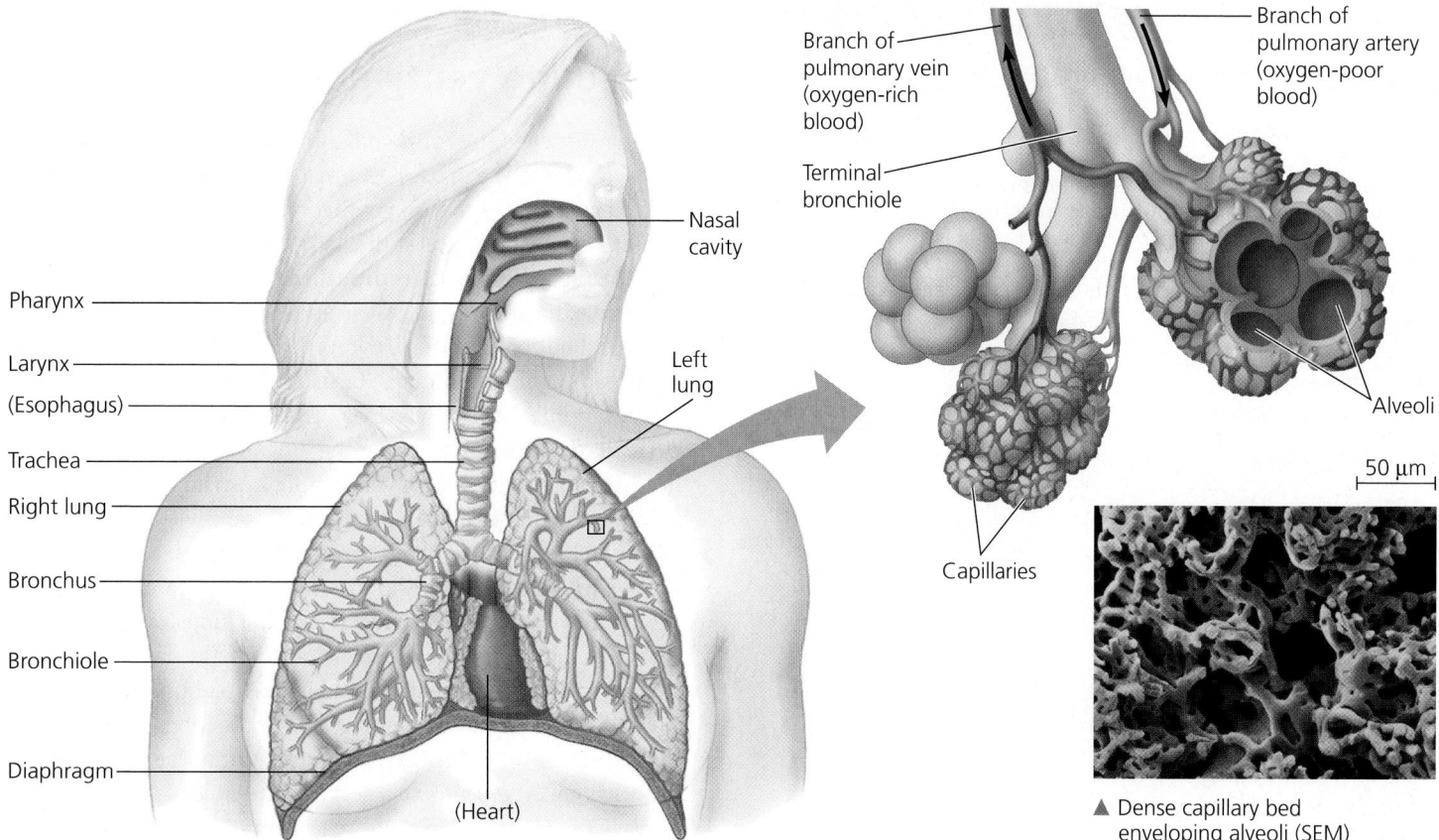

Branch of pulmonary vein (oxygen-rich blood)

Branch of pulmonary artery (oxygen-poor blood)

Terminal bronchiole

Nasal cavity

Pharynx

Larynx

(Esophagus)

Trachea

Right lung

Bronchus

Bronchiole

Diaphragm

(Heart)

Left lung

Alveoli

Capillaries

50 μm

▲ Dense capillary bed enveloping alveoli (SEM)

▲ **Figure 34.20 The mammalian respiratory system.** From the nasal cavity and pharynx, inhaled air passes through the larynx, trachea, and bronchi to the bronchioles, which end in microscopic alveoli lined by a thin, moist epithelium. Branches of the pulmonary arteries convey oxygen-poor blood to the alveoli; branches of the pulmonary veins transport oxygen-rich blood from the alveoli back to the heart.

repeatedly into finer and finer tubes called *bronchioles*. The entire system of air ducts has the appearance of an inverted tree, the trunk being the trachea. The epithelium lining the major branches of this respiratory tree is covered by cilia and a thin film of mucus. The mucus traps dust, pollen, and other particulate contaminants, and the beating cilia move the mucus upward to the pharynx, where it can be swallowed into the esophagus. This process, sometimes referred to as the "mucus escalator," plays a crucial role in cleansing the respiratory system.

Gas exchange in mammals occurs in **alveoli** (singular, *alveolus*; see Figure 34.20), air sacs clustered at the tips of the tiniest bronchioles. Human lungs contain millions of alveoli, which together have a surface area of about 100 m², about 50 times that of the skin. Oxygen in the air entering the alveoli dissolves in the moist film lining their inner surfaces and rapidly diffuses across the epithelium into a web of capillaries that surrounds each alveolus. Net diffusion of carbon dioxide occurs in the opposite direction, from the capillaries across the epithelium of the alveolus and into the air space.

Lacking cilia or significant air currents to remove particles from their surface, alveoli are highly susceptible to contamination. White blood cells patrol the alveoli, engulfing foreign particles. However, if too much particulate matter reaches the alveoli, the defenses can be overwhelmed, leading to inflammation and irreversible damage. For example, particulates from cigarette smoke that enter alveoli can cause a permanent reduction in lung capacity. For coal miners, inhalation of large amounts of coal dust can lead to silicosis, a disabling, irreversible, and sometimes fatal lung disease.

The film of liquid that lines alveoli is subject to surface tension, an attractive force that has the effect of minimizing a liquid's surface area (see Concept 2.5). Given their tiny diameter (about 0.25 mm), why don't alveoli collapse under high surface tension? It turns out that alveoli produce a mixture of phospholipids and proteins called **surfactant**, for *surface-active* agent, which coats the alveoli and reduces surface tension.

In the late 1950s, Mary Ellen Avery did the first experiment linking a lack of surfactant to *respiratory distress syndrome* (RDS), a disease common in early preterm infants—those born 6 weeks or more before their due dates **(Figure 34.21)**. (The average full-term human pregnancy is 38 weeks.) Later studies revealed that surfactant typically appears in the lungs after 33 weeks of development. In the 1950s, RDS killed 10,000 infants annually in the United States, but artificial surfactants are

## ▼ Figure 34.21 Inquiry

### What causes respiratory distress syndrome?

**Experiment** Mary Ellen Avery, a research fellow at Harvard University, hypothesized that a lack of surfactant caused respiratory distress syndrome (RDS) in preterm infants. To test this idea, she obtained autopsy samples of lungs from infants that had died of RDS or from other causes. She extracted material from the samples and let it form a film on water. Avery then measured the tension (in dynes per centimeter) across the water surface and recorded the lowest surface tension observed for each sample.

**Results** Avery noted a pattern when she grouped the samples based on the body mass of the infants: less than 1,200 g (2.7 lb) and 1,200 g or greater.

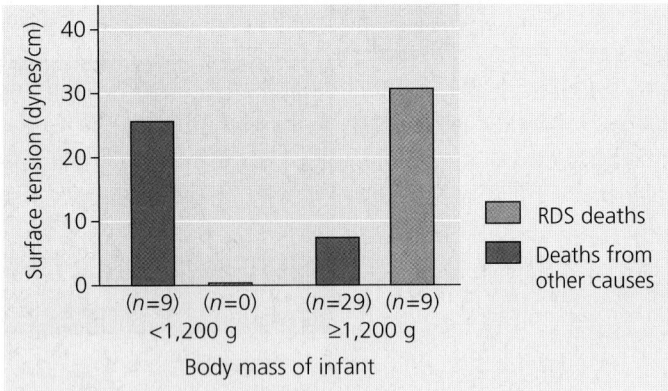

**Conclusion** For infants with a body mass of 1,200 g or greater, the material from those that had died of RDS exhibited much higher surface tension than the material from those that had died from other causes. Avery inferred that infants' lungs normally contain a surface tension–reducing substance (now called surfactant) and that a lack of this substance was a likely cause of RDS. The results from infants with a body mass less than 1,200 g were similar to those of infants who had died from RDS, suggesting that surfactant is not normally produced until a fetus reaches this size.

**Data from** M. E. Avery and J. Mead, Surface properties in relation to atelectasis and hyaline membrane disease, *American Journal of Diseases of Children* 97:517–523 (1959).

**WHAT IF?** If the researchers had measured the amount of surfactant in lung samples from the infants, what relationship would you expect between the amount of surfactant and infant body mass?

now used successfully to treat early preterm infants. Treated babies with a body mass over 900 g (2 pounds) at birth usually survive without long-term health problems. For her contributions, Avery received the National Medal of Science in 1991.

### CONCEPT CHECK 34.5

1. Why is an internal location for gas exchange tissues advantageous for terrestrial animals?
2. After a heavy rain, earthworms come to the surface. How would you explain this behavior in terms of an earthworm's requirements for gas exchange?
3. **MAKE CONNECTIONS** Describe similarities in the countercurrent exchange that facilitates respiration in fish and thermoregulation in geese (see Figure 32.15).

For suggested answers, see Appendix A.

## CONCEPT 34.6

# Breathing ventilates the lungs

Having surveyed the route that air follows when we breathe, we turn now to the process of breathing itself. Like fishes, terrestrial vertebrates rely on ventilation to maintain high $O_2$ and low $CO_2$ concentrations at the gas exchange surface. The process that ventilates lungs is **breathing**, the alternating inhalation and exhalation of air. A variety of mechanisms for moving air in and out of lungs have evolved, as we will see by considering breathing in amphibians, birds, and mammals.

An amphibian such as a frog ventilates its lungs by **positive pressure breathing**, inflating the lungs with forced airflow. During each cycle of ventilation, fresh air is first drawn through the nostrils into a specialized oral cavity. Next, stale air in the lungs is forced out through the mouth and nostrils. Finally, with the nostrils and mouth closed, the floor of the oral cavity moves upward, forcing air into the lungs.

When a bird breathes, it passes air over the gas exchange surface in only one direction, making ventilation highly efficient. Air sacs situated on either side of the lungs act as bellows that direct air flow through the lungs. Within the lungs, tiny channels called *parabronchi* serve as the sites of gas exchange. Passage of air through the entire system—air sacs and lungs—requires two cycles of inhalation and exhalation.

## How a Mammal Breathes

To understand how a mammal breathes, think about filling a syringe. By pulling back on the plunger, you lower the pressure in the syringe, drawing gas or fluid through the needle into the syringe chamber. Similarly, mammals employ **negative pressure breathing**—pulling, rather than pushing, air into their lungs **(Figure 34.22)**. Using muscle contraction to actively expand the thoracic cavity, mammals lower air pressure in their lungs below that of the air outside their body. Because

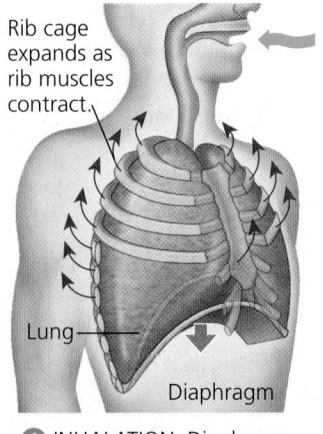
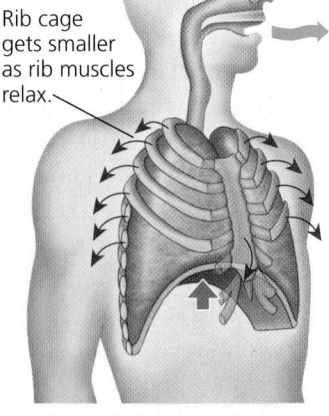

① INHALATION: Diaphragm contracts (moves down).

② EXHALATION: Diaphragm relaxes (moves up).

▲ **Figure 34.22 Negative pressure breathing.** A mammal breathes by changing the air pressure within its lungs relative to the pressure of the outside atmosphere.

gas flows from a region of higher pressure to a region of lower pressure, air rushes through the nostrils and mouth and down the breathing tubes to the alveoli.

Expanding the thoracic cavity during inhalation involves the animal's rib muscles and the **diaphragm**, a sheet of skeletal muscle that forms the bottom wall of the cavity. Contracting one set of rib muscles expands the rib cage, the front wall of the thoracic cavity, by pulling the ribs upward and the sternum outward. At the same time, the diaphragm contracts, expanding the thoracic cavity downward. It is this descending diaphragm that is analogous to a plunger being drawn out of a syringe.

Whereas inhalation is always active and requires work, exhalation is usually passive. During exhalation, the muscles controlling the thoracic cavity relax, and the volume of the cavity is reduced. The increased air pressure in the alveoli forces air up the breathing tubes and out of the body.

Within the thoracic cavity, a double membrane surrounds the lungs. The inner layer of this membrane adheres to the outside of the lungs, and the outer layer adheres to the wall of the thoracic cavity. A thin space filled with fluid separates the two layers. Surface tension in the fluid causes the two layers to stick together like two plates of glass separated by a film of water: The layers can slide smoothly past each other, but they cannot be pulled apart easily. Consequently, the volume of the thoracic cavity and the volume of the lungs change in unison.

The volume of air inhaled and exhaled with each breath is called **tidal volume**. It averages about 500 mL in resting adult humans. The tidal volume during maximal inhalation and exhalation is the **vital capacity**, which is about 3.4 L and 4.8 L for college-age women and men, respectively. The air that remains after a forced exhalation is called the **residual volume**. With age, the lungs lose their resilience, and residual volume increases at the expense of vital capacity.

Because the lungs in mammals do not completely empty with each breath, and because inhalation occurs through the same airways as exhalation, each inhalation mixes fresh air with oxygen-depleted residual air. As a result, the maximum $P_{O_2}$ in alveoli is always considerably less than in the atmosphere. This is one reason mammals function less well than birds at high altitude. For example, humans have great difficulty obtaining enough $O_2$ when climbing at high elevations, such as those in the Himalayas. However, bar-headed geese and several other bird species easily fly through high Himalayan passes during their migrations.

## Control of Breathing in Humans

Although you can voluntarily hold your breath or breathe faster and deeper, most of the time your breathing is regulated by involuntary mechanisms. These control mechanisms ensure that gas exchange is coordinated with blood circulation and with metabolic demand.

The neurons mainly responsible for regulating breathing are in the medulla oblongata, near the base of the brain **(Figure 34.23)**. Neural circuits in the medulla form a pair of breathing control centers that establish the breathing rhythm. When you breathe deeply, a negative-feedback mechanism prevents the lungs from overexpanding: During inhalation, sensors that detect stretching of the lung tissue send nerve impulses to the control circuits in the medulla, inhibiting further inhalation.

In regulating breathing, the medulla uses the pH of the surrounding tissue fluid as an indicator of blood $CO_2$ concentration. The reason pH can be used in this way is that blood $CO_2$ is the main determinant of the pH of cerebrospinal fluid, the fluid surrounding the brain and spinal cord. Carbon dioxide diffuses from the blood to the cerebrospinal fluid, where it reacts with water and forms carbonic acid ($H_2CO_3$). The $H_2CO_3$ can then dissociate into a bicarbonate ion ($HCO_3^-$) and a hydrogen ion ($H^+$):

$$CO_2 + H_2O \rightleftharpoons H_2CO_3 \rightleftharpoons HCO_3^- + H^+$$

Consider what happens if metabolic activity increases, such as during exercise. Increased metabolism raises the concentration of $CO_2$ in the blood and cerebrospinal fluid. Through the reactions shown above, the higher $CO_2$ concentration leads to an increase in the concentration of $H^+$, lowering pH. Sensors in the medulla as well as in major blood vessels detect this pH change. In response, the medulla's control circuits increase the depth and rate of breathing (see Figure 34.23). Both remain high until the excess $CO_2$ is eliminated in exhaled air and pH returns to a normal value.

The blood $O_2$ level usually has little effect on the breathing control centers. However, when the $O_2$ level drops very low (at high altitudes, for instance), $O_2$ sensors in the aorta and the

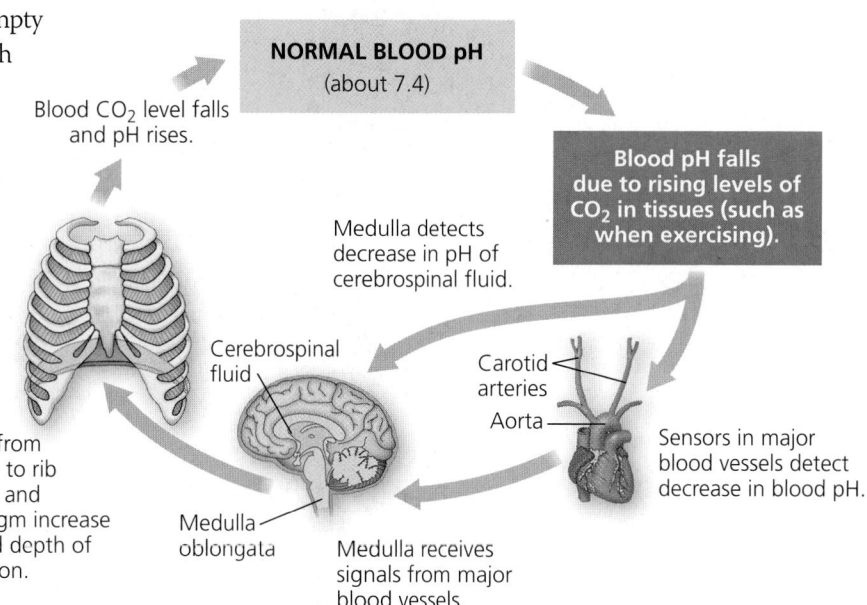

**NORMAL BLOOD pH**
(about 7.4)

Blood $CO_2$ level falls and pH rises.

Blood pH falls due to rising levels of $CO_2$ in tissues (such as when exercising).

Medulla detects decrease in pH of cerebrospinal fluid.

Cerebrospinal fluid

Carotid arteries

Aorta

Sensors in major blood vessels detect decrease in blood pH.

Signals from medulla to rib muscles and diaphragm increase rate and depth of ventilation.

Medulla oblongata

Medulla receives signals from major blood vessels.

▲ **Figure 34.23 Homeostatic control of breathing.**

**WHAT IF?** *Suppose a person began breathing very rapidly while resting. Describe the effect on blood $CO_2$ levels and the steps by which the negative-feedback circuit would restore homeostasis.*

carotid arteries in the neck send signals to the breathing control centers, which respond by increasing the breathing rate.

Breathing control is effective only if ventilation is matched to blood flow through alveolar capillaries. During exercise, for instance, such coordination couples an increased breathing rate, which enhances $O_2$ uptake and $CO_2$ removal, with an increase in cardiac output.

### CONCEPT CHECK 34.6

1. How does an increase in the $CO_2$ concentration in the blood affect the pH of cerebrospinal fluid?
2. A drop in blood pH causes an increase in heart rate. What is the function of this control mechanism?
3. **WHAT IF?** If an injury tore a small hole in the membranes surrounding your lungs, what effect on lung function would you expect?

   **For suggested answers, see Appendix A.**

---

### CONCEPT 34.7

# Adaptations for gas exchange include pigments that bind and transport gases

The high metabolic demands of many animals necessitate the exchange of large quantities of $O_2$ and $CO_2$. Here we'll examine how blood molecules called respiratory pigments facilitate this exchange through their interaction with $O_2$ and $CO_2$. As a basis for exploring respiratory pigment function, we'll first summarize the basic gas exchange circuit in humans.

## Coordination of Circulation and Gas Exchange

To appreciate how the gas exchange and circulatory systems function together, let's track the variation in partial pressure for $O_2$ and shed $CO_2$ across these systems **(Figure 34.24)**. ❶ During inhalation, fresh air mixes with air remaining in the lungs. ❷ The resulting mixture formed in the alveoli has a lower $P_{O_2}$ than the blood flowing through the alveolar capillaries. Consequently, there is a net diffusion of $O_2$ down its partial pressure gradient from the air in the alveoli to the blood. Meanwhile, the presence of a $P_{CO_2}$ in the alveoli that is higher in the capillaries than in the air drives the net diffusion of $CO_2$ from blood to air. ❸ By the time the blood leaves the lungs in the pulmonary veins, its $P_{O_2}$ and $P_{CO_2}$ match the values for the alveolar air. After returning to the heart, this blood is pumped through the systemic circuit.

❹ In the systemic capillaries, gradients of partial pressure favor the net diffusion of $O_2$ out of the blood and $CO_2$ into the blood. These gradients exist because cellular respiration in the mitochondria of cells near each capillary removes $O_2$ from and adds $CO_2$ to the surrounding interstitial fluid. ❺ After the blood unloads $O_2$ and loads $CO_2$, it is returned to the heart and pumped to the lungs again. ❻ There, exchange occurs across the alveolar capillaries, resulting in exhaled air enriched in $CO_2$ and partially depleted of $O_2$.

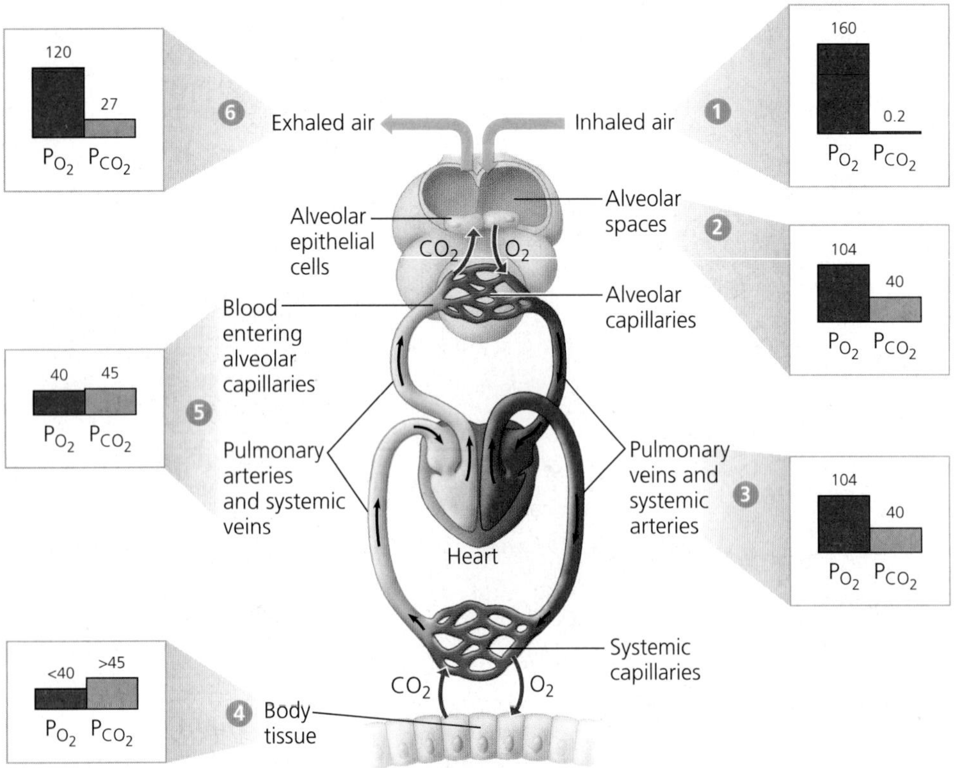

## Respiratory Pigments

The low solubility of $O_2$ in water (and thus in blood) poses a problem for animals that rely on the circulatory system to deliver $O_2$. For example, a person requires almost 2 L of $O_2$ per minute during intense exercise, and all of it must be carried in the blood from the lungs to the active tissues. At normal body temperature and air pressure, however, only 4.5 mL of $O_2$ can dissolve into a liter of blood in the lungs. Even if 80% of the dissolved $O_2$ were delivered to the tissues, the heart would still need to pump 555 L of blood per minute!

In fact, animals transport most of their $O_2$ bound to proteins called **respiratory pigments**. Respiratory pigments circulate with the blood or hemolymph and are often contained within specialized cells. The pigments greatly increase the amount of $O_2$ that can be carried in the circulatory fluid (from 4.5 to about 200 mL of $O_2$

▲ **Figure 34.24 Loading and unloading of respiratory gases.**

**WHAT IF?** *If you consciously forced more air out of your lungs each time you exhaled, how would that affect the values shown in the figure?*

per liter in mammalian blood). In our example of an exercising human with an $O_2$ delivery rate of 80%, the presence of a respiratory pigment reduces the cardiac output necessary for $O_2$ transport to a manageable 12.5 L of blood per minute.

A variety of respiratory pigments have evolved in animals. With a few exceptions, these molecules have a distinctive color (hence the term *pigment*) and consist of a metal bound to a protein. One example is the blue pigment hemocyanin, which has copper as its oxygen-binding component and is found in arthropods and many molluscs.

The respiratory pigment of many invertebrates and almost all vertebrates is hemoglobin. In vertebrates, it is contained in the erythrocytes and has four subunits (polypeptide chains), each with a cofactor called a heme group that has an iron atom at its center (**Figure 34.25**). Each iron atom binds one molecule of $O_2$; hence, a single hemoglobin molecule can carry four $O_2$ molecules. Like all respiratory pigments, hemoglobin binds $O_2$ reversibly, loading $O_2$ in the lungs or gills and unloading it elsewhere in the body. This process is enhanced by cooperativity between the hemoglobin subunits (see Concept 6.5). When $O_2$ binds to one subunit, the others change shape slightly, increasing their affinity for $O_2$. When four $O_2$ molecules are bound and one subunit unloads its $O_2$, the other three subunits more readily unload $O_2$, as an associated shape change lowers their affinity for $O_2$.

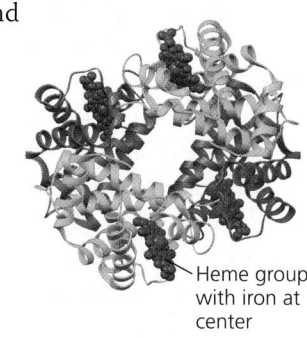

Heme group with iron at center

▲ **Figure 34.25 Hemoglobin.**

Cooperativity in $O_2$ binding and release is evident in the dissociation curve for hemoglobin (**Figure 34.26a**). Over the range of $P_{O_2}$ where the dissociation curve has a steep slope, even a slight change in $P_{O_2}$ causes hemoglobin to load or unload a substantial amount of $O_2$. The steep part of the curve corresponds to the range of $P_{O_2}$ found in body tissues. When cells in a particular location begin working harder—during exercise, for instance—$P_{O_2}$ dips in their vicinity as the $O_2$ is consumed in cellular respiration. Because of the effect of subunit cooperativity, a slight drop in $P_{O_2}$ causes a relatively large increase in the amount of $O_2$ the blood unloads.

Hemoglobin is especially efficient at delivering $O_2$ to tissues actively consuming $O_2$. However, this increased efficiency results not from $O_2$ consumption, but rather from $CO_2$ production. As tissues consume $O_2$ in cell respiration, they also produce $CO_2$. As we have seen, $CO_2$ reacts with water, forming carbonic acid, which lowers the pH of its surroundings. Low pH, in turn, decreases the affinity of hemoglobin for $O_2$, an effect called the **Bohr shift** (**Figure 34.26b**). Thus, where $CO_2$ production is greater, hemoglobin releases more $O_2$, which can then be used to support more cellular respiration.

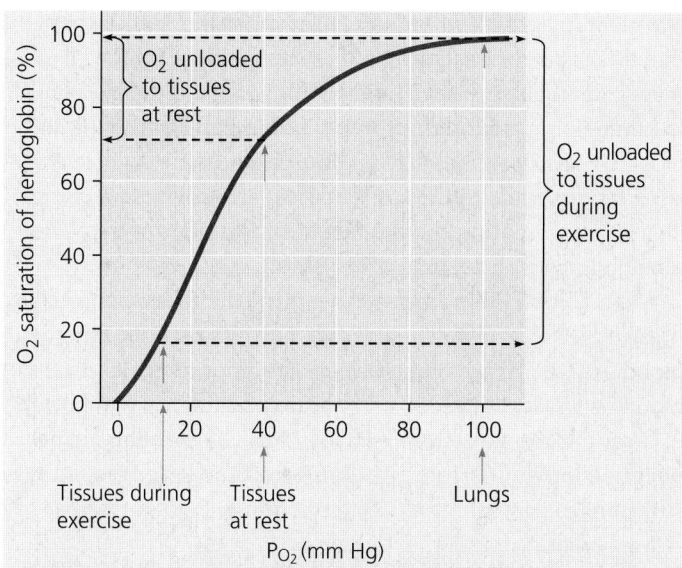

**(a) $P_{O_2}$ and hemoglobin dissociation at pH 7.4.** The curve shows the relative amounts of $O_2$ bound to hemoglobin exposed to solutions with different $P_{O_2}$. At a $P_{O_2}$ of 100 mm Hg, typical in the lungs, hemoglobin is about 98% saturated with $O_2$. At a $P_{O_2}$ of 40 mm Hg, common in resting tissues, hemoglobin is about 70% saturated, having unloaded nearly a third of its $O_2$. As shown in the above graph, hemoglobin can release much more $O_2$ to metabolically very active tissues, such as muscle tissue during exercise.

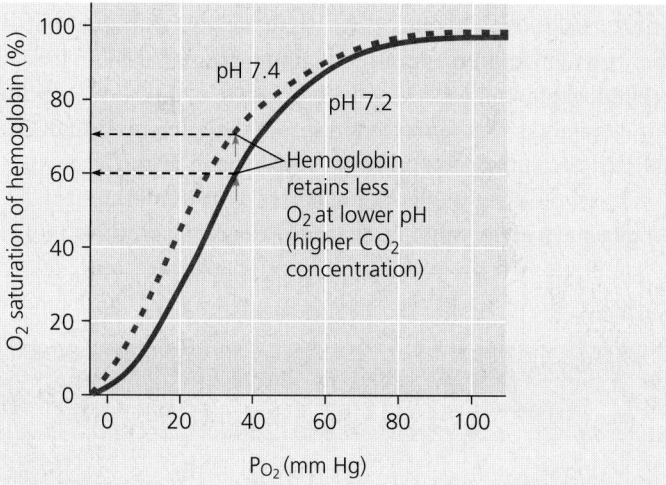

**(b) pH and hemoglobin dissociation.** In very active tissues, $CO_2$ from cellular respiration reacts with water to form carbonic acid, decreasing pH. Because hydrogen ions affect hemoglobin shape, a drop in pH shifts the $O_2$ dissociation curve toward the right (the Bohr shift). For a given $P_{O_2}$, hemoglobin releases more $O_2$ at a lower pH, supporting increased cellular respiration.

▲ **Figure 34.26 Dissociation curves for hemoglobin at 37°C.**

Hemoglobin also assists in buffering the blood—that is, preventing harmful changes in pH. In addition, it has a minor role in $CO_2$ transport, the topic we'll explore next.

## Carbon Dioxide Transport

Only about 7% of the $CO_2$ released by respiring cells is transported in solution in blood plasma. The rest diffuses from the plasma into erythrocytes and reacts with water (assisted by the enzyme carbonic anhydrase), forming $H_2CO_3$. The $H_2CO_3$ readily dissociates into $H^+$ and $HCO_3^-$. Most of the $H^+$ binds to hemoglobin and other proteins, minimizing the change in blood pH. Most of the $HCO_3^-$ diffuses out of the erythrocytes and is transported to the lungs in the plasma. The remaining $HCO_3^-$, representing about 5% of the $CO_2$, binds to hemoglobin and is transported in erythrocytes.

When blood flows through the lungs, the relative partial pressures of $CO_2$ favor the net diffusion of $CO_2$ out of the blood. As $CO_2$ diffuses into alveoli, the amount of $CO_2$ in the blood decreases. This decrease shifts the chemical equilibrium in favor of the conversion of $HCO_3^-$ to $CO_2$, enabling further net diffusion of $CO_2$ into alveoli. Overall, the $P_{CO_2}$ gradient is sufficient to drive about a 15% reduction in $P_{CO_2}$ during passage of blood through the lungs.

## Respiratory Adaptations of Diving Mammals

**EVOLUTION** Animals vary greatly in their ability to spend time in environments in which there is no access to their normal respiratory medium—for example, when an air-breathing mammal swims underwater. Whereas most humans, even expert divers, cannot hold their breath longer than 2 or 3 minutes or swim deeper than 20 m, the Weddell seal of Antarctica routinely plunges to 200–500 m and remains there for 20 minutes to more than an hour. Another diving mammal, the Cuvier's beaked whale, can reach depths of more than 2,900 m—nearly 2 miles—and stay submerged for more than 2 hours! What enables these animals to perform such amazing feats?

One evolutionary adaptation of diving mammals to prolonged stays underwater is a capacity to store large amounts of $O_2$ in their bodies. The Weddell seal has about twice the volume of blood per kilogram of body mass as a human. Furthermore, the muscles of seals and other diving mammals contain a high concentration of an oxygen-storing protein called **myoglobin**. As a result, the Weddell seal can store about twice as much $O_2$ per kilogram of body mass as can a human.

Diving mammals not only have a relatively large $O_2$ stockpile but also have adaptations that conserve $O_2$. They swim with little muscular effort and glide passively for prolonged periods. During a dive, most blood is routed to the brain, spinal cord, eyes, adrenal glands, and, in pregnant seals, the placenta. Blood supply to the muscles is restricted or, during the longest dives, shut off altogether. During dives of more than about 20 minutes, a Weddell seal's muscles deplete the $O_2$ stored in myoglobin and then derive their ATP from fermentation instead of respiration (see Concept 7.5).

The unusual abilities of the Weddell seal and other air-breathing divers to power their bodies during long dives showcase two related themes in our study of organisms—the response to environmental challenges over the short term by physiological adjustments and over the long term as a result of natural selection.

### CONCEPT CHECK 34.7

1. What determines whether $O_2$ or $CO_2$ undergoes net diffusion into or out of capillaries? Explain.
2. How does the Bohr shift help deliver $O_2$ to very active tissues?
3. **WHAT IF?** A doctor might give bicarbonate ($HCO_3^-$) to a patient who is breathing very rapidly. What is the doctor assuming about the patient's blood chemistry?

For suggested answers, see Appendix A.

---

## 34 | Chapter Review

Go to **MasteringBiology®** for Assignments, the eText, and the Study Area with Animations, Activities, Vocab Self-Quiz, and Practice Tests.

## SUMMARY OF KEY CONCEPTS

VOCAB SELF-QUIZ
goo.gl/gbai8v

### CONCEPT 34.1

**Circulatory systems link exchange surfaces with cells throughout the body (pp. 707–710)**

- In animals with simple body plans, a **gastrovascular cavity** mediates exchange between the environment and body cells. Because net **diffusion** is slow over long distances, most complex animals have a circulatory system that moves fluid between cells and the organs that carry out exchange with the environment. Arthropods and most molluscs have an **open circulatory system**, in which **hemolymph** bathes organs directly. Vertebrates have a **closed circulatory system**, in which **blood** circulates in a closed network of pumps and vessels.
- The closed circulatory system of vertebrates consists of blood, blood vessels, and a two- to four-chambered **heart**. Blood pumped by a heart **ventricle** passes to **arteries** and then to the

**capillaries**, sites of chemical exchange between blood and interstitial fluid. **Veins** return blood from capillaries to an **atrium**, which passes blood to a ventricle.

- Fishes, rays, and sharks have a single pump in their circulation. Air-breathing vertebrates have two pumps combined in a single heart. Variations in ventricle number and separation reflect adaptations to different environments and metabolic needs.

**?** *How does the flow of a fluid in a closed circulatory system differ from the movement of molecules between cells and their environment with regard to distance traveled, direction traveled, and driving force?*

### CONCEPT 34.2

**Coordinated cycles of heart contraction drive double circulation in mammals (pp. 710–712)**

- The right ventricle pumps blood to the lungs, where it loads $O_2$ and unloads $CO_2$. Oxygen-rich blood from the lungs enters the heart at the left atrium and is pumped to the body tissues by the left ventricle. Blood returns to the heart through the right atrium.

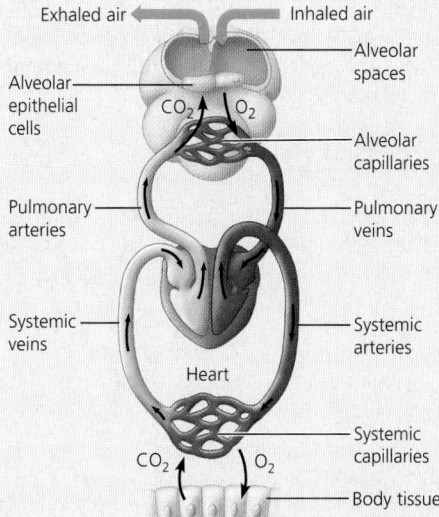

- The **cardiac cycle**, a complete sequence of the heart's pumping and filling, consists of a period of contraction, called **systole**, and a period of relaxation, called **diastole**. The heartbeat originates with impulses at the **sinoatrial (SA) node** (pacemaker) of the right atrium. They trigger atrial contraction, are delayed at the **atrioventricular (AV) node**, and then are conducted into the ventricles, triggering contraction. The nervous system, hormones, and body temperature affect pacemaker activity.

**?** *What changes in cardiac function might you expect after surgical replacement of a defective heart valve?*

CONCEPT 34.3

### Patterns of blood pressure and flow reflect the structure and arrangement of blood vessels (pp. 712–716)

- Blood vessels have structures well adapted to function. Capillaries have narrow diameters and thin walls that facilitate exchange. The velocity of blood flow is lowest in the capillary beds as a result of their large total cross-section area. Arteries contain thick elastic walls that maintain blood pressure. Veins contain one-way valves that contribute to the return of blood to the heart. Blood pressure is altered by changes in cardiac output and by variable constriction of arterioles.
- Fluid leaks out of capillaries and is returned to blood by the **lymphatic system**, which also defends against infection.

**?** *If you placed your forearm on your head, how, if at all, would the blood pressure in that arm change? Explain.*

CONCEPT 34.4

### Blood components function in exchange, transport, and defense (pp. 717–720)

- Whole blood consists of cells and cell fragments (**platelets**) suspended in a liquid matrix called **plasma**. Plasma proteins influence blood pH, osmotic pressure, and viscosity, and they function in lipid transport, immunity (antibodies), and blood clotting (fibrinogen). Red blood cells, or **erythrocytes**, transport $O_2$. White blood cells, or **leukocytes**, function in defense against microbes and foreign substances. Platelets function in blood clotting.

- A variety of diseases impair function of the circulatory system. In **sickle-cell disease**, an aberrant form of **hemoglobin** disrupts erythrocyte shape and function. In cardiovascular disease, inflammation of the arterial lining enhances deposition of lipids and cells, resulting in the potential for life-threatening damage to the heart or brain.

**?** *In the absence of infection, what percentage of cells in human blood are leukocytes?*

CONCEPT 34.5

### Gas exchange occurs across specialized respiratory surfaces (pp. 720–726)

- At all sites of **gas exchange**, a gas undergoes net diffusion from where its **partial pressure** is higher to where it is lower. Air is more conducive to gas exchange than water because air has a higher $O_2$ content, lower density, and lower viscosity.
- Gills are outfoldings of the body surface specialized for gas exchange in water. The effectiveness of gas exchange in some gills, including those of fishes, is increased by **ventilation** and **countercurrent exchange** between blood and water. Gas exchange in insects relies on a **tracheal system**, a branched network of tubes that bring $O_2$ directly to cells. Spiders, land snails, and most terrestrial vertebrates have **lungs**. In mammals, inhaled air passes through the pharynx into the **trachea, bronchi,** bronchioles, and dead-end **alveoli**, where gas exchange occurs.

**?** *Why does altitude have almost no effect on an animal's ability to rid itself of $CO_2$ through gas exchange?*

CONCEPT 34.6

### Breathing ventilates the lungs (pp. 726–728)

- Breathing mechanisms vary substantially among vertebrates. An amphibian ventilates its lungs by **positive pressure breathing**, which forces air down the trachea. Birds use a system of air sacs as bellows to keep air flowing through the lungs in one direction only. Mammals ventilate their lungs by **negative pressure breathing**, which pulls air into the lungs. Lung volume increases when the rib muscles and **diaphragm** contract. Incoming and outgoing air mix, decreasing the efficiency of ventilation.
- Sensors detect the pH of cerebrospinal fluid (reflecting $CO_2$ concentration in the blood), and a control center in the brain adjusts breathing rate and depth to match metabolic demands. Additional input to the control center is provided by sensors in the aorta and carotid arteries that monitor blood levels of $O_2$ as well as $CO_2$ (via blood pH).

**?** *How does air in the lungs differ from the fresh air that enters the body during inspiration?*

CONCEPT 34.7

### Adaptations for gas exchange include pigments that bind and transport gases (pp. 728–730)

- In the lungs, gradients of partial pressure favor the net diffusion of $O_2$ into the blood and $CO_2$ out of the blood. The opposite situation exists in the rest of the body. **Respiratory pigments**, such as hemoglobin, bind $O_2$, greatly increasing the amount of $O_2$ transported by the circulatory system.
- Evolutionary adaptations enable some animals to satisfy extraordinary $O_2$ demands. Deep-diving mammals stockpile $O_2$ in blood and other tissues and deplete it slowly.

**?** *How is the role of a respiratory pigment like that of an enzyme?*

# TEST YOUR UNDERSTANDING

## Level 1: Knowledge/Comprehension

**1.** Which of the following respiratory systems is *not* closely associated with a blood supply?
(A) the lungs of a vertebrate
(B) the gills of a fish
(C) the tracheal system of an insect
(D) the skin of an earthworm

**2.** Blood returning to the mammalian heart in a pulmonary vein drains first into the
(A) left atrium.
(B) right atrium.
(C) left ventricle.
(D) right ventricle.

**3.** Pulse is a direct measure of
(A) blood pressure.
(B) stroke volume.
(C) cardiac output.
(D) heart rate.

**4.** When you hold your breath, which of the following blood gas changes first leads to the urge to breathe?
(A) rising $O_2$
(B) falling $O_2$
(C) rising $CO_2$
(D) falling $CO_2$

**5.** One feature that amphibians and humans have in common is
(A) the number of heart chambers.
(B) a complete separation of circuits for circulation.
(C) the number of circuits for circulation.
(D) a low blood pressure in the systemic circuit.

## Level 2: Application/Analysis

**6.** If a molecule of $CO_2$ released into the blood in your left toe is exhaled from your nose, it must pass through all of the following *except*
(A) the pulmonary vein.
(B) the trachea.
(C) the right atrium.
(D) the right ventricle.

**7.** Compared with the interstitial fluid that bathes active muscle cells, blood reaching these cells in arteries has a
(A) higher $P_{O_2}$.
(B) higher $P_{CO_2}$.
(C) greater bicarbonate concentration.
(D) lower pH.

## Level 3: Synthesis/Evaluation

**8.** **DRAW IT** Plot blood pressure against time for one cardiac cycle in humans, drawing separate lines for the pressure in the aorta, the left ventricle, and the right ventricle. Below the time axis, add a vertical arrow pointing to the time when you expect a peak in atrial blood pressure.

**9. SCIENTIFIC INQUIRY**
**INTERPRET THE DATA**
The hemoglobin of a human fetus differs from adult hemoglobin. Compare the dissociation curves of the two hemoglobins in the graph. Describe how they differ, and propose a hypothesis to explain the benefit of this difference.

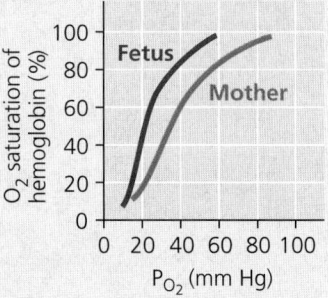

**10. FOCUS ON EVOLUTION**
One of the opponents of the movie monster Godzilla is Mothra, a giant mothlike creature with a wingspan of several dozen meters. The largest known insects were Paleozoic dragonflies with half-meter wingspans. Focusing on respiration and gas exchange, explain why giant insects are improbable.

**11. FOCUS ON INTERACTIONS**
Some athletes prepare for competition at sea level by sleeping in a tent in which $P_{O_2}$ is kept artificially low. When climbing high peaks, some mountaineers breathe from bottles of pure $O_2$. In a short essay (100–150 words), relate these behaviors to the mechanism of $O_2$ transport in the human body and to physiological interactions with our gaseous environment.

**12. SYNTHESIZE YOUR KNOWLEDGE**

The diving bell spider (*Argyroneta aquatica*) stores air underwater in a net of silk. Explain why this adaptation could be more advantageous than having gills, taking into account differences in gas exchange media and gas exchange organs among animals.

*For selected answers, see Appendix A.*

**KEY CONCEPTS**

**35.1** In innate immunity, recognition and response rely on traits common to groups of pathogens

**35.2** In adaptive immunity, receptors provide pathogen-specific recognition

**35.3** Adaptive immunity defends against infection of body fluids and body cells

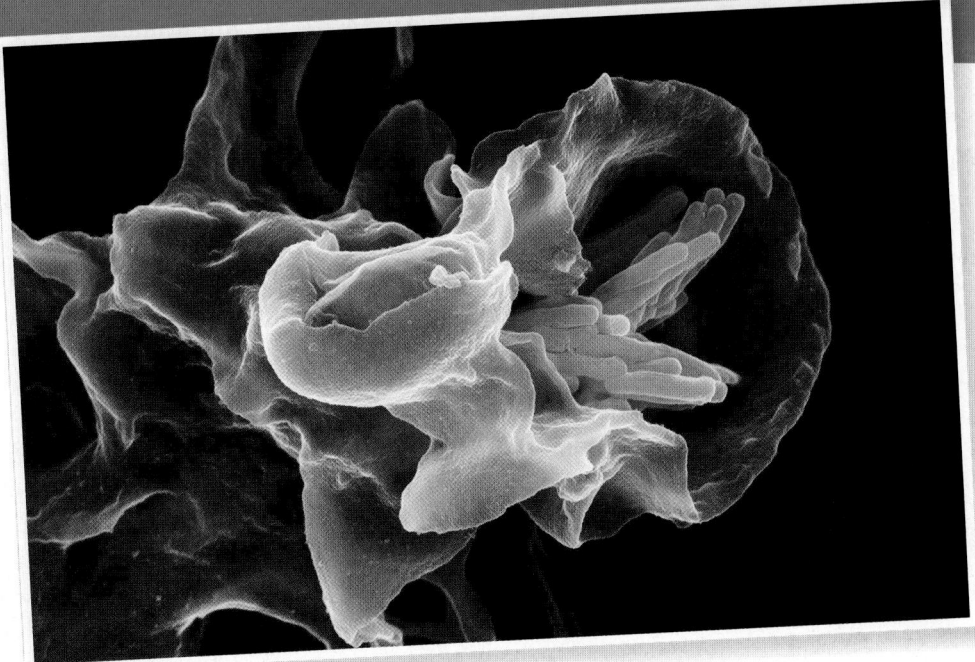

▲ **Figure 35.1** What triggered this attack by an immune cell on a clump of rod-shaped bacteria?

# Recognition and Response

For a **pathogen**—a bacterium, fungus, virus, or other disease-causing agent—the internal environment of an animal is a nearly ideal habitat. The animal body offers a ready source of nutrients, a protected setting for growth and reproduction, and a means of transport to new environments. From the perspective of a cold or flu virus, we are wonderful hosts. From our vantage point, the situation is not so ideal. Fortunately, adaptations have arisen over the course of evolution that protect animals against many pathogens.

Dedicated immune cells in the body fluids and tissues of most animals specifically interact with and destroy pathogens. In **Figure 35.1**, for example, an immune cell called a macrophage (brown) is engulfing rod-shaped bacteria (green). Immune cells also release defense molecules into body fluids, including proteins that punch holes in bacterial membranes or block viruses from entering body cells. Together, the body's defenses make up the **immune system**, which enables an animal to avoid or limit many infections. A foreign molecule or cell doesn't have to be pathogenic (disease-causing) to elicit an immune response, but we'll focus in this chapter on the immune system's role in defending against pathogens.

The first lines of defense offered by immune systems help prevent pathogens from gaining entrance to the body. For example, an outer covering, such as a skin or shell, blocks entry by many pathogens. Sealing off the entire body surface is impossible, however, because gas exchange, nutrition, and reproduction require openings to the environment. Secretions that trap or kill microbes guard the body's entrances and exits, while the linings of the digestive tract, airway, and other exchange surfaces provide additional barriers to infection.

If a pathogen breaches barrier defenses and enters the body, the problem of how to fend off attack changes substantially. Housed within body fluids and tissues, the invader is no longer an outsider. To fight infections, an animal's immune system must detect foreign particles and cells within the body. In other words, a properly functioning immune system distinguishes nonself from self. How? Immune cells produce receptor molecules that bind specifically to molecules from foreign cells or viruses and activate defense responses. The specific binding of immune receptors to foreign molecules is a type of *molecular recognition* and is the central event in identifying nonself molecules, particles, and cells.

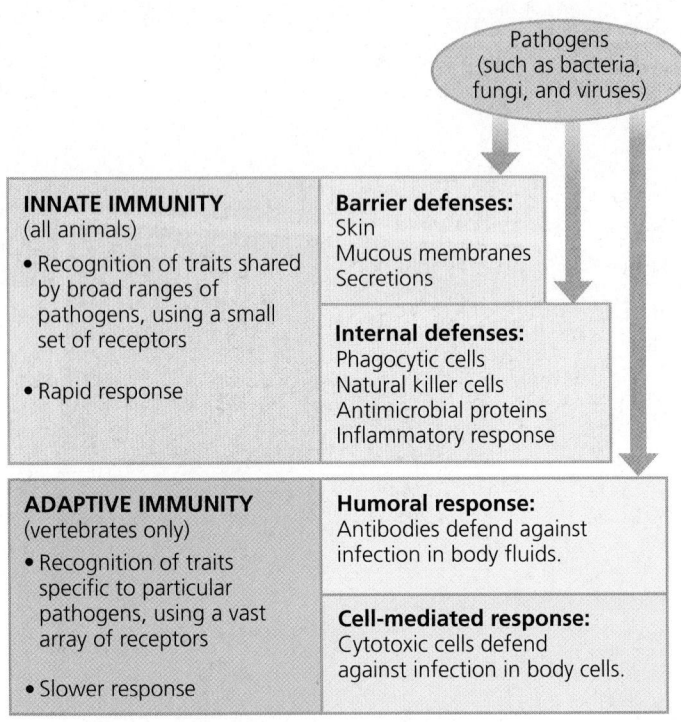

**▲ Figure 35.2 Overview of animal immunity.** Immune responses can be divided into innate and adaptive immunity. Some components of innate immunity help activate adaptive immune defenses.

Two types of molecular recognition provide the basis for the two types of immune defense found among animals: innate immunity, which is common to all animals, and adaptive immunity, which is found only in vertebrates. **Figure 35.2** summarizes these two types of immunity, highlighting fundamental similarities and differences.

In **innate immunity**, which includes barrier defenses, molecular recognition relies on a small set of receptor proteins that bind to molecules or structures that are absent from animal bodies but common to a group of viruses, bacteria, or other microbes. Binding of an innate immune receptor to a foreign molecule activates internal defenses, enabling responses to a very broad range of pathogens.

In **adaptive immunity**, molecular recognition relies on a vast arsenal of receptors, each of which recognizes a feature typically found only on a particular part of a particular molecule in a particular pathogen. As a result, recognition and response in adaptive immunity occur with remarkable specificity.

The adaptive immune response, also known as the acquired immune response, is activated after the innate immune response and develops more slowly. As reflected by the names *adaptive* and *acquired*, this immune response is enhanced by previous exposure to the infecting pathogen. Examples of adaptive responses include the synthesis of proteins that inactivate a bacterial toxin and the targeted killing of a virus-infected body cell.

In this chapter, we'll examine how each type of immunity protects animals from disease. We'll also investigate how pathogens can avoid or overwhelm the immune system and how defects in the immune system can imperil an animal's health.

# In innate immunity, recognition and response rely on traits common to groups of pathogens

Innate immunity is found in all animals (as well as in plants). In exploring innate immunity, we'll begin with invertebrates, which repel and fight infection with only this type of immunity. We'll then turn to vertebrates, in which innate immunity serves both as an immediate defense against infection and as the foundation for adaptive immune defenses.

## Innate Immunity of Invertebrates

The great success of insects in terrestrial and freshwater habitats teeming with diverse pathogens highlights the effectiveness of invertebrate innate immunity. One part of this defense system is the insect exoskeleton, which acts as a physical barrier against infection. Within the digestive system, **lysozyme**, an enzyme that breaks down bacterial cell walls, acts as a chemical barrier against pathogens ingested with food.

Any pathogen that breaches an insect's barrier defenses encounters internal immune defenses. Immune cells of insects recognize pathogens by binding to molecules specific to viruses or microorganisms. These molecules include double-stranded viral RNAs and components of fungal or bacterial cell walls that are not normally found in animal cells. Such macromolecules serve as "identity tags" in the process of pathogen recognition. Each recognition protein of insects binds to a macromolecule characteristic of a broad class of pathogens. Once bound to a macromolecule, the recognition protein triggers an innate immune response specific for that class.

The major immune cells of insects are called *hemocytes*. Like amoebas, some hemocytes ingest and break down microorganisms, a process known as **phagocytosis (Figure 35.3)**. Many hemocytes release antimicrobial peptides, which inactivate or kill bacteria or fungi by disrupting their plasma membranes. One class of hemocytes produces another type of defense molecule that helps entrap larger pathogens, such as *Plasmodium*, the single-celled parasite of mosquitoes that causes malaria in humans.

## Innate Immunity of Vertebrates

In jawed vertebrates, innate immune defenses coexist with the more recently evolved system of adaptive immunity. Because most of the recent discoveries regarding vertebrate innate immunity have come from studies of mice and humans, we'll focus here on mammals. In this section we'll consider first the innate defenses that are similar to those found among invertebrates: barrier defenses, phagocytosis, and antimicrobial peptides. We'll then examine some unique aspects of vertebrate innate immunity, such as natural killer cells, interferons, and the inflammatory response.

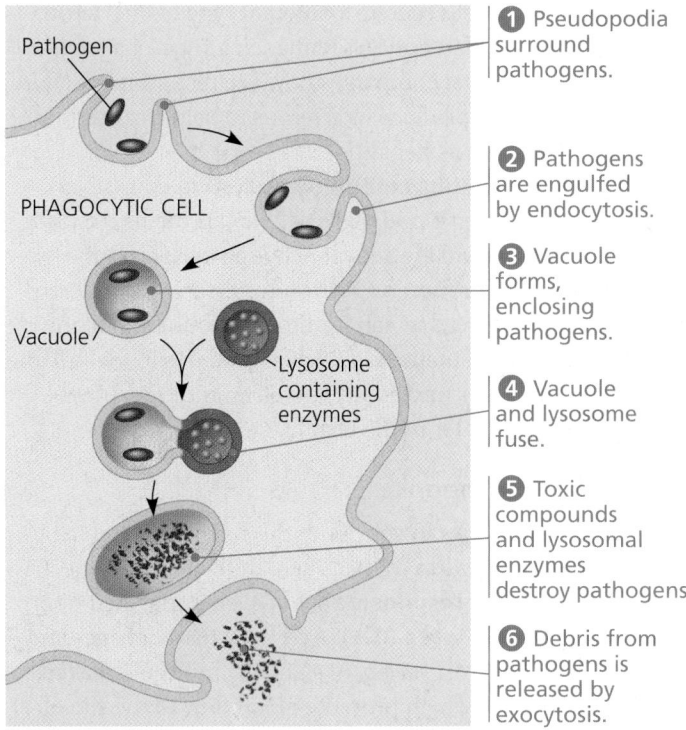

Pathogen

PHAGOCYTIC CELL

Vacuole

Lysosome containing enzymes

**1** Pseudopodia surround pathogens.

**2** Pathogens are engulfed by endocytosis.

**3** Vacuole forms, enclosing pathogens.

**4** Vacuole and lysosome fuse.

**5** Toxic compounds and lysosomal enzymes destroy pathogens.

**6** Debris from pathogens is released by exocytosis.

▲ **Figure 35.3 Phagocytosis.** This schematic depicts events in the ingestion and destruction of a microbe by a typical phagocytic cell.

## Barrier Defenses

The barrier defenses of mammals, which block the entry of many pathogens, include the skin and the mucous membranes. The mucous membranes that line the digestive, respiratory, urinary, and reproductive tracts produce *mucus*, a viscous fluid that traps pathogens and other particles.

Beyond their physical role in inhibiting microbial entry, body secretions create an environment that is hostile to many pathogens. Lysozyme in tears, saliva, and mucous secretions destroys the cell walls of susceptible bacteria as they enter the openings around the eyes or the upper respiratory tract. Microbes in food or water and those in swallowed mucus must also contend with the acidic environment of the stomach, which kills most of them before they can enter the intestines. Similarly, secretions from oil and sweat glands give human skin a pH ranging from 3 to 5, acidic enough to prevent the growth of many bacteria.

## Cellular Innate Defenses

In mammals, as in insects, there are innate immune cells dedicated to detecting, devouring, and destroying invading pathogens. In recognizing viral, fungal, or bacterial components, phagocytic mammalian cells rely on several types of receptors. Some are very similar to Toll, a key activator of innate immunity in insects, a discovery recognized with the Nobel Prize in Physiology or Medicine in 2011. Each mammalian **Toll-like receptor (TLR)** binds to fragments of molecules normally absent from the vertebrate body but characteristic of a set of

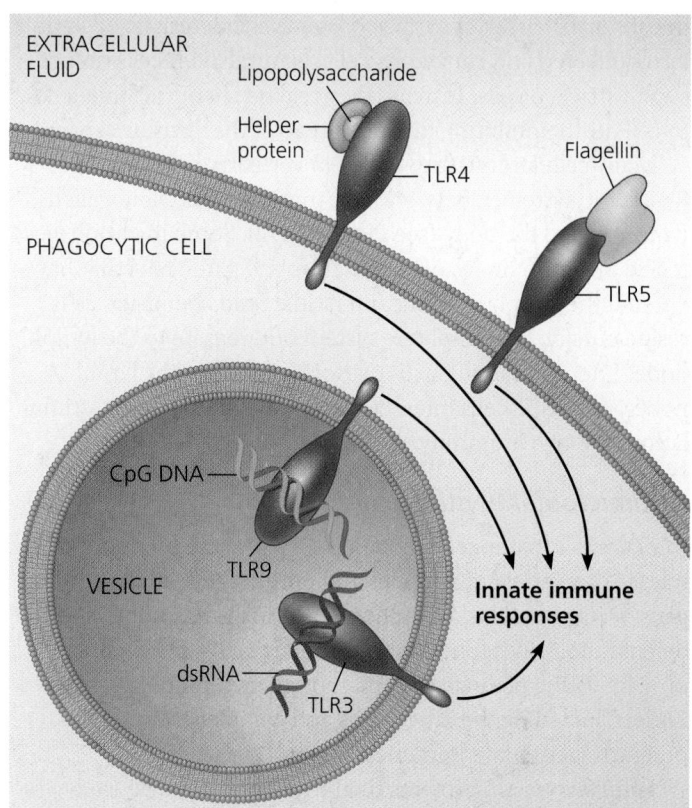

▲ **Figure 35.4 TLR signaling.** Each mammalian Toll-like receptor (TLR) recognizes a molecular pattern characteristic of a group of pathogens. Lipopolysaccharide, flagellin, CpG DNA (DNA containing unmethylated CG sequences), and double-stranded RNA (dsRNA) are all found in bacteria, fungi, or viruses, but not in animal cells. Together with other recognition and response factors, TLR proteins trigger internal innate immune defenses.

**?** *Some TLR proteins are on the cell surface, whereas others are inside vesicles. Suggest a possible benefit of this distribution.*

pathogens **(Figure 35.4)**. For example, TLR3 binds to double-stranded RNA, a form of nucleic acid characteristic of certain viruses. Similarly, TLR4 recognizes lipopolysaccharide, a molecule found on the surface of many bacteria.

The two main types of phagocytic cells in the mammalian body are neutrophils and macrophages. **Neutrophils**, which circulate in the blood, are attracted by signals from infected tissues and then engulf and destroy the infecting pathogens. **Macrophages** ("big eaters"), like the one shown in Figure 35.1, are larger phagocytic cells.

Two other types of cells—dendritic cells and eosinophils—also have roles in innate defense. *Dendritic cells* mainly populate tissues, such as skin, that contact the environment. They stimulate adaptive immunity against pathogens they encounter and engulf, as we'll explore shortly. *Eosinophils*, often found in tissues underlying an epithelium, are important in defending against multicellular invaders, such as parasitic worms. Upon encountering such parasites, eosinophils discharge destructive enzymes.

Cellular innate defenses in vertebrates also involve **natural killer cells**. These cells circulate through the body and detect the abnormal array of surface proteins characteristic of some virus-infected and cancerous cells. Natural killer cells do not engulf stricken cells. Instead, they release chemicals that lead to cell death, inhibiting further spread of the virus or cancer.

Many cellular innate defenses in vertebrates involve the lymphatic system, a network that transports the fluid called lymph within the body (see Figure 34.12). Some macrophages reside in lymph nodes, where they engulf pathogens that have entered the lymph from the interstitial fluid. Dendritic cells reside outside the lymphatic system but migrate to the lymph nodes after interacting with pathogens. Within the lymph nodes, dendritic cells interact with other immune cells, stimulating adaptive immunity.

### Antimicrobial Peptides and Proteins

In mammals, pathogen recognition triggers the production and release of a variety of peptides and proteins that attack pathogens or impede their reproduction. As in insects, some of these defense molecules function as antimicrobial peptides, killing or inactivating pathogens by disrupting membrane integrity. Others, including the interferons and complement proteins, are unique to vertebrate immune systems.

**Interferons** are proteins that provide innate defense by interfering with viral infections. Virus-infected body cells secrete interferon proteins that induce nearby uninfected cells to produce substances that inhibit viral replication. In this way, these interferons limit the cell-to-cell spread of viruses in the body, helping control viral infections such as colds and influenza.

Some white blood cells secrete a different type of interferon that helps activate macrophages, enhancing their phagocytic ability. Pharmaceutical companies now use recombinant DNA technology to mass-produce interferons to help treat certain viral infections, such as hepatitis C.

The infection-fighting **complement system** consists of roughly 30 proteins in blood plasma. These proteins circulate in an inactive state and are activated by substances on the surface of many microbes. Activation results in a cascade of biochemical reactions that can lead to lysis (bursting) of invading cells. The complement system also functions in the inflammatory response, our next topic, as well as in the adaptive defenses discussed later in the chapter.

### Inflammatory Response

When a splinter lodges in your skin, the surrounding area becomes swollen and warm to the touch. Both changes reflect a local **inflammatory response**, a set of events triggered upon injury or infection **(Figure 35.5)**. Activated macrophages and neutrophils discharge a variety of signaling molecules called **cytokines**, some of which promote blood flow to the site of injury or infection. In addition, *mast cells*, found in connective tissue, release the signaling molecule **histamine** at sites of damage. Histamine triggers nearby blood vessels to dilate and become more permeable. The resulting increase in local blood supply produces the redness and increased skin temperature typical of the inflammatory response (from the Latin *inflammare*, to set on fire).

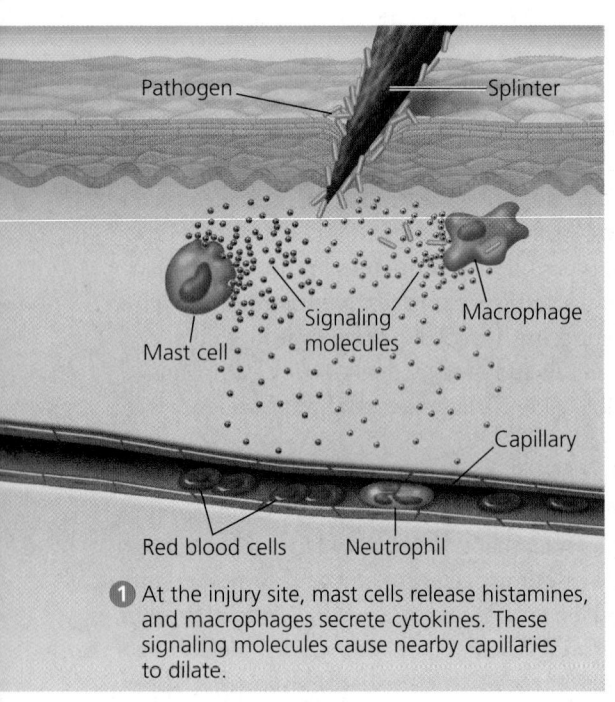

① At the injury site, mast cells release histamines, and macrophages secrete cytokines. These signaling molecules cause nearby capillaries to dilate.

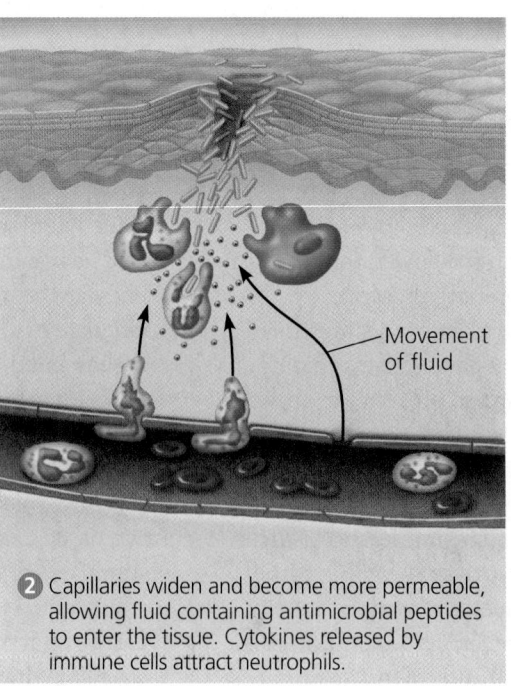

② Capillaries widen and become more permeable, allowing fluid containing antimicrobial peptides to enter the tissue. Cytokines released by immune cells attract neutrophils.

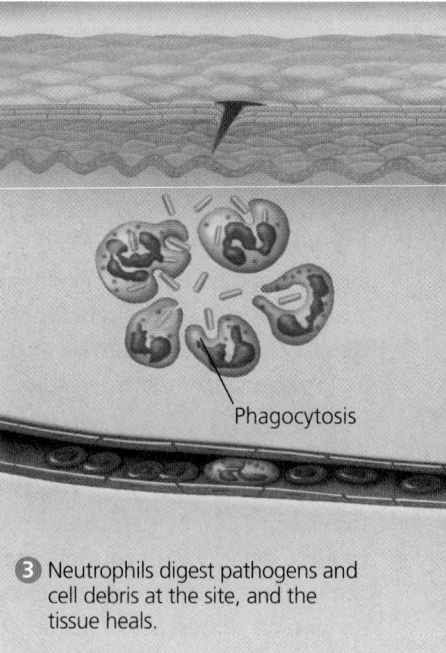

③ Neutrophils digest pathogens and cell debris at the site, and the tissue heals.

▲ **Figure 35.5 Major events in a local inflammatory response.**

**?** *From your experience with splinters, deduce whether the signals mediating an inflammatory response are short- or long-lived. Explain your answer.*

During inflammation, cycles of signaling and response transform the site of injury and infection. Activated complement proteins promote further release of histamine, enabling more phagocytic cells to enter the site of injury. At the same time, enhanced blood flow to the site helps deliver antimicrobial peptides. The result is an accumulation of pus, a fluid rich in white blood cells, dead pathogens, and debris from damaged tissue.

A minor injury or infection causes a local inflammatory response, but more extensive tissue damage or infection may lead to a response that is systemic (throughout the body). Cells in injured or infected tissue often secrete molecules that stimulate the release of additional neutrophils from the bone marrow. In the case of a severe infection, such as meningitis or appendicitis, the number of white blood cells in the bloodstream may increase severalfold within only a few hours.

A systemic inflammatory response sometimes involves fever. In response to certain pathogens, substances released by activated macrophages cause the body's thermostat to reset to a higher temperature (see Concept 32.3). The benefits of the resulting fever are still a subject of debate. One hypothesis is that an elevated body temperature may enhance phagocytosis and, by speeding up chemical reactions, accelerate tissue repair.

Certain bacterial infections can induce an overwhelming systemic inflammatory response, leading to a life-threatening condition called *septic shock*. Characterized by very high fever, low blood pressure, and poor blood flow through capillaries, septic shock occurs most often in the very old and the very young. It is fatal in more than one-third of cases and kills more than 90,000 people each year in the United States alone.

Chronic (ongoing) inflammation can also threaten human health. For example, millions of individuals worldwide suffer from Crohn's disease and ulcerative colitis, often debilitating disorders in which an unregulated inflammatory response disrupts intestinal function.

## Evasion of Innate Immunity by Pathogens

Adaptations have evolved in some pathogens that enable them to avoid destruction by phagocytic cells. For example, the outer capsule that surrounds certain bacteria interferes with molecular recognition and phagocytosis. One such bacterium, *Streptococcus pneumoniae*, is a major cause of pneumonia and meningitis in humans (see Concept 13.1).

Some bacteria are recognized, but are resistant to breakdown after being engulfed by a host cell. One example is *Mycobacterium tuberculosis*, the bacterium shown in Figure 35.1. Rather than being destroyed, this bacterium grows and reproduces within host cells, effectively hidden from further attack by the body's immune defenses. The result of this infection is tuberculosis (TB), a disease that attacks the lungs and other tissues. Worldwide, TB kills more than 1 million people a year.

### CONCEPT CHECK 35.1

1. Pus is both a sign of infection and an indicator of immune defenses in action. Explain.
2. **MAKE CONNECTIONS** How do the molecules that activate the vertebrate TLR signal transduction pathway differ from the ligands in most other signaling pathways (see Concept 5.6)?
3. **WHAT IF?** Parasitic wasps inject their eggs into host larvae of other insects. If the host immune system doesn't kill the wasp egg, the egg hatches and the wasp larva devours the host larva as food. Why might insect species differ in whether they can initiate an innate immune response to a wasp egg?

For suggested answers, see Appendix A.

## CONCEPT 35.2

# In adaptive immunity, receptors provide pathogen-specific recognition

Vertebrates are unique in having both adaptive and innate immunity. The adaptive response relies on T cells and B cells, which are types of white blood cells called **lymphocytes**. Like all blood cells, lymphocytes originate from stem cells in the bone marrow. Some migrate from the bone marrow to the **thymus**, an organ in the thoracic cavity above the heart (see Figure 34.12). These lymphocytes mature into **T cells**. Lymphocytes that remain and mature in the bone marrow develop as **B cells**.

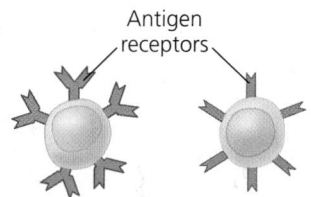

Antigen receptors

Mature B cell    Mature T cell

Any substance that elicits a B or T cell response is called an **antigen**. In adaptive immunity, recognition occurs when a B cell or T cell binds to an antigen, such as a bacterial or viral protein, via a protein called an **antigen receptor**. Each antigen receptor binds to just one part of one molecule from a particular pathogen, such as a species of bacteria or strain of virus.

The cells of the immune system produce millions of different antigen receptors. Any particular lymphocyte, however, produces just one variety; all of the antigen receptors made by a single B or T cell are identical. Infection by a virus, bacterium, or other pathogen triggers activation of B and T cells with antigen receptors specific for parts of that pathogen. B and T cells are shown in this text with only a few antigen receptors, but there are actually about 100,000 antigen receptors on the surface of a single B or T cell.

Antigens are usually foreign and are typically large molecules, either proteins or polysaccharides. The small, accessible portion of an antigen that binds to an antigen receptor is called an **epitope**. An example is a group of amino acids in a particular protein. A single antigen usually has several epitopes, each binding a receptor with a different specificity. Because all antigen receptors produced by a single B cell or T cell are identical, they bind to the same epitope. Each B or T cell thus displays *specificity* for a particular epitope, enabling it to respond to any pathogen that produces molecules containing that epitope.

The antigen receptors of B cells and T cells have similar components, but they encounter antigens in different ways. We'll consider the two processes in turn.

## Antigen Recognition by B Cells and Antibodies

Each B cell antigen receptor is a Y-shaped protein consisting of four polypeptide chains: two identical **heavy chains** and two identical **light chains**. Disulfide bridges link the chains together **(Figure 35.6)**.

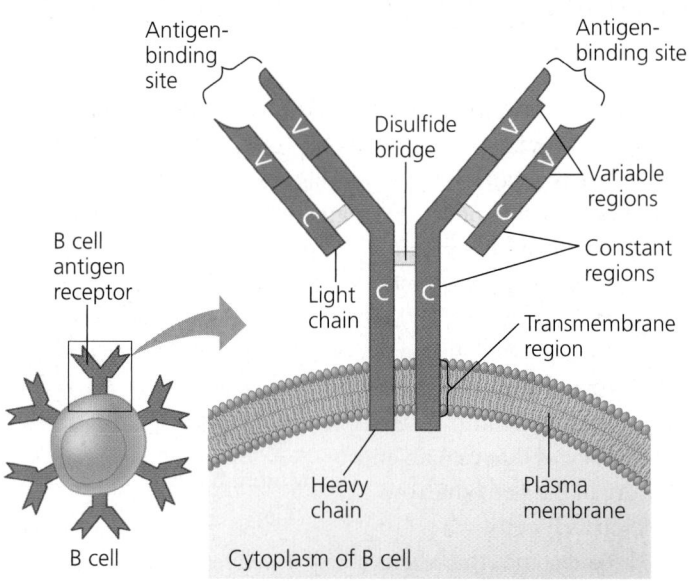

▲ **Figure 35.6 The structure of a B cell antigen receptor.**

The light and heavy chains each have a *constant (C) region*, where amino acid sequences vary little among the receptors on different B cells. The constant region of heavy chains contains a transmembrane region, which anchors the receptor in the cell's plasma membrane. As shown in Figure 35.6, each heavy and light chain also has a *variable (V) region*, so named because its amino acid sequence varies extensively from one B cell to another. Together, parts of a heavy-chain V region and a light-chain V region form an asymmetric binding site for an antigen. As shown in Figure 35.6, each B cell antigen receptor has two identical antigen-binding sites.

Binding of a B cell antigen receptor to an antigen is an early step in B cell activation, leading eventually to formation of cells that secrete a soluble form of the receptor **(Figure 35.7a)**. This secreted protein is called an **antibody**, also known as an

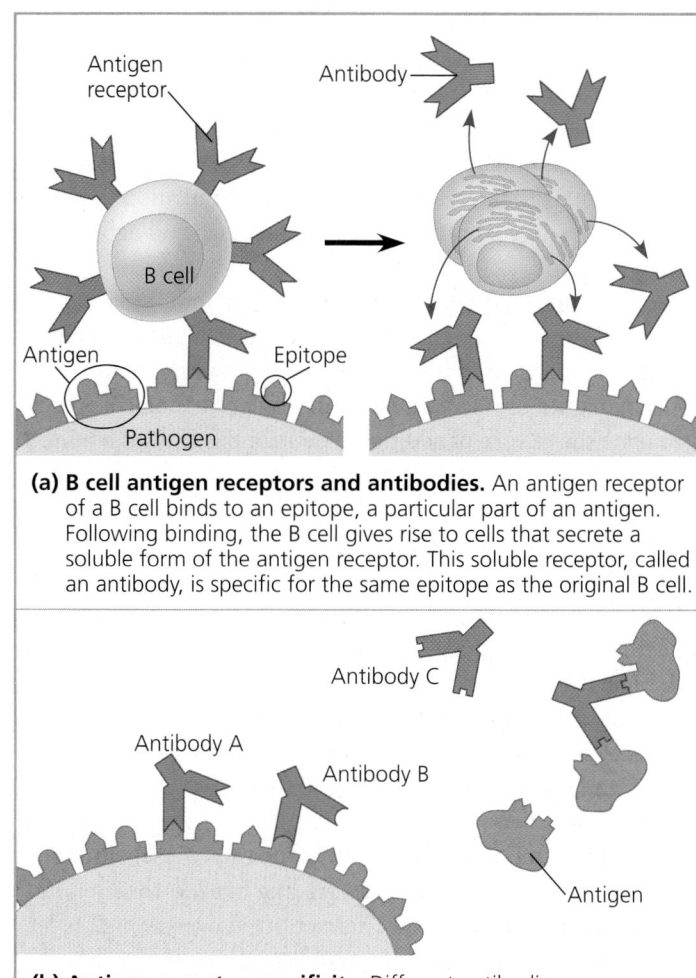

**(a) B cell antigen receptors and antibodies.** An antigen receptor of a B cell binds to an epitope, a particular part of an antigen. Following binding, the B cell gives rise to cells that secrete a soluble form of the antigen receptor. This soluble receptor, called an antibody, is specific for the same epitope as the original B cell.

**(b) Antigen receptor specificity.** Different antibodies can recognize distinct epitopes on the same antigen. Furthermore, antibodies can recognize free antigens as well as antigens on a pathogen's surface.

▲ **Figure 35.7 Antigen recognition by B cells and antibodies.**

**MAKE CONNECTIONS** *The interactions depicted here involve a highly specific binding between antigen and receptor (see also Figure 3.20). How is this similar to an enzyme-substrate interaction (see Figure 6.14)?*

**immunoglobulin (Ig)**. Antibodies have the same Y-shaped structure as B cell antigen receptors, but they are secreted rather than membrane-bound. As we'll discuss, antibodies provide a direct defense against pathogens in body fluids.

The antigen-binding site of a membrane-bound receptor or antibody has a unique shape that provides a lock-and-key fit for a particular epitope. Many noncovalent bonds between an epitope and the surface of the binding site provide a stable and specific interaction. Differences in the amino acid sequences of variable regions provide the variation in binding surfaces that enables this highly specific binding.

B cell antigen receptors and antibodies bind to intact antigens in the blood and lymph. As illustrated in **Figure 35.7b** for antibodies, they can bind to antigens on the surface of pathogens or free in body fluids.

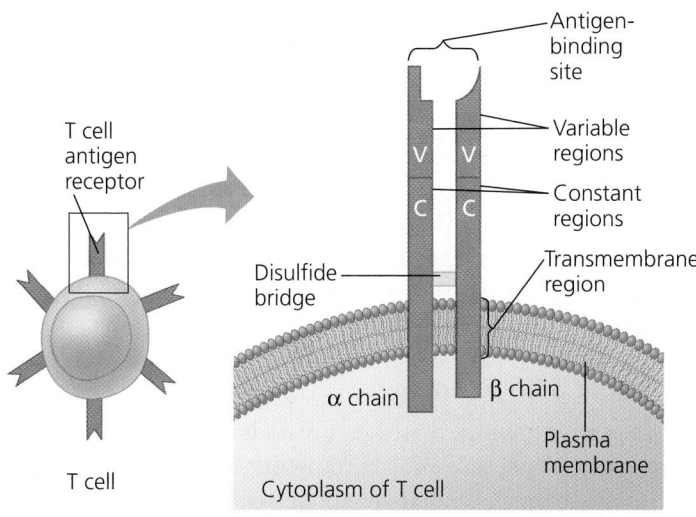

▲ Figure 35.8 **The structure of a T cell antigen receptor.**

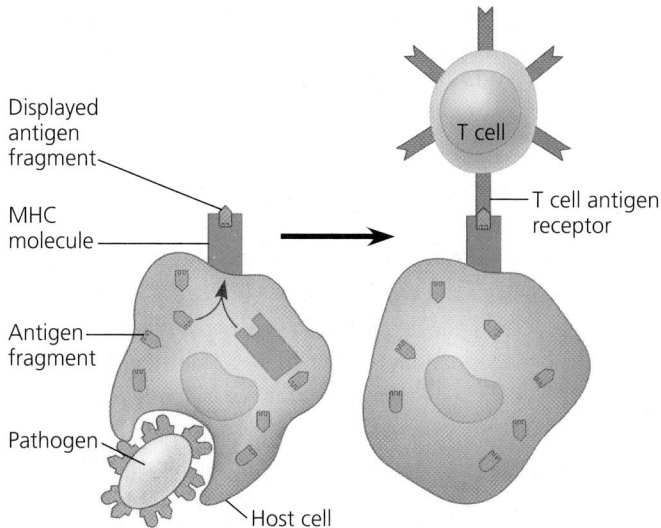

▲ Figure 35.9 **Antigen recognition by T cells.** Inside the host cell, an antigen fragment from a pathogen binds to an MHC molecule and is brought to the cell surface, where it is displayed. The combination of MHC molecule and antigen fragment is recognized by a T cell.

## Antigen Recognition by T Cells

For a T cell, the antigen receptor consists of two different poly-peptide chains, an α *chain* and a β *chain*, linked by a disulfide bridge **(Figure 35.8)**. Near the base of the T cell antigen receptor (often called simply a T cell receptor) is a transmembrane region that anchors the molecule in the cell's plasma membrane. At the outer tip of the molecule, the variable (V) regions of the α and β chains together form a single antigen-binding site. The remainder of the molecule is made up of the constant (C) regions.

Whereas the antigen receptors of B cells bind to epitopes of intact antigens protruding from pathogens or circulating free in body fluids, those of T cells bind only to fragments of antigens that are displayed, or presented, on the surface of host cells. The host protein that displays the antigen fragment on the cell surface is called a **major histocompatibility complex (MHC) molecule**.

Recognition of protein antigens by T cells begins when a pathogen or part of a pathogen either infects or is taken in by a host cell **(Figure 35.9)**. Inside the host cell, enzymes in the cell cleave the antigen into smaller peptides. Each peptide, called an *antigen fragment*, binds to an MHC molecule inside the cell. The MHC molecule and bound peptide are then transported to the cell surface. The result is **antigen presentation**, the display of the antigen fragment in an exposed groove of the MHC protein.

In effect, antigen presentation advertises the fact that a host cell contains a foreign substance. If the cell displaying an antigen fragment encounters a T cell with the right specificity, the interaction of an MHC molecule, an antigen fragment, and an antigen receptor triggers an adaptive immune response, as we'll explore in Concept 35.3.

## B Cell and T Cell Development

Now that you know how B cells and T cells recognize antigens, let's consider four major characteristics of adaptive immunity.

First, the immense repertoire of lymphocytes and receptors enables detection of antigens and pathogens never before encountered. Second, adaptive immunity normally has self-tolerance, the lack of reactivity against an animal's own molecules and cells. Third, cell proliferation triggered by activation greatly increases the number of B and T cells specific for an antigen. Fourth, there is a stronger and more rapid response to an antigen encountered previously, due to a feature known as *immunological memory*, which we'll discuss later in the chapter.

Receptor diversity and self-tolerance arise as a lymphocyte matures. Cell proliferation and the formation of immunological memory occur later, after a mature lymphocyte encounters and binds to a specific antigen. We'll consider these four characteristics in the order in which they develop.

### Generation of B Cell and T Cell Diversity

Each person makes more than 1 million different B cell antigen receptors and 10 million different T cell antigen receptors. Yet there are only about 20,000 protein-coding genes in the human genome. How, then, do we generate so many different antigen receptors? The answer lies in combinations. Think of a menu offering a choice of nine dinners and five desserts. There are 45 (9 × 5) combinations to consider. Similarly, by combining variable elements, the immune system assembles millions of different receptors from a very small collection of parts.

To understand the origin of receptor diversity, let's consider an immunoglobulin (Ig) gene that encodes the light chain of both membrane-bound B cell antigen receptors and secreted antibodies (immunoglobulins). Although we'll analyze only a single Ig light-chain gene, B and T cell antigen receptor genes undergo very similar transformations.

The capacity to generate diversity is built into the structure of Ig genes. A receptor light chain is encoded by three gene segments: a variable (*V*) segment, a joining (*J*) segment, and a constant (*C*) segment. The *V* and *J* segments together encode the variable region of the receptor chain, while the *C* segment encodes the constant region. The light-chain gene contains a single *C* segment, 40 different *V* segments, and 5 different *J* segments. These alternative copies of the *V* and *J* segments are arranged within the gene in a series (**Figure 35.10**). Because a functional gene is built from one copy of each type of segment, the pieces can be combined in 200 different ways ($40\ V \times 5\ J \times 1\ C$). The number of different heavy-chain combinations is even greater, resulting in even more diversity.

Assembling a functional Ig gene requires rearranging the DNA. Early in B cell development, an enzyme complex called *recombinase* links one light-chain *V* gene segment to one *J* gene segment. This recombination event eliminates the long stretch of DNA between the segments, forming a single exon that is part *V* and part *J*.

Recombinase acts randomly, linking any one of the 40 *V* gene segments to any one of the 5 *J* gene segments. Heavy-chain genes undergo a similar rearrangement. In any given cell, however, only one allele of a light-chain gene and one allele of a heavy-chain gene are rearranged. Furthermore, the rearrangements are permanent and are passed on to the daughter cells when the lymphocyte divides.

After both a light-chain and a heavy-chain gene have been rearranged, antigen receptors can be synthesized. The rearranged genes are transcribed, and the transcripts are processed for translation. Following translation, the light chain and heavy chain assemble together, forming an antigen receptor (see Figure 35.10). Each pair of randomly rearranged heavy and light chains results in a different antigen-binding site. For the total population of B cells in a human body, the number of such combinations has been calculated as $3.5 \times 10^6$. Furthermore, mutations introduced during *VJ* recombination add additional variation, making the number of antigen-binding specificities even greater.

### Origin of Self-Tolerance

In adaptive immunity, how does the body distinguish self from nonself? Because antigen receptor genes are randomly rearranged, some immature lymphocytes produce receptors specific for epitopes on the organism's own molecules. If these self-reactive lymphocytes were not eliminated or inactivated, the immune system could not distinguish self from nonself and would attack body proteins, cells, and tissues. Instead, as lymphocytes mature in the bone marrow or thymus, their antigen receptors are tested for self-reactivity. Some B and T cells with receptors specific for the body's own molecules are destroyed by programmed cell death (see Concept 16.1). The remaining self-reactive lymphocytes are typically rendered nonfunctional, leaving only those lymphocytes that react to foreign molecules. Since the body normally lacks mature lymphocytes that can react against its own components, the immune system is said to exhibit self-tolerance.

▶ **Figure 35.10 Immunoglobulin (antibody) gene rearrangement.** The joining of randomly selected *V* and *J* gene segments ($V_{39}$ and $J_5$ in this example) results in a functional gene that encodes the light-chain polypeptide of a B cell antigen receptor. Transcription, splicing, and translation result in a light chain that combines with a polypeptide produced from an independently rearranged heavy-chain gene to form a functional receptor. Mature B cells (and T cells) are exceptions to the generalization that all nucleated cells in the body have exactly the same DNA.

**MAKE CONNECTIONS** *Both alternative splicing and joining of V and J segments by recombination generate diverse gene products from a limited set of gene segments. How do these processes differ (see Figure 15.12)?*

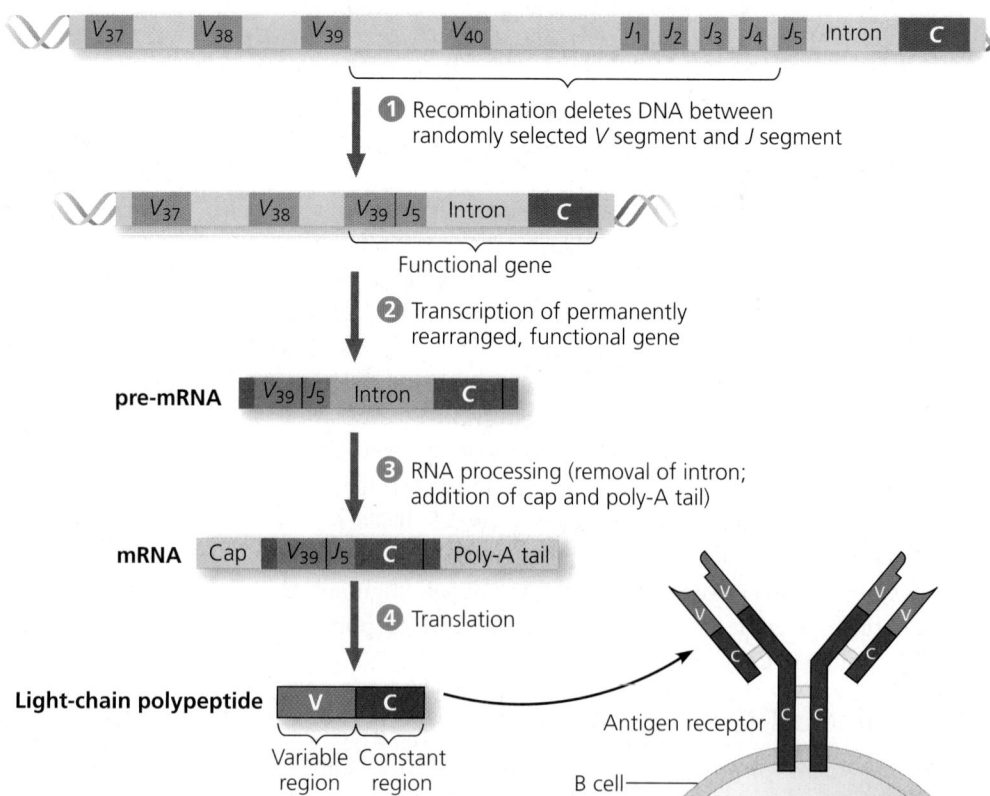

### Proliferation of B Cells and T Cells

Despite the enormous variety of antigen receptors, only a tiny fraction are specific for a given epitope. How, then, does an effective adaptive response develop? To begin with, an antigen is presented to a steady stream of lymphocytes in the lymph nodes (see Figure 34.12) until a match is made. A successful match between an antigen receptor and an epitope initiates events that activate the lymphocyte bearing the receptor.

Once activated, a B cell or T cell undergoes multiple cell divisions. For each activated cell, the result of this proliferation is a clone, a population of cells that are identical to the original cell. Some cells from this clone become **effector cells**, mostly short-lived cells that take effect immediately against the antigen and any pathogens producing that antigen. For B cells, the effector forms are **plasma cells**, which secrete antibodies. The For T cells, the effector forms are helper T cells and cytotoxic T cells, whose roles we'll explore in Concept 35.3. The remaining cells in the clone become **memory cells**, long-lived cells that can give rise to effector cells if the same antigen is encountered later in the animal's life.

**Figure 35.11** summarizes the proliferation of a lymphocyte into a clone of cells in response to binding to an antigen, using B cells as an example. This process is called **clonal selection** because an encounter with an antigen *selects* which

lymphocyte will divide to produce a *clonal* population of thousands of cells specific for a particular epitope.

### Immunological Memory

Immunological memory is responsible for the long-term protection that a prior infection provides against many diseases, such as chickenpox. This type of protection was noted almost 2,400 years ago by the Greek historian Thucydides. He observed that individuals who had recovered from the plague could safely care for those who were sick or dying, "for the same man was never attacked twice—never at least fatally."

Prior exposure to an antigen alters the speed, strength, and duration of the immune response. The production of effector cells from a clone of lymphocytes during the first exposure to an antigen is the basis for the **primary immune response**. The primary response peaks about 10–17 days after the initial exposure. During this time, selected B cells and T cells give rise to their effector forms. If an individual is exposed again to the same antigen, the response is faster (typically peaking only 2–7 days after exposure), of greater magnitude, and more prolonged. This is the **secondary immune response**, a hallmark of adaptive, or acquired, immunity. Because selected B cells give rise to antibody-secreting effector cells, measuring the concentrations

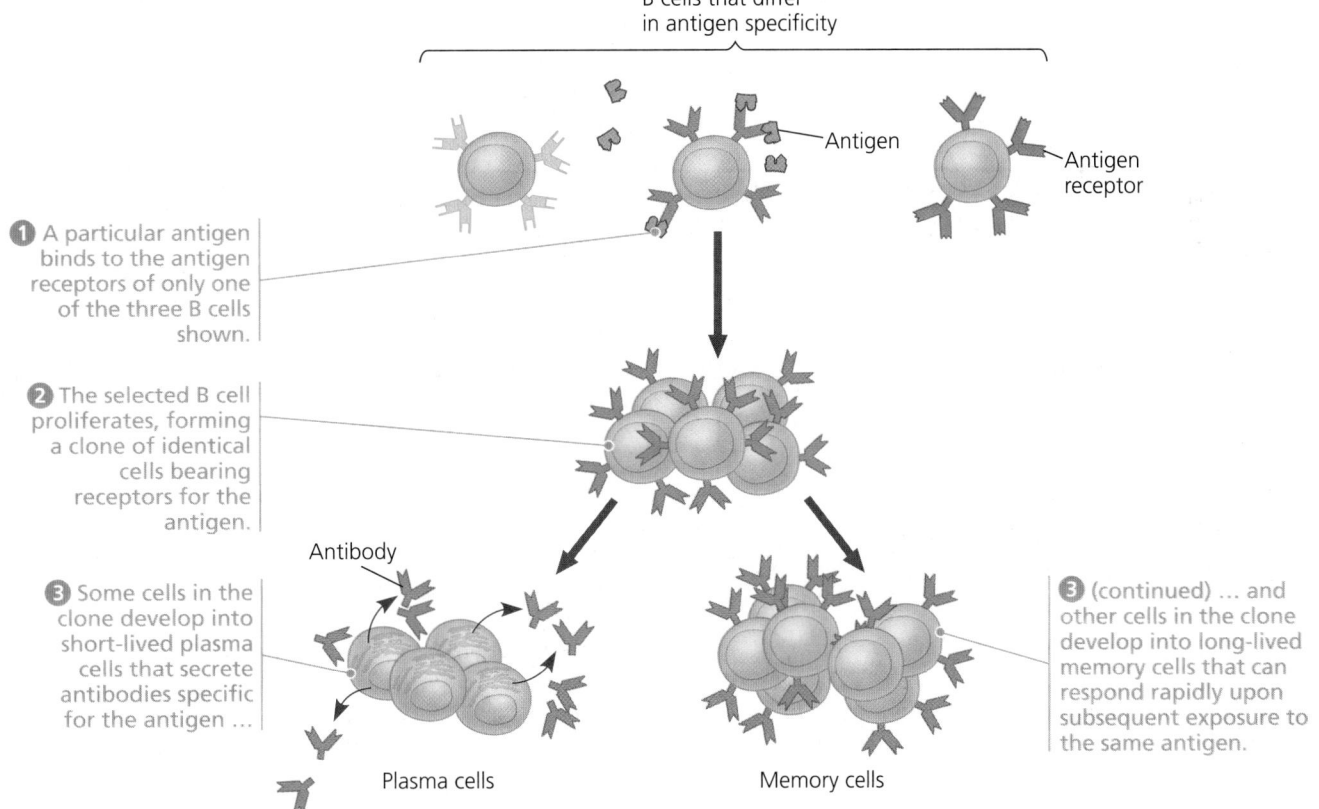

B cells that differ in antigen specificity

Antigen

Antigen receptor

❶ A particular antigen binds to the antigen receptors of only one of the three B cells shown.

❷ The selected B cell proliferates, forming a clone of identical cells bearing receptors for the antigen.

Antibody

❸ Some cells in the clone develop into short-lived plasma cells that secrete antibodies specific for the antigen ...

❸ (continued) ... and other cells in the clone develop into long-lived memory cells that can respond rapidly upon subsequent exposure to the same antigen.

Plasma cells

Memory cells

▲ **Figure 35.11 Clonal selection.** This figure illustrates clonal selection, using B cells as an example. In response to a specific antigen and to immune cell signals (not shown), one B cell divides and forms a clone of cells. The remaining B cells, which have antigen receptors specific for other antigens, do not respond. The clone of cells formed by the selected B cell gives rise to antibody-secreting plasma cells and memory B cells. T cells also undergo clonal selection, generating effector T cells (cytotoxic T cells and helper T cells) as well as memory T cells.

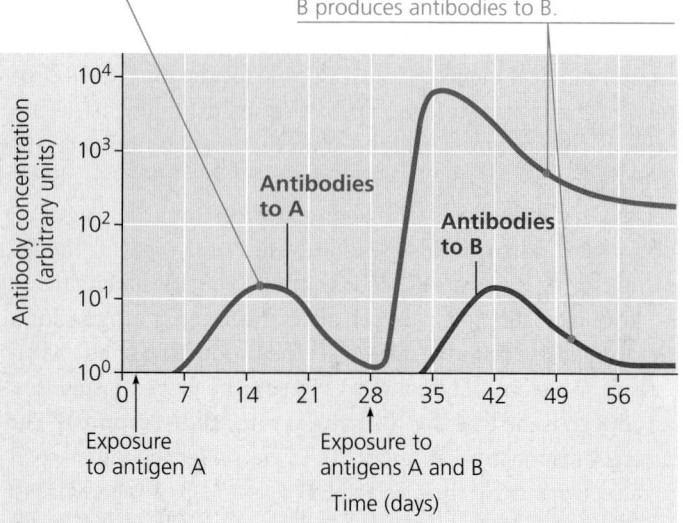

▲ **Figure 35.12 The specificity of immunological memory.**
Long-lived memory cells generated in the primary response to antigen A
give rise to a heightened secondary response to the same antigen but
do not affect the response to a different antigen (antigen B).

**INTERPRET THE DATA** *Assume that on average one out of every
$10^5$ B cells in the body is specific for antigen A on day 16 and that
the number of B cells producing a specific antibody is proportional
to the concentration of that antibody. What would you predict is the
frequency of B cells specific for antigen A on day 36?*

of specific antibodies in blood over time distinguishes the primary and secondary immune responses **(Figure 35.12)**.

The secondary immune response relies on the reservoir of T and B memory cells generated upon initial exposure to an antigen. Because these cells are long-lived, they provide the basis for immunological memory, which can span many decades. (Most effector cells have much shorter life spans, which is why the immune response diminishes after an infection is overcome.) If an antigen is encountered again, memory cells specific for that antigen enable the rapid formation of clones of thousands of effector cells also specific for that antigen, thus generating a greatly enhanced immune defense.

Although the processes for antigen recognition, clonal selection, and immunological memory are similar for B cells and T cells, these two classes of lymphocytes fight infection in different ways and in different settings, as we'll explore in Concept 35.3.

### CONCEPT CHECK 35.2

1. **DRAW IT** Sketch a B cell antigen receptor. Label the V and C regions. Where are the antigen-binding sites, disulfide bridges, and transmembrane region located relative to these regions?
2. Explain how memory cells strengthen the immune response when a pathogen is encountered for a second time.
3. **WHAT IF?** If both copies of a light-chain gene and a heavy-chain gene recombined in each (diploid) B cell, how would this affect B cell development and function?

For suggested answers, see Appendix A.

# Adaptive immunity defends against infection of body fluids and body cells

Having considered how clones of lymphocytes arise, we now explore how these cells help fight infections and minimize damage by pathogens. The defenses provided by B and T lymphocytes can be divided into humoral and cell-mediated immune responses. The **humoral immune response** protects the blood and lymph (once called body *humors*, or fluids). In this response, antibodies help neutralize or eliminate toxins and pathogens in these fluids. In the **cell-mediated immune response**, specialized T cells destroy infected host cells. Both humoral and cellular immunity can include a primary and a secondary immune response, with memory cells enabling the secondary response.

## Helper T Cells: Activating Adaptive Immunity

A type of T cell called a **helper T cell** activates humoral and cell-mediated immune responses. Before this can happen, however, two conditions must be met. First, a foreign molecule must be present that can bind specifically to the antigen receptor of the helper T cell. Second, this antigen must be displayed on the surface of an **antigen-presenting cell**. The antigen-presenting cell can be a dendritic cell, macrophage, or B cell.

When animal cells are infected, they, too, display foreign antigens on their surface. What, then, distinguishes an antigen-presenting cell? The answer lies in the existence of two classes of MHC molecules. Most body cells have only class I MHC molecules, but antigen-presenting cells have both class I and class II MHC molecules. The class II molecules provide a molecular signature by which an antigen-presenting cell is recognized.

A helper T cell and the antigen-presenting cell displaying its specific epitope have a complex interaction **(Figure 35.13)**. The antigen receptors on the surface of the helper T cell bind to the antigen fragment and to the class II MHC molecule displaying that fragment on the antigen-presenting cell. At the same time, an accessory protein called CD4 on the helper T cell surface binds to the class II MHC molecule, helping keep the cells joined. As the two cells interact, signals in the form of cytokines are exchanged.

Antigen-presenting cells interact with helper T cells in several different contexts. Antigen presentation by a dendritic cell or macrophage activates a helper T cell, which then proliferates, forming a clone of activated cells. B cells present antigens to *already* activated helper T cells, which in turn activate the B cells themselves. Activated helper T cells also help stimulate cytotoxic T cells, as we'll discuss next.

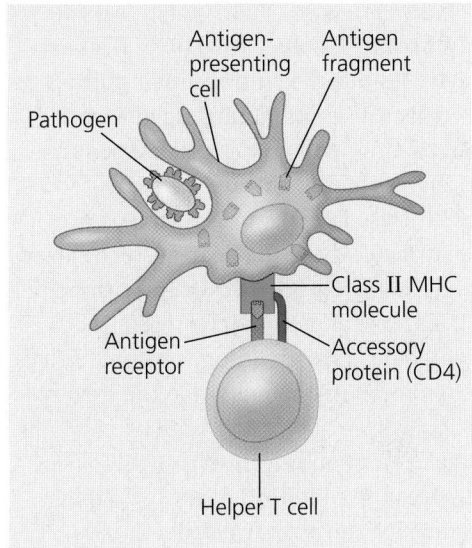

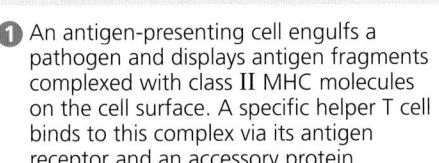

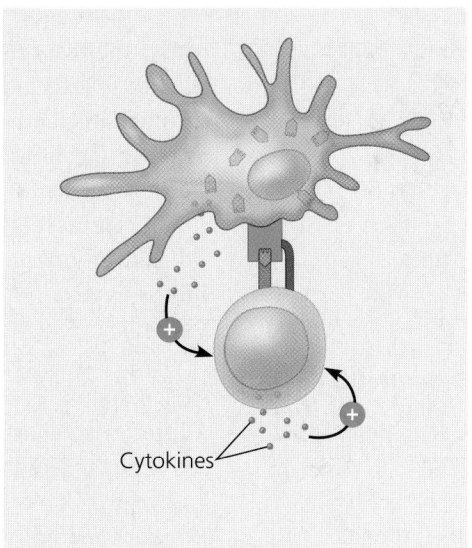

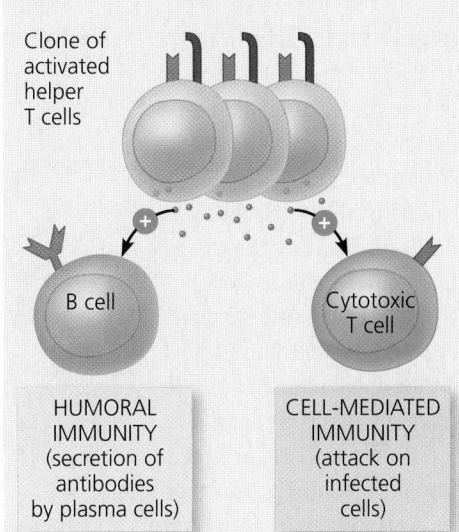

**①** An antigen-presenting cell engulfs a pathogen and displays antigen fragments complexed with class II MHC molecules on the cell surface. A specific helper T cell binds to this complex via its antigen receptor and an accessory protein.

**②** Binding of the helper T cell promotes secretion of cytokines by the antigen-presenting cell. These cytokines, along with cytokines from the helper T cell itself, activate the helper T cell and stimulate its proliferation.

**③** Cell proliferation produces a clone of activated helper T cells. All cells in the clone have receptors with the same antigen specificity. These cells secrete other cytokines, which help activate B cells and cytotoxic T cells.

▲ **Figure 35.13 The central role of helper T cells in humoral and cell-mediated immune responses.**

## B Cells and Antibodies: A Response to Extracellular Pathogens

The secretion of antibodies by clonally selected B cells is the hallmark of the humoral immune response. As illustrated in **Figure 35.14**, activation of B cells involves both helper T cells and proteins on the surface of pathogens. Stimulated by both an antigen and cytokines, the B cell proliferates and differentiates into memory B cells and antibody-secreting plasma cells.

A single activated B cell gives rise to thousands of identical plasma cells. Each plasma cell secretes approximately 2,000 antibodies every second during its 4- to 5-day life span, nearly a trillion antibody molecules in total. Antibodies do not directly kill pathogens, but by binding to antigens, they interfere with pathogen activity or mark pathogens for inactivation or destruction. Consider, for example, *neutralization*, a process in which antibodies bind to proteins on the surface of a virus

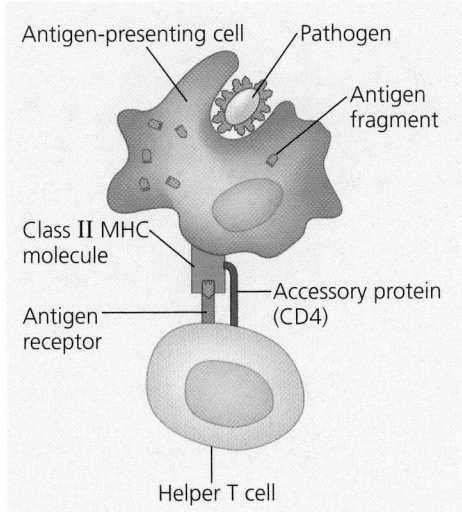

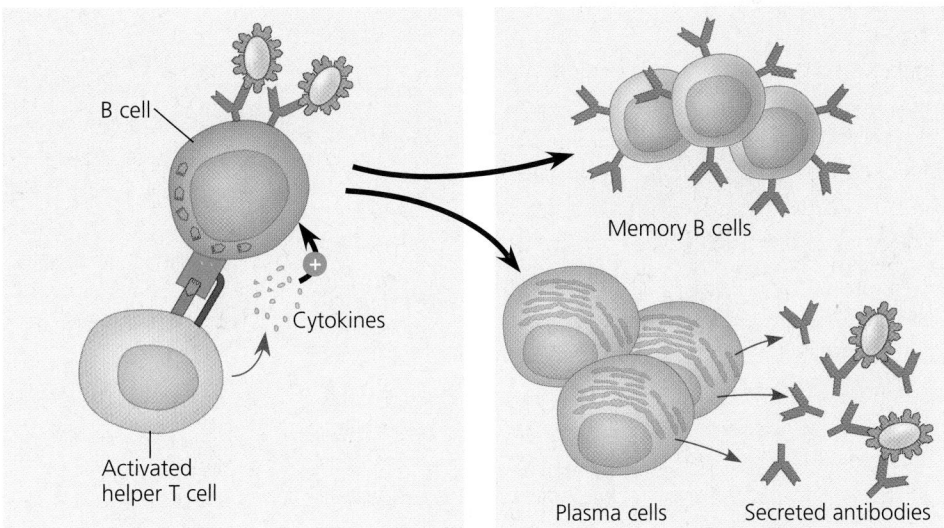

**①** After an antigen-presenting cell (here, a macrophage) engulfs and degrades a pathogen, it displays an antigen fragment complexed with a class II MHC molecule. A helper T cell that recognizes the complex is then activated.

**②** A B cell with receptors for the same epitope internalizes the antigen and displays a fragment on the cell surface in a complex with a class II MHC molecule. An activated helper T cell specific for the displayed fragment activates the B cell.

**③** The activated B cell proliferates and differentiates into memory B cells and antibody-secreting plasma cells. The secreted antibodies are specific for the same antigen that initiated the response.

▲ **Figure 35.14 Activation of a B cell in the humoral immune response.**

**(Figure 35.15).** The bound antibodies prevent infection of a host cell, thus neutralizing the virus. Similarly, antibodies sometimes bind to toxins released in body fluids, preventing the toxins from entering body cells. Because each antibody has two antigen-binding sites, antibodies can also facilitate phagocytosis by linking bacterial cells, viruses, or other foreign substances into aggregates.

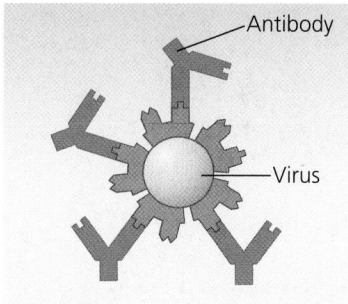

▲ **Figure 35.15 Neutralization.** Antibodies bind to antigens on the surface of a virus, blocking its ability to bind to a host cell.

Antibodies sometimes work together with the proteins of the complement system. (The name *complement* reflects the fact that these proteins add to the effectiveness of antibody-directed attacks on bacteria.) Binding of a complement protein to an antigen-antibody complex on a foreign cell triggers events leading to formation of a pore in the membrane of the cell. Ions and water rush into the cell, causing it to swell and lyse.

B cells can express five different types of immunoglobulin. For a given B cell, each type has an identical antigen-binding specificity but a distinct heavy-chain C region. One type of Ig, the B cell antigen receptor, is membrane bound. The other four Ig types consist of soluble antibodies, including those found in blood, tears, saliva, and breast milk.

## Cytotoxic T Cells: A Response to Infected Host Cells

In the absence of an immune response, pathogens can reproduce in and kill infected cells. In the cell-mediated immune response, **cytotoxic T cells** use toxic proteins to kill cells infected by viruses or other intracellular pathogens. To become active, cytotoxic T cells require signals from helper T cells and interaction with an antigen-presenting cell. Fragments of foreign proteins produced in infected host cells associate with class I MHC molecules and are displayed on the cell surface, where they can be recognized by activated cytotoxic T cells **(Figure 35.16).** As with helper T cells, cytotoxic T cells have an accessory protein that binds to the MHC molecule. This accessory protein, called CD8, helps keep the two cells in contact while the cytotoxic T cell is activated.

The targeted destruction of an infected host cell by a cytotoxic T cell involves the secretion of proteins that disrupt membrane integrity and trigger cell death (see Figure 35.16). The death of the infected cell not only deprives the pathogen of a place to multiply, but also exposes cell contents to circulating antibodies, which mark released antigens for disposal.

## Summary of the Humoral and Cell-Mediated Immune Responses

As noted earlier, both humoral and cell-mediated immunity can include primary as well as secondary immune responses. Memory cells of each type—helper T cell, B cell, and cytotoxic T cell—enable the secondary response. For example, when body fluids are reinfected by a pathogen encountered previously, memory B cells and memory helper T cells initiate a secondary humoral response. **Figure 35.17** summarizes adaptive immunity, reviews the events that initiate humoral and cell-mediated immune responses, and highlights the central role of the helper T cell.

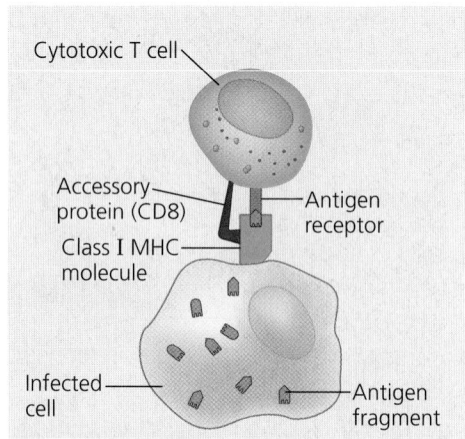

❶ An activated cytotoxic T cell binds to a class I MHC–antigen fragment complex on an infected cell via its antigen receptor and an accessory protein.

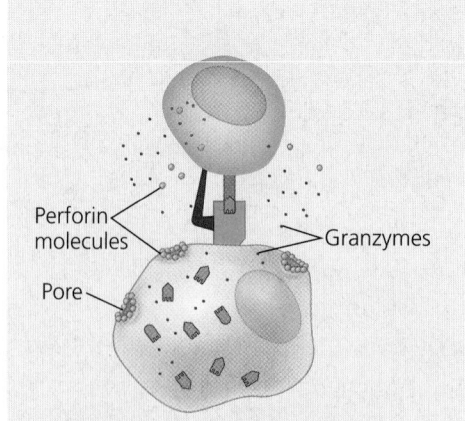

❷ The T cell releases perforin molecules, which form pores in the infected cell membrane, and granzymes, enzymes that break down proteins. Granzymes enter the infected cell by endocytosis.

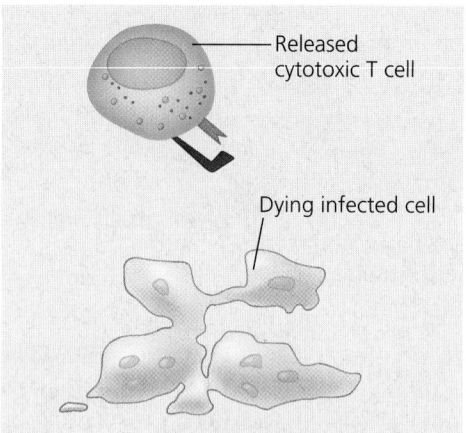

❸ The granzymes initiate apoptosis within the infected cell, leading to fragmentation of the nucleus and cytoplasm and eventual cell death. The released cytotoxic T cell can attack other infected cells.

▲ **Figure 35.16 The killing action of cytotoxic T cells on an infected host cell.** An activated cytotoxic T cell releases molecules that make pores in an infected cell's membrane and enzymes that break down proteins, promoting the cell's death.

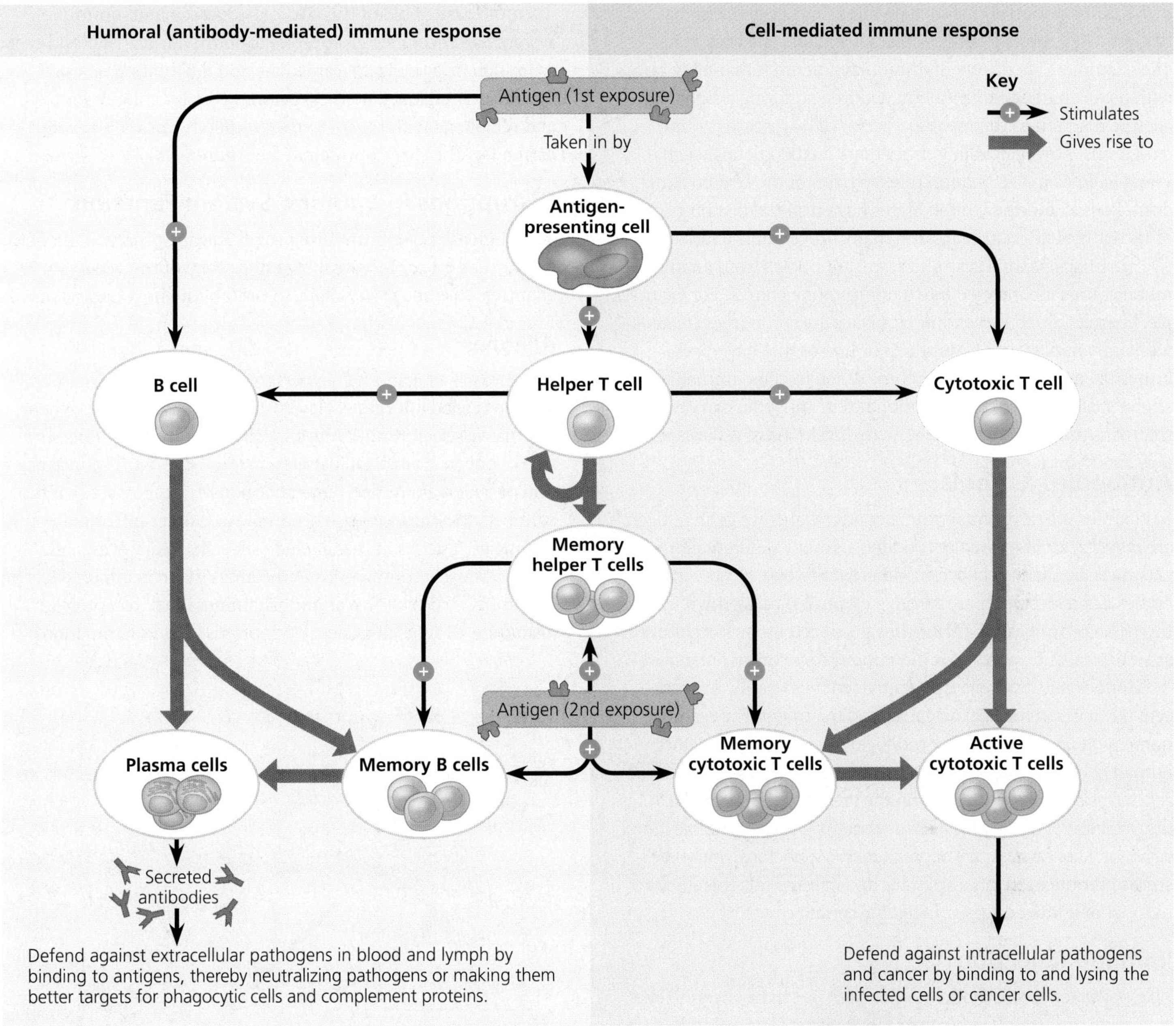

## Figure (Figure 35.17)

**Humoral (antibody-mediated) immune response** | **Cell-mediated immune response**

Key
+ → Stimulates
→ Gives rise to

Antigen (1st exposure)

Taken in by

Antigen-presenting cell

B cell ← Helper T cell → Cytotoxic T cell

Memory helper T cells

Antigen (2nd exposure)

Plasma cells ← Memory B cells ← → Memory cytotoxic T cells → Active cytotoxic T cells

Secreted antibodies

Defend against extracellular pathogens in blood and lymph by binding to antigens, thereby neutralizing pathogens or making them better targets for phagocytic cells and complement proteins.

Defend against intracellular pathogens and cancer by binding to and lysing the infected cells or cancer cells.

▲ **Figure 35.17 An overview of the adaptive immune response.**

**?** *Identify each black or brown arrow as representing part of the primary or secondary response.*

**ANIMATION** Visit the Study Area in **MasteringBiology** for the BioFix® 3-D Animation on Immunology.

## Active and Passive Immunity

Our discussion of adaptive immunity has focused to this point on *active immunity*, the defenses that arise when a pathogen infects the body. A different type of immunity results when, for example, antibodies in the blood of a pregnant female cross the placenta to her fetus. This protection is called *passive immunity* because the antibodies in the recipient (in this case, the fetus) are produced by another individual (the mother). Antibodies present in breast milk provide additional passive immunity to the infant's digestive tract while the infant's immune system develops. Because passive immunity does not involve the recipient's B and T cells, it

persists only as long as the transferred antibodies last (a few weeks to a few months).

Both active immunity and passive immunity can be induced artificially. Active immunity is induced when antigens are introduced into the body in *vaccines*, which may be made from inactivated bacterial toxins, killed or weakened pathogens, or even genes encoding microbial proteins. This process, called **immunization** (or vaccination), induces a primary immune response and immunological memory. As a result, any subsequent encounter with the pathogen from which the vaccine was derived triggers a rapid and strong secondary immune response (see Figure 35.12).

Misinformation about vaccine safety and disease risk has led to a growing public health problem. Consider measles as just one example. Side effects of immunization are remarkably rare, with fewer than one in a million children suffering a significant allergic reaction to the measles vaccine. The disease is quite dangerous, however, killing more than 200,000 people each year. Declines in vaccination rates in parts of the United Kingdom, Russia, and the United States have resulted in a number of recent measles outbreaks and many preventable deaths.

In artificial passive immunization, antibodies from an immune animal are injected into a nonimmune animal. For example, humans bitten by venomous snakes are sometimes treated with antivenin, serum from sheep or horses that have been immunized against a snake venom. When injected immediately after a snakebite occurs, the antibodies in antivenin can neutralize toxins in the venom before the toxins do massive damage.

## Antibodies as Tools

Antibodies that an animal produces after exposure to an antigen are *polyclonal*: They are the products of many different clones of plasma cells, each specific for a different epitope. However, antibodies can also be prepared from a clone of B cells grown in culture. The **monoclonal antibodies** produced by such a culture are identical and specific for the same epitope on an antigen.

Monoclonal antibodies have provided the basis for many recent advances in medical diagnosis and treatment. For example, home pregnancy test kits use monoclonal antibodies to detect human chorionic gonadotropin (hCG). Because hCG is produced as soon as an embryo implants in the uterus (see Concept 36.4), the presence of this hormone in a woman's urine is a reliable indicator for a very early stage of pregnancy. Monoclonal antibodies are also produced in large amounts and injected as a therapy for a number of human diseases, including certain cancers.

## Immune Rejection

Like pathogens, cells from another person can be recognized as foreign and attacked by immune defenses. For example, skin transplanted from one person to a genetically nonidentical person will look healthy for a week or so but will then be destroyed (rejected) by the recipient's immune response. It turns out that MHC molecules are a primary cause of rejection. Why? Each of us expresses MHC proteins from more than a dozen different MHC genes. Furthermore, there are more than 100 different versions, or alleles, of human MHC genes. As a consequence, the sets of MHC proteins on cell surfaces are likely to differ between any two people, except identical twins. Such differences can stimulate an immune response in the recipient of a transplant or graft, causing rejection.

To minimize rejection of a transplant or graft, surgeons use donor tissue bearing MHC molecules that match those of the recipient as closely as possible. In addition, the recipient takes medicines that suppress immune responses (but as a result leave the recipient more susceptible to infections).

In the case of blood transfusions, the recipient's immune system can recognize glycoproteins on the surface of blood cells as foreign, triggering an immediate and devastating reaction. To avoid this danger, medical personnel match blood donors and recipients with regard to the so-called A,B, and O groups defined by these glycoproteins (see Figure 11.11).

## Disruptions in Immune System Function

Although adaptive immunity protects against many pathogens, it is not fail-safe. Here we'll examine some of the ways the adaptive immune system fails to protect the host organism.

### Allergies

Allergies are exaggerated (hypersensitive) responses to certain antigens called **allergens**. Hay fever, for instance, occurs when plasma cells secrete antibodies specific for antigens on the surface of pollen grains, as illustrated in **Figure 35.18**. The interaction of pollen grains and these antibodies triggers mast cells in connective tissue to release histamine and other inflammatory chemicals. The results can include sneezing, teary eyes, and smooth muscle contractions in the lungs that can inhibit effective breathing. Drugs known as antihistamines block receptors for histamine, diminishing allergy symptoms (and inflammation).

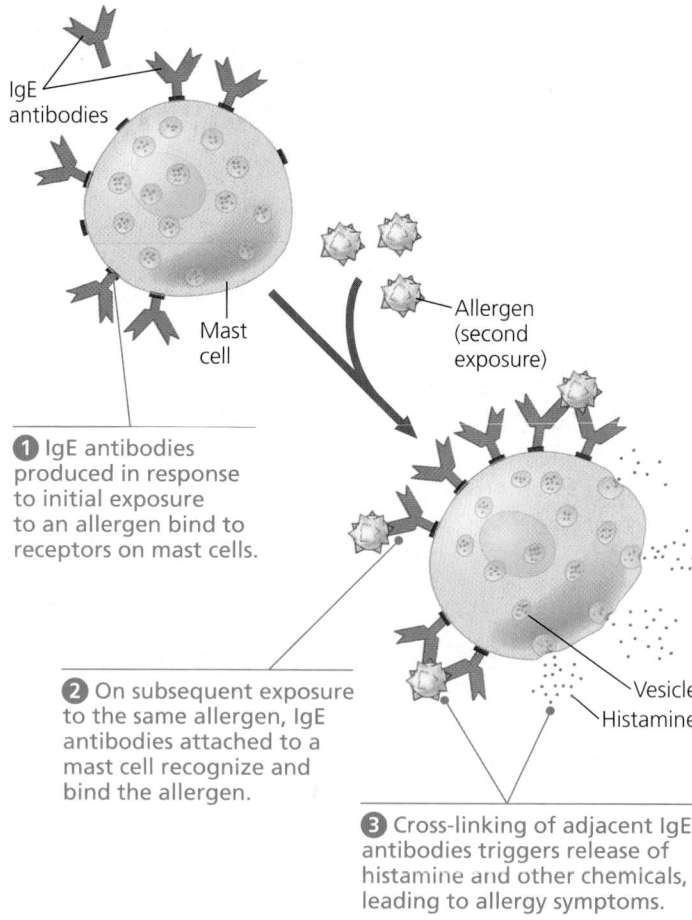

❶ IgE antibodies produced in response to initial exposure to an allergen bind to receptors on mast cells.

❷ On subsequent exposure to the same allergen, IgE antibodies attached to a mast cell recognize and bind the allergen.

❸ Cross-linking of adjacent IgE antibodies triggers release of histamine and other chemicals, leading to allergy symptoms.

▲ **Figure 35.18 Mast cells and the allergic response.** In this example, pollen grains act as the allergen, and the immunoglobulins that mediate the response are of a type called IgE.

An acute allergic response sometimes leads to a life-threatening reaction called *anaphylactic shock*. Inflammatory chemicals released from immune cells trigger constriction of bronchioles and sudden dilation of peripheral blood vessels, which causes a precipitous drop in blood pressure. Death may occur within minutes due to the inability to breathe and lack of blood flow. Substances that can cause anaphylactic shock in allergic individuals include bee venom, penicillin, peanuts, and shellfish. People with severe hypersensitivities often carry syringes containing the hormone epinephrine, which counteracts this allergic response.

## Autoimmune Diseases

In some people, the immune system is active against particular molecules of the body, causing an **autoimmune disease**. In systemic lupus erythematosus, commonly called lupus, the immune system generates antibodies against histones and DNA. Other targets of autoimmunity include the insulin-producing beta cells of the pancreas (in type 1 diabetes) and the myelin sheaths that encase many neurons (in multiple sclerosis).

Gender, heredity, and environment all influence susceptibility to autoimmune disorders. For example, many autoimmune diseases afflict females more often than males. Women are nine times as likely as men to suffer from lupus and two to three times as likely to develop *rheumatoid arthritis*, a damaging and painful inflammation of the cartilage and bone in joints **(Figure 35.19)**. The causes of this sex bias, as well as of the rise in autoimmune disease frequency in industrialized countries, are areas of active research and debate.

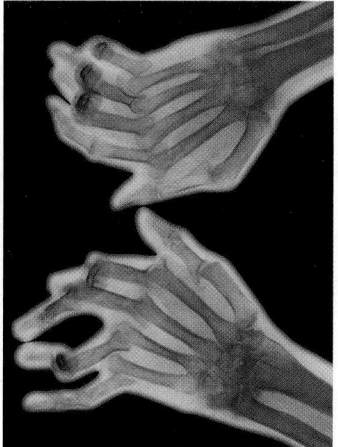

▲ **Figure 35.19 Colored X-ray of hands deformed by rheumatoid arthritis.**

## Immune System Avoidance

**EVOLUTION** Just as immune systems that ward off pathogens have evolved in animals, mechanisms that thwart immune responses have evolved in pathogens. In one such mechanism, a pathogen alters how it appears to the immune system. If a pathogen changes the epitopes it expresses to ones that a host has not previously encountered, it can reinfect or remain in the host without triggering the rapid and robust response mediated by memory cells. Such changes in epitope expression are called *antigenic variation*. The parasite that causes sleeping sickness provides an extreme example, periodically switching at random among 1,000 different versions of the protein found over its entire surface. In the **Scientific Skills Exercise**, you will interpret data on this form of antigenic variation and the body's response.

Antigenic variation is the main reason the influenza, or "flu," virus remains a major public health problem. As it replicates in one human host after another, the human flu virus undergoes frequent mutations. Because any change that lessens recognition by the immune system provides a selective advantage, the virus steadily accumulates mutations that change its surface proteins, reducing the effectiveness of the host immune response. As a result, a new flu vaccine must be developed, produced, and distributed each year. In addition, the human flu virus occasionally forms new strains by exchanging genes with influenza viruses that infect domesticated animals, such as pigs or chickens. When this occurs, the new strain may not be recognized by any of the memory cells in the human population. The resulting outbreak can be deadly: The 1918–1919 influenza outbreak killed more than 20 million people.

Some viruses avoid an immune response by infecting cells and then entering a largely inactive state called *latency*. In latency, the production of most viral proteins and free viruses ceases; as a result, latent viruses do not trigger an adaptive immune response. Latency typically persists until conditions arise that are favorable for viral transmission or unfavorable for host survival.

Herpes simplex viruses provide a good example of latency. The type 1 virus causes most oral herpes infections, whereas the sexually transmitted type 2 virus is responsible for most cases of genital herpes. These viruses remain latent in sensory neurons until a stimulus such as fever, emotional stress, or menstruation reactivates the viruses. Activation of the type 1 virus can result in blisters around the mouth that are inaccurately called "cold" sores. Infections of the type 2 virus pose a serious threat to the babies of infected mothers and can increase transmission of HIV.

The **human immunodeficiency virus (HIV)**, the pathogen that causes AIDS, both escapes and attacks the adaptive immune response. Once introduced into the body, HIV infects helper T cells with high efficiency. Although the body responds to HIV with an immune response sufficient to eliminate most viral infections, some HIV invariably escapes. One reason HIV persists is that it has a very high mutation rate. Altered proteins on the surface of some mutated viruses reduce interaction with antibodies and cytotoxic T cells. Such viruses replicate and mutate further. HIV thus evolves within the body. The continued presence of HIV is also helped by latency.

Over time, an untreated HIV infection not only avoids the adaptive immune response but also abolishes it. Viral replication and cell death triggered by the virus lead to loss of helper T cells, impairing both humoral and cell-mediated immune responses. The eventual result is *acquired immune deficiency syndrome* (AIDS), an impairment in immune responses that leaves the body susceptible to infections and cancers that a healthy immune system would usually defeat. For example, *Pneumocystis carinii*, a common fungus that does not cause disease in healthy individuals, can result in severe pneumonia in people with AIDS. Such opportunistic diseases, as well as nerve damage and wasting, are the primary causes of death from AIDS, not HIV itself.

# Comparing Two Variables on a Common x-Axis

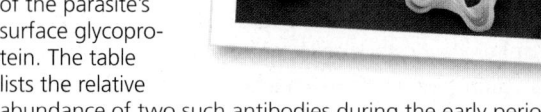

**How Does the Immune System Respond to a Changing Pathogen?** Natural selection favors parasites that are able to maintain a low-level infection in a host for a long time. *Trypanosoma*, the unicellular parasite that causes sleeping sickness, is one example. The glycoproteins covering a trypanosome's surface are encoded by a gene that is duplicated more than a thousand times in the organism's genome. Each copy is slightly different. By periodically switching among these genes, the trypanosome can display a series of surface glycoproteins with different molecular structures. In this exercise, you will interpret two data sets to explore hypotheses about the benefits of the trypanosome's ever-shifting surface glycoproteins and the host's immune response.

**Part A: Data from a Study of Parasite Levels** This study measured the abundance of parasites in the blood of one human patient during the first few weeks of a chronic infection.

| Day | Number of Parasites (in millions) per mL of Blood |
|---|---|
| 4 | 0.1 |
| 6 | 0.3 |
| 8 | 1.2 |
| 10 | 0.2 |
| 12 | 0.2 |
| 14 | 0.9 |
| 16 | 0.6 |
| 18 | 0.1 |
| 20 | 0.7 |
| 22 | 1.2 |
| 24 | 0.2 |

### PART A: INTERPRET THE DATA

1. Plot the data in the table in part A as a line graph. Which column is the independent variable, and which is the dependent variable? Put the independent variable on the *x*-axis. (For additional information about graphs, see the Scientific Skills Review in Appendix F and in the Study Area in MasteringBiology.)

2. Visually displaying data in a graph can help make patterns in the data more noticeable. Describe any patterns revealed by your graph.

3. Assume that a drop in parasite abundance reflects an effective immune response by the host. Formulate a hypothesis to explain the pattern you described in question 2.

**Part B: Data from a Study of Antibody Levels** Many decades after scientists first observed the pattern of *Trypanosoma* abundance over the course of infection, researchers identified antibodies specific

to different forms of the parasite's surface glycoprotein. The table lists the relative abundance of two such antibodies during the early period of chronic infection, using an index ranging from 0 (absent) to 1.

| Day | Antibody Specific to Glycoprotein Variant A | Antibody Specific to Glycoprotein Variant B |
|---|---|---|
| 4 | 0 | 0 |
| 6 | 0 | 0 |
| 8 | 0.2 | 0 |
| 10 | 0.5 | 0 |
| 12 | 1 | 0 |
| 14 | 1 | 0.1 |
| 16 | 1 | 0.3 |
| 18 | 1 | 0.9 |
| 20 | 1 | 1 |
| 22 | 1 | 1 |
| 24 | 1 | 1 |

**Data from** L. J. Morrison et al., Probabilistic order in antigenic variation of *Trypanosoma brucei*, *International Journal for Parasitology* 35:961–972 (2005) and L. J. Morrison et al., Antigenic variation in the African trypanosome: molecular mechanisms and phenotypic complexity, *Cellular Microbiology* 1:1724–1734 (2009).

### PART B: INTERPRET THE DATA

4. Note that the data in the table in part B were collected over the same period of infection (days 4–24) as the parasite abundance data in part A. Therefore, you can incorporate these new data into your first graph, using the same *x*-axis. However, since the antibody level data are measured in a different way than the parasite abundance data, add a second set of *y*-axis labels on the right side of your graph. Then, using different colors or sets of symbols, graph the data for the two antibody types. Labeling the *y*-axis two different ways enables you to compare how two dependent variables change relative to a shared independent variable.

5. Describe any patterns you observe by comparing the two data sets over the same period. Do these patterns support your hypothesis from part A? Do they prove that hypothesis? Explain.

6. Scientists can now also distinguish the abundance of trypanosomes recognized specifically by antibodies type A and type B. How would incorporating such information change your graph?

(MB) A version of this Scientific Skills Exercise can be assigned in MasteringBiology.

Transmission of HIV requires the transfer of virus particles or infected cells via body fluids such as semen, blood, or breast milk. Unprotected sex (that is, without using a condom) and transmission via HIV-contaminated needles (often among intravenous drug users) cause the vast majority of HIV infections. People infected with HIV can transmit the disease in the first few weeks of infection, *before* they express HIV-specific antibodies that can be detected in a blood test. Currently, 10–50% of all new HIV infections appear to be caused by recently infected individuals. Although no cure has been found

for HIV infection, drugs that can significantly slow HIV replication and the progression to AIDS have been developed.

## Cancer and Immunity

When adaptive immunity is inactivated, the frequency of certain cancers increases dramatically. For example, the risk of developing Kaposi's sarcoma is 20,000 times greater for untreated AIDS patients than for healthy people. This observation was initially puzzling. If the immune system recognizes only nonself, it should fail to recognize the uncontrolled growth of self cells that is the hallmark of cancer. It turns out, however, that viruses are involved in about 15–20% of all human cancers. Because the immune system can recognize viral proteins as foreign, it can act as a defense against viruses that can cause cancer and against cancer cells that harbor viruses.

Scientists have identified six viruses that can cause cancer in humans. The Kaposi's sarcoma herpesvirus is one such virus. Hepatitis B virus, which can trigger liver cancer, is another. A vaccine introduced in 1986 for hepatitis B virus was the first vaccine shown to help prevent a specific human cancer. Rapid progress on developing vaccines for virus-induced cancers continues. In 2006, the release of a vaccine specific for human papillomavirus (HPV) marked a major victory against cervical cancer, as well as other cancers that can affect sexually active men and women.

# 35 Chapter Review

Go to **MasteringBiology**® for Assignments, the eText, and the Study Area with Animations, Activities, Vocab Self-Quiz, and Practice Tests.

## SUMMARY OF KEY CONCEPTS

**VOCAB SELF-QUIZ**

goo.gl/gbai8v

### CONCEPT 35.1

**In innate immunity, recognition and response rely on traits common to groups of pathogens (pp. 734–737)**

- In both invertebrates and vertebrates, **innate immunity** is mediated by physical and chemical barriers as well as cell-based defenses. Activation of innate immune responses relies on recognition proteins specific for broad classes of **pathogens**. Microbes that penetrate barrier defenses are ingested by phagocytic cells, which in vertebrates include **macrophages** and **dendritic cells**. In the **inflammatory response**, **histamine** and other chemicals released at the injury site promote changes in blood vessels that enhance immune cell access and action.

**?** *In what ways does innate immunity protect the mammalian digestive tract?*

### CONCEPT 35.2

**In adaptive immunity, receptors provide pathogen-specific recognition (pp. 737–742)**

- **Adaptive immunity** relies on two types of **lymphocytes** that arise from stem cells in the bone marrow: **T cells** and **B cells**. Lymphocytes have cell-surface **antigen receptors** for foreign molecules (**antigens**). Some T cells help other lymphocytes; others kill infected host cells. B cells called **plasma cells** produce soluble receptor proteins called **antibodies**, which bind to foreign molecules and cells. Activated T and B lymphocytes called **memory cells** defend against future infections by the same pathogen.
- Recognition of foreign molecules involves the binding of variable regions of receptors to an **epitope**, a small region of an antigen. B cells and antibodies recognize epitopes on the surface of antigens. T cells recognize protein epitopes in small antigen fragments (peptides) that are presented on the surface of host cells

by proteins called **major histocompatibility complex (MHC) molecules**.
- The four major characteristics of B and T cell development are the generation of cell diversity, self-tolerance, proliferation, and immunological memory. Proliferation and memory are both based on **clonal selection**, illustrated here for B cells.

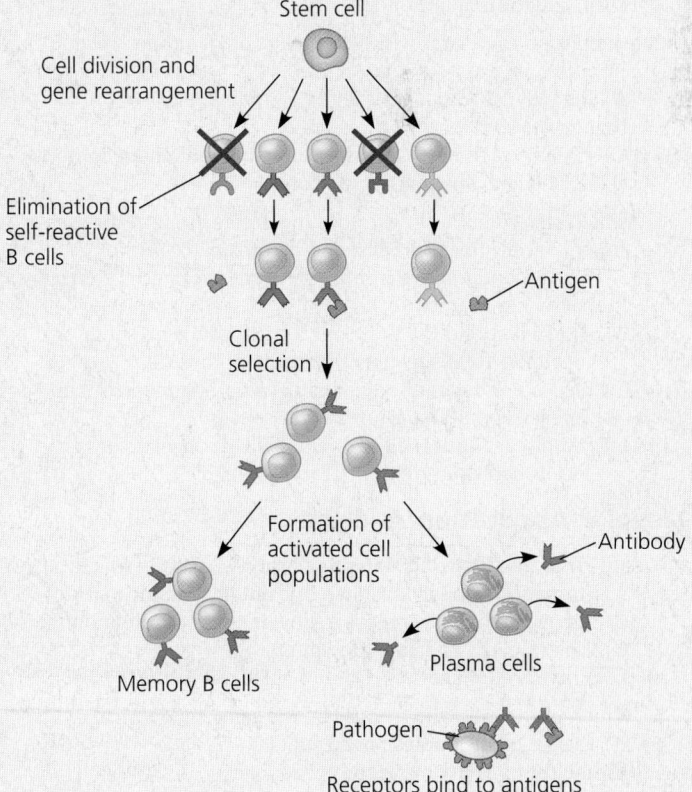

Stem cell

Cell division and gene rearrangement

Elimination of self-reactive B cells

Antigen

Clonal selection

Formation of activated cell populations

Memory B cells

Antibody

Plasma cells

Pathogen

Receptors bind to antigens

**?** *Why is the adaptive immune response to an initial infection slower than the innate response?*

## Adaptive immunity defends against infection of body fluids and body cells (pp. 742–749)

- **Helper T cells** interact with antigen fragments displayed by class II MHC molecules on the surface of **antigen-presenting cells:** dendritic cells, macrophages, and B cells. Activated helper T cells secrete **cytokines** that stimulate other lymphocytes. In the **humoral immune response**, antibodies help eliminate antigens by facilitating phagocytosis and complement-mediated lysis. In the **cell-mediated immune response**, activated **cytotoxic T cells** trigger destruction of infected cells.
- Active immunity develops in response to infection or to **immunization**. The transfer of antibodies in passive immunity provides immediate, short-term protection.
- In organ transplants, MHC molecules stimulate immune rejection.
- In allergies, the interaction of antibodies and **allergens** triggers immune cells to release histamine and other mediators that cause vascular changes and allergic symptoms. Loss of self-tolerance can lead to **autoimmune diseases**, such as multiple sclerosis.
- Antigenic variation, latency, and direct assault on the immune system allow some pathogens to thwart immune responses. **HIV** infection destroys helper T cells.

> **?** *Do natural infection and immunization result in different types of immunological memory? Explain.*

# TEST YOUR UNDERSTANDING

## Level 1: Knowledge/Comprehension

**PRACTICE TEST**

goo.gl/gbai8v

1. Which of these is NOT part of insect immunity?
   (A) antibacterial digestive enzymes
   (B) activation of natural killer cells
   (C) phagocytosis by hemocytes
   (D) production of antimicrobial peptides

2. An epitope associates with which part of an antigen receptor or antibody?
   (A) the tail
   (B) heavy-chain constant regions only
   (C) variable regions of a heavy chain and light chain combined
   (D) light-chain constant regions only

3. Which statement best describes the difference in responses of effector B cells (plasma cells) and cytotoxic T cells?
   (A) B cells confer active immunity; cytotoxic T cells confer passive immunity.
   (B) B cells respond the first time a pathogen is present; cytotoxic T cells respond subsequent times.
   (C) B cells secrete antibodies against a pathogen; cytotoxic T cells kill pathogen-infected host cells.
   (D) B cells carry out the cell-mediated response; cytotoxic T cells carry out the humoral response.

## Level 2: Application/Analysis

4. Which of the following statements is NOT true?
   (A) An antibody has more than one antigen-binding site.
   (B) A lymphocyte has receptors for multiple different antigens.
   (C) An antigen can have different epitopes.
   (D) A liver cell makes one class of MHC molecule.

5. Which of the following should be the same in identical twins?
   (A) the set of antibodies produced
   (B) the set of MHC molecules produced
   (C) the set of T cell antigen receptors produced
   (D) the set of immune cells eliminated as self-reactive

## Level 3: Synthesis/Evaluation

6. Vaccination increases the number of
   (A) different receptors that recognize a pathogen.
   (B) lymphocytes with receptors that can bind to the pathogen.
   (C) epitopes that the immune system can recognize.
   (D) MHC molecules that can present an antigen.

7. Which of the following would NOT help a virus avoid triggering an adaptive immune response?
   (A) having frequent mutations in genes for surface proteins
   (B) infecting cells that produce very few MHC molecules
   (C) producing proteins very similar to those of other viruses
   (D) infecting and killing helper T cells

8. **DRAW IT** Consider a pencil-shaped protein with two epitopes, Y (the "eraser" end) and Z (the "point" end). They are recognized by antibodies A1 and A2, respectively. Draw and label a picture showing the antibodies linking proteins into a complex that could trigger endocytosis by a macrophage.

9. **MAKE CONNECTIONS** Contrast clonal selection with Lamarck's idea for the inheritance of acquired characteristics (see Concept 19.1).

10. **SCIENTIFIC INQUIRY**
    The presence of bacterial lipopolysaccharide (LPS) in the blood is a major cause of septic shock. Suppose you have available purified LPS and several strains of mice, each with a mutation that inactivates a particular TLR gene. Explain how you might use these mice to test the feasibility of treating septic shock with a drug that blocks TLR signaling.

11. **FOCUS ON EVOLUTION**
    Describe one invertebrate defense mechanism and discuss how it is an evolutionary adaptation retained in vertebrates.

12. **FOCUS ON INFORMATION**
    Among all nucleated body cells, only B and T cells lose DNA during their development and maturation. In a short essay (100–150 words), discuss the relationship between this loss and the theme of DNA as heritable biological information, focusing on similarities between cellular and organismal generations.

13. **SYNTHESIZE YOUR KNOWLEDGE**

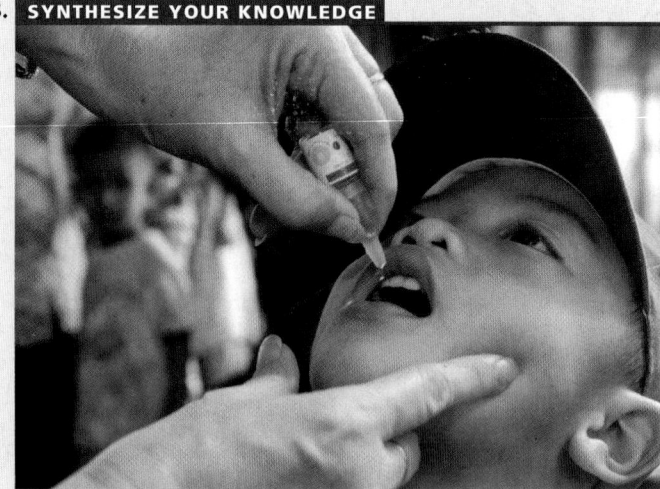

This child is receiving an oral vaccine against polio, a disease caused by a virus that infects neurons. Given that the body cannot replace most neurons, why must a polio vaccine stimulate both a cell-mediated and a humoral response?

*For selected answers, see Appendix A.*

▲ **Figure 36.1** How can each of these sea slugs be both male and female?

## Let Me Count the Ways

The sea slugs, or nudibranchs (*Nembrotha chamberlaini*), in **Figure 36.1** are mating. If not disturbed, these marine molluscs may remain joined for hours as sperm are transferred and eggs are fertilized. A few weeks later, new individuals will hatch, and sexual reproduction will be complete—but which parent is the mother of these offspring? The answer is simple yet probably unexpected: both. Each sea slug produces eggs *and* sperm and therefore will be both a mother and a father to the next generation.

As humans, we tend to think of reproduction in terms of the mating of males and females and the fusion of sperm and eggs. Across the animal kingdom, however, reproduction takes many forms. In some species, individuals change their sex during their lifetime; in other species, such as sea slugs, an individual is both male and female. There are animals that can fertilize their own eggs, as well as others that can reproduce

without any form of sex. For example, a sea anemone can simply split in two. In certain species, such as honeybees, only a few members of a large population reproduce.

A population outlives its members only by reproduction, the generation of new individuals from existing ones. In this chapter, we'll compare the diverse reproductive mechanisms that have evolved in the animal kingdom. Then we'll examine details of reproduction in mammals, with particular emphasis on the intensively studied example of humans. Lastly, we'll explore fundamental events in the earliest stages in the development of an animal from a fertilized egg.

▶ **Sea anemone dividing into two offspring**

# Both asexual and sexual reproduction occur in the animal kingdom

There are two modes of animal reproduction—sexual and asexual. In **sexual reproduction**, the fusion of haploid gametes forms a diploid cell, the **zygote**. The animal that develops from a zygote can in turn give rise to gametes by meiosis (see Figure 10.8). The female gamete, the **egg**, is large and nonmotile, whereas the male gamete, the **sperm**, is generally much smaller and motile. In **asexual reproduction**, new individuals are generated without the fusion of egg and sperm. For most asexual animals, reproduction relies entirely on mitosis. Asexual and sexual reproduction are both common in nature **(Figure 36.2)**. We'll consider each of them in turn.

## Mechanisms of Asexual Reproduction

Among animals, several simple forms of asexual reproduction are found exclusively in invertebrates. One of these is *budding*, in which new individuals arise from outgrowths of existing ones (see Figure 10.2). In stony corals, for example, buds form and remain attached to the parent. The eventual result is a colony more than 1 m across, consisting of thousands of connected individuals. Also common among invertebrates is *fission*, the splitting of a parent organism into two individuals of approximately equal size, as in sea anemones.

One form of asexual reproduction is a two-step process: *fragmentation*, the breaking of the body into several pieces, followed by *regeneration*, the regrowth of lost body parts. If more than one piece grows and develops into a complete animal, the net effect is reproduction. For example, certain annelid worms can split their body into several fragments, each regenerating a complete worm in less than a week. Many sponges, cnidarians, bristle worms, sea stars, and sea squirts also reproduce by fragmentation and regeneration.

▲ **Figure 36.2 Varied modes of reproduction among coral reef animals.** Some reef animals reproduce sexually, including the orange and white clown fish. Many others can reproduce either sexually or asexually, including the sea star at the lower left, the finger-like sea anemones, and the coral itself.

A particularly intriguing form of asexual reproduction is **parthenogenesis**, in which an egg develops without being fertilized. Among invertebrates, parthenogenesis occurs in certain species of bees, wasps, and ants. The offspring can be either haploid or diploid. If haploid, they develop into adults that produce eggs or sperm without meiosis.

Among vertebrates, parthenogenesis is thought to be a rare response to low population density. For example, for both the Komodo dragon and hammerhead shark, zookeepers observed that a female produced offspring when kept apart from males of its species. In 2015, DNA analysis revealed an example of vertebrate parthenogenesis in the wild, a group of female sawfish that were genetically identical to one another.

## Sexual Reproduction: An Evolutionary Enigma

**EVOLUTION** Although our species and many others reproduce sexually, the existence of sexual reproduction is actually puzzling. To see why, imagine an animal population in which half the females reproduce sexually and half reproduce asexually **(Figure 36.3)**. We'll assume that the number of offspring per female is a constant, two in this case. The offspring of an asexual female will both be daughters that will each give birth to two more daughters that can reproduce. In contrast, on average, half of a sexual female's offspring will be male. The number of sexual offspring will remain the same at each generation, because both a male and a female are required to reproduce. Thus, the asexual condition will increase in frequency at each generation. Yet despite this "twofold cost," sex is maintained even in animal species that can also reproduce asexually.

What advantage does sex provide that counteracts its twofold cost? The answer remains elusive. Most hypotheses focus on the unique combinations of parental genes formed during meiotic recombination and fertilization. By producing offspring of varied genotypes, sexual reproduction may enhance the reproductive success of parents when environmental factors, such as pathogens, change relatively rapidly. In contrast, we would expect asexual reproduction to be most advantageous in stable, favorable environments because it perpetuates successful genotypes precisely.

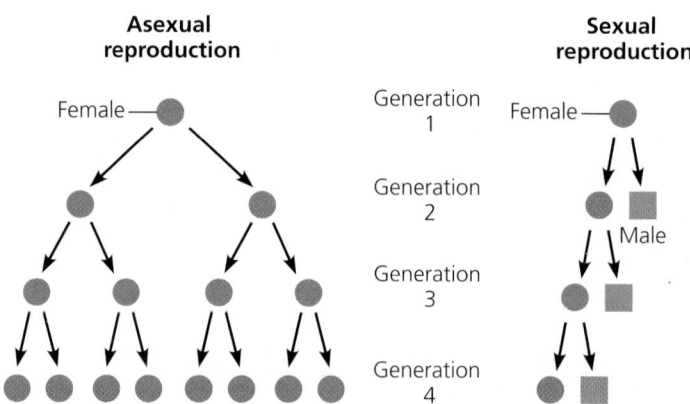

▲ **Figure 36.3 The "reproductive handicap" of sex.** These diagrams contrast asexual versus sexual reproduction over four generations, assuming two surviving offspring per female.

## Reproductive Cycles

Most animals, whether asexual or sexual, exhibit cycles in reproductive activity, often related to changing seasons. These cycles are controlled by hormones, whose secretion is in turn regulated by environmental cues. In this way, animals conserve resources, reproducing only when sufficient energy sources or stores are available and when environmental conditions favor the survival of offspring.

Because seasonal temperature is often an important cue for reproduction, climate change can decrease reproductive success. Researchers have discovered such an effect on caribou (wild reindeer) in Greenland. In spring, caribou migrate to calving grounds to eat sprouting green plants, give birth, and care for their calves. Prior to 1993, the arrival of caribou at the calving grounds coincided with the brief period during which the plants were nutritious and digestible. By 2006, however, average spring temperatures in the calving grounds had increased by more than 4°C, and the plants sprouted two weeks earlier. Because caribou migration is triggered by day length, not temperature, there is a mismatch between the timing of new plant growth and caribou birthing. Without adequate nutrition for the nursing females, production of caribou offspring has declined by 75% since 1993.

For some asexual animals, a cycle of reproductive behavior appears to reflect a sexual evolutionary past. In certain parthenogenetic lizard species of the genus *Aspidoscelis*, reproduction is only asexual, and there are no males. However, these lizards have courtship and mating behaviors very similar to those of sexual species of *Aspidoscelis*. In breeding season, one female of each mating pair mimics a male **(Figure 36.4a)**. Each member of the pair alternates roles two or three times during the season. An individual adopts female behavior prior to ovulation, when the level of the female sex hormone estradiol is high, and then switches to male-like behavior after ovulation, when the level of progesterone is high **(Figure 36.4b)**. Ovulation is more likely to occur if the individual is mounted during the critical time of the hormone cycle; isolated lizards lay fewer eggs than those that go through the motions of sex. These observations support the hypothesis that these parthenogenetic lizards evolved from species having two sexes and require sexual stimuli for maximum reproductive success.

## Variation in Patterns of Sexual Reproduction

For many sexual animals, finding a partner for reproduction can be challenging. Adaptations that arose during the evolution of some species meet this challenge in a novel way—by blurring the strict distinction between male and female. One such adaptation arose among sessile (stationary) animals, such as barnacles; burrowing animals, such as clams; and some parasites, including tapeworms. Largely lacking mobility, these animals have little opportunity to find a mate. The evolutionary solution in this case is **hermaphroditism**, in which each individual has both male and female reproductive

**(a)** Both lizards in this photograph are *A. uniparens* females. The one on top is playing the role of a male. Individuals switch sex roles two or three times during the breeding season.

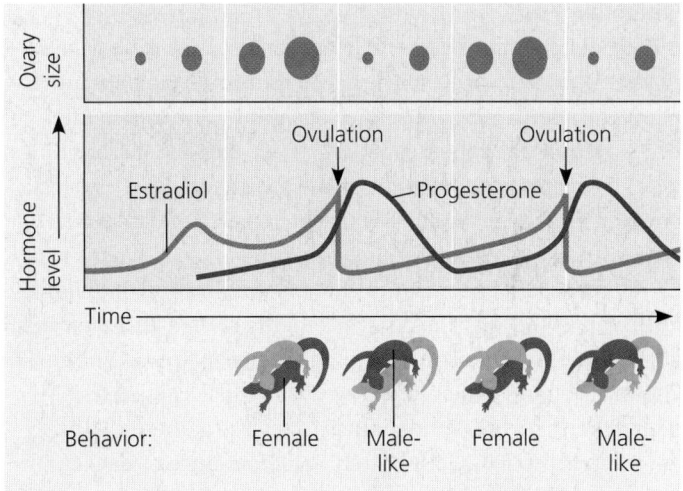

**(b)** The changes in sexual behavior of *A. uniparens* individuals are correlated with the cycles of ovulation and changing levels of the sex hormones estradiol and progesterone. These drawings track the changes in ovary size, hormone levels, and sexual behavior of one female lizard (shown in brown).

▲ **Figure 36.4 Sexual behavior in parthenogenetic lizards.** The desert-grassland whiptail lizard (*Aspidoscelis uniparens*) is an all-female species. These reptiles reproduce by parthenogenesis, the development of an unfertilized egg, but ovulation is stimulated by mating behavior.

**INTERPRET THE DATA** *If you plotted hormone levels for the lizard shown in gray, how would your graph differ from the graph in (b)?*

systems (the term *hermaphrodite* merges the names Hermes and Aphrodite, a Greek god and goddess). Because each hermaphrodite reproduces as both a male and a female, *any* two individuals can mate. Each animal donates and receives sperm during mating, as the sea slugs in Figure 36.1 are doing. In some species, hermaphrodites are also capable of self-fertilization, allowing a form of sexual reproduction that doesn't require any partner.

The bluehead wrasse (*Thalassoma bifasciatum*) provides an example of a quite different variation in sexual reproduction. These coral reef fish live in harems, each consisting of a single male and several females. When the lone male dies, the opportunity for sexual reproduction would appear lost. Instead, the largest female in the harem transforms into a male and within a week begins to produce sperm instead of eggs. What selective pressure in the evolution of the bluehead wrasse resulted in sex reversal for that female with the largest body? Because it is the male wrasse that defends a harem against intruders, a larger size may be particularly important for a male in ensuring successful reproduction.

## External and Internal Fertilization

Sexual reproduction requires **fertilization**, the union of sperm and egg. In species with *external fertilization*, the female releases eggs into the environment, where the male then fertilizes them. In species with *internal fertilization*, sperm deposited in or near the female reproductive tract fertilize eggs within the tract.

A moist habitat is almost always required for external fertilization, both to prevent the gametes from drying out and to allow the sperm to swim to the eggs. Many aquatic invertebrates simply shed their eggs and sperm into the surroundings, and fertilization occurs without the parents making physical contact. However, timing is crucial to ensure that mature sperm and eggs encounter one another.

Among some species with external fertilization, individuals clustered in the same area release their gametes into the water at the same time, a process known as *spawning*. In some cases, chemical signals generated by one individual as it releases gametes trigger other individuals to release gametes. In other cases, environmental cues, such as temperature or day length, cause a whole population to release gametes at one time. For example, the palolo worm, native to coral reefs of the South Pacific, coordinates its spawning to both the season and the lunar cycle. Sometime in spring when the moon is in its last quarter, palolo worms break in half, releasing tail segments engorged with sperm or eggs. These packets rise to the ocean surface and burst in such vast numbers that the sea appears milky with gametes. The sperm quickly fertilize the floating eggs, and within hours, the palolo's once-a-year reproductive frenzy is complete.

When external fertilization is not synchronous across a population, individuals may exhibit specific "courtship" behaviors leading to the fertilization of the eggs of one female by one male. By triggering the release of both sperm and eggs, these behaviors increase the probability of successful fertilization.

Internal fertilization is an adaptation that enables sperm to reach an egg efficiently, even when the external environment is dry. It typically requires sophisticated and compatible reproductive systems, as well as cooperative behavior that leads to copulation. The male copulatory organ delivers sperm, and the female reproductive tract often has receptacles for storage and delivery of sperm to mature eggs **(Figure 36.5)**.

No matter how fertilization occurs, the mating animals may make use of *pheromones*, chemicals released by one organism that can influence the physiology and behavior of other

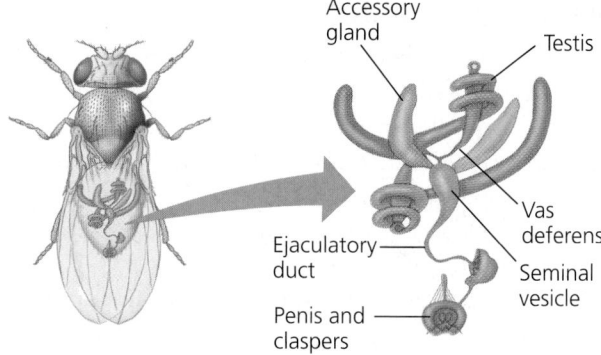

**(a) Male fruit fly.** Sperm form in the testes, pass through a sperm duct (vas deferens), and are stored in the seminal vesicles. The male ejaculates sperm along with fluid from the accessory glands. (Males of some species of insects and other arthropods have appendages called claspers that grasp the female during copulation.)

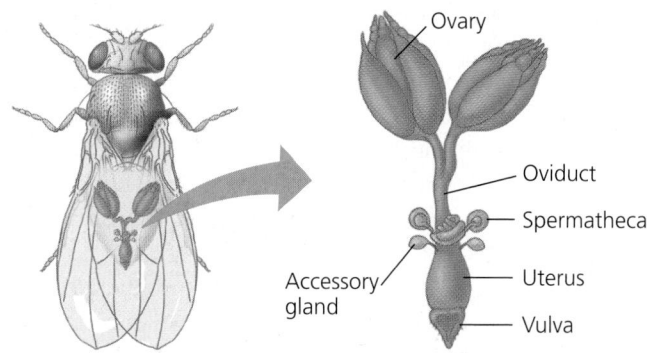

**(b) Female fruit fly.** Eggs develop in the ovaries and then travel through the oviducts to the uterus. After mating, sperm are stored in the spermathecae, which are connected to the uterus by short ducts. The female uses a stored sperm to fertilize each egg as it enters the uterus before she passes the egg out through the vulva.

▲ **Figure 36.5 Reproductive anatomy for internal fertilization.** Sexual reproduction that requires internal fertilization often relies on specialized tissues, organs, and ducts, as shown in *Drosophila melanogaster*.

individuals of the same species. Pheromones are small, volatile or water-soluble molecules that disperse into the environment and, like hormones, are active at very low concentrations. Many pheromones function as mate attractants, enabling some female insects to be detected by males more than a kilometer away.

## Ensuring the Survival of Offspring

Animals that fertilize eggs internally typically produce fewer gametes than species with external fertilization, but a higher fraction of their zygotes survive. Better zygote survival is due in part to the fact that eggs fertilized internally are sheltered from potential predators. However, internal fertilization is also more often associated with mechanisms that provide greater protection of the embryos and parental care of the young. For example, the eggs of birds and other reptiles have calcium- and protein-containing shells and internal membranes that protect against water loss and physical damage (see Figure 27.28). In

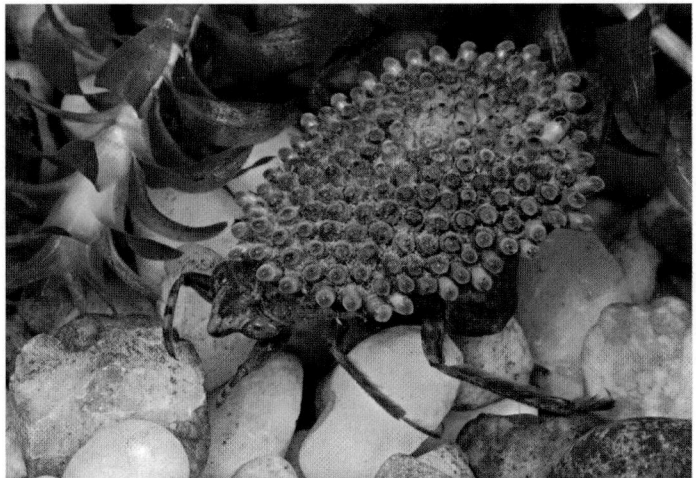

▲ **Figure 36.6 Parental care in an invertebrate.** Compared with many other insects, giant water bugs of the genus *Belostoma* produce relatively few offspring, but offer much greater parental protection. Following internal fertilization, the female glues her fertilized eggs to the back of the male. The male (shown here) carries the eggs for days, frequently fanning water over them to keep them moist, aerated, and free of parasites.

contrast, the eggs of fishes and amphibians have only a gelatinous coat and lack internal membranes.

Rather than secreting a protective eggshell, some animals retain the embryo for a portion of its development within the female's reproductive tract. Embryos of marsupial mammals, such as kangaroos and opossums, spend only a short period in the uterus; the embryos then crawl out and complete fetal development attached to a mammary gland in the mother's pouch. Embryos of eutherian (placental) mammals, such as sheep and humans, remain in the uterus throughout fetal development. There they are nourished by the mother's blood supply through a temporary organ, the placenta. The embryos of some fishes and sharks also complete development internally.

When an eagle hatches out of an egg or when a human is born, the newborn is not yet capable of independent existence. Instead, adult birds feed their young and adult mammals nurse their offspring. Examples of parental care are in fact widespread among animals, including some invertebrates **(Figure 36.6)**.

### CONCEPT CHECK 36.1

1. How does internal fertilization facilitate life on land?
2. **WHAT IF?** If a hermaphrodite self-fertilizes, will the offspring be identical to the parent? Explain.
3. **MAKE CONNECTIONS** What examples of plant reproduction are most similar to asexual reproduction in animals? (See Concept 30.2.)

For suggested answers, see Appendix A.

# Reproductive organs produce and transport gametes

Sexual reproduction in animals relies on sets of cells that are precursors for eggs and sperm. A group of cells dedicated to this function is often established early in the formation of the embryo. After the body takes shape, these cells undergo cycles of growth and mitosis that increase, or *amplify*, the number of cells available for making eggs or sperm.

## Variation in Reproductive Systems

In producing gametes and making them available for fertilization, animals employ a variety of reproductive systems. **Gonads**, organs that produce gametes, are found in many but not all animals. Exceptions include the palolo worm, discussed earlier. The palolo and most other polychaete worms (phylum Annelida; see Figure 27.11) have separate sexes but lack distinct gonads; rather, the eggs and sperm develop from undifferentiated cells lining the coelom (body cavity). As the gametes mature, they are released from the body wall and fill the coelom. Depending on the species, mature gametes in these worms may be shed through the excretory opening, or the swelling mass of eggs may split a portion of the body open, spilling the eggs into the environment.

More elaborate reproductive systems include sets of accessory tubes and glands that carry, nourish, and protect the gametes and sometimes the developing embryos. Most insects, for example, have separate sexes with complex reproductive systems (see Figure 36.5). In many insect species, the female reproductive system includes one or more *spermathecae*, sacs in which sperm may be kept alive for extended periods, a year or more in some species. Because the female releases male gametes from the spermathecae only in response to the appropriate stimuli, fertilization occurs under conditions likely to be well suited to survival of offspring.

Vertebrate reproductive systems display limited but significant variations. In many nonmammalian vertebrates, the digestive, excretory, and reproductive systems have a common opening to the outside, the **cloaca**, a structure probably present in the ancestors of all vertebrates. Lacking a well-developed penis, males of these species release sperm by turning the cloaca inside out. In contrast, mammals generally lack a cloaca and have a separate opening for the digestive tract. Most female mammals also have separate openings for the excretory and reproductive systems. In some vertebrates, the uterus is divided into two chambers; in others, including humans and birds, it is a single structure.

Having surveyed some general features of animal reproduction, we turn now to human reproduction, beginning with the reproductive anatomy of males.

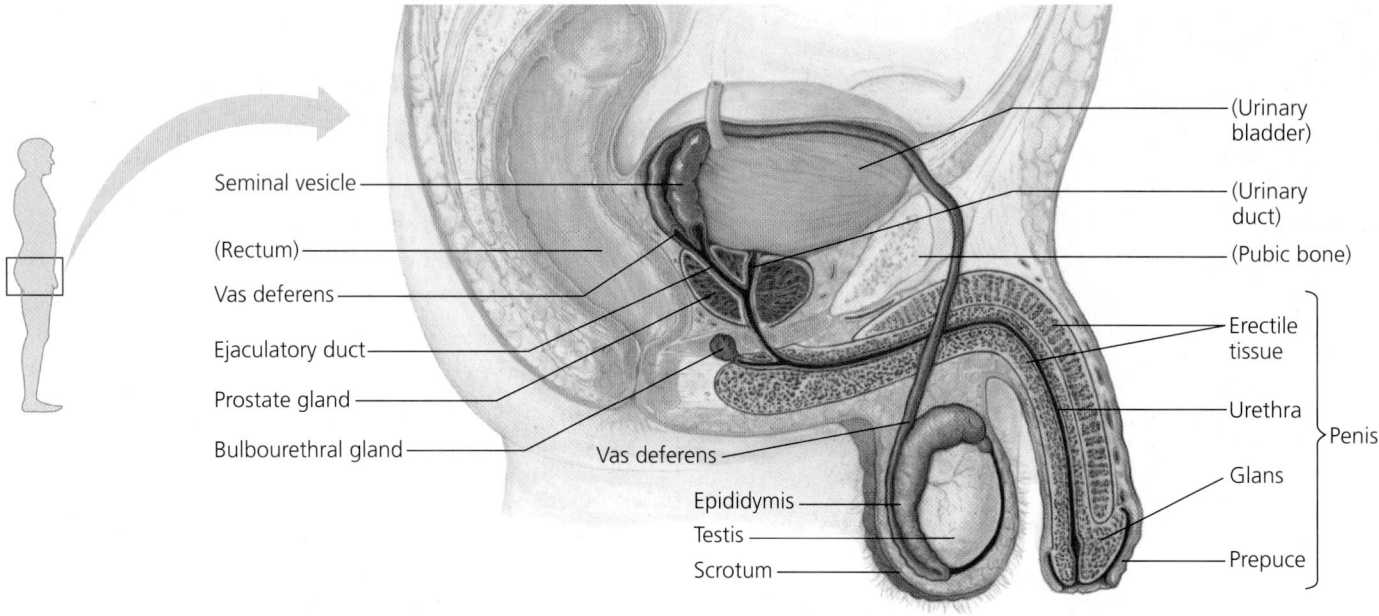

▲ **Figure 36.7 Reproductive anatomy of the human male.** Some nonreproductive structures are labeled in parentheses for orientation purposes.

## Human Male Reproductive Anatomy

The human male's external reproductive organs are the scrotum and penis. The internal reproductive organs consist of a pair of gonads that produce both sperm and reproductive hormones, accessory glands that secrete products essential to sperm movement, and ducts that carry the sperm and glandular secretions **(Figure 36.7)**.

### Testes

The male gonads, or **testes** (singular, *testis*), produce sperm in highly coiled tubes called **seminiferous tubules**. Most mammals produce sperm properly only when the testes are cooler than the rest of the body. In humans and many other mammals, the **scrotum**, a fold of the body wall, maintains testis temperature about 2°C below the core body temperature. The testes develop in the abdominal cavity and descend into the scrotum just before birth (a testis within a scrotum is a *testicle*).

In many rodents, the testes are drawn back into the cavity between breeding seasons, interrupting sperm maturation. Some mammals whose body temperature is low enough to allow sperm maturation—such as whales and elephants—retain the testes in the abdominal cavity at all times.

### Ducts

From the seminiferous tubules of a testis, the sperm pass into the coiled duct of an **epididymis**, where they complete maturation and become motile. During **ejaculation**, the sperm are propelled from each epididymis through a muscular duct, the **vas deferens**. A vas deferens from each epididymis extends around and behind the urinary bladder, where it joins a duct from the seminal vesicle, forming a short **ejaculatory duct**. The ejaculatory ducts open into the **urethra**, the outlet tube

for both the excretory system and the reproductive system. The urethra runs through the penis and opens to the outside at the tip of the penis.

### Accessory Glands

Three sets of accessory glands—the seminal vesicles, the prostate gland, and the bulbourethral glands—produce secretions that combine with sperm to form **semen**, the fluid that is ejaculated. Two **seminal vesicles** contribute about 60% of the volume of semen. The fluid from the seminal vesicles is thick, yellowish, and alkaline. It contains mucus, the sugar fructose (which provides most of the sperm's energy), a coagulating enzyme, ascorbic acid, and local regulators called prostaglandins.

The **prostate gland** secretes its products into the urethra through small ducts. This fluid is thin and milky; it contains anticoagulant enzymes and citrate (a sperm nutrient). The *bulbourethral glands* are a pair of small glands along the urethra below the prostate. Before ejaculation, they secrete clear mucus that neutralizes any acidic urine remaining in the urethra. Bulbourethral fluid also carries some sperm released before ejaculation, which is one reason for the high failure rate of the withdrawal method of birth control (coitus interruptus).

### Penis

The human **penis** contains the urethra as well as three cylinders of spongy erectile tissue. During sexual arousal, the erectile tissue, which is derived from modified veins and capillaries, fills with blood from the arteries. As this tissue fills, the increasing pressure seals off the veins that drain the penis, causing it to engorge with blood. The resulting erection enables the penis to be inserted into the vagina. Alcohol consumption, certain drugs, emotional issues, and aging all can cause an inability to achieve an erection (erectile dysfunction). For individuals with long-term erectile dysfunction, drugs such as Viagra promote the vasodilating action of nitric oxide; the resulting relaxation of smooth muscles in the blood vessels of the penis enhances blood flow into the erectile tissues. Although all mammals rely on penile erection for mating, the penis of dogs, raccoons, polar bears, and several other mammals also contains a bone, the baculum, which is thought to further stiffen the penis for mating.

The main shaft of the penis is covered by relatively thick skin. The head, or **glans**, of the penis has a much thinner covering and is consequently more sensitive to stimulation. The human glans is covered by a fold of skin called the *prepuce*, or foreskin, which is removed if a male is circumcised.

## Human Female Reproductive Anatomy

The human female's external reproductive structures are the clitoris and two sets of labia, which surround the clitoris and vaginal opening. The internal organs are the gonads, which produce both eggs and reproductive hormones, and a system of ducts and chambers, which receive and carry gametes and house the embryo and fetus **(Figure 36.8)**.

### Ovaries

The female gonads are a pair of **ovaries** that flank the uterus and are held in place in the abdominal cavity by ligaments. The outer layer of each ovary is packed with **follicles**, each consisting of an **oocyte**, a partially developed egg, surrounded by support cells. The surrounding cells nourish and protect the oocyte during much of its formation and development.

### Oviducts and Uterus

An **oviduct**, or fallopian tube, extends from the uterus toward a funnel-like opening at each ovary. The dimensions of this tube vary along its length, with the inside diameter near the uterus being as narrow as a human hair. Upon **ovulation**, the release of a mature egg, cilia on the epithelial lining of the oviduct help collect the egg by drawing fluid from the body cavity into the oviduct. Together with wavelike contractions of the oviduct, the cilia convey the egg down the duct to the **uterus**, also known as the womb. The uterus is a thick, muscular organ that can expand during pregnancy to accommodate a 4-kg fetus. The inner lining of the uterus, the **endometrium**, is richly supplied with blood vessels. The neck of the uterus, called the **cervix**, opens into the vagina.

### Vagina and Vulva

The **vagina** is a muscular but elastic chamber that is the site for insertion of the penis and deposition of sperm during copulation. The vagina, which also serves as the birth canal through which a baby is born, opens to the outside at the **vulva**, the collective term for the external female genitalia.

The **labia majora**, a pair of thick, fatty ridges, enclose and protect the rest of the vulva. The vaginal opening and the separate opening of the urethra are located within a cavity bordered by a pair of slender skin folds, the **labia minora**. A thin piece of tissue called the **hymen** partly covers the vaginal opening in humans at birth, but becomes thinner over time and typically wears away through physical activity. Located at the top of the labia minora, the **clitoris** consists of erectile tissue supporting a rounded glans, or head, covered by a small hood of skin, the prepuce. During sexual arousal, the clitoris, vagina, and labia minora all engorge with blood and enlarge. Richly supplied with nerve endings, the clitoris is one of the most sensitive points of sexual stimulation. Sexual arousal also induces the vestibular glands near the vaginal opening to secrete lubricating mucus, thereby facilitating intercourse.

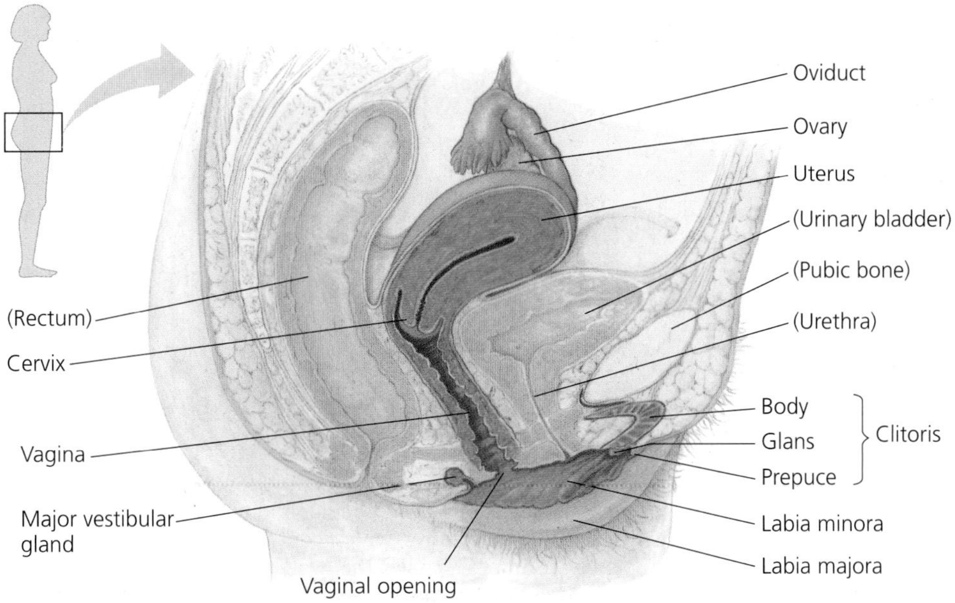

### Mammary Glands

The **mammary glands** are present in both sexes, but normally produce milk only in females. Though not part of the reproductive system, the female mammary glands are important to reproduction. Within the glands, small sacs of epithelial tissue secrete milk, which drains into a series of ducts that open at the nipple. The breasts contain connective and fatty (adipose) tissue in addition to the mammary glands.

## Gametogenesis

With this overview of reproductive anatomy in mind, we turn now to **gametogenesis**, the production of gametes. **Figure 36.9** compares this process in human males and females, highlighting the close relationship between the gonads' structure and their function.

▲ **Figure 36.8 Reproductive anatomy of the human female.**
Some nonreproductive structures are labeled in parentheses for orientation purposes.

Labels: Oviduct, Ovary, Uterus, (Urinary bladder), (Pubic bone), (Urethra), Body, Glans, Prepuce } Clitoris, Labia minora, Labia majora, Vaginal opening, Major vestibular gland, Vagina, Cervix, (Rectum)

## Spermatogenesis

Stem cells that give rise to sperm are situated near the outer edge of the seminiferous tubules. Their progeny move inward as they pass through the spermatocyte and spermatid stages, and sperm are released into the lumen (fluid-filled cavity) of the tubule. The sperm travel along the tubule into the epididymis, where they become motile.

The stem cells arise from division and differentiation of primordial germ cells in the embryonic testes. In mature testes, they divide mitotically to form **spermatogonia**, which in turn generate spermatocytes by mitosis. Each spermatocyte gives rise to four spermatids through meiosis, reducing the chromosome number from diploid ($2n = 46$ in humans) to haploid ($n = 23$). Spermatids undergo extensive changes in differentiating into sperm.

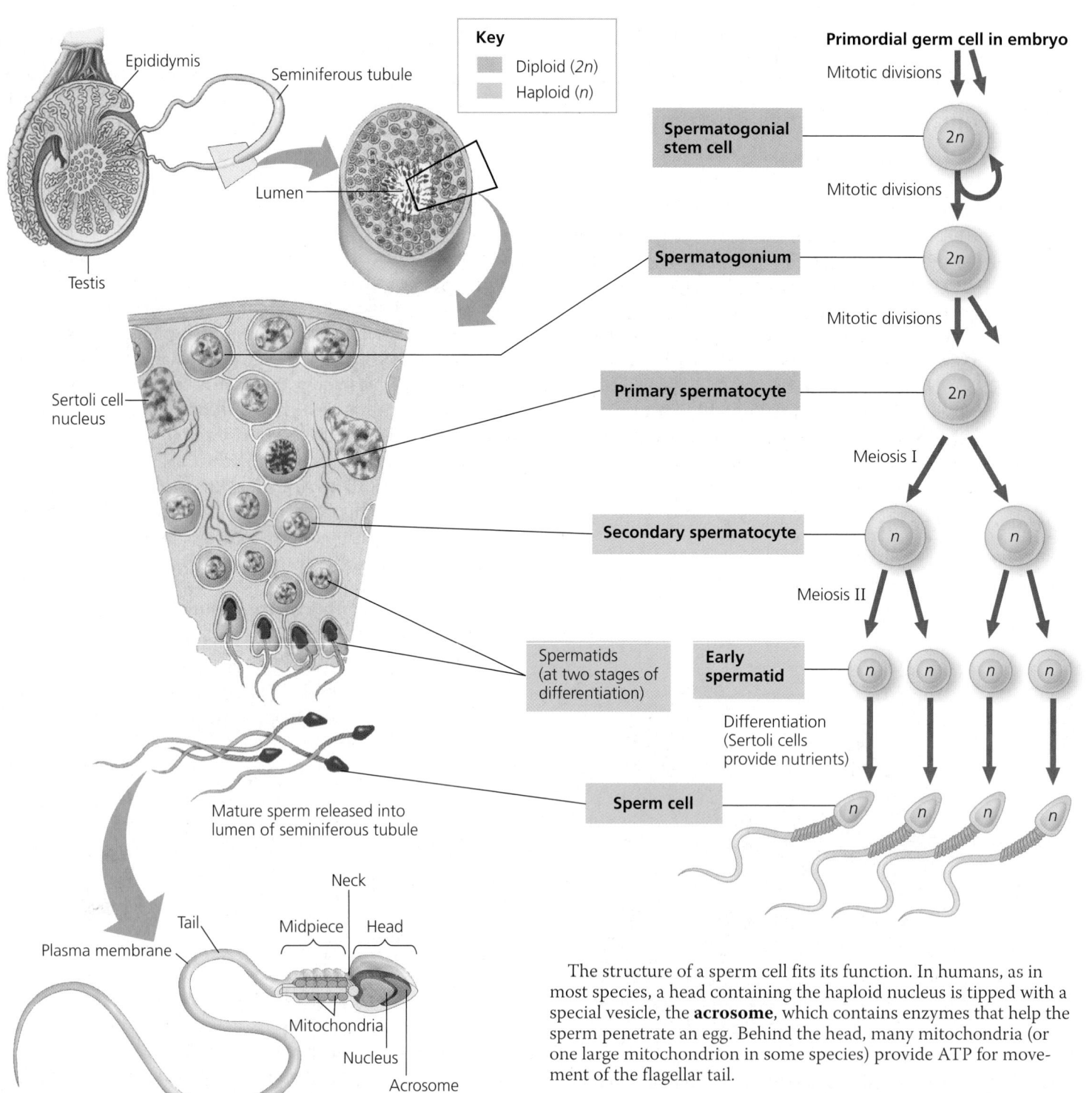

The structure of a sperm cell fits its function. In humans, as in most species, a head containing the haploid nucleus is tipped with a special vesicle, the **acrosome**, which contains enzymes that help the sperm penetrate an egg. Behind the head, many mitochondria (or one large mitochondrion in some species) provide ATP for movement of the flagellar tail.

## Oogenesis

Oogenesis begins in the female embryo with the production of **oogonia** from primordial germ cells. The oogonia divide by mitosis to form cells that begin meiosis, but stop the process at prophase I before birth. These developmentally arrested cells, which are **primary oocytes**, each reside within a small follicle, a cavity lined with protective cells. At birth, the ovaries together contain about 1–2 million primary oocytes, of which about 500 fully mature between puberty and menopause.

To the best of our current knowledge, women are born with all the primary oocytes they will ever have. It is worth noting, however, that a similar conclusion regarding most other mammals was overturned in 2004 when researchers discovered that the ovaries of adult mice contain multiplying oogonia that develop into oocytes. If the same turned out to be true of humans, it might be that the marked decline in fertility that occurs as women age results from both a depletion of oogonia and the degeneration of aging oocytes.

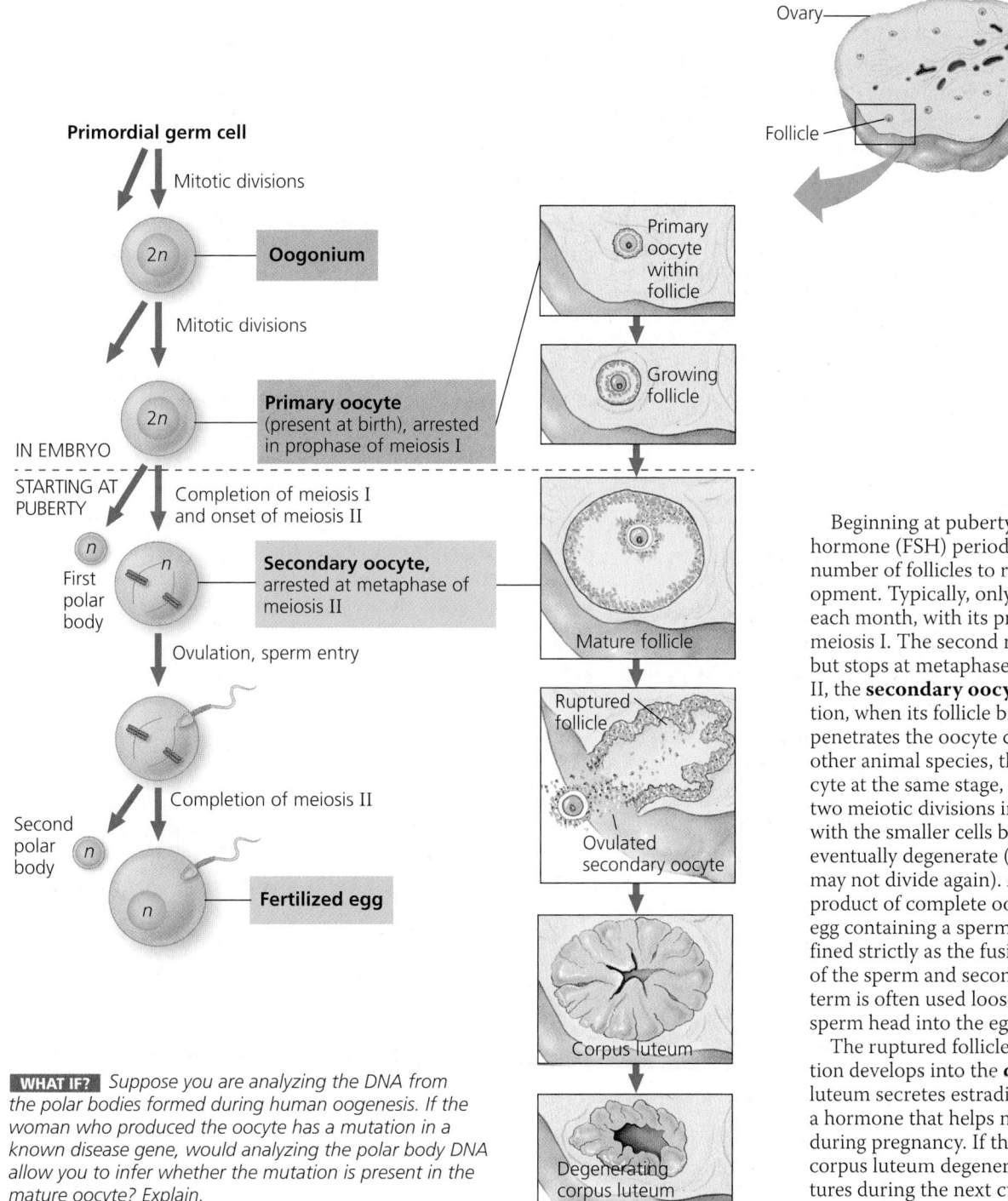

Beginning at puberty, follicle-stimulating hormone (FSH) periodically stimulates a small number of follicles to resume growth and development. Typically, only one follicle fully matures each month, with its primary oocyte completing meiosis I. The second meiotic division begins, but stops at metaphase. Thus arrested in meiosis II, the **secondary oocyte** is released at ovulation, when its follicle breaks open. Only if a sperm penetrates the oocyte does meiosis II resume. (In other animal species, the sperm may enter the oocyte at the same stage, earlier, or later.) Each of the two meiotic divisions involves unequal cytokinesis, with the smaller cells becoming polar bodies that eventually degenerate (the first polar body may or may not divide again). As a result, the functional product of complete oogenesis is a single mature egg containing a sperm head. Fertilization is defined strictly as the fusion of the haploid nuclei of the sperm and secondary oocyte, although the term is often used loosely to mean the entry of the sperm head into the egg.

The ruptured follicle left behind after ovulation develops into the **corpus luteum**. The corpus luteum secretes estradiol as well as progesterone, a hormone that helps maintain the uterine lining during pregnancy. If the egg is not fertilized, the corpus luteum degenerates, and a new follicle matures during the next cycle

**WHAT IF?** *Suppose you are analyzing the DNA from the polar bodies formed during human oogenesis. If the woman who produced the oocyte has a mutation in a known disease gene, would analyzing the polar body DNA allow you to infer whether the mutation is present in the mature oocyte? Explain.*

**Spermatogenesis**, the formation and development of sperm, is continuous and prolific in adult males. To produce hundreds of millions of sperm each day, cell division and maturation occur throughout the seminiferous tubules coiled within the two testes. For a single sperm, the process takes about 7 weeks from start to finish.

**Oogenesis**, the development of mature oocytes (eggs), is a prolonged process in the human female. Immature eggs form in the ovary of the female embryo but do not complete their development until years, and often decades, later.

In humans, spermatogenesis differs from oogenesis in three significant ways:

- Only in spermatogenesis do all four products of meiosis develop into mature gametes. In oogenesis, cytokinesis during meiosis is unequal, with almost all the cytoplasm segregated to a single daughter cell. This large cell is destined to become the egg; the other products of meiosis, smaller cells called polar bodies, degenerate.
- Spermatogenesis occurs throughout adolescence and adulthood. In contrast, the mitotic divisions of oogenesis in human females are thought to be complete before birth, and the production of mature gametes ceases at about age 50.
- Spermatogenesis produces mature sperm from precursor cells in a continuous sequence, whereas oogenesis has long interruptions.

## CONCEPT CHECK 36.2

1. In what ways are a second polar body and an early spermatid similar? In what ways are they dissimilar?
2. Why might using a hot tub frequently make it harder for a couple to conceive a child?
3. **MAKE CONNECTIONS** How are the uterus of an insect and the ovary of a flowering plant similar in function? How are they different? (See Figure 30.7.)
4. **WHAT IF?** If each vas deferens in a male was surgically sealed off, what changes would you expect in sexual response and ejaculate composition?

For suggested answers, see Appendix A.

---

CONCEPT 36.3

# The interplay of tropic and sex hormones regulates reproduction in mammals

Mammalian reproduction is governed by the coordinated actions of hormones from the hypothalamus, anterior pituitary, and gonads (see Figure 32.5).

Endocrine control of reproduction begins with the hypothalamus, which secretes *gonadotropin-releasing hormone (GnRH)*. This hormone directs the anterior pituitary to secrete **follicle-stimulating hormone (FSH)** and **luteinizing hormone (LH)**. Both are **tropic hormones**, meaning that

they act on endocrine tissues to trigger the release of other hormones. They are called *gonadotropins* because the endocrine tissues they act on are in the gonads. There, FSH and LH control sex hormone production.

The gonads produce and secrete three major types of steroid sex hormones: *androgens*, principally **testosterone**; *estrogens*, principally **estradiol**; and **progesterone**. All three hormones are found in both males and females, but at quite different concentrations. Testosterone levels in the blood are about 10 times higher in males than in females, whereas estradiol levels are about 10 times higher in females than in males; peak progesterone levels are also much higher in females. Although the gonads are the major source of sex hormones, the adrenal gland secretes sex hormones in small amounts.

In mammals, sex hormone function in reproduction begins in the embryo. In particular, androgens produced in male embryos direct the appearance of the primary sex characteristics of males, the structures directly involved in reproduction. These include the seminal vesicles and associated ducts, as well as external reproductive structures. In the **Scientific Skills Exercise**, you can interpret the results of an experiment investigating the development of reproductive structures in mammals.

During sexual maturation, sex hormones induce formation of secondary sex characteristics, the physical and behavioral features that are not directly related to the reproductive system. Secondary sex characteristics often lead to sexual dimorphism, the difference in appearance between the male and female adults of a species **(Figure 36.10)**. When human males enter puberty, androgens cause the voice to deepen, facial and pubic hair to develop, and muscles to grow (by stimulating protein synthesis). Androgens also promote specific sexual behaviors and sex drive, as well as an increase in general aggressiveness. Estrogens similarly have multiple effects in females.

▲ **Figure 36.10 Androgen-dependent male anatomy and behavior in a moose.** The male and female in a mating pair of moose (*Alces alces*) differ in both anatomy and physiology. High levels of testosterone in the male are responsible for the appearance of secondary sex characteristics, such as antlers, and for male courtship and territorial behavior.

# Making Inferences and Designing an Experiment

► Human sex chromosomes

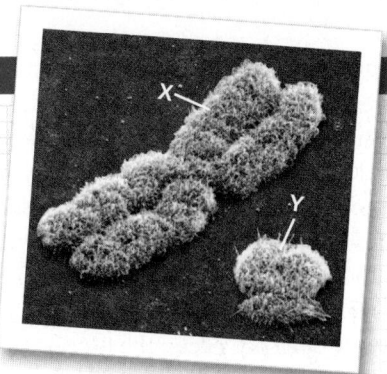

**What Role Do Hormones Play in Making a Mammal Male or Female?** In non-egg-laying mammals, females have two X chromosomes, whereas males have one X chromosome and one Y chromosome. In the 1940s, French physiologist Alfred Jost wondered whether development of mammalian embryos as female or male in accord with their chromosome set requires instructions in the form of hormones produced by the gonads. In this exercise, you will interpret the results of an experiment that Jost performed to answer this question.

**How the Experiment Was Done** Working with rabbit embryos still in the mother's uterus and at a stage before sex differences are observable, Jost surgically removed the portion of each embryo that would form the ovaries or testes. When the baby rabbits were born, he made note of their chromosomal sex and whether their genital structures were male or female.

**Data from the Experiment**

| | Appearance of Genitalia | |
|---|---|---|
| Chromosome Set | No Surgery | Embryonic Gonad Removed |
| XY (male) | Male | Female |
| XX (female) | Female | Female |

**Data from** A. Jost, Recherches sur la differenciation sexuelle de l'embryon de lapin (Studies on the sexual differentiation of the rabbit embryo), *Archives d'Anatomie Microscopique et de Morphologie Experimentale* 36:271–316 (1947).

**INTERPRET THE DATA**

1. This experiment is an example of a research approach in which scientists infer how something works normally based on what happens when the normal process is blocked. What normal process was blocked in Jost's experiment? From the results, what inference can you make about the role of the gonads in controlling the development of mammalian genitalia?

2. The data in Jost's experiment could be explained if some aspect of the surgery other than gonad removal caused female genitalia to develop. If you were to repeat Jost's experiment, how might you test the validity of such an explanation?

3. What result would Jost have obtained if female development also required a signal from the gonad?

4. Design another experiment to determine whether the signal that controls male development is a hormone. Be sure to identify your hypothesis, prediction, data collection plan, and controls.

(MB) A version of this Scientific Skills Exercise can be assigned in MasteringBiology.

At puberty, estradiol stimulates breast and pubic hair development. Estradiol also influences female sexual behavior, induces fat deposition in the breasts and hips, increases water retention, and alters calcium metabolism.

Once mammals reach sexual maturity, the sex hormones, as well as the gonadotropins, have essential roles in gametogenesis. In exploring this hormonal control of reproduction, we'll begin with the relatively simple system found in males.

## Hormonal Control of the Male Reproductive System

In directing spermatogenesis, FSH and LH act on two types of cells in the testes **(Figure 36.11)**. **Sertoli cells**, located within the seminiferous tubules, respond to FSH by nourishing developing sperm (see Figure 36.9). **Leydig cells**, scattered in connective tissue between the tubules, respond to LH by producing testosterone and other androgens, which promote spermatogenesis in the tubules.

Two negative-feedback mechanisms control sex hormone production in males (see Figure 36.11). Testosterone regulates blood levels of GnRH, FSH, and LH through inhibitory effects on the hypothalamus and anterior pituitary. In addition, *inhibin*, a hormone that in males is produced by Sertoli cells, acts on the anterior pituitary gland to reduce FSH secretion. Together, these negative-feedback circuits maintain androgen production at optimal levels.

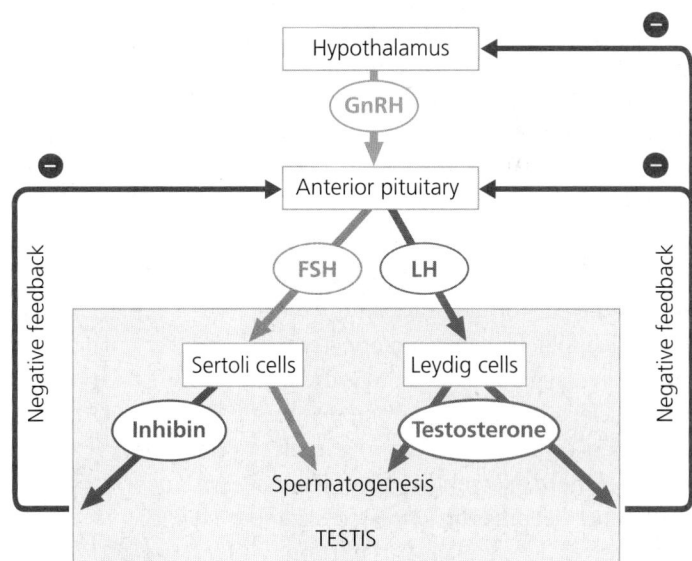

▲ **Figure 36.11 Hormonal control of the testes.**

## Hormonal Control of Female Reproductive Cycles

Whereas sperm are produced continuously in human males, there are two closely linked reproductive cycles in human females. Both are controlled by cyclic patterns of endocrine signaling.

Cyclic events in the ovaries define the **ovarian cycle**: Once per cycle a follicle matures and an oocyte is released. Changes in the uterus define the **menstrual cycle**, also called the **uterine cycle**.

During each cycle, the endometrium (lining of the uterus) thickens and develops a rich blood supply. By linking the ovarian and uterine cycles, hormone activity synchronizes ovulation with the establishment of a uterine lining that can support embryo implantation and development.

If an oocyte is not fertilized and pregnancy does not occur, the uterine lining is sloughed off, and another pair of ovarian and uterine cycles begins. The cyclic shedding of the blood-rich endometrium from the uterus, a process that occurs in a flow through the cervix and vagina, is called **menstruation**. Menstrual cycles average 28 days but can range from about 20 to 40 days.

**Figure 36.12** outlines the major events of the female reproductive cycles, illustrating the close coordination across different tissues in the body.

### The Ovarian Cycle

In human females, as in males, the hypothalamus has a central role in regulating reproduction. The ovarian cycle begins ❶ with the release from the hypothalamus of GnRH, which stimulates the anterior pituitary to ❷ secrete small amounts of FSH and LH. ❸ Follicle-stimulating hormone (as its name implies) stimulates follicle growth, aided by LH, and ❹ the cells of the growing follicles start to make estradiol. There is a slow rise in estradiol secreted during most of the **follicular phase**, the part of the ovarian cycle during which follicles grow and oocytes mature. (Several follicles begin to grow with each cycle, but usually only one matures; the others disintegrate.) The low levels of estradiol inhibit secretion of the pituitary hormones, keeping the levels of FSH and LH relatively low. During this portion of the cycle, regulation of the hormones controlling reproduction closely parallels the regulation observed in males.

❺ When estradiol secretion by the growing follicle begins to rise steeply, ❻ the FSH and LH levels in the blood increase markedly. Whereas a low level of estradiol inhibits the secretion of pituitary gonadotropins, a high concentration has the opposite effect: It stimulates gonadotropin secretion by acting on

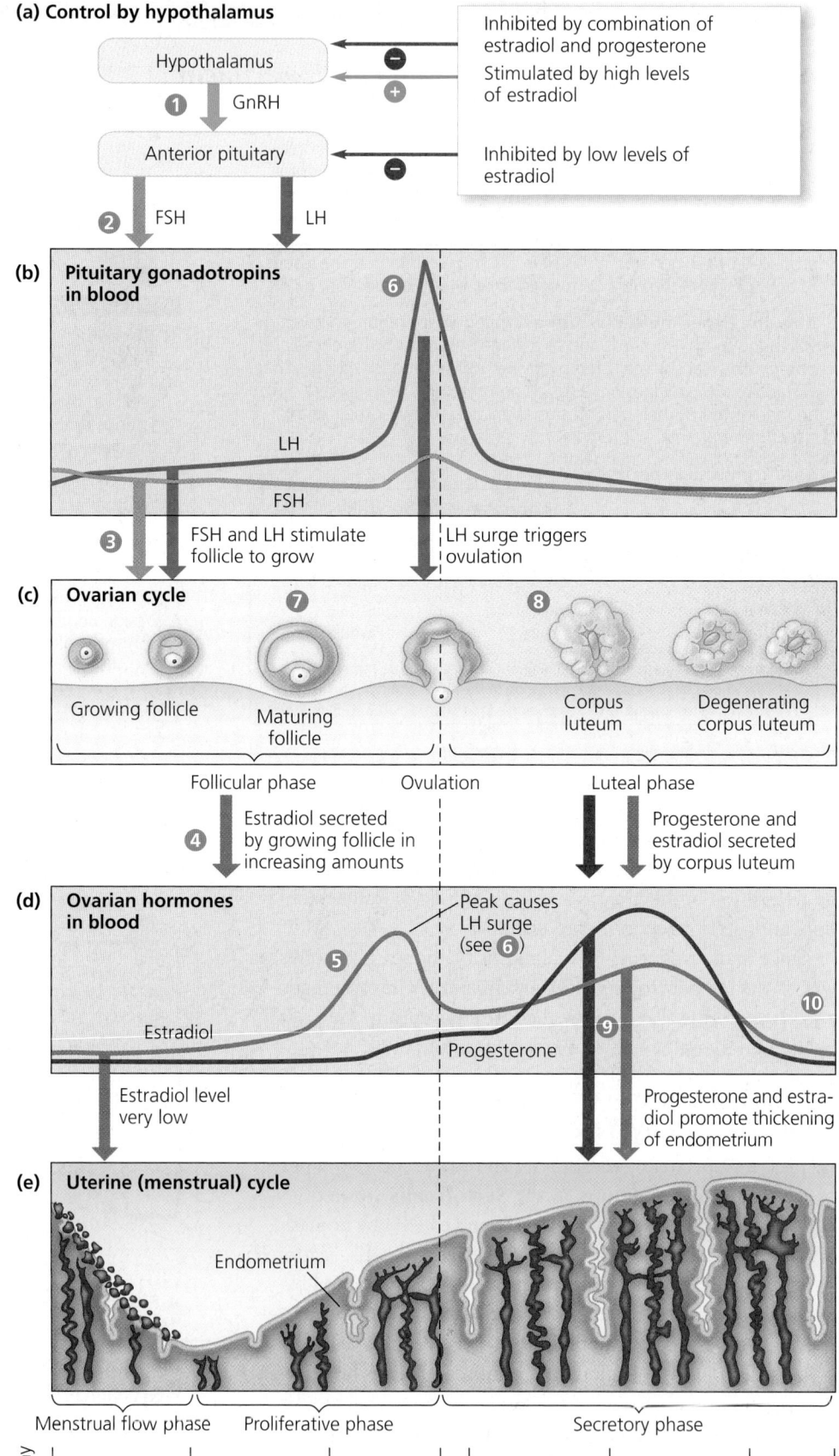

**(a) Control by hypothalamus**

Inhibited by combination of estradiol and progesterone

Stimulated by high levels of estradiol

Inhibited by low levels of estradiol

Hypothalamus

❶ GnRH

Anterior pituitary

❷ FSH    LH

**(b) Pituitary gonadotropins in blood**

❻

LH

FSH

❸ FSH and LH stimulate follicle to grow

LH surge triggers ovulation

**(c) Ovarian cycle**    ❼    ❽

Growing follicle    Maturing follicle    Corpus luteum    Degenerating corpus luteum

Follicular phase    Ovulation    Luteal phase

❹ Estradiol secreted by growing follicle in increasing amounts

Progesterone and estradiol secreted by corpus luteum

**(d) Ovarian hormones in blood**

Peak causes LH surge (see ❻)

❺

Estradiol

Progesterone

❾    ❿

Estradiol level very low

Progesterone and estradiol promote thickening of endometrium

**(e) Uterine (menstrual) cycle**

Endometrium

Menstrual flow phase    Proliferative phase    Secretory phase

Day    0    5    10    14 15    20    25    28

▲ **Figure 36.12 The reproductive cycles of the human female.** This figure shows how **(c)** the ovarian cycle and **(e)** the uterine (menstrual) cycle are regulated by changing hormone levels in the blood, depicted in parts **(a)**, **(b)**, and **(d)**. The time scale at the bottom of the figure applies to parts **(b)–(e)**.

the hypothalamus to increase its output of GnRH. The effect is greater for LH because the high concentration of estradiol increases the GnRH sensitivity of LH-releasing cells in the pituitary. In addition, follicles respond more strongly to LH at this stage because more of their cells have receptors for this hormone.

The increase in LH concentration caused by increased estradiol secretion from the growing follicle is an example of positive feedback. The result is final maturation of the follicle. ❼ The maturing follicle, containing a fluid-filled cavity, enlarges, forming a bulge near the surface of the ovary. The follicular phase ends at ovulation, about a day after the LH surge. In response to the peak in LH levels, the follicle and adjacent wall of the ovary rupture, releasing the secondary oocyte. There is sometimes a distinctive pain in the lower abdomen at or near the time of ovulation; this pain is felt on the left or right side, corresponding to whichever ovary has matured a follicle during that cycle.

The *luteal phase* of the ovarian cycle follows ovulation. ❽ LH stimulates the follicular tissue left behind in the ovary to transform into a corpus luteum, a glandular structure. Under continued stimulation by LH, the corpus luteum secretes progesterone and estradiol, which in combination exert negative feedback on the hypothalamus and pituitary. This feedback reduces the secretion of LH and FSH to very low levels, preventing another egg from maturing when a pregnancy may already be under way.

Near the end of the luteal phase, low gonadotropin levels cause the corpus luteum to disintegrate, triggering a sharp decline in estradiol and progesterone concentrations. The decreasing levels of ovarian steroid hormones liberate the hypothalamus and pituitary from the negative-feedback effect of these hormones. The pituitary can then begin to secrete enough FSH to stimulate the growth of new follicles in the ovary, initiating the next ovarian cycle.

### The Uterine (Menstrual) Cycle

Prior to ovulation, ovarian steroid hormones stimulate the uterus to prepare for support of an embryo. Estradiol secreted in increasing amounts by growing follicles signals the endometrium to thicken. In this way, the follicular phase of the ovarian cycle is coordinated with the *proliferative phase* of the uterine cycle. After ovulation, ❾ the estradiol and progesterone secreted by the corpus luteum stimulate maintenance of the uterine lining, as well as further development, including enlargement of arteries and growth of endometrial glands. These glands secrete a nutrient fluid that can sustain an early embryo even before it implants in the uterine lining. Thus, the luteal phase of the ovarian cycle is coordinated with what is called the *secretory phase* of the uterine cycle.

Once the corpus luteum has disintegrated, ❿ the rapid drop in ovarian hormone levels causes arteries in the endometrium to constrict. Deprived of its circulation, the uterine lining largely disintegrates, and the uterus, in response to prostaglandin secretion, contracts. Small endometrial blood vessels constrict, releasing blood that is shed along with endometrial tissue and fluid. The result is menstruation—the *menstrual*

*flow phase* of the uterine cycle. During this phase, which usually lasts a few days, a new set of ovarian follicles begin to grow. By convention, the first day of flow is designated day 1 of the new uterine (and ovarian) cycle.

Overall, the hormonal cycles in females coordinate egg maturation and release with changes in the uterus, the organ that must accommodate an embryo if the egg cell is fertilized. If an embryo has not implanted in the endometrium by the end of the secretory phase, a new menstrual flow commences, marking the start of the next cycle. Later in the chapter, you'll learn about override mechanisms that prevent disintegration of the endometrium in pregnancy.

### Menopause

After about 500 cycles, a woman undergoes **menopause**, the cessation of ovulation and menstruation. Menopause usually occurs between the ages of 46 and 54. During this interval, the ovaries lose their responsiveness to FSH and LH, resulting in a decline in estradiol production.

Menopause is an unusual phenomenon. In most other species, females and males retain their reproductive capacity throughout life. Is there an evolutionary explanation for menopause? One intriguing hypothesis proposes that during early human evolution, undergoing menopause after bearing several children allowed a mother to provide better care for her children and grandchildren, thereby increasing the survival of individuals who share much of her genetic makeup.

### Menstrual Versus Estrous Cycles

In all female mammals, the endometrium thickens before ovulation, but only humans and some other primates have menstrual cycles. Other mammals have **estrous cycles**, in which in the absence of a pregnancy, the uterus reabsorbs the endometrium and no extensive fluid flow occurs. Whereas human females may engage in sexual activity throughout the menstrual cycle, mammals with estrous cycles usually copulate only during the period surrounding ovulation. This period, called estrus (from the Latin *oestrus*, frenzy, passion), is the only time the female is receptive to mating. It is often called "heat," and the female's body temperature does increase slightly.

The length, frequency, and nature of estrous cycles vary widely among mammals. Bears and wolves have one estrous cycle per year; elephants typically have multiple cycles lasting 14–16 weeks each. Rats have estrous cycles throughout the year, each lasting only 5 days. Their nemesis, the household cat, ovulates only upon mating.

## Human Sexual Response

In humans, the arousal of sexual interest is complex, involving a variety of psychological as well as physical factors. Although many reproductive structures in the male and female are quite different in appearance, they often serve similar functions in arousal, reflecting their shared developmental origin. For example, the same embryonic tissues give rise to the glans of the penis and the clitoris, to the scrotum and the labia majora, and to the skin on the penis and the labia minora. Furthermore, the general pattern of human sexual response is similar in males and females. Two types of physiological reactions predominate in both sexes: *vasocongestion*, the filling of a tissue with blood, and *myotonia*, increased muscle tension.

The sexual response cycle can be divided into four phases: excitement, plateau, orgasm, and resolution. An important function of the excitement phase is to prepare the vagina and penis for *coitus* (sexual intercourse). During this phase, vasocongestion is particularly evident in erection of the penis and clitoris and in enlargement of the testicles, labia, and breasts. The vagina becomes lubricated and myotonia may occur, resulting in nipple erection or tension of the arms and legs.

In the plateau phase, these responses continue as a result of direct stimulation of the genitalia. In females, the outer third of the vagina becomes vasocongested, while the inner two-thirds slightly expands. This change, coupled with the elevation of the uterus, forms a depression for receiving sperm at the back of the vagina. Breathing increases and heart rate rises, sometimes to 150 beats per minute—not only in response to the physical effort of sexual activity, but also as an involuntary response to stimulation of the autonomic nervous system (see Chapter 38).

*Orgasm* is characterized by rhythmic, involuntary contractions of the reproductive structures in both sexes. Male orgasm has two stages. The first, emission, occurs when the glands and ducts of the reproductive tract contract, forcing semen into the urethra. Expulsion, or ejaculation, occurs when the urethra contracts and the semen is expelled. During female orgasm, the uterus and outer vagina contract, but the inner two-thirds of the vagina does not. Orgasm is the shortest phase of the sexual response cycle, usually lasting only a few seconds. In both sexes, contractions occur at about 0.8-second intervals and may also involve the anal sphincter and several abdominal muscles.

The resolution phase completes the cycle and reverses the responses of the earlier stages. Vasocongested organs return to their normal size and color, and muscles relax. Most of the changes of resolution are completed within 5 minutes, but some may take as long as an hour. Following orgasm, the male typically enters a refractory period, lasting anywhere from a few minutes to hours, during which erection and orgasm cannot be achieved. Females do not have a refractory period, making possible multiple orgasms within a short period of time.

## CONCEPT CHECK 36.3

1. FSH and LH get their names from events of the female reproductive cycle, but they also function in males. How are their functions in females and males similar?
2. How does an estrous cycle differ from a menstrual cycle, and in what animals are the two types of cycles found?
3. **WHAT IF?** If a human female were to take estradiol and progesterone immediately after the start of a new menstrual cycle, how would ovulation be affected? Explain.

For suggested answers, see Appendix A.

# Development of an egg into a mature embryo requires fertilization, cleavage, gastrulation, and organogenesis

Sperm meets egg. What's next? Across a range of animal species, development involves common stages that occur in a set order. As shown in **Figure 36.13**, the process begins with fertilization, the formation of a zygote. Development proceeds with the cleavage stage, during which a series of mitoses divide, or cleave, the zygote into a many-celled *embryo*. Over the course of these cleavage divisions, which typically are rapid and lack accompanying cell growth, the embryo is transformed into a hollow ball of cells called a blastula. Next, the blastula folds in on itself, rearranging into a three-layered embryo, the gastrula, in a process called gastrulation. During organogenesis, the last major stage of embryonic development, local changes in cell shape and large-scale changes in cell location generate the rudimentary organs from which adult structures grow.

With this overview of embryonic development in mind, let's take a brief look at the early events—fertilization, cleavage, and gastrulation—in sea urchins (phylum Echinodermata; see Figure 27.11). Why sea urchins? Their gametes are easy to collect, and they have external fertilization; as a result, researchers can observe fertilization and subsequent events simply by combining eggs and sperm in seawater. Sea urchins are one example of a model organism, a species chosen for in-depth research in part because it is easy to study in the laboratory.

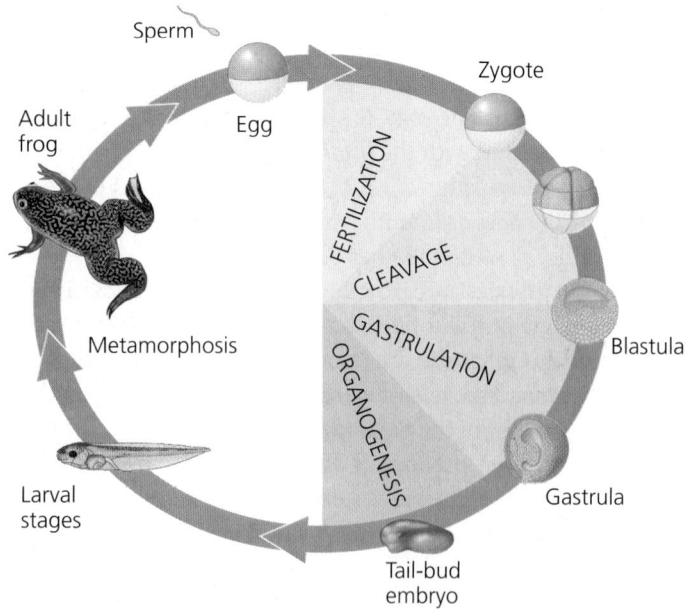

▲ **Figure 36.13 Developmental events in the life cycle of a frog.**

## Fertilization

Molecules and events at the egg surface play a crucial role in each step of fertilization (**Figure 36.14**). First, sperm dissolve or penetrate any protective layer surrounding the egg to reach the plasma membrane. Next, molecules on the sperm surface bind to receptors on the egg, helping ensure that a sperm of the same species fertilizes the egg. Finally, changes at the surface of the egg prevent *polyspermy*, the entry of multiple sperm nuclei into the egg. If polyspermy were to occur, the resulting abnormal number of chromosomes in the embryo would be lethal. Instead, both a fast and a slow block to polyspermy ensure that only one sperm nucleus crosses the egg plasma membrane.

Fertilization initiates and speeds up metabolic reactions that bring about the onset of embryonic development. There is, for example, a marked increase in the rates of cellular respiration and protein synthesis in the egg following entry of the sperm nucleus. Soon thereafter, the egg and sperm nuclei fully fuse, and cycles of DNA synthesis and cell division begin. What triggers activation of the egg? Studies show that sperm entry causes release of internal $Ca^{2+}$ stores into the egg cytoplasm. Furthermore, injecting $Ca^{2+}$ into an unfertilized egg activates egg metabolism. Additional experiments indicate that the rise in $Ca^{2+}$ concentration triggers not only egg activation but also the cortical reaction responsible for the slow block to polyspermy (see Figure 36.14).

Fertilization in other species is similar to the process in sea urchins. However, the timing of events differs, as does the stage of meiosis the egg has reached when it is fertilized. Sea urchin eggs have already completed meiosis when they are released from the female. In other species, eggs are arrested at a specific stage of meiosis and complete the meiotic divisions only after fertilization. Human eggs, for example, arrest at metaphase of meiosis II prior to fertilization (see Figure 36.9).

## Cleavage and Gastrulation

Once fertilization is complete, the zygotes of many animal species undergo a series of rapid cell divisions that characterize the **cleavage** stage of early development. During cleavage, the cell cycle consists primarily of the S (DNA synthesis) and M (mitosis) phases (for a review of the cell cycle, see Figure 9.6). Cells essentially skip the $G_1$ and $G_2$ (gap) phases, and little or no protein synthesis occurs. As a result, cleavage partitions the cytoplasm of the large fertilized egg into many smaller cells. The first five to seven cleavage divisions produce a hollow

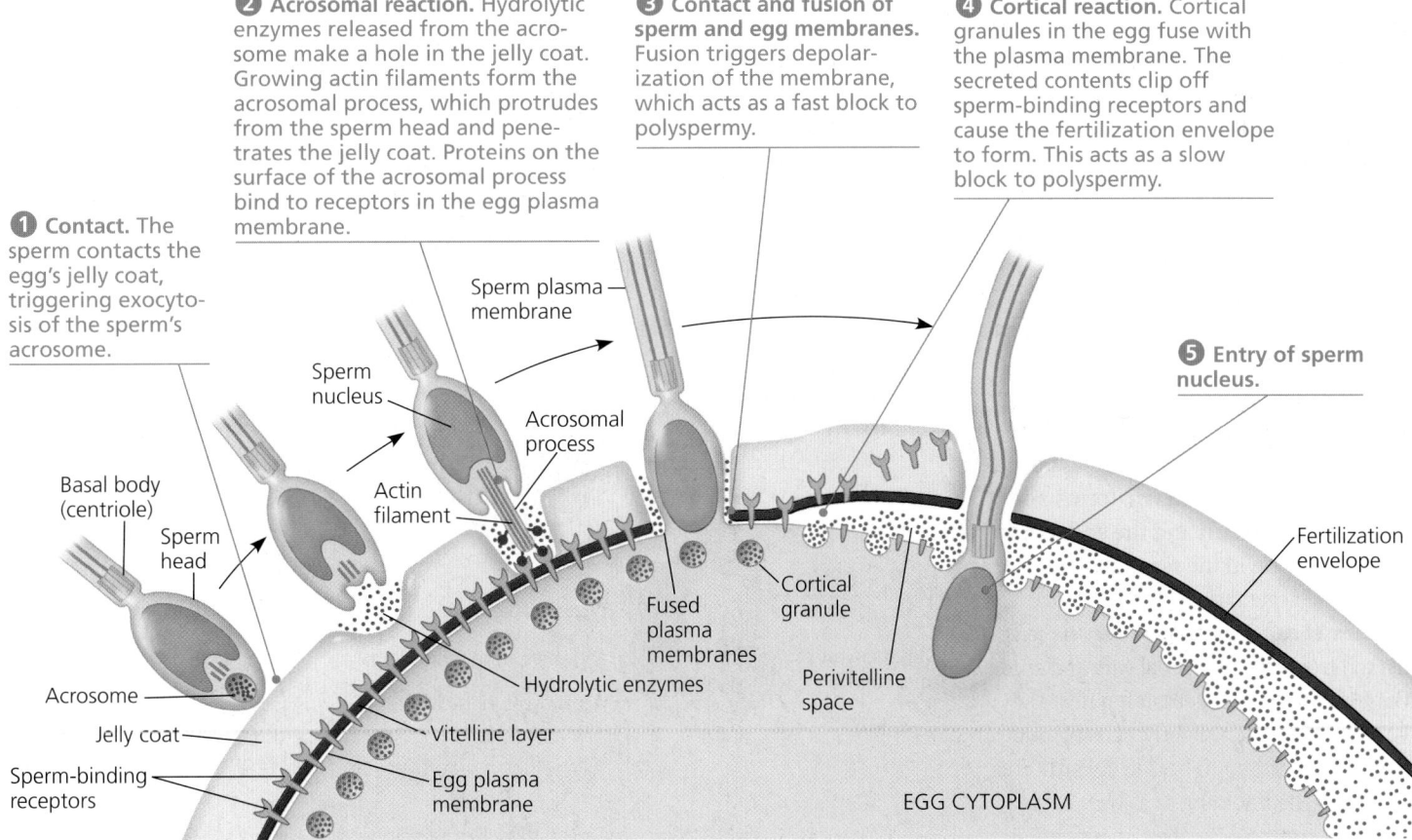

**1 Contact.** The sperm contacts the egg's jelly coat, triggering exocytosis of the sperm's acrosome.

**2 Acrosomal reaction.** Hydrolytic enzymes released from the acrosome make a hole in the jelly coat. Growing actin filaments form the acrosomal process, which protrudes from the sperm head and penetrates the jelly coat. Proteins on the surface of the acrosomal process bind to receptors in the egg plasma membrane.

**3 Contact and fusion of sperm and egg membranes.** Fusion triggers depolarization of the membrane, which acts as a fast block to polyspermy.

**4 Cortical reaction.** Cortical granules in the egg fuse with the plasma membrane. The secreted contents clip off sperm-binding receptors and cause the fertilization envelope to form. This acts as a slow block to polyspermy.

**5 Entry of sperm nucleus.**

Basal body (centriole)
Sperm head
Acrosome
Jelly coat
Sperm-binding receptors
Sperm nucleus
Actin filament
Acrosomal process
Sperm plasma membrane
Hydrolytic enzymes
Vitelline layer
Egg plasma membrane
Fused plasma membranes
Cortical granule
Perivitelline space
EGG CYTOPLASM
Fertilization envelope

▲ **Figure 36.14 The acrosomal and cortical reactions during sea urchin fertilization.** The events following contact of a single sperm and egg ensure that the nucleus of only one sperm enters the egg cytoplasm.

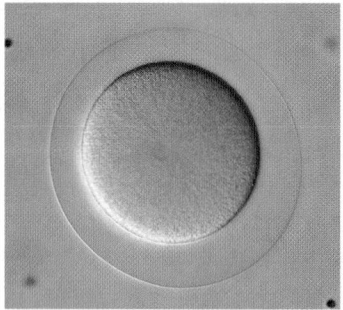

**(a) Fertilized egg.** Shown here is the zygote shortly before the first cleavage division, surrounded by the fertilization envelope.

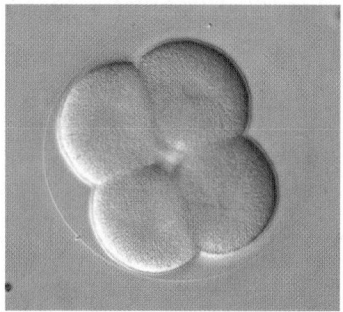

**(b) Four-cell stage.** Remnants of the mitotic spindle can be seen between the two pairs of cells that have just completed the second cleavage division.

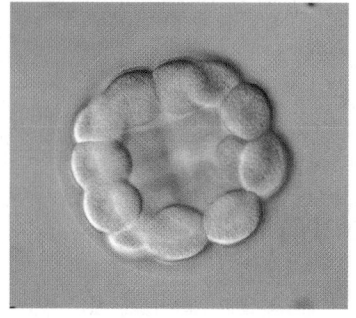

**(c) Early blastula.** After further cleavage divisions, the embryo is a multicellular ball that is still surrounded by the fertilization envelope. The blastocoel has begun to form in the center.

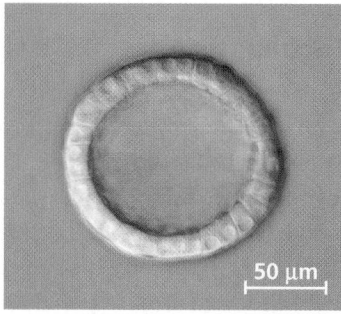

**(d) Later blastula.** A single layer of cells surrounds a large blastocoel. Although not visible here, the fertilization envelope is present; the embryo will soon hatch from it and begin swimming.

50 μm

▲ **Figure 36.15 Cleavage in an echinoderm embryo.** Cleavage is a series of mitotic cell divisions that transform the zygote into a blastula, a hollow ball of cells called blastomeres. These light micrographs show the embryonic stages of a sand dollar, which are virtually identical to those of a sea urchin.

ball of cells, the **blastula**, surrounding a fluid-filled cavity called the **blastocoel** (Figure 36.15).

After cleavage, the rate of cell division slows as the normal cell cycle is restored. The remaining stages of embryonic development bring about **morphogenesis**, the cellular and tissue-based processes by which the animal body takes shape.

During **gastrulation**, a set of cells at or near the surface of the blastula moves to an interior location, cell layers are established, and a primitive digestive tube is formed. Gastrulation reorganizes the hollow blastula into a two-layered or three-layered embryo called a **gastrula**. The cell layers produced by gastrulation are called the embryonic *germ layers* (from the Latin *germen*, to sprout or germinate). In the late gastrula, **ectoderm** forms the outer layer and **endoderm** forms the lining of the embryonic digestive tract. In vertebrates and other animals with bilateral symmetry, a third germ layer, the **mesoderm**, forms between the ectoderm and the endoderm.

Gastrulation in the sea urchin begins at the vegetal pole of the blastula (**Figure 36.16**). There, *mesenchyme cells* detach from the blastocoel wall and enter the blastocoel. Cells remaining near the vegetal pole flatten slightly, causing that end of the embryo to buckle inward. This process—the infolding of a sheet of cells into the embryo—is called *invagination*. Extensive rearrangement of cells transforms the shallow depression into a

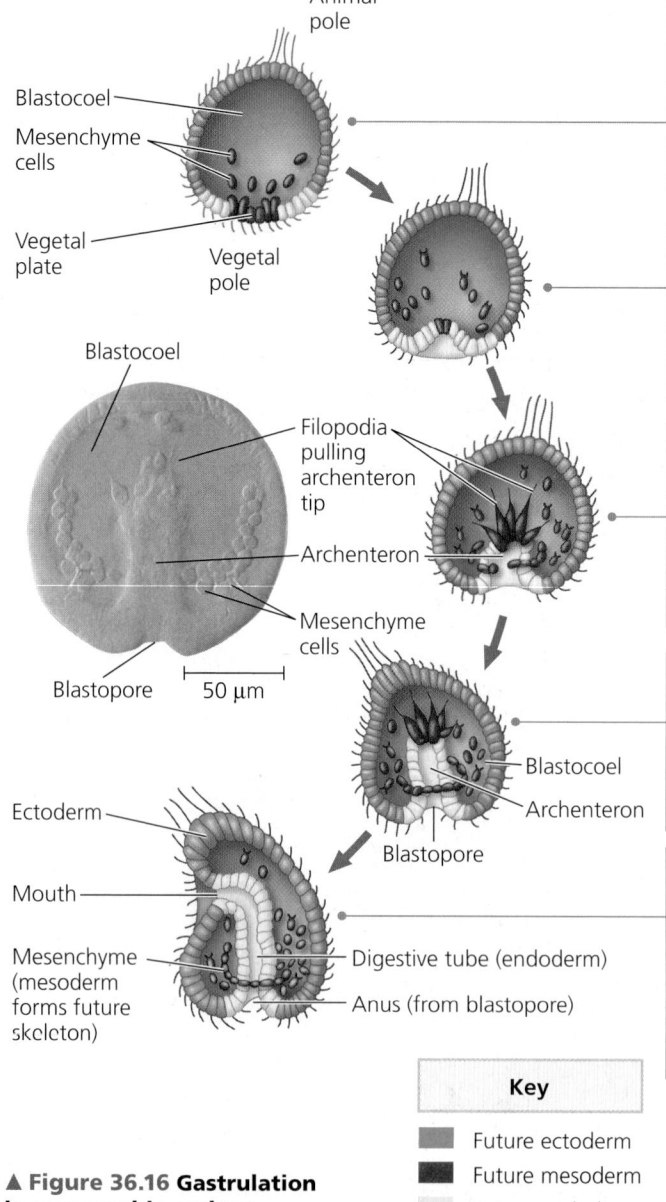

❶ A group of *mesenchyme* cells migrates from the vegetal pole of the blastula into the blastocoel. Some of these cells will eventually secrete calcium carbonate and form a simple internal skeleton.

❷ Cells at the vegetal plate flatten slightly, causing the vegetal pole of the embryo to buckle inward. This infolding of a sheet of cells is called *invagination*.

❸ Endoderm cells form the archenteron, the future digestive tube. New mesenchyme cells at the tip of the archenteron send out thin extensions (filopodia) toward the blastocoel wall (left, LM).

❹ The filopodia contract, dragging the archenteron across the blastocoel. The open end of the archenteron, which will become the anus, is called the blastospore.

❺ Fusion of the archenteron with the blastocoel wall forms the digestive tube, which now has a mouth and an anus. The gastrula has three germ layers and is covered with cilia, which will function in feeding and movement.

▲ **Figure 36.16 Gastrulation in a sea urchin embryo.**

| Key |
| --- |
| Future ectoderm |
| Future mesoderm |
| Future endoderm |

deeper, narrower, blind-ended tube called the *archenteron*. The open end of the archenteron, which will become the anus, is called the *blastopore*. A second opening, which will become the mouth, forms when the other end of the archenteron touches the inside of the ectoderm and the two layers fuse, producing a rudimentary digestive tube.

The cell movements and interactions that form the germ layers vary considerably among species. One basic distinction is whether the mouth develops from the first opening that forms in the embryo (protostomes) or the second (deuterostomes). The mouth develops from the second opening in sea urchins and in humans and other vertebrates.

## Organogenesis

During **organogenesis**, regions of the embryonic germ layers develop into the rudiments of body organs. Each germ layer contributes to a distinct set of structures in the adult animal, as shown for vertebrates in **Figure 36.17**. Note that some organs and many organ systems of the adult derive from more than one germ layer. For example, the adrenal glands have both ectodermal and mesoderm tissue.

Having introduced the stages of fertilization, cleavage, gastrulation, and organogenesis, we now return to our consideration of human reproduction.

## Conception and Embryo Implantation in Humans

During human copulation, the male delivers hundreds of millions of sperm in 2–5 mL of semen. When first ejaculated, the semen coagulates, which may keep the ejaculate in place until sperm reach the cervix. Soon after, anticoagulants liquefy the semen, and sperm swim through the cervix and oviducts.

Fertilization—also called **conception** in humans—occurs when a sperm fuses with an egg (mature oocyte) in the oviduct

**ECTODERM (outer layer of embryo)**

- Epidermis of skin and its derivatives (including sweat glands, hair follicles)
- Nervous and sensory systems
- Pituitary gland, adrenal medulla
- Jaws and teeth
- Germ cells

**MESODERM (middle layer of embryo)**

- Skeletal and muscular systems
- Circulatory and lymphatic systems
- Excretory and reproductive systems (except germ cells)
- Dermis of skin
- Adrenal cortex

**ENDODERM (inner layer of embryo)**

- Epithelial lining of digestive tract and associated organs (liver, pancreas)
- Epithelial lining of respiratory, excretory, and reproductive tracts and ducts
- Thymus, thyroid, and parathyroid glands

▲ **Figure 36.17 Major derivatives of the three embryonic germ layers in vertebrates.**

(Figure 36.18, ❶ and ❷). As in sea urchin fertilization, sperm binding triggers a cortical reaction, which results in a slow block to polyspermy. (No fast block to polyspermy has been found in mammals.) The zygote begins cleavage ❸ about 24 hours after fertilization and produces a blastocyst ❹ after an additional 4 days. A few days later, the embryo implants into the endometrium of the uterus ❺.

The condition of carrying one or more embryos in the uterus is called *pregnancy*, or **gestation**. Human pregnancy averages 266 days (38 weeks) from fertilization of the egg, or 40 weeks from the start of the last menstrual cycle. In comparison, gestation averages 21 days in many rodents, 280 days in cows, and more than 600 days in elephants.

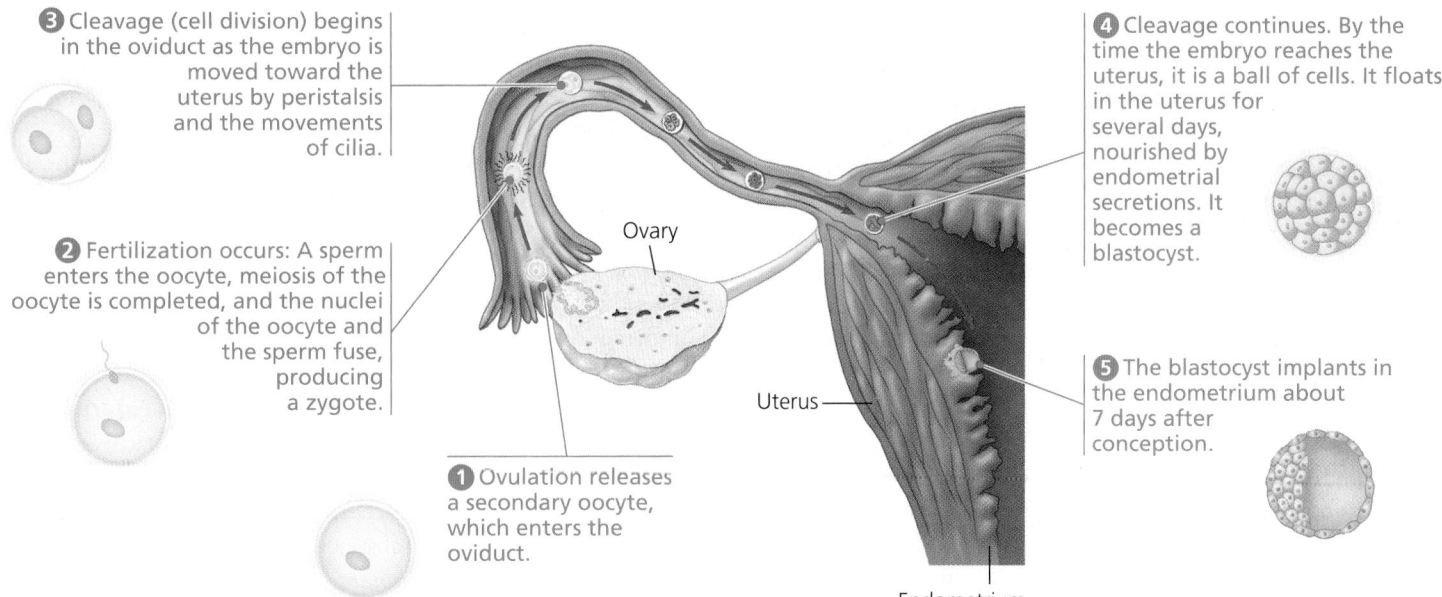

❸ Cleavage (cell division) begins in the oviduct as the embryo is moved toward the uterus by peristalsis and the movements of cilia.

❷ Fertilization occurs: A sperm enters the oocyte, meiosis of the oocyte is completed, and the nuclei of the oocyte and the sperm fuse, producing a zygote.

❶ Ovulation releases a secondary oocyte, which enters the oviduct.

Ovary

❹ Cleavage continues. By the time the embryo reaches the uterus, it is a ball of cells. It floats in the uterus for several days, nourished by endometrial secretions. It becomes a blastocyst.

❺ The blastocyst implants in the endometrium about 7 days after conception.

Uterus

Endometrium

▲ **Figure 36.18 Formation of a human zygote and early postfertilization events.**

Human gestation can be divided into three *trimesters* of about three months each. During the first trimester, the implanted embryo secretes hormones that signal its presence and regulate the mother's reproductive system. One hormone, *human chorionic gonadotropin (hCG)*, acts like LH in maintaining secretion of progesterone and estrogens by the corpus luteum through the first few months of pregnancy. Some hCG passes from the maternal blood to the urine; detecting hCG in the urine is the basis of a common pregnancy test.

Occasionally, the embryo splits during the first month of development, resulting in identical, or *monozygotic* (one-egg), twins. Fraternal, or *dizygotic*, twins arise in a very different way: Two follicles mature in a single cycle, followed by independent fertilization and implantation of two genetically distinct embryos.

Not all embryos complete development. Many spontaneously stop developing as a result of chromosomal or developmental abnormalities. Such spontaneous abortion, or *miscarriage*, occurs in as many as one-third of all pregnancies, often before the woman is even aware she is pregnant.

## Human Development and Birth

During its first 2–4 weeks of development, the embryo obtains nutrients directly from the endometrium. Meanwhile, the outer layer of the embryo, called the **trophoblast**, grows outward and mingles with the endometrium (see step 5 in Figure 36.18), eventually helping form the **placenta**. This disk-shaped organ, containing both embryonic and maternal blood vessels, can weigh close to 1 kg. Material diffusing between the maternal and embryonic circulatory systems supplies nutrients, provides immune protection, exchanges respiratory gases, and disposes of metabolic wastes for the embryo.

The first trimester is the main period of organogenesis **(Figure 36.19a)**. During this stage, the embryo is particularly susceptible to damage. For example, alcohol that passes through the placenta and reaches the developing nervous system of the embryo can cause fetal alcohol syndrome, a disorder that can result in mental retardation and other serious birth defects. At 8 weeks, all the major structures of the adult are present in rudimentary form, and the embryo is called a **fetus**. The heart begins beating by the fourth week; a heartbeat can be detected at 8–10 weeks. At the end of the first trimester (13 weeks), the fetus is well differentiated but only 5 cm long.

During the second trimester **(Figure 36.19b)**, the fetus grows to about 30 cm in length. Development continues, including formation of fingernails, external sex organs, and outer ears. The mother may feel fetal movements as early as one month into this trimester. Growth to nearly 20 cm in length requires adoption of the fetal position (head at knees)

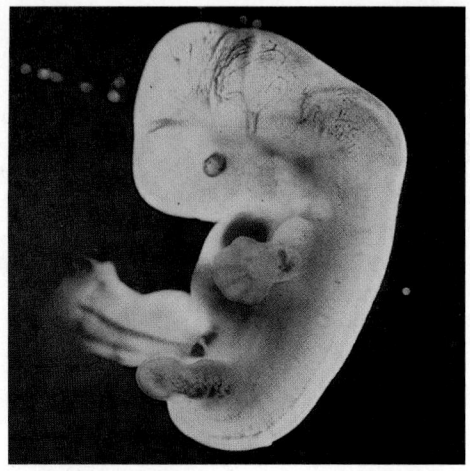
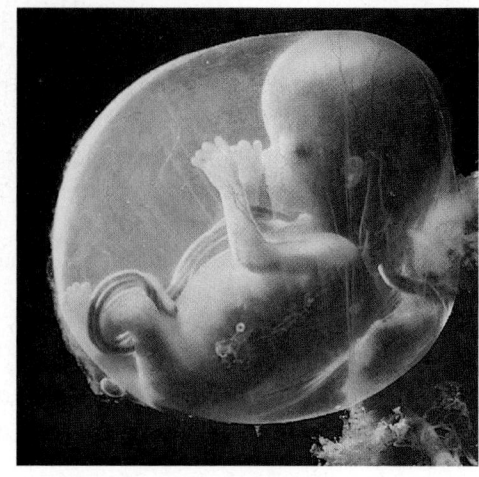

(a) **5 weeks.** Limb buds, eyes, the heart, the liver, and rudiments of all other organs have started to develop in the embryo, which is only about 1 cm long.

(b) **14 weeks.** Growth and development of the offspring, now called a fetus, continue during the second trimester. This fetus is about 6 cm long.

▲ **Figure 36.19 Snapshots of human development.**

due to the limited space available. During the third trimester, the fetus grows to about 3–4 kg in weight and 50 cm in length. Fetal activity may decrease as the fetus fills the available space.

Childbirth begins with *labor*, a series of strong, rhythmic uterine contractions that push the fetus and placenta out of the body. Once labor begins, local regulators (prostaglandins) and hormones (chiefly estradiol and oxytocin) induce and regulate further contractions of the uterus. A positive-feedback loop (see Concept 32.2) is central to this regulation: Uterine contractions stimulate secretion of oxytocin, which in turn stimulates further contractions.

One aspect of postnatal care unique to mammals is *lactation*, the production of mother's milk. In response to suckling by the newborn and changes in estradiol levels, the hypothalamus signals the anterior pituitary to secrete prolactin, which stimulates the mammary glands to produce milk. Suckling also stimulates secretion of oxytocin from the posterior pituitary, which triggers milk release (see Figure 32.7).

## Contraception

We'll look now at **contraception**, the deliberate prevention of pregnancy. Some contraceptive methods prevent gamete development or release from female or male gonads; others prevent fertilization by keeping sperm and egg apart; and still others prevent implantation of an embryo. For complete information on contraceptive methods, you should consult a health-care provider. The following brief introduction to common methods and the corresponding diagram in **Figure 36.20** make no pretense of being a contraception manual.

Fertilization can be prevented by abstinence from sexual intercourse or by any of several barriers that keep live sperm from contacting the egg. Temporary abstinence, often called the *rhythm method* or *natural family planning*, depends on refraining from intercourse when conception is most likely. Because the egg can survive in the oviduct for 24–48 hours and sperm for up to 5 days,

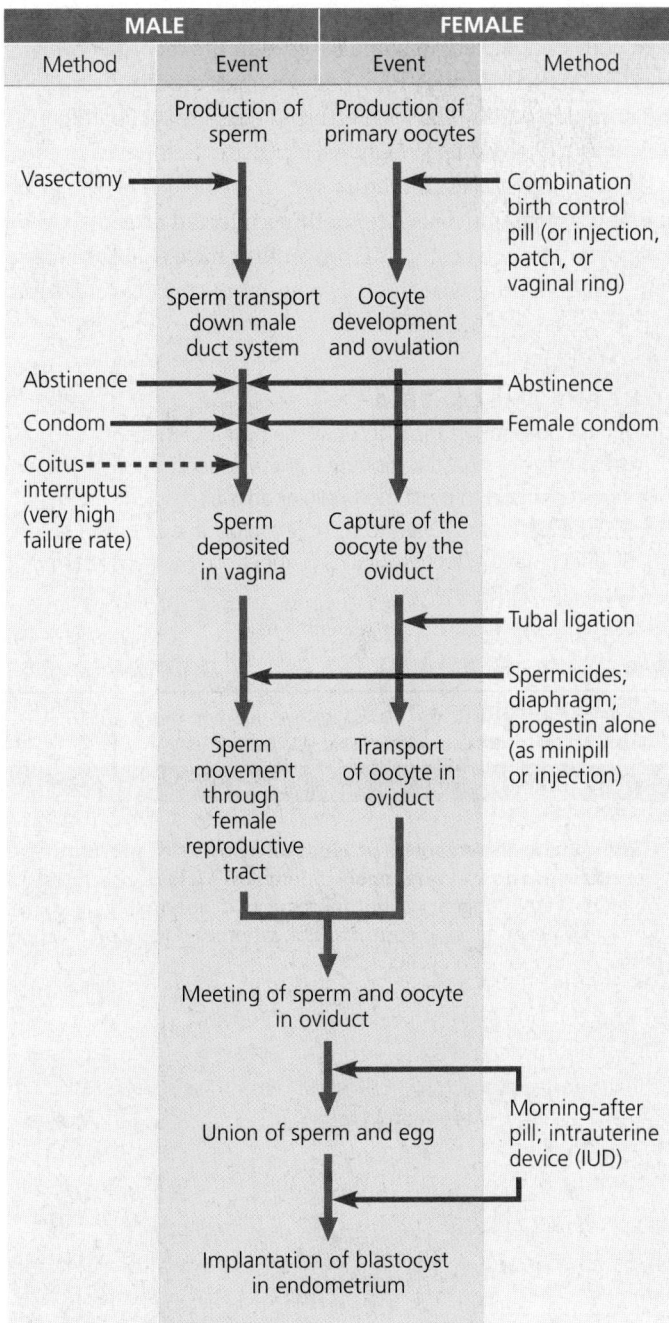

| MALE | | FEMALE | |
|---|---|---|---|
| Method | Event | Event | Method |
| | Production of sperm | Production of primary oocytes | |
| Vasectomy → | | | ← Combination birth control pill (or injection, patch, or vaginal ring) |
| | Sperm transport down male duct system | Oocyte development and ovulation | |
| Abstinence → | | | ← Abstinence |
| Condom → | | | ← Female condom |
| Coitus interruptus (very high failure rate) ⤑ | Sperm deposited in vagina | Capture of the oocyte by the oviduct | |
| | | | ← Tubal ligation |
| | | | ← Spermicides; diaphragm; progestin alone (as minipill or injection) |
| | Sperm movement through female reproductive tract | Transport of oocyte in oviduct | |

Meeting of sperm and oocyte in oviduct

Union of sperm and egg ← Morning-after pill; intrauterine device (IUD)

Implantation of blastocyst in endometrium

▲ **Figure 36.20 Mechanisms of several contraceptive methods.** Red arrows indicate where these methods, devices, or products interfere with events from the production of sperm and primary oocytes to implantation of a developing embryo.

a couple practicing temporary abstinence should not engage in intercourse for a significant number of days before and after ovulation. Contraceptive methods based on fertility awareness require that the couple understand physiological indicators associated with ovulation, such as changes in cervical mucus. A pregnancy rate of 10–20% is typically reported for natural family planning. (Pregnancy rate is the average number of women who become pregnant during a year for every 100 women using a particular pregnancy prevention method, expressed as a percentage.)

*Coitus interruptus*, or withdrawal (removal of the penis from the vagina before ejaculation), is unreliable. Sperm from a previous ejaculate may be transferred in secretions that precede ejaculation. Also, a split-second lapse in timing or willpower can result in millions of sperm being transferred before withdrawal.

Several barrier methods that block sperm from meeting the egg have pregnancy rates of less than 10%. The *condom* is a thin sheath that fits over the penis to collect the semen. For sexually active individuals, latex condoms are the only contraceptives that are highly effective in preventing the spread of sexually transmitted diseases, including AIDS. (This protection is, however, not absolute.) Another device is the *diaphragm*, a dome-shaped rubber cap inserted into the upper portion of the vagina before intercourse. Both devices have lower pregnancy rates when used with a spermicidal (sperm-killing) foam or jelly. Another barrier device is the vaginal pouch, or "female condom."

Except for complete abstinence, the most effective means of birth control are sterilization, intrauterine devices (IUDs), and hormonal contraceptives. Sterilization (vasectomy in males or tubal ligation in females) is almost 100% effective. The IUD has a pregnancy rate of 1% or less and is the most common reversible method of birth control outside the United States. Placed in the uterus by a doctor, the IUD interferes with fertilization and implantation. Hormonal contraceptives, usually *birth control pills*, also have pregnancy rates of 1% or less.

The most commonly prescribed hormonal contraceptives are a combination of a synthetic estrogen and a synthetic progesterone-like hormone (progestin). This combination mimics negative feedback in the ovarian cycle, stopping the release of GnRH by the hypothalamus and thus of FSH and LH by the pituitary. The prevention of LH release blocks ovulation. In addition, the inhibition of FSH secretion by the low dose of estrogens in the pills prevents follicles from developing. Such combination birth control pills can also act as "morning-after" pills. Taken within 3 days after unprotected intercourse, they prevent fertilization or implantation with an effectiveness of about 75%.

A different type of hormonal contraceptive contains only progestin. Progestin causes a woman's cervical mucus to thicken so that it blocks sperm from entering the uterus. Progestin also decreases the frequency of ovulation and causes changes in the endometrium that may interfere with implantation if fertilization occurs. Progestin can be administered as injections that last for three months or as a tablet ("minipill") taken daily. Pregnancy rates for progestin treatment are very low.

Hormonal contraceptives have both beneficial and harmful side effects. Women who regularly smoke cigarettes face a three to ten times greater risk of dying from cardiovascular disease if they also use oral contraceptives. Among nonsmokers, birth control pills slightly raise a woman's risk of abnormal blood clotting, high blood pressure, heart attack, and stroke. Although oral contraceptives increase the risk for these cardiovascular disorders, they eliminate the dangers of pregnancy; women on birth control pills have mortality rates about one-half those of pregnant women.

## Infertility and *In Vitro* Fertilization

Infertility—an inability to conceive offspring—is quite common, affecting about one in ten couples both in the United States and worldwide. The causes of infertility are varied, and the likelihood of a reproductive defect is nearly the same for men and women. Among preventable causes of infertility, the most significant is *sexually transmitted disease (STD)*. In women 15–24 years old, approximately 700,000 cases of chlamydia and gonorrhea are reported annually in the United States. The actual number infected is considerably higher because most women with these STDs have no symptoms and are therefore unaware of their infection. Up to 40% of women who remain untreated for chlamydia or gonorrhea develop an inflammatory disorder that can lead to infertility or to potentially fatal complications during pregnancy.

Some forms of infertility are treatable. Hormone therapy can sometimes increase sperm or egg production, and surgery can often correct ducts that have failed to form properly or have become blocked. In some cases, doctors recommend *in vitro* **fertilization (IVF)**, which involves mixing oocytes and sperm in culture dishes. Fertilized eggs are incubated until they have formed at least eight cells and are then transferred to the woman's uterus. If mature sperm are defective or low in number, a sperm nucleus is sometimes injected directly into an oocyte. Though costly, IVF procedures have enabled more than a million otherwise infertile couples to conceive children.

### CONCEPT CHECK 36.4

1. Where does fertilization occur in the human female?
2. What role does gastrulation play in the specialization of cell types common to most multicellular animals?
3. **WHAT IF?** If an STD led to complete blockage of both oviducts, what effect would you expect on the menstrual cycle and on fertility?

For suggested answers, see Appendix A.

---

# 36 Chapter Review

Go to **Mastering**Biology® for Assignments, the eText, and the Study Area with Animations, Activities, Vocab Self-Quiz, and Practice Tests.

## SUMMARY OF KEY CONCEPTS

**VOCAB SELF-QUIZ**

goo.gl/gbai8v

### CONCEPT 36.1

**Both asexual and sexual reproduction occur in the animal kingdom (pp. 752–755)**

- **Sexual reproduction** requires the fusion of male and female gametes, forming a diploid **zygote**. **Asexual reproduction** is the production of offspring without gamete fusion. Mechanisms of asexual reproduction include budding, fission, and fragmentation with regeneration. Variations on the mode of reproduction are achieved through **parthenogenesis**, **hermaphroditism**, and sex reversal. Hormones and environmental cues control reproductive cycles.
- **Fertilization**, whether external or internal, requires coordinated timing, which may be mediated by environmental cues, pheromones, or courtship behavior. Internal fertilization is typically often associated both with fewer offspring and with greater protection of offspring by the parents.

? *Would a pair of haploid offspring produced by parthenogenesis be genetically identical? Explain.*

### CONCEPT 36.2

**Reproductive organs produce and transport gametes (pp. 755–760)**

- Systems for gamete production and delivery range from undifferentiated cells in the body cavity to complex **gonads** with accessory tubes and glands that carry and protect gametes and embryos. In human males, **sperm** are produced in **testes**, which are suspended outside the body in the **scrotum**. Ducts connect the testes to internal accessory glands and to the **penis**. The reproductive system of the human female consists principally of the **ovaries**, **oviducts**, **uterus**, and **vagina** internally and the **labia** and the **glans** of the **clitoris** externally. **Eggs** are produced in the ovaries and upon fertilization develop in the uterus.

- **Gametogenesis**, or gamete production, consists of **spermatogenesis** in males and **oogenesis** in females. Meiosis generates one large egg in oogenesis, but four sperm in spermatogenesis. In humans, sperm develop continuously, whereas oocyte maturation is discontinuous and cyclic.

**Human gametogenesis**

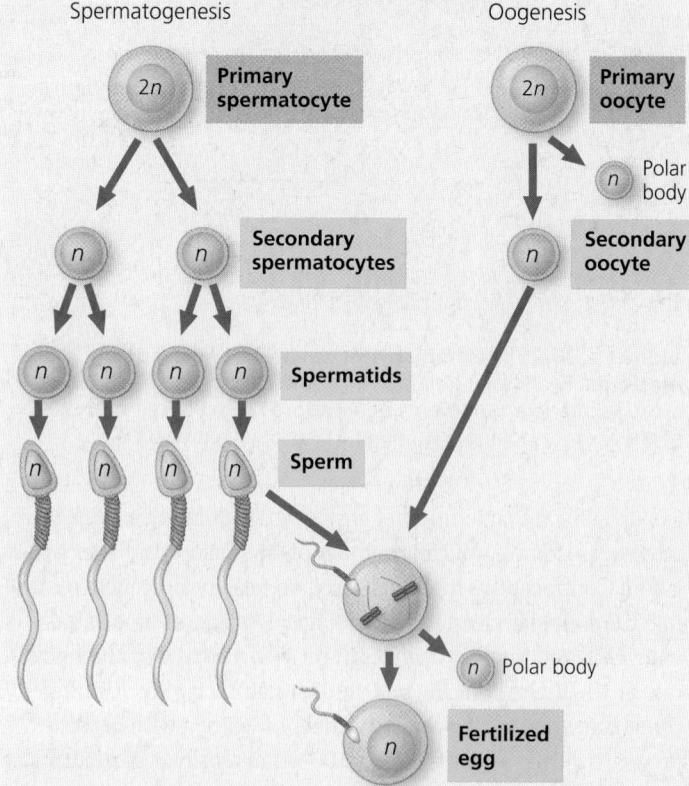

Spermatogenesis — Oogenesis

Primary spermatocyte ($2n$) — Primary oocyte ($2n$)

Polar body ($n$)

Secondary spermatocytes ($n$) — Secondary oocyte ($n$)

Spermatids ($n$)

Sperm ($n$)

Polar body ($n$)

Fertilized egg ($n$)

? *How does the difference in size and cellular contents between sperm and eggs relate to their specific functions?*

## The interplay of tropic and sex hormones regulates reproduction in mammals (pp. 760–764)

- GnRH from the hypothalamus regulates release of **FSH** and **LH** from the anterior pituitary, which orchestrates gametogenesis. Secretion of androgens (chiefly **testosterone**) and sperm production are controlled by FSH and LH. In the **menstrual cycle**, cyclic secretion of FSH and LH causes changes in the ovary and uterus via estrogens, primarily **estradiol**, and **progesterone**. The follicle and corpus luteum also secrete hormones, with feedback coordinating the uterine and ovarian cycles.
- In **estrous cycles**, the endometrial lining is reabsorbed, and sexual receptivity is limited to a heat period.

**?** *Why do anabolic steroids lead to reduced sperm count?*

## Development of an egg into a mature embryo requires fertilization, cleavage, gastrulation, and organogenesis (pp. 764–770)

- Fertilization brings together the nuclei of sperm and egg, forming a diploid zygote, and activates the egg, initiating embryonic development. Changes at the egg surface triggered by sperm entry help block polyspermy in many animals.
- Fertilization is followed by **cleavage**, a period of rapid cell division without growth, which generates a large number of cells. In many species, cleavage creates a multicellular ball called the **blastula**, which contains a fluid-filled cavity, the **blastocoel**.
- **Gastrulation** converts the blastula to a **gastrula**, which has a primitive digestive cavity and three germ layers: **ectoderm**, **mesoderm**, and **endoderm**.
- The mammalian zygote becomes a blastocyst before implanting in the **endometrium**. Major organs start developing by 8 weeks.
- **Contraception** may prevent release of gametes from the gonads, fertilization, or embryo implantation. Infertile couples may be helped by hormonal methods or *in vitro* **fertilization**.

**?** *How does the fertilization envelope form in sea urchins? What is its function?*

# TEST YOUR UNDERSTANDING

**PRACTICE TEST**

goo.gl/CRZjvS

## Level 1: Knowledge/Comprehension

1. Which of these refers to parthenogenesis?
   (A) An individual may change its sex.
   (B) Groups of cells grow into new individuals.
   (C) An organism is first a male and then a female.
   (D) An egg develops without being fertilized.

2. The cortical reaction of sea urchin eggs functions directly in
   (A) the formation of a fertilization envelope.
   (B) the production of a fast block to polyspermy.
   (C) the generation of an electrical impulse by the egg.
   (D) the fusion of egg and sperm nuclei.

3. Which of the following is *not* properly paired?
   (A) seminiferous tubule—cervix      (C) testosterone—estradiol
   (B) vas deferens—oviduct            (D) scrotum—labia majora

4. Peaks of LH and FSH production occur during
   (A) the menstrual flow phase of the uterine cycle.
   (B) the beginning of the follicular phase of the ovarian cycle.
   (C) the period just before ovulation.
   (D) the secretory phase of the menstrual cycle.

5. During human gestation, rudiments of all organs develop
   (A) in the first trimester.        (C) in the third trimester.
   (B) in the second trimester.       (D) during the blastocyst stage.

## Level 2: Application/Analysis

6. Which of the following statements is true?
   (A) All mammals have menstrual cycles.
   (B) The endometrial lining is shed in menstrual cycles but reabsorbed in estrous cycles.
   (C) Estrous cycles are more frequent than menstrual cycles.
   (D) Ovulation occurs before the endometrium thickens in estrous cycles.

7. For which is the number the same in males and females?
   (A) interruptions in meiotic divisions
   (B) functional gametes produced by meiosis
   (C) meiotic divisions required to produce each gamete
   (D) different cell types produced by meiosis

8. Which statement about human reproduction is false?
   (A) Fertilization occurs in the oviduct.
   (B) Spermatogenesis and oogenesis require different temperatures.
   (C) An oocyte completes meiosis after a sperm penetrates it.
   (D) The earliest stages of spermatogenesis occur closest to the lumen of the seminiferous tubules.

## Level 3: Synthesis/Evaluation

9. **DRAW IT** In human spermatogenesis, mitosis of a stem cell gives rise to a stem cell and a spermatogonium. (a) Draw four rounds of mitosis for a stem cell, and label the daughter cells. (b) For one spermatogonium, draw the cells it would produce from one round of mitosis followed by meiosis. Label the cells, and label mitosis and meiosis. (c) Explain what would happen if stem cells divided like spermatogonia.

10. **SCIENTIFIC INQUIRY**
    Suppose that you discover a new egg-laying worm species. You dissect four adults and find both oocytes and sperm in each. Cells outside the gonad contain five chromosome pairs. Lacking genetic variants, explain how you would determine whether the worms can self-fertilize.

11. **FOCUS ON EVOLUTION**
    Hermaphroditism is more often found in animals that are fixed to a surface than in motile species. Explain why.

12. **FOCUS ON ENERGY AND MATTER**
    In a short essay (100–150 words), discuss how energy investments by a female frog, which lays many eggs, and a human female contribute to reproductive success.

13. **SYNTHESIZE YOUR KNOWLEDGE**

A female Komodo dragon kept isolated in a zoo had offspring. The offspring were not identical, but each one had two identical copies of every gene in its genome. Make a hypothesis to explain these observations, referring to parthenogenesis and meiosis.

*For selected answers, see Appendix A.*

**KEY CONCEPTS**

**37.1** Neuron structure and organization reflect function in information transfer

**37.2** Ion pumps and ion channels establish the resting potential of a neuron

**37.3** Action potentials are the signals conducted by axons

**37.4** Neurons communicate with other cells at synapses

▲ **Figure 37.1** What makes this snail such a deadly predator?

## Lines of Communication

The tropical cone snail (*Conus geographus*) in **Figure 37.1** is small and slow moving, yet it is a dangerous predator. A carnivore, this marine snail hunts, kills, and devours fish. Injecting venom with a hollow, harpoon-like tooth, the cone snail paralyzes its free-swimming prey almost instantaneously. The venom is so deadly that unlucky scuba divers have died from just a single injection. What component renders it so fast acting and lethal? The answer is a mixture of toxin molecules: Each has a specific mechanism of disabling **neurons**, the nerve cells that transfer information within the body. Because an animal attacked by the cone snail loses neuronal control of locomotion and respiration, it cannot escape, defend itself, or otherwise survive.

Communication by neurons largely consists of long-distance electrical signals and short-distance chemical signals. The specialized structure of neurons allows them to use pulses of electrical current to receive, transmit, and regulate the flow of information over long distances within the body.

In transferring information from one cell to another, neurons often rely on chemical signals that act over very short distances. Cone snail venom is particularly potent because it interferes with both electrical signaling and chemical signaling by neurons.

All neurons transmit electrical signals within the cell in an identical manner. Thus a neuron that detects an odor transmits information along its length in the same way as a neuron that controls the movement of a body part. The particular connections made by the active neuron are what distinguish the type of information being transmitted. Interpreting nerve impulses involves sorting those signals according to neuronal paths and connections. In more complex animals, this higher-order processing is carried out largely in groups of neurons organized into a **brain** or into simpler clusters called **ganglia**.

In this chapter, we look closely at the structure of a neuron and explore the molecules and physical principles that govern signaling by neurons. In the remainder of this unit, we'll examine nervous, sensory, and motor systems before exploring how their functions are integrated in producing behavior.

# Neuron structure and organization reflect function in information transfer

Our starting point for exploring the nervous system is the neuron, a cell type exemplifying the close fit of form and function that often arises over the course of evolution.

## Neuron Structure and Function

The ability of a neuron to receive and transmit information is based on a highly specialized cellular organization **(Figure 37.2)**. Most of a neuron's organelles, including its nucleus, are located in the **cell body**. In a typical neuron, the cell body is studded with highly branched extensions called **dendrites** (from the Greek *dendron*, tree). Together with the cell body, the dendrites *receive* signals from other neurons. A neuron also has a single **axon**, an extension that *transmits* signals to other cells. Axons are often much longer than dendrites, and some, such as those that reach from the spinal cord of a giraffe to the muscle cells in its feet, are over a meter long. The cone-shaped base of an axon, called the axon hillock, is

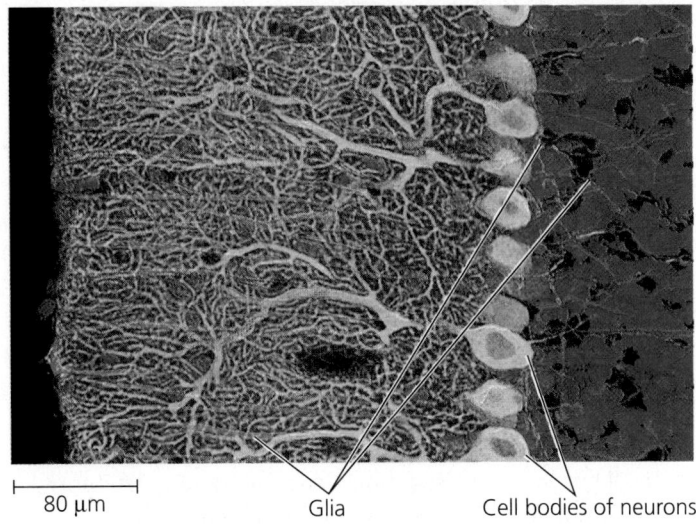

80 μm          Glia          Cell bodies of neurons

▲ **Figure 37.3 Glia in the mammalian brain.** This micrograph (a fluorescently labeled laser confocal image) shows a region of the rat brain packed with glia and interneurons. Glia are labeled red, DNA in nuclei is labeled blue, and dendrites of neurons are labeled green.

typically where signals that travel down the axon are generated. Near its other end, an axon usually divides into many branches.

Each branched end of an axon transmits information to another cell at a junction called a **synapse**. The part of each axon branch that forms this specialized junction is a *synaptic terminal*. At most synapses, chemical messengers called **neurotransmitters** pass information from the transmitting neuron to the receiving cell (see Figure 37.2). In describing a synapse, we refer to the transmitting neuron as the *presynaptic cell* and the neuron, muscle, or gland cell that receives the signal as the *postsynaptic cell*.

The neurons of vertebrates and most invertebrates require supporting cells called **glial cells**, or **glia** (from a Greek word meaning "glue") **(Figure 37.3)**. Overall, glia outnumber neurons in the mammalian brain 10- to 50-fold. Glia nourish neurons, insulate the axons of neurons, and regulate the extracellular fluid surrounding neurons. In addition, glia sometimes function in replenishing certain groups of neurons and in transmitting information (as we'll discuss in Chapter 38).

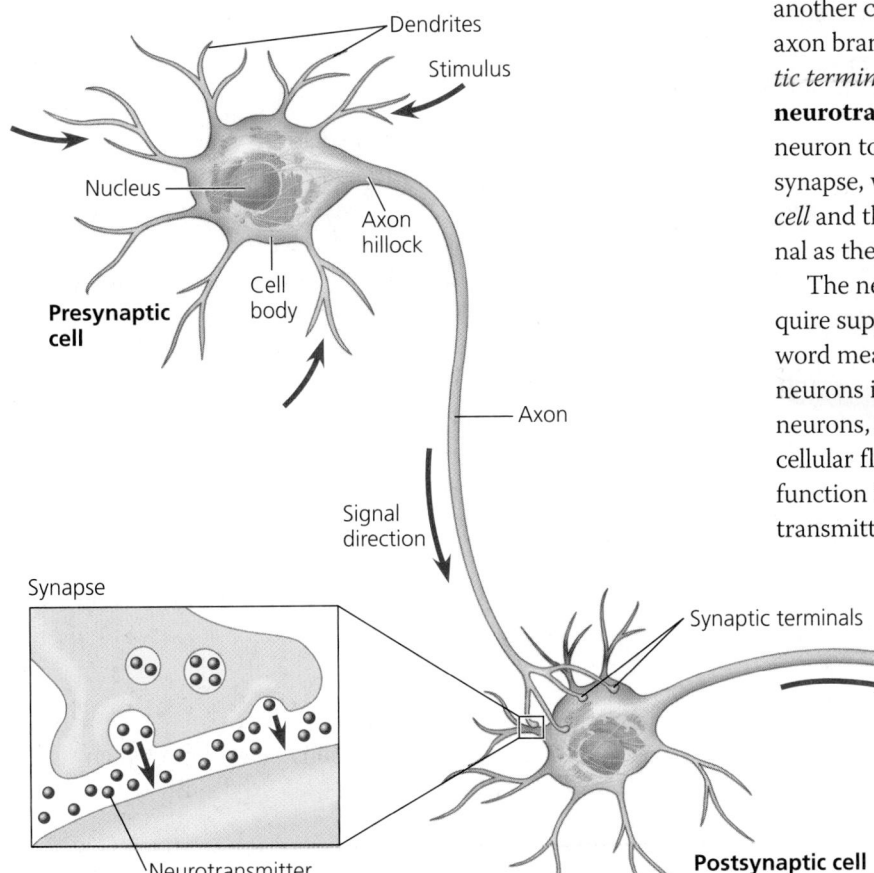

▲ **Figure 37.2 Neuron structure.** Arrows illustrate the flow of signals into, along, and out of neurons. The cells are labeled presynaptic and postsynaptic with reference to the synapse shown in the inset.

## Introduction to Information Processing

Information processing by a nervous system occurs in three stages: sensory input, integration, and motor output. As an example, let's consider how a cone snail like the one in Figure 37.1 identifies and attacks its prey. To generate sensory input to the nervous system, the snail surveys its environment with its tubelike siphon, sampling scents that might reveal a nearby fish **(Figure 37.4)**. During the integration stage, networks of neurons process this information to determine if a fish is in fact present and, if so, where the fish is located. Motor output from the processing center then initiates attack, activating neurons that trigger release of the harpoon-like tooth toward the prey.

In all but the simplest animals, specialized populations of neurons handle each stage of information processing.

- **Sensory neurons**, like those in the snail's siphon, transmit information about external stimuli—such as light, touch, or smell—or internal conditions—such as blood pressure or muscle tension.
- **Interneurons** form the local circuits connecting neurons in the brain or ganglia. Interneurons are responsible for the integration (analysis and interpretation) of sensory input.
- **Motor neurons** transmit signals to muscle cells, causing them to contract. Additional neurons that extend out of the processing centers trigger gland activity.

The neurons that carry out integration are often organized in a **central nervous system (CNS)**. The neurons that carry information into and out of the CNS constitute the **peripheral nervous system (PNS)**. When bundled together, PNS neurons form **nerves**.

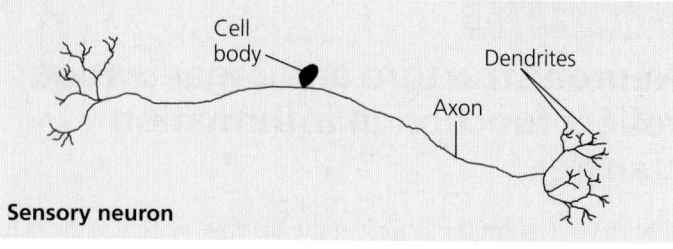

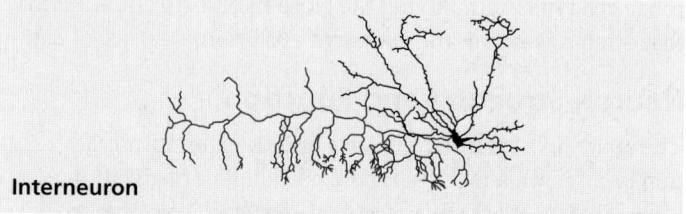

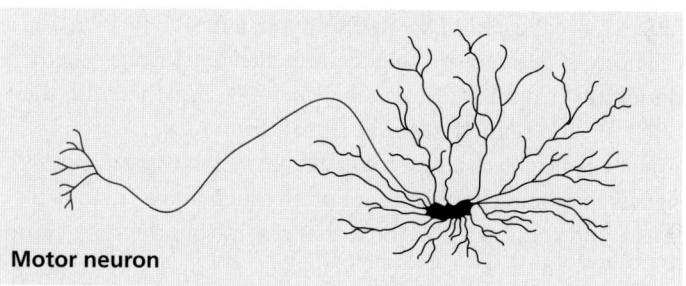

▲ **Figure 37.5 Structural diversity of neurons.** In these drawings of neurons, cell bodies and dendrites are black and axons are red.

Depending on its role in information processing, the shape of a neuron can vary from simple to quite complex **(Figure 37.5)**. Neurons that have highly branched dendrites, such as some interneurons, can receive input through tens of thousands of synapses. Similarly, neurons that transmit information to many target cells do so through highly branched axons.

Whatever their type and structure, neurons do not function alone. Instead, they form circuits, transferring information within a neuron and from a neuron to another cell. In exploring how neurons transmit information, we will begin with the remarkable properties that arise when proteins facilitate and regulate the movement of ions across a plasma membrane.

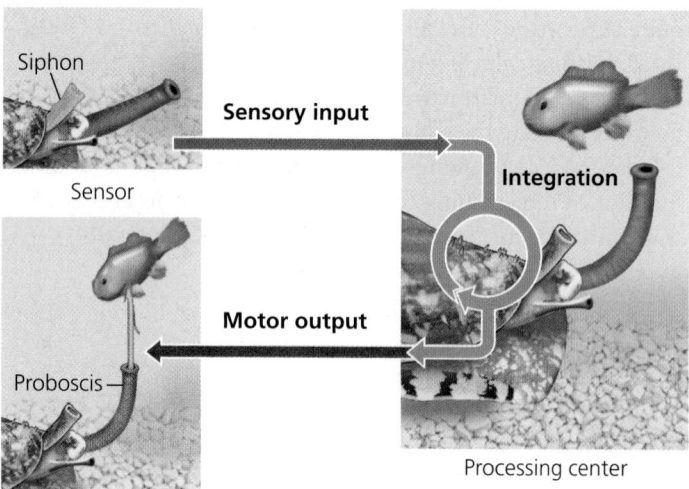

▲ **Figure 37.4 Summary of information processing.** The cone snail's siphon acts as a sensor, transferring information to neuronal circuits in the snail's head. If, upon processing the input, prey is detected, these circuits issue motor commands, triggering release of a harpoon-like tooth from the proboscis.

### CONCEPT CHECK 37.1

1. Compare and contrast the structure and function of axons and dendrites.
2. Describe the basic pathway of information flow through neurons that causes you to turn your head when someone calls your name.
3. **WHAT IF?** How might increased branching of an axon help coordinate responses to signals communicated by the nervous system?

For suggested answers, see Appendix A.

# Ion pumps and ion channels establish the resting potential of a neuron

We turn now to the essential role of ions in neuronal signaling. In neurons, as in other cells, ions are unequally distributed between the interior of cells and the surrounding fluid (see Concept 5.4). As a result, the inside of a cell is negatively charged relative to the outside. This charge difference, or *voltage*, across the plasma membrane is called the **membrane potential**, reflecting the fact that the attraction of opposite charges is a source of potential energy. For a resting neuron—one that is not sending a signal—the membrane potential is called the **resting potential** and is typically between $-60$ and $-80$ mV (millivolts).

When a neuron receives a stimulus, the membrane potential changes. Rapid shifts in membrane potential are what enable us to see the pattern of a spiderweb, remember a song, or ride a bicycle. To understand how changes in membrane potential convey information, we first need to explore the ways in which membrane potentials are formed, maintained, and altered.

## Formation of the Resting Potential

Potassium ions ($K^+$) and sodium ions ($Na^+$) play an essential role in the formation of the resting potential. These ions each have a concentration gradient across the plasma membrane of a neuron **(Table 37.1)**. For most neurons, the concentration of $K^+$ is higher inside the cell, while the concentration of $Na^+$ is higher outside. These $Na^+$ and $K^+$ gradients are maintained by the **sodium-potassium pump** (see Figure 5.14). This pump uses the energy of ATP hydrolysis to actively transport $Na^+$ out of the cell and $K^+$ into the cell **(Figure 37.6)**. (There are also concentration gradients for chloride ions [$Cl^-$] and other anions, as shown in Table 37.1, but we can ignore these for now.)

The sodium-potassium pump transports three $Na^+$ out of the cell for every two $K^+$ that it transports in. Although this pumping generates a net export of positive charge, the resulting change in the membrane potential is only a few millivolts. Why, then, is there a membrane potential of $-60$ to $-80$ mV in a resting neuron? The answer lies in ion movement through **ion channels**, pores formed by clusters of specialized proteins

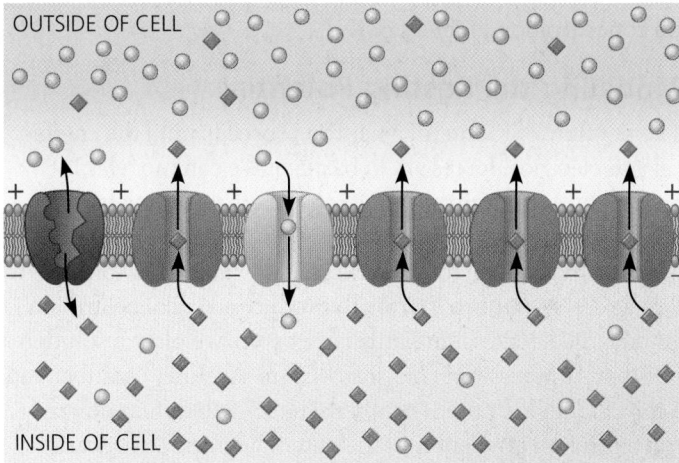

**Key**

○ $Na^+$    ◆ $K^+$    Sodium-potassium pump    Potassium channel    Sodium channel

▲ **Figure 37.6 The basis of the membrane potential.** The sodium-potassium pump generates and maintains the ionic gradients of $Na^+$ and $K^+$ shown in Table 37.1. The $Na^+$ gradient results in very little net diffusion of $Na^+$ in a resting neuron because very few sodium channels are open. In contrast, the many open potassium channels allow a significant net outflow of $K^+$. Because the membrane is only weakly permeable to chloride and other anions, this outflow of $K^+$ results in a net negative charge inside the cell.

that span the membrane. Ion channels allow ions to diffuse back and forth across the membrane. As ions diffuse through channels, they carry with them units of electrical charge. Any resulting *net* movement of positive or negative charge will generate a membrane potential across the membrane.

Concentration gradients of ions across a plasma membrane represent a chemical form of potential energy that can be used for cellular processes. In neurons, the ion channels that convert this chemical potential energy into electrical potential energy can do so because they have *selective permeability*—they allow only certain ions to pass. For example, a potassium channel allows $K^+$ to diffuse freely across the membrane, but not other ions, such as $Na^+$ or $Cl^-$.

Diffusion of $K^+$ through potassium channels that are always open (sometimes called *leak channels*) is critical for establishing the resting potential. The $K^+$ concentration is 140 m$M$ inside the cell, but only 5 m$M$ outside. The chemical concentration gradient thus favors a net outflow of $K^+$. Furthermore, a resting neuron has many open potassium channels, but very few open sodium channels. Because $Na^+$ and other ions can't

**Table 37.1 Ion Concentrations Inside and Outside of Mammalian Neurons**

| Ion | Intracellular Concentration (m$M$) | Extracellular Concentration (m$M$) |
|---|---|---|
| Potassium ($K^+$) | 140 | 5 |
| Sodium ($Na^+$) | 15 | 150 |
| Chloride ($Cl^-$) | 10 | 120 |
| Large anions ($A^-$) inside cell, such as proteins | 100 | Not applicable |

readily cross the membrane, K⁺ outflow leads to a net negative charge inside the cell. This buildup of negative charge within the neuron is the major source of the membrane potential.

What stops the buildup of negative charge? The excess negative charges inside the cell exert an attractive force that opposes the flow of additional positively charged potassium ions out of the cell. The separation of charge (voltage) thus results in an electrical gradient that counterbalances the chemical concentration gradient of K⁺.

## Modeling the Resting Potential

The net flow of K⁺ out of a neuron proceeds until the chemical and electrical forces are in balance. We can model this process by considering a pair of chambers separated by an artificial membrane. To begin, imagine that the membrane contains many open ion channels, all of which allow only K⁺ to diffuse across **(Figure 37.7a)**. To produce a K⁺ concentration gradient like that of a mammalian neuron, we place a solution of 140 m*M* potassium chloride (KCl) in the inner chamber and 5 m*M* KCl in the outer chamber. The K⁺ will diffuse down its concentration gradient into the outer chamber. But because the chloride ions (Cl⁻) lack a means of crossing the membrane, there will be an excess of negative charge in the inner chamber.

When our model neuron reaches equilibrium, the electrical gradient will exactly balance the chemical gradient, so that no further net diffusion of K⁺ occurs across the membrane. The magnitude of the membrane voltage at equilibrium for a particular ion is called that ion's **equilibrium potential** ($E_{ion}$). For a membrane permeable to a single type of ion, $E_{ion}$ can be calculated using a formula called the *Nernst equation*. At human body temperature (37°C) and for an ion with a net charge of 1+, such as K⁺ or Na⁺, the Nernst equation is

$$E_{ion} = 62 \text{ mV}\left(\log \frac{[\text{ion}]_{outside}}{[\text{ion}]_{inside}}\right)$$

Plugging the K⁺ concentrations into the Nernst equation reveals that the equilibrium potential for K⁺ ($E_K$) is −90 mV (see Figure 37.7a). The minus sign indicates that K⁺ is at equilibrium when the inside of the membrane is 90 mV more negative than the outside.

Whereas the equilibrium potential for K⁺ is −90 mV, the resting potential of a mammalian neuron is somewhat less negative. This difference reflects the small but steady movement of Na⁺ across the few open sodium channels in a resting neuron. The concentration gradient of Na⁺ has a direction opposite to that of K⁺ (see Table 37.1). Na⁺ therefore diffuses into the cell, making the inside of the cell less negative. If we model a membrane in which the only open channels are selectively permeable to Na⁺, we find that a tenfold higher concentration of Na⁺ in the outer chamber results in an equilibrium potential ($E_{Na}$) of +62 mV **(Figure 37.7b)**. In an actual neuron, the resting potential (−60 to −80 mV) is much closer to $E_K$ than to $E_{Na}$ because there are many open potassium channels but only a small number of open sodium channels.

Because neither K⁺ nor Na⁺ is at equilibrium in a resting neuron, there is a net flow of each ion across the membrane. The resting potential remains steady, which means that these K⁺ and Na⁺ currents are equal and opposite. Ion concentrations on either side of the membrane also remain steady. Why? The resting potential arises from the net movement of far fewer ions than would be required to alter the concentration gradients.

If Na⁺ is allowed to cross the membrane more readily, the membrane potential will move toward $E_{Na}$ and away from $E_K$. As you'll see shortly, this is precisely what happens during the generation of a nerve impulse.

A neuron is just one example of a cell that relies on ion movement across membranes for its specialized function. **Figure 37.8** illustrates how ion movement contributes to a range of physiological processes in animals, plants, and microorganisms.

▶ **Figure 37.7 Modeling a mammalian neuron.** In this model of the membrane potential of a resting neuron, an artificial membrane divides each container into two chambers. Ion channels allow free diffusion for particular ions, resulting in the net ion flow represented by arrows. **(a)** The presence of open potassium channels makes the membrane selectively permeable to K⁺, and the inner chamber contains a 28-fold higher concentration of K⁺ than the outer chamber; at equilibrium, the inside of the membrane is −90 mV relative to the outside. **(b)** The membrane is selectively permeable to Na⁺, and the inner chamber contains a tenfold lower concentration of Na⁺ than the outer chamber; at equilibrium, the inside of the membrane is +62 mV relative to the outside.

**WHAT IF?** *How would adding potassium or chloride channels to the membrane in (b) affect the membrane potential?*

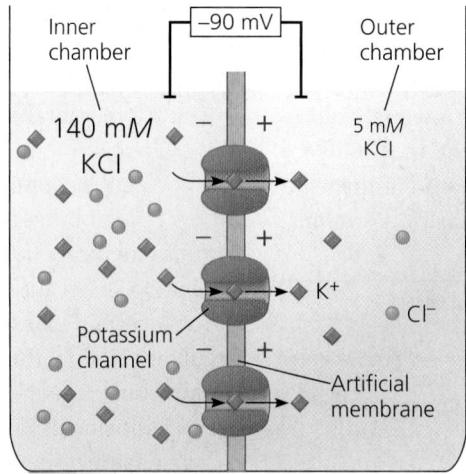

**(a) Membrane selectively permeable to K⁺**

Nernst equation for K⁺ equilibrium potential at 37°C:

$$E_K = 62 \text{ mV}\left(\log \frac{5 \text{ m}M}{140 \text{ m}M}\right) = -90 \text{ mV}$$

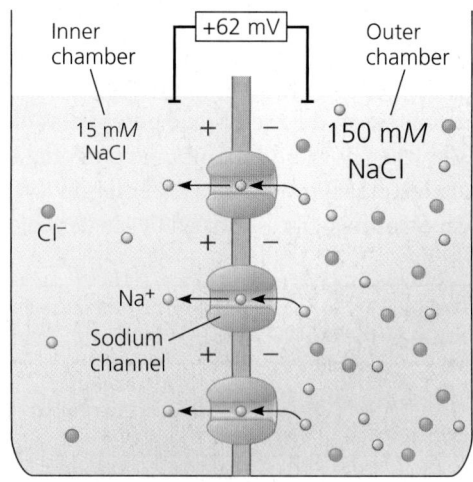

**(b) Membrane selectively permeable to Na⁺**

Nernst equation for Na⁺ equilibrium potential at 37°C:

$$E_{Na} = 62 \text{ mV}\left(\log \frac{150 \text{ m}M}{15 \text{ m}M}\right) = +62 \text{ mV}$$

# Ion Movement and Gradients

The transport of ions across the plasma membrane of a cell is a fundamental activity of all animals, and indeed of all living things. By generating ion gradients, ion transport provides the potential energy that powers processes ranging from an organism's regulation of salts and gases in internal fluids to its perception of and locomotion through its environment.

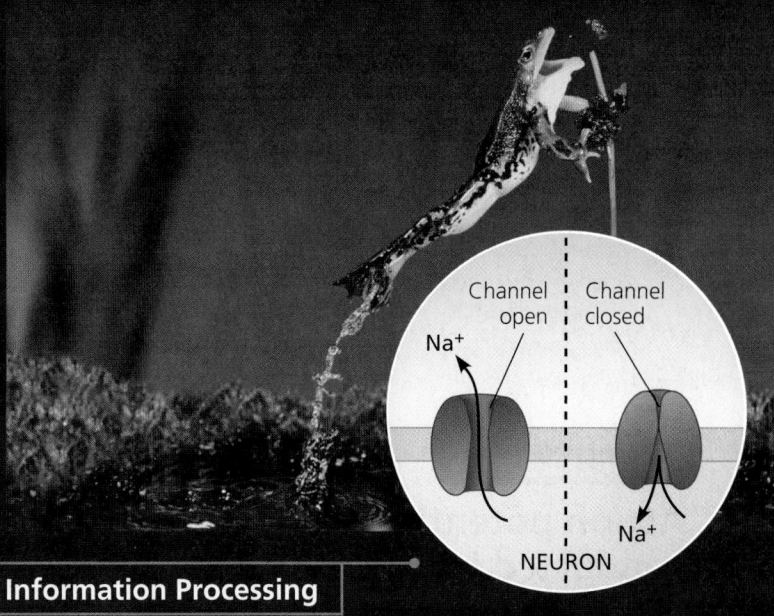

## Information Processing

In neurons, transmission of information as nerve impulses is made possible by the opening and closing of channels selective for sodium or other ions. These signals enable nervous systems to receive and process input and to direct appropriate output, such as this leap of a frog capturing prey. (See Concepts 37.3 and 39.1.)

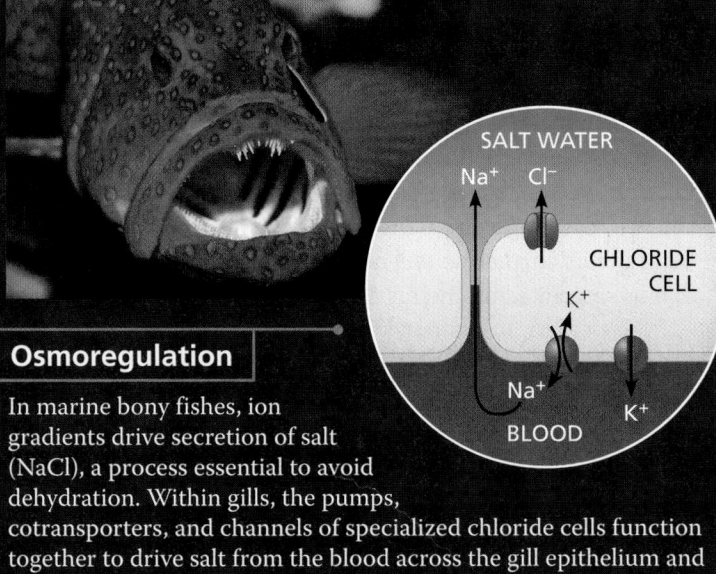

## Osmoregulation

In marine bony fishes, ion gradients drive secretion of salt (NaCl), a process essential to avoid dehydration. Within gills, the pumps, cotransporters, and channels of specialized chloride cells function together to drive salt from the blood across the gill epithelium and into the surrounding salt water. (See Figure 32.17.)

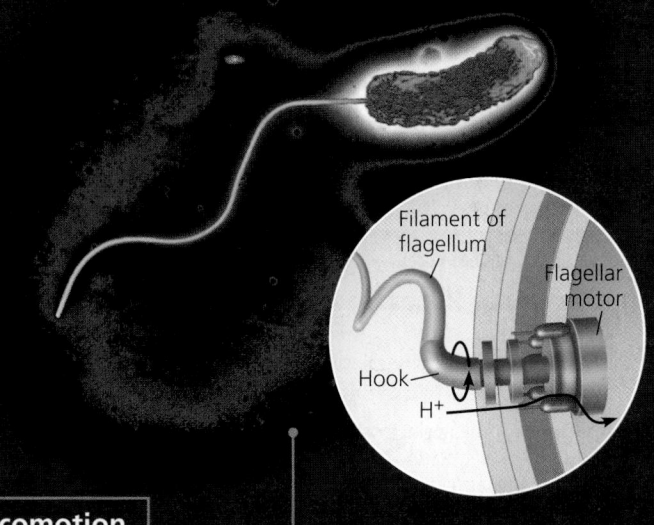

## Locomotion

A gradient of $H^+$ ions powers the bacterial flagellum. An electron transport chain generates this gradient, establishing a higher concentration of $H^+$ outside the bacterial cell. Protons reentering the cell provide a force that causes the flagellar motor to rotate. The rotating motor turns the curved hook, causing the attached filament to propel the cell. (See Concept 7.4 and Figure 24.10.)

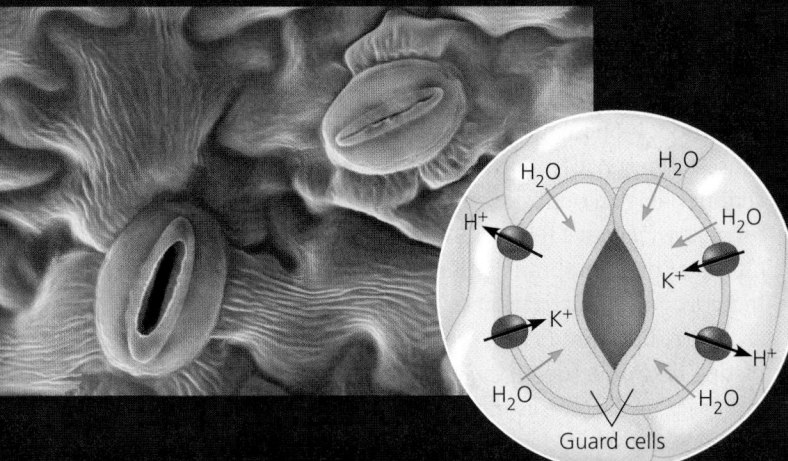

## Gas Exchange

Ion gradients provide the basis for the opening of plant stomata by surrounding guard cells. Active transport of $H^+$ out of a guard cell generates a voltage (membrane potential) that drives inward movement of $K^+$ ions. This uptake of $K^+$ by guard cells triggers an osmotic influx of water that changes cell shape, bowing the guard cells outward and thereby opening the stoma. (See Concept 29.6.)

**MAKE CONNECTIONS** *Explain why the set of forces driving ion movement across the plasma membrane of a cell are described as an electrochemical (electrical and chemical) gradient. (See Concept 5.4.)*

**ANIMATION** Visit the Study Area in **MasteringBiology** for the BioFlix 3-D Animation on Membrane Transport (Chapter 5).

1. Under what circumstances could ions flow through an ion channel from a region of lower ion concentration to a region of higher ion concentration?

2. **WHAT IF?** Suppose a cell's membrane potential shifts from −70 mV to −50 mV. What changes in the cell's permeability to K⁺ or Na⁺ could cause such a shift?

3. **MAKE CONNECTIONS** Review Figure 5.9, which illustrates diffusion of dye molecules across a membrane. Could diffusion eliminate the concentration gradient of a dye that has a net charge? Explain.

*For suggested answers, see Appendix A.*

## CONCEPT 37.3

# Action potentials are the signals conducted by axons

When a neuron responds to a stimulus, such as the scent of fish detected by a hunting cone snail, the membrane potential changes. Using intracellular recording **(Figure 37.9)**, researchers can graph these changes as a function of time. As you will see, such graphs are extremely helpful in analyzing information transfer by neurons.

How does a stimulus bring about a change in the membrane potential? It turns out that some of the ion channels in a neuron are **gated ion channels**, ion channels that open or close in response to stimuli. When a gated ion channel opens or closes, it alters the membrane's permeability to particular ions **(Figure 37.10)**. This in turn alters the membrane potential.

### ▼ Figure 37.9 Research Method

#### Intracellular Recording

**Application** Electrophysiologists use intracellular recording to measure the membrane potential of neurons and other cells.

**Technique** A microelectrode is made from a glass capillary tube filled with an electrically conductive salt solution. One end of the tube tapers to an extremely fine tip (diameter < 1 µm). While looking through a microscope, the experimenter uses a micropositioner to insert the tip of the microelectrode into a cell. A voltage recorder (usually an oscilloscope or a computer-based system) measures the voltage between the microelectrode tip inside the cell and a reference electrode placed in the solution outside the cell.

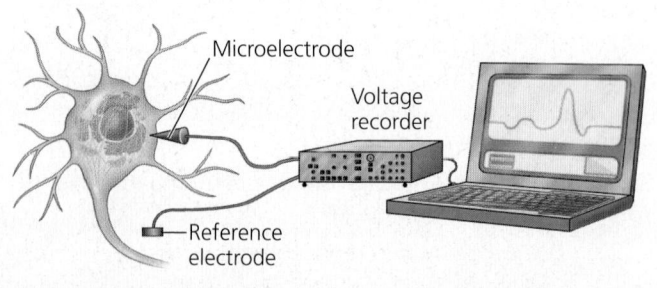

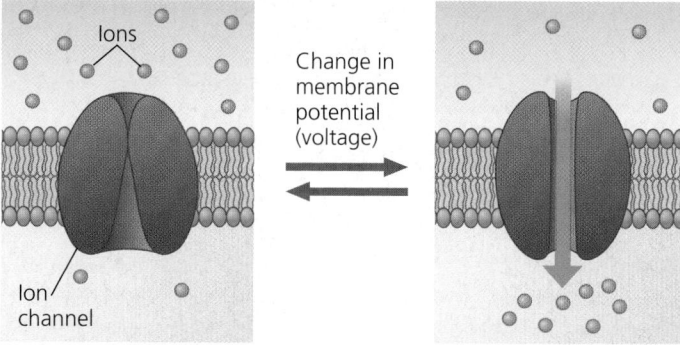

**Gate closed:** No ions flow across membrane.

**Gate open:** Ions flow through channel.

▲ **Figure 37.10 Voltage-gated ion channel.** A change in the membrane potential in one direction (right-pointing arrow) opens the channel. The opposite change (left-pointing arrow) closes the channel.

Particular types of gated channels respond to different stimuli. For example, Figure 37.10 illustrates a **voltage-gated ion channel**, a channel that opens or closes in response to a shift in the voltage across the plasma membrane of the neuron. Later in this chapter we will discuss gated channels in neurons that are regulated by chemical signals.

## Hyperpolarization and Depolarization

Let's consider now what happens when a stimulus causes closed voltage-gated ion channels to open. If gated potassium channels in a resting neuron open, the membrane's permeability to K⁺ increases. As a result, net diffusion of K⁺ out of the neuron increases, shifting the membrane potential toward $E_K$ (−90 mV at 37°C). This shift, called a **hyperpolarization**, makes the inside of the membrane more negative and increases the magnitude of the membrane potential **(Figure 37.11a)**. In a resting neuron, a hyperpolarization results from any stimulus that increases the outflow of positive ions or the inflow of negative ions.

Although opening potassium channels in a resting neuron causes hyperpolarization, opening some other types of ion channels has an opposite effect, making the inside of the membrane less negative **(Figure 37.11b)**. A reduction in the magnitude of the membrane potential is called a **depolarization**. In neurons, depolarization often involves gated sodium channels. If a stimulus causes the gated sodium channels in a resting neuron to open, the membrane's permeability to Na⁺ increases. Na⁺ diffuses into the cell along its concentration gradient, causing a depolarization as the membrane potential shifts toward $E_{Na}$ (+62 mV at 37°C).

## Graded Potentials and Action Potentials

Sometimes, as shown in the example in Figure 37.11a and b, the response to hyperpolarization or depolarization is simply a shift in the membrane potential. This shift, called a **graded potential**, has a magnitude that varies with the strength of the stimulus: A larger stimulus causes a greater

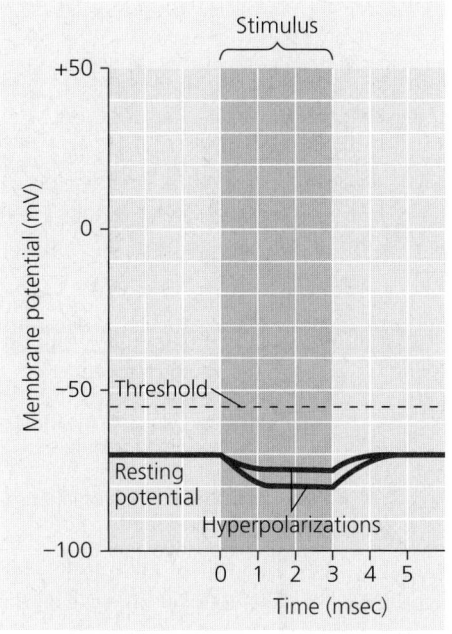

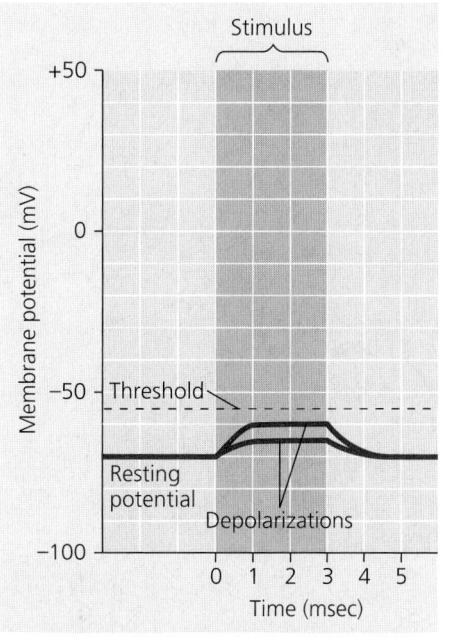

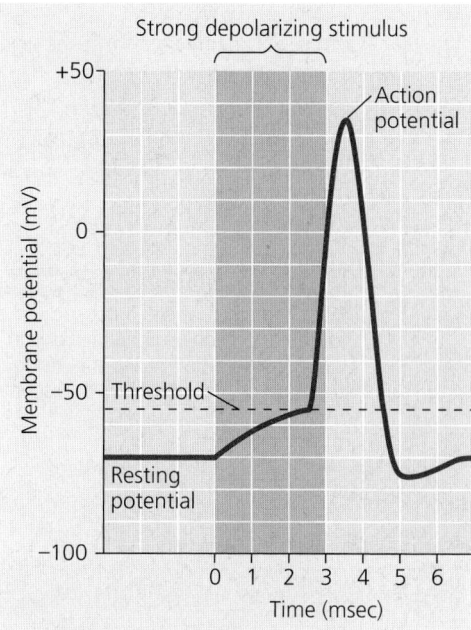

**(a) Graded hyperpolarizations produced by two stimuli that increase membrane permeability to K⁺.** The larger stimulus produces a larger hyperpolarization.

**(b) Graded depolarizations produced by two stimuli that increase membrane permeability to Na⁺.** The larger stimulus produces a larger depolarization.

**(c) Action potential triggered by a depolarization that reaches the threshold.**

▲ **Figure 37.11 Graded potentials and an action potential in a neuron.**

`DRAW IT` *Redraw the graph in part (c), extending the y-axis. Then label the positions of $E_K$ and $E_{Na}$.*

change in the membrane potential. Graded potentials induce a small electrical current that dissipates as it flows along the membrane. Graded potentials thus decay with time and with distance from their source.

If a depolarization shifts the membrane potential sufficiently, the result is a massive change in membrane voltage called an **action potential (Figure 37.11c)**. Unlike graded potentials, action potentials have a constant magnitude and can regenerate in adjacent regions of the membrane. Action potentials can therefore spread along axons, making them well suited for transmitting a signal over long distances.

The fact that some of the ion channels in neurons are voltage-gated is central to the formation of action potentials. As shown in Figure 37.10, such channels open or close when the membrane potential passes a particular level. If a depolarization opens voltage-gated sodium channels, the resulting flow of Na⁺ into the neuron results in further depolarization. Because the sodium channels are voltage gated, the increased depolarization causes more sodium channels to open, leading to an even greater flow of current. The result is a process of positive feedback that triggers a very rapid opening of many voltage-gated sodium channels and the marked temporary change in membrane potential that defines an action potential.

Action potentials occur whenever a depolarization increases the membrane voltage beyond a particular value, called the **threshold** (see Figure 37.11c). Once initiated, the action

potential has a magnitude that is independent of the strength of the triggering stimulus. Because action potentials either occur fully or do not occur at all, they represent an *all-or-none* response to stimuli. This all-or-none property reflects the fact that depolarization opens voltage-gated sodium channels, causing further depolarization. The positive-feedback loop of depolarization and channel opening triggers an action potential whenever the membrane potential reaches threshold, about −55 mV for many mammalian neurons.

## Generation of Action Potentials: *A Closer Look*

The characteristic shape of the graph of an action potential (see Figure 37.11c) reflects changes in membrane potential resulting from ion movement through voltage-gated sodium and potassium channels. Depolarization opens both types of channels, but they respond independently and sequentially. Sodium channels open first, initiating the action potential. As the action potential proceeds, sodium channels become *inactivated*: A loop of the channel protein moves, blocking ion flow through the opening. Sodium channels remain inactivated until after the membrane returns to the resting potential and the channels close. Potassium channels open more slowly than sodium channels, but remain open and functional until the end of the action potential.

To understand further how voltage-gated channels shape the action potential, consider the process as a series of stages,

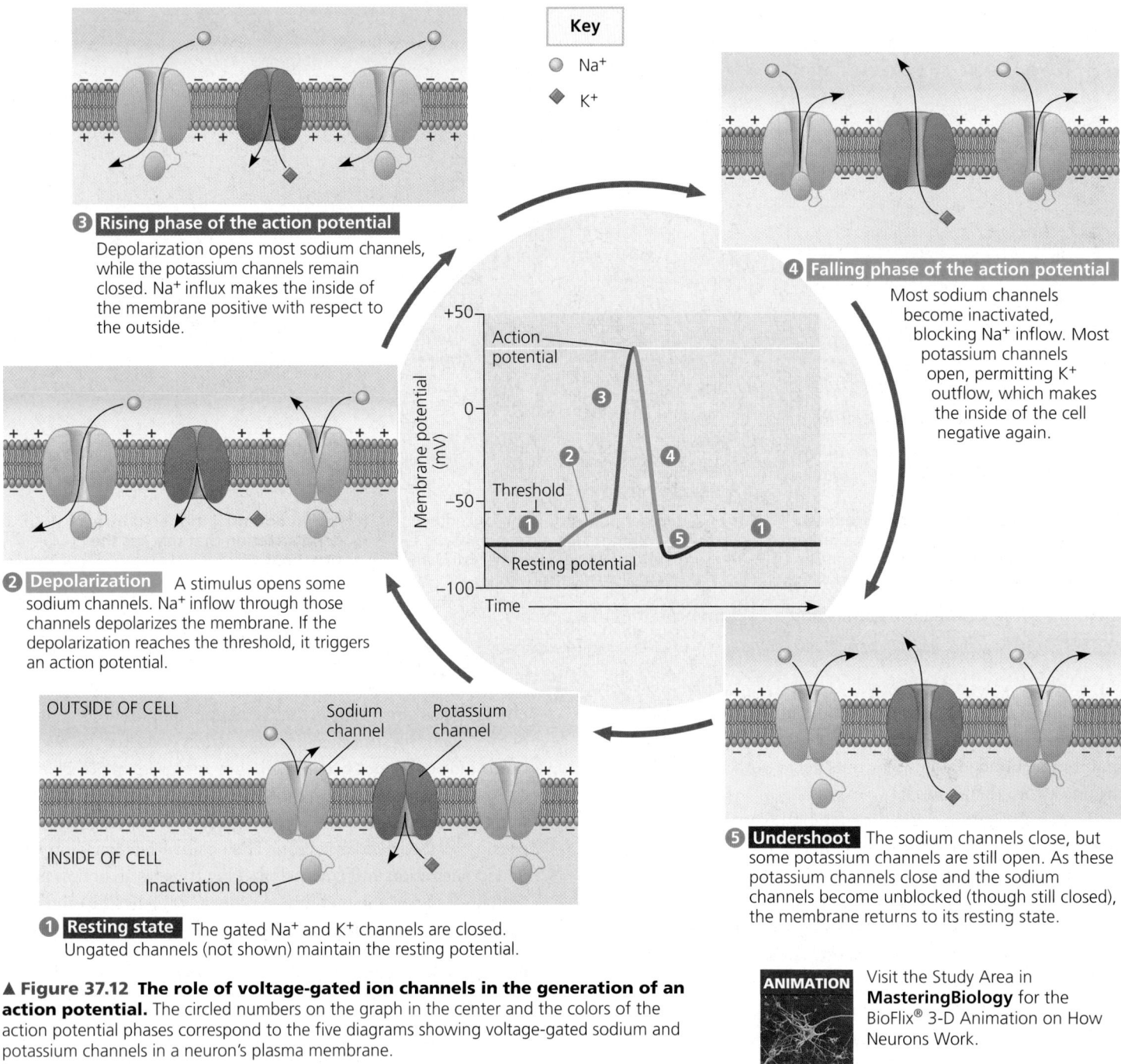

**③ Rising phase of the action potential**
Depolarization opens most sodium channels, while the potassium channels remain closed. Na⁺ influx makes the inside of the membrane positive with respect to the outside.

**④ Falling phase of the action potential**
Most sodium channels become inactivated, blocking Na⁺ inflow. Most potassium channels open, permitting K⁺ outflow, which makes the inside of the cell negative again.

**Key**
○ Na⁺
◆ K⁺

**② Depolarization** A stimulus opens some sodium channels. Na⁺ inflow through those channels depolarizes the membrane. If the depolarization reaches the threshold, it triggers an action potential.

OUTSIDE OF CELL

Sodium channel

Potassium channel

INSIDE OF CELL

Inactivation loop

**① Resting state** The gated Na⁺ and K⁺ channels are closed. Ungated channels (not shown) maintain the resting potential.

**⑤ Undershoot** The sodium channels close, but some potassium channels are still open. As these potassium channels close and the sodium channels become unblocked (though still closed), the membrane returns to its resting state.

▲ **Figure 37.12 The role of voltage-gated ion channels in the generation of an action potential.** The circled numbers on the graph in the center and the colors of the action potential phases correspond to the five diagrams showing voltage-gated sodium and potassium channels in a neuron's plasma membrane.

**ANIMATION** Visit the Study Area in **MasteringBiology** for the BioFlix® 3-D Animation on How Neurons Work.

as depicted in **Figure 37.12**. ❶ When the membrane of the axon is at the resting potential, most voltage-gated sodium channels are closed. Some potassium channels are open, but most voltage-gated potassium channels are closed. ❷ When a stimulus depolarizes the membrane, some gated sodium channels open, allowing more Na⁺ to diffuse into the cell. The Na⁺ inflow causes further depolarization, which opens still more gated sodium channels, allowing even more Na⁺ to diffuse into the cell. ❸ Once the threshold is crossed, the positive-feedback cycle rapidly brings the membrane potential close to $E_{Na}$. This stage of the action potential is called the *rising phase*. ❹ Two events prevent the membrane potential from actually

reaching $E_{Na}$: Voltage-gated sodium channels inactivate soon after opening, halting Na⁺ inflow, and most voltage-gated potassium channels open, causing a rapid outflow of K⁺. Both events quickly bring the membrane potential back toward $E_K$. This stage is called the *falling phase*. ❺ In the final phase of an action potential, called the *undershoot*, the membrane's permeability to K⁺ is higher than at rest, so the membrane potential is closer to $E_K$ than it is at the resting potential. The gated potassium channels eventually close, and the membrane potential returns to the resting potential.

The sodium channels remain inactivated during the falling phase and the early part of the undershoot. As a result, if a second depolarizing stimulus occurs during this period, it will be unable to trigger an action potential. The "downtime" when a second action potential cannot be initiated is called the **refractory period**. One consequence of the refractory period is to limit the maximum frequency at which action potentials can be generated. As we will discuss shortly, the refractory period also ensures that all signals in an axon travel in one direction, from the cell body to the axon terminals.

Note that the refractory period is due to the inactivation of sodium channels, not to a change in the ion gradients across the plasma membrane. The flow of charged particles during an action potential involves far too few ions to change the concentration on either side of the membrane significantly.

## Conduction of Action Potentials

Having described the events of a single action potential, we'll explore next how a series of action potentials moves a signal along an axon. At the site where an action potential is initiated (usually the axon hillock), Na$^+$ inflow during the rising phase creates an electrical current that depolarizes the neighboring region of the axon membrane **(Figure 37.13)**. The depolarization is large enough to reach threshold, causing an action potential in the neighboring region. This process is repeated many times along the length of the axon. Because an action potential is an all-or-none event, the magnitude and duration of the action potential are the same at each position along the axon. The net result is the movement of a nerve impulse from the cell body to the synaptic terminals, much like the cascade of events triggered by knocking over the first domino in a line.

An action potential that starts at the axon hillock moves along the axon only toward the synaptic terminals. Why? Immediately behind the traveling zone of depolarization the axon membrane is in the refractory period, and the sodium channels therefore remain inactivated. Consequently, the inward current that depolarizes the axon membrane *ahead* of the action potential cannot produce another action potential *behind* it. This is the reason action potentials do not travel back toward the cell body.

In many neurons, action potentials last less than 2 milliseconds (msec) and can therefore be produced as frequently as hundreds per second. Furthermore, action potential frequency conveys information: The rate at which action potentials are produced in a particular neuron is proportional to input signal strength. In hearing, for example, louder sounds trigger more frequent action potentials in neurons linking the ear to the brain. Differences in the number of action potentials in a given time are in fact the only variable in how information is encoded and transmitted along an axon.

Gated ion channels and action potentials have a central role in nervous system activity. As a consequence, mutations in genes that encode ion channel proteins can cause disorders affecting the nerves, brain, muscles, or heart, depending largely on where in the body the gene for the ion channel protein is expressed. For example, mutations affecting sodium channels in skeletal muscle can cause myotonia, a periodic spasming of those muscles. Mutations affecting sodium channels in the brain can cause epilepsy, in which groups of nerve cells fire simultaneously and excessively, producing seizures.

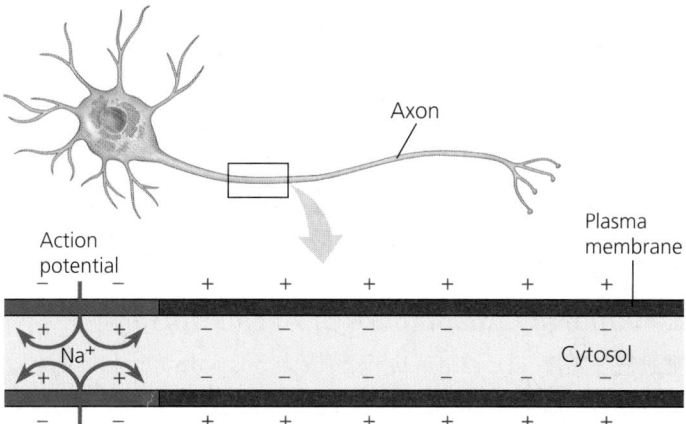

**1** An action potential is generated as Na$^+$ flows inward across the membrane at one location.

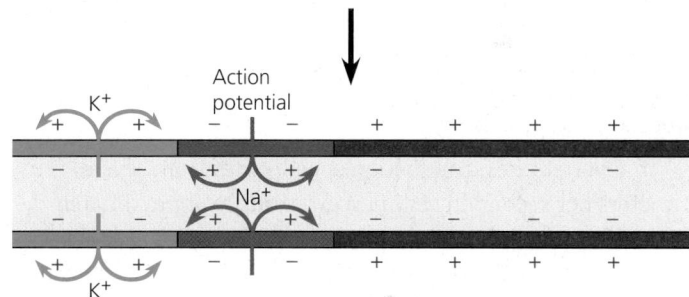

**2** The depolarization of the action potential spreads to the neighboring region of the membrane, reinitiating the action potential there. To the left of this region, the membrane is repolarizing as K$^+$ flows outward.

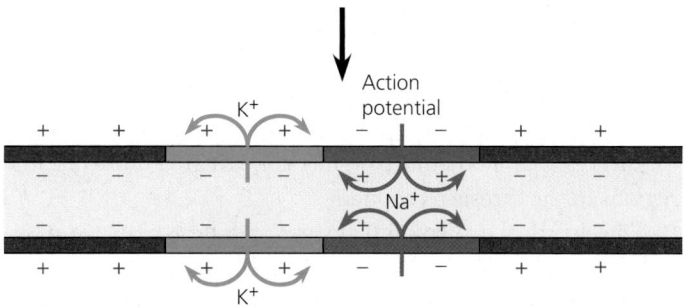

**3** The depolarization-repolarization process is repeated in the next region of the membrane. In this way, local currents of ions *across* the plasma membrane cause the action potential to be propagated *along* the length of the axon.

▲ **Figure 37.13 Conduction of an action potential.** This figure shows events at three successive times as an action potential passes from left to right. At each point along the axon, voltage-gated ion channels go through the sequence of changes shown in Figure 37.12. Membrane colors correspond to the action potential phases in Figure 37.12.

**DRAW IT** *For the axon segment shown, consider a point at the left end, a point in the middle, and a point at the right end. Draw a graph for each point showing the change in membrane potential over time at that point as a single nerve impulse moves from left to right across the segment.*

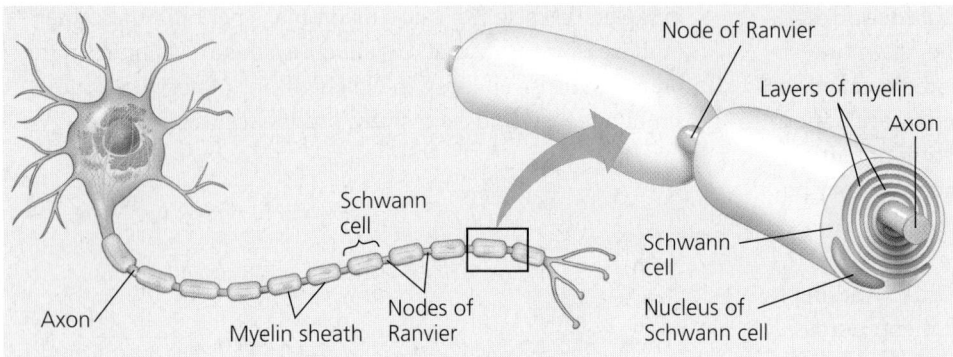
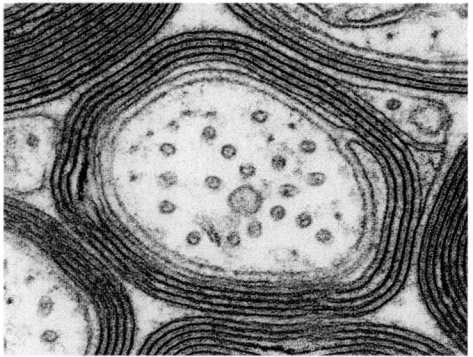

0.1 µm

▲ **Figure 37.14 Schwann cells and the myelin sheath.** In the PNS, glia called Schwann cells wrap themselves around axons, forming layers of myelin. Gaps between adjacent Schwann cells are called nodes of Ranvier. The TEM shows a cross section through a myelinated axon.

### Evolutionary Adaptations of Axon Structure

**EVOLUTION** The rate at which the axons within nerves conduct action potentials governs how rapidly an animal can react to danger or opportunity. As a consequence, natural selection often results in anatomical adaptations that increase conduction speed. One such adaptation is a wider axon. In the same way that a wide hose offers less resistance to the flow of water than does a narrow hose, a wide axon provides less resistance to the current associated with an action potential than does a narrow axon.

In invertebrates, conduction speed varies from several centimeters per second in very narrow axons to approximately 30 m/sec in the giant axons of some arthropods and molluscs. These giant axons (up to 1 mm wide) function in rapid behavioral responses, such as the muscle contraction that propels a hunting squid toward its prey.

Vertebrate axons have narrow diameters but can still conduct action potentials at high speed. How is this possible? The evolutionary adaptation that enables fast conduction in vertebrate axons is electrical insulation, analogous to the plastic insulation that encases many electrical wires. Insulation causes the depolarizing current associated with an action potential to travel farther along the axon interior, bringing more distant regions to the threshold sooner.

The electrical insulation that surrounds vertebrate axons is called a **myelin sheath (Figure 37.14)**. Myelin sheaths are produced by glia—**oligodendrocytes** in the CNS and **Schwann cells** in the PNS. During development, these specialized glia wrap axons in many layers of membrane. The membranes forming these layers are mostly lipid, which is a poor conductor of electrical current and thus a good insulator.

In myelinated axons, voltage-gated sodium channels are restricted to gaps in the myelin sheath called **nodes of Ranvier** (see Figure 37.14). An action potential propagating along a myelinated axon appears to jump from one such node to another, a process called **saltatory conduction** (from the Latin *saltare*, to leap) **(Figure 37.15)**. Why? The answer lies in the fact that the extracellular fluid contacts the axon membrane only at the nodes. As a result, action potentials are not generated in the regions between the nodes. Rather, the inward current produced during the rising phase of the action potential at a node travels within the axon all the way to the next node. There, the current depolarizes the membrane and regenerates the action potential.

Action potentials propagate more rapidly in myelinated axons because the time-consuming process of opening and closing ion channels occurs at only a limited number of positions along the axon. A myelinated axon 20 µm in diameter has a conduction speed faster than that of a squid giant axon with a diameter 40 times greater. For this reason, myelination is very space efficient: More than 2,000 of those myelinated axons can be packed into the space occupied by just one giant axon.

For any axon, myelinated or not, the conduction of an action potential to the end of the axon sets the stage for the next step in neuronal signaling—the transfer of information to another cell. This information handoff occurs at synapses, the topic of the next section.

▶ **Figure 37.15 Saltatory conduction.** In a myelinated axon, the depolarizing current during an action potential at one node of Ranvier spreads along the interior of the axon to the next node (blue arrows), where voltage-gated sodium channels enable reinitiation. Therefore, the action potential appears to jump from node to node as it travels along the axon (red arrows).

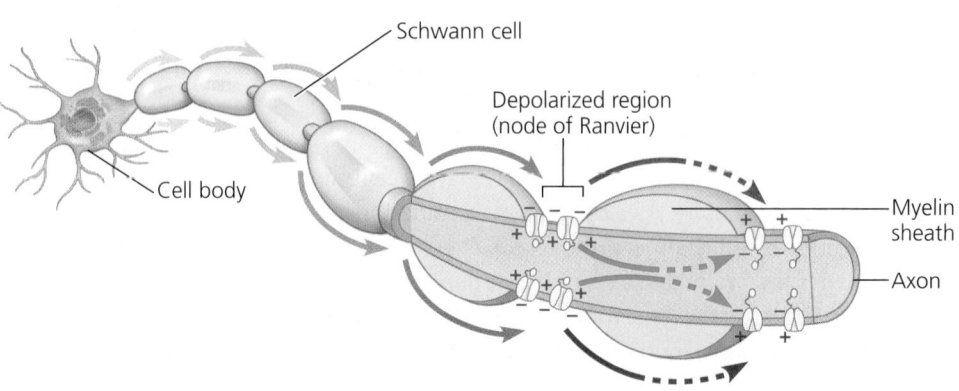

1. How do action potentials and graded potentials differ?

2. In multiple sclerosis (from the Greek *skleros*, hard), a person's myelin sheaths harden and deteriorate. How would this affect nervous system function?

3. In what ways do both positive and negative feedback contribute to the shape of an action potential?

4. **WHAT IF?** Suppose a mutation caused gated sodium channels to remain inactivated longer after an action potential. How would this affect the frequency at which action potentials could be generated? Explain.

For suggested answers, see Appendix A.

## CONCEPT 37.4

# Neurons communicate with other cells at synapses

In most cases, action potentials are not transmitted from neurons to other cells. However, information is transmitted, and this transmission occurs at synapses.

The majority of synapses are *chemical synapses*, which rely on the release of a chemical neurotransmitter by the presynaptic neuron to transfer information to the target cell. While at rest, the presynaptic neuron synthesizes the neurotransmitter at each synaptic terminal, packaging it in multiple membrane-enclosed compartments called *synaptic vesicles*. When an action potential arrives at the synapse, it depolarizes the plasma membrane at the synaptic terminal, opening voltage-gated channels that allow $Ca^{2+}$ to diffuse in (**Figure 37.16**). The $Ca^{2+}$ concentration in the terminal rises, causing synaptic vesicles to fuse with the terminal membrane and release the neurotransmitter.

Once released, the neurotransmitter diffuses across the *synaptic cleft*, the gap that separates the presynaptic neuron from the postsynaptic cell. Diffusion time is very short because the gap is less than 50 nm across. Upon reaching the postsynaptic membrane, the neurotransmitter binds to and activates a specific receptor in the membrane, which in turn triggers a response in the postsynaptic cell.

After a response is triggered, a chemical synapse returns to its resting state. How does this happen? The key step is clearing

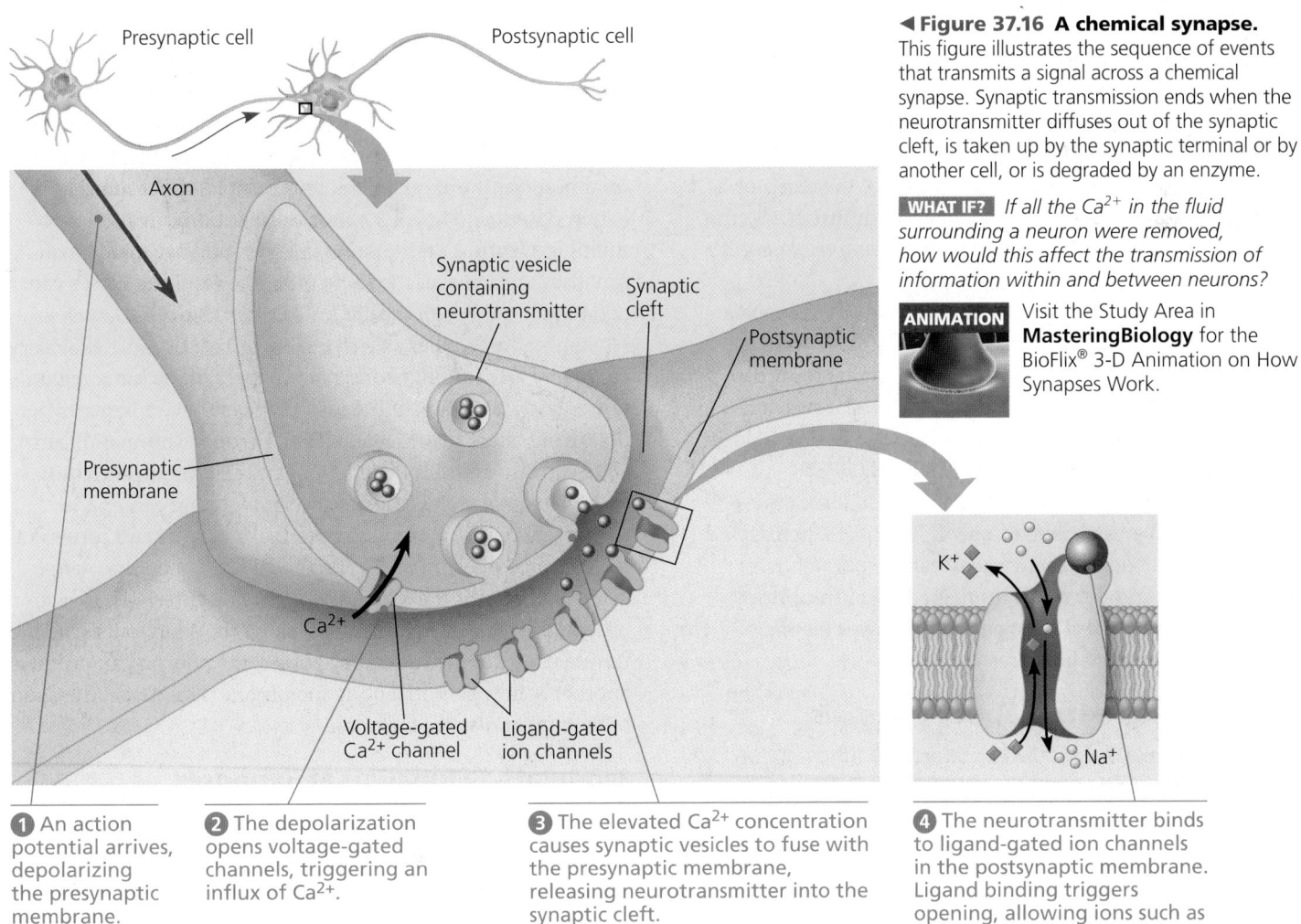

◄ **Figure 37.16 A chemical synapse.**
This figure illustrates the sequence of events that transmits a signal across a chemical synapse. Synaptic transmission ends when the neurotransmitter diffuses out of the synaptic cleft, is taken up by the synaptic terminal or by another cell, or is degraded by an enzyme.

**WHAT IF?** *If all the $Ca^{2+}$ in the fluid surrounding a neuron were removed, how would this affect the transmission of information within and between neurons?*

**ANIMATION** Visit the Study Area in **MasteringBiology** for the BioFlix® 3-D Animation on How Synapses Work.

Presynaptic cell          Postsynaptic cell

Axon

Synaptic vesicle containing neurotransmitter

Synaptic cleft

Postsynaptic membrane

Presynaptic membrane

$Ca^{2+}$

$K^+$

$Na^+$

Voltage-gated $Ca^{2+}$ channel

Ligand-gated ion channels

**1** An action potential arrives, depolarizing the presynaptic membrane.

**2** The depolarization opens voltage-gated channels, triggering an influx of $Ca^{2+}$.

**3** The elevated $Ca^{2+}$ concentration causes synaptic vesicles to fuse with the presynaptic membrane, releasing neurotransmitter into the synaptic cleft.

**4** The neurotransmitter binds to ligand-gated ion channels in the postsynaptic membrane. Ligand binding triggers opening, allowing ions such as $Na^+$ and $K^+$ to diffuse through.

neurotransmitter molecules from the synaptic cleft. Some neurotransmitters leave the synaptic cleft by diffusion or are re-captured into presynaptic neurons or glia for recycling. Others are cleaved by an enzyme into inactive fragments. Blocking this step can have severe consequences: The nerve gas sarin triggers paralysis and death by inhibiting the enzyme that breaks down the neurotransmitter controlling skeletal muscles.

Information transfer at chemical synapses can be modi-fied by altering the amount of neurotransmitter released by the presynaptic cell or the responsiveness of the postsynaptic cell. Such modifications underlie an animal's ability to alter its behavior in response to change and form the basis for learning and memory (as discussed in Chapter 38).

Some synapses are *electrical synapses*, which rely on the movement of electric current rather than a neurotransmitter. In electrical synapses, current flows from one neuron to an-other via gap junctions (see Figure 4.27). Such synapses, which are less readily modified than chemical synapses, are common in rapid and unvarying neural pathways. For example, electri-cal synapses associated with the giant axons of squids and lob-sters facilitate swift escapes from danger. Electrical synapses are also found in the vertebrate heart and brain.

## Generation of Postsynaptic Potentials

At many chemical synapses, the receptor protein that binds and responds to neurotransmitters is a **ligand-gated ion channel**, often called an *ionotropic receptor*. These receptors are clustered in the membrane of the postsynaptic cell, directly opposite the synaptic terminal. Binding of the neurotransmit-ter (the receptor's ligand) to a particular part of the receptor opens the channel and allows specific ions to diffuse across the postsynaptic membrane. The result is a *postsynaptic potential*, a graded potential in the postsynaptic cell.

At some chemical synapses, the ligand-gated ion channels are permeable to both $K^+$ and $Na^+$ (see Figure 37.16). When these channels open, the membrane potential depolarizes toward a value roughly midway between $E_K$ and $E_{Na}$. Because such a de-polarization brings the membrane potential toward threshold, it is called an **excitatory postsynaptic potential (EPSP)**.

At other chemical synapses, the ligand-gated ion chan-nels are selectively permeable for only $K^+$ or $Cl^-$. When such channels open, the postsynaptic membrane hyperpolarizes. A hyperpolarization produced in this manner is an **inhibitory postsynaptic potential (IPSP)** because it moves the mem-brane potential further from threshold.

## Summation of Postsynaptic Potentials

The interplay between multiple excitatory and inhibitory in-puts is the essence of integration in the nervous system. The cell body and dendrites of a given postsynaptic neuron may re-ceive inputs from chemical synapses formed with hundreds or even thousands of synaptic terminals **(Figure 37.17)**. How do so many synapses contribute to information transfer?

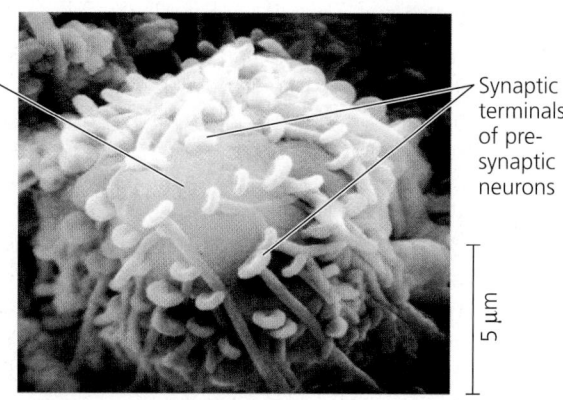

▲ **Figure 37.17  Synaptic terminals on the cell body of a postsynaptic neuron** (colorized SEM).

The input from an individual synapse is typically insuf-ficient to trigger a response in a postsynaptic neuron. To see why, consider an EPSP arising at a single synapse. As a graded potential, the EPSP becomes smaller as it spreads from the synapse. Therefore, by the time a particular EPSP reaches the axon hillock, it is usually too small to trigger an action poten-tial **(Figure 37.18a)**.

On some occasions, individual postsynaptic potentials combine to produce a larger postsynaptic potential, a process called **summation**. For instance, two EPSPs may occur at a single synapse in rapid succession. If the second EPSP arises before the postsynaptic membrane potential returns to its resting value, the EPSPs add together through *temporal sum-mation*. If the summed postsynaptic potentials depolarize the membrane at the axon hillock to threshold, the result is an action potential **(Figure 37.18b)**. Summation can also involve multiple synapses on the same postsynaptic neuron. If such synapses are active at the same time, the resulting EPSPs can add together through *spatial summation* **(Figure 37.18c)**.

Summation applies as well to IPSPs: Two or more IPSPs oc-curring nearly simultaneously at synapses in the same region or in rapid succession at the same synapse have a larger hyper-polarizing effect than a single IPSP. Through summation, an IPSP can also counter the effect of an EPSP **(Figure 37.18d)**.

The axon hillock is the neuron's integrating center, the region where the membrane potential at any instant represents the summed effect of all EPSPs and IPSPs. When the mem-brane potential at the axon hillock reaches threshold, an action potential is generated and travels along the axon to its synaptic terminals. After the refractory period, the neuron can produce another action potential if the membrane potential at the axon hillock again reaches threshold.

## Modulated Signaling at Synapses

So far, we have focused on chemical synapses where a neu-rotransmitter binds directly to an ion channel, causing the channel to open. However, there are also chemical synapses in which the receptor for the neurotransmitter is *not* part of an

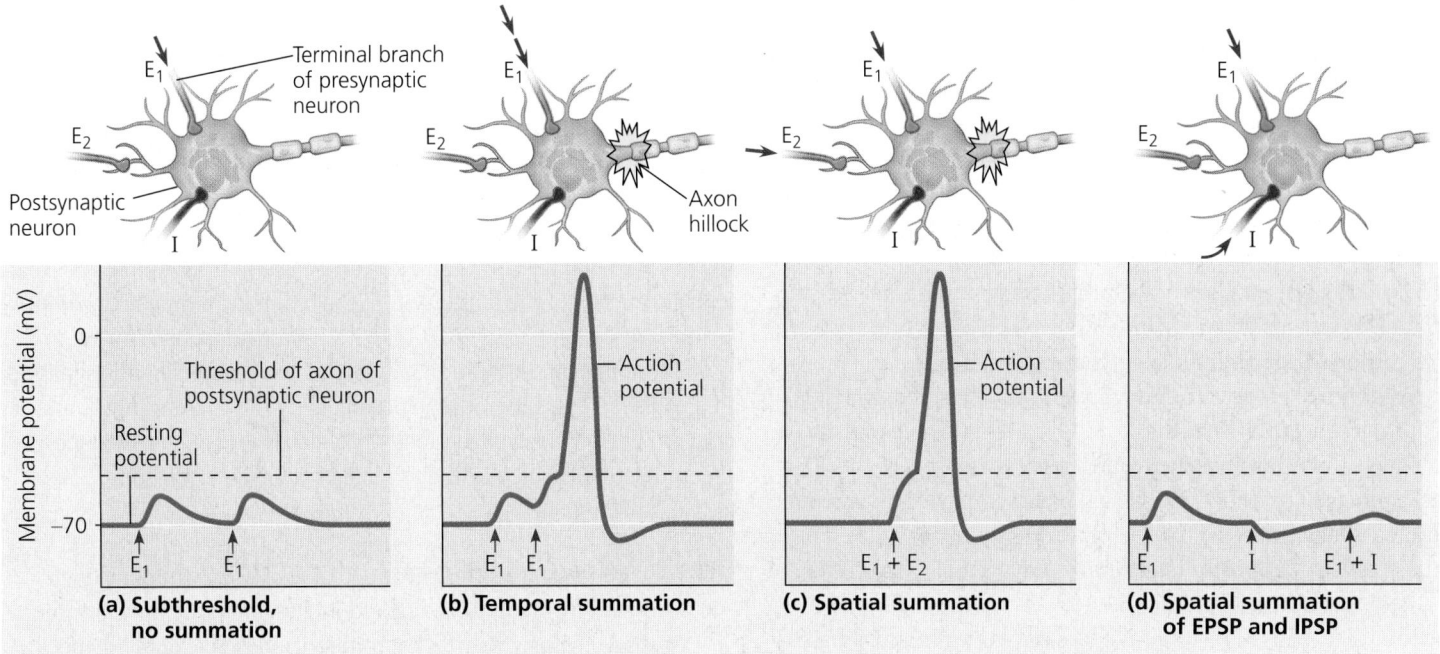

▲ **Figure 37.18 Summation of postsynaptic potentials.** These graphs trace changes in the membrane potential at a postsynaptic neuron's axon hillock. The arrows indicate times when postsynaptic potentials occur at two excitatory synapses ($E_1$ and $E_2$, green in the diagrams above the graphs) and at one inhibitory synapse (I, red). Like most EPSPs, those produced at $E_1$ or $E_2$ do not reach the threshold at the axon hillock without summation.

ion channel. At these synapses, the neurotransmitter binds to a G protein-coupled receptor, activating a signal transduction pathway involving a second messenger (see Concept 5.6). Because the resulting opening or closing of ion channels depends on one or more metabolic steps, these G protein-coupled receptors are also called *metabotropic receptors*.

G protein-coupled receptors modulate the responsiveness and activity of postsynaptic neurons in diverse ways. Consider, for example, the metabotropic receptor for the neurotransmitter norepinephrine. Binding of norepinephrine to its G protein-coupled receptor activates a G protein, which in turn activates adenylyl cyclase, the enzyme that converts ATP to cyclic AMP (see Figure 5.25). Cyclic AMP activates protein kinase A, which phosphorylates specific ion channel proteins in the postsynaptic membrane, causing them to open or close. Because this signal transduction pathway has an amplifying effect, one norepinephrine molecule can trigger the opening or closing of many channels.

Many neurotransmitters have both ionotropic and metabotropic receptors. Compared with the postsynaptic potentials produced by ligand-gated channels, the effects of G protein pathways typically have a slower onset but last longer.

## Neurotransmitters

Signaling at a chemical synapse brings about a response that depends on both the neurotransmitter released from the presynaptic membrane and the receptor produced at the postsynaptic membrane. A single neurotransmitter may bind to more than a dozen different receptors. Indeed, a particular neurotransmitter

can excite postsynaptic cells expressing one receptor and inhibit postsynaptic cells expressing a different receptor. As an example, let's examine **acetylcholine**, a common neurotransmitter in both invertebrates and vertebrates.

### Acetylcholine

Acetylcholine is vital for nervous system functions that include muscle stimulation, memory formation, and learning. In vertebrates, there are two major classes of acetylcholine receptor. One type is a ligand-gated ion channel. We know the most about its function at the vertebrate *neuromuscular junction*, the site where a motor neuron forms a synapse with a skeletal muscle cell. When acetylcholine released by motor neurons binds this receptor, the ion channel opens, producing an EPSP. This excitatory activity is soon terminated by acetylcholinesterase, an enzyme in the synaptic cleft that hydrolyzes the neurotransmitter into an inactive form.

A G protein-coupled receptor for acetylcholine is found at locations that include the vertebrate CNS and heart. In heart muscle, acetylcholine released by neurons activates a signal transduction pathway. The G proteins in the pathway inhibit adenylyl cyclase and open potassium channels in the muscle cell membrane. Both effects reduce the rate at which the heart pumps. Thus, the effect of acetylcholine in heart muscle is inhibitory rather than excitatory.

Several chemicals with profound effects on the nervous system mimic or alter the function of acetylcholine. Nicotine, a chemical found in tobacco and tobacco smoke, acts as a stimulant by binding to an ionotropic acetylcholine receptor in the CNS. As discussed earlier, the nerve gas sarin blocks enzymatic cleavage of acetylcholine. A third example is botulinum toxin, which inhibits presynaptic release of acetylcholine. The result is a form of food poisoning called botulism. Because muscles required for breathing fail to contract when acetylcholine release is blocked, untreated botulism is typically fatal. Today, injections of the botulinum toxin, known by the trade name Botox, are used cosmetically to minimize wrinkles around the eyes or mouth by inhibiting synaptic transmission to particular facial muscles.

Although acetylcholine has many roles, it is just one of more than 100 known neurotransmitters. As shown by the examples in **Table 37.2**, the rest fall into four classes: amino acids, biogenic amines, neuropeptides, and gases. In the remainder of this chapter, we will use examples from vertebrates to illustrate some functions and properties of each of these types of neurotransmitters.

## Amino Acids

*Glutamate*, one of several amino acids that can act as a neurotransmitter, is the most common neurotransmitter in the CNS. Synapses at which glutamate is the neurotransmitter have a key role in the formation of long-term memory (as we will discuss in Chapter 38).

Two amino acids act as inhibitory neurotransmitters in the CNS. *Glycine* acts at inhibitory synapses in parts of the CNS that lie outside of the brain. Within the brain, the amino acid *gamma-aminobutyric acid (GABA)* is the neurotransmitter at most inhibitory synapses. Binding of GABA to receptors in postsynaptic cells increases membrane permeability to $Cl^-$, resulting in an IPSP. The widely prescribed drug diazepam (Valium) reduces anxiety through binding to a site on a GABA receptor.

## Biogenic Amines

The neurotransmitters grouped as *biogenic amines* are synthesized from amino acids and include *norepinephrine*, which is made from tyrosine. Norepinephrine is an excitatory neurotransmitter in the autonomic nervous system, a branch of the PNS. Outside the nervous system, norepinephrine has distinct but related functions as a hormone, as does the chemically similar biogenic amine *epinephrine* (see Chapter 32).

The biogenic amines *dopamine*, made from tyrosine, and *serotonin*, made from tryptophan, are released at many sites in the brain and affect sleep, mood, attention, and learning. Some psychoactive drugs, including LSD and mescaline, apparently produce their hallucinatory effects by binding to brain receptors for these neurotransmitters.

Biogenic amines have a central role in a number of nervous system disorders and treatments. The degenerative illness Parkinson's disease is associated with a lack of dopamine in the brain. In addition, depression is often treated with drugs that increase the brain concentrations of biogenic amines. Prozac, for instance, enhances the effect of serotonin by inhibiting its reuptake after release.

## Neuropeptides

Several **neuropeptides**, relatively short chains of amino acids, serve as neurotransmitters that operate via G protein-coupled receptors. Such peptides are typically produced by cleavage of much larger protein precursors. The neuropeptide *substance P*, a key excitatory neurotransmitter, mediates our perception of pain. Neuropeptides called **endorphins** function as natural analgesics, decreasing pain perception.

Endorphins are produced in the brain during times of physical or emotional stress, such as childbirth. In addition to relieving pain, they reduce urine output, decrease respiration, and produce euphoria, as well as other emotional effects. Opiates (drugs such as morphine and heroin) mimic endorphins and produce many of the same physiological effects (see Figure 2.14). In the **Scientific Skills Exercise**, you can

| Table 37.2 Major Neurotransmitters | |
|---|---|
| **Neurotransmitter** | **Structure** |
| **Acetylcholine** | |
| **Amino Acids** | |
| Glutamate | |
| Glycine | |
| GABA (gamma-aminobutyric acid) | |
| **Biogenic Amines** | |
| Norepinephrine | 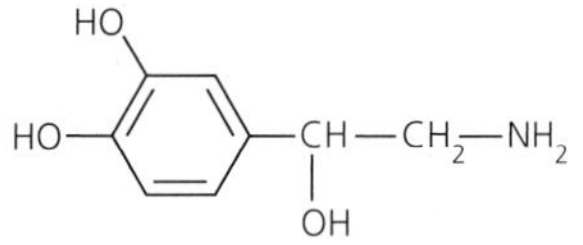 |
| Dopamine | 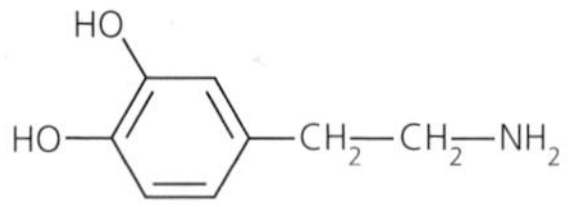 |
| Serotonin | 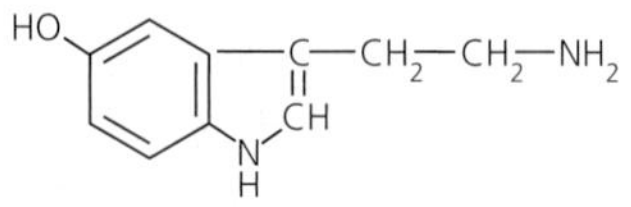 |
| **Neuropeptides** (a very diverse group, only two of which are shown) | |
| Substance P | Arg—Pro—Lys—Pro—Gln—Gln—Phe—Phe—Gly—Leu—Met |
| Met-enkephalin (an endorphin) | Iyr—Gly—Gly—Phe—Met |
| **Gases** | |
| Nitric oxide | $N=O$ |

# Interpreting Data Values Expressed in Scientific Notation

**Does the Brain Have Specific Protein Receptors for Opiates?**
A team of researchers was looking for opiate receptors in the mammalian brain. Knowing that the drug naloxone blocks the analgesic effect of opiates, they hypothesized that naloxone acts by binding tightly to brain opiate receptors without activating them. In this exercise, you will interpret the results of an experiment that the researchers conducted to test their hypothesis.

**How the Experiment Was Done** The researchers added radioactive naloxone to a protein mixture prepared from rodent brains. If the mixture contained opiate receptors or other proteins that could bind naloxone, the radioactivity would stably associate with the mixture. To determine whether the binding was due to specific opiate receptors, they tested other drugs, opiate and non-opiate, for their ability to block naloxone binding.

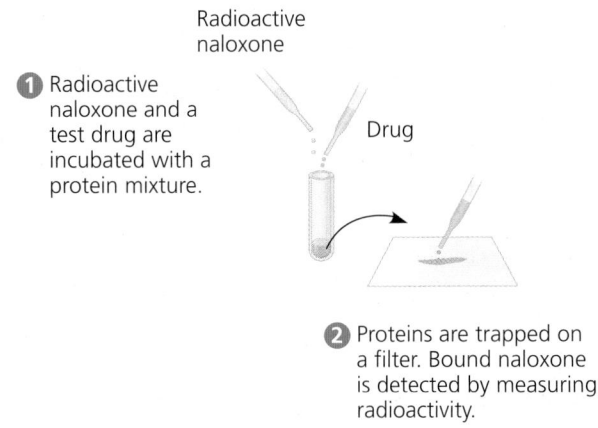

① Radioactive naloxone and a test drug are incubated with a protein mixture.

Radioactive naloxone

Drug

② Proteins are trapped on a filter. Bound naloxone is detected by measuring radioactivity.

**Data from the Experiment**

| Drug | Opiate | Lowest Concentration That Blocked Naloxone Binding |
|---|---|---|
| Morphine | Yes | $6 \times 10^{-9}\ M$ |
| Methadone | Yes | $2 \times 10^{-8}\ M$ |
| Levorphanol | Yes | $2 \times 10^{-9}\ M$ |
| Phenobarbital | No | No effect at $10^{-4}\ M$ |
| Atropine | No | No effect at $10^{-4}\ M$ |
| Serotonin | No | No effect at $10^{-4}\ M$ |

**Data from** C. B. Pert and S. H. Snyder, Opiate receptor: Demonstration in nervous tissue, *Science* 179:1011–1014 (1973).

**INTERPRET THE DATA**

1. The data from this experiment are expressed using scientific notation: a numerical factor times a power of 10. Remember that a negative power of 10 means a number less than 1. For example, $10^{-1}\ M$ (molar) can also be written as 0.1 M. Write the concentrations in the table above for morphine and atropine in this alternative format.

2. Compare the concentrations listed in the table for methadone and phenobarbital. Which concentration is higher? By how much?

3. Would phenobarbital, atropine, or serotonin have blocked naloxone binding at a concentration of $10^{-5}\ M$? Explain why or why not.

4. Which drugs blocked naloxone binding in this experiment? What do these results indicate about the brain receptors for naloxone?

5. When the researchers instead used tissue from mammalian intestinal muscles rather than brains, they found no naloxone binding. What does that suggest about opiate receptors in mammalian muscle tissue?

(MB) A version of this Scientific Skills Exercise can be assigned in MasteringBiology.

---

interpret data from an experiment designed to search for opiate receptors in the brain.

## Gases

Some vertebrate neurons release dissolved gases as neurotransmitters. In human males, for example, certain neurons release nitric oxide (NO) into the erectile tissue of the penis during sexual arousal. The resulting relaxation of smooth muscle in the blood vessel walls of the spongy erectile tissue allows the tissue to fill with blood, producing an erection. The erectile dysfunction drug Viagra works by inhibiting an enzyme that terminates the action of NO.

Unlike most neurotransmitters, NO is not stored in cytoplasmic vesicles but is instead synthesized on demand. NO diffuses into neighboring target cells, produces a change, and is broken down—all within a few seconds. In many of its targets, including smooth muscle cells, NO works like many hormones, stimulating an enzyme to synthesize a second messenger that directly affects cellular metabolism.

Although inhaling air containing the gas carbon monoxide (CO) can be deadly, the vertebrate body uses the enzyme heme oxygenase to produce small amounts of CO, some of which acts as a neurotransmitter.

In the next chapter, we'll consider how the cellular and biochemical mechanisms we have discussed contribute to nervous system function on the system level.

### CONCEPT CHECK 37.4

1. How is it possible for a particular neurotransmitter to produce opposite effects in different tissues?

2. Organophosphate pesticides work by inhibiting acetylcholinesterase, the enzyme that breaks down the neurotransmitter acetylcholine. Explain how these toxins would affect EPSPs produced by acetylcholine.

3. **MAKE CONNECTIONS** Name one or more membrane activities that occur both in fertilization of an egg and in neurotransmission across a synapse (see Figure 36.14).

For suggested answers, see Appendix A.

Go to **MasteringBiology**® for Assignments, the eText, and the Study Area with Animations, Activities, Vocab Self-Quiz, and Practice Tests.

## SUMMARY OF KEY CONCEPTS

VOCAB SELF-QUIZ

goo.gl/gbai8v

### CONCEPT 37.1

**Neuron structure and organization reflect function in information transfer (pp. 773–774)**

- Most neurons have branched **dendrites** that receive signals from other neurons and an **axon** that transmits signals to other cells at **synapses**. Neurons rely on **glia** for functions that include nourishment, insulation, and regulation.

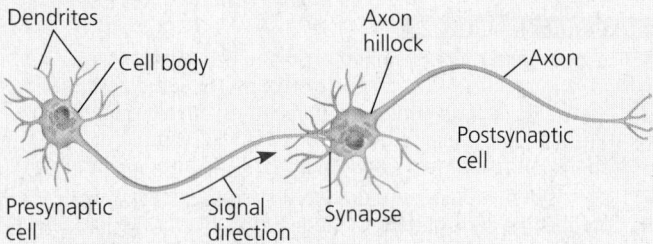

- A **central nervous system (CNS)** and a **peripheral nervous system (PNS)** process information in three stages: sensory input, integration, and motor output to effector cells.

? *How would severing an axon affect the flow of information in a neuron?*

### CONCEPT 37.2

**Ion pumps and ion channels establish the resting potential of a neuron (pp. 775–778)**

- Ionic gradients generate a voltage difference, or **membrane potential**, across the plasma membrane of cells. The concentration of $Na^+$ is higher outside than inside; the reverse is true for $K^+$. In resting neurons, the plasma membrane has many open potassium channels but few open sodium channels. Diffusion of ions, principally $K^+$, through channels generates a **resting potential**, with the inside more negative than the outside.

? *Suppose you placed an isolated neuron in a solution similar to extracellular fluid and later transferred the neuron to a solution lacking any sodium ions. What change would you expect in the resting potential?*

### CONCEPT 37.3

**Action potentials are the signals conducted by axons (pp. 778–783)**

- Neurons have gated ion channels that open or close in response to stimuli, leading to changes in the membrane potential. An increase in the magnitude of the membrane potential is a **hyperpolarization**; a decrease is a **depolarization**. Changes in membrane potential that vary continuously with the strength of a stimulus are known as **graded potentials**.

- An **action potential** is a brief, all-or-none depolarization of a neuron's plasma membrane. When a graded depolarization brings the membrane potential to **threshold**, many **voltage-gated ion channels** open, triggering an inflow of $Na^+$ that rapidly brings the membrane potential to a positive value. A negative

membrane potential is restored by the inactivation of sodium channels and by the opening of many voltage-gated potassium channels, which increases $K^+$ outflow. A **refractory period** follows, corresponding to the interval when the sodium channels remain inactivated.

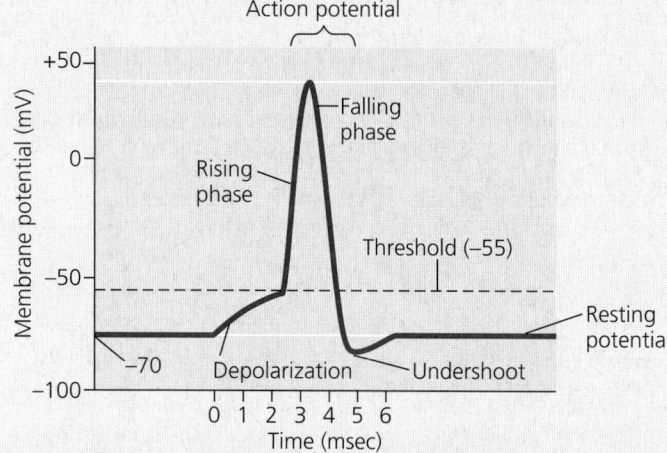

- A nerve impulse travels from the axon hillock to the synaptic terminals by propagating a series of action potentials along the axon. The speed of conduction increases with the diameter of the axon and, in many vertebrate axons, with myelination. Action potentials in axons insulated by myelination appear to jump from one **node of Ranvier** to the next, a process called **saltatory conduction**.

**INTERPRET THE DATA** *Assuming that the refractory period does not extend past the end of the action potential (see graph above), what is the maximum frequency per unit time at which a neuron could fire action potentials?*

### CONCEPT 37.4

**Neurons communicate with other cells at synapses (pp. 783–787)**

- In an electrical synapse, electrical current flows directly from one cell to another. In a chemical synapse, depolarization causes synaptic vesicles to fuse with the terminal membrane and release **neurotransmitter** into the synaptic cleft.

- At many synapses, the neurotransmitter binds to **ligand-gated ion channels** in the postsynaptic membrane, producing an **excitatory** or **inhibitory postsynaptic potential** (**EPSP** or **IPSP**). The neurotransmitter then diffuses out of the cleft, is taken up by surrounding cells, or is degraded by enzymes. A single neuron has many synapses on its dendrites and cell body. Temporal and spatial **summation** of EPSPs and IPSPs at the axon hillock determines whether a neuron generates an action potential.

- Different receptors for the same neurotransmitter produce different effects. Some neurotransmitter receptors activate signal transduction pathways, which can produce long-lasting changes in postsynaptic cells. Major neurotransmitters include **acetylcholine**; the amino acids glutamate, glycine, and GABA; biogenic amines; **neuropeptides**; and gases such as NO.

? *Why are many drugs that are used to treat nervous system diseases or to affect brain function targeted to specific receptors rather than particular neurotransmitters?*

# TEST YOUR UNDERSTANDING

PRACTICE TEST
goo.gl/CRZjvS

## Level 1: Knowledge/Comprehension

1. What happens when a resting neuron's membrane depolarizes?
   (A) There is a net diffusion of $Na^+$ out of the cell.
   (B) The equilibrium potential for $K^+$ ($E_K$) becomes more positive.
   (C) The neuron's membrane voltage becomes more positive.
   (D) The cell's inside is more negative than the outside.

2. A common feature of action potentials is that they
   (A) cause the membrane to hyperpolarize and then depolarize.
   (B) can undergo temporal and spatial summation.
   (C) are triggered by a depolarization that reaches threshold.
   (D) move at the same speed along all axons.

3. Where are neurotransmitter receptors located?
   (A) the nuclear membrane
   (B) the nodes of Ranvier
   (C) the postsynaptic membrane
   (D) synaptic vesicle membranes

## Level 2: Application/Analysis

4. Why are action potentials usually conducted in one direction?
   (A) Ions can flow along the axon in only one direction.
   (B) Voltage-gated $Na^+$ channels are inactivated during the refractory period.
   (C) The axon hillock has a higher membrane potential than the terminals of the axon.
   (D) Voltage-gated channels for both $Na^+$ and $K^+$ open in only one direction.

5. Which of the following is a *direct* result of depolarizing the presynaptic membrane of an axon terminal?
   (A) Voltage-gated calcium channels in the membrane open.
   (B) Synaptic vesicles fuse with the membrane.
   (C) Ligand-gated channels open, allowing neurotransmitters to enter the synaptic cleft.
   (D) An EPSP or IPSP is generated in the postsynaptic cell.

6. Suppose a particular neurotransmitter causes an IPSP in postsynaptic cell X and an EPSP in postsynaptic cell Y. A likely explanation is that
   (A) the threshold value in the postsynaptic membrane is different for cell X and cell Y.
   (B) the axon of cell X is myelinated, but that of cell Y is not.
   (C) only cell Y produces an enzyme that terminates the activity of the neurotransmitter.
   (D) cells X and Y express different receptor molecules for this particular neurotransmitter.

## Level 3: Synthesis/Evaluation

7. **WHAT IF?** Ouabain, a plant substance used in some cultures to poison hunting arrows, disables the sodium-potassium pump. What change in the resting potential would you expect to see if you treated a neuron with ouabain? Explain.

8. **WHAT IF?** If a drug mimicked the activity of GABA in the CNS, what general effect on behavior might you expect? Explain.

9. **DRAW IT** Suppose a researcher inserts a pair of electrodes at two different positions along the middle of an axon dissected out of a squid. By applying a depolarizing stimulus, the researcher brings the plasma membrane at both positions to threshold. Using the drawing below as a model, create one or more drawings that illustrate where each action potential would terminate.

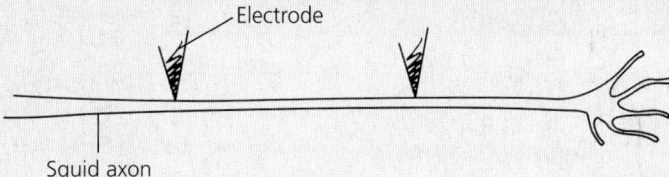

Electrode

Squid axon

10. **SCIENTIFIC INQUIRY**
    From what you know about action potentials and synapses, propose two hypotheses for how various anesthetics might block pain.

11. **FOCUS ON EVOLUTION**
    An action potential is an all-or-none event. This on/off signaling is an evolutionary adaptation of animals that must sense and act in a complex environment. It is possible to imagine a nervous system in which the action potentials are graded, with the amplitude depending on the size of the stimulus. Describe what evolutionary advantage on/off signaling might have over a graded (continuously variable) kind of signaling.

12. **FOCUS ON ORGANIZATION**
    In a short essay (100–150 words), describe how the structure and electrical properties of vertebrate neurons reflect similarities and differences with other animal cells.

13. **SYNTHESIZE YOUR KNOWLEDGE**

The rattlesnake alerts enemies to its presence with a rattle—a set of modified scales at the tip of its tail. Describe the distinct roles of gated ion channels in initiating and moving a signal along the nerve from the snake's head to its tail and then from that nerve to the muscle that shakes the rattle.

*For selected answers, see Appendix A.*

# 38 Nervous and Sensory Systems

## KEY CONCEPTS

**38.1** Nervous systems consist of circuits of neurons and supporting cells

**38.2** The vertebrate brain is regionally specialized

**38.3** The cerebral cortex controls voluntary movement and cognitive functions

**38.4** Sensory receptors transduce stimulus energy and transmit signals to the central nervous system

**38.5** In hearing and equilibrium, mechanoreceptors detect moving fluid or settling particles

**38.6** The diverse visual receptors of animals depend on light-absorbing pigments

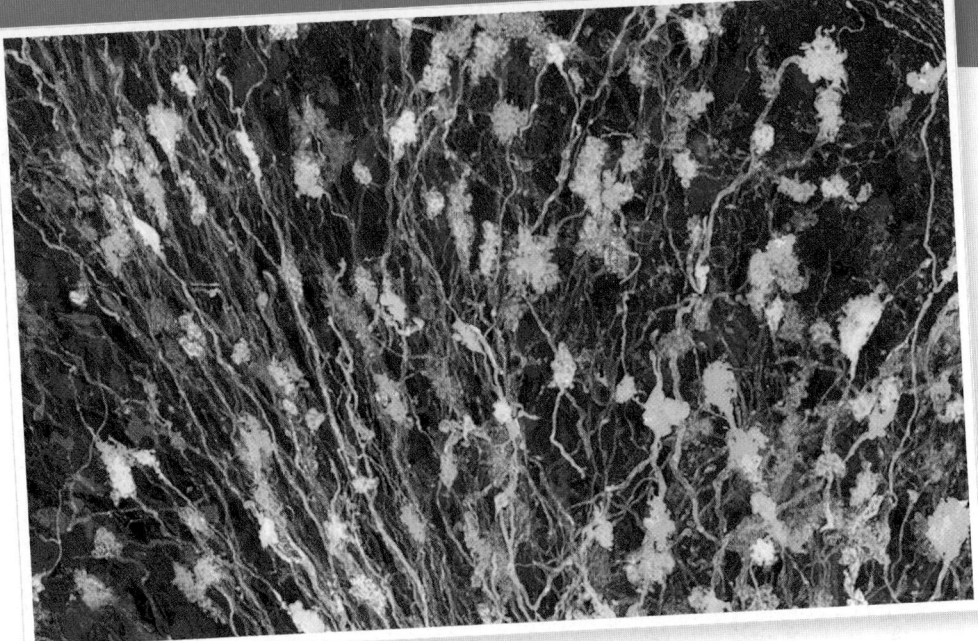

▲ **Figure 38.1** How do scientists identify individual neurons in the brain?

## Command and Control Center

What happens in your brain when you solve a math problem or listen to music? Answering such a question was for a long time nearly unimaginable. The human brain contains an estimated $10^{11}$ (100 billion) neurons organized into circuits more complex than those of even the most powerful supercomputers. However, thanks in part to several exciting new technologies, scientists have begun to sort out the cellular mechanisms for information processing in the brain, thus delineating the processes that underlie thought and emotion.

One breakthrough came with the development of powerful imaging techniques that reveal activity in the working brain. Researchers record events in multiple areas of the human brain while a subject is performing various tasks, such as speaking, looking at pictures, or forming a mental image of a person's face. The investigators then analyze these data to determine if there is a correlation between a particular task and activity detected in specific brain areas.

A more recent advance in exploring the brain relies on a method for expressing random combinations of colored proteins in brain cells—such that each cell shows up in a different color. The result is a "brainbow" like the one in **Figure 38.1**.

In this image, each neuron in the brain of a genetically engineered mouse expresses one of more than 90 different color combinations of four fluorescent proteins. Using the brainbow technology, neuroscientists hope to develop detailed maps of the neuronal connections that transfer information between particular regions of the brain.

Gathering, processing, and organizing information are essential functions of all nervous systems. In this chapter, we'll begin by examining the basic organization of nervous systems. Next, we'll consider specialization in regions of the vertebrate brain and how brain activity makes information storage and organization possible. Finally, we'll investigate the sensory processes that convey information about an animal's external and internal environments to its brain.

### CONCEPT 38.1

## Nervous systems consist of circuits of neurons and supporting cells

Hydras and other cnidarians are the simplest animals with nervous systems. In most cnidarians, interconnected neurons form a diffuse **nerve net (Figure 38.2a)**, which controls the contraction and expansion of the gastrovascular cavity. In more complex animals, the axons of multiple neurons are often bundled together to form **nerves**, fibrous structures that channel and organize information flow.

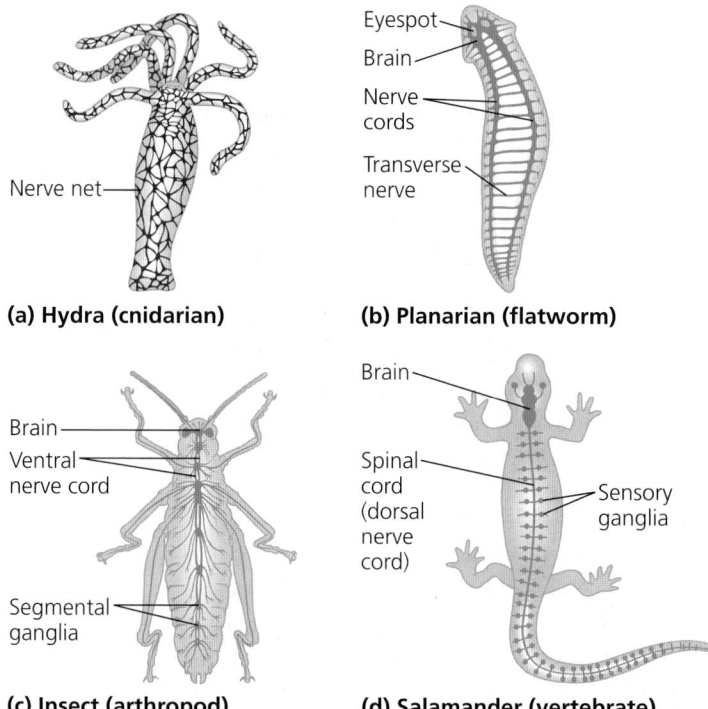

**(a) Hydra (cnidarian)**

Nerve net

Eyespot
Brain
Nerve cords
Transverse nerve

**(b) Planarian (flatworm)**

Brain
Ventral nerve cord
Segmental ganglia

**(c) Insect (arthropod)**

Brain
Spinal cord (dorsal nerve cord)
Sensory ganglia

**(d) Salamander (vertebrate)**

▲ **Figure 38.2 Nervous system organization. (a)** In a hydra's nervous system, individual neurons (purple) form a diffuse nerve net. **(b–d)** More complex nervous systems have groups of neurons (blue) organized into nerves and often ganglia and a brain.

Animals with elongated, bilaterally symmetric bodies have even more specialized nervous systems. Neuron organization in such animals reflects *cephalization*, an evolutionary trend toward a clustering of sensory neurons and interneurons at the anterior (front) end of the body. Nerves that extend toward the posterior (rear) end enable these clustered anterior neurons to communicate with cells elsewhere in the body.

In many animals, neurons that carry out integration form a **central nervous system (CNS)**, and neurons that carry information into and out of the CNS form a **peripheral nervous system (PNS)**. Nonsegmented worms, such as the planarian shown in **Figure 38.2b**, have a small brain and longitudinal nerve cords. This arrangement constitutes the simplest clearly defined CNS. More complex invertebrates, such as insects and other arthropods **(Figure 38.2c)**, have more complicated brains. In addition, their ventral nerve cords contain *ganglia*, segmentally arranged clusters of neurons that act as relay points in transmitting information. In vertebrates **(Figure 38.2d)**, the brain and spinal cord together form the CNS; nerves and ganglia are the key elements of the PNS.

## Glia

As discussed in Concept 37.1, the nervous systems of vertebrates and most invertebrates include not only neurons but also **glial cells**, or **glia**. These include, for example, microglia, which provide immune defense in the CNS. **Figure 38.3** illustrates the major types of glia in the adult vertebrate and provides an overview of the ways in which they nourish, support, and regulate the functioning of neurons.

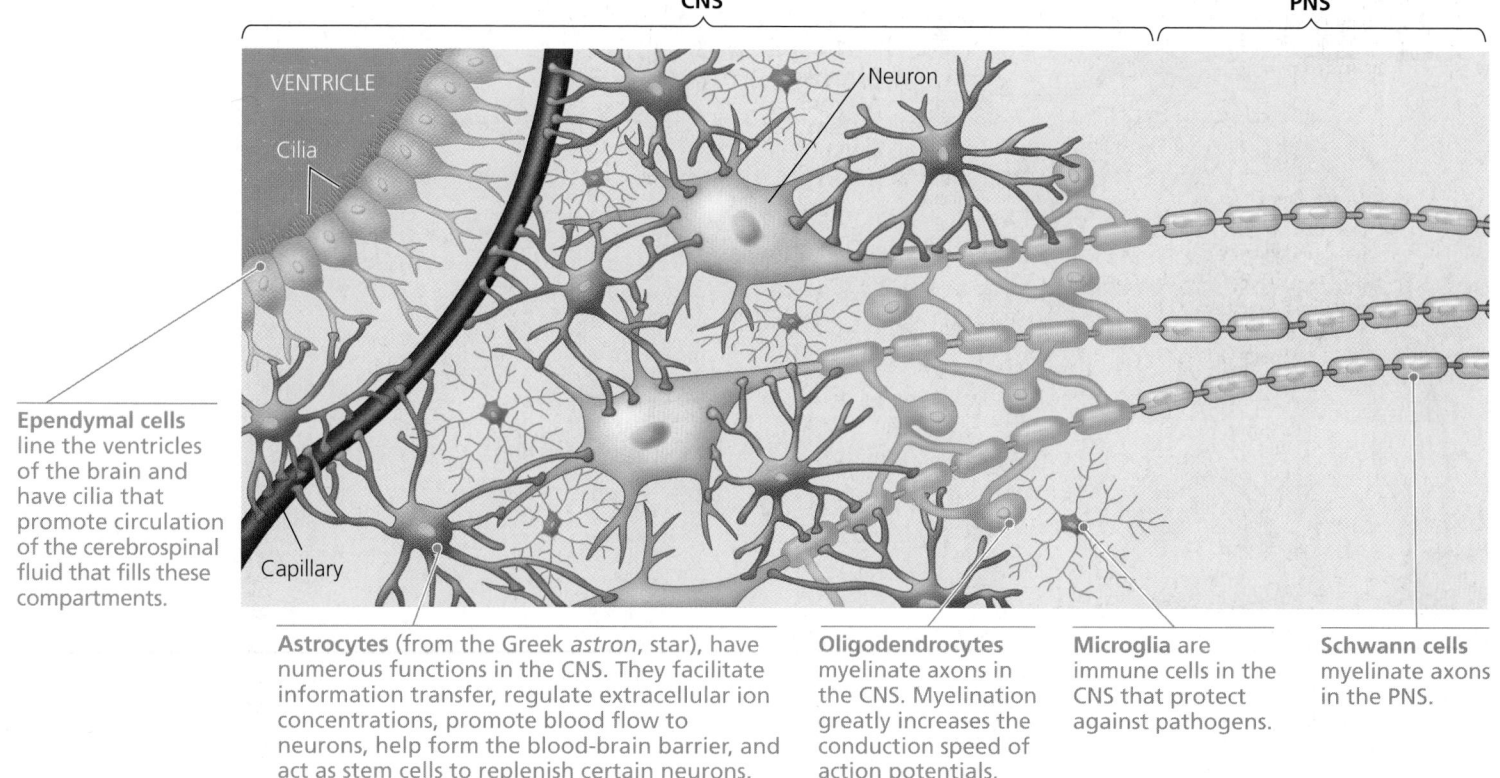

CNS

PNS

VENTRICLE
Cilia
Neuron

**Ependymal cells** line the ventricles of the brain and have cilia that promote circulation of the cerebrospinal fluid that fills these compartments.

Capillary

**Astrocytes** (from the Greek *astron*, star), have numerous functions in the CNS. They facilitate information transfer, regulate extracellular ion concentrations, promote blood flow to neurons, help form the blood-brain barrier, and act as stem cells to replenish certain neurons.

**Oligodendrocytes** myelinate axons in the CNS. Myelination greatly increases the conduction speed of action potentials.

**Microglia** are immune cells in the CNS that protect against pathogens.

**Schwann cells** myelinate axons in the PNS.

▲ **Figure 38.3 Glia in the vertebrate nervous system.**

In the vertebrate embryo, two types of glia play an essential role in the development of the nervous system. *Radial glia* form tracks along which newly formed neurons migrate from the neural tube, the structure that gives rise to the CNS. Later, star-shaped glia called *astrocytes*, located alongside capillaries in the brain, participate in forming the *blood-brain barrier*, a filtering mechanism that restricts the entry of most substances from the blood into the CNS.

Both radial glia and astrocytes can act as stem cells, which undergo unlimited cell divisions to self-renew and to form more specialized cells. Studies using mice reveal that stem cells in the brain give rise to neurons that mature, migrate to particular locations, and become incorporated into the circuitry of the adult nervous system. Researchers are now exploring approaches to use these stem cells as a means of replacing brain tissue that has ceased to function properly.

## Organization of the Vertebrate Nervous System

In vertebrates, the spinal cord runs lengthwise inside the vertebral column, known as the spine **(Figure 38.4)**. The spinal cord conveys information to and from the brain and generates basic patterns of locomotion. It also acts independently of the brain as part of the simple nerve circuits that produce certain **reflexes**, the body's automatic responses to particular stimuli.

For example, the spinal cord controls the reflex that jerks your hand away if you accidently touch a hot frying pan.

Both the brain and the spinal cord contain gray and white matter. **Gray matter** is primarily made up of neuron cell bodies. **White matter** consists of bundled axons. In the spinal cord, white matter makes up the outer layer, consistent with its function in linking the CNS to sensory and motor neurons of the PNS. In the brain, white matter is predominantly located in the interior, where signaling between neurons functions in learning, feeling emotions, processing sensory information, and generating commands.

The CNS also contains fluid-filled spaces, called the *central canal* in the spinal cord and *ventricles* in the brain (see Figure 38.3). The fluid inside, called *cerebrospinal fluid*, is formed in the brain by filtering arterial blood. It supplies the CNS with nutrients and hormones and carries away wastes, circulating through the ventricles and central canal before draining into the veins.

## The Peripheral Nervous System

The PNS transmits information to and from the CNS and plays a large role in regulating an animal's movements and its internal environment **(Figure 38.5)**. Sensory information reaches the CNS along PNS neurons designated as *afferent* (from the Latin, meaning "to carry toward"). Following information processing within the CNS, instructions travel to muscles, glands, and endocrine cells along PNS neurons designated as *efferent* (meaning "to carry away"). Note that most nerves contain both afferent and efferent neurons.

The PNS has two efferent components: the motor system and the autonomic nervous system (see Figure 38.5). The

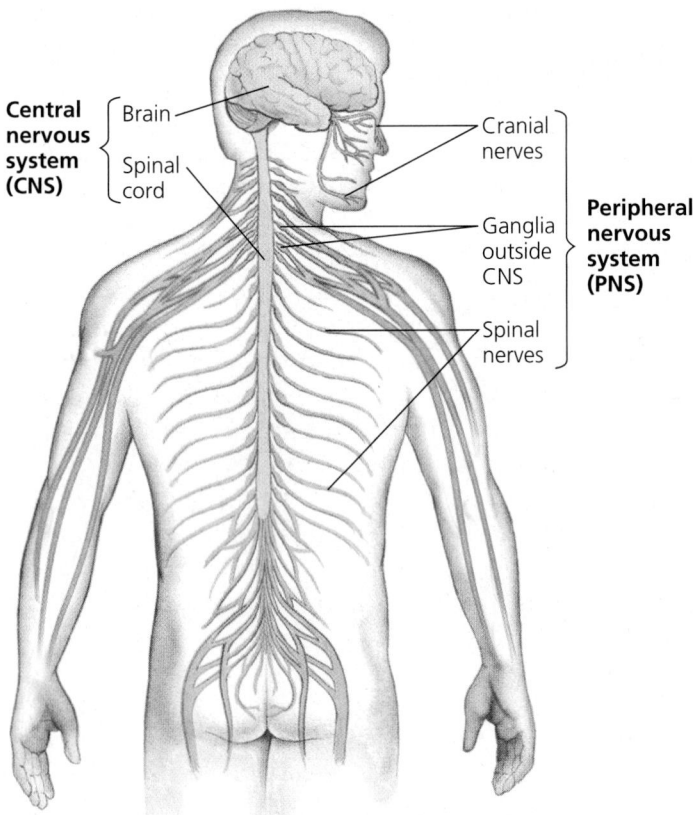

▲ **Figure 38.4 The vertebrate nervous system.** The central nervous system consists of the brain and spinal cord (yellow). Left-right pairs of cranial nerves, spinal nerves, and ganglia make up most of the peripheral nervous system (dark gold).

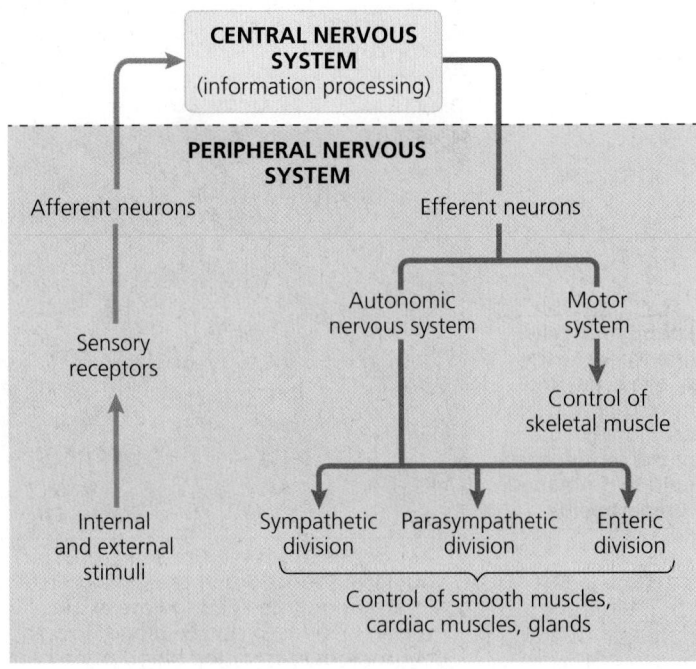

▲ **Figure 38.5 Functional hierarchy of the vertebrate peripheral nervous system.**

neurons of the **motor system** carry signals to skeletal muscles. Motor control can be voluntary, as when you raise your hand to ask a question, or involuntary, as in a reflex. In contrast, regulation of smooth and cardiac muscles by the **autonomic nervous system** is generally involuntary. The three divisions of the autonomic nervous system—enteric, sympathetic, and parasympathetic—together control the organs of the digestive, cardiovascular, excretory, and endocrine systems. For example, networks of neurons that form the **enteric division** of the autonomic nervous system are active in the digestive tract, pancreas, and gallbladder.

The sympathetic and parasympathetic divisions of the autonomic nervous system have largely antagonistic (opposite) functions in regulating organ function. Activation of the **sympathetic division** is responsible for the "fight-or-flight" response, a state of hyperarousal with which we and other animals respond to a threat. In mammals, the heart beats faster, digestion slows or stops, and the adrenal medulla secretes more epinephrine (adrenaline). Activation of the **parasympathetic division** generally causes opposite responses that promote calming and a return to self-maintenance functions ("rest and digest").

As we have seen, each component of the PNS has functions specific to particular locations in the body. We therefore describe the PNS as having regional specialization, a property also apparent in the brain, our next topic.

### CONCEPT CHECK 38.1

1. Which division of the autonomic nervous system would likely be activated if a student learned that an exam she had forgotten about would start in 5 minutes? Explain.
2. **WHAT IF?** Suppose a person had an accident that severed a small nerve required to move some of the fingers of the right hand. Would you also expect an effect on sensation from those fingers?

For suggested answers, see Appendix A.

---

### CONCEPT 38.2

# The vertebrate brain is regionally specialized

We'll now discuss the vertebrate brain, returning to the question raised at the beginning of this chapter: What happens in your brain when you are processing and responding to sensory input? One way to explore this question is by using technology called functional imaging to visualize brain activity.

## Functional Imaging of the Brain

Functional imaging is transforming our understanding of normal and diseased brains, allowing doctors to monitor patient recovery from stroke, map abnormalities in migraine headaches, and increase the effectiveness of brain surgery.

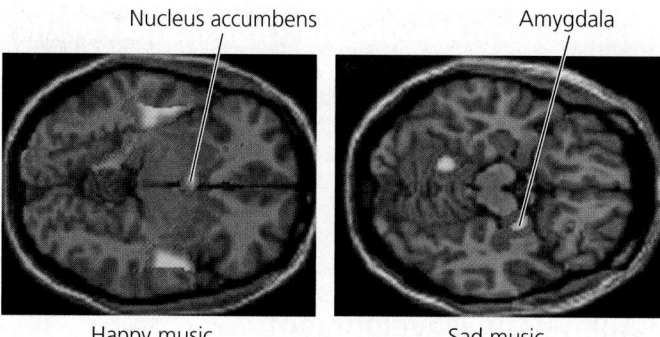

Nucleus accumbens      Amygdala

Happy music      Sad music

▲ **Figure 38.6 Functional brain imaging in the working brain.** Functional magnetic resonance imaging (fMRI) was used to reveal brain activity associated with happy or sad music.

**WHAT IF?** *In the experiment that produced the images shown above, some regions of the brain were active under all conditions. What function might such regions carry out?*

The first widely used technique was positron-emission tomography (PET), in which an injection of radioactive glucose enables a display of metabolic activity. Today, many studies use functional magnetic resonance imaging (fMRI), in which a subject lies with his or her head in the center of a large, doughnut-shaped magnet. Brain activity in a region is detected by changes in the local oxygen concentration. By scanning the brain while the subject performs a task, such as forming a mental image of a face, researchers can correlate particular tasks with activity in specific brain areas.

In one experiment using fMRI, researchers mapped brain activity while subjects listened to music that they described as happy or sad **(Figure 38.6)**. The findings were striking: Different regions of the brain were associated with the experience of these contrasting emotions. Listening to happy music led to increased activity in the nucleus accumbens, a brain structure important for the perception of pleasure. In contrast, subjects who heard sad music had increased activity in the amygdala, the focus for emotional memory.

To learn more about regional specialization of the brain, let's begin with **Figure 38.7**, which explores the overall architecture of the brain. Use this figure to trace how brain structures arise during embryonic development; as a reference for their size, shape, and location in the adult brain; and as an introduction to their best-understood functions.

In examining specific examples of brain organization and function, we'll consider first the structures that control when we are awake or asleep.

## Arousal and Sleep

If you've ever drifted off to sleep while listening to a lecture (or reading a textbook), you know that your attentiveness and mental alertness can change rapidly. Such transitions are regulated by the brainstem and cerebrum, which control arousal and sleep. Arousal is a state of awareness of the external world. Sleep is a state in which external stimuli are received but not consciously perceived.

The brain is the most complex organ in the human body. Surrounded by the thick bones of the skull, the brain is divided into a set of distinctive structures, some of which are visible in the magnetic resonance image (MRI) of an adult's head shown at right. The diagram below traces the development of these structures in the embryo. Their major functions are explained in the main text of the chapter.

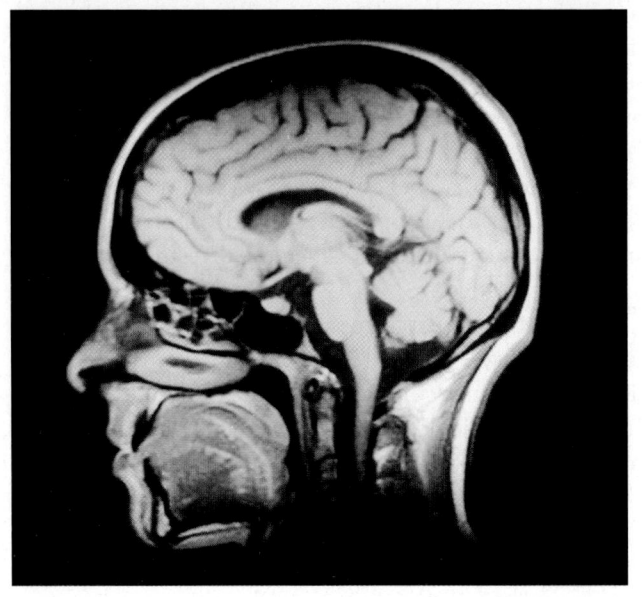

## Human Brain Development

As a human embryo develops, the neural tube forms three anterior bulges—the forebrain, midbrain, and hindbrain—that together produce the adult brain. The midbrain and portions of the hindbrain give rise to the **brainstem,** a stalk that joins with the spinal cord at the base of the brain. The rest of the hindbrain gives rise to the **cerebellum**, which lies behind the brainstem. Meanwhile, the forebrain develops into the diencephalon, including the neuroendocrine tissues of the brain, and the telencephalon, which becomes the **cerebrum**. Rapid, expansive growth of the telencephalon during the second and third months causes the outer portion, or cortex, of the cerebrum to extend over and around much of the rest of the brain.

**Embryonic brain regions**

**Brain structures in child and adult**

| Forebrain | Telencephalon | Cerebrum (includes cerebral cortex, basal nuclei) |
| | Diencephalon | Diencephalon (thalamus, hypothalamus, epithalamus) |
| Midbrain | Mesencephalon | Midbrain (part of brainstem) |
| Hindbrain | Metencephalon | Pons (part of brainstem),  cerebellum |
| | Myelencephalon | Medulla oblongata (part of brainstem) |

Midbrain
Hindbrain
Forebrain

**Embryo at 1 month**

Mesencephalon
Metencephalon
Diencephalon
Myelencephalon
Telencephalon
Spinal cord

**Embryo at 5 weeks**

Cerebrum
Diencephalon
Midbrain
Pons
Medulla oblongata
Cerebellum
Spinal cord
Brainstem

**Child**

## The Cerebrum

The cerebrum controls skeletal muscle contraction and is the center for learning, emotion, memory, and perception. It is divided into right and left *cerebral hemispheres*. The outer layer of the cerebrum is called the **cerebral cortex** and is vital for perception, voluntary movement, and learning. The left side of the cerebral cortex receives information from, and controls the movement of, the right side of the body, and vice versa. A thick band of axons known as the **corpus callosum** enables the right and left cerebral cortices to communicate. Deep within the white matter, clusters of neurons called *basal nuclei* serve as centers for planning and learning movement sequences. Damage to these sites during fetal development can result in cerebral palsy, a disorder resulting from a disruption in the transmission of motor commands to the muscles.

## The Cerebellum

The cerebellum coordinates movement and balance and helps in learning and remembering motor skills. The cerebellum receives sensory information about the positions of the joints and the lengths of the muscles, as well as input from the auditory (hearing) and visual systems. It also monitors motor commands issued by the cerebrum. The cerebellum integrates this information as it carries out coordination and error checking during motor and perceptual functions.

Hand-eye coordination is an example of cerebellar control; if the cerebellum is damaged, the eyes can follow a moving object, but they will not stop at the same place as the object. Hand movement toward the object will also be erratic.

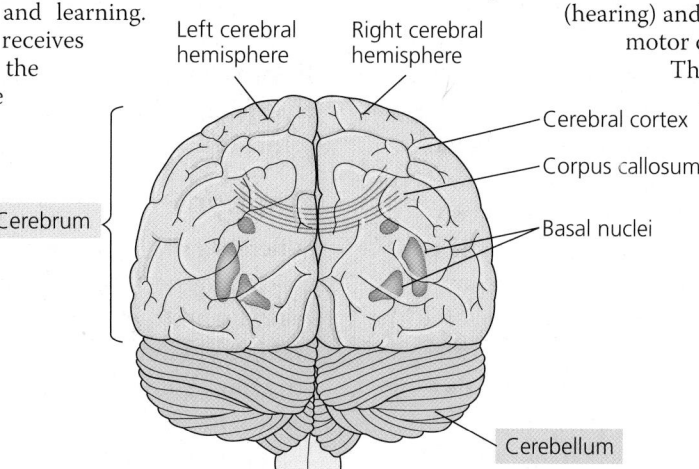

**Adult brain viewed from the rear**

Left cerebral hemisphere · Right cerebral hemisphere · Cerebral cortex · Corpus callosum · Basal nuclei · Cerebrum · Cerebellum

## The Diencephalon

The diencephalon gives rise to the thalamus, hypothalamus, and epithalamus. The **thalamus** is the main input center for sensory information going to the cerebrum. Incoming information from all the senses, as well as from the cerebral cortex, is sorted in the thalamus and sent to the appropriate cerebral centers for further processing. The thalamus is formed by two masses, each roughly the size and shape of a walnut. A much smaller structure, the **hypothalamus**, constitutes a control center that includes the body's thermostat as well as the central biological clock. Through its regulation of the pituitary gland, the hypothalamus regulates hunger and thirst, plays a role in sexual and mating behaviors, and initiates the fight-or-flight response. The hypothalamus is also the source of posterior pituitary hormones and of releasing hormones that act on the anterior pituitary. The *epithalamus* includes the pineal gland, the source of melatonin.

Diencephalon · Thalamus · Pineal gland · Hypothalamus · Pituitary gland · Spinal cord

## The Brainstem

The brainstem consists of the midbrain, the **pons**, and the **medulla oblongata** (commonly called the *medulla*). The midbrain receives and integrates several types of sensory information and sends it to specific regions of the forebrain. All sensory axons involved in hearing either terminate in the midbrain or pass through it on their way to the cerebrum. In addition, the midbrain coordinates visual reflexes, such as the peripheral vision reflex: The head turns toward an object approaching from the side without the brain having formed an image of the object.

A major function of the pons and medulla is to transfer information between the PNS and the midbrain and forebrain. The pons and medulla also help coordinate large-scale body movements, such as running and climbing. Most axons that carry instructions about these movements cross from one side of the CNS to the other in the medulla. As a result, the right side of the brain controls much of the movement of the left side of the body, and vice versa. An additional function of the medulla is the control of several automatic, homeostatic functions, including breathing, heart and blood vessel activity, swallowing, vomiting, and digestion. The pons also participates in some of these activities; for example, it regulates the breathing centers in the medulla.

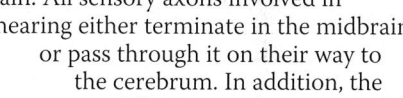

Brainstem · Midbrain · Pons · Medulla oblongata

Contrary to appearances, sleep is an active state, at least for the brain. By placing electrodes at multiple sites on the scalp, researchers can record patterns of electrical activity—brain waves—in an electroencephalogram (EEG). These recordings reveal that brain wave frequencies change as the brain progresses through distinct stages of sleep.

Within the brainstem, clusters of neurons regulating arousal and sleep are found in the midbrain and pons (see Figure 38.7). These neurons control the timing of sleep periods characterized by rapid eye movements (REM) and by vivid dreams. Sleep is also regulated by the biological clock, discussed next, and by regions of the forebrain that regulate sleep intensity and duration.

Some animals have evolutionary adaptations that allow for substantial activity during sleep. Bottlenose dolphins, for example, swim while sleeping, rising to the surface on a regular basis to breathe air. How is this possible? As in other mammals, the forebrain of dolphins is divided into two halves, the right and left hemispheres. Noting that dolphins sleep with one eye open and one closed, researchers hypothesized that only one side of the dolphin brain is asleep at a time. EEG recordings from each hemisphere of sleeping dolphins support this hypothesis (Figure 38.8).

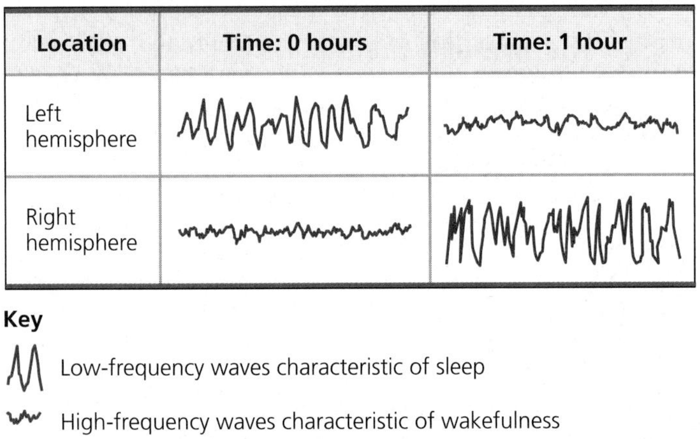

| Location | Time: 0 hours | Time: 1 hour |
|---|---|---|
| Left hemisphere | ∿∿∿∿∿ | ∿∿∿∿ |
| Right hemisphere | ∿∿∿∿ | ∿∿∿∿∿ |

**Key**

∿ Low-frequency waves characteristic of sleep

∿ High-frequency waves characteristic of wakefulness

▲ **Figure 38.8 Dolphins can be asleep and awake at the same time.** EEG recordings were made separately for the two sides of a dolphin's brain. At each time point, low-frequency activity was recorded in one hemisphere while higher-frequency activity typical of being awake was recorded in the other hemisphere.

## Biological Clock Regulation

Cycles of sleep and wakefulness are an example of a circadian rhythm, a daily cycle of biological activity. Such cycles, which occur in organisms ranging from bacteria to humans, rely on a **biological clock**, a molecular mechanism that directs periodic gene expression and cellular activity. Although biological clocks are typically synchronized to the cycles of light and dark in the environment, they can maintain a roughly 24-hour cycle even in the absence of environmental cues. For example, humans studied in a constant environment exhibit a sleep/wake cycle of 24.2 hours, with very little variation among individuals.

What normally links the biological clock to cycles of light and dark in an animal's surroundings? In mammals, circadian rhythms are coordinated by clustered neurons in the hypothalamus (see Figure 38.7). These neurons form a structure called the **SCN**, which stands for **suprachiasmatic nucleus**. (Certain clusters of neurons in the CNS are referred to as "nuclei.") In response to sensory information from the eyes, the SCN acts as a pacemaker, synchronizing the biological clock in cells throughout the body to the natural cycles of day length. In the **Scientific Skills Exercise**, you can interpret data from an experiment and propose additional experiments to test the role of the SCN in the circadian rhythms of a mammal.

## Emotions

Whereas a single structure in the brain controls the biological clock, generating and experiencing emotions depend on many brain structures, including the amygdala, hippocampus, and parts of the thalamus. As shown in **Figure 38.9**, these structures border the brainstem in mammals and are therefore termed the *limbic system* (from the Latin *limbus*, border).

One way the limbic system contributes to our emotions is by storing emotional experiences as memories that can be recalled by similar circumstances. This is why, for example, a situation that causes you to remember a frightening event can trigger a faster heart rate, sweating, or fear, even if there is currently nothing scary or threatening in your surroundings. Such storage and recall of emotional memory are especially dependent on the amygdala (see Figure 38.6).

Often, generating and experiencing emotion require interaction between different regions of the brain. For example, both laughing and crying involve the limbic system interacting with sensory areas of the forebrain. Similarly, structures in the forebrain attach emotional "feelings" to survival-related functions controlled by the brainstem, including aggression, feeding, and sexuality.

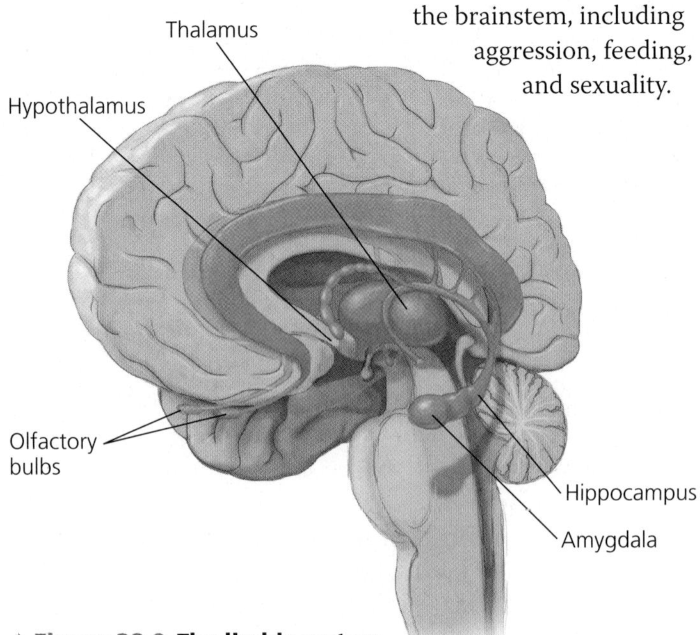

▲ **Figure 38.9 The limbic system.**

# Designing an Experiment Using Genetic Mutants

**Does the SCN Control the Circadian Rhythm in Hamsters?** By surgically removing the SCN from laboratory mammals, scientists demonstrated that the SCN is required for circadian rhythms. But these studies did not reveal whether circadian rhythms originate in the SCN. To answer this question, researchers performed an SCN transplant experiment on wild-type and mutant hamsters (*Mesocricetus auratus*). Whereas for wild-type hamsters the period between cyclic peaks in activity in the absence of external cues is 24 hours, hamsters homozygous for the τ (tau) mutation have a period lasting only about 20 hours. In this exercise, you will evaluate the design of this experiment and propose additional experiments to gain further insight.

**How the Experiment Was Done** The researchers surgically removed the SCN from wild-type and τ hamsters. Several weeks later, each of these hamsters received a transplant of an SCN from a hamster of the opposite genotype. The researchers then measured the period length for the transplant recipients.

**Data from the Experiment** In 80% of the hamsters from which the SCN had been removed, transplanting an SCN from another hamster restored rhythmic activity. For hamsters in which an SCN transplant restored a circadian rhythm, the net effect of the two procedures (SCN removal and replacement) on the circadian cycle is graphed. Each red line represents the change in the measured period for an individual hamster.

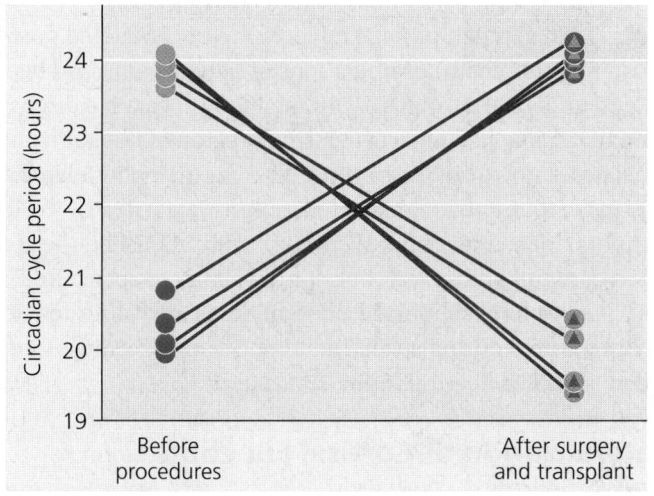

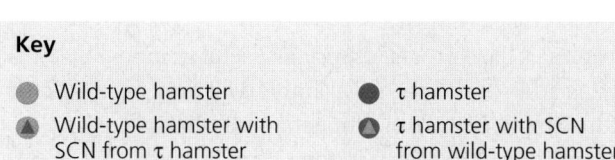

**Key**

- Wild-type hamster
- Wild-type hamster with SCN from τ hamster
- τ hamster
- τ hamster with SCN from wild-type hamster

**Data from** M. R. Ralph et al., Transplanted suprachiasmatic nucleus determines circadian period, *Science* 247:975–978 (1990).

**INTERPRET THE DATA**

1. In a controlled experiment, researchers manipulate one variable at a time. What was the experimental variable in this study? Why did the researchers use more than one hamster for each procedure? What traits of the individual hamsters would likely have been controlled among the treatment groups?

2. To interpret the effects of transplanting the SCN from τ hamsters to the wild-type hamsters, what would be an appropriate control?

3. What general trends does the graph reveal about the period of the circadian rhythm in transplant recipients? Do the trends differ for the wild-type and τ recipients? Based on these data, what can you conclude about the role of the SCN in determining the period length?

4. In 20% of the hamsters, there was no restoration of rhythmic activity following the SCN transplant. What are some possible reasons for this finding? Do you think you can be confident of your conclusion about the role of the SCN based on data from 80% of the hamsters? Explain.

5. Suppose that researchers identified a mutant hamster that lacked rhythmic activity; that is, its circadian cycle had no regular pattern. Propose SCN transplant experiments using such a mutant along with (a) wild-type and (b) τ hamsters. Predict the results of those experiments in light of your conclusion in question 3.

(MB) A version of this Scientific Skills Exercise can be assigned in MasteringBiology.

## The Brain's Reward System and Drug Addiction

Emotions are strongly influenced by a neural circuit in the brain called the *reward system*. The reward system provides motivation for activities that enhance survival and reproduction, such as eating in response to hunger, drinking when thirsty, and engaging in sexual activity when aroused. Inputs to the reward system are received by neurons in a region near the base of the brain called the *ventral tegmental area (VTA)*. When activated, these neurons release the neurotransmitter dopamine from their synaptic terminals within specific regions of the cerebrum. Targets of this dopamine signaling include the prefrontal cortex, discussed shortly, and the nucleus accumbens (see Figure 38.6).

The brain's reward system is dramatically affected by drug addiction, a disorder characterized by compulsive consumption of a drug and loss of control in limiting intake. Addictive drugs range from sedatives to stimulants and include alcohol, cocaine, nicotine, and heroin. All enhance the activity of the

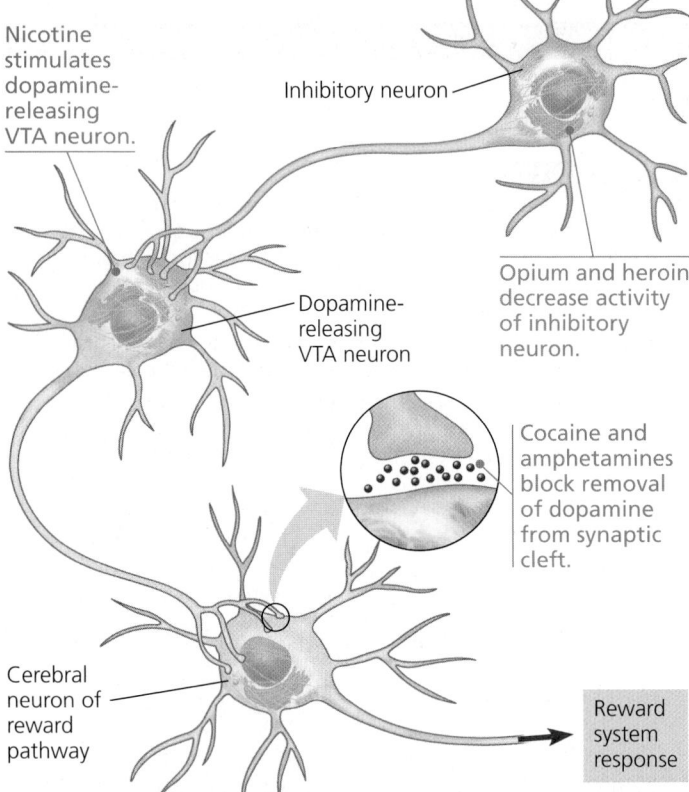

Nicotine stimulates dopamine-releasing VTA neuron.

Inhibitory neuron

Dopamine-releasing VTA neuron

Opium and heroin decrease activity of inhibitory neuron.

Cocaine and amphetamines block removal of dopamine from synaptic cleft.

Cerebral neuron of reward pathway

Reward system response

▲ **Figure 38.10 Effects of addictive drugs on the reward system of the mammalian brain.** Addictive drugs alter the transmission of signals in the pathway formed by neurons of the ventral tegmental area (VTA).

**MAKE CONNECTIONS** *Review depolarization in Concept 37.3. What effect would you expect if you depolarized the neurons in the VTA? Explain.*

dopamine pathway **(Figure 38.10)**. As addiction develops, there are also long-lasting changes in the reward circuitry. The result is a craving for the drug independent of any pleasure associated with consuming it.

Laboratory animals have proved especially useful in modeling and studying addiction. Rats, for example, will provide themselves with cocaine, heroin, or amphetamine when given a dispensing system linked to a lever in their cage. Furthermore, they exhibit addictive behavior in such circumstances, continuing to self-administer the drug rather than seek food, even to the point of starvation.

### CONCEPT CHECK 38.2

1. When you wave your right hand, what part of your brain initiates the action?
2. People who are inebriated have difficulty touching their nose with their eyes closed. Based on this observation, name one of the brain regions impaired by alcohol.
3. **WHAT IF?** Two groups of individuals have CNS damage. In one group, the damage has resulted in a coma (a prolonged state of unconsciousness). In the other group, it has caused total paralysis (a loss of skeletal muscle function throughout the body). Relative to the position of the midbrain and pons, where is the likely site of damage in each group? Explain.

For suggested answers, see Appendix A.

# The cerebral cortex controls voluntary movement and cognitive functions

We turn now to the cerebrum, the part of the brain essential for language, cognition, memory, consciousness, and awareness of our surroundings. For the most part, cognitive functions reside in the cerebral cortex, the outer layer of the cerebrum. Within this cortex, sensory areas receive and process sensory information, association areas integrate the information, and motor areas transmit instructions to other parts of the body. In discussing the cortex, neurobiologists often use four regions, or *lobes*, as physical landmarks. Each lobe—frontal, temporal, occipital, and parietal—is named for a nearby bone of the skull and each is the focus of specific brain activities **(Figure 38.11)**.

## Language and Speech

The mapping of cognitive functions within the cortex began in the 1800s when physicians studied the effects of damage to particular regions of the cortex by injuries, strokes, or tumors. The French physician Pierre Broca conducted postmortem (after death) examinations of patients who had been able to understand language but unable to speak. He discovered that many had defects in a small region of the left frontal lobe, now known as *Broca's area*. The German physician Karl Wernicke found that damage to a posterior portion of the left temporal lobe, now called *Wernicke's area*, abolished the ability to comprehend speech but not the ability to speak. Recent PET studies have supported these findings, revealing activity in Broca's area during speech generation and Wernicke's area when speech is heard **(Figure 38.12)**.

## Lateralization of Cortical Function

Both Broca's area and Wernicke's area reside in the left cortical hemisphere, reflecting a greater role with regard to language for the left side of the cerebrum than for the right side. The left hemisphere is also more adept at math and logical operations. In contrast, the right hemisphere appears to be dominant in the recognition of faces and patterns, spatial relations, and nonverbal thinking. This difference in function between the right and left hemispheres is called **lateralization**.

The cortical hemispheres are not entirely independent, but instead exchange information through the fibers of the corpus callosum (see Figure 38.7). If this connection is severed, the two hemispheres function independently. The affected individual cannot read even a familiar word that appears in their left field of vision: The sensory information travels from the left field of vision to the right hemisphere, but cannot then reach the language centers in the left hemisphere.

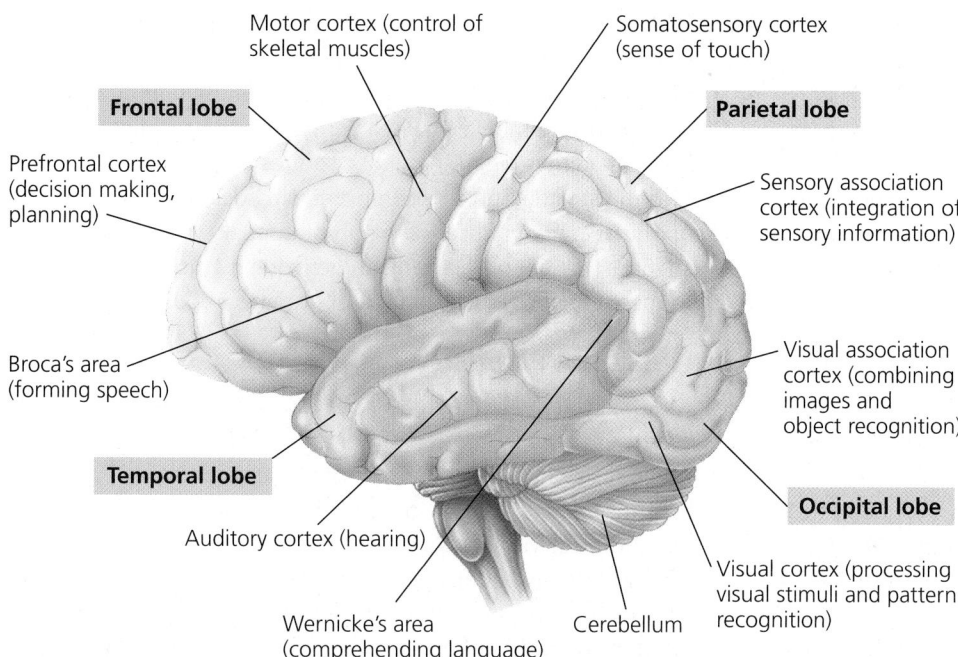

Motor cortex (control of skeletal muscles)

Somatosensory cortex (sense of touch)

**Frontal lobe**

**Parietal lobe**

Prefrontal cortex (decision making, planning)

Sensory association cortex (integration of sensory information)

Broca's area (forming speech)

Visual association cortex (combining images and object recognition)

**Temporal lobe**

**Occipital lobe**

Auditory cortex (hearing)

Visual cortex (processing visual stimuli and pattern recognition)

Wernicke's area (comprehending language)

Cerebellum

▲ **Figure 38.11 The human cerebral cortex.** Each of the four lobes of the cerebral cortex has specialized functions, some of which are listed here. Some areas on the left side of the brain (shown here) have different functions from those on the right side (not shown).

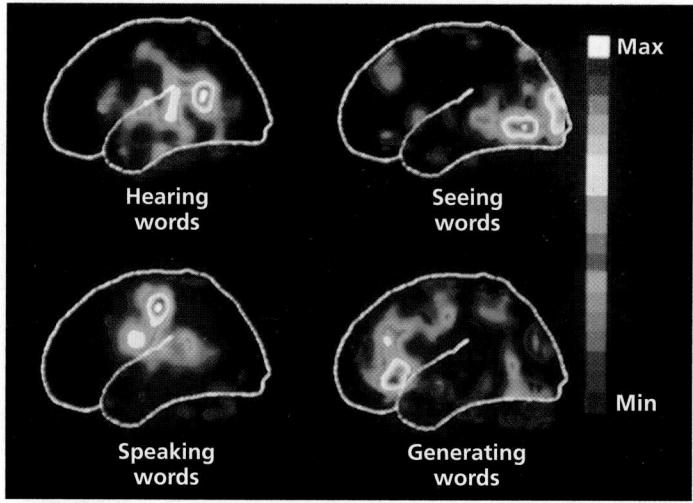

Max

Hearing words

Seeing words

Speaking words

Generating words

Min

▲ **Figure 38.12 Mapping language areas in the cerebral cortex.** These PET images show activity levels on the left side of one person's brain during four activities related to speech. Hearing words activates Wernicke's area, speaking words activates Broca's area, seeing words activates the visual cortex, and generating words (without reading them) activates parts of the frontal lobe.

## Information Processing

The cerebral cortex receives sensory input from groups of receptors in dedicated sensory organs, such as the eyes and nose. Additional input comes from individual receptors in the hands, scalp, and elsewhere in the body. These *somatosensory* receptors (from the Greek *soma*, body) provide information about touch, pain, pressure, temperature, and the position of muscles and limbs.

When sensory information reaches the cortex, most is directed by the thalamus to primary sensory areas within the brain lobes. This information is then passed along to nearby association areas, which process particular features of the sensory input. In the occipital lobe, for instance, some groups of neurons in the primary visual area are specifically sensitive to rays of light oriented in a particular direction. In the visual association area, information related to such features is combined in a region dedicated to recognizing complex images, such as faces.

Once processed, sensory information passes to the prefrontal cortex, which helps plan actions. The cerebral cortex may then generate motor commands that cause particular behaviors, such as waving a hand.

## Frontal Lobe Function

In 1848, a railroad worker named Phineas Gage had a horrific accident: An explosion drove a wide iron rod through his skull from just below his left eye to the top of his head (**Figure 38.13**), damaging large portions of his frontal lobe. Remarkably, Gage recovered, but his personality changed dramatically. He became emotionally detached, impatient, and erratic in his behavior, providing evidence of the role of the prefrontal cortex in temperament and decision-making.

Additional observations support the hypothesis that the frontal lobe helps shape personality and guide decisions. First, frontal lobe tumors cause similar symptoms: Intellect and memory seem intact, but decision making is flawed and emotional responses are diminished. Second, the same problems arise when the connection between the prefrontal cortex and the limbic system is surgically severed. (This procedure, called a frontal lobotomy, was once a common treatment for severe behavioral disorders, but it is no longer in use.) Together, these observations provide evidence that the frontal lobes have a substantial influence on what are called "executive functions."

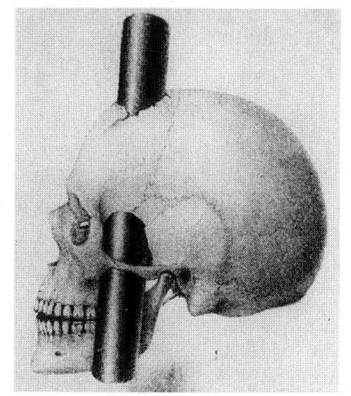

▲ **Figure 38.13 Phineas Gage's skull injury.**

## Evolution of Cognition in Vertebrates

**EVOLUTION** In nearly all vertebrates, the brain has the same basic structures. Given this uniform organization, what changed during evolution that provided certain species with a capacity for higher-order reasoning? One hypothesis is that advanced *cognition*, the perception and reasoning that constitute knowledge, required evolution of a highly convoluted cerebral cortex, such as is found in humans, other primates, and cetaceans (whales, dolphins, and porpoises). Indeed, in humans, the cerebral cortex accounts for about 80% of total brain mass.

Birds lack a convoluted cerebral cortex and were long thought to have much lower intellectual capacity than primates and cetaceans. However, recent experiments have refuted this idea. Western scrub jays (*Aphelocoma californica*) can remember the relative period of time that has passed since they hid a particular cache of food items. Furthermore, African gray parrots (*Psittacus erithacus*) understand numerical and abstract concepts, such as "same," "different," and "none."

The anatomical basis for sophisticated information processing in birds appears to be a nuclear (clustered) organization of neurons within the *pallium*, the top or outer portion of the brain **(Figure 38.14)**. This arrangement is different from that seen in the human cerebral cortex, where six parallel layers of neurons are arranged tangential to the brain surface. Thus, evolution has resulted in two different types of outer brain organization in vertebrates, each of which supports complex and flexible brain function.

How did the bird pallium and human cerebral cortex arise during evolution? The current consensus is that the common ancestor of birds and mammals had a pallium in which neurons were organized into nuclei, as is still found in birds.

Early in mammalian evolution, this nuclear organization was transformed into a layered one. However, connectivity was maintained such that, for example, the pallium in birds and the cerebral cortex in mammals each receive sensory input relayed by the thalamus regarding sights, sounds, and touch.

Sophisticated information processing depends not only on the overall organization of a brain but also on small-scale changes that enable learning and encode memory. We'll turn to these changes in the context of humans in the next section.

## Neuronal Plasticity

Much of the reshaping of the nervous system occurs at synapses. Synapses in circuits that link information in useful ways are maintained, whereas those that convey bits of information lacking any context may be lost. Specifically, when the activity of a synapse coincides with that of other synapses, changes may occur that reinforce that synaptic connection. Conversely, when the activity of a synapse fails to coincide with that of other synapses, the synaptic connection sometimes becomes weaker.

**Figure 38.15a** illustrates how activity-dependent events can trigger the gain or loss of a synapse. If you think of signals in the nervous system as traffic on a highway, such changes are comparable to adding or removing an entrance ramp. The net effect is to increase signaling between particular pairs of neurons and decrease signaling between other pairs. Signaling at a synapse can also be strengthened or weakened, as shown in

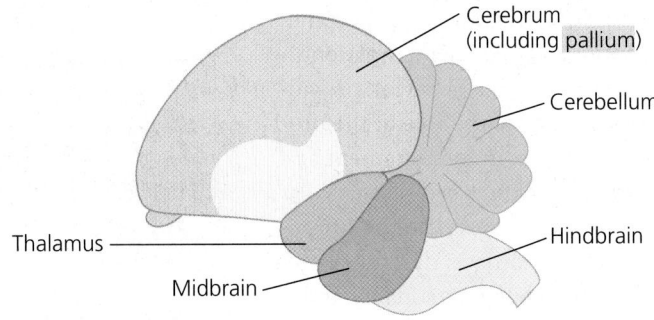

**(a)** High-level activity at the synapse neuron $N_1$ with the postsynaptic neuron leads to recruitment of additional axon terminals from that neuron. Lack of activity at the synapse with neuron $N_2$ leads to loss of functional connections with that neuron.

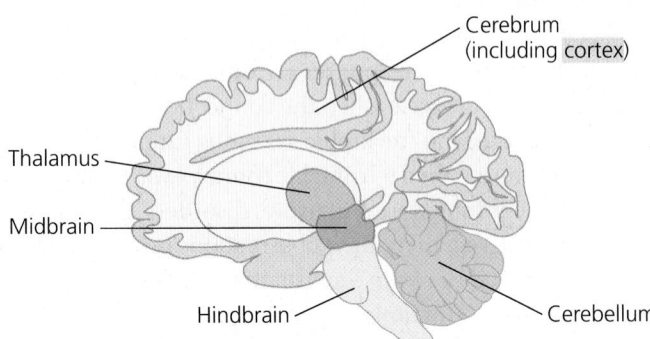

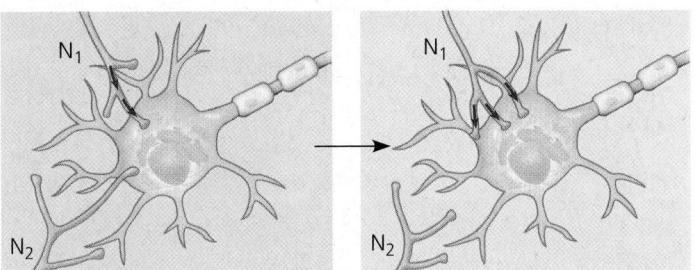

**(b)** If two synapses on a postsynaptic cell are often active at the same time, the strength of both responses may increase.

▲ **Figure 38.15 Neural plasticity.** Synaptic connections can strengthen or weaken in response to activity at the synapse.

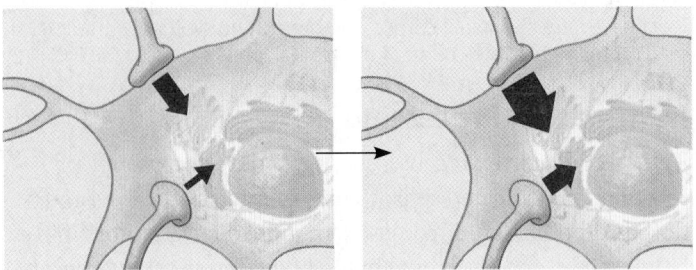

**(a) Songbird brain** (cross section)

Cerebrum (including pallium)
Cerebellum
Thalamus
Hindbrain
Midbrain

**(b) Human brain** (cross section)

Cerebrum (including cortex)
Thalamus
Midbrain
Hindbrain
Cerebellum

▲ **Figure 38.14 Comparison of regions for higher cognition in avian and human brains.** Although structurally different, the pallium of a songbird brain **(a)** and the cerebral cortex of the human brain **(b)** play similar roles in higher cognitive activities and make many similar connections with other brain structures.

**Figure 38.15b**. In our traffic analogy, this would be equivalent to widening or narrowing an entrance ramp.

This capacity for the nervous system to be remodeled, especially in response to its own activity, is called **neuronal plasticity**. A defect in neuronal plasticity may underlie the disorder *autism*, which results in impaired communication and social interaction, as well as restricted and repetitive behaviors, beginning in early childhood. There is now growing evidence that autism involves a disruption of activity-dependent remodeling at synapses.

## Memory and Learning

Neuronal plasticity is essential to the formation of memories. We are constantly checking what is happening against what just happened. We hold information for a time in **short-term memory** and then release it if it becomes irrelevant. If we wish to retain knowledge of a name, phone number, or other fact, the mechanisms of **long-term memory** are activated. If we later need to recall the name or number, we fetch it from long-term memory and return it to short-term memory.

Short-term and long-term memory both involve the storage of information in the cerebral cortex. In short-term memory, this information is accessed via temporary links formed in the hippocampus. When information is transferred to long-term memory, these links are replaced by connections within the cerebral cortex itself. Some of this memory consolidation is thought to occur during sleep. Furthermore, the reactivation of the hippocampus that is required for memory consolidation likely forms the basis for at least some of our dreams.

What evolutionary advantage might be offered by organizing short-term and long-term memories differently? One hypothesis is that the delay in forming connections in the cerebral cortex allows long-term memories to be integrated gradually into the existing store of knowledge and experience, providing a basis for more meaningful associations. Consistent with this hypothesis, the transfer of information from short-term to long-term memory is enhanced by the association of new data with data previously learned and stored in long-term memory. For example, it's easier to learn a new card game if you already have "card sense" from playing other card games.

# Sensory receptors transduce stimulus energy and transmit signals to the central nervous system

When a stimulus is detected by a sensory receptor, the resulting change in membrane potential alters the transmission of action potentials to the CNS. When this information is decoded within the CNS, a sensation results. In this section, we'll examine these steps in more detail.

## Sensory Reception and Transduction

A sensory pathway begins with **sensory reception**, the detection of a stimulus by sensory cells. Each sensory cell is either a specialized neuron or a non-neuronal cell that regulates a neuron **(Figure 38.16)**. Some sensory cells exist singly; others are collected in sensory organs, such as an eye.

The term **sensory receptor** is used to describe a sensory cell or organ, as well as the subcellular structure that detects stimuli. Some sensory receptors detect stimuli from outside the body, such as heat, light, pressure, or chemicals. Other receptors respond to internal stimuli such as blood pressure and body position. Activating a sensory receptor does not necessarily require a large amount of stimulus energy. Most light receptors, for example, can detect a single quantum (photon) of light.

Although animals use a range of sensory receptors to detect widely varying stimuli, the effect in all cases is to open or close ion channels. The resulting flow of ions across the membrane

▼ **Figure 38.16** Neuronal and non-neuronal sensory receptors.

| Neuronal receptors: Receptor *is* afferent neuron. | Non-neuronal receptors: Receptor *regulates* afferent neuron. |
|---|---|

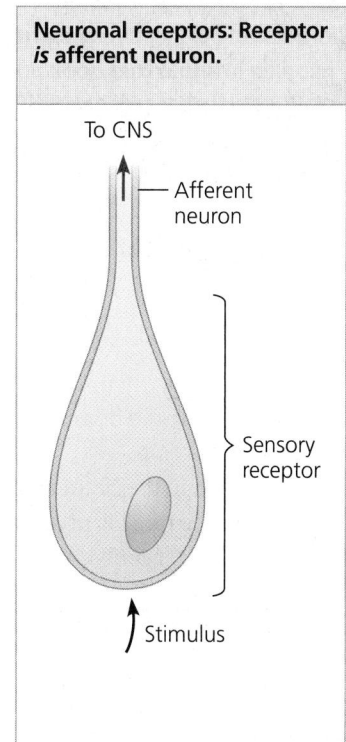

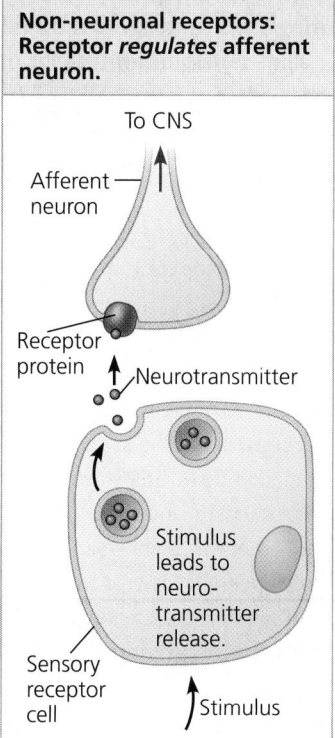

alters the membrane potential, a change called a **receptor potential**. Note that receptor potentials are graded: their magnitude varies with the strength of the stimulus. The conversion of a stimulus to a change in the membrane potential of a sensory receptor is called **sensory transduction**.

## Transmission

Sensory information travels through the nervous system as nerve impulses, or action potentials. Neurons that act directly as sensory receptors have an axon that extends into the CNS and transmit information directly as action potentials (see Figure 38.16). Non-neuronal sensory receptor cells form chemical synapses with sensory (afferent) neurons; these receptors generate a response to stimuli by altering the rate at which the afferent neurons produce action potentials.

The size of a receptor potential increases with stimulus intensity. If the receptor is a sensory neuron, a larger receptor potential results in more frequent action potentials **(Figure 38.17)**. If the receptor is not a sensory neuron, a larger receptor potential usually causes the receptor to release more neurotransmitter.

Many sensory neurons spontaneously generate action potentials at a low rate. In these neurons, a stimulus changes *how often* an action potential is produced, alerting the nervous system to changes in stimulus intensity.

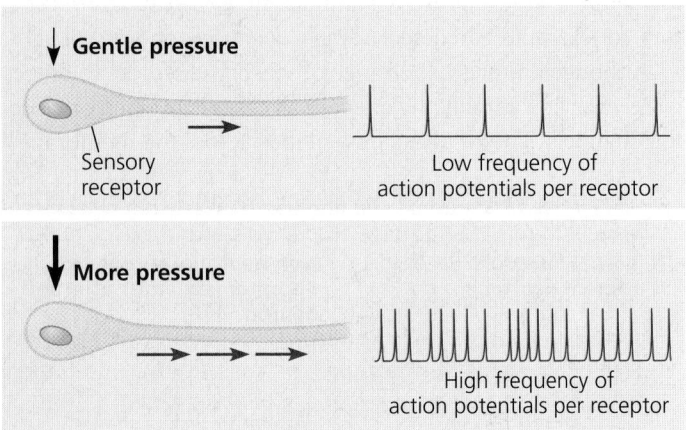

▲ **Figure 38.17 Coding of stimulus intensity by a single sensory receptor.**

## Perception

When action potentials reach the brain via sensory neurons, circuits of neurons process this input, generating the brain's **perception** of the stimulus. Perceptions—such as colors, smells, and sounds—are constructions formed in the brain and do not exist outside it. So, if a tree falls and no animal is present to hear it, is there a sound? The falling tree certainly produces pressure waves in the air, but there is no sound unless an animal senses the waves and its brain perceives them.

An action potential triggered by light striking the eye has the same properties as an action potential triggered by air vibrating in the ear. How, then, do we distinguish sights, sounds, and other stimuli? The answer lies in the connections that link

sensory receptors to the brain. Action potentials from sensory receptors travel along neurons that are dedicated to a particular stimulus; these dedicated neurons form synapses with particular neurons in the brain or spinal cord. As a result, the brain distinguishes stimuli such as sight and sound solely by the path along which the action potentials have arrived.

## Amplification and Adaptation

The sensitivity of the sensory system to stimuli is strongly influenced by the processes of amplification and sensory adaptation. **Amplification**—the strengthening of a sensory signal during transduction—enables animals to detect and respond to weak stimuli. The effect of amplification can be considerable: For example, an action potential conducted from the eye to the human brain has about 100,000 times as much energy as the few photons of light that triggered it.

Amplification that occurs in sensory receptor cells often requires signal transduction pathways involving enzyme-catalyzed reactions (see Figure 5.24). Because a single enzyme molecule catalyzes the formation of many product molecules, these pathways amplify signal strength considerably. Amplification may also take place in accessory structures of a sense organ. For example, accessory structures in the ear enhance the pressure associated with sound waves more than 20-fold before the stimulus reaches receptors in the innermost part of the ear.

Upon continued stimulation, many receptors undergo a decrease in responsiveness termed **sensory adaptation** (not to be confused with the evolutionary term *adaptation*). Without sensory adaptation, you would be constantly aware of feeling every beat of your heart and every bit of clothing on your body. Adaptation also enables you to see, hear, and smell changes in the environment that vary widely in stimulus intensity.

## Types of Sensory Receptors

We can classify sensory receptors based on the nature of the stimuli they transduce: mechanoreceptors, electromagnetic receptors, thermoreceptors, pain receptors, and chemoreceptors.

### Mechanoreceptors

**Mechanoreceptors** sense physical deformation (shape change) caused by forms of mechanical energy such as pressure, touch, stretch, motion, and sound. Mechanoreceptors typically consist of ion channels that are linked to structures that extend outside the cell, such as "hairs" (cilia). Bending or stretching of the external structure generates tension that alters the permeability of the ion channels. This change in ion permeability alters the membrane potential, resulting in a depolarization or hyperpolarization (see Figure 37.11a and b).

Some animals use mechanoreceptors to literally get a feel for their environment. For example, cats and many rodents have extremely sensitive mechanoreceptors at the base of their whiskers. Because deflection of different whiskers triggers action potentials that reach different cells in the brain, an animal's whiskers provide detailed information about nearby objects.

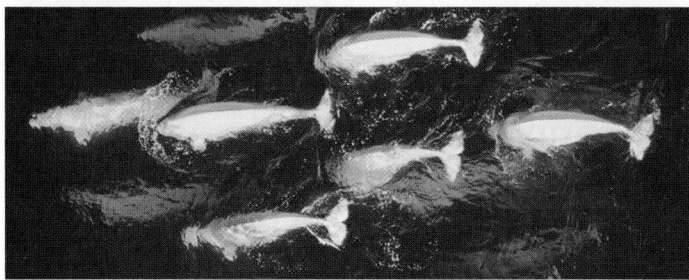

**(a)** Some migrating animals, such as these beluga whales, apparently sense Earth's magnetic field and use the information, along with other cues, for orientation.

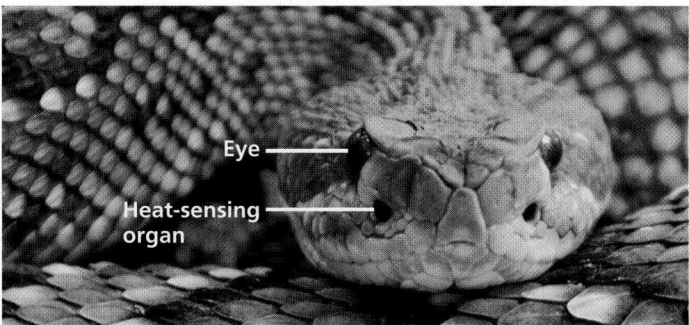

Eye

Heat-sensing organ

**(b)** This rattlesnake and other pit vipers have a pair of heat-sensing pit organs, one anterior to and just below each eye. These organs are sensitive enough to detect the infrared radiation emitted by a warm prey a meter away. The snake moves its head from side to side until the radiation is detected equally by the two pit organs, indicating that the prey is straight ahead.

▲ **Figure 38.18 Specialized electromagnetic receptors.**

## Electromagnetic Receptors

**Electromagnetic receptors** detect electromagnetic energy in forms such as light, electricity, and magnetism. For example, the platypus has electroreceptors on its bill that are thought to detect the electric field generated by the muscles of crustaceans, small fish, and other prey. In a few cases, the animal detecting the stimulus is also its source: Some fishes generate electric currents and then use electroreceptors to locate prey or other objects that disturb those currents.

Many animals use Earth's magnetic field lines to orient during migration **(Figure 38.18a)**; the iron-containing mineral magnetite may be responsible for this ability. Magnetite is found in many vertebrates (including salmon, pigeons, sea turtles, and humans), in bees, and in some molluscs, as well as in certain protists and prokaryotes that orient to Earth's magnetic field.

## Thermoreceptors

**Thermoreceptors** detect heat and cold. For example, certain venomous snakes rely on thermoreceptors to detect the infrared radiation emitted by warm prey **(Figure 38.18b)**. In humans, thermoreceptors located in the skin and in the anterior hypothalamus send information to the body's thermostat in the posterior hypothalamus.

Our understanding of thermoreceptors in humans has been greatly enhanced by scientists with an appreciation for fiery foods. It turns out that exposing sensory neurons to capsaicin—the molecule that makes jalapeno peppers taste "hot"—triggers calcium ion influx. When scientists identified the receptor protein that binds capsaicin, they made a fascinating discovery: The receptor opens a calcium channel in response not only to capsaicin, but also to high temperatures (42°C or higher). In essence, spicy foods taste "hot" because they activate the same receptors as hot soup and coffee. What is more, the receptor for temperatures below 28°C is activated by menthol, a plant product that we perceive to have a "cool" flavor.

## Pain Receptors

Extreme pressure or temperature, as well as certain chemicals, can damage animal tissues. To detect stimuli that reflect such noxious (or harmful) conditions, animals rely on **nociceptors** (from the Latin *nocere*, to hurt), also called **pain receptors**. By triggering defensive reactions, such as withdrawal from danger, the perception of pain serves an important function.

In humans, certain naked dendrites act as nociceptors by detecting noxious thermal, mechanical, or chemical stimuli. The capsaicin receptor is thus a nociceptor as well as a thermoreceptor. Although nociceptor density is highest in skin, some pain receptors are associated with other organs, enabling us to detect, for example, a sore throat or stomachache.

Chemicals produced in an animal's body sometimes enhance the perception of pain. For example, damaged tissues produce prostaglandins, local regulators that worsen pain by increasing nociceptor sensitivity to noxious stimuli. Because aspirin and ibuprofen inhibit the synthesis of prostaglandins, they have a pain-reducing, or analgesic, effect.

## Chemoreceptors

**Chemoreceptors** fall into two broad categories, those that transmit information about total solute concentration and those that respond to individual kinds of molecules. Osmoreceptors in the mammalian brain, for example, detect changes in the solute concentration of the blood and stimulate thirst when osmolarity increases. Most animals also have receptors for specific molecules, including glucose, oxygen, carbon dioxide, and amino acids.

The perceptions of **olfaction** (smell) and **gustation** (taste) both depend on chemoreceptors. In the case of terrestrial animals, smell is the detection of **odorants** that are carried through the air, and taste is the detection of chemicals called **tastants** that are present in a solution. Humans can distinguish thousands of different odors, each caused by a structurally distinct odorant. This level of sensory discrimination requires many different olfactory receptors. In 2004, Richard Axel and Linda Buck shared a Nobel Prize for their discovery of olfactory receptor genes, which in humans encode nearly 400 receptors specific for particular odorants.

With regard to tastants, humans (and other mammals) recognize just five types: sweet, sour, salty, bitter, and umami. Umami (Japanese for "delicious") is elicited by the amino acid glutamate. Often used as a flavor enhancer, monosodium glutamate (MSG) occurs naturally in foods such as meat and aged cheese, imparting a quality described as savory.

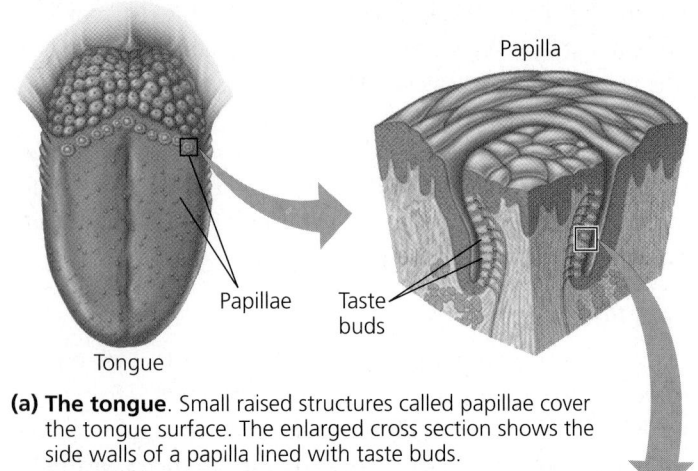

Papilla

Papillae
Taste buds
Tongue

(a) **The tongue**. Small raised structures called papillae cover the tongue surface. The enlarged cross section shows the side walls of a papilla lined with taste buds.

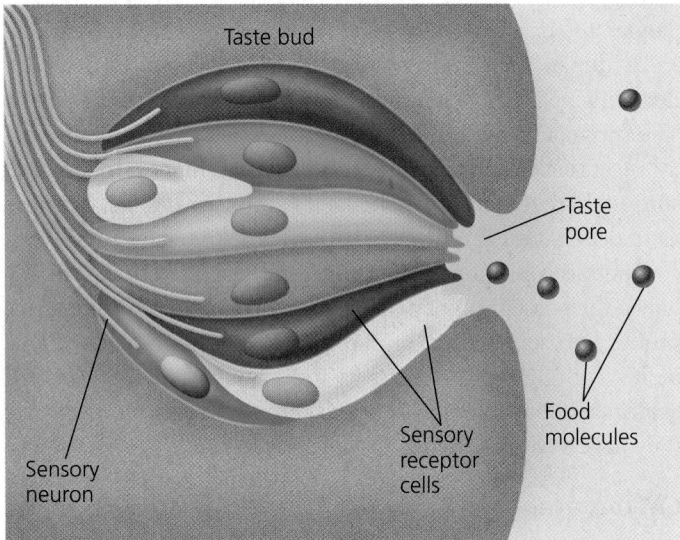

Taste bud

Taste pore

Sensory neuron

Sensory receptor cells

Food molecules

**Key**

Sweet   Bitter   Umami
Salty   Sour

(b) **A taste bud**. Taste buds in all regions of the tongue contain sensory receptor cells specific for each of the five taste types.

▲ **Figure 38.19 Human taste receptors.**

Taste receptor cells are organized into **taste buds**, most of which are found in projections called papillae (**Figure 38.19**). In contrast to the common misconception, any region of the tongue with taste buds can detect any of the five types of taste. Each taste cell, however, has just a single receptor type, programming the cell to detect only one of the five tastes.

**CONCEPT CHECK 38.4**

1. Which one of the five categories of sensory receptors is primarily dedicated to external stimuli?
2. Why can eating "hot" peppers cause a person to sweat?
3. **WHAT IF?** If you stimulated a sensory neuron of an animal electrically, how would that stimulation be perceived?

For suggested answers, see Appendix A.

## CONCEPT 38.5

# In hearing and equilibrium, mechanoreceptors detect moving fluid or settling particles

In most animals, hearing is closely related to the perception of bodily equilibrium, the sense that underlies balance. For both senses, mechanoreceptor cells produce receptor potentials when settling particles or moving fluid causes deflection of cell-surface structures.

## Sensing of Gravity and Sound in Invertebrates

To sense gravity and maintain equilibrium, most invertebrates rely on mechanoreceptors located in organs called **statocysts**. In a typical statocyst, *statoliths*, granules formed by grains of sand or other dense materials, sit freely in a chamber lined with ciliated receptor cells (**Figure 38.20**). Each time an animal repositions itself, the statoliths resettle, stimulating mechanoreceptors at the low point in the chamber.

How did researchers test the hypothesis that resettling of statoliths provides information about body position relative to Earth's gravity? In one key experiment, statoliths in crayfish were replaced with metal shavings. Researchers then "tricked" the crayfish into swimming upside down by using magnets to pull the shavings to the upper end of statocysts at the base of their antennae.

Many (perhaps most) insects have body hairs that vibrate in response to sound waves. Hairs differing in stiffness and length vibrate at different frequencies. Many insects also detect sound by means of vibration-sensitive organs. Such organs are what allow cockroaches to sense and react to air movement, such as that caused by a descending human foot.

## Hearing and Equilibrium in Mammals

In mammals, as in most other terrestrial vertebrates, the sensory organs for hearing and equilibrium are closely associated. **Figure 38.21** explores the structure and function of these organs in the human ear.

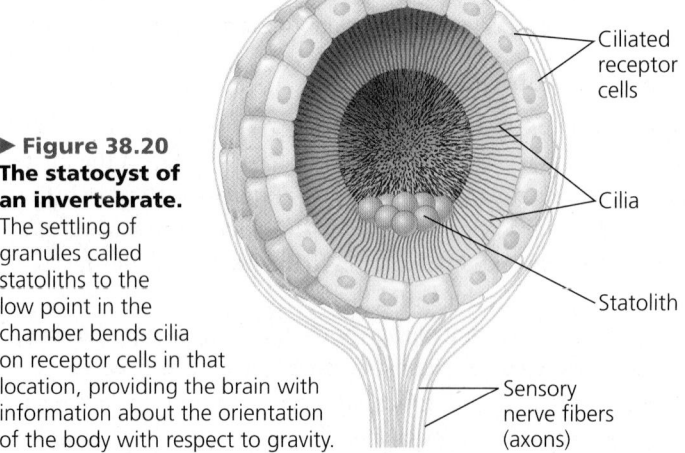

► **Figure 38.20 The statocyst of an invertebrate.** The settling of granules called statoliths to the low point in the chamber bends cilia on receptor cells in that location, providing the brain with information about the orientation of the body with respect to gravity.

Ciliated receptor cells

Cilia

Statolith

Sensory nerve fibers (axons)

## 1 Overview of Ear Structure

The **outer ear** consists of the external pinna and the auditory canal, which collect sound waves and channel them to the **tympanic membrane** (eardrum), a thin tissue that separates the outer ear from the **middle ear**. In the middle ear, three small bones—the malleus (hammer), incus (anvil), and stapes (stirrup)—transmit vibrations to the **oval window**, which is a membrane beneath the stapes. The middle ear also opens into the **Eustachian tube**, a passage that connects to the pharynx and equalizes pressure between the middle ear and the atmosphere. The **inner ear** consists of fluid-filled chambers, including the **semicircular canals**, which function in equilibrium, and the coiled **cochlea** (from the Latin meaning "snail"), a bony chamber that is involved in hearing.

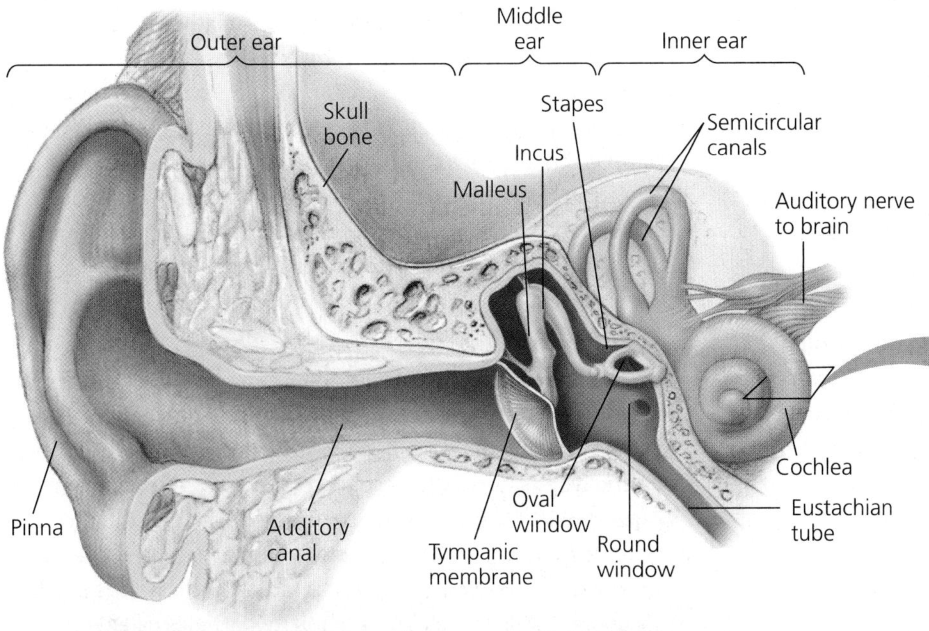

## 2 The Cochlea

The cochlea has two large canals—an upper vestibular canal and a lower tympanic canal—separated by a smaller cochlear duct. Both canals are filled with fluid.

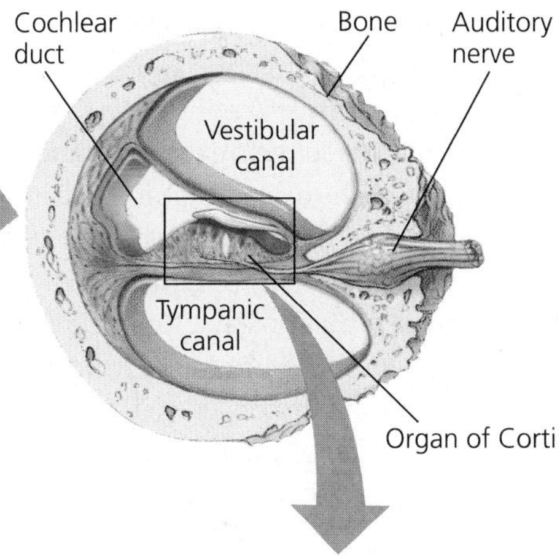

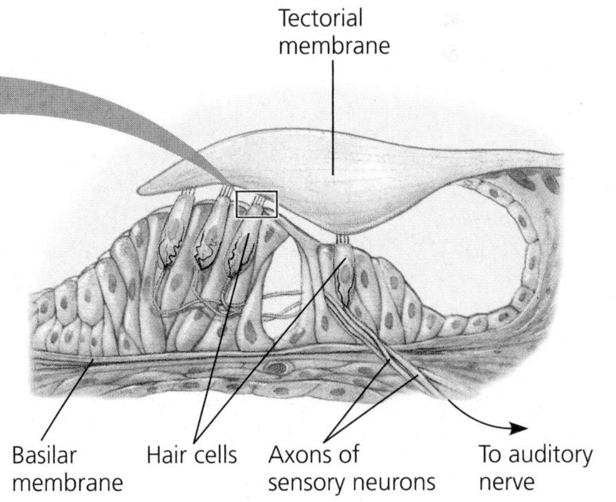

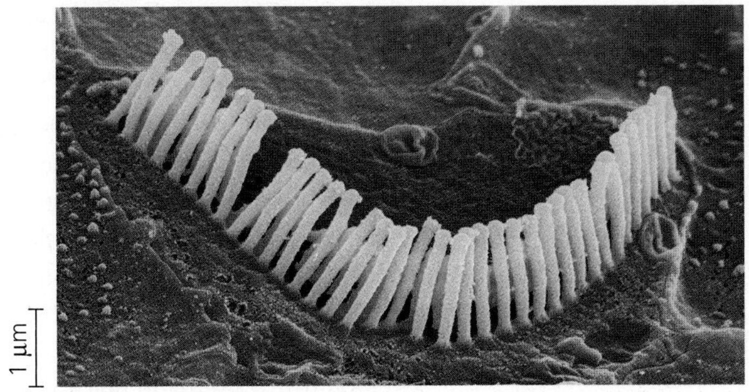

▲ Bundled hairs projecting from a single mammalian hair cell (SEM). Two shorter rows of hairs lie behind the tall hairs in the foreground.

## 4 Hair Cell

Projecting from each hair cell is a bundle of rod-shaped "hairs," within which lie a core of actin filaments. Vibration of the basilar membrane in response to sound raises and lowers the hair cells, bending the hairs against the surrounding fluid and the tectorial membrane. When the hairs are displaced, mechanoreceptors are activated, changing the membrane potential of the hair cell.

## 3 The Organ of Corti

The floor of the cochlear duct, the basilar membrane, bears the **organ of Corti**, which contains the mechanoreceptors of the ear— hair cells with hairs projecting into the cochlear duct. Many of the hairs are attached to the tectorial membrane, which hangs over the organ of Corti like an awning. Sound waves make the basilar membrane vibrate, which results in bending of the hairs and depolarization of the hair cells.

## Hearing

Vibrating objects, such as a plucked guitar string or the vocal cords of a person who is speaking, create pressure waves in the surrounding air. In hearing, the ear transduces this mechanical stimulus (pressure waves) into nerve impulses that the brain perceives as sound. To hear music, speech, or other sounds in our environment, we rely on **hair cells**, sensory cells with hair-like projections that detect motion.

Before vibration waves reach the hair cells, they are amplified and transformed by accessory structures. The process begins when moving air that reaches the outer ear causes the tympanic membrane to vibrate. Next, the three bones of the middle ear transmit these vibrations to the oval window, a membrane on the cochlea's surface. When one of those bones, the stapes, vibrates against the oval window, it creates pressure waves in the fluid inside the cochlea.

Upon entering the vestibular canal, fluid pressure waves push down on the cochlear duct and basilar membrane. In response, the basilar membrane and attached hair cells vibrate up and down. The hairs projecting from the moving hair cells are deflected by the fixed tectorial membrane, which lies above (see Figure 38.21). With each vibration, the hairs bend first in one direction and then the other, causing ion channels in the hair cells to open or close **(Figure 38.22)**. The result is a change in auditory nerve sensations that the brain interprets as sound.

What prevents pressure waves from reverberating within the ear and causing prolonged sensation? After propagating through the vestibular canal, pressure waves pass around the apex (tip) of the cochlea and dissipate as they strike the **round window**. This damping of sound waves resets the apparatus for the next vibrations that arrive.

The ear captures information about two important sound variables: volume and pitch. *Volume* (loudness) is determined by the amplitude, or height, of the sound wave. A large-amplitude wave causes more vigorous vibration of the basilar membrane, greater bending of the hairs on hair cells, and more action potentials in the sensory neurons. *Pitch* is determined by a sound wave's frequency, the number of vibrations per unit time. The detection of sound wave frequency takes place in the cochlea and relies on the asymmetric structure of that organ.

The cochlea can distinguish pitch because the basilar membrane is not uniform along its length: It is relatively narrow and stiff near the oval window and wider and more flexible at the apex at the base of the cochlea. Each region of the basilar membrane is tuned to a different vibration frequency. Furthermore, each region is connected by axons to a different location in the cerebral cortex. Consequently, when a sound wave causes vibration of a particular region of the basilar membrane, a specific site in our cortex is stimulated and we perceive sound of a particular pitch.

## Equilibrium

Several organs in the inner ear of humans and most other mammals detect body movement, position, and equilibrium. For example, the chambers called the **utricle** and **saccule** allow us to perceive position with respect to gravity or linear movement **(Figure 38.23)**. Each of these chambers, which are situated in a vestibule behind the oval window, contains hair cells that project into a gelatinous material. Embedded in this gel are small calcium carbonate particles called otoliths ("ear stones"). When you tilt your head, the otoliths shift position, contacting a different set of hairs protruding into the gel. The hair cell receptors transform this deflection into a change in the output of sensory neurons, signaling the brain that your head is at an angle. The otoliths are also responsible for your ability to perceive acceleration, as, for example, when a stationary car in which you are sitting pulls forward.

Three fluid-filled semicircular canals connected to the utricle detect turning of the head and other rotational acceleration. Because the three canals are arranged in the three spatial planes, they can detect angular motion of the head in any direction. If you spin in place, the fluid in each canal eventually comes to equilibrium and remains in that state until you stop. At that point, the moving fluid encounters a stationary cupula (see Figure 38.23), triggering the false sensation of angular motion that we call dizziness.

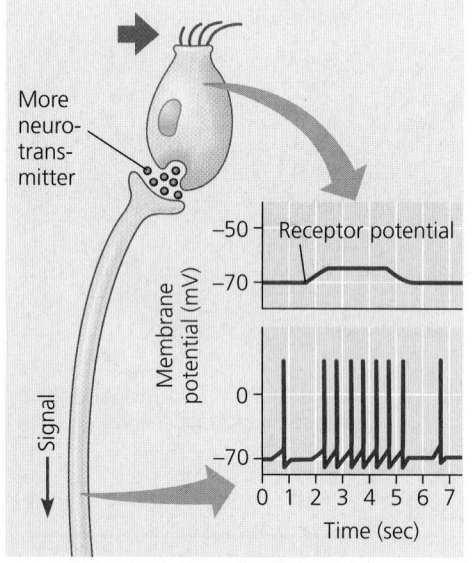

**(a) Bending of hairs in one direction**

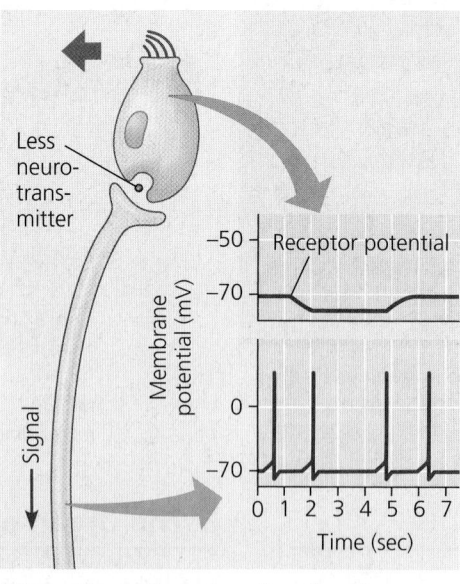

**(b) Bending of hairs in other direction**

▲ **Figure 38.22 Sensory reception by hair cells.** Bending of a hair cell bundle in one direction depolarizes the hair cell. This increases neurotransmitter release, resulting in more frequent action potentials in the sensory neuron. Bending in the other direction has the opposite effect.

**DRAW IT** *Draw a graph of the membrane potentials you would expect to observe in the hair cell and sensory neuron when there is no stimulation.*

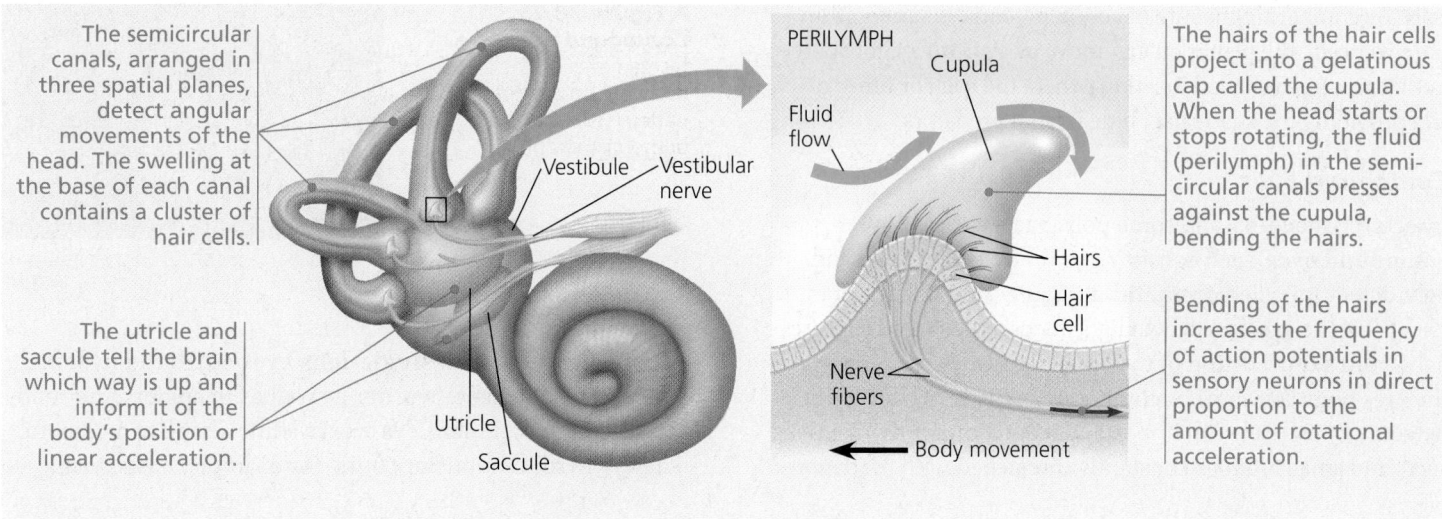

The semicircular canals, arranged in three spatial planes, detect angular movements of the head. The swelling at the base of each canal contains a cluster of hair cells.

The utricle and saccule tell the brain which way is up and inform it of the body's position or linear acceleration.

Vestibule    Vestibular nerve

Utricle

Saccule

PERILYMPH    Cupula

Fluid flow

The hairs of the hair cells project into a gelatinous cap called the cupula. When the head starts or stops rotating, the fluid (perilymph) in the semicircular canals presses against the cupula, bending the hairs.

Hairs

Hair cell

Nerve fibers

Bending of the hairs increases the frequency of action potentials in sensory neurons in direct proportion to the amount of rotational acceleration.

← Body movement

▲ Figure 38.23 Organs of equilibrium in the inner ear.

CONCEPT CHECK 38.5

1. How are otoliths adaptive for burrowing mammals, such as the star-nosed mole?

2. **WHAT IF?** Suppose a series of pressure waves in your cochlea caused a vibration of the basilar membrane that moves gradually from the apex toward the base. How would your brain interpret this stimulus?

3. **WHAT IF?** If the stapes became fused to the other middle ear bones or to the oval window, how would this condition affect hearing? Explain.

4. **MAKE CONNECTIONS** Plants use statoliths to detect gravity (see Figure 31.19). How do plants and animals differ with regard to the type of compartment in which statoliths are found and the physiological mechanism for detecting their response to gravity?

For suggested answers, see Appendix A.

## CONCEPT 38.6

# The diverse visual receptors of animals depend on light-absorbing pigments

The ability to detect light has a central role in the interaction of nearly all animals with their environment. Although the organs used for vision vary considerably among animals, the underlying mechanism for capturing light is the same, suggesting a common evolutionary origin.

## Evolution of Visual Perception

**EVOLUTION** Light detectors in the animal kingdom range from simple clusters of cells that detect only the direction and intensity of light to complex organs that form images. These diverse light detectors all contain **photoreceptors**, sensory cells that contain light-absorbing pigment molecules. Furthermore, the

genes that specify where and when photoreceptors arise during embryonic development are shared among flatworms, annelids, arthropods, and vertebrates. It is thus very probable that the genetic underpinnings of all photoreceptors were already present in the earliest bilaterian animals.

### Light-Detecting Organs

Most invertebrates have some kind of light-detecting organ. One of the simplest is that of planarians **(Figure 38.24)**. A pair of ocelli (singular, *ocellus*), sometimes called eyespots, are located in the head region. Photoreceptors in each ocellus receive light only through an opening where there are no pigmented

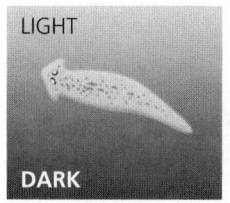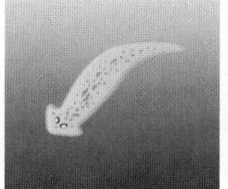

LIGHT

DARK

**(a)** The planarian's brain directs the body to turn until the sensations from the two ocelli are equal and minimal, causing the animal to move away from light.

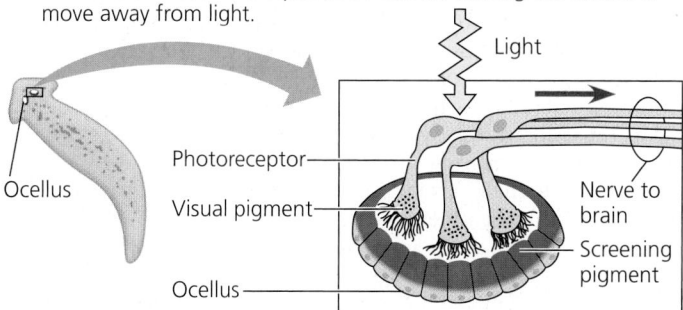

Light

Ocellus

Photoreceptor

Visual pigment

Ocellus

Nerve to brain

Screening pigment

**(b)** Whereas light striking the front of an ocellus excites the photoreceptors, light striking the back is blocked by the screening pigment. In this way, the ocelli indicate the direction of a light source, enabling the light avoidance behavior.

▲ Figure 38.24 Ocelli and orientation behavior of a planarian.

cells. By comparing the rate of action potentials coming from the two ocelli, the planarian can move away from a light source until it reaches a shaded location, where the rock or other object providing the shade may hide it from predators.

### Compound Eyes

Insects, crustaceans, and some polychaete worms have **compound eyes**, each consisting of up to several thousand light detectors called **ommatidia (Figure 38.25)**. Each of these "facets" of the eye has its own light-focusing lens that captures light from a tiny portion of the visual field (the area seen when the eyes point forward). A compound eye is very effective at detecting movement, an important adaptation for flying insects and small animals constantly threatened with predation.

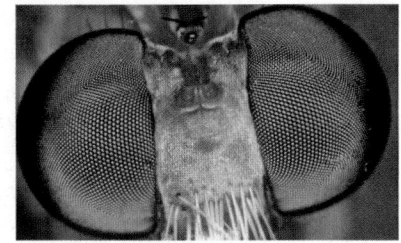

▶ **Figure 38.25**
**Compound eyes.** The faceted eyes on the head of a fly form a repeating pattern visible in this photomicrograph.

2 mm

### Single-Lens Eyes

Among invertebrates, **single-lens eyes** are found in some jellies and polychaete worms, as well as in spiders and many molluscs. A single-lens eye works somewhat like a camera. The eye of an octopus or squid, for example, has a small

---

▼ **Figure 38.26**   **Exploring the Structure of the Human Eye**

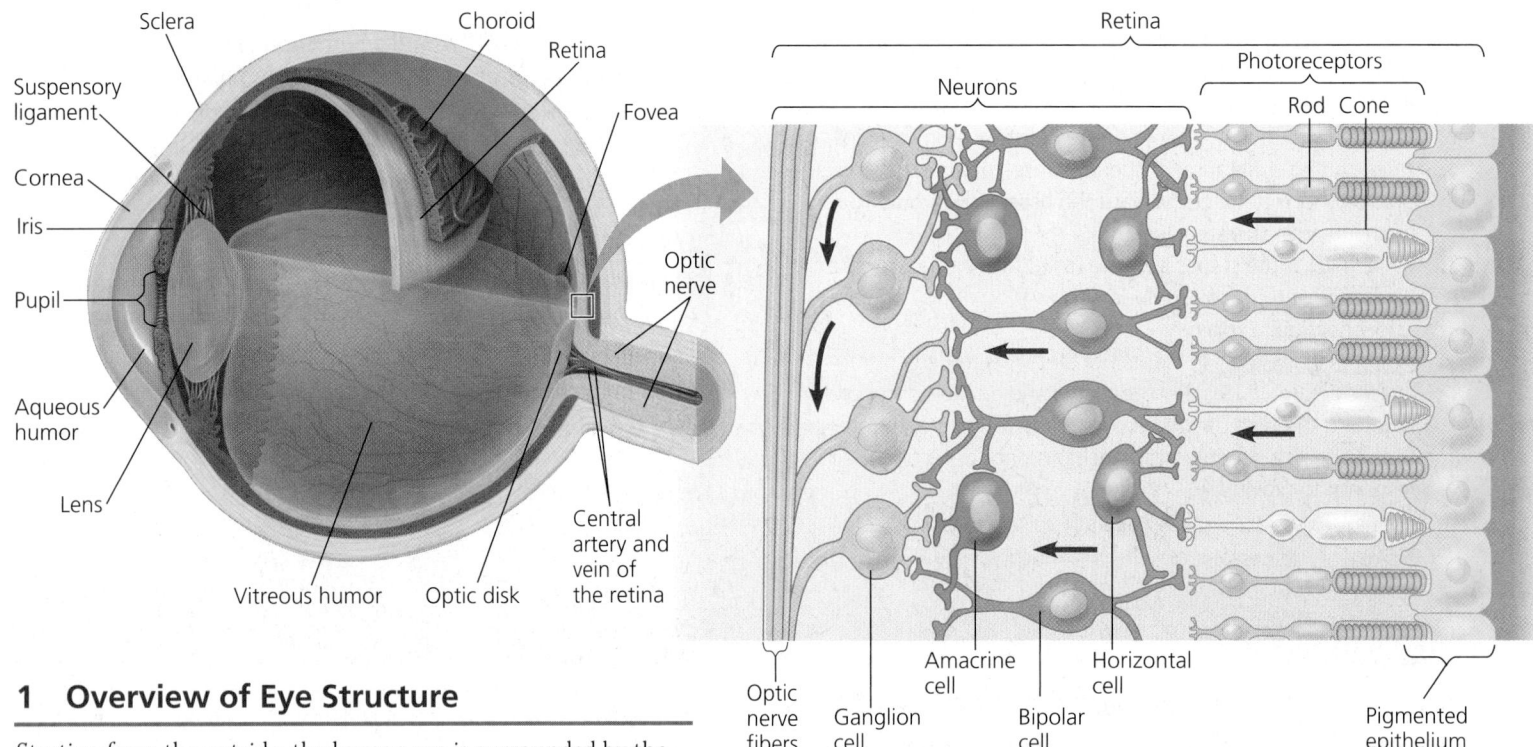

### 1   Overview of Eye Structure

Starting from the outside, the human eye is surrounded by the conjunctiva, a mucous membrane (not shown); the sclera, a connective tissue; and the choroid, a thin, pigmented layer. At the front, the sclera forms the transparent *cornea* and the choroid forms the colored *iris*. By changing size, the iris regulates the amount of light entering the pupil, the hole in the center of the iris. Just inside the choroid, the neurons and photoreceptors of the **retina** form the innermost layer of the eyeball. The optic nerve exits the eye at the optic disk.

The **lens**, a transparent disk of protein, divides the eye into two cavities. In front of the lens lies the *aqueous humor*, a clear watery substance. Blockage of ducts that drain this fluid can produce glaucoma, a condition in which increased pressure in the eye damages the optic nerve, causing vision loss. Behind the lens lies the jellylike *vitreous humor* (illustrated here in the lower portion of the eyeball).

### 2   The Retina

Light (coming from left in the above view) strikes the retina, passing through largely transparent layers of neurons before reaching the rods and cones, two types of photoreceptors that differ in shape and in function. The neurons of the retina then relay visual information captured by the photo receptors to the optic nerve and brain along the pathways shown with red arrows. Each *bipolar cell* receives information from several rods or cones, and each *ganglion cell* gathers input from several bipolar cells. *Horizontal* and *amacrine cells* integrate information across the retina.

The optic disk, where the optic nerve exits the retina, lacks photoreceptors. As a result, this region forms a "blind spot" where light is not detected.

opening, the **pupil**, through which light enters. Like a camera's adjustable aperture, the **iris** expands or contracts, changing the diameter of the pupil to let in more or less light. Behind the pupil, a single lens directs light on a layer of photoreceptors. Similar to a camera's focusing action, muscles in an invertebrate's single-lens eye move the lens forward or backward, focusing on objects at different distances.

The eyes of all vertebrates have a single lens. In fishes, focusing occurs as in invertebrates, with the lens moving forward or backward. In other species, including mammals, focusing is achieved by changing the shape of the lens.

## The Vertebrate Visual System

The human eye will serve as our model of vision in vertebrates. As described in **Figure 38.26**, vision begins when photons of light enter the eye and strike the rods and cones. There the energy of each photon is captured by a shift in configuration of a single chemical bond in retinal.

Although light detection in the eye is the first stage in vision, remember that it is actually the brain that "sees." Thus, to understand vision, we must examine how the capture of light by retinal changes the production of action potentials and then follow these signals to the visual centers of the brain, where images are perceived.

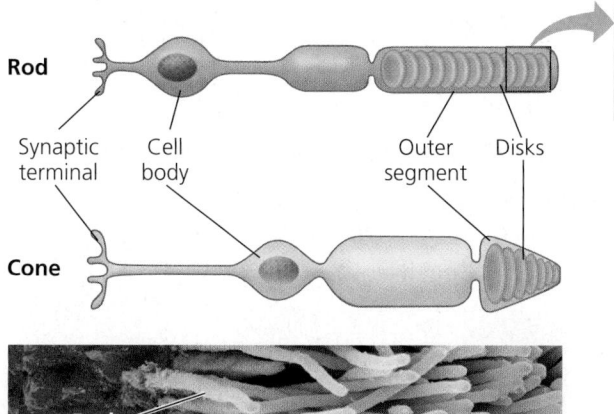

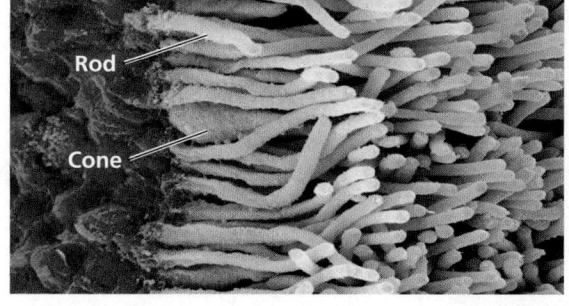

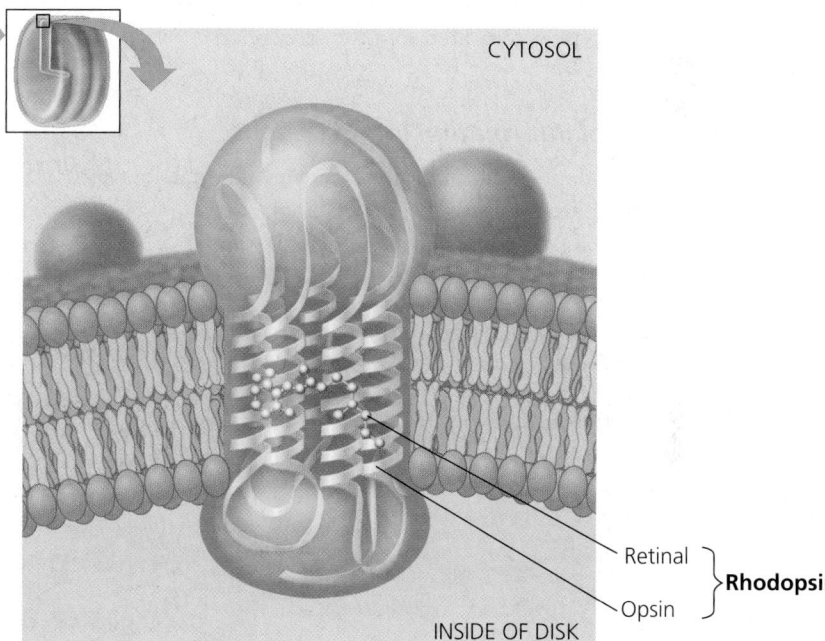

## 3  Photoreceptor Cells

Humans have two main types of photoreceptor cells: rods and cones. Within the outer segment of a rod or cone is a stack of membranous disks in which *visual pigments* are embedded. **Rods** are more sensitive to light than cones are, but they do not distinguish colors; rods enable us to see at night, but only in black and white. **Cones** provide color vision, but, being less sensitive, contribute very little to night vision. There are three types of cones. Each has a different sensitivity across the visible spectrum, providing an optimal response to red, green, or blue light.

In the colorized SEM shown above, cones (green), rods (light tan), and adjacent neurons (red) are visible. The pigmented epithelium, which was removed in this preparation, would be to the right.

## 4  Visual Pigments

Vertebrate visual pigments consist of a light-absorbing molecule called **retinal** (a derivative of vitamin A) bound to a membrane protein called an **opsin**. Seven α helices of each opsin molecule span the disk membrane. The visual pigment of rods, shown here, is called **rhodopsin**.

Retinal exists as two isomers. Absorption of light shifts one bond in retinal from a *cis* to a *trans* arrangement, converting the molecule from an angled shape to a straight shape. This change in configuration activates the opsin protein to which retinal is bound.

## Sensory Transduction in the Eye

The transduction of visual information to the nervous system begins with the light-induced conversion of *cis*-retinal to *trans*-retinal. As shown in **Figure 38.27**, this conversion activates rhodopsin, which activates a G protein, which in turn activates an enzyme called phosphodiesterase. The substrate for this enzyme is cyclic GMP, which in the dark binds to sodium ion ($Na^+$) channels and keeps them open. When phosphodiesterase hydrolyses cyclic GMP, $Na^+$ channels close, and the cell becomes hyperpolarized. The signal transduction pathway in photoreceptor cells then shuts off as enzymes convert retinal back to the *cis* form, returning rhodopsin to its inactive state.

In very bright light, rhodopsin remains active and the response in the rods becomes saturated. If the amount of light entering the eyes decreases abruptly, the rods do not regain full responsiveness for several minutes. This is why you are temporarily blinded if you pass quickly from bright sunshine into a movie theater or other dark environment. (Because light activation changes the color of rhodopsin from purple to yellow, rods in which the light response is saturated are often described as "bleached.")

### Processing of Visual Information in the Retina

The processing of visual information begins in the retina itself, where both rods and cones form synapses with bipolar cells (see Figure 38.25). In the dark, rods and cones are depolarized and continually release the neurotransmitter glutamate at these synapses. Some bipolar cells depolarize in response to glutamate, whereas others hyperpolarize. When light strikes the eye, the rods and cones hyperpolarize, shutting off their release of glutamate. In response, the bipolar cells that are depolarized by glutamate hyperpolarize, and those that are hyperpolarized by glutamate depolarize.

Signals from rods and cones can follow several different pathways in the retina. Some information passes directly from photoreceptors to bipolar cells to ganglion cells. In other cases, horizontal cells carry signals from one rod or cone to other photoreceptors and to several bipolar cells. Amacrine cells also contribute to signal processing, distributing information from one bipolar cell to several ganglion cells.

A single ganglion cell receives information from an array of rods and cones, each of which responds to light coming from a particular location. Together, the rods and cones that are feeding information to one ganglion cell define a *receptive field*—the part of the visual field to which that ganglion cell can respond. The fewer rods or cones that supply a single ganglion cell, the smaller the receptive field. A smaller receptive field typically results in a sharper image because the information about where light has struck the retina is more precise.

### Processing of Visual Information in the Brain

Axons of ganglion cells form the optic nerves that transmit sensations from the eyes to the brain. The two optic nerves meet at the *optic chiasm* near the center of the base of the

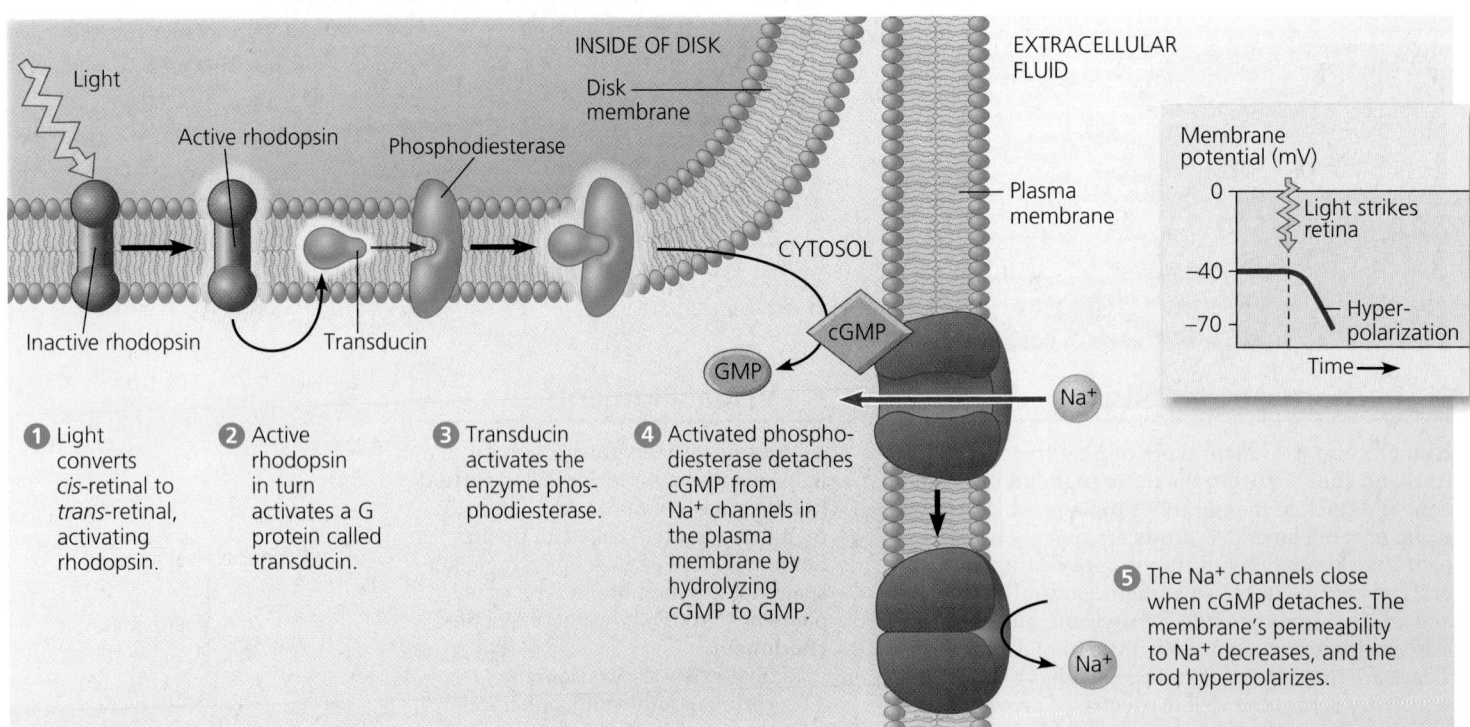

▲ **Figure 38.27 Production of the receptor potential in a rod cell.** In rods (and cones), the receptor potential triggered by light is a hyperpolarization, not a depolarization.

**?** *Which steps in the cascade of events shown here provide an opportunity for amplification, the strengthening of the sensory signal?*

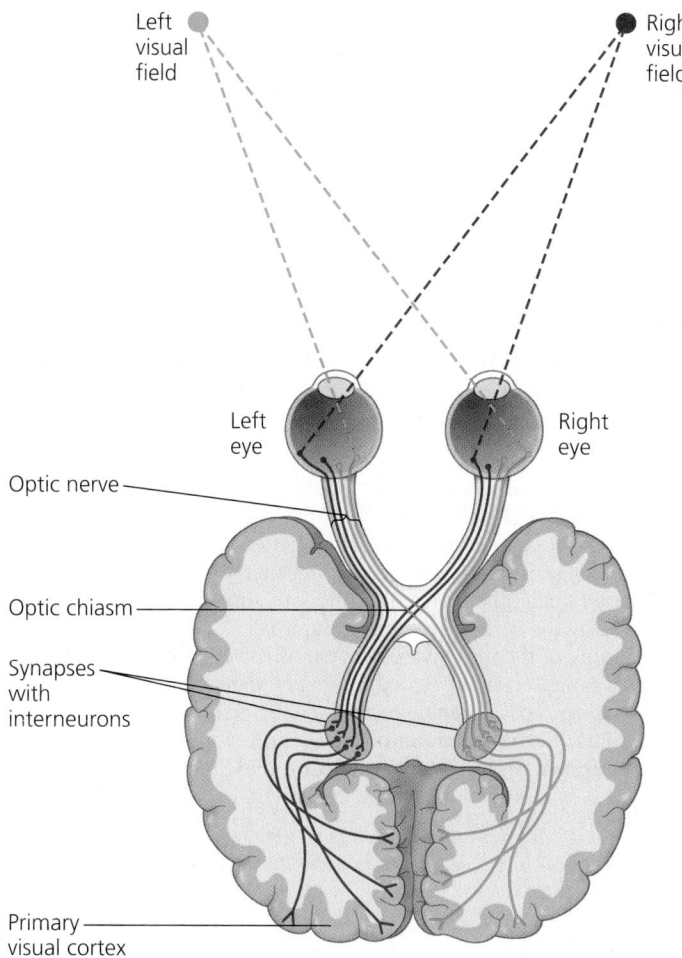

Left visual field

Right visual field

Left eye

Right eye

Optic nerve

Optic chiasm

Synapses with interneurons

Primary visual cortex

▲ **Figure 38.28 Neural pathways for vision.** Each optic nerve contains about a million axons that form synapses with interneurons in the brain. These interneurons relay sensations to the primary visual cortex, one of many brain centers that cooperate in constructing visual perceptions.

cerebral cortex **(Figure 38.28)**. Axons in the optic nerves are routed at the optic chiasm such that sensations from the left visual field are transmitted to the right side of the brain, and sensations from the right visual field are transmitted to the left side of the brain. (Note that each visual field, whether right or left, involves input from both eyes.)

Researchers estimate that at least 30% of the cerebral cortex, comprising hundreds of millions of neurons in perhaps dozens of integrating centers, takes part in formulating what we actually "see." Determining how these centers integrate such components of our vision as color, motion, depth, shape, and detail is the focus of much exciting research.

## Color Vision

Among vertebrates, most fishes, amphibians, and reptiles, including birds, have very good color vision. Humans and other primates also see color well, but are among the minority of mammals with this ability.

In humans, the perception of color is based on three types of cones, each with a different visual pigment—red, green, or blue. The three visual pigments, called *photopsins*, are formed from the binding of retinal to three distinct opsin proteins. Slight differences in the opsin proteins cause each photopsin to absorb light optimally at a different wavelength.

Abnormal color vision typically results from mutations in the genes for one or more photopsin proteins. Because the human genes for the red and green pigments are located on the X chromosome, a mutation in one copy of either gene can disrupt color vision in males. For this reason, color blindness is more common in males than in females (5–8% of males but fewer than 1% of females) and nearly always affects perception of red or green. (The human gene for the blue pigment is on chromosome 7.)

Unlike primates, many mammals are nocturnal. For such animals, a high proportion of rods in the retina is an evolutionary adaptation that provides keen night vision. Cats, for instance, have limited color vision; when they are awake during the day, they probably see a pastel world.

### The Visual Field

The brain not only processes visual information but also controls what information is captured. One important control is focusing, which, as noted earlier, in humans occurs by changing the shape of the lens. When you focus your eyes on a close object, your lenses become almost spherical. When you view a distant object, your lenses are flattened.

Although our peripheral vision allows us to see objects over a nearly 180° range, the distribution of photoreceptors across the eye limits both what we see and how well we see it. Overall, the human retina contains about 125 million rods and 6 million cones. At the **fovea**, the center of the visual field, there are no rods but a very high density of cones—about 150,000 cones per square millimeter. The ratio of cones to rods decreases with distance from the fovea, with the peripheral regions having only rods. In daylight, you achieve your sharpest vision by looking directly at an object, such that light shines on the tightly packed cones in your fovea. At night, looking directly at a dimly lit object is ineffective, since the rods—the more sensitive light receptors—are absent from the fovea. For this reason, you see a dim star best by focusing on a point just to one side of it.

### CONCEPT CHECK 38.6

1. Contrast the light-detecting organs of planarians and flies. How is each organ adaptive for the lifestyle of the animal?
2. **WHAT IF?** Our brain receives more action potentials when our eyes are exposed to light even though our photoreceptors release more neurotransmitter in the dark. Propose an explanation.
3. **MAKE CONNECTIONS** Compare the function of retinal in the eye with that of the pigment chlorophyll in a plant photosystem (see Concept 8.2).

For suggested answers, see Appendix A.

# 38 Chapter Review

Go to **MasteringBiology**® for Assignments, the eText, and the Study Area with Animations, Activities, Vocab Self-Quiz, and Practice Tests.

## SUMMARY OF KEY CONCEPTS

**VOCAB SELF-QUIZ**

goo.gl/gbai8v

### CONCEPT 38.1

**Nervous systems consist of circuits of neurons and supporting cells (pp. 790–793)**

- Nervous systems range in complexity from simple **nerve nets** to highly centralized nervous systems. In vertebrates, the **central nervous system (CNS)**, consisting of the brain and the spinal cord, integrates information, while the **nerves** of the **peripheral nervous system (PNS)** transmit sensory and motor signals between the CNS and the rest of the body. The simplest circuits in the vertebrate nervous system control **reflex** responses, in which sensory input is linked to motor output without involvement of the brain. Vertebrate neurons are supported by **glia**, which nourish, support, and regulate neuron function.

- Afferent neurons carry sensory signals to the CNS. Efferent neurons function in either the **motor system**, which carries signals to skeletal muscles, or the **autonomic nervous system**, which regulates smooth and cardiac muscles. The **sympathetic division** and **parasympathetic division** of the autonomic nervous system have antagonistic effects on a diverse set of target organs, while the **enteric division** controls the activity of many digestive organs.

? *In what ways do different types of glia support the growth and survival of neurons?*

### CONCEPT 38.2

**The vertebrate brain is regionally specialized (pp. 793–798)**

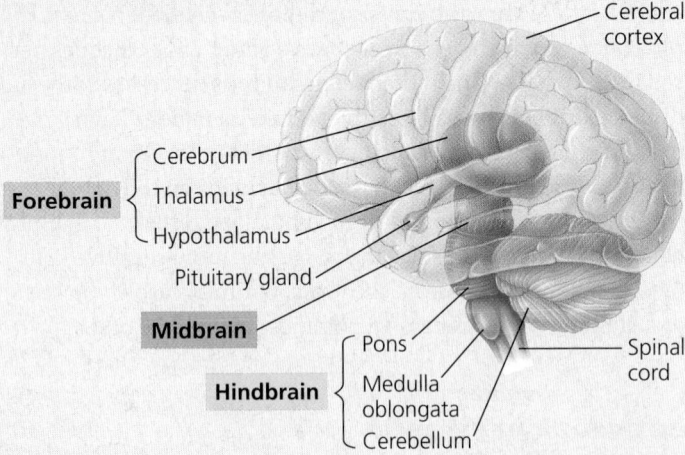

Forebrain { Cerebrum, Thalamus, Hypothalamus }
Pituitary gland
Cerebral cortex
**Midbrain**
**Hindbrain** { Pons, Medulla oblongata, Cerebellum }
Spinal cord

- The cerebrum has two hemispheres, each of which consists of cortical **gray matter** overlying **white matter** and basal nuclei. The basal nuclei are important in planning and learning movements. The **pons** and **medulla oblongata** are relay stations for information traveling between the PNS and the cerebrum. The pons, together with the **brainstem** and parts of the forebrain, also regulate sleep and arousal. The **cerebellum** helps coordinate motor, perceptual, and cognitive functions. The **thalamus** is the main center through which sensory information passes to the **cerebrum**. The **hypothalamus** regulates homeostasis and basic survival behaviors. Within the hypothalamus, the **suprachiasmatic nucleus (SCN)** acts as the pacemaker for

circadian rhythms. The amygdala plays a key role in recognizing and recalling a number of emotions.

? *What roles do the midbrain, cerebellum, thalamus, and cerebrum play in vision and responses to visual input?*

### CONCEPT 38.3

**The cerebral cortex controls voluntary movement and cognitive functions (pp. 798–801)**

- Each side of the **cerebral cortex** has four lobes that contain primary sensory areas and association areas. Specific types of sensory input enter the primary sensory areas. Association areas integrate information from different sensory areas. Broca's area and Wernicke's area are essential for generating and understanding language. In birds, a brain region called the pallium contains clustered nuclei that carry out functions similar to those performed by the cerebral cortex of mammals.

- Reshaping of the nervous system can involve the loss or addition of synapses or the strengthening or weakening of signaling at synapses. This capacity for remodeling is termed **neuronal plasticity**. **Short-term memory** relies on temporary links in the hippocampus. In **long-term memory**, these temporary links are replaced by connections within the cerebral cortex.

? *After an accident, a patient has trouble with language and has paralysis on one side of the body. Which side would you expect to be paralyzed? Why?*

### CONCEPT 38.4

**Sensory receptors transduce stimulus energy and transmit signals to the central nervous system (pp. 801–804)**

- The detection of a stimulus precedes **sensory transduction**, the change in the membrane potential of a **sensory receptor** in response to a stimulus. The resulting **receptor potential** controls transmission of action potentials to the CNS, where sensory information is integrated to generate **perceptions**. The frequency of action potentials in an axon and the number of axons activated determine stimulus strength. The identity of the axon carrying the signal encodes the nature or quality of the stimulus.

- **Mechanoreceptors** respond to stimuli such as pressure, touch, stretch, motion, and sound. **Electromagnetic receptors** detect different forms of electromagnetic radiation. **Thermoreceptors** signal surface and core temperatures of the body. Pain is detected by a group of **nociceptors** that respond to excess heat, pressure, or specific classes of chemicals. **Chemoreceptors** detect either total solute concentration or specific molecules, as in smell (**olfaction**) and taste (**gustation**). Humans can distinguish five taste perceptions and make use of 400 different receptors to distinguish among odorants.

? *To simplify sensory receptor classification, why might it make sense to eliminate nociceptors as a distinct class?*

### CONCEPT 38.5

**In hearing and equilibrium, mechanoreceptors detect moving fluid or settling particles (pp. 804–807)**

- Most invertebrates sense their orientation with respect to gravity by means of **statocysts**. Specialized **hair cells** form the basis for hearing and balance in mammals. In mammals, the **tympanic**

**membrane** (eardrum) transmits sound waves to bones of the **middle ear**, which transmit the waves through the **oval window** to the fluid in the coiled **cochlea** of the **inner ear**. Pressure waves in the fluid vibrate the basilar membrane, depolarizing hair cells and triggering action potentials that travel via the auditory nerve to the brain. Receptors in the inner ear function in balance and equilibrium.

**?** *How are music volume and pitch encoded in signals sent to the brain?*

**CONCEPT 38.6**

### The diverse visual receptors of animals depend on light-absorbing pigments (pp. 807–811)

- Invertebrates have varied light detectors, including simple light-sensitive eyespots, image-forming **compound eyes**, and **single-lens eyes**. In the vertebrate eye, a single **lens** is used to focus light on **photoreceptors** in the **retina**. Both **rods** and **cones** contain a pigment, **retinal**, bonded to a protein (**opsin**). Absorption of light by retinal triggers a signal transduction pathway that hyperpolarizes the photoreceptors, causing them to release less neurotransmitter. Synapses transmit information from photoreceptors to cells that integrate information and convey it to the brain along axons that form the optic nerve.

**?** *How does the processing of sensory information sent to the vertebrate brain in vision differ from that in hearing or olfaction?*

## TEST YOUR UNDERSTANDING

**PRACTICE TEST**

goo.gl/CRZjvS

### Level 1: Knowledge/Comprehension

1. Patients with damage to Wernicke's area have difficulty
   (A) coordinating limb movement.
   (B) generating speech.
   (C) recognizing faces.
   (D) understanding language.

2. The cerebral cortex does *not* play a major role in
   (A) short-term memory.
   (B) long-term memory.
   (C) circadian rhythm.
   (D) breath holding.

3. The middle ear converts
   (A) air pressure waves to fluid pressure waves.
   (B) air pressure waves to nerve impulses.
   (C) fluid pressure waves to nerve impulses.
   (D) pressure waves to hair cell movements.

4. If the following events are arranged in the order in which they occur for an animal hiding and holding still in response to seeing a predator, which is the fourth event in the series?
   (A) signaling by an afferent PNS neuron
   (B) signaling by an efferent PNS neuron
   (C) information processing in the CNS
   (D) activation of a sensory receptor

### Level 2: Application/Analysis

5. Injury to just the hypothalamus would most likely disrupt
   (A) regulation of body temperature.
   (B) short-term memory.
   (C) executive functions, such as decision making.
   (D) sorting of sensory information.

6. Which sensory distinction is NOT encoded by a difference in which axon transfers the information to the brain?
   (A) white and red
   (B) red and green
   (C) loud and faint
   (D) salty and sweet

### Level 3: Synthesis/Evaluation

7. Although some sharks close their eyes just before they bite, their bites are on target. Researchers have noted that sharks often misdirect their bites at metal objects and that they can find batteries buried under sand. This evidence suggests that sharks keep track of their prey during the split second before they bite in the same way that
   (A) a rattlesnake finds a mouse in its burrow.
   (B) an insect avoids being stepped on.
   (C) a crayfish senses the directions up and down.
   (D) a platypus locates its prey in a muddy river.

8. **SCIENTIFIC INQUIRY**
   Consider an individual who had been fluent in American Sign Language before suffering an injury to his left cerebral hemisphere. After the injury, he could still understand signs but could not readily generate signs that represented his thoughts. Propose two hypotheses that could explain this finding. How might you distinguish between them?

9. **FOCUS ON EVOLUTION**
   Scientists often use measures of "higher-order thinking" to assess intelligence in other animals. For example, birds are judged to have sophisticated thought processes because they can use tools and make use of abstract concepts. Identify problems you see in defining intelligence in these ways.

10. **FOCUS ON ORGANIZATION**
    In a short essay (100–150 words), describe three ways in which the structure of the lens of the human eye is well adapted to its function in vision.

11. **SYNTHESIZE YOUR KNOWLEDGE**

Bloodhounds, which are adept at following a scent trail even days old, have 1,000 times as many olfactory receptor cells as humans have. How might this difference contribute to the tracking ability of these dogs? What differences in brain organization would you expect in comparing a bloodhound and a human?

*For selected answers, see Appendix A.*

# CHAPTER

# 39 Motor Mechanisms and Behavior

## KEY CONCEPTS

**39.1** The physical interaction of protein filaments is required for muscle function

**39.2** Skeletal systems transform muscle contraction into locomotion

**39.3** Discrete sensory inputs can stimulate both simple and complex behaviors

**39.4** Learning establishes specific links between experience and behavior

**39.5** Selection for individual survival and reproductive success can explain diverse behaviors

**39.6** Genetic analyses and the concept of inclusive fitness provide a basis for studying the evolution of behavior

▲ **Figure 39.1** What prompts a male fiddler crab to display his giant claw?

## The How and Why of Animal Activity

Unlike most animals, male fiddler crabs (genus *Uca*) are highly asymmetric. One claw grows to giant proportions, half the mass of the crab's entire body **(Figure 39.1)**. The name *fiddler* reflects the asymmetry as well as the crab's behavior as it feeds on algae from the mudflats where it lives: The smaller front claw moves to and from the mouth in front of the enlarged claw. At times, however, the male waves his large claw in the air. What triggers this behavior? What purpose does it serve?

Claw-waving behavior by a male fiddler crab has two functions. Waving the claw, which can be used as a weapon, helps the crab *repel* other males wandering too close to his burrow. Vigorous claw waving also helps him *attract* females who wander through the crab colony in search of a mate. After the male fiddler crab lures a female to his burrow, he seals her in with mud or sand in preparation for mating.

Animal behaviors, whether solitary or social, fixed or variable, are based on physiological systems and processes. An individual **behavior** is an action carried out by muscles under control of the nervous system. Examples include an animal using its throat muscles to produce a song, releasing a scent to mark its territory, or simply waving a claw. Behavior is an essential part of acquiring nutrients and finding a partner for sexual reproduction. Behavior also contributes to homeostasis, as when honeybees huddle to conserve heat. In short, all of physiology contributes to behavior, and behavior influences all of physiology.

Many behaviors, especially those involved in recognition and communication, rely on specialized body structures or form. For instance, the enormous claw of a male fiddler crab enables recognition by rival males and by potential mates. Similarly, having eyes on stalks high above his head enables the male crab to identify other crabs from far away. As these examples illustrate, the process of natural selection that shapes behaviors also influences the evolution of animal anatomy.

In this chapter, we'll begin by considering the structure and function of muscles and skeletons, as well as mechanisms of animal movement. These topics will lead us naturally to the questions of how behavior is controlled, how it develops during an animal's life, and how it is influenced by genes and the environment. Finally, we'll investigate the ways in which behavior evolves over many generations. Moving our study from an animal's inner workings to its interactions with the outside world will set the stage for exploring ecology (the focus of Unit Seven).

CONCEPT 39.1

## The physical interaction of protein filaments is required for muscle function

The whisker-guided maneuvers of a mouse, the upside-down swimming of a crayfish with manipulated statocysts, and the light-avoiding maneuvers of planarians are examples of specific behaviors triggered by sensory inputs to the nervous system (see Chapter 38). Underlying these diverse behaviors are common fundamental mechanisms—walking, swimming, and crawling all require muscle activity in response to nervous system input.

Muscle cell contraction relies on the interaction between protein structures called thin and thick filaments. The major component of **thin filaments** is the globular protein actin. In thin filaments, two strands of polymerized actin are coiled around one another; similar actin structures called microfilaments function in cell motility (see Concept 4.6). The **thick filaments** are staggered arrays of myosin molecules. Muscle contraction is the result of filament movement powered by chemical energy; muscle extension occurs only passively. To understand how filaments bring about muscle contraction, we will begin by examining the skeletal muscle of vertebrates.

### Vertebrate Skeletal Muscle

Vertebrate **skeletal muscle**, which moves individual bones and the whole body, has a hierarchy of smaller and smaller units **(Figure 39.2)**. Within a typical skeletal muscle is a bundle of long muscle fibers running along the length of the muscle. Each individual fiber is a single cell. Within are multiple nuclei, each derived from one of the embryonic cells that fused to form the fiber. Surrounding these nuclei are many longitudinal **myofibrils**, which consist of bundles of thin and thick filaments.

The myofibrils in muscle fibers are made up of repeating sections called **sarcomeres**, which are the basic contractile units of skeletal muscle. The borders of the sarcomeres line up in adjacent myofibrils, forming a pattern of light and dark bands (striations) visible with a microscope. For this reason,

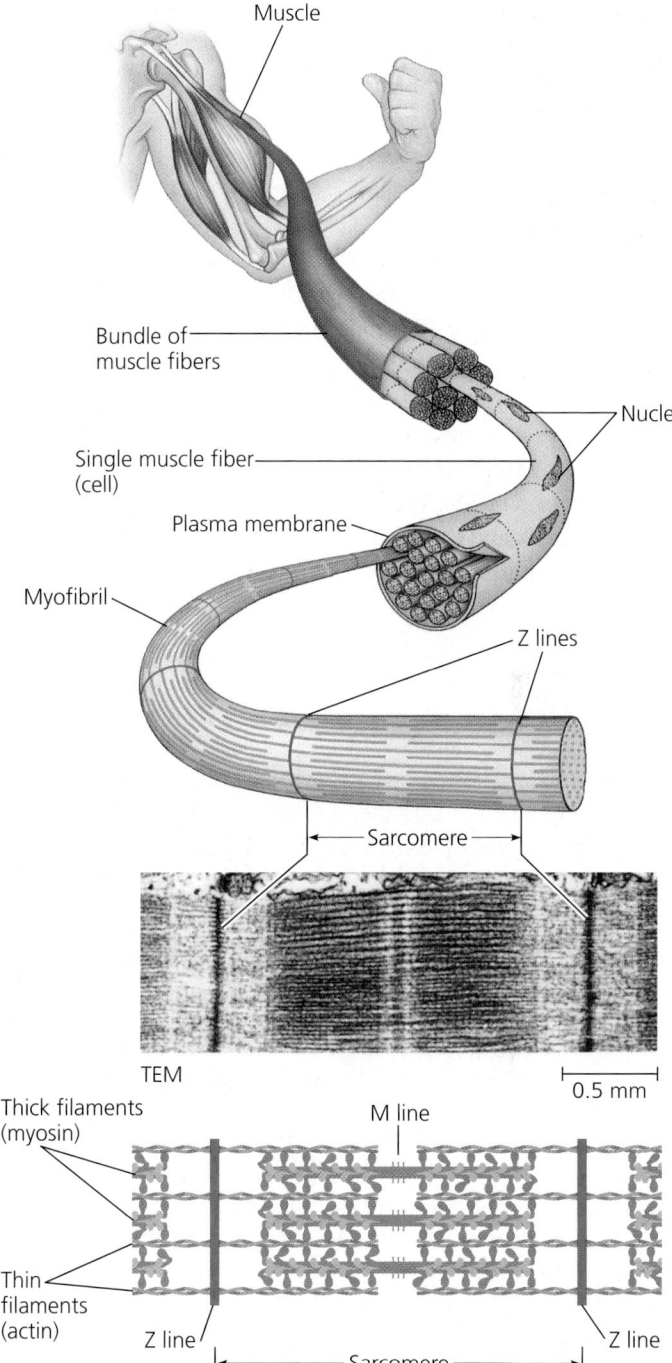

▲ Figure 39.2 **The structure of skeletal muscle.**

skeletal muscle is also sometimes called *striated muscle*. As shown in Figure 39.2, thin filaments attach at the Z lines at the sarcomere ends, while thick filaments are anchored in the middle of the sarcomere (M line).

In a resting (relaxed) myofibril, thick and thin filaments partially overlap. Near the edge of the sarcomere there are only thin filaments, whereas the zone in the center contains only thick filaments. This arrangement is the key to how the sarcomere, and hence the whole muscle, contracts.

### *The Sliding-Filament Model of Muscle Contraction*

A contracting muscle shortens, but the filaments that bring about contraction stay the same length. To explain this apparent paradox, we'll focus first on a single sarcomere. As shown in **Figure 39.3**, the filaments slide past each other, much like the segments of a telescoping support pole. According to the well-accepted **sliding-filament model**, the thin and thick filaments ratchet past each other, powered by myosin molecules.

**Figure 39.4** illustrates the cycles of change in myosin that convert the chemical energy of ATP into the longitudinal sliding of thick and thin filaments in a contracting muscle. Each myosin molecule has a long "tail" region and a globular "head" region. The tail adheres to the tails

Relaxed muscle
Z
M
Z
Sarcomere
Z
M
Z
Contracting muscle
Fully contracted muscle
0.5 μm
Contracted sarcomere


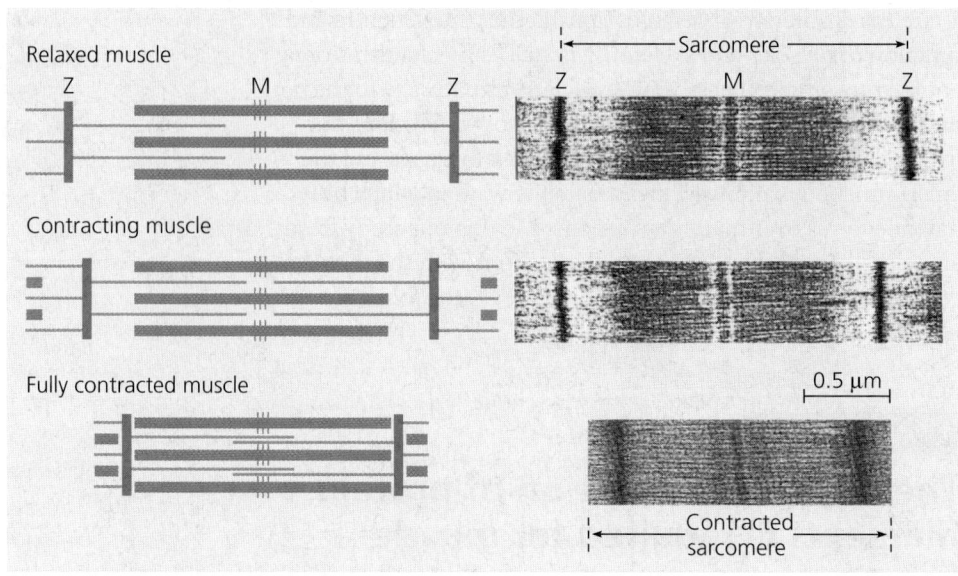

▲ **Figure 39.3 The sliding-filament model of muscle contraction.** The drawings on the left show that the lengths of the thick (myosin) filaments (purple) and thin (actin) filaments (orange) remain the same as a muscle fiber contracts.

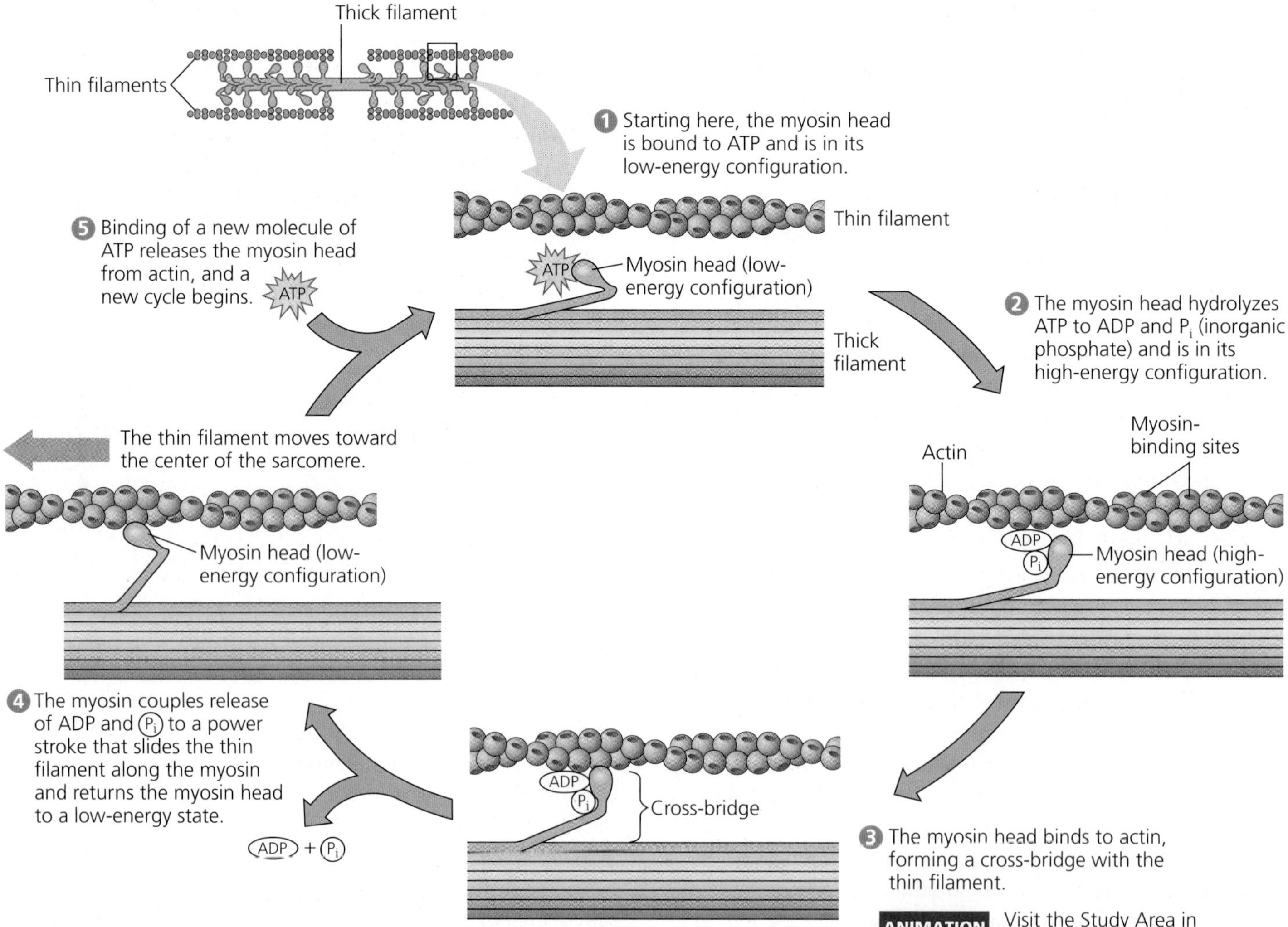

Thick filament
Thin filaments
1 Starting here, the myosin head is bound to ATP and is in its low-energy configuration.
Thin filament
ATP
Myosin head (low-energy configuration)
Thick filament
2 The myosin head hydrolyzes ATP to ADP and $P_i$ (inorganic phosphate) and is in its high-energy configuration.
Actin
Myosin-binding sites
ADP
$P_i$
Myosin head (high-energy configuration)
3 The myosin head binds to actin, forming a cross-bridge with the thin filament.
ADP
$P_i$
Cross-bridge
4 The myosin couples release of ADP and $P_i$ to a power stroke that slides the thin filament along the myosin and returns the myosin head to a low-energy state.
ADP + $P_i$
The thin filament moves toward the center of the sarcomere.
Myosin head (low-energy configuration)
5 Binding of a new molecule of ATP releases the myosin head from actin, and a new cycle begins.
ATP


▲ **Figure 39.4 Myosin-actin interactions underlying muscle fiber contraction.**

**?** *When ATP binds, what prevents the filaments from sliding back into their original positions?*

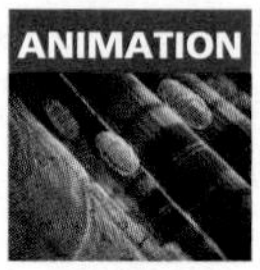

ANIMATION


Visit the Study Area in **MasteringBiology** for the BioFlix® 3-D Animation on Muscle Contraction.

of other myosin molecules, forming the thick filament. The head, jutting to the side, can bind ATP. Hydrolysis of bound ATP converts myosin to a high-energy form that binds to actin, forming a cross-bridge between the myosin and the thin filament. The myosin head then returns to its low-energy form as it helps to pull the thin filament toward the center of the sarcomere. When a new ATP molecule binds to the myosin head, the cross-bridge is broken, thereby releasing the myosin head from the thin filament.

Muscle contraction requires repeated cycles of binding and release. During each cycle of each myosin head, the head is freed from a cross-bridge, cleaves the newly bound ATP, and binds again to actin. Because the thin filament moved toward the center of the sarcomere in the previous cycle, the myosin head now attaches to a new binding site farther along the thin filament. A thick filament contains approximately 350 myosin heads, each of which forms and re-forms about five cross-bridges per second, driving the thick and thin filaments past each other.

At rest, most muscle fibers contain only enough ATP for a few contractions. Powering repetitive contractions requires two other storage compounds: creatine phosphate and glycogen. Transfer of a phosphate group from creatine phosphate to ADP in an enzyme-catalyzed reaction synthesizes additional ATP. In this way, the resting supply of creatine phosphate can sustain contractions for about 15 seconds. ATP stores are also replenished when glycogen is broken down to glucose. During light or moderate muscle activity, this glucose is metabolized by aerobic respiration. This highly efficient metabolic process yields enough power to sustain contractions for nearly an hour.

### The Role of Calcium and Regulatory Proteins

Proteins bound to actin play crucial roles in controlling muscle contraction. In a muscle fiber at rest, **tropomyosin**, a regulatory protein, and the **troponin complex**, a set of additional regulatory proteins, are bound to the actin strands of thin filaments. Tropomyosin covers the myosin-binding sites along the thin filament, preventing actin and myosin from interacting **(Figure 39.5a)**.

Motor neurons enable actin and myosin to interact by triggering a release of calcium ions ($Ca^{2+}$) into the cytosol. There, $Ca^{2+}$ binds to the troponin complex, causing the myosin-binding sites on actin to be exposed **(Figure 39.5b)**. Note that the effect of $Ca^{2+}$ is indirect: Binding to $Ca^{2+}$ causes the troponin complex to change shape, dislodging tropomyosin from the myosin-binding sites.

When the $Ca^{2+}$ concentration rises in the cytosol, the thin and thick filaments slide past each other, and the muscle fiber contracts. When the $Ca^{2+}$ concentration falls, the binding sites are covered, and contraction stops.

Motor neurons cause muscle contraction by triggering the release of $Ca^{2+}$ into the cytosol of muscle cells with which they form synapses **(Figure 39.6)**. The process that regulates this release of $Ca^{2+}$ begins with the arrival of an action potential at the synaptic terminal of a motor neuron. The synaptic terminal

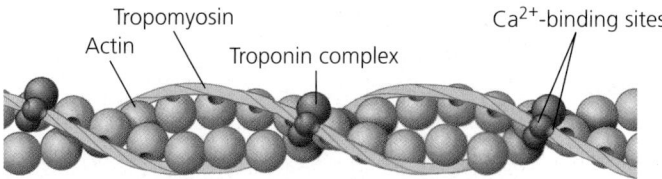

**(a) Myosin-binding sites blocked by tropomyosin**

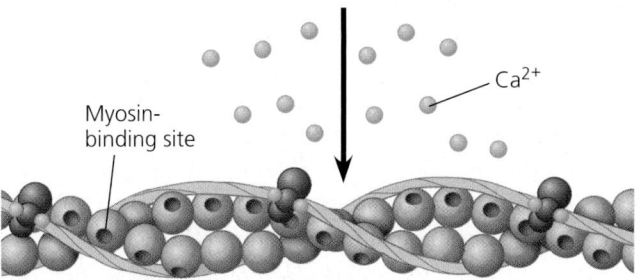

**(b) Myosin-binding sites exposed**

▲ **Figure 39.5 The role of regulatory proteins and calcium in muscle fiber contraction.** Each thin filament consists of two strands of actin and associated regulatory proteins: two long molecules of tropomyosin and multiple copies of the troponin complex.

releases the neurotransmitter acetylcholine, which binds to receptors on the muscle fiber, leading to a depolarization that initiates an action potential. Within the muscle fiber, the action potential spreads deep into the interior, following infoldings of the plasma membrane called **transverse (T) tubules**. These transverse tubules make close contact with the **sarcoplasmic reticulum (SR)**, a specialized endoplasmic reticulum. As the action potential spreads along the T tubules, it triggers changes in the SR, opening $Ca^{2+}$ channels. Calcium ions stored in the

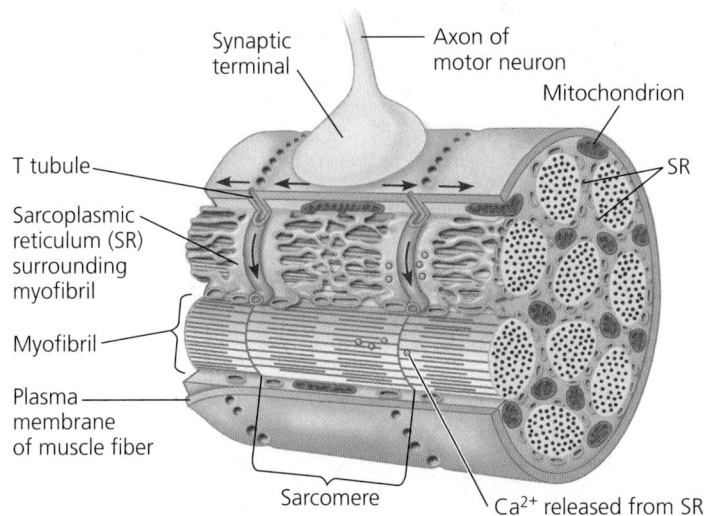

▲ **Figure 39.6 The roles of the sarcoplasmic reticulum and T tubules in muscle fiber contraction.** The synaptic terminal of a motor neuron releases acetylcholine, which depolarizes the plasma membrane of the muscle fiber. The depolarization causes action potentials (red arrows) to sweep across the muscle fiber and deep into it along the transverse (T) tubules. The action potentials trigger the release of calcium (green dots) from the sarcoplasmic reticulum into the cytosol. Calcium initiates the sliding of filaments by allowing myosin to bind to actin.

interior of the SR flow through open channels into the cytosol and bind to the troponin complex, initiating muscle fiber contraction.

When motor neuron input stops, the filaments slide back to their starting position as the muscle relaxes. Relaxation begins as proteins in the SR pump $Ca^{2+}$ back in from the cytosol. When the $Ca^{2+}$ concentration in the cytosol drops to a low level, the regulatory proteins bound to the thin filament shift back to their starting position, once again blocking the myosin-binding sites. At the same time, the $Ca^{2+}$ pumped from the cytosol accumulates in the SR, providing the stores needed to respond to the next action potential. These events involved in skeletal muscle contraction are summarized in **Figure 39.7**.

Several diseases cause paralysis by interfering with the excitation of skeletal muscle fibers by motor neurons. In amyotrophic lateral sclerosis (ALS), motor neurons in the spinal cord and brainstem degenerate, and muscle fibers atrophy. ALS is progressive and usually fatal. In myasthenia gravis, a person produces antibodies to the acetylcholine receptors of skeletal muscle. As the disease progresses and the number of receptors decreases, transmission between motor neurons and muscle fibers declines. Myasthenia gravis can generally be controlled with drugs that inhibit acetylcholinesterase or suppress the immune system.

### Nervous Control of Muscle Tension

Whereas contraction of a single skeletal muscle fiber is a brief all-or-none twitch, contraction of a whole muscle, such as the biceps in your upper arm, is graded; you can voluntarily alter the extent and strength of its contraction. The nervous system

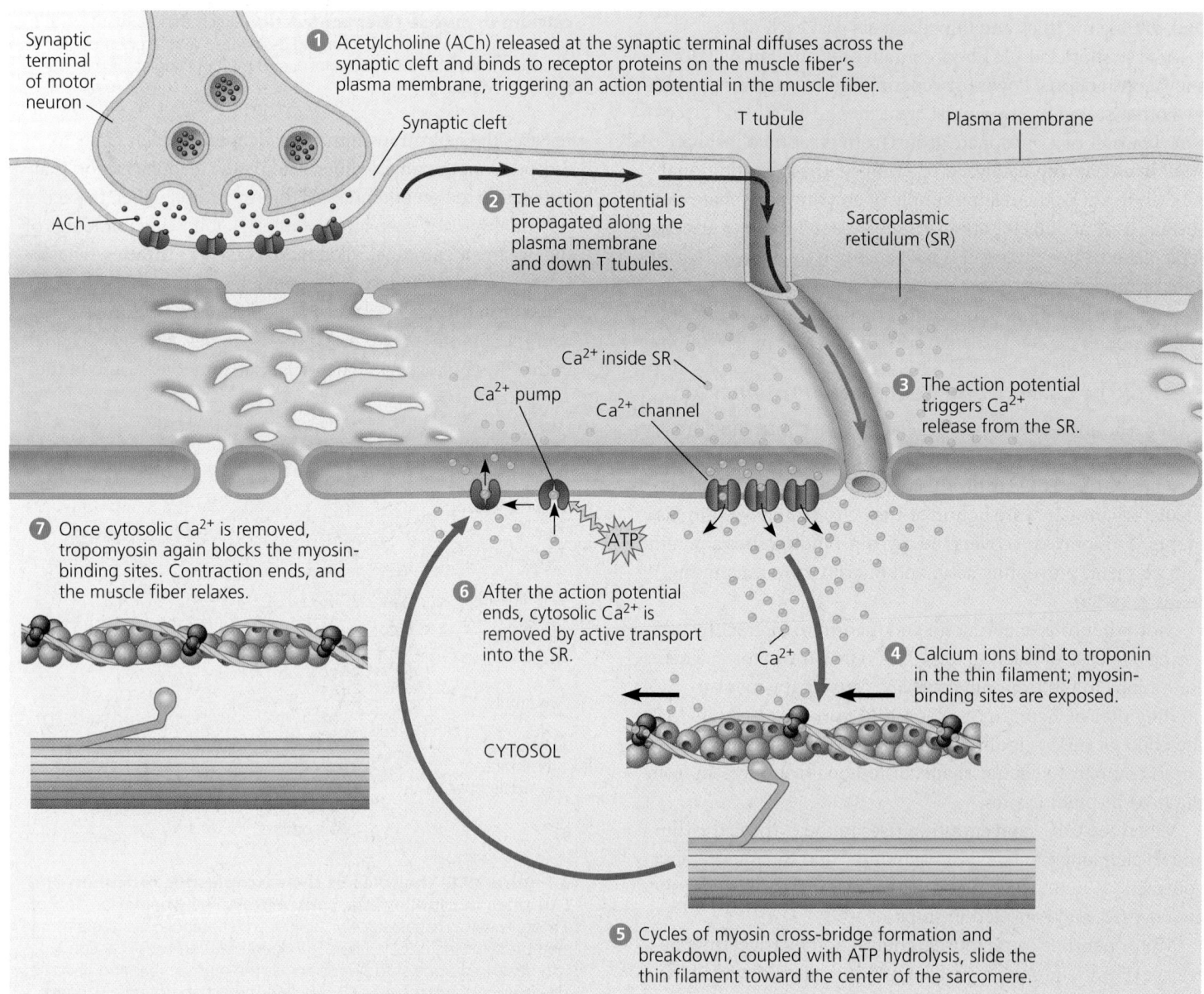

**1** Acetylcholine (ACh) released at the synaptic terminal diffuses across the synaptic cleft and binds to receptor proteins on the muscle fiber's plasma membrane, triggering an action potential in the muscle fiber.

Synaptic terminal of motor neuron

ACh

Synaptic cleft

**2** The action potential is propagated along the plasma membrane and down T tubules.

T tubule

Plasma membrane

Sarcoplasmic reticulum (SR)

$Ca^{2+}$ inside SR

$Ca^{2+}$ pump

$Ca^{2+}$ channel

**3** The action potential triggers $Ca^{2+}$ release from the SR.

ATP

**7** Once cytosolic $Ca^{2+}$ is removed, tropomyosin again blocks the myosin-binding sites. Contraction ends, and the muscle fiber relaxes.

**6** After the action potential ends, cytosolic $Ca^{2+}$ is removed by active transport into the SR.

$Ca^{2+}$

**4** Calcium ions bind to troponin in the thin filament; myosin-binding sites are exposed.

CYTOSOL

**5** Cycles of myosin cross-bridge formation and breakdown, coupled with ATP hydrolysis, slide the thin filament toward the center of the sarcomere.

▲ **Figure 39.7 Review of contraction in a skeletal muscle fiber.**

produces graded contractions of whole muscles by varying (1) the number of muscle fibers that contract and (2) the rate at which muscle fibers are stimulated. Let's consider each mechanism in turn.

A vertebrate motor neuron typically synapses with many muscle fibers. A **motor unit** consists of a single motor neuron and all the muscle fibers it controls **(Figure 39.8)**. When a motor neuron produces an action potential, all the muscle fibers in its motor unit contract as a group. The strength of the resulting contraction depends on how many muscle fibers the motor neuron controls.

Although in vertebrates each muscle fiber is controlled by one motor neuron, there may be hundreds of motor units in a whole muscle. As more and more of the motor neurons controlling the muscle are activated, a process called *recruitment*, the force (tension) developed by a muscle progressively increases. Depending on the number of motor neurons your brain recruits and the size of their motor units, you can lift a fork or something much heavier, like your textbook. Some muscles, especially those that hold up the body and maintain posture, are almost always partially contracted. In such muscles, the nervous system may alternate activation among the motor units, reducing the length of time any one set of fibers is contracted.

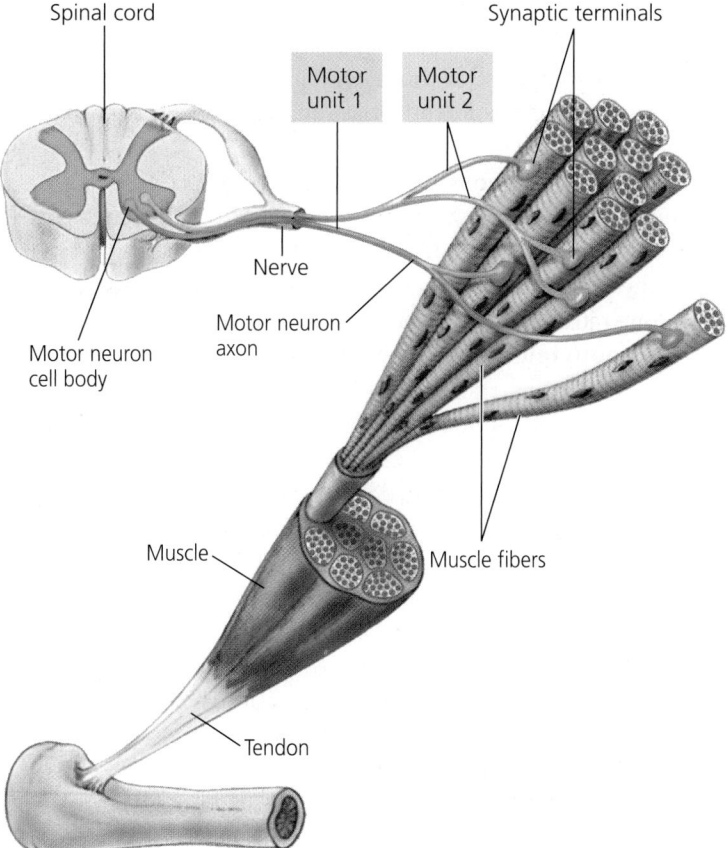

▲ **Figure 39.8 Motor units in a vertebrate skeletal muscle.** Each muscle fiber (cell) forms synapses with only one motor neuron, but each motor neuron typically synapses with many muscle fibers. A motor neuron and all the muscle fibers it controls constitute a motor unit.

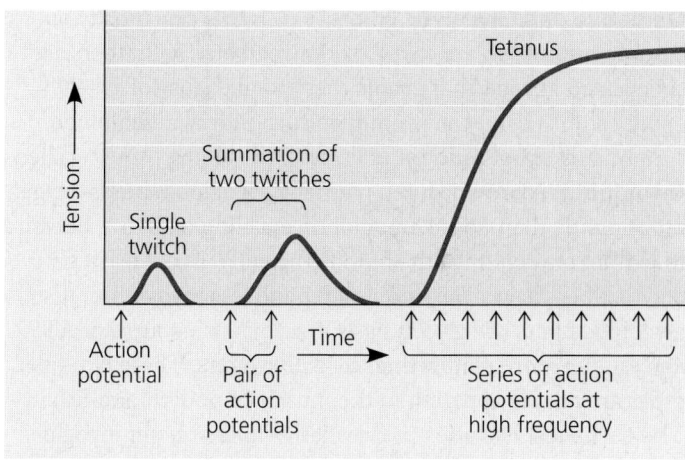

▲ **Figure 39.9 Summation of twitches.** This graph illustrates how the number of action potentials during a short period of time influences the tension developed in a muscle fiber.

**?** *How could the nervous system cause a skeletal muscle to produce the most forceful contraction it is capable of?*

The nervous system regulates muscle contraction not only by controlling which motor units are activated but also by varying the rate of muscle fiber stimulation. A single action potential produces a twitch lasting about 100 milliseconds or less. If a second action potential arrives before the muscle fiber has completely relaxed, the two twitches add together, resulting in greater tension **(Figure 39.9)**. Further summation occurs as the rate of stimulation increases. When the rate is so high that the muscle fiber cannot relax at all between stimuli, the twitches fuse into one smooth, sustained contraction called **tetanus**. (Note that tetanus is also the name of a disease of uncontrolled muscle contraction caused by a bacterial toxin.)

### Types of Skeletal Muscle Fibers

Our discussion to this point has focused on the general properties of vertebrate skeletal muscles. There are, however, several distinct types of skeletal muscle fibers, each of which is adapted to a particular set of functions. We typically classify these varied fiber types both by the source of ATP used to power their activity and by the speed of their contraction **(Table 39.1)**.

**Table 39.1 Types of Skeletal Muscle Fibers**

|  | Slow-Twitch | Fast-Twitch | |
|---|---|---|---|
|  | Oxidative | Oxidative | Glycolytic |
| Major ATP source | Aerobic respiration | Aerobic respiration | Glycolysis |
| Rate of fatigue | Slow | Intermediate | Fast |
| Mitochondria | Many | Many | Few |
| Myoglobin content | High (red muscle) | High (red muscle) | Low (white muscle) |

**Oxidative and Glycolytic Fibers** Fibers that rely mostly on aerobic respiration are called oxidative fibers. Such fibers are specialized in ways that enable them to make use of a steady energy supply: They have many mitochondria, a rich blood supply, and a large amount of an oxygen-storing protein called **myoglobin**. A brownish red pigment, myoglobin binds oxygen more tightly than does hemoglobin, enabling oxidative fibers to extract oxygen from the blood efficiently. In contrast, glycolytic fibers have a larger diameter and less myoglobin. Also, glycolytic fibers use glycolysis as their primary source of ATP and fatigue more readily than oxidative fibers. These two fiber types are readily apparent in the muscle of poultry and fish: The dark meat is made up of oxidative fibers rich in myoglobin, and the light meat is composed of glycolytic fibers.

**Fast-Twitch and Slow-Twitch Fibers** Muscle fibers vary in the speed with which they contract: **Fast-twitch fibers** develop tension two to three times faster than **slow-twitch fibers**. Fast fibers enable brief, rapid, powerful contractions. Compared with a fast fiber, a slow fiber has less sarcoplasmic reticulum and pumps $Ca^{2+}$ more slowly. Because $Ca^{2+}$ remains in the cytosol longer, a muscle twitch in a slow fiber lasts about five times as long as one in a fast fiber.

The difference in contraction speed between slow-twitch and fast-twitch fibers mainly reflects the rate at which their myosin heads hydrolyze ATP. However, there isn't a one-to-one relationship between contraction speed and ATP source. Whereas all slow-twitch fibers are oxidative, fast-twitch fibers can be either glycolytic or oxidative.

Most human skeletal muscles contain both fast-twitch and slow-twitch fibers, although the muscles of the eye and hand are exclusively fast-twitch. In a muscle that has a mixture of fast and slow fibers, the relative proportions of each are largely genetically determined. However, if such a muscle is used repeatedly for activities requiring high endurance, some fast glycolytic fibers can develop into fast oxidative fibers. Because fast oxidative fibers fatigue more slowly than fast glycolytic fibers, the result will be a muscle that is more resistant to fatigue.

Some vertebrates have skeletal muscles that twitch at rates far faster than any human muscle. For example, superfast muscles produce a rattlesnake's rattle and a dove's coo. Even faster are the muscles surrounding the gas-filled swim bladder of the male toadfish **(Figure 39.10)**. In producing its "boat whistle" mating call, the toadfish can contract and relax these muscles more than 200 times per second.

## Other Types of Vertebrate Muscle

Although all muscles share the same fundamental mechanism of contraction—actin and myosin filaments sliding past each other—there are many different types of muscle. Vertebrates, for example, have cardiac muscle and smooth muscle in addition to skeletal muscle.

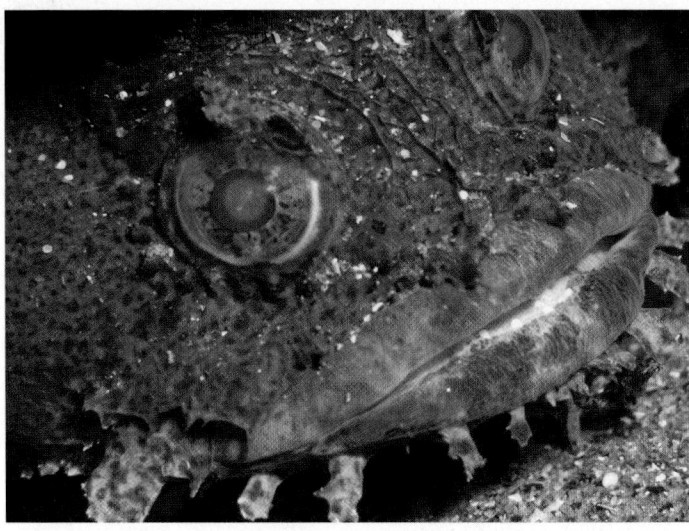

▲ **Figure 39.10 Specialization of skeletal muscle.** The male toadfish (*Opsanus tau*) uses superfast muscles surrounding its swim bladder to produce its mating call.

Vertebrate **cardiac muscle** is found only in the heart. Like skeletal muscle, cardiac muscle is striated. However, skeletal and cardiac muscle fibers differ in their electrical and membrane properties. Whereas skeletal muscle fibers require motor neuron input, ion channels in the plasma membrane of cardiac muscle cells cause rhythmic depolarizations that trigger action potentials without nervous system input. Furthermore, these action potentials last up to 20 times longer than those of skeletal muscle fibers.

Specialized regions called *intercalated disks* electrically couple adjacent cardiac muscle cells. This coupling enables the action potential generated by specialized cells in one part of the heart to spread, causing the whole heart to contract. A long refractory period prevents summation and tetanus.

**Smooth muscle** in vertebrates is found mainly in the walls of hollow organs, such as blood vessels and organs of the digestive tract. Smooth muscle cells lack striations because their actin and myosin filaments are not regularly arrayed along the length of the cell. Instead, the thick filaments are scattered throughout the cytoplasm, and the thin filaments are attached to structures called dense bodies, some of which are tethered to the plasma membrane. Some smooth muscle cells contract only when stimulated by neurons of the autonomic nervous system. Others are electrically coupled to one another and can generate action potentials without input from neurons. Smooth muscles contract and relax more slowly than striated muscles.

Although $Ca^{2+}$ regulates smooth muscle contraction, the mechanism for regulation is different from that in skeletal and cardiac muscle. Smooth muscle cells have no troponin complex or T tubules, and their sarcoplasmic reticulum is not well developed. During an action potential, $Ca^{2+}$ enters the cytosol mainly through the plasma membrane. Calcium ions cause contraction by binding to the protein calmodulin, which activates an enzyme that phosphorylates the myosin head, enabling cross-bridge activity.

## Invertebrate Muscle

Invertebrates have muscle cells similar to vertebrate skeletal and smooth muscle cells, and arthropod skeletal muscles are nearly identical to those of vertebrates. However, because the flight muscles of insects are capable of independent, rhythmic contraction, the wings of some insects can actually beat faster than action potentials can arrive from the central nervous system. Another interesting evolutionary adaptation has been discovered in the muscles that hold a clam's shell closed. The thick filaments in these muscles contain a protein called paramyosin that enables the muscles to remain contracted for as long as a month with only a low rate of energy consumption.

### CONCEPT CHECK 39.1

1. Contrast the role of $Ca^{2+}$ in the contraction of a skeletal muscle fiber and a smooth muscle cell.
2. **WHAT IF?** Why are the muscles of an animal that has recently died likely to be stiff?
3. **MAKE CONNECTIONS** How does the activity of tropomyosin and troponin in muscle contraction compare with the activity of a competitive inhibitor in enzyme action? (See Figure 6.17.)

For suggested answers, see Appendix A.

---

### CONCEPT 39.2

# Skeletal systems transform muscle contraction into locomotion

Converting muscle contraction to movement requires a skeleton—a rigid structure to which muscles can attach. An animal changes its shape or location by contracting muscles connecting two parts of its skeleton.

Because muscles exert force only during contraction, moving a body part back and forth typically requires two muscles attached to the same section of the skeleton. We can see such an arrangement of muscles in the upper portion of a human arm or grasshopper leg **(Figure 39.11)**. Although we call such muscles an antagonistic pair, their function is actually cooperative, coordinated by the nervous system. For example, when you extend your arm, motor neurons trigger your triceps muscle to contract while the absence of neuronal input allows your biceps to relax.

Vital for movement, the skeletons of animals also function in support and protection. Most land animals would collapse if they had no skeleton to support their mass. Even an animal living in water would be formless without a framework to maintain its shape. In many animals, a hard skeleton also protects soft tissues. For example, the vertebrate skull protects the brain, and the ribs of terrestrial vertebrates form a cage around the heart, lungs, and other internal organs.

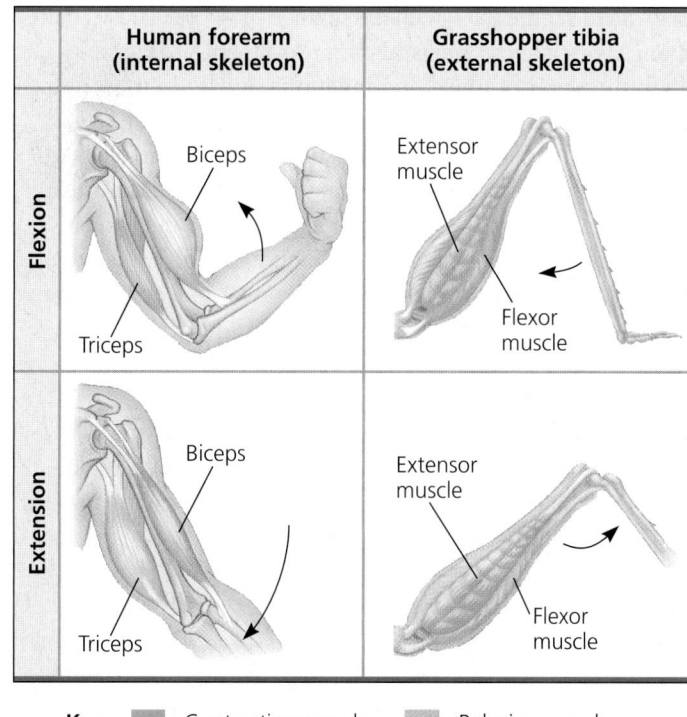

**Key** ■ Contracting muscle ■ Relaxing muscle

▲ **Figure 39.11 The interaction of muscles and skeletons in movement.** Back-and-forth movement of a body part is generally accomplished by antagonistic muscles. This arrangement works with either an internal skeleton, as in mammals, or an external skeleton, as in insects.

## Types of Skeletal Systems

Although we tend to think of skeletons only as interconnected sets of bones, skeletons come in many different forms. Hardened support structures can be external (as in exoskeletons), internal (as in endoskeletons), or even absent (as in fluid-based, or hydrostatic, skeletons).

### Hydrostatic Skeletons

A **hydrostatic skeleton** consists of fluid held under pressure in a closed body compartment. This is the main type of skeleton in most cnidarians, flatworms, nematodes, and annelids. These animals control their form and movement by using muscles to change the shape of fluid-filled compartments. Among the cnidarians, for example, a hydra elongates its body by closing its mouth and constricting its central gastrovascular cavity. using contractile cells in its body wall. Because water maintains its volume under pressure, the cavity must elongate when its diameter is decreased.

Worms carry out locomotion in a variety of ways. In planarians and other flatworms, body movement results mainly from muscles in the body wall exerting localized forces against the interstitial fluid. In nematodes (roundworms), longitudinal muscles contracting around the fluid-filled body cavity move the animal forward by wavelike motions called undulations. In earthworms and many other annelids, circular and longitudinal muscles act together to change the shape of individual fluid-filled segments, which are divided by septa. These shape changes bring about **peristalsis**, a movement produced by rhythmic waves of muscle contractions passing from front to back **(Figure 39.12)**.

Hydrostatic skeletons are well suited for life in aquatic environments. On land, they provide support for crawling and burrowing, but not for walking or running, in which an animal's body is held off the ground. In humans and some other mammals, there is a hydrostatic skeleton for one portion of the male body—the penis.

### Exoskeletons

The clamshell you find on a beach once served as an **exoskeleton**, a hard covering deposited on an animal's surface. The shells of clams and most other molluscs are made of calcium carbonate secreted by the mantle, a sheetlike extension of the body wall. Clams and other bivalves close their hinged shell using muscles attached to the inside of this exoskeleton. As the animal grows, it enlarges its shell by adding to the outer edge.

Insects and other arthropods have a jointed exoskeleton called a *cuticle,* a coat secreted by the epidermis. About 30–50% of the arthropod cuticle is **chitin**, a polysaccharide similar to cellulose. Fibrils of chitin are embedded in a protein matrix, forming a composite material that combines strength and flexibility. In body parts requiring rigidity, such as biting mouthparts or armored regions, the cuticle is hardened with organic compounds and, in some cases, calcium salts. In body parts that must be flexible, such as leg joints, the cuticle remains unhardened. Muscles are attached to knobs and plates of the cuticle that extend into the interior of the body. With each growth spurt, an arthropod must shed its exoskeleton (molt) and produce a larger one.

### Endoskeletons

Animals ranging from sponges to mammals have a hardened internal skeleton, or **endoskeleton**, buried within their soft tissues. In sponges, the endoskeleton often consists of hard needlelike structures of inorganic material or fibers made of protein. Echinoderms' bodies are reinforced by ossicles, hard plates composed of magnesium carbonate and calcium carbonate crystals. Whereas the ossicles of sea urchins are tightly bound, the ossicles of sea stars are more loosely linked, allowing a sea star to change the shape of its arms.

Chordates have an endoskeleton consisting of cartilage, bone, or some combination of these materials. The mammalian skeleton is built from more than 200 bones **(Figure 39.13)**, some fused together and others connected at joints by ligaments that allow freedom of movement **(Figure 39.14)**.

### Size and Scale of Skeletons

How thick does an endoskeleton need to be? We can begin to answer this question by applying ideas from civil engineering. The weight of a building increases with the cube of its dimensions. However, the strength of a support depends on its cross-sectional area, which only increases with the square of its diameter. We can thus predict that if we scaled up a mouse to the size of an elephant, the legs of the giant mouse would be too thin to support its weight. Indeed, large animals do have very different body proportions from small ones.

In applying the building analogy, we might also predict that the size of leg bones should be directly proportional to the strain imposed by its body weight. Animal bodies, however, are complex and nonrigid. In supporting body weight, it turns out that body posture—the position of the legs relative to the main body—is more important than leg size, at least in mammals and birds. In addition, muscles and tendons, which hold

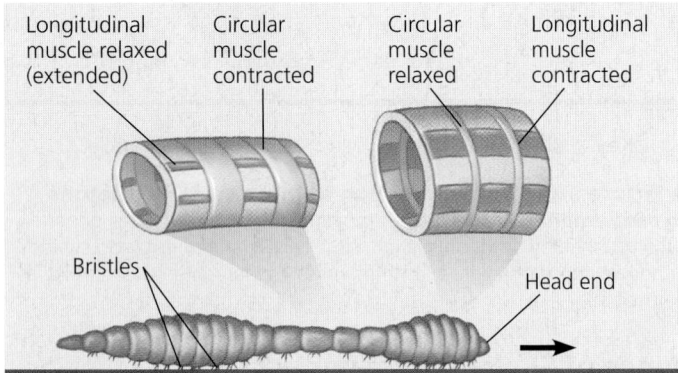

① At the moment depicted, body segments at the earthworm's head end and just in front of the rear end are short and thick (longitudinal muscles contracted; circular muscles relaxed) and are anchored to the ground by bristles. The other segments are thin and elongated (circular muscles contracted; longitudinal muscles relaxed).

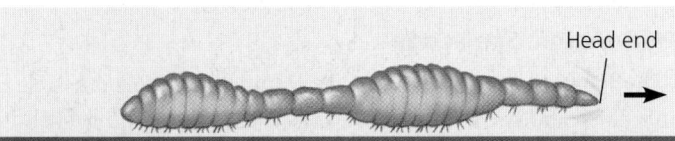

② The head has moved forward because circular muscles in the head segments have contracted. Segments behind the head and at the rear are now thick and anchored, thus preventing the worm from slipping backward.

③ The head segments are thick again and anchored in their new positions. The rear segments have released their hold on the ground and have been pulled forward.

▲ **Figure 39.12 Crawling by peristalsis.** Contraction of the longitudinal muscles thickens and shortens the earthworm; contraction of the circular muscles constricts and elongates it.

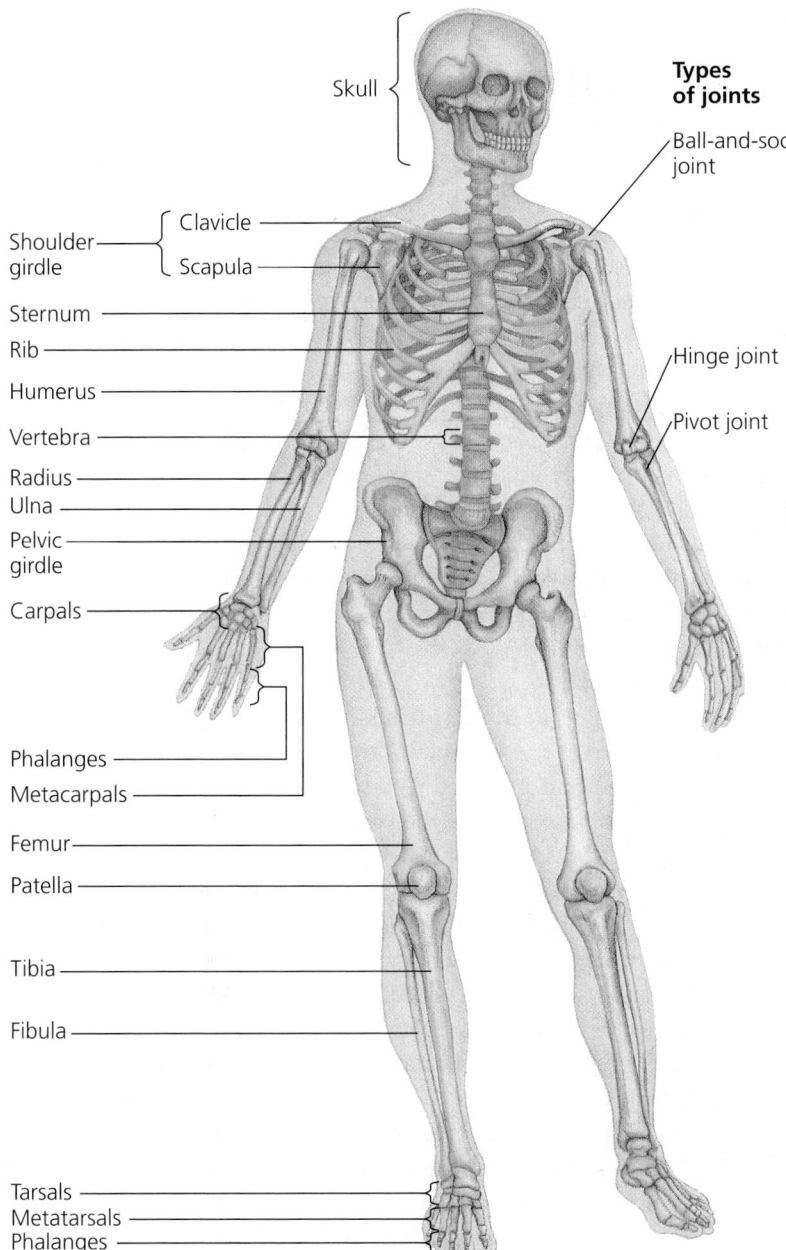

Skull

Shoulder girdle
- Clavicle
- Scapula

Sternum

Rib

Humerus

Vertebra

Radius

Ulna

Pelvic girdle

Carpals

Phalanges

Metacarpals

Femur

Patella

Tibia

Fibula

Tarsals

Metatarsals

Phalanges

**Types of joints**

Ball-and-socket joint

Hinge joint

Pivot joint

▲ **Figure 39.13 Bones and joints of the human skeleton.**

▼ **Figure 39.14 Types of joints.**

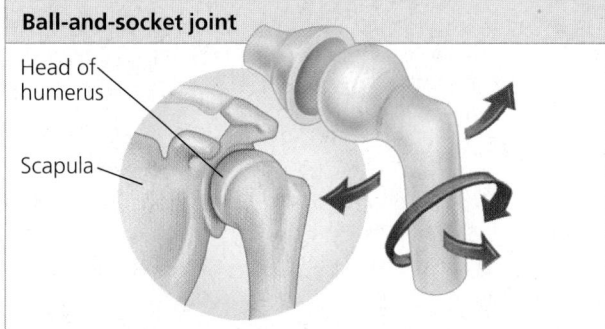

**Ball-and-socket joint**

Head of humerus

Scapula

Ball-and-socket joints are found where the humerus contacts the shoulder girdle and where the femur contacts the pelvic girdle. These joints enable the arms and legs to rotate and move in several planes.

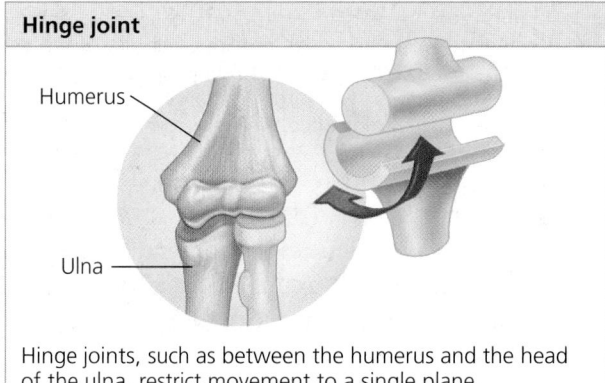

**Hinge joint**

Humerus

Ulna

Hinge joints, such as between the humerus and the head of the ulna, restrict movement to a single plane.

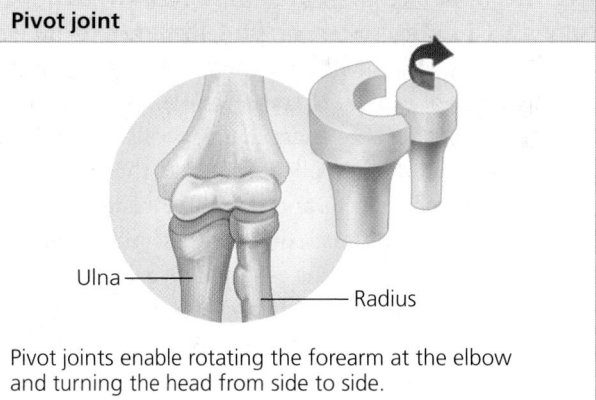

**Pivot joint**

Ulna

Radius

Pivot joints enable rotating the forearm at the elbow and turning the head from side to side.

the legs of large mammals relatively straight and positioned under the body, actually bear most of the load.

## Types of Locomotion

For most animals, activities such as obtaining food, avoiding danger, and finding a mate involve **locomotion**—active travel from place to place. To move, an animal must expend energy to overcome two forces: friction and gravity. As we will see next, the amount of energy required to oppose friction or gravity is often reduced by an animal body plan adapted for movement in a particular environment.

### Locomotion on Land

On land, a walking, running, hopping, or crawling animal must be able to support itself and move against gravity. When an animal walks, runs, or hops, its leg muscles expend energy both to propel it and to keep it from falling down. With each step, the leg muscles must overcome inertia by accelerating a leg from a standing start. At the same time, air provides relatively little resistance to movement, at least when locomotion is at moderate speeds. Therefore, powerful muscles and strong skeletal support are more important than a streamlined shape.

▲ **Figure 39.15 Energy-efficient locomotion on land.**
Members of the kangaroo family travel from place to place mainly by leaping on their large hind legs. Kinetic energy momentarily stored in tendons after each leap provides a boost for the next leap. In fact, a large kangaroo hopping at 30 km/hr uses no more energy per minute than it does at 6 km/hr. The large tail provides extra force for leaping and helps to balance the kangaroo when it leaps as well as when it sits.

Diverse adaptations for traveling on land have evolved in various vertebrates. For example, kangaroos have large, powerful muscles in their hind legs, suitable for hopping **(Figure 39.15)**. As a kangaroo lands after each leap, tendons in its hind legs momentarily store energy. The farther the animal hops, the more energy its tendons store. Like the energy in a compressed spring, the energy stored in the tendons is available for the next jump, reducing the total amount of energy the animal must expend to travel. The legs of an insect, dog, or human also retain some energy during walking or running, although a considerably smaller share than those of a kangaroo.

Maintaining balance is another prerequisite for walking, running, or hopping. A cat, dog, or horse maintains balance by keeping three feet on the ground when walking. Illustrating the same principle, bipedal animals, such as humans and birds, keep part of at least one foot on the ground when walking. At running speeds, momentum more than foot contact keeps the body upright, enabling all the feet to be off the ground briefly. A kangaroo's large tail forms a stable tripod with its hind legs when the animal sits or moves slowly and helps balance its body during leaps. Recent studies reveal that the tail also generates significant force during hopping, helping to propel the kangaroo's forward movement.

Crawling poses a very different challenge. Having much of its body in contact with the ground, a crawling animal must exert considerable effort to overcome friction. As you have read, earthworms crawl by peristalsis. In contrast, many snakes crawl by undulating their entire body from side to side. Assisted by large, movable scales on their underside, such snakes use their body to push against the ground, propelling the animal forward. Some other snakes, such as boa constrictors and pythons, creep straight forward, driven by muscles that

lift belly scales off the ground, tilt the scales forward, and then push them backward against the ground.

## Swimming

Because most animals are reasonably buoyant in water, overcoming gravity is less of a problem for swimming than for movement on land or through the air. On the other hand, water is a much denser and more viscous medium than air, and thus drag (friction) is a major problem for aquatic animals. A sleek, fusiform (torpedo-like) shape is a common adaptation of fast swimmers, such as tuna.

Swimming occurs in diverse ways. Many insects and four-legged vertebrates use their legs as oars to push against the water. Squids, scallops, and some cnidarians are jet-propelled, taking in water and squirting it out in bursts. Sharks and bony fishes swim by moving their body and tail from side to side, while whales and dolphins move by undulating their body and tail up and down.

## Flying

Active flight (in contrast to gliding downward from a tree) has evolved in only a few animal groups: insects, reptiles (including birds), and, among the mammals, bats. One group of flying reptiles, the pterosaurs, died out millions of years ago, leaving birds and bats as the only flying vertebrates.

Gravity poses a major problem for a flying animal because its wings must develop enough lift to overcome gravity's downward force. The key to flight is wing shape. All wings act as airfoils—structures whose shape alters air currents in a way that helps animals or airplanes stay aloft. As for the body to which the wings attach, a fusiform shape helps reduce drag in air as it does in water.

Flying animals are relatively light, with body masses ranging from less than a gram for some insects to about 20 kg for the largest flying birds. Many flying animals have structural adaptations that contribute to low body mass. Birds, for example, have no urinary bladder or teeth and have relatively large bones with air-filled regions that help lessen the bird's mass.

Running, swimming, and flying each impose different energetic demands on animals. In the **Scientific Skills Exercise**, you can interpret a graph that compares the relative energy costs of these three forms of locomotion.

## CONCEPT CHECK 39.2

1. In what way are septa an important feature of the earthworm skeleton?
2. Contrast swimming and flying in terms of the main problems they pose and the adaptations that allow animals to overcome those problems.
3. **WHAT IF?** When using your arms to lower yourself into a chair, you bend your arms without using your biceps. Explain how this is possible. (*Hint:* Think about gravity as a force antagonistic to the force exerted by a body muscle.)

For suggested answers, see Appendix A.

# Interpreting a Graph with Log Scales

**What Are the Energy Costs of Locomotion?** In the 1960s, animal physiologist Knut Schmidt-Nielsen, at Duke University, wondered whether general principles govern the energy costs of different forms of locomotion among diverse animal species. To answer this question, he drew on his own experiments as well as those of other researchers. In this exercise, you will analyze the combined results of these studies and evaluate the rationale for plotting the experimental data on a graph with logarithmic scales.

**How the Experiments Were Done** Researchers measured the rate of oxygen consumption or carbon dioxide production in animals that ran on treadmills, swam in water flumes, or flew in wind tunnels. For example, a tube connected to a plastic face mask collected gases exhaled by a parakeet during flight (see photo at upper right). From these measurements, Schmidt-Nielsen calculated the amount of energy each animal used to transport a given amount of body mass over a given distance (in calories per kilogram · meter).

**Data from the Experiments** Schmidt-Nielsen plotted the cost of running, flying, and swimming versus body mass on a single graph with logarithmic (log) scales for the axes. He then drew a best-fit straight line through the data points for each form of locomotion. (On the graph, the individual data points are not shown.)

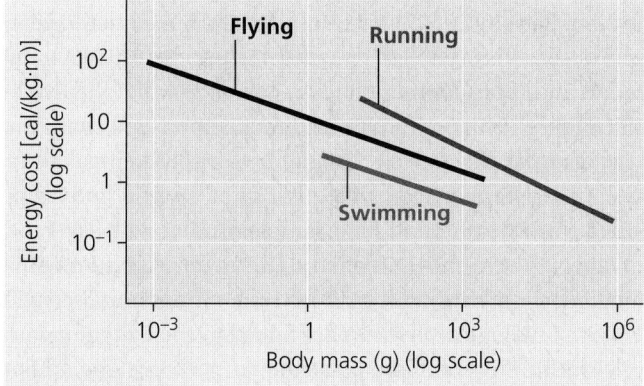

**Data from** K. Schmidt-Nielsen, Locomotion: Energy cost of swimming, flying, and running, *Science* 177:222–228 (1972)

### INTERPRET THE DATA

1. The body masses of the animals used in these experiments ranged from about 0.001 g to 1,000,000 g, and their rates of energy use ranged from about 0.1 cal/(kg·m) to 100 cal/(kg·m). If you were to plot these data on a graph with linear instead of log scales for the axes, how would you draw the axes so that all of the data would be visible? What is the advantage of using log scales for plotting data with a wide range of values? (For additional information about graphs, see the Scientific Skills Review in Appendix F and in the Study Area in MasteringBiology.)

2. Based on the graph, how much greater is the energy cost of flying for an animal that weighs $10^{-3}$ g than for an animal that weighs 1 g? For any given form of locomotion, which travels more efficiently, a larger animal or a smaller animal?

3. The slopes of the flying and swimming lines are very similar. Based on your answer to question 2, if the energy cost of a 2-g swimming animal is 1.2 cal/(kg·m), what is the estimated energy cost of a 2-kg swimming animal?

4. Considering animals with a body mass of about 100 g, rank the three forms of locomotion from highest energy cost to lowest energy cost. Were these the results you expected, based on your own experience? What could explain the energy cost of running compared with that of flying or swimming?

5. Schmidt-Nielson calculated the swimming cost for a mallard duck and found that it was nearly 20 times as high as the swimming cost for a salmon of the same body mass. What could explain the greater swimming efficiency of salmon?

(MB) A version of this Scientific Skills Exercise can be assigned in MasteringBiology.

---

# Discrete sensory inputs can stimulate both simple and complex behaviors

So far we have been discussing the mechanics of animal behaviors—how the animal body produces the movements that make up a particular behavior. In the rest of the chapter, we'll take a broader look at the function of animal behaviors as well as their evolution.

What approach do biologists use to determine how behaviors arise and what functions they serve? The Dutch scientist Niko Tinbergen, a pioneer in the study of animal behavior, suggested that understanding any behavior requires answering four questions, which can be summarized as follows:

1. What stimulus elicits the behavior, and what physiological mechanisms mediate the response?
2. How does the animal's experience during growth and development influence the response?
3. How does the behavior make the animal more likely to survive and reproduce?
4. What is the behavior's evolutionary history?

Tinbergen's first two questions ask about *proximate causation*: "how" a behavior occurs or is modified. The last two questions ask about *ultimate causation*: "why" a behavior occurs in the context of natural selection.

Studies on proximate causation by Tinbergen earned him a share of a Nobel Prize awarded in 1973. The concept of ultimate causation is central to **behavioral ecology**, the study of the ecological and evolutionary basis for animal behavior. We'll explore these topics in turn.

▲ **Figure 39.16 Sign stimuli in a classic fixed action pattern.** A male stickleback fish attacks other male sticklebacks that invade its nesting territory. The red belly of the intruding male (left) acts as a sign stimulus that releases the aggressive behavior.

**?** *Suggest an explanation for why this behavior evolved (its ultimate causation).*

## Fixed Action Patterns

In addressing Tinbergen's first question, the nature of the stimuli that trigger behavior, we'll begin with behavioral responses to well-defined stimuli, starting with an example from one of Tinbergen's own experiments.

As part of his research, Tinbergen kept fish tanks containing three-spined sticklebacks (*Gasterosteus aculeatus*), a species in which males, but not females, have red bellies. Male sticklebacks attack other males that invade their nesting territories **(Figure 39.16)**. Tinbergen noticed that his male sticklebacks also behaved aggressively when a red truck passed within view of their tank. Inspired by this chance observation, he carried out experiments showing that the red color of an intruder's underside is the proximate cause of the attack behavior. A male stickleback will not attack a fish lacking red coloration, but will attack even unrealistic models if they contain areas of red color.

The territorial response of male sticklebacks is an example of a **fixed action pattern**, a sequence of unlearned acts directly linked to a simple stimulus. Fixed action patterns are essentially unchangeable and, once initiated, usually carried to completion. The trigger for the behavior is an external cue called a **sign stimulus**, such as a red object that prompts the male stickleback's aggressive behavior.

## Migration

Environmental stimuli not only trigger behaviors but also provide cues that animals use to carry out those behaviors. For example, a wide variety of birds, fishes, and other animals use environmental cues to guide **migration**—a regular, long-distance change in location. In the course of migration, many animals pass through environments they have not previously encountered. How, then, do they find their way in these foreign settings?

Some migrating animals track their position relative to the sun, even though the sun's position relative to Earth changes throughout the day. Animals can adjust for these changes by means of a *circadian clock*, an internal mechanism that maintains a 24-hour activity rhythm or cycle. For example,

experiments have shown that migrating birds orient differently relative to the sun at distinct times of the day.

Although the sun as well as stars can provide clues for navigation, clouds can obscure these landmarks. How do migrating animals overcome this problem? A simple experiment with homing pigeons provides one answer. On an overcast day, placing a small magnet on the head of a homing pigeon prevents it from returning efficiently to its roost. Researchers concluded that pigeons sense their position relative to Earth's magnetic field and can thereby navigate without solar or celestial cues.

## Behavioral Rhythms

Although the circadian clock plays a small but significant role in navigation by some migrating species, it has a major role in the daily activity of all animals. The clock is responsible for a circadian rhythm, a daily cycle of rest and activity (see Concept 38.2). The clock is normally synchronized with the light and dark cycles of the environment but can maintain rhythmic activity even under constant environmental conditions, such as during hibernation.

Some behaviors, such as migration and reproduction, reflect biological rhythms with a longer cycle, or period, than the circadian rhythm. Behavioral rhythms linked to the yearly cycle of seasons are called *circannual rhythms*. Although migration and reproduction typically correlate with food availability, these behaviors are not a direct response to changes in food intake. Instead, circannual rhythms, like circadian rhythms, are influenced by the periods of daylight and darkness in the environment. For example, studies with several bird species have shown that an artificial environment with extended daylight can induce out-of-season migratory behavior.

Not all biological rhythms are linked to the light and dark cycles in the environment. Consider, for instance, the fiddler crab shown in Figure 39.1. The male's claw-waving courtship behavior is linked to the timing of the new and full moon. This timing helps the development of offspring. Fiddler crabs begin their lives as larvae settling in the mudflats. The tides disperse the larvae to deeper waters, where they complete early development in relative safety before returning to the tidal flats. By courting at the time of the new or full moon, crabs link their reproduction to the times of greatest tidal movement.

## Animal Signals and Communication

Claw waving by fiddler crabs during courtship is an example of one animal (the male crab) generating the stimulus that guides the behavior of another animal (the female crab). A stimulus transmitted from one organism to another is called a **signal**. The transmission and reception of signals between animals constitute **communication**, which often has a role in the proximate causation of behavior.

### Forms of Animal Communication

Let's consider the courtship behavior of the fruit fly, *Drosophila melanogaster*, as an introduction to the four common modes of animal communication: visual, chemical, tactile, and auditory.

Fruit fly courtship constitutes an example of a *stimulus-response chain*, in which the response to each stimulus is itself the stimulus for the next behavior. In the first step, a male detects a female in his field of vision and orients his body toward hers. To confirm she belongs to his species, he uses his olfactory system to detect chemicals she releases into the air. The male then approaches and touches the female with a foreleg **(Figure 39.17)**. This touching, or tactile communication, alerts the female to the male's presence. In the third stage of courtship, the male extends and vibrates one of his wings, producing a courtship song. This auditory communication informs the female whether the male is of the same species. Only if all of these forms of communication are successful will the female allow the male to attempt copulation.

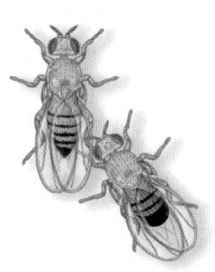

▲ **Figure 39.17 Male fruit fly tapping female with foreleg.**

In general, the form of communication that evolves is closely related to an animal's lifestyle and environment. For example, most terrestrial mammals are nocturnal, which makes visual displays relatively ineffective. Instead, these species use olfactory and auditory signals, which work as well in the dark as in the light. In contrast, most birds are diurnal (active mainly in daytime) and communicate primarily by visual and auditory signals. Humans are also diurnal and use primarily visual and auditory communication. We can thus appreciate the songs and bright colors used by birds to communicate but miss many chemical cues on which other mammals base their behavior.

The information content of animal communication varies considerably. One of the most remarkable examples is the symbolic language of the European honeybee (*Apis mellifera*), discovered in the early 1900s by Austrian researcher Karl von Frisch. Using glass-walled observation hives, he and his students spent several decades observing honeybees. Methodical recordings of bee movements enabled von Frisch to decipher a "dance language" that returning foragers use to inform other bees about the distance and direction of travel to food sources **(Figure 39.18)**. When the other bees then exit the hive, they fly almost directly to the area indicated by the returning foragers.

### Pheromones

Animals that communicate through odors or tastes emit chemical substances called **pheromones**. Pheromones are especially common among mammals and insects and often relate to reproductive behavior. For example, pheromones are the basis for the chemical communication in fruit fly courtship. Pheromones are not limited to short-distance signaling, however. Male silkworm moths have receptors that can detect the pheromone from a female moth from several kilometers away.

**(a) Worker bees cluster around a recently returned bee.**

**(b) The round dance indicates that food is near.**

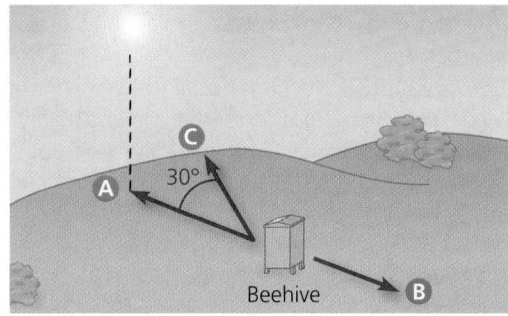

Beehive

Location **A**: Food source is in same direction as sun.

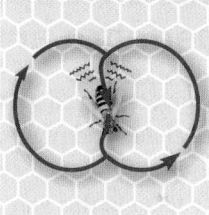

Location **B**: Food source is in direction opposite sun.

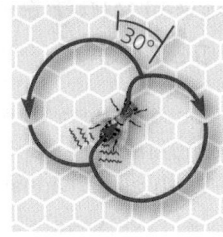

Location **C**: Food source is 30° to right of sun.

**(c) The waggle dance is performed when food is distant.** The waggle dance resembles a figure eight. Distance is indicated by the number of abdominal waggles performed in the straight-run part of the dance. Direction is indicated by the angle (in relation to the vertical surface of the hive) of the straight run.

▲ **Figure 39.18 Honeybee dance language.** Honeybees returning to the hive communicate the location of food sources through the symbolic language of a dance.

In a honeybee colony, pheromones produced by the queen and her daughters, the workers, maintain the hive's complex social order. One pheromone (once called the queen substance) has a particularly wide range of effects. It attracts workers to the queen, inhibits development of ovaries in workers, and attracts males (drones) to the queen during her mating flights out of the hive.

Pheromones can also serve as alarm signals. For example, when a minnow or catfish is injured, a substance released from the fish's skin disperses in the water, causing nearby minnows to become more vigilant and seek safety near the lake bottom.

So far, we have explored the types of stimuli that elicit behaviors—the first part of Tinbergen's first question. The second part of that question—the physiological mechanisms that mediate responses—involves the nervous, muscular, and skeletal systems: Stimuli activate sensory systems, are processed in the central nervous system, and result in motor outputs that constitute behavior. Thus, we are ready to focus on Tinbergen's second question—how experience influences behavior.

## CONCEPT CHECK 39.3

1. If an egg rolls out of the nest, a mother greylag goose will retrieve it by nudging it with her beak and head. If researchers remove the egg or substitute a ball during this process, the goose continues to bob her beak and head while she moves back to the nest. Explain how and why this behavior occurs.
2. **MAKE CONNECTIONS** How is the lunar-linked rhythm of fiddler crab courtship similar in mechanism and function to the seasonal timing of plant flowering? (See Concept 31.2.)

   **For suggested answers, see Appendix A.**

## CONCEPT 39.4

# Learning establishes specific links between experience and behavior

For some behaviors—such as a fixed action pattern, a courtship stimulus-response chain, and pheromone signaling—nearly all individuals in a population behave alike. Behavior that is developmentally fixed in this way is known as **innate behavior**. Other behaviors, however, vary with experience and thus differ between individuals.

## Experience and Behavior

Tinbergen's second question asks how an animal's experiences during growth and development influence the response to stimuli. One informative approach to this question is a **cross-fostering study**, in which the young of one species are placed in the care of adults from another species. The extent to which the offspring's behavior changes in such a situation provides a measure of how the social and physical environment influences behavior.

Certain mouse species have behaviors well suited for cross-fostering studies. Male California mice (*Peromyscus californicus*) are highly aggressive toward other mice and provide extensive parental care. In contrast, male white-footed mice (*Peromyscus leucopus*) are less aggressive and engage in little parental care. When the pups of each species were placed in the nests of the other species, the cross-fostering altered some behaviors of both species **(Table 39.2)**. For instance, male California mice raised by white-footed mice were less aggressive toward intruders. Thus, experience

**Table 39.2 Influence of Cross-Fostering on Male Mice***

| Species | Aggression Toward an Intruder | Aggression in Neutral Situation | Paternal Behavior |
|---|---|---|---|
| California mice fostered by white-footed mice | Reduced | No difference | Reduced |
| White-footed mice fostered by California mice | No difference | Increased | No difference |

*Comparisons are with mice raised by parents of their own species.

during development can strongly influence aggressive behavior in these rodents.

One of the most important findings of the cross-fostering experiments with mice was that the influence of experience on behavior can be passed on to progeny: When the cross-fostered California mice became parents, they spent less time retrieving offspring who wandered off than did California mice raised by their own species. Thus, experience during development can modify physiology in a way that alters parental behavior, extending the influence of environment to a subsequent generation.

For humans, the influence of genetics and environment on behavior can be explored by a **twin study**, in which researchers compare the behavior of identical twins raised apart with the behavior of those raised in the same household. Twin studies have been instrumental in studying disorders that alter human behavior, such as anxiety disorders, schizophrenia, and alcoholism.

## Learning

One powerful way that an animal's environment can influence its behavior is through **learning**, the modification of behavior as a result of specific experiences. The capacity for learning depends on nervous system organization established during development following instructions encoded in the genome. Learning itself involves the formation of memories by specific changes in neuronal connectivity (see Concept 38.3). Therefore, the essential challenge for research into learning is not to decide between nature (genes) and nurture (environment), but rather to explore the contributions of *both* nature and nurture in shaping learning and, more generally, behavior.

### Imprinting

In some species, the ability of offspring to recognize and be recognized by a parent is essential for survival. In the young, this learning often takes the form of **imprinting**, the establishment of a long-lasting behavioral response to a particular individual or object. Imprinting can take place only during a specific time period in development, called the **sensitive period**. Among gulls, for instance, the sensitive period for a parent to bond with its young lasts one to two days. During the sensitive period, the young imprint on their parent and learn basic behaviors, while the parent learns to recognize its offspring. If bonding does not occur, the parent will not care for the offspring, leading to the death of the offspring and a decrease in the reproductive success of the parent.

How do the young know on whom—or what—to imprint? Experiments with many species of waterfowl indicate that young birds have no innate recognition of "mother." Rather, they identify with the first object they encounter that has certain key characteristics. In the 1930s, experiments showed that the principal *imprinting stimulus* in greylag geese (*Anser anser*) is a nearby object that is moving away from the young. When incubator-hatched goslings spent their first few hours with a person rather than with a goose, they imprinted on the human and steadfastly followed the person from then on **(Figure 39.19a)**. Furthermore, they showed no recognition of their biological mother.

Imprinting has become an important component of efforts to save endangered species, such as the whooping crane (*Grus americana*). Scientists tried raising whooping cranes in captivity by using sandhill cranes (*Grus canadensis*) as foster parents. However, because the whooping cranes imprinted on their foster parents, none formed a *pair-bond* (strong attachment) with a whooping crane mate. To avoid such problems, captive breeding programs now isolate young cranes, exposing them to the sights and sounds of members of their own species.

Scientists have made further use of imprinting to teach cranes born in captivity to migrate along safe routes. Young whooping cranes are imprinted on humans in "crane suits" and then are allowed to follow these "parents" as they fly ultralight aircraft along selected migration routes **(Figure 39.19b)**. Importantly, these cranes still pair-bond with other whooping cranes, indicating that the crane costumes have the features required to direct "normal" imprinting.

### Spatial Learning and Cognitive Maps

Every natural environment has spatial variation, as in locations of nest sites, hazards, food, and prospective mates. Therefore, an organism's fitness may be enhanced by the capacity for **spatial learning**, the establishment of a memory that reflects the environment's spatial structure.

The idea of spatial learning intrigued Tinbergen while he was a graduate student in the Netherlands. At that time, he was studying the female digger wasp (*Philanthus triangulum*), which nests in small burrows dug into sand dunes. When a wasp leaves her nest to go hunting, she hides the entrance to the burrow from potential intruders by covering it with sand.

**(a)** These young greylag geese imprinted on a man.

**(b)** A pilot wearing a crane suit and flying an ultralight plane acts as a surrogate parent to direct the migration of whooping cranes.

▲ **Figure 39.19 Imprinting.** Imprinting can be altered to **(a)** investigate animal behavior or **(b)** direct animal behavior.

**WHAT IF?** *Suppose the geese shown in (a) were bred to each other. How might their having imprinted on a human affect their offspring? Explain.*

When she returns, however, she flies directly to her hidden nest, despite the presence of hundreds of other burrows in the area. How does she accomplish this feat? Tinbergen hypothesized that a wasp locates her nest by learning its position relative to visible landmarks. To test his hypothesis, he carried out an experiment in the wasps' natural habitat **(Figure 39.20)**. By manipulating objects around nest entrances, he demonstrated that digger wasps engage in spatial learning.

In some animals, spatial learning involves formulating a **cognitive map**, a representation in an animal's nervous system of the spatial relationships between objects in its surroundings. One striking example is found in the Clark's nutcracker (*Nucifraga columbiana*), a relative of ravens, crows, and jays. In the fall, nutcrackers hide pine seeds for retrieval during the winter. By experimentally varying the distance between landmarks in the birds' environment, researchers discovered that birds used the halfway point between landmarks, rather than a fixed distance, to find their hidden food stores.

### Associative Learning

Learning often involves making associations between experiences. Consider, for example, a blue jay (*Cyanocitta cristata*) that ingests a brightly colored monarch butterfly (*Danaus plexippus*). Substances that the monarch accumulates from milkweed plants cause the blue jay to vomit almost immediately **(Figure 39.21)**. Following such experiences, blue jays avoid attacking monarchs and similar-looking butterflies. The ability to associate one environmental feature (such as a color) with another (such as a foul taste) is called **associative learning**.

Studies reveal that animals can learn to link many pairs of features of their environment, but not all. For example, pigeons can learn to associate danger with a sound but not with a color. However, they can learn to associate a color with food. What does this mean? The development and organization of the pigeon's nervous system apparently restrict the associations that can be formed. Moreover, such restrictions are not limited to birds. Rats, for example, can learn to avoid illness-inducing foods on the basis of smells, but not on the basis of sights or sounds.

If we consider how behavior evolves, the fact that some animals can't learn to make particular associations appears logical. The associations an animal can readily form typically reflect relationships likely to occur in nature. Conversely, associations that can't be formed are those unlikely to be of selective advantage in a native environment. In the case of a rat's diet in the wild, for example, a harmful food is far more likely to have a certain odor than to be associated with a particular sound.

### Cognition and Problem Solving

The most complex forms of learning involve **cognition**—the process of knowing that involves awareness, reasoning, recollection, and judgment. Although it was once argued that only primates and certain marine mammals have high level thought processes, many other groups of animals, including insects, appear to exhibit cognition in controlled laboratory studies. For example, an experiment using Y-shaped mazes

### Does a digger wasp use landmarks to find her nest?

**Experiment** A female digger wasp covers the entrance to her underground nest while foraging long distances for food, but she can return later to the exact location and uncover her hidden nest. Niko Tinbergen hypothesized that the female, before flying off, learns visual landmarks that mark her nest location. To test this hypothesis, he first marked one nest with a ring of pinecones while the wasp was in the nest. After leaving the nest to forage, the wasp returned to the nest successfully.

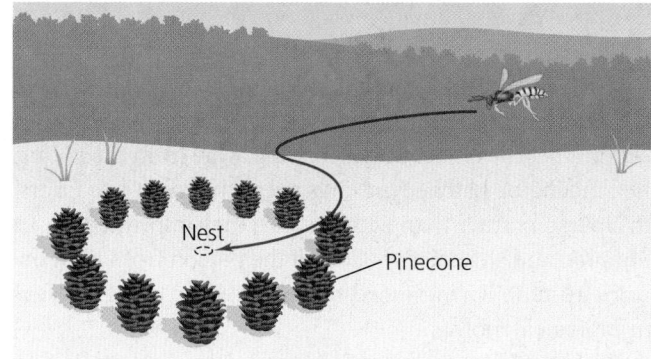

Two days later, after the wasp had again left, Tinbergen shifted the ring of pinecones away from the nest. Then he waited to observe the wasp's behavior.

**Results** When the wasp returned, she flew to the center of the pinecone circle instead of to the nearby nest. Repeating the experiment with many wasps, Tinbergen obtained the same results.

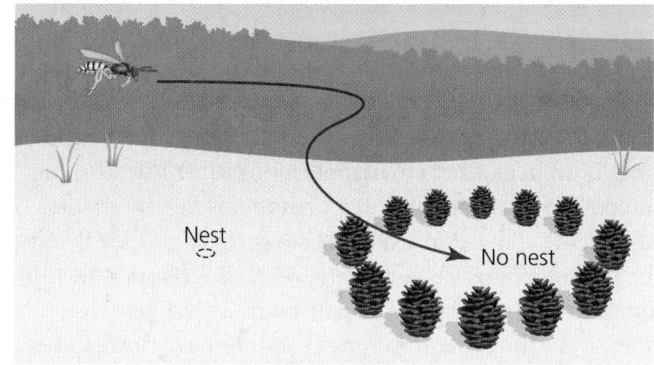

**Conclusion** The experiment supported the hypothesis that digger wasps use visual landmarks to keep track of their nests.

**Data from** N. Tinbergen, The study of instinct, Clarendon Press, Oxford (1951).

**WHAT IF?** Suppose the digger wasp had returned to her original nest site, despite the pinecones having been moved. What alternative hypotheses might you propose regarding how the wasp finds her nest and why the pinecones didn't misdirect the wasp?

demonstrated that honeybees can distinguish between "same" and "different," concepts requiring abstract thinking.

The information-processing ability of a nervous system can also be revealed in **problem solving**, the cognitive activity of

▲ **Figure 39.21 Associative learning.** Having eaten and vomited a monarch butterfly, a blue jay has probably learned to avoid this species.

devising a method to proceed from one state to another in the face of real or apparent obstacles. For example, if a chimpanzee is placed in a room with several boxes on the floor and a banana hung high out of reach, the chimpanzee can assess the situation and stack the boxes, enabling it to reach the food. Problem-solving behavior is highly developed in some mammals, especially primates and dolphins. Notable examples have also been observed in some bird species, especially corvids. In one study, ravens were confronted with food hanging from a branch by a string. After failing to grab the food in flight, one raven flew to the branch and alternately pulled up and stepped on the string until the food was within reach. A number of other ravens eventually arrived at similar solutions. Nevertheless, some ravens failed to solve the problem, indicating that problem-solving success in this species, as in others, varies with individual experience and abilities.

Many animals learn to solve problems by observing the behavior of other individuals. Young wild chimpanzees, for example, learn how to crack open oil palm nuts with two stones by copying experienced chimpanzees. This type of learning through observing others is called **social learning**. Social learning forms the roots of **culture**, which can be defined as a system of information transfer through social learning or teaching that influences the behavior of individuals in a population. Cultural transfer of information can alter behavioral phenotypes and thereby influence the fitness of individuals.

### CONCEPT CHECK 39.4

1. How might associative learning explain why different species of distasteful or stinging insects have similar colors?
2. **WHAT IF?** How might you position and manipulate a few objects in a lab to test whether an animal can use a cognitive map to remember the location of a food source?
3. **MAKE CONNECTIONS** How might a learned behavior contribute to speciation? (See Concept 22.1.)

*For suggested answers, see Appendix A.*

# Selection for individual survival and reproductive success can explain diverse behaviors

**EVOLUTION** We turn now to Tinbergen's third question—how behavior enhances survival and reproduction in a population. Our focus thus shifts from proximate causation—the "how" questions—to ultimate causation—the "why" questions. We'll begin by considering the activity of gathering food. Food-obtaining behavior, or **foraging**, includes not only eating but also any activities an animal uses to search for, recognize, and capture food items.

## Evolution of Foraging Behavior

The fruit fly allows us to examine one way that foraging behavior might have evolved. Variation in a gene called *forager* (*for*) dictates how far *Drosophila* larvae travel when foraging. On average, larvae carrying the *for*$^R$ ("Rover") allele travel nearly twice as far while foraging as do larvae with the *for*$^s$ ("sitter") allele.

Both the *for*$^R$ and *for*$^s$ alleles are present in natural populations. What circumstances might favor one or the other allele? The answer became apparent in experiments that maintained flies at either low or high population densities for many generations. Larvae in populations kept at a low density foraged over shorter distances than those in populations kept at high density **(Figure 39.22)**. Furthermore, the *for*$^s$ allele increased in frequency in the low-density populations, whereas the *for*$^R$ allele increased in frequency in the high-density group. These changes make sense. At a low population density, short-distance foraging

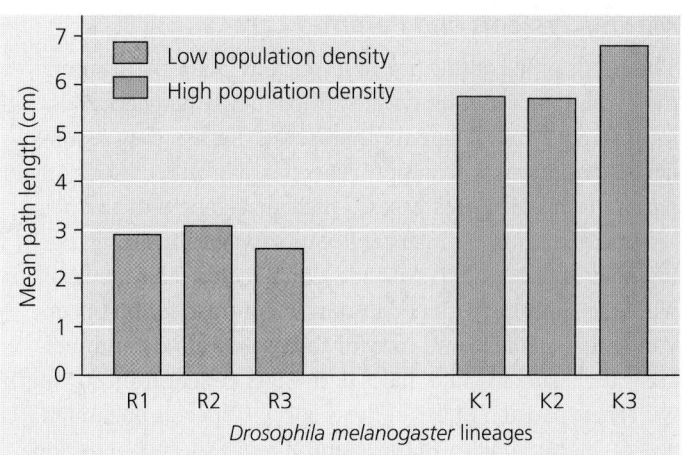

▲ **Figure 39.22 Evolution of foraging behavior by laboratory populations of *Drosophila melanogaster*.** After 74 generations of living at low population density, *Drosophila* larvae (populations R1–R3) followed foraging paths significantly shorter than those of *Drosophila* larvae that had lived at high density (populations K1–K3).

**INTERPRET THE DATA** *What alternative hypothesis is ruled out by having three R and K lines, rather than one of each?*

yields sufficient food, while long-distance foraging would result in unnecessary energy expenditure. Under crowded conditions, long-distance foraging could enable larvae to move beyond areas depleted of food. Thus, an interpretable evolutionary change in behavior occurred in the course of the experiment.

## Mating Behavior and Mate Choice

Just as foraging is crucial for individual survival, mating behavior and mate choice play a major role in determining reproductive success. These behaviors include seeking or attracting mates, choosing among potential mates, competing for mates, and caring for offspring.

### Mating Systems and Sexual Dimorphism

Although we tend to think of mating simply as the union of a male and female, species vary greatly with regard to *mating systems,* the length and number of relationships between males and females. In some animal species, mating is *promiscuous,* with no strong pair-bonds. In others, mates form a relationship of some duration that is **monogamous** (one male mating with one female) or **polygamous** (an individual of one sex mating with several of the other). Polygamous relationships involve either *polygyny,* a single male and many females, or *polyandry,* a single female and multiple males.

The extent to which males and females differ in appearance, a characteristic known as *sexual dimorphism,* typically varies with the type of mating system **(Figure 39.23)**. Among monogamous species, males and females often look very similar. In contrast, among polygamous species, the sex that attracts multiple mating partners is typically showier and larger than the opposite sex. We'll discuss the evolutionary basis of these differences shortly.

### Mating Systems and Parental Care

The needs of the young are an important factor constraining the evolution of mating systems. Most newly hatched birds, for instance, cannot care for themselves. Rather, they require a large, continuous food supply, a need that is difficult for a single parent to meet. In such cases, a male that stays with and helps a single mate may ultimately have more viable offspring than if it would by going off to seek additional mates. This may explain why many birds are monogamous. In contrast, for birds with young that can feed and care for themselves almost immediately after hatching, the males derive less benefit from staying with their partner. Males of these species, such as pheasants and quail, can maximize their reproductive success by seeking other mates, and polygyny is relatively common in such birds. In the case of mammals, the lactating female is often the only food source for the young, and males usually play no role in raising the young. In mammalian species where males protect the females and young, such as lions, a male or small group of males typically cares for a harem of many females.

Another factor influencing mating behavior and parental care is *certainty of paternity.* Young born to or eggs laid by a

**(a) Monogamy** (one male, one female)

In monogamous species, such as these western gulls (*Larus occidentalis*), males and females are difficult to distinguish using external characteristics only.

**(b) Polygyny** (one male, multiple females)

Among polygynous species, such as elk (*Cervus canadensis*), the male (right) is often highly ornamented.

**(c) Polyandry** (one female, multiple males)

In polyandrous species, such as these red-necked phalaropes (*Phalaropus lobatus*), females (right) are generally more ornamented than males.

▲ **Figure 39.23 Relationship between mating system and male and female forms.**

▲ **Figure 39.24 Paternal care by a male jawfish.** The male jawfish, which lives in tropical marine environments, holds the eggs it has fertilized in its mouth, keeping them aerated and protecting them from egg predators until the young hatch.

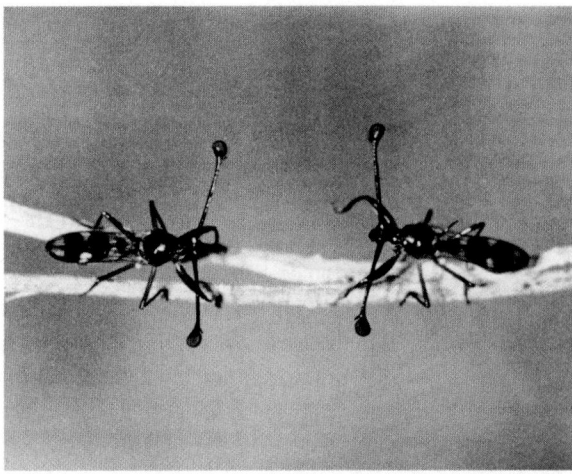

▲ **Figure 39.25 A face-off between male stalk-eyed flies for female attention.** In such ritual showdowns, the male whose eyestalk length is smaller usually retreats peacefully.

female definitely contain that female's genes. However, even within a normally monogamous relationship, a male other than the female's usual mate may have fathered that female's offspring. The certainty of paternity is relatively low in most species with internal fertilization because the acts of mating and birth (or mating and egg laying) are separated over time. This could explain why exclusively male parental care is rare in bird and mammal species. However, the males of many species with internal fertilization engage in behaviors that appear to increase their certainty of paternity. These behaviors include guarding females, removing any sperm from the female reproductive tract before copulation, and introducing large quantities of sperm that displace the sperm of other males.

Certainty of paternity is high when egg laying and mating occur together, as in external fertilization. This may explain why parental care in aquatic invertebrates, fishes, and amphibians, when it occurs at all, is at least as likely to be by males as by females **(Figure 39.24)**. Among fishes and amphibians, parental care occurs in fewer than 10% of species with internal fertilization but in more than half of species with external fertilization.

It is important to note that certainty of paternity does not mean that animals are aware of those factors when they behave a certain way. Parental behavior correlated with certainty of paternity exists because it has been reinforced over generations by natural selection. The intriguing relationship between certainty of paternity and male parental care remains an area of active research.

### Sexual Selection and Mate Choice

Sexual dimorphism results from sexual selection, a form of natural selection in which differences in reproductive success among individuals are a consequence of differences in mating success (see Concept 21.4). Sexual selection can take the form of *intersexual selection*, in which members of one sex choose mates on the basis of characteristics of the other sex, such as

courtship songs, or *intrasexual selection*, which involves competition between members of one sex for mates.

Mate preferences of females may play a central role in the evolution of male behavior and anatomy through intersexual selection. Consider, for example, the courtship behavior of stalk-eyed flies. The eyes of these insects are at the tips of stalks, which are longer in males than in females. During courtship, a male approaches the female headfirst. Researchers have shown that females are more likely to mate with males that have relatively long eyestalks. Why would females favor this seemingly arbitrary trait? Ornaments such as long eyestalks in these flies and bright coloration in birds correlate in general with health and vitality. A female whose mate choice is a healthy male is likely to produce more offspring that survive to reproduce. As a result, males may compete with each other in ritualized contests to attract female attention **(Figure 39.25)**.

Our consideration of stalk-eyed flies illustrates how female choice can select for one best type of male in a given situation, resulting in low variation among males. This example also shows how male competition for mates can reinforce the tendency for reduced variation among males. The length of the eyestalks of the male flies in Figure 39.25 is an important factor in conflict between males. This competition takes the form of an *agonistic behavior*, an often-ritualized contest that determines which competitor gains access to a resource, such as food or a mate.

### CONCEPT CHECK 39.5

1. Why does the mode of fertilization correlate with the presence or absence of male parental care?
2. **MAKE CONNECTIONS** Balancing selection can maintain variation at a locus (see Concept 21.4). Based on the foraging experiments described in this chapter, devise a simple hypothesis to explain the presence of both $for^R$ and $for^s$ alleles in natural fly populations.

For suggested answers, see Appendix A.

# Genetic analyses and the concept of inclusive fitness provide a basis for studying the evolution of behavior

**EVOLUTION** We'll now explore issues related to Tinbergen's fourth question—the evolutionary history of behaviors. We will first look at the genetic control of a behavior. Next, we will examine the genetic variation underlying the evolution of particular behaviors. Finally, we will see how expanding the definition of fitness beyond individual survival can help explain "selfless" behavior.

## Genetic Basis of Behavior

It is now well established that differences in behavior can arise from variation in the activity or amount of a single gene product. One striking example comes from the study of two related species of voles, which are small, mouse-like rodents. Male meadow voles (*Microtus pennsylvanicus*) are solitary and do not form lasting relationships with mates. Following mating, they pay little attention to their pups. In contrast, male prairie voles (*Microtus ochrogaster*) form a pair-bond with a single female after they mate **(Figure 39.26)**. Male prairie voles hover over their young pups, licking them and carrying them, while acting aggressively toward intruders.

A peptide neurotransmitter is critical for the partnering and parental behavior of male voles. Known as **antidiuretic hormone (ADH)**, or **vasopressin** (see Figure 32.5), this peptide is released during mating and binds to a specific receptor in the central nervous system. When male prairie voles are given a drug that inhibits the brain receptor for vasopressin, they fail to form pair-bonds after mating.

► **Figure 39.26 A pair of prairie voles (*Microtus ochrogaster*) huddling.** Male North American prairie voles associate closely with their mates, as shown here, and contribute substantially to the care of young.

The vasopressin receptor gene is much more highly expressed in the brain of prairie voles than in the brain of meadow voles. Testing the hypothesis that vasopressin receptor levels in the brain regulate postmating behavior, researchers inserted the vasopressin receptor gene from prairie voles into meadow voles. The male meadow voles carrying this gene not only developed brains with higher levels of the vasopressin receptor, but also showed many of the same mating behaviors as male prairie voles, such as pair-bonding. Thus, although many genes influence pair-bonding and parenting among voles, a change in vasopressin receptor levels is sufficient to alter the development of these behaviors.

## Genetic Variation and the Evolution of Behavior

Behavioral differences between closely related species, such as meadow and prairie voles, are common. Significant differences in behavior can also be found *within* a species, but are often less obvious. When behavioral variation between populations of a species correlates with variation in environmental conditions, it may reflect natural selection.

## Case Study: *Variation in Prey Selection*

An example of genetically based behavioral variation within a species involves prey selection by the western garter snake (*Thamnophis elegans*). The natural diet of this species differs widely across its range in California. Coastal populations feed predominantly on banana slugs (*Ariolimax californicus*) **(Figure 39.27)**. Inland populations feed on frogs, leeches, and fish, but not banana slugs. In fact, banana slugs are rare or absent in the inland habitats.

When researchers offered banana slugs to snakes from each wild population, most coastal snakes readily ate them, whereas inland snakes tended to refuse. To what extent does genetic variation contribute to a fondness for banana slugs? To answer

▲ **Figure 39.27 Western garter snake from a coastal habitat eating a banana slug.** Experiments indicate that the preference of these snakes for banana slugs may be influenced more by genetics than by environment.

this question, researchers collected pregnant snakes from each wild population and housed them in separate cages in the laboratory. While still very young, the offspring were offered a small piece of banana slug on each of ten days. More than 60% of the young snakes from coastal mothers ate banana slugs on eight or more of the ten days. In contrast, fewer than 20% of the young snakes from inland mothers ate a piece of banana slug even once. Perhaps not surprisingly, banana slugs thus appear to be a genetically acquired taste.

How did a genetically determined difference in feeding preference come to match the snakes' habitats so well? It turns out that the coastal and inland populations also vary with respect to their ability to recognize and respond to odor molecules produced by banana slugs. Researchers hypothesize that when inland snakes colonized coastal habitats more than 10,000 years ago, some of them could recognize banana slugs by scent. Because these snakes took advantage of this food source, they had higher fitness than snakes in the population that ignored the slugs. Over hundreds or thousands of generations, the capacity to recognize the slugs as prey increased in frequency in the coastal population. The marked variation in behavior observed today between the coastal and inland populations may be evidence of this past evolutionary change.

## Altruism

We typically assume that behaviors are selfish; that is, they benefit the individual at the expense of others, especially competitors. For example, superior foraging ability by one individual may leave less food for others. The problem comes with "unselfish" behaviors. How can such behaviors arise through natural selection? To answer this question, let's look more closely at some examples of unselfish behavior and consider how they might arise.

In discussing selflessness, we will use the term **altruism** to describe a behavior that reduces an animal's individual fitness but increases the fitness of other individuals in the population. Consider, for example, the Belding's ground squirrel, which lives in the western United States and is vulnerable to predators such as coyotes and hawks. A squirrel that sees a predator approach often gives a high-pitched alarm call that alerts unaware individuals to retreat to their burrows. Note that for the squirrel that warns others, the conspicuous alarm behavior increases the risk of being killed because it brings attention to the caller's location.

Altruism is also observed in naked mole rats (*Heterocephalus glaber*), highly social rodents that live in underground chambers and tunnels in southern and northeastern Africa. The naked mole rat, which is almost hairless and nearly blind, lives in colonies of 20 to 300 individuals **(Figure 39.28)**. Each colony has only one reproducing female, the queen, who mates with one to three males, called kings. The rest of the colony consists of nonreproductive females and males who at times sacrifice themselves to protect the queen or kings from snakes or other predators that invade the colony.

▲ **Figure 39.28 Naked mole rats, a species of colonial mammal that exhibits altruistic behavior.** Pictured here is a queen nursing offspring while surrounded by other members of the colony.

## Inclusive Fitness

With these examples from ground squirrels and mole rats in mind, let's return to the question of how altruistic behavior arises during evolution. The easiest case to consider is that of parents sacrificing for their offspring. When parents sacrifice their own well-being to produce and aid offspring, this act actually increases the fitness of the parents because it maximizes their genetic representation in the population. By this logic, altruistic behavior can be maintained by evolution even though it does not enhance the survival and reproductive success of the self-sacrificing individuals.

What about circumstances when individuals help others who are not their offspring? By considering a broader group of relatives than just parents and offspring, biologist William Hamilton found an answer. He began by proposing that an animal could increase its genetic representation in the next generation by helping close relatives other than its own offspring. Like parents and offspring, full siblings have half their genes in common. Therefore, selection might also favor helping siblings or helping one's parents produce more siblings. This thinking led Hamilton to the idea of **inclusive fitness**, the total effect an individual has on proliferating its genes by producing its own offspring *and* by providing aid that enables other close relatives to produce offspring.

### Hamilton's Rule and Kin Selection

The power of Hamilton's hypothesis was that it provided a way to measure, or quantify, the effect of altruism on fitness. According to Hamilton, the three key variables in an act of altruism are the benefit to the recipient, the cost to the altruist, and the coefficient of relatedness. The benefit, $B$, is the average number of *extra* offspring that the recipient of an altruistic act produces. The cost, $C$, is how many *fewer* offspring the altruist produces. The **coefficient of relatedness**, $r$, equals the fraction of genes that, on average, are shared. Natural selection favors altruism when the benefit to the recipient multiplied by the coefficient of relatedness exceeds the cost to the altruist—in other words, when $rB > C$. This statement is called **Hamilton's rule**.

To better understand Hamilton's rule, let's apply it to a human population in which the average individual has two children. We'll imagine that a young man is close to drowning in heavy surf, and his sister risks her life to swim out and pull her sibling to safety. If the young man had drowned, his reproductive output would have been zero; but now, if we use the average, he can father two children. The benefit to the man is thus two offspring ($B = 2$). What cost does his sister incur? Let's say that she has a 25% chance of drowning in attempting the rescue. The cost of the altruistic act to the sister is then

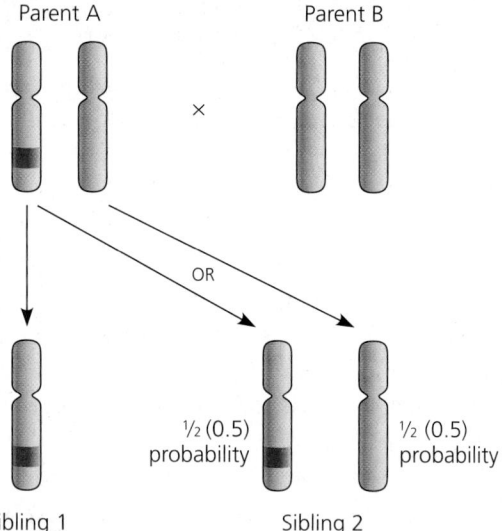

▲ **Figure 39.29 The coefficient of relatedness between siblings.** The red band indicates a particular allele (version of a gene) present on one chromosome, but not its homolog, in parent A. Sibling 1 has inherited the allele from parent A. There is a probability of ½ that sibling 2 will also inherit this allele from parent A. Any allele present on one chromosome of either parent will behave similarly. The coefficient of relatedness between the two siblings is thus ½, or 0.5.

**WHAT IF?** *The coefficient of relatedness of an individual to a full (nontwin) sibling or to either parent is the same: 0.5. Does this value also hold true in cases of polyandry and polygyny?*

0.25 times 2, the number of offspring she would be expected to have if she had stayed on shore ($C = 0.25 \times 2 = 0.5$). Finally, we note that a brother and sister share half their genes on average ($r = 0.5$). One way to see this is in terms of the segregation of homologous chromosomes that occurs during meiosis of gametes (**Figure 39.29**).

We can use our values of $B$, $C$, and $r$ to evaluate whether natural selection would favor the altruistic act in our imaginary scenario. For the surf rescue, $rB = 0.5 \times 2 = 1$, whereas $C = 0.5$. Because $rB$ is greater than $C$, Hamilton's rule is satisfied; thus, natural selection would favor this altruistic act.

Averaging over many individuals and generations, any particular gene in a sister faced with the situation described will be passed on to more offspring if she risks the rescue than if she does not. Among the genes propagated in this way may be some that contribute to altruistic behavior. Natural selection that thus favors altruism by enhancing the reproductive success of relatives is called **kin selection**.

Kin selection weakens with hereditary distance. Consequently, natural selection would not favor rescuing a cousin unless the surf was less treacherous. British geneticist J. B. S. Haldane appears to have anticipated these ideas when he jokingly stated that he would not lay down his life for one brother, but would do so for two brothers or eight cousins.

**CONCEPT CHECK 39.6**

1. Explain why geographic variation in garter snake prey choice might indicate that the behavior evolved by natural selection.
2. **WHAT IF?** Suppose you applied Hamilton's logic to a situation in which one individual is past reproductive age. Could there still be a selection for an altruistic act? Explain.

   **For suggested answers, see Appendix A.**

---

# 39 | Chapter Review

Go to **MasteringBiology**® for Assignments, the eText, and the Study Area with Animations, Activities, Vocab Self-Quiz, and Practice Tests.

## SUMMARY OF KEY CONCEPTS

**VOCAB SELF-QUIZ**

goo.gl/gbai8v

### CONCEPT 39.1

**The physical interaction of protein filaments is required for muscle function (pp. 815–821)**

- The muscle cells (fibers) of vertebrate **skeletal muscle** contain **myofibrils** composed of **thin filaments** of (mostly) actin and **thick filaments** of myosin. These filaments are organized into repeating units called **sarcomeres**. Myosin heads, energized by the hydrolysis of ATP, bind to the thin filaments, forming cross-bridges, and then release upon binding ATP anew. As this cycle repeats, the thick and thin filaments slide past each other, contracting the muscle fiber.

- Motor neurons release acetylcholine, triggering action potentials in muscle fibers that stimulate $Ca^{2+}$ release from the **sarcoplasmic reticulum**. When $Ca^{2+}$ binds the **troponin complex**, **tropomyosin** moves, exposing the myosin-binding sites on actin and thus initiating cross-bridge formation. A **motor unit** consists of a motor neuron and the muscle fibers it controls. Skeletal muscle fibers can be **slow-** or **fast-twitch** and oxidative or glycolytic.
- **Cardiac muscle**, found in the heart, consists of striated cells that are electrically connected by intercalated disks and can generate action potentials without neuronal input. In **smooth muscle**, contractions are initiated by the muscles themselves or by stimulation from the autonomic nervous system.

? *How do oxidative and glycolytic muscle fibers differ?*

## CONCEPT 39.2

### Skeletal systems transform muscle contraction into locomotion (pp. 821–825)

- Skeletal muscles, often in antagonistic pairs, pull against the skeleton. Skeletons may be **hydrostatic** and maintained by fluid pressure, as in worms; hardened into **exoskeletons**, as in insects; or internal **endoskeletons**, as in vertebrates. Both exoskeletons and endoskeltons contain unhardened connections, or joints, that allow freedom of movement.
- Each form of **locomotion**—movement on land, swimming, and flying—presents particular challenges. For example, swimmers face more of a challenge from friction and less from gravity than do animals that walk or fly.

? *Explain how microscopic and macroscopic anchoring of muscle filaments enables you to bend your elbow.*

## CONCEPT 39.3

### Discrete sensory inputs can stimulate both simple and complex behaviors (pp. 825–828)

- Behavior is the sum of an animal's responses to external and internal stimuli. In behavior studies, proximate, or "how," questions focus on the stimuli that trigger a behavior and on the genetic, physiological, and anatomical underpinnings of a behavioral act. Ultimate, or "why," questions address evolutionary significance.
- A **fixed action pattern** is a largely invariant behavior triggered by a simple cue known as a **sign stimulus**. Migratory movements involve navigation, which can be based on orientation relative to the sun, the stars, or Earth's magnetic field. Animal behavior is often synchronized to the circadian cycle of light and dark in the environment or to cues that cycle over the seasons.
- The transmission and reception of **signals** constitute animal **communication**. Animals use visual, auditory, chemical, and tactile signals. Chemical substances called **pheromones** transmit information between members of a species.

? *How is migration based on circannual rhythms poorly suited for adaptation to global climate change?*

## CONCEPT 39.4

### Learning establishes specific links between experience and behavior (pp. 828–831)

- **Cross-fostering studies** can be used to measure the influence of social environment and experience on behavior.
- **Learning**, the modification of behavior based on experience, can take many forms:

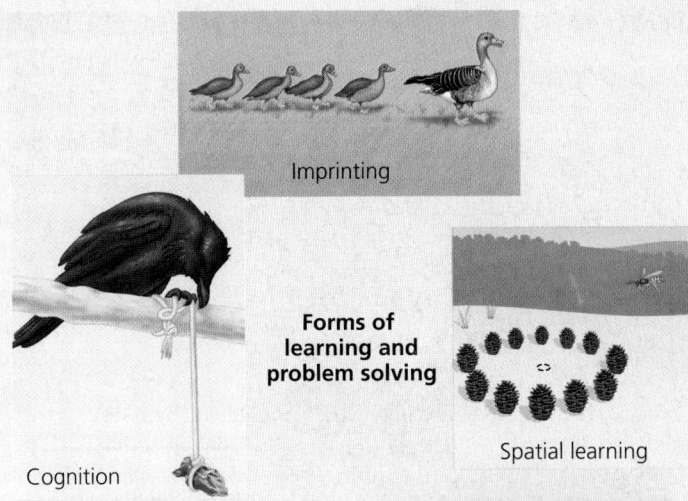

Imprinting

Cognition

**Forms of learning and problem solving**

Spatial learning

**Associative learning**

Social learning

? *Pigeons can learn to associate color with a food but not with danger. How might this observation be explained in terms of the selective forces acting during evolution?*

## CONCEPT 39.5

### Selection for individual survival and reproductive success can explain diverse behaviors (pp. 831–833)

- Controlled experiments in the laboratory can give rise to interpretable evolutionary changes in behavior.
- Sexual dimorphism correlates with the types of mating relationships, which include **monogamous** and **polygamous** mating systems. Variations in mating system and mode of fertilization affect certainty of paternity, which has a significant influence on mating behavior and parental care.

? *In some spider species, the female eats the male immediately after copulation. How might you explain this behavior from an evolutionary perspective?*

## CONCEPT 39.6

### Genetic analyses and the concept of inclusive fitness provide a basis for studying the evolution of behavior (pp. 834–836)

- Research on voles illustrates how variation in a single gene can determine differences in complex behaviors.
- Behavioral variation within a species that corresponds to environmental variation may be evidence of past evolution.
- **Altruism** can be explained by the concept of **inclusive fitness**, the effect an individual has on proliferating its genes by producing offspring *and* by providing aid that enables close relatives to reproduce. **Kin selection** favors altruistic behavior by enhancing the reproductive success of relatives.

? *If an animal were unable to distinguish close from distant relatives, would the concept of inclusive fitness still be applicable? Explain.*

# TEST YOUR UNDERSTANDING

## Level 1: Knowledge/Comprehension

1. During the contraction of a vertebrate skeletal muscle fiber, calcium ions
   (A) break cross-bridges by acting as a cofactor in the hydrolysis of ATP.
   (B) bind with troponin, changing its shape so that the myosin-binding sites on actin are exposed.
   (C) transmit action potentials from the motor neuron to the muscle fiber.
   (D) spread action potentials through the T tubules.

2. Which of the following is true of innate behaviors?
   (A) Their expression is only weakly influenced by genes.
   (B) They occur with or without environmental stimuli.
   (C) They are expressed in most individuals in a population.
   (D) They occur in invertebrates and some vertebrates but not mammals.

3. According to Hamilton's rule,
   (A) natural selection does not favor altruistic behavior that causes the death of the altruist.
   (B) natural selection favors altruistic acts when the resulting benefit to the recipient, corrected for relatedness, exceeds the cost to the altruist.
   (C) natural selection is more likely to favor altruistic behavior that benefits an offspring than altruistic behavior that benefits a sibling.
   (D) the effects of kin selection are larger than the effects of direct natural selection on individuals.

4. The binding of calcium to the troponin complex
   (A) disrupts cross-bridges, allowing filaments to slide past each other.
   (B) allows tropomyosin to bind actin.
   (C) opens ion channels, allowing sodium to rush into the muscle cells.
   (D) causes tropomyosin to shift position, exposing myosin bind sites on actin.

## Level 2: Application/Analysis

5. Curare, a substance that blocks the acetylcholine receptors on skeletal muscle, will cause
   (A) rapid muscle twitches.
   (B) sustained muscle contraction (tetanus).
   (C) muscle relaxation.
   (D) no effect in the absence of acetylcholinesterase.

6. Although many chimpanzees live in environments with oil palm nuts, members of only a few populations use stones to crack open the nuts. The likely explanation is that
   (A) the behavioral difference is caused by genetic differences between populations.
   (B) members of different populations have different nutritional requirements.
   (C) the cultural tradition of using stones to crack nuts has arisen in only some populations.
   (D) members of different populations differ in learning ability.

7. Which of the following is *not* required for a behavioral trait to evolve by natural selection?
   (A) The form of the behavior is determined entirely by genes.
   (B) The behavior varies among individuals.
   (C) An individual's reproductive success depends in part on how the behavior is performed.
   (D) Some component of the behavior is genetically inherited.

## Level 3: Synthesis/Evaluation

8. **SCIENTIFIC INQUIRY**
   Propose a hypothesis to explain how paramyosin enables clamshell muscles to remain contracted for long periods. Explain how you would test your hypothesis.

9. **SCIENTIFIC INQUIRY**
   Scientists studying scrub jays found that "helpers" often assist mated pairs of birds by gathering food for their offspring. Propose a hypothesis to explain what advantage there might be for the helpers to engage in this behavior instead of seeking their own territories and mates. How would you test your hypothesis? If it is correct, what results would you expect your tests to yield?

10. **FOCUS ON EVOLUTION**
    We often explain our behavior in terms of subjective feelings, motives, or reasons, but evolutionary explanations are based on reproductive fitness. How would you describe the relationship between the two kinds of explanation? For instance, is an explanation for behavior such as "falling in love" incompatible with an evolutionary explanation?

11. **FOCUS ON INFORMATION**
    Learning is defined as a change in behavior based on experience. In a short essay (100–150 words), describe how heritable information contributes to the acquisition of learning, using examples from imprinting and associative learning.

12. **SYNTHESIZE YOUR KNOWLEDGE**

Acorn woodpeckers (*Melanerpes formicivorus*) stash acorns in storage holes they drill in trees. When these woodpeckers breed, the offspring from previous years often help with parental duties. Activities of these nonbreeding helpers include incubating eggs and defending stashed acorns. What questions about the proximate and ultimate causation of these behaviors might a biologist ask?

*For selected answers, see Appendix A.*

# Unit 7 Ecology

### 40 Population Ecology and the Distribution of Organisms

Ecologists study the interactions of organisms and the environment to understand the **distribution of species. Populations** of a species may be relatively stable in size or fluctuate greatly, driven by ecological and evolutionary factors.

### 41 Species Interactions

Populations of different species interact in ecological **communities** through processes such as competition, predation, and mutualism.

### 42 Ecosystems and Energy

Energy flow and chemical cycling occur in an **ecosystem**, the community of organisms living in an area and the physical factors with which they interact. Within an ecosystem, organisms transfer **energy** through trophic levels and are characterized by their main source of nutrition and energy.

### 43 Global Ecology and Conservation Biology

Human activities are changing climate patterns, trophic structures, energy flow, and disturbance patterns throughout the biosphere. Efforts to sustain ecosystem processes and to preserve biodiversity from habitat loss and other threats comprise the fields of **global ecology** and **conservation biology**.

Ecology

40
41
42
43

Population Ecology and the Distribution of Organisms
Species Interactions
Ecosystems and Energy
Global Ecology and Conservation Biology

Animals
Plants
History of Life
Evolution
Genetics
Chemistry and Cells

# CHAPTER

# 40 Population Ecology and the Distribution of Organisms

## KEY CONCEPTS

**40.1** Earth's climate influences the distribution of terrestrial biomes

**40.2** Aquatic biomes are diverse and dynamic systems that cover most of Earth

**40.3** Interactions between organisms and the environment limit the distribution of species

**40.4** Biotic and abiotic factors affect population density, dispersion, and demographics

**40.5** The exponential and logistic models describe the growth of populations

**40.6** Population dynamics are influenced strongly by life history traits and population density

▲ **Figure 40.1** What limits the distribution of this tiny frog?

## Discovering Ecology

Kneeling by a stream in Papua New Guinea in 2008, Cornell University undergraduate Michael Grundler heard a series of clicks. He first thought that the sounds must be coming from a nearby cricket. Turning to look, he saw instead a tiny frog inflating its vocal sac to call for a mate. Grundler would later learn that he had just discovered the first of two new frog species, *Paedophryne swiftorum* and *Paedophryne amauensis* **(Figure 40.1)**. The entire *Paedophryne* genus is known only from the Papuan Peninsula in eastern New Guinea. Adult frogs of both species are typically only 8 mm (0.3 inch) long and may be the smallest adult vertebrates on Earth.

What environmental factors limit the geographic distribution of *Paedophryne* frogs? How do variations in their food supply or interactions with other species, such as pathogens, affect the size of their population?

Questions like these are the subject of **ecology** (from the Greek *oikos*, home, and *logos*, study), the scientific study of the interactions between organisms and the environment. The interactions studied by ecologists can be organized into a hierarchy that ranges in scale from single organisms to the globe **(Figure 40.2)**.

Ecology is a rigorous experimental science that requires a breadth of biological knowledge. Ecologists observe nature, generate hypotheses, manipulate environmental variables, and observe outcomes. In this chapter, we'll first consider how Earth's climate and other factors determine the location of major life zones on land and in the oceans. We'll then examine how ecologists investigate what controls the distribution of species and the density and size of populations. The next three chapters focus on community, ecosystem, and global ecology, as we explore how ecologists apply biological knowledge to predict the global consequences of human activities and to conserve Earth's biodiversity.

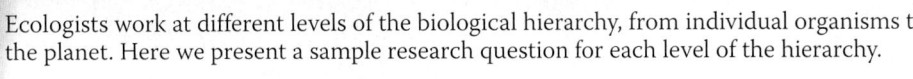

Ecologists work at different levels of the biological hierarchy, from individual organisms to the planet. Here we present a sample research question for each level of the hierarchy.

## Organismal Ecology

**Organismal ecology**, which includes the subdisciplines of physiological, evolutionary, and behavioral ecology, is concerned with how an organism's structure, physiology, and behavior meet the challenges posed by its environment.

◄ How do flamingos select a mate?

## Population Ecology

A **population** is a group of individuals of the same species living in an area. **Population ecology** analyzes factors that affect population size and how and why it changes through time.

◄ What environmental factors affect the reproductive rate of flamingos?

## Community Ecology

A **community** is a group of populations of different species in an area. **Community ecology** examines how interactions between species, such as predation and competition, affect community structure and organization.

◄ What factors influence the diversity of species that interact at this African lake?

## Ecosystem Ecology

An **ecosystem** is the community of organisms in an area and the physical factors with which those organisms interact. **Ecosystem ecology** emphasizes energy flow and chemical cycling between organisms and the environment.

◄ What factors control photosynthetic productivity in this aquatic ecosystem?

## Landscape Ecology

A **landscape** (or seascape) is a mosaic of connected ecosystems. Research in **landscape ecology** focuses on the factors controlling exchanges of energy, materials, and organisms across multiple ecosystems.

◄ To what extent do nutrients from terrestrial ecosystems affect organisms in the lake?

## Global Ecology

The **biosphere** is the global ecosystem—the sum of all the planet's ecosystems and landscapes. Global ecology examines how the regional exchange of energy and materials influences the functioning and distribution of organisms across the biosphere.

◄ How do global patterns of air circulation affect the distribution of organisms?

## Latitudinal Variation in Sunlight Intensity

Earth's curved shape causes latitudinal variation in the intensity of sun light. Because sunlight strikes the tropics (those regions that lie between 23.5° north latitude and 23.5° south latitude) most directly, more heat and light per unit of surface area are delivered there. At higher latitudes, sunlight strikes Earth at an oblique angle, and thus the light energy is more diffuse on Earth's surface.

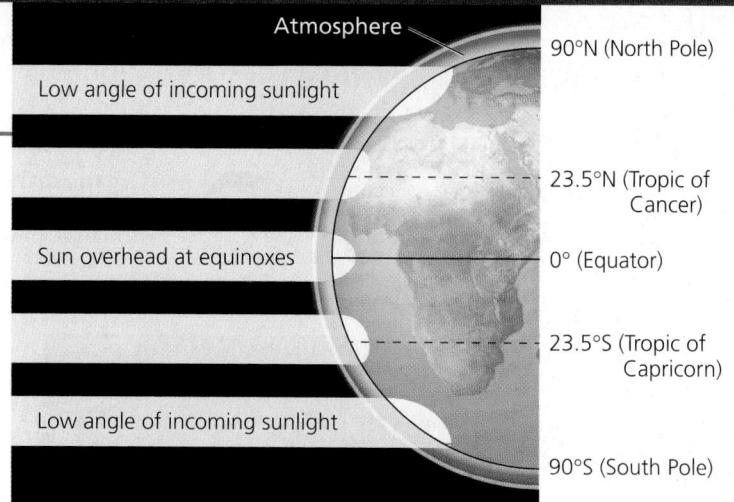

## Global Air Circulation and Precipitation Patterns

Intense solar radiation near the equator initiates a global pattern of air circulation and precipitation. High temperatures in the tropics evaporate water from Earth's surface and cause warm, wet air masses to rise (blue arrows) and flow toward the poles. As the rising air masses expand and cool, they release much of their water content, creating abundant precipitation in tropical regions. The high-altitude air masses, now dry, descend (tan arrows) toward Earth around 30° north and south, absorbing moisture from the land and creating an arid climate conducive to the development of the deserts that are common at those latitudes. Some of the descending air then flows toward the poles. At latitudes around 60° north and south, the air masses again rise and release abundant precipitation (though less than in the tropics). Some of the cold, dry rising air then flows to the poles, where it descends and flows back toward the equator, absorbing moisture and creating the comparatively rainless and bitterly cold climates of the polar regions.

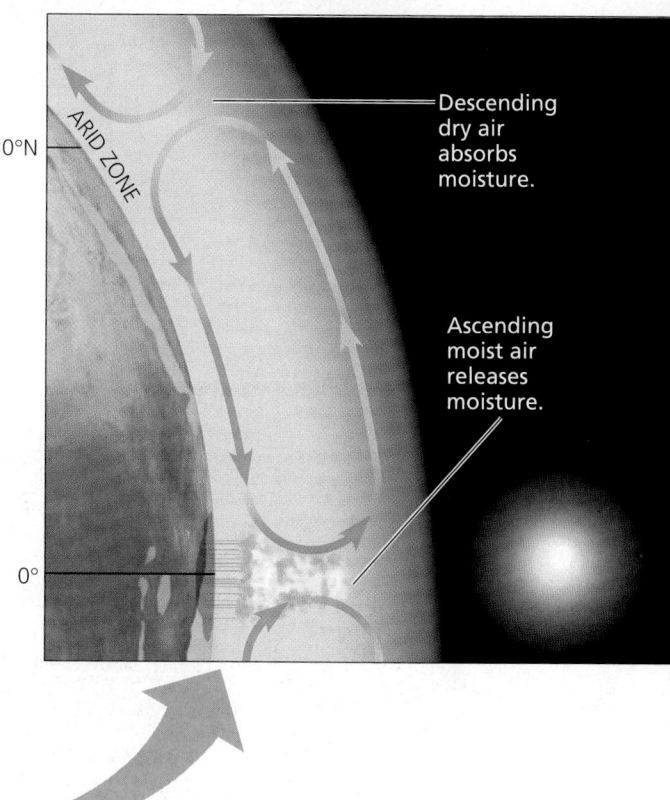

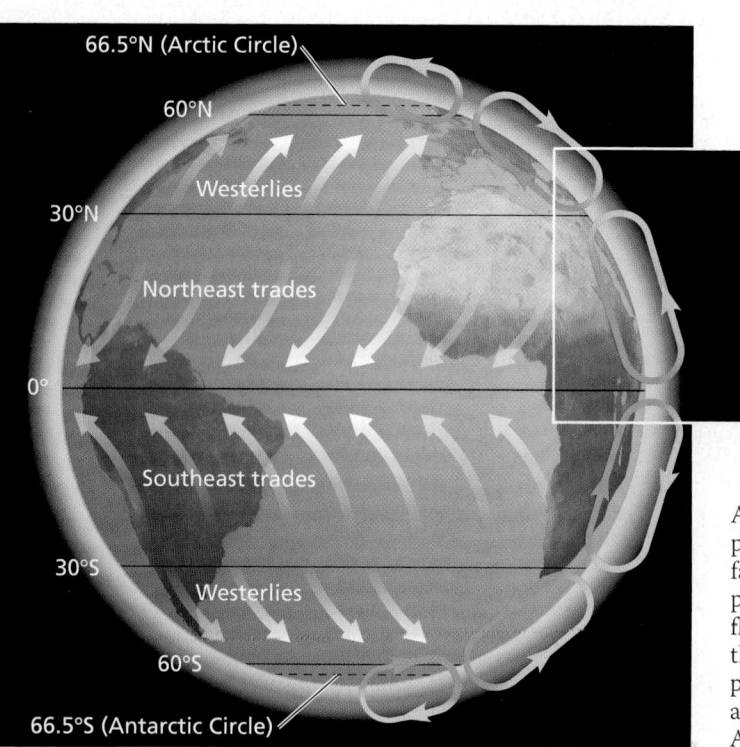

Air flowing close to Earth's surface creates predictable global wind patterns. As Earth rotates on its axis, land near the equator moves faster than that at the poles, deflecting the winds from the vertical paths shown above and creating the more easterly and westerly flows shown at left. Cooling trade winds blow from east to west in the tropics; prevailing westerlies blow from west to east in the temperate zones, defined as the regions between the Tropic of Cancer and the Arctic Circle and between the Tropic of Capricorn and the Antarctic Circle.

# Earth's climate influences the distribution of terrestrial biomes

The most significant influence on the distribution of organisms on land is **climate**, the long-term, prevailing weather conditions in a given area. Four physical factors—temperature, precipitation, sunlight, and wind—are particularly important components of climate. Such **abiotic**, or nonliving, factors are the chemical and physical attributes of the environment that influence the distribution and abundance of organisms. **Biotic**, or living, factors—the other organisms that are part of an individual's environment—also influence the distribution and abundance of life.

We begin by describing patterns in climate that can be observed at the global, regional, and landscape levels.

## Global Climate Patterns

Global climate patterns are determined largely by the input of solar energy and Earth's movement in space. The sun warms the atmosphere, land, and water. This warming establishes the temperature variations, movements of air and water, and evaporation of water that cause dramatic latitudinal variations in climate. **Figure 40.3** summarizes Earth's climate patterns and how they are formed.

## Regional Effects on Climate

Climate varies seasonally and can be modified by other factors, such as large bodies of water and mountain ranges.

### Seasonality

In middle to high latitudes, Earth's tilted axis of rotation and its annual passage around the sun cause strong seasonal cycles in day length, solar radiation, and temperature **(Figure 40.4)**. The changing angle of the sun over the course of the year also affects local environments. For example, the belts of wet and dry air on either side of the equator move slightly northward and southward as the sun's angle changes; this produces marked wet and dry seasons around 20° north and 20° south latitude, where many tropical deciduous forests grow. In addition, seasonal changes in wind patterns alter ocean currents, sometimes causing the upwelling of cold water from deep ocean layers. This nutrient-rich water stimulates the growth of surface-dwelling phytoplankton and the organisms that feed on them. These upwelling zones make up only a small percentage of ocean area but are responsible for more than a quarter of fish caught globally.

### Bodies of Water

Ocean currents influence climate along the coasts of continents by heating or cooling overlying air masses that pass across the land. Coastal regions are also generally wetter than inland areas at the same latitude. The cool, misty climate

▼ **Figure 40.4 Seasonal variation in sunlight intensity.** Because Earth is tilted on its axis relative to its plane of orbit around the sun, the intensity of solar radiation varies seasonally. This variation is smallest in the tropics and increases toward the poles.

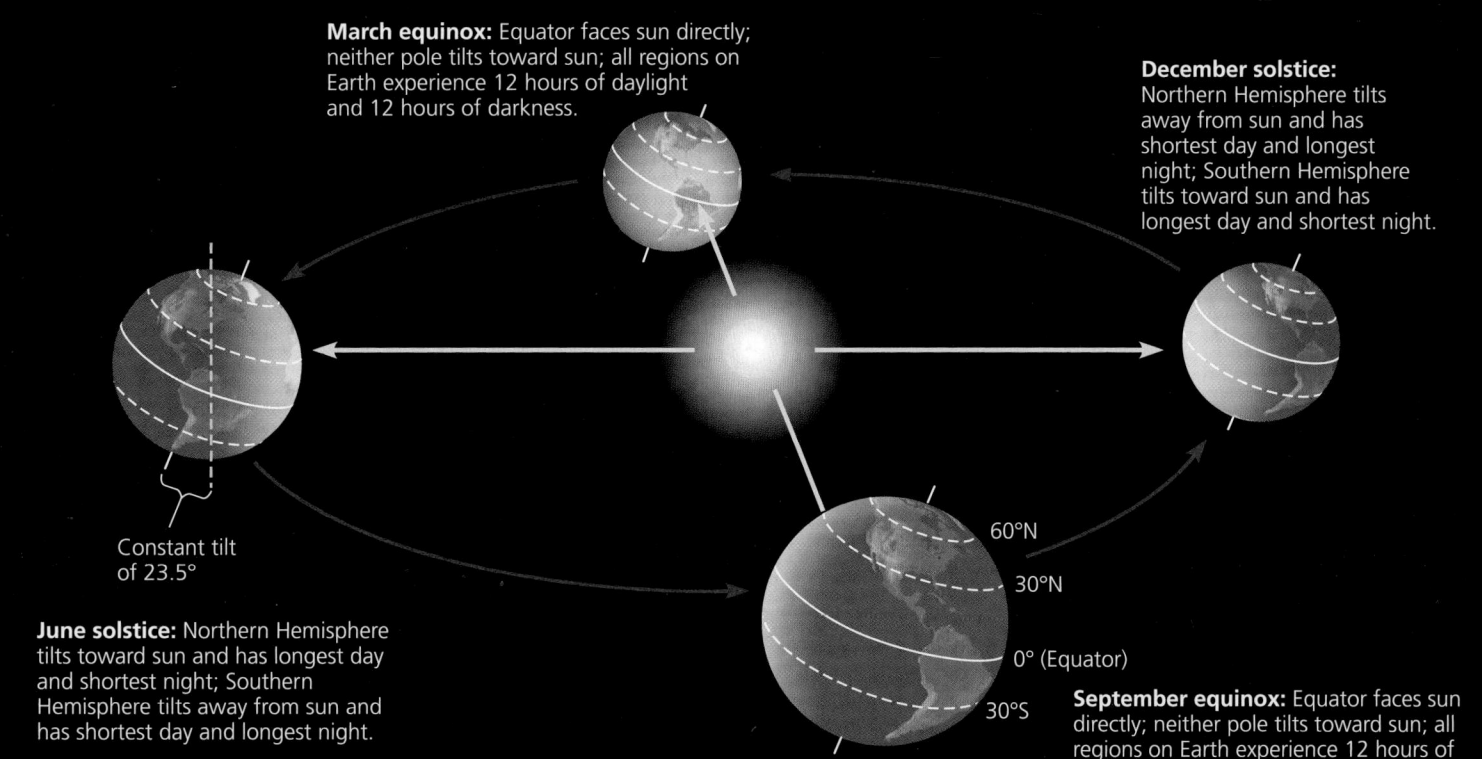

**March equinox:** Equator faces sun directly; neither pole tilts toward sun; all regions on Earth experience 12 hours of daylight and 12 hours of darkness.

**December solstice:** Northern Hemisphere tilts away from sun and has shortest day and longest night; Southern Hemisphere tilts toward sun and has longest day and shortest night.

Constant tilt of 23.5°

**June solstice:** Northern Hemisphere tilts toward sun and has longest day and shortest night; Southern Hemisphere tilts away from sun and has shortest day and longest night.

60°N
30°N
0° (Equator)
30°S

**September equinox:** Equator faces sun directly; neither pole tilts toward sun; all regions on Earth experience 12 hours of

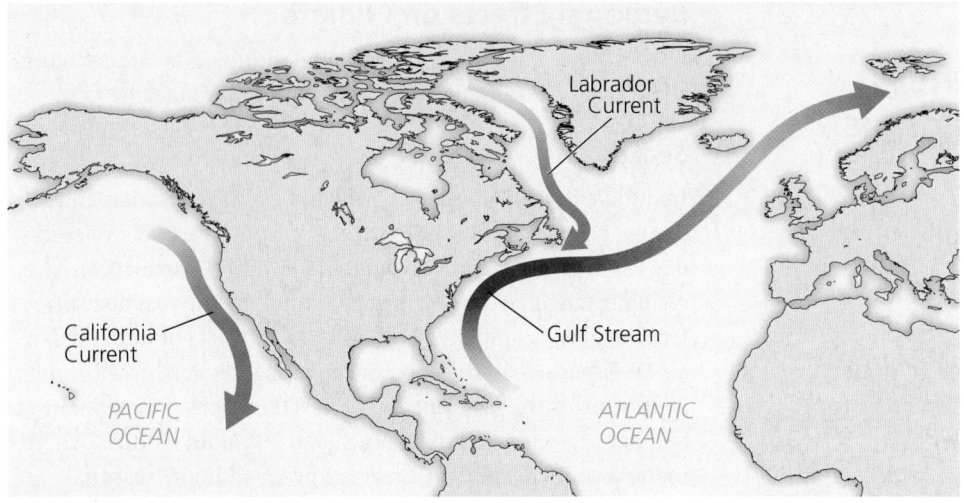

▲ **Figure 40.5 Circulation of surface water in the oceans around North America.** The California Current carries cold water southward along the western coast of North America. Along the eastern coast, the Labrador Current carries cold water southward, and the Gulf Stream moves warm water toward northern Europe.

produced by the cold California Current that flows southward along western North America supports a coniferous rain forest ecosystem along much of the continent's Pacific coast **(Figure 40.5)**. Conversely, the west coast of northern Europe has a mild climate because the Gulf Stream carries warm water from the equator to the North Atlantic. As a result, northwestern Europe is warmer during winter than southeastern Canada, which is farther south but is cooled by the Labrador Current flowing south from the coast of Greenland.

Because of the high specific heat of water (see Concept 2.5), oceans and large lakes tend to moderate the climate of nearby land. During a hot day, when land is warmer than the water, air over the land heats up and rises, drawing a cool breeze from the water across the land **(Figure 40.6)**. In contrast, because temperatures drop more quickly over land than over water at night, air over the now warmer water rises, drawing cooler air from the land back out over the water and replacing it with warmer air from offshore.

### Mountains

Like large bodies of water, mountains influence air flow over land. When warm, moist air approaches a mountain, the air expands and cools as it rises, releasing moisture on the windward side of the peak (see Figure 40.6). On the leeward side, cooler, dry air descends, absorbing moisture and producing a "rain shadow." Such leeward rain shadows determine where many deserts are found, including the Mojave Desert of western North America and the Gobi Desert of Asia.

Mountains also affect the amount of sunlight reaching an area and thus the local temperature and rainfall. South-facing slopes in the Northern Hemisphere receive more sunlight than north-facing slopes and are therefore warmer and drier. These physical differences influence species distributions locally. In many mountains of western North America, spruce and other conifers grow on the cooler north-facing slopes, but shrubby, drought-resistant plants inhabit the south-facing slopes. In addition, every 1,000-m increase in elevation produces an average temperature drop of 6°C, equivalent to that produced by an 880-km increase in latitude. This is one reason that high-elevation communities near the equator, for example, can be similar to lower-elevation communities that are far from the equator.

## Climate and Terrestrial Biomes

We turn now to the role of climate in determining the nature and location of Earth's **biomes**, major life zones characterized by vegetation type (in terrestrial biomes) or by the physical environment (in aquatic biomes, which we will survey in Concept 40.2).

▼ **Figure 40.6 How large bodies of water and mountains affect climate.** This figure illustrates what can happen on a hot summer day.

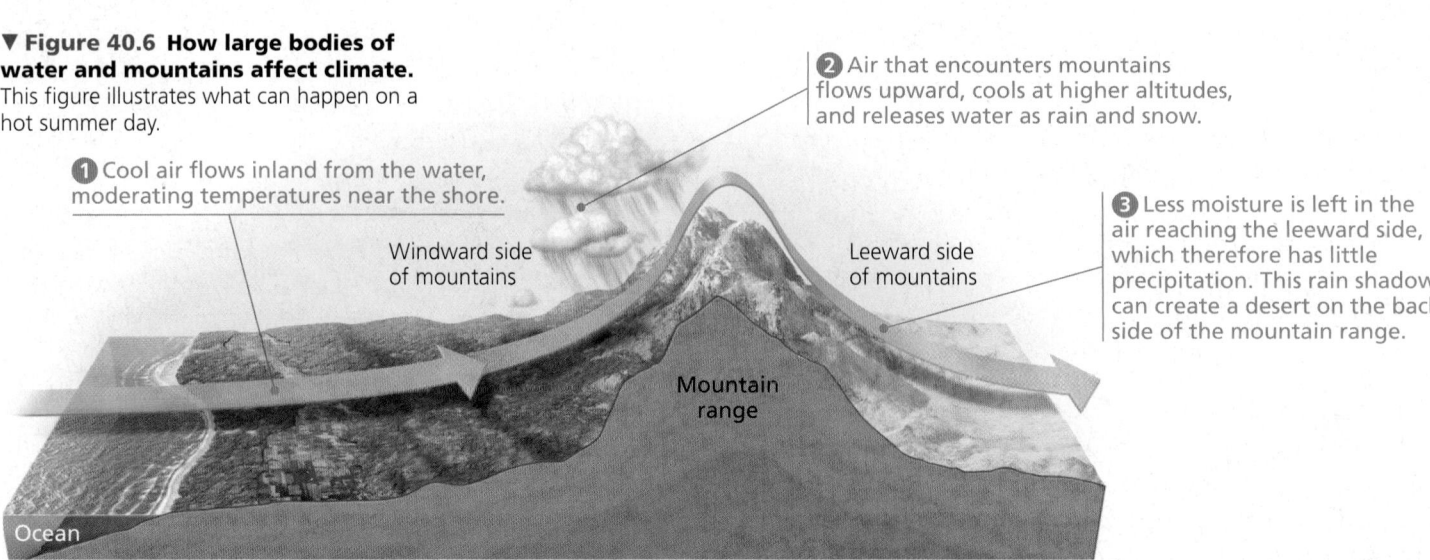

❷ Air that encounters mountains flows upward, cools at higher altitudes, and releases water as rain and snow.

❶ Cool air flows inland from the water, moderating temperatures near the shore.

❸ Less moisture is left in the air reaching the leeward side, which therefore has little precipitation. This rain shadow can create a desert on the back side of the mountain range.

Windward side of mountains

Leeward side of mountains

Mountain range

Ocean

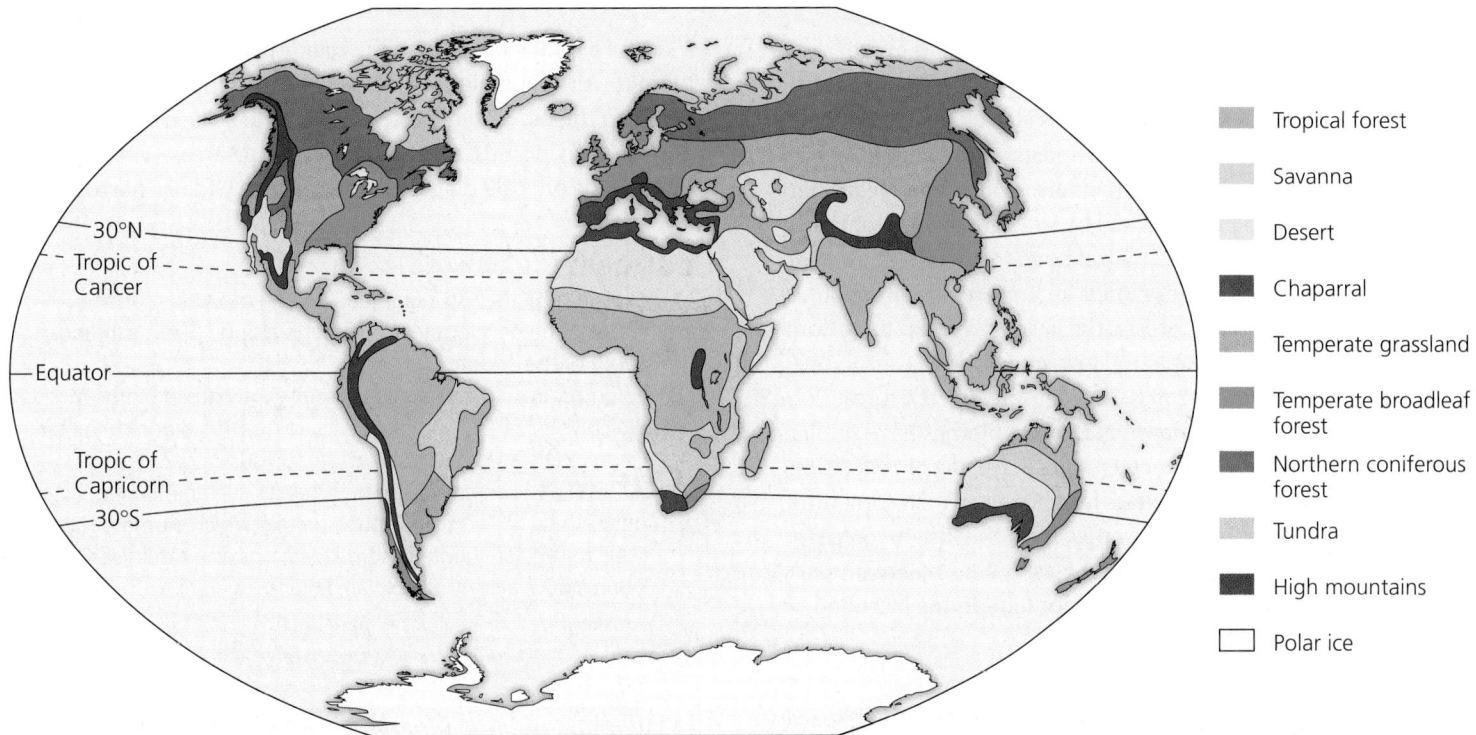

▲ **Figure 40.7 The distribution of major terrestrial biomes.**

Because of the latitudinal patterns of climate described in Figure 40.3, the locations of terrestrial biomes also show strong latitudinal patterns **(Figure 40.7)**. For example, the dry air that descends at 30° north and south often leads to the formation of deserts at those latitudes. We can highlight the importance of climate on the distribution of biomes is by constructing a **climograph**, a plot of the annual mean temperature and precipitation in a particular region **(Figure 40.8)**. Notice, for instance, that grasslands in North America are typically drier than forests and that deserts are drier still.

Factors other than mean temperature and precipitation also play a role in determining where biomes exist. For example, some areas in North America with a particular combination of temperature and precipitation support a temperate broadleaf forest, but other areas with similar values for these variables support a coniferous forest (see the overlap in Figure 40.8). One reason for this variation is that climographs are based on annual *averages*, but the *pattern* of climatic variation is often as important as the average climate. For example, some areas may receive regular precipitation throughout the year, whereas other areas may have distinct wet and dry seasons.

Natural and human-caused disturbances also alter the distribution of biomes. A **disturbance** is an event such as a storm, fire, or human activity that changes a community, re-moving organisms from it and altering resource availability. For instance, frequent fires can kill woody plants and keep a sa-vanna from becoming the woodland that climate alone would support. Hurricanes and other storms create openings for new species in many tropical and temperate forests. Human-caused disturbances have altered much of Earth's surface, replacing natural communities with urban and agricultural ones.

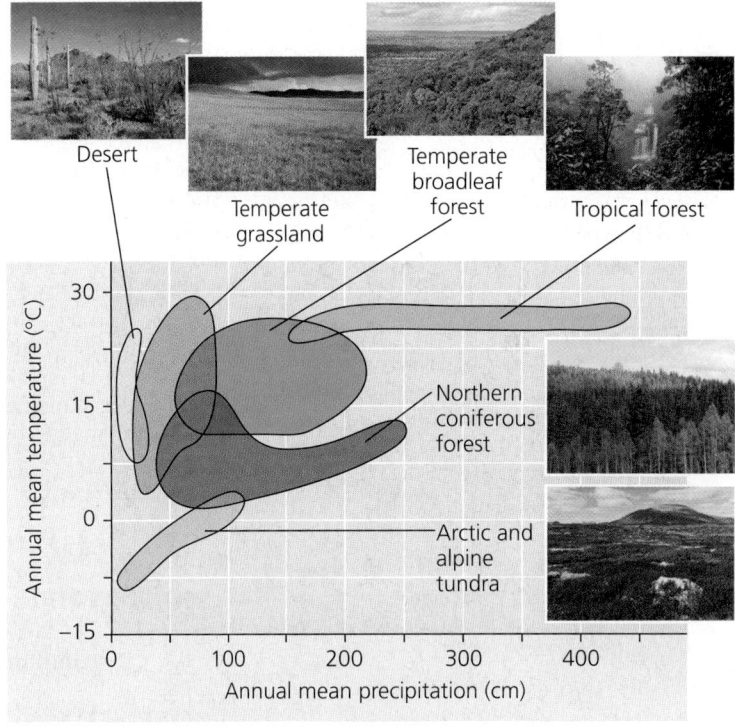

▲ **Figure 40.8 A climograph for some major types of biomes in North America.** The areas plotted here encompass the ranges of annual mean temperature and precipitation in the biomes.

**INTERPRET THE DATA** *Some arctic tundra ecosystems receive as little rainfall as deserts but have much denser vegetation. What climatic factor might explain this difference? Explain.*

## General Features of Terrestrial Biomes

Most terrestrial biomes are named for major physical or cli-matic features and for their predominant vegetation. Tem-perate grasslands, for instance, are generally found in middle latitudes, where the climate is more moderate than in the tropics or polar regions, and are dominated by various grass species (see Figure 40.7). Each biome is also characterized by microorganisms, fungi, and animals adapted to that particular environment. Temperate grasslands are usually more likely than temperate forests to be populated by large grazing mammals and to have arbuscular mycorrhizal fungi (see Figure 29.14).

Vertical layering of vegetation is an important feature of terrestrial biomes. In many forests, the layers from top to bottom consist of the upper **canopy**, the low-tree layer, the shrub understory, the ground layer of herbaceous plants, the forest floor (litter layer), and the root layer. Nonforest biomes have similar, though usually less pronounced, layers. Layering of vegetation provides many different habitats for animals, which sometimes exist in well-defined feeding groups, from the insectivorous birds and bats that feed above canopies to the small mammals, worms, and arthropods that search for food in the litter and root layers below.

**Figure 40.9** summarizes the major features of terrestrial biomes. Although Figure 40.7 shows distinct boundaries between the biomes, terrestrial biomes usually grade into neighboring biomes, sometimes over large areas. The area of intergradation, called an **ecotone**, may be wide or narrow.

## CONCEPT CHECK 40.1

1. Explain how the sun's unequal heating of Earth's surface leads to the development of deserts around 30° north and south of the equator.
2. Using Figures 40.7 and 40.9, identify the natural biome in which you live and summarize its abiotic and biotic characteristics. Do these reflect your actual surroundings? Explain.
3. **WHAT IF?** If global warming increases average temperatures on Earth by 4°C in this century, predict which biome is most likely to replace tundra in some locations as a result (see Figures 40.7 and 40.8). Explain your answer.

For suggested answers, see Appendix A.

---

▼ **Figure 40.9**  **Exploring Terrestrial Biomes**

## Tropical Forest

**Distribution** Equatorial and subequatorial regions

**Climate** Temperature is usually high, averaging 25–29°C with little seasonal variation. In **tropical rain forests**, rainfall is relatively constant, about 200–400 cm annually. In **tropical dry forests**, precipitation averages about 150–200 cm annually, with a six-to seven-month dry season.

**Organisms** Tropical forests are vertically layered, and plants compete strongly for light. Broadleaf evergreen trees are dominant in rain forests, whereas many dry forest trees drop their leaves during the dry season. Tropical forests are home to millions of animal species, including an estimated 5–30 million still undescribed species of insects, spiders, and other arthropods. Animal diversity is higher than in any other terrestrial biome. The animals are adapted to the vertically layered environment and are often inconspicuous.

**Human Impact** Humans long ago established thriving communities in tropical forests. Rapid population growth leading to agriculture and development is now destroying many tropical forests.

A tropical rain forest in Costa Rica

## Savanna

**Distribution** Equatorial and subequatorial regions

**Climate** Rainfall averages 30–50 cm per year in **savannas** and is seasonal, with a dry season that can last up to nine months. Temperature averages 24–29°C but varies seasonally more than in tropical forests.

**Organisms** Scattered trees often are thorny and have small leaves, an apparent adaptation to the relatively dry conditions. Fires are common in the dry season, and the dominant plant species are fire-adapted and tolerant of seasonal drought.

Grasses and small nonwoody plants called forbs make up most of the ground cover. Large plant-eating mammals, such as wildebeests and zebras, and predators, including lions and hyenas, are common inhabitants. However, the dominant herbivores are insects, especially termites.

**Human Impact** The earliest humans may have lived in savannas. Overly frequent fires set by humans reduce tree regeneration by killing the seedlings and saplings. Cattle ranching and overhunting have led to declines in large-mammal populations.

A savanna in Kenya

## Desert

**Distribution** **Deserts** occur in bands near 30° north and south latitude or at other latitudes in the interior of continents (for instance, the Gobi Desert of north-central Asia).

**Climate** Precipitation is low and highly variable, generally less than 30 cm per year. Temperature varies seasonally and daily. It may exceed 50°C in hot deserts and fall below −30°C in cold deserts.

**Organisms** Desert landscapes are dominated by low, widely scattered vegetation. Common plants include succulents such as cacti or euphorbs, deeply rooted shrubs, and herbs that grow during the infrequent moist periods. Desert plant adaptations include tolerance to heat and desiccation, water storage, reduced leaf surface area, and physical defenses such as spines and toxins in leaves. Many desert plants carry out C4 or CAM photosynthesis. Common desert animals include scorpions, ants, beetles, snakes, lizards, migratory and resident birds, and seed-eating rodents. Many species in hot deserts are active at night, when the air is cooler. Water conservation is a common adaptation, and some animals can obtain all their water by breaking down carbohydrates in seeds.

**Human Impact** Long-distance transport of water and deep groundwater wells have allowed humans to maintain substantial populations in deserts. Urbanization and conversion to irrigated agriculture have reduced the natural biodiversity of some deserts.

Organ Pipe Cactus National Monument, Arizona

## Chaparral

**Distribution** Midlatitude coastal regions on several continents

**Climate** Annual precipitation is typically 30–50 cm and is highly seasonal, with rainy winters and dry summers. Fall, winter, and spring are cool, with average temperatures of 10–12°C. Average summer temperature can reach 30°C.

**Organisms** **Chaparral** is dominated by shrubs and small trees adapted to frequent fires. Some fire-adapted shrubs produce seeds that will germinate only after a hot fire; food reserves stored in their roots enable them to resprout quickly and use nutrients released by the fire. Adaptations to drought include tough, evergreen leaves, which reduce water loss. Animals include browsers, such as deer and goats, that feed on twigs and buds of woody vegetation; there are also many species of insects, amphibians, small mammals, and birds.

**Human Impact** Chaparral areas have been heavily settled and reduced through conversion to agriculture and urbanization. Humans contribute to the fires that sweep across the chaparral.

An area of chaparral in California

## Temperate Grassland

**Distribution** Typically at midlatitudes, often in the interior of continents

**Climate** Annual precipitation in **temperate grasslands** generally averages 30 to 100 cm and can be highly seasonal, with relatively dry winters and wet summers. Average temperatures frequently are below −10°C in winter and reach 30°C in summer.

**Organisms** Dominant plants are grasses and forbs, which vary in height from a few centimeters to 2 m in tallgrass prairie. Many grassland plants have adaptations that help them survive periodic, protracted droughts and fire. Grazing by large mammals such as bison and wild horses helps prevent establishment of woody shrubs and trees. Burrowing mammals, such as prairie dogs in North America, are also common.

**Human Impact** Because of their deep, fertile soils, temperate grasslands in North America and Eurasia have frequently been converted to farmland. In some drier grasslands, cattle and other grazers have turned parts of the biome into desert.

A grassland in Mongolia

## Northern Coniferous Forest

**Distribution** In a broad band across northern North America and Eurasia to the edge of the arctic tundra, the **northern coniferous forest**, or *taiga*, is the largest terrestrial biome.

**Climate** Annual precipitation generally ranges from 30 to 70 cm. Winters are cold. Some areas of coniferous forest in Siberia typically range in temperature from −50°C in winter to over 20°C in summer.

**Organisms** Cone-bearing trees (conifers), such as pine, spruce, fir, and hemlock, are common, and some species depend on fire to regenerate. The conical shape of many conifers prevents snow from accumulating and breaking their branches, and their needlelike or scalelike leaves reduce water loss. Plant diversity in the shrub and herb layers is lower than in temperate broadleaf forests. Many migratory birds nest in northern coniferous forests. Mammals include moose, brown bears, and Siberian tigers. Periodic outbreaks of insects can kill vast tracts of trees.

**Human Impact** Although they have not been heavily settled by human populations, northern coniferous forests are being logged at a fast rate, and old-growth stands may soon disappear.

A coniferous forest in Norway

## Temperate Broadleaf Forest

A temperate broadleaf forest in New Jersey

**Distribution** Midlatitudes in the Northern Hemisphere, with smaller areas in Chile, South Africa, Australia, and New Zealand

**Climate** Precipitation averages about 70 to 200 cm annually. Significant amounts fall during all seasons, with winter snow in some forests. Winter temperatures average around 0°C. Summers are humid, with maximum temperatures near 35°C.

**Organisms** The dominant plants of **temperate broadleaf forests** in the Northern Hemisphere are deciduous trees, which drop their leaves before winter, when low temperatures would reduce photosynthesis. In Australia, evergreen eucalyptus trees are common. In the Northern Hemisphere, many mammals hibernate in winter, while many bird species migrate to areas with warmer climates.

**Human Impact** Temperate broadleaf forests have been heavily settled globally. Logging and land clearing for agriculture and urban development have destroyed virtually all the original deciduous forests in North America, but these forests are returning over much of their former range.

## Tundra

**Distribution** **Tundra** covers expansive areas of the Arctic, amounting to 20% of Earth's land surface. High winds and low temperatures produce alpine tundra on very high mountaintops at all latitudes, including the tropics.

**Climate** Precipitation averages 20 to 60 cm annually in arctic tundra but may exceed 100 cm in alpine tundra. Winters are cold, with average temperatures in some areas below −30°C. Summer temperatures generally average less than 10°C.

**Organisms** The vegetation of tundra is mostly herbaceous, typically a mixture of mosses, grasses, and forbs, with some dwarf shrubs, trees, and lichens. A permanently frozen soil layer called permafrost restricts the growth of plant roots. Large grazing musk oxen are resident, while caribou and reindeer are migratory. Predators include bears, wolves, foxes, and snowy owls. Many bird species migrate to the tundra for summer nesting.

**Human Impact** Tundra is sparsely settled but has become the focus of significant mineral and oil extraction in recent years.

Dovrefjell -Sunndalsfjella National Park, Norway

## CONCEPT 40.2

# Aquatic biomes are diverse and dynamic systems that cover most of Earth

Unlike terrestrial biomes, aquatic biomes are characterized primarily by their physical and chemical environment. For example, marine biomes generally have salt concentrations that average 3%, whereas freshwater biomes such as lakes and streams typically have a salt concentration of less than 0.1%.

Another important feature of many aquatic biomes is that they are divided into vertical and horizontal zones, as illustrated for a lake in **Figure 40.10**. Light is absorbed by water and by photosynthetic organisms, so its intensity decreases rapidly with depth. The upper **photic zone** is where there is sufficient light for photosynthesis, and the lower **aphotic zone** is where little light penetrates. These two zones together make up the **pelagic zone**. At the bottom of these zones, deep or shallow, is the **benthic zone**, which consists of organic and inorganic sediments and is occupied by communities of organisms called the **benthos**. In a lake, aquatic biomes can be divided horizontally into the **littoral zone**, waters close to shore that are shallow enough for rooted plants, and the **limnetic zone**, waters farther from shore that are too deep to support plants with roots.

Thermal energy from sunlight warms surface waters, but the deeper waters remain cold. In the ocean and in most lakes, a narrow layer of abrupt temperature change called a **thermocline** separates the more uniformly warm upper layer from more uniformly cold deeper waters.

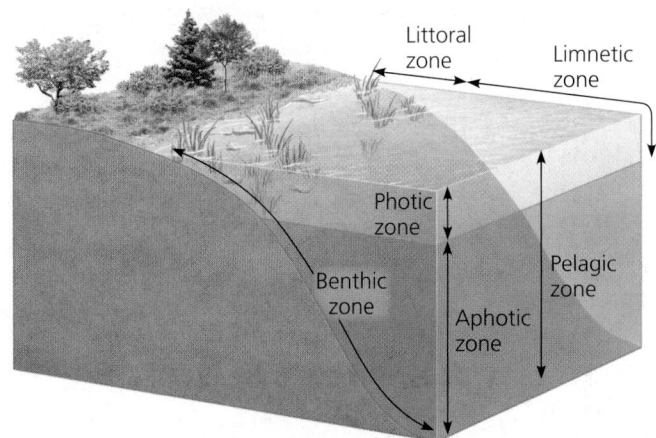

▲ **Figure 40.10 Zonation in a lake.** A lake environment is generally classified on the basis of three physical criteria: light penetration (photic and aphotic zones), distance from shore and water depth (littoral and limnetic zones), and whether the environment is open water (pelagic zone) or bottom (benthic zone).

Aquatic biomes show far less latitudinal variation than terrestrial biomes, with all types found across the globe **(Figure 40.11)**. The oceans make up the largest marine biome, covering about 75% of Earth's surface. Because of their vast size, they greatly impact the biosphere. Water evaporated from the oceans provides most of the planet's rainfall. Marine algae and photosynthetic bacteria supply much of the world's oxygen and consume large amounts of atmospheric carbon dioxide. Ocean temperatures have a major effect on global climate and wind patterns (see Figure 40.3), and along with large lakes, oceans tend to moderate the climate of nearby land.

---

▼ **Figure 40.11** | **Exploring Aquatic Biomes**

## Wetlands and Estuaries

**Physical and Chemical Environment**
**Wetlands** are inundated by water at least sometimes and support plants adapted to water-saturated soil. In an **estuary,** the transition zone between a river and the sea, seawater flows up and down the estuary channel during the changing tides. Nutrients from upstream make wetlands and estuaries among the most productive habitats on Earth. Because of high organic production by plants and decomposition by microorganisms, the water and soils are often low in dissolved oxygen. Both habitats filter dissolved nutrients and chemical pollutants.

**Geologic Features** Wetlands develop in diverse habitats, including shallow basins, the flooded banks of rivers and streams, and lake coasts. Along seacoasts, sediments from rivers and tidal waters create channels, islands, and mudflats in estuaries.

**Organisms** Water-saturated soils favor the growth of plants, such as cattails and sedges in wetlands and saltmarsh grasses in estuaries, that can grow in water or in soil that is anaerobic at times. In freshwater wetlands, herbivores may include crustaceans, aquatic insect larvae, and muskrats, and carnivores may include dragonflies, frogs, alligators, and herons. Estuaries support an abundance of oysters, crabs, and fish species that humans eat. Many marine invertebrates and fishes use estuaries as breeding grounds.

A basin wetland in the United Kingdom

**Human Impact** Draining and filling have destroyed up to 90% of wetlands in Europe. Filling, dredging, and upstream pollution have disrupted estuaries worldwide.

An oligotrophic lake in Alberta, Canada

## Lakes

**Physical and Chemical Environment**
Standing bodies of water range from ponds a few square meters in area to lakes covering thousands of square kilometers. Light decreases with depth, creating photic and aphotic zones. Temperate lakes may have a seasonal thermocline; tropical lowland lakes have a thermocline year-round. **Oligotrophic lakes** are nutrient-poor and generally oxygen-rich; **eutrophic lakes** are nutrient-rich and often depleted of oxygen in the deepest zone in summer and if covered with ice in winter. High rates of decomposition in deeper layers of eutrophic lakes cause periodic oxygen depletion.

**Geologic Features** Oligotrophic lakes may become more eutrophic over time as runoff adds sediments and nutrients. They tend to have less surface area relative to their depth than eutrophic lakes.

**Organisms** Rooted and floating aquatic plants live in the littoral zone, the shallow, well-lit waters close to shore. The limnetic zone, where water is too deep to support rooted aquatic plants, is inhabited by a variety of phytoplankton, including cyanobacteria, and small drifting heterotrophs, or zooplankton, that graze on the phytoplankton. The benthic zone is inhabited by assorted invertebrates whose species composition depends partly on oxygen levels. Fishes live in all zones with sufficient oxygen.

**Human Impact** Runoff from fertilized land and dumping of wastes lead to nutrient enrichment, which can produce algal blooms, oxygen depletion, and fish kills.

## Streams and Rivers

**Physical and Chemical Environment**
Headwater streams are generally cold, clear, turbulent, and swift. Downstream in larger rivers, the water is generally warmer and more turbid because of suspended sediment. The salt and nutrient content of streams and rivers increases from the headwaters to the mouth, but oxygen content typically decreases.

**Geologic Features** Headwater stream channels are often narrow, have a rocky bottom, and alternate between shallow sections and deeper pools. Rivers are generally wide and meandering. River bottoms are often silty from sediments deposited through time.

**Organisms** Headwater streams that flow through grasslands or deserts may be rich in phytoplankton or rooted aquatic plants. Diverse fishes and invertebrates inhabit unpolluted rivers and streams. In streams flowing through forests, organic matter from terrestrial vegetation is the primary source of food for aquatic consumers.

**Human Impact** Municipal, agricultural, and industrial pollution degrade water quality and can kill aquatic organisms. Dams impair the natural flow of streams and rivers and threaten migratory species such as salmon.

A headwater stream in Washington

## Intertidal Zones

A rocky intertidal zone on the Oregon coast

**Physical and Chemical Environment**
An **intertidal zone** is periodically submerged and exposed by the tides, twice daily on most marine shores. Upper strata experience longer exposures to air and greater variations in temperature and salinity, conditions that limit the distributions of many organisms to particular strata. Oxygen and nutrient levels are generally high and are renewed with each turn of the tides.

**Geologic Features** The rocky or sandy substrates of intertidal zones select for particular behavior and anatomy among intertidal organisms. The configuration of bays or coastlines influences the magnitude of tides and the exposure of intertidal zones to waves.

**Organisms** Diverse and plentiful marine algae grow on rocks in intertidal zones. Sandy intertidal zones exposed to waves generally lack attached plants or algae, while those in protected bays or lagoons often support rich beds of seagrass and algae. Some animals have structural adaptations that enable them to attach to rocks. Many animals in sandy or muddy intertidal zones, such as worms, clams, and predatory crustaceans, bury themselves and feed as the tides bring food. Other common animals are sponges, sea anemones, and small fishes.

**Human Impact** Oil pollution has disrupted many intertidal areas. Rock walls and barriers built to reduce erosion from waves and storm surges disrupt some areas.

## Coral Reefs

A coral reef in the Red Sea

**Physical and Chemical Environment**
**Coral reefs** are formed largely from the calcium carbonate skeletons of corals. Shallow reef-building corals live in the clear photic zone of tropical oceans, primarily near islands and along the edge of some continents. They are sensitive to temperatures below about 18–20°C and above 30°C. Deep-sea coral reefs are found at a depth of 200–1,500 m. Corals require high oxygen levels and can be harmed or killed by high inputs of fresh water and nutrients.

**Geologic Features** Corals require a solid substrate for attachment. A typical coral reef begins as a fringing reef on a young, high island, forms an offshore barrier reef later, and becomes a coral atoll as the older island submerges.

**Organisms** Unicellular algae live within the tissues of the corals in a mutualism that provides the corals with organic molecules. Diverse multicellular red and green algae also contribute substantial amounts of photosynthesis. Corals are the predominant animals on coral reefs, but fish and invertebrate diversity is also exceptionally high. Animal diversity on coral reefs rivals that of tropical forests.

**Human Impact** Collecting of coral skeletons and overfishing have reduced populations of corals and reef fishes. Global warming and pollution may be contributing to large-scale coral death. Development of coastal mangroves for aquaculture has also reduced spawning grounds for many species of reef fishes.

## Oceanic Pelagic Zone

**Physical and Chemical Environment**
The **oceanic pelagic zone** is a vast realm of open blue water, whose surface is constantly mixed by wind-driven currents. Because of higher water clarity, the photic zone extends to greater depths than in coastal marine waters. Oxygen content is generally high. Nutrient levels are generally lower than in coastal waters. Mixing of surface and deeper waters in fall and spring renews nutrients in the photic zones of temperate and high-latitude ocean areas.

**Geologic Features** This biome covers approximately 70% of Earth's surface. It has an average depth of nearly 4,000 m and a maximum depth of more than 10,000 m.

**Organisms** The dominant photosynthetic organisms are bacteria and other phytoplankton, which drift with the currents and account for half of global productivity. Zooplankton, including protists, worms, krill, jellies, and small larvae of invertebrates and fishes, eat the phytoplankton. Free-swimming animals include large squids, fishes, sea turtles, and marine mammals.

**Human Impact** Overfishing has depleted fish stocks in all oceans, which have also been affected by climate warming and pollution.

Open ocean near Iceland

## Marine Benthic Zone

A deep-sea hydrothermal vent community

**Physical and Chemical Environment**
The **marine benthic zone** consists of the seafloor. Except for shallow, near-coastal areas, the marine benthic zone is dark. Water temperature declines with depth, while pressure increases. Organisms in the very deep benthic, or abyssal, zone are adapted to continuous cold (about 3°C) and high water pressure. Oxygen concentrations are generally sufficient to support diverse animal life.

**Geologic Features** Soft sediments cover most of the benthic zone, but there are areas of rocky substrate on reefs, submarine mountains, and new oceanic crust.

**Organisms** Photosynthetic organisms, mainly seaweeds and filamentous algae, live in shallow benthic areas with sufficient light. In the dark, hot environments near **deep sea hydrothermal vents**, the food producers are chemoautotrophic prokaryotes. Coastal benthic communities include numerous invertebrates and fishes. Below the photic zone, most consumers depend entirely on organic matter raining down from above. Among the animals of the deep-sea hydrothermal vent communities are giant tube worms (pictured at left), some more than 1 m long. They are nourished by chemoautotrophic prokaryotes that live as symbionts within their bodies. Many other invertebrates, including arthropods and echinoderms, are also abundant around the hydrothermal vents.

**Human Impact** Overfishing has decimated important benthic fish populations, such as the cod of the Grand Banks off Newfoundland. Dumping of organic wastes has created oxygen-deprived benthic areas.

Freshwater biomes are closely linked to the soils and biotic components of the surrounding terrestrial biome. Freshwater biomes are also influenced by the patterns and speed of water flow and the climate to which the biome is exposed.

In both freshwater and marine environments, communities are distributed according to water depth, degree of light penetration, distance from shore, and whether they are found in open water or near the bottom. Plankton and many fish species occur in the relatively shallow photic zone (see Figure 40.10). Most of the deep ocean is virtually devoid of light (the aphotic zone) and harbors relatively little life.

## CONCEPT CHECK 40.2

The first two questions refer to Figure 40.11.
1. Why are phytoplankton, and not benthic algae or rooted aquatic plants, the dominant photosynthetic organisms of the oceanic pelagic zone?
2. **MAKE CONNECTIONS** Many organisms living in estuaries experience both freshwater and saltwater conditions each day with the rising and falling of tides. Explain how these changing conditions challenge the survival of these organisms (see Concept 32.3).
3. **WHAT IF?** Water leaving a reservoir behind a dam is often taken from deep layers of the reservoir. Would you expect fish found in a river below a dam in summer to be species that prefer colder or warmer water than fish found in an undammed river? Explain.

For suggested answers, see Appendix A.

---

CONCEPT 40.3

# Interactions between organisms and the environment limit the distribution of species

So far in this chapter, we've examined Earth's climate and the characteristics of terrestrial and aquatic biomes. We've also introduced the range of biological levels at which ecologists work (see Figure 40.2). In this section, we'll examine how ecologists determine what factors control the distribution of species, such as the *Paedophryne* frog shown in Figure 40.1.

Species distributions are a consequence of both ecological factors and evolutionary history. Consider kangaroos, which are found in Australia and nowhere else in the world. Fossil evidence indicates that kangaroos and their close relatives originated in Australia, roughly 5 million years ago. By that time, Australia had moved (by continental drift; see Concept 23.2) close to its present location and it was not connected to other landmasses. Thus, kangaroos occur only in Australia in part because of an accident of history: The kangaroo lineage originated there at a point in time when the continent was geographically isolated.

But ecological factors are also important. To date, kangaroos have not dispersed (on their own) to other continents; hence, they are restricted to the continent on which they originated. And within Australia, kangaroos are found in some habitats but not in others. The red kangaroo, for example, occurs in the arid grasslands of central Australia, but not in the tall, open forests of eastern Australia. Such observations can lead us to ask whether food availability, predators, temperature, or other factors cause red kangaroos to be found in some regions but not others.

As our discussion of kangaroos suggests, ecologists ask not only *where* species occur, but also *why* species occur where they do: What factors determine their distribution? Ecologists generally need to consider multiple factors and alternative hypotheses when attempting to explain the distribution of a species. We'll focus here on the ecological factors highlighted by the questions in the flowchart in **Figure 40.12**.

## Dispersal and Distribution

One factor that contributes greatly to the global distribution of organisms is **dispersal**, the movement of individuals or gametes away from their area of origin or from centers of high population density. For example, while land-bound kangaroos have not reached North America under their own power, other organisms that disperse more readily, such as some birds, have. The dispersal of organisms is critical to understanding the

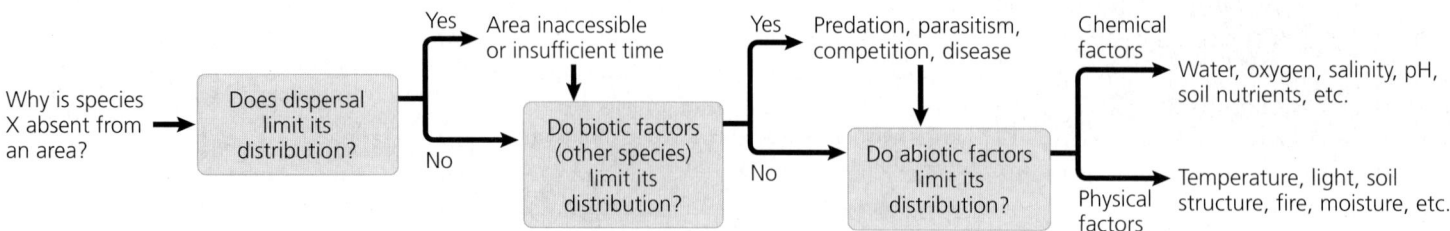

▲ **Figure 40.12 Flowchart of factors limiting geographic distribution.** An ecologist studying factors limiting a species' distribution might consider questions like these. As suggested by the arrows leading from the "yes" responses, the ecologist would answer all of these questions because more than one of these factors can limit a species' distribution.

❓ *How might the importance of various abiotic factors differ for aquatic and terrestrial ecosystems?*

role of geographic isolation in evolution (see Concept 22.2) as well as the broad patterns of species distribution that we see around the world today.

To determine if dispersal is a key factor limiting the distribution of a species, ecologists may observe the results of intentional or accidental transplants of the species to areas where it was previously absent. For a transplant to be successful, some of the organisms must not only survive in the new area but also reproduce there sustainably. If a transplant is successful, then we can conclude that the *potential* range of the species is larger than its *actual* range; in other words, the species *could* live in certain areas where it currently does not.

Species introduced to new geographic locations can disrupt the communities and ecosystems to which they have been introduced (see Concept 43.1). Consequently, ecologists rarely move species to new geographic regions. Instead, they document the outcome when a species has been transplanted for other purposes, as when a predator is introduced to control a pest species, or when a species has been moved to a new region accidentally.

## Biotic Factors

Our next question is whether biotic factors—other species—limit the distribution of a species. Often, the ability of a species to survive and reproduce is reduced by its interactions with other species, such as predators (organisms that kill their prey) or herbivores (organisms that eat plants or algae). **Figure 40.13** describes a specific case in which an herbivore, a sea urchin, has the potential to limit the distribution of a food species, a seaweed.

In addition to predation and herbivory, the presence or absence of pollinators, food resources, parasites, pathogens, and competing organisms can act as a biotic limitation on species distribution. As you will see in Chapter 41, such biotic limitations are common in nature.

## Abiotic Factors

The last question in the flowchart in Figure 40.12 considers whether abiotic factors—such as temperature, water, oxygen, salinity, sunlight, or soil—might be limiting a species' distribution. If the physical conditions at a site do not allow a species to survive and reproduce, then the species will not be found there.

---

### ▼ Figure 40.13 Inquiry

### Does feeding by sea urchins limit seaweed distribution?

**Experiment** W. J. Fletcher, of the University of Sydney, Australia, reasoned that if sea urchins are a limiting biotic factor in a particular ecosystem, then more seaweeds should invade an area from which sea urchins have been removed. To isolate the effect of sea urchins from that of a seaweed-eating mollusc, the limpet, he removed only urchins, only limpets, or both from study areas adjacent to a control site.

**Results** Fletcher observed a large difference in seaweed growth between areas with and without sea urchins.

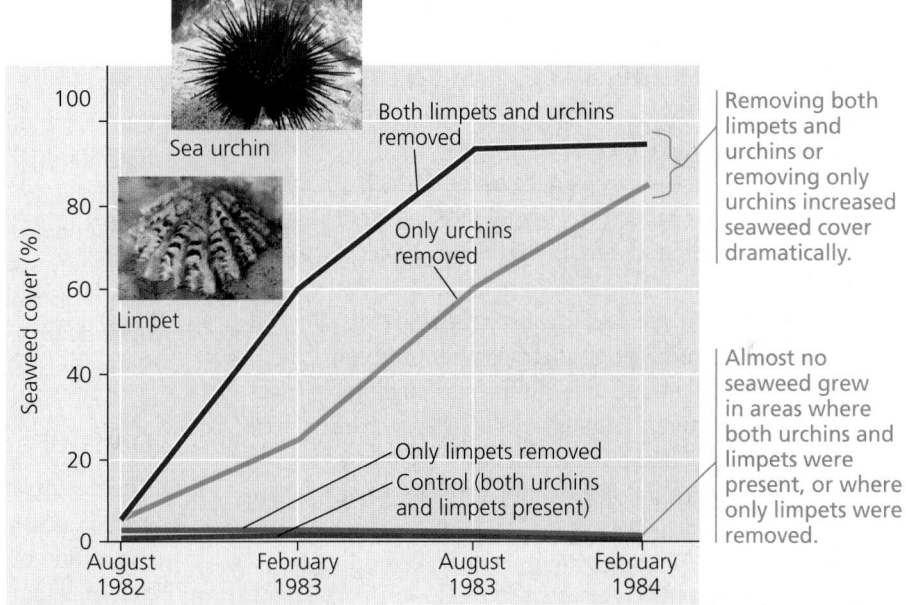

**Conclusion** Removing both limpets and urchins resulted in the greatest increase in seaweed cover, indicating that both species have some influence on seaweed distribution. But since removing only urchins greatly increased seaweed growth whereas removing only limpets had little effect, Fletcher concluded that sea urchins have a much greater effect than limpets in limiting seaweed distribution.

**Data from** W. J. Fletcher, Interactions among subtidal Australian sea urchins, gastropods, and algae: Effects of experimental removals, *Ecological Monographs* 57:89–109 (1987).

**INTERPRET THE DATA** Removing only limpets had little effect on seaweed growth compared to the control. Given this result, suggest a reason why removing both urchins and limpets resulted in greater seaweed growth than removing only urchins.

---

- **Temperature** Environmental temperature is an important factor in the distribution of organisms because of its effect on biological processes. Cells may rupture if the water they contain freezes (at temperatures below 0°C), and the proteins of most organisms denature at temperatures above 45°C. Most organisms function best within a specific range of environmental temperature.

- **Water and Oxygen** The dramatic variation in water availability among habitats is another important factor in species distribution. Species living at the seashore or in tidal wetlands can desiccate (dry out) as the tide recedes. Terrestrial organisms face a nearly constant threat of desiccation, and the distribution of terrestrial species reflects their ability to obtain and retain water. Many amphibians, such as the

*Paedophryne* frog in Figure 40.1, are particularly vulnerable to drying because they use their moist, delicate skin for gas exchange.

Water affects oxygen availability in aquatic environments and in flooded soils. Oxygen diffuses slowly in water. As a result, its concentration can be low in certain aquatic systems and soils, limiting cellular respiration and other physiological processes. Oxygen concentrations can be particularly low in both deep ocean and deep lake waters and sediments where organic matter is abundant.

- **Salinity** The salt concentration of water in the environment affects the water balance of organisms through osmosis. Most aquatic organisms are restricted to either freshwater or saltwater habitats by their limited ability to osmoregulate (see Concept 32.4). Although most terrestrial organisms can excrete excess salts from specialized glands or in feces or urine, salt flats and other high-salinity habitats typically have few species of plants or animals.

- **Sunlight** Sunlight provides the energy that drives most ecosystems, and too little sunlight can limit the distribution of photosynthetic species. In forests, shading by leaves makes competition for light especially intense, particularly for seedlings growing on the forest floor. In aquatic environments, most photosynthesis occurs near the surface, where sunlight is more available.

- **Rocks and Soil** On land, the pH, mineral composition, and physical structure of rocks and soil limit the distribution of plants and therefore of the animals that feed on them. The pH of soil can limit the distribution of organisms directly, through extreme acidic or basic conditions, or indirectly, by affecting the solubility of nutrients and toxins.

In a river, the composition of rocks and soil that make up the substrate (riverbed) can affect water chemistry, which in turn influences the resident organisms. In freshwater and marine environments, the structure of the substrate determines the organisms that can attach to it or burrow into it.

So far in this chapter, you have seen how the distributions of biomes and organisms depend on abiotic and biotic factors. In the rest of the chapter, we'll focus on how abiotic and biotic factors influence the ecology of populations.

### CONCEPT CHECK 40.3

1. Give examples of human actions that could expand a species' distribution by changing its (a) dispersal or (b) biotic interactions.

2. **WHAT IF?** You suspect that deer are restricting the distribution of a tree species by preferentially eating the seedlings of the tree. How might you test this hypothesis?

For suggested answers, see Appendix A.

# Biotic and abiotic factors affect population density, dispersion, and demographics

Population ecology explores how biotic and abiotic factors influence the density, distribution, and size of populations. A population is a group of individuals of a single species living in the same general area. Members of a population rely on the same resources, are influenced by similar environmental factors, and are likely to interact and breed with one another.

Populations are often described by their boundaries and size (the number of individuals living within those boundaries). Ecologists usually begin investigating a population by defining boundaries appropriate to the organism under study and to the questions being asked. A population's boundaries may be natural ones, as in the case of an island or a lake, or they may be arbitrarily defined by an investigator—for example, a specific county in Minnesota for a study of oak trees.

## Density and Dispersion

The **density** of a population is the number of individuals per unit area or volume: the number of oak trees per square kilometer in the Minnesota county or the number of *Escherichia coli* bacteria per milliliter in a test tube. **Dispersion** is the pattern of spacing among individuals within the boundaries of the population.

### Density: A Dynamic Perspective

In some cases, population size and density can be determined by counting all individuals within the boundaries of the population. We could count all the sea stars in a tide pool, for instance. Large mammals that live in herds, such as elephants, can sometimes be counted accurately from airplanes. In most cases, however, it is impractical or impossible to count all individuals in a population. Instead, ecologists use various sampling techniques to estimate densities and total population sizes. They might count the number of oak trees in several randomly located 100 × 100 m plots, calculate the average density in the plots, and then extend the estimate to the population size in the entire area. Such estimates are most accurate when there are many sample plots and when the habitat is fairly homogeneous. In other cases, instead of counting single organisms, ecologists estimate density from an indicator of population size, such as the number of nests, burrows, tracks, or fecal droppings.

Density is not a static property but changes as individuals are added to or removed from a population **(Figure 40.14)**. Additions occur through birth (which we define here to include all forms of reproduction) and **immigration**, the influx of new individuals from other areas. The factors that remove

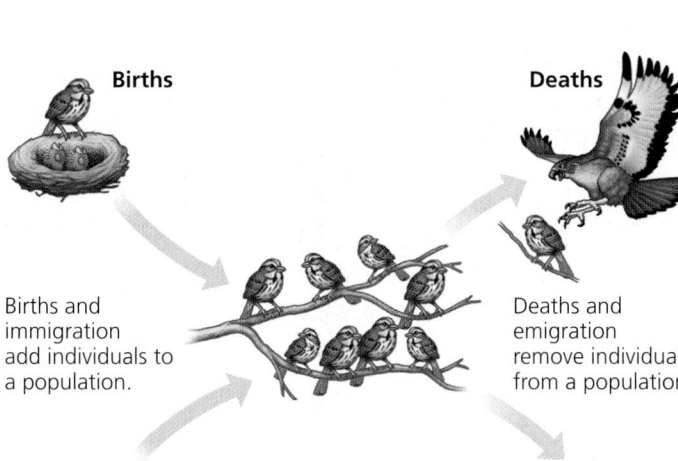

Births

Deaths

Births and
immigration
add individuals to
a population.

Deaths and
emigration
remove individuals
from a population.

Immigration

Emigration

▲ **Figure 40.14 Population dynamics.**

individuals from a population are death (mortality) and
**emigration**, the movement of individuals out of a population
and into other locations.

While birth and death rates influence the density of all
populations, immigration and emigration also alter the density
of many populations. Long-term studies of Belding's ground
squirrels (*Spermophilus beldingi*) in the vicinity of Tioga Pass,
in the Sierra Nevada of California, showed that some of the
squirrels moved nearly 2 km from where they were born. This
long-distance movement made them immigrants to other populations. In fact, immigrants made up 1–8% of the males and
0.7–6% of the females in the study population. Such immigration is a meaningful biological exchange between populations
over time.

### Patterns of Dispersion

Within a population's geographic range, local densities may
differ substantially, creating contrasting patterns of dispersion.
Differences in local density are among the most important
characteristics for a population ecologist to study, since they
provide insight into the environmental associations and social
interactions of individuals in the population.

The most common pattern of dispersion is *clumped*, in
which individuals are aggregated in patches. Plants and fungi
are often clumped where soil conditions and other environmental factors favor germination and growth. Mushrooms, for
instance, may be clumped within and on top of a rotting log.
Insects and salamanders may be clumped under the same log
because of the higher humidity there. Clumping of animals
may also be associated with mating behavior. Sea stars group
together in tide pools, where food is readily available and
where they can breed successfully **(Figure 40.15a)**. Forming
groups may also increase the effectiveness of predation or defense; for example, a wolf pack is more likely than a single wolf

**(a) Clumped.** Sea stars group together where food is abundant.

**(b) Uniform.** Nesting king penguins exhibit uniform spacing,
maintained by aggressive interactions between neighbors.

**(c) Random.** Dandelions grow from windblown seeds that land
at random and later germinate.

▲ **Figure 40.15 Patterns of dispersion within a population's
geographic range.**

to subdue a moose, and a flock of birds is more likely than a
single bird to warn of a potential attack.

A *uniform*, or evenly spaced, pattern of dispersion may
result from direct interactions between individuals in the
population. Some plants secrete chemicals that inhibit the
germination and growth of nearby individuals that could
compete for resources. Animals often exhibit uniform dispersion as a result of antagonistic social interactions, such
as **territoriality**—the defense of a bounded physical space
against encroachment by other individuals **(Figure 40.15b)**.

In *random* dispersion (unpredictable spacing), the position of each individual in a population is independent of other individuals. This pattern occurs in the absence of strong attractions or repulsions among individuals or where key physical or chemical factors are relatively constant across the study area. Plants established by windblown seeds, such as dandelions, may be randomly distributed in a fairly uniform habitat **(Figure 40.15c)**.

## Demographics

The factors that influence population density and dispersion patterns—ecological needs of a species, environmental conditions, and interactions among individuals within the population—also influence other characteristics of populations. **Demography** is the study of the vital statistics of populations and how they change over time. Of particular interest to demographers are birth rates and death rates. A useful way to summarize some of the vital statistics of a population is to make a life table.

### Life Tables

A **life table** provides a summary of the age-specific survival and reproductive rates of individuals in a population. When it is possible to do so, the best way to construct a life table is to follow the fate of a **cohort**, a group of individuals of the same age, from birth until all of the individuals are dead. To build such a life table, we need to determine the number of individuals that die in each age-group and to calculate the proportion of the cohort surviving from one age class to the next. We also need to keep track of the number of offspring produced by females in each age-group.

Demographers who study sexually reproducing species often ignore the males and concentrate on the females in a population because only females produce offspring. When this is done, a population is viewed in terms of females giving rise to new females. **Table 40.1** illustrates this approach for female Belding's ground squirrels from a population located in the Sierra Nevada mountains of California. Next, we'll take a closer look at some of the data in a life table, beginning with a discussion of survivorship curves.

### Survivorship Curves

The survival rate data in a life table can be represented graphically as a **survivorship curve**, a plot of the proportion or numbers in a cohort still alive at each age. Often, a survivorship curve begins with a cohort of a convenient size—say, 1,000 individuals. Though diverse, survivorship curves can be classified into three general types **(Figure 40.16)**.

A Type I curve is flat at the start, reflecting low death rates during early and middle life, and then drops steeply as death rates increase among older age-groups. Many large mammals, including humans, that produce few offspring but provide them with good care exhibit this kind of curve. In contrast, a Type III curve drops sharply at the start, reflecting very high

**Table 40.1 Life Table for Female Belding's Ground Squirrels (Tioga Pass, in the Sierra Nevada of California)**

| Age (years) | Number Alive at Start of Year | Proportion Alive at Start of Year[†] | Death Rate[‡] | Average Number of Female Offspring |
|---|---|---|---|---|
| 0–1 | 653 | 1.000 | 0.614 | 0.00 |
| 1–2 | 252 | 0.386 | 0.496 | 1.07 |
| 2–3 | 127 | 0.197 | 0.472 | 1.87 |
| 3–4 | 67 | 0.106 | 0.478 | 2.21 |
| 4–5 | 35 | 0.054 | 0.457 | 2.59 |
| 5–6 | 19 | 0.029 | 0.526 | 2.08 |
| 6–7 | 9 | 0.014 | 0.444 | 1.70 |
| 7–8 | 5 | 0.008 | 0.200 | 1.93 |
| 8–9 | 4 | 0.006 | 0.750 | 1.93 |
| 9–10 | 1 | 0.002 | 1.00 | 1.58 |

**Data from** P. W. Sherman and M. L. Morton, Demography of Belding's ground squirrel, *Ecology* 65:1617–1628 (1984).

[†]Indicates the proportion of the original cohort of 653 individuals that are still alive at the start of a time interval.

[‡]The death rate is the proportion of individuals alive at the start of a time interval that die during that time interval.

death rates for the young, but flattens out as death rates decline for those few individuals that survive the early period of die-off. This type of curve is usually associated with organisms that produce very large numbers of offspring but provide little or no care, such as long-lived plants, many fishes, and most marine invertebrates. An oyster, for example, may release millions of eggs, but most larvae hatched from fertilized eggs die from predation or other causes. Those few offspring that survive long enough to attach to a suitable substrate and begin growing a hard shell tend to survive for a relatively long time.

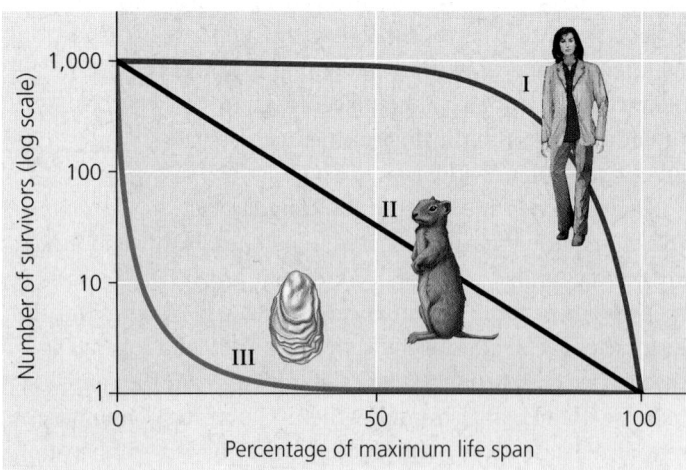

▲ **Figure 40.16 Idealized survivorship curves: Types I, II, and III.** The *y*-axis is logarithmic and the *x*-axis is on a relative scale so that species with widely varying life spans can be presented together on the same graph.

Type II curves are intermediate, with a constant death rate over the organism's life span. This kind of survivorship occurs in some rodents (including Belding's ground squirrel), invertebrates, lizards, and annual plants.

Many species fall somewhere between these basic types of survivorship or show more complex patterns. In birds, mortality is often high among the youngest individuals (as in a Type III curve) but fairly constant among adults (as in a Type II curve). Some invertebrates, such as crabs, may show a "stair-stepped" curve, with brief periods of increased mortality during molts, followed by periods of lower mortality when their protective exoskeleton is hard.

In populations not experiencing immigration or emigration, survivorship is one of the two key factors determining changes in population size. The other key factor determining population trends is reproductive rate.

### Reproductive Rates

As mentioned above, demographers often ignore the males and concentrate on the females in a population because only females produce offspring. Therefore, demographers view populations in terms of females giving rise to new females. The simplest way to describe the reproductive pattern of a population is to ask how reproductive output varies with the ages of females.

Reproductive output for sexual organisms such as birds and mammals is typically measured as the average number of female offspring for each female in a given age-group. For Belding's ground squirrels, which begin to reproduce at age 1 year, reproductive output rises to a peak at 4–5 years of age and then gradually falls off in older females (see Table 40.1).

Age-specific reproductive rates vary considerably by species. Squirrels, for example, have a litter of two to six young once a year for less than a decade, whereas oak trees drop thousands of acorns each year for tens or hundreds of years. Mussels and other invertebrates may release millions of eggs and sperm in a spawning cycle. However, a high reproductive rate will not lead to rapid population growth unless conditions are near ideal for the growth and survival of offspring, as you'll learn in the next section.

### CONCEPT CHECK 40.4

1. **DRAW IT** Each female of a particular fish species produces millions of eggs per year. Draw and label the most likely survivorship curve for this species, and explain your choice.
2. Imagine that you are constructing a life table for a different population of Belding's ground squirrels (see Table 40.1). If 485 individuals are alive at the start of year 0–1 and 218 are still alive at the start of year 1–2, what is the proportion alive at the start of each of these years (see column 3 in Table 40.1)?
3. **MAKE CONNECTIONS** A male stickleback fish attacks other males that invade its nesting territory (see Figure 39.16). Predict the likely pattern of dispersion for male sticklebacks, and explain your reasoning.

For suggested answers, see Appendix A.

# The exponential and logistic models describe the growth of populations

Populations of all species have the potential to expand greatly when resources are abundant. To appreciate the potential for population increase, consider a bacterium that can reproduce by fission every 20 minutes under ideal laboratory conditions. There would be two bacteria after 20 minutes, four after 40 minutes, and eight after 60 minutes. If reproduction continued at this rate for a day and a half without mortality, there would be enough bacteria to form a layer 30 cm deep over the entire globe! But unlimited growth does not occur for long in nature, where individuals typically have access to fewer resources as a population grows. Ecologists study population growth in idealized conditions and in the more realistic conditions where different factors limit growth. We'll examine both scenarios in this section.

## Changes in Population Size

Imagine a population consisting of a few individuals living in an ideal, unlimited environment. Under these conditions, there are no external limits on the abilities of individuals to harvest energy, grow, and reproduce. The population will increase in size with every birth and with the immigration of individuals from other populations, and it will decrease in size with every death and with the emigration of individuals out of the population. We can thus define a change in population size during a fixed time interval with the following verbal equation:

$$\begin{matrix} \text{Change in} \\ \text{population} \\ \text{size} \end{matrix} = \text{Births} + \begin{matrix} \text{Immigrants} \\ \text{entering} \\ \text{population} \end{matrix} - \text{Deaths} - \begin{matrix} \text{Emigrants} \\ \text{leaving} \\ \text{population} \end{matrix}$$

For now, we will simplify our discussion by ignoring the effects of immigration and emigration.

We can use mathematical notation to express this simplified relationship more concisely. If $N$ represents population size and $t$ represents time, then $\Delta N$ is the change in population size and $\Delta t$ is the time interval (appropriate to the life span or generation time of the species) over which we are evaluating population growth. (The Greek letter delta, $\Delta$, indicates change, such as change in time.) Using $B$ for the number of births in the population during the time interval and $D$ for the number of deaths, we can rewrite the verbal equation:

$$\frac{\Delta N}{\Delta t} = B - D$$

Typically, population ecologists are most interested in changes in population size—the number of individuals that are added to or subtracted from a population during a given time interval, symbolized by $R$. Here, $R$ represents the *difference* between the number of births ($B$) and the number of deaths ($D$)

that occur in the time interval. Thus, $R = B - D$, and we can simplify our equation by writing:

$$\frac{\Delta N}{\Delta t} = R$$

Next, we can convert our model to one in which changes in population size are expressed on a per individual (per capita) basis. The *per capita* change in population size ($r_{\Delta t}$) represents the contribution that an average member of the population makes to the number of individuals added to the population during the time interval $\Delta t$. If, for example, a population of 1,000 individuals increases by 16 individuals per year, then on a per capita basis, the annual change in population size is 16/1,000, or 0.016. If we know the annual per capita change in population size, we can use the formula $R = r_{\Delta t}N$ to calculate how many individuals will be added to a population each year. For example, if $r_{\Delta t} = 0.016$ and the population size is 500,

$$R = r_{\Delta t}N = 0.016 \times 500 = 8 \text{ per year}$$

Since the number of individuals added to the population ($R$) can be expressed on a per capita basis as $R = r_{\Delta t}N$, we can revise our population growth equation to take this into account:

$$\frac{\Delta N}{\Delta t} = r_{\Delta t}N$$

Remember that our equation is for a specific time interval (often one year). However, many ecologists prefer to use differential calculus to express population growth as a rate of change *at each instant in time*:

$$\frac{dN}{dt} = rN$$

In this case, $r$ represents the per capita change in population size that occurs at each instant in time (whereas $r_{\Delta t}$ represented the per capita change that occurred during the time interval $\Delta t$). If you have not yet studied calculus, don't be intimidated by the last equation; it is similar to the previous one, except that the time intervals $\Delta t$ are very short and are expressed in the equation as $dt$.

## Exponential Growth

Earlier we described a population whose members all have access to abundant food and are free to reproduce at their physiological capacity. In some cases, a population that experiences such ideal conditions increases in size by a constant proportion at each instant in time. When this occurs, the pattern of growth that results is called **exponential population growth**. The equation for exponential growth is the one presented at the end of the previous section, namely:

$$\frac{dN}{dt} = rN$$

In this equation, $dN/dt$ represents the rate at which the population is increasing in size at each moment in time, akin to how a glance at the speedometer of a car reveals the speed at that instant in time. As seen in the equation, $dN/dt$ equals the current population size, $N$, multiplied by a constant, $r$. Ecologists

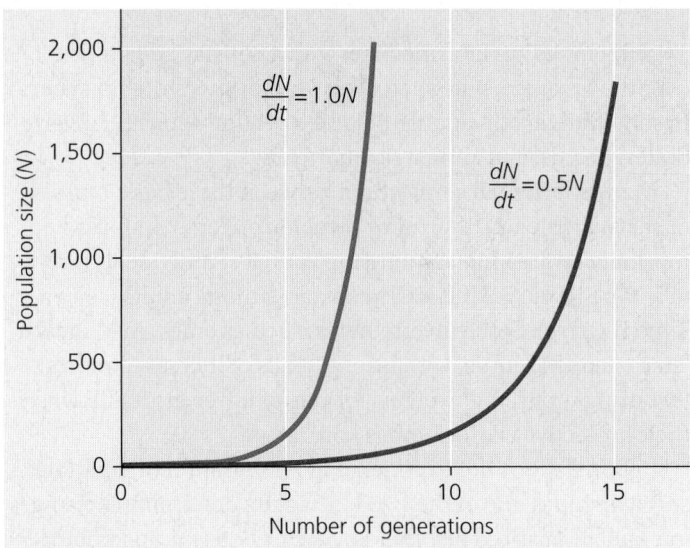

▲ **Figure 40.17 Population growth predicted by the exponential model.** This graph compares growth in a population with $r = 1.0$ (blue curve) to growth in a population with $r = 0.5$ (red curve).

**?** *How many generations does it take each of these populations to reach a size of 1,500 individuals?*

refer to $r$ as the **intrinsic rate of increase**, the per capita rate at which an exponentially growing population increases in size at each instant in time.

On a per capita basis, the size of a population that is growing exponentially increases at a constant rate, resulting eventually in a J-shaped growth curve when population size is plotted over time **(Figure 40.17)**. Although the per capita rate of population growth is constant (and equals $r$), more new individuals are added per unit of time when the population is large than when it is small; thus, the curves in Figure 40.17 get progressively steeper over time. This occurs because population growth depends on $N$ as well as $r$, and hence more individuals are added to larger populations than to small ones growing at the same per capita rate. It is also clear from Figure 40.17 that a population with a higher intrinsic rate of increase ($dN/dt = 1.0N$) will grow faster than one with a lower intrinsic rate of increase ($dN/dt = 0.5N$).

Exponential growth can occur in populations that are introduced into a new environment or whose numbers were drastically reduced by a catastrophic event and are rebounding. For example, the population of elephants in Kruger National Park, South Africa, grew exponentially for approximately 60 years after they were first protected from hunting **(Figure 40.18)**. The increasingly large number of elephants eventually caused enough damage to vegetation that a collapse in their food supply was likely. To protect other species and the park ecosystem before that happened, park managers began limiting the elephant population by using birth control and exporting elephants to other countries.

## Carrying Capacity

The exponential growth model assumes that resources remain abundant, which is rarely the case in the real world. Instead, as the size of a population increases, each individual has access to fewer resources. Ultimately, there is a limit to the number

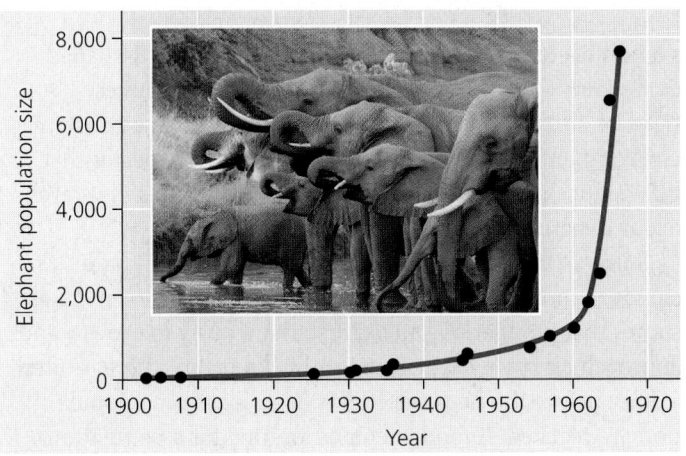

**▲ Figure 40.18 Exponential growth in the African elephant population of Kruger National Park, South Africa.**

**Table 40.2 Logistic Growth of a Hypothetical Population (K = 1,500)**

| Population Size (N) | Intrinsic Rate of Increase (r) | $\dfrac{K - N}{K}$ | Per Capita Rate of Increase $r\dfrac{(K - N)}{K}$ | Population Growth Rate* $rN\dfrac{(K - N)}{K}$ |
|---|---|---|---|---|
| 25 | 1.0 | 0.983 | 0.983 | +25 |
| 100 | 1.0 | 0.933 | 0.933 | +93 |
| 250 | 1.0 | 0.833 | 0.833 | +208 |
| 500 | 1.0 | 0.667 | 0.667 | +333 |
| 750 | 1.0 | 0.500 | 0.500 | +375 |
| 1,000 | 1.0 | 0.333 | 0.333 | +333 |
| 1,500 | 1.0 | 0.000 | 0.000 | 0 |

*Rounded to the nearest whole number.

of individuals that can occupy a habitat. Ecologists define the **carrying capacity**, symbolized by $K$, as the maximum population size that a particular environment can sustain. Carrying capacity varies over space and time with the abundance of limiting resources. Energy, shelter, refuge from predators, nutrient availability, water, and suitable nesting sites can all be limiting factors. For example, the carrying capacity for bats may be high in a habitat with abundant flying insects and roosting sites, but lower where there is abundant food but fewer suitable shelters.

Crowding and resource limitation can have a profound effect on population growth rate. If individuals cannot obtain sufficient resources to reproduce, then the per capita birth rate will decline. Similarly, if starvation or disease increases with density, the per capita death rate may increase. Falling per capita birth rates or rising per capita death rates will cause the per capita rate of population growth to drop—a very different situation from the constant per capita growth rate ($r$) seen in a population that is growing exponentially.

## The Logistic Growth Model

In the **logistic population growth** model, the per capita rate of population growth approaches zero as the population size nears the carrying capacity ($K$). To construct the logistic model, we start with the exponential population growth model and add an expression that reduces the per capita rate of population growth as $N$ increases. If the carrying capacity is $K$, then $K - N$ is the number of additional individuals the environment can support, and $(K - N)/K$ is the fraction of $K$ that is still available for population growth. By multiplying the exponential rate of population growth $rN$ by $(K - N)/K$, we modify the change in population size as $N$ increases:

$$\frac{dN}{dt} = rN\frac{(K - N)}{K}$$

When $N$ is small compared to $K$, the term $(K - N)/K$ is close to 1. When this occurs, the per capita rate of increase, $r(K - N)/K$, will be close to (but slightly less than) $r$, the intrinsic rate of increase seen in exponential population growth. But when $N$ is large and resources are limiting, then $(K - N)/K$ is close to 0,

and the per capita rate of increase is small. When $N$ equals $K$, the population stops growing. **Table 40.2** shows calculations of population growth rate for a hypothetical population growing according to the logistic model, with $r = 1.0$ per individual per year. Notice that the overall population growth rate is highest, +375 individuals per year, when the population size is 750, or half the carrying capacity. At a population size of 750, the per capita rate of increase remains relatively high (one-half the value of $r$), but there are more reproducing individuals ($N$) in the population than at lower population sizes.

As shown in **Figure 40.19**, the logistic model of population growth produces a sigmoid (S-shaped) growth curve when $N$ is plotted over time (the red line). New individuals are added to

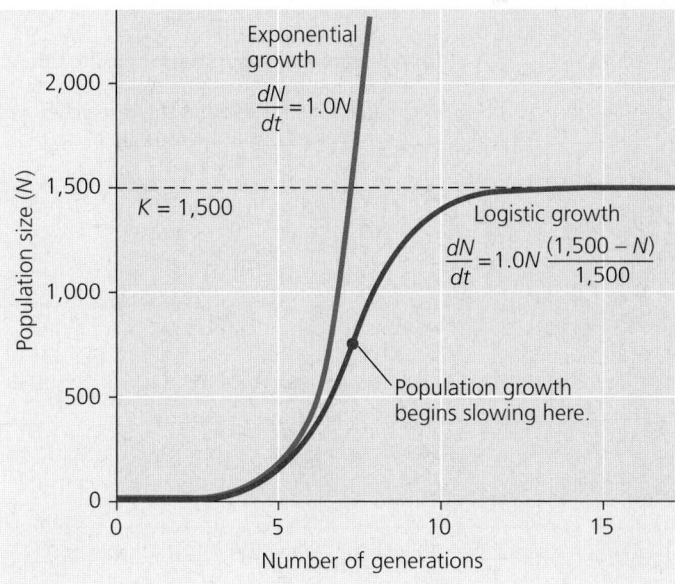

**▲ Figure 40.19 Population growth predicted by the logistic model.** The rate of population growth decreases as population size ($N$) approaches the carrying capacity ($K$) of the environment. The red line shows logistic growth in a population where $r = 1.0$ and $K = 1,500$ individuals. For comparison, the blue line illustrates a population continuing to grow exponentially with the same $r$.

the population most rapidly at intermediate population sizes, when there is not only a breeding population of substantial size, but also lots of available space and other resources in the environment. The number of individuals added to the population decreases dramatically as $N$ approaches $K$. As a result, the population growth rate ($dN/dt$) also decreases as $N$ approaches $K$.

Note that we haven't said anything yet about *why* the population growth rate decreases as $N$ approaches $K$. For a population's growth rate to decrease, the birth rate must decrease, the death rate must increase, or both. Later in the chapter, we'll consider some of the factors affecting these rates, including the presence of disease, predation, and limited amounts of food and other resources.

## The Logistic Model and Real Populations

The growth of laboratory populations of some small animals, such as beetles and crustaceans, and of some microorganisms, such as bacteria, *Paramecium*, and yeasts, fits an S-shaped curve fairly well under conditions of limited resources **(Figure 40.20a)**. These populations are grown in a constant environment lacking predators and competing species that may reduce growth of the populations, conditions that rarely occur in nature.

Some of the assumptions built into the logistic model clearly do not apply to all populations. The logistic model assumes that populations adjust instantaneously to growth and approach carrying capacity smoothly. In reality, there is often a delay before the negative effects of an increasing population are realized. If food becomes limiting for a population, for instance, reproduction will decline eventually, but females may use their energy reserves to continue reproducing for a short time. This may cause the population to overshoot its carrying capacity temporarily, as shown for the water fleas in **Figure 40.20b**. In the **Scientific Skills Exercise**, you can

▶ **Figure 40.20 How well do these populations fit the logistic growth model?** In each graph, the black dots plot the measured growth of the population, and the red curve is the growth predicted by the logistic model.

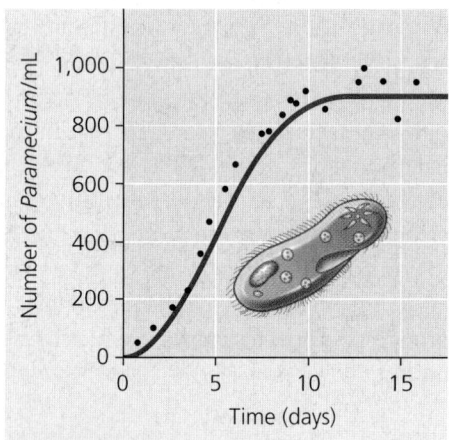

**(a) A *Paramecium* population in the lab.** Growth in a small culture closely approximates logistic growth if the researcher maintains a constant environment.

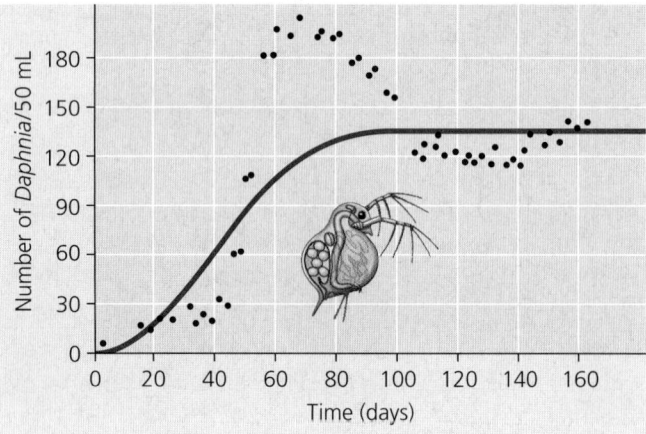

**(b) A *Daphnia* (water flea) population in the lab.** Growth in a small culture does not correspond well to the logistic model. This population overshoots the carrying capacity of its artificial environment before reaching an approximately stable size.

model what can happen to such a population when $N$ becomes *greater* than $K$. Other populations fluctuate greatly, making it difficult even to define carrying capacity. We'll examine some possible reasons for such fluctuations later in the chapter.

**CONCEPT CHECK 40.5**

1. Explain why a constant rate of increase ($r$) for a population produces a growth curve that is J-shaped.
2. Explain why a population that fits the logistic growth model increases more rapidly at intermediate size than at relatively small and large sizes.
3. **MAKE CONNECTIONS** Many viruses are pathogens of animals and plants (see Concept 17.3). How might the presence of pathogens alter the carrying capacity of a population? Explain.

For suggested answers, see Appendix A.

## CONCEPT 40.6

# Population dynamics are influenced strongly by life history traits and population density

**EVOLUTION** Natural selection favors traits that improve an organism's chances of survival and reproductive success. In every species, there are trade-offs between survival and reproductive traits such as frequency of reproduction, number of offspring (number of seeds produced by plants; litter or clutch size for animals), and investment in parental care. The traits that affect an organism's schedule of reproduction and survival make up its **life history**. A life history entails three main variables: when reproduction begins (the age at first reproduction or age at maturity), how often the organism reproduces, and how many offspring are produced per reproductive episode.

## "Trade-offs" and Life Histories

No organism could produce unlimited numbers of offspring *and* provision them well. There is a trade-off between reproduction and survival. For instance, researchers in Scotland found that female red deer that reproduced in a given summer were more likely to die the next winter than were females that did not reproduce.

Selective pressures also influence trade-offs between the number and size of offspring. Plants and animals whose young are more likely to die often produce many small offspring. Plants that colonize disturbed environments, for example, usually produce many small seeds, only a few of which may reach a suitable habitat. Small size may also increase the chance of seedling establishment by enabling the seeds to be carried longer distances to a broader range of habitats **(Figure 40.21a)**. Animals that suffer high predation rates, such as quail, sardines, and mice, also tend to produce many offspring.

In other organisms, extra investment on the part of the parent greatly increases the offspring's chance of survival. Walnut

**(a)** Dandelions grow quickly and release a large number of tiny fruits, each containing a single seed. Producing numerous seeds increases the chance that at least some will grow into plants that eventually produce seeds themselves.

**(b)** Some plants, such as the Brazil nut tree (right), produce a moderate number of large seeds in pods (above). Each seed's large endosperm provides nutrients for the embryo, an adaptation that helps a relatively large fraction of offspring survive.

▲ **Figure 40.21 Variation in the number and size of seeds in plants.**

and Brazil nut trees produce large seeds packed with nutrients that help the seedlings become established **(Figure 40.21b)**. Primates generally bear only one or two offspring at a time; parental care and an extended period of learning in the first several years of life are very important to offspring fitness. Such provisioning and extra care can be especially important in habitats with high population densities.

Ecologists have attempted to connect differences in favored traits at different population densities with the logistic growth model discussed in Concept 40.5. Selection for traits that are sensitive to population density and are favored at high densities is known as **K-selection**, or density-dependent selection. In contrast, selection for traits that maximize reproductive success in uncrowded environments (low densities) is called **r-selection**, or density-independent selection. These names follow from the variables of the logistic equation. *K*-selection is said to operate in populations living at a density near the limit imposed by their resources (the carrying capacity, $K$), where competition among individuals is stronger. Mature trees growing in an old-growth forest are an example of *K*-selected organisms.

In contrast, *r*-selection is said to maximize *r*, the intrinsic rate of increase, and occurs in environments in which population densities are well below carrying capacity or individuals face little competition. Such conditions are often found in disturbed habitats that are being recolonized. Weeds growing in an abandoned agricultural field are an example of *r*-selected organisms.

## Population Change and Population Density

Similar to the case of *r*-selection, a birth rate or death rate that does *not* change with population density is said to be **density independent**. In a classic study of population regulation, Andrew Watkinson and John Harper, of the University of Wales, found that the mortality of dune fescue grass (*Vulpia fasciculata*) is mainly due to physical factors that kill similar proportions of a local population, regardless of its density. For example, drought stress that arises when the roots of the grass are uncovered by shifting sands is a density-independent factor. In contrast, a death rate that increases with population density or a birth rate that falls with rising density is said to be **density dependent**, a situation similar to *K*-selection. Watkinson and Harper found that reproduction by dune fescue declines as population density increases, in part because water or nutrients become more scarce. Thus, the key factors regulating birth rate in this population are density dependent, while death rate is largely regulated by density-independent factors. **Figure 40.22** shows how the combination of density-dependent reproduction and density-independent mortality can stop population growth, leading to an equilibrium population density in species such as dune fescue.

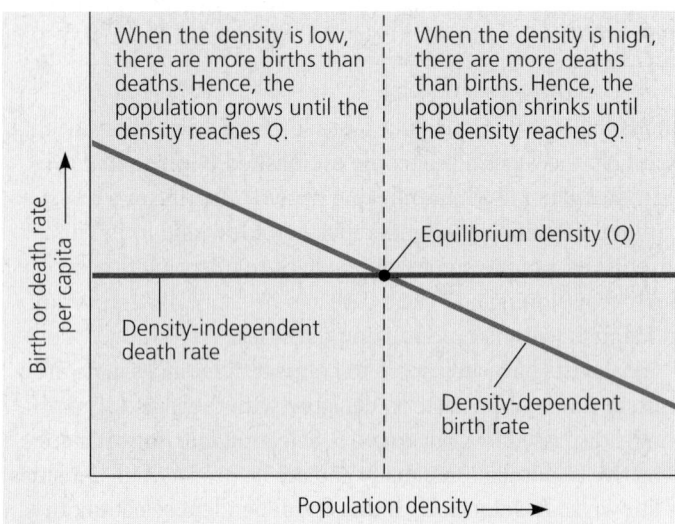

When the density is low, there are more births than deaths. Hence, the population grows until the density reaches Q.

When the density is high, there are more deaths than births. Hence, the population shrinks until the density reaches Q.

Equilibrium density (Q)

Density-independent death rate

Density-dependent birth rate

Birth or death rate per capita

Population density ⟶

▲ **Figure 40.22 Determining equilibrium for population density.** This simple model considers only birth and death rates. (Immigration and emigration rates are assumed to be either zero or equal.) In this example, the birth rate changes with population density, while the death rate is constant. At the equilibrium density (Q), the birth and death rates are equal.

**DRAW IT** *Redraw this figure for the case where the birth and death rates are both density dependent, as occurs for many species.*

## Mechanisms of Density-Dependent Population Regulation

Without some type of negative feedback between population density and the rates of birth and death, a population would never stop growing. But no population can increase in size indefinitely. Ultimately, at large population sizes, negative feedback is provided by density-dependent regulation, which halts population growth through mechanisms that reduce birth rates or increase death rates. Several mechanisms of density-dependent population regulation are described in **Figure 40.23**.

These various examples of population regulation by negative feedback show how increased densities cause population growth rates to decline by affecting reproduction, growth, and survival. But although negative feedback helps explain why populations stop growing, it does not address why some populations fluctuate dramatically while others remain relatively stable. That is the topic we address next.

## Population Dynamics

All populations show some fluctuation in size. Such population fluctuations from year to year or place to place, called **population dynamics**, are influenced by many factors and in turn affect other species. For example, fluctuations in fish populations influence seasonal harvests of commercially important species. The study of population dynamics focuses on the complex interactions between biotic and abiotic factors that cause variation in population sizes.

### *Stability and Fluctuation*

Populations of large mammals were once thought to remain relatively stable, but long-term studies have challenged that idea. For instance, the moose population on Isle Royale in Lake Superior has fluctuated substantially since around 1900. At that time, moose from the Ontario mainland (25 km away) colonized the island, perhaps by walking across the lake when it was frozen. Wolves, which rely on moose for most of their food, reached the island around 1950 by walking across the frozen lake. The lake has not frozen over in recent years, and both populations appear to have been isolated from immigration and emigration since then. Despite this isolation, the

As population density increases, many density-dependent mechanisms slow or stop population growth by decreasing birth rates or increasing death rates.

## Competition for Resources

Increasing population density intensifies competition for nutrients and other resources, reducing reproductive rates. Farmers minimize the effect of resource competition on the growth of grains such as wheat (*Triticum aestivum*) and other crops by applying fertilizers to reduce nutrient limitations on crop yield.

## Toxic Wastes

Yeasts, such as the brewer's yeast *Saccharomyces cerevisiae*, are used to convert carbohydrates to ethanol in winemaking. The ethanol that accumulates in the wine is toxic to yeasts and contributes to density-dependent regulation of yeast population size. The alcohol content of wine is usually less than 13% because that is the maximum concentration of ethanol that most wine-producing yeast cells can tolerate.

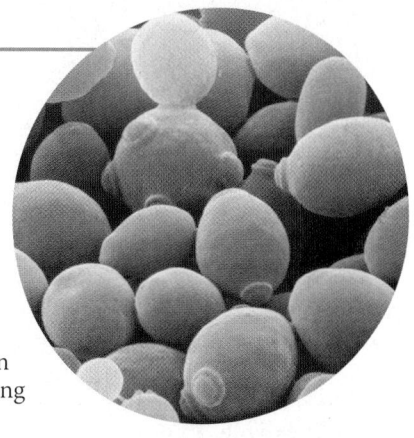

5 μm

## Territoriality

Territoriality can limit population density when space becomes the resource for which individuals compete. Cheetahs (*Acinonyx jubatus*) use a chemical marker in urine to warn other cheetahs of their territorial boundaries. The presence of surplus, or nonbreeding, individuals is a good indication that territoriality is restricting population growth.

## Predation

Predation can be an important cause of density-dependent mortality if a predator captures more food as the population density of the prey increases. As a prey population builds up, predators may also feed preferentially on that species. Population increases in the collared lemming (*Dicrostonyx groenlandicus*) lead to density-dependent predation by several predators, including the snowy owl (*Bubo scandiacus*).

## Disease

If the transmission rate of a disease increases as a population becomes more crowded, then the disease's impact is density dependent. In humans, the respiratory diseases influenza (flu) and tuberculosis are spread through the air when an infected person sneezes or coughs. Both diseases strike a greater percentage of people in densely populated cities than in rural areas.

## Intrinsic Factors

Intrinsic physiological factors (those operating within an individual organism) sometimes regulate population size. Reproductive rates of white-footed mice (*Peromyscus leucopus*) in a field enclosure can drop even when food and shelter are abundant. This drop in reproduction at high population density is associated with aggressive interactions and hormonal changes within individual mice that delay sexual maturation and depress the immune system.

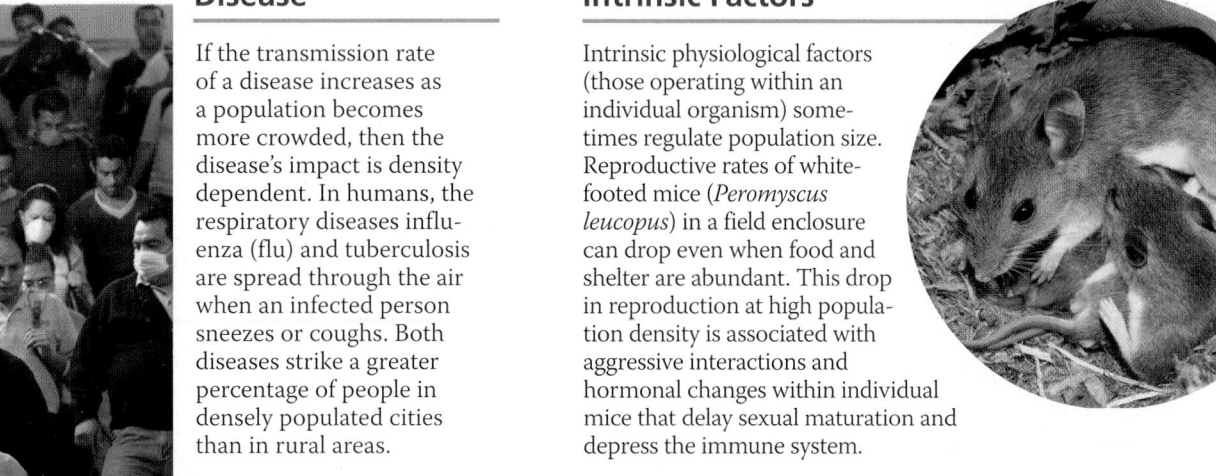

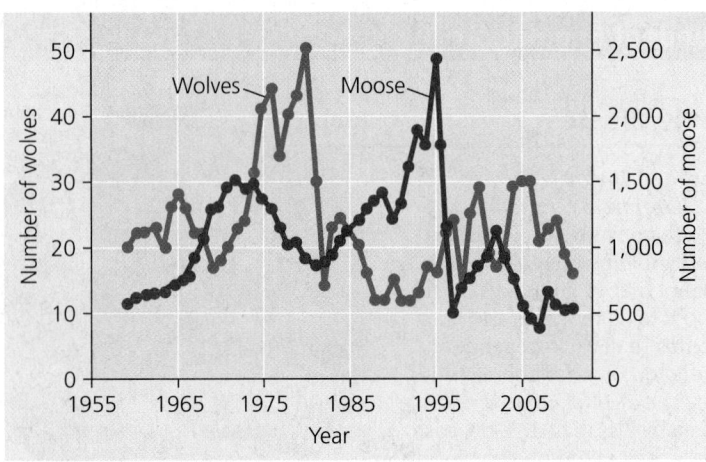

▲ **Figure 40.24 Fluctuations in moose and wolf populations on Isle Royale, 1959–2011.**

 **ANIMATION** Visit the Study Area in **MasteringBiology** for the BioFlix® 3-D Animation on Population Ecology.

moose population experienced two major increases and collapses during the last 50 years **(Figure 40.24)**.

What factors cause the size of the moose population to change so dramatically? Harsh weather, particularly cold winters with heavy snowfall, can weaken moose and reduce food availability, decreasing the population size. When moose numbers are low and the weather is mild, food is readily available and the population grows quickly. Conversely, when moose numbers are high, factors such as predation and an increase in the density of ticks and other parasites cause the population to shrink. The effects of some of these factors can be seen in Figure 40.24. The first collapse coincided with a peak in the numbers of wolves from 1975 to 1980. The second major collapse, around 1995, coincided with harsh winter weather, which increased the energy needs of the animals and made it harder for the moose to find food under the deep snow.

### Immigration, Emigration, and Metapopulations

So far, our discussion of population dynamics has focused mainly on the contributions of births and deaths. However, immigration and emigration also influence populations. When a population becomes crowded and resource competition increases, emigration often increases.

Immigration and emigration are particularly important when a number of local populations are linked, forming a **metapopulation**. Local populations in a metapopulation can be thought of as occupying discrete patches of suitable habitat in a sea of otherwise unsuitable habitat. Such patches vary in size, quality, and isolation from other patches, factors that influence how many individuals move among the populations. If one population becomes extinct, the patch it occupied may be recolonized by immigrants from another population.

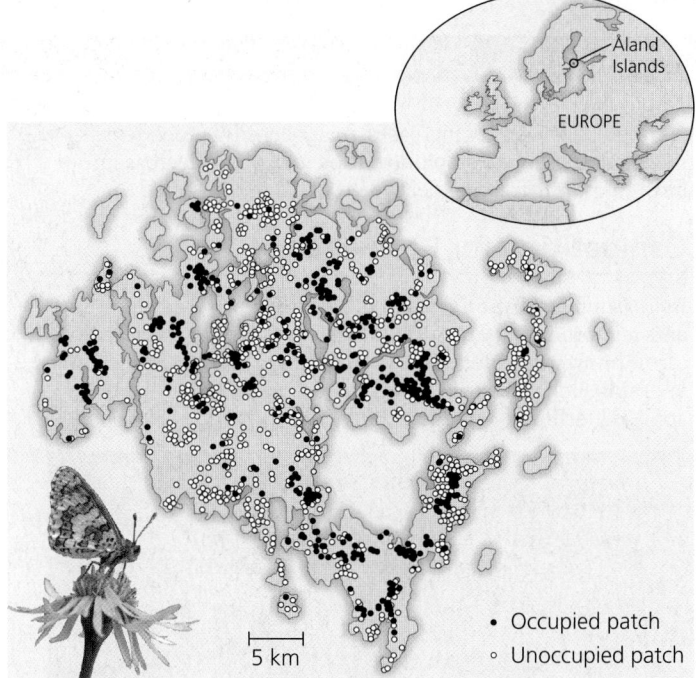

▲ **Figure 40.25 The Glanville fritillary: a metapopulation.** On the Åland Islands, local populations of this butterfly (filled circles) are found in only a fraction of the suitable habitat patches (open circles) at any given time. Individuals can move between local populations and colonize unoccupied patches.

The Glanville fritillary (*Melitaea cinxia*) illustrates the movement of individuals between populations. This butterfly is found in about 500 meadows across the Åland Islands of Finland, but its potential habitat in the islands is much larger, approximately 4,000 suitable patches. New populations of the butterfly regularly appear and existing populations become extinct, constantly shifting the locations of the 500 colonized patches **(Figure 40.25)**. The species persists in a balance of extinctions and recolonizations.

The metapopulation concept underscores the significance of immigration and emigration for the Glanville fritillary and many other species. It also helps ecologists understand population dynamics and gene flow in patchy habitats, providing a framework for the conservation of species living in a network of habitat fragments and reserves.

### CONCEPT CHECK 40.6

1. In the fish called the peacock wrasse (*Symphodus tinca*), females disperse some of their eggs widely and lay other eggs in a nest. Only the latter receive parental care. Explain the trade-offs in reproduction that this behavior illustrates.

2. **WHAT IF?** Mice that experience stress such as a food shortage will sometimes abandon their young. Explain how this behavior might have evolved in the context of reproductive trade-offs and life history.

3. **MAKE CONNECTIONS** Negative feedback is a process that regulates biological systems (see Concept 32.2). Explain how the density-dependent birth rate of dune fescue grass exemplifies negative feedback.

For suggested answers, see Appendix A.

Go to **MasteringBiology**® for Assignments, the eText, and the Study Area with Animations, Activities, Vocab Self-Quiz, and Practice Tests.

## SUMMARY OF KEY CONCEPTS

**VOCAB SELF-QUIZ**
goo.gl/gbai8v

### CONCEPT 40.1

**Earth's climate influences the distribution of terrestrial biomes (pp. 843–848)**

- Global **climate** patterns are largely determined by the input of solar energy and Earth's revolution around the sun.
- The changing angle of the sun over the year, bodies of water, and mountains exert seasonal, regional, and local effects on climate.
- **Climographs** show that temperature and precipitation are correlated with **biomes**. Other factors also affect biome location.
- Terrestrial biomes are often named for major physical or climatic factors and for their predominant vegetation. Vertical layering is an important feature of terrestrial biomes.

❓ *Suppose global air circulation suddenly reversed, with most air ascending at 30° north and south latitude and descending at the equator. At what latitude would you most likely find deserts?*

### CONCEPT 40.2

**Aquatic biomes are diverse and dynamic systems that cover most of Earth (pp. 849–852)**

- Aquatic biomes are characterized primarily by their physical environment rather than by climate and are often layered with regard to light penetration, temperature, and community structure.
- In the ocean and in most lakes, an abrupt temperature change called a **thermocline** separates a more uniformly warm upper layer from more uniformly cold deeper waters.

❓ *In which aquatic biomes might you find an aphotic zone?*

### CONCEPT 40.3

**Interactions between organisms and the environment limit the distribution of species (pp. 852–854)**

- Ecologists want to know not only *where* species occur but also *why* those species occur where they do.
- The distribution of species may be limited by **dispersal**, **biotic** (living) factors, and **abiotic** (chemical and physical) factors, such as temperature extremes, salinity, and water availability.

❓ *If you were an ecologist studying the chemical and physical limits to the distributions of species, how might you rearrange the flowchart in Figure 40.12?*

### CONCEPT 40.4

**Biotic and abiotic factors affect population density, dispersion, and demographics (pp. 854–857)**

- Population **density**—the number of individuals per unit area or volume—reflects the interplay of births, deaths, immigration, and emigration. Environmental and social factors influence the **dispersion** of individuals.
- Populations increase from births and **immigration** and decrease from deaths and **emigration**. **Life tables** and **survivorship curves** summarize specific trends in **demography**.

❓ *Gray whales (Eschrichtius robustus) gather each winter near Baja California to give birth. How might such behavior make it easier for ecologists to estimate birth and death rates for the species?*

### CONCEPT 40.5

**The exponential and logistic models describe the growth of populations (pp. 857–861)**

- If immigration and emigration are ignored, a population's growth rate (the per capita rate of increase) equals its birth rate minus its death rate.
- The **exponential population growth** equation $dN/dt = rN$ represents a population's growth when resources are relatively abundant, where $r$ is the (per capita) **intrinsic rate of increase** and $N$ is the number of individuals in the population.

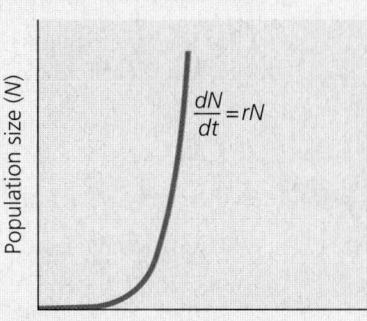

- Exponential growth cannot be sustained for long in any population. A more realistic population model limits growth by incorporating **carrying capacity** ($K$), the maximum population size the environment can support. According to the **logistic population growth** equation $dN/dt = rN(K - N)/K$, growth levels off as population size approaches the carrying capacity.

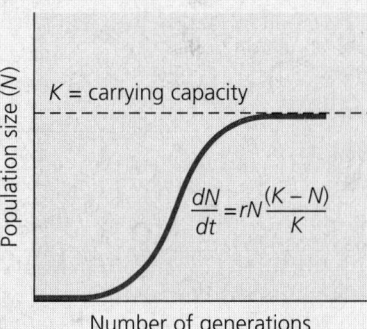

- The logistic model fits few real populations perfectly, but it is useful for estimating possible growth.

❓ *As an ecologist who manages a wildlife preserve, you want to increase the preserve's carrying capacity for a particular endangered species. How might you go about accomplishing this?*

### CONCEPT 40.6

**Population dynamics are influenced strongly by life history traits and population density (pp. 861–864)**

- **Life history** traits are evolutionary outcomes reflected in the development, physiology, and behavior of organisms.
- **Density-dependent** changes in birth and death rates curb population increase through negative feedback and can eventually stabilize a population near its carrying capacity. Density-dependent limiting factors include intraspecific competition for limited food or space, increased predation, disease, intrinsic physiological factors, and buildup of toxic substances.
- All populations exhibit some size fluctuations, and many undergo substantial changes in size that are influenced by complex interactions between biotic and abiotic factors. A **metapopulation** is a group of populations linked by immigration and emigration.

❓ *Name one biotic and one abiotic factor that could contribute to yearly fluctuations in the size of the human population in a given region.*

# TEST YOUR UNDERSTANDING

**PRACTICE TEST**

goo.gl/CRZjvS

## Level 1: Knowledge/Comprehension

1. Which of the following biomes is correctly paired with the description of its climate?
   (A) savanna—low temperature, precipitation uniform during the year
   (B) tundra—long summers, mild winters
   (C) temperate broadleaf forest—relatively short growing season, mild winters
   (D) tropical forests—nearly constant day length and temperature

2. A population's carrying capacity
   (A) may change as environmental conditions change.
   (B) can be accurately calculated using the logistic model.
   (C) generally remains constant over time.
   (D) increases as the per capita growth rate decreases.

## Level 2: Application/Analysis

3. When climbing a mountain, we can observe transitions in biological communities that are analogous to the changes
   (A) in different depths in the ocean.
   (B) in biomes at different latitudes.
   (C) in a community through different seasons.
   (D) in an ecosystem as it evolves over time.

4. According to the logistic growth equation

$$\frac{dN}{dt} = rN\frac{(K - N)}{K}$$

   (A) the number of individuals added per unit time is greatest when $N$ is close to zero.
   (B) the per capita growth rate increases as $N$ approaches $K$.
   (C) population growth is zero when $N$ equals $K$.
   (D) the population grows exponentially when $K$ is small.

## Level 3: Synthesis/Evaluation

5. **WHAT IF?** If the direction of Earth's rotation reversed, the most predictable effect would be
   (A) the elimination of ocean currents.
   (B) a big change in the length of the year.
   (C) winds blowing from west to east along the equator.
   (D) a loss of seasonal variation at high latitudes.

6. **INTERPRET THE DATA** After examining Figure 40.13, you decide to study feeding relationships among sea otters, sea urchins, and kelp. You know that sea otters prey on sea urchins and that urchins eat kelp. At four coastal sites, you measure kelp abundance. Then you spend one day at each site and mark whether otters are present or absent every 5 minutes during the day. Graph kelp abundance (on the $y$-axis) versus otter density (on the $x$-axis), using the data below. Then formulate a hypothesis to explain the pattern you observed.

| Site | Otter Density (# sightings per day) | Kelp Abundance (% cover) |
|------|-------------------------------------|--------------------------|
| 1 | 98 | 75 |
| 2 | 18 | 15 |
| 3 | 85 | 60 |
| 4 | 36 | 25 |

7. **SCIENTIFIC INQUIRY**
   Jens Clausen and colleagues, at the Carnegie Institution of Washington, studied how the size of yarrow plants (*Achillea lanulosa*) growing on the slopes of the Sierra Nevada varied with elevation. They found that plants from low elevations were generally taller than plants from high elevations, as shown below:

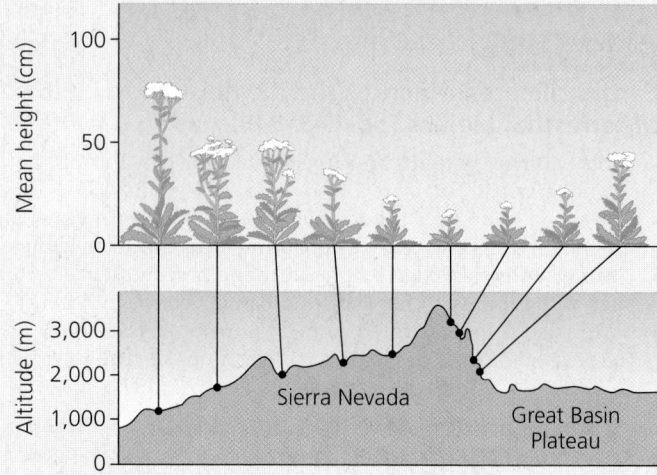

**Data from** J. Clausen et al., Experimental studies on the nature of species. III. Environmental responses of climatic races of *Achillea*, Carnegie Institution of Washington Publication No. 581 (1948).

Clausen and colleagues proposed two hypotheses to explain this variation within a species: (1) There are genetic differences between populations of plants found at different elevations. (2) The species has developmental flexibility and can assume tall or short growth forms, depending on local abiotic factors. If you had seeds from yarrow plants found at low and high elevations, describe the experiments you would perform to test these hypotheses.

8. **FOCUS ON EVOLUTION**
   Discuss how the distribution of a species can be affected both by its evolutionary history and by ecological factors.

9. **FOCUS ON INTERACTIONS**
   In a short essay (100–150 words), identify the factor or factors in Figure 40.23 that you think may ultimately be most important for density-dependent population regulation in humans, and explain your reasoning.

10. **SYNTHESIZE YOUR KNOWLEDGE**

Locusts (a type of grasshopper) undergo cyclic population outbreaks. Of the mechanisms of density-dependent regulation shown in Figure 40.23, choose the two that you think most apply to locust swarms, and explain why.

*For selected answers, see Appendix A.*

# 41

# Species Interactions

## KEY CONCEPTS

**41.1** Interactions within a community may help, harm, or have no effect on the species involved

**41.2** Diversity and trophic structure characterize biological communities

**41.3** Disturbance influences species diversity and composition

**41.4** Biogeographic factors affect community diversity

**41.5** Pathogens alter community structure locally and globally

▲ **Figure 41.1** Which species benefits from this interaction?

## Communities in Motion

Deep in the Lembeh Strait of Indonesia, a carrier crab scuttles across the ocean floor using its modified rear legs to hold a large sea urchin on its back **(Figure 41.1)**. When a predatory fish arrives, the crab quickly settles into the sediments and puts its living shield to use. The fish darts in and tries to bite the crab. In response, the crab tilts the spiny sea urchin toward whichever side the fish attacks. The fish eventually gives up and swims away. Carrier crabs use many organisms to protect themselves, including jellies (see the small photo).

The crab in Figure 41.1 clearly benefits from having the sea urchin on its back. But how does the sea urchin fare in this relationship? Its association with the crab might harm it, help it, or have no effect on its survival and reproduction. For example, the sea urchin may be harmed if the crab sets it down in an unsuitable habitat or in a place where it is vulnerable to predators. On the other hand, the crab may also protect the sea urchin from predators while carrying it. Additional observations or experiments would be needed before ecologists could answer this question.

In Chapter 40, you learned how individuals within a population can affect other individuals of the same species. This chapter will examine ecological interactions between populations of different species. A group of populations of different species living close enough to interact is called a biological **community**. Ecologists define the boundaries of a particular community to fit their research questions: They might study the community of decomposers and other organisms living on a rotting log, the benthic community in Lake Superior, or the community of trees and shrubs in Sequoia National Park in California.

We begin this chapter by exploring the kinds of interactions that occur between species in a community, such as the crab and sea urchin in Figure 41.1. We'll then consider several of the factors that are most significant in structuring a community—in determining how many species there are, which particular species are present, and the relative abundance of these species. Finally, we'll apply some of the principles of community ecology to the study of human disease.

# Interactions within a community may help, harm, or have no effect on the species involved

Some key relationships in the life of an organism are its interactions with individuals of other species in the community. These **interspecific interactions** include competition, predation, herbivory, parasitism, mutualism, and commensalism. In this section, we'll define and describe each of these interactions, grouping them according to whether they have positive (+) or negative (−) effects on the survival and reproduction of the two species engaged in the interaction.

For example, predation is a +/− interaction, with a positive effect on the survival and reproduction of the predator population and a negative effect on that of the prey population. Mutualism is a +/+ interaction because the survival and reproduction of both species are increased in the presence of the other. A 0 indicates that a population is not affected by the interaction in any known way. We'll consider three broad categories of ecological interactions: competition (−/−), exploitation (+/−), and positive interactions (+/+ or +/0).

## Competition

**Interspecific competition** is a −/− interaction that occurs when individuals of different species compete for a resource that limits the survival and reproduction of each species. Weeds growing in a garden compete with garden plants for nutrients and water. Lynx and foxes in the northern forests of Alaska and Canada compete for prey such as snowshoe hares. In contrast, some resources, such as oxygen, are rarely in short supply, at least on land; most terrestrial species use this resource, but they do not usually compete for it.

### Competitive Exclusion

What happens in a community when two species compete for limited resources? In 1934, Russian ecologist G. F. Gause studied this question using laboratory experiments with two closely related protist species, *Paramecium aurelia* and *Paramecium caudatum*. He cultured the species under stable conditions, adding a constant amount of food each day. When Gause grew the two species separately, each population increased rapidly in number and then leveled off at the apparent carrying capacity of the culture (see Figure 40.20a for an illustration of the logistic growth of a *Paramecium* population). But when Gause grew the two species together, *P. caudatum* became extinct. Gause inferred that *P. aurelia* had a competitive edge in obtaining food. More generally, he concluded that two species competing for the same limiting resources cannot coexist permanently in the same place. In the absence of disturbance, one species will use the resources more efficiently and reproduce more rapidly than the other. Even a slight reproductive

advantage will eventually lead to local elimination of the inferior competitor, an outcome called **competitive exclusion**.

### Ecological Niches and Natural Selection

**EVOLUTION** The specific set of biotic and abiotic resources that an organism uses in its environment is called its **ecological niche**. American ecologist Eugene Odum used the following analogy to explain the niche concept: If an organism's habitat is its "address," the niche is the organism's "profession." The niche of a tropical tree lizard, for instance, includes the temperature range it tolerates, the size of branches on which it perches, the time of day when it is active, and the sizes and kinds of insects it eats. Such factors define the lizard's niche, or ecological role—how it fits into an ecosystem.

We can use the niche concept to restate the principle of competitive exclusion: Two species cannot coexist permanently in a community if their niches are identical. However, ecologically similar species *can* coexist in a community if one or more significant differences in their niches arise through time. Evolution by natural selection can result in one of the species using a different set of resources or similar resources at different times of the day or year. The differentiation of niches that enables similar species to coexist in a community is called **resource partitioning (Figure 41.2)**.

As a result of competition, a species' *fundamental niche*, which is the niche potentially occupied by that species, is often

A. distichus perches on fence posts and other sunny surfaces.

A. insolitus usually perches on shady branches.

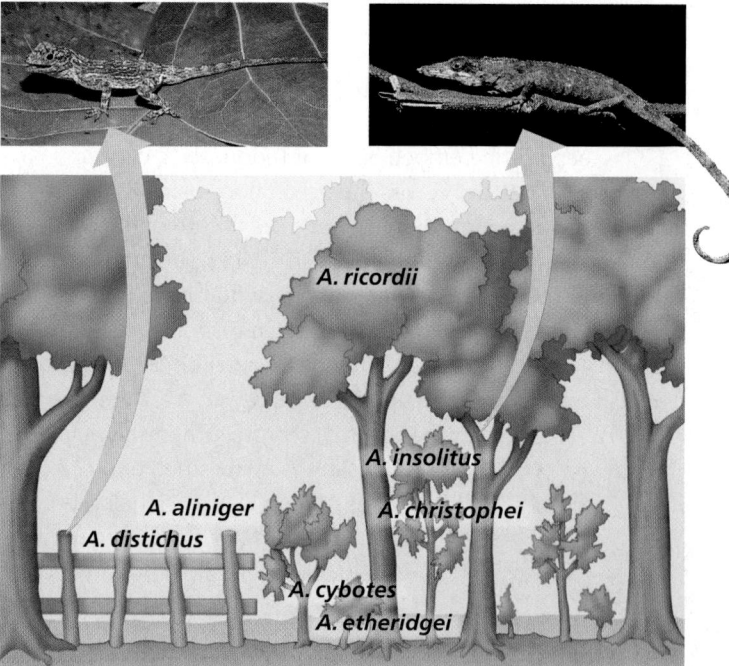

▲ **Figure 41.2 Resource partitioning among Dominican Republic lizards.** Seven species of *Anolis* lizards live in close proximity, and all feed on insects and other small arthropods. However, competition for food is reduced because each lizard species has a different preferred perch, thus occupying a distinct niche.

different from its *realized niche*, the portion of its fundamental niche that it actually occupies. Ecologists can identify the fundamental niche of a species by testing the range of conditions in which it grows and reproduces in the absence of competitors. They can also test whether a potential competitor limits a species' realized niche by removing the competitor and seeing if the first species expands into the newly available space. The classic experiment depicted in **Figure 41.3** clearly showed that competition between two barnacle species kept one species from occupying part of its fundamental niche.

▼ **Figure 41.3** Inquiry

### Can a species' niche be influenced by interspecific competition?

**Experiment** Ecologist Joseph Connell studied two barnacle species—*Chthamalus stellatus* and *Balanus balanoides*—that have a stratified distribution on rocks along the coast of Scotland. *Chthamalus* is usually found higher on the rocks than *Balanus*. To determine whether the distribution of *Chthamalus* is the result of interspecific competition with *Balanus*, Connell removed *Balanus* from the rocks at several sites.

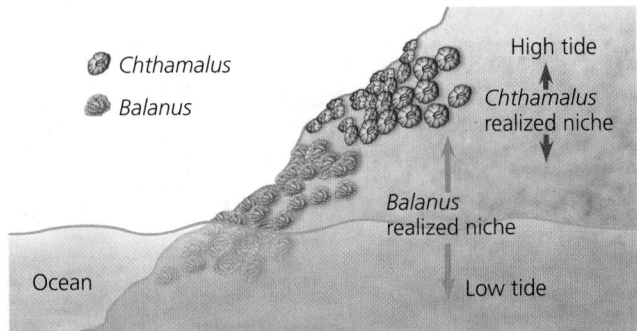

**Results** *Chthamalus* spread into the region formerly occupied by *Balanus*.

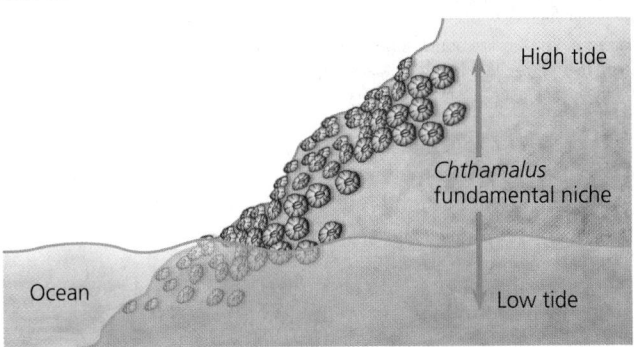

**Conclusion** Interspecific competition makes the realized niche of *Chthamalus* much smaller than its fundamental niche.

**Data from** J. H. Connell, The influence of interspecific competition and other factors on the distribution of the barnacle *Chthamalus stellatus*, *Ecology* 42:710–723 (1961).

(MB) See the related Experimental Inquiry Tutorial in MasteringBiology.

**WHAT IF?** Other observations showed that *Balanus* cannot survive high on the rocks because it dries out during low tides. How would *Balanus*'s realized niche compare with its fundamental niche?

### Character Displacement

Closely related species whose populations are sometimes allopatric (geographically separate; see Concept 22.2) and sometimes sympatric (geographically overlapping) provide more evidence for the importance of competition in structuring communities. In some cases, the allopatric populations of such species are morphologically similar and use similar resources. By contrast, sympatric populations, which would potentially compete for resources, show differences in body structures and in the resources they use. This tendency for characteristics to diverge more in sympatric than in allopatric populations of two species is called **character displacement**. An example of character displacement in Galápagos finches is shown in **Figure 41.4**.

## Exploitation

All nonphotosynthetic organisms must eat, and all organisms are at risk of being eaten. Thus, much of the drama in nature involves **exploitation**, a term for any type of $+/-$ interaction in which one species benefits by feeding on the other species, which in turn is harmed by the interaction. Exploitative interactions include predation, herbivory, and parasitism.

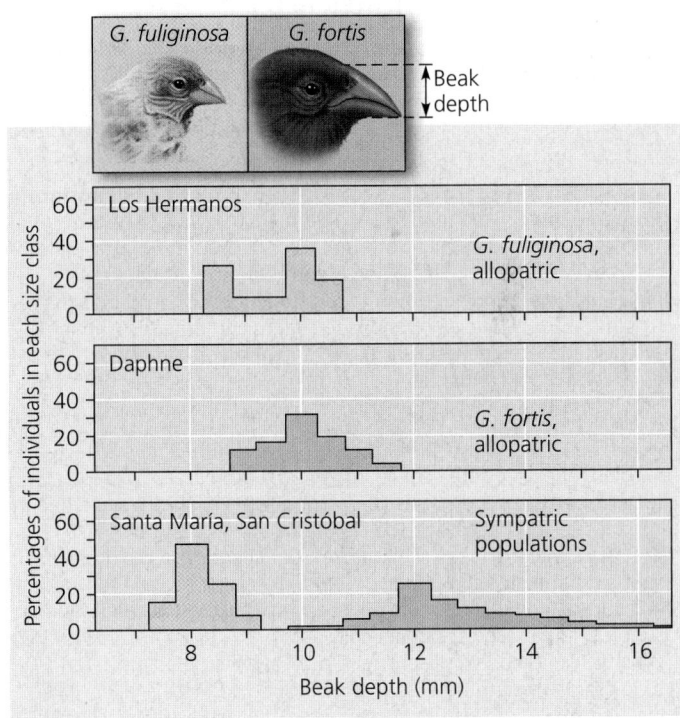

▲ **Figure 41.4 Character displacement: indirect evidence of past competition.** Allopatric populations of *Geospiza fuliginosa* and *Geospiza fortis* on Los Hermanos and Daphne Islands have similar beak morphologies (top two graphs) and presumably eat similarly sized seeds. However, where the two species are sympatric on Santa María and San Cristóbal, *G. fuliginosa* has a shallower, smaller beak and *G. fortis* a deeper, larger one (bottom graph), adaptations that favor eating different-sized seeds.

**INTERPRET THE DATA** *If the beak length of* G. fortis *is typically 12% longer than the beak depth, what is the predicted beak length of* G. fortis *individuals with the smallest beak depths observed on Santa María and San Cristóbal Islands?*

## Predation

**Predation** refers to a +/− interaction between species in which one species, the predator, kills and eats the other, the prey. Though the term *predation* generally elicits such images as a lion attacking and eating an antelope, it applies to a wide range of interactions. A rotifer (a tiny aquatic animal that is smaller than many protists) that kills a unicellular alga by eating it can also be considered a predator. Because eating and avoiding being eaten are prerequisites to reproductive success, the adaptations of both predators and prey tend to be refined through natural selection. In the **Scientific Skills Exercise**, you can interpret data from an experiment investigating a specific predator-prey interaction.

Many important feeding adaptations of predators are obvious and familiar. Most predators have acute senses that enable them to find and identify potential prey. Rattlesnakes and other pit vipers, for example, find their prey with a pair of heat-sensing organs located between their eyes and nostrils (see Figure 38.17b). Many predators also have adaptations such as claws, fangs, or poison that help them catch and subdue their food. Predators that pursue their prey are generally fast and agile, whereas those that lie in ambush are often disguised in their environments.

Just as predators possess adaptations for capturing prey, potential prey animals have adaptations that help them avoid being eaten. Some common behavioral defenses are hiding, fleeing, and forming herds or schools. Active self-defense is less common, though some large grazing mammals vigorously defend their young from predators such as lions.

Animals also display a variety of morphological and physiological defensive adaptations. **Cryptic coloration**, or camouflage, makes prey difficult to see **(Figure 41.5a)**. Mechanical or chemical defenses protect species such as porcupines and skunks. Some animals, such as the European fire salamander, can synthesize toxins; others accumulate toxins passively from the plants they eat. Animals with effective chemical defenses often exhibit bright **aposematic coloration**, or warning coloration, such as that of poison dart frogs **(Figure 41.5b)**. Such coloration seems to be adaptive because predators often avoid brightly colored prey.

Some prey species are protected by their resemblance to other species. For example, in **Batesian mimicry**, a palatable

---

## Scientific Skills Exercise

### Using Bar Graphs and Scatter Plots to Present and Interpret Data

**Can a Native Predator Species Adapt Rapidly to an Introduced Prey Species?** Cane toads (*Bufo marinus*) were introduced to Australia in 1935 in a failed attempt to control an insect pest. Since then, the toads have spread across northeastern Australia, with a population of over 200 million today. Cane toads have glands that produce a toxin that is poisonous to snakes and other potential predators. In this exercise, you will graph and interpret data from a two-part experiment conducted to determine whether native Australian predators have developed resistance to the cane toad toxin.

**How the Experiment Was Done** In part 1, researchers collected 12 black snakes (*Pseudechis porphyriacus*) from areas where cane toads had existed for 40–60 years and another 12 from areas free of cane toads. They recorded the percentage of snakes from each area that ate either a freshly killed native frog (*Limnodynastes peronii*, a species the snakes commonly eat) or a freshly killed cane toad from which the toxin gland had been removed (making the toad nonpoisonous). In part 2, researchers collected snakes from areas where cane toads had been present for 5–60 years. To assess how cane toad toxin affected the physiological activity of these snakes, they injected small amounts of the toxin into the snakes' stomachs and measured the snakes' swimming speed in a small pool.

**Data from the Experiment, Part 1**

| Type of Prey Offered | Percentage of Snakes from Each Area That Ate the Native Frog vs. Cane Toad | |
|---|---|---|
| | Area with Cane Toads Present for 40–60 Years | Area with No Cane Toads |
| Native frog | 100 | 100 |
| Cane toad | 0 | 50 |

**Data from the Experiment, Part 2**

| Number of Years Cane Toads Were Present in the Area | 5 | 10 | 10 | 20 | 50 | 60 | 60 | 60 | 60 | 60 |
|---|---|---|---|---|---|---|---|---|---|---|
| Percent Reduction in Snake Swimming Speed | 52 | 19 | 30 | 30 | 5 | 5 | 9 | 11 | 12 | 22 |

**Data from** B. L. Phillips and R. Shine, An invasive species induces rapid adaptive change in a native predator: cane toads and black snakes in Australia, *Proceedings of the Royal Society B* 273:1545–1550 (2006).

**INTERPRET THE DATA**

1. Make a bar graph of the data in part 1. (For additional information about graphs, see the Scientific Skills Review in Appendix F and in the Study Area in MasteringBiology.)

2. What do the data represented in the graph suggest about the effects of cane toads on the predatory behavior of black snakes in areas where the toads are and are not currently found?

3. Suppose a novel enzyme that deactivates the cane toad toxin evolved in a black snake population exposed to cane toads. If the researchers repeated part 1 of this study, predict how the results would change.

4. Identify the dependent and independent variables in part 2 and make a scatter plot. What conclusion would you draw about whether exposure to cane toads is having a selective effect on black snakes? Explain.

5. Explain why a bar graph is appropriate for presenting the data in part 1 and a scatter plot is appropriate for the data in part 2.

(MB) A version of this Scientific Skills Exercise can be assigned in MasteringBiology.

**(a) Cryptic coloration**

▶ Canyon tree frog

**(b) Aposematic coloration**

▶ Poison dart frog

**(c) Batesian mimicry: A harmless species mimics a harmful one.**

◀ Nonvenomous hawkmoth larva

▼ Venomous green parrot snake

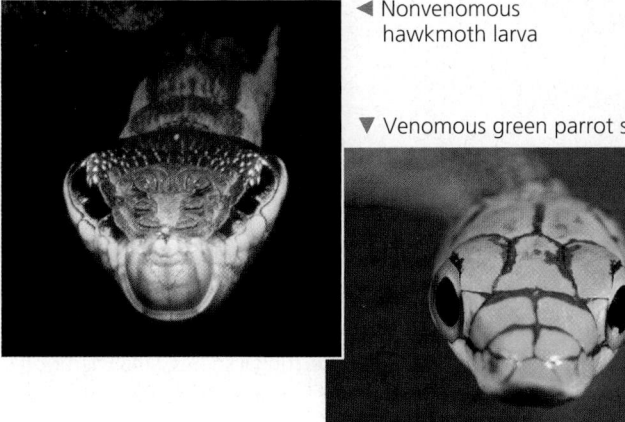

▲ **Figure 41.5 Examples of defensive adaptations in animals.**

**MAKE CONNECTIONS** *Explain how natural selection could increase the resemblance of a harmless species to a distantly related harmful species. Along with selection, what else could account for a harmless species resembling a closely related harmful species? (See Concept 19.2.)*

or harmless species mimics an unpalatable or harmful species to which it is not closely related. The larva of the hawkmoth *Hemeroplanes ornatus* puffs up its head and thorax when disturbed, looking like the head of a small venomous snake **(Figure 41.5c)**. In this case, the mimicry even involves behavior; the larva weaves its head back and forth and hisses like a snake. In Batesian mimicry, the resemblance of a prey species to a distantly related unpalatable or harmful species is thought to have resulted from natural selection.

Many predators also use mimicry. The alligator snapping turtle has a tongue that resembles a wriggling worm, which is used to lure small fish. Any fish that tries to eat the "bait" is itself quickly consumed as the turtle's strong jaws snap closed.

▲ **Figure 41.6 A marine herbivore.** This West Indian manatee (*Trichechus manatus*) in Florida is grazing on *Hydrilla*, an introduced plant.

### Herbivory

Ecologists use the term **herbivory** to refer to a +/− interaction in which an organism—an herbivore—eats parts of a plant or alga, thereby harming it. While large mammalian herbivores such as cattle, sheep, and water buffalo may be most familiar, most herbivores are actually invertebrates, such as grasshoppers, caterpillars, and beetles. In the ocean, herbivores include sea urchins, some tropical fishes, and certain mammals, including the manatee **(Figure 41.6)**.

Like predators, herbivores have many specialized adaptations. Many herbivorous insects have chemical sensors on their feet that enable them to distinguish between plants based on their toxicity or their nutritional value. Some mammalian herbivores, such as goats, use their sense of smell to examine plants, rejecting some and eating others. They may also eat just a specific part of a plant, such as the flowers. Many herbivores also have specialized teeth or digestive systems adapted for processing vegetation (see Concept 33.4).

Unlike prey animals, plants cannot run away to avoid being eaten. Instead, a plant's arsenal against herbivores may feature chemical toxins or structures such as spines and thorns. Among the plant compounds that serve as chemical defenses are the poison strychnine, produced by the tropical vine *Strychnos toxifera*, and nicotine, from the tobacco plant. Compounds that are not toxic to humans but may be distasteful to many herbivores are responsible for the familiar flavors of cinnamon, cloves, and peppermint.

### Parasitism

**Parasitism** is a +/− exploitative interaction in which one organism, the **parasite**, derives its nourishment from another organism, its **host**, which is harmed in the process. Parasites that live within the body of their host, such as tapeworms, are called **endoparasites**; parasites that feed on the external surface of a host, such as ticks and lice, are called **ectoparasites**. Some ecologists have estimated that at least one-third of all species on Earth are parasites. In one particular type of parasitism, parasitoid insects—usually small wasps—lay eggs on or in living hosts, such as the braconid wasp parasitizing a tobacco hornworm (*Manduca sexta*) in the photo.

Many parasites have complex life cycles involving multiple hosts. The blood fluke, which currently infects approximately 200 million people around the world, requires two hosts at different times in its development: humans and freshwater snails. Some parasites change the behavior of their current host in ways that increase the likelihood that the parasite will reach its next host. For instance, crustaceans that are parasitized by acanthocephalan (spiny-headed) worms leave protective cover and move into the open, where they are more likely to be eaten by the birds that are the second host in the worm's life cycle.

Parasites can significantly affect the survival, reproduction, and density of their host population, either directly or indirectly. For example, ticks that feed as ectoparasites on moose can weaken their hosts by withdrawing blood and causing hair breakage and loss. In their weakened condition, the moose have a greater chance of dying from cold stress or predation by wolves (see Figure 40.24).

## Positive Interactions

While nature abounds with dramatic and gory examples of exploitive interactions, ecological communities are also heavily influenced by **positive interactions**, a term that refers to a +/+ or +/0 interaction in which at least one species benefits and neither is harmed. Positive interactions include mutualism and commensalism. As we'll see, they can affect the diversity of species found in ecological communities.

### Mutualism

**Mutualism** is an interspecific interaction that benefits both species (+/+). Mutualisms are common in nature, as illustrated by examples seen in previous chapters, including cellulose digestion by microorganisms in the digestive systems of termites and ruminant mammals, animals that pollinate flowers or disperse seeds, nutrient exchange between fungi and plant roots in mycorrhizae, and photosynthesis by unicellular algae in corals. In the acacia-ant example shown in **Figure 41.7**, each species depends on the other for their survival and reproduction. However, in other mutualisms—including some other acacia-ant interactions—both species can survive on their own.

Typically, both partners in a mutualism incur costs as well as benefits. In mycorrhizae, for example, the plant often

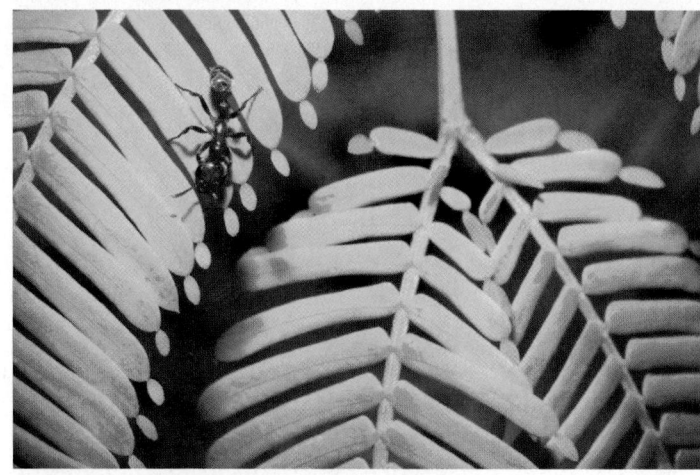

**(a)** Certain species of acacia trees in Central and South America have hollow thorns (not shown) that house stinging ants of the genus *Pseudomyrmex*. The ants feed on nectar produced by the tree and on protein-rich swellings (yellow in the photograph) at the tips of leaflets.

**(b)** The acacia benefits because the pugnacious ants, which attack anything that touches the tree, remove fungal spores, small herbivores, and debris. They also clip vegetation that grows close to the acacia.

▲ **Figure 41.7 Mutualism between acacia trees and ants.**

transfers carbohydrates to the fungus, while the fungus transfers limiting nutrients, such as phosphorus. Each partner benefits, but each partner also experiences a cost: It transfers materials that it could have used to support its own growth and metabolism. The key point is that for an interaction to be a mutualism, the benefits to each partner must exceed the costs.

### Commensalism

An interaction between species that benefits one of the species but neither harms nor helps the other (+/0) is called **commensalism**. Like mutualism, commensal interactions are common in nature. For instance, many wildflowers that live on the forest floor depend entirely on the trees that tower above them—the trees provide the habitat in which they live. Yet the survival and reproduction of the trees are not affected by these wildflowers. Thus, these species are involved in a +/0 interaction in which the wildflowers benefit and the trees are not affected.

**▲ Figure 41.8 Commensalism between cattle egrets and an African buffalo.**

In another example of a commensal association, cattle egrets feed on insects flushed out of the grass by grazing bison, cattle, and other herbivores **(Figure 41.8)**. Because the birds typically find more prey when they follow herbivores, they clearly benefit from the association. Much of the time, the herbivores are not affected by the birds. At times, however, they, too, may derive some benefit; the birds occasionally remove and eat ticks and other ectoparasites from the herbivores or may warn the herbivores of a predator's approach. This example illustrates another key point about ecological interactions: Their effects can change. In this case, an interaction whose effects are typically +/0 (commensalism) may at times become +/+ (mutualism).

Positive interactions can have large effects on ecological communities. For instance, the black rush *Juncus gerardii* makes the soil more hospitable for other plant species in some zones of New England salt marshes **(Figure 41.9a)**. *Juncus* helps prevent salt buildup in the soil by shading the soil surface,

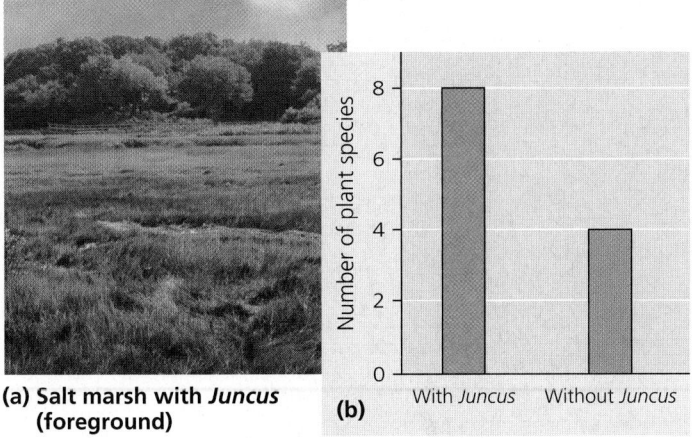

**(a) Salt marsh with *Juncus* (foreground)**

**(b)**

**▲ Figure 41.9 Facilitation by black rush (*Juncus gerardii*) in New England salt marshes.** Black rush increases the number of plant species that can live in the upper middle zone of the marsh.

which reduces evaporation. *Juncus* also prevents the salt marsh soils from becoming oxygen depleted as it transports oxygen to its belowground tissues. In one study, when *Juncus* was removed from areas in the upper middle intertidal zone, those areas supported 50% fewer plant species **(Figure 41.9b)**.

In fact, as is true for positive interactions, competition and exploitation (predation, herbivory, and parasitism) also can have large effects on ecological communities. You'll see examples of how this can occur throughout this chapter.

### CONCEPT CHECK 41.1

1. Explain how interspecific competition, predation, and mutualism differ in their effects on the interacting populations of two species.
2. According to the principle of competitive exclusion, what outcome is expected when two species with identical niches compete for a resource? Why?
3. **MAKE CONNECTIONS** Figure 22.13 illustrates how a hybrid zone can change over time. Imagine that two finch species colonize a new island and are capable of hybridizing (mating and producing viable offspring). The island contains two plant species, one with large seeds and one with small seeds, growing in isolated habitats. If the two finch species specialize in eating different plant species, would reproductive barriers be reinforced, weakened, or unchanged in this hybrid zone? Explain.

For suggested answers, see Appendix A.

---

### CONCEPT 41.2

# Diversity and trophic structure characterize biological communities

Along with the specific interactions described in the previous section, communities are also characterized by more general attributes, including how diverse they are and the feeding relationships of their species. In this section, you'll see why such ecological attributes are important. You'll also learn how a few species sometimes exert strong control on a community's structure, particularly on the composition, relative abundance, and diversity of its species.

## Species Diversity

The **species diversity** of a community—the variety of different kinds of organisms that make up the community—has two components. One is **species richness**, the number of different species in the community. The other is the **relative abundance** of the different species, the proportion each species represents of all individuals in the community.

Imagine two small forest communities, each with 100 individuals distributed among four tree species (A, B, C, and D) as follows:

Community 1: 25A, 25B, 25C, 25D

Community 2: 80A, 5B, 5C, 10D

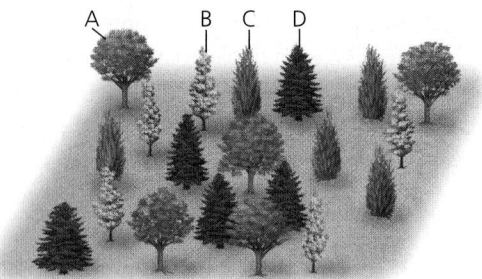

**Community 1**
A: 25%  B: 25%  C: 25%  D: 25%

**Community 2**
A: 80%  B: 5%  C: 5%  D: 10%

▲ **Figure 41.10  Which forest is more diverse?** Ecologists would say that community 1 has greater species diversity, a measure that includes both species richness and relative abundance.

The species richness is the same for both communities because they both contain four species of trees, but the relative abundance is very different **(Figure 41.10)**. You would easily notice the four types of trees in community 1, but without looking carefully, you might see only the abundant species A in the second forest. Most observers would intuitively describe community 1 as the more diverse of the two communities.

Ecologists use many tools to compare the diversity of communities across time and space. They often calculate indexes of diversity based on species richness and relative abundance. One widely used index is the **Shannon diversity index** (*H*):

$$H = -(p_A \ln p_A + p_B \ln p_B + p_C \ln p_C + \ldots)$$

where A, B, C . . . are the species in the community, *p* is the relative abundance of each species, and ln is the natural logarithm; the ln of each value of *p* can be determined using the "ln" key on a calculator. A higher value of *H* indicates a more diverse community. Let's use this equation to calculate the Shannon diversity index of the two communities in Figure 41.10. For community 1, *p* = 0.25 for each species, so

$$H = -4(0.25 \ln 0.25) = 1.39$$

For community 2,

$$H = -[0.8 \ln 0.8 + 2(0.05 \ln 0.05) + 0.1 \ln 0.1] = 0.71$$

These calculations confirm our intuitive description of community 1 as more diverse.

Determining the number and relative abundance of species in a community can be challenging. Because most species in a community are relatively rare, it may be hard to obtain a

sample size large enough to be representative. It can also be difficult to census highly mobile or less visible members of communities, such as microorganisms, insects, and nocturnal species. The small size of microorganisms makes them particularly difficult to sample, so ecologists now use molecular tools to help determine microbial diversity **(Figure 41.11)**.

▼ **Figure 41.11   Research Method**

## Determining Microbial Diversity Using Molecular Tools

**Application**  Ecologists are increasingly using molecular techniques to determine microbial diversity and richness in environmental samples. One such technique produces a DNA profile for microbial taxa based on sequence variations in the DNA that encodes the small subunit of ribosomal RNA. Noah Fierer and Rob Jackson, of Duke University, used this method to compare the diversity of soil bacteria in 98 habitats across North and South America to help identify environmental variables associated with high bacterial diversity.

**Technique**  Researchers first extract and purify DNA from the microbial community in each sample. They use the polymerase chain reaction (PCR; see Figure 13.27) to amplify the ribosomal DNA and label it with a fluorescent dye. Restriction enzymes then cut the amplified, labeled DNA into fragments of different lengths, which are separated by gel electrophoresis. The number and abundance of these fragments characterize the DNA profile of the sample. Based on their analysis, Fierer and Jackson calculated the Shannon diversity index (*H*) of each sample. They then looked for a correlation between *H* and several environmental variables, including vegetation type, mean annual temperature and rainfall, and soil acidity.

**Results**  The diversity of the sampled bacteria was related almost exclusively to soil pH, with the Shannon diversity index being highest in neutral soils and lowest in acidic soils. Amazonian rain forests, which have extremely high plant and animal diversity, had the most acidic soils and the lowest bacterial diversity of the samples tested.

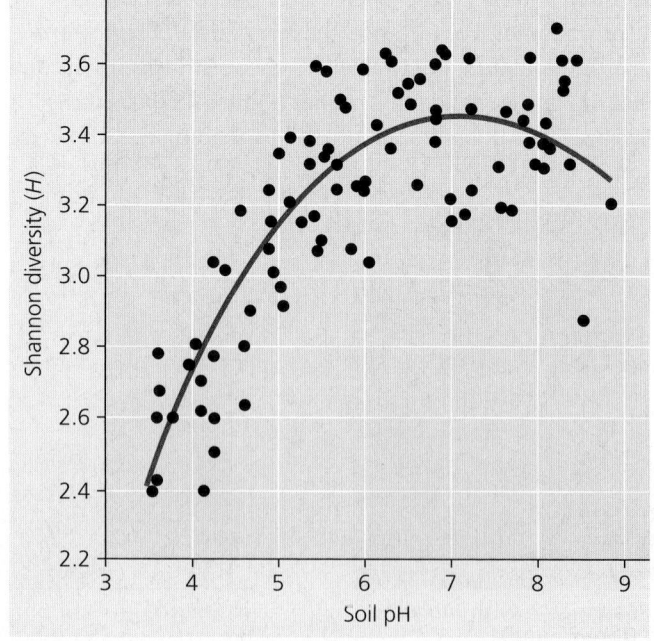

**Data from**  N. Fierer and R. B. Jackson, The diversity and biogeography of soil bacterial communities, *Proceedings of the National Academy of Sciences USA* 103:626–631 (2006).

▲ **Figure 41.12 Study plots at the Cedar Creek Ecosystem Science Reserve, site of long-term experiments in which researchers have manipulated plant diversity.**

## Diversity and Community Stability

In addition to measuring species diversity, ecologists manipulate diversity in experimental communities in nature and in the laboratory. They do this to examine the potential benefits of diversity, including increased productivity and stability of biological communities.

Researchers at the Cedar Creek Ecosystem Science Reserve, in Minnesota, have been manipulating plant diversity in experimental communities for more than two decades **(Figure 41.12)**. Higher-diversity communities generally are more productive and are better able to withstand and recover from environmental stresses, such as droughts. More diverse communities are also more stable year to year in their productivity. In one decade-long experiment, for instance, researchers at Cedar Creek created 168 plots, each containing 1, 2, 4, 8, or 16 perennial grassland species. The most diverse plots consistently produced more **biomass** (the total mass of all organisms in a habitat) than the single-species plots each year.

Higher-diversity communities are often more resistant to **invasive species**, which are organisms that become established outside their native range. Scientists working in Long Island Sound, off the coast of Connecticut, created communities with different levels of diversity consisting of sessile marine invertebrates, including tunicates (see Figure 27.15b). They then examined how vulnerable these experimental communities were to invasion by an exotic tunicate. They found that the exotic tunicate was four times more likely to survive in lower-diversity communities than in higher-diversity ones. The researchers concluded that relatively diverse communities captured more of the resources available in the system, leaving fewer resources for the invader and decreasing its survival.

## Trophic Structure

Experiments like the ones just described often examine the importance of diversity within one trophic level. The structure and dynamics of a community also depend on the feeding

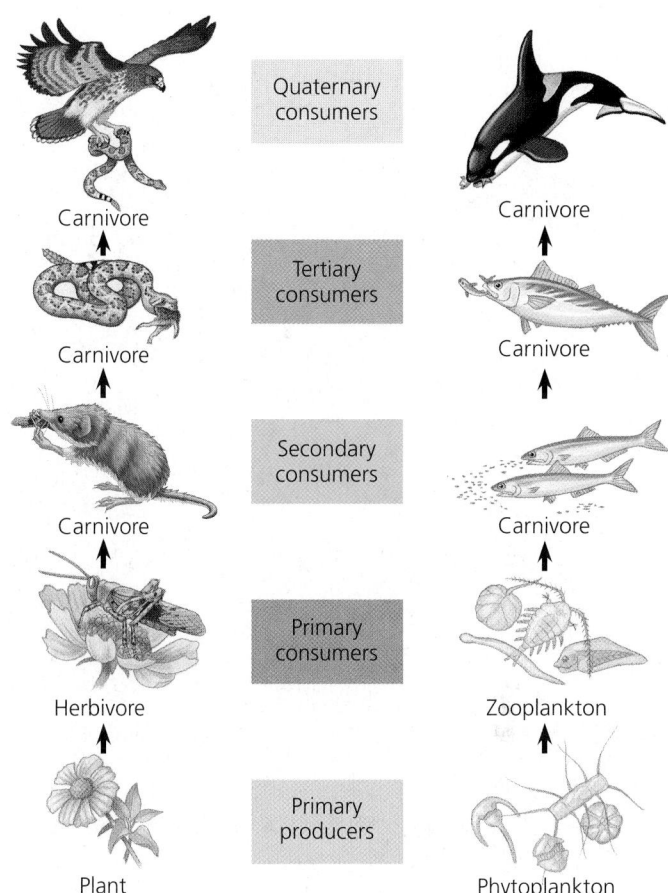

A terrestrial food chain          A marine food chain

▲ **Figure 41.13 Examples of terrestrial and marine food chains.** The arrows trace energy and nutrients that pass through the trophic levels of a community when organisms feed on one another. Decomposers, which feed on the remains of organisms from all trophic levels, are not shown here.

**?** *Suppose the abundance of carnivores that eat zooplankton increased greatly. How might that affect phytoplankton abundance?*

relationships between organisms—the **trophic structure** of the community. The transfer of food energy up the trophic levels from its source in plants and other autotrophs (primary producers) through herbivores (primary consumers) to carnivores (secondary, tertiary, and quaternary consumers) and eventually to decomposers is referred to as a **food chain** **(Figure 41.13)**.

In the 1920s, Oxford University biologist Charles Elton recognized that food chains are not isolated units but are linked together in **food webs**. Ecologists diagram the trophic relationships of a community using arrows that link species according to who eats whom. In an Antarctic pelagic community, for example, the primary producers are phytoplankton, which serve as food for the dominant grazing zooplankton, especially krill and copepods, both of which are crustaceans. These zooplankton species are in turn eaten by various carnivores, including other plankton, penguins, seals, fishes, and baleen whales. Squids, which are carnivores that feed on fish and zooplankton,

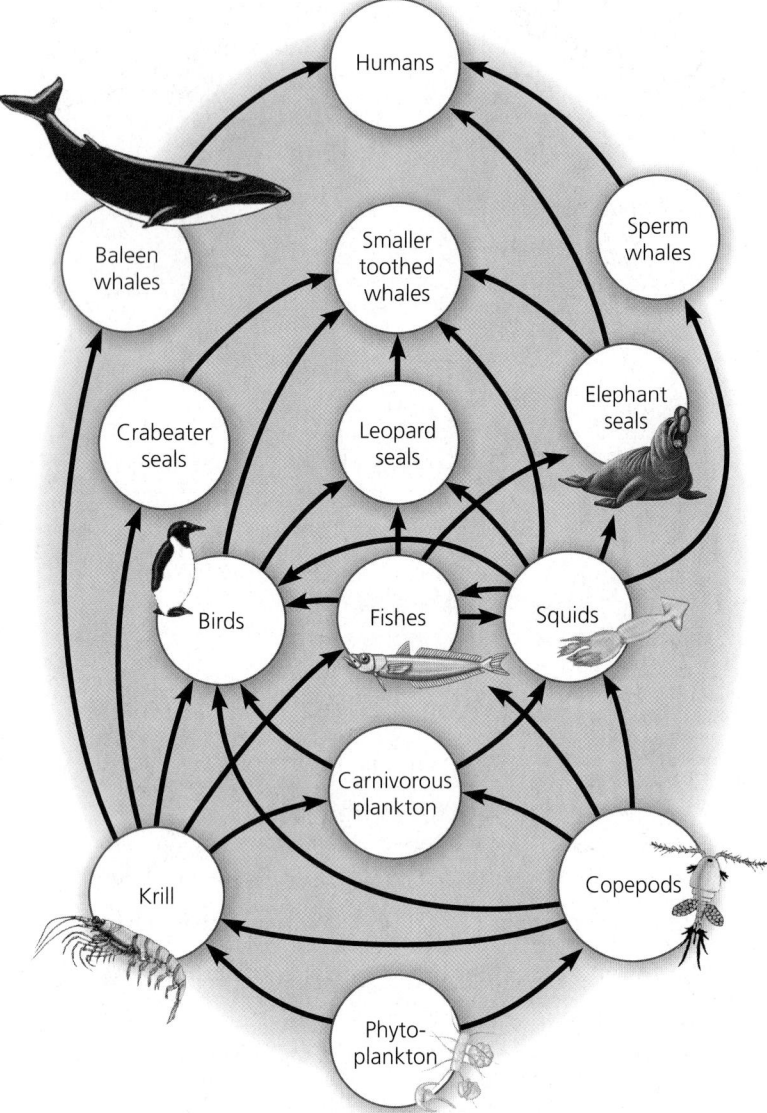

▲ **Figure 41.14 An Antarctic marine food web.** Arrows follow the transfer of food from the producers (phytoplankton) up through the trophic levels. For simplicity, this diagram omits decomposers.

❓ *In the food web shown here, indicate the number of organism types that each group eats. Which two groups are both predator and prey for each other?*

are another important link in these food webs, as they are in turn eaten by seals and toothed whales **(Figure 41.14)**.

Note that a given species may weave into the web at more than one trophic level. For example, in the food web shown in Figure 41.14, krill feed on phytoplankton as well as on other grazing zooplankton, such as copepods.

## Species with a Large Impact

Certain species have an especially large impact on the structure of entire communities because they are highly abundant or play a pivotal role in community dynamics. The impact of these species occurs through trophic interactions and their influence on the physical environment.

**Dominant species** in a community are the species that are the most abundant or that collectively have the highest

biomass. There can be different explanations for why different species become dominant. One hypothesis suggests that dominant species are competitively superior in exploiting limited resources such as water or nutrients. Another hypothesis is that dominant species are most successful at avoiding predation or the impact of disease. The latter idea could explain the high biomass attained in some environments by invasive species. Such species may not face the natural predators or parasites that would otherwise hold their populations in check.

In contrast to dominant species, **keystone species** are not usually abundant in a community. They exert strong control on community structure not by numerical might but by their pivotal ecological roles, or niches. **Figure 41.15** highlights the importance of a keystone species, a sea star, in maintaining the diversity of an intertidal community.

Still other organisms exert their influence on a community not through trophic interactions but by changing their physical environment. Species that dramatically alter their environment are called **ecosystem engineers** or, to avoid implying

▼ **Figure 41.15** **Inquiry**

### Is *Pisaster ochraceus* a keystone predator?

**Experiment** In rocky intertidal communities of western North America, the relatively uncommon sea star *Pisaster ochraceus* preys on mussels such as *Mytilus californianus*, a dominant species and strong competitor for space.

Robert Paine, of the University of Washington, removed *Pisaster* from an area in the intertidal zone and examined the effect on species richness.

**Results** In the absence of *Pisaster*, species richness declined as mussels monopolized the rock face and eliminated most other invertebrates and algae. In a control area where *Pisaster* was not removed, species richness changed very little.

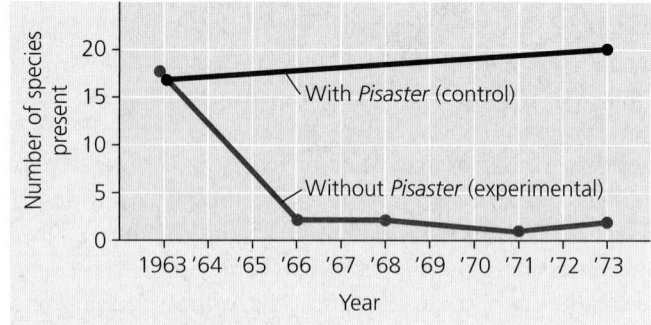

**Conclusion** *Pisaster* acts as a keystone species, exerting an influence on the community that is not reflected in its abundance.

**Data from** R. T. Paine, Food web complexity and species diversity, *American Naturalist* 100:65–75 (1966).

**WHAT IF?** Suppose that an invasive fungus killed most individuals of *Mytilus* at these sites. Predict how species richness would be affected if *Pisaster* were then removed.

▲ **Figure 41.16 Beavers as ecosystem engineers.** By felling trees, building dams, and creating ponds, beavers can transform large areas of forest into flooded wetlands.

conscious intent, "foundation species." A familiar ecosystem engineer is the beaver **(Figure 41.16)**. The effects of ecosystem engineers on other species can be positive or negative, depending on the needs of the other species.

## Bottom-Up and Top-Down Controls

Simplified models based on relationships between adjacent trophic levels are useful for describing community organization. Let's consider the three possible relationships between plants ($V$ for vegetation) and herbivores ($H$):

$$V \rightarrow H \qquad V \leftarrow H \qquad V \leftrightarrow H$$

The arrows indicate that a change in the biomass of one trophic level causes a change in the other trophic level. $V \rightarrow H$ means that an increase in vegetation will increase the numbers or biomass of herbivores, but not vice versa. In this situation, herbivores are limited by vegetation, but vegetation is not limited by herbivory. In contrast, $V \leftarrow H$ means that an increase in herbivore biomass will decrease the abundance of vegetation, but not vice versa. A double-headed arrow indicates that each trophic level is sensitive to changes in the biomass of the other.

Two models of community organization are common: the bottom-up model and the top-down model. The $V \rightarrow H$ linkage suggests a **bottom-up model**, which postulates a unidirectional influence from lower to higher trophic levels. In this case, the presence or absence of mineral nutrients ($N$) controls plant ($V$) numbers, which control herbivore ($H$) numbers, which in turn control predator ($P$) numbers. The simplified bottom-up model is thus $N \rightarrow V \rightarrow H \rightarrow P$. To change the community structure of a bottom-up community, you need to alter biomass at the lower trophic levels, allowing those changes to propagate up through the food web. If you add mineral nutrients to stimulate plant growth, then the higher trophic levels should also increase in biomass. If you change predator abundance, however, the effect should not extend down to the lower trophic levels.

In contrast, the **top-down model** postulates the opposite: Predation mainly controls community organization because predators limit herbivores, herbivores limit plants, and plants limit nutrient levels through nutrient uptake. The simplified top-down model, $N \leftarrow V \leftarrow H \leftarrow P$, is also called the *trophic cascade model*. In a lake community with four trophic levels, the model predicts that removing the top carnivores will increase the abundance of primary carnivores, in turn decreasing the number of herbivores, increasing phytoplankton abundance, and decreasing concentrations of mineral nutrients. The effects thus move down the trophic structure as alternating $+/-$ effects.

Ecologists have applied the top-down model to improve water quality in lakes with high abundances of algae. This approach, called **biomanipulation**, attempts to prevent algal blooms by altering the density of higher-level consumers. In lakes with three trophic levels, removing fish should improve water quality by increasing zooplankton density, thereby decreasing algal populations. In lakes with four trophic levels, adding top predators should have the same effect **(Figure 41.17)**.

Ecologists in Finland used biomanipulation to help purify Lake Vesijärvi, a large lake that was polluted with city sewage and industrial wastewater until 1976. After pollution controls reduced these inputs, the water quality of the lake began to improve. By 1986, however, massive blooms of cyanobacteria started to occur in the lake. These blooms coincided with an increase in the population of roach, a fish species that eats zooplankton, which otherwise keep the cyanobacteria and algae in check. To reverse these changes, ecologists removed nearly a million kilograms of fish from the lake between 1989 and 1993, reducing roach abundance by about 80%. At the same time, they added a fourth trophic level by stocking the lake with pike perch, a predatory fish that eats roach. The water became clear, and the last cyanobacterial bloom was in 1989. Ecologists continue to monitor the lake for evidence of cyanobacterial blooms and low oxygen availability, but the lake has remained clear, even though roach removal ended in 1993.

As these examples show, communities vary in their degree of bottom-up and top-down control. To manage agricultural landscapes, parks, reservoirs, and fisheries, we need to understand each particular community's dynamics.

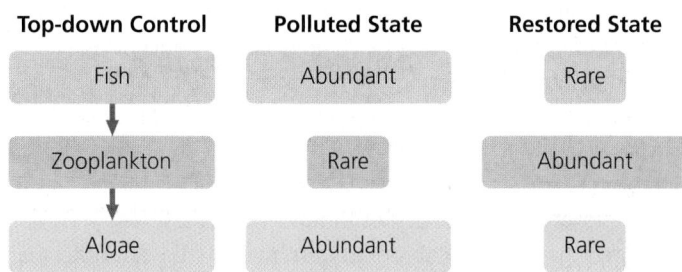

▲ **Figure 41.17 Results of biomanipulation in a lake with top-down control of community organization.** Decreasing the abundance of fish that ate zooplankton results in a decrease in the biomass of algae, improving water quality.

1. What two components contribute to species diversity? Explain how two communities with the same number of species can differ in species diversity.

2. How is a food chain different from a food web?

3. **WHAT IF?** Consider a grassland with five trophic levels: grasses, mice, snakes, raccoons, and bobcats. If you released additional bobcats into the grassland, how would grass biomass change if the bottom-up model applied? If the top-down model applied? Explain.

4. **MAKE CONNECTIONS** Rising atmospheric $CO_2$ levels lead to ocean acidification (see Figure 2.24) and global warming, both of which can reduce krill abundance. Predict how a drop in krill abundance might affect other organisms in the food web shown in Figure 41.14. Which organisms are particularly at risk? Explain.

For suggested answers, see Appendix A.

# CONCEPT 41.3

# Disturbance influences species diversity and composition

Decades ago, most ecologists favored the traditional view that biological communities are at equilibrium, a more or less stable balance, unless seriously disturbed by human activities. The "balance of nature" view focused on interspecific competition as a key factor determining community composition and maintaining stability in communities. *Stability* in this context refers to a community's tendency to reach and maintain a relatively constant composition of species.

One of the earliest proponents of this view, F. E. Clements, of the Carnegie Institution of Washington, argued in the early 1900s that the community of plants at a site had only one stable equilibrium, a *climax community* controlled solely by climate. According to Clements, biotic interactions caused the species in the community to function as an integrated unit—in effect, as a superorganism. His argument was based on the observation that certain species of plants are consistently found together, such as the oaks, maples, birches, and beeches in deciduous forests of the northeastern United States.

Other ecologists questioned whether most communities were at equilibrium or functioned as integrated units. A. G. Tansley, of Oxford University, challenged the concept of a climax community, arguing that differences in soils, topography, and other factors created many potential communities that were stable within a region. H. A. Gleason, of the University of Chicago, saw communities not as superorganisms but as chance assemblages of species found together because they happen to have similar abiotic requirements—for example, for temperature, rainfall, and soil type. Gleason and other ecologists also realized that disturbance keeps many communities from reaching a state of equilibrium in species diversity or

composition. A **disturbance** is an event—such as a storm, fire, flood, drought, or human activity—that changes a community by removing organisms from it or altering resource availability.

This emphasis on change has led to the formulation of the **nonequilibrium model**, which describes most communities as constantly changing after disturbance. Even relatively stable communities can be rapidly transformed into nonequilibrium communities. Let's examine some of the ways that disturbances influence community structure and composition.

## Characterizing Disturbance

The types of disturbances and their frequency and severity vary among communities. Storms disturb almost all communities, even those in the oceans through the action of waves. Fire is a significant disturbance; in fact, chaparral and some grassland biomes require regular burning to maintain their structure and species composition. Many streams and ponds are disturbed by spring flooding and seasonal drying. A high level of disturbance is generally the result of frequent *and* intense disturbance, while low disturbance levels can result from either a low frequency or low intensity of disturbance.

The **intermediate disturbance hypothesis** states that moderate levels of disturbance foster greater species diversity than do high or low levels of disturbance. High levels of disturbance reduce diversity by creating environmental stresses that exceed the tolerances of many species or by disturbing the community so often that slow-growing or slow-colonizing species are excluded. At the other extreme, low levels of disturbance can reduce species diversity by allowing competitively dominant species to exclude less competitive ones. Meanwhile, intermediate levels of disturbance can foster greater species diversity by opening up habitats for occupation by less competitive species. Such intermediate disturbance levels rarely create conditions so severe that they exceed the environmental tolerances or recovery rates of potential community members.

The intermediate disturbance hypothesis is supported by many terrestrial and aquatic studies. In one study, ecologists in New Zealand compared the richness of invertebrates living in the beds of streams exposed to different frequencies and intensities of flooding (**Figure 41.18**). When floods occurred either very frequently or rarely, invertebrate richness was low. Frequent floods made it difficult for some species to become established in the streambed, while rare floods resulted in species being displaced by superior competitors. Invertebrate richness peaked in streams that had an intermediate frequency or intensity of flooding, as predicted by the hypothesis.

Although moderate levels of disturbance appear to maximize species diversity in some cases, small and large disturbances also can have important effects on community structure. Small-scale disturbances can create patches of different habitats across a landscape, which help maintain diversity in a community. Large-scale disturbances are also a natural part of many communities. Much of Yellowstone National

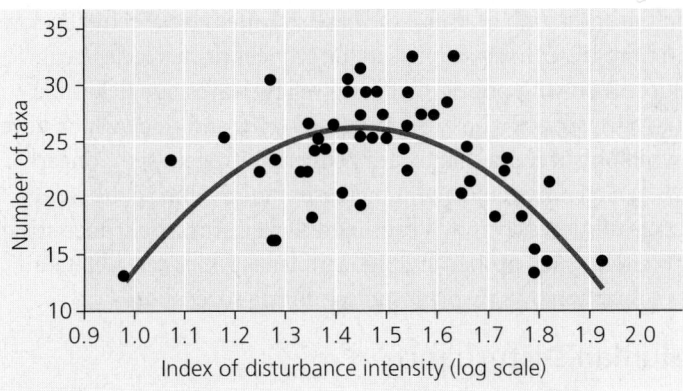

▲ **Figure 41.18 Testing the intermediate disturbance hypothesis.** Researchers identified the taxa (species or genera) of invertebrates at two locations in each of 27 New Zealand streams. They assessed the intensity of flooding at each location using an index of streambed disturbance. The number of invertebrate taxa peaked where the intensity of flooding was at intermediate levels.

Park, for example, is dominated by lodgepole pine, a tree species that requires the rejuvenating influence of periodic fires. Lodgepole pine cones remain closed until exposed to intense heat. When a forest fire burns the trees, the cones open and the seeds are released. The new generation of lodgepole pines can then thrive on nutrients released from the burned trees and in the sunlight that is no longer blocked by taller trees.

In the summer of 1988, extensive areas of Yellowstone burned during a severe drought. By 1989, burned areas in the park were largely covered with new vegetation, suggesting that the species in this community are adapted to rapid recovery after fire **(Figure 41.19)**. In fact, large-scale fires have periodically swept through the lodgepole pine forests of Yellowstone and other northern areas for thousands of years.

Studies of the Yellowstone forest community and many others indicate that they are nonequilibrium communities, changing continually because of natural disturbances and the internal processes of growth and reproduction. Mounting evidence suggests that nonequilibrium conditions are in fact the norm for most communities.

## Ecological Succession

Changes in the composition and structure of terrestrial communities are most apparent after a severe disturbance, such as a volcanic eruption or a glacier, strips away all the existing vegetation. The disturbed area may be colonized by a variety of species, which are gradually replaced by other species, which are in turn replaced by still other species—a process called **ecological succession**. When this process begins in a virtually lifeless area where soil has not yet formed, such as on a new volcanic island or on the rubble (moraine) left by a retreating glacier, it is called **primary succession**. Another type of succession, **secondary succession**, occurs when an existing community has been cleared by a disturbance that leaves the soil intact, as in Yellowstone following the 1988 fires (see Figure 41.19).

**(a) Soon after fire.** While all trees in the foreground of this photograph were killed by the fire, unburned trees can be seen in other locations.

**(b) One year after fire.** The community has begun to recover. Herbaceous plants, different from those in the former forest, cover the ground.

▲ **Figure 41.19 Recovery following a large-scale disturbance.** The 1988 Yellowstone National Park fires burned large areas of forests dominated by lodgepole pines.

During primary succession, the only life-forms initially present are often prokaryotes and protists. Lichens and mosses, which grow from windblown spores, are commonly the first macroscopic photosynthesizers to colonize such areas. Soil develops gradually as rocks weather and organic matter accumulates from the decomposed remains of the early colonizers. Once soil is present, the lichens and mosses are usually overgrown by grasses, shrubs, and trees that sprout from seeds blown in from nearby areas or carried in by animals. Eventually, an area is colonized by plants that become the community's dominant form of vegetation. Producing such a community through primary succession may take hundreds or thousands of years.

Early-arriving species and later-arriving ones may be linked by one of three key processes. The early arrivals may *facilitate* the appearance of the later species by making the environment more favorable—for example, by increasing the fertility of the soil. Alternatively, the early species may *inhibit* establishment of the later species, so that successful colonization by later species occurs in spite of, rather than because of, the activities of the early species. Finally, the early species may be completely independent of the later species, which *tolerate* conditions created early in succession but are neither helped nor hindered by early species.

Ecologists have conducted some of the most extensive research on primary succession at Glacier Bay in southeastern Alaska, where glaciers have retreated more than 100 km since 1760 **(Figure 41.20)**. By studying the communities at different distances from the mouth of the bay, ecologists can examine different stages in succession. ❶ The exposed glacial moraine is colonized first by pioneering species that include liverworts, mosses, fireweed, scattered *Dryas* (a mat-forming shrub), and willows. ❷ After about three decades, *Dryas* dominates the plant community. ❸ A few decades later, the area is invaded by alder, which forms dense thickets up to 9 m tall. ❹ In the next two centuries, these alder stands are overgrown first by Sitka spruce and later by western hemlock and mountain hemlock. In areas of poor drainage, the forest floor of this spruce-hemlock forest is invaded by sphagnum moss, which holds water and acidifies the soil, eventually killing the trees. Thus, by about 300 years after glacial retreat, the vegetation consists of sphagnum bogs on the poorly drained flat areas and spruce-hemlock forest on the well-drained slopes.

Succession on glacial moraines is related to changes in soil nutrients and other environmental factors caused by transitions in the vegetation. Because the bare soil after glacial retreat is low in nitrogen content, almost all the pioneer plant species begin succession with poor growth and yellow leaves due to limited nitrogen supply. The exceptions are *Dryas* and alder, which have symbiotic bacteria that fix atmospheric nitrogen (see Concept 29.4). Soil nitrogen content increases quickly during the alder stage of succession and keeps increasing during the spruce stage. By altering soil properties, pioneer plant species can facilitate colonization by new plant species during succession.

## Human Disturbance

Ecological succession is a response to disturbance of the environment, and the strongest disturbances are human activities. Agricultural development has disrupted what were once the vast grasslands of the North American prairie. Tropical rain forests are quickly disappearing as a result of clear-cutting for lumber, cattle grazing, and farmland. Centuries of overgrazing and agricultural disturbance have contributed to famine in parts of Africa by turning seasonal grasslands into vast barren areas.

Humans disturb marine ecosystems as well as terrestrial ones. The effects of ocean trawling, in which boats drag weighted nets across the seafloor, are similar to those of clear-cutting a forest or plowing a field **(Figure 41.21)**. The trawls scrape and scour corals and other life on the seafloor. In a typical year, ships trawl an area about the size of South America, 150 times larger than the area of forests that are clear-cut annually.

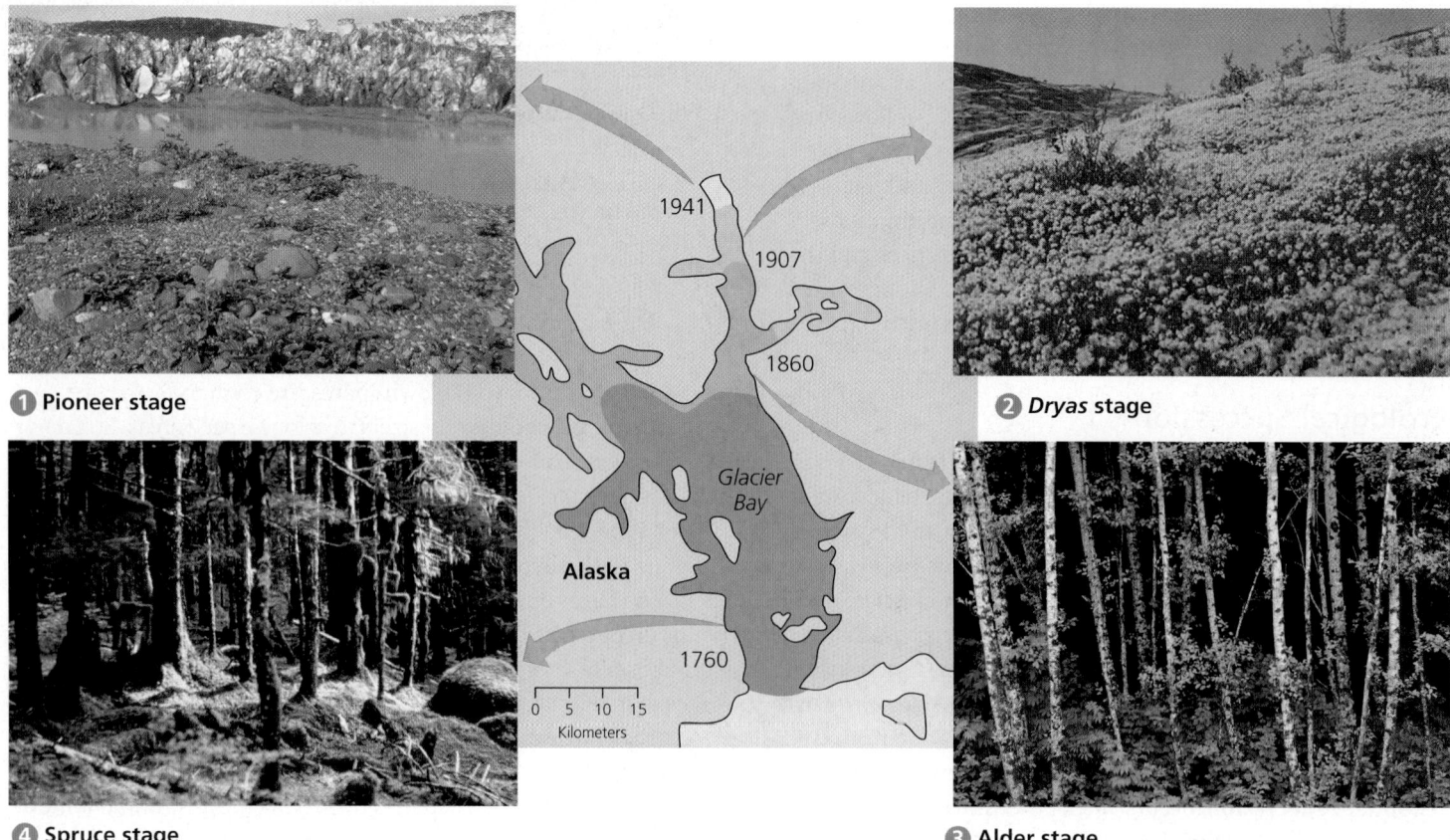

❶ **Pioneer stage**

❷ *Dryas* **stage**

1941
1907
1860
Glacier Bay
Alaska
0  5  10  15
Kilometers
1760

❹ **Spruce stage**

❸ **Alder stage**

▲ **Figure 41.20 Glacial retreat and primary succession at Glacier Bay, Alaska.** The different shades of blue on the map show retreat of the glacier since 1760, based on historical descriptions.

◄ Before trawling

◄ After trawling

▲ **Figure 41.21 Disturbance of the ocean floor by trawling.** These photos show the seafloor off northwestern Australia before (top) and after (bottom) deep-sea trawlers have passed.

Because disturbance by human activities is often severe, it reduces species diversity in many communities. In Chapter 43, we'll take a closer look at how human-caused disturbance is affecting the diversity of life.

### CONCEPT CHECK 41.3

1. Why do high and low levels of disturbance usually reduce species diversity? Why does an intermediate level of disturbance promote species diversity?

2. During succession, how might the early species facilitate the arrival of other species?

3. **WHAT IF?** Most prairies experience regular fires, typically every few years. If these disturbances were relatively modest, how would the species diversity of a prairie likely be affected if no burning occurred for 100 years? Explain your answer.

For suggested answers, see Appendix A.

---

## CONCEPT 41.4

# Biogeographic factors affect community diversity

So far, we have examined relatively small-scale or local factors that influence the diversity of communities, including the effects of species interactions, dominant species, and many types of disturbances. Ecologists also recognize that large-scale biogeographic factors contribute to the tremendous range of diversity observed in biological communities. The contributions of two biogeographic factors in particular—the latitude of a community and the area it occupies—have been investigated for more than a century.

## Latitudinal Gradients

In the 1850s, both Charles Darwin and Alfred Wallace pointed out that plant and animal life was generally more abundant and diverse in the tropics than in other parts of the globe. Since

that time, many researchers have confirmed this observation. One study found that a 6.6-hectare (1 ha = 10,000 m²) plot in tropical Malaysia contained 711 tree species, while a 2-ha plot of deciduous forest in Michigan typically contained just 10 to 15 tree species. Many groups of animals show similar latitudinal gradients. For instance, there are more than 200 species of ants in Brazil, but only 7 in Alaska.

Two key factors that can affect latitudinal gradients of species richness are evolutionary history and climate. Over the course of evolution, species richness may increase in a community as more speciation events occur (see Concept 22.2). Tropical communities are generally older than temperate or polar communities, which have repeatedly "started over" after major disturbances such as glaciations. As a result, species diversity may be highest in the tropics simply because there has been more time for speciation to occur in tropical communities than in temperate or polar communities.

Climate is another key factor thought to affect latitudinal gradients of richness and diversity. In terrestrial communities, the two main climatic factors correlated with diversity are sunlight and precipitation, both of which occur at high levels in the tropics. These factors can be considered together by measuring a community's rate of **evapotranspiration**, the evaporation of water from soil and plants. Evapotranspiration, a function of solar radiation, temperature, and water availability, is much higher in hot areas with abundant rainfall than in areas with low temperatures or low precipitation. *Potential evapotranspiration*, a measure of potential water loss that assumes that water is readily available, is determined by the amount of solar radiation and temperature and is highest in regions where both are plentiful. The species richness of plants and animals correlates with both measures, as shown for vertebrates and potential evapotranspiration in **Figure 41.22**.

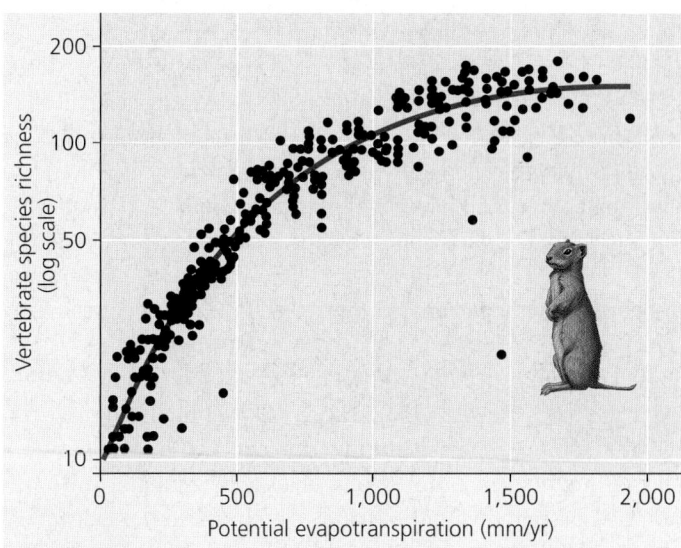

▲ **Figure 41.22 Energy, water, and species richness.** Vertebrate species richness in North America increases predictably with potential evapotranspiration, expressed as rainfall equivalents (mm/yr).

## Area Effects

In 1807, naturalist and explorer Alexander von Humboldt described one of the first patterns of species richness to be recognized, the **species-area curve**: All other factors being equal, the larger the geographic area of a community, the more species it has, in part because larger areas offer a greater diversity of habitats and microhabitats. The basic concept of diversity increasing with increasing area applies in many situations, from surveys of ant diversity in New Guinea to studies of plant species richness on islands of different sizes.

Because of their isolation and limited size, islands provide excellent opportunities for studying the biogeographic factors that affect the species diversity of communities. By "islands," we mean not only oceanic islands, but also habitat islands on land, such as lakes, mountain peaks separated by lowlands, or habitat fragments—any patch surrounded by an environment not suitable for the "island" species. American ecologists Robert MacArthur and E. O. Wilson developed a general model of island biogeography, identifying the key determinants of species diversity on an island with a given set of physical characteristics.

Consider a newly formed oceanic island that receives colonizing species from a distant mainland. Two factors that determine the number of species on the island are the rate at which new species immigrate to the island and the rate at which species on the island become extinct. At any given time, an island's immigration and extinction rates are affected by the number of species already present. As the number of species on the island increases, the immigration rate of new species decreases, because any individual reaching the island is less likely to represent a species that is not already present. At the same time, as more species inhabit an island, extinction rates on the island increase because of the greater likelihood of competitive exclusion.

Two physical features of the island further affect immigration and extinction rates: its size and its distance from the mainland. Small islands generally have lower immigration rates because potential colonizers are less likely to reach a small island than a large one. Small islands also have higher extinction rates because they generally contain fewer resources, have less diverse habitats, and have smaller population sizes. Distance from the mainland is also important; a closer island generally has a higher immigration rate and a lower extinction rate than one farther away. Arriving colonists help sustain the presence of a species on a near island and prevent its extinction.

MacArthur and Wilson's model is called the *island equilibrium model* because an equilibrium will eventually be reached where the rate of species immigration equals the rate of species extinction. The number of species at this equilibrium point is correlated with the island's size and distance from the mainland. Like any ecological equilibrium, this species equilibrium is dynamic; immigration and extinction continue, and the exact species composition may change over time.

---

▼ **Figure 41.23** **Inquiry**

### How does species richness relate to area?

**Field Study** Ecologists Robert MacArthur and E. O. Wilson studied the number of plant species on the Galápagos Islands in relation to the area of the different islands.

**Results**

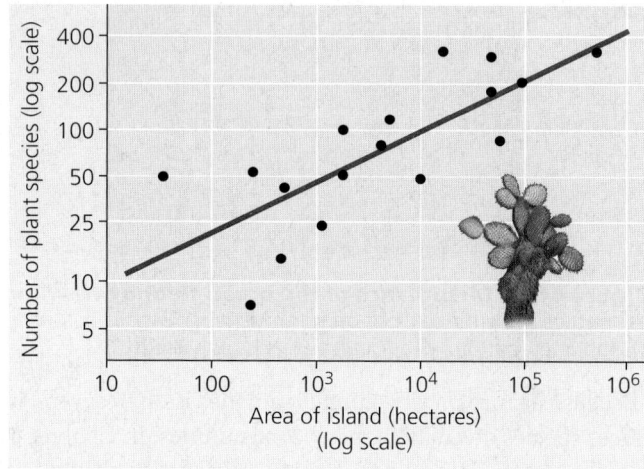

**Conclusion** Plant species richness increases with island size, supporting the island equilibrium model.

**Data from** R. H. MacArthur and E. O. Wilson, *The theory of island biogeography*, Princeton University Press, Princeton, NJ (1967).

**WHAT IF?** Five islands in this study ranging in area from about 40 ha to 10,000 ha each contained about 50 plant species. What does such variation tell you about the simple assumptions of the island equilibrium model?

---

MacArthur and Wilson's studies of the diversity of plants and animals on island chains support the prediction that species richness increases with island size, in keeping with the island equilibrium model **(Figure 41.23)**. Species counts also fit the prediction that the number of species decreases with increasing remoteness of the island.

Over long periods, disturbances such as storms, adaptive evolutionary changes, and speciation generally alter the species composition and community structure on islands. Nonetheless, the island equilibrium model is widely applied in ecology. Conservation biologists in particular use it when designing habitat reserves or establishing a starting point for predicting the effects of habitat loss on species diversity.

### CONCEPT CHECK 41.4

1. Describe two hypotheses that explain why species diversity is greater in tropical regions than in temperate and polar regions.
2. Describe how an island's size and distance from the mainland affect the island's species richness.
3. **WHAT IF?** Based on MacArthur and Wilson's island equilibrium model, how would you expect the richness of birds on islands to compare with the richness of snakes and lizards? Explain.

For suggested answers, see Appendix A.

## CONCEPT 41.5

# Pathogens alter community structure locally and globally

Now that we have examined several important factors that structure biological communities, we will finish the chapter by examining community interactions involving **pathogens**—disease-causing microorganisms and viruses. Scientists have only recently come to appreciate how universal the effects of pathogens are in structuring ecological communities.

## Effects on Community Structure

Pathogens produce especially clear effects on community structure when they are introduced into new habitats. Coral reef communities, for example, are increasingly susceptible to the influence of newly discovered pathogens. White-band disease, caused by an unknown pathogen, has resulted in dramatic changes in the structure and composition of Caribbean reefs. The disease kills corals by causing their tissue to slough off in a band from the base to the tip of the branches. Because of the disease, staghorn coral (*Acropora cervicornis*) has virtually disappeared from the Caribbean since the 1980s. Populations of elkhorn coral (*Acropora palmata*) have also been decimated. Such corals provide key habitat for lobsters as well as snappers and other fish species. When the corals die, they are quickly overgrown by algae. Surgeonfish and other herbivores that feed on algae come to dominate the fish community. Eventually, the corals topple because of damage from storms and other disturbances. The complex, three-dimensional structure of the reef disappears, and diversity plummets.

Pathogens also influence community structure in terrestrial ecosystems. In the forests and savannas of California, trees of several species are dying from sudden oak death (SOD). This recently discovered disease is caused by the protist *Phytophthora ramorum* (see Concept 25.4). SOD was first described in California in 1995, when hikers noticed trees dying around San Francisco Bay. By 2014, it had spread more than 1,000 km, from the central California coast to southern Oregon, and it had killed more than a million oaks and other trees. The loss of the oaks has led to the decreased abundance of at least five bird species, including the acorn woodpecker and the oak titmouse, that rely on the oaks for food and habitat. Although there is currently no cure for SOD, scientists recently sequenced the genome of *P. ramorum* in hopes of finding a way to fight the pathogen.

## Community Ecology and Zoonotic Diseases

Three-quarters of emerging human diseases and many of the most devastating diseases are caused by **zoonotic pathogens**—those that are transferred to humans from other animals, either through direct contact with an infected animal or by means of an intermediate species, called a **vector**. The

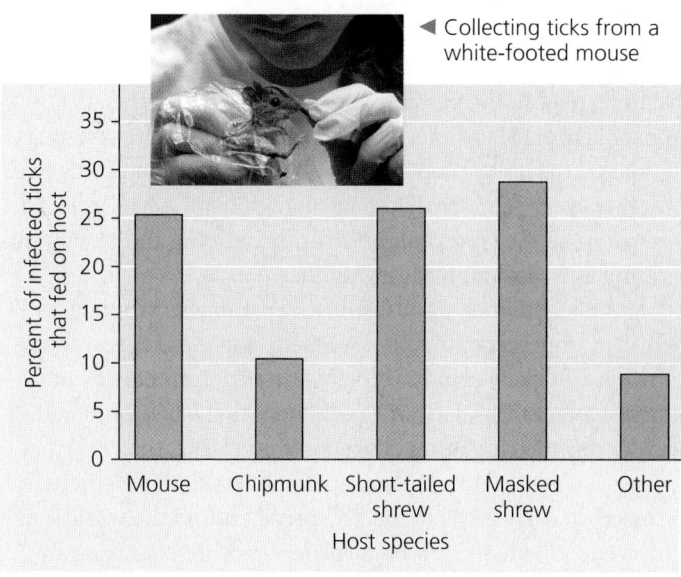

◄ Collecting ticks from a white-footed mouse

**▲ Figure 41.24 Unexpected hosts of the Lyme disease pathogen.** A combination of ecological data and genetic analyses enabled scientists to show that more than half of ticks carrying the Lyme pathogen became infected by feeding on the short-tailed shrew (*Blarina brevicauda*) or the masked shrew (*Sorex cinereus*).

**MAKE CONNECTIONS** *Concept 21.1 describes genetic variation between populations. How might genetic variation between shrew populations in different locations affect the number of infected ticks?*

vectors that spread zoonotic diseases are often parasites, including ticks, lice, and mosquitoes.

Identifying the community of hosts and vectors for a pathogen can help prevent illnesses such as Lyme disease, which is spread by ticks. For years, scientists thought that the primary host for the Lyme pathogen was the white-footed mouse because mice are heavily parasitized by young ticks. When researchers vaccinated mice against Lyme disease and released them into the wild, however, the number of infected ticks hardly changed. Further investigation in New York revealed that two inconspicuous shrew species were the hosts of more than half the infected ticks collected in the field **(Figure 41.24)**. Identifying the dominant hosts for a pathogen provides information that may be used to control the hosts most responsible for spreading diseases.

Ecologists also use their knowledge of community interactions to track the spread of zoonotic diseases. One example, avian flu, is caused by highly contagious viruses transmitted through the saliva and feces of birds (see Concept 17.3). Most of these viruses affect wild birds mildly, but they often cause stronger symptoms in domesticated birds, the most common source of human infections. Since 2003, one particular viral strain, called H5N1, has killed hundreds of millions of poultry and more than 300 people.

Control programs that quarantine domestic birds or monitor their transport may be ineffective if avian flu spreads naturally through the movements of wild birds. From 2003 to 2006, the H5N1 strain spread rapidly from southeast Asia into Europe and Africa. By 2015, the virus had not appeared in Australia or South America, but one human case had occurred in North America; this took place in Canada when a person returning from China became ill with the virus and later died. With respect to the possible spread of H5N1 by birds, the most likely place for infected wild birds to enter the Americas is Alaska, the entry point for ducks, geese, and shorebirds that migrate every year across the Bering Sea from Asia. Ecologists are studying the spread of the virus by trapping and testing migrating and resident birds in Alaska.

Human activities are transporting pathogens around the world at unprecedented rates. Genetic analyses suggest that *P. ramorum* likely came to North America from Europe in nursery plants. Similarly, the pathogens that cause human diseases are spread by our global economy. H1N1, the virus that causes "swine flu" in humans, was first detected in Veracruz, Mexico, in early 2009. It quickly spread around the world when infected individuals flew on airplanes to other countries. By 2010, this flu outbreak had a confirmed death toll of more than 18,000 people. The actual number may have been significantly

higher since many people who died with flu-like symptoms were not tested for H1N1.

While our emphasis here has been on community ecology, pathogens are also greatly influenced by changes in the physical environment. To control pathogens and the diseases they cause, scientists need an ecosystem perspective—an intimate knowledge of how the pathogens interact with other species and with all aspects of their environment. Ecosystems are the subject of Chapter 42.

### CONCEPT CHECK 41.5

1. What are pathogens?
2. **WHAT IF?** Rabies, a viral disease in mammals, is not currently found in the British Isles. If you were in charge of disease control there, what practical approaches might you employ to keep the rabies virus from reaching these islands?

For suggested answers, see Appendix A.

---

## 41 | Chapter Review

Go to **MasteringBiology®** for Assignments, the eText, and the Study Area with Animations, Activities, Vocab Self-Quiz, and Practice Tests.

### SUMMARY OF KEY CONCEPTS

**VOCAB SELF-QUIZ**

goo.gl/gbai8v

#### CONCEPT 41.1

**Interactions within a community may help, harm, or have no effect on the species involved (pp. 868–873)**

- As shown in the table, ecological interactions can be grouped into three broad categories: competition, exploitation, and positive interactions.

| Interaction | Description |
|---|---|
| Competition (−/−) | Two or more species compete for a resource that is in short supply. |
| Exploitation (+/−) | One species benefits by feeding upon the other species, which is harmed. Exploitation includes: |
| Predation | One species, the predator, kills and eats the other, the prey. |
| Herbivory | An herbivore eats part of a plant or alga. |
| Parasitism | The **parasite** derives its nourishment from a second organism, its **host**, which is harmed. |
| Positive interactions (+/+ or +/0) | One species benefits, while the other species benefits or is not harmed. Positive interactions include: |
| Mutualism (+/+) | Both species benefit from the interaction. |
| Commensalism (+/0) | One species benefits, while the other is not affected. |

- **Competitive exclusion** states that two species competing for the same resource cannot coexist permanently in the same place. **Resource partitioning** is the differentiation of **ecological niches** that enables species to coexist in a community.

**?** *For each interaction listed in the table above, give an example of a pair of species that exhibit the interaction.*

#### CONCEPT 41.2

**Diversity and trophic structure characterize biological communities (pp. 873–878)**

- **Species diversity** is affected by both the number of species in a community—its **species richness**—and their **relative abundance**. A community with similar abundances of species is more diverse than one in which one or two species are abundant and the remainder are rare.
- **Trophic structure** is a key factor in community dynamics. **Food chains** link the trophic levels from producers to top carnivores. Branching food chains and complex trophic interactions form **food webs**.
- **Dominant species** are the most abundant species in a community. **Keystone species** are usually less abundant species that exert a disproportionate influence on community structure. **Ecosystem engineers** influence community structure through their effects on the physical environment.
- The **bottom-up model** proposes a unidirectional influence from lower to higher trophic levels, in which nutrients and other abiotic factors primarily determine community structure. The **top-down model** proposes that control of each trophic level comes from the trophic level above, with the result that predators control herbivores, which in turn control primary producers.

**?** *Based on indexes such as Shannon diversity, is a community of higher species richness always more diverse than a community of lower species richness? Explain.*

#### CONCEPT 41.3

**Disturbance influences species diversity and composition (pp. 878–881)**

- Increasing evidence suggests that **disturbance** and lack of equilibrium, rather than stability and equilibrium, are the norm for most communities. According to the **intermediate disturbance hypothesis**, moderate levels of disturbance can foster higher species diversity than can low or high levels of disturbance.

- **Ecological succession** is the sequence of community and ecosystem changes after a disturbance. **Primary succession** occurs where no soil exists when succession begins; **secondary succession** begins in an area where soil remains after a disturbance.
- Humans are the most widespread agents of disturbance, and their effects on communities often reduce species diversity.

> **?** *Is the disturbance pictured in Figure 41.21 more likely to initiate primary or secondary succession? Explain.*

CONCEPT 41.4

### Biogeographic factors affect community diversity (pp. 881–882)

- Species richness generally declines along a latitudinal gradient from the tropics to the poles. Climate influences the diversity gradient through energy (heat and light) and water. The greater age of tropical environments also may contribute to their greater species richness.
- Species richness is directly related to a community's geographic size, a principle formalized in the **species-area curve**. The island equilibrium model maintains that species richness on an ecological island reaches an equilibrium where new immigrations are balanced by extinctions.

> **?** *How have periods of glaciation influenced latitudinal patterns of diversity?*

CONCEPT 41.5

### Pathogens alter community structure locally and globally (pp. 883–884)

- Recent work has highlighted the role that **pathogens** play in structuring terrestrial and marine communities.
- **Zoonotic pathogens** are transferred from other animals to humans. Community ecology provides the framework for identifying key species interactions associated with such pathogens and for helping us track and control their spread.

> **?** *Suppose a pathogen attacks a keystone species. Explain how this could alter the structure of the community.*

## TEST YOUR UNDERSTANDING

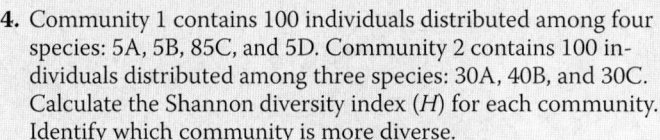

PRACTICE TEST

goo.gl/CRZjvS

### Level 1: Knowledge/Comprehension

1. The feeding relationships among the species in a community determine the community's
   (A) secondary succession.
   (B) ecological niche.
   (C) species richness.
   (D) trophic structure.

2. Based on the intermediate disturbance hypothesis, a community's species diversity is increased by
   (A) frequent massive disturbance.
   (B) stable conditions with no disturbance.
   (C) moderate levels of disturbance.
   (D) human intervention to eliminate disturbance.

### Level 2: Application/Analysis

3. Which of the following could qualify as a top-down control on a grassland community?
   (A) limitation of plant biomass by rainfall amount
   (B) influence of temperature on competition among plants
   (C) influence of soil nutrients on the abundance of grasses versus wildflowers
   (D) effect of grazing intensity by bison on plant species diversity

4. Community 1 contains 100 individuals distributed among four species: 5A, 5B, 85C, and 5D. Community 2 contains 100 individuals distributed among three species: 30A, 40B, and 30C. Calculate the Shannon diversity index (*H*) for each community. Identify which community is more diverse.

### Level 3: Synthesis/Evaluation

5. **DRAW IT** In the Chesapeake Bay, the blue crab is an omnivore, eating eelgrass and other primary producers as well as clams. It is also a cannibal. In turn, the crabs are eaten by humans and by the endangered Kemp's Ridley sea turtle. Based on this information, draw a food web that includes the blue crab. Assuming that the top-down model holds for this system, describe what would happen to the abundance of eelgrass if humans stopped eating blue crabs.

6. **SCIENTIFIC INQUIRY**
   An ecologist studying plants in the desert performed the following experiment. She staked out two identical plots, containing sagebrush plants and small annual wildflowers. She found the same five wildflower species in roughly equal numbers on both plots. She then enclosed one of the plots with a fence to keep out kangaroo rats, the most common grain-eaters of the area. After two years, four of the wildflower species were no longer present in the fenced plot, but one species had increased drastically. The control plot had not changed in species diversity. Using the principles of community ecology, propose a hypothesis to explain her results. What additional evidence would support your hypothesis?

7. **FOCUS ON EVOLUTION**
   Explain why adaptations of particular organisms to interspecific competition may not necessarily represent instances of character displacement. What would a researcher have to demonstrate about two competing species to make a convincing case for character displacement?

8. **FOCUS ON INFORMATION**
   In Batesian mimicry, a palatable species gains protection by mimicking an unpalatable one. Imagine that individuals of a palatable, brightly colored fly species are blown to three remote islands. The first island has no predators of that species; the second has predators but no similarly colored, unpalatable species; and the third has both predators and a similarly colored, unpalatable species. In a short essay (100–150 words), predict what might happen to the coloration of the palatable species on each island over time if coloration is a genetically controlled trait. Explain your predictions.

9. **SYNTHESIZE YOUR KNOWLEDGE**

Describe two types of interspecific interactions that appear to be occurring between the three species shown in this photo. What morphological adaptation can be seen in the species that is at the highest trophic level in this scene?

*For selected answers, see Appendix A.*

▲ **Figure 42.1** How can a fox transform a grassland into tundra?

## Transformed to Tundra

The arctic fox (*Vulpes lagopus*) is a predator native to arctic regions of North America, Europe, and Asia **(Figure 42.1)**. Valued for its fur, it was introduced onto hundreds of islands between Alaska and Russia a century ago in an effort to establish populations that could be easily harvested. The introduction had a surprising effect—it transformed many habitats on the islands from grassland to tundra.

How did this remarkable change come about? The foxes fed on seabirds, decreasing their density almost 100-fold compared to that on fox-free islands. Fewer seabirds meant less bird guano, the primary source of nutrients for plants on the islands. The reduction in nutrient availability in turn favored slower-growing forbs and shrubs typical of tundra instead of grasses and sedges, which require more nutrients. To test this explanation, researchers added fertilizer to plots of tundra on one of the fox-infested islands in 2001. Three years later, the fertilized plots had turned back into grassland.

Each of these islands and the community of organisms on it make up an **ecosystem**, the sum of all the organisms living in a given area and the abiotic factors with which they interact. An ecosystem can encompass a large area, such as a lake, forest, or island, or a microcosm, such as a small desert spring or the space under a fallen log **(Figure 42.2)**. As with populations and communities, the boundaries of ecosystems are not always discrete. Many ecologists view the entire biosphere as a global ecosystem, a composite of all the local ecosystems on Earth.

Regardless of an ecosystem's size, two key ecosystem processes cannot be fully described by population or community phenomena: energy flow and chemical cycling. Energy enters most ecosystems as sunlight. It is converted to chemical energy by autotrophs, passed to heterotrophs in the organic compounds of food, and dissipated as heat. Chemical elements, such as carbon and nitrogen, are cycled among abiotic and biotic components of the ecosystem. Photosynthetic and chemosynthetic organisms take up these elements in inorganic form from the air, soil, and water and incorporate them into

▲ **Figure 42.2** **A desert spring ecosystem.**

their biomass, some of which is consumed by animals. The elements are returned in inorganic form to the environment by the metabolism of plants and animals, and by bacteria and fungi that break down organic wastes and dead organisms.

Both energy and chemical elements are transformed in ecosystems through photosynthesis and feeding relationships. But unlike chemicals, energy cannot be recycled. An ecosystem must be powered by a continuous influx of energy from an external source—in most cases, the sun. As we'll see, energy flows through ecosystems, whereas chemicals cycle within them.

Resources critical to human survival and welfare—from the food we eat to the oxygen we breathe—are products of ecosystem processes. In this chapter, we'll explore the dynamics of energy flow and chemical cycling, emphasizing the results of ecosystem experiments. We'll also consider how human activities have affected energy flow and chemical cycling. Finally, we'll examine the science of restoration ecology, which focuses on returning degraded ecosystems to a more natural state.

CONCEPT 42.1

# Physical laws govern energy flow and chemical cycling in ecosystems

Cells transform energy and matter, subject to the laws of thermodynamics (see Concept 6.1). Cell biologists study these transformations within organelles and cells and measure the amounts of energy and matter that cross the cell's boundaries. Ecosystem ecologists do the same thing, except in their case the "cell" is a complete ecosystem. By studying how populations change in size and by grouping the species in a community into trophic levels of feeding relationships (see Concept 41.2), ecologists can follow the transformations of energy in an ecosystem and map the movements of chemical elements.

## Conservation of Energy

Because ecosystem ecologists study the interactions of organisms with the physical environment, many ecosystem approaches are based on laws of physics and chemistry. The first law of thermodynamics states that energy cannot be created or destroyed but only transferred or transformed (see Concept 6.1). Plants and other photosynthetic organisms convert solar energy to chemical energy, but the total amount of energy does not change: The energy stored in organic molecules must equal the total solar energy intercepted by the plant minus the amounts reflected and dissipated as heat. Ecosystem ecologists measure transfers within and across ecosystems to understand how many organisms a habitat can support and the amount of food humans can harvest from a site.

One implication of the second law of thermodynamics, which states that every exchange of energy increases the entropy of the universe, is that energy conversions are inefficient. Some energy is always lost as heat. As a result, each unit of energy that enters an ecosystem eventually exits as heat. Thus, energy flows through ecosystems—it does not cycle within them for long periods of time. Because energy flowing through ecosystems is ultimately lost as heat, most ecosystems would vanish if the sun were not continuously providing energy to Earth.

## Conservation of Mass

Matter, like energy, cannot be created or destroyed. This **law of conservation of mass** is as important for ecosystems as are the laws of thermodynamics. Because mass is conserved, we can determine how much of a chemical element cycles within an ecosystem or is gained or lost by that ecosystem over time.

Unlike energy, chemical elements are continually recycled within ecosystems. For example, a carbon atom in $CO_2$ is released from the soil by a decomposer, taken up by a blade of grass through photosynthesis, consumed by a grazing animal, and returned to the soil in the animal's waste. This process of chemical cycling is a key feature of how ecosystems work.

Although few elements are gained or lost from Earth, they can be gained by or lost from a particular ecosystem. In a forest, most mineral nutrients—the essential elements that plants obtain from soil—typically enter as dust or as solutes dissolved in rainwater or leached from rocks in the ground. Nitrogen is also supplied through the biological process of nitrogen fixation (see Figure 29.12). In terms of losses, some elements return to the atmosphere as gases, and others are carried out of the ecosystem by moving water or by wind. Like organisms, ecosystems are open systems, absorbing energy and mass and releasing heat and waste products.

In nature, most gains and losses to ecosystems are small compared to the amounts recycled within them. Still, the balance between inputs and outputs determines whether an ecosystem is a source or a sink for a given element. If a mineral nutrient's outputs exceed its inputs, it will eventually limit production in that system. Human activities often change the balance of inputs and outputs considerably, as we'll see later in this chapter and in Chapter 43.

## Energy, Mass, and Trophic Levels

Ecologists group species into trophic levels based on their main source of nutrition and energy (see Concept 41.2). The trophic level that ultimately supports all others consists of autotrophs, also called the **primary producers** of the ecosystem. Most autotrophs are photosynthetic organisms that use light energy to synthesize sugars and other organic compounds, which they use as fuel for cellular respiration and as building material for growth. The most common autotrophs are plants, algae, and photosynthetic prokaryotes, although chemosynthetic prokaryotes are the primary producers in ecosystems such as deep-sea hydrothermal vents (see Figure 40.11) and places deep underground or beneath ice.

Organisms in trophic levels above the primary producers are heterotrophs, which depend directly or indirectly on the primary producers for their source of energy. Herbivores, which eat plants and other primary producers, are **primary consumers**. Carnivores that eat herbivores are **secondary consumers**, and carnivores that eat other carnivores are **tertiary consumers**.

Another group of heterotrophs is the **detritivores**, or **decomposers**, terms used synonymously in this text to refer to consumers that get their energy from detritus. **Detritus** is nonliving organic material, such as the remains of dead organisms, feces, fallen leaves, and wood. Although some animals (such as earthworms) feed on detritus, the most important detritivores are prokaryotes and fungi **(Figure 42.3)**. These organisms secrete enzymes that digest organic material; they then absorb the breakdown products. Many detritivores are in turn eaten by secondary and tertiary consumers. In a forest, for instance, birds eat earthworms that have been feeding on leaf litter and its associated prokaryotes and fungi. As a result, chemicals originally synthesized by plants pass from plants to leaf litter to detritivores to birds.

Detritivores also play a critical role in recycling chemical elements to primary producers. Detritivores convert organic matter from all trophic levels to inorganic compounds usable by primary producers. When the detritivores excrete waste products or die, those inorganic compounds are returned

▲ **Figure 42.3 Fungi decomposing a dead tree.**

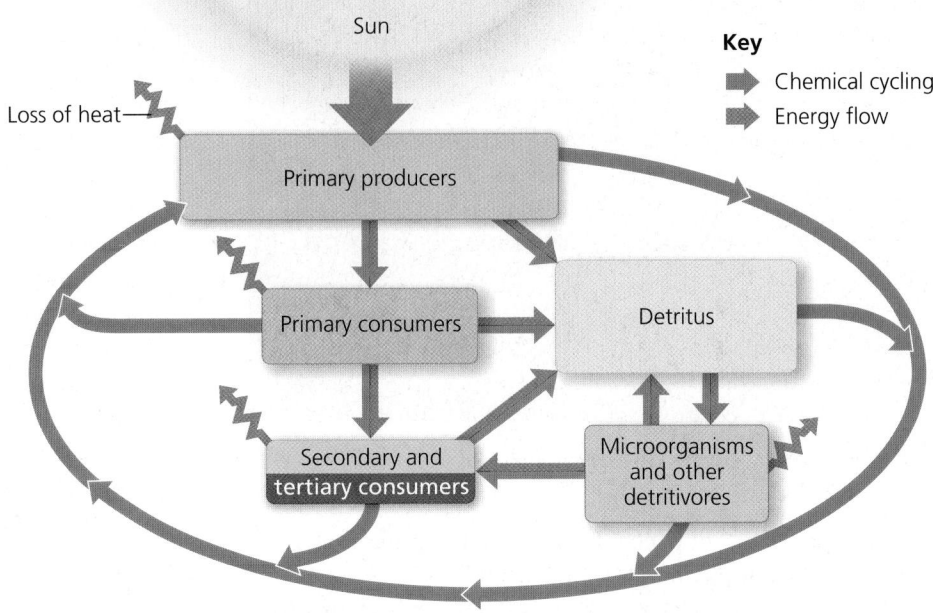

▲ **Figure 42.4 An overview of energy and nutrient dynamics in an ecosystem.** Energy enters, flows through, and exits an ecosystem, whereas chemical nutrients cycle within it. Energy (dark orange arrows) entering from the sun as radiation is transferred as chemical energy through the food web; each of these units of energy ultimately exits as heat radiated into space. Most transfers of nutrients (blue arrows) through the food web lead eventually to detritus; the nutrients then cycle back to the primary producers.

to the soil. Producers can then recycle these elements into organic compounds. If decomposition stopped, life as we know it would cease as detritus piled up and the supply of ingredients needed to synthesize organic matter was exhausted. **Figure 42.4** summarizes the trophic relationships in an ecosystem.

### CONCEPT CHECK 42.1

1. Why is the transfer of energy in an ecosystem referred to as energy flow, not energy cycling?
2. **WHAT IF?** You are studying nitrogen cycling on the Serengeti Plain in Africa. During your experiment, a herd of migrating wildebeests grazes through your study plot. What would you need to know to measure their effect on nitrogen balance in the plot?
3. **MAKE CONNECTIONS** Use the second law of thermodynamics to explain why an ecosystem's energy supply must be continually replenished. (See Concept 6.1.)

For suggested answers, see Appendix A.

### CONCEPT 42.2

## Energy and other limiting factors control primary production in ecosystems

The theme of energy transfer underlies all biological interactions (see Concept 1.1). In most ecosystems, the amount of light energy converted to chemical energy—in the form of organic compounds—by autotrophs during a given time period is the ecosystem's **primary production**. These photosynthetic products are the starting point for most studies of ecosystem metabolism and energy flow. In ecosystems where the primary producers are chemoautotrophs, the initial energy input is chemical, and the initial products are the organic compounds synthesized by the microorganisms.

## Ecosystem Energy Budgets

In most ecosystems, primary producers use light energy to synthesize energy-rich organic molecules, and consumers acquire their organic fuels secondhand (or even third- or fourth-hand) through food webs (see Figure 41.14). Therefore, the total amount of photosynthetic production sets the spending limit for the entire ecosystem's energy budget.

### The Global Energy Budget

Each day, Earth's atmosphere is bombarded by a total of about $10^{22}$ joules of solar radiation (1 J = 0.239 cal). This is enough energy to supply the demands of the entire human population for 19 years at 2013 energy consumption levels. The intensity of the solar energy striking Earth varies with latitude, with the tropics receiving the greatest input (see Figure 40.3). About 50% of incoming solar radiation is absorbed, scattered, or reflected by clouds and dust in the atmosphere. The amount of solar radiation that ultimately reaches Earth's surface limits the possible photosynthetic output of ecosystems.

Only a small fraction of the sunlight that reaches Earth's surface is actually used in photosynthesis. Much of the radiation strikes materials that don't photosynthesize, such as ice and soil. Of the radiation that does reach photosynthetic organisms, only certain wavelengths are absorbed by photosynthetic pigments (see Figure 8.9); the rest is transmitted, reflected, or lost as heat. As a result, only about 1% of the visible light that strikes photosynthetic organisms is converted to chemical energy. Nevertheless, Earth's primary producers create about 150 billion metric tons ($1.50 \times 10^{14}$ kg) of organic material each year.

### Gross and Net Production

Total primary production in an ecosystem is known as that ecosystem's **gross primary production (GPP)**—the amount of energy from light (or chemicals, in chemoautotrophic systems) converted to the chemical energy of organic molecules per unit time. Not all of this production is stored as organic material in the primary producers because they use some of the molecules as fuel in their own cellular respiration. **Net primary production (NPP)** is equal to gross primary production minus the energy used by the primary producers for their "autotrophic respiration" ($R_a$):

$$NPP = GPP - R_a$$

On average, NPP is about one-half of GPP. To ecologists, NPP is the key measurement because it represents the storage of

chemical energy that will be available to consumers in the ecosystem.

Net primary production can be expressed as energy per unit area per unit time [$J/(m^2 \cdot yr)$] or as biomass (mass of vegetation) added per unit area per unit time [$g/(m^2 \cdot yr)$]. (Note that biomass is usually expressed in terms of the dry mass of organic material.) An ecosystem's NPP should not be confused with the total biomass of photosynthetic autotrophs present, a measure called the *standing crop*. The net primary production is the amount of *new* biomass added in a given period of time. Although a forest has a large standing crop, its NPP may actually be less than that of some grasslands; grasslands do not accumulate as much biomass as forests because animals consume the plants rapidly and because grasses and herbs decompose more quickly than trees do.

Satellites provide a powerful tool for studying global patterns of primary production **(Figure 42.5)**. Images produced from satellite data show that different ecosystems vary considerably in their NPP. Tropical rain forests are among the

---

▼ **Figure 42.5**  **Research Method**

## Determining Primary Production with Satellites

**Application** Because chlorophyll captures visible light, photosynthetic organisms absorb more light at visible wavelengths (about 380–750 nm) than at near-infrared wavelengths (750–1,100 nm) (see Figure 8.6). Scientists use this difference in absorption to estimate the rate of photosynthesis in different regions of the globe using satellites.

**Technique** Most satellites determine what they "see" by comparing the ratios of wavelengths reflected back to them. Vegetation reflects much more near-infrared radiation than visible radiation, producing a reflectance pattern very different from that of snow, clouds, soil, and liquid water.

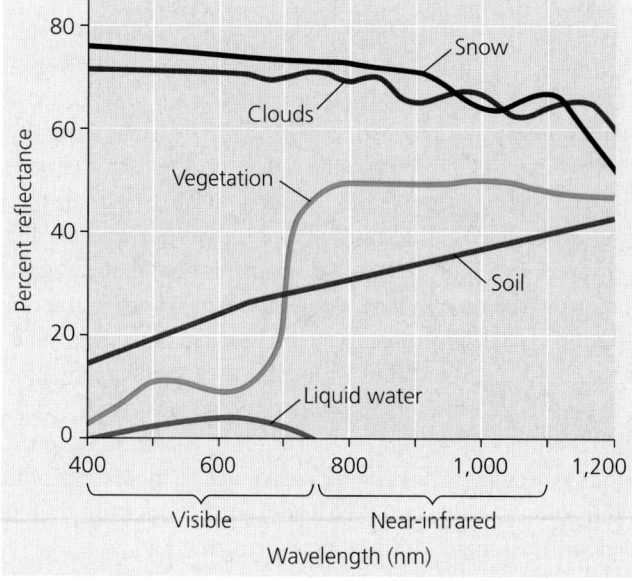

**Results** Scientists use the satellite data to help produce maps of primary production like the one in Figure 42.6.

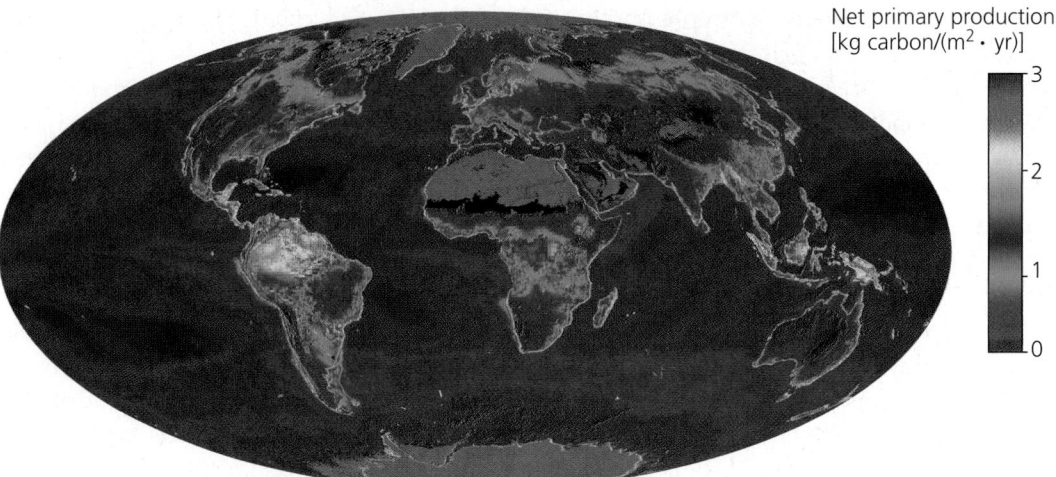

► **Figure 42.6 Global net primary production.** This map is based on satellite-collected data, such as amount of sunlight absorbed by vegetation. Note that tropical land areas have the highest rates of production (yellow and red on the map).

**?** *Does this map accurately reflect the importance of some highly productive habitats, such as wetlands, coral reefs, and coastal zones? Explain.*

Net primary production [kg carbon/(m² · yr)]

most productive terrestrial ecosystems and contribute a large portion of the planet's NPP. Estuaries and coral reefs also have very high NPP, but their contribution to the global total is smaller because these ecosystems cover only about one-tenth the area covered by tropical rain forests. In contrast, while the open oceans are relatively unproductive **(Figure 42.6)**, their vast size means that together they contribute as much global NPP as terrestrial systems do.

Whereas NPP can be stated as the amount of new biomass added by producers in a given period of time, **net ecosystem production (NEP)** is a measure of the *total biomass accumulation* during that time. NEP is defined as gross primary production minus the total respiration of all organisms in the system ($R_T$)—not just primary producers, as for the calculation of NPP, but decomposers and other heterotrophs as well:

$$NEP = GPP - R_T$$

NEP is useful to ecologists because its value determines whether an ecosystem is gaining or losing carbon over time. A forest may have a positive NPP but still lose carbon if heterotrophs release it as $CO_2$ more quickly than primary producers incorporate it into organic compounds.

The most common way to estimate NEP is to measure the net flux (flow) of $CO_2$ or $O_2$ entering or leaving the ecosystem. If more $CO_2$ enters than leaves, the system is storing carbon. Because $O_2$ release is directly coupled to photosynthesis and respiration (see Figure 7.2), a system that is giving off $O_2$ is also storing carbon. On land, ecologists typically measure only the net flux of $CO_2$ from ecosystems because detecting small changes in $O_2$ flux in a large atmospheric $O_2$ pool is difficult. In the oceans, researchers use both approaches.

What limits production in ecosystems? To ask this question another way, what factors could we change to increase production for a given ecosystem? We'll address this question first for aquatic ecosystems.

## Primary Production in Aquatic Ecosystems

In aquatic (marine and freshwater) ecosystems, both light and nutrients are important in controlling primary production.

### Light Limitation

Because solar radiation drives photosynthesis, you would expect light to be a key variable in controlling primary production in oceans. Indeed, the depth of light penetration affects primary production throughout the photic zone of an ocean or lake (see Figure 40.10). About half of the solar radiation is absorbed in the first 15 m of water. Even in "clear" water, only 5–10% of the radiation may reach a depth of 75 m.

If light were the main variable limiting primary production in the ocean, you would expect production to increase along a gradient from the poles toward the equator, which receives the greatest intensity of light. However, you can see in Figure 42.6 that there is no such gradient. Another factor must strongly influence primary production in the ocean.

### Nutrient Limitation

More than light, nutrients limit primary production in most oceans and lakes. A **limiting nutrient** is the element that must be added for production to increase. The nutrients that most often limit marine production are nitrogen and phosphorus. Concentrations of these nutrients are typically low in the photic zone because they are rapidly taken up by phytoplankton and because detritus tends to sink.

For example, as detailed in **Figure 42.7**, nutrient enrichment experiments found that nitrogen was limiting phytoplankton growth off the south shore of Long Island, New York. One practical application of this work is in preventing algal "blooms" caused by excess nitrogen runoff that fertilizes the phytoplankton. Prior to this research, phosphate contamination was thought to cause many such blooms in the ocean, but eliminating phosphates alone may not help unless nitrogen pollution is also controlled.

The macronutrients nitrogen and phosphorus are not the only nutrients that limit aquatic production. Several large areas of the ocean have low phytoplankton densities despite relatively high nitrogen concentrations. The Sargasso Sea, a subtropical region of the Atlantic Ocean, has some of the clearest water in the world because of its low phytoplankton

## ▼ Figure 42.7 Inquiry

### Which nutrient limits phytoplankton production along the coast of Long Island?

**Experiment** Pollution from duck farms concentrated near Moriches Bay adds both nitrogen and phosphorus to the coastal water off Long Island, New York. To determine which nutrient limits phytoplankton growth in this area, John Ryther and William Dunstan, of the Woods Hole Oceanographic Institution, cultured the phytoplankton *Nannochloris atomus* with water collected from several sites, identified as A through G. They added either ammonium ($NH_4^+$) or phosphate ($PO_4^{3-}$) to some of the cultures.

**Results** The addition of ammonium caused heavy phytoplankton growth in the cultures, but the addition of phosphate did not.

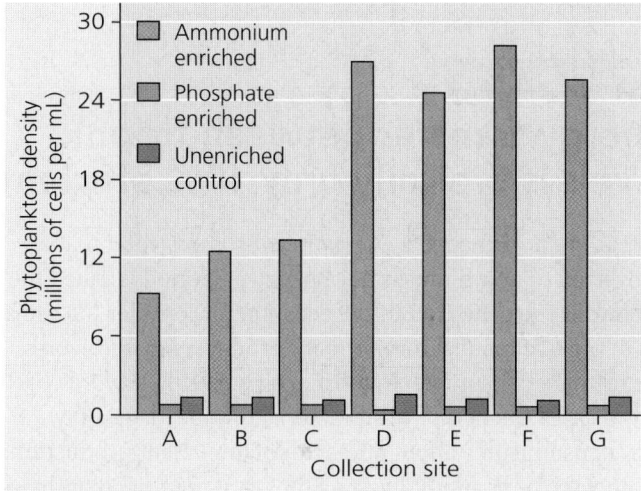

**Conclusion** The researchers concluded that nitrogen is the nutrient that limits phytoplankton growth in this ecosystem because adding phosphorus did not increase *Nannochloris* growth, whereas adding nitrogen increased phytoplankton density dramatically.

**Data from** J. H. Ryther and W. M. Dunstan, Nitrogen, phosphorus, and eutrophication in the coastal marine environment, *Science* 171:1008–1013 (1971).

**WHAT IF?** Predict how the results would change if water samples were drawn from areas where new duck farms had greatly increased the amount of pollution in the water. Explain.

---

### Table 42.1 Nutrient Enrichment Experiment for Sargasso Sea Samples

| Nutrients Added to Experimental Culture | Relative Uptake of $^{14}C$ by Cultures* |
|---|---|
| None (controls) | 1.00 |
| Nitrogen (N) + phosphorus (P) only | 1.10 |
| N + P + metals, excluding iron (Fe) | 1.08 |
| N + P + metals, including Fe | 12.90 |
| N + P + Fe | 12.00 |

*$^{14}C$ uptake by cultures measures primary production.

**Data from** D. W. Menzel and J. H. Ryther, Nutrients limiting the production of phytoplankton in the Sargasso Sea, with special reference to iron, *Deep Sea Research* 7:276–281 (1961).

**INTERPRET THE DATA** *The element molybdenum (Mo) is another micronutrient that can limit primary production in the oceans. If the researchers found the following results for additions of Mo, what would you conclude about its relative importance for growth?*

| | |
|---|---|
| N + P + Mo | 6.0 |
| N + P + Fe + Mo | 72.0 |

and in the coastal waters off Peru, California, and parts of western Africa.

Nutrient limitation is also common in freshwater lakes. During the 1970s, scientists showed that sewage and fertilizer runoff from farms and lawns adds considerable nutrients to lakes, promoting the growth of primary producers. When the primary producers die, detritivores decompose them, depleting the water of much or all of its oxygen. The ecological impacts of this process, known as **eutrophication** (from the Greek *eutrophos*, well nourished), include the loss of many fish species from the lakes (see Figure 40.11).

To control eutrophication, scientists need to know which nutrient is responsible. While nitrogen rarely limits primary production in lakes, whole-lake experiments showed that phosphorus availability limited cyanobacterial growth. This and other ecological research led to the use of phosphate-free detergents and other water quality reforms.

## Primary Production in Terrestrial Ecosystems

At regional and global scales, temperature and moisture are the main factors controlling primary production in terrestrial ecosystems. Tropical rain forests, with their warm, wet conditions that promote plant growth, are the most productive terrestrial ecosystems (see Figure 42.6). In contrast, low-productivity systems are generally hot and dry, like many deserts, or cold and dry, like arctic tundra. Between these extremes lie the temperate forest and grassland ecosystems, with moderate climates and intermediate productivity.

The climate variables of moisture and temperature are very useful for predicting NPP in terrestrial ecosystems. Primary production is greater in wetter ecosystems, as shown for the

density. Nutrient enrichment experiments have revealed that the availability of the micronutrient iron limits primary production there **(Table 42.1)**. Windblown dust from land supplies most of the iron to the oceans but is relatively scarce in the Sargasso Sea and certain other regions compared to the oceans as a whole.

Areas of *upwelling*, where deep, nutrient-rich waters circulate to the ocean surface, have exceptionally high primary production. This fact supports the hypothesis that nutrient availability determines marine primary production. Because upwelling stimulates growth of the phytoplankton that form the base of marine food webs, upwelling areas typically host highly productive, diverse ecosystems and are prime fishing locations. The largest areas of upwelling occur in the Southern Ocean (also called the Antarctic Ocean), along the equator,

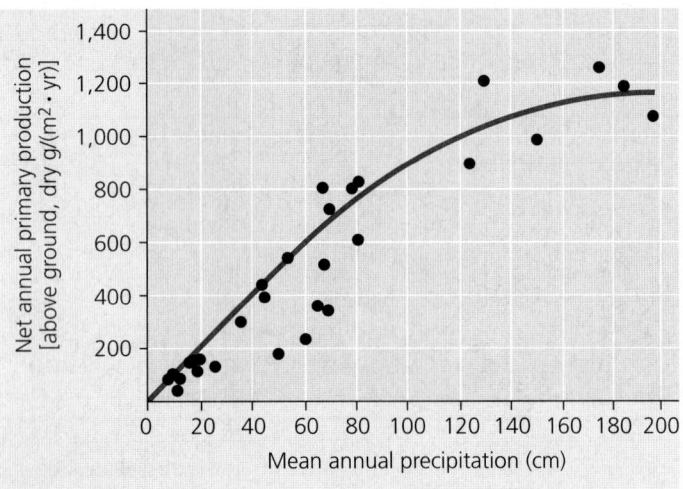

▲ **Figure 42.8  A global relationship between net primary production and mean annual precipitation for terrestrial ecosystems.**

plot of NPP and annual precipitation in **Figure 42.8**. Along with mean annual precipitation, a second useful predictor is *evapotranspiration*, the total amount of water transpired by plants and evaporated from a landscape. Evapotranspiration increases with the temperature and amount of solar energy available to drive evaporation and transpiration.

### Nutrient Limitations and Adaptations That Reduce Them

**EVOLUTION**  Soil nutrients also limit primary production in terrestrial ecosystems. As in aquatic systems, nitrogen and phosphorus are the nutrients that most commonly limit terrestrial production. Globally, nitrogen limits plant growth most. Phosphorus limitations are common in older soils where phosphate molecules have been leached away by water, such as in many tropical ecosystems. Phosphorus availability is also often low in soils of deserts and other ecosystems with a basic pH, where some phosphorus precipitates and becomes unavailable to plants. Adding a nonlimiting nutrient, even one that is scarce, will not stimulate production. Conversely, adding more of the limiting nutrient will increase production until some other nutrient becomes limiting.

Various adaptations have evolved in plants that can increase their uptake of limiting nutrients. One important adaptation is the symbiosis between plant roots and nitrogen-fixing bacteria. Another is the mycorrhizal association between plant roots and fungi that supply phosphorus and other limiting elements to plants (see Concept 29.4). Plant roots also have hairs and other anatomical features that increase their surface area and, hence, the area of soil in contact with the roots (see Figure 28.4). Many plants release enzymes and other substances into the soil that increase the availability of limiting nutrients; such substances include phosphatases, which cleave a phosphate group from larger molecules, and certain molecules (called chelating agents) that make micronutrients such as iron more soluble in the soil.

Studies relating nutrients to terrestrial primary production have practical applications in agriculture. Farmers maximize their crop yields by using fertilizers with the right balance of nutrients for the local soil and type of crop. This knowledge of limiting nutrients helps feed billions of people.

### CONCEPT CHECK 42.2

1. Why is only a small portion of the solar energy that strikes Earth's atmosphere stored by primary producers?
2. How can ecologists experimentally determine the factor that limits primary production in an ecosystem?
3. **MAKE CONNECTIONS**  Explain how nitrogen and phosphorus, the nutrients that most often limit primary production, are necessary for the Calvin cycle to function in photosynthesis (see Concept 8.3).

For suggested answers, see Appendix A.

### CONCEPT 42.3

# Energy transfer between trophic levels is typically only 10% efficient

The amount of chemical energy in consumers' food that is converted to new biomass during a given period is called the **secondary production** of the ecosystem. Consider the transfer of organic matter from primary producers to herbivores, the primary consumers. In most ecosystems, herbivores eat only a small fraction of plant material produced; globally, they consume only about one-sixth of total plant production. Moreover, they cannot digest all the plant material that they *do* eat, as anyone who has walked through a field where cattle have been grazing will attest. Most of an ecosystem's production is eventually consumed by detritivores. Next, we'll look at how such processes affect the transfer of energy in ecosystems.

## Production Efficiency

We'll begin by examining secondary production in one organism—a caterpillar. When a caterpillar feeds on a leaf, only about 33 J out of 200 J, or one-sixth of the potential energy in the leaf, is used for secondary production, or growth **(Figure 42.9)**. The caterpillar stores some of the remaining energy in organic compounds that will be used for cellular respiration and passes the rest in its feces. The energy in the feces remains in the ecosystem temporarily, but most of it is lost as heat after the feces are consumed by detritivores. The energy used for the caterpillar's respiration is also eventually lost from the ecosystem as heat. This is why energy is said to flow through, not cycle within, ecosystems. Only the chemical energy stored by herbivores as biomass, through growth or the production of offspring, is available as food to secondary consumers.

We can measure the efficiency of animals as energy transformers using the following equation:

$$\text{Production efficiency} = \frac{\text{Net secondary production} \times 100\%}{\text{Assimilation of primary production}}$$

Net secondary production is the energy stored in biomass represented by growth and reproduction. Assimilation consists

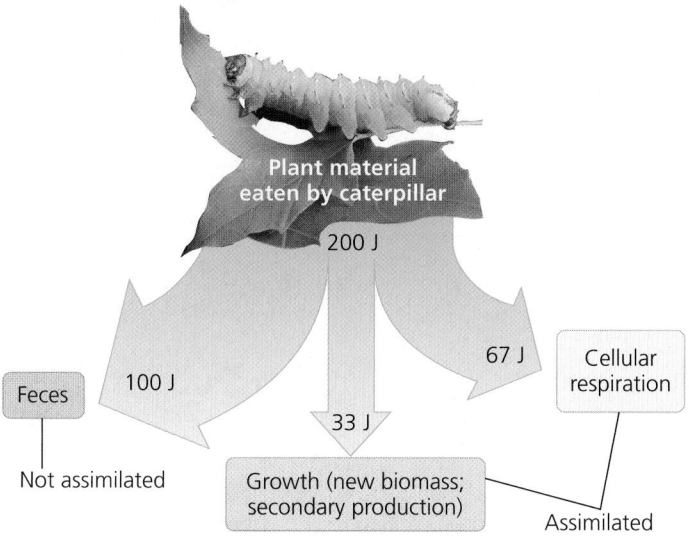

Plant material eaten by caterpillar

200 J

Feces — 100 J — Not assimilated

33 J — Growth (new biomass; secondary production)

67 J — Cellular respiration

Assimilated

▲ **Figure 42.9 Energy partitioning within a link of the food chain.**

**INTERPRET THE DATA** *What percentage of the energy in the caterpillar's food is actually used for secondary production (growth)?*

of the total amount of energy an organism has consumed and used for growth, reproduction, and respiration. **Production efficiency**, therefore, is the percentage of energy stored in assimilated food that is used for growth and reproduction, *not* respiration. For the caterpillar in Figure 42.9, production efficiency is 33%; 67 J of the 100 J of assimilated energy is used for respiration. (The 100 J of energy lost as undigested material in feces does not count toward assimilation.) Birds and mammals typically have low production efficiencies, in the range of 1–3%, because they use so much energy in maintaining a constant, high body temperature. Fishes, which are mainly ectothermic (see Concept 32.3), have production efficiencies around 10%. Insects and microorganisms are even more efficient, with production efficiencies averaging 40% or more.

## Trophic Efficiency and Ecological Pyramids

Let's scale up now from the production efficiencies of individual consumers to the flow of energy through trophic levels.

**Trophic efficiency** is the percentage of production transferred from one trophic level to the next. Trophic efficiencies must always be less than production efficiencies because they take into account not only the energy contained in feces and the energy lost through respiration, but also the energy converted to new biomass in a lower trophic level but not consumed by the next trophic level. Trophic efficiencies range from roughly 5% to 20% in different ecosystems, but on average are only about 10%. In other words, 90% of the energy available at one trophic level typically is *not* transferred to the next. This loss is multiplied over the length of a food chain. If 10% of available energy is transferred from primary producers to primary consumers, such as caterpillars, and 10% of that energy is transferred to secondary consumers (carnivores), then only 1% of net primary production is available to secondary consumers (10% of 10%). In the **Scientific Skills Exercise**, you

## Interpreting Quantitative Data in a Table

### How Efficient Is Energy Transfer in a Salt Marsh Ecosystem?
In a classic experiment, John Teal studied the flow of energy through the producers, consumers, and detritivores in a salt marsh. In this exercise, you will use the data from this study to calculate some measures of energy transfer between trophic levels in this ecosystem.

**How the Study Was Done** Teal measured the amount of solar radiation entering a salt marsh in Georgia over a year. He also measured the aboveground biomass of the dominant primary producers, which were grasses, as well as the biomass of the dominant consumers, including insects, spiders, and crabs, and of the detritus that flowed out of the marsh to the surrounding coastal waters. To determine the amount of energy in each unit of biomass, he dried the biomass, burned it in a calorimeter, and measured the amount of heat produced.

**Data from the Study**

| Form of Energy | $kcal/(m^2 \cdot yr)$ |
|---|---|
| Solar radiation | 600,000 |
| Gross grass production | 34,580 |
| Net grass production | 6,585 |
| Gross insect production | 305 |
| Net insect production | 81 |
| Detritus leaving marsh | 3,671 |

**Data from** J. M. Teal, Energy flow in the salt marsh ecosystem of Georgia, *Ecology* 43:614–624 (1962).

**INTERPRET THE DATA**

1. What proportion of the solar energy that reaches the marsh is incorporated into gross primary production? Into net primary production? (A proportion is the same as a percentage divided by 100. Both measures are useful for comparing relative efficiencies across different ecosystems.)

2. How much energy is lost by primary producers as respiration in this ecosystem? How much is lost as respiration by the insect population?

3. If all of the detritus leaving the marsh is plant material, what proportion of all net primary production leaves the marsh as detritus each year?

(MB) A version of this Scientific Skills Exercise can be assigned in MasteringBiology.

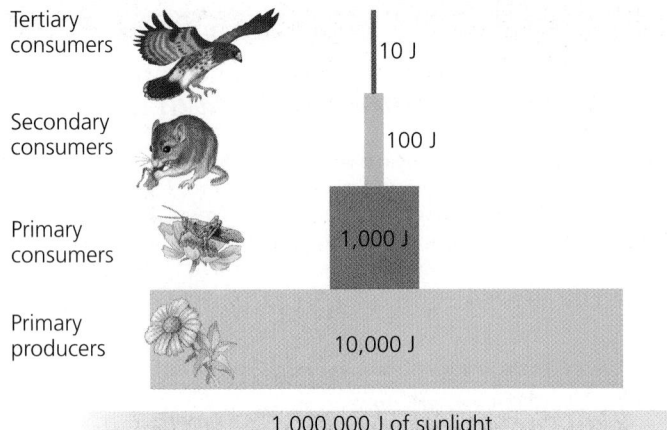

| Tertiary consumers | 10 J |
| Secondary consumers | 100 J |
| Primary consumers | 1,000 J |
| Primary producers | 10,000 J |

1,000,000 J of sunlight

▲ **Figure 42.10 An idealized pyramid of energy.** This example assumes a trophic efficiency of 10% for each link in the food chain. Notice that primary producers convert only about 1% of the energy available to them to net primary production.

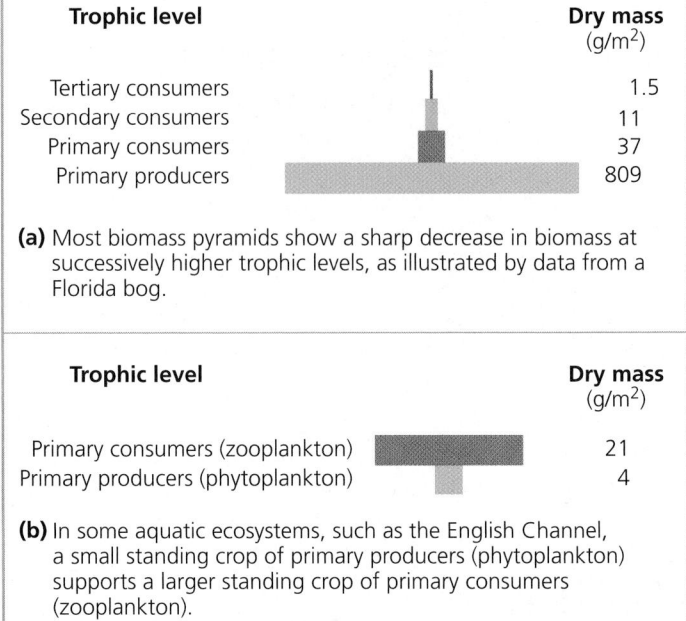

| Trophic level | Dry mass (g/m²) |
|---|---|
| Tertiary consumers | 1.5 |
| Secondary consumers | 11 |
| Primary consumers | 37 |
| Primary producers | 809 |

**(a)** Most biomass pyramids show a sharp decrease in biomass at successively higher trophic levels, as illustrated by data from a Florida bog.

| Trophic level | Dry mass (g/m²) |
|---|---|
| Primary consumers (zooplankton) | 21 |
| Primary producers (phytoplankton) | 4 |

**(b)** In some aquatic ecosystems, such as the English Channel, a small standing crop of primary producers (phytoplankton) supports a larger standing crop of primary consumers (zooplankton).

▲ **Figure 42.11 Pyramids of biomass.** Numbers denote the dry mass of all organisms at each trophic level.

can calculate trophic efficiency and other measures of energy flow in a salt marsh ecosystem.

The progressive loss of energy along a food chain limits the abundance of top-level carnivores that an ecosystem can support. Only about 0.1% of the chemical energy fixed by photosynthesis can flow all the way through a food web to a tertiary consumer, such as a snake or a shark. This explains why most food webs include only about four or five trophic levels (see Concept 41.2).

The loss of energy with each transfer in a food chain can be represented by an *energy pyramid*, in which the net productions of different trophic levels are arranged in tiers **(Figure 42.10)**. The width of each tier is proportional to the net production, expressed in joules, of each trophic level. The highest level, which represents top-level predators, contains relatively few individuals. The small population size typical of top predator species is one reason they tend to be vulnerable to extinction (and to the evolutionary consequences of small population size, discussed in Concept 21.3).

One important ecological consequence of low trophic efficiencies is represented in a *biomass pyramid*, in which each tier represents the standing crop (the total dry mass of all organisms) in one trophic level. Most biomass pyramids narrow sharply from primary producers at the base to top-level carnivores at the apex because energy transfers between trophic levels are so inefficient **(Figure 42.11a)**. Certain aquatic ecosystems, however, have inverted biomass pyramids: Primary consumers outweigh the producers **(Figure 42.11b)**.

Such inverted biomass pyramids occur because the producers—phytoplankton—grow, reproduce, and are consumed so quickly by the zooplankton that their total biomass remains at comparatively low levels. However, because the phytoplankton continually replace their biomass at such a rapid rate, they can support a biomass of zooplankton bigger than

their own biomass. Likewise, because phytoplankton reproduce so quickly and have much higher production than zooplankton, the pyramid of *energy* for this ecosystem is still bottom-heavy, like the one in Figure 42.10.

The dynamics of energy flow through ecosystems have important implications for human consumers. Eating meat is a relatively inefficient way of tapping photosynthetic production. The same pound of soybeans that a person could eat for protein produces only a fifth of a pound of beef or less when fed to a cow. Worldwide agriculture could, in fact, feed many more people and require less land if we all fed more efficiently—as primary consumers, eating plant material.

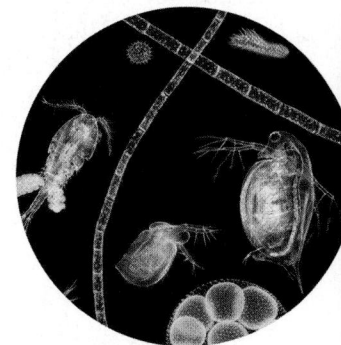

▲ Phytoplankton and zooplankton

### CONCEPT CHECK 42.3

1. If an insect that eats plant seeds containing 100 J of energy uses 30 J of that energy for respiration and excretes 50 J in its feces, what is the insect's net secondary production? What is its production efficiency?

2. Tobacco leaves contain nicotine, a poisonous compound that is energetically expensive for the plant to make. What advantage might the plant gain by using some of its resources to produce nicotine?

3. **WHAT IF?** Detritivores are consumers that obtain their energy from detritus. How many joules of energy are potentially available to detritivores in the ecosystem represented in Figure 42.10?

For suggested answers, see Appendix A.

# Biological and geochemical processes cycle nutrients and water in ecosystems

Although most ecosystems receive abundant solar energy, chemical elements are available only in limited amounts. Life therefore depends on the recycling of essential chemical elements. Much of an organism's chemical stock is replaced continuously as nutrients are assimilated and waste products are released. When the organism dies, the atoms in its body are returned to the atmosphere, water, or soil by decomposers. Decomposition replenishes the pools of inorganic nutrients that plants and other autotrophs use to build new organic matter.

## Decomposition and Nutrient Cycling Rates

Decomposers are heterotrophs that get their energy from detritus. Their growth is controlled by the same factors that limit primary production in ecosystems, including temperature, moisture, and nutrient availability. Decomposers usually grow faster and decompose material more quickly in warmer ecosystems **(Figure 42.12)**. In tropical rain forests, most organic material decomposes in a few months to a few years, whereas in temperate forests, decomposition takes four to six years, on average. The difference is largely the result of the higher temperatures and more abundant precipitation in tropical rain forests.

Because decomposition in a tropical rain forest is rapid, relatively little organic material accumulates as leaf litter on the forest floor; about 75% of the ecosystem's nutrients is present in the woody trunks of trees, and only about 10% is contained in the soil. Thus, the relatively low concentrations of some nutrients in the soil of tropical rain forests result from a short cycling time, not from a lack of these elements in the ecosystem. In temperate forests, where decomposition is much slower, the soil may contain as much as 50% of all the organic material in the ecosystem. The nutrients that are present in temperate forest detritus and soil may remain there for years before plants assimilate them.

Decomposition on land is also slower when conditions are either too dry for decomposers to thrive or too wet to supply them with enough oxygen. Ecosystems that are both cold and wet, such as peatlands, store large amounts of organic matter. Decomposers grow poorly there, and net primary production greatly exceeds the rate of decomposition.

In aquatic ecosystems, decomposition in anaerobic muds can take 50 years or longer. Bottom sediments are comparable to the detritus layer in terrestrial ecosystems, but algae and aquatic plants usually assimilate nutrients directly from the water. Thus, the sediments often constitute a nutrient sink, and aquatic ecosystems are very productive only when there is exchange between the bottom layers of water and surface waters (as occurs in the upwelling regions described earlier).

### How does temperature affect litter decomposition in an ecosystem?

**Experiment** Researchers with the Canadian Forest Service placed identical samples of organic material—litter—on the ground in 21 sites across Canada (marked by letters on the map below). Three years later, they returned to see how much of each sample had decomposed.

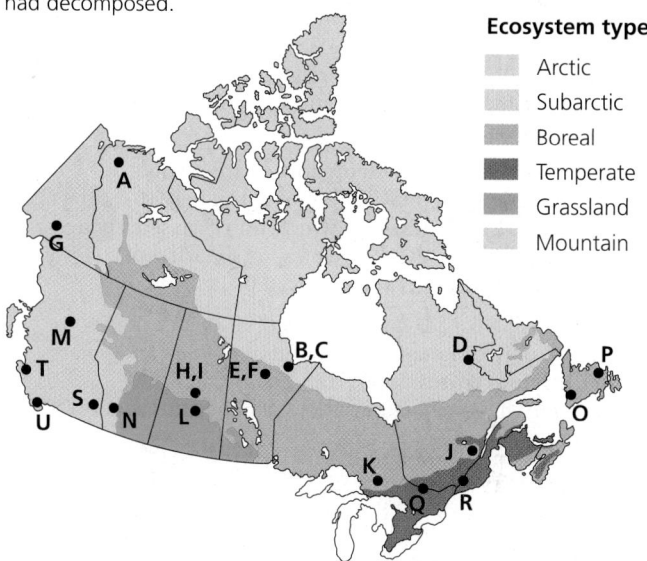

**Results** The mass of litter in the warmest ecosystem decreased four times faster than in the coldest ecosystem.

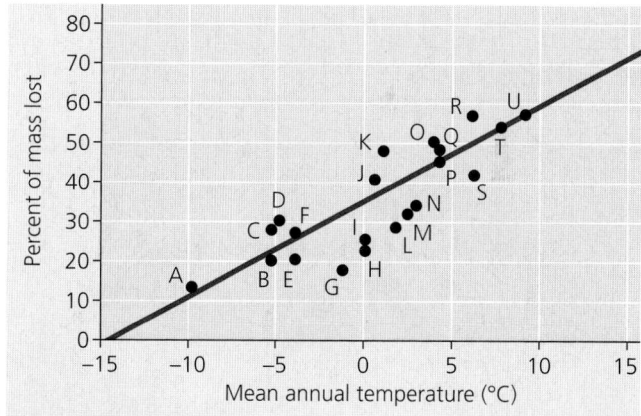

**Conclusion** Decomposition rate increases with temperature across much of Canada.

**Data from** T. R. Moore et al., Litter decomposition rates in Canadian forests, *Global Change Biology* 5:75–82 (1999).

**WHAT IF?** What other factors might have varied across these sites? How might this variation have affected the interpretation of the data?

## Biogeochemical Cycles

Because nutrient cycles involve both biotic and abiotic components, they are called **biogeochemical cycles (Figure 42.13)**. For convenience, we can recognize two general categories of biogeochemical cycles: global and local. Gaseous forms of carbon, oxygen, sulfur, and nitrogen occur in the atmosphere, and cycles of these elements are essentially

Examine each cycle closely, considering the major reservoirs of water, carbon, nitrogen, and phosphorus and the processes that drive each cycle. The widths of the arrows in the diagrams approximately reflect the relative contribution of each process to the movement of water or a nutrient in the biosphere.

## The Water Cycle

**Biological importance**  Water is essential to all organisms, and its availability influences the rates of ecosystem processes, particularly primary production and decomposition in terrestrial ecosystems.

**Forms available to life**  All organisms are capable of exchanging water directly with their environment. Liquid water is the primary physical phase in which water is used, though some organisms can harvest water vapor. Freezing of soil water can limit water availability to terrestrial plants.

**Reservoirs**  The oceans contain 97% of the water in the biosphere. Approximately 2% is bound in glaciers and polar ice caps, and the remaining 1% is in lakes, rivers, and groundwater, with a negligible amount in the atmosphere.

**Key processes**  The main processes driving the water cycle are evaporation of liquid water by solar energy, condensation of water vapor into clouds, and precipitation. Transpiration by terrestrial plants also moves large volumes of water into the atmosphere. Surface and groundwater flow can return water to the oceans, completing the water cycle.

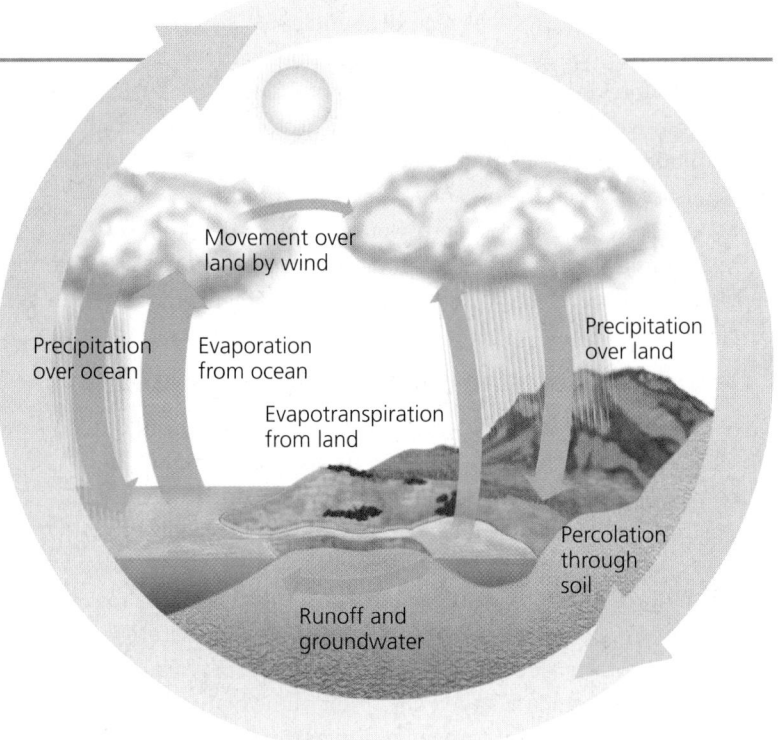

## The Carbon Cycle

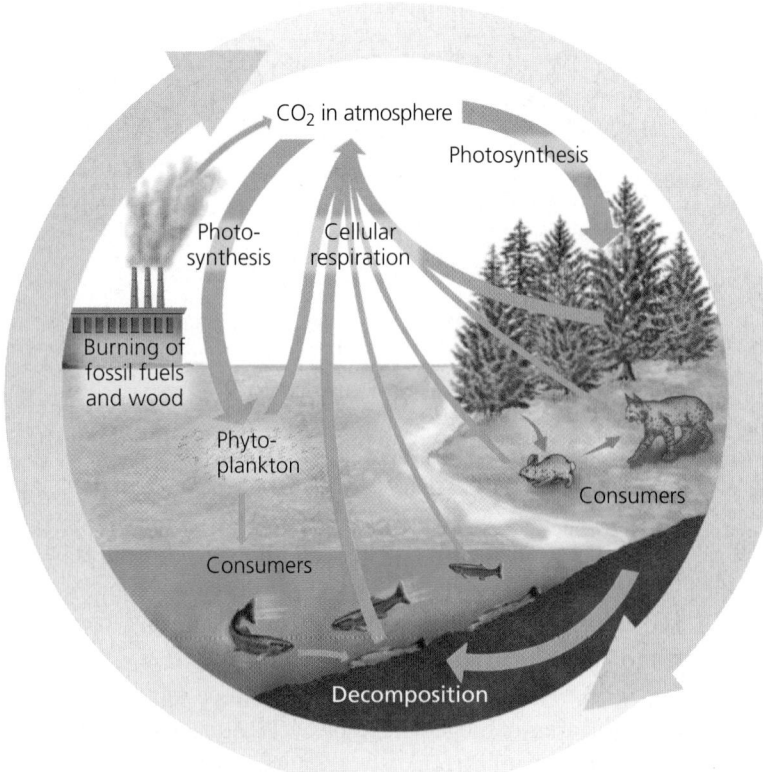

**Biological importance**  Carbon forms the framework of the organic molecules essential to all organisms.

**Forms available to life**  Photosynthetic organisms utilize $CO_2$ during photosynthesis and convert the carbon to organic forms that are used by consumers, including animals, fungi, and heterotrophic protists and prokaryotes.

**Reservoirs**  The major reservoirs of carbon include fossil fuels, soils, the sediments of aquatic ecosystems, the oceans (dissolved carbon compounds), plant and animal biomass, and the atmosphere ($CO_2$). The largest reservoir is sedimentary rocks such as limestone; however, carbon remains in this pool for long periods of time. All organisms are capable of returning carbon directly to their environment in its original form ($CO_2$) through respiration.

**Key processes**  Photosynthesis by plants and phytoplankton removes substantial amounts of atmospheric $CO_2$ each year. This quantity is approximately equaled by $CO_2$ added to the atmosphere through cellular respiration by producers and consumers. The burning of fossil fuels and wood is adding significant amounts of additional $CO_2$ to the atmosphere. Over geologic time, volcanoes are also a substantial source of $CO_2$.

 Visit the Study Area in **MasteringBiology** for the BioFlix® 3-D Animation on The Carbon Cycle.

# The Nitrogen Cycle

**Biological importance** Nitrogen is part of amino acids, proteins, and nucleic acids and is often a limiting plant nutrient.

**Forms available to life** Plants can assimilate (use) two inorganic forms of nitrogen—ammonium ($NH_4^+$) and nitrate ($NO_3^-$)—and some organic forms, such as amino acids. Various bacteria can use all of these forms as well as nitrite ($NO_2^-$). Animals can use only organic forms of nitrogen.

**Reservoirs** The main reservoir of nitrogen is the atmosphere, which is 80% free nitrogen gas ($N_2$). The other reservoirs of inorganic and organic nitrogen compounds are soils and the sediments of lakes, rivers, and oceans; surface water and groundwater; and the biomass of living organisms.

**Key processes** The major pathway for nitrogen to enter an ecosystem is via *nitrogen fixation*, the conversion of $N_2$ to forms that can be used to synthesize organic nitrogen compounds. Certain bacteria, as well as lightning and volcanic activity, fix nitrogen naturally. Nitrogen inputs from human activities now outpace natural inputs on land. Two major contributors are industrially produced fertilizers and legume crops that fix nitrogen via bacteria in their root nodules. Other bacteria in soil convert nitrogen to different forms. Examples include nitrifying bacteria, which convert ammonium to nitrate, and denitrifying bacteria, which convert nitrate to nitrogen gas. Human activities also release large quantities of reactive nitrogen gases, such as nitrogen oxides, to the atmosphere.

# The Phosphorus Cycle

**Biological importance** Organisms require phosphorus as a major constituent of nucleic acids, phospholipids, and ATP and other energy-storing molecules and as a mineral constituent of bones and teeth.

**Forms available to life** The most biologically important inorganic form of phosphorus is phosphate ($PO_4^{3-}$), which plants absorb and use in the synthesis of organic compounds.

**Reservoirs** The largest accumulations of phosphorus are in sedimentary rocks of marine origin. There are also large quantities of phosphorus in soil, in the oceans (in dissolved form), and in organisms. Because soil particles bind $PO_4^{3-}$, the recycling of phosphorus tends to be quite localized in ecosystems.

**Key processes** Weathering of rocks gradually adds $PO_4^{3-}$ to soil; some leaches into groundwater and surface water and may eventually reach the sea. Phosphate taken up by producers and incorporated into biological molecules may be eaten by consumers. Phosphate is returned to soil or water by either decomposition of biomass or excretion by consumers. Because there are no significant phosphorus-containing gases, only relatively small amounts of phosphorus move through the atmosphere, usually in the forms of dust and sea spray.

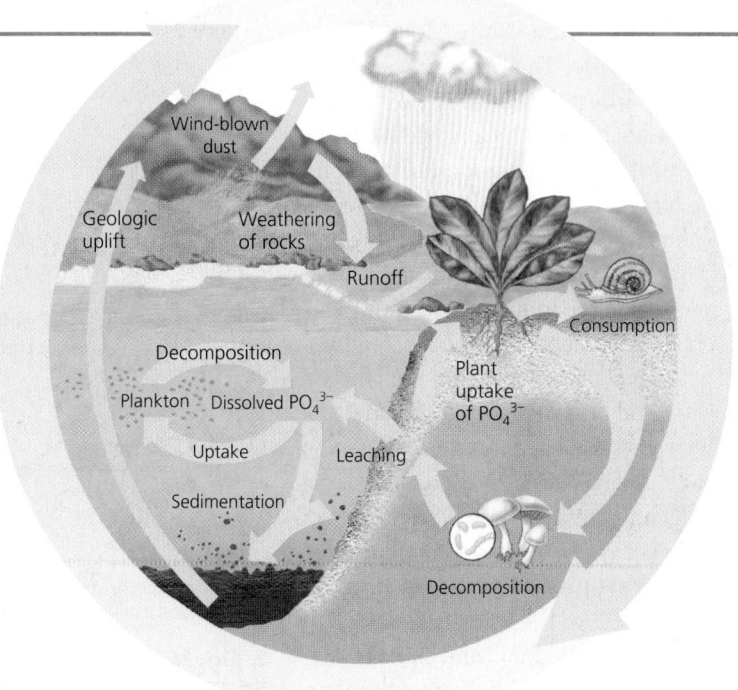

global. Other elements, including phosphorus, potassium, and calcium, are too heavy to occur as gases at Earth's surface. They cycle locally in terrestrial ecosystems and more broadly in aquatic ecosystems.

Figure 42.13 provides a detailed look at the cycling of water, carbon, nitrogen, and phosphorus. When you study each cycle, consider which steps are driven primarily by biological processes. For the carbon cycle, for instance, plants, animals, and other organisms control most of the key steps, including photosynthesis and decomposition. For the water cycle, however, purely physical processes control many key steps, such as evaporation from the oceans.

How have ecologists worked out the details of chemical cycling in various ecosystems? One common method is to follow the movement of naturally occurring, nonradioactive isotopes through the biotic and abiotic components of an ecosystem. Another method involves adding tiny amounts of radioactive isotopes of specific elements and tracing their progress. Scientists have also been able to make use of the radioactive carbon ($^{14}$C) released into the atmosphere during atom bomb testing in the 1950s and early 1960s. This "spike" of $^{14}$C can reveal where and how quickly carbon flows into ecosystem components, including plants, soils, and ocean water.

## *Case Study*: Nutrient Cycling in the Hubbard Brook Experimental Forest

Since 1963, ecologist Gene Likens and colleagues have been studying nutrient cycling at the Hubbard Brook Experimental Forest in the White Mountains of New Hampshire. Their research site is a deciduous forest that grows in six small valleys, each drained by a single creek. Impermeable bedrock underlies the soil of the forest.

The research team first determined the mineral budget for each of six valleys by measuring the input and outflow of several key nutrients. They collected rainfall at several sites to measure the amount of water and dissolved minerals added to the ecosystem. To monitor the loss of water and minerals, they constructed a small concrete dam with a V-shaped spillway across the creek at the bottom of each valley **(Figure 42.14a)**. They found that about 60% of the water added to the ecosystem as rainfall and snow exits through the stream, and the remaining 40% is lost by evapotranspiration.

Preliminary studies confirmed that internal cycling conserved most of the mineral nutrients in the system. For example, only about 0.3% more calcium ($Ca^{2+}$) leaves a valley via its creek than is added by rainwater, and this small net loss is probably replaced by chemical decomposition of the bedrock. During most years, the forest even registers small net gains of a few mineral nutrients, including nitrogen.

Experimental deforestation of a watershed dramatically increased the flow of water and minerals leaving the watershed **(Figure 42.14b)**. Over three years, water runoff from the newly deforested watershed was 30–40% greater than in a control watershed, apparently because there were no plants to absorb and transpire water from the soil. Most remarkable was the loss of nitrate, whose concentration in the creek increased 60-fold, reaching levels considered unsafe for drinking water **(Figure 42.14c)**. The Hubbard Brook deforestation study showed that the amount of nutrients leaving an intact forest

**(a)** Concrete dams and weirs built across streams at the bottom of watersheds enabled researchers to monitor the outflow of water and nutrients from the ecosystem.

**(b)** One watershed was clear-cut to study the effects of the loss of vegetation on drainage and nutrient cycling. All of the original plant material was left in place to decompose.

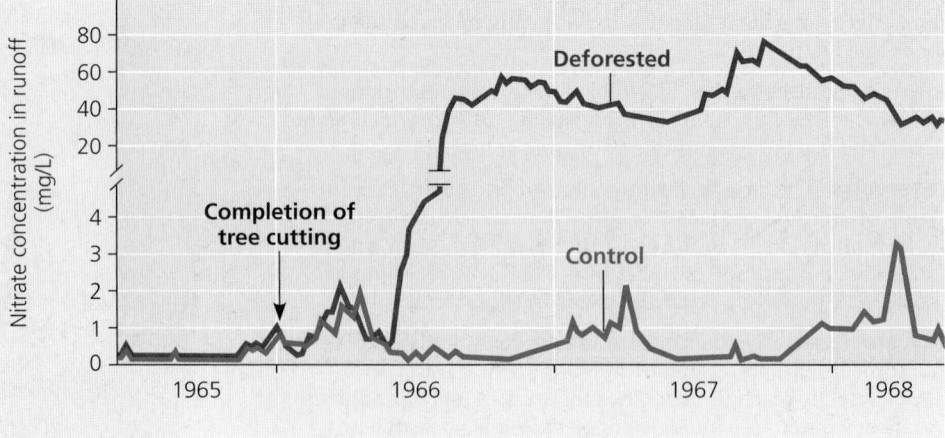

**(c)** The concentration of nitrate in runoff from the deforested watershed was 60 times greater than in a control (unlogged) watershed.

▲ **Figure 42.14 Nutrient cycling in the Hubbard Brook Experimental Forest: an example of long-term ecological research.**

(MB) A related Experimental Inquiry Tutorial can be assigned in MasteringBiology.

ecosystem is controlled mainly by the plants. Retaining nutrients in an ecosystem helps to maintain the productivity of the system, as well as to avoid algal "blooms" and other problems caused by excess nutrient runoff (see Figure 42.7).

## CONCEPT CHECK 42.4

1. **DRAW IT** For each of the four biogeochemical cycles in Figure 42.13, draw a simple diagram that shows one possible path for an atom of that chemical from abiotic to biotic reservoirs and back.
2. Why does deforestation of a watershed increase the concentration of nitrates in streams draining the watershed?
3. **WHAT IF?** Why is nutrient availability in a tropical rain forest particularly vulnerable to logging?

For suggested answers, see Appendix A.

## CONCEPT 42.5

# Restoration ecologists return degraded ecosystems to a more natural state

Ecosystems can recover naturally from most disturbances (including the experimental deforestation at Hubbard Brook) through the stages of ecological succession (see Concept 41.3). Sometimes that recovery takes centuries, though, particularly when humans have degraded the environment. Tropical areas that are cleared for farming may quickly become unproductive because of nutrient losses. Mining activities may last for several decades, and the lands are often abandoned in a degraded state. Ecosystems can also be damaged by salts that build up in soils from irrigation and by toxic chemicals or oil spills. Biologists increasingly are called on to help restore and repair damaged ecosystems.

Restoration ecologists seek to initiate or speed up the recovery of degraded ecosystems. One of the basic assumptions is that environmental damage is at least partly reversible. This optimistic view must be balanced by a second assumption—that ecosystems are not infinitely resilient. Restoration ecologists therefore work to identify and manipulate the processes that most limit recovery of ecosystems from disturbances. Where disturbance is so severe that restoring all of a habitat is impractical, ecologists try to reclaim as much of a habitat or ecological process as possible, within the limits of the time and money available to them.

In extreme cases, the physical structure of an ecosystem may need to be restored before biological restoration can occur. If a stream was straightened to channel water quickly through a suburb, ecologists may reconstruct a meandering channel to slow down the flow of water eroding the stream bank. To restore an open-pit mine, engineers may first grade the site with heavy equipment to reestablish a gentle slope, spreading topsoil when the slope is in place **(Figure 42.15)**.

Once physical reconstruction of the ecosystem is complete—or when it is not needed—biological restoration is the next step. The long-term objective of restoration is to return an ecosystem as much as possible to its predisturbance state. **Figure 42.16** explores four ambitious and successful restoration projects.

There are many such projects throughout the world, and they often employ two key strategies: bioremediation and biological augmentation.

## Bioremediation

Using organisms—usually prokaryotes, fungi, or plants—to detoxify polluted ecosystems is known as **bioremediation**. Some plants and lichens adapted to soils containing heavy metals can accumulate high concentrations of toxic metals such as lead and cadmium in their tissues. Restoration ecologists can introduce such species to sites polluted by mining and other human activities and then harvest these organisms to remove

**(a)** In 1991, before restoration

**(b)** In 2000, near the completion of restoration

▲ **Figure 42.15  A gravel and clay mine site in New Jersey before and after restoration.**

The examples highlighted on this page are just a few of the many restoration ecology projects taking place around the world.

## Kissimmee River, Florida

In the 1960s, the Kissimmee River was converted from a meandering river to a 90-km canal to control flooding. This channelization diverted water from the floodplain, causing the wetlands to dry up, threatening many fish and wetland bird populations. Kissimmee River restoration has filled 12 km of drainage canal and reestablished 24 km of the original 167 km of natural river channel. Pictured here is a section of the Kissimmee canal that has been plugged (wide, light strip on the right side of the photo), diverting flow into remnant river channels (center of the photo). The project will also restore natural flow patterns, which will foster self-sustaining populations of wetland birds and fishes.

## Succulent Karoo, South Africa

In the Succulent Karoo desert region of southern Africa, as in many arid regions, overgrazing by livestock has damaged vast areas. Private landowners and government agencies in South Africa are restoring large areas of this unique region, revegetating the land and employing more sustainable resource management. The photo shows a small sample of the exceptional plant diversity of the Succulent Karoo; its 5,000 plant species include the highest diversity of succulent plants in the world.

## Maungatautari, New Zealand

Weasels, rats, pigs, and other introduced species pose a serious threat to New Zealand's native plants and animals, including kiwis, a group of flightless, ground-dwelling bird species. The goal of the Maungatautari restoration project is to exclude all exotic mammals from a 3,400-ha reserve located on a forested volcanic cone. A specialized fence around the reserve eliminates the need to continue setting traps and using poisons that can harm native wildlife. In 2006, a pair of critically endangered takahe (a species of flightless rail) were released into the reserve with the hope of reestablishing a breeding population of this colorful bird on New Zealand's North Island.

## Coastal Japan

Seaweed and seagrass beds are important nursery grounds for a wide variety of fishes and shellfish. Once extensive but now reduced by development, these beds are being restored in the coastal areas of Japan. Techniques include constructing suitable seafloor habitat, transplanting seaweeds and seagrasses from natural beds using artificial substrates, and hand seeding (shown in this photograph).

the metals from the ecosystem. For instance, researchers in the United Kingdom have discovered a lichen species that grows on soil polluted with uranium dust left over from mining. The lichen concentrates uranium in a dark pigment, making it useful as a biological monitor and potentially as a remediator.

Ecologists already use the abilities of many prokaryotes to carry out bioremediation of soils and water. Scientists have sequenced the genomes of at least ten prokaryotic species specifically for their bioremediation potential. One of the species, the bacterium *Shewanella oneidensis*, appears particularly promising. It can metabolize a dozen or more elements under aerobic and anaerobic conditions. In doing so, it converts soluble forms of uranium, chromium, and nitrogen to insoluble forms that are less likely to leach into streams or groundwater. Researchers at Oak Ridge National Laboratory, in Tennessee, stimulated the growth of *Shewanella* and other uranium-reducing bacteria by adding ethanol to groundwater contaminated with uranium; the bacteria can use ethanol as an energy source. In just five months, the concentration of soluble uranium in the ecosystem dropped by 80% **(Figure 42.17)**.

## Biological Augmentation

In contrast to bioremediation, which is a strategy for removing harmful substances from an ecosystem, **biological augmentation** uses organisms to *add* essential materials to a degraded ecosystem. To augment ecosystem processes, restoration ecologists need to determine which factors, such as chemical nutrients, have been lost from a system and are limiting its recovery.

Encouraging the growth of plants that thrive in nutrient-poor soils often speeds up succession and ecosystem recovery. In alpine ecosystems of the western United States, nitrogen-fixing plants such as lupines are often planted to raise nitrogen concentrations in soils disturbed by mining and other activities. Once these nitrogen-fixing plants become established, other native species are better able to obtain enough soil nitrogen to survive. In other systems where the soil has been severely disturbed or where topsoil is missing entirely, plant roots may lack the mycorrhizal symbionts that help them meet their nutritional needs (see Concept 26.2). Ecologists restoring a tallgrass prairie in Minnesota recognized this limitation and enhanced the recovery of native species by adding mycorrhizal symbionts to the soil they seeded.

Restoring the physical structure and plant community of an ecosystem does not always ensure that animal species will recolonize a site and persist there. Because animals provide critical ecosystem services, including pollination and seed dispersal, restoration ecologists sometimes help wildlife to reach and use restored ecosystems. They might release animals at a site or establish habitat corridors that connect a restored site to places where the animals are found. They may build perches for birds to use. These and other efforts can increase the biodiversity of restored ecosystems and help the community persist.

**(a)** Wastes containing uranium were dumped in these four unlined pits for more than 30 years, contaminating soils and groundwater.

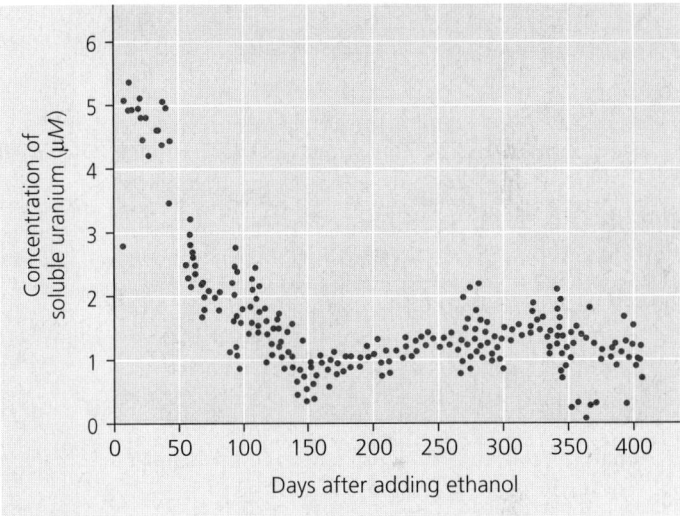

**(b)** After ethanol was added, microbial activity decreased the concentration of soluble uranium in groundwater near the pits.

▲ **Figure 42.17 Bioremediation of groundwater contaminated with uranium at Oak Ridge National Laboratory, Tennessee.**

### CONCEPT CHECK 42.5

1. Identify the main goal of restoration ecology.
2. How do bioremediation and biological augmentation differ?
3. **WHAT IF?** In what way is the Kissimmee River project a more complete ecological restoration than the Maungatautari project (see Figure 42.16)?

For suggested answers, see Appendix A.

## Ecosystems: *A Review*

**Figure 42.18** illustrates energy transfer, nutrient cycling, and other key processes for an arctic tundra ecosystem. Note the conceptual similarities between this figure and Make Connections Figure 8.20 (The Working Cell). The scale of the two figures is different, but the physical laws and biological rules that govern life apply equally to both systems.

## MAKE CONNECTIONS

# The Working Ecosystem

This arctic tundra ecosystem teems with life in the short two-month growing season each summer. In ecosystems, organisms interact with each other and with the environment around them in diverse ways, including those illustrated here.

**Caribou**

**Snow geese**

**Herbivory**

### Populations Are Dynamic (Chapter 40)

**1** Populations change in size through births and deaths and through immigration and emigration. Caribou migrate across the tundra to give birth at their calving grounds each year. (See Figure 40.14.)

**2** Snow geese and many other species migrate to the Arctic each spring for the abundant food found there in summer. (See Concept 39.3.)

**3** Birth and death rates influence the density of all populations. Death in the tundra comes from many causes, including predation, competition for resources, and lack of food in winter. (See Figure 40.23.)

**Arctic fox**

**3**

**4**

**Predation**

**Snow goose**

### Species Interact in Diverse Ways (Chapter 41)

**4** In predation, an individual of one species kills and eats another. (See Concept 41.1.)

**5** In herbivory, an individual of one species eats part of a plant or other primary producer, such as a caribou eating a lichen. (See Concept 41.1.)

**6** In mutualism, two species interact in ways that benefit each other. In some mutualisms, the partners live in direct contact, forming a symbiosis; for example, a lichen is a symbiotic mutualism between a fungus and an alga or cyanobacterium. (See Concept 41.1 and Figure 26.29.)

**7** In competition, individuals seek to acquire the same limiting resources. For example, snow geese and caribou both eat cottongrass. (See Concept 41.1.)

**Nitrogen cycle**

N₂

Denitrification

Organisms  N fixation

**Carbon cycle**

CO₂

Cellular respiration

Photosynthesis

## Organisms Transfer Energy and Matter in Ecosystems (Chapter 42)

**8** Primary producers convert the energy in sunlight to chemical energy through photosynthesis. Their growth is often limited by abiotic factors such as low temperatures, scarce soil nutrients, and lack of light in winter. (See Figure 8.5, Figure 40.9, and Figure 42.4.)

**9** Food chains are typically short in the tundra because primary production is lower than in most other ecosystems. (See Figure 41.13.)

**10** When one organism eats another, the transfer of energy from one trophic level to the next is usually less than 10%. (See Figure 42.10.)

**11** Detritivores recycle chemical elements back to primary producers. (See Figures 42.3 and 42.4.)

**12** Chemical elements such as carbon and nitrogen move in cycles between the physical environment and organisms. (See Figure 42.14.)

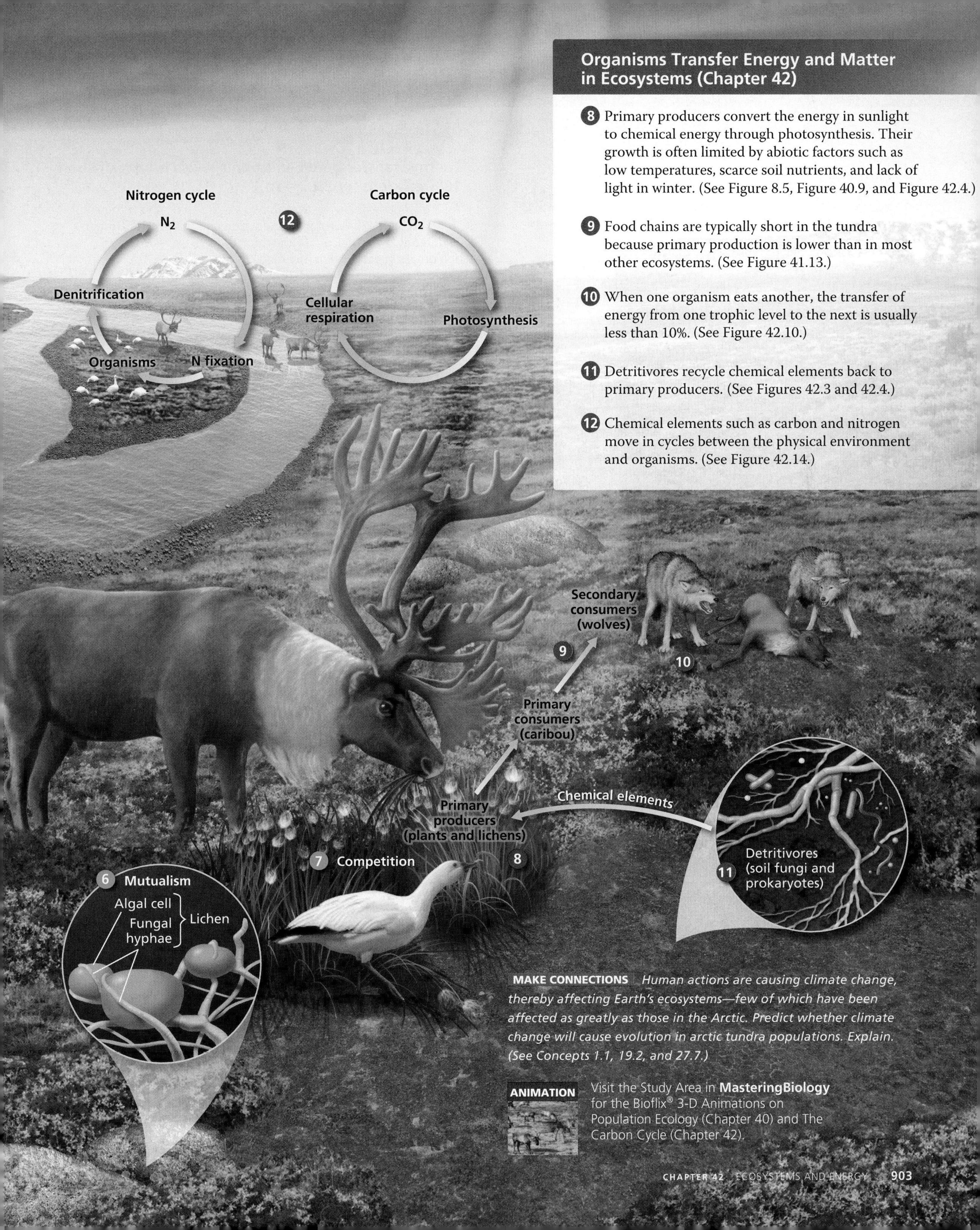

Secondary consumers (wolves)

Primary consumers (caribou)

Chemical elements

Primary producers (plants and lichens)

**7** Competition

**6** Mutualism

Algal cell

Fungal hyphae ⟩ Lichen

Detritivores (soil fungi and prokaryotes)

**MAKE CONNECTIONS** *Human actions are causing climate change, thereby affecting Earth's ecosystems—few of which have been affected as greatly as those in the Arctic. Predict whether climate change will cause evolution in arctic tundra populations. Explain. (See Concepts 1.1, 19.2, and 27.7.)*

**ANIMATION** Visit the Study Area in **MasteringBiology** for the Bioflix® 3-D Animations on Population Ecology (Chapter 40) and The Carbon Cycle (Chapter 42).

# 42 Chapter Review

Go to **MasteringBiology**® for Assignments, the eText, and the Study Area with Animations, Activities, Vocab Self-Quiz, and Practice Tests.

## SUMMARY OF KEY CONCEPTS

**VOCAB SELF-QUIZ**

goo.gl/gbai8v

### CONCEPT 42.1

**Physical laws govern energy flow and chemical cycling in ecosystems (pp. 887–888)**

- An **ecosystem** consists of all the organisms in a community and the abiotic factors with which they interact. Energy is conserved but degraded to heat during ecosystem processes. As a result, energy flows through ecosystems (rather than being recycled).
- Chemical elements enter and leave an ecosystem and cycle within it, subject to the **law of conservation of mass**. Inputs and outputs are generally small compared to recycled amounts, but their balance determines whether the ecosystem gains or loses an element over time.

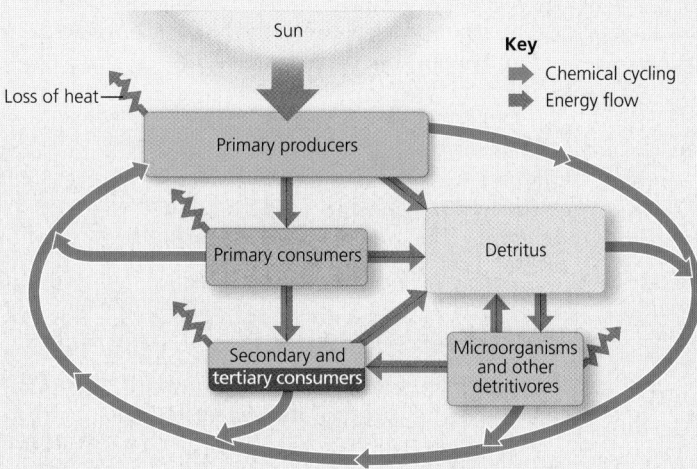

**Key**
- ➡ Chemical cycling
- ➡ Energy flow

? *Considering the second law of thermodynamics, would you expect the typical biomass of primary producers in an ecosystem to be greater than or less than the biomass of secondary producers in the system? Explain your reasoning.*

### CONCEPT 42.2

**Energy and other limiting factors control primary production in ecosystems (pp. 888–892)**

- **Primary production** sets the spending limit for the global energy budget. **Gross primary production** is the total energy assimilated by an ecosystem in a given period. **Net primary production**, the energy accumulated in autotroph biomass, equals gross primary production minus the energy used by the primary producers for respiration. **Net ecosystem production** is the total biomass accumulation of an ecosystem, defined as the difference between gross primary production and total ecosystem respiration.
- In aquatic ecosystems, light and nutrients limit primary production. In terrestrial ecosystems, climatic factors such as temperature and moisture affect primary production at large scales, but a soil nutrient is often the limiting factor locally.

? *If you know NPP, what additional variable do you need to know to estimate NEP? Why might measuring this variable be difficult, for instance, in a sample of ocean water?*

### CONCEPT 42.3

**Energy transfer between trophic levels is typically only 10% efficient (pp. 892–894)**

- The amount of energy available to each trophic level is determined by the net primary production and the **production efficiency**, the efficiency with which food energy is converted to biomass at each link in the food chain.
- The percentage of energy transferred from one trophic level to the next, called **trophic efficiency**, is typically 10%. Pyramids of net production and biomass reflect low trophic efficiency.

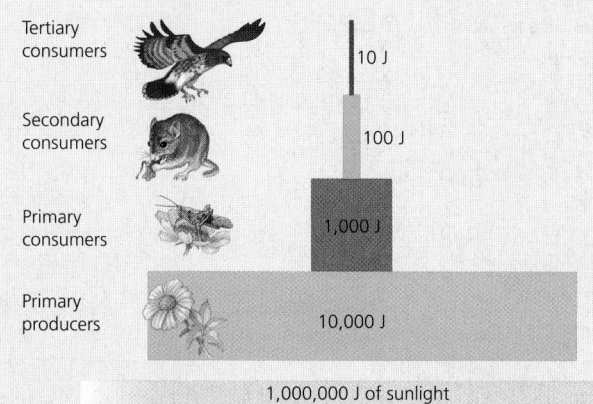

Tertiary consumers — 10 J
Secondary consumers — 100 J
Primary consumers — 1,000 J
Primary producers — 10,000 J

1,000,000 J of sunlight

? *Why would runners have a lower production efficiency when running a long-distance race than when they are sedentary?*

### CONCEPT 42.4

**Biological and geochemical processes cycle nutrients and water in ecosystems (pp. 895–899)**

- Water moves in a global cycle driven by solar energy. The carbon cycle primarily reflects the reciprocal processes of photosynthesis and cellular respiration. Nitrogen enters ecosystems through atmospheric deposition and nitrogen fixation by prokaryotes.
- The proportion of a nutrient in a particular form and its cycling in that form vary among ecosystems, largely because of differences in the rate of decomposition.
- Nutrient cycling is strongly regulated by vegetation. The Hubbard Brook case study showed that logging increases water runoff and can cause large losses of minerals.

? *If decomposers usually grow faster and decompose material more quickly in warmer ecosystems, why is decomposition in hot deserts so slow?*

### CONCEPT 42.5

**Restoration ecologists return degraded ecosystems to a more natural state (pp. 899–901)**

- Restoration ecologists harness organisms to detoxify polluted ecosystems through the process of **bioremediation**.
- In **biological augmentation**, ecologists use organisms to add essential materials to ecosystems.

? *In preparing a site for surface mining and later restoration, why would engineers separate the topsoil from the deeper soil, rather than removing all soil at once and mixing it in a single pile?*

# TEST YOUR UNDERSTANDING

PRACTICE TEST

goo.gl/CRZjvS

## Level 1: Knowledge/Comprehension

1. Which of the following organisms is *incorrectly* paired with its trophic level?
   (A) cyanobacterium—primary producer
   (B) grasshopper—primary consumer
   (C) zooplankton—primary producer
   (D) fungus—detritivore

2. Which of these ecosystems has the *lowest* net primary production per square meter?
   (A) a salt marsh     (C) a coral reef
   (B) an open ocean     (D) a tropical rain forest

3. The discipline that applies ecological principles to returning degraded ecosystems to a more natural state is known as
   (A) restoration ecology.
   (B) thermodynamics.
   (C) eutrophication.
   (D) biogeochemistry.

## Level 2: Application/Analysis

4. Nitrifying bacteria participate in the nitrogen cycle mainly by
   (A) converting nitrogen gas to ammonia.
   (B) releasing ammonium from organic compounds, thus returning it to the soil.
   (C) converting ammonium to nitrate, which plants absorb.
   (D) incorporating nitrogen into amino acids and organic compounds.

5. Which of the following has the greatest effect on the rate of chemical cycling in an ecosystem?
   (A) the rate of decomposition in the ecosystem
   (B) the production efficiency of the ecosystem's consumers
   (C) the trophic efficiency of the ecosystem
   (D) the location of the nutrient reservoirs in the ecosystem

6. The Hubbard Brook watershed deforestation experiment yielded all of the following results *except*:
   (A) Most minerals were recycled within a forest ecosystem.
   (B) Calcium levels remained high in the soil of deforested areas.
   (C) Deforestation increased water runoff.
   (D) The nitrate concentration in waters draining the deforested area became dangerously high.

7. Which of the following would be considered an example of bioremediation?
   (A) adding nitrogen-fixing microorganisms to a degraded ecosystem to increase nitrogen availability
   (B) using a bulldozer to regrade a strip mine
   (C) reconfiguring the channel of a river
   (D) adding seeds of a chromium-accumulating plant to soil contaminated by chromium

8. If you applied a fungicide to a cornfield, what would you expect to happen to the rate of decomposition and net ecosystem production (NEP)?
   (A) Both decomposition rate and NEP would decrease.
   (B) Neither would change.
   (C) Decomposition rate would increase and NEP would decrease.
   (D) Decomposition rate would decrease and NEP would increase.

## Level 3: Synthesis/Evaluation

9. **INTERPRET THE DATA** Draw a simplified global water cycle showing ocean, land, atmosphere, and runoff from the land to the ocean. (a) Label your drawing with these annual water fluxes: ocean evaporation, 425 km$^3$; ocean evaporation that returns to the ocean as precipitation, 385 km$^3$; ocean evaporation that falls as precipitation on land, 40 km$^3$; evapotranspiration from plants and soil that falls as precipitation on land, 70 km$^3$; runoff to the oceans, 40 km$^3$. (b) What is the ratio of ocean evaporation that falls as precipitation on land compared with runoff from land to the oceans? (c) How would this ratio change during an ice age, and why?

10. **SCIENTIFIC INQUIRY**
    Using two neighboring ponds in a forest as your study site, design a controlled experiment to measure the effect of falling leaves on net primary production in a pond.

11. **FOCUS ON EVOLUTION**
    Some biologists have suggested that ecosystems are emergent, "living" systems capable of evolving. One manifestation of this idea is environmentalist James Lovelock's Gaia hypothesis, which views Earth itself as a living, homeostatic entity—a kind of superorganism. Are ecosystems capable of evolving? If so, would this be a form of Darwinian evolution? Why or why not? Explain.

12. **FOCUS ON ENERGY AND MATTER**
    Decomposition typically occurs quickly in moist tropical forests. However, waterlogging in the soil of some moist tropical forests results over time in a buildup of organic matter called peat. In a short essay (100–150 words), discuss the relationship of net primary production, net ecosystem production, and decomposition for such an ecosystem. Are NPP and NEP likely to be positive? What do you think would happen to NEP if a landowner drained the water from a tropical peatland, exposing the organic matter to air?

13. **SYNTHESIZE YOUR KNOWLEDGE**

This dung beetle (genus *Scarabaeus*) is burying a ball of dung it has collected from a large mammalian herbivore in Kenya. Explain why this process is important for the cycling of nutrients and for primary production.

*For selected answers, see Appendix A.*

# 43 Global Ecology and Conservation Biology

## KEY CONCEPTS

**43.1** Human activities threaten Earth's biodiversity

**43.2** Population conservation focuses on population size, genetic diversity, and critical habitat

**43.3** Landscape and regional conservation help sustain biodiversity

**43.4** Earth is changing rapidly as a result of human actions

**43.5** The human population is no longer growing exponentially but is still increasing rapidly

**43.6** Sustainable development can improve human lives while conserving biodiversity

▲ **Figure 43.1** What will be the fate of this newly described lizard species?

## Psychedelic Treasure

Scurrying across a rocky outcrop, a lizard stops abruptly in a patch of sunlight. A conservation biologist senses the motion and turns to find a gecko splashed with rainbow colors, its bright orange legs and tail blending into a blue body, its head and neck splotched with yellow and green. The psychedelic rock gecko (*Cnemaspis psychedelica*) was discovered in 2010 during an expedition to the Greater Mekong region of southeast Asia **(Figure 43.1)**. Its known habitat is restricted to an island of just 8 km² (3 square miles) in southern Vietnam. Other new species found during the same series of expeditions include the Elvis monkey, which sports a hairdo like that of a certain legendary singer. Between 2000 and 2010, biologists identified more than 1,000 new species in the Greater Mekong region alone.

To date, scientists have described and formally named about 1.8 million species of organisms. In addition to these named species, many others remain to be discovered: Estimates for the number of species that currently exist range from 5 million to 100 million. Some of the greatest concentrations of species are in the tropics. Unfortunately, tropical forests are being cleared at an alarming rate to make room

for and support a burgeoning human population. In Vietnam, rates of deforestation are among the highest in the world **(Figure 43.2)**. What will become of the psychedelic rock gecko and other newly discovered species if such activities continue unchecked?

Throughout the biosphere, human activities are altering natural disturbances, trophic structures, energy flow, and chemical cycling—ecosystem processes on which we and all other species depend (see Chapter 42). We have physically altered nearly half of Earth's land surface, and we use over half of all accessible surface fresh water. In the oceans, stocks of most major fisheries are shrinking because of overharvesting. By some estimates, we may be pushing more species toward extinction than did the large asteroid that triggered the mass extinctions at the close of the Cretaceous period 66 million years ago (see Figure 23.12).

In this chapter, we'll examine changes happening across Earth, focusing on **conservation biology**, a discipline that integrates ecology, physiology, molecular biology, genetics, and evolutionary biology to conserve biological diversity at all levels. Efforts to sustain ecosystem processes and stem the loss of biodiversity also connect the life sciences with the social sciences, economics, and humanities.

▲ **Figure 43.2 Tropical deforestation in Vietnam.**

We'll begin by taking a closer look at the biodiversity crisis and examining some of the conservation strategies being adopted to slow the rate of species loss. We'll also examine how human activities are altering the environment through climate change and other global processes, and we'll discuss how the growing human population drives these changes. Finally, we'll consider how current decisions about long-term conservation priorities could affect life on Earth.

## CONCEPT 43.1

# Human activities threaten Earth's biodiversity

Species are disappearing at an alarming rate. More than 1,000 species have become extinct in the last 400 years, a rate that is 100 to 1,000 times the "background," or typical, extinction rate seen in the fossil record (see Concept 23.2). This comparison suggests that the extinction rate today is high and that human activities threaten Earth's biodiversity at all levels.

## Three Levels of Biodiversity

Biodiversity—short for biological diversity—can be considered at three main levels: genetic diversity, species diversity, and ecosystem diversity **(Figure 43.3)**.

### Genetic Diversity

Genetic diversity comprises not only the individual genetic variation *within* a population, but also the genetic variation *between* populations that is often associated with adaptations to local conditions (see Chapter 21). If one population becomes extinct, then a species may have lost some of the genetic diversity that makes microevolution possible. This erosion of genetic diversity in turn reduces the adaptive potential of the species.

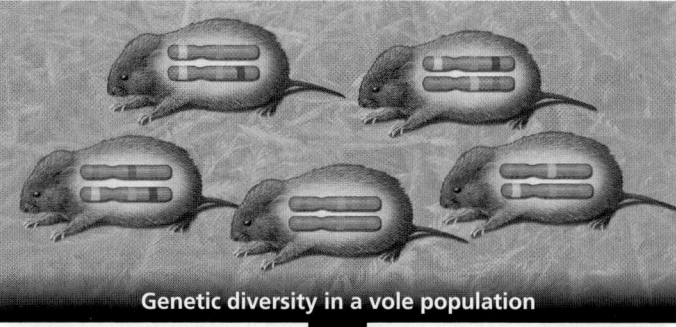

Genetic diversity in a vole population

Species diversity in a coastal redwood ecosystem

Community and ecosystem diversity across the landscape of an entire region

▲ **Figure 43.3 Three levels of biodiversity.** The oversized chromosomes in the top diagram symbolize the genetic variation within the population.

### Species Diversity

Public awareness of the biodiversity crisis centers on species diversity—the number of species in an ecosystem or across the biosphere. Of particular concern are species that are endangered or threatened. An **endangered species** is in danger of extinction throughout all or much of its range, while a **threatened species** is considered likely to become endangered in the near future. The following are just a few statistics that illustrate the problem of species loss:

- According to the International Union for Conservation of Nature and Natural Resources (IUCN), 12% of the 10,000 known species of birds and 21% of the 5,500 known species of mammals are threatened.
- A survey by the Center for Plant Conservation showed that of the nearly 20,000 known plant species in the United States, 200 have become extinct since such records have been kept, and 730 are endangered or threatened.

**Philippine eagle**

**Yangtze River dolphin**

▲ **Figure 43.4 A hundred heartbeats from extinction.** These are two members of what Harvard biologist E. O. Wilson calls the Hundred Heartbeat Club, species with fewer than 100 individuals remaining on Earth. The Yangtze River dolphin may be extinct, but a few individuals were reportedly sighted in 2007.

**?** *To document that a species has actually become extinct, what factors would you need to consider?*

- In North America, at least 123 freshwater animal species have become extinct since 1900, and hundreds more species are threatened. The extinction rate for North American freshwater fauna is about five times as high as that for terrestrial animals.

Extinction of species may also be local; for example, a species may be lost in one river system but survive in an adjacent one. Global extinction of a species means that it is lost from *all* the ecosystems in which it lived, leaving them permanently impoverished **(Figure 43.4)**.

### Ecosystem Diversity

The variety of ecosystems on Earth is a third level of biological diversity. Because of the many interactions between populations of different species in an ecosystem, the local extinction of one species can have a negative impact on other species in the ecosystem (see Figure 41.15). For instance, bats called "flying foxes" are important pollinators and seed dispersers in the Pacific Islands, where they are increasingly hunted as a luxury food **(Figure 43.5)**. Conservation biologists fear that the extinction of flying foxes would also harm the native plants of the Samoan Islands, where four-fifths of the tree species depend on flying foxes for pollination or seed dispersal.

Some ecosystems have already been heavily affected by humans, and others are being altered at a rapid pace. Since European colonization, more than half of the wetlands in the contiguous United States have been drained and converted to agricultural and other uses. In California, Arizona, and

▲ **Figure 43.5 The endangered Marianas "flying fox" bat (*Pteropus mariannus*), an important pollinator.**

New Mexico, roughly 90% of native riparian (streamside) communities have been affected by overgrazing, flood control, water diversions, lowering of water tables, and invasion by non-native plants.

## Biodiversity and Human Welfare

Why should we care about the loss of biodiversity? One reason concerns *biophilia*, our sense of connection to nature and all life. Moreover, the belief that other species are entitled to life is a pervasive theme of many religions and the basis of a moral argument that we should protect biodiversity. There is also a concern for future human generations. Paraphrasing an old proverb, G. H. Brundtland, a former prime minister of Norway, said: "We must consider our planet to be on loan from our children, rather than being a gift from our ancestors." In addition to such philosophical and moral justifications, species and genetic diversity bring us many practical benefits.

### Benefits of Species and Genetic Diversity

Many species that are threatened could potentially provide medicines, food, and fibers for human use, making biodiversity a crucial natural resource. Products from aspirin to antibiotics were originally derived from natural sources. In food production, if we lose wild populations of plants closely related to agricultural species, we lose genetic resources that could be used to improve crop qualities, such as disease resistance. For instance, in the 1970s, plant breeders responded to devastating outbreaks of the grassy stunt virus in rice (*Oryza sativa*) by screening 7,000 populations of this species and its close relatives for resistance to the virus. One population of a single relative, Indian rice (*Oryza nivara*), was found to be resistant to the virus, and scientists succeeded in breeding the resistance trait into commercial rice varieties. Today, the original disease-resistant population has apparently become extinct in the wild.

In the United States, about 25% of the prescriptions dispensed from pharmacies contain substances originally derived from plants. In the 1970s, researchers discovered that the rosy periwinkle, which grows in Madagascar, contains alkaloids that inhibit cancer cell growth. This discovery led to treatments for two deadly forms of cancer, Hodgkin's lymphoma and childhood leukemia, resulting in remission in most cases. Madagascar is also home to five other species of periwinkles, one of which is approaching extinction. Losing these species would mean the loss of any medicinal benefits they might offer.

**Rosy periwinkle**

Each loss of a species means the loss of unique genes, some of which may code for enormously useful proteins. The enzyme Taq polymerase was first extracted from a bacterium, *Thermus aquaticus*, found in hot springs at Yellowstone National Park. This enzyme is essential for the polymerase chain reaction (PCR) because it is stable at the high temperatures required for automated PCR (see Figure 13.27). DNA from many other species of prokaryotes, living in a variety of environments, is used in the mass production of proteins for new medicines, foods, petroleum substitutes, industrial chemicals, and other products. However, because many species of prokaryotes and other organisms may become extinct before we discover them, we stand to lose the valuable genetic potential held in their unique libraries of genes.

### Ecosystem Services

The benefits that individual species provide to humans are substantial, but saving individual species is only part of the reason for preserving ecosystems. Humans evolved in Earth's ecosystems, and we rely on these systems and their inhabitants for our survival. **Ecosystem services** encompass all the processes through which natural ecosystems help sustain human life. Ecosystems purify our air and water. They detoxify and decompose our wastes and reduce the impacts of extreme weather and flooding. The organisms in ecosystems pollinate our crops, control pests, and create and preserve our soils. Moreover, these diverse services are provided for free.

If we had to pay for them, how much would the services of natural ecosystems be worth? In 1997, scientists estimated the value of Earth's ecosystem services at $33 trillion per year, nearly twice the gross national product of all the countries on Earth at the time ($18 trillion). It may be more realistic to do the accounting on a smaller scale. In 1996, New York City invested more than $1 billion to buy land and restore habitat in the Catskill Mountains, the source of much of the city's fresh water. This investment was spurred by increasing pollution of the water by sewage, pesticides, and fertilizers. By harnessing ecosystem services to purify its water naturally, the city saved $8 billion it would have otherwise spent to build a new water treatment plant and $300 million a year to run the plant.

There is growing evidence that the functioning of ecosystems, and hence their capacity to perform services, is linked to biodiversity. As human activities reduce biodiversity, we are reducing the capacity of the planet's ecosystems to perform processes critical to our own survival.

## Threats to Biodiversity

Many different human activities threaten biodiversity on local, regional, and global scales. The threats posed by these activities include four major types: habitat loss, introduced species, overharvesting, and global change.

### Habitat Loss

Human alteration of habitat is the single greatest threat to biodiversity throughout the biosphere. Habitat loss has been brought about by factors such as agriculture, urban development, forestry, mining, and pollution. As discussed later in this chapter, global climate change is already altering habitats today and will have an even larger effect later this century. When no alternative habitat is available or a species is unable to move, habitat loss may mean extinction. The IUCN implicates destruction of habitat for 73% of the species that have become extinct, endangered, vulnerable, or rare in the last few hundred years.

Habitat loss and fragmentation may occur over immense regions. Approximately 98% of the tropical dry forests of Central America and Mexico have been cut down. The clearing of tropical rain forest in the state of Veracruz, Mexico, mostly for cattle ranching, has resulted in the loss of more than 90% of the original forest, leaving relatively small, isolated patches of forest. Many other natural habitats have also been fragmented by human activities **(Figure 43.6)**.

▲ **Figure 43.6 Habitat fragmentation in the foothills of Los Angeles.** Development in the valleys may confine the organisms that inhabit the narrow strips of hillside.

In almost all cases, habitat fragmentation leads to species loss because the smaller populations in habitat fragments have a higher probability of local extinction. Prairie covered about 800,000 hectares (ha) of southern Wisconsin when Europeans first arrived in North America but occupies only 800 ha today; most of the original prairie in this area is now used to grow crops. Plant diversity surveys of 54 Wisconsin prairie remnants conducted in 1948–1954 and 1987–1988 showed that the remnants lost 8–60% of their plant species in the time between the two surveys.

Habitat loss is also a major threat to aquatic biodiversity. About 70% of coral reefs, among Earth's most species-rich aquatic communities, have been damaged by human activities. At the current rate of destruction, 40–50% of the reefs, home to one-third of marine fish species, could disappear in the next 30 to 40 years. Freshwater habitats are also being lost, often as a result of the dams, reservoirs, channel modification, and flow regulation now affecting most of the world's rivers. For example, the more than 30 dams and locks built along the Mobile River basin in the southeastern United States changed river depth and flow. While providing the benefits of hydroelectric power and increased ship traffic, these dams and locks also helped drive more than 40 species of mussels and snails to extinction.

### Introduced Species

**Introduced species**, also called non-native or exotic species, are those that humans move intentionally or accidentally from the species' native locations to new geographic regions. Human travel by ship and airplane has accelerated the transplant of species. Free from the predators, parasites, and pathogens that limit their populations in their native habitats, such transplanted species may spread rapidly through a new region.

Some introduced species disrupt their new community, often by preying on native organisms or outcompeting them for resources. The brown tree snake was accidentally introduced to the island of Guam from other parts of the South Pacific after World War II, as a "stowaway" in military cargo. Since then, 12 species of birds and 6 species of lizards that the snakes ate have become extinct on Guam. The devastating zebra mussel, a suspension-feeding mollusc, was introduced into the Great Lakes of North America in 1988, most likely in the ballast water of ships arriving from Europe. Zebra mussels form dense colonies and have disrupted freshwater ecosystems, threatening native aquatic species. They have also clogged water intake structures, causing billions of dollars in damage to domestic and industrial water supplies.

Humans have deliberately introduced many species with good intentions but disastrous effects. An Asian plant called kudzu, which the U.S. Department of Agriculture once introduced in the southern United States to help control erosion, has taken over large areas of the landscape there **(Figure 43.7)**.

Introduced species are a worldwide problem, contributing to approximately 40% of the extinctions recorded since 1750

▲ **Figure 43.7 Kudzu, an introduced species, thriving in South Carolina.**

and costing billions of dollars each year in damage and control efforts. There are more than 50,000 introduced species in the United States alone.

### Overharvesting

The term *overharvesting* refers generally to the harvesting of wild organisms at rates exceeding the ability of their populations to rebound. Species with restricted habitats, such as small islands, are particularly vulnerable to overharvesting. One such species was the great auk, a large, flightless seabird found on islands in the North Atlantic Ocean. By the 1840s, humans had hunted the great auk to extinction to satisfy demand for its feathers, eggs, and meat.

Also susceptible to overharvesting are large organisms with low reproductive rates, such as elephants, whales, and rhinoceroses. The decline of Earth's largest terrestrial animals, the African elephants, is a classic example of the impact of overhunting. Largely because of the trade in ivory, elephant populations have been declining in most of Africa for the last 50 years. An international ban on the sale of new ivory resulted in increased poaching (illegal hunting), so the ban had little effect in much of central and eastern Africa. Only in South Africa, where once-decimated herds have been well protected for nearly a century, have elephant populations been stable or increasing (see Figure 40.18).

Conservation biologists increasingly use the tools of molecular genetics to track the origins of tissues harvested from endangered species. For example, researchers constructed a DNA reference map for the African elephant using DNA isolated from elephant dung. By comparing this reference map with DNA isolated from ivory harvested legally or by poachers, they can determine to within a few hundred kilometers where the elephants were killed **(Figure 43.8)**. Such work in Zambia suggested that poaching rates were 30 times higher than previously estimated, leading to improved anti-poaching efforts by the Zambian government. Similarly, biologists using phylogenetic analyses of mitochondrial DNA (mtDNA) showed that some whale meat sold in Japanese fish markets

▲ **Figure 43.8 Ecological forensics and elephant poaching.**
These tusks were part of an illegal shipment of ivory intercepted on its way from Africa to Singapore in 2002. DNA evidence showed that the thousands of elephants killed for the tusks came from a relatively narrow east-west band centered in Zambia rather than from across Africa.

came from illegally harvested species, including fin and humpback whales, which are endangered (see Figure 20.6).

Many commercially important fish populations, once thought to be inexhaustible, have been decimated by overfishing. Demands for protein-rich food from an increasing human population, coupled with new harvesting technologies, such as long-line fishing and modern trawlers, have reduced these fish populations to levels that cannot sustain further exploitation. Until the past few decades, the North Atlantic bluefin tuna was considered a sport fish of little commercial value—just a few cents per pound for use in cat food. In the 1980s, however, wholesalers began airfreighting fresh, iced bluefin to Japan for sushi and sashimi. In that market, the fish now brings up to $100 per pound **(Figure 43.9)**. With increased harvesting spurred by such high prices, it took just ten years to reduce the western North Atlantic bluefin population to less than 20% of its 1980 size.

### Global Change

The fourth threat to biodiversity, global change, alters the fabric of Earth's ecosystems at regional to global scales. Global change includes alterations in climate, atmospheric chemistry, and broad ecological systems that reduce the capacity of Earth to sustain life.

One of the first types of global change to cause concern was *acid precipitation*, which is rain, snow, sleet, or fog with a pH less than 5.2. The burning of wood and fossil fuels releases oxides of sulfur and nitrogen that react with water in air, forming sulfuric and nitric acids. The acids eventually fall to Earth's surface, harming some aquatic and terrestrial organisms.

In the 1960s, ecologists determined that lake-dwelling organisms in eastern Canada were dying because of air pollution from factories in the Midwestern United States. Newly hatched lake trout, for instance, die when the pH drops below 5.4. Lakes and streams in southern Norway and Sweden were losing fish because of pollution generated in Great Britain and central Europe. By 1980, the pH of precipitation in large areas of North America and Europe averaged 4.0–4.5 and sometimes dropped as low as 3.0. (To review pH, see Concept 2.5.)

Environmental regulations and new technologies have enabled many countries to reduce sulfur dioxide emissions in recent decades. In the United States, sulfur dioxide emissions decreased more than 40% between 1993 and 2009, gradually reducing the acidity of precipitation **(Figure 43.10)**. However, ecologists estimate that it will take decades for aquatic ecosystems to recover. Meanwhile, emissions of nitrogen oxides are increasing in the United States, and emissions of sulfur dioxide and acid precipitation continue to damage forests in Central and Eastern Europe.

We'll explore the importance of global change for Earth's biodiversity in more detail in Concept 43.4, where we examine such factors as climate change.

▲ **Figure 43.9 Overharvesting.** North Atlantic bluefin tuna are auctioned in a Japanese fish market.

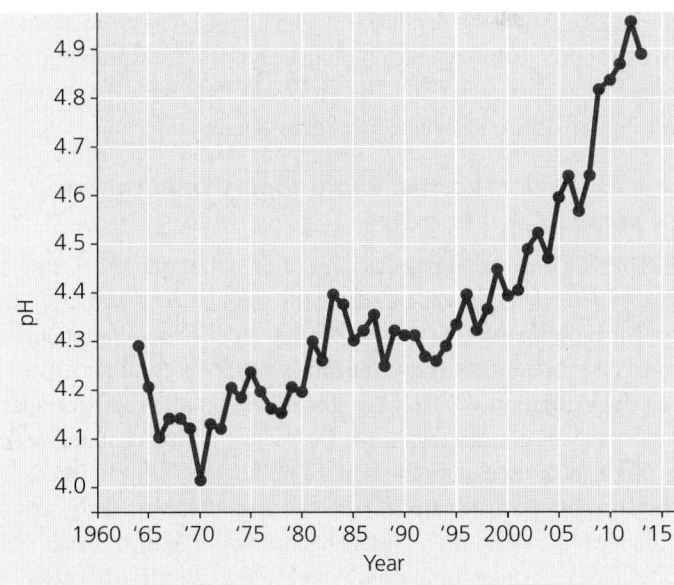

▲ **Figure 43.10 Changes in the pH of precipitation at Hubbard Brook Experimental Forest, New Hampshire.**

**MAKE CONNECTIONS** *Describe the relationship between pH and acidity. (See Concept 2.5.) Overall, is the precipitation in this forest becoming more acidic or less acidic?*

1. Explain why it is too narrow to define the biodiversity crisis as simply a loss of species.

2. Identify the four main threats to biodiversity and explain how each damages diversity.

3. **WHAT IF?** Imagine two populations of a fish species, one in the Mediterranean Sea and one in the Caribbean Sea. Now imagine two scenarios: (1) The populations breed separately, and (2) adults of both populations migrate yearly to the North Atlantic to interbreed. Which scenario would result in a greater loss of genetic diversity if the Mediterranean population were harvested to extinction? Explain your answer.

*For suggested answers, see Appendix A.*

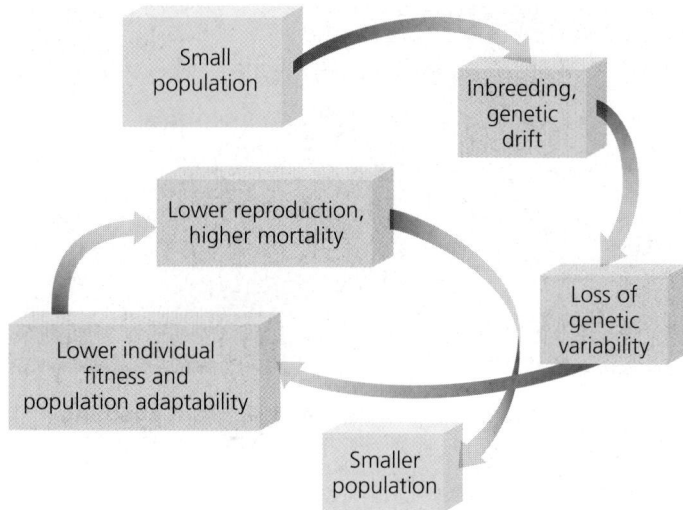

▲ **Figure 43.11 Processes driving an extinction vortex.**

## CONCEPT 43.2

# Population conservation focuses on population size, genetic diversity, and critical habitat

Biologists who work on conservation at the population and species levels use two main approaches. One approach focuses on populations that are small and hence often vulnerable. The other emphasizes populations that are declining rapidly, even if they are not yet small.

## Small-Population Approach

Small populations are particularly vulnerable to overharvesting, habitat loss, and the other threats to biodiversity that you read about in Concept 43.1. After such factors have reduced a population's size, the small size itself can push the population to extinction. Conservation biologists who adopt the small-population approach study the processes that cause extinctions once population sizes have been greatly reduced.

### The Extinction Vortex: Evolutionary Implications of Small Population Size

**EVOLUTION** A small population is vulnerable to inbreeding and genetic drift, which can draw the population down an **extinction vortex** toward smaller and smaller population size until no individuals survive **(Figure 43.11)**. A key factor driving the extinction vortex is the loss of the genetic variation that enables evolutionary responses to environmental change, such as the appearance of new strains of pathogens. Both inbreeding and genetic drift can cause a loss of genetic variation (see Concept 21.3), and their effects become more harmful as a population shrinks. Inbreeding often reduces fitness because offspring are more likely to be homozygous for harmful recessive traits.

Not all small populations are doomed by low genetic diversity, and low genetic variability does not automatically lead to permanently small populations. For instance,

overhunting of northern elephant seals in the 1890s reduced the species to only 20 individuals—clearly a bottleneck with reduced genetic variation. Since that time, however, the northern elephant seal populations have rebounded to about 150,000 individuals today, though their genetic variation remains relatively low. Thus, low genetic diversity does not always impede population growth.

### Case Study: *The Greater Prairie Chicken and the Extinction Vortex*

When Europeans arrived in North America, the greater prairie chicken (*Tympanuchus cupido*) was common from New England to Virginia and across the western prairies of the continent. Land cultivation for agriculture fragmented the populations of this species, and its abundance decreased rapidly (see Figure 21.11). Illinois had millions of greater prairie chickens in the 19th century but fewer than 50 by 1993. Researchers found that the decline in the Illinois population was associated with a decrease in fertility. As a test of the extinction vortex hypothesis, scientists increased genetic variation by importing 271 birds from larger populations elsewhere **(Figure 43.12)**. The Illinois population rebounded, confirming that it had been on its way to extinction until rescued by the transfusion of genetic variation.

### Minimum Viable Population Size

How small does a population have to be before it starts down an extinction vortex? The answer depends on the type of organism and other factors. Large predators that feed high on the food chain usually require extensive individual ranges, resulting in low population densities. Therefore, not all rare species concern conservation biologists. All populations, however, require some minimum size to remain viable.

The minimal population size at which a species is able to sustain its numbers is known as the **minimum viable population (MVP)**. MVP is usually estimated for a

▼ **Figure 43.12** Inquiry

## What caused the drastic decline of the Illinois greater prairie chicken population?

**Experiment** Researchers had observed that the population collapse of the greater prairie chicken was mirrored in a reduction in fertility, as measured by the hatching rate of eggs. Comparison of DNA samples from the Jasper County, Illinois, population with DNA from feathers in museum specimens showed that genetic variation had declined in the study population (see Figure 21.11). In 1992, Ronald Westemeier and colleagues began translocating prairie chickens from neighboring states in an attempt to increase genetic variation.

**Results** After translocation (black arrow), the viability of eggs rapidly increased, and the population rebounded.

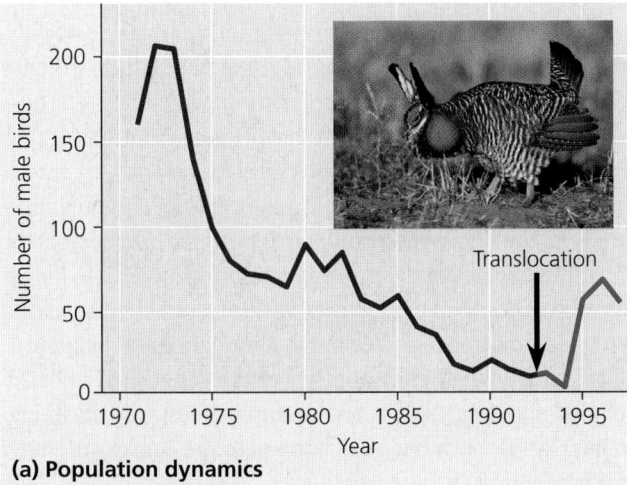

**(a) Population dynamics**

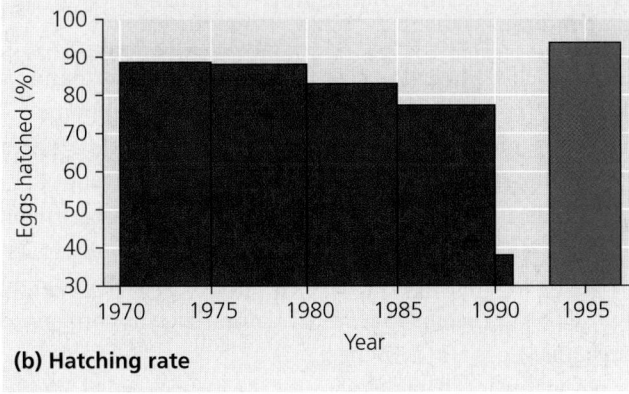

**(b) Hatching rate**

**Conclusion** Reduced genetic variation had started the Jasper County population of prairie chickens down the extinction vortex.

**Data from** R. L. Westemeier et al., Tracking the long-term decline and recovery of an isolated population, *Science* 282:1695–1698 (1998).

**Inquiry in Action** Read and analyze the original paper in *Inquiry in Action: Interpreting Scientific Papers.*

**WHAT IF?** Given the success of using transplanted birds as a tool for increasing the percentage of hatched eggs in Illinois, why wouldn't you immediately transplant additional birds to Illinois?

given species using computer models that integrate many factors. The calculation may include, for instance, an estimate of how many individuals in a small population are likely to be killed by a natural catastrophe such as a storm. Once in the extinction vortex, two or three consecutive years of bad weather could finish off a population that is already below its MVP.

### Effective Population Size

Genetic variation is a key issue in the small-population approach. The *total* size of a population may be misleading because only certain members of the population breed successfully and pass their alleles on to offspring. Therefore, a meaningful estimate of MVP requires the researcher to determine the **effective population size**, which is based on the breeding potential of the population.

The following formula illustrates one way to estimate the effective population size, abbreviated $N_e$:

$$N_e = \frac{4N_fN_m}{N_f + N_m}$$

where $N_f$ and $N_m$ are, respectively, the number of females and the number of males that successfully breed. If we apply this formula to an idealized population whose total size is 1,000 individuals, $N_e$ will also be 1,000 if every individual breeds and the sex ratio is 500 females to 500 males. In this case, $N_e = (4 \times 500 \times 500)/(500 + 500) = 1,000$. Any deviation from these conditions (not all individuals breed or there is not a 1:1 sex ratio) reduces $N_e$. For instance, if the total population size is 1,000 but only 400 females and 400 males breed, then $N_e = (4 \times 400 \times 400)/(400 + 400) = 800$, or 80% of the total population size. Numerous factors can influence $N_e$. Alternative formulas for estimating $N_e$ take into account factors such as age at maturation, genetic relatedness among population members, the effects of gene flow, and population fluctuations.

In actual study populations, $N_e$ is always some fraction of the total population. Thus, simply determining the total number of individuals in a small population does not provide a good measure of whether the population is large enough to avoid extinction. Whenever possible, conservation programs attempt to sustain total population sizes that include at least the minimum viable number of *reproductively active* individuals. The conservation goal of sustaining effective population size ($N_e$) above MVP stems from the concern that populations retain enough genetic diversity to adapt as their environment changes.

### Case Study: *Analysis of Grizzly Bear Populations*

One of the first population viability analyses was conducted in 1978 by Mark Shaffer as part of a long-term study of grizzly bears in Yellowstone National Park and its surrounding areas

▲ **Figure 43.13 Long-term monitoring of a grizzly bear population.** The ecologist is fitting this tranquilized bear with a radio collar so that the bear's movements can be compared with those of other grizzlies in the Yellowstone National Park population.

**(Figure 43.13)**. A threatened species in the United States, the grizzly bear (*Ursus arctos horribilis*) is currently found in only 4 of the 48 contiguous states. Its populations in those states have been drastically reduced and fragmented. In 1800, an estimated 100,000 grizzlies ranged over about 500 million ha of habitat, while today only 1,000 individuals in six relatively isolated populations range over less than 5 million ha.

Shaffer attempted to determine viable sizes for the Yellowstone grizzly population. Using life history data obtained for individual Yellowstone bears over a 12-year period, he simulated the effects of environmental factors on survival and reproduction. His models predicted that, given a suitable habitat, a Yellowstone grizzly bear population of 70–90 individuals would have about a 95% chance of surviving for 100 years. A slightly larger population of only 100 bears would have a 95% chance of surviving for twice as long, about 200 years.

How does the actual size of the Yellowstone grizzly population compare with Shaffer's predicted MVP? A current estimate puts the total grizzly bear population in the greater Yellowstone ecosystem at about 400 individuals. The relationship of this estimate to the effective population size, $N_e$, depends on several factors. Usually, only a few dominant males breed, and it may be difficult for them to locate females, since individuals inhabit such large areas. Moreover, females may reproduce only when there is abundant food. As a result, $N_e$ is only about 25% of the total population size, or about 125 bears.

Because small populations tend to lose genetic variation over time, researchers have used proteins, mitochondrial DNA, and short tandem repeats (see Concept 18.4) to assess genetic variability in the Yellowstone grizzly bear population. All results to date indicate that the Yellowstone population has less genetic variability than other grizzly bear populations in North America.

How might conservation biologists increase the effective size and genetic variation of the Yellowstone grizzly bear population? Migration between isolated populations of grizzlies could increase both effective and total population sizes. Computer models predict that introducing only two unrelated bears each decade into a population of 100 individuals would reduce the loss of genetic variation by about half. For the grizzly bear, and probably for many other species with small populations, finding ways to promote dispersal among populations may be one of the most urgent conservation needs.

This case study and that of the greater prairie chicken bridge small-population models and practical applications in conservation. Next, we look at an alternative approach to understanding the biology of extinction.

## Declining-Population Approach

The declining-population approach focuses on threatened and endangered populations that show a downward trend, even if the population is far above its minimum viable population. The distinction between a declining population, which may not be small, and a small population, which may not be declining, is less important than the different priorities of the two approaches. While the small-population approach emphasizes smallness itself as an ultimate cause of a population's extinction, the declining-population approach emphasizes the environmental factors that caused a population decline in the first place. If, for instance, an area is deforested, then species that depend on trees will decline in abundance and become locally extinct, whether or not they retain genetic variation. The following case study is one example of how the declining-population approach has been applied to the conservation of an endangered species.

### Case Study: *Decline of the Red-cockaded Woodpecker*

The red-cockaded woodpecker (*Picoides borealis*) is found only in the southeastern United States. It requires mature pine forests, preferably ones dominated by the longleaf pine, for its habitat. Most woodpeckers nest in dead trees, but the red-cockaded woodpecker drills its nest holes in mature, living pine trees. It also drills small holes around the entrance to its nest cavity, which causes resin from the tree to ooze down the trunk. The resin seems to repel predators, such as corn snakes, that eat bird eggs and nestlings.

Another critical habitat factor for the red-cockaded woodpecker is that the undergrowth of plants around the pine trunks must be low **(Figure 43.14a)**. Breeding birds tend to abandon nests when vegetation among the pines is thick and higher than about 4.5 m **(Figure 43.14b)**. Apparently, the birds need a clear flight path between their home trees and the neighboring feeding grounds. Periodic fires have historically swept through longleaf pine forests, keeping the undergrowth low.

One factor leading to the decline of the red-cockaded woodpecker has been the destruction or fragmentation of suitable habitats by logging and agriculture. By recognizing key habitat factors, protecting some longleaf pine forests, and using controlled fires to reduce forest undergrowth,

Red-cockaded woodpecker

**(a)** Forests that can sustain red-cockaded woodpeckers have low undergrowth.

**(b)** Forests that cannot sustain red-cockaded woodpeckers have high, dense undergrowth that interferes with the woodpeckers' access to feeding grounds.

▲ **Figure 43.14 A habitat requirement of the red-cockaded woodpecker.**

**?** *How is habitat disturbance absolutely necessary for the long-term survival of the woodpecker?*

conservation managers have helped restore habitat that can support viable populations.

Sometimes conservation managers also help species colonize restored habitats. Because red-cockaded woodpeckers take months to excavate nesting cavities, researchers performed an experiment to see whether providing cavities for the birds would make them more likely to use a site. The researchers constructed cavities in pine trees at 20 sites. The results were dramatic. Cavities in 18 of the 20 sites were colonized by red-cockaded woodpeckers, and new breeding groups formed only in those sites. Based on this experiment, conservationists initiated a habitat maintenance program that included controlled burning and excavation of new nesting cavities, enabling this endangered species to begin to recover.

## Weighing Conflicting Demands

Determining population numbers and habitat needs is only part of a strategy to save species. Scientists also need to weigh a species' needs against other conflicting demands. Conservation biology often highlights the relationship between science, technology, and society. For example, an ongoing, sometimes bitter debate in the western United States pits habitat preservation for wolf, grizzly bear, and bull trout populations against job opportunities in the grazing and resource extraction industries. Programs that restocked wolves in Yellowstone National Park remain controversial for people concerned about human safety and for many ranchers concerned with potential loss of livestock outside the park.

Large, high-profile vertebrates are not always the focal point in such conflicts, but habitat use is almost always the issue. Should work proceed on a new highway bridge if it destroys the only remaining habitat of a species of freshwater mussel? If you owned a coffee plantation growing varieties that thrive in bright sunlight, would you be willing to change to shade-tolerant varieties that produce less coffee per hectare but can grow beneath trees that support large numbers of songbirds?

Another important consideration is the ecological role of a species. Because we cannot save every endangered species, we must determine which species are most important for conserving biodiversity as a whole. Identifying keystone species and finding ways to sustain their populations can be central to maintaining communities and ecosystems. In most situations, conservation biologists must also look beyond single species and consider the whole community and ecosystem as an important unit of biodiversity.

### CONCEPT CHECK 43.2

1. How does the reduced genetic diversity of small populations make them more vulnerable to extinction?
2. If there were 100 greater prairie chickens in a population, and 30 females and 10 males bred, what would be the effective population size ($N_e$)?
3. **WHAT IF?** In 2005, at least ten grizzly bears in the greater Yellowstone ecosystem were killed through contact with people. Most of these deaths resulted from three things: collisions with automobiles, hunters (of other animals) shooting when charged by a female grizzly bear with cubs nearby, and conservation managers killing bears that attacked livestock repeatedly. If you were a conservation manager, what steps might you take to minimize such encounters in Yellowstone?

For suggested answers, see Appendix A.

### CONCEPT 43.3

# Landscape and regional conservation help sustain biodiversity

Although conservation efforts have historically focused on saving individual species, efforts today often seek to sustain the biodiversity of entire communities, ecosystems, and landscapes (see Figure 40.2). Such a broad view requires applying not just ecological principles, but aspects of human population dynamics and economics as well.

## Landscape Structure and Biodiversity

The biodiversity of a given landscape is heavily influenced by its physical features, or *structure*. Understanding landscape structure is critically important in conservation because many species use more than one kind of ecosystem, and many live on the borders between ecosystems.

### Fragmentation and Edges

The boundaries, or *edges*, between ecosystems—such as between a lake and the surrounding forest or between cropland and suburban housing tracts—are defining features of landscapes **(Figure 43.15)**. An edge has its own set of physical conditions, which differ from those on either side of it. The soil surface of an edge between a forest patch and a burned area receives more sunlight and is usually hotter and drier than the forest interior, but it is cooler and wetter than the soil surface in the burned area.

Some organisms thrive in edge communities because they gain resources from both adjacent areas. The ruffed grouse (*Bonasa umbellus*) is a bird that needs forest habitat for nesting, winter food, and shelter, but it also needs forest openings with dense shrubs and herbs for summer food.

Ecosystems in which edges arise from human alterations often have reduced biodiversity and a preponderance of edge-adapted species. For example, white-tailed deer thrive in edge habitats, where they can browse on woody shrubs; deer populations often expand when forests are logged and more

▲ **Figure 43.16 Amazon rain forest fragments created as part of the Biological Dynamics of Forest Fragments Project.**

edges are generated. The brown-headed cowbird (*Molothrus ater*) is an edge-adapted species that lays its eggs in the nests of other birds, often migratory songbirds. Cowbirds need forests, where they can parasitize the nests of other birds, and open fields, where they forage on seeds and insects. Consequently, their populations are growing where forests are being cut and fragmented, creating more edge habitat and open land. Increasing cowbird parasitism and habitat loss are correlated with declining populations of several of the cowbird's host species.

The influence of fragmentation on the structure of communities has been explored since 1979 in the long-term Biological Dynamics of Forest Fragments Project. Located in the heart of the Amazon River basin, the study area consists of isolated fragments of tropical rain forest separated from surrounding continuous forest by distances of 80–1,000 m **(Figure 43.16)**. Numerous researchers working on this project have clearly documented the effects of this fragmentation on organisms ranging from bryophytes to beetles to birds. They have consistently found that species adapted to forest interiors show the greatest declines when patches are the smallest, suggesting that landscapes dominated by small fragments will support fewer species.

### Corridors That Connect Habitat Fragments

In fragmented habitats, the presence of a **movement corridor**, a narrow strip or series of small clumps of habitat connecting otherwise isolated patches, can be extremely important for conserving biodiversity. Riparian habitats often serve as corridors, and in some nations, government policy prohibits altering these habitats. In areas of heavy human use, artificial corridors are sometimes constructed. Bridges or tunnels, for instance, can reduce the number of animals killed trying to cross highways **(Figure 43.17)**.

▲ **Figure 43.15 Edges between ecosystems in Yellowstone National Park.**

**?** *What edges between ecosystems do you see in this photo?*

▲ **Figure 43.17** **An artificial corridor.** This highway overpass in the Netherlands helps animals cross a human-created barrier.

Movement corridors can also promote dispersal and reduce inbreeding in declining populations. Corridors have been shown to increase the exchange of individuals among populations of many organisms, including butterflies, voles, and aquatic plants. Corridors are especially important to species that migrate between different habitats seasonally. However, a corridor can also be harmful—for example, by allowing the spread of disease. In a 2003 study, a scientist at the University of Zaragoza, Spain, showed that habitat corridors facilitate the movement of disease-carrying ticks among forest patches in northern Spain. All the effects of corridors are not yet understood, and their impact is an area of active research in conservation biology.

## Establishing Protected Areas

Currently, governments have set aside about 7% of the world's land in various forms of reserves. Choosing where to place nature reserves and how to design them poses many challenges. Should the reserve be managed to minimize the risks of fire and predation to a threatened species? Or should the reserve be left as natural as possible, with such processes as fires ignited by lightning allowed to play out on their own? This is just one of the debates that arise among people who share an interest in the health of national parks and other protected areas.

### Preserving Biodiversity Hot Spots

In deciding which areas are of highest conservation priority, biologists often focus on hot spots of biodiversity. A **biodiversity hot spot** is a relatively small area with numerous endemic species (species found nowhere else in the world) and a large number of endangered and threatened species **(Figure 43.18)**. Nearly 30% of all bird species can be found in hot spots that make up only about 2% of Earth's land area.

Together, the "hottest" of the terrestrial biodiversity hot spots total less than 1.5% of Earth's land but are home to more than a third of all species of plants, amphibians, reptiles (including birds), and mammals. Aquatic ecosystems also have hot spots, such as coral reefs and certain river systems.

Biodiversity hot spots are good choices for nature reserves, but identifying them is not always simple. One problem is that a hot spot for one taxonomic group, such as butterflies, may not be a hot spot for some other taxonomic group, such as birds. Designating an area as a biodiversity hot spot is often biased toward saving vertebrates and plants, with less attention paid to invertebrates and microorganisms. Some biologists are also concerned that the hot-spot strategy places too much emphasis on such a small fraction of Earth's surface.

Global change makes the task of preserving hot spots even more challenging because the conditions that favor a particular community may not be found in the same location in the future. The biodiversity hot spot in the southwest corner of Australia (see Figure 43.18) holds thousands of species of endemic plants and numerous endemic vertebrates. Researchers recently concluded that between 5% and 25% of the plant species they examined may become extinct by 2080 because the plants will be unable to tolerate the increased dryness predicted for this region.

### Philosophy of Nature Reserves

Nature reserves are protected "islands" of high biodiversity in a sea of habitat altered or degraded by human activity. An earlier policy—that protected areas should be set aside to remain unchanged forever—was based on the concept that ecosystems are balanced, self-regulating units. However, disturbance is common in all ecosystems (see Concept 41.3). Management policies that ignore natural disturbances or attempt to prevent them have generally failed. For instance, setting aside an area of a fire-dependent community, such as a portion of a tallgrass prairie, chaparral, or dry pine forest, with the intention of

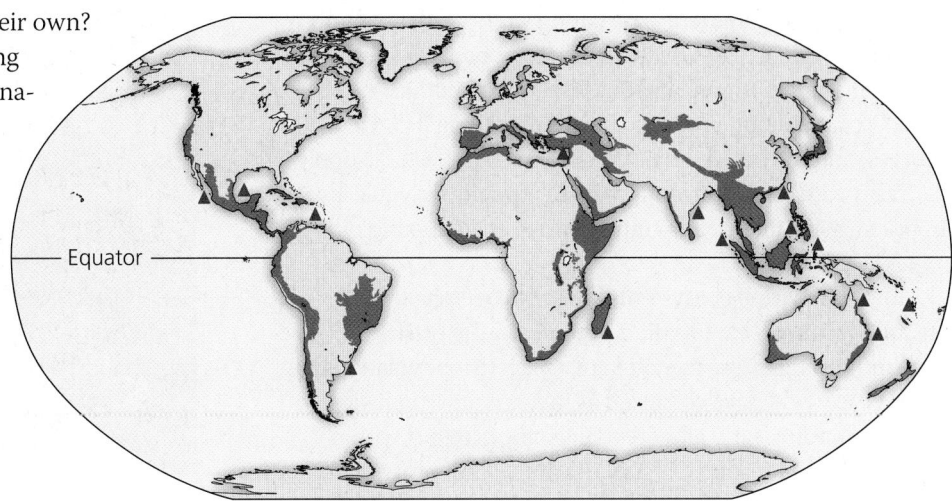

▲ **Figure 43.18** **Earth's terrestrial (■) and marine (▲) biodiversity hot spots.**

saving it is unrealistic if periodic burning is excluded. Without the dominant disturbance, the fire-adapted species are usually outcompeted and biodiversity is reduced.

An important conservation question is whether to create numerous small reserves or fewer large reserves. Small, unconnected reserves may slow the spread of disease between populations. One argument for large reserves is that large, far-ranging animals with low-density populations, such as the grizzly bear, require extensive habitats. Large reserves also have proportionately smaller perimeters than small reserves and are therefore less affected by edges.

As conservation biologists have learned more about the requirements for achieving minimum viable populations for endangered species, they have realized that most national parks and other reserves are far too small. The area needed for the long-term survival of the Yellowstone grizzly bear population, for instance, is more than 11 times the area of Yellowstone National Park. Areas of private and public land surrounding reserves will likely have to contribute to biodiversity conservation.

### Zoned Reserves

Several nations have adopted a zoned reserve approach to landscape management. A **zoned reserve** is an extensive region that includes areas relatively undisturbed by humans surrounded by areas that have been changed by human activity and are used for economic gain. The key challenge of the zoned reserve approach is to develop a social and economic climate in the surrounding lands that is compatible with the long-term viability of the protected core. These surrounding areas continue to support human activities, but regulations prevent the types of extensive alterations likely to harm the protected area. As a result, the surrounding habitats serve as buffer zones against further intrusion into the undisturbed area.

The Central American nation of Costa Rica has become a world leader in establishing zoned reserves. An agreement initiated in 1987 reduced Costa Rica's international debt in return for land preservation there. The country is now divided into 11 Conservation Areas, which include national parks and other protected areas, both on land and in the ocean **(Figure 43.19)**. Costa Rica is making progress toward managing its zoned reserves, and the buffer zones provide a steady, lasting supply of forest products, water, and hydroelectric power while also supporting sustainable agriculture and tourism, both of which employ local people.

Although marine ecosystems have also been heavily affected by human exploitation, reserves in the ocean are far less common than reserves on land. Many fish populations around the world have collapsed as increasingly sophisticated equipment puts nearly all potential fishing grounds within human reach. In response, scientists at the University of York, England, have proposed establishing marine reserves around the world that would be off limits to fishing. They present

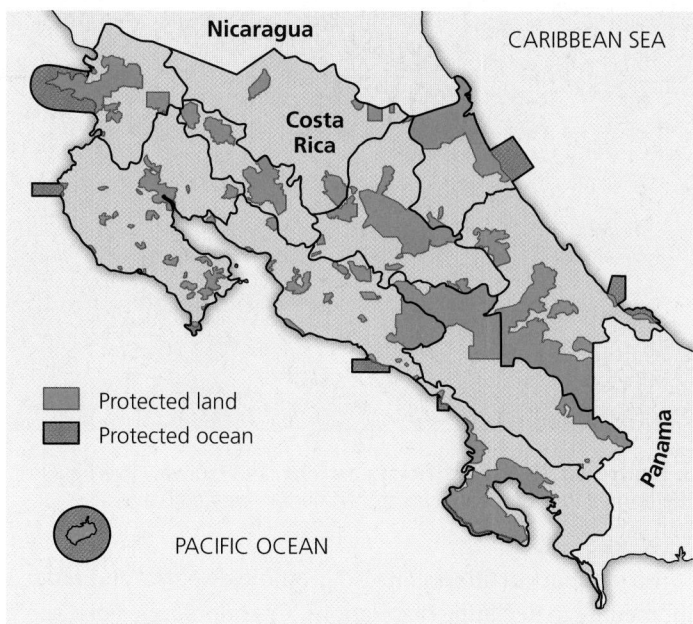

▲ **Figure 43.19 Protected areas in Costa Rica.** Boundaries of the Conservation Areas are indicated by black outlines.

strong evidence that a patchwork of marine reserves can serve as a means of both increasing fish populations within the reserves and improving fishing success in nearby areas. Their proposed system is a modern application of a centuries-old practice in the Fiji Islands in which some areas have historically remained closed to fishing—a traditional example of the zoned reserve concept.

The United States adopted such a system in creating a set of 13 national marine sanctuaries, including the Florida Keys National Marine Sanctuary, which was established in 1990 **(Figure 43.20)**. Populations of marine organisms, including fishes and lobsters, recovered quickly after harvests were banned in the 9,500-km² reserve. Larger and more abundant

▲ **Figure 43.20 A diver measuring coral in the Florida Keys National Marine Sanctuary.**

fish now produce larvae that help repopulate reefs and improve fishing outside the sanctuary. The increased marine life within the sanctuary also makes it a favorite for recreational divers, increasing the economic value of this zoned reserve.

**CONCEPT CHECK 43.3**

1. What is a biodiversity hot spot?
2. How do zoned reserves provide economic incentives for long-term conservation of protected areas?
3. **WHAT IF?** Suppose a developer proposes to clear-cut a forest that serves as a corridor between two parks. To compensate, the developer also proposes to add the same area of forest to one of the parks. As a professional ecologist, how might you argue for retaining the corridor?

For suggested answers, see Appendix A.

**CONCEPT 43.4**

# Earth is changing rapidly as a result of human actions

As we've discussed, landscape and regional conservation help protect habitats and preserve species. However, environmental changes that result from human activities are creating new challenges. As a consequence of human-caused climate change, for example, the place where a vulnerable species is found today may not be the same place that is needed for preservation in the future. What would happen if *many* habitats on Earth changed so quickly that the locations of preserves today were unsuitable for their species in 10, 50, or 100 years? Such a scenario is increasingly possible.

The rest of this section describes three types of environmental change that threaten biodiversity: nutrient enrichment, toxin accumulation, and climate change. The impacts of these and other changes are evident not just in human-dominated ecosystems, such as cities and farms, but also in the most remote ecosystems on Earth.

## Nutrient Enrichment

Human activity often removes nutrients from one part of the biosphere and adds them to another. Someone eating broccoli in Washington, DC, consumes nutrients that only days before were in the soil in California; a short time later, some of these nutrients will be in the Potomac River, having passed through the person's digestive system and a local sewage treatment facility. Likewise, nutrients in farm soil may run off into streams and lakes, depleting nutrients in one area, increasing them in another, and altering chemical cycles in both.

Farming illustrates how human activities can lead to nutrient enrichment. After vegetation is cleared from an area, the existing reserve of nutrients in the soil is depleted because many of these nutrients are exported from the area in crop biomass. Consider nitrogen, the main nutrient lost through agriculture (see Figure 42.14). Plowing mixes the soil and speeds up decomposition of organic matter, releasing nitrogen that is then removed when crops are harvested. Fertilizers containing nitrates and other forms of nitrogen that plants can absorb are used to replace the nitrogen that is lost. However, after crops are harvested, few plants remain to take up nitrates from the soil. As we saw in Figure 42.15, without plants to absorb them, nitrates are often leached from the ecosystem.

Recent studies indicate that human activities have more than doubled Earth's supply of fixed nitrogen available to primary producers. Industrial fertilizers provide the largest additional nitrogen source. Fossil fuel combustion also releases nitrogen oxides, which enter the atmosphere and dissolve in rainwater; the nitrogen ultimately enters ecosystems as nitrate. Increased cultivation of legumes, with their nitrogen-fixing symbionts, is a third way in which humans increase the amount of fixed nitrogen in the soil.

A problem arises when the nutrient level in an ecosystem exceeds the **critical load**, the amount of added nutrient, usually nitrogen or phosphorus, that can be absorbed by plants without damaging ecosystem integrity. For example, nitrogenous minerals in the soil that exceed the critical load eventually leach into groundwater or run off into freshwater and marine ecosystems, contaminating water supplies and killing fish. Nitrate concentrations in groundwater are increasing in most agricultural regions, sometimes reaching levels that are unsafe for drinking.

Many rivers contaminated with nitrates and ammonium from agricultural runoff and sewage drain into the Atlantic Ocean, with the highest inputs coming from northern Europe and the central United States. The Mississippi River carries nitrogen pollution to the Gulf of Mexico, fueling a phytoplankton bloom each summer. When the phytoplankton die, their decomposition by oxygen-using organisms creates an extensive "dead zone" of low oxygen levels along the coast **(Figure 43.21)**.

▲ **Figure 43.21 A phytoplankton bloom arising from nitrogen pollution in the Mississippi basin that leads to a dead zone.** In this satellite image from 2004, red and orange represent high concentrations of phytoplankton in the Gulf of Mexico.

Fish and other marine animals disappear from some of the most economically important waters in the United States. To reduce the size of the dead zone, farmers have begun using fertilizers more efficiently, and managers are restoring wetlands in the Mississippi watershed, two changes stimulated by the results of ecosystem experiments.

Nutrient runoff can also lead to the eutrophication of lakes (see Concept 42.2). The bloom and subsequent die-off of algae and cyanobacteria and the ensuing depletion of oxygen are similar to what occurs in a marine dead zone. Such conditions threaten the survival of many organisms. For example, eutrophication of Lake Erie coupled with overfishing wiped out commercially important fishes such as blue pike, whitefish, and lake trout by the 1960s. Since then, tighter regulations on the dumping of sewage into the lake have enabled some fish populations to rebound, but many native species of fish and invertebrates have not recovered.

## Toxins in the Environment

Humans release an immense variety of toxic chemicals, including thousands of synthetic compounds previously unknown in nature, with little regard for the ecological consequences. Organisms acquire toxic substances from the environment along with nutrients and water. Some of the poisons are metabolized or excreted, but others accumulate in specific tissues, often fat. One of the reasons accumulated toxins are particularly harmful is that they become more concentrated in successive trophic levels of a food web. This phenomenon, called **biological magnification**, occurs because the biomass at any given trophic level is produced from a much larger biomass ingested from the level below (see Concept 42.3). Thus, top-level carnivores tend to be most severely affected by toxic compounds in the environment.

Chlorinated hydrocarbons are a class of industrially synthesized compounds that have demonstrated biological magnification. Chlorinated hydrocarbons include the industrial chemicals called PCBs (polychlorinated biphenyls) and many pesticides, such as DDT. Current research implicates many of these compounds in endocrine system disruption in a large number of animal species, including humans. Biological magnification of PCBs has been found in the food web of the Great Lakes, where the concentration of PCBs in herring gull eggs, at the top of the food web, is nearly 5,000 times that in phytoplankton, at the base of the food web **(Figure 43.22)**.

An infamous case of biological magnification that harmed top-level carnivores involved DDT, a chemical used to control insects such as mosquitoes and agricultural pests. In the decade after World War II, the use of DDT grew rapidly; its ecological consequences were not yet fully understood. By the 1950s, scientists were learning that DDT persists in the environment and is transported by water to areas far from where it is applied. One of the first signs that DDT was a serious

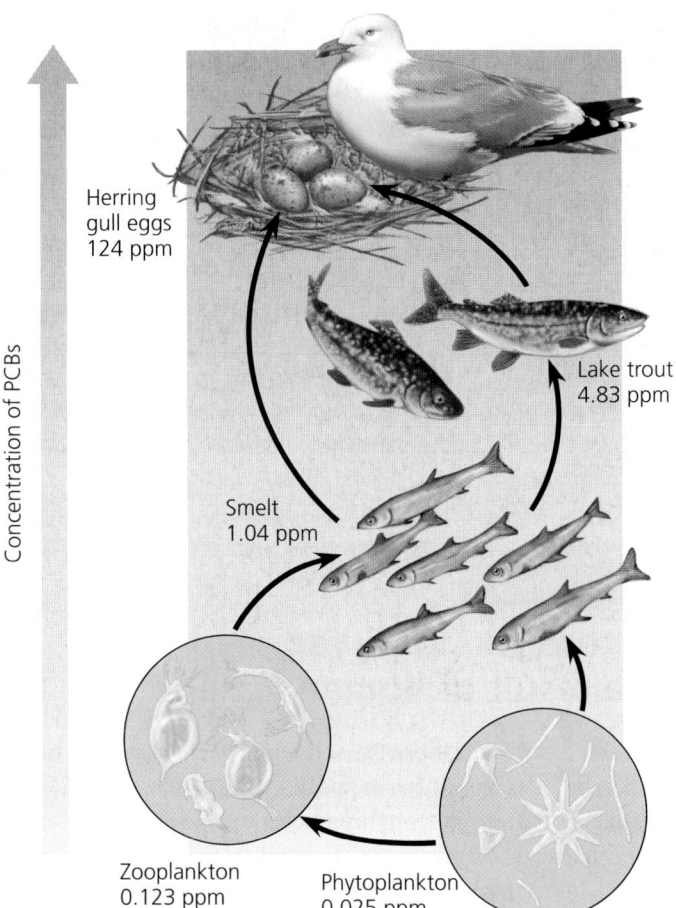

Herring gull eggs 124 ppm

Lake trout 4.83 ppm

Smelt 1.04 ppm

Concentration of PCBs

Zooplankton 0.123 ppm

Phytoplankton 0.025 ppm

▲ **Figure 43.22 Biological magnification of PCBs in a Great Lakes food web.** (ppm = parts per million)

**?** *Calculate how much the PCB concentration increased at each step in the food web.*

environmental problem was a decline in the populations of pelicans, ospreys, and eagles, birds that feed at the top of food webs. The accumulation of DDT (and DDE, a product of its breakdown) in the tissues of these birds interfered with the deposition of calcium in their eggshells. When the birds tried to incubate their eggs, the weight of the parents broke the shells of affected eggs, resulting in catastrophic declines in the birds' reproduction rates. Rachel Carson's book *Silent Spring* helped bring the problem to public attention in the 1960s **(Figure 43.23)**, and DDT was banned in the United States in 1971. A dramatic recovery in populations of the affected bird species followed.

In much of the tropics, DDT is still used to control the mosquitoes that spread malaria and other diseases. Societies there face a trade-off between saving human lives and protecting other species. The best approach seems to be to apply DDT sparingly and to couple its use with mosquito netting and other low-technology solutions. The complicated history of DDT illustrates the importance of understanding the ecological connections between diseases and communities (see Concept 41.5).

► Figure 43.23
**Rachel Carson.**
Through her writing and her testimony before the U.S. Congress, biologist and author Rachel Carson helped promote a new environmental ethic. Her efforts led to a ban on DDT use in the United States and stronger controls on the use of other chemicals.

Pharmaceuticals make up another group of toxins in the environment, one that is a growing concern among ecologists. The use of over-the-counter and prescription drugs has risen in recent years, particularly in industrialized nations. People who consume such products excrete residual chemicals in their waste and may also dispose of unused drugs improperly, such as in their toilets or sinks. Drugs that are not broken down in sewage treatment plants may then enter rivers and lakes with the material discharged from these plants. Growth-promoting drugs given to farm animals can also enter rivers and lakes with agricultural runoff. As a consequence, many pharmaceuticals are spreading in low concentrations across the world's freshwater ecosystems **(Figure 43.24)**.

Among the pharmaceuticals that ecologists are studying are the sex steroids, including forms of estrogen used for birth control. Some fish species are so sensitive to certain estrogens that concentrations of a few parts per trillion in their water can alter sexual differentiation and shift the female-to-male sex ratio toward females. Researchers in Ontario, Canada, conducted a seven-year experiment in which they applied the synthetic estrogen used in contraceptives to a lake in very low concentrations (5–6 ng/L). They found that chronic exposure

of the fathead minnow (*Pimephales promelas*) to the estrogen led to feminization of males and a near extinction of the species from the lake.

Many toxins cannot be degraded by microorganisms and persist in the environment for years or even decades. In other cases, chemicals released into the environment may be relatively harmless but are converted to more toxic products by reaction with other substances, by exposure to light, or by the metabolism of microorganisms. Mercury, a by-product of plastic production and coal-fired power generation, has been routinely expelled into rivers and the sea in an insoluble form. Bacteria in the bottom mud convert the waste to methylmercury $(CH_3Hg^+)$, an extremely toxic water-soluble compound that accumulates in the tissues of organisms, including humans who consume fish from the contaminated waters.

## Greenhouse Gases and Climate Change

Human activities release a variety of gaseous waste products. People once thought that the vast atmosphere could absorb these materials indefinitely, but we now know that such additions can lead to **climate change**, a directional change to the global climate that lasts for three decades or more (as opposed to short-term changes in the weather).

To see how human actions can cause climate change, consider atmospheric $CO_2$ levels. Over the past 150 years, the concentration of $CO_2$ in the atmosphere has been increasing as a result of the burning of fossil fuels and deforestation. Scientists estimate that the average $CO_2$ concentration in the atmosphere before 1850 was about 274 ppm. In 1958, a monitoring station began taking accurate measurements on Hawaii's Mauna Loa peak, a location far from cities and high enough for the atmosphere to be well mixed. At that time, the average $CO_2$ concentration was 316 ppm **(Figure 43.25)**. Today, it exceeds 400 ppm, an increase of more than 45% since the mid-19th century. In the **Scientific Skills Exercise**, you can graph and interpret changes in $CO_2$ concentration that occur during the course of a year and over longer periods.

The marked increase in the concentration of atmospheric $CO_2$ over the last 150 years concerns scientists because of its link to increased global temperature. Much of the solar radiation that warms Earth's surface is emitted toward space as infrared radiation (known informally as "heat radiation"). Although $CO_2$, methane, water vapor, and other greenhouse gases in the atmosphere are transparent to visible light, they intercept and absorb much of the infrared radiation that Earth emits, radiating most of it back toward Earth. This process,

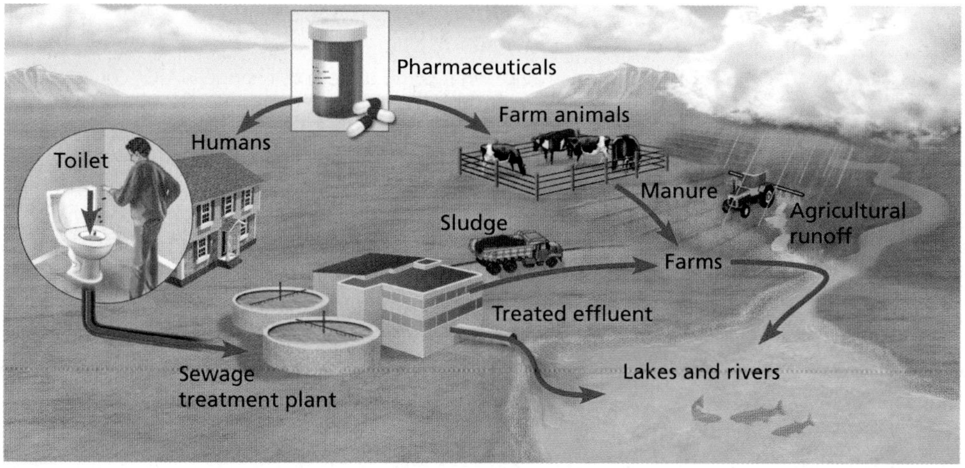

▲ **Figure 43.24 Sources and movements of pharmaceuticals in the environment.**

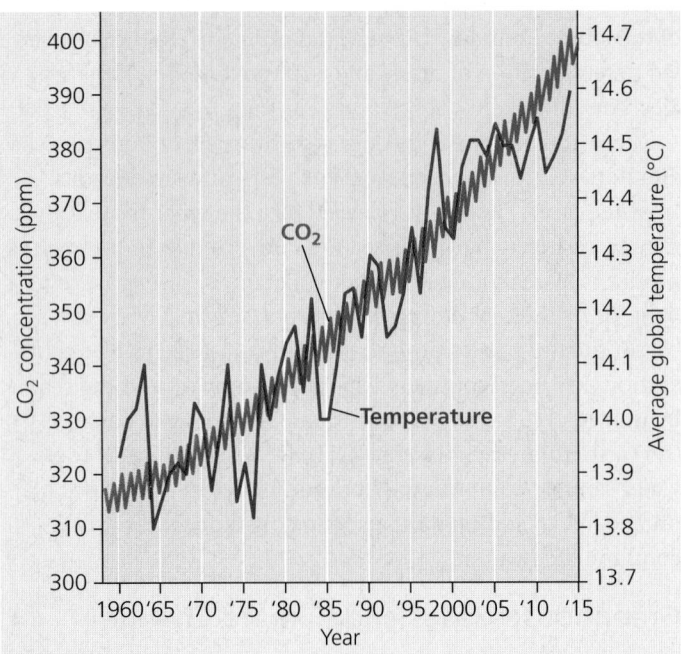

▲ **Figure 43.25 Increase in atmospheric carbon dioxide concentration at Mauna Loa, Hawaii, and average global temperatures.** Aside from normal seasonal fluctuations, the $CO_2$ concentration (blue curve) has increased steadily from 1958 to 2015. Though average global temperatures (red curve) fluctuated a great deal over the same period, there is a clear warming trend.

called the **greenhouse effect**, retains some of the solar heat **(Figure 43.26)**. If it were not for this greenhouse effect, the average air temperature at Earth's surface would be a frigid −18°C (−0.4°F), and most life as we know it could not exist.

As the concentrations of $CO_2$ and other greenhouse gases rise, more solar heat is retained, thereby increasing the temperature of our planet. So far, Earth has warmed by an average of 0.9°C (1.6°F) since 1900. At the current rates that $CO_2$ and other gases are being added to the atmosphere, global models predict an additional rise of at least 3°C (5°F) by the end of the 21st century.

As our planet warms, the climate is changing in other ways as well: Wind and precipitation patterns are shifting, and extreme weather events (such as droughts and storms) are occurring more often. What are likely consequences of such changes to Earth's climate?

### Biological Effects of Climate Change

Many organisms, especially plants that cannot disperse rapidly over long distances, may not be able to survive the rapid climate change projected to result from global warming. Furthermore, many habitats today are more fragmented than ever (see Concept 43.3), further limiting the ability of many organisms to migrate.

---

## Scientific Skills Exercise

### Graphing Cyclic Data

**How Does the Atmospheric $CO_2$ Concentration Change During a Year and from Decade to Decade?** The blue curve in Figure 43.25 shows how the concentration of $CO_2$ in Earth's atmosphere has changed over a span of more than 50 years. For each year in that span, two data points are plotted, one in May and one in November. A more detailed picture of the change in $CO_2$ concentration can be obtained by looking at measurements made at more frequent intervals. In this exercise, you'll graph monthly $CO_2$ concentrations for each of three one-year periods.

**Data from the Study** The data in the table below are average $CO_2$ concentrations (in parts per million) at the Mauna Loa monitoring station for each month in 1990, 2000, and 2010.

| Month | 1990 | 2000 | 2010 |
|---|---|---|---|
| January | 353.79 | 369.25 | 388.45 |
| February | 354.88 | 369.50 | 389.82 |
| March | 355.65 | 370.56 | 391.08 |
| April | 356.27 | 371.82 | 392.46 |
| May | 359.29 | 371.51 | 392.95 |
| June | 356.32 | 371.71 | 392.06 |
| July | 354.88 | 369.85 | 390.13 |
| August | 352.89 | 368.20 | 388.15 |
| September | 351.28 | 366.91 | 386.80 |
| October | 351.59 | 366.91 | 387.18 |
| November | 353.05 | 366.99 | 388.59 |
| December | 354.27 | 369.67 | 389.68 |

**Data from** National Oceanic & Atmospheric Administration, Earth System Research Laboratory, Global Monitoring Division.

**INTERPRET THE DATA**

1. Plot the data for each of the three years on one graph (producing three curves). Select a type of graph that is appropriate for these data, and choose a vertical-axis scale that allows you to clearly see the patterns of $CO_2$ concentration changes, both during each year and from decade to decade. (For additional information about graphs, see the Scientific Skills Review in Appendix F and in the Study Area in MasteringBiology.)

2. Within each year, what is the pattern of change in $CO_2$ concentration? Why might this pattern have occurred?

3. The measurements taken at Mauna Loa represent average atmospheric $CO_2$ concentrations for the Northern Hemisphere. Suppose you could measure $CO_2$ concentrations under similar conditions in the Southern Hemisphere. What pattern would you expect to see in those measurements over the course of a year? Explain.

4. In addition to the changes within each year, what changes in $CO_2$ concentration occurred between 1990 and 2010? Calculate the average $CO_2$ concentration for the 12 months of each year. By what percentage did this average change from 1990 to 2000 and from 1990 to 2010?

(MB) A version of this Scientific Skills Exercise can be assigned in MasteringBiology.

▲ A researcher sampling the air at the Mauna Loa monitoring station, Hawaii.

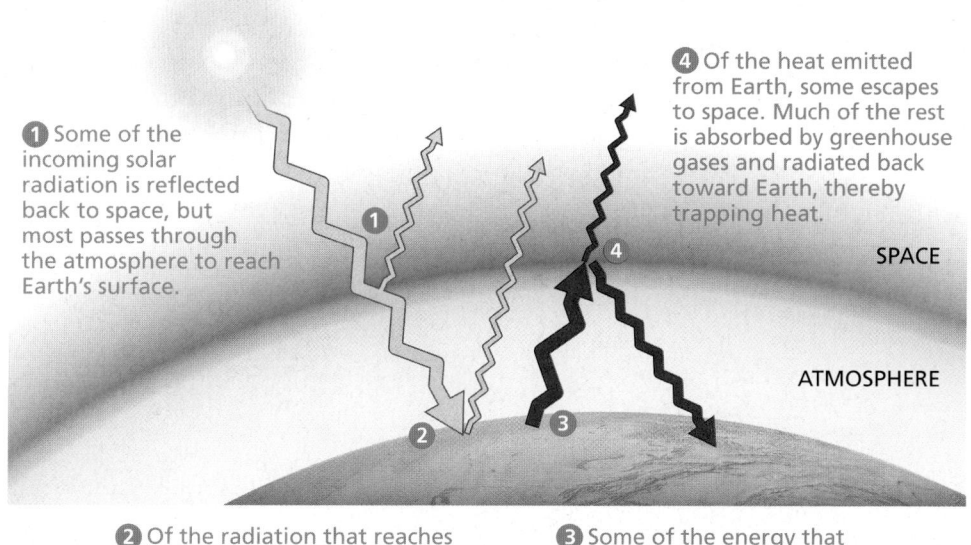

① Some of the incoming solar radiation is reflected back to space, but most passes through the atmosphere to reach Earth's surface.

④ Of the heat emitted from Earth, some escapes to space. Much of the rest is absorbed by greenhouse gases and radiated back toward Earth, thereby trapping heat.

SPACE

ATMOSPHERE

② Of the radiation that reaches Earth's surface, some is reflected back to space. Much of the rest is absorbed, warming Earth's surface.

③ Some of the energy that warms the surface is then emitted from Earth as heat.

▲ **Figure 43.26 The greenhouse effect.** Carbon dioxide and other greenhouse gases in the atmosphere absorb heat emitted from Earth's surface and then radiate much of that heat back to Earth.

One way to predict the possible effects of future climate change on geographic ranges is to look back at the changes that have occurred since the last ice age ended. Until about 16,000 years ago, continental glaciers covered vast regions across much of North America and Eurasia. As the climate warmed and the glaciers retreated, tree distributions expanded northward. Fossil pollen shows that while some species moved northward rapidly, others moved more slowly, with their range expansion lagging several thousand years behind the shift in suitable habitat.

Will plants and other species be able to keep up with the much more rapid warming projected for this century? Consider the American beech, *Fagus grandifolia*. Ecological models predict that the northern limit of the beech's range may move 700–900 km northward in the next century, and its southern range limit will shift even more. The current and predicted ranges of this species under two different climate-change scenarios are shown in **Figure 43.27**. If these predictions are even approximately correct, the beech's range must shift 7–9 km northward per year to keep pace with the warming climate. However, since the end of the last ice age, the beech has moved at a rate of only 0.2 km per year. Without human help in moving to new habitats, species such as the American beech may have much smaller ranges or even become extinct.

In fact, the climate change that has *already* occurred has had wide-ranging biological effects. For example, hundreds of terrestrial, marine, and freshwater organisms have shifted their geographic ranges as the climate has changed. In Europe, for instance, the northern range limits of 22 of 35 butterfly species studied have shifted farther north by 35–240 km as the climate has warmed over the last 100 years. In western

North America, many plant species have moved to lower elevations, in this case, apparently in response to decreased rain and snow at higher elevations. Other research shows that a Pacific diatom, *Neodenticula seminae*, recently has colonized the Atlantic Ocean for the first time in 800,000 years. As Arctic sea ice has receded in the past decade, the increased flow of water from the Pacific has swept these diatoms around Canada and into the Atlantic, where they quickly became established. In these and many other such cases, when climate change causes a species to expand its range into a new geographic area, other organisms living there may be harmed. **Figure 43.28** shows an example of this and other effects of climate change.

The ecosystems where the climate has changed the most are those in the far north, particularly northern coniferous forests and tundra. As snow and ice melt and uncover darker, more absorptive surfaces, these systems reflect less radiation back to the atmosphere and warm further (see Figure 43.26). Arctic sea ice in the summer of 2012 covered the smallest area on record. Climate models suggest that there may be no summer ice there within a few decades, decreasing habitat for polar bears, seals, and seabirds. In addition, over the past 30 years, some Arctic regions have switched from being a $CO_2$ *sink* (absorbing more $CO_2$ from the atmosphere than they release to the atmosphere) to a $CO_2$ *source* (releasing more $CO_2$ than they absorb)—a worrisome change that could contribute to further climate warming.

Coniferous forests in western North America have also been hard hit, in this case by a combination of higher temperatures, decreased winter snowfall, and a lengthening of the summer dry

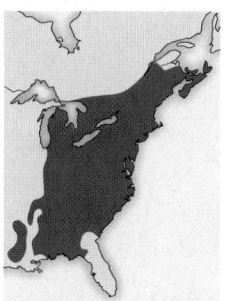 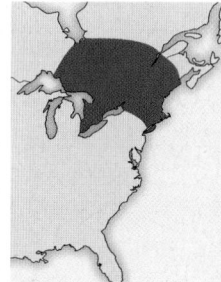 

**(a) Current range**

**(b) 4.5°C warming over next century**

**(c) 6.5°C warming over next century**

▲ **Figure 43.27 Current range and predicted range for the American beech under two climate-change scenarios.**

[?] *The predicted range in each scenario is based on climate factors alone. What other factors might alter the distribution of this species?*

MAKE CONNECTIONS

# Climate Change Has Effects at All Levels of Biological Organization

The burning of fossil fuels by humans has caused atmospheric concentrations of carbon dioxide and other greenhouse gases to rise dramatically (see Figure 43.25). This, in turn, is changing Earth's climate: The planet's average temperature has increased by about 1°C since 1900, and extreme weather events are occurring more often in some regions of the globe. How are these changes affecting life on Earth today?

## Effects on Cells

Temperature affects the rates of enzymatic reactions (see Figure 6.16), and as a result, the rates of DNA replication, cell division, and other key processes in cells are affected by rising temperatures.

Global warming and other aspects of climate change have also impaired some organisms' defense responses at the cellular level. For example, in the vast coniferous forests of western North America, climate change has reduced the ability of pine trees to defend themselves against attack by the mountain pine beetle (*Dendroctonus ponderosae*).

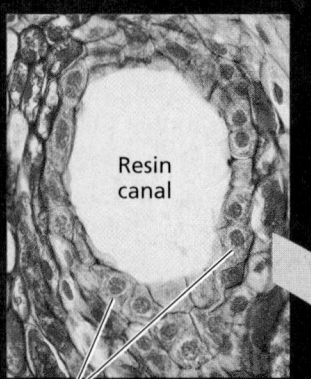

◄ Pine defenses include specialized resin cells that secrete a sticky substance (resin) that can entrap and kill mountain pine beetles. Resin cells produce less resin in trees that are stressed by rising temperatures and drought conditions.

Resin canal

Resin cells   |—100 μm—|

► When beetles overwhelm a tree's cellular defenses, they produce large numbers of offspring that tunnel through the wood, causing extensive damage. Rising temperatures have shortened how long it takes beetles to mature and reproduce, resulting in even more beetles. The beetles can also infect the tree with a harmful fungus, which appears as blue stains on the wood.

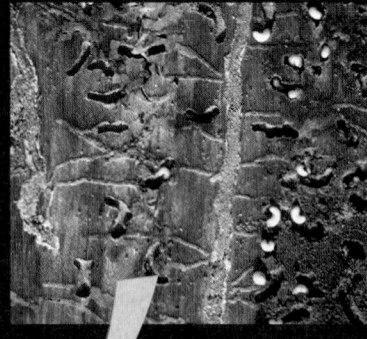

◄ This aerial view shows the scope of destruction in one North American forest due to mountain pine beetles; dead trees appear orange and red.

## Effects on Individual Organisms

Organisms must maintain relatively constant internal conditions (see Concept 32.3); for example, an individual will die if its body temperature becomes too high. Global warming has increased the risk of overheating in some species, leading to reduced food intake and reproductive failure.

For instance, an American pika (*Ochotona princeps*) will die if its body temperature rises just 3°C above its resting temperature—and this can happen quickly in regions where climate change has already caused significant warming.

► As summer temperatures have risen, American pikas are spending more time in their burrows to escape the heat. Thus, they have less time to forage for food. Lack of food has caused mortality rates to increase and birth rates to drop. Pika populations have dwindled, some to the point of extinction. (See Figure 1.11 for another example.)

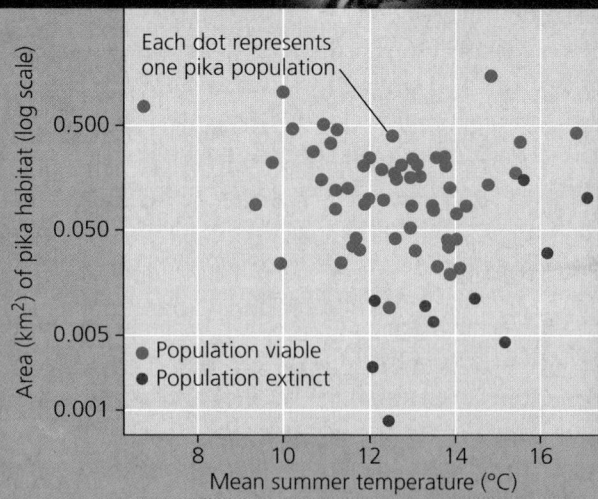

▲ This graph represents conditions in 2015 at 67 sites that previously supported a pika population; the populations at 10 of these sites had become extinct. Most extinctions occurred at sites with high summer temperatures and a small area of pika habitat. As temperatures continue to increase, more extinctions are expected.

Climate change has caused some populations to increase in size, while others have declined (see Concepts 1.1 and 36.1). In particular, as the climate has changed, some species have adjusted when they grow, reproduce, or migrate—but others have not, causing their populations to face food shortages and reduced survival or reproductive success.

In one example, researchers have documented a link between rising temperatures and declining populations of caribou (*Rangifer tarandus*) in the Arctic.

Climate affects where species live (see Figure 40.8). Climate change has caused hundreds of species to move to new locations, in some cases leading to dramatic changes in ecological communities. Climate change has also altered primary production (see Figure 25.24) and nutrient cycling in ecosystems.

In the example we discuss here, rising temperatures have enabled a sea urchin to invade southern regions along the coast of Australia, causing catastrophic changes to marine communities there.

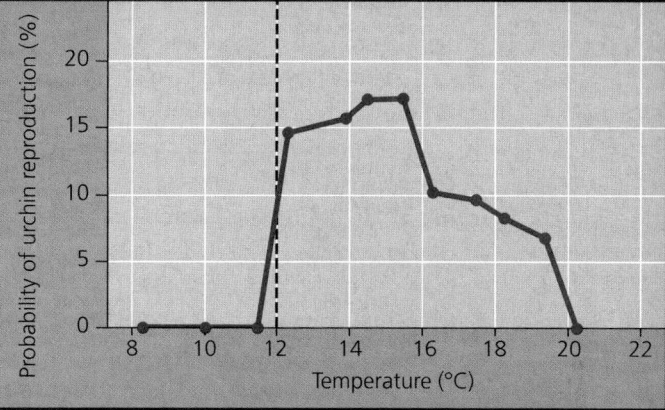

▲ The sea urchin *Centrostephanus rodgersii* requires water temperatures above 12˚C to reproduce successfully, as shown in this graph. As ocean waters rise above this critical temperature, the urchin has been able to expand its range to the south, destroying kelp beds as it moves into new regions.

▲ Caribou populations migrate north in the spring to give birth and to eat sprouting plants.

▶ Alpine chickweed is an early-flowering plant on which caribou depend.

▲ As the urchin has expanded its range to the south, it has destroyed high-diversity kelp communities, leaving bare regions called "urchin barrens" in its wake.

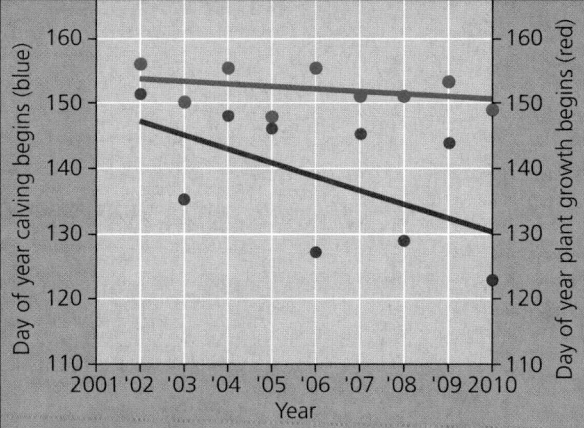

▲ As the climate has warmed, the plants on which caribou depend have emerged earlier in the spring. Caribou have not made similar changes in the timing of when they migrate and give birth. As a result, there is a shortage of food, and caribou offspring production has dropped fourfold.

**MAKE CONNECTIONS** *In addition to causing climate change, rising concentrations of $CO_2$ are causing ocean acidification (see Figure 2.24). Explain how ocean acidification can affect individual organisms, and how that, in turn, can cause dramatic changes in ecological communities.*

period. As a result, since the later half of the 20th century, otherwise healthy forests have experienced a steady increase in the percentage of trees that die each year. Higher temperatures and more frequent droughts also increase the likelihood of fires. In boreal forests of western North America and Russia, for example, fires have burned twice the usual area in recent decades, again leading to widespread tree mortality. As the climate continues to warm, this will likely bring other changes in the geographic distribution of precipitation, such as making agricultural areas of the central United States much drier.

Climate change has affected many other ecosystems as well. In Europe and Asia, for example, plants are producing leaves earlier in the spring, while in tropical regions, the growth and survival of some species of coral have declined as water temperatures have warmed. Still other effects of climate change are discussed in Figure 43.28. A key take-home message from these examples is that a given effect of climate change may, in turn, cause a series of other biological changes. The exact nature of such cascading effects can be hard to predict, but it is clear that the more our planet warms, the more severely its ecosystems will be affected.

### Climate Change Solutions

We will need many approaches to slow global warming and other aspects of climate change. Quick progress can be made by using energy more efficiently and by replacing fossil fuels with renewable solar and wind power and, more controversially, with nuclear power. Today, coal, gasoline, wood, and other organic fuels remain central to industrialized societies and cannot be burned without releasing $CO_2$. Stabilizing $CO_2$ emissions will require concerted international effort and changes in both personal lifestyles and industrial processes. International negotiations have yet to reach a global consensus on how to reduce greenhouse gas emissions.

Another important approach to slowing climate change is to reduce deforestation around the world, particularly in the tropics. Deforestation currently accounts for about 10% of greenhouse gas emissions. Recent research shows that paying countries *not* to cut forests could decrease the rate of deforestation by half within 10 to 20 years. Reduced deforestation would not only slow the buildup of greenhouse gases in our atmosphere, but also sustain native forests and preserve biodiversity, a positive outcome for all.

#### CONCEPT CHECK 43.4

1. How can the addition of excess mineral nutrients to a lake threaten its fish population?
2. **MAKE CONNECTIONS** There are vast stores of organic matter in the soils of northern coniferous forests and tundra around the world. Suggest an explanation for why scientists who study global warming are closely monitoring these stores (see Figures 7.2 and 42.13).

**For suggested answers, see Appendix A.**

# The human population is no longer growing exponentially but is still increasing rapidly

Global environmental problems, such as climate change, arise from the intersection of two factors. One is the growing amount of goods and resources that each of us consumes. The other is the increasing size of the human population, which has grown at an unprecedented rate in the last few centuries. No population can grow indefinitely, however. In this section, we'll apply ecological concepts to the specific case of the human population.

## The Global Human Population

The human population has grown explosively over the last four centuries **(Figure 43.29)**. In 1650, about 500 million people inhabited Earth. Our population doubled to 1 billion within the next two centuries, doubled again to 2 billion by 1930, and doubled still again by 1975 to more than 4 billion. Notice that the time it took our population to double in size decreased from 200 years in 1650 to just 45 years in 1930. Thus, historically our population has grown even *faster* than in exponential growth, which assumes a constant rate of increase (see Figure 40.19).

The global population is now more than 7.2 billion people and is increasing by about 78 million each year. This translates into more than 200,000 people each day, the equivalent of adding a city the size of Amarillo, Texas. At this rate, it takes only about four years to add the equivalent of another United States to the world population. Ecologists predict a population of 8.1–10.6 billion people on Earth by the year 2050.

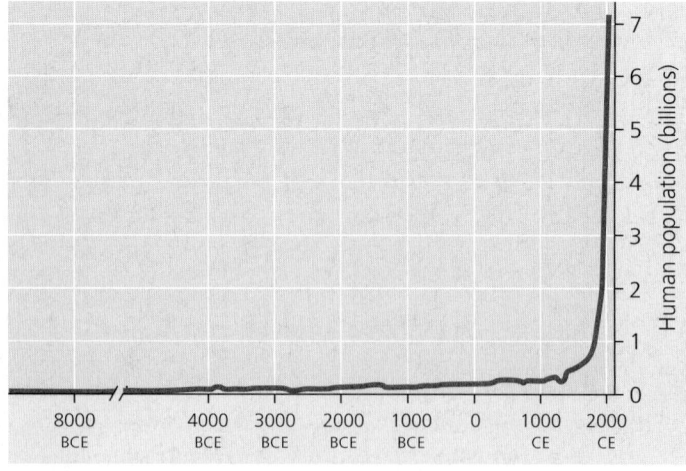

▲ Figure 43.29 **Human population growth (data as of 2014).** The global human population has grown almost continuously throughout history, but it skyrocketed after the Industrial Revolution. Though it is not apparent at this scale, the rate of population growth has slowed in recent decades, mainly as a result of decreased birth rates throughout the world.

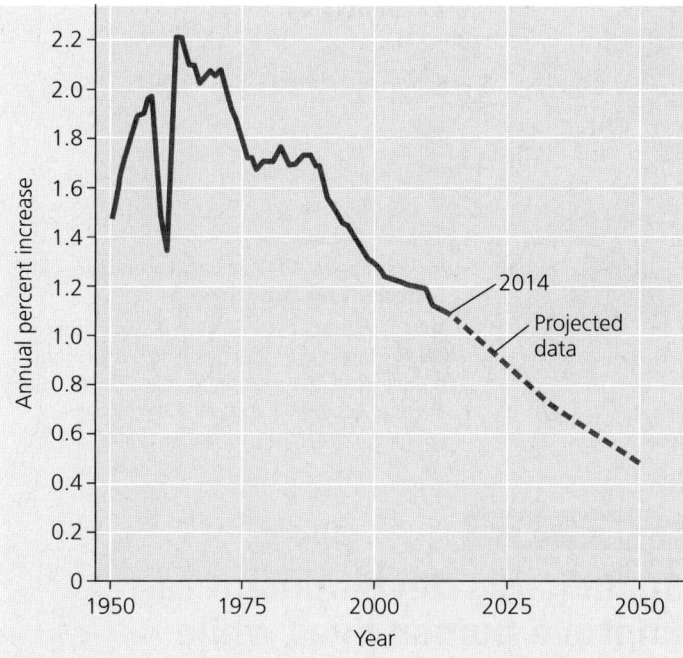

▲ **Figure 43.30 Annual percent increase in the global human population (data as of 2014).** The sharp dip in the 1960s is due mainly to a famine in China in which about 60 million people died.

Though the global population is still growing, the *rate* of growth began to slow during the 1960s **(Figure 43.30)**. The annual rate of increase in the global population peaked at 2.2% in 1962 but was only 1.1% in 2014. Current models project a growth rate of 0.5% by 2050, which would still add 45 million more people per year if the population climbs to a projected 9 billion. The reduction in growth rate over the past four decades shows that the human population is now growing more slowly than expected in exponential growth. This change resulted from fundamental shifts in population dynamics due to diseases, including AIDS, and to voluntary population control.

The growth rates of individual nations vary with their degree of industrialization. In industrialized nations, populations are near equilibrium, with growth rates of about 0.1% per year and reproductive rates near the replacement level (total fertility rate = 2.1 children per female). In countries such as Canada, Germany, Japan, and the United Kingdom, total reproductive rates are in fact *below* replacement. These populations will eventually decline if there is no immigration and if the birth rate does not change. In fact, the population is already declining in many eastern and central European countries.

In contrast, most of the current global population growth occurs in less industrialized nations, where about 80% of the world's people now live. Countries such as Afghanistan, Uganda, and Jordan had populations that grew by more than 3% *per year* between 2005 and 2010. Although death rates have declined rapidly since 1950 in many less industrialized countries, birth rates have declined substantially in only some of

them. The fall in birth rate has been most dramatic in China. Largely because of the Chinese government's strict one-child policy, the expected total fertility rate (children per woman per lifetime) decreased from 5.9 in 1970 to 1.6 in 2011. The transition to lower birth rates has also been rapid in some African countries, though birth rates remain high in most of sub-Saharan Africa.

A unique feature of human population growth is our ability to control family sizes through planning and voluntary contraception. Social change and the rising educational and career aspirations of women in many cultures encourage women to delay marriage and postpone reproduction. Delayed reproduction helps to decrease population growth rates and to move a society toward zero population growth under conditions of low birth rates and low death rates. However, there is a great deal of disagreement as to how much support should be provided for global family planning efforts.

## Global Carrying Capacity

No ecological question is more important than the future size of the human population. As we noted earlier, population ecologists project a global population of approximately 8.1–10.6 billion people by the year 2050. Thus, an estimated 1–4 billion people will be added to the population in the next four decades because of the momentum of population growth. But just how many humans can the biosphere support? Will the world be overpopulated in 2050? Is it *already* overpopulated?

### Estimates of Carrying Capacity

Estimates of the human carrying capacity of Earth have varied from less than 1 billion to more than 1,000 billion (1 trillion), with an average of 10–15 billion. Carrying capacity is difficult to estimate, and scientists use different methods to produce their estimates. Some researchers use curves like that produced by the logistic equation (see Figure 40.19) to predict the future maximum of the human population. Others generalize from existing "maximum" population density and multiply this number by the area of habitable land. Still others base their estimates on a single limiting factor, such as food, and consider variables such as the amount of farmland, the average yield of crops, the prevalent diet—vegetarian or meat based—and the number of calories needed per person per day.

### Limits on Human Population Size

A more comprehensive approach to estimating the carrying capacity of Earth is to recognize that humans have multiple constraints: We need food, water, fuel, building materials, and other resources, such as clothing and transportation. The **ecological footprint** concept summarizes the aggregate land and water area required by each person, city, or nation to produce all the resources it consumes and to absorb all the waste it generates. What is a sustainable ecological footprint for the human population? One way to estimate this is to add up all

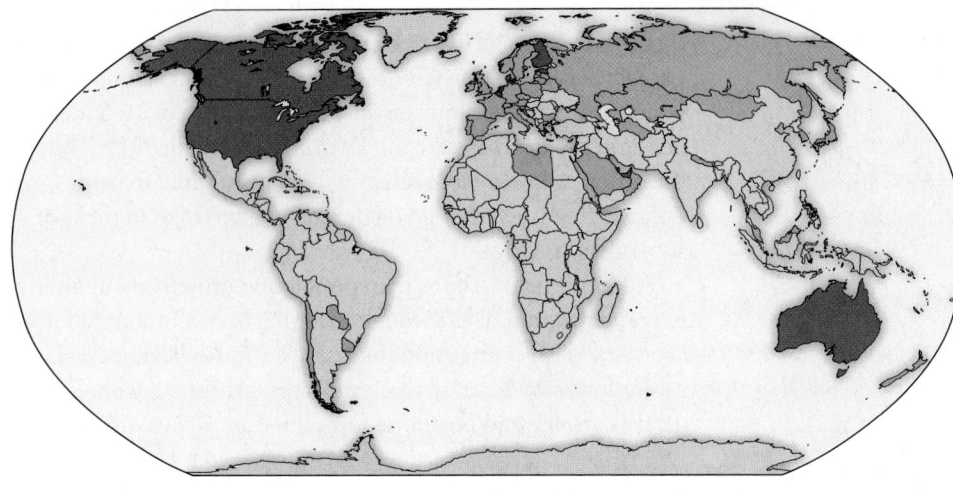

Per capita ecological
footprint (gha)

0–3

3–6

>6

Insufficient data

◄ **Figure 43.31 Per capita ecological
footprint by country.**

**?** *Earth has a total of 11.9 billion global
hectares of productive land. How many
people could Earth support sustainably if the
average ecological footprint were 8 global
hectares per person (as in the United States)?*

the ecologically productive land on the planet and divide by
the population. Typically, this estimate is made using *global
hectares*, where a global hectare (gha) is defined as a hectare
of land or water that is of world-average biological productivity (1 hectare = 2.47 acres). This calculation yields an allot-
ment of 1.7 gha per person—the benchmark for comparing
actual ecological footprints. Anyone who consumes resources
that require more than 1.7 gha to produce is using an unsustainable share of Earth's resources, as is the case for the citizens of many countries **(Figure 43.31)**. For example, a typical
ecological footprint for a person in the United States is 8 gha.

Our impact on the planet can also be assessed using currencies
other than area, such as energy use. Average energy use differs
greatly in developed and developing nations. A typical person in
the United States, Canada, or Norway consumes roughly 30 times
the energy that a person in central Africa does. Moreover, fossil
fuels, such as oil, coal, and natural gas, are the source of 80% or
more of the energy used in most developed nations. This unsustainable reliance on fossil fuels is changing Earth's climate and increasing the amount of waste that each of us produces. Ultimately,
the combination of resource use per person and population density determines our global ecological footprint.

How many people our planet can sustain depends on the
quality of life each of us enjoys and the distribution of wealth
across people and nations, topics of great concern and political
debate. We can decide whether zero population growth will be
attained through social changes based on human choices or,
instead, through increased mortality due to resource limitation, plagues, war, and environmental degradation.

**CONCEPT CHECK 43.5**

1. How has the growth rate and number of people added to the
   human population each year changed in recent decades?
2. **WHAT IF?** Type "personal ecological footprint calculator"
   into a search engine and use one of the resulting calculators
   to estimate your footprint. Is your current lifestyle sustainable?
   If not, what choices can you make to reduce your footprint?

For suggested answers, see Appendix A.

CONCEPT 43.6

# Sustainable development can improve human lives while conserving biodiversity

With the increasing loss and fragmentation of habitats,
changes in Earth's physical environment and climate, and
increasing human population, we face difficult trade-offs in
managing the world's resources. Preserving all habitat patches
isn't feasible, so biologists must help societies set conservation priorities by identifying which habitat patches are most
crucial, while likewise improving the quality of life for local
people. Ecologists use the concept of *sustainability* as a tool to
establish long-term conservation priorities.

## Sustainable Development

We need to understand the interconnections of the biosphere
if we are to protect species from extinction and improve the
quality of human life. To this end, many nations, scientific
societies, and other groups have embraced the concept of
**sustainable development**, economic development that meets
the needs of people today without limiting the ability of future
generations to meet their needs.

Achieving sustainable development is an ambitious goal. To
sustain ecosystem processes and stem the loss of biodiversity,
we must connect life science with the social sciences, economics, and the humanities. We must also reassess our personal
values. Those of us living in wealthier nations have a larger
ecological footprint than do people living in developing nations. By considering the long-term costs of consumption, we
can learn to value the natural processes that sustain us.

## Case Study: *Sustainable Development in Costa Rica*

The success of conservation in Costa Rica discussed earlier
has required a partnership between the national government,
nongovernment organizations (NGOs), and private citizens.
Many nature reserves established by individuals have been

recognized by the government as national wildlife reserves and given significant tax benefits. However, conservation and restoration of biodiversity make up only one facet of sustainable development; the other is improving the human condition.

How have the living conditions of the Costa Rican people changed as the country has pursued its conservation goals? Two fundamental indicators of living conditions are infant mortality rate and life expectancy. From 1930 to 2010, the infant mortality rate in Costa Rica declined from 170 to 9 per 1,000 live births; over the same period, life expectancy increased from about 43 years to 79 years. Another indicator is the literacy rate. The 2011 literacy rate in Costa Rica was 96%, compared to an average of 82% in the other six Central American countries. Such statistics show that living conditions in Costa Rica have improved greatly over the period in which the country has dedicated itself to conservation and restoration. While this result does not prove that conservation *causes* an improvement in human welfare, we can say with certainty that development in Costa Rica has attended to both nature *and* people.

## The Future of the Biosphere

Our modern lives are very different from those of early humans, who hunted and gathered to survive. Their reverence for the natural world is evident in early cave paintings of wildlife **(Figure 43.32a)** and in the stylized visions of life they sculpted from bone and ivory **(Figure 43.32b)**.

Our lives reflect remnants of our ancestral attachment to nature and the diversity of life—the concept of *biophilia*. We evolved in natural environments rich in biodiversity, and we still have an affinity for such settings **(Figure 43.32c and d)**. Indeed, our biophilia may be innate, an evolutionary product of natural selection acting on a brainy species whose survival depended on a close connection to the environment and a practical appreciation of plants and animals.

Our appreciation of life guides the field of biology today. We celebrate life by deciphering the genetic code that makes each species unique. We embrace life by using fossils and DNA to chronicle evolution through time. We preserve life through our efforts to classify and protect the millions of species on Earth. We respect life by using nature responsibly and reverently to improve human welfare.

Biology is the scientific expression of our desire to know nature. We are most likely to protect what we appreciate, and we are most likely to appreciate what we understand. By learning about the processes and diversity of life, we also become more aware of ourselves and our place in the biosphere. We hope this text has served you well in this lifelong adventure.

**(a)** Detail of animals in a 17,000-year-old cave painting, Lascaux, France

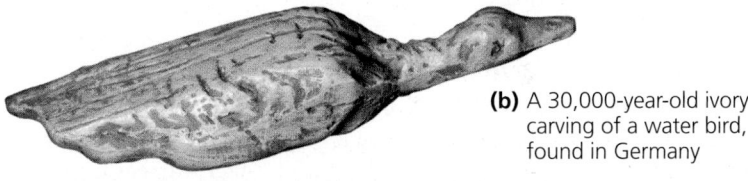

**(b)** A 30,000-year-old ivory carving of a water bird, found in Germany

**(c)** Nature lovers on a wildlife-watching expedition

▲ **Figure 43.32 Biophilia, past and present.**

### CONCEPT CHECK 43.6

1. What is meant by the term *sustainable development*?
2. How might biophilia influence us to conserve species and restore ecosystems?
3. **WHAT IF?** Suppose a new fishery is discovered, and you are put in charge of developing it sustainably. What ecological data might you want on the fish population? What criteria would you apply for the fishery's development?

For suggested answers, see Appendix A.

**(d)** A young biologist holding a songbird

# 43 Chapter Review

Go to **MasteringBiology**® for Assignments, the eText, and the Study Area with Animations, Activities, Key Terms, and Self-Tests.

## SUMMARY OF KEY CONCEPTS

### CONCEPT 43.1

**Human activities threaten Earth's biodiversity (pp. 907–912)**

• Biodiversity can be considered at three main levels:

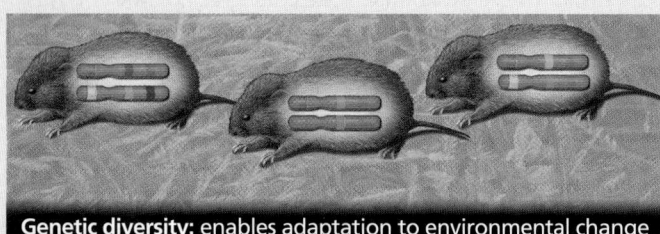

**Genetic diversity:** enables adaptation to environmental change

**Species diversity:** maintains communities and food webs

**Ecosystem diversity:** provides life-sustaining services

• Our biophilia enables us to recognize the value of biodiversity for its own sake. Other species also provide humans with food, fiber, medicines, and **ecosystem services**.
• Four major threats to biodiversity are habitat loss, **introduced species**, overharvesting, and global change.

**?** *Give at least three examples of key ecosystem services that nature provides for people.*

### CONCEPT 43.2

**Population conservation focuses on population size, genetic diversity, and critical habitat (pp. 912–915)**

• When a population drops below a **minimum viable population (MVP)** size, its loss of genetic variation due to nonrandom mating and genetic drift can trap it in an **extinction vortex**.
• The declining-population approach focuses on the environmental factors that cause decline, regardless of absolute population size. It follows a step-by-step conservation strategy.
• Conserving species often requires resolving conflicts between the habitat needs of endangered species and human demands.

**?** *Why is the minimum viable population size smaller for a genetically diverse population than for a less genetically diverse population?*

### CONCEPT 43.3

**Landscape and regional conservation help sustain biodiversity (pp. 915–919)**

• As habitat fragmentation increases and edges become more extensive, biodiversity tends to decrease. **Movement corridors** can promote dispersal and help sustain populations.
• **Biodiversity hot spots** are also hot spots of extinction and thus prime candidates for protection. The **zoned reserve** model recognizes that conservation efforts often involve working in landscapes that are greatly affected by human activity.

**?** *Give two examples that show how habitat fragmentation can harm species in the long term.*

### CONCEPT 43.4

**Earth is changing rapidly as a result of human actions (pp. 919–926)**

• Agriculture removes plant nutrients from ecosystems, so large supplements are usually required. The nutrients in fertilizer can pollute groundwater and surface water, where they can stimulate excess algal growth (eutrophication).
• The release of toxic wastes and pharmaceuticals has polluted the environment with harmful substances that often persist for long periods and become increasingly concentrated in successively higher trophic levels of food webs (**biological magnification**).
• Because of the burning of fossil fuels and other human activities, the atmospheric concentration of $CO_2$ and other greenhouse gases has been steadily increasing. These increases have caused **climate change**, including significant global warming and changing patterns of precipitation. Climate change has already affected many ecosystems.

**?** *In the face of biological magnification of toxins, is it healthier to feed at a lower or higher trophic level? Explain.*

### CONCEPT 43.5

**The human population is no longer growing exponentially but is still increasing rapidly (pp. 926–928)**

• Since about 1650, the global human population has grown exponentially, but within the last 50 years, the rate of growth has fallen by half. While some nations' populations are growing rapidly, those of others are stable or declining in size.
• The carrying capacity of Earth for humans is uncertain. **Ecological footprint** is the aggregate land and water area needed to produce all the resources a person or group of people consume and to absorb all of their waste. With a world population of more than 7 billion people, we are already using many resources in an unsustainable manner.

**?** *How do humans differ from other species in the ability to "choose" a carrying capacity for their environment?*

### CONCEPT 43.6

**Sustainable development can improve human lives while conserving biodiversity (pp. 928–929)**

• The goal of the Sustainable Biosphere Initiative is to acquire the ecological information needed for the development, management, and conservation of Earth's resources.

- Costa Rica's success in conserving tropical biodiversity has involved a partnership among the government, other organizations, and private citizens. Human living conditions in Costa Rica have improved along with ecological conservation.
- By learning about biological processes and the diversity of life, we become more aware of our close connection to the environment and the value of other organisms that share it.

**?** *Why is sustainability such an important goal for conservation biologists?*

## TEST YOUR UNDERSTANDING

**SELF-TEST**

### Level 1: Knowledge/Comprehension

goo.gl/CRZjvS

**1.** One characteristic that distinguishes a population in an extinction vortex from most other populations is that
   (A) it is a rare, top-level predator.
   (B) its effective population size is much lower than its total population size.
   (C) its genetic diversity is very low.
   (D) it is not well adapted to edge conditions.

**2.** The main cause of the increase in the amount of $CO_2$ in Earth's atmosphere over the past 150 years is
   (A) increased worldwide primary production.
   (B) increased worldwide standing crop.
   (C) an increase in the amount of infrared radiation absorbed by the atmosphere.
   (D) the burning of larger amounts of wood and fossil fuels.

**3.** What is the single greatest threat to biodiversity?
   (A) overharvesting of commercially important species
   (B) habitat alteration, fragmentation, and destruction
   (C) introduced species that compete with native species
   (D) pollution of Earth's air, water, and soil

### Level 2: Application/Analysis

**4.** Which of the following is a consequence of biological magnification?
   (A) Toxic chemicals in the environment pose greater risk to top-level predators than to primary consumers.
   (B) Populations of top-level predators are generally smaller than populations of primary consumers.
   (C) The biomass of producers in an ecosystem is generally higher than the biomass of primary consumers.
   (D) Only a small portion of the energy captured by producers is transferred to consumers.

**5.** Which of the following strategies would most rapidly increase the genetic diversity of a population in an extinction vortex?
   (A) Establish a reserve that protects the population's habitat.
   (B) Introduce new individuals transported from other populations of the same species.
   (C) Sterilize the least fit individuals in the population.
   (D) Control populations of the endangered population's predators and competitors.

**6.** Of the following statements about protected areas that have been established to preserve biodiversity, which one is *not* correct?
   (A) About 25% of Earth's land area is now protected.
   (B) National parks are one of many types of protected areas.
   (C) Management of a protected area should be coordinated with management of the land surrounding the area.
   (D) It is especially important to protect biodiversity hot spots.

### Level 3: Synthesis/Evaluation

**7.** **DRAW IT** (a) Estimate the average $CO_2$ concentration in 1975 and in 2012 using the data provided in Figure 43.25. (b) On average, how rapidly did $CO_2$ concentrations increase (ppm/yr) from 1975 to 2012? (c) Estimate the approximate $CO_2$ concentration in 2100, assuming that the $CO_2$ concentration continues to rise as fast as it did from 1975 to 2012. (d) Draw a graph of average $CO_2$ concentration from 1975 to 2012 and then use a dashed line to extend the graph to the year 2100. (e) Identify the ecological factors and human decisions that might influence the actual rise in $CO_2$ concentration. (f) Discuss how additional scientific data could help societies predict this value.

**8.** **SCIENTIFIC INQUIRY**
   **DRAW IT** Suppose that you are managing a forest reserve, and one of your goals is to protect local populations of woodland birds from parasitism by the brown-headed cowbird. You know that female cowbirds usually do not venture more than about 100 m into a forest and that nest parasitism is reduced when woodland birds nest away from forest edges. The reserve you manage extends about 6,000 m from east to west and 3,000 m from north to south. It is surrounded by a deforested pasture on the west, an agricultural field for 500 m in the southwest corner, and intact forest everywhere else. You must build a road, 10 m by 3,000 m, from the north to the south side of the reserve and construct a maintenance building that will take up 100 $m^2$ in the reserve. Draw a map of the reserve, showing where you would put the road and the building to minimize cowbird intrusion along edges. Explain your reasoning.

**9.** **FOCUS ON EVOLUTION**
   The fossil record indicates that there have been five mass extinction events in the past 500 million years (see Concept 23.2). Many ecologists think we are currently entering a human-caused sixth mass extinction event. Briefly discuss the history of mass extinctions and the length of time it typically takes for species diversity to recover through the process of evolution. Explain why this should motivate us to slow the loss of biodiversity today.

**10.** **FOCUS ON INTERACTIONS**
   In a short essay (100–150 words), identify the factor or factors that you think may ultimately have the greatest effect in regulating the human population, and explain your reasoning.

**11.** **SYNTHESIZE YOUR KNOWLEDGE**

Big cats, such as the snow leopard (*Panthera uncia*) shown here and on the cover of the book, are one of the most endangered groups of mammals in the world. Based on what you've learned in this chapter, discuss some of the approaches you would use to help preserve them.

*For selected answers, see Appendix A.*

**NOTE:** Answers to Scientific Skills Exercises, Interpret the Data questions, and short-answer essay questions are available only for instructors in the Instructor Resources area of MasteringBiology.

## Chapter 1

### Concept Check 1.1
**1.** Examples: A molecule consists of *atoms* bonded together. Each organelle has an orderly arrangement of *molecules*. Photosynthetic plant cells contain *organelles* called chloroplasts. A tissue consists of a group of similar *cells*. Organs such as the heart are constructed from several *tissues*. A complex multicellular organism, such as a plant, has several types of *organs*, such as leaves and roots. A population is a set of *organisms* of the same species. A community consists of *populations* of the various species inhabiting a specific area. An ecosystem consists of a biological *community* along with the nonliving factors important to life, such as air, soil, and water. The biosphere is made up of all of Earth's *ecosystems*. **2.** (a) New properties emerge at successive levels of biological organization: Structure and function are correlated. (b) Life's processes involve the expression and transmission of genetic information. (c) Life requires the transfer and transformation of energy and matter. **3.** Some possible answers: *Organization (emergent properties)*: The ability of a human heart to pump blood requires an intact heart; it is not a capability of any of the heart's tissues or cells working alone. *Organization (structure and function)*: The strong, sharp teeth of a wolf are well suited to grasping and dismembering its prey. *Information*: Human eye color is determined by the combination of genes inherited from the two parents. *Energy and matter*: A plant, such as a grass, absorbs energy from the sun and transforms it into molecules that act as stored fuel. Animals can eat parts of the plant and use the food for energy to carry out their activities. *Interactions*: A mouse eats food, such as nuts or grasses, and deposits some of the food material as wastes (feces and urine). Construction of a nest rearranges the physical environment and may hasten degradation of some of its components. The mouse may also act as food for a predator. *Evolution*: All plants have chloroplasts, indicating their descent from a common ancestor.

### Concept Check 1.2
**1.** An address pinpoints a location by tracking from broader to narrower categories—a state, city, zip code, street, and building number. This is analogous to the groups-subordinate-to-groups structure of biological taxonomy. **2.** The naturally occurring heritable variation in a population is "edited" by natural selection because individuals with heritable traits better suited to the environment survive and reproduce more successfully than others. Over time, better-suited individuals persist and their percentage in the population increases, while less well-suited individuals become less prevalent—a type of population editing.
**3.**

Ancestral eukaryotes
— Plants
— Fungi
— Animals

### Concept Check 1.3
**1.** Inductive reasoning derives generalizations from specific cases; deductive reasoning predicts specific outcomes from general premises. **2.** Mouse coat color matches the environment for both beach and inland populations. **3.** Compared to a hypothesis, a scientific theory is usually more general and substantiated by a much greater amount of evidence. Natural selection is an explanatory idea that applies to all kinds of organisms and is supported by vast amounts of evidence. **4.** Science aims to understand natural phenomena and how they work, while technology involves application of scientific discoveries for a particular purpose or to solve a specific problem.

### Summary of Key Concepts Questions
**1.1** Finger movements rely on the coordination of the many structural components of the hand (muscles, nerves, bones, etc.), each of which is composed of elements from lower levels of biological *organization* (cells, molecules). The development of the hand relies on the genetic *information* encoded in chromosomes found in cells throughout the body. To power the finger movements that result in a text message, muscle and nerve cells require chemical *energy* that they transform in powering muscle contraction or in propagating nerve impulses. Texting is in essence communication, an *interaction* that conveys information between organisms, in this case of the same species. Finally, all of the anatomical and physiological features that allow the activity of texting are the outcome of a process of natural selection that resulted in the *evolution* of hands and of the mental facilities for use of language. **1.2** Ancestors of the beach mouse may have exhibited variations in their coat color. Because of the prevalence of visual predators, the better camouflaged (lighter) mice may have survived longer and been able to produce more offspring. Over time, a higher and higher proportion of individuals in the population would have had the adaptation of lighter fur that acted to camouflage the mouse. **1.3** Gathering and interpreting data are core activities in the scientific process, and they are affected by, and affect in turn, three other arenas of the

scientific process: exploration and discovery, community analysis and feedback, and societal benefits and outcomes.

### Test Your Understanding
**1.** B **2.** C **3.** B **4.** C **5.** D
**6.** Your figure should show the following: (1) For the biosphere, Earth with an arrow coming out of a tropical ocean; (2) for the ecosystem, a distant view of a coral reef; (3) for the community, a collection of reef animals and algae, with corals, fishes, some seaweed, and any other organisms you can think of; (4) for the population, a group of fish of the same species; (5) for the organism, one fish from your population; (6) for the organ, the fish's stomach (see Chapter 33 for help); (7) for a tissue, a group of similar cells from the stomach; (8) for a cell, one cell from the tissue, showing its nucleus and a few other organelles; (9) for an organelle, the nucleus, where most of the cell's DNA is located; and (10) for a molecule, a DNA double helix. Your sketches can be very rough!

## Chapter 2

### Figure Questions
**Figure 2.6** Atomic number = 12; 12 protons, 12 electrons; 3 electron shells; 2 valence electrons **Figure 2.15** The plant is submerged in water ($H_2O$), in which the $CO_2$ is dissolved. The sun's energy is used to make sugar, which is found in the plant and can act as food for the plant itself, as well as for animals that eat the plant. The oxygen ($O_2$) is present in the bubbles. **Figure 2.16** One possible answer is the following:

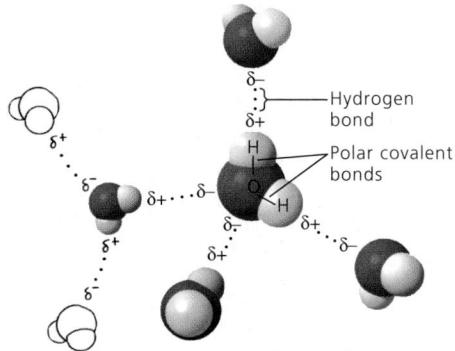

**Figure 2.20** Without hydrogen bonds, water would behave like other small molecules, and the solid phase (ice) would be denser than liquid water. The ice would sink to the bottom and would no longer insulate the body of water. All the water would eventually freeze because the average annual temperature at the South Pole is –50°C. The krill could not survive. **Figure 2.21** Heating the solution would cause the water to evaporate faster than it is evaporating at room temperature. At a certain point, there wouldn't be enough water molecules to dissolve the salt ions. The salt would start coming out of solution and re-forming crystals. Eventually, all the water would evaporate, leaving behind a pile of salt like the original pile. **Figure 2.24** By causing the loss of coral reefs, a decrease in the ocean's carbonate concentration would have a ripple effect on noncalcifying organisms. Some of these organisms depend on the reef structure for protection, while others feed on species associated with reefs.

### Concept Check 2.1
**1.** Yes, because an organism requires trace elements, even though only in small amounts **2.** A person with an iron deficiency will probably show fatigue and other effects of a low oxygen level in the blood. (The condition is called anemia and can also result from too few red blood cells or abnormal hemoglobin.)

### Concept Check 2.2
**1.** $^{15}_{7}N$ **2.** 9 electrons; two electron shells; 1 electron is needed to fill the valence shell. **3.** The elements in a row all have the same number of electron shells. In a column, all the elements have the same number of electrons in their valence shells.

### Concept Check 2.3
**1.** Each carbon atom has only three covalent bonds instead of the required four. **2.** The attraction between oppositely charged ions, forming ionic bonds **3.** If you could synthesize molecules that mimic these shapes, you might be able to treat diseases or conditions caused by the inability of affected individuals to synthesize such molecules.

### Concept Check 2.4
**1.** At equilibrium, the forward and reverse reactions occur at the same rate.
**2.** $C_6H_{12}O_6 + 6 O_2 \rightarrow 6 CO_2 + 6 H_2O + Energy$. Glucose and oxygen react to form carbon dioxide and water, releasing energy. We breathe in oxygen because we need it for this reaction to occur, and we breathe out carbon dioxide because it is a product of this reaction. (This reaction is called cellular respiration, and you will learn more about it in Chapter 7.)

## Concept Check 2.5

**1.** Hydrogen bonds hold neighboring water molecules together. This cohesion helps the chains of water molecules move upward against gravity in water-conducting cells as water evaporates from the leaves. Adhesion between water molecules and the walls of the water-conducting cells also helps counter gravity. **2.** As water freezes, it expands because water molecules move farther apart in forming ice crystals. When there is water in a crevice of a boulder, expansion due to freezing may crack the boulder. **3.** $10^5$, or 100,000 **4.** The covalent bonds of water molecules would not be polar, and water molecules would not form hydrogen bonds with each other. Water would therefore not have the unusual properties described in this chapter—such as cohesion, surface tension, high specific heat, high heat of vaporization, and versatility as a solvent.

### Summary of Key Concepts Questions

**2.1** Iodine (part of a thyroid hormone) and iron (part of hemoglobin in blood) are both trace elements, required in minute quantities. Calcium and phosphorus (components of bones and teeth) are needed by the body in much greater quantities.
**2.2**

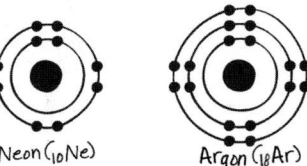

Neon ($_{10}$Ne)    Argon ($_{18}$Ar)

Both neon and argon are unreactive because they have completed valence shells. They do not have unpaired electrons that could participate in chemical bonds. **2.3** Electrons are shared equally between the two atoms in a nonpolar covalent bond. In a polar covalent bond, the electrons are drawn closer to the more electronegative atom. In the formation of ions, one or more electrons are completely transferred from one atom to a much more electronegative atom. **2.4** The concentration of products would increase as the added reactants were converted to products. Eventually, an equilibrium would again be reached in which the forward and reverse reactions were proceeding at the same rate and the relative concentrations of reactants and products returned to where they were before the addition of more reactants. **2.5** The polar covalent bonds of a water molecule allow it to form hydrogen bonds with other water molecules and other polar molecules as well. The sticking together of water molecules, called cohesion, and the sticking of water to other molecules, called adhesion, help water rise from the roots of plants to their leaves, among other biological benefits. Hydrogen bonding between water molecules is responsible for water's high specific heat (resistance to temperature change), which helps moderate temperature on Earth. Hydrogen bonding is also responsible for water's high heat of vaporization, which makes water useful for evaporative cooling. A lattice of stable hydrogen bonds in ice makes it less dense than liquid water, so that it floats, creating an insulating surface on bodies of water that allows organisms to live underneath. Finally, the polarity of water molecules resulting from their polar covalent bonds makes water an excellent solvent; polar and ionic atoms and molecules that are needed for life can exist in a dissolved state and participate in chemical reactions.

### Test Your Understanding

**1.** B  **2.** D  **3.** D  **4.** D  **5.** B  **6.** C  **7.** D  **8.** D
**9.**

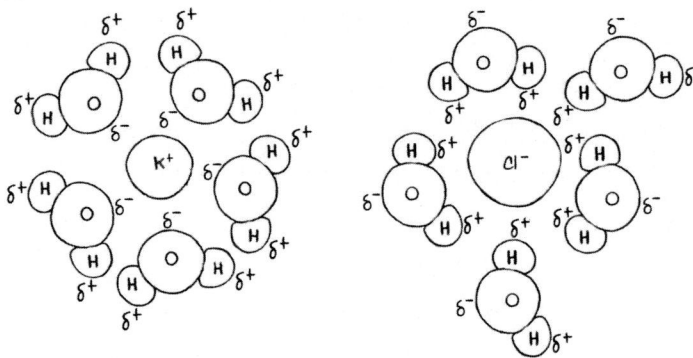

**10.** Both global warming and ocean acidification are caused by increasing levels of carbon dioxide in the atmosphere, the result of burning fossil fuels.

## Chapter 3

### Figure Questions
**Figure 3.5**

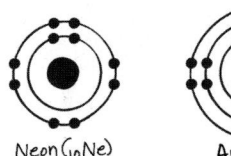

---

**Figure 3.9**

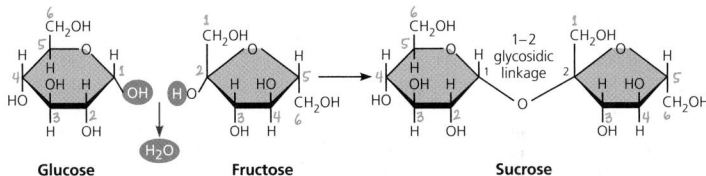

Note that the oxygen on carbon 5 lost its proton and that the oxygen on carbon 2, which used to be the carbonyl oxygen, gained a proton. Four carbons are in the fructose ring, and two are not. (The latter two carbons are attached to carbons 2 and 5, which are in the ring.) The fructose ring differs from the glucose ring, which has five carbons in the ring and one that is not. (Note that the orientation of this fructose molecule is flipped horizontally relative to that of the one in Figure 3.10.)
**Figure 3.10**

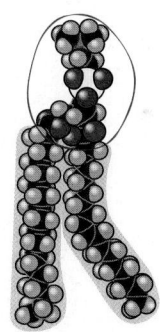

In sucrose, the linkage is called a 1–2 glycosidic linkage because the number 1 carbon in the left monosaccharide (glucose) is linked to the number 2 carbon in the right monosaccharide (fructose). (Note that the fructose molecule is oriented differently from the glucose molecules in Figures 3.9 and 3.10, and from the fructose shown in the answer for Figure 3.9, above. In Figure 3.10 and here, carbon 2 of fructose is close to carbon 1 of glucose.)
**Figure 3.15**

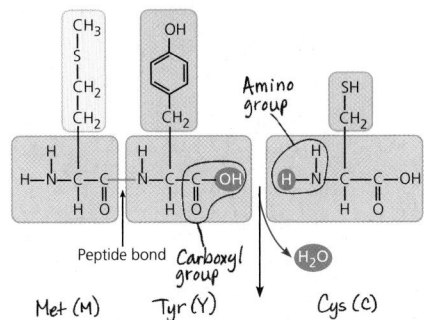

**Figure 3.19**

**Figure 3.23** The R group of glutamic acid is acidic and hydrophilic, whereas that of valine is nonpolar and hydrophobic. Therefore, it is unlikely that valine can participate in the same intramolecular interactions that glutamic acid can. A change in these interactions would be expected to (and does) cause a disruption of molecular structure. **Figure 3.30** Using a genomics approach allows us to use gene sequences to identify species and to learn about evolutionary relationships among any two species. This is because all species are related by their evolutionary history, and the evidence is

in the DNA sequences. Proteomics—looking at proteins that are expressed—allows us to learn about how organisms or cells are functioning at a given time or in an association with another species.

**Concept Check 3.1**
**1.** Both consist largely of hydrocarbon chains, which provide fuel—gasoline for engines and fats for plant embryos and animals. Reactions of both types of molecule release energy.   **2.** The forms of $C_4H_{10}$ in (b) are structural isomers, as are the butenes (forms of $C_4H_8$) in (c).   **3.** It has both an amino group (—$NH_2$), which makes it an amine, and a carboxyl group (—COOH), which makes it a carboxylic acid.   **4.** A chemical group that can act as a base (by picking up $H^+$) has been replaced with a group that can act as an acid, increasing the acidic properties of the molecule. The shape of the molecule would also change, likely changing the molecules with which it can interact.

$$\begin{array}{c} O \quad\quad H \quad\quad O \\ \parallel \quad\quad | \quad\quad \parallel \\ C-C-C \\ / \quad | \quad \backslash \\ HO \quad CH_2 \quad OH \\ | \\ SH \end{array}$$

The original cysteine molecule has an asymmetric carbon in the center. After replacement of the amino group with a carboxyl group, this carbon is no longer asymmetric.

**Concept Check 3.2**
**1.** Nine, with one water molecule required to hydrolyze each connected pair of monomers   **2.** The amino acids in the fish protein must be released in hydrolysis reactions and incorporated into other proteins in dehydration reactions.

**Concept Check 3.3**
**1.** $C_3H_6O_3$   **2.** $C_{12}H_{22}O_{11}$   **3.** The antibiotic treatment is likely to have killed the cellulose-digesting prokaryotes in the cow's stomach. The absence of these prokaryotes would hamper the cow's ability to obtain energy from food and could lead to weight loss and possibly death. Thus, prokaryotic species are reintroduced, in appropriate combinations, in the gut culture given to treated cows.

**Concept Check 3.4**
**1.** Both have a glycerol molecule attached to fatty acids. The glycerol of a fat has three fatty acids attached, whereas the glycerol of a phospholipid is attached to two fatty acids and one phosphate group.   **2.** Human sex hormones are steroids, a type of compound that is hydrophobic and thus classified as a lipid.   **3.** The oil droplet membrane could consist of a single layer of phospholipids rather than a bilayer, because an arrangement in which the hydrophobic tails of the membrane phospholipids were in contact with the hydrocarbon regions of the oil molecules would be more stable.

**Concept Check 3.5**
**1.** The function of a protein is a consequence of its specific shape, which is lost when a protein becomes denatured.   **2.** Secondary structure involves hydrogen bonds between atoms of the polypeptide backbone. Tertiary structure involves interactions between atoms of the side chains of the amino acid monomers.   **3.** These are all nonpolar, hydrophobic amino acids, so you would expect this region to be located in the interior of the folded polypeptide, where it would not contact the aqueous environment inside the cell.

**Concept Check 3.6**
**1.**

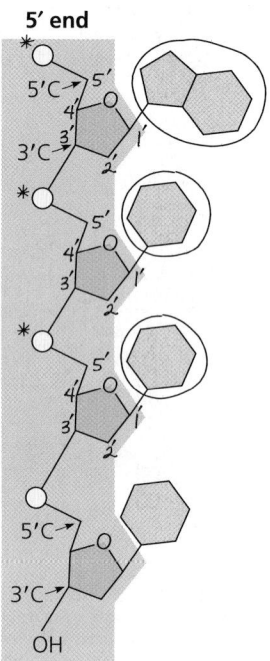

**5′ end**

**3′ end**

**2.**

$$5'-T\ A\ G\ G\ C\ C\ T-3'$$
$$3'-A\ T\ C\ C\ G\ G\ A-5'$$

**Concept Check 3.7**
**1.** The DNA of an organism encodes all of its proteins, and proteins are the molecules that carry out the work of cells, whether an organism is unicellular or multicellular. By knowing the DNA sequence of an organism, scientists would be able to catalog the protein sequences as well.   **2.** Ultimately, the DNA sequence carries the information necessary to make the proteins that determine the traits of a particular species. Because the traits of the two species are similar, you would expect the proteins to be similar as well, and therefore the gene sequences should also have a high degree of similarity.

**Summary of Key Concepts Questions**
**3.1** The methyl group is nonpolar and not reactive. The other six groups are called functional groups because they can participate in chemical reactions. Except for the sulfhydryl group, these functional groups are hydrophilic; they increase the solubility of organic compounds in water.   **3.2** The polymers of large carbohydrates, proteins, and nucleic acids are built from three different types of monomers (monosaccharides, amino acids, and nucleotides, respectively).   **3.3** Both starch and cellulose are polymers of glucose, but the glucose monomers are in the α configuration in starch and the β configuration in cellulose. The glycosidic linkages thus have different geometries, giving the polymers different shapes and thus different properties. Starch is an energy-storage compound in plants; cellulose is a structural component of plant cell walls. Humans can hydrolyze starch to provide energy but cannot hydrolyze cellulose. Cellulose aids in the passage of food through the digestive tract.   **3.4** Lipids are not polymers because they do not exist as a chain of linked monomers. They are not considered macromolecules because they do not reach the giant size of many polysaccharides, proteins, and nucleic acids.   **3.5** A polypeptide, which may consist of hundreds of amino acids in a specific sequence (primary structure), has regions of coils and pleats (secondary structure), which are then folded into irregular contortions (tertiary structure) and may be noncovalently associated with other polypeptides (quaternary structure). The linear order of amino acids, with the varying properties of their side chains (R groups), determines what secondary and tertiary structures will form to produce a protein. The resulting unique three-dimensional shapes of proteins are key to their specific and diverse functions.   **3.6** The complementary base pairing of the two strands of DNA makes possible the precise replication of DNA every time a cell divides, ensuring that genetic information is faithfully transmitted. In some types of RNA, complementary base pairing enables RNA molecules to assume specific three-dimensional shapes that facilitate diverse functions.   **3.7** You would expect the human gene sequence to be most similar to that of the mouse (another mammal), and then to that of the fish (another vertebrate), and least similar to that of the fruit fly (an invertebrate).

**Test Your Understanding**
**1.** A   **2.** B   **3.** C   **4.** C   **5.** B   **6.** A   **7.** B   **8.** C   **9.** B
**10.**

|  | Monomers or Components | Polymer or larger molecule | Type of linkage |
|---|---|---|---|
| Carbohydrates | Monosaccharides | Polysaccharides | Glycosidic linkages |
| Fats | Fatty acids | Triacylglycerols | Ester linkages |
| Proteins | Amino acids | Polypeptides | Peptide bonds |
| Nucleic acids | Nucleotides | Polynucleotides | Phosphodiester linkages |

**11.**

Original strand    Complementary strand

# Chapter 4

## Figure Questions

**Figure 4.5** A phospholipid is a lipid consisting of a glycerol molecule joined to two fatty acids and one phosphate group. Together, the glycerol and phosphate end of the phospholipid form the "head," which is hydrophilic, while the hydrocarbon chains on the fatty acids form hydrophobic "tails." The presence in a single molecule of both a hydrophilic and a hydrophobic region makes the molecule ideal as the main building block of a cell or organelle membrane, which is a phospholipid bilayer. In this bilayer, the hydrophobic regions can associate with each other on the inside of the membrane, while the hydrophilic region of each can be in contact with the aqueous solution on either side. **Figure 4.8** The DNA in a chromosome dictates synthesis of a messenger RNA (mRNA) molecule, which then moves out to the cytoplasm. There, the information is used for the production, on ribosomes, of proteins that carry out cellular functions. **Figure 4.9** Any of the bound ribosomes (attached to the endoplasmic reticulum) could be circled, because any could be making a protein that will be secreted. **Figure 4.22** Each centriole has 9 sets of 3 microtubules, so the entire centrosome (two centrioles) has 54 microtubules. Each microtubule consists of a helical array of tubulin dimers (as shown in Table 4.1).

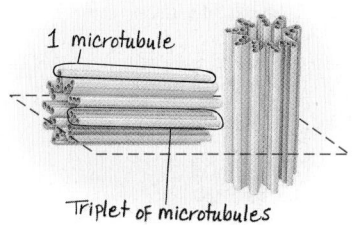

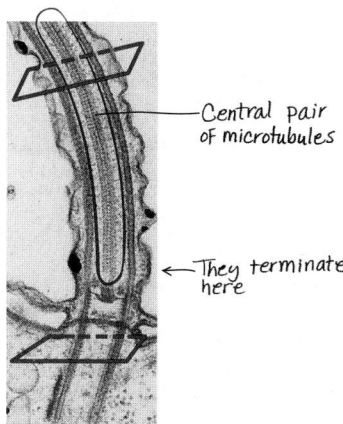

**Figure 4.23** The two central microtubules terminate above the basal body, so they aren't present at the level of the cross section through the basal body, indicated by the lower red rectangle in **(a)**.

## Concept Check 4.1

**1.** Stains used for light microscopy are colored molecules that bind to cell components, affecting the light passing through, while stains used for electron microscopy involve heavy metals that affect the beams of electrons. **2.** (a) Light microscope, (b) scanning electron microscope

## Concept Check 4.2

**1.** See Figure 4.7. **2.** This cell would have the same volume as the cells in columns 2 and 3 but proportionally more surface area than that in column 2 and less than that in column 3. Thus, the surface-to-volume ratio should be greater than 1.2 but less than 6. To obtain the surface area, you would add the area of the six sides (the top, bottom, sides, and ends): 125 + 125 + 125 + 125 + 1 + 1 = 502. The surface-to-volume ratio equals 502 divided by a volume of 125, or 4.0.

## Concept Check 4.3

**1.** Ribosomes in the cytoplasm translate the genetic message, carried from the DNA in the nucleus by mRNA, into a polypeptide chain. **2.** Nucleoli consist of DNA and the ribosomal RNA (rRNA) made according to its instructions, as well as proteins imported from the cytoplasm. Together, the rRNA and proteins are assembled into large and small ribosomal subunits. (These are exported through nuclear pores to the cytoplasm, where they will participate in polypeptide synthesis.) **3.** Each chromosome consists of one long DNA molecule attached to numerous protein molecules, a combination called chromatin. As a cell begins division, each chromosome becomes "condensed" as its diffuse mass of chromatin coils up.

## Concept Check 4.4

**1.** The primary distinction between rough and smooth ER is the presence of bound ribosomes on the rough ER. Both types of ER make phospholipids, but membrane proteins and secretory proteins are all produced by the ribosomes on the rough ER. The smooth ER also functions in detoxification, carbohydrate metabolism, and storage of calcium ions. **2.** Transport vesicles move membranes and the substances they enclose between other components of the endomembrane system. **3.** The mRNA is synthesized in the nucleus and then passes out through a nuclear pore to be translated on a bound ribosome, attached to the rough ER. The protein is synthesized into the lumen of the ER and perhaps modified there. A transport vesicle carries the protein to the Golgi apparatus. After further modification in the Golgi, another transport vesicle carries it back to the ER, where it will perform its cellular function.

## Concept Check 4.5

**1.** Both organelles are involved in energy transformation, mitochondria in cellular respiration and chloroplasts in photosynthesis. They both have multiple membranes that separate their interiors into compartments. In both organelles, the innermost membranes—cristae, or infoldings of the inner membrane, in mitochondria, and the thylakoid membranes in chloroplasts—have large surface areas with embedded enzymes that carry out their main functions. **2.** Yes. Plant cells are able to make their own sugar by photosynthesis, but mitochondria in these eukaryotic cells are the organelles that are able to generate energy from sugars, a function required in all cells. **3.** Mitochondria and chloroplasts are not derived from the ER, nor are they connected physically or via transport vesicles to organelles of the endomembrane system. Mitochondria and chloroplasts are structurally quite different from vesicles derived from the ER, which are bounded by a single membrane.

## Concept Check 4.6

**1.** Dynein arms, powered by ATP, move neighboring doublets of microtubules relative to each other. Because they are anchored within the organelle and with respect to one another, the doublets bend instead of sliding past each other. Synchronized bending of the nine microtubule doublets brings about bending of both cilia and flagella. **2.** Such individuals have defects in the microtubule-based movement of cilia and flagella. Thus, the sperm can't move because of malfunctioning or nonexistent flagella, and the airways are compromised because cilia that line the trachea malfunction or don't exist, and so mucus cannot be cleared from the lungs.

## Concept Check 4.7

**1.** The most obvious difference is the presence of direct cytoplasmic connections between cells of plants (plasmodesmata) and animals (gap junctions). These connections result in the cytoplasm being continuous between adjacent cells. **2.** The cell would not be able to function properly and would probably soon die, as the cell wall or ECM must be permeable to allow the exchange of matter between the cell and its external environment. Molecules involved in energy production and use must be allowed entry, as well as those that provide information about the cell's environment. Other molecules, such as products synthesized by the cell for export and the byproducts of cellular respiration, must be allowed to exit. **3.** The parts of the protein that face aqueous regions would be expected to have polar or charged (hydrophilic) amino acids, while the parts that go through the membrane would be expected to have nonpolar (hydrophobic) amino acids. You would predict polar or charged amino acids at each end (tail), in the region of the cytoplasmic loop, and in the regions of the two extracellular loops. You would predict nonpolar amino acids in the four regions that go through the membrane between the tails and loops.

## Summary of Key Concepts Questions

**4.1** Both light and electron microscopy allow cells to be studied visually, thus helping us understand internal cellular structure and the arrangement of cell components. Cell fractionation techniques separate out different groups of cell components, which can then be analyzed biochemically to determine their function. Performing microscopy on the same cell fraction helps to correlate the biochemical function of the cell with the cell component responsible. **4.2** The separation of different functions in different organelles has several advantages. Reactants and enzymes can be concentrated in one area instead of spread throughout the cell. Reactions that require specific conditions, such as a lower pH, can be compartmentalized. And enzymes for specific reactions are often embedded in the membranes that enclose or partition an organelle. **4.3** The nucleus contains the genetic material of the cell in the form of DNA, which codes for messenger RNA, which in turn provides instructions for the synthesis of proteins (including the proteins that make up part of the ribosomes). DNA also codes for ribosomal RNA, which is combined with proteins in the nucleolus into the subunits of ribosomes. Within the cytoplasm, ribosomes join with mRNA to build polypeptides, using the genetic information in the mRNA. **4.4** Transport vesicles move proteins and membrane synthesized by the rough ER to the Golgi for further processing and then to the plasma membrane, lysosomes, or other locations in the cell, including back to the ER. **4.5** According to the endosymbiont theory, mitochondria originated from an oxygen-using prokaryotic cell that was engulfed by an ancestral eukaryotic cell. Over time, the host and endosymbiont evolved into a single unicellular organism. Chloroplasts originated when at least one of these eukaryotic cells containing mitochondria engulfed and then retained a photosynthetic prokaryote. **4.6** Inside the cell, motor proteins interact with components of the cytoskeleton to move cellular parts. Motor proteins may "walk" vesicles along microtubules. The movement of cytoplasm within a cell involves interactions of the motor protein myosin and microfilaments (actin filaments). Whole cells can be moved by the rapid bending of flagella or cilia, which is caused by the motor-protein-powered sliding of microtubules within these structures. Some cells move by amoeboid movement, which involves interactions of microfilaments with myosin. Interactions of motor proteins and microfilaments in muscle cells cause muscle contraction that can propel whole organisms (for example, by walking or swimming). **4.7** A plant cell wall is primarily composed of microfibrils of cellulose embedded in other polysaccharides and proteins. The ECM of animal cells is primarily composed of collagen and other protein fibers, such as fibronectin and other glycoproteins. These fibers are embedded in a network of carbohydrate-rich proteoglycans. A plant cell wall provides structural support for the cell and, collectively, for the plant body. In addition to giving support, the ECM of an animal cell allows for communication of environmental changes into the cell.

**Test Your Understanding**
**1.** B **2.** C **3.** B **4.** A **5.** A **6.** C **7.** C **8.** See Figure 4.7.

## Chapter 5

### Figure Questions
**Figure 5.3**

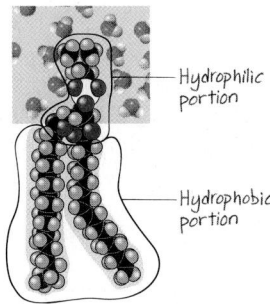

The hydrophilic portion is in contact with an aqueous environment (cytosol or extracellular fluid), and the hydrophobic portion is in contact with the hydrophobic portions of other phospholipids in the interior of the bilayer.    **Figure 5.4** You couldn't rule out movement of proteins within membranes of the same species. You might propose that the membrane lipids and proteins from one species weren't able to mingle with those from the other species because of some incompatibility.    **Figure 5.7** A transmembrane protein like the dimer in (f) might change its shape upon binding to a particular extracellular matrix (ECM) molecule. The new shape might enable the interior portion of the protein to bind to a second, cytoplasmic protein that would relay the message to the inside of the cell, as shown in (c).
**Figure 5.8**

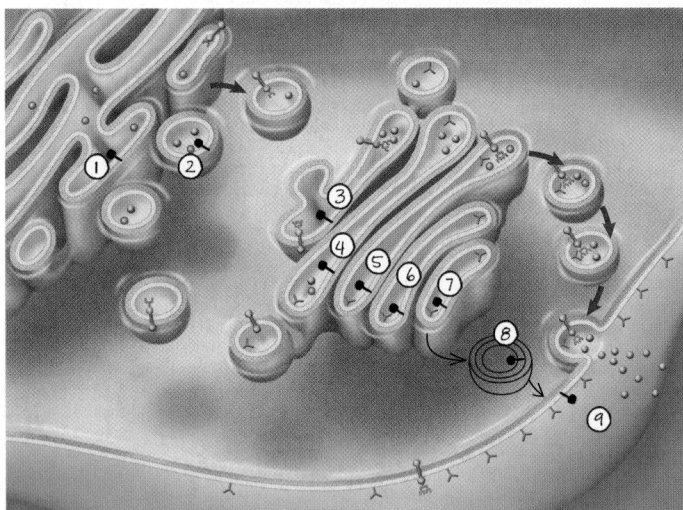

The protein would contact the extracellular fluid.    **Figure 5.10** The orange dye would be evenly distributed throughout the solution on both sides of the membrane. The solution levels would not be affected because the orange dye can diffuse through the membrane and equalize its concentration. Thus, no additional osmosis would take place in either direction.    **Figure 5.15** The diamond solutes are moving into the cell (down), and the round solutes are moving out of the cell (up); each is moving against its concentration gradient.    **Figure 5.23** The aldosterone molecule is hydrophobic and can therefore pass directly through the lipid bilayer of the plasma membrane into the cell. (Hydrophilic molecules cannot do this.)    **Figure 5.24** The active form of protein kinase 1

### Concept Check 5.1
**1.** They are on the inner side of the transport vesicle membrane.    **2.** The grasses living in the cooler region would be expected to have more unsaturated fatty acids in their membranes because those fatty acids remain fluid at lower temperatures. The grasses living immediately adjacent to the hot springs would be expected to have more saturated fatty acids, which would allow the fatty acids to "stack" more closely, making the membranes less fluid and therefore helping them to stay intact at higher temperatures. (Cholesterol could not moderate the effects of temperature on membrane fluidity in this case because it is not found within plant cell membranes.)

### Concept Check 5.2
**1.** $O_2$ and $CO_2$ are both nonpolar molecules that can easily pass through the hydrophobic interior of a membrane.    **2.** Water is a polar molecule, so it cannot pass very rapidly through the hydrophobic region in the middle of a phospholipid bilayer.    **3.** The hydronium ion is charged, while glycerol is not. Charge is probably more significant than size as a basis for exclusion by the aquaporin channel.

### Concept Check 5.3
**1.** $CO_2$ is a nonpolar molecule that can diffuse through the plasma membrane. As long as it diffuses away so that the concentration remains low outside the cell, it will continue to exit the cell in this way. (This is the opposite of the case for $O_2$, described in this section of the text.)    **2.** The activity of *Paramecium caudatum*'s contractile vacuole will decrease. The vacuole pumps out excess water that accumulates in the cell; this accumulation occurs only in a hypotonic environment.

### Concept Check 5.4
**1.** The pump uses ATP. To establish a voltage, ions have to be pumped against their gradients, which requires energy.    **2.** Each ion is being transported against its electrochemical gradient. If either ion were transported down its electrochemical gradient, this process *would* be considered cotransport.    **3.** The internal environment of a lysosome is acidic, so it has a higher concentration of $H^+$ than does the cytoplasm. Therefore, you might expect the membrane of the lysosome to have a proton pump such as that shown in Figure 5.16 to pump $H^+$ into the lysosome.

### Concept Check 5.5
**1.** Exocytosis. When a transport vesicle fuses with the plasma membrane, the vesicle membrane becomes part of the plasma membrane.
**2.**

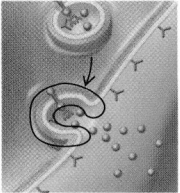

**3.** The glycoprotein would be synthesized in the ER lumen, move through the Golgi apparatus, and then travel in a vesicle to the plasma membrane, where it would undergo exocytosis and become part of the ECM.

### Concept Check 5.6
**1.** Glycogen phosphorylase acts in the third stage, the response to epinephrine signaling.    **2.** Protein phosphatases reverse the effects of the kinases, and unless the signaling molecule is at a high enough concentration that it is continuously rebinding the receptor, the kinase molecules will be returned to their inactive states by phosphatases.    **3.** At each step in a cascade of sequential activations, one molecule or ion may activate numerous molecules functioning in the next step.

### Summary of Key Concepts Questions
**5.1** Plasma membranes define the cell by separating the cellular components from the external environment. This allows conditions inside cells to be controlled by membrane proteins, which regulate entry and exit of molecules and even cell function (see Figure 5.7). The processes of life can be carried out inside the controlled environment of the cell, so membranes are crucial. In eukaryotes, membranes also function to subdivide the cytoplasm into different compartments where distinct processes can occur, even under differing conditions such as low or high pH.    **5.2** Aquaporins are channel proteins that greatly increase the permeability of a membrane to water molecules, which are polar and therefore do not readily diffuse through the hydrophobic interior of the membrane.    **5.3** There will be a net diffusion of water out of a cell into a hypertonic solution. The free water concentration is higher inside the cell than in the solution (where many water molecules are not free, but are clustered around the higher concentration of solute particles).    **5.4** One of the solutes moved by the cotransporter is actively transported against its concentration gradient. The energy for this transport comes from the concentration gradient of the other solute, which was established by an electrogenic pump that used energy (usually provided by ATP) to transport the other solute across the membrane.    **5.5** In receptor-mediated endocytosis, specific molecules bind to receptors on the plasma membrane in a region where a coated pit develops. The cell can acquire bulk quantities of those specific molecules when the coated pit forms a vesicle and carries the bound molecules into the cell.    **5.6** A cell is able to respond to a hormone only if it has a receptor protein on the cell surface or inside the cell that can bind to the hormone. The response to a hormone depends on the specific signal transduction pathway within the cell, which will lead to the specific cellular response. The response can vary for different types of cells.

### Test Your Understanding
**1.** B **2.** A **3.** C **4.** B **5.** C **6.** B

## Chapter 6

### Figure Questions
**Figure 6.9** Glutamic acid (Glu) has a carboxyl group at the end of its R group. Glutamine (Gln) has exactly the same structure as glutamic acid, except that there is an amino group in place of the —OH on the R group. Thus, in this figure, Gln is drawn as a Glu with an attached $NH_2$.

**Figure 6.12**

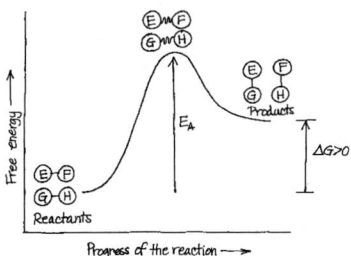

**Figure 6.16**

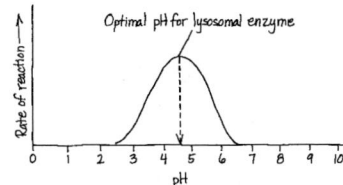

## Concept Check 6.1

**1.** The second law is the trend toward randomization, or increasing entropy. When the concentrations of a substance on both sides of a membrane are equal, the distribution is more random than when they are unequal. Diffusion of a substance to a region where it is initially less concentrated increases entropy, making it an energetically favorable (spontaneous) process as described by the second law. (This explains the process seen in Figure 5.9.)   **2.** The apple has potential energy in its position hanging on the tree, and the sugars and other nutrients it contains have chemical energy. The apple has kinetic energy as it falls from the tree to the ground. Finally, when the apple is digested and its molecules broken down, some of the chemical energy is used to do work, and the rest is lost as thermal energy.

## Concept Check 6.2

**1.** Cellular respiration is a spontaneous and exergonic process. The energy released from glucose is used to do work in the cell or is lost as heat.   **2.** Catabolism breaks down organic molecules, releasing their chemical energy and resulting in smaller products with more entropy, as when moving from the top to the bottom of part (c). Anabolism consumes energy to synthesize larger molecules from simpler ones, as when moving from the bottom to the top of part (c).   **3.** The reaction is exergonic because it releases energy—in this case, in the form of light. (This is a nonbiological version of the bioluminescence seen in Figure 6.1.)

## Concept Check 6.3

**1.** ATP usually transfers energy to endergonic processes by phosphorylating (adding phosphate groups to) other molecules. (Exergonic processes phosphorylate ADP to regenerate ATP.)   **2.** A set of coupled reactions can transform the first combination into the second. Since this is an exergonic process overall, $\Delta G$ is negative and the first combination must have more free energy (see Figure 6.9).   **3.** Active transport: The solute is being transported against its concentration gradient, which requires energy, provided by ATP hydrolysis.

## Concept Check 6.4

**1.** A spontaneous reaction is a reaction that is exergonic. However, if it has a high activation energy that is rarely attained, the rate of the reaction may be low.   **2.** Only the specific substrate(s) will fit properly into the active site of an enzyme, the part of the enzyme that carries out catalysis.   **3.** In the presence of malonate, increase the concentration of the normal substrate (succinate) and see whether the rate of reaction increases. If it does, malonate is a competitive inhibitor.

## Concept Check 6.5

**1.** The activator binds in such a way that it stabilizes the active form of an enzyme, whereas the inhibitor stabilizes the inactive form.

## Summary of Key Concepts Questions

**6.1** The process of "ordering" a cell's structure is accompanied by an increase in the entropy, or disorder, of the universe. For example, an animal cell takes in highly ordered organic molecules as the source of matter and energy used to build and maintain its structures. In the same process, however, the cell releases heat and the simple molecules of carbon dioxide and water to the surroundings. The increase in entropy of the latter process offsets the entropy decrease in the former.   **6.2** Spontaneous reactions supply the energy to perform cellular work.   **6.3** The free energy released from the hydrolysis of ATP may drive endergonic reactions through the transfer of a phosphate group to a reactant molecule, forming a more reactive phosphorylated intermediate. ATP hydrolysis also powers the mechanical and transport work of a cell, often by powering shape changes in the relevant motor proteins. Cellular respiration, the catabolic breakdown of glucose, provides the energy for the endergonic regeneration of ATP from ADP and $P_i$.   **6.4** Activation energy barriers prevent the complex molecules of the cell, which are rich in free energy, from spontaneously breaking down to less ordered, more stable molecules. Enzymes permit a regulated metabolism by binding to specific substrates and forming enzyme-substrate complexes that selectively lower the $E_A$ for the chemical reactions in a cell.   **6.5** A cell tightly regulates its metabolic pathways in response to fluctuating needs for energy and materials. The binding of

activators or inhibitors to regulatory sites on allosteric enzymes stabilizes either the active or inactive form of the subunits. For example, the binding of ATP to a catabolic enzyme in a cell with excess ATP would inhibit that pathway. Such types of feedback inhibition preserve chemical resources within a cell. If ATP supplies are depleted, binding of ADP to the regulatory site of catabolic enzymes will activate that pathway, generating more ATP.

**Test Your Understanding**
**1.** B   **2.** C   **3.** B   **4.** A   **5.** C   **6.** D   **7.** C
**8.**

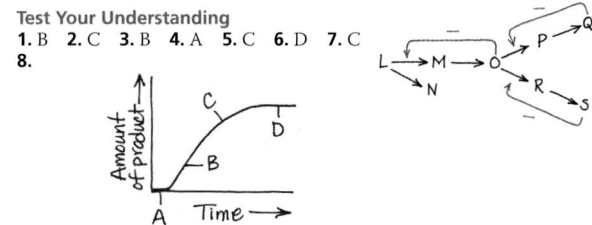

**A.** The substrate molecules are entering the cells, so no product is made yet.
**B.** There is sufficient substrate, so the reaction is proceeding at a maximum rate.
**C.** As the substrate is used up, the rate decreases (the slope is less steep).
**D.** The line is flat because no new substrate remains and thus no new product appears.

## Chapter 7

### Figure Questions
**Figure 7.7** Because there is no external source of energy for the reaction, it must be exergonic, and the reactants must be at a higher energy level than the products.
**Figure 7.9** The removal would probably stop glycolysis, or at least slow it down, since it would push the equilibrium for step 5 toward the bottom (toward DHAP). If less (or no) glyceraldehyde 3-phosphate were available, step 6 would slow down (or be unable to occur).
**Figure 7.13**

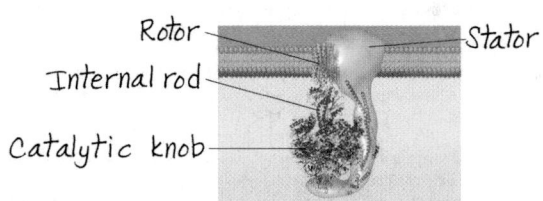

**Figure 7.14** At first, some ATP could be made, since electron transport could proceed as far as complex III, and a small $H^+$ gradient could be built up. Soon, however, no more electrons could be passed to complex III because it could not be reoxidized by passing its electrons to complex IV.   **Figure 7.15** First, there are 2 NADH from the oxidation of pyruvate plus 6 NADH from the citric acid cycle (CAC); 8 NADH × 2.5 ATP/NADH = 20 ATP. Second, there are 2 $FADH_2$ from the CAC; 2 $FADH_2$ × 1.5 ATP/$FADH_2$ = 3 ATP. Third, the 2 NADH from glycolysis enter the mitochondrion through one of two types of shuttle. They pass their electrons either to 2 FAD, which become $FADH_2$ and result in 3 ATP, or to 2 $NAD^+$, which become NADH and result in 5 ATP. Thus, 20 + 3 + 3 = 26 ATP or 20 + 3 + 5 = 28 ATP from all NADH and $FADH_2$.

### Concept Check 7.1
**1.** Both processes include glycolysis, the citric acid cycle, and oxidative phosphorylation. In aerobic respiration, the final electron acceptor is molecular oxygen ($O_2$); in anaerobic respiration, the final electron acceptor is a different substance.   **2.** Substrate-level phosphorylation, which occurs during glycolysis and the citric acid cycle, involves the direct transfer of a phosphate group from an organic substrate to ADP by an enzyme. The process of oxidative phosphorylation occurs during the third stage of cellular respiration, which is called oxidative phosphorylation. In this process, the synthesis of ATP from ADP and inorganic phosphate ($P_i$) is powered by the redox reactions of the electron transport chain.   **3.** $C_4H_6O_5$ would be oxidized, and $NAD^+$ would be reduced.

### Concept Check 7.2
**1.** $NAD^+$ acts as the oxidizing agent in step 6, accepting electrons from glyceraldehyde 3-phosphate (G3P), which thus acts as the reducing agent.

### Concept Check 7.3
**1.** NADH and $FADH_2$; they will donate electrons to the electron transport chain.
**2.** $CO_2$ is released from the pyruvate that is the end product of glycolysis, and $CO_2$ is also released during the citric acid cycle.

### Concept Check 7.4
**1.** Oxidative phosphorylation would eventually stop entirely, resulting in no ATP production by this process. Without oxygen to "pull" electrons down the electron transport chain, $H^+$ would not be pumped into the mitochondrion's intermembrane space and chemiosmosis would not occur.   **2.** Decreasing the pH means the addition of $H^+$. This would establish a proton gradient even without the function of the electron transport chain, and we would expect ATP synthase to function and synthesize ATP. (In fact, it was experiments like this that provided support for chemiosmosis as an energy-coupling mechanism.)   **3.** One of the components of the electron transport chain, ubiquinone (Q), must be able to diffuse within the membrane. It could not do so if the membrane components were locked rigidly into place.

## Concept Check 7.5

**1.** A derivative of pyruvate, such as acetaldehyde during alcohol fermentation, or pyruvate itself during lactic acid fermentation; oxygen during aerobic respiration
**2.** The cell would need to consume glucose at a rate about 16 times the consumption rate in the aerobic environment (2 ATP are generated by fermentation versus up to 32 ATP by cellular respiration).

## Concept Check 7.6

**1.** The fat is much more reduced; it has many —$CH_2$— units, and in all these bonds the electrons are equally shared. The electrons present in a carbohydrate molecule are already somewhat oxidized (shared unequally in bonds), as quite a few of them are bound to oxygen. Electrons that are equally shared, as in fat, have a higher energy level than electrons that are unequally shared, as in carbohydrates. Thus, fats are much better fuels than carbohydrates.   **2.** When you consume more food than necessary for metabolic processes, your body synthesizes fat as a way of storing energy for later use.
**3.** When oxygen is present, the fatty acid chains containing most of the energy of a fat are oxidized and fed into the citric acid cycle and the electron transport chain. During intense exercise, however, oxygen is scarce in muscle cells, so ATP must be generated by glycolysis alone. A very small part of the fat molecule, the glycerol backbone, can be oxidized via glycolysis, but the amount of energy released by this portion is insignificant compared with that released by the fatty acid chains. (This is why moderate exercise, staying below 70% maximum heart rate, is better for burning fat—because enough oxygen remains available to the muscles.)

## Summary of Key Concepts Questions

**7.1** Most of the ATP produced in cellular respiration comes from oxidative phosphorylation, in which the energy released from redox reactions in an electron transport chain is used to produce ATP. In substrate-level phosphorylation, an enzyme directly transfers a phosphate group to ADP from an intermediate substrate. All ATP production in glycolysis occurs by substrate-level phosphorylation; this form of ATP production also occurs at one step in the citric acid cycle.   **7.2** The oxidation of the three-carbon sugar glyceraldehyde 3-phosphate yields energy. In this oxidation, electrons and $H^+$ are transferred to $NAD^+$, forming NADH, and a phosphate group is attached to the oxidized substrate. ATP is then formed by substrate-level phosphorylation when this phosphate group is transferred to ADP.   **7.3** The release of six molecules of $CO_2$ represents the complete oxidation of glucose. During the processing of two pyruvates to acetyl CoA, the fully oxidized carboxyl groups (—$COO^-$) are given off as 2 $CO_2$. The remaining four carbons are released as $CO_2$ in the citric acid cycle as citrate is oxidized back to oxaloacetate.   **7.4** The flow of $H^+$ through the ATP synthase complex causes the rotor and attached rod to rotate, exposing catalytic sites in the knob portion that produce ATP from ADP and $Ⓟ_i$. ATP synthases are found in the inner mitochondrial membrane, the plasma membrane of prokaryotes, and membranes within chloroplasts.   **7.5** Anaerobic respiration yields more ATP. The 2 ATP produced by substrate-level phosphorylation in glycolysis represent the total energy yield of fermentation. NADH passes its "high-energy" electrons to pyruvate or a derivative of pyruvate, recycling $NAD^+$ and allowing glycolysis to continue. In anaerobic respiration, the NADH produced during glycolysis, as well as additional molecules of NADH produced as pyruvate is oxidized, are used to generate ATP molecules. An electron transport chain captures the energy of the electrons in NADH via a series of redox reactions; ultimately, the electrons are transferred to an electronegative molecule other than oxygen. Also, additional molecules of NADH are produced in anaerobic respiration as pyruvate is oxidized.   **7.6** The ATP produced by catabolic pathways is used to drive anabolic pathways. Also, many of the intermediates of glycolysis and the citric acid cycle are used in the biosynthesis of a cell's molecules.

## Test Your Understanding

**1.** C   **2.** C   **3.** B   **4.** A   **5.** D   **6.** A   **7.** B
**8.**

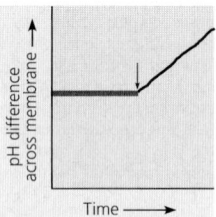

$H^+$ would continue to be pumped across the membrane into the intermembrane space, increasing the difference between the matrix pH and the intermembrane space pH. $H^+$ would not be able to flow back through ATP synthase, since the enzyme is inhibited by the poison, so rather than maintaining a constant difference across the membrane, the difference would continue to increase. (Ultimately, the $H^+$ concentration in the intermembrane space would be so high that no more $H^+$ would be able to be pumped against the gradient, but this isn't shown in the graph.)

## Chapter 8

### Figure Questions

**Figure 8.15** You would (a) decrease the pH outside the mitochondrion (thus increasing the $H^+$ concentration) and (b) increase the pH in the chloroplast stroma (thus decreasing the $H^+$ concentration). In both cases, this would generate an $H^+$ gradient across the membrane that would cause ATP synthase to synthesize ATP.

**Figure 8.17**

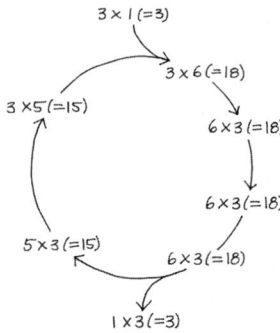

Three carbon atoms enter the cycle, one by one, as individual $CO_2$ molecules and leave the cycle in one three-carbon molecule (G3P) per three turns of the cycle.
**Figure 8.19** Yes, plants can break down the sugar (in the form of glucose) by cellular respiration, producing ATPs for various cellular processes such as endergonic chemical reactions, transport of substances across membranes, and movement of molecules in the cell.   **Figure 8.20** The gene encoding hexokinase is part of the DNA of a chromosome in the nucleus. There, the gene is transcribed into mRNA, which is transported to the cytoplasm where it is translated on a free ribosome into a polypeptide. The polypeptide folds into a functional protein with secondary and tertiary structure. Once functional, it carries out the first reaction of glycolysis in the cytoplasm.

### Concept Check 8.1

**1.** $CO_2$ enters the leaves via stomata, and water enters the plant via roots and is carried to the leaves through veins.   **2.** Using $^{18}O$, a heavy isotope of oxygen, as a label, researchers were able to confirm van Niel's hypothesis that the oxygen produced during photosynthesis comes from water, not from carbon dioxide.   **3.** The light reactions could *not* keep producing NADPH and ATP without the $NADP^+$, ADP, and $Ⓟ_i$ that the Calvin cycle generates. The two cycles are interdependent.

### Concept Check 8.2

**1.** Green, because green light is mostly transmitted and reflected—not absorbed— by photosynthetic pigments   **2.** Water ($H_2O$) is the initial electron donor; $NADP^+$ accepts electrons at the end of the electron transport chain, becoming reduced to NADPH.   **3.** The rate of ATP synthesis would slow and eventually stop. Because the added compound would not allow a proton gradient to build up across the membrane, ATP synthase could not catalyze ATP production.

### Concept Check 8.3

**1.** The more energy and reducing power a molecule stores, the more energy and reducing power are required for formation of that molecule. Glucose is a valuable energy source because it is highly reduced, storing lots of potential energy in its electrons. To reduce $CO_2$ to glucose, a large amount of energy and reducing power are required in the form of large numbers of ATP and NADPH molecules, respectively.   **2.** The light reactions require ADP and $NADP^+$, which would not be formed in sufficient quantities from ATP and NADPH if the Calvin cycle stopped.   **3.** Photorespiration decreases photosynthetic output by adding oxygen, instead of carbon dioxide, to the Calvin cycle. As a result, no sugar is generated (no carbon is fixed), and $O_2$ is used rather than generated.

### Summary of Key Concepts Questions

**8.1** $CO_2$ and $H_2O$ are the products of respiration; they are the reactants in photosynthesis. In respiration, glucose is oxidized to $CO_2$ and electrons are passed through an electron transfer chain from glucose to $O_2$, producing $H_2O$. In photosynthesis, $H_2O$ is the source of electrons, which are energized by light, temporarily stored in NADPH, and used to reduce $CO_2$ to carbohydrate.   **8.2** The action spectrum of photosynthesis shows that some wavelengths of light that are not absorbed by chlorophyll *a* are still effective at promoting photosynthesis. The light-harvesting complexes of photosystems contain accessory pigments, such as chlorophyll *b* and carotenoids, which absorb different wavelengths and pass the energy to chlorophyll *a*, broadening the spectrum of light usable for photosynthesis.
**8.3**

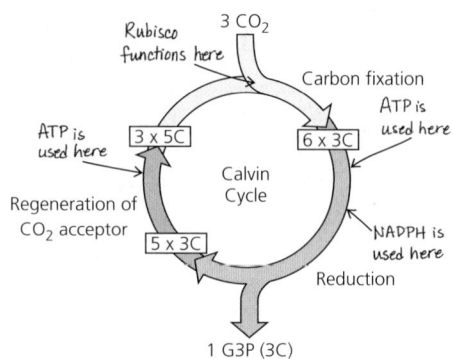

In the reduction phase of the Calvin cycle, ATP phosphorylates a three-carbon compound, and NADPH then reduces this compound to G3P. ATP is also used in the regeneration phase, when five molecules of G3P are converted to three molecules of the five-carbon compound RuBP. Rubisco catalyzes the first step of carbon fixation—the addition of $CO_2$ to RuBP.

**Test Your Understanding**
**1.** D   **2.** B   **3.** C   **4.** A   **5.** C   **6.** B   **7.** C   **8.** 6; 18; 12
**10.**

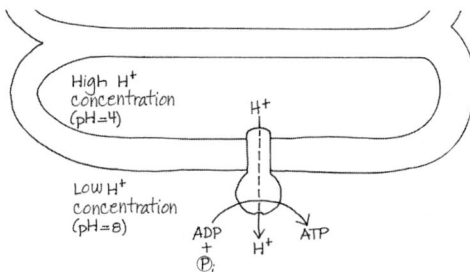

The ATP would end up outside the thylakoid. The thylakoids were able to make ATP in the dark because the researchers set up an artificial proton concentration gradient across the thylakoid membrane; thus, the light reactions were not necessary to establish the $H^+$ gradient required for ATP synthesis by ATP synthase.

## Chapter 9

**Figure Questions**
**Figure 9.4**

Circling the other chromatid instead would also be correct.   **Figure 9.5** The chromosome has four chromatid arms. The single (duplicated) chromosome in step 2 becomes two (unduplicated) chromosomes in step 3. The duplicated chromosome in step 2 is considered one single chromosome.   **Figure 9.7** 12; 2; 2; 1
**Figure 9.8**

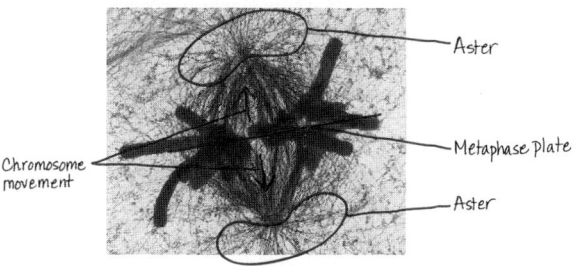

**Figure 9.9** The mark would have moved toward the nearer pole. The lengths of fluorescent microtubules between that pole and the mark would have decreased, while the lengths between the chromosomes and the mark would have remained the same.
**Figure 9.14** In both cases, the $G_1$ nucleus would have remained in $G_1$ until the time it normally would have entered the S phase. Chromosome condensation and spindle formation would not have occurred until the S and $G_2$ phases had been completed.
**Figure 9.16** The cell would divide under conditions where it was inappropriate to do so. If the daughter cells and their descendants also ignored either of the checkpoints and divided, there would soon be an abnormal mass of cells. (This type of inappropriate cell division can contribute to the development of cancer.)   **Figure 9.17** The cells in the vessel with PDGF would not be able to respond to the growth factor signal and thus would not divide. The culture would resemble that without the added PDGF.

**Concept Check 9.1**
**1.** 1; 1; 2   **2.** 39; 39; 78

**Concept Check 9.2**
**1.** 6 chromosomes; they are duplicated; 12 chromatids   **2.** Following mitosis, cytokinesis results in two genetically identical daughter cells in both plant cells and animal cells. However, the mechanism of dividing the cytoplasm is different in animals and plants. In an animal cell, cytokinesis occurs by cleavage, which divides the parent cell in two with a contractile ring of actin filaments. In a plant cell, a cell plate forms in the middle of the cell and grows until its membrane fuses with the plasma membrane of the parent cell. A new cell wall grows inside the cell plate.   **3.** From the end of S phase in interphase through the end of metaphase in mitosis   **4.** During eukaryotic cell division, tubulin is involved in spindle formation and chromosome movement, while actin functions during cytokinesis. In bacterial binary fission, it's the opposite: Tubulin-like molecules are thought to act in daughter cell separation, and actin-like molecules are thought to move the daughter bacterial chromosomes to opposite ends of the cell.

**Concept Check 9.3**
**1.** The nucleus on the right was originally in the $G_1$ phase; therefore, it had not yet duplicated its chromosomes. The nucleus on the left was in the M phase, so it had already duplicated its chromosomes.   **2.** Most body cells are in a nondividing state called $G_0$.   **3.** Both types of tumors consist of abnormal cells, but their characteristics are different. A benign tumor stays at the original site and can usually be surgically removed; the cells have some genetic and cellular changes from normal, non-tumor cells. Cancer cells from a malignant tumor have more significant genetic and cellular changes, can spread from the original site by metastasis, and may impair the functions of one or more organs.   **4.** The cells might divide even in the absence of PDGF. In addition, they would not stop when the surface of the culture vessel was covered; they would continue to divide, piling on top of one another.

**Summary of Key Concepts Questions**
**9.1** The DNA of a eukaryotic cell is packaged into structures called *chromosomes*. Each chromosome is a long molecule of DNA, which carries hundreds to thousands of genes, with associated proteins that maintain chromosome structure and help control gene activity. This DNA-protein complex is called *chromatin*. The chromatin of each chromosome is long and thin when the cell is not dividing. Prior to cell division, each chromosome is duplicated, and the resulting sister *chromatids* are attached to each other by proteins at the centromeres and, for many species, all along their lengths (a phenomenon called sister chromatid cohesion).   **9.2** Chromosomes exist as single DNA molecules in $G_1$ of interphase and in anaphase and telophase of mitosis. During S phase, DNA replication produces sister chromatids, which persist during $G_2$ of interphase and through prophase, prometaphase, and metaphase of mitosis.
**9.3** Checkpoints allow cellular surveillance mechanisms to determine whether the cell is prepared to go to the next stage. Internal and external signals move a cell past these checkpoints. The $G_1$ checkpoint, called the "restriction point" in mammalian cells, determines whether a cell will complete the cell cycle and divide or switch into the $G_0$ phase. The signals to pass this checkpoint often are external—such as growth factors. Regulation of the cell cycle is carried out by a molecular system, including proteins called cyclins and other proteins that are kinases. The signal to pass the M phase checkpoint is not activated until all chromosomes are attached to kinetochore fibers and are aligned at the metaphase plate. Only then will sister chromatid separation occur.

**Test Your Understanding**
**1.** B   **2.** A   **3.** B   **4.** A   **5.** D
**6.**

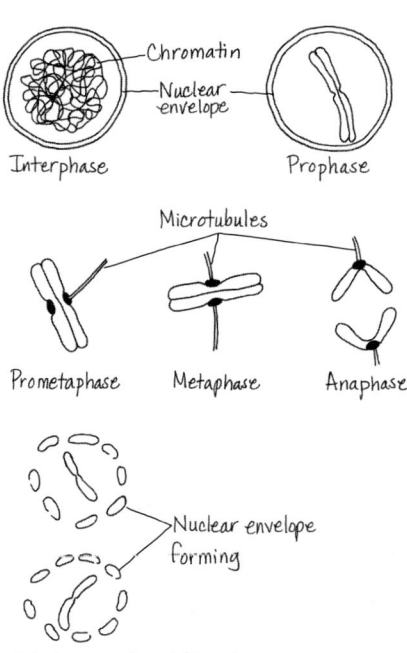

**7.** See Figure 9.7 for a description of major events.

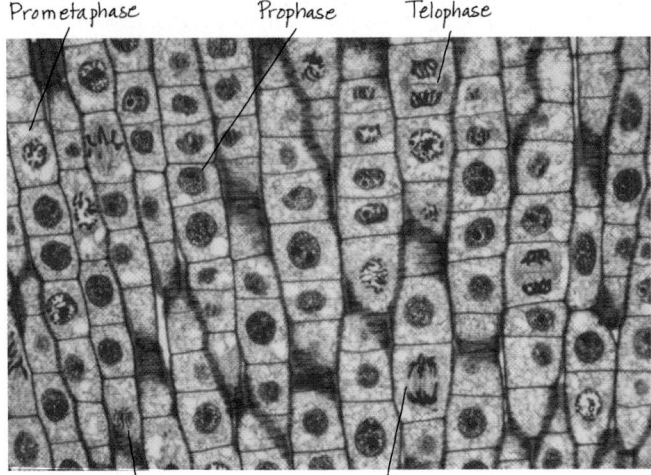

Prometaphase  Prophase  Telophase

Metaphase  Anaphase

Only one cell is indicated for each stage, but other correct answers are also present in this micrograph.

## Chapter 10

### Figure Questions
**Figure 10.4** Two sets of chromosomes are present. Three pairs of homologous chromosomes are present.
**Figure 10.7**

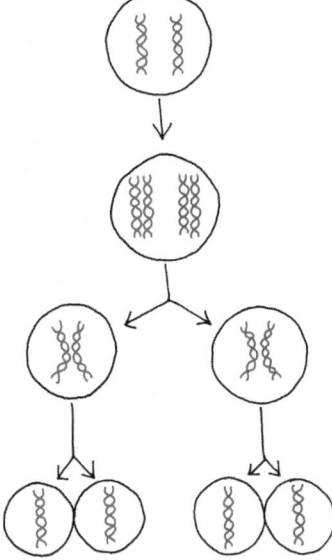

(A short strand of DNA is shown here for simplicity, but each chromosome or chromatid contains a very long coiled and folded DNA molecule.)  **Figure 10.8** If a cell with six chromosomes undergoes two rounds of mitosis, each of the four resulting cells will have six chromosomes, while the four cells resulting from meiosis in Figure 10.8 each have three chromosomes. In mitosis, DNA replication (and thus chromosome duplication) precedes each prophase, ensuring that daughter cells have the same number of chromosomes as the parent cell. In meiosis, in contrast, DNA replication occurs only before prophase I (not before prophase II). Thus, in two rounds of mitosis, the chromosomes duplicate twice and divide twice, while in meiosis, the chromosomes duplicate once and divide twice.  **Figure 10.10** Yes. Each of the six chromosomes (three per cell) shown in telophase I has one nonrecombinant chromatid and one recombinant chromatid. Therefore, eight possible sets of chromosomes can be generated for the cell on the left and eight for the cell on the right.

### Concept Check 10.1
**1.** Parents pass genes to their offspring; by dictating the production of messenger RNAs (mRNAs), the genes program cells to make specific enzymes and other proteins, whose cumulative action produces an individual's inherited traits.  **2.** Such organisms reproduce by mitosis, which generates offspring whose genomes are exact copies of the parent's genome (in the absence of mutations).  **3.** She should clone it. Crossbreeding it with another plant would generate offspring that have additional variation, which she no longer desires now that she has obtained her ideal orchid.

### Concept Check 10.2
**1.** Each of the six chromosomes is duplicated, so each contains two DNA double helices. Therefore, there are 12 DNA molecules in the cell. The haploid number, $n$, is 3. One set is always haploid.  **2.** There are 23 pairs of chromosomes and two sets.  **3.** The haploid number ($n$) is 7; the diploid number ($2n$) is 14.  **4.** This organism has the life cycle shown in Figure 10.6c. Therefore, it must be a fungus or a protist, perhaps an alga.

### Concept Check 10.3
**1.** The chromosomes are similar in that each is composed of two sister chromatids, and the individual chromosomes are positioned similarly at the metaphase plate. The chromosomes differ in that in a mitotically dividing cell, sister chromatids of each chromosome are genetically identical, but in a meiotically dividing cell, sister chromatids are genetically distinct because of crossing over in meiosis I. Moreover, the chromosomes in metaphase of mitosis can be a diploid set or a haploid set, but the chromosomes in metaphase of meiosis II always consist of a haploid set.  **2.** If crossing over did not occur, the two homologs would not be associated in any way; each sister chromatid would be either all maternal or all paternal, and would only be attached to its sister, not to a nonsister chromatid. This might result in incorrect arrangement of homologs during metaphase I and ultimately in formation of gametes with an abnormal number of chromosomes.

### Concept Check 10.4
**1.** Mutations in a gene lead to the different versions (alleles) of that gene.  **2.** If the segments of the maternal and paternal chromatids that undergo crossing over are genetically identical and thus have the same two alleles for every gene, then the recombinant chromosomes will be genetically equivalent to the parental chromosomes. Crossing over contributes to genetic variation only when it involves the rearrangement of different alleles.

### Summary of Key Concepts Questions
**10.1** Genes program specific traits, and offspring inherit their genes from each parent, accounting for similarities in their appearance to one or the other parent. Humans reproduce sexually, which ensures new combinations of genes (and thus traits) in the offspring. Consequently, the offspring are not clones of their parents (which would be the case if humans reproduced asexually).  **10.2** Animals and plants both reproduce sexually, alternating meiosis with fertilization. Both have haploid gametes that unite to form a diploid zygote, which then goes on to divide mitotically, forming a diploid multicellular organism. In animals, haploid cells become gametes and don't undergo mitosis, while in plants, the haploid cells resulting from meiosis undergo mitosis to form a haploid multicellular organism, the gametophyte. This organism then goes on to generate haploid gametes. (In plants such as trees, the gametophyte is quite reduced in size and not obvious to the casual observer.)  **10.3** At the end of meiosis I, the two members of a homologous pair end up in different cells, so they cannot pair up and undergo crossing over.  **10.4** First, during independent assortment in metaphase I, each pair of homologous chromosomes lines up independent of every other pair at the metaphase plate, so a daughter cell of meiosis I randomly inherits either a maternal or paternal chromosome. Second, due to crossing over, each chromosome is not exclusively maternal or paternal, but includes regions at the ends of the chromatid from a nonsister chromatid (a chromatid of the other homolog). (The nonsister segment can also be in an internal region of the chromatid if a second crossover occurs beyond the first one before the end of the chromatid.) This provides much additional diversity in the form of new combinations of alleles. Third, random fertilization ensures even more variation, since any sperm of a large number containing many possible genetic combinations can fertilize any egg of a similarly large number of possible combinations.

### Test Your Understanding
**1.** A  **2.** B  **3.** D  **4.** C
**5.** (a)

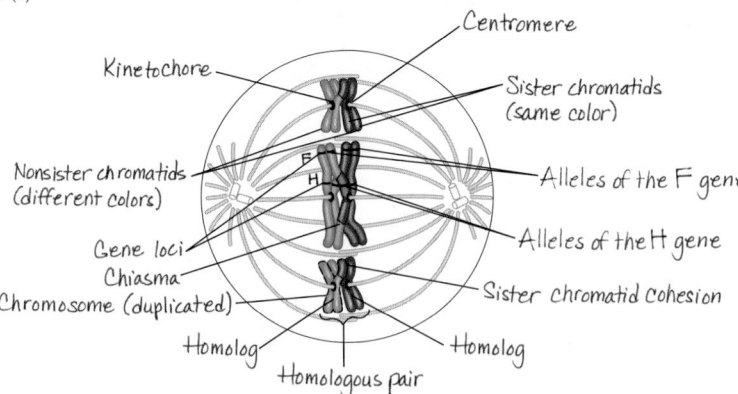

Kinetochore
Centromere
Sister chromatids (same color)
Nonsister chromatids (different colors)
Alleles of the F gene
Alleles of the H gene
Gene loci
Chiasma
Chromosome (duplicated)
Sister chromatid cohesion
Homolog
Homolog
Homologous pair

(b) Metaphase I (c) A haploid set is made up of one long, one medium, and one short chromosome, no matter what combination of colors. For example, one red long, one blue medium, and one red short chromosome make up a haploid set. (In cases where crossovers have occurred, a haploid set of one color may include segments of chromatids of the other color.) All red and blue chromosomes together make up a diploid set.
**6.** This cell must be undergoing meiosis because homologous chromosomes are associated with each other at the metaphase plate; this does not occur in mitosis.

## Chapter 11

### Figure Questions

**Figure 11.3** All offspring would have purple flowers. (The ratio would be 1 purple : 0 white.) The P generation plants are true-breeding, so mating two purple-flowered plants produces the same result as self-pollination: All the offspring have the same trait.

**Figure 11.8**

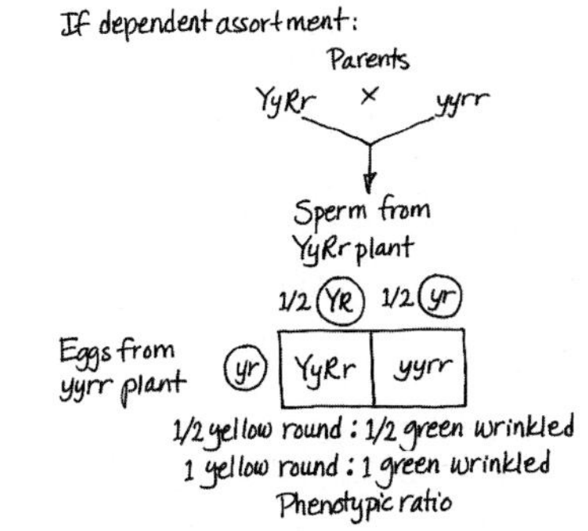


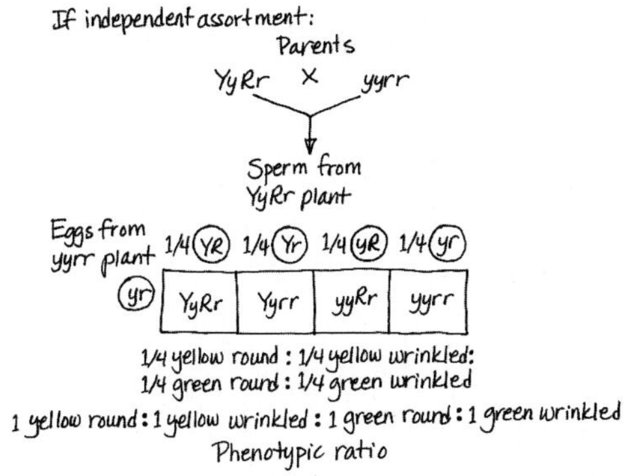


Yes, this cross would also have allowed Mendel to make different predictions for the two hypotheses, thereby allowing him to distinguish the correct one. **Figure 11.10** Your classmate would probably point out that the $F_1$ generation hybrids show an intermediate phenotype between those of the homozygous parents, which supports the blending hypothesis. You could respond that crossing the $F_1$ hybrids results in the reappearance of the white phenotype, rather than identical pink offspring, which fails to support the idea of traits blending during inheritance, in which case the white trait would have been lost after the $F_1$ generation. **Figure 11.11** Both the $I^A$ and $I^B$ alleles are dominant to the *i* allele, which is recessive and results in no attached carbohydrate. The $I^A$ and $I^B$ alleles are codominant; both are expressed in the phenotype of $I^AI^B$ heterozygotes, who have type AB blood. **Figure 11.15** In the Punnett square, two of the three individuals with normal coloration are carriers, so the probability is ⅔. (Note that you must take into account everything you know when you calculate probability: You know she is not *aa*, so there are only three possible genotypes to consider.)

### Concept Check 11.1

**1.** According to the law of independent assortment, 25 plants (1/16 of the offspring) are predicted to be *aatt*, or recessive for both characters. The actual result is likely to differ slightly from this value.

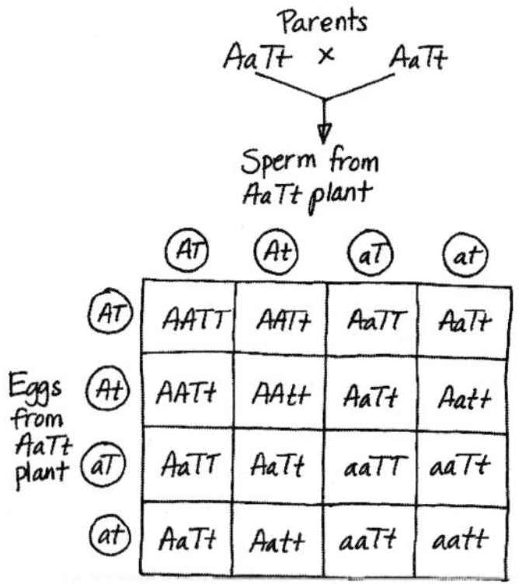


**2.** The plant could make eight different gametes (*YRI, YRi, YrI, Yri, yRI, yRi, yrI,* and *yri*). To fit all the possible gametes in a self-pollination, a Punnett square would need 8 rows and 8 columns. It would have spaces for the 64 possible unions of gametes in the offspring. **3.** Self-pollination is sexual reproduction because meiosis is involved in forming gametes, which unite during fertilization. As a result, the offspring in self-pollination are genetically different from the parent.

### Concept Check 11.2

**1.** ½ homozygous dominant (*AA*), 0 homozygous recessive (*aa*), and ½ heterozygous (*Aa*) **2.** ¼ *BBDD*; ¼ *BbDD*; ¼ *BBDd*; ¼ *BbDd* **3.** The genotypes that fulfill this condition are *ppyyIi, ppYyii, Ppyyii, ppYYii,* and *ppyyii*. Use the multiplication rule to find the probability of getting each genotype and then use the addition rule to find the overall probability of meeting the conditions of this problem:

| | | |
|---|---|---|
| ppyyIi | ½ (probability of pp) × ¼ (yy) × ½ (ii) | $= \frac{1}{16}$ |
| ppYyii | ½ (pp) × ½ (Yy) × ½ (ii) | $= \frac{2}{16}$ |
| Ppyyii | ½ (Pp) × ¼ (yy) × ½ (ii) | $= \frac{1}{16}$ |
| ppYYii | ½ (pp) × ¼ (YY) × ½ (ii) | $= \frac{1}{16}$ |
| ppyyii | ½ (pp) × ¼ (yy) × ½ (ii) | $= \frac{1}{16}$ |
| Fraction predicted to have at least two recessive traits | | $= \frac{6}{16}$ or 3/8 |

### Concept Check 11.3

**1.** Incomplete dominance describes the relationship between two alleles of a single gene, whereas epistasis relates to the genetic relationship between two genes (and the respective alleles of each). **2.** Half of the children would be expected to have type A blood and half type B blood. **3.** The black and white alleles are incompletely dominant, with heterozygotes being gray in color. A cross between a gray rooster and a black hen should yield approximately equal numbers of gray and black offspring.

### Concept Check 11.4

**1.** 1/9 (Since cystic fibrosis is caused by a recessive allele, Beth and Tom's siblings who have CF must be homozygous recessive. Therefore, each parent must be a carrier of the recessive allele. Since neither Beth nor Tom has CF, this means they each have a ⅔ chance of being a carrier. If they are both carriers, there is a ¼ chance that they will have a child with CF. ⅔ × ⅔ × ¼ = 1/9); 0 (Both Beth and Tom would have to be carriers to produce a child with the disease.) **2.** In the monohybrid cross involving flower color, the ratio is 3.15 purple : 1 white, while in the human family in the pedigree, the ratio in the third generation is 1 can taste PTC : 1 cannot taste PTC. The difference is due to the small sample size (two offspring) in the human family. If the second-generation couple in this pedigree were able to have 929 offspring as in the pea plant cross, the ratio would likely be closer to 3:1. (Note that none of the pea plant crosses in Table 11.1 yielded *exactly* a 3:1 ratio.)

### Summary of Key Concepts Questions

**11.1** Alternative versions of genes, called alleles, are passed from parent to offspring during sexual reproduction. In a cross between purple- and white-flowered homozygous parents, the $F_1$ offspring are all heterozygous, each inheriting a purple allele from one parent and a white allele from the other. Because the purple allele is dominant, it determines the phenotype of the $F_1$ offspring to be purple, and the expression of the recessive white allele is masked. Only in the $F_2$ generation is it possible for a white allele to exist in the homozygous state, which causes the white trait to be expressed.

**11.2**

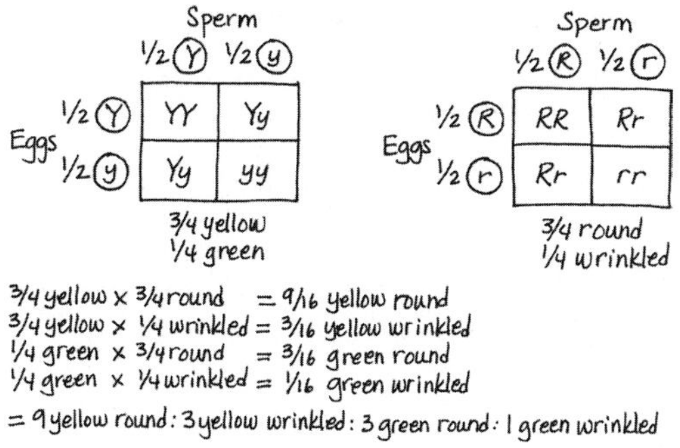


**11.3** The ABO blood group is an example of multiple alleles because this single gene has more than two alleles ($I^A$, $I^B$, and *i*). Two of the alleles, $I^A$ and $I^B$, exhibit codominance, since both carbohydrates (A and B) are present when these two alleles exist together in a genotype. $I^A$ and $I^B$ each exhibit complete dominance over the *i* allele. This situation is not an example of incomplete dominance because each allele affects the phenotype in a distinguishable way, so the result is not intermediate between the two phenotypes. Because this situation involves a single gene, it is not an example of epistasis or polygenic inheritance. **11.4** The chance of the fourth child having cystic fibrosis is ¼, as it was for each of the other children, because each birth is an independent event. We already know that both parents are carriers, so whether their first three children are carriers or not has no bearing on the probability that their next child will have the disease. The parents' genotypes provide the only relevant information.

**1.**

Parents
GgIi × GgIi

Sperm

| Eggs | GI | Gi | gI | gi |
|------|-----|-----|-----|-----|
| GI | GGII | GGIi | GgII | GgIi |
| Gi | GGIi | GGii | GgIi | Ggii |
| gI | GgII | GgIi | ggII | ggIi |
| gi | GgIi | Ggii | ggIi | ggii |

9 green inflated : 3 green constricted :
3 yellow inflated : 1 yellow constricted

**2.** Man $I^A i$; woman $I^B i$; child $ii$. Genotypes for future children are predicted to be $\frac{1}{4} I^A I^B$, $\frac{1}{4} I^A i$, $\frac{1}{4} I^B i$, $\frac{1}{4} ii$. **3.** $\frac{1}{2}$ **4.** A cross of $Ii \times ii$ would yield offspring with a genotypic ratio of 1 $Ii$ : 1 $ii$ (2:2 is an equivalent answer) and a phenotypic ratio of 1 inflated : 1 constricted (2:2 is equivalent).

Parents
Ii × ii

Sperm from
ii plant

| Eggs from Ii plant | i | i |
|------|-----|-----|
| I | Ii | Ii |
| i | ii | ii |

Genotypic ratio 1 Ii : 1 ii
(2:2 is equivalent)

Phenotypic ratio 1 inflated : 1 constricted
(2:2 is equivalent)

**5.** (a) $\frac{1}{64}$; (b) $\frac{1}{64}$; (c) $\frac{1}{8}$; (d) $\frac{1}{32}$ **6.** (a) $\frac{3}{4} \times \frac{3}{4} \times \frac{3}{4} = \frac{27}{64}$; (b) $1 - \frac{27}{64} = \frac{37}{64}$; (c) $\frac{1}{4} \times \frac{1}{4} \times \frac{1}{4} = \frac{1}{64}$; (d) $1 - \frac{1}{64} = \frac{63}{64}$ **7.** (a) $\frac{1}{256}$; (b) $\frac{1}{16}$; (c) $\frac{1}{256}$; (d) $\frac{1}{64}$; (e) $\frac{1}{128}$ **8.** (a) 1; (b) $\frac{1}{32}$; (c) $\frac{1}{8}$; (d) $\frac{1}{2}$ **9.** $\frac{1}{9}$ **10.** Matings of the original mutant cat with true-breeding noncurl cats will produce both curl and noncurl $F_1$ offspring if the curl allele is dominant, but only noncurl offspring if the curl allele is recessive. You would obtain some true-breeding offspring homozygous for the curl allele from matings between the $F_1$ cats resulting from the original curl × noncurl crosses. If dominant, you wouldn't be able to tell true-breeding, homozygous offspring from heterozygotes without further crosses whether the curl trait is dominant or recessive. You know that cats are true-breeding when curl × curl matings produce only curl offspring for several generations. As it turns out, the allele that causes curled ears is dominant. **11.** 25%, or $\frac{1}{4}$, will be cross-eyed; all (100%) of the cross-eyed offspring will also be white. **12.** The dominant allele $I$ is epistatic to the $P/p$ locus, and thus the genotypic ratio for the $F_1$ generation will be 9 $I–P–$ (colorless) : 3 $I–pp$ (colorless) : 3 $iiP–$ (purple) : 1 $iipp$ (red). Overall, the phenotypic ratio is 12 colorless : 3 purple : 1 red. **13.** Recessive. All affected individuals (Arlene, Tom, Wilma, and Carla) are homozygous recessive $aa$. George is $Aa$, since some of his children with Arlene are affected. Sam, Ann, Daniel, and Alan are each $Aa$, since they are all unaffected children with one affected parent. Michael also is $Aa$, since he has an affected child (Carla) with his heterozygous wife Ann. Sandra, Tina, and Christopher can each have either the $AA$ or $Aa$ genotype. **14.** $\frac{1}{6}$

## Chapter 12

### Figure Questions
**Figure 12.2** The ratio would be 1 yellow round : 1 green round : 1 yellow wrinkled : 1 green wrinkled. **Figure 12.4** About $\frac{3}{4}$ of the $F_2$ offspring would have red eyes and about $\frac{1}{4}$ would have white eyes. About $\frac{1}{4}$ of the white-eyed flies would be female and half would be male; similarly, about half of the red-eyed flies would be female and half would be male. (Note that the homologs with the eye-color alleles would be the same shape in the Punnett square, and each offspring would inherit two alleles. The sex of the flies would be determined separately by inheritance of the sex chromosomes. Thus your Punnett square would have four possible combinations in sperm and four in eggs; it would have 16 squares altogether.) **Figure 12.7** All the males would be color-blind, and all the females would be carriers. (Another way to say this is that $\frac{1}{2}$ the offspring would be color-blind males, and $\frac{1}{2}$ the offspring would be carrier females.) **Figure 12.9** The two largest classes would still be the offspring with the phenotypes of the true-breeding P generation flies, but now they would be gray vestigial and black normal because those were the specific allele combinations in the P generation. **Figure 12.10** The two chromosomes on the left side of the sketch are like the two chromosomes inherited by the $F_1$ female, one from each P generation fly. They are

passed by the $F_1$ female intact to the offspring and thus could be called "parental" chromosomes. The other two chromosomes result from crossing over during meiosis in the $F_1$ female. Because they have combinations of alleles not seen in either of the $F_1$ female's chromosomes, they can be called "recombinant" chromosomes. (Note that in this example, the alleles on the recombinant chromosomes, $b^+ vg^+$ and $b\ vg$, are the allele combinations that were on the parental chromosomes in the cross shown in Figures 12.9 and 12.10. The basis for calling them parental chromosomes is that they have the combination of alleles that was present on the P generation chromosomes.)

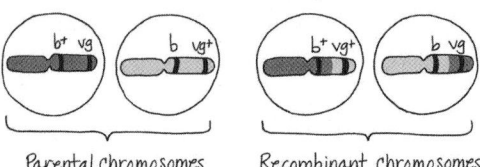

Parental chromosomes          Recombinant chromosomes

### Concept Check 12.1
**1.** The law of segregation relates to the inheritance of alleles for a single character. The law of independent assortment of alleles relates to the inheritance of alleles for two characters. **2.** The physical basis for the law of segregation is the separation of homologs in anaphase I. The physical basis for the law of independent assortment is the alternative arrangements of all the different homologous chromosome pairs in metaphase I. **3.** To show the mutant phenotype, a male needs to possess only one mutant allele. If this gene had been on a pair of autosomes, the two alleles would both have had to be mutant for an individual to show the recessive mutant phenotype, a much less probable situation.

### Concept Check 12.2
**1.** Because the gene for this eye-color character is located on the X chromosome, all female offspring will be red-eyed and heterozygous ($X^{w^+}X^w$); all male offspring will inherit a Y chromosome from the father and be white-eyed ($X^wY$). (Another way to say this is that $\frac{1}{2}$ the offspring will be red-eyed, heterozygous [carrier] females, and $\frac{1}{2}$ will be white-eyed males.) **2.** $\frac{1}{4}$ ($\frac{1}{2}$ chance that the child will inherit a Y chromosome from the father and be male × $\frac{1}{2}$ chance that he will inherit the X carrying the disease allele from his mother); if the child is a boy, there is a $\frac{1}{2}$ chance he will have the disease; a female would have zero chance (but $\frac{1}{2}$ chance of being a carrier). **3.** With a disorder caused by a dominant allele, there is no such thing as a "carrier," since those with the allele have the disorder. Because the allele is dominant, the females lose any "advantage" in having two X chromosomes, since one disorder-associated allele is sufficient to result in the disorder. All fathers who have the dominant allele will pass it along to *all* their daughters, who will also have the disorder. A mother who has the allele (and thus the disorder) will pass it to half of her sons and half of her daughters.

### Concept Check 12.3
**1.** Crossing over during meiosis I in the heterozygous parent produces some gametes with recombinant genotypes for the two genes. Offspring with a recombinant phenotype arise from fertilization of the recombinant gametes by homozygous recessive gametes from the double-mutant parent. **2.** In each case, the alleles contributed by the female parent (in the egg) determine the phenotype of the offspring because the male in this cross contributes only recessive alleles. Thus, identifying the phenotype of the offspring tells you what alleles were in its mother's (the dihybrid female's) egg. **3.** No. The order could be $A$-$C$-$B$ or $C$-$A$-$B$. To determine which possibility is correct, you need to know the recombination frequency between $B$ and $C$.

### Concept Check 12.4
**1.** In meiosis, a combined 14-21 chromosome will behave as one chromosome. If a gamete receives the combined 14-21 chromosome and a normal copy of chromosome 21, trisomy 21 will result when this gamete combines with a normal gamete during fertilization. **2.** No. The child can be either $I^A I^A i$ or $I^A ii$. A sperm of genotype $I^A I^A$ could result from nondisjunction in the father during meiosis II, while an egg with the genotype $ii$ could result from nondisjunction in the mother during either meiosis I or meiosis II. **3.** Activation of this gene could lead to the production of too much of this kinase. If the kinase is involved in a signaling pathway that triggers cell division, too much of it could trigger unrestricted cell division, which in turn could contribute to the development of a cancer (in this case, a cancer of one type of white blood cell). **4.** The inactivation of two X chromosomes in XXX women would leave them with one genetically active X, as in women with the normal number of chromosomes. Microscopy should reveal two Barr bodies in XXX women.

### Summary of Key Concepts Questions
**12.1** Because the sex chromosomes are different from each other and because they determine the sex of the offspring, Morgan could use the sex of the offspring as a phenotypic character to follow the parental chromosomes. (He could also have followed them under a microscope, as the X and Y chromosomes look different.) At the same time, he could record eye color to follow the eye-color alleles. **12.2** Males have only one X chromosome, along with a Y chromosome, while females have two X chromosomes. The Y chromosome has very few genes on it, while the X has about 1,000. When a recessive X-linked allele that causes a disorder is inherited by a male on the X from his mother, there isn't a second allele present on the Y (males are hemizygous), so the male has the disorder. Because females have two X chromosomes, they must inherit two recessive alleles in order to have the disorder, a rarer occurrence. **12.3** Crossing over results in new combinations of alleles. Crossing over is a random occurrence, and the more distance there is between two genes, the more chances there are for crossing over to occur, leading to a new allele combination. **12.4** In inversions and reciprocal translocations, the same genetic material

is present in the same relative amount but just organized differently. In aneuploidy, duplications, deletions, and nonreciprocal translocations, the balance of genetic material is upset, as large segments are either missing or present in more than one copy. Apparently, this type of imbalance is very damaging to the organism. (Although it isn't lethal in the developing embryo, the reciprocal translocation that produces the Philadelphia chromosome can lead to a serious condition, cancer, by altering the expression of important genes.)

**Test Your Understanding**
**1.** 0; ½; ¹⁄₁₆    **2.** Recessive; if the disorder were dominant, it would affect at least one parent of a child born with the disorder. The disorder's inheritance is sex-linked because it is seen only in boys. For a girl to have the disorder, she would have to inherit recessive alleles from *both* parents. This would be very rare, since males with the recessive allele on their X chromosome die in their early teens.    **3.** Between $T$ and $A$, 12%; between $A$ and $S$, 5%    **4.** Between $T$ and $S$, 18%; sequence of genes is $T$–$A$–$S$.
**5.** ¼ for each daughter (½ chance that the child will be female × ½ chance of a homozygous recessive genotype); ½ for first son    **6.** About one-third of the distance from the vestigial-wing locus to the brown-eye locus    **7.** 6%; wild-type heterozygous for normal wings and red eyes × recessive homozygous for vestigial wings and purple eyes
**8.** Fifty percent of the offspring will show phenotypes resulting from crossovers. These results would be the same as those from a cross where $A$ and $B$ were *not* on the same chromosome, and you would interpret the results to mean that the genes are unlinked. (Further crosses involving other genes on the same chromosome would reveal the genetic linkage and map distances.)    **9.** 450 each of blue-oval and white-round (parentals) and 50 each of blue-round and white-oval (recombinants)    **10.** (a) For each pair of genes, you had to generate an $F_1$ dihybrid fly; let's use the $A$ and $B$ genes as an example. You obtained homozygous parental flies, either the first with dominant alleles of the two genes ($AABB$) and the second with recessive alleles ($aabb$), or the first with dominant alleles of gene $A$ and recessive alleles of gene $B$ ($AAbb$) and the second with recessive alleles of gene $A$ and dominant alleles of gene $B$ ($aaBB$). Breeding either of these pairs of P generation flies gave you an $F_1$ dihybrid, which you then testcrossed with a doubly homozygous recessive fly ($aabb$). You classed the offspring as parental or recombinant, based on the genotypes of the P generation parents (either of the two pairs described above). You added up the number of recombinant types and then divided by the total number of offspring. This gave you the recombination percentage (in this case, 8%), which you can translate into map units (8 map units) to construct your map.
(b)

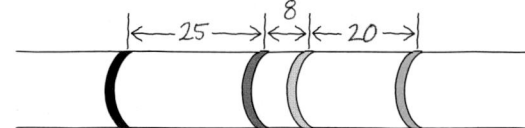

## Chapter 13

**Figure Questions**
**Figure 13.3** The living S cells found in the blood sample were able to reproduce to yield more S cells, indicating that the S trait is a permanent, heritable change, rather than just a one-time use of the dead S cells' capsules.    **Figure 13.5** The radioactivity would have been found in the pellet when proteins were labeled (batch 1) because proteins would have had to enter the bacterial cells to program them with genetic instructions. It's hard for us to imagine now, but the DNA might have played a structural role that allowed some of the proteins to be injected while it remained outside the bacterial cell (thus no radioactivity in the pellet in batch 2).    **Figure 13.13** The tube from the first replication would look the same, with a middle band of hybrid $^{15}$N - $^{14}$N DNA, but the second tube would not have the upper band of two light blue strands. Instead, it would have a bottom band of two dark blue strands, like the bottom band in the result predicted after one replication in the conservative model.    **Figure 13.15** In the bubble at the top of the micrograph in (b), arrows should be drawn pointing left and right to indicate the two replication forks.    **Figure 13.16** Looking at any of the DNA strands, we see that one end is called the 5′ end and the other the 3′ end. If we proceed from the 5′ end to the 3′ end on the left-most strand, for example, we list the components in this order: phosphate group → 5′ C of the sugar → 3′ C → phosphate → 5′ C → 3′ C. Going in the opposite direction on the same strand, the components proceed in the reverse order: 3′ C → 5′ C → phosphate. Thus, the two directions are distinguishable, which is what we mean when we say that the strands have directionality. (Review Figure 13.6 if necessary.)
**Figure 13.19**

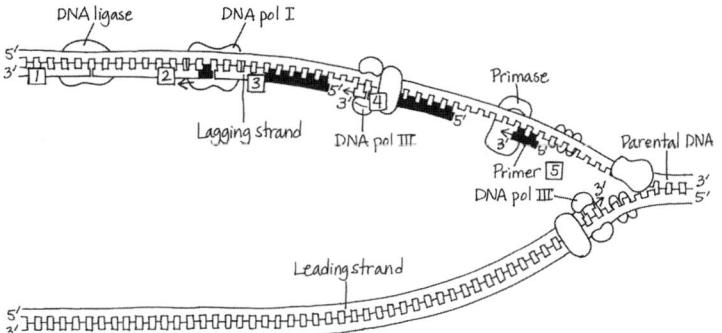

**Figure 13.20**

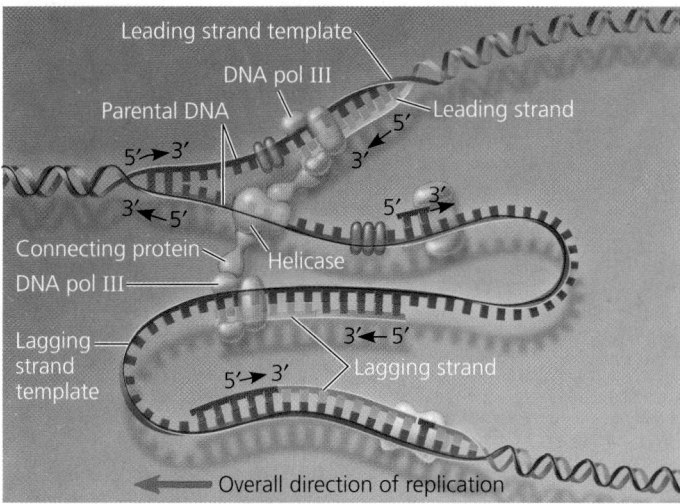

**Figure 13.25**

**Concept Check 13.1**
**1.** In order to tell which end is the 5′ end, you need to know which end has a phosphate group on the 5′ carbon (the 5′ end) or which end has an —OH group on the 3′ carbon (the 3′ end).    **2.** Griffith expected that the mouse injected with the mixture of heat-killed S cells and living R cells would survive, since neither type of cell alone could kill the mouse.

**Concept Check 13.2**
**1.** Complementary base pairing ensures that the two daughter molecules are exact copies of the parental molecule. When the two strands of the parental molecule separate, each serves as a template on which nucleotides are arranged, by the base-pairing rules, into new complementary strands.
**2.**

| Protein | Function |
|---|---|
| Helicase | Unwinds parental double helix at replication forks |
| Single-strand binding protein | Binds to and stabilizes single-stranded DNA until it can be used as a template |
| Topoisomerase | Relieves "overwinding" strain ahead of replication forks by breaking, swiveling, and rejoining DNA strands |
| Primase | Synthesizes an RNA primer at 5′ end of leading strand and at 5′ end of each Okazaki fragment of lagging strand |
| DNA pol III | Using parental DNA as a template, synthesizes new DNA strand by covalently adding nucleotides to 3′ end of a preexisting DNA strand or RNA primer |
| DNA pol I | Removes RNA nucleotides of previous fragment's primer from its 5′ end and replaces them with DNA nucleotides attached to 3′ end of next fragment |
| DNA ligase | Joins 3′ end of DNA that replaces primer to rest of leading strand and joins Okazaki fragments of lagging strand |

**3.** In the cell cycle, DNA synthesis occurs during the S phase, between the $G_1$ and $G_2$ phases of interphase. DNA replication is therefore complete before the mitotic phase begins.

**Concept Check 13.3**
**1.** A nucleosome is made up of eight histone proteins, two each of four different types, around which DNA is wound. Linker DNA runs from one nucleosome to the next.
**2.** Euchromatin is chromatin that becomes less compacted during interphase and is accessible to the cellular machinery responsible for gene activity. Heterochromatin, on the other hand, remains quite condensed during interphase and contains genes that are largely inaccessible to this machinery.

## Concept Check 13.4

**1.** The covalent sugar-phosphate bonds of the DNA strands   **2.** Yes, *Pvu*I will cut the molecule (at the position indicated by the dashed red line).

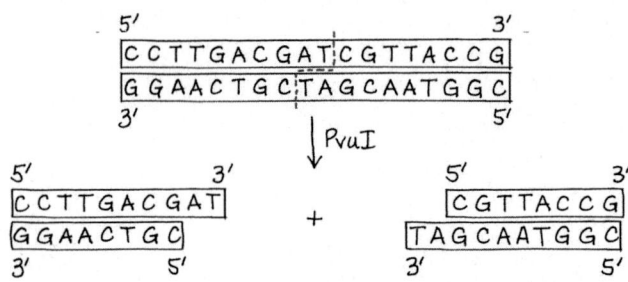

**3.** Cloning requires joining two pieces of DNA—a cloning vector, such as a bacterial plasmid, and a gene or DNA fragment from another source. Both pieces must be cut with the same restriction enzyme, creating sticky ends that will base-pair with complementary ends on other fragments. (The sugar-phosphate backbones will then be ligated together by ligase.) In PCR, the primers must base-pair with their target sequences in the DNA mixture, bracketing one specific region among many, and complementary base pairing is the basis for the building of the new strand during the extension step. In DNA sequencing, the primers base-pair to the template, allowing DNA synthesis to start, and then nucleotides are added to the growing strand based on complementarity of base pairing. Using the CRISPR-Cas9 system to edit genes involves an RNA used as a guide that is complementary to the sequence to be edited. Complementary base pairing anchors the Cas9 protein to the target DNA sequence it will then cut.

### Summary of Key Concepts Questions

**13.1** Each strand in the double helix has polarity, the end with a phosphate group on the 5′ carbon of the sugar being called the 5′ end, and the end with an —OH group on the 3′ carbon of the sugar being called the 3′ end. The two strands run in opposite directions, one running 5′ → 3′ and the other running 3′ → 5′. Thus, each end of the molecule has both a 5′ and a 3′ end. This arrangement is called "antiparallel." If the strands were parallel, they would both run 5′ → 3′ in the same direction, so an end of the molecule would have either two 5′ ends or two 3′ ends.   **13.2** On both the leading and lagging strands, DNA polymerase adds onto the 3′ end of an RNA primer synthesized by primase, synthesizing DNA in the 5′ → 3′ direction. Because the parental strands are antiparallel, however, only on the leading strand does synthesis proceed continuously into the replication fork. The lagging strand is synthesized bit by bit in the direction away from the fork as a series of shorter Okazaki fragments, which are later joined together by DNA ligase. Each fragment is initiated by synthesis of an RNA primer by primase as soon as a given stretch of single-stranded template strand is opened up. Although both strands are synthesized at the same rate, synthesis of the lagging strand is delayed because initiation of each fragment begins only when sufficient template strand is available.   **13.3** Most of the chromatin in an interphase nucleus is fairly uncondensed. Much is present as the 30-nm fiber, with some in the form of the 10-nm fiber and some as looped domains of the 30-nm fiber. (These different levels of chromatin packing may reflect differences in gene expression occurring in these regions.) Also, a small percentage of the chromatin, such as that at the centromeres and telomeres, is highly condensed heterochromatin.   **13.4** A plasmid vector and a source of foreign DNA to be cloned are both cut with the same restriction enzyme, generating restriction fragments with sticky ends. These fragments are mixed together, ligated, and reintroduced into bacterial cells, which can then make many copies of the foreign DNA or its product.

### Test Your Understanding

**1.** C   **2.** A   **3.** B   **4.** D   **5.** C   **6.** B   **7.** D   **8.** B   **9.** A   **10.** Like histones, the *E. coli* proteins would be expected to contain many basic (positively charged) amino acids, such as lysine and arginine, which can form weak bonds with the negatively charged phosphate groups on the sugar-phosphate backbone of the DNA molecule.

**11.**

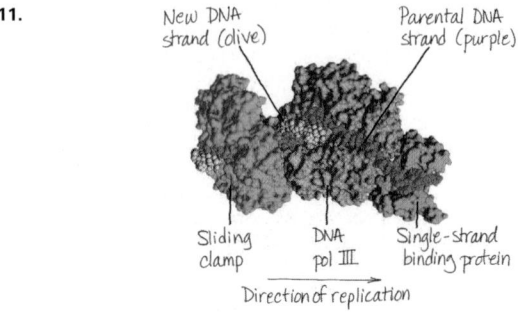

## Chapter 14

### Figure Questions

**Figure 14.5** The mRNA sequence (5′-UGGUUUGGCUCA-3′) is the same as the nontemplate DNA strand sequence (5′-TGGTTTGGCTCA-3′), except there is a U

in the mRNA wherever there is a T in the DNA.   **Figure 14.8** The processes are similar in that polymerases form polynucleotides complementary to an antiparallel DNA template strand. In replication, however, both strands act as templates, whereas in transcription, only one DNA strand acts as a template.   **Figure 14.9** The RNA polymerase would bind directly to the promoter, rather than being dependent on the previous binding of other factors.

**Figure 14.12**

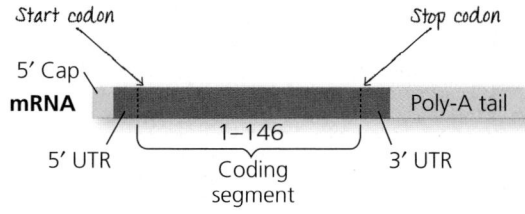

**Figure 14.23** The mRNA farthest to the right (the longest one) started being transcribed first. The ribosome at the top, closest to the DNA, started translating first and thus has the longest polypeptide.

### Concept Check 14.1

**1.** Recessive   **2.** A polypeptide made up of 10 Gly (glycine) amino acids

**3.**

> Nontemplate sequence in 3′ to 5′ order:   3′–ACGACTGAA–5′
>
> mRNA sequence:   5′–UGCUGACUU–3′
>
> Translated:   Cys–STOP

If the nontemplate sequence could have been used as a template for transcribing the mRNA, the protein translated from the mRNA would have a completely different amino acid sequence and would most likely be nonfunctional. (It would also be shorter because of the UGA stop signal shown in the mRNA sequence above—and possibly others earlier in the mRNA sequence.)

### Concept Check 14.2

**1.** A promoter is the region of DNA to which RNA polymerase binds to begin transcription. It is at the upstream end of the gene (transcription unit).   **2.** In a bacterial cell, part of the RNA polymerase recognizes the gene's promoter and binds to it. In a eukaryotic cell, transcription factors mediate the binding of RNA polymerase to the promoter. In both cases, sequences in the promoter bind precisely to the RNA polymerase, so the enzyme is in the right location and orientation.   **3.** The transcription factor that recognizes the TATA sequence would be unable to bind, so RNA polymerase could not bind and transcription of that gene probably would not occur.

### Concept Check 14.3

**1.** Due to alternative splicing of exons, each gene can result in multiple different mRNAs and can thus direct synthesis of multiple different proteins.   **2.** In watching a show recorded with a DVR, you watch segments of the show itself (exons) and fast-forward through the commercials, which are thus like introns. However, unlike introns, commercials remain in the recording, while the introns are cut out of the RNA transcript during RNA processing.   **3.** Once the mRNA has exited the nucleus, the cap prevents it from being degraded by hydrolytic enzymes and facilitates its attachment to ribosomes. If the cap were removed from all mRNAs, the cell would no longer be able to synthesize any proteins and would probably die.

### Concept Check 14.4

**1.** First, each aminoacyl-tRNA synthetase specifically recognizes a single amino acid and attaches it only to an appropriate tRNA. Second, a tRNA charged with its specific amino acid binds only to an mRNA codon for that amino acid.   **2.** The structure and function of the ribosome seem to depend more on the rRNAs than on the ribosomal proteins. Because it is single-stranded, an RNA molecule can hydrogen-bond with itself and with other RNA molecules. RNA molecules make up the interface between the two ribosomal subunits, so presumably RNA-RNA binding helps hold the ribosome together. The binding site for mRNA in the ribosome includes rRNA that can bind the mRNA. Also, complementary hydrogen bonding within an RNA molecule allows it to assume a particular three-dimensional shape and, along with the RNA's functional groups, presumably enables rRNA to catalyze peptide bond formation during translation.   **3.** A signal peptide on the leading end of the polypeptide being synthesized is recognized by a signal-recognition particle that brings the ribosome to the ER membrane. There the ribosome attaches and continues to synthesize the polypeptide, depositing it in the ER lumen.   **4.** Because of wobble, the tRNA could bind to either 5′-GCA-3′ or 5′-GCG-3′, both of which code for alanine (Ala). Alanine would be attached to the tRNA (see diagram, upper right).

## Concept Check 14.5

**1.** In the mRNA, the reading frame downstream from the deletion is shifted, leading to a long string of incorrect amino acids in the polypeptide, and in most cases, a stop codon will arise, leading to premature termination. The polypeptide will most likely be nonfunctional. **2.** Heterozygous individuals, said to have sickle-cell trait, have a copy each of the wild-type allele and the sickle-cell allele. Both alleles will be expressed, so these individuals will have both normal and sickle-cell hemoglobin molecules. Apparently, having a mix of the two forms of β-globin has no effect under most conditions, but during prolonged periods of low blood oxygen (such as at higher altitudes), these individuals can show some signs of sickle-cell disease.

**3.**

Normal DNA sequence
(template strand is on top):

3′–TACTTGTCCGATATC–5′
5′–ATGAACAGGCTATAG–3′

mRNA sequence:

5′–AUGAACAGGCUAUAG–3′

Amino acid sequence:

Met-Asn-Arg-Leu-STOP

---

Mutated DNA sequence
(template strand is on top):

3′–TACTTGTCCAATATC–5′
5′–ATGAACAGGTTATAG–3′

mRNA sequence:

5′–AUGAACAGGUUAUAG–3′

Amino acid sequence:

Met-Asn-Arg-Leu-STOP

No effect: The amino acid sequence is Met-Asn-Arg-Leu both before and after the mutation because the mRNA codons 5′-CUA-3′ and 5′-UUA-3′ both code for Leu. (The fifth codon is a stop codon.)

### Summary of Key Concepts Questions

**14.1** A gene contains genetic information in the form of a nucleotide sequence. The gene is first transcribed into an RNA molecule, and a messenger RNA molecule is ultimately translated into a polypeptide. The polypeptide makes up part or all of a protein, which performs a function in the cell and contributes to the phenotype of the organism. **14.2** Both bacterial and eukaryotic genes have promoters, regions where RNA polymerase ultimately binds and begins transcription. In bacteria, RNA polymerase binds directly to the promoter; in eukaryotes, transcription factors bind first to the promoter, and then RNA polymerase binds to the transcription factors and promoter together. **14.3** Both the 5′ cap and the 3′ poly-A tail help the mRNA exit from the nucleus and then, in the cytoplasm, help ensure mRNA stability and allow it to bind to ribosomes. **14.4** In the context of the ribosome, tRNAs function as translators between the nucleotide-based language of mRNA and the amino-acid-based language of polypeptides. A tRNA carries a specific amino acid, and the anticodon on the tRNA is complementary to the codon on the mRNA that codes for that amino acid. In the ribosome, the tRNA binds to the A site. Then the polypeptide being synthesized (currently on the tRNA in the P site) is joined to the new amino acid, which becomes the new (C-terminal) end of the polypeptide. Next, the tRNA in the A site moves to the P site. After the polypeptide is transferred to the new tRNA, thus adding the new amino acid, the now empty tRNA moves from the P site to the E site, where it exits the ribosome. **14.5** When a nucleotide base is altered chemically, its base-pairing characteristics may be changed. When that happens, an incorrect nucleotide is likely to be incorporated into the complementary strand during the next replication of the DNA, and successive rounds of replication will perpetuate the mutation. Once the gene is transcribed, the mutated codon may code for a different amino acid that inhibits or changes the function of a protein. If the chemical change in the base is detected and repaired by the DNA repair system before the next replication, no mutation will result.

### Test Your Understanding

**1.** B **2.** C **3.** A **4.** A **5.** B **6.** C **7.** D **8.** No. Transcription and translation are separated in space and time in a eukaryotic cell, as a result of the eukaryotic cell's nuclear membrane.

**9.**

| Type of RNA | Functions |
|---|---|
| Messenger RNA (mRNA) | Carries information specifying amino acid sequences of proteins from DNA to ribosomes |
| Transfer RNA (tRNA) | Serves as translator molecule in protein synthesis; translates mRNA codons into amino acids |
| Ribosomal RNA (rRNA) | Plays catalytic (ribozyme) roles and structural roles in ribosomes |
| Primary transcript | Is a precursor to mRNA, rRNA, or tRNA, before being processed; some intron RNA acts as a ribozyme, catalyzing its own splicing |
| Small RNAs in spliceosome | Play structural and catalytic roles in spliceosomes, the complexes of protein and RNA that splice pre-mRNA |

## Chapter 15

### Figure Questions

**Figure 15.3** As the concentration of tryptophan in the cell falls, eventually there will be none bound to *trp* repressor molecules. These will then change into their inactive shapes and dissociate from the operator, allowing transcription of the operon to resume. The enzymes for tryptophan synthesis will be made, and they will again synthesize tryptophan in the cell. **Figure 15.11** In both types of cell, the albumin gene enhancer has the three control elements colored yellow, gray, and red. The sequences in the liver and lens cells would be identical, since the cells are in the same organism.

### Concept Check 15.1

**1.** Binding by the *trp* corepressor (tryptophan) activates the *trp* repressor, allowing it to bind to the *trp* operator, shutting off transcription of the *trp* operon. Binding by the *lac* inducer (allolactose) inactivates the *lac* repressor so it can no longer bind to the *lac* operator, leading to transcription of the *lac* operon. **2.** When glucose is scarce, cAMP is bound to CRP and CRP is bound to the *lac* promoter, favoring the binding of RNA polymerase. However, in the absence of lactose, the *lac* repressor is bound to the operator, blocking RNA polymerase binding to the promoter. Therefore, the *lac* operon genes are not transcribed. **3.** The cell would continuously produce β-galactosidase and the two other enzymes for using lactose, even in the absence of lactose, thus wasting cell resources.

### Concept Check 15.2

**1.** Histone acetylation is generally associated with gene expression, while DNA methylation is generally associated with lack of expression. **2.** General transcription factors function in assembling the transcription initiation complex at the promoters for all genes. Specific transcription factors bind to control elements associated with a particular gene and, once bound, either increase (activators) or decrease (repressors) transcription of that gene. **3.** The three genes should have some similar or identical sequences in the control elements of their enhancers. Because of this similarity, the same specific transcription factors in muscle cells could bind to the enhancers of all three genes and stimulate their expression coordinately.

### Concept Check 15.3

**1.** The mRNA would persist and be translated into the cell division–promoting protein, and the cell would probably divide. If the intact miRNA is necessary for inhibition of cell division, then division of this cell might be inappropriate. Uncontrolled cell division could lead to formation of a mass of cells (tumor) that prevents proper functioning of the organism and could contribute to the development of cancer. **2.** The *XIST* RNA is transcribed from the *XIST* gene on the X chromosome that will be inactivated. It then binds to that chromosome and induces heterochromatin formation. A likely model is that the *XIST* RNA somehow recruits chromatin modification enzymes that lead to formation of heterochromatin.

### Concept Check 15.4

**1.** In RT-PCR, the primers must base-pair with their target sequences in the DNA mixture, locating one specific region among many. Also, the DNA polymerase (e.g., Taq polymerase) used in PCR relies on complementary base pairing to the template strand to add new nucleotides during synthesis of the fragments. In DNA microarray analysis, the labeled probe binds only to the specific target sequence due to complementary nucleic acid hybridization (DNA-DNA hybridization). **2.** As a researcher interested in cancer, you would want to study genes represented by spots that are green or red because these are genes for which the expression level differs between the two types of tissues. Some of these genes may be expressed differently as a result of cancer, while others might play a role in causing cancer, so both would be of interest.

### Summary of Key Concepts Questions

**15.1** A corepressor and an inducer are both small molecules that bind to the repressor protein in an operon, causing the repressor to change shape. In the case of a corepressor (like tryptophan), this shape change allows the repressor to bind to the operator, blocking transcription. In contrast, an inducer causes the repressor to dissociate from the operator, allowing transcription to begin. **15.2** In that specific type of cell, the chromatin must not be tightly condensed because it must be accessible to transcription factors. The appropriate specific transcription factors (activators), which are made in that type of cell, must bind to the control elements in the enhancer of the gene, while repressors must not be bound. The DNA must be bent by a bending protein so the activators can contact the mediator proteins and form a complex with general transcription factors at the promoter. Then RNA polymerase must bind and begin transcription. **15.3** miRNAs do not "code" for the amino acids of a protein—they are never translated. Each miRNA associates with a group of proteins to form a complex. Binding of the complex to an mRNA with a complementary sequence causes that mRNA to be degraded or blocks its translation. This is considered gene regulation because it controls the amount of a particular mRNA that can be translated into a functional protein. **15.4** The genes that are expressed in a given tissue or cell type determine the proteins (and noncoding RNAs) that are the basis of the structure and functions of that tissue or cell type. Understanding which groups of interacting genes establish particular structures and carry out certain functions will help us learn how the parts of an organism form and are maintained. We will also be better able to treat diseases that occur when faulty gene expression leads to malfunctioning tissues.

### Test Your Understanding

**1.** C **2.** B **3.** C **4.** D **5.** C **6.** B

**7.**

(a)

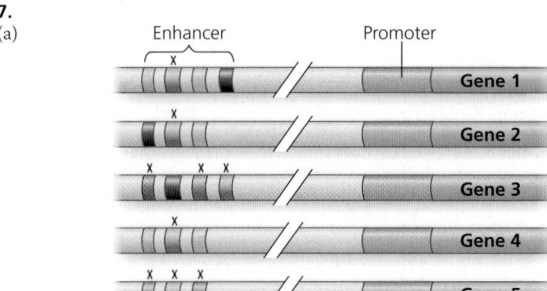

The purple, blue, and red activator proteins would be present.

(b)

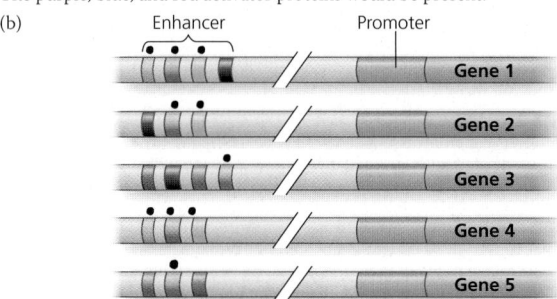

Only gene 4 would be transcribed. (c) In nerve cells, the orange, blue, green, and black activators would have to be present, thus activating transcription of genes 1, 2, and 4. In skin cells, the red, black, purple, and blue activators would have to be present, thus activating genes 3 and 5.

## Chapter 16

### Figure Questions

**Figure 16.4** Even if the mutant MyoD protein couldn't activate the *myoD* gene, it could still turn on genes for the other proteins in the pathway (other transcription factors, which would turn on the genes for muscle-specific proteins, for example). Therefore, some differentiation would occur. But unless there were other activators that could compensate for the loss of the MyoD protein's activation of the *myoD* gene, the cell would not be able to maintain its differentiated state. **Figure 16.10** Normal Bicoid protein would be made in the anterior end and compensate for the presence of mutant *bicoid* mRNA put into the egg by the mother. Development should be normal, with a head present. (This is what was observed.) **Figure 16.11** None of the eggs with the transplanted nuclei from the four-cell embryo at the upper left would have developed into a tadpole. Also, the resulting samples might include only some of the tissues of a tadpole. The tissues that develop might differ from treatment to treatment, depending on which of the four nuclei was transplanted. (This assumes that there was some way to tell the four cells apart, as one can in some frog species.) **Figure 16.16** Using converted iPS cells would not carry the same risk, which is its major advantage. Because the donor cells would come from the patient, they would be perfectly matched. The patient's immune system would recognize them as "self" cells and would not mount an attack (which is what leads to rejection). **Figure 16.21** Cancer is a disease in which cell division occurs without its usual regulation. Cell division can be stimulated by growth factors (see Figure 9.17), which bind to cell-surface receptors (see Figure 5.20). Cancer cells evade these normal controls and can often divide in the absence of growth factors (see Figure 9.18). This suggests that the receptor proteins or some other components in a signaling pathway are abnormal in some way (see, for example, the mutant Ras protein in Figure 16.18) or are expressed at abnormal levels, as seen for the receptors in this figure. Under some circumstances in the mammalian body, steroid hormones such as estrogen and progesterone can also promote cell division. These molecules also use cell-signaling pathways, as described in Concept 5.6 (see Figure 5.23). Because signaling receptors are involved in triggering cells to undergo cell division, it is not surprising that altered genes encoding these proteins might play a significant role in the development of cancer. Genes might be altered either through a mutation that changes the function of the protein product or through a mutation that causes the gene to be expressed at abnormal levels that disrupt the overall regulation of the signaling pathway.

### Concept Check 16.1

**1.** Cells undergo differentiation during embryonic development, becoming different from each other. Therefore, the adult organism is made up of many highly specialized cell types. **2.** By binding to a receptor on the receiving cell's surface and triggering a signal transduction pathway involving intracellular molecules such as second messengers and transcription factors that affect gene expression **3.** The products of maternal effect genes, made and deposited into the egg by the mother, determine the head and tail ends, as well as the back and belly, of the egg and embryo (and eventually the adult fly).

### Concept Check 16.2

**1.** The state of chromatin modification in the nucleus from the intestinal cell was undoubtedly less similar to that of a nucleus from a fertilized egg, explaining why many fewer of these nuclei were able to be reprogrammed. In contrast, the chromatin

in a nucleus from a cell at the four-cell stage would have been much more like that of a nucleus in a fertilized egg and therefore much more easily programmed to direct development. **2.** Dolly's face matched that of the donor of the nucleus, because the nucleus contained the genetic information that determined Dolly's phenotype. No genetic information was contributed by either the egg donor (the egg was enucleated) or the surrogate mother. **3.** A technique would have to be worked out for turning a human iPS cell into a pancreatic cell (probably by inducing expression of pancreas-specific regulatory genes in the cell).

### Concept Check 16.3

**1.** Apoptosis is signaled by p53 protein when a cell has extensive DNA damage, so apoptosis plays a protective role in eliminating a cell that might contribute to cancer. If mutations in the genes in the apoptotic pathway blocked apoptosis, a cell with such damage could continue to divide and might lead to tumor formation. **2.** When an individual has inherited an oncogene or a mutant allele of a tumor-suppressor gene **3.** A cancer-causing mutation in a proto-oncogene usually makes the gene product overactive, whereas a cancer-causing mutation in a tumor-suppressor gene usually makes the gene product nonfunctional.

### Summary of Key Concepts Questions

**16.1** The first process involves cytoplasmic determinants, including mRNAs and proteins, placed into specific locations in the egg by the mother. The cells that are formed from different regions in the egg during early cell divisions will have different proteins in them, which will direct different programs of gene expression. The second process involves the cell in question responding to signaling molecules secreted by neighboring cells. The signaling pathway in the responding cells also leads to a different pattern of gene expression. The coordination of these two processes results in each cell following a unique pathway in the developing embryo. **16.2** Cloning a mouse involves transplanting a nucleus from a differentiated mouse cell into a mouse egg cell that has had its own nucleus removed. Activating the egg cell and promoting its development into an embryo in a surrogate mother results in a mouse that is genetically identical to the mouse that donated the nucleus. In this case, the differentiated nucleus has been reprogrammed by factors in the egg cytoplasm. Mouse ES cells are generated from inner cells in mouse blastocysts, so in this case the cells are "naturally" reprogrammed by the process of reproduction and development. (Cloned mouse embryos can also be used as a source of ES cells.) iPS cells can be generated without the use of embryos from a differentiated adult mouse cell by adding certain transcription factors into the cell. In this case, the transcription factors are reprogramming the cells to become pluripotent. **16.3** The protein product of a proto-oncogene is usually involved in a pathway that stimulates cell division. The protein product of a tumor-suppressor gene is usually involved in a pathway that inhibits cell division.

### Test Your Understanding

**1.** A **2.** C **3.** C **4.** A **5.** D

## Chapter 17

### Figure Questions

**Figure 17.3** Top vertical arrow: Infection. Left upper arrow: Replication. Right upper arrow: Transcription. Right middle arrow: Translation. Lower left and right arrows: Self-assembly. Bottom middle arrow: Exit. **Figure 17.8** There are many steps that could be interfered with: binding of the virus to the cell, reverse transcriptase function, integration into the host cell chromosome, genome synthesis (in this case, transcription of RNA from the integrated provirus), assembly of the virus inside the cell, and budding of the virus. (Many of these, if not all, are targets of actual medical strategies to block progress of the infection in HIV-infected people.) **Figure 17.9** The shape of a protein on the HIV surface is likely to be complementary to the shape of the receptor (CD4) and also to that of the co-receptor (CCR5). A molecule with a shape similar to the HIV surface protein could bind CCR5, blocking HIV binding. (Another answer would be a molecule that bound to CCR5 and changed the shape of CCR5 so it could no longer bind to HIV.)

### Concept Check 17.1

**1.** TMV consists of one molecule of RNA surrounded by a helical array of proteins. The influenza virus has eight molecules of RNA, each surrounded by a helical array of proteins, similar to the arrangement of the single RNA molecule in TMV. Another difference between the viruses is that the influenza virus has an outer envelope and TMV does not. **2.** The T2 phages were an excellent choice for use in the Hershey-Chase experiment because they consist of only DNA surrounded by a protein coat, and DNA and protein were the two candidates for macromolecules that carried genetic information. Hershey and Chase were able to radioactively label each type of molecule alone and follow it during separate infections of *E. coli* cells with T2. Only the DNA entered the bacterial cell during infection, and only labeled DNA showed up in some of the progeny phage. Hershey and Chase concluded that the DNA must carry the genetic information necessary for the phage to reprogram the cell and produce progeny phages.

### Concept Check 17.2

**1.** Lytic phages can only carry out lysis of the host cell, whereas lysogenic phages may either lyse the host cell or integrate into the host chromosome. In the latter case, the viral DNA (prophage) is simply replicated along with the host chromosome. Under certain conditions, a prophage may exit the host chromosome and initiate a lytic cycle. **2.** Both the viral RNA polymerase and the cellular RNA polymerase in Figure 14.10 synthesize an RNA molecule complementary to a template strand. However, the cellular RNA polymerase in Figure 14.10 uses one of the strands of the DNA double helix as a template, whereas the RNA of the viral RNA polymerase uses the RNA of the viral genome as a template. **3.** HIV is called a retrovirus because it synthesizes DNA using its RNA genome as a template. This is the reverse ("retro") of the usual DNA → RNA information flow. **4.** Both the CRISPR system and miRNAs involve RNA molecules bound in a protein complex and acting as "homing devices" that enable the complex to bind a complementary sequence, but miRNAs are involved in regulating gene expression (by affecting mRNAs) and the CRISPR system protects bacterial cells from foreign invaders (infecting phages). Thus the CRISPR system is more like an immune system than are miRNAs.

### Concept Check 17.3

**1.** Mutations can lead to a new strain of a virus that can no longer be effectively fought by the immune system, even if an animal had been exposed to the original strain; a virus can jump from one species to a new host; and a rare virus can spread if a host population becomes less isolated.   **2.** In horizontal transmission, a plant is infected from an external source of the virus, which enters through a break in the plant's epidermis due to damage by herbivores or other agents. In vertical transmission, a plant inherits viruses from its parent either via infected seeds (sexual reproduction) or via an infected cutting (asexual reproduction).   **3.** Humans are not within the host range of TMV, so they can't be infected by the virus. (TMV can't bind to receptors on human cells and infect them.)

### Summary of Key Concepts Questions

**17.1** Viruses are generally considered nonliving because they are not capable of replicating outside of a host cell and are unable to carry out the energy-transforming reactions of metabolism. To replicate and carry out metabolism, they depend completely on host enzymes and resources.   **17.2** Single-stranded RNA viruses require an RNA polymerase that can make RNA using an RNA template. (Cellular RNA polymerases make RNA using a DNA template.) Retroviruses require reverse transcriptases to make DNA using an RNA template. (Once the first DNA strand has been made, the same enzyme can promote synthesis of the second DNA strand.)   **17.3** The mutation rate of RNA viruses is higher than that of DNA viruses because RNA polymerase has no proofreading function, so errors in replication are not corrected. Their higher mutation rate means that RNA viruses change faster than DNA viruses, leading to their being able to have an altered host range and to evade immune defenses in possible hosts.

### Test Your Understanding

**1.** C   **2.** D   **3.** C   **4.** D   **5.** B
**6.** As shown in the sketch, the viral genome would be translated into capsid proteins and envelope glycoproteins directly, rather than after a complementary RNA copy was made. A complementary RNA strand would still be made, however, that could be used as a template for many new copies of the viral genome.

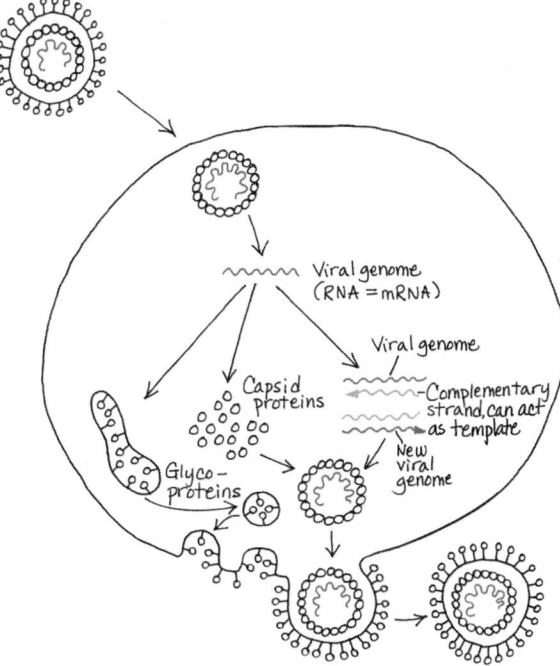

## Chapter 18

### Figure Questions

**Figure 18.2** In stage 2 of this figure, the order of the fragments relative to each other is not known and will be determined later by computer. The unordered nature of the fragments is reflected by their scattered arrangement in the diagram.   **Figure 18.7** The transposon would be cut out of the DNA at the original site rather than copied, so the figure would show the original stretch of DNA without the transposon after the mobile transposon had been cut out.   **Figure 18.9** The RNA transcripts extending from the DNA in each transcription unit are shorter on the left and longer on the right. This means that RNA polymerase must be starting on the left end of the unit and moving toward the right.
**Figure 18.12**

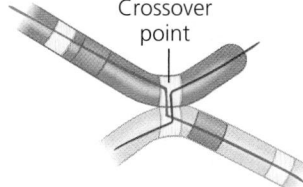

Crossover point

**Figure 18.13** Pseudogenes are nonfunctional. They could have arisen by any mutations in the second copy that made the gene product unable to function. Examples would be base changes that introduce stop codons in the sequence, alter amino acids, or change a region of the gene promoter so that the gene can no longer be expressed.   **Figure 18.14** Let's say a transposable element (TE) existed in the intron to the left of the indicated EGF exon in the EGF gene, and the same TE was present in the intron to the right of the indicated F exon in the fibronectin gene. During meiotic recombination, these TEs could cause nonsister chromatids on homologous chromosomes to pair up incorrectly, as seen in Figure 18.12. One gene might end up with an F exon next to an EGF exon. Further mistakes in pairing over many generations might result in these two exons being separated from the rest of the gene and placed next to a single or duplicated K exon. In general, the presence of repeated sequences in introns and between genes facilitates these processes because it allows incorrect pairing of nonsister chromatids, leading to novel exon combinations.

### Concept Check 18.1

**1.** In the whole-genome shotgun approach, short fragments are generated by cutting the genome with multiple restriction enzymes. These fragments are cloned, sequenced, and then ordered by computer programs that identify overlapping regions (see Figure 18.2).

### Concept Check 18.2

**1.** The Internet allows centralization of databases such as GenBank and software resources such as BLAST, making them freely accessible. Having all the data in a central database, easily accessible on the Internet, minimizes the possibility of errors and of researchers working with different data. It streamlines the process of science, since all researchers are able to use the same software programs, rather than each having to obtain their own, possibly different, software. It speeds up dissemination of data and ensures as much as possible that errors are corrected in a timely fashion. These are just a few answers; you can probably think of more.   **2.** Cancer is a disease caused by multiple factors. To focus on a single gene or a single defect would ignore other factors that may influence the cancer and even the behavior of the single gene being studied. The systems approach, because it takes into account many factors at the same time, is more likely to lead to an understanding of the causes and most useful treatments for cancer.   **3.** Some of the transcribed region is accounted for by introns. The rest is transcribed into noncoding RNAs, including small RNAs, such as microRNAs (miRNAs) or siRNAs. These RNAs help regulate gene expression by blocking translation, causing degradation of mRNA, binding to the promoter and repressing transcription, or causing remodeling of chromatin structure. The longer noncoding RNAs may also contribute to gene regulation or to chromatin remodeling.

### Concept Check 18.3

**1.** Alternative splicing of RNA transcripts from a gene and post-translational processing of polypeptides   **2.** At the top of the web page, you can see the number of genomes completed and those considered permanent drafts in a bar graph by year. Scrolling down, you can see the number of complete and incomplete sequencing projects by year, the number of projects by domain by year (the genomes of viruses and metagenomes are counted too, even though these are not "domains"), the phylogenetic distribution of bacterial genome projects, and projects by sequencing center. Finally, near the bottom, you can see a pie chart of the "Project Relevance of Bacterial Genome Projects," which shows that about 47% have medical relevance. The web page ends with another pie chart showing the sequencing centers for archaeal and bacterial projects.
**3.** Prokaryotes are generally smaller cells than eukaryotic cells, and they reproduce by binary fission. The evolutionary process involved is natural selection for more quickly reproducing cells: The faster they can replicate their DNA and divide, the more likely they will be able to dominate a population of prokaryotes. The less DNA they have to replicate, then, the faster they will reproduce.

### Concept Check 18.4

**1.** The number of genes is higher in mammals, and the amount of noncoding DNA is greater. Also, the presence of introns in mammalian genes makes them larger, on average, than prokaryotic genes.   **2.** In the copy-and-paste transposon mechanism and in retrotransposition   **3.** In the rRNA gene family, identical transcription units for all three different RNA products are present in long arrays, repeated one after the other. The large number of copies of the rRNA genes enables organisms to produce the rRNA for enough ribosomes to carry out active protein synthesis, and the single transcription unit for the three rRNAs ensures that the relative amounts of the different rRNA molecules produced are correct—every time one rRNA is made, a copy of each of the other two is made as well. Rather than including many identical units, each globin gene family consists of a relatively small number of nonidentical genes. The differences in the globin proteins encoded by these genes result in production of hemoglobin molecules adapted to particular developmental stages of the organism.

### Concept Check 18.5

**1.** If meiosis is faulty, two copies of the entire genome can end up in a single cell. Errors in crossing over during meiosis can lead to one segment being duplicated while another is deleted. During DNA replication, slippage backward along the template strand can result in segment duplication.   **2.** For either gene, a mistake in crossing over during meiosis could have occurred between the two copies of that gene, such that one ended up with a duplicated exon. (The other copy would have ended up with a deleted exon.) This could have happened several times, resulting in the multiple copies of a particular exon in each gene.   **3.** Homologous transposable elements scattered throughout the genome provide sites where recombination can occur between different chromosomes. Movement of these elements into coding or regulatory sequences may change expression of genes. Transposable elements also can carry genes with them, leading to dispersion of genes and in some cases different patterns of expression. Transport of an exon during transposition and its insertion into a gene may add a new functional domain to the originally encoded protein, a type of exon shuffling. (For any of these changes to be heritable, they must happen in germ cells, cells that will give rise to gametes.)

## Concept Check 18.6

**1.** Because both humans and macaques are primates, their genomes are expected to be more similar than the macaque and mouse genomes are. The mouse lineage diverged from the primate lineage before the human and macaque lineages diverged. **2.** Homeotic genes differ in their *non*homeobox sequences, which determine the interactions of homeotic gene products with other transcription factors and hence which genes are regulated by the homeotic genes. These nonhomeobox sequences differ in the two organisms, as do the expression patterns of the homeobox genes.

### Summary of Key Concepts Questions

**18.1** One focus of the Human Genome Project was to improve sequencing technology in order to speed up the process. During the project, many advances in sequencing technology allowed faster reactions, which were therefore less expensive. **18.2** The most significant finding was that more than 75% of the human genome appears to be transcribed at some point in at least one of the cell types studied. Also, at least 80% of the genome contains an element that is functional, participating in gene regulation or maintaining chromatin structure in some way. The project was expanded to include other species to further investigate the functions of these transcribed DNA elements. It is necessary to carry out this type of analysis on the genomes of species that are able to be used in laboratory experiments. **18.3** (a) In general, bacteria and archaea have smaller genomes, lower numbers of genes, and higher gene density than eukaryotes. (b) Among eukaryotes, there is no apparent systematic relationship between genome size and phenotype. The number of genes is often lower than would be expected from the size of the genome—in other words, the gene density is often lower in larger genomes. (Humans are an example.) **18.4** Transposable elements can move from place to place in the genome, and some of these sequences make a new copy of themselves when they do so. Thus, it is not surprising that they make up a significant percentage of the genome, and this percentage might be expected to increase over evolutionary time. **18.5** Chromosomal rearrangements within a species lead to some individuals having different chromosomal arrangements. Each of these individuals could still undergo meiosis and produce gametes, and fertilization involving gametes with different chromosomal arrangements could result in viable offspring. However, during meiosis in the offspring, the maternal and paternal chromosomes might not be able to pair up, causing gametes with incomplete sets of chromosomes to form. Most often, when zygotes are produced from such gametes, they do not survive. Ultimately, a new species could form if two different chromosomal arrangements became prevalent within a population and individuals could mate successfully only with other individuals having the same arrangement. **18.6** Comparing the genomes of closely related species can reveal information about more recent evolutionary events, perhaps events that resulted in the distinguishing characteristics of the species. Comparing the genomes of very distantly related species can tell us about evolutionary events that occurred a very long time ago. For example, genes that are shared between distantly related species must have arisen before those species diverged.

### Test Your Understanding

**1.** B **2.** A **3.** C **4.**

1. ATETI…PKSSD…TSSTT…NARRD

2. ATETI…PKSS[E]…TSS[TT]…[N]ARRD

3. ATETI…PKSSD…TSSTT…NARRD

4. ATETI…PKSSD…TSS[NT]…[S]ARRD

5. ATETI…PKSSD…TSSTT…NARRD

6. VTETI…PKSSD…TSSTT…NARRD

(a) Lines 1, 3, and 5 are the C, G, R species. (b) Line 4 is the human sequence. See the above figure for the difference between the human and C, G, R sequences—the underlined amino acids at which the human sequence has an N where the C, G, R sequences have a T, and an S where C, G, R have an N. (c) Line 6 is the orangutan sequence. (d) See the figure. There is one amino acid difference between the mouse (the circled E on line 2) and the C, G, R species (which have a D in that position). There are three amino acid differences between the mouse and the human. (The boxed E, T, and N in the mouse sequence are instead D, N, and S, respectively, in the human sequence.) (e) Because only one amino acid difference arose during the 60–100 million years since the mouse and C, G, R species diverged, it is somewhat surprising that two additional amino acid differences resulted during the 6 million years since chimpanzees and humans diverged. This indicates that the *FOXP2* gene has been evolving faster in the human lineage than in the lineages of other primates.

## Chapter 19

### Figure Questions

**Figure 19.6** The common cactus finch is more closely related to the large ground finch; Figure 1.17 shows that they share a more recent common ancestor than the common cactus finch shares with the green warbler finch. **Figure 19.9** The common ancestor lived more than 5.5 million years ago. **Figure 19.13** The colors and body forms of these mantises allow them to blend into their surroundings, providing an example of how organisms are well suited for life in their environments. The mantises also share features with one another (and with all other mantises), such as six legs, grasping forelimbs, and large eyes. These shared features illustrate another key observation about life: the unity of life that results from descent from a common ancestor. Over time, as these mantises diverged from a common ancestor, they accumulated different adaptations that made them well suited for life in their different environments. Eventually, as enough differences accumulated between mantis populations, new species were formed, thus contributing to the great diversity of life. **Figure 19.14** These

results show that being reared from the egg stage on one plant species or the other did not result in the adult having a beak length appropriate for that host; instead, adult beak lengths were determined primarily by the population from which the eggs were obtained. Because an egg from a balloon vine population likely had long-beaked parents, while an egg from a goldenrain tree population likely had short-beaked parents, these results indicate that beak length is an inherited trait. **Figure 19.20** Hind limb structure changed first. *Rodhocetus* lacked flukes, but its pelvic bones and hind limbs had changed substantially from how those bones were shaped and arranged in *Pakicetus*. For example, in *Rodhocetus*, the pelvis and hind limbs appear to be oriented for paddling, whereas they were oriented for walking in *Pakicetus*.

### Concept Check 19.1

**1.** Hutton and Lyell proposed that geologic events in the past were caused by the same processes operating today, at the same gradual rate. This principle suggested that Earth must be much older than a few thousand years, the age that was widely accepted in the early 19th century. Hutton's and Lyell's ideas also stimulated Darwin to reason that the slow accumulation of small changes could ultimately produce the profound changes documented in the fossil record. In this context, the age of Earth was important to Darwin, because unless Earth was very old, he could not envision how there would have been enough time for evolution to occur. **2.** By this criterion, Cuvier's explanation of the fossil record and Lamarck's hypothesis of evolution are both scientific. Cuvier thought that species did not evolve over time. He also suggested that sudden, catastrophic events caused extinctions in particular areas. These assertions can be tested against the fossil record. With respect to Lamarck, his principle of use and disuse can be used to make testable predictions for fossils of groups such as whale ancestors as they adapted to a new habitat. Lamarck's principle of use and disuse and his associated principle of the inheritance of acquired characteristics can also be tested directly in living organisms.

### Concept Check 19.2

**1.** Organisms share characteristics (the unity of life) because they share common ancestors; the great diversity of life occurs because new species have repeatedly formed when descendant organisms gradually adapted to different environments, becoming different from their ancestors. **2.** The fossil mammal species (or its ancestors) would most likely have colonized the Andes from within South America, whereas ancestors of mammals currently found in Asian mountains would most likely have colonized those mountains from other parts of Asia. As a result, the Andes fossil species would share a more recent common ancestor with South American mammals than with mammals in Asia. Thus, for many of its traits, the fossil mammal species would probably more closely resemble mammals that live in South American jungles than mammals that live on Asian mountains. It is also possible, however, that the fossil mammal species could resemble the Asian mountain mammals because similar environments selected for similar adaptations (even though they were only distantly related to one another). **3.** As long as the white phenotype (encoded by the genotype *pp*) continues to be favored by natural selection, the frequency of the *p* allele will likely increase over time in the population. If the proportion of white individuals increases relative to purple individuals, the frequency of the recessive *p* allele will also increase relative to that of the *P* allele, which only appears in purple individuals (some of which also carry a *p* allele).

### Concept Check 19.3

**1.** An environmental factor such as a drug does not *create* new traits, such as drug resistance, but rather *selects for* traits among those that are already present in the population. **2.** (a) Despite their different functions, the forelimbs of different mammals are structurally similar because they all represent modifications of a structure found in the common ancestor. (b) This is a case of convergent evolution. The similarities between the sugar glider and flying squirrel indicate that similar environments selected for similar adaptations despite different ancestry. **3.** At the time that dinosaurs originated, Earth's landmasses formed a single large continent, Pangaea. Because many dinosaurs were large and mobile, it is likely that early members of these groups lived on many different parts of Pangaea. When Pangaea broke apart, fossils of these organisms would have moved with the rocks in which they were deposited. As a result, we would predict that fossils of early dinosaurs would have a broad geographic distribution (this prediction has been upheld).

### Summary of Key Concepts Questions

**19.1** Darwin thought that descent with modification occurred as a gradual, steplike process. The age of Earth was important to him because if Earth were only a few thousand years old (as conventional wisdom suggested), there wouldn't have been sufficient time for major evolutionary change. **19.2** All species have the potential to overreproduce—that is, to produce more offspring than can be supported by the environment. This ensures that there will be what Darwin called a "struggle for existence" in which many of the offspring are eaten, starved, diseased, or unable to reproduce for a variety of other reasons. Members of a population exhibit a range of heritable variations, some of which make it likely that their bearers will leave more offspring than other individuals (for example, the bearer may escape predators more effectively or be more tolerant of the physical conditions of the environment). Over time, natural selection resulting from factors such as predators, lack of food, or the physical conditions of the environment can increase the proportion of individuals with favorable traits in a population (evolutionary adaptation). **19.3** The hypothesis that cetaceans originated from a terrestrial mammal and are closely related to even-toed ungulates is supported by several lines of evidence. For example, fossils document that early cetaceans had hind limbs, as expected for organisms that descended from a land mammal; these fossils also show that cetacean hind limbs became reduced over time. Other fossils show that early cetaceans had a type of ankle bone that is otherwise found only in even-toed ungulates, providing strong evidence that even-toed ungulates are the land mammals to which cetaceans are most closely related. DNA sequence data also indicate that even-toed ungulates are the land mammals to which cetaceans are most closely related.

### Test Your Understanding

**1.** B **2.** D **3.** C **4.** B **5.** A

**6.** (a)

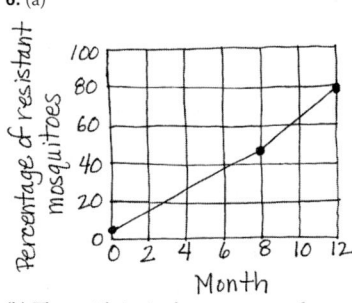

(b) The rapid rise in the percentage of mosquitoes resistant to DDT was most likely caused by natural selection in which mosquitoes resistant to DDT could survive and reproduce while other mosquitoes could not. (c) In India—where DDT resistance first appeared—natural selection would have caused the frequency of resistant mosquitoes to increase over time. If resistant mosquitoes then migrated from India (for example, transported by wind or in planes, trains, or ships) to other parts of the world, the frequency of DDT resistance would increase there as well. In addition, if resistance to DDT were to arise independently in mosquito populations outside of India, those populations would also experience an increase in the frequency of DDT resistance.

## Chapter 20

### Figure Questions

**Figure 20.4** The branching pattern of the tree indicates that the badger and the wolf share a common ancestor that is more recent than the ancestor these two animals share with the leopard.   **Figure 20.5** The new version (shown below) does not alter any of the evolutionary relationships shown in Figure 20.5. For example, B and C remain sister taxa, taxon A is still as closely related to taxon B as it is to taxon C, and so on.

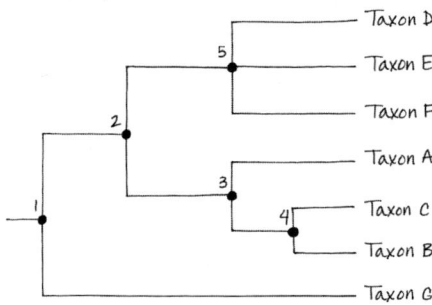

**Figure 20.6** Unknown mtDNA sample #1b (a portion of sample #1) and unknown mtDNA samples #9–13 all would have to be located on the branch of the gene tree that currently leads to Minke (Southern Hemisphere) mtDNA and unknown mtDNA #1a and 2–8.   **Figure 20.9** There are four possible bases (A, C, G, T) at each nucleotide position. If the base at each position depends on chance, not common descent, we would expect roughly one out of four (25%) of them to be the same. **Figure 20.11** You should have circled the branch point that is drawn farthest to the left (the common ancestor of all taxa shown). Both cetaceans and seals descended from terrestrial lineages of mammals, indicating that the cetacean-seal common ancestor had legs and lacked a streamlined body form. As a result, that ancestor would not be part of the cetacean-seal group.   **Figure 20.12** You should have circled the frog, turtle, and leopard lineages, along with their most recent common ancestor. **Figure 20.16** The lizard and snake lineage is the most basal taxon shown (closest to the root of the tree).   **Figure 20.21** This tree indicates that the sequences of rRNA and other genes in mitochondria are most closely related to those of proteobacteria, while the sequences of chloroplast genes are most closely related to those of cyanobacteria. These gene sequence relationships are what would be predicted from endosymbiont theory, which posits that both mitochondria and chloroplasts originated as engulfed prokaryotic cells.

### Concept Check 20.1

**1.** We are classified the same from the domain level to the class level; both the leopard and human are mammals. Leopards belong to order Carnivora, whereas humans do not.   **2.** The tree in (c) shows a different pattern of evolutionary relationships. In (c), C and B are sister taxa, whereas C and D are sister taxa in (a) and (b).
**3.**

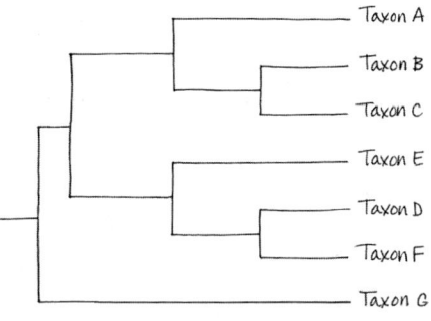

### Concept Check 20.2

**1.** (a) Analogy, since porcupines and cacti are not closely related and since most other animals and plants do not have similar structures; (b) homology, since cats and humans are both mammals and have homologous forelimbs, of which the hand and paw are the lower part; (c) analogy, since owls and hornets are not closely related and since the structure of their wings is very different   **2.** Species B and C are more likely to be closely related. Small genetic changes (as between species B and C) can produce divergent physical appearances, but if many genes have diverged greatly (as in species A and B), then the lineages have probably been separate for a long time.

### Concept Check 20.3

**1.** No; hair is a shared ancestral character common to all mammals and thus is not helpful in distinguishing different mammalian subgroups.   **2.** The principle of maximum parsimony states that the hypothesis about nature we investigate first should be the simplest explanation found to be consistent with the facts. Actual evolutionary relationships may differ from those inferred by parsimony owing to complicating factors such as convergent evolution.   **3.** The traditional classification provides a poor match to evolutionary history, thus violating the basic principle of cladistics—that classification should be based on common descent. Both birds and mammals originated from groups traditionally designated as reptiles, making reptiles (as traditionally delineated) a paraphyletic group. These problems can be addressed by removing *Dimetrodon* and cynodonts from the reptiles and by regarding birds as a group of reptiles (specifically, as a group of dinosaurs).

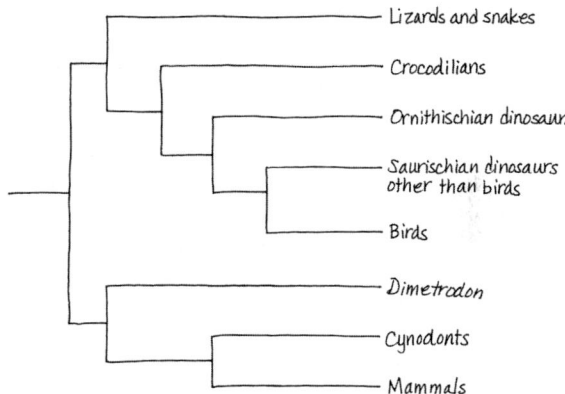

### Concept Check 20.4

**1.** A molecular clock is a method of estimating the actual time of evolutionary events based on numbers of base changes in genes that are related by descent. It is based on the assumption that the regions of genomes being compared evolve at constant rates. **2.** There are many portions of the genome that do not code for genes; mutations that alter the sequence of bases in such regions could accumulate without affecting an organism's survival and reproduction. Even in coding regions of the genome, some mutations may not have a critical effect on genes or proteins.   **3.** The gene (or genes) used for the molecular clock may have evolved more slowly in these two taxa than in the species used to calibrate the clock; as a result, the clock would underestimate the time at which the taxa diverged from each other.

### Concept Check 20.5

**1.** The kingdom Monera included bacteria and archaea, but we now know that these organisms are in separate domains. Kingdoms are subsets of domains, so a single kingdom (like Monera) that includes taxa from different domains is not valid.   **2.** Because of horizontal gene transfer, some genes in eukaryotes are more closely related to bacteria, while others are more closely related to archaea; thus, depending on which genes are used, phylogenetic trees constructed from DNA data can yield conflicting results.
**3.**

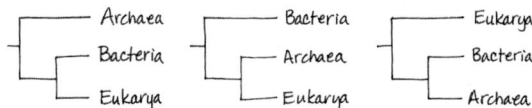

The fossil record indicates that prokaryotes originated long before eukaryotes. This suggests that the third tree, in which the eukaryotic lineage diverged first, is not accurate and hence is not likely to support from genetic data.

### Summary of Key Concepts Questions

**20.1** The fact that humans and chimpanzees are sister species indicates that we share a more recent common ancestor with chimpanzees than we do with any other living primate species. But that does not mean that humans evolved from chimpanzees, or vice versa; instead, it indicates that both humans and chimpanzees are descendants of that common ancestor.   **20.2** Homologous characters result from shared ancestry. As organisms diverge over time, some of their homologous characters will also diverge. The homologous characters of organisms that diverged long ago typically differ more than do the homologous characters of organisms that diverged more recently. As a result, differences in homologous characters can be used to infer phylogeny. In contrast, analogous characters result from convergent evolution, not shared ancestry, and hence can give misleading estimates of phylogeny.   **20.3** All features of organisms arose at some point in the history of life. In the group in which a new feature first arose, that feature is a shared derived character that is unique to that clade. The group in which each shared derived character first appeared can be determined, and the resulting nested pattern can be used to infer evolutionary history. **20.4** A key assumption of molecular clocks is that nucleotide substitutions occur at fixed rates, and hence the number of nucleotide differences between two DNA sequences is proportional to the time since the sequences diverged from each other. Some limitations of molecular clocks: No gene marks time with complete precision; natural selection can favor certain DNA changes over others; nucleotide substitution rates can change over long periods of time (causing molecular clock estimates of when events in the distant past occurred to be highly uncertain); and the same gene can evolve at different rates in different organisms. **20.5** Genetic data indicated that many prokaryotes differed as much from each other as they did from eukaryotes. This indicated that organisms should be grouped into three "super-kingdoms," or domains (Archaea, Bacteria, Eukarya). These data also indicated that the previous kingdom Monera (which had contained all the prokaryotes) did not make biological sense and should be abandoned. Later genetic and morphological data also indicated that the former kingdom Protista (which had primarily contained single-celled organisms) should be abandoned because some protists are more closely related to plants, fungi, or animals than they are to other protists.

## Test Your Understanding

**1.** A  **2.** C  **3.** B  **4.** D  **5.** D  **6.** A  **7.** D

**8.**

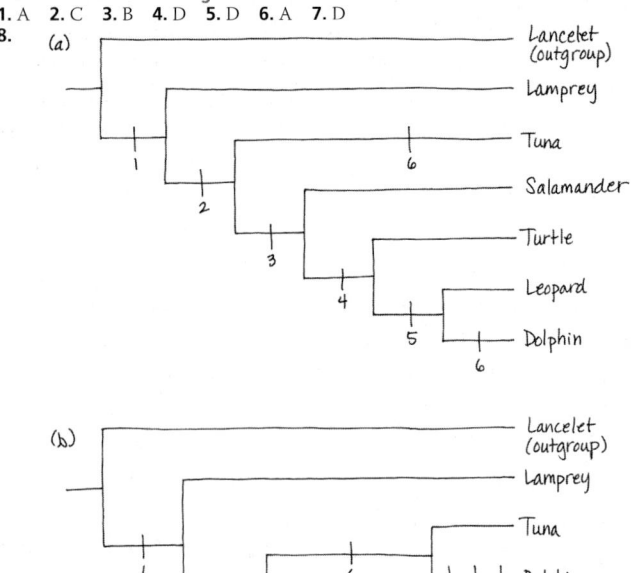

(c) The tree in (a) requires seven evolutionary changes, while the tree in (b) requires nine evolutionary changes. Thus, the tree in (a) is more parsimonious, since it requires fewer evolutionary changes.

## Chapter 21

### Figure Questions

**Figure 21.4** The genetic code is redundant, meaning that more than one codon can specify the same amino acid. As a result, a substitution at a particular site in a coding region of the *Adh* gene might change the codon but not the translated amino acid, and thus not the resulting protein encoded by the gene. One way an insertion in an exon would not affect the gene produced is if it occurs in an untranslated region of the exon. This is the case for the insertion at location 1,703. **Figure 21.8** The predicted frequencies are 36% $C^R C^R$, 48% $C^R C^W$, and 16% $C^W C^W$. **Figure 21.13** Directional selection. Goldenrain tree has smaller fruit than does the native host, balloon vine. Thus, in soapberry bug populations feeding on goldenrain tree, bugs with shorter beaks had an advantage, resulting in directional selection for shorter beak length. **Figure 21.15** Under prolonged low-oxygen conditions, some of the red blood cells of a heterozygote may sickle, leading to harmful effects. This does not occur in individuals with two wild-type hemoglobin alleles, suggesting that there may be selection against heterozygotes in malaria-free regions (where there is no heterozygote advantage). However, since heterozygotes are healthy under most conditions, selection against them is unlikely to be strong. **Figure 21.18** Crossing a single female's eggs with both an SC and an LC male's sperm allowed the researchers to directly compare the effects of the males' contribution to the next generation, since both batches of offspring had the same maternal contribution. This isolation of the male's impact enabled researchers to draw conclusions about differences in genetic "quality" between the SC and LC males.

### Concept Check 21.1

**1.** Within a population, genetic differences among individuals provide the raw material on which natural selection and other mechanisms can act. Without such differences, allele frequencies could not change over time—and hence the population

could not evolve. **2.** Many mutations occur in somatic cells, which do not produce gametes and so are lost when the organism dies. Of mutations that do occur in cell lines that produce gametes, many do not have a phenotypic effect on which natural selection can act. Others have a harmful effect and are thus unlikely to increase in frequency because they decrease the reproductive success of their bearers. **3.** Its genetic variation (whether measured at the level of the gene or at the level of nucleotide sequences) would probably drop over time. During meiosis, crossing over and the independent assortment of chromosomes produce many new combinations of alleles. In addition, a population contains a vast number of possible mating combinations, and fertilization brings together the gametes of individuals with different genetic backgrounds. Thus, via crossing over, independent assortment of chromosomes, and fertilization, sexual reproduction reshuffles alleles into fresh combinations each generation. Without sexual reproduction, the rate of forming new combinations of alleles would be vastly reduced, causing the overall amount of genetic variation to drop.

### Concept Check 21.2

**1.** Each individual has two alleles, so the total number of alleles is 1,400. To calculate the frequency of allele A, note that each of the 85 individuals of genotype AA has two A alleles, each of the 320 individuals of genotype Aa has one A allele, and each of the 295 individuals of genotype aa has zero A alleles. Thus, the frequency (p) of allele A is

$$p = \frac{(2 \times 85) + (1 \times 320) + (0 \times 295)}{1,400} = 0.35$$

There are only two alleles (A and a) in our population, so the frequency of allele a must be q = 1 − p = 0.65. **2.** Because the frequency of allele a is 0.45, the frequency of allele A must be 0.55. Thus, the expected genotype frequencies are $p^2 = 0.3025$ for genotype AA, 2pq = 0.495 for genotype Aa, and $q^2 = 0.2025$ for genotype aa. **3.** There are 120 individuals in the population, so there are 240 alleles. Of these, there are 124 V alleles—32 from the 16 VV individuals and 92 from the 92 Vv individuals. Thus, the frequency of the V allele is p = 124/240 = 0.52; hence, the frequency of the v allele is q = 0.48. Based on the Hardy-Weinberg equation, if the population were not evolving, the frequency of genotype VV should be $p^2 = 0.52 \times 0.52 = 0.27$; the frequency of genotype Vv should be 2pq = 2 × 0.52 × 0.48 = 0.5; and the frequency of genotype vv should be $q^2 = 0.48 \times 0.48 = 0.23$. In a population of 120 individuals, these expected genotype frequencies lead us to predict that there would be 32 VV individuals (0.27 × 120), 60 Vv individuals (0.5 × 120), and 28 vv individuals (0.23 × 120). The actual numbers for the population (16 VV, 92 Vv, 12 vv) deviate from these expectations (fewer homozygotes and more heterozygotes than expected). This indicates that the population is not in Hardy-Weinberg equilibrium and hence may be evolving at this locus.

### Concept Check 21.3

**1.** Natural selection is more "predictable" in that it alters allele frequencies in a nonrandom way: It tends to increase the frequency of alleles that increase the organism's reproductive success in its environment and decrease the frequency of alleles that decrease the organism's reproductive success. Alleles subject to genetic drift increase or decrease in frequency by chance alone, whether or not they are advantageous. **2.** Genetic drift results from chance events that cause allele frequencies to fluctuate at random from generation to generation; within a population, this process tends to decrease genetic variation over time. Gene flow is the transfer of alleles between populations, a process that can introduce new alleles to a population and hence may increase its genetic variation (albeit slightly, since rates of gene flow are often low). **3.** Selection is not important at this locus; furthermore, the populations are not small, and hence the effects of genetic drift should not be pronounced. Gene flow is occurring via the movement of pollen and seeds. Thus, allele and genotype frequencies in these populations should become more similar over time as a result of gene flow.

### Concept Check 21.4

**1.** Zero, because fitness includes reproductive contribution to the next generation, and a sterile mule cannot produce offspring. **2.** Although both gene flow and genetic drift can increase the frequency of advantageous alleles in a population, they can also decrease the frequency of advantageous alleles or increase the frequency of harmful alleles. Only natural selection *consistently* results in an increase in the frequency of alleles that enhance survival or reproduction. Thus, natural selection is the only mechanism that consistently leads to adaptive evolution. **3.** The three modes of natural selection (directional, stabilizing, and disruptive) are defined in terms of the selective advantage of different *phenotypes*, not different genotypes. Thus, the type of selection represented by heterozygote advantage depends on the phenotype of the heterozygotes. In this question, because heterozygous individuals have a more extreme phenotype than either homozygote, heterozygote advantage represents directional selection.

### Summary of Key Concepts Questions

**21.1** Much of the nucleotide variability at a genetic locus occurs within introns. Nucleotide variation at these sites typically does not affect the phenotype because introns do not code for the protein product of the gene. (Note: In certain circumstances, it is possible that a change in an intron could affect RNA splicing and ultimately have some phenotypic effect on the organism, but such mechanisms are not covered in this introductory text.) There are also many variable nucleotide sites within exons. However, most of the variable sites within exons reflect changes to the DNA sequence that do not change the sequence of amino acids encoded by the gene (and hence may not affect the phenotype). **21.2** No, this is not an example of circular reasoning. Calculating p and q from observed genotype frequencies does not imply that those genotype frequencies must be in Hardy-Weinberg equilibrium. For example, consider a population that has 195 individuals of genotype AA, 10 of genotype Aa, and 195 of genotype aa. Calculating p and q from these values yields

$p = q = 0.5$. Using the Hardy-Weinberg equation, the predicted equilibrium frequencies are $p^2 = 0.25$ for genotype $AA$, $2pq = 0.5$ for genotype $Aa$, and $q^2 = 0.25$ for genotype $aa$. Since there are 400 individuals in the population, these predicted genotype frequencies indicate that there should be 100 $AA$ individuals, 200 $Aa$ individuals, and 100 $aa$ individuals—numbers that differ greatly from the values that we used to calculate $p$ and $q$. **21.3** It is unlikely that two such populations would evolve in similar ways. Since their environments are very different, the alleles favored by natural selection would probably differ between the two populations. Although genetic drift may have important effects in each of these small populations, drift causes unpredictable changes in allele frequencies, so it is unlikely that drift would cause the populations to evolve in similar ways. Both populations are geographically isolated, suggesting that little gene flow would occur between them (again making it less likely that they would evolve in similar ways). **21.4** Compared to males, it is likely that the females of such species would be larger, more colorful, endowed with more elaborate ornamentation (for example, a large morphological feature such as the peacock's tail), and more apt to engage in behaviors intended to attract mates or prevent other members of their sex from obtaining mates.

### Test Your Understanding
**1.** D  **2.** C  **3.** D  **4.** B  **5.** A  **6.** C

## Chapter 22

### Figure Questions
**Figure 22.7** If this had not been done, the strong preference of "starch flies" and "maltose flies" to mate with like-adapted flies could have occurred simply because the flies could detect (for example, by sense of smell) what their potential mates had eaten as larvae—and they preferred to mate with flies that had a similar smell to their own. **Figure 22.12** In murky waters where females distinguish colors poorly, females of each species might mate often with males of the other species. Hence, since hybrids between these species are viable and fertile, the gene pools of the two species could become more similar over time. **Figure 22.13** The graph suggests there has been gene flow of some fire-bellied toad alleles into the range of the yellow-bellied toad. Otherwise, all individuals located to the left of the hybrid zone portion of the graph would have allele frequencies close to 1. **Figure 22.14** Because the populations had only just begun to diverge from one another at this point in the process, it is likely that any existing barriers to reproduction would weaken over time. **Figure 22.18** Over time, the chromosomes of the experimental hybrids came to resemble those of *H. anomalus*. This occurred even though conditions in the laboratory differed greatly from conditions in the field, where *H. anomalus* is found, suggesting that selection for laboratory conditions was not strong. Thus, it is unlikely that the observed rise in the fertility of the experimental hybrids was due to selection for life under laboratory conditions. **Figure 22.19** The presence of *M. cardinalis* plants that carry the *M. lewisii yup* allele would make it more likely that bumblebees would transfer pollen between the two monkey flower species. As a result, we would expect the number of hybrid offspring to increase.

### Concept Check 22.1
**1.** (a) All except the biological species concept can be applied to both asexual and sexual species because they define species on the basis of characteristics other than the ability to reproduce. In contrast, the biological species concept can be applied only to sexual species. (b) The easiest species concept to apply in the field would be the morphological species concept because it is based only on the appearance of the organism. Additional information about its ecological habits, evolutionary history, and reproduction is not required. **2.** Because these birds live in fairly similar environments and can breed successfully in captivity, the reproductive barrier in nature is probably prezygotic; given the species' differences in habitat preference, this barrier could result from habitat isolation.

### Concept Check 22.2
**1.** In allopatric speciation, a new species forms while in geographic isolation from its parent species; in sympatric speciation, a new species forms in the absence of geographic isolation. Geographic isolation greatly reduces gene flow between populations, whereas ongoing gene flow is more likely in sympatric populations. As a result, sympatric speciation is less common than allopatric speciation. **2.** Allopatric speciation would be less likely to occur on an island near a mainland than on a more isolated island of the same size. We expect this result because continued gene flow between mainland populations and those on a nearby island reduces the chance that enough genetic divergence will take place for allopatric speciation to occur. **3.** If all of the homologs failed to separate during anaphase I of meiosis, some gametes would end up with an extra set of chromosomes (and others would end up with no chromosomes). If a gamete with an extra set of chromosomes fused with a normal gamete, a triploid would result; if two gametes with an extra set of chromosomes fused with each other, a tetraploid would result.

### Concept Check 22.3
**1.** Hybrid zones are regions in which members of different species meet and mate, producing some offspring of mixed ancestry. Such regions can be viewed as "natural laboratories" in which to study speciation because scientists can directly observe factors that cause (or fail to cause) reproductive isolation. **2.** (a) If hybrids consistently survived and reproduced poorly compared with the offspring of intraspecific matings, reinforcement could occur. If it did, natural selection would cause prezygotic barriers to reproduction between the parent species to strengthen over time, decreasing the production of unfit hybrids and leading to a completion of the speciation process. (b) If hybrid offspring survived and reproduced as well as the offspring of intraspecific matings, indiscriminate mating between the parent species would lead to the production

of large numbers of hybrid offspring. As these hybrids mated with each other and with members of both parent species, the gene pools of the parent species could fuse over time, reversing the speciation process.

### Concept Check 22.4
**1.** The time between speciation events includes (1) the length of time that it takes for populations of a newly formed species to begin diverging reproductively from one another and (2) the time it takes for speciation to be complete once this divergence begins. Although speciation can occur rapidly once populations have begun to diverge from one another, it may take millions of years for that divergence to begin. **2.** Investigators transferred alleles at the *yup* locus (which influences flower color) from each parent species to the other. *M. lewisii* plants with an *M. cardinalis yup* allele received many more visits from hummingbirds than usual; hummingbirds usually pollinate *M. cardinalis* but avoid *M. lewisii*. Similarly, *M. cardinalis* plants with an *M. lewisii yup* allele received many more visits from bumblebees than usual; bumblebees usually pollinate *M. lewisii* and avoid *M. cardinalis*. Thus, alleles at the *yup* locus can influence pollinator choice, which in these species provides the primary barrier to interspecific mating. Nevertheless, the experiment does not prove that the *yup* locus alone controls barriers to reproduction between *M. lewisii* and *M. cardinalis*; other genes might enhance the effect of the *yup* locus (by modifying flower color) or cause entirely different barriers to reproduction (for example, gametic isolation or a postzygotic barrier). **3.** Crossing over. If crossing over did not occur, each chromosome in an experimental hybrid would remain as in the $F_1$ generation: composed entirely of DNA from one parent species or the other.

### Summary of Key Concepts Questions
**22.1** According to the biological species concept, a species is a group of populations whose members interbreed and produce viable, fertile offspring; thus, gene flow occurs between populations of a species. In contrast, members of different species do not interbreed, and hence no gene flow occurs between their populations. Overall, then, in the biological species concept, species can be viewed as designated by the *absence* of gene flow—making gene flow of central importance to the biological species concept. **22.2** Sympatric speciation can be promoted by factors such as polyploidy, habitat shifts, and sexual selection, all of which can reduce gene flow between the subpopulations of a larger population. But such factors can also occur in allopatric populations and hence can also promote allopatric speciation. **22.3** If the hybrids are selected against, the hybrid zone could persist if individuals from the parent species regularly travel into the zone, where they mate to produce hybrid offspring. If hybrids are not selected against, there is no cost to the continued production of hybrids, and large numbers of hybrid offspring may be produced. However, natural selection for life in different environments may keep the gene pools of the two parent species distinct, thus preventing the loss (by fusion) of the parent species and once again causing the hybrid zone to be stable over time. **22.4** As the goatsbeard plant, Bahamas mosquitofish, and apple maggot fly illustrate, speciation continues to happen today. A new species can begin to form whenever gene flow is reduced between populations of the parent species. Such reductions in gene flow can occur in many ways: A new, geographically isolated population may be founded by a few colonists; some members of the parent species may begin to utilize a new habitat; and sexual selection may isolate formerly connected populations or subpopulations. These and many other such events are happening today.

### Test Your Understanding
**1.** B  **2.** C  **3.** D  **4.** A  **5.** D  **6.** C
**7.** Here is one possibility:

## Chapter 23

### Figure Questions
**Figure 23.4** Because uranium-238 has a half-life of 4.5 billion years, the *x*-axis would be relabeled (in billions of years) as 4.5, 9, 13.5, and 18. **Figure 23.9** The Australian plate's current direction of movement is roughly similar to the northeasterly direction the continent traveled over the past 66 million years. **Figure 23.18** In this bat, the ratio of the length of the longest set of hand and finger bones to the length of the radius is approximately equal to 2. Although answers will vary from person to person, the corresponding ratio is typically less than 1 in humans. **Figure 23.21** The coding sequence of the *Pitx1* gene would differ between the marine and lake populations, but patterns of gene expression would not.

## Concept Check 23.1

**1.** 22,920 years (four half-lives: 5,730 × 4)   **2.** The fossil record shows that different groups of organisms dominated life on Earth at different points in time and that many organisms once alive are now extinct; specific examples of these points can be found in Figure 23.3. The fossil record also indicates that new groups of organisms can arise via the gradual modification of previously existing organisms, as illustrated by fossils that document the origin of mammals from their cynodont ancestors.   **3.** A fossil record of life today would include many organisms with hard body parts (such as vertebrates and many marine invertebrates), but might not include some species we are very familiar with, such as those that have small geographic ranges and/or small population sizes (for example, endangered species such as the giant panda, tiger, and several rhinoceros species).   **4.** The discovery of such a (hypothetical) fossil organism would indicate that aspects of our current understanding of the origin of mammals are not correct because mammals are thought to have originated much more recently (see Figure 23.5). For example, such a discovery could suggest that the dates of previous fossil discoveries are not correct or that the lineages shown in Figure 23.5 shared features with mammals but were not their direct ancestors. Such a discovery would also suggest that radical changes in multiple aspects of the skeletal structure of organisms could arise suddenly—an idea that is not supported by the known fossil record.

## Concept Check 23.2

**1.** The theory of plate tectonics describes the movement of Earth's continental plates, which alters the physical geography and climate of Earth, as well as the extent to which organisms are geographically isolated. Because these factors affect extinction and speciation rates, plate tectonics has a major impact on life on Earth.   **2.** In each of the five mass extinctions documented in the fossil record, 50% or more of marine species became extinct, as did large numbers of terrestrial species. As a result, a mass extinction alters the course of evolution dramatically, removing many evolutionary lineages and reducing the diversity of life on Earth for millions of years. A mass extinction can also change ecological communities by changing the types of organisms that live in them. **3.** Mass extinctions; major evolutionary innovations; the diversification of another group of organisms (which can provide new sources of food); migration to new locations where few competitor species exist   **4.** In theory, fossils of both common and rare species would be present right up to the time of the catastrophic event, then disappear. Reality is more complicated because the fossil record is not perfect. So the most recent fossil for a species might be a million years before the mass extinction—even though the species did not become extinct *until* the mass extinction. This complication is especially likely for rare species because few of their fossils will form and be discovered. Hence, for many rare species, the fossil record would not document that the species was alive immediately before the extinction (even if it was).

## Concept Check 23.3

**1.** Heterochrony can cause a variety of morphological changes. For example, if the onset of sexual maturity changes, a retention of juvenile characteristics (paedomorphosis) may result. Paedomorphosis can be caused by small genetic changes that result in large changes in morphology, as seen in the axolotl salamander.   **2.** In animal embryos, *Hox* genes influence the development of structures such as limbs and feeding appendages. As a result, changes in these genes—or in the regulation of these genes—are likely to have major effects on morphology.   **3.** From genetics, we know that gene regulation is altered by how well transcription factors bind to noncoding DNA sequences called control elements. Thus, if changes in morphology are often caused by changes in gene regulation, portions of noncoding DNA that contain control elements are likely to be strongly affected by natural selection.

## Concept Check 23.4

**1.** Complex structures do not evolve all at once, but in increments, with natural selection selecting for adaptive variants of the earlier versions.   **2.** Although the myxoma virus is highly lethal, initially some of the rabbits are resistant (0.2% of infected rabbits are not killed). Thus, assuming resistance is an inherited trait, we would expect the rabbit population to show a trend for increased resistance to the virus. We would also expect the virus to show an evolutionary trend toward reduced lethality. We would expect this trend because a rabbit infected with a less lethal virus would be more likely to live long enough for a mosquito to bite it and hence potentially transmit the virus to another rabbit. (A virus that kills its rabbit host before a mosquito transmits the virus to another rabbit dies with its host.)

## Summary of Key Concepts Questions

**23.1** One challenge is that radioisotopes with very long half-lives are not used by organisms to build their bones or shells. As a result, fossils older than 75,000 years cannot be dated directly. Fossils are often found in sedimentary rock, but those rocks typically contain sediments of different ages, again posing a challenge when trying to date old fossils. To circumvent these challenges, geologists use radioisotopes with long half-lives to date layers of volcanic rock that surround old fossils. This approach provides minimum and maximum estimates for the ages of fossils sandwiched between two layers of volcanic rock.   **23.2** The broad evolutionary changes documented by the fossil record reflect the rise and fall of major groups of organisms. In turn, the rise or fall of any particular group results from a balance between speciation and extinction rates: A group increases in size when the rate at which its members produce new species is greater than the rate at which its member species are lost to extinction, while a group shrinks in size if extinction rates are greater than speciation rates. **23.3** A change in the sequence or regulation of a developmental gene can produce major morphological changes. In some cases, such morphological changes may enable organisms to perform new functions or live in new environments—thus potentially leading to an adaptive radiation and the formation of a new group of organisms. **23.4** Evolutionary change results from interactions between organisms and their current environments. No goal is involved in this process. As environments change over time, the features of organisms favored by natural selection may also change.

When this happens, what once may have seemed like a "goal" of evolution (for example, improvements in the function of a feature previously favored by natural selection) may cease to be beneficial or may even be harmful.

**Test Your Understanding**
**1.** D   **2.** B   **3.** C   **4.** C   **5.** A

## Chapter 24

**Figure Questions**
**Figure 24.3** Proteins are almost always composed of the same 20 amino acids shown in Figure 3.18. However, many other amino acids could potentially form in this or any other experiment. For example, any molecule that had a different R group than those listed in Figure 3.18 in the 20 most common amino acids (yet still contained an α carbon, an amino group, and a carboxyl group) would be an amino acid—yet it would not be one of the 20 amino acids commonly found in nature today.   **Figure 24.4** The hydrophobic regions of such molecules are attracted to one another and excluded from water, whereas the hydrophilic regions have an affinity for water. As a result, the molecules can form a bilayer in which the hydrophilic regions are on the outside of the bilayer (facing water on each side of the bilayer) and the hydrophobic regions point toward each other (that is, toward the inside of the bilayer).   **Figure 24.14** It is likely that the expression or sequence of genes that affect glucose metabolism may have changed; genes for metabolic processes no longer needed by the cell also may have changed.   **Figure 24.15** Transduction results in horizontal gene transfer when the host and recipient cells are members of different species.   **Figure 24.18** Eukarya **Figure 24.20** Thermophiles live in very hot environments, so it is likely that their enzymes can continue to function normally at much higher temperatures than can the enzymes of other organisms. At low temperatures, however, the enzymes of thermophiles may not function as well as the enzymes of other organisms. **Figure 24.22** From the graph, plant uptake can be estimated as 0.7, 0.6, and 0.95 mg $K^+$ for strains 1, 2, and 3, respectively. These values average to 0.75 mg $K^+$. If bacteria had no effect, the average plant uptake of $K^+$ for strains 1, 2, and 3 should be close to 0.5 mg $K^+$, the value observed for plants grown in bacteria-free soil.

## Concept Check 24.1

**1.** The hypothesis that conditions on early Earth could have permitted the synthesis of organic molecules from inorganic ingredients   **2.** In contrast to random mingling of molecules in an open solution, segregation of molecular systems by the membranes of protocells could concentrate organic molecules, assisting biochemical reactions.   **3.** The earliest prokaryotic fossils are of stromatolites that lived in shallow marine environments 3.5 billion years ago. By 3.1 billion years ago, stromatolites had diversified into two different morphological types, and by 2.8 billion years ago, they had expanded to live in salty lakes as well as marine environments. Fossils of individual prokaryotic cells have also been found, the earliest dating to 3.4 billion years ago. By 2.5 billion years ago, diverse communities of photosynthetic cyanobacteria lived in the oceans. These cyanobacteria released oxygen to Earth's atmosphere during the water-splitting step of photosynthesis. As a result, the composition of the atmosphere changed, and many prokaryotic groups were driven to extinction—thus altering the course of evolution.   **4.** Today, genetic information usually flows from DNA to RNA, as when the DNA sequence of a gene is used as a template to synthesize the mRNA encoding a particular protein. However, the life cycle of retroviruses such as HIV shows that genetic information can flow in the reverse direction (from RNA to DNA). In these viruses, the enzyme reverse transcriptase uses RNA as a template for DNA synthesis, suggesting that a similar enzyme could have played a key role in the transition from an RNA world to a DNA world.

## Concept Check 24.2

**1.** Prokaryotic cells lack the complex compartmentalization associated with the membrane-enclosed organelles of eukaryotic cells. Prokaryotic genomes have much less DNA than eukaryotic genomes, and most of this DNA is contained in a single ring-shaped chromosome located in the nucleoid rather than within a true membrane-enclosed nucleus. In addition, many prokaryotes also have plasmids, small ring-shaped DNA molecules containing a few genes.   **2.** A phototroph derives its energy from light, while a chemotroph gets its energy from chemical sources. An autotroph derives its carbon from $CO_2$, $HCO_3^-$, or related compounds, while a heterotroph gets its carbon from organic nutrients such as glucose. Thus, there are four nutritional modes: photoautotrophic, photoheterotrophic (unique to prokaryotes), chemoautotrophic (unique to prokaryotes), and chemoheterotrophic.   **3.** Plastids such as chloroplasts are thought to have evolved from an endosymbiotic photosynthetic prokaryote. More specifically, the phylogenetic tree shown in Figure 20.21 indicates that plastids are closely related to cyanobacteria. Hence, we can hypothesize that the thylakoid membranes of chloroplasts resemble those of cyanobacteria because chloroplasts evolved from a cyanobacterial endosymbiont.   **4.** If humans could fix nitrogen, we could build proteins using atmospheric $N_2$ and hence would not need to eat high-protein foods such as meat, fish, or soy. Our diet would, however, need to include a source of carbon, along with minerals and water. Thus, a typical meal might consist of carbohydrates as a carbon source, along with fruits and vegetables to provide essential minerals (and additional carbon).

## Concept Check 24.3

**1.** Prokaryotes can have extremely large population sizes, in part because they often have short generation times. The large number of individuals in prokaryotic populations makes it likely that in each generation there will be many individuals that have new mutations at any particular gene, thereby adding considerable genetic diversity to the population.   **2.** In transformation, naked, foreign DNA from the environment is taken up by a bacterial cell. In transduction, phages carry bacterial genes from one bacterial cell to another. In conjugation, a bacterial cell directly transfers plasmid or chromosomal DNA to another cell via a mating bridge that temporarily connects the two cells.   **3.** The population that includes individuals capable of conjugation would probably be more successful, since some of its members could form recombinant cells whose new gene combinations might be advantageous in a novel environment. **4.** Yes. Genes for antibiotic resistance could be transferred (by transformation, transduction, or conjugation) from the nonpathogenic bacterium to a pathogenic bacterium; this could make the pathogen an even greater threat to human health. In general, transformation, transduction, and conjugation tend to increase the spread of resistance genes.

### Concept Check 24.4
**1.** Molecular systematic studies indicate that some organisms once classified as bacteria are more closely related to eukaryotes and belong in a domain of their own: Archaea. Metagenomic studies have added many new branches to the prokaryotic tree of life, highlighting the extensive genetic diversity of these organisms. Genomic studies have also shown that horizontal gene transfer is common and plays an important role in the evolution of prokaryotes. **2.** At present, all known methanogens are archaea in the clade Euryarchaeota; this suggests that this unique metabolic pathway probably arose in ancestral species within Euryarchaeota. Since Bacteria and Archaea have been separate evolutionary lineages for billions of years, the discovery of a methanogen from the domain Bacteria would suggest that adaptations that enabled the use of $CO_2$ to oxidize $H_2$ may have evolved twice—once in Archaea (within Euryarchaeota) and once in Bacteria. (It is also possible that a newly discovered bacterial methanogen could have acquired the genes for this metabolic pathway by horizontal gene transfer from a methanogen in domain Archaea. However, horizontal gene transfer is not a likely explanation because of the large number of genes involved and because gene transfers between species in different domains are rare.)

### Concept Check 24.5
**1.** Although prokaryotes are small, their large numbers and metabolic abilities enable them to play key roles in ecosystems by decomposing wastes, recycling chemicals, and affecting the concentrations of nutrients available to other organisms. Prokaryotes also play a key role in ecological interactions such as mutualism and parasitism. **2.** No. If the poison is secreted as an exotoxin, live bacteria could be transmitted to another person. But the same is true if the poison is an endotoxin—only in this case, the live bacteria that are transmitted may be descendants of the (now-dead) bacteria that produced the poison. **3.** Cyanobacteria produce oxygen when water is split in the light reactions of photosynthesis. The Calvin cycle incorporates $CO_2$ from the air into organic molecules, which are then converted to sugars. **4.** Some of the many different species of prokaryotes that live in the human gut compete with one another for resources (from the food that you eat). Because different prokaryotic species have different adaptations, a change in diet may alter which species can grow most rapidly, thus altering species abundance.

### Summary of Key Concepts Questions
**24.1** Particles of montmorillonite clay may have provided surfaces on which organic molecules became concentrated and hence were more likely to react with one another. Montmorillonite clay particles may also have facilitated the transport of key molecules, such as short strands of RNA, into vesicles. These vesicles can form spontaneously from simple precursor molecules, "reproduce" and "grow" on their own, and maintain internal concentrations of molecules that differ from those in the surrounding environment. These features of vesicles represent key steps in the emergence of protocells and (ultimately) the first living cells. **24.2** Specific structural features that enable prokaryotes to thrive in diverse environments include their cell walls (which provide shape and protection), flagella (which function in directed movement), and ability to form capsules or endospores (both of which can protect against harsh conditions). Prokaryotes also have an exceptionally broad range of metabolic adaptations, enabling them to thrive in many different environments. **24.3** Many prokaryotic species can reproduce extremely rapidly, and their populations can number in the trillions. As a result, even though mutations are rare, every day many offspring are produced that have new mutations at particular gene loci. In addition, even though prokaryotes reproduce asexually and hence the vast majority of offspring are genetically identical to their parent, the genetic variation of their populations can be increased by transduction, transformation, and conjugation. Each of these (nonreproductive) processes can increase genetic variation by transferring DNA from one cell to another—even among cells that are of different species. **24.4** Phenotypic criteria such as shape, motility, and nutritional mode do not provide a clear picture of the evolutionary history of the prokaryotes. In contrast, molecular data have revealed that prokaryotes form two domains (Bacteria and Archaea), and they have elucidated relationships among major groups of prokaryotes. Molecular data have also allowed researchers to sample genes directly from the environment; using such genes to construct phylogenies has led to the discovery of major new groups of prokaryotes. **24.5** Prokaryotes play key roles in the chemical cycles on which life depends. For example, prokaryotes are important decomposers, breaking down corpses and waste materials, thereby releasing nutrients to the environment, where they can be used by other organisms. Prokaryotes also convert inorganic compounds to forms that other organisms can use. With respect to their ecological interactions, many prokaryotes form life-sustaining mutualisms with other species. For example, human well-being depends on our associations with mutualistic prokaryotes, such as the many species that live in our intestines and digest food that we cannot. In some cases, such as hydrothermal vent communities, the metabolic activities of prokaryotes provide an energy source on which hundreds of other species depend; in the absence of the prokaryotes, the community collapses.

### Test Your Understanding
**1.** D  **2.** C  **3.** A  **4.** B  **5.** C  **6.** D  **7.** A

## Chapter 25

### Figure Questions
**Figure 25.4** Four. The first (and primary) genome is the DNA located in the chlorarachniophyte nucleus. A chlorarachniophyte also contains remnants of a green alga's nuclear

DNA, located in the nucleomorph. Finally, mitochondria and chloroplasts contain DNA from the (different) bacteria from which they evolved. These two prokaryotic genomes comprise the third and fourth genomes contained within a chlorarachniophyte. **Figure 25.7** As described in observations 1 and 2, choanoflagellates and several groups of animals have collar cells. Since collar cells have never been observed in plants, fungi, or non-choanoflagellate protists, this suggests that choanoflagellates may be more closely related to animals than to other eukaryotes. If choanoflagellates are more closely related to animals than to any other group of eukaryotes, choanoflagellates and animals should share other traits that are not found in other eukaryotes. The data described in observation 3 are consistent with this prediction. **Figure 25.9** Based on the age of the oldest taxonomically resolved fossil eukaryote, a red alga that lived 1.2 billion years ago, we can conclude that the supergroups must have begun to diverge no later than 1.2 billion years ago. **Figure 25.21** If the assumption is correct, then their results indicate that the fusion of the genes for DHFR and TS may be a derived trait shared by members of three supergroups of eukaryotes (Excavata, SAR, and Archaeplastida). However, if the assumption is not correct, the presence or absence of the gene fusion may tell little about phylogenetic history. For example, if the genes fused multiple times, groups could share the trait because of convergent evolution rather than common descent. If instead the genes were secondarily split, a group with such a split could be placed (incorrectly) in Unikonta rather than its correct placement in one of the other three supergroups. **Figure 25.26** The apicoplast is a modified plastid, and hence it descended from a cyanobacterium. Thus, the metabolic pathway that the apicoplast uses to synthesize this essential chemical would likely differ from pathways found in humans—and hence drugs that target this pathway would probably not harm humans.

### Concept Check 25.1
**1.** The earliest fossil eukaryotes date to 1.8 billion years ago. By 1.3 billion years ago, the fossil record documents a moderate diversity of unicellular and simple multicellular eukaryotes, some of which had asymmetric forms indicating the presence of a well-developed cytoskeleton. Fossil organisms that lived from 1.3 billion to 635 million years ago include those with complex multicellularity, sexual life cycles, and eukaryotic photosynthesis. Large, multicellular eukaryotes first appeared about 600 million years ago. **2.** Eukaryotes are considered "combination" organisms because some of their genes and cellular characteristics are derived from archaea, while others are derived from bacteria. Strong evidence shows that eukaryotes acquired mitochondria after a host cell (either an archaean or a cell with archaeal ancestors) first engulfed and then formed an endosymbiotic association with an alpha proteobacterium. Similarly, chloroplasts in red and green algae appear to have descended from a photosynthetic cyanobacterium that was engulfed by an ancient heterotrophic eukaryote. Secondary endosymbiosis also played an important role: Various protist lineages acquired plastids by engulfing unicellular red or green algae. **3.** Photosynthetic eukaryotes can trace their ancestry to the primary endosymbiotic event that gave rise to plastids. Thus, such a discovery would suggest that eukaryotic photosynthesis arose at least twice, in two separate primary endosymbiotic events in which a cyanobacterium was engulfed by a heterotrophic eukaryote.

### Concept Check 25.2
**1.** Morphologically, choanoflagellates are almost indistinguishable from the collar cells of sponges, a basal animal lineage. Other animals also have collar cells, whereas such cells have never been observed in fungi, plants, or protists other than choanoflagellates. Finally, DNA sequence comparisons indicate that choanoflagellates are the sister group of animals. **2.** The evolution of proteins that attach animal cells to one another was a key step in the origin of multicellularity in animals. Choanoflagellates encode many of the domains found in one such group of animal attachment proteins, the cadherins. Other eukaryotes do not encode these domains; thus, animal cadherin proteins appear to have descended from proteins found in choanoflagellates. Evidence for modification is also clear: As seen in Figure 25.8, the protein domains found in animal cadherins differ in type, number, and location from those found in the ancestral choanoflagellate protein. **3.** Multicellularity originated independently in *Volvox*, plants, and fungi. Since each of these groups arose from different single-celled ancestors, it is likely that their cell-to-cell attachments form using different molecules. (Data from recent molecular studies are consistent with this prediction.)

### Concept Check 25.3
**1.** Many members of the supergroup Excavata have unique cytoskeletal features, and some have an "excavated" feeding groove on one side of their cells; two major clades of excavates are characterized by having reduced mitochondria. The SAR supergroup contains three large clades—stramenopiles, alveolates, and rhizarians—which collectively include diatoms and other key photosynthetic species, protists that move using cilia, and amoebas with threadlike pseudopodia. The supergroup Archaeplastida contains clades that descended from a protist ancestor that engulfed a cyanobacterium, including red algae, green algae, and plants. Finally, the supergroup Unikonta includes a large clade of amoebas that have lobe- or tube-shaped pseudopodia, as well as animals, fungi, and their close protist relatives. **2.** During photosynthesis, aerobic algae produce $O_2$ and use $CO_2$. $O_2$ is produced as a by-product of the light reactions, while $CO_2$ is used as an input to the Calvin cycle (the end products of which are sugars). Aerobic algae also perform cellular respiration, which uses $O_2$ as an input and produces $CO_2$ as a waste product. **3.** Since the unknown protist is more closely related to diplomonads than to euglenozoans, it must have originated after the diplomonads and parabasalids diverged from the euglenozoans. In addition, since the unknown species has fully functional mitochondria—yet both diplomonads and parabasalids do not—it is likely that it originated *before* the last common ancestor of the diplomonads and parabasalids.

### Concept Check 25.4
**1.** Because photosynthetic protists constitute the base of aquatic food webs, many aquatic organisms depend on them for food, either directly or indirectly. In addition, a substantial percentage of the oxygen produced by photosynthesis is made by

photosynthetic protists. **2.** Protists form mutualistic and parasitic associations with other organisms. Examples include photosynthetic dinoflagellates that form a mutualistic symbiosis with coral polyps; parabasalids that form a mutualistic symbiosis with termites; and the stramenopile *Phytophthora ramorum*, a parasite of oak trees.
**3.** Corals depend on their dinoflagellate symbionts for nourishment, so coral bleaching would probably cause the corals to die. As the corals died, less food would be available for fishes and other species that eat coral. As a result, populations of these species might decline, and that, in turn, might cause populations of their predators to decline.

### Summary of Key Concepts Questions

**25.1** All eukaryotes have mitochondria or remnants of these organelles, but not all eukaryotes have plastids. **25.2** Two such examples are described in this chapter: the evolution of multicellularity in *Volvox* and the evolution of multicellularity in animals. In each case, structures or genes present in unicellular ancestors were co-opted and used for new purposes in the multicellular lineage. In *Volvox*, cells are attached to one another using proteins that are homologous to proteins in the cell wall of their closest unicellular relative, *Chlamydomonas*. Likewise, in animals the cadherin proteins that function in cell attachment represent modified versions of proteins that served other purposes in the choanoflagellates, the protists that are the sister group of animals.
**25.3** Kingdom Protista has been abandoned because some protists are more closely related to plants, fungi, or animals than they are to other protists. In addition, biologists once hypothesized that a collection of eukaryotes that seemed to lack mitochondria belonged to a lineage that diverged from all other eukaryotes early in the history of the eukaryotes. That hypothesis, known as the "amitochondriate hypothesis," has also been abandoned for two reasons: Species previously thought to lack mitochondria have since been shown to have reduced mitochondria, and DNA sequence data have shown that some of these organisms are not closely related to one another. Finally, morphological studies and DNA sequence analyses suggest that the vast diversity of eukaryotes alive today can be grouped into four very large clades, the eukaryotic "supergroups." **25.4** Sample response: Ecologically important protists include photosynthetic dinoflagellates that provide essential sources of energy to their symbiotic partners, the corals that build coral reefs. Other important protistan symbionts include those that enable termites to digest wood and *Plasmodium*, the pathogen that causes malaria. Photosynthetic protists such as diatoms are among the most important producers in aquatic communities; as such, many other species in aquatic environments depend on them for food.

### Test Your Understanding

**1.** A **2.** D **3.** D **4.** B **5.** B **6.** C **7.** The two approaches differ in the evolutionary changes they may bring about. A strain of *Wolbachia* that confers resistance to infection by *Plasmodium* and does not harm mosquitoes would spread rapidly through the mosquito population. In this case, natural selection would favor any *Plasmodium* individuals that could overcome the resistance to infection conferred by *Wolbachia*. If insecticides are used, mosquitoes that are resistant to the insecticide would be favored by natural selection. Hence, use of *Wolbachia* could cause evolution in *Plasmodium* populations, while using insecticides could cause evolution in mosquito populations.
**8.**

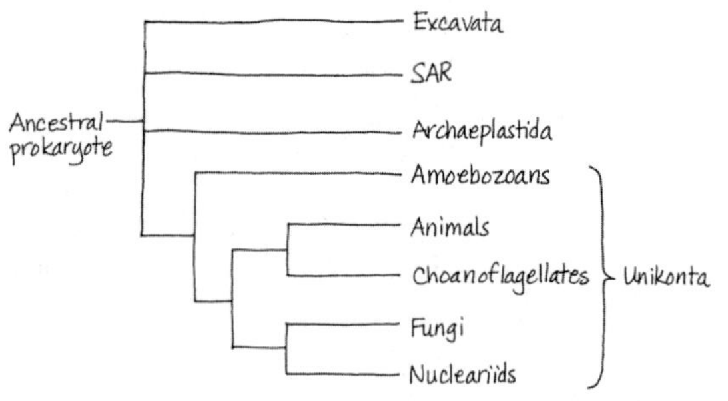


Pathogens that share a relatively recent common ancestor with humans will likely also share metabolic and structural characteristics with humans. Because drugs target the pathogen's metabolism or structure, developing drugs that harm the pathogen but not the patient should be most difficult for pathogens with which we share the most recent evolutionary history. Working backward in time, we can use the phylogenetic tree to determine the order in which humans shared a common ancestor with pathogens in different taxa. This process leads to the prediction that it should be hardest to develop drugs to combat animal pathogens, followed by choanoflagellate pathogens, fungal and nucleariid pathogens, amoebozoans, other protists, and finally prokaryotes.

## Chapter 26

### Figure Questions

**Figure 26.6** The life cycle of plants and some algae, shown in Figure 10.6b, has alternation of generations; the others do not. Unlike in the animal life cycle (Figure 10.6a), in the plant/algal life cycle, meiosis produces spores, not gametes. These spores then divide repeatedly by mitosis, ultimately forming a multicellular haploid individual that produces gametes. There is no multicellular haploid stage in the animal life cycle. An alternation of generations life cycle also has a multicellular diploid stage, whereas the life cycle of most fungi and some protists shown in Figure 10.6c does not.
**Figure 26.13** DNA from each of these mushrooms would be identical if each mushroom is part of a single hyphal network, as could well be the case.

**Figure 26.14** Possible examples include the Golgi apparatus (flattening; increases area for receiving and transporting proteins), the endoplasmic reticulum (folding; increases area for biosynthesis), the cardiovascular system (branching; increases area for materials exchange in tissues), and root hairs (projections; increase area for absorption). **Figure 26.23** It contains cells from three generations: (1) the current sporophyte (cells of ploidy $2n$, found in the seed coat and in the megasporangium remnant that surrounds the spore wall); (2) the female gametophyte (cells of ploidy $n$, found in the food supply); and (3) the sporophyte of the next generation (cells of ploidy $2n$, found in the embryo). **Figure 26.31** Two possible controls would be E−P− and E+P−. Results from an E−P− control could be compared with results from the E−P+ experiment, and results from an E+P− control could be compared with results from the E+P+ experiment. Together, these two comparisons would indicate whether the addition of the pathogen causes an increase in leaf mortality. Results from an E−P− experiment could also be compared with results from the second control (E+P−) to determine whether adding the endophytes has a negative effect on the plant.

### Concept Check 26.1

**1.** Plants share some key traits only with charophytes, including rings of cellulose-synthesizing complexes and similarity in sperm structure. Comparisons of nuclear, chloroplast, and mitochondrial DNA sequences also indicate that certain groups of charophytes (such as *Zygnema* and *Coleochaete*) are the closest living relatives of plants. **2.** Possible answers include: spore walls toughened by sporopollenin (protects against harsh environmental conditions); multicellular, dependent embryos (provides nutrients and protection to the developing embryo); cuticle (reduces water loss); stoma (supports photosynthesis by allowing the exchange of $CO_2$ and $O_2$ between the outside air and the plant body; stoma close during dry conditions, reducing water loss).
**3.** The earliest fossil evidence of plants comes from spores that date to 470 million years ago. These spores had a chemical composition similar to that in plant spores but different from the spores of other organisms; the walls of these spores also had structural features found only in spore walls of certain plants. Larger plant structures appear in the fossil record by 425 million years ago. By 400 million years ago, fossil evidence shows that a diverse assemblage of plants lived on land; collectively, these plants had key traits not found in their algal ancestors, such as specialized tissues for water transport, stomata, and branched sporophytes. **4.** The multicellular diploid stage of the life cycle would not produce gametes. Instead, both males and females would produce haploid spores by meiosis. These spores would give rise to multicellular male and female haploid stages—a major change from the single-celled haploid stages (sperm and eggs) that we actually have. The multicellular haploid stages would produce gametes and reproduce sexually. An individual at the multicellular haploid stage of the human life cycle might look like a present-day human or might look completely different.

### Concept Check 26.2

**1.** Both a fungus and a human are heterotrophs. Many fungi digest their food externally by secreting enzymes into the food and then absorbing the small molecules that result from digestion. Other fungi absorb such small molecules directly from their environment. In contrast, humans (and most other animals) ingest relatively large pieces of food and digest the food within their bodies. **2.** Mycorrhizae form extensive networks of hyphae through the soil, enabling nutrients to be absorbed more efficiently than a plant can do on its own; this is true today, and similar associations were probably very important for the earliest plants (which lacked roots). Evidence for the antiquity of mycorrhizal associations includes fossils showing arbuscular mycorrhizae in the early plant *Aglaophyton* and molecular results showing that genes required for the formation of mycorrhizae are present in liverworts and other basal plant lineages. **3.** Carbon that enters the plant through stomata is fixed into sugar through photosynthesis. Some of these sugars are absorbed by the fungus that partners with the plant to form mycorrhizae; others are transported within the plant body and used in the plant. Thus, the carbon may be deposited in either the body of the plant or the body of the fungus.

### Concept Check 26.3

**1.** Both seedless vascular plants and bryophytes have flagellated sperm that require moisture for fertilization; this shared similarity poses challenges for these species in arid regions. With respect to key differences, seedless vascular plants have lignified, well-developed vascular tissue, a trait that enables the sporophyte to grow tall and that has transformed life on Earth (via the formation of forests). Seedless vascular plants also have true leaves and roots, which, when compared with bryophytes, provide increased surface area for photosynthesis and improve their ability to extract nutrients from soil. **2.** Plants, vascular plants, and seed plants are monophyletic because each of these groups includes the common ancestor of the group and all of the descendants of that common ancestor. The other two categories of plants, the nonvascular plants and the seedless vascular plants, are paraphyletic: These groups do not include all of the descendants of the group's most recent common ancestor. **3.** The phylogeny in Figure 26.18 shows that while monilophytes and lycophytes are all seedless vascular plants, monilophytes share a more recent common ancestor with seed plants than with lycophytes. Therefore, we would expect key traits (such as megaphylls) that arose *after* monilophytes diverged from lycophytes but *before* monilophytes diverged from seed plants should be found in the most recent common ancestor of monilophytes and seed plants. The concept of descent with modification indicates that key traits found in the common ancestor of monilophytes and seed plants would likely also be found in that ancestor's descendants, the monilophytes and the seed plants.

### Concept Check 26.4

**1.** The reduced gametophytes of seed plants are nurtured by sporophytes and protected from stress, such as drought conditions and UV radiation. Pollen grains, with walls containing sporopollenin, provide protection during transport by wind or animals; because the sperm-producing male gametophytes are contained within pollen grains, the sperm of seed plants do not require water to reach the eggs. The ovule has a layer of tissue called integument that protects the female gametophyte as it develops from a megaspore. When mature, the ovule forms a seed, which has a thick layer of protective tissue, the seed coat. Seeds also contain a stored supply of food, which provides nourishment for growth after dormancy is broken and the embryo emerges as a seedling. **2.** Based on fossils known during his lifetime, Darwin was troubled by the relatively sudden and geographically widespread appearance of angiosperms in the fossil record. Recent fossil evidence shows that angiosperms arose and began to diversify over a period of 20–30 million years, a less rapid event than was suggested by the fossils known during Darwin's lifetime. Fossil discoveries have also uncovered extinct lineages of woody seed plants that may have been closely related to angiosperms; one such group, the Bennettitales, had flowerlike structures that may have been pollinated by insects. Phylogenetic analyses have identified *Amborella* as the most basal angiosperm lineage; *Amborella* is woody, and hence its basal position supports the conclusion (from fossils) that the common ancestor of angiosperms was likely woody. **3.** All taxa in this tree are vascular plants; you should have circled the lycophytes, which was the first group of vascular plants to have diverged from all other vascular plant lineages.

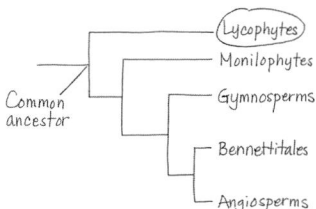

### Concept Check 26.5

**1.** Lichens, symbiotic associations between fungi and photosynthetic microorganisms (algae or cyanobacteria), break down bare rock surfaces by physically penetrating and chemically altering them. This influences the formation of soil and enables a succession of plants to grow. Plants also affect the formation of soil: Their roots hold soil in place, and leaf litter and other decaying plant parts add nutrients to the soil. Plants also affect the composition of Earth's atmosphere by releasing oxygen to the air and by their impact on the atmospheric concentration of $CO_2$. **2.** Mutualistic fungi absorb nutrients from their host organism but reciprocate by providing benefits to the host. Important examples include mycorrhizal associations with plant roots (in which fungal hyphae increase the efficiency with which the plant can absorb nutrients such as phosphorous from the soil) and symbiotic endophytes (fungi that live within leaves or other plant parts and provide the plant with benefits such as increased resistance to disease or increased tolerance of heat, drought, or heavy metals). Parasitic fungi also absorb nutrients from host cells but provide no benefits in return. Examples include the ascomycete fungus *Cryphonectria parasitica* (which causes chestnut blight, a disease that has virtually eliminated the once-common chestnut tree from forests of the northeastern United States). **3.** You should have circled steps that represent light energy absorption, photosynthesis, consumers eating producers, uptake of nutrients by plants, and decomposition. **4.** By focusing on cases in which a radial clade shared an immediate common ancestor with a bilateral clade, the researchers could control for effects of time; that is, each radial clade had the same amount of time over which new species could form as did the bilateral clade to which it was compared. As a result, differences in the number of species between the two clades could be attributed to flower shape (rather than to differences in the length of time over which new species could form).

### Summary of Key Concepts Questions

**26.1** The earliest fossil evidence of plants comes from spores that date to 470 million years ago. These spores have a chemical composition that matches that found in the spores of extant plants, yet differs from the spores of other organisms. Furthermore, the structure of the walls of these spores is only found in the spores of certain plants (liverworts). Finally, similar spores dating to 450 million years ago have been found embedded in plant cuticle material. **26.2** The body of a multicellular fungus typically consists of thin filaments called hyphae. These filaments form an interwoven mass (mycelium) that penetrates the substrate on which the fungus grows and feeds. Because the individual filaments are thin, the surface-to-volume ratio of the mycelium is maximized, making nutrient absorption highly efficient. Furthermore, fungi that form mycorrhizal associations with plant roots have specialized hyphae through which they can exchange nutrients with their host plant. The high efficiency with which fungal mycelia absorb nutrients, together with the ability of mycorrhizae to exchange nutrients through specialized hyphae, may have provided early plants (which lacked roots) with greater access to soil nutrients—thus aiding the colonization of land by plants. **26.3** Lignified vascular tissue provided the strength needed to support a tall plant against gravity, as well as a means to transport water and nutrients to plant parts located high above ground. Roots were another key trait, anchoring the plant to the ground and providing additional structural support for plants that grew tall. Tall plants could shade shorter plants, thereby outcompeting them for light. Because the spores of a tall plant disperse farther than the spores of a short plant, it is also likely that tall plants could colonize new habitats more rapidly than short plants. **26.4** The Bennettitales is one of several groups of fossil seed plants that are thought to be more closely related to extant angiosperms than to extant gymnosperms. All of the species in the Bennettitales and other such fossil seed plants were woody. The most basal lineage of extant angiosperms (*Amborella*) is also woody. The fact that both the seed plant ancestors of angiosperms and the most basal taxon of extant angiosperms were woody suggests that the angiosperm common ancestor was woody. **26.5** During photosynthesis, plants convert light energy to the chemical energy of food; that chemical energy supports all life on land, either directly (as when an herbivore eats a plant) or indirectly (as when a predator eats an herbivore that ate a plant). Large animals, such as vertebrate herbivores and their predators, could not survive on land in the absence

of plants, so the presence of plants on land has enabled the myriad biotic interactions that occur among large animal species today. Similarly, plants extract nutrients from the soil and capture carbon (in the form of $CO_2$) from the air; as a result, those nutrients become available to terrestrial animals. Fungi also play an essential role in increasing the availability of nutrients to other terrestrial organisms. As decomposers, fungi break down the bodies of dead organisms, thereby recycling chemical nutrients to the physical environment. If plants and fungi had not colonized land, photosynthesis and decomposition would still occur—but all terrestrial life would be microbial, and hence biotic interactions among terrestrial organisms would occur on a much smaller scale than they do today.

### Test Your Understanding

**1.** B  **2.** D  **3.** (a) diploid; (b) haploid; (c) haploid; (d) diploid  **4.** C  **5.** D
**6.**

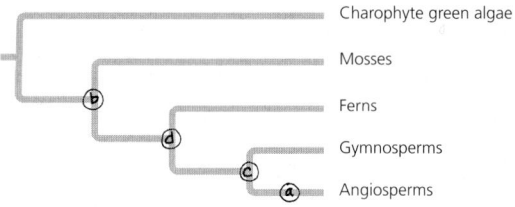

## Chapter 27

### Figure Questions

**Figure 27.5** You should have circled the node shown in the tree diagram at approximately 635 million years ago (mya), which leads to the echinoderm/chordate lineage and to the lineage that gave rise to brachiopods, annelids, molluscs, and arthropods. Although the 635 mya date is shown in the figure, this common ancestor must be at least as old as any of its descendants. Since fossil molluscs date to about 560 mya, the common ancestor represented by the circled branch point must be at least 560 million years old. **Figure 27.10** Cnidaria is the sister phylum in this figure.
**Figure 27.15** Such a result would be consistent with the origin of the *Ubx* and *abd-A Hox* genes having played a major role in the evolution of increased body segment diversity in arthropods. However, note that such a result would simply show that the presence of the *Ubx* and *abd-A Hox* genes was *correlated with* an increase in body segment diversity in arthropods; it would not provide direct experimental evidence that the origin of the *Ubx* and *abd-A* genes *caused* an increase in arthropod body segment diversity. **Figure 27.19** You would expect the vertebrate groups Actinopterygii, Actinistia, Dipnoi, and Tetrapoda to have lungs or lung derivatives. All of these groups originate to the right of (evolved after) the hatch mark indicating the appearance of this derived character in their lineage. **Figure 27.26** Between 370 mya and 340 mya. We can infer this because amphibians must have originated after the most recent common ancestor of *Tulerpeton* and living tetrapods (and that ancestor originated 370 mya), but no later than the date of the earliest known fossils of amphibians (shown in the figure as 340 mya). **Figure 27.29** Pterosaurs did not descend from the common ancestor of all dinosaurs; hence, pterosaurs are not dinosaurs. However, birds are descendants of the common ancestor of the dinosaurs. As a result, a monophyletic clade of dinosaurs must include birds. In that sense, birds are dinosaurs. **Figure 27.31** In a catabolic pathway, like the aerobic processes of cellular respiration, water is released as a by-product when an organic compound such as glucose is mixed with oxygen. The kangaroo rat can retain and use that water, decreasing its need to drink water.
**Figure 27.32**

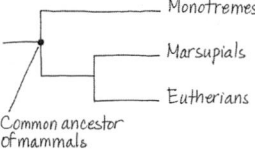

**Figure 27.40** Since the cod are adapting to the pressure of fishing by reproducing at younger ages, the overall number of offspring they produce each year will be lower. This may cause the population to decline as time goes on, thereby further reducing the population's ability to recover.

### Concept Check 27.1

**1.** The earliest evidence of animal life comes from 710-million-year-old sediments containing steroids indicative of sponges. This fossil biochemical evidence is consistent with molecular clock results indicating that animals originated 770 million years ago (mya), sponges originated 700 mya, and cnidarians originated 680 mya. The oldest fossils of large animals date to about 560 mya; these fossils are of sponges, cnidarians, and molluscs. Thus, by 560 mya at the latest, both sponges and cnidarians had diverged from other animal groups. **2.** Evolution is not goal oriented; hence, it would not be correct to argue that sponges were not "highly evolved" simply because they lack tissues (unlike nearly all other animals). Instead, the fact that sponges have persisted for hundreds of millions of years indicates that their lineage is a highly successful one.

### Concept Check 27.2

**1.** The "Cambrian explosion" refers to a relatively short interval of time (535–525 million years ago) during which large forms of many present-day animal phyla first appear in the fossil record. The evolutionary changes that occurred during this time, such as the appearance of large predators and well-defended prey, were important because they set the stage for many of the key events in the history of life over the last 500 million years. **2.** Following such a change, predators that were best able to kill or catch these well-defended prey might leave more offspring than would other (less capable) predators. As a result, evolution by natural selection in the predator population would likely improve the ability of the predators to eat these prey. If that took place, prey individuals with new defensive adaptations would be favored by natural selection, potentially leading to further changes in predator populations, and so on.

## Concept Check 27.3

**1.** A body plan is a set of morphological and developmental traits, integrated into a functional whole (the living animal). One key feature is the type of symmetry (or absence of symmetry): Sponges lack symmetry, some animals exhibit radial symmetry, and others are bilaterally symmetric. Another key feature is the way tissues are organized. Sponges and a few other animal groups lack true tissues; the tissues of cnidarians and ctenophores originate from two embryonic germ layers, while the tissues of most animals (bilaterians) originate from three germ layers. A third feature found in most bilaterians is a body cavity, a fluid- or air-filled space located between the digestive tract and the outer body wall. **2.** It would be accurate to describe the Cambrian explosion as consisting of three "explosions." To see why, note that the phylogeny in Figure 27.10 indicates that molluscs are members of Lophotrochozoa, one of the three main groups of bilaterians (the others being Deuterostomia and Ecdysozoa). As discussed in Concept 27.2, the fossil record shows that molluscs were present tens of millions of years before the Cambrian explosion. Thus, long before the Cambrian explosion, the lophotrochozoan clade had formed and was evolving independently of the evolutionary lineages leading to Deuterostomia and Ecdysozoa. Based on the phylogeny in Figure 27.10, we can also conclude that the lineages leading to Deuterostomia and Ecdysozoa were independent of one another before the Cambrian explosion. Thus, since the lineages leading to the three main clades of bilaterians were evolving independently of one another prior to the Cambrian explosion, that explosion could be viewed as consisting of three "explosions," not one. With respect to summarizing the major steps in animal evolution, the phylogeny in Figure 27.10 indicates that all animals share a common ancestor, that sponges are basal animals, that Eumetazoa is a clade of animals with true tissues, and that most phyla belong to the clade Bilateria.

## Concept Check 27.4

**1.** The four characters are a notochord; a dorsal, hollow nerve chord; pharyngeal slits or clefts; and a muscular, post-anal tail. **2.** Two key adaptations of aquatic gnathostomes are their jaws (an adaptation for feeding) and their paired fins and a tail (adaptations for swimming). **3.** During the time period covered by this question, a broad range of invertebrate phyla diversified in marine environments. Invertebrates in one of these phyla—Chordata—gave rise to early vertebrates, and those early vertebrates diversified further into two lineages of jawless vertebrates and three lineages of jawed vertebrates. One lineage of jawed vertebrates would ultimately give rise to the tetrapods, the vertebrate lineage that colonized land. But the other lineages of jawed vertebrates—along with the many lineages of invertebrates—continued to diversify in aquatic environments, making it hard to argue that the evolutionary changes that took place were directed toward the origin of terrestrial vertebrates.

## Concept Check 27.5

**1.** The arthropod exoskeleton, which had already evolved in the ocean, allows terrestrial species to retain water and support their bodies on land. Wings allow insects to disperse quickly to new habitats and to find food and mates. The tracheal system allows for efficient gas exchange despite the presence of an exoskeleton. **2.** Descent with modification—the process by which organisms gradually accumulate differences from their ancestors—occurred in the colonization of land by plants as well as the colonization of land by animals. The modifications over time, however, were more extensive in plants than in animals. This was because plants arose from a small alga with few features that were suitable for life on land. Animals, in contrast, colonized land repeatedly; in each of these events, the animals that colonized land arose from aquatic animals that typically had well-developed skeletal, muscle, digestive, circulatory, respiratory, and nervous systems—all of which facilitated the colonization of land.

## Concept Check 27.6

**1.** The amniotic egg provides protection to the embryo and allows the embryo to develop on land, eliminating the necessity of a watery environment for reproduction. Another key adaptation is rib cage ventilation, which improves the efficiency of air intake and may have allowed early amniotes to dispense with breathing through their skin. Finally, not breathing through their skin allowed amniotes to develop relatively impermeable skin, thereby conserving water. **2.** As illustrated in Figure 27.33, chimpanzees and humans represent the tips of separate branches of evolution. As such, the human and chimpanzee lineages have evolved independently since they diverged from their common ancestor—an event that took place between 6 million and 7 million years ago. Hence, it is incorrect to say that humans evolved from chimpanzees (or vice versa). **3.** The egg came first. The amniotic egg, which all reptiles (including chickens) and all mammals have, arose more than 310 million years ago, long before the first chicken (or any other bird).

## Concept Check 27.7

**1.** The oceans had cloudy waters and low oxygen levels for more than a billion years after the origin of eukaryotes; throughout this time, cyanobacteria were the dominant producers. By the early Cambrian period, the ocean waters were clearer and had higher oxygen levels; in addition, cyanobacteria were less abundant and algae had become the dominant producers. By removing large quantities of cyanobacteria, early suspension-feeding animals would have made the waters less cloudy, a change that favored algae (which require more light for photosynthesis than do cyanobacteria). By about 530 million years ago, a variety of large animals were present, leading to dramatic changes in feeding relationships as formidable predators pursued well-defended prey. **2.** Before animals colonized land, terrestrial communities had a simple structure, the main elements of which consisted of producers (early plants) and decomposers. The colonization of land by animals introduced new types of biotic interactions that involved herbivorous animals that ate plants, detritivores such

as millipedes that consumed decaying organic matter, and predators. **3.** Gene flow occurs more readily between nearby populations than between distant populations; hence, we would predict that gene flow would be higher in the original population than in the remnant populations. And since genetic drift has more pronounced effects in small populations, we would predict that the role of genetic drift would be more pronounced in the remnant populations. Finally, since genetic drift can lead to the fixation of harmful alleles, we would predict that the risk of extinction would be higher in the remnant populations than in the original populations.

### Summary of Key Concepts Questions

**27.1** Sponge choanocyte cells are similar morphologically to the cells of choanoflagellates; DNA sequences of sponges and choanoflagellates are also very similar. These observations are consistent with the hypothesis that animals descended from a lineage of single-celled eukaryotes similar to present-day choanoflagellates. **27.2** Current hypotheses about the cause of the Cambrian explosion include new predator-prey relationships, an increase in atmospheric oxygen, and an increase in developmental flexibility provided by the origin of *Hox* genes and other genetic changes.

**27.3**

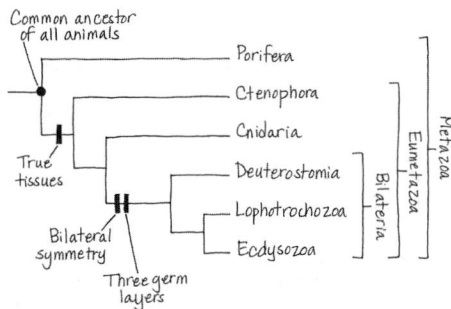

**27.4** The origin of jaws altered how early aquatic vertebrates obtained food, which in turn would have had large effects on ecological interactions. Predators could use their jaws to grab prey or remove chunks of flesh, stimulating the evolution of increasingly sophisticated means of defense in prey species. Evidence for these changes can be found in the fossil record, which includes fossils of 10-m-long predators with remarkably powerful jaws, as well as lineages of well-defended prey species whose bodies were covered by armored plates. **27.5** Tetrapods are thought to have originated about 365 million years ago when the fins of some lobe-fins evolved into the limbs of tetrapods. Steps in this process are illustrated by *Tiktaalik*. Like a fish, this species had fins, gills, and lungs, and its body was covered in scales. But unlike a fish, *Tiktaalik* had ribs and a neck, and the bones of its front fin had the same basic pattern as those in a tetrapod limb. In addition to their four limbs with digits—a key derived trait for which the group is named—other derived traits of tetrapods include a neck (consisting of vertebrae that separate the head from the rest of the body) and a pelvic girdle that is fused to the backbone. **27.6** The major derived character of the amniotes is the amniotic egg, which contains four specialized membranes: the amnion, the chorion, the yolk sac, and the allantois. The amniotic egg provides protection to the embryo and allows the embryo to develop on land, eliminating the necessity of a watery environment for reproduction. As a result, the amniotes were able to expand into a wider range of terrestrial habitats than were other tetrapods (such as early tetrapods and amphibians). **27.7** As organisms interact over time with other organisms and the physical environment, their populations can evolve. The activities of animals have altered the physical structure (for example, the water clarity) of the ocean and fundamentally changed biotic interactions in the sea and on land—thus potentially causing evolutionary change in a wide range of species. Examples include the effects animals have had on evolutionary radiations in parasites and plants, as well as the ongoing evolutionary changes that are taking place in populations that humans hunt for sport or food.

### Test Your Understanding

**1.** B **2.** D **3.** C **4.** A **5.** C **6.** B **7.** (a) Because brain size tends to increase consistently in such lineages, we can conclude that natural selection favored the evolution of larger brains and hence that the benefits outweighed the costs. (b) As long as the benefits of brains that are large relative to body size are greater than the costs, large brains can evolve. Natural selection might favor the evolution of brains that are large relative to body size because such brains confer an advantage in obtaining mates and/or an advantage in survival.

(c)

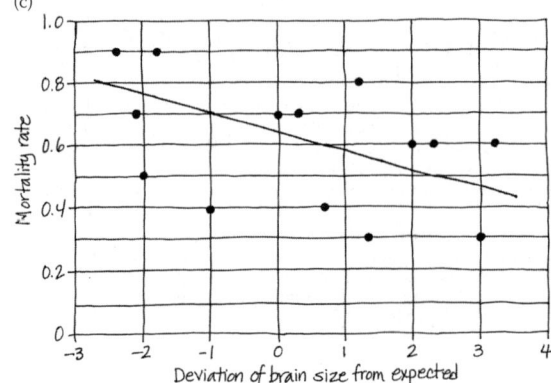

Adult mortality tends to be lower in birds with larger brains.   **8.** The circled clade should include birds, the two dinosaur lineages, and the common ancestor of the dinosaurs. The phylogeny shows that dinosaurs other than birds are nested between crocodilians and birds. Since crocodilians and birds differ with respect to whether they are endothermic, we cannot use phylogenetic bracketing to predict whether dinosaurs other than birds were endothermic (or not). However, we can conclude that the dinosaur that gave rise to birds was endothermic, as are all living birds.

## Chapter 28

### Figure Questions

**Figure 28.12** Every root epidermal cell would develop a root hair.
**Figure 28.18** Pith and cortex are defined, respectively, as ground tissue that is internal and ground tissue that is external to vascular tissue. Since vascular bundles of monocot stems are scattered throughout the ground tissue, there is no clear distinction between internal and external relative to the vascular tissue.   **Figure 28.19** The vascular cambium produces growth that increases the diameter of a stem or root. The tissues that are exterior to the vascular cambium cannot keep pace with the growth because their cells no longer divide. As a result, these tissues rupture.

### Concept Check 28.1

**1.** The vascular tissue system connects leaves and roots, allowing sugars to move from leaves to roots in the phloem and allowing water and minerals to move to the leaves in the xylem.   **2.** To get sufficient energy from photosynthesis, we would need lots of surface area exposed to the sun. This large surface-to-volume ratio, however, would create a new problem—evaporative water loss. We would have to be permanently connected to a water source—the soil, also our source of minerals. In short, we would probably look and behave very much like plants.   **3.** As plant cells enlarge, they typically form a huge central vacuole that contains a dilute watery sap. Central vacuoles enable plant cells to become large with only a minimal investment of new cytoplasm. The orientation of the cellulose microfibrils in plant cell walls affects the growth pattern of cells.

### Concept Check 28.2

**1.** Primary growth arises from apical meristems and involves production and elongation of organs. Secondary growth arises from lateral meristems and adds to the girth of roots and stems.   **2.** The largest, oldest leaves would be lowest on the shoot. Since they would probably be heavily shaded, they would not photosynthesize much regardless of their size.   **3.** No. The carrot roots will probably be smaller at the end of the second year because the food stored in the root will be used to produce flowers, fruit, and seeds.

### Concept Check 28.3

**1.** In roots, primary growth occurs in three successive stages, moving away from the tip of the root: the zones of cell division, elongation, and differentiation. In shoots, it occurs at the tip of apical buds, with leaf primordia arising along the sides of an apical meristem. Most growth in length occurs in older internodes below the shoot tip.
**2.** No. Because vertically oriented leaves, such as maize, can capture light equally on both sides of the leaf, you would expect them to have mesophyll cells that are not differentiated into palisade and spongy layers. This is typically the case. Also, vertical leaves usually have stomata on both leaf surfaces.   **3.** Root hairs are cellular extensions that increase the surface area of the root epidermis, thereby enhancing the absorption of minerals and water. Microvilli are extensions that increase the absorption of nutrients by increasing the surface area of the gut.

### Concept Check 28.4

**1.** The sign will still be 2 m above the ground because this part of the tree is no longer growing in length (primary growth); it is now growing only in thickness (secondary growth).   **2.** Since there is little temperature variation in the tropics, the growth rings of a tree from the tropics would be difficult to discern unless the tree came from an area that had pronounced wet and dry seasons.   **3.** The tree would die slowly. Girdling removes an entire ring of secondary phloem (part of the bark), completely preventing transport of sugars and starches from the shoots to the roots. After several weeks, the roots would have used all of their stored carbohydrate reserves and would die.

### Summary of Key Concepts Questions

**28.1** Here are a few examples: The cuticle of leaves and stems protects these structures from desiccation. Collenchyma and sclerenchyma cells have thick walls that provide support for plants. Strong, branching root systems help anchor the plant in the soil.
**28.2** All plant organs and tissues are ultimately derived from meristematic activity.
**28.3** Lateral roots emerge from the pericycle and destroy plant cells as they emerge. In stems, branches arise from axillary buds and do not destroy any cells.   **28.4** With the evolution of secondary growth, plants were able to grow taller and shade competitors.

### Test Your Understanding

**1.** D   **2.** C   **3.** C   **4.** D   **5.** D
**6.**

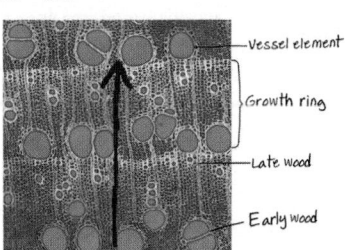

## Chapter 29

### Figure Questions

**Figure 29.2** Cellular respiration is occurring in all parts of a growing plant at all times, with mitochondria continuously releasing $CO_2$ and consuming $O_2$. In photosynthetic cells, the $CO_2$ produced by mitochondria during the day is consumed by chloroplasts, which also consume $CO_2$ from the air. Meanwhile, the mitochondria obtain $O_2$ from the chloroplasts, which also release $O_2$ into the air. At night, when photosynthesis does not occur, the mitochondria must exchange gases with the air rather than with the chloroplasts. As a result, at night photosynthetic cells are *releasing* $CO_2$ into the air and *consuming* $O_2$ from the air, the opposite of what happens during the day.
**Figure 29.3** The leaves are being produced in a counterclockwise spiral.
**Figure 29.9** Anions. Because cations are bound to soil particles, they are less likely to be lost from the soil following heavy rains.   **Figure 29.10** Some other examples of mutualism are the following relationships. *Sea turtles and various fish:* The fish feed on small organisms living on the turtle's shell, while the turtle benefits by the removal of these parasites. *Flashlight fish and bioluminescent bacteria:* The bacteria gain nutrients and protection from the fish, while the bioluminescence attracts prey and mates for the fish. *Flowering plants and pollinators:* Animals distribute the pollen and are rewarded by a meal of nectar or pollen. *Vertebrate herbivores and some bacteria in the digestive system:* Microorganisms in the alimentary canal break down cellulose to glucose and, in some cases, provide the animal with vitamins or amino acids. Meanwhile, the microorganisms have a steady supply of food and a warm environment. *Humans and some bacteria in the digestive system:* Some bacteria provide humans with vitamins, while the bacteria get nutrients from the digested food.   **Figure 29.13** The legume plants benefit because the bacteria fix nitrogen that is absorbed by their roots. The bacteria benefit because they acquire photosynthetic products from the plants.
**Figure 29.16** The Casparian strip blocks water and minerals from moving between endodermal cells or moving around an endodermal cell via the cell's wall. Therefore, water and minerals must pass through an endodermal cell's plasma membrane.

### Concept Check 29.1

**1.** Vascular plants must transport minerals and water absorbed by the roots to all the other parts of the plant. They must also transport sugars from sites of production to sites of use.   **2.** Many features of plant architecture affect self-shading, including leaf arrangement and the orientations of stems and leaves.   **3.** Increased stem elongation would raise the plant's upper leaves. Erect leaves and reduced lateral branching would make the plant less subject to shading by the encroaching neighbors.

### Concept Check 29.2

**1.** The cell's $\Psi_P$ is 0.7 MPa. In a solution with a $\Psi$ of −0.4 MPa, the cell's $\Psi_P$ at equilibrium would be 0.3 MPa.   **2.** The cells would still adjust to changes in their osmotic environment, but their responses would be slower. Although aquaporins do not affect the water potential gradient across membranes, they allow for more rapid osmotic adjustments.   **3.** The protoplasts would burst. Because the cytoplasm has many dissolved solutes, water would enter the protoplast continuously without reaching equilibrium. (When present, the cell wall prevents rupturing by excessive expansion of the protoplast.)

### Concept Check 29.3

**1.** No. Even though macronutrients are required in greater amounts, all essential elements are necessary for the plant to complete its life cycle.   **2.** No. The fact that the addition of an element results in an increase in the growth rate of a plant does not mean that the element is strictly required for the plant to complete its life cycle.
**3.** Waterlogging displaces air from the soil, leading to low $O_2$ conditions. Such conditions promote the anaerobic process of alcoholic fermentation in plants, the end product of which is ethanol.

### Concept Check 29.4

**1.** The rhizosphere is a narrow zone in the soil immediately adjacent to living roots. This zone is especially rich in both organic and inorganic nutrients and has a microbial population that is many times greater than the bulk of the soil.   **2.** Soil bacteria and mycorrhizae enhance plant nutrition by making certain minerals more available to plants. For example, many types of soil bacteria are involved in the nitrogen cycle, and the hyphae of mycorrhizae provide a large surface area for the absorption of nutrients, particularly phosphate ions.   **3.** Saturating rainfall may deplete the soil of oxygen. A lack of soil oxygen would inhibit nitrogen fixation by the soybean root nodules and decrease the nitrogen available to the plant. Alternatively, heavy rain may leach nitrate from the soil. A symptom of nitrogen deficiency is yellowing of older leaves.

### Concept Check 29.5

**1.** The endodermis regulates the passage of water-soluble solutes by requiring all such molecules to cross a selectively permeable membrane. Presumably, the inhibitor never reaches the plant's photosynthetic cells.   **2.** Perhaps greater root mass helps compensate for the lower water permeability of the plasma membranes.   **3.** The Casparian strip and tight junctions both prevent movement of fluid between cells.

### Concept Check 29.6

**1.** The activation of the proton pumps of stomatal cells would cause the guard cells to take up $K^+$. The increased turgor of the guard cells would lock the stomata open and lead to extreme evaporation from the leaf.   **2.** After the flowers are cut, transpiration from any leaves and from the petals (which are modified leaves) will continue to draw water up the xylem. If cut flowers are transferred directly to a vase, air pockets in xylem vessels prevent delivery of water from the vase to the flowers. Cutting stems again underwater, a few centimeters from the original cut, will sever the xylem above the air pocket. The water droplets prevent another air pocket from forming while placing the flowers in a vase.   **3.** Water molecules are in constant motion, traveling at different rates. The average speed of these particles depends on the water's temperature. If water molecules gain enough energy, the most energetic molecules near the liquid's surface will impart sufficient speed, and therefore sufficient kinetic energy, to cause water molecules to propel away from the liquid in the form of gaseous molecules or, more simply, as water vapor. As the particles with the highest kinetic energy levels evaporate, the average kinetic energy of the remaining liquid decreases. Because a liquid's temperature is directly related to the average kinetic energy of its molecules, the liquid cools as it evaporates.

### Concept Check 29.7
**1.** The main sugar sources are fully grown leaves (by photosynthesis) and fully developed storage organs (by breakdown of starch). Roots, buds, stems, expanding leaves, and fruits are powerful sugar sinks because they are actively growing. A storage organ may be a sugar sink in the summer when accumulating carbohydrates, but a sugar source in the spring when breaking down starch into sugar for growing shoot tips. **2.** Positive pressure in the sieve-tube elements of the phloem requires active transport. Most long-distance transport in the xylem depends on bulk flow driven by negative pressure potential generated ultimately by the evaporation of water from the leaf and does not require living cells. **3.** The spiral slash prevents optimal bulk flow of the phloem sap to the root sinks. Therefore, more phloem sap can move from the source leaves to the fruit sinks, making them sweeter.

### Summary of Key Concepts Questions
**29.1** Plants with tall shoots and elevated leaf canopies generally had an advantage over shorter competitors. A consequence of the selective pressure for tall shoots was the further separation of leaves from roots. This separation created problems for the transport of materials between root and shoot systems. Plants with xylem cells were more successful at supplying their shoot systems with soil resources (water and minerals). Similarly, those with phloem cells were more successful at supplying sugar sinks with carbohydrates. **29.2** Xylem sap is pulled up the plant by transpiration much more often than it is pushed up the plant by root pressure. **29.3** No. Plants can complete their life cycle when grown hydroponically, that is, in aerated salt solutions containing the proper ratios of all the minerals needed by plants. **29.4** No. Some parasitic plants obtain their energy by siphoning off carbon nutrients from other organisms. **29.5** Hydrogen bonds are necessary for the cohesion of water molecules to each other and for the adhesion of water to other materials, such as cell walls. Both adhesion and cohesion of water molecules are involved in the ascent of xylem sap under conditions of negative pressure. **29.6** Although stomata account for most of the water lost from plants, they are necessary for exchange of gases—for example, for the uptake of carbon dioxide needed for photosynthesis. **29.7** Although the movement of phloem sap depends on bulk flow, the pressure gradient that drives phloem transport depends on the osmotic uptake of water in response to the loading of sugars into sieve-tube elements at sugar sources. Phloem loading depends on $H^+$ cotransport processes that ultimately depend on $H^+$ gradients established by active $H^+$ pumping.

### Test Your Understanding
**1.** B  **2.** A  **3.** B  **4.** B  **5.** C  **6.** A  **7.** C  **8.**

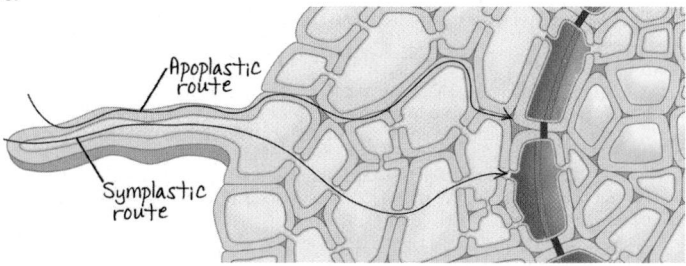

## Chapter 30

### Figure Questions
**Figure 30.3** Another example of a homeotic gene mutation is the mutation in a *Hox* gene that causes legs to form in place of antennae in *Drosophila* (depicted in Figure 16.8). **Figure 30.4** The flower would consist of nothing but carpels. **Figure 30.8** In addition to having a single cotyledon, monocots generally have leaves with parallel leaf venation, scattered vascular tissue in their stems, a fibrous root system, pollen grains with only one opening, and floral organs in multiples of three. In contrast, eudicots have two cotyledons and generally netlike leaf venation, vascular tissue in a ring, taproots, pollen grains with three openings, and floral organs in multiples of four or five. **Figure 30.9** Beans use a hypocotyl hook to push through the soil. The delicate leaves and shoot apical meristem are also protected by being sandwiched between two large cotyledons. The coleoptile of maize seedlings helps protect the emerging leaves.

### Concept Check 30.1
**1.** Long styles help to weed out pollen grains that are genetically inferior and not capable of successfully growing long pollen tubes. **2.** Hypothetically, tepals could arise if *B* gene activity were present in all three of the outer whorls of the flower. **3.** No. The haploid (gametophyte) generation of plants is multicellular and arises from spores. The haploid phase of the animal life cycle is a single-celled gamete (egg or sperm) that arises directly from meiosis: There are no spores.

### Concept Check 30.2
**1.** Flowering plants can avoid self-fertilization by being dioecious, by having different flowers with reproductive parts of different lengths, or by self-incompatibility. **2.** Asexually propagated crops lack genetic diversity. Genetically diverse populations are less likely to become extinct in the face of an epidemic because there is a greater

likelihood that a few individuals in the population are resistant. **3.** In the short term, selfing may be advantageous in a population that is so dispersed and sparse that pollen delivery is unreliable. In the long term, however, selfing is an evolutionary dead end because it leads to a loss of genetic diversity that may preclude adaptive evolution.

### Concept Check 30.3
**1.** Traditional breeding and genetic engineering both involve artificial selection for desired traits. However, genetic engineering techniques facilitate faster gene transfer and are not limited to transferring genes between closely related varieties or species. **2.** *Bt* maize suffers less insect damage; therefore, *Bt* maize plants are less likely to be infected by fumonisin-producing fungi that infect plants through wounds. **3.** In such species, engineering the transgene into the chloroplast DNA would not prevent its escape in pollen; such a method requires that the chloroplast DNA be found only in the egg. An entirely different method of preventing transgene escape would therefore be needed, such as male sterility, apomixis, or self-pollinating closed flowers.

### Summary of Key Concepts Questions
**30.1** After pollination, a flower typically changes into a fruit. The petals, sepals, and stamens typically fall off the flower. The stigma of the pistil withers and the ovary begins to swell. The ovules (embryonic seeds) inside the ovary begin to mature. **30.2** Asexual reproduction can be advantageous in a stable environment because individual plants that are well suited to that environment pass on all their genes to offspring. Also, asexual reproduction generally results in offspring that are less fragile than the seedlings produced by sexual reproduction. However, sexual reproduction offers the advantage of dispersal of tough seeds. Moreover, sexual reproduction produces genetic variety, which may be advantageous in an unstable environment. It is more likely that at least one offspring of sexual reproduction will survive in a changed environment. **30.3** "Golden Rice" has been engineered to produce more vitamin A, thereby raising the nutritional value of rice. A protoxin gene from a soil bacterium has been engineered into *Bt* maize. This protoxin is lethal to invertebrates but harmless to vertebrates. *Bt* crops require less pesticide spraying and have lower levels of fungal infection. Genetic engineering has increased the nutritional value of cassava by boosting the amount of iron and beta-carotene (a vitamin A precursor) and almost eliminating cyanide-producing chemicals from the roots.

### Test Your Understanding
**1.** C  **2.** A  **3.** C  **4.** D  **5.** C  **6.** D  **7.** D  **8.** C  **9.** D
**10.**

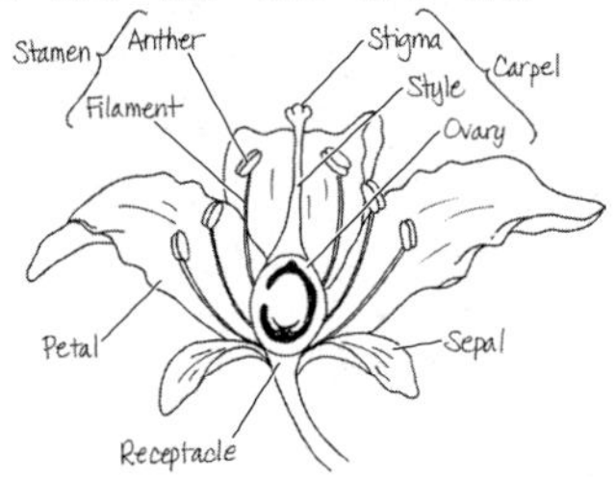

Stamen, Anther, Filament, Stigma, Style, Carpel, Ovary, Petal, Sepal, Receptacle

## Chapter 31

### Figure Questions
**Figure 31.2** To determine which wavelengths of light are most effective in phototropism, you could use a glass prism to split white light into its component colors and see which colors cause the quickest bending (the answer is blue; see Figure 31.12). **Figure 31.3** More auxin would move down the side without the TIBA-containing agar bead, causing greater elongation on this side and, consequently, bending of the coleoptile toward the side with the bead. **Figure 31.4** No. Polar auxin transport depends on the distribution of auxin transport proteins at the basal ends of cells. **Figure 31.13** Yes. The white light, which contains red light, would stimulate seed germination in all treatments. **Figure 31.17** The short-day plant would not flower. The long-day plant would flower. **Figure 31.18** If this were true, florigen would be an inhibitor of flowering, not an inducer.

### Concept Check 31.1
**1.** Fusicoccin's ability to cause an increase in plasma $H^+$ pump activity would have an auxin-like effect and promote stem cell elongation. **2.** The plant will exhibit a constitutive triple response. Because the kinase that normally prevents the triple response is dysfunctional, the plant will undergo the triple response regardless of whether ethylene is present or the ethylene receptor is functional. **3.** Since ethylene often stimulates its own synthesis, it is under positive-feedback regulation.

### Concept Check 31.2
**1.** Not necessarily. Many environmental factors, such as temperature and light, change over a 24-hour period in the field. To determine whether the enzyme is under circadian control, a scientist would have to demonstrate that its activity oscillates even when environmental conditions are held constant. **2.** It is impossible to say. To establish that this species is a short-day plant, it would be necessary to establish the critical night length for flowering and that this species only flowers when the night is longer than the critical night length. **3.** According to the action spectrum, red and blue light are the most effective in photosynthesis. Thus, it is not surprising that plants assess their light environment using blue- and red-light-absorbing photoreceptors.

### Concept Check 31.3
**1.** A plant that overproduces ABA would undergo less evaporative cooling because its stomata would not open as widely. **2.** Plants close to the aisles may be more subject to mechanical stresses caused by passing workers and air currents. The plants nearer

to the center of the bench may also be taller as a result of shading and less evaporative stress. **3.** No. Because root caps are involved in sensing gravity, roots that have their root caps removed are almost completely insensitive to gravity.

### Concept Check 31.4
**1.** Some insects increase plants' productivity by eating harmful insects or aiding in pollination. **2.** Mechanical damage breaches a plant's first line of defense against infection, its protective dermal tissue. **3.** Perhaps the breeze blows away a volatile defense compound that the plants produce.

### Summary of Key Concepts Questions
**31.1** Yes, there is truth to the old adage that one bad apple spoils the whole bunch. Ethylene, a gaseous hormone that stimulates ripening, is produced by damaged, infected, or overripe fruits. Ethylene can diffuse to healthy fruit in the "bunch" and stimulate their rapid ripening. **31.2** Plant physiologists proposed the existence of a floral-promoting factor (florigen) based on the fact that a plant induced to flower could induce flowering in a second plant to which it was grafted, even though the second plant was not in an environment that would normally induce flowering in that species. **31.3** Plants subjected to drought stress are often more resistant to freezing stress because the two types of stress are quite similar. Freezing of water in the extracellular spaces causes free water concentrations outside the cell to decrease. This, in turn, causes free water to leave the cell by osmosis, leading to the dehydration of cytoplasm, much like what is seen in drought stress. **31.4** In response to herbivory, plants suffer a loss in biomass. They divert energy away from growth and toward defense. The wounds associated with herbivore damage can create portals for pathogen invasion.

### Test Your Understanding
**1.** B **2.** C **3.** D **4.** B **5.** B **6.** B
**7.**

|  | Control | Ethylene added | Ethylene synthesis inhibitor |
|---|---|---|---|
| Wild-type | | | |
| Ethylene insensitive (*ein*) | | | |
| Ethylene overproducing (*eto*) | | | |
| Constitutive triple response (*ctr*) | | | |

## Chapter 32

### Figure Questions
**Figure 32.3** In hot environments, both plants and animals experience evaporative cooling as a result of transpiration (in plants) or bathing, sweating, and panting (in animals); both plants and animals synthesize heat-shock proteins, which protect other proteins from heat stress; and animals also use various behavioral responses to minimize heat absorption. In cold environments, both plants and animals increase the proportion of unsaturated fatty acids in their membrane lipids and use antifreeze proteins that prevent or limit the formation of intracellular ice crystals; plants increase cytoplasmic levels of specific solutes that help reduce the loss of intracellular water during extracellular freezing; and animals increase metabolic heat production and use insulation, circulatory adaptations such as countercurrent exchange, and behavioral responses to minimize heat loss. **Figure 32.12** The stimuli (gray boxes) are the room temperature increasing in the top loop or decreasing in the bottom loop. The responses could include the heater turning off and the temperature decreasing in the top loop and the heater turning on and the temperature increasing in the bottom loop. The sensor/control center is the thermostat. The air conditioner would form a second control circuit, cooling the house when air temperature exceeded the set point. Such opposing, or antagonistic, pairs of control circuits increase the effectiveness of a homeostatic mechanism. **Figure 32.14** Convection is occurring as the movement of the fan passes air over your skin. Evaporation may also be occurring if your skin is damp with sweat. You also radiate a small amount of heat to the surrounding air at all times. **Figure 32.22** Furosemide increases urine volume. The absence of ion transport in the ascending limb that results from this drug leaves the filtrate too concentrated for substantial volume reduction in the distal tubule and collecting duct.

### Concept Check 32.1
**1.** All types of epithelia consist of cells that line a surface, are tightly packed, are situated on top of a basal lamina, and form an active and protective interface with the external environment. **2.** Sheets of connective tissue support many of the body's organs, and connective tissue forms an integral part of most organs.

### Concept Check 32.2
**1.** Yes, the response can differ if the pathway regulated by the receptor is different in the two cell types. **2.** If the function of the pathway is to provide a transient response, a short-lived stimulus would be less dependent on negative feedback. **3.** Epinephrine in animals and auxin in plants act as hormones that trigger specific cellular responses that vary among different tissues of the organism.

### Concept Check 32.3
**1.** No; an animal's internal environment fluctuates within a normal range or around set points. Homeostasis is a dynamic state. Furthermore, there are sometimes programmed changes in set points, such as those resulting in radical increases in hormone levels at particular times in development. **2.** In thermoregulation, the product of the pathway (a change in temperature) decreases pathway activity by reducing the stimulus. In an enzyme-catalyzed biosynthetic process, the product of the pathway (in this case, isoleucine) inhibits the pathway that generated it. **3.** The ice water would cool tissues in your head, including blood, that would then circulate throughout your body. This effect would accelerate the return to a normal body temperature. If, however, the ice water reached the eardrum and cooled the blood vessel that supplies the hypothalamus, the hypothalamic thermostat would respond by inhibiting sweating and constricting blood vessels in the skin, slowing cooling elsewhere in the body.

### Concept Check 32.4
**1.** Filtration produces a fluid for exchange processes that is free of cells and large molecules, which are of benefit to the animal and could not readily be reabsorbed. **2.** Because uric acid is largely insoluble in water, it can be excreted as a semisolid paste, thereby reducing an animal's water loss. **3.** Without a layer of insulating fur, the camel must use the cooling effect of evaporative water loss to maintain body temperature, thus linking thermoregulation and osmoregulation.

### Concept Check 32.5
**1.** The consumption of a large amount of water in a very short period of time, coupled with an absence of solute intake, can reduce sodium levels in the blood below tolerable levels. This condition, called hyponatremia, leads to disorientation and, sometimes, respiratory distress. It has occurred in some marathon runners who drink water rather than sports drinks. (It has also caused the death of a fraternity pledge as a consequence of a water hazing ritual and the death of a contestant in a water-drinking competition.) **2.** The kidney medulla would absorb less water; consequently, the drug would increase the amount of water lost in the urine. **3.** A decline in blood pressure in the afferent arteriole would reduce the force driving water and solutes across the membranes of glomerular capillaries and would therefore reduce the filtration rate.

### Summary of Key Concepts Questions
**32.1** The epithelium lining the inner surface of the stomach secretes mucus, which lubricates and protects the surface, as well as digestive juices. In addition, the tight packing of the epithelial cells provides a protective barrier. **32.2** Because receptors for water-soluble hormones are located on the cell surface, facing the extracellular space, injecting the hormone into the cytoplasm would not trigger a response. **32.3** Heat exchange across the skin is a primary mechanism for the regulation of body core temperature, with the result that the skin is cooler than the body core.
**32.4**

| Waste Attribute | Ammonia | Urea | Uric Acid |
|---|---|---|---|
| Toxicity | High | Very low | Low |
| Energy content | Low | Moderate | High |
| Water loss in excretion | High | Moderate | Low |

**32.5** Kangaroo rats, but not birds, have loops of Henle that extend deep into the renal medulla, enabling the production of urine that is more concentrated than the blood.

### Test Your Understanding
**1.** B **2.** C **3.** A **4.** B **5.** C **6.** C **7.** B
**8.**

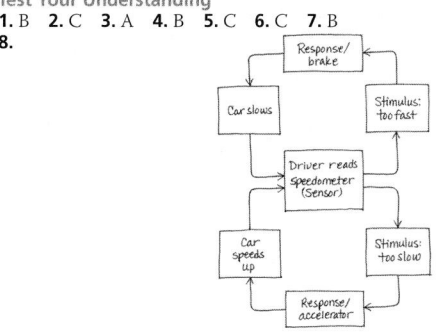

## Chapter 33

### Figure Questions
**Figure 33.9** Since enzymes are proteins, and proteins are hydrolyzed in the small intestine, the digestive enzymes in that compartment need to be resistant to enzymatic cleavage other than the cleavage required to activate them. **Figure 33.10** None. Since digestion is completed in the small intestine, tapeworms simply absorb predigested nutrients through their large body surface. **Figure 33.18** The transport of nutrients across membranes and the synthesis of RNA and protein are coupled to ATP hydrolysis. These processes proceed spontaneously because there is an overall drop in free energy, with the excess energy given off as heat. Similarly, less than half of the free energy in glucose is captured in the coupled reactions of cellular respiration. The remainder of the energy is released as heat. **Figure 33.19** Both insulin and glucagon are involved in negative-feedback circuits.

## Concept Check 33.1
**1.** The only essential amino acids are those that an animal cannot synthesize from other molecules. **2.** Many vitamins serve as enzyme cofactors, which, like enzymes themselves, are unchanged by the chemical reactions in which they participate. Therefore, only very small amounts of vitamins are needed. **3.** To identify the essential nutrient missing from an animal's diet, a researcher could supplement the diet with individual nutrients and determine which nutrient eliminates the signs of malnutrition.

## Concept Check 33.2
**1.** A gastrovascular cavity is a digestive pouch with a single opening that functions in both ingestion and elimination; an alimentary canal is a digestive tube with a separate mouth and anus at opposite ends. **2.** As long as nutrients are within the cavity of the alimentary canal, they are in a compartment that is continuous with the outside environment via the mouth and anus and have not yet crossed a membrane to enter the body. **3.** Just as food remains outside the body in a digestive tract, gasoline moves from the fuel tank to the engine, and waste products exit through the exhaust without ever entering the passenger compartment of the car. In addition, gasoline, like food, is broken down in a specialized compartment, so that the rest of the car (or body) is protected from disassembly. In both cases, high-energy fuels are consumed, complex molecules are broken down into simpler ones, and waste products are eliminated.

## Concept Check 33.3
**1.** By peristalsis, which can squeeze food through the esophagus even without the help of gravity **2.** Because parietal cells in the stomach pump $H^+$ to produce HCl, a proton pump inhibitor reduces the acidity of chyme and thus the irritation that occurs when chyme enters the esophagus. **3.** Proteins would be denatured and digested into peptides. Further digestion, to individual amino acids, would require enzymatic secretions found in the small intestine. No digestion of carbohydrates or lipids would occur.

## Concept Check 33.4
**1.** The increased time for transit through the alimentary canal allows for more extensive processing, and the increased surface of the canal area provides greater opportunity for absorption. **2.** A mammal's digestive system provides mutualistic microorganisms an environment protected against other microorganisms by saliva and gastric juice, held at a constant temperature conducive to enzyme action, and that provides a steady source of nutrients. **3.** For the yogurt treatment to be effective, the bacteria from yogurt would have to establish a mutualistic relationship with the small intestine, where disaccharides are broken down and sugars are absorbed. Conditions in the small intestine are likely to be very different from those in a yogurt culture. The bacteria might be killed before they reach the small intestine, or they might not be able to grow there in sufficient numbers to aid in digestion.

## Concept Check 33.5
**1.** Over the long term, the body stores excess calories in fat, whether those calories come from fat, carbohydrate, or protein in food. **2.** Since each gram of the mouse requires more calories than each gram of the elephant, the metabolic rate *per gram* must be higher in the mouse than in the elephant. **3.** Excess insulin production will cause blood glucose level to decrease below normal levels and also trigger glycogen synthesis in the liver, further decreasing blood glucose. However, low blood glucose level will stimulate the release of glucagon from alpha cells in the pancreas, which will trigger glycogen breakdown. Thus, there will be antagonistic effects on the liver.

## Summary of Key Concepts Questions
**33.1** Since the cofactor is necessary in all animals, those animals that do not require it in their diet must be able to synthesize it from other organic molecules. **33.2** A liquid diet containing glucose, amino acids, and other building blocks could be ingested and absorbed without the need for mechanical or chemical digestion. **33.3** The small intestine has a much larger surface area than the stomach. **33.4** The assortment of teeth in our mouth and the short length of our cecum suggest that our ancestors' digestive systems were not specialized for digesting plant material. **33.5** When mealtime arrives, nervous inputs from the brain signal the stomach to prepare to digest food through secretions and churning.

## Test Your Understanding
**1.** C  **2.** B  **3.** C  **4.** B  **5.** D  **6.** B
**7.**

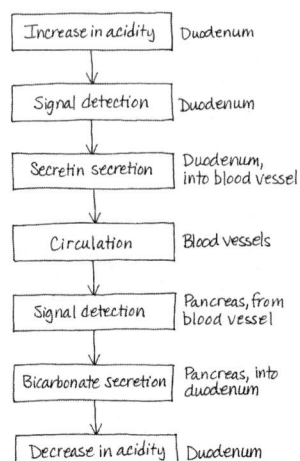

# Chapter 34
## Figure Questions
**Figure 34.8** Each feature of the ECG recording, such as the sharp upward spike, occurs once per cardiac cycle. Using the *x*-axis to measure the time in seconds between successive spikes and dividing that number into 60 would yield the heart rate as the number of cycles per minute. **Figure 34.21** The reduction in surface tension results from the presence of surfactant. Therefore, for all the infants who had died of RDS, you would expect the amount of surfactant to be near zero. For infants who had died of other causes, you would expect the amount of surfactant to be near zero for body masses less than 1,200 g but much greater than zero for body masses above 1,200 g. **Figure 34.23** Breathing at a rate greater than that needed to meet metabolic demand (hyperventilation) would lower blood $CO_2$ levels. Sensors in major blood vessels and the medulla would signal the breathing control center to decrease the rate of contraction of the diaphragm and rib muscles, decreasing the breathing rate and restoring normal $CO_2$ levels in the blood and other tissues. **Figure 34.24** The resulting increase in tidal volume would enhance ventilation within the lungs, increasing $P_{O_2}$ and decreasing $P_{CO_2}$ in the alveoli.

## Concept Check 34.1
**1.** In both an open circulatory system and a fountain, fluid is pumped through a tube and then returns to the pump after collecting in a pool. **2.** The ability to shut off blood supply to the lungs when the animal is submerged **3.** The $O_2$ content would be abnormally low because some oxygen-depleted blood returned to the right atrium from the systemic circuit would mix with the oxygen-rich blood in the left atrium.

## Concept Check 34.2
**1.** The pulmonary veins carry blood that has just passed through capillary beds in the lungs, where it accumulated $O_2$. The venae cavae carry blood that has just passed through capillary beds in the rest of the body, where it lost $O_2$ to the tissues. **2.** The delay allows the atria to empty completely, filling ventricles fully before they contract. **3.** The heart, like any other muscle, becomes stronger through regular exercise. You would expect a stronger heart to have a greater stroke volume, which would allow for the decrease in heart rate.

## Concept Check 34.3
**1.** The large total cross-sectional area of the capillaries **2.** An increase in blood pressure and cardiac output combined with the diversion of more blood to the skeletal muscles would increase the capacity for action by increasing the rate of blood circulation and delivering more $O_2$ and nutrients to the skeletal muscles. **3.** Additional hearts could be used to improve blood return from the legs. However, it might be difficult to coordinate the activity of multiple hearts and to maintain adequate blood flow to hearts far from the gas exchange organs.

## Concept Check 34.4
**1.** An increase in the number of white blood cells (leukocytes) may indicate that the person is combating an infection. **2.** Clotting factors do not initiate clotting but are essential steps in the clotting process. Also, the clots that form a thrombus typically result from an inflammatory response to an atherosclerotic plaque, not from clotting at a wound site. **3.** The chest pain results from inadequate blood flow in coronary arteries. Vasodilation promoted by nitric oxide from nitroglycerin increases blood flow, providing the heart muscle with additional oxygen and thus relieving the pain. **4.** Embryonic stem cells can give rise to many rather than just a few different cell types.

## Concept Check 34.5
**1.** Their interior position helps them stay moist. If the respiratory surfaces of lungs extended out into the terrestrial environment, they would quickly dry out, and diffusion of $O_2$ and $CO_2$ across these surfaces would stop. **2.** Earthworms need to keep their skin moist for gas exchange, but they need air outside this moist layer. If they stay in their waterlogged tunnels after a heavy rain, they will suffocate because they cannot get as much $O_2$ from water as from air. **3.** In fish, water passes over the gills in the direction opposite to that of blood flowing through the gill capillaries, maximizing the extraction of oxygen from the water along the length of the exchange surface. Similarly, in the extremities of some vertebrates, blood flows in opposite directions in neighboring veins and arteries; this countercurrent arrangement maximizes the recapture of heat from blood leaving the body core in arteries, which is important for thermoregulation in cold environments.

## Concept Check 34.6
**1.** An increase in blood $CO_2$ concentration causes an increase in the rate of $CO_2$ diffusion into the cerebrospinal fluid, where the $CO_2$ combines with water to form carbonic acid. Dissociation of carbonic acid releases hydrogen ions, decreasing the pH of the cerebrospinal fluid. **2.** Increased heart rate increases the rate at which $CO_2$-rich blood is delivered to the lungs, where $CO_2$ is removed. **3.** A hole would allow air to enter the space between the inner and outer layers of the double membrane, resulting in a condition called a pneumothorax. The two layers would no longer stick together, and the lung on the side with the hole would collapse and cease functioning.

## Concept Check 34.7
**1.** A difference in partial pressure between the capillaries and the surrounding tissues or medium; the net diffusion of a gas occurs from a region of higher partial pressure to a region of lower partial pressure. **2.** The Bohr shift causes hemoglobin to release more $O_2$ at a lower pH, such as found in the vicinity of tissues with high rates of cellular respiration and $CO_2$ release. **3.** The doctor is assuming that the rapid breathing is the body's response to low blood pH. Metabolic acidosis, the lowering of blood pH as a result of metabolism, can have many causes, including complications of certain types of diabetes, shock (extremely low blood pressure), and poisoning.

## Summary of Key Concepts Questions
**34.1** In a closed circulatory system, an ATP-driven muscular pump generally moves fluids in one direction on a scale of millimeters to meters. Exchange between cells

and their environment relies on diffusion, which involves random movements of molecules. Concentration gradients of molecules across exchange surfaces can drive rapid net diffusion on a scale of 1 mm or less. **34.2** Replacement of a defective valve should increase stroke volume. A lower heart rate would therefore be sufficient to maintain the same cardiac output. **34.3** Blood pressure in the arm would fall by 25 to 30 mm Hg, the same difference as is normally seen between your heart and your brain. **34.4** One microliter of blood contains about 5 million erythrocytes and 5,000 leukocytes, so leukocytes make up only about 0.1% of the cells in the absence of infection. **34.5** Because $CO_2$ is such a small fraction of atmospheric gas (0.29 mm Hg/760 mm Hg, or less than 0.04%), the partial pressure gradient of $CO_2$ between the respiratory surface and the environment always strongly favors the release of $CO_2$ to the atmosphere. **34.6** Because the lungs do not empty completely with each breath, incoming and outgoing air mix. Lungs thus contain a mixture of fresh and stale air. **34.7** An enzyme speeds up a reaction without changing the equilibrium and without being consumed. Similarly, a respiratory pigment speeds up the exchange of gases between the body and the external environment without changing the equilibrium state and without being consumed.

**Test Your Understanding**
**1.** C  **2.** A  **3.** D  **4.** C  **5.** C  **6.** A  **7.** A
**8.**

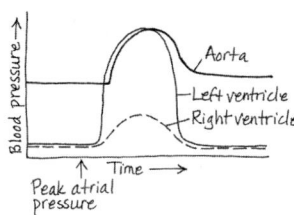

## Chapter 35

### Figure Questions
**Figure 35.4** Cell-surface TLRs recognize pathogens identifiable by surface molecules, whereas TLRs in vesicles recognize pathogens identifiable by internal molecules after the pathogens are broken down. **Figure 35.5** Because the pain of a splinter stops almost immediately when you remove it from the skin, you can deduce that the signals that mediate the inflammatory response are quite short-lived. **Figure 35.7** Part of the enzyme or antigen receptor provides a structural "backbone" that maintains overall shape, while interaction occurs at a surface with a close fit to the substrate or antigen. The combined effect of multiple noncovalent interactions at the active site or binding site is a high-affinity interaction of tremendous specificity. **Figure 35.10** After gene rearrangement, a lymphocyte and its daughter cells make a single version of the antigen receptor. In contrast, alternative splicing is not heritable and can give rise to diverse gene products in a single cell. **Figure 35.17** Primary response: arrows extending from Antigen (1st exposure), Antigen-presenting cell, Helper T cell, B cell, Plasma cells, Cytotoxic T cell, and Active cytotoxic T cells; secondary response: arrows extending from Antigen (2nd exposure), Memory helper T cells, Memory B cells, and Memory cytotoxic T cells

### Concept Check 35.1
**1.** Because pus contains white blood cells, fluid, and cell debris, it indicates an active and at least partially successful inflammatory response against invading microbes. **2.** Whereas the ligand for the TLR receptor is a foreign molecule, the ligand for many signal transduction pathways is a molecule produced by the animal itself. **3.** Mounting an immune response would require recognition of some molecular feature of the wasp egg not found in the host. It might be that only some potential hosts have a receptor with the necessary specificity.

### Concept Check 35.2
**1.** See Figure 35.6. The transmembrane regions lie within the C regions, which also form the disulfide bridges. In contrast, the antigen-binding sites are in the V regions. **2.** Generating memory cells ensures both that a receptor specific for a particular epitope will be present and that there will be more lymphocytes with this specificity than in a host that had never encountered the antigen. **3.** If each B cell produced two different light and heavy chains for its antigen receptor, different combinations would make four different receptors. If any one was self-reactive, the lymphocyte would be eliminated in the generation of self-tolerance. For this reason, many more B cells would be eliminated, and those that could respond to a foreign antigen would be less effective at doing so due to the variety of receptors (and antibodies) they express.

### Concept Check 35.3
**1.** It is considered an autoimmune disease because the immune system produces antibodies against self molecules (certain receptors on muscle cells). **2.** A child lacking a thymus would have no functional T cells. Without helper T cells to help activate B cells, the child would be unable to produce antibodies against extracellular bacteria. Furthermore, without cytotoxic T cells or helper T cells, the child's immune system would be unable to kill virus-infected cells. **3.** These receptors enable memory cells to present antigen on their cell surface to a helper T cell. Antigen presentation is required to activate memory cells in a secondary immune response. **4.** If the handler developed immunity to proteins in the antivenin, another injection could provoke a severe immune response. The handler's immune system might also now produce antibodies that could neutralize the venom in the absence of antivenin.

## Summary of Key Concepts Questions
**35.1** Lysozyme in saliva destroys bacterial cell walls; the viscosity of mucus helps trap bacteria; acidic pH in the stomach kills many bacteria; and the tight packing of cells lining the gut provides a physical barrier to infection. **35.2** Sufficient numbers of cells to mediate an innate immune response are always present, whereas an adaptive response requires selection and proliferation of an initially very small cell population specific for the infecting pathogen. **35.3** No. Immunological memory after a natural infection and immunological memory after immunization are very similar. There may be minor differences in the particular antigens that can be recognized in a subsequent infection.

## Test Your Understanding
**1.** B  **2.** C  **3.** C  **4.** B  **5.** B  **6.** B  **7.** C
**8.** One possible answer:

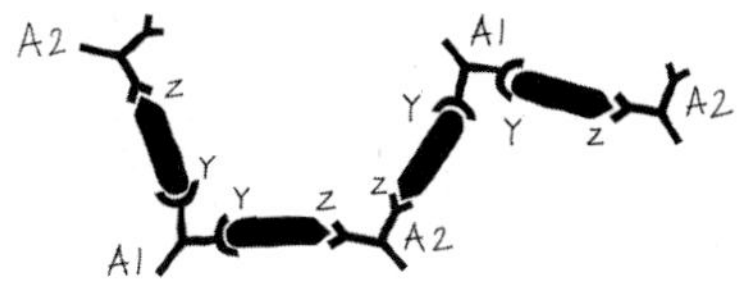

## Chapter 36

### Figure Question
**Figure 36.9** The analysis would be informative because the polar bodies contain all of the maternal chromosomes that don't end up in the mature egg. For example, finding two copies of the disease gene in the polar bodies would indicate its absence in the egg. This method of genetic testing is sometimes carried out when oocytes collected from a female are fertilized with sperm in a laboratory dish.

### Concept Check 36.1
**1.** Internal fertilization allows the sperm to reach the egg without either gamete drying out. **2.** No. Owing to random assortment of chromosomes during meiosis, the offspring may receive the same copy or different copies of a particular parental chromosome from the sperm and the egg. Furthermore, genetic recombination during meiosis will result in reassortment of genes between pairs of parental chromosomes. **3.** Both fragmentation and budding in animals have direct counterparts in the asexual reproduction of plants.

### Concept Check 36.2
**1.** Both have a haploid DNA content and very little cytoplasm. However, the early spermatid will develop into a functional gamete, whereas a polar body is a by-product of oocyte production. **2.** Spermatogenesis occurs normally only when the testicles are cooler than the rest the body. Extensive use of a hot tub (or of very tight-fitting underwear) can cause a decrease in sperm quality and number. **3.** Like the uterus of an insect, the ovary of a plant is the site of fertilization. Unlike the plant ovary, the uterus is not the site of egg production, which occurs in the insect ovary. In addition, the fertilized insect egg is expelled from the uterus, whereas the plant embryo develops within a seed in the ovary. **4.** The only effect of sealing off each vas deferens is an absence of sperm in the ejaculate. Sexual response and ejaculate volume are unchanged. The cutting and sealing off of these ducts, a *vasectomy*, is a common surgical procedure for men who do not wish to produce any (more) offspring.

### Concept Check 36.3
**1.** In both females and males, FSH encourages the growth of cells that support and nourish developing gametes (follicle cells in females and Sertoli cells in males), and LH stimulates the production of sex hormones that promote gametogenesis (estrogens, primarily estradiol, in females and androgens, especially testosterone, in males). **2.** In estrous cycles, which occur in most female mammals, the endometrium is reabsorbed (rather than shed) if fertilization does not occur. Estrous cycles often occur just once or a few times a year, and the female is usually receptive to copulation only during the period around ovulation. Menstrual cycles are about four weeks in length, do not restrict receptivity to copulation to a particular interval, and are found only in humans and some other primates. **3.** The combination of estradiol and progesterone would have a negative-feedback effect on the hypothalamus, blocking release of GnRH. This would interfere with LH secretion by the pituitary, thus preventing ovulation. This is in fact one basis of action of the most common hormonal contraceptives.

### Concept Check 36.4
**1.** Fertilization occurs in one of the two oviducts. **2.** Gastrulation places layers of cells with specific functions in particular locations in the developing embryo. Therefore, the process is essential for development of an organism in which specialized cells, tissues, and organs have shapes and connections that enable basic physiological functions, such as circulation, gas exchange, and reproduction. **3.** The menstrual cycle would be unaffected because it is controlled by hormones, which circulate in the bloodstream. However, the woman could not become pregnant naturally because the oviduct blockage would prevent sperm from reaching her eggs.

### Summary of Key Concepts Questions
**36.1** No. Because parthenogenesis involves meiosis, the mother would pass on to each offspring a random and therefore typically distinct combination of the chromosomes she inherited from her mother and father. **36.2** The small size and lack of cytoplasm characteristic of a sperm are adaptations well suited to its function as a delivery vehicle for DNA. The large size and rich cytoplasmic contents of eggs support the growth and development of the embryo. **36.3** Circulating anabolic steroids mimic the feedback regulation of testosterone, turning off pituitary signaling to the testes and thereby blocking the release of signals required for spermatogenesis. **36.4** The fertilization envelope forms after cortical granules release their contents outside the egg, causing the vitelline membrane to rise and harden. The fertilization envelope serves as a barrier to fertilization by more than one sperm.

## Test Your Understanding

**1.** D  **2.** A  **3.** A  **4.** C  **5.** A  **6.** B  **7.** C  **8.** D
**9.**

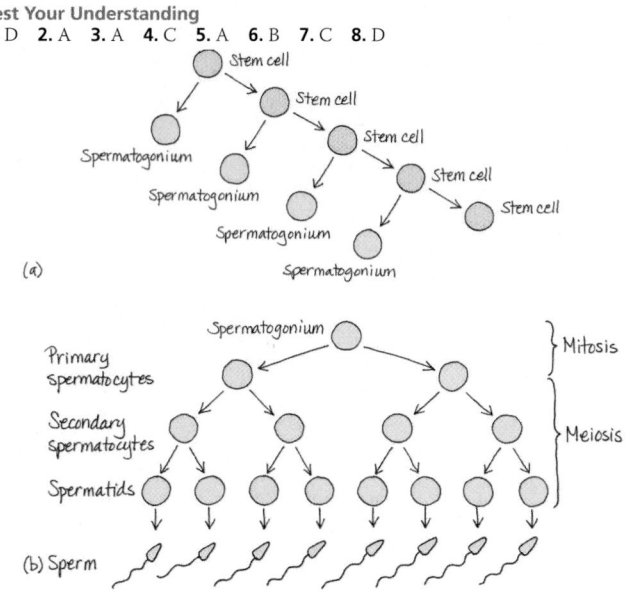

(a)

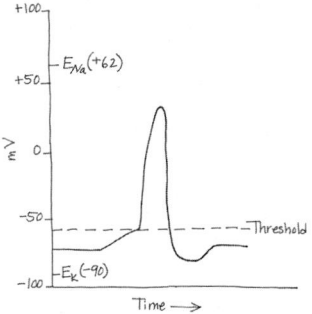

(b) Sperm

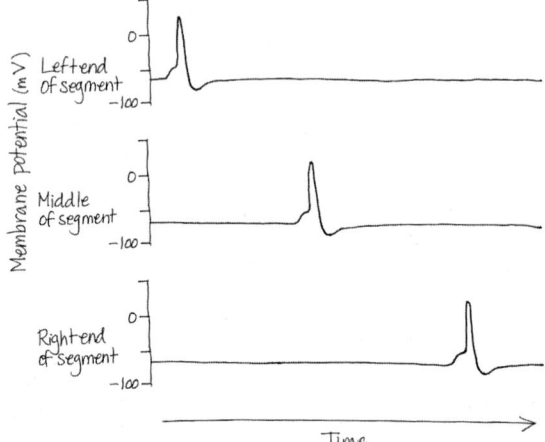

(c) The supply of stem cells would be used up, and spermatogenesis would not be able to continue.

# Chapter 37

## Figure Questions

**Figure 37.7** Adding chloride channels would make the membrane potential less positive. Adding sodium or potassium channels would have no effect because sodium ions are already at equilibrium and there are no potassium ions present.
**Figure 37.8** When the concentration of an ion differs across a plasma membrane, the difference in the concentration of ions inside and outside represents chemical potential energy, while the resulting difference in charge inside and outside represents electrical potential energy.
**Figure 37.11**

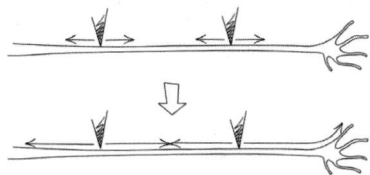

**Figure 37.13**

**Figure 37.16** The production and transmission of action potentials would be unaffected. However, action potentials arriving at chemical synapses would be unable to trigger release of neurotransmitter. Signaling at such synapses would thus be blocked.

## Concept Check 37.1

**1.** Axons and dendrites extend from the cell body and function in information flow. Dendrites transfer information to the cell body, whereas axons transmit information from the cell body. A typical neuron has multiple dendrites and one axon.  **2.** Sensors in your ear transmit information to your brain. There, the activity of interneurons in processing centers enables you to recognize your name. In response, signals transmitted via motor neurons cause contraction of muscles that turn your neck.  **3.** The increased branching would allow control of a greater number of postsynaptic cells, enhancing coordination of responses to nervous system signals.

## Concept Check 37.2

**1.** Ions can flow against a chemical concentration gradient if there is an opposing electrical gradient of greater magnitude.  **2.** A decrease in permeability to $K^+$, an increase in permeability to $Na^+$, or both  **3.** Charged dye molecules could equilibrate only if other charged molecules could also cross the membrane. If not, a membrane potential would develop that would counterbalance the chemical gradient.

## Concept Check 37.3

**1.** A graded potential has a magnitude that varies with stimulus strength, whereas an action potential has an all-or-none magnitude that is independent of stimulus strength.  **2.** Loss of the insulation provided by myelin sheaths leads to a disruption of action potential propagation along axons. Voltage-gated sodium channels are restricted to the nodes of Ranvier, and without the insulating effect of myelin, the inward current produced at one node during an action potential cannot depolarize the membrane to the threshold at the next node.  **3.** Positive feedback is responsible for the rapid opening of many voltage-gated sodium channels, causing the rapid outflow of sodium ions responsible for the rising phase of the action potential. As the membrane potential becomes positive, voltage-gated potassium channels open in a form of negative feedback that helps bring about the falling phase of the action potential.  **4.** The maximum frequency would decrease because the refractory period would be extended.

## Concept Check 37.4

**1.** It can bind to different types of receptors, each triggering a specific response in postsynaptic cells.  **2.** These toxins would prolong the EPSPs that acetylcholine produces because the neurotransmitter would remain longer in the synaptic cleft.  **3.** Membrane depolarization, exocytosis, and membrane fusion each occur in fertilization and in neurotransmission.

## Summary of Key Concepts Questions

**37.1** It would prevent information from being transmitted away from the cell body along the axon.  **37.2** There are very few open sodium channels in a resting neuron, so the resting potential either would not change or would become slightly more negative (hyperpolarization).  **37.4** A given neurotransmitter can have many receptors that differ in their location and activity. Drugs that target receptor activity rather than neurotransmitter release or stability are therefore likely to exhibit greater specificity and potentially have fewer undesirable side effects.

## Test Your Understanding

**1.** C  **2.** C  **3.** C  **4.** B  **5.** A  **6.** D  **7.** The activity of the sodium-potassium pump is essential to maintain the resting potential. With the pump inactivated, the sodium and potassium concentration gradients would gradually disappear, resulting in a greatly reduced resting potential.  **8.** Since GABA is an inhibitory neurotransmitter in the CNS, this drug would be expected to decrease brain activity. A decrease in brain activity might be expected to slow down or reduce behavioral activity. Many sedative drugs act in this fashion.  **9.** As shown in this pair of drawings, a pair of action potentials would move outward in both directions from each electrode. (Action potentials are unidirectional only if they begin at one end of an axon.) However, because of the refractory period, the two action potentials between the electrodes both stop where they meet. Thus, only one action potential reaches the synaptic terminals.

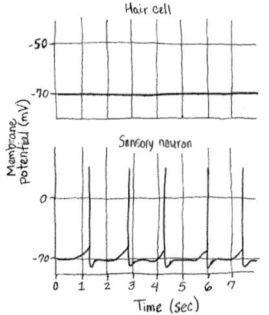

# Chapter 38

## Figure Questions

**Figure 38.6** Regions you would expect to be active regardless of the type of music played would include ones that are important for processing and interpreting sounds.
**Figure 38.10** If the depolarization brings the membrane potential to or past threshold, it should initiate action potentials that cause dopamine release from the VTA neurons. This should mimic natural stimulation of the brain reward system, resulting in positive and perhaps pleasurable sensations.
**Figure 38.22**

**Figure 38.27** In step 4, the activity of phosphodiesterase can bring about signal amplification, because each enzyme molecule can cleave many molecules of cyclic GMP. In step 5, the Na⁺ channels can contribute to amplification, since large numbers of ions can cross the membrane through a single open channel.

### Concept Check 38.1

**1.** The sympathetic division would likely be activated. It mediates the "fight-or-flight" response in stressful situations. **2.** Nerves contain bundles of axons: some that belong to motor neurons, which send signals outward from the CNS, and some that belong to sensory neurons, which bring signals into the CNS. Therefore, you would expect effects on both motor control and sensation.

### Concept Check 38.2

**1.** The cerebral cortex on the left side of the brain initiates voluntary movement of the right side of the body. **2.** Alcohol diminishes function of the cerebellum. **3.** A coma reflects a disruption in the cycles of sleep and arousal regulated by communication between the midbrain and pons and the cerebrum. You would expect this group to have damage to the midbrain, pons, cerebrum, or any part of the brain between these structures. Paralysis reflects an inability to carry out motor commands transmitted from the cerebrum to the spinal cord. You would expect this group to have damage to the portion of the CNS extending from the spinal cord up to but not including the midbrain and pons.

### Concept Check 38.3

**1.** There can be an increase in the number of synapses between the neurons or an increase in the strength of existing synaptic connections. **2.** Broca's area, which is active during the generation of speech, is located near the motor cortex, which controls skeletal muscles, including those in the face. Wernicke's area, which is active when speech is heard, is located in the posterior part of the temporal lobe, which is involved in hearing. **3.** Each cerebral hemisphere is specialized for different parts of this task—the right for face recognition and the left for language. Without an intact corpus callosum, neither hemisphere can take advantage of the other's processing abilities.

### Concept Check 38.4

**1.** Electromagnetic receptors in general detect only external stimuli. Other sensory receptors, such as chemoreceptors or mechanoreceptors, can act as either internal or external sensors. **2.** The capsaicin in the peppers activates the thermoreceptor for high temperatures. In response to the perceived high temperature, the nervous system triggers sweating to achieve evaporative cooling. **3.** The electrical stimulus would be perceived as if the sensory receptors that regulate that neuron had been activated. For example, electrical stimulation of a sensory neuron that forms synapses with a thermoreceptor activated by menthol would likely be perceived as a local cooling.

### Concept Check 38.5

**1.** Otoliths in the utricle and saccule enable a mammal to detect its orientation with respect to gravity, providing information that is essential in environments where light cues are absent. **2.** As a sound that changed gradually from a very low to a very high pitch **3.** The stapes and the other middle ear bones transmit vibrations from the tympanic membrane to the oval window. Fusion of these bones (as occurs in a disease called otosclerosis) would block this transmission and result in hearing loss. **4.** In animals, the statoliths are extracellular. In contrast, the statoliths of plants are found within an intracellular organelle. The methods for detecting their location also differ. In animals, detection is by means of mechanoreceptors on ciliated cells. In plants, the mechanism appears to involve calcium signaling.

### Concept Check 38.6

**1.** Planarians have ocelli that cannot form images but can sense the intensity and direction of light, providing enough information to enable the animals to find protection in shaded places. Flies have compound eyes that form images and excel at detecting movement. **2.** In the light, the photoreceptors hyperpolarize, shutting off their release of the neurotransmitter glutamate. In response, the bipolar cells that are depolarized by glutamate hyperpolarize, and those that are hyperpolarized by glutamate depolarize. If the bipolar cells that hyperpolarize in the light release a neurotransmitter that inhibits ganglion cells, then those ganglion cells will generate more action potentials in the light. Similarly, if the bipolar cells that depolarize in the light release a neurotransmitter that excites ganglion cells, then those ganglion cells will also generate more action potentials in the light. **3.** Absorption of light by retinal converts retinal from its *cis* isomer to its *trans* isomer, initiating the process of light detection. In contrast, a photon absorbed by chlorophyll does not bring about isomerization, but instead boosts an electron to a higher-energy orbital, initiating the electron flow that generates ATP and NADPH.

### Summary of Key Concepts Questions

**38.1** Glia have diverse functions. For example, ependymal cells help circulate cerebrospinal fluid, which carries nutrients, hormones, and waste products. In contrast, astrocytes promote increased blood flow to active neurons, and microglia defend against pathogens within the nervous system. **38.2** The midbrain coordinates visual reflexes; the cerebellum controls coordination of movement that depends on visual input; the thalamus serves as a routing center for visual information; and the cerebrum is essential for converting visual input to a visual image. **38.3** You would expect the right side of the body to be paralyzed because it is controlled by the left cerebral hemisphere, where language generation and interpretation are localized. **38.4** Nociceptors overlap with other classes of receptors in the type of stimulus they detect. They differ from other receptors in how a particular stimulus is perceived. **38.5** Volume is encoded by the frequency of action potentials transmitted to the brain; pitch is encoded by which axons are transmitting action potentials. **38.6** The major difference is that neurons in the retina integrate information from multiple sensory receptors (photoreceptors) before transmitting information to the central nervous system.

### Test Your Understanding

**1.** D **2.** C **3.** A **4.** B **5.** A **6.** C **7.** D

## Chapter 39

### Figure Questions

**Figure 39.4** Hundreds of myosin heads participate in sliding each pair of thick and thin filaments past each other. Because cross-bridge formation and breakdown are not synchronized, many myosin heads are exerting force on the thin filaments at all times during muscle contraction. **Figure 39.9** By causing all of the motor neurons that control the muscle to generate action potentials at a rate high enough to produce tetanus in all of the muscle fibers. **Figure 39.16** The fixed action pattern based on the sign stimulus of a red belly ensures that the male will chase away any invading males of his species. By chasing away such males, the defender decreases the chance that another male will fertilize eggs laid in his nesting territory. **Figure 39.19** There should be no effect. Imprinting is an innate behavior that is carried out anew in each generation. Assuming that the nest was not disturbed, the offspring of the geese imprinted on a human would imprint on the mother goose. **Figure 39.20** Perhaps the wasp doesn't use visual cues. It might also be that wasps recognize objects native to their environment, but not foreign objects, such as the pinecones. Tinbergen addressed these ideas before carrying out the pinecone study. When he swept away the pebbles and sticks around the nest, the wasps could no longer find their nests. If he shifted the natural objects in their natural arrangement, the shift in the landmarks caused a shift in the site to which the wasps returned. Finally, if natural objects around the nest site were replaced with pinecones while the wasp was in the burrow, the wasp nevertheless found her way back to the nest site. **Figure 39.29** It holds true for some, but not all, individuals. If a parent has more than one reproductive partner, the offspring of different partners will have a coefficient of relatedness less than 0.5.

### Concept Check 39.1

**1.** In a skeletal muscle fiber, $Ca^{2+}$ binds to the troponin complex, which moves tropomyosin away from the myosin-binding sites on actin and allows cross-bridges to form. In a smooth muscle cell, $Ca^{2+}$ binds to calmodulin, which activates an enzyme that phosphorylates the myosin head and thus enables cross-bridge formation. **2.** *Rigor mortis*, a Latin phrase meaning "stiffness of death," results from the complete depletion of ATP in skeletal muscle. Since ATP is required to release myosin from actin and to pump $Ca^{2+}$ out of the cytosol, muscles become chronically contracted beginning about 3–4 hours after death. **3.** A competitive inhibitor binds to the same site as the substrate for the enzyme. In contrast, the troponin and tropomyosin complex masks, but does not bind to, the myosin-binding sites on actin.

### Concept Check 39.2

**1.** Septa provide the divisions of the coelom that allow for peristalsis, a form of locomotion requiring independent control of different body segments. **2.** The main problem in swimming is drag; a fusiform body minimizes drag. The main problem in flying is overcoming gravity; wings shaped like airfoils provide lift, and adaptations such as air-filled bones reduce body mass. **3.** When you grasp the sides of the chair, you are using a contraction of the triceps to keep your arms extended against the pull of gravity on your body. As you lower yourself slowly into the chair, you gradually decrease the number of motor units in the triceps that are contracted. Contracting your biceps would jerk you down, since you would no longer be opposing gravity.

### Concept Check 39.3

**1.** The proximate explanation for this fixed action pattern might be that nudging and rolling are released by the sign stimulus of an object outside the nest, and the behavior is carried to completion once initiated. The ultimate explanation might be that ensuring that eggs remain in the nest increases the chance of producing healthy offspring. **2.** In both cases, the detection of periodic variation in the environment results in a reproductive cycle timed to environmental conditions that optimize the opportunity for success.

### Concept Check 39.4

**1.** Natural selection would tend to favor convergence in color because a predator learning to associate a color with a sting would avoid all other individuals with that same color, regardless of species. **2.** You might move objects around to establish an abstract rule, such as "past landmark A, the same distance as A is from the starting point," while maintaining a minimum of fixed metric relationships, that is, avoiding having the food directly adjacent to or a set distance from a landmark. As you might surmise, designing an informative experiment of this kind is not easy. **3.** Learned behavior, just like innate behavior, can contribute to reproductive isolation and thus to speciation. For example, learned bird songs contribute to species recognition during courtship, thereby helping ensure that only members of the same species mate.

### Concept Check 39.5

**1.** Certainty of paternity is higher with external fertilization. **2.** Balancing selection could maintain the two alleles at the *forager* locus if population density fluctuated from one generation to another. At times of low population density, the energy-conserving sitter larvae (carrying the *for^s* allele) would be favored, while at higher population density, the more mobile Rover larvae (*for^R* allele) would have a selective advantage.

### Concept Check 39.6

**1.** Because this geographic variation corresponds to differences in prey availability between two snake habitats, it seems likely that snakes with characteristics enabling them to feed on the abundant prey in their locale would have had increased survival and reproductive success. In this way, natural selection would have resulted in the divergent foraging behaviors. **2.** The older individual cannot be the beneficiary because he or she cannot have extra offspring. However, the cost is low for an older individual performing the altruistic act because he or she has already reproduced (but perhaps is still caring for a child or grandchild). There can therefore be selection for an altruistic act by a postreproductive individual that benefits a young relative.

## Summary of Key Concepts Questions

**39.1** Oxidative fibers rely mostly on aerobic respiration and have many mitochondria, a rich blood supply, and a large amount of myoglobin. Glycolytic fibers use glycolysis as their primary source of ATP. They have a larger diameter and less myoglobin than oxidative fibers and fatigue more readily. **39.2** In response to nervous system motor output, the formation and breakdown of cross-bridges between myosin heads and actin cause the thin and thick filaments to slide past each other within each sarcomere. Because the thick filaments are anchored in the center of the sarcomeres and the thin filaments are anchored at the ends of the sarcomeres, this sliding movement shortens the sarcomeres and the muscle fibers that contain them. Furthermore, because the fibers themselves are part of muscles anchored at each end to bones, this shortening results in the movement of bones, as in the bending of an elbow. **39.3** Circannual rhythms are typically based on the cycles of light and dark in the environment. As the global climate changes, animals that migrate in response to these rhythms may shift to a location before or after local environmental conditions are optimal for reproduction and survival. **39.4** Because many foods have a distinctive color, associating a color with food can provide a selective advantage in foraging. However, the environment of a pigeon is unlikely to differ in color when a threat is present. Consequently, there would be no selective forces favoring the ability to associate color with danger. **39.5** Because feeding the female is likely to improve her reproductive success, the genes from the sacrificed male are likely to appear in a greater number of progeny. **39.6** Yes. Kin selection does not require any recognition or awareness of relatedness.

### Test Your Understanding
**1.** B   **2.** C   **3.** B   **4.** D   **5.** C   **6.** C   **7.** A

## Chapter 40

### Figure Questions
**Figure 40.12** Some factors, such as fire, are relevant only for terrestrial systems. At first glance, water availability is primarily a terrestrial factor, too. However, species living along the intertidal zone of oceans or along the edge of lakes also suffer desiccation (drying out). Salinity stress is important for species in some aquatic and terrestrial systems. Oxygen availability is an important factor primarily for species in some aquatic systems and in soils and sediments. **Figure 40.17** The population with $r = 1.0$ (blue curve) reaches 1,500 individuals in about 7.5 generations, whereas the population with $r = 0.5$ (red curve) reaches 1,500 individuals in about 14.5 generations.
**Figure 40.22**

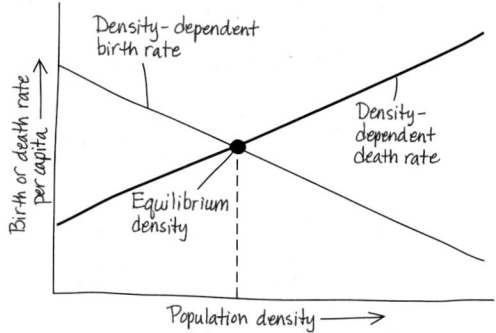

### Concept Check 40.1
**1.** In the tropics, high temperatures evaporate water and cause warm, moist air to rise. The rising air cools and releases much of its water as rain over the tropics. The remaining dry air descends at approximately 30° north and south, causing deserts to occur in those regions.   **2.** Answers will vary by location but should be based on the information and maps in Figures 40.7 and 40.9. How much your local area has been altered from its natural state will influence how much it reflects the expected characteristics of your biome, particularly the expected plants and animals.   **3.** Northern coniferous forest is likely to replace tundra along the boundary between these biomes. To see why, note that northern coniferous forest is adjacent to tundra throughout North America, northern Europe, and Asia (see Figure 40.7) and that the temperature range for northern coniferous forest is just above that for tundra (see Figure 40.8).

### Concept Check 40.2
**1.** In the oceanic pelagic zone, the ocean bottom lies below the photic zone, so there is too little light to support benthic algae or rooted plants.   **2.** Aquatic organisms either gain or lose water by osmosis if the osmolarity of their environment differs from their internal osmolarity. Water gain can cause cells to swell, and water loss can cause them to shrink. To avoid excessive changes in cell volume, organisms that live in estuaries must be able to compensate for both water gain (under freshwater conditions) and water loss (under saltwater conditions).   **3.** In a river below a dam, the fish are more likely to be species that prefer colder water. In summer, the deep layers of a reservoir are colder than the surface layers, so a river below a dam will be colder than an undammed river.

### Concept Check 40.3
**1.** (a) Humans might transplant a species to a new area that it could not previously reach because of a geographic barrier. (b) Humans might eliminate a predator or herbivore species, such as sea urchins, from an area.   **2.** One test would be to build a fence around a plot of land in an area that has trees of that species, excluding all deer from the plot. You could then compare the abundance of tree seedlings inside and outside the fenced plot over time.

### Concept Check 40.4
**1.**

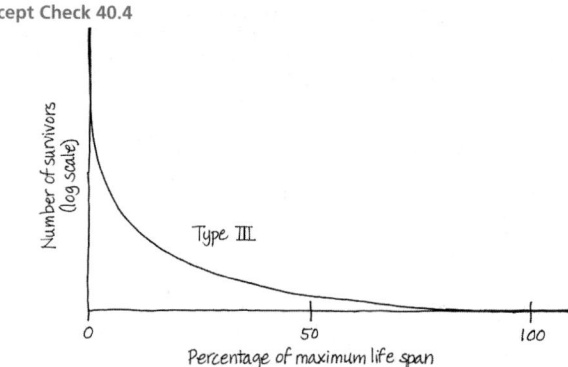

A Type III survivorship curve is most likely because very few of the young probably survive.   **2.** The proportion alive at the start of year 0–1 is 485/485 = 1.0. The proportion alive at the start of year 1–2 is 218/485 = 0.449.   **3.** Male sticklebacks would likely have a uniform pattern of dispersion, with antagonistic interactions maintaining a relatively constant spacing between them.

### Concept Check 40.5
**1.** Though $r$ is constant, $N$, the population size, is increasing. As $r$ is applied to an increasingly large $N$, population growth ($rN$) accelerates, producing the J-shaped curve. **2.** When $N$ (population size) is small, there are relatively few individuals producing offspring. When $N$ is large, near the carrying capacity, the per capita growth rate is relatively small because it is limited by available resources. The steepest part of the logistic growth curve corresponds to a population with a number of reproducing individuals that is substantial but not yet near carrying capacity.   **3.** If a population becomes too crowded, the likelihood of disease and mortality may increase because of the effects of pathogens. Thus, pathogens can reduce the long-term carrying capacity of a population.

### Concept Check 40.6
**1.** By preferentially investing in the eggs it lays in the nest, the peacock wrasse increases the chance those eggs will survive. The eggs it disperses widely and does not provide care for are less likely to survive, at least some of the time, but require a lower investment by the adults. (In this sense, the adults avoid the risk of placing all their eggs in one basket.)   **2.** If a parent's survival is compromised greatly by bearing young during times of stress, the animal's fitness may increase if it abandons its current young and survives to produce healthier young at a later time.   **3.** In negative feedback, the output, or product, of a process slows that process. In populations that have a density-dependent birth rate, such as dune fescue grass, an accumulation of product (more individuals, resulting in a higher population density) slows the process (population growth) by decreasing the birth rate.

### Summary of Key Concepts Questions
**40.1** Because dry air would descend at the equator instead of at 30° north and south (where deserts exist today), deserts would be more likely to exist along the equator (see Figure 40.3).   **40.2** An aphotic zone is most likely to be found in the deep waters of a lake, the oceanic pelagic zone, or the marine benthic zone.   **40.3** You might arrange a flowchart that begins with abiotic limitations—first determining the physical and chemical conditions under which a species could survive—and then moves through the other factors listed in the flowchart.   **40.4** Ecologists can potentially estimate birth rates by counting the number of young whales born each year, and they can estimate death rates by seeing how the number of adults changes each year. **40.5** There are many things you can do to increase the carrying capacity of the species, including increasing its food supply, protecting it from predators, and providing more sites for nesting or reproduction.   **40.6** An example of a biotic factor would be disease caused by a pathogen; natural disasters, such as earthquakes and floods, are examples of abiotic factors.

### Test Your Understanding
**1.** D   **2.** A   **3.** B   **4.** C   **5.** C

## Chapter 41

### Figure Questions
**Figure 41.3** Its realized and fundamental niches would be similar, unlike those of *Chthamalus*.   **Figure 41.5** Individuals of a harmless species that resembled a distantly related harmful species might be attacked by predators less often than other individuals of the harmless species that did not resemble the harmful species. As a result, individuals of the harmless species that resembled a harmful species would tend to contribute more offspring to the next generation than would other individuals of the harmless species. Over time, as natural selection by predators continued to favor those individuals of the harmless species that most closely resembled the harmful species, the resemblance of the harmless species to the harmful species would increase. Selection is not the only process that could cause a harmless species to resemble a closely related harmful species. In this case, the two species also could resemble each other because they descended from a recent common ancestor and hence share many traits (including a resemblance to one another).   **Figure 41.13** An increase in the abundance of carnivores that ate zooplankton might cause zooplankton abundance to drop, thereby causing phytoplankton abundance to increase.   **Figure 41.14** The number of other organism types eaten is zero for phytoplankton; one for copepods, crabeater seals, baleen whales, and sperm whales; two for krill, carnivorous plankton, and elephant seals; three for squids, fishes, leopard seals, and humans; and five for birds and smaller toothed whales. The

two groups that both consume and are consumed by each other are fishes and squids. **Figure 41.15** The death of individuals of *Mytilus*, a dominant species, should open up space for other species and increase species richness even in the absence of *Pisaster*. **Figure 41.23** Other factors not included in the model must contribute to the number of species. **Figure 41.24** Shrew populations in different locations and habitats might show substantial genetic variation in their susceptibility to the Lyme pathogen. As a result, there might be fewer infected ticks where shrew populations were less susceptible to the Lyme pathogen and more infected ticks where shrews were more susceptible.

### Concept Check 41.1

**1.** Interspecific competition has negative effects on both species ($-/-$). In predation, the predator population benefits at the expense of the prey population; this is an example of exploitative interaction ($+/-$). Mutualism is an interaction in which both species benefit ($+/+$).  **2.** One of the competing species will become locally extinct because of the greater reproductive success of the more efficient competitor.  **3.** By specializing in eating seeds of different plant species, individuals of the two finch species may be less likely to come into contact in the separate habitats, reinforcing a barrier to hybridization.

### Concept Check 41.2

**1.** Species richness (the number of species in the community) and relative abundance (the proportions of the community represented by the various species) both contribute to species diversity. Compared to a community with a very high proportion of one species, one with a more even proportion of species is considered more diverse.
**2.** A food chain presents a set of one-way transfers of food energy up to successively higher trophic levels. A food web documents how food chains are linked together, with many species weaving into the web at more than one trophic level.  **3.** According to the bottom-up model, adding extra predators would have little effect on lower trophic levels, particularly vegetation. If the top-down model applied, increased bobcat numbers would decrease raccoon numbers, increase snake numbers, decrease mouse numbers, and increase grass biomass.  **4.** A decrease in krill abundance might increase the abundance of organisms that krill eat (phytoplankton and copepods), while decreasing the abundance of organisms that eat krill (baleen whales, crabeater seals, birds, fishes, and carnivorous plankton); baleen whales and crabeater seals might be particularly at risk because they only eat krill. However, many of these possible changes could lead to other changes as well, making the overall outcome hard to predict. For example, a decrease in krill abundance could cause an increase in copepod abundance—but an increase in copepod abundance could counteract some of the other effects of decreased krill abundance (since like krill, copepods eat phytoplankton and are eaten by carnivorous plankton and fishes).

### Concept Check 41.3

**1.** High levels of disturbance are generally so disruptive that they eliminate many species from communities, leaving the community dominated by a few tolerant species. Low levels of disturbance permit competitively dominant species to exclude other species from the community. On the other hand, moderate levels of disturbance can facilitate coexistence of a greater number of species in a community by preventing competitively dominant species from becoming abundant enough to eliminate other species from the community.  **2.** Early successional species can facilitate the arrival of other species in many ways, including increasing the fertility or water-holding capacity of soils or providing shelter to seedlings from wind and intense sunlight.  **3.** The absence of fire for 100 years would represent a change to a low level of disturbance. According to the intermediate disturbance hypothesis, this change should cause diversity to decline as competitively dominant species gain sufficient time to exclude less competitive species.

### Concept Check 41.4

**1.** Ecologists propose that the greater species richness of tropical regions is the result of their longer evolutionary history and the greater solar energy input and water availability in tropical regions.  **2.** Immigration to islands declines with distance from the mainland and increases with island area. Extinction of species is lower on larger islands and on less isolated islands. Since the number of species on islands is largely determined by the difference between rates of immigration and extinction, the number of species will be highest on large islands near the mainland and lowest on small islands far from the mainland.  **3.** Because of their greater mobility, birds disperse to islands more often than snakes and lizards, so birds should have greater richness.

### Concept Check 41.5

**1.** Pathogens are microorganisms or viruses that cause disease.  **2.** To keep the rabies virus out, you could ban imports of all mammals, including pets. Potentially, you could also attempt to vaccinate all dogs in the British Isles against the virus. A more practical approach might be to quarantine all pets brought into the country that are potential carriers of the disease, the approach the British government actually takes.

### Summary of Key Concepts Questions

**41.1** Note: Sample answers follow; other answers could also be correct. Competition: a fox and a bobcat competing for prey. Predation: an orca eating a sea otter. Herbivory: a bison grazing in a prairie. Parasitism: a parasitoid wasp that lays its eggs on a caterpillar. Mutualism: a fungus and an alga that make up a lichen. Commensalism: a beetle (that feeds upon wildflowers growing in a maple forest) and a maple tree.  **41.2** Not necessarily if the more species-rich community is dominated by only one or a few species. **41.3** Similar to clear-cutting a forest or plowing a field, some species would be present initially. As a result, the disturbance would initiate secondary succession in spite of its severe appearance.  **41.4** Glaciations are major disturbances that can completely destroy communities found in temperate and polar regions. As a result, tropical communities are older than temperate or polar communities. This can cause species diversity to be high in the tropics simply because there has been more time for speciation to occur.  **41.5** A keystone species is one with a pivotal ecological role. Hence, a pathogen that reduces the abundance or otherwise harms a keystone species could greatly alter the structure of the community. For example, if a novel pathogen drove a keystone species to local extinction, drastic changes in species diversity could occur.

### Test Your Understanding

**1.** D  **2.** C  **3.** D  **4.** Community 1: $H = -(0.05 \ln 0.05 + 0.05 \ln 0.05 + 0.85 \ln 0.85 + 0.05 \ln 0.05) = 0.59$. Community 2: $H = -(0.30 \ln 0.30 + 0.40 \ln 0.40 + 0.30 \ln 0.30) = 1.1$. Community 2 is more diverse.  **5.** Crab numbers should increase, reducing the abundance of eelgrass.

## Chapter 42

### Figure Questions

**Figure 42.6** The map does not accurately reflect the productivity of wetlands, coral reefs, and coastal zones because these habitats cover areas that are too small to show up clearly on global maps.  **Figure 42.7** New duck farms would add extra nitrogen and phosphorus to the water samples used in the experiment. We would expect that the extra phosphorus from these new duck farms would not alter the results (because in the original experiment, phosphorus levels were *already* so high that adding phosphorus did not increase phytoplankton growth). However, the new duck farms might increase nitrogen levels to the point where adding extra nitrogen in an experiment would not increase phytoplankton density.  **Figure 42.12** Water availability is probably another factor that varied across the sites. Such factors not included in the experimental design could make the results more difficult to interpret. Multiple factors can be correlated to each other in nature, so ecologists must be careful that the factor they are studying is actually causing the observed response and is not just correlated with it.  **Figure 42.18** Populations evolve as organisms interact with each other and with the physical and chemical conditions of their environment. As a result, any human action that alters the environment has the potential to cause evolutionary change. In particular, since climate change has greatly affected Arctic ecosystems, we would expect that climate change will cause evolution in Arctic tundra populations.

### Concept Check 42.1

**1.** Energy passes through an ecosystem, entering as sunlight and leaving as heat. It is not recycled within the ecosystem.  **2.** You would need to know how much biomass the wildebeests ate from your plot and how much nitrogen was contained in that biomass. You would also need to know how much nitrogen they deposited in urine or feces.  **3.** The second law states that in any energy transfer or transformation, some of the energy is dissipated to the surroundings as heat. For the ecosystem to remain intact, this "escape" of energy from the ecosystem must be offset by the continuous influx of solar radiation.

### Concept Check 42.2

**1.** Only a fraction of solar radiation strikes plants or algae, only a portion of that fraction is of wavelengths suitable for photosynthesis, and much energy is lost as a result of reflection or heating of plant tissue.  **2.** By manipulating the level of the factors of interest, such as phosphorus availability or soil moisture, and measuring responses by primary producers  **3.** The enzyme rubisco, which catalyzes the first step in the Calvin cycle, is the most abundant protein on Earth. Like all proteins, rubisco contains nitrogen, and rubisco is so abundant that photosynthetic organisms require considerable nitrogen to make it. Phosphorus is also needed as a component of several metabolites in the Calvin cycle and as a component of both ATP and NADPH (see Figure 8.17).

### Concept Check 42.3

**1.** 20 J; 40%  **2.** Nicotine protects the plant from herbivores.  **3.** Total net production is $10,000 + 1,000 + 100 + 10$ J = 11,110 J. This is the amount of energy theoretically available to detritivores.

### Concept Check 42.4

**1.** For example, for the carbon cycle:

Cycling of a carbon atom

**2.** Removal of the trees stops nitrogen uptake from the soil, allowing nitrate to accumulate there. The nitrate is washed away by precipitation and enters the streams.
**3.** Most of the nutrients in a tropical rain forest are contained in the trees, so removing the trees by logging rapidly depletes nutrients from the ecosystem. The nutrients that remain in the soil are quickly carried away into streams and groundwater by the abundant precipitation.

### Concept Check 42.5

**1.** The main goal is to restore degraded ecosystems to a more natural state.
**2.** Bioremediation uses organisms, generally prokaryotes, fungi, or plants, to detoxify or remove pollutants from ecosystems. Biological augmentation uses organisms, such

as nitrogen-fixing plants, to add essential materials to degraded ecosystems. **3.** The Kissimmee River project returns the flow of water to the original channel and restores natural flow, a self-sustaining outcome. Ecologists at the Maungatautari reserve will need to maintain the integrity of the fence indefinitely, an outcome that is not self-sustaining in the long term.

### Summary of Key Concepts Questions

**42.1** Because energy conversions are inefficient, with some energy inevitably lost as heat, you would expect that a given mass of primary producers would support a smaller biomass of secondary producers. **42.2** For estimates of NEP, you need to measure the respiration of all organisms in an ecosystem, not just the respiration of primary producers. In a sample of ocean water, primary producers and other organisms are usually mixed together, making their respective respirations hard to separate. **42.3** Runners use much more energy in respiration when they are running than when they are sedentary, reducing their production efficiency. **42.4** Factors other than temperature, including a shortage of water and nutrients, slow decomposition in hot deserts. **42.5** If the topsoil and deeper soil are kept separate, the engineers could return the deeper soil to the site first and then apply the more fertile topsoil to improve the success of revegetation and other restoration efforts.

### Test Your Understanding

**1.** C **2.** B **3.** A **4.** C **5.** A **6.** B **7.** D **8.** D

## Chapter 43

### Figure Questions

**Figure 43.4** You would need to know the complete range of the species and that it is missing across all of that range. You would also need to be certain that the species isn't hidden, as might be the case for an animal that is hibernating underground or a plant that is present in the form of seeds or spores. **Figure 43.10** The higher the pH, the lower the acidity. Thus, the precipitation in this forest is becoming less acidic. **Figure 43.12** Because the population of Illinois birds has a different genetic makeup than birds in other regions, you would want to maintain to the greatest extent possible the frequency of beneficial genes or alleles found only in that population. In restoration, preserving genetic diversity in a species is as important as increasing organism numbers. **Figure 43.14** The natural disturbance regime in this habitat includes frequent fires that clear undergrowth but do not kill mature pine trees. Without these fires, the undergrowth quickly fills in and the habitat becomes unsuitable for red-cockaded woodpeckers. **Figure 43.15** The photo shows edges between forest and grassland ecosystems, grassland and river ecosystems, and grassland and lake ecosystems. **Figure 43.22** The PCB concentration increased by a factor of 4.9 from phytoplankton to zooplankton, 41.6 from phytoplankton to smelt, 8.5 from zooplankton to smelt, 4.6 from smelt to lake trout, 119.2 from smelt to herring gull eggs, and 25.7 from lake trout to herring gull eggs. **Figure 43.27** Dispersal limitations, activities of people (such as a broad-scale conversion of forests to agriculture or selective harvesting), interactions with other species, soil conditions, and many other factors (including those discussed in Concept 40.3) **Figure 43.28** Ocean acidification reduces the availability of carbonate ions ($CO_3^{2-}$). Corals and many other marine organisms require carbonate ions to build their shells. Since shell-building organisms depend upon their shells for survival, scientists have predicted that ocean acidification will cause many shell-building organisms to die. In turn, increased mortality rates of organisms that build shells would cause many other changes to ecological communities. For example, increased mortality rates of corals would harm the many other species that seek protection in coral reefs or that feed upon the species living there. **Figure 43.31** If the average ecological footprint were 8 global hectares per person, Earth could support about 1.5 billion people in a sustainable fashion. This estimate is obtained by dividing the total amount of Earth's productive land (11.9 billion gha) by the number of global hectares used per person (8 gha/person), which yields 1.49 billion people.

### Concept Check 43.1

**1.** In addition to species loss, the biodiversity crisis includes the loss of genetic diversity within populations and species and the degradation of entire ecosystems. **2.** Habitat destruction, such as deforestation, channelizing of rivers, or conversion of natural ecosystems to agriculture or cities, deprives species of places to live. Introduced species, which are transported by humans to regions outside their native range, where they are not controlled by their natural pathogens or predators, often reduce the population sizes of native species through competition or predation. Overharvesting has reduced populations of plants and animals or driven them to extinction. Finally, global change is altering the environment to the extent that it reduces the capacity of Earth to sustain life. **3.** If both populations breed separately, then gene flow between the populations would not occur and genetic differences between them would be greater. As a result, the loss of genetic diversity would be greater than if the populations interbreed.

### Concept Check 43.2

**1.** Reduced genetic variation decreases the capacity of a population to evolve in the face of change. **2.** The effective population size, $N_e$, would be $(4 \times 30 \times 10)/(30 + 10) = 30$ birds. **3.** Because millions of people use the greater Yellowstone ecosystem each year, it would be impossible to eliminate all contact between people and bears. Instead, you might try to reduce the kinds of encounters where bears are killed. You might recommend lower speed limits on roads in the park, adjust the timing or location of hunting seasons (where hunting is allowed outside the park) to minimize contact with mother bears and cubs, and provide financial incentives for livestock owners to try alternative means of protecting livestock, such as using guard dogs.

### Concept Check 43.3

**1.** A small area supporting numerous endemic species as well as a large number of endangered and threatened species **2.** Zoned reserves may provide sustained supplies of forest products, water, hydroelectric power, educational opportunities, and income from tourism. **3.** Habitat corridors can increase the rate of movement or dispersal of organisms between habitat patches and thus the rate of gene flow between subpopulations. They thus help prevent a decrease in fitness attributable to inbreeding. They can also minimize interactions between organisms and humans as the organisms disperse; in cases involving potential predators, such as bears or large cats, minimizing such interactions is desirable.

### Concept Check 43.4

**1.** Adding nutrients causes population explosions of algae and the organisms that feed on them. Increased respiration by algae and consumers, including detritivores, depletes the lake's oxygen, which the fish require. **2.** Decomposers are consumers that use nonliving organic matter as fuel for cellular respiration, which releases $CO_2$. Because higher temperatures lead to faster decomposition, organic matter in these soils could be decomposed to $CO_2$ more rapidly, thereby speeding up global warming.

### Concept Check 43.5

**1.** The growth rate of Earth's human population has dropped by half since the 1960s, from 2.2% in 1962 to 1.1% today. Nonetheless, the yearly increase in population size has not slowed as much because the smaller growth rate is counterbalanced by increased population size; hence, the number of additional people on Earth each year remains enormous—approximately 78 million. **2.** Each student will calculate his or her own ecological footprint. Each of us influences our ecological footprint by how we live—what we eat, how much energy we use, and the amount of waste we generate—as well as by how many children we have. Making choices that reduce our demand for resources makes our ecological footprint smaller.

### Concept Check 43.6

**1.** Sustainable development is an approach to development that works toward the long-term prosperity of human societies and the ecosystems that support them, which requires linking the biological sciences with the social sciences, economics, and humanities. **2.** Biophilia, our sense of connection to nature and all forms of life, may act as a significant motivation for the development of an environmental ethic that resolves not to allow species to become extinct or ecosystems to be destroyed. Such an ethic is necessary if we are to become more attentive and effective custodians of the environment. **3.** At a minimum, you would want to know the size of the population and the average reproductive rate of the individuals in it. To develop the fishery sustainably, you would seek a harvest rate that maintains the population near its original size and maximizes its harvest in the long term rather than the short term.

### Summary of Key Concepts Questions

**43.1** Nature provides us with many beneficial services, including a supply of reliable, clean water, the production of food and fiber, and the dilution and detoxification of our pollutants. **43.2** A more genetically diverse population is better able to withstand pressures from disease or environmental change, making it less likely to become extinct over a given period of time. **43.3** Habitat fragmentation can isolate populations, leading to inbreeding and genetic drift, and it can make populations more susceptible to local extinctions resulting from the effects of pathogens, parasites, or predators. **43.4** It's healthier to feed at a lower trophic level because biological magnification increases the concentration of toxins at higher levels. **43.5** We are unique in our potential ability to reduce global population through contraception and family planning. We also are capable of consciously choosing our diet and personal lifestyle, and these choices influence the number of people Earth can support. **43.6** One goal of conservation biology is to preserve as many species as possible. Sustainable approaches that maintain the quality of habitats are required for the long-term survival of organisms.

### Test Your Understanding

**1.** C **2.** D **3.** B **4.** A **5.** B **6.** A **7.** (a) The average $CO_2$ concentration was approximately 330 ppm in 1975 and 394 ppm in 2012. (b) The rate of $CO_2$ concentration increase was (394 ppm – 330 ppm)/(2012 – 1975) = 64 ppm/37 years = 1.73 ppm/yr. (c) If this rate of increase continues, the concentration in 2100 will be about 550 ppm (1.73 ppm/yr × 88 yr = 152 ppm for the increase from 2012–2100 + 394 ppm in 2012 = 546 ppm, rounded off to 550 ppm).
(d)

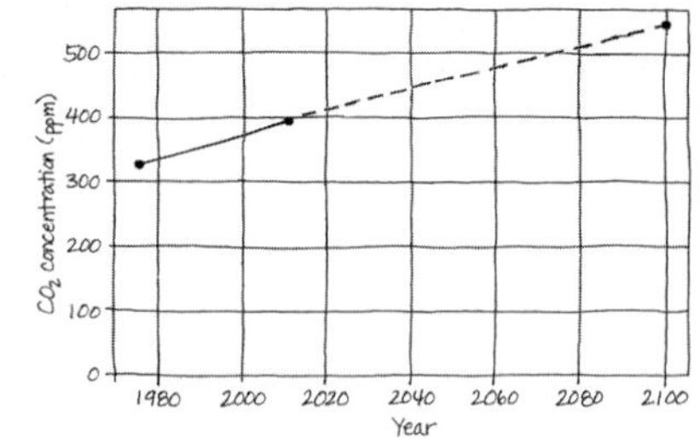


(e) The actual rise could be larger or smaller, depending on Earth's human population, per capita energy use, and the extent to which societies take steps to reduce $CO_2$ emissions, including replacing fossil fuels with renewable or nuclear fuels. (f) Additional data will be important for many reasons, including determining how quickly greenhouse gases such as $CO_2$ are removed from the atmosphere by the biosphere.
**8.**

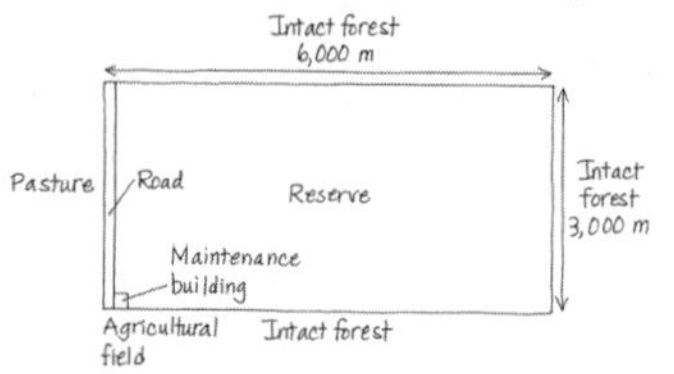


To minimize the area of forest into which the cowbirds penetrate, you should locate the road along the west edge of the reserve (since that edge abuts deforested pasture and an agricultural field). Any other location would increase the area of affected habitat. Similarly, the maintenance building should be in the southwest corner of the reserve to minimize the area susceptible to cowbirds.

# Periodic Table of the Elements

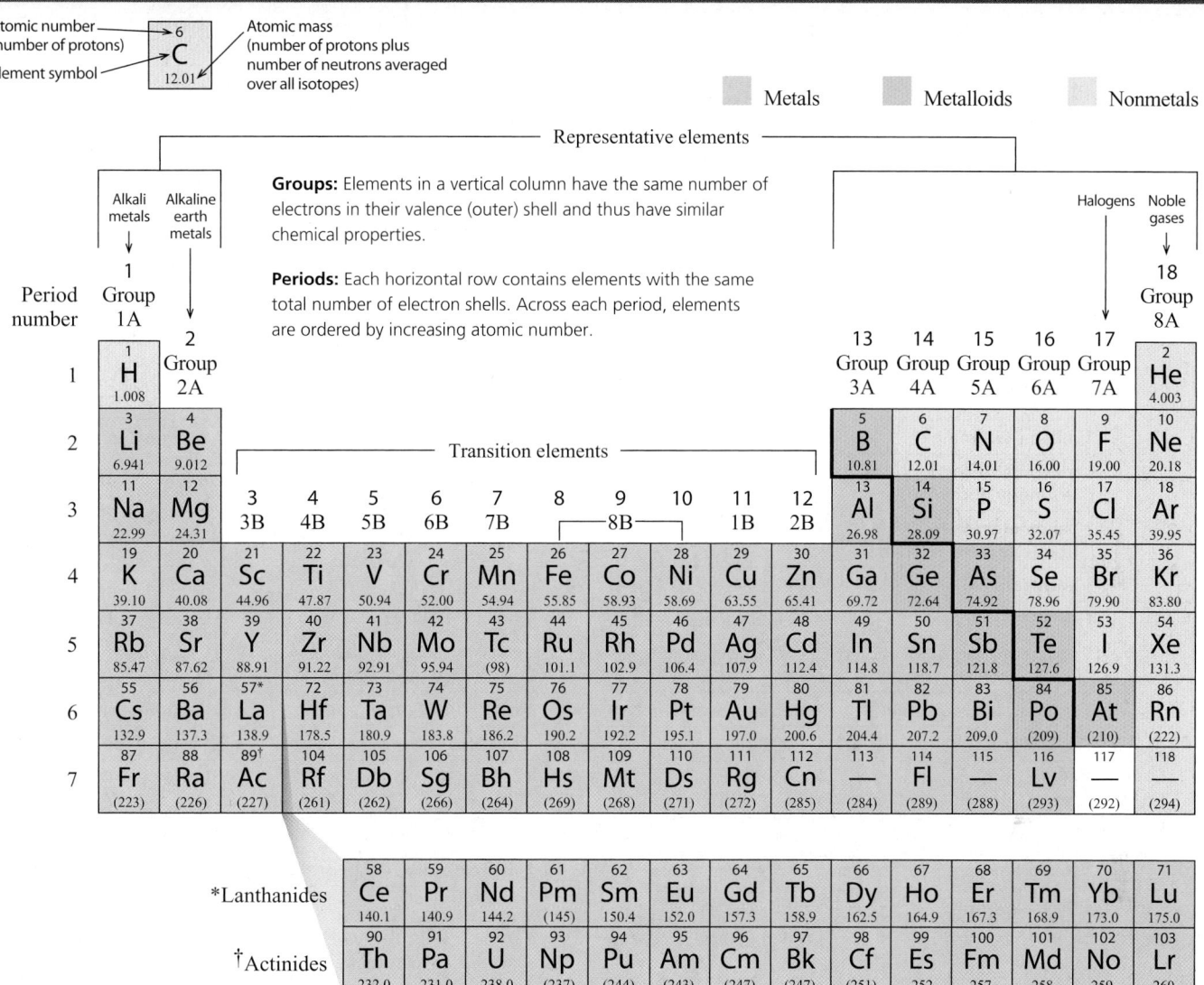

Atomic number (number of protons)
Element symbol
6 C 12.01

Atomic mass (number of protons plus number of neutrons averaged over all isotopes)

Metals   Metalloids   Nonmetals

Representative elements

**Groups:** Elements in a vertical column have the same number of electrons in their valence (outer) shell and thus have similar chemical properties.

**Periods:** Each horizontal row contains elements with the same total number of electron shells. Across each period, elements are ordered by increasing atomic number.

Alkali metals   Alkaline earth metals

Halogens   Noble gases

Transition elements

| Name (Symbol) | Atomic Number | Name (Symbol) | Atomic Number | Name (Symbol) | Atomic Number | Name (Symbol) | Atomic Number | Name (Symbol) | Atomic Number |
|---|---|---|---|---|---|---|---|---|---|
| Actinium (Ac) | 89 | Copernicium (Cn) | 112 | Iodine (I) | 53 | Osmium (Os) | 76 | Silicon (Si) | 14 |
| Aluminum (Al) | 13 | Copper (Cu) | 29 | Iridium (Ir) | 77 | Oxygen (O) | 8 | Silver (Ag) | 47 |
| Americium (Am) | 95 | Curium (Cm) | 96 | Iron (Fe) | 26 | Palladium (Pd) | 46 | Sodium (Na) | 11 |
| Antimony (Sb) | 51 | Darmstadtium (Ds) | 110 | Krypton (Kr) | 36 | Phosphorus (P) | 15 | Strontium (Sr) | 38 |
| Argon (Ar) | 18 | Dubnium (Db) | 105 | Lanthanum (La) | 57 | Platinum (Pt) | 78 | Sulfur (S) | 16 |
| Arsenic (As) | 33 | Dysprosium (Dy) | 66 | Lawrencium (Lr) | 103 | Plutonium (Pu) | 94 | Tantalum (Ta) | 73 |
| Astatine (At) | 85 | Einsteinium (Es) | 99 | Lead (Pb) | 82 | Polonium (Po) | 84 | Technetium (Tc) | 43 |
| Barium (Ba) | 56 | Erbium (Er) | 68 | Lithium (Li) | 3 | Potassium (K) | 19 | Tellurium (Te) | 52 |
| Berkelium (Bk) | 97 | Europium (Eu) | 63 | Livermorium (Lv) | 116 | Praseodymium (Pr) | 59 | Terbium (Tb) | 65 |
| Beryllium (Be) | 4 | Fermium (Fm) | 100 | Lutetium (Lu) | 71 | Promethium (Pm) | 61 | Thallium (Tl) | 81 |
| Bismuth (Bi) | 83 | Flerovium (Fl) | 114 | Magnesium (Mg) | 12 | Protactinium (Pa) | 91 | Thorium (Th) | 90 |
| Bohrium (Bh) | 107 | Fluorine (F) | 9 | Manganese (Mn) | 25 | Radium (Ra) | 88 | Thulium (Tm) | 69 |
| Boron (B) | 5 | Francium (Fr) | 87 | Meitnerium (Mt) | 109 | Radon (Rn) | 86 | Tin (Sn) | 50 |
| Bromine (Br) | 35 | Gadolinium (Gd) | 64 | Mendelevium (Md) | 101 | Rhenium (Re) | 75 | Titanium (Ti) | 22 |
| Cadmium (Cd) | 48 | Gallium (Ga) | 31 | Mercury (Hg) | 80 | Rhodium (Rh) | 45 | Tungsten (W) | 74 |
| Calcium (Ca) | 20 | Germanium (Ge) | 32 | Molybdenum (Mo) | 42 | Roentgenium (Rg) | 111 | Uranium (U) | 92 |
| Californium (Cf) | 98 | Gold (Au) | 79 | Neodymium (Nd) | 60 | Rubidium (Rb) | 37 | Vanadium (V) | 23 |
| Carbon (C) | 6 | Hafnium (Hf) | 72 | Neon (Ne) | 10 | Ruthenium (Ru) | 44 | Xenon (Xe) | 54 |
| Cerium (Ce) | 58 | Hassium (Hs) | 108 | Neptunium (Np) | 93 | Rutherfordium (Rf) | 104 | Ytterbium (Yb) | 70 |
| Cesium (Cs) | 55 | Helium (He) | 2 | Nickel (Ni) | 28 | Samarium (Sm) | 62 | Yttrium (Y) | 39 |
| Chlorine (Cl) | 17 | Holmium (Ho) | 67 | Niobium (Nb) | 41 | Scandium (Sc) | 21 | Zinc (Zn) | 30 |
| Chromium (Cr) | 24 | Hydrogen (H) | 1 | Nitrogen (N) | 7 | Seaborgium (Sg) | 106 | Zirconium (Zr) | 40 |
| Cobalt (Co) | 27 | Indium (In) | 49 | Nobelium (No) | 102 | Selenium (Se) | 34 | | |

# The Metric System

Metric Prefixes:
$10^9$ = giga (G)  $10^{-2}$ = centi (c)  $10^{-9}$ = nano (n)
$10^6$ = mega (M)  $10^{-3}$ = milli (m)  $10^{-12}$ = pico (p)
$10^3$ = kilo (k)  $10^{-6}$ = micro (μ)  $10^{-15}$ = femto (f)

| Measurement | Unit and Abbreviation | Metric Equivalent | Metric-to-English Conversion Factor | English-to-Metric Conversion Factor |
|---|---|---|---|---|
| **Length** | 1 kilometer (km) | = 1,000 ($10^3$) meters | 1 km = 0.62 mile | 1 mile = 1.61 km |
| | 1 meter (m) | = 100 ($10^2$) centimeters<br>= 1,000 millimeters | 1 m = 1.09 yards<br>1 m = 3.28 feet<br>1 m = 39.37 inches | 1 yard = 0.914 m<br>1 foot = 0.305 m |
| | 1 centimeter (cm) | = 0.01 ($10^{-2}$) meter | 1 cm = 0.394 inch | 1 foot = 30.5 cm<br>1 inch = 2.54 cm |
| | 1 millimeter (mm) | = 0.001 ($10^{-3}$) meter | 1 mm = 0.039 inch | |
| | 1 micrometer (μm)<br>(formerly micron, μ) | = $10^{-6}$ meter ($10^{-3}$ mm) | | |
| | 1 nanometer (nm)<br>(formerly millimicron, mμ) | = $10^{-9}$ meter ($10^{-3}$ μm) | | |
| | 1 angstrom (Å) | = $10^{-10}$ meter ($10^{-4}$ μm) | | |
| **Area** | 1 hectare (ha) | = 10,000 square meters | 1 ha = 2.47 acres | 1 acre = 0.405 ha |
| | 1 square meter ($m^2$) | = 10,000 square centimeters | 1 $m^2$ = 1.196 square yards<br>1 $m^2$ = 10.764 square feet | 1 square yard = 0.8361 $m^2$<br>1 square foot = 0.0929 $m^2$ |
| | 1 square centimeter ($cm^2$) | = 100 square millimeters | 1 $cm^2$ = 0.155 square inch | 1 square inch = 6.4516 $cm^2$ |
| **Mass** | 1 metric ton (t) | = 1,000 kilograms | 1 t = 1.103 tons | 1 ton = 0.907 t |
| | 1 kilogram (kg) | = 1,000 grams | 1 kg = 2.205 pounds | 1 pound = 0.4536 kg |
| | 1 gram (g) | = 1,000 milligrams | 1 g = 0.0353 ounce<br>1 g = 15.432 grains | 1 ounce = 28.35 g |
| | 1 milligram (mg) | = $10^{-3}$ gram | 1 mg = approx. 0.015 grain | |
| | 1 microgram (μg) | = $10^{-6}$ gram | | |
| **Volume (solids)** | 1 cubic meter ($m^3$) | = 1,000,000 cubic centimeters | 1 $m^3$ = 1.308 cubic yards<br>1 $m^3$ = 35.315 cubic feet | 1 cubic yard = 0.7646 $m^3$<br>1 cubic foot = 0.0283 $m^3$ |
| | 1 cubic centimeter ($cm^3$ or cc) | = $10^{-6}$ cubic meter | 1 $cm^3$ = 0.061 cubic inch | 1 cubic inch = 16.387 $cm^3$ |
| | 1 cubic millimeter ($mm^3$) | = $10^{-9}$ cubic meter<br>= $10^{-3}$ cubic centimeter | | |
| **Volume (liquids and gases)** | 1 kiloliter (kL or kl) | = 1,000 liters | 1 kL = 264.17 gallons | |
| | 1 liter (L or l) | = 1,000 milliliters | 1 L = 0.264 gallon<br>1 L = 1.057 quarts | 1 gallon = 3.785 L<br>1 quart = 0.946 L |
| | 1 milliliter (mL or ml) | = $10^{-3}$ liter<br>= 1 cubic centimeter | 1 mL = 0.034 fluid ounce<br>1 mL = approx. ¼ teaspoon<br>1 mL = approx. 15–16 drops (gtt.) | 1 quart = 946 mL<br>1 pint = 473 mL<br>1 fluid ounce = 29.57 mL<br>1 teaspoon = approx. 5 mL |
| | 1 microliter (μL or μl) | = $10^{-6}$ liter ($10^{-3}$ milliliter) | | |
| **Pressure** | 1 megapascal (MPa) | = 1,000 kilopascals | 1 MPa = 10 bars | 1 bar = 0.1 MPa |
| | 1 kilopascal (kPa) | = 1,000 pascals | 1 kPa = 0.01 bar | 1 bar = 100 kPa |
| | 1 pascal (Pa) | = 1 newton/$m^2$ ($N/m^2$) | 1 Pa = $1.0 \times 10^{-5}$ bar | 1 bar = $1.0 \times 10^5$ Pa |
| **Time** | 1 second (s or sec) | = 1/60 minute | | |
| | 1 millisecond (ms or msec) | = $10^{-3}$ second | | |
| **Temperature** | Degrees Celsius (°C) (0 K [Kelvin] = −273.15°C) | | °F = 9/5°C + 32 | °C = 5/9 (°F − 32) |

# A Comparison of the Light Microscope and the Electron Microscope

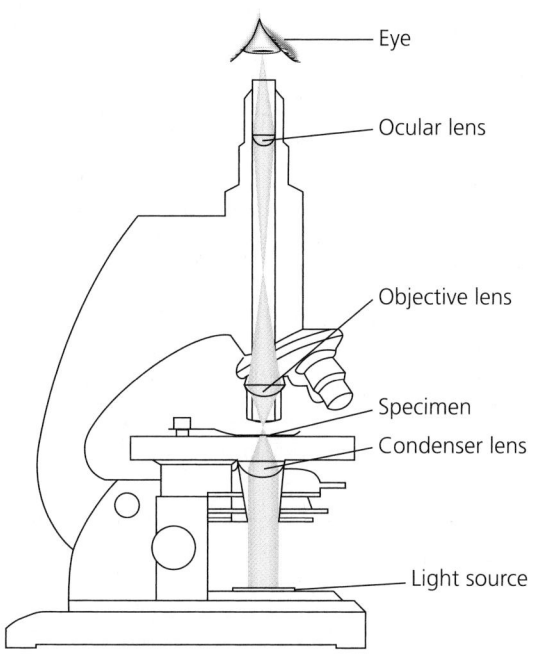

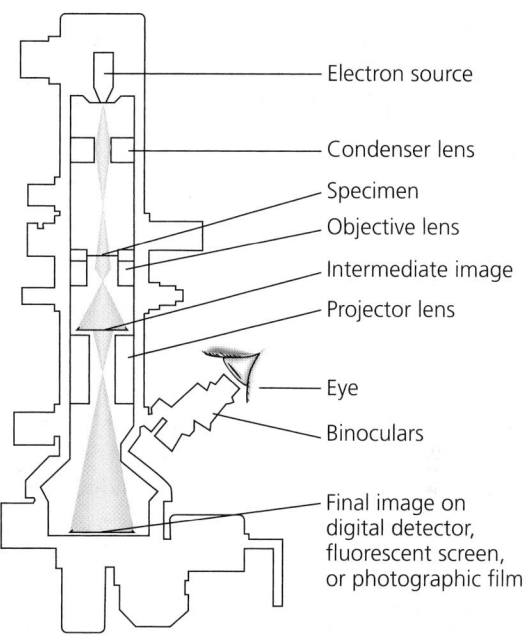

## Light Microscope

In light microscopy, light is focused on a specimen by a glass condenser lens; the image is then magnified by an objective lens and an ocular lens for projection on the eye, digital camera, digital video camera, or photographic film.

## Electron Microscope

In electron microscopy, a beam of electrons (top of the microscope) is used instead of light, and electromagnets are used instead of glass lenses. The electron beam is focused on the specimen by a condenser lens; the image is magnified by an objective lens and a projector lens for projection on a digital detector, fluorescent screen, or photographic film.

# Classification of Life

This appendix presents a taxonomic classification for the major extant groups of organisms discussed in this text; not all phyla are included. The classification presented here is based on the three-domain system, which assigns the two major groups of prokaryotes, bacteria and archaea, to separate domains (with eukaryotes making up the third domain).

Various alternative classification schemes are discussed in Unit Four of the text. The taxonomic turmoil includes debates about the number and boundaries of kingdoms and about the alignment of the Linnaean classification hierarchy with the findings of modern cladistic analysis. In this review, asterisks (*) indicate currently recognized phyla thought by some systematists to be paraphyletic.

## DOMAIN BACTERIA

- **Proteobacteria**
- **Chlamydia**
- **Spirochetes**
- **Cyanobacteria**
- **Gram-positive bacteria**

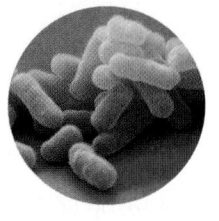

## DOMAIN ARCHAEA

- **Korarchaeota**
- **Euryarchaeota**
- **Crenarchaeota**
- **Nanoarchaeota**

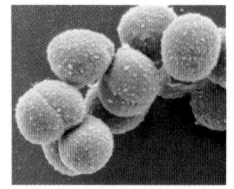

## DOMAIN EUKARYA

In the phylogenetic hypothesis we present in Chapter 25, major clades of eukaryotes are grouped together in the four "supergroups" listed in green type. Formerly, all the eukaryotes generally called protists were assigned to a single kingdom, Protista. However, advances in systematics have made it clear that some protists are more closely related to plants, fungi, or animals than they are to other protists. As a result, the kingdom Protista has been abandoned.

### Excavata
- Diplomonadida (diplomonads)
- Parabasala (parabasalids)
- Euglenozoa (euglenozoans)
  Kinetoplastida (kinetoplastids)
  Euglenophyta (euglenids)

### SAR
- Stramenopila (stramenopiles)
  Bacillariophyta (diatoms)
  Phaeophyta (brown algae)
- Alveolata (alveolates)
  Dinoflagellata (dinoflagellates)
  Apicomplexa (apicomplexans)
  Ciliophora (ciliates)

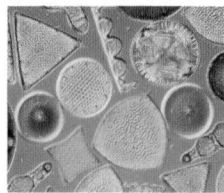

- Rhizaria
  Foraminifera (forams)
  Cercozoa (cercozoans)

### Archaeplastida
- Rhodophyta (red algae)
- Chlorophyta (green algae: chlorophytes)
- Charophyta (green algae: charophytes)
- Plantae
  Phylum Hepatophyta (liverworts) ⎱
  Phylum Bryophyta (mosses)          ⎰ Nonvascular plants (bryophytes)
  Phylum Anthocerophyta (hornworts)
  Phylum Lycophyta (lycophytes)      ⎱ Seedless vascular
  Phylum Monilophyta (ferns, horsetails, whisk ferns) ⎰ plants
  Phylum Ginkgophyta (ginkgo)        ⎱
  Phylum Cycadophyta (cycads)        ⎰ Gymnosperms ⎱ Seed
  Phylum Gnetophyta (gnetophytes)                   ⎰ plants
  Phylum Coniferophyta (conifers)
  Phylum Anthophyta (flowering plants) ⎰ Angiosperms

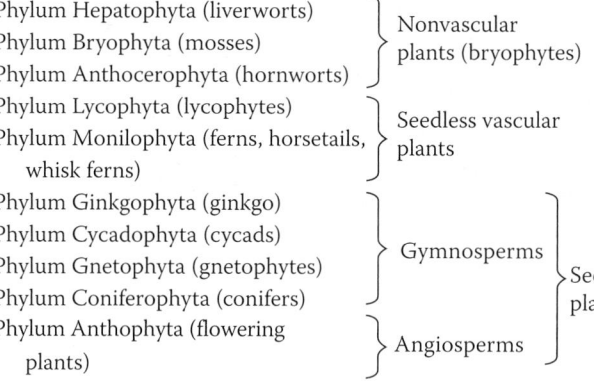

## Unikonta

- Amoebozoa (amoebozoans)
  - Tubulinea (tubulinids)
  - Dictyostelida (cellular slime molds)
- Nucleariida (nucleariids)
- Fungi
  - *Phylum Chytridiomycota (chytrids)
  - *Phylum Zygomycota (zygomycetes)
  - Phylum Glomeromycota (glomeromycetes)
  - Phylum Ascomycota (sac fungi)
  - Phylum Basidiomycota (club fungi)

- Choanoflagellata (choanoflagellates)
- Animalia
  - Phylum Porifera (sponges)
  - Phylum Ctenophora (comb jellies)
  - Phylum Cnidaria (cnidarians)
  - Lophotrochozoa (lophotrochozoans)
    - Phylum Platyhelminthes (flatworms)
    - Phylum Ectoprocta (ectoprocts)
    - Phylum Brachiopoda (brachiopods)
    - Phylum Rotifera (rotifers)
    - Phylum Mollusca (molluscs)
    - Phylum Annelida (segmented worms)

Ecdysozoa (ecdysozoans)
- Phylum Nematoda (roundworms)
- Phylum Arthropoda (This survey groups arthropods into a single phylum, but some zoologists now split the arthropods into multiple phyla.)
  - Chelicerata (horseshoe crabs, arachnids)
  - Myriapoda (millipedes, centipedes)
  - Pancrustacea (crustaceans, insects)

Deuterostomia (deuterostomes)
- Phylum Hemichordata (hemichordates)
- Phylum Echinodermata (echinoderms)
- Phylum Chordata (chordates)
  - Cephalochordata (lancelets)
  - Urochordata (tunicates)
  - Cyclostomata (cyclostomes)
    - Myxini (hagfishes)
    - Petromyzontida (lampreys)
  - Gnathostomata (gnathostomes)
    - Chondrichthyes (sharks, rays, chimaeras)
    - Actinopterygii (ray-finned fishes)
    - Actinistia (coelacanths)
    - Dipnoi (lungfishes)
    - Amphibia (amphibians)
    - Reptilia (tuataras, lizards, snakes, turtles, crocodilians, birds)
    - Mammalia (mammals)

> Vertebrates

## Graphs

Graphs provide a visual representation of numerical data. They may reveal patterns or trends in the data that are not easy to recognize in a table. A graph is a diagram that shows how one variable in a data set is related (or perhaps not related) to another variable. The **independent variable** is the factor that is manipulated or changed by the researchers. The **dependent variable** is the factor that the researchers are measuring in relationship to the independent variable. The independent variable is typically plotted on the *x*-axis and the dependent variable on the *y*-axis. Types of graphs that are frequently used in biology include scatter plots, line graphs, bar graphs, and histograms.

▶ A **scatter plot** is used when the data for all variables are numerical and continuous. Each piece of data is represented by a point. In a **line graph**, each data point is connected to the next point in the data set with a straight line, as in the graph to the right. (To practice making and interpreting scatter plots and line graphs, see the Scientific Skills Exercises in Chapters 2, 5, 6, 8, 10, 17, 22, 35, 38, 39, 41, and 43.)

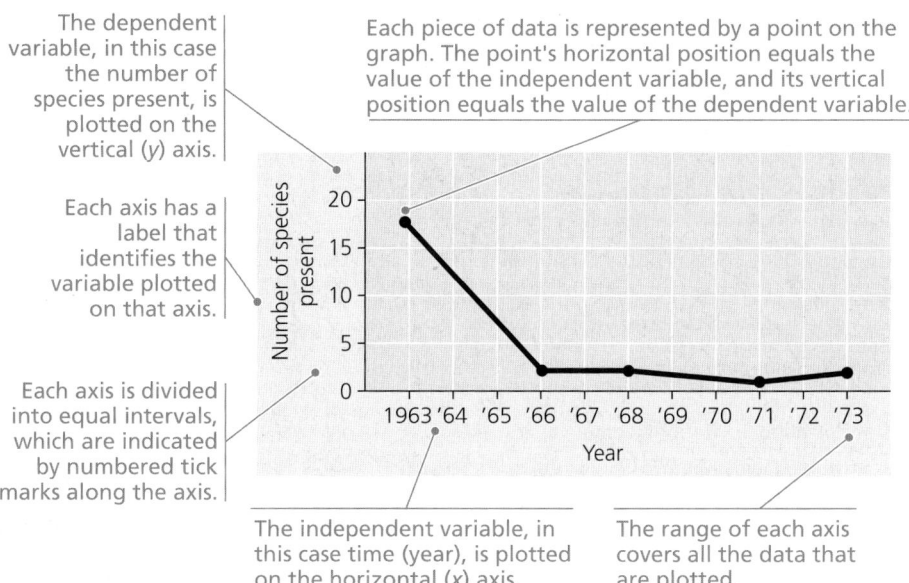

The dependent variable, in this case the number of species present, is plotted on the vertical (*y*) axis.

Each axis has a label that identifies the variable plotted on that axis.

Each axis is divided into equal intervals, which are indicated by numbered tick marks along the axis.

Each piece of data is represented by a point on the graph. The point's horizontal position equals the value of the independent variable, and its vertical position equals the value of the dependent variable.

The independent variable, in this case time (year), is plotted on the horizontal (*x*) axis.

The range of each axis covers all the data that are plotted.

▼ Two or more data sets can be plotted on the same line graph to show how two dependent variables are related to the same independent variable. (To practice making and interpreting line graphs with two or more data sets, see the Scientific Skills Exercises in Chapters 5, 35, 38, 39, and 43.)

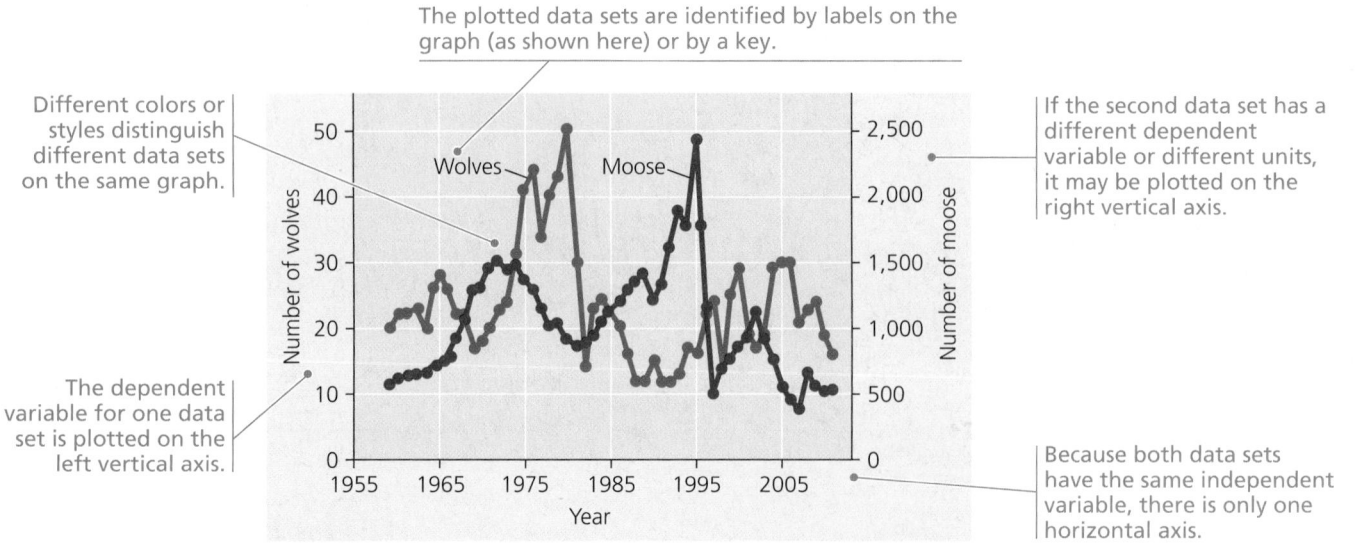

The plotted data sets are identified by labels on the graph (as shown here) or by a key.

Different colors or styles distinguish different data sets on the same graph.

The dependent variable for one data set is plotted on the left vertical axis.

If the second data set has a different dependent variable or different units, it may be plotted on the right vertical axis.

Because both data sets have the same independent variable, there is only one horizontal axis.

▼ In some scatter plot graphs, a straight or curved line is drawn through the entire data set to show the general trend in the data. A straight line that mathematically best fits the data is called a *regression line.* Alternatively, a mathematical function that best fits the data may describe a curved line, often termed a *best-fit curve.* (To practice making and interpreting regression lines, see the Scientific Skills Exercises in Chapters 2 and 8.)

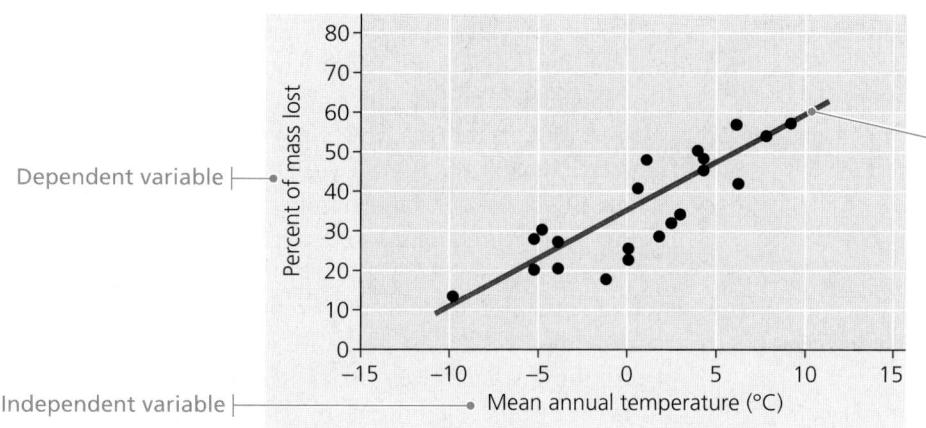

Dependent variable ⊢

Independent variable ⊢

The regression line can be expressed as a mathematical equation. It allows you to predict the value of the dependent variable for any value of the independent variable within the range of the data set and, less commonly, beyond the range of the data.

▼ A **bar graph** is a kind of graph in which the independent variable represents groups or non-numerical categories and the values of the dependent variable(s) are shown by bars. (To practice making and interpreting bar graphs, see the Scientific Skills Exercises in Chapters 1, 7, 15, 16, 19, 23, 24, 27, 28, 31, and 41.)

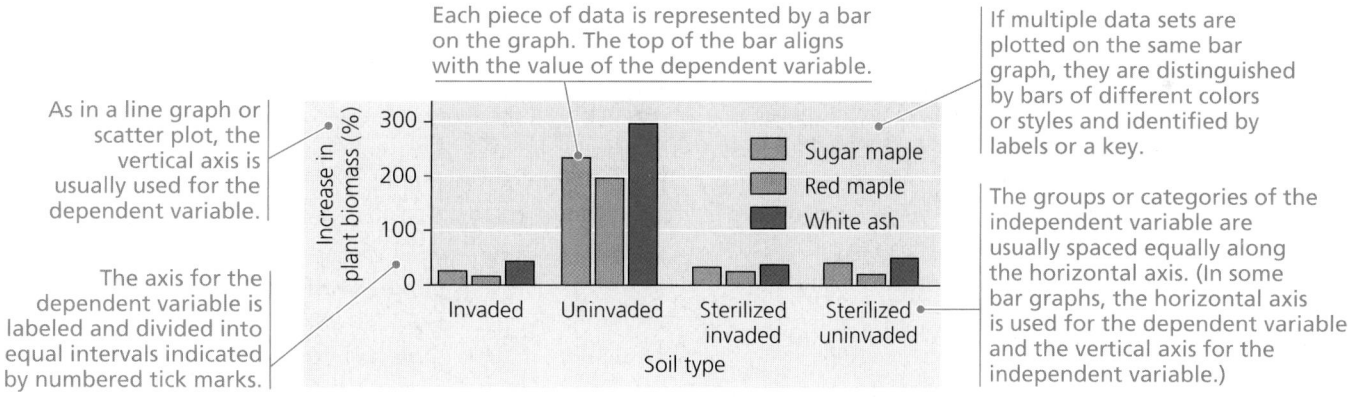

Each piece of data is represented by a bar on the graph. The top of the bar aligns with the value of the dependent variable.

If multiple data sets are plotted on the same bar graph, they are distinguished by bars of different colors or styles and identified by labels or a key.

As in a line graph or scatter plot, the vertical axis is usually used for the dependent variable.

The axis for the dependent variable is labeled and divided into equal intervals indicated by numbered tick marks.

The groups or categories of the independent variable are usually spaced equally along the horizontal axis. (In some bar graphs, the horizontal axis is used for the dependent variable and the vertical axis for the independent variable.)

▶ A variant of a bar graph called a **histogram** can be made for numeric data by first grouping, or "binning," the variable plotted on the *x*-axis into intervals of equal width. The "bins" may be integers or ranges of numbers. In the histogram at right, the intervals are 25 mg/dL wide. The height of each bar shows the percent (or, alternatively, the number) of experimental subjects whose characteristics can be described by one of the intervals plotted on the *x*-axis. (To practice making and interpreting histograms, see the Scientific Skills Exercises in Chapters 9, 11, and 34.)

The height of this bar shows the percent of individuals (about 4%) whose plasma LDL cholesterol levels are in the range indicated on the *x*-axis.

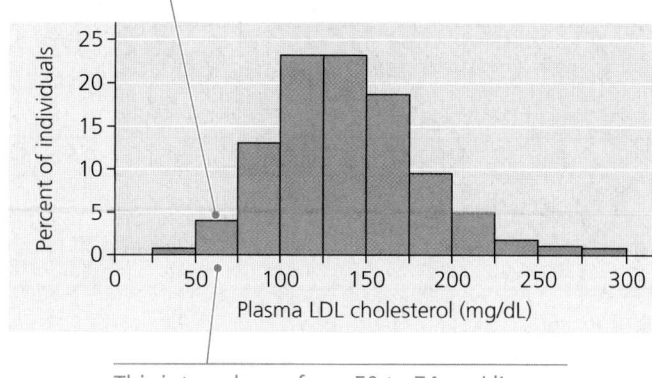

This interval runs from 50 to 74 mg/dL.

# Glossary of Scientific Inquiry Terms

See Concept 1.3 for more discussion of the process of scientific inquiry.

**control group** In a controlled experiment, a set of subjects that lacks (or does not receive) the specific factor being tested. Ideally, the control group should be identical to the experimental group in other respects.

**controlled experiment** An experiment designed to compare an experimental group with a control group; ideally, the two groups differ only in the factor being tested.

**data** Recorded observations.

**deductive reasoning** A type of logic in which specific results are predicted from a general premise.

**dependent variable** A factor whose value is measured in an experiment to see whether it is influenced by changes in another factor (the independent variable).

**experiment** A scientific test. Often carried out under controlled conditions that involve manipulating one factor in a system in order to see the effects of changing that factor.

**experimental group** A set of subjects that has (or receives) the specific factor being tested in a controlled experiment. Ideally, the experimental group should be identical to the control group for all other factors.

**hypothesis** A testable explanation for a set of observations based on the available data and guided by inductive reasoning. A hypothesis is narrower in scope than a theory.

**independent variable** A factor whose value is manipulated or changed during an experiment to reveal possible effects on another factor (the dependent variable).

**inductive reasoning** A type of logic in which generalizations are based on a large number of specific observations.

**inquiry** The search for information and explanation, often focusing on specific questions.

**model** A physical or conceptual representation of a natural phenomenon.

**prediction** In deductive reasoning, a forecast that follows logically from a hypothesis. By testing predictions, experiments may allow certain hypotheses to be rejected.

**theory** An explanation that is broader in scope than a hypothesis, generates new hypotheses, and is supported by a large body of evidence.

**variable** A factor that varies during an experiment.

# Chi-Square ($\chi^2$) Distribution Table

To use the table, find the row that corresponds to the degrees of freedom in your data set. (The degrees of freedom is the number of categories of data minus 1.) Move along that row to the pair of values that your calculated $\chi^2$ value lies between. Move up from those numbers to the probabilities at the top of the columns to find the probability range for your $\chi^2$ value. A probability of 0.05 or less is generally considered significant. (To practice using the chi-square test, see the Scientific Skills Exercise in Chapter 12.)

| Degrees of Freedom (df) | Probability | | | | | | | | | | |
|---|---|---|---|---|---|---|---|---|---|---|---|
| | 0.95 | 0.90 | 0.80 | 0.70 | 0.50 | 0.30 | 0.20 | 0.10 | 0.05 | 0.01 | 0.001 |
| 1 | 0.004 | 0.02 | 0.06 | 0.15 | 0.45 | 1.07 | 1.64 | 2.71 | 3.84 | 6.64 | 10.83 |
| 2 | 0.10 | 0.21 | 0.45 | 0.71 | 1.39 | 2.41 | 3.22 | 4.61 | 5.99 | 9.21 | 13.82 |
| 3 | 0.35 | 0.58 | 1.01 | 1.42 | 2.37 | 3.66 | 4.64 | 6.25 | 7.82 | 11.34 | 16.27 |
| 4 | 0.71 | 1.06 | 1.65 | 2.19 | 3.36 | 4.88 | 5.99 | 7.78 | 9.49 | 13.28 | 18.47 |
| 5 | 1.15 | 1.61 | 2.34 | 3.00 | 4.35 | 6.06 | 7.29 | 9.24 | 11.07 | 15.09 | 20.52 |
| 6 | 1.64 | 2.20 | 3.07 | 3.83 | 5.35 | 7.23 | 8.56 | 10.64 | 12.59 | 16.81 | 22.46 |
| 7 | 2.17 | 2.83 | 3.82 | 4.67 | 6.35 | 8.38 | 9.80 | 12.02 | 14.07 | 18.48 | 24.32 |
| 8 | 2.73 | 3.49 | 4.59 | 5.53 | 7.34 | 9.52 | 11.03 | 13.36 | 15.51 | 20.09 | 26.12 |
| 9 | 3.33 | 4.17 | 5.38 | 6.39 | 8.34 | 10.66 | 12.24 | 14.68 | 16.92 | 21.67 | 27.88 |
| 10 | 3.94 | 4.87 | 6.18 | 7.27 | 9.34 | 11.78 | 13.44 | 15.99 | 18.31 | 23.21 | 29.59 |

# Credits



## Photo Credits

**Cover image, Title Page, Ready-to-Go Teaching Modules** Klaus Honal/age fotostock.

**Endpapers** **Unit 1** Albert Tousson; **Unit 2** Andrey Prokhorov/Getty Images; **Unit 3** Sylvain Cordier/Science Source; **Unit 4** Lisa-Ann Gershwin; **Unit 5** Dennis Frates; **Unit 6** Oliver Wien/epa; European Pressphoto Agency/Alamy; **Unit 7** Juan Carlos Munoz/age fotostock.

**Preface** Gerard Lacz/FLPA.

**Organization and New Content** **Chapter 1** J.B. Miller/Florida Park Service; **Unit 1** Albert Tousson; **Unit 2** Andrey Prokhorov/Getty Images; **Unit 3** Sylvain Cordier/Science Source; **Unit 4** Lisa-Ann Gershwin; **Unit 5** Dennis Frates; **Unit 6** Oliver Wien/epa; European Pressphoto Agency/Alamy; **Unit 7** Juan Carlos Muñoz/age fotostock.

**Detailed Contents** **ants** Paul Quagliana/BNPS; **gecko** Martin Harvey/Photolibrary/Getty Images; **cell** Michael W. Davidson/Molecular Expressions; **geyser** Jack Dykinga/NaturePL.com; **kelp** Lawrence Naylor/Science Source; **dividing cells** Steve Gschmeissner/Science Source; **chromosome** Peter Menzel/Science Source; **protein** Barry Stoddard; **embryo** Stephen W. Paddock; **adenovirus** Linda M. Stannard, University of Cape Town/Science Source; **flower mantid** Lighthouse/UIG/age fotostock; **fossils** B. O'Kane/Alamy; ***Pandorina*** De Agostini Picture Library/Science Source; **mycelium** Hecker/blickwinkel/Alamy; **fern** John Martin/Alamy; **turtle** Juniors Bildarchiv/age fotostock; **flytrap** Chris Mattison/NaturePL.com; **bobcat** Thomas Kitchin & Victoria Hurst/All Canada Photos/age fotostock; **capillary** Ed Reschke/Getty Images; **spider** Stefan Hetz Supplied by WENN.com/Newscom; **locusts** Stringer/Reuters; **urchins** Scott Ling, Institute for Marine & Antarctic Studies, University of Tasmania.

**Unit Openers** **Unit 1: 2** Paul Quagliana/BNPS; **3** Mark J. Winter/Science Source; **4** Don W. Fawcett/Science Source; **5** Used by permission of B. L. de Groot. Related to work done for: Water Permeation Across Biological Membranes: Mechanism and Dynamics of Aquaporin-1 and GlpF. BL de Groot, H Grubmüller. *Science* 294:2353–2357 (2001); **6** Doug Perrine/Nature Picture Library; **7** Uryadnikov Sergey/Shutterstock; **8** Aflo/Nature Picture Library; **9** George von Dassow; **Unit 2: 10** Mango Productions/Corbis; **11** John Swithinbank/age fotostock; **12** Peter Menzel/Science Source; **13** Andrey Prokhorov/Getty Images; **14** Alessandro Bianchi/Reuters; **15** Andreas Werth; **16** Stephen W. Paddock; **17** Richard Bizley/Science Source; **18** Karen Huntt/Corbis; **Unit 3: 19** William Mullins/Alamy; **20** Pierson Hill; **21** Sylvain Cordier/Science Source; **22** Joel Sartore/Getty Images; **23** Juergen Ritterbach/Alamy; **Unit 4: 24** B Christopher/Alamy; **25** Biophoto Associates/Science Source; **26** Christopher Meder/Shutterstock; **27** Scott Linstead/Science Source; **Unit 5: 28** John Walker; **29** Dennis Frates/Alamy; **30** Ch'ien Lee; **31** Ernesto Gianoli, Departamento de Biología, Universidad de La Serena, Chile; **Unit 6: 32** Matthias Wittlinger; **33** Jeff Foott/Discovery Channel Images/Getty Images; **34** Jane Burton/BRUCE COLEMAN/Alamy; **35** SPL/Science Source; **36** Colin Marshall/FLPA; **37** Franco Banfi/Science Source; **38** Tamily Weissman; **39** Manamana/Shutterstock; **Unit 7: 40** Christopher Austin; **41** Jurgen Freund/Nature Picture Library; **42** Jordi Bas Casas/Photoshot; **43** Phung My Trung, vncreatures.net.

**Chapter 1** **1.1** J.B. Miller/Florida Park Service; **1.2** Shawn P. Carey, Migration Productions; **1.UN1 hummingbird** Jim Zipp/Science Source; **1.3.1** Leonello Calvetti/Stocktrek Images/Getty Images; **1.3.2** Terry Donnelly/Alamy; **1.3.3, 1.3.4** Floris van Breugel/Nature Picture Library; **1.3.5** Greg Vaughn/Alamy; **1.3.6** Pat Burner, Pearson Education; **1.3.7** Jeremy Burgess/Science Source; **1.3.8** Texas A & M University; **1.3.9** Jeremy Burgess/Science Source; **1.4a** Steve Gschmeissner/Science Source; **1.4b** A. Barry Dowsett/Science Source; **1.5** Conly Rieder; **1.6** Gelpi/fotolia; **1.7a** Photodisc/Getty Images; **1.8a.1** Carol Yepes/Moment/Getty Images; **1.8a.2** Ralf Dahm/Max Planck Institute of Neurobiology; **1.10** WaterFrame/Alamy; **1.11** Rod Williams/Nature Picture Library; **1.12a** Oliver Meckes and Nicole Ottawa/Eye of Science/Science Source; **1.12b** Eye of Science/Science Source; **1.12c1** Kunst & Scheidulin/age fotostock; **1.12c2** daksel/fotolia; **1.12c3** Anup Shah/Nature Picture Library; **1.12c4** M. I. Walker/Science Source; **1.13** Dede Randrianarisata/Macalester College; **1.14a** From: *Origin of Species* by Charles Darwin, 1859. Murray edition; **1.14b** ARCHIV/Science Source; **1.15a** zhaoyan/Shutterstock; **1.15b** Sebastian Knight/Shutterstock; **1.15c** Volodymyr Goinyk/Shutterstock; **1.18 inset** Karl Ammann/Corbis; **1.18** Tim Ridley/Dorling Kindersley; **1.19 clockwise from top** Martin Shields/Alamy; Maureen Spuhler, Pearson Education; Rolf Hicker Photography/All Canada Photos/Alamy; xPacifica/The Image Bank/Getty Images; **1.20a** From: Darwin to DNA: The Genetic Basis of Color Adaptations. From *In the Light of Evolution: Essays from the Laboratory and Field*, by J. Losos, Roberts and Co. Photo by Sacha Vignieri; **1.20a inset** Courtesy of Hopi Hoekstra, Harvard University; **1.20b** Sacha Vignieri; **1.20b inset** Shawn P. Carey, Migration Productions; **1.21** From: The selective advantage of cryptic coloration in mice. S. N. Vignieri, J. Larson, and H. E. Hoekstra. 2010. *Evolution* 64:2153–2158. Fig. 1; **1 SSE** Imagebroker/FLPA; **Chapter Review eukaryotic cell** Steve Gschmeissner/Science Source; **prokaryotic cell** A. Barry Dowsett/Science Source; **turtle** WaterFrame/Alamy; **SYK** Chris Mattison/Alamy.

**Chapter 2** **2.1** Paul Quagliana/BNPS; **2.2a** Chip Clark; **2.2bc** Stephen Frisch/Pearson Education; **2.4** National Library of Medicine; **2.11** Stephen Frisch/Pearson Education; **2.UN01 gecko** Martin Harvey/Photolibrary/Getty Images; **2.15** Nigel Cattlin/Science Source; **2.17** N.C. Brown Center for Ultrastructure Studies, SUNY, Syracuse, NY; **2.18** Alasdair Thomson/E+/Getty Images; **2.20** Jan van Franeker, Alfred-Wegener Institute fur Polar-und-Meeresforschung; **2.23a** Paulista/fotolia; **2.23b** Fotofermer/fotolia; **2.23c** SciePro/SPL/age fotostock; **2.23d** Beth Van Trees/Shutterstock; **2 SSE** From: Coral reefs under rapid climate change and ocean acidification. O. Hoegh-Guldberg et al. *Science*. 2007 Dec 14; 318(5857): 1737–42. Fig. 5a; **Chapter Review moth** Rolf Nussbaumer/Nature Picture Library; **SYK** Eric Guilloret/Biosphoto/Science Source.

**Chapter 3** **3.1** Mark J. Winter/Science Source; **3.11 clockwise from top left** Dougal Waters/Getty Images; John N.A. Lott/Biological Photo Service; Paul Lazarow; Biophoto Associates/Science Source; John Durham/Science Source; **3.14a** Jim Forrest/Dorling Kindersley; **3.14b** David Murray/Dorling Kindersley; **3.17a** Andrey Stratilatov/Shutterstock; **3.17bc** Nina Zanetti/Pearson Education; **3.20ab** PDB ID 2LYZ, From: Real-space refinement of the structure of hen egg-white lysozyme. R. Diamond. *Journal of Molecular Biology* 82(3):371–91 (Jan 25, 1974); **3.20c** Clive Freeman, The Royal Institution/Science Source; **3.21** Peter Colman; **3.22 spider** Dieter Hopf/Imagebroker/age fotostock; **blood cells** SciePro/SPL/age fotostock; **3.23** Eye of Science/Science Source; **3.25 top** CC-BY-3.0 Photo: Dsrjsr; **3.25 bottom** Laguna Design/Science Source; **3.29** P. Morris/Garvan Institute of Medical Research; **3.30 DNA** Alfred Pasieka/Science Source; **hippo** Frontline Photography/Alamy; **whale** WaterFrame/Alamy; **elephants** Imagebroker/FLPA; **Neanderthal** Illustration by Viktor Deak; **doctor** Chassenet/BSIP SA/Alamy; **fungus** David Read, Department of Animal and Plant Sciences, University of Sheffield; **3 SSE.1** lanych/Shutterstock; **3 SSE.2** David Bagnall/Alamy; **3 SSE.3** Eric Isselée/Shutterstock; **Chapter Review butter** Jim Forrest/Dorling Kindersley; **oil** David Murray/Dorling Kindersley; **SYK** Africa Studio/Shutterstock.

**Chapter 4** **4.1** Don W. Fawcett/Science Source; **4.UNCO paramecium** M.I. Walker/Science Source; **4.3.1–4.3.4** Elisabeth Pierson, Pearson Education; **4.3.5** Michael W. Davidson/Molecular Expressions; **4.3.6** Karl Garsha; **4.3.7a** J.L. Carson/Custom Medical Stock Photo; **4.3.7b** William Dentler/Biological Photo Service; **4.4** CNRI/Science Source; **4.5** Daniel S. Friend; **4.7 left to right** S. Cinti/Science Source; SPL/Science Source; A. Barry Dowsett/Science Source; Biophoto Associates/Science Source; SPL/Science Source; The University of Kansas Center for Bioinformatics; **4 SSE** Kelly Tatchell; **4.UN01 nucleus** Thomas Deerinck and Mark Ellisman, NCMIR; **4.8.1** Reproduced with permission from Freeze-Etch Histology, by L. Orci and A. Perrelet, Springer-Verlag, Heidelberg, 1975, Plate 25, page 53. Copyright © 1975 by Springer-Verlag GmbH & Co KG; **4.8.2** Don W. Fawcett/Science Source; **4.8.3** Ueli Aebi; **4.9.1** Don W. Fawcett/Science Source; **4.9.2** Harry Noller; **4.10** Don Fawcett; R. Bolender/Science Source; **4.11** Don W. Fawcett/Science Source; **4.12–4.13** Daniel S. Friend; **4.14** Eldon H. Newcomb; **4.17a** Daniel S. Friend; **4.17b** From: The shape of mitochondria and the number of mitochondrial nucleoids during the cell cycle of *Euglena gracilis*. Y. Hayashi and K. Ueda. *Journal of Cell Science*, 93:565–570, Copyright © 1989 by Company of Biologists; **4.18a** Jeremy Burgess/Science Source; **4.18b** Franz Golig, Philipps-University, Marburg, Germany; **4.19** Eldon H. Newcomb; **4.20** Albert Tousson; **4.21b** Bruce J. Schnapp; **Table 4.1 left to right** Mary Osborn; Frank Solomon; Mark Ladinsky; **4.23a** Omikron/Science Source; **4.23b** Dartmouth Electron Microscope Facility, Dartmouth College; **4.23c** Richard W. Linck; **4.24** From: Cross-linker system between neurofilaments, microtubules, and membranous organelles in frog axons revealed by the quick-freeze, deep-etching method. Hirokawa Nobutaka. *Journal of Cell Biology* 94(1): 129–142, 1982. Reproduced by permission of Rockefeller University Press; **4.25** G. F. Leedale/Science Source; **4.27.1** Reproduced with permission from *Freeze-Etch Histology*, by L. Orci and A. Perrelet, Springer-Verlag, Heidelberg, 1975. Plate 32. Page 68. Copyright © 1975 by Springer-Verlag GmbH & Co KG; **4.27.2** From: Fine structure of desmosomes, hemidesmosomes, and an adepidermal globular layer in developing newt epidermis. D. E. Kelly. *J Cell Biol.* 1966 Jan; 28(1):51–72. Fig. 7; **4.27.3** From: Low resistance junctions in crayfish. Structural changes with functional uncoupling. C. Peracchia and A. F. Dulhunty. *The Journal of Cell Biology*. 1976 Aug; 70(2 pt 1):419–39. Fig. 6. Reproduced by permission of Rockefeller University Press; **4.28** Eye of Science/Science Source; **Chapter Review vacuole, peroxisome** Eldon H. Newcomb; **SYK** Susumu Nishinaga/Science Source.

**Chapter 5** **5.1** Used by permission of B. L. de Groot. Related to work done for: Water Permeation Across Biological Membranes: Mechanism and Dynamics of Aquaporin-1 and GlpF. B. L. de Groot and H. Grubmüller. *Science* 294:2353–2357 (2001); **5.12** Michael Abbey/Science Source; **5 SSE** Photo Fun/Shutterstock; **5.18 phagocytosis** Biophoto Associates/Science Source; **pinocytosis** Don W. Fawcett/Science Source; **endocytosis** From M. M. Perry and A. B. Gilbert, *Journal of Cell Science* 39: 257–272, Fig. 11 (1979). © 1979 The Company of Biologists Ltd; **Chapter Review cheetah** Federico Veronesi/Alamy; **SYK** Kristoffer Tripplaar/Alamy.

**Chapter 6** **6.1** Doug Perrine/Nature Picture Library; **6.2** Stephen Simpson/Photolibrary/Getty Images; **6.3** Robert N. Johnson; **6.4a** Image Quest Marine; **6.4b** asharkyu/Shutterstock; **6.14** Thomas A. Steitz; **6 SSE** Fer Gregory/Shutterstock; **6.16** Jack Dykinga/Nature Picture Library; **6.20** Nicolae Simionescu; **Chapter Review SYK** PayPal/Flickr/Moment Select/Getty Images.

**Chapter 7** **7.1** Sergey Uryadnikov/Shutterstock; **7.4** Dionisvera/fotolia; **7.13b** The image of the ATP synthase is reproduced with the kind permission of Professor Sir John Walker and Dr. Martin Montgomery of the Medical Research Council's Mitochondrial Biology Unit; **7 SSE** Kitchin and Hurst/All Canada Photos/age fotostock; **Chapter Review SYK** Stephen Rees/Shutterstock.

**Chapter 8** **8.1** Aflo/Nature Picture Library; **8.2a** Jean-Paul Nacivet/age fotostock; **8.2b** Lawrence Naylor/Science Source; **8.2c** M. I. Walker/Science Source; **8.2d** Susan M. Barns; **8.2e** Heide Schulz; **8.3a** Andreas Holzenburg and Stanislav Vitha, Dept. of Biology and Microscopy & Imaging Center, Texas A&M University; **8.3b** Jeremy Burgess/Science Source; **8.11b** Christine L. Case; **8.18a** Doukdouk/Alamy; **8.18b** Keysurfing/Shutterstock; **8 SSE** Dennis Barnes/Agstockusa/age fotostock; **8.19** Andreas Holzenburg and Stanislav Vitha, Dept. of Biology and Microscopy & Imaging Center, Texas A&M University; **Chapter Review SYK** Gary Yim/Shutterstock.

**Chapter 9** **9.1** George von Dassow; **9.2a** Biophoto Associates/Science Source; **9.2b** Biology Pics/Science Source; **9.2c** Biophoto/Science Source; **9.3** J. M. Murray, Univ. of Pennsylvania Medical School; **9.4, 9.5b** Biophoto/Science Source; **9.7** Conly L. Rieder; **9.8.1 top** J. Richard McIntosh; **9.8.2** From: The kinetochore fiber structure in the acentric spindles of the green alga *Oedogonium*. M. J. Schibler and J. D. Pickettheaps. *Protoplasma*, 1987; 137(1): 29–44. Fig. 17. Springer-Verlag GmbH & Co. KG; **9.10a** Don W. Fawcett/Science Source; **9.10b** Eldon H. Newcomb; **9.11** Elizabeth Pierson, Pearson Education; **9.17** Guenter Albrecht-Buehler, Northwestern University; **9.18** Lan Bo Chen; **9.19** Anne Weston/Wellcome Institute Library; **9 SSE** Mike Davidson; **Chapter Review mitosis** J.L. Carson/Custom Medical Stock Photo; **SYK** Steve Gschmeissner/Science Source.

**Chapter 10** **10.1** Mango Productions/Comet/Corbis; **10.2a** Roland Birke/OKAPIA/Science Source; **10.2b** George Ostertag/SuperStock; **10.3a** Ermakoff/Science Source; **10.3c** CNRI/Science Source; **10 SSE** SciMAT/Science Source; **10.12** Mark Petronczki and Maria Siomos; **Chapter Review SYK** Randy Ploetz.

**Chapter 11** **11.1** John Swithinbank/age fotostock; **11 SSE** Alberto Pomares/Getty Images; **11.14** Barbara Bowman, Pearson Education; **11.15** Rick Guidotti, Positive Exposure; **11.17** Michael Ciesielski Photography; **Chapter Review hairlines** Barbara Bowman, Pearson Education; **cat** Breeder/owner Patricia Speciale, photographer Norma JubinVille; **SYK** Rene Maltete/RAPHO/Gamma-Rapho via Getty Images.

**Chapter 12** **12.1** Peter Menzel/Science Source; **12.3** From: Learning to Fly: Phenotypic Markers in *Drosophila*. A poster of common phenotypic markers used in *Drosophila* genetics. Jennifer Childress, Richard Behringer, and Georg Halder. 2005. *Genesis* 43(1). Cover illustration; **12.5** Andrew Syred/Science Source; **12.8** Jagodka/Shutterstock; **12 SSE** Oliver911119/Shutterstock; **12.15a** SPL/Science Source; **12.15b** denys_kuvaiev/fotolia; **Chapter Review SYK** James K. Adams, Biology, Dalton State College, Dalton, Georgia.

**coelacanth** Laurent Ballesta/www.blancpain-ocean-commitment.com/www.andromede-ocean.com and iSimangaliso Wetland Park Authority; **lungfish** Tom McHugh/Science Source; **giraffes** Malcolm Schuyl/FLPA; **27.20** The Field Museum; **27.23 top inset** Stuart Wilson/Science Source; **top right** James Laurie/Shutterstock; **middle** John Cancalosi/Nature Picture Library; **bottom** Danté Fenolio/Science Source; **27.24** Meul/ARCO/Nature Picture Library; **27.25 head, ribs, scales** Ted Daeschler/Academy of Natural Sciences/VIREO; **fin skeleton** Kalliopi Monoyios Studio; **27.27a** Alberto Fernández/age fotostock; **27.27b** Paul A. Zahl/Science Source; **27.27c** Zeeshan Mirza/ephotocorp/Alamy; **27.29 crocodile** defpicture/Shutterstock; **hummingbird** Mariusz Blach/fotolia; **turtle** Juniors Bildarchiv/age fotostock; **tuatara** Heather Angel/Natural Visions/Alamy; **snake** Nick Garbutt/Nature Picture Library; **27.30a** visceralimage/fotolia; **27.30b** The Natural History Museum/Alamy; **27.32 top** clearviewstock/Shutterstock; **egg** Mervyn Griffiths, Commonwealth Scientific and Industrial Research Organization; **middle** Will & Deni McIntyre/Science Source; **bottom** Anup Shah/Image State/Alamy; **27.34** T. White/David L. Brill Photography; **27.35** David L Brill/David L. Brill Photography; **27.36** Erik Trinkaus; **27.38** R.L. Jefferies; **27.39** W. Barthlott and W. Rauh, Nees Institute for Biodiversity of Plants; **27 SSE** Christophe Courteau/Nature Picture Library; **27.40** Hecker/Blickwinkel/Alamy; **27.41 snail** Dave Clarke/The Zoological Society of London; **workers** The U.S. Bureau of Fisheries (1919); **Chapter Review SYK** Lucy Arnold.

**Chapter 28** **28.1** John Walker; **28.4** Jeremy Burgess/Science Source; **28.5.1** Geoff Tompkinson/SPL/Science Source; **28.5.2** Rob Walls/Alamy; **28.5.3** Dana Tezarr/Photodisc/Getty Images; **28.6.1** D. G. Clever; **28.6.2** Dorling Kindersley; **28.6.3** Aflo Foto Agency/Alamy; **28.7.1** Neil Cooper/Alamy; **28.7.2** Martin Ruegner/Photodisc/Getty Images; **28.7.3** Gusto Production/SPL/Science Source; **28.7.4** Jerome Wexler/Science Source; **28.9.1** M. I. (Spike) Walker/Alamy; **28.9.2** Clouds Hill Imaging/Lastrefuge.co.uk; **28.9.3ab** Graham Kent, Pearson Education; **28.9.4** N.C. Brown Center for Ultrastructure Studies; **28.9.5a** Brian Gunning; **28.9.5b** Ray F. Evert; **28.9.5c** Graham Kent, Pearson Education; **28 SSE** Matthew Ward/Dorling Kindersley; **28.12** John Schiefelbein, American Society of Plant Biologists; **28.13** The Roslin Institute, The University of Edinburgh; **28.14a1** Ed Reschke; **28.14a2** Chuck Brown/Science Source; **28.14b** Ed Reschke; **28.15 28.16** Michael Clayton; **28.17 28.18** Ed Reschke; **28.19b1 left** Michael Clayton; **28.19b2** Alison W. Roberts; **Chapter Review woody eudicot** From: Anatomy of the vessel network within and between tree rings of *Fraxinus lanuginosa* (Oleaceae). Peter B. Kitin, Tomoyuki Fujii, Hisashi Abe and Ryo Funada. *American Journal of Botany*. 2004; 91:779–788; **SYK** Biophoto Associates/Science Source.

**Chapter 29** **29.1** Dennis Frates/Alamy; **29.UNCO leaf** Bjorn Svensson/Science Source; **29.3** Rolf Rutishauser; **29.6** Nigel Cattlin/Holt Studios International/Science Source; **29.8 clockwise from top left** View Stock RF/age fotostock; Guillermo Roberto Pugliese, International Plant Nutrition Institute (IPNI); M. K. Sharma and P. Kumar, IPNI; C. Witt, IPNI; **29.10 Peltigera** David T. Webb, University of Montana; **layers** Ralf Wagner; **puffer** Andrey Nekrasov/Pixtal/age fotostock; **Azolla** Daniel L Nickrent; **ants** Alex Wild, alexanderwild.com; **root** USDA/Science Source; **cut leaf** Tim Flach/The Image Bank/Getty Images; **nest** Martin Dohrn/Nature Picture Library; **29.11** Sarah Lydia Lebeis; **29.13** Scimat/Science Source; **29.14a left** Hugues B. Massicotte, University of Northern British Columbia Ecosystem and Management Program, Prince George, BC, Canada; **29.14a right, 29.14b** Mark Brundrett (http://mycorrhizas.info); **29.15 staghorn** David Wall/Alamy; **mistletoe** Peter Lane/Alamy; **dodder** Emilio Ereza/Alamy; **Indian pipe** Martin Shields/Alamy; **sundew** Fritz Polking/FLPA; **pitcher plants** Philip Blenkinsop/Dorling Kindersley; **ant on plant** Paul Zahl/Science Source; **Venus flytrap** Chris Mattison/Nature Picture Library; **29.20 leafless** Gerald Holmes, California Polytechnic State University at San Luis Obispo, Bugwood.org; **with leaves** Kate Shane, Southwest School of Botanical Medicine; **leaves close-up** Steven Baskauf, Nature Conservancy Tennessee Chapter Headquarters; **cross section** Natalie Bronstein; **flowers** Andrew de Lory/Dorling Kindersley; **cactus** Rob Tilley/Danita Delimont/Alamy; **Chapter Review SYK** Catalin Petolea.

**Chapter 30** **30.1** Ch'ien Lee; **30.3** From: Genetic interactions among floral homeotic genes of *Arabidopsis*. J. L. Bowman et al. *Development*. 1991 May; 112(1):1–20; Fig. 1A; **30.6 staminate** Wildlife GmbH/Alamy; **carpellate** Friedhelm Adam/Imagebroker/Getty Images; **dandelion and UV** Bjorn Rorslett; **bat** Rolf Nussbaumer/Imagebroker/age fotostock; **hummingbird** Rolf Nussbaumer/Nature Picture Library; **30.11 coconut** Kevin Schafer/Alamy; **A. macrocarpa** Aquiya/fotolia; **dandelion** Steve Bloom Images/Alamy; **maple** Chrispo/fotolia; **tumbleweed** Nurlan Kalchinov/Alamy; **puncture vine** California Department of Food and Agriculture's Plant Health and Pest Prevention Services; **squirrel** Alan Williams/Alamy; **scat** Kim A. Cabrera; **ant** Benoit Guénard; **30.12** Dennis Frates/Alamy; **30.13a** Marcel Dorken; **30.13b** Nobumitsu Kawakubo, Gifu University; **30 SSE** Toby Bradshaw; **30.14** Meriel G. Jones, University of Liverpool School of Biological Sciences; **30.15** Andrew McRobb/Dorling Kindersley; **30.16** Ton Koene/Lineair/Still Pictures/Robert Harding World Imagery; **Chapter Review SYK** Dartmouth College Electron Microscope Facility.

**Chapter 31** **31.1, 31.UNCO Boquila/Ranunculus** Ernesto Gianoli, Departamento de Biología, Universidad de La Serena, Chile; **31.4** From: Regulation of polar auxin transport by AtPIN1 in *Arabidopsis* vascular tissue. L. Gälweiler et al. *Science*. 1998 Dec 18; 282(5397):2226–30; Fig. 4ac; **31.6a** Richard Amasino; **31.6b** Fred Jensen; **31.8.1 left** Mia Molvray; **31.8.2** Karen E. Koch; **31.9a** Kurt Stepnitz, DOE Plant Research Lab, Michigan State Univ.; **31.9b** Joseph J. Kieber, Univ. of North Carolina; **31.10** Ed Reschke; **31.11** Natalie Bronstein; **31.13** Nigel Cattlin/Alamy; **31.15** Martin Shields/Alamy; **31.19** Michael Evans, Ohio State University; **31.20** From the cover of *Cell*, Volume 60, Issue 3, 9 February 1990. Janet Braam, Ronald W. Davis. Used by permission, Copyright © 1990 Cell Press. Image courtesy of Elsevier Sciences, Ltd; **31.21** Martin Shields/Science Source; **31 SSE** sarsmis/fotolia; **31.22** J. L. Basq and M. C. Drew; **31.23a** Susumu Nishinaga/Science Source; **31.23b** Johan De Meester/Arterra Picture Library/Alamy; **31.23c** Custom Life Science Images/Alamy; **31.24** New York State Agricultural Experiment Station (NYSAES) Cornell University; **Chapter Review SYK** Gary Crabbe/Alamy.

**Chapter 32** **32.1** Matthias Wittlinger; **32.2 epithelial** Steve Downing/Pearson Education; **muscle, connective** Nina Zanetti, Pearson Education; **blood** Gopal Murti/SPL/Science Source; **brain** Thomas Deerinck/National Center for Microscopy and Imaging Research, University of California, San Diego; **32.3 sunflowers** Phil_Good/fotolia; **fly** WildPictures/Alamy; **leaves** irin-k/Shutterstock; **bobcat** Thomas Kitchin & Victoria Hurst/All Canada Photos/age fotostock; **sprouts** Bogdan Wankowicz/Shutterstock; **gecko** Nature's Images/Science Source; **tubes** Last Refuge/Robert Harding Picture Library/Alamy; **vessels** Susumu Nishinaga/Science Source; **peas** Scott Rothstein/Shutterstock; **pigs** Walter Hodges/Lithium/age fotostock; **root** Rosanne Quinnell © The University of Sydney. eBot http://hdl.handle.net/102.100.100/1463; **villi** David M. Martin/Science Source; **spongy mesophyll** Rosanne Quinnell © The University of Sydney. eBot http://hdl.handle.net/102.100.100/2574; **lung** David M. Phillips/Science Source; **32.10.1** F. Hecker/blickwinkel/Alamy; **32.10.2** Jurgen & Christine Sohns/FLPA; **32.11.1 top** Jeffrey Lepore/Science Source; **32.11.2** Neil McNicoll/Alamy; **32.13a** John Shaw/Science Source; **32.13b** Matt T. Lee, Pearson Education; **32 SSE** Jiri Lochman/Lochman Transparencies; **32.18.1** George Peters/Getty Images; **32.18.2** Eric Isselée/fotolia; **32.18.3** Maksym Gorpenyuk/Shutterstock; **32.23** Michael Lynch/Shutterstock; **Chapter Review SYK** Oliver Wien/epa European Pressphoto Agency creative account/Alamy.

**Chapter 33** **33.1** Jeff Foott/Discovery Channel Images/Getty Images; **33.3** Stefan Huwiler/Russ Nussbaumer Photography/Alamy; **33.4** Gunter Ziesler/Photolibrary/Getty Images; **33.14.1** Fritz Polking/Peter Arnold/Alamy; **33.14.2** Tom Brakefield/Stockbyte/Getty Images; **33.15** Juergen Berger/Science Source; **33 SSE** The Jackson Laboratory, Bar Harbor, Maine; **Chapter Review SYK** Jack Moskovita.

**Chapter 34** **34.1** Jane Burton/Bruce Coleman/Alamy; **34.2** Eric Grave/Science Source; **34.9.1 top** From: Human Histology Photo CD, Indigo Instruments; **34.9.2** Ed Reschke/Getty Images; **34.15** Eye of Science/Science Source; **34.16** Image Source/Exactostock/SuperStock; **34 SSE** fotolia; **34.17a** Peter Batson/Image Quest Marine; **34.17b** Jez Tryner/Image Quest Marine; **34.19c** Prepared by Dr. Hong Y. Yan, University of Kentucky and Dr. Peng Chai, University of Texas; **34.20** Motta & Macchiarelli/Science Source; **Chapter Review SYK** Stefan Hetz, Supplied by WENN.com/Newscom.

**Chapter 35** **35.1** SPL/Science Source; **35.19** CNRI/Science Source; **35 SSE** Eye of Science/Science Source; **Chapter Review SYK** Tatan Syuflana/AP Images.

**Chapter 36** **36.1** Colin Marshall/FLPA; **36.UNCO anemone** David Wrobel; **36.2** Anastasia Pyryeva/fotolia; **36.4a** P. de Vries and David Crews; **36.6** John Cancalosi/Getty Images; **36.10** Philippe Henry/Design Pics/Alamy; **36 SSE** Andrew Syred/Science Source; **36.15** George von Dassow; **36.16** Charles A. Ettensohn; **36.19** Lennart Nilsson/Scanpix; **Chapter Review SYK** Dave Thompson/AP Images.

**Chapter 37** **37.1** Franco Banfi/Science Source; **37.3** Thomas Deerinck/National Center for Microscopy and Imaging Research, University of California, San Diego; **37.8 fish** Roger Steene/Image Quest Marine; **frog** F. Rauschenbach/F1online digitale Bildagentur GmbH/Alamy; **stomata** Power and Syred/Science Source; **bacteria** Eye of Science/Science Source; **37.14** Alan Peters; **37.17** Edwin R. Lewis; **Chapter Review SYK** B.A.E./Alamy.

**Chapter 38** **38.1** Tamily Weissman; **38.6** From: A functional MRI study of happy and sad affective states induced by classical music. M. T. Mitterschiffthaler et al. *Hum Brain Mapp.* 2007 Nov; 28(11):1150–62. Fig. 1; **38.7** Larry Mulvehill/Corbis; **38.12** Marcus E. Raichle, M.D.; **38.13** Images from the History of Medicine (NLM); **38.18a** Michael Nolan/Robert Harding World Imagery; **38.18b** Grischa Georgiew/Panther Media/age fotostock; **38.21** SPL/Science Source; **38.25** USDA/APHIS Animal and Plant Health Inspection Service; **38.26** Science Photo Library/age fotostock; **Chapter Review SYK** Dogs/fotolia.

**Chapter 39** **39.1** Manamana/Shutterstock; **39.2** Clara Franzini-Armstrong; **39.3** H. E. Huxley; **39.10** George Cathcart Photography; **39.15** Dave Watts/NHPA/Science Source; **39 SSE** Vance A. Tucker; **39.18a** Kenneth Lorenzen; **39.19a** Thomas D. McAvoy/Time Life Pictures/Getty Images; **39.19b** Operation Migration - USA; **39.21** Lincoln Brower, Sweet Briar College; **39.23a** Matt T. Lee; **39.23b** David Osborn/Alamy; **39.23c** David Tipling/FLPA; **39.24** James D Watt/Image Quest Marine; **39.25** Gerald S. Wilkinson; **39.26** Lowell Getz; **39.27** Rory Doolin; **39.28** Jennifer Jarvis; **Chapter Review SYK** William Leaman/Alamy.

**Chapter 40** **40.1 and 40.UNCO frogs** Christopher Austin; **40.2.1** Peter Blackwell/Nature Picture Library; **40.2.2** Barrie Britton/Nature Picture Library; **40.2.3** Oleg Znamenskiy/fotolia; **40.2.4** Juan Carlos Muñoz/age fotostock; **40.2.5** John Downer/Nature Picture Library; **40.2.6** 1xpert/fotolia; **40.8.1** Joseph Sohm/age fotostock; **40.8.2** David Halbakken/age fotostock; **40.8.3** gary718/Shutterstock; **40.8.4** Siepmann/Imagebroker/Alamy; **40.8.5** Bent G. Nordeng/Shutterstock; **40.8.6** Juan Carlos Muñoz/Nature Picture Library; **40.9.1** Siepmann/Imagebroker/Alamy; **40.9.2** Robert Harding Picture Library/Alamy; **40.9.3** Joseph Sohm/age fotostock; **40.9.4** The California Chaparral Institute; **40.9.5** David Halbakken/age fotostock; **40.9.6** Bent G. Nordeng/Shutterstock; **40.9.7** gary718/Shutterstock; **40.9.8** Juan Carlos Muñoz/Nature Picture Library; **40.11.1** David Tipling/Nature Picture Library; **40.11.2 40.11.3** Susan Lee Powell; **40.11.4** Stuart Westmorland/Alamy; **40.11.5** Digital Vision/Photodisc/Getty Images; **40.11.6** Tatonka/Shutterstock; **40.11.7** William Lange, Woods Hole Oceanographic Institute; **40.13.1 top** Scott Ling, Institute for Marine & Antarctic Studies, University of Tasmania; **40.13.2** Timothy J. Alexander; **40.15a** Bernard Castelein/Alamy; **40.15b** Michael S. Nolan/age fotostock; **40.15c** Alexander Chaikin/Shutterstock; **40.18** Villiers Steyn/Shutterstock; **40 SSE** Laguna Design/Science Source; **40.21a** Steve Bloom Images/Alamy; **40.21b1** Fernanda Preta/Alamy; **40.21b2** Edward Parker/Alamy; **40.UN02 moose** DLILLC/Cardinal/Corbis; **40.23 wheat** fotoVoyager/E+/Getty Images; **yeast** Andrew Syred/Science Source; **owl** Hellio & Van Ingen/NHPA/Photoshot; **cheetah** Gregory G. Dimijian/Science Source; **subway** Jorge Dan, Mexico/Reuters; **mice** Nicholas Bergkessel, Jr./Science Source; **Chapter Review SYK** Stringer/Reuters.

**Chapter 41** **41.1 and 41.UNCO crabs** Jurgen Freund/Nature Picture Library; **41.2.1** Joseph T. Collins/Science Source; **41.2.2** National Museum of Natural History/Smithsonian Institution; **41 SSE** Johan Larson/Shutterstock; **41.5a** Barry Mansell/Nature Picture Library; **41.5bc1** Danté Fenolio/Science Source; **41.5c2** Robert Pickett/Papilio/Alamy; **41.6** Doug Perrine/Nature Picture Library; **41.UN01 hornworm** Custom Life Science Images/Alamy; **41.7a** Bazzano Photography/Alamy; **41.7b** Nicholas Smythe/Science Source; **41.8** Daryl Balfour/NHPA/Photoshot; **41.9a** Sally D. Hacker; **41.12** Cedar Creek Ecosystem Science Reserve, University of Minnesota; **41.15** Genny Anderson; **41.16** Adam Welz; **41.19** NPS Photos by Bob Stevenson; **41.20.1** Charles D. Winters/Science Source; **41.20.2** Douglas W Veltre; **41.20.3** Terry Donnell and Mary Liz Austin; **41.20.4** Glacier Bay National Park and Preserve; **41.21.1 top** R. Grant Gilmore, Dynamac Corporation; **41.21.2** Lance Horn, National Undersea Research Center, University of North Carolina-Wilmington; **41.24** Nelish Pradhan, Bates College, Lewiston, ME; **Chapter Review SYK** Jacques Rosès/Science Source.

**Chapter 42** **42.1** Jordi Bas Casas/Photoshot; **42.2** Stone Nature Photography/Alamy; **42.3** Justus de Cuveland/Imagebroker/age fotostock; **42.6** Image by Reto Stöckli, based on data provided by the MODIS Science Team/Earth Observatory/NASA; **42.9** Matt Meadows/Photolibrary/Getty Images; **42 SSE** David R. Frazier Photolibrary/Science Source; **42.UN01 plankton** Laguna Design/Science Source; **42.14** Hubbard Brook Research Foundation/USDA Forest Service; **42.15** Mark Gallagher, Princeton Hydro, LLC, Ringoes, NJ; **42.16 Florida** Photo provided by Kissimmee Division staff, South Florida Water Management District; **South Africa** Jean Hall/Holt Studio/Science Source; **New Zealand** Tim Day, Xcluder Pest Proof Fencing Company; **Japan** Kenji Morita/Environment Division, Tokyo Kyuei Co., Ltd; **42.17a** U.S. Department of Energy; **Chapter Review SYK** Eckart Pott/Photoshot.

**Chapter 43** **43.1** Phung My Trung, vncreatures.net; **43.2** Trinh Le Nguyen/Shutterstock; **43.4.1 top** Neil Lucas/Nature Picture Library; **43.4.2** Mark Carwardine/Getty Images; **43.5** Merlin D. Tuttle/Science Source; **43.UN01 periwinkle** Scott Camazine/Science Source; **43.6** Michael Edwards/The Image Bank/Getty Images; **43.7** Robert Ginn/PhotoEdit; **43.8** Benezeth Mutayoba, University of Washington Center for Conservation Biology; **43.9** Pictures Colour Library/Travel Picture/Alamy; **43.12a** William Ervin/Science Source; **43.13** Craighead Institute; **43.14a inset** William Leaman/Alamy; **43.14a** Chuck Bargeron, University of Georgia, Bugwood.org; **43.14b** William D. Boyer, USDA Forest Service, Bugwood.org; **43.15** Guido Alberto Rossi/AGF RM/age fotostock; **43.16** R. O. Bierregaard, Jr., Biology Department, University of North Carolina, Charlotte; **43.17** frans lemmens/Alamy; **43.20** Mark Chiappone

37:4767–4772; **18 SSE1 and 2** Compiled using data from the National Center for Biotechnology Information (NCBI); **18.16b** Data from W. Shu et al., Altered ultrasonic vocalization in mice with a disruption in the *Foxp2* gene, *Proceedings of the National Academy of Sciences USA* 102:9643–9648 (2005); **18.17** Adapted from *The Homeobox: Something Very Precious That We Share with Flies, From Egg to Adult* by Peter Radetsky. Copyright © 1992 by William McGinnis. Image provided by William McGinnis; **18.18** Adapted from "*Hox* genes and the evolution of diverse body plans" by M. Akam, *Philosophical Transactions of the Royal Society B: Biological Sciences*, 349(1329):313–319 (9/29/95). Copyright © 1995 by the Royal Society. Reprinted with permission. Permission conveyed through Copyright Clearance Center, Inc.

**Chapter 19** **p.383 Quote** ("Your words have . . .") Darwin, Charles (1887:116), Darwin, F., ed., The life and letters of Charles Darwin, including an autobiographical chapter, London: John Murray (*The Autobiography of Charles Darwin*). Retrieved on 2006-12-15; **19.9** Adapted from artwork by U. Kikutani, as appeared in "What can make a four-ton mammal a most sensitive beast?", *Natural History* 106(1), 36–45 (11/97). Copyright ©1997 by Utako Kikutani. Reprinted with permission of the artist; **19.14b** Data from S. P. Carroll and C. Boyd, Host race radiation in the soapberry bug: natural history with the history, *Evolution* 46:1052–1069 (1992); **19.15a** Courtesy of B.A. Diep for author Michael Cain; **19.15b** From www.hcup.ahrq.gov: A. Elixhauser & C. Steiner, Statistical Brief #35. Infections with methicillin-resistant *Staphylococcus aureus* (MRSA) in U.S. hospitals, 1993–2005, Fig. 1. Healthcare Cost and Utilization Project. Agency for Healthcare Research and Quality, Rockville, MD (2007); **p.393 Quote** ("There is grandeur . . .") Charles Darwin, *The Origin of Species*, 1859; **19 SSE.1** Data from J. A. Endler, Natural selection on color patterns in *Poecilia reticulata*, *Evolution* 34:76–91 (1980); **19 EOC Table** Data from C. F. Curtis et al., Selection for and against insecticide resistance and possible methods of inhibiting the evolution of resistance in mosquitoes, *Ecological Entomology* 3:273–287 (1978).

**Chapter 20** **20.6** Data from C. S. Baker and S. R. Palumbi, Which whales are hunted? A molecular genetic approach to monitoring whaling, *Science* 265:1538–1539 (1994); **20.13** Adapted from "The Evolution of the Hedgehog Gene Family in Chordates; Insights from Amphioxus Hedgehog" by S.M. Shimeld, *Developmental Genes and Evolution* 209(1): 40–47 (1/1/99). Copyright © 1999 by Springer. Reprinted with kind permission from Springer Science+Business Media B.V.; **20.19** Based on *Molecular Markers, Natural History, and Evolution*, 2nd ed., by J.C. Advise. Sinauer Associates, 2004; **20.20** Adapted from "Timing the Ancestor of the HIV-1 Pandemic Strains" by B. Korber et al., *Science* 288(5472):1789–1796 (6/9/00). Copyright © 2000 by AAAS. Reprinted with permission from AAAS; **20.22** "Phylogenetic Classification and the Universal Tree" by W.F. Doolittle, *Science* 284(5423):2124–2128 (6/25/99). Copyright © 1999 by AAAS. Reprinted with permission from AAAS; **20 SSE.1** Data from Nancy A. Moran, Yale University. See N. A. Moran and T. Jarvik, Lateral transfer of genes from fungi underlies carotenoid production in aphids, *Science* 328:624–627 (2010).

**Chapter 21** **21.4** Data from M. Kreitman, Nucleotide polymorphism at the alcohol dehydrogenase locus of *Drosophila melanogaster*, *Nature* 304:412–416 (1983); **21.11** *Discover Biology*, 2nd ed., by M.L. Cain and H. Damman, by R.A. Lue and C.K. Loon (eds.). W. W. Norton & Company, Inc., 2002; **21.12** Data from "Gene Flow Maintains a Large Genetic Difference in Clutch Size at a Small Spatial Scale" by E. Postma and A.J. van Noordwijk, *Nature* 433(7021): 65–68 (1/6/05); **21.14** Adapted from many sources including D.J. Futuyma, *Evolution*, Fig. 11.3, Sinauer Associates, Sunderland, MA (2005) and from R.L. Carroll, *Vertebrate Paleontology and Evolution*, W.H. Freeman & Co. (1988); **21.15** Adapted from A.C. Allison. Abnormal hemoglobin and erythrocyte enzyme-deficiency traits. In *Genetic Variation in Human Populations*, by G.A. Harrison (ed.), Oxford: Elsevier Science (1961), and from S. I. Hay et al., A world malaria map: *Plasmodium falciparum* endemicity in 2007, *PLoS Medicine* 6:291 (2009), fig. 3; **21.16** Adapted from "Frequency-Dependent Natural Selection in the Handedness of Scale-Eating Cichlid Fish" by M. Hori, *Science* 260(5105):216–219 (4/9/93). Copyright © 1993 by AAAS. Reprinted with permission from AAAS; **21.18** Data from A. M. Welch et al., Call duration as an indicator of genetic quality in male gray tree frogs, *Science* 280:1928–1930 (1998); **21 EOC02** Data from R. K. Koehn and T. J. Hilbish, The adaptive importance of genetic variation, *American Scientist* 75:134–141 (1987).

**Chapter 22** **p.434 Quote** ("Both in space . . .") Charles Darwin, *The Voyage of the Beagle* (1839), chapter VIII: "Excursion to Colonia del Sacramiento, etc." (second edition, 1845), entry for 19 November 1833, pages 147–148; **22.6** R. Brian Langerhans, an original unpublished graph; **22.7** Data from D. M. B. Dodd, Reproductive isolation as a consequence of adaptive divergence in *Drosophila pseudoobscura*, *Evolution* 43:1308–1311 (1989); **22.12** Data from O. Seehausen and J. J. M. van Alphen, The effect of male coloration on female mate choice in closely related Lake Victoria cichlids (*Haplochromis nyererei* complex), *Behavioral Ecology and Sociobiology* 42:1–8 (1998); **22.13** Based on "Analysis of Hybrid Zones with *Bombina*" by J.M. Szymura, by R.G. Harrison (ed.), *Hybrid Zone and the Evolutionary Process*. Oxford University Press, 1993; **22 SSE.1** Data from S. G. Tilley, P. A. Verrell, and S. J. Arnold, Correspondence between sexual isolation and allozyme differentiation: A test in the salamander *Desmognathus ochrophaeus*, *Proceedings of the National Academy of Sciences USA*. 87:2715–2719 (1990); **22.18** Data from L. H. Rieseberg et al., Role of gene interactions in hybrid speciation: Evidence from ancient and experimental hybrids, *Science* 272:741–745 (1996).

**Chapter 23** **23.4** Figure 6.1 adapted from *Geologic Time* by D.L. Eicher, 1976. Copyright © 1976 by Pearson Education, Inc. Reprinted and electronically reproduced by permission of Pearson Education, Inc. Upper Saddle River, NJ; **23.5c-f** Adapted from many sources including D.J. Futuyma, *Evolution*, fig. 4.10, Sunderland, MA: Sinauer Associates, Sunderland, MA (2005) and from R.L. Carroll, *Vertebrate Paleontology and Evolution*. W.H. Freeman & Co. (1988); **23.5g** Adapted from Z. Luo et al., A new mammalia form from the Early Jurassic and evolution of mammalian characteristics, *Science* 292:1535 (2001); **23.8** Adapted from http://geology.er.usgs.gov/eastern/plates.html; **23.10** Adapted from Figure 1 of "Amphibians as Indicators of Early Tertiary "Out-of-India" Dispersal of Vertebrates" by F. Bossuyt and M.C. Milinkovitch, *Science* 292(5514):93–95 (4/01). Copyright © 2001 by AAAS. Reprinted with permission from AAAS. Permission conveyed through Copyright Clearance Center, Inc.; **23.11** Graph created from D.M. Raup & J.J. Sepkoski, Jr., Mass extinctions in the marine fossil record, *Science* 215:1501–1503 (1982); J.J. Sepkoski, Jr., A kinetic model of Phanerozoic taxonomic diversity. III. Post-Paleozoic families and mass extinctions, *Paleobiology* 10:246–267 (1984); and D.J. Futuyma, The Evolution of Biodiversity, p. 143, fig. 7.3a and p. 145, fig. 7.6, Sinauer Associates, Sunderland, MA; **23.13** Based on data from "A Long-Term Association Between Global Temperature and Biodiversity, Origination and Extinction in the Fossil Record" by P.J. Mayhew, G.B. Jenkins and T.G. Benton, *Proceedings of the Royal Society B: Biological Sciences* 275(1630):47–53. The Royal Society, 2008; **23.14** Adapted from "Anatomical and ecological constraints on Phanerozoic animal diversity in the marine realm" by R.K. Bambach, A.H. Knoll and J.J. Sepkoski, Jr., *Proceedings of the National Academy of Sciences* 99(10):6854–6859 (5/14/02). Copyright © 2002 by National Academy of Sciences, U.S.A. Reprinted with permission; **23.15** Based on Hickman, Roberts, & Larson, *Zoology*, 10th ed., Wm. C. Brown, fig. 31.1 (1997); **23.20** Based on "Hox Protein Mutation and Macroevolution of the Insect Body Plan" by M. Ronshaugen, N. McGinnis and W. McGinnis, *Nature* 415(6874):914–917 (2/6/02); **23.21** Data from M. D. Shapiro et al., Genetic and developmental basis of evolutionary pelvic reduction in three-spine sticklebacks, *Nature* 428:717–723 (2004); **23.22** Based on M. Strickberger, *Evolution*, Jones & Bartlett, Boston, MA (1990); **23 SSE.1** Data from T. Hansen, Larval dispersal and species longevity in Lower Tertiary gastropods, *Science* 199:885–887 (1978).

**Chapter 24** **24.3** Based on data from A.P. Johnson et al., The Miller volcanic spark discharge experiment, *Science* 322:404 (2008); **24.4a** Data from "Experimental Models of Primitive Cellular Compartments: Encapsulation, Growth, and Division" by M.M. Hanczyc, S.M. Fujikawa, and J.W. Szostak, *Science* 302(5645):618–622 (10/24/03). American Association for the Advancement of Science (AAAS), 2003; **24.14** Data from V. S. Cooper and R. E. Lenski, The population genetics of ecological specialization in evolving *Escherichia coli* populations, *Nature* 407: 736–739 (2000); **24.22a** Based on data from C. Calvaruso et al., Root-associated bacteria contribute to mineral weathering and to mineral nutrition in trees: A budgeting analysis, *Applied and Environmental Microbiology* 72:1258–1266 (2006); **24 SSE.1** Data from R. Mendes, et al. Deciphering the rhizosphere for disease-suppressive bacteria, *Science* 332:1097–1100 (2011); **24 EOC Table** Data from J. J. Burdon et al., Variation in the effectiveness of symbiotic associations between native rhizobia and temperate Australian *Acacia*: Within-species interactions, *Journal of Applied Ecology* 36:398–408 (1999).

**Chapter 25** **25.4** Adaptation of Figure 2 from "The Number, Speed, and Impact of Plastid Endosymbioses in Eukaryotic Evolution" by Patrick J. Keeling, from *ANNUAL REVIEW OF PLANT BIOLOGY*, April 2013, Volume 64. Copyright © 2002 by Annual Reviews, Inc. Permission to reprint conveyed through Copyright Clearance Center, Inc.; **25.8** Based on H. Oda and M. Takeichi, Structural and functional diversity of cadherin at the adherens junction, *The Journal of Cell Biology* 193(7): 1137–1146 (2011); and M. Abedin and N. King, The premetazoan ancestry of cadherins, *Science* 319:946–948 (2011); **25.17** Adapted from an illustration by K. X. Probst in *Microbiology* by R.W. Bauman, 2004:350. Copyright © by Kenneth X. Probst. Reprinted with permission; **25.21** Data from A. Stechmann and T. Cavalier-Smith, Rooting the eukaryote tree by using a derived gene fusion, *Science* 297:89–91 (2002); **25.24** Data from D.G. Boyce et al., Global phytoplankton decline over the past century, *Nature* 466:591–596 (2010) and from personal communication; **25 SSE.1** Data from D. Yang, et al., Mitochondrial origins, *Proceedings of the National Academy of Sciences USA* 82:4443–4447 (1985).

**Chapter 26** **26.27** Adapted from: "A revision of *Williamsoniella*" by T.M. Harris, *Philosophical Transactions of the Royal Society B: Biological Sciences* 231(583):313–328 (10/10/44). Copyright © 1944 by the Royal Society. Reprinted with permission, and Figure 11a from "Phylogenetic Analysis of Seed Plants and the Origin of Angiosperms" by P.R. Crane, *Annals of the Missouri Botanical Garden* 72:738 (1985). Copyright © 1985 by P.R. Crane. Reprinted with permission by Missouri Botanical Garden Press; **26.28a** Adapted from *Phylogeny and Evolution of Angiosperms*, by P. Soltis et al., Sinauer Associates, Inc., 2005; **26 SSE.1** Data from F. Martin et al., The genome of *Laccaria bicolor* provides insights into mycorrhizal symbiosis, *Nature* 452:88–93 (2008); **26.31** Data from A. E. Arnold et al., Fungal endophytes limit pathogen damage in a tropical tree, *Proceedings of the National Academy of Sciences USA* 100:15649–15654 (2003); **26 EOC Table** Data from R. S. Redman et al., *Science* 298:1581 (2002).

**Chapter 27** **27.15** Data from J. K. Grenier et al., Evolution of the entire arthropod *Hox* gene set predated the origin and radiation of the onychophoran/arthropod clade, *Current Biology* 7:547–553 (1997); **27.26 Left** Adapted from "The Devonian tetrapod *Acanthostega gunnari Jarvik*: postcranial anatomy, basal tetrapod relationships and patterns of skeletal evolution" by M.I. Coates, *Transaction of the Royal Society of Edinburgh: Earth Sciences* 87: 363–421 (1996). Copyright © 1996 by M.I. Coates. Reprinted with permission of the Royal Society of Edinburgh; **27.26 Right** Adapted from "The Pectoral Fin of *Tiktaalik roseae* and the Origin of the Tetrapod Limb" by N.H. Shubin, E.B. Daeschler, and F.A. Jenkins, Jr., *Nature* 440(7085):764–771 (4/6/06); **27.33** Based on N.J. Butterfield, Animals and the invention of the Phanerozoic Earth system, *Trends in Ecology and Evolution* 26:81–87, fig. 1 (2011); **27.40** Adapted from "Correlates of recovery for Canadian Atlantic cod" by J.A. Hutchings and R.W. Rangeley, *Canadian Journal of Zoology* 89(5):386–400 (5/1/11). Copyright © 2011 by NRC Research Press. Reprinted with permission; **27.41** Adapted from "The Global Decline of Nonmarine Mollusks" by C. Lydeard, et al., *Bioscience* 54(4):321–330 (4/04). (Updated data are from International Union for Conservation of Nature, 2008.) **27 SSE.1** Data from R. Rochette, S. P. Doyle, and T. C. Edgell, Interaction between an invasive decapod and a native gastropod: Predator foraging tactics and prey architectural defenses, *Marine Ecology Progress Series* 330:179–188 (2007); **27 EOC Table** Data from D. Sol et al., Big-brained birds survive better in nature, *Proceedings of the Royal Society B* 274:763–769 (2007).

**Chapter 28** **28 SSE.1** Data from D. L. Royer et al., Phenotypic plasticity of leaf shape along a temperature gradient in *Acer rubrum*, *PLoS ONE* 4(10): e7653 (2009).

**Chapter 29** **29 SSE.1** Data from J.D. Murphy and D.L. Noland, Temperature effects on seed imbibition and leakage mediated by viscosity and membranes, *Plant Physiology* 69:428–431 (1982); **29.11b** Data from D.S. Lundberg et al., Defining the core *Arabidopsis thaliana* root microbiome, *Nature* 488:86–94 (2012).

**Chapter 30** **30 SSE.1** Data from S. Sutherland and R. K. Vickery, Jr., Trade-offs between sexual and asexual reproduction in the genus *Mimulus*, *Oecologia* 76:330–335 (1988).

**Chapter 31** **31.2** Data from C. R. Darwin, *The Power of Movement in Plants*, John Murray, London (1880). P. Boysen-Jensen, Concerning the performance of phototropic stimuli on the *Avena* coleoptile, *Berichte der Deutschen Botanischen Gesellschaft* (*Reports of the German Botanical Society*) 31:559–566 (1913); **31.3** Data from F. Went, A growth substance and growth, *Recueils des Travaux Botaniques Néerlandais* (*Collections of Dutch Botanical Works*) 25:1–116 (1928); **31.4** Data from L. Gälweiler et al., Regulation of polar auxin transport by AtPIN1 in *Arabidopsis* vascular tissue, *Science* 282:2226–2230 (1998); **31.12a** Based on M. Wilkins, *Plant Watching*, Facts on File Publ. (1988); **31.13** Data from H. Borthwick et al., A reversible photoreaction controlling seed germination, *Proceedings of the National Academy of Sciences, USA* 38:662–666 (1952); **31 SSE.1** Data from O. Falik et al., Rumor has it . . .: Relay communication of stress cues in plants, *PLoS ONE* 6(11): e23625 (2011).

**Chapter 32** **32.18** Adapted from *Zoology*, by L.G. Mitchell, J.A. Mutchmor, and W.D. Dolphin, 1998. Copyright © 1998 by Pearson Education, Inc. Reprinted and electronically reproduced by permission of Pearson Education, Inc. Upper Saddle River, NJ; **32.21** Figure 25.3b adapted from *Human Anatomy and Physiology*, 8th ed., by E.N. Marieb and K.N. Hoehn, 2010. Copyright © 2010 by Pearson Education, Inc. Reprinted and electronically reproduced by permission of Pearson Education, Inc. Upper Saddle River, NJ; **32.22** Figure 25.3b adapted from *Human Anatomy and Physiology*, 8th ed., by E.N. Marieb and K.N. Hoehn, 2010. Copyright © 2010 by Pearson Education, Inc. Reprinted and electronically reproduced by permission of Pearson Education, Inc. Upper Saddle River, NJ; **32 SSE.1** Data from R.E. MacMillen et al., Water economy and energy metabolism of the sandy inland mouse, *Leggadina hermannsburgensis*, *Journal of Mammalogy* 53:529–539 (1972).

**Chapter 33** **p.689 Quote** ("A vitamin is . . .") Albert Szent-Györgyi, Nobel Prize in Physiology or Medicine, 1937; **33.7** Adapted from *Human Anatomy and Physiology*, 8th ed., by E.N. Marieb and K.N. Hoehn, 2010. Copyright © 2010 by Pearson Education, Inc. Reprinted and electronically reproduced by permission of Pearson Education, Inc. Upper Saddle River, NJ; **33 SSE.1** Data from D. L. Coleman, Effects of parabiosis of obese mice with diabetes and normal mice, *Diabetologia* 9:294–298 (1973).

**Chapter 34** **34.12** Adapted from *Human Anatomy and Physiology*, 8th ed., by E.N. Marieb and K.N. Hoehn, 2010. Copyright © 2010 by Pearson Education, Inc. Reprinted and electronically reproduced by permission of Pearson Education, Inc. Upper Saddle River, NJ; **34.21** Data from M. E. Avery and J. Mead, Surface properties in relation to atelectasis and hyaline membrane disease, *American Journal of Diseases of Children* 97:517–523 (1959); **34.24** Adapted from *Human Anatomy and Physiology*, 4th ed., by E.N. Marieb. Copyright © 1998 by Pearson Education, Inc. Reprinted and electronically reproduced by permission of Pearson Education, Inc., Upper Saddle River, NJ; **34.25** PDB ID 2HHB: G. Fermi, M.F. Perutz, B. Shaanan, and R. Fourme. The crystal structure of human deoxyhaemoglobin at 1.74 Å resolution. *J. Mol. Biol.* 175:159–174 (1984); **34 SSE.1** Data from J. C. Cohen et al., Sequence variations in PCSK9, low LDL, and protection against coronary heart disease, *New England Journal of Medicine* 354:1264–1272 (2006).

**Chapter 35** **35.5** Adapted from *Microbiology: An Introduction*, 6th ed., by G. J. Tortora, B.R. Funke, and C.L. Case. Copyright © 1998 by Pearson Education, Inc. Reprinted and electronically reproduced by permission of Pearson Education, Inc. Upper Saddle River, NJ; **p.741 Quote** ("for the same . . .") Thucydides quoted in *The History of the Peloponnesian War*, by Thucydides, written 431 B.C.E., translated by Richard Crawley (Floating Press, © 2008); **35 SSE.1** Data from L. J. Morrison, et al., Probabilistic order in antigenic variation of *Trypanosoma brucei*, *International Journal for Parasitology* 35:961–972 (2005), and L. J. Morrison, et al., Antigenic variation in the African trypanosome: molecular mechanisms and phenotypic complexity, *Cellular Microbiology* 1: 1724–1734 (2009).

**Chapter 36** **36 SSE.1** Data from A. Jost, Recherches sur la differenciation sexuelle de l'embryon de lapin (Studies on the sexual differentiation of the rabbit embryo), *Archives d'Anatomie Microscopique et de Morphologie Experimentale* 36:271–316 (1947).

**Chapter 37** **37.12** Based on *Cellular Physiology of Nerve and Muscle*, 4th ed., by G. Matthews. Wiley-Blackwell, 2002; **37 SSE.1** Data from "Opiate Receptor: Demonstration in Nervous Tissue" by C.B. Pert and S.H. Snyder, *Science* 179(4077):1011–1014 (3/9/73). American Association for the Advancement of Science (AAAS), 1973.

**Chapter 38** **38.8** Based on L.M. Mukhametov. Sleep in marine mammals. In *Sleep Mechanisms*, by A.A. Borbély and J.L. Valatx (eds.), pp. 227–238, Springer-Verlag, Munich (1984); **38.14** Based on E.D. Jarvis et al., Avian brains and a new understanding of vertebrate brain evolution, *Nature Reviews Neuroscience* 6:151–159, fig. 1c (2005); **38.19** Figure 15.23 adapted from *Human Anatomy and Physiology*, 8th ed., by E.N. Marieb and K.N. Hoehn, 2010. Copyright © 2010 by Pearson Education, Inc. Reprinted and electronically reproduced by permission of Pearson Education, Inc. Upper Saddle River, NJ; **38.23** Adapted from *Human Anatomy and Physiology*, 8th ed., by E.N. Marieb, 2001. Copyright © 2001 by E.N. Marieb. Reprinted and electronically reproduced with permission by Pearson Education, Inc. Upper Saddle River, NJ; **38.26a** Adapted from *Human Anatomy and Physiology*, 8th ed., by E.N. Marieb and K.N. Hoehn, 2010. Copyright © 2010 by Pearson Education, Inc. Reprinted and electronically reproduced by permission of Pearson Education, Inc. Upper Saddle River, NJ; **38.26c,d** Adapted from *Human Anatomy and Physiology*, 8th ed., by E.N. Marieb and K.N. Hoehn, 2010. Copyright © 2010 by Pearson Education, Inc. Reprinted and electronically reproduced by permission of Pearson Education, Inc. Upper Saddle River, NJ; **38 SSE.1** Data from M. R. Ralph et al., Transplanted suprachiasmatic nucleus determines circadian period, *Science* 247:975–978 (1990).

**Chapter 39** **39.2** Adapted from *Human Anatomy and Physiology*, 4th ed., by E.N. Marieb. Copyright © 1998 by Pearson Education, Inc. Reprinted and electronically reproduced by permission of Pearson Education, Inc., Upper Saddle River, NJ; **39.6** Adapted from *Human Anatomy and Physiology*, 4th ed., by E.N. Marieb. Copyright © 1998 by Pearson Education, Inc. Reprinted and electronically reproduced by permission of Pearson Education, Inc., Upper Saddle River, NJ; **39.11** Grasshopper based on Hickman et al., *Integrated Principles of Zoology*, 9th ed., p. 518, fig. 22.6, McGraw-Hill Higher Education, NY (1993); **39.17** Figure 1 in "*Drosophila*: Genetics Meets Behavior" by Marla B. Sokolowski, *Nature Reviews: Genetics*, November 2001, Volume 2(11); **39.20a,b** Data from N. Tinbergen, *The Study of Instinct*, Clarendon Press, Oxford (1951); **39.22** Adapted from "Evolution of foraging behavior in *Drosophila* by density-dependent selection" by M.B. Sokolowski, H.S. Pereira, and K. Hughes, *Proceedings of the National Academy of Sciences* 94(14):7375 (7/8/97). Copyright © 1997 by National Academy of Sciences, U.S.A. Reprinted with permission; **39 SSE.1** Data from K. Schmidt-Nielsen, Locomotion: Energy cost of swimming, flying, and running, *Science* 177:222–228 (1972).

**Chapter 40** **40.7** Based on H. Walter & S. Walter-Breckle, *Walter's Vegetation of the Earth*, p. 36, fig. 16, Springer-Verlag, Munich (2003); **40.13** Data from W.J. Fletcher, Interactions among subtidal Australian sea urchins, gastropods and algae: Effects of experimental removals, *Ecological Monographs* 57:89–109 (1987); **40.22** Based on J.T. Enright, Climate and population regulation: The biogeographer's dilemma, *Oecologia* 24:295–310 (1976); **40.24** Data courtesy of Rolf O. Peterson, Michigan Technological University; **Table 40.1** Data from P. W. Sherman and M. L. Morton, Demography of Belding's ground squirrel, *Ecology* 65:1617–1628 (1984); **40.EOC07** Data from J. Clausen et al., Experimental studies on the nature of species. III. Environmental responses of climatic races of *Achillea*, Carnegie Institution of Washington Publication No. 581 (1948).

**Chapter 41** **41.2** Adapted from "The Anoles of La Palma: Aspects of Their Ecological Relationships" by A.S. Rand and E.E. Williams, *Breviora* 327:1–19. Copyright © 1969 by Museum of Comparative Zoology, Harvard University. Reprinted with permission; **41.3** Based on J. H. Connell, The influence of interspecific competition and other factors on the distribution of the barnacle *Chthamalus stellatus*, *Ecology* 42:710–723 (1961); **41.4** Data adapted from David Lambert Lack, *Darwin's Finches*, p. 57, Figure 9, and p. 82, Figure 17. Cambridge: Cambridge University Press, 1947; **41 SSE.1** Data from B. L. Phillips and R. Shine, An invasive species induces rapid adaptive change in a native predator: cane toads and black snakes in Australia, *Proceedings of the Royal Society B* 273:1545–1550 (2006); **41.9b** Data from S.D. Hacker and M.D. Bertness, Experimental evidence for factors maintaining plant species diversity in a New England salt marsh, *Ecology* 80:2064–2073(1999); **41.11** Data from Figure 1A in "The diversity and biogeography of soil bacterial communities" by N. Fierer and R.B. Jackson, *Proceedings of the National Academy of Sciences* 103(3):626–631 (1/17/06); **41.14** Based on E.A. Knox, Antarctic marine ecosystems. In *Antarctic Ecology* by M. W. Holdgate (ed.), 69–96, Academic Press, London (1970); **41.15b** Data from R.T. Paine, Food web complexity and species diversity, *American Naturalist* 100:65–75 (1966); **41.18** Data from Figure 2a in "The intermediate disturbance hypothesis, refugia, and biodiversity in streams" by C.R. Townsend, M.R. Scarsbrook, and S. Doledec, *Limnology and Oceanography* 42:938–949 (1997); **41.20** Based on R.L. Crocker & J. Major, Soil development in relation to vegetation and surface age at Glacier Bay, Alaska, *Journal of Ecology* 43:427–448 (1955); **41.22** Data from "Energy and Large-Scale Patterns of Animal- and Plant-Species Richness" by D.J. Currie, *American Naturalist* 137(1): 27–49 (1/91); **41.23** Data from R. H. MacArthur and E. O. Wilson, *The Theory of Island Biogeography*, Princeton University Press, Princeton, NJ (1967); **41.24** Data for mouse, chipmunk, short-tailed shrew, and squirrel (part of "other") from D. Brisson and D.E. Dykhuizen, ospC diversity in *Borrelia burgdorferi*: different hosts are different niches, *Genetics* 168: 713–722 (2004); data for mouse, chipmunk, short-tailed shrew, and "other" from K. LoGiudice et al., The ecology of infectious disease: Effects of host diversity and community composition on Lyme disease risk, *PNAS* 100(2):567–571 (2003); data for masked shrew from D. Brisson, D.E. Dykhuizen, and R.S. Ostfeld, Conspicuous impacts of inconspicuous hosts on the Lyme disease epidemic, *Proceedings of the Royal Society B* 275:227–235 (2008).

**Chapter 42** **42.4** Based on D.L. DeAngelis, *Dynamics of Nutrient Cycling and Food Webs*, Chapman & Hall, New York (1992); **42.7** Data from J. H. Ryther and W. M. Dunstan, Nitrogen, phosphorus, and eutrophication in the coastal marine environment, *Science* 171:1008–1013 (1971); **42.8** Data from Fig. 4.1, p. 82, in *Communities and Ecosystems*, 1e by R.H. Whittaker, Macmillan, New York (1970); **42.12a** Figure based on "The Canadian Intersite Decomposition Experiment: Project and Site Establishment Report" (Information Report BC-X-378) by J. A. Trofymow and the CIDET Working Group. Natural Resources Canada, Canadian Forest Service, Pacific Forestry Centre, 1998; **42.12b** Data from "Litter Decomposition Rates in Canadian Forests" by T. R. Moore, et al., from *Global Change Biology*, Volume 5(1): 75–82; **42.13a** Based on *The Economy of Nature*, 5th ed., by R.E. Ricklefs. Copyright © 2001 by W.H. Freeman and Company. Reprinted with permission; **42.17** Data from "Pilot-Scale in Situ Bioremediation of Uranium in a Highly Contaminated Aquifer. 2. Reduction of U(VI) and Geochemical Control of U(VI) Bioavailability" by Wei-Min Wu et al., *Environmental Science Technology* 40(12):3986–3995 (5/13/06); **42.Table01** Based on data from D.W. Menzel & J.H. Ryther, Nutrients limiting the production of phytoplankton in the Sargasso Sea, with special reference to iron, *Deep Sea Research* 7:276–281 (1961); **42 SSE.1** Data from J. M. Teal, Energy flow in the salt marsh ecosystem of Georgia, *Ecology* 43:614–624 (1962).

**Chapter 43** **p.908 Quote** ("We must consider . . .") G. H. Brundtland quoted in Editorial in *Science*, Vol. 277, July (1997); **43.10** Data from Gene Likens; **43.11** Adapted from Figure 19.1 in *Ecology: The Experimental Analysis of Distribution and Abundance*, 5th Edition, by C.J. Krebs, 2001. Copyright © 2001 by C.J. Krebs. Reprinted and electronically reproduced by permission of Pearson Education, Inc. Upper Saddle River, NJ; **43.12** Data from R. L. Westemeier et al., Tracking the long-term decline and recovery of an isolated population, *Science* 282:1695–1698 (1998); **43.18** Based on Ocean Biogeographic Information System; http://www.iobis.org/node/591; Conservation International Foundation, 2011. Reprinted with permission; **43.25** Based on $CO_2$ data from www.esrl.noaa.gov/gmd/ccgg/trends. Temperature data from www.giss.nasa.gov/gistemps/graphs/Fig.A.lrg.gif; **43.27** Adapted from "How Fast Can Trees Migrate?" by L. Roberts, *Science* 243(4892):735–737 (2/10/89). Copyright © 1989 by AAAS. Reprinted with permission from AAAS. Permission conveyed through Copyright Clearance Center, Inc.; **43.28a (pikas)** Data from J. A. E. Stewart et al., Revisiting the past to foretell the future: summer temperature and habitat area predict pika extirpations in California, *Journal of Biogeography* 42:880–890 (2015). doi:10.1111/jbi.12466; **43.28b (caribou)** Data from Post, *Ecology of climate change—the importance of biotic interactions*, Princeton University Press, Princeton, NJ (2013); **43.28c (sea urchins)** Based on S. D. Ling et al., Reproductive potential of a marine ecosystem engineer at the edge of a newly expanded range, *Global Change Biology* 14:907–915 (2008). doi:10.1111/j.1365-2486.2008.01543.x; **43.29** Based on data from U.S. Census Bureau International Database; **43.30** Based on data from U.S. Census Bureau International Database; **43.31** Ewing B., D. Moore, S. Goldfinger, A. Oursler, A. Reed, and M. Wackernagel. 2010. *The Ecological Footprint Atlas* 2010. Oakland: Global Footprint Network, p. 33 (www.footprintnetwork.org); **43 SSE.1** Data from National Oceanic & Atmospheric Administration, Earth System Research Laboratory, Global Monitoring Division.

**Appendix A** **3.Ans01** © Pearson Education, Inc.; **3.Ans04** Adapted from *Biology: The Science of Life*, 3rd ed., by R. Wallace, G. Sanders, and R. Ferl, 1991. Copyright © 1991 by Pearson Education, Inc. Reprinted and electronically reproduced by permission of Pearson Education, Inc. Upper Saddle River, NJ; **11.Ans01** Data from G. Mendel, Experiments in plant hybridization, *Proceedings of the Natural History Society of Brünn* 4:3–47 (1866); **14.Ans01** Adapted from *The World of the Cell*, 3rd Edition, by W.M. Becker, J.B. Reece, and M.F. Poenie, 1996. Copyright © 1996 by Pearson Education, Inc. Reprinted and electronically reproduced by permission of Pearson Education, Inc. Upper Saddle River, NJ; **37.Ans02** Based on *Cellular Physiology of Nerve and Muscle*, 4th ed., by G. Matthews. Wiley-Blackwell, 2002; **40.Ans01** Based on J. T. Enright, Climate and population regulation: The biogeographer's dilemma, *Oecologia* 24:295–310 (1976).

**Test Bank** **T34.02** © Pearson Education, Inc.; **T37.03** Data from http://www.sciencemag.org/content/342/6157/441; **T37.04** Data from http://www.sciencemag.org/content/342/6157/441; **T37.05** Data from http://www.sciencemag.org/content/342/6157/441.

# Glossary

## Pronunciation Key

Pronounce

| | | |
|---|---|---|
| ā | as in | ace |
| a/ah | | ash |
| ch | | chose |
| ē | | meet |
| e/eh | | bet |
| g | | game |
| ī | | ice |
| i | | hit |
| ks | | box |
| kw | | quick |
| ng | | song |
| ō | | robe |
| o | | ox |
| oy | | boy |
| s | | say |
| sh | | shell |
| th | | thin |
| ū | | boot |
| u/uh | | up |
| z | | zoo |

′ = primary accent
′ = secondary accent

**5′ cap** A modified form of guanine nucleotide added onto the end of a pre-mRNA molecule.

**A site** One of a ribosome's three binding sites for tRNA during translation. The A site holds the tRNA carrying the next amino acid to be added to the polypeptide chain. (A stands for aminoacyl tRNA.)

**ABC hypothesis** A model of flower formation identifying three classes of organ identity genes that direct formation of the four types of floral organs.

**abiotic** (ā′-bī-ot′-ik) Nonliving; referring to the physical and chemical properties of an environment.

**abscisic acid (ABA)** (ab-sis′-ik) A plant hormone that slows growth, often antagonizing the actions of growth hormones. Two of its many effects are to promote seed dormancy and facilitate drought tolerance.

**absorption** The third stage of food processing in animals: the uptake of small nutrient molecules by an organism's body.

**absorption spectrum** The range of a pigment's ability to absorb various wavelengths of light; also a graph of such a range.

**accessory fruit** A fruit, or assemblage of fruits, in which the fleshy parts are derived largely or entirely from tissues other than the ovary.

**acetyl CoA** Acetyl coenzyme A; the entry compound for the citric acid cycle in cellular respiration, formed from a two-carbon fragment of pyruvate attached to a coenzyme.

**acetylcholine** (as′-uh-til-kō′-lēn) One of the most common neurotransmitters; functions by binding to receptors and altering the permeability of the postsynaptic membrane to specific ions, either depolarizing or hyperpolarizing the membrane.

**acid** A substance that increases the hydrogen ion concentration of a solution.

**acrosome** (ak′-ruh-sōm) A vesicle in the tip of a sperm containing hydrolytic enzymes and other proteins that help the sperm reach the egg.

**actin** (ak′-tin) A globular protein that links into chains, two of which twist helically about each other, forming microfilaments (actin filaments) in muscle and other kinds of cells.

**action potential** An electrical signal that propagates (travels) along the membrane of a neuron or other excitable cell as a nongraded (all-or-none) depolarization.

**action spectrum** A graph that profiles the relative effectiveness of different wavelengths of radiation in driving a particular process.

**activation energy** The amount of energy that reactants must absorb before a chemical reaction will start; also called free energy of activation.

**activator** A protein that binds to DNA and stimulates gene transcription. In prokaryotes, activators bind in or near the promoter; in eukaryotes, activators generally bind to control elements in enhancers.

**active site** The specific region of an enzyme that binds the substrate and that forms the pocket in which catalysis occurs.

**active transport** The movement of a substance across a cell membrane against its concentration or electrochemical gradient, mediated by specific transport proteins and requiring an expenditure of energy.

**adaptation** Inherited characteristic of an organism that enhances its survival and reproduction in a specific environment.

**adaptive evolution** A process in which traits that enhance survival or reproduction tend to increase in frequency in a population over time.

**adaptive immunity** A vertebrate-specific defense that is mediated by B lymphocytes (B cells) and T lymphocytes (T cells) and that exhibits specificity, memory, and self-nonself recognition; also called acquired immunity.

**adaptive radiation** Period of evolutionary change in which groups of organisms form many new species whose adaptations allow them to fill different ecological roles in their communities.

**addition rule** A rule of probability stating that the probability of any one of two or more mutually exclusive events occurring can be determined by adding their individual probabilities.

**adenosine triphosphate** *See* ATP (adenosine triphosphate).

**adhesion** The clinging of one substance to another, such as water to plant cell walls, by means of hydrogen bonds.

**aerobic respiration** A catabolic pathway for organic molecules, using oxygen ($O_2$) as the final electron acceptor in an electron transport chain and ultimately producing ATP. This is the most efficient catabolic pathway and is carried out in most eukaryotic cells and many prokaryotic organisms.

**aggregate fruit** A fruit derived from a single flower that has more than one carpel.

**AIDS (acquired immunodeficiency syndrome)** The symptoms and signs present during the late stages of HIV infection, defined by a specified reduction in the number of T cells and the appearance of characteristic secondary infections.

**alcohol fermentation** Glycolysis followed by the reduction of pyruvate to ethyl alcohol, regenerating $NAD^+$ and releasing carbon dioxide.

**alga** (plural, **algae**) A general term for any species of photosynthetic protist, including both unicellular and multicellular forms. Algal species are included in three eukaryote supergroups (Excavata, SAR, and Archaeplastida).

**alimentary canal** (al′-uh-men′-tuh-rē) A complete digestive tract, consisting of a tube running between a mouth and an anus.

**allele** (uh-lē′-ul) Any of the alternative versions of a gene that may produce distinguishable phenotypic effects.

**allergen** An antigen that triggers an exaggerated immune response.

**allopatric speciation** (al′-uh-pat′-rik) The formation of new species in populations that are geographically isolated from one another.

**allopolyploid** (al′-ō-pol′-ē-ployd) A fertile individual that has more than two chromosome sets as a result of two different species interbreeding and combining their chromosomes.

**allosteric regulation** The binding of a regulatory molecule to a protein at one site that affects the function of the protein at a different site.

**alpha (α) helix** (al′-fuh hē′-liks) A coiled region constituting one form of the secondary structure of proteins, arising from a specific pattern of hydrogen bonding between atoms of the polypeptide backbone (not the side chains).

**alternation of generations** A life cycle in which there is both a multicellular diploid form, the sporophyte, and a multicellular haploid form, the gametophyte; characteristic of plants and some algae.

**alternative RNA splicing** A type of eukaryotic gene regulation at the RNA-processing level in which different mRNA molecules are produced from the same primary transcript, depending on which RNA segments are treated as exons and which as introns.

**altruism** (al′-trū-iz-um) Selflessness; behavior that reduces an individual's fitness while increasing the fitness of another individual.

**alveolates** (al-vē'-uh-lets) One of the three major subgroups for which the SAR eukaryotic supergroup is named. This clade arose by secondary endosymbiosis, and its members have membrane-enclosed sacs (alveoli) located just under the plasma membrane.

**alveolus** (al-vē'-uh-lus) (plural, **alveoli**) One of the dead-end air sacs where gas exchange occurs in a mammalian lung.

**amino acid** (uh-mēn'-ō) An organic molecule possessing both a carboxyl and an amino group. Amino acids serve as the monomers of polypeptides.

**amino group** A chemical group consisting of a nitrogen atom bonded to two hydrogen atoms; can act as a base in solution, accepting a hydrogen ion and acquiring a charge of $1+$.

**aminoacyl-tRNA synthetase** An enzyme that joins each amino acid to the appropriate tRNA.

**ammonia** A small, toxic molecule ($NH_3$) produced by nitrogen fixation or as a metabolic waste product of protein and nucleic acid metabolism.

**amniote** (am'-nē-ōt) Member of a clade of tetrapods named for a key derived character, the amniotic egg, which contains specialized membranes, including the fluid-filled amnion, that protect the embryo. Amniotes include mammals as well as birds and other reptiles.

**amniotic egg** An egg that contains specialized membranes that function in protection, nourishment, and gas exchange. The amniotic egg was a major evolutionary innovation, allowing embryos to develop on land in a fluid-filled sac, thus reducing the dependence of tetrapods on water for reproduction.

**amoeba** (uh-mē'-buh) A member of one of several groups of unicellular eukaryotes that have pseudopodia.

**amoebocyte** (uh-mē'-buh-sīt') An amoeba-like cell that moves by pseudopodia and is found in most animals. Depending on the species, it may digest and distribute food, dispose of wastes, form skeletal fibers, fight infections, or change into other cell types.

**amoebozoan** (uh-mē'-buh-zō'-an) A protist in a clade that includes many species with lobe- or tube-shaped pseudopodia.

**amphibian** Member of the tetrapod class Amphibia, including salamanders, frogs, and caecilians.

**amphipathic** (am'-fē-path'-ik) Having both a hydrophilic region and a hydrophobic region.

**amplification** The strengthening of stimulus energy during transduction.

**amylase** (am'-uh-lās') An enzyme that hydrolyzes starch (a glucose polymer from plants) and glycogen (a glucose polymer from animals) into smaller polysaccharides and the disaccharide maltose.

**anabolic pathway** (an'-uh-bol'-ik) A metabolic pathway that consumes energy to synthesize a complex molecule from simpler molecules.

**anaerobic respiration** (an-er-ō'-bik) A catabolic pathway in which inorganic molecules other than oxygen accept electrons at the "downhill" end of electron transport chains.

**analogous** Having characteristics that are similar because of convergent evolution, not homology.

**analogy** (an-al'-uh-jē) Similarity between two species that is due to convergent evolution rather than to descent from a common ancestor with the same trait.

**anaphase** The fourth stage of mitosis, in which the chromatids of each chromosome have separated and the daughter chromosomes are moving to the poles of the cell.

**anatomy** The structure of an organism.

**anchorage dependence** The requirement that a cell must be attached to a substratum in order to initiate cell division.

**aneuploidy** (an'-yū-ploy'-dē) A chromosomal aberration in which one or more chromosomes are present in extra copies or are deficient in number.

**angiosperm** (an'-jē-ō-sperm) A flowering plant, which forms seeds inside a protective chamber called an ovary.

**anion** (an'-ī-on) A negatively charged ion.

**anterior** Pertaining to the front, or head, of a bilaterally symmetric animal.

**anterior pituitary** A portion of the pituitary that develops from non-neural tissue; consists of endocrine cells that synthesize and secrete several tropic and nontropic hormones.

**anther** In an angiosperm, the terminal pollen sac of a stamen, where pollen grains containing sperm-producing male gametophytes form.

**antibody** A protein secreted by plasma cells (differentiated B cells) that binds to a particular antigen; also called immunoglobulin. All antibodies have the same Y-shaped structure and in their monomer form consist of two identical heavy chains and two identical light chains.

**anticodon** (an'-tī-kō'-don) A nucleotide triplet at one end of a tRNA molecule that base-pairs with a particular complementary codon on an mRNA molecule.

**antidiuretic hormone (ADH)** (an'-tī-dī-yū-ret'-ik) A peptide hormone, also known as vasopressin, that promotes water retention by the kidneys. Produced in the hypothalamus and released from the posterior pituitary, ADH also functions in the brain.

**antigen** (an'-ti-jen) A substance that elicits an immune response by binding to receptors of B cells, antibodies, or T cells.

**antigen presentation** The process by which an MHC molecule binds to a fragment of an intracellular protein antigen and carries it to the cell surface, where it is displayed and can be recognized by a T cell.

**antigen receptor** The general term for a surface protein, located on B cells and T cells, that binds to antigens, initiating adaptive immune responses. The antigen receptors on B cells are called B cell receptors, and the antigen receptors on T cells are called T cell receptors.

**antigen-presenting cell** A cell that upon ingesting pathogens or internalizing pathogen proteins generates peptide fragments that are bound by class II MHC molecules and subsequently displayed on the cell surface to T cells. Macrophages, dendritic cells, and B cells are the primary antigen-presenting cells.

**antiparallel** Referring to the arrangement of the sugar-phosphate backbones in a DNA double helix (they run in opposite $5' \rightarrow 3'$ directions).

**aphotic zone** (ā'-fō'-tik) The part of an ocean or lake beneath the photic zone, where light does not penetrate sufficiently for photosynthesis to occur.

**apical bud** (ā'-pik-ul) A bud at the tip of a plant stem; also called a terminal bud.

**apical dominance** (ā'-pik-ul) Tendency for growth to be concentrated at the tip of a plant shoot, because the apical bud partially inhibits axillary bud growth.

**apical meristem** (ā'-pik-ul mār'-uh-stem) Embryonic plant tissue in the tips of roots and buds of shoots. The dividing cells of an apical meristem enable the plant to grow in length.

**apicomplexan** (ap'-ē-kom-pleks'-un) A group of alveolate protists, this clade includes many species that parasitize animals. Some apicomplexans cause human disease.

**apomixis** (ap'-uh-mik'-sis) The ability of some plant species to reproduce asexually through seeds without fertilization by a male gamete.

**apoplast** (ap'-ō-plast) Everything external to the plasma membrane of a plant cell, including cell walls, intercellular spaces, and the space within dead structures such as xylem vessels and tracheids.

**apoptosis** (ā-puh-tō'-sus) A type of programmed cell death that is brought about by activation of enzymes that break down many chemical components in the cell.

**aposematic coloration** (ap'-ō-si-mat'-ik) The bright warning coloration of many animals with effective physical or chemical defenses.

**appendix** A small, finger-like extension of the vertebrate cecum; contains a mass of white blood cells that contribute to immunity.

**aquaporin** A channel protein in the plasma membrane of a plant, animal, or microorganism cell that specifically facilitates osmosis, the diffusion of free water across the membrane.

**aqueous solution** (ā'-kwē-us) A solution in which water is the solvent.

**arbuscular mycorrhizae** (ar-bus'-kyū-lur mī'-kō-rī'-zē) Associations of a fungus with a plant root system in which the fungus causes the invagination of the host (plant) cells' plasma membranes.

**arbuscular mycorrhizal fungus** A symbiotic fungus whose hyphae grow through the cell wall of plant roots and extend into the root cell (enclosed in tubes formed by invagination of the root cell plasma membrane).

**Archaea** (ar'-kē'-uh) One of two prokaryotic domains, the other being Bacteria.

**Archaeplastida** (ar'-kē-plas'-tid-uh) One of four supergroups of eukaryotes proposed in a current hypothesis of the evolutionary history of eukaryotes. This monophyletic group, which includes red algae, green algae, and plants, descended from an ancient protistan ancestor that engulfed a cyanobacterium. *See also* Excavata, SAR, and Unikonta.

**artery** A vessel that carries blood away from the heart to organs throughout the body.

**arthropod** A segmented, molting bilaterian animal with a hard exoskeleton and jointed appendages. Familiar examples include insects, spiders, millipedes, and crabs.

**artificial selection** The selective breeding of domesticated plants and animals to encourage the occurrence of desirable traits.

**asexual reproduction** The generation of offspring from a single parent that occurs without the fusion of gametes (by budding, division of a single cell, or division of the entire organism into two or more parts). In most cases, the offspring are genetically identical to the parent.

**associative learning** The acquired ability to associate one environmental feature (such as a color) with another (such as danger).

**aster** A radial array of short microtubules that extends from each centrosome toward the plasma membrane in an animal cell undergoing mitosis.

**atherosclerosis** A cardiovascular disease in which fatty deposits called plaques develop in the inner walls of the arteries, obstructing the arteries and causing them to harden.

**atom** The smallest unit of matter that retains the properties of an element.

**atomic mass** The total mass of an atom, numerically equivalent to the mass in grams of 1 mole of the atom. (For an element with more than one isotope, the atomic mass is the average mass of the naturally occurring isotopes, weighted by their abundance.)

**atomic nucleus** An atom's dense central core, containing protons and neutrons.

**atomic number** The number of protons in the nucleus of an atom, unique for each element and designated by a subscript.

**ATP (adenosine triphosphate)** (a-den′-ō-sēn trī-fos′-fāt) An adenine-containing nucleoside triphosphate that releases free energy when its phosphate bonds are hydrolyzed. This energy is used to drive endergonic reactions in cells.

**ATP synthase** A complex of several membrane proteins that functions in chemiosmosis with adjacent electron transport chains, using the energy of a hydrogen ion (proton) concentration gradient to make ATP. ATP synthases are found in the inner mitochondrial membranes of eukaryotic cells and in the plasma membranes of prokaryotes.

**atrioventricular (AV) node** A region of specialized heart muscle tissue between the left and right atria where electrical impulses are delayed for about 0.1 second before spreading to both ventricles and causing them to contract.

**atrioventricular (AV) valve** A heart valve located between each atrium and ventricle that prevents a backflow of blood when the ventricle contracts.

**atrium** (ā′-trē-um) (plural, **atria**) A chamber of the vertebrate heart that receives blood from the veins and transfers blood to a ventricle.

**autoimmune disease** An immunological disorder in which the immune system turns against self.

**autonomic nervous system** (ot′-ō-nom′-ik) An efferent branch of the vertebrate peripheral nervous system that regulates the internal environment; consists of the sympathetic, parasympathetic, and enteric divisions.

**autopolyploid** (ot′-ō-pol′-ē-ployd) An individual that has more than two chromosome sets that are all derived from a single species.

**autosome** (ot′-ō-sōm) A chromosome that is not directly involved in determining sex; not a sex chromosome.

**autotroph** (ot′-ō-trōf) An organism that obtains organic food molecules without eating other organisms or substances derived from other organisms. Autotrophs use energy from the sun or from oxidation of inorganic substances to make organic molecules from inorganic ones.

**auxin** (ôk′-sin) A term that primarily refers to indoleacetic acid (IAA), a natural plant hormone that has a variety of effects, including cell elongation, root formation, secondary growth, and fruit growth.

**axillary bud** (ak′-sil-ār-ē) A structure that has the potential to form a lateral shoot, or branch. The bud appears in the angle formed between a leaf and a stem.

**axon** (ak′-son) A typically long extension, or process, of a neuron that carries nerve impulses away from the cell body toward target cells.

**B cells** The lymphocytes that complete their development in the bone marrow and become effector cells for the humoral immune response.

**Bacteria** One of two prokaryotic domains, the other being Archaea.

**bacteriophage** (bak-tēr′-ē-ō-fāj) A virus that infects bacteria; also called a phage.

**bacteroid** A form of the bacterium *Rhizobium* contained within the vesicles formed by the root cells of a root nodule.

**balancing selection** Natural selection that maintains two or more phenotypic forms in a population.

**bar graph** A graph in which the independent variable represents groups or nonnumerical categories. Each piece of data is represented by a bar, whose height (or length) represents the value of the independent variable for the group or category indicated.

**bark** All tissues external to the vascular cambium, consisting mainly of the secondary phloem and layers of periderm.

**Barr body** A dense object lying along the inside of the nuclear envelope in cells of female mammals, representing a highly condensed, inactivated X chromosome.

**basal body** (bā′-sul) A eukaryotic cell structure consisting of a "9 + 0" arrangement of microtubule triplets. The basal body may organize the microtubule assembly of a cilium or flagellum and is structurally very similar to a centriole.

**basal taxon** In a specified group of organisms, a taxon whose evolutionary lineage diverged early in the history of the group.

**base** A substance that reduces the hydrogen ion concentration of a solution.

**Batesian mimicry** (bāt′-zē-un mim′-uh-krē) A type of mimicry in which a harmless species resembles an unpalatable or harmful species to which it is not closely related.

**behavior** Individually, an action carried out by muscles or glands under control of the nervous system in response to a stimulus; collectively, the sum of an animal's responses to external and internal stimuli.

**behavioral ecology** The study of the evolution of and ecological basis for animal behavior.

**benign tumor** A mass of abnormal cells with specific genetic and cellular changes such that the cells are not capable of surviving at a new site and generally remain at the site of the tumor's origin.

**benthic zone** The bottom surface of an aquatic environment.

**benthos** (ben′-thōz) The communities of organisms living in the benthic zone of an aquatic biome.

**beta (β) pleated sheet** One form of the secondary structure of proteins in which the polypeptide chain folds back and forth. Two regions of the chain lie parallel to each other and are held together by hydrogen bonds between atoms of the polypeptide backbone (not the side chains).

**beta oxidation** A metabolic sequence that breaks fatty acids down to two-carbon fragments that enter the citric acid cycle as acetyl CoA.

*bicoid* A maternal effect gene that codes for a protein responsible for specifying the anterior end in *Drosophila melanogaster*.

**bilateral symmetry** Body symmetry in which a central longitudinal plane divides the body into two equal but opposite halves.

**bilaterian** (bī′-luh-ter′-ē-uhn) Member of a clade of animals with bilateral symmetry and three germ layers.

**bile** A mixture of substances that is produced in the liver and stored in the gallbladder; enables formation of fat droplets in water as an aid in the digestion and absorption of fats.

**binary fission** A method of asexual reproduction in single-celled organisms in which the cell grows to roughly double its size and then divides into two cells. In prokaryotes, binary fission does not involve mitosis, but in single-celled eukaryotes that undergo binary fission, mitosis is part of the process.

**binomial** A common term for the two-part, latinized format for naming a species, consisting of the genus and specific epithet; also called a binomen.

**biodiversity hot spot** A relatively small area with numerous endemic species and a large number of endangered and threatened species.

**bioenergetics** (1) The overall flow and transformation of energy in an organism. (2) The study of how energy flows through organisms.

**biofilm** A surface-coating colony of one or more species of prokaryotes that engage in metabolic cooperation.

**biofuel** A fuel produced from biomass.

**biogeochemical cycle** Any of the various chemical cycles that involve both biotic and abiotic components of ecosystems.

**biogeography** The scientific study of the past and present geographic distributions of species.

**bioinformatics** The use of computers, software, and mathematical models to process and integrate biological information from large data sets.

**biological augmentation** An approach to restoration ecology that uses organisms to add essential materials to a degraded ecosystem.

**biological clock** An internal timekeeper that controls an organism's biological rhythms. The biological clock marks time with or without environmental cues but often requires signals from the environment to remain tuned to an appropriate period. *See also* circadian rhythm.

**biological magnification** A process in which retained substances become more concentrated at each higher trophic level in a food chain.

**biological species concept** Definition of a species as a group of populations whose members have the potential to interbreed in nature and produce viable, fertile offspring, but do not produce viable, fertile offspring with members of other such groups.

**biology** The scientific study of life.

**biomanipulation** An approach that applies the top-down model of community organization to alter ecosystem characteristics. For example, ecologists can prevent algal blooms and eutrophication by altering the density of higher-level consumers in lakes instead of by using chemical treatments.

**biomass** The total mass of organic matter comprising a group of organisms in a particular habitat.

**biome** (bī'-ōm) Any of the world's major ecosystem types, often classified according to the predominant vegetation for terrestrial biomes and the physical environment for aquatic biomes and characterized by adaptations of organisms to that particular environment.

**bioremediation** The use of organisms to detoxify and restore polluted and degraded ecosystems.

**biosphere** The entire portion of Earth inhabited by life; the sum of all the planet's ecosystems.

**biotic** (bī-ot'-ik) Pertaining to the living factors—the organisms—in an environment.

**blade** (1) A leaflike structure of a seaweed that provides most of the surface area for photosynthesis. (2) The flattened portion of a typical leaf.

**blastocoel** (blas'-tuh-sēl) The fluid-filled cavity that forms in the center of a blastula.

**blastula** (blas'-tyū-luh) A hollow ball of cells that marks the end of the cleavage stage during early embryonic development in animals.

**blood** A connective tissue with a fluid matrix called plasma in which red blood cells, white blood cells, and cell fragments called platelets are suspended.

**blue-light photoreceptor** A type of light receptor in plants that initiates a variety of responses, including phototropism and slowing of hypocotyl elongation.

**body cavity** A fluid- or air-filled space located between the digestive tract and the outer body wall; also called a coelom.

**body plan** In multicellular eukaryotes, a set of morphological and developmental traits that are integrated into a functional whole—the living organism.

**Bohr shift** A lowering of the affinity of hemoglobin for oxygen, caused by a drop in pH. It facilitates the release of oxygen from hemoglobin in the vicinity of active tissues.

**bolus** A lubricated ball of chewed food.

**bottleneck effect** Genetic drift that occurs when the size of a population is reduced, as by a natural disaster or human actions. Typically, the surviving population is no longer genetically representative of the original population.

**bottom-up model** A model of community organization in which mineral nutrients influence community organization by controlling plant or phytoplankton numbers, which in turn control herbivore numbers, which in turn control predator numbers.

**Bowman's capsule** (bō'-munz) A cup-shaped receptacle in the vertebrate kidney that is the initial, expanded segment of the nephron where filtrate enters from the blood.

**brain** Organ of the central nervous system where information is processed and integrated.

**brainstem** A collection of structures in the vertebrate brain, including the midbrain, the pons, and the medulla oblongata; functions in homeostasis, coordination of movement, and conduction of information to higher brain centers.

**branch point** The representation on a phylogenetic tree of the divergence of two or more taxa from a common ancestor. A branch point is usually shown as a dichotomy in which a branch representing the ancestral lineage splits (at the branch point) into two branches, one for each of the two descendant lineages.

**brassinosteroid** A steroid hormone in plants that has a variety of effects, including inducing cell elongation, retarding leaf abscission, and promoting xylem differentiation.

**breathing** Ventilation of the lungs through alternating inhalation and exhalation.

**bronchus** (brong'-kus) (plural, **bronchi**) One of a pair of breathing tubes that branch from the trachea into the lungs.

**brown alga** A multicellular, photosynthetic protist with a characteristic brown or olive color that results from carotenoids in its plastids. Most brown algae are marine, and some have a plantlike body.

**bryophyte** (brī'-uh-fīt) An informal name for a moss, liverwort, or hornwort; a nonvascular plant that lives on land but lacks some of the terrestrial adaptations of vascular plants.

**buffer** A solution that contains a weak acid and its corresponding base. A buffer minimizes changes in pH when acids or bases are added to the solution.

**bulk feeder** An animal that eats relatively large pieces of food.

**bulk flow** The movement of a fluid due to a difference in pressure between two locations.

**C₃ plant** A plant that uses the Calvin cycle for the initial steps that incorporate $CO_2$ into organic material, forming a three-carbon compound as the first stable intermediate.

**C₄ plant** A plant in which the Calvin cycle is preceded by reactions that incorporate $CO_2$ into a four-carbon compound, the end product of which supplies $CO_2$ for the Calvin cycle.

**callus** A mass of dividing, undifferentiated cells growing in culture.

**calorie (cal)** The amount of heat energy required to raise the temperature of 1 g of water by 1°C; also the amount of heat energy that 1 g of water releases when it cools by 1°C. The Calorie (with a capital C), usually used to indicate the energy content of food, is a kilocalorie.

**Calvin cycle** The second of two major stages in photosynthesis (following the light reactions), involving fixation of atmospheric $CO_2$ and reduction of the fixed carbon into carbohydrate.

**CAM plant** A plant that uses crassulacean acid metabolism, an adaptation for photosynthesis in arid conditions. In this process, carbon dioxide entering open stomata during the night is converted to organic acids, which release $CO_2$ for the Calvin cycle during the day, when stomata are closed.

**Cambrian explosion** A relatively brief time in geologic history when many present-day phyla of animals first appeared in the fossil record. This burst of evolutionary change occurred about 535–525 million years ago and saw the emergence of the first large, hard-bodied animals.

**canopy** The uppermost layer of vegetation in a terrestrial biome.

**capillary** (kap'-il-ār'-ē) A microscopic blood vessel that penetrates the tissues and consists of a single layer of endothelial cells that allows exchange between the blood and interstitial fluid.

**capillary bed** A network of capillaries in a tissue or organ.

**capsid** The protein shell that encloses a viral genome. It may be rod-shaped, polyhedral, or more complex in shape.

**capsule** (1) In many prokaryotes, a dense and well-defined layer of polysaccharide or protein that surrounds the cell wall and is sticky, protecting the cell and enabling it to adhere to substrates or other cells. (2) The sporangium of a bryophyte (moss, liverwort, or hornwort).

**carbohydrate** (kar'-bō-hī'-drāt) A sugar (monosaccharide) or one of its dimers (disaccharides) or polymers (polysaccharides).

**carbon fixation** The initial incorporation of carbon from $CO_2$ into an organic compound by an autotrophic organism (a plant, another photosynthetic organism, or a chemoautotrophic prokaryote).

**carbonyl group** (kar-buh-nēl') A chemical group present in aldehydes and ketones and consisting of a carbon atom double-bonded to an oxygen atom.

**carboxyl group** (kar-bok'-sil) A chemical group present in organic acids and consisting of a single carbon atom double-bonded to an oxygen atom and also bonded to a hydroxyl group.

**cardiac cycle** (kar'-dē-ak) The alternating contractions and relaxations of the heart.

**cardiac muscle** A type of striated muscle that forms the contractile wall of the heart. Its cells are joined by intercalated disks that relay the electrical signals underlying each heartbeat.

**cardiovascular system** A closed circulatory system with a heart and branching network of arteries, capillaries, and veins. The system is characteristic of vertebrates.

**carnivore** An animal that mainly eats other animals.

**carotenoid** (kuh-rot′-uh-noyd′) An accessory pigment, either yellow or orange, in the chloroplasts of plants and in some prokaryotes. By absorbing wavelengths of light that chlorophyll cannot, carotenoids broaden the spectrum of colors that can drive photosynthesis.

**carpel** (kar′-pul) The ovule-producing reproductive organ of a flower, consisting of the stigma, style, and ovary.

**carrier** In genetics, an individual who is heterozygous at a given genetic locus for a recessively inherited disorder. The heterozygote is generally phenotypically normal for the disorder but can pass on the recessive allele to offspring.

**carrying capacity** The maximum population size that can be supported by the available resources, symbolized as $K$.

**Casparian strip** (ka-spär′-ē-un) A water-impermeable ring of wax in the endodermal cells of plants that blocks the passive flow of water and solutes into the stele by way of cell walls.

**catabolic pathway** (kat′-uh-bol′-ik) A metabolic pathway that releases energy by breaking down complex molecules to simpler molecules.

**catalysis** (kuh-ta′-luh-sis) A process by which a chemical agent called a catalyst selectively increases the rate of a reaction without being consumed by the reaction.

**catalyst** (kat′-uh-list) A chemical agent that selectively increases the rate of a reaction without being consumed by the reaction.

**cation** (cat′-ī′-on) A positively charged ion.

**cation exchange** A process in which positively charged minerals are made available to a plant when hydrogen ions in the soil displace mineral ions from the clay particles.

**cecum** (sē′-kum) (plural, **ceca**) The blind pouch forming one branch of the large intestine.

**cell body** The part of a neuron that houses the nucleus and most other organelles.

**cell cycle** An ordered sequence of events in the life of a cell, from its origin in the division of a parent cell until its own division into two. The eukaryotic cell cycle is composed of interphase (including $G_1$, S, and $G_2$ subphases) and M phase (including mitosis and cytokinesis).

**cell cycle control system** A cyclically operating set of molecules in the eukaryotic cell that both triggers and coordinates key events in the cell cycle.

**cell division** The reproduction of cells.

**cell fractionation** The disruption of a cell and separation of its parts by centrifugation at successively higher speeds.

**cell plate** A membrane-bounded, flattened sac located at the midline of a dividing plant cell, inside which the new cell wall forms during cytokinesis.

**cell wall** A protective layer external to the plasma membrane in the cells of plants, prokaryotes, fungi, and some protists. Polysaccharides such as cellulose (in plants and some protists), chitin (in fungi), and peptidoglycan (in bacteria) are important structural components of cell walls.

**cell-mediated immune response** The branch of adaptive immunity that involves the activation of cytotoxic T cells, which defend against infected cells.

**cellular respiration** The catabolic pathways of aerobic and anaerobic respiration, which break down organic molecules and use an electron transport chain for the production of ATP.

**cellulose** (sel′-yū-lōs) A structural polysaccharide of plant cell walls, consisting of glucose monomers joined by β glycosidic linkages.

**central nervous system (CNS)** The portion of the nervous system where signal integration occurs; in vertebrate animals, the brain and spinal cord.

**central vacuole** In a mature plant cell, a large membranous sac with diverse roles in growth, storage, and sequestration of toxic substances.

**centriole** (sen′-trē-ōl) A structure in the centrosome of an animal cell composed of a cylinder of microtubule triplets arranged in a 9 + 0 pattern. A centrosome has a pair of centrioles.

**centromere** (sen′-trō-mēr) In a duplicated chromosome, the region on each sister chromatid where it is most closely attached to its sister chromatid by proteins that bind to specific DNA sequences; this close attachment causes a constriction in the condensed chromosome. (An uncondensed, unduplicated chromosome has a single centromere, identified by its DNA sequence.)

**centrosome** (sen′-trō-sōm) A structure present in the cytoplasm of animal cells that functions as a microtubule-organizing center and is important during cell division. A centrosome has two centrioles.

**cercozoan** An amoeboid or flagellated protist that feeds with threadlike pseudopodia.

**cerebellum** (sār′-ruh-bel′-um) Part of the vertebrate hindbrain located dorsally; functions in unconscious coordination of movement and balance.

**cerebral cortex** (suh-rē′-brul) The surface of the cerebrum; the largest and most complex part of the mammalian brain, containing nerve cell bodies of the cerebrum; the part of the vertebrate brain most changed through evolution.

**cerebrum** (suh-rē′-brum) The dorsal portion of the vertebrate forebrain, composed of right and left hemispheres; the integrating center for memory, learning, emotions, and other highly complex functions of the central nervous system.

**cervix** (ser′-viks) The neck of the uterus, which opens into the vagina.

**chaparral** A scrubland biome of dense, spiny evergreen shrubs found at midlatitudes along coasts where cold ocean currents circulate offshore; characterized by mild, rainy winters and long, hot, dry summers.

**character** An observable heritable feature that may vary among individuals.

**character displacement** The tendency for characteristics to be more divergent in sympatric populations of two species than in allopatric populations of the same two species.

**checkpoint** A control point in the cell cycle where stop and go-ahead signals can regulate the cycle.

**chemical bond** An attraction between two atoms, resulting from a sharing of outer-shell electrons or the presence of opposite charges on the atoms. The bonded atoms gain complete outer electron shells.

**chemical energy** Energy available in molecules for release in a chemical reaction; a form of potential energy.

**chemical equilibrium** In a chemical reaction, the state in which the rate of the forward reaction equals the rate of the reverse reaction, so that the relative concentrations of the reactants and products do not change with time.

**chemical reaction** The making and breaking of chemical bonds, leading to changes in the composition of matter.

**chemiosmosis** (kem′-ē-oz-mō′-sis) An energy-coupling mechanism that uses energy stored in the form of a hydrogen ion gradient across a membrane to drive cellular work, such as the synthesis of ATP. Under aerobic conditions, most ATP synthesis in cells occurs by chemiosmosis.

**chemoreceptor** A sensory receptor that responds to a chemical stimulus, such as a solute or an odorant.

**chiasma** (plural, **chiasmata**) (kī-az′-muh, kī-az′-muh-tuh) The X-shaped, microscopically visible region where crossing over has occurred earlier in prophase I between homologous nonsister chromatids. Chiasmata become visible after synapsis ends, with the two homologs remaining associated due to sister chromatid cohesion.

**chitin** (kī′-tin) A structural polysaccharide, consisting of amino sugar monomers, found in many fungal cell walls and in the exoskeletons of all arthropods.

**chlorophyll** (klōr′-ō-fil) A green pigment located in membranes within the chloroplasts of plants and algae and in the membranes of certain prokaryotes. Chlorophyll *a* participates directly in the light reactions, which convert solar energy to chemical energy.

**chlorophyll *a*** A photosynthetic pigment that participates directly in the light reactions, which convert solar energy to chemical energy.

**chlorophyll *b*** An accessory photosynthetic pigment that transfers energy to chlorophyll *a*.

**chloroplast** (klōr′-ō-plast) An organelle found in plants and photosynthetic protists that absorbs sunlight and uses it to drive the synthesis of organic compounds from carbon dioxide and water.

**choanocyte** (kō-an′-uh-sīt) A flagellated feeding cell found in sponges. Also called a collar cell, it has a collar-like ring that traps food particles around the base of its flagellum.

**cholesterol** (kō-les′-tuh-rol) A steroid that forms an essential component of animal cell membranes and acts as a precursor molecule for the synthesis of other biologically important steroids, such as many hormones.

**chondrichthyan** (kon-drik′-thē-an) Member of the class Chondrichthyes, vertebrates with skeletons made mostly of cartilage, such as sharks and rays.

**chordate** Member of the phylum Chordata, animals that at some point during their development have a notochord; a dorsal, hollow nerve cord; pharyngeal slits or clefts; and a muscular, post-anal tail.

**chromatin** (krō'-muh-tin) The complex of DNA and proteins that makes up eukaryotic chromosomes. When the cell is not dividing, chromatin exists in its dispersed form, as a mass of very long, thin fibers that are not visible with a light microscope.

**chromosome** (krō'-muh-sōm) A cellular structure consisting of one DNA molecule and associated protein molecules. (In some contexts, such as genome sequencing, the term may refer to the DNA alone.) A eukaryotic cell typically has multiple, linear chromosomes, which are located in the nucleus. A prokaryotic cell often has a single, circular chromosome, which is found in the nucleoid, a region that is not enclosed by a membrane. *See also* chromatin.

**chromosome theory of inheritance** A basic principle in biology stating that genes are located at specific positions (loci) on chromosomes and that the behavior of chromosomes during meiosis accounts for inheritance patterns.

**chylomicron** (kī'-lō-mī'-kron) A lipid transport globule composed of fats mixed with cholesterol and coated with proteins.

**chyme** (kīm) The mixture of partially digested food and digestive juices formed in the stomach.

**ciliate** (sil'-ē-it) A type of protist that moves by means of cilia.

**cilium** (sil'-ē-um) (plural, **cilia**) A short appendage containing microtubules in eukaryotic cells. A motile cilium is specialized for locomotion or moving fluid past the cell; it is formed from a core of nine outer doublet microtubules and two inner single microtubules (the "9 + 2" arrangement) ensheathed in an extension of the plasma membrane. A primary cilium is usually nonmotile and plays a sensory and signaling role; it lacks the two inner microtubules (the "9 + 0" arrangement).

**circadian rhythm** (ser-kā'-dē-un) A physiological cycle of about 24 hours that persists even in the absence of external cues.

**cis-trans isomer** One of two or more compounds that have the same molecular formula and covalent bonds between atoms but differ in the spatial arrangements of their atoms owing to the inflexibility of double bonds; formerly called a geometric isomer.

**citric acid cycle** A chemical cycle involving eight steps that completes the metabolic breakdown of glucose molecules begun in glycolysis by oxidizing acetyl CoA (derived from pyruvate) to carbon dioxide; occurs within the mitochondrion in eukaryotic cells and in the cytosol of prokaryotes; together with pyruvate oxidation, the second major stage in cellular respiration.

**clade** (klād) A group of species that includes an ancestral species and all of its descendants.

**cladistics** (kluh-dis'-tiks) An approach to systematics in which organisms are placed into groups called clades based primarily on common descent.

**class** In Linnaean classification, the taxonomic category above the level of order.

**cleavage** (1) The process of cytokinesis in animal cells, characterized by pinching of the plasma membrane. (2) The succession of rapid cell divisions without significant growth during early embryonic development that converts the zygote to a ball of cells.

**cleavage furrow** The first sign of cleavage in an animal cell; a shallow groove around the cell in the cell surface near the old metaphase plate.

**climate** The long-term prevailing weather conditions at a given place.

**climate change** A directional change in temperature, precipitation, or other aspect of the global climate that lasts for three decades or more.

**climograph** A plot of the temperature and precipitation in a particular region.

**clitoris** (klit'-uh-ris) An organ at the upper intersection of the labia minora that engorges with blood and becomes erect during sexual arousal.

**cloaca** (klō-ā'-kuh) A common opening for the digestive, urinary, and reproductive tracts found in many nonmammalian vertebrates but in few mammals.

**clonal selection** The process by which an antigen selectively binds to and activates only those lymphocytes bearing receptors specific for the antigen. The selected lymphocytes proliferate and differentiate into a clone of effector cells and a clone of memory cells specific for the stimulating antigen.

**clone** (1) A lineage of genetically identical individuals or cells. (2) In popular usage, an individual that is genetically identical to another individual. (3) As a verb, to make one or more genetic replicas of an individual or cell. *See also* gene cloning.

**cloning vector** In genetic engineering, a DNA molecule that can carry foreign DNA into a host cell and replicate there. Cloning vectors include plasmids.

**closed circulatory system** A circulatory system in which blood is confined to vessels and is kept separate from the interstitial fluid.

**cochlea** (kok'-lē-uh) The complex, coiled organ of hearing that contains the organ of Corti.

**codominance** The situation in which the phenotypes of both alleles are exhibited in the heterozygote because both alleles affect the phenotype in separate, distinguishable ways.

**codon** (kō'-don) A three-nucleotide sequence of DNA or mRNA that specifies a particular amino acid or termination signal; the basic unit of the genetic code.

**coefficient of relatedness** The fraction of genes that, on average, are shared by two individuals.

**coenzyme** (kō-en'-zīm) An organic molecule serving as a cofactor. Most vitamins function as coenzymes in metabolic reactions.

**coevolution** The joint evolution of two interacting species, each in response to selection imposed by the other.

**cofactor** Any nonprotein molecule or ion that is required for the proper functioning of an enzyme. Cofactors can be permanently bound to the active site or may bind loosely and reversibly, along with the substrate, during catalysis.

**cognition** The process of knowing that may include awareness, reasoning, recollection, and judgment.

**cognitive map** A neural representation of the abstract spatial relationships between objects in an animal's surroundings.

**cohesion** The linking together of like molecules, often by hydrogen bonds.

**cohesion-tension hypothesis** The leading explanation for the ascent of xylem sap. It states that transpiration exerts pull on xylem sap, putting the sap under negative pressure or tension, and that the cohesion of water molecules transmits this pull along the entire length of the xylem from shoots to roots.

**cohort** A group of individuals of the same age in a population.

**coleoptile** (kō-lē-op'-tul) The covering of the young shoot of the embryo of a grass seed.

**coleorhiza** (kō-lē-uh-rī'-zuh) The covering of the young root of the embryo of a grass seed.

**collagen** A glycoprotein in the extracellular matrix of animal cells that forms strong fibers, found extensively in connective tissue and bone; the most abundant protein in the animal kingdom.

**collecting duct** The location in the kidney where processed filtrate, called urine, is collected from the renal tubules.

**collenchyma cell** (kō-len'-kim-uh) A flexible plant cell type that occurs in strands or cylinders that support young parts of the plant without restraining growth.

**colon** (kō'-len) The largest section of the vertebrate large intestine; functions in water absorption and formation of feces.

**commensalism** (kuh-men'-suh-lizm) A +/0 ecological interaction in which one organism benefits but the other is neither helped nor harmed.

**communication** In animal behavior, a process involving transmission of, reception of, and response to signals. The term is also used in connection with other organisms, as well as individual cells of multicellular organisms.

**community** All the organisms that inhabit a particular area; an assemblage of populations of different species living close enough together for potential interaction.

**community ecology** The study of how interactions between species affect community structure and organization.

**companion cell** A type of plant cell that is connected to a sieve-tube element by many plasmodesmata and whose nucleus and ribosomes may serve one or more adjacent sieve-tube elements.

**competitive exclusion** The concept that when populations of two similar species compete for the same limited resources, one population will use the resources more efficiently and have a reproductive advantage that will eventually lead to the elimination of the other population.

**competitive inhibitor** A substance that reduces the activity of an enzyme by entering the active site in place of the substrate, whose structure it mimics.

**complement system** A group of about 30 blood proteins that may amplify the inflammatory response, enhance phagocytosis, or directly lyse extracellular pathogens.

**complementary DNA (cDNA)** A double-stranded DNA molecule made *in vitro* using mRNA as a template and the enzymes reverse transcriptase and DNA polymerase. A cDNA molecule corresponds to the exons of a gene.

**complete dominance** The situation in which the phenotypes of the heterozygote and dominant homozygote are indistinguishable.

**complete flower** A flower that has all four basic floral organs: sepals, petals, stamens, and carpels.

**compound** A substance consisting of two or more different elements combined in a fixed ratio.

**compound eye** A type of multifaceted eye in insects and crustaceans consisting of up to several thousand light-detecting, focusing ommatidia.

**concentration gradient** A region along which the density of a chemical substance increases or decreases.

**conception** The fertilization of an egg by a sperm in humans.

**cone** A cone-shaped cell in the retina of the vertebrate eye, sensitive to color.

**conformer** An animal for which an internal condition conforms to (changes in accordance with) changes in an environmental variable.

**conifer** Member of the largest gymnosperm phylum. Most conifers are cone-bearing trees, such as pines and firs.

**conjugation** (kon′-jū-gā′-shun) (1) In prokaryotes, the direct transfer of DNA between two cells that are temporarily joined. When the two cells are members of different species, conjugation results in horizontal gene transfer. (2) In ciliates, a sexual process in which two cells exchange haploid micronuclei but do not reproduce.

**connective tissue** Animal tissue that functions mainly to bind and support other tissues, having a sparse population of cells scattered through an extracellular matrix.

**conservation biology** The integrated study of ecology, evolutionary biology, physiology, molecular biology, and genetics to sustain biological diversity at all levels.

**contraception** The deliberate prevention of pregnancy.

**contractile vacuole** A membranous sac that helps move excess water out of certain freshwater protists.

**control element** A segment of noncoding DNA that helps regulate transcription of a gene by serving as a binding site for a transcription factor. Multiple control elements are present in a eukaryotic gene's enhancer.

**control group** In a controlled experiment, a set of subjects that lacks (or does not receive) the specific factor being tested. Ideally, the control group should be identical to the experimental group in other respects.

**controlled experiment** An experiment designed to compare an experimental group with a control group; ideally, the two groups differ only in the factor being tested.

**convergent evolution** The evolution of similar features in independent evolutionary lineages.

**cooperativity** A kind of allosteric regulation whereby a shape change in one subunit of a protein caused by substrate binding is transmitted to all the other subunits, facilitating binding of additional substrate molecules to those subunits.

**coral reef** Typically a warm-water, tropical ecosystem dominated by the hard skeletal structures secreted primarily by corals. Some coral reefs also exist in cold, deep waters.

**corepressor** A small molecule that binds to a bacterial repressor protein and changes the protein's shape, allowing it to bind to the operator and switch an operon off.

**cork cambium** (kam′-bē-um) A cylinder of meristematic tissue in woody plants that replaces the epidermis with thicker, tougher cork cells.

**corpus callosum** (kor′-pus kuh-lō′-sum) The thick band of nerve fibers that connects the right and left cerebral hemispheres in mammals, enabling the hemispheres to process information together.

**corpus luteum** (kor′-pus lū′-tē-um) A secreting tissue in the ovary that forms from the collapsed follicle after ovulation and produces progesterone.

**cortex** (1) The outer region of cytoplasm in a eukaryotic cell, lying just under the plasma membrane, that has a more gel-like consistency than the inner regions due to the presence of multiple microfilaments. (2) In plants, ground tissue that is between the vascular tissue and dermal tissue in a root or eudicot stem.

**cortical nephron** In mammals and birds, a nephron with a loop of Henle located almost entirely in the renal cortex.

**cotransport** The coupling of the "downhill" diffusion of one substance to the "uphill" transport of another against its own concentration gradient.

**countercurrent exchange** The exchange of a substance or heat between two fluids flowing in opposite directions. For example, blood in a fish gill flows in the opposite direction of water passing over the gill, maximizing diffusion of oxygen into and carbon dioxide out of the blood.

**countercurrent multiplier system** A countercurrent system in which energy is expended in active transport to facilitate exchange of materials and generate concentration gradients.

**covalent bond** (kō-vā′-lent) A type of strong chemical bond in which two atoms share one or more pairs of valence electrons.

**crassulacean acid metabolism (CAM)** (crass-yū-lā′-shen) An adaptation for photosynthesis in arid conditions, first discovered in the family Crassulaceae. In this process, a plant takes up $CO_2$ at night when stomata are open and incorporates it into a variety of organic acids; during the day, when stomata are closed, $CO_2$ is released from the organic acids for use in the Calvin cycle.

**crista** (plural, **cristae**) (kris′-tuh, kris′-tē) An infolding of the inner membrane of a mitochondrion. The inner membrane houses electron transport chains and molecules of the enzyme catalyzing the synthesis of ATP (ATP synthase).

**critical load** The amount of added nutrient, usually nitrogen or phosphorus, that can be absorbed by plants without damaging ecosystem integrity.

**cross-fostering study** A behavioral study in which the young of one species are placed in the care of adults from another species.

**crossing over** The reciprocal exchange of genetic material between nonsister chromatids during prophase I of meiosis.

**cryptic coloration** Camouflage that makes a potential prey difficult to spot against its background.

**culture** A system of information transfer through social learning or teaching that influences the behavior of individuals in a population.

**cuticle** (kyū′-tuh-kul) (1) A waxy covering on the surface of stems and leaves that prevents desiccation in terrestrial plants. (2) The exoskeleton of an arthropod, consisting of layers of protein and chitin that are variously modified for different functions. (3) A tough coat that covers the body of a nematode.

**cyclic AMP (cAMP)** Cyclic adenosine monophosphate, a ring-shaped molecule made from ATP that is a common intracellular signaling molecule (second messenger) in eukaryotic cells. It is also a regulator of some bacterial operons.

**cyclostome** (sī′-cluh-stōm) Member of one of the two main clades of vertebrates; cyclostomes lack jaws and include lampreys and hagfishes. *See also* gnathostome.

**cystic fibrosis** (sis′-tik fī-brō′-sis) A human genetic disorder caused by a recessive allele for a chloride channel protein; characterized by an excessive secretion of mucus and consequent vulnerability to infection; fatal if untreated.

**cytochrome** (sī′-tō-krōm) An iron-containing protein that is a component of electron transport chains in the mitochondria and chloroplasts of eukaryotic cells and the plasma membranes of prokaryotic cells.

**cytogenetic map** A map of a chromosome that locates genes with respect to chromosomal features distinguishable in a microscope.

**cytokine** (sī′-tō-kīn′) Any of a group of small proteins secreted by a number of cell types, including macrophages and helper T cells, that regulate the function of other cells.

**cytokinesis** (sī′-tō-kuh-nē′-sis) The division of the cytoplasm to form two separate daughter cells immediately after mitosis, meiosis I, or meiosis II.

**cytokinin** (sī′-tō-kī′-nin) Any of a class of related plant hormones that retard aging and act in concert with auxin to stimulate cell division, influence the pathway of differentiation, and control apical dominance.

**cytoplasm** (sī′-tō-plaz′-um) The contents of the cell enclosed by the plasma membrane;

in eukaryotes, the portion exclusive of the nucleus.

**cytoplasmic determinant** A maternal substance, such as a protein or RNA, that when placed into an egg influences the course of early development by regulating the expression of genes that affect the developmental fate of cells.

**cytoskeleton** A network of microtubules, microfilaments, and intermediate filaments that extends throughout the cytoplasm and serves a variety of mechanical, transport, and signaling functions.

**cytosol** (sī'-tō-sol) The semifluid portion of the cytoplasm.

**cytotoxic T cell** A type of lymphocyte that, when activated, kills infected cells as well as certain cancer cells and transplanted cells.

**dalton** A measure of mass for atoms and subatomic particles; the same as the atomic mass unit, or amu.

**data** Recorded observations.

**day-neutral plant** A plant in which flower formation is not controlled by photoperiod or day length.

**decomposer** An organism that absorbs nutrients from nonliving organic material such as corpses, fallen plant material, and the wastes of living organisms and converts them to inorganic forms; a detritivore.

**deductive reasoning** A type of logic in which specific results are predicted from a general premise.

**deep-sea hydrothermal vent** A dark, hot, oxygen-deficient environment associated with volcanic activity on or near the seafloor. The producers in a vent community are chemoautotrophic prokaryotes.

**de-etiolation** The changes a plant shoot undergoes in response to sunlight; also known informally as greening.

**dehydration reaction** A chemical reaction in which two molecules become covalently bonded to each other with the removal of a water molecule.

**deletion** (1) A deficiency in a chromosome resulting from the loss of a fragment through breakage. (2) A mutational loss of one or more nucleotide pairs from a gene.

**demography** The study of changes over time in the vital statistics of populations, especially birth rates and death rates.

**denaturation** (dē-nā'-chur-ā'-shun) In proteins, a process in which a protein loses its native shape due to the disruption of weak chemical bonds and interactions, thereby becoming biologically inactive; in DNA, the separation of the two strands of the double helix. Denaturation occurs under extreme (noncellular) conditions of pH, salt concentration, or temperature.

**dendrite** (den'-drīt) One of usually numerous, short, highly branched extensions of a neuron that receive signals from other neurons.

**density** The number of individuals per unit area or volume.

**density dependent** Referring to any characteristic that varies with population density.

**density independent** Referring to any characteristic that is not affected by population density.

**density-dependent inhibition** The phenomenon observed in normal animal cells that causes them to stop dividing when they come into contact with one another.

**deoxyribonucleic acid (DNA)** (dē-ok'-sē-rī'-bō-nū-klā'-ik) A nucleic acid molecule, usually a double-stranded helix, in which each polynucleotide strand consists of nucleotide monomers with a deoxyribose sugar and the nitrogenous bases adenine (A), cytosine (C), guanine (G), and thymine (T); capable of being replicated and determining the inherited structure of a cell's proteins.

**deoxyribose** (dē-ok'-si-rī'-bōs) The sugar component of DNA nucleotides, having one fewer hydroxyl group than ribose, the sugar component of RNA nucleotides.

**dependent variable** A factor whose value is measured in an experiment to see whether it is influenced by changes in another factor (the independent variable).

**depolarization** A change in a cell's membrane potential such that the inside of the membrane is made less negative relative to the outside. For example, a neuron membrane is depolarized if a stimulus decreases its voltage from the resting potential of −70 mV in the direction of zero voltage.

**dermal tissue system** The outer protective covering of plants.

**desert** A terrestrial biome characterized by very low precipitation.

**desmosome** A type of intercellular junction in animal cells that functions as a rivet, fastening cells together.

**determinate growth** A type of growth characteristic of most animals and some plant organs, in which growth stops after a certain size is reached.

**determination** The progressive restriction of developmental potential whereby the possible fate of each cell becomes more limited as an embryo develops. At the end of determination, a cell is committed to its fate.

**detritivore** (deh-trī'-tuh-vōr) A consumer that derives its energy and nutrients from nonliving organic material such as corpses, fallen plant material, and the wastes of living organisms; a decomposer.

**detritus** (di-trī'-tus) Dead organic matter.

**diabetes mellitus** (dī'-uh-bē'-tis mel'-uh-tus) An endocrine disorder marked by an inability to maintain glucose homeostasis. The type 1 form results from autoimmune destruction of insulin-secreting cells; treatment usually requires daily insulin injections. The type 2 form most commonly results from reduced responsiveness of target cells to insulin; obesity and lack of exercise are risk factors.

**diaphragm** (dī'-uh-fram') (1) A sheet of muscle that forms the bottom wall of the thoracic cavity in mammals. Contraction of the diaphragm pulls air into the lungs. (2) A dome-shaped rubber cup fitted into the upper portion of the vagina before sexual intercourse. It serves as a physical barrier to the passage of sperm into the uterus.

**diastole** (dī-as'-tō-lē) The stage of the cardiac cycle in which a heart chamber is relaxed and fills with blood.

**diatom** A photosynthetic protist in the stramenopile clade; diatoms have a unique glasslike wall made of silicon dioxide embedded in an organic matrix.

**differential gene expression** The expression of different sets of genes by cells with the same genome.

**differentiation** The process by which a cell or group of cells becomes specialized in structure and function.

**diffusion** The random thermal motion of particles of liquids, gases, or solids. In the presence of a concentration or electrochemical gradient, diffusion results in the net movement of a substance from a region where it is more concentrated to a region where it is less concentrated.

**digestion** The second stage of food processing in animals: the breaking down of food into molecules small enough for the body to absorb.

**dihybrid** (dī'-hī'-brid) An organism that is heterozygous with respect to two genes of interest. All the offspring from a cross between parents doubly homozygous for different alleles are dihybrids. For example, parents of genotypes *AABB* and *aabb* produce a dihybrid of genotype *AaBb*.

**dihybrid cross** A cross between two organisms that are each heterozygous for both of the characters being followed (or the self-pollination of a plant that is heterozygous for both characters).

**dinoflagellate** (dī'-nō-flaj'-uh-let) Member of a group of mostly unicellular photosynthetic algae with two flagella situated in perpendicular grooves in cellulose plates covering the cell.

**dioecious** (dī-ē'-shus) In plant biology, having the male and female reproductive parts on different individuals of the same species.

**diploid cell** (dip'-loyd) A cell containing two sets of chromosomes (2*n*), one set inherited from each parent.

**diplomonad** A protist that has modified mitochondria and multiple flagella.

**directional selection** Natural selection in which individuals at one end of the phenotypic range survive or reproduce more successfully than do other individuals.

**disaccharide** (dī-sak'-uh-rīd) A double sugar, consisting of two monosaccharides joined by a glycosidic linkage formed by a dehydration reaction.

**dispersal** The movement of individuals or gametes away from their parent location. This movement sometimes expands the geographic range of a population or species.

**dispersion** The pattern of spacing among individuals within the boundaries of a population.

**disruptive selection** Natural selection in which individuals on both extremes of a phenotypic range survive or reproduce more successfully than do individuals with intermediate phenotypes.

**distal tubule** In the vertebrate kidney, the portion of a nephron that helps refine filtrate and empties it into a collecting duct.

**disturbance** A natural or human-caused event that changes a biological community and usually removes organisms from it. Disturbances, such as fires and storms, play a pivotal role in structuring many communities.

**disulfide bridge** A strong covalent bond formed when the sulfur of one cysteine monomer bonds to the sulfur of another cysteine monomer.

**DNA (deoxyribonucleic acid)** (dē-ok′-sē-rī′-bō-nū-klā′-ik) A nucleic acid molecule, usually a double-stranded helix, in which each polynucleotide strand consists of nucleotide monomers with a deoxyribose sugar and the nitrogenous bases adenine (A), cytosine (C), guanine (G), and thymine (T); capable of being replicated and determining the inherited structure of a cell's proteins.

**DNA ligase** (lī′-gās) A linking enzyme essential for DNA replication; catalyzes the covalent bonding of the 3′ end of one DNA fragment (such as an Okazaki fragment) to the 5′ end of another DNA fragment (such as a growing DNA chain).

**DNA methylation** The presence of methyl groups on the DNA bases (usually cytosine) of plants, animals, and fungi. (The term also refers to the process of adding methyl groups to DNA bases.)

**DNA microarray assay** A method to detect and measure the expression of thousands of genes at one time. Tiny amounts of a large number of single-stranded DNA fragments representing different genes are fixed to a glass slide and tested for hybridization with samples of labeled cDNA.

**DNA polymerase** (puh-lim′-er-ās) An enzyme that catalyzes the elongation of new DNA (for example, at a replication fork) by the addition of nucleotides to the 3′ end of an existing chain. There are several different DNA polymerases; DNA polymerase III and DNA polymerase I play major roles in DNA replication in *E. coli.*

**DNA replication** The process by which a DNA molecule is copied; also called DNA synthesis.

**DNA sequencing** Determining the order of nucleotide bases in a gene or DNA fragment.

**domain** (1) A taxonomic category above the kingdom level. The three domains are Archaea, Bacteria, and Eukarya. (2) A discrete structural and functional region of a protein.

**dominant allele** An allele that is fully expressed in the phenotype of a heterozygote.

**dominant species** A species with substantially higher abundance or biomass than other species in a community. Dominant species exert a powerful control over the occurrence and distribution of other species.

**dormancy** A condition typified by extremely low metabolic rate and a suspension of growth and development.

**dorsal** Pertaining to the top of an animal with radial or bilateral symmetry.

**double bond** A double covalent bond; the sharing of two pairs of valence electrons by two atoms.

**double circulation** A circulatory system consisting of separate pulmonary and systemic circuits, in which blood passes through the heart after completing each circuit.

**double fertilization** A mechanism of fertilization in angiosperms in which two sperm cells unite with two cells in the female gametophyte (embryo sac) to form the zygote and endosperm.

**double helix** The form of native DNA, referring to its two adjacent antiparallel polynucleotide strands wound around an imaginary axis into a spiral shape.

**Down syndrome** A human genetic disease usually caused by the presence of an extra chromosome 21; characterized by developmental delays and heart and other defects that are generally treatable or non-life-threatening.

**Duchenne muscular dystrophy** (duh-shen′) A human genetic disease caused by a sex-linked recessive allele; characterized by progressive weakening and a loss of muscle tissue.

**duodenum** (dū′-uh-dēn′-um) The first section of the small intestine, where chyme from the stomach mixes with digestive juices from the pancreas, liver, and gallbladder as well as from gland cells of the intestinal wall.

**duplication** An aberration in chromosome structure due to fusion with a fragment from a homologous chromosome, such that a portion of a chromosome is duplicated.

**dynein** (dī′-nē-un) In cilia and flagella, a large motor protein extending from one microtubule doublet to the adjacent doublet. ATP hydrolysis drives changes in dynein shape that lead to bending of cilia and flagella.

**E site** One of a ribosome's three binding sites for tRNA during translation. The E site is the place where discharged tRNAs leave the ribosome. (E stands for exit.)

**ecological footprint** The aggregate land and water area required by a person, city, or nation to produce all of the resources it consumes and to absorb all of the waste it generates.

**ecological niche** (nich) The sum of a species' use of the biotic and abiotic resources in its environment.

**ecological species concept** Definition of a species in terms of ecological niche, the sum of how members of the species interact with the nonliving and living parts of their environment.

**ecological succession** Transition in the species composition of a community following a disturbance; establishment of a community in an area virtually barren of life.

**ecology** The study of how organisms interact with each other and their environment.

**ecosystem** All the organisms in a given area as well as the abiotic factors with which they interact; one or more communities and the physical environment around them.

**ecosystem ecology** The study of energy flow and the cycling of chemicals among the various biotic and abiotic components in an ecosystem.

**ecosystem engineer** An organism that influences community structure by causing physical changes in the environment.

**ecosystem service** A function performed by an ecosystem that directly or indirectly benefits humans.

**ecotone** The transition from one type of habitat or ecosystem to another, such as the transition from a forest to a grassland.

**ectoderm** (ek′-tō-durm) The outermost of the three primary germ layers in animal embryos; gives rise to the outer covering and, in some phyla, the nervous system, inner ear, and lens of the eye.

**ectomycorrhizae** (ek′-tō-mī′-kō-rī′-zē) Associations of a fungus with a plant root system in which the fungus surrounds the roots but does not cause invagination of the host (plant) cell's plasma membrane.

**ectomycorrhizal fungus** A symbiotic fungus that forms sheaths of hyphae over the surface of plant roots and also grows into extracellular spaces of the root cortex.

**ectoparasite** A parasite that feeds on the external surface of a host.

**ectothermic** Referring to organisms for which external sources provide most of the heat for temperature regulation.

**Ediacaran biota** (ē′-dē-uh-keh′-run bī-ō′-tuh) An early group of macroscopic, soft-bodied, multicellular eukaryotes known from fossils that range in age from 635 million to 535 million years old.

**effective population size** An estimate of the size of a population based on the numbers of females and males that successfully breed; generally smaller than the total population.

**effector** A pathogen-encoded protein that cripples a plant's innate immune system.

**effector cell** (1) A muscle cell or gland cell that performs the body's response to stimuli as directed by signals from the brain or other processing center of the nervous system. (2) A lymphocyte that has undergone clonal selection and is capable of mediating an adaptive immune response.

**egg** The female gamete.

**ejaculation** The propulsion of sperm from the epididymis through the muscular vas deferens, ejaculatory duct, and urethra.

**ejaculatory duct** In mammals, the short section of the ejaculatory route formed by the convergence of the vas deferens and a duct from the seminal vesicle. The ejaculatory duct transports sperm from the vas deferens to the urethra.

**electrocardiogram (ECG or EKG)** A record of the electrical impulses that travel through heart muscle during the cardiac cycle.

**electrochemical gradient** The diffusion gradient of an ion, which is affected by both the concentration difference of an ion across a membrane (a chemical force) and the ion's tendency to move relative to the membrane potential (an electrical force).

**electrogenic pump** An active transport protein that generates voltage across a membrane while pumping ions.

**electromagnetic receptor** A receptor of electromagnetic energy, such as visible light, electricity, or magnetism.

**electromagnetic spectrum** The entire spectrum of electromagnetic radiation, ranging in wavelength from less than a nanometer to more than a kilometer.

**electron** A subatomic particle with a single negative electrical charge and a mass about 1/2,000 that of a neutron or proton. One or more electrons move around the nucleus of an atom.

**electron microscope (EM)** A microscope that uses magnets to focus an electron beam on or through a specimen, resulting in a practical resolution 100-fold greater than that of a light microscope using standard techniques. A transmission electron microscope (TEM) is used to study the internal structure of thin sections of cells. A scanning electron microscope (SEM) is used to study the fine details of cell surfaces.

**electron shell** An energy level of electrons at a characteristic average distance from the nucleus of an atom.

**electron transport chain** A sequence of electron carrier molecules (membrane proteins) that shuttle electrons down a series of redox reactions that release energy used to make ATP.

**electronegativity** The attraction of a given atom for the electrons of a covalent bond.

**element** Any substance that cannot be broken down to any other substance by chemical reactions.

**elimination** The fourth and final stage of food processing in animals: the passing of undigested material out of the body.

**embryo sac** (em′-brē-ō) The female gametophyte of angiosperms, formed from the growth and division of the megaspore into a multicellular structure that typically has eight haploid nuclei.

**embryonic lethal** A mutation with a phenotype leading to death of an embryo or larva.

**embryophyte** Alternate name for plants that refers to their shared derived trait of multicellular, dependent embryos.

**emergent properties** New properties that arise with each step upward in the hierarchy of life, owing to the arrangement and interactions of parts as complexity increases.

**emigration** The movement of individuals out of a population.

**enantiomer** One of two compounds that are mirror images of each other and that differ in shape due to the presence of an asymmetric carbon.

**endangered species** A species that is in danger of extinction throughout all or a significant portion of its range.

**endemic** (en-dem′-ik) Referring to a species that is confined to a specific geographic area.

**endergonic reaction** (en′-der-gon′-ik) A nonspontaneous chemical reaction, in which free energy is absorbed from the surroundings.

**endocrine gland** (en′-dō-krin) A gland that secretes hormones directly into the interstitial fluid, from which they diffuse into the bloodstream.

**endocrine system** The internal system of communication involving hormones, the ductless glands that secrete hormones, and the molecular receptors on or in target cells that respond to hormones; functions in concert with the nervous system to effect internal regulation and maintain homeostasis.

**endocytosis** (en′-dō-sī-tō′-sis) Cellular uptake of biological molecules and particulate matter via formation of vesicles from the plasma membrane.

**endoderm** (en′-dō-durm) The innermost of the three primary germ layers in animal embryos; lines the archenteron and gives rise to the liver, pancreas, lungs, and the lining of the digestive tract in species that have these structures.

**endodermis** In plant roots, the innermost layer of the cortex that surrounds the vascular cylinder.

**endomembrane system** The collection of membranes inside and surrounding a eukaryotic cell, related either through direct physical contact or by the transfer of membranous vesicles; includes the plasma membrane, the nuclear envelope, the smooth and rough endoplasmic reticulum, the Golgi apparatus, lysosomes, vesicles, and vacuoles.

**endometrium** (en′-dō-mē′-trē-um) The inner lining of the uterus, which is richly supplied with blood vessels.

**endoparasite** A parasite that lives within a host.

**endophyte** A fungus that lives inside a leaf or other plant part without causing harm to the plant.

**endoplasmic reticulum (ER)** (en′-dō-plaz′-mik ruh-tik′-yū-lum) An extensive membranous network in eukaryotic cells, continuous with the outer nuclear membrane and composed of ribosome-studded (rough) and ribosome-free (smooth) regions.

**endorphin** (en-dōr′-fin) Any of several hormones produced in the brain and anterior pituitary that inhibit pain perception.

**endoskeleton** A hard skeleton buried within the soft tissues of an animal.

**endosperm** In angiosperms, a nutrient-rich tissue formed by the union of a sperm with two polar nuclei during double fertilization. The endosperm provides nourishment to the developing embryo in angiosperm seeds.

**endospore** A thick-coated, resistant cell produced by some bacterial cells when they are exposed to harsh conditions.

**endosymbiont theory** The theory that mitochondria and plastids, including chloroplasts, originated as prokaryotic cells engulfed by host cells. The engulfed cell and its host cell then evolved into a single organism. *See also* endosymbiosis.

**endosymbiosis** A mutually beneficial relationship between two species in which one organism lives inside the cell or cells of another organism.

**endothelium** (en′-dō-thē′-lē-um) The simple squamous layer of cells lining the lumen of blood vessels.

**endothermic** Referring to organisms that are warmed by heat generated by their own metabolism. This heat usually maintains a relatively stable body temperature higher than that of the external environment.

**endotoxin** A toxic component of the outer membrane of certain gram-negative bacteria that is released only when the bacteria die.

**energy** The capacity to cause change, especially to do work (to move matter against an opposing force).

**energy coupling** In cellular metabolism, the use of energy released from an exergonic reaction to drive an endergonic reaction.

**enhancer** A segment of eukaryotic DNA containing multiple control elements, usually located far from the gene whose transcription it regulates.

**enteric division** One of three divisions of the autonomic nervous system; consists of networks of neurons in the digestive tract, pancreas, and gallbladder; normally regulated by the sympathetic and parasympathetic divisions of the autonomic nervous system.

**entropy** A measure of disorder, or randomness.

**enzyme** (en′-zīm) A macromolecule serving as a catalyst, a chemical agent that increases the rate of a reaction without being consumed by the reaction. Most enzymes are proteins.

**enzyme-substrate complex** A temporary complex formed when an enzyme binds to its substrate molecule(s).

**epicotyl** (ep′-uh-kot′-ul) In an angiosperm embryo, the embryonic axis above the point of attachment of the cotyledon(s) and below the first pair of miniature leaves.

**epidemic** A general outbreak of a disease.

**epidermis** (1) The dermal tissue system of nonwoody plants, usually consisting of a single layer of tightly packed cells. (2) The outermost layer of cells in an animal.

**epididymis** (ep′-uh-did′-uh-mus) A coiled tubule located adjacent to the mammalian testis where sperm are stored.

**epigenetic inheritance** Inheritance of traits transmitted by mechanisms not directly involving the nucleotide sequence of a genome.

**epinephrine** (ep′-i-nef′-rin) A catecholamine that, when secreted as a hormone by the adrenal medulla, mediates "fight-or-flight" responses to short-term stresses; also released by some neurons as a neurotransmitter; also known as adrenaline.

**epiphyte** (ep′-uh-fīt) A plant that nourishes itself but grows on the surface of another plant for support, usually on the branches or trunks of trees.

**epistasis** (ep′-i-stā′-sis) A type of gene interaction in which the phenotypic expression of one gene alters that of another independently inherited gene.

**epithelial tissue** (ep′-uh-thē′-lē-ul) Sheets of tightly packed cells that line organs and body cavities as well as external surfaces; also called epithelium.

**epithelium** An epithelial tissue.

**epitope** A small, accessible region of an antigen to which an antigen receptor or antibody binds; also called an antigenic determinant.

**equilibrium potential ($E_{ion}$)** The magnitude of a cell's membrane voltage at equilibrium, calculated using the Nernst equation.

**erythrocyte** (eh-rith'-ruh-sīt) A blood cell that contains hemoglobin, which transports oxygen; also called a red blood cell.

**esophagus** (eh-sof'-uh-gus) A muscular tube that conducts food, by peristalsis, from the pharynx to the stomach.

**essential amino acid** An amino acid that an animal cannot synthesize itself and must be obtained from food in prefabricated form.

**essential element** A chemical element required for an organism to survive, grow, and reproduce.

**essential fatty acid** An unsaturated fatty acid that an animal needs but cannot make.

**essential nutrient** A substance that an organism cannot synthesize from any other material and therefore must absorb in preassembled form.

**estradiol** (es'-truh-dī'-ol) A steroid hormone that stimulates the development and maintenance of the female reproductive system and secondary sex characteristics; the major estrogen in mammals.

**estrogen** (es'-trō-jen) Any steroid hormone, such as estradiol, that stimulates the development and maintenance of the female reproductive system and secondary sex characteristics.

**estrous cycle** (es'-trus) A reproductive cycle characteristic of female mammals except humans and certain other primates, in which the nonpregnant endometrium is reabsorbed rather than shed, and sexual response occurs only during mid-cycle at estrus.

**estuary** The area where a freshwater stream or river merges with the ocean.

**ethylene** (eth'-uh-lēn) A gaseous plant hormone involved in responses to mechanical stress, programmed cell death, leaf abscission; and fruit ripening.

**etiolation** Plant morphological adaptations for growing in darkness.

**euchromatin** (yū-krō'-muh-tin) The less condensed form of eukaryotic chromatin that is available for transcription.

**euglenozoan** Member of a diverse clade of flagellated protists that includes predatory heterotrophs, photosynthetic autotrophs, and pathogenic parasites.

**Eukarya** (yū-kar'-ē-uh) The domain that includes all eukaryotic organisms.

**eukaryotic cell** (yū'-ker-ē-ot'-ik) A type of cell with a membrane-enclosed nucleus and membrane-enclosed organelles. Organisms with eukaryotic cells (protists, plants, fungi, and animals) are called eukaryotes.

**eumetazoan** (yū'-met-uh-zō'-un) Member of a clade of animals with true tissues. All animals except sponges and a few other groups are eumetazoans.

**Eustachian tube** (yū-stā'-shun) The tube that connects the middle ear to the pharynx.

**eutherian** (yū-thēr'-ē-un) Placental mammal; mammal whose young complete their embryonic development within the uterus, joined to the mother by the placenta.

**eutrophic lake** (yū-trōf'-ik) A lake that has a high rate of biological productivity supported by a high rate of nutrient cycling.

**eutrophication** A process by which nutrients, particularly phosphorus and nitrogen, become highly concentrated in a body of water, leading to increased growth of organisms such as algae or cyanobacteria.

**evaporative cooling** The process in which the surface of an object becomes cooler during evaporation, a result of the molecules with the greatest kinetic energy changing from the liquid to the gaseous state.

**evapotranspiration** The total evaporation of water from an ecosystem, including water transpired by plants and evaporated from a landscape, usually measured in millimeters and estimated for a year.

**evo-devo** Evolutionary developmental biology; a field of biology that compares developmental processes of different multicellular organisms to understand how these processes have evolved and how changes can modify existing organismal features or lead to new ones.

**evolution** Descent with modification; the idea that living species are descendants of ancestral species that were different from the present-day ones; also defined more narrowly as the change in the genetic composition of a population from generation to generation.

**Excavata** One of four supergroups of eukaryotes proposed in a current hypothesis of the evolutionary history of eukaryotes. Excavates have unique cytoskeletal features, and some species have an "excavated" feeding groove on one side of the cell body. *See also* SAR, Archaeplastida, and Unikonta.

**excitatory postsynaptic potential (EPSP)** An electrical change (depolarization) in the membrane of a postsynaptic cell caused by the binding of an excitatory neurotransmitter from a presynaptic cell to a postsynaptic receptor; makes it more likely for a postsynaptic cell to generate an action potential.

**excretion** The disposal of nitrogen-containing metabolites and other waste products.

**exergonic reaction** (ek'-ser-gon'-ik) A spontaneous chemical reaction, in which there is a net release of free energy.

**exocrine gland** (ek'-sō-krin) A gland that secretes substances through a duct onto a body surface or into a body cavity.

**exocytosis** (ek'-sō-sī-tō'-sis) The cellular secretion of biological molecules by the fusion of vesicles containing them with the plasma membrane.

**exon** A sequence within a primary transcript that remains in the RNA after RNA processing; also refers to the region of DNA from which this sequence was transcribed.

**exoskeleton** A hard encasement on the surface of an animal, such as the shell of a mollusc or the cuticle of an arthropod, that provides protection and points of attachment for muscles.

**exotoxin** (ek'-sō-tok'-sin) A toxic protein that is secreted by a prokaryote or other pathogen and that produces specific symptoms, even if the pathogen is no longer present.

**expansin** Plant enzyme that breaks the cross-links (hydrogen bonds) between cellulose microfibrils and other cell wall constituents, loosening the wall's fabric.

**experiment** A scientific test. Often carried out under controlled conditions that involve manipulating one factor in a system in order to see the effects of changing that factor.

**experimental group** A set of subjects that has (or receives) the specific factor being tested in a controlled experiment. Ideally, the experimental group should be identical to the control group for all other factors.

**exploitation** A +/− ecological interaction in which one species benefits by feeding on the other species, which is harmed. Exploitative interactions include predation, herbivory, and parasitism.

**exponential population growth** Growth of a population in an ideal, unlimited environment, represented by a J-shaped curve when population size is plotted over time.

**extinction vortex** A downward population spiral in which inbreeding and genetic drift combine to cause a small population to shrink and, unless the spiral is reversed, become extinct.

**extracellular matrix (ECM)** The meshwork surrounding animal cells, consisting of glycoproteins, polysaccharides, and proteoglycans synthesized and secreted by the cells.

**extreme halophile** An organism that lives in a highly saline environment, such as the Great Salt Lake or the Dead Sea.

**extreme thermophile** An organism that thrives in hot environments (often 60–80°C or hotter).

**extremophile** An organism that lives in environmental conditions so extreme that few other species can survive there. Extremophiles include extreme halophiles ("salt lovers") and extreme thermophiles ("heat lovers").

**F factor** In bacteria, the DNA segment that confers the ability to form pili for conjugation and associated functions required for the transfer of DNA from donor to recipient. The F factor may exist as a plasmid or be integrated into the bacterial chromosome.

**F plasmid** The plasmid form of the F factor.

**F$_1$ generation** The first filial, hybrid (heterozygous) offspring arising from a parental (P generation) cross.

**F$_2$ generation** The offspring resulting from interbreeding (or self-pollination) of the hybrid F$_1$ generation.

**facilitated diffusion** The passage of molecules or ions down their electrochemical gradient across a biological membrane with the assistance of specific transmembrane transport proteins, requiring no energy expenditure.

**facultative anaerobe** (fak'-ul-tā'-tiv an'-uh-rōb) An organism that makes ATP by aerobic respiration if oxygen is present but that switches to anaerobic respiration or fermentation if oxygen is not present.

**family** In Linnaean classification, the taxonomic category above genus.

**fast-twitch fiber** A muscle fiber used for rapid, powerful contractions.

**fat** A lipid consisting of three fatty acids linked to one glycerol molecule; also called a triacylglycerol or triglyceride.

**fatty acid** A carboxylic acid with a long carbon chain. Fatty acids vary in length and in the number and location of double bonds; three fatty acids linked to a glycerol molecule form a fat molecule, also known as a triacylglycerol or triglyceride.

**feces** (fē'-sēz) The wastes of the digestive tract.

**feedback inhibition** A method of metabolic control in which the end product of a metabolic pathway acts as an inhibitor of an enzyme within that pathway.

**fermentation** A catabolic process that makes a limited amount of ATP from glucose (or other organic molecules) without an electron transport chain and that produces a characteristic end product, such as ethyl alcohol or lactic acid.

**fertilization** (1) The union of haploid gametes to produce a diploid zygote. (2) The addition of mineral nutrients to the soil.

**fetus** (fē'-tus) A developing mammal that has all the major structures of an adult. In humans, the fetal stage lasts from the 9th week of gestation until birth.

**fiber** A lignified cell type that reinforces the xylem of angiosperms and functions in mechanical support; a slender, tapered sclerenchyma cell that usually occurs in bundles.

**fibronectin** An extracellular glycoprotein secreted by animal cells that helps them attach to the extracellular matrix.

**filter feeder** An animal that feeds by using a filtration mechanism to strain small organisms or food particles from its surroundings.

**filtrate** Cell-free fluid extracted from the body fluid by the excretory system.

**filtration** In excretory systems, the extraction of water and small solutes, including metabolic wastes, from the body fluid.

**fimbria** (plural, **fimbriae**) A short, hairlike appendage of a prokaryotic cell that helps it adhere to the substrate or to other cells.

**first law of thermodynamics** The principle of conservation of energy: Energy can be transferred and transformed, but it cannot be created or destroyed.

**fixed action pattern** In animal behavior, a sequence of unlearned acts that is essentially unchangeable and, once initiated, usually carried to completion.

**flaccid** (flas'-id) Limp. Lacking turgor (stiffness or firmness), as in a plant cell in surroundings where there is a tendency for water to leave the cell. (A walled cell becomes flaccid if it has a higher water potential than its surroundings, resulting in the loss of water.)

**flagellum** (fluh-jel'-um) (plural, **flagella**) A long cellular appendage specialized for locomotion. Like motile cilia, eukaryotic flagella have a core with nine outer doublet microtubules and two inner single microtubules (the "9 + 2" arrangement) ensheathed in an extension of the plasma membrane. Prokaryotic flagella have a different structure.

**florigen** A flowering signal, probably a protein, that is made in leaves under certain conditions and that travels to the shoot apical meristems, inducing them to switch from vegetative to reproductive growth.

**flower** In an angiosperm, a specialized shoot with up to four sets of modified leaves, bearing structures that function in sexual reproduction.

**fluid feeder** An animal that lives by sucking nutrient-rich fluids from another living organism.

**fluid mosaic model** The currently accepted model of cell membrane structure, which envisions the membrane as a mosaic of protein molecules drifting laterally in a fluid bilayer of phospholipids.

**follicle** (fol'-uh-kul) A microscopic structure in the ovary that contains the developing oocyte and secretes estrogens.

**follicle-stimulating hormone (FSH)** A tropic hormone that is produced and secreted by the anterior pituitary and that stimulates the production of eggs by the ovaries and sperm by the testes.

**follicular phase** That part of the ovarian cycle during which follicles are growing and oocytes maturing.

**food chain** The pathway along which food energy is transferred from trophic level to trophic level, beginning with producers.

**food vacuole** A membranous sac formed by phagocytosis of microorganisms or particles to be used as food by the cell.

**food web** The interconnected feeding relationships in an ecosystem.

**foot** One of the three main parts of a mollusc; a muscular structure usually used for movement. *See also* mantle and visceral mass.

**foraging** The seeking and obtaining of food.

**foram (foraminiferan)** An aquatic protist that secretes a hardened shell containing calcium carbonate and extends pseudopodia through pores in the shell.

**fossil** A preserved remnant or impression of an organism that lived in the past.

**founder effect** Genetic drift that occurs when a few individuals become isolated from a larger population and form a new population whose gene pool composition is not reflective of that of the original population.

**fovea** (fō'-vē-uh) The place on the retina at the eye's center of focus, where cones are highly concentrated.

**fragmentation** A means of asexual reproduction whereby a single parent breaks into parts that regenerate into whole new individuals.

**frameshift mutation** A mutation occurring when nucleotides are inserted in or deleted from a gene and the number inserted or deleted is not a multiple of three, resulting in the improper grouping of the subsequent nucleotides into codons.

**free energy** The portion of a biological system's energy that can perform work when temperature and pressure are uniform throughout the system. The change in free energy of a system ($\Delta G$) is $G_{\text{final state}} - G_{\text{initial state}}$. It can be calculated by the equation $\Delta G = \Delta H - T\Delta S$, where $\Delta H$ is the change in enthalpy (in biological systems, equivalent to total energy), $T$ is the absolute temperature, and $\Delta S$ is the change in entropy.

**frequency-dependent selection** Selection in which the fitness of a phenotype depends on how common the phenotype is in a population.

**fruit** A mature ovary of a flower. The fruit protects dormant seeds and often aids in their dispersal.

**functional group** A specific configuration of atoms commonly attached to the carbon skeletons of organic molecules and involved in chemical reactions.

**fusion** In evolutionary biology, a process in which gene flow between two species that can form hybrid offspring weakens barriers to reproduction between the species. This process causes their gene pools to become increasingly alike and can cause the two species to fuse into a single species.

**G protein** A GTP-binding protein that relays signals from a plasma membrane signal receptor, known as a G protein-coupled receptor, to other signal transduction proteins inside the cell.

**G protein-coupled receptor (GPCR)** A signal receptor protein in the plasma membrane that responds to the binding of a signaling molecule by activating a G protein. Also called a G protein-linked receptor.

**$G_0$ phase** A nondividing state occupied by cells that have left the cell cycle, sometimes reversibly.

**$G_1$ phase** The first gap, or growth phase, of the cell cycle, consisting of the portion of interphase before DNA synthesis begins.

**$G_2$ phase** The second gap, or growth phase, of the cell cycle, consisting of the portion of interphase after DNA synthesis occurs.

**gallbladder** An organ that stores bile and releases it as needed into the small intestine.

**gamete** (gam'-ēt) A haploid reproductive cell, such as an egg or sperm. Gametes unite during sexual reproduction to produce a diploid zygote.

**gametogenesis** The process by which gametes are produced.

**gametophyte** (guh-mē'-tō-fīt) In organisms (plants and some algae) that have alternation of generations, the multicellular haploid form that produces haploid gametes by mitosis. The haploid gametes unite and develop into sporophytes.

**ganglion** (gang'-glē-uhn) (plural, **ganglia**) A cluster (functional group) of nerve cell bodies.

**gap junction** A type of intercellular junction in animal cells, consisting of proteins surrounding a pore that allows the passage of materials between cells.

**gas exchange** The uptake of molecular oxygen from the environment and the discharge of carbon dioxide to the environment.

**gas exchange circuit** The branch of the circulatory system that supplies the organs where gases are exchanged with the environment; in many amphibians, it supplies the lungs and skin and is called a *pulmocutaneous circuit*, whereas in birds and mammals, it supplies only the lungs and is called a *pulmonary circuit*.

**gastric juice** A digestive fluid secreted by the stomach.

**gastrovascular cavity** A central cavity with a single opening in the body of certain animals, including cnidarians and flatworms, that functions in both the digestion and distribution of nutrients.

**gastrula** (gas'-trū-luh) An embryonic stage in animal development encompassing the formation of three layers: ectoderm, mesoderm, and endoderm.

**gastrulation** (gas'-trū-lā'-shun) In animal development, a series of cell and tissue movements in which the blastula-stage embryo folds inward, producing a three-layered embryo, the gastrula.

**gated channel** A transmembrane protein channel that opens or closes in response to a particular stimulus.

**gated ion channel** A gated channel for a specific ion. The opening or closing of such channels may alter a cell's membrane potential.

**gel electrophoresis** (ē-lek'-trō-fōr-ē'-sis) A technique for separating nucleic acids or proteins on the basis of their size and electrical charge, both of which affect their rate of movement through an electric field in a gel made of agarose or another polymer.

**gene** A discrete unit of hereditary information consisting of a specific nucleotide sequence in DNA (or RNA, in some viruses).

**gene cloning** The production of multiple copies of a gene.

**gene expression** The process by which information encoded in DNA directs the synthesis of proteins or, in some cases, RNAs that are not translated into proteins and instead function as RNAs.

**gene flow** The transfer of alleles from one population to another, resulting from the movement of fertile individuals or their gametes.

**gene pool** The aggregate of all copies of every type of allele at all loci in every individual in a population. The term is also used in a more restricted sense as the aggregate of alleles for just one or a few loci in a population.

**genetic drift** A process in which chance events cause unpredictable fluctuations in allele frequencies from one generation to the next. Effects of genetic drift are most pronounced in small populations.

**genetic engineering** The direct manipulation of genes for practical purposes.

**genetic map** An ordered list of genetic loci (genes or other genetic markers) along a chromosome.

**genetic profile** An individual's unique set of genetic markers, detected most often today by PCR.

**genetic recombination** General term for the production of offspring with combinations of traits that differ from those found in either parent.

**genetic variation** Differences among individuals in the composition of their genes or other DNA segments.

**genetics** The scientific study of heredity and hereditary variation.

**genome** (jē'-nōm) The genetic material of an organism or virus; the complete complement of an organism's or virus's genes along with its noncoding nucleic acid sequences.

**genomics** (juh-nō'-miks) The study of whole sets of genes and their interactions within a species, as well as genome comparisons between species.

**genotype** (jē'-nō-tīp) The genetic makeup, or set of alleles, of an organism.

**genus** (jē'-nus) (plural, **genera**) A taxonomic category above the species level, designated by the first word of a species' two-part scientific name.

**geologic record** A standard time scale dividing Earth's history into time periods grouped into four eons—Hadean, Archaean, Proterozoic, and Phanerozoic—and further subdivided into eras, periods, and epochs.

**gestation** (jes-tā'-shun) Pregnancy; the state of carrying developing young within the female reproductive tract.

**gibberellin** (jib'-uh-rel'-in) Any of a class of related plant hormones that stimulate growth in the stem and leaves, trigger the germination of seeds and breaking of bud dormancy, and (with auxin) stimulate fruit development.

**glans** The rounded structure at the tip of the clitoris or penis that is involved in sexual arousal.

**glia (glial cells)** Cells of the nervous system that support, regulate, and augment the functions of neurons.

**glomerulus** (glō-mār'-yū-lus) A ball of capillaries surrounded by Bowman's capsule in the nephron and serving as the site of filtration in the vertebrate kidney.

**glucagon** (glū-kuh-gon) A hormone secreted by the pancreas that raises blood glucose levels. It promotes glycogen breakdown and release of glucose by the liver.

**glyceraldehyde 3-phosphate (G3P)** (glis'-er-al'-de-hīd) A three-carbon carbohydrate that is the direct product of the Calvin cycle; it is also an intermediate in glycolysis.

**glycogen** (glī'-kō-jen) An extensively branched glucose storage polysaccharide found in the liver and muscle of animals; the animal equivalent of starch.

**glycolipid** A lipid with one or more covalently attached carbohydrates.

**glycolysis** (glī-kol'-uh-sis) A series of reactions that ultimately splits glucose into pyruvate. Glycolysis occurs in almost all living cells, serving as the starting point for fermentation or cellular respiration.

**glycoprotein** A protein with one or more covalently attached carbohydrates.

**glycosidic linkage** A covalent bond formed between two monosaccharides by a dehydration reaction.

**gnathostome** (na'-thu-stōm) Member of one of the two main clades of vertebrates; gnathostomes have jaws and include sharks and rays, ray-finned fishes, coelacanths, lungfishes, amphibians, reptiles, and mammals. *See also* cyclostome.

**Golgi apparatus** (gol'-jē) An organelle in eukaryotic cells consisting of stacks of flat membranous sacs that modify, store, and route products of the endoplasmic reticulum and synthesize some products, notably noncellulose carbohydrates.

**gonad** (gō'-nad) A male or female gamete-producing organ.

**graded potential** In a neuron, a shift in the membrane potential that has an amplitude proportional to signal strength and that decays as it spreads.

**gram-negative** Describing the group of bacteria that have a cell wall that is structurally more complex and contains less peptidoglycan than the cell wall of gram-positive bacteria. Gram-negative bacteria are often more toxic than gram-positive bacteria.

**gram-positive** Describing the group of bacteria that have a cell wall that is structurally less complex and contains more peptidoglycan than the cell wall of gram-negative bacteria. Gram-positive bacteria are usually less toxic than gram-negative bacteria.

**granum** (gran'-um) (plural, **grana**) A stack of membrane-bounded thylakoids in the chloroplast. Grana function in the light reactions of photosynthesis.

**gravitropism** (grav'-uh-trō'-pizm) A response of a plant or animal to gravity.

**gray matter** Regions of dendrites and clustered neuron cell bodies within the CNS.

**green alga** A photosynthetic protist, named for green chloroplasts that are similar in structure and pigment composition to the chloroplasts of plants. Green algae are a paraphyletic group; some members are more closely related to plants than they are to other green algae.

**greenhouse effect** The warming of Earth due to the atmospheric accumulation of carbon dioxide and certain other gases, which absorb reflected infrared radiation and reradiate some of it back toward Earth.

**gross primary production (GPP)** The total primary production of an ecosystem.

**ground tissue system** Plant tissues that are neither vascular nor dermal, fulfilling a variety of functions, such as storage, photosynthesis, and support.

**growth factor** (1) A protein that must be present in the extracellular environment (culture medium or animal body) for the growth and normal development of certain types of cells. (2) A local regulator that acts on nearby cells to stimulate cell proliferation and differentiation.

**guard cells** The two cells that flank the stomatal pore and regulate the opening and closing of the pore.

**gustation** The sense of taste.

**gymnosperm** (jim'-nō-sperm) A vascular plant that bears naked seeds—seeds not enclosed in protective chambers.

**hair cell** A mechanosensory cell that alters output to the nervous system when hairlike projections on the cell surface are displaced.

**half-life** The amount of time it takes for 50% of a sample of a radioactive isotope to decay.

**Hamilton's rule** The principle that for natural selection to favor an altruistic act, the benefit to the recipient, devalued by the coefficient of relatedness, must exceed the cost to the altruist.

**haploid cell** (hap'-loyd) A cell containing only one set of chromosomes ($n$).

**Hardy-Weinberg equilibrium** The state of a population in which frequencies of alleles and genotypes remain constant from generation to generation, provided that only Mendelian segregation and recombination of alleles are at work.

**heart** A muscular pump that uses metabolic energy to elevate the hydrostatic pressure of the circulatory fluid (blood or hemolymph). The fluid then flows down a pressure gradient through the body and eventually returns to the heart.

**heart attack** The damage or death of cardiac muscle tissue resulting from prolonged blockage of one or more coronary arteries.

**heart murmur** A hissing sound that most often results from blood squirting backward through a leaky valve in the heart.

**heat** Thermal energy in transfer from one body of matter to another.

**heat of vaporization** The quantity of heat a liquid must absorb for 1 g of it to be converted from the liquid to the gaseous state.

**heat-shock protein** A protein that helps protect other proteins during heat stress. Heat-shock proteins are found in plants, animals, and microorganisms.

**heavy chain** One of the two types of polypeptide chains that make up an antibody molecule and B cell receptor; consists of a variable region, which contributes to the antigen-binding site, and a constant region.

**helicase** An enzyme that untwists the double helix of DNA at replication forks, separating the two strands and making them available as template strands.

**helper T cell** A type of T cell that, when activated, secretes cytokines that promote the response of B cells (humoral response) and cytotoxic T cells (cell-mediated response) to antigens.

**hemoglobin** (hē′-mō-glō′-bin) An iron-containing protein in red blood cells that reversibly binds oxygen.

**hemolymph** (hē′-mō-limf′) In invertebrates with an open circulatory system, the body fluid that bathes tissues.

**hemophilia** (hē′-muh-fil′-ē-uh) A human genetic disease caused by a sex-linked recessive allele, resulting in the absence of one or more blood-clotting proteins; characterized by excessive bleeding following injury.

**hepatic portal vein** A large vessel that conveys nutrient-laden blood from the small intestine to the liver, which regulates the blood's nutrient content.

**herbivore** (hur′-bi-vōr′) An animal that mainly eats plants or algae.

**herbivory** A +/− interaction in which an organism eats parts of a plant or alga.

**heredity** The transmission of traits from one generation to the next.

**hermaphroditism** (hur-maf′-rō-dī-tizm) A condition in which an individual has both female and male gonads and functions as both a male and female in sexual reproduction by producing both sperm and eggs.

**heterochromatin** (het′-er-ō-krō′-muh-tin) Eukaryotic chromatin that remains highly compacted during interphase and is generally not transcribed.

**heterochrony** (het′-uh-rok′-ruh-nē) Evolutionary change in the timing or rate of an organism's development.

**heterocyst** (het′-er-ō-sist) A specialized cell that engages in nitrogen fixation in some filamentous cyanobacteria; also called a heterocyte.

**heterotroph** (het′-er-ō-trōf) An organism that obtains organic food molecules by eating other organisms or substances derived from them.

**heterozygote** An organism that has two different alleles for a gene (encoding a character).

**heterozygote advantage** Greater reproductive success of heterozygous individuals compared with homozygotes; tends to preserve variation in a gene pool.

**heterozygous** (het′-er-ō-zī′-gus) Having two different alleles for a given gene.

**high-density lipoprotein (HDL)** A particle in the blood made up of thousands of cholesterol molecules and other lipids bound to a protein. HDL scavenges excess cholesterol.

**histamine** (his′-tuh-mēn) A substance released by mast cells that causes blood vessels to dilate and become more permeable in inflammatory and allergic responses.

**histogram** A variant of a bar graph in which a numerical independent variable is divided into equal intervals (or groups called "bins"). The height (or length) of each bar represents the value of the dependent variable for a particular interval.

**histone** (his′-tōn) A small protein with a high proportion of positively charged amino acids that binds to the negatively charged DNA and plays a key role in chromatin structure.

**histone acetylation** The attachment of acetyl groups to certain amino acids of histone proteins.

**HIV (human immunodeficiency virus)** The infectious agent that causes AIDS. HIV is a retrovirus.

**holdfast** A rootlike structure that anchors a seaweed.

**homeobox** (hō′-mē-ō-boks′) A 180-nucleotide sequence within homeotic genes and some other developmental genes that is widely conserved in animals. Related sequences occur in plants and yeasts.

**homeostasis** (hō′-mē-ō-stā′-sis) The steady-state physiological condition of the body.

**homeotic gene** (hō-mē-o′-tik) Any of the master regulatory genes that control placement and spatial organization of body parts in animals, plants, and fungi by controlling the developmental fate of groups of cells.

**hominin** A group consisting of humans and the extinct species that are more closely related to us than to chimpanzees.

**homologous chromosomes** (hō-mol′-uh-gus) A pair of chromosomes of the same length, centromere position, and staining pattern that possess genes for the same characters at corresponding loci. One homologous chromosome is inherited from the organism's father, the other from the mother. Also called homologs, or a homologous pair.

**homologous structures** Structures in different species that are similar because of common ancestry.

**homologs** *See* homologous chromosomes.

**homology** (hō-mol′-ō-jē) Similarity in characteristics resulting from a shared ancestry.

**homoplasy** (hō′-muh-play′-zē) A similar (analogous) structure or molecular sequence that has evolved independently in two species.

**homozygote** An organism that has a pair of identical alleles for a gene (encoding a character).

**homozygous** (hō′-mō-zī′-gus) Having two identical alleles for a given gene.

**horizontal gene transfer** The transfer of genes from one genome to another through mechanisms such as transposable elements, plasmid exchange, viral activity, and perhaps fusions of different organisms.

**hormone** In multicellular organisms, one of many types of secreted chemicals that are formed in specialized cells, travel in body fluids, and act on specific target cells in other parts of the body, changing the target cells' functioning. Hormones are thus important in long-distance signaling.

**host** The larger participant in a symbiotic relationship, often providing a home and food source for the smaller symbiont.

**host range** The limited number of species whose cells can be infected by a particular virus.

**Human Genome Project** An international collaborative effort to map and sequence the DNA of the entire human genome.

**human immunodeficiency virus (HIV)** The pathogen that causes AIDS (acquired immune deficiency syndrome).

**humoral immune response** (hyū′-mer-ul) The branch of adaptive immunity that involves the activation of B cells and that leads to the production of antibodies, which defend against bacteria and viruses in body fluids.

**humus** (hyū′-mus) Decomposing organic material that is a component of topsoil.

**Huntington's disease** A human genetic disease caused by a dominant allele; characterized by uncontrollable body movements and degeneration of the nervous system; usually fatal 10 to 20 years after the onset of symptoms.

**hybrid** Offspring that results from the mating of individuals from two different species or from two true-breeding varieties of the same species.

**hybrid zone** A geographic region in which members of different species meet and mate, producing at least some offspring of mixed ancestry.

**hybridization** In genetics, the mating, or crossing, of two true-breeding varieties.

**hydration shell** The sphere of water molecules around a dissolved ion.

**hydrocarbon** An organic molecule consisting of only carbon and hydrogen.

**hydrogen bond** A type of weak chemical bond that is formed when the slightly positive hydrogen atom of a polar covalent bond in one molecule is attracted to the slightly negative atom of a polar covalent bond in another molecule or in another region of the same molecule.

**hydrogen ion** A single proton with a charge of 1+. The dissociation of a water molecule ($H_2O$) leads to the generation of a hydroxide ion ($OH^-$) and a hydrogen ion ($H^+$); in water, $H^+$ is not found alone but associates with a water molecule to form a hydronium ion.

**hydrolysis** (hī-drol′-uh-sis) A chemical reaction that breaks bonds between two molecules by the addition of water; functions in disassembly of polymers to monomers.

**hydronium ion** A water molecule that has an extra proton bound to it; $H_3O^+$, commonly represented as $H^+$.

**hydrophilic** (hī′-drō-fil′-ik) Having an affinity for water.

**hydrophobic** (hī′-drō-fō′-bik) Having no affinity for water; tending to coalesce and form droplets in water.

**hydrophobic interaction** A type of weak chemical interaction caused when molecules that do not mix with water coalesce to exclude water.

**hydroponic culture** A method in which plants are grown in mineral solutions rather than in soil.

**hydrostatic skeleton** A skeletal system composed of fluid held under pressure in a closed body compartment; the main skeleton of most cnidarians, flatworms, nematodes, and annelids.

**hydroxide ion** A water molecule that has lost a proton; $OH^-$.

**hydroxyl group** (hī-drok′-sil) A chemical group consisting of an oxygen atom joined to a hydrogen atom. Molecules possessing this group are soluble in water and are called alcohols.

**hymen** A thin membrane that partly covers the vaginal opening in the human female. The hymen is ruptured by sexual intercourse or other vigorous activity.

**hyperpolarization** A change in a cell's membrane potential such that the inside of the membrane becomes more negative relative to the outside. Hyperpolarization reduces the chance that a neuron will transmit a nerve impulse.

**hypersensitive response** A plant's localized defense response to a pathogen, involving the death of cells around the site of infection.

**hypertension** A disorder in which blood pressure remains abnormally high.

**hypertonic** Referring to a solution that, when surrounding a cell, will cause the cell to lose water.

**hypha** (plural, **hyphae**) (hī′-fuh, hī′-fē) One of many connected filaments that collectively make up the mycelium of a fungus.

**hypocotyl** (hī′-puh-cot′-ul) In an angiosperm embryo, the embryonic axis below the point of attachment of the cotyledon(s) and above the radicle.

**hypothalamus** (hī′-pō-thal′-uh-mus) The ventral part of the vertebrate forebrain; functions in maintaining homeostasis, especially in coordinating the endocrine and nervous systems; secretes hormones of the posterior pituitary and releasing factors that regulate the anterior pituitary.

**hypothesis** (hī-poth′-uh-sis) A testable explanation for a set of observations based on the available data and guided by inductive reasoning. A hypothesis is narrower in scope than a theory.

**hypotonic** Referring to a solution that, when surrounding a cell, will cause the cell to take up water.

**imbibition** The physical adsorption of water by a seed or other structure, resulting in swelling.

**immigration** The influx of new individuals into a population from other areas.

**immune system** An animal body's system of defenses against agents that cause disease.

**immunization** The process of generating a state of immunity by artificial means. In active immunization, also called vaccination, an inactive or weakened form of a pathogen is administered, inducing B and T cell responses and immunological memory. In passive immunization, antibodies specific for a particular microbe are administered, conferring immediate but temporary protection.

**immunoglobulin (Ig)** (im′-yū-nō-glob′-yū-lin) *See* antibody.

**imprinting** In animal behavior, the formation at a specific stage in life of a long-lasting behavioral response to a specific individual or object.

***in situ* hybridization** A technique using nucleic acid hybridization with a labeled probe to detect the location of a specific mRNA in an intact organism.

***in vitro* fertilization (IVF)** (vē′-trō) Fertilization of oocytes in laboratory containers followed by artificial implantation of the early embryo in the mother's uterus.

**inclusive fitness** The total effect an individual has on proliferating its genes by producing its own offspring and by providing aid that enables other close relatives to increase production of their offspring.

**incomplete dominance** The situation in which the phenotype of heterozygotes is intermediate between the phenotypes of individuals homozygous for either allele.

**incomplete flower** A flower in which one or more of the four basic floral organs (sepals, petals, stamens, or carpels) are either absent or nonfunctional.

**independent variable** A factor whose value is manipulated or changed during an experiment to reveal possible effects on another factor (the dependent variable).

**indeterminate growth** A type of growth characteristic of plants, in which the organism continues to grow as long as it lives.

**induced fit** Caused by entry of the substrate, the change in shape of the active site of an enzyme so that it binds more snugly to the substrate.

**inducer** A specific small molecule that binds to a bacterial repressor protein and changes the repressor's shape so that it cannot bind to an operator, thus switching an operon on.

**induction** The process in which one group of embryonic cells influences the development of another, usually by causing changes in gene expression.

**inductive reasoning** A type of logic in which generalizations are based on a large number of specific observations.

**inflammatory response** An innate immune defense triggered by physical injury or infection of tissue involving the release of substances that promote swelling, enhance the infiltration of white blood cells, and aid in tissue repair and destruction of invading pathogens.

**inflorescence** A group of flowers tightly clustered together.

**ingestion** The first stage of food processing in animals: the act of eating.

**ingroup** A species or group of species whose evolutionary relationships are being examined in a given analysis.

**inhibitory postsynaptic potential (IPSP)** An electrical change (usually hyperpolarization) in the membrane of a postsynaptic neuron caused by the binding of an inhibitory neurotransmitter from a presynaptic cell to a postsynaptic receptor; makes it more difficult for a postsynaptic neuron to generate an action potential.

**innate behavior** Animal behavior that is developmentally fixed and under strong genetic control. Innate behavior is exhibited in virtually the same form by all individuals in a population despite internal and external environmental differences during development and throughout their lifetimes.

**innate immunity** A form of defense common to all animals that is active immediately upon exposure to pathogens and that is the same whether or not the pathogen has been encountered previously.

**inner ear** One of three main regions of the vertebrate ear; includes the cochlea (which in turn contains the organ of Corti) and the semicircular canals.

**inquiry** The search for information and explanation, often focusing on specific questions.

**insertion** A mutation involving the addition of one or more nucleotide pairs to a gene.

**insulin** (in′-suh-lin) A hormone secreted by pancreatic beta cells that lowers blood glucose levels. It promotes the uptake of glucose by most body cells and the synthesis and storage of glycogen in the liver.

**integral protein** A transmembrane protein with hydrophobic regions that extend into and often completely span the hydrophobic interior of the membrane and with hydrophilic regions in contact with the aqueous solution on one or both sides of the membrane (or lining the channel in the case of a channel protein).

**integrin** In animal cells, a transmembrane receptor protein with two subunits that interconnects the extracellular matrix and the cytoskeleton.

**integument** (in-teg′-yū-ment) Layer of sporophyte tissue that contributes to the structure of an ovule of a seed plant.

**interferon** (in′-ter-fēr′-on) A protein that has antiviral or immune regulatory functions.

For example, interferons secreted by virus-infected cells help nearby cells resist viral infection.

**intermediate disturbance hypothesis** The concept that moderate levels of disturbance can foster greater species diversity than low or high levels of disturbance.

**intermediate filament** A component of the cytoskeleton that includes filaments intermediate in size between microtubules and microfilaments.

**internal fertilization** The fusion of eggs and sperm within the female reproductive tract. The sperm are typically deposited in or near the tract.

**interneuron** An association neuron; a nerve cell within the central nervous system that forms synapses with sensory and/or motor neurons and integrates sensory input and motor output.

**internode** A segment of a plant stem between the points where leaves are attached.

**interphase** The period in the cell cycle when the cell is not dividing. During interphase, cellular metabolic activity is high, chromosomes and organelles are duplicated, and cell size may increase. Interphase often accounts for about 90% of the cell cycle.

**interspecific competition** Competition for resources between individuals of two or more species when resources are in short supply.

**interspecific interaction** A relationship between individuals of two or more species in a community.

**interstitial fluid** The fluid filling the spaces between cells in most animals.

**intertidal zone** The shallow zone of the ocean adjacent to land and between the high- and low-tide lines.

**intrinsic rate of increase (*r*)** In population models, the per capita rate at which an exponentially growing population increases in size at each instant in time.

**introduced species** A species moved by humans, either intentionally or accidentally, from its native location to a new geographic region; also called a non-native or exotic species.

**intron** (in'-tron) A noncoding, intervening sequence within a primary transcript that is removed from the transcript during RNA processing; also refers to the region of DNA from which this sequence was transcribed.

**invasive species** A species, often introduced by humans, that takes hold outside its native range.

**inversion** An aberration in chromosome structure resulting from reattachment of a chromosomal fragment in a reverse orientation to the chromosome from which it originated.

**invertebrate** An animal without a backbone. Invertebrates make up 95% of animal species.

**ion** (ī'-on) An atom or group of atoms that has gained or lost one or more electrons, thus acquiring a charge.

**ion channel** A transmembrane protein channel that allows a specific ion to diffuse across the membrane down its concentration or electrochemical gradient.

**ionic bond** (ī-on'-ik) A chemical bond resulting from the attraction between oppositely charged ions.

**ionic compound** A compound resulting from the formation of an ionic bond; also called a salt.

**iris** The colored part of the vertebrate eye, formed by the anterior portion of the choroid.

**isomer** One of two or more compounds that have the same numbers of atoms of the same elements but different structures and hence different properties.

**isotonic** (ī'-sō-ton'-ik) Referring to a solution that, when surrounding a cell, causes no net movement of water into or out of the cell.

**isotope** (ī'-sō-tōp') One of several atomic forms of an element, each with the same number of protons but a different number of neutrons, thus differing in atomic mass.

**iteroparity** Reproduction in which adults produce offspring over many years; also known as repeated reproduction.

**joule (J)** A unit of energy: 1 J = 0.239 cal; 1 cal = 4.184 J.

**juxtaglomerular apparatus (JGA)** (juks'-tuh-gluh-mār'-yū-ler) A specialized tissue in nephrons that releases the enzyme renin in response to a drop in blood pressure or volume.

**juxtamedullary nephron** In mammals and birds, a nephron with a loop of Henle that extends far into the renal medulla.

**karyogamy** (kār'-ē-og'-uh-mē) In fungi, the fusion of haploid nuclei contributed by the two parents; occurs as one stage of sexual reproduction, preceded by plasmogamy.

**karyotype** (kār'-ē-ō-tīp) A display of the chromosome pairs of a cell arranged by size and shape.

**keystone species** A species that is not necessarily abundant in a community yet exerts strong control on community structure by the nature of its ecological role or niche.

**kidney** In vertebrates, one of a pair of excretory organs where blood filtrate is formed and processed into urine.

**kilocalorie (kcal)** A thousand calories; the amount of heat energy required to raise the temperature of 1 kg of water by 1°C.

**kin selection** Natural selection that favors altruistic behavior by enhancing the reproductive success of relatives.

**kinetic energy** (kuh-net'-ik) The energy associated with the relative motion of objects. Moving matter can perform work by imparting motion to other matter.

**kinetochore** (kuh-net'-uh-kōr) A structure of proteins attached to the centromere that links each sister chromatid to the mitotic spindle.

**kingdom** A taxonomic category, the second broadest after domain.

***K*-selection** Selection for life history traits that are sensitive to population density; also called density-dependent selection.

**labia majora** A pair of thick, fatty ridges that encloses and protects the rest of the vulva.

**labia minora** A pair of slender skin folds that surrounds the openings of the vagina and urethra.

**lacteal** (lak'-tē-ul) A tiny lymph vessel extending into the core of an intestinal villus and serving as the destination for absorbed chylomicrons.

**lactic acid fermentation** Glycolysis followed by the reduction of pyruvate to lactate, regenerating NAD$^+$ with no release of carbon dioxide.

**lagging strand** A discontinuously synthesized DNA strand that elongates by means of Okazaki fragments, each synthesized in a 5' → 3' direction away from the replication fork.

**landscape** An area containing several different ecosystems linked by exchanges of energy, materials, and organisms.

**landscape ecology** The study of how the spatial arrangement of habitat types affects the distribution and abundance of organisms and ecosystem processes.

**larynx** (lār'-inks) The portion of the respiratory tract containing the vocal cords; also called the voice box.

**lateral meristem** (mār'-uh-stem) A meristem that thickens the roots and shoots of woody plants. The vascular cambium and cork cambium are lateral meristems.

**lateral root** A root that arises from the pericycle of an established root.

**lateralization** Segregation of functions in the cortex of the left and right cerebral hemispheres.

**law of conservation of mass** A physical law stating that matter can change form but cannot be created or destroyed. In a closed system, the mass of the system is constant.

**law of independent assortment** Mendel's second law, stating that each pair of alleles segregates, or assorts, independently of each other pair during gamete formation; applies when genes for two characters are located on different pairs of homologous chromosomes or when they are far enough apart on the same chromosome to behave as though they are on different chromosomes.

**law of segregation** Mendel's first law, stating that the two alleles in a pair segregate (separate from each other) into different gametes during gamete formation.

**leading strand** The new complementary DNA strand synthesized continuously along the template strand toward the replication fork in the mandatory 5' → 3' direction.

**leaf** The main photosynthetic organ of vascular plants.

**leaf primordium** A finger-like projection along the flank of a shoot apical meristem, from which a leaf arises.

**learning** The modification of behavior based on specific experiences.

**lens** The structure in an eye that focuses light rays onto the photoreceptors.

**lenticel** (len'-ti-sel) A small raised area in the bark of stems and roots that enables gas exchange between living cells and the outside air.

**leukocyte** (lū'-kō-sīt') A blood cell that functions in fighting infections; also called a white blood cell.

**Leydig cell** (lī'-dig) A cell that produces testosterone and other androgens and is located between the seminiferous tubules of the testes.

**lichen** A mutualistic association between a fungus and a photosynthetic alga or cyanobacterium.

**life cycle** The generation-to-generation sequence of stages in the reproductive history of an organism.

**life history** The traits that affect an organism's schedule of reproduction and survival.

**life table** A summary of the age-specific survival and reproductive rates of individuals in a population.

**ligand** (lig'-und) A molecule that binds specifically to another molecule, usually a larger one.

**ligand-gated ion channel** A transmembrane protein containing a pore that opens or closes as it changes shape in response to a signaling molecule (ligand), allowing or blocking the flow of specific ions; also called an ionotropic receptor.

**light chain** One of the two types of polypeptide chains that make up an antibody molecule and B cell receptor; consists of a variable region, which contributes to the antigen-binding site, and a constant region.

**light microscope (LM)** An optical instrument with lenses that refract (bend) visible light to magnify images of specimens.

**light reactions** The first of two major stages in photosynthesis (preceding the Calvin cycle). These reactions, which occur on the thylakoid membranes of the chloroplast or on membranes of certain prokaryotes, convert solar energy to the chemical energy of ATP and NADPH, releasing oxygen in the process.

**light-harvesting complex** A complex of proteins associated with pigment molecules (including chlorophyll *a*, chlorophyll *b*, and carotenoids) that captures light energy and transfers it to reaction-center pigments in a photosystem.

**lignin** (lig'-nin) A strong polymer embedded in the cellulose matrix of the secondary cell walls of vascular plants that provides structural support in terrestrial species.

**limiting nutrient** An element that must be added for production to increase in a particular area.

**limnetic zone** In a lake, the well-lit, open surface waters far from shore.

**linear electron flow** A route of electron flow during the light reactions of photosynthesis that involves both photosystems (I and II) and produces ATP, NADPH, and $O_2$. The net electron flow is from $H_2O$ to $NADP^+$.

**line graph** A two-dimensional graph in which each data point is connected to the next point in the data set with a straight line.

**linkage map** A genetic map based on the frequencies of recombination between markers during crossing over of homologous chromosomes.

**linked genes** Genes located close enough together on a chromosome that they tend to be inherited together.

**lipid** (lip'-id) Any of a group of large biological molecules, including fats, phospholipids, and steroids, that mix poorly, if at all, with water.

**littoral zone** In a lake, the shallow, well-lit waters close to shore.

**liver** A large internal organ in vertebrates that performs diverse functions, such as producing bile, maintaining blood glucose level, and detoxifying poisonous chemicals in the blood.

**loam** The most fertile soil type, made up of roughly equal amounts of sand, silt, and clay.

**lobe-fin** Member of a clade of osteichthyans having rod-shaped muscular fins. The group includes coelacanths, lungfishes, and tetrapods.

**local regulator** A secreted molecule that influences cells near where it is secreted.

**locomotion** Active motion from place to place.

**locus** (plural, **loci**) (lō'-kus, lō'-sī) A specific place along the length of a chromosome where a given gene is located.

**logistic population growth** Population growth that levels off as population size approaches carrying capacity.

**long-day plant** A plant that flowers (usually in late spring or early summer) only when the light period is longer than a critical length.

**long-term memory** The ability to hold, associate, and recall information over one's lifetime.

**loop of Henle** The hairpin turn, with a descending and ascending limb, between the proximal and distal tubules of the vertebrate kidney; functions in water and salt reabsorption.

**low-density lipoprotein (LDL)** A particle in the blood made up of thousands of cholesterol molecules and other lipids bound to a protein. LDL transports cholesterol from the liver for incorporation into cell membranes.

**lung** An infolded respiratory surface of a terrestrial vertebrate, land snail, or spider that connects to the atmosphere by narrow tubes.

**luteinizing hormone (LH)** (lū'-tē-uh-nī'-zing) A tropic hormone that is produced and secreted by the anterior pituitary and that stimulates ovulation in females and androgen production in males.

**lycophyte** (lī'-kuh-fīt) An informal name for a member of the phylum Lycophyta, a group of seedless vascular plants that includes club mosses and their relatives.

**lymph** The colorless fluid, derived from interstitial fluid, in the lymphatic system of vertebrates.

**lymph node** An organ located along a lymph vessel. Lymph nodes filter lymph and contain cells that attack viruses and bacteria.

**lymphatic system** A system of vessels and nodes, separate from the circulatory system, that returns fluid, proteins, and cells to the blood.

**lymphocyte** A type of white blood cell that mediates immune responses. The two main classes are B cells and T cells.

**lysogenic cycle** (lī'-sō-jen'-ik) A type of phage replicative cycle in which the viral genome becomes incorporated into the bacterial host chromosome as a prophage, is replicated along with the chromosome, and does not kill the host.

**lysosome** (lī'-suh-sōm) A membrane-enclosed sac of hydrolytic enzymes found in the cytoplasm of animal cells and some protists.

**lysozyme** (lī'-sō-zīm) An enzyme that destroys bacterial cell walls; in mammals, found in sweat, tears, and saliva.

**lytic cycle** (lit'-ik) A type of phage replicative cycle resulting in the release of new phages by lysis (and death) of the host cell.

**macroclimate** Large-scale patterns in climate; the climate of an entire region.

**macroevolution** Evolutionary change above the species level. Examples of macroevolutionary change include the origin of a new group of organisms through a series of speciation events and the impact of mass extinctions on the diversity of life and its subsequent recovery.

**macromolecule** A giant molecule formed by the joining of smaller molecules, usually by a dehydration reaction. Polysaccharides, proteins, and nucleic acids are macromolecules.

**macronutrient** An essential element that an organism must obtain in relatively large amounts. *See also* micronutrient.

**macrophage** (mak'-rō-fāj) A phagocytic cell present in many tissues that functions in innate immunity by destroying microbes and in acquired immunity as an antigen-presenting cell.

**major histocompatibility complex (MHC) molecule** A host protein that functions in antigen presentation. Foreign MHC molecules on transplanted tissue can trigger T cell responses that may lead to rejection of the transplant.

**malignant tumor** A cancerous tumor containing cells that have significant genetic and cellular changes and are capable of invading and surviving in new sites. Malignant tumors can impair the functions of one or more organs.

**mammal** Member of the class Mammalia, amniotes that have hair and mammary glands (glands that produce milk).

**mammary gland** An exocrine gland that secretes milk to nourish the young. Mammary glands are characteristic of mammals.

**mantle** One of the three main parts of a mollusc; a fold of tissue that drapes over the mollusc's visceral mass and may secrete a shell. *See also* foot and visceral mass.

**map unit** A unit of measurement of the distance between genes. One map unit is equivalent to a 1% recombination frequency.

**marine benthic zone** The ocean floor.

**marsupial** (mar-sū'-pē-ul) A mammal, such as a koala, kangaroo, or opossum, whose young complete their embryonic development inside a maternal pouch.

**mass extinction** The elimination of a large number of species throughout Earth, the result of global environmental changes.

**mass number** The total number of protons and neutrons in an atom's nucleus.

**maternal effect gene** A gene that, when mutant in the mother, results in a mutant phenotype in the offspring, regardless of the offspring's genotype. Maternal effect genes, also called egg-polarity genes, were first identified in *Drosophila melanogaster*.

**matter** Anything that takes up space and has mass.

**maximum parsimony** A principle that states that when considering multiple explanations for an observation, one should first investigate the simplest explanation that is consistent with the facts.

**mechanoreceptor** A sensory receptor that detects physical deformation in the body's environment associated with pressure, touch, stretch, motion, or sound.

**medulla oblongata** (meh-dul'-uh ob'-long-go'-tuh) The lowest part of the vertebrate brain, commonly called the medulla; a swelling of the hindbrain anterior to the spinal cord that controls autonomic, homeostatic functions, including breathing, heart and blood vessel activity, swallowing, digestion, and vomiting.

**megapascal (MPa)** (meg'-uh-pas-kal') A unit of pressure equivalent to about 10 atmospheres of pressure.

**megaphyll** A leaf with a highly branched vascular system, found in almost all vascular plants other than lycophytes.

**megaspore** A spore from a heterosporous plant species that develops into a female gametophyte.

**meiosis** (mī-ō'-sis) A modified type of cell division in sexually reproducing organisms consisting of two rounds of cell division but only one round of DNA replication. It results in cells with half the number of chromosome sets as the original cell.

**meiosis I** The first division of a two-stage process of cell division in sexually reproducing organisms that results in cells with half the number of chromosome sets as the original cell.

**meiosis II** The second division of a two-stage process of cell division in sexually reproducing organisms that results in cells with half the number of chromosome sets as the original cell.

**membrane potential** The difference in electrical charge (voltage) across a cell's plasma membrane due to the differential distribution of ions. Membrane potential affects the activity of excitable cells and the transmembrane movement of all charged substances.

**memory cell** One of a clone of long-lived lymphocytes, formed during the primary immune response, that remains in a lymphoid organ until activated by exposure to the same antigen that triggered its formation. Activated memory cells mount the secondary immune response.

**menopause** The cessation of ovulation and menstruation, marking the end of a human female's reproductive years.

**menstrual cycle** (men'-strū-ul) In humans and certain other primates, a type of reproductive cycle in which the nonpregnant endometrium is shed through the cervix into the vagina; also called the uterine cycle.

**menstruation** The shedding of portions of the endometrium during a uterine (menstrual) cycle.

**meristem** (mar'-uh-stem) Plant tissue that remains embryonic as long as the plant lives, allowing for indeterminate growth.

**mesoderm** (mez'-ō-derm) The middle primary germ layer in a triploblastic animal embryo; develops into the notochord, the lining of the coelom, muscles, skeleton, gonads, kidneys, and most of the circulatory system in species that have these structures.

**mesophyll** (mez'-ō-fil) Leaf cells specialized for photosynthesis. In $C_3$ and CAM plants, mesophyll cells are located between the upper and lower epidermis; in $C_4$ plants, they are located between the bundle-sheath cells and the epidermis.

**messenger RNA (mRNA)** A type of RNA, synthesized using a DNA template, that attaches to ribosomes in the cytoplasm and specifies the primary structure of a protein. (In eukaryotes, the primary RNA transcript must undergo RNA processing to become mRNA.)

**metabolic pathway** A series of chemical reactions that either builds a complex molecule (anabolic pathway) or breaks down a complex molecule to simpler molecules (catabolic pathway).

**metabolic rate** The total amount of energy an animal uses in a unit of time.

**metabolism** (muh-tab'-uh-lizm) The totality of an organism's chemical reactions, consisting of catabolic and anabolic pathways, which manage the material and energy resources of the organism.

**metagenomics** The collection and sequencing of DNA from a group of species, usually an environmental sample of microorganisms. Computer software sorts partial sequences and assembles them into genome sequences of individual species making up the sample.

**metaphase** The third stage of mitosis, in which the spindle is complete and the chromosomes, attached to microtubules at their kinetochores, are all aligned at the metaphase plate.

**metaphase plate** An imaginary structure located at a plane midway between the two poles of a cell in metaphase on which the centromeres of all the duplicated chromosomes are located.

**metapopulation** A group of spatially separated populations of one species that interact through immigration and emigration.

**metastasis** (muh-tas'-tuh-sis) The spread of cancer cells to locations distant from their original site.

**methanogen** (meth-an'-ō-jen) An organism that produces methane as a waste product of the way it obtains energy. All known methanogens are in domain Archaea.

**methyl group** A chemical group consisting of a carbon bonded to three hydrogen atoms. The methyl group may be attached to a carbon or to a different atom.

**microevolution** Evolutionary change below the species level; change in the allele frequencies in a population over generations.

**microfilament** A cable composed of actin proteins in the cytoplasm of almost every eukaryotic cell, making up part of the cytoskeleton and acting alone or with myosin to cause cell contraction; also known as an actin filament.

**micronutrient** An essential element that an organism needs in very small amounts. *See also* macronutrient.

**microphyll** A small, usually spine-shaped leaf supported by a single strand of vascular tissue, found only in lycophytes.

**microRNA (miRNA)** A small, single-stranded RNA molecule, generated from a hairpin structure on a precursor RNA transcribed from a particular gene. The miRNA associates with one or more proteins in a complex that can degrade or prevent translation of an mRNA with a complementary sequence.

**microspore** A spore from a heterosporous plant species that develops into a male gametophyte.

**microtubule** A hollow rod composed of tubulin proteins that makes up part of the cytoskeleton in all eukaryotic cells and is found in cilia and flagella.

**microvillus** (plural, **microvilli**) One of many fine, finger-like projections of the epithelial cells in the lumen of the small intestine that increase its surface area.

**middle ear** One of three main regions of the vertebrate ear; in mammals, a chamber containing three small bones (the malleus, incus, and stapes) that convey vibrations from the eardrum to the oval window.

**middle lamella** (luh-mel'-uh) In plants, a thin layer of adhesive extracellular material, primarily pectins, found between the primary walls of adjacent young cells.

**migration** A regular, long-distance change in location.

**mineral** In nutrition, a simple nutrient that is inorganic and therefore cannot be synthesized in the body.

**minimum viable population (MVP)** The smallest population size at which a species is able to sustain its numbers and survive.

**mismatch repair** The cellular process that uses specific enzymes to remove and replace incorrectly paired nucleotides.

**missense mutation** A nucleotide-pair substitution that results in a codon that codes for a different amino acid.

**mitochondrial matrix** The compartment of the mitochondrion enclosed by the inner membrane and containing enzymes and substrates for the citric acid cycle, as well as ribosomes and DNA.

**mitochondrion** (mī'-tō-kon'-drē-un) (plural, **mitochondria**) An organelle in eukaryotic cells that serves as the site of cellular respiration; uses oxygen to break down organic molecules and synthesize ATP.

**mitosis** (mī-tō'-sis) A process of nuclear division in eukaryotic cells conventionally divided into five stages: prophase, prometaphase, metaphase, anaphase, and telophase. Mitosis conserves chromosome number by allocating replicated chromosomes equally to each of the daughter nuclei.

**mitotic (M) phase** The phase of the cell cycle that includes mitosis and cytokinesis.

**mitotic spindle** An assemblage of microtubules and associated proteins that is involved in the movement of chromosomes during mitosis.

**mixotroph** An organism that is capable of both photosynthesis and heterotrophy.

**model** A physical or conceptual representation of a natural phenomenon.

**model organism** A particular species chosen for research into broad biological principles because it is representative of a larger group and usually easy to grow in a lab.

**molarity** A common measure of solute concentration, referring to the number of moles of solute per liter of solution.

**mole (mol)** The number of grams of a substance that equals its molecular weight in daltons and contains Avogadro's number of molecules.

**molecular clock** A method for estimating the time required for a given amount of evolutionary change, based on the observation that some regions of genomes evolve at constant rates.

**molecular mass** The sum of the masses of all the atoms in a molecule; sometimes called molecular weight.

**molecule** Two or more atoms held together by covalent bonds.

**monilophyte** An informal name for a member of the phylum Monilophyta, a group of seedless vascular plants that includes ferns and their relatives.

**monoclonal antibody** (mon'-ō-klōn'-ul) Any of a preparation of antibodies that have been produced by a single clone of cultured cells and thus are all specific for the same epitope.

**monogamous** (muh-nog'-uh-mus) Referring to a type of relationship in which one male mates with just one female.

**monohybrid** An organism that is heterozygous with respect to a single gene of interest. All the offspring from a cross between parents homozygous for different alleles are monohybrids. For example, parents of genotypes *AA* and *aa* produce a monohybrid of genotype *Aa*.

**monohybrid cross** A cross between two organisms that are heterozygous for the character being followed (or the self-pollination of a heterozygous plant).

**monomer** (mon'-uh-mer) The subunit that serves as the building block of a polymer.

**monophyletic** (mon'-ō-fī-let'-ik) Pertaining to a group of taxa that consists of a common ancestor and all of its descendants. A monophyletic taxon is equivalent to a clade.

**monosaccharide** (mon'-ō-sak'-uh-rīd) The simplest carbohydrate, active alone or serving as a monomer for disaccharides and polysaccharides. Also known as simple sugars, monosaccharides have molecular formulas that are generally some multiple of $CH_2O$.

**monosomic** Referring to a diploid cell that has only one copy of a particular chromosome instead of the normal two.

**monotreme** An egg-laying mammal, such as a platypus or echidna. Like all mammals, monotremes have hair and produce milk, but they lack nipples.

**morphogen** A substance, such as Bicoid protein in *Drosophila*, that provides positional information in the form of a concentration gradient along an embryonic axis.

**morphogenesis** (mōr'-fō-jen'-uh-sis) The development of the form of an organism and its structures.

**morphological species concept** Definition of a species in terms of measurable anatomical criteria.

**motor neuron** A nerve cell that transmits signals from the brain or spinal cord to muscles or glands.

**motor protein** A protein that interacts with cytoskeletal elements and other cell components, producing movement of the whole cell or parts of the cell.

**motor system** An efferent branch of the vertebrate peripheral nervous system composed of motor neurons that carry signals to skeletal muscles in response to external stimuli.

**motor unit** A single motor neuron and all the muscle fibers it controls.

**movement corridor** A series of small clumps or a narrow strip of quality habitat (usable by organisms) that connects otherwise isolated patches of quality habitat.

**mucus** A viscous and slippery mixture of glycoproteins, cells, salts, and water that moistens and protects the membranes lining body cavities that open to the exterior.

**multifactorial** Referring to a phenotypic character that is influenced by multiple genes and environmental factors.

**multigene family** A collection of genes with similar or identical sequences, presumably of common origin.

**multiple fruit** A fruit derived from an entire inflorescence.

**multiplication rule** A rule of probability stating that the probability of two or more independent events occurring together can be determined by multiplying their individual probabilities.

**muscle tissue** Tissue consisting of long muscle cells that can contract, either on its own or when stimulated by nerve impulses.

**mutagen** (myū'-tuh-jen) A chemical or physical agent that interacts with DNA and can cause a mutation.

**mutation** (myū-tā'-shun) A change in the nucleotide sequence of an organism's DNA or in the DNA or RNA of a virus.

**mutualism** (myū'-chū-ul-izm) A +/+ ecological interaction that benefits each of the interacting species.

**mycelium** (mī-sē'-lē-um) The densely branched network of hyphae in a fungus.

**mycorrhiza** (plural, **mycorrhizae**) (mī'-kō-rī'-zuh, mī'-kō-rī'-zē) A mutualistic association of plant roots and fungus.

**myelin sheath** (mī'-uh-lin) Wrapped around the axon of a neuron, an insulating coat of cell membranes from Schwann cells or oligodendrocytes. It is interrupted by nodes of Ranvier, where action potentials are generated.

**myofibril** (mī'-ō-fī'-bril) A longitudinal bundle in a muscle cell (fiber) that contains thin filaments of actin and regulatory proteins and thick filaments of myosin.

**myoglobin** (mī'-uh-glō'-bin) An oxygen-storing, pigmented protein in muscle cells.

**myosin** (mī'-uh-sin) A type of motor protein that associates into filaments that interact with actin filaments, causing cell contraction.

**NAD⁺** Nicotinamide adenine dinucleotide, a coenzyme that cycles easily between oxidized ($NAD^+$) and reduced (NADH) states, thus acting as an electron carrier.

**NADP⁺** Nicotinamide adenine dinucleotide phosphate, an electron acceptor that, as NADPH, temporarily stores energized electrons produced during the light reactions.

**natural killer cell** A type of white blood cell that can kill tumor cells and virus-infected cells as part of innate immunity.

**natural selection** A process in which individuals that have certain inherited traits tend to survive and reproduce at higher rates than other individuals *because of* those traits.

**negative feedback** A form of regulation in which accumulation of an end product of a process slows the process; in physiology, a primary mechanism of homeostasis, whereby a change in a variable triggers a response that counteracts the initial change.

**negative pressure breathing** A breathing system in which air is pulled into the lungs.

**nephron** (nef'-ron) The tubular excretory unit of the vertebrate kidney.

**nerve** A fiber composed primarily of the bundled axons of neurons.

**nerve net** A weblike system of neurons, characteristic of radially symmetric animals, such as hydras.

**nervous system** The fast-acting internal system of communication involving sensory receptors, networks of nerve cells, and connections to muscles and glands that respond to nerve signals; functions in concert with the endocrine system to effect internal regulation and maintain homeostasis.

**nervous tissue** Tissue made up of neurons and supportive cells.

**net ecosystem production (NEP)** The gross primary production of an ecosystem minus the energy used by all autotrophs and heterotrophs for respiration.

**net primary production (NPP)** The gross primary production of an ecosystem minus the energy used by the producers for respiration.

**neuron** (nyūr'-on) A nerve cell; the fundamental unit of the nervous system, having structure and properties that allow it to conduct signals by taking advantage of the electrical charge across its plasma membrane.

**neuronal plasticity** The capacity of a nervous system to change with experience.

**neuropeptide** A relatively short chain of amino acids that serves as a neurotransmitter.

**neurotransmitter** A molecule that is released from the synaptic terminal of a neuron at a chemical synapse, diffuses across the synaptic cleft, and binds to the postsynaptic cell, triggering a response.

**neutral variation** Genetic variation that does not provide a selective advantage or disadvantage.

**neutron** A subatomic particle having no electrical charge (electrically neutral), with a mass of about $1.7 \times 10^{-24}$ g, found in the nucleus of an atom.

**neutrophil** The most abundant type of white blood cell. Neutrophils are phagocytic and tend to self-destruct as they destroy foreign invaders, limiting their life span to a few days.

**nitrogen cycle** The natural process by which nitrogen, either from the atmosphere or from decomposed organic material, is converted by soil bacteria to compounds assimilated by plants. This incorporated nitrogen is then taken in by other organisms and subsequently released, acted on by bacteria, and made available again to the nonliving environment.

**nitrogen fixation** The conversion of atmospheric nitrogen ($N_2$) to ammonia ($NH_3$). Biological nitrogen fixation is carried out by certain prokaryotes, some of which have mutualistic relationships with plants.

**nociceptor** (nō′-si-sep′-tur) A sensory receptor that responds to noxious or painful stimuli; also called a pain receptor.

**node** A point along the stem of a plant at which leaves are attached.

**node of Ranvier** (ron′-vē-ā′) Gap in the myelin sheath of certain axons where an action potential may be generated. In saltatory conduction, an action potential is regenerated at each node, appearing to "jump" along the axon from node to node.

**nodule** A swelling on the root of a legume. Nodules are composed of plant cells that contain nitrogen-fixing bacteria of the genus *Rhizobium*.

**noncompetitive inhibitor** A substance that reduces the activity of an enzyme by binding to a location remote from the active site, changing the enzyme's shape so that the active site no longer effectively catalyzes the conversion of substrate to product.

**nondisjunction** An error in meiosis or mitosis in which members of a pair of homologous chromosomes or sister chromatids fail to separate properly from each other.

**nonequilibrium model** A model that maintains that communities change constantly after being buffeted by disturbances.

**nonpolar covalent bond** A type of covalent bond in which electrons are shared equally between two atoms of similar electronegativity.

**nonsense mutation** A mutation that changes an amino acid codon to one of the three stop codons, resulting in a shorter and usually nonfunctional protein.

**northern coniferous forest** A terrestrial biome characterized by long, cold winters and dominated by cone-bearing trees.

**notochord** (nō′-tuh-kord′) A longitudinal, flexible rod that runs along the anterior-posterior axis of a chordate in the dorsal part of the body.

**nuclear envelope** In a eukaryotic cell, the double membrane that surrounds the nucleus, perforated with pores that regulate traffic with the cytoplasm. The outer membrane is continuous with the endoplasmic reticulum.

**nuclear lamina** A netlike array of protein filaments that lines the inner surface of the nuclear envelope and helps maintain the shape of the nucleus.

**nucleariid** Member of a group of unicellular, amoeboid protists that are more closely related to fungi than they are to other protists.

**nuclease** An enzyme that cuts DNA or RNA, either removing one or a few bases or hydrolyzing the DNA or RNA completely into its component nucleotides.

**nucleic acid** (nū-klā′-ik) A polymer (polynucleotide) consisting of many nucleotide monomers; serves as a blueprint for proteins and, through the actions of proteins, for all cellular activities. The two types are DNA and RNA.

**nucleic acid hybridization** The process of base pairing between a gene and a complementary sequence on another nucleic acid molecule.

**nucleic acid probe** In DNA technology, a labeled single-stranded nucleic acid molecule used to locate a specific nucleotide sequence in a nucleic acid sample. Molecules of the probe hydrogen-bond to the complementary sequence wherever it occurs; radioactive, fluorescent, or other labeling of the probe allows its location to be detected.

**nucleoid** (nū′-klē-oyd) A non-membrane-enclosed region in a prokaryotic cell where its chromosome is located.

**nucleolus** (nū-klē′-ō-lus) (plural, **nucleoli**) A specialized structure in the nucleus consisting of chromosomal regions containing ribosomal RNA (rRNA) genes along with ribosomal proteins imported from the cytoplasm; site of rRNA synthesis and ribosomal subunit assembly. *See also* ribosome.

**nucleosome** (nū′-klē-ō-sōm′) The basic, bead-like unit of DNA packing in eukaryotes, consisting of a segment of DNA wound around a protein core composed of two copies of each of four types of histone.

**nucleotide** (nū′-klē-ō-tīd′) The building block of a nucleic acid, consisting of a five-carbon sugar covalently bonded to a nitrogenous base and one to three phosphate groups.

**nucleotide excision repair** A repair system that removes and then correctly replaces a damaged segment of DNA using the undamaged strand as a guide.

**nucleotide-pair substitution** A type of point mutation in which one nucleotide in a DNA strand and its partner in the complementary strand are replaced by another pair of nucleotides.

**nucleus** (1) An atom's central core, containing protons and neutrons. (2) The organelle of a eukaryotic cell that contains the genetic material in the form of chromosomes, made up of chromatin. (3) A cluster of neurons.

**nutrition** The process by which an organism takes in and makes use of food substances.

**obligate anaerobe** (ob′-lig-et an′-uh-rōb) An organism that only carries out fermentation or anaerobic respiration. Such organisms cannot use oxygen and in fact may be poisoned by it.

**ocean acidification** The process by which the pH of the ocean is lowered (made more acidic) when excess $CO_2$ dissolves in seawater and forms carbonic acid ($H_2CO_3$).

**oceanic pelagic zone** Most of the ocean's waters far from shore, constantly mixed by ocean currents.

**odorant** A molecule that can be detected by sensory receptors of the olfactory system.

**Okazaki fragment** (ō′-kah-zah′-kē) A short segment of DNA synthesized away from the replication fork on a template strand during DNA replication. Many such segments are joined together to make up the lagging strand of newly synthesized DNA.

**olfaction** The sense of smell.

**oligodendrocyte** A type of glial cell that forms insulating myelin sheaths around the axons of neurons in the central nervous system.

**oligotrophic lake** A nutrient-poor, clear lake with few phytoplankton.

**ommatidium** (ōm′-uh-tid′-ē-um) (plural, **ommatidia**) One of the facets of the compound eye of arthropods and some polychaete worms.

**omnivore** An animal that regularly eats animals as well as plants or algae.

**oncogene** (on′-kō-jēn) A gene found in viral or cellular genomes that is involved in triggering molecular events that can lead to cancer.

**oocyte** A cell in the female reproductive system that differentiates to form an egg.

**oogenesis** (ō′-uh-jen′-uh-sis) The process in the ovary that results in the production of female gametes.

**oogonium** (ō′-uh- gō′-nē-em) (plural, **oogonia**) A cell that divides mitotically to form oocytes.

**open circulatory system** A circulatory system in which fluid called hemolymph bathes the tissues and organs directly and there is no distinction between the circulating fluid and the interstitial fluid.

**operator** In bacterial and phage DNA, a sequence of nucleotides near the start of an operon to which an active repressor can attach. The binding of the repressor prevents RNA polymerase from attaching to the promoter and transcribing the genes of the operon.

**operon** (op′-er-on) A unit of genetic function found in bacteria and phages, consisting of a promoter, an operator, and a coordinately regulated cluster of genes whose products function in a common pathway.

**opisthokont** (uh-pis′-thuh-kont′) Member of an extremely diverse clade of eukaryotes that includes fungi, animals, and several closely related groups of protists.

**opposable thumb** A thumb that can touch the ventral surface (fingerprint side) of the fingertip of all four fingers of the same hand with its own ventral surface.

**opsin** A membrane protein bound to a light-absorbing pigment molecule.

**oral cavity** The mouth of an animal.

**order** In Linnaean classification, the taxonomic category above the level of family.

**organ** A specialized center of body function composed of several different types of tissues.

**organ of Corti** The actual hearing organ of the vertebrate ear, located in the floor of the cochlear duct in the inner ear; contains the receptor cells (hair cells) of the ear.

**organ system** A group of organs that work together in performing vital body functions.

**organelle** (ōr-guh-nel′) Any of several kinds of membrane-enclosed structures with specialized functions, suspended in the cytosol of eukaryotic cells.

**organic compound** A chemical compound containing carbon.

**organism** An individual living thing, consisting of one or more cells.

**organismal ecology** The branch of ecology concerned with the morphological, physiological, and behavioral ways in which individual organisms meet the challenges posed by their biotic and abiotic environments.

**organogenesis** (ōr-gan′-ō-jen′-uh-sis) The process in which organ rudiments develop from the three germ layers after gastrulation.

**origin of replication** Site where the replication of a DNA molecule begins, consisting of a specific sequence of nucleotides.

**osmoconformer** An animal that is isoosmotic with its environment.

**osmolarity** (oz′-mō-lār′-uh-tē) Solute concentration expressed as molarity.

**osmoregulation** Regulation of solute concentrations and water balance by a cell or organism.

**osmoregulator** An animal that controls its internal osmolarity independent of the external environment.

**osmosis** (oz-mō′-sis) The diffusion of free water molecules across a selectively permeable membrane.

**osteichthyan** (os′-tē-ik′-thē-an) Member of a vertebrate clade with jaws and mostly bony skeletons.

**outer ear** One of three main regions of the ear in reptiles (including birds) and mammals; made up of the auditory canal and, in many birds and mammals, the pinna.

**outgroup** A species or group of species from an evolutionary lineage that is known to have diverged before the lineage that contains the group of species being studied. An outgroup is selected so that its members are closely related to the group of species being studied, but not as closely related as any study-group members are to each other.

**oval window** In the vertebrate ear, a membrane-covered gap in the skull bone, through which sound waves pass from the middle ear to the inner ear.

**ovarian cycle** (ō-vār′-ē-un) The cyclic recurrence of the follicular phase, ovulation, and the luteal phase in the mammalian ovary, regulated by hormones.

**ovary** (ō′-vuh-rē) (1) In flowers, the portion of a carpel in which the egg-containing ovules develop. (2) In animals, the structure that produces female gametes and reproductive hormones.

**oviduct** (ō′-vuh-duct) A tube passing from the ovary to the vagina in invertebrates or to the uterus in vertebrates, where it is also known as a fallopian tube.

**ovulation** The release of an egg from an ovary. In humans, an ovarian follicle releases an egg during each uterine (menstrual) cycle.

**ovule** (ō′-vyūl) A structure that develops within the ovary of a seed plant and contains the female gametophyte.

**oxidation** The complete or partial loss of electrons from a substance involved in a redox reaction.

**oxidative phosphorylation** (fos′-fōr-uh-lā′-shun) The production of ATP using energy derived from the redox reactions of an electron transport chain; the third major stage of cellular respiration.

**oxidizing agent** The electron acceptor in a redox reaction.

**oxytocin** (ok′-si-tō′-sen) A hormone produced by the hypothalamus and released from the posterior pituitary. It induces contractions of the uterine muscles during labor and causes the mammary glands to eject milk during nursing.

**P generation** The true-breeding (homozygous) parent individuals from which $F_1$ hybrid offspring are derived in studies of inheritance; P stands for "parental."

**P site** One of a ribosome's three binding sites for tRNA during translation. The P site holds the tRNA carrying the growing polypeptide chain. (P stands for peptidyl tRNA.)

***p53* gene** A tumor-suppressor gene that codes for a specific transcription factor that promotes the synthesis of proteins that inhibit the cell cycle.

**paedomorphosis** (pē′-duh-mōr′-fuh-sis) The retention in an adult organism of the juvenile features of its evolutionary ancestors.

**pain receptor** A sensory receptor that responds to noxious or painful stimuli; also called a nociceptor.

**paleontology** (pā′-lē-un-tol′-ō-jē) The scientific study of fossils.

**pancreas** (pan′-krē-us) A gland with exocrine and endocrine tissues. The exocrine portion functions in digestion, secreting enzymes and an alkaline solution into the small intestine via a duct; the ductless endocrine portion functions in homeostasis, secreting the hormones insulin and glucagon into the blood.

**pandemic** A global epidemic.

**Pangaea** (pan-jē′-uh) The supercontinent that formed near the end of the Paleozoic era, when plate movements brought all the landmasses of Earth together.

**parabasalid** A protist, such as a trichomonad, with modified mitochondria.

**paraphyletic** (pār′-uh-fī-let′-ik) Pertaining to a group of taxa that consists of a common ancestor and some, but not all, of its descendants.

**parasite** (pār′-uh-sīt) An organism that feeds on the cell contents, tissues, or body fluids of another species (the host) while in or on the host organism. Parasites harm but usually do not kill their host.

**parasitism** (pār′-uh-sit-izm) A +/− ecological interaction in which one organism, the parasite, benefits by feeding upon another organism, the host, which is harmed; some parasites live within the host (feeding on its tissues), while others feed on the host's external surface.

**parasympathetic division** One of three divisions of the autonomic nervous system; generally enhances body activities that gain and conserve energy, such as digestion and reduced heart rate.

**parenchyma cell** (puh-ren′-ki-muh) A relatively unspecialized plant cell type that carries out most of the metabolism, synthesizes and stores organic products, and develops into a more differentiated cell type.

**parental type** An offspring with a phenotype that matches one of the true-breeding parental (P generation) phenotypes; also refers to the phenotype itself.

**parthenogenesis** (par′-thuh-nō′-jen′-uh-sis) A form of asexual reproduction in which females produce offspring from unfertilized eggs.

**partial pressure** The pressure exerted by a particular gas in a mixture of gases (for instance, the pressure exerted by oxygen in air).

**passive transport** The diffusion of a substance across a biological membrane with no expenditure of energy.

**pathogen** An organism or virus that causes disease.

**pathogen-associated molecular pattern (PAMP)** A molecular sequence that is specific to a certain pathogen.

**pattern formation** The development of a multicellular organism's spatial organization, the arrangement of organs and tissues in their characteristic places in three-dimensional space.

**PCR** *See* polymerase chain reaction.

**pedigree** A diagram of a family tree with conventional symbols, showing the occurrence of heritable characters in parents and offspring over multiple generations.

**pelagic zone** The open-water component of aquatic biomes.

**penis** The copulatory structure of male mammals.

**pepsin** An enzyme present in gastric juice that begins the hydrolysis of proteins. Pepsin is synthesized as an inactive precursor form, pepsinogen.

**peptide bond** The covalent bond between the carboxyl group on one amino acid and the amino group on another, formed by a dehydration reaction.

**peptidoglycan** (pep′-tid-ō-glī′-kan) A type of polymer in bacterial cell walls consisting of modified sugars cross-linked by short polypeptides.

**perception** The interpretation of sensory system input by the brain.

**pericycle** The outermost layer in the vascular cylinder, from which lateral roots arise.

**periderm** (pār′-uh-derm′) The protective coat that replaces the epidermis in woody plants during secondary growth, formed of the cork and cork cambium.

**peripheral nervous system (PNS)** The sensory and motor neurons that connect to the central nervous system.

**peripheral protein** A protein loosely bound to the surface of a membrane or to part of an integral protein and not embedded in the lipid bilayer.

**peristalsis** (păr′-uh-stal′-sis) (1) Alternating waves of contraction and relaxation in the smooth muscles lining the alimentary canal that push food along the canal. (2) A type of movement on land produced by rhythmic waves of muscle contractions passing from front to back, as in many annelids.

**peritubular capillary** One of the tiny blood vessels that form a network surrounding the proximal and distal tubules in the kidney.

**peroxisome** (puh-rok′-suh-sōm′) An organelle containing enzymes that transfer hydrogen atoms from various substrates to oxygen ($O_2$), producing and then degrading hydrogen peroxide ($H_2O_2$).

**petal** A modified leaf of a flowering plant. Petals are the often colorful parts of a flower that advertise it to insects and other pollinators.

**petiole** (pet′-ē-ōl) The stalk of a leaf, which joins the leaf to a node of the stem.

**pH** A measure of hydrogen ion concentration equal to $-\log [H^+]$ and ranging in value from 0 to 14.

**phage** (fāj) A virus that infects bacteria; also called a bacteriophage.

**phagocytosis** (fag′-ō-sī-tō′-sis) A type of endocytosis in which large particulate substances or small organisms are taken up by a cell. It is carried out by some protists and by certain immune cells of animals (in mammals, mainly macrophages, neutrophils, and dendritic cells).

**pharyngeal cleft** (fuh-rin′-jē-ul) In chordate embryos, one of the grooves that separate a series of pouches along the sides of the pharynx and may develop into a pharyngeal slit.

**pharyngeal slit** (fuh-rin′-jē-ul) In chordate embryos, one of the slits that form from the pharyngeal clefts and communicate to the outside, later developing into gill slits in many vertebrates.

**pharynx** (făr′-inks) (1) An area in the vertebrate throat where air and food passages cross. (2) In flatworms, the muscular tube that protrudes from the ventral side of the worm and ends in the mouth.

**phenotype** (fē′-nō-tīp) The observable physical and physiological traits of an organism, which are determined by its genetic makeup.

**pheromone** (făr′-uh-mōn) In animals and fungi, a small molecule released into the environment that functions in communication between members of the same species. In animals, it acts much like a hormone in influencing physiology and behavior.

**phloem** (flō′-em) Vascular plant tissue consisting of living cells arranged into elongated tubes that transport sugar and other organic nutrients throughout the plant.

**phloem sap** The sugar-rich solution carried through a plant's sieve tubes.

**phosphate group** A chemical group consisting of a phosphorus atom bonded to four oxygen atoms; important in energy transfer.

**phospholipid** (fos′-fō-lip′-id) A lipid made up of glycerol joined to two fatty acids and a phosphate group. The hydrocarbon chains of the fatty acids act as nonpolar, hydrophobic tails, while the rest of the molecule acts as a polar, hydrophilic head. Phospholipids form bilayers that function as biological membranes.

**phosphorylated intermediate** A molecule (often a reactant) with a phosphate group covalently bound to it, making it more reactive (less stable) than the unphosphorylated molecule.

**photic zone** (fō′-tic) The narrow top layer of an ocean or lake, where light penetrates sufficiently for photosynthesis to occur.

**photomorphogenesis** Effects of light on plant morphology.

**photon** (fō′-ton) A quantum, or discrete quantity, of light energy that behaves as if it were a particle.

**photoperiodism** (fō′-tō-pēr′-ē-ō-dizm) A physiological response to photoperiod, the relative lengths of night and day. An example of photoperiodism is flowering.

**photophosphorylation** (fō′-tō-fos′-fōr-uh-lā′-shun) The process of generating ATP from ADP and phosphate by means of chemiosmosis, using a proton-motive force generated across the thylakoid membrane of the chloroplast or the membrane of certain prokaryotes during the light reactions of photosynthesis.

**phosphorylation cascade** A series of chemical reactions during cell signaling mediated by enzymes (kinases), in which each kinase in turn phosphylates and activates another, ultimately leading to phosphorylation of many proteins.

**photoreceptor** An electromagnetic receptor that detects the radiation known as visible light.

**photorespiration** A metabolic pathway that consumes oxygen and ATP, releases carbon dioxide, and decreases photosynthetic output. Photorespiration generally occurs on hot, dry, bright days, when stomata close and the $O_2/CO_2$ ratio in the leaf increases, favoring the binding of $O_2$ rather than $CO_2$ by rubisco.

**photosynthesis** (fō′-tō-sin′-thi-sis) The conversion of light energy to chemical energy that is stored in sugars or other organic compounds; occurs in plants, algae, and certain prokaryotes.

**photosystem** A light-capturing unit located in the thylakoid membrane of the chloroplast or in the membrane of some prokaryotes, consisting of a reaction-center complex surrounded by numerous light-harvesting complexes. There are two types of photosystems, I and II; they absorb light best at different wavelengths.

**photosystem I (PS I)** One of two light-capturing units in a chloroplast's thylakoid membrane or in the membrane of some prokaryotes; it has two molecules of P700 chlorophyll *a* at its reaction center.

**photosystem II (PS II)** One of two light-capturing units in a chloroplast's thylakoid membrane or in the membrane of some prokaryotes; it has two molecules of P680 chlorophyll *a* at its reaction center.

**phototropism** (fō′-tō-trō′-pizm) Growth of a plant shoot toward or away from light.

**phyllotaxy** (fil′-uh-tak′-sē) The pattern of leaf attachment to the stem of a plant.

**phylogenetic species concept** Definition of a species as the smallest group of individuals that share a common ancestor, forming one branch on the tree of life.

**phylogenetic tree** A branching diagram that represents a hypothesis about the evolutionary history of a group of organisms.

**phylogeny** (fī-loj′-uh-nē) The evolutionary history of a species or group of related species.

**phylum** (fī′-lum) (plural, **phyla**) In Linnaean classification, the taxonomic category above class.

**physiology** The processes and functions of an organism.

**phytochrome** (fī′-tuh-krōm) A type of light receptor in plants that mostly absorbs red light and regulates many plant responses, such as seed germination and shade avoidance.

**pilus** (plural, **pili**) (pī′-lus, pī′-lī) In bacteria, a structure that links one cell to another at the start of conjugation; also known as a sex pilus or conjugation pilus.

**pinocytosis** (pī′-nō-sī-tō′-sis) A type of endocytosis in which the cell ingests extracellular fluid and its dissolved solutes.

**pistil** A single carpel (a simple pistil) or a group of fused carpels (a compound pistil).

**pith** Ground tissue that is internal to the vascular tissue in a stem; in many monocot roots, parenchyma cells that form the central core of the vascular cylinder.

**pituitary gland** (puh-tū′-uh-tār′-ē) An endocrine gland at the base of the hypothalamus; consists of a posterior lobe, which stores and releases two hormones produced by the hypothalamus, and an anterior lobe, which produces and secretes many hormones that regulate diverse body functions.

**placenta** (pluh-sen′-tuh) A structure in the pregnant uterus for nourishing a viviparous fetus with the mother's blood supply; formed from the uterine lining and embryonic membranes.

**plasma** (plaz′-muh) The liquid matrix of blood in which the blood cells are suspended.

**plasma cell** The antibody-secreting effector cell of humoral immunity. Plasma cells arise from antigen-stimulated B cells.

**plasma membrane** The membrane at the boundary of every cell that acts as a selective barrier, regulating the cell's chemical composition.

**plasmid** (plaz′-mid) A small, circular, double-stranded DNA molecule that carries accessory genes separate from those of a bacterial chromosome; in DNA cloning, can be used as a vector carrying up to about 10,000 base pairs (10 kb) of DNA.

**plasmodesma** (plaz′-mō-dez′-muh) (plural, **plasmodesmata**) An open channel through the cell wall that connects the cytoplasm of adjacent plant cells, allowing water, small solutes, and some larger molecules to pass between the cells.

**plasmogamy** (plaz-moh′-guh-mē) In fungi, the fusion of the cytoplasm of cells from two individuals; occurs as one stage of sexual reproduction, followed later by karyogamy.

**plasmolysis** (plaz-mol′-uh-sis) A phenomenon in walled cells in which the cytoplasm shrivels and the plasma membrane pulls away from the cell wall; occurs when the cell loses water to a hypertonic environment.

**plastid** One of a family of closely related organelles that includes chloroplasts, chromoplasts, and amyloplasts. Plastids are found in the cells of photosynthetic eukaryotes.

**plate tectonics** The theory that the continents are part of great plates of Earth's crust that float on the hot, underlying portion of the mantle. Movements in the mantle cause the continents to move slowly over time.

**platelet** A pinched-off cytoplasmic fragment of a specialized bone marrow cell. Platelets circulate in the blood and are important in blood clotting.

**pleiotropy** (plī′-o-truh-pē) The ability of a single gene to have multiple effects.

**pluripotent** Describing a cell that can give rise to many, but not all, parts of an organism.

**point mutation** A change in a single nucleotide pair of a gene.

**polar covalent bond** A covalent bond between atoms that differ in electronegativity. The shared electrons are pulled closer to the more electronegative atom, making it slightly negative and the other atom slightly positive.

**polar molecule** A molecule (such as water) with an uneven distribution of charges in different regions of the molecule.

**pollen grain** In seed plants, a structure consisting of the male gametophyte enclosed within a pollen wall.

**pollen tube** A tube that forms after germination of the pollen grain and that functions in the delivery of sperm to the ovule.

**pollination** (pol′-uh-nā′-shun) The transfer of pollen to the part of a seed plant containing the ovules, a process required for fertilization.

**poly-A tail** A sequence of 50–250 adenine nucleotides added onto the 3′ end of a pre-mRNA molecule.

**polygamous** Referring to a type of relationship in which an individual of one sex mates with several of the other.

**polygenic inheritance** (pol′-ē-jen′-ik) An additive effect of two or more genes on a single phenotypic character.

**polymer** (pol′-uh-mer) A long molecule consisting of many similar or identical monomers linked together by covalent bonds.

**polymerase chain reaction (PCR)** (puh-lim′-uh-rās) A technique for amplifying DNA *in vitro* by incubating it with specific primers, a heat-resistant DNA polymerase, and nucleotides.

**polynucleotide** (pol′-ē-nū′-klē-ō-tīd) A polymer consisting of many nucleotide monomers in a chain. The nucleotides can be those of DNA or RNA.

**polypeptide** (pol′-ē-pep′-tīd) A polymer of many amino acids linked together by peptide bonds.

**polyphyletic** (pol′-ē-fī-let′-ik) Pertaining to a group of taxa that includes distantly related organisms but does not include their most recent common ancestor.

**polyploidy** (pol′-ē-ploy′-dē) A chromosomal alteration in which the organism possesses more than two complete chromosome sets. It is the result of an accident of cell division.

**polysaccharide** (pol′-ē-sak′-uh-rīd) A polymer of many monosaccharides, formed by dehydration reactions.

**polytomy** (puh-lit′-uh-mē) In a phylogenetic tree, a branch point from which more than two descendant taxa emerge. A polytomy indicates that the evolutionary relationships between the descendant taxa are not yet clear.

**pons** A portion of the brain that participates in certain automatic, homeostatic functions, such as regulating the breathing centers in the medulla.

**population** A group of individuals of the same species that live in the same area and interbreed, producing fertile offspring.

**population dynamics** The study of how complex interactions between biotic and abiotic factors influence variations in population size.

**population ecology** The study of populations in relation to their environment, including environmental influences on population density and distribution, age structure, and variations in population size.

**positional information** Molecular cues that control pattern formation in an animal or plant embryonic structure by indicating a cell's location relative to the organism's body axes. These cues elicit a response by genes that regulate development.

**positive feedback** A form of regulation in which an end product of a process speeds up that process; in physiology, a control mechanism in which a change in a variable triggers a response that reinforces or amplifies the change.

**positive interactions** A +/+ or +/0 ecological interaction in which at least one of the interacting species benefits and neither is harmed; positive interactions include mutualism and commensalism.

**positive pressure breathing** A breathing system in which air is forced into the lungs.

**posterior** Pertaining to the rear, or tail end, of a bilaterally symmetric animal.

**posterior pituitary** An extension of the hypothalamus composed of nervous tissue that secretes oxytocin and antidiuretic hormone made in the hypothalamus; a temporary storage site for these hormones.

**postzygotic barrier** (pōst′-zī-got′-ik) A reproductive barrier that prevents hybrid zygotes produced by two different species from developing into viable, fertile adults.

**potential energy** The energy that matter possesses as a result of its location or spatial arrangement (structure).

**predation** An interaction between species in which one species, the predator, eats the other, the prey.

**prediction** In deductive reasoning, a forecast that follows logically from a hypothesis. By testing predictions, experiments may allow certain hypotheses to be rejected.

**prepuce** (prē′-pyūs) A fold of skin covering the head of the clitoris or penis.

**pressure potential** ($\Psi_P$) A component of water potential that consists of the physical pressure on a solution, which can be positive, zero, or negative.

**prezygotic barrier** (prē′-zī-got′-ik) A reproductive barrier that impedes mating between species or hinders fertilization if interspecific mating is attempted.

**primary cell wall** In plants, a relatively thin and flexible layer that surrounds the plasma membrane of a young cell.

**primary consumer** An herbivore; an organism that eats plants or other autotrophs.

**primary electron acceptor** In the thylakoid membrane of a chloroplast or in the membrane of some prokaryotes, a specialized molecule that shares the reaction-center complex with a pair of chlorophyll *a* molecules and that accepts an electron from them.

**primary growth** Growth produced by apical meristems, lengthening stems and roots.

**primary immune response** The initial adaptive immune response to an antigen, which appears after a lag of about 10–17 days.

**primary oocyte** (ō′-uh-sīt) An oocyte prior to completion of meiosis I.

**primary producer** An autotroph, usually a photosynthetic organism. Collectively, autotrophs make up the trophic level of an ecosystem that ultimately supports all other levels.

**primary production** The amount of light energy converted to chemical energy (organic compounds) by the autotrophs in an ecosystem during a given time period.

**primary structure** The level of protein structure referring to the specific linear sequence of amino acids.

**primary succession** A type of ecological succession that occurs in an area where there were originally no organisms present and where soil has not yet formed.

**primary transcript** An initial RNA transcript from any gene; also called pre-mRNA when transcribed from a protein-coding gene.

**primase** An enzyme that joins RNA nucleotides to make a primer during DNA replication, using the parental DNA strand as a template.

**primer** A short stretch of RNA with a free 3′ end, bound by complementary base pairing to the template strand and elongated with DNA nucleotides during DNA replication.

**prion** An infectious agent that is a misfolded version of a normal cellular protein. Prions appear to increase in number by converting correctly folded versions of the protein to more prions.

**problem solving** The cognitive activity of devising a method to proceed from one state to another in the face of real or apparent obstacles.

**producer** An organism that produces organic compounds from $CO_2$ by harnessing light energy (in photosynthesis) or by oxidizing inorganic chemicals (in chemosynthetic reactions carried out by some prokaryotes).

**product** A material resulting from a chemical reaction.

**production efficiency** The percentage of energy stored in assimilated food that is not used for respiration or eliminated as waste.

**progesterone** A steroid hormone that contributes to the menstrual cycle and prepares the uterus for pregnancy.

**prokaryote** An organism that has a prokaryotic cell; an informal term for an organism in either domain Bacteria or domain Archaea.

**prokaryotic cell** (prō′-kār′-ē-ot′-ik) A type of cell lacking a membrane-enclosed nucleus and membrane-enclosed organelles. Organisms with prokaryotic cells (bacteria and archaea) are called prokaryotes.

**prometaphase** The second stage of mitosis, in which the nuclear envelope fragments and the spindle microtubules attach to the kinetochores of the chromosomes.

**promoter** A specific nucleotide sequence in the DNA of a gene that binds RNA polymerase, positioning it to start transcribing RNA at the appropriate place.

**prophage** (prō′-fāj) A phage genome that has been inserted into a specific site on a bacterial chromosome.

**prophase** The first stage of mitosis, in which the chromatin condenses into discrete chromosomes visible with a light microscope, the mitotic spindle begins to form, and the nucleolus disappears but the nucleus remains intact.

**prostate gland** (pros′-tāt) A gland in human males that secretes an acid-neutralizing component of semen.

**protease** An enzyme that digests proteins by hydrolysis.

**protein** (prō′-tēn) A biologically functional molecule consisting of one or more polypeptides folded and coiled into a specific three-dimensional structure.

**protein kinase** An enzyme that transfers phosphate groups from ATP to a protein, thus phosphorylating the protein.

**protein phosphatase** An enzyme that removes phosphate groups from (dephosphorylates) proteins, often functioning to reverse the effect of a protein kinase.

**proteoglycan** (prō′-tē-ō-glī′-kan) A large molecule consisting of a small core protein with many carbohydrate chains attached, found in the extracellular matrix of animal cells. A proteoglycan may consist of up to 95% carbohydrate.

**proteome** The entire set of proteins expressed by a given cell, tissue, or organism.

**proteomics** (prō′-tē-ō′-miks) The systematic study of the full protein sets (proteomes) encoded by genomes.

**protist** An informal term applied to any eukaryote that is not a plant, animal, or fungus. Most protists are unicellular, though some are colonial or multicellular.

**protocell** An abiotic precursor of a living cell that had a membrane-like structure and that maintained an internal chemistry different from that of its surroundings.

**proton** (prō′-ton) A subatomic particle with a single positive electrical charge, with a mass of about $1.7 \times 10^{-24}$ g, found in the nucleus of an atom.

**proton pump** An active transport protein in a cell membrane that uses ATP to transport hydrogen ions out of a cell against their concentration gradient, generating a membrane potential in the process.

**proton-motive force** The potential energy stored in the form of a proton electrochemical gradient, generated by the pumping of hydrogen ions ($H^+$) across a biological membrane during chemiosmosis.

**proto-oncogene** (prō′-tō-on′-kō-jēn) A normal cellular gene that has the potential to become an oncogene.

**protoplast** The living part of a plant cell, which also includes the plasma membrane.

**provirus** A viral genome that is permanently inserted into a host genome.

**proximal tubule** In the vertebrate kidney, the portion of a nephron immediately downstream from Bowman's capsule that conveys and helps refine filtrate.

**pseudogene** (sū′-dō-jēn) A DNA segment that is very similar to a real gene but does not yield a functional product; a DNA segment that formerly functioned as a gene but has become inactivated in a particular species because of mutation.

**pseudopodium** (sū′-dō-pō′-dē-um) (plural, **pseudopodia**) A cellular extension of amoeboid cells used in moving and feeding.

**pulse** The rhythmic bulging of the artery walls with each heartbeat.

**punctuated equilibria** In the fossil record, long periods of apparent stasis, in which a species undergoes little or no morphological change, interrupted by relatively brief periods of sudden change.

**Punnett square** A diagram used in the study of inheritance to show the predicted genotypic results of random fertilization in genetic crosses between individuals of known genotype.

**pupil** The opening in the iris, which admits light into the interior of the vertebrate eye. Muscles in the iris regulate its size.

**purine** (pyū′-rēn) One of two types of nitrogenous bases found in nucleotides, characterized by a six-membered ring fused to a five-membered ring. Adenine (A) and guanine (G) are purines.

**pyrimidine** (puh-rim′-uh-dēn) One of two types of nitrogenous bases found in nucleotides, characterized by a six-membered ring. Cytosine (C), thymine (T), and uracil (U) are pyrimidines.

**quantitative character** A heritable feature that varies continuously over a range rather than in an either-or fashion.

**quaternary structure** (kwot′-er-nār′-ē) The particular shape of a complex, aggregate protein, defined by the characteristic three-dimensional arrangement of its constituent subunits, each a polypeptide.

**R plasmid** A bacterial plasmid carrying genes that confer resistance to certain antibiotics.

**radial symmetry** Symmetry in which the body is shaped like a pie or barrel (lacking a left side and a right side) and can be divided into mirror-imaged halves by any plane through its central axis.

**radicle** An embryonic root of a plant.

**radioactive isotope** An isotope (an atomic form of a chemical element) that is unstable; the nucleus decays spontaneously, giving off detectable particles and energy.

**radiometric dating** A method for determining the absolute age of rocks and fossils, based on the half-life of radioactive isotopes.

**ras gene** A gene that codes for Ras, a G protein that relays a growth signal from a growth factor receptor on the plasma membrane to a cascade of protein kinases, ultimately resulting in stimulation of the cell cycle.

**ray-finned fish** Member of the class Actinopterygii, aquatic osteichthyans with fins supported by long, flexible rays, including tuna, bass, and herring.

**reabsorption** In excretory systems, the recovery of solutes and water from filtrate.

**reactant** A starting material in a chemical reaction.

**reaction-center complex** A complex of proteins associated with a special pair of chlorophyll *a* molecules and a primary electron acceptor. Located centrally in a photosystem, this complex triggers the light reactions of photosynthesis. Excited by light energy, the pair of chlorophylls donates an electron to the primary electron acceptor, which passes an electron to an electron transport chain.

**reading frame** On an mRNA, the triplet grouping of ribonucleotides used by the translation machinery during polypeptide synthesis.

**receptacle** The base of a flower; the part of the stem that is the site of attachment of the floral organs.

**reception** The binding of a signaling molecule to a receptor protein, activating the receptor by causing it to change shape. *See also* sensory reception.

**receptor potential** An initial response of a receptor cell to a stimulus, consisting of a change in voltage across the receptor membrane proportional to the stimulus strength.

**receptor-mediated endocytosis** (en′-dō-sī-tō′-sis) The movement of specific molecules into a cell by the inward budding of vesicles containing proteins with receptor sites specific to the molecules being taken in; enables a cell to acquire bulk quantities of specific substances.

**recessive allele** An allele whose phenotypic effect is not observed in a heterozygote.

**recombinant chromosome** A chromosome created when crossing over combines DNA from two parents into a single chromosome.

**recombinant DNA molecule** A DNA molecule made *in vitro* with segments from different sources.

**recombinant type (recombinant)** An offspring whose phenotype differs from that of the true-breeding P generation parents; also refers to the phenotype itself.

**rectum** The terminal portion of the large intestine, where the feces are stored prior to elimination.

**red alga** A photosynthetic protist, named for its color, which results from a red pigment that masks the green of chlorophyll. Most red algae are multicellular and marine.

**redox reaction** (rē'-doks) A chemical reaction involving the complete or partial transfer of one or more electrons from one reactant to another; short for **red**uction-**ox**idation reaction.

**reducing agent** The electron donor in a redox reaction.

**reduction** The complete or partial addition of electrons to a substance involved in a redox reaction.

**reflex** An automatic reaction to a stimulus, mediated by the spinal cord or lower brain.

**refractory period** (rē-frakt'-ōr-ē) The short time immediately after an action potential in which the neuron cannot respond to another stimulus, owing to the inactivation of voltage-gated sodium channels.

**regression line** A line drawn through a scatter plot that shows the general trend of the data. It represents an equation that is calculated mathematically to best fit the data and can be used to predict the value of the dependent variable for any value of the independent variable.

**regulator** An animal for which mechanisms of homeostasis moderate internal changes in a particular variable in the face of external fluctuation of that variable.

**regulatory gene** A gene that codes for a protein, such as a repressor, that controls the transcription of another gene or group of genes.

**reinforcement** In evolutionary biology, a process in which natural selection strengthens prezygotic barriers to reproduction, thus reducing the chances of hybrid formation. Such a process is likely to occur only if hybrid offspring are less fit than members of the parent species.

**relative abundance** The proportional abundance of different species in a community.

**relative fitness** The contribution an individual makes to the gene pool of the next generation, relative to the contributions of other individuals in the population.

**renal cortex** The outer portion of the vertebrate kidney.

**renal medulla** The inner portion of the vertebrate kidney, beneath the renal cortex.

**renal pelvis** The funnel-shaped chamber that receives processed filtrate from the vertebrate kidney's collecting ducts and is drained by the ureter.

**repetitive DNA** Nucleotide sequences, usually noncoding, that are present in many copies in a eukaryotic genome. The repeated units may be short and arranged tandemly (in series) or long and dispersed in the genome.

**replication fork** A Y-shaped region on a replicating DNA molecule where the parental strands are being unwound and new strands are being synthesized.

**repressor** A protein that inhibits gene transcription. In prokaryotes, repressors bind to the DNA in or near the promoter. In eukaryotes, repressors may bind to control elements within enhancers, to activators, or to other proteins in a way that blocks activators from binding to DNA.

**reproductive isolation** The existence of biological factors (barriers) that impede members of two species from producing viable, fertile offspring.

**reptile** Member of the clade of amniotes that includes tuataras, lizards, snakes, turtles, crocodilians, and birds.

**residual volume** The amount of air that remains in the lungs after forceful exhalation.

**resource partitioning** The division of environmental resources by coexisting species such that the niche of each species differs by one or more significant factors from the niches of all coexisting species.

**respiratory pigment** A protein that transports oxygen in blood or hemolymph.

**response** (1) In cellular communication, the change in a specific cellular activity brought about by a transduced signal from outside the cell. (2) In feedback regulation, a physiological activity triggered by a change in a variable.

**resting potential** The membrane potential characteristic of a nonconducting excitable cell, with the inside of the cell more negative than the outside.

**restriction enzyme** An endonuclease (type of enzyme) that recognizes and cuts DNA molecules foreign to a bacterium (such as phage genomes). The enzyme cuts at specific nucleotide sequences (restriction sites).

**restriction fragment** A DNA segment that results from the cutting of DNA by a restriction enzyme.

**restriction site** A specific sequence on a DNA strand that is recognized and cut by a restriction enzyme.

**retina** (ret'-i-nuh) The innermost layer of the vertebrate eye, containing photoreceptor cells (rods and cones) and neurons; transmits images formed by the lens to the brain via the optic nerve.

**retinal** The light-absorbing pigment in rods and cones of the vertebrate eye.

**retrotransposon** (re'-trō-trans-pō'-zon) A transposable element that moves within a genome by means of an RNA intermediate, a transcript of the retrotransposon DNA.

**retrovirus** (re'-trō-vī'-rus) An RNA virus that replicates by transcribing its RNA into DNA and then inserting the DNA into a cellular chromosome; an important class of cancer-causing viruses.

**reverse transcriptase** (tran-skrip'-tās) An enzyme encoded by certain viruses (retroviruses) that uses RNA as a template for DNA synthesis.

**reverse transcriptase–polymerase chain reaction (RT-PCR)** A technique for determining expression of a particular gene. It uses reverse transcriptase and DNA polymerase to synthesize cDNA from all the mRNA in a sample and then subjects the cDNA to PCR amplification using primers specific for the gene of interest.

**rhizarians** (rī-za'-rē-uhns) One of the three major subgroups for which the SAR eukaryotic supergroup is named. Many species in this clade are amoebas characterized by threadlike pseudopodia.

**rhizobacterium** A soil bacterium whose population size is much enhanced in the rhizosphere, the soil region close to a plant's roots.

**rhizoid** (rī'-zoyd) A long, tubular single cell or filament of cells that anchors bryophytes to the ground. Unlike roots, rhizoids are not composed of tissues, lack specialized conducting cells, and do not play a primary role in water and mineral absorption.

**rhizosphere** The soil region close to plant roots and characterized by a high level of microbiological activity.

**rhodopsin** (rō-dop'-sin) A visual pigment consisting of retinal and opsin. Upon absorbing light, the retinal changes shape and dissociates from the opsin.

**ribonucleic acid (RNA)** (rī'-bō-nū-klā'-ik) A type of nucleic acid consisting of a polynucleotide made up of nucleotide monomers with a ribose sugar and the nitrogenous bases adenine (A), cytosine (C), guanine (G), and uracil (U); usually single-stranded; functions in protein synthesis, gene regulation, and as the genome of some viruses.

**ribose** The sugar component of RNA nucleotides.

**ribosomal RNA (rRNA)** (rī'-buh-sō'-mul) RNA molecules that, together with proteins, make up ribosomes; the most abundant type of RNA.

**ribosome** (rī'-buh-sōm') A complex of rRNA and protein molecules that functions as a site of protein synthesis in the cytoplasm; consists of a large subunit and a small subunit. In eukaryotic cells, each subunit is assembled in the nucleolus. *See also* nucleolus.

**ribozyme** (rī'-buh-zīm) An RNA molecule that functions as an enzyme, such as an intron that catalyzes its own removal during RNA splicing.

**RNA interference (RNAi)** A technique used to silence the expression of selected genes. RNAi uses synthetic double-stranded RNA molecules that match the sequence of a particular gene to trigger the breakdown of the gene's messenger RNA.

**RNA polymerase** An enzyme that links ribonucleotides into a growing RNA chain during transcription, based on complementary binding to nucleotides on a DNA template strand.

**RNA processing** Modification of RNA primary transcripts, including splicing out of introns, joining together of exons, and alteration of the 5' and 3' ends.

**RNA splicing** After synthesis of a eukaryotic primary RNA transcript, the removal of portions of the transcript (introns) that will not be included in the mRNA and the joining together of the remaining portions (exons).

**rod** A rodlike cell in the retina of the vertebrate eye, sensitive to low light intensity.

**root** An organ in vascular plants that anchors the plant and enables it to absorb water and minerals from the soil.

**root cap** A cone of cells at the tip of a plant root that protects the apical meristem.

**root hair** A tiny extension of a root epidermal cell, growing just behind the root tip and increasing surface area for absorption of water and minerals.

**root system** All of a plant's roots, which anchor it in the soil, absorb and transport minerals and water, and store food.

**rooted** Describing a phylogenetic tree that contains a branch point (often, the one farthest to the left) representing the most recent common ancestor of all taxa in the tree.

**rough ER** That portion of the endoplasmic reticulum with ribosomes attached.

**round window** In the mammalian ear, the point of contact where vibrations of the stapes create a traveling series of pressure waves in the fluid of the cochlea.

**r-selection** Selection for life history traits that maximize reproductive success in uncrowded environments; also called density-independent selection.

**rubisco** (rū-bis′-kō) Ribulose bisphosphate (RuBP) carboxylase, the enzyme that catalyzes the first step of the Calvin cycle (the addition of $CO_2$ to RuBP).

**ruminant** (rū′-muh-nent) A cud-chewing animal, such as a cow or a sheep, with multiple stomach compartments specialized for an herbivorous diet.

**S phase** The synthesis phase of the cell cycle; the portion of interphase during which DNA is replicated.

**saccule** (sack′-yū-uhl) In the vertebrate ear, a chamber in the vestibule behind the oval window that participates in the sense of balance.

**salicylic acid** (sal′-i-sil′-ik) A signaling molecule in plants that may be partially responsible for activating systemic acquired resistance to pathogens.

**salivary gland** A gland associated with the oral cavity that secretes substances that lubricate food and begin the process of chemical digestion.

**salt** A compound resulting from the formation of an ionic bond; also called an ionic compound.

**saltatory conduction** (sol′-tuh-tōr′-ē) Rapid transmission of a nerve impulse along an axon, resulting from the action potential jumping from one node of Ranvier to another, skipping the myelin-sheathed regions of membrane.

**SAR clade** One of four supergroups of eukaryotes proposed in a current hypothesis of the evolutionary history of eukaryotes. This supergroup contains a large, extremely diverse collection of protists from three major subgroups: stramenopiles, alveolates, and rhizarians. *See also* Excavata, Archaeplastida, and Unikonta.

**sarcomere** (sar′-kō-mēr) The fundamental, repeating unit of striated muscle, delimited by the Z lines.

**sarcoplasmic reticulum (SR)** (sar′-kō-plaz′-mik ruh-tik′-yū-lum) A specialized endoplasmic reticulum that regulates the calcium concentration in the cytosol of muscle cells.

**saturated fatty acid** A fatty acid in which all carbons in the hydrocarbon tail are connected by single bonds, thus maximizing the number of hydrogen atoms attached to the carbon skeleton.

**savanna** A tropical grassland biome with scattered trees and large herbivores and maintained by occasional fires and drought.

**scanning electron microscope (SEM)** A microscope that uses an electron beam to scan the surface of a sample, coated with metal atoms, to study details of its topography.

**scatter plot** A graph in which each piece of data is represented by a point, but individual points are not connected by lines.

**Schwann cell** A type of glial cell that forms insulating myelin sheaths around the axons of neurons in the peripheral nervous system.

**science** An approach to understanding the natural world.

**scion** (sī′-un) The twig grafted onto the stock when making a graft.

**sclereid** (sklār′-ē-id) A short, irregular sclerenchyma cell in nutshells and seed coats. Sclereids are scattered throughout the parenchyma of some plants.

**sclerenchyma cell** (skluh-ren′-kim-uh) A rigid, supportive plant cell type usually lacking a protoplast and possessing thick secondary walls strengthened by lignin at maturity.

**scrotum** A pouch of skin outside the abdomen that houses the testes; functions in maintaining the testes at the lower temperature required for spermatogenesis.

**second law of thermodynamics** The principle stating that every energy transfer or transformation increases the entropy of the universe. Usable forms of energy are at least partly converted to heat.

**second messenger** A small, nonprotein, water-soluble molecule or ion, such as a calcium ion ($Ca^{2+}$) or cyclic AMP, that relays a signal to a cell's interior in response to a signaling molecule bound by a signal receptor protein.

**secondary cell wall** In plant cells, a strong and durable matrix that is often deposited in several laminated layers around the plasma membrane and that provides protection and support.

**secondary consumer** A carnivore that eats herbivores.

**secondary endosymbiosis** A process in eukaryotic evolution in which a heterotrophic eukaryotic cell engulfed a photosynthetic eukaryotic cell, which survived in a symbiotic relationship inside the heterotrophic cell.

**secondary growth** Growth produced by lateral meristems, thickening the roots and shoots of woody plants.

**secondary immune response** The adaptive immune response elicited on second or subsequent exposures to a particular antigen. The secondary immune response is more rapid, of greater magnitude, and of longer duration than the primary immune response.

**secondary oocyte** (ō′-uh-sīt) An oocyte that has completed the first of the two meiotic divisions.

**secondary production** The amount of chemical energy in consumers' food that is converted to their own new biomass during a given time period.

**secondary structure** Regions of repetitive coiling or folding of the polypeptide backbone of a protein due to hydrogen bonding between constituents of the backbone (not the side chains).

**secondary succession** A type of succession that occurs where an existing community has been cleared by some disturbance that leaves the soil or substrate intact.

**secretion** (1) The discharge of molecules synthesized by a cell. (2) The discharge of wastes from the body fluid into the filtrate.

**seed** An adaptation of some terrestrial plants consisting of an embryo packaged along with a store of food within a protective coat.

**seed coat** A tough outer covering of a seed, formed from the outer coat of an ovule. In a flowering plant, the seed coat encloses and protects the embryo and endosperm.

**seedless vascular plant** An informal name for a plant that has vascular tissue but lacks seeds. Seedless vascular plants form a paraphyletic group that includes the phyla Lycophyta (club mosses and their relatives) and Monilophyta (ferns and their relatives).

**selective permeability** A property of biological membranes that allows them to regulate the passage of substances across them.

**self-incompatibility** The ability of a seed plant to reject its own pollen and sometimes the pollen of closely related individuals.

**semelparity** Reproduction in which an organism produces all of its offspring in a single event; also known as big-bang reproduction.

**semen** (sē′-mun) The fluid that is ejaculated by the male during orgasm; contains sperm and secretions from several glands of the male reproductive tract.

**semicircular canals** A three-part chamber of the inner ear that functions in maintaining equilibrium.

**semiconservative model** Type of DNA replication in which the replicated double helix consists of one old strand, derived from the parental molecule, and one newly made strand.

**semilunar valve** A valve located at each exit of the heart, where the aorta leaves the left ventricle and the pulmonary artery leaves the right ventricle.

**seminal vesicle** (sem′-i-nul ves′-i-kul) A gland in males that secretes a fluid component of semen that lubricates and nourishes sperm.

**seminiferous tubule** (sem′-i-nif′-er-us) A highly coiled tube in the testis in which sperm are produced.

**senescence** (se-nes′-ens) The growth phase in a plant or plant part (as a leaf) from full maturity to death.

**sensitive period** A limited phase in an animal's development when learning of particular behaviors can take place; also called a critical period.

**sensor** In homeostasis, a receptor that detects a stimulus.

**sensory adaptation** The tendency of sensory neurons to become less sensitive when they are stimulated repeatedly.

**sensory neuron** A nerve cell that receives information from the internal or external environment and transmits signals to the central nervous system.

**sensory reception** The detection of a stimulus by sensory cells.

**sensory receptor** An organ, cell, or structure within a cell that responds to specific stimuli from an organism's external or internal environment.

**sensory transduction** The conversion of stimulus energy to a change in the membrane potential of a sensory receptor cell.

**sepal** (sē′-pul) A modified leaf in angiosperms that helps enclose and protect a flower bud before it opens.

**serial endosymbiosis** A hypothesis for the origin of eukaryotes consisting of a sequence of endosymbiotic events in which mitochondria, chloroplasts, and perhaps other cellular structures were derived from small prokaryotes that had been engulfed by larger cells.

**Sertoli cell** A support cell of the seminiferous tubule that surrounds and nourishes developing sperm.

**set point** In homeostasis in animals, a value maintained for a particular variable, such as body temperature or solute concentration.

**sex chromosome** A chromosome responsible for determining the sex of an individual.

**sex-linked gene** A gene located on either sex chromosome. Most sex-linked genes are on the X chromosome and show distinctive patterns of inheritance; there are very few genes on the Y chromosome.

**sexual dimorphism** (dī-mōr′-fizm) Differences between the secondary sex characteristics of males and females of the same species.

**sexual reproduction** A type of reproduction in which two parents give rise to offspring that have unique combinations of genes inherited from both parents via the gametes.

**sexual selection** A process in which individuals with certain inherited characteristics are more likely than other individuals of the same sex to obtain mates.

**Shannon diversity index** An index of community diversity symbolized by $H$ and represented by the equation $H = -(p_A \ln p_A + p_B \ln p_B + p_C \ln p_C + \ldots)$, where A, B, C . . . are species, $p$ is the relative abundance of each species, and ln is the natural logarithm.

**shared ancestral character** A character that is shared by members of a particular clade but that originated in an ancestor that is not a member of that clade.

**shared derived character** An evolutionary novelty that is unique to a particular clade.

**shoot system** The aerial portion of a plant body, consisting of stems, leaves, and (in angiosperms) flowers.

**short tandem repeat (STR)** Simple sequence DNA containing multiple tandemly repeated units of two to five nucleotides. Variations in STRs act as genetic markers in STR analysis, used to prepare genetic profiles.

**short-day plant** A plant that flowers (usually in late summer, fall, or winter) only when the light period is shorter than a critical length.

**short-term memory** The ability to hold information, anticipations, or goals for a time and then release them if they become irrelevant.

**sickle-cell disease** A recessively inherited human blood disorder in which a single nucleotide change in the β-globin gene causes hemoglobin to aggregate, changing red blood cell shape and causing multiple symptoms in afflicted individuals.

**sieve plate** An end wall in a sieve-tube element, which facilitates the flow of phloem sap in angiosperm sieve tubes.

**sieve-tube element** A living cell that conducts sugars and other organic nutrients in the phloem of angiosperms; also called a sieve-tube member. Connected end to end, they form sieve tubes.

**sign stimulus** An external sensory cue that triggers a fixed action pattern by an animal.

**signal** In animal behavior, transmission of a stimulus from one animal to another. The term is also used in the context of communication in other kinds of organisms and in cell-to-cell communication in all multicellular organisms.

**signal peptide** A sequence of about 20 amino acids at or near the leading (amino) end of a polypeptide that targets it to the endoplasmic reticulum or other organelles in a eukaryotic cell.

**signal transduction pathway** A series of steps linking a mechanical, chemical, or electrical stimulus to a specific cellular response.

**signal-recognition particle (SRP)** A protein-RNA complex that recognizes a signal peptide as it emerges from a ribosome and helps direct the ribosome to the endoplasmic reticulum (ER) by binding to a receptor protein on the ER.

**silent mutation** A nucleotide-pair substitution that has no observable effect on the phenotype; for example, within a gene, a mutation that results in a codon that codes for the same amino acid.

**simple fruit** A fruit derived from a single carpel or several fused carpels.

**simple sequence DNA** A DNA sequence that contains many copies of tandemly repeated short sequences.

**single bond** A single covalent bond; the sharing of a pair of valence electrons by two atoms.

**single circulation** A circulatory system consisting of a single pump and circuit, in which blood passes from the sites of gas exchange to the rest of the body before returning to the heart.

**single nucleotide polymorphism (SNP)** A single base-pair site in a genome where nucleotide variation is found in at least 1% of the population.

**single-lens eye** The camera-like eye found in some jellies, polychaete worms, spiders, and many molluscs.

**single-strand binding protein** A protein that binds to the unpaired DNA strands during DNA replication, stabilizing them and holding them apart while they serve as templates for the synthesis of complementary strands of DNA.

**sinoatrial (SA) node** A region in the right atrium of the heart that sets the rate and timing at which all cardiac muscle cells contract; the pacemaker.

**sister chromatids** Two copies of a duplicated chromosome attached to each other by proteins at the centromere and, sometimes, along the arms. While joined, two sister chromatids make up one chromosome. Chromatids are eventually separated during mitosis or meiosis II.

**sister taxa** Groups of organisms that share an immediate common ancestor and hence are each other's closest relatives.

**skeletal muscle** A type of striated muscle that is generally responsible for the voluntary movements of the body.

**sliding-filament model** The idea that muscle contraction is based on the movement of thin (actin) filaments along thick (myosin) filaments, shortening the sarcomere, the basic unit of muscle organization.

**slow-twitch fiber** A muscle fiber that can sustain long contractions.

**small interfering RNA (siRNA)** One of multiple small, single-stranded RNA molecules generated by cellular machinery from a long, linear, double-stranded RNA molecule. The siRNA associates with one or more proteins in a complex that can degrade or prevent translation of an mRNA with a complementary sequence. In some cases, siRNA can also block transcription by promoting chromatin modification.

**small intestine** The longest section of the alimentary canal, so named because of its small diameter compared with that of the large intestine; the principal site of the enzymatic hydrolysis of food macromolecules and the absorption of nutrients.

**smooth ER** That portion of the endoplasmic reticulum that is free of ribosomes.

**smooth muscle** A type of muscle lacking the striations of skeletal and cardiac muscle because of the uniform distribution of myosin filaments in the cells; responsible for involuntary body activities.

**social learning** Modification of behavior through the observation of other individuals.

**sodium-potassium pump** A transport protein in the plasma membrane of animal cells that actively transports sodium out of the cell and potassium into the cell.

**solute** (sol′-yūt) A substance that is dissolved in a solution.

**solute potential ($\Psi_S$)** A component of water potential that is proportional to the molarity of a solution and that measures the effect of solutes on the direction of water movement; also called osmotic potential, it can be either zero or negative.

**solution** A liquid that is a homogeneous mixture of two or more substances.

**solvent** The dissolving agent of a solution. Water is the most versatile solvent known.

**somatic cell** (sō-mat′-ik) Any cell in a multicellular organism except a sperm or egg or their precursors.

**spatial learning** The establishment of a memory that reflects the environment's spatial structure.

**speciation** (spē'-sē-ā'-shun) An evolutionary process in which one species splits into two or more species.

**species** (spē'-sēz) A population or group of populations whose members have the potential to interbreed in nature and produce viable, fertile offspring, but do not produce viable, fertile offspring with members of other such groups.

**species-area curve** The biodiversity pattern that shows that the larger the geographic area of a community, the more species it has.

**species diversity** The number and relative abundance of species in a biological community.

**species richness** The number of species in a biological community.

**specific heat** The amount of heat that must be absorbed or lost for 1 g of a substance to change its temperature by 1°C.

**spectrophotometer** An instrument that measures the proportions of light of different wavelengths absorbed and transmitted by a pigment solution.

**sperm** The male gamete.

**spermatogenesis** The continuous and prolific production of mature sperm cells in the testis.

**spermatogonium** (plural, **spermatogonia**) A cell that divides mitotically to form spermatocytes.

**sphincter** (sfink'-ter) A ringlike band of muscle fibers that controls the size of an opening in the body, such as the passage between the esophagus and the stomach.

**spliceosome** (spli'-sō-sōm) A large complex made up of proteins and RNA molecules that splices RNA by interacting with the ends of an RNA intron, releasing the intron and joining the two adjacent exons.

**spontaneous process** A process that occurs without an overall input of energy; a process that is energetically favorable.

**sporangium** (spōr-an'-jē-um) (plural, **sporangia**) A multicellular organ in fungi and plants in which meiosis occurs and haploid cells develop.

**spore** (1) In the life cycle of a plant or alga undergoing alternation of generations, a haploid cell produced in the sporophyte by meiosis. A spore can divide by mitosis to develop into a multicellular haploid individual, the gametophyte, without fusing with another cell. (2) In fungi, a haploid cell, produced either sexually or asexually, that produces a mycelium after germination.

**sporophyte** (spō-ruh-fīt') In organisms (plants and some algae) that have alternation of generations, the multicellular diploid form that results from the union of gametes. The sporophyte produces haploid spores by meiosis that develop into gametophytes.

**sporopollenin** (spōr-uh-pol'-eh-nin) A durable polymer that covers exposed zygotes of charophyte algae and forms the walls of plant spores, preventing them from drying out.

**stability** In evolutionary biology, a term referring to a hybrid zone in which hybrids continue to be produced; this causes the hybrid zone to be "stable" in the sense of persisting over time.

**stabilizing selection** Natural selection in which intermediate phenotypes survive or reproduce more successfully than do extreme phenotypes.

**stamen** (stā'-men) The pollen-producing reproductive organ of a flower, consisting of an anther and a filament.

**starch** A storage polysaccharide in plants, consisting entirely of glucose monomers joined by α glycosidic linkages.

**start point** In transcription, the nucleotide position on the promoter where RNA polymerase begins synthesis of RNA.

**statocyst** (stat'-uh-sist') A type of mechanoreceptor that functions in equilibrium in invertebrates by use of statoliths, which stimulate hair cells in relation to gravity.

**statolith** (stat'-uh-lith') (1) In plants, a specialized plastid that contains dense starch grains and may play a role in detecting gravity. (2) In invertebrates, a dense particle that settles in response to gravity and is found in sensory organs that function in equilibrium.

**stele** (stēl) The vascular tissue of a stem or root.

**stem** A vascular plant organ consisting of an alternating system of nodes and internodes that support the leaves and reproductive structures.

**stem cell** Any relatively unspecialized cell that can produce, during a single division, one identical daughter cell and one more specialized daughter cell that can undergo further differentiation.

**steroid** A type of lipid characterized by a carbon skeleton consisting of four fused rings with various chemical groups attached.

**sticky end** A single-stranded end of a double-stranded restriction fragment.

**stigma** (plural, **stigmata**) The sticky part of a flower's carpel, which receives pollen grains.

**stimulus** In feedback regulation, a fluctuation in a variable that triggers a response.

**stipe** A stemlike structure of a seaweed.

**stock** The plant that provides the root system when making a graft.

**stoma** (stō'-muh) (plural, **stomata**) A microscopic pore surrounded by guard cells in the epidermis of leaves and stems that allows gas exchange between the environment and the interior of the plant.

**stomach** An organ of the digestive system that stores food and performs preliminary steps of digestion.

**stramenopiles** One of the three major subgroups for which the SAR eukaryotic supergroup is named. This clade arose by secondary endosymbiosis and includes diatoms and brown algae.

**stratum** (strah'-tum) (plural, **strata**) A rock layer formed when new layers of sediment cover older ones and compress them.

**stroke** The death of nervous tissue in the brain, usually resulting from rupture or blockage of arteries in the head.

**stroma** (strō'-muh) The dense fluid within the chloroplast surrounding the thylakoid membrane and containing ribosomes and DNA; involved in the synthesis of organic molecules from carbon dioxide and water.

**stromatolite** Layered rock that results from the activities of prokaryotes that bind thin films of sediment together.

**structural isomer** One of two or more compounds that have the same molecular formula but differ in the covalent arrangements of their atoms.

**style** The stalk of a flower's carpel, with the ovary at the base and the stigma at the top.

**substrate** The reactant on which an enzyme works.

**substrate feeder** An animal that lives in or on its food source, eating its way through the food.

**substrate-level phosphorylation** The enzyme-catalyzed formation of ATP by direct transfer of a phosphate group to ADP from an intermediate substrate in catabolism.

**sugar sink** A plant organ that is a net consumer or storer of sugar. Growing roots, shoot tips, stems, and fruits are examples of sugar sinks supplied by phloem.

**sugar source** A plant organ in which sugar is being produced by either photosynthesis or the breakdown of starch. Mature leaves are the primary sugar sources of plants.

**sulfhydryl group** A chemical group consisting of a sulfur atom bonded to a hydrogen atom.

**summation** A phenomenon of neural integration in which the membrane potential of the postsynaptic cell is determined by the combined effect of EPSPs or IPSPs produced in rapid succession at one synapse or simultaneously at different synapses.

**suprachiasmatic nucleus (SCN)** A group of neurons in the hypothalamus of mammals that functions as a biological clock.

**surface tension** A measure of how difficult it is to stretch or break the surface of a liquid.

**surfactant** A substance secreted by alveoli that decreases surface tension in the fluid that coats the alveoli.

**survivorship curve** A plot of the number of members of a cohort that are still alive at each age; one way to represent age-specific mortality.

**suspension feeder** An animal that feeds by removing suspended food particles from the surrounding medium by a capture, trapping, or filtration mechanism.

**sustainable development** Development that meets the needs of people today without limiting the ability of future generations to meet their needs.

**symbiont** (sim'-bē-ont) The smaller participant in a symbiotic relationship, living in or on the host.

**symbiosis** An ecological relationship between organisms of two different species that live together in direct and intimate contact.

**sympathetic division** One of three divisions of the autonomic nervous system; generally increases energy expenditure and prepares the body for action.

**sympatric speciation** (sim-pat'-rik) The formation of new species in populations that live in the same geographic area.

**symplast** In plants, the continuum of cytoplasm connected by plasmodesmata between cells.

**synapse** (sin'-aps) The junction where a neuron communicates with another cell across a narrow gap via a neurotransmitter or an electrical coupling.

**synapsid** Member of an amniote clade distinguished by a single hole on each side of the skull. Synapsids include the mammals.

**synapsis** (si-nap'-sis) The pairing and physical connection of duplicated homologous chromosomes during prophase I of meiosis.

**synaptonemal complex** A zipper-like structure composed of proteins, which connects two homologous chromosomes tightly along their lengths during part of prophase I of meiosis.

**systematics** A scientific discipline focused on classifying organisms and determining their evolutionary relationships.

**systemic acquired resistance** A defensive response in infected plants that helps protect healthy tissue from pathogenic invasion.

**systemic circuit** The branch of the circulatory system that supplies oxygenated blood to and carries deoxygenated blood away from organs and tissues throughout the body.

**systems biology** An approach to studying biology that aims to model the dynamic behavior of whole biological systems based on a study of the interactions among the system's parts.

**systole** (sis'-tō-lē) The stage of the cardiac cycle in which a heart chamber contracts and pumps blood.

**T cells** The class of lymphocytes that mature in the thymus; they include both effector cells for the cell-mediated immune response and helper cells required for both branches of adaptive immunity.

**taproot** A main vertical root that develops from an embryonic root and gives rise to lateral (branch) roots.

**tastant** Any chemical that stimulates the sensory receptors in a taste bud.

**taste bud** A collection of modified epithelial cells on the tongue or in the mouth that are receptors for taste in mammals.

**TATA box** A DNA sequence in eukaryotic promoters crucial in forming the transcription initiation complex.

**taxis** (tak'-sis) An oriented movement toward or away from a stimulus.

**taxon** (plural, **taxa**) A named taxonomic unit at any given level of classification.

**taxonomy** (tak-son'-uh-mē) A scientific discipline concerned with naming and classifying the diverse forms of life.

**Tay-Sachs disease** A human genetic disease caused by a recessive allele for a dysfunctional enzyme, leading to accumulation of certain lipids in the brain. Seizures, blindness, and degeneration of motor and mental performance usually become manifest a few months after birth, followed by death within a few years.

**technology** The application of scientific knowledge for a specific purpose, often involving industry or commerce but also including uses in basic research.

**telomere** (tel'-uh-mēr) The tandemly repetitive DNA at the end of a eukaryotic chromosome's DNA molecule. Telomeres protect the organism's genes from being eroded during successive rounds of replication. *See also* repetitive DNA.

**telophase** The fifth and final stage of mitosis, in which daughter nuclei are forming and cytokinesis has typically begun.

**temperate broadleaf forest** A biome located throughout midlatitude regions where there is sufficient moisture to support the growth of large, broadleaf deciduous trees.

**temperate grassland** A terrestrial biome that exists at midlatitude regions and is dominated by grasses and forbs.

**temperate phage** A phage that is capable of replicating by either a lytic or lysogenic cycle.

**temperature** A measure in degrees of the average kinetic energy (thermal energy) of the atoms and molecules in a body of matter.

**template strand** The DNA strand that provides the pattern, or template, for ordering, by complementary base pairing, the sequence of nucleotides in an RNA transcript.

**terminator** In bacteria, a sequence of nucleotides in DNA that marks the end of a gene and signals RNA polymerase to release the newly made RNA molecule and detach from the DNA.

**territoriality** A behavior in which an animal defends a bounded physical space against encroachment by other individuals, usually of its own species.

**tertiary consumer** (ter'-shē-ār'-ē) A carnivore that eats other carnivores.

**tertiary structure** The overall shape of a protein molecule due to interactions of amino acid side chains, including hydrophobic interactions, ionic bonds, hydrogen bonds, and disulfide bridges.

**test** In foram protists, a porous shell that consists of a single piece of organic material hardened with calcium carbonate.

**testcross** Breeding an organism of unknown genotype with a homozygous recessive individual to determine the unknown genotype. The ratio of phenotypes in the offspring reveals the unknown genotype.

**testis** (plural, **testes**) The male reproductive organ, or gonad, in which sperm and reproductive hormones are produced.

**testosterone** A steroid hormone required for development of the male reproductive system, spermatogenesis, and male secondary sex characteristics; the major androgen in mammals.

**tetanus** (tet'-uh-nus) The maximal, sustained contraction of a skeletal muscle, caused by a very high frequency of action potentials elicited by continual stimulation.

**tetrapod** Member of a vertebrate clade characterized by limbs with digits. Tetrapods include mammals, amphibians, and birds and other reptiles.

**thalamus** (thal'-uh-mus) An integrating center of the vertebrate forebrain. Neurons with cell bodies in the thalamus relay neural input to specific areas in the cerebral cortex and regulate what information goes to the cerebral cortex.

**theory** An explanation that is broader in scope than a hypothesis, generates new hypotheses, and is supported by a large body of evidence.

**thermal energy** Kinetic energy due to the random motion of atoms and molecules; energy in its most random form. *See also* heat.

**thermocline** A narrow stratum of abrupt temperature change in the ocean and in many temperate-zone lakes.

**thermodynamics** (ther'-mō-dī-nam'-iks) The study of energy transformations that occur in a collection of matter. *See also* first law of thermodynamics; second law of thermodynamics.

**thermoreceptor** A receptor stimulated by either heat or cold.

**thermoregulation** The maintenance of internal body temperature within a tolerable range.

**thick filament** A filament composed of staggered arrays of myosin molecules; a component of myofibrils in muscle fibers.

**thigmomorphogenesis** A response in plants to chronic mechanical stimulation, resulting from increased ethylene production. An example is thickening stems in response to strong winds.

**thigmotropism** (thig-mo'-truh-pizm) A directional growth of a plant in response to touch.

**thin filament** A filament consisting of two strands of actin and two strands of regulatory protein coiled around one another; a component of myofibrils in muscle fibers.

**threatened species** A species that is considered likely to become endangered in the foreseeable future.

**threshold** The potential that an excitable cell membrane must reach for an action potential to be initiated.

**thrombus** A fibrin-containing clot that forms in a blood vessel and blocks the flow of blood.

**thylakoid** (thī'-luh-koyd) A flattened, membranous sac inside a chloroplast. Thylakoids often exist in stacks called grana that are interconnected; their membranes contain molecular "machinery" used to convert light energy to chemical energy.

**thymus** (thī'-mus) A small organ in the thoracic cavity of vertebrates where maturation of T cells is completed.

**tidal volume** The volume of air a mammal inhales and exhales with each breath.

**tight junction** A type of intercellular junction between animal cells that prevents the leakage of material through the space between cells.

**tissue** An integrated group of cells with a common structure, function, or both.

**Toll-like receptor (TLR)** A membrane receptor on a phagocytic white blood cell that recognizes fragments of molecules common to a set of pathogens.

**tonicity** The ability of a solution surrounding a cell to cause that cell to gain or lose water.

**top-down model** A model of community organization in which predation influences community organization by controlling herbivore numbers, which in turn control plant or phytoplankton numbers, which in turn control nutrient levels; also called the trophic cascade model.

**topoisomerase** A protein that breaks, swivels, and rejoins DNA strands. During DNA replication, topoisomerase helps to relieve strain in the double helix ahead of the replication fork.

**totipotent** (tō′-tuh-pōt′-ent) Describing a cell that can give rise to all parts of the embryo and adult, as well as extraembryonic membranes in species that have them.

**trace element** An element indispensable for life but required in extremely minute amounts.

**trachea** (trā′-kē-uh) The portion of the respiratory tract that passes from the larynx to the bronchi; also called the windpipe.

**tracheal system** In insects, a system of branched, air-filled tubes that extends throughout the body and carries oxygen directly to cells.

**tracheid** (trā′-kē-id) A long, tapered water-conducting cell found in the xylem of nearly all vascular plants. Functioning tracheids are no longer living.

**trait** One of two or more detectable variants in a genetic character.

**transcription** The synthesis of RNA using a DNA template.

**transcription factor** A regulatory protein that binds to DNA and affects transcription of specific genes.

**transcription initiation complex** The completed assembly of transcription factors and RNA polymerase bound to a promoter.

**transcription unit** A region of DNA that is transcribed into an RNA molecule.

**transduction** (1) A process in which phages (viruses) carry bacterial DNA from one bacterial cell to another. When these two cells are members of different species, transduction results in horizontal gene transfer. (2) In cellular communication, the conversion of a signal from outside the cell to a form that can bring about a specific cellular response; also called signal transduction.

*trans* **fats** An unsaturated fat, formed artificially during hydrogenation of oils, containing one or more *trans* double bonds.

**transfer RNA (tRNA)** An RNA molecule that functions as a translator between nucleic acid and protein languages by picking up a specific amino acid and carrying it to the ribosome, where the tRNA recognizes the appropriate codon in the mRNA.

**transformation** (1) The process by which a cell in culture acquires the ability to divide indefinitely, similar to the division of cancer cells. (2) A change in genotype and phenotype due to the assimilation of external DNA by a cell. When the external DNA is from a member of a different species, transformation results in horizontal gene transfer.

**transgenic** Pertaining to an organism whose genome contains DNA introduced from another organism of the same or a different species.

**translation** The synthesis of a polypeptide using the genetic information encoded in an mRNA molecule. There is a change of "language" from nucleotides to amino acids.

**translocation** (1) An aberration in chromosome structure resulting from attachment of a chromosomal fragment to a nonhomologous chromosome. (2) During protein synthesis, the

third stage in the elongation cycle, when the RNA carrying the growing polypeptide moves from the A site to the P site on the ribosome. (3) The transport of organic nutrients in the phloem of vascular plants.

**transmission electron microscope (TEM)** A microscope that passes an electron beam through very thin sections stained with metal atoms and is primarily used to study the internal ultrastructure of cells.

**transpiration** The evaporative loss of water from a plant.

**transport epithelium** One or more layers of specialized epithelial cells that carry out and regulate solute movement.

**transport protein** A transmembrane protein that helps a certain substance or class of closely related substances to cross the membrane.

**transport vesicle** A small membranous sac in a eukaryotic cell's cytoplasm carrying molecules produced by the cell.

**transposable element** A segment of DNA that can move within the genome of a cell by means of a DNA or RNA intermediate; also called a transposable genetic element.

**transposon** A transposable element that moves within a genome by means of a DNA intermediate.

**transverse (T) tubule** An infolding of the plasma membrane of skeletal muscle cells.

**triacylglycerol** (trī-as′-ul-glis′-uh-rol) A lipid consisting of three fatty acids linked to one glycerol molecule; also called a fat or triglyceride.

**triple response** A plant growth maneuver in response to mechanical stress, involving slowing of stem elongation, thickening of the stem, and a curvature that causes the stem to start growing horizontally.

**triplet code** A genetic information system in which a series of three-nucleotide-long words specifies a sequence of amino acids for a polypeptide chain.

**trisomic** Referring to a diploid cell that has three copies of a particular chromosome instead of the normal two.

**trophic efficiency** The percentage of production transferred from one trophic level to the next.

**trophic structure** The different feeding relationships in an ecosystem, which determine the route of energy flow and the pattern of chemical cycling.

**trophoblast** The outer epithelium of a mammalian blastocyst. It forms the fetal part of the placenta, supporting embryonic development but not forming part of the embryo proper.

**tropic hormone** A hormone that has an endocrine gland or endocrine cells as a target.

**tropical dry forest** A terrestrial biome characterized by relatively high temperatures and precipitation overall but with a pronounced dry season.

**tropical rain forest** A terrestrial biome characterized by relatively high precipitation and temperatures year-round.

**tropics** Latitudes between 23.5° north and south.

**tropism** A growth response that results in the curvature of whole plant organs toward or

away from stimuli due to differential rates of cell elongation.

**tropomyosin** The regulatory protein that blocks the myosin-binding sites on actin molecules.

**troponin complex** The regulatory proteins that control the position of tropomyosin on the thin filament.

**true-breeding** Referring to organisms that produce offspring of the same variety over many generations of self-pollination.

**tumor-suppressor gene** A gene whose protein product inhibits cell division, thereby preventing the uncontrolled cell growth that contributes to cancer.

**tundra** A terrestrial biome at the extreme limits of plant growth. At the northernmost limits, it is called arctic tundra, and at high altitudes, where plant forms are limited to low shrubby or matlike vegetation, it is called alpine tundra.

**turgid** (ter′-jid) Swollen or distended, as in plant cells. (A walled cell becomes turgid if it has a lower water potential than its surroundings, resulting in entry of water.)

**turgor pressure** The force directed against a plant cell wall after the influx of water and swelling of the cell due to osmosis.

**twin study** A behavioral study in which researchers compare the behavior of identical twins raised apart with that of identical twins raised in the same household.

**tympanic membrane** Another name for the eardrum, the membrane between the outer and middle ear.

**Unikonta** (yū′-ni-kon′-tuh) One of four supergroups of eukaryotes proposed in a current hypothesis of the evolutionary history of eukaryotes. This clade, which is supported by studies of myosin proteins and DNA, consists of amoebozoans and opisthokonts. *See also* Excavata, SAR, and Archaeplastida.

**unsaturated fatty acid** A fatty acid that has one or more double bonds between carbons in the hydrocarbon tail. Such bonding reduces the number of hydrogen atoms attached to the carbon skeleton.

**urea** A soluble nitrogenous waste produced in the liver by a metabolic cycle that combines ammonia with carbon dioxide.

**ureter** (yū-rē′-ter) A duct leading from the kidney to the urinary bladder.

**urethra** (yū-rē′-thruh) A tube that releases urine from the mammalian body near the vagina in females and through the penis in males; also serves in males as the exit tube for the reproductive system.

**uric acid** A product of protein and purine metabolism and the major nitrogenous waste product of insects, land snails, and many reptiles. Uric acid is relatively nontoxic and largely insoluble.

**urinary bladder** The pouch where urine is stored prior to elimination.

**uterine cycle** The changes that occur in the uterus during the reproductive cycle of the human female; also called the menstrual cycle.

**uterus** A female organ where eggs are fertilized and/or development of the young occurs.

**utricle** (yū'-trick-uhl) In the vertebrate ear, a chamber in the vestibule behind the oval window that opens into the three semicircular canals.

**vaccine** A harmless variant or derivative of a pathogen that stimulates a host's immune system to mount defenses against the pathogen.

**vacuole** (vak'-yū-ōl') A membrane-bounded vesicle whose specialized function varies in different kinds of cells.

**vagina** Part of the female reproductive system between the uterus and the outside opening; the birth canal in mammals. During copulation, the vagina accommodates the male's penis and receives sperm.

**valence** The bonding capacity of a given atom; the number of covalent bonds an atom can form, which usually equals the number of unpaired electrons in its outermost (valence) shell.

**valence electron** An electron in the outermost electron shell.

**valence shell** The outermost energy shell of an atom, containing the valence electrons involved in the chemical reactions of that atom.

**van der Waals interactions** Weak attractions between molecules or parts of molecules that result from transient local partial charges.

**variable** A factor that varies during an experiment.

**variation** Differences between members of the same species.

**vas deferens** In mammals, the tube in the male reproductive system in which sperm travel from the epididymis to the urethra.

**vasa recta** The capillary system in the kidney that serves the loop of Henle.

**vascular cambium** A cylinder of meristematic tissue in woody plants that adds layers of secondary vascular tissue called secondary xylem (wood) and secondary phloem.

**vascular plant** A plant with vascular tissue. Vascular plants include all living plant species except liverworts, mosses, and hornworts.

**vascular tissue** Plant tissue consisting of cells joined into tubes that transport water and nutrients throughout the plant body.

**vascular tissue system** A transport system formed by xylem and phloem throughout a vascular plant. Xylem transports water and minerals; phloem transports sugars, the products of photosynthesis.

**vasoconstriction** A decrease in the diameter of blood vessels caused by contraction of smooth muscles in the vessel walls.

**vasodilation** An increase in the diameter of blood vessels caused by relaxation of smooth muscles in the vessel walls.

**vasopressin** *See* antidiuretic hormone (ADH).

**vector** An organism that transmits pathogens from one host to another.

**vegetative propagation** Cloning of plants by humans.

**vegetative reproduction** Cloning of plants in nature.

**vein** (1) In animals, a vessel that carries blood toward the heart. (2) In plants, a vascular bundle in a leaf.

**ventilation** The flow of air or water over a respiratory surface.

**ventral** Pertaining to the underside, or bottom, of an animal with radial or bilateral symmetry.

**ventricle** (ven'-tri-kul) (1) A heart chamber that pumps blood out of the heart. (2) A space in the vertebrate brain, filled with cerebrospinal fluid.

**vernalization** The use of cold treatment to induce a plant to flower.

**vertebrate** A chordate animal with a backbone. Vertebrates include sharks and rays, ray-finned fishes, coelacanths, lungfishes, amphibians, reptiles, and mammals.

**vesicle** (ves'-i-kul) A membranous sac in the cytoplasm of a eukaryotic cell.

**vessel** A nonliving, water-conducting tube found in most angiosperms and a few nonflowering vascular plants that is formed by the end-to-end connection of vessel elements.

**vessel element** A short, wide, water-conducting cell found in the xylem of most angiosperms and a few nonflowering vascular plants. Dead at maturity, vessel elements are aligned end to end to form vessels.

**vestigial structure** A feature of an organism that is a historical remnant of a structure that served a function in the organism's ancestors.

**villus** (plural, **villi**) (1) A finger-like projection of the inner surface of the small intestine. (2) A finger-like projection of the chorion of the mammalian placenta. Large numbers of villi increase the surface areas of these organs.

**viral envelope** A membrane, derived from membranes of the host cell, that cloaks the capsid, which in turn encloses a viral genome.

**virulent phage** A phage that replicates only by a lytic cycle.

**virus** An infectious particle incapable of replicating outside of a cell, consisting of an RNA or DNA genome surrounded by a protein coat (capsid) and, for some viruses, a membranous envelope.

**visceral mass** One of the three main parts of a mollusc; the part containing most of the internal organs. *See also* foot and mantle.

**visible light** That portion of the electromagnetic spectrum that can be detected as various colors by the human eye, ranging in wavelength from about 380 nm to about 750 nm.

**vital capacity** The maximum volume of air that a mammal can inhale and exhale with each breath.

**vitamin** An organic molecule required in the diet in very small amounts. Many vitamins serve as coenzymes or parts of coenzymes.

**voltage-gated ion channel** A specialized ion channel that opens or closes in response to changes in membrane potential.

**vulva** Collective term for the female external genitalia.

**water potential (Ψ)** The physical property predicting the direction in which water will flow, governed by solute concentration and applied pressure.

**wavelength** The distance between crests of waves, such as those of the electromagnetic spectrum.

**wetland** A habitat that is inundated by water at least some of the time and that supports plants adapted to water-saturated soil.

**white matter** Tracts of axons within the CNS.

**whole-genome shotgun approach** Procedure for genome sequencing in which the genome is randomly cut into many overlapping short segments that are sequenced; computer software then assembles the complete sequence.

**wild type** The phenotype most commonly observed in natural populations; also refers to the individual with that phenotype.

**wilting** The drooping of leaves and stems that occurs when plant cells become flaccid.

**wobble** Flexibility in the base-pairing rules in which the nucleotide at the 5′ end of a tRNA anticodon can form hydrogen bonds with more than one kind of base in the third position (3′ end) of a codon.

**xerophyte** A plant adapted to an arid climate.

**X-linked gene** A gene located on the X chromosome; such genes show a distinctive pattern of inheritance.

**X-ray crystallography** A technique used to study the three-dimensional structure of molecules. It depends on the diffraction of an X-ray beam by the individual atoms of a crystallized molecule.

**xylem** (zī'-lum) Vascular plant tissue consisting mainly of tubular dead cells that conduct most of the water and minerals upward from the roots to the rest of the plant.

**xylem sap** The dilute solution of water and dissolved minerals carried through vessels and tracheids.

**yeast** Single-celled fungus. Yeasts reproduce asexually by binary fission or by the pinching of small buds off a parent cell. Many fungal species can grow both as yeasts and as a network of filaments; relatively few species grow only as yeasts.

**zero population growth (ZPG)** A period of stability in population size, when additions to the population through births and immigration are balanced by subtractions through deaths and emigration.

**zoned reserve** An extensive region that includes areas relatively undisturbed by humans surrounded by areas that have been changed by human activity and are used for economic gain.

**zoonotic pathogen** A disease-causing agent that is transmitted to humans from other animals.

**zygote** (zī'-gōt) The diploid cell produced by the union of haploid gametes during fertilization; a fertilized egg.

# Index

Blindness, 422, 489f, 690
color, 811
Blind spot, 808f
Blood, 707f, **708**. See also Blood pressure; Blood
    vessels
    ABO blood groups for human, 104, 224, 225f, 746
    apoptosis of human white blood cells of, 326f
    cell division of bone marrow cells and, 182, 183f
    in closed circulatory systems, 707f–709f, 710
    clotting of, 241, 718, 719f
    components of, 717f
    composition and function of, 717f–720f
    connective tissue of, 665f
    countercurrent exchange and fish, 723f
    flow velocity of, 713, 714f
    gas exchange adaptations of, 728f–729f, 730
    hemophilia and clotting of, 241
    immune rejection of transfusions of, 746
    osmolarity of, 684
    pH of human, 38f, 39
    processing of filtrate from, by kidneys, 682, 683f
    sickle-cell disease and, 428f–429f
    volume and pressure of, in kidney regulation, 685
Blood-brain barrier, 792
Blood doping, 718
Blood flukes, 872
Bloodhound, 813
Blood poisoning, 488f
Blood pressure
    in cardiovascular systems, 714, 715f
    in circulatory systems, 708–709
    hypertension and, 720
    kidney homeostasis and, 685
    temporal changes in, 732
Blood proteins, 60f–61f
Blood types, human, 104, 224, 225f, 746
Blood vessels
    adaptations of, for thermoregulation, 675, 676f
    blood flow velocity in, 713, 714f
    blood pressure in, 714, 715f
    capillary function, 715
    in circulatory systems, 707f, 708
    diseases of, 719, 720f
    structure and function of, 713f
Blooms, 510, 515
    algal, 890, 920
    phytoplankton, 919f, 920
Blowflies, 619
Blueberry maggot fly, 436f
Blue crab, 885
Blue-footed boobies, 436f
Bluehead wrasse, 753–754
Blue jays, 830, 831f
Blue-light photoreceptors, **649**, 651
Blue whales, 398f
BMR. See Basal metabolic rate
Bobcat, 666f
Bodies, animal, 664t, 665f, 822
Body cavities, animal, **550**
Body fat, 703
Body hairs, insect, 804
Body plans, **549**
    animal, 549
    apoptosis and, 324, 326f
    arthropod, 553
    fungal, 526f
    homeotic genes and, 374f–375f
    macroevolution of, from changes in
        developmental genes, 465f–468f
    pattern formation and, 326, 327f–329f
    symmetry, tissues, and body cavities in animal,
        549, 550f
    unity and diversity of bird, 11f
Bohr shift, **729f**
Boletus edulis, 525f
Bolting, 644, 645f
Bolus, **693–694**
Bombina, 444f, 445–446
Bonasa umbellus, 916
Bonding, parental, 829
Bone cells, 332f
Bone marrow, 182, 183f, 332f, 737

Bones. See also Skeletal systems
    endoskeletons of, 822, 823f
    for flight, 565f
    human middle ear, 805f, 806
    of human skeleton, 822, 823f
Bonneia, 499f
Bonnemaisonia hamifera, 512f
Bonobo, 373
Boquila trifoliata, 639f
Borisy, Gary, 189f
Borrelia burgdorferi, 489f, 492f
Botox, 786
Bottleneck effect, 422f–423f
Bottlenose dolphins, 796f
Bottom-up model, **877**
    of trophic control, 877
Botulism, 116, 346, 489f, 492, 786
Boundaries
    community, 867
    ecosystem, 886, 887f, 916f
    population, 854
Bound ribosomes, 82f, 295f
Boveri, Theodor, 236
Bowman's capsule, **681f**, 682
Boysen-Jensen, Peter, 640, 641f
Brachiopods, 548f, 551f
Braconid wasp, 872
Bradybaena, 436f
Brainbow technology, 790f
Brain cancer, 196, 361
Brains, **772**
    arousal and sleep functions of, 793, 796f
    biological clock regulation by, 796
    breathing control centers in human, 727f, 728
    in central nervous system, 791, 792f
    cerebral cortex functions in, 798, 799f–801f
    drug addiction and reward system of, 797, 798f
    evolution of cognition in, 799, 800f
    frontal lobe function in, 799f
    functional brain imaging of, 793f
    information processing by, 799
    language and speech functions of, 798, 799f
    lateralization of cortical function in, 798
    limbic system of, and emotions, 796f
    mammalian, 773f
    memory and learning in, 801
    neuronal plasticity of, 800f, 801
    neurons in, 772, 773f–774f. See also Neurons
    opiate receptors in mammalian, 786–787
    organization of human, 794f–795f
    sensory systems and, 801–811. See also Sensory
        systems
    size of, 573
    songbird vs. human, 800f
    stem cells of, 332
    strokes in, 720
    structure of human, 794f–795f, 800f
    vertebrate, 793–801
    visual information processing in, 810, 811f
Brainstem, 794f–795f
Brain waves, 796f
Branching
    carbon skeleton, 45f
    plant stem, 524, 586, 595
    surface area and, 526f
Branching evolution, 471
Branch length, phylogenetic tree, 402f–404f, 405
Branch points, phylogenetic tree, **397f**, 398
Brassinosteroids, 642t, 645
Brazil nut trees, 861f
BRCA1 and BRCA2 genes, 337, 338f–339f, 340
Bread mold, 279f–280f
Breakdown pathways, 123
Breast cancer, 195f, 337, 338f–339f, 340
Breasts, human, 757
Breathing, **726f**–727f, 728
    gait and, 573
Breathing control centers, 727f, 728
Breeding, 385f, 386, 619, 633f, 634
Brewer's yeast, 863f. See also Saccharomyces cerevisiae
Briggs, Robert, 330
Brightfield microscopy, 74f
Brine shrimp, 375f, 467f

Broadleaf tree, 161f
Broca, Pierre, 798
Broca's area, 798, 799f
Bronchi, **724**, 725f
Bronchioles, 725f
Brooding, 405f–406f
Brown algae, **509f**, 510
Brown bears, 124f
Brown fat, 154
Brown-headed cowbird, 916, 931
Brown tree snake, 910
Brundtland, G. H., 908
Brush border, 696
Bryophytes, **530f**–531f
Bryozoans, 552f
Bt toxin, 634
Bubo scandiacus, 863f
Buchloe dactyloides, 595
Buck, Linda, 803
Budding, 78f, 201f, 210, 752
Buffalo grass, 595
Buffers, **39**
Bufo marinus, 870
Bugs, 559f
Bulbourethral glands, 756f
Bulk feeders, 691f
Bulk flow, vascular plant, **599**, 610, 611f, 612, 616
Bulk transport, 112, 113f
Bumblebees, 449f
Bundle-sheath cells, 175f, 587f
Burgess Shale fossil bed, 453f
Burkholderia glathei, 491f
Burkitt's lymphoma, 340
Butane, 45f
Butterflies, 252, 472, 559f, 624f, 635–636, 830, 831f, 864f

## C

C. See Calorie
C₃ plants, **175**
C₄ plants, **175f**
Cactus, 614f, 624f, 847f
Cactus-eater finches, 383f
Caddisflies, 472
Cadherins, 504, 505f
Caecilians, 561, 562f
Caenorhabditis elegans, 326, 360, 362t
cal. See Calorie
Calcification, 40
Calcitonin, 669f
Calcium, 23t
    dietary requirements for, 690
Calcium carbonate, 39f, 40
Calcium ions, 83–84, 118, 783, 817f–818f, 820–821
California, chaparral biome in, 847f
California Current, 844f
California mouse, 828t
Callus, **632**
Callyspongia plicifera, 546f
Calmodulin, 820–821
Calorie (cal), **34**, 701. See also Kilocalorie
Calvin, Melvin, 164–165
Calvin cycle, **164**. See also Photosynthesis
    evolution of alternative mechanisms of, 175f, 176
    overview of, 165f, 174f, 177f
    phases of, 173, 174f
    as stage of photosynthesis, 164, 165f
Cambrian explosion, 466, 499f, **547**–557
Camouflage, 16f–17f, 386f, 870, 871f
CAM (crassulacean acid metabolism) plants, 175f,
    **176**, 614f, 615
cAMP (cyclic adenosine monophosphate), **118**, 119f,
    **307f**, 785
cAMP receptor protein (CRP), 307f
Campylobacter, 488f
Canada goose, 676f
Canadian Forest Service, 895f
Cancer
    biodiversity benefits for treatment of, 909
    brain cancer, 196, 361
    breast cancer, 195f, 337, 338f–339f, 340
    carcinogen screening and, 300
    cell-signaling pathways, 335f–336f